CW00725623

Jane's Space Systems and Industry 2008-2009

Edited by Peter Bond

Twenty-fourth Edition

Bookmark jsd.janes.com today!

The title is also available on online/CD-rom (New Gen), JDS (Jane's Data Services), EIS (Electronic Information Service), and Intel centres. Online gives the capability of real-time editing, permitting frequent updating. We trust our readers will use these facilities to keep abreast of the latest changes as and when they occur.
Updates online: Any update to the content of this product will appear online as it occurs.

Jane's Space Systems and Industry online site gives you details of the additional information that is unique to online subscribers and the many benefits of upgrading to an online subscription. Don't delay, visit jsd.janes.com today and view the list of the latest updates to this online service.

ISBN-13 978 0 7106 28 57 2
"Jane's" is a registered trademark

Copyright © 2008 by Jane's Information Group Limited, Sentinel House, 163 Brighton Road, Coulsdon, Surrey, CR5 2YH, UK

In the US and its dependencies
Jane's Information Group Inc, 110 N. Royal Street, Suite 200, Alexandria, Virginia 22314, US

Copyright enquiries
e-mail: copyright@janes.com

All rights reserved. No part of this publication may be reproduced, stored in retrieval systems or transmitted in any form or by any means, electronic, mechanical, photocopying, recording or otherwise, without the prior written permission of the Publishers. Licences, particularly for use of the data in databases or local area networks are available on application to the Publishers.
Infringements of any of the above rights will be liable to prosecution under UK or US civil or criminal law.
Whilst every care has been taken in the compilation of this publication to ensure its accuracy at the time of going to press, the Publishers cannot be held responsible for any errors or omissions or any loss arising therefrom.

This book was produced using FSC certified paper

Printed and bound in Great Britain by Cambridge University Press

Front cover image: The International Space Station seen from Space Shuttle Discovery as the two spacecraft moved apart on 5 November 2007. STS-120 delivered the European-built Harmony Node 2, which was later installed to the forward end of the Destiny laboratory. Node 2 provides attachment points for the European Columbus laboratory and the Japanese Kibo laboratry. During STS-120, the Port 6 truss strucuture was also moved from atop the Station's Z1 truss to the end of the Port 5 truss structure. 1330369

Contents

Jane's Space Systems and Industry website: jsd.janes.com

Jane's Defence Equipment Intelligence

With a global network of defence experts, Jane's provides unrivalled accurate and authoritative information on commercial and military aerospace systems, ground-based military equipment, naval vessels and weapon systems – a complete resource for market intelligence, threat assessment and recognition.

janes.com

Jane's An IHS Company
Intelligence and Insight You Can Trust

101540205

EDITORIAL AND ADMINISTRATION

Chief Content Officer: Ian Kay, e-mail: ian.kay@janes.com

Group Publishing Director: Sean Howe, e-mail: sean.howe@janes.com

Publisher: Sara Morgan, e-mail: sara.morgan@janes.com

Compiler/Editor: Welcomes information and comments from users who should send material to:
Information Collection
Jane's Information Group, Sentinel House, 163 Brighton Road, Coulsdon, Surrey CR5 2YH, UK
Tel: (+44 20) 87 00 37 00 Fax: (+44 20) 87 00 39 00
e-mail: yearbook@janes.com

SALES OFFICES

Europe and Africa
Jane's Information Group, Sentinel House, 163 Brighton Road, Coulsdon, Surrey CR5 2YH, UK
Tel: (+44 20) 87 00 37 50 Fax: (+44 20) 87 00 37 51
e-mail: customer.servicesuk@janes.com

North/Central/South America
Jane's Information Group, 110 N Royal Street, Suite 200, Alexandria, Virginia 22314, US
Tel: (+1 703) 683 21 34 Fax: (+1 703) 836 02 97
Tel: (+1 800) 824 07 68 Fax: (+1 800) 836 02 97
e-mail: customer.servicesus@janes.com

Asia
Jane's Information Group, 78 Shenton Way, #10-02, Singapore 079120, Singapore
Tel: (+65) 63 25 08 66 Fax: (+65) 62 26 11 85
e-mail: asiapacific@janes.com

Oceania
Jane's Information Group, Level 3, 33 Rowe Street, Eastwood, NSW 2122, Australia
Tel: (+61 2) 85 87 79 00 Fax: (+61 2) 85 87 79 01
e-mail: oceania@janes.com

Middle East
Jane's Information Group, PO Box 502138, Dubai, United Arab Emirates
Tel: (+971 4) 390 23 36 Fax: (+971 4) 390 88 48
e-mail: mideast@janes.com

Japan
Jane's Information Group, CERA51 Bldg, 1-21-8 Ebisu, Shibuya-ku, Tokyo 150-0013, Japan
Tel: (+81 3) 57 91 96 63 Fax: (+81 3) 54 20 64 02
e-mail: japan@janes.com

ADVERTISEMENT SALES OFFICES

(Head Office)
Jane's Information Group
Sentinel House, 163 Brighton Road,
Coulsdon, Surrey CR5 2YH, UK
Tel: (+44 20) 87 00 37 00 Fax: (+44 20) 87 00 38 59/37 44
e-mail: defadsales@janes.com

Janine Boxall, Global Advertising Sales Director,
Tel: (+44 20) 87 00 38 52 Fax: (+44 20) 87 00 38 59/37 44
e-mail: janine.boxall@janes.com

Richard West, Senior Key Accounts Manager
Tel: (+44 1892) 72 55 80 Fax: (+44 1892) 72 55 81
e-mail: richard.west@janes.com

Carly Litchfield, Advertising Sales Manager
Tel: (+44 20) 87 00 39 63 Fax: (+44 20) 87 00 37 44
e-mail: carly.litchfield@janes.com

(US/Canada office)
Jane's Information Group
110 N Royal Street, Suite 200,
Alexandria, Virginia 22314, US
Tel: (+1 703) 683 37 00 Fax: (+1 703) 836 55 37
e-mail: defadsales@janes.com

US and Canada
Sean Fitzgerald, Southeast Region Advertising Sales Manager
Tel: (+1 703) 836 24 46 Fax: (+1 703) 836 55 37
e-mail: sean.fitzgerald@janes.com

Linda Hewish, Northeast Region Advertising Sales Manager
Tel: (+1 703) 836 24 13 Fax: (+1 703) 836 55 37
e-mail: linda.hewish@janes.com

Janet Murphy, Central Region Advertising Sales Manager
Tel: (+1 703) 836 31 39 Fax: (+1 703) 836 55 37
e-mail: janet.murphy@janes.com

Richard L Ayer
127 Avenida del Mar, Suite 2A, San Clemente, California 92672, US
Tel: (+1 949) 366 84 55 Fax: (+1 949) 366 92 89
e-mail: ayercomm@earthlink.com

Rest of the World
Australia: *Richard West* (UK Head Office)

Benelux: *Carly Litchfield* (UK Head Office)

Eastern Europe (excl. Poland): MCW Media & Consulting Wehrstedt
Dr Uwe H Wehrstedt
Hagenbreite 9, D-06463 Ermsleben, Germany
Tel: (+49 03) 47 43/620 90 Fax: (+49 03) 47 43/620 91
e-mail: info@Wehrstedt.org

France: Patrice Février
BP 418, 35 avenue MacMahon,
F-75824 Paris Cedex 17, France
Tel: (+33 1) 45 72 33 11 Fax: (+33 1) 45 72 17 95
e-mail: patrice.fevrier@wanadoo.fr

Germany and Austria: *MCW Media & Consulting Wehrstedt* (see Eastern Europe)

Greece: *Carly Litchfield* (UK Head Office)

Hong Kong: *Carly Litchfield* (UK Head Office)

India: *Carly Litchfield* (UK Head Office)

Israel: *Oreet International Media*
15 Kinneret Street, IL-51201 Bene Berak, Israel
Tel: (+972 3) 570 65 27 Fax: (+972 3) 570 65 27
e-mail: admin@oreet-marcom.com
Defence: Liat Heiblum
e-mail: liat_h@oreet-marcom.com

Italy and Switzerland: *Ediconsult Internazionale Srl*
Piazza Fontane Marose 3, I-16123 Genoa, Italy
Tel: (+39 010) 58 36 84 Fax: (+39 010) 56 65 78
e-mail: genova@ediconsult.com

Japan: *Carly Litchfield* (UK Head Office)

Middle East: *Carly Litchfield* (UK Head Office)

Pakistan: *Carly Litchfield* (UK Head Office)

Poland: *Carly Litchfield* (UK Head Office)

Russia: Anatoly Tomashevich
1/3, appt 108, Zhivopisnaya Str, Moscow, 123103, Russia
Tel/Fax: (+7 495) 942 04 65
e-mail: to-anatoly@tochka.ru

Scandinavia: *Falsten Partnership*
23, Walsingham Road, Hove, East Sussex BN41 2XA, UK
Tel: (+44 1273) 77 10 20 Fax: (+ 44 1273) 77 00 70
e-mail: sales@falsten.com

Singapore: *Richard West* (UK Head Office)

South Africa: *Richard West* (UK Head Office)

Spain: Macarena Fernandez
VIA Exclusivas S.L., Virato, 69 - Sotano C, E-28010, Madrid, Spain
Tel: (+34 91) 448 76 22 Fax: (+34 91) 446 02 14
e-mail: macarena@viaexclusivas.com

Turkey: *Richard West* (UK Head Office)

ADVERTISING COPY
Kate Gibbs (UK Head Office)
Tel: (+44 20) 87 00 37 42 Fax: (+44 20) 87 00 38 59/37 44
e-mail: kate.gibbs@janes.com

For North America, South America and Caribbean only:
Tel: (+1 703) 683 37 00 Fax: (+1 703) 836 55 37
e-mail: us.ads@janes.com

How to Use: Jane's Space Systems and Industry

Content

This is the second year of publication of Jane's Space Systems and Industry , the successor to *Jane's Space Directory*. By giving the book a new structure and focus, *Jane's Space Systems and Industry* is intended to better address the growing space industry's current information requirements regarding spacecraft, operators, space organisations, contractors and facilities.

Structure

Jane's Space Systems and Industry is divided into seven main sections, detailed below, each representing major space industry sectors, spacecraft or launches. Subdivisions organised under each main section are further divided by country.

Commercial Space Services — This section is divided into two parts: Launch Facilities and Satellite Operators. These subdivisions are further divided by operator or facility country.

Civil Space Organisations — The Civil Space Organisations section consists of five subdivisions: Launch Facilities; Multi-national Agencies; National Agencies; Research and Development; and Satellite Operators. It is also further divided by country.

Space Defence — The Space Defence section consists of three subdivisions: Bases; Operators; and Research and Development. The Bases subsection includes military-operated launch facilities. Each subsection is divided by country.

Space Industry — This section contains three subdivisions: Joint Ventures; Major Sub-contractors; and Prime Contractors. These subsections are also divided by country.

Space Launch Vehicles — The Launch Vehicles section is divided into two subdivisions: Orbital and Sub-orbital, both of which are further divided by country.

Spacecraft — The Spacecraft section is comprised of seven subdivisions: Communications; Defence (including early warning and signals intelligence); Earth Observation (including meteorology); Human; Navigation; Scientific; and Technology Demonstration. All of these are subdivided by country.

Space Logs — This final main section contains numerous logs related to human spaceflight missions and launches, the most historic of which are published in the online version only, which can be found at: jsd.janes.com.

Record Structure

The following main headings are used within most records:

Current Status — Current events or recent changes or developments impacting an organisation or spacecraft can be found under this heading.

Background — A detailed development history and background information for each organisation, launch vehicle or or spacecraft can be found under this heading.

Specifications — The main technical details of space systems are listed under this heading, including launch history, dimensions, weight and operating parameters (where available).

Images

Photographs are provided for each space system or facility wherever possible. Line drawings and graphics are also provided in some cases. Images are annotated with a seven digit number which uniquely identifies them in the *Jane's* image database.

Other Information

Contractor List — Space-related manufacturers and organisations, with their contact details, are listed alphabetically here. Web site and e-mail details are included. URLs are generally written in lower case using the Roman alphabet; e-mail addresses are sometimes case sensitive.

General index — Finally, there is an alphabetical list of the book's contents in the General Index, reflecting section and page number.

DISCLAIMER This publication is based on research, knowledge and understanding, and to the best of the author's ability the material is current and valid. While the authors, editors, publishers and Jane's Information Group have made reasonable effort to ensure the accuracy of the information contained herein, they cannot be held responsible for any errors found in this publication. The authors, editors, publishers and Jane's Information Group do not bear any responsibility or liability for the information contained herein or for any uses to which it may be put.

This publication is provided for informational purposes only. Users may not use the information contained in this publication for any unlawful purpose. Without limiting the generality of the foregoing, users must comply with all applicable laws, rules and regulations with regard to the transmission of facsimiles.

While reasonable care has been taken in the compilation and editing of this publication, it should be recognised that the contents are for information purposes only and do not constitute any guidance to the use of the equipment described herein. Jane's Information Group cannot accept any responsibility for any accident, injury, loss or damage arising from the use of this information.

Jane's Libraries

To assist your information gathering and to save you money, Jane's has grouped some related subject matter together to form 'ready-made' libraries, which you can access in whichever way suits you best – online, on CD-ROM, via Jane's EIS or through Jane's Data Service.

The entire contents of each library can be cross-searched, to ensure you find every reference to the subjects you are looking for. All Jane's libraries are updated according to the delivery service you choose and can stand alone or be networked throughout your organisation.

Jane's Defence Equipment Library

Aero-Engines
Air-Launched Weapons
Aircraft Upgrades
All the World's Aircraft
Ammunition Handbook
Armour and Artillery
Armour and Artillery Upgrades
Avionics
C4I Systems
Electro-Optic Systems
Explosive Ordnance Disposal
Fighting Ships
Infantry Weapons
Land-Based Air Defence
Military Communications
Military Vehicles and Logistics
Mines and Mine Clearance
Naval Weapon Systems
Nuclear, Biological and Chemical Defence
Radar and Electronic Warfare Systems
Strategic Weapon Systems
Underwater Warfare Systems
Unmanned Aerial Vehicles and Targets

Jane's Defence Magazines Library

Defence Industry
Defence Weekly
Foreign Report
Intelligence Digest
Intelligence Review
International Defence Review
Islamic Affairs Analyst
Missiles and Rockets
Navy International
Terrorism and Security Monitor

Jane's Market Intelligence Library

Aircraft Component Manufacturers
All the World's Aircraft
Defence Industry
Defence Weekly
Electronic Mission Aircraft
Fighting Ships
Helicopter Markets and Systems
International ABC Aerospace Directory
International Defence Directory
Marine Propulsion
Naval Construction and Retrofit Markets
Police and Homeland Security Equipment
Simulation and Training Systems
Space Systems and Industry
Underwater Security Systems and Technology
World Armies
World Defence Industry

Jane's Security Library

Amphibious and Special Forces
Chemical-Biological Defense Guidebook
Facility Security
Fighting Ships
Intelligence Digest
Intelligence Review
Intelligence Watch Report
Islamic Affairs Analyst
Police and Homeland Security Equipment
Police Review
Terrorism and Security Monitor
Terrorism Watch Report
World Air Forces
World Armies
World Insurgency and Terrorism

Jane's Sentinel Library

Central Africa
Central America and the Caribbean
Central Europe and the Baltic States
China and Northeast Asia
Eastern Mediterranean
North Africa
North America
Oceania
Russia and the CIS
South America
South Asia
Southeast Asia
Southern Africa
The Balkans
The Gulf States
West Africa
Western Europe

Jane's Transport Library

Aero-Engines
Air Traffic Control
Aircraft Component Manufacturers
Aircraft Upgrades
Airport Review
Airports and Handling Agents –
  Central and Latin America (inc. the Caribbean)
  Europe
  Far East, Asia and Australasia
  Middle East and Africa
  United States and Canada
Airports, Equipment and Services
All the World's Aircraft
Avionics
High-Speed Marine Transportation
Marine Propulsion
Merchant Ships
Naval Construction and Retrofit Markets
Simulation and Training Systems
Transport Finance
Urban Transport Systems
World Airlines
World Railways

nes.com

Jane's Electronic Solutions

Jane's online service

For sheer timeliness, accuracy and scope, nothing matches Jane's online service

www.janes.com is the most comprehensive open-source intelligence resource on the Internet. It is your ultimate online facility for security, defence, aerospace, transport, and related business information, providing you with easy access, extensive content and total control.

Jane's online service is subscription based and gives you instant access to Jane's information and expert analysis 24 hours a day, 7 days a week, 365 days a year, wherever you have access to the Internet.

To see what is available online in your specialist area simply go to **www.janes.com** and click on the **Intel Centres** tab

Once you have entered the **Intel Centres** page, choose the link that most suits your requirements from the following list:

- Defence Intelligence Centre
- Transport Intelligence Centre
- Aerospace Intelligence Centre
- Security Intelligence Centre
- Business Intelligence Centre

Jane's offers you information from over 200 sources covering areas such as:

- Market forecasts and trends
- Risk analysis
- Industry insight
- Worldwide news and features
- Country assessments
- Equipment specifications

As a Jane's Online subscriber you have instant access to:

- *Accurate and impartial* information
- *Archives* going back five years
- *Additional reference content*, source data, analysis and high-quality images
- *Multiple search tools* providing browsing by section, by country or by date, plus an optional word search to narrow results further
- *Jane's text and images* for use in internal presentations
- *Related information* using active interlinking

Jane's An IHS Company
Intelligence and Insight You Can Trust

janes.com

Executive Overview: Space

On 4 October 1957, the grass-covered steppes surrounding the secret missile base near Tyuratam, Kazakhstan, reverberated to the sound of the Soviet Union's R-7 vehicle as it blasted off on a journey into history. Located in its nose cone was a small aluminium ball named Sputnik. For the next 21 days, the world's first artificial satellite circled the Earth every 96 minutes, transmitting a beeping signal that transfixed the citizens of every country it passed over. Sputnik marked the dawn of the Space Age, and the world has never been the same since.

50 years later, our modern technological society relies upon numerous space-based applications, ranging from satellite navigation to instantaneous global communications, weather forecasting and Earth observation. Dozens of military spacecraft are now an indispensable adjunct to the global armed forces. Automated spacecraft have ventured to every planet in the Solar System and explored the furthest reaches of the universe. More than 450 citizens of planet Earth have experienced the wonders of weightlessness, and 12 men have walked on the Moon.

Since the space-based economy pervades so much of our lives, it is hardly surprising that the value of the global space economy passed USD250 billion in 2007, an 11 per cent increase over the previous year, according to a report by the Space Foundation. US government spending alone accounted for USD62.6 billion (25 per cent) of the total market, with other space-faring governments accounting for another USD14.7 billion.

As in previous years, the largest chunk of NASA's USD17.3 billion budget was devoted to construction of the International Space Station (ISS) and keeping the Space Shuttle flying. The past 12 months have generally been favourable to the human spaceflight programme. Five Shuttles and the first European Automated Transfer Vehicle successfully docked with the ISS, delivering major pieces of hardware from several different nations. With the long-delayed arrival of additional US truss segments, the European-built Harmony Node, Columbus and ATV, together with the first element of Japan's Kibo laboratory and the Canadian Dextre robotic manipulator system, all of the major ISS partners could at last claim their own corner of the Station - almost 10 years after its construction began.

ESA's Columbus laboratory being lifted from the Shuttle payload bay prior to attachment to the Harmony node on the ISS in February 2008 (NASA/ESA) 1343783

Another major step towards the completion of the ISS came in June when the third of four pairs of massive solar arrays were successfully deployed. Five months later, another potential stumbling block was overcome when a truss and its solar arrays were successfully relocated to their permanent position on the left side of the station. Once again, the importance of human intervention was displayed when NASA astronaut Scott Parazynski used his surgeon's skills to stitch together a torn array, thus enabling its full deployment. The only down side came with the discovery of damage to a critical solar array rotating mechanism, which threatened to limit the station's planned power output and full utilisation.

With only 10 Shuttle missions to the ISS, plus a Hubble Space Telescope servicing mission, remaining on the manifest before the world's only reusable orbiter ends its operational life in 2010, the time is nearing when there will be a hiatus of perhaps five years in US human spaceflight. With the prospect of the US and other ISS partners becoming totally dependent on Russia for crew access to the Station, it is hardly surprising that some US politicians and industry officials are arguing for a limited extension of the Shuttle programme alongside accelerated development of the Ares I launcher and Orion crew vehicle. Meanwhile, in Europe the desire for independent access to space is driving proposals for development of a joint crew transportation system endeavour with Russia, or even a home-grown system based on a human-rated Ariane 5.

NASA remains focused on the Vision for Space Exploration (VSE) with more than USD3 billion devoted to Exploration in the FY08 budget, with a jump to over USD7 billion foreseen for 2011 onwards, after the Shuttle is retired and construction of the ISS is completed. Over the past year, NASA has been assembling the team of contractors that will build the Ares I launch vehicle and the Orion Crew Exploration Vehicle. Concerns have arisen over a vibration in the Ares I first stage that spreads to affect the remainder of the stack, although NASA officials insist that this will be fairly straightforward to resolve. Meanwhile, the date of the Orion's debut is likely to slip to 2015 as NASA's budget fails to grow sufficiently in the next few years to meet demands on its resources.

In February 2008, NASA selected Orbital Sciences to provide the Cygnus cargo spacecraft and the Taurus II launch vehicle as part of the Commercial Orbital Transportation Services (COTS) project (OSC) 1166951

Unwilling to rely solely upon Europe, Russia and Japan to supply the ISS beyond 2010, NASA is still pursuing a Commercial Orbital Transportation Services (COTS) option. This approach met with mixed success in 2007, when funding was withdrawn from Rocketplane Kistler after it was unable to meet agreed-upon milestones in its effort to develop and demonstrate commercial transportation capabilities to low Earth orbit. Its place was taken by Orbital Sciences, which has committed itself to provide an automated supply craft, known as Cygnus, and the new Taurus II medium-lift expendable rocket by the end of 2010. The other main COTS beneficiary, SpaceX, has yet to achieve a successful launch of its Falcon rocket, but hopes remain high that the company's Falcon 9 vehicle will be able to demonstrate its ability to launch the Dragon capsule in 2009.

The last 12 months saw some heated discussions in which politicians and scientists expressed their concern over a perceived drop in funding in NASA's Space Science programmes, particularly in the area of astrophysics, where complex projects such as the James Web Space Telescope threatened to consume much of the shrinking budget allocation. The agency's Director of Science, Alan Stern, made fiscal responsibility his priority, threatening to cancel overspending programmes, such as the Kepler mission to search for extrasolar planets. However, after only one year in office, Stern handed in his notice when

Administrator Michael Griffin countermanded an order to trim USD4 million from the budget of the long-lived Mars Exploration Rovers, Spirit and Opportunity.

Nevertheless, 2007 was a good year for Solar System science, with spacecraft from various countries returning new data on the Sun, Mercury, Venus, Mars, Jupiter and Saturn. The longevity of the Mars Rovers continued to astound, while the Cassini mission - now extended until 2010 - sent back more revelations about smog-shrouded Titan and the icy geysers of Enceladus. The renewed enthusiasm for all things lunar resulted in Japan launching Kaguya, the largest spacecraft sent to the Moon since the days of Apollo, and in the first Chinese deep space mission, Chang'e-1. NASA's much-delayed Dawn spacecraft finally set off for an historic rendezvous with two of the largest asteroids, while Phoenix was launched towards Mars for another pioneering mission to explore the planet's frozen arctic wastes.

Japan also continued to play a leading role in space science. New insights into the nature of solar flares and the mysterious superheating of the Sun's corona were provided by the Hinode satellite. The Akari satellite produced the most detailed infrared map of the heavens before its supply of helium coolant was depleted in August 2007.

The numerous programmes funded under the US Department of Defense (DoD) budget continued to exhibit a mixture of some successes alongside many setbacks. 2007 saw the launch of an unusually high number of US military satellites, including various innovative and successful technology demonstration missions, such as NFIRE and Orbital Express. Following the launch of the first Wideband Global Satellite in October 2007, Australia agreed to fund an identical spacecraft, boosting the order book for this next generation of military communications satellites to six. The SBIRS missile early warning system also got off to a promising start with the launch of two payloads into HEO. The first of these payloads completed its initial checkout and its performance reportedly met or exceeded specifications. However, the cost of the yet-to-be-launched SBIRS GEO satellites keeps climbing and their scheduled launch date keeps moving backwards. A major modification of the SBIRS GEO software was also required following the complete failure of the control computer on USA 193, a National Reconnaissance Office (NRO) satellite, soon after launch in December 2006.

The embarrassing loss of NRO's USA 193, brought into the global spotlight by the subsequent launch of a missile to bring about its destruction before it tumbled to Earth (see image), was symptomatic of the deeply troubled Future Imagery Architecture (FIA) programme. The failed satellite is believed to be the first of the FIA radar imaging spacecraft designed to replace the aging Lacrosse/Vega/Onyx family. Meanwhile, plans for a USAF/NRO Space Radar constellation that could detect moving targets on the ground have been scrapped due to its technical complexity and spiralling costs. A simplified, lower cost programme is now being proposed.

Anti-SATellite (ASAT) technologies and policies, came to the forefront once more on 21 February 2008 when a US Aegis-class cruiser fired an SM-3 missile that destroyed the malfunctioning USA 193 in low Earth orbit. The DoD justified this demonstration of American missile defence and ASAT capabilities by explaining there was a (faint) possibility of a satellite loaded with toxic hydrazine plummeting to Earth in a densely populated area.

This incident also highlighted once more the threat posed by orbital space debris. US authorities were at pains to point out that the amount of debris created by the obliteration of USA 193 was only a fraction of that produced when China destroyed its aged Fenyung 1C weather satellite in January 2007. Whereas most of the Chinese debris will be in orbit 100 years from now, almost all of the fragments of the NRO satellite re-entered in six months. Meanwhile, the ramifications of the Chinese action rumbled on, with military leaders and politicians openly discussing ways to protect America's invaluable, but vulnerable, orbital assets. This problem was further illustrated when Libya was identified as the source of jamming that interrupted services on a Thuraya communications satellite, while Intelsat had to take action to end the hijacking of a vacant Ku-band transponder by the Tamil Tiger rebels in Sri Lanka.

With national security and global terrorism high on the agenda, it was not surprising to see the launch of various satellites that could provide high resolution Earth imaging from orbit. One of the most innovative approaches involved DigitalGlobe's WorldView-1, which began to deliver extremely detailed imagery to the National Geospatial-Intelligence Agency (NGA) as part of the NextView programme, in return for partial funding. WorldView-1 and GeoEye-1, scheduled for launch in summer 2008, are the first non-military imaging satellites to offer resolution below 0.5 m.

On 24 February 2008, an SM-3 missile launched from the cruiser Lake Erie destroyed a non-functioning NRO satellite (US Navy) 1294384

At the same time, Europe became a serious player in the reconnaissance/Earth observation arena with the debut of Germany's TerraSar-X, the first launches in Italy's Cosmo-Skymed constellation and three more deployments of Germany's SAR-Lupe radar reconnaissance satellites. Israel also climbed on board with the launches of its Ofeq-7 optical reconnaissance satellite and TECSAR multimode X-band radar imaging satellite. Secure, high speed, military communications for NATO members were also a priority with the introduction of the UK's first two Skynet 5 satellites, together with XTAR-Eur and Spainsat.

In 2007, commercial satellite products and services continued to dominate the space market, accounting for USD138.8 billion or 55 per cent of activity worldwide, according to the Space Foundation's Space Report 2008. Furthermore, the commercial sector demonstrated a healthy expansion, with a 20 per cent increase over 2006. Much of this growth was accounted for by digital satellite/direct-to-home television and Global Positioning System (GPS) equipment, the two largest sectors of the commercial space industry.

In recent years, the ongoing efforts to consolidate the major satellite operators have been offset by the emergence of various newcomers, including some national or regional flag carriers, such as NigcomSat and Rascomstar-QAF. Among those placing orders for the first time in 2007 were YahSat of the United Arab Emirates and S2M, a Dubai-based start-up.

The expansion of the fixed service satellite sector was fed by high demand in Europe, the Middle East, North Africa and some North American sectors. Companies such as DIRECTV were responsible for the introduction of large, powerful satellites that provide consumer broadband services and high definition TV.

Another area of future growth is likely to be associated with the introduction of satellite-terrestrial hybrid networks as growing demand for new multimedia services such as mobile TV pushes the satellite and terrestrial communications industries together. In the US, Mobile Satellite Ventures has already been granted the first FCC license to provide hybrid satellite-terrestrial services, and the company has signed a contract for two next-generation satellites that will be launched in 2009 - 2010. The satellites feature 22 m diameter, elliptical mesh

Four of Germany's SAR-Lupe radar reconnaissance satellites were launched by Russian Kosmos-3M vehicles between December 2006 and March 2008 (OHB-System)
1329936

reflectors that will support L-band communication with conventional handsets through a network based on MSV's patented ancillary terrestrial component technology. Whereas the current MSAT satellites are capable of handling thousands of simultaneous voice calls and low speed data transmissions, the new satellites will provide more than 10 times this capacity. More importantly, they will be optimised for mobile broadband services, including IP-based voice and dispatch, audio and video. Where MSV is paving the way, other companies seem sure to follow.

Meanwhile, the low earth orbit/medium earth orbit (LEO/MEO) mobile satellite sector has continued its re-emergence from the doldrums, with the launch of 8 Globalstar satellites during 2007 as an insurance against degrading signals from the existing constellation. These will eventually be merged into the Globalstar second-generation. A similar upgrade of the Orbcomm system will be taking place in 2008, with subsequent orders for up to 100 satellites expected from that company and Iridium in the next 24 months.

Other growth areas included emerging space programmes in China, which tripled its spending between 2006 and 2007, and India, which approved a 29 per cent increase in spending for FY07. A priority for both countries was a national satellite navigation system, while India began to consider a human space endeavour that could eventually compete with the Chinese Shenzhou programme. Satellite navigation was also a priority elsewhere, with the US continuing to upgrade its GPS fleet and Russia starting the long process of replenishing its Glonass constellation. In contrast, the sometimes fractious debate over Galileo continued throughout the year in Europe, ending with an agreement to fund the full system so that it may be deployed by 2013.

The number of orbital launches attempted in 2007 reached 68, the second consecutive year of growth and the highest number of launches since 2000, when 85 missions were attempted worldwide. The 2007 launch manifest included 118 satellites, only the second time in the past decade that the number of payloads had increased year on year. However, the increase in demand for launches, delays in satellite deliveries and a shortage of launch vehicles resulted in a continuation of the tight commercial launch market.

The steady rise in the number of launches during the last couple of years is partly attributable to an increase in the number of Chinese Long March rockets being launched. In 2007, 10 Long Marches were successfully launched, the highest number since the programme began in 1970. In the US, the Delta II remained the stalwart of the government and commercial launch market, while the Atlas V finally began to launch satellites for the DoD - albeit with an early engine shutdown in June 2007 that caused two NRO ocean surveillance spacecraft to be placed in the wrong orbits. Worse setbacks involved the Sea Launch Zenit-3SL, which was grounded for the entire year following a launch pad explosion in January, and the Russian Proton, which suffered a launch failure in September 2007, followed by another in March 2008.

The Sea Launch Zenit-3SL was grounded for the whole of 2007 following a January launch platform explosion (Sea Launch)
0137129

The revival in the commercial fortunes of China's Long March has coincided with the policy of Thales Alenia to bypass US ITAR restrictions by manufacturing comsats that avoid the use of American components. This has led to some dissatisfaction over loss of market share and security concerns on both sides of the Atlantic, with complaints from Arianespace officials as well as US industry and members of Congress. It has also directed attention once more to what many in the industry regard as over-strict enforcement of ITAR, with its harmful impact on US sales of satellites and components and the consequent encouragement of foreign competitors to develop their own capabilities. One small example of the desire for change is the petition from Bigelow Aerospace to the US government to change the export licensing jurisdiction for the inflatable technology used in its Genesis satellites.

More than four decades after Gagarin became the first human to leave the planet, the day when modestly wealthy private citizens can experience the wonder of spaceflight is creeping ever closer. In June 2007, EADS Astrium announced that it had begun preliminary work on a winged vehicle designed to carry four passengers 100 km into space. In January 2008, Virgin Galactic unveiled the designs of the SpaceShipTwo suborbital spaceplane and the White Knight Two carrier aircraft, although progress was slowed after Scaled Composites was hit six months earlier by a fatal accident while testing components. Other companies are also beavering away to fulfil the perceived demand for brief excursions in microgravity.

Finally, the success of the X Prize in inspiring the private development of SpaceShipOne and White Knight One has spawned an even more ambitious successor. In 2007, the X Prize Foundation combined with Google to offer up to USD25 million to the first privately funded team to successfully place a roving vehicle on the lunar surface and send back to Earth high definition video, still images and data. By March 2008, the contest had attracted 567 expressions of interest from 53 nations. It seems that space still has the ability to enthuse the next generation of scientists and engineers, 50 years after Sputnik soared across the night sky.

Peter Bond

Acknowledgements

The Editor would like to offer his warm thanks to Rebecca Davies, Senior Content Editor, Kimberley Ebner, Assistant Editor until 27 July 2007, and Johanna Catena, Research Assistant, for their assistance, diligence, competence and attention to detail during his first year in the post. Thanks also to Jonathan McDowell for his continued support in giving us permission to reprint his worldwide launch logs, and to Andrew Wilson of ESA, former editor of *Jane's Spaceflight Directory*, who provided valuable advice.

Peter Bond

Peter Bond is a lifelong space and astronomy enthusiast. He has written 10 books, including the award-winning "*DK Guide to Space*", and has contributed to or acted as consultant/editor for many others. His latest book, "*Distant Worlds: Milestones in Planetary Exploration*", was published by Springer-Copernicus-Praxis in February 2007. Two books scheduled for publication in 2008 are "*Jane's Space Recognition Guide*" (HarperCollins) and "*Pop Up Space*" (Templar).

Peter has also written hundreds of articles on space and astronomy for British and American newspapers and magazines, as well as material for the Sunday Times "*Window On The Universe*" CD-ROM, the Nature-IOP "*Encyclopedia of Astronomy and Astrophysics*", the "*Encarta Reference Library*" and the Philip's "*Encyclopedia of Astronomy and Astrophysics*". He was the Space Science Advisor/Press Officer for the Royal Astronomical Society 1995-2007, frequently appearing on TV and radio to comment on the latest space discoveries and events. Peter has also been a consultant/writer for the European Space Agency for many years, writing and editing brochures and reports, as well as writing articles for the ESA Bulletin and several of the Agency's web sites. He received a certificate for an "outstanding contribution" from ESA in 2005, and a Group Achievement Award from NASA in 2004. He is a Fellow of both the Royal Astronomical Society and the British Interplanetary Society. Peter was appointed Consultant Editor for *Jane's Space Systems and Industry* on 1 May 2007.

Jane's Users' Charter

This publication is brought to you by Jane's Information Group, a global company with more than 100 years of innovation and an unrivalled reputation for impartiality, accuracy and authority.

Our collection and output of information and images is not dictated by any political or commercial affiliation. Our reportage is undertaken without fear of, or favour from, any government, alliance, state or corporation.

We publish information that is collected overtly from unclassified sources, although much could be regarded as extremely sensitive or not publicly accessible.

Our validation and analysis aims to eradicate misinformation or disinformation as well as factual errors; our objective is always to produce the most accurate and authoritative data.

In the event of any significant inaccuracies, we undertake to draw these to the readers' attention to preserve the highly valued relationship of trust and credibility with our customers worldwide.

If you believe that these policies have been breached by this title, you are invited to contact the editor.

A copy of Jane's Information Group's Code of Conduct for its editorial teams is available from the publisher.

janes.com

Jane's
An IHS Company
Intelligence and Insight You Can Trust

Glossary: Jane's Space Systems and Industry

The metric system of measurement is employed throughout *Jane's Space Systems and Industry*. Times and dates are quoted in Greenwich Mean Time (GMT) or Universal Time (UT) unless otherwise noted. Abbreviations and acronyms are usually explained in the text; the most common appear below, along with some unit conversions not given elsewhere.

µg: microgravity
µm: 10^{-6} m (micron)
Å: Angstrom, 10^{-10} m
ACS: Attitude Control System
A/D: Analogue/Digital
AEB: Agência Espacial Brasileira; Brazilian Space Agency
AF: Air Force
AFB: Air Force Base
AIT: Assembly, Integration and Testing
AKM: Apogee Kick Motor
AO: Announcement of Opportunity
AOCS: Attitude and Orbit Control System
AS: Air Station
ASEAN: Association of South East Asia Nations
ASI: Agenzia Spaziale Italiana; Italian Space Agency
ASIC: Application Specific Integrated Circuit
atm: Earth atmospheric pressure (1.0336 kg/cm^2; 14.7 lb/in^2)
ATV: Automated Transfer Vehicle
AU: ESA Accounting Unit (equivalent to the ECU)
AU: Astronomical Unit (149.598 million km); billion = 1,000 million
BNSC: British National Space Centre
BOL: Beginning Of Life
BOM: Beginning Of Mission
C3: Command, Control and Communications
C4: Command, Control, Communications and Computers
C4I: Command, Control, Communications, Computers and Intelligence
CAD: Computer-Aided Design
CCD: Charge Coupled Device
CCT: Computer-Compatible Tape
CDR: Critical Design Review
cleanliness: *Jane's Space Systems and Industry* adopts the US system specified by the numerical limit of 0.5 particles per ft^3 (0.028 m^3). Thus a class 10,000 environment has fewer than 10,000 particles of 0.5 in 1 ft^3 (0.028 m^3). Most satellite and launcher integration and testing facilities are class 100,000. Class 10 and class 100 facilities are used for assembly of sensitive items such as gyros
cm: centimetre (1 inch = 2.54 cm)
CNC: Computer Numerical Control
CNES: Centre National d'Études Spatiales; French Space Agency
CNSA: China National Space Administration
Conus: Continental United States
COTS: Commercial-Off-The-Shelf
CPV: Common Pressure Vessel
CSA: Canadian Space Agency
DAMA: Demand Assigned Multiple Access
DBS: Direct Broadcast Satellite
DGPS: Differential GPS
DLR: Deutsches Zentrum für Luft- und Raumfahrt; German Aerospace Centre
DoD: US Department of Defense
DoE: US Department of Energy
DoF: Degrees of Freedom
DRAM: Dynamic Random Access Memory
DSN: NASA's Deep Space Network
DTH: Direct To Home
EAFB: Edwards Air Force Base, California
ECLSS: Environment Control and Life Support System
ECU: European Currency Unit
EDO: Extended Duration Orbiter
EGSE: Electrical Ground Support Equipment
EIRP: Equivalent Isotropically Radiated Power
ELINT: Electronic Intelligence
ELV: Expendable Launch Vehicle
EM: Engineering Model
EOL: End Of Life
EOM: End Of Mission
EPC: Electronic Power Conditioner
ESA: European Space Agency
eV: electron Volt
EVA: Extravehicular Activity
fl: focal length
FM: Flight Model
FoV: Field of View
FSA: Russian Federal Space Agency; Roscosmos
FSD: Full Scale Development
FWHM: Full Width Half Maximum
FY: Fiscal Year
g: gravitational acceleration (9.806 m/s)
GaAs: Gallium Arsenide
GAS: Get Away Special
GDP: Gross Domestic Product
GEO: Geostationary Orbit; altitude 35,786 km, 0° inclination, period 23.934 h
GH_2: Gaseous Hydrogen
GIS: Geographic Information System
GMDSS: Global Maritime Distress and Safety System
GMT: Greenwich Mean Time (also UT)
GN&C: Guidance, Navigation and Control
GNP: Gross National Product
GOX: Gaseous Oxygen
GPS: Global Positioning System
GSE: Ground Support Equipment
GSM: Groupe Spéciale Mobile
GSO: Geosynchronous Orbit
GTO: Geostationary Transfer Orbit
HES: Health, Environment and Safety
HGA: High Gain Antenna
HOOD: Hierarchical Object Oriented Design
HPA: High-Power Amplifier
HRVIR: High-Resolution Visible Infra-Red
HTPB: Hydroxyl Terminated Polybutadiene
IC: Integrated Circuit
ICBM: InterContinental Ballistic Missile (range >5,500 km)
IFOG: Interferometric Fibre Optic Gyro
IMINT: Image Intelligence
IMU: Inertial Measurement Unit
INS: Inertial Navigation System
INU: Inertial Navigation Unit
IOC: Initial Operational Capability
IRBM: Intermediate Range Ballistic Missile (range 500-5,500 km)
IRU: Inertial Reference Unit
ISAS: Japanese Institute of Space and Astronautical Science
ISDN: Integrated Services Digital Network
ISRO: Indian Space Research Organisation
ISS: International Space Station
IUS: Inertial Upper Stage
JAXA: Japan Aerospace Exploration Agency
km: kilometre (0.6214 mile)
KSC: Kennedy Space Center
L: Launched
LEO: Low Earth Orbit, typically up to 1,500 km
LEOP: Launch and Early Orbit Phase
LH_2: Liquid Hydrogen
LHCP: Left-Hand Circular Polarisation
LiOH: Lithium Hydroxide
LITVC: Liquid Injection Thrust Vector Control
LNA: Low-Noise Amplifier
LOX: Liquid Oxygen
m: metre (39.37 inch)
MAU: Million Accounting Units
MEOP: Maximum Expected Operating Pressure
MeV: Mega-electronvolt
MGSE: Mechanical Ground Support Equipment
MHz: Megahertz
micron: 10^{-6} m

MIR: Mid Infra-Red
MMH: Monomethyl Hydrazine
MMIC: Monolithic Microwave Integrated Circuit
MON: Mixed Oxides of Nitrogen
MoU: Memorandum of Understanding
MPD: Magnetoplasmadynamic
MW: Momentum Wheel
N: Newton (0.225 lbf)
NAL: National Aerospace Laboratory of Japan
NASA: US National Aeronautics and Space Administration
NASDA: National Space Development Agency of Japan
Ni/Cd: Nickel Cadmium
Ni/H_2: Nickel Hydrogen
NiMH: Nickel Metal Hydride
NIR: Near Infra-Red
NOAA: US National Oceanic and Atmospheric Administration
NORAD: North American Aerospace Defense Command
NSSK: North-South Station-Keeping
NTO, N_2O_4: Nitrogen Tetroxide
OBDH: Onboard Data Handling
OEM: Original Equipment Manufacturer
PAM: Payload Assisted Module
PAN: Panchromatic
PDR: Preliminary Design Review
PI: Principal Investigator
PKM: Perigee Kick Motor
PMD: Propellant Management Device
ppm: parts per million
PROM: Programmable Read-Only Memory
QM: Qualification Model
R&D: Research and Development
RAM: Random Access Memory
RCS: Reaction Control System
RDT&E: Research, Development, Test and Evaluation
RF: Radio Frequency
RHCP: Right Hand Circular Polarisation
RKA: Russian Aviation and Space Agency
RLG: Ring Laser Gyro
RLV: Reusable Launch Vehicle
RTG: Radioisotope Thermoelectric Generator
RW: Reaction Wheel
SAR: Synthetic Aperture Radar
SCADA: Supervisory Control and Data Acquisition
SCOE: Special Check-Out Equipment
SDI: Stategic Defense Initiative
SEP: Spherical Error Probability
SEU: Single Event Upset
SGLS: US DoD's Space Ground Link Subsystem
SIGINT: Signals Intelligence
SL: Sea Level
SLBM: Submarine-Launched Ballistic Missile
SNG: Satellite News Gathering
SOS: Sapphire On Silicon
SRB: Solid Rocket Booster
SSC: Swedish Space Corporation
SSPA: Solid-State Power Amplifier
SSR: Solid-State Recorder
STDN: NASA's Spacecraft Tracking and Data Network
STS: Space Transportation System
SW: Space Wing
SWIR: Short Wave Infra-Red
t: metric tonne (1,000 kg; 0.984 ton; 1.102 US ton)
TC: Telecommand
TCR: Telemetry, Command and Ranging
TDRS: Tracking and Data Relay Satellite
TT&C: Tracking, Telemetry and Command
TVC: Thrust Vector Control
TVRO: TV Receive Only
TWTA: Travelling Wave Tube Amplifier
USN: US Navy
UT: Universal Time (also GMT)
UTM: Universal Transverse Mercator grid projection
VHSIC: Very High-Speed Integrated Circuit
VLBI: Very Long Baseline Interferometry
VNIR: Very Near Infra-Red
VTIR: Visible/Thermal Infra-Red
W: Watt
WWW: World Wide Web
XS: multispectral

COMMERCIAL SPACE SERVICES

Launch Facilities
Satellite Operators

LAUNCH FACILITIES

International

Odyssey Launch Platform (Sea Launch)

Location: 0°/154°W

Current Status

The Sea Launch consortium, the only launch operator to launch large GTO satellites from a floating platform, launched four successful missions in 2005 and has completed four (as of August) of its six launches booked for 2006. In 2005, the group booked the first mission for its Land Launch branded service from Baikonur, due to begin launch services in 2007.

For further information on the Sea Launch Zenit-3SL launch vehicle system, please see the Zenit-3SL (Sea Launch) entry in the Launch Vehicles, International Section.

Background

Studies of a sea-based launching system, designed to launch from a point on the Equator for maximum performance, started in Russia at the end of 1991. The first meeting between NPO Energia and Boeing took place in March 1993, and three partners – Boeing, Energia and Norway's Kvaerner shipbuilding – reached a framing agreement in November 1993. The Ukraine's NPO Yuzhnoye and PO Yuzhmash, designer and manufacturer respectively of the Zenit rocket, joined discussions in March 1994 and a Joint Venture (JV) was established in April 1995. Sea Launch is a limited liability company with headquarters in Long Beach, California. The original partners remain the same, although Kvaerner is now part of Aker ASA. Hughes signed the first order for the service in December 1995.

In December 1996, the command ship Sea Launch Commander was launched in Scotland, and in the following May the launch platform, named Odyssey, sailed from Norway to Russia. In June 1998, with modifications complete, the two marine platforms set sail from St Petersburg and Vyborg to the Sea Launch home port of Long Beach, California. The first launch took place on 27 March 1999.

The Sea Launch system includes two marine craft. The Assembly and Command Ship (ACS) Sea Launch Commander acts as an assembly facility in port at Long Beach and provides mission control facilities at sea. The 201 m long, 30,830 ton ship, based on a roll-on, roll-off design, was built in Scotland and modified in Russia. It features an internal vehicle assembly bay, a rear loading and transfer door and extending scissor ramp, and accommodates 40 ship crew and up to 200 rocket and payload operators, customer representatives and VIPs invited to view the launch.

The Launch Platform (LP) Odyssey was modified from a semi-submersible, self-propelled North Sea oil drilling platform. It was refurbished at the Rosenberg Shipyard in Stavanger, Norway, then transferred to Russia for the installation of launch equipment. With a submerged (transit) displacement of 50,600 tons, it is one of the largest platforms of its type. It has a diesel-electric propulsion system with generators on the main deck and motors in its submerged pontoons.

Aerial view of the Sea Launch Odyssey Platform with the Assembly and Command Ship in the background (Yuzhnoye Design Office) 1182775

Launch equipment on the LP is an almost exact replica of the launch equipment installed at Baikonur. The main deck is fitted with railroad tracks, supporting the rail-mounted transporter-erector vehicle. (The tracks are the only Russian-gauge tracks in the US). During transit, the rocket is housed in an environmentally protected hangar. The tracks extend to a launch pad, also based on the Baikonur design. It has an exhaust deflector below the rocket, which is cooled by fresh water. Umbilical connectors retract behind steel protective doors as the motor ignites. The LP also carries the fuel supply for the rocket.

The launch campaign begins with the delivery of inert components to Long Beach. The spacecraft is tested and sealed in its fairing in an on-land facility. The rocket is assembled aboard the ACS. The ACS then docks with the LP, and the rocket is rolled on to the ramp. Cranes on the LP then lift the rocket 45 m upwards, move it aft and lower it on to the tracks. The complete system then sails to the launch site, around 370 km from Kiritimati (Christmas Island).

The LP is unmanned at launch and all operations are automatic, controlled by redundant radio link from the ACS. (The control room on the ACS is divided into two sections, one for payload crew and one for the Russian launch crew). The last 5 h of the countdown can be completed without human intervention. In preparation for launch, the platform is ballasted to its lowest depth. The rocket is moved out of the hangar and erected on the pad as propellant chill-down begins. The fuel and LOX are loaded 2.5 h before launch and the transporter-erector arm is lowered at launch minus 17 min and the TEL is rolled back into the protected hangar.

The launch pad and other equipment on the LP are designed and protected so that they can be reused without any refurbishment. The original plan was to perform up to three launches per voyage. The ACS has space to accommodate components for three vehicles, and the ACS-to-LP docking mechanism is designed to transfer completed rockets at sea. However, due to low demand, this capability has not been exercised.

The Sea Launch system was successfully demonstrated on 27 March 1999, placing an instrumented demonstration spacecraft into GTO. By the end of 2005, Sea Launch had established a routine pace of operations of around six flights per year. The only satellite loss to date has been ICO F-1, on the second commercial launch attempt on 12 March 2000.

In October 2003, Sea launch announced plans to offer commercial launches from Baikonur under the Land Launch brand name, using both two-stage and three-stage versions (Zenit-3SLB) of the Zenit launcher. PanAmSat signed a contract in July 2005 covering the first commercial Land Launch mission, plus several options. At that time, the contract called for a Zenit-3SLB to launch the PAS-11 satellite to GTO by the end of the second quarter of 2007. Israel Aircraft Industries has also signed up for Land Launch, with a contract to orbit the Amos-3 communications satellite in the last quarter of 2007.

United States

California Spaceport

Location: 34.4°N/120.35°W

Current Status

California Spaceport serves commercial and government customers. In September 2002, SSI signed the interface control document (ICD) for National Reconnaissance Office (NRO) mission NRO L-22. The launch was a success and an NRO satellite was carried into space aboard a Delta IV on 28 June 2006 at 03:33 GMT from Space Launch Complex 6 (SLC-6). SSI also provides space vehicle processing, including Delta IV processing for Vandenberg launches, to NRO and is contracted through 2011.

Background

California Spaceport is operated by Spaceport Systems International (SSI), a limited partnership between ITT Federal Service Corporation and California Commercial Spaceport Incorporated (CCSI). The Spaceport is located on Vandenberg Air Force Base (see separate entry) and was the first federally licensed commercial launch site operator in the US. California Spaceport's first licence, since renewed, was issued by the Office of Commercial Space Transportation on 19 September 1996. SSI leases a small piece of land that includes the USD300 million SLC-6 Payload Preparation Room originally built for the Space Shuttle and renovated by SSI into the Integrated Processing Facility (IPF) as well as the Commercial Launch Facility (CLF), or SLC-8, located just south of SLC-6.

California Spaceport's original 25-year lease was signed by the Western Commercial Space Center (WCSC) and the US Air Force (USAF). WCSC merged in 2001, with the California Space and Technology Alliance (CSTA), formed in 1996, to become the California Space Authority (CSA). CSA, a non-profit organisation, was appointed in 2003 as California's Spaceport Authority, a responsibility formerly held by CSTA. The Spaceport's annual fee is USD70,000 plus about USD10,000 monthly for operations costs and utilities. The US government provided USD10 million, primarily as Air Force grant money. ITT committed USD30 million. The State of California also provides grant funds to the facility.

Most vehicles depart due south on an 180° azimuth, and do not overfly population centres. 168 to 220° azimuths are available without impacting the local community or off-shore oil rigs. A wide range between 150–270° launches are possible.

Kodiak Launch Complex (KLC)

Location: 57°26'09"N/152°20'16"W

Current Status

Kodiak Launch Complex (KLC) is both a sub-orbital and orbital launch site. KLC's only successful orbital launch so far, took place in September 2001. The Kodiak Star mission launched four satellites, Starshine 3; Picosat; PCsat; and Sapphire, into LEO aboard a Lockheed Athena I rocket.

The Alaska Aerospace Development Corporation (AADC) and the US Missile Defense Agency (MDA) signed a five-year launch support contract in 2003. Three launches have taken place between December 2004 and February 2006 under this contract. As of August 2006, KLC had conducted a total of nine launches.

Background

The Alaska Aerospace Development Corporation (AADC) was established in July 1991 by Alaska's State Legislature to promote development of the aerospace industry in Alaska. As the spaceport authority for the State of Alaska, it constructed the commercial USD18 million to USD20 million KLC on Narrow Cape, Kodiak Island. KLC is situated on 15 km^2 of state-owned land that has been provided free for 30 years. The site is about 45 miles south of Kodiak and 250 miles south of Anchorage. KLC's construction was completed in 2000. It accepts commercial, civil and military launch customers.

KLC has two all-weather launch pads and can accommodate Castor 120 launched payloads into LEO ±40° polar and Molniya orbits, taking advantage of the wide launch azimuth over the Pacific Ocean. Indoor facilities include the Launch Control Centre, the Payload Processing Facility and the Integration Processing Facility. The Launch Service Structure (LSS) services pad 1, which can accommodate orbital launches, and the Spacecraft and Assemblies Transfer Facility services the second launch pad, used for sub-orbital flights. Both pads are enclosed.

Launch pad 1 can accommodate a variety of launch vehicles, including the Minuteman series, Taurus and Athena; it's flame trench can handle up to about 544,000 kg of rocket thrust.

The first launch from KLC took place on 5 November 1998 when the US Air Force launched the AIT-1 atmospheric intercepter test rocket. AIT consisted of two solid rocket notes: the Thiokol Castor IV with a flexseal as first stage and the M57 as second stage.

KLC users can find accommodation in the Narrow Cape Lodge, which is exclusively contracted to serve the site.

Mid-Atlantic Regional Spaceport (MARS)

Location: 37.8°N/75.5°W

Current Status

Two commercial launches have taken place from the Mid-Atlantic Regional Spaceport (MARS). On both occasions, Minotaur I launch vehicles lifted off successfully, carrying US Air Force payloads. The first, on 16 December 2006, delivered TacSat-2 into orbit, while the second, on 24 April 2007, inserted the Near Field Infrared Experiment (NFIRE). The December 2006 launch was the Minotaur's maiden flight from MARS, as well as the first orbital launch from Wallops to take place in a number of years.

The Minotaur is a Minuteman ICMB derivative rocket; its first two stages are taken from the decommissioned missiles, and its third and fourth stages are Pegasus and Taurus boosters. Orbital Sciences manufactures and assembles the Minotaur. Launch cost is roughly USD20 million. The next Minotaur to launch from MARS is scheduled for December 2007; it will carry TacSat-3.

Background

Formerly the Virginia Space Flight Center, the Mid-Atlantic Regional Spaceport (MARS) is located on the Southern point of NASA's Wallops Flight Facility (see separate entry), on Wallops Island, off the Virginia coast. The facility is about 145 km north of Norfolk, Virginia. The Main Base offices and launch pads are located on separate segments of land. MARS operates from Wallops launch pads 0-A and 0-B; it leases the launch pads, as well as the range control center, clean room, assembly and payload processing buildings from NASA, under a 30-year agreement, originating in 1997. Launch pad 0-B can support launches carrying up to 4,500 kg into LEO.

Since 2003, both Virginia and Maryland have partnered on the spaceport project. Funding to found the commercial spaceport was raised from federal, state and private sources. Management work and fundraising was initiated by the Virginia Commercial Space Flight Authority (VCSFA), which remains Virginia's spaceport authority.

Mojave Air and Spaceport

Location: 35°03'N/118°09' W

Current Status

Mojave Air and Spaceport hosted all three SpaceShipOne (see separate entry) test and Ansari X Prize-winning flights from June through October 2004. The reusable SpaceShipOne was the first private spacecraft to achieve an 100 km altitude twice in two weeks, and was also the first private craft to exceed Mach 2 and Mach 3. Previous to that time, Scaled Composites, SpaceShipOne's developer, was a Mojave Air and Spaceport tenant, using the facility for test purposes. Scaled Composites, as well as XCOR Aerospace, the EZ-Rocket developer, continue to use the Spaceport as a test base. The Spaceship Company, Virgin Galactic, Scaled Composites, and Mojave Aerospace Ventures will conduct numerous flight tests of SpaceShipTwo and its carrier aircraft, White Knight, from the Mojave Spaceport, prior to SpaceShipTwo carrying its first paying passengers in 2009, most likely from Spaceport America (see separate entry).

Background

The 3,300 acre Mojave Air and Spaceport, near the town of Mojave, in Southern California, is about 10 km northwest of Edwards Air Force Base and about 130 km northeast of Los Angeles. From 1942–1961, the Air and Spaceport was owned and operated by both the US Marine Corps and US Navy as an auxiliary air station. Kern County authorities then obtained the airfield from the military, and the site has remained a County-operated commercial airport since that time. In June 2004, Mojave was the first site to be granted commercial spaceport status by the Federal Aviation Administration (FAA) licensing authority.

BAE Systems and Orbital Sciences also conduct operations out of Mojave. Mojave Air and Spaceport is linked to Southern California and beyond by rail, as well as highway infrastructure.

Spaceport America (Southwest Regional Spaceport, New Mexico)

Current Status

On 25 September 2006, UP Aerospace conducted Spaceport America's inaugural launch. A SpaceLoft XL sounding rocket carrying 50 experiments from institutions worldwide lifted off at 14:14 local time. The flight was postponed three times before the 25 September launch date was set. The launch was originally scheduled to occur at 07:30, and so ran nearly seven hours late. Once launched, the rocket failed to reach its planned sub-orbital apogee of about 112 km, and rose to only about 12 km. The rocket then crashed outside its planned drop area. Further launches from Spaceport America's temporary pad are planned for site, the next is scheduled for October 2006.

New Mexico's government is hoping that Spaceport America's business will grow over several years into a multi-million dollar industry for the State. Futron Corporation and New Mexico State University studies have projected that the Spaceport will possibly generate up to USD750 million in revenue by 2020 and USD300 million in payroll generation by around 2010 through the Spaceport's construction period.

Background

Dubbed the "world's first purpose-built commercial spaceport," with the goal to provide direct access to space for the commercial sector, Spaceport America is located near Upham, New Mexico, about 70 km north of Las Cruces and just under 50 km east of Truth or Consequences. The site covers about 60 km^2 of State Trust Lands granted by New Mexico's Commissioner of Public Lands.

The New Mexico Economic Development Department played a large part in the Spaceport's initial concept, land acquisition and development, and is a crucial Spaceport partner, with its Cabinet Secretary acting as the Chairperson of New Mexico's Spaceport Authority. In December 2005, Virgin Galactic signed an agreement with the Spaceport Authority to operate future sub-orbital flights out of Spaceport America as well as locating its headquarters and mission control centre there.

The New Mexico Economic Development Department estimates that the Spaceport will cost USD225 million to build. In January 2006, New Mexico committed USD100 million to the project over the next three years, as well as USD10 million to cover initial planning costs. The Spaceport Authority plans to complete Spaceport construction by 2010.

In 2005, Spaceport America was named home to the annual X Prize Cup competition, the next of which is scheduled for 17–21 October 2006. Other companies that have expressed interest in operating out of Spaceport America are Starchaser and Rocket Racing League.

Spaceport America was first on record as the Southwest Regional Spaceport. The name change to Spaceport America was announced in July 2006 at the Farnborough International Air Show. In June 2006, New Mexico selected DMJM to lead the architecture and engineering of the Spaceport project. New Mexico-based design consultants Dekker Perich Sabatini, and Molzen-Corbin & Associates were also selected. Spaceport Design and construction is to take place in two Phases and be fully completed by 2010.

The New Mexico Spaceport Authority expects to be granted a commercial launch site operator licence by the Federal Aviation Administration (FAA) by sometime in 2007.

West Texas Spaceport

Location: 31°N/105°W

Current Status

Blue Origin conducted its first sub-orbital test launch on 13 November 2006, from its purpose built spaceport in a remote area of West Texas. The spacecraft, called Goddard, is part of Blue Origin's ambitions to develop New Shepard, a sub-orbital spacecraft that will provide transportation to the edge of space for those wishing to experience weightlessness and a view of Earth. There are few details available regarding either the spaceport or Blue Origin spacecraft. The company has been relatively secretive regarding its aerospace engineering and infrastructure. Goddard is a conical, capsule type, vertical launch and land spacecraft.

Background

Blue Origin founder and Amazon.com billionaire, Jeff Bezos purchased the 165,000 acre Corn Ranch, about 32 km north of Van Horn, off Highway 54 in West Texas, with the intention of building the Spaceport to support Blue Origin launches. Construction at the site began in 2006, six years after Blue Origin was established in Kent, Washington, near Seattle. The Spaceport facility remains basic, however, Blue Origin plans to add to the infrastructure as New Shepard spacecraft development progresses.

The Texas state government is also considering the establishment of spaceport infrastructure in Willacy and Brazos Counties: both these locations are situated on the East Texas coastline, which would enable launches over the sea.

SATELLITE OPERATORS

Australia

Optus Networks Pty Ltd

Current Status
On 14 October 2006, the Optus D1 communications satellite was launched successfully from Centre Spatial Guyanais (CSG), aboard an Ariane 5. Optus D1 was built by Orbital Sciences and has a 15-year lifespan. The satellite is operated by Optus' Sydney ground station. The Optus B1 satellite will be retired as a result of Optus D1's successful operation. Optus D2 is scheduled for launch from CSG in 2007.

The only Optus C series satellite, C1, was launched in June 2003 and is operated equally by Optus and the Australian Ministry of Defence. At the time, it was the largest hybrid commercial/military communications satellite launched. Built by Space Systems/Loral, it has a design life of 15 years.

Background
Optus is a satellite communications service provider that supplies Australians and New Zealanders with mobile and terrestrial telecommunications services, including internet and television subscription services. The company operates the largest fleet of Australian satellites, with five in orbit as of 2006. In 2001, Optus became a wholly owned subsidiary of SingTel, the Singapore based communications corporation. Optus owns several of its own branded subsidiaries, including Virgin Mobile Australia.

Optus was formed in 1992, after changes were made to the Australian telecommunications industry. The national carrier, Aussat, when privatised, was transformed into Optus, and Optus then became the second carrier to Telstra (see separate entry). Aussat, Australia's national satellite operator, was first established by the government in 1981. The company was legislated under the 1984 Aussat Act and was 75 per cent owned by the Commonwealth of Australia, with the remaining portion allocated to the Australian Telecommunications Commission (Telecom Australia).

The Australian government announced its plans, in November 1990, to de-regulate the telecommunications industry. It sold Aussat to establish a second carrier to compete with the merged Overseas Telecommunications Commission (OTC)/Telecom Australia: Telstra. It immediately removed Telecom Australia's 25 per cent and lifted all legislative restrictions imposed upon Aussat's operation. The successful bidder was announced 19 November 1991: Optus Communications Pty Ltd, at the time, a 51 per cent Australian-owned company, with BellSouth of Atlanta and Cable and Wireless of the UK holding the remainder. Optus acquired Aussat for AUD800 million. The government paid out Aussat's debt and terminated the leverage lease financing arrangements for the satellites at a total cost of about AUD740 million. The Optus ownership arrangements came into operation 31 January 1992.

Brazil

Empresa Brasileira de Telecomunicações (Embratel)

Current Status
Star One, Embratel's subsidiary and satellite operator, anticipated the launch of its Star One C1 (also called Simon Bolivar F1) satellite in 2006 to expand its market throughout South America. However, the satellite's flight, contracted with Arianespace to occur from Guiana Space Center, has been repeatedly delayed since 2004. It may launch in 2007 or 2008. Alcatel Alenia Space built the spacecraft, which is to replace Brasilsat B1 in orbit. Alcatel contracted again with Star One to build Star One C2, which may launch in 2007, to replace Brasilsat B2. However, noting the delays that have already dogged C1, C2 may wait for a firm launch date as well.

Background
Empresa Brasileira de Telecomunicações (Embratel) was created in September 1965 as the operational arm of the Ministry of Communication's Telecomunicações Brasileiras (Telebras) component, and was responsible for the domestic Brazilian communications network and links through the Intelsat and Inmarsat systems (see separate entries). Embratel was privatised with the Telebras telecommunications monopoly split in July 1998. Embratel Participações (Embrapar) is Embratel's parent company, owning a 99 per cent stake in Embratel. In 1998, the US telecommunications corporation, WorldCom/MCI purchased a 19.26 per cent controlling share of Embrapar. In 2004, the bankrupt MCI sold its Embrapar interests to Teléfonos de México (Telmex). In 2006, Telmex purchased a large portion of Embrapar common and preferred stock and as of early 2007 owned 96.4 per cent of Embrapar. It had made a similar move in 2004, that time achieving 72.3 per cent ownership. At the end of 2005, Embratel and its subsidiaries employed 13,888 staff.

Star One SA
Star One manages and operates Embratel's satellites. Founded in 2000 out of Embratel's satellite department, it is 80 per cent owned by Embratel; SES Global (see separate entry) owns 20 percent.

Star One operates Brasilsat B1, B2, B3 and B4, the last of which was launched in August 2000. Brasilsat B1 and B2 were launched in 1994 and 1995 respectively; their need of replacement by C1 and C2 is based upon the satellites' 12-year lifespans.

BolivarSat was formed by the South American Andesat consortium and Alcatel in 1999; Alcatel later sold its shares to Star One. BolivarSat will operate Star One C1, once it is launched. As of December 2005, Star One employed 195 staff. The company also leases capacity on AMC-12, sometimes known as Star One 12.

The satellites are controlled from Star One's Ground Control Center, located in Guaratiba near Rio de Janeiro, comprising the Spacial Segment Control Center and the Communications Operation and Control Center. There is a backup station in Tanguá.

Canada

Telesat Canada

Current Status
In 2006, Loral Space & Communications Inc and the Public Sector Pension Investment Board (PSP Investments) formed an acquisition company to purchase Telesat from BCE for CAD3.42 billion. The acquisition was announced by the organisations in December 2006. Telesat will retain its name and Ottawa headquarters location. At the acquisition's start, Loral will hold a 64 per cent interest and PSP a 36 per cent interest and majority voting control in Telesat. If the deal is successfully completed, Loral Skynet, Loral Space & Communications' satellite operator, will merge with Telesat, creating the world's fourth largest geostationary communications satellite operator.

As of January 2007, Telesat operated a seven satellite fleet, including two Nimiq satellites leased from DirecTV. Telesat anticipates the launch of its Anik F3 satellite in early 2007 and its Nimiq 4 in 2008. The Nimiq 5 contract was awarded to Space Systems/Loral; it will be launched in 2009.

Background
A private company, incorporated by an act of Parliament in 1969, Telesat was originally 53.7 per cent owned by the Canadian government and the remainder by a consortium of common carriers and Canadian National-Canadian Pacific (CNCP). In 1992, the government sold its shares to Alouette Telecommunications Inc, a consortium of Canadian telephone companies, and Spar Aerospace Ltd. In 1998, Bell Canada Enterprises Inc (BCE) increased its majority stake in Telesat to 100 per cent by acquiring all outstanding common shares in Alouette Telecommunications Inc. BCE will remain Telesat's sole owner, until the Telesat acquisition, announced in December 2006, is complete.

Telesat Communications Services, Telesat's US subsidiary, was founded in 2000, following Telesat's US Federal Communication Commission (FCC) listing of Aniks E1 and E2, a move that allowed Telesat's Canadian-owned satellites to service US customers. In 2001, Telesat acquired Infosat Communications, also a BCE-owned company, and consolidated all BCE satellite operations and services under Telesat. In 2005, Telesat acquired the US company, The SpaceConnection, also a satellite services provider. The company's Telesat Brasil Limitada provides satellite services to South American customers. Telesat owns 3484203 Canada Inc, the company that owns 100 per cent of TMI Communications and Company, Limited Partnership's (TMI) limited partner units. TMI, in turn, owns an interest in Mobile Satellite Ventures (MSV).

Telesat reported revenues of CAD475 million at the end of 2005. Telesat's broadcast services account for 44 per cent of its revenues, with its business network contributing 26 per cent, its subsidiaries contributing 19 per cent, and consulting and carrier businesses contributing six per cent of revenues each.

Telesat currently operates seven satellites. Aniks F1, F2, F1R, and Nimiqs 1, 2, 3 and 4i. Nimiq 3 and Nimiq 4i are DirecTV satellites, leased and operated by Telesat and positioned in Telesat's orbital slots. The Anik FSS satellites are used for broadcast and business telecommunications services. Telesat's Nimiq Direct Broadcast Satellites (DBS) are used in the delivery of Direct To Home (DTH) services. Telesat also operates eight satellites on behalf of several customers, including the MSAT1 and MSAT2 mobile communications satellites for MSV, and the Rock and Roll satellites for XM Satellite Radio.

Egypt

Nilesat

Current Status
Nilesat is a relatively new, and thriving satellite communications company; its satellites provide over five million North African and Middle Eastern homes with TV and radio broadcasting. It has increased its number of channels steadily over the years, now offering over 350, including the US Broadcasting Board of Governors' owned Middle East Broadcasting Networks' (MBN) Alhurra and Alhurra-Iraq channels. Additionally, Nilesat, and Arabsat, transmit the formerly Iraqi and now Syrian-based al-Zawraa channel, which airs material that, according to some, broadcasts insurgent propaganda. As of early 2007, the US and Iraqi governments have requested that the Egyptian government's Nilesat cease carrying the al-Zawraa channel. Egypt has refused, citing commercial reasons.

Background
Nilesat, headquartered in Cairo, was founded in 1996, and is the first Egyptian-owned satellite operator. The Egyptian Radio & Television Union (ERTU), an arm of Egypt's Ministry of Information, owns a 40 per cent interest in Nilesat; 14 per cent of Nilesat's shares are publicly owned. The Arabian Organization of Industrialization (AOI) and the Egyptian Company for Investment Projects

(ECIP) own 10 and nine per cent respectively; the remainder of shares are held by Egyptian financial institutions and other investors.

Nilesat operates two satellites, Nilesats 101 and 102, launched in 1998 and 2000 respectively. The satellites provide Direct To Home (DTH) broadcasting and data services to North African and Middle Eastern customers. The company offers over 350 Egyptian, Arabic and international TV and radio channels. Nilesat leases capacity on Nilesat 103, also known as Eutelsat's (see separate entry) Atlantic Bird 4, formerly Hot Bird 4.

Nilesat operates two ground stations, a primary station in 6th of October City, near Cairo, and a secondary station in Alexandria. EADS Astrium built both ground stations as well as Nilesats 101 and 102. Nilesat 101 is Egypt's first satellite.

France

Centre National d'Études Spatiales (CNES)

Current Status

Centre National d'Études Spatiales (CNES) and its partners, along with Spot Image, are planning the Pleiades (see separate entry) microsatellite series, which CNES expects to launch beginning in 2008; Pleiades will eventually replace the Spot Earth imaging series.

Background

CNES and their subsidiary, Spot Image, operate the Spot satellite series and archive imagery collected by the satellites, which currently include Spot 2, Spot 4 and Spot 5. For further information on CNES, please see the separate CNES entry in the National Agencies section.

Spot 1 was launched in February 1986, followed by Spot 2 in January 1990, Spot 3 in September 1993, and Spot 4 in March 1998. Spot 3 ceased functioning in 1997. Spot 5 was launched by Ariane 4 on 3 May 2002 and began operations on 12 July 2002. As of early 2007, Spot 5 data was responsible for attracting much of Spot Image's revenues, due to its higher resolution images compared to the others in the series.

Spot Image SA

Spot Image, with CNES and Matra as the majority shareholders, was established in 1982 to distribute Earth remote sensing data products from the Spot series satellites. As of early 2007, CNES; EADS Astrium; the Swedish Space Corporation (SSC); Alcatel; Institut Géographique National (IGN); and the state of Belgium were the principal shareholders in the company, with CNES and EADS holding 41 and 40 per cent respectively.

Spot Image also markets and distributes the Taiwanese National Space Organization's (NSPO) Formosat-2 and the Korean Aerospace Research Institute's (KARI), Korean Multi-purpose Satellite (Kompsat-2) imagery, as well as distributing imagery from numerous other Earth imaging spacecraft such as the high resolution Quickbird, Eros and Ikonos satellites. The company's Main Receiving Stations are in Toulouse, France and Kiruna, Sweden. Clients can also arrange to licence Direct Receiving Stations (DRS), to download images directly from the satellite, circumventing the need to route data through Toulouse or Kiruna, however DRS can only receive imagery within their 2,500 km radius areas. Spot Image currently provides multi-resolution imagery from 2.5 to 20 m from the Spot series and 1 and 2 m resolution imagery from Kompsat-2 and Formosat-2, respectively. Spot images are archived at the Spot Image Archiving and Processing Centre (CAP); between 1986 and 2003, CAP had amassed about 130,000 Giga bytes of data. The company has archived more than 10 million images.

Spot Image is headquartered in Toulouse, France; the company also has five subsidiaries located in Australia, China, Japan, Singapore and the US, as well as offices in Brazil and Mexico. The company's customer base includes government clients such as disaster relief organisations and the defence and security communities; Spot Image products also have forestry, agriculture and maritime applications.

Indonesia

PT Indonesian Satellite Corporation Tbk (Indosat)

Current Status

Indosat announced in January 2007, that it will launch the first in the Palapa D series by 2009. Palapa D will replace Palapa C2, which is nearing the end of its design life. The satellite cost is projected at about USD200 million.

As of June 2002, PT Satelit Palapa Indonesia (PT Satelindo) was a wholly owned entity within PT Indosat Tbk, with Indosat having purchased for USD325 million, the remaining 25 per cent of Satelindo once held by Deutsche Telekom. In 2001, Indosat had paid USD111.6 million for 61.5 per cent of the company, raising its stake to 75 per cent. Indosat, previously a state-owned telecommunications organisation, became Indonesia's second largest telecommunications provider, upon the merger of Satelindo and two other telecommunications companies into Indosat. The privatisation of the Indonesian telecommunications industry was precipitated in 2000 when Telecommunications Law number 36 was passed, and the government disentangled itself from the Indonesian telecom market. The government in 2002, sold its remaining 42 per cent stake in Indosat to Singapore Technologies Telemedia Pte (STT). The largest telecommunications organisations then restructured to become more competitive. Previous to these activities, PT Satelindo was Indonesia's largest telecommunications satellite operator.

Background

Indosat was created in 1967 to provide international telecommunications services. In 1980 the Indonesian government took over the company and it became a state-owned enterprise. In 1994, Indosat was first listed on the local Jakarta and Surabaya Stock Exchanges as well as the New York Stock Exchange. After privatisation, Indosat focused its efforts on becoming Indonesia's primary cellular telecommunications services provider.

PT Satelit Palapa Indonesia (PT Satelindo) was created on 29 January 1993 to finance and operate the new USD240 million Palapa C1 and C2 communications satellites. PT Pasifik Satelit Nusantara (see separate entry) now operates the Palapa C1 satellite, and Satelindo operates the Palapa C2 satellite.

Satelindo shareholders at the company's founding were PT Bimagraha Telekomindo (60 per cent), PT Telkom (30 per cent) and PT Indosat (10 per cent), until the DeTeMobil subsidiary of Deutsche Telekom agreed in March 1995 to pay USD566 million for a 25 per cent interest. PT Satelindo also took over the Palapa Bs from PT Telkom before they were retired.

PT Pasifik Satelit Nusantara (PSN)

Current Status

PT Pasifik Satelit Nusantara (PSN) encountered financial problems beginning with the Indonesian finance crisis of 1998. Unable to make a profit, the company went through a debt restructure process and the Indonesian Bank Restructuring Agency (IBRA) sold PSN for USD205 million in January 2002, along with its Asian Cellular Satellite System (ACeS) joint venture. Once a publicly listed company, PSN was delisted from the NASDAQ in 2001 following its financial difficulties. Several mobile satellite service providers experienced great difficulties during this period, as terrestrial cellular services infrastructure grew and became increasingly popular.

Background

Founded in 1991, PT Pasifik Satelit Nusantara (PSN) was Indonesia's first private telecommunications satellite company. PSN is headquartered in Bekasi, Java and provides mobile and fixed voice and data satellite services to the Asia-Pacific region via its Byru (Satellite GSM), Pasti (Portable Fixed Satellite), and Bina (Integrated Data Communication Network) products. Previous to its debt restructure, the company's major shareholders included PT Telekomunikasi Indonesia (Telkom), Hughes Space and Communications (the Boeing Satellite Development Center since 2000) and Telesat Canada.

PSN led the development of the USD650 million Asian Cellular Satellite System (ACes) joint venture, with the Philippines Long Distance Telephone Co (PLDT) and Jasmine International Public Company Ltd of Thailand. An agreement was also signed with Lockheed Martin in May 1995 for exclusive negotiations; the contract was announced 6 July 1995. The Garuda 1 satellite, launched aboard a Proton on 12 February 2000 from Baikonur, can handle 10,000 simultaneous calls from the 2 million subscribers via the 150-beam phased array antenna. Subsequent problems with the satellite's L-band system reduced operating capability and PSN recovered some finances by winning an insurance claim related to the satellite's malfunction. PSN also operates the Aguila-II and Palapa C1 satellites.

International

Eutelsat Communications

Current Status

Eutelsat anticipates the launch of Hot Bird™ 9 and W2M in 2008, as well as Hot Bird™ 10 and W2A in 2009. Hot Bird™ 9 and W2M will initially provide HDTV broadcast, data networks and broadband services to the Middle East, North Africa and Europe. W2A will provide telecommunications, broadband and TV services to Middle East, North Africa and Europe, while Hot Bird™ 10 will provide redundancy for the Hot Bird™ 13° E neighbourhood, including Hot Birds™ 2, 6, 7A and 8. Hot Bird™ 7A was launched on 11 March 2006 and Hot Bird™ 8 was launched on 5 August 2006.

Eutelsat currently operates a fleet of 23 geostationary satellites, 19 of which are fully owned, including four Hot Birds™; six W series; four Eurobirds™ (some of which were formerly Hot Birds™); four Atlantic Birds™ and the SESAT 1. Eutelsat also leases capacity on Telstar 12; Express A3; Telecom 2D and SESAT 2/EXPRESS AM 22. Eutelsat is the largest European commercial satellite operator, and ranks at number three worldwide.

Background

On 2 July 2001, Eutelsat completed corporate restructuring to become a private company under French law: Eutelsat SA. In April 2005, Eutelsat shareholders created a new corporate entity: Eutelsat Communications. Eutelsat Communications is Eutelsat SA's primary shareholder, owning 95.2 per cent. As of FY 2005–2006, the organisation had 490 staff members and a revenue of EUR791.1 million; it is headquartered in Paris.

The European Telecommunications Satellite Organization (Eutelsat) was founded in 1977 as an intergovernmental entity established to operate Europe's regional satellite system for 47 member states; it launched its first communications satellite in 1983, and has launched 30 more since then. Eutelsat was the first European satellite operator to deliver television broadcast services direct to the domestic market, with coverage beginning in Western Europe, but expanding to Central and Eastern Europe by the end of the 1980s. Although four of its seven subsidiary offices are in Europe, Eutelsat now provides services in Asia and the Americas as well, and maintains offices in China, Brazil and the US.

Two thirds of Eutelsat business activities are dedicated to audiovisual services, including interactive TV, HDTV, and Satellite News Gatherers (SNG) delivery to cable headends, Digital Terrestrial Television (DTT) transmitters, and domestic satellite reception systems. The remainder of Eutelsat's services include broadband, data networks, telecommunications, and terrestrial and maritime mobile services. Eutelsat provides customer access to 2,100 TV stations and 970 radio stations.

ImageSat International NV

Current Status

ImageSat International successfully launched it's EROS B high resolution imaging satellite on 25 April 2006, from Svobodny Cosmodrome in Russia, aboard a Start-1 launch vehicle. EROS A was launched in 2000 from the same site, and remains operational as of 2007. EROS C is expected to launch around 2009. The Russian firm, ZAO Puskovie Uslugi acted as launch provider; the EROS B launch cost was reported to be about USD30 million.

Background

ImageSat International, based in the Netherlands Antilles, also maintains offices in Tel Aviv, Israel and Limassol, Cyprus. The company is majority owned by Israel Aerospace Industries (IAI); Elbit Systems Electro-Optics ElOp Ltd (El-Op), also based in Israel, owns a further 14 per cent. Core Software Technology in the US, and other investors, hold the remainder of interests in the joint venture. ImageSat has expressed interest in issuing an Initial Public Offering (IPO) in the future.

ImageSat was founded in 1997, in the Cayman Islands, as West Indian Space Ltd. It was renamed and re-domiciled in the Netherlands Antilles in June 2000, prior to the EROS A launch. Upon EROS A's successful launch, ImageSat became the first non-US commercial entity to offer high resolution satellite imagery to its customers; it was the second operational company of its kind.

ImageSat expects to grow and establish a constellation of its Earth Remote Observation Satellites (EROS) to serve its customers, who can purchase exclusive use of the satellite's imagery and independently control the satellite's camera over a specific area of interest, excluding ImageSat from their intelligence loop. The EROS series is the commercial version of Israel's Ofeq reconnaissance satellite series (see separate entries). IAI builds the spacecraft bus for both series, and El-Op develops and manufactures the optics for both satellite series. The company primarily markets its products to the defence and homeland security sectors, however EROS imagery also has civil applications such as mapping and environmental disaster response.

Intelsat Ltd

Current Status

Announced in August 2005, Intelsat's acquisition of PanAmSat in July 2006, created the world's largest commercial geostationary communications satellite operator. As of January 2007, Intelsat commanded a 51 satellite fleet. Intelsat plans to launch the Space Systems/Loral manufactured IA-9 in 2007 from Sea Launch's Pacific platform; Space Systems/Loral will also build Intelsat 14, set to replace IS-1R upon its launch in 2009.

In connection with the PanAmSat acquisition, Intelsat, as of 1 February 2007, has renamed 16 satellites in its fleet. Please see the table below for specific details.

On 23 June 2005, Intelsat launched IA-8 aboard a Zenit-3SL, using Sea Launch's Pacific launch services. In March 2004 Intelsat acquired four operational satellites from Loral giving immediate entry to the North American video and corporate data markets. On 15 June 2004 the second Intelsat 10 satellite was successfully launched by Proton from Baikonur.

Background

Established on 20 August 1964 as an intergovernmental organisation, the International Telecommunications Satellite Consortium (Intelsat) was a not-for-profit commercial cooperative of more than 140 member nations, prior to its July 2001 privatisation. In January 2005, Zeus Holdings, Ltd, now known as Intelsat Holdings, Ltd, a consortium of funds, acquired Intelsat Ltd. Domiciled in Bermuda, Intelsat maintains offices around the world. Its US headquarters is in Washington, DC and the corporation employs 1,300 staff worldwide. Intelsat claimed a pro forma USD2 billion in revenue as of 31 March 2006, previous to its final acquisition of PanAmSat.

Intelsat provides international telecommunications services to more than 200 countries, territories and dependencies, as well as directly to the domestic market, in the forms of telecommunications; data; Internet Protocol (IP); broadband; mobile; TV and HDTV services. More than 1,800 international financial networks, multinational corporations, international news services, governments, and TV and radio broadcasters rely on the system for day-to-day global communications. Intelsat customers operate their own Earth-based receiving stations.

Intelsat has a rich history; on 6 April 1965, it launched Early Bird, or Intelsat I, the first commercial communications satellite. On 20 July 1969 Intelsat satellites relayed globally, Neil Armstrong's first steps on the Moon. Beginning in 1974, Intelsat also enabled the White House – Kremlin 'Hot Line'.

Previous to its acquisition by Intelsat, PanAmSat operated a fleet of about 25 geostationary communications satellites. Like Intelsat, PanAmSat provided network, TV, HDTV and IP services. PanAmSat assets are being fully integrated into Intelsat's structure.

2007 Intelsat Satellite Name Changes

Launch Date	Old Name	New Name	New Acronym
		Galaxy Series:	
24 May 1997	Telstar 5/Intelsat America IA-5	Galaxy 25	G-25
15 February 1999	Telstar 6/IA-6	Galaxy 26	G-26
25 September 1999	Telstar 7/IA-7	Galaxy 27	G-27
7 August 2003	Echostar IX / Telstar 13/IA-13	Galaxy 23	G-23
23 June 2005	IA-8	Galaxy 28	G-28
		Intelsat Series:	
8 July 1994	PAS-2	Intelsat 2	IS-2
2 August 1995	PAS-4	Intelsat 4	IS-4
12 January 1996	PAS-3R	Intelsat 3R	IS-3R
28 August 1997	PAS-5/Arabsat 2C/Badr C	Intelsat 5/Arabsat 2C/Badr C	IS-5
16 September 1998	PAS-7	Intelsat 7	IS-7
4 November 1998	PAS-8	Intelsat 8	IS-8
22 December 1998	PAS-6B	Intelsat 6B	IS-6B
28 July 2000	PAS-9	Intelsat 9	IS-9
29 October 2000	Europe*Star 1/PAS-12	Intelsat 12 / Europe*Star 1	IS-12
16 November 2000	PAS-1R	Intelsat 1R	IS-1R
15 May 2001	PAS-10	Intelsat 10	IS-10

International Mobile Satellite Organization (Inmarsat)

Current Status

Inmarsat operates a commercial fleet of 10 geostationary telecommunications satellites. The Inmarsat constellation includes second, third and fourth generation Inmarsat satellites: three I-2 (F1, F2 and F4); five I-3 (F1-F5); and two I-4 (F1 and F2). I-2 F3 was retired in 2006. The first two satellites in the I-4 series were launched in March and November 2005; Inmarsat plans to launch the third and final I-4 in late 2007, to cover services in the Pacific region. It will likely use the services of Sea Launch (see separate entry) or International Launch Services at Cape Canaveral. EADS Astrium built the I-4 series, which provide broadband voice and data services. The I-4 launches marked the beginning of Inmarsat's Broadband Global Area Network (BGAN) service. Inmarsat E Emergency Position-Indicating Radio Beacons (EPIRB) services ended on 1 December 2006. Inmarsat A analogue services will close toward the end of 2007, marking the full transition to Inmarsat's digital products.

Inmarsat is part of Europe's Galileo (see separate entry) satellite navigation system (see separate entry) joint venture along with seven other organisations that make up the Galileo Operating Company (GOC). Inmarsat is responsible for the oversight of the Galileo Operations Company (OpCo), including security and performance management. Galileo will eventually be comprised of a 30 satellite constellation.

Background

Headquartered in London, the International Mobile Satellite Organization (founded as the International Maritime Satellite Organization) was established as an intergovernmental organisation in July 1979. It was privatised in 1999, and is now a commercial company operating a satellite system to provide mobile telecommunications, broadband and data services to the shipping, aviation, offshore, and land mobile industries. At the end of 2005, Inmarsat had achieved USD491.1 million in revenue and a pre-tax profit of USD95.5 million.

When the 1999 privatisation process created Inmarsat plc, a separate, residual, intergovernmental organisation, also known as the International Mobile Satellite Organisation, but using the acronym 'IMSO' was formed. IMSO oversees Inmarsat's remaining civil responsibilities, namely Inmarsat's operation of the Global Maritime Distress and Safety System (GMDSS), the principal reason for Inmarsat's 1979 creation. While Inmarsat remains the sole GMDSS operator, IMSO anticipates that services will be expanded to include other companies and/or civil organisations in the future. IMSO has also determined that Inmarsat's privatisation has not affected the level of GMDSS service.

Inmarsat's Satellite Control Centre (SCC) and Network Operations Centre (NOC) are located at its London headquarters. Tracking, Telemetry and Control (TT&C) facilities are located in Fucino, Italy; Beijing, China; Lake Cowichan, Canada; and Pennant Point, Canada. Secondary stations are located in Eik, Norway and Auckland, New Zealand. 29 Land Earth Stations (LESs; also known as Coast Earth Stations in maritime circles and as Ground Earth Stations for aeronautical applications) located in 29 different nations link the satellites and international telecommunications networks. BGAN has a separate access route; communications travel through the BGAN Satellite Access Station (SAS) located in Fucino.

Inmarsat offers a wide variety of service systems to its maritime, aviation and land-based customers. Commercial and government products include BGAN; Global Area Network (GAN); and Mini-M services. Inmarsat offers Fleet F77; Fleet F55; Fleet F33; and Inmarsat B; C; Mini-C; D+; E+ M and Mini-M, as well as Crew Calling products to the maritime industry. Aviation products include Swift64 Mobile; Aero C; Aero H; Aero H+; Aero I; Aero L; and Aero M.

The Inmarsat B terminal is the updated digital equivalent of the analogue Inmarsat A, and continues to dominate the maritime market. Inmarsat B supports voice, data, fax, telex and distress and safety services. Inmarsat's large product range supports varying communications services and speeds, and includes Internet Protocol (IP), GPS, store and forward messaging, GMDSS and non-GMDSS compatible service units.

Until the 1990's Inmarsat 2 debut, the organisation operated with leased satellites. The second generation was followed by the five Inmarsat 3 series launched between 1996 and 1998, providing multiple spot beams. The Inmarsat 3 series was built by Lockheed Martin AstroSpace, now Lockheed Martin Missiles & Space, with European Matra Marconi Space, now Astrium, developing and building the communication payload. Please also see the separate Inmarsat 2, Inmarsat 3 and Inmarsat 4 entries.

Inmarsat has strenuously marketed new business opportunities including leased transponders for military communications, and open channels carrying unencrypted or uncoded signals embracing voice, data and TV.

Mobile Satellite Ventures (MSV)

Current Status

MSV announced in January 2006 that it had ordered three new large communications satellites from Boeing to deliver mobile satellite services across the Americas. The first two satellites are scheduled for launch in mid-2009 and 2010. For further technical information on MSV's current and future satellites, please see the separate MSV Satellite Series entry.

Background
Headquartered in Reston, Virginia, MSV was formed in 2001 by renaming the former American Mobile Satellite Corporation (AMSC). AMSC had been founded in the late 1980s to provide mobile communications services across North America, in collaboration with Canada's Telesat Mobile Inc (TMI). The two companies launched twin satellites, AMSC-1 and TMI's MSAT, in 1995 and 1996. Still in service in early 2007, the spacecraft provide L-band telephone, data, fax and Push-To-Talk (PTT) two-way radio service across North America via MSV and its joint-venture partner and subsidiary, MSV Canada. The satellites are expected to continue operations through 2009.

MSV is seeking to deliver hybrid satellite communications services to North America, integrating both terrestrial and space-based mobile communications technologies. These services will begin after the company launches two new replacement satellites beginning in 2009. MSV's current coverage includes North and Central America, northern South America, and the Caribbean and Hawaiian islands. As of early 2007, the company employed 120 personnel; MSV expects to hire more employees as it launches its hybrid services.

SkyTerra Communications owns the controlling stake in MSV, holding 100 per cent of MSV's general partner shares and about 83 per cent of its outstanding limited partnership interests. In January 2007, SkyTerra acquired Bell Canada Enterprises' (BCE) limited partnership interests in the company, exchanging 22.5 million non-voting common stock shares in the deal.

Telesat Canada (see separate entry), in 2003, loaned MSV USD1 million to secure and build ground station premises; Telesat also operates MSAT; an arrangement that may terminate upon the launch of two new MSV satellites in 2009 and 2010.

In 2005, TerreStar (see separate entries) spun off from MSV as a separate company. TerreStar will provide communications services in S-band rather than L-band.

SES Global

Current Status
Comprised primarily of SES Astra; SES Americom; and SES New Skies, SES Global operates a combined fleet of about 44 geostationary communications satellites. SES Global also owns interests in several other satellite operators and service providers. The corporation plans to launch 10 satellites by mid-2009; these will both replace some of the more aged of the fleet as well as adding capacity.

SES Astra expects to launch Astra 1L in early 2007, and successfully launched Astra 1KR in April 2006, replacing Astra 1B, which was retired. SES Americom launched AMC-18 on 8 December 2006 and anticipates the launch of AMC-21 in 2008.

In April 2005 Astra announced that it will replace Astra 1H with Astra 1M and provide in-orbit backup capacity for the Astra fleet at 19.2°E. Astra 1M will carry 32 active Ku-band transponders and is being procured ahead of scheduled replacement requirements to provide additional security to support Astra's growth at 28.2°E. This will be achieved by relieving Astra 2C so that it can be moved to that location from its present position at 19.2°E. This event is scheduled to occur after the launch of Astra 1KR and Astra 1L, and will provide added capacity into the UK and Irish markets.

Background
Société Européenne des Satellites (SES) was incorporated as Europe's first private satellite communications company in March 1985, to establish a medium-power satellite system for TV distribution from one of the geostationary slots assigned to the Grand Duchy of Luxembourg. The Duchy continues to retain a 5.8 per cent interest in SES Global.

In 2001, SES, operator of the Astra satellite series, and GE Capital, formed SES Global; it then became the parent company of several corporations: SES Astra, which traces its history back to 1985 through SES; SES Americom, formerly GE American Communications Inc (including its Latin American and Asian interests); SES Multimedia; AsiaSat; Nordic Satellite AB; and Star One. SES Astra serves the European market, while SES Americom, acquired by SES during the same 2001 period, serves the North American market. At the time of SES Global's creation in 2001, it was the largest communications satellite operator in the world. SES Global is a public company listed on the Luxembourg and Euronext Paris Stock Exchanges. As of December 2006, the US's GE Capital owned a 19.4 per cent interest in SES Global, however GE Capital may gradually divest its interest in the communications satellite giant completely. About 53 per cent of SES Global shares are available for public trading. As of June 2006, SES Global reported EUR710.5 million in revenue and a net profit of EUR215.6 million.

SES Astra
Astra, headquartered in Betzdorf, Luxembourg along with SES Global, is one of the largest direct-to-home satellite broadcast and broadband service providers in the world. It serves the European market, operating a 12 satellite fleet. Astra provides more than 1,700 TV and radio channels to 107 million European homes.

As of January 2007, Astra's constellation was comprised of: Astra 1E, 1F, 1G, 1H, 2C and 1KR at 19.2°E; Astra 3A and 1D at 23.5°E; and Astra 2A, 2B and 2D at 28.2°E. Astra 1C, launched in 1993, will soon be retired, most likely after the launch of Astra 1L in 2007.

Astra satellites are manufactured by Boeing Satellite Systems, Lockheed Martin and Astrium. The company uses Arianespace and International Launch Services for launch services, typically launching satellites from Kourou and Baikonur aboard Ariane and Proton launch vehicles, although the Astra 1KR was launched from Cape Canaveral aboard an Atlas V.

SES Americom
SES Americom was formerly known as GE American Communications Inc (Americom), and RCA Americom before that. RCA Americom was created in 1975 as a wholly owned RCA subsidiary to own and operate the parent company's domestic satellite system. Following GE's 1986 RCA acquisition, GE Americom began operating as part of GE's Communications & Services Group.

SES Americom primarily serves the North American market, providing TV, radio, internet and network services via its 19 satellite constellation. As of January 2007, Americom's fleet was comprised of: AMC-1 (GE-1), AMC-2 (GE-2), AMC-3 (GE-3), AMC-4 (GE-4), AMC-5 (GE-5), AMC-6 (GE-6), AMC-7 (GE-7), AMC-8 (GE-8), AMC-9 Gstar 4 (formerly GE Gstar 4), AMC-10, AMC-11, AMC-12, AMC-15, AMC-16, and AMC-23, as well as AAP-1, Satcom C3 (GE Satcom C3) and Satcom C4 (GE Satcom C4). AMC-18 was launched from Kourou in December 2006, and AMC-21 is scheduled to launch toward the end of 2008.

SES New Skies
New Skies Satellites (NSS) was founded in November 1998 during a partial privatisation move by Intelsat (see separate entry), which transferred five satellites, about a quarter of its fleet at the time, to the fledgeling company. New Skies' 1998 Pro forma revenue was approximately USD117 million. By 2005's end, New Skies reported a revenue of USD250.5 million. SES Global acquired New Skies Satellites on 30 March 2006. Headquartered in the Netherlands, New Skies provides global access to TV, internet and telecommunications services; New Skies operates six of SES Global's total satellite fleet: NSS-5, NSS-6, NSS-7, NSS-703, NSS-806 and IS-603. A seventh satellite, NSS-8, was scheduled to launch in January 2007 from Sea Launch's Odyssey platform aboard a Zenit-3SL , however that scheduled flight ended in catastrophic failure, destroying the satellite and damaging the Odyssey platform. New Skies has said that it will continue to operate NSS-703 until 2009 due to the accident, replacing that satellite with NSS-9. Launched in 1990 and primarily a back-up satellite, IS-603 is likely to be fully retired soon.

Further SES Holdings
SES Global also owns large interests in many other satellite operators, notably Sirius AB (75 per cent); AsiaSat (34.1 per cent); Ciel Satellite Group (70 per cent); Quetzsat (49 per cent); and Star One (19.99 per cent).

The SES parent company also wholly or partially owns seven satellite service providers: Americom Government Services (100 per cent); APS (100 per cent); Satlynx (100 per cent); ND Satcom (100 per cent); SES Astra Techcom (100 per cent); Accelon (43.5 per cent); and Netsystem (15.02 per cent).

SES Sirius AB
Sirius, which until 2005 was known as Nordic Satellite AB (NSAB), provides TV and multimedia services to the Scandinavian and Baltic regions, as well as to segments of Central and Eastern Europe. Astra owns 75 per cent of Sirius, with the remaining shares owned by the Swedish Space Corporation (SSC). Astra and SSC began in 2000 as equal shareholders in NSAB, however Astra, in 2003, acquired a further 25 per cent of the company following the formation of SES Global.

As of 2007, Sirius will operate three satellites: Sirius 2 and Sirius 3, launched in 1997 and 1998 respectively, and Sirius 4, which Sirius plans to launch in mid-2007. Sirius 2 was built by Aerospatiale, now Alcatel, and was the largest satellite of its kind to be built and to serve Europe when it was launched. It has a lifespan of 15 years. Hughes Space & Communications in California built Sirius 3. It is expected to remain in service through about 2010. Lockheed Martin is completing Sirius 4, using its A2100AX bus; it will be the largest in the Sirius constellation. It has a 15-year design life.

Asia Satellite Telecommunications Company Ltd (AsiaSat)
Hong Kong based AsiaSat, founded in 1988, was Asia's first commercial communications satellite operator. Since January 1999, SES Global (then SES) has owned 34.1 per cent of the company. AsiaSat operates a fleet of three satellites, providing TV, radio, telecommunications, IP and network services to the Asian and Pacific regions. AsiaSat 2, AsiaSat 3S and AsiaSat 4 are operational; AsiaSat 5 will launch in late 2008, set to replace the Lockheed Martin built AsiaSat 2, which has been in service since November 1995. AsiaSat 3S was launched in March 1999, replacing AsiaSat 1; it was built by Boeing. Also built by Boeing, AsiaSat 4 has been in service since April 2003. The AsiaSat 5 contract was awarded to Space Systems/Loral in April 2006; it will be launched aboard one of the first Land Launch Zenit-2SLB flights from Baikonur.

Ciel Satellite Group
A 2004 start-up founded by a consortium including SES Global and other financial backers and satellite professionals, Ciel provides TV and radio

SES Astra's Betzdorf control facility 0517408

broadcast services, and will, in the future, provide direct-to-home, HDTV and broadband services to the Canadian market. In 2006, Ciel awarded Alcatel Alenia Space the Ciel-2 satellite contract. Ciel expects to launch the spacecraft in late 2008; it has a design life of 16 years and will be the company's first new geostationary satellite in orbit. In November 2006, Ciel filed nine applications for orbital slots. Its first slot was awarded in 2004 at 129°W, where Ciel-2 will be placed. Since August 2005 Ciel has leased EchoStar-V (also known as Ciel-1 and Echo-V) capacity; the crippled (as of 2001) satellite was launched aboard an Atlas IIAS in September 1999 by the US company EchoStar Communications (see separate entry); it also provides some back-up services to EchoStar's Dish Network customers. The satellite's lifespan has been shortened due to the loss of its momentum wheel and some solar array capacity. EchoStar-V will most likely be retired soon after Ciel-2's launch and positioning. Ciel-2 will continue to provide some Dish Network services.

Quetzsat SRL de CV

Quetzsat is a start-up that was founded by SES Global via SES Americom, and the Mexican government in 2004, to provide satellite broadcast services to the Mexican market. The company, like Ciel, also uses a damaged satellite, EchoStar-IV, courtesy of EchoStar Communications. EchoStar-IV, launched in 1998, provides some services to Dish Network customers, as will Quetzsat 1, once it is launched. The contract to build Quetzsat 1 will be awarded in the future, and the satellite will probably launch around the 2009 time frame.

Star One SA

Star One, headquartered in Brazil, serves the Brazilian TV, radio, telecommunications and network satellite market. Founded in 2000, It is 80 per cent owned by Embratel (see separate entry); SES Global owns 20 percent. Star One manages and operates Embratel's satellites. The company anticipated the launch of its Star One C1 (also called Simon Bolivar F1) satellite in 2006 to expand its market throughout South America, however, the satellite's flight, contracted with Arianespace to occur from Guiana Space Center, has been repeatedly delayed since 2004. It may launch in 2007 or 2008. Alcatel Alenia Space built the spacecraft, which is to replace Brasilsat B1 in orbit. Alcatel contracted again with Star One to build Star One C2, which may launch in 2007, to replace Brasilsat B2. However, noting the delays that have already dogged C1, C2 may wait for a firm launch date as well.

Star One operates Brasilsat B1, B2, B3 and B4, the last of which was launched in August 2000. Brasilsat B1 and B2 were launched in 1994 and 1995 respectively; their need of replacement by C1 and C2 is based upon the satellites' 12-year lifespans. BolivarSat was formed by the South American Andesat consortium and Alcatel in 1999; Alcatel later sold its shares to Star One. BolivarSat will operate Star One C1, once it is launched. As of December 2005, Star One employed 195 staff. The company also leases capacity on AMC-12, sometimes known as Star One 12.

WorldSpace Inc

Current Status

Beginning in December 2007, WorldSpace will temporarily suspend its broadcast services to Northern and Western Africa and Western Europe. Those customers previously reached by Afristar's Western beam will be unable to access services. WorldSpace claims it is upgrading its technology and will resume services to these areas at a later, unannounced date.

WorldSpace's 2006 third quarter results reported that the company had achieved a total 176,831 subscribers, 138,065 of which are in India. WorldSpace also reported USD3.3 million in revenues for that period, however the company is still reporting net losses: these totalled USD28.9 million at the end of the third quarter 2006. Financial analysts are not optimistic regarding WorldSpace's present and future health.

Background

WorldSpace was founded in 1990 with audio services to Africa and the Middle East beginning in October 1999 via AfriStar. One of the company's initial goals was to offer direct audio broadcasting to help bridge the digital divide and provide better access to information, thereby helping to prevent the spread of AIDS and other diseases. Today, WorldSpace offers a range of audio services to the people of Africa and Asia, providing 100 channels of programming, including over 30 self-produced channels.

WorldSpace is headquartered in Silver Spring, Maryland and also runs offices and studios in Africa, the Middle East, Asia, Europe and the US. As of 2005, WorldSpace was a NASDAQ listed company; the same year, XM Satellite Radio invested USD25 million in WorldSpace. Noah Samara, WorldSpace's CEO, was also involved in the formation of XM Satellite Radio, and WorldSpace originally owned part of the company, as well as the satellite technology with which XM's receivers operate; WorldSpace sold its XM stake in 1999. XM, which broadcasts to the US, carries four WorldSpace channels. XM and Sirius Satellite Radio announced in 2007 their intention to merge, however some industry analysts foresee potential problems with the completion of this deal. As of early 2007, WorldSpace, XM and Sirius were the only operational satellite radio companies in the world.

WorldSpace originally planned to launch three 1467–1492 MHz L-band satellites, each beaming high sound quality radio channels to Africa, the Middle East, Asia, the Caribbean and South and Central America. A contract was signed 21 January 1995 with Alcatel Espace to build, launch and insure the three satellites (plus a ground spare) for USD600 million. Alcatel Espace signed the Ariane launch contract 18 March 1996. More than USD750 million financing was secured, with investment banker Morgan Stanley of New York serving as financial advisor, by a group of international investors.

AfriStar was launched by Ariane 44L on 28 August 1998 and has a lifetime of 15 years. AsiaStar was launched by Ariane 5 on 20 March 2000. The launch of AmeriStar has been greatly delayed and WorldSpace has not announced a launch date for the satellite, although its planned coverage includes South and Central America, as well as part of the Caribbean. A conflict exists with the US Department of Defense (DoD) regarding AmeriStar, because the US DoD also broadcasts in the L band and claims that AmeriStar's broadcasts would overlap with those they use for telemetry. WorldSpace and the US DoD could not come to an agreement and AmeriStar has remained in storage since 1999.

A WorldSpace receiver, initially costing USD150 to USD250, can now be procured for about USD70. Radios can be used for stationary or portable applications. The radio chipset has also been integrated onto computer interface cards for data downloads and audio programmes via PC.

Japan

Broadcasting Satellite System Corporation (B-SAT)

Current Status

B-SAT is planning its next generation of communications satellites, and awarded Lockheed Martin, in 2005, with the contract to provide B-SAT with BSAT-3a. Expected to launch from Kourou in 2007, the satellite will replace BSAT 1a.

B-SAT operates satellites that, through service providers, broadcast direct audio-visual programming to Japanese homes and businesses.

Background

The Broadcasting Satellite System Corporation (B-SAT) was incorporated in Tokyo on 13 April 1993 as a consortium of Japan Broadcasting Corporation (NHK), WOWOW (formerly Japan Satellite Broadcasting Inc), five other broadcasters and eight banks, to procure and operate the BS-4 generation of broadcasting satellites. NHK and WOWOW remain the majority shareholders, at 49.9 and 19.6 per cent respectively. Tokyo Broadcasting System Inc, TV Asahi Corporation, BS Nippon Corporation, BS Fuji Inc, and BS Japan Corporation also own B-SAT shares at about 5 per cent each. As of 2007, B-SAT employed 66 staff.

As of early 2007, B-SAT operated the BS-3N (Yuri-3N) satellite, launched in 1994, and the BSAT 1 and BSAT 2 series of broadcast satellites. The BS-4 generation satellites were redesignated as the BSAT series; BSAT-1a, launched in April 1997, and BSAT-1b, launched in April 1998. BSAT-2a launched in March 2001 and BSAT-2c launched in June 2003.

BSAT-2b was launched by Ariane 5 on 12 July 2001 but an upper stage failure left the satellite stranded in low Earth orbit.

Japan Satellite Systems Inc (JSAT)

Current Status

As of May 2007, JSAT operated a 10-satellite fleet, including one backup satellite. Horizons-2, the follow-on to the joint JSAT-Intelsat owned Horizons-1, launched in 2003 from Sea Launch's Odyssey platform, is expected to launch from Kourou in late 2007. JSAT and Intelsat have also reached a transponder purchase agreement regarding Intelsat's IS-15, expected to launch in 2009.

Background

Japan Communications Satellite Company Inc (JCSat) was formed in April 1985 to provide the first commercial Japanese satellite communication system. Satellite Japan Inc (SAJAC, licensed in May 1991 to launch two satellites in 1994 and 1995) was forced by financing problems to merge with JCSat in 1993, creating JSAT. JSAT is now a wholly owned subsidiary of Sky Perfect JSAT Corporation and was first listed on the Tokyo Stock Exchange in 2000. It employs roughly 230 staff and maintains offices in Tokyo and Osaka; JSAT also runs three ground stations in Yokohama and Gunma, Japan, and at Kapolei on Oahu Island, Hawaii. JSAT International, in El Segundo California, is a wholly owned JSAT subsidiary founded to market JSAT programming broadcast via the JSAT-Intelsat Horizons satellites.

JSAT provides direct to home (DTH) broadcast services and business network services via its satellite system. Sky PerfecTV! and Sky PerfecTV! 110 DTH services broadcasted more than 300 channels to over 4 million subscribers as of mid-2006. JSAT took a USD56 million loss in FY05 (March 2006 year end); the reasons for this were primarily associated with JCSAT-1B technical issues, the sale of its JC-HITS subsidiary, Horizons-associated North American service expansion, and the launch of JCSAT-9. The company's revenues for the same period totalled about USD363 million. JSAT's revenues for FY04 were very similar, however the company was profitable during that period and all other periods dating back through at least 2001.

JSAT operates satellites in the JCSAT and Horizons series, as well as the N-STARb telecommunications satellite, all of which were launched between 1995 and 2006; Horizons-2 is set to launch in 2007. JCSAT-3A, launched in 2006, will eventually replace JCSAT-3, launched in 1995. JCSAT-R is the company's in-orbit back up. Launched in 1997, it is nearing the end of its design life, and may not be able to provide adequate back up should a more modern satellite in the series fail.

Sky Perfect JSAT should not be confused with British Sky Broadcasting (BSkyB), which broadcasts its programming via leased transponders aboard SES and Eutelsat satellites.

Space Communications Corporation (SCC)

Current Status

Superbird-D, launched in 2000, was the last of SCC's satellites to achieve operational orbit. Mitsubishi Electric Corporation, SCC's part owner, is developing and building Superbird-7, which is expected to launch in early 2008; it will replace Superbird-C at 144°E.

Launched in April 2004, Superbird-A2 (Superbird-6), did not achieve its target orbit. As a result, it suffered fuel and power losses and was unable to enter operational service.

Background

Established on 22 March 1985, Space Communications Corporation (SCC) is wholly owned by the Mitsubishi Group. The Mitsubishi Corporation and Mitsubishi Electric Corporation

(MELCO) are the majority shareholders, at 28.41 per cent and 18.94 per cent respectively; Mitsubishi Heavy Industries Inc owns a 10 per cent stake in SCC. Another 25 Mitsubishi Group Companies own a 42.65 per cent aggregate. SCC is based in Tokyo and employs about 200 personnel. Its satellite control centres are located in Ibaraki and Yamaguchi. As of 2006, SCC's revenues totalled JPY17,985 million (about USD153 million).

SCC's Superbird satellites offer telecommunications services, including audio-visual digital broadcast, commercial data, and telephony services, to the Asia-Pacific region, and not only cover Japan, but also South Korea, Taiwan and parts of China. As of early 2007, SCC operated a four-satellite fleet, including Superbirds A1, B2, C and D (D is also known as Superbird-5, N-SAT-110, and JCSAT 110).

The first Superbirds A and B were planned and built by Space Systems/Loral, based on the FS-1300 platform; programme cost was JPY70 billion (USD636 million), covering two satellites, launch services and insurance at premiums of about 20 per cent. The launch of Superbird A on 5 June 1989 by Ariane to 158°E was successful, but most of its stationkeeping oxidiser was lost in December 1990 and commercial operations were ended; some customers were then transferred to the rival JCSat. The insurance claim was reportedly USD170 million. Superbird B, delayed from a December 1989 launch because of transponder problems, was lost in the Ariane 4 accident of 22 February 1990. Insurance cover was USD94.3 million.

The launch contract for the replacement satellite Superbird B1 was signed with Arianespace in November 1990; SS/L delivered the satellite within 24 months and the satellite was launched on 27 February 1992. Superbird C was ordered from Hughes in March 1995 to expand coverage to the Asia Pacific region, including HDTV and 150 Mbit/s data.

Russian Federation

JSC Gazkom

Current Status

Developed by RSC Energia, Gazkom's Yamal 200 series satellites were successfully launched in November 2003. Yamal 201 and 202 have been operational since that time. Yamal 102 also continues to function.

The Yamal 300 series, also comprised of a satellite pair, is planned for launch in 2008. It will provide HDTV and data services to Russia and Eastern Europe. Gazkom is also planning a four-satellite constellation called the Smotr (Inspection) Remote Sensing System, which will provide surveillance and monitoring services to protect Gazprom's infrastructure. The Smotr constellation will likely launch during the 2009–2010 time frame.

Background

RSC Energia, Gazprombank and Gazprom, Russia's state natural gas monopoly (90 per cent of Russia's extraction and 20 per cent of the world's), created Gazkom in 1992 to develop the Yamal communications satellite system to support Gazprom's transcontinental transport and communications through remote station relays and terminals.

Gazkom owns and operates the Yamal satellites and provides corporate communications services to Gazprom as well as providing broadcast and data services to other government and commercial users in the Russian Federation, Eastern Europe and the Middle East; the latter services were developed and enabled subsequent to the Yamal 200 series entering service. Gazprom uses the satellites to communicate with its gas fields, particularly in Siberia. Gazprom were already using 30 stations, operating through Gorizont by 1995, with another 75 over 1995–96. By June 2004 there were more than 120 Earth stations operating through three teleports.

Satellite control is maintained through a mission complex in Moscow. Yamal 101 was launched on 6 September 1999 along with Yamal 102. Yamal 101 reportedly malfunctioned soon after its placement in orbit and was never a fully operational satellite within the Gazkom fleet.

Thailand

Shin Satellite Public Company Ltd

Current Status

Shin Satellite operates the four-satellite Thaicom fleet, the last of which, Thaicom 5, was launched in May 2006. Thaicom 1A, 2 and 4 (IPSTAR) were also operational as of early 2007. Thaicom 3 was deorbited in October 2006; it suffered a final power loss in its ninth year of service. Customer broadcast services had been transferred to Thaicom 5 previous to the Thaicom 3 failure.

At a meeting of its investors on 17 August 2004, an agreement was met to proceed with a public offering of 208 million new shares following the launch of the IPSTAR (Thaicom 4) satellite in August 2005.

Background

First known as the Shinawatra Satellite Public Company Ltd, Shin Satellite was founded in 1991 as Thailand's and Asia's first commercial satellite company. In 1987, the Thai Ministry of Transport and Communications invited tenders for a domestic satellite system; the winner would receive a 30-year concession to lease transponders to government and private sector users, with excess capacity offered to neighbouring countries. Shinawatra Computer & Communications Group (SC&C), the largest group of Thai companies operating telecommunications and broadcasting service concessions, was selected in 1991, and given the 30-year concession. It had an eight-year monopoly to supply satellite capacity to domestic users and was required to pay existing users of other satellites to transfer to Thaicom. The concession will expire in 2021.

Shin Satellite provides telecommunications, internet broadband, and television broadcasting services to business and residential customers in Thailand, Asia, Australia, Africa, the Middle East and much of Europe. It is publicly listed on the Stock Exchange of Thailand. Shin Satellite is headquartered in Bangkok and also operates satellite control facilities north of Bangkok. The company took a THB46 million loss in 2006, however it achieved a net profit in 2005 of THB1,337 million. Shin Satellite assigns blame for the 2006 loss on the malfunction and deorbit of Thaicom 3, which did not operate to the end of its 14-year design life, as well as the new depreciation costs associated with Thaicom 5, launched in 2006. Shin Satellite claims that if Thaicom 3 had not failed, the company would have earned a THB629 million net profit.

Shinawatra began formal negotiations in June 1991 with Hughes, signing a contract on 8 October 1991, at around USD100 million for two satellites, ground equipment and training. Deliveries were due within 24 to 28 months. The first launch contract was signed with Arianespace 6 December 1991. As lighter versions of Hughes' HS-376 bus, Thaicom satellites were the second to come within Ariane's 1 tonne SDS (Spelda Dedicated Satellite) class. Shinawatra signed a letter of intent on 1 November 1993 to negotiate with Hughes for a third satellite, but Aerospatiale's SpaceBus 3000A platform was selected December 1994 for the more powerful Thaicom 3; negotiations for Thaicom 4 continued. Total cost for Thaicom 3 on station was estimated at USD240 million.

Previous to the Thaicom system, Thailand leased 2¼ transponders aboard AsiaSat 1 for telecommunications services. Thaicom 1 was intended to be placed at 101°E and Thaicom 2 at 78.5°E but an agreement was reached with AsiaSat in September 1993 to co-locate at the more westerly slot to avoid interference with AsiaSat 2 at 100.5°E. Thaicom 1 was launched on 17 December 1993 followed by Thaicom 2 on 2 October 1994 and Thaicom 3 on 16 April 1997.

Shin Satellite ordered the broadband satellite, IPSTAR, from Space Systems/Loral for launch in 2003 by Arianespace, but the launch was delayed to August 2005. When launched, its 45 Gbps capability equalled that of all the existing satellites serving Asia. IPSTAR is a high powered satellite with 87 Ku-band transponders and a mass of about 6,500 kg. It is based on the Space Systems/Loral LS-1300SX bus. Thaicom 5 was launched on 27 May 2006. It serves direct to home (DTH) broadcast customers in Asia, Australia, Africa, the Middle East and Europe.

United States

DirecTV Group, Inc

Current Status

As of December 2006, DirecTV operated a nine satellite constellation. The company launched DirecTV 9S in October 2006, and expects to launch DirecTVs 10 and 11 in 2007. DirecTV 12 will remain in storage as a ground spare and the company expects to launch DirecTV 13 in the future, as a replacement for the DirecTV 5 satellite.

Background

DirecTV US and DirecTV Latin America are the two main units of the DirecTV Group Inc. It is currently one of the world's largest providers of digital TV entertainment, broadband satellite networks and services, together with video and data broadcasting.

DirecTV is headquartered in El Segundo, California; as of 2005's end, the company employed about 9,200 staff. DirecTV serves more than 15.5 million US customers, providing them with about 1,500 HDTV, TV and radio channels. The company reported a total revenue of USD13.2 billion at the end of 2005, USD12.2 billion derived from the DirecTV US unit.

DirecTV traces its history to the Hubbard Broadcasting subsidiary United States Satellite Broadcasting (USSB), founded in the early 1980s, and the General Motors subsidiary, Hughes Electronics Corporation. The demise of the Sky Cable consortium of Hughes, Cablevision Systems, NBC and The News Corporation was followed in June 1991 by Hughes' creation of a Direct Broadcast Services business unit. Hughes unveiled its DirecTV DTH distribution system at the same time and simultaneously sold five of the 16 transponders aboard the first satellite to USSB for USD100+ million. Before that, in 1991 the pair had agreed to operate a domestic US Direct to Home (DTH) system using Ku-band HS-601s. The partner companies launched their first satellite in December 1993 and the DTH service in 1994. DirecTV became an independent GM subsidiary in November 1993. In 1999, DirecTV acquired USSB, the same year it acquired PrimeStar and its Tempo satellites.

In December 2003, the Fox Entertainment Group acquired a 34 per cent interest in DirecTV, which is listed on the New York Stock Exchange, along with the Fox Entertainment Group and Fox's owner, News Corporation. As of April 2006, Fox owned a 36.8 per cent interest in the company. In 2004, DirecTV sold all interest in PanAmSat and Hughes Network Systems, and after further restructuring, its former holding company, Hughes Electronics Corporation, changed its corporate name and stock listing to DirecTV Group.

DirecTV Latin America, LLC is an 86 per cent owned subsidiary of the DirecTV Group. It provides DTH services to about 1.5 million customers, located primarily in South America and Puerto Rico.

DirecTV is licenced to operate satellites from the 101°W, 103°W, 110°W and 119°W orbital slots. Its current fleet includes: DirecTV 1 (formerly Direct Broadcast Satellite or DBS 1), DirecTV 1R, DirecTV 4S, DirecTV 5 (PrimeStar's Tempo 1), DirecTV 7S, DirecTV 8, DirecTV 9S, and Spaceway 1 and Spaceway 2, both launched in 2005. Additionally, DirecTV leases an orbital slot from Telesat Canada at 72.5°W, and leases capacity on a satellite located at 95°W. DirecTV leases DirecTV 2 (formerly DBS 2 and now Nimiq 4i) and DirecTV 3 (formerly DBS 3 and Nimiq 2i and now Nimiq 3) to Telesat Canada; Nimiqs 3 and 4i occupy Telesat orbital slots.

EchoStar Communications

Current Status

EchoStar X was successfully launched in February 2006; EchoStar IX was launched in August 2003. Both took flight from Sea Launch's Pacific-based Odyssey platform. It is likely that Space Systems/Loral, will complete EchoStar XI and deliver it for launch in 2007; this spacecraft is likely to also be sent into orbit by Sea Launch (see separate entry). EchoStar is planning to launch five more satellites during the 2008–2011 time frame.

Virgin America, a new Virgin group low-cost airline, will begin US operations in 2007 and will offer EchoStar's Dish Network satellite programming to its in-flight customers.

Background

EchoStar Communications Corporation (ECC) owns 100 per cent of EchoStar Satellite Operating Corporation (ESOC), the company's satellite operator. EchoStar Communications is a publicly traded company, listed on NASDAQ. ECC reported, as of December 2005, an annual revenue of USD8.4 billion and USD1.5 billion in earnings; as of early 2007, ECC and its subsidiaries employed approximately 21,000 people.

EchoStar is the Dish Network's parent company; the Dish Network is EchoStar's single source satellite service provider. EchoStar founded the Dish Network brand name upon EchoStar I's launch in 1995. By December 2006, the Dish Network was providing TV, radio and internet services to over 13 million North American customers.

EchoStar launched its first communications satellite, EchoStar I, in 1995 from Xichang, China, eight years after it filed for its Direct Broadcast Satellite (DBS) licence with the US Federal Communications Commission (FCC), and three years after the FCC granted EchoStar its 119°W orbital slot. EchoStar now operates satellites in the 61.5°W, 148°W and 110°W orbital positions as well. Directsat merged with EchoStar in January 1994 and both organisations had FCC approval to merge their DBS licences to occupy the 119° W slot. Directsat originally had been allocated 10 Broadcasting Satellite Service (BSS) channels at 119° W by FCC in December 1993.

ESOC currently operates a fleet of 14 satellites, including EchoStars I – X and EchoStar 12 (previously known as Rainbow 1 or Cablevision 1, and purchased from Rainbow DBS Co in 2005), all of which EchoStar owns, as well as three leased satellites: AMC2, AMC-15 and AMC-16. The AMC satellites belong to SES Global (see separate entry), which also broadcasts via the satellites through SES Americom. EchoStar IX is also known as Telstar 13, Intelsat Americas 13, and Galaxy 23: Intelsat (see separate entry) leases C-band capacity from EchoStar on the satellite. EchoStar also plans to lease capacity on AMC-14 and Telesat's Anik F3 when they are launched in 2007, and SES Global's Ciel 2 when it is launched in 2008.

EchoStar Europe is comprised of EchoStar International, with offices in the Netherlands and Spain, and Eldon Technology Ltd in the UK. Together, these companies develop EchoStar software and hardware product technologies.

GeoEye, Inc

Current Status

Orbimage and Space Imaging announced in September 2005 that Orbimage would acquire Space Imaging; the consolidation created a new company, GeoEye, in January 2006.

Orbview 3 was launched by Orbital Sciences Corporation aboard a Pegasus rocket on 26 June 2003. It offers 1 m resolution panchromatic and 4 m multispectral digital imagery. In September 2004, Orbimage was awarded a contract by the US National Imaging and Mapping Agency (NIMA), now the US National Geospatial-Intelligence Agency (NGA), for the high resolution Orbview 5 satellite. Now known as GeoEye 1, the first satellite in the NextView programme, it will join Orbview 3 in orbit in 2007. NGA will share the satellite's cost, financing USD237 million out of the anticipated USD502 million total cost.

Background

Orbital Sciences Corporation (OSC) founded its Orbital Imaging Corporation (Orbimage) subsidiary in 1992 to build and operate OSC's commercial Earth observation system, and to market the company's remote sensing products and services. Lockheed Martin Space Systems, Raytheon Systems Company, Mitsubishi and Hyundai founded Space Imaging in 1994. In November 1996, Space Imaging acquired Earth Observation Satellite Co (EOSAT), which in 1995 experienced a USD3 million loss on revenues of USD30 million.

GeoEye was founded in 2006, after Orbimage acquired Space Imaging Inc, rebranded the combined companies, and formed GeoEye. Headquartered in Dulles, Virginia, it is the world's largest commercial satellite imaging company. GeoEye is a publicly traded company listed on NASDAQ. As of 2006, it employed 290 staff in three US locations.

In 1995, Orbimage decided to pursue the launch of the small, low-cost Orbview satellite, derived from OSC's MicroLab design. At the time, the cost of deploying one satellite and establishing the ground system was estimated at less than USD100 million. Orbview 1 was launched in 1995, followed by Orbview 2 in 1997. The first of Orbimage's second generation, Orbview 4, launched in September 2001 but was lost in a Taurus launcher failure. Orbview 3 was launched on 26 June 2003 and offers 1 m panchromatic and 4 m multispectral digital imagery. Orbview 4 would have offered the first hyperspectral imagery.

Following Orbview 4's loss in 2001, Orbimage filed for financial restructuring and sought protection from bankruptcy under US Chapter 11 in September of that year. The company did not emerge from that status until December 2003, following the successful launch of Orbview 3. At that time, it changed its name to Orbimage, no longer calling itself Orbital Imaging Corporation.

GeoEye's constellation consists of three satellites: Space Imaging's IKONOS, launched in September 1999; Orbview 2, and Orbview 3. GeoEye's NextView programme will launch its first satellite, GeoEye 1 (formerly Orbview 5), in 2007; it claims a .41 panchromatic resolution. GeoEye 1 was built by General Dynamics/C4 Systems; ITT built the electro-optical camera payload. It is scheduled to take flight from Vandenberg aboard a Delta II rocket.

Additionally, since 1994, Space Imaging, has worked with the Indian Space Research Organization's (ISRO) Antrix Corporation to deliver Indian Remote Sensing (IRS) products. Carrying on this relationship following the Space Imaging acquisition, the IRS products GeoEye now distributes originate from IRS 1C, IRS 1D, IRS-P5 (or Cartosat) and IRS-P6 (or RESOURCESAT).

NGA is GeoEye's largest customer; the agency purchases commercial satellite imagery to satisfy some of its intelligence requirements. NGA has first rights to all GeoEye 1 imagery, and will likely use approximately 50 per cent of the satellite's capacity once it is operational.

Globalstar Inc

Current Status

Globalstar offers satellite data and voice communications services, primarily to business and government clients. Their satellites serve 120 countries, with expansion anticipated as the company launches new spacecraft; Globalstar's second generation satellites are planned for launch beginning in 2009 and 2010.

By late 2006 Globalstar claimed to be serving more than 250,000 active voice and data units, with service offered over most of the world's land masses, with the major exceptions of the polar regions, sub-Saharan Africa and South Asia, including India.

Background

Globalstar was one of a number of 'big LEO' or 'global mobile' communications satellite projects started during the 1990s. The objective was to provide direct satellite communications to users with hand-held terminals, in virtually any location in the world, at a cost that would be affordable by civilian users. Originally dubbed Loral Qualcomm Satellite Services Inc (LQSS), the project was started by Loral and Qualcomm in 1991 and formally launched in March 1994 with Alcatel (now Alcatel Alenia Space) and Deutsche Aerospace (now part of EADS) as partners; after the 1994 launch, the company was renamed Globalstar.

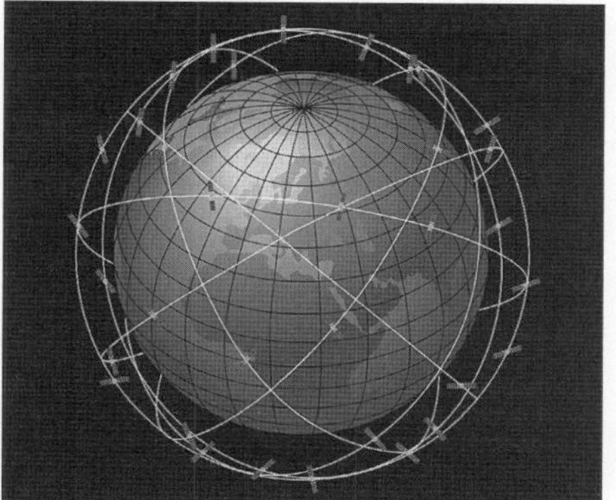

The 48-satellite Globalstar constellation was completed by early 1999 (Globalstar) 0005570

Like its fellow 'big LEOs', Globalstar experienced serious business difficulties because of weak demand, much of the market having been addressed by the unexpectedly rapid growth of conventional mobile telephones and the advent of global roaming services. In February 2002 Globalstar filed for Chapter 11 bankruptcy protection. An attempted takeover by ICO Global Communications failed in late 2003, and in October of that year Globalstar warned that it would have to de-orbit its satellites without a rescue plan.

In December 2003, private investors Thermo Capital Partners stepped in with a USD43 million bid and acquired 81 per cent of Globalstar. The company emerged from Chapter 11 in April 2004 and embarked on plans to improve its services – adding gateways in Alaska and the Caribbean, for example – and to offer new services tailored to specific markets, including maritime users, homeland defence and the oil and gas extraction industries. General Electric has selected Globalstar as the medium for its asset-management product, using the satellite system to report the location and status of commercial and transportation assets worldwide.

Headquartered in Milpitas, California, Globalstar is a NASDAQ listed company; it announced an Initial Public Offering (IPO) of 7.5 million shares in November 2006. The primary objectives given by Globalstar for the IPO, were to increase funding to support the development and purchase of Globalstar's second generation of satellites, which the company intends to launch beginning in 2009. Funds will also finance facilities upgrades and the launch of eight on-orbit spare satellites in 2007.

Globalstar's stock quickly dropped in February 2007. According to some sources, Globalstar's satellites, although approaching the end of their design-lives, are degrading more quickly than planned, leaving the possibility of inadequate two-way transmission performance due to crippled amplifiers. The company's ability to continue to provide sufficient services before the launch of their second generation spacecraft, may therefore be jeopardised, since spacecraft performance could degrade to an inoperable point by 2008.

ICO Global Communications

Current Status

On 14 April 2008, ICO successfully launched ICO G1, a geosynchronous satellite designed to provide S-band 2 GHz Mobile Satellite Services (MSS) in the US. Space Systems/Loral constructed ICO G1, one of the largest commercial satellites ever launched, and the largest ever launched by an Atlas V launch vehicle. It is also the first commercial satellite to utilise Ground Based Beam Forming in both the satellite-receive and satellite-transmit modes. ICO has a 15 May 2008 deadline to meet its 12th and last milestone set by the US Federal Communications Commission (FCC) to certify that ICO G1 is operational. ICO's application to the FCC to utilise an Ancillary Terrestrial Component (ATC) for its network is in the public comment phase (March 2008). ICO G1 is capable of delivering a variety of services to mobile and handheld devices. Trials of its ICO mim™ (mobile interactive media) service will be conducted in mid-2008 in Raleigh-Durham, North Carolina and Las Vegas, Nevada, with the company expecting to offer commercial service in 2009. In April 2008, ICO announced an exclusive three year partnership with Delphi Corp to produce DVB-SH equipment for automobiles in the United States.

Background

Headquartered in Reston, Virginia, ICO Global Communications was formed in January 1995 to provide global hand-held telephone services using a Medium Earth Orbit (MEO) satellite system. Four million subscribers were projected by 2010 and original plans anticipated operations starting in 2000. ICO's first launch was scheduled for the second half of 1998, with services beginning in late 1999. The original 47 investors provided USD1.5 billion total capitalisation, including USD150 million from Inmarsat (see separate entry). At the time, at least 70 per cent was to have been owned by Inmarsat and its signatories. Hughes Electronics Corporation also became a strategic investor, paying out USD93.8 million. The rest was to be financed by outside strategic investors and bank loans.

To operate a viable system, ICO required a constellation of 10 satellites plus two spares in two planes at MEO. An order for 12 Hughes (now Boeing) HS601 satellites, worth USD1.4 billion, was announced 20 July 1995; the final contract was signed 5 October 1995. An additional USD925 million launch services contract was signed 7 December 1995. This was to have been the first HS-601 application in medium orbits; the design would have employed GaAs cells, direct injection by the launch vehicle, an enhanced thermal radiator and a modified attitude control system.

Due to financial difficulties, the slow acquisition of business contracts and the unanticipated rapid expansion of terrestrial cellular networks, ICO Global Communications filed for US Chapter 11 bankruptcy protection in August 1999. In October 1999, telecommunications entrepreneur Craig McCaw and his affiliates, Teledesic and Eagle River Investments LLC agreed to lead a group of investors that provided up to USD1.2 billion to ICO to enable it to emerge from bankruptcy, thus allowing the first satellite launch to proceed. As a result of the new financial backing, the company that emerged from Chapter 11 on 17 May 2000 was known as New ICO. Craig McCaw currently serves as the Chairman of ICO, J. Timothy Bryan is Chief Executive Officer.

New ICO has modified its services concept to make its system more attractive to a wider range of users. Applications include maritime markets in commercial shipping; fishing and recreational vessels; transportation for freight companies that carry goods over distances greater than 480 km; government services such as data encryption, restricted user groups and global coverage; the oil and gas industry; sets to link professionals in remote locations; and equipment for individuals and small businesses where there is a need for fixed and mobile networks. The ICO mim™ service will feature 10–15 channels of live mobile television, interactive navigation and emergency satellite communications.

ICO's F1 HS 601 satellite was to have been placed in a medium orbit of 10,104 km by direct injection with an inclination of 45°. However, the satellite was destroyed during a second stage launch failure after liftoff from the Sea Launch Odyssey platform in March 2000. The F-1 was the testbed satellite for the 10 further ICO satellites that were designed to form the company's first constellation. Five satellites were to have been positioned in each of two orthogonal planes providing overlapping coverage of the Earth, orbiting every six hours.

After its financial tribulations and the loss of its first satellite, New ICO's F2 satellite was successfully launched on 19 June 2001, passing in-orbit tests in September 2001. ICO F2 essentially acts as a place holder, protecting the company's orbital slot and frequency licence. The remaining 10 satellites are in storage and in various stages of near completion, awaiting the outcome of the litigation between ICO and Boeing, the contractor selected to provide ICO with its original satellite constellation. The companies have been in litigation since 2004, with the trial scheduled to start in May 2008.

In 2005, ICO selected Space Systems/Loral to build a geostationary satellite, ICO G1, in order to provide US customers with mobile services. In February 2007, the US Federal Communications Commission (FCC) granted ICO a milestone extension for the manufacture and launch of ICO G1. ICO is again a NASDAQ listed company, after being delisted following its financial troubles in 1999.

Iridium Satellite LLC

Current Status

Iridium Satellite LLC is now operating profitably and is predominantly supported by civil users. Late in 2006, the company announced plans to develop a replacement constellation of satellites.

Iridium's commercial service commenced in March 2001 following a contract from the US Department of Defense (DoD) which effectively rescued the system. It gives unlimited air time to thousands of US government users for mobile voice, data and paging.

Background

Iridium LLC provides remote communications solutions for heavy construction, defence and government agencies, emergency services, maritime operators, mining operators, forestry agencies, the oil and gas industry, aviation (commercial, corporate and leisure) and leisure and adventure pursuits.

On 20 November 2000, newly formed Iridium Satellite LLC received approval from a New York bankruptcy court to buy Iridium Inc's satellites for USD25 million. Subsequently, the Defense Information Systems Agency confirmed a two-year contract to Iridium Satellite LLC for USD72 million, with options for extending satellite utilisation through 2008 at a value of USD252 million. Boeing Service Company, a subsidiary of Boeing Integrated Defense Systems, was tasked with maintaining the network. Following the terrorist attacks of 11 September 2001, the Department of Defense made increasing use of Iridium for voice and data communications, generating income for the new Iridium company and returning the system to profit.

Underpinned by military contracts, Iridium Satellite rolled out commercial voice and data services in March and June 2001 respectively, targeting industrial and transportation users rather than individual consumers. Since then, Iridium and its partners have developed installations for applications such as small commercial or private vessels and private aircraft, taking advantage of the system's ability to meet basic voice and data requirements worldwide with very small and inexpensive user equipment. In July 2006, Iridium reported that the number of airborne terminals connected to the system had more than doubled in the preceding 12 months to 7,500 units and that more than 300 new units were being added each month. An Iridium datalink is also being built into the new Eclipse 500 small business jet, transmitting real-time operational data and engine diagnostics to the company's support system.

Motorola originally unveiled its Iridium mobile communications system proposal in June 1990 after 2½ years of internal work. Iridium planned a system of 66 small Low Earth Orbit (LEO) satellites by end-1998 providing global pocket mobile digital 4.8 kbit/s telephone services via onboard switching and inter-satellite links for USD3 per minute to the user. Cost was projected at USD3.45 billion for establishing the operational system of 66 satellites and 14 in-orbit spares by 1998, followed by another USD2.8 billion over the next five years for operations and maintenance. 800,000 users were required within five years for viability. Motorola projected 1.824 million by 2001, broken down as 42 per cent pager, 42 per cent portable phone and 7 per cent mobile. After 10 years, in 2006, it projected 3.224 million.

Iridium, Inc was incorporated 14 June 1991 to own and operate Motorola's Iridium system, with Motorola eventually holding 15 per cent. The company's original owner organisations were: Motorola, Lockheed Martin, Iridium Africa Corporation, Iridium Canada Inc, Iridium India Telecom Private Ltd, Iridium Middle East Corporation, Iridium SudAmerica Corporation, Nippon Iridium Corporation, Korea Mobile Telecommunications Corporation, Raytheon, Sprint Corporation, STET (Italy), Khrunichev (Russia), China Great Wall Industry Corporation, Pacific Electric Wire & Cable Co (Taiwan), Thai Satellite Telecommunications Corporation and Vebacom GmbH (Germany). A total of USD1.888 billion was raised by early 1996; the remainder came from debt financing.

The first cluster of five Iridium satellites was launched by Delta II in May 1997 and it took less than two years to orbit 66 operational satellites for the prime constellation. In December 1998, a CZ-2C/SD launch from Taiyuan completed the 72 satellite (66 prime, 6 spares) Iridium constellation. The service was launched immediately. Iridium's predicted subscriber levels were far higher than actually achieved, however, and by 1 April 1999, only 10,294 subscribers had signed up to the system, far short of the 500,000–600,000 needed to begin paying debt on the USD4.85 billion programme. In the first three months of 1999 the company announced revenues of USD1.4 million and a net loss of USD505 million, bringing down stock from USD72 in 1998 to USD15 by late April 1999, and casting doubts on the viability of high-end satellite telephone services. Analysts did, however, point to the high pricing of Iridium, which, with 1.5 billion minutes of capacity per year, had to charge USD3+ per minute for calls. Iridium needed to recoup at the rate of USD1.4 a minute just to break even. At the time, competing systems such as Globalstar, with 7 billion minutes of capacity, could wholesale at 21 cents while ICO, with 5 billion minutes, could do it for 23 cents. Iridium's telephone handset price of USD2,700 was set to plummet during 1999, a year in which the

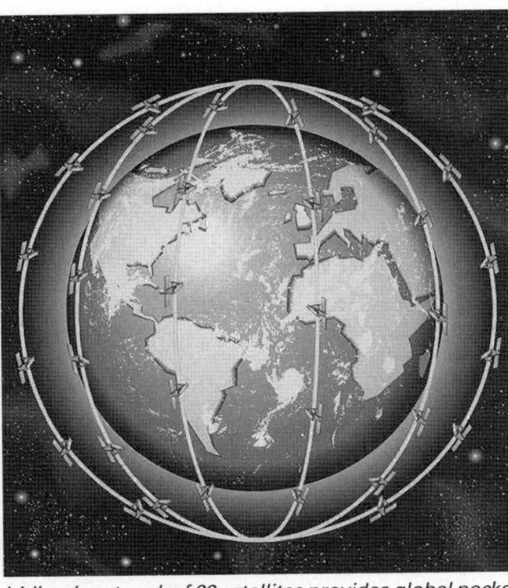

Iridium's network of 66 satellites provides global pocket mobile telephone services via the L/S-band phased array antennas. Iridium is the 77th element, reflecting the original concept's constellation size 0517224

company still expected to sign up a total 500,000 subscribers. On 13 August 1999, Iridium filed for US Chapter 11 bankruptcy protection. In the same month the company filed a bankruptcy petition. In common with other 'global mobile' services, Iridium had failed to anticipate the rapid growth of conventional mobile phone networks or the speed with which they rolled out worldwide roaming services and multi-band handsets.

Following the Iridium Satellite LLC acquisition, the company gradually returned to health, and by late 2006, Iridium could report seven successive quarters of profitability, with commercial users accounting for 70 per cent of revenues. That same year, the company claimed 169,000 subscribers–still only one-twentieth of Motorola's initial projections.

Loral Skynet

Current Status

Loral Space & Communications Inc, Loral Skynet's parent company, and the Canadian Public Sector Pension Investment Board (PSP Investments) announced in December 2006, their intention to acquire Telesat (see separate entry) from Bell Canada Enterprises (BCE) for CAD3.42 billion. Once the acquisition is completed in mid-2007, Loral Space & Communications will hold a 64 per cent interest, and PSP a 36 per cent interest and majority voting control in Telesat. To complete the joint venture, Loral Skynet will then merge its assets with Telesat to create the world's fourth largest geostationary communications satellite operator. The new entity will continue operations under the corporate name, 'Telenet', and will be headquartered in Ottawa, Canada.

Background

Loral Skynet is a subsidiary of Loral Space & Communications (see separate entry). In March 1997, Loral completed its acquisition from AT&T, of Skynet, a leading US domestic satellite service provider. Headquartered in Bedminster, New Jersey, Loral Skynet operates Loral Space & Communications' geostationary communications satellite fleet. As of December 2006, the company employed about 260 staff in 15 locations, and reported revenues of USD161 million.

Loral Skynet provides fixed satellite services, direct to home broadcast, network, and IP satellite services. In March 2004, Loral Skynet sold half of its satellite fleet, along with its North American coverage, to Intelsat (see separate entry), at a time when Loral Space & Communications was restructuring after having filed for US Chapter 11 bankruptcy protection in 2003. As a result of the satellite sales to Intelsat and related agreements between the two companies, Loral Skynet did not serve the North American market for two years, beginning US service once again in March 2006.

As of 2007, Loral Skynet operated a five satellite fleet: Telstar 10 (Apstar IIR), Telstar 11 (Orion 1), Telstar 12, Telstar 14 (Estrela do Sul 1) and Telstar 18. Loral expects to launch Telstar 11N in 2008 to replace the aged Telstar 11. The company also leases capacity on seven further satellites. Satmex 6, partially owned by Loral, and built by Space Systems/Loral, helped reintroduce Loral Skynet to North American customers in 2006.

At the end of 2005, Loral owned approximately 49 per cent of Satelites Mexicanos SA de CV (Satmex); Satmex, however, underwent restructuring in 2006 and as a result, Loral now owns about five per cent interest.

Loral Skynet owns and operates Telemetry, Tracking and Control (TT&C) facilities in Hawley, Pennsylvania and Mount Jackson, Virginia. It also leases and operates TT&C centres in Rio de Janeiro, Brazil; Quito, Ecuador; and Utive, Panama.

Orbcomm Inc

Current Status

One of a number of 'global mobile' satellite companies to have emerged from bankruptcy by establishing new markets, Orbcomm Inc specializes in 'Machine to Machine' (M2M) data messaging services, monitoring mobile or remote industrial equipment.

Background

Orbcomm was created as a subsidiary of Orbital Sciences Corporation (OSC) at the beginning of 1990. The company filed an application with the US Federal Communications Commission (FCC) on 28 February 1990 for a constellation of Low Earth Orbit (LEO) satellites. Orbcomm planned to provide for two-way communications and geolocation services with low-cost alphanumeric data communications and position determination (375 m) for emergency assistance, data acquisition and messaging services. OSC and Canada's Teleglobe Inc signed an agreement on 26 July 1993 to jointly develop and operate the system. Teleglobe provided USD10 million of the USD55 million cost of the two first phase satellites. For the second phase, which required 34 more satellites (including eight ground spares), Teleglobe increased its equity share in Orbcomm Global to 50 per cent.

After launching one unsuccessful experimental satellite (Orbcomm-X) by Ariane on 17 July 1991, Orbcomm launched a total of 36 satellites using OSC's own Taurus and Pegasus boosters between 1995 and 1999. However, like other global-mobile services, Orbcomm had vastly over-estimated demand – forecasting more than 5 million users after seven years of operation – and had failed to predict the rapid expansion of conventional mobile-phone coverage and worldwide roaming services. Revenues were inadequate to cover payments on the launch costs and Orbcomm filed for US Chapter 11 bankruptcy protection in September 2000. The company's assets were purchased in 2001 by a group of investors including Pacific Corporate Group, Ridgewood Satellite LLC and OHB Technology AG of Germany. Orbcomm has slowly returned to health.

The new owners have focused marketing activities on the M2M customer base. An important step forward came in March 2002 when Volvo Trucks North America announced that Orbcomm would provide two-way communication, tracking and monitoring of all its trucks in North America. Typical current applications include the monitoring of widely dispersed fixed assets such as oil and gas pipelines and storage facilities. Oil and gas levels, pressures and flow rates can be metered and monitored without land lines or human attendants. Heavy equipment fleet operators can use Orbcomm services to monitor the operation of their assets worldwide, checking location (referred to as geo-fencing), mechanical condition and usage rates. Truck operators can also use geo-fencing and make sure that drivers are complying with operating restrictions, speed limits and designated routes. Orbcomm customers include General Electric, Caterpillar, Volvo Group and Komatsu and Wal-Mart uses a GE-provided Orbcomm service to monitor its trucking fleet.

After filing a registration statement with the US Securities and Exchange Commission (SEC) in May 2006 for an Initial Public Offering (IPO) announced in October 2006, consisting of more than 11 million shares, Orbcomm began trading on the NASDAQ as of November 2006. The company is raising funds to finance its next generation satellite system.

The Orbcomm fleet consists of 30 operational satellites in LEO and five on-orbit spares. Orbcomm will begin replacing its aging fleet with six satellites, scheduled for launch in 2007; 18 more will follow beginning in 2008. Headquartered in Fort Lee, New Jersey, the company also operates 12 Gateway Earth Stations (GES) around the world, along with its Network Control Center (GCC) in Dulles, Virginia and six Gateway Control Centers in the US (at the GCC), South America, Asia and Europe.

For further technical information on Orbcomm's fleet, please see the separate Orbcomm Constellation entry.

TerreStar Networks Inc

Current Status

TerreStar Networks anticipates that its space-based mobile communications services for the North American market will be operational before the end of 2008.

Background

TerreStar Networks became an independent company when it spun off from Mobile Satellite Ventures (MSV – see separate entry) in May 2005. Its headquarters are located in Reston, Virginia. At the time of TerreStar's separation from MSV, Motient Corporation owned a 61 per cent controlling interest in the company; as of December 2006, it owned 70 per cent. SkyTerra Communications and Telesat Mobile Inc (TMI) also own some interest in TerreStar at 15.4 and 14.7 per cent respectively.

TerreStar's goal is to offer mobile voice and data communications services to cell-phone-sized user devices across North America, using a combination of its satellite and an Ancillary Terrestrial Component (ATC) to provide service in almost any location. Its satellite will operate in the S-band (2–2.2 GHz) range rather than the L-band (1.5–1.6 GHz) band used by MSV; the company expects to launch TerreStar-1 in November 2007. For further satellite technical details, please see the separate TerreStar Series entry.

While TerreStar sees a large commercial market for its services, the largest growth prospects in the next few years will likely be in government services, and the company will seek a government 'anchor tenant' for the network.

CIVIL SPACE ORGANISATIONS

Launch Facilities
Multi-national Agencies
National Agencies
Research & Development
Satellite Operators

LAUNCH FACILITIES

Brazil

Alcântara

Location: 2°28'S/44°38'W

Current Status
Brazil continues to develop its indigenous launch vehicle, the *Veiculo Lançador de Satelite* (VLS). On 22 August 2003, the third VLS launch vehicle exploded while being prepared for flight three days hence when one of the four solid propellant strap-on boosters ignited unexpectedly resulting in the death of 21 people. This explosion, which could be heard 18 km away, completely destroyed the launch pad. The Brazilian Space Agency (AEB) hopes to make a fourth successful attempt to launch the VLS, which could occur around 2007. Brazilian space officials continue to negotiate several agreements with other nations, including the Ukraine, Russia, China and the US, to construct and provide launch facilities, as well as to exchange launch technology.

Background
The Alcântara Launch Center (CLA) was expanded on a 520 km² site on the Atlantic coast outside Sao Luis to handle the VLS orbital launcher. The facility also had pads for the Sonda 3 and Sonda 4 sounding rockets, meteorological rockets and other science vehicles (NASA, for example, launched four Nike Orions there in 1994). The launch facility's position near the equator offers 25 per cent greater initial launch energy due to the Earth's rotation compared to Cape Canaveral.

Brazilian government and military officials held a formal opening for Alcântara in February 1990. The fifth and last Sonda 4 was launched from this site on 21 February 1990. The first two unsuccessful VLS flights were attempted on 2 November 1997 and 11 December 1999.

China

Jiuquan satellite launch centre

Location: 41.10°N/100.30°E

Current Status
On 12 October 2005, Chinese astronauts Fei Junlong and Nei Haishung rode aboard Shenzou 6, and maintained their craft in LEO for five days. Shenzou 6 was launched from Jiuquan as a human-rated CZ-2F payload.

The manned space flight plan that began in 1992 and grew out of the 1996 to 2000 Five Year Plan saw development of a new launch complex at Jiuquan to support flight operations with the CZ-2F, developed from the CZ-2E first flown in 1990. The first flight of the CZ-2F carrying an unmanned precursor took place on 20 November 1999. Further unmanned test flights of the Shenzhou spacecraft took place on 9 January 2001, 25 March 2002 and 29 December 2002. It is from this site that the first Chinese astronaut, Yang Liwei, aboard Shenzhou 5, was sent into space, an event that took place on 15 October 2003 when the astronaut was placed in a 343 km circular orbit and returned to Earth later the same day.

The China Great Wall Industry Corporation (CGWIC), an arm of the nationally owned China Aerospace Science and Technology Corporation (CASC), markets and provides Jiuquan launch services to international commercial customers. Up to the end of 2004, Jiuquan had hosted at least 28 orbital launch attempts.

Background
The Chinese built Jiuquan in the late 1950s in the Gobi desert 1,600 km west of Beijing near the Soviet border, north of Jiuquan city in Gansu province, although later the area was redistricted and now belongs to China's Inner Mongolia region. The first launch took place on 1 September 1960; it was also the first Chinese launch of a Chinese R-2, and was followed on 5 November 1960 by an indigenous Chinese R-2. The first attempt at an orbited launch on 1 November 1969 was a failure but the first successful launch came on 24 April 1970. Hundreds of missile tests conducted at Jiuquan have been reported, including into the Lop Nor nuclear test site.

Formerly known in the West as Shuang Cheng Tzu (the former US Space Command/Strategic Command designator), Jiuquan was China's first launch site but it is limited to southeastern launches into 57° to 70° orbits to avoid over-flying Russia and Mongolia. The site is also used to launch sounding rockets. It consists of two pads 416 m apart, with a shared mobile service tower that can be employed separately for CZ-2/FB-1 and CZ-1 space vehicles. In the early 1980s, the latter added the capability for CZ-2C missions. CZ-2D flew from this pad as well, beginning in 1992.

The site's base area (called Dong Feng, 'East Wind') houses the Huxi Xincun range control centre, power station and residential accommodation. The technical centre, where vehicles are assembled and checked out, is 18 km north. Supporting facilities BL 1 and BL 2 can each integrate two launchers and several satellites simultaneously. BL 1, an enormous 140 m long building has an 18 × 90 m test and assembly hall, where the launch vehicles are delivered on rail from the connected transfer hall. There are 25 test rooms down each side of the main hall, including a class 100,000 satellite assembly and test room. The pads are another 30 km further north.

Within the chamber, 41.6 m towers share a 55 m mobile service tower on 17 m span rails for preparation and delivery of the launch vehicle. The mobile tower's maximum speed is about 15 m/min. It has 15/5 tonne-capacity main/auxiliary cranes. To prepare the satellite, commercial users have a class 100,000 clean room which employs a 6 × 6.5 m 13.1 m high cabin to protect the vehicle's upper section. Temperature can be maintained at 20 ±5°C, with 50 to 70 per cent humidity.

A launch control centre is 200 m from pad 1. Both road and rail systems link the facility to the nearest airport, 94 km to the south. A 270 km line connects to the main railway network so satellites can be delivered for launch from virtually any point in China. Jiuquan will continue to be used primarily for recoverable Earth observation and microgravity missions but, because of its geographical constraints, greater commercial activities may be focused China's other launch bases.

Taiyuan satellite launch centre

Location: 37.8°N/111.5°E

Current Status
Taiyuan satellite launch centre is located in northern Shanxi Province, nearly 300 km away from Taiyuan city, the Provincial capital. The site is used as a missile testing range as well as for placing satellites into polar orbit. Taiyuan is equipped with a command and control centre as well as tracking and telemetry facilities.

Taiyuan Satellite Launch Centre in 2004 launched the second Double Star satellite, an ESA-Chinese joint project designed to study the Earth's magnetosphere.

The China Great Wall Industry Corporation (CGWIC), an arm of the nationally owned China Aerospace Science and Technology Corporation (CASC), markets and provides Taiyuan launch services to international commercial customers.

Background
Taiyuan was inaugurated on 6 September 1988 with the first of the CZ-4A designed to place satellites into polar orbits for remote sensing, meteorological and reconnaissance missions. The second launch was made during September 1990.

Taiyuan was initially operated for missile testing as an extension of Jiuquan for larger vehicles, including the DF-5. The first photographs of the site were released in 1990. CZ-2C launches began with a load of Iridium satellites. US Strategic Command (which absorbed US Space Command) continues to call the site Wuzhai.

Jiuquan provides two pads for CZ-1/2 launchers but the site's location limits missions to 57 to 70° 0517071

Long March 4 at the Taiyuan launch site used for Sun-synchronous missions (Theo Pirard) 0054384

Wenchang Satellite Launch Center, Hainan

Location: 19.5°N/110.5°E

Current Status
According to People's Daily and Xinhua, China's state-owned news agencies, development plans for Wenchang Satellite Launch Center, located on Hainan Island, off China's Southern cost, are complete and awaiting central government approval.The launch site is located in the Wenchang district, on Hainan's Northeast coast, about 60 km East of Hainan's capital, Haikou; it will reportedly cover an area of about 20 km². The current plan, as outlined by the media, includes the launch centre, the mission control centre, launch vehicle assembly facilities, and a science theme park.

Background
Discussions regarding the development of a Satellite Launch Centre located on Hainan have continued for many years. The launch site, once built, will be capable of launching the next generation Chang Zheng, or Long March rocket series and may serve as an additional site from which the country will operate its burgeoning manned spaceflight programme. Long term development and construction plans could last several years before the site is active; media reported in 2007 that the site could conduct its first launch by 2010. China has operated a sounding rocket launch site on Hainan island since the 1980s. The island's proximity to the Equator makes it an advantageous spot from which to launch satellites into geosynchronous orbit, as well as LEO, increasing launch efficiency.

The sounding rocket site on Hainan Island was inaugurated in late 1988 with the launches of four Zhinui 1 ('Weaver Girl') vehicles over Beibu (Tonkin) Gulf. A Zhinui 3 attained 120 km in January 1991. In the previous Five Year Plan, the Chinese began to upgrade the facilities at Hainan to handle 300 km-altitude vehicles.

Xichang launch site used by CZ-2E and CZ-3 rockets for geosynchronous transfer (Theo Pirard) 0054385

Xichang: the main launcher processing facility can handle three vehicles simultaneously. A CZ-2E requires all three tracks because of the strap-ons 0517072

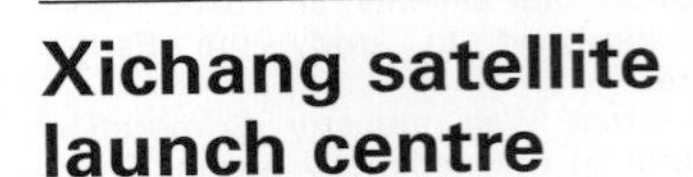

Xichang satellite launch centre

Location: 28.25°N/102.0°E

Current Status
Xichang Satellite Launch Centre is located in southern Sichuan Province, in southwestern China, north of Xichang city. The site has been used for several commercial launches, so far dominating China's other launch centres in this respect. Xichang's position in southwest China makes it more suitable for placing satellites into geosynchronous orbit.

CZ-2E ready for launch at Xichang. The older pad is in the background 0517250

In 2003, Xichang Satellite Launch Centre launched the first of the Double Star satellites, a joint ESA-Chinese project designed to study the Earth's magnetosphere. Taiyuan Satellite Launch Centre launched the other in 2004. By the end of 2005, the Xichang launch site had hosted nearly 40 orbital launch attempts.

Xichang was also the launch site for China's three Beidou GPS satellites in 2000 and 2003.

The China Great Wall Industry Corporation (CGWIC), an arm of the nationally owned China Aerospace Science and Technology Corporation (CASC), markets and provides Xichang launch services to international commercial customers.

Background
Xichang was selected from a shortlist of 16 sites, of 81 surveyed, for a more favourable GEO mission base than Jiuquan. Construction work in the canyon site at the foot of Mt Liang Shan, 65 km north of Xichang city, began in 1978 and resulted in the first launch in January 1984. The site hosted its first commercial mission when it launched the Hong Kong AsiaSat 1 communications satellite on a CZ-3 in April 1990, followed by Australia's two Optus Bs in 1992.

The local population has traditionally been allowed to live close to the pads. When the maiden CZ-3B crashed on 14 February 1996, into a hillside 1.5 km from the newer complex, the official death toll was six plus 57 injured, and 250 homes were damaged or destroyed. When the CZ-2E of 25 January 1995 exploded at 51 seconds, the debris killed six and injured 23 others in a village 7 km downrange.

Xichang's Mission Command and Control Centre 0517251

The dry season, running from October to May, provides the most suitable time for launch campaigns, but the site enjoys 320 days of sunshine annually. Annual temperature range is –10 to +33°C, because the site sits at 1,839 m above sea level. Xichang's main airport, upgraded to handle B-747 and C-130s, is 50 km south of Xichang city. The site lies within a PLA military installation and administration of the site includes co-operation with security interests of customers selecting China's launch vehicles.

The single CZ-3 pad can support up to five missions annually. It has a fixed 77 m high/11-level gantry. In order to set the launch azimuth a technician hand cranks a firing table. The firing room is buried in the hillside 60 m away. Originally, the Chinese stacked vehicles on the pad and mated the payload in the open air but, as part of the CZ-3 commercial marketing, a stage 3/payload class 100,000 clean room was added.

Vehicle stacking requires about two days, followed by 14 days of checkout with a launch rehearsal on day 15 aiming for the actual launch on day 20. Stage ½ UDMH loading from underground reservoirs begins 16 hours before launch, followed by N_2O_4. Stage 3's cryogenic loading about 5½ hours before launch. 36 days are allowed for launcher and payload checkout.

The second complex is sited 300 m distant for CZ-2E/3A/3B vehicles with add-on boosters. It was first used on 16 July 1990. The pad is at 28.25° N/102.02° E, 1,826 m above sea level. Like the original pad, this one has a fixed 74 m 1,025 tonne tower plus 98 m 4,300 tonne mobile tower. Just before launch the ground crew withdraws the tower 130 m.

Payloads are encapsulated within their fairings in the class 100,000 preparation facilities and delivered to the pad for vehicle mating. Vehicle erection begins 11 weeks before launch, the encapsulated payload is mated at six weeks, launch rehearsal concluded at three weeks and fuelling completed at one week.

The Mission Command and Control Centre (MCCC) lies 7 km southeast of the pads and was upgraded 1994–95. At this site, a communications centre includes an 11 m Intelsat B Earth station. Vehicle/payload processing facilities are 2.2 km from the pads in the technical centre.

Launcher stages are received by rail directly inside a 31 × 14 m transit hall and transferred to an associated 92 × 28 m assembly room capable of processing three complete vehicles. The launchers are assembled, checked out and then broken up for transfer by road for stacking in stages on the pad. The Non-hazardous Operations Building (BS2) houses the main 756 m² 42 × 18 × 18 m satellite test hall, class 100,000, climate-controlled (20 to 25°C, humidity 35 to 55 per cent) facility. Personnel access it by an airlock with 5.4 m wide 12.5 m high door. The building includes offices, test rooms and storage.

The associated Hazard Operations & Fuelling Building (BS3) is a 324 m² spacecraft fuelling and assembly hall, with a class 100,000 clean room. The door is 5.4 m wide 13 m high. Solid kick motors are installed in here and the payload encapsulated in its fairing.

After the stack begins its preparation, the Checkout/Preparation Building (BM) is used to inspect the solid motors. It is a 108 m² 12 × 9 × 9.5 m hall (door 3.6 m wide 4.2 m high). Next door sits the SPM x-ray Building (BMX), a solid motor test facility with an x-ray hall sized 125 m² 12.5 × 10 × 15.4 m high (the door is 3.2 m wide 4.5 m high). In the same building the Cold soak chamber, 9.6 m² 3.3 × 3.0 × 4.0 m, door 3.2 m wide, 3.5 m high, provides 0 to 15°C.

India

Balasore Rocket Launching Station

Location: 21°25′N/87°00′E

Current Status

Balasore is a sub-orbital launch site in Chandipur in the state of Orissa, on India's northeast coast. The site is used to launch meteorological sounding rockets and is also in service as a military missile test range.

On 9 July 2006, an Agni-III IRBM was launched from the Wheeler Island Integrated Test Range, also known as launch complex IV. However, the missile developed second stage problems, and dropped into the Bay of Bengal, failing to hit its test target.

Background

Balasore was set up by Indian Space Research Organisation (ISRO) and has been in operation since 1989 as a meteorological research station supporting sounding rocket flights from facilities adjacent to weather balloon release sites. Balasore has been considered for due-south departures into polar paths, enhancing PSLV's Sun-synchronous capability by 500 kg. Balasore currently supports RH-200 sounding rockets.

Satish Dhawan Space Centre (SDSC), Sriharikota Range (SHAR)

Location: 13°41′N/80°14′E

Current Status

Launch facilities are now available for sounding rockets, the Satellite Launch Vehicle (SLV-3), the Augmented Satellite Launch Vehicle (ASLV), the Polar Satellite Launch Vehicle (PSLV) and the Geosynchronous Satellite Launch Vehicle (GSLV). Modifications for GSLV, the first of which was launched on 18 April 2001, have provided facilities for the eventual commercialisation of the launcher. The second development GSLV was launched on 8 May 2003 carrying the 1,825 kg GSAT-2 experimental satellite into GTO. The third GSLV successfully launched Edusat on 20 September 2004 by the first operational launcher in this series, designated F01.

On 10 July 2006, 17:38 IST, GSLV-F02, the GSLV's second operational launch from the Satish Dhawan Space Centre (SDSC), and the first GSLV launch since 2004, ended in catastrophic failure. The launch vehicle, which carried the 2,168 kg INSAT 4C commercial broadcast communications satellite developed an anomaly during the booster separation stage and had to be destroyed by mission control over the Bay of Bengal. Within two days of the failure, a 15-member Failure Analysis Committee was organised to delve into the detected launch anomalies and reasons behind the failure. Initial reports indicated that one of the GSLV's four strap-on motors lost pressure. ISRO expects to launch its next GSLV as soon as mid-2007.

Background

Originally known as the Sriharikota Range (SHAR) and later named after Satish Dhawan, India's primary orbital launch site is housed at the Indian Space Research Organisation's (ISRO) SDSC on Sriharikota Island, on the east coast of Andhra Pradesh, approximately 100 km north of Chennai. SHAR began launching rockets in October 1971, however, ISRO's first rocket firings were made from the Thumba Equatorial Rocket Launching Station (see separate entry) in 1963. Together with the northerly Balasore Rocket Launching Station, the facilities are operated under the ISRO Range Complex (IREX) headquartered at SHAR. SHAR launches the full range of Indian vehicles from sub-orbital up to the GSLV but range safety considerations limit SHAR to less than 140° launch azimuths. The Centre has two operational orbital launch pads.

SHAR is ISRO's satellite launching base and additionally provides launch facilities for the full range of Rohini sounding rockets. The Vehicle Assembly, Static Test and Evaluation Complex (VAST, previously STEX) and the Solid Propellant Space Booster Plant (SPROB) are located at SHAR for casting and testing solid motors.

The ASLV orbital launcher was integrated vertically, beginning with motor and subassembly preparations in the Vehicle Integration Building (VIB) and completed on the pad within the 40 m tall Mobile Service Structure. The PSLV launch complex was commissioned in 1990. It has a 3,000 tonne 76.5 m high Mobile Service Tower (MST) which provides the SP-3 payload clean room. The MST is rolled back 180 m for launch. SHAR also has a 52 m umbilical tower, an open steel structure next to the launch pedestal. Modifications were completed to allow the complex to handle the GSLV. The Launch Control Centre (LCC) is 5.6 km distant and the CTR Checkout Terminal Room is next to the pad.

Liquid stage fuelling is done remotely from the LCC. All solid motors are processed in the SMPF Solid Motor Preparation Facility. As a launch campaign nears completion, stage 1's segments are stacked on the pad. Satellite preparation/integration is undertaken in three SP clean room facilities. SP-1 at the LCC has three bays: class 100,000 17 × 15 × 8 m high bay, materials airlock bay and dynamic balancing room. It provides satellite inspection, subsystem tests, RCS leak tests and solar panel deployment tests. SP-2 next to the pad is used for RCS filling and checking compatibility with PSLV's adaptor. SP-3 is part of MST for integrating payloads with the launcher.

PSLV's Mobile Service Tower provides payload clean room facilities. At bottom is the Checkout Terminal Room for vehicle testing before the more distant Launch Control Centre takes over (ISRO) 0517080

The VAST was originally designed to handle motors up to 1 m diameter, 10 m length and 980 kN. ISRO modified it for PSLV for 2.8 m diameter and 4.9 MN levels (first PSLV stage 1 test was made in 1989). The horizontal firing stands provide 1-/6-component facilities. The vibration platform was upgraded from 13.5 tonnes to a 30 tonne dual shaker operating over 5–2,000 Hz with maximum acceleration of 100 g. Specimens up to 2 tonnes can be shock tested to 80 g by free-fall to simulate motor ignition shocks, and upper stage motors are fired in a 50 km-equivalent high-altitude test facility. The 25 km² SPROB was commissioned in March 1977 for SLV-3's motors but it was then augmented to cope with PSLV's 2.8 m diameter 25 tonne segments, requiring a 2.5 tonne vertical mixer and 15 MeV linear accelerator for the radiographic facility in addition to the previous 8 MeV machine.

Approximately 2,400 people were employed at the show centre in 2005.

Thumba equatorial rocket launching station

Location: 8°32′N/76°52′E

Current Status

The Thumba Equatorial Rocket Launching Station (TERLS) is located in southern India, in Kerala, on the Arabian Sea and near Thiruvananthapuram, Kerala's capital. The facility retains three launch pads and remains true to its roots as a suborbital launch centre.

India's Vikram Sarabhai Space Center (VSSC – see separate entry) was built on the TERLS site and includes some of the original TERLS infrastructure. The Liquid Propulsion Systems Center (LPSC) is also nearby.

Background

TERLS supports RH-200/300 sounding rockets. Centaure, Nike Apache and Judi-Dart rockets were launched there through the 1960s, followed by the RH series and other types. Thumba was the site of India's first sub-orbital Nike Apache launch in 1963. In 1968 TERLS was dedicated to the United Nations.

Indonesia

Pameungpeuk

Location: 1.0°S/136°E

Current Status

The Pameungpeuk rocket launching station is a sub-orbital launch site located in Garut district, West Java. As of 2005, the launch station employed 51 personnel. Indonesia's island of Java is located within the Pacific Rim 'Ring of Fire' and was affected by the 2004 Asian Tsunami, as well as the Java Tsunami in July 2006.

Background

Pameungpeuk launched its first experimental rocket, a Kartika I, on 14 August 1964. The Indonesian National Institute of Aeronautics and Space (LAPAN) launches its one- and two-stage solid propellant sounding rockets for upper atmospheric research from the Pameungpeuk range. Indonesia has unsuccessfully sought to attract foreign finance for developing the site as an international research station. These efforts have attracted potential foreign investors but there is no interest from the Indonesian government in supporting such an initiative. China considered contracting with Indonesia to build an equatorial launch site nearby for China's CZ-3 during 1987. One possible location under consideration at the time was Biak Island. Such plans have now been abandoned.

International

Baikonur Cosmodrome (Tyuratam)

Location: 45.6°N/63.4°E

Current Status

On 26 July 2006 a Dnepr (see separate entry) rocket that was scheduled to have launched 18 medium- and small-satellites into LEO, failed. The accident generated up to three crash sites across Kazakhstan and renewed Kazakh concerns over possible environmental damage caused by Baikonur's launch accidents. The Dnepr, adapted from R-36M2 ICBM technology, was to have launched Belarus' first remote-sensing satellite. The rocket also carried science satellites built by the US, Norway, Korea, Italy and Russia. Kazakh officials have suspended Dnepr launches out of Baikonur for the foreseeable future due to the accident.

On 22 December 2004, the Russian Federation and Kazakhstan signed an agreement to jointly fund a launch complex for the new Angara series (see separate entry) launch vehicle at Baikonur. The Angara remains in development and the first test flights are due to occur, most likely at Plesetsk Cosmodrome, in 2010–2011. The Baikonur launch system has been dubbed the Baiterek project and will provide Angara with processing and launch facilities. Four years of construction work is planned, and Kazakhstan has committed a USD225 million loan to the project over 19 years, beginning in 2006. In 2005, Kazakhstan agreed to continue leasing Baikonur to the Russian Federation through 2050.

In January 2006, Russian Space Forces Commander, Colonel General Vladimir Popovkin, announced that troops would permanently relocate from Baikonur to Plesetsk Cosmodrome as soon as 2007–2008. Popovkin later expanded upon this statement in August 2006 by stating that 4,224 troops would be affected: 707 Baikonur personnel will be retained and moved to Plesetsk, and 3,517 will transfer to reserve service. The downsize is part of a broad decrease within the Russian Federation armed forces. The Space Forces' exit from Baikonur will mark the end of defence related launches at the Cosmodrome.

Sea Launch LLC (see separate entries), the successful Boeing, RSC Energia, SDO Yuzhnoye/PO Yuzhmash, Aker ASA joint venture, announced in 2003 that the company would expand to provide launch services at Baikonur, beginning in 2007. The new concept is known as Land Launch and is operated in partnership with the subcontractor, Space International Services, headquartered in Moscow. The Group will use the Zenit-3SLB, based upon Sea Launch's successful Zenit-3SL, to launch medium-weight satellites into GTO from Baikonur. Heavy-lift services will also be made available for LEO payloads. Commercial customers' satellites will launch from the existing Area 45 Zenit complex, as well as using the existing Baikonur Right Flank facilities, known as Areas 42 and 31 for launch vehicle integration and payload processing. Land Launch also uses Area 254 in Baikonur's Centre section for upper stage processing, and plans in the future to use it as the payload processing facility.

Background

All Russian Federation manned flights and planetary missions are launched from Baikonur. The site was originally designed and built as an operational missile launch and test site and still retains those duties, however, with the imminent departure of Russian Space Forces from the site, it is likely that military activity will cease at Baikonur. Due-east azimuth launches (the most efficient) are prohibited out of Baikonur because the lower stages would impact in China.

The Baikonur Cosmodrome has traditionally been split into three sectors: the Right and Left Flanks and the Centre. Nine launch complexes with a total of 15 pads are offered between all sectors, to accommodate space and missile activities. It is the only cosmodrome supporting Proton, and Zenit launches. It also provides services for Soyuz, Molniya (launched primarily from Plesetsk), Tsyklon, Rockot (launched primarily from Plesetsk), Kosmos and various missile and missile-derived rocket launches. It will also conduct Angara launches after the completion of that planned launch complex.

The initial R-7 launch complex, Area 1, in the Centre sector was designed and built by a team led by VP Barmin, to house a site for missile and rocket tests which began in the area in 1955. The first pad built at Baikonur launched both Sputnik 1 and Yuri Gagarin into orbit. Today the refurbished pad hosts Soyuz launches.

Baikonur's Gagarin pad has supported more than 400 orbital launches. Vehicles are moved to the pad horizontally by rail – the transporter is visible at right 0517074

A Japanese astronomer pinpointed the site's location in 1957 as being near Tyuratam in Kazakhstan, east of the Aral Sea and about 370 km southwest of the original Baikonur. However, the Soviets continued for at least 17 years to give its latitude/longitude as that of the original town of Baikonur. The Cosmodrome continues to be known as Baikonur, and the town that grew alongside it was later renamed as such, although the site is actually located in a large expanse north of the Tyuratam rail station. Not until Soyuz-T 6 in June 1982 was a Western journalist admitted, but US reconnaissance satellites and U2 flights provided detailed information for the military on the site's whereabouts, while Landsat and Spot images brought the sprawl of pads into the public domain in 1974.

Baikonur town, previously known as Leninsk, and before that as Zarya (Dawn), is 2,100 km from Moscow and was founded 5 May 1955 as Baikonur Cosmodrome's residential area. It has an official population of about 60,000, down from the 100,000 peak of the mid-1980s, and grew up alongside the launch facilities as the administrative centre of the 1,560 km² Cosmodrome. 46,011 km² is allocated for downrange stage impacts. The Cosmodrome has several hotels and residential areas, closer to launch facilities, and separate from the main town. The climate can be extreme, +60°C in summer, while suffering violent snowstorms and –40°C frosts in winter.

Soyuz and Progress spacecraft are integrated and launched in the Centre sector. The outdated N1/Energia pads are also located in the Centre. Proton, Tsyklon and Rockot launch and processing facilities are on the Left Flank, and Zenit and Kosmos facilities are located on the Right Flank. Also on the Left Flank, is the 4,500 m long, 84 m wide Yubileiniy landing strip on a northeast-southwest orientation. It supported the Buran landing and jet-powered test and training flights. Operations were controlled from a nearby building. Silos supporting various missile types (most no longer in service) are spread over all three Sectors.

Centre Spatial Guyanais (Kourou)

Location: 5°14'N/52°46'W

Current Status

The Centre Spatial Guyanais (CSG), or the Guiana Space Centre, in partnership with France's Centre National d'Études Spatiales (CNES), Arianespace, Starsem, the Russian Federal Space Agency, TsSKB Progress, KBOM and NPO-Lavotchkin has planned, since 2002, to launch Soyuz rockets from CSG. Beginning in 2004, European Space Agency (ESA) member states Austria, Belgium, France, Germany, Italy, Spain and Switzerland have financed the Soyuz launch complex construction project as well as the technical adaptation of the Soyuz for successful launch in Kourou's equatorial and tropical conditions. France and Russia have, since 2003, had an international agreement to allow Soyuz launches from the site, including those of human-rated launch vehicles.

The Ariane 5 pad. The 90 m tower holds 1,500 m³ of water for trench cooling and sound suppression. Four pylons provide lightning protection (ESA) 0517428

The 1.2 km² Soyuz launch site falls under Sinnamary town (about 18 km away) jurisdiction and is 27 km northwest of Kourou; it is around 20 km distance from the Ariane 5 Launch Complex (LC). The new Soyuz LC, Ensemble de Lancement Soyuz (ELS), will include payload and launch processing facilities, the launch pad and a control centre. The LC's projected cost is EUR135 million. CSG expects that the first flight from the site will occur sometime in 2008, after construction is complete.

Vega, ESA's small satellite launcher, will also be launched from CSG. ELV SpA, a joint venture formed by Avio and Agenzia Spaziale Italiana (ASI), the Italian Space Agency, is developing the Vega, which will launch from CSG at the Ensemble de Lancement Ariane 1 (ELA 1) complex. Originally built and used to launch first Europa rockets, and then Arianes 1-3, the LC is being refurbished to accommodate the Vega. The complex will be renamed Zone de Lancement Vega (ZLV). Vega operations will share launch control facilities with the Ariane 5 complex. The first Vega launch is expected to take place toward the end of 2007.

Arianespace will be the launch operator for both Soyuz and Vega flights as well as continuing to launch the Ariane 5. The three launchers will enable ESA to launch small, medium and large payloads aboard separate launch vehicles. As of 2006, Kourou's only operational orbital launch pad is ELA 3, used to launch Ariane 5.

Background

CSG is 'Europe's Spaceport' and is used by ESA to launch Ariane 5 Launch Vehicles (LV). On 15 February 2003, the last Ariane 4 was launched from the site, making it the 159th Ariane series flight from CSG.

Although owned by CNES (see separate entry), CSG is made available to ESA (see separate entry) and Arianespace under a governmental agreement that guarantees access to the ESA-owned Ariane launch facilities. CSG is the most favourable major site for GEO launches, its near equatorial position providing a 15 per cent payload advantage over Cape Canaveral for eastward launches. French Guiana's coastline permits launches into both equatorial and polar/Sun-synchronous paths, with inclinations up to 100.5° possible.

ESA funding makes up about two thirds of the CSG annual budget, although ESA's share includes

A MAN tractor pulls the Ariane 5 launch table from the Final Assembly Building (ESA) 0054382

French contributions to the Space Agency. In 2005, ESA provided more than EUR82.7 million in operations funding to Kourou, and in 2004 it supplied nearly EUR87 million. CNES provides the remaining funds for the spaceport. ESA also supplies a large amount of resources toward the construction and maintenance of CSG's infrastructure. As of 2004, ESA had contributed EUR1.6 billion to various CSG ground facilities projects. CSG employees work under secondment to CNES if they are initially employed by other agencies or companies. As of 2004, CSG staff totalled 1,360. A large number of contractors also work at the site.

CSG covers 750 km^2 and extends more than 30 km along the French Guiana coast from Kourou town to Sinnamary. Ariane launch facilities are about 18 km northwest of Kourou. CSG's Technical Centre (TC), close to Kourou, incorporates Spaceport administration, logistics and operations functions, including the Jupiter Control Centre. The first Ensemble de Préparation Charges Utiles (ECPU), or satellite preparation facility, in service, the S1 complex, is also located at the TC but is presently inactive. Payload processing takes place at a separate four-building complex called ECPU S5, completed in 2002, on the Ariane 5 ELA 3 complex.

The Jupiter 2 Control Centre handled its first campaign in January 1996. The preceding Jupiter Control Centre had been operational since 1968. Radar and telemetry stations are sited at Montagne des Pères , 20 km southeast of the Technical Centre, with downrange stations for easterly trajectories in Natal, Brazil; Ascension Island; N'Koltang, near Libreville in Gabon; and Malindi in Kenya.

Detailed study of 14 potential launch sites worldwide resulted in the selection of Cayenne, French Guiana, as the CNES spaceport site in 1964. Kourou town, northwest of Cayenne, the spaceport, and air and sea port facilities were constructed by more than 2,500 workers between 1965 and 1968. CSG became operational in April 1968; eight flights of France's Diamant B were conducted from the site from March 1970, followed in 1971 by the Europa launch vehicle. The site has also witnessed more than 60 balloon flights (the last occurring in 1980).

CSG's meteorological conditions reflect its tropical location. The mean daytime temperature is 26.4°C, with maximums of 34°C, and minimums of 18.1°C. The windspeed has an annual mean of 3.3 m/s, a maximum of 23 m/s. Annual rainfall is 2.9 m, 438 h/year over 235 days.

Ariane 5 launch requirements preclude launches if storms exist within 10 km at T–10 minutes or if convective clouds exist at 5,000 to 6,500 m altitude at T-0. These criteria have been used since Ariane 4, flight 19.

Integration of Ariane 5.501 in the launcher integration building

0084630

Ariane 5 emerges from the BAF final assembly building (ESA) 0054383

The Launcher Integration Building, in the preparation area (ESA) 0054386

For details of the latest updates to ***Jane's Space Systems and Industry*** online and to discover the additional information available exclusively to online subscribers please visit
jsd.janes.com

The main entrance of the Centre Spatial Guyanais, Kourou, French Guiana 0084626

CSG continues to maintain its four-pad sounding rocket launch site, although it is seldom used for launches; its buildings are occupied by payload preparation personnel. CSG's orbital LCs are:

ELA 1/ZLV
ELA 1, once used to launch Ariane versions 1–3, has been redesignated for Vega launches. The LC is undergoing reconstruction and the first Vega launch is scheduled to occur in 2007.

ELA 2
ELA 2 was used to launch Ariane 4 rockets but has since been deactivated. The final Ariane 4 to fly from CSG lifted off on 15 February 2003.

ELA 3
ELA 3 is CSG's sole active orbital launch complex. It is equipped with two pads and is capable of launching eight to 10 Ariane 5s per year. Launch campaigns last 20 days. The 20 km² LC includes separate vehicle launch and preparation zones. The preparation area buildings include the ECPU S5, the Batiment d'Intégration Lanceur (BIL) launcher integration building where the core is erected on the mobile launch platform and the boosters added, and the Batiment d'Assemblage Final (BAF) final assembly building for completing the vehicle, including payload installation. Launch vehicle main stage cryogenic fueling occurs at the launch zone. Ariane 5 operations facilities are located in the number 3 launch control centre (CDL3), on the ELA 3 site.

From concept and design phases through to ELA 3 completion, the Ariane 5 LC cost a total of EUR800 million to build.

ELS
Groundbreaking and earthworks began in 2005 for the future Ensemble de Lancement Soyuz (ELS). LC construction will continue into 2008, when the first Soyuz launch is expected to take place.

Japan

Tanegashima Space Center (TNSC)

Location: 30°24'N; 130°58'E.

Current Status
Two H-IIA (see separate entry) flights were launched from Tanegashima in 2003; the November launch carrying the remote sensing satellite IGS-2 was a failure when one of its solid rocket boosters failed to separate as planned. After several launch delays, the next H-IIA was launched from Tanegashima in February 2005, its payload the Space Systems/Loral FS-1300 series MTSAT-1R, or Himawari-6 satellite. Two further successful satellite launches, ALOS and MTSAT-2 followed in January and February 2006 respectively.

Background
The Japan Aerospace Exploration Agency's (JAXA) 8,600,000 m² TNSC orbital launch centre is sited on the southeast tip of Tanegashima Island, Kagoshima Prefecture, 1,000 km southwest of Tokyo. TNSC's northern Osaki Range includes the Yoshinobu Launch Complex, from which H-IIA rockets are launched, and a second pad from which the J-I, launches took place. By the end of 2005, TNSC had conducted 38 orbital launches. The Takesaki Range to the south handles sounding rockets and provides facilities for H-II solid booster static firings and the H-II Range Control Centre.

With few exceptions, departures from both ranges are normally restricted to 45-day January-February/August-September 'launch seasons' because of range safety procedures and the influential fishing industry lobby. The solid strap-ons of JAXA's H-I formerly impacted some 19 to 29 km out, but H-II's fall beyond the 190 km range safety limit. Rocket Systems Corporation may continue to negotiate to increase launch opportunities for the H-IIA. Although there may be potential for year-round firings, launch failures, such as the IGS-2 failure, would still be potentially hazardous.

The Takesaki Range at 30°22'25"N/130°57'56"E, sits in the southeast section of Tanegashima and is used to launch small suborbital solid propellant sounding rockets. Its pad is activated as required for tests related to JAXA programmes. Sounding rockets flown from Takesaki include the TR-1A, flown from here since 1991, as well as the LS-C, JCR, and TT-500A.

The H-II's 150,000 m² Yoshinobu complex was built from 1985 to 1992 for launching the larger vehicle from the headland west of the J-I site. H-IIs are integrated in the Vehicle Assembly Building on a mobile platform. A single platform is adequate for two annual launches, but increasing the rate to four would require a second platform.

The pad at 30°23'45"N/130°58'22"E handled its last H-I launch in February 1992 and was then modified for the heavier J-I. J-I F1 was launched in 1996 and carried the Hypersonic Flight Experiment test plane (HYFLEX) on a suborbital flight. J-I F2 was originally scheduled to depart in late 2001. The J-I was JAXA's proposed small satellite rocket, capable of lifting 1,000 kg. It reportedly proved cost prohibitive to launch and the programme was suspended by the National Space Development Agency (NASDA) previous to the J-I F2 flight. The J-I programme has remained inactive; NASDA was later folded into JAXA, upon that agency's creation in 2003. The J-I used the combined technologies of the H-II solid rocket booster and the M-3S II's upper stage.

Satellites are processed in the class 100,000 3,930 m² Second Spacecraft Test & Assembly (STA2) building and then moved to the 4,010 m² Spacecraft & Fairing Assembly (SFA) building, where propellants are loaded, pyrotechnics and solid motors inserted, and the fairing is added. The complete payload assembly is then transferred to the pad to be attached to the waiting H-IIA.

H-II's stages and solid rocket boosters (SRB) are received in the vehicle assembly building's (VAB)' 32 × 46 m 19.1 m high low bay before hoisting on to the mobile launcher (ML) platform in the 32 × 27 m 66 m high bay in solids/stage 1/stage 2 order. The 800 tonne 18 × 22 m 4 m high ML is powered by electric motors on four dollies at 0 to 0.48 km/h for the 500 m rail transfer to the pad. ML provides LOX/LH_2 to stage 1 and is protected by a water deluge system. The exhaust opening is 4 × 9 m.

The Pad Service Tower (PST) comprises a fixed tower plus two rotating service structures (RSS). The 8 × 15 m 67 m high 1,500 tonne tower provides 10 floors, H-II umbilicals and payload assembly installation. The 2,500 tonne main RSS provides 12 floors, the top two adjustable, and a 20 tonne crane. The 800 m² launch blockhouse, submerged 4.8 m, is next to the VAB. Control after lift-off, and up to about 30 minutes later when the payload separates, is the responsibility of the Takesaki Range Control Center (RCC), built because the existing Osaki Center is within the launch danger zone.

Uchinoura Space Center (Kagoshima)

Location: 31.2°N/131.1°E

Current Status
Uchinoura Space Center (USC), in Kimotsuki District, Kagoshima Prefecture, is primarily a suborbital sounding rocket launch site, however USC also conducts Japan's scientific satellite and probe launches. Hayabusa (see separate entry) was launched from the site in 2003, followed by Suzaku (Astro-E2) and Akari (Astro-F) in 2005 and 2006, respectively. More than 360 launches have been carried out from this site since 1970.

Background
Uchinoura Space Center (formerly Kagoshima Space Center) was originally established in 1962

Orbital launch facilities at JAXAs Tanegashima site. The static test stand at centre right is used for LE-7 firings. The H-II is stacked on its mobile platform in the Vehicle Assembly Building at right and moved the 500 m by rail to the pad. At top centre is the J-I pad. Beyond it, the old Osaki Range Control Center stands unused. Launcher and payload processing buildings are at top right (NASDA) 0517427

USC's protected pad is used for sounding rockets. This is the S-520 single stage vehicle, capable of delivering 150 kg to 350 km 0517082

as part of the University of Tokyo's Institute of Industrial Science. In 1964 it became a part of the Institute of Space and Aeronautical Science (ISAS), and in 1981 became an independent research organisation, the Kagoshima Space Center (KSC), although it remained linked to ISAS. When the Japan Aerospace Exploration Agency (JAXA) was formed in 2003, by joining Japan's space organisations, Kagoshima was merged into the space agency and renamed Uchinoura Space Center.

Japan launched its first satellite, Osumi, from the site in 1970. The Sakigake and Suisei Halley's Comet probes of 1985–1988 provided Japan's first experience of beyond Earth orbit.

USC has been restricted, by government directive, to all-solid vehicles of 1.4 m diameter but approval for the 2.5 m diameter M-V was made in 1989 to undertake modest deep space missions.

Japan's first six satellites were launched from these levelled hilltops facing the Pacific Ocean at Uchinoura (Kimotsuki, since 2005) on the southern tip of Kyushu Island. Construction of a sounding rocket site began in February 1962, with extensions for science satellite launchings by M rockets completed in 1966. The first Lambda 4 was fired in 1964 but it was another six years before the first successful orbital launch was achieved.

USC uses the M-V rocket (see separate entry) to launch its science satellites. A satellite launch control room is linked with the Mu Center pad and satellite assembly room.

The KS Center launches S-310 and S-520 sounding rockets as well as the MT-135 meteorological rocket. The annual launch rate is typically three to four for the larger sounding rockets.

Departures from this site are normally restricted to 45 day January-February/August-September 'launch seasons' because of range safety procedures and the influential fishing industry lobby.

In addition to Usuda's 60 m station, JAXA operates a 34 m dish at USC, as well as, 20 and 30 m dishes used for telemetry and tracking purposes.

Norway

Andøya Rocket Range

Location: 69°17′N/16°01′E

Current Status

The Andøya Rocket Range (ARR) is located 280 km north of the Arctic Circle, and is the world's northernmost, permanent suborbital launch site. Andøya is also capable of launching small orbital polar satellites, however this option has yet to be exercised by the European Space Agency (ESA) or any other organisation.

Background

ARR has conducted more than 800 launches since the first sounding rocket, a Nike Cajun, was launched from the site on 18 August 1962. The Norwegian Space Centre owns 90 per cent of ARR, with Kongsberg Defense and Aerospace owning the remaining 10 per cent. As of 2006, Andøya employed about 50 staff. Sounding rockets launched from the site are primarily used to investigate the upper and middle atmosphere in the polar region and to study ionospheric and magnetospheric processes at high latitudes.

The range has been supported since 1972 by ESA through the Special Project Agreement, under which it is maintained by and made available to some ESA states. It is also operated for commercial and bilateral projects. The range has a very large vacant impact area offering users a wide range of possible rocket configurations and launch vehicles without guidance, reducing costs and expanding the user market. The largest launch facility can handle vehicles up to 20,000 kg. Vehicles currently handled include Rohini, Viper, Loki, Orion variants and Black Brant. There are eight pads, including three universal launchers (two with 3 tonne and one with 20 tonne load limits) and a small rail launcher (over-slung type). The universal launcher 3 (U3) with 20.5 m rail length and 20 tonne safety weight limit became available from late 1993.

A vehicle is prepared horizontally in the protective 36 m long 5.5 m wide 5.2 m high concrete housing and raised shortly before launch. At launch, the launch elevation can be set from 65 to 90° (nominally 85°), the azimuth 260 to 20° (nominally 340°). Four-stage rockets up to 23 m long, 1.2 m diameter, 520 kN thrust (Black Brant 12 class) can be launched into an impact area about 1,900 km downrange.

The 272 m^2 Main Assembly Hall accommodates vehicles up to 15 tonnes. The User Science Operations Centre is located 300 m northeast of the control centre with two laboratories and a recording room. Payload recovery is performed routinely from the Norwegian Sea using ships or helicopters. New tracking facilities using radar and slant range systems are designed to improve impact prediction to 500 m for sea-recovered capsules.

The Arctic Lidar Observatory for Middle Atmosphere Research (ALOMAR) is responsible for Andøya's ground based instrumentation. It is owned and operated by ARR.

A downrange telemetry station and a sounding rocket mobile launch facility became operational at Ny- Ålesund (79° N) on Svalbard archipelago during 1998. Climatically, Svalbard is an arctic desert, warmed by the Gulf Stream (mean monthly winter low of –14°C) with 200 to 300 mm annual precipitation. Longyearbyen is the administrative centre, with 1,000 residents and an international-standard airport. The SvalRak launch facility allows four-tonne rockets to climb parallel and perpendicular to the magnetic field, as well as, directly into the polar cusp and cleft regions.

There is a 20 m VLBI antenna to measure continental drift, and the Norwegian Mapping Authority also operates GPS receivers as part of the systems' core network to determine precisely the satellites' orbits.

ARR can handle balloons over 100,000 m^3 and the site has seen about 500 launches. A mobile tracking telemetry unit allows flight times of about 72 h. The balloons drift east in winter and west in summer.

Russian Federation

Kapustin Yar Cosmodrome

Location: 48.4°N/45.8°E

Current Status

Kapustin Yar is primarily a tactical and strategic missile test range. The site has a rich history dating back through the Cold War to the post-Second World War era, however, its launch facilities are now little used for orbital purposes. The last orbital launch to occur at Kapustin Yar was on 28 April 1999, from launch complex 107, when ABRIXAS, a German X-ray astronomy satellite, was launched aboard a Kosmos 3-M rocket.

Kapustin Yar has been used in the past to launch only small satellites and therefore it is unlikely that the site could be used at this time as a Baikonur replacement should the Russian Federation require indigenous facilities such as those it retains by agreement with Kazakhstan.

Background

Founded in 1946 as the Fourth State Central Test Range (GTsP-4), Kasputin Yar, located near the Kazakh border, conducted its first missile launch in 1947. It became a source of renewed interest when it was used in June 1982 for Kosmos 1374, the first test of the subscale spaceplane flights. Its two launches in 1984 included the fourth spaceplane test. Kapustin Yar in later years handled only occasional orbital missions, possibly for radar calibration. Missile testing activities and sub-orbital launches still occur at the site. The facility remains capable of handling Kosmos launch vehicles. Kapustin Yar's first orbital launch was Kosmos 1 in 1962. In 1975, a Landsat 2 image revealed publicly that its facilities extend over a 96 × 72 km area northeast of the River Volga and 965 km southeast of Moscow.

During its early years, Kapustin Yar tested captured German V-2 (A-4) missiles and conducted sounding rocket experiments carrying dogs and other animals up to 500 km. By 1980, there had been 70 orbital launches, mostly small Kosmos science satellites but, as work was switched to Plesetsk, the annual launch rate fell to an average of only one.

Plesetsk Cosmodrome

Location: 62.8°N/40.1°E

Current Status

Plesetsk is an orbital launch site as well as a missile range. ICBM tests are still conducted at Plesetsk regularly.

In January 2006, Space Troops Commander Colonel General Vladimir Popovkin announced that Russian Space Forces would leave Baikonur by 2007–2008 to be permanently based in Plesetsk.

In July 2006, a launch platform large enough to accommodate the Angara (see separate entry in the Launch Vehicles section) was completed and delivered to Plesetsk. Test flights for the launch vehicle, which have slipped several years, are now set to begin in 2010–2011. As of 2006, Plesetsk remains the only orbital launch site on the continent of Europe.

Since 2000, Eurockot, which is 51 per cent owned by EADS Space Transportation and 49 per cent owned by Khrunichev State Research and Production Space Center, has provided Rockot launch services at Plesetsk. On 28 July 2006, Eurockot successfully launched Korea's KOMPSAT-2.

Background

Plesetsk was, for a long time, by far the world's busiest spaceport but has now been overtaken by Baikonur. Orbital launches have declined for several years, falling from a steady 62 per year from 1983 to 1985, to only six in 2005. Plesetsk's

An RS-12M Topol ICBM is launched from the northern Plesetsk Cosmodrome on 1 October 1999 (Empics) 1132551

total orbital record of 1,500 launches by the end of 2005 is still higher than Baikonur's total orbital launch record of 1,170.

The site's 1,752 km^2 base is located 800 km northeast of Moscow and about 200 km south of Archangel, on the Barents Sea coast, which enables communications and intelligence satellites to be placed in polar (but never retrograde) and highly elliptical orbits. Range safety restrictions limit most flights to 62.8°, 67.1°, 73–74°, 82–83°.

Plesetsk's existence was established publicly by Geoffrey Perry, a physics teacher at Kettering Grammar School in England, and founder of the Kettering Group, following Plesetsk's first orbital launch, Kosmos 112, aboard a Vostok rocket on 17 March 1966 from pad 1. The Soviet Union officially acknowledged it in 1983.

Originally intended as an ICBM site, construction of the first of several pads for the R7-A class vehicle began in April 1957, and was completed in December 1959. The pad and ICBM began active duty in January 1960. Four pads were active by July 1961 and 15 were in service by 1964, with the range capable of launching R-9A, R-16 and R-16A missiles. Landsat imagery taken in 1973 showed Plesetsk to be about 100 km long, housing at least four complexes and defended by surface-to-air missiles.

There are nine operational orbital pads at Plesetsk: three for the Kosmos-3M, one converted for the Rockot SS-19; four for Soyuz/Molniya launchers; and two for the Tsyklon-3. The Angara series launch vehicle, still under development, will use the complex that was originally under construction for the Ukranian Zenit, which will not fly from Plesetsk, as originally planned in the early 1990s, previous to the complete collapse of the Soviet Union. The pads remain under construction, however the Russian Federation is committed to completion of the project, since operational pads for the indigenous rocket will free the Russian Federation of further reliance upon Baikonur, in Kazakhstan, to conduct orbital launches.

The Soviet Union acknowledged two major accidents in 1989. Nine technicians and soldiers were killed on 26 June 1973 (SL-8 Kosmos) and a further 50 on 18 March 1980 (SL-3 Vostok). On both occasions a vehicle exploded during fuelling operations. Collection of the hundreds of strap-ons and first stages discarded over the area started in 1991 in response to complaints from local residents over environmental contamination; compensation was also claimed. The junk was processed to recover useful materials.

The town of Mirny (Peaceful), about 40 km from the launch facilities, has grown alongside to house staff and their families; its population is about 40,000. Eurockot guests can find accommodation in the Rockot Hotel in Mirny. Climatic conditions are extreme, ranging from –35°C to +35°C, with up to 85 cm of snow cover in Winter.

Svobodny Cosmodrome

Location: 51°47'N/128°11'E'

Current Status

On 25 April 2006 the Israeli imaging satellite, EROS-B1, was successfully launched aboard a Start-1 launch vehicle from Svobodny.

Since 2005, there has been speculation that Svobodny will cease operations in the coming years. The Russian Federation's government had announced in 2005 that it will pull Svobodny's funding, while honouring previously scheduled launches. Russian Space Forces may also abandon the site, relocating to Plesetsk by 2007–2008.

United Start has an agreement to use the Svobodny Cosmodrome for commercial customers, however, there have been few Start-1 rocket (based on the RS-12M Topol ICBM) launches from Svobodny. The last orbital launch to occur at the site before April 2006 was in 2001, totalling five orbital launches altogether since the first two took place in 1997.

Background

Svobodny, in Amur Oblast, on the northeastern Chinese border, was activated in 1968 as a missile test range and was off limits to civilians during the Soviet era. Proposals for a new Cosmodrome at the decommissioned missile site at Svobodny surfaced during Russia's negotiations with Kazakhstan over Baikonur's future after the Soviet Union's collapse. This was widely perceived as a negotiating tool, but President Yeltsin signed a decree on 1 March 1996 approving the Cosmodrome's creation.

Temperatures at Svobodny range from –38°C to +32°C with up to 30 cm of snow cover in winter. Facilities include Spacecraft and launch vehicle Assembly and Test Buildings (ATBs), launch pads and storage buildings.

United Start has been the only company to provide launch services at Svobodny, employing its own transportable launcher. Rockot was supposed to have followed, employing one of five Svobodny silos, but so far this proposal has not come to fruition. The largest investment discussed in the past was the concept for a two-pad Angara complex, however, since Svobodny has been plagued with budget woes, it is unlikely that the Angara will ever fly from the site, launching instead, from both Plesetsk and Baikonur.

60 UR-100 (SS-11/Sego) ICBMs were silo based at the site, however when Svobodny was deactivated as a missile base in 1993, all these were dismantled by 1996, in compliance with Strategic Arms Reduction Treaty (START) agreements.

The maiden orbital launch for Svobodny took place on 4 March 1997 when Zeya was placed into orbit by a Start-1 flown from a mobile launch platform. A second Start-1 flight on 24 December 1997 carried the Early Bird 1 commercial satellite for EarthWatch Inc. Another Start-1 launched the Israeli satellite EROS A1 from Svobodny in December 2000 followed by a Start-1 carrying Sweden's Odin satellite in February 2001.

South Africa

Overberg test range (Overberg Toetsbaan)

Location: 34°35'S/20°19'E

Current Status

Overberg Test Range, or Overberg Toetsbaan (OTB) is a suborbital test site, owned and operated by Denel (Pty) Ltd. OTB specialises in missile testing, however, the site has been qualified, since March 1991, to provide LEO launch services. As of 2006, no organisation has yet taken advantage of this service. OTB also engages in flight and long-range artillery testing.

Background

Developed by the Houwteq organisation in the 1980s, then a subsidiary of Denel, OTB is a 43,000 ha site along 70 km of southeast Western Cape coastline. OTB is about 200 km from Cape Town. Its position near Africa's southernmost tip is ideal for launches into Sun-synchronous, polar and Earth-synchronous orbits down to about 38° inclination. The site is a business unit within Denel's Aerospace Division. Denel is wholly owned by the South African government.

Employment at the site peaked at more than 500 in the early 1990s, while South Africa aspired to indigenous nuclear and satellite launch capabilities, however OTB now has a much smaller permanent staff to handle its launch services.

Construction of the test range and launch facility began in 1983, allowing the first simple tests in 1987. A first stage test of South Africa's first planned orbital vehicle was launched in June 1989, followed by stage 1 and 2 in July 1989 and November 1990, leading to the facility's formal qualification by the end of March 1991. South Africa cancelled its satellite launch programme here in 1994, but OTB's core capability is maintained through regular use on a wide variety of flight tests. South Africa joined the Missile Technology Control Regime (MTCR) as a full member in 1995, allowing OTB to be used by other parties subject to non-proliferation controls.

OTB's central control room (Denel-OTB) 0517430

The range, close to the villages of Bredasdorp and Waenhuiskrans, is divided into east/west sectors by the De Hoop Nature Reserve, with most of the instrumentation and infrastructure in the western half. Adjacent to OTB, South African Air Force (SAAF) Air Force base Overberg, of which the Test Flight and Development Centre (TFDC) is a part, includes a 3,000 m runway and can accommodate a Boeing 747 and is available for OTB clients. A good quality highway between Cape Town and Bredasdorp allows the 200 km to be covered by car in two hours. Nearby Die Herberg accommodation caters to 120 people.

The area has a mild Mediterranean climate with a temperature of +5/25°C, 40 cm annual rainfall (mostly May to September) and 80 per cent humidity).

OTB has numerous measurements sites spread over 20 × 60 km. Sites are linked with the command and control centre by fibre optic and microwave. Some instruments can be moved outside the Range for a 300 km baseline. A fully self-contained telemetry station with stabilised antennas can be deployed on land or onboard ships. All instrumentation is linked to a central computer in the Range control centre, and can be slaved to the tracked object's state vector, computed by the central computer and based on prioritised pointing and direction information from the instruments and telemetry. Instrumentation includes, three tracking radars, four mobile cinetheodolites, one mobile optical high-speed cameras track mount (with film development laboratory and film/video readers), two telemetry stations, one downrange telemetry station (ship or land), four Doppler (two mobile); Satellite Ground Station (10 m S/X band, 2 m S-band TT&C), Command Destruct System; electronic sky screens.

There is also a fully equipped meteorological station, with rocket sondes that can probe up to 65 km.

Sweden

Esrange Space Centre

Location: 67°56'N/21°04'E

Current Status

Esrange, 45 km east of Kiruna, is the Swedish Space Corporation's (SSC) sounding rocket, stratospheric balloon and satellite operations range, and is located about 200 km north of the Arctic Circle. In June 2006, ESA reselected Esrange to operate the Galileo constellation's northern telemetry, tracking & command (TT&C) station. EuroLaunch, a cooperative organisation founded by SSC and the German Aerospace Centre, or *Deutschen Zentrums für Luft- und Raumfahrt* (DLR), provide launch services from Esrange.

Background

Esrange was founded in 1966 by the European Space Research Organisation (ESRO), a forerunner of ESA. SSC has operated and owned the site since 1972. More than 500 sounding rocket launches have been conducted from the Esrange since 1966. The site includes seven permanent launch facilities able to handle most types of sounding rocket (including Terrier, Black Brant and Skylark 12). The seventh became operational in 1991 to handle the large Maxus vehicle. Sweden has been studying the commercial use of Esrange for orbiting small polar satellites, but no launches have yet taken place.

Esrange has a 75 × 120 km impact area available for land recovery. Payload and control facilities were upgraded, partly to accommodate the Maxus long-duration sounding rocket's first launch in 1991. Maxus uses a Castor 4B motor, developed jointly by SSC/DASA. The 23 m high Maxus building has a 15 m-high umbilical mast. A 145 m^2 storage building can house two Castors at 20±2°C. Maxus 7, carrying five ESA microgravity experiment modules, was successfully launched on 2 May 2006.

Stratospheric balloons of up to 1,200,000 m^3 are launched to 45 km and regularly recovered within Sweden, Finland and Russia. Esrange can handle balloon payloads of up to 2,000 kg. French and US balloons are launched from the Zodiac and Raven areas, respectively. Since 1974, Esrange has launched more than 500 balloons.

The North European Aerospace Test range (NEAT), founded by SSC and the Swedish Defence Materiel Administration (FMV), has the ability to link the Esrange and Vidsel Missile Test Range for some users, and in doing so can provide a 20,000 km² overland aerospace test range.

Esrange has provided satellite control services since 1989. Since 1978, the site has received, processed and disseminated data from Earth observation satellites such as Landsat, SPOT, MOS and JERS. The satellite management and operations division employs roughly 25 people.

Esrange provides on-site accommodation for its launch site users. Estrange visitors must travel by road from Kiruna, about a 45 minute journey by private vehicle or taxi.

United States

Kennedy Space Center (KSC)

Location: 28°36'33"N/80°36'15"W

Current Status

The Space Shuttle is due to retire in 2010, concurrent with International Space Station (ISS) completion, according to NASA's Vision for Space Exploration. Lockheed Martin's Orion (NASA's Crew Exploration Vehicle) development and testing work will continue. KSC will remain NASA's primary launch site during the exploration programme. If all goes according to NASA's plan, KSC could once again send astronauts to the Moon by 2020.

In September 2006, Lockheed Martin took over a part of KSC's Operations and Checkout Building in order to complete the assembly and testing of the Orion vehicle. NASA announced on 31 August 2006 that Lockheed had won the competition with Boeing to design and manufacture the capsule formerly known as the Crew Exploration Vehicle (CEV).

The Orion capsule will be launched into space aboard an Ares I rocket, formerly known as the Crew Launch Vehicle (CLV), from KSC. ATK Thiokol is designing and building Ares' first stage for NASA; the LV's first test flight is scheduled to occur at Space Launch Complex (SLC) 39 in 2009. NASA will renovate SLC 39, the current Shuttle launch site, in phases, to accommodate the Ares. Launch tests for the Ares V heavy lift vehicle will follow afterward.

Background

KSC is NASA's site for processing, launching and landing the Space Shuttle and its payloads, including International Space Station (ISS) components. Located on Merritt Island adjacent to the US Air Force (USAF) launch facilities on Cape Canaveral Air Station (CCAS), KSC was originally built to support the Apollo lunar landing programme of the 1960s. After the last Apollo lunar launch in 1972, Launch Complex (LC) 39 supported Skylab in 1973–74, Apollo-Soyuz in 1975 and the Shuttle from the late 1970s. The installation covers 567 km². All non-operational areas are part of the Merritt Island National Wildlife Refuge. KSC extends some 55 km north to south and 16 km across its widest point.

Land acquisition to build the KSC began in 1962, and clearing for the Vehicle Assembly Building (VAB) began in November 1962. The VAB was structurally completed in mid-1965. The first LC 39 launch was Apollo 4 in November 1967. Shuttle processing and launch facilities are in the LC 39 area. About 8 km to the south, office buildings, payload processing facilities (including the Space Station Processing Facility) and flight crew quarters are in the Industrial Area. NASA also uses some buildings on neighbouring CCAS. The USAF Atlas V LC 41 lies inside KSC within a 1.44 km² piece of land assigned to the USAF.

2002 data reflects that a total of 11,747 personnel worked at KSC at the time the data was collected; 1,796 were civil service employees, and 9,951 were contractor employed.

United Space Alliance (USA) is NASA's primary Space Shuttle contractor and is responsible for Shuttle fleet management and operations. USA was formed in 1995 by merging Lockheed Martin Space Operations Company's and Rockwell International's Shuttle contract interests; today the Alliance is made up of Lockheed Martin and Boeing Company concerns, due to Boeing's purchase of Rockwell's space entities in 1996. The USA/NASA Space Flight Operations Contract (SFOC) became active in 1996 and USA will remain the prime contractor until the Shuttle fleet is retired in 2010. Other contractors on site at KSC include Boeing, Lockheed Martin, Orbital Sciences Corporation and Space Gateway Support.

Columbia occupies pad 39A in the foreground during STS-35 preparations, while Discovery is on 39B for STS-41. The white Payload Changeout Room in the Rotating Service Structure is used for installing vertically processed payloads on the pad (NASA)
0517254

Final Shuttle assembly takes place in the VAB and the stack is then transported either 5,535 m east to pad 39A or 6,828 m northeast to 39B. LC 39's design reversed the fixed launch concept of earlier missions, under which assembly, checkout and launch were conducted at the pad. KSC's first director, Dr Kurt Debus, developed the mobile launch concept.

The Shuttle's primary processing and launch facilities are:

Vehicle Assembly Building (VAB)

The VAB is 160 m high, 218 m long, 158 m wide, divided into four 160 m high bays and a 64 m low bay and transected by a north to south transfer aisle. One of the world's largest buildings by volume, it covers 32,400 m². A maintenance shop for Shuttle main engines and holding areas for the solid rocket booster (SRB) forward assembly and aft skirt is in the low bay. External Tanks are stored and checked out in high bays 2 and 4, with 4 also used for payload canister handling. High bays 1 and 3, facing east, are used for assembly of the three main Shuttle elements (two SRBs, external tank and Orbiter) on the Mobile Launch Platform. The stack is transported from here to the pad.

Launch Control Center (LCC)

A four-storey building adjacent to the VAB on the east side, housing four firing rooms. Firing rooms 1 and 3 provide full control of launch and orbiter operations; number 3 was previously secured for classified DoD departures. Number 2 is usually used for software development and testing. Number 4's refurbishment was completed in 2006, and it is the most technologically advanced firing room. Computer and software equipment comprising the Launch Processing System (LPS) is also housed in the LCC. Designed at KSC for the Shuttle programme, the LPS automatically controls many of the procedures formerly performed mechanically, while also requiring fewer personnel (about 230). Shuttle checkout, countdown and launch are conducted from the LCC using the LPS. A countdown begins typically about two days before scheduled lift-off, at the T-43 h mark and including an average of 28 hours of built-in holds. As soon as the two solids ignite at T-0, control of the mission shifts from KSC to Mission Control at Johnson Space Center in Houston. KSC again assumes responsibility after the crew exits the Shuttle (usually within an hour of touchdown) and ground cooling has been established.

Mobile Launcher Platform (MLP)

Final Shuttle assembly, transport to the pad and launch are conducted on the MLP, a two-storey steel structure, 7.6 m high, 49 m long, 41 m wide, and weighing 4,190 tonnes. MLP-3, previously used for Apollo 11 and manned Skylab missions, was added to the existing pair in late 1989 and first used by STS-32.

Crawler-transporter

Two crawlers, each weighing 2,700 tonnes, were in service by early 1967 for the Apollo lunar programme. About 40 m long, 35 m wide and height adjustable from 6 to 8 m, with an operator control cab at each end. There are eight tracks, each with 57 0.9 tonne shoes. Two 2,750 hp diesel engines drive four 1 MW DC generators powering 26 traction motors. Two 750 kW AC generators, driven by two 1,065 hp diesels, provide jacking, steering, lighting and ventilation. Two 150 kW AC generators provide MLP power. Maximum speed

Vehicle Assembly Building and Shuttle landing strip (background). The Launch Control Center is the low building angled at the VAB's base. Note Mobile Launcher Platform 3 at right, awaiting modification from its Apollo configuration (NASA)
0517256

is about 3.2 km/h unloaded, 1.5 km/h loaded, consuming about 350 litres of diesel oil per km. It requires between 6 and 8 hours to move Shuttle from the VAB to 'hard down' at the pad.

Launch Pads 39A and 39B

39A and 39B are of an identical size and roughly octagonal design. The central structure on the Pad's hard stand is the 106 m tall Fixed Service Structure (FSS), including the 24.4 m lightning mast. The FSS has three service arms, the External Tank Gaseous Oxygen Vent Arm (with 'beanie cap' hood), the Orbiter Access Arm, the External Tank Hydrogen Vent Line and access Arm. Attached to the FSS is the Rotating Service Structure (RSS), which rotates 120° on a track to encase the Orbiter so that its Payload Changeout Room fits flush with the cargo bay for vertical payload installation.

All manned flights after Apollo 7 have been launched from SLC 39, and all except Apollo 10 during the lunar programme from 39A. The first 24 Shuttles also went from 39A; the fateful Challenger 51L mission was the first to use 39B. During the post-Challenger hiatus, 39B underwent extensive weather protection modifications. Other refinements included safety improvements to the crew's emergency escape system. The Orbiter Access Arm was covered with solid panels for fire protection and a water spray system was incorporated. Two slidewire baskets were added to the existing five, along with brakes; and improvements were made to the emergency shelter bunker near the end of the slidewires. Modifications to both Shuttle pads are scheduled periodically. The ill-fated Columbia also launched on 16 January 2003 from Pad 39B.

Orbiter Processing Facility (OPF)

The original OPF is capable of handling two Orbiters in three identical high bays 29 m high, 60 m long, 46 m wide. The bulk of Orbiter processing between missions is performed in the OPF. Equipped with movable work platforms that completely envelop an Orbiter, they are equipped to fuel the Auxiliary Power Units and thrusters with hypergolic propellants, in addition to draining them after missions. Horizontal payloads such as Spacelab are installed in the Orbiter within the OPF; vertical payloads are installed on the pad. Before transfer to the VAB for final assembly with the External Tank/SRBs, the Orbiter is weighed and its centre of gravity located.

Thermal Protection System Facility

The two-storey 4,088 m² Thermal Protection System Facility is opposite OPF bays one and two. The facility is used for final manufacture and repair of thermal tiles and gap fillers.

Rotation Processing and Surge Facility

One of several facilities devoted to SRB processing. Includes four buildings north of the VAB. New and reloaded SRB segments shipped by manufacturer Thiokol in Utah are received here. Recovered casings, refurbished at the Solid Rocket Booster Disassembly Facility, are installed on rail cars for shipping to Utah. Inspection, rotation and aft booster build-up occur in the processing building. Completed aft skirt assemblies, brought from the ARF (see below), are integrated here with aft segments. Two nearby surge buildings store up to eight segments each (two flight sets). Final assembly and stacking into complete SRBs is performed in the VAB.

SRB Assembly & Refurbishment Facility (ARF)

Managed by NASA Marshall for refurbishment and subassembly of inert SRB hardware, including the forward and aft skirt assemblies, frustrum, thrust vector controls, recovery systems and electronic controls takes place in this seven-building complex. The three-storey Manufacturing Building includes an ordnance area for installing separation devices; 24.4 × 61 m high bay; class 100,000 area for building up aft skirt elements; two 13.6 tonne bridge cranes and three of the world's largest overhead gantry industrial robots for painting and spraying on insulation. A hot fire test stand on the southeast corner of the 178,000 m² site supports testing of the SRB's hydrazine TVC system. Completed aft skirt assemblies are transported to the RPF for integration with the motor segments.

Shuttle Landing Facility (SLF)

The SLF is 4.8 km northwest of the VAB on a Northwest/Southeast alignment. The concrete paved runways are 4,572 m long, with 305 m hardened asphalt overruns at each end, 91 m wide, with additional 15 m asphalt shoulders, and 40.6 cm thick at the centre. The 61 cm centre to edge slope facilitates drainage. Post-Challenger, a section 1,067 m long at each end was ground down to create a smoother surface without cross grooves to reduce Orbiter tyre wear. The 6.3 mm grooves provide a skid-resistant surface and rapid water runoff. The centre's surface was also smoothed off (but without reducing the grooves) in September 1994. The landing strip consists of two runways: Runway 15 with approach from northwest and Runway 33 from southeast. The Tactical Air Navigation (TACAN) system provides range/bearing measurements from 44.2 km altitude; the Microwave Scanning Beam Landing System provides more precise slant range, azimuth and elevation from 5.5 to 6.1 km. A 149 × 168 m aircraft parking apron is at the southeast end.

The Mate/De-mate Device is on the northeast corner of the ramp. It attaches the Orbiter to or lifts it from the Shuttle Carrier Aircraft during ferry operations. There are also movable platforms for access to some Orbiter components. The device can lift 104.3 tonnes and withstand winds up to 200 km/h.

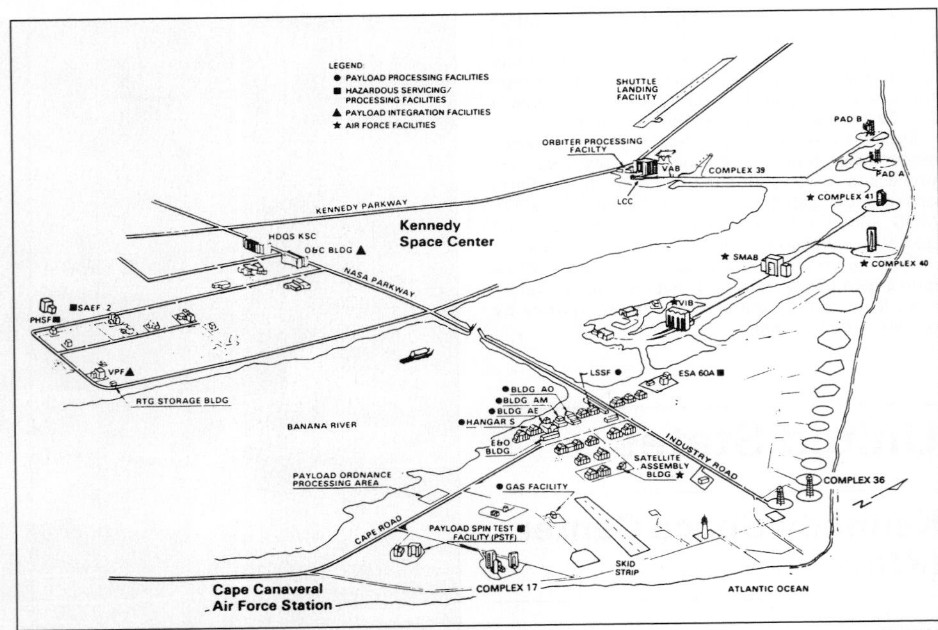

Principal launch and payload handling facilities at KSC and CCAS (NASA) 0517088

Logistics Facility

The Logistics Facility is a 30,159 m² main building south of the VAB. It houses some 150,000 Shuttle parts and uses an automated storage-retrieval system. The separate NASA Shuttle Logistics Depot maintains and repairs 95 per cent of Shuttle's 4,000 Line Replaceable Units and parts.

Operations Support Buildings

The Operations Support Building is a 27,700 m², six-storey structure, southwest of the VAB. It houses the technical documentation centre, library and photo analysis area. Operations Support Building II opened in 2006; it is a 17,560 m², five-storey office building housing office spaces, a conference centre with an observation deck, training rooms and technical libraries. The second support building replaced temporary office and technical facilities.

Processing Control Center (PCC)

The three-storey 9,200 m² PCC between the OPF and Operations Support Building, began operations in 1992 for Orbiter testing. Three control rooms monitor Shuttle processing in the three OPF high bays. Launch team training and LPS maintenance also occur here.

Shuttle Launch Processing

A KSC-led team prepares a returning Orbiter for turnaround whether it lands at KSC or Edwards Air Force Base. It is returned from Edwards atop the Shuttle Carrier Aircraft (a modified 747) to the Shuttle Landing Facility, where it is removed using the Mate/De-mate Device. Processing for the next flight then begins in the OPF, including ordnance safing, payload removal, payload bay reconfiguration, removal and installation of the three main engines, horizontal payload insertion and any required maintenance. During this work, the SRBs are stacked in the VAB atop the MLP and the external tank (ET) attached. The Orbiter is then transferred on a transporter, lifted vertically and mated to the ET. The Crawler moves the stack to the pad, where connections and hookups are made and checkout begins. Equipment in a room below the pad hard stand provides the links to the Launch Control Centre. If the payload is not installed in the OPF, it is inserted on the pad.

NASA Payload Processing Facilities

Shuttle payloads can be processed horizontally or vertically. Facilities are located in the Industrial Area and across the river on CCAS. NASA keeps several buildings devoted to payload processing for the Shuttle, ISS and satellites launched aboard expendable vehicles, some of which are on CCAS; brief descriptions of several facilities follow:

Operations and Checkout Building (O&C)

Lockheed Martin occupies a segment of the building and will use it to assemble, test and checkout the Orion Crew Exploration Vehicle.

Horizontal Shuttle payloads are received, assembled and tested in the O&C, the Industrial Area's largest building. They are then transferred to the OPF for installation in the cargo bay; returned payloads follow a reverse procedure. The 55,742 m² O&C not only contains payload processing facilities, but also astronaut quarters, office space and laboratories.

Multi-Payload Processing Facility (MPPF)

Located in the KSC Industrial Area, this 1,825 m² facility can process, as its name suggests, multiple Shuttle payloads simultaneously. The MPPF also houses the Raffaello Multi-Purpose Logistic Module (MLPM) Access Certification training unit.

Hangar L

Hangar L is also known as the Life Sciences Support Facility (LSSF). Live specimens are received here; it is also a life sciences research facility.

Spacecraft Assembly and Checkout Building (Hangar AE)

This hangar is in the CCAS Industrial Area and contains high bay clean rooms for processing deployable satellites aboard expendable launch vehicles (ELV). The Building also houses the Launch Vehicle Data Center and telemetry facilities.

Vertical Processing Facility (VPF)

Processing of vertical payloads, including upper stages, takes place in the Vertical Processing Facility south of the O&C in the Industrial Area. The VPF houses Cargo Integration Test Equipment (CITE) to verify Orbiter/payload interfaces. It is equipped with a 32 m high bay, airlock and single storey support facilities along the sides of the high bay. The entire 2,503 m² facility is a class 100,000 clean room.

Spacecraft Assembly & Encapsulation Facility 2 (SAEF-2)

The 1,588 m² SAEF-2, also in the southern part of the Industrial Area, is used for assembly, testing, encapsulation and sterilisation of spacecraft, particularly large payloads. It is a class 100,000 clean room.

Payload Hazardous Servicing Facility (PHSF)

The PHSF is a class 100,000 clean area used for large payload processing, integration of solid motors and fuelling operations. The west end of the high service bay has a sloping area for fuelling. PHSF operations are handled from the separate Multi-Operations Support Building (MOSB), with two 112 m^2 control rooms.

The Canister Rotation Facility (CRF)
The CRF, with its 670 m^2 high bay was officially activated in December 1992 for maintaining the Shuttle's two environmentally controlled 19.8 m long payload canisters. Previously, rotation had to be performed in the VAB.

Space Station Processing Facility (SSPF)
In order to prepare for the "Space Station era," NASA built the SSPF, a three-storey 42,455 m^2 facility east of the O&C which receives, processes and checks out Space Station flight hardware as it arrives at KSC. In fact, SSPF is a generic facility capable of handling any Shuttle payload. The 5,853 m^2 class 100,000 clean room includes a high bay and an intermediate bay, accessed by a 465 m^2 airlock.

The 21 km^2 launch area contains five launchers, two movable launcher enclosures and separate facilities for payload assembly, rocket motor assembly, motor storage, scientific support, telemetry, radar and data communication (Evelyn Trabant/Poker Flats) 0008102

Poker Flat Research Range (PFRR)

Location: 65°07'N, 147°28'W

Current Status
Due to its remote location and large flight area, Poker Flat is being considered, as of 2006, as a UAV test site, in part to determine whether it is advisable that UAVs share US airspace with other aircraft and spacecraft. By March 2005, Poker Flat had conducted over 300 sub-orbital launches.

Background
Set up in the early 1960s by Neil Davis of the Geophysical Institute, Poker Flat Research Range (PFRR), primarily a sounding rocket launch facility dedicated to auroral and middle to upper atmospheric research, is 50 km northeast of Fairbanks, Alaska. Operated by the University of Alaska Geophysical Institute since 1968, it is the world's only university-owned launch range. The site fired off its first rocket in 1969. PFRR's 21 km^2, site includes telemetry stations, optical observing facilities, five launch pads, a blockhouse, rocket storage and assembly buildings, maintenance and communication facilities, a payload assembly building, including a Class-100 cleanroom, radar and the range office. PFRR has been operating under contract with NASA's Wallops Flight Facility (see separate entry) and agreements with the Geophysical Institute since 1979.

With flight zones covering over 67,000 km^2 of northern Alaska, it is the largest land-based launch facility in the world and, as such, permits extensive payload recovery programmes along with established down range instrument sites.

Poker Flat is funded through contracts with DoD, NASA, NOAA and the National Science Foundation. NASA currently funds the USD1.1 million annual operation cost. Five employees work year-round directing design and construction of new facilities, coordinating science projects, maintaining the physical plant, obtaining the various waivers, approvals and agreements necessary for each launch, and providing launch services. Additional employees are added during major campaigns.

An average of two to 10 major sounding rockets are launched annually. Poker flat has launched several sounding rocket types, including Nike-Tomahawk, Black Brant, Aries, Nike-Black Brant, Terrier-Malemute, Orion, Nike-Orion, Taurus-Orion, Taurus-Tomohawk, Viper-Dart and Taurus-Nike-Tomahawk.

Most missions require simultaneous ground-based data from a worldwide network of stations or are in coordination with satellite experiments. The T. Neil Davis Science Operations Center (SOC) houses magnetometers and riometer data displays, all-sky and narrow field auroral TV cameras, meridian scanning photometers and other optical and RF observing instruments for auroral and upper atmospheric research. A large worldwide data display is available at the SOC for launch team use.

The range operates five launch pads. Pads one and two each have a 3,400 kg Missile Rail Launcher (MRL). Pad three and four have 9,000 kg rail launchers and are sheltered from severe weather by moveable launch enclosures. Pad five has a 1,800 kg twin boom launcher.

Range telemetry support is provided by two S-band autotrack systems, incorporating 2.44, 4.9 and 11 m dishes, provided by NASA and located in a 15.3 m dome at the top of the hill overlooking the complex. A NASA C-band radar unit was established at Poker Flat in 1983 and was replaced with a new C-band system in 1995.

Authorisation for overflight, landing and recovery of rockets and payloads roughly northeast of the range head to the coast of the Beaufort Sea, is granted by the Bureau of Land Management, the US Fish and Wildlife Service, the State of Alaska Department of Natural Resources (Division of Lands), the Venetie Tribal government, the Traditional Councils of Arctic Village and Venetie and Doyon, Ltd (the Interior Alaska Regional Native Corporation). The Federal Aviation Administration approves rocket flight zones and coordinates air space use during launches, and the Alaska Department of Transportation and Public Facilities authorises range personnel to stop road traffic on the Steese Highway during launches. In addition to the normal Notices to Airmen (NOTAMS) procedure, each launch is coordinated with NORAD.

Wallops Flight Facility, National Aeronautics and Space Administration

Current Status
Wallops Flight Facility (WFF), on Wallops Island, on Virginia's Eastern Shore, has been managed by NASA's Goddard Space Flight Center (see separate entry) since 1982. It is primarily a sub-orbital launch site but retains orbital launch capabilities. WFF is comprised of three separate areas: the Mainland, the Main Base and Wallops Island. The site was founded in 1945 as a National Advisory Committee for Aeronautics (NACA) facility. It employs about 1,000 civil service and contractor staff. NASA Wallops also hosts four tenant organisations: the US Navy's Surface Combat Systems Center (SCSC), the US Coast Guard, the National Oceanic and Atmospheric Administration (NOAA) and the Mid-Atlantic Regional Spaceport (MARS). MARS is located on the south end of Wallops Island. It is a licensed commercial spaceport.

WFF primarily focuses on NASA sub-orbital missions and also partners with commercial, academic and governmental organisations that wish to use WFF's tracking capabilities, scientific balloon, sounding rocket and aircraft research resources. Wallops' launch range now maintains six launch pads. Two Wallops facilities, the National Scientific Balloon Facility in Palestine, Texas and the Scientific Balloon Flight Facility in Fort Sumner, New Mexico, launch balloons used for atmospheric and other scientific applications at a rate of roughly 25 per year.

NASA's Sounding Rocket Program (NSRP) is based at WFF. Some 20 launches are made each year from a range of sites, including Wallops itself, Poker Flat Research Range, Alaska and White Sands Missile Range, New Mexico. Wallops conducted up to 11 sounding rocket launches in 2003, but has completed only a few in each of the years hence. NASA currently employs 14 sounding rocket types, including the Black Brant and Terrier series. The site can accommodate launch azimuths between 90 and 160°.

Two US Air Force Space and Missile Systems Center (SMC), Detachment 12 orbital launches are scheduled to take place at WFF during 2006–2007. The first, due to occur in November 2006, will launch from the MARS pad and carry the TacSat 2 satellite aboard a Minotaur I launch vehicle. Additionally, SMC has arranged to launch the Near-Field Infrared Experiment (NFIRE) satellite, also using the Minotaur, from the site sometime in 2007.

Background
Wallops became the third US orbital site on 16 February 1961 with the launch of the Explorer 9 aboard a Scout launch vehicle. The Scout was retired in 1994. The first sub-orbital firing made from Wallops was a Tiamat on 4 July 1945. Through the mid to late 1990s, Wallops conducted orbital launches at the rate of one or two per year, but has not hosted any at all since that time. The facility retains its orbital launch capabilities for occasional projects. The State of Virginia hopes to grow the site into a thriving commercial spaceport, while preserving its defence and civil space infrastructure and duties.

NASA, the US Air Force and public/private sector partnerships have built extensive launch facilities on the island over the years and the active maintenance of these facilities reflects a commitment on the part of the US and State governments to maintain this infrastructure. Operational buildings include:

Six launch areas, including two commercial launch complexes

Payload Assembly and processing facilities

Dynamic Balance Facility

The Wallops Ground Station was established to provide NASA tracking support for various LEO projects. High-speed data transfer to NASA Goddard in Greenbelt, Maryland is provided by a satellite link. WFF utilises three telemetry receiving and command antennas for LEO satellite support. WFF's three chief tracking antennas are: the five m Low Earth Orbiter-Terminal (LEO-T), the eight m Transportable Orbital Tracking System (TOTS), and the 11.3 m X- and S-Band tracking system. Wallops is also equipped with a nine m system as well as a number of other systems, including mobile tracking stations and a dedicated Meteosat station.

MULTI-NATIONAL AGENCIES

International

European Space Agency (ESA)

Current Status

ESA has continued to invest in science, navigation and human spaceflight projects. While its budget is dwarfed next to NASA's, ESA is steadily building upon its successes, although the agency has encountered some launch delay problems.

GIOVE-A, ESA's first Galileo navigation system technology demonstrator, has been in orbit since December 2005. The agency expects to launch GIOVE-B by the end of 2007, a delay already well over one year from the date ESA originally intended; the delay could stretch to two years, threatening to throw off the Galileo project plan to have 30 satellites in space by the end of 2010.

CNES and ESA launched the COROT telescope on 27 December 2006. It has begun returning images and normal science operations began in February 2007. ESA is an active International Space Station (ISS) participant, with astronauts Thomas Reiter and Christer Fuglesang contributing to ISS and Space Shuttle missions in 2006, and Paolo Nespoli, Hans Schlegel and Léopold Eyharts scheduled to fly in 2007.

Background

Headquartered in Paris, France, the European Space Agency (ESA) is comprised of 17 Member States: Austria, Belgium, Denmark, Finland, France, Germany, Greece, Ireland, Italy, Luxembourg, the Netherlands, Norway, Portugal, Spain, Sweden, Switzerland and the United Kingdom. Canada, the Czech Republic and Hungary also participate co-operatively in some ESA programmes. ESA has no affiliation with the European Union (EU), and ESA members need not necessarily be EU member states. However, ESA and the EU work together to develop European space-related goals, strategies and policies.

Ten European states created ESA on 30 May 1975 by signing the ESA Convention. The Agency's charter members were: Belgium, Denmark, France, (West) Germany, Italy, the Netherlands, Spain, Sweden, Switzerland and the UK. Ireland became a signatory in December 1975. While the Agency commenced operations after the agreement was signed, ESA did not 'officially' exist until five years later, following formal Convention ratification on 30 October 1980. It replaced its parent organisations, the European Launcher Development Organisation (ELDO) and the European Space Research Organisation (ESRO). Canada has participated in ESA programmes as a Co-operating State since December 1978.

Member states contribute to mandatory ESA general and space science budgets on a gross national product basis; these activities account for approximately 25 per cent of ESA expenditures. States can contribute further funding on an optional basis, depending upon voluntary programme participation. As of early 2005, Germany and the UK were the largest mandatory programme contributors, followed by France and Italy. France however, was by far the largest optional programme contributor, followed by Germany, Italy and Belgium. ESA's estimated 2006 budget is EUR2,904 million. As of early 2005, ESA employed approximately 1,900 total personnel in offices across Europe, Australia and the Americas, a level that has remained relatively consistent since 2002.

ESA is comprised of 10 Directorates charged with space programme and organisational management. Besides the Space Science programme and administrative Directorates, to which member states must contribute funds, ESA has four other programme Directorates: Launcher Programmes; Earth Observation Programmes; Human Spaceflight, Microgravity and Exploration Programmes; and EU and Industrial Programmes. Member states contribute to these based upon programme participation.

ESA policy formulation activities take place at the Paris headquarters, however ESA maintains several Centres in locations across Europe. The largest, the European Space Research and Technology Centre (ESTEC) is located in Noordwijk, the Netherlands. ESTEC scientists and other personnel are largely responsible for ESA's programme management (except launchers), and spacecraft engineering, design and testing. ESTEC staff design and run ESA's many space science programmes as well as Earth observation, satellite navigation, telecommunications, and human space flight projects.

The European Space Operations Centre (ESOC), located in Darmstadt, Germany, is responsible for ESA's in-flight satellite systems operations, mission control and management. ESOC also runs ESA's ESTRACK ground and tracking stations, based in several sites across Europe and Australia and in Kourou, French Guiana. The Centre provides spare capacity mission management and tracking services to the private sector and co-operates with other national space agencies in this endeavour. It also manages the Database and Information System Characterising Objects in Space (DISCOS), which uses data assembled by USSPACECOM (see separate entry), to track and prevent possible ESA spacecraft collision risks with space debris.

The ESA Centre for Earth Observation (ESRIN), ESA's Earth observation headquarters since 2004, is in Frascati, south of Rome, Italy. The Centre collects, stores and distributes Earth Observation (EO) satellite data to ESA's users in both the public and private sectors, including Envisat data. ESRIN's EO archive collection is the largest in Europe. ESRIN also manages various other ESA projects, including ESA's Vega small launcher programme and administration of the Agency's technical space records.

The European Astronauts Centre (EAC) is located in Cologne, Germany. The facility houses ESA's astronauts and is where they train and prepare for human space flight missions. EAC also co-ordinates co-operative training and mission activities with ESA's International Space Station (see separate entry) partners.

ESA maintains liaison offices in Brussels; Houston, Texas; and Moscow; and with France, maintains the Centre Spatial Guyanais (CSG) launch base in Kourou, French Guiana (see separate entry).

ESA Space Programme

The European Space Agency (ESA) has been designing and implementing space programmes for roughly 30 years and Europe's space presence is growing. ESA has been particularly successful in consistently planning and launching Science and EO satellites over the past 15 years and the Agency will continue to build on its experience with such programmes. ESA will also continue to develop its Launch Vehicle and Human Space flight programmes in preparation for missions to the International Space Station (ISS), the Moon, Mars and other deep space destinations.

ESA has recently been seeking ways to co-operate with potential military clients. With its present EO and future navigational capabilities in mind, ESA is aware that it would be able to equip national and co-operative military organisations with geospatial intelligence and navigational products, providing capabilities that are becoming increasingly important to the world's armed forces in order to build situational awareness in both terrestrial and space environments.

Current In-Orbit ESA Missions

Mission Name	Launch Date	Mission Type	Mission Notes
Hubble Space Telescope	24.04.1990	Science	Joint NASA programme; 2013 anticipated mission end
Ulysses	06.10.1990	Science	Joint NASA programme; mission extended to March 2008
Meteosat-5 (MOP-2)	02.03.1991	Meteorology	Administered by EUMETSAT; near the end of its service
Meteosat-6 (MOP-3)	20.11.1993	Meteorology	Administered by EUMETSAT; near the end of its service
ERS-2	21.04.1995	Earth Observation	Has outlived its expected minimum 10-year service-life and remains operational
SOHO	02.12.1995	Science	Joint NASA programme; 2009 scheduled mission end
Meteosat-7 (MTP)	02.09.1997	Meteorology	Administered by EUMETSAT; 2008 scheduled mission end
Cassini-Huygens	15.10.1997	Science	Joint NASA programme; Huygens descent to Titan 14.01.2005; Huygens segment completed
Pastel Spot-4	24.04.1998	Telecommunications and Earth Observation	Joint CNES programme
XMM-Newton	10.12.1999	Science	Mission extended to 2010
Cluster II	16.07.2000	Science	2009 scheduled de-orbit
Cluster II	09.08.2000	Science	2009 scheduled de-orbit
Proba-1	22.10.2001	Earth Observation	Technology demonstrator microsatellite; remains operational
Artemis	12.07.2001	Telecommunications	2011 anticipated mission end
Envisat	01.03.2002	Earth Observation	Mission extended to 2010
Meteosat-8 (MSG-1)	28.08.2002	Meteorology	Administered by EUMETSAT
Integral	17.10.2002	Science	Mission extended to 2010
Mars Express	02.06.2003	Science	Mission extended through 10.2007 (Beagle-2 Lander failure 25.12.2003)
SMART-1	28.09.2003	Science	Mission end upon impact with the Moon 3 September 2006
Double Star	29.12.2003	Science	Joint CNSA mission; mission extended through September 2007
Rosetta	02.03.2004	Science	2015 scheduled mission end
Double Star	25.07.2004	Science	Joint CNSA mission; mission extended through September 2007
Venus Express	09.11.2005	Science	2007 scheduled mission end
Meteosat-9 (MSG-2)	21.12.2005	Meteorology	Administered by EUMETSAT
GIOVE-A	28.12.2005	Navigation	First Galileo constellation validator satellite
MetOp-A	19.10.2006	Meteorology	Joint US programme; Administered by EUMETSAT
COROT	27.12.2006	Science	Joint CNES mission; Planned 2.5 year mission duration

Planned ESA Missions

Mission Name	Launch Date	Mission Type	Mission Notes
Columbus ISS Module	2007	Science; Human Spaceflight and Exploration	Space Shuttle transport and ISS assembly
GOCE	2007	Earth Observation	Planned 2 year mission duration
GIOVE-B	2007	Navigation	Second Galileo constellation validator satellite
Jules Verne Automated Transfer Vehicle (ATV)	2007	Human Spaceflight and Exploration	Manufacture and launch of seven total vehicles anticipated
SMOS	2007	Earth Observation	Planned minimum 3 year mission duration
Proba-2	2007	Science and Earth Observation	Technology demonstrator microsatellite
Herschel Space Observatory	2008	Science	Planned 3–4 year mission duration
Planck	2008	Science	Planned 2 year mission duration
ADM-Aeolus	2008	Earth Observation	Planned 3 year mission duration
LISA-Pathfinder (SMART-2)	2009	Science	Planned 1 year mission duration
CryoSat-2	2009	Earth Observation	CryoSat replacement
Swarm	2010	Earth Observation	Constellation of 3 satellites
Microscope	2010	Earth Observation	Joint CNES mission
Gaia	2011	Science	Planned 9 year mission
EarthCARE	2012	Earth Observation	Joint JAXA mission
ExoMars	2013	Science/Exploration	Aurora programme Mars exploration mission
James Webb Space Telescope (JWST)	2013	Science	Joint NASA programme
BepiColombo	2013	Science	Planned 7–8 year mission
Darwin	2015	Science	4 spacecraft telescope constellation
Solar Orbiter	2015	Science	9 year planned mission
XEUS	2015	Science	XMM-Newton replacement
LISA	2017	Science	Joint NASA mission; 3 spacecraft
Hyper	2020	Science	Planned 2 year mission

ESA's core activities are organised into Science; Launcher; Human Spaceflight, Microgravity and Exploration; Earth Observation; and EU and Industrial Programmes, which includes ESA's Telecommunications programmes. With the exception of ESA's Science sector, which requires mandatory member contributions, ESA technical programmes are funded by members on an optional basis.

Much of ESA's 2005 budget was spent on the Launcher; Human Spaceflight; Science; and Earth Observation programmes, together representing EUR1,906.2 million out of a total EUR2,951.8 million 2005 expenditures. The Launcher and Human Spaceflight programmes alone represented about 40 per cent of ESA's budget. The Agency spent EUR244.3 million on the Galileo Navigation programme in 2005 – about 8 per cent of total outlays. ESA reports that it expended EUR159.5 million on Telecommunications programmes, EUR81.1 million on Microgravity programmes and EUR92.7 million on Technology and Exploration programmes in 2005.

Science

Space science and research have played mandatory ESA roles, with the agency spending EUR345.9 million in 2005 on its obligatory Science programme. As of early 2006, ESA had sponsored or partnered in over 1,660 space science and microgravity experiments. Examination of ESA's science goals resulted, as of 2001, in the organisation of ESA's science activities into six disciplines, each of which contains specific ESA Research Cornerstones. The ESA science disciplines are: Fundamental Physics; Fluid and Combustion Physics; Material Sciences; Biology; Physiology; and Astro/exobiology and Planetary Exploration.

ESA's recent Science missions include Rosetta, Venus Express, and Mars Express. Rosetta, the first long-term comet research mission of its kind, was launched in 2004, and is making its journey to 67P/Churyumov-Gerasimenko. Rendezvous is scheduled for 2014. It will then orbit the comet and release its 100 kg lander onto the surface. Rosetta will revolve around 67P/Churyumov-Gerasimenko for about two years, collecting data with its 11 on-board instruments. The lander is equipped with nine instruments designed to analyse the comet's surface and composition. Rosetta's mission is anticipated to end in December 2015. The mission's cost, including launch and operations, totals at nearly EUR1,000 million and is the most ambitious comet study to date.

Venus Express, launched in November 2005, arrived and entered Venusian orbit in April 2006. The initial mission went according to plan and as of early 2007, Venus Express is on an extended mission. Venus Express uses some of the same technology that was developed for Mars Express and Rosetta. Seven Instruments collect data on Venus' atmosphere and surface, in more detail than former US and Soviet-era Venusian missions.

Launched in 2003, the Mars Express mission is scheduled to come to a close in October 2007. The orbiter is designed to detect subsurface water sources and has returned streams of data on the planet's exterior and geological composition. Mars Express' Lander, Beagle-2, failed after its release from the main craft and crashed on the planet's surface after its uncontrolled, rapid descent into the Martian atmosphere. A Commission of Inquiry investigated the failure and subsequent findings indicated that organisational and project management problems within ESA as well as spacecraft technical problems existed previous to the Beagle's project approval by ESA, but that problems were not appropriately addressed by the Agency.

ESA also counts the XMM-Newton X-ray space observatory; the Cluster II Earth plasma research vessels; the Ulysses solar polar probe; the Solar Heliospheric Observatory (SOHO) and the International Gamma-Ray Astrophysics Laboratory (Integral) among its continuing shared and independent missions. ESA participation in Hubble should also continue, currently planned through 2013.

ESA is planning several future joint and independent Science missions as well. Among these are the Herschel Space Observatory; and Planck. COROT was launched in December 2006 and is a CNES led project. The space telescope will detect minute changes in star brightness, thereby assisting with the identification of new extra-solar system planets. The Herschel Observatory is heralded as a pioneer in the field of infra-red astronomy and will cover the full far infra-red and sub-millimetre waveband; it will be the largest space observatory yet launched. Planck will ride along with Herschel in a delayed, joint 2008 Ariane 5 launch. Both the Planck radiation collector and Herschel Observatory should assist in promoting further scientific understanding of the universe's origins.

Launch Vehicles

ESA and the French Centre Nationale d'Etudes Spatiales (CNES), with Aerospatiale as industrial architect, were responsible for developing the Ariane 1-5 range of orbital launchers. The vehicles are now procured, marketed and launched by Arianespace. The Centre Spatial Guyanais (CSG) in Kourou, French Guiana, will continue as Europe's launch site. Launch facilities are being added and upgraded to accommodate future Vega and Soyuz-2 launches.

In 2003 ESA and the EU initiated the European Guaranteed Access to Space (EGAS) policy. The policy's main aim is to ensure that Ariane 5 launchers (see separate entry) are available to ESA through 2009, facilitating Europe's independent, continued access to space. The multinational agreement also brings stability to Arianespace and the Ariane 5 programme in a slowing launcher market; ESA has agreed to conduct at least six Ariane 5 launches per year from 2005 to 2009, and will therefore require a significant number of the heavy-lift launch vehicles over the 5-year EGAS period.

Beginning in 2008, Arianespace and Europe's Spaceport in Kourou will also begin operating the Vega and Soyuz-2 launchers (see separate entries). Vega and Soyuz build upon the ESA/EU EGAS policy and progress the organisations' collective aspiration to establish a strong, long-term European launch vehicles industrial base.

The Vega programme, developed and led by an Italian team comprised of ELV SpA, a joint-venture of FiatAvio and the Italian Space Agency (ASI), has

ESA's ESTEC in Noordwijk, the Netherlands (European Space Agency) 1133870

ESA spacecraft Mars Express in final assembly and checkout (European Space Agency) 0572491

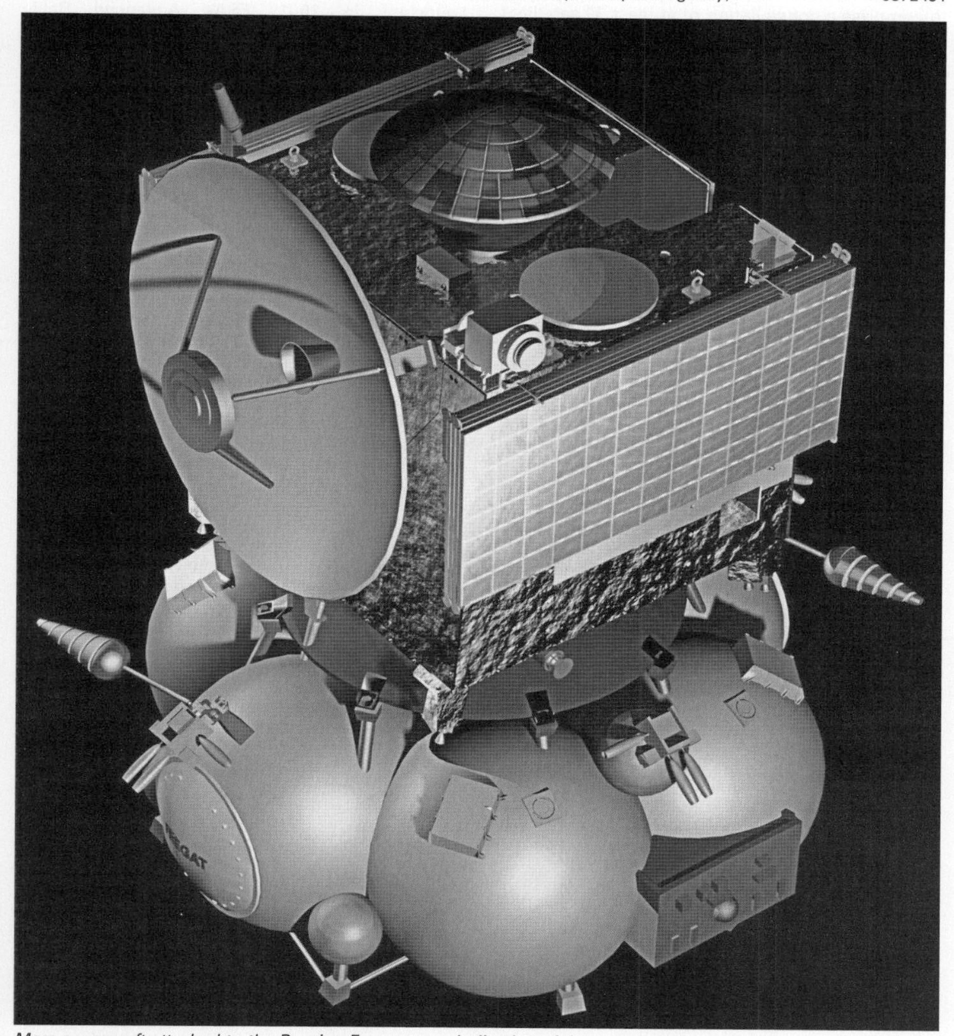

Mars spacecraft attached to the Russian Fregat stage indicative of the co-operation that exists between ESA and the Russian Space Agency (European Space Agency) 0572490

provided ESA with its first small launcher; Vega is capable of placing multiple payloads up to a total of 2,000 kg into Low Earth and Sun Synchronous Orbits (LEO, SSO). It will enable ESA to conduct more cost effective and possibly more frequent launches, as well as take advantage of the emerging small satellite market.

Starsem, a European-Russian joint venture founded by EADS, Arianespace, the Russian Federal Space Agency, and the Samara Space Center (TsSKB-Progress) will supply ESA with Soyuz-2 launch vehicles and services. Soyuz will add medium-class launch capabilities to ESA's launch services palette, allowing the organisation to place payloads of up to 3,000 kg into Geostationary Transfer Orbit (GTO).

ESA's Future Launchers Preparatory Programme (FLPP), initiated in February 2004, has as its objective, to provide Europe with the Next Generation Launcher (NGL) by approximately 2020. Studies on both reusable and expendable launchers have commenced and will likely continue through 2009 and beyond. ESA requirements dictate that future launch vehicles must be affordable, reliable and flexible in order to meet EGAS policy goals. Part of FLPP, ESA's Intermediate eXperimental Vehicle (IXV) is in development and is expected to reach space in 2010, atop a Vega launcher. The vehicle will be used as an in-orbit validator. IXV will test developing aerothermodynamics, materials and guidance re-entry technologies.

Human Spaceflight

Exploitation of the ISS is aimed at developing European operational capabilities for long-term manned space exploration, to build up the necessary know-how to master operations with complex manned outposts in space and to encourage support of the ISS by the European user community. Technical content includes operations and maintenance of the ESA ISS elements and ground segments such as the Columbus module, the Jules Verne Automated Transfer Vehicle (ATV), and the European Robotic Arm (ERA), including microgravity facilities for Columbus as well as respective ATV and Columbus control centres. Other commitments include the fulfilment of ISS common-systems operation, including the Ariane 5-ATV launcher, refuel and reboost services to the ISS, and support and co-ordination of ISS utilisation by European users including payload integration, logistics, data transmission and astronauts.

The Columbus Module, a major European ISS contribution, is planned for 2006 departure to the ISS, depending upon the US Space Shuttle's future flight readiness and launch schedule. It will add a further 75 m^3 of area to the ISS with the express purpose of conducting microgravity experimentation. The 4.5 m diameter pressurised cylinder will be linked to ISS Node 2. It can support three crew members and will contain ten internal racks to accommodate numerous experimental payloads, as well as four external mounting areas for space environment exposure and observation experiments. Columbus will house five specialised lab facilities: the Biological Experimentation Laboratory or BIOLAB; the Fluid Science Laboratory (FSL); and the Material Science Laboratory Electromagnetic Levitator (MSL-EML), in addition to the European Drawer Rack (EDR), and European Physiology Modules (EPM).

The ATV is intended for ISS support and servicing. Launched by the newly developed version Ariane-5-ATV, the 20 tonne ballistic vehicle is automated and will annually deliver cargo weighing up to 7,500 kg to the ISS crew. Upon arrival, the ATV will dock with the Russian service module and remain linked for approximately six months. When released, it will transport ISS waste toward Earth, where both vehicle and contents will be incinerated upon atmospheric re-entry. Both ATV and Columbus build upon existing Multi-Purpose Logistics Module (MPLM) technology.

The ERA is a joint ESA-Russian Federal Space Agency project and will be used on the Russian segment of ISS to support assembly and servicing tasks such as the installation and exchange of external equipment. The 11 m long symmetrical arm will have two 'wrists' and be attached at either end, thus being able to relocate itself, operating with either wrist acting as the manipulator's base. ERA can lift and transport cargo up to 8,000 kg. The programme was approved in 1994 and Dutch Space is leading the ERA international consortium. ERA is a direct descendant of the Hermes Robotic Arm (HERA), originally intended for that long-cancelled programme. Launch and journey to the ISS is anticipated in November 2007.

ESA's developing Aurora Exploration Programme, initiated in 2001, aspires to send European astronauts to Mars by 2030. Europe's first robotic mission to Mars, ExoMars, is planned for launch in 2013, a delay of two years from the initial launch date. Spacecraft for the mission will consist of the launcher and transfer stage; an orbiter; a descent module; an ascent module; one rover; and a re-entry vehicle. The descent module, carrying the rover, will be released toward the planet once the main orbiter craft enters low-altitude orbit. After collecting geological samples, the rover will transfer these to the ascent vehicle, which will rendezvous with the re-entry craft, and then return to Earth.

Earth Observation

ESA's Earth Observation (EO) programme is now better known as the Living Planet Programme. EO programmes are split into two main elements: the Earth Explorer Component, which includes mission definition, development, launch and operations of EO missions covering Earth's interior, oceans, atmosphere, cryosphere and land surface; and the Earth Watch component, which comprises ESA's meteorological missions, including those administered by the Organisation for the Exploitation of Meteorological Satellites (EUMETSAT), as well as satellite data collection and dissemination for operational use. More recent missions under the joint ESA-European Commission Global Monitoring for Environment and Security (GMES) initiative are also included.

The Earth Explorer component is comprised of Core and Opportunity Missions. Current Core Missions include: the Gravity and Ocean Circulation Explorer (GOCE); ADM-Aeolus Doppler Wind Lidar; and the Clouds, Aerosols and Radiation Explorer (EarthCARE). Current Opportunity

ESA Headquarters, Paris (European Space Agency) 1133869

GIOVE-A, the first validator satellite for Europe's Galileo system, with solar arrays collapsed (European Space Agency) 1133871

Missions are: the Soil Moisture and Ocean Salinity (SMOS); and the Swarm magnetic field constellation. The Cryosat ice sheet measurement satellite was an Opportunity Mission, however, the satellite was lost on 8 October 2005 due to a launch failure. ESA's 2005 Call for Earth Explorer Proposals capped proposed project costs at EUR300 million.

The GMES programme is still in development. The programme has begun with a five-year initiative called the GMES Services Element (GSE), although ESA has also started investigations into a new generation of satellites called Sentinels. GSE focuses upon end-user data needs and is targeted at streamlining ESA's EO product delivery. GMES EO priorities fall into the following categories: land and vegetation cover, forest monitoring, the marine, coastal and polar environment, maritime security, food security, atmospheric monitoring, humanitarian aid and risk management.

ERS-2 and Envisat remain operational, although ERS-2 has outlived its planned lifespan. ESA expects Envisat to remain in service until at least 2010. The Meteosat series has served ESA well, and will continue to do so well into the future: Meteosat 9 (MSG-2) was launched in 2005; Meteosat 11 will be the final satellite in that series, and the first Meteosat Third Generation (MTG) satellite is planned for 2015 launch.

Microgravity

ESA aims to continue with the technological development of microgravity research, implementing microgravity missions involving the Columbus module; Foton capsules; Maser, Mini Texus, and Maxus sounding rockets; Airbus A-300 Zero-G parabolic flights; the ZARM drop tower; and other ground-based simulation facilities. Further details on Columbus can be found in the section on human spaceflight, above.

Navigation

The joint ESA-European Commission Galileo European Satellite Navigation System, is expected to comprise a constellation of at least 30 satellites in Medium Earth Orbit (MEO) at 23,616 km altitude. Galileo began in-orbit test operations in late 2005 with the launch of GIOVE-A. Surrey Satellite Technology's GIOVE-A is the first of four trial satellites orbited as part of the Galileo System Test Bed (GSTB). Following a successful In-Orbit Validation (IOV) period, 26 further satellites will be launched, with Galileo achieving Full Operational Capability (FOC) by 2010. On the ground, ESA will maintain two European Galileo Control Centres (GCC) as well as 20 Galileo Sensor Stations (GSS), located across the globe. Additionally, Five S-band up-link stations and 10 C-band up-link stations will be built to facilitate data exchange between GCC and operational Galileo satellites. Galileo will also include a Search and Rescue (SAR) service featuring user transponders and a two way user communication system. GIOVE-B, the second Galileo validator satellite, developed by Galileo Industries, a consortium including EADS Astrium, is set to launch in 2007.

Galileo, the civilian controlled European Global Navigation Satellite System (GNSS) segment, will become a central feature of all European transport systems. It provides worldwide navigation coverage and is compatible with US GPS and Russian GLONASS GNSS navigation services. Consumers will be able to use one receiver to employ any Global Navigation Satellite System. Future Galileo users previously dependent upon the GPS or GLONASS systems, will enjoy independent, uninterrupted, European-based navigation services. Galileo's development has required substantial international co-operation; not simply among the European nations themselves, but also with the US and Russia, to ensure system interoperability for the benefit of the end user. Galileo will not only be operable independent of GPS and GLONASS, but will also be capable of augmenting GPS or GLONASS services, providing users with highly defined positioning data – accurate to within one metre.

The goal to form a complete European GNSS has first provided Europe with the European Geostationary Navigation Overlay Service (EGNOS), which became operational in 2005. EGNOS uses existing European Geostationary Orbit (GSO) infrastructure and a ground station network as well as GPS and GLONASS satellites to provide navigation services to users in European nations and potentially in other geographic areas as well. Providing accurate positioning to within five metres, EGNOS was considered the first step to developing an independent European System; Galileo, at the fully operational stage, represents the second step's culmination.

Telecommunications

Launched in 2001, ESA's Advanced Relay Technology Mission (Artemis) satellite is equipped with mobile telecommunications, data relay, and navigation instruments. Its telecommunications capabilities facilitate mobile communications among users in remote areas of Europe and Africa. ESA also provided the Amazonas satellite, launched in 2004, with the AmerHis switchboard instrument. Its AlphaBus 12-18 kW satellite platform is currently in development and promises to provide Europe with a European-developed high-powered telecommunications satellite.

The ESA Telecommunications Programme is based on a portfolio of activities that reflect the need to adapt swiftly to market conditions and consists of several elements covering different projects. ESA's telecommunications department seeks to strengthen and promote growth in the European telecom industry. Telecommunications programmes are organised into eight 'Programme Lines': Programme Development, Technology, User Segment, Multimedia Systems, Mobility, Large Platform Mission, In-Orbit Demonstrations, and Inter-Satellite Links (ISL). The Telecommunications department previously managed their portfolio using a framework called ARTES (Advanced Research in Telecommunications). Programme Lines are intended to organise telecom programmes into more practical, functional sectors.

NATIONAL AGENCIES

Argentina

Comision Nacional de Actividades Espaciales (CONAE)

Current Status
The Comision Nacional de Actividades Espaciales (CONAE), or the National Commission for Space Activities, in partnership with the Argentine Air Force, is leading the indigenous Argentine launcher project, called Tronador. The first vehicle in development, Tronador I, is ballistic in nature. The development of Tronador II, a more sophisticated launch vehicle, will follow, and will pave the way for the development of an Argentine satellite launch vehicle.

CONAE continues to work on several satellite programmes. SAC-D, in partnership with NASA, and SAC-E, a joint project with Brazil, are in development; SAC-D is scheduled for launch in 2008. SAOCOM 1-A, in co-operation with the Italian Space Agency, is also in development. All are Earth observation satellites.

Background
CONAE plans and executes Argentina's national civil space programme. Argentina's National Space Plan was last updated in 2004, and projects through to the year 2015. CONAE's estimated direct national funding is USD66 million for 2006, and USD56 million for 2007 (in 2002 USD). The plan includes goals associated with Argentina's ground support infrastructure, satellite missions, and access to space.

Besides its Buenos Aires headquarters, CONAE runs the Teófilo Tabanera Space Center, about 30 km southwest of Córdoba, which houses the Córdoba Ground Station, responsible for command, control and tracking of Argentina's satellites; the satellite Mission Control Centre; the satellite Testing and Integration Facility; and The Mario Gulich Institute for Advanced Space Studies.

CONAE, in co-operation with other Argentine and external national organisations, has played a part in satellite development and launch activities for many years, developing the Satélite de Aplicaciones Científicas (SAC) series, beginning in 1981. Argentina's imaging satellite, SAC-C launched from Vandenberg in 2000 aboard a Delta 2.

In May 1991, the Comision Nacional de Investigaciones Espaciales (CNIE), or National Commission for Space Research, was transferred from the Argentine Air Force to become the civil CONAE, under the direct supervision of Argentina's President. CNIE directed Argentina's space science activities from the 1960s. It began with development of a two-stage Castor sounding rocket able to carry small payloads up to 500 km.

Australia

Australian Centre for Remote Sensing (ACRES)

Current Status
The Australian Centre for Remote Sensing (ACRES) operates its Earth observation ground stations to collect and distribute imagery for geoscientific and environmental impact monitoring purposes, primarily to benefit the Australian public. ACRES provides imagery to support such projects as bushfire and water quality monitoring, agricultural monitoring, mapping, land use assessment, resource exploration, and regional and urban planning.

Background
Established in 1979, ACRES is a unit within the Geospatial and Earth Monitoring Division of Geoscience Australia, a part of the Australian Department of Industry, Tourism and Resources. It is the data acquisition, archiving, processing and distribution facility for ERS, Landsat, Radarsat, ALOS, Terra Modis, Aqua Modis, and NOAA satellite imagery. Processed data is used to generate remote sensing products on either magnetic or photographic media.

ACRES has facilities at Canberra, the home of the Data Processing Facility, and Alice Springs, where the Data Acquisition Facility and Ground Station are located. ACRES ground station coverage includes all of Australia, Papua New Guinea and parts of Indonesia. ACRES also operates a Hobart, Tasmania facility: the Tasmanian Earth Resources Satellite Station (TERSS), whose coverage extends across the Southern Ocean to Antarctica, including New Zealand.

Austria

Aeronautics and Space Agency

Current Status
In March 2005, the Austrian Space Agency (ASA) became the Agentur für Luft-und Raumfahrt (ALR), or the Aeronautics and Space Agency. ALR is a division of Die Österreichische Forschungsförderungsgesellschaft mbH (FFG): the Austrian Research Promotion Agency, founded in September 2004.

Background
ALR designs and executes the Austrian space programme and is responsible for developing Austrian space policy. It also represents the nation's interests to international space organisations, such as the European Space Agency (ESA), of which Austria is a member. ALR's primary goal is to serve as an institutional foundation working for the promotion and development of Austria's aerospace industry, including space-related products and services, and technical demonstrations.

Now managed by ALR, the Austrian national space programme grew from two national space programmes which, in 2002, first fell under ASA and Ministry for Transport, Innovation and Technology management: the Austrian Space Applications Programme (ASAP), and the Austrian Radionavigation Technology and Integrated Satnav services and products Testbed (ARTIST). ASAP supports research and commercial organisations developing space-related products, services and technologies linked with space science and exploration projects. ARTIST is Austria's satellite navigation testbed and is linked to Austria's involvement in ESA's Galileo programme. ALR is Austria's Galileo programme point of contact.

ALR's main roles and responsibilities are:
- Formulate Austrian national aeronautics and space policy
- Act as a co-ordinating partner in all national aerospace programmes
- Act as the official Austrian representative to international space-related organisations and committees, such as ESA and EUMETSAT, as well as the European Union
- Support and advise the Ministry of Foreign Affairs on space-related topics and act as Austria's official representative to space-related programmes run by organisations such as the UN Committee on the Peaceful Uses of Outer Space (UNCUOPOS) and the European Commission
- Organise and manage Austria's participation in bilateral and international aerospace programmes
- Provide stability and a solid structure for the Austrian space industry
- Organise and manage aerospace events and aerospace training events and activities.

ASA, the organisation from which ALR evolved, was established by the Federal authorities in Vienna in 1972. It co-ordinated national space activities and was the Austrian representative for international space activities. ASA formerly assisted the Federal Ministry for Transport, Innovation and Technology and was responsible for establishing the programmatic framework and strategic guidance to promote industry and scientific institutions to play a significant role in international space-related undertakings. ASA placed particular emphasis on the established teams in Austria dedicated to remote sensing to integrate international data for Austria's benefit.

Brazil

Agência Espacial Brasileira (AEB)

Current Status
The Agência Espacial Brasileira (AEB), the Brazilian Space Agency, sent its first astronaut into space aboard a Soyuz, with the International Space Station (ISS) Expedition 13 crew, in March 2006. Mission Specialist Marcos Pontes worked aboard the ISS and returned to Earth about nine days later with the Expedition 12 crew. He received both NASA and Russian Federal Space Agency (FSA) training previous to his flight.

Background
AEB, headquartered in Brasília, was founded in February 1994, as an autonomous agency under the Ministério da Ciência e Tecnologia (MCT), the Ministry of Science and Technology, to co-ordinate, plan and execute Brazil's space programme. The Agency replaced the Brazilian Commission for Space Activities, Comissão Brasileira de Atividades Espaciais (COBAE), which led Brazil's space activities from the 1970s. Several other space-related entities reporting to separate government agencies were also eliminated. Previous to AEB's creation, Brazil's military primarily led the country's space programme. AEB can trace its history back to 1941 with the creation of the Ministry of Aeronautics.

AEB is currently working under its 2005–2014 National Space Activities Programme (PNAE). The PNAE outlines Brazil's goals and space activities over a ten-year period and stresses the peaceful use of space. The Programme's activities fulfil the requirements of Brazil's National Policy for the Development of Space Activities (PNDAE). Traditionally, the country's satellite activity has concentrated upon Earth observation applications, including remote sensing, meteorology and oceanography; the programme also stresses telecommunications and scientific applications as well as Brazil's access to space via the development of indigenous launch capabilities, of which Brazil aims to develop both small satellite and heavy lift versions.

AEB is comprised of four directorates: The Directorate for Space Policy and Strategic Investment; the Directorate for Satellite Applications and Development; the Directorate for Space Transportation and Licensing; and the Directorate for Planning, Budget and Administration. Brazil's Ministry of Defence operates the Alcântara Launch Centre, Centro de Lançamento de Alcântara (CLA), the site from which AEB tests its small satellite orbital launch vehicle, the Veículo Lançador de Satélites (VLS).

AEB works in partnership with the Instituto Nacional de Pesquisas Espaciais (INPE), the National Institute for Space Research, which reports to the Ministry of Science and Technology. INPE is responsible for Brazil's satellite development (see separate entry).

AEB has worked in co-operation with other countries on various programmes, including China and the Ukraine. Brazil and Argentina are also involved in the joint Earth observation satellite programme, SAC-E. Brazil launched its first satellite, Satélite de Coleta de Dados (SCD-1), from Kennedy Space Center (KSC), aboard a Pegasus in 1993.

Canada

Canadian Space Agency

Current Status
The Canadian Space Agency (CSA) expects to be engaged in a number of activities in 2007, including the launch of the hybrid telecommunications and science satellite, Cassiope, and the launch and installation on the International Space Station (ISS) of the Dextre robotic arm. CSA has also contributed the meteorological (MET) package to NASA's Phoenix mission (see separate entry), due also to launch in 2007. CSA actively partners with both NASA and ESA, contributing especially to science and human spaceflight programmes.

The Canadian Space Strategy stresses space science, Earth observation, satellite communications, exploration and the advancement of Canadian space expertise as the primary areas for Canadian space development and participation.

Background
CSA is responsible for the management, planning and policy development of the Canadian space programme, including co-ordination of the space activities of other federal government agencies. It was created in March 1989, drawing together the space activities of the Ministry of State for Science, the Department of Communications, the Department of Energy, Mines and Resources and the National Research Council. Parliament, on 14 December 1990, passed the Canadian Space Agency Act, which, besides giving formal approval of the agency's creation, also allowed CSA to present its own budget and appropriate its own funds. The CSA President reports to the Minister of Industry.

CSA is headquartered in Longueuil, Quebec at the John H Chapman Space Centre. The agency employs roughly 635 personnel as well as 170 contractors. The David Florida Laboratory (see separate entry) is located in Ottawa. CSA also maintains liaison offices in Longueuil, Ottawa, Washington, Houston and Paris.

CSA focuses its efforts on five core operations: the Canadian Astronaut Office; Space Technologies, Space Science, Space Programs and Space Operations. The agency's 2005–2006 expenditures totalled CAD288 million and included activities in the Earth Observation; Space Science and Exploration; Satellite Communications; and Space Awareness and Learning programmes.

Canada is an active International Space Station (ISS) partner. CSA developed and deployed the ISS Mobile Servicing System (MSS), consisting of Canadarm 2, the Mobile Base System (MBS) and the Special Purpose Dexterous Manipulator (Dextre). Canadarm 2, the remote manipulator system, is permanently attached to the ISS. It is capable of joint activity with the shuttle orbiter-attached Canadarm, when the Shuttle is docked to the station. Canadarm 2 was launched and attached to the ISS in 2001. The MBS was transported and installed on the ISS in 2002. It stores and carries equipment while astronauts carry out work during spacewalks. CSA and NASA expect Dextre installation to take place in late 2007, launching aboard the Shuttle Endeavour mission STS-123 for ISS assembly flight 1J/A. Dextre will add to the ISS's independent maintenance and servicing capabilities. As of early 2007, Canada's astronaut corps included six astronauts, all of whom have flown on Space Shuttle and ISS missions.

China

China National Space Administration

Current Status
In October 2006, China released its white paper on space activities. Its primary work in the future, as outlined in the paper, will be in the areas of launch vehicle development; human spaceflight; Earth observation and remote sensing systems; communications and navigation satellites; ground segment development; and space science, including materials science, astrobiology, microgravity experimentation, astronomy and space physics. The stated goals of China's space programme are to support the country's economic and sustainable development strategies; to advance indigenous science and technology capabilities; to encourage social progress for all mankind; and to strengthen the national defence.

China shocked the world in January 2007, when it successfully destroyed one of its own aged Fengyun (FY-1C) meteorological satellites (see separate entry) with an anti-satellite (ASAT) weapon, creating over 1,000 pieces of orbiting debris. Although the 2006 white paper states that China will explore and use space for peaceful purposes, this latest weapons demonstration has made other space faring nations wary of China's aspirations; it has confirmed suspicions that the nation was funding a thriving ASAT development programme and that China will compete with other nations for space supremacy.

China became the third nation to independently launch a human into space, when Yang Liwei took flight aboard Shenzhou 5 in 2003. Shenzhou 6, with Fei Junlong and Nie Haisheng aboard, was successfully launched, and returned to Earth in 2005. Please see the separate Shenzhou entry for more details on China's human spaceflight programme.

Background
Although the Chinese space programme can be traced back to 1956 under the Chinese military, The China National Space Administration (CNSA) was founded much more recently, in 1993, upon the approval of the Eighth National People's Congress. It is subordinate to the State Council and China's Commission of Science, Technology and Industry for National Defense (COSTIND). CNSA is comprised of four divisions: General Planning; System Engineering; Science, Technology and Quality Control; and Foreign Affairs. The agency is responsible for China's civil space policy, as well as the country's space-related intergovernmental agreements and relationships. There is no full separation between China's civil and military space programmes, although COSTIND holds mainly civilian administrative responsibilities following 1999 Chinese defence industry reforms. This lack of segregation between the defence and civil programmes fuels suspicion among other nations regarding Chinese access to and use of space.

The State Council and COSTIND also oversee the China Aerospace Science and Technology Corporation (CASC) and the China Aerospace Science and Industry Corporation (CASIC). CASC and CASIC are responsible for launch vehicle, missile and spacecraft research, development and manufacture.

The Chinese space programme budget for 2007 was about one-tenth that of NASA's, according to CNSA's Administrator in 2006. The nation still spends quite a bit on its space programme for a developing nation: somewhere under USD1.7 billion. USD2.4 billion was expended on Shenzhous 1 – 5, and under USD125 million on Shenzhou 6. The agency's budget for the three-step Chang'e programme is USD170 million, with the Chang'e-1 first phase costing about USD125 million.

China is not an International Space Station (ISS) member nation, however it has sought ISS access; this has so far been rejected by ISS member nations. The country has announced ambitions to develop and build its own space station, as well as to send its astronauts (sometimes known as taikonauts) to the Moon. CNSA anticipates the launch of its Chang'e-1 Lunar orbiter from Xichang in 2007. It also has plans to launch Lunar lander and Lunar sample collector spacecraft previous to sending humans to the Moon, possibly by 2020.

Despite the barriers to obtaining ISS access, China has fostered and lead many other multinational and bilateral co-operative relationships, and will continue to do so. Examples include the CBERS project with Brazil; Double Star with ESA; and membership in the Asia-Pacific Space Co-operation Organization (APSCO).

China launched its first satellite, Dongfanghong (the East is Red), in 1970 from Jiuquan. It was a communications satellite that transmitted the patriotic song of the same name. Although the satellite is infamous for this function, examination of the payload suggests that it also carried communications transponders similar to the US Telstar 1 and 2. This raises the possibility that it might have had a limited communications repeater role. In 1981, China achieved its first multiple satellite launch from a single launch vehicle. Since Dongfanghong, China has launched over 100 satellites for both government and commercial customers.

XSSC-Xian Satellite Control Centre

Background
China's launch and satellite TT&C network is centred on the XSSC Xian Satellite Control Centre in Shaanxi province, controlling six fixed, three mobile and three ship (Yuanwang 1-3) stations. Yuanwang 3 was delivered in March 1995. The fixed sites are Weinan (near Xian, and the lead site), Min-Xi (Fujian province), Xiamen, Changchun (Jilin), Karshi (Xinjiang) and Nanning (Guanxi), with the first three providing GEO control. XSSC issues de-orbit commands to FSW recoverable satellites through the two mobile stations, while the third locates the re-entry modules during descent. XSSC's hardware covers three NCI 2780, two VAX 8700 and several VAX-II computers. Improved and updated tracking facilities. New telemetry stations.

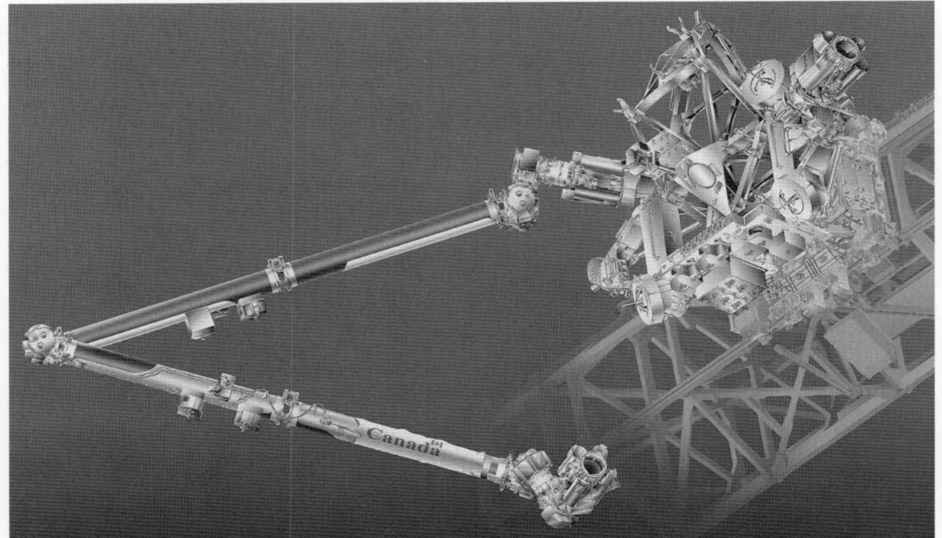

Canada has built decades of space engineering expertise on a wide range of space-related products, not least of which is the Shuttle remote manipulator system, seen here in its refined and redesigned form as the Space Station Remote Manipulator System
0572493

Denmark

Danish National Space Centre

Current Status
On 1 January 2007, the Danish National Space Centre (DNSC) absorbed other Danish space-related

organisations and then merged with the Technical University of Denmark (DTU), along with several other institutions, becoming a University research facility. The Danish government anticipates that extra funding will be diverted to the University to facilitate the mergers during the 2007 and 2008 fiscal years, however, DNSC's director will not hold a position in the Board of Governors, the body that will oversee merger activity.

Background

DNSC was formed in January 2005 by combining the Danish Space Research Institute, founded in 1966, and the National Survey and Cadastre of Denmark's geodesy division. DNSC remains subordinate to the Danish Ministry of Science, Technology and Innovation, as does DTU. Although it is part of DTU, DNSC is Denmark's lead space organisation, heading the Danish Space Consortium, which includes all major Danish space organisations in the commercial and public sectors. In this role, it is responsible for advising the Ministry on Space issues, as well as acting as Denmark's ESA liaison. DNSC is headed by a director; the Centre employs roughly 110 scientific, administration and student personnel.

DNSC's primary areas of research are aerospace Instrumentation, astrophysics, geodesy, geodynamics, remote sensing, solar system physics, and Sun-Earth climate studies. The Centre contributes to high profile science programmes such as Integral, the James Webb Space Telescope (JWST) and Planck, and provides instrumentation for the scientific satellites, ensuring Danish access to the programme's scientific databases.

France

Centre National d'Études Spatiales (CNES)

Current Status

The heads of CNES and the Russian Federal Space Agency (FSA), Roskosmos, have agreed to work together on the Oural future launcher project. Reportedly, CNES has devoted EUR250 million to the Oural programme over five years. The project promises to deliver a heavy-lift launcher by 2020 that will be used to lift payloads of up to 7,000 kg into GEO. The new launcher, once in active service, will replace the Ariane.

For a number of years, cooperative agreements have existed that will allow Soyuz launches from the Centre Spatial Guyanais (CSG – see separate entry). CNES anticipates that the first Soyuz will launch from CSG in 2008.

Background

CNES, the French Space Agency, was established in 1961 to execute the French national space programme. The Agency is a body that is considered to be a "Public Establishment of an Industrial and Commercial Character (EPIC)." CNES underwent a leadership change and directorate reorganisation in 2003 after a EUR35 million budget deficit in fiscal year 2002 and a EUR90 million shortfall in project funding during the same period. It is headed by a Director General and a President, who is appointed by the French Council of Ministers; CNES is comprised of 13 Directorates. In 1993 CNES was also allocated responsibility for the design and development of French military space projects, under the Ministry of Defence. As of 2006, personnel totalled 2,400, including 1,800 engineers, located at the Paris Headquarters, the Evry Launcher Directorate (DLA), the Toulouse Space Centre and the Centre Spatial Guyanais (CSG) launch site in Kourou, French Guiana (please see separate entries).

CNES has nine subsidiaries, four Economic Interest Groups (GIE), and five Public Interest Groups (GIP) to exploit and market its space activities. The subsidiaries are: Arianespace, which produces and markets Ariane launchers; Intespace, which performs space environment tests and maintains facilities for the tests; Spot Image, a provider of geographical and vegetation satellite imagery; SCOT, an Earth observation consultancy; Collecte Localisation Satellites (CLS), Argos System operator and provider of environmental, oceanic and location satellite services; Novespace, a microgravity and space technology transfer organisation; the European centre for research and advanced training in scientific computation (CERFACS), which develops advanced numerical simulation and algorithmic solution research methods; Cité de l'Espace (Space City), a space education park in Toulouse; and Novacom Services, a CLS subsidiary and satellite tracking service provider. The GIEs and GIPs are also made up of space-related groups, with specialisations ranging from remote sensing to communications. Additionally, CNES does not maintain in-house space science laboratories but does support a number of principal French research facilities.

CNES engages in satellite projects with research, civil, environmental and defence applications goals. Primarily, the satellites carry out Earth and environmental observation, navigation and meteorological missions. CNES supports space science and planetary missions, as well as the International Space Station (ISS). It also funds the Ariane, Soyuz and Vega launcher programmes, fulfilling France's goal of assured access to space.

Centre Spatial de Toulouse (CST)

Current Status

CST at Toulouse is the agency's principal engineering and technology facility and exploits the operational systems. It has developed several engineering activities designed to provide research, evaluation and test of small and medium satellites. CST has conducted research activities in the design and development of small to medium-class satellites. It developed nanotechnology concepts for the micro-miniaturisation of satellite and spacecraft systems for application in small or medium size bus designs. CST also manages the balloon launching sites at Aire-sur-l 'Adour and Gap-Tallard. The centre comprises 20 laboratories, including the spacecraft test facility managed by CNES subsidiary Intespace. The operation centres located at Toulouse include: two satellite launch/early orbital phase control rooms (second opened 1991), Sarsat-Cospas control/mission centre; the Telecom 1/2 dedicated control centre; the TDF 1/2 dedicated control centre; the Spot 1–3 control centre and the Spot imagery rectification centre; the Helios 1/Spot 4 Fresnel building; the computer centre; and the 2 GHz network control centre, with stations at Issus Aussaguel (15 km southwest of Toulouse), Kourou, Hartebeesthoek (South Africa) and Kerguelen Island (3,000 km southeast of South Africa). CST has conducted integrated mission control studies for multinational projects. This work has expanded into the development of designs for small satellite applications and for the development of integrated nanotechnologies on large platforms.

Centre Spatial d'Evry

Current Status

Evry is CNES' launcher development centre, responsible for the Ariane series for the European Space Agency. It has a mandate to continue development of the Ariane 5 and had authority to provide technical direction for the FESTIP and FTLP activities. Responsibility for upper stage development and for propulsion systems analysis. Instrumental in resolving Ariane 5 failure determination. On the continuing search for a reusable launch vehicle Evry has played a very minor role in providing consultative recommendations on the impact a reusable fly-back booster may have on launch complex design and resource requirements. No formal programme has been authorised for such a vehicle, but consistent interest from government agencies across Europe have encouraged a sustained study and analysis programme. In recent months there have been unofficial discussions with the Russian Space Agency about the lift requirements for a new proposed successor to the Russian Soyuz manned spacecraft. Any such consideration would concern the possibility of using Ariane to lift a Russo-European manned space vehicle.

Germany

Deutsche Forschungsanstalt für Luft und Raumfahrt

Current Status

In close co-operation with DARA, the German Aerospace Research Establishment's responsibility is to execute the research element of the German space programme. It was formed in 1969 and currently employs a total staff of about 4,400, including 1,550 scientists in five research areas: Telecommunications Technology and Remote Sensing, Materials and Structures, Energetics, Fluid Mechanics, and Flight Mechanics/Guidance and Control. The annual budget is about DM700 million. Research centres are located in Berlin, Braunschweig, Göttingen, Köln, Lampoldshausen, Oberpfaffenhofen and Stuttgart, with local branches in Hamburg, Bonn, Trauen, Neustrelitz and Weilheim.

Test facilities include subsonic-to-hypersonic wind tunnels, rocket propulsion test stands, spacecraft TT&C stations and space simulators. Space mission operations, particularly those for Spacelab, Columbus and Eureca, are centred at the Köln-Porz and Oberpfaffenhofen sites. The Crew Training Complex (for ESA's astronaut cadre) and Microgravity User Support Centre in Köln-Porz, and the German Space Operations Centre (GSOC), Manned Space Laboratories Control Centre, User Data Centre and Automation In Orbit Centre at Oberpfaffenhofen have performed manned/unmanned missions for more than 20 years. Change options for development of on-orbit mission tasks and payload monitoring. This allows crew elements to control general modular housekeeping studies and simultaneously track progress with experiments.

The DFD German Remote Sensing Data Centre is the national PAF Processing and Archiving Facility for ERS and other satellites. Köln is the home base for federal astronauts and seat of the EAC European Astronaut Centre of ESA. The Materials and Structures department includes the Institute of Space Simulation, which operates the Microgravity User Support Centre and was most recently primarily concerned with preparing users, astronaut crew and equipment for Spacelab D2. Under the Department of Energetics, the Space Propulsion Division at Lampoldshausen operates test facilities for rocket motors such as Ariane 5's Vulcain. A new stand was completed in 1995 under a bilateral programme with France.

India

Indian Space Research Organisation (ISRO)

Current Status

ISRO has notched up a creditable series of flights and anticipates a robust period for development of both satellite and launch vehicle technologies. The INSAT 3A multipurpose satellite was launched by Ariane 5 on 10 April 2003 and will augment the present INSAT capacity for telecommunication and broadcasting as well as providing meteorological services along with INSAT 2E and KALPANA 1. The second development test of the GSLV launch vehicle took place on 8 May 2003 from the Satish Dhawan Space Centre, revalidating vehicle systems and carrying the 1.8 tonne GSAT-2 satellite into orbit. The first operational flight of the GSLV took place on 20 September 2004 and is expected to be followed by five more GSLV flights in 2006 to 2008. The advanced GSLV-III is scheduled for its first flight in 2007 to 2008. Success with the PSLV launcher saw the KALPANA-1 satellite orbited on 12 September 2002 by PSLV-C4 followed by Resourcesat-1 (IRS-P6) by PSLV-C5 on 17 October 2003. Cartosat and Hamsat were launched by PSLV-C6 on 5 May 2005. Five more PSLV flights are scheduled for the period 2005 to 2008. ISRO was established in 1969 as India's primary space R&D organisation, responsible for developing launcher and propulsion systems, launch sites, satellites and their tracking networks. This was followed in 1972 by the Space Commission and Department of Space. Personnel totals rose from 13,488 in 1986 to 16,800 in 1996 and stood at 16,400 in 2002, accommodated in eight main centres and units:

Vikram Sarabhai Space Centre (VSSC). ISRO's single largest facility (4,700 personnel), near Trivandrum, providing the technology base for launcher and propulsion development.

Liquid Propulsion Systems Centre (LPSC). Development wings in Bangalore and Trivandrum are supported by major test facilities at Mahendragiri for a wide spectrum of liquid motors, from reaction control system thrusters to the 720 kN Vikas and cryogenic engines. Personnel total 1,600.

ISRO Satellite Centre (ISAC). ISRO's lead centre for the design, fabrication and testing of science, technology and applications satellites. Staff strength is 2,400.

ISRO Inertial Systems Unit (IISU). Provides inertial systems/components for satellites and launchers.

SHAR Centre. ISRO's orbital launch site and largest solid motor production and test facility with 2,300 employees.

ISRO Telemetry, Tracking and Command Network (Istrac). Headquartered in Bangalore, Istrac operates a network of ground stations to provide TTC support for launcher and satellite operations. Personnel total is 460.

Space Applications Centre (SAC). Located at Ahmedabad, SAC is ISRO's applications R&D centre, including communications, remote sensing and geodesy. For example, it developed IRS 1C's cameras. 2,000 personnel.

Development and Educational Communications Unit (DECU) at Ahmedabad.

Insat Master Control Facility (MCF) at Hassan, 180 km from Bangalore. 295 personnel.

National Remote Sensing Agency Ground facilities for the reception, processing and dissemination of IRS, Landsat and ERS data, as well as the development of remote sensing techniques and applications.

Regional Remote Sensing Service Centre located at Bangalore, Naqpur, Kharagpur, Jodhpur and Dehradun.

Physical Research Laboratory covering research into astronomy and astrophysics, planetary atmospheres and aeronomy.

National Mesosphere/Stratosphere Troposphere Radar Facility at Gadankinear Tirupati.

ISTRAC-ISRO Telemetry Tracking and Command network

Background

Istrac operates a network of TT&C stations for ISRO's launch vehicle and satellite activities. The facility employs 465. Stations are sited at Trivandrum (8 m), Sriharikota (10 m), Carnicobar (8 m, principally for launcher downrange TT&C), Lucknow (10 m) and Bangalore (10 m), which also accommodates the network's Spacecraft Control Facility. Its two 10 m Cassegrain dishes (20.5 dB/K, 2.200–2.300/2.025–2.120 GHz down/up link, 9°/s slew rate, 0.8° beamwidth).

ISTRAC has two VAX 11/985 computers to currently support the IRS remote sensing satellite (imagery is routed via the National Remote Sensing Agency's Hyderabad station) with a back-up VHF command system. Insat is controlled from its own dedicated station at Hassan.

Three satellites can be controlled simultaneously, one undergoing orbital checkout and two operational. SSC has a main control room, mission analysis room, two dedicated control rooms and a data control room, the first two operating until orbital checkout is completed and the satellite is handed over to a dedicated room. The data control room acts as a node for all the network stations and SCC. Added data handling and telemetry processing centre. New data analysis section.

The network also operates an S-band receive-only TT&C station near Port Louis, Mauritius. Optical/laser tracking is performed at the Satellite Tracking & Ranging Station (STARS) established at Kavalur in co-operation with Russia.

Indonesia

National Institute of Aeronautics and Space (LAPAN)

Current Status

LAPAN (Lembaga Penerbangan dan Antariksa Nasional), founded in 1963, is a non-departmental government agency and is directly responsible to the President. Its chief activities are remote sensing and probing of the upper atmosphere by balloon and rocket sondes. LAPAN pioneered methods of seeding rain clouds and used meteorological satellite data to co-ordinate rain-making activities to water rice fields. It also performs telecommunications research, notably on signal attenuation caused by heavy tropical rainfall. First and second stage solid propellant sounding rockets designed and built by LAPAN are launched up to 100 km from the range at Pameungpeuk some 150 km south of Bandung. The agency maintains Earth stations at Jakarta and Biak Island (Irian Jaya) for reception of GMS, NOAA and Landsat MSS/TM imagery. The Multimission Remote Sensing Satellite Ground Station was commissioned in October 1993 at Pare-Pare on Sulawesi island to receive Spot, ERS and Landsat 5 (Landsat 6 ETM was also to have been included); Japan's JERS was added December 1995 and the facility is being upgraded to take advantage of new commercial remote sensing systems. LAPAN has secured optical and radar imaging to integrate with sounding rocket data for national and regional evaluations.

Israel

Israel Space Agency (ISA)

Current Status

ISA was established in 1983 within the Ministry of Science and Technology to direct national space research and R&D activities. The Ofeq programme is conducted under its auspices. The ISA has broad responsibilities for both civil and military technologies but only on the same basis as that of NASA in the USA whereby common technologies and experiments are frequently hosted by the civilian agency. However, unlike US civil and military space organisations, the ISA conducts research and technology studies directly applicable to both interests. This has proved an effective use of limited resources. R and D includes space physics research, space systems engineering and nano-technologies. Liaison with other government agencies for policy planning. ISA actively conducts international projects and provides a point of convergence for military and non-military projects. ISA has conducted new initiatives to link industries in Israel with joint venture offerings throughout Europe and the Far East.

Italy

Agenzia Spaziale Italiana (ASI)

Current Status

In April 2007, the AGILE satellite was launched into space aboard the Indian PSLV-C8 vehicle. AGILE was designed and developed by Carlo Gavazzi Space, in association with several industrial and research institutions for the Agenzia Spaziale Italiana (ASI). AGILE is equipped with scientific instruments capable of imaging distant celestial objects in the X-ray and Gamma ray regions of the electromagnetic spectrum. At lift-off, AGILE weighed 352 kg.

Also in 2007, the first and second of the COSMO-SkyMed (COnstellation of small Satellites for the Mediterranean Observation) satellites were launched into orbit. COSMO-SkyMed is an Earth observation satellite system funded by the Italian Ministry of Research and Ministry of Defence, which was conducted by ASI. The system is intended for both military and civilian use. The completed system will consist of four medium-sized satellites, each equipped with Synthetic Aperture Radar (SAR) sensors with global coverage of the planet. The four satellites are planned for sun-synchronous polar orbits, phased at 90° and at an altitude of 619 km with an orbit of 97 minutes. The satellites main components consist of two solar array for 3.8 kW at 42 V DC, stabilisation, navigation and GPS systems, SAR working in X band, 300 Gbit on-board memory and 310 Mbit/s datalink with ground segments. The expected operating life expectancy for each satellite is five years. The third satellite in the series is expected to launch in mid-2008 and the launch date for the fourth is yet to be confirmed.

The ASI and ESA also developed the expendable launch system Vega (see separate entry), which was mostly funded by the ASI and is now planned for launch in 2008.

Background

ASI was formally established 25 August 1988 under MURST to manage the PSN national space plan (Piano Spaziale Nazionale), previously directed from within the National Research Council (CNR, Consiglio Nazionale delle Ricerche). ASI is responsible for Italy's ESA involvement and manages the four principal national satellite projects: Italsat, Tethered Satellite, SAX and Lageos 2.

The ASI has been instrumental in establishing a strong contribution to the European effort supporting the NASA led International Space Station. Negotiations with NASA over the possible development of a habitation module for the ISS began when the Bush administration cancelled the US element in early 2001. Little progress has been made on this and further discussion will

Bangalore's Istrac complex. Centre: main control centre; right: TT&C station. The twin 10 m dishes are controlling the IRS remote sensing satellites (ISRO) 0517081

await resolution of the changing nature of the ISS programme and the future of the Shuttle.

The agency's annual budget is described at the beginning of the national entry. The Cabinet elects chairman of the Board, proposed by the MURST Minister. The DG is appointed by MURST after consultations with the Board of Directors. Board Members, elected by the Cabinet, sit for five years.

The CGS at Matera is part of the worldwide integrated geodetic network. Returns are regularly made using the principal laser ranging satellites, Lageos 1/2, Starlette, Ajisai and Stella. AlliedSignal Technical Services Corp of the US installed a Mobile Laser Ranging System in 1994. The station includes a 20 m VLBI dish. ASI's Milo base in Sicily (38°01′ north/12°35′ east) launches stratospheric balloons along the Mediterranean to Spain's west coast.

In November 2004, Dr. Tachikawa, the President of Japan Aerospace Exploration Agency (JAXA) and Professor Vetrella from the ASI signed a joint statement outlining plans for increased co-operation between the twin agencies.

Japan

Japan Aerospace Exploration Agency (JAXA)

Current Status

The Japan Aerospace Exploration Agency (JAXA) was created on 1 October 2003, when three separate aerospace organisations merged to form one agency. Although the Institute of Space and Astronautical Science (ISAS), the National Aerospace Laboratory of Japan (NAL) and the National Space Development Agency of Japan (NASDA) no longer exist as separate entities, their legacy continues in JAXA's work. JAXA is organised into five key divisions and institutes: the Office of Spaceflight and Operations; the Office of Space Applications; the Aviation Program Group (APG); the Institute of Aerospace Technology (IAT) and the Institute of Space and Astronautical Science (ISAS).

JAXA is headquarted in Tokyo, Japan. The Agency maintains over 20 Japan-based field centre and tracking station sites as well as offices in Washington, DC; Houston, Texas; Paris, France; Bangkok, Thailand and a liaison office at NASA's Kennedy Space Center, Florida. Staff total nearly 2000, including graduate student and foreign researchers. JAXA's FY2003 budget was JPY98 billion.

ISAS retained its name after JAXA was formed and continues to focus upon space science, observation and planetary research. ISAS launched Japan's first satellite, OHSUMI, in 1970; it is the lead organisation on the troubled Hayabusa (MUSES-C) asteroid research project. ISAS works in co-operation with Japanese universities wishing to conduct space science research.

The Office of Space Applications oversees telecommunications and information satellite development and research, including satellites used for environmental applications, such as the Greenhouse Gasses Observing Satellite (GOSAT) and the Advanced Land Observing Satellite (ALOS).

The Office of Spaceflight and Operations has largely inherited NASDA's work and is responsible for conducting satellite launch and command and control operations as well as launch vehicle systems development and all International Space Station (ISS) activities. The office runs JAXA's Usuda Deep Space Center as well as its satellite tracking and communications stations.

IAT is charged with aerospace research and development for both atmospheric and space-based applications. Its space-related activities include research into propulsion systems, rocket engines, space structures engineering and space power systems.

APG leads aviation research and continues to build on NAL's past achievements. Projects include the Next Generation Supersonic Transport (SST) craft, a supersonic passenger plane similar to the Concorde, and Unmanned Aerial Vehicle (UAV) development.

Tsukuba Space Center (TKSC)

Current Status

Japan's contribution to the International Space Station, the Japanese Experiment Module (JEM), or Kibo, will be controlled and supported from the Space Station Integration & Promotion (SSIP) facility at Tsukuba. NASA's ambitious Space Shuttle launch schedule has Kibo slated for an early 2008 launch aboard Atlantis' STS-124 ISS assembly flight 1J mission.

TKSC also houses an astronaut training facility. As of early 2007, Japan had certified a corps of eight astronauts, two of whom have been selected to accompany Kibo when it makes its journey to the ISS.

Background

Located in Tsukuba Science City 50 km northeast of Tokyo, TKSC is the Japan Aerospace Exploration Agency's (JAXA) largest technical facility, with seven sections responsible for launcher and satellite R&D, design and testing. TKSC began operations in 1972 as the National Space Development Agency of Japan's (NASDA) primary facility. NASDA was merged into JAXA in 2003 and TKSC is now a JAXA field centre.

The Office of Space Flight and Operations, and the Office of Space Application are located at TKSC. Tsukuba's SSIP provides the Space Experiment Lab (SEL); the user experiment preparation and testing building; the Space Station Test (SST) building for JEM testing, training, payload integration and operations simulation; the Astronaut Training Facility (ATF) for crew operations; and the Weightless Environment Test (WET) building (a 16 m diameter, 11 m deep, water tank) for procedure verification and crew training. It acts as the Space Station Operation Facility (SSOF) for JEM operation.

The Tsukuba Tracking and Control Centre is the command hub for all Japanese-launched satellites, including those of ISAS and commercial organisations. JAXA maintains a network of ground facilities in and outside of Japan, including Okinawa, Masuda and Usuda. The Institute of Space Technology and Aeronautics as well as part of the Institute of Space and Astronautical Science are located at TKSC.

Netherlands

Space Research Organisation Netherlands

Current Status

Space Research Organisation Netherlands' (SRON) main activities are currently focused within manufacturing secondary space hardware for use on other spacecraft, and analysing returned data from experiments from other international space agencies. SRON share contracts with ESA are based on the money return policy agreed by ESA member states. SRON also has partnerships with other national space agencies, notably ASI and NASA.

SRON is the primary contractor on the Heterodyne Instrument for the Far-Infrared (HIFI), selected by ESA as one of the three main instruments on the Herschel Space Observatory. HIFI is a high resolution spectrometer capable of mapping the chemical composition of nearby stars and galaxies. HIFI is designed to identify atomic and molecular spectral lines in the submilimeter category, operating at frequencies between 480–1,250 GHz and 1,410–1,910 GHz, it is the first heterodyne instrument to cover such a broad range. SRON's role is to lead the international consortium building the instrument and to verify its integration with the satellite structure. The instrument is expected to return its first data shortly after launch of the Herschel Space Observatory, which is planned for mid-2008.

ESA and the United States are currently in a joint venture to monitor the climate and improve weather forecasting using the European built MetOp satellites. One of the European instruments on MetOp is the improved Global Ozone Monitoring Experiment (GOME-2) and SRON's primary activities on the project concern monitoring instrument performance, radiometric calibration algorithms and the interpretation of the measurements, with the main focus on the retrieval of tropospheric ozone and aerosol properties. The MetOp satellites are being launched sequentially over 14 years, with the first launched in 2005.

Other future projects involving SRON include the Gravity Field and Steady-State Ocean Circulation Explorer (GOCE), which is due for launch in May 2008 and the LISA Pathfinder, due for launch at the end of 2009. The GOCE satellite will provide data to help determine global and regional models of the Earth's quasi-static gravity and the geoid. The principle aim of the project is to further understanding of the Earth's climate and to do this they are focusing in the fields of steady-state ocean circulation, physics of the Earth's interior, geodesy, surveying and sea-level change. SRON has the lead, under EAS contract, for the definition study of the Calibration and Monitoring Facility (CMF), which is part of the GOCE Ground Segment. The CMF will be responsible for the continuous monitoring of the Space Segment, the calibration and the observational data quality, the High level Processing Facility (HPF). SRON is also responsible for the scientific pre-processing and the external calibration as well as hosting the project management office. SRON co-ordinates the scientific use of GOCE data in the Netherlands in the fields of Oceanography and Geodynamics.

With regards to the LISA Pathfinder programme, SRON is responsible for the development of the Inertial Sensor Special Check-out Equipment (IS-SCOE). This unit replaces the Inertial Sensor core in the ground tests of the LISA Pathfinder system.

With SRON currently involved in ESA's Earth Gravity mission GOCE is preparing for a role in a future Earth gravity mission using LISA technology. In this way, the participation in LISA Pathfinder serves a dual purpose.

Background

The Stichting Ruimteonderzoek Nederland (SRON) was founded 10 June 1983 within the Netherlands Organization for Scientific Research (NWO) in The Hague, by which it is financed, to manage/execute space science projects. SRON comprises two laboratories, in Utrecht and Groningen, with specific interest in X/V-ray and IR/sub-mm astronomy. The Institute acts as the national agency for space research and as the national point of contact for ESA programmes. SRON operates in an advisory role for government and also conducts analysis of capabilities and makes recommendations to the government regarding future programmes. Projects include detectors for NASA's Compton Observatory and AXAF, two wide-field cameras for SAX, the short wavelength spectrometer for ESA's ISO, the SCIAMACHY instrument on ESA's ENVISAT and a reflection grating spectrometer for ESA's XMM cornerstone mission. SRON also manages Dutch μg, radiation, exobiology and Earth-oriented research. Some 10 per cent of its EUR20 million annual budget derives from commissioned research. This is set to increase with an involvement in major international research programmes sought by SRON.

Norway

Norsk Romsenter

Current Status

The Norwegian Space Centre is an independent organisation established by the Ministry of Industry 5 June 1987 as the immediate successor to NTNF, Space Activity Division. Including daughter companies there are about 80 personnel. Its mandate includes compiling and submitting to the Ministry recommendations for co-ordinated long-term programmes. In particular, it represents Norway in ESA and co-ordinates the space-related activities of Norwegian companies, universities and research institutes. Maintains policy planning and future mission recommendations for government. The Andoya Rocket Range and Tromso Satellite Station are both limited companies, with the Norwegian Space Centre having 100 per cent and 50 per cent interests. The organisation also conducts research into international projects and programmes likely to benefit from Norwegian science. Norsk Romsenter has been integrated with several levels of government in an attempt to maximise the potential economic benefit of space-related activities.

Pakistan

Space and Upper Atmosphere Research Commission (SUPARCO)

Current Status
The Space and Upper Atmosphere Research Commission (SUPARCO) is Pakistan's national space agency. It oversees the nation's satellite and launch vehicle development programmes. Pakistan's space technology and capabilities development efforts are focused upon the desire to achieve independent, indigenous satellite manufacture and independent space access via a small satellite launcher. SUPARCO's work is dedicated to research and development programmes that exploit the peaceful uses of space for the national benefit.

Background
Headquartered in Karachi, SUPARCO is Pakistan's national space agency and executes the country's atmospheric science; communications, science, Earth observation and remote sensing satellite programmes; its sounding rocket and launch vehicle development programmes; and its telemetry, tracking and ground station development, management and operations. It was established in 1961 under the chairmanship of Professor Abdus Salam and re-organised in 1981 as an autonomous commission under the aegis of the 10-member Space Research Council (SRC) under the Cabinet Division. Following the 1999 military coup, the new government set up a new structure, to which SUPARCO is subordinate. In December 2000, SUPARCO's former reporting chain was abolished and today, SUPARCO reports to the Development Control Committee (DCC), part of Pakistan's National Command Authority (NCA), the body that oversees Pakistan's nuclear programme. The DCC is chaired by Pakistan's President. SUPARCO is comprised of a Chairman, who is SUPARCO's head, and five members; the members each head a division.

SUPARCO is Pakistan's co-ordinator and international liaison for remote sensing activities. The Commission has reported that it will open a National Center for Remote Sensing (RS) and Geographic Information Systems (GIS) in Karachi in 2007. SUPARCO is also responsible for conducting national and regional land and water use studies using SPOT and Landsat satellite resources.

The Commission's multi-satellite ground station is located outside Islamabad. It is equipped with a 10 m S and X-band Cassegrain antenna to receive SPOT and Landsat series data and an L-Band antenna to receive NOAA data. The ground station was established in 1989 and upgraded in 2004. The ground station continuously acquired and processed SPOT data to support relief efforts following the Major earthquake northern Pakistan suffered in October 2005.

SUPARCO managed the development of the Badr-1 and Badr-B satellites (see separate entry) and is now engaged in the development of its Remote Sensing Satellite System (RSSS). The Commission expects to launch two RSSS satellites sometime before 2010, preferably upon an indigenous Satellite Launch Vehicle (SLV), if the plan keeps to schedule. SUPARCO considers the satellites, and their development, manufacture and launch, to be important to Pakistani sustainable and technology development progress. RSSS will be used primarily for environmental, land use and mapping applications. A Telemetry, Tracking and Command (TT&C) ground station is being established to support RSSS and any subsequent Pakistani satellites. SUPARCO will engineer RSSS-2 and the Italian M/S Space Engineering SPA, which partnered with SUPARCO on the satellite feasibility study, will develop RSSS-2. The Pakistan Remote Sensing Satellite (PRSS), developed by SUPARCO, is the RSSS prototype. It is equipped with a 3 m ground resolution panchromatic camera and has a three-year design life; revisit time is 2–4 days.

SUPARCO also developed the prototype PakSat 1R C-band communications satellite. PakSat-1, a C and Ku-band communications satellite located at 38°, is operated by SUPARCO and administered by PakSat International Ltd.

SUPARCO operates the Sonmiani Bay launch site, North of Karachi; it is a sub-orbital site used primarily to launch sounding rockets as well as for missile testing. An SLV launch date has not been announced. However, Pakistan's missile technology is well-developed (please see *Jane's Strategic Weapon Systems*), and the nation may soon demonstrate that it has achieved the ability to independently launch satellites.

Russian Federation

Russian Space Agency

Current Status
RSA was created by President Yeltsin's decree in February 1992 to draw up and manage Russia's civil space programme. Its director, Yuri Koptev, was formerly MOM deputy minister and an NPO Lavotchkin engineer. It has 300 HQ employees and a further 600 eslewhere.The nine principal divisions are: state programmes, manned projects and launch facilities, implementation of state programmes, science and commercial satellites, international; ground infrastructure, external relations and legal affairs and resources and business affairs. All rocket engine organisations, such as Energomash and NII TP, now come under RSA. In 1999, the Aviation Industry Ministry became part of the State Committee on the Defence Industry and then under the Defence Industry Ministry before becoming part of the Russian Space Agency.

Controls the Russian side of the International Space Station partnership, negotiates with the other ISS partners and serves as the government agent on the procurement of Soyuz and Progress vehicles. The RSA had developed strategies for maintaining the ISS during the hiatus in Shuttle operations pending a return to flight no earlier than September 2003. NASA has decided to fly two Shuttle missions to demonstrate its compliance with the recommendations of the Accident Investigation Board which would put the earliest resumption of full ISS support at early 2005. As of June 2004 those plans had slipped considerably and the RSS now anticipates maintaining the ISS until late 2005 following the first Shuttle flight earlier that year.

Spain

Instituto Nacional de Técnica Aeroespacial (INTA)

Current Status
The defence ministry established the Instituto Nacional de Técnica Aeroespacial (INTA) in 1942. It remains an autonomous R&D body reporting to the Ministry of Defence. Personnel total about 1,200 in four divisions, namely, Materials and Structures, Aerodynamics, Energy and Propulsion, and Avionics. It also operates the Aeronautical/Space Documentation and Information Centre (CIDAE). In the space field, it provides management and technical expertise and is currently managing the Minisat and Capricornio programmes. It was responsible for Spain's first satellite (Intasat), conducts INTA 100/300 sounding rocket firings, provides technical management for Spain's Helios contribution, developed S/Ku-band TT&C antennas used on satellites such as Olympus, Hipparcos, Telecom 2, Eureca and Hispasat, and operates the ESA Spasolab Laboratory for testing and certifying photovoltaic cells. INTA's INSA subsidiary operates and maintains the Villafranca tracking sites for ESA and the Madrid site for NASA. Its own 10 m S-band antenna at Maspalomas receives Landsat and Spot imagery, which it distributes as ESRIN's National Point of Contact. Also receives ERS data and handles tape data supplied by ESA. Special applications to government, defence and national security needs. Co-ordinates the use of earth observation images for military applications and has conducted studies of commercial applications for civil satellites.

Sweden

Swedish Space Corporation (SSC)

Current Status
In January 2007, the Swedish Space Corporation (SSC) signed an agreement with Virgin Galactic to carry out initial spaceport feasibility studies, effectively inaugurating Spaceport Sweden. The new spaceport, founded by SSC, Icehotel, LFV Group (Kiruna Airport) and Progressum, is located at the Esrange Space Centre (see separate entry) near Kiruna, and aspires to be the first European commercial, suborbital space tourist launch site.

SSC expects to launch its two Prisma satellites in 2009. The technology demonstrator programme consists of Main and Target satellites, and will validate satellite formation flying and rendezvous guidance and navigation strategies and technologies. The Prisma satellites have been developed in co-operation with German, French and Danish partners.

Background
Originally formed in 1961 as the Space Technology Group, and renamed in 1972, SSC is a government-owned limited company under the Ministry of Enterprise, Energy and Communications. It is responsible for the technical implementation of Sweden's space and remote sensing programmes, including feasibility studies, operational applications, systems engineering and research and development; it also supports the Swedish National Space Board with technical interpretation and is Sweden's ESA liaison. SSC's headquarters are located in Solna, Stockholm; it also operates the Esrange sounding rocket range (see separate entry) launch, satellite control and data reception facilities. The agency employs more than 300 personnel, about 180 of whom are located at Esrange, near Kiruna.

SSC designs and operates spacecraft and provides launch services for a range of clients. Its suborbital launch vehicles include Maser, Maxus, Black Brandt, Orion and Nike-Orion. SCC is also engaged in the design and manufacture of satellites, having been involved in about 10 science and communications satellite programmes since the 1980s.

SSC owns three subsidiaries: LSE, ECAPS and NanoSpace, as well as being part-owner (25 per cent) of the communications satellite operator SES Sirius (please see separate entry listed as SES Global). LSE is a German space consulting and engineering firm; ECAPS works on ecological propellent systems, and NanoSpace a space microsystems developer.

SSC employees operate the Sirius 2 and 3 satellites from Esrange, as well as the Odin science satellite.

Ukraine

Ukrainian Space Agency

Current Status
The agency and commission were established in November 1992 to manage and supervise national space activities while licensing national companies. The Agency has extended negotiations with several

foreign governments and an agreement with NASA in 1994 resulted in a Ukrainian cosmonaut flying on a shuttle mission in 1997. USA has been gradually moving toward a policy alignment to ESA and NASA organisations in an attempt to increase participation in space science and engineering projects. The USA has attempted to forge a unique set of skill capabilities for harnessing national assets from government and semi-private sectors. These have increased contacts and projects participation with several institutes, universities and research centres outside the country.

United Kingdom

British National Space Centre (BNSC)

Current Status
The British National Space Centre (BNSC), created in 1985, is comprised of 11 partner agencies and research institutions with vested interests in space science, exploration, exploitation and Earth observation. BNSC focuses and co-ordinates the civil space interests of its member organisations and represents UK civil space interests to the European Space Agency (ESA) and other space-related consortiums and organisations. BNSC partners, as of 2005, are: the Department of Trade and Industry (DTI), the Department for Environment, Food and Rural Affairs (DEFRA), the Ministry of Defence (MoD), the Department for Transport (DfT), the Foreign and Commonwealth Office (FCO), the Department for Education and Skills (DfES), The Meteorological Office (Met Office), the Natural Environment Research Council (NERC), the Office of Science and Technology (OST), the Particle Physics and Astronomy Research Council (PPARC) and the Central Laboratory for the Research Councils' (CCLRC) Rutherford Appleton Laboratory (RAL). BNSC staff operate from the London Headquarters and technical centres such as CCLRC. DTI's Parliamentary Under-Secretary of State for Science and Innovation oversees BNSC as well as OST and Research Councils UK (RCUK). BNSC is headed by a Director General.

BNSC partners spent, in 2004/2005, a total of nearly GBP200 million on space-related programmes. Civil expenditures have grown again after a sharp down turn in spending from 2001 to 2003. The Space Centre co-ordinates partner organisation space resources; approximately 65 per cent of these finances are earmarked to ESA programmes. BNSC has recently undergone reorganisation to streamline its financial and policy decision-making processes. The UK Space Board, formed in 2005, is composed of the five largest BNSC financial partners: the MoD, DTI, PPARC, NERC and the Met Office. The Board plays a role in BNSC's financial goals by assisting and enabling financial negotiation between member organisations. The Space Advisory Council, also formed in 2005, advises the Space Board on issues between BNSC member organisations, and is made up of BNSC partners.

United States

Department of Energy, Office of Radioisotope Power Systems

Current Status
NASA has teamed with the Department of Energy (DoE) on NASA's Prometheus Nuclear Systems and Technology programme. DoE's Naval Reactors (NR) Programme was also originally to assist NASA with nuclear power for its Jupiter Icy Moons mission, however that programme was cancelled. Prometheus aspires to provide NASA with a nuclear reactor powered space exploration craft.

DoE offices have supplied NASA with the radioisotopes necessary to provide electrical power to deep space craft since 1961. The Office of Radioisotope Power Systems is subordinate to DoE's Office of Nuclear Energy.

Background
DoE's Office of Radioisotope Power Systems developed the Radioisotope Thermoelectric Generators (RTGs) for the Cassini mission launched in 1997. Previous to Cassini, it supplied US national security and civil space missions with RTGs, which provide spacecraft with the necessary power to explore deep space, beginning in 1961 with the Transit 4A satellite programme. DoE has also provided Radioisotope Heater Units (RHU) for such spacecraft as the Mars Rovers and Pathfinder; heating allows the craft to remain operational in cold surroundings, where solar heat generation is not possible.

In 2005, the Office of Radioisotope Power Systems supplied NASA with the necessary approval and plutonium-238 to power the RTG for the New Horizons Pluto exploration craft, launched in January 2006. Idaho National Laboratory (INL), operated by Battelle Energy Alliance for DoE, manufactures the RTGs. INL anticipates its business growth potential for the design and manufacture of space power systems, estimated at USD18 million in 2005, to expand to USD70 million over the next ten years.

Federal Aviation Administration, Office of the Associate Administrator for Commercial Space Transportation

Current Status
Title 49 United States Code, Subtitle IX, Chapter 701 (formerly known as the Commercial Space Launch Act of 1984) and Executive Order 12465 gave the Department of Transportation (DoT), and subsequently the Federal Aviation Administration (FAA), the authority to regulate US commercial space launch and re-entry activities. The Commercial Space Launch Amendments Act of 2004 further defines DoT responsibilities and authority, most notably with regard to commercial human space flight. AST previously reported directly to DoT's Office of the Secretary but was transferred intact on 16 November 1995 to the FAA.

Commercial orbital and suborbital firings are listed on the Office of the Associate Administrator for Commercial Space Transportation's (FAA/AST) website and in their Quarterly Launch Reports, published by AST's Space Systems Developments Division. The first licensed suborbital launch was of the CONSORT-1 satellite in March 1989 by a Starfire launch vehicle.

Licences are granted based on launch providers presenting evidence they are in compliance with all safety regulations and other requirements, and have sufficient insurance or financial resources to cover any probable losses from a launch mishap. The office is also responsible for the regulation of any future commercial launch sites and commercial re-entry vehicles oversight. It is also the compliance monitor for the Chinese and Russian commercial launch agreements. The Space Systems Developments Division studies environmental impacts of new launch sites, develops long range commercial launch forecasts and develops space transportation policies. AST Divisions also evaluate privately owned rocket launcher projects, license entrepreneurial ventures and compile launch forecasts for government agency analysis of future requirements.

The Commercial Space Transportation Advisory Committee (COMSTAC) provides recommendations and advice to the FAA Administrator, as well as policy guidance to commercial organisations.

Florida Space Authority

Current Status
Created in 1989 by Florida's Governor and Legislature, and governed by a nine member, Governor appointed board of supervisors, the Florida Space Authority (FSA – formerly the Spaceport Florida Authority) engages in space policy activities and aims to foster development of Florida's space enterprise, including industry, research and education. FSA is a public corporation and is authorised to own, operate and construct spaceport infrastructure and facilities in the State. The Authority reports to the Governor through Florida's Office of Tourism, Trade, and Economic Development.

In 2005, Futron Corporation conducted a Florida Spaceport feasibility study for FSA. The study determined that a Florida-based commercial spaceport, separate from Cape Canaveral facilities, would be both feasible and desirable for the State, especially to serve the developing sub-orbital space tourist market. FSA will, in the future acquire the land for a commercial spaceport, possibly at two split sites to separately accommodate horizontal and vertical sub-orbital launch vehicles. The Authority is investigating possible areas for the spaceport, including the Florida Keys.

SFA's offices are located at Cape Canaveral and in Tallahassee. As of 2006, FSA employed about 20 staff.

Background
FSA actively partners with civil, military, academic and commercial organisations. Currently, FSA's launch facilities are located at Cape Canaveral Air Station (please see the separate Cape Canaveral entry for launch complex details) and include Space Launch Complex (SLC) 20, SLC 46, and SLC 47. FSA has partially funded both Delta IV and Atlas V EELV facilities at Cape Canaveral Air Station (CCAS). The Authority also manages the Space Life Sciences Lab, the Reusable Launch Vehicle Flight Support Facility, the FSA Conference Center and the Space Operations Control Center. The International Space Research Park (ISRP) is being developed jointly by FSA and NASA and will be located at Kennedy Space Center (see separate entry). Facility design for the site began in 2006, and FSA anticipates ISRP construction will commence in 2008.

During the 1990s, FSA financed USD30 million of the total USD41 million Apollo-Saturn V Visitor Complex project for NASA, and built a USD28 million 6,094 m^2 Titan IV Solid Rocket Motor Unit Long Term Storage Facility for the US Air Force at Camp Blanding in northeast Florida. Lockheed Martin has leased the facility until September 2007.

The Authority frequently hosts visits from students and foreign space delegates while playing a major role in disseminating information about Florida's space related activities.

National Aeronautics and Space Administration, Headquarters

Current Status
At least two high profile NASA science missions are due for launch in 2007. The first, Dawn (see separate entry), a project that was one of the first casualties of Vision for Exploration related budget cuts, was reinstated by NASA and is scheduled for launch in June. The second, Phoenix (see separate entry), is the first in the Mars Scout group, and will launch in August 2007.

Separate to the Vision for Space Exploration budget, NASA is funding its Commercial Orbital Transportation Services project. Two contractors, Space Exploration Technologies (SpaceX) and Rocketplane Kistler, are competing in a bid to eventually provide NASA with commercial cargo, and also human transportation systems, for trips to the International Space Station (ISS). The two companies have shared USD485 million in NASA funding to develop their cargo systems. NASA is also working with Transformational Space Corporation (t/Space) and PlanetSpace Inc, but has not provided funds to these contractors. NASA anticipates that future commercial spacecraft purchases will be more cost effective than using government designed and managed craft. This will allow NASA to continue to dedicate resources to Moon and Mars exploration programmes.

Background
NASA was formally established on 1 October 1958 to plan and execute the US civil space programme. It recently has undergone reorganisation to

better define organisational responsibilities and accountability as well as to address requirements for the Vision for Space Exploration, announced in January 2004. NASA now comprises four Mission Directorates and numerous Mission Support and administrative offices. About a dozen major research centres and facilities (see separate entries) report to the Headquarters Associate Administrators who lead the four specialised Mission Directorates, which are listed below. NASA employs more than 18,000 civil servants, over half of which are scientists and engineers. Roughly 1,100 personnel are located at NASA Headquarters in Washington, DC.

NASA's four Mission Directorates are:

- Exploration Systems: This Directorate's main objective is to support the Vision for Space Exploration. It is composed of eight Divisions and has absorbed the former Office of Biological and Physical Research. It is responsible for developing the Crew Exploration Vehicle (CEV) and Crew Launch Vehicle (CLV).
- Space Operations: NASA's Johnson, Kennedy, Marshall and Stennis Centers are organised under this Directorate. It is charged with the management of NASA's current low-Earth orbit programmes, the International Space Station (ISS) and the Space Shuttle, as well as launch services and space communications.
- Science: NASA's Ames and Goddard Centers, and the Jet Propulsion Laboratory (JPL) are organised under this Directorate. It manages solar and Earth systems research and a range of space science research as well as NASA's satellite systems research and development.
- Aeronautics Research: NASA's Dryden, Glenn and Langley Centers are organised under this Directorate. It is primarily responsible for atmospheric aeronautics transportation and systems research and development.

Other important NASA Headquarters Offices are:

- Office of Program Analysis and Evaluation (PA&E): This office performs NASA planning activities, cost analyses, programme assessments and evaluates whether NASA's combination of programmes is appropriate.
- Office of Safety and Mission Assurance: This office is responsible for NASA-wide quality assurance, mission risk assessment methodologies and safety standards compliance.
- Office of Inspector General: This independent office is responsible for management and programme audits, and the detection of possible waste, fraud and abuse within NASA.

The NASA Space Programme

In September 2005, NASA released the results of its Exploration Systems Architecture Study (ESAS), recommending the development of a heavy-lift vehicle, a smaller human-rated crew lift vehicle and a capsule-type, six-person Crew Exploration Vehicle to shift the US human spaceflight programme away from Space Shuttle dependency as soon as possible. The new launch systems – based on modified Shuttle propulsion systems – are expected to support a return to the Moon by 2020, in accordance with the Vision for Space Exploration (VSE) unveiled by President Bush in January 2004. The FY2007 NASA budget focuses on this new goal, over the protests of scientists who see cuts on the horizon in unmanned space missions and fundamental space technology development.

Boeing and Lockheed Martin, participants in the Evolved Expendable Launch Vehicle (EELV) programme providing expendable launchers for US military and NASA missions, announced in 2005 that they intended to merge their activities into a joint venture: the United Launch Alliance (ULA). The joint venture was approved in 2006. ULA provides Lockheed Atlas V and Boeing Delta IV, as well as Delta II launch vehicles and services to NASA and the US military, conducting launches from facilities at Cape Canaveral Air Force Station and Vandenberg Air Force Base.

Current In-Flight NASA Missions

Mission Name	Launch Date	Mission Type	Mission Notes
Voyager 2	20.08.1977	Science	Remains operational and departing the solar system; on the Voyager Interstellar Mission (VIM) until at least 2020
Voyager 1	05.09.1977	Science	Operational and departing the solar system; on the Voyager Interstellar Mission (VIM) until at least 2020; now the farthest human-made object from Earth
Hubble Space Telescope	24.04.1990	Science	Joint ESA programme; 2010 anticipated mission end
Ulysses	06.10.1990	Science	Joint ESA programme; mission extended to March 2008
Geotail	24.07.1992	Science/Earth Observation	Joint JAXA mission and IASTP mission; has far surpassed its nominal mission duration
WIND	01.11.1994	Science/Earth Observation	IASTP mission; remains operational; nominal mission duration surpassed
SOHO	02.12.1995	Science	Joint ESA programme and IASTP mission; 2007 scheduled mission end
RXTE	30.12.1995	Science	Initial mission duration surpassed; remains operational
Polar	24.02.1996	Earth Observation	IASTP mission; nominal mission duration far surpassed
FAST	21.08.1996	Science/Earth Observation	Has survived long past its expected service-life; remains operational
Mars Global Surveyor	07.11.1996	Science	Lost communications with craft in late 2006; likely mission has ended
Advanced Composition Explorer (ACE)	25.08.1997	Science	Most instruments functioning; ability to maintain orbit to 2019
Cassini-Huygens	15.10.1997	Science	Joint ESA programme; Huygens descent to Titan 14.01.2005; Huygens segment completed
TRMM	27.11.1997	Earth Observation	Joint JAXA mission; nominal mission completed in 2000; remains operational
TRACE	01.04.1998	Science	Planned operations through at least 2008
ISS – Zarya Control Module	20.11.1998	Science	ISS Assembly Mission 1 A/R – the first ISS component in orbit; construction completion due in 2010; ISS is a joint RSA, ESA, JAXA, CSA, AEB mission; planned operations through at least 2016
Landsat 7	15.04.1999	Science	Nominal mission through 2004; craft remains operational; USGS managed
QuikSCAT	19.06.1999	Earth Observation	Nominal 2-year mission; remains operational
FUSE	24.06.1999	Science	Extended mission since 2003
Chandra X-Ray Observatory (AXAF)	23.07.1999	Science	Launched from Space Shuttle Columbia; could be mission capable up to 2016
Terra (EOS AM-1)	19.12.1999	Earth Observation	Nominal mission through 2006; remains operational
ACRIMSAT	20.12.1999	Earth Observation	Primary mission completed in 2005
CHAMP	15.07.2000	Earth Observation	Primary mission completed in 2005; continues operations
Cluster II	16.07.2000	Science	Joint ESA programme; mission end 2009
Cluster II	09.08.2000	Science	Joint ESA programme; mission end 2009
HETE-2	10.09.2000	Science	2-year design life; remains fully operational
NMP EO-1	21.11.2000	Earth Observation	Nominal mission completed; extended mission
Mars Odyssey	07.04.2001	Science	On extended mission period; serves also as a communications relay unit for Mars Rovers
WMAP	30.06.2001	Science	Scheduled to remain in operational L2 orbit through at least 2007
GOES-M	23.07.2001	Meteorology	Remains operational as of early 2007
Jason-1	07.12.2001	Earth Observation	Five-year design life; launched with TIMED; operational
TIMED	07.12.2001	Science	Launched with Jason; nominal mission complete; remains operational
METEOR-3M	10.12.2001	Earth Observation	Joint Russian Federal Space Agency mission; NASA provided the SAGE III instrument
RHESSI	05.02.2002	Science	Nominal mission complete; remains operational
GRACE	17.03.2002	Earth Observation	2-spacecraft; joint Deutsche Forschungsanstalt fur Luft und Raumfahrt (DLR) mission
Aqua (EOS PM)	04.05 2002	Earth Observation	First in the Afternoon Constellation/A-Train EO satellite series
NOAA-M (POES)	24.06.2002	Earth Observation	Nominal mission through 2004; remains operational
Integral	17.10.2002	Science	ESA mission; NASA operates one ground station; operational until at least 2010
ADEOS II	14.12.2002	Earth Observation	Joint JAXA programme; NASA provided the SeaWinds instrument
ICESat	12.01.2003	Earth Observation	3-year nominal mission; remains operational; launched with CHIPSat
CHIPSat	12.01.2003	Science	Nominal mission complete; remains operational; launched with ICESat
SORCE	25.01.2003	Science	Planned operation through 2008
GALEX	28.04.2003	Science	Nominal mission complete; remains operational
Mars Express	02.06.2003	Science	Joint ESA mission; NASA provided the ASPERA instrument; extended mission through 2007
Mars Rover Spirit	10.06.2003	Science	Operational as of early 2007; years beyond nominal mission completion
Mars Rover Opportunity	07.07.2003	Science	Operational as of early 2007; years beyond nominal mission completion
Spitzer Space Telescope (SIRTF)	25.08.2003	Science	Completed its nominal mission; on an extended mission period through at least 2008
Rosetta	02.03.2004	Science	ESA mission; US sources provided the ALICE, MIRO and IES instruments; mission duration to at least 2015
Gravity Probe-B	20.04.2004	Science	Initial mission complete; remains operational
Aura	15.07.2004	Earth Observation	Second in the Afternoon Constellation/A-Train EO satellite series
MESSENGER	03.08.2004	Science	Due in Mercury's orbit March 2011
Swift Gamma-Ray Burst Explorer	20.11.2004	Science	2-year nominal mission

Mission Name	Launch Date	Mission Type	Mission Notes
PARASOL	18.12.2004	Earth Observation	Third in the Afternoon Constellation/A-Train EO satellite series; joint CNES programme
NOAA-N (POES)	23.06.2005	Earth Observation	Minimum mission through 2007
Suzaku (Astro-E2)	10.07.2005	Science	JAXA mission developed in collaboration with NASA; minimum 2-year service-life
Mars Reconnaissance Orbiter	12.08.2005	Science	Primary mission to November 2008
New Horizons	19.01.2006	Science	Pluto arrival due July 2015
NMP ST5	22.03.2006	Science	3-microsat technology demonstrator constellation; primary mission ended June 2006
CloudSat	28.04.2006	Earth Observation	Fourth in the Afternoon Constellation/A-Train EO satellite series; launched with CALIPSO
CALIPSO (PICASSO-CENA)	28.04.2006	Earth Observation	Fifth in the Afternoon Constellation/A-Train EO satellite series; launched with CloudSat
GOES-N	24.05.2006	Meteorology	Minimum 5-year mission
Solar-B	22.09.2006	Science	Joint JAXA mission; nominal 3-year planned service
MetOp-A	19.10.2006	Meteorology	Joint ESA/EUMETSAT mission
STEREO	26.10.2006	Science	2-year nominal mission
NMP ST6	16.12.2006	Science	Compass technology demonstrator segment launched aboard TacSat-2
THEMIS	17.02.2007	Science	five-satellite, 2-year nominal mission

Notes: Joint international missions with significant NASA input are included in this table.
Space Shuttle and ISS missions are not included in these tables.

Planned NASA Missions

Mission Name	Launch Date	Mission Type	Mission Notes
AIM	2007	Science/Earth Observation	2–3 year expected service-life
Dawn	2007	Science	Mission cancelled and then re-instated by NASA, March 2006
GLAST	2007	Science	Minimum 5-year mission duration
NOAA N' (POES)	2007	Earth Observation	2-year minimum mission
Phoenix Mars Scout	2007	Science	First Mars Scout programme mission; Mars Exploration programme
GOES-O	2008	Meteorology	Minimum 5-year mission
Kepler Telescope	2008	Science	Minimum 5-year mission
OSTM	2008	Earth Observation	5-year mission
GOES-P	2008	Meteorology	Minimum 5-year mission
OCO	2008	Earth Observation	Sixth and final in the Afternoon Constellation/A-Train EO satellite series; minimum 2-year mission
Mars Science Laboratory	2009	Science	Mars Exploration programme
Aquarius	2009	Earth Observation	Minimum 3-year mission
WISE (NGSS)	2009	Science	Minimum 7-month mission
NPP	2009	Earth Observation	5-year minimum mission; launch date slip from 2006
LDCM	2009	Earth Observation	Minimum 7-year mission
LISA-Pathfinder	2009	Science	Joint ESA mission; 5-year minimum service-life
NMP ST7	2009	Science	DRS technology demonstrator to be launched with LISA Pathfinder; joint ESA programme
NMP ST8	2009	Science	Technology demonstrator
NMP ST9	2010	Science	Technology demonstrator
GPM	2010	Earth Observation	5-year mission
Mars Scout	2011	Science	Second Mars Scout mission; Mars Exploration programme
Mars Sample Return	2011	Science	Mars Exploration programme
Astrobiology Field Laboratory	2011	Science	Mars Exploration programme
Mars Deep Drill Lander	2011	Science	Mars Exploration programme
GOES-R	2012	Meteorology	Minimum 8-year mission
NPOESS	2013	Earth Observation	Programme/launch delayed
James Webb Space Telescope	2013	Science	Minimum 5-year mission
Space Interferometry Mission (SIM)	2015	Science	Minimum 5-year mission

Note: Missions currently lacking projected or announced launch years are not included in this table.

The Vision for Space Exploration

The current centrepiece of US civilian space policy is the Vision for Space Exploration, announced by President Bush on 14 January 2004. The Vision asserts four overarching objectives: to develop and realise a viable, budget-conscious human and robotic space exploration programme, with astronauts first visiting destinations within the Solar System, and then beyond; to expand human presence across the solar system, sending astronauts to the Moon by 2020 and to Mars thereafter; to develop space-related technology and infrastructure to accomplish the first two goals as well as to support future exploration decisions and expeditions; and to foster international and commercial space exploration co-operation in order to advance US science, security and economic interests.

The Vision charges NASA with International Space Station (ISS) completion by 2010; safe Space Shuttle operation and retirement by 2010; development and full operation of new Crew Exploration and Crew Launch Vehicles by 2014; delivery of astronauts to the Moon by 2020; development and implementation of a plan to carry astronauts to Mars following the human Moon mission; and continuing execution of other advanced science and planetary missions to help develop and support human spaceflight and exploration goals and aspirations.

NASA created the Exploration Systems Missions Directorate (ESMD) in 2004 to support the Exploration Vision's goals, manage exploration

NASA plans to retire the Shuttle in 2010 and develop a new class of vehicles allowing a return to the moon and the beginning of human Mars visits (NASA)
0572488

For details of the latest updates to ***Jane's Space Systems and Industry*** online and to discover the additional information available exclusively to online subscribers please visit
jsd.janes.com

Technicians test components on the Mars Exploration Rover launched in 2003 (JPL) 1047539

NASA envisions a mission to Mars for human colonisation some time before the middle of this century (NASA) 1047544

programmes, and attempt to draw clear lines of accountability for achieving the Vision's goals (see also the NASA Centers entries). ESMD in 2004, announced its Exploration Systems Interim Strategy. ESMD's Interim Strategy, NASA's 2005 Exploration Systems Architecture Study (ESAS) Report, and NASA's 2006 Strategic Plan, which expresses goals reaching through 2016, are all consistent with the Vision for Space Exploration's objectives and further detail NASA's technological goals, science and human spaceflight programmes, and the means by which the organisation will achieve the Vision's goals and objectives.

NASA's plan to retire the Space Shuttle by 2010 is coupled with its charge to develop and successfully fly a new human rated vessel initially called the Crew Exploration Vehicle (CEV) by 2014. Current CEV plans reveal, among widespread criticism, that it is a capsule vehicle, similar to Apollo, rather than a spaceplane like the Shuttle. NASA must also develop launch vehicles suitable for projecting the CEV into Low Earth Orbit (LEO) as well as propelling payloads beyond LEO. Development of the initial CEV, Crew Launch Vehicle (CLV), and other Exploration equipment and systems fall into a programme dubbed "Project Constellation."

NASA has completed its Great Observatories programme with the SIRTF Spitzer telescope (NASA) 1047540

NASA's technological focus relies upon reusable vehicles or reusable technology that can be applied to multiple missions.

In August 2006, NASA selected Lockheed Martin as the lead CEV contractor over a Northrop Grumman/Boeing consortium. The CEV spacecraft has been given the name Orion. At approximately 5 m in outer diameter, Orion will be able to carry up to six astronauts to the ISS – double the number of crew the original Apollo design could carry- but will only carry four astronauts to the Moon. It will be equipped with an abort system giving the crew an opportunity to eject the vehicle from the launch system in case of catastrophic launch vehicle failure. Orion will have ISS docking capabilities and a dedicated propulsion system. It will be designed for Earth re-entry and dry land touch down but will also be capable of water touch down and survivability. NASA plans to have Orion ready for its first automated test flight by 2008. Test and evaluation work began in 2006 and will continue through 2013 at an estimated cost of USD3.9 billion. Spacecraft orders between 2009 and 2019 are projected to cost USD3.5 billion. The craft is scheduled to commence fully operational human flights by 2014.

NASA has plans to develop and fly two separate launchers: the four stage CLV and the Heavy Lift Launch Vehicle (HLLV), dubbed in 2006, Ares I and Ares V. Ares I and Ares V must be fully operational by 2014 and 2020 respectively. Ares I will have both CEV and cargo launch capabilities up to 32,000 kg if a fifth stage is built into the vehicle. Ares V, as envisaged in its current design iteration, will first carry only cargo but will be powerful enough to propel payloads up to 130,000 kg beyond LEO. Both launch vehicles are currently designed to use Space Shuttle rocket engine and booster technologies. The Ares I preliminary design review is due in mid-2008.

NASA's ISS construction completion target year is 2010, in order to both honour international agreements as well as to ensure completion by the Space Shuttle's retirement date. Crew and cargo missions to the ISS will be separate, to the largest extent possible thereafter. Different spacecraft will play support roles, depending upon mission objectives. Between 2005 and 2010, the Space Shuttle will continue to service the ISS and deliver components for assembly, such as the European Space Agency's (ESA) Columbus module. Afterwards, Orion and Soyuz vehicles will carry crews to the ISS, while automated craft, such as the Russian Progress, ESA's Automated Transfer Vehicle (ATV), and the Japan Aerospace Exploration Agency's (JAXA) H-II Transfer Vehicle (HTV) will handle ISS cargo missions. NASA's Exploration schedule also demands that ISS research supporting human exploration and healthy human survival in space be complete by 2016.

According to NASA's plan, US astronauts will return to the Moon by 2020 – possibly before that date if NASA exceeds its initial expectations of Exploration Vision goal timeframes. Robotic Moon missions, including the Lunar Reconnaissance Orbiter (LRO), will begin by 2008 and continue at least through 2011. Initial data indicates that the Moon's South Pole might be a productive human landing target. Robotic missions will help to validate or reject this hypothesis. NASA projects that the first human Moon missions will last approximately seven days. Astronauts will build upon each Moon exploration experience and gradually lengthen stays, leading ultimately to the construction of a permanent Lunar Base.

According to the present plan, the crewed Lunar flight plan and Lunar landing are broken into several mission segments. Orion and Lunar Lander crafts will be launched into LEO separately. The two spacecrafts will then dock, and afterwards journey to the Moon. Once in Lunar orbit, the Lander will separate and descend to the Moon's surface, while Orion remains in flight. Astronauts will spend about a week on the surface; they will use an ascent module to return to Orion, still in Lunar orbit. The Lander will be discarded, and Orion will approach and re-enter Earth's atmosphere.

NASA's Lunar missions will enable the agency to further develop the technologies necessary to eventually carry humans to Mars. If NASA keeps to its schedule, humans could potentially make the first trip sometime after 2020. ESA aspires to send European astronauts to Mars as soon as 2030; this may create some competition between the agencies. Successful robotic missions have already begun: the Spirit and Opportunity Rovers have been transmitting data streams for years. By 2010, NASA will have sent orbiters, rovers, and landers to the planet. More advanced, dedicated robotic missions with a view to landing humans on Mars will commence in 2011.

Science missions connected with the Space Exploration Vision are in development, and NASA expects to conduct robotic exploration to other solar system destinations. Some science programmes have lost funding and have been cancelled, however, for example, the Jupiter Icy Moons Orbiter (JIMO), that was supposed to launch around 2017. Also in danger of cancellation is the Terrestrial Planet Finder observatory, an observatory duo that was to launch between 2014 and 2020. The telescopes were to have found Earth-like planets outside of this solar system,

The Spitzer telescope launched in 2003 (NASA) 1047541

and also would have been capable of examining planetary chemical composition so scientists could have speculated about any evidence of life that might have been found. The Prometheus Project, a segment of the Exploration Technology Development Program (ETDP), will continue in its endeavour to develop spacecraft nuclear power and propulsion systems to fuel more advanced, deep space missions beyond 2020.

When the Vision was initially publicised in 2004, there was no concrete information released on what the Exploration programme would cost over time. NASA was already spending approximately USD15.4 billion annually on its programmes and operations. The Bush Administration then announced the plan to increase NASA's budget by USD1 billion over five years, first in five per cent increments until 2007, and then in one per cent increments until 2009. NASA's budget would then increase at the rate of inflation thereafter. Most likely, much of the ongoing funding directed toward NASA's existing programmes will simply continue, providing existing NASA science and exploration programmes that are deemed 'Vision-related' with sustained resources. However, some programmes have already suffered in favour of those that are considered better Vision-aligned. NASA's 2007 budget request totalled at USD16.8 billion; a 3.2 per cent increase over 2006. Its 2008 budget request comes to USD17.3 billion; a 3.1 per cent increase over the 2007 fiscal year request. Requests are clearly topping the one per cent increments originally planned for the 2007–2009 period.

NASA projected in its 2005 budget request, that during the years 2004–2009, it will spend more than USD14 billion, of a projected total USD84 billion during that five-year period, on projects supporting Exploration Vision objectives. By 2010, NASA projects that Vision-related expenditures would total about USD4 billion annually; that figure would more than double by 2020, to USD9 billion, providing that the Shuttle retires by 2010 and the ISS is completed by 2017.

NASA estimates that by 2020, it will have spent about USD66 billion to develop and implement the technology required to land humans on the moon. This represents about one-quarter of the total NASA budget up to that time. Robotic Lunar missions are separate to this figure and will cost about USD29 billion through 2020, or 11 per cent of NASA's total outlays. A 2004 Congressional Budget Office (CBO) analysis indicates, however, that NASA's programme estimates may be overly optimistic. CBO projects, at highest, that crewed Lunar mission costs by 2020 could be nearly double that of NASA's estimates. CBO also projects that at NASA's planned expenditure rates, Exploration programmes could fall behind schedule and delay NASA's human Moon mission by as much as seven years – putting humans on the Moon by 2027 instead of 2020, as the Exploration Vision requires.

RESEARCH & DEVELOPMENT

Argentina

Asociacion Argentina de Tecnologia Espacial (AATE)

Current Status
The Asociacion Argentina de Tecnologia Espacial (AATE) is a non-profit institution engaged in space science activities. Since its inception it has conducted more than one hundred sub-orbital and orbital launches.

AATE co-founder and President, Pablo de León, represented Argentina and his organisation Pablo de León and Associates, with the Gauchito vehicle, in the Ansari X Prize competition in 2004.

Background
AATE, located in Buenos Aires, was founded in 1987 from a membership that consisted of institutions, which beforehand, were conducting separate space science experiments and launches. In 1989, Argentina granted AATE, non-profit, non-governmental organisation status. AATE works in partnership with other Argentine institutions, as well as national and international organisations.

Beginning in 1997, AATE campaigned to develop a set of experiments that could be lifted into space in NASA's Getaway Special (GAS) programme and brought together a wide range of potential experiments from several research institutes in Argentina. Eventually, it obtained a reservation for Getaway Special G-761, aboard US shuttle flight STS-108, which took place in December 2001. Known as the Argentine Experiments Package, or Paquete Argentino de Experimentos (PADE), its seven experiments were used to study fluid movement, the reaction of seeds exposed to space, and other microgravity effects. GAS G-761 was the first set of experiments launched by NASA on behalf of a Latin American country.

AATE has conducted detailed scientific studies of technical innovations capable of providing zero-gravity time for low-cost experiments developed at technical and academic institutions.

Instituto de Astronomia y Fisica del Espacio (IAFE)

Current Status
The Instituto de Astronomia y Fisica del Espacio (IAFE), or the Astronomy and Space Physics Institute, was founded in 1971 and is an organisation within Argentina's Consejo Nacional de Investigaciones Científicas y Técnicas (CONICET), or National Research and Technology Council; it is also associated with the Universidad de Buenos Aires. IAFE's main fields of research are astronomy, theoretical astrophysics, upper atmospheric physics and atomic collisions.

Background
IAFE was the Principal Investigator institute for SAC-B's Hard X-Ray Spectrometer (HXRS) instrument. The failed SAC-B remained attached to the third stage of its Pegasus vehicle following launch from Wallops Flight Facility on 4 November 1996.

In the 1990s, IAFE raised plans for a science satellite for aeronomy and space physics studies and conducted discussions with other countries in South America regarding international participation. This attracted interest from some groups in the Middle East although no agreements were reached. Further possibilities have been suggested whereby CONICET would serve as the organising body for a new South American effort to focus on future astronomy research. Considerable effort was expended to facilitate this, but no firm commitments were secured. Approaches to other astronomy related organisations in other South American countries has stimulated a wider debate about regional co-operation, which stands a greater possibility of success.

Austria

Space Research Institute

Current Status
The Institut für Weltraumforschung (IWF), Austria's Space Research Institute, provides technical expertise and tracking for many international science satellite programmes. NASA's Solar TErrestrial RElations Observatory (STEREO), and the joint CNES – Observatoire de Paris, Meudon Convection, Rotation and Planetary Transits (COROT) missions are among recent examples. IWF participated in the STEREO/WAVES (SWAVES) instrument team, and developed the Boîtes Extracteur (BEX) image data computer system for COROT.

Background
Founded in 1970, the Space Research Institute is a division of the Österreichische Akademie der Wissenschaften (ÖAW), Austria's Academy of Sciences (AAS); it is Austria's principal space science research facility. IWF places research emphasis on Satellite Geodesy, Experimental Space Research and Extraterrestrial physics, and maintains three departments that carry out studies in these areas. As of 2006, IWF employed about 70 staff members. Its offices are located in Graz, in the ÖAW Research Centre.

IWF has strong connections throughout Europe and is a major contributor to international projects both inside and outside Europe. IWF teams work with the world's space agencies, including ESA, NASA, JAXA and CNES. The Institute has provided experiments, instruments and scientific expertise, such as modelling and calibration teams for high profile missions such as Mars Express, Venus Express, Cassini/Huygens and Rosetta, among others.

IWF also provides one of the world's Satellite Laser Ranging (SLR) and GPS stations, Observatory Lustbühel Graz (OLG), a joint venture with the Office of Meteorology and Surveying.

Belgium

Centre Spatial de Liège (CSL)

Current Status
The Centre Spatial de Liège (CSL) designs, tests and builds optical instruments for scientific satellites. Examples include optic elements for CNES' COnvection, ROtation and planetary Transits (COROT), NASA's Solar TErrestrial RElations Observatory (STEREO), and ESA's Herschel space observatory.

CSL also develops satellite remote sensing processors, algorithms and instruments through its Synthetic Aperture Radar (SAR) work. It is developing SAR processors for Argentina's SAOCOM.

Background
Employing about 100 staff members, CSL is an autonomous unit within the Université de Liège; it evolved from the University's Astrophysical Institute and has been engaged in space research activities since the 1960s. CSL works closely with ESA and provides space science services and satellite instrumentation to ESA as well as other space agencies, such as NASA and CNES. The Centre also collaborates with and provides services to large laboratories and the commercial space industry.

CSL is an ESA Test Facility specialising in optical instruments. Its space environment test facilities are based in two class 10,000 cleanrooms and include four thermal vacuum optical test space simulation chambers. Focal 1.5, Focal 2, Focal 3 and Focal 5 range from 1.5 to five m in diameter and from 1 m^3 to 190 m^3 in capacity. Thermal tests ranging from –268°C to 120°C can be conducted. Pressure within the chambers can reach to levels lower than 10-3 Pa. Each simulation chamber is equipped with an optical bench. CSL also maintains vibration and offgassing test facilities.

The Centre undertakes testing and calibration of camera tubes, image intensifiers, Charge-Coupled Devices (CCD), photomultipliers and photon counting systems, in addition to the design and manufacture of meteorological instruments. CSL has developed new systems for multifunctional satellite concepts involving platforms capable of conducting remote sensing surveys and data store and dump. It is also involved in graduate and post-graduate studies and supports the development of new instruments for weather satellites. CNL leases its test facilities to government agencies and to international groups involved in Belgian and international space research activities.

Canada

David Florida Laboratory, Canadian Space Agency

Current Status
The David Florida Laboratory (DFL) participated in the Assembly, Integration and Testing (AIT) of Radarsat-2, scheduled for launch in 2007. It also integrated and tested the Mobile Servicing System (MSS), Canada's contribution to the International Space Station (ISS).

Background
Operated on government funding and collocated with the Communications Research Centre (CRC), CSA's David Florida Laboratory (DFL) is the national facility for spacecraft Assembly, Integration and Testing (AIT); its facilities are available to other organisations on a fee for service basis. Built in 1971 and opened in 1972 for the Canadian-US Hermes Communications Technology Satellite project, it has since supported the development of Aniks C – E, the Shuttle remote manipulator, Brasilsat, Olympus, MSAT, Indostar and BSAT 2A and 2B, among other projects. DFL facilities include:

Integration and Assembly Areas/Clean Rooms
The integration and assembly areas comprise two temperature (22±2°C) and humidity (40±5% Winter and 45±5% Summer) controlled clean rooms: high bay 2 at 315 m^2 and high bay 3 at 1,080 m^2. The clean rooms provide class 100,000 conditions, with class 10,000 available when required, and are traversed by travelling bridge cranes with 9, 11 and 16 m hook heights.

Thermal Qualification Facility
The thermal facilities are comprised of eight chambers, with three principal cylindrical vacuum chambers, TV2, TV3 and TV5, measuring 1×1, 2.5×2.5 and 7×10 m respectively. TV4 was retired from service in 2005.

The facility is also equipped with a 3×4 m thermal chamber, a 1×1 thermal humidity chamber, two 2.7×3.7 m and 3.7×3.7 m thermal Passive Intermodulation (PIM) chambers, and a 1×1 bakeout chamber.

Structural Qualification Facilities
DFL's structural qualification facilities include, among its extensive equipment range, UD 4000 and MB C-150 electrodynamic shakers as well as

Radarsat in the CSA's David Florida Laboratory (CSA) 1047515

An artist's depiction of Canada's Radarsat 2 in orbit (CSA) 1047516

a Ling A395 shaker, a Burnsco Thermal Chamber, and a MTS shock machine.

DFL's Static Load Test Facility was first used in 1992 by the Radarsat bus structural model, it incorporates a 12 channel Cyber Fatigue Master 7000 Digital Control System combined with a new data processing/acquisition system accommodating 200 strain gauge channels and up to 40 LVDTs.

Radio Frequency Test Facility

The RF facility comprises five anechoic chambers, a rooftop range, an Electromagnetic Compatibility (EMC) facility, a Passive Intermodulation (PIM)/ Multipaction facility and antenna test facilities. The chambers and rooftop turntable provide indoor and outdoor ranges for automatic antenna measurements; the three-axis positioners and sources are controlled via a fibre optic link between the 62 m tower and control room. DFL has added a spherical near field antenna measurement facility for performing near field to far field transformations. The Lab is designated by Inmarsat as both an Authorised Antenna Test House (AATH), authorised to conduct antenna tests to Inmarsat specifications, as well as a Designated Inmarsat Representative (DIR), providing the DFL representative with the authority to approve antennas on Inmarsat's behalf.

Canadarm 2 end effector developed from the Shuttle remote manipulator system but destined for the ISS (CSA) 1047517

China

China Aerospace Science and Technology Corporation (CASC)

Current Status

China launched its fourth Beidou navigation satellite on 2 February 2007 from Xichang, the first of the Beidou second generation satellites. The China Academy of Space Technology (CAST), an organisation subordinate to the China Aerospace Science and Technology Corporation (CASC), designed Beidou 1D (also known as Beidou 2A). The satellite did not achieve geostationary orbit, and has remained in transfer orbit, signifying a malfunction.

Background

CASC specialises in the development and manufacture of spacecraft and launch vehicles as well as a variety of ballistic missiles. It is a satellite, launcher and propellant technology provider and also performs work in automated control and systems integration.

On 1 July 1999, in an endeavour to force China's defence and technology industries to be more competitive, the State Council approved reforms affecting the nation's top corporations in these sectors. The reforms also brought about the establishment of CASC from the former China Aerospace Corporation (CAC). With some 270,000 employees at the time, CAC was divided into CASC and the China Aerospace Machinery and Electronics Corporation (CAMEC). In 2001, CAMEC became the China Aerospace Science and Industry Corporation (CASIC – see separate entry). Both CASC and CASIC report to the Commission of Science, Technology and Industry for National Defence (COSTIND), primarily a civilian administrative arm after the 1999 reforms, and the State Council.

CAC, founded in 1993, traces its roots back to 1956 with the establishment of the Number 5 Research Academy of the Ministry of National

Defense, which specialised in rocket and ballistic missile development. Renamed Ministry 7 in 1964, it was subsequently known as the Ministry of Aerospace Industry, from which grew a variety of research and development institutes. A series of facilities numbered 061–068 possessed a wide range of research, development and manufacturing capabilities and although they were set up to comprise third-tier manufacturing centres they quickly became first-line plants.

More than 130 separate organisations are subordinate to CASC including seven major research academies and production facilities: the Chinese Academy of Space Technology (CAST); the China Academy of Launch Vehicle Technology (CALT); the Shanghai Academy of Space Flight Technology (SAST); the Academy of Space Solid Propulsion Technology; the Academy of Space Liquid Propulsion Technology; the Aerospace Times Instrument Corporation; and the Sichuan Aerospace Industry Corporation.

CASC employs approximately 110,000 personnel, of which nearly 40 per cent are technical staff. It has primary responsibility for the development and manufacture of launch vehicles as well as ballistic missiles; it also provides commercial launch services, through its China Great Wall Industry Corporation (CGWIC). CASC has played a major role in the adaptation of CZ launchers for human space flight and in the development of the Shenzhou manned spacecraft, which carried the first Chinese astronaut into orbit in 2003.

France

Intespace

Current Status
Intespace released its DynaWorks version 5.1.D software testing product in March 2007. ESA will use DynaWorks to store mechanical and thermal testing data.

Background
Intespace (Ingéniére Tests en Environnement Spatial) has been based in Toulouse, France and in operation since 1963. Sopemea divested its interest in Intespace in 2000, as did CNES at a later date. As of 2007, EADS Astrium was Intespace's majority shareholder, owning 87 per cent. Thales Alenia Space owns the 13 per cent remainder, having inherited its shares from Alcatel Space. Intespace employs a workforce of 155, including staff located in its Toulouse and European Test Services' (ETS) location in the Netherlands.

Intespace is part of the Co-ordinated European Test Facilities (CETeF) group, along with Germany's Industrieanlagen-Betriebsgesellschaft mbH (IABG) and Belgium's Centre Spatial de Liège, organised by ESA's ESTEC (see separate entries). The group ensures that ESA's programmes have adequate test facilities available.

Intespace's facilities cover a range of thermal and mechanical space environment testing capabilities, including thermal vacuum, acoustic, amagnetic and climatic chambers; electrodynamic vibrators; anechoic faraday cages; and a centrifuge machine.

MPLM during acoustic testing at Intespace 0008101

Office National d'Etudes et de Recherches Aérospatiales (ONERA)

Current Status
ONERA created the first two-way optical communications link between ESA's Artemis satellite and an aircraft cruising at 30,000 ft, successfully tested in 2006. The technology will potentially impact UAV command, control and communications. The organisation has sustained development programmes for Ariane 5 booster performance improvement and is engaging in emerging technology spacecraft research such as formation flight, as well as space surveillance and materials science work.

Background
The Office National d'Etudes et de Recherches Aéronautiques (ONERA) was founded in 1946 as a national aeronautical research centre under the French MoD's Délégation Générale pour l'Armement (DGA). In 1963 its name was changed to the Office National d'Etudes et de Recherches Aérospatiales to reflect its growing involvement in space-related research. 2,000 staff, about three-quarters of whom are scientists and technical personnel, are employed at eight major centres specialising in aerodynamics, fluid mechanics, flight mechanics, structures, materials, optics, acoustics, electronics, radars, computer sciences, and ground, wind tunnel and flight testing. The French government subsidises ONERA, contributing about 40 per cent of the organisation's 2006 EUR188 million budget. The remainder is derived from research contracts between ONERA and its clients.

ONERA's Châtillon site, outside Paris, houses the Headquarters, operational departments and laboratories; Modane-Avrieux and Fauga-Mauzac house research wind tunnels; Palaiseau specialises in materials science, optics, instrumentation and energetics; Meudon in aerodynamics; Lille in aerodynamics and flight mechanics; Salon de Provence in system control and electo-optics; and Toulouse accommodates the Center d'Etudes et de Recherches de Toulouse (CERT) as well as l'École Nationale Supérieure de l'Aéronautique et de l'Espace (SUPAERO). The largest test facilities are operated at Modane-Avrieux and le Fauga-Mauzac. Modane-Avrieux maintains the S1MA continuous sonic wind tunnel (8 m diameter, 14 m long), the S2MA continuous transonic/supersonic (1.8 × 1.75 m), the S3MA transonic/supersonic blowdown (0.8 m high test sections), the S4MA hypersonic (M12.0) and R4 transonic/supersonic cascade tunnel. The S4B 15 mbar pressure chamber (2.5 m diameter, 10 m long), is primarily used today for the calibration of nacelles and air-powered engine simulators. Fauga-Mauzac operates the subsonic F1 (pressurised) and F2, and the high enthalpy F4 wind tunnels.

Germany

Industrieanlagen-Betriebsgesellschaft GmbH

Background
IABG was established in 1961 to provide aircraft and spacecraft test capabilities; it is now an element of the Co-ordinated European Space Test Facilities group, along with Intespace (France), Centre Spatial de Liège (Belgium) and ESTEC. For space, IABG offers thermal vacuum, thermo-mechanical, thermo-environmental, vibration, acoustic, modal, static, magnetic and EMC testing on launcher and satellite components and assemblies. IABG also conducts structural and mechanical tests and electromagnetic compatibility evaluation. It has tested more than 100 satellites and major satellite subsystems; recent activities include Huygens, Cassini, Artemis, ISO, Cluster, Soho, Freja and Ariane 5 stages and major components. IABG employs approximately 1,200 people and has diversified into several discrete business units: Information and Communications; Defence; Aerospace; Environmental and Management Services; Transportation, and Automotive. Facilities include:

Electromagnetic (EMC) Test Facility
The facility comprises a shielded test chamber, fitted with RF absorbers and providing class 100,000 conditions, and EMC measuring equipment. For emission testing, a computer-controlled 10 Hz to 40 GHz spectrum surveillance system, current probes and antennas are employed. For susceptibility measurements, high-power broadband amplifiers provide up to 600 V/m field strength from 10 kHz to 18 GHz. Special pulse generators and current probes allow various susceptibility tests, including electrical tests for power supply simulation.
Usable dimensions: 10.5 m length, 7 m width, and 8 m height.
Access door: 4 m width, 6 m height.
Shielding attenuation: H-field >60 dB for 10 kHz, >100 dB above 1 MHz; E-field and plane waves 100 dB for 1 MHz to 10 GHz.
Internal reflectivity: >20 dB above 100 MHz, >40 dB for 300 MHz to 10 GHz.

Vibration Test Facility
A 300 kN vibration system can operate horizontally or vertically under class 100,000 conditions, powered by four coupled electrodynamic shakers. Specimen response information is fed into a 360-channel digital data acquisition system. Sensors include 500 accelerometers with graded performance ranges. There are also several smaller single-shaker facilities.
Mounting table: 3 × 3 m (80 × 80 mm M10 hole pattern).
Maximum specimen mass: 15,000 kg.
Maximum acceleration: 15 g without load, 4.5 g with 5.0 t load.
Maximum displacement: ±25 mm.
Frequency ranges: 4–2,000 Hz sine low level, 4–300 Hz sine high level, 10–2,000 Hz random high level.

Modal Test Facilities
The systems provide tuned sinusoidal and non-tuned broadband excitation. The largest is housed in a transportable container; the two smaller ones are easily transported. The software undertakes test control, data acquisition/processing, modal analyses and result presentation. A separate hall is available for testing a larger range of structures, under clean room conditions if needed.
Excitation: sine, random 8 channels.
Vibration exciters: 26, 10–7,000 N.
Measurement: 400 accelerometers, plus force and displacement transducers.
Data acquisition: 882 channels (in 7 blocks) up to 2 kHz; 384 channels (in 8 blocks) up to 4(20) kHz; 16 channels up to 25 kHz

Acoustic Test Facility
Acoustic environments are simulated by electro-pneumatic noise generators that can reproduce overall sound pressure levels and spectra of operational fields. The total acoustic power available is 80 kW AC. The recording system includes 20 microphone channels, 128 accelerometer channels and 24 strain gauge channels. The facility comprises several reverberation chambers, control/computer rooms, air supply systems (5 kg/s) and 9.6 m height × 15.4 × 11.5 m preparation hall. The 800 m^3 chamber is an irregular pentagon 5.7 m high, 13.0 m wide, 10.8 m deep and can create 150 dB levels in class 100,000 conditions. The 206 m^3 chamber, 4.7 m high, 8.0 m wide, 5.5 m deep, can attain a maximum of 162 dB. The Progressive Wave Tube provides up to 170 dB in a 0.8 × 1.2 m test section. A thermoacoustic facility can simulate combined environments such as surface temperatures up to 1,300°C on a specimen plus acoustic excitation up to 158 dB.

Magnetic Field Simulation
The facility consists of a square, triaxial coil system with four coils per axis and with an edge length of 15 m. The test volume measures 10 × 10 × 10 m with free access of 4 × 4 m. It is used for measuring the magnetic cleanliness of objects, recording magnetic moments and eddy current fields, attitude control testing of magnetically stabilised spacecraft, calibrating magnetometers, and magnetising or demagnetising objects to determine permanent, remnant and induced fields. DC field values up to 75,000 nT can be produced,

corresponding to Earth's magnetic field level at the poles. The limit for uniform AC fields is 25 kHz. Hall probes, fluxgate magnetometers and search coil magnetometers are available; for precise measurements, proton spin magnetometers and optically- pumped magnetometers are employed.

Thermal-Vacuum Facilities
A 6.8 m diameter chamber (6.2 m diameter test volume) providing $<10^{-8}$atm vacuum conditions and a 3.6 m diameter or 3 × 4.5 m rectangular solar beam for test objects up to 2.5 tonnes, 4 m diameter/5 m high was introduced in 1983. The solar beam, with ±2° collimation angle, provides up to 1.4 solar constants. The cryogenic shroud creates 100 to 385 K conditions within the 13 m long chamber. Test specimens can be mounted on a two-axis motion simulator, providing up to 10 rpm continuous spin and ±200° attitude excursions and levelling capability.

Other Thermal-Vacuum Chambers
These chambers are used for subsystem and component trials. 3 m chamber: 3.2 m diameter, 3.8 m long shroud, 10^{-8} atm, 100 to 400 K, infra-red radiators; 1.3 m chamber: 1.35 m diameter, 2.2 m high shroud, 10^{-8} atm, 100 to 400 K. Clean room conditions from class 100,000 to 100.

Thermo-Mechanical Vacuum Facility
This is a 1.5 m diameter cylindrical vacuum chamber, with two-zone heating by graphite elements up to 1,600°C, mechanical loading by vacuum-tight feed-through and hot load introduction. Displacement measurement in hot zones relative to fixation points. 600 × 600 × 100 mm maximum sample.

Thermo-Mechanical Facilities
Mechanical loading is applied to shingle-type test samples (up to 350 × 350 mm) by pressure difference in a cylindrical pressure chamber. Seven infra-red modules with total 252 kW_e. Maximum temperature is 1,450°C applied thermal loading. Maximum pressure difference is 500 mbar. Reaction forces, temperature, pressure and displacements can be measured and visualised online. In addition, cover furnaces with resistance heaters up to 1,600°C in air or inert gas atmosphere; metallic and carbon heater up to 2,000°C in vacuum or inert gas available. Samples can be loaded and measured by extensometer and/or displacement by LVDT.

Thermo-Environmental Test Facility
A 1.5 m diameter cylinder of nominal 1.5 m length, extendible by modules. Houses a water-cooled reference system for, for example, displacement measurements. Several heating systems can be installed; maximum sample temperature =1,800°C. Different atmospheres and pressure profiles can be simulated. Mechanical loads can be applied by a hydraulic external loading device.

Institut für Raumfahrtsysteme

Current Status
The Space Systems Institute (IRS) was founded in 1970 and plasma thruster development has always been a principal interest. Facilities allow MW-class testing at 0.5–2 g/s propellant flow rates (argon) under selected 10^{-6}–10^{-3} atmosphere conditions. 0.5–1 MW MPD stationary thrusters of different geometries have been developed, mainly under AFOSR grants. Based on this experience, IRS has designed and built a range of thermal arcjet thrusters. Current arcjet projects include: Atos 600 to 700 W ammonia, Artus 2 1 to 2 kW NSSK hydrazine (with DASA; DARA contract); 10 kW and 100 kW devices. Atos is a simplified version of Artus 2 for flight on Amsat-Deutschland's amateur satellite aboard Ariane 502. The satellite was launched by 502 in October 1997.

Four plasma wind tunnels are used to investigate re-entry vehicle thermal protection materials and to validate aerothermodynamic CFD codes. Two have MPD plasma generators (0.10 to 1 MW) and are especially suited for high specific enthalpy (up to 150 MJ/kg) and low total pressure conditions. One tunnel has an inductive plasma generator, specially used for catalycity investigations, and the fourth has a thermal plasma generator for higher total pressure and low specific enthalpy areas. Sustained research into plasma generators for research and test activity with the facility available for hire by research institutes and government organisations and agencies.

IRS' Mission and System Analysis division studies development and numerical simulation and design tools for space transportation systems, mission and system optimisation, and performance assessment of air-breathing launchers. The Space Technology and Utilization division encompasses space station design, numerical flow field and simulation methods, and space systems safety. Developed software fail-safe checkout systems technology for mapping fault tree analysis. Has developed software and fault-tree mapping for advanced automated systems that the IRS believes would be fundamental to operational efficiency with the ESA Aurora programmes.

ZARM

Current Status
ZARM (Zentrum für angewandte Raumfahrttechnologie und Mikrogravitation) was established in September 1985 in the Faculty for Production Technology at the University of Bremen. ZARM consists of eight departments covering hydrodynamic stability, space technology, rotating fluids, aerodynamics, interface phenomena, gravitation physics, ferrofluids and combustion. μg research interests include fluid mechanics, combustion and thermodynamics, and technical assistance in its industrial exploitation. ZARM's DM30 million 146 m Drop Tower Bremen was commissioned in 1990, providing 10^{-5} *g* for up to 5 seconds. ZARM-FAB mbH (ZARM Fallturm Betriebsgesellschaft mbH: ZARM-Drop Tower Operation and Service Co) co-ordinates drop tower operations. ZARM-Lab GmbH provides technology/science support for users of Columbus and other flight facilities. ZARM-Förderverein eV (ZARM Promoting Association) is a non-profit organisation to support space and μg activities. Conducts new and off-nominal test evaluations and has created new rationales for generating accurate results. Has designed an automated, multifunctional, tribology research facility for encapsulation in existing test structures on manned habitats such as the ISS and research tests have resulted in a new series of proposed experiments involving suborbital rocket flights.

ZARM has played a substantial role in developing microgravity experiments for the ISS.

A 0.25 m hypersonic wind tunnel (ZARM-HHK) became operational in 1993 within the Hypersonic Technology research division. ZARM was also responsible for the university's Bremsat Shuttle satellite.

The 3.5 m diameter 110 m long drop shaft provides a 4.74 second free-fall through a 10^{-5} atmosphere vacuum (pumping requires 1½ h). The pressurised capsule is decelerated over 6 m by styropore granules. A larger capsule became operational in 1995. This is being adapted for unmanned remotely operated platforms.

Specifications
Capsule accommodation: 700 mm diameter, 1,200 mm high; 150 kg mass; 28 V 10 A (100 A short term) power; 1.6 Mbit/s telemetry; peak deceleration 350 m/s^2 (design must allow for safety factor of 2).

India

Indian Space Research Organisation Satellite Centre (ISRO-ISAC)

Current Status
In July 2006, a meeting was held at the ISRO Satellite Centre (ISAC) during which Indian leaders agreed that the Indian satellite navigation programme is of great importance to the Indian economy and people. India will embark first on a GPS compatible system as a test bed using the GSAT-4 satellite. Once that segment has proved successful, India will build and implement the Indian Regional Navigation Satellite System (IRNSS), which will be comprised of an independent 7-satellite system, supported by an extensive ground segment. ISAC is bound to play a large role in the development of the system, due to the Centre's satellite expertise and experience in developing the GSAT series. India anticipates that IRNSS will begin operation around 2012.

ISAC, in partnership with the UK's Rutherford Appleton Laboratory, contributed to the development of Chandrayaan-1's Imaging X-Ray Spectrometer (CIXS).

Background
The ISRO Satellite Centre (ISAC) in Bangalore is the lead centre for ISRO's satellite development and construction programmes. It grew from the Indian Scientific Satellite Project, established in 1972 to build the Aryabhata satellite. ISAC employs a staff of about 2,500, responsible for the design, fabrication and testing of Indian national programme satellites. The Centre has developed at least 23 communications; meteorology; remote sensing; and scientific spacecraft, including CARTOSAT-1 and the IRS, INSAT and GSAT series satellites.

ISAC is organised into five main functional Areas: Mechanical Systems (MSA); Control and Mission (CMA); Digital and Communications (DCA); Integration and Power (IPA); and Reliability and Components (RCA). The Centre also houses three separate Project Offices for the GEOSAT; IRS and small satellites; and ASTROSAT programmes. ISAC's independent Programme Planning & Evaluation (PP&E) division provides technical and managerial planning, resource allocation and progress monitoring on each project. The Laboratory for Electro-Optic Sensors (LEOS) is also organised under ISAC.

Physical Research Laboratory (PRL)

Current Status
India's principal centre for space science activities, covering ground-based telescopes and satellite/sounding rocket instrumentation as well as physics experiments carried aboard foreign satellites. Latterly an enhanced emphasis on international ventures with Europe, Russia and the USA. Astrophysical experiments and a continuing programme of co-operation with Russia on space science experiments. The PRL has a wide range of test and laboratory facilities and is linked with other national assets to provide a national resource which conducts co-operative programmes with foreign equivalents. The PRL has conducted a study of international ventures which it proposes to join. These are determined to be favourable to the future direction of India's physics and astronomy programme.

Vikram Sarabhai Space Centre (VSSC)

Current Status
The Vikram Sarabhai Space Centre (VSSC) continues its space systems and sub-systems research and development work, contributing, with the Swedish Institute of Space Physics, the Sub keV Atom Reflecting Analyser (SARA) for India's first Lunar exploration mission, Chandrayaan-1. The spacecraft is anticipated to launch in 2008, and will be carried into space by the Polar Satellite Launch Vehicle (PSLV), also developed by VSSC.

Background
Vikram Sarabhai Space Centre (VSSC), located around the village of Thumba near Trivandrum and with a personnel total of about 5,000, is the Indian Space Research Organisation's (ISRO) largest centre. It is the development base for the country's indigenous orbital launchers, including the SLV3, ASLV, PSLV and GSLV, as well as sounding rockets such as Rohini. Trivandrum's Liquid Propulsion Systems Centre operates as a separate entity. A Space Physics Laboratory carries out research in atmospheric and related space sciences.

Specialised research and development groups support VSSC programmes in avionics and mission dynamics; solid propulsion; propellants and chemicals; Materials and Mechanical Systems (MMS); systems reliability; computers and information systems; and programme planning and management. Extension facilities are located at Valiamala for the PSLV launcher project, at Vattiyoorkavu for composite development, and an experimental ammonium perchlorate plant is based at Aluvaye. A space physics laboratory conducts research in atmospheric and related space sciences.

VSCC Liquid Propulsion Systems Centre (LPSC)

Current Status

VSCC is responsible for the development of launcher liquid and cryogenic propulsion stages and auxiliary propulsion for launch vehicles and spacecraft. Main research activities cover avionics, aeronautics, materials and mechanical engineering, solid propulsion and composites, propellants, polymers and chemicals. It also conducts research on systems reliability and computer integration. It is growing in importance as main liquid engines are introduced to India's orbital launchers and with the increasing size of indigenous satellites. Current main projects cover the 720 kN Vikas PSLV stage 2 engine, PSLV's stage 4 dual 7.5 kN system, Insat 2's 440 N Liquid Apogee Motor and unified network of 22 N ACS thrusters, and the 680 kN Vikas GSLV strap-on. Seven Russian 76 kN cryogenic engines are being provided for the initial GSLVs, the first of which flew on 18 April 2001, but India's own engine will complete development by 2006. India signed a deal for the Russian engines in 1990 incorporating technical support on cryogenic development. Facilities development of cryogenic handling structures assisted by Russian aid.

VSSC's test facilities are sited southeast of Trivandrum at Mahendragiri. Vikas' Principal Test Stand (PTS) was commissioned during 1987 and used in January 1988 for the engine's first full-duration 150-second firing. Altitude facilities are also available for PSLV's 7.5 kN motor, Insat 2's LAM and smaller thrusters. Cryogenic engine and stage facilities were commissioned in 1997. The site for liquid propulsion and cryogenic stages is located at Thiravananthapuram while test facilities are located at Mahendragiri in Tamil Nadu. An ammonium perchlorate experimental plant is operated by VSSC at Aluva.

VSSC Solid Propulsion Group

Current Status

India's first 75 mm diameter solid motor was produced in 1967, followed by a 125 to 560 mm range for sounding rocket applications and, in the early 1970s, motors for the SLV-3 satellite launcher. Work on the most powerful so far began in 1984: the 2.8 m diameter 3,500 kN thrust solid powers PSLV's first stage. Although the Solid Propulsion Group is an element of VSSC, the primary Solid Propellant Space Booster Plant (SPROB) is sited at SHAR Centre on Sriharikota Island, along with the Vehicle Assembly Static Test and Evaluation Complex (VAST) for solids. At VSSC the motor cases are produced in the Mechanical Engineering Facility and the propellant binders in the centre's Propellant Fuel Complex. Hybrid propellant combinations studied for possible adoption at primary stage rocket motors for future launches. VSSC has begun development of hybrid systems for advanced high-energy propulsion requirements. These support new proposals for a heavy lift development of the GSLV and studies carried out to date suggest a significant increase in capability with a reduction in the costs of kilogramme/payload weight to orbit.

International

European Space Research and Technology Centre

Current Status

ESTEC has reorganised ESA astronaut assignments in the wake of the Columbia accident. It initially agreed to defer flights aboard scheduled Soyuz missions to the ISS in April and October 2003, relieving these spacecraft to carry two-people crews to the ISS for housekeeping duties. Subsequent to the release of the Columbia Accident Investigation Board report on the loss of the Shuttle these flights are deferred to a schedule yet to be determined.

Background

ESTEC is ESA's largest single establishment, with about 1,600 employees. It provides project management for science, communications, earth observation, ügravity and space station programmes, executes the space science programme, performs future satellite programme studies, and undertakes design, development and testing of components and complete space vehicles. Half of the on-site staff belongs to the spacecraft project teams reporting to the Programme Directors at ESA Headquarters; the majority of the remainder belong to the specialised technical divisions based at ESTEC.

Capital investment is about US$55 million for buildings and US$150 million for the technical facilities. The development laboratories include Mechanical Systems, Propulsion, RF Systems, European Space Battery Test Centre, Materials/Processes, Components, Onboard Data Processing, Simulation and Electrical Facilities for Automated and Manned Missions, including a robotics testbed. The Satellite Communications Building supports special telecom services such as videoconferencing, mobile links and other in-orbit testing. The Fuel Cell Test Facility is an annexe to the Space Battery Centre.

The largest laboratory, the ESTEC Test Centre, operates a wide range of environmental test facilities. Working with Intespace (Toulouse), IABG (Ottobrunn) and Centre Spatial de Liège, ESTEC heads the Co-ordinated European Test Facilities to ensure that national facilities can be used for agency projects. All test areas are air-conditioned (19 to 23°C; 40 to 60 per cent RH), class 100,000.

ESTEC's test facilities include:

Large Space Simulator (LSS)

Europe's largest high-performance solar/vacuum simulator became operational in 1986 and was inaugurated January 1987 for Ariane 4-class payload testing. The first major test was on Alenia's IRIS upper stage test model. The LSS consists of a main 10 m diameter chamber with a removable lid for ease of access and a 5 m side port, with an auxiliary chamber containing the collimating mirror. Payloads can be supported on a vibration-isolated (-10^{-3} g) 3.2 × 3.2 m platform, or suspended from the upper volume, or mounted on a two-axis motion simulator. An array of 19 unfiltered 25 kW high-pressure Xenon lamps provides a 6 m diameter collimated beam, with a maximum 1.3 solar constant intensity. The LSS is depressurised using both the test centre's central pumping system and a dedicated high-vacuum system of turbo-molecular pumps and a liquid helium cryo-pump to attain 3×10^{-7} mbar in about 10 hours. Both chambers incorporate stainless steel shrouds operating on LN_2 or GN_2.

Chamber capacity: 10 m diameter × 15 m high, auxiliary chamber 8 to 11.5 m diameter × 14.5 m long. The contractor is BSL.

Vacuum conditions: 3×10^{-7} mbar in 10 hours; repressurised by GN_2 to 100 mbar followed by filtered air over 4–24 hours.

Illumination: 1 solar constant (1,360 W/m^2) by 12 of 19 × 25 kW Xe lamps. 32 kW lamps can be used if needed. 6 m diameter horizontal beam with ±4% uniformity; 7.2 m collimation mirror consists of 121 hexagonal segments, collimation angle 1.9°. Contractor Carl Zeiss.

Shroud temperature: <100 K by LN_2 circulation, 150–350 K by GN_2. Cool/warm-up period ±2 hours. Shrouds supplied by Leybold-Heraeus.

Motion simulator: spinbox (1–6 rev/min + 1–24 rev/day), turntable (1–24 rev/day). The contractor is SIGRI.

Thermal Facilities

HBF3 thermal vacuum chamber. The chamber is employed for subsystem vacuum temperature cycling in a 3 m diameter usable test volume. The facility is equipped with a demountable infra-red rig for solar panel testing, comprising a LN_2 shroud system with two compartments 3.85 m long, 2 m deep each fitted with 38 500 W infra-red lamps. Maximum temperature range on carbon fibre-backed solar arrays is 170/+90°C. The basic chamber incorporates five shrouds temperature controlled by LN_2 or GN_2. Loading is via removable top lid.

Vacuum conditions: $2–6 \times 10^{-6}$ mbar in 5 h (0.02 mbar in 1 hour) from atmospheric pressure.

Shroud temperature: <100 K by LN_2 circulation, 200–373 K by GN_2. 2 K/min cooling; 3 K/min heating.

Small thermal vacuum chambers: VTC 1.5 usable volume 1.5 m diameter × 2.5 m high, up to 500 kg test subjects, 1.3×10^{6} mbar vacuum, <100 and 125–425 K using LN_2/GN_2. Corona horizontal cylinder for up to 400 kg masses in 1.8 m diameter × 3.2 m long test volume. The LN_2 shroud with two 1.21 m deep, 2.9 m long compartments each equipped with 32 500 W infra-red lamps for solar panels testing, featuring rapid de-pressurisation (down to 1.3×10^{-4} mbar in 7 min), down to 6.6×10^{-6} mbar, 170/+110°C achievable on carbon fibre-backed solar arrays (the chamber can also be used for outgassing tests); Ultra-High Vacuum Chamber UHV 0.5 0.120 m^3 test volume in 45 cm diameter × 76 cm high cylinder, attaining $<1.3 \times 10^{-10}$ mbar with bake-out facility up to 523 K; Accelerated Thermal Cycling Chamber ATC II for rapid automatic thermal cycling of small lightweight items at atmospheric pressure, T range 93–403 K using dry N_2 in 48 cm high × 50 cm long × 25 cm wide usable volume, typical cycle +100 to −100°C and back in 3 minutes.

Electrodynamic Shakers

Located in the same complex as the Large Space Simulator, the centre is equipped with a 280 kN multishaker and a 70 kN shaker. The larger system incorporates two Ling Dynamics Systems 984 LS vibrators, each capable of 144 kN thrust. The 70 kN system uses one Ling Dynamics 964 LS exciter. All systems generate 52,000 Hz sinusoidal or random vibrations in horizontal and vertical modes. The data acquisition system allows for 250 channels, processed on-line.

HYDRA Hydraulic Shaker the 5.5 m diameter 22 tonne octagonal table (flush with the floor) of the 6-DOF shaker is driven by four actuators vertically and two in each lateral direction. Each actuator has a ±70 mm stroke, max 0.8 m/s piston velocity and 630 kN force rating, providing about 5 g for a 5 tonne test payload and 3.5 g for a 15 tonne mass in the vertical axis. Frequency range 0.1–100 Hz. In addition to the traditional sine dwell and sine sweep tests, HYDRA can provide transient excitations in 6-DOF. The 6-DOF allows the specimen to be sine tested along the vertical/lateral axes with one single set-up. Adding transients was a major HYDRA objective: multidirectional transients at the launcher/spacecraft interface produces a more realistic structural response. The shaker is supplied by Mannesmann-Rexroth.

Large European Acoustic Facility (LEAF)

The 154.5 dB 1624 m^3 LEAF, the largest in Europe, became operational in 1990. It provides an internal height of 16.4 m, allowing testing of Ariane 4/5 payloads. The noise generation system, employing pressurised N_2, comprises four horns with cut-off frequencies of 25, 35, 80 and 160 Hz. The overall noise level may be increased to 158.5 dB in the future.

Chamber size: 1624 m^3, 9 × 11 × 16.4 m (W × L × H), accessed by 7 × 16.4 m (W × H) door.

Cleanliness level: class 100,000.

Temperature range: 20 ±2°C during test.

Suspension points: 9 of 80 kN capacity, 35 of 15 kN; max crane load 160 kN.

Overall sound pressure level range: 125–154.5 dBL.

Specified octave band pressure levels in empty chamber at 154.5 dBL: 136.5 dBL at 31.5 Hz centre frequency, 141.5 at 63, 147.5 at 125, 150.5 at 250, 147.5 at 500, 144.5 at 1,000, 137.5 at 2,000, 131.5 at 4,000, 125.5 at 8,000.

Field homogeneity in test volume: ±2 dBL.

Control tolerance of sound field: ±1.5 dBL overall.

Noise measurement/data acquisition: 16 microphones, 250 accelerometers, and 50 strain gauges.

Compact Payload Test Range (CPTR)

The CPTR became operational in mid-1992 primarily for measuring the electrical performance characteristics of Ariane 4-class RF-radiating payloads in their operational configurations. It comprises a shielded anechoic chamber with a feed scanner room and control room together with a test preparation area. The PWZ Plane Wave Zone test volume is obtained using two large reflectors in an offset Cassegrain configuration. Payloads of up to 5 tonnes are located at the PWZ centre by means of a positioner providing movement in azimuth, elevation and polar. Test data can be presented in a wide range of formats, including 2-D radiation patterns, contour plots and projections in 3-D. System performance parameters can also be measured. By locating feed horns at different positions in the focal plane (in the scanner room), the direction of the electrical boresight can be changed. In this way, the performance of the satellite

transmit /rx from stations at different locations is verified. Similarly, Earth coverage contours or in-orbit reconfiguration can be verified. As the chamber is shielded, it can undertake a range of electromagnetic compatibility measurements.
Internal size: 10.9 × 9.6 × 24.5 m.
Frequency range: 1.5–40 GHz.
Plane wave zone: 7 × 5 × 5 m.
PWZ performance: ±0.2 dB amplitude ripple, ±4° phase ripple, <0.4 dB taper, <–40 dB cross polar.
Reflectors: 9.2 × 8.0 m subreflector, 10.2 × 7.6 m main reflector, less than 100 μm peak-to-peak surface accuracy.
Environmental: class 100,000, 20±2°C, 50±10% humidity.

EMC/ESD Facility
The electromagnetic compatibility/electrostatic discharge facility comprises a shielded anechoic chamber and two operating rooms, including a complete range of automated test equipment. The 20 Hz 40 GHz range satisfies all test requirements for science satellites (low magnetic) and communications satellites (high RF power). All emission and susceptibility measurements are fully automated with on-line data reduction narrow/broadband identification. Data output is corrected for probe antenna factors and compensated for resonance effects.
Test area: 6 × 6 × 4.5 m, accessed by 3.5 × 4.5 m door.
Environmental: class 10,000, 20 to 24°C, 610 to 775 mm pressure, 40 to 55 per cent humidity.

European Space Research Institute (ESRIN)

Background
Based in Frascati, Italy, ESRIN is one of four ESA establishments and operates with a permanent staff of 140 and a contractor staff of about 250. Its main focus is Earth observation satellite data handling, including management of the ground segment to acquire, pre-process and archive data, and handling distribution, either directly or via a distributor. ESRIN handles missions for non-ESA clients as well including Landsat, Tiros, MOS, JERS 1 and Spot but the principal emphasis is on ERS and preparations for Envisat launched in 2002.

The ESRIN ERS Central Facility (EECF) maintains constant links with the ERS Mission Management and Control Centre at ESOC. ESOC executes the mission operation plan prepared by the EECF, with ERS ground stations for the scheduling of near-realtime distribution of Fast Delivery products. The Processing and Archiving Facilities handle data product orders, providing users with access to the online worldwide catalogue of ERS data. To this end, ESRIN is responsible for stations in Europe and Canada, ERS PAFs in Italy, Germany, France and UK, and has contracts with worldwide national stations for ERS data acquisition. ESRIN manages more than 30 ground stations that receive data from ESA satellites.

ESRIN also operates the ESA-IRS Information Retrieval Service which provides online access to bibliographic and factual databases covering most fields of science and technology. The ESIS European Space Information System caters for the astronomy and space physics community.

Israel

Asher Space Research Institute

Current Status
The institute was established in 1984 to provide expertise in space sciences and engineering, including astrophysics, spacecraft propulsion, control, materials, structural design and remote sensing. It employs 32 full time engineers and 24 academic members drawn from the aerospace engineering and physics faculties. Its TechSat 1 satellite was launched on 28 March 1995 along with two other satellites by Start launcher. The Gurwin-II Techsat satellite was launched from Baikonur by Zenit 2 launcher on 10 July 1998. The Institute continues to carry out research on physics experiments in planning for a new generation of Israeli science satellites. These are envisaged as multipurpose platforms for diverse users and experimenters. Integrated space support systems packages would ensure maximum efficiency for minimum mass/volume allowed. Several potential missions have been defined from new nanotech designs for ultra-low mass properties. Research programmes include broadband laser inter-satellite technology, hyperspectral data acquisition and retrieval for micro-satellite remote sensing, integrated optical inertial space navigation systems, strap-down stellar star-track sensors, magnetic attitude control systems for small satellites and diagnostic techniques for plasma thrusters in flight.

Views of ESA's main establishments: HQ, ESTEC, ESOC, ESRIN 0019686

Japan

Institute for Unmanned Space Experiment Free Flyer (USEF)

Current Status
The Institute for Unmanned Space Experiment Free Flyer (USEF), located in Tokyo, was established in 1986 to promote the development and utilisation of unmanned space systems.

USEF is a non-profit organisation under the direction of the Ministry of Economy, Trade and Industry (METI) and endowed by 20 companies: Denso, Fuji Heavy Industries, Fujitsu, Hitachi, Furukawa Battery Company, IHI Aerospace, Ishikawajima-Harima Heavy Industries, Itochu Corporation, Kawasaki Heavy Industries, Keyware Solutions, Mitsubishi Company, Mitsubishi Electric Corporation, Mitsubishi Heavy Industries, Mitsubishi Precision, Mitsubishi Space Software Company, Mitsui Bussan Aerospace, NEC Toshiba Space Systems, Sharp, Sony Corporation and Sumitomo Electric Industries.

Background
USEF company products were stimulated by US interest in free-flyers for manned space stations, assembled through a series of modules co-orbiting microgravity structures, achieving lower levels of vibration and accelerations than possible through manned, pressurised, modules. The Space Flyer Unit (SFU) carried nine experiments on its first flight before retrieval by NASA's Shuttle in January 1996. The Express and Space Robotics projects began in 1992. The Unmanned Space Experiment Recovery System (USERS) project began in 1995 and the technology has been applied to the development of concepts for unmanned platforms. This in turn has provided further research and development opportunities for advanced SFU design concepts. USERS was recovered successfully after its 2002–2003 flight. The Space Environment Reliability Verification Integrated System (SERVIS) project began in 1999 and will continue through 2007. The launch of the second SERVIS spacecraft is expected to take place, although the spacecraft is not yet scheduled for launch by the Japan Aerospace Exploration Agency (JAXA). SERVIS' primary objective is to test and validate the ability of COTS technology to withstand the severe radiation satellite parts are exposed to in the space environment. The free-flyer concept has returned to favour in recent years as a relatively low economic means of gaining zero-gravity experience and the unexpected increase in costs for the International Space Station makes this option a more promising opportunity.

Microgravity Laboratory of Japan (MGLAB)

Current Status
MGLAB began operations in 1993, providing a 1.50 m diameter drop tube with 100 m free-drop zone. The 900 mm diameter 2.280 m capsule is decelerated over 50 m and provides two payload racks to users. Test equipment has been utilised for development of microgravity equipment for the JEM. This has expanded to include offers for Japan's metallurgical industries to invest in new research programmes for materials science. Under government reorganisation of nationally funded research projects MGLAB has developed a set of design initiatives which may expand Japan's ability to carry out materials processing tests in space. Stalled plans to get JEM operational due largely to the hiatus in Shuttle missions has encouraged MGLAB to develop new concepts for man-tended and unmanned platform research lifting total reliance from the manned elements formerly believed to be in the vanguard of this research.

In the aftermath of the reorganisation of NASDA and ISAS the JAXA has integrated several strands of work linking the space centres with industry and universities. This has been benefited by the smoother integration of the disparate policies and plans of separate organisations, now integrated into one cohesive body.

Telecommunications Advancement Organisation of Japan (TAO)

Current Status
The Telecommunications Satellite Corp of Japan (TSCJ) was established in August 1979 to operate the CS and BS satellite series procured and launched by NASDA. The name changed to TAO in October 1992 as it introduced new services such as R&D. Japan's first two operational communications satellites, CS-2a/2b, each with six Ka-band and two C-band transponders, were launched in 1983 principally for maintaining communications with the more scattered islands of the Japanese archipelago. Both were still operational when they were replaced by CS-3a/3b in 1988. NTT's NStar took over in 1995. Now operated by TAO, NStar A was launched in August 1995 followed by NStar B in February 1996, both built by SpaceSystems Loral, and NStar C in July 2002, a satellite built by Lockheed Martin Commercial Space Systems. The BS-2 series introduced DBS services in 1984 using 100+W transponders but two of BS-2a's three amplifiers failed soon after launch on 23 January 1984. BS-2b brought the system to full capacity in 1986 but a BS-2x was purchased from GE Astro Space by NHK for an Ariane launch to ensure system continuity before the first of the BS-3 series could be launched in 1990. However, BS-2x was lost in the Ariane failure of 22 February 1990; the replacement BS-3H was then lost in its April 1991 launch. Both would have been controlled through TSCJ. BS-3a, launched 28 August 1990, replaced BS-2b but it operates on marginal power. 3b completed the system in 1991, but NHK procured BS-3N in late 1992 (launched July 1994) to guarantee services. The 'BS-4' generation is operated by B-SAT through its BSAT-1 series, BSAT-1a launched April 1997 and BSAT-1b launched April 1998. Manufactured by Orbital Sciences, BSAT-2a was launched by Ariane 5 on 8 March 2001 followed by BSAT-2b on 12 July 2001, the latter placed in an incorrect orbit due to a terminal stage failure.

Research and development activity includes the test and evaluation of new fail-safe operating concepts.

Netherlands

National Aerospace Laboratory (NLR)

Current Status
In mid-2006, the National Aerospace Laboratory (NLR) presented the Galileo satellite navigation system at various conferences and symposiums. Galileo deploys 30 satellites to determine the position of receivers on the ground. It is intended for use by transport enterprises, electricity companies and by governments. It is also designed for use in relation to car navigation systems and personal navigation devices integrated into mobile telephones and PDAs. The purpose in the development of Galileo is to reduce the current dependency on the Global Positioning System (GPS). The two systems will work in tandem with one another. Galileo is more accurate then GPS and is being developed mainly for the civil market. The Galileo programme is being conducted by the European space industry for the EU and ESA, with NLR's role to fully test the functionality of the system, its part and the receivers. Galileo is expected to be fully operational by 2013.

Background
The NLR is the central institute for aerospace research in the Netherlands. Since 1937 NLR had been an independent non-profit organisation that provides technological support to aerospace industries, operators, authorities and international organisations concerned with aviation and spaceflight, all over the world. NLR participates in programmes of the European Space Agency (ESA) and the European Union. NLR closely co-operates with national industrial partners in space-related projects supported by the Netherlands Agency for Aerospace Programme (NIVR).

NLR employs a staff of about 900 in two main establishments, one in Amsterdam and one in the Noordoostpolder, Flevoland. Over two-thirds of the staff are graduates from universities or technical colleges. NLR holds an ISO 9001; 2000/AQAP-110 quality assurance certificate and several calibration accreditations.

NLR owns several wind tunnels, including a transonic (DNW-HST) and a supersonic (DNW-SST) wind tunnel extensively used in test programmes for the development of the ARIANE launchers, ARD and CTV/CRV capsules and the X-38, as well as heat protection systems. Both of these wind tunnels are capable of being used for the validation of the aerodynamic performance of future space transport systems. NLR's wind tunnels are operated by the German-Dutch Wind Tunnels organisation, DNW, along with similar German facilities. NLR develops wind tunnel models and test equipment.

NLR operates facilities for research in the area of structures and materials used in space projects. Expertise in the area of loads and use monitoring is combined with structural response monitoring and materials characterisation to perform damage tolerance assessment. Advanced dynamic and non-linear analysis and test systems are applied to solving problems associated with design and verification of spacecraft structures, including fatigue and crack growth analysis. Structural optimisation with regard to strength, active vibration control or stability performance is also executed. Materials science and engineering projects include the evaluation of properties of metallic, composite and hybrid materials, also at elevated temperatures, and the testing of full-size structures.

NLR's extensive computer network, featuring a 64 GFlops NEC SX-5/8 supercomputer, is used for theoretical research and simulation, especially in the fields of computational fluid dynamics, computational solid mechanics, robotics, computational electromagnetics, and structural design and analysis, among other things.

NLR's moving-base research flight simulators can be configured not only for transport aircraft, fighter aircraft and helicopters but also for shuttle or spaceplane-type vehicles.

Test and verification equipment is available for use in the development of space avionics systems. Various environmental testing facilities and an antenna test range are available.

Spaceflight specialisations include the simulation and testing of satellite subsystems, the development of two-phase flow heat control systems, and research on liquid dynamics problems in space. In addition, NLR develops electronic control systems for biological experiments in micro-gravity. NLR activities in the area of satellite navigation are focused on validation and validation. To support mission preparation, execution and evaluation, NLR works in the area of space robotics, telescience, and utilisation support.

NLR is developing operational systems for the management of natural resources using remote sensing data. NLR acts as National Point of Contact for the dissemination of remote sensing data, and has co-developed a mobile receiving station. To enable Dutch users to access a wide range of Earth observation products and to allow Dutch service providers to make their products and services available, NLR is working on the development of a Netherlands Earth Observation Network (NEONET), in correct with the European CEO initiative. Optical and microwave remote systems are developed, and operated from NLR's Cessna Citation and Fairchild Metro research aircraft. NLR is one of the developers and the ESA focal point of the NASA crack growth program NASGRO.

Prins Maurits Laboratory (PML) TNO

Current Status
PML is one of the three defence research institutes of the Netherlands Organization for Applied Scientific Research (TNO). PML has about 270 employees in two research and development departments that includes propulsion technology. It developed the gas generator igniter, thrust chamber igniter and turbine pump starter for Ariane 5's Vulcain cryogenic engine, in addition to undertaking research and development of high-performance solid-propellants based on phase-stabilised ammonium nitrate and hydrazinium nitroformate for future spacecraft. Facilities include an indoor test stand for solids, a connected pipe facility for ramjets, mixers for propellant and ramjet fuel manufacturing, and equipment for determining the chemical, mechanical and burning properties of propellants. The laboratory seeks to perform, and conduct, ground breaking research into new and exotic propellants that show results in tests commensurate with substantial improvements in launch vehicle performance. Research into advanced propellant combinations and tri-propellants supporting high performance and high Isp propulsion concepts. Has specialised in matching new and innovative propulsion concepts with projected launcher and in-space applications.

Technisch Physische Dienst TNO-TU DELFT

Current Status
TPD, the TNO Institute of Applied Physics, is an element of Netherlands Organization for Applied Scientific Research (TNO). Space personnel total 24. TPD began space instrumentation research and development in 1964 with the S59 UV stellar spectrometer for Europe's TD-1A astronomy satellite. It now specialises in attitude sensors, science instrumentation and Earth observation sensors. For example, the department provided Giotto's star mapper and Eureca's Sun acquisition sensors. Science instruments cover the wavelength range from hard X-ray up to the far-IR/mm. Recent projects include Hipparcos' modulation grid and refocusing mechanism, and ISO's cryogenic Short Wavelength Spectrometer. Remote sensing activities are concentrating on instruments for Meteosat Second Generation, Envisat and Metop, including optical systems, detection techniques, focal plane layout and precision mechanisms. Examples are the MIPAS cryogenic focal plane assembly (4 to 15 μm) and SCIAMACHY's optical bench. It was intensively involved in the development of the GOME Global Ozone Monitoring Experiment for ERS 2. TPD supplied the onboard calibration unit and performed the whole instrument's ground calibration under ESA contract. For that, TPD developed a calibration facility unique in Europe; it will also be used for the calibration of SCIAMACHY and other atmospheric science instruments, such as MERIS. Development of new sensor instruments and research into alternative sensor suite packages for remote sensing, meteorological and earth observation payload requirements. Has taken a lead role in projecting a new generation of requirements for next-generation atmospheric sciences satellites.

Russian Federation

Izmiran

Current Status
The Institute of Terrestrial Magnetism, Ionosphere and Radio Wave Propagation of the Academy of Sciences is Russia's principal investigator of the magnetosphere, ionosphere, Sun and Sun-Earth phenomena. It operated the Coronas-I solar observatory launched in 1994 from Plesetsk, the first combined Russian/Ukraine helioseismology satellite. It was to be followed by Coronas F originally planned for launch in 1997 and then for 2001. Izmiran also provides instruments, particularly fields and particles measurement devices for satellites and is developing devices which it believes it can sell to ESA countries. Proposes major set of small scale upper atmosphere sounding rocket programmes to accommodate incremental payload packages.

Has proposed a series of internationally funded near-earth space physics experiments but has not yet received approval.

Keldysh Research Centre

Current Status
The Keldysh Research Centre – previously known as RNII, NII-1 and then NIITP – was established in 1933 as the first institute for research and development of liquid and solid propellant rocket engines. It currently develops, manufactures and tests advanced prototype of rocket engines, space power systems, high-energy beam generators and particle accelerators of various types. Its products have been used in the development of launch vehicles, spacecraft and orbital stations. Before and during the Second World War the centre concentrated on ground-based and air-launched weapons, the major achievement at that time being the development and manufacture of the first batch of Katyusha missiles, for which the centre received the Order of the Red Star. During its history the centre boasted having M V Keldysh, S P Korolev and V P Glushko working within its organisation.

In rocket motor development, KRC has developed staged combustion cycle motors and pioneered work on highly reliable engines of this type for Proton, Zenit and Energia launchers. Hybrid rocket motors are being developed utilising solid fuel and liquid or gaseous oxidiser intermediate in Isp between solid rocket motors and cryogenic, high-energy, motors. New moveable nozzles are being designed in the Nozzle Differential Facility as are ultralight uncooled composite nozzles reducing engine weight by 12 to 15 per cent.

In general, space activity KRC has researched energy transport in space and from space to earth and into solar dynamics power systems on a Brayton cycle gas turbine with 10 kW output power. Also, research into large solar electric propulsion systems for powering satellites and orbital transfer vehicles. Also support systems include the droplet-coolant radiator for powerplant weight reduction applications.

KRC also conducts research into electron beam generators of 35–1,000 kW, 30–300 KeV, negative ions producing a beam current of 1 mA, 1–20 KeV.

OKB Fakel

Current Status
Fakel is a propulsion laboratory which was established in the Academy of Sciences in 1955 and re-organised in 1972 as OKB Fakel ('Torch'), specialising in spacecraft attitude control thrusters, ion engines and plasma sources. The first EPS flew in 1972 aboard a Meteor-1 satellite and further models was subsequently flown on the later Meteor generation satellites, including the Meteor-Priroda series. Research and development work on Stationary Plasma Thrusters (SPT) started at Fakel in 1964. Starting in 1982, EPS units based upon STP-70 and the K-10 hydrazine thermal-catalytic thrusters (TCT) have flown on communications satellites such as Geizer and Luch (including flights within the Kosmos programme). In 1994, a third generation EPS was introduced, SPT-100 aboard Gals and Ekspress satellites. In 1992, Fakel together with Space Systems/Loral and the Moscow Scientific Research Institute of Applied Mechanics and Electrodynamics (RIAME) established a joint venture – International Space Technology Inc (ISTI) – for promoting, marketing and selling the EPS outside Russia. Subsequently, SEP and ARC joined the company. In 1996 the SPT-100 was certified in accordance with western standards: 7,440 hours of firing tests were successfully conducted at Fakel and 5,000 hours were accomplished at the Jet Propulsion Laboratory in Pasadena. Fakel proposed a joint venture through a commercial company whereby the SPT-100 concept could be applied to US projects for space applications.

Its SPT-70 and SPT-100 (Stationary Plasma Thruster) Hall electric thrusters have flown on more than 50 Meteor polar meteorological satellites since 1972 to provide orbit control. The numerical designators indicate beam diameter in mm. SPTs employ a DC-gas discharge in an annular chamber, in which a radial magnetic field traps the electrons. These Hall currents ionise the Xe propellant, the ions of which are accelerated inside a quasi-neutral plasma, without grids, and by the discharge voltage itself. The advantages are their rugged simplicity (no grid system or high voltage supply), but they provide lower efficiencies, lower exhaust velocities and increased Xe consumption than ion thrusters such as the UK-10. Thrusters utilised on Zord-3, Meteor and Meteor-Priroda satellites, Luch, Gals, Express, Yamal, Arkos and Kupon satellites. In 1994 OKB Fakel became part of the Russian Space Agency. Co-operative exchange system with ESA and has extended several international opportunities to countries in the Far East and South East Asia.

Space Research Institute

Current Status
Space Research Institute, part of the Russian Academy of Sciences since its creation in 1965, is Russia's premier space science centre and leads missions such as Spektrum and Relikt 2. Bureaux in Frunze and Tarusa support it. Employment declined from around 1,600 at the collapse of the USSR in 1991 to 1,300 in 1994 and about 750 today. Responsible for international co-operation between numerous establishments in Europe and the USA. Conducts high-level academic discussions on international space science policy and research trends. Has continued to provide a focus for Russian research and technology development programmes related to space activities. Emphasis on obtaining joint ventures and co-operative projects with foreign organisations. Since 2001 IKI has developed several long-term strategies for resuming the unmanned exploration of the solar system and is working several concepts for deep space missions to the outer solar system and flight beyond the heliosphere to interstellar space. This is one of several long-term mission goals and objectives being worked as theoretical possibilities for government guidance.

TsAGI

Current Status
The Central Aerohydrodynamics Institute (TsAEI), was formed in 1918 in Moscow under the leadership of N I Zhukovsky. It maintains aircraft and satellite test facilities employing about 5,900 personnel. TsAGI worked with British Aerospace on interim Hotol research. In 1993, it worked with ESA on development of a Buran real-time entry flight simulator. Buran pilots trained on TsAGI's 6-DOF simulator. Has conducted extensive research into hypersonic flight and conducted tests with scramjet engine designs. Some co-operation with the French. Emphasis in space related research has focused on the potential development of trans-atmospheric vehicles, a technology which has historically interested Russian engineers and their former Soviet colleagues and some effort has been made in recent years to interest European companies in an agreement to focus research on hypersonic sramjet technology. The Russians are building a credible research base on these applications and have attracted the interest of Indian companies. Refocusing on exoatmospheric boundary vehicles (capable of flying just inside or outside the earth's atmosphere) TsAGI has designed several hypersonic vehicle configurations for scramjet applications, extending original work conducted during the 1960s and again during the 1980s.

VNII TransMash

Current Status
Established in 1930 TransMash specialises in the design and fabrication of transport vehicles and tanks. In the space field, it developed the Mars-98 rover (Lavotchkin integrates the payload) and was responsible for the Lunokhods, small Mars 3 rover and the Phobos hopper. It has experience in the design and initial breadboard testing of systems layouts for semi-automated planetary exploration vehicles, capable of significantly closing the gap between the conventional rolling rover, an independently self-controlled robotic systems, with the capability to climb, roll across rough but level surfaces and crossing small ravines. Continues to research the technology for advanced planetary rovers and remotely controlled vehicles. Some work in this direction has been contracted by Energia and Khrunichev. Employees total 1,750.

Sweden

Swedish Institute of Space Physics

Current Status
The Swedish Institute of Space Physics (Institutet för rymdfysik, IRF) was established in 1957. It has participated in more than 40 rocket and 25 satellite investigations since 1964, primarily targeted at magnetospheric and ionospheric studies. IRF provided Promics particle detectors for Prognoz 7/8, hot plasma and low-frequency wave experiments onboard the Swedish satellites Viking and Freja, the Aspera hot ion composition spectrometres onboard both Phobos spacecraft, the Promics-3 hot plasma detectors for both Interball orbiters, the Electric Fields and Waves instrument for Cluster and Cluster-2, the Aspera-C ion/neutral particle imager for Mars-96, the 0.5 eV–40 keV/q ion spectrograph on Japan's Nozomi/Planet-B Mars orbiter launched in July 1998, and in November 2000 launched its own nanosatellite Munin with MEDUSA, a combined electron and ion spectrometer, and DINA, a neutral particle detector. IRTF assembled instruments for two major ESA missions: an energetic atoms analyser Aspera-3 for Mars Express launched in June 2003, and an ion composition analyser RPC-ICA for the spacecraft Rosetta launched to the comet Churyumov-Gerasimenko in February 2004. The institute has carried out several analyses on possible future science packages for comet landers and sample return missions. This has been stimulated by new directions in deep space design concepts pioneered by the Institute. These directions include self-test and repair, all electric spacecraft (including propulsion) and new autonomous navigation systems.

The research facility's head office is at Kiruna and is primarily concerned with upper atmosphere phenomena, the ionosphere and the planetary magnetosphere. Since 1996 IRF has also had an atmospheric research programme in Kiruna specialising in the physics of the middle and upper atmosphere. IRF's headquarters in Kiruna and its division in Uppsala focus on studies of the ionosphere and magnetosphere. It also has researchers based in Umeå and Lund specialising respectively in the propagation of infrasonic waves and solar terrestrial physics.

United Kingdom

Central Laboratory of the Research Councils

Background
The Central Laboratory was created 1 April 1995 as an independent entity from the merged Daresbury and Rutherford Appleton Labs, reporting to the Office of Science & Technology and exercised by the Department of Trade and Industry. It is operated by the Council for the Central Laboratory of the Research Councils (abbreviated to CCLRC). Its space funding is mostly from PPARC and NERC but some 20 per cent comes from commercial contracts. RAL's Space Science Department is a partner member of the British National Space Centre. RAL is the focus for the UK's space science projects and directs both ESA and bilateral activities, such as the Spectrum-X astronomical satellite collaboration with the Soviets. It hosts 200 personnel. Policy recommendations made to various UK government

bodies including DTI and BNSC. CLRC directs external university participation.

Centralised facilities for university research groups include a satellite operations centre, data processing facilities, and the Starlink data analysis network, the British Atmospheric Data Centre and assembly/testing of space payloads. Projects include:

Badr-B
CCD camera for cloud monitoring. Chilton will act as back-up to Lahore ground station. The same camera is used for Morocco.

Cassini
Participation in Saturn orbiter's dust analyser and plasma electron spectrometer, and Titan lander's surface science package.

Cluster
Participated in the development of the four satellites' 16 plasma sensors.

SOHO
Led development of Coronal Diagnostic Spectrometer.

ISO
RAL managed development of the Long Wavelength Spectrometer.

Spektrum-X
Participation in the JET-X grazing incidence multimirror x-ray telescope.

IUE
Managed UK participation until shutdown in 1996.

Rosat
Participated in development of the Wide Field Camera.

Minisat
Built star tracker for Legri gamma telescope.

UARS
Joint developer of the Improved Stratospheric & Mesospheric Sounder, incorporating a Stirling cycle cooler now available from Matra Marconi Space.

ERS 1/2
ATSR Along-Track Scanning Radiometer developed by RAL-led consortium. Advanced ATSR calibration carried out at RAL. The centre provides data processing for all three.

IRAS
Controlled from RAL's Satellite Control Centre, via the 12 m antenna. The IRAS database is available at RAL for the UK astronomical community.

Ginga
Leicester University/RAL provided the Large Area Counter to Japan's X-ray satellite (re-entered 1991).

Solar Max
Major collaboration in the X-ray Polychromator for studying solar flares.

Spacelab 2
CHASE Coronal Helium Abundance Spectrometer Experiment developed by RAL/Mullard Space Science Laboratory.

UKS
RAL and MSSL satellite and sub-satellite.

Yohkoh
Developed Bragg Crystal Spectrometer with MSSL and US Naval Research Laboratory.

Rosetta
Management and a system engineering for sample analysis instrument on the lander and mother spacecraft to visit comet Wirtanen.

EOS
Management of High-Resolution Dynamics Limb Sounder to measure global temperature and atmospheric chemical species.

Polar
RAL provided hardware on two instruments to study Earth's magnetosphere.

Facilities include:

Clean rooms
A 13 × 8 m room containing 4 × 3 m Class 100 tunnel and three class 100 laminar flow benches; a 90 m² room housing 7.5 × 3.5 m Class 100 working area in form of vertical laminar flow downdraft units.

Thermal vacuum
5.5 × 3.0 m diameter tank with cryopumps down to 10^{-10} atmosphere −190/100°C, plus 1.7 × 1.0 m diameter with turbomolecular pump down to 10^{-10} atmosphere, −170/100°C. Bakeout Tank: 90 × 60 cm diameter with LN_2 traps on diffusion pumps down to 10^{-10} atmosphere. Ambient temperature to +150°C.

Vibration facility
Electromagnetic vibrator with 90 × 90 cm slip table capable of vibrating a 35 kg mass to ESA/NASA specs in three-axes. Max displacement ±19 mm, max rate 1.78 m/s. The facility has a cryostat to fit on the vibrator for testing 30 kg objects at down to 4 K. A second electromagnetic vibrator can test a 5 kg mass to ESA/NASA specs. Maximum displacement ±22 mm, max rate 1.2 m/s.

United States

National Aeronautics and Space Administration, Ames Research Center

Current Status
Ames Research Center (ARC) was founded in 1939 as a National Advisory Committee on Aeronautics (NACA) facility and became part of NASA upon that agency's creation in 1958. ARC is located about 10 miles north of San Jose, at Moffett Field. The Center is named after Joseph S Ames, NACA's former Chair.

ARC's primary research areas and expertise take advantage of its Silicon Valley location; the centre actively partners with high-tech companies and educational institutions in the region. ARC is NASA's lead for both information technology and astrobiology research and development projects. Ames also engages in nanotechnology, biotechnology, aerospace operations systems and thermal protection systems research and development. Through its activities in these diverse spheres, Ames supports NASA's space missions and the Bush Administration's Vision for Space Exploration by developing technologies that make space study and exploration feasible.

Ames maintains the ARC Jet Complex, a system of bays with various facilities that simulate aerothermodynamic heating; the Complex is used to conduct spacecraft thermal protection and heat shield component tests. The ARC Jet Complex has been involved in the development of every NASA planetary and transportation spacecraft Thermal Protection System (TPS) over the last 40 years, including those of the Space Shuttle and Apollo crafts. Other examples of Ames' research and development efforts include investigating robotic inspection vehicles used to examine spacecraft exteriors *in situ*; spacecraft materials radiation hardiness testing, using high-end computing facilities and simulation software specially developed to test and model the radiation damage spacecraft components may suffer while operating in space; and *in-situ* Resource Utilization (ISRU) technologies that will support NASA's mission to Mars, such as the development of atmosphere acquisition equipment.

Ames employs nearly 4,000 personnel, including members of academia. NASA is building a new, NASA Research Park (NRP) on the former 1,500 acre Naval Air Station, Moffett Field site that the agency now controls, while retaining Ames' current facilities. The agency has been seeking and working with partners in this endeavour since 2002.

National Aeronautics and Space Administration, Dryden Flight Research Center

Current Status
Dryden Flight Research Center (DFRC) is located 130 km north of Los Angeles, California on the edge of the 114 km² Rogers Dry Lake at the south end

Artist's illustration of the X-37 ATLV Advanced Technology Demonstrator in flight (NASA) 1151055

of an 800 km high-speed flight corridor at the US Air Force's (USAF) Edwards Air Force Base (AFB). It has 51,800 km² of restricted airspace to conduct aeronautical research with high performance aircraft such as the F-15B, SR-71 Blackbird, F-16XL and the X-36 tail-less fighter.

Dryden is NASA's lead centre for aeronautical research. DFRC supports the X-37 Approach and Landing Test Vehicle (ATLV) and X-43 hypersonic research aircraft programmes. It operates NASA's ER-2 High-altitude Science Aircraft (the U-2S aircraft's civil version) and acts as the Space Shuttle's alternate landing site, using the 8 km to 17 km runways on the dry lakebed and Edwards's main 4,600 m concrete runway.

DFRC is active in Unmanned Aerial Vehicle (UAV) research and development, including NASA's High-altitude, Long-Endurance Remotely Operated Aircraft in the National Airspace System (HALE ROA in the NAS) programme.

Background
DFRC activities began in 1946 with the arrival of five National Advisory Committee on Aeronautics (NACA) X-1 programme personnel at Muroc Army Airfield, the present Edwards AFB, California. The site gained permanent status in 1947 as the Muroc Flight Test Unit and was renamed the Flight Research Center under NASA in 1959. It was renamed in honour of NACA's Hugh Dryden in 1976. DFRC merged in 1981 with NASA Ames, located at Moffett Field, California, but again became a fully independent centre in 1994.

Dryden led the Pegasus Hypersonic Experiment in 1998 and was involved in the development of the X-33 and X-34 reusable launcher programmes until cancellation by the Bush administration in 2001. DFRC supported development of the X-38 Crew Return Vehicle (CRV), which completed its third and final free flight on 30 March 2000. The Apollo Lunar Landing Research Vehicle and Lifting Body family operated out of Dryden in the 1960s to 1970s.

National Aeronautics and Space Administration, Goddard Space Flight Center

Current Status
Goddard Space Flight Center (GSFC) is located in Greenbelt, Maryland, in the Washington, DC area. GSFC was established in May 1959 around a core staff of 157, transferred from the Naval Research Laboratory's (NRL) Vanguard team. It is the only US national facility that can develop, fabricate, test, launch and analyse data from its own space science missions. Goddard's total workforce is about 8,200, including approximately 3,200 civil servants and 5,000 contract personnel. GSFC also manages the Wallops Flight Facility (see separate entry), the Independent Verification & Validation Facility, the Goddard Institute for Space Sciences and the White Sands Complex in New Mexico. Civil service and contract employees at these facilities total about 1,500.

GSFC has managed the development of more than 160 satellites for NASA and the National Oceanic and Atmospheric Administration (NOAA), including the Explorer series, Cosmic Background Explorer (COBE), Compton Observatory, Solar Max, Earth Radiation Budget Satellite (ERBS), Spartan, the Geostationary Operational Environmental Satellites (GOES) series and Upper Atmosphere Research Satellite (UARS). The Center has managed approximately 275 Science missions since its inception.

GSFC's focus centres on Earth science and environmental observation and monitoring missions including the Earth Observing Satellites, the first of which was launched on 21 November 2000. GSFC is responsible for Hubble Space Telescope project management. It also manages programmes such as the Thermosphere, Ionosphere, Mesophere, Energetics and Dynamics (TIMED) mission to study a region of Earth's atmosphere 60 to180 km above the Earth's surface. GSFC developed the Galaxy Evolution Explorer (GALEX) satellite, now managed by NASA's Jet Propulsion Laboratory (JPL), and the Swift satellite launched in 2004.

Goddard directs the operation of NASA's Tracking and Data Relay Satellite Project (TDRS). TDRS' in-orbit segment which uses six Tracking and Data Relay Satellites. The TDRS Ground Network (GN) is in White Sands, New Mexico. Goddard also manages further separate Space and Ground Networks, responsible for NASA spacecraft tracking and operation.

The 8,082 m^2 USD16 million Spacecraft Systems Development and Integration Facility opened in June 1990 for handling Shuttle payloads of up to 27 tonnes. The 1,161 m^2 class 10,000 laminar flow cleanroom is one of the world's largest and can accommodate two full size Shuttle payloads (up to 27,216 kg, 4.572 m diameter) simultaneously. Other facilities include: Large Area Pulsed Solar Simulator (test complete panels), Space Simulation Test Facility, Vibration Facility, Battery Test Facility, High Voltage Test Facility, Magnetic Field Component Test Facility (calibrate/align magnetometers), Spacecraft Magnetic Test Facility (entire craft and sounding rockets), High Capacity Centrifuge, Acoustic Facility, Electromagnetic Interference Facility, Static/Dynamic Balance Facility, Optical Thin Film Deposition Facility, Material Properties and Analysis Laboratories.

NASA's Goddard runs the Getaway Special Program (GAS) which accommodates small, autonomous payloads in the Shuttle's cargo bay utilising standardised hardware. Hitchhiker (HH) offers two carriers providing capabilities beyond GAS. Hitchhiker Junior entered service 1995, using HH avionics and GAS cans but offering satellite ejection (previously from GAS) and limited Orbiter power and services such as pointing.

The canister is an aluminium cylinder provided with a standard experiment mounting plate which, while it may not be altered, does carry provisions for a variety of attachments. GAS can be evacuated and/or pressurised and includes an insulated exterior on the bottom/sides for passive thermal control (an insulated top end cap is available). Operations are independent of the Shuttle Orbiter other than three on/off controls activated by the crew. The experimenter is responsible for providing electrical power, heating/cooling and data acquisition systems. A Motorised Door Assembly is available (satellite ejection no longer is).

National Aeronautics and Space Administration, Jet Propulsion Laboratory

Current Status

The Jet Propulsion Laboratory (JPL) is a federally funded facility operated by the California Institute of Technology (CalTech) in Pasadena, California. JPL has been under contract to NASA since 1958. The Laboratory has been associated with US government research since the 1930s; previous to NASA, JPL reported to the US Army. Although the Laboratory conducted much research in jet propulsion in its early days, JPL has not focused on this type of work for many years. The laboratory's main areas of expertise are planetary exploration spacecraft and research, Earth science research using satellite-borne instruments, space science and observation and space communications. NASA is JPL's primary client, however the Laboratory also has performed work for other US government agencies, such as DoD. JPL's work force totals approximately 4,500 personnel, including contractors. The Laboratory's FY05 total operating budget was about USD1.6 billion.

JPL is responsible for most of NASA's recent deep space missions, including the Mars Exploration Rovers, Spirit and Opportunity; the Mars Odyssey; the Spitzer Space Telescope, and Deep Impact. It has also been responsible for many NASA missions launched in the past, some of which remain in operation. Examples include, Voyager, Galileo, Cassini, the Mars Surveyor and the Mars Pathfinder. It manages NASA's portion of the joint European Space Agency Ulysses solar probe, the US/French Topex/Poseidon oceanographic satellite project, and developed the Wide Field/Planetary Camera for Hubble as well as many Space Shuttle-borne research instruments. JPL supports the Vision for Space Exploration through its numerous Mars and other planetary exploration missions.

NASA's first planetary launch, Mariner 1, lifts off from Cape Canaveral on 22 July 1962. Intended for a fly-by of Venus, the spacecraft was destroyed when the launch vehicle failed due to an error in flight guidance equations (NASA/JPL) 0569815

JPL developed and manages NASA's worldwide Deep Space Network (DSN) for providing links with spacecraft above 10,000 km. DSN comprises three complexes: near Canberra, Australia; Madrid, Spain; and Goldstone, near Barstow, California, (opened in 1958). JPL also runs a launch support facility at the NASA Kennedy Center, and network control and spacecraft compatibility test facilities at JPL. DSN can supplement Goddard's tracking system for Shuttle and selected Earth satellites. The DSN Deep Space Communications Complexes, run by host governments under agreement, each operate a 70 m antenna (S/X-band downlink; 20 kW and 400 kW S-band uplink); a 34 m high efficiency system (full downlink; X-band uplink), a 26 m antenna (S-band up/down) and a 34 m beam waveguide X-band antenna (Goldstone has three of these and Madrid plans to add a second). To boost data reception rates, the Goldstone complex can be arrayed with the 27 25 m Very Large Array dishes of the US National Radio Astronomy Observatory in New Mexico, and Canberra linked with the 64 m Parkes radio telescope. JPL also runs the astronomical observatory at the Table Mountain Facility near Wrightwood, California.

JPL's Spacecraft Assembly Facility (SAF) includes an 890 m^2 high bay area, class 100,000, and a 16.8 m very high bay class 10,000 area of 390 m^2. A flight system testbed opened in 1993 for development-phase integration and testing of small spacecraft prototypes and subsystems, in a 150 m^2 lab near the high bays. Environmental test facilities provide vibration and spin balancing. The 6.1 m diameter $\times$ 7.63 m (working volume) space/solar simulator was built in 1961 and reopened in early 1994 after refurbishment. It provides 5×10^{-6} torr, −196°C to +93°C, solar intensity to 2.7 Suns (5.6 m diameter beam) or 12 Suns (2.2 m beam). A 3 m simulator, added in 1965, and a 93 m^2 class 10,000 assembly room supplement it.

Mission data facilities include the Multimission Image Processing Laboratory (MIPL) based on the digital image processing technology begun at JPL in the early 1960s. The Planetary Data System archives and distributes digital data from planetary missions, ground observations and lab measurements.

National Aeronautics and Space Administration, John H Glenn Research Center

Current Status

Founded in 1941 as the National Advisory Committee for Aeronautics (NACA) Aircraft Engine Research Laboratory, the John H Glenn Research Center at Lewis Field, Cleveland, Ohio, is NASA's lead organisation for aeronautics propulsion, aeronautics technology, space propulsion, space power and microgravity science research and development. It contributes to NASA's Space Exploration Systems priorities primarily through its expertise in these areas. Glenn Research Center is the lead for ion propulsion development; it participated in the Deep Space 1 project and leads the Electric Propulsion Segment (EPS) of the Prometheus programme. Glenn also conducts turbomachinery research. Until 1998 Glenn Research Center managed NASA's Atlas/Centaur launch activities; these responsibilities were then transferred to Kennedy Space Center (see separate entry). Glenn continues to contribute its expertise to International Space Station (ISS) electrical power system design and development. Staff total approximately 2000, including civil service employees and support services contractors.

In 1948 the centre was named the Lewis Flight Propulsion Laboratory, in honor of George W Lewis, the former Director of Aeronautical Research for NACA, NASA's precursor organisation. When NASA was established in 1958, the NASA Lewis Research Center was created. In 1999 the Lewis Research Center (LeRC) was renamed the John H Glenn Research Center at Lewis Field to honour the first American to orbit the Earth while retaining the Lewis heritage. Lewis Field encompasses approximately 350 acres and includes over 150 buildings.

National Aeronautics and Space Administration, Johnson Space Center

Current Status

The Johnson Space Center (JSC), in Houston Texas, was founded in 1961 as NASA's centre for design, development and testing of manned spacecraft. When the facility first began operations, it was known as the Manned Spacecraft Center; it was named after President Lyndon B Johnson in 1973. JSC remains NASA's lead centre for human space flight. Its many capabilities include management of the Space Shuttle Program and the International Space Station Program, studies in advanced human support technology, biomedical research and countermeasures, space medicine, space operations/communications management, Extra Vehicular Activity (EVA) and curatorial care and study of Lunar and Planetary materials. JSC is also home to NASA's astronaut corps, where they train for space flight and support NASA's human space flight programs. JSC's staff is comprised of nearly 3,000 civil servants and approximately 13,000 contractors. The Center is divided into Directorates responsible for specific functions: Flight Crew Operations, Mission Operations, Engineering, Space Life Sciences, Information Resources and Center Operations. The Space Shuttle and International Space Station Program Offices manage those specific programmes. JSC also operates the White Sands Test Facility in New Mexico, a pre-eminent resource for testing and evaluating potentially hazardous materials, space flight components and rocket propulsion.

Most of JSC's 100 buildings on the 6.5 km^2 Houston site provide office space and laboratories. Its research facilities support life, planetary and Earth sciences, artificial intelligence and lunar sample analysis. Specialised training facilities include the Shuttle simulators, a Shuttle Orbiter trainer in the Space Vehicle Mock-up Facility, Precision Air-Bearing Facility and Space Station mock-ups, Space Environment Simulations and acoustic test facilities.

NASA's Johnson Space Center has directed all US manned space flights from the early Gemini missions in 1965 including historic moon walks in the early 1970s (NASA)
1047482

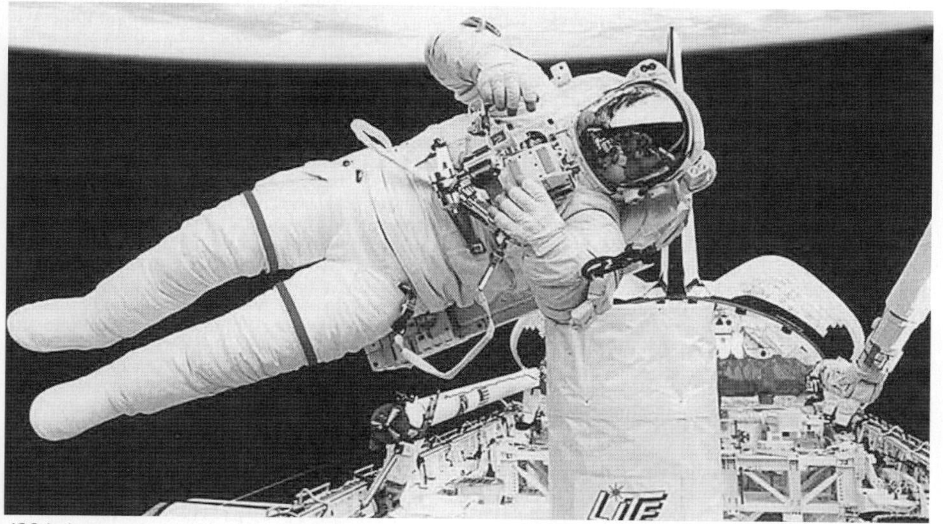

JSC in-house mission support includes crew equipment and aids such as this SAFER system for EVAs (NASA)
1047488

JSC has overall responsibility for the ISS seen here in its May 2004 configuration of partial assembly (NASA)
1047537

NASA's Mission Control Center (MCC) has controlled all NASA manned flights since Gemini IV in 1965. The Flight Control Room (FCR), on the third floor is now an historical landmark. A new wing was added to facilitate new technology for Mission Control – adding two new FCRs, one for the Shuttle and one for the International Space Station (ISS) to facilitate new technology for Mission Control. The new workstations are more efficient requiring less manpower to control the combined Shuttle and Station operations.

JSC trains astronaut candidates on simulators, both fixed and motion based, to prepare crew for space flight. The motion flight deck simulator uses hydraulic actuators for 6 Degrees Of Freedom (DOF) to simulate launch, landing and other dynamic manoeuvers. A mid-deck might be added for entire crews to train together. Integrated simulations with MCC begin some 10 weeks before a flight, concluding with a fifth ascent/entry session two days before the crew flies to Kennedy Space Center (KSC). The two fixed-base simulators provide on-orbit operations training. The detailed aft flight decks cover activities such as Remote Manipulator System (RMS), EVA, rendezvous and payload deployment. Experiments, meals and housekeeping chores are practised in the mid-deck. There are also high-fidelity ISS simulators.

The Space Vehicle Mock-up Facility houses two high-fidelity Orbiter forward sections (flight deck and mid-deck), which can be rotated into launch attitude. Astronauts use it to learn the cabin's layout and engineers make fit-checks of new equipment. A separate full Fuselage Trainer cargo bay mock-up allows a fit-check of payload mock-ups. The Mission Development Facility trains Mission Specialists for RMS operations. Crew also use the Shuttle Engineering Simulator (SES) and the Shuttle Mission Simulator (SMS) for robotic arm training.

The Sonny Carter Neutral Buoyancy Laboratory houses a 23.47 million litre, 31 × 10 × 61.59 × 12.20 m pool capable of accommodating a full-scale Orbiter payload bay and working RMS for crews to simulate microgravity. Space station assembly missions are also simulated in the NBL.

The Lunar Sample Building holds most of Apollo's Lunar samples, as well as meteorite and cosmic dust samples. JSC is the designated curatorial center for all of NASA's extraterrestrial samples.

National Aeronautics and Space Administration, Langley Research Center

Current Status

Located in Hampton, on Virginia's coast, adjacent to Langley Air Force Base (AFB), Langley Research Center (LaRC) was established in 1917 as the Langley Memorial Aeronautical Laboratory under the National Advisory Committee for Aeronautics (NACA). It became one of the four original NASA facilities (with Ames, Edwards, and Lewis) when the agency was created in 1958. LaRC has undergone reorganisation but continues to focus on research in aeronautics, atmospheric science, space technology, and structures and materials science. The Center employs about 3,300 civil service and contractor employees. Langley covers 3.2 km^2 in its West area, plus 0.08 km^2 on its East side and has more than 220 buildings. The NASA Engineering and Safety Center (NESC), founded following the Columbia accident, is based at LaRC. Langley's Atmospheric Sciences Data Center (ASDC) is an Earth Observing System (EOS) Distributed Active Archive Center (DAAC), which carries data from about 35 NASA atmospheric field missions and satellites.

LaRC is primarily a research centre for advanced aerospace technology. Major research fields include aerodynamics, materials, structures, flight controls, information systems, acoustics, aeroelasticity, atmospheric sciences and non-destructive evaluation. About 35 per cent of LaRC's work supports space activities, including technology and instrumentation for advanced space transportation and large space structures, such as the Space Shuttle, International Space Station (ISS), and the Clouds and the Earth's Radiant Energy System (CERES). The Center plays a strong role in the Cloud-Aerosol Lidar and Infrared Pathfinder Satellite

Observations (CALIPSO), leading NASA's systems engineering and payload mission operation responsibilities for the joint Centre National d'Etudes Spatiales (CNES) French-US programme. LaRC is developing the Aerial Regional-Scale Environmental Survey (ARES) of Mars aircraft, an Unmanned Aerial Vehicle (UAV) designed to withstand Martian atmospheric conditions and relay environmental data back to NASA.

Background

LaRC has played an active role in NASA's space technology research and development efforts for many years. The US manned space programme began at LaRC in 1959 as the Space Task Group, which completed its transfer to new dedicated facilities in Houston in mid-1962. The Center made contributions to the highly successful Lunar Orbiters and Viking Mars orbiters/landers.

Langley's Light Detection and Ranging (Lidar) In-Space Technology Experiment (LITE) during STS-64 in September 1994 detected stratospheric and tropospheric aerosols, probed the planetary boundary layer and measured cloud top heights. The 1 m diameter telescope with three-wavelength neodymium doped yttrium aluminum garnet (Nd: YAG) laser was a testbed for future operational spaceborne lidars. The Cloud-Aerosol Lidar with Orthogonal Polarization (CALIOP) aboard CALIPSO uses similar technology. The Lidar Atmospheric Sensing Experiment (LASE), the first autonomous aircraft-based lidar to measure water vapour, completed final validation studies aboard NASA's ER 2 in September 1995.

The centre manages the Halogen Occultation Experiment (HALOE) aboard the Upper Atmosphere Research Satellite (UARS) to monitor the vertical distributions of ozone and key trace gases. The Measurement of Air Pollution from Satellites (MAPS) instrument flew several times on the Shuttle to measure the global distribution of carbon monoxide in the free troposphere; it operated for one year aboard Mir from June 1996. Langley's Earth Radiation Budget Experiment continues operations on the ERBS satellite (which also carries LaRC's Stratospheric Aerosol Gas Experiment (SAGE) instrument) and NOAA 9 and 10.

LaRC leads NASA in developing high energy, high efficiency, long-life solid-state lasers for advanced lidar and other applications requiring stable laser sources; it also develops new technologies for future systems, including solid-state lasers, detectors, lightweight optics, deployable telescopes, miniature sensors and onboard data processing.

LaRC provided several experiments during the 1990s for Russia's Mir station. The Mir Environmental Effects Payload (MEEP) was attached to the docking module by STS-76 in March 1996, where it remained until late 1997 collecting micrometeoroid particles and the effects of strikes on Station Alpha materials. The Enhanced Dynamic Load Sensors set up carried Mir's Priroda module to measure the effects of crew movement. The Materials in Devices As Superconductors (MIDAS) experiment remained aboard for several months after delivery by STS-79 in August 1996. Four circuit boards, including one with Russian samples, measured effects on the electrical and magnetic properties of high temperature superconductor materials.

LaRC had the lead in developing the composite primary structures and the advanced composite thermal protection systems for the cancelled X-33 and X-34 programmes. It helped the industry teams to develop reusable cryogenic tanks and advanced propulsion, in addition to vehicle systems analysis, aerodynamic and aerothermodynamic testing/analysis, and flight control development. It performed wind tunnel testing of an X-34 model mated to a Boeing 737.

National Aeronautics and Space Administration, Marshall Space Flight Center

Current Status

Marshall Space Flight Center (MSFC) was established in 1960 around a rocket team headed by Dr Wernher von Braun, the team which, as part of the US Army Ballistic Missile Agency (ABMA), orbited the Explorer 1, the first US satellite.

Rocket engines developed at the Marshall Space Flight Center and elsewhere, tested in giant cells like these, were the lifeblood of the US civilian launch industry since the early 1960s (MSFC) 1047489

Named after General George C Marshall, the creator of the "Marshall Plan," MSFC is located on a 7.3 km^2 tract within the US Army's Redstone Arsenal in Huntsville, Alabama. Also operated by Marshall is the Michoud Assembly Facility in Louisiana, occupying a 3.37 km^2 site where Shuttle external fuel tanks are manufactured and shipped to Florida by barge. Marshall's employees total approximately 2,400 plus over 3,000 contractors engaged in mission support work.

MSFC is the lead NASA centre for space transportation and propulsion systems research. It is charged with technology maturation, design, development and integration of space transportation and propulsion systems. This includes both reusable and expendable launch vehicles, as well as vehicles for orbital transfer and deep space missions. MSFC has reorganised and hopes to play a significant role in the Bush Administration's Vision for Space Exploration, announced in January 2004. Predicated to a large extent by the need to retire the Shuttle, the Vision's objectives appeal to the core duties of MSFC and the work for which the NASA Center was originally established. Marshall developed and continues to provide the Space Shuttle's main engines, solid rocket boosters and external tank. MSFC provides support to the International Space Station programme with engineering personnel and facilities to develop pressurised modules, support equipment and assigned payloads.

Marshall's Science and Technology Directorate conducts Earth science, space science and microgravity research. MSFC has lead responsibility for the Chandra X-Ray Observatory programme and partners with the Chandra X-Ray Center (CXC) at the Smithsonian Astrophysical Observatory (SAO) to operate the observatory and examine and archive the data Chandra collects. Marshall also manages the Gravity Probe B and other smaller scientific payloads and instruments and conducts research in high-energy astrophysics, solar magnetic fields, and low-energy space plasma physics.

MSFC supports the Prometheus programme's spacecraft nuclear propulsion and power research as well as NASA's Crew Exploration Vehicle (CEV) development. The Center manages the Demonstration of Autonomous Rendezvous Technology (DART) programme, and is set to investigate the failure of the unmanned rendezvous craft to complete its mission after its successful launch in April 2005.

Historic programmes in which MSFC played a large role include Skylab; Saturn V, the launch vehicle that carried the Apollo missions; Mercury-Redstone; and Spacelab.

National Aeronautics and Space Administration, Stennis Space Center

Current Status

Stennis Space Center (SSC), on the Mississippi Gulf Coast, is NASA's primary rocket propulsion test facility. Previously known as the National Space Technology Laboratories (NSTL), SSC was named after US Senator John C Stennis in 1988. SSC's total land area is 562 km^2, of which 55.8 km^2 constitute the operational base and the remainder is held as an acoustic buffer zone. The site was selected in 1961 to accommodate Apollo's F-1 and J-2 engines and Saturn V stages, testing a total of 27 stages. The NASA-related workforce totals about 250. However, 30 other federal and state agencies, private firms and educational institutions occupy the Centre as tenants, including the Naval Meteorology and Oceanography Command and a Department of Energy (DoE) Strategic Petroleum Reserve (SPR) office, bringing the workforce to about 4,000.

All Space Shuttle Main Engines (SSME) undergo acceptance testing at Stennis before flight. The first SSME test was made during June 1975. Two test stands handle single SSME trials. Stand A-1 evaluates sea level performance; A-2 simulates firing between 16.4 and 21.3 km altitude, operating with the engine 18° from the vertical to mimic the installation on the launch pad. A-1 and A-2 were both built in the 1960s for Saturn V second stage testing and can each handle 4.89 MN force, 10.07 m diameter test articles. Stand B-1 was modified and is used to test the Boeing-Rocketdyne RS-68 engine that powers the Delta IV launch vehicle. The B-2 stand was where three-engine SSME cluster testing was performed; it originally handled Saturn V's first stage and as such can cope with 48.9 MN force, 10.07 m diameter test articles. It is the largest rocket test stand of its kind in the US, however it has not been used for several years.

The three-cell E-1 Test Facility can handle engines using liquid hydrogen LH_2 or hydrocarbons, solid or hybrid, high flow rates, high pressure cryogenics and ultra-high pressure gases. The state of the art complex is used for testing hardware for future launch vehicles. The two-cell E-2 Test

A technician prepares an SSME for tests (NASA) 0572501

Local crowds gather to experience the exhilaration of a live Shuttle engine test (NASA) 0572502

Facility supports component testing similar to E-1 but at about 1/10 scale. The two-cell E-3 facility supports even smaller scale component tests utilizing hydrogen peroxide, hydrocarbon or liquid oxygen propellants.

Stennis also plays a large role in NASA's Earth Science Enterprise and is home to the Applied Sciences Directorate. The Directorate uses NASA's unique Earth science research results, data, remote sensing and other technical capabilities to analyse and solve practical problems, allowing managers to make more informed decisions on such national applications as disaster management, national security issues, coastal management, ecological forecasting and agricultural efficiency.

SATELLITE OPERATORS

Australia

Telstra Corp Ltd

Current Status
In 2005, the Australian government announced its plan to offer up for sale much of its 51.8 per cent interest in Telstra. After legislative approval of the plan, the sale to investors to further privatise the company, by purchasing AUD8 billion worth of the Australian government's AUD22.5 billion stake in Telstra, was set to begin in November 2006. The remainder of the government's Telstra shares will be migrated to the Australian Future Fund and sold over time.

Background
Telstra is Australia's majority share government-owned telecommunications company. It was formed in 1992 by the merger of Overseas Telecommunications Commission (OTC) and Telecom Australia and began trading under the name Telstra in 1993. Telstra was Australia's signatory to Intelsat, owning roughly 1.7 per cent in FY2004. It was also Australia's Inmarsat investor. In May 2000, Telstra sold its global satellite business to Xantic for an AUD170 million stake in the company; this included Telstra's Inmarsat shares, which in FY1999 amounted to about 2 per cent. Telstra sold its Intelsat investment in January 2005, for AUD69 million.

Brazil

Instituto Nacional de Pesquisas Espaciais (INPE)

Current Status
The National Institute for Space Research (INPE) plans to launch its joint China-Brazil Earth Resources Satellite, CBERS-2B, from China aboard a Long March 4B (see separate entry) in 2006 or 2007; its CBERS-3 satellite will follow in 2008 and CBERS-4 in 2011. CBERS-2 remains operational and continues to provide INPE with imagery. NASA and the US Geological Survey have expressed interest in receiving CBERS imagery to augment Landsat data.

Background
Headquartered in São José dos Campos, São Paulo, INPE was created in 1971 under the Ministry of Aeronautics, to succeed the National Space Activities Commission created in 1961. It became an organisation within the newly created Ministry of Science and Technology in 1985, and is responsible for the development of the ground and space segments of Brazil's satellite programmes. As such, it is developing the five Brazilian Complete Space Mission (MECB) satellites and the Brazilian contribution to the four joint Sino-Brazil CBERS remote sensing satellites, along with several other satellite projects. INPE has expanded remote sensing activities and integrates satellite operations with ground truth data. The Institute operates its Centro de Rastreio e Controle de Satélites (CRC) Satellite Tracking and Control Centre at São José dos Campos (23° S/46° W); its mission centre for satellite data collection (SCD) in Cachoeira Paulista (23° S/45° W, 80 km north of São Paulo); Brazil's Landsat and CBERS station in Cuiaba (15° S/56° W); and an installation at Comando-Geral de Tecnologia Aeroespacial's (CTA – see separate entry) sounding rocket launching station, Centro de Lançamento da Barreira do Inferno, in Natal (5° S/35° W). CBERS satellites are controlled from the CRC, with tracking stations provided in Alcântara and Cuiaba, which also receives and records the CBERS images.

INPE's research and management areas include satellite tracking and control, atmospheric and space science, space technology development, Earth observation, and spacecraft integration and testing. Further INPE interests include propulsion (including bipropellant and monopropellant catalytic thrusters), space materials, solar cells, sensors, computer sciences, meteorology, remote sensing and science scanning technology development. INPE has formed many co-operative partnerships with space agencies and organisations of various countries including Argentina, China, France, Germany and Japan. It also works with the Brazilian Space Agency (see separate entry) and numerous Brazilian institutions.

INPE's 10,000 m^2 Laboratory for Test and Integration (LIT) opened in 1988 as the largest such facility in the southern hemisphere. It includes a 1,600 m^2 work hall, a 450 m^2 integration room, a control room, 13 and 80 kN shakers, 22 m^3 thermal vacuum chamber and electromagnetic test chambers; Intespace added an acoustic chamber in 1997. CBERS-2 was assembled here. All system qualification testing of Argentina's SAC-B satellite was also undertaken here; contact with the satellite was maintained through the Comisión Nacional de Actividades Espaciales (CONAE's – see separate entry) ground station at San Miguel, Argentina.

INPE began operating a 14.2 m Very Long Baseline Interferometry (VLBI) antenna in 1993 at Eusebio, in northeastern Brazil to link South America geodetically with other continents as part of the International Interferometry Survey, co-ordinated by the International Earth Rotation Service. It also has a GPS station to compare results with VLBI-derived parameters.

China

China Telecommunications Broadcast Satellite Co

Current Status
ChinaSat has organised and set up China Space Moblie Satellite Telecommunications Co Ltd for operating global satellite mobile telecommunications services and Beijing Spacenet Information Telecommunications Pte Ltd for operating satellite broadcast services. These plans have run into problems likely to delay the initiative, not least because of uncertainty in technical standards.

Background
China Telecommunications Broadcast Satellite Co or 'ChinaSat' is the state satellite operator and a wholly owned subsidiary of the Ministry of Posts and Telecommunications. Capacity is also provided by AsiaSat, which is partly owned by China. ChinaSat was established under the Ministry of Radio, Film and Television in 1983 to provide satellite based TV services to rural China.

Official estimates claimed more than 40,000 TVROs by 1993. The government in October 1993 announced restrictions on the manufacture, sale and installation of terminals, but it remained to be seen how strictly the regulations would be enforced in such a huge market. A 1994 estimate claimed 500,000 home dishes. There were 2,600 VSATs in operation by 1992, rising to more than 8,000 by 2002. Hughes Network Systems in 1993 provided 300 VSATs to the Bank of China, working through AsiaSat 1. MPT lost its monopoly in January 1994.

MPT is China's signatory to Intelsat, and the Beijing Marine Communications and Navigation Co to Inmarsat.

Preparations for the STTW-T1 included a 1983 agreement to employ Italy's Sirio 1 for communications trials, and a USD20 million March 1984 contract to Spar Aerospace for 26 Earth stations. 15 were in operation by October 1986 at Lhasa, Tibet, Guangzhou and Hohhot, linking off-shore oil/gas production facilities by voice/data. Spar received a further USD36 million contract mid-1991 for 13 m Intelsat A stations at 10 new locations and enhancements at seven existing sites. Spar signed a letter of intent 8 November 1994 with CASC, China National Railway Signal and Communications Corp, and Space Wit Communications Co Ltd to develop Earth stations and terminal networks for voice/data. A new Shanghai station opened August 1992 to provide European links via Indian Ocean satellites. A new Intelsat TT&C and Monitoring station was inaugurated at Beijing in October 1992 as part of the organisation's worldwide network. Scientific-Atlanta in September 1995 was awarded a USD3.7 million contract by MPT for an emergency satellite communication network to support relief operations. The initial phase of the Skylinx DDS Digital DAMA telephony network became operational in early 1996 using AsiaSat 2 Ku-band capacity. It employs a 4.5 m master station in Shanghai working with 3.6 m stations in seven provinces and 37 mobile 1.2/1.8 m VSATs. The service will eventually be expanded to include a 3.6 m station in all 22 provinces, with at least five mobile units per province. A USD9.9 million contract for the Ministry of Foreign Trade and Economic Co-operation installed hubs in Beijing and Shenzhen working to almost 60 VSATs in trading locations for financial and customs communications.

Studies are under way of a GMSIS Global Mobile Satellite Information System, which would provide personal hand-held communications via 18 to 24 satellites in medium orbits. Three or four of the 500 kg satellites could be delivered by a single CZ3A. The Twin Star proposal would use two satellites some 40° apart in the GEO arc to provide positioning, messaging and timing services, after the fashion of OmniTracs and EutelTracs.

India

Indian National Satellite System

Current Status
The Insat system uniquely provides geostationary platforms for simultaneous domestic communications and Earth observation functions. The four first-generation Insat 1 satellites were all US-built, but the advanced Insat 2 and Insat 3 series is indigenously produced. The Insat 2 series were launched between July 1992 and April 1999 while the first Insat 3 was launched in March 2000. The Insat system is a joint venture of India's Department of Space, Dept of Telecommunications, India Meteorological Dept, All India Radio and Doordarshan. Overall co-ordination and management rests with the inter-ministerial Insat Coordination Committee (ICC); the DOS has direct responsibility for establishing and operating the space segment. Insat 3B was launched in March 2000 followed by Insat 3C in January 2002, both launched by Ariane 5. They have design lifetimes of 14 years. These were followed by Insat 3A on 10 April 2003 and Insat 3E on 28 September 2003 both launched by Ariane 5 and will be followed in 2005 to 2006 by Insat 4A and 4B, advanced meteorological satellites. Insats 4C and 4D are scheduled for launch in 2006 to 2007 followed by 4E by the end of 2008.

The satellites are handled from the Insat Master Control Facility (MCF) at Hassan, Karnataka through two Satellite Control Earth Stations (one with a 14 m fully steerable antenna and the other with a 7.5 m limited steering dish), one additional 14 m fully steerable antenna and an Insat 1 Satellite Control Centre (SCC) with associated TT&C equipment, on-orbit checkout equipment, computer facilities and auxiliary power services. MCF was upgraded with the addition of the Insat 2 SCC and two associated 11 m Satellite Control Earth Stations.

International

ArabSat

Current Status
Arabsat expects to launch Arabsat-4C, also known as Badr-6, in mid-2008 from the Guiana Space Center. The organisation has also announced

its intention to launch three Arabsat-5, or fifth generation satellites, beginning in 2009. Badr-5, with its anticipated launch in 2009 will also be placed in Arabsat's 26°E slot. Arabsat-5A and Arabsat-5C, also planned for 2009 flights, will be placed in 30.5°E and 20°E locations, respectively. EADS Astrium and Alcatel Space built the Arabsat fourth generation.

Background

The Arab Satellite Communications Organization was established as an intergovernmental organisation in 1976 by the 20 member Arab League. It provides telecommunications, data, TV, radio and broadband services to the Middle East and North African regions for 130 million viewers. Arabsat is directed by three internal departments: the General Assembly, the Board of Directors and the Management Committee. As of 2007, Arabsat, and the Arab League, had 22 members and one observer state (Eritrea), each represented in the organisation by its national communications ministry. Arabsat's capital, paid by its member states, amounted to USD163 million in 2006. Saudi Arabia, Kuwait, Libya and Qatar contribute the most financially to Arabsat, at 36.66, 14.59, 11.28 and 9.81 (72.34) per cent of the total, respectively. The United Arab Emirates (UAE), Jordan, Lebanon, Bahrain and Syria collectively contribute 17.07. Iraq, Algeria, Yemen, Egypt, Oman, Tunisia, Morocco, Mauritania, Sudan, Palestine, Somalia and Djibouti contribute the remaining 10.59 per cent. According to Arabsat, at present Comoros, a member since 1993, and Eritrea do not contribute funding.

Arabsat operates a five-satellite fleet: Arabsat-2B; Badr-C (also known as Arabsat-2C, Intelsat 5, IS-5 and PAS-5), Badr-2 (formerly known as Hot Bird 5, Eurobird 2 and Arabsat-2D); Badr-3 (formerly known as Arabsat-3A and Arabsat-2-BSS); and Badr-4 (Arabsat-4B). The latest fleet member, Badr-4, was successfully launched aboard a Proton from Baikonur in November 2006. It should serve Arabsat for 15 years. Arabsat-4A, also known as Badr-1, was launched in February 2006, but failed to reach its designated orbit; the satellite was declared a total loss, and was deorbited in March 2006. Arabsat-2B and Badr-C will be replaced by Badr-4 and Badr-6, respectively.

Arabsat launched its first two satellites in 1985; both were removed from service in 1992. The third satellite, the last of Arabsat's first generation, was launched in 1992, extending operations until the first in the second generation, Arabsat-2A, could appear in July 1996, launched by an Ariane 44L. It was followed by Arabsat-2B in November 1996, and Arabsat-3A (known before that as Arabsat-2-BSS, and now known as Badr-3) in February 1999. Arabsat-2C was launched by PanAmSat in August 1997 and was known as PAS-5 at the time; this satellite is leased by Arabsat and is now known as Badr-C, Intelsat 5 or IS-5. Arabsat leases Badr-2 from Eutelsat. 26°E is Arabsat's primary orbital slot, however Arabsat-2B operates from 30.5°E. While 20°E is reserved for Arabsat, it has not yet placed a satellite in this location, but plans to operate Arabsat-5C from the slot beginning around 2009.

Arabsat's Badr series should not be confused with Pakistan's Badr science microsatellite series (see separate entry).

Argos

Current Status

Argos is a joint programme of CNES, NASA and NOAA, operated under an MoU signed in 1974 and extended in 1986. It provides global data telemetry and geo-positioning services. Presently a candidate for expanded global environmental data monitoring activity with potential for integrating advanced position fixing. It was first developed under a Memorandum of Understanding between Centre National d'Etudes Spatiales (CNES), NASA in the USA and NOAA also in the USA. System operation and promotion is the responsibility of CLS, established in April 1986 as a subsidiary of CNES (55 per cent), IFREMER (French Marine Institute; 15 per cent) and a pool of banks. CNES operates the French Global Processing Center (FRGPS) in Toulouse and handles all user relations outside North America. US subsidiary Service Argos Inc is responsible for all US/Canadian users; it operates the identical USGPC, which provides redundancy. North American CLS is concerned with developing new services and products. The Australasian regional processing centre began operations in December 1988.

The system consists of independent user Platform Transmitter Terminals (PTTs, down to 25 g and the size of a matchbox), two NOAA polar satellites with Argos Data Collection and Location System packages to receive PTT messages on a random access basis for separation, time coding, formatting and retransmission to ground stations, and the ground stations and two Global Processing Centers in Toulouse and Landover, where data are retrieved, processed and distributed to users. Location of PTTs can be determined to within typically 150 m, if required, by Doppler techniques.

The PTTs transmit data from up to 32 sensors encoded in 8-bit words on 401.650 MHz with typically 200 mW power at intervals of 60 to 120 seconds for location PTTs and >200 s for data collection-only platforms. Message length is 360 to 920 ms and contains the PTT identification number.

The principal Argos system processes and disseminates space and terrestrial environmental data received by dedicated packages aboard US NOAA polar meteorological satellite from fixed/mobile platforms anywhere in the world. More than 5,000 transmitters are operating; the past five years has seen a 15 per cent growth in activity. More than 23,000 transmitters have used the system.

The NOAA Argos packages receive all messages within a 5,000 km diameter visibility circle at any instant; four can be processed simultaneously by NOAA A-J and eight will be handled by the third generation system aboard NOAA K and its successors from 1996. NOAA 11 can handle data at 960 bit/s, NOAA J 1,200 bit/s and NOAA K-N 2,560 bit/s. NOAA K-N increase bandwidth from 24 kHz to 80 kHz. The data are formatted/stored, then dumped each time the satellite moves within reach of one of the three ground stations. VHF/S-band transmitters also perform real-time relay for any user station within the visibility circle. Current onboard packages can process up to 1,400 data-only PTTs, or 415 requiring the location service, but from NOAA K on the capacity will be quadrupled.

Under an agreement signed in April 1990 with EUMETSAT, CLS manages all non-meteorological applications using the Meteosat data collection function. The MAEDS Multisatellite Applications Extended Dissemination Service adds Meteosat and GOES to the existing Argos system to provide high data collection capacity (649 byte messages up to 24 times daily), real-time access to data, and continuous coverage from the Pacific to the Urals. The first MAEDS element became operational in October 1990, followed in 1991 by the GOES satellites covering America and the Pacific. CLS also operates the Doris location system and is involved in the Starsys system.

Global Processing Centers NOAA's tape recorders are read out every 100 minutes over one of the Gilmore Creek (Alaska), Wallops Island (Virginia) and Lannion (France) stations. All of the satellite's data are routed to NOAA's National Environment Satellite, Data and Information Service (NESDIS) in Maryland, where the Argos component is extracted and relayed to CLS in Toulouse and Landover. Here, PTT locations are determined and the sensor data processed. About two-thirds of the processed messages are provided to users within 3 hours of the uplink. Location is derived from measuring the Doppler shift in the carrier frequency of received messages during the 10 minute overflight. An accuracy of 150 m is typical but for PTTs at fixed positions or on slow-moving carriers 100 m is possible. The transmit frequency must remain stable throughout the pass. Each frequency measurement yields a set of possible positions and with a known elevation only four messages are required on each pass to generate a unique solution. The satellites' own positions are known to within 300 m at any instant from a reference system of 11 beacons worldwide at precisely known geodesic positions.

Cospas-Sarsat

Current Status

Cospas-Sarsat is phasing out emergency beacons operating in the 121.5/243 MHz range in favour of the 406 MHz Emergency Position-Indicating Radio Beacons (EPIRB), which provide more accurate and reliable alert data. There are an estimated 429,000 406 MHz EPIRB in use. After 1 February 2009, the 121.5 MHz analogue beacons will be obsolete, and all users should have transferred by that time to registered, digital 406 MHz EPIRB. Cospas-Sarsat estimates that there are approximately 556,000 121.5 MHz units in use. The organisation has promised to ensure that technology will also be available for potential interface with the future Galileo (see separate entry) satellite navigation system for accurate position fixing information modulated on to the carrier signal. Guidance originating from the two United Nations organisations, the International Civil Aviation Organization (ICAO) and the International Maritime Organization (IMO), prompted Cospas-Sarsat to impose the change.

Background

Cospas-Sarsat, the International Satellite System For Search and Rescue (SAR), was established in 1979 by Canada, France, the US and the USSR (the membership of which has been assumed by the Russian Federation). The system was governed by a Memorandum of Understanding (MoU) from October 1984 until the International Cospas-Sarsat Program Agreement was signed by the four original member states on 1 July 1988 in Paris. Cospas-Sarsat headquarters are located in Montreal, Canada. As of 2007, 40 participating states contributed resources to and used the co-operative SAR ground and space infrastructure.

The Cospas-Sarsat (Space System for the Search of Vessels in Distress – Search and Rescue Satellite-aided Satellite Tracking System) system was originally based on the detection of distress beacons relayed by four polar satellites. As of 2007, Cospar-Sarsat was operating SAR system payloads aboard a 13 satellite fleet. Eight of these are polar Low Earth Orbit (LEO) and five are geostationary (GEO), comprising the LEOSAR and GEOSAR constellations, respectively. Several of the polar satellites are only partially operational. The satellites are capable of locating an activated beacon to within 5 km and the system was credited with helping to save more than 20,531 lives by December 2005.

The US NOAA; EUMETSAT; and India provided the GEOSAR satellites. Having launched in 1995, GOES-9 is scheduled for retirement in 2007. GEO satellites cannot pinpoint a beacon source but must rely on position information encoded in distress signals. GOES 7 detected its first real mayday 15 June 1988, 75 minutes before the LEO satellites responded. 406 MHz GEO relays are included aboard the GEO satellites; 121.5 MHz is not supported.

Current and Planned Cospas-Sarsat Satellites and Payloads

Constellation	Satellite	Payload	Launch	Status
LEOSAR	Nadezhda-1	Cospas-4	July 1989	Partially operational
LEOSAR	Nadezhda-6	Cospas-9	June 2000	Operational
LEOSAR	NOAA-14	Sarsat-6	December 1994	Operational
LEOSAR	NOAA-15	Sarsat-7	May 1998	Partially operational
LEOSAR	NOAA-16	Sarsat-8	September 2000	Partially operational
LEOSAR	NOAA-17	Sarsat-9	June 2002	Operational
LEOSAR	NOAA-18	Sarsat-10	May 2005	Operational
LEOSAR	METOP-A	Sarsat-11	October 2006	Operational
LEOSAR	Dedicated satellite	Cospas-11	2007 (planned)	Pre-launch
LEOSAR	Dedicated satellite	Cospas-12	2008 (planned)	Pre-launch
LEOSAR	NOAA-N	Sarsat-12	2008 (planned)	Pre-launch
GEOSAR	GOES-9	406 MHz relay	May 1995	Operational; 2007 decommission
GEOSAR	GOES-10	406 MHz relay	April 1997	In-orbit spare
GEOSAR	GOES-11/ GOES-West	406 MHz relay	May 2000	Operational
GEOSAR	GOES-12/ GOES-East	406 MHz relay	July 2001	Operational
GEOSAR	INSAT-3A	406 MHz relay	April 2003	Operational
GEOSAR	MSG-1	406 MHz relay	August 2002	Operational
GEOSAR	MSG-2	406 MHz relay	December 2005	MSG-1 back-up
GEOSAR	INSAT-3D	406 MHz relay	2007 (planned)	Pre-launch
GEOSAR	Electro-L 1	406 MHz relay	2008 (planned)	Pre-launch

The satellites in the fleet relay distress signals to Local User Terminals (LUT) for processing to determine a distress beacon location. The information is passed to a regional Mission Control Centre (MCC) to alert local rescue authorities, or Rescue Co-ordination Centres (RCC). Many 121.5 MHz beacons are still operational, but these are restricted to real-time relay by the satellite and therefore require a LUT to be within about 2,500 km range for position determination to within 10 to 15 km. The 406.025 MHz units include user identification codes in the message but they still operate on the Doppler shift concept: this provides two locations for each distress signal, an ambiguity resolved by allowing for the Earth's rotation.

If the beacon's frequency stability is sufficient, as with the 406 MHz devices, the true solution is determined on a single satellite pass. The 406 MHz information is not only relayed in real-time but also time-tagged and stored for dumping as each LUT comes into view. This frequency therefore provides a global service with an average waiting time of 44 minutes, with 94 per cent detected in more than 90 minutes (results from 1990 exercise). All satellite downlinks operate at 1.5445 GHz.

The 406 MHz beacons emit a 5 W RF burst of about 500 m/s duration every 50 seconds. The improved frequency stability 10^{-9}/min (a factor of two relaxation was permitted until 1991) assures more precise location accuracy (90 per cent within 5 km), while the high peak power increases the probability of detection on a single pass to more than 98 per cent. Typical delay before detection is 1 to 2 hours near the equator (less at the poles).

Each satellite can handle 90 simultaneous signals. The digital encoded message conveys the country of origin and identification of the vessel or aircraft. Position information from systems such as GPS, Glonass and Galileo are encoded in the messages of new 406 MHz beacons. Depending on the beacon type (maritime, aviation or land), the unit can be activated manually or automatically by shock or immersion. An optional homing device is usually collocated with the 406 MHz beacon for the SAR services but it is not a system specification. Beacons are required to function within a –20/55°C thermal range for a minimum 24 hours and automatically limit any inadvertent continuous transmission to more than 45 seconds.

The USSR began deploying the Cospas-Sarsat space segment with the launch of Kosmos 1383 in 1982. Designated Cospas 1 ('Space System for the Search of Vessels in Distress'), the 121.5 MHz band remained operational until Mar 1988, with 406 MHz utilized primarily for interference monitoring. The other relays to bring the system up to its full operational complement of four were provided by packages on the US NOAA meteorological satellites. Sarsat 1 (S&R Satellite-aided Tracking) was implemented with NOAA 8 in 1983.

European Organisation for the Exploitation of Meteorological Satellites (EUMETSAT)

A EUMETSAT MSG satellite (EUMETSAT) 1047513

Current Status

The first Meteosat Second Generation (MSG) geostationary satellite, Meteosat-8 (formerly MSG-2), was successfully launched on 28 August 2002, by Ariane 5 from Kourou, French Guiana. MSG satellites, of which there will be four in all, are manufactured by Alcatel Alenia Space. The MSG satellites are expected to remain operational until at least 2018 when the Meteosat Third Generation (MTG) satellites will be introduced. Meteosat-9 (formerly MSG-2) was launched on 21 December 2005; it is Meteosat-8's back-up satellite. MSG-3 will launch in 2009, followed by MSG-4 in 2011. Each MSG has a nominal lifespan of seven years. Meteosats 5–7, belonging to the Meteosat First Generation, remain operational as of early 2007, although the satellites no longer orbit at 0°; MSG satellites have replaced First Generation satellites in the equatorial latitude. All three First Generation satellites will soon close out their service; Meteosat-7, the last in the series, is expected to remain functional through about 2008, when the First Generation will have been completely replaced by the MSG series.

Developed in co-operation with ESA, the Meteorological Operational (MetOp) satellite series, or the EUMETSAT Polar System (EPS), are polar orbiting satellites providing 'morning' coverage. The US National Oceanic and Atmospheric Administration (NOAA) co-operates with EUMETSAT and ESA in this programme, known as the Initial Joint Polar System (IJPS), by providing 'afternoon' satellite coverage. MetOp-A (formerly MetOp-2) was launched on 19 October 2006 from Baikonur; It will be followed by MetOp-B in 2010 (formerly MetOp-1), currently in storage; MetOp-3 will launch around 2015. EADS Astrium builds the MetOp satellites, which each have a nominal lifespan of five years. MetOps observe global meteorological, climatological and environmental features and transmit regional data to user stations throughout the world; stored global data is sent to central facilities in Europe for further processing and distribution.

EUMETSAT has also teamed with the French CNES, NOAA and NASA on the Ocean Surface Topography Mission (OSTM) Jason-2 marine meteorology satellite. Jason-2, set to launch in June 2008 from Vandenberg, will replace Jason-1, a joint NASA-CNES programme.

Background

EUMETSAT provides meteorological data to its member states, co-operative partner states, and other users; it is an intergovernmental organisation created through an international convention agreed by 18 original member states: Austria, Belgium, Denmark, Finland, France, Germany, Greece, Ireland, Italy, Luxembourg, the Netherlands, Norway, Portugal, Spain, Sweden, Switzerland, Turkey and the UK. Since that time, two more states have become full members, increasing membership to 20: Croatia and Slovakia. EUMETSAT also works with ten co-operating states: Bulgaria, the Czech Republic, Estonia, Hungary, Iceland, Latvia, Lithuania, Poland, Romania, and Slovenia. Co-operating states retain the same data usage rights as full members, however their financial contribution is 50 per cent that of full member fees and they have no organisation governing rights.

The EUMETSAT Convention came into force 19 June 1986, with the organisation assuming overall and financial control of the Meteosat Operational Program (MOP) on 12 January 1987. Its headquarters are located in Darmstadt, Germany. A Council providing one seat per member state controls EUMETSAT. The Council represents the national meteorological services of all member European states. Three major departments make up EUMETSAT: Programme Development, Operations, and Administration.

EUMETSAT staff are drawn from member nations. The majority of staff are German (28.3 per cent), British (16.4 per cent), French (15.5 per cent), and Italian (11.9 per cent). As of 2005, overall, Germany contributes 22.73 per cent of the total EUMETSAT budget, followed by the UK at 16.23, France at 15.81 per cent, Italy at 12.78 per cent, Spain at 6.47 per cent, and the Netherlands at 4.33 per cent. Other member nations and co-operating states contribute to the balance. EUMETSAT's 2005 budget was EUR291.67 million, 92 per cent of which was dedicated to its satellite programmes.

The MSG Mission Control Centre (MCC) is located at the EUMETSAT headquarters in Darmstadt. The centre comprises four main MSG facilities, the Image Processing Facility (IMPF); the Meteorological Products Extraction Facility

For details of the latest updates to ***Jane's Space Systems and Industry*** online and to discover the additional information available exclusively to online subscribers please visit **jsd.janes.com**

Eumetsat headquarters buildings (EUMETSAT) 1047514

(MPEF); the Data Acquisition and Dissemination Facility (DADF); and the Unified Meteorological Archive and Retrieval Facility (U-MARF). The MPEF completes full data processing and generates a range of meteorological products for the end-users. The U-MARF allows users to access historical data from the satellites. Since 1978 the Unified Meteorological Archive and Retrieval Facility (U-MARF) has been the single repository for all Meteosat image data and derived products. Live data is continuously archived and historic data was copied. EUMETSAT's Primary Ground Station (PGS) is located at Fucino, Italy; the back-up station is at Cheia, Romania.

With the launch and placement of MSG Meteosats 8 and 9 at 0°, EUMETSAT WEFAX analogue dissemination services were terminated. The same is true of Meteosat 0° High Resolution Image (HRI) data services. EUMETCast digital dissemination services and the Direct dissemination scheme have replaced prior data access methods. Operation of EUMETCast for access to Meteosat, MetOp and external provider data typically requires an independent reception station and a EUMETCast Key Unit (EKU) to decrypt the transmitted data. Direct dissemination, with data directly downloaded from Meteosat-7, is available to Primary Data User Stations (PDUS) equipped with a Meteosat Key Unit (MKU) to decrypt the data. Meteosat-9 Low Rate Information Transmission (LRIT) data access requires a Low Rate User Station (LRUS) and Station Key Unit (SKU) for data decryption. EUMETSAT also disseminates some types of data via the World Meteorological Organization's (WMO) Global Telecommunication System (GTS) and the internet.

Please also see the separate Meteosat, MetOp and Jason entries.

Intersputnik

Current Status

Intersputnik operates 3 Gorizont satellites at 50° E, 130°E and 142.5°E, 2 Express satellites at 14°W and 80°E and LM-1 at 75°E. Express 3A will operate from 11°W. Intersputnik membership has now grown to 24 countries with the latest LMI-1 satellite manufactured under the management of the Lockheed Martin Intersputnik joint venture. In July 2002 Intersputnik and Eutelsat signed a distribution agreement allowing Intersputnik to sell Eutelsat's satellite capacity. LMI-1 commenced service in November 1999 from 75°E over the Indian Ocean providing C-band and Ku-band services to Eastern Europe, Asia, Africa and Australia.

Background

Intersputnik was established in 1971 to offer satellite-based TV, radio, telephony and datalinks. Membership is Afghanistan, Belarus, Bulgaria, Cuba, Czech Republic, Georgia, Germany, Hungary, India, Kazakhstan, Kyrgyzstan, Laos, Mongolia, Poland, Republic of Korea, Republic of Nicaragua, Romania, Russia, Syria, Tajikistan, Turkmenistan, Ukraine, Vietnam and Yemen. A Board composed of one representative from each member governs the organisation; sessions are held at least annually. The Directorate is a permanent executive and administrative body, headed by a Director General. Financial activities are controlled by an Auditing Committee elected for a three-year term. On joining Intersputnik, a member makes a minimum 1 per cent contribution to the Statutory Fund; profits are distributed in proportion to these contributions.

As of 1 January 1996, Intersputnik provided its services through 26 36 MHz C-band transponders on Express 1 at 14° W and eight Gorizont satellites at 40°E, 70°E, 80°E, 85°E, 96.5°E, 103°E, 130°E and 142.5°E, plus two 27 MHz Ku-band transponders on Gals 1 at 70° E. Four transponders (#7/8/10 at 14°W + #8 at 80°E) provide voice/data using digital IDR technology. The others are used for TV, three of which (#7/9 14°W + #10 80°E) are for occasional use. The even-numbered transponders (except for #6) provide minimum 31.0 dBW EIRP hemispheric coverage, the odd numbers are global, at 28.0 dBW minimum. 14°W/80°E are the principal Intersputnik locations.

C-band Earth stations in the system utilise 12 m diameter main dishes of more than 31 dB/K G/T at 3.800 GHz 5° elevation. Transmitter output power is 1 to 3 kW. The stations operate at 3.700 to 4.150 GHz rx, 6.025–6.475 GHz transmit. Ku-band 10 m dishes provide 38.0 dB/K. The network comprises 42 standard Earth stations. TV can be carried in PAL, SECAM and NTSC formats. 6.5 to 7.5 m C-band and Ku-band stations also operate. 2.0 to 3.0 and 3.5 to 4.5 m are used for telephony, fax, telex, data, videoconferencing (3.5–4.5 m only) and TV/radio reception.

Bulgaria, Hungary, German Democratic Republic, Cuba, Mongolia, Poland, Romania, Czechoslovakia and USSR signed the Intersputnik agreement 15 November 1971. The first installations to enter service were those of the USSR and Cuba's Caribe ground station, used for TV transmissions via Molniya 1 of Leonid Brezhnev's 1973 Cuban visit. The system formally became operational in 1974; by 1980 there were eight stations and 11 member countries. By end-1984, the organisation was claiming that 60 per cent of all TV transmissions between member states was being relayed on its channels. Germany's DBP Telkom inherited the GDR membership in German unification. In 2005, Intersputnik has 24 member countries. These have formed a new alliance for co-operating with national and regional governments to improve the links between systems.

Intersputnik has approved the development of a series of small/medium satellites designated Intersputnik-100 m. Development of micro-dish antennas for special user needs and selected frequency allocations. Frames user needs into operational specifications.

Japan

Japan Meteorological Agency (JMA)

Current Status

The Multi-functional Transport Satellite-2 (MTSAT-2) was successfully launched in February 2006 aboard an H-IIA vehicle from Tanegashima Space Center (see separate entries). The MTSAT series follows Japan's Geostationary Meteorological Satellite (GMS) series and covers East Asian and Western Pacific weather patterns. MTSAT-1R was launched in February 2005. MTSAT-2 will remain at in-orbit test status and in standby until about 2010.

The Japan Meteorology Agency (JMA) employs approximately 6,000 personnel and had a budget, as of FY2003, of about JPY72.8 billion.

Background

Renamed the JMA when it became an organ of the Ministry of Transport in 1956, JMA can trace its origins to the Tokyo Meteorological Observatory (TMO), set up in 1875. JMA, located in central Tokyo, is responsible for providing space-based meteorological services. The Meteorological Satellite Centre (MSC), established April 1977 as an auxiliary organ of JMA, is situated in Kiyose, a northwestern Tokyo suburb. MSC undertakes satellite control and image data acquisition through its Command and Data Acquisition Station (CDAS) at Hatoyama, 35 km northwest of Tokyo, as well as data analysis for MTSAT and GMS. JMA also oversees aviation and marine weather services as well as meteorological observatories and weather stations.

JMA played a leading role in the conceptualisation, design and development plans for the first Japanese meteorological satellites and has played a vital role in subsequent years developing and refining successors to the first satellites. The National Space Development Agency (NASDA), now absorbed into the Japan Aerospace Exploration Agency – JAXA, procured and launched the satellites under JMA funding and requirements. Raw imagery, data collection platform signals and spacecraft telemetry are received on the two CDAS 18 m Cassegrain dishes; they are then relayed by 2.0 GHz microwave link to the Data Processing Centre (DPC). The DPC maintains four mainframe computers: two Fujitsu M-1600/2 (one primary and one redundant) for spacecraft control and operations, and two Fujitsu M-1600/10R for image processing and subsequent extraction of parameters such as sea temperature and cloud top height. The smaller machines can assume some of the M-1600/10R processing duties in the event of failure. The Stretched-VISSR and WEFAX data are passed to the CDAS for uplinking and broadcast to users: Medium Data Utilization Stations (MDUS) can receive S-VISSR but the Smaller Data Utilization Stations (SDUS) receive the lower-resolution (7 to 8.5 km) WEFAX. Two Turn-Around Ranging Stations (TARS) are operated on Ishigaki Island (Japan) and Crib Point (Australia) in conjunction with CDAS for satellite orbit determination and prediction.

Malaysia

Binariang SDN BHD

Current Status

Binariang signed an MoU in November 1991 with Hughes for two HS-376 satellites to provide domestic and regional services. The first MEASAT (Malaysian-East Asian Satellite) launch was then planned for 1994 aboard Ariane. Before MEASAT appeared, Malaysia spent some US$30 million annually on Intelsat and Palapa leases. A contract was not signed with Hughes Communications International Inc until 17 May 1994 following a four-month international competition. The agreement covered one satellite to be launched by late 1995, an option for a second (authorised

January 1995), the ground control station and training for the Malaysian operators. This HS-376 is the first to carry GaAs solar cells (increased output), a lightweight shaped antenna (improves gain and eliminates multiple feedhorns) and a bipropellant stationkeeping propulsion system. Binariang also holds a licence for international gateway services. MEASAT 1 was launched by Ariane 44L in January 1996 to 91.5°E followed by MEASAT 2 to 148°E in November 1996 on another Ariane 44L. Plans for MEASAT 3 envisage C-band services to a wide area from central Africa to Australia and Japan, with Ku-band coverage from Pakistan to Taiwan. Binariang selected Boeing Satellite Systems to build the satellite around a Boeing 601 HP bus and will employ 24 C-band and 24 Ku-band transponders, each providing 36 MHz of bandwidth over a 15-year service life, using a Proton Breeze M launcher from ILS to put MEASAT 3 in orbit during 2005, co-located with MEASAT 1 at 91.5° E. The MEASAT 1 replacement is planned for launch no later than the fourth quarter of 2007. Binariang is also planning a regional expansion which could see a need for additional capacity. This has stimulated further research into high powered satellite systems and the optimum capacity required for domestic versus leased requirements.

United States

National Oceanic and Atmospheric Administration

Current Status

The National Oceanic and Atmospheric Administration (NOAA), an element of the US Department of Commerce (DoC) since 1970, is currently second only to NASA as the US government agency most closely focused on environmental monitoring and climate change. NOAA has approximately 12,500 personnel and had an annual budget, in 2004, of about USD3.9 billion.

Satellite data are used to monitor weather conditions, to issue warnings of severe weather such as hurricanes, prepare charts and coastal maps, and improve assessment and conservation of marine life. They are also employed to assess the impact of natural factors and human activities on global food and fuel supplies and on environmental quality. NOAA offices participating directly in space projects include the National Environmental Satellite, Data, and Information Service (NESDIS), the Systems Acquisition Office (SAO), the National Weather Service (NWS), the National Marine Fisheries Service, the Office of Oceanic and Atmospheric Research, and the National Ocean Service. NOAA offices collect, interpret and archive meteorological data that is used by civil, defence and commercial organisations.

In 1994, changes in the discrimination and budgeting of national meteorological and environmental satellites unified the once separate civilian and defence systems. NESDIS manages the National Polar-orbiting Operational Environmental Satellite System (NPOESS) through its Integrated Program Office (IPO), which is jointly staffed by DoD, DoC and NASA personnel. NESDIS also oversees the Geostationary Operating Environmental Satellite (GOES) and Polar Operational Environmental Satellite (POES) systems, gathering, archiving and distributing environmental data. NESDIS personnel gather and archive Defense Meteorological Satellite Program (DMSP) data, although that programme is officially managed by the USAF Space and Missile Systems Center (SMC).

Background

NOAA was federal overseer until October 1992 of the commercial Landsat 4/5 remote sensing satellite system, a role it repeated for Landsat 7. NOAA is the US signatory to the International Satellite System For Search and Rescue Cospas-Sarsat agreement. Its Space Environment Center (see separate entry under National Weather Service) in Boulder, Colorado provides real-time monitoring and forecasting services of solar disturbances and their effects on Earth's environment.

In co-ordination and co-operation with other appropriate organisations, SAO has 'responsibility for designated major systems acquisition', such as the GOES series. The NOAA is integral to the convergence of civil and military meteorlogical services now that DoD has decided to obtain its weather information from civil satellites.

An NOAA polar satellite shows Hurricane Katrina making landfall on the US Gulf Coast on 29 August 2005 (US National Oceanic and Atmospheric Administration, NESDIS) 1133740

Increasingly, organisations like the NOAA collect and assimilate data from NASA earth science and observation missions such as Aqua, see here (NASA) 1047497

SPACE DEFENCE

Bases
Operators
Research & Development

BASES

Australia

Woomera Test Facility

Location: 31.1°S/136.8°E

Current Status
Rocketplane Kistler plans to launch its K-1 out of Woomera as soon as 2008. Leighton Contractors are building Kistler's Woomera launch facilities. The K-1 is being developed as part of NASA's Commercial Orbital Transportation Services programme, a plan to test and build reusable space vehicles to supply the International Space Station (ISS) after the Space Shuttle retires. SpaceX could also launch its Falcon 9 vehicle from Woomera, a rocket that is also being developed, along with its Dragon supply vehicle, under the same NASA programme; however it is more likely that SpaceX will fly its spacecraft out of Kwajalein or Cape Canaveral.

Despite many years' talk of establishing launch facilities in Cape York, Darwin, and Christmas Island, Woomera remains Australia's single launch facility. Orbital launches have not taken place at Woomera since 1971. The facility has largely fallen into disuse, but retains some sub-orbital launch capabilities. Japan's asteroid probe, Hayabusa (see separate entry) will probably use Woomera as its landing site when it returns to Earth in 2010. The Australian Department of Defence continues to allow foreign government and commercial access to Woomera for various testing purposes.

Background
About 430 km north of Adelaide, running eastwards for 2,000 km across the desert and comprising about 127,000 km^2, the Woomera range in South Australia was established jointly by the UK and Australia in 1946. Separate parts of the range are owned by the South Australian and Commonwealth governments as well as the Maralinga Tjarutja and Pitjantjatjara Aboriginal communities, however Woomera is administered by the Australian Department of Defence and remains in use as a weapons test site. The Department of Defence stated in 2006 that commercial launches will only be allowed if the flights do not conflict with Defence activities at Woomera.

Woomera was originally a test range for ballistic missiles and sounding rockets, as well as a target practice range for pilotless aircraft, and there are continued Australian aspirations to evolve the site into a commercial satellite launch centre; these aspirations have not yet been completely fulfilled. However, if Rocketplane Kistler, after securing its lease and building it spaceport, manages to successfully orbit its K-1, then a full Woomera orbital launch revival could occur.

The first Skylark sounding rocket launched from Woomera was flown on 13 February 1957. The first Black Knight launched 7 September 1958 and 10 Europa flights took place between 1964 and 1970. During the period when the UK's Blue Streak was under development as the main stage of the pre-ESA European Launcher Development Organisation's (ELDO) Europa rocket, a township of 4,500, with more than 500 houses supplied with power and water across 160 km of desert, was created. However, Woomera was not suitable for equatorial launches and France preferred to develop its own centre at Kourou. Europa research and development flights were conducted between 5 June 1964 and 12 June 1970, the last three were unsuccessful attempts to place a satellite in orbit, before launches switched to the Kourou launch site in French Guiana. After successive cuts in its contributions to the range, the UK announced there would be no more work for Woomera after 1976.

NASA also decided to close down its Carnarvon tracking station, which had once played a major part in manned and unmanned space flights. The small Wresat test satellite, launched into polar orbit by a US Redstone in November 1967, and the UK's Prospero in October 1971 remain the only two satellites orbited from Woomera. The range reopened in August 1987 for the launch of a West German-funded Skylark sounding rocket to undertake about 5 minutes of observations of Supernova 1987A.

The South Australian government has long wanted to revitalise Woomera as a commercial site for aircraft and aerospace systems testing, but rejected initial proposals in 1990 for relying excessively on the use of the Australian armed forces. Woomera's advantages include polar orbit access, a sparsely inhabited downrange area, an existing infrastructure and largely cloud-free weather.

Israel

Palmachim Air Force Base

Location: 31.52°N/34.45°E

Current Status
Palmachim Air Force Base is located south of Tel Aviv, near the town of Yavne. It is Israel's military satellite launch site. The Palmachim launch facility can accommodate Shavit and LK-1 launch vehicles. Israel launches its Ofeq (also Ofek) satellites from this location.

On 6 September 2004, Israel attempted to launch Ofeq 6 from Palmachim, however the launch failed during the third stage, and the remote sensing satellite fell into the Mediterranean Sea near Ashdod. Ofeq 6 was designed to augment intelligence data received from Ofeq 5, and to eventually replace the older satellite, which remains operational.

Background
The Israelis inaugurated Palmachim as an orbital site with the launch of the first Shavit rocket, carrying the Ofeq 1 satellite into low Earth orbit (LEO) on 19 September 1988. The facilities are classified, although they are visible from the coast road. Located at the eastern end of the Mediterranean, the site is restricted to retrograde launches due to range safety and security considerations. Although the trajectory still passes over southern Europe, Israel does not launch toward the west, over neighbouring Arab countries.

Israel's second Ofeq orbital launch was made in April 1990, and the third in April 1995. Reportedly, several launch failures occurred between 1991 and 1994. Ofeq 4 was launched from Palmachim in January 1998, but it failed to reach orbit.

The Ofeq 5 imaging satellite was successfully launched on 28 May 2002. It is equipped with a high-resolution camera, which is likely to observe military movements inside Iran, Syria and Iraq for intelligence, surveillance and reconnaissance (ISR) purposes.

Israeli commercial and research satellites are placed into orbit by launch vehicles outside Israel.

Russian Federation

Dombarovsky (Yasny)

Location: 50.75°N/59.50°E

Current Status
The Russian Federation launched its first orbital flight from Dombarovsky on 12 July 2006. Although the site is a military base, the Dnepr launch was conducted commercially, for Bigelow Aerospace. The Dnepr is a converted, Ukrainian-built R-36M, or SS-18 "Satan" ICBM. The payload, Genesis I, was an experimental, inflatable structure, used to validate concepts to build commercial leisure space station modules, such as hotels. Once in orbit, it inflated and deployed its solar arrays successfully. Kosmotras, a joint Russian, Ukranian and Kazakh endeavour, converts, markets and launches the Dnepr rocket for commercial customers.

Background
Dombarovsky is the home of the Russian Strategic Missile Forces' (RVSN) 13th Missile Division. The base began construction in the 1960s, and still houses dozens of active R-36M and R-36M2 ICBM missile silos. It is located in Russia's Orenburg region, near the Northern Kazakhstan border. Yasny, the civilian name for the site, is a nearby town where the 13th Missile Division is quartered.

The military began conducting launch tests beginning in 2004, to trial the SS-18 for possible space launches. The long-term objective, if the launches remain successful, will be to permanently move Dnepr launches from Baikonur to Dombarovsky. The catastrophic Dnepr launch failure at Baikonur on 26 July 2006, makes this aspiration especially urgent, since the Kazakhs have suspended all Dnepr launches out of Baikonur for the foreseeable future, primarily due to the cleanup costs and hazards to the environment and the public that Dnepr failures cause, which stem from the rocket's toxic nitrogen tetroxide/unsymmetrical dimethylhydrazine (N_2O_4/UDMH) liquid propellant.

The Russian military and Orenburg authorities are examining the possible environmental impacts of Dnepr launches in the region. However, commercial Dnepr launches could potentially reap millions of dollars in profits for the Russians, if the plan to permanently launch from Dombarovsky moves forward.

United States

Cape Canaveral Air Station

Location: 28°26′48.73″N/80°33′54.58″W

Current Status
The final Titan to fly from Cape Canaveral carried a National Reconnaissance Office (NRO) satellite into orbit on 30 April 2005, GMT. Cape Canaveral Air Station (CCAS) supports Evolved Expendable Launch Vehicle (EELV) programme launches, and both Atlas V and Delta IV heavy launch vehicles have flown from the site since 2002.

Background
A unit of the US Air Force (USAF) Space Command, the 45th Space Wing (see separate entry) is the CCAS and Patrick Air Force Base (AFB) host organisation and is responsible for DoD Eastern Range orbital launches and operations, as well as supporting NASA launches and other government and commercial customers. CCAS is located adjacent to and East of Merritt Island's Kennedy Space Center (KSC), on the Florida Atlantic Coast, about 85 km East of Orlando. Patrick AFB lies some 35 km South of CCAS and Cocoa Beach, Florida. CCAS and Patrick AFB origins date back to September 1948 when the Banana River Naval Air Station was transferred to the USAF as a base for operations as a joint-service missile range. The Headquarters opened on 10 June 1949, but joint operations proved unwieldy and it was replaced by the USAF's Long Range Proving Ground Division on 16 May 1950. On 30 June 1951, in keeping with policy to designate intermediate headquarters as 'centres', the Long Range Proving Ground Division became the USAF Missile Test Center. It was renamed the USAF Eastern Test Range in 1964. The 45th Space Wing comprises four major Groups: Operations, Launch, Mission Support and Medical.

The 45th Space Wing provides extensive support for Space Shuttle launch operations; the USAF Eastern Range provides weather forecasting, range safety, tracking and preparation of DoD payloads. CCAS's principal current launch activities take place at Space Launch Complex (SLC) 41 for the Atlas V, SLC 37 for the Delta IV, and SLC 17 for the Delta II. The Atlas V SLC 41 is actually north of the CCAS boundary, on KSC, on a 1.44 km^2 site allocated to the USAF.

The Eastern Range tracking network stretches 16,000 km over the Antigua and Ascension Island stations and into the Indian Ocean, where it meets

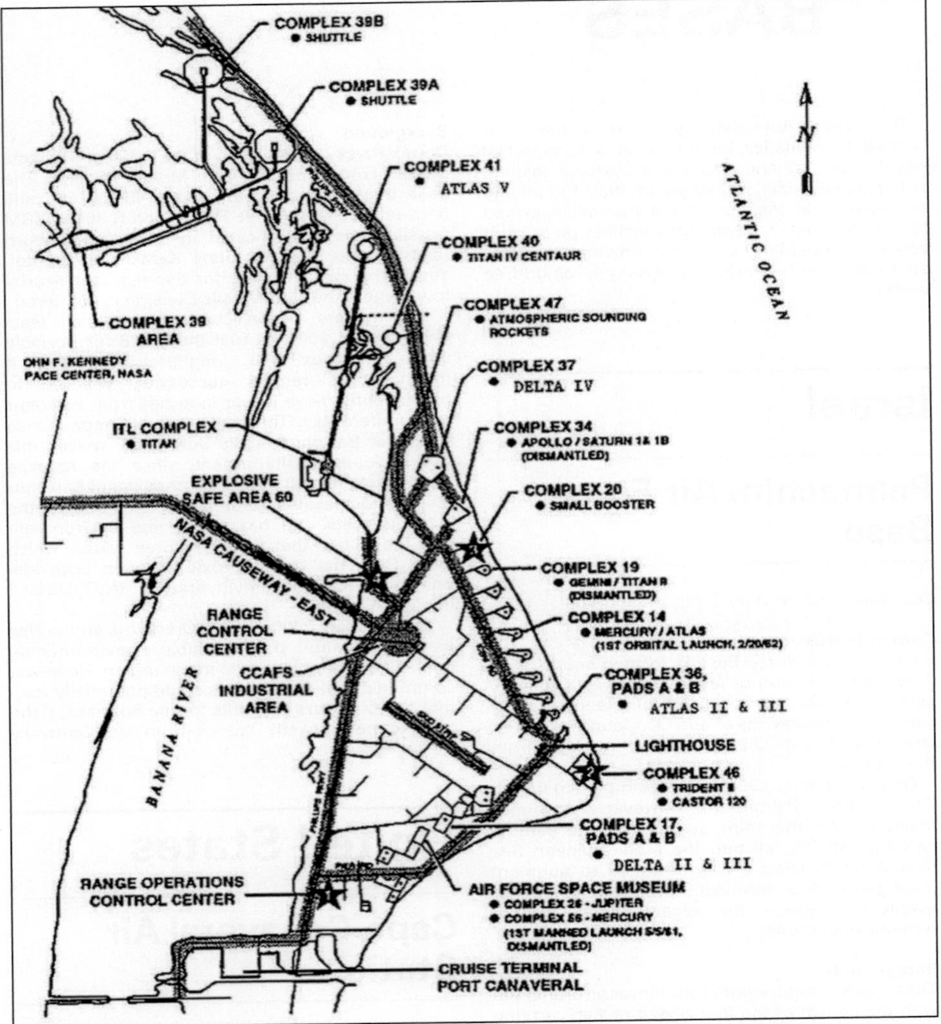

Cape Canaveral Air Station launch complex map (45th Space Wing History Office, Patrick Air Force Base) 1133877

the Western Range. Mainland Florida stations are maintained at CCAS, Jonathan Dickinson and Malabar annexes, augmented by the fleet of advanced range instrumentation aircraft (ARIA). In addition, the Eastern Range can use instrumentation operated by NASA on Bermuda and Wallops Island. Another station is available at Argentina in Newfoundland for high inclination launches on northeastern flights of about 37°.

CCAF has operated numerous Launch Complexes (LC) over the years, many having hosted historic programmes. Below is a listing of each LC's status:

1–4 Inactive. Used for Snark, Bomarc, Bumper (first major Cape launch on 24 July 1950 from Complex 3), Redstone, Jason, X-17, Polaris launches; the sites also conducted 'Fat Albert' tethered aerostat tests and radar testing of air defence equipment to be used at other locations.

5–6 Dismantled. The first US manned launches were conducted here: Mercury Redstone 3 and 4, the Shepard/Grissom sub-orbital flights in May and July 1961; Jupiter C, Explorer and Pioneer. The complexes have been a part of the CCAS USAF Space Museum since January 1964.

9–10 Dismantled. Used for Navaho launches. After demolition, LC 31 and 32 were built on the same site.

11 Dismantled. Used for Atlas ballistic missiles.

12 Dismantled. Used for Atlas and Atlas Agena carrying Mariner and Ranger.

13 Demolished on 6 August 2005 after being declared an environmental and safety hazard. Used for Atlas and Atlas Agena carrying NASA and DoD missions.

14 Demolished in 1976 due to its dangerous condition caused by salt corrosion. Mercury Atlas 6–9 manned orbital flights were launched from this SLC, starting with John Glenn, February 1962.

15 Dismantled. Used for Titan I and II ballistic missiles.

16 Inactive. Previously used for Titan I and II ballistic missiles, Pershing 1, 1A and 2, and Apollo tests

17 Active. CCAS's Delta II and III launch site. Plans were made to close 17 down at the end of 1986, but the selection of Delta II to launch GPS satellites ensured SLC 17's survival. 17B was modified to accommodate the cryogenic Delta III.

18 Dismantled. Used for Vanguard, Blue Scout, Thor.

19 Dismantled. Titan I and II; all 10 Gemini Titan II manned flights 1965–66.

20 Active. The Florida Space Authority refurbished the complex to use it for sub-orbital commercial launches and reactivated the site in 1999. The pad had been used sporadically from 1964 through the early 1990s, prior to the Space Authority take-over. Historically it was used for the Titan I and modified for Titan IIIA testing; it was then renovated to launch Starbird suborbital vehicles; Joust 1 and Strategic Defense Initiative (SDI) Red Tigress launches were also conducted.

21–22 Dismantled. Used for Mace and Bull Goose.

25 Dismantled. Used for Trident I, Poseidon and Polaris.

26 Deactivated. The site is now the USAF Space Museum. During the 1950s, the two pads launched 24 Jupiter IRBMs, two Jupiter bio monkey flights, two Redstone IRBMs, two Jupiter C (Explorer 1 and 2) and six Juno 2 (Explorer 8 and 11) flights.

29 Deactivated. Used for Polaris and UK's Chevaline.

30 Dismantled. Used for Pershing tests.

31–32 Dismantled. See 9-10. Used for Minuteman and Pershing launches; built on the dismantled 9–10 site. Challenger debris has been stored in a missile silo at LC 31 since 1987.

34 Dismantled. Used for Apollo Saturn 1 and 1B.

36 Deactivated. Until 2005 SLC 36 was used for Atlas II and III launches. Pioneer, Mariner and Intelsat launches were conducted from the site. About 145 Atlas Centaurs were launched from Complex 36.

37 Active. SLC 37 is CCAS's Delta IV launch site; the first Delta IV lifted off from the site in November 2002. Historically, the site was used for Apollo Saturn 1 and 1B launches.

39A–39B Active. NASA launches the Space Shuttle from these KSC SLCs. Please see the KSC entry.

40 Deactivated. Previously used for Titan IV launches, until the vehicle's final launch on 30 April 2005 GMT.

41 Active. CCAS's Atlas V launch site. SLC 41 was formerly used for Titan IV launches until the Titan's final launch from the site in April 1999.

43 Dismantled. Previously used to launch weather rockets. The site was deactivated to make room to construct LC 46.

45 Deactivated. The site was built to test Roland mobile missiles, although none were ever launched.

46 Active. The Florida Space Authority uses the site for commercial launches that require small orbital launch vehicles. Athena I and II have been launched from the site. The Space Authority won USD4.89 million in USAF grants and spent an additional USD2 million in industry matching funds to refurbish the site. SLC 46 was previously used by the US Navy for the Trident, and that Service still retains the right to use the site.

47 Active. The Florida Space Authority took over the site on 5 June 2003 for commercial launches, upon licence with the USAF. LC 47 facilities are used primarily for small rocket launches conducted by academic institutions and other customers. The site was previously used to launch weather rockets.

Ronald Reagan Ballistic Missile Defense Test Site, Kwajalein Atoll

Location: 9°N/167°E

Current Status

Space Exploration Technologies (SpaceX) has, as of early 2007, conducted two Falcon 1 satellite launch vehicle flights from Kwajalein. Falcon 1 is billed as the cheapest small satellite launch vehicle yet, costing around USD6.7 million to launch. SpaceX leased Omelek launch facilities from the US Army to conduct the commercial launch tests. The company has organised a launch manifest, including US government and commercial customers, that includes Kwajalein, through 2010. SpaceX also expects to launch its Falcon series from Vandenberg and Cape Canaveral (see separate entries). The first Falcon 1 demonstration flights took place on 24 March 2006 and 20 March 2007 from the Omelek site.

Background

The Ronald Reagan Ballistic Missile Defense Test Site (RTS) is operated by the US Army Kwajalein Atoll (USAKA), a unit of the US Army Space and Missile Defense Command/U.S. Army Forces Strategic Command (SMDC/ARSTRAT), and is located in The Republic of the Marshall Islands (RMI). The US Army leases 11 islands, of the roughly 100 that make up the Atoll, from the RMI. Kwajalein is the Southernmost island in the Kwajalein Atoll; mission control facilities are located here. Missile tests and orbital launches are conducted from Omelek, Meck and Roi-Namur Islands, as well as other Pacific-based facilities at Wake Island and Aur Atoll. RTS constitutes about 1,942,500 km², although its land area is only about 181 km²; the range primarily encompasses open ocean.

RTS is also home to the Advanced Research Project Agency (ARPA) Long-Range Tracking and Instrumentation Radar (ALTAIR) and the ARPA Lincoln C-Band Observables Radar (ALCOR) on Roi-Namur, part of the Space Surveillance Network (SSN). USAKA provides tracking services for the SSN and also communicates with the Defense Satellite Communications System (DSCS). The site also detects non-US Eastern Hemisphere launches and supports US space operations, including manned missions.

Vandenberg Air Force Base

Location: 34.4°N/120.35°W

Current Status

Vandenberg is a missile testing and orbital launch site. The base supports the US Air Force's Evolved Expendable Launch Vehicle (EELV) Programme, and is capable of launching Boeing Delta IV and Lockheed Martin Atlas V heavy lift vehicles. In 2005, Boeing and Lockheed announced that they would enter into a joint venture called the United Launch Alliance (ULA). Spurred by the small demand for EELV flights since the launch vehicles entered

SLC-4 and SLC-3 complexes. From left: SLC-4W (former Titan II), SLC-4E (former Titan IV), SLC-3E and SLC-3W (Atlas) (USAF) 0517084

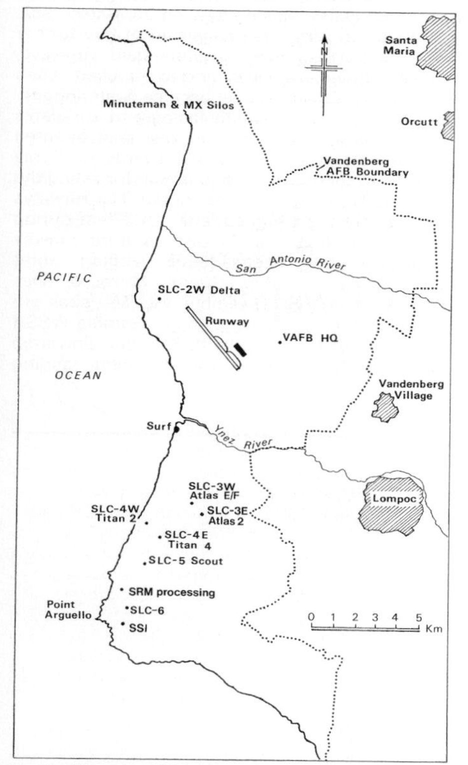

Vandenberg provides access to polar orbits. The northern area is generally employed for missile development and operational test launches; most are de-activated but Minuteman and Peacekeeper ICBM silos and pads remain in use. Apart from the SLC-2W Delta pad, the orbital pads are clustered in the Point Arguello area: SLC-3E and 3W for the Atlas, SLC-4W and 4E previously for the Titan family, and SLC-5 for the Scout (retired 1994). SLC-6 is now used for Delta IV launches. SLC-8, California Spaceport's pad, was built at 34.576° North/120.631° West, South of SLC-6, for commercial operations, which began in 1997 0517426

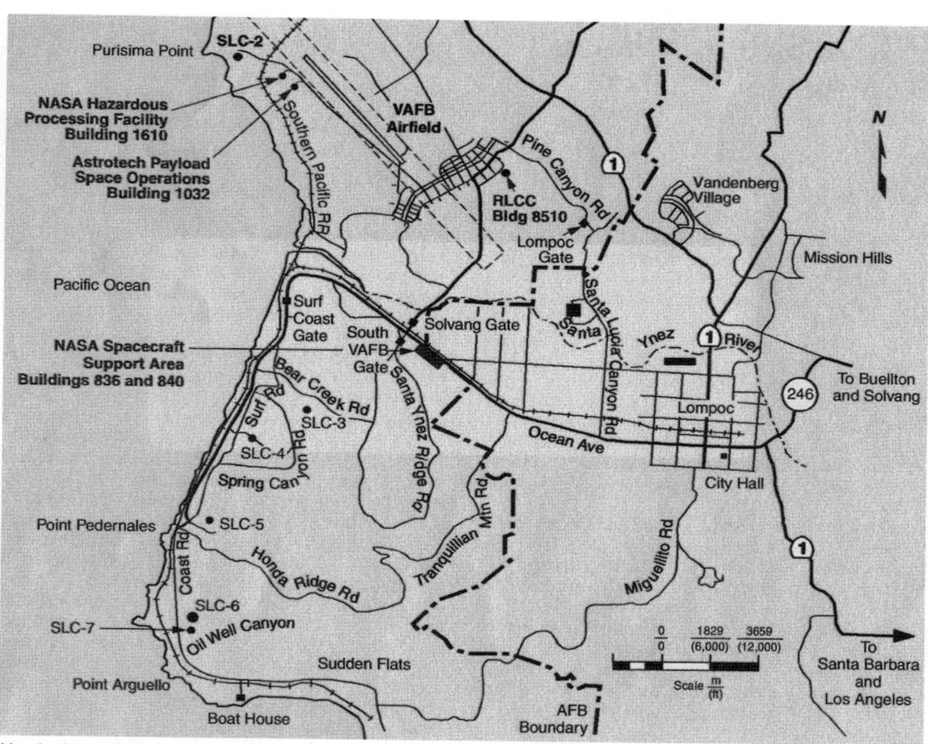

Vandenberg Air Force Base facilities 0132804

NASA Facts

Vandenberg Space Shuttle Launch Complex

1. MOBILE SERVICE TOWER
2. LAUNCH PAD
3. ACCESS TOWER
4. PAYLOAD CHANGEOUT ROOM, ET & ORBITER ERECTION
5. PAYLOAD PREPARATION ROOM
6. LO_2 STORAGE
7. LH_2 STORAGE
8. SRB DUCTS
9. SSME EXHAUST DUCT
10. GAS STORAGE AREA
11. ESCAPE WIRE SYSTEM
12. LAUNCH CONTROL CENTER
13. SHUTTLE ASSEMBLY BUILDING

SLC-6 was created in the 1960s for the Manned Orbiting Laboratory (MOL), before it was converted for Shuttle operations (shown here). The Payload Preparation Room (5) is being leased to California Spaceport's Spaceport Systems International as a processing facility and launch control centre 0517086

service in 2002, the proposed ULA structure purportedly will save EELV programme money by merging Boeing and Lockheed capabilities and expertise. Vandenberg's first Atlas V was scheduled to fly in 2006 carrying a National Reconnaissance Office payload. The launch has been delayed several times, and is anticipated in 2007. All Atlas Vs have thus far been flown out of Cape Canaveral (see separate entry).

Background

Since 1956 Vandenberg has been responsible for missile and space launches on the US West Coast and operates the Western Range tracking network extending into the Indian Ocean, where it meets the Eastern Range. Vandenberg provides the US with access to polar orbits using due South launches and at one time was to have provided a base for Space Shuttle departures on high-inclination missions. Delta missions are supported by the Space Launch Complex (SLC)-2W pad, and Atlas by SLC-3. Titan vehicles were supported by SLC-4 until the final Titan launch from Vandenberg occurred on 19 October 2005.

Vandenberg is home to the 30th Space Wing (see separate entry). The 30th Operations Group is responsible for Western Range Operations. Vandenberg is an important base for development and operational ICBM flight testing. Above-ground pads and underground silos support Peacekeeper and Minuteman launches.

The 399 km² base is located along 40 km of Central California coastline, near the towns of Lompoc and Santa Maria. The 1,100 buildings house 53 host organisations, 37 associate units and 49 contractor organisations, supporting more than 13,000 military, civil and contractor personnel and service families. Classified satellites are launched and missiles are test-fired under operational conditions over the Western Range, extending 8,000 km across the Pacific and into the Indian Ocean.

From its southern tip, Point Arguello, spacecraft are launched into polar orbits for both NASA and DoD. California Spaceport (see separate entry) is also located at the southern end of the base. The Spaceport is a commercial facility, however, it counts DoD and NASA among its customers.

A host of missile launch and testing areas and space launch complexes (SLCs) are located at Vandenberg. The major space launch facilities are:

SLC-2

Built in 1958 as part of a group of seven launch installations for the Thor IRBM. Pad 2W was upgraded in 1995 to accommodate the Delta II/7000 Series. (Radarsat was the first launch). The second Delta pad, 2E, is no longer operational. Pad 2W can cope with six launches annually.

SLC-3

SLC-3E attained initial launch capability in October 1996 following a three year, USD300 million modification project to launch the Atlas family of launch vehicles. Pad 3E again underwent upgrades, completed in 2005, in order to accommodate the EELV heavy launch vehicle, Atlas V. After several delays, the first Atlas V launch from Vandenberg is slated to occur in 2007.

SLC-4

SLC-4 previously supported Titan launches, until that vehicle was retired. The last Titan to launch from Vandenberg lifted off on 19 October 2005 and carried a classified NRO payload.

SLC-6

SLC-6 was selected as the launch site for the Delta IV and is used to launch satellite payloads for both commercial and government customers. The first Delta IV launched from SLC-6 flew on 28 June 2006 and carried an NRO payload. SLC-6 has a chequered history; it once was supposed to have been the West Coast launch site for the Space Shuttle, with the launch complex project costing as much as USD3 billion into the 1980s. It is widely believed that those plans were scrapped due to the 1986 Challenger tragedy. California Spaceport (see separate entry) now leases the site's former Payload Preparation Room, renamed the Integrated Processing Facility (IPF); the IPF accommodates military, civil and commercial customers.

White Sands Missile Range

Location: 32.23°N/106.28°W

Current Status

Predominately a sub-orbital missile testing launch site, White Sands Missile Range (WSMR) is located in southern New Mexico. In addition to its own facilities and activities, WSMR also houses NASA's White Sands Test Facility (WSTF) and White Sands Space Harbor (WSSH).

The 8,200 km² WSMR is operated by the US Army. The site headquarters is about 30 km east of Las Cruces, New Mexico and is serviced by the El Paso, Texas Airport some 70 km to the south. WSMR is about 160 km in length, from north to south, and is roughly 65 km wide. As of 2004, WSMR retained a total staff of 6,237, including those from all Services, civil government employees and contractors.

WSMR provides missile, rocket, Unmanned Aerial Vehicle (UAV) and space vehicle test facilities, among other capabilities, to all the US Services and DoD, as well as to NASA, foreign militaries, and academic and commercial customers. The Range houses several tenant organisations, including NASA and the National Geospatial-Intelligence Agency. Navy and Air Force Deputies are assigned to represent those services at the White Sands facilities. WSMR plans to provide support to New Mexico's Southwest Regional Spaceport, or Spaceport America, which is being constructed on land near Upham, New Mexico, adjacent to the White Sands facilities.

Background

WSMR was the first major US rocket firing site after the Second World War. The first V-2 flight from US territory took place on 16 January 1946 from White Sands, south of Route 70. It was also the site of the first atomic explosion on 16 July 1945; the Trinity Site is on the northern part of the range. White Sands is still a major sounding rocket firing base and during FY 2004 launched 770 missiles and rockets. Up to September 2004, WSMR had conducted a total of 44,506 launches.

NASA's White Sands Test Facility (WSTF) is located on WSMR's west side and occupies an area of roughly 243 km². Johnson Space Center chose the site in 1962 to test Apollo propulsion systems; WSTF has been active since 1964. The site has supported testing and evaluation efforts for Apollo, Skylab, Delta Clipper, X-34 and Space Shuttle projects. The Tracking and Data Relay Satellite System (TDRSS) ground station is located on the site as well. WSTF facilities and services are available to other governmental and commercial customers.

WSTF has conducted over 300 engine tests, for a total of over two million firings since it began operations. The Facility has two separate propulsion test sites, designated areas 300 and 400, originally established for trials of Apollo Command and Service Modules (CSM) and Lunar Module (LM) main engines and thrusters. NASA now maintains nine test stands at the site: six altitude test stands to provide vacuum test capability and three ambient test stands. All are located 1,100 m above sea level.

Managed by WSTF, White Sands Space Harbor (WSSH) is also on the WSMR and located about 50 km west of Alamogordo, New Mexico. It is one of the primary training areas for Space Shuttle pilots, who develop approaches and landings skills in the Shuttle Training Aircraft (STA). Shuttle pilots are required to complete approximately 1,000 simulated landings, using a modified Grumman Gulfstream II, prior to a Shuttle mission. NASA selected the site in early 1976 within WSMR; the original 3 km Northrup gypsum lakebed strip was lengthened to 4.6 km in time for the first Shuttle Training Aircraft flight in August 1976.

White Sands Space Harbor provides the third Shuttle landing site in the US after Kennedy Space Center (KSC) and Edwards Air Force Base. The third mission, STS-3, landed on the Northrup Strip, now called Akalai Flats, on 30 March 1982. The Northrup Strip then consisted of two further lengthened 11 km long gypsum-sand runways, 1,200 m above sea level, and were used when the primary Edwards site became waterlogged. One runway is north-south facing and simulates the KSC runway; the other is east-west oriented and simulates the Runway at Edwards Air Force Base. A full set of convoy equipment for safing the Shuttle is kept at the Space Harbor. The runways were laser-levelled and widened to 275 m during the post-Challenger standdown. A third runway to simulate the transatlantic landing abort (TAL) runway at Ben Guerir, Morocco, was constructed in 1987. The other two TAL sites are located at Moron, Spain and Banjul, Gambia. WSSH is also one of the three (with KSC and Edwards) US 'Abort Once Around' (AOA) Shuttle landing areas.

OPERATORS

France

Etat Major des Armées

Current Status
The Armed Forces Headquarters is responsible for developing France's multiyear military space plan and is also involved in planning military space surveillance systems. It has had a significant role in the Syracuse military communications system. Optional projects have been analysed for their viability, including electronic surveillance satellite systems and advanced imaging reconnaissance systems. It has organised a military space projects development plan under, the aegis of France's government, and formulated plans for technical and operational activities necessary for expanded satellite programmes driven by perceived national security needs. Considerable emphasis has been placed upon integrated space capabilities for military space-based systems and these have been accelerated in the period 2003 to 2005. Several studies have been conducted for third parties interested in participating projects and co-operative ventures.

Ministère de la Défense

Current Status
This ministry is the customer through the Délégation Générale pour l'Armement (DGA, military procurement agency) and CNES for the Syracuse military communication element of the Telecom satellite system and the Helios reconnaissance satellite. Has assembled a portfolio of missions capable of giving France's defence interests independent access to surveillance and reconnaissance information. Alcatel Space won the contract to build the new Syracuse 3 military communications system, the contract being signed on 30 November 2000. The Syracuse 3 series comprises three satellites all to be launched by Ariane 5 beginning late 2005 or early 2006. The ministry has evaluated other defence related satellite projects and is examining optional spy and reconnaissance concepts.

United States

21st Space Wing

Current Status
In 2004, the US Air Force took charge of the Naval Space Surveillance System, or the 'Navy Fence', from the US Navy and the Alternate Space Control Center (ASCC), components of the US Space Surveillance Network (SSN). The Air Force created the 20th Space Control Squadron's Detachment (DET) 1, under the 21st Space Wing, to handle the responsibility of space control and object tracking over the US. The SSN is comprised of about 20 stations worldwide, the mission of which is to track manmade space objects 10 cm or larger, including space debris that could endanger military or civil spacecraft launches and in-orbit missions. The ASCC directs the SSN. The Naval Space Surveillance System is now known as the Air Force Space Surveillance System (AFSS), however the Command (and DET 1) location remains in Dahlgren, Virginia, as it was under the Navy. 20th Space Control Squadron command is located at Eglin Air Force Base in Florida.

The Joint Space Operations Center (JSpOC), the US military's primary space control entity, is located at Vandenberg AFB, having been transferred in 2005 from Cheyenne Mountain Directorate.

Background
The 21st Space Wing, a 14th Air Force Unit, and subordinate to the Air Force Space Command (AFSPC – see separate entry), was activated on 15 May 1992. It was formed from the assets of the decommissioned 1st Space Wing and 3rd Space Support Wing, both formerly located at Peterson AFB, Colorado, to operate and maintain a system of space and land-based sensors to detect and track ballistic missile launches, detect space launches, and provide data on foreign ballistic missile nuclear detonations. Space-based early warning is provided by the Defense Support Program (DSP – see separate entry) satellites, the catalyst of the US early warning system as it is the first to detect missile launches. 21st Space Wing DSP, ground-based early warning system and space control squadrons send launch and nuclear detonation, as well as space control reports to the HQ AFSPC, North American Aerospace Defense Command (NORAD) and US Strategic Command (USSTRATCOM – please see separate entries for each command) centres at Cheyenne Mountain Directorate and combatant commands and deployed forces around the world. Daily Cheyenne Mountain operations are transferring to Petersen AFB over the next few years.

The Wing's ground-based early warning systems units are comprised of Ballistic Missile Early Warning System (BMEWS) and Sea-Launched Ballistic Missile (SLBM) warning systems squadrons. The BMEWS radar units are operated by the 12th Space Warning Squadron (SWS) at Thule AB in Greenland and the 13th SWS at Clear Air Force Station (AFS) in Alaska; the Wing maintains a liaison with the Royal Air Force, Fylingdales, UK, also considered part of the early warning system. The Wing's SLBM warning system units are the 6th SWS at Cape Cod AFS, Massachusetts, and the 7th SWS at Beale AFB, California, although all early warning units are able to detect both ICBMs and SLBMs. The 10th SWS at Cavalier Air Force Station, North Dakota also detects and tracks both SLBM and ICBM launches using the Perimeter Acquisition Radar Characterization System (PARCS).

The 21st Operations Group directs the 21st Space Wing's operational units. The Group is also the lead for the AF's eventual transition from DSP to the Space-Based Infrared Radar System (SBIRS – see separate entry).

The 21st Space Wing's 1st, 4th and 20th Space Control Squadrons' and 21st Operations Group's domestic and foreign located Detachments are responsible for detecting, tracking and cataloguing more than 9,500 manmade objects in space from LEO to deep space, and make up the US Space Surveillance Network. The Ground-based Electro-Optical Deep Space Surveillance (GEODSS) system, operated by the 21st Operations Group, is comprised of three telescopes at Socorro, New Mexico; Diego Garcia, British Indian Ocean Territory; and Maui, Hawaii. GEODSS is able to track objects more than 32,000 km away from Earth. Besides being employed by the US to inform NORAD for security and informational purposes, the space surveillance data gathered by the 21st Space Wing's units are used to help generate the UN register of space objects.

As host wing for the Peterson AFB Complex, the 21st Space Wing provides base support services for Peterson AFB Complex, including HQ Northern Command (NORTHCOM), HQ NORAD, HQ AFSPC, HQ US Army Space and Missile Defense Command (USASMDC/ARSTRAT – see separate entry) and the 302nd Airlift Wing (Reserve), as well as Cheyenne Mountain Directorate, and Schriever AFB (Formerly Falcon AFB). Worldwide, personnel total more than 4,000 military, 1000 AF civilian and 2,600 contractor employees.

30th Space Wing

Current Status
The 30th Space Wing is an element of the 14th Air Force and is headquartered at Vandenberg Air Force Base (AFB), California, Air Force Space Command's (AFSPC) largest location. The Wing is comprised of four groups: the 30th Operations Group, the 30th Launch Group, the 30th Mission Support Group and the 30th Medical Group.

The 30th Operations Group is responsible for Western Range operations, including satellite launch, missile test launch and space surveillance duties. The 2nd Range Operations Squadron is responsible for Western Range launch planning and scheduling, and command and control. It also oversees ballistic missile and aeronautical test and evaluation operations. The 30th Weather Squadron provides Vandenberg's units with various weather data, including forecasts, in support of Western Range launches and other base operations.

The final Titan IV launches from Space Launch Complex-4 at Vandenberg AFB on 19 October 2005 carrying a classified National Reconnaissance Office payload (US Air Force) 1129978

The Launch Group is responsible for Western Range space launch systems. The 4th Space Launch Squadron oversees Delta and Atlas Evolved Expendable Launch Vehicle (EELV) operations. The squadron was created in December 2003 to take the lead on EELV medium/heavy lift launches. The 1st Air and Space Test Squadron tests and evaluates missile weapons systems and space launch vehicles. The Wing's 2nd Space Launch Squadron was responsible for Titan launches; the final of which was conducted in October 2005. The Squadron was deactivated as a result of the Titan retirement. The 30th Launch Support Squadron was activated in October 2005, replacing the 2nd Space Launch Squadron; it is charged with launch infrastructure management and payload processing support.

Several associate units also reside alongside the 30th Space Wing at Vandenberg AFB. One of these, the 576th Flight Test Squadron, reports directly to the Space Warfare Center (SWC), at Schriever AFB, Colorado, and handles Intercontinental Ballistic Missile (ICBM) test and evaluation operations. The squadron reports its test data and results to US Strategic Command (USSTRATCOM – see separate entry), the Joint Staff, Air Force Space Command (AFSPC) and Air Force Air Staff.

The 148th Space Operations Squadron is an Air National Guard (ANG) unit, also stationed at Vandenberg AFB. The 148th is responsible for Milstar satellite constellation operations, facilitating communications across the US military and services.

45th Space Wing

Current Status
An Air Force Space Command unit, and subordinate to the 14th Air Force, the 45th Space Wing is the host organisation for Cape Canaveral Air Force Station (AFS), Florida, Patrick Air Force Base (AFB), Florida and the two Eastern Range island station Detachment locations, Ascension and Antigua. The Wing is comprised of four major Groups: Operations, Launch, Mission Support and Medical. Besides managing DoD Eastern Range launches, the Wing provides launch support and services to NASA and International commercial customers.

The 45th Operations Group provides command and control of DoD space launches on the Eastern Range, and oversees the 1st Range Operations Squadron, as well as Detachment 1 at Antigua

A Delta IV EELV lifts off from Cape Canaveral AFS, on a demonstration mission (US Air Force) 1129979

AFS and Detachment 2 at Ascension Auxiliary Air Field (AAF). The 1st Range Operations Squadron manages Eastern Range launch scheduling, operations and operations support. Detachments 1 and 2 support Eastern Range space launches by providing telemetry and radar tracking data. The Detachments are also a part of the Space Surveillance Network, locating and cataloguing space objects as a secondary mission. The Operations Group also leads Operation Support, Range Management, Space Communications and Weather Squadrons.

The Launch Group is responsible for satellite launch operations and is comprised of three squadrons. The 1st Space Launch Squadron was activated on 1 October 1990 to assume all responsibility for Delta operations, its current primary mission being the launch of DoD's Global Positioning System (GPS) payloads. The 3rd Space Launch Squadron was deactivated after Cape Canaveral's final Titan IV launch on 29 April, 2005; the 45th Launch Support Squadron (LCSS) has been formed from those assets and is tasked with Evolved Expendable Launch Vehicle (EELV) and Delta payload processing and launch support duties. The 5th Space Launch Squadron, activated on 14 April 1994, now oversees all EELV launches.

Background

The 45th Space Wing designation in November 1991 replaced the Eastern Space and Missile Center organisation, established on 1 October 1979. Its origins date back to 1950 with the Army/Air Force establishment of the Joint Long Range Proving Ground. The Air Force assumed full control in 1951, redesignating it the Air Force Missile Test Center. It was renamed the Air Force Eastern Test Range in 1964. The eastern and western launch ranges and bases were managed by Air Force Systems Commands' Space Systems Division until 1 October 1990.

A single story 11,800 m² Range Operations Control Center (ROCC) became operational in March 1995 as the Eastern Range control facility under a USD134 million Harris Corporation contract. The goal was to allow flights every 12 hours. Operationally, it serves the Eastern Range by processing worldwide radar metric data, generating inter-range vectors to aid in acquisition, calculating state vectors for navigation uplink, displaying telemetry and orbital navigation information and functioning as Lead Range Control Center (LRCC) to co-ordinate radar tracking and retrieval.

50th Space Wing

Current Status

Host unit at Schriever Air Force Base (AFB), Colorado, the 50th Space Wing is a component of Air Force Space Command at 18 km-distant Peterson AFB, Colorado. A 14th Air Force unit, the wing was originally established 8 July 1985 as the 2nd Space Wing and then redesignated 30 January 1992 as the 50th Space Wing. The facility at Schriever (formerly Falcon AFB) was built under the air force programme known as the CSOC Consolidated Space Operations Center. Its mission is to command and control operational DoD satellites and to manage the worldwide Air Force Satellite Control Network (AFSCN) in support of US Strategic Command (USSTRATCOM – see separate entry) missions.

The Wing is responsible for the command and control of over 140 satellite systems, including Defense Support Program (DSP), Milstar communications constellation, NATO III, Navstar Global Positioning System (GPS) and the Defense Satellite Communications System (DSCS).

The Wing has 3,200 military and civil personnel assigned to its worldwide locations. It is comprised of three Groups: 50th Operations Group, 50th Network Operations Group and 50th Mission Support Group at Schriever AFB. The 50th Operations and Network Operations Groups command and control the satellites, and through the 50th Operations Support Squadron and Space Operations Squadrons (SOPS), train space operations crews and provide operational support and evaluation for managing the satellite operations centres and ground stations. The Groups' Space Operations Squadrons are:

1st SOPS offers support during launch, early orbit, mission and anomaly phases of GPS, DSP and Midcourse Space Experiment (MSX). This includes routinely operating older, more troublesome, satellites and eventually de-orbiting them. Once satellites are checked out, day to day control is turned over to the other SOPS.

2nd SOPS performs day to day GPS command and control. The Navstar GPS Master Control Station (MCS) took command of the first GPS operational satellite following launch on 14 February 1989. Detachment 1 at Cape Canaveral Air Force Station (AFS) operates a ground antenna for prelaunch compatibility testing and on-orbit control.

3rd SOPS performs day to day command and control of DoD, NATO and British MOD communications satellites, including DSCS III, the management of which was assumed from the deactivated 5th SOPS. In 2005 the Squadron will also assume control of Dod's Wideband Gapfiller System.

4th SOPS is responsible for day to day operations of the Milstar communications constellation, including satellite command and control, communications management and ground segment maintenance from both fixed and mobile Constellation Control Stations (CCS). The 148th SOPS (an Air National Guard unit), located at Vandenberg AFB, California operates one of the fixed CCS.

21st SOPS is a unit reporting to the 50th Network Operations Group, rather than those above, which report to the 50th Operations Group. 21st SOPS performs its duties from Onizuka AFS, Sunnyvale, California; the Squadron provides AFSCN resource scheduling and support, and network communications to a full range of international users, as well as supporting all space shuttle missions.

22nd SOPS also reports to the 50th Network Operations Group; the Squadron develops, publishes and executes the network tasking order and co-ordinates launch and on-orbit operations of over 140 DoD satellites. The 22nd SOPS is also responsible for the operation of worldwide AFSCN remote tracking stations, and maintains Detachments (DET) at each. Locations are: DET 1, Vandenberg AFB, California; DET 2, Diego Garcia, British Indian Ocean Territory; DET 3, Thule Air Base, Greenland; DET. 4, Kaena Point, Hawaii; DET 5, Andersen AFB, Guam; Colorado Tracking Station, Schriever AFB, Colorado and a liaison at RAF Oakhanger, United Kingdom.

23rd SOPS has been a 50th Network Operations Group subordinate since 2004. Operating from New Boston AFS, New Hampshire, the unit operates the largest of the 50th Network Operations Group's eight AFSCN tracking stations. The squadron provides command and control capability for over 140 satellites.

The 50th Network Operations Group also commands two Space Communications Squadrons: the 50th and the 850th. These squadrons support the Wing by ensuring AFSCN ground communications and computer systems operability and integrity. The 50th Space Communications Squadron also operates one Milstar and two DSCS ground terminals and provides integrated tactical warning and attack assessment data to NORAD and USSTRATCOM.

Department of Defense, Defense Information Systems Agency (DISA)

Current Status

DISA is a Department of Defense (DoD) combat support agency whose main objective is to anticipate and respond to the needs of its customers, the warfighters, by providing them with joint, network-centric and innovative information services, which provide a fused picture of the battlefield. Core mission areas incorporate the Defense Information System Network (DISN), Global Combat Support System (GCSS) and the Global Command and Control System (GCCS). DISA is the DoD's executive agent for information standards and provides the department with Information Assurance (IA) services and joint interoperability support. It is responsible for planning, developing and supporting command, control, communications and computers (C4) and information systems that serve the needs of the National Command Authorities (NCA) under all conditions of peace and war as well as providing guidance and support on technical and operational C4 and information systems issues. DISA holds contract authority for the Satellite Transmission Service-Global (DSTS-G), which provides global fixed satellite bandwidth to DoD users, the Iridium Enhanced Mobile Satellite Service (EMSS), providing encrypted satellite telecommunications services to field-based DoD personnel and International Maritime Satellite (INMARSAT), which provides secure and non-secure mobile telecommunications services to authorised users. DISA has worked with DARPA on geolocation systems. The agency is also authorised to work on creative concept evaluation and analysis for potential new systems capabilities in supporting defence requirements in space and application of new technologies to evolving defence strategies.

Department of Defense, National Geospatial-Intelligence Agency (NGA)

Current Status

Headquartered in Bethesda, Maryland, the National Geospatial-Intelligence Agency's (NGA) self-described mission is to provide timely, relevant and accurate geospatial intelligence to US defence and civil authorities. The agency was created as the National Imaging and Mapping Agency (NIMA) on 1 October, 1996 to co-ordinate imagery collection, processing, exploitation, analysis and distribution. It is a DoD combat support agency and was formed by consolidating the Defense Mapping Agency (DMA), Central Imagery Office (CIO), Defense Dissemination Program Office (DDPO), National Photographic Interpretation Center (NPIC) and other DoD imagery organisation assets. In November 2003 NIMA changed its name to NGA to reflect its evolving military and civil activities. It is a US Intelligence Community member.

NGA partners with international, government and commercial imagery suppliers in order to provide geospatial information to its users. The agency provides such services as Global Positioning System (GPS) data and analysis, ballistic missile trajectory and target miss error analysis and geodetic surveys to DoD and other Intelligence Community members.

NGA worked with NASA on the Shuttle Radar Topography Mission (SRTM).

Department of Defense, National Reconnaissance Office (NRO)

Current Status

A member of the US Intelligence Community, the National Reconnaissance Office (NRO) was established in September 1961, although its existence was not officially acknowledged by DoD until 18 September 1992, to co-ordinate overhead reconnaissance operations, including those from satellites. The NRO oversees the development and operations of US reconnaissance satellites in support of US national security interests. It is composed of approximately 3,000 DoD and Central Intelligence Agency (CIA) military and civilian staff. NRO has been headquartered in Chantilly, Virginia since 1994; previously it was housed in the Pentagon.

NRO receives its budget through the National Reconnaissance Program portion of the National Foreign Intelligence Program. The office reports to the Secretary of Defense, who appoints NRO's Director, and the Director of National Intelligence, who holds management responsibility for the US Intelligence Community.

In September 2005 the NRO declassified its joint Naval Research Laboratory (NRL) POPPY SIGINT satellite system programme. The POPPY programme operated during the Cold War from 1962 to 1977. In February 1995 it declassified the Corona programme and 800,000 images taken between 1960 and 1972 were transferred to the National Archives and Records Administration.

NRO has previously launched some satellites using Titan launch vehicles, which were retired from service in October 2005. The final Titan launch at Vandenberg Air Force Base, California, carried an undisclosed NRO payload. NRO launches also take place from Cape Canaveral Air Force Station, Florida.

Department of Defense, Office of the Secretary of Defense (OSD) Space-Related Offices

Current Status
In May 2005, DoD further defined military space policy and acquisition responsibilities, placing some authority under newly formed OSD offices. With the creation of the Assistant Secretary of Defense office for Networks & Information Integration/DoD Chief Information Officer (ASD-NII/DoD CIO), non-intelligence space responsibilities were realigned under the Deputy Assistant Secretary of Defence (DASD) for Command, Control and Communications (C3) Policies & Programs & Space Programs. A Space Programs office, led at the Director level, oversees policy and programmes associated with issues such as space access and control, satellite operations and space technology. ASD-NII/DoD CIO responsibilities and functions are enumerated in DoD Directive 5144.1. The ASD-NII/DoD CIO structure has replaced the Deputy Under Secretary of Defense for Space Acquisition and Technology Programs (DUSD SA&TP); that office is now defunct. The DISA (see separate entry) Director reports to ASD-NII/DoD CIO.

ASD-NII/DoD CIO works in conjunction with both the Under Secretary of Defense (USD) for Acquisition,Technology and Logistics (AT&L) and the USD for Intelligence (I) to formulate space-related policies and procedures and engage in activities related to the Space Major Defense Acquisition Program. The USD-I provides intelligence-related space policy guidance to US intelligence agencies, including the National Geospatial-Intelligence Agency (NGA) and the National Reconnaissance Office (NRO – see separate entries).

The Secretary of the Air Force, in June 2003, was appointed DoD's Executive Agent (EA) for Space. The EA for Space holds responsibility for co-ordinating and integrating space planning, budgeting, space systems programmes and space acquisition for Space Major Defense Acquisition Programs. The Space EA is also responsible for providing operational space forces and capabilities, including the development of a professional space cadre. The post works with OSD units, primarily USD-AT&L, USD-I and ASD-NII/DoD CIO to develop and disseminate DoD's space policies. It is also responsible for co-ordinating and integrating space-related requirements for DoD Components.

The National Security Space Office (NSSO) was created in May 2004, by combining the assets of the former National Security Space Architect (NSSA) and the former National Security Space Integration Office (NSSI). The office is composed of both military and Intelligence Community (IC) personnel. Presently, NSSO reports to DoD's EA for Space and the National Reconnaissance Office (NRO) Director. Its role is chiefly influential; NSSO works with players across the space community, including DoD, IC and joint organisations, to encourage national security space programme and planning integration and collaboration.

North American Aerospace Defense Command (NORAD)

Current Status
In July 2006, NORAD and USNORTHCOM announced that the Cheyenne Mountain Operations Center would become the Cheyenne Mountain Directorate, and the facility would, over the course of the next few years, effectively stand down to 'warm stand-by' status, and only be maintained for use on short notice. All personnel currently occupying Cheyenne Mountain will be transferred to the NORAD and NORTHCOM headquarters at nearby Peterson AFB. Joint Space Operations Center (JSpOC) personnel are relocating to Vandenberg AFB in California.

Background
NORAD is the bi-national US-Canadian command charged with safeguarding the sovereign airspace of the two countries. The NORAD commander is accountable to both the US President and the Canadian Prime Minister and also commands the newest unified combatant command, USNORTHCOM, founded in 2002.

The Cheyenne Mountain Operations Center is where NORAD had historically received and co-ordinated data generated by the Space Surveillance Network, operated by the 21st Space Wing (see separate entry), in order to fulfil its North American airspace protection mission. NORAD will continue with this mission, but it will be carried out at Peterson AFB headquarters in the future.

As of 2006, NORAD also has been charged with US and Canadian maritime approaches and North American waterways security.

Along with Cheyenne Mountain and the Petersen AFB HQ, NORAD commands regional centres located at Elmendorf AFB, Alaska; Canadian Forces Base in Winnipeg, Manitoba; and Tyndall AFB, Florida.

The Cheyenne Mountain facility was built by the Army Corps of Engineers, and is located more than 730 m underground. It began operations in 1966.

US Air Force Space Command (AFSPC)

Current Status
AFSPC is the Air Force (AF) Service Component Command for US Strategic Command (USSTRATCOM – see separate entry). Headquartered at Peterson Air Force Base (AFB), Colorado, this major command reports to the AF Chief of Staff and oversees US Air Force space operations. It is responsible for equipping and training its personnel, as well as providing space technology, expertise and operational services to USSTRATCOM. AFSPC maintains two numbered Air Forces and two centres under its command:

- 14th Air Force, Vandenberg AFB, California
- 20th Air Force, FE Warren AFB, Wyoming
- Space and Missile Systems Center (SMC), Los Angeles AFB, California
- Space Warfare Center, Schriever AFB, Colorado

The 14th and 20th Air Forces command numerous Space Wings, Groups and Squadrons (see separate entries) important to US military satellite, launch vehicle, ICBM strategic forces and ballistic missile early warning systems operations.

As of 2005, the Space Warfare Center will undergo integration with the USAF Air Warfare Center, Nellis AFB, Nevada, to form the US Air Force Warfare Center. Although neither existing unit will relocate or close, the Air Force Warfare Center will headquarter at Nellis AFB and will be commanded by USAF Air Combat Command. The SMC is responsible for AF missile and satellite research, development and acquisition.

US Army, Space and Missile Defense Command/Army Forces Strategic Command (USASMDC/ARSTRAT)

Current Status
USASMDC/ARSTRAT (formerly Army Space Command, or ARSPACE), is the Army Service Component Command (ASCC) for US Strategic Command (USSTRATCOM – see separate entry), and is an Army major command (MACOM). As such, USASMDC/ARSTRAT provides space and missile defence related US Army planning and co-ordination, and engages in Army research, development and acquisition activities in support of Army and STRATCOM missions.

USASMDC/ARSTRAT commands and controls the US Army Space and Ballistic Missile Defense Forces (SBMDF), composed of the 1st Space Brigade and 100th Missile Defense Brigade (Ground-based Midcourse Defense). The command is also responsible for Army Astronaut Detachment, Technical Center, Future Warfare Center and Technical Interoperability and Matrix Center (TIMC) activities.

The 1st Space Brigade commands three subordinate Battalions:

- The 53rd Signal Battalion
- The 1st Space Battalion
- The 193rd Space Support Battalion

The 53rd Signal Battalion (formerly the 1st Satellite Control (SATCON) Battalion) is charged with Defense Satellite Communications System (DSCS) operations and management, including five DSCS facilities, located around the world. The Battalion provides around-the-clock communications, command, control, intelligence and logistics information for joint tactical use, to US, joint and coalition soldiers and sailors worldwide. The 1st Space Battalion's Army Space Support Company (ARSST) supports the military's space operations by providing deployed forces with technology, capabilities and expertise. Its Theater Missile Warning Company operates five Joint Tactical Ground Stations (JTAGS) and provides theatre commanders with a transportable system for processing and disseminating near-real-time warning of theatre ballistic missile launches. JTAGS ties directly to theatre communications systems to transmit launch point information, impact area prediction, time of flight and positional information. The 193rd Space Support Battalion is a Colorado Army National Guard unit providing deployed forces with tactical satellite imagery analysis and information.

Ground-based missile defence is the 100th Missile Defense Brigade's responsibility, including

The US Army has been involved with launching satellites since the late 1950s. Here a Juno 2 derivative of the Jupiter missile is prepared for launch (DoD) 0572492

their 49th Missile Defence Battalion in Fort Greely, Alaska. Duties are carried out by Colorado and Alaska Army National Guard soldiers, tasked to protect the US from ballistic missile attack as well as accidental launches projected to enter US airspace. 100th Missile Defense Brigade (Ground-based Midcourse Defense) soldiers are trained to maintain and fire Ground-Based Interceptors (GBI), designed to destroy ballistic missiles while in midcourse phase.

The Space and Missile Defense Technical Center conducts space technology and missile defence research and development for the Army and the US Missile Defense Agency (MDA – see separate entry). It is also the organisation in charge of test and evaluations at the High Energy Laser Systems Test Facility (HELSTF), White Sands Missile Range and the US Army's Kwajalein Atoll/Ronald Reagan Missile Defense Test Site.

The Future Warfare Center is charged with developing space and missile defence doctrine for the Army, while TIMC, on behalf of USASMDC/ARSTRAT, interfaces with technology and equipment developers to guide joint and coalition space and missile defence interoperability and technology requirements. The Army Astronaut Detachment is composed of Army astronauts and trainees and is located at NASA's Johnson Space Center, Houston, Texas.

US Strategic Command (USSTRATCOM)

Current Status

USSTRATCOM is a Unified Combatant Command, headquartered at Offut Air Force Base, Nebraska, tasked with managing US forces and data relevant to the nation's nuclear deterrence and space control missions. Through its service components, USSTRATCOM operates a worldwide network of missile warning sensors providing tactical warning and attack assessment of Inter-Continental Ballistic Missile (ICBM) and Sea-Launched Ballistic Missile (SLBM) launches posing potential threats against North America. They also provide tactical warning of ballistic missile attacks to US Commanders worldwide. The data is also employed by NORAD at Cheyenne Mountain AFS. STRATCOM's ground and space-based missile warning sensors are used for space surveillance. Together with dedicated space surveillance sensors operated by the 21st Space Wing (see separate entry) and contributing US Army Space and Missile Defense Command/Army Strategic Command (USASMDC/ARSTRAT – see separate entry) and civil assets, they form the Space Surveillance Network, which provides timely and accurate detection, tracking and identification of space objects and events. The network maintains a catalogue of more than 9,500 orbiting space objects.

Air Force Space Command (AFSPC – see separate entry), formed in September 1982, is the largest service component, with approximately 25,000 military and civilian personnel and 13,700 contractors worldwide. AFSPC provides forces for strategic and tactical ballistic missile warning, space control, satellite operations, DoD satellite launchings and, since 1 July 1993, ICBM missile operations. The 14th Air Force, activated 1 July 1993, is the day-to-day manager of AFSPC's space forces. The 20th Air Force has daily operational control over ICBMs. AFSPC's 50th Space Wing (see separate entry) at Schriever AFB (formerly Falcon AFB) is the focal point for the daily command and control of more than 140 operational satellites. Responsibility for the space launch bases at Cape Canaveral AFS, Florida and Vandenberg AFB, California fall to the 45th and 30th Space Wings, respectively (see separate entries). The Wings' space launch Squadrons are responsible for Delta, Titan, Atlas and Evolved Expendable Launch Vehicle (EELV) launch operations.

The Army component, USASMDC/ARSTRAT was created in 1997 with the redesignation and responsibilities upgrade of the U.S. Army Space and Strategic Defense Command. The former Army Space Command (ARSPACE) has been absorbed into the USASMDC/ARSTRAT command structure. Naval Fleet Forces Command at Norfolk, Virginia is the Navy's USSTRATCOM Component. Naval Network and Space Operations Command (NNOC), headquartered at Dahlgren, Virginia, formed in 2002 from Naval Space Command and Naval Network Operations Command, and Naval Space and Warfare Systems Command (SPAWAR), headquartered in San Diego, California perform space command, control and acquisition functions for the Navy. Both commands are subordinate to Naval Network Warfare Command, located at Norfolk, Virginia. The Marines also maintain a USSTRATCOM component: Marine Corps Forces, U.S. Strategic Command (MARFORSTRAT).

STRATCOM oversees four Functional Component Commands (FCC) as well as the service commands, and anticipates forming a 5th command to manage

USSTRATCOM Headquarters, General Curtis E. LeMay Building (US Strategic Command) 1116752

and support the US combat of weapons of mass destruction mission, responsibilities for which were assigned to STRATCOM by the US Secretary of Defense in 2005. STRATCOM's FCCs are: Integrated Missile Defense (JFCC-IMD), Intelligence, Surveillance and Reconnaissance (JFCC-ISR), Network Warfare (JFCC-NW) and Space & Global Strike (JFCC-SGS), which oversees the Joint Information Operations Center. FCC commanders and executives also lead other major military and intelligence organisations. Additionally, STRATCOM manages several task forces that work with its service and functional components to meet the command's mission requirements.

Background

In October 2002 Secretary of Defense Rumsfeld merged US Space Command (USSPACECOM) and USSTRATCOM, with USSTRATCOM absorbing USSPACECOM personnel and duties. USSPACECOM, a Unified Combatant Command comprising the Army, Naval and Air Force (AF) Space Commands, was authorised by President Reagan in November 1984 and activated in September 1985. Its singular responsibility was to ensure the US had continuous access to the use of space for military, civil and commercial purposes. Army, Naval and AF space operations had previously functioned separately since the DoD's space programme began in the 1950s. USSPACECOM conducted joint space operations in accordance with Unified Command Plan assigned missions in space forces support, space force enhancement, space force application and space force control.

RESEARCH & DEVELOPMENT

Brazil

Comando-Geral de Tecnologia Aeroespacial (CTA)

Current Status
Comando-Geral de Tecnologia Aeroespacial (CTA), the Aerospace Technology General Command, created in January 2006, oversees much of Brazil's aerospace defence research and development activities. CTA is a command within the Brazilian Air Force, under the Ministry of Defence.

Background
CTA is headquartered in São José dos Campos, São Paulo. Although the Command is, since 2006, the dominant aerospace research command, it evolved from an organisation (and maintains its original CTA acronym), known since 1969, as the Centro Técnico Aeroespacial, the Aerospace Technical Centre. This entity will remain in existence through 2006, until the command restructure is complete. The Aerospace Technology General Command replaces the Departamento de Pesquisas e Desenvolvimento (DEPED), the Air Force's Research and Development Department, to which the Aerospace Technical Centre formerly was subordinate. Since 2006, CTA oversees four aerospace institutes, among other groups, as well as operating Brazil's Alcântara Launch Centre (see separate entry).

Instituto de Aeronáutica e Espaço, (IAE), the Institute of Aeronautics and Space (see separate entry), is one of CTA's institutes. IAE is responsible for developing Sonda and VS sounding rockets, as well as Brazil's Veículos Lançadores de Satélites (VLS) orbital launcher (see separate entry) for the national space programme.

Instituto de Aeronáutica e Espaço (IAE)

Current Status
As of 2006, the Instituto de Aeronáutica e Espaço (IAE), the Aeronautics and Space Institute, is subordinate to the newly formed Comando-Geral de Tecnologia Aeroespacial (CTA), the Brazilian Air Force's Aerospace Technology General Command (see separate entry).

Background
IAE is located on the CTA campus in São José dos Campos, São Paulo. The Institute is responsible for developing the Sonda and VS sounding rockets as well as the Veículos Lançadores de Satélites (VLS) orbital launcher (see separate entry) for the national space programme. IAE conducts research in a number of disciplines, including aeronautics, materials science, and space and defence systems. The Institute is responsible for VLS guidance equipment and has conducted tests with advanced navigation and guidance equipment. IAE has conducted an intensive investigation of the VLS assembly and launch preparation safety criteria, examining a wide range of potential fail points from sneak circuits to spontaneous combustion of propellants. This technology has been adapted for other aerospace applications. IAE also conducts microgravity experiments along with its sounding rocket launches, and partners with a number of organisations and countries, including Germany, with which it has developed the VS sounding rocket series.

On 2 November 1997, the VLS's first launch from Alcântara Launch Centre (see separate entry) ended in catastrophic failure. Due to a first stage ignition network anomaly, the VLS exploded about 3,800 m into its flight. Its second launch on 11 December 1999, also ended in failure after the second stage failed to ignite. A third launch attempt was scheduled for August 2003; this also ended in failure when the vehicle exploded on the pad during launch preparations, several days before its scheduled liftoff, killing 21 people. Satellite payloads were also destroyed in the explosion. Brazil may make a fourth VLS launch attempt some time in 2007.

China

China Aerospace Science and Industry Corporation (CASIC)

Current Status
The China Aerospace Science and Industry Corporation's (CASIC) Aerospace Solid Launch Vehicle Company is developing the Kaitouzhe (KT) launch vehicle series (see separate entry), a DF-21 missile derivative. A KT launched from Xichang on 11 January 2007 is thought to have carried the anti-satellite (ASAT) kill vehicle that China used to destroy its own in-orbit Fengyun meteorological satellite (see separate entry). The test and subsequent explosion created over 1,000 pieces of orbiting debris, and attracted international condemnation.

Background
CASIC took on its present name in 2001. Previous to that, it was known as the China Aerospace Machinery and Electronics Corporation (CAMEC). It became CAMEC when the China Aerospace Corporation (CAC) was split, during 1999 state-owned defence industry reforms, into two separate state owned organisations: CAMEC and the China Aerospace Science and Technology Corporation (CASC – see separate entry). Both organisations report to the Commission of Science, Technology and Industry for National Defence (COSTIND), primarily a civilian administrative arm following the 1999 reforms, and the State Council. CAC was founded in 1993, having evolved from a succession of organisations, beginning with the Number 5 Research Academy of the Ministry of National Defense in 1956.

CASIC employs about 100,000 personnel, 40 per cent of whom are technical staff. It is responsible for roughly 180 subordinate organisations, including six academic institutions.

CASIC is primarily engaged in defence work; it is China's leading missile developer, working alongside CASC in this endeavour. CASIC scientists also design and build satellites as well as guidance systems, among a variety of aerospace electronics products. The organisation also offers IT and systems integration solutions and services.

United Kingdom

DERA – Space Department

Current Status
DERA is a government-owned UK research and development organisation providing independent services to civil and military organisations and industries world-wide. DERA draws upon well-equipped test and production facilities, and state-of-the-art satellite control and ground receiving stations. DERA is involved in most space activities ranging from concept and feasibility studies through to project management, design and hardware construction. Space Department also supports the satellite communications work of the Strategic Communications and Networks Department at DERA Defford.

Activities include analysis of future satellite and ground system requirements, procurement support, post launch mission analysis and resolution of spacecraft operating anomalies. Experience in the UK Skynet military communications programme is used to support overseas customers in the initial studies and procurement of new civil and military satellite systems. DERA provides expertise and operational guidance on control systems and ground station electronics architecture.

In the last two decades DERA has had a responsibility to develop remote sensing applications. Capabilities include SAR and visible/infra-red instrument design, and research into the exploitation of satellite remote sensing data for both civil and military use. Capabilities include digital imaging, radar and wet film photoanalysis and interpretation. Two ground stations for operational reception of meteorological data and high resolution radar data and near real-time delivery of application-specific products support the activities. DERA teams with industry to promote and develop commercially self-sustaining satellite remote sensing markets. Examples of collaboration includes Radarsolutions, a consortium to receive and distribute RADARSAT data across the British Isles and Denmark. DERA works with privately run and publicly owned remote sensing organisations supporting data interpretation and analysis.

DERA's research, development and consultancy is offered on all aspects of space technology, including spacecraft subsystems, attitude and orbit control systems, power systems, onboard data handling, materials and structures, thermal modelling and space debris. In addition, DERA is an acknowledged leader in the field of space environmental effects and ion propulsion. DERA has worked on Ion Thruster technology for more than 20 years, leading the development of the T5 Thruster which is employed in the UK10 Ion Propulsion system. Considerable research work conducted at Culham laboratory on electric propulsion technology and engineering.

DERA also maintains a comprehensive range of test facilities available to industry and academia for qualification testing of space hardware, instruments, components and satellites.

DERA runs a micro-satellite development programme offering a complete service to organisations wishing to take advantage of the comprehensive DERA space capability. This includes complete systems, satellite design and predictive modelling, comprehensive testing, manufacture and the complete provision of ground segment facilities.

Facilities
West Freugh satellite ground station
Acquisition of radar data
ERS-1,2; RADARSAT
Reception downlink at X band
Lasham satellite ground station
Acquisition of meteorological data:
NOAA, Meteosat, GOES
COSPAS/SARSAT data dissemination
L&S band TT&C
Image Data Facility
Image and data transcription facilities
Space Test Facility
Thermal vacuum chambers
Solar simulator
Vibration testing
Mass properties facility
Solar cell measurement and testing.

United States

Air Force Research Laboratory (AFRL)

Current Status
TacSat-2, AFRL's tactical communications technology demonstration satellite, launched in December 2006, was a programme requiring joint participation from both AFRL's Space Vehicles and Propulsion Directorates. Space Vehicles designed the spacecraft

and Propulsion designed the craft's Hall effect thruster, its electrical station-keeping system. The 370 kg micro-satellite will be in service demonstrating responsive space technologies and processes for one year, however its propulsion system has a three-year design life. AFRL expects to launch TacSat-3 during the last part of 2007.

In 2005, the Base Realignment and Closure (BRAC) Commission decided to move AFRL's Hanscom Air Force Base (AFB) assets to Wright-Patterson AFB and Kirtland AFB, to merge with existing AFRL facilities. The consolidation will be completed between 2007–2012. Roughly 200 Space Vehicles Directorate personnel will transfer from Hanscom to Kirtland.

Background

In 1997, the US Air Force (USAF) further consolidated its research assets, including the Air Force Office of Scientific Research (AFOSR), into one entity: the Air Force Research Laboratory (AFRL). AFRL absorbed what was then known as the Phillips Laboratory, as well as three other 'Superlabs': Armstrong, Rome and Wright-Patterson. A former element of the USAF Space and Missile Systems Center, the Phillips Laboratory was founded in December 1990 as part of the USAF's reorganisation and consolidation of its 14 existing laboratories and research centres into the four former 'Super Laboratories'. Phillips studied military space, missile, directed energy, propulsion and geophysics research and development. It replaced the USAF Space Technology Center, created in October 1982, at Kirtland Air Force Base (AFB); the USAF Weapons Laboratory at Kirtland; USAF Geophysics Lab at Hanscom AFB; and the USAF Astronautics Lab at Edwards AFB.

In total, AFRL employs about 5,400 personnel in facilities across the US. The Lab oversees a USAF science and technology budget worth about USD2 billion, as well as a further USD1.7 billion in customer revenue.

AFRL Directorates

AFRL is comprised of ten Directorates, charged with investigating a wide range of space and aeronautical subjects.

Space Vehicles Directorate

The Space Vehicles directorate is based at Kirtland AFB in New Mexico, with further facilities at Hanscom AFB in Massachusetts. It is the USAF lead for space vehicle structure and technology research and development and was formed from the Phillips Lab's Space Technology, Space Experiments and Geophysics Directorates. The Directorate controls an annual budget of about USD378 million and employs roughly 940 personnel, including contractors. It is comprised of three divisions: Spacecraft Technology; Battlespace Environment; and Integrated Experiments and Evaluation.

The Space Vehicle Directorate focuses upon developing and pursuing spacecraft and space superiority technologies and processes, including space structures; control systems and sensors; power generation; electrical components; launch vehicle and spacecraft mechanisms; microsystems; infra-red, Space-Based Radar (SBR) and cryogenic technologies; specialised software; space weather data collection and analysis; and integration and testing.

Propulsion Directorate

AFRL's Propulsion Directorate is headquartered at Wright-Patterson AFB in Ohio and also maintains facilities at Edwards AFB in California. It conducts research and development associated with four separate technological sectors: Space and Missile propulsion; Aerospace; Power; and Turbine Engines.

The Space and Missile Propulsion Division undertakes the research and development of advanced solid, liquid, hybrid and electric missile and space propulsion on a 170 km² Edwards AFB site. Facilities include test stands accommodating engines up to 45 MN thrust and altitude chambers in which attitude control thrusters can fire up to 7 hours continuously. The National Hover Test Facility to support the Missile Defence Agency (MDA) was inaugurated in November 1988 by the first hovering trials of a space-based interceptor demonstration model. The Aerospace Division conducts research into hypersonic vehicles. The Directorate teamed with the Space Vehicles Directorate on AFRL's Military SpacePlane (MSP) project.

Directed Energy Directorate

Formerly the Phillips Lab's Lasers and Imaging Directorate, the Directed Energy Directorate is also located at Kirtland AFB. Elements of the Directorate support the USAF's space control mission, operating sophisticated experimental telescopes. The Starfire Optical Range at Kirtland is one of these. The Directorate also manages telescope facilities at the Maui Space Surveillance and White Sands Missile Range Sites.

The Directed Energy Directorate performs and directs research and development of technologies for directed energy weapons (microwave, laser and particle beam) and related advanced weapon concepts, space systems survivability and ElectroMagnetic (EM) effects. It measures and models directed energy effects on air and space systems and determines lethality to threat targets, develops hardening countermeasures and transitions survivability technology to product and logistic centres.

Air Vehicles Directorate

The Air Vehicles Directorate is based at Wright-Patterson AFB. It engages in some space-related activities, primarily related to responsive access to space, the transition of air vehicle technologies to space vehicle technologies or applications and global strike capabilities.

Sensors Directorate

The Sensors Directorate is based at Wright-Patterson AFB. Among its responsibilities, the Directorate develops electro-optical sensors for space applications, including threat detection, targeting, and countermeasures sensors.

Munitions Directorate

Headquartered at Eglin Air Force Base, Florida, the Munitions Directorate develops conventional air-launched weapons, primarily for air applications.

Materials and Manufacturing Directorate

Based at Wright-Patterson AFB, with facilities at Tyndall AFB in Florida, the Materials and Manufacturing Directorate conducts materials research and development for both aircraft and spacecraft. The Directorate's research reaches into survivability and sensors materials research as well as air and spacecraft construction. It also houses a Nonstructural Materials for Space Systems section that looks into new materials such as thermal coatings and lubricated systems that can increase spacecraft design life, efficiency, survivability and effectiveness. The section has also used the International Space Station as a base to conduct materials experiments.

Information Directorate

The Information Directorate is Headquartered at Wright-Patterson, however its primary facilities are located on a site that was once part of Griffiss AFB in Rome, New York; it engages in several space-related Information Technology (IT) pursuits and manages facilities relevant to satellite Intelligence, Surveillance and Reconnaissance (ISR) and Command and Control (C2) endeavours. Among these are the Advanced Ka Band SATCOM Test Facility and the Electronic Intelligence (ELINT) Development Facility.

Human Effectiveness Directorate

The Human Effectiveness Directorate is based at Wright-Patterson and also maintains research facilities at Mesa, on the former Williams AFB site, and at the USAF's Brooks City Base. The Directorate studies human factors in the air and space warfighting environments as well as force protection technologies and processes.

Air Force Office of Scientific Research

The Air Force Office of Scientific Research (AFOSR) is included within the AFRL structure. AFOSR is based in Arlington, Virginia and works with the academic community, as well as government, to plan and co-ordinate the broad range of research projects in which it invests.

Applied Physics Laboratory of The Johns Hopkins University

Current Status

In January 2003, APL announced that NASA had awarded a contract to formulate, implement and operate multiple spacecraft for the agency's Geospace missions to study the effect of the Sun on earth's magnetic field.

Background

Founded in 1942, APL is a not-for-profit R&D laboratory and independent division of The Johns Hopkins University. It operates primarily under a contract with the US Naval Sea Systems Command. The DOD sponsors some three-quarters of APL's effort. Full-time staff total 2,700, of which 400 work within the Space Department's two branches, named Engineering and Technology and Science and Analysis.

APL is a major centre for satellite development, construction and operation. Through 1996, it had built 57 satellites (45 of them 50–180 kg). The Laboratory gained early experience by putting instruments on captured German V2 rockets after World War II. Based on its analysis of Sputnik 1's Doppler signals, APL conceived and built the Transit satellite navigation system for USN. It built and operated Geosat, provided the altimeters on Geos-C, Geosat, Seasat and Topex, and led the AMPTE project for NASA. Topex also includes APL's laser retro-reflectors and frequency reference unit, the most stable oscillator ever orbited. Earlier satellites include Dodge, Anna, Triad, Traac, Magsat, Nova and three small astronomy satellites, Injun, Hilat, Polar Bear. The HUT Hopkins UV Telescope (a joint project with the Department of Physics & Astronomy) flew on Shuttle's 1990 Astro 1 mission and March 1995's Astro 2. Voyager, Galileo and Ulysses carry particle radiation detectors. Japan's Geotail includes the EPIC Energetic Particles and Ion Composition experiment. Sweden's Freja and NASA's UARS carry APL magnetic field instruments.

Current projects include the NEAR Near-Earth Asteroid Rendezvous craft (launched February 1996) successfully orbiting the asteroid Eros since February 2000 until it was put to the surface in February 2001. Mid-course Space Experiment (MSX, launched April 1996), the ACE Advanced Composition Explorer, the TIMED spacecraft and the Magnetospheric Imaging Instrument for Cassini. Facilities include:

Richard B Kershner Space Integration and Test Facility

The test facility, 7,350 m², was dedicated in 1983. It houses seven clean rooms: three class 100,000 (139 + 2 × 93 m²); two 93 m² class 10,000; two 22 m² class 100 rooms are suitable for assembling small space instruments; they can also be held sterile for assembling implantable biomedical devices.

The Space Simulation Laboratory

The SSL includes two 2.4 m diameter 3 m high vertical thermal vacuum chambers (–196/+121°C; 10^{-7} torr cryopump and diffusion pump), a 0.91 m diameter 0.91 m long space simulation chamber (–78/+126°C; 10^{-8} torr) and a 1.52 × 1.83 × 2.13 m T/humidity/altitude chamber (–100/+125°C; 35 to 95 per cent humidity; 0 to 61 km altitude), in its 447 m² area.

The Vibration Test Laboratory

A facility that offers: two Unholtz-Dickie shakers (T4000: 178 kN, 152 × 152 cm Team bearing-line slip table; T1000: 89 kN, 4.4 cm peak-to-peak displacement, 61 × 61 cm combined slip table assembly) and an MRC mass properties machine (2,270 kg capacity).

The Solar Array Laboratory and Simulator Control Room

A Large Area Pulsed Solar Simulator (LAPSS) characterises and calibrates solar panel and coupon performance using a Xe source. LAPSS can calibrate panels up to 1.83 m across in 1 ms. Smaller laboratories serve the Satellite Reliability Group, where new parts are inspected and radiation tested.

Satellite Communications Facility

Originally developed as a dedicated Transit station, the 623 m² facility has been upgraded for S/L/X-band station for a wide range of NASA and DoD satellites. It is used for continuous MSX operations (a 450 m² addition houses a new command/control centre and a data-processing centre for MSX). Antennas: 18.3 m parabolic, 2.2 to 2.3 GHz autotrack feed, receive G/T 23 dB/K, simultaneous RH/LHCP. A 5 m parabolic, 2.2 to 2.3 GHz + 1.650 to 1.750 GHz feeds with selectable RH/LHCP, rx G/T 14 dB/K at 2.2 GHz; 10 m parabolic X/S/L-band simultaneous coverage, fully operational SGLS site with 0.2/2 kW transmit, associated 1 m acquisition antenna. Real-time computers steer the dishes and redundant back-up control and configure the 10 m dish.

Time and Frequency Standards Laboratory

Supports APL projects requiring highly stable time (1 pps), frequency (1, 5, 10 MHz) and time code signals (IRIG-A/B). Other activities include measurement of frequency standards stability, design of atomic hydrogen masers, spaceborne oscillators and synthesisers, and superconducting oscillators.

Anechoic Chamber
7.62 × 7.62 m internal, accessed by 1.83 × 2.44 m door, attenuates outside interference by >70 dB over 0.1 to 10 GHz.

Arnold Engineering Development Center (AEDC)

Current Status
Alliant Techsystems (ATK) announced on 15 Feb 2008 that AEDC completed an acceptance test of the Combustion Air Heater (CAH). Under contract with AEDC the CAH was designed by ATK at the company's facilities in Ronkonkoma New York and Florida. The CAH was installed at AEDC's Aerodynamics and Propulsion Test Unit facility and will be used for hypersonic engine development, where air flow speeds of up to mach 8 can be simulated using a combustor and rocket engine-like injector array within the CAH.

AEDC announced a successful aerodynamic test of NASA's Mars Science Laboratory's (MSL) parachute on 2 February 2008 at the Center's National Full-Scale Aerodynamics Complex, where Martian atmospheric subsonic entry conditions were simulated in preparation for the MSL launch scheduled in 2009. The parchute was built by Pioneer Aerospace, Connecticut and supports 36,000 kg (80,000 pds) loads and measures 65 ft in length and 55 ft in diameter.

Arnold Engineering Development Center (AEDC) announced on 9 September 2007 aerothermal testing of the heat shield candidate materials for NASA's next generation Orion Crew Exploration Vehicle (CEV) has been conducted at the Center's H2 test facility. Re-entry conditions were reproduced with temperatures of 13,000 degrees Rankine. Heat is created through arc heaters which is then confined to a high pressure water-cooled channel. The heat flows through the nozzle in an evacuated chamber over the arc-heated sample. This was following on from preproduction tests on the sample heat shield that was part of the calibration facility validation tests in June 2007. Further aerothermal tests are planned. The Mars Science Laboratory has also undergone H2 testing at AEDC.

The large main fuel tank incorporates a Thermal Protection System (TPS) and the materials used prevent any ice or frost forming during the launch phase. A foam sample was used to evaluate the TPS performance under simulated flight conditions similar to the ascent of the space shuttle, which was conducted at ADEC's von Karman Facility Hypsersonic Wind Tunnel C during the second half of 2007. The test originally began in March but had to be halted due to cooling problems associated with the tunnel. A team of 200 specialists were assembled to tackle the problem, which ensured deadlines with the Center's customer NASA were met.

In June 2007 AEDC conducted tests at the von Karman Gas Dynamics Facility's Wind Tunnel B for NASA's Hypersonic Boundary Layer Transition (HyBoLT) experiment and ATK's ALV X-1 launch vehicle, set to launch from Wallops Flight Facility in 2008. AECD tested the vehicle at Mach 6 and Mach 8 conditions. This marks the first wind tunnel test for the HyBoLT/ALV X-1 stability. Further testing for supersonic machs will be conducted at NASA Langley Research Center, where the test article has been moved to and eventually transonic testing at NASA Ames Research Center. The HyBoLT payload will reside on the nose of the launch vehicle and an internal payload SOAREX will be ejected to collect atmospheric re-entry technology data.

Also in June the Blended Body Wing (BBW), a two per cent scale model of a Boeing flying wing, was tested at AEDC's 16-foot transonic wind tunnel (16T). The test was sponsored by the Air Force Research Laboratory (AFRL) in a co-operative venture between NASA, Boeing and the US Air Force.

AEDC conducted an aerothermal test of a full scale model of the Terminal High Altitude Area Defense (THAAD) missile system nosecone, as announced on 8 May 2007. The thermal structure of the new infra-red seeker window was tested beyond the final stages of a missile flight at the Hypervelocity Wind Tunnel 9 facility.

In a configuration upgrade, the first system checkout has been conducted for AEDC's Characterization of Combined Orbital Surface Effects (CCOSE) chamber, announced in March 2007. In addition to testing real-time orbital simulation with respect to low earth orbit and geosynchronous Earth orbit, the CCOSE is able to test environmental effects on space hardware and materials which occur in space. In May 2006 the first test was run using thermal control paint. The checkout, which involved organisations located within AFRL's Directed Energy Directorate, who provided the thermal paint and solar cell, was completed in December 2006.

A scale model of the NASA's CEV has undergone Mach 8 and 10 tests at AEDC's Hypervelocity Wind Tunnel 9 to obtain surface temperature data using a Temperature Sensitive Paint (TSP) system. To assess the heat transfer of the CEV a sensitive charge coupled device camera, part of the TSP system, as well as a special paint and ultra violet illumination source, provides experimental data for the Orion database to validate Computational Fluid Dynamic (CFD) models.

Background
AEDC is located on Arnold Air Force Base in Tennessee, on the former Camp Forrest site, and is part of the US Air Force Materiel Command. It is the Command's aircraft, missile and spacecraft systems test centre, and has also provided test services for Department of Defense, the US Army, the US Navy and NASA. AEDC was founded toward the end of the Second World War, by General Henry H 'Hap' Arnold and Dr Theodore von Karman, as an Air Force facility responsible for the Service's long term aeronautical technology research and development goals. The facility took its present name in 1951. AEDC also operates facilities in Silver Spring, Maryland and at NASA's Ames Research Center in California.

For many years AEDC has operated the world's largest and most advanced complex of aerospace simulation test facilities. AEDC is home to 58 wind tunnels, space environment chambers and engine and propulsion test units. Divided into AEDC's product capabilities, Test Squadrons (718th, 717th, and 716th), are responsible for the center's test facilities. With regard to space missiles, the 718th Squadron conducts space simulation, hypersonics, rocket propulsion and missile signature testing. Test units include the deep space environmental 7 V (20 Kelvin (K) cryogenically-cooled liner and optical bench giving a background of low infra-red) and the 10 V (20 K low infra-red background using LightTight GHe Cryoliner) Space Chamber, both with simulated altitudes of 200 miles, testing interceptors and surveillance sensors. The thermal vacuum 12 V Space Chamber provides for 77 K conditions with a shroud of liquid nitrogen lining the chamber, additionally optional helium inner liner to reduce the temperature a further 10 K can be applied. The 12 V Space Chamber also has a solar simulation capability. Other units include the Aerodynamic and Propulsion Test Unit, High-Enthalpy Arc-Heated Facilities, Hypervelocity Gun Range, J-6 Large Rocket Test Facility, Mark 1 Test Facility and the Rocket Development Test Cell J-4. The latter facility is currently inactive. The 717th Test Squadron is responsible for aeropropulsion and the 716th deals with flight systems-aerodynamics. Facilities under 716th Test Squadron include: Aerodynamic Wind Tunnel 4T (operated from 0.2 to 2.0 Mach); Hypervelocity Wind Tunnel 9 (from Mach 7 to 16.5); Propulsion Wind Tunnels 16S and 16T (from 0.06 to 1.60 Mach); and the von Karman Gas Dynamics Facility Wind Tunnels A (supersonic), B and C (hypersonic) operated from 7 to 16.5 Mach. The tunnels are supported by the Aerodynamic Measurement Laboratory (ATMLab), although the 16 foot supersonic (16-S) tunnel is currently not active.

AEDC performs engineering tests and mechanical evaluations relating to government programmes and commercial projects. For example in March 2006 AEDC was contracted by NASA with regard to CEV testing to provide Pressure Sensitive Paint (PSP) system for NASA Ames' 4T and Tunnel A. A portable system was specially developed and was installed at NASA facilities. The PSP system enables surface temperature data to be assimilated, providing CFD validation. AEDC has also performed tests for high profile space programmes as the Navstar GPS and GOES-M satellites; the Atlas and Titan launch vehicles; and for human spaceflight programmes including the Space Shuttle, International Space Station and Apollo.

Specifications
Mark 1 Test Facility
Mission: Space environment simulation test chamber.
Thermal: Chamber lined with liquid nitrogen shroud provided thermal conditions down to 77 degrees Kelvin.
Vacuum Range: ≤5 × 10^{-7} Torr to local atmospheric pressure.
Working Volume: 11 metre × 21 metres (36 foot diameter × 70 foot tall)
Support Structure: 200, 000 pds/12-foot-square seismic mass provides vibration isolated of test articles.
Pumping system: Mechanical roughing pumps, turbo molecular and cryogenic high vacuum pumps, all with cryogenic traps.
Loading: Vertical: 20 foot opening on the top chamber, via a 15-ton precision crane.
Horizontal: 8-foot diameter manway in the chamber side near the bottom.
Cold Wall: Lined with 77 degrees Kelvin liquid nitrogen Cryogenic shroud.
Special features: Laminar flow optical quality clean room (400 square foot, better than Class 1000). 15-ton precision crane.
Data: Time lagged and archived. 1,500 channel test data system available.
Work Areas: Test customer offices available. Limited hardware storage. For test article build-up two large bay areas available.
Solar simulation: Infra-red lamp arrays available for thermal cycling.
Other: 15 degrees Kelvin gaseous helium cyropumping capability available.

Department of Defense, Defense Advanced Research Projects Agency (DARPA)

Current Status
DARPA was established in February 1958 as the Advanced Research Projects Agency, partially as a response to the technology shock of the Sputnik I launch. On 23 March 1972 the agency's name was changed to the Defense Advanced Research Projects Agency, and it was established as a separate agency under the US Department of Defense (DoD). DARPA has no laboratories and few other fixed assets, and operates in collaboration with industry, universities, the armed forces, intelligence agencies and NASA. It is charged with pursuing very ambitious technical developments, often with a cross-service application. For example, it funded development of the Pegasus and Taurus launch vehicles, several small satellites and numerous advanced technologies for satellite subsystems and payloads.

In 1994 DARPA officially withdrew from space activities, returning to structure the Discoverer II programme at the end of that decade, a joint initiative with the US Air Force and the National Reconnaissance Office to develop an affordable space-based radar for tactical geolocation. Since that time, DARPA's space activities have steadily expanded. Programme managers and resources from its seven principal offices – Advanced Technology, Defense Sciences, Information Processing Technology, Information Exploitation, Microsystems Technology, Special Projects and Tactical Technology – have been formed into a Virtual Space Office (VSO). VSO is "virtual" insofar as it draws resources from the other offices. VSO's overriding goal is to develop new and less costly ways to meet military requirements for space access and on-orbit missions.

Dr. Gary Graham, deputy director of DARPA's Tactical Technology Office (TTO) stated at the agency's 2005 DARPATech conference in Anaheim in August 2005 that DARPA's FY2005 space budget was USD400 million, representing a factor-of-40 increase in four years.

Department of the Navy, Office of Naval Research and Naval Research Laboratory

Current Status
The Office of Naval Research (ONR) Science and Technology (S&T) Directorate conducts space-related activities through its Ocean, Atmosphere

A 50-foot radio telescope dish atop NRL's main administration building is the unofficial symbol of the Laboratory (Naval Research Laboratory) 1133738

and Space (OAS) Department, however a great deal of ONR's space-related research and development takes place at Naval Research Laboratory (NRL) installations. OAS maintains two divisions, Sensing and Systems, and Processes and Prediction. Sensing and Systems conducts remote sensing research and development relevant to atmospheric and marine applications and operations, such as the investigation of radio frequency (including Global Positioning Satellite frequencies) and space weather phenomena that affect Naval communications and intelligence systems. ONR reports to the Secretary of the Navy and advises the Secretary and the Chief of Naval Operations on technical matters.

NRL was established in July 1923 as the Naval Experimental and Research Laboratory. Now an ONR element, it is the navy's corporate laboratory. Staff, mostly civilian, operate primarily in Washington DC, Stennis Space Center, Mississippi, Monterey, California and Maryland Detachments. NRL's Naval Center for Space Technology (NCST) maintains its space technology expertise within two departments: Space Systems Development (SSDD) and Spacecraft Engineering (SED). The Remote Sensing Division (RSD) and the Space Science Division (SSD) fall under NRL's Ocean and Atmospheric Science and Technology Directorate.

SSDD and SED work together to satisfy spacecraft mission requirements for the US Navy, DoD, and other national security interests. The two departments are jointly responsible for spacecraft research, design, development, construction and testing. SSDD and SED are involved in the development and assembly of satellite, launch vehicle and propulsion systems, as well as on-board communications and power systems. NCST facilities include cleanrooms, anechoic RF chambers, shock/vibration chambers, an acoustic reverberation chamber, thermal/vacuum chambers and long term testing of satellite clock time/frequency standards.

RSD undertakes a broad programme in sensing applications over frequencies from ultra-violet to radio. Sensor systems include Real Aperture Radar (RAR), scatterometers, lidars, optical/radio interferometers and passive microwave imagers. Its middle atmosphere sensors include the Polar Ozone and Aerosol Monitor (POAM) on Spot 3, the Millimetre-wave Atmospheric Sounder (MAS) on Shuttle Atlas missions (three flown), and a suite of water vapour and ozone monitors as part of the Network for Detection of Stratospheric Change. RSD's facilities include digital image processing, a tactical environmental visualisation centre, an aerosol measurement facility, 25.6/25.9 m radio antennas at Maryland Point Observatory and an optical interferometer at Mount Wilson Observatory for monitoring background environmental emissions at high angular resolution.

SSD maintains facilities for designing, constructing, assembling, calibrating and analysing space experiments, principally upper atmosphere UV sensing, solar atmosphere spectrometry and celestial radiation over UV to cosmic rays. For example, it provided the Solar UV Spectral Irradiance Monitor (SUSIM) for the Upper Atmosphere Research Satellite (UARS) and Atlas, the Remote Atmospheric and Ionospheric Detection System (RAIDS) to measure airglow, the Oriented Scintillation Spectrometer Experiment (OSSE) gamma instrument on NASA's Compton Observatory, solar X-ray measurements using a Bragg crystal on Japan's Yohkoh and the Large Angle Spectrometric Coronograph (LASCO) on Soho. SSD in 1999 provided three of the main Advanced Research and Global Observation Satellite (ARGOS) instruments: Unconventional Stellar Aspect (USA), High Resolution Airglow and Aurora spectroscopy (HIRAAS) and Global Imaging Monitor of the Ionosphere (GIMI).

NRL directed the Vanguard programme of the mid-1950s, America's first publicly revealed satellite programme. Since 1960 it has developed more than 80 satellites, including the Clementine (also known as the Deep Space Program Science Experiment) technology demonstrator lunar/asteroid probe for the former Ballistic Missile Defense Organisation (now the Missile Defense Agency) launched in January 1994. It has flown numerous payloads and experiments, such as the Hercules unit carried by Space Shuttle STS-53 in December 1992 to allow astronauts to photograph surface features while automatically recording positions within 2 km. It flew again on STS-56 in April 1993 and on STS-70 in July 1995.

Los Alamos National Laboratory (LANL)

Current Status

Los Alamos National Laboratory's (LANL) Cibola Flight Experiment (CFE) was launched aboard an Atlas V in March 2007. Cibola is designed to detect Earth-based nuclear detonations, like the Fast On-orbit Recording of Transient Events (FORTE) satellite before it. Surrey Satellite Technology, Ltd (SSTL) built the CFESat micro-satellite platform.

Background

Los Alamos National Laboratory (LANL), founded in 1943, is a multidisciplinary research facility operated by Los Alamos National Security, LLC for the Department of Energy's (DoE) National Nuclear Security Administration (NNSA). Los Alamos National Security LLC is a consortium of organisations comprised of Bechtel National, the University of California, BWX Technologies, and Washington Group International. LANL undertakes research in nuclear and non-nuclear defence programmes, nuclear safeguards and security, biomedical sciences, computational sciences, materials science, and environmental cleanup. LANL's first director was J Robert Oppenheimer, who helped found the Laboratory to develop the first atomic weapons under the Manhattan Project. As its primary mission, LANL, along with Sandia National Laboratories (see separate entry), manufactures and oversees the US's nuclear weapons arsenal.

LANL encompasses about 93 km^2 of mesas and canyons in northern New Mexico. The Lab manages an annual budget of about USD2.2 billion and employs a workforce of 9,000, in addition to about 650 contractors who work at the site.

LANL is responsible for developing instruments that monitor Earth-based nuclear explosions from space. Its detectors are in orbit aboard the Defense Support Program (DSP) early warning, and lower altitude GPS navigation satellites.

LANL also participates in science satellite and planetary projects. Its Alexis satellite was launched in 1993 to conduct an all-sky survey of the soft X-ray background glow. LANL and Sandia National Laboratory launched their Fast On-orbit Recording of Transient Events (FORTE) satellite on a Pegasus launch vehicle to demonstrate technology to look for pulses from low-technology nuclear explosions. LANL provided the X-ray detectors for the High Energy Transient Explorer (HETE) mission, three instruments on NASA's Polar, four imaging spectrometers on ESA's Cluster, two gamma-ray detectors on Mars-96 and a gamma-ray spectrometer on Rosetta's comet lander. It collaborated with Finland's VTT to provide Cassini's Ion Beam Spectrometer.

LANL operates the only US facility capable of fabricating Pu-238 into ceramic form as a component in the manufacture of Radioisotope Thermoelectric Generators (RTGs): New Horizons has one, Cassini has three, Galileo had two and Ulysses one. The laboratory provided technical direction for the Strategic Defense Initiative's (SDI) Beam Experiments Aboard a Rocket (BEAR) flight of July 1989, which fired the first neutral particle beam in space.

Sandia National Laboratories (SNL)

Current Status

Among its many nuclear technology and security responsibilities, Sandia develops and builds space-based proliferation detection technologies, such as the Global Burst Detector (GBD) aboard the Navstar Block IIR GPS Space Vehicle Number (SVN) 47 satellite launched in December 2003. The GBD, which detects atmospheric and near-space nuclear detonations, is part of the US Nuclear Detonation Detection (NUDET) system, components of which are present on all US Air Force Navstar GPS satellites. Sandia has also developed plug-in software that enhances commercial software capabilities used for multispectral and thermal remote sensing.

Background

Sandia was established in 1949 and is now operated for the US Department of Energy's (DoE) National Nuclear Security Administration (NNSA) by Lockheed Martin's Sandia Corporation. The Laboratories are active in defence and energy national security programmes, with a principal emphasis on nuclear weapons research and development. Headquartered in Albuquerque, New Mexico on Kirtland Air Force Base, Sandia's personnel total 8,500. The laboratory also employs staff at facilities in Livermore, California, the Tonopah test range in Nevada, operated by Sandia in co-operation with the USAF, and the Kauai Test Range in Hawaii. Sandia manages a programme portfolio worth about USD2.3 billion.

Sandia's aerospace research areas include flight control and telemetry systems, Synthetic Aperture Radar (SAR) systems, robotics, materials, high-velocity atmospheric entry vehicles and aerothermal designs for hypersonic aerospace vehicles, and computerised and physical laboratory testing of space debris impact on satellites; Sandia's Hypervelocity Launcher gas gun accelerates particles up to 12 km/s. It also develops and operates suborbital launch vehicles for the DoE, the military and other users. Tonopah Test Range at the north end of Nellis Air Force Base, Nevada, houses test facilities for small rockets.

Since 1962, Sandia has operated the Kauai Test Facility on the US Navy Pacific Missile Range Facility (PMRF) at Barking Sands. Sandia maintains its own launch, handling and launch control equipment and operates its Strategic Target System (STARS) and sounding rocket programmes from this location. On a typical target mission, STARS is launched from Kauai to impact in the Kwajalein Atoll. The DoE's Strypi family of solid propellant rockets has been launched from Kauai since the 1960's.

Starfire Optical Range

Current Status

According to some reports, the Starfire Optical Range has been engaged in anti-satellite weapons related research since at least January 2006; reports claim the Range is researching the use of ground-based directed energy weapons to disable Low Earth Orbiting (LEO) satellites. Starfire has studied ground-based laser beam control technologies for many years, and reportedly in 1973, for the first time, disabled a remote-controlled aerial target using a ground-based high-energy CO_2 laser.

The Range has worked for a number of years to develop the means to reliably track space-based objects using adaptive optics, as well as the ability to focus a laser beam upon a moving target at increasingly remote altitudes, for satellite communications applications. The first such demonstrations using Starfire's 3.5 m observatory were completed in 1994–1995, when the Range successfully trained a ruby laser and a frequency-doubled YAG laser upon several LEO satellites located in orbits between 1,000–6,000 km.

Background

Formerly the Sandia Optical Range, a unit of the Air Force Weapons Laboratory (AFWL), the Starfire Optical Range took its present name in the early 1970s. Following the US Air Force's 1997 reorganisation of its laboratories into the Air Force Research Laboratory (AFRL – see separate entry),

Starfire became subordinate to AFRL's Directed Energy Directorate. Starfire is located on Kirtland Air Force Base in New Mexico.

Starfire's facilities include a 3.5 m satellite tracking telescope equipped with adaptive optics, as well as a 1.5 m adaptive optics telescope, which was the first of its kind, and a 1.0 m auxiliary beam director. The 3.5 m telescope one of the largest in the US, joined the existing 1.5 m telescope and 1 m laser director for research into space object tracking and atmospheric compensation techniques. The 3.5 m, USD27 million optical observatory was installed in 1993 and began operations in 1994. The 1.5 m mirror carries 341 actuators 7 mm apart on its 1 mm-thick glass. The system samples atmospheric distortion 1,700 times per second by observing laser scattering off molecules in the lower atmosphere, and the continuously adjusts its deformable mirror.

During the 1980s the Starfire Optical Range did much of the pioneering work on adaptive optics and compensated imaging technology development, now used in the US Air Force's Airborne Laser (ABL) programme. In 1993, it produced the first important astronomical observation via adaptive optics: ionised hydrogen clouds around Orion nebula stars. The Starfire facility routinely tracks objects in space and others descending through the atmosphere. Two technical workers used the re-entering Shuttle Columbia as a test target during its tragic descent on 1 February 2003, but those images were not taken by the facility's main telescope.

SPACE INDUSTRY

Joint Ventures
Major Sub-contractors
Prime Contractors

JOINT VENTURES

International

Eurockot Launch Services GmbH

Current Status
Eurockot will launch the second Space Environment Reliability Verification Integrated System (SERVIS) in 2009 for the Institute of Unmanned Space Experiment Free Flyer (USEF), Tokyo, Japan. USEF is under contract to the New Energy and Industrial Technology Organisation (NEDO), under the Ministry of Economy Trade and Industry (METI) of Japan, to build the satellite. SERVIS-2 is part of Japan's space environment verification of commercial off the shelf parts and technologies.

The European Space Agency has signed an agreement with Eurockot to launch the Gravity field and steady-state Ocean Circulation Explorer (GOCE) spacecraft into a Sun-Synchronous Orbit (SSO), at approximately between 270/300 km altitude, in the first half 2008. The GOCE is part of the European Space Agency's (ESA) Earth Observation Envelope Programme (EOEP) and represents one of the four scheduled Earth Explorers (EE) under the living planet mission. The Soil Moisture and Ocean Salinity (SMOS) is another EE mission and will also be launched by Eurockot in early 2008. An adapter system built by Khrunichev Space Centre (KSC) is being used for the SMOS spacecraft alongside PROBA-2 (Project for On-Board Autonomy), a secondary payload built by Verhaert Space, Belgium, for the ESA. PROBA-2 was successfully integrated to the adapter interface ring on 24 July, 2007. The SMOS satellite underwent fit checks in December 2007 with the adapter system, releasing and separating successfully during the test.

Background
Eurockot Launch Services was formed in 1995 and is a joint venture between the European Aeronautic Defence and Space Company (EADS) and Khrunichev State Research and Production Space Centre. EADS holds 51 per cent shares in Eurockot and Khrunichev the remaining 49 per cent. The company is located in Bremen, Germany and provides international launch services for Low Earth Orbit (LEO) satellites at Plestsk Cosmodrome, north Russia at 63 degrees latitude and 40 degrees longitude. The company began commercial launches in 2000 and has performed eight successful launches delivering 16 spacecrafts to their required orbits. Eurockot procures the initial launch services and finalises the launch contract with the customer, which is then sub-contracted to Khrunichev to perform the launch mission.

Additional Eurockot customer obligations include providing a dedicated mission team and operational onsite management during the pre-launch phase, as well as mission management services. Customers include government agencies, satellite operators and research institutes.

The Rockot is a three stage liquid fuelled launch vehicle with a re-ignitable Breeze KM upper stage engine. The launch vehicle is capable of lifting 1,950 kg multiple, dual or single stage LEO satellites for earth observation, remote sensing, communications and scientific disciplines into SSO, near polar and highly inclined orbits. The Rockot is based on SS-19 inter-continental ballistic missile. Larger satellites can be accommodated as well as smaller payloads for planetary missions with an additional propulsion module.

Eurockot provides payload adapter systems for mounting spacecrafts, which can be customised for the customer. Either a Clamp Band Separation Systems (CBSS) or a point fixation system incorporating a Mechanical Lock Systems (MLS) can be used. The adapter or dispenser systems include the electrical interface between the payload and ground support. The company uses CBSS from EADS Construcciones Aeronáuticas SA (CASA), Spain with interface diameters of 937 mm and 1194 mm; and SAAB from Sweden with 600 mm, 937 mm and 1194 mm interface diameters. Both differ in the CBSS tension. A CBSS was used on Servis-1 and Kompsat-2. Compatible payload adapters are provided in 600 mm, 937 mm, 1194 mm and 1664 mm available in either aluminium or carbon fibre structures. For example a cylindrical aluminium payload adapter was specifically designed for the GOCE spacecraft. Russian MLS are flight proven and an alternative to traditional CBSS using an interface ring. The payload adapter using a MLS is customised to the shape and mass of the spacecraft. For example CyroSat, a European Space Agency mission, used a single adapter MLS. However due to an error in the flight control system the CryoSat launch aboard the Rockot/Breeze KM failed during October 2005. The company also provides Multi-Satellite Dispenser systems for either side or base mounted satellites, for example two GRACE satellites were side mounted, incorporating a MLS onto the dispenser system.

To safely follow the launch, the remote mission centre is located 30 km from Plestsk. The centre includes, a launch countdown display, real-time video featuring the launch and mission data, as well as audio coverage from the launch site and for worldwide communications.

Europropulsion SA

Current Status
Europropulsion is the prime contractor for the new Vega launch vehicle's P80 Solid Rocket Motor (SRM). Its partners include Avio, SNPE Matériaux Energétiques, Regulus, Snecma Propulsion Solide and Sabca. As with the Ariane, Vega launch vehicles are assembled at Kourou. Vega's first flight is expected to occur in 2008.

Background
Europropulsion was established in 1991 by France's Société Européenne de Propulsion (SEP) and Italy's BPD Difesa e Spazio to bid for solid motor development contracts within Europe's civil space programme. SEP merged with Snecma in 1997, and Fiat Avio (now Avio) acquired BPD in 1994. Snecma and Avio now control equal shares in Europropulsion. Europropulsion's main operations are located in Kourou, French Guiana at the Centre Spatial Guyanais (CSG).

Europropulsion's principal responsibilities lie with the production, transportation, integration and final checkout of Ariane 5's Moteur Propergol Solide (MPS) solid rocket motors. The company leads a team of sub-contractors that contribute MPS components. Individually, Avio provides the lead for the igniter development, thermal insulation and the insulation liner; Snecma Propulsion Solide contributes the nozzle assembly; German sub-contractors MAN AG provide the motor casings; and Regulus (the SNPE-Avio partnership) the two large segment casting operations. The small forward segment is completed by Avio in

Table of Launches of the Rockot/Breeze KM Launch Vehicle from Plestsk Cosmodrome 40 degrees east and 60 degrees north by Eurockot Launch Services

Date of launch	Contractor/Operator	Payload	Discipline	Mass	Orbit
28 July 2006	Korea Aerospace Research Institute (KARI).	KOMPSAT-2	Earth Observation.	798 kg	SSO 685 km
30 October 2003	USEF on behalf of NEDO under METI of Japan.	SERVIS-1	Space Science.	850 kg	SSO 1000 km
30 June 2003	Czech Astronomical Institute.	MINOSA part of the MOM (Multi Orbit Mission).	Earth Observation.	66 kg	Elliptical orbit of 320 × 820 km
	Canadian Space Agency.	MOST MOM	Astronomy.	51 kg	SSO 820 km
	UTIAS/SFL, Canada.	NLS-1 MOM (3 nano-satellites) CanX-1AAU/Danish – CubeSat/DTUsat.	Educational.	3 kg	SSO 820 km
		NLS-2 MOM Cube-sat XI-IV (Tokoyo University) and CUTE-1 (Tokoyo Institute of Technology).	Educational.	2 kg	SSO 820 km
	Space and Systems Development Laboratory in the Department of Aeronautics and Astronautics at Stanford University (US)/ Quakefinder (US).	QuakeSat-1	Earth Sciences.	3 kg	SSO 820 km
	KSC, Russia.	MONITOR Mock-up.	Mass and Frequency Simulator.	250 kg	Remains on Breeze Upper Stage
20 July 2002	IRIDIUM Satellite LLC.	2 Iridium spacecraft (IS-2).	Communication.	1360 kg	650 km
17 March 2002	German Aerospace Centre/NASA.	2 Gravity Recovery and Climate Experiment (GRACE) spacecrafts.	Space Science.	974.4 kg	near polar orbit 500 km
16 May 2000	KSC, Russia.	Commercial Demonstration Flight (CDF) SIMSAT1/-2.	Flight qualification of the Rockot-KM.	1380 kg	540 km

For details of the latest updates to *Jane's Space Systems and Industry* online and to discover the additional information available exclusively to online subscribers please visit **jsd.janes.com**

Italy. SNPE Propulsion produces the ammonium perchlorate required for the solid rocket boosters at its Toulouse site.

International Launch Services (ILS)

Current Status
After the merger of Lockheed and Martin Marietta in 1995, ILS was created to handle the joint commercial marketing of Proton and Atlas for the Lockheed Khrunichev Energia venture and Lockheed Martin Commercial Launch Services (LMCLS). LKE was formed in early 1993 as a joint venture to market Russia's Proton launch vehicle and associated launch services. LMCLS has been doing business since 1988 as the commercial marketing company for the Atlas launch vehicles. ILS provides integrated marketing of both vehicles, with the concept of mutual back-up.

The ILS services cover spacecraft integration, Proton and Atlas supply, mission management, insurance brokering, launch site support, post-mission support and customer support. Backlog of US$3 billion representing more than 30 launches. Over USD1 billion in contracts signed in 2000 (16 launches). 14 launches in 2000 including first flight of Breeze M upper stage.

By the end of August 2002, ILS had accomplished 52 Atlas launches and 24 commercial Proton launches. Khrunichev has in addition launched 26 Proton variants for the Russian government since the end of 1995. During 2003 ILS launched six times, won contracts for 11 new missions and received nine additional assignments for future US government launches. During 2003, five launches were with various configurations of the Atlas and one with the Proton using a Breeze M upper stage.

Regulus

Current Status
Regulus continues to supply Ariane 5 rockets with filled solid propellant sections.

Background
Regulus is a partnership between France's SNPE (40 per cent) and Italy's BPD Difesa e Spazio, now Avio (60 per cent), formed to build and operate Kourou's Ariane 5 French Guiana Propellant Plant, or Usine Propergols en Guyane (UPG). The plant comprises 40 buildings, totalling 26,000 m^2 on a 300 hectare site and was commissioned 24 October 1991.

Each Regulus-supplied strap-on booster contains more than 200 tonnes of ammonium perchlorate solid propellant. Of the three rocket sections, Regulus provides the propellant required for the mid and aft units, while the forward element is shipped from Avio. When the facility became fully operational in 1995, its 150 personnel were able to support eight flights a year, requiring 3,300 tonnes of propellant for 32 segments. The first segment was poured November 1991 for the B1 battleship motor, tested 16 February 1993. The first flight of a full segment booster on Ariane 5 took place 4 June 1996.

Sea Launch LLC

Current Status
In January 2007, a Sea Launch Zenit-3SL failed at the first stage, destroying the SES New Skies NSS-8 satellite payload and damaging the Odyssey launch platform. Odyssey was transported to the Sea Launch Home Port in Long Beach, California for repairs and Sea Launch began an investigation into what caused the failure, which the company expected to complete around June 2007. Sea Launch aspires to relaunch its services in October 2007.

Sea Launch was planning to inaugurate its Land Launch services in 2007, operating from the Baikonur Cosmodrome's launch complex 45. The company offers the two-stage Zenit-2SLB and the three-stage Zenit-3SLB, the first launch of which was scheduled for mid-2007. The Space International Services Ltd (SIS) joint venture, headquartered in Moscow, sub-contracts with Sea Launch to provide the Zenit systems, mission integration and launch services. The January 2007 Sea Launch accident has caused further Land Launch delays, since the launch vehicle requires recertification subsequent to the investigation that the company is carrying out. Several companies opted to take their business elsewhere after the January failure. Both Land Launch and Sea Launch experienced launch delays prior to the January accident, however, reportedly due to launch vehicle technology supply difficulties. It is not clear when the first Land Launch mission will lift off.

Background
The Sea Launch joint venture was created in 1995 by a consortium comprising the Boeing Commercial Space Company (40 per cent), Russia's RSC Energia (25 per cent), Norway's Aker ASA (20 per cent) and the Ukraine's SDO Yuzhnoye and PO Yuzhmash (15 per cent). The company produces the three-stage Zenit-3SL, ships the launch vehicle to its Long Beach Home Port headquarters in the US, then assembles and transports the launch vehicle and payload into international waters to Sea Launch's Odyssey Platform for a near-Equatorial launch. The Zenit-3SL components are manufactured in the Ukraine, Russia and Seattle, Washington.

The first launch of the Zenit-3SL with a demonstration payload took place on 27 March 1999, followed by its first commercial launch on 9 October 1999, with the DirecTV satellite. Sea Launch had a very good launch record prior to their January 2007 failure, at 91.3 per cent. Even after the accident, the company holds an 87.5 per cent success rate, where three out of its 24 launches ended in failure. One of the failures, the Telstar 18 mission in June 2004, can actually be counted as a partial failure, since the satellite was deployed in good health, although about 14,000 km short of its intended orbit; Telstar 18 completed the journey under its own steam, with no significant fuel shortages expected in its future.

Kværner Rosenberg (now Aker Kværner) converted a 30,000 tonne oil drilling rig into the ocean-going launch platform, Odyssey (see separate entry), in Stavanger under a USD78 million contract; Kværner Govan (also now Aker Kværner) in Scotland built the 34,000-tonne, 200 m Assembly and Command Ship (ACS), Sea Launch Commander, for USD93 million. Both vessels were modified in Russia prior to final delivery.

A Russian Proton launch vehicle lifts off from Baikonur with an Inmarsat payload (ILS) 0572503

Starsem SA

Current Status
Soyuz launch services will commence at ESA's Centre Spatial Guyanais in Kourou in 2008, however Arianespace, rather than Starsem, will provide the launch services lead for that location. Starsem will continue to provide Soyuz production and institutional support for Kourou-based Soyuz launches.

Background
Starsem is a joint-venture company, founded in 1996, to market and provide commercial launch services for the Soyuz family of launch vehicles. Starsem's founding partners are EADS (35 per cent), the Russian Federal Space Agency (25 per cent), the Samara Space Centre/TsSKB-Progress (25 per cent) and Arianespace (15 per cent). The agreement for the creation of Starsem was signed by the partners on 17 July 1996 in Moscow, and on 6 August the statutes of the company were signed in Paris. The company is a French-registered corporation and was originally set up with FRF500,000. Starsem is headquartered in Evry, France and employs about 25 specialists supporting sales, finance, programme development, launch campaigns and payload processing and integration.

The first Starsem-marketed Soyuz flight took place on 9 February 1999, placing four Globalstar satellites in orbit. As of the end of 2006, the company had conducted 17 commercial launches, all successful. Starsem operates from the Baikonur Cosmodrome's Sites 112 and 31.

MAJOR SUB-CONTRACTORS

France

Aerospatiale

Current Status
Aerospatiale has developed a range of rigid solar arrays optimised for 1 to 10 kW outputs. The first generation GSR-1 is employed on Arabsat 1, Italsat 1, Telecom 1, TDF/TV-Sat and Tele-X. The GSR-3 array offers 40 to 45 W/kg and 115 W/m² (12-year EOL). The arrays are flying on Spacebus 2000/3000, such as Italsat 2 (5.3 kW) and Turksat (3 kW). The GSR-3 on Spot 4/Helios 1 use ADELE frictionless hinges. Up to seven panels, each a maximum 3 × 6 m, can be accommodated in a variety of 2-D/3-D configurations. Standard or ultra-thin (50 μm) cells can be employed, up to 6 × 8 cm. The panels comprise an aluminium honeycomb sandwich faced by 2-D carbon skins and insulated by a single Kapton sheet. Final panel mass density is 1.050 kg/m². The GSR-4 family was qualified in 1996–97. The panels of co-cured aluminium sandwich with 50 μm Kapton reinforced carbon fibre face skins (max density 1 kg/m²) can host advanced Si and GaAs cells. ILSA is a deployable flexible Kapton array for 4 to 7 kW missions such as Spot 1 to 3 and ERS 1/2. A 55 kW system has been studied for ESA and larger systems with high-power SSPA for station modules and free-flying laboratories. Developing the concept of the unfurlable semi-rigid array, company has designed a 10 to100 kW output array for satellite and ISS applications.

The company operates an 875 m² class 100,000 solar array integration area at La Frayère near its major Cannes facility, housing a 35 m³ vacuum chamber, a Spectrolab flasher for solar cell testing and two deployment jigs.

SAGEM SA

Current Status
SAGEM, one of Europe's principal inertial systems manufacturers, comprises three divisions: Defence and Security; Terminals and Telecommunications; and Electronics and Industry. Research and Development, prototype design and system integration are concentrated in four sites (Argenteuil, Eragny, Saint-Christophe, Cergy), with four manufacturing centres (Montluçon, Saint-Etienne-du-Rouvray, Fougères and Poitiers). Total floor area of 200,000 m² of class 10–10,000 cleanrooms. Principal specialisations for satellites and launchers cover gyros, linear and rotating mechanisms (with associated electronics), bubble memories, optics, microelectronics and navigation/guidance systems. The linear and rotating mechanisms and the mechanical systems used in telescopes and optical instruments, mirror refocusing, optical systems and structures for spectrometers include: Spacelab spectrometer, mirror scanning system for Metop's IASI instrument, Mir MIRAS spectrometer, and Helios refocusing mechanism. Combined linear and rotating systems have been developed for adaptive optics in advanced astronomy telescopes. Adaptive optics have been specified for the Next Generation Telescope and Sagem has conducted primary level research for that programme.

Regys 3S, the newest member of the Regys family of dry-tuned two-axis inertial reference units developed for space applications, it incorporates the Gildas 3 gyro. A fully autonomous unit, it provides error-compensated angular rates in digital form ready for AOCS computer processing. No thermal control is needed as there is built-in compensation for thermal-induced inertial errors. The separate gyro and electronics units are connected by a flexible harness; the electronics units can be stacked. Regys 35 has been selected for the following programmes: ODIN, XMM, Integral, Stentor and Proteus.
Size: 103 × 96 × 95 mm sensor unit;
283 × 164 × 45 mm electronic unit; less than 2.3 kg total
Reliability: 0.99 for a three-axis redundant configuration on a 15-year telecom application
Power requirement: 15 W maximum, at 22–51 V DC
Output interfaces: OBDH/RU, 1553, MACS (optional)
Range (°/s): ±2 fine, ±40 coarse
Resolution: less than 0.1 arcsec
Bandwidth: more than 5 Hz (up to 20 Hz by software adaptation)
Scale factor stability: 10^{-3} BOL; 10^{-2} EOL
g-insensitive drift stability: 0.15° max over 1 h, with ΔT±2ιC
Environmental tolerance: –10/50°C; vibration 22 g rms 0–2 kHz random, 20 g 0–100 Hz 0-peak; radiation 15 year GEO

The Regys 10 rate gyro package is a digital-output package for sensing satellite angular rates about two orthogonal axes, incorporating a Gildas 1/4 dynamically tuned gyro adapted from an air navigation system, and mounted with a shock absorber in a cradle that can be adjusted for required axis orientation. Regys 10 is used by Spot 4, Helios 1 and Envisat, and the newer version aboard Spot 5 and Helios 2. The computed angular displacement is delivered on request at least every 200 ms. If the measured rate is approximately 2°/s, the OBC can order the gyropackage to operate at quarter speed in security mode and provide coarse data (0.400 arcsec/impulse in contrast to 0.0288 arcsec/impulse in fine mode).
Size: 16 × 16 × 16 cm; 2.5 kg
Power requirement: 21 W null speed, 29 W at 8°/s
MTBF: 200,000 h
Range (°/s): ±0.6 fine, ±2 coarse, ±8 security
Scale factor (arcsec/impulse): 0.0288 fine, 0.0972 coarse, 0.400 security
Scale factor stability: 10^{-4} short-term; 10^{-3} long-term
Environmental tolerance: 0/50°C operation, –20/50°C transient; 40 g/4 ms shock: 100 krad radiation

SAGEM began developing a space-qualified brushless DC motor under ESA/CNES contracts in 1975, producing a high resolution unit offering 1,200 steps/rev for application in: stepper mode for solar array drive mechanisms on LEO/GEO spacecraft, synchronous mode for de-spinning onboard structures (as on Giotto), precision positioning mode for mirror pointing (as on Spot) and mini-stepping mode for manipulator articulation. 100 μrad accuracy has been demonstrated for the stepper version, with a 6 per cent maximum variation at 0.01 rpm in the synchronous variant. SAGEM's 18EM brushless DC motor drives the rotating mirror of Sodern's STD 12 Earth horizon sensor aboard Spot.

Sydem is a motorised unfolding and pointing system designed to actuate and precisely position various appendages including large antennae on satellites. Typical use includes also single-shot movement with final locking.

Sydem (Type 42) is qualified for space applications, and represents the top of the Sydem range. A smaller and lighter Sydem (Type 32) supports less demanding applications.

Snecma Moteurs

Current Status
The principal space business of Snecma Moteurs is the design, development and construction of propulsion systems for launch vehicles and satellites, including Ariane's Viking, HM-7, Vulcain and P230 (partnered with BPD in Europropulsion), motors. In satellite work, SM produces Mage apogee kick motors, hydrazine thrusters for attitude control, bipropellant liquid engines and electric/ion engines. The company also produces solar array drive mechanisms, μg furnaces, surface tension tanks, magnetic bearings, transducers and flow regulators. In 2002, personnel totalled 3,800, with a turnover of FFr7,200 million. Aerospatiale in January 1994 sold its 13.6 per cent interest for FFr178.9 million. SM and Spain's Empresarios Agrupados established Iberespacio in October 1989 to undertake joint development of space propulsion systems. SM and its SNECMA parent established Hyperspace in January 1990 to co-ordinate propulsion activities for hypersonic aircraft and launchers. Snecma has studied air breathing hypersonic propulsion systems and bench tests of components have been performed. Snecma continues to study these economical propulsion concepts for reusable launch vehicles.

Other interests include: Arianespace 8 per cent, Europropulsion (50/50 with BPD), G2P 75 per cent (large solid-propulsion systems for military applications), Spot Image 12 per cent, Carbone-Industrie 100 per cent (carbon brakes), TECHLAM 50 per cent (laminated elastomeric components) and S2M 54 per cent (magnetic bearings).

Test/Assembly Facilities: SM maintains a 4,500 m² Ariane propulsion assembly hall containing 14 Viking engine assembly stands, five for HM-7 and nine for stage 1 propulsion systems. There is also a 400 m² class 100,000 cleanroom for cryogenic engine component assembly. TDF propulsion was installed in an integration building provided with two 300 m² class 100,000 rooms. The Vulcain Engine Assembly Building can produce eight engines annually. The 1,220 m² class 100,000 cleanroom accommodates 12 assembly cells and 25 class 100 laminar flow booths/hoods for storage before assembly. Also in the building are a 700 m² warehouse, a 1,400 m² engineering and mechanical area and 1,150 m² of office space.

Engine test facilities are concentrated at the 1.5 km² Vernon site but some are located at the 24,300 m² Villaroche area. For storable propellant motors, there are four component test stands (PF1-3, A48, F22) and one engine stand (PF2). Cryogenic testing is undertaken on three stands: the two of Vernon's PF41 (one with altitude simulation) and one at Villaroche (with altitude simulation). Eleven existing cryogenic component stands were joined in 1988 by PF52 for Ariane 5 Vulcain gas generator and turbopump development. The PF50 stand for complete Vulcain engine tests was inaugurated September 1990 at Vernon. SM inaugurated its 11,000 m² production plant for Ariane 5's P230 nozzles in October 1990 at Haillan. A FFr180 million investment, it was fabricating 10 nozzles annually by year 2000. Assembly space is 650 m², plus 825 m² class 100,000 cleanrooms.

Solid rocket motor development and production is the responsibility of Europropulsion. A new research department into hybrid propellants has been proposed to respond to initiatives from Russian engineers for an international development programme on hybrid rocket motors.

Société Nationale des Poudres et Explosifs (SNPE)

Current Status
The Société Nationale des Poudres et Explosifs SNPE Group, created as a state-owned company in 1971 and now with approximately 5,600 personnel and FFr5.7 billion 2004 turnover (FFr132 million profit), manufactures liquid rocket propellants, with emphasis on UDMH, and develops/produces solid propellants. Has conducted research into exotic and tripropellant combinations as well as hybrid solid-liquid propellant combinations for exotic and specific propulsion requirements. Along with BPD Difesa e Spazio it participates in the Regulus partnership. The company provides the propellant for Ariane 5's P230 solid boosters using facilities at Kourou constructed by subsidiary SNPE Ingénierie. SNPE maintains facilities at Saint-Médard en Jalles near Bordeaux (solid propellant grains), Toulouse (UDMH, AP), Toulon (pyrotechnics) and Le Bouchet near Paris (HQ and R&D). The Toulouse site manages the production of ammonium perchlorate with a capacity of 6,000 tonnes annually to ensure Ariane 5 supplies. SNPE formed a second partnership with BPD, Société Européenne de perchlorate d'ammonium (EUPERA; SNPE 66 per cent, BPD 34 per cent), to handle production. Toulouse also provides the UDMH for Ariane 4 and the MMH for Ariane 5's upper stage. Subsidiary Pyromeca manufactures detonating cords and flexible linear cutting charges and ultra-fast pyrotechnic valves for Ariane 4 release from the launch pad. Propellant management test rig and flow evaluation equipment capable of handling rapid fuelling and dump capability.

In cryogenic propulsion, SNPE provides the main igniter and turbine starter grain for Ariane 4's HM-7B stage 3 engine and the shaped ducts for the Ariane 5 Vulcain igniter and starter.

SODERN

Current Status

The Société Anonyme d'Etudes et Réalisations Nucléaires (SODERN) specialises in instrumentation for particle radiation detection, Sun/Earth/star sensors for satellite attitude control systems, and spaceborne rendezvous/docking cameras. Personnel total more than 300. More than 130 SODERN attitude sensors have been operated in space since the first launch of a SODERN sensor in 1974 (Symphonie). Sensor development began with the STR 01 Earth horizon-crossing and STA 01 Earth static sensing devices for the GEO Symphonies, launched 1974/75. All the Earth-sensing devices operate at 14 to 16 μm IR. An improved STA 03 followed on OTS and Marecs, and four STR 03s on each Meteosat. A slightly improved STA 03A is employed by the ECS/Eutelsats. 35 STR 04 horizon-crossing sensors were provided for Intelsat 5 and the self-redundant STA 04 was introduced on Telecom 1 in addition to successor derivatives. The STD 12 two-axis conical scan sensor was developed for Spot and ERS, and has been further developed as the STD 15/16 for operations in all orbits from low altitude to GEO. The SED 04 star tracker was employed by Spacelab 2's Instrument Pointing System. In 1985, CNES awarded SODERN the contract for developing the SED 12 CCD star tracker for the Franco-Soviet Granat X/γ observatory. SODERN also developed the Earth-imaging cameras for Spot 1–4 and Helios, and Envisat's MERIS spectrometer and has developed the Spot 5 and Helios II cameras. New sensor development undertaken by an expanded product research and technology development facility. SODERN provides the capability to integrate star sensors with earth limb scanning sensors in a unique unit slaved to directional target sensors for attitude orientation and control.

SED 04 star/sun tracker

The SED 04 was developed to update the reference gyros of ESA's Instrument Pointing System, first flown aboard Spacelab 2 in 1985.

SED 12 CCD star tracker

The SED 12 was designed and built to provide attitude data to CNES's control system for the French Sigma telescope aboard the Soviet Granat astronomical observatory. The SED 12D model is used by France's Helios 1 reconnaissance satellite and has been adapted for Helios 2, Envisat and Sweden's Odin. The tracker is based on Thomson's TH7861 CCD area array, each element measuring 23 × 23 μm and providing 40 mV/lx sensitivity. Acquisition probability is more than 0.997 for stars of magnitude +8 to –1. Simultaneous tracking of five stars over the 7.5 × 10° rectangular Field Of View (FOV) is possible. From an initial standby mode, the sensor can operate in search mode for up to 30 stars over the whole FOV (requiring minimum detectable magnitude, number of stars to be tracked after search, and apparent star motion), or in tracking mode for a maximum of five local fields (requiring co-ordinates of five stars, expected magnitude and motion). For both modes, the sensor outputs the located stellar co-ordinates, magnitudes, time delay between charge collection and read-out, detector/optical head/electronics temperatures.
Field Of View: 7.5 × 10° rectangular
Accuracy (arcsec): 3 after one observation, 0.5 after filtering/calibration (1 σ); 17 on two axes after one observation (3 σ), 5 after filtering/calibration
Acquisition time: less than 8 seconds overall field; less than 1.5 seconds in 0.25° square field. Probability of single star acquisition more than 0.997
Maximum star rate: 0.1°/s for magnitude brighter than 9, 1.0°/s for magnitude brighter than 3
Size: 141 mm diameter, 230 mm length/2.5 kg optical head without baffle; 271 × 222 × 217/7.2 kg electronics unit
Input power: 2 W optical unit with cooler off; 2–8 W optical unit with cooler on; 40 W electronics unit typical; 20–50 V DC (PWM 1525)
Temperature range: –20 to +50°C
Memory capacity: 40–64 kbyte ROM; 16–32 kbyte RAM depending on mission
Failure rate: 3,800 × 10⁻⁹/h

SED 15 CCD star tracker

The SED 15 CCD StarTracker is under development for GEO satellites to provide pitch/roll data. One version will track only the Pole Star. 30 arcsec accuracy in 5 × 7° FOV. By adding barycentric signal processing, the accuracy holds for a larger Field of View (FoV), such as 25 × 35°.

STD 15 scanning earth sensor

The STD 15 scanning earth sensor derived from the STD 12 of Spot and ERS 1/2. Its first flight was aboard Telecom 2 in December 1991; 12 are now flying on Telecom 2A/2B, Hispasat 1A/1B and Hot Bird 2. The sensor has been selected by NPO PM (Russia) for a new telecommunications platform. The sensor permits the attitude control subsystem to generate roll/pitch angular deviation data by using the four Earth-space/space-Earth transitions of the two scan traces (generated by the rotating mirror and two fixed mirrors); an optical encoder measures the angular position of each transition with respect to the spacecraft reference axes. Data rate 1.25 Hz in normal mode, 5 Hz in stationkeeping mode.
Field Of View: 1.5° roll/12° pitch; 14° roll/30° pitch in acquisition mode. Instantaneous FOV 1.8 × 1.8°
Accuracy: 0.025° 1 second
Mass: 3.4 kg
Input power: 7.5 W at 22 to 50 V DC
Temperature range: –20 to +50°C operating; –40 to +60°C storage
Vibration tolerance: 20 g rms over 20 to 2,000 Hz
Lifetime: 15 years; 0.9 × 10⁻⁶/h failure rate

STD 16 scanning earth sensor

STD 16 scanning earth sensor is Sodern's latest sensor, designed for LEO satellites: Helios 1, Spot 4, Envisat, ADEOS and ETS 7. Operation principle is as STD 15. Nine were supplied to Matra for Spot 4/Helios and Envisat: six three were supplied 1992–93 to Toshiba for ADEOS and ETS 7.
Field Of View: effective scan for one trace is 152°
Altitude range: 200 to 2,000 km
Accuracy: 0.045° bias, 0.03° noise; 3 seconds
Mass: 3.5 kg; length 386 mm; height 175 mm
Input power: 7.5 W at 22 to 52 V DC
Temperature range: –20 to +50°C; –40 to +65°C storage
Failure rate: 1.3 × 10⁻⁶/h
Lifetime: 10 years

DTA 01 high-resolution imager

Spot 1–3 incorporate Sodern's DTA 01 high-resolution visible push broom imaging unit. Each of the four spectral bands employs four Thomson TH7801A 1728-element CCD arrays (Spot 1's utilised Fairchild's 122 DC detectors). PAN passes directly through the prisms; its spectral range is determined by absorption and interference filters on the CCDs. Green band B1 is reflected off to the side by a dichroic mirror; red band B2 and near-IR B3 are separated by a second mirror. The rigid primary titanium structure holds the beam splitter and the CCD assembly of four butted lines; each line of four CCDs is straight to within 2 μm. A mechanically isolated secondary structure carries the front-end electronics. Resistance heaters help to maintain CCD temperature to within 2°C.
Spectral response (μm, at half-ht): 0.495–0.580 B1; 0.610–0.665 B2; 0.775–0.785 B3, 0.490–0.715 PAN
Number of pixels: 3,000 (1,326 μm) B1-3; 6,000 (1,313 μm) PAN
Resolution (from 800 km): 20 m B1-3; 10 m PAN
Typical peak beam splitter transmittance: 0.8 B1; 0.75 B2; 0.8 B3; 0.85 PAN
Saturation spectral irradiance (W/m²/μm): 0.004 B1; 0.005 B2; 0.003 B3; 0.006 PAN
S/N for max/min illuminance: 700/90 B1; 650/40 B2; 800/50 B3; 500/70 PAN
Response uniformity: 5% B1-3; 3% PAN
Size: 211 mm wide, mounting flange 255 mm diameter; 2.5 kg
Power consumption: 2.7 W
Thermal: 18 to 22°C operating, –20 to +60°C storage.

DTA 03 high-resolution imager

The DTA 03 high-resolution imager, derived from DTA 01 for Spot 4, enhanced by a 1.6 μm SWIR channel and by using the B2 channel (instead of PAN) for 10 m resolution. The SWIR scanline is formed by 3,000 30 × 30 μm InGaAs Thomson detectors in two staggered rows of 1,500 each with a centre to centre pitch of 26 μm. The three B channels use TH7811 detectors, derived from the TH7801. Short-term stability is 0.01°C at 5°C for SWIR and 0.1°C at 20°C for the CCDs.
Spectral response (μm, at half-ht): 0.495–0.580 B1; 0.610–0.665 B2; 0.775–0.785 B3, SWIR 1.54–1.75
Typical peak beam splitter transmittance: 0.8 B1; 0.7 B2; 0.8 B3; 0.7 SWIR
S/N for max/min illuminance: 900/250 B1; 350/60 B2; 1,000/100 B3; 650/3 SWIR
Response uniformity: 4% B1/2; 4% B3; 10% SWIR
Size: 200 × 250 mm, mounting flange 250 mm diameter; 7.4 kg
Power consumption: 10.5 W
Thermal: 18 to 22°C operating, –15 to +50°C storage.

DTA 02/04 very high-resolution imagers

SODERN developed the first prototype of the imager (DTA 02) for the Samro military satellite. When the programme re-started as Helios, the company became responsible for the whole DTA 04 detection chain, including both focal plane and video electronics. An improved new detection chain DTA 06 is being developed for the new-generation Helios II. The programme is classified.

DTA 05 multispectral cameras

The DTA 05 multispectral cameras were developed for Spot 4's Vegetation payload, this imaging subassembly comprises separate focal plane detection units: 1,728-pixel CCDs for 0.49, 0.58, 0.64, 0.83 μm visible and a 1,500-pixel AsGaAs hybrid detector for the 1.6 μm SWIR channel. The front-end electronics are linked to a common video electronic unit, each CCD being read sequentially in series. Each detection unit is mounted on optics (developed by SODERN subsidiary Cerco) with an interposed dedicated interference filter. Optics are telecentric, ±50.5° FOV with registration less than 0.3 pixel regardless of incident angle. Global MTF 0.3 along track. 23 W power consumption.

DTA 10 spectro imager

The DTA 10 spectro imager was developed for Envisat 1's MERIS. Five identical cameras each analyse a 1,200 km swath (300 m resolution) with a spectrometer. Up to 15 bands can be programmed in position and width within the swath. Optics developed by Cerco.
Spectral range: 400 to 1,050 nm
Spectral resolution: 2.5 nm
Radiometric accuracy: less than 2%

Other imagers

A new imager is being developed for Spot 5. The detection unit is built around a five spectral band beam splitter. Four bands are fitted with a Thomson TH7834 12,000 element array and the fifth band (SWIR) is fitted with a 30,000 element GoAo detector line. The PA channel is designed to achieve a ground resolution of 5 m.

ALICE BOE/BOS

The ALICE BOE/BOS was developed for the ALICE 1 and ALICE 2 instruments and used aboard the Mir Complex for microgravity experiments, the ALICE BOE and ALICE BOS achieve the optical diagnosis of cells which are filled with liquid subjected to critical conditions. Cells are temperature-regulated by high-precision thermostats and are analysed during the physical phase transitions by the optic-electronic devices of BOE and BOS. The ALICE optical capabilities include: microscopic inspection at various magnifications, interferometry, optical transmission, index gradient visualisation and optical diffusion. The total volume is 0.15 m³ and the mass is 55 kg.

POLDER-BOM

The POLDER-BOM was developed for the POLDER instrument aboard Japan's ADEOS 1 and 2 spacecraft, the optical-mechanical sub-assembly POLDER-BOM comprises a wide angle lens, a rotating filter wheel and a CCD focal plane. The optics have a 114° field of view, low distortion and constant illumination over the local plane. The filters and polarisers achieve spectral analysis from violet to the near-IR in different directions of polarisation. Eight spectral channels are available, three of which have polarisers.
Spatial resolution: 244 × 276 pixels
Temperature: 15 to 20°C operating
–20 to +60°C non-operating.

SIS 01 very broad band 1-axis seismometer

The SIS 01 very broad band 1-axis seismometer, a vertical-axis seismometer, was developed for the Russian Mars 8 (Mars-96) mission for operations on the Martian surface.
Band pass: 0.03 to 8 Hz
Acceleration resolution: 10⁻⁸ m/s²
Power consumption: 60 mW
Dimensions: 90 × 90 × 90 mm
Mass: 350 g

SIS 02 very broad band 3-axis seismometer

The SIS 02 very broad band 3-axis seismometer is an improved-technology development of SIS 01 with three seismic masses on a pyramid corner. Developed for planetary seismology.
Band pass: 0 to 5 Hz
Acceleration resolution: 2 × 10⁻⁸ m/s²
Power consumption: 1 mW
Dimensions: sphere, diameter 22 cm
Mass: less than 1.6 kg

Germany

Carl Zeiss

Current Status
Carl Zeiss provides optical equipment with separate divisions providing instruments for astronomy, industrial meterology, microscopy, opthalmology, photogrammetry, semiconductor technology, surgical operations and surveying. Carl Zeiss specialises, in the space field, in high-quality optical systems such as telescopes and microscopes. It will provide Abrixas's X-ray optics system. Hipparcos ROSAT and SELEX and ESTEC's large space simulator employ the company's optics. Research into satellite imaging optics technology. New optical folding mirror mechanisms for remote sensing satellites and planetary spacecraft. CZ has developed folding optic arrays for adaptive lens systems on telescopes for Earth observation and astronomical purposes. Has designed adaptive optical mirrors for large observational devices.

Kayser-Threde GmbH (KT)

Current Status
Kayser-Threde is one of seven companies in the SEKAS group and is active in telematics, process control, data management, Earth observation, scientific payloads and specialised optical systems. KT provides expertise in the design, build and integration of space experiment payloads, covering sounding rockets, Spacelab, SPAS, Foton, Shuttle and re-entry capsules. Sales in 1995 were DM51.5 million (DM80 million including subsidiaries); staff totals about 240 people. Experience covers: 43 sounding rocket payloads (7 Aries, 33 Texus, 3 Maxus), vestibular experiments on Spacelab 1 and Mir-92, venous pressure measurements on Spacelab 1/D1, high-precision thermostat on Eureca 1, Spacelab D1/D2, Cryostat protein growth facility on IML-1, main contractor for Astro-SPAS' Orfeus telescope and Holop holographic lab for Spacelab D2. It was system integrator for all German experiments on Mir-92. The Energia Deutschland GmbH partnership with NPO Energia exploits an exclusive right to distribute Earth observation data from Mir.

KT built Temisat (plus a spare) under a USD10 million contract to Telespazio for piggyback launch on 31 August 1993 with a Meteor 2 aboard a Tsyklon, demonstrating relay from environmental data collection platforms. Temisat was the first of KT's Blackbird family of micro-satellites, although no others are under development. KT also acted as launch agent for the Technical University of Berlin's Tubsat B in January 1994. It built the GFZ 1 laser geodetic satellite for Germany's National Centre of Earth Sciences for an April 1995 launch on a Russian Progress Mir station ferry.

KT developed a range of DGPS stations for DARA and is involved in the pre-development of an advanced combined GPS/Glonass receiver for ESA. The company has been marketing Glonass receivers since June 1991. It provided the MOMSNAV orbit/attitude determination package for Germany's MOMS-02P imaging system on Mir's Priroda module. This uses two GPS receivers and a high-precision gyro package for 2 m/3 arcsec accuracy. A Precise Range & Range Rate Experiment (PRARE) package was provided under DARA contract for January 1994's Meteor 3 meteorological satellites. KT developed the transputer-based multiprocessor for Soho's SUMER instrument and the camera control processor for Mars-96's High Resolution Stereo Camera.

KT developed the Biopan biological experiment carrier for the exterior of Foton and Resurs descent capsules under ESA contract. The hinged lid on the 37 cm diameter 22 cm high pan-shaped container opens through 180° to expose samples mounted on the lid and fixed bottom plate, totaling 910 cm^2 for experiments. Its own 2 Mbyte microcontroller commands the power supply for the hinge drive and heaters (maintaining 0–37°C) and performs temperature sensor switching. Solar aspect angle and 0.25–4.5 μm ultra-violet radiation are also monitored. A 3.8 cm thick layer of glass fibre and phenol provides re-entry protection. The test flight was made aboard Foton 5 in October 1992. ESA has approved two further flights; first was June 1994 on Foton 6 carrying six experiments, using the refurbished ground qualification unit. The next was in September 1996.

The company signed an agreement with Glavcosmos in December 1987 for options on Foton and Resurs missions, acting as marketing agent. Four missions flew 1989 to 1991, carrying Intospace's Cosima 2–4, Casimir and LZZ experiments. A further contract was signed in autumn 1991 for up to six flights. The next crystal growth experiments flew on Foton 6 in June 1994. Foton 7 in February 1995 carried seven KT materials processing experiments under DARA contract in the Zona 4M and Konstanta 2M furnaces. MIRKA (Micro Re-entry Capsule) was launched aboard the Foton 8 satellite in October 1997. The 150 kg MIRKA was mounted on the outside of the Foton descent capsule where it remained until after the main spacecraft's de-orbit manoeuvre had taken place. It came down 110 km southeast of Orsk on 23 October 1997.

Kayser-Threde was awarded the contract from DARA to be responsible for the logistical and technical preparation for the MIR-97 visit of Reinhold Ewald to Mir – a role which the company had played for the MIR-92 mission. Progress-M33 launched in November 1996 carried 158 kg of German scientific equipment to Mir in advance of the Soyuz-TM 25 launch. Kayser-Threde was responsible for a MIR-97 experiment involving real-time transmission between Mir and the German control centre at Oberpfaffenhofen. The company is now participating in the Columbus experiment racking and experiment schedule. Development work has been carried out on long-term science experiment support aboard the International Space Station with a view to producing cost effective logistics and resupply operations. Kayser-Threde is a partner in the Rapideye commercial satellite-based geo-information service set up in 1998 with products and services for agriculture, cartography, natural disaster assessment and 3D visualisation and so forth. The concept envisages five Earth orbiting imaging satellites with S-band TT&C and X-band data downloads. From an evenly distributed polar orbit the five satellites provide multispectral images with a resolution of about 6.5 metres and an imaging area of 80 × 1,500 km.

MT Aerospace

Current Status
In July 2006, Mitsubishi Heavy Industries (MHI) awarded MT Aerospace with a contract to supply components for Japan's H-II Transfer Vehicle (HTV).

In June 2005, the MAN Group sold MAN Technologie AG to OHB Technology AG. At that time, the company changed its name to MT Aerospace.

Background
MT Aerospace, like its predecessor, MAN Technologie, is headquartered in Augsberg, Germany. The company also operates facilities in Mainz, Germany; Wolverhampton, UK; and Kourou, French Guiana. MAN Technologie was founded in 1986, however the company can trace its roots back to 1969 with the formation of MAN Neue Technologie as a branch of MAN AG. As of 2007, MT Aerospace employed 500 personnel and claimed annual revenues of about EUR100 million. The company owns an eight per cent stake in Arianespace.

MT Aerospace is responsible for the manufacture of about 10 per cent of the Ariane 5 launch vehicle's components. The company builds the flow-turned steel booster casings, the main tank's front skirt, the Vulcain engine's GAT and GAM high-pressure vessels for the engine Thrust Vector Control (TVC) and booster nozzles, Vulcain's gimbal joint, and the steering mechanism's heat shields. In December 1996, MT took over the production of the main tank bulkheads. MT is also responsible for Ariane 5 erection and maintenance services at the launch vehicle's ELA 2, ELA 3, and payload preparation facility S5 launch facilities in Kourou. Additionally, MT manufactures fuel tank domes for Japan's H-IIA launch vehicle.

Apart from launch vehicle components, MT Aerospace manufactures orbital transfer systems and satellite components. MT supplies the propellant, water and gas tanks for Europe's Automated Transfer Vehicle (ATV), as well as the thruster module. Under a contract signed with MHI in July 2006, MT will continue to supply Japan with HTV components through 2009.

MT has manufactured satellite propellant tanks and Meteosat high pressure tanks for a number of years. The company's propellant tanks have been used aboard science and commercial spacecraft, including Cluster; XMM-Newton; the Mars and Venus Express craft; Hot Bird; and Inmarsat.

MT Aerospace also manufactures antennas; radio and optical telescopes; and missile components.

TELDIX GmbH

Current Status
TELDIX GmbH was founded in 1960 as a subsidiary of TELEFUNKEN GmbH and the BENDIX Corporation, USA. In 1973, the Bosch GmbH acquired the BENDIX interest and in 1981 also that of AEG-TELEFUNKEN.

In February 1996, Litton Industries Inc, Woodland Hills, California, USA, acquired TELDIX GmbH from Bosch and made TELDIX a 'Center of Excellence' for high-technology products in the fields of computer technology, displays and space products.

In April 2001 Northrop Grumman Corporation, Los Angeles, California, USA acquired Litton Industries. TELDIX now belongs to the Navigation Systems Division within Northrop Grumman's Electronic Systems sector. Space products include high-precision satellite mechanisms, ball-bearing momentum and reaction wheels with integrated of external wheel drive electronics, low-noise magnetic bearing stabilisation wheels for advanced satellites, solar array drive assemblies, antenna pointing mechanisms and Systems for Optical Satellite Communication (SILX).

TELDIX is DIN-EN ISO-9001, AQAP-110, aqap-150, JAR 21 and JAR 145 certified.

For more than 30 years TELDIX has been successful in the field of gyroscopic actuators such as Momentum and Reaction Wheels.

TELDIX Momentum Wheels DRALLRAD® are utilised for the stabilisation of all kinds of satellites. As of June 2002, 411 wheels installed in 182 satellites are in orbit representing nearly 1,350 years of flawless operation.

Range of products include:

High-precision mechanisms for satellites, for example:

- Ball bearing momentum and reaction wheels with Wheel Drive Electronics (WDE)
- Low-noise magnetic bearing stabilisation wheels
- Solar Array Drive Assemblies (SADA)
- Antenna pointing mechanisms
- Systems for optical satellite communication SILEX.

Ball bearing momentum and reaction wheels
TELDIX is the sole commercial manufacturer of ball bearing momentum and reaction wheels in Germany and leading manufacturer in Europe.

TELDIX ball bearing momentum and reaction wheels are the ultimate choice for advanced satellite stabilisation.

Reference projects
European satellite programmes:

Abrixas, Artemis, Astra-2B, -1K, Beppo-SAX, Demeter, DFS, ECS, ESSAIM, Eurasiasat 1, Europe*Star, EUROSTAR 2000+, EUROSTAR 3000, Eutelsat II, Eutelsat W, GE-1E/Sirius 2, FBM, Hispasat 1A, 1B, 1C, Hot Bird, Inspector, ISO, Italsat, MARECS, MAROTS, Mars Express, Microscope, OTS, Parasol, Picard, Proba, Proteus/Jason, ROSAT, Skynet 4, Spacebus, Stentor, Symphonie A/B, TDF-1, TDF-2, Telecom-1, Telecom II, TELE-X, TUBSAT-B, Turksat 1, TV-SAT and TV-SAT-2.

International Satellite Programmes
Agila 2, Amos, Apple, Aqua (PM-1), Arabsat II, Arabsat III, AsiaStar, Aura (Chemistry), Beidou 1A, 1B, BSat 2a, 2b, Chandra (CXO), Chinasat 8,22, DFH3, DFH4, Echostar VI, ETS-V, FBM, GE 5, GOES, Inmarsat II, Insat-1D, Insat-2, Insat-3, Intelsat V, Intelsat VII, IRAS, iSKY (KaStar), KaistSat, KitSat 3, METSAT, MOS-1, MS-T5, MT Sat, Nahuel, NATO IV, Nilestar, N-Star, NSS-6 (K-TV), OmegaSat, Orbcomm, Orion 1, Orion 2, PanAmSat 6, 7, 8, Pioneer, Planet-A, SBIRS Low, Sinosat, Sirius 1-3, Sky-1, Sky-2, ST-1, Step 4, Superbird, Telstar 5, Telstar 6-12, Tempo, Thaicom, Worldstar.

In 1990 TELDIX started the development of an experimental low-cost reaction/momentum wheel with integrated WDE for use in very small satellites. The wheels of this type were delivered in 1993 for the satellites TUBSAT B and C, built by the Technical University of Berlin and were also selected and built for KITSAT-3 (Korea) and a small spacecraft called Inspector built by DASA, Bremen. The more stringent requirements of small commercially used satellites require an adaptation of the experimental design to a highly reliable, medium-life space-qualified reaction wheel as for the ORBCOMM constellation.

Magnetic bearing momentum wheel MWX
The TELDIX MWX, a 5 DOF magnetic bearing momentum wheel means a real breakthrough with regard to vibrations exerted on the satellite, and a quantum step forward in wheel technology towards high-accuracy space missions. The MWX combines the following highly attractive features:

- extremely low vibrational noise due to active vibration suppression feature
- capability to tilt the rotor actively allowing very high pointing control
- capability to damp structural vibrations generated by other sources in the spacecraft
- and, as a general advantage of magnetic bearing wheels: higher speed capability, no lubrication, no stiction and low friction.

Thus the MWX is the most appropriate wheel for:

- scientific missions, μ-gravity
- optical communication links
- space telescopes.

Drive assemblies and pointing mechanisms
As a highly experienced mechanism manufacturer, TELDIX has been working for many years in the field of low-speed mechanisms as well. Within the framework of various contracts, pointing mechanisms for different applications were designed and built:

- Antenna pointing mechanisms
- Fine Pointing and Flip-Flop Mechanisms for SILEX (Semiconductor Intersatellite Laser Link Experiment)
- Pancake-type stepper motors
- Gimbal mechanisms
- Solar Array Drive Assemblies (SADA).

These SADA's are being used on DFS-Kopernikus and the Chinese DFH-3 satellites.

Israel

Elbit Systems Ltd

Current Status
Elbit Systems' business areas are mainly concentrated in the Intelligence, Surveillance and Reconnaissance (ISR); Command, Control, Communications, Computers and Intelligence (C4I); space-based defence; and Unmanned Aerial Vehicle (UAV) industries. The company provides electronic and electro-optic upgrades and solutions to its government and prime contractor customers.

Elbit Systems owns several subsidiaries based in both Israel and the US. Elbit Systems Electro Optics Elop Ltd (El-Op) has been a wholly owned subsidiary of Elbit Systems Ltd, since the companies' 2000 merger. Please see the separate El-Op entry for further details on that company's space-related activities.

Background
Elbit Systems was founded in 1966 as Elbit Computers Ltd. It has since grown into the largest, wholly commercial defence company in Israel, and is publicly traded on the Nasdaq National Market (NASDAQ) and the Tel-Aviv Stock Exchange (TASE).

Elbit began its participation in the space industry in September 1989 when it was awarded a contract by General Dynamics Space Systems for Atlas launch vehicle avionics boxes. Elbit also provided TechSat-1's On-Board Computer (OBC). It is one of Israel's largest electronics systems houses, developing control systems for Israeli satellites, including communications and data handling and advanced reconnaissance and surveillance systems. Elbit provides high-resolution panchromatic and multispectral earth observation systems (civil and military), science payloads and a wide range of optical equipment for space and ground-based systems. The company also develops interlinked radar and optical devices with software convergence solutions. Elbit Systems Ltd has developed a wide range of satellite and space vehicle programmes for foreign clients and agencies.

Elbit Systems Electro-Optics Elop Ltd (El-Op)

Current Status
El-Op provides electro-optic systems, including satellite-based imagery intelligence (IMINT) equipment to its prime contractor, commercial and public sector customers. The company's Earth imaging systems are especially suited to mini and micro-satellites.

El-Op provided the electro-optic imaging payloads for ImageSat International's high-resolution Earth Resources Observation System satellites, EROS-A and EROS-B; it will also provide the imagery payload for EROS-C. Launched in December 2000, EROS-A had a resolution of 1.9 m; EROS-B, launched in April 2006, bettered that, with a resolution of 0.7 m. At its time of launch, EROS-A was the only non-US high-resolution commercial imaging satellite in orbit. EROS-B imagery quality is high enough for it to be a dual use satellite.

El-Op provided the Korea Multi-Purpose Satellite 2 (Kompsat-2) with its Multi-Spectral Camera (MSC), and has also developed the Earth Resources Monitoring System (ERMS).

El-Op supplied the Israeli Ministry of Defense with the Ofeq 3 and Ofeq 5 high-resolution remote sensing payloads, launched in 1995 and 2002 respectively. As of 2007, Ofeq 5 remained operational. Ofeq 7 is under development and is likely to be launched during 2007.

Background
El-Op is based in Rehovot, Israel and employs about 1,200 personnel, of which, 40 per cent are engineers and scientists. The company was founded in 1937 and in 1960 became a government-owned business. Its public-sector status ended 18 years later, when it was sold to Tadiran and the Federmann Group. El-Op has been a wholly owned subsidiary of Elbit Systems Ltd (see separate entry) since the companies' 2000 merger.

El-Op's Remote Sensing Operation is principally concerned with airborne and ground electro-optical observation systems, warning systems and optical processing. El-Op has developed key elements of Israel's space-based reconnaissance and surveillance defence system and has developed advanced optical systems for the EROS series.

El-Op's civil research and development activities cover remote sensing; astronomy; optical sensing instruments for high-resolution telescope systems applicable to ground-, air- and space-based systems; Earth resources monitoring; and electro-optical navigation sensors. It developed Israel's TAUVEX UV camera in co-operation with Tel Aviv University for Russia's Spektr-RG (cancelled, but possibly to be resurrected); the VENμS micro-satellite for CNES; and the static horizon sensor and low resolution camera for Technion's TechSat-1 (Gurwin-1).

Israel Military Industries Ltd (IMI)

Current Status
IMI is a government-owned company comprising a HQ group and four operating groups of 10 divisions and 19 plants. There are about 4,400 personnel. The Rocket Systems Division (RSD) of the Systems Group develops/manufactures solid rocket motors, weapon systems and metal/composite material products. For example, it provides the propulsion systems for the Gabriel surface-to-surface and Shafrir air-to-air missiles. Conducts basic R&D in exotic propellant combinations and mixed liquid/solid combinations for hybrid motors. In the space field, it developed/produces the stage 1/2 motors for the Shavit/NEXT launchers. These are HTPB cylindrical grains in filament wound composite cases. Steering options are jet vane, LITVC and flexseal. The motors are marketed as the ASTM family in the US by Atlantic Research Corporation. RSD is deeply involved in both Israel's space programme and the SD10 Arrow. The propulsion systems and the pyrotechnics for these programmes are IMI developed and produced. IMI has researched technology requirements for NEXT launcher derivatives using HTPB grains. No development commitment has yet been made and plans for further development have been shelved.

Rafael Manor

Current Status
Rafael's Manor, the Propulsion & Explosive Systems Division, specialises in the research and development, and production of propulsion systems for mini and micro-satellites.

Background
Integrated Systems and Propulsion
Rafael has experience in the design, development, qualification and production of complete propulsion systems for satellite applications and applies a variety of propulsion concepts to its systems, including:

- Monopropellant (hydrazine)
- Electric (Hall-effect based)
- Cold Gas Reaction Control System

Rafael's propulsion systems are based on indigenous components, customized to meet specific requirements. The company's Attitude-Control and Orbital-Correction Hydrazine Propulsion Modules (HPM) have flown on-board Ofeq and Eros satellites.

Rafael is experienced in the field of space propulsion and has in-house capabilities for design, development, qualification, integration and manufacturing of advanced propulsion systems. Its infrastructure consists of clean rooms, space simulation testing facilities, environmental testing facilities and manufacturing facilities for metallic and composite-material components.

Micro-satellites
Rafael has joined with Israel Aircraft Industries (IAI) to develop and build micro-satellites based on IAI and Rafael's heritage in the mini-satellite industry. This research and development activity is aimed at designing a low-cost, efficient satellite bus, with incorporated state-of-the-art technologies and microelectromechanical systems (MEMS) components. The bus is intended for use in formation flying and Low Earth Orbit (LEO) observation missions. The LEO bus is supported by an electric propulsion system. In response to the Mission on Demand requirement, Rafael is developing an air-launch concept.

Italy

Alenia Difesa

Current Status
The Optics and Space Division (DIOS) was established in 1983, with a staff of 270 by early 2002, to complement the activities of the Military Systems and Environment divisions. It is concerned principally with onboard equipment and instrumentation for spacecraft, including electro-optical sensors for attitude determination (IR Earth, Sun and star sensors), CCD video systems, optical sensors for robotic rendezvous and docking, remote sensing electro-optical instruments (visible and near-infra-red), and mechanisms and components for optical payloads (lightweight mirrors, baffles and optics). DIOS has provided more than 250 attitude sensors for Eureca, Turksat, Soho, SAX, Inmarsat, Telecom, Hipparcos, Olympus, Eutelsat, Italsat, Tethered Satellite, ISO, DFH 3, Cassini, Skynet, Orion, Hispasat, Arabsat 2 and Nahuelsat. The light baffle and aluminium mirror for Giotto's multicolour camera were produced by DIOS, as were the CDD monitoring cameras of Eureca's Protein Crystallisation Facility. Other instruments have been delivered or are under development for astrophysics research and Earth observation projects. UVCS/Soho precision mechanisms for Soho's three-channel UV coronograph spectrometer. JET-X + XMM Optical Monitor ancillary telescopes to detect faint stars in visible/near-UV close to the optical axis. Cassini VIMS camera visible channel on the orbiter; Huygen's Atmospheric Structure Experiment to probe Titan's atmosphere; optical heads for star reference unit. IASI IR atmospheric sounding interferometer for Metop 1 polar satellite for temperature and humidity soundings and trace gases. GOME (Global Ozone Monitoring Experiment) for ERS 2. MIPAS (Michelson Interferometer for Passive Atmospheric Sounding) for Envisat 1. Sounding

rocket μg: MITE (Measuring Interfacial Tension Experiment); INEX MAM (Interactive Experiment for Marangoni Migration).

The MSS, Modular Star Sensor, designed for three-axis satellites such as ISO, Soho and SAX, is a two-axis star tracker incorporating a catadioptric head providing attitude control data in addition to the precise pointing of astronomical instrumentation. In the latter role, it also provides magnitude measurements of stars brighter than 9^{mv}. It first flew aboard STS-52 in October 1992 for ESA's In-Orbit Technology Demonstration Programme. With a few modifications (10 × 10° FOV), it was developed for the Cassini spacecraft launched to the planet Saturn in 1997. Advanced navigation systems development for very long duration space missions and complex navigation requirements in deep space where limb surfaces are not available.

Field of view: 4 × 3°
Accuracy: 6 arcsec uncalibrated
Size: 137 × 134 × 227 mm/3.5 kg (sensor head)
137 × 228 × 187 mm/2.5 kg (electronics unit)
Detector: silicon CCD matrix of 288 × 384 pixels
Input power: 10.5 W

The IFPSS, Integrated Focal Plane Star Sensor is a focal plane star tracker developed for ISO but is suitable for operation in any IR astronomical satellite with provision for cooling down to 2 K.

Accuracy: 1 arcsec rms
Size: 70 × 70 × 30 mm/1 kg (optical head at cryogenic temperatures)
200 × 150 × 70 mm/1.5 kg (electronics unit at –30 to +60°C)

The FDSS, Fine Digital Sun Sensor, is a two-axis wide angle static precise Sun sensor for application aboard three-axis satellites such as Eureca, Italsat and TSS.

Number of optical heads: two or more
Field of view: ±45°
Accuracy: 0.02° rms
Size: 170 × 94 × 88 mm/1.0 kg (optical head)
204 × 152 × 121 mm/2.7 kg (electronics)
Input power: 5 W

The CASS, Coarse Analogue Sun Sensor is a two-axis sensor designed for coarse determination of Sun angle aboard three-axis satellites such as Italsat. The sensor head incorporates four redundant Si cells assembled on a truncated pyramid with light baffle.

Field of view: 2π steradians
Accuracy: 0.1° rms
Size: 86 × 86 × 53 mm/0.230 kg

The ESS, Earth Sun Elevation Sensor, incorporates two 15 μm IR telescopes and two visible light solar split sensors to measure Earth/Sun elevation angles for spin-stabilised satellites. Users include ECS, Hipparcos, Giotto and Inmarsat.

Field of view: ±55° Sun
Accuracy: 0.03° both Earth/Sun, rms
Size: 166 × 150 × 127 mm/1.4 kg total package
Input power: 0.7 W

LACES, the Low Altitude Conical Earth Sensor is a two-axis IR (15 μm thermistor bolometer) Earth horizon sensor designed for pitch/roll determination of three-axis LEO satellites. Users include Eureca and TSS; it also flew October 1992 aboard STS-52 with MSS.

Number of sensors: 2 optical heads with conical scanning equipment
Field of view: 360° pitch, 90° roll
Accuracy: 0.1° rms
Size: 170 × 100 × 116 mm/1.55 kg (optical head)
204 × 172 × 107 mm/2.5 kg (electronics)
Input power: 6 W

The IRES, Infra-red Earth Sensor, is a two-axis IR (15 μm thermistor bolometer) Earth horizon sensor designed for pitch/roll determination of three-axis GEO satellites. Users include ECS, Inmarsat, Olympus, Skynet, Eutelsat, DFH-3, Turksat, Italsat, Orion, Arabsat 2 and Nahuelsat.

Field of view: ±11° pitch, ±15° roll
Accuracy: 0.02° rms
Size: 137 × 190 × 158 mm/3 kg integrated unit
Input power: 2.2 W

IRES with new electronics is replacing the original, improving accuracy to less than 0.02°rms and failure rate to 1,700 × 10^{-9}/h, while reducing cost and mass. Qualified for Sinosat and Eutelsat 3.

The CCD onboard camera is a 288 × 384 pixel CCD TV-standard camera (625 lines/50 Hz) is designed for space operations such as rendezvous/docking, robotic vision, ERA teleoperations and routine monitoring. The optics, processing electronics and LED illumination system are integrated in a single housing.

Field of view: 30 × 40°, with 1.8 × 10^{-13} W lower detection limit
Size: 106 × 100 × 95 mm/1.5 kg
Power input: 2 W camera head, 10 W processing electronics and illumination system.

Avio SpA

Current Status
In September 2003, FiatAvio was acquired from the Fiat Group by Avio Holding SpA, which was 70 per cent owned by the US investment firm, The Carlyle Group and 30 per cent owned by Italy's Finmeccanica. FiatAvio then became simply, Avio. In August 2006, Carlyle and Finmeccanica agreed to sell Avio, valued at EUR2.57 billion, to the European equity group, Cinven Ltd. Finmeccanica reinvested its funds in Avio and retains a 15 per cent stake in the company; Cinven owns the 85 per cent majority share. Finmeccanica sold its original 30 per cent share to Cinven for EUR432 million and bought back its 15 per cent stake for EUR130 million.

Through its joint Venture with the Italian Space Agency, ELV, Avio is leading the development of the small to medium four-stage Vega launch vehicle, which is scheduled for a maiden flight in 2008. Avio will produce Vega's three solid rocket motors: the P80, the Zefiro 23 and the Zefiro 9. Vega's final stage is the bi-propellant liquid upper stage Attitude and Vernier Upper Module (AVUM). Avio will produce the AVUM's propulsion system and is also Vega's system integrator.

Background
Avio traces its history back to the founding of Fiat SpA in 1908. The company is headquartered in Rivalta di Torino, Italy. 2,000 personnel work in this location, which also houses industrial facilities; Avio employs a total of 4,800 workers in 16 locations throughout the world. 2005 consolidated revenues for Avio Holding were reported at EUR1, 281 million.

FiatAvio entered the space propulsion field in 1985 as an extension of its primary aero gas turbine business. Under a Société Européenne De Propulsion (SEP – now a part of Snecma) contract, it designed, developed and manufactured the solid fuel strap-on boosters and turbopumps for Ariane's Vulcain 1 and 2 cryogenic engines. Avio's primary space technology activities remain in the launch vehicle engine and propulsion and satellite propulsion areas. Avio, along with its joint venture companies, is Europe's sole source for solid rocket motors. The company has produced solid propellant motors for every Ariane version. Avio also manufactures solid propellant motors for missiles and air-launched rockets.

Avio holds interests in Regulus, Europropulsion, ELV SpA, The Italian Aerospace Research Centre (CIRA), Arianespace and several other aerospace companies. In January 1994, FiatAvio acquired BPD Difesa e Spazio, which was founded in 1913. BPD Difesa e Spazio was involved with propulsion systems, including the study and production of solid, liquid and gaseous propellants, as well as studies on magneto-plasma dynamic systems. BPD designed and produced the Ariane 3 and 4 solid propellant strap-on boosters.

Japan

IHI Corporation

Current Status
Ishikawajima-Harima Heavy Industries Company Ltd formally changed its name to IHI Corporation in 2007.

Background
IHI traces its history back to 1853, when the Ishikawajima Shipyard was founded. In 1960, the company merged with Harima Shipbuilding and Engineering Company Ltd, which was itself established in 1907, to form Ishikawajima-Harima Heavy Industries Company Ltd (IHI). IHI is headquartered in Tokyo and employed 6,866 staff in 41 offices worldwide as of early 2006. The company achieved a net income of JPY5,283 million in FY2006. Its aerospace divisions were responsible for 22 per cent of net sales during the same period. IHI is listed on the Tokyo, Osaka, Nagoya, Fukuoka and Sapporo stock exchanges.

IHI has 193 subsidiaries and affiliates doing business in a diverse range of industries. Its Soma, Tanashi and Mizuho aero-engine plants engage in space systems development and manufacture. The IHI Aerospace Company Ltd (formerly Nissan Motor's Aerospace and Defense Divisions), founded in 2000, and IHI Aerospace Engineering Company Ltd, are IHI's primary space technology businesses. The companies develop and manufacture space launch vehicle and satellite propulsion systems, including the solid propellant M-V and H-IIA rockets, as well as International Space Station (ISS), and science satellite systems, including Japanese Experiment Module (JEM) systems and liquid apogee satellite engines. Research, development and systems manufacture and assembly take place at IHI Aerospace's Tomioka and Kawagoe plants. IHI Corporation and IHI Aerospace also own shares in the Galaxy Express Corporation (GALEX), developer of the medium lift GX Launch Vehicle.

Kawasaki Heavy Industries Ltd (KHI)

Current Status
KHI became the first company in Japan to successfully develop and produce the next generation of Unmanned Space Experiment Recovery System (USERS) Re-Entry Module, for use after a satellites atmospheric re-entry from Earth's orbit. This technology makes it possible to fly a large unmanned airship and hover it in the stratosphere (at an altitude of about 20 kilometres) via computer control. The airship is used for new broadcast and communications systems, digital broadcasting and mobile communications.

Background
KHI was involved in the construction of launch facilities for the N, H-1, H2 and H2A at NASDA's Tanegashima Space Centre. It has participated in system development of the berthing mechanism, airlock and environmental control systems for Space Station's Japanese Experiment Module (JEM) since 1985. It has been involved with the HOPE project and spaceplane research since 1986. It is system integrator of the ALFLEX Auto Landing Flight Experiment and rear body integrator for the HYFLEX Hypersonic Flight Experiment. Both are key steps in developing HOPE and possible crew return vehicle for the ISS.

KHI has developed and manufactured the experimental Geodetic Satellite launched by H1 in 1986.

Space robotics and docking mechanisms with advanced proximity sensors are being studied on ground testbeds for future in-orbit servicing. KHI developed the docking systems for 1997's ETS-7. In 1994, it completed the WETS Weightless Environment Training System neutral buoyancy facility at NASDA's Tsukuba Space Center for crew basic training, procedure development for on-orbit rack replacement and JEM system development. It was contracted in 1994 for a crew Isolation Chamber at Tsukuba completed in 1996. In 2000, KHI developed and manufactured the ablative heat shield and shell structure for the re-entry module of the Unmanned Space Experiment Recovery System. Adapted this technology to unmanned recovery systems designs and to potential thermal protection systems for unmanned or manned reusable space vehicles.

NOF Corporation

Current Status
NOF continues to supply solid propellant for Japan's M-V and S-310 rockets, as well as the Solid Rocket Booster (SRB-A) propellant for the H-IIA launch vehicle.

Background
Nippon Oil and Fats (NOF) was founded in June 1937. The company has engaged in solid rocket propellant production since 1954 and is now JAXA's sole solid propellant supplier. NOF is based in Tokyo and employs 1,588 personnel. The company's FY2006 net income was JPY6,456. Its FY2006 sales were JPY143,157, of which JPY31,398 were from its explosives and propulsion industry.

NOF's Taketoyo Plant in Aichi Prefecture produces M-V and S-310 propellant, while its Tanegashima Works in Kagoshima Prefecture manufactures the H-IIA SRB-A propellant using the largest solid propellant mixer in the world. NOF's subsidiary,

Nichiyu Giken Kogyo Company Ltd manufactures rocket pyrotechnic components. These range from igniters and small pyro devices to string pyros capable of severing attachment adaptors securing rocket stages and satellite mounting adaptors.

Netherlands

Black Holes B.V.

Current Status

Black Holes are contributing to a University College of London (UCL) study regarding satellite monitoring as a legal compliance tool within the environmental sector. This involves monitoring national laws and regulations, European law and international treaties dealing with environmental degradation and the risks thereof. The project from a legal perspective deals with issues relating to the proper usage of satellite data in court proceedings, and how to enhance such benefits further by legal and regulatory means. It is scheduled to be completed in 2008.

Black Holes with GMV SA have been working on the 'Humanitarian Aid, Emergency Management and Law Enforcement user applications of Galileo' (HARMLESS project), for the Galileo Joint Undertaking. In the context of future Galileo services, emerging issues such as liability, privacy and governmental involvement in downstream terrestrial services, require further investigation and analysis with respect to humanitarian aid and emergency management, as well as law enforcement. For example virtual prisons and zoning prohibitions. The project is scheduled for completion in 2008.

Black Holes monitors the legal aspects for Finance Innovation Network Addressing New Commercial Enterprises using Space, referred to as FinanceSpace, which is investigating financial tools for undertaking private and commercial space activities. FinanceSpace is co-ordinated through a consortium led by the European Space Agency (ESA), consisting of Europe Unlimited, Assystem UK, the International Space University and MST Aerospace. A key role for Black Holes is to analyse international and trade issues, to ensure all relevant legal and regulatory considerations are taken into account. The project began in January 2006 and will be completed in August 2008.

A study on the regulatory framework for space tourist vehicles flying from Sweden, entitled 'International Liability Issues in Private Spaceflight', was commissioned to Black Holes by the Swedish Space Corporation and Virgin Galactic. The project was initiated to address international liability and certification issues with regard to Virgin Galactic, a UK company, using US licensed technology offering space tourist flights from a Swedish Spaceport. The study outlined the need to reconcile existing Swedish and US regulatory frameworks for spaceflight, as well as aviation from the Swedish side in co-operation with the Swedish Civil Aviation Authority. The project set out relevant recommendations to be initiated by the companies to enable future space tourism flights from a Swedish Spaceport in Kiruna.

A study was undertaken by Black Holes in 2007 for a commercial operator interested in the tentative possibilities of operating space tourist flights from the Dutch Antilles. The project focused on the legal parameters and ramifications of such space flights in particular from the perspective of international, US and Dutch space law.

Black Holes with Risk & Policy Analysts Ltd (RPA) completed a study of the economic and governance evolution of space in Europe for the European Commission in 2007. The study addressed key legal and governance issues with regard to the multinational character of the European 'spacescape'. Related issues included dual-use export controls, the International Telecommunications Union regime, the World Trade Organisation framework, space debris and liability, and the Internal Market for satellite communications.

The Global Navigation Satellite System (GNSS) Supervisory Authority is preparing for the introduction of GNSS services, notably Galileo, into a number of downstream markets, where Location-Based Services (LBS) is one of the most important markets. In 2007 Black Holes with Logica CMG and Thales Alenia Space for the Galileo Joint Undertaking completed a study on 'Application of Galileo in the LBS Environment' referred to as the AGILE project. Black Holes consultancy services outlined and analysed data privacy and protection issues, information communications technology convergence, Intellectual Property Rights (IPR) and liabilities as well as service guarantees that will impact on GNSS services and subsequently upon overall market perspectives of relevant satellite systems.

With regard to the International Space Station (ISS), Black Holes along with European Science Foundation contributed to a study regarding humans in space in 2007. The project involved assessing the ISS legal framework in respect to liability, IPR and criminal law.

The International Institute of Air and Space Law (IIASL), Netherlands funded a Black Holes 2007 study to analyse the legal aspects of the Chinese ASAT-test regarding space debris and associated risks from the destroyed weather satellite by the 'killer-vehicle.' This resulted in an article in the Korean Journal of Air and Space Law in 2007.

Black Holes along with the Aerospace Corporation and the Association of Space Explorers (ASE) completed a study on planetary defence in 2007. The project involved legal issues on how to mitigate future disasters such as near-Earth objects hitting Earth.

Background

Black Holes BV, was incorporated on 4 July 2007 by Professor Dr Frans G. von der Dunk to provide consultancy services in space law and policy. The company is headquartered in the university-town of Leiden, Netherlands. Dr von der Dunk, who is company director, earned his Doctorate in International Space Law at the Universiteit Leiden in September 1998 and has been Director of Space Law Research and Co-Director of the IIASL at Universiteit Leiden since 1990. Dr von der Dunk is also advisor to the Dutch Government as well as Honorary Professor from 2004 at the Gujarat National Law University, Gandhinagar, India. On 2 January 2008 Dr von der Dunk was appointed Professor of Space Law at the Law Faculty of the University of Nebraska, Lincoln.

Dr von der Dunk is presently Director and Treasurer of the International Institute of Space Law (IISL) and member of the International Academy of Astronautics (IAA). IISL, founded in 1960 alongside IAA, is an associate of the International Astronautical Federation (IAF). IAF and its associates host yearly conferences entitled the International Astronautical Congress (IAC) in which space law colloquiums are held by IISL. The IISL proceedings are published by American Institute of Aeronautics and Astronautics (AIAA), where Dr von der Dunk is a senior member.

The Black Holes director is also a member of the Board of the European Centre for Space Law, a ESA initiative which was founded in 1989. Other legal memberships include the International Bar Association's Section on Business Law – Committee Z on Outer Space Law; International Law Association's Committee on Space Law for the Netherlands and is also a member of the editorial board of 'Space Policy'. Lastly he is a corresponding member of Centro de Investigacion y Difusion Aeronautico-Espacial.

During October 2004 at the 55th IAC held in Vancouver Dr von der Dunk was awarded the distinguished Service Award by the IISL. He was also awarded a Social Science Award by the IAA during the 57th IAC held in Valencia, Spain in October 2006.

Black Holes consultancy services include international legal, policy and the political framework for all major space projects. Customers range from, the European Commission, ESA, the United Nations, the Organisation for Economic Co-operation and Development (OECD), the Nederlands Instituut voor Vliegtuigontwikkeling en Ruimtevaart (Netherlands Agency for Aerospace Programs), the Deutsches Zentrum für Luft und Raumfahrt eV (German Aerospace Centre), the Agência Espacial Brasileira (Brazilian Space Agency), and the Centre for Strategic and International Studies, to various private and commercial companies.

Black Holes provides international and national legal, regulatory, policy and political framework for any space activity as well as downstream terrestrial applications for space tourism, satellite communications and satellite navigation. The range of consultancy services includes: researching,

Black Holes B.V. A Selection of Space Law Commissioned Studies Undertaken From 2000–2007

Name of Project	Area of Expertise	Commissioned By	Date
Support to the implementation of impending Dutch national space legislation.	Space tourism and private spaceflight.	Ministry of Economic Affairs of the Netherlands.	2006
Cluster of major projects on GNSS Introduction 1) in the Aviation Sector (GIANT); 2) in the RAIL sector (GRAIL).	Satellite navigation systems.	INECO/Galileo Joint Undertaking.	2005 to 2007
Study on the Moon Agreement and the prospect of commercial exploitation of lunar resources.	Legal status of the moon and other celestial bodies.	IISL	2006–2007
Study on a European legal regime for commercial utilisation of the International Space Station.	Privatisation and commercialisation of space activities; ISS.	Meijers Research Institute & European Centre for Space Law.	2006
GOSIS. Project on GMES Organisation and System Integration Scenarios.	Military/dual use space.	ESYS for the European Commission.	2005 to 2006
Study on legal aspects of geospatial data-gathering in space.	Satellite earth observation and remote sensing.	Technical University Delft, Netherlands.	2005
Study on cost effective earth observation missions.	Launching and space transportation.	IAA	2004 to 2005
Global and European challenges for air and space law at the edge of the 21st Century (Project 2001 Plus).	International and national space law and policy developments.	Deutsches Zentrum für Luft und Raumfahrt eV (German Aerospace Centre) and the University of Cologne.	2001 to 2005
The commercialisation of space and the development of space infrastructure: the role of public and private actors.	International co-operation.	OECD	2004
Study on development of the European Radio-Navigation Plan (ERNP), with Helios technology.	Satellite communications; multi national.	European Commission.	2003 to 2004
Property rights on the Moon.	Planetary defence and planetary protection.	IISL	2002 to 2003
GALILEI. Study on further definition of the future GALILEO system. Study on sovereignty	Trade issues regarding space products and services; internal financing.	With Galileo Industries for the European Commission.	2001 to 2003
and outer space. Study on international	Colonisation of space and celestial bodies.	IIASL	2002
legal aspects of the establishment of international radio quiet zones.	Radio astronomy and space science.	OECD	2001

analysing and evaluating results into a final report and or presentation; preparing 'legal reality checks' for the initial activity; organising workshops and meetings to address issues within the framework; and providing advice and recommendations for future actions in order to help facilitate the space activity and or application. Other services provided by Black Holes include developing, monitoring and providing tailor-made or self-study tutorials for professionals and to address audiences and issues within the framework.

Black Holes area of expertise include: international and national space law and policy developments; international co-operation; multinational developments (such as ESA or the European Union); privatisation and commercialisation of space activities; space tourism and private spaceflight as well as issues relating to liability and insurance issues with regard to space transportation. The company provides for the long term future analysis of colonising outer space as well as the legal status of the moon and celestial bodies including planetary defence and protection. Ballistic missile defence is also dealt with under military space systems, along with dual use systems such as Global Monitoring for Environment and Security (GMES). Trading issues with slot, frequency, allocations and assignments as well as intellectual property rights and liability issues forms part of Black Holes expertise in satellite communications.

Black Holes also provides expert knowledge in Earth observation and remote sensing satellites such as IPR, data policy and archiving issues; satellite navigation systems such as Galileo; ISS; radio astronomy and space science; trade issues regarding space products and services; and international financing of space activities.

Dutch Space

Current Status

Work on the flight model of the Herschel Solar Array and Sunshade (HSS) has been completed. HSS was delivered to European Space Research and Technology Centre (ESTEC), Noordwijk Netherlands on 5 November 2007 and is set to be integrated into the spacecraft. Hershchel is a European Space Agency mission which is scheduled for a 2008 launch.

Dutch Space are developing solar arrays for the following forthcoming missions: the European Space Agency's (ESA's) Gravity field and steady-state Ocean Circulation Explorer (GOCE), Herschel Mission and Planck Mission, NASA's Dawn, and also for the joint co-operative European and US Metop mission.

A new type of solar array called 'Thin Solar Arrays' is under development, with the aim of a price reduction of up to 50 per cent. It is based upon CIGS thin film technology, which is applied to a matrix concept thereby reducing the mass of current solar arrays. Dutch Space, in co-operation with Delft Technical University, plan to test this new solar array design on the Dephi C3 Cubesat, (a nano – satellite).

Dutch Space contributes to the booster recovery system of Ariane 5, as well as the main engine frame. They are also working on aluminium conical ½ inter-stages for the new small launcher, Vega, which is not due for launch until 2008.

BepiColombo is a future mission to Mercury and Dutch Space is supplying its thermal control experience to the antenna pointing system.

They are also responsible for the development of Real-Time Simulator (RTS) for the ESA's GIAI mission, due for launch in December 2011 using a Soyuz-Fregat Launcher.

Dutch Space are in collaboration with Italy, Switzerland and the United States regarding the Automated Transfer Vehicle for the International Space Station (ISS). In addition, they are a prime contractor for the European Robotic Arm (ERA) which they are developing for the ISS' Russian segment.

The ConeXpress Orbital Life Extension Vehicle, which is designed by Dutch Space, is intended to extend the life of a satellite by up to ten years. It is expected to become operational in 2008.

Dutch Space are part of the Contraves Consortium for developing the Dexterous Robotic Arm (DEXARM), which will have various applications, for example planetary exploration. The DEXARM is a design blueprint for the future Eurobot Arms, which Dutch Space is involved with as part of the Alenia Spazio consortium.

Background

Dutch Space has over 45 years experience in the space industry. Originally called Fokker Space and Systems, after Anthony Herman Gerard Fokker (1890–1939) who specialised in aircraft construction and was the founder of the aircraft company NV Fokker, they set their sights on space in 1968.

Despite Daimler-Benz AG and subsidiary Deutsche Aerospace AG (DASA) purchasing shares in Fokker NV in 1993, the company filed for bankruptcy in 1995. Fokker Space and Systems were not affected by the bankruptcy and became an independent company in that same year, dropping Systems from their company name. In April 2002 the company changed its name to Dutch Space and in 2005 a merger with EADS Space took place and was completed on 28 July 2006. They are a subsidiary of Astrium N.V. which in turn is a subsidiary of EADS.

Historically, in the 1970's, Dutch Space founded the Industrial Consortium IRAS, where in joint collaboration with NASA and the British Research Council they participated in the InfraRed Astronomic Satellite programme. The satellite was launched in 1983 and using four infrared bands identified a substantial amount of infrared and astronomical sources, as well as discovering the galaxy core – the milky way. The Infrared Space Observatory (ISO) was a follow-on from IRAS and Dutch Space designed the Short Wavelength Spectrometer (SWS).

They have over 55 solar arrays in orbit, 90 per cent of which are supplied to the European Space Agency and to commercial satellites, for example Rosetta, XMM-Newton and SMART-1 among others, and their solar arrays are based upon two designs:

a) the Advanced Rigid Array (ARA), serving telecommunications satellites, science and Earth observation applications. A special variety of ARA known as 'Fred' has also been designed to give extra stiffness and strength in the arrays. The Automated Transfer Vehicle for the ISS is one such example. The ARAMkIII is the latest in the array family. Eight different configurations of the ARAMkIII have been designed with varying differences to give performance and structural flexibility. Approximately twenty have been tested and delivered.

b) Flatpack Solar arrays are flexible and very thin, being approximately 10 mm wide as they are stowed during launch. They provide up to 10 kW of power to satellites with large platforms, usually Earth observations satellites in polar orbit. To prevent any interferences, particularly during the launch phase, the Kevlar cables that keep the panels in place are fired upon by 'Thermal Knives' (patented by Dutch Space), in order to release the panels. This principle has been applied to numerous (Dutch) solar arrays as well as to other technologies.

With regard to the family of Ariane Launchers 1 to 4, Dutch Space designed and manufactured inter-stages (1/2) and (2/3). During its development phase, Ariane 4 historically yielded a 23 per cent mass reduction due to the carbon fibre (previously replacing the aluminium structure) 2/3 inter-stage. Ariane 4 production ended in 2002.

Dutch Space have successfully contributed to planetary explorations, such as Cassini Huygens, Ulysses and Giotto, supplying thermal control systems and/or solar arrays.

Utilising its solar array experience, Dutch Space provided a flatpack solar array and built the radiant coolers for the Envisat Satellite. Alongside Germany's Dornier Satellite Systems (now EADS) they were prime contractors in developing the imaging spectrometre, the SCanning Imaging Absorption SpectroMeter for Atmospheric CHartographY (SCIAMACHY) instrument for Envisat, which has been in working order since the launch of Envisat in 2002.

The Ozone Monitoring Instrument (OMI) for the NASA EOS Aura satellite was built in partnership between the Netherlands and Finland. The Netherlands Agency for Aerospace Programmes and the Finnish Meteorological Institute were responsible for mangaging the project and the OMI was built by Dutch Space, TNO (alongside Finnish Patria Finavitec Oy Systems and VTT Automation) and EEV in England. Dutch Space also contributed to the Ground Data Processing software for the EOS Satellite.

Metosat is another Dutch Space project as they contributed to the thermal design for the Second Generation Passive Cooler Assembly, including the mechanical arrangements for the Assembly.

In respect to microgravity research, they have contributed to the design of scientific experiments for Mission Delta (2004), which was tested on board the ISS.

Norway

Kongsberg Gruppen A/S

Current Status

The Kongsberg Group is Norway's largest defence products manufacturer, established as an independent company in 1987; 51 per cent is owned by the Ministry of Industry. Its space-related activities are managed by the Space and Avionics Dept of Kongsberg Defence and Aerospace and began with development of booster attach and release mechanisms for Ariane 5: structures and mechanisms; SCOE and simulators; electro-optical systems; ground stations for Earth observation satellites (Kongsberg Spacetec). Principal space contracts cover development of Ariane 5's booster attach/separation system; SCOEs and simulators for Soho, Huygens and Ariane 5; MIPAS optical differential system. Set up in 2002, Kongsberg Satellite Services employs around 55 people.

Konsberg Informasionskontroll A/S or IK is a subsidiary of Kongsberg Defense. Principal space capabilities cover system development and integration, software engineering computer systems, signal processing and electronics. Main products include software and computer systems for processing and operational analysis of Earth observation data, and EGSE. Contracts include SAR processing systems for ERS and Radarsat, a system for automatic ship detection in SAR images, VICOS Verification Integration Check-Out Software for Columbus Ground System Software, and EGSE (hard/software) for Soho. Kongsberg has developed new checkout and housekeeping system software package for ISS modules based on a pattern originally designed for scientific satellites. Further development work has focused on man-tended systems for platform builders.

Kongsberg Spacetec is an information technology company, a subsidiary of Kongsberg Defence and Aerospace, providing system engineering services in electronics, computers and software. 75 staff. Located in the Tromsø Satellite Station, it maintains the station under contract to the Norwegian Space Centre. It offers Earth observation ground stations, meteorological systems for Meteosat PDUS/DCP and NOAA AVHRR systems, and value added products and services based on satellite data, including vegetation maps and pollution monitoring. MEOS (Multimission Earth Observation System) is Spacetec's UNIX-based concept for satellite data acquisition and processing. Increased applications for scientific tasks in addition to support of national Earth observation and remote sensing programmes.

Russian Federation

GNPP Kvant

Current Status

Set up in 1987 Kvant builds solar cells and arrays for space applications. The origins of the company go back decades and it has been involved in power production systems since the late 1950s. They have been involved in producing solar cell arrays for almost every Russian satellite and spacecraft so equipped. They conducted research into new forms of deployment for array configurations and were instrumental in developing the arrays for the early Soyuz spacecraft concept. Scientific facilities include electron microscopy, micro x-ray analysis, secondary ion spectroscopy and photo cathode electro-fluorimetric analysis. Expanding research in large solar array configurations for high-power requirements. These have included orbiting solar power satellites capable of beaming electrical energy to rectennas on Earth using microwave frequencies. Developed concepts for unfurlable antenna arrays as a spin-off application. Kvant has conducted several engineering studies for companies in India.

KB Khimautomatiki

Current Status
The bureau of Semyin A Kosberg (1903–65) was formed in 1941, and began investigating liquid propellant rocket propulsion for aircraft in the early 1950s. It later became responsible for some Soviet space launcher upper stages. In 1974 it was reorganised as KB Khimautomatiki. Developed hybrid propulsion systems using mixed solid-liquid propellant combinations, tri-propellant systems and long duration motors using high temperature burning cycles.

Responsible for Energia's RD-0120 cryogenic core engine, Proton's stage 2/3 engines and Soyuz/Molniya stage 2 engines. Formed from old Kosberg upper stage propulsion bureau. Hybrid propellant development. New chemical research facilities for supporting launch vehicle and rocket motor development.

KB Khimmach

Current Status
A bureau headed by Alexei m Isayev was established in 1944 for research into storable liquid propellant rocket engines. Whereas Glushko's GDL concentrated on major launcher engines and Kosberg's bureau on upper stage propulsion systems, Isayev's produced smaller spacecraft engines, including manned, Earth orbit and deep space systems. A number of designations have been adopted but 'KTDU' (Korrektyroushchaya-Tormorznaya Dvigatelnaya Ustanovka, or Corrective-Braking Rocket Motor) is most frequent, with the others variations on this theme.

Information released in early 1991 indicated the bureau is part of NPO Soyuz in Kaliningrad. It is responsible for the 73.58 kN cryogenic engine of the new Proton km version, originally designed for the L-3M manned lunar mission, and sold to India for its GSLV vehicle. A newly-developed DMT-600 600 N NTO/UDMH thruster was referred to in 1994. Khimmach developed all of the engines for the Heavy KosmosTKS transport craft. Continued with research into new and improved forms of hybrid propulsion systems for special requirements set out by the Ministry of Defence. Produces the 11D49 engine for the second stage of the Kosmos 3 launch vehicle.

Starting in 1971 and at the instigation of Isayev, the bureau started the development of low-thrust engines, which could be used for attitude control of spacecraft. A series of 11 engines with thrusts up to 2,206 kN (225 kg) and eight with thrusts up to 49 kN (5 kg) were developed. They are capable of burns lasting from hundredths of a second through to hundreds and thousands of seconds. Considerable progress has been made with this technology since the mid-1990s. Aspirations to extend research activities into ion propulsion induced some early efforts in this area but these have been halted.

Korolev Design Bureau

Current Status
One of the most famous engineering design organisations in Russia, Sergei Korolev's bureau, founded in 1946, developed Proton's Block D/DM and Molniya's Block L engines. The organisation is responsible for the development of approximately 250 separate rocket motor designs and 40 patents for innovative technologies. Mikhail Melnikov designed the Proton engine.

Korolev has evolved into one of Russia's premier space systems contractors, SP Korolev Rocket and Space Corporation Energia (see separate entry).

Lyulka-Saturn OAO

Current Status
Set up in 1944 as a gas turbine design bureau under Arkhip m Lulka, the bureau specialises in military aircraft engines. It also provides turbomachinery and gas generators for space applications, including Buran's gas generators. The 400 kN LOX/LH_2 D57 engine was developed in the 1960s for the lunar N1 launcher's third and fourth stages. Lunar project work was cancelled in 1972 and sometime in that decade the bureau was reorganised into the Saturn Science and Production Association (NPO). After privatisation in the 1990s it became the A Lyulka-Saturn Joint-Stock Company. Lyulka teamed with Aerojet Propulsion in 1993 to improve and market the engine. During this period the company successfully tested combinations of propellant that it had been unable to apply to proven designs and this stimulated dialogue between Russian and American rocket propulsion engineers.

NPO cryogenic engine, which was been described in the 1990s as the D-57, would appear to be the 11D57 gimballing engine. This was developed during the early 1970s for use on the Block S of the modified N-1 launch vehicle, which would have been flown for the dual-launch L-3M manned lunar missions. The bureau was also tasked to develop the 11D54 fixed-chamber cryogenic engine with the same thrust level of 40 tonnes (390 kN). This engine would have been used in a cluster of six to eight on a modified Block V of the N-1 vehicle. The propulsion system was marketed as a potential export for interested clients and some derivatives were tested in support of that. Higher power versions of these engines were proposed for a commercial version that failed to receive development funds. Following the cancellation of Buran and the lack of support for the Energia launcher, Lyulka-Saturn OAO seems to have had no space business but has retained its skill base funded by other activities.

NII KHIMMASH

Current Status
NII Khimmash was founded in 1948 as a facility of NII-88 and conducted its first firing in December 1949. Known as Branch 2 it was located 15 km north of Zagorsk and, in 1961, separated from its parent body to become NII-229. From 1965 it was under the Ministry of General Machine Building but when that organisation was disbanded in 1991 it was embraced by the Neptun corporation until absorbed into the Russian Space Agency in April 1992. From meagre beginnings as Russia's oldest motor test facility dating back to the R-7 ballistic missile, it is Russia's largest rocket engine test facilities, covering 25 km^2 and capable of testing complete stages up to 9 m diameter and 12 MN. Unsuccessful efforts to gain political approval for a new generation of single-chamber rocket motors operating on cryogenic propellants. The major test stands are, Stand V1: 100 kN cryogenic; V2: 2 positions for 100 to 2,000 kN cryogenic (such as RD-0120); V3: stages to 490 kN. Stand 102: 11.8 MN LOX/kerosene. Six thermal vacuum chambers of 0.3 to 100 m^3 can simulate missions of up to nine months for 5 N to 6 kN engines. 900 m^3 and 8,300 m^3 chambers test complete satellites. Employees total 3,800. SEP heads a European team that began testing the RD-0120 at the site in 1995. NII Khimmash is developing several initiatives for foreign participation with the SEP tests stimulating new joint deals and international ventures.

NII Machinostroenye R&DIME (Research and Development Institute of Mechanical Engineering)

Current Status
NII Machinostroenye R&DIME (Scientific Research Institute of Mechanical Engineering) was established in 1958 as part of NIIThermal Processes, but became independent in the early 1980s. Development and production of 0.8 to 400 N spacecraft thrusters (implying connection with old Isayev bureau), including Soyuz, Progress and Mir. Development and small batch production of engines up to 6 kN can be accommodated. Pioneering work on nano-thrust propulsion systems for long duration firing and low acceleration. Test facilities can handle hydrogen (up to 3 kN), methane and kerosene fuels. The Elektro Earth observation satellite carries 16 DEN-16 electrothermal thrusters. Note: LTRE means Low Thrust Rocket Engine, or RDMT in Russian (Raketnay Dvigatel Maloi Tyagi). Employs fewer than 2,000 people and further work is based more on continued production of extant hardware than development of new technologies. Diversified into applications within defence industries. Has engineered new variants of existing small scale engines and has completed tests with hybrid reaction control systems.

NII Parachutostroenye

Current Status
The NIIPS Institute of Parachute Construction provides recovery parachutes for Soyuz, Buran, Energia boosters and Progress return capsule. Developed new lock-up control links to satisfy safety evaluation recommendations. It is also supplying the recovery parachutes for ESA's Ariane 5 and has conducted experiments with the recovery of very heavy booster stages. New, phased deceleration deployment for masses in excess of 25 tonnes. Experimental development of very heavy load retardation in support of a new generation of Russian launch vehicles proposed by Energomash but not funded. Work has begun on deceleration systems for recoverable launch vehicle stages and on re-shaped parafoils. Some manufacturers would like to develop land recovery for reusable launch vehicle stages.

NII PM (Prikladnoi Mekhaniki)

Current Status
NII Applied Mechanics is a major designer and developer of launcher and spacecraft navigation instruments, including laser gyros and has produced guidance equipment for planetary spacecraft. Has also produced several different inertial navigation systems and co-operative and interactive navigation sets for ballistic missiles and test rockets. Developed research programmes and technology efforts resulting in radical changes to guidance platform design during the 1970s and 1980s. Ring gyro systems for deepspace operations and stop-restart functions following dormancy. NII PM has conducted research into integrated guidance and navigation systems for deep space systems.

NII Radio

Current Status
Established in the 1930s, NIIR is the main research centre of the State Committee on Communications and Information of the Russian Federation subordinated to the State Committee on Communications and Informatisation of the Russian Federation. Controls standards on frequencies and makes appropriate recommendations on boundaries within which to allocate band spread. NII Radio has carried out tests with unusual frequency distribution networks design but these are related to civilian applications outside the space industry. Carries out several military related programmes for frequency allocation. Research projects paid for by industrial manufacturers have been carried out over the last several years.

NII Thermal Processes

Current Status
NII Thermal Processes performs research in all forms of launcher and space propulsion, including ramjet, electric and nuclear. BMDO took delivery in October 1993 of a T-100 1.5 kW Hall-type Stationary Plasma Thruster for evaluation. N_2O_4/UDMH thrusters include 10 N and 1,000 N models. NII TP developed the Phobos spacecraft's network of hydrazine thrusters for attitude control and minor orbital modifications. 28 (24 × 50 N, 4 × 10 N) were mounted on four spherical tanks. A set of 0.5 N hydrazine microthrusters of an unusual design was also included to avoid the need for separate N_2 cold

gas jets; the propellant was burned internally and cooled inside an expansion tank before passing to the thrusters. Successfully completed extended life cycle tests showing operability after lengthy period of dormancy in cryogenic environment. NIITP has attempted to engage with other European countries to enhance the development of microthruster technology. Responding to a direction of effort in the United States toward nuclear propulsion for deep space transportation, NII TP has attempted to link its own (largely historic) development of nuclear and nuclear-electric in an effort to combine research and provide a base for international co-operation. High thermal demand propulsion systems are being researched and some limited testing carried out.

NPO AP

Current Status
NPO Automatics and Instrument Engineering originated in 1946 as the Department of Automation in the NII-885 department of the Ministry of Communication Means. In 1963 the Ministry of Industry of Communications Means separated to become an independent entity. NPO AP was run by Nikolai created in 1978 as part of NII AP to develop control, navigation and guidance systems for launchers and spacecraft. NPO AP was run by Nikolai A Pilyugin (1908–82; Lapigin succeeded him in 1962) and provided the flight control systems for the Vostok launcher family, Proton (including Block D/DM), N1, Zenit and Buran. In 1978 it formally received its present title and in 1994 it was absorbed by the Russian Space Agency. Areas of diversification included automated control systems, robotics and 'intelligent' machine technology. Its capabilities remain intact and have been dispersed into other departments. Prior to dispersing the assets of the organisation, NPO AP developed automated flight control systems, that were designed to remotely control spacecraft and space vehicles originally designed for man-tended operations and this technology was adapted to automotive engineering processes in ground facilities unrelated to the space programme.

NPO Elas

Current Status
Elas builds communications and imaging sensor payloads. It is providing the payloads for the Kupon and Courier communications satellites; it is probably responsible for the military Geizer satellite's payload and has developed advanced military communications payloads for future milsats. Elas conducted extensive research into the development of commercial by-products from its work on military communications satellite systems and has derived some business from that transfer, although more revenue accrued from diverse civilian applications than work for the military. Expanded sensor development for multispectral instruments on remote sensing satellites.

NPO Energomash

Current Status
The history of NPO Energomash can be traced back to 15 May 1929 when the Group for the Development of Rocket Engines was organised at the Gas Dynamics Laboratory (GDL) in Leningrad (now St Petersburg). For more than 60 years the organisation repeatedly changed its name and location, but remained at the forefront of rocket engine design. The founder and the permanent head of the GDL as Academician Valentin P Glushko, considered to be the initiator of the Russian rocket engine industry.

In May 1974, the GDL merged with Sergei Korolev's design bureau, the new organisation being re-named NPO Energia, with Glushko as its head. Following Glushko's death in January 1989, the former GDL left Energia to become NPO Energomash, continuing to concentrate on the development of rocket engines: the bureau was headed by Vitali P Radovsky during 1974–1994. The current general director and chief designer is Boris I Katorgin. In 1994, Energomash was transferred to the Russian Space Agency.

NPO Energomash incorporates the design bureau, plant and test facilities, with the capability to test engines with thrusts of up to 10 MN either as components or complete. An agreement was signed with Pratt and Whitney in October 1992 providing exclusive US marketing rights for Energomash's LOX/kerosene and tri-propellant products and technology.

In 1998, Energomash began transforming itself from a state enterprise to an open-stock holding company but with majority shareholding held by the state. Rocket motors under development include RD-161, RD-180, RD-120K, RD-701, RD-704.

Energomash designed and built the RD-180 which now powers the most powerful variant of the Atlas launch vehicle, the Atlas 5 series. The first flight took place on 21 August 2002.

NPO Kompozit

Current Status
Advanced metallic, non-metallic and composite material research in Russia, including Buran thermal protection, takes place at Kompozit. Research work has been expanded in high thermal stress requirements for multi-use re-entry systems for large structures. Other products include microgravity research equipment, including semiconductor and optical materials units. Spacecraft and launcher instrumentation. Thermal protection research provided base for development of ceramic insulation for boilers and gas turbines. Employment stood at 7,500 in 2003. NPO Kompozit has completed joint ventures with Canada, Germany, Italy and Switzerland.

NPO Kraznaya Zvezda (Red Star)

Current Status
Established in 1972 Red Star specialises in nuclear power systems. It developed the 2 kW Bouk thermionic reactor for the Rorsat programme, and the twice-flown (Kosmos 1818/1867) Topaz 1 5 to 6 kW unit. Redstar has developed plans for the design of high power reactors on long duration planetary space flights and has conceptually speculated on the requirements of a nuclear power source for planetary surface basis. Limited research and development on a small nuclear power source for remote regions of Earth. Brayton cycle converters. The company has also developed commercial applications and has specialised in heat-pipe technologies. Developed high-thermal transfer systems for oil driven powerplants. Space-related applications have continued despite little work other than theoretical research in that area.

NPO Precision Instruments

Current Status
Gonets-D1M 1 is a Russian low-altitude communications satellite that was launched by a Kosmos-3M rocket from Plesetsk at 18:34 UT on 21 December 2005. It is the first of the new generation Gonets satellites. The 250 kg, 40 W craft is expected to serve some 30 Russian agencies and organisations with e-mail and short messages. An earlier fleet of six Gonets-D1 satellites had served that role in the 1990s. The initial orbital parameters of Gonets-D1M 1 were period 114.7 min, apogee 1,424 km, perigee 1,414 km, and inclination 82.5°. It is currently unclear what involvement NPO Precision Instruments has in the production of the new generation of Gonets satellites.

Background
Set up in 1963 as Department 4 of NII-885 the Scientific Research Institute of Precision Instruments NPO TP specialises in manned and unmanned spacecraft payloads and electronics, including control systems, telemetry, reception stations and communications. The organisation served as one of the most important suppliers and design bureau for small instrumentation and for subsystems and components. NPO Precision Instruments provided guidance systems for SS-11 to SS-19 ICBMs, Buran orbiter, Luna, Mars, Venera and Vega spacecraft and SS-25/Start 1/Start launchers. It developed the Salyut/Mir Igla and Kurs rendezvous and docking systems and the Vostok telecommand system. Prior to the collapse of the USSR they developed a set of sensor suites for autonomous planetary explorers. It is part of the Smolsat consortium developing the Gonets satellite system. Prime contractor is Applied Mechanics.

NPO Vega-M

Current Status
Headed by the MNIIP Moscow Science and Research Institute of Instrument Engineering, Vega provides space SAR radars, including those of the Almaz remote sensing satellites. These have been applied to both military and civilian projects. Some space-based radar designs have been adapted for airborne applications. Formerly developed technologies for military reconnaissance and surveillance systems. Pursued development of SAR and other space-based radar systems for the Ministry of Defence. This work led to more advanced SAR designs for Almaz vehicles. Developed advanced radar surveillance equipment for military applications to domestic clients. The company has developed various technology applications which has added revenue value greater than that accruing from space business.

EDB Fakel

Current Status
In 2003, Snecma and EDB Fakel signed an agreement to collaborate on the more powerful plasma PPS-5000 propulsion system for ESA's AlphaBus programme. Fakel works with European and US, as well as Russian, customers. As early as 1993, Loral Space and Communications and Atlantic research Corporation (the propulsion business of which was sold to American Pacific Corporation in 2004) teamed with Snecma and Fakel in the joint venture, International Space Technology Inc to produce licensed Fakel SPT-100 plasma engines, and afterwards, PPS-1350 systems. The PPS-1350 is on ESA's SMART-1, launched in 2003, as well as other geostationary satellites.

Background
Experimental Design Bureau (EDB) Fakel was founded in 1955 in Kaliningrad, on the Baltic coast, to develop the Soviet Union's spacecraft propulsion systems. First established as an arm of the Russian Academy of Sciences, in 1962 it transitioned into an EDB, complete with manufacturing facilities. Fakel supports the Russian Federal Space Agency, Roskosmos, with their propulsion technology. The organisation employs roughly 960 personnel.

Fakel specialises in spacecraft Electric Propulsion Thrusters (EPT) and systems including, Stationary Plasma Thrusters (SPT), ThermoCatalytic Thrusters (TCT), attitude control thrusters, ion engines and plasma sources. Fakel began its research and development work in the 1950s and tested a considerable range of designs in the early 1970s. Its Hall-effect Stationary Plasma Thrusters (SPT) have flown on more than 200 satellites to provide orbit control. SPTs employ a DC-gas discharge in an annular chamber, in which a radial magnetic field traps the electrons. These Hall currents ionise the Xe propellant, the ions of which are accelerated inside a quasi-neutral plasma, without grids, and by the discharge voltage itself. The advantages are their rugged simplicity (no grid system or high voltage supply), but they provide lower efficiencies, lower exhaust velocities and increased Xe consumption.

PO Samara Frunze Engine Building

Current Status
Originated by the French in Moscow in 1912, becoming Aviation Plant 24 named after M V Frunze and from 1932 produced first indigenously

manufactured aircraft engines. Redirected for serial production of R-7 ICBM engines in 1957. One of Russia's principal aerospace engine production plants, Samara builds the Soyuz/Molniya RD-107/108 engines for NPO Energomash. After privatisation, became Open Joint-Stock Company named Motorostroitel. Manufacture of propulsion components for the Ministry of Defence and contracts let through the Russian Space Agency. Has received some funding for innovative hybrid propulsion systems designed for long life storage and quick reaction operation. Demonstrated innovative methods of assembling propulsion systems for low cost production methods. This has extended to a new series of small upper-stage propulsion systems for vernier and attitude control applications.

TsNII Mashinostroenye

Current Status
TsNIIMash central research establishment of Russia's Ministry of General Machine Building (previously the USSR's MOM) is now part of RKA. The TsUP Space Operations Centre is one of its departments. Originally an artillery factory, it became NII-88 (Science Research Institute 88) in May 1946 and, with the help of interned German V2 specialists, under Korolev developed the first Soviet rockets. Korolev's OKB-1 branched off from NII-88 in August 1956. Utkin was General Director of NPO Yuzhnoye until he moved to TsNIIMash in 1990. TsNIIMash-Export was created in 1991 to market TsNIIMash capabilities. It is working with the US BMDO on TAL (thruster with anode layer) D-55 1.5 kW and TAL D-100 5 kW Hall-type Stationary Plasma Thrusters. TsNIIMash employs about 5,400 people. Has provided research development work for the Russian Ministry of Defence on ABM technologies. Has developed new methods of creating high-energy plasma directors which are mobile and could theoretically be used in orbiting platforms. Several direct applications into civilian, earth-based, activity has resulted in some commercial success for projects that had little direct application in space projects.

Sweden

Saab Ericsson Space

Current Status
In June 2006, Saab Ericsson Space received an order for 10 X-band helix antennas for Surrey Satellite Technology's RapidEye satellite. RapidEye is a commercial Earth observation satellite system built under the leadership of the Canadian Macdonald Dettweiler group. The system will include five 175 kg satellites built by Surrey Satellite Technology Ltd (SSTL). The satellites will carry a high resolution optical instrument developed by Jena Optronik and will be capable of seeing features on ground larger then 6.5 m from a 620 km orbit. The satellites are scheduled for launch in 2008 aboard a DNEPR rocket in Kazakhstan.

Background
Saab Ericsson (SE) Space was founded 1 January 1992 by the merger of Saab Space AB and the Space Department of Ericsson Radar Electronics AB. Saab Combitech AB owns 60 per cent and Ericsson Microwave Systems 40 per cent. The company supplies launcher and spacecraft onboard subsystems and equipment, including fault-tolerant computers, data handling systems, antenna systems, microwave electronics, satellite separation systems and guidance systems. Each Ariane 5 carries two new-generation parallel guidance and navigation computers; the equivalent continue to be supplied for Ariane 4. The Columbus Polar Platform employs SE's Cassegrain antenna (including DRS pointing control electronics) and an OBC similar to the Spot 4 fault tolerant concept. For Envisat, SE will deliver control units for two instruments, central RF electronics and a radar signal processing subsystem for the ASAR radar. Meteosat Second Generation will carry a data handling system, UFH receiver/transmitter and TT&C antennas. Thaicom 3, Agila, Nahuel, Sirius 2 and Eutelsat W24 will have SE's data handling systems. Earlier deliveries include computers and microwave antennas for ERS ½, computers for Spot 1-3, data handling system, antennas, RF distribution network and central software for Soho, and attitude control computer, antennas and cover release mechanism for ISO. Data handling systems and antennas were supplied for Eutelsat 2, Arabsat 2 and Turksat.

Volvo Aero

Current Status
Pratt & Whitney Rocketdyne (PWR), have sub-contracted Volvo Aero to produce the nozzle for the J-2X engine: stage I of NASA's Ares launch vehicle, successor to the Space Shuttle. The J-2X is a J-2 derivative, the engine that powered the Saturn V of the Apollo programme. The nozzle that Volvo Aero will produce will resemble the nozzle the company manufactures for the Ariane 5 Vulcain 2 engine.

Background
Volvo Aero has been a subsidiary of the Volvo Group since 1941, when it was known as Svenska Flygmotor AB, or the Swedish Aero-Engine Company Ltd. In the 1970's, the company entered the space age when it began producing Viking motor components for the Joint European Space Programme. In addition to aero-engines, Volvo Aero's Component Development and Production division specialises in the design, development and production of rocket engine turbines, rocket combustion chambers and nozzles for space propulsion systems. Component Development and Production facilities are located in Trollhättan, Sweden; Kongsberg, Norway; and Newington, Connecticut. Volvo Aero employs roughly 3,500 personnel; in 2006 the company had net sales of SEK8,048 million, representing about 3 per cent of the Volvo Group's combined net sales. The company's strongest financial performance originated in North America, with European business a close second.

Volvo Aero produces the turbines and nozzles for the Ariane 5's Vulcain and Vulcain 2 engines, as well as its Vinci upper stage engine turbines. The company also produced Ariane 4's Viking 4 and Viking 5 engine combustion chambers and nozzles. Volvo Aero manufactures the sandwich nozzle for Pratt & Whitney's Advanced Rocket Technology Demonstrator Engine (ARTDE) cryogenic RL60, an RL10 upgrade.

Zarlink Semiconductors

Current Status
Zarlink semiconductors provides wireless, medical and optical communications and analog foundary for the healthcare, consumer, industrial, security, military and avionics markets. They specialise in TDM voice and data switching chips, as well as timing and synchronisation chips, RF and digital design. Optoelectronic components and modules are produced at Zarlink's Järfälla operations in Sweden.

Background
ABB Hafo, of Järfälla, Sweden, was acquired by Mitel Corporation in March 1996, and operated as a subsidiary called Mitel Semiconductors AB. ABB Hafo specialised in radiation hardened Complementary Metal Oxide Semiconductors on Silicon On Sapphire (CMOS/SOS) Application Specific Integrated Circuits (ASICs) as well as optoelectronics. CMOS/SOS technology products, originally part of Mitel's Lincoln UK operations was sold in 2000 to Dynex Power Inc, a Canadian company, who subsequently established a subsidiary company called Dynex Semiconductors. Please refer to the Dynex Semiconductors entry for CMOS/SOS product information.

Mitel Corporation was founded in 1973 in Ottawa Canada and began its semi-conductor business in 1992. Mitel acquired from GEC PLC the semiconductor business entitled GEC Plessey in 1998. The US division operated as Mitel Americas Inc. After the ABB Hafo acquisition the company merged the subsidiaries under one division, entitled the Semiconductor Division. The subsidiaries continued to work independently. In January 2001 Mitel Corporation put its communications business up for sale, including the use of the name Mitel to Dr Terence H Matthews. This sale was concluded in February 2001. The company re-emerged as Mitel Networks Corporation. After the sale of its telephony based business Mitel re-branded under a new global corporate identity, entitled Zarlink Semiconductor.

For wireless communications Zarlink Semiconductors provides GPS chipsets for 32 Bit ARM6 RISC Processors such as the P610ARM-B and the P60ARM-B. The company also provides GPS correlators such as the GP1020 Six-Channel Parallel Correlator Circuit, GP2021 12 Channel GPS Correlator and GP4020 GPS Receiver Baseband Processor. The GP1020 is a 120-pin plastic quad flatpack with six independent correlation channel and is used for tracking GPS Course Acquisition (C/A) code or Glonass Receivers using integrated circuits, for example the GP1010 L1-channel down-converter. Most 16 or 32 bit microprocessors interface with the GP1020 and ares compatible with GP1010 GPS Receiver Front-End. Other applications for GP1020 include High Integrity Combined Receivers, GPS Geodetic Receivers and GPS time reference.

The GP2021, a low voltage 12 independent channel correlator, incorporates a low-current power-down mode with battery backup voltage of 2.2 V min and is used with NAVSTAR GPS satellite navigation receivers. GP2021 is compatible with most 16 or 32 bit microprocessors which together support Dual UART, a Real Time Clock and Memory Control functions where the latter can be configured to support an ARM system, in particular Zarlink's P60ARM-B, or under a standard interface node. It is also compatible with GP2015 and 2010 Radio Frequency (RF) Front-End and can also be used in GPS Geodetic Receivers and Time Transfer Receivers.

GP4020 is a low-power 3.3. V receiver which operates with a Zarlink Firefly MF1 Core, which includes ARM7TDMI (thumb) microprocessor. GPS4020 with 12 independent channels is compatible with GP2015 GPS Front End device and is a baseband processor that produces NAVSTAR GPS C/A code. Zarlink does not offer any upgrades to GP4020 from GP2021 with ARM60 users. The company suggests Novatel for GPS OEM receivers.

Other products are the GP2010 GPS Receiver Radio Frequency (RF) Front-End and GP2015 Miniature GPS Receiver RF Front-End. The GP2010 is a 44-pin lead or matte tin Surface Mount Quad Flat-Pack Package (10 × 10 mm) low voltage (3 V-5 V) C/A code compatible GPS Front-End RF Receiver. The input signal, received via an antenna and LNA, is GPS L1 (1575.42 MHz). Using a triple down conversion the signal is converted to 4.309 MHz IF and is subsequently sampled to produce a 2-bit digital output. The GP2021 CMOS correlator can be used in conjunction with GP2010. The GP2015 is similar to the GP2010 except that the Front-End RF Receiver is provided in 48-Lead (7 × 7 × 1.4 mm) Surface Mount Quad Flat-Pack Package which is also compatible with GP2021 as well as the GP4020 Baseband Processor. The RF Front-End Receivers are used within surveying, timing, C/A Code Global Positioning by Satellite Receivers and navigation applications.

Switzerland

Contraves

Current Status
Contraves' Space Division is part of Oerlikon-Bührle Holding Ltd and is responsible for that organisation's space activities. Provides special services in fields of engineering, design, structural analysis, thermal analysis, materials and processes, optics, electronics, computer aided engineering and testing. The company provided the spacecraft structures for ESRO 1, GEOS, Giotto, ISEE 2, Exosat, Giotto and Ulysses, in addition to deployment mechanisms for scientific instruments, antennas and solar arrays. Other space interests include the production of Ariane 4/5 payload fairings, thermal control and EODP. It also provides fairings for Lockheed Martin's commercial Titan 3 and Titan 4. Has available designs for lightweight oversize fairings and payload protection envelopes, also for massive payloads with load support structures as an integral part of the fairing and enclosure.

Contraves has developed fairing and payload shroud designs capable of addressing very large volume requirements and specifications for oversize structures.

RST Raumfahrt Systemtechnik AG

Current Status
In January 2008, OHB Technology AG acquired a 50 per cent ownership of RST Raumfahrt Systemtechnik AG. RST was already a radar partner of OHB System AG in the joint project SAR-Lupe (the satellite-based reconnaissance system) for a number of years, assuming project responsibility for the radar conceptual design and Synthetic Aperture Radar (SAR) processing.

The merger enables increased access to radar technologies for OHB Technology AG, especially within the concepts of spaceborne reconnaissance on the basis of SAR-Lupe. As of early 2008, OHB and RST are in the development stages of strategising future enhancements for the second generation of SAR-Lupe, scheduled for deployment in 2017.

Background
From its offices in Switzerland and Germany, RST specialises in the design and development of SAR, altimeter and ground penetration radars. Provides consultancy, engineering services and space products: space systems, space-related ground systems for Earth observation, navigation and communications. Studies completed on integrated Earth observation systems. In 1994 it introduced a gyro-stabilised (less than 40 arcmin accuracy elevation/azimuth) 1.2 m antenna for shipboard reception of 1 to 13 GHz satellite TV. In 1996 a new forward-looking CW radar was introduced. RST is the radar system architect for the German SAR-Lupe satellite reconnaissance system and has developed the wideband altimeter for ASIRAS, an ESA project. RST is capable of providing high-level digitised imagery from SAR systems and can integrate altimetry and SAR swaths. Also new SAR applications for small satellites with limited power capabilities.

Ukraine

NPO Khartron/ Elektropribor

Current Status
With facilities in Kiev, Moscow and Zhaporje, Khartron supplies onboard spacecraft computers, including Mir and Buran and it provides Zenit SL-16 and Kosmos 3 SL-8 avionics and has developed guidance equipment for autonomous control of launchers. Has conducted software analysis for trajectory changes supporting off-nominal mission requirements. Guidance equipment upgrades and complete replacement of control computer systems. The company has started a new open architecture plug-in module for launch vehicle applications. Intention is to sell the unit to a variety of LV designers, manufacturers and operators.

United Kingdom

British Aerospace Defence Co Ltd

Royal Ordnance RMD provides the principal UK capability in propellants and rocket motors, including double base/composite solids and mono/bipropellant liquid engines. Solid rocket performances cover 0.008–600 s burning time, 0.012–425 kg charge mass and 0.8–200 kN thrust.

Liquid Propellant Engines
The 500 N Leros liquid apogee engine represents the first phase of RO's expansion into civil space propulsion, with the long-term objective of developing a comprehensive range of mono and bipropellant engines. Leros 1 was the first to be qualified; Leros 1b, 2, 20 and 20H have been developed since. RO is improving performance by using advanced materials in thrust chambers for prolonged operation at elevated temperatures.

Olin Aerospace's MR 103C 0.5 N mono thruster is marketed in Europe under licence by RO. It is being supplied to MMS for Skynet 4. RO has a long history of work with catalytic thrusters for attitude control applications or stage attitude control systems. For example, in 1974 RO was appointed as Technical Authority for Chevaline's Hydrazine Actuation System.

Leros 1
RO began development of Leros in 1986 to satisfy the orbital manoeuvring requirements of large telecom satellites using dual mode propulsion, which provides a high performance bipropellant MON/hydrazine liquid apogee engine combined with mono hydrazine thrusters for attitude control and stationkeeping. The design approach was based on proven design features, optimisation of chamber length for smooth combustion over a wide operating envelope, a long combustion chamber to reduce thermal soakback and to ensure a cool running injector assembly and stable film conditions, conventional fabrication and assembly, and in-process injector screening in a slave chamber for acceptable performance.

Leros 1 design verification testing was completed in spring 1989, with the qualification programme at Boeing Tulalip for Martin Marietta Astro Space demonstrating by mid-1990 that the engine can handle a 3 h burn (double the worse-case mission duration identified). Tests covered oxidant depletion, bubble ingestion, multiple hot restarts, hard vacuum ignition and helium-saturated propellants. The 140 starts accumulated 23,000 s and accounted for 3.7 t propellant throughput. Lockheed Martin Astro Space has purchased 29 Leros 1 engines for use in telecom platforms. Leros 1 became the world's first flight-proven dual mode LAE when a pair placed Astra 1B into GEO in Mar 1991. Other users are Telstar 4, AsiaSat 2, Echostar and Intelsat 8/8A.
Applications: dual mode satellite propulsion systems
Dry mass: 4.2 kg
Length: 610 mm
Max diameter: 288 mm (nozzle)
Oxidiser: MON3
Fuel: hydrazine
Propellant feed method: pressure, by helium
Inlet pressure: 15 atm
Valves: Moog model 53–177 torque motor valve, dual coil single seat configuration, based on the model 53–135 unit used on Rocketdyne's Peacekeeper attitude control bipropellant thrusters. >1,500 produced
Thrust: 500±25 N vacuum
Specific impulse: 314 s vacuum nominal
Response time: <10 ms
Mixture ratio (O/F): 0.8
Expansion ratio: 150:1 with 8.5° exit plane half angle
Expansion bell: disilicide-coated niobium C103, radiatively cooled
Chamber pressure: 7.0 atm
Chamber combustion stability: ±12% above 100 Hz, ±3% below 100 Hz
Chamber materials: disilicide-coated niobium C103, T<=1,360°C
Injector: 6Al/4V titanium alloy, jet impingement type using eight unlike doublets for the core around which are 16 fuel film coolant orifices. A backing plate encloses the injector elements and provides the welded interface with the engine mounting flange and mounting points for the propellant control valves
Duty cycle: unlimited, but qualified to 3 h

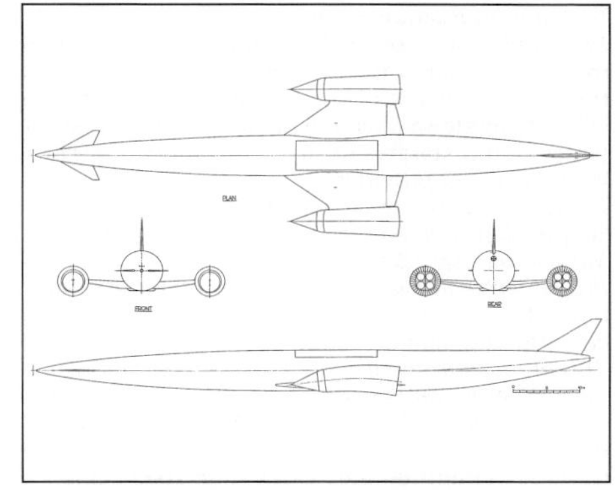
Skylon configuration C1 (Reaction Engines) 0517322

Leros 1b
This improved version delivers 318 s Isp. Design/construction is identical in all respects to Leros 1 with the exception of the propellant valve, which has been replaced by two Moog solenoid valves (model 53–200). Leros 1b was qualified in 1995 for Lockheed Martin and flight engines have been developed.

Leros 1c
Under development for a major US company using high T materials for the thrust chamber. Target vacuum SI is 325 s.

Leros 2
The Leros 2 apogee engine is designed for use in unified MON/MMH bipropellant systems. It differs from Leros 1 only in its injector configuration and combustion chamber geometry. An alternative valve option is available. The engine has completed development and is being used as a springboard for the development of the higher performance (320 s Isp) **Leros 2A** version under ESA contract employing an advanced high temperature material chamber capable of withstanding prolonged operation at up to 2,000°C. Design verification testing is planned for 3Q 1997. **Leros 2B** features conventional columbium/disilicide technology for the thrust chamber and the 300:1 expansion cone. Demonstrated vacuum SI >317 s.
Dry mass: 3.40 kg
Inlet pressure: 16.0 atm
Thrust: 556 N
Mixture ratio (O/F): 1.65
Specific impulse: 312 s

Leros 20
This second generation bipropellant attitude control thruster was developed for MMH propulsion systems.
Dry mass: 0.73 kg
Length: 202 mm
Max diameter: 60 mm (nozzle)
Oxidiser: MON3 at 4.8 gm/s
Fuel: MMH at 2.9 gm/s
Propellant feed method: pressure, by helium
Inlet pressure: 10–20 atm
Valves: Moog single or dual seat
Thrust: 22 N nominal vacuum
Specific impulse: 294 s vacuum nominal
Mixture ratio (O/F): 1.65
Expansion ratio: 180:1
Expansion bell materials: niobium C103 or titanium
Chamber pressure: 7.5 atm
Chamber materials: niobium C103, T<=1,400°C
Injector: multiple impinging jet element with three unlike doublets and six fuel film cooling orifices
Duty cycle: unlimited
Minimum impulse bit: 25 mNs

The **Leros 20H** MON/hydrazine engine is aimed at dual mode systems under ESA contract: SI 300 s, mixture ratio 0.7–0.8, minimum impulse bit 15 mNs. Performance characterisation testing is scheduled for 3Q 1996.

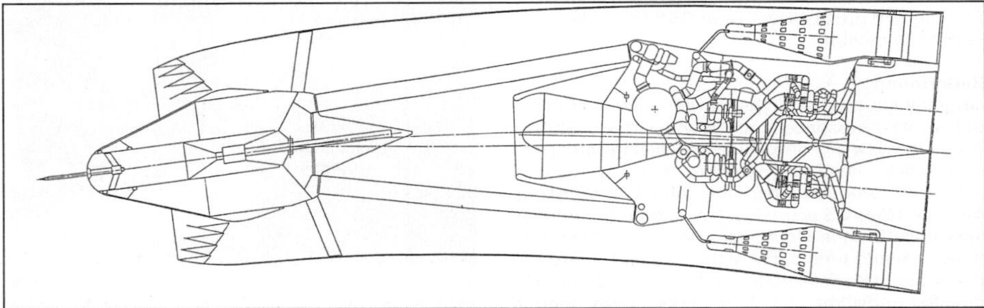
Sabre nacelle vertical cross section. Each nacelle carries two Sabre engines (Reaction Engines) 0517321

Test and Production Facilities

The company most recently commissioned two major high altitude test facilities for its liquid thrusters. Westcott's new facility handles attitude control thrusters, while the Boeing site at Marysville (Washington state) is used for the apogee motors.

Westcott's unit provides:

Pumping system: two stage ejector, 52 km at 10 gm/s (airflow) and 0.57 mbar for Leros 20

Vessel volume: 3.5 m^3

Propellant feed system: 30 l oxidant tank, 20 l fuel tank, 35 bar working pressure, 4–45°C thermal conditioning, helium saturation

Thrust measurement: single axis, high response, piezo electric transducer

Instrumentation: 64 channels, 120 kHz, analogue recorders, Agema thermal imaging system, Cyclops IR thermometer, CCTV high speed ciné.

The Boeing facility provides:

Pumping system: 5-stage steam ejectors with condensors and engine diffuser

Performance: 55 km at 195 gm/s (airflow) and 0.22 mbar for Leros 1

Vessel volume: 285 m^3

Propellant feed system: 450 l oxidant tank, 450 l fuel tank, 20 bar working pressure, ambient to 30°C thermal conditioning

Thrust measurement: six component, multi-axis fast response

Instrumentation: 50 channels, 1 kHz, analogue recorders, Agema thermal imaging system, Cyclops IR thermometer, CCTV high speed ciné.

Westcott provides other firing sites for mono/bi system testing at SL/altitude. The altitude site utilises three interconnected chambers totalling 600 m^3; two positive displacement pumps reduce the pressure in a single chamber to about 4 mbar, and to 6 mbar with all three connected. The 60 l hydrazine expulsion rig can supply propellant at up to 45 atm. The sea level facilities accommodate all storable propellants and thrusts up to 100 kN.

Westcott also offers:

Climatic chambers six chambers covering –55/90°C; seven cold chambers for long-term storage testing down to –60°C, with additional hot air and hot water long-term chambers

Vibration two tables, one covering 5–2,000 Hz, 144 kN sine thrust (±25.6 mm amplitude) and the other 10–3,000 Hz, 27 kN sine thrust (±12.7 mm amplitude)

Shock 1,000 kg test items subjected to 250 g

Centrifuge two centrifuges for testing solid propellant motors under varying g conditions. One provides up to 100 g for up to 33.4 kN motors on a 3 m arm, the second 60 g for 800 kN motors on a 5 m arm

Dynamic balancing up to 1,000 kg items on 56-300 rpm 1 m dia table; minimum measurable out-of-balance 0.3 kg-mm at 270–300rpm.

Westcott's **Liquid Motor Processing Facility** was established to process propellant and gas storage assemblies for Chevaline's post boost propulsion system and has handled >1,000 tanks and gas storage assemblies. The LMPF processes, fills, tests and performs final closure welds on aluminium and stainless steel tanks, with class 100,000 clean rooms. Propellants in use include IRFNA, MAF 1 and 4 and hydrazine.

Smiths Aerospace

Current Status

In May 2007, General Electric (GE) completed its acquisition of the aerospace division of the Smiths Group for USD4.8 billion. Smiths revealed the increased investment requirements of the industry was the primary reason for its decision to sell the aerospace division to General Electric.

For the 53 week period ending 5 August 2006, Smiths Aerospace recorded operating profits of GBP156 million on a turnover of GBP1.3 billion. Total aerospace net assets before intercompany funding and gross assets were GBP808 million and GBP1,254 million.

Background

Originally formed from the merger between Smiths Industries Aerospace and Dowty Group, Smiths Aerospace is a UK company emphasising aerospace electronics, medical systems and specialised industrial products, deriving about half of its revenue from US activities. The company provides six Space Shuttle elements: engine interface unit, mission/event timers, FM signal processor, ground command interface logic controller and attitude director indicator. Its fibre optic gyro inertial measurement unit (0.4 kg, 10.7 × 7.1 × 6.6 cm) is used by Boeing's LEAP BMDO demonstration kinetic kill vehicle. System and subsystem upgrades for critical space elements and development of new layout concepts for satellite design engineering.

UK Ion Propulsion

UK ion thruster development began in the 1960s with a 10 mN 10 cm mercury Kaufman thruster at RAE/Culham. The 10 cm work was re-activated in 1985 using Xe because of projected satcom growth in the 1990s. UK-10 in its qualified form provides a thrust of typically 10–25 mN but can be smoothly throttled from 0.2 to 30 mN. Culham has also performed extensive trials on the UK-25 25 cm version, capable of delivering a throttlable 50–300 mN. ESA's Artemis will carry two 18 mN UK-10 versions for NSSK and elements of the UK-25 are being adapted for the ESA-XX (qv) interplanetary thruster. A thruster of intermediate size, in the 50 mN category is also under development by the DRA (ex RAE).

DRA provides: programme technical lead, thruster design/manufacture, testing, electronics and hollow cathode development. Culham: thruster testing (including diagnostic life tests), plasma physics, ion beam extraction modelling, UK-25 development/testing. Matra Marconi Space (MMS): development of the UK-10 system for Artemis, including the power conditioning and control equipment and the propellant supply/monitoring equipment and the manufacture and qualification of the flight hardware for Artemis. An element of AEA Technology, Culham houses Europe's largest Xe propellant ion thruster test chamber: 5.8 m long, 1.3 m dia, equipped with helium cryopumps for handling the inert gas. During 1996 MMS commissioned a new test facility, capable of handling two 25 mN thrusters simultaneously.

UK-10 ion thruster development in the UK began in the late 1960s at RAE/Culham, concentrating on a 10 cm dia thruster with a nominal 10 mN thrust and mercury propellant. The T4A engineering model achieved very high efficiency, stable operation and a long operational lifetime. The propellant was changed to Xe in 1985 and new versions of the T5 flight model thruster have been manufactured/tested in a collaboration between DRA, Culham and Matra Marconi Space. 0.2–70 mN thrust has been demonstrated; qualification testing at 18 mN is underway at MMS and DRA, specifically for operational NSSK application on ESA's Artemis. This will be extended to 25 mN during 1997. A major experimental programme was completed in 1995 at the Aerospace Corp in Los Angeles, using USAF funding and a thruster supplied by DRA. This concentrated on a detailed characterisation of the ion beam. A lower level effort continues at Aerospace funded by DRA, MMS and the USAF. The thruster comprises a cylindrical discharge chamber closed at one end by a soft iron backplate and with a set of closely-spaced grids at the other. A magnetic field is generated by six equispaced peripheral solenoids. The field lines inside the chamber link an inner cylindrical soft iron pole and a larger diameter outer pole. Propellant gas is introduced through the axial hollow cathode and a bypass distributor on the backplate. A DC discharge is set up between the cathode and the cylindrical anode. This ionises the gas, the efficiency being enhanced by the magnetic field and the correct design of the inner pole/baffle disc arrangement. The positive ions are extracted and accelerated by a high electric field between the grids, attaining typically 30–60 km/s. A triple grid design (see diagram) is used to minimise damage caused by

The UK-25 xenon ion thruster was performance-mapped in Culham's vacuum facility (UKAEA Culham) 0516923

DRA's UK-10 thruster under test at the Aerospace Corp. Detailed characterisation of the ion beam was completed in 1995 (The Aerospace Corp) 0003458

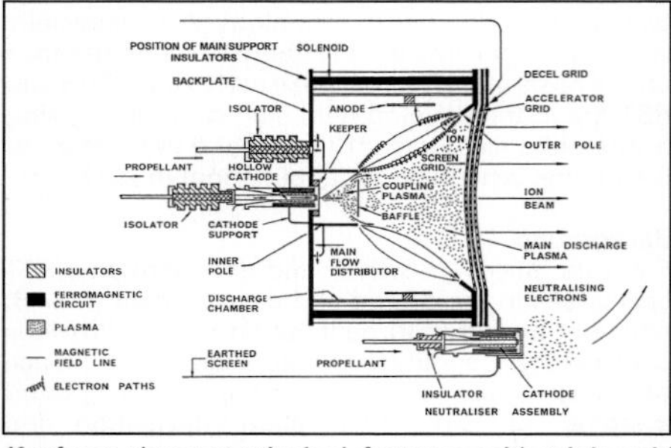

Kaufman thruster principal features, with triple grid (DRA) 0517338

the impact of charge-exchange ions. The ion beam's charge is neutralised by electrons from an external cathode, and the whole thruster is surrounded by an earthed screen to prevent those electrons from reaching other parts of the device. MMS UK is responsible for the electronics and propellant feed systems, and commercial exploitation of the fully qualified operational system. Philips Components Ltd previously manufactured the hollow cathode and neutraliser; the work has now been transferred to DRA. DRA/Culham in 1994 produced a low cost version of the complete system for experimental applications. This was accepted for flight on Johns Hopkins Univ's NEPSTP Nuclear Electric Propulsion Space Test Program, since cancelled, which was intended to test the Russian Topaz 2 nuclear reactor. Other flight opportunities are being sought.

ESA's Artemis carries two 18 mN versions. Another application may be ESA's Gravity Explorer Mission; DRA/Dornier studied the possibility in 1995/6 with encouraging results. DRA's STRV-1A (see UK National), launched in Jun 1994, included an experiment to allow the hollow cathode assembly to demonstrate spacecraft electrostatic discharging. Although the cathode assembly operated correctly, charging was not observed. The nominal operating parameters given below are in 18/25 mN order.

Thrust: 18/25 mN

Beam voltage: 1,100/1,100 V

Accelerator voltage: –250/–250 V

Decelerator voltage: –50/–50 V

Beam current: 0.329/0.457 A

Exhaust velocity: 40,233/40,233 m/s

Specific impulse: 3,084/3,131 s

Discharge power: 80/106 W

Beam power: 362/503 W

Power/thrust ratio: 26.4/25.7 W/mN

Electrical efficiency: 76/78%

Total efficiency: 57/60%

UK-25 design of a laboratory model 25 cm thruster began in early 1986 using scaling laws formulated during the UK-10 programme. Culham testing began Oct 1986; nominal thrust was 200 mN (highest obtained was 316 mN). Following the successful conclusion of the lab model work, an engineering model was manufactured at the beginning of 1989 and completed a comprehensive test/evaluation programme in 1992. Included were studies of life-limiting factors. Investigation of other propellants was completed in 1995 at Southampton Univ (although an ideal inert propellant, xenon is costly). Development of a power conditioning system was undertaken at Birmingham Univ and high current hollow cathodes were developed by DRA/Philips Components. Further work on the ESA-XX hybrid design, using features from this thruster and the German RIT-35, is underway with ESA funding. The first prototype of this thruster, intended for interplanetary missions, has been successfully tested at the Univ of Giessen in Germany.

United States

AEC-ABLE Engineering Company Inc

Current Status

Founded in 1975, and acquired in 1999 by Pressure Systems Inc, ABLE provides deployable structures and special-purpose space mechanisms. In addition to deployable coilable and articulated lattice masts and tubular columns, ABLE provides: composite structures; honeycomb panels; wire deployers; solar array deployment and tracking/drive mechanisms; payload separation devices; brushless DC motors; linear/rotary drive systems; hinge devices; structural and fluid quick-connect; atomic oxygen protection and thermal control blankets. Composite, high tensile, umbrella antenna structures extending from compact folded configuration to full deployment of up to 100 m. ABLE is providing the bearing assembly and solar array deployer for Space Station. The failed Meteor 1 satellite in 1995 carried ABLE's SCARLET solar wing, with four concentrator panels and two Si planar panels. New panel assemblies from twin Si arrays. Other space projects include Deep Space 1, Mars Surveyor, Cassini, SRTM and Galileo.

ABLE offers a range of solar arrays, including:

- The UltraFlex Solar Array which provides 142 W/kg, 10.6 kg at 1,500 W BOL with Si; 166 W/kg, 9.0 kg at 1,500 W BOL with GaAs/Ge cells. 0.0794 m³ stowed volume for 1,500 W wing.
- The PUMA Solar Array which provides 47 W/kg for 1,300 W BOL; self-synchronising deployment, high stiffness. Provided for MSTI 3.

The deployable masts and columns include:

- The CoilABLE Boom which is a triangular cross section with three continuous longerons, or rectangular cross section with four, for use with any of three deployment systems: nut; lanyard; carousel. Booms up to 45 m have been deployed in space. A Nut-Deployed CoilABLE boom was first used to deploy the large solar array on Shuttle STS-41D in September 1984. Others include LACE, UARS and Galileo.
- The ABLE Articulated Retractable Mast or AARM employing four articulated longerons for greater bending strength and stiffness than above. Deployed by Nut Deployer. The FASTMast is the strongest version, stowing in 2 to 5 per cent of its deployed length. Side length 18 to 274 cm. Italy's TSS Tethered Satellite employs FASTMast, as does ISSA for solar array deployment. These designs have formed the basis for new terrestrial antenna design construction facilitating many new design concepts for commercial applications.
- The Tubular Masts which complement the lattice structures consist of a line of tubular masts of 0.6 to 10 cm diameter, including BESTMast and STALK. Flights include six on Transit (SOOS) and a pair for Cornell University's WISP II.

Aerojet

Current Status

Aerojet has been awarded a contract by the Air Force Research Laboratory for computational capability improvements in advanced technology upper stage engines. The three-phase programme begins with the development of a conceptual engine design followed by a computational tools phase and concluding with a hardware and computational tools validation phase. The contract was awarded on 2 February 2004 and will run for nine months.

Background

A segment of GenCorp of Akron, Ohio, Aerojet was founded in 1942 by astronautics pioneer Dr Theodore von Karman and initially developed JATO jet-assist take-off rockets for aircraft. It was the first US company to produce storable bipropellant rockets. The propulsion plant, as Aerojet TechSystems (formerly Liquid Rocket Co), provided liquid engines for Aerobee, Bomarc, Nike, Vanguard, Delta, Atlas Able, Titan 1 and Apollo, and continues to manufacture Titan, Delta and Shuttle engines (see individual entries below). Aerojet, Rocketdyne and P&W formed the Space Transportation Propulsion Team in 1990 for work on the STME Space Transportation Main Engine.

Aerojet manufactured the solid motors for MX, Minuteman, Small ICBM, Standard Missile (rolling out the 10,000th motor in Jun 1989) and Hawk, and successfully bid in partnership as Aerojet Space Boosters (becoming Aerojet ASRM Division) with Lockheed for NASA's Shuttle Advanced Solid Rocket Motor contract. Congress cancelled ASRM in Oct 1993. Propulsion activities are performed at the 52.6 km² Sacramento plant, founded 1951 and capable of testing cryogenic and storable propellant motors up to 6,670 kN. The Mach 8/30 km-altitude hypersonic Hytest facility was added for National AeroSpace Plane engines and components tests of up to 35 s, and the company has access to the Hypulse M25 facility at subsidiary General Applied Sciences Laboratories Inc (Ronkonkoma, NY), which it acquired in 1989.

Aerojet teamed with Lyulka in 1993 to improve and market the 400 kN LOX/LH_2 D57 engine, originally developed in the 1960s for the improved N1-L3M manned lunar mission. It is a candidate for an SSTO demonstrator. A teaming agreement was signed Jul 1993 with Kuznetsov/NPO Trud (qv) to use the NK series exN1 engines and their technology in the US market. Aerojet made five tests totalling 410 s with an NK-33 in OctNov 1995. If it had been selected for a US vehicle as the AJ26-NK33A (it lost to the RD-180 for Lockheed Martin's new Atlas 2AR and 5 launch vehicles), a US production line would have been established. Unit cost would have been about $4 million. See the CIS Propulsion section for further information. A similar agreement was signed Oct 1994 with KB Khimautomatiki for the RD0120, principally developing it into a tripropellant engine. Work is being performed under a $17.2 million NASA Marshall contract.

Aerojet presently employs about 500 people.

AJ10-118K

This pressure-fed engine, optimised for altitude operation, has flown as the stage 2 propulsion system for McDonnell Douglas' Delta and the related Japanese N vehicle, in addition to flying paired (as the AJ10138) on Titan 3's Transtage upper stage. Contracts continue through 2000, building about 10 annually.

Applications: Delta stage 2
First flown: Aug 1982, Delta 164
Dry mass: 124.7 kg
Length: 269 cm
Mounting: fixed
Engine cycle: pressure-fed
Oxidiser: nitrogen tetroxide at 9.1 kg/s
Fuel: Aerozine-50 at 4.76 kg/s
Mixture ratio: 1.9:1
Thrust: 43.38 kN vacuum
Specific impulse: 320.5 s vacuum
Expansion ratio: 65:1
Combustion chamber pressure: 8.84 atm
Cooling method: ablative chamber, radiative skirt
Burn time: qualified up to 500 s (unlimited starts)

Orbital Maneuvering System (OMS)

NASA's Shuttle Orbiter carries two OMS pods, each housing a single Aerojet OMS engine for orbit insertion, manoeuvring and reentry initiation. The engines are capable of 100 missions and 500 starts in space. 12 were originally delivered for the Shuttle fleet (including four reserves), with two more added following *Challenger's* loss (#102/112) in 1986 (these were completed as #115/116 in summer 1989 and first flown on STS50 in Jun 1992). Aerojet was awarded a $3.6 million 38month contract in Nov 1986 to continue development of an uprated OMS featuring increased chamber pressure through the addition of a pumpfeed system. Successful testing beyond 600 s was completed in autumn 1990 at NASA's White Sands altitude simulation facility. OMS is derived from Aerojet's Apollo Service Propulsion System, designed with high reliability for manned lunar missions. The first OMS prototype and demonstration tests were completed 19723 and the company was awarded the full development contract in 1974. The thrust chamber was successfully tested for the first time in 1976 and the first prototype development engine was delivered to NASA in Feb 1977 for extensive White Sands testing to begin the following Aug. A further engine, used to qualify the system as part of the Orbiter's pod, was delivered in Jan 1979 and fired 270 times that year, accumulating 10,817 s. The first two production engines, for *Columbia*, were delivered Mar 1979, and testing of the final qualification engine, delivered Oct 1979, was completed in Jun 1980. In the first 24 Shuttle missions, nine of the 12 engines were carried and fired 254 times for a total of 21,875 s. In the 27 missions following *Challenger* to end 1992, eight engines were used for 377 firings totalling 39,995 s without anomaly.

Applications: Space Shuttle orbit/de orbit insertion, circularisation
First flown: 12 Apr 1981, Orbiter
Columbia Number flown: 14, to end-1995 (103/104 have never flown)
Dry mass: 118 kg
Length: 195.6 cm
Maximum diameter: 116.8 cm
Mounting: gimballed ±7° yaw/±6° pitch by two electromechanical actuators for thrust vector control
Engine cycle: pressure-fed (improvement programme underway for pumpfeeding)
Oxidiser: 6,743 kg nitrogen tetroxide in each pod (pods can be cross linked)
Fuel: 4,087 kg of monomethyl hydrazine in each pod (pods can be cross linked)
Mixture ratio: 1.65:1
Thrust: 26.7 kN vacuum
Specific impulse: 316 s vacuum
Expansion ratio: 55:1
Combustion chamber pressure: 8.62 atm

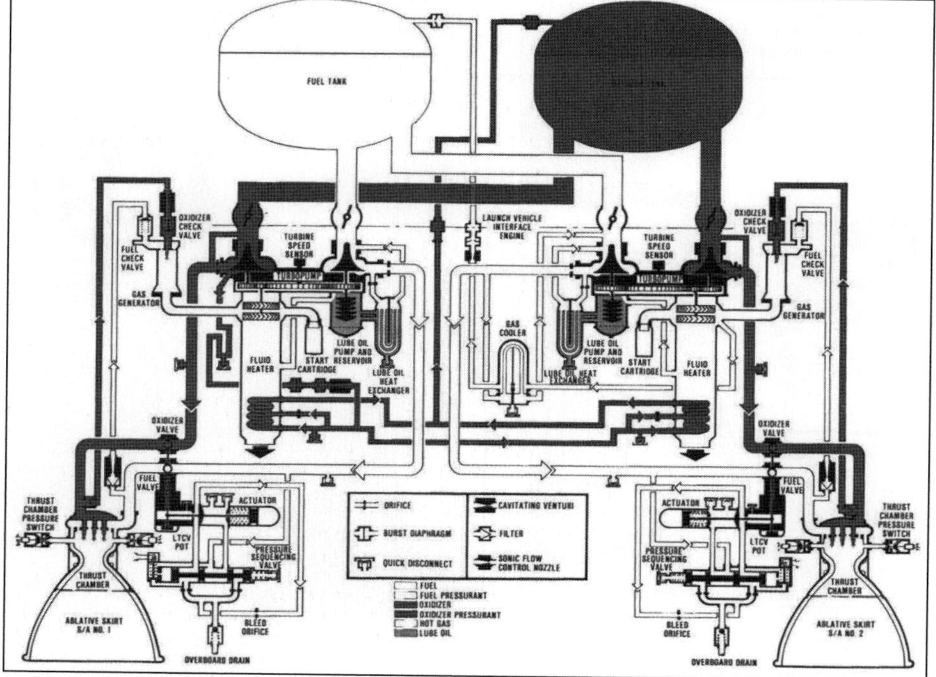

LR-87-AJ-11 schematic (Aerojet)

0517368

Aerojet's paired LR-87 AJ11 engines power the Titan 3/4 first stages 0517369

Cooling method: fuel regenerative for chamber, radiative for nozzle
Burn time: qualified for 500 starts, 15 h/100 mission life, longest firing 1,250 s, de orbit burn typically 150–250 s.

Titan Space Launcher Engines
Aerojet has developed and manufactured the engines for four generations of USAF/NASA Titan ICBMs and space launchers since 1958, attaining a peak annual production rate of 270. Each Titan core carries a dualchamber engine in the first stage and a scaleddown single-chamber version to power stage 2. The company has refurbished 14 sets of Titan 2 engines for orbital launcher applications, and is manufacturing new 11family engines for Titan 4. The Titan 4 contracts, worth some $350 million, call for 41 USAF and four NASA engine sets.

AJ-5 Titan 2 engines
Aerojet began testing the first stage 1/2 experimental engines in 1960 and completed production of the 474 chambers by end1967. The ICBMs were overhauled 1974–82 and the company subsequently refurbished 14 sets for the space launcher version. The Gemini Titan 2 manned version flown 1964–66 utilised manrated 7-type engines.
Designation: Aerojet LR-87-AJ-5
Configuration: twin fixed motors with individual turbopump assemblies
Applications: Titan 2 stage 1
First flown: 1962 ICBM; Sep 1988 orbital
Dry mass: 1,266 kg for full assembly
Length: 2.3 m
Maximum diameter: 1.1 m
Engine cycle: gas generator

LR-91-AJ-11 schematic (Aerojet) 0517370

Propellants: hypergolic nitrogen tetroxide and Aerozine50, delivered at 754 kg/s
Mixture ratio: 1.93:1
Thrust: 1,913 kN sea level
Specific impulse: 259 s sea level
Expansion ratio: 8:1
Combustion chamber pressure: 53.7 atm
Burn time: about 165 s
Designation: Aerojet LR-91-AJ-5
Configuration: scaled-down version of stage 1 engine, featuring fixed single chamber
Applications: Titan 2 stage 2
First flown: as stage 1 engine above
Dry mass: 472 kg
Length: 2.80 m
Maximum diameter: 1.68 m
Engine cycle: gas generator
Propellants: as stage 1, at 146.5 kg/s
Thrust: 444.8 kN vacuum
Specific impulse: 312 s vacuum
Expansion ratio: 49.2:1
Combustion chamber pressure: 56.2 atm
Burn time: about 185 s

LR-87-AJ-11
Titan 4's first stage (and Titan 3's if it flies again) are powered by paired AJ-11 engines. The 9 model was first tested in 1963 and 476 were produced before being replaced by the 11 in late 1968, first tested earlier that year. Together with its stage 2 derivative, it is the only US engine to be operated on storable N_2O_4/A-50 *and* tested on LOX/RP1. Ablative skirts with 12:1 and 15:1 expansion ratios are available, producing 254 s (sea level) and 302 s (vacuum) SI. USAF awarded a contract to Martin Marietta in 1990 for a modified Titan 4B nozzle. This requires an increased expansion ratio for ignition at greater altitude when Hercules' higher performance strap ons are carried.
Applications: Titan 3/4 stage 1
First flown: 1968 Titan 3, 1989 Titan 4
Dry mass: 1,874 kg (paired), 758 kg (single)
Length: 3.84 m to top of thrust structure, 3.23 m to top of turbopump assembly
Maximum diameter: 1.6 m
Mounting: gimballed pair
Engine cycle: gas generator
Oxidiser: nitrogen tetroxide at 513 kg/s
Fuel: Aerozine-50 at 268 kg/s
Mixture ratio (O/F): 1.91:1
Thrust: 2,340 kN vacuum paired
Specific impulse: 301 s vacuum
Expansion ratio: 15:1
Combustion chamber pressure: 55 atm
Cooling method: fuel regenerative and ablative skirt
Burn time: about 200 s

LR-91-AJ-11
Development of the current Titan 3/4 stage 2 engines followed that of the stage 1 version. It has flown without failure on every operational Titan 3 mission.
Applications: Titan 3/4 stage 2
First flown: late 1968 Titan 3, 1989 Titan 4
Dry mass: 589 kg

Aerojet's LR-91-AJ-11 Titan 3/4 stage 2 engine 0517371

Length: 281 cm
Maximum diameter: 163 cm (skirt outer diameter)
Mounting: gimballed, turbine exhaust utilised for roll control
Engine cycle: gas generator
Oxidiser: nitrogen tetroxide at 97.0 kg/s
Fuel: Aerozine-50 at 54.7 kg/s
Mixture ratio: 1.86:1
Thrust: 467 kN vacuum
Specific impulse: 316 s vacuum
Expansion ratio: 49.2:1
Combustion chamber pressure: 58.5 atm
Cooling method: fuel regenerative thrust chamber, with separate ablative skirt
Burn time: about 247 s

Transtar
Transtar was developed as an upper stage engine using injector, chamber and nozzle derived from OMS. However, the propellants are pump-fed, increasing chamber pressure and SI, and permitting use of lowpressure lightweight tankage.
Applications: upper stage
First flight: not flown
Dry mass: 67 kg
Length: 127 cm
Mounting: gimballed ±10° by two electromechanical actuators
Propellants: N_2O_4/MMH
Mixture ratio: 1.8:1
Thrust: 16.680 kN vacuum
Specific impulse: 328 s vacuum
Expansion ratio: 132:1
Combustion chamber pressure: 23.8 atm
Cooling method: fuel regenerative for chamber, radiative for extension
Burn time: not available, 15 starts

Aerojet Satellite Engines
Aerojet also produces 2, 21, 62 and 445 N bipropellant (NTO/MMH) thrusters for satellite orbitadjust and attitude control systems. Qualification of a 21 N thruster is being completed for a USAF satellite. Principal characteristics are:

Thrust (N, vac)	*2.00*	*21.35*	*62*	*445*
Specific impulse (s, vac)	265	285	287	309
Dry mass (kg)	0.27	0.57	1.13	1.86
Expansionratio	150	150	75	150
Mixtureratio (O/F)	1.65	1.60	1.65	1.65

Alliant Techsystems, Inc (ATK)

Current Status
ATK has consolidated its space propulsion capabilities with the acquisition of Hercules Aerospace Company in 1995, Thiokol Propulsion in 2001 and Composite Optics in 2003, a specialist composite space structures company which ATK has applied to innovative design concepts.

Background
Alliant Techsystems acquired Hercules Aerospace Co in Mar 1995 for USD296 million. Hercules' rocket motor, weapons and ordnance businesses totalled USD660 million revenue in 1993, for a USD105 million operating profit. Hercules was selected in Oct 1987 as the propulsion contractor for Titan's Solid Rocket Motor Upgrade programme, potentially worth USD725 million for the development, qualification and production of 15 sets of Titan 4 solid boosters. Lightweight graphite composite materials replaced the steel used in current motor cases, and a high performance propellant similar to that developed for the Delta 2 strapon is employed. PQM-1 (Preliminary Qualification Motor) was the first full scale test, but the motor failed after a few seconds at Edwards AFB 1 Apr 1991. The added PQM1' firing was successful 12 Jun 1992. The QM2 third test, run at 2.5°C, was successful 21 Feb 1993. QM-3 at the 41°C upper limit was successful 2 June. Delivery of the first flight set was made 1Q 1994. Lockheed Martin awarded the company a contract forTitan launch vehicles support which extended to 2003 and was followed by contract extensions.

The company was awarded a McDonnell Douglas contract in 1987 to develop the stretched solid rocket GEM graphite epoxy strapon motors for Delta 2; they first flew Nov 1990. The initial contract for 144 motors (16 flight sets) was followed in 1991 by a second for 117 (13 sets) beginning production in 1993, and a third in Jun 1995 for 144 (16 sets). GEM production rate is 6/month. For Delta 2, from 1996, the expansion ratio for the airlit nozzles will be increased from 10.6 to 16.3 by lengthening the nozzle by 30 cm and increasing its dia by 19.5 cm. The GEMVN version incorporates a vectorable nozzle; the qualification firing was made May 1994. The TVC system is provided by AlliedSignal and the nozzle by BP-Hitco as part of the strategic partnership. Delta 3's nine strapons will be lengthened by 1.22 m and their diameters increased to 116.8 cm. Three of the six groundlit motors will carry gimballed nozzles for TVC, and the three airlit will improve performance with extended nozzles. The first of 16 sets ordered Jun 1995 were delivered by early 1998.

Alliant produces Pegasus' Orion solid motors, derived from GEM, and the fairing. Taurus also uses the Orion motors. Hercules also designs/fabricates composite structures: spars, struts and optical benches used in satellites.

Analytical Graphics Inc (AGI)

Current Status

AGI was awarded its 14th US patent in eight years in June 2007 for designing attitude and position software that includes a spherical interface. Among other applications, the software can be used to calculate the effectiveness of a communications satellite based upon its antenna orientation.

Background

Analytical Graphics Inc (AGI) was founded in 1989. Headquartered in Exton, Pennsylvania, AGI operates nine facilities, including small sales offices in the UK and Spain, and the Center for Space Standards and Innovation (CSSI) in Colorado Springs, Colorado. It employs about 250 personnel and in 2002 reported sales of USD27 million.

AGI develops and markets Commercial Off The Shelf (COTS) analysis and visualisation software for the space and defence communities. AGI software can be used to represent space superiority; space situational awareness; spacecraft mission design and operation; and geospatial intelligence applications. AGI developed the Satellite Tool Kit (STK) software family, an interactive 3-D graphical software package to access, manage, display and manipulate space environment data. As of 2007, STK 8 was the current version.The STK family also includes the Navigational Tool Kit (NavTK) and the Orbital Determination Tool Kit (ODTK), which are used to analyse satellite navigation and tracking systems. As of 2007, AGI software had been deployed to over 32,000 space and defence community terminals.

Atlantic Research Corp

Current Status

ARC, founded in 1949, is a manufacturer of solid propulsion motors and gas generators, and in 1987 expanded into liquid propulsion with the acquisition of Bell Aerospace Textron. The company's Virginia and Arkansas facilities take solid motor concepts from design through to high volume production; the company's annual propellant production capacity is 9,000 t. ARC produces propulsion units for missiles such as Stinger, Tomahawk, Peacekeeper and Trident 2, but also undertakes space-related projects: the company teamed with Hercules in Dec 1987 to bid (unsuccessfully) for NASA's Shuttle Advanced Solid Rocket Motor. ASRM was subsequently cancelled by NASA. Israel's 1.3 m and AUS51 'Marble' motor, flown as Shavit's stage 3, is marketed by ARC/Rafael. ARC markets the ASTM family of Shavit and NEXT launcher stage 1/2 motors in the US under agreement with TAAS Israel Industries Ltd.

International Space Technology's plasma thruster is flight proven on Russian Meteor satellites
(Space Systems/Loral) 0517183

ARC also manufactures rocket motor components, including igniters, initiators, casings and nozzles using advanced materials and filament winding/braiding techniques. Nov 1987's acquisition of Bell Aerospace Textron added liquid bipropellant attitude control and apogee satellite propulsion systems, and their associated tanks & valves. ARC's 22 N (5 lbf) bipropellant thruster was developed for Intelsat 6's attitude control system and has demonstrated 400,000 pulses, 32 h aggregate firing and 748 kg total throughput. This thruster is carried by Optus B, UFO and Intelsat 7. As Bell, ARC/LP developed the Agena propulsion system that flew on >370 missions. The engine remains available for 31.1120 kN thrust.

ARC is developing the Agena 2000, providing the thrusters for the X-37 and is a partner of International Space Technology, Inc.

Charles Stark Draper Laboratory (CSDL) Inc

Current Status

In March 2007 a second Zero Propellant Manoeuvre (ZPM) was conducted on the International Space Station (ISS). A 180-° rotation was completed using mach three in two hours and 45 minutes. To complete the ZPM 80 combined attitude and rate commands were transmitted and separated by 125 seconds to the Control Moment Gyro (CMG).

During mid-2007 a sensor web demonstration was completed using NASA's Ikhana, a Predator B aircraft, which tested new imaging and real-time communications technology. The Ikhana captured thermal infrared images of wildfires over the west coast of US using the sensor payload on the remotely piloted aircraft. Images were transferred in real time through a satellite data link to NASA Ames Research Center. Sensing platforms include EO-1, Terra and Aqua. The flight was conducted through NASA's Dryden Flight Research Center.

Background

The Charles Stark Draper Laboratory (CSDL) was established in 1932 by Charles Stark, a professor and pilot, as a Massachusetts Institute ofTechnology (MIT) aeronautics instrumentation research facility. The CSDL was named after its founder in 1970, and in 1973 separated from MIT to become an independent, non-profit Research & Development (R&D) organisation, primarily serving the public sector to practice science and engineering in the US national interest. The Laboratory is headquartered in Cambridge, Massachusetts, runs five field offices, and as of 2007, employed a workforce of approximately 1,025, 750 of which are scientists and technical staff. It manages approximately USD20 million per year devoted entirely to R&D programmes.

CSDL conducts research and development work in five main business areas: strategic systems; space systems; tactical systems; special operations; and biomedical engineering. Capabilities include Guidance, Navigation and Control (GN&C); embedded, real-time software; microelectronics and packaging; autonomous systems; distributed systems; MicroElectroMechanical Systems (MEMS); biomedical engineering; and prototyping system solutions. CSDL provide specialist laboratory facilities and class 100 and 1,000 clean rooms.

Previous examples for CSDL's strategic systems include successfully testing with Northrop Gumman the GPS/Inertial Navigation System instrumentation package for the Minuteman III re-entry vehicles. To accurately read a Silicon Oscillating Accelerometer the CSDL developed a radiation hardened electric circuit card in support of the US Air Force Research Laboratory (AFRL) advance strategic upgrade accelerometer technology. The CSDL is developing pro-types and testing Flexured Mass Accelerometer solid-state designs.

CSDL's space systems develop technologies for space science, space exploration, spaceflight and military space. Space science is divided into four disciplines; advanced GN&C, command and control, science instruments and mission integration. For example semi-autonomous rendezvous algorithms were studied extensively both by the Jet Propulsion Laboratory (JPL) and CSDL for robotic sample returns of Mars, developing GN&C sensors for precision landing. Georgia Institute of Technology Space Systems Design Laboratory was sponsored by CSDL to conduct research regarding "Pinpoint Landing Assessment for Autonomous Space Exploration Systems", involving terrain navigation for future robotic landers on Mars, Titan and Venus. CSDL alongside JPL and LAngley Research Center (LARC) is supporting a Johnson Space Center (JSC) project which involves developing an autonomous lunar ascent and landing GN&C technologies for the Autonomous Landing and Hazard Avoidance Technology programme, incorporating Precision Landing System. With regard to the Hubble Robotic Servicing Deorbit Mission CSDL is leading the GN&C aspect.

The Laboratory is also a flight system partner to NASA's Mars scout mission, called the Aerial Regional-scale Environmental Survey (ARES), an autonomous powered airplane which will fold compactly into a protective aeroshell on route to Mars. CSDL alongside NASA LARC as well as other team partners will provide system engineering, navigation and guidance engineering, and embedded flight software for data and flight control of ARES. The Hardware-In-The-Loop Simulator (HILSIM) was developed by CSDL which incorporates two simulation frameworks, CSDL's C-language-based Simulator (CSIM) and NASA LARC's Standard Real-Time Simulation in C to support the development of ARES flight codes. A series of wind tunnel tests at NASA's LARC Transonic Dynamics Tunnel was concluded in 2006. The ejection system to ARES was also tested in 2004 using a one-third scale model. Simulating conditions on Mars a high altitude research balloon released a 50 per cent prototype model of ARES at an altitude of 103,500 ft to test the autonomous deployment of the model in 2002. After following pre-programmed flight patterns for 90 minutes the aeroplane returned to Earth and landed successfully.

Through NASA's New Millennium Programme (NMP) CSDL has also developed a full 3-axis stellar inertial attitude determination system, called Inertial Stellar Compass (ISC) with an accuracy of better than 0.1°.

The ISC is a low power (~3.6 W) and low mass (~2.5 kg) real-time system that incorporates two miniaturised devices; firstly a wide-of-field view star camera with an active pixel sensor plus a MEMS gyro and secondly a microprocessor. The devices form two separate but interconnected assemblies called the Camera-Gyro Assembly and the Data Processor Assembly. Both devices communicate with each other to ensure the position and motion is periodically updated to the control system so that the spacecraft is steady and facing in the correct direction.

Under the NMP, Space Technology 6 (ST6) selected the ISC technology to conduct a flight validation experiment. The ISC was mounted aboard TacSat-2 micro-satellite and was launched aboard Minotaur launch vehicle on 16 December 2006. TacSat-2 is a joint project with the AFRL, US Air Force Space Command, US Army Space and Missile Defense Command, the Department of Defense Operationally Responsive Space Office and the Office of Naval Research.

The purpose of the ST6 programme is to reduce the risk and cost of future space science NASA missions. The performance of the ISC is evaluated through ground and space validation tests. Detailed tests of the ISC were conducted from January to March 2007 with the AFRL TacSat-2 team. The validation test will aid the development of the next-generation ISC, which is being investigated by CSDL with the possibility of using ISC for a future TacSat mission. Also the CSDL is developing an advanced ISC for landing or as a deep-space navigation sensor.

For command and control systems CSDL under the NMP designed the mission planning software, the Earth Phenomena Observing System (EPOS), for the Earth Observing-1 (EO-1) spacecraft. EO-1 was launched aboard Delta 7320 from Vandenberg Air Force Base on 21 November 2000. The EPOS server provides cloud cover data, including 8 hour forecasts from the Air Force Weather Agency to predict the cloud cover at chosen imaging targets.

CSDL was one of 28 proposals awarded a NASA Earth Science Technology Office grant in 2005 under the advanced information systems technology programme for Sensor Web Dynamic Re-planning, which consists of space-based remote sensors, in-situ sensors on unmanned, surface vessels, and remote sensors on unmanned air vehicles. CSDL plan to make enhancements and extensions to EPOS re-planning sensors to provide analysis and monitoring a variety of environmental events. For example incorporating new science models such as prediction models for fire and hurricane forecasting into the EPOS functional architecture.

The Laboratory is exploring wet chemistry sensor research to develop an instrument for life detection on a Mars 2016 mission. CSDL is also exploring lander technologies that will enable the deployment of multiple probes on the Moon and Mars. With regard to mission integration services CSDL offers complete programme management which includes mission design, system engineering and integration.

CSDL is partnered with MIT assisting NASA's Project Constellation. Under Space Exploration both CSDL with MIT provided conceptual designs for the Crew Exploration Vehicle (CEV) under a 2004 NASA study contract. In 2005 CSDL joined a Northrop Grumman-Boeing led industry team in competition with a Lockheed Martin led team that included Honeywell, Orbital Sciences Corporation, United Space Alliance and Hamilton Sundstrand. The two phase selection process was finalised in 2006 where Lockheed Martin was selected as prime contractor to design, develop and build Orion CEV. CSDL is acting as consultant to Lockheed Martin and NASA's JSC for GN&C systems for the CEV.

For human spaceflight CSDL provides support for the NASA Space Shuttle as well as the International Space Station (ISS) and historically designed the GN&C system for the Apollo spacecraft. NASA awarded a USD34 million contract to CSDL to provide engineering support to the Space Shuttle and ISS, supporting GN&C plus integrated avionics systems for the Engineering Directorate at NASA's JSC, Houston on 18 January 2006. The contract will last with options exercised for 4 years.

CSDL assessed various control issues to perfect the Orbital Repair Manoeuvre where the Space Shuttle is rotated onto the belly surface whilst held in the arm of ISS. In a 360° pitch rotation, called the R-Bar Pitch Rotation Manoeuvre, CSDL developed and analysed the control algorithms. The Laboratory was also involved in software development and algorithms certification for the Orbiter Boom Sensor System, essential for inspecting the Shuttle's surface before re-entry and landing. The Zero Propellant Manoeuvre (ZPM), developed by CSDL, enables large angle rotations without using any modified flight software or the ISS thrusters. A 90° angle rotation using the ZPM was conducted in November 2006, which in total took 2 hours to complete using 80 combined attitude and rate commands transmitted to the CMG, which were spaced 90 seconds apart.

CSDL redesigned the Shuttle's on-orbit flight control system for docking with large structures, and developed the International Space Station's (ISS) Timeliner software.

Under military space CSDL designed the mission management Timeliner software for DoD's Orbital Express (OE) satellites servicing mission, which was successfully launched on 9 March 2007. Servicing operations using CSDL's software will be performed by OE autonomously capturing two satellites ASTRO and NextSat for refuelling and recovery supplies. CSDL also designed the autonomous rendezvous and proximity operations guidance systems for the US Air Force's Experimental Satellite System 11 (XSS-11), launched in 2005.

Goodrich Corporation

Current Status

Goodrich's Intelligence, Surveillance and Reconnaissance (ISR) division's Electro-Optical Systems team supplied the satellite attitude control systems for the Missile Defense Agency's (MDA) Near Field Infrared Experiment (NFIRE) satellite, launched in April 2007.

The US Naval Research Laboratory, in June 2007, awarded Goodrich with a contract to adapt its U2 high altitude aircraft electro-optical sensor for use aboard spacecraft.

Background

BF Goodrich changed its name to the Goodrich Corporation in 2001. Headquartered in Charlotte, North Carolina, Goodrich reported revenues of nearly USD6 billion, and profits of USD482 million in 2006. The company employs a workforce of approximately 23,000 in 90 locations across 16 countries. Its defence and space sector nets about 25 per cent of Goodrich's sales.

Goodrich's space systems manufacturing divisions are primarily headed by the company's Intelligence, Surveillance and Reconnaissance (ISR) Systems group. ISR Systems is comprised of the divisions: Electro-Optical Systems (EOS); EOS-Ithaco Space Systems; Space Flight Systems (SFS); Surveillance & Reconnaissance Systems; and SUI (Formerly Sensors Unlimited).

Goodrich's space-related divisions manufacture satellite and launch vehicle airframe, electronic and engine components and systems, as well as satellite Attitude Determination and On-orbit Control Systems (ADOCS). The company supplies the Space Shuttle's rubber-coated strut covers, a component that protects the wiring between the Shuttle's External Tank (ET) and its Solid Rocket Boosters (SRBs). They also manufacture the Shuttle's landing gear, wheels, carbon brakes, airborne and ground based lighting systems, and some Shuttle sensor systems. Goodrich also manufactures acquisition, command, and control electronics for Atlas and Delta launch vehicles, as well as Trident missiles.

Rosemount Aerospace, acquired by BF Goodrich in 1993, is now part of Goodrich's Sensors Systems division. Rosemount manufactured more than 12,500 onboard- and ground-support sensors for the Space Shuttle, and also supplied sensors for every other US human spaceflight mission previous to the Space Shuttle's service.

BF Goodrich purchased Raytheon Optical Systems (ROS) in December 2000, now Goodrich's Electro-Optical Systems business unit, based in Danbury, Connecticut. Formerly known as Hughes Danbury Optical Systems, Raytheon Optical Systems had been in business for more than 30 years previous to its acquisition by Goodrich. ROS was responsible for supplying electro-optics systems aboard both the Hubble Telescope, including its optical telescope assembly and 2.4 m main and 30 cm secondary mirrors, and the three fine guidance sensors, as well as optical systems aboard the Chandra X-Ray Observatory. Goodrich's Electro-Optical Systems has carried on the tradition, supplying the beryllium optics and telescope structure for the Spitzer Space Telescope (SIRTF).

Hamilton Sundstrand

Current Status

Hamilton Sundstrand remains an important player in US human spaceflight programmes. It recently contributed the P5 truss to the International Space Station (ISS). Rocketdyne Space Power and Energy designed the P5; that Rocketdyne business unit was merged into Hamilton Sundstrand upon United Technologies Corporation's acquisition of Rocketdyne from Boeing in 2005.

Background

Headquartered in Windsor Locks, Connecticut, Hamilton Sundstrand is a subsidiary of United Technologies Corporation (UTC). Hamilton Sundstrand was formed when Hamilton Standard and the Sundstrand Corporation merged in 1999. As of 2006, the company reported USD5 billion in annual sales, and employed a total workforce of 17,500 engaged in aerospace and industrial technologies production and services delivery. Hamilton Sundstrand's Space, Land and Sea business unit is largely responsible for the company's space-related industry, which encompass their Missile, Satellite and Launch Systems; and Human Space Systems enterprises.

Hamilton Sundstrand's space interests focus on power systems, environmental control, thermal control and life support systems. Hamilton Standard provided Apollo's lunar suit life-support system and the Lunar Module's environmental and thermal control system. Hamilton Sundstrand and its predecessors have, since 1981, continued to manufacture environmental control systems for the Space Shuttle and also provide the Shuttle's Extravehicular Mobility Unit (EMU) space suit. Hamilton Sundstrand provides on-orbit replaceable units for the EMU including the displays and controls module, the primary life support system, the secondary oxygen pack and enhancements such as the heated gloves, improved hard upper torso, the EMU battery and arms and legs.

Hamilton Sundstrand provides many Space Shuttle subsystems, including seven major Orbiter environmental and life support systems. Among these are the Atmosphere Revitalisation System (ARS), which provides the shirtsleeve environment, avionics bays cooling air and a water coolant loop to collect and transport cabin heat to the external Freon Coolant Loop (FCL); the Freon Coolant Loop, which collects heat from the ARS and other equipment in the Orbiter mid-body and aft areas and delivers it to one of three heat rejection devices; the Flash Evaporator System, which rejects all FCL heat during ascent and entry when the Orbiter is at approximately 30 km altitude and supplements radiators on-orbit and rejects excess water collected onboard; the Water Spray Boiler, which cools the hydraulic fluid and lubricating oil during launch and descent; the Ambient Temperature Catalytic Oxidiser system, which control temperature and convert CO to CO_2; and the Waste Collector Subsystem, which collects and stores crew solid and liquid body wastes.

For the International Space Station (ISS), Hamilton Sundstrand provides temperature and humidity control systems to circulate air between the modules, remove airborne particulates and cool and dehumidify cabin air; similar components to circulate air to cool avionics and other heat loads; a waste water processor; active thermal control systems, including pumps, heat exchangers and cold plates for the external loop.

Harris Corporation

Current Status

Harris Corporation was awarded a US Air Force USD410 million Network and Space Operations and Maintenance (NSOM) contract on 15 January 2008. Operation, maintenance and logistic support will be provided by Harris Corporation to the 50th Space Wing's Air Force Satellite Control Network (AFSCN) including the remote tracking AFSCN stations around the world. The NSOM contract also includes software analysis and support for GPS including maintenance and support for GPS ground antennas and monitoring stations as well as performing orbital analysis of military satellites. Harris Corporation is leading the NOSM team, comprised of Lockheed Martin Information Technology, L-3 Communications, Faith Enterprises, ASRC Aerospace, Arctic Slope World Services, Nortel Government Solutions and Gunther Douglas. From previous contracts Harris Corporation has seven years of experience in providing maintenance support to the 50th Space wing.

The Missile Defense Agency's Ground-Based Midcourse Defense System advanced terminal underwent a successful test during September 2007, which is part of USD200 million Boeing contract support by Harris and Northrop Grumman Mission System, for the design and development of IFICS Data terminals. The interceptor was launched from Vandenburg Air Force Base, California successfully and communicated with the terminal that was developed by Harris, to destroy the missile target.

A potential five year USD10.7 million follow-on ground systems engineering and depot services contract was awarded to Harris on 25 September 2007. Under the Defense Meteorological Satellite Programme (DMSP) Harris will provide technical support and engineering services for the DMSP ground system.

Harris has assembled a team to compete for a USD600 million contract to develop global military satellite terminals for the US Army Modernisation of Enterprise Terminals (MET) programme. The MET terminals will be used for military communications and missile defence. General Dynamics Satcom Technologies was added to the team on 9 August 2007 and O'Neil and Associates was added on 4 October 2007. Harris plan to develop X- and Ka-band terminals that will operate on new and legacy satellite systems. The contract is for 15 years and is scheduled to be awarded in 2008 and completed by 2020. The first terminal will be ready for fielding in 2010.

Franklin Van Rensselaer is the vice president and senior executive account manager for NASA programmes, as announced by Harris 24 July 2007.

Harris donated product licenses of the company's satellite command and control software OS/COMET as well as technical support to the US Air Force Academy Cadets, Department of Astronautics on 11 April 2007. The OS/COMET software will be used to support the department's FalconSAT programme.

Harris is competing for the National Oceanic and Atmospheric Administration's Geostationary Operational Environmental Satellite-R (GOES-R) satellite ground segment contract. Operations are scheduled to run until 2029. Harris is bidding as prime contractor and systems integrator for the programme. The contract is scheduled to be awarded in the second half of 2008. GOES-R is scheduled for launch in 2014.

A contract was awarded to Harris by Thales Alenia Space in March 2007 to supply an S-band unfurlable reflector antenna for the Eutelsat W2A communications satellite. W2A, serving Europe, Africa and the Middle East, will provide Ku-band and C-band broadcasting services and mobile devices. The satellite is scheduled for launch in the first quarter of 2009.

The company's Large Aperture Multiband Deployable Antenna (LAMDA) terminals was given Intelsat's Type Approval in March 2007. Harris delivered nine LAMDA terminals to the US Marines and three to the US Air Force in March 2007 under a 2004 USD42 million contract with the US Army. Harris will provide 39 LAMDA terminals for the US Marine and US Air Force to be used for multiband, mobile satellite comunications. The terminals work with Defense Satellite Communications System, the Wideband Gapfiller System, NATO, Skynet, XTAR, and Intelsat satellite constellations.

On 28 February 2007 Harris Corporation was awarded an USD25.6 million contract by Space Superiority Systems Wing, Los Angles Air Force Base Station, to develop a new Counter Communication System (CCS) with dual Tactical Multi-band SATCOM Antennas or dual apertures capability. The contract also includes systems verification testing, maintenance, configuration and data management. The CCS contract is projected to end in 2010.

Background
Harris is headquartered in Melbourne, Florida. The company reported revenues of USD4.2 billion for Fiscal Year 2007 (FY07), which is a 22 percent increase from USD3.5 billion in FY06 and a net income of USD480 million in FY07, compared to USD238 million in FY06. Harris employs approximately 14,000 personnel, including more than 6,000 engineers and scientists, in 150 countries. Harris is a publicly traded company, listed on the New York Stock Exchange.

Harris is an international communications and information technology equipment company focused on providing product, system and service solutions to commercial, government and defence customers. The company's four operating divisions serve markets for microwave, broadcast, Radio Frequency (RF) and government communications systems. Key space related technologies manufactured by Harris include satellite antennas, including unfurlable mesh antennas, ground satellite command, control and communications terminals, and geospatial imagery asset management technologies.

Harris is working with Boeing on the Ground-Based Midcourse Defense (GMD) Communications Network (GCN), and also providing Boeing with the Ka-band antennae for the Wideband Gapfiller Satellite (WGS), set to launch in 2009. Harris participated in the GOES-R project in a USD2.8 million Program Definition and Risk Reduction (PDRR) contract from Boeing to provide design concepts for the ground-based segment for the GOES-R satellite. The contract ended in April 2007. Harris teamed with the Defense Advanced Research Projects Agency (DARPA) and the US Air Force Research Laboratory (AFRL) on the Innovative Space-Based Radar Antenna Technology (ISAT) programme and was the lead organisation on the project, charged with developing large, light-weight, deployable space-based radar antennae. Harris is providing Lockheed Martin with the Mobile User Objective System (MUOS) satellite's unfurlable mesh reflectors; MUOS is scheduled for launch around 2010.

Honeywell International Inc

Current Status
The present company was formed in late 1999 with the merger of AlliedSignal and Honeywell Inc, but it can trace its origins to 1885. Aerospace Solutions and services has sales exceeding US$10.5 billion and employs 49,000 people at 20 sites. Honeywell International Inc serves the space industry through Honeywell Aerospace and Honeywell Specialty Materials.

Under Aerospace Services, the space related activities of the Defense and Space division concentrate on: inertial systems; onboard processors; altitude and control; control-moment gyroscopes; reaction wheel and momentum wheel assemblies; communications/data handling; and isolation/pointing. This includes the development and spacecraft application of fibre optic rotation sensors to supplement the presently manufactured ring laser and dry tuned gyro inertial measurement units. Integrated satellite attitude-control systems and their components are also available, including star trackers, reaction wheels and control-moment gyroscopic effectors, phased array momentum control systems and their operational software.

Data-management systems, in selectable configurations, employing fibre-optic networks, flat-panel displays and signal processors, are ready for spacecraft incorporation and are being provided for Space Station. Integrated components are also available to include optical mass storage, upgraded Shuttle standard multiplexers/demultiplexers, network and bus interface units and onboard data processors.

Honeywell, under contract to the USAF Phillips Laboratory, developed the GVSC Generic VHSIC Spaceborne Computer, which has a high-performance general purpose 1750A CPU, a local-memory controller and 256 + 8 static RAM. A 32-bit processor and single-board computer is in production and has been ordered for various military and commercial space platforms. All elements are radiation hardened. HII produces radiation-hardened computers for meteorological satellite imaging sensor and inertial navigation units for Atlas 5.

Advanced payload pointing systems with magnetic and passive isolation have been developed to attain sub-arcsecond accuracy with low bandwidth isolation. To support a wide range of satellite pointing applications, high-gain modular antenna pointing systems, solar array drives with integrated power transfer, gimbal actuators and spin/de-spin payload control systems are being manufactured. On Shuttle, Honeywell provides the main engine controllers, hand controllers, flight control system, radar altimeter and autoland software. It is subcontracted (US$70 million) by Rockwell for the Crew Multifunction Electronic Display System (MEDS 'glass cockpit') to replace the Orbiter's old cockpit displays; this was completed in 1999 and is installed in all orbiters on a rotation basis.

HII provides range safety tracking systems for the Kodiak launch complex and the company has developed a new range of optical and radar tracking instruments for fast accelerating rockets and missiles.

ITT Corporation

Current Status
Spaceport Systems International (in partnership with ITT since 1993) has been awarded a NASA contract (from John F. Kennedy Space Centre) to provide Commercial Payload Processing Services for Expendable Launch Vehicles at Vandenberg Air Force Base, California.

ITT Systems Division has been awarded a contract worth approximately USD14 million from the Launch and Range Systems Wing, Los Angeles Air Force base, to provide a new CT-4 system in Building 1 of Pillar Point at San Mateo, California.

ITT Advanced Engineering and Sciences (AES), was awarded a contract in 2006 that had an initial value of USD147 million, for Electromagnetic Spectrum Engineering Services by Defense Information Systems Agency (DISA) for the Joint Spectrum Centre (JSC), providing, for example, analysis, modelling, simulation, testing and evaluation for radio frequencies. AES has assembled a team to work on this particular contract that include large to small businesses. AES were also awarded a Cost Plus Award Fee Contract, allowing for a base year value plus four optional one year add-ons by US Space and Missile Defense Command in support of the Missile Defense Agency for Lethality Testing and Criteria Development. The contract has an estimated total value of USD57.3 million. This work relates to target configurations for use by the Ballistic Missile Defense Systems.

ITT Space Systems Division (SSD), is contributing to the next series of GOES satellites N-P, providing the imager and sounder for the NASA and the National Oceanic and Atmospheric Administration (NOAA) project. On 24 May 2006, Goes-N was launched using a Boeing Delta IV Launch Vehicle, Goes-O is scheduled for April 2008 launch and Goes-P is scheduled for October 2008. Goes-R follows on from N/O/P and is scheduled for 2012 launch. SSD is designing the Advanced Baseline Imager (ABI) instrument for GOES-R and earlier in 2007 SSD passed a critical design review where NASA has approved the ABI, moving SSD into the next stage of the mission, which is module integration.

Also, earlier in 2007, SSD completed work for GeoEye, delivering to prime contractors General Dynamics the sensor system for high resolution satellite GeoEye-1. SSD will be working on the next generation GeoEye satellite. The division was awarded a contract to develop the camera for GeoEye-2 in mid October 2007.

SSD provided the sensor system for the WorldView-1 high resolution satellite for DigitalGlobe, which is part of the National Geospatial-Intelligence Agency's (NGA) NextView programme. Worldview-1 was launched on 18 September 2007. Three images, which were captured by SSD's advanced sensor system from WorldView-1, were released 16 October 2007. SSD and partners Technologies Corp and Ball Aerospace are also developing Worldview-2 for DigitalGlobe where work is scheduled to be completed by the end of 2008.

SSD is working with prime contractors Northrop Grumman Space Technologies in an international collaboration with NASA, (which will be managed through the Goddard Space flight Centre), the European Space Agency and Canadian Space Agency regarding the James Webb Space Telescope (JWST) programme, the successor to the Hubble Space Telescope. SSD will be providing cost effective glass mirror architecture, (forming the optical part of the telescope) as well as the Focal Plane Assembly for Near- Infrared Spectrograph (NIRSpec) instrument. Launch for the JWST is scheduled for 2013.

ITT Visual Information Solutions (VIS) has provided software to Kongsberg Satellite Services (KSAT), that will enable satellite images from Envisat (European Space Agency) and Radarsat-1 (Canada) to be transmitted over a low bandwidth for Arctic Ocean expeditions during the International Polar Year (March 2007 to March 2009). A new version of IDL, The Data Visualization & Analysis Platform 6.4, was released in July 2007 with additional features plus enhancements. ITT VIS have entered into what has been described as a "business relationship" with Environmental Systems Research Institute (ESRI) to extend ESRI's ARCGIS platform with a certified National Imagery Transmission Format (NITF) for users within the defence and intelligence communities.

Background
ITT Defense and Electronics is one of three technology segments under ITT Corporation, (previously ITT Industries). All financial reports are prepared under ITT Defense and Electronics including its sub-divisions. Defense and Electronics generates approximately 47 per cent sales revenue for ITT Corporation, accounting for almost 50 per cent of the income in respect to the other two segments. Headquartered in McLean, Virginia, 40,000 people work within ITT Defense and Electronics under President Steven F. Gaffney. They have 60 years of experience in providing solutions to numerous markets.

Defense and Electronics is divided into two areas, Systems and Services and Defense Electronics. Systems and Services is further subdivided into Systems and Advanced Engineering and Science. Defense Electronics is subdivided into Aerospace and Communications, Space Systems, Night Vision and Electronic Systems.

The Systems Division serves the following markets, Space Ground and Range Systems offering 43 years worth of experience in supporting launch and range services. Through their venture partnership with SSI they were awarded the lease in 1995 of Vandenberg Air Force base for 25 years, where launch services include, polar orbiting satellites requiring a medium-lift capability. Examples of launches include Jawsat and Mightysat, both launched in 2000, and XSS-11 and STP-R1 launched in 2005, as well as the Cosmic Mission, a Taiwanese satellite that was launched in 2006. SSI also offer payload processing services. The Systems Division contributes to deep space missions, for example NASA's Jet Propulsion Laboratory (JPL) for Deep Space Network, providing support for tracking and communications, as well as Command, Control, and Communication (C3) Networks and electro-optics. They also offer range services, which include operation, planning, tracking instrumentation, space surveillance including sensor for Missile Defense and sensor support for phased array and X-Band radars as well as electro-optic systems. The Systems Division serves the

Communications and Information Systems (CIS) market, offering engineering services, integration for either commercial or military users in single, multi-band satellite communications, providing communications operation and maintenance for the US Department of Defense. The Systems Division employs over 7,000 people and its headquarters are based in Colorado Springs, US. Product and service users of this division include government, commercial and international customers. A mission example is Mars Reconnaissance Orbiter. Through the Deep Space Network in 2006, ITT, based at Goldstone Deep Space Communications Complex, supported a successful Mars orbit insertion. The Deep Space Network is contracted through NASA's Jet Propulsion Laboratory (JPL). Other NASA Deep Space facilities supported by ITT include Madrid, Spain and Canberra, Australia.

AES employs 1,300 people all over the US, where their headquarters are located, in Herndon, Virginia. They provide services to either commercial, industrial or governmental customers. Areas of expertise include 'Advanced Communications and Information Systems' serving national and commercial enterprises where the Space Technology Department of Advanced Engineering and Science provides 35 years of solid experience for Low Earth Orbit (LEO) applications. They also have 25 years of experience in supporting, for example, NASA Goddard with radio frequency communications systems engineering in respect to Tracking and Data Relay Systems (TDRSS). Spectrum management is another expertise of AES, ensuring that NASA requirements are met for missions requiring optimum radio frequency quality. AES performs spectrum analysis, for example, interference, sharing, distribution, risk and coverage, thus ensuring no degradation or loss of communication links.

Under Defense and Electronics, the Aerospace and Communications Division (ACD) offers GPS products such as navigation, time transfer, two-way transfer, survey and navigation receivers, for example the Selective Availability Anti-Spoofing Module (SAASM) embedded GPS Receiver, GPS SAASM Kit and GPS Fanout. ACD are also leading providers in worldwide wireless communication for the defence industry and headquarters are located in Fort Wayne, Indiana, US.

SSD headquarters are located in Rochester, New York. 2,600 employees work for the SSD. Within the space sector they have built up 50 years of experience, where their expertise lies within remote sensing and SSD's capabilities include image and data collection as well as processing and disseminating data. They serve a range of applications offering a variety of solutions, from the design to the testing of individual components to complete payloads including support services for interpreting images and data. Examples of applications served include intelligence, surveillance and reconnaissance systems, GPS, Navigation, Meteorological, Earth and Space Science Systems, High Resolution Imaging and Decision Support Solutions. The SSD works with defence and intelligence departments to provide, for example, space-qualified remote sensing payloads, Spaceborne RF receiver systems including anti-jam, data encryption as well as support services in Synthentic Aperture Radar (SAR) for high azimuth resolution images.

SSD began working on GPS in 1974 for NAVSTAR GPS, which is currently operated by the US Air Force and managed at the Space and Missile Systems Center (SMC), Air Force Space Command, Los Angeles Air Force Base, California, making up the Global Positioning System Wing for the Department of Defense who are key to developing NAVSTAR GPS systems. The SSD have contributed to 50 payloads which have been in orbit, without any operational failures, for 490 orbit years. Lockheed Martin Space Systems are prime contractors, with SSD, for a modernising project for Block IIR-M. SSD are contributing to the navigation payload for Block IIR-M.

SSD has been working with NOAA and NASA developing key instruments for weather forecasting. They have gained considerable experience from working with geostationary and LEO programmes. In a partnership with Space System/Loral (SS/L) they developed the Geostationary Operational Environmental Satellite (GOES) series I-M, part of the early warning weather system in the US, where SSD contributed the imager and sounder to the project. In respect to LEO applications, SSD has designed Advanced Very High Resolution Radiometer (AVHRR) which is part of the Polar Operational Environmental Satellites (POES) projects and Metop, which is a joint European Space Agency and the European Organisation for the Exploitation of Meteorological Satellites (EUMETSAT) programme. With regard to the Multi-Function Transport Satellite (MTSAT), a Japanese Advanced Meteorological Imager (JAMI), SSD contributed to MTSAT-2's imager.

In 2004, SSD acquired from Eastman Kodak Company its Remote Sensing Systems (RSS), business which was part of their Government and Systems operation. Customers of RSS have included NASA, for example, they have been involved in historical space missions such as Apollo 11, as well as other contributions such as Chandra and Hobby-Eberly Telescope and commercial high resolution Earth Observation/Imaging Satellites, Quickbird and IKONOS Satellite cameras. SSD was one of 22 major suppliers contributing optical components to the Southern African Large Telescope (SALT) programme.

With regard to Earth Science Systems, SSD have contributed to advanced climate prediction utilising latest technologies such as Light Detection and Ranging (LIDAR) and Imaging Spectroscopy. They have a strong history in designing Optical Components and Systems for ground- and space-based applications. For high resolution imaging, SSD offer a complete service from building custom imaging systems to capturing and processing images including supplying sensor technology, for example Multi-Spectral and Advanced Sensors and Electronics Technology (ASET). Under Decision Support solutions ITT VIS, part of the SSD, provides complete expertise with respect to spectrum analysis offering software solutions through ENVI – The Remote Sensing Exploitation Platform, that is used in a variety of applications, for example environment, pharmaceuticals and defence and intelligence. Customers that use ENVI include major space agencies and government agencies as well as commercial entities. Other Software solutions include The Data Visualization & Analysis Platform (IDL), supplied to NASA for use in the Mars Rover Project and SeaWiFS, and the NOAA. The SSD also offers solutions related to specific applications as well as comprehensive support services.

Kearfott Guidance & Navigation Corporation

Current Status

Kearfott is a subsidiary of the Astronautics Corp of America, specialising in space-qualified inertial guidance and inertial reference systems. Employees total almost 2,000. The company has now produced more than 100,000 inertial devices and more than four million precision components. Principal space hardware covers one-, two- and three-axis attitude control and pointing systems, star trackers, rotary components and cryogenic systems. In its New Jersey facility, Kearfott also produces a line of gimballed and strapdown GNC systems, using both ring lasers and dynamically tuned gyros (DTGs), test equipment and inertial sensors. In North Carolina and Mexico, it designs and manufactures precision rotary components, actuators and electronics. Developing new nano-sized guidance sensors and tracking equipment. This technology underpins a new generation of precision guidance packages which Kerfott have been researching since the early 1990s.

Kearfott initially furnished floated rate integrating gyros to Vela Hotel, OAO, Mariner, Nimbus 2–4, Surveyor, Viking, Skylab and a three-axis rate system (GRA) for classified satellites. These programmes involved approximately 180 gyros, accruing more than 360,000 hours in space without failure.

In the early 1970s, Kearfott developed the first of the Space-qualified Kearfott Inertial reference Unit (SKIRU) three-axis attitude reference systems employing the Gyroflex DTG, which has 6.2 million hours of data on 82 life test units. SKIRU is used on Galileo, DSCS III, Voyager, TDRS, Milstar, XTE, TRMM, EOS AM 1 and AXAF-I, plus classified satellites. 70 SKIRUs have been delivered. Some systems have 45,000 to 110,000 hours in space without failures; total is 1.9 million operating hours. SKIRU has also been developed into a fine pointing system, utilising the same basic components with temperature control and circuit modifications to permit very accurate space pointing over long periods. SKIRU I to V are two-gyro single-string units; the D-II is a three-gyro internally redundant unit that is interchangeable with NASA's DRIRU II Dry Gyro Inertial Reference Unit.

Specifications

Performance: 0.002–0.009°/h, 3σ, 8 h
Size (cm): 17.8 × 22.8 × 9.5 IV; 17.8 × 22.8 × 14.2 IV-H; 17.8 × 22.8 × 15.0 V; 17.8 × 35.5 × 17.8 D-II
Mass (kg): 3.2 IV; 5.0 IV-H; 5.0 V; 12.7 D-II
Power (max quiescent;w): 25 (typical 17) IV; 25 (typical 22.5) IV-H; 26 (typical 17) V; 26 D-II

TARA, the Two-Axis Rate Assembly utilises an IRU comprising a two-axis CONEX MOD I/S DTG. 15,000 CONEX gyros have been built. TARA's modular concept allows two packages to be used for three-axis control, while three would provide complete redundancy. It was developed and qualified by early 1993 under a company-funded programme. Users are TOMS-EP, Lewis, MSTI 2/3, Rocsat 1 and Komsat. The MOD I/S CONEX is also being delivered as part of a three-axis IRU for Hughes' HS-601.

Performance: 0.05°/h
Size (cm): 16.0 × 14.3 × 8.6
Mass: 1.8 kg
Power (max quiescent; W): 11 (typical 7) operating from unregulated DC input.

L-3 Communications

Current Status

Alliant Techsystems (ATK) has sub-contracted L-3's subsidiary Cincinnati Electronics (CE) to develop subsystems and components for the Ares 1 rocket, the future two-stage launch vehicle for human spaceflight. ATK is prime contractor for NASA, developing the first stage of the Ares 1 rocket which will carry the next generation human spacecraft, Orion Crew Exploration Vehicle. It is part of a USD1.8 billion continuation contract which was awarded to ATK in August 2007 for the Design, Development, Test and Evaluation (DDT&E) phase of the first stage, lasting until 2014. The first stage is based upon the Space Shuttle's solid rocket motors four-segment design. The Ares 1 will have a single, five-segment reusable solid rocket booster in the first stage. L-3 CE will be responsible for the avionics subsystem line replaceable units which includes; booster control, power distribution, data acquisition, recorder and recovery units, as well as the ignition separation controller and hydraulic power unit controller. Under the DDT&E L-3 CE are scheduled to perform flight tests in 2009 and the contract is extendable to 2014.

In May 2007 L-3 finalised its acquisition of Global Communications Solutions Inc. The company is now an L-3 Division and has been renamed L-3 GCS. GCS is a satellite communications systems integrator; at the time of its acquisition, the company was achieving about USD90 million in annual sales.

Background

L-3 Communications is headquartered in New York City, and maintains 98 business units across North America, Europe and Australia. The company was founded in 1997 by spinning off 10 Loral Corporation divisions acquired by Lockheed Martin in the previous year. L-3 issued its first Initial Public Offering (IPO) in 1998, after which it grew and began to acquire further companies and divisions at a rapid rate.

L-3 is organised into four primary Business Segments: Command, Control, Communications, Intelligence, Surveillance and Reconnaissance (C3ISR); Government Services; Aircraft Modernization and Maintenance (AM&M); and Specialized Products. In 2006, the company achieved total sales of nearly USD12.5 billion and employed a total workforce of roughly 63,000.

L-3 sub-contracts to provide secure satellite communications; ground station; guidance; and Telemetry, Tracking and Control (TT&C) systems and subsystems to prime contractor customers. The company supplies its space and satellite systems and services through several of its divisions and companies, including Space and Navigation; Narda Satellite Networks; L-3 Communication Systems-East (CS-E); Telemetry-West; and Datron Advanced Technologies (obtained upon L-3's 2005 Titan acquisition). L-3 Analytics Corporation, formerly one of L-3's space products and services divisions, was folded into L-3 Communications' Government Services in 2003.

In 2003, L-3's Storm Control Systems Ltd (L-3 Storm), now a subdivision of L-3 Telemetry-West, signed a USD7.5 million agreement with Astrium to provide command and control software for the UK MoD's Skynet 4 and Skynet 5 satellite control centres, operated by Paradigm Secure Communications. In 2006, L-3 Storm also agreed to supply Lockheed Martin with test software for the US Air Force's Advanced Extremely High Frequency (AEHF) satellite constellation, the successor to the Milstar satellites.

L-3 and its legacy companies have contributed to many high-profile programmes and have provided, among many other systems, the International Space Station's (ISS) communication and tracking system; the Saturn launch vehicle's inertial guidance system; the Hubble Telescope's pointing and attitude control system; Iridium satellites' gyro and momentum wheel assemblies; as well as providing fixed and mobile satellite terminals and stations to DoD.

Olin Aerospace Company

Current Status

Primex Technologies was acquired by General Dynamics in 2001 and in 2002 Aerojet acquired GD Space Propulsion Systems. In 1985, Rocket Research Corporation was acquired by the Olin Corporation and Olin Aerospace was born. Primex Technologies spun off from Olin in 1997 and Primex Aerospace was created. Specialises in monopropellant hydrazine engines and gas generators for spacecraft and upper stage applications; >9,300 assemblies and 100 propulsion systems have been produced since the first system was qualified in 1964. There are 600 personnel at the Redmond and Moses Lake sites. The MR508 hydrazine arcjet was qualified in 1991 and Telstar 4 in 1993 became the first commercial satellite to carry an arcjet for stationkeeping; see p295 1995–96 for specifications. The model is superseded by the MR-509/510. OAC has delivered >250 gas generators for Shuttle's Auxiliary Power Unit. Olin is working on designs for high temperature rocket motors and on microthruster applications with extended life.

Facilities: OAC's 0.29 km² site at the Grant County Airport in Moses Lake houses 2,840 m² of solid propellant manufacturing plant, a hazardous device assembly area, storage magazines, a shortrange flight test facility, a sea level large liquid engine test complex, ballistic test bays and chemistry lab.

MR-50 22.2N

Applications: attitude control for SMS, Viking, Meteosat, GOES, Voyager, GPS, Intelsat 5, Scatha, Lockheed Martin 5000, Delta Star, Magellan, Wind/Polar, EOS AM 1
First flown: SMS metsat (May 1974)
Flown: 695 (748 produced) to end 1995
Dry mass (g): 680
Length (cm): 18.3
Max diameter (cm): 6.6
Mounting method: bolted three places
Propellant: hydrazine at 4.67 17.33 g/s (Shell 405 catalytic decomposition)
Feed method: GN_2 or GHe at 4.832.7 atm, through 29 W solenoid dual seat valve
Thrust (N): 9.79–38.7 (22.2 nominal)
Specific impulse (s, vacuum): 215 228
Time to full thrust (ms): 150 to 90% PC
Nozzle area ratio: 40:1
Nozzle length (cm): 3.95
Nozzle cooling method: radiative
Combustion chamber:
pressure (atm): 2.9 11.3
temperature (°C): 800
cooling method: radiative
Duty cycle (s): 0.020 min to 5,400 max; 471,000 pulses
Total impulse (kNs): 459.4

MR-103 0.89N

Applications: attitude control thruster for Voyager, GPS, Intelsat 5, Lockheed Martin 3000, 4000, 5000 & 7000, Mars Observer, ACTS, Magellan, Cassini, Iridium
First flown: ATS 6 (May 1974)
Flown: 1,260 (2,705 produced) to end 1995
Dry mass (g): 332
Length (cm): 14.7

Magellan thruster module: 2 × 445 N MR 104; 3 × 0.89 N MR 103; 1 × 22.2 N MR-50 (OAC) 0516943

Max diameter (cm): 3.43
Mounting method: bolted three places
Propellant: hydrazine at 0.091 0.499 g/s (Shell 405 catalytic decomposition)
Feed method: GN_2 at 4.8–28.6 atm through 9W solenoid dual seat valve
Thrust (N): 0.19–1.12 (0.89 nominal)
Specific impulse (s, vacuum): 206227
Time to full thrust (ms): 150 to 90% PC
Nozzle area ratio: 100:1
Nozzle length (cm): 0.98
Nozzle cooling method: radiative
Combustion chamber:
pressure (atm): 4.35–23.9 atm
temperature (°C): 800
cooling method: radiative
Duty cycle (s): 0.008 min to unlimited max burn; 750,000 pulses
Total impulse (kNs): 158.46

MR-104 445N

Applications: attitude control and {d} V corrections, Voyager, Magellan, DMSP, Tiros N, Landsat
First flown: Voyager 2 (Aug 1977)
Flown: 68 (101 produced) to end 1995
Dry mass (g): 1,860
Length (cm): 46.0
Max diameter (cm): 15.2
Mounting method: bolted three places
Propellant: hydrazine at 91–290 g/s (Shell 405/LCH202 catalytic decomposition)
Feed method: GN_2 or GHe at 6.8 28.6 atm through Wright Components 30 W single seat solenoid valve
Thrust (N): 205–572 (445 nominal)
Specific impulse (s, vacuum): 228239
Time to full thrust (ms): 50 to 90% PC
Expansion ratio: 53:1 (dia 15.2 cm)
Nozzle length (cm): 17.8
Nozzle cooling method: radiative
Combustion chamber:
pressure (atm): 3.8–19.0
temperature (°C): 800
cooling method: radiative
Duty cycle (s): 0.022–2,000 single firing; 2,654 *cumulative;* 1,742 pulses
Total impulse (kNs): 693.9

MR-106 27N

Applications: spacecraft and upper stage attitude control and z {d} V corrections, PAM A/S, Radarsat, GPS Block 2R, HAS/Peace Courage, Titan Centaur, Atlas Centaur, A2100, Lunar Prospector, NEAR
First flown: HAS/Peace Courage
Flown: 396 (2,381 produced) end 1995
Dry mass (g): 476
Length (cm): 17.8
Max diameter (cm): 6.4
Mounting method: bolted three places
Propellant: hydrazine at 4.0811.79 g/s (LCH 227/202 catalytic decomposition)
Feed method: GN_2 or GHe at 6.8 30.6 atm through Wright Components 27 W solenoid valve
Thrust (N): 8.9–26.7 (26.7 nominal)
Specific impulse (s, vacuum): 218–232
Time to full thrust (ms): 200 to 90% PC
Expansion ratio: 61:1
Nozzle length (cm): 4.72
Nozzle cooling method: radiative

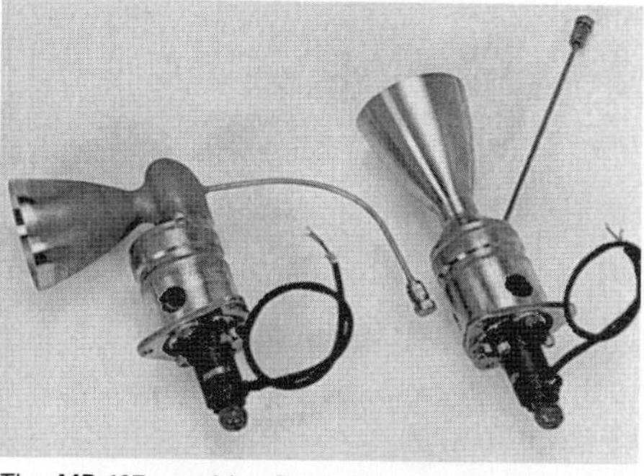

The MR-107 provides Delta, Titan and Atlas attitude control (Olin Aerospace Co) 0516944

Combustion chamber:
pressure (atm): 6.12–10.8
temperature (°C): 800
cooling method: radiative
Duty cycle (s): 0.016 min to 2,000 max; 12,397 pulses
Total impulse (kNs): 167

MR-107 178N

Applications: spacecraft and upper stage attitude control and {d} V corrections, Delta 2, Titan 2, Commercial Titan, PAM D, Small ICBM, HAS/Peace Courage, Atlas roll control module, STEP, Pegasus, LMLV/OAM
First flown: HAS/Peace Courage
Flown: 328 (1,898 produced) to end 1995
Dry mass (g): 885
Length (cm): 21.8
Max diameter (cm): 6.6
Mounting method: bolted three places
Propellant: hydrazine at 24–113 g/s (LCH 207/LCH 202 catalytic decomposition)
Feed method: GN_2 or GHe at 6.8 34 atm through 50 W solenoid valve
Thrust (N): 51.2–257.9 (178 nominal)
Specific impulse (s, vacuum): 217 236
Time to full thrust (ms): 200 to 90% PC
Expansion ratio: 21:1
Nozzle length (cm): 4.72
Nozzle cooling method: radiative
Combustion chamber:
pressure (atm): 1.9–9.5
temperature (°C): 800
cooling method: radiative
Duty cycle (s): 0.016 min to 2,137 max; 7,005 pulses
Total impulse (kNs): 332.3

MR-111 2.2–4.4N

Applications: attitude control, Intelsat 5, ERBS, ACTS, Radarsat, Lockheed Martin 4000, 5000, 7000, Wind/Polar, Landsat, Mars Observer, Radarsat, ACE, Skynet 4, MGS, EOS AM 1, MSTI 2/3, Mars Pathfinder, Mars 98, Clementine 1
First flown: Intelsat 5 (Dec 1980)
Flown: 410 (627 produced) to end 1995
Dry mass (g): 345
Length (cm): 16.7
Max diameter (cm): 35.6
Mounting method: bolted three places
Propellant: hydrazine at 0.19 2.4 g/s (Shell 405 catalytic decomposition)
Feed method: GN_2 or He at 4.08 27.2 atm through Wright Components 9 W dual seat solenoid valve
Thrust (N): 0.44–5.34 (2.2–4.4 nominal)
Specific impulse (s, vacuum): 213 229
Time to full thrust (ms): 150 to 90% PC
Expansion ratio: 200:1
Nozzle length (cm): 2.18
Nozzle cooling method: radiative
Combustion chamber:
pressure (atm): 3.1–13.9
temperature (°C): 800
cooling method: radiative
Duty cycle: 0.020 s min to 15 h max; 420,000 pulses
Total impulse (kNs): 260.21

MR-501

Electrothermal Hydrazine Thruster
Applications: communications satellite NSSK, Lockheed Martin 4000
First flown: Satcom 1R (Apr 1983)
Flown: 72 (92 produced) to end 1995
Dry mass (g): 816

For details of the latest updates to ***Jane's Space Systems and Industry*** online and to discover the additional information available exclusively to online subscribers please visit
jsd.janes.com

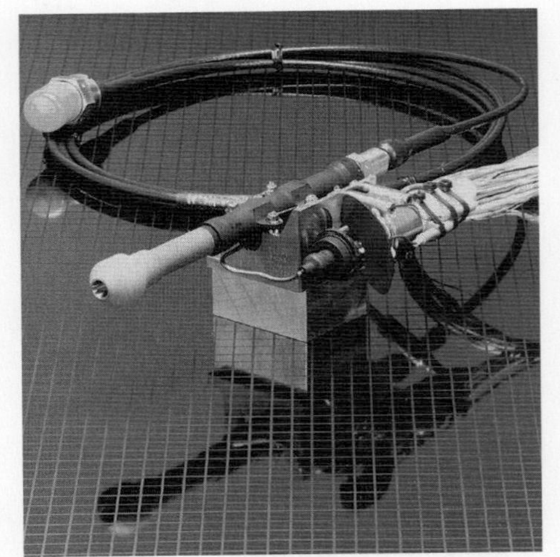

OAC's MR-509 arcjet was introduced on AsiaSat 2 0517356

Length (cm): 19.4
Max diameter (cm): 8.84
Mounting method: bolted four places
Propellant: hydrazine at 0.059 0.127 g/s over Shell 405 catalyst and 350510 W electric heater
Feed method: GN_2 or GHe at 6.8 23.8 atm through Wright Components 9 W dual seat solenoid valve
Thrust (N): 0.18–0.33
Specific impulse (s, vacuum): 280 304
Time to full thrust (ms): not applicable
Expansion ratio: 100:1
Nozzle length (cm): 0.879
Nozzle cooling method: radiative
Combustion chamber:
pressure (atm): 3.4–6.5
temperature (°C): 800
cooling method: radiative
Duty cycle: 300 s min to 1.7 h max; 500,000 pulses
Total impulse (kNs): 311.4

MR-502
Applications: communications satellite NSSK, Lockheed Martin 5000
First flown: Astra 1B (Mar 1991)
Flown: 16 (32 produced) to end 1995
Dry mass (g): 848
Length (cm): 19.4
Max diameter (cm): 8.84
Mounting method: bolted four places
Propellant: hydrazine at 0.119 0.167 g/s over Shell 405 catalyst and 610839.6 W electric heater
Feed method: GN_2 or GHe at 21.8 23.8 atm through Wright Components 9 W dual seat solenoid valve
Thrust (N): 0.360.50
Specific impulse (s, vacuum): 280304
Time to full thrust (ms): not applicable
Expansion ratio: 100:1
Nozzle length (cm): 0.879
Nozzle cooling method: radiative
Combustion chamber:
pressure (atm): 3.4–6.5
temperature (°C): 800
cooling method: radiative
Duty cycle: 300 s min to 2.0 h max; 430,000 pulses
Total impulse (kNs): 525

MR-509 Low Power Hydrazine Arcjet
Applications: communications satellite NSSK, Lockheed Martin 7000
First flight: AsiaSat 2 Nov 1995
Flown: 8 (36 produced) to end 1995
Dry mass (g): 1,338
Length (cm): 24.4
Max diameter (cm): 9.27
Mounting method: bolted four places
Propellant: hydrazine at 0.068 0.077 g/s over Shell 405 catalyst and 1.8 kW (PCU input) electric arc
Feed method: GN_2 or GHe at 14.920.4 atm through Wright Components 9 W dual seat solenoid valve
Thrust (N): 0.21–0.25
Specific impulse (s, vacuum): 511
Time to full thrust (ms): not applicable
Expansion ratio: 100:1
Nozzle length (cm): 0.879
Nozzle cooling method: radiative
Combustion chamber:
pressure (atm): 4.1–4.7
temperature (°C): 800
cooling method: radiative
Duty cycle: 5 min minimum to 65 h max
Total impulse (kNs): 883

MR-510 Low Power Hydrazine Arcjet
Applications: communications satellite NSSK, Lockheed Martin A2100
First flight: planned GE 1 1996
Flown: 0 (44 produced) to end 1995
Dry mass (g): 1,338
Length (cm): 24.4
Max diameter (cm): 9.27
Mounting method: bolted four places
Propellant: hydrazine at 0.068 0.077 g/s over Shell 405 catalyst and 2.2 kW (PCU input) electric arc
Feed method: GN_2 or GHe at 14.9 20.4 atm through Wright Components 9 W dual seat solenoid valve
Thrust (N): 0.210.25
Specific impulse (s, vacuum): 586
Time to full thrust (ms): not applicable
Expansion ratio: 100:1
Nozzle length (cm): 0.879
Nozzle cooling method: radiative
Combustion chamber:
pressure (atm): 4.1–4.7
temperature (°C): 800
cooling method: radiative
Duty cycle: 5 min minimum to 65 h max
Total impulse (kNs): 812

Parker Aerospace

Air/Space Division
18321 Jamboree Road, Irvine, California 92715.
Tel: +1 714 833-3000
Fax: +1 714 851-3341

Senior staff
General Manager: Jim Sabin

Control Systems Division
14300 Alton Parkway, Irvine, California 92718.
Tel: +1 714 833-3000
Telex: 678427
Fax: +1 714 586-8456

Senior staff
VP/General Manager: Robert Barker
VP Marketing: Jim Ryder
1425 West 2675 North, Ogden, Utah 84404.
Tel: +1 801 782-3100
Telex: 388424
Fax: +1 801 786-3045

Senior staff
Plant Manager: Mike Romito

Parker Aerospace, part of the Parker Hannifin Corp, designs, manufactures and services fluid systems, components and related electronic control systems for aerospace applications. Personnel total 5,300. The space related products of its divisions are specified below.

Air/Space This division specialises in the design/manufacture of valves and coupling devices to handle gaseous, liquid and cryogenic propellants, including hydrogen, nitrogen, oxygen, helium, NTO and MMH. Such systems fly on Shuttle, RL10 engine, Peacekeeper and Pegasus.

High Pressure Helium Regulator
Applications: ullage pressure control in a lightweight 680 atm blowdown propellant feed systems for missile and spacecraft propulsion systems. The regulator controls ullage pressure within ±10% with a 10:1 variation in inlet pressure and from lockup to rated flow. It is a welded

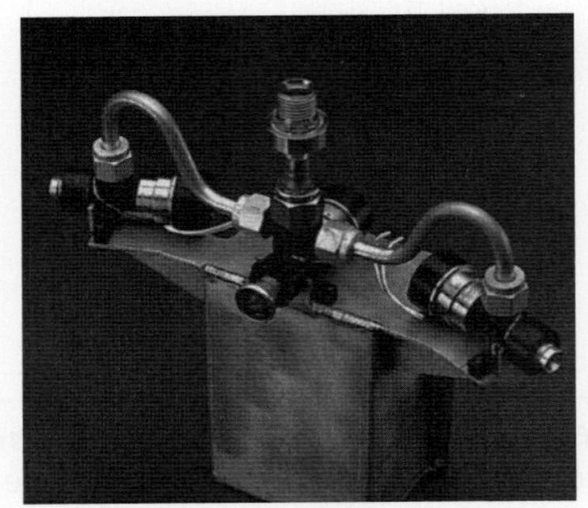

Parker Bertea Aerospace produces the nitrogen thruster packages for Pegasus 0516942

Through to Feb 1997 Pressure Systems delivered 60 titanium hydrazine tanks for the Iridium satellites. (PSI) 0517165

assembly constructed entirely from titanium alloys. The basic regulator design can be used for a variety of fluids/gases.
Inlet pressure: 680–102 atm
Regulated pressure: 87 atm ±10%
Flow rate: 0.16 kg/s helium min at 102 atm inlet
Inlet/outlet ports: 6.35 mm dia × 0.89 mm wall tube stubs
Startup transient: 95.8 atm max regulated pressure at 680 atm inlet pressure
Mass: 0.077 kg

Cold Gas Thruster Module
This module is designed to provide 3-axis control for small launch vehicles; it is flown on Pegasus. The module comprises three cold gas thrusters operated by integral independent solenoid valves. The central thruster generates 111.2 N and the two lateral thrusters 55.6 N when operated with a 136 atm N_2 gas supply. SI of each is 68 s. The solenoid valves are pilot operated and have a response time of <5 ms at max gas inlet pressure.
Operating pressure range: 136–13.6 atm
Thrust at 136 atm: 4.5–111.2 N + 8.9 55.6 N
Voltage: 28 ± 4 VDC
Response time: 8 ms max
Mass: 0.086 kg per thruster
Size: 25.4 mm dia × 63.5 mm per thruster.

Control SystemsThe division, with 1,700 employees, designs/manufactures flight control actuators and associated electronics for space applications that provide system functions of proportional control, performance enhancement and availability enhancement utilising available hydromechanical, electrohydraulic, electromechanical and electro-optic technologies. CSD provides the stage 2 3axis fin control actuator system for Pegasus and the stage 2/3TVC actuator system.

Rockwell Collins Optronics

Current Status
Rockwell Collins Optronics was, until September 2005, Kaiser Electro-Optics Inc (KEO). Rockwell Collins acquired KEO in December 2000.

Background
Rockwell Collins Optronics is based in Carlsbad, California. Originally established as the optical department for Kaiser Electronics in 1983, Kaiser Electro-Optics Inc (KEO) specialised in refractive and reflective optical and electro-optical assemblies for space, surface, airborne and underwater military and commercial applications. Optronics' expertise covers production, assembly and testing of spherical, flat, cylindrical and aspheric optics and systems for visible to infra-red wavelengths, as well as electro-optical systems for high revolution imaging cameras and Charge-Coupled Device (CCD) elements in visible and non-visible portions of the spectrum.

KEO delivered, in 1984, the first of 12 Cassegrain telescopes for the Defense Meteorological Satellites Program's (DMSP) Operational Linescan System imagers. The 20 cm f/1.0 primary and hyperboloid secondary nickel-coated beryllium mirrors were mounted in a beryllium structure. Eight 15 cm aperture telescopes (f/0.5 parabola primary and hyperbolic secondary) were delivered for the National Oceanic and Atmospheric Administration's (NOAA) High Resolution Infra-Red Sounder, in addition to 20 cm Cassegrain systems for NOAA's Advanced Very High

Resolution Radiometer. KEO and Rockwell Collins Optronics have also produced the High Resolution Infrared Radiation Sounder for NOAA; the Block 5 Telescope for the DMSP; the NASA Lightning Imaging Sensor; the NASA Lightning Mapper Sensor; the National Polar-orbiting Operational Environmental Satellite System (NPOESS) Visible/ Infrared Imager/Radiometer Suite; the JPL MUSES-CN Nanorover Lens; the 92 mm and 200 mm Star Tracker lenses; nine camera lenses each for the Mars Rovers, Spirit and Opportunity, as well as one lens each for the Rover's Landers; and star trackers lens assemblies on Deep Impact's impactor and flyby spacecraft.

Textron Inc

Current Status

United Industrial Corporation (UIC) and its subsidiary AAI Corporation are on course to become part of Textron for an agreed sale price of USD1.1 billion, where the majority of AAI's businesses are set to be picked up by Textron Systems, (which is under the Bell segment of Textron).

As part of the High Energy Liquid Laser Area Defence System (HELLADS) programme, Textron Defense Systems has been awarded a contract to design a Unit Cell Module for 150 kW Laser Weapon System (LWS).

In May 2007, NASA Ames Research Centre awarded Textron Defence Systems a contract to design an alternative heat shield for the Orion Spacecraft. The contract has a value of USD24 million and the company will be investigating two heat ablative materials, Avocat (previously used in the Apollo missions) and Dual Layer (which was designed for re-entry vehicles). This contract forms part of the Alternative Block 2 Thermal Protection System (TPS) Materials and Heat Shield Systems Advanced Development programme and work is expected to be completed by August 2008.

The US Department of Energy, operators of the Idaho National Laboratory, Battelle Energy Alliance (BEA), awarded Textron Defence Systems a three-year contract on 5 January 2006. The company will be designing Fine Weave Pierced Fabric (FWPF) billets for NASA missions and is scheduled to provide 18 3-D carbon-carbon FWPF billets.

The US Space and Missile Defense Command (SMDC) and the Army Forces Strategic Command (ARSTRAT) signed a contract with Textron Systems in December 2005 to develop a 100-kW High Power Solid State Laser. The contract is for three years and is valued with all options at USD30 million.

Background

Textron was founded in 1923 by Royal Little, a young entrepreneur, and has grown into a multi-industry operation with 2006 revenues of USD11.4 billion. Third quarter revenues for 2007 has seen a 15 per cent increase from USD2.8 billion in 2006 to USD3.3 billion. In 1952 Textron branched out into new industry areas, acquiring companies to diversify its brand of products and later, in 1963, selling its textile operations. Some of the acquisitions are still a part of Textron, for example, E-Z-GO, which is part of their Industrial Segment, and Bell Aerospace (including Bell Helicopters), which was acquired in 1960.

Textron's world headquarters is located in Providence, Rhode Island, United States. It has a presence in 32 countries with approximately 40,000 employees, under Chief Executive Officer and Director of Textron Inc, Lewis B. Campbell. Textron has four core business segments; Bell, Cessna, Finance, and Industrial (which includes Greenlee, E-Z-GO, Jacobsen, Kautex and Fluid & Power Group plus other smaller units).

The Bell segment includes Bell Helicopters, Textron Systems and Lycoming Engines under President and Chief Executive Officer Richard Millman. Textron Systems Corporation, headquarted in Wilmington, Massachusetts, United States, has five operating units under President Frank Tempesta. These units are Textron Defense Systems, HR Textron, Textron Marine and Land and Overwatch Systems. Textron Systems is trading under Textron Defence Systems and is a division of the Avco Corporation, alongside Lycoming Engines and the Ocean Acquisition Company, who owns Overwatch Systems. Avco Corporation is a subsidiary of Textron alongside HR Textron and Textron Marine and Land Systems.

The origin of Textron Defense Systems began in 1955 under Avco Research and Advanced Development (RAD), which was acquired by Textron in 1984, and has since been providing products for delivery systems, submunitions, airborne surveillance, automated test equipment, munitions, sensors, laser systems and Adaptable Radiation Area Monitor (ARAM) systems to the defence, homeland security and aerospace industry. Avco Corporation (before it was acquired by Textron Inc) designed the heat shield for the Apollo space programme. Once the Avocat material has charred from burning a protective layer is created around the outer surface of the spacecraft thus preventing any heat penetration. This breakthrough led to various technologies that could be used in many industrial applications, for example fire protection. Textron Defense Systems has designed the FWPF, a heat ablative carbon-carbon composite material that makes up the graphite impact shell and aeroshell of the General Purpose Heat Source (GPHS) module and was used, for example, in NASA's Cassini-Huygen mission, as well as US Air Force Inter-Continental Ballistic Missile (ICBM) programme for re-entry and test flights.

HR Textron was formerly the Hydraulic Research and Manufacturing Division of Bell Helicopters before its acquisition by Textron. Their products include photonic controls (Fly-By-Light (FBL)) and servo technology, for example high pressure servo-valves and advanced systems for launch vehicles.

Overwatch Systems was purchased by Textron Systems in December 2006. The Overwatch Geospatial Operations family of products include geospatial analysis software, for example RemoteView, that encompasses both imagery and analysis solutions for remote sensing applications and ELT, (a commercial imaging software) which is under the Overwatch Boston operations, (formerly Paragon Imaging). Also under Visual Learning Systems (purchased by Overwatch Systems in April 2006) they provide feature extraction tools such as Feature Analyst (which acts as a plug-in to ArcGIS, that is an Environmental Systems Research Institute (ESRIN) product), LIDAR Analyst (an extension to ArcGIS plus other GIS software), and Urban Analyst (which is an extension to ArcGIS 9.2). Interscope software, providing 2-D/3-D interactive maps, is also part of Overwatch Geospatial's product line. Their operations have relocated from Washington DC to Sterling Virginia, home to Overwatch Geospatial Operations, where they are working together to update product lines as well as developing extensions for existing products.

All-Source Imaging Systems, Northampton, Massachusetts, as of 24 June 2006 has become a part of Overwatch Geospatial Operations enabling them to develop product integration, as well as fuse interests within the geospatial and intelligence communities.

United Technologies Corporation Pratt & Whitney

Current Status

UTC combined the space propulsion activities of Chemical Systems Division (CSD), USBI and P&W in 1990 within P&W's Government Engines & Space Propulsion unit based at West Palm Beach in Florida. USBI is under contract to NASA Marshall to provide assembly, test and refurbishment of the non-motor segments of Shuttle's solid rocket boosters. NASA extended the contract in 1994 for recovering and refurbishing boosters to September 1997, worth US$1.8 billion. The previous US$1 billion award in January 1989 ran through September 1994. It was extended in 1996 through September 1999. The contracts have been suspended following the loss of Shuttle Columbia on 1 February 2003 pending a determination on utilisation rates at the end of 2004 prior to resumption of Shuttle missions no earlier than the second quarter 2005.

P&W produces the Orbus 21 solid propulsion system for applications such as the first stage of the IUS, PKM for Intelsat 6 and the TOS and provides the Orbus 21D for the Athena series of launchers, the RD-180 and the RL-10 in addition to high-pressure turbopumps for the SSME. Also produces the booster separation motors for the Shuttle SRB, eight being provided for each booster. P&W is developing a series of engines for the NASA Space Launch Initiative in co-operation with Rocketdyne division of Boeing.

CSD's Coyote facility in San Jose includes the world's largest solid rocket vertical test stand: 29 m high, it can handle 26.7 MN thrust motors and was most recently used for qualification firings of the seven segment USAF Titan 4 strap-on motor. Small/ medium- sized horizontal and vertical test stands are controlled from a central complex and structural test facilities can apply omni-axial simulated flight loads of up to 4.9 MN to motor cases and attach skirts. CSD's integral rocket/ramjet test complex can simulate booster, transition and ramjet operations in a single sequence under sea level to high altitude conditions up to M6.0. The Coyote site also houses a 5,400-tonne annual capacity solid propellant mixing facility comprising two 2,270 litre, one 2,840 litre and one 1,510 litre mixers.

Wyle Laboratories

Current Status

Founded in 1949 as an engineering consultancy, Wyle Laboratories now serve aerospace, telecommunications, space science, electronics, transportation, nuclear, meteorological and airports industries. Specialists in engineering, testing and support services, small payload integration and small experiment payload engineering. Can integrate several national experiments in an international package for launch or support in a third country. Wyle is setting up life sciences facilities and extending its activities into the field of bio-astronautics. Wyle can provide analytical surveys and has studied a broad range of space related projects and programmes including civil and military activities. Has conducted research into new technologies for advanced military space missions and innovative operational tasks. Wyle also carries out conceptual solutions to specified requirements.

PRIME CONTRACTORS

Canada

Bristol Aerospace Limited

Current Status
Bristol Aerospace continues to build the Black Brant sounding rocket; it has been one of the most-launched family of sounding rockets since the introduction of the Black Brant I in the early 1960s. The Black Brant has been used by NASA, the Canadian Space Agency and many other public sector research institutions and universities for upper atmospheric and microgravity research.

Background
A unit of the Magellan Aerospace Corporation since 1997, Bristol Aerospace, as it has been known since 1967, operates at three sites in Winnipeg, Manitoba. The company can trace its roots back to 1914.

Bristol's Defence and Space business unit is largely responsible for the company's space-related development activities. In the late 1950s, Bristol began developing the solid propellant Black Brant sounding rocket (see separate entry), of which, by 2006, over 800 had been launched. Two Black Brant launcher types are usually employed: a rail-type typically providing 6 to 8 m of guided rail travel and a 30 m travel tower. Bristol also operates facilities for vibration; temperature and humidity; salt fog; and mechanical testing.

In addition to the supply of basic rocket vehicle hardware, Bristol provides a range of self-contained vehicle and payload support technology, including thrust termination, parachute recovery, payload separation, despin, telemetry and fairing deployment systems. The company also produces CRV7 unguided rockets; pyrophoric infrared flares; and the GyroWheel spacecraft attitude control system.

Bristol Aerospace has manufactured small satellites since 1999, when it was awarded the SCISAT-1 contract by the Canadian Space Agency. The satellite was launched in August 2003, and as of 2007, remained operational.

Dynex Power, Inc

Current Status
On 19 November 2007 third quarter results for Dynex Power were announced, showing profits of USD915,000 and profits of USD1.1 million were announced for that year to date.

Background
In 2000, Dynex Power Inc, a Canadian company, acquired Mitel Corporation's Lincoln UK operations who specialise in radiation hardened Complementary Metal Oxide Semiconductors on Silicon On Sapphire (CMOS/SOS) Application Specific Integrated Circuits (ASICs). Dynex Semiconductors were subsequently established as a subsidiary of Dynex Power.

Dynex Semiconductors provide products to aerospace and telecommunications industries, for example CMOS/SOS technologies which are radiation hardened space components that can provide resistance to single event and transient upset and total radiation dosage.

Products with the Silicon On Sapphire (SOS) technology Integrated Circuits (IC) databus include MA1916 Radiation Hard Reed Solomon and Convolution Encoder. The Reed-Solomon Encoder guards against message error bursts and the Convolution Encoder provides for noise immunity. Interleave depths at 1, 4 and 5 are provided to improve performance. The MA1916 is ideal for satellite communication technology applications as it is manufactured in radiation hard low power CMOS technology. The MA1916 Encoder uses international Consultative Committee for Space Data Systems (CCSDS) standards code RS (255, 223).

The MA28139 On-Board Data Handling (OBDH) Bus Terminal is a five-channel modem where the Litton Bus drivers are operated by digital wave forms that are provided on the Bus Side and receive outputs from the Litton Bus detector. The OBI signals and the internal user bus levels also have a multiplexing and demultiplexing function.

The MA28140 Packet Telecommand Decoder provides a high performance radiation hardened CMOS/SOS 1.5 μm technology that implements the entire Telecommand Decoder core into a single chip. It is compliant with European Spacy Agency (ESA) Procedures, Standards and Specifications (PSS) for packet Telecommand (ESA PSS-04-107 and decoder specification ESA PSS-04-151).

SOS IC Logics include, a 16-bit parallel 54HSC Radiation Hard Error Detection and Correction circuit that is compatible with TTL or CMOS, which performs a 6-bit, check word. 54HSC/T Series Radiation Hard High Speed CMOS/SOS Logic circuit has latch up and single event immunity and is compatible with the 54 L/S circuits pin for pin.

SOS IC Memories include a high performance MA7001 512 × 9-bit FIFO radiation hardened 3 μm technology, MA5114 Radiation Hard 1,024 × 4-bit Static RAM and the MA9264 Radiation Hard 8,192 × 8-bit Static RAM are radiation latch-up free with a six transistor cell with full static operation. They require no clock or timing strobe.

SOS IC microprocessors include the MAS281 Mil-Std-1750 Microprocessor, a 16-bit central processor which incorporates three large scale integration chips. The MAS281 unit that is interconnected by a 64-pin dual-in line ceramic substrate with components MA17501 Radiation Hardened Mil-Std-1750A Execution Unit, MA17502 Radiation Hardened Mil-Std-1750A Control Unit and the MA17503 Radiation Hardened Mil-Std-1750A Control Unit, which makes up the MIL-STD-1750A Instruction Set. It is also available without the ceramic substrate. The single chip MA31750 Mil-Std-1750 Microprocessor has a 32-bit internal bus structure with a 24 × 24 bit multiplier and 32-bit ALU. Compared to the MAS281 the MA31750 provides increased performance and incorporates the MIL-STD-1750A instruction set architecture, or option 2 of Draft MIL-STD-1750B.

SOS IC peripherals and support circuits includes 24 programmable I/O pins MA28155 Radiation Hard Programmable Peripheral Interface that is used for the MAS281 microprocessor. Conforming to the MIL-STD-1750A and 1750B standard is the MA31751 Memory Management and Block Protection Unit, which is an optional chip that can be used to expand MAS1750 and is user configurable for the 1750B.

The four independent prioritised channels MA31753 Direct Memory Access Controller (DMAC) supports a MIL-STD-1750A or B Operation in an MA31750 System, where three basic transfer nodes are performed Direct Memory to I/O peripheral transfers, direct I/O to memory transfers, memory to memory transfers, I/O to I/O transfers. The MA31755 16-bit Feedthrough Error Detection and Correction Unit (EDAC) is positioned in the data bus located between the processor and the memory, which provides added protection for the unit. EDAC is used when retrieving data from the memory monitors and corrects those data values.

Macdonald, Dettwiler and Associates Ltd (MDA)

Current Status
Macdonald, Dettwiler and Associates (MDA) were the prime contractor on the next generation observational satellite RADARSAT-2, which after a delay was successfully launched in December 2007. RADARSAT-2 is an Earth observation satellite jointly developed by MDA and the Canadian Space Agency (CSA). The satellite will replace its predecessor RADARSAT-1 and offers enhanced range of radar imagery. The RADARSAT-2 captures precision images from anywhere on the globe, day or night, in all weather conditions, with the information relayed back to various ground stations.

MDA are also performing work for the CAD75.8 million NASA contract awarded to Canadian Commercial Corporation for continued hardware and software robotics support for the Space Shuttle Programme and the International Space Station (ISS). MDA will sustain engineering services to the Space Shuttle Remote Manipulator System robotic arm, known as Canadarm, the inspection boom assembly and robotic work station.

Background
In 2001 MDA filed total revenues of CAD481.275 million compared with CAD401.125 million in 2000. The former majority shareholder Orbital Sciences Corporation reduced from 67 per cent in January 2000 to zero in July 2001.

MDA specialises in computer-based systems for aerospace, defence and resource management applications, exporting 75 per cent of its products. In November 1995 it became part of the Orbital Sciences Corporation group. Company employs 1,800 people worldwide.

MDA's product line includes: space-qualified software; MIL-SPEC (including DOD-STD-2167A) systems and software; turnkey remote sensing satellite ground stations; image mapping system for generating maps from digital images; optical/radar image analysis systems for space/airborne sensor data; meteorological data analysis systems. For Canada's Space Station Mobile Servicing System, it is contributing to development of software, data processing subsystems and artificial intelligence applications.

MDA is a leading supplier of ground receiving/processing systems for remote sensing satellites, working as prime contractor on 18 turnkey satellite ground stations worldwide and as major subcontractor on 10. It provides systems for all optical and SAR missions. Its SAR team was the first to digitally process SAR data and to develop a commercial digital processor, and is the world's leading supplier of spaceborne SAR processing systems. MDA is active in the development of analytical software for RADARSAT data.

In 2001, the company was selected to build the first QuickBird ground station for receiving and processing high resolution imagery from QuickBird satellites. It has contributed information about optimised hybrid satellite technologies and software options for optimised systems. The company is processing high resolution images for use with several companies and organisations related to security concerns and for organisations involved in planning regional ultra-large development schemes.

In October 2005, MDA announced an agreement with IHI Aerospace of Japan to help provide information solutions for helping to load pallets containing equipment and supplies from the International Space Station (ISS) into Japan's H2 Transfer Vehicle.

Since then, MDA have also agreed a long term agreement with Rocketplane Kistler for a joint participation in the initial and future phases of the Commercial Orbital Transportation Services (COTS) programme. The agreement is expected to evolve as financing, technical and customer targets are met.

Germany

OHB-System GmbH

Current Status
In 2007, OHB-Systems GmbH were awarded a contract by the European Space Agency (ESA) to develop the next generation of small geostationary satellites. Small geostationary satellites have their origins in technical specifications developed by OHB-System. The small geostationary satellites are considerably smaller then previous conventional versions offering reduced risk and greater cost effectiveness for the customer.

Background
Orbital- und Hydrotechnologie Bremen-System GmbH, founded in 1958, undertakes system engineering, hardware development/production and project management for space and environmental programmes. 1995 sales were DM41 million. Principal activities are: design/development of mechanical and electronic equipment; μgravity

systems, experiment facilities/equipment; manned/unmanned space systems and mission operations, re-entry technology and aerodynamics/aerothermodynamics; small satellites. Its μgravity activities include industrial lead in the Mikroba programme, parabolic flights in chartered aircraft (NASA's KC-135 and CNES' Caravelle), collaboration in the construction and utilisation of ZARM's drop tower at Bremen University (providing the drop capsule), development of the Holop holographic camera for Germany's Spacelab D2 mission and the High Speed Centrifuge for ESA's Anthrorack. The company also developed the electronic control unit for Intospace's Cosima crystal growth facility, first flown aboard a recoverable Chinese platform in August 1988. OHB has continued to develop software controlled electronic processing units for microgravity research on recoverable modules.

OHB manufactures the flight harness for the BPDU Spacelab rack, it developed the video control for IML 2's NIZemi, and it was responsible for the design of ergometry, blood and urine monitoring experiments flown on Germany's Mir-92 mission. On Spacelab D2, OHB provided the Gravitational Biology element of incubators and centrifuges. OHB is MGSE prime contractor for ESA's Polar Platform, it was awarded the Phase A study for Columbus' Anthrolab facility and within the Europayload Consortium, it is working on the Fluid Science Laboratory and Automated Bio Laboratory for Columbus integrated as Europe's contribution to the ISS. OHB-System reconfigured these elements following an ESA decision on the future of European policy on the reduced ISS programme plan reducing microgravity research time.

OHB developed Bremsat for the University of Bremen and its SAFIR-R1 demonstration package was launched November 1994 attached to a Russian Resurs-O1, preceding 1997's SAFIR 2 small satellite to provide global two-way data services. Instrumental in developing joint programme studies for international projects involving ESA and Russian activity in microgravity research.

The creation of Space Business International was announced in June 1993 by OHB, Space Industries International, Inc and Novespace to co-operate on space contracts.

International

Arianespace

Current Status

In April 2007, Arianespace and Mitsubishi Heavy Industries (MHI – see separate entry) agreed to jointly propose European and Japanese launch services to better support customers' satellite launch requirements. This expands upon the original Launch Services Alliance agreement, created in 2003 between Arianespace, MHI and Boeing Launch Services Inc, which provides back up launch services to customers should an originally planned launch prove impractical or improbable due to launch schedule conflicts and delays, or launch pad damage caused by a previous launch failure. MHI offers the H-IIA and Boeing offers Zenit-3SL services through its Sea Launch joint venture (see separate entries).

Background

Arianespace was founded in 1980 by European financial, civil and space technology interests and is headquartered in Evry, France. The company also maintains offices in Washington, DC; Tokyo, Japan; and Singapore. Arianespace's shareholders represent ten European countries: Belgium, Denmark, Germany, France, Italy, Netherlands, Norway, Spain, Sweden and Switzerland. The French Centre National d'Etudes Spatiales and Astrium SAS remain the company's two largest shareholders at 33 per cent and 16 per cent respectively. Other major shareholders include Astrium GmbH, the Safran group, Avio and MT Aerospace AG. As of 2006, Arianespace employed about 270 personnel and had achieved EUR985 million in annual sales. As of early 2007, the company had agreed to 46 further launches, including those that will be conducted by Starsem (see separate entry), in which Arianespace holds a 15 per cent interest. Arianespace has launched about two thirds of the commercial satellites in service.

Arianespace markets and provides commercial launch services for the Ariane 5 launch vehicle, which launches from the Centre Spatial Guyanais in Kourou, French Guiana (see separate entry). Soyuz and Vega rockets will also lift off from Kourou as soon as 2008; Arianespace will be responsible for launch services for these launch vehicles as well.

While ESA is the design authority and developer for the Ariane launch vehicle series, Arianespace is responsible for validated Ariane rocket production financing and oversight; CNES manages for ESA, the total technical and financial aspects of the Ariane programme. These various roles and responsibilities have been in effect since the first commercial launch of the Ariane I, in 1984, although Ariane I's first flight was in December 1979.

EADS Astrium

Current Status

In June 2007, EADS Astrium announced that it was joining the space tourism field of competitors by developing a four-passenger spaceplane designed to take crew and tourists on sub-orbital flights 100 km above the Earth to experience about three minutes of weightlessness. Tickets are expected to cost between EUR150,000 and EUR200,000 per flight. Astrium expects that the craft will rocket its first paying passengers into sub-orbital space around 2012, and project development is slated to cost approximately EUR1 billion.

On 24 October 2003, Astrium announced the award of the Skynet 5 contract to its subsidiary, Paradigm Secure Communications for the UK MoD's next generation Skynet 5 satellite system. Under this contract Astrium built three Skynet 5 satellites, and under the Skynet 5 Private Finance Initiative (PFI) the MOD procured satellite communications services from Paradigm for 15 years. In order to provide a seamless transition, Skynet 4 satellites, the last of which was launched in February 2001, are now operated and managed by Paradigm as well as the Skynet 5A and 5B satellites. Skynet 5A launched in March 2007 and Skynet 5B is scheduled to launch in late 2007; Skynet 5C, the in-orbit spare, will launch in 2008.

Background

Astrium was formed in May 2000 with the merger of Matra Marconi Space in France and the UK, and the space division of DaimlerChrysler Aerospace (DASA) in Germany. In 2003, the European Aeronautic Defence and Space Company (EADS) bought out BAE Systems' 25 per cent stake in Astrium, including the Paradigm Secure Communications subsidiary, and became Astrium's sole owner, creating EADS Astrium. Construcciones Aeronauticas SA (CASA) Espacio was already wholly owned by EADS in 2000 (CASA merged into EADS with Aerospatiale Matra and DaimlerChrysler Aerospace creating the EADS parent company in July 2000), when Astrium was formed. At that time, EADS also had a Launch Vehicles sector, separate to Astrium, CASA Espacio and Space Services: all these comprised EADS' space division. Astrium and Launch Vehicles were soon afterward combined. Infoterra had been a subsidiary of Astrium since 2001 and Astrium acquired Dutch Space BV (formerly part of the Fokker group) in 2005. In 2006, the EADS space groups were restructured to better organise all the company's space pursuits; this created three Business Units to bring together the space assets of the several company divisions descended from EADS' merger and acquisition activities. As of 2007, EADS Astrium was comprised of the three following space Business Units: Astrium Space Transportation; Astrium Satellites; and Astrium Services. Each of these existed separately for a while within the EADS business structure, previous to the introduction of the umbrella Astrium identity.

EADS Astrium is Europe's largest space company. Headquartered in Paris, it maintains locations in France, Germany, the Netherlands, Spain and the UK. As of 2006, Astrium employed just under 12,000 personnel, nearly half of whom are based in France, and that year achieved revenues of EUR3.2 billion. During the same period, EADS as a whole achieved EUR39.4 billion in revenues and employed roughly 116,000 personnel. Astrium owns stakes in Arianespace and London Satellite Exchange, and counts Computadoras, Redes e Ingeniería, SA (CRISA); Tesat-Spacecom; and Dutch Space, BV, as wholly owned subsidiaries.

EADS Astrium, through its three Business Units, plays a role in every space market. The Space Transportation unit designs and manufactures launch vehicles and propulsion systems; the Satellite unit not only develops satellites, but also develops and builds ground stations; and the Services unit provides satellite operations and telecommunications services.

Astrium is the prime contractor for the Ariane 5 launch vehicle, and is responsible for the rocket's delivery to Arianespace, the launch services provider which EADS in part owns. EADS is playing a role in ESA's Future Launchers Preparatory Programme (FLPP), and has designed vehicles such as Phoenix and Pre-X, which remain in demonstrator stages. The company also owns stakes in the Eurockot Launch Services and Starsem, Russian joint ventures that provide launch services for the Rockot and Soyuz launch vehicles.

Astrium designs and builds satellites, including equipment and subsystems, for a wide range of applications and has served the science, Earth observation, private communications and defence sectors. It can name ESA, Inmarsat, Intelsat, NASA and the UK MoD among its customers.

In addition to Astrium Services, Paradigm Secure Communications; MilSat Services GmbH; and Infoterra, also EADS Astrium owned subsidiaries, provide military satellite communications and Earth observation and imagery services to EADS' customers.

Telespazio

Current Status

In April 2007, Finmeccanica and Thales formed the New Space Alliance, a group composed of Telespazio and Thales Alenia Space (see separate entry), which was created from the former Alcatel Alenia Space. As of April 2007, Telespazio was 67 per cent owned by Finmeccanica and 33 per cent owned by Thales; Thales Alenia Space was 67 per cent owned by Thales and 33 per cent owned by Finmeccanica.

Background

Telespazio, headquartered in Rome, is a satellite services provider. It operates in 22 locations worldwide and employs about 1,400 personnel. The company achieved a revenue of EUR340 million in 2005. Telespazio provides business and public sector satellite telecommunications and broadband internet services, as well as offering secure networks for military customers who have high data security requirements. The company works closely with the Agenzia Spaziale Italiana (ASI) and ESA, to manage Italian satellite services linked with national and European space programmes.

Telespazio contributes to European telecommunications, navigation and Earth imaging programmes, including the Galileo, and Global Monitoring for Environment and Security (GMES) programmes, primarily by designing, building and operating programme ground segments; providing in-orbit operations services; providing mission management services; designing and building ground segment simulation and testing equipment and managing data acquisition operations. It operates five space centres, including the Fucino Space Centre, and also has facilities at the Centre Spatial Guyanais in Kourou, French Guiana (see separate entry).

Telespazio is active in the Earth observation field; it receives, processes and distributes Landsat, Quickbird, ERS, Envisat and IRS imagery via its Fucino facility. In 2000, Telespazio established E-Geos, a joint venture with the Italian Space Agency, to found Italy's first national Earth observation centre. In 2001 Telespazio managed the early orbit phase of the SICRAL military telecommunications satellite. In 2003, the company acquired the German remote sensing services firm, GAF.

Thales Alenia Space

Current Status

Following the transfer to Thales of Alcatel-Lucent's space sector company shares in Alcatel Alenia Space (67 per cent) and Telespazio (33 per cent), an agreement was reached in 2007 between Finmeccanica and Thales to form the New Space Alliance, a group that includes Thales Alenia Space and Telespazio. Thales Alenia Space evolved from the former Alcatel Alenia Space, formed in 2005 as a result of the Alenia Spazio and Alcatel Space merger. After the transaction, as a Thales investor, Alcatel-Lucent owned a 21 per cent stake in Thales, having acquired 25 million stock shares and EUR710 million in cash. The French State remains Thales' primary single shareholder at 27 per cent. As of April 2007, Thales Alenia Space was 67 per cent owned by Thales and 33 per cent owned by Finmeccanica; Telespazio is 67 per cent owned by Finmeccanica and 33 per cent owned by Thales.

Background

Formerly Alcatel Alenia Space, (and the Alenia Spazio and Alcatel Space companies before that), Thales Alenia Space is 67 per cent owned by Thales and 33 per cent owned by Finmeccanica. Thales Alenia Space is Europe's largest satellite and space systems manufacturer, employing over 7,200 personnel in 13 locations worldwide, including subsidiary offices.

Company expertise covers complete space systems, satellites for telecommunications, remote sensing, meteorology, navigation and science applications, manned systems and space infrastructures, optics systems, transport, launch and re-entry systems, telemetry and control stations and specialist space software. Some of the company's recent high profile projects include: the International Space Station's (ISS) three Multi-Purpose Logistics Modules (MPLM), Leonardo, Raffaello and Donatello; the thermal-mechanical segment for the ISS Columbus module; and Cargo Carrier development and integration, as well as thermal control studies for the ISS's Jules Verne Automated Transfer Vehicle (ATV). Thales Alenia Space owns a 38 per cent interest in Galileo Industries, the Galileo navigation system's prime contractor. The company has contributed to many of Europe's premier space science programmes, including Cassini-Huygens, COROT and Venus Express.

Israel

Israel Aerospace Industries Ltd (IAI)

Current Status

Built by IAI, the Israeli military remote sensing satellite, Ofeq 7, was launched aboard a Shavit rocket in June 2007, returning its first images on 14 June. It is expected to remain operational for at least four years. OpSat 3000, unveiled at the Paris Air Show in June 2007, is IAI's latest generation observation satellite.

The Earth Resource Observation Satellite, Eros B, launched on 25 April 2006. The IAI/MBT-built Eros series has proved a successful commercial descendent of IAI's Ofeq military reconnaissance series. Eros A launched from Svobodny, Russia on 5 December 2000.

Background

First established as the Bedek Aviation Company in 1953, and later known as Israel Aircraft Industries (IAI), the company changed its name again, in 2006, to Israel Aerospace Industries (IAI) to better reflect its expanded role in the aerospace industry. IAI is headquartered at Ben Gurion International Airport, southeast of Tel Aviv. The company employs a workforce of about 15,000 and in FY2006 reported USD130 million in net profits. IAI issued a well-received USD250 million bond offering in June 2007.

Systems, Missiles and Space Group

Much of IAI's space-related development work is conducted from within the Systems, Missiles and Space Group, which is comprised of several divisions. IAI's Elta Systems group also develops space systems and products.

The IAI/MBT Space Division houses Israel's major space centre. It was responsible for developing Israel's first satellite, the Ofeq, as well as the all-solid fuel Shavit launcher, the Eros imaging satellite series, and the Amos commercial communications satellite (see separate entries).

The MBT Space Division space technology centre includes a satellite integration and test centre; an S/X-band Tracking, Telemetry and Control (TT&C) station; laboratories; image processing facilities; and fabrication facilities. The satellite centre covers 3,400 m², including a 2,000 m² class 100,000 clean room, and a class 10,000 30 m² portable laminar flow booth. Class 100 facilities can also be provided.

The MLM Division was founded in 1964 and specialises in systems integration; telemetry systems; Shavit small satellite launch vehicle systems; ballistic missile systems; test range instrumentation; and solar arrays. MLM employs about 1,070 personnel.

IAI's Tamam Division, in operation since 1964, specialises in inertial navigation, space attitude control systems, and Electro-Optic (EO) products and systems.

The MBT Missiles Division develops and builds a range of missile systems, including naval and ground-based air defence systems and naval attack systems.

Established in 1967, Elta Systems Ltd is an IAI subsidiary, independent of the other IAI groups. Elta participates in the development and fabrication of Israeli satellites including remote sensing and surveillance satellites. Elta manufactures a range of products, including military electronics systems; radars; active/passive early warning systems; secure communications; signal processing systems; microwave components; tubes and antennas; high-voltage power s upplies; and microelectronics components.

Japan

Fuji Heavy Industries Limited (FHI)

Current Status

FHI participated in studies of Japan's HOPE unmanned shuttle and the initial design of a horizontal takeoff and landing spaceplane. It built the two ALFLEX models first flown in 1996. Redirected efforts to changed requirements statement for fly-back recovery vehicle and now developing concepts for ISS rescue vehicles. FHI has researched possible variants of the HOPE concept with conceptual designs similar to the revised NASA re-entry vehicle being considered as an ISS rescue vehicle. These studies have now shifted emphasis as Japan's plans for orbital space planes have changed. Early plans for a reusable Hermes-like shuttlecraft have been moderated in favour of continued research for technology development with the eventual goal of a reusable, affordable, system for spaceplane applications. This has resulted in renewed interest in a reusable delivery system capable of carrying astronauts if necessary, but this is not crucial to the craft's operation. Technology development supported by FHI includes wind tunnel, CFD, composites and structures and materials science and engineering.

Mitsubishi Electric Corporation (MELCO)

Current Status

In 2003, Mitsubishi Electric Corporation (MELCO) became the first Japanese satellite maker to be a prime contractor for a commercial satellite, when it was chosen for the Optus C1 communications satellite for SingTel Optus Pty. Ltd. This was followed in 2005 when the company was chosen to build the Superbird-7 satellite for Japanese satellite service provider Space Communications Corp. (SCC).

MELCO announced that the satellite would weigh 5 tons and be launched into a geostationary orbit at 144°E in 2008, and will carry 28 Ku-band transponders.

Since winning the Optus C1 contract, MELCO had been struggling to break into the commercial market, and so the Superbird-7 contract was a significant success for the company.

When operational, the Superbird-7 satellite will replace the Boeing 601 designed Superbird-C which was launched in 1997.

Background

Established in 1921, MELCO provides telecom and Earth observation satellites, remote sensing systems and satellite communication systems. It acted as prime for Japan's CS telecom satellites, the ETS-2/4/5 engineering test satellites, the JERS 1 synthetic aperture radar Earth observation satellite and the ISAS/USEF Space Flyer Unit. It is working with Space Systems/Loral and Aerospatiale as contractors for the Intelsat 7 series and with Lockheed Martin Astro Space on Inmarsat 3. Its xenon ion thruster and inter-satellite communications system flew on ETS-6. MELCO owns the Space Communications Corp of Japan, operators of the Superbird communications satellite system. Subsystem expertise covers AOCS (electronics, wheel drive assembly, Earth/Sun sensors, nutation dampers), communications/data handling (transponders, telemetry/command units, central processing unit), electrical power (battery assemblies, converters, shunt, control units), antennas (shaped beam, multibeam, phased array and omni antennas, feeds, multiplexers), solar arrays (GFRP, KFRP, CFRP substrates, semi-rigid/rigid, solar cells modules, GaAs cells), propulsion (mono hydrazine thrusters, ion and MPD arc jet engines), thermal (louvre, heat pipe, shield, multilayer insulation) and structures.

MELCO provided the microwave scanning radiometer for the MOS satellites and developed the SAR for JERS 1. The company also offers turnkey Earth stations, small Earth stations for SNG, VSAT and USAT, and was responsible for ISAS' 64 m Usuda deep space tracking facility. It is a major supplier of mobile Earth terminals for the US MSAT system.

The class IB spacecraft manufacture/test facility at Kamakura can fabricate/test subsystems and components, and integrate/test the entire spacecraft system in one building. The Satellite Integration and Test Facility includes a centre of gravity machine, mass measuring machine, dynamic balancing machine and a data acquisition system. The Subsystem Test Facility houses an anechoic chamber, antenna range test tower, EMC test system, Sun/Earth simulators, three-axis air bearing and gimballed servo tables, infra-red reflectometer and three-axis co-ordinates measuring machine. The environment test facility accommodates a space simulation chamber, spin table, vibration machine, rotary accelerator, shock test machine and outgassing measuring machine. The manufacturing facility allows fabrication of solar cell modules and heat pipes. MELCO employs 103,000 people in its main organisation and subsidiary companies. Net sales in June 2007 were reported as approximately USD33 billion with equity of approximately USD9 billion.

Mitsubishi Heavy Industries Ltd (MHI)

Current Status

In 2007, Mitsubishi Heavy Industries (MHI) and Arianespace (see separate entry) formed an agreement to augment their back-up satellite launch services for the H-IIA and Ariane 5 launch vehicles. MHI is the H-IIA launch vehicle systems integrator and launch services provider, operating from Tanegashima Space Center (see separate entry). Customers will be offered a choice between the two launch vehicles to better accommodate their launch requirements.

Background

Mitsubishi Heavy Industries (MHI) was first founded when the Mitsubishi Mail Steamship Company began leasing the Nagasaki Shipyard and Machinery Works from the Japanese government in 1884; the modern MHI corporate entity was established in 1950. The company began R&D work on liquid rocket engines in 1954; much of the subsequent early technologies were derived from the US Delta rocket family development programme.

As of 2006, the company employed 32,552 workers in 31 locations. MHI achieved a consolidated net income of JPY48.8 billion in FY2006. Its aerospace businesses represented 16.6 per cent of the company's contracts, worth JPY543.3 billion.

MHI's aerospace development and manufacturing work principally takes place at its Guidance and Propulsion Systems Works and Aerospace Works, both located in Nagoya, as well as its Kobe and Nagasaki Shipyard and Machinery Works plants. The Kobe plant houses the Kobe Space Center, which is equipped with clean rooms. Both Nagoya plants are also equipped with clean rooms. Nagoya also has a space environmental test laboratory, a rocket engine firing test stand and a materials test laboratory, among other R&D and manufacturing facilities.

MHI provides Japan's primary capability in liquid rocket engines, developing the cryogenic oxygen/hydrogen LE-7A stage 1 and LE-5B stage 2 engines for JAXA's H-IIA launcher. MHI has also developed and manufactured Japanese Experiment Module (JEM) and H-II Transfer Vehicle (HTV) components as part of Japan's contribution to the International Space Station (ISS) and its servicing missions. The company is also developing the cryogenic MB-XX upper stage engine with Pratt & Whitney Rocketdyne. MHI has prime industrial responsibility for the H-IIA, as it did for the H-II. The 10-year USD108 million LE-5 development programme produced Japan's first indigenous operational LOX/LH_2 engine, providing the H-I's stage 2 with a restart capability. The technology formed the basis for developing the larger and more advanced LE-7 and the LE-5A direct outgrowth.

Mitsui & Co Ltd

Current Status
Established in 1947, Mitsui has capital of ¥192 billion and employs more than 10,000 people of which 3,500 are overseas in 43 subsidiaries with facilities in 88 countries. Hardware, software and services for satellites, launchers and Space Station. Has committed some modest resources to integrated launch vehicle automation processes sought by national and company efforts to reduce costs, limit risks, remove excess monitoring and control equipment installation expansion. This work continues with emphasis on the new generation of reusable spacecraft sought by the new Japanese aerospace agency JAXA. A readjustment of priorities has stimulated the company to direct its efforts at improving performance, manufacturing reliability and test criteria. It has also developed new cost-effective mechanisms for production efficiencies. The company holds a 24.5 per cent interest in Japan Satellite Systems (covered in the Communications section), which began operating its first satellite in 1988.

Russian Federation

AKO Polyot

Current Status
Polyot designs and builds the Kosmos 3M satellite launcher (marketed by the Cosmos USA joint venture) and has been instrumental in the development of several specialised spacecraft types. It has built the Glonass, Tsikada, Cospas, Parus, Strela, Gonets and Informator satellites in conjunction with designer NPO PM. Some of these satellites were hybrid, multi-purpose and carried several separate control functions required by different payloads. It built the RD-170 engine for NPO Energomash, but production has halted and the plant is now dismantled.

Polyot was created in 1941 to build aircraft. It employed 20,000 in 1994 and approximately 9,000 as of 2005. Production of the An-74 cargo aircraft resulted in the PO designation changing to AKO.

KB Arsenal

Current Status
Collocated with PO Arsenal, historically a major producer of artillery, KB Arsenal provides the bus for the military Eorsats to TsNPO Kometa. They now offer science and commercial satellite design and are developing a new family of bus designs applicable to civil and military applications. Conceptual work on secondary military piggy-back payloads operating as autonomous vehicles and hybrid satellite and spacecraft bus configurations which, due to a downturn in global market uptake, have not yet succeeded in orders. Much of this work has been abandoned in favour of more traditional engineering contracts.

KB Makeyev

Current Status
V P Makeyev (1925–1985) founded the bureau in 1955 as an offshoot of Korolev's bureau and produced Russia's first ballistic missile to use storable propellants, the R11, a development of the German Wasserfall SAM. It developed the SS-N-6 Zyb, SS-N-8 Vysota, SS-N-18 Volna, SS-N-20 and SS-N-23 Shtil SLBM. KB Makeyev was involved in the development of the Surf/Priboy satellite launcher under an agreement between the US Sea launch organisation and Russia's Ramcon Association. The bureau was responsible for several unsuccessful designs and its research department developed a wide range of propellant combinations, chemical matrices and motor designs that characterised rocket development in Russia during the 1960s. Some of this effort was directed toward the use of storable propellants for rapid-response launches to replace satellites destroyed in time of war. The absence of an effective anti-satellite system in the west brought an end to this work.

The Makeyev bureau offers satellite launch services using its converted SLBMs. A Volna was launched 7 June 1995 from the 'Kalmar' Delta 4 boat submerged in the Barents Sea on a 20 minute suborbital µg flight carrying a 120 kg experiment from Germany's ZARM (electrically-driven thermal convection in a fluid shell between two concentric spheres). Recovery was 5,600 km downrange on Kamchatka. There were apparently three SLBM demonstrations 1991-93. A Shtil 2N orbital demonstration launch was reportedly planned carrying a 100 kg Izmiran satellite. Major activity by the end of the 1990s was conversion of SLBMs into potential satellite launches.

Discussions with potential customers for modified Surf/Priboy variants. A wide range of options continues to be developed with few indications of solid customer orders.

Khrunichev State Research and Production Space Centre

Current Status
The history of GKNPTs Khrunichev began in 1916 when the factory was established in Fili for building cars by the Russian-Baltic Carriage Factory Joint Stock Co. It began aircraft construction in the 1920s, initially leased to Junkers. In 1927 Plant 22 was established to build metal aircraft for domestic use including types designed by Tupolev, Arkhangelskiy and Petlyakov. In 1941 after the outbreak of war with Germany Plant 22 was moved to Kazan where it remained and is now known as the Kayan Aviation Production Association. In Moscow, on the original site of Plant 22, a new plant, designated No 23, was set up to build bombers, including the turbojet Tu-12 and Tu-14 of 1947–49. In 1951 Plant 23 built Myasishchev bombers from which date it became OKB-23 under the Moscow Council of People's Economy. The plant was renamed after the late Minister of Aviation Industry, Mikhail Vasilievich Khrunichev, on 3 July 1961. On 30 October 1960 OKB-23 had been transferred to OKB-52 headed by V N Chelomei for the production of rockets and missiles and, later in the decade, the development of the Almaz spacecraft. In 1965 the Khrunichev Plant came under the Ministry of General Machine-building of the USSR but after this structure was disbanded in 1991 it became part of the Kompomash association. In 1997 it reported to the Ministry of Economy and, in May 1998, control was transferred to the Russian Space Agency. This move encouraged research and test activity in hybrid propulsion systems. In 1994 Khrunichev merged with KB Salyut to become the M V Khrunichev State Space Science and Production Centre (GKNPTs) after which it received a role in development of spacecraft and launchers. At this point Khrunichev accessed several technologies for exotic propellants and built its own research facilities for high-pressure propellant and combustion systems. These funded new research and development structures for potential competition with parallel organisations. Serial Proton production has been undertaken since the 1960s, followed by the SS-19 missile (now available as the Rockot satellite launcher) and space station modules (Salyut, Mir, Almaz, Kvant, Kristall, Priroda, Spektr, FGB and Service Module), these latter elements for the ISS. Khrunichev is actively developing the Angara series of modular launch vehicles and the Baikal reusable launch vehicle which could fly on Angara stages. The ambitious Baikal incorporates a deployable wing for lifting entry followed by deployable air-breathing engines for controlled flight in the atmosphere preceding touchdown.

Moscow Institute of Thermal Technology (MITT)

Current Status
In May 2007, Russia tested the RS-24 missile, which was developed at the Moscow Institute of Thermal Technology. Fired from a mobile launcher, very little information was released about the missile in contrast to the usual degree of relative openness about strategic missile programmes. However experts immediately declared it as a Russian response to the US plans to deploy elements of a missile defence system in Poland and the Czech Republic, a plan Russia has expressed strong opposition to.

Background
MITT was founded in 1948 by Alexander Nadiradze's (1914–87) design bureau and was responsible for the design and development of the RS-12/SS-13 silo-based, RS-14/SS-16, RSD-10/SS-20 and RS-12M/SS-25 road mobile and RS-22/SS-24 rail mobile solid propellant ICBMs. Its Science and Technical Centre Complex was created in 1991 to develop the Start orbital launchers based on the SS-20/SS-25; the Start-1/Start test launches were made 25 March 1993/28 March 1995 from Plesetsk. Responsible for enhanced, boosted, propellants capable of propelling high-acceleration rockets and ballistic missiles. In 1998 the MITT was subordinated to the Russian Space Agency but its expertise remains intact and it has played a leading role in developing the Start launcher. The company has designed a new range of small launchers based on Start technology but lack of financial support from potential investors and a general glut in the launch vehicle market has conspired to prevent further development.

NPO Lavochkin

Current Status
NPOL specialises in planetary and deep space craft, science satellites, including Luna, Venera, Zond 1–3, Mars, Phobos, Spektrum, Astron and Granat. The Babakin centre, which emerged from Korolev's OKB-1, is solely responsible for planetary probes and science satellites until its associated NPO Lavochkin was identified openly in 1989. Georgi Babakin (1914-1971) established his own OKB in 1965 and at the same time took over the satellite arm of OKB Lavochkin (founded 1937 as an aircraft manufacturer). Lunar and deep space work was transferred from Korolev's OKB-1 in 1965. Sergei Kryukov was chief designer August 1971 to December 1977. Vyacheslav M Kovtunenko then took over (died 11 July 1995, aged 73), succeeded in turn by Stanislav D Kulikov from 1996. Lavochkin was embraced by the Russian Space Agency from 1994. Specialises in autonomous operation on pre-programmed command and control systems with software uplink changes.

Lavochkin is providing the Kupon telecom satellite for Global Information Systems. It provides the bus for the Oko and Prognoz early warning satellites to TsNPO Kometa, and Fregat as a launcher upper stage. Has made industrial connections with companies in the Ukraine and with several ESA industrial consortia with a view to international cooperation and technical deals.

NPO Mashinostroyenia

Current Status
Originates from the Design Bureau Factory No.51 set up in October 1944 as Special Design Group (SKG) under the Ministry of Aviation Industry and headed by Vladimir N Chelomei to develop a naval cruise

missile. Reorganised into OKB-52 in 1955 and from 1958 expanded to ballistic missile development absorbing several major manufacturing plants. Developed anti-missile system 1963–64 until Krunichev dismissed and developed lunar fly-by mission using LK spacecraft and UR-500 Proton. In 1965 the OKB was placed under the administration of the new Ministry of General Machine-building. Vladimir Chelomei's bureau was divided after his death in December 1984 into KB Salyut and NPO Machinostroenye under Gerbert A Yefremou. Almaz radar satellites, and probably others, are built at the Khrunichev factory. It operates the Almaz foreign trade company to market Almaz remote sensing data. Technology acquired in SAR development has been used in development projects for a proposed new generation of remote sensing satellites. Has made a major contribution to the development of military SAR systems and platforms. In 2000 NPO Mashinostroyenia had sales of US$24.2 million, an increase of 46 per cent over 1999. The company states that where once all its business came from the Russian government, now only 10 per cent originates from that source and 75 per cent of its income is generated from work with the Indian government on missiles and rockets.

NPO Molniya

Current Status
By merging several design bureau the Molniya Scientific Production Association was set up in 1976 specifically to develop the Buran reusable spaceplane. As such it was responsible for its aerodynamic design, flew BOR Kosmos subscale missions, and conducted hypersonic vehicle research. Dr Gleb E Lozino-Lozinsky was Director until February 1993. Employment declined from 5,000 in 1992 to 3,500 in 1995 and to 2,800 in 2000. Responsible for proposed MAKS concept whereby a reusable orbiter is carried aloft on the back of an Antonov An-225 from where it is released for flight to orbit. Design teams from NPO Molniya continued to refine spaceplane concepts but have no realistic funding agent for tests. Molniya are responding to renewed interest from the Russian government in cost-effective launch vehicle programmes through enhanced reusability and recovery techniques.

NPO PM (Prikladnoi Mekhaniki)

Current Status
Set up in 1959 as a separate branch of the OKB-1, the M F Reshetnyov Science and Production Association of Applied Mechanics (NPO PM) was dedicated to serial production of ballistic missiles. Production on licence from major missile design bureau.

Since 1961 it switched exclusively to satellites and had been responsible for 900 satellites by the end of 2002 (13 in 1995), including the Express, Gals, Ekran, Gorizont, Raduga and most other major telecom satellites. Designed large military consats offered as a constellation of high-powered geostationary satellites. Its other responsibilities included the Strela 3 military sextets, Glonass and Kosmos navigation and GEO-IK geodetic satellites (the NIIKP Institute of Space Device Engineering provides the payloads). Also responsible for Molniya satellites. It remains a state enterprise, reporting to RKA, and has not yet decided if it will privatise. Employees total about 4,900 in 2000. Its business has been divided equally between civil and military, but the military element is now sharply declining. As a result, it is moving into other areas, such as ground antennas and mobile antenna structures with autonomous power sources. Space vehicle assembly has stabilised and is no longer in decline.

Since January 1994, NPO-PM controls the Gals and Express satellites itself through its Persei subsidiary, instead of relying on the military. TT&C stations in Krasnoyarsk and Gouss Khroustalni allow control over 20° W-154° E GEO. Its Mercury subsidiary offers transponder leasing. From inception NPO PM has been led by Mikhail F Reshetnev (died 26 January 1996) succeeded by his first Deputy Albert Eavrilovich Kozlov. It developed Yangel's R14/SS-5 Skean missile into the Kosmos space launcher. NPO PM made a commercial breakthrough in July 1995 when Eutelsat selected it to provide Sesat (Siberia Europe Satellite). The satellite bus has been adapted for other applications and missions.

SP Korolev Rocket and Space Corporation Energia (RSC Energia)

Current Status
Energia traces its history back to 1946, with the establishment of the Korolev Design Bureau (see separate entry). Set up in 1956 as a separate long-range missile development department within NII-88, it was tasked with launching an artificial earth satellite under the jurisdiction of Sergei Korolev. It quickly became responsible for all manned spacecraft, planetary flights and orbiting space stations. RSC Energia (also known as RKK Energia, after the Russian) is Russia's largest space company, responsible for the design, development and manufacture of the manned and man-related vehicles: Soyuz, Progress, Mir, Buran, and Energia. President Yeltsin signed an order in early 1994 to sell 49 per cent of the company with the government retaining its 51 per cent interest for at least 3 years.

RSC Energia's predecessor, NPO Energia was formed May 1974 by merger of Sergei Korolev's bureau and the Gas Dynamics Laboratory of Valentin Glushko. Korolev's Bureau had been renamed Central Construction Bureau of Experimental Machine Building (TsKBEM) after his death in January 1966. The Gas Dynamics Laboratory left Energia after Glushko's death in January 1989, becoming NPO Energomash (see separate entry). Mir management and foreign cosmonaut missions are handled by Energia, reporting to the Russian Ministry of Industry.

Soyuz and Progress craft are assembled in Energia's KIS integration and test facility, and shipped to Baikonur in rail containers. KIS is also used for Proton stage 4. Energia has developed plans for an advanced manned Mars mission using elements which it would like to test and evaluate on the International Space Station. These plans envisage some level of international co-operation and limited numbers of hardware elements have been built but only small-scale engineering fixtures for test.

Energia negotiates for non-Russian trade deals and the export of Russian space technology and has a broad base of expertise for dealing with foreign companies.

State Research and Production Space Rocket Centre (TsSKB) Progress

Current Status
The Federal State Unitary Enterprise, State Research and Production Space Rocket Center (TsSKB) Progress, was created in 1996 by merging the Central Specialised Design Bureau (CSDB) and the Progress manufacturing plant in Samara. The headquarters are in Moscow, however TsSKB Progress also maintains offices at the Baikonur and Plestsk launch facilities. TsSKB Progress is subordinate to the Russian Federal Space Agency.

Background
The Central Specialised Design Bureau (CSDB), was founded in 1959 as a division of Korolev's OKB-1 Moscow design bureau to provide design support for launcher manufacturing at the Progress factory, where the Soyuz launch vehicles and Foton satellites, among others, are built. It is the parent of KB Foton. Employees, including KB Foton, totalled about 2,000 in 2005. TsSKB has reduced manufacturing levels as a response to fluctuations in demand, and cost saving measures were introduced to eliminate outages of resources and personnel.

TsNII Kometa

Current Status
Kometa is responsible for several high value tactical and strategic space systems, including the ASAT anti-satellite, Oko/Prognoz early warning, Eorsat and Rorsat. TsNII evolved from the KB-1 in the late 1950s which specialised in anti-satellite and ocean reconnaissance and became an independent bureau under A I Savin. It also played a leading role in monitoring the development of US nuclear weapons and in the development of space-based early warning systems. Charged with projecting an enhanced ABM deployment concept. Has shadowed US ABM technologies and tested warhead entry trajectories to evade ABM missiles using space-based sensors. Has evaluated new methods of in-orbit detection, examination and analysis of in-orbit satellites as a simple inspection capability. Some new work on satellite interception has been conducted since the renewed US anti-ballistic missile programme. This has been matched by interest from the Russian government investing new research funds to carry out tests on counter systems.

VNII Elektromekaniki (VNIIEM)

Current Status
Founded in 1941 to build electrical equipment with post-war emphasis on the development of military equipment including missile electronics and guidance systems, VNIIEM is responsible for the Elektro, Resurs-O and Meteor Earth observation satellites. It contributes instruments to other projects, including Almaz and Mir. VNIIEM has diversified into non-space related applications and in 1991 teamed with General Electric to produce x-ray computer tomography scanners for medical purposes. Performing detailed evaluation of large earth environment monitoring platforms similar to those envisaged by NASA's Mission to Planet Earth programmes of the early 1990s. Continues to maintain a close working relationship with ESA. Has worked with several companies in western Europe to determine common specifications for earth observation instruments. VNIIEM employs about 2,300 people.

Ukraine

NPO Yuzhnoye

Current Status
Based at the Yuzhmash factory, Yuzhnoye is the world's largest integrated missile production facility. NPO Yuzhnoye (formerly OKB-586) is a major launcher, missile and satellite design, development and manufacture bureau. Formed in 1954 under the directorship of M K Yangel until his death in 1971, NPO Yuzhnoye has a 15 per cent interest in Sea launch. Origin of the company goes back to 1951 when what was a large automobile factory was transformed into a missile factory in Dniepropetrovsk and it developed and tested the R-1 (SS-1), R-2 (SS-2) and R-5M (SS-3) missiles and was made responsible for serial production. It is responsible for the SL-11/14 Tsyklon and SL-16 Zenit space launchers, the SS-4 Sandal (basis of SL-7), SS-5 Skean (basis of SL-8 Kosmos), SS-9 Scarp (basis of SL-11/14 Tsyklon), SS-18 Satan (now offered as a satellite launcher) and SS-N-5 Sark missiles, the Okean-O/Sich satellites, Tselina global Elints, and about 400 Kosmos and Interkosmos satellites (including those based on the AUOS bus family). It also probably handles the military Rorsats. Employees totalled about 39,000 (including 14,000 on R&D), with 5,800 involved in space work by 1991. Facilities total about 185,000 m², including titanium forging and propellant tank fabrication (also undertaken for other organisations). Test facilities now open for leasing to other Ukrainian companies.

Yuzhmash was producing automobiles and aircraft at the end of the Second World War. In 1951,

Stalin allocated it to Sergei Korolev. In about 1954, Korolev's deputy Mikhail KYangel (25 October 1911 – 25 October 1971) assumed control for production of the SS-6 Sapwood, which formed the basis of the current SL-4 Soyuz and SL-6 Molniya launchers (later switched to the Central specialised Design Bureau). OKB Yangel was also responsible for the manned lunar lander's propulsion system (led by Boris Gubanov, who later became responsible for the Energia launcher), and proposed the R56 manned lunar launcher. NPO Energia allocated it in 1974. These were developed in parallel as the SL-16 Zenit 2. NPO Yuzhnoye was formed out of OKB Yangel in 1974.

United Kingdom

Reaction Engines Ltd

Current Status

Reaction Engines continues its research and development work on space transportation and hypersonic engines and vehicles. UK government financial support for the company's projects has not materialised.

Background

Reaction Engines has been headquartered in Abingdon, UK, since 2001. The company was formed in 1989 to continue research on the RB545 engines and Horizontal Take-Off and Landing (HOTOL) spaceplane that Reaction Engines' founders had begun developing while working for Rolls-Royce. Rolls-Royce cancelled the project in 1988 when the UK government withdrew funding. Reaction Engines employs a small staff, numbering less than 20. The company funds its development efforts through investors, loans and consulting work.

Reaction Engines has continued its development work on advanced space transportation propulsion systems, primarily through its Skylon spaceplace and Sabre engines projects. Skylon and Sabre are effectively the successors to HOTOL and the RB545. The unmanned single stage Skylon is designed to lift a payload of 12 tonnes into Low Earth Orbit (LEO). The spacecraft's hybrid Sabre engines operate in both closed cycle rocket and airbreathing supersonic modes. Reaction Engines' Abingdon location is equipped with a cryogenic engine test facility used to test engine components, especially heat exchangers.

50 per cent funded by the European Union, Reaction Engines has also been conducting studies for the Long-Term Advanced Propulsion Concepts and Technologies (LAPCAT) project. LAPCAT proposes to develop a hypersonic commercial vehicle designed to carry long distance passengers half-way around the globe in two to four hours. The Scimitar Engine, a Sabre derivative, would power the hypersonic vehicle, called the A2. The A2 has some commonality of design with the Skylon, however, it would of course not be required to re-enter the atmosphere, but demands a design that could withstand Mach 5 flights several hours in duration.

Reaction Engines also works in partnership with AEA Technology, and the Bristol, Kingston and York Universities on aerothermodynamics, propulsion, structures and control technologies research and development.

Surrey Satellite Technology Ltd (SSTL)

Current Status

University of Surrey began its micro-satellite research with the UoSAT programme of 1979. UoSAT-1 was launched by Delta 2910 from Vandenberg AFB in October 1981.

SSTL was formed in 1985 by the University of Surrey to support technology transfer to industry and commercial development of research activities, in addition to undertaking industrial contracts. SSTL is located within the Surrey Space Centre supporting 35 post-graduate students and research fellows. The company's expertise includes designing, fabricating and operating low-cost (around GBP2 million each) micro-satellites: UoSAT 1-12, Kitsat 1/2, S80/T, PoSAT 1, HealthSat 2, Cerise, FASat satellites, Thai-Phutt, TiungSat-1, SNAP-1, PicoSAT, Clementine and Tsinghua-1 launched 1984–99. SSTL is located within the purpose-built 1,500 m² Surrey Space Centre, operating as an autonomous business with management and administration separate from the University. There are three varieties of cleanrooms (which can accommodate AIT of five micro-satellites simultaneously): 25 m² class 10,000 instrument assembly, 80 m² flight subsystem/payload assembly and 80 m² flight model AIT. The 300 m² R&D laboratories cover RF and ground station support, OBDH and digital systems, power and sensors, and an MCAD/ECAD design office. SSTL's Mission Operations & Control Groundstation was improved and relocated to CSER in March 1994. The three independent tracking antennas (VHF, UHF, S-band) frequently track as many as eight satellites, supporting more than 110 daily passes.

Research studies cover: power systems; OBCs and data-handling networks; modulation/demodulation schemes; RF transmitters/receivers; signal processing; attitude determination; stabilisation and control; imaging; radiation effects on components; failure-resilient architectures; VLSI subsystem design and store/forward techniques from LEO satellites. CSER is developing a 350 N hybrid motor. Using hydrogen peroxide as oxidiser and a solid fuel, it was successfully tested at Royal Ordnance's Westcott facility in February 1995. Under an ESA contract, SSTL is developing a GPS receiver for providing micro-satellite position and attitude; it will be carried by FASat-Bravo and TMSAT.

The modular multipurpose bus forming the core of the UoSat-12 mini-satellite provides S-band communications and enhanced OBDH for 180 kg payloads. The first will carry RWs, a 40 m resolution Earth camera and propulsion system for orbit control.

SSTL contracts have included supply of UoSAT ground stations to South Korea, Pakistan and Portugal; ESTEC spacecraft studies; mission studies for the Swedish Space Corporation; licensing of UoSAT ground station software; manufacture of flight hardware and launch and integration services on UoSAT missions; and supply of platforms for integration of dedicated payloads for commercial customers Matra and Aerospatiale (S80/T and Cerise). SSTL signed an agreement with Arianespace in August 1988 to act as the prime customer for the Ariane Structure for Auxiliary Payloads (ASAP) project. Of the six spacecraft slots on the first ASAP mission in January 1990, two were filled by UoSAT D/E, and four by micro-satellites from Amsat of North America. As a result, SSTL undertook a number of contracts for integration and launch services for experiments from ESTEC, RAE and Volunteers in Technical Assistance (VITA, US). In July 1990, SSTL signed a further Ariane agreement for UoSAT 5 launch aboard ERS 1's vehicle in 1991. The SSTL-built Kitsat 1 was launched in 1992 (SSTL supplied parts for the South Koreans to build 1993's Kitsat 2 based on SSTL's design), PoSAT 1 in 1993 and Healthsat 2 in 1993. A GBP3 million contract was signed in May 1994 with the Chilean Air Force for the FASat-Alfa satellite, ground station and training. The GBP1.3 million contract was signed in January 1996 for the identical FASat-Bravo. A GBP3 million contract was signed in November 1995 with the Thai MicroSatellite Company for the first national Thai micro-satellite, TMSAT, in 1997. At the beginning of 1996, contracts in progress totalled GBP11 million. A GBP1.35 million contract was signed in March 1996 for a communications payload aboard UoSAT 2. Photographs of SSTL-built satellites are included in the South Korea, Portugal, UK, and Oscar entries.

The GBP6 million demonstration mission of SSTL's multipurpose MiniBus platform will be carried free on the fourth and last test launch of Russia's Rockot. MiniBus can support 50 to 180 kg LEO payloads, including SSTL standard module boxes and a payload bay for heavier instruments or those with nadir-pointing apertures. Payload data will be distributed via the Internet.

During 1996, SSTL became part of a team selected by ESA to conduct a Lunar orbiter Mission study including the Technical University of Munich and the Swedish Institute of Space Physics. Known as the Lunar Academic Research Satellite (LunARSat) the study sought to encourage young pupils in schools and universities in a belief that they could participate in useful science. In October 1998, SSTL announced that it had broken into China's tightly controlled internal satellite industry with formation of a collaborative venture company in Beijing to develop micro-satellites. The first GBP3 million contract had been signed between SSTL and Tsinghua for a 50 kg micro-satellite to be called Tsinghua-1, the demonstrator for a constellation of seven micro-satellites which will provide daily worldwide high-resolution imagery for disaster monitoring. Launched 29 June 2000 by Kosmos.

First in a constellation of five disaster monitoring satellites, AlSAT-1 was successfully launched by Kosmos 3M from the Plesetsk Cosmodrome on 28 November 2002. This is Algeria's first satellite, joined in orbit by four more disaster monitoring satellites in 2003. SSTL has achieved a placing in the Sunday Times Tech Track 100, companies with the fastest growing technology. It has seen an increase in sales from GBP5.4 million in 2000 to GBP12.2 million in 2002.

Specifications

UoSat Bus

Configuration: octagonal prism, 1.1 m (diameter) × 80 cm (height), accommodating standard SSTL module boxes
Mass: 300 kg on-orbit (180 kg payload)
Power supply: 9 GaAs panels provide 150 W orbit average processed power; 21 Ah Ni/Cd battery supports eclipse operations
OBCs: 186 and 386 computer operating under a multitasking system provide data handling. Specialist computing resources provided by TMS320C32 and Transputer-T800 systems. Standard SSTL TM/TC protocol. Inherent in the store/forward architecture is a 72 h autonomy period with automatic fault reporting and data downloading
Attitude control: 3-axis by MWs, magnetorquers and cold gas thrusters
Communications: data rates of 9.6-76.8 kbit/s on 400 MHz using redundant UHF and 1 Mbit/s on redundant S-band. 9.6 kbit/s VHF command uplink
TT&C: by SSTL's Mission Control and Operations Centre

MiniSat Bus

Total mass: 50 kg, including launcher fittings
Configuration: 600 × 345 × 345 mm box-shaped body with solar panels on four sides, 6 m boom deployed from top face for gravity gradient stabilisation. Each housekeeping system or payload is housed in a standard module box, mass-produced by a computer-controlled process. The boxes are then stacked to form the main spacecraft structure; their number can be varied to meet mission requirements
Power supply: 4 GaAs panels totalling 4,992 cm² providing 18 W (orbit average processed); 6 Ah Ni/Cd battery supporting eclipse operations
OBC/Data Handling: the 80C186 payload experiment computer is used for high-level data manipulation. 80C188 version provides secondary OBC. On UoSAT 5 the payload Transputers are employed for high-performance data processing and attitude control
Attitude control: Earth pointing is maintained by combination of passive gravity gradient stabilisation and active magnetorquing; Sun/Earth horizon sensors and 3-axis fluxgate magnetometer with ±2 nT resolution over ±64 μT range provide attitude information. Gravity gradient supplied by 3 kg tip mass on 6 m boom deployed from top face, magnetorquing by six 150-turn aluminium wire coils (one on each spacecraft facet)
Communications: data rate of 9.6 or 38.4 kbit/s, 3-channel receiver (2 communications, 1 command), redundant transmitter switchable to 1/2.5/10 W RF output, 4-element turnstile VHF antenna, one monopole UHF antenna.

United States

AeroAstro Corporation

Current Status

AeroAstro was formed in 1988 to design and build small low-cost space systems and components. There are about 65 personnel. Its first major contract was spacecraft development, test and payload integration and ground station for the Los Alamos National Laboratory Alexis satellite. MIT's 1996 HETE satellite and ground station was developed under a similar contract, awarded in February 1991. The company built Boston University's Terriers satellite to study the upper atmosphere and ionosphere. Terriers was launched in May 1999 and is part of the Student Explorer Demonstration

Initiative sponsored by the Universities Space Research Association. AeroAstro is planning a low-cost small satellite launcher. In September 2001, AeroAstro was contracted to develop the DoD Space Test Programme (STP) Satellite Mission 1 (STPSat-1) for the USAF Space Command. A successful Preliminary Design Review was held in July 2002. STPSat-1 will offer more than 200 W of orbit power with pointing accuracy better than 0.1°. Launch is planned for 2006 on Delta 4.

AeroAstro provided the 1 Gbit solid-state digital mass memory system for the NASA/Los Alamos MOXE x-ray detector on Russia's Spektrum-X satellite in 1998. The same mission carries a 0.4 Gbit memory for the Canadian Space Agency's Euvita experiment.

The company also designs and fabricates spacecraft RF equipment, torque coils, Sun sensors (0.1° accuracy over 10°), solar panels, integrated control systems, radiation-hardened flight computers, software, ground stations and other elements for small satellite projects. It provided Ni/Cd batteries for Sweden's 1992 Freja satellite under a USD340,000 contract.

Co-developed with the USAF, AeroAstro's Bitsy micro-satellite provides an autonomous three-axis satellite weighing only 1 kg. Its simple, single circuit board design is easily tailored to specific mission requirements, such as store/forward, asset tracking, and hosting remote sensing instruments. Cost is USD100,000 for the basic satellite, or less than USD1 million for a turnkey system, including launch. Bitsy can be operated by the customer's PC or Macintosh-based ground station, or via AeroAstro on Internet. The debut Bitsy will determine its attitude using a star camera, image the Earth and transmit the compressed image.

Specifications

Availability: 6 months after order
Mass: 1.0 kg
Size: 5 × 15 × 15 cm with attached solar panel
AOCS: 0.5 mrad (2σ) using star camera, 50 mrad (2σ) attitude control by cold gas thrusters, 1 mrad/s stability (target, orbit dependent)
Power: 2 to 10W (increase optional), 8V unregulated; 4Wh lithium ion battery
Communications: 8 Mbit minimum, up to 1 Gbit, 5 Mbit/s payload raw data max output; UHF or S-band, 2 kbit/s up/down (10 to 100 kbit/s down option)

The Aerospace Corporation

Current Status

As of early 2008, Aerospace Corporation currently has 14 satellites in orbit, which are a combination of observational, communication, weather and military orientated models. The company also provided support on various other programmes, such as Atlas II and III, Delta II and the Titan projects. The Aerospace Corporation has preliminary plans for a further 10 satellites to be launched over the next 5 years.

Background

The Aerospace Corporation was founded in 1960 and operates as a private, non-profit federally funded research and development centre dedicated to national security needs, specialising in space systems and related technologies, and providing general system engineering and integration. It provided detailed analytical surveys on defence procurement requirements to meet perceived threats from anti-satellite systems and was instrumental in conducting background research for the Saint programme. Played a major role in the development of the Navstar GPS and was awarded the Robert J. Collier Award for this in 1992. The Aerospace Corporation provides the Space and Missile Systems Center of the Air Force Space Command with risk reduction analysis in launch vehicle operations. Technology development for picosatellites, remote sensing and hyperspectral imaging. Aerospace provides technical oversight for the national security programme, including launchers and satellites such as Milstar, DSCS, GPS, DMSP, DSP and future missile launch detection. For example, it wrote the DSCS II programme plan and provided technical oversight from 1965's study contract through the continuing operations phase. The Aerospace Corporation has full technical support facilities and tracking and control facilities for civilian and military satellite constellations.

Ball Aerospace & Technologies Corporation

Current Status

Ball Aerospace has been awarded a system architecture contract for the next generation GOES-R system planned for launch in 2012. Ball will develop end-to-end configuration for the satellites and associated ground system for the NOAA. A subsidiary of The Ball Corporation, BATC reported 1998 sales of USD363 million. The company was founded in 1956 and comprises the aerospace systems and telecommunications products divisions. It was prime contractor for NASA's ERBS satellite, SDI's Relay Mirror Experiment, and the CRRES science satellite. Ball was awarded the USD46 million contract in 1992 to build the Geosat Follow-On radar altimeter satellite for the US Navy. It provided technical assistance to Spain's small satellite programme under a four-year agreement signed in October 1992.

In 1996, Ball Aerospace Australia was set up with operations in Brisbane, Canberra, Sydney, Melbourne and Adelaide, Australia. BATC built two instruments for the Chandra X-ray observatory. Past programmes include Brilliant Pebbles (1988), Compton GRO (1991), Chandra X-ray observatory (1999), COSTAR for HST (1993), CRRES (1990), Darpasat (1994), Geosat Follow-on (1998), GHRS (1990), HST (1990), IRAS (1983), MTI (2000), NICMOS for HST (1997), OSO (1962), OSSE (1991), QuickSCAT (1999), Radarsat (1995), RME (1990), SIR-C (1994), SBUV-2 (1996), SRTM (2000), STIS (1997), and SWAS (1998). Current programmes include CALIPSO, CloudSat, Deep Impact, ICESat, JWST, Kepler, Mars Rovers, NPOESS, QuickBird, SAGE III, SIRTF and THEL.

BATC employs 2,200 people and reported sales of USD363 million for 2000. Ball Corporation reported sales of USD3.9 billion in 2002.

Bigelow Aerospace

Current Status

On 28 June 2007 the Genesis II pathfinder space habitat structure was successfully launched aboard a Dnepr rocket from Dombarovsky (ISC Kosmotras Yasny Cosmodrome in the Orenburg region of Russia) the same location from which Genesis I was launched. Both of the experimental pathfinder spacecraft have expandable structures, alongside their solar arrays. Bigelow Aerospace was able to maintain communication links with each spacecraft soon after launch and orbital insertion. Both Genesis I and II are one-third scale models of the company's ultimate full-scale flexible space habitat structures. Bigelow Aerospace has booked ISC Kosmotras launch services for 2008 to continue their space structure testing.

Background

Founded in 1999, Bigelow Aerospace is headquartered in Las Vegas, Nevada, with other facilities in Houston and Washington DC. As of

Ball built the Combined Radiation Release Effects Satellite as a joint NASA/USAF endeavour (NASA) 1047493

2007, billionaire Robert Bigelow had invested over USD100 million of his personal fortune in the company and claims to be committed to investing up to USD500 million by 2015.

Bigelow Aerospace is developing expandable space habitat structures, primarily for commercial use. Part of the company's mission objective is to develop Low Earth Orbit (LEO) manned or unmanned space complexes for commercial experimentation and habitation purposes. Bigelow Aerospace launched Genesis I, an experimental pathfinder, with an anticipated lifespan between three to 13 years, on 12 July 2006. A small NASA experiment regarding weightlessness, called the Genebox, was attached to the internal structure of Genesis I.

Bigelow Aerospace plans to launch its Galaxy module in late 2008. The Sundancer and BA 330, sized to a human habitable scale, will follow by 2012 at the latest. By 2015, with USD500 million invested, Bigelow Aerospace aspires for a fully operational and habitable LEO spacecraft, which was originally announced in 2000 by Robert Bigelow.

The company plans to achieve between six to 10 pathfinder demonstrator spacecraft flight missions by 2010 and, depending on whether the space transportation is fully operational, Bigelow Aerospace may be ready to deploy its first habitable structure in 2010/2012, three to five years ahead of schedule. Furthermore the company will be providing ground infrastructure for low cost space missions to commercial and civil communities.

Bigelow Aerospace entered into an agreement with Lockheed Martin in September 2006 to study the possibility of developing the Atlas V launch vehicle into a human qualified space rocket that is capable of launching a passenger spacecraft.

Bigelow Aerospace is the sponsor for 'America's Space Prize'. The prize is a USD50 million award to any US spacecraft developer who can deliver a reusable spacecraft with the capability to fly at least five passengers to a Bigelow space structure and successfully link with that habitat module by January 2010. The winner must demonstrate two successful missions with at least five passengers aboard the transfer spacecraft.

Blue Origin

Current Status

Blue Origin is a privately held, Seattle-based company started in September 2000 by Jeff Bezos, founder of the Amazon online retail empire. Its initial goal is to develop a reusable sub-orbital vehicle for space tourism.

Background

Blue Origin (the company's formal name is Blue Operations LLC) has disclosed little about its activities. The company says that its "initial research efforts are focused on reusable liquid propulsion systems, low cost operations, life support, abort systems and human factors. We are currently working to develop a crewed, suborbital launch system that emphasizes safety and low cost of operations."

In 2005, Blue Origin announced plans to establish a spaceport close to Van Horn, Texas. At the same time, it was revealed that the company was planning a Vertical-Take-off, Vertical Landing (VTVL) vehicle fuelled by kerosene and hydrogen peroxide. (However, Blue Origin is also recruiting engineers with LOX experience.) It has also been reported that the company has already tested a turbojet-powered VTVL vehicle, probably to help develop flight-control systems.

Late in 2005, Blue Origin was preparing to move into a larger facility in Kent, south of Seattle – suited to manufacturing with some provision for propulsion testing – and was in the process of increasing its staff from 40 to 70–100 people. Some reports suggest that an early rocket demonstrator could be flying by late 2006.

Boeing Integrated Defense Systems

Current Status

Boeing Integrated Defense Systems (IDS) is one of the largest and most prosperous defence and space companies in the world. It runs, or contributes to, some of the most complex defence and civil space programmes in existence. Aside from its programme involvement in the International Space Station (ISS) and the Space Shuttle, IDS is developing the GPS Block IIF and Wideband Gapfiller (also known as the Wideband Global Satcom – WGS) satellites; competing with Lockheed Martin for the Transformational Satellite Communications (TSAT) programme; and pioneering the world's first in-orbit autonomous robotic satellite servicing system, validating the concept using the NextSat and Autonomous Space Transport Robotic Operations (ASTRO) satellites of IDS' Orbital Express programme.

Background

Boeing Integrated Defense Systems (IDS) is based in St. Louis, Missouri. One of five Boeing Company Business Units, IDS employs about 72,000 personnel worldwide and achieved revenues in 2006 of USD32.4 billion. IDS elements specific to space systems are NASA Systems division based at Houston, Texas, Space and Intelligence Systems division at Seal Beach, California and Launch and Satellite Systems division also located at Seal Beach.

The Boeing Company's space industry experience stretches back to the beginning of the space age and US manned spaceflight. Boeing managed the overall systems integration for the Apollo project and also produced lunar orbiters, the astronauts' lunar rover and the Apollo-Saturn V launch vehicle's first stage S-1C engine. The company has been building interplanetary exploration spacecraft and satellites since the 1960s and 1970s.

In 1996, Boeing merged with Rockwell International Corporation's aerospace and defense businesses, which at that time became the Boeing North American Inc subsidiary. The acquisition provided Boeing with Rockwell's Space Shuttle operations and satellite manufacturing expertise, particularly GPS satellite contracts, as well as Rocketdyne's engine production facilities. Nine years later, in 2005, Boeing quit manufacturing rocket propulsion systems and sold its Rocketdyne Propulsion and Power division to United Technologies Corporation (UTC), which was then folded into UTC's Pratt & Whitney Business Unit.

In 1997, Boeing acquired the McDonnell Douglas Corporation; the McDonnell and Douglas companies having merged thirty years before, in 1967. McDonnell Douglas, had, since before the two companies' merger, developed substantial missile and rocket expertise, marking among their technology developments the Nike and the Thor, precursor to the Delta launch vehicle.

In 2000, Boeing acquired the Hughes Space and Communications Company, along with Hughes Electron Dynamics; Hughes' Spectrolab Inc; and Hughes' HRL laboratory shares. The Hughes divisions were merged into an entity that was called at that time, Boeing Satellite Systems. Boeing Satellite Systems later became Boeing Integrated Defense Systems' Satellite Development Center, the manufacturing site for the current Boeing Space and Intelligence Systems division.

Since 2006, Boeing IDS has been comprised of four Business Centers: Advanced Systems; Network and Space Systems; Precision Engagement and Mobility Systems; and Support Systems. The company's space-related work is primarily housed in the Network and Space Systems Center, with development projects also originating from IDS' Advanced Systems division and the independent Phantom Works Business Unit. The Network and Space Systems Business Center is comprised of five divisions: Combat Systems; Command, Control and Communications (C3) Networks; Missile Defense Systems; Space and Intelligence Systems; and Space Exploration.

Boeing has been the prime contractor for the International Space Station (ISS) since 1993; the contract, valued at USD5.63 billion, was signed with NASA on 13 January 1995. At that time Boeing's Defense and Space Group led the ISS work, supported by the Missiles and Space Division. IDS also subcontracts with the United Space Alliance (USA) to provide support for Space Shuttle operations; Boeing has supported Space Shuttle operations, maintenance, upgrades and payload integration since 1981.

Boeing owns stakes in three joint venture launch services companies: United Launch Alliance (ULA), founded in 2006; United Space Alliance (USA), founded in 1995; and Sea Launch, also founded in 1995. Both ULA and USA were jointly established by Boeing and Lockheed Martin. ULA offers Boeing Delta and Lockheed Atlas series launch vehicle services to US government customers; USA is NASA's Space Shuttle prime contractor. Sea Launch is an international joint venture established to primarily serve the commercial launch market. See separate entries for further details.

E'Prime Aerospace Corporation (EPAC)

Current Status

In March 2007, E'Prime's Board of Directors authorised an internal investigation into the company and its founders, who made alleged statements that E'Prime Aerospace Corporation's (EPAC's) ESM-9 Peacekeeper-derived motor technology was eliminated from the Strategic Arms Reduction Treaty (START). The question remains as to whether Peacekeeper technology can be used for commercial launch purposes.

The company has struggled to obtain funding and in 2007 ordered a reverse stock split in the hope that the move would raise further funds to allow a launch to take place to lift up to four M Star Global Communications satellites into GEO. M Star is a company with which EPAC has signed an Memorandum of Understanding (MoU). It remains to be seen whether this customer is itself a profitable communications company and whether EPAC will ever be capable of delivering any M-Star satellites into orbit.

EPAC and Space Propulsion Systems Incorporated (SPS), in June 2007, signed a strategic alliance agreement allowing EPAC to use SPS's propulsion systems aboard EPAC's Eagle launch vehicle series. The companies anticipate that SPS's solid-fueled Micro-Fuel Cell (MFC) motors will decrease launch-related negative environmental impacts and enable a EPAC to provide enhanced customised launch services tailored to the customer's launch requirements.

Background

EPAC was established in 1987 and is headquartered in Titusville, Florida. The company was founded to develop a commercial satellite launch vehicle derived from Peacekeeper ICBM technology.

EPAC claims to have developed the most advanced launchers on the market. Although EPAC has been developing its Eagle launch vehicle series for 20 years, no commercial launches of the rocket family have taken place as of 2007. EPAC was hampered in part by the 1990 Strategic Arms Reduction Treaty (START), which prohibited the practical use of Peacekeeper first stage technology. In 1998, the company's founders claimed that EPAC had changed the rocket's technology and customer base sufficiently for its launchers to be eliminated from START: a statement that, as of early 2007, was under investigation.

EPAC is proposing to launch from either Kennedy Space Center or the Mid-Atlantic Regional Spaceport (MARS) on Wallops Island, and Kodiak Launch Complex. Formerly it proposed to launch from Ascension Island or Vandenberg Air Force Base.

EPAC's Eagle S-series solid propellant orbital launchers employ Peacekeeper stage 1 motors. The company is in some stage of developing its Eagle S-I through Eagle S-VII versions, as well as the smaller Eagle and Eaglet rockets. EPAC claims that all apart from the two smaller versions are capable of lifting payloads into GTO, and all but the S-VII are capable of launching Polar orbiting payloads. Eagle and Eaglet would launch LEO or Polar orbiting payloads.

The company's debut Loft 1 was launched on 17 November 1988 from Canaveral's complex 47 to 4.6 km altitude; it carried a 15.5 kg educational payload.

General Dynamics Information Systems and Technology Group (Spectrum Astro, Inc)

Current Status

In May 2007 General Dynamics Advanced Information Systems (GDAIS) released an updated version of its Geographical Information Systems (GIS) based Global Maritime Boundaries Database (GMBD). In this latest version of GMBD a list of international maritime contacts and co-ordinate lists for boundary agreements is included and maritime areas and boundary lines among others have been updated. GDAIS was awarded a NASA Goddard Spaceflight contract in May 2007 for the Landsat Data Continuity Mission (LDCM) spacecraft accommodation study. Under the LDCM study GDAIS is responsible for highlighting methods of achieving all mission objectives, as well as providing design concepts that will incorporate all scientific instruments on board the Landsat spacecraft. Contract offers for building the LDCM observatory will begin once the study is completed.

In April 2007 the Near Field Infrared Experiment (NFIRE) satellite, a Missile Defense Agency mission, was launched successfully from Wallops Island, Virginia. GDAIS was responsible for building the satellite and payload integration. Operational on-orbit support for one year will be provided by GDAIS.

GDAIS is working together with NASA to develop the Gamma-ray Large Area Space Telescope (GLAST). GLAST is also an international collaboration of governmental agencies and educational institutions from France, Germany, Italy, Japan, Sweden and the US. The terms of the contract include design and fabrication of the GLAST observatory, payload integration, observatory-level testing, and on-orbit observatory check-out. The GLAST launch is scheduled for 2008.

GDAIS is developing the spacecraft bus for Earth imaging satellite GeoEye-1, which is scheduled to be launched in 2008. Also GDAIS is providing payload integration, environmental testing and on-orbit checkout services for GeoEye-1.

In 2006 General Dynamics (GD) C4 Systems was awarded a contract by the Air Force Headquarters Cryptologic Systems Group of San Antonio, Texas to modernise encryption and decryption for satellite telemetry tracking and control. The contract is part of the US Air Force's Aerospace Vehicle Equipment-Increment One programme. The encryption/decryption products will come in two configurations – the embedded solution and the end cryptographic unit. Delivery and product demonstrations will commence in mid 2008. Also in 2006 a joint contract was awarded to GD C4 Systems and Northrop Gruman Space Technology by the Space and Missiles Systems Centre, Los Angeles Air Force Base. Both companies will research and develop the Alternative Infrared Satellite System (AIRSS) as an alternative to the Space Based Infrared High (SBIRS-High) GEO-3 satellite. The GD C4 Systems contract is worth USD23.4 million. A two day AIRSS System Requirements Review (SRR) was recently completed in front of industry and governmental participates. GDAIS was awarded a USD6.5 million contract modification to perform system design work. A system design review is planned and work is set to complete mid-2008.

Background

GD Corporation headquarters is located, Falls Church, Virginia, US, employing approximately 82,900 people under Chairman and Chief Executive Officer Nicholas D. Chabraja.

Spectrum Astro Inc was acquired by GD in 2004 and became part of GD C4 Systems unit under the Information Technology Group. Historically, Spectrum Astro acted as a prime contractor in building spacecraft buses. For example, Spectrum Astro manufactured three Miniature Sensor Technology Integration satellites (MSTI) for the US Air Force in order to develop detection and tracking capabilities for Ballistic Missile Defence (BMD) systems. MSTI-1, MSTI-2 and MSTI-3 were launched in 1992, 1994 and 1996 respectively.

Prior to its acquisition by GD, Spectrum Astro was awarded the Rapid Spacecraft Acquisition contract by the NASA Goddard Space Flight Centre to build the Coriolis Spacecraft Bus and provide payload integration services. The mission was jointly sponsored by the Department of Defense (DoD) space test programme and the US Navy. The Coriolis spacecraft was launched in 2003. Other contributions from Spectrum Astro include, Mightysat 2.1 (a US Air Force Phillips Laboratory multimission satellite) and New Millenium DS-1 (a NASA Jet Propulsion Laboratory, California Institute of Technology mission satellite which was launched in 1998 and retired in 2001). Spectrum Astro were also contracted for the Reuven Ramaty High Energy Solar Spectroscopic Imager (launched in 2002) and Swift (launched in 2004 with an operational life of seven years) satellites, both of which were NASA Space Goddard Flight Centre missions. Spectrum Astro also designed the SA-200 High Performance (HP), SA-200 Standard (S), SA-200 Basic (B), which are now part of GD family of products.

There are four subdivisions under GD: aerospace; combat systems; information systems and technology group; and marine systems. Under the information technology group a further four business units are subdivided into advanced information systems, C4 Systems, information technology and UK Ltd.

Five key areas fall within GDAIS (formerly GTE Government Systems Corporation, electronic systems division): homeland security; integrated space systems; intelligence, surveillance and reconnaissance; information assurance; and maritime combat systems.

Under integrated space systems, GDAIS capabilities include spacecraft development and integration, space mission system integration, space computing, and geospatial intelligence and imaging. Spacecraft development and integration incorporates GDAIS Accommodating Bus approach, which allows flexibility for specific payloads and missions. For example GDAIS (continuing a contract awarded under Spectrum Astro in 2001) was responsible for building the Communications/Navigation Outage Forecasting System (C/NOFS) satellite. The C/NOFS is a joint project between DoD Space Test Program and the Air Force Research Laboratory, which is set be launched onboard a Pegasus XL Launch Vehicle at the Ronald Reagan BMD Test Site, Kwajalein Atoll, Republic of the Marshall Islands.

Space mission integration provides command and control systems for the defence, intelligence and civil sectors. GDAIS provides imagery exploitation support system and information warfare planning capability services, as well as communication subsystems, electronic solutions, and configurable subsystems. Also with regard to restricted space programmes GDAIS offers specialised payload integration.

GDAIS space computing business provides radio frequency, electro-optical and radar processors as well as restricted payloads, plus any planning, targeting and assessment tools. Furthermore, GDAIS designs, manufactures and installs various subsystem electronics. Subsystem components include AstroRT, modular power control electronics, charge control unit, and payload and attitude control interface, forming part of GDAIS capabilities in command and data handling as well as data acquisition. The space qualified redundant array of independent disk drives, a data storage solution, is also provided through the business.

Other product examples include: Astro Accel (a sensitive accelerometer) and the GPS constellation crosslink transponders, which provide a communication link between satellites within the GPS system. They come in two models – the crosslink transponder subsystem and the crosslink transponder data unit. Under the GPS system GDAIS also provides spaceborne receivers that acquire and track GPS signals. For example the Monarch precise positioning service receiver, Viceroy GPS spaceborne receiver and the Viceroy geosynchronous receiver all of which facilitate real time, position and velocity. The Starlight space-based configurable subsystem is capable of acting as a single module transponder in multi-applications for next generation spacecrafts. Tracking, telemetry and control transponders include the Cassini X-band transponder, small deep space transponder and the multimode S-Band transceiver. Other GDAIS space communication products are the HRT150 high rate transmitter for delivering large amounts of data, an uplink/downlink communications board and an X-band solid state power amplifier that has a range of 7.8 to 8.8 GHz.

Under geospatial intelligence and imaging, GDAIS capabilities include image processing, as well as data storage, mining and archiving. Products include; Geoworx, Oceanworx, Eagle Vision (EV) programs, Geospatial Knowledge Discovery (GKD), Multimedia Analysis and Archive System, and Mobile Integrated Geospatial Intelligence System. Additionally, GDAIS is developing and providing sensor technology such as GPS Anomaly Monitoring Equipment Suite (GAMES) and Electro-Optical / Infrared Remote (EO/IR) sensing systems.

Geoworx is an image-processing commercial software and produces orthorectified images as well as advanced workflow management. The software supports commercial sensors such as French Systeme Pour l'Observation La Terre (SPOT), Landsat-4, -5 and -7, and Canadian SAR Satellite (RADAR-1), among others. GDAIS also provides upgraded Tadpole notebooks that support Geoworx software as well as Environmental Systems Research Institute's Arcview among other applications. Data can be uploaded into EV systems as well as downloaded from EV using a docked Tadpole notebook. Oceanworx provides environmental coastal information from remote sensing satellites, offering spatial grids and maps of costal wind and wave fields, as well as spatial location of ships and coastal fronts.

The EV ground stations one through five enable direct access to broad area and multispectral imagery supporting topographic applications, in particular military operations. For example EV-2 is a transportable image processing ground station, which can also be loaded onto C-130H or C-141 transport aircraft where satellites such as SPOT and RADAR-1 (through the EV-2 imagery downlink) enable direct access for commercial imaging to military commanders. Operation and management support as well training and system upgrades are also supplied.

GKD is a data mining system and provides a problem solving technique for numerical, text and geospatial datasets. By placing one dataset on top of another, factual data can be extracted and, for example, interpreted into patterns and trends or used to identify search locations.

GAMES provide military and civilians users with GPS warning reports, as well as monitoring GPS signals for jamming and interference. GAMES sensors are usually strategically placed, for example at airfields and ports, so that information is as accurate as possible. All files are catalogued and archived for post mission analysis.

EO/IR incorporates spectral sensing, phase diversity imaging, polarimetry sensing, EO/IR modelling, design and analysis, and experiment design and execution. GDAIS is developing new algorithms for multi-spectral and hyperspectral systems to extract information in respect to spectral sensing. With regard to phase diversity imaging GDAIS is working to exploit data as well as conduct field experiments by performing modelling and simulation analysis. Polarimetry sensing involves identifying the shape of an object that is either built or natural when, for example, the background is highly cluttered. EO/IR system modelling, design and analysis incorporate two models. Firstly the Images-Based Sensor Model (IBSM) and integrated tools, which provides support for image intensified, multispectral, broadband and ladar sensors. Secondly the IBSM extension Hyperspectral System Image Model (HySIM), which ensures data is collected to meet quality standards. GDAIS plans to advance EO/IR technologies including their applications.

GD C4 Systems headquarters is located in Arizona, US and employs approximately 11,000 people worldwide under President Chris Marzilli. Customers are mainly from the defence sector and government, as well as select US or international commercial entities. GD C4 Systems merged with General Dynamics Decision Systems (GDDS) in 2003, (that also included the 2001 acquisition of Motorola's Integrated Information Systems Group) to form a single business unit. GD C4 Systems capabilities include command and control; communications and networking; information assurance; intelligence, surveillance and reconnaissance; platform integration; RF networking; ruggedised computing; satellite/wireless and ground-based communication; space systems and services; switching systems; training and simulation.

Under space systems and services, GD C4 Systems provides satellite communication products and services, as well as information assurance for space systems. GD C4 Systems range of services to the customer also includes offering satellite operations. Satellite communications incorporates the Motorola satellite series 9505A, Motorola satellite series pager 9501, Reachback inverse multiplexer, Reachback Voyager multichannel data and voice multiplexer, and a tactical secure fly away kit for on the move communications (comprised of Iridium accessories), plus various other satellite communications accessories such as batteries, data kits, chargers, antennas and security modules. With regard to the Motorola satellite series 9505A which is a portable satellite secure phone, US government subscribes can attach Type 1 Iridium Security Module (ISM) with module compatible STU-IIIs for secure communications. Customers include defence and federal, state and local governments where satellite communication products have been utilised in disaster and humanitarian relief, emergency management, homeland security, combat search and rescue and asset tracking using GPS.

Information assurance incorporates space-based encryption services and products for satellite-based systems and communications. Products include Type-1 certified encryptors with corresponding ground encryptors, such as the AVE Optimised

SWap, which is CMMI level 5 certified; KGR GOE (supports the AVE), and KG-STE (supports the KG T and KG R products). Other encryption services include the Sectéra Black Digital Interface (BDI) terminal. In April 2006 the National Security Agency certified the software upgrade for the security module through their test services facility. The Sectéra BDI terminal connected to an ISDN adaptor provides secure communications that has data rates of 128 Kb/s over landline digital networks with ATI compatible phones and mobile phones over Irridium, Inmarsat, Globalstar and Thuraya networks. Also available is the Sectéra remote access security solution for voice and data communications over the public switched telephone network. Any communications that are top secret go through the Secure Communication Interoperability Protocol (SCIP) signalling which is certified to protect secure conversations and transfer of data using NSA certified Type 1 encryption.

GD SATCOM Technologies (part of GD C4 Systems), specialises in satellite and wireless communications. Leading brand names such as VertexRSI, Prodelin and Gabriel are among GD SATCOM Technologies products. VertexRSI products are for fixed and mobile satellite communications supplying for example SATCOM antennas, controls, electronics systems and microwave components. Prodelin provides antennas for the very small aperture terminal as well as terrestrial microwave antenna products. Gabriel specialises in backhaul wireless products that have a variety of applications such as government agencies and scientific communities.

Kelly Space and Technology

Current Status

Kelly Space and Technology Inc (KST) is a technology development company which was incorporated in April 1993 under the laws of the State of California. The company's current focus is the development of the 'Eclipse' family of low-cost, reusable, commercial orbital and suborbital space launch vehicles which are intended to make access to space both affordable and routine. KST has focused initial development activities on solving this problem by offering what could become a significantly more operationally efficient launch system at substantially lower cost, with plans to reduce costs even further by incrementally incorporating new technology into improved operational vehicles as it becomes available. The Eclipse launch vehicles are designed to address this need in the near-term by combining the best of available, proven technology to mitigate development risk and improve operational flexibility while defining an implementation plan that minimises non-recurring investment to ensure sound financial performance.

The Eclipse concept utilises a Boeing 747 to tow the Eclipse winged launch vehicle from a conventional runway to the launch altitude of approximately 12 km. At this altitude, the Eclipse vehicle's rocket engine is ignited, the tow line is released and the vehicle climbs to the payload separation altitude of approximately 120 km. Following ejection from the Eclipse vehicle, the upper stages are ignited to deliver the payload to the specified destination while the Eclipse vehicle descends as a glider. For final descent from approximately 9 km, the Eclipse deploys two turboprop engines located aft to support powered approach and landing.

Importantly, KST has mitigated its development risk by employing only existing, proven technology in its launch vehicle configurations. No technological advances or breakthroughs are required.

The Eclipse tow launch concept is currently in a flight demonstration programme being conducted by KST under a United States Air Force Phase II Small Business Innovation Research (SBIR) contract. Under this cost-share contract KST will demonstrate the key features of the Eclipse launch concept using a QF-106 jet powered supersonic aircraft towed behind a C141A aircraft. The USAF has provided the Company with two QF 106 aircraft under the SBIR contract to serve as scaled representatives of the Eclipse launch vehicle. The USAF Flight Test Centre at Edwards AFB is supporting flight test operations by providing a C141A aircraft for use as the tow plane. Flight tests are being supported by and conducted at NASA's Dryden Flight Research Center (DFRC) at Edwards Air Force Base. Flight test activities continue using the QF-106 aircraft throughout the remaining useful life of the aircraft.

The Eclipse family of reusable launch vehicles consists of three basic, sequentially developed, flight vehicle configurations: Eclipse Sprint, Eclipse Express and Eclipse Astroliner. Each vehicle supports the development of the next and is placed into commercial use following testing to service incrementally larger segments of the space launch market. Adopting this approach enables each vehicle to serve initially as a test vehicle and to then generate revenues concurrent with the development of the next vehicle.

The first vehicle in the family, a small launch vehicle named Eclipse Sprint, will be constructed to serve both as a development test bed for the Eclipse Astroliner and as a commercial suborbital launch vehicle to generate early revenue. The Eclipse Sprint vehicle will be used to develop flight operations procedures to ensure safety, evaluate and demonstrate various recovery options, gain operational flight experience with the Eclipse tow launch method, support test activities of the Eclipse Astroliner, and generate early revenues. Eclipse Space Lines Inc, will be established and licensed by KST to commercially provide Eclipse Sprint suborbital launch services. The Eclipse Sprint vehicle, a small-scale rocket powered glider which is representative of the Eclipse Astroliner, will support successive Eclipse development activities and provide suborbital launches to service approximately half of the existing sounding rocket market. Sounding rockets provide payloads with up to 20 minutes of exposure to the space environment supporting microgravity experiments, Earth observation, atmospheric sampling and numerous US Department of Defense applications. The Eclipse Sprint service offers significant price advantages with reliability not provided by any of the expendable rockets currently servicing this market.

The Eclipse Express is a follow-on and enhanced version of the Eclipse Sprint vehicle which incorporates upper stages that could be ejected with a payload to service the balance of the existing suborbital market as well as being capable of launching small satellite payloads (under 90 kg) into low-earth orbit. Development of the Eclipse Express will be initiated as a follow-on to Eclipse Sprint operations in order to escalate the development of the Eclipse Astroliner. The Eclipse Sprint vehicle will serve as the test vehicle to support Eclipse Astroliner development activities. As currently envisioned, Eclipse Express would be developed under the same financing approach and business model as the Eclipse Sprint and is expected to be deployed before flight testing of the Eclipse Astroliner.

The flagship of KST's family of launch vehicles will be the Eclipse Astroliner. This vehicle will be capable of launching payloads of up to 1,600 kg into a 465 km polar orbit or up to 2,300 kg into a 465 km equatorial orbit. Commercial operations are projected to commence around mid-2000. The Eclipse Astroliner recoverable flight vehicle consists of the airframe, propulsion systems, thermal protection systems, avionics packages (including software), attitude control system, ordnance initiation system, and landing gear. The Eclipse Astroliner airframe will be a low to mid-wing delta monoplane with a single vertical stabiliser. The airframe will include an articulated nose door similar to a 747 Freighter or C-5 aircraft for both ground loading of the payload and payload deployment at altitude. Access doors servicing the payload and all other units requiring inspection and maintenance are also incorporated into the airframe.

In 1998, Kelly successfully demonstrated the Eclipse concept under a USAF small business innovative research (SBIR) programme. In September 1999, Kelly received a US$2.1 million NASA contract to perform a long-term transportation architecture study. On 7 January 2000 NASA awarded a risk reduction contract worth US$3.1 million.

In August 2000, NASA awarded Kelly a US$3.1 million contract for risk reduction and analysis of possible shuttle replacement contenders for initial service in 2010.

On 23 January 2001, Kelly Space and Technology Inc and Vought Aircraft Industries Inc announced a teaming agreement to refine 2nd generation RLV architectures and further develop the Eclipse Astroliner concept. Studies have responded to NASA's future launch vehicle studies enshrined within the newly re-launched Space Launch Initiative. The architecture has been redesigned to match potential opportunities.

Kistler Aerospace Corporation

Current Status

Kistler Aerospace Corporation is proposing the development of a two-stage, fully reusable launch vehicle, designated the K-1. The vehicle will use Russian engines designed and built by NPO Trud (previously the Kuznetsov design bureau) for the aborted Soviet manned lunar programme.

The company completed the phase 1 preliminary design in September 1995 and has secured the right of first refusal for all NK33 and NK43 engines which were built for the N-1 manned lunar landing vehicle. Phase 2 development and testing was completed in 1996 and Kistler is currently undertaking phase 3 development and testing. Phase 3 calls for completion of the detailed design of the launch vehicle, the establishment of an advisory group comprising contractors and Kistler supporters, finalisation of the contract team, fund raising, assembly and test of the first K-1 vehicle, conduction of flight tests and the initial commercial operations. Two launch sites are being established at the Woomera range in South Australia.

Each of the two stages of the K-1 vehicle would be recovered using a combination of parachutes and airbags. Ideally, both stages would automatically return to the launch site, the second stage after initially entering orbit with the payload.

When complete a fleet of five K-1 vehicles is proposed, with each vehicle designed for 100 flights. Each vehicle would have a turnaround time of nine days, with each stage being returned to the launch site within 24 hours of launch. The launch vehicle would be assembled horizontally, transported and erected on a launch stand using a wheeled mobile launcher. A Memorandum of Understanding with the State of Nevada to establish a launch site was signed as part of the phase 1 work for the vehicle.

Kistler is forecasting a price of US$1,700 per lb (or about US$770 per kg) to a 28.5°/km LEO. Since 1998, KAC has been re-financing and technical development work has slowed.

NASA has awarded Kistler a US$135 million contract under the Space Launch Initiative for 13 embedded technology evaluations on the first four flights of the K-1.

Redirecting its efforts from a reusable launch vehicle to a low cost conventional launcher, deferring reusability to a later development.

Lockheed Martin Space Systems Company

Current Status

Under a memorandum of understanding, Lockheed Martin and the United Space Alliance (USA) are to pursue NASA's Johnson's Space Center Facilities Development and Operations Contract (FDOC) announced on 8 January 2008. FDOC will incorporate mission support operations (which is currently executed by Lockheed Martin) and space programme operations (executed by USA); it provides engineering, maintenance, training, software development, flight design and planning services. Additionally, for International Space Station (ISS) and Space Shuttle human spaceflight missions, FDOC will provide real-time operations services.

On 14 January 2008, Lockheed Martin announced it had been awarded an USD40.4 million contract to provide the Aegis Ballistic Missile Defense (BMD) capability for *JS Myoko*. *JS Myoko* is the third Japanese destroyer to be equipped with the weapon system. Work is underway to install the Aegis BMD system in the second destroyer *JS Chokai*.

Lockheed Martin announced it will be building the Gravity Recovery And Interior Laboratory (GRAIL) spacecraft, which is part of NASA's Discovery Programme on 19 December. The GRAIL spacecraft is a lunar mission and is scheduled for launch in 2011.

On 14 December 2007, Lockheed Martin delivered to Vandenburg Air Force Base the Defense Meteorological Satellite Programme (DMSP) F-18 Block 5D-3 spacecraft. Three other DMSP satellites have yet to be launched and remain at Lockheed Martin's Sunnyvale facilities. Operational use of the Lockheed Martin DMSP F-17 Block 5D-3 spacecraft was handed over to the National Oceanic and Atmospheric Administration (NOAA) by Lockheed

Martin in a ceremony held at the Air Force Weather Agency at Offutt Air Force Base in Omaha, Nebraska on 29 January 2007. The spacecraft was launched from Vandenburg Air Force base in 2006.

Dedicated to support NASA's Project Orion and Constellation Programme, a new Exploration Development Laboratory (EDL) funded by Lockheed Martin, United Launch Alliance and Honeywell was unveiled in an opening ceremony held on 12 December 2007. The laboratory, adjacent to NASA's Johnson's Space Centre in Houston is part of an integrated EDL network consisting of three other facilities in Colorado, Arizona and Virginia, which will enable early performance tests for the next-generation human spaceflight vehicle Orion.

On 3 December 2007 Lockheed Martin announced it is working on the Demand-Based Geospatial intelligence programme with the National Geospatial Intelligence Agency. This will deal with the timely delivery of maps and imagery to worldwide users.

The targets and countermeasures single integration capability, which deals with targets for BMD systems, was inaugurated at newly built facilities in Courtland, Alabama, on 15 October 2007 by Lockheed Martin and the US Missile Defense Agency (MDA). It will also be home to the assembly and integration of the Multiple Kill Vehicle (MKV) programme. Lockheed Martin is the prime contractor for targets and countermeasures, a deal which was awarded in 2003.

A USD178 million Geostationary Operational Environmental Satellites (GOES)-R Series contract was awarded to Lockheed Martin by NASA in co-ordination with the National Oceanic and Atmospheric Administration on 24 September 2007. Under the contract Lockheed Martin will provide one Solar Ultra Violet Imager with three options for additional instruments for the GOES-R Series, scheduled for launch in 2014. Work will be conducted by Lockheed Martin's Solar and Astrophysics Laboratory (LMSAL) at the Space Systems Advanced Technology Centre in Palo Alto. LMSAL has previously provided the Solar X-Ray Imager for GOES-N which was launched in 2006. Lockheed Martin was selected to build the Geostationary Lightning Mapper (GLM) for the GOES-R Series in a USD96.7 million contract awarded on 19 December 2007. The GLM will provide new insight into storm evolution. The University of Alabama are part of Lockheed Martin's GLM team.

Under a task order, Lockheed Martin will provide management and infrastructure framework regarding current and future missions for the Multi-Mission Satellite Operations Centre (MMSOC), Schriever Air Force Base. The task order awarded on 18 September 2007 is valued at USD2.4 million and will increase incrementally for the next two years.

A critical subsystem test for the MKV payload of the ground-based mid-course BMD was conducted at Edwards Air Force National Hover Test facility on 27 August 2007. The performance of the divert and attitude control system, built by Pratt and Whitney Rocketdyne, met with the MDA performance objectives. The MKV will have a carrier system holding smaller hit-to-kill vehicles which can be used for cluster targets. Using the carrier's onboard sensors and ground-based radars, the MKV will destroy itself on impact with the target. Lockheed Martin was awarded the USD27 million contract for the MKV in 2004; originally called the Miniature Kill Vehicle. Pratt and Whitney Rocketdyne is Lockheed Martin's subcontractor for the MKV payload system.

The Phoenix Mars Lander, the first in the NASA Scout programme, was successfully launched on 4 August 2007 at Cape Canaveral (CC) aboard a Delta II 7925 rocket. The spacecraft, based upon the cancelled 2001 Mars Surveyor Lander, was built by Lockheed Martin.

NASA has awarded Lockheed Martin a USD25 million contract for ground mission support for the Stardust-NExT (New Exploration of Tempel) programme, which will use the Stardust spacecraft, built by Lockheed Martin and launched in 1999, to fly by the Temp 1 comet on 14 Feb 2011.

Lockheed Martin dedicated the new USD3 million Milsatcom Centre of Excellence on 10 July 2007. The Rockville facility will consolidate all ground segment activities as well as conduct software development and testing. Preparing for the next follow on Advanced Extremely High Frequency (AEHF) satellites, the Milstar five satellite constellation underwent on-orbit reconfiguration on 16 July 2007, which was conducted jointly by the US Air Force and Lockheed Martin.

B-Sat-3a was launched on 14 August 2007 aboard an Ariane 5 ECA launch vehicle from Kourou, French Guiana. B-Sat-3a is a Japanese GEO communication satellite built by Lockheed Martin for the Broadcasting Satellite System Corporation using an A2100A bus. The spacecraft was declared operational after successful on-orbit checkouts on 1 October 2007. JCSAT-11, a telecommunication satellite, was also built by Lockheed Martin for JSAT Corporation using an A2100AX spacecraft bus. An attempt to launch JCSAT-11 on 5 September 2007 aboard a Proton M rocket from Baikonur Cosmodrome, Kazakhstan was unsuccessful. In light of the launch failure JSAT Corporation, on 16 September 2007, placed another order for the A2100AX bus from Lockheed Martin. The spacecraft will be called JCSAT-12 and the company has agreed with Arianespace to conduct the launch aboard an Ariane 5 rocket from Guiana Space Centre during mid 2009.

The US Navy and MDA certified Aegis BMD Weapon System version 3.6 equipped on the USS Lake Erie (CG 70) intercepted two simultaneous targets on 26 April 2007. The Aegis BMD Weapon System guided two standard missiles to firstly intercept outside the Earth's atmosphere a short-range non-separating ballistic missile target and secondly a low altitude cruise missile intercept. During a mid-course flight on 22 June 2007 the Aegis BMD System intercepted and destroyed a ballistic missile with a separating re-entry vehicle outside the Earth's atmosphere. This marks the ninth successful test out of 11 intercept attempts. The test also demonstrated the exchange of tracking data between Terminal High Altitude Area Defense (THAAD) and the Ageis BMD system. This was also the ninth intercept test for the Standard Missile 3 (SM-3), produced by Raytheon who are leading an industry team of Alliant Techsystems, Boeing and Aerojet.

The SM-3 Block IA which will be deployed on Aegis cruisers is designed to intercept short to mid-course range BM threats. A further exo-atmosphere test was completed on 6 November 2007 when the SM-3, guided by the Aegis BMD Weapon System 3.6 successfully intercepted two short-range non separating BM. On 17 December 2007, Japanese Kongo Class Destroyers equipped with Lockheed Martin's Aegis BMD system successfully guided a Raytheon SM-3 Block IA to intercept, in space, a separating medium range BM target. THAAD also participated in the test.

Lockheed Martin is working on the Solar Dynamics Observatory (SDO) with Stanford University, who have built together the Helioseismic Magnetic Imager, which was delivered to NASA's Goddard Space Flight Centre (GSFC) on 15 November 2007. Lockheed Martin has also built the Atmospheric Imaging Assembly (AIA), which will provide views of the solar Corona. The AIA was delivered to NASA's GSFC on 17 December 2007 to prepare SDO for a December 2008 launch.

A production review was completed for the Mobile User Objective System (MUOS) on 20 June 2007. MUOS is managed by the US Navy's programme Executive Office for Space Systems. Following the review the MUOS team can begin developing two MUOS satellites as well as the ground system. Lockheed Martin is prime contractor for the MUOS programme.

A baseline integrated system test to prepare the first Advanced Extremely High Frequency (AEHF) satellite for launch was conducted between 27 July and 23 October 2007, allowing a move to the next critical environment test phase. On 18 June 2007 the propulsion structure and payload module was integrated with the AEHF A2100 spacecraft bus built by Lockheed Martin; it is scheduled for launch in 2008. Lockheed Martin has completed the core propulsion module for the second AEHF satellite which is scheduled for a 2009 launch.

The STS-117 Mission Crew over 11 days, in a mission to the ISS launched on the Space Shuttle Atlantis 8 June 2007, delivered, installed and connected the third of four pairs of solar arrays, as well as the second Solar Array Rotary Joint, built by Lockheed Martin.

Lockheed Martin was awarded a USD47 million contract on 6 June 2007 by the US Air Force Space Command to continue to support the Space Command's weather receiving stations, using the MARK IV-B system to provide near-real time weather, a component to Global Information Grid.

Software developed by Lockheed Martin for the spacecraft GEO-1 has been completed three months ahead of schedule on 5 June 2007. GEO-1 satellite is the first satellite that will support the High Space-Based Infra-red System (SBIRS-High). Thermal vacuum testing on the GEO-1 payload was successfully completed between 11 March and 15 June 2007, which was conducted at Northrop Grumman's Azusa facilities, who are the payload subcontractors. Thermal vacuum testing was conducted inside the Dual Entry Large Thermal Altitude Chamber of SBIRS-High GEO-1 spacecraft bus between 16 January and 2 February at the company's Sunnyvale California facilities. The performance of the core structure in extreme temperatures was successfully tested. GEO-1 is scheduled for launch in 2008. Lockheed Martin is also developing GEO-2, which is ready for propulsion subsystem integration after a successful pyroshock test. GEO-2 is scheduled for a late 2009 launch.

The High Elliptical Orbit (HEO-1) sensor has successfully completed on-orbit checkouts and is due to become operational in 2008. A critical end-to-end test, simulating a command and control test between SBIRS-High ground segments with GEO-1 spacecraft, which will prepare GEO-1 for launch and subsequent on-orbit operations, was successfully completed on 1 August 2007. The GEO payload by Northrop Grumman arrived at Sunnyvale for spacecraft integration on 6 August 2007. Integration was completed on 10 September 2007. The ground-based information processing segment of SBIRS has been operational since 2001.

Broadcasting satellite Astra 1L, built by Lockheed Martin for SES ASTRA, was successfully launched from Kourou on 4 May 2007 aboard an Ariane 5ECA rocket. ASTRA 1L was declared operational on 11 July 2007 after successful on-orbit checkouts. AMC-14 is currently being built by Lockheed Martin for SES AMERICOM. It is scheduled for launch in the first quarter of 2008. Sirius 4 is a communications satellite built by Lockheed Martin for SES Sirius and was successfully launched on 17 November 2007 aboard a Proton M/Breeze rocket from Baiknour Cosmodrome, Kazakhstan. It will provide across Africa, Europe and Scandinavia direct-to-home broadcasting plus interactive services.

On 27 October 2007, in an exo-atmosphere test, THAAD successfully detected, tracked and intercepted a unitary target in a third flight test at Pacific Missile Range Facility (PMRF) on Kauai Hawaii. The THAAD weapon system also successfully intercepted a unitary target in a mid endo-atmosphere test on 6 April 2007 at PMRF. This was the second flight test conducted both by Lockheed Martin and US MDA. A simulated Aegis BMD ship was used to communicate via datalink to the THAAD fire and control communication unit using a satellite link with the Navy's Space and Naval Warfare Systems Command. The first flight test of the THAAD weapon system was conducted on 26 January 2007 at the PMRF, which successfully intercepted a unitary target in the high endo-atmosphere. A low endo-atmosphere THAAD test conducted on 27 June 2007 successfully tested the interceptor at White Sands Missile Range. Previous flight tests, which began in 2005, were held at White Sands Missile Range, where three successful tests were conducted, including a unitary target interception in July 2006.

Lockheed Martin with Northorp Grumman is developing the Space Radar system. The space segment involves a constellation of nine radar satellites, which will provide synthetic radar imagery, surface moving target indication, open ocean surveillance and high resolution terrain information as well as geospatial intelligence products. Both Lockheed Martin and Northrop Grumman were awarded in 2004 a Phase A concept and development USD220 million contract for the Space Based Radar, which was renamed Space Radar following a restructure in January 2005. An integrated baseline review for the Space Radar programme was completed ahead of schedule in April 2007. Decision to proceed to the design phase of the Space Radar will take place at the latest, FY2009.

On 21 March 2007 the eighth and final GPS Block IIR (designated GPS IIR-M) satellite was delivered to the US Air Force by Lockheed Martin, prime contractors for the GPS IIR modernisation programme. The fourth GPS IIR-M satellite (GPS IIR-17M), was launched on 17 October 2007 aboard a Delta 2 rocket. The satellite was declared operational on 5 November 2007 after successful on-orbit checkouts conducted by the US Air Force and Lockheed Martin. GPS IIR-18M was successfully launched aboard a Delta 2 rocket at CC on 20 December 2007, which marks the fifth GPS IIR-M satellite developed by Lockheed Martin. Completing on-orbit checks in three days GPS IIR-18M was declared operational on 2 January 2008 for civil and military users. Lockheed Martin is modernising the GPS constellation on behalf of the Global Positioning Systems Wing, which is a joint service effort directed by the US Air Force and managed at the Space and Missile Systems Centre (SMSC), Air Force Space Command, Los Angeles Air Force Base, California.

The new GPS IIR-M series will include a modernised antenna, two new military signals and a second civil signal. A USD6 million contract was awarded to Lockheed Martin earlier in 2007 to develop a payload that will transmit an additional frequency, located at 1176.45 MHz. ITT Corporation is Lockheed Martin's navigation payload provider, delivering the SV 09 demonstration payload on 20 December 2007, two months ahead of schedule and only nine months after the contract was awarded. The SV 09 payload is ready for integration with the GPS IIR-M spacecraft series at Lockheed Martin's facilities at Valley Forge, Philadelphia. Lockheed Martin is leading the GPS III space segment team to build the next generation GPS and ITT Corporation and General Dynamics are part of the team. The GPS III space segment team is under a Phase A GPS III concept development contract. The system design review for GPS III was completed on 5 April 2007. A single contractor is yet to be selected; this is scheduled for announcement during the first quarter of 2008.

A new facility was allocated to Lockheed Martin on 7 February 2007, to support the Transformational Satellite Communications Systems (TSAT) missions' operations system programme. The facility will be used to test various systems before being put to operational use. The TSAT Mission Operations System (TMOS) is part of a 10 year USD2 billion 2006 contract awarded to Lockheed Martin to link the global information grid with TSAT, which is scheduled to become operational in 2014. The space segment to TSAT includes, space laser communications (which will have five satellites in constellation) ground terminals for radio users and TMOS, where a single network mode will consist of satellite, ground and user segments. A network design review was held five months prior to the system design review for the TMOS architecture, which was completed successfully on 4 June 2007.

A design review for the TSAT Space Segment (TSAT-SS) system was successfully completed on 7 May 2007 by the TSAT-SS team, consisting of Lockheed Martin and Northrop Grumman. Government representatives and key stakeholders attended the design review. This is part of a USD514 million systems definition and risk reduction 2004 contract which could lead to a further multi-billion contract for the satellite development and production phase. This is set to be awarded to a single contractor in the first quarter of 2008, subject to the Defense Acquisition Space Board approval in the last quarter of 2007. A proposal detailing the design and development of TSAT was jointly submitted on 30 July 2007 by Lockheed Martin and the Northrop Grumman team, who are together competing to act as prime contractor for the TSAT-SS. Juniper Networks and ViaSat are also part of the team. The Military Satellite Communications Systems Wing at the SMSC, Los Angeles Air Force Base in California will be awarding the contract.

On 5 February 2007 Lockheed Martin delivered a new missile warning system to Cheyenne Mountain Directorate Command Centre as part of an Integrated Space Command and Control (ISC2) contract. Forty Combatant Commander's Integrated and Control System Spiral 2 upgrades have been delivered and will be used by US Strategic Command, US Northern Command and North American Aerospace Defence commanders.

A USD36.5 million two year contract was awarded by Inmarsat to Lockheed Martin to install and deliver three network control centres and gateways for public terrestrial networks on 15 January 2007. Implementation of the terminals, which will operate via existing Inmarsat satellites one to four and dual-mode Global System for Mobile Communications as well as satellite handheld terminals, is planned for 2008.

Background

The Lockheed Martin Space Systems Company (SSC) is headquartered in Littleton, Colorado. SSC represents about 20 per cent of Lockheed's total business; in 2006, SSC achieved USD7.9 billion in sales. Lockheed's satellite enterprise produced the highest yields, representing 66 per cent of SSC's sales, with Strategic and Defensive Missile Systems, and Space Transportation following, at 18 and 16 per cent, respectively. Lockheed Martin SSC employed about 16,320 personnel as of 2006.

Lockheed Martin, via its heritage companies, has been involved in the space industry since the early days of missile, rocket and satellite technology development, as well as the dawn of human spaceflight. The Martin and Lockheed companies, as well as other companies which Lockheed Martin later absorbed, were responsible for the Vanguard rocket project, which successfully launched the US' second satellite in 1958; the Pershing and Bold Orion missiles; the Titan I ICBM, the forerunner to the Titan launch vehicle; the Scout and Atlas launch vehicles; and contributing to the Mercury, Gemini, and Apollo projects.

The Lockheed and Martin Marietta companies merged in 1995, the Martin and Marietta organisations having combined 34 years previously, in 1961. With the 1995 merger, Lockheed Martin became one of the largest space, defence and information technology companies in the world. The company now has four core Business Areas: Aeronautics; Electronic Systems; Information Systems and Global Services; and Space Systems, which of course, houses Lockheed's primary space industry efforts.

The Lockheed Missiles & Space Company, became with the 1995 merger, the Lockheed Martin Missiles & Space Company, a part of Lockheed Martin's Space & Strategic Missiles sector. Lockheed's Astronautics division was folded into its Space Systems and Missile Systems divisions at the end of 1990. Lockheed Martin Missiles & Space is now the Sunnyvale Operations facility within the larger Lockheed Martin Space Systems Company (SSC) structure. SSC is comprised of six operating units: the Advanced Technology Center; Commercial Space; Military Space; NASA and Civil Space Systems; Missile Defense Systems; and Submarine Ballistic Missile Systems.

Lockheed Martin's Commercial Space Group, since the early 1990s, has manufactured the A2100 satellite bus. The platform continues to be widely used for commercial communications satellites.

SSC is equal partners with the Boeing Company in two joint ventures: the United Space Alliance (USA), established in 1995; and the United Launch Alliance (ULA), founded in 2006, to provide Atlas V and Delta series launch vehicle services. In 2006, Lockheed Martin divested its interests in the US-Russian International Launch Services (ILS) joint venture. ILS continues to provide Proton launch vehicle services without SSC involvement.

SSC develops and manufactures satellite communications, navigation, remote sensing and space science systems for the defence, civil and commercial markets, as well as missile defence systems and launch vehicles. SSC was the systems integrator and prime contractor for NASA's Hubble Space Telescope. Responsible to NASA for Hubble's structure and support, SSC designed the shroud that encloses the telescope assembly and the scientific instruments.

Lockheed Martin built the Focal Plane Package (FPP) which resides on the Solar Optical Telescope that is onboard the Hinode (Sunrise) satellite, formerly known as Solar-B, which was launched on 23 September 2006. Onboard instruments have provided detailed views of the Sun giving new insights regarding changes to the Sun's magnetic fields. Hinode images were released by NASA on 21 March 2007. Hinode is an international collobration between ESA, JAXA, NASA and the Science and Technology Facilities Council (STFC). Lockheed Martin also provided the Extreme Ultraviolet Imager (EUVI) within the Sun Earth Connection Coronal and Heliospheric Investigation (SECCHI) for the NASA Solar TErrestrial RElations Observatory (STEREO) mission. This was the third mission in NASA's Solar Terrestrial Probes (STP) programme. Two STEREO spacecrafts were launched in 2006 aboard a Delta II rocket on different trajectories, one ahead of Earth's orbit and one behind for a two year mission to study coral mass injections on the surface of the Sun. The first 3D images from the SECCHI instruments, EUVI and 3D counterparts were taken on 17–27 March, 2007.

SSC also built the Iridium satellites for the Iridium LEO system; developed and manufactures the silica fibre insulation tiles for the Shuttle Orbiters; and assembles the Space Shuttle's External Tank. Lockheed has been responsible for more military satellites than any other US or European company. It is prime contractor for the Milstar satellite constellation; the GPS IIR Block; the Space Based Infra-red System (SBIRS); and, with Northrop Grumman Space Technology, the AEHF programme.

The SBIRS contract was originally awarded to Lockheed Martin in 1996 by the US Air Force to develop five geosynchronous satellites, a ground system and production of two HEO payloads, to replace the Defense Support Programme. Deployment for SBIRS was originally scheduled for 2004 but following a 25 per cent increase in costs and delays in the schedule, a Nunn McCurdy breach was triggered in December 2001 and a review was implemented. A certification was issued in May 2002 by the Under Secretary of Defense for Acquisition, Technology, and Logistics (USD AT&L). Revised schedules for the launch of GEO-1 and -2 and the procurement of GEO-3 to -5 were implemented. However, in May 2004 a second Nunn McCurdy breach occurred, which was signed by USD AT&L in a letter to Congress in June 2004. A 15 per cent unit increase was reported. The allotted timescale to deliver GEO-1 and -2 had passed by one year adding to further cost delays. The contract was also extended beyond FY2010. Production and development costs for SBIRS were revised again, including the addition of GEO-3 to -5 production costs causing a third breach to occur in December 2005. Total costs of the programme were estimated at USD10.4 billion. The programme was restructured again with contract modifications to include three GEO satellites at cost of USD3.5 billion plus two additional HEO sensors, although it is envisioned four GEO satellites are required to meet operational needs. An option to procure a fourth GEO satellite is provided for, however procurement may be dependant upon the next Defense Executive Review.

The AEHF is part of a USD2.6 billion contract awarded in 2001 to both Lockheed Martin and Northrop Grumman by the SMSC, Los Angeles Air Force Base, California. The contract includes two AEHF satellites the mission control segment as well as the upgrade of the Department of Defense (DoD) Milstar satellite ground segment, payloads and other components, digital processor and RF equipment. In 2006 Lockheed Martin was awarded a USD491 million contract by the US Air Force to build the third AEHF spacecraft. Lockheed Martin is also contracted to provide ground support for existing and new AEHF terminals.

Lockheed Martin was awarded USD619 million contract to produce fire unit fielding and support equipment for the THAAD system. Forty eight interceptors, six launchers and two fire control and communications units are included in the contract. The THAAD system is scheduled to be fielded in 2009. Lockheed Martin's Camden, Arkansas operations facilities will be building the THAAD launcher and the fire control communications unit.

Lockheed Martin was awarded USD980 million 2006 contract by Naval Sea Systems Command for the design and testing of the AEGIS BMD, which involves computer programme development. Deployment of the Aegis BMD in Aegis cruisers and destroyers is scheduled for 2012.

In 2006, NASA selected Lockheed Martin Space Systems Company (SSC) as the Orion Crew Exploration Vehicle's (see separate entry) prime contractor. The contract was initially valued at USD3.9 billion.

The DoD awarded SSC USD2.1 billion in 2004 for the manufacture of the first two satellites and the ground system for the Mobile User Objective System (MUOS). The constellation, managed by the US Navy, could comprise up to five satellites; SSC has the option to supply a further three under the current contract. If selected, SSC could net up to USD3.26 billion for the five satellite and ground station delivery. SSC expects to orbit the first MUOS satellite in 2010.

Lockheed Martin Space Systems Company – Michoud Operations

NASA's Michoud Assembly Facility is located in New Orleans, Louisiana; Lockheed Martin SSC – Michoud Operations assembles the Space Shuttle's External Tanks (ET) at this location. SSC also engages in other spacecraft assembly activities at Michoud, and future uses of the site will include Orion assembly.

Now called SSC's Michoud Operations, Martin Marietta's Manned Space Systems was formed in 1973 to design, assemble and test External Tanks (ET) for NASA's Space Shuttle. It was originally part of Martin Marietta's Information and Technology Services Sector. As of 2006, Michoud staff totalled about 2,200. The first external tank, the main propulsion test article, was delivered to NASA on 7 September 1977. Michoud is one of the world's largest manufacturing plants – spanning 174,000 m^2 under a single roof. The facility also has a port with deep water access for barging the ETs to Florida. Three ocean-going barges are available for the five-day journey.

Loral Space & Communications

Current Status

Loral Space & Communications Inc, Loral Skynet's (see separate entry) parent company, and the Canadian Public Sector Pension Investment Board

(PSP Investments) announced in December 2006, their intention to acquire Telesat (see separate entry) from Bell Canada Enterprises (BCE) for CAD3.42 billion. Once the acquisition is completed in mid-2007, Loral Space & Communications will hold a 64 per cent interest, and PSP a 36 per cent interest and majority voting control (according to Canadian law) in Telesat. To complete the deal, Loral Skynet will then merge its assets with Telesat to create the world's fourth largest geostationary communications satellite operator. The new entity will continue operations under the new corporate name, 'Telenet', and will be headquartered in Ottawa, Canada.

Background
Loral Space & Communications Inc, headquartered in New York City, was spun off and became a separate entity from the Loral Corporation in April 1996, as a result of the sale of Loral Corporation's defence electronics and systems integration businesses to the Lockheed Martin Corporation. Many of the Loral Corporation's businesses, one year later, separated again from Lockheed Martin, to become L-3 Communications (see separate entry). Loral Space & Communications is a publicly traded company listed on the NASDAQ, reporting an FY2006 combined revenue of USD797 million.

Loral Space and Communications is the parent company of both Space Systems/Loral and Loral Skynet (see separate entries), which form Loral Space & Communications' satellite manufacturing and satellite communications services and operations businesses, respectively. Loral Skynet will merge, in 2007, with Canada's Telesat to become Telenet, a new Canadian company part-owned by Loral Space and Communications.

On 15 July 2003, Loral Space & Communications announced that it had filed for reorganisation under Chapter 11 of US bankruptcy law. Formerly registered as a Bermuda Company, Loral Space & Communications is now domiciled in Delaware following its Chapter 11 declaration and subsequent restructure. At the time of its Chapter 11 announcement, Loral Space & Communications reached a definitive agreement to sell Loral Skynet's six North American telecommunications satellites to Intelsat Ltd for up to USD1.1 billion in cash. Loral expected to use USD959 million of this to pay off its outstanding secured bank debt. Under its reorganisation plan, Loral Space & Communications would continue to own and operate Space Systems/Loral and Loral Skynet.

On 22 September 2003, Loral Skynet declared the Telstar 4 satellite, launched in 1995, a total loss following a short circuit in the primary power bus three days earlier. Nearly all Telstar 4 customers were shifted to the Telstar 5, 6 and 7 satellites. Telstar 4 was insured for USD141 million and under the agreement to sell the six satellites covering North America to Intelsat Ltd, the purchase price was adjusted less the receipt of the insured sum.

Loral Space & Communications emerged from Chapter 11 on 21 November 2005, after successful reorganisation.

Affiliate Companies
Loral Skynet currently holds a 56 per cent interest in XTAR, along with the Spanish partner company, Hisdesat Servicios Estrategicos SA. XTAR owns and operates the XTAR-EUR communications satellite, and also leases capacity on the Hisdesat-owned Spainsat. Following Loral Skynet's transformation into Telenet, XTAR ownership interests will transfer to Loral Space & Communications.

As of December 2006, Loral owned 1,168,934 Globalstar (see separate entry) shares, following Globalstar's November 2006 Initial Public Offering (IPO). Loral's shares represent about one-seventh of all the initial shares offered by Globalstar. Loral has agreed not to divest 70 per cent of its shares until at least 180 days following Globalstar's IPO. Since Globalstar could be in danger of ceasing its satellite services, it is uncertain whether Loral will continue its investment in the company, should Globalstar's in-orbit back-up satellite plan fail. Globalstar launched four back-up satellites in May 2007, and will launch a further four before 2007's end, in order to continue services to its customers. It plans to begin launch of a second generation constellation in 2009.

Loral has gradually divested its interests in Satmex; it once held a 49 per cent investment in the company but as of December 2006, held only 1.3 per cent. Loral Space & Communications will, however, retain end of life satellite transponder use rights on Satmex 5 and 6.

Microcosm Inc

Current Status
Microcosm continues work on the Scorpius expendable launch vehicle family, which includes the Sprite and Exodus orbital vehicles (see separate entry). Funding for rocket development has originated mainly from the US Air Force Research Laboratory (AFRL), the Ballistic Missile Defense Organization (BMDO) and NASA. The goal is to provide an affordable, rapidly available launch vehicle that will meet USAF responsive space requirements. Sprite's (360 kg to LEO) launch cost projection is USD3 million, and Exodus (5,900 kg to LEO) is slated to cost USD9 million.

Microcosm successfully launched the sub-orbital Scorpius sounding rocket versions, the SR-S and SR-XM in 1999 and 2001 from White Sands Missile Range. The latest version, the SR-M sounding rocket, is expected to launch in 2007.

Background
Established in 1984, Microcosm developed a pressure-fed LOX/kerosene 22.2 kN engine with a claimed manufacturing cost, at the time, of less than USD5,000 each, excluding the injector. A 200-second engine test run was successful on 20 November 1995. Microcosm proposed the first generation of Scorpius family of vehicles based on the same engine: the SR-S sounding rocket ; the SR-1 sounding rocket; the Liberty Light Lift; and the Exodus Medium Lift. Development was halted in the late 1990s due to funding problems and a surfeit of systems in the launch vehicle market. Development was rejuvenated with the USAF Falcon programme, however Microcosm was unsuccessful in advancing to Falcon's second phase. Presently, Microcosm's work is associated with the USAF Launch Vehicle Technology (LVT) programme.

Microcosm also Provides mission operating software, analysis and design together with hybrid engineering solutions. The technology development conducted during the formative evolution of this concept was fed across into other technology applications the company is using to develop new and efficient versions of the propulsion system.

Microcosm founded the Scorpius Space Launch Company in 1999 to provide Scorpius launch services. The companies are co-located in Hawthorne California.

Northrop Grumman Space Technology

Current Status
Northrop Grumman acquired TRW in December 2002 and absorbed part of the company into what is now Northrop Grumman Space Technology (NGST), while other TRW business areas were merged into Northrop's Information Technology and Mission Systems sectors.

TRW was selected by NASA in September 2002 to build the James Webb Space Telescope at an initial contract value of USD824.8 million. NGST inherited the project and NASA expects to launch the Hubble successor in 2013. The James Webb telescope's projected total cost is estimated at USD4.5 billion.

Background
Based in Redondo Beach, California, Northrop Grumman Space Technology (NGST) employs roughly 9,300 personnel. The company reported gross profits of USD5,488 million at 2006's year end. NGST, with its TRW legacy, has engaged in space technology development for 50 years. TRW was responsible for developing Pioneer 1, launched in 1958; it was the first commercially-built US satellite. Among other past achievements, TRW also developed and built the bipropellant Lunar Module Descent Engine (LMDE) that was responsible for landing 12 astronauts on the Moon from 1969–1972.

NGST develops and manufactures rocket engines, mono and bipropellant propulsion systems, communications satellites, Earth observation, early warning and surveillance satellites, and space observatories, as well as developing missile defence and directed energy systems and technologies. NGST is the prime contractor for the James Webb Space Telescope, and also for the Chandra X-Ray Observatory. It built the US Defense Support Program (DSP) satellites and is developing, for the US Missile Defense Agency (MDA), the Space Tracking and Surveillance System (STSS), formerly known as the SBIRS-Low programme, the first two satellites of which are set to launch in 2007. According to a 2005 US Government Accountability Office (GAO) report, SBSS programme costs between 2002 and 2011 have swelled to USD4.5 billion.

NGST also develops gel propellant propulsion, micropropulsion and electronic propulsion systems such as Hall effect thrusters and pulsed plasma thrusters. The company is one of NASA's Space Launch Initiative (SLI) contractors, developing liquid oxygen/ethanol and liquid oxygen/liquid hydrogen thrusters as well as the TR107 main engine.

Propulsion test facilities are centred at NGST's 8 km^2 Capistrano Test Site (CTS), 108 km south of Redondo Beach. CTS has been in operation since 1963; it houses a number of facilities and is equipped for thermal vacuum, vibration, shock, cold flow calibration, pressure/leak, functional and electrical, and accurate hot fire performance and life testing. Engine designs burning storable, cryogenic, gelled and high energy bipropellants can be evaluated under sea level and simulated altitude conditions. The High-Energy Propellant Test Stand (HEPTS), designed to test propulsion, laser and gas generator systems, has nine test positions, seven of which can simulate altitude firings.

Orbital Sciences Corporation

Current Status
OSC has joined the team led by Lockheed Martin competing to develop the NASA Orbital Space Plane. Also including Northrop Grumman, the team is competing against Boeing and will capitalise on OSC's work with the DART (Demonstration for Advanced Rendezvous Technology) programme. OSC's role on the OSP will be autonomous rendezvous and docking and with ISS integration. NASA has accelerated the OSP programme by targeting a return ISS crew capability by 2008 and a two-way crew transfer capability by 2012.

Background
OSC was established in 1982 to develop Space Shuttle upper stages, resulting in the solid propellant Transfer Orbit Stage. In June 1988, OSC announced a partnership with motor manufacturer Hercules Aerospace (now Alliant Techsystems) to develop the Pegasus air-launched small orbital delivery system. OSC was contracted in 1989 by DARPA for the Taurus rapid response orbital launcher based on Pegasus. Space Data Corp of Tempe, Arizona became a subsidiary in November 1988, extending OSC's capabilities to a wide range of sounding rockets: it has manufactured/launched approximately 600 suborbital boosters in 35 different configurations, in addition to producing up to 1,000 meteorological rockets annually. The division has participated in approximately 60 Minuteman 1 re-entry tests since 1971. A series of winged air-launched suborbital vehicles was announced September 1992. OSC's L-1011 aircraft can carry three 1- or 2-stage suborbital vehicles derived from the company's other carriers, totalling 11,340 kg with payloads up to 1,130 kg. The OSC/Rockwell team was selected by NASA in March 1995 to develop the semi reusable X34 launcher, the same month the OSC/McDonnell Douglas team was awarded NASA's MedLite launch contract, involving Taurus for the smaller satellites. At this date the X-34 was a two-stage, small, reusable winged launcher, partially reusable, on which NASA would spend about US$70 million. Moving with uncharacteristic haste, NASA selected an Orbital Sciences Corporation (OSC)/Rockwell team to develop the two-stage X-34 aiming for a suborbital flight in 1997 and orbital flights beginning a year later. The team formed American Space Lines as an 'Arianespace' lookalike for commercial development but in early 1996 it pulled out, believing the costs could not produce a viable commercial return. NASA re-contracted with OSC for a single-stage suborbital demonstrator, still designated X-34. Three X-34 test vehicles were scheduled to support flight trials planned to begin in 2000 with the winged vehicle launched from the under-fuselage cradle of a converted L-1011 operated by OSC. Capable of a maximum speed of Mach 8, the X-34 was to have demonstrated all-weather capability and pilotless landings as a technology precursor to

An artist's impression of the DART flight test demonstrator of the space manoeuvring vehicle (OSC) 1047530

fully reusable launch systems of the future. X-34 was scheduled for flight drop tests in 2001–2002. Frustrated by repeated delays and prompted by pressure from the White House, NASA terminated the X-34 programme in February 2001. Active in the development of missile defence systems and its Orbital Boost Vehicle supplying interceptor boosters for 'kill' vehicles. On 24 May 2001 OSC announced that it had been selected for two contracts totalling USD53 million as part of NASA's Space Launch Initiative (SLI). These consisted of a USD47 million contract for the Demonstration of Autonomous Rendezvous Technology (DART) programme where OSC would build, test and launch a space manoeuvring vehicle derived from the Pegasus rocket to demonstrate autonomous rendezvous and close proximity operations. The second contract was for a USD6 million study effort to design and develop the Crew Transfer Vehicle (CTV) concept that the company proposed in 2000 under NASA's Space Transportation Architecture Study (STAS) programme. The DART flight demonstration was scheduled for be flown aboard a Pegasus rocket in 2004 to conduct a series of rendezvous manoeuvres with an orbiting satellite to demonstrate autonomous proximity operations using a modified Pegasus upper stage and avionics system. The space taxi CTV was designed to perform a variety of future NASA missions with significant improvements in cost and safety over the current Space Shuttle. The multifunctional Space Taxi vehicle would serve as an emergency Crew Return Vehicle (CRV) for the ISS, a two-way human space transportation system, a small cargo delivery and return vehicle and a platform for performing satellite servicing or in-orbit construction.

Scaled Composites, LLC

Current Status

Scaled Composites is building what purports to be the world's first feasible, privately funded, vehicle for human space flight. The plan involves a three-person spaceship (SpaceShipOne) carried by a turbojet powered aircraft (White Knight) to an altitude of 50,000ft (15,240m) where it separates, ignites its integral rocket motor reaching a speed of 402.25km/hr coasting to a peak altitude of 100 km (62 miles). During the coast and fall back to earth the spaceship is weightless and converts to a high-drag configuration to allow safe, stable atmospheric entry allowing a 17 min glide from peak deceleration at 80,000ft (24.4km). Concept design began in 1996 and preliminary development started in 1999. In 2000 Scaled Composites conducted a study of rocket motor technologies and selected a hybrid combination using nitrous oxide (liquid N2O) and rubber (HTPB) propellants with a motor mounted to the spacecraft by skirt flanges on the oxidiser tank. The nitrous oxide tank consists of a composite liner laid up on to titanium flanges with a graphite overwrap provided by Thiokol. The unified fuel casket and nozzle components use a high-temperature composite insulator with a graphite/epoxy structure laid up on to an ablative nozzle supplied by AAE Aerospace. The ground firing development programme started in November 2002 with a 15 second run by the SpaceDev team. White Knight made its first flight on 1 August 2002. The first captive carry flight took place on 20 May 2003 with SpaceShipOne mated to White Knight. The first glide flight took place on 7 August 2003 when SpaceShipOne was released from an altitude of 47,000ft (14,325m). The first powered flight, the eighth flight of SpaceShipOne, took place on 17 December 2003 when it was released from the White Knight carrier aircraft with burnout at Mach 1.2. No first-flight date to space has been announced and the company has said that it has no urgency in achieving interim milestones.

Background

Founded in 1982 by Burt Rutan, Scaled Composites provides air vehicle design, tooling and manufacture utilising specialist composite structures design, analysis and test.

Scaled Composites provides complete composite aerospace vehicle design, manufacture and test services. It builds the fins and wing for Orbital Sciences Corporation's Pegasus launcher, and provided the aeroshell for the DC-X demonstrator. It is currently working on the K-1 vehicle for Kistler Aerospace Corporation and the HMX Roton. SCI has produced new composite structures for replacement elements on new launch vehicles proposed by Kistler and OSC.

Space Exploration Technologies (SpaceX)

Current Status

SpaceX launched its second Falcon 1 (see separate entry) demonstrator from Omelek Island, Kwajalein Atoll on 21 March 2007 UT. The launch was considered a partial success on the part of SpaceX, which managed to develop and launch its first rocket (in 2006) in a very swift time frame, considering the company was only established in 2002. The second launch reportedly achieved nearly all of its demonstration objectives, however it failed due to a second stage anomaly, with the second stage falling short of orbit after reaching a 300 km altitude; the rocket did not carry a payload on this test flight. SpaceX launched its maiden demonstrator on 24 March 2006, also from Kwajalein, however that launch ended in failure when the rocket, carrying the 19.5 kg DARPA/USAF Academy student-designed FalconSAT-2, exploded soon after launch, due to a first stage fault. SpaceX claimed that the lessons learned from its first test

Falcon 1 lifts off for its second demonstration launch, 21 March 2007 (SpaceX) 1133896

launch assisted in making its second launch a better success, providing more test data.

The first fully operational Falcon 1 flights are scheduled to launch in the third and fourth quarters of 2007, two years later than originally planned. SpaceX's customers are the US Naval Research Laboratory (NRL), flying the TacSat 1 payload; and the Malaysian company, Astronautic Technology (Malaysia) Sdn Bdh (ATSB) with the Malaysian Space Agency, flying Razaksat (MACSat).

SpaceX is developing its Falcons 1 and 9. Falcon 9 is the heavy lift version, and may also be human rated at a later time in its evolution. The company had plans also to develop a medium lift vehicle, Falcon 5, but that version was cancelled as there is little customer demand so far for that rocket size.

In August 2006, NASA awarded SpaceX, along with Rocketplane-Kistler, the Commercial Orbital Transportation Services (COTS) programme contract; the two companies will develop cargo spacecraft to service the International Space Station (ISS). The SpaceX craft is called Dragon. The contract is worth USD500 million, and is shared by both companies. Falcon 9 is expected to launch Dragon on its first demonstration, as early as 2008. The contract also includes an option to design a human-rated crew carrier.

Background

SpaceX was founded in 2002 by Elon Musk, the PayPal billionaire. The company is headquartered in El Segundo, California, with test facilities in Texas and launch sites at the Ronald Reagan Ballistic Missile Defense Test Site (RTS), Kwajalein Atoll, in the Marshall Islands, Cape Canaveral Air Station and Vandenberg Air Force Base.

The SpaceX launch manifest already stretches through 2010 and includes all three launch sites.

An engineering model of SpaceX's Dragon cargo capsule (Chris Thompson/SpaceX) 1133897

XCOR's Rocket Engines

Name	Spacecraft/Subsystem	Thrust	Propellant	Status
XR-2PI (Tea Cart engine)	Engine	67 N (15 lbf)	Nitrous oxide and ethan. Propellants and third fluid injection are variable.	Experimental. Supports Research and Development (R & D) and education.
XR-3B4 (NRO)	Engine	222 N (50 lbf)	Nitrous oxide and 70/30 mixture of isopropyl alcohol and water, (non-toxic storable propellants).	Satellite stationkeeping.
XR-3M9 (LNG-1)	Reaction control systems and satellite manoeuvring systems.	222 N (50 lbf)	LOX and methane (storable).	R & D. Stationkeeping.
XR-3A2	None	712N (160 lbf)	LOX and isopropanol.	Testbed for XR-4A3
XR-4A3	Auxiliary and primary propulsion to the EZ Rocket.	1,779 N (400 lbf)	LOX and anhydrous isopropyl alcohol.	Supports R & D.
XR-4K14	Primary propulsion, Rocket Racer (X-Racer).	6,672 N (1,500 lbf)	LOX and kerosene.	Supports R & D.
XR-4K5	Sub-orbital main propulsion for Xerus and other future applications.	8,007 N (1,800 lbf)	LOX and kerosene.	Prototype. Supports R & D.
XR-5M15 (ATK-XCOR)	Lunar main ascent engine.	33,361 N (7,500 lbf)	LOX and Methane (storable).	Prototype. Under R & D.

Despite its two initial launch failures, and a two year delay in conducting its first operational launch, US government, international, and commercial customers are queuing up to contract SpaceX as a launch provider. This is not surprising when one considers that the Falcon 1 is the least expensive satellite launch option on the market, coming in at around USD6.9 million per rocket. Still costly, this figure actually represents less than one-third of the cost it takes to get other comparable launch vehicles off the ground.

In 2005 SpaceX bought a 10 per cent stake in Surrey Satellite Technology Ltd (SSTL), a leader in small satellite development, based in the UK.

Space Systems/Loral

Current Status

In May 2007, SES New Skies awarded Space Systems/Loral (SS/L) with the contract to build SES' NSS-12 communication satellite. SS/L expects NSS-12 to be completed in 2009.

SS/L also manufactured the Globalstar (see separate entry) satellite ground spares that were launched in 2007 to backup Globalstar's older, failing satellites, until the company can orbit a new satellite generation. Other recent customers include EchoStar and Intelsat, a company for which SS/L has built 43 other satellites over 40 years. Intelsat 14 will be delivered in 2009.

Background

Headquartered in Palo Alto, California, Space Systems/Loral (SS/L) is a subsidiary of Loral Space & Communications (see separate entry) and is responsible for the satellite manufacturing side of Loral Space & Communications' satellite businesses. SS/L reported revenues of USD637 million in FY2006 and employs a workforce of more than 2,200 on its Palo Alto campus.

In 1996, Loral Space & Communications made a strategic decision to increase its ownership in SS/L to 100 per cent by acquiring the 18.3 per cent interest held by the Lehman Partnerships in August 1996, and in February 1997, by acquiring the remaining 49 per cent minority interest held by SS/L's four European aerospace partners: Aerospatiale, Alcatel, DaimlerChrysler Aerospace (DASA) and Finmeccanica. Aerospatiale Matra and DASA as of 2000, merged with Construcciones Aeronáuticas SA (CASA) to form the European Aeronautic Defence and Space Company (EADS). Space Systems/Loral was originally formed from the Loral Corporation's Space Division, when the parent company took over Ford Aerospace Corporation in October 1990 in an acquisition worth USD715 million. Previous to that time, it was a Philco laboratory and then Philco-Ford.

Specialising especially in civil and commercial communications and meteorological spacecraft, SS/L and its corporate predecessors have built more than 220 satellites over 50 years in the business. As of June 2007, 53 SS/L-manufactured satellites were in orbit. The company's satellites have delivered a total of more than 1,300 years of on-orbit service, the majority of that time from its tri-axis stabilised 1300 series. The company is a full service manufacturer that develops, produces and delivers satellites, payloads and complete satellite systems, while also offering risk management, insurance, mission control and launch services.

SS/L manufactures the 1300 series satellite platform, first offered in the 1980s. The 1300 series is in a continuous state of evolution. Its structure is composed of graphite composite and the platform and can provide satellite power from 5 kW to 25 kW. The 1300 can accommodate up to 150 transponders.

In addition to its primary satellite business, SS/L supplies on-board electrical power systems to Boeing for the International Space Station (ISS).

XCOR Aerospace

Current Status

A new 249N (56 lbf) thrust copper engine with a lightweight aluminium cooling jacket was successfully tested by XCOR Aerospace at their facilities in Mojave Air and Space Port on 12 December 2007. The engine, designated XR-3E17, was financed through private investment and is a descendant of the XR-2P1 Tea Cart engine. Self pressurising propellants ethane and nitrous oxide fuel the engine.

XCOR Aerospace was ranked 446 in the top 500 fastest growing US companies by Inc. Magazine in August 2007.

The Boston Harbour Angels, consisting of 36 investors on 7 June 2007 have invested in the development of XCOR Aerospace sub-orbital vehicle.

XCOR Aerospace is researching and designing sub-orbital rocket powered vehicles that can reach 200,000 ft altitude using supersonic speeds. This is part of a Small Business Innovative Research (SBIR) Air Force Research Laboratory's (AFRL) Phase 1 contract, which was awarded in April 2007. XCOR Aerospace will be utilising existing product experience and capabilities in respect to rocket engines, piston propellant pumps and valves. The contract is part of the Air Vehicles Directorate mission to deliver Operationally Responsive Space (ORS). The company plans to match the US Air Force grant with private funding during vehicle fabrication, which will be Phase II of the contract.

XCOR Aerospace, in a USD3.3 million contract with Alliant Techsystems on 16 January 2007, performed a series of successful test firings of the new workhorse 33,361 N (7,500 lbf) Liquid Oxygen (LOX) and Methane engine, in the Mojave desert. This is an early stage development of the engine and the results will help determine the future development of the first flight weight version of the engine.

Background

XCOR Aerospace was established in 1999 by Jeff Greason, Dan DeLong, Aleta Jackson, and Doug Jones. It is a privately owned company with capital provided by investors. The company continues to actively seek investors, who are US Securities and Exchange Commission accredited. Consultancy services, commercial and government contracts also contribute to XCOR Aerospace's revenues.

The company is headquartered in Mojave, California, at the Mojave Spaceport and Civilian Aerospace Test Centre. The site received the first commercial spaceport licence from the Federal Aviation Administration Office of Commercial

The XCOR XR-5M15 LOX/methane rocket engine is tested at the Mojave Spaceport in Mojave, California (Mike Masse/XCOR) 1166504

For details of the latest updates to *Jane's Space Systems and Industry* online and to discover the additional information available exclusively to online subscribers please visit **jsd.janes.com**

Space Transportation (FAA AST), thereby providing the facilities for XCOR to test reusable rocket horizontal launch vehicles for commercial human spaceflight purposes. XCOR Aerospace was the second company to be issued a mission licence by the FAA AST on 23 April 2004, which incorporated 35 flight tests for an intermediate sub-orbital demonstration test vehicle called Sphinx. This also included, after all tests were completed, initial payload revenue flights from Mojave Air and Space Port. However no test flights took place and the licence expired in 2006.

XCOR Aerospace will initially focus on sub-orbital vehicles for human spaceflight as well as launching payloads into low Earth orbit. In the long term the company's goal is to develop orbital space vehicles. XCOR Aerospace capabilities include designing, testing and manufacturing rocket vehicles, engines, propulsion systems and components.

The Tea Cart engine was used in a Defense Advanced Research Projects Agency (DARPA) contract, which was awarded to XCOR Aerospace and ATK in 2006. The contract supported research regarding third fluid cooled liquid engine, which was used to improve overall performances in liquid engines. Tests have shown that superheated steam can be generated at the same time the coolant flow is reduced.

The XR-3B4 was developed in 2000 through a National Reconnaissance Office (NRO) contract to design a long-life rocket that does not use toxic propellants. Tests showed an efficient regenerative cooling system, transferring for example waste heat to the oxidiser thereby enhancing the lifetime performance of the rocket. Shortly after tests were completed on the 712 N (160 lbf) XR-3A2 engine in November 2000 XCOR Aerospace began constructing the XR-4A3, which is the main propulsion engine to the EZ-Rocket, also built by the company.

The EZ-Rocket has twin engines giving a total thrust power of 3,558 N (800 lbf) and in its first test flight the rocket reached an altitude of 9,000 ft for a duration of 10 minutes. A total of 26 flights have been conducted, two of which were performed at Wisconsin's EAA AirVenture Air Show and three were performed in New Mexico in 2005 at the Countdown to the X-Prize Cup. The X-Prize Cup event is in partnership with the X-Prize Foundation and New Mexico to promote and advance private spaceflights by holding various educational activities and test demonstrations of future space vehicles. The National Aeronautical Association (NAA) certified the 25th flight of the EZ-Rocket, which flew just under 10 miles (16 km), the longest distance for a ground launched rocket powered vehicle. The maximum altitude the EZ-Rocket has attained is 11,500 ft. Currently the rocket is retired.

The development of the XR-3M9 engine was privately financed and in part continued under a Phase 1 SBIR contract. The AFRL propulsion directorate at Edwards Air Force Base contracted XCOR Aerospace in 2005 to test alternative, liquid oxygen and methane-fuelled rocket engines. The testing of XR-3MP revealed a high performance capability for human missions to the moon and Mars.

XCOR Aerospace is contracted by the Rocket Racing league to develop the X-Racer, the next generation EZ-Rocket. The Mark-1 X-Racer will be a piloted, one engine vehicle. The XR-4K14 will be incorporated into the airframe, which will give a maximum speed of up to 230 mph (370 kph). Peter Diamandis (X-Prize founder) and Granger Whitelaw established the Rocket Racing League to introduce rocketry as a competition and sport. The Mark-1 X-Racer, now called the Thunderhawk, was unveiled during the X-Prize Cup in 2006 and also featured in the October 2007 Wirefly X-Prize Cup in Las Cruces, New Mexico.

The XR-4K5 engine is currently under development. Future designs of the engine will form part of the main propulsion system of sub-orbital vehicle Xerus. XCOR Aerospace plan to use piston pumps as the propellant feed system for Xerus, a single stage horizontal take-off rocket plane that will be capable of reaching an altitude of 100 km. Fuel will be stored in the wing and strakes, and LOX will be stored in the cylindrical fuselage tank. The space plane will be targeted towards three markets: space tourism, sub-orbital payloads and microsatellite delivery. The company plans to use Xerus to launch microgravity research experiments, which are usually carried aboard sounding rockets, to reduce launch costs. XCOR Aerospace also plans to attach an expendable stage with its own separate engine to the first stage of their rocket plane, which will be capable of placing a microsatellite in low Earth orbit upon stage separation. It is estimated the weight of the microsatellite would be 10 kg with launch costs of approximately USD500,000. Xerus is currently in design phase.

ATK was awarded a USD10.4 million contract by NASA on 8 May 2006 to design, develop and test a 33,361 N (7,500 lbf) engine for NASA's Crew Exploration Vehicle using non-toxic propellants, LOX and methane. XCOR Aerospace was subcontracted in a USD3.3 million contract by ATK on 9 May 2006 to help develop a workhorse version of the engine. XCOR Aerospace was previously sponsored by DARPA to design a 44,482N (10,000 lbf) LOX and methane engine alongside other XCOR Aerospace engines the XR-3M9, XR-4K5 and XR-4A3, which will be built from work on the 33,361N (7000 lbf) workhorse engine.

XCOR Aerospace have designed and manufactured a Three Cylinder Rocket Propellant Piston Pump. Tests have shown faster operational speeds and quick starts and stops compared to traditional turbo pumps. The pump is designed for LOX and Kersone propellants and can be adapted for use with other common propellants.

XCOR Aerospace also fabricates composite materials. In a USD7.7 million 12 month contract entitled the Long-Life, Light Weight Oxidation Resistant Cryogen Tank awarded by NASA to XCOR Aerospace, the company is developing and manufacturing a LOX composite cryogenic tank. XCOR Aerospace will be fabricating new fire resistant resin and fibre materials suitable for use with a LOX tank. The contract is part of the NASA's Exploration Systems Research Technology (ESRT) programme for future human explorations of the moon, Mars and beyond. The company began using a fluoropolymer composite developing sample coupons where test results showed their suitability for the tank's inner skin. This led to the construction and testing of skin-foam-skin sandwich coupons to also demonstrate their tank suitability. Using the new material to build cylindrical test banks XCOR Aerospace were able to test the lifespan of the material from pressurising and cryogenically cycling the test banks. The themoplastic fluoropolymer resin is called Nonburnite and is a non-combustible material which could have a variety of potential applications, for example life support systems and attitude control thrusters or secondary propulsion systems. In 2006 the Nonburnite was tested at NASA's White Sands Test Facility, New Mexico, in an oxygen enhanced atmosphere alongside a graphite composite using a blowtorch flame. The result of which led to graphite composite burning in contrast to the Nonburnite, which was heat resistant.

XCOR Aerospace provides custom built and modified fluid control valves. They manufacture two different manually-actuated ball valves, allowing a greater custom flange for one of the manual valves. The company also develops the interlocked fuel and oxidiser valve, which is the main propellant valve for the XR-4K15 engine, which controls 1 kg/sec of kerosene and 2.5 kg/sec LOX. Other valves include the three quarters inch spool valve and the solenoid controlled, pilot actuated poppet valve, which is servo configurable. A cryogenic version is also available.

SPACE LAUNCH VEHICLES

Orbital
Sub-orbital

ORBITAL

Brazil

Sonda and VLS

Current Status

On 23 August 2003, the third prototype VLS-1 launcher was destroyed in a pad explosion at the Alacantra launch site in which 21 people were killed. It is believed the first stage prematurely ignited causing the S-43 propulsion system to fire, an event possibly caused by electrical discharge, electromagnetic pulse or a piece of metal in the propellant according to the Brazilian space agency. This launch, designated V3, was carrying the Brazilian Satec-1 and Unosat satellites. VLS-1 and VLS-2 were launched in 1997 and 1999 respectively and both were destroyed by the range safety officer when they failed during ascent.

Background

Sonda 1 was a two-stage vehicle first launched 1964, and flown more than 200 times before retirement. Sonda 2's development then took over in 1966 and has flown about 60 times, offering 44 kg capacity to 80 km. Single stage, 300 mm diameter, 4.1 m long, 360 kg launch mass. Three versions are still employed, principally for technology-proving flights. Sonda 3 development began 1969 to provide a two-stage vehicle capable of delivering 50 kg to 500 km, providing three-axis payload control and sea recovery. The S3 basic version handles 50 to 80 kg to high altitudes using the Sonda 2 as the S20 stage 2. A reduced Sonda two-stage 2 (S23) is carried for S3-M1 missions with 130 to 160 kg payloads destined for lower altitudes. (The M2 version, using the smaller S24, has been used once, 23 August 1979.)

Preliminary studies of a Sonda 4, began in 1974 and produced the decision to launch five prototypes for vehicle qualification and as technology demonstrators for the VLS (Velculo Lancador de Satelites) satellite launcher, which incorporates clustered Sonda 4 stage 1 motors. Four launches have been made: 21 November 1984, 19 November 1985, 8 October 1987, and 28 April 1989. There have been no flights since the first and last Sonda 4 on 21 February 1990 from the new launch site Centro de Lancamento de Foguetes in Alacantra. Stage 2 failed to separate in 1987's flight. The programme has developed 300M steel for the motor casings, digital control system (although Sonda has so far carried an analogue system), payload fairing ejection and three-axis control (including liquid injection TVC and movable nozzle TVC). MBB provided technical assistance in developing the LITVC system. #4 demonstrated movable nozzle TVC for the first time on stage 2.

IAE, under the aegis of the Ministry of Aeronautics, has developed the Veiculo Lancador

A 1/3 VLS model was launched May 1989 to demonstrate strap-on separation (IAE/CTA) 0516861

Specifications
VLS motor

Motor	S-43	S-43TM	S-40TM	S-44
Motor mass (kg)	8,210	8,400	5,340	917
Propellant mass (kg)	7,180	7,180	4,450	810
Burn time (web, s)	58.9	58.9	56.4	67.9
Action time (s)	62.0	62.0	62.0	72.0
Average pressure (atmospheres)	56.2	55.3	57.2	39.5
Average thrust (kN, vacuum)	309.0	327.0	212.5	33.9
Total vacuum imp (MNs)	18.42	19.52	12.00	2.31
SI (s, vacuum)	260.0*	277.0	274.9	281.6
Expansion ratio	12.82	37.68	25.97	66.0
Nozzle exit diameter (mm)	700	1,200	800	602

*sea level SI 230.5 s. S-44 employs Kevlar epoxy casing; others are steel.

de Satelites. Originally intended to launch the MECB satellites. But the programme ran afoul of the Missile Technology Control Regime signed by the G-7 group of the world's seven most industrialised nations. Many components such as the liquid roll control package planned forward of stage 2 and replaced by solid motors (but is now again liquid), had to be reduced to less capable technology.

The strap-ons and two of the three core stages are derived from Sonda 4's stage 1. The four strap-ons ignite on the pad, employing nozzle flexure for steering. The core stage 1 is nested within the strap-on cluster and ignites at 20 km altitude. A liquid stage 1 is under consideration as a future upgrading. Stage 2 is a shortened version of the S-43 motor. The stage 3 orbit injection motor is newly developed and spin-stabilised with a fixed nozzle. By end-1995, two of four planned strap-on static firings had been made, two of three stage 1, four of five stage 2 and four of six stage 3. The 6.6 tonne 9.52 m 'VS-40' vehicle was launched 2 April 1993 on a 24 minute, 1,248 km altitude flight testing VLS's stages 2/3. A 9.76 tonne 12.0 m VS-43 vehicle using VLS stage 1/3 could deliver 1.2 m diameter 200 to 500 kg suborbital payloads to 1,000–2,000 km.

Brazil had major problems acquiring the inertial guidance platform and the vehicle has yet to be launched. The four flights of the original MECB were planned in 1989 to begin 1992, followed by the second in 1993 and the remaining two during 1994. This never materialised but the first launch attempt for VLS-1 took place on 2 November 1997. It failed to reach orbit. The 50 tonne four-stage all-solid vehicle stands 19 m high, capable of placing 200 kg into a 750 km circular 25° orbit or a 450 km Sun-synchronous orbit.

The second launch attempt took place on 11 December 1999 and again the VLS failed to reach orbit, the SACI-2 satellite being destroyed in the atmosphere.

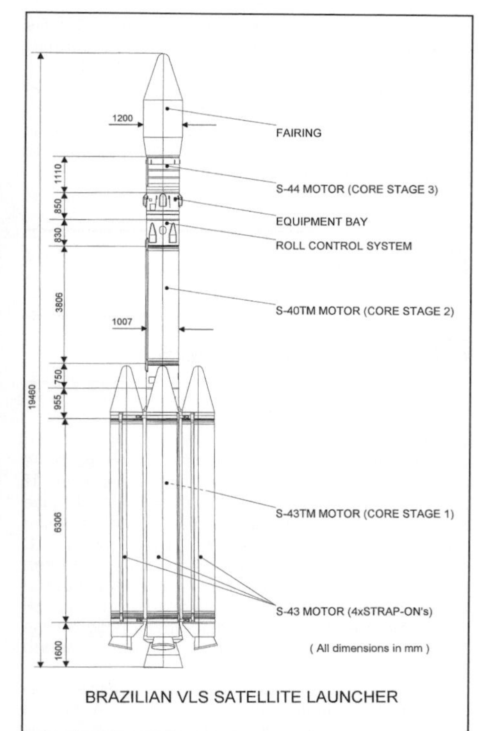

VLS principal features (IAE/CTA) 0003408

Sonda 3/3-M1

First launch: 26 February 1976
Number launched: 28 (2 failures: 31 October/ 14 November 1983) to end-1996; M1 4 flights, successful
Number of stages: 2 (solid propellant)
Overall length: 8.0 m both versions
Principal diameter: 557 mm
Launch mass: 1,521 kg (M1: 1,527 kg), excluding payload
Typical performance: 60 kg to 600 km (M1: 140 kg to 275 km).

Sonda 3/3-M1 stage 1

Overall length: 373 cm
Principal diameter: 557 mm
Stage mass: 1,205 kg both versions
Propellant mass: 860 kg both versions
Average thrust: 102 kN sea level both versions
Burn time: 24 s.

Sonda 3/3-M1 stage 2

Designation: S20 (M1: S23)
Overall length: 290 cm (M1: 160 cm), excluding payload
Principal diameter: 300 mm both versions
Stage mass: 316 kg (M1: 182 kg)
Propellant mass: 229 kg (M1: 113 kg)
Average thrust: 33 kN vacuum (M1: 18 kN)
Burn time: 15 s (M1: 15 s).

Sonda 4

First launch: 21 November 1984
Number of launches: 4, to end-1995 (8 November 1987 stage separation failure)
Number of stages: 2 (solid propellant)
Overall length: 11.0 m
Principal diameter: 1,008 mm
Launch mass: 6,800 kg, excluding payload
Typical performance: 500 kg to 650 km
Guidance: the inertial platform and control system are housed in the equipment bay atop stage 2. Three-axis control during stage 1 burn is provided by LITVC (2.5° max vector deflection; 81.6 atmospheres injection by N_2; 9 ms valve response time; Three valves per quadrant) for pitch/yaw and thrusters for roll.

Sonda 4 stage 1

Overall length: about 537 cm
Principal diameter: 1,008 mm
Stage mass: 5,670 kg
Propellant mass: 4,220 kg
Average thrust: 203 kN sea level
Burn time: about 60 s

Sonda 4 stage 2

Overall length: about 325 cm, excluding payload
Principal diameter: 555 mm
Stage mass: 1,130 kg
Propellant mass: 869 kg
Average thrust: 95 kN vacuum
Burn time: 28 s.

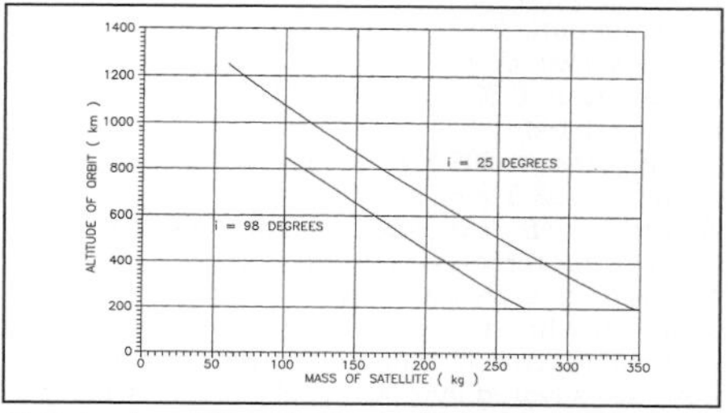

VLS circular orbit performance (IAE/CTA) 0516860

The VS-40 tested VLS stages 2/3 (IAE/CTA) 0517188

End view of Sonda 4 rocket motor propellant grain, for which AGQuimica manufactures the ammonium perchlorate 0516862

VLS
First launch: 1997 (fail); second launch 11 December 1999 (fail)
Launch site: Alcantara
Principal uses: delivery of small payloads to LEO
Schedule of missions: SCD 2A fourth quarter 1997?; first half 1998 SCD 3; SSR 1 1999; SSR 2 2000
Typical performance: 200 kg into 750 km, 25°
Number of stages: 3 + 4 strap-ons (all solid)
Overall length: 19.46 m
Principal diameter: 1.00 m
Launch mass: 50 t
Guidance: inertial, mounted in equipment bay below stage 3
Launch sequence (min:s):

0.00	strap-ons ignite
0.55	stage 1 ignites
1.07	strap-ons burn out/separate
1.58	stage 1 burnout/separation and stage 2 ignition
2.07	heatshield separates
3.00	stage 2 burnout
3.08	stage 2 separates
7.12	equipment bay separates/stage 3
–	Ignites
8.23	stage 3 burnout/separate

VLS strap-ons
Designation: S-43
Overall length: 8.92 m
Principal diameter: 1,006.6 mm
Stage mass: 8,550 kg each (7,180 kg propellant + consumables)
Average thrust: 309 kN vacuum each
Burn time: 58.9 s (62 s. action time)
Attitude control: movable nozzles provide up to 3.0° TVC for three-axis control from four motors
Separation: each strap-on is attached by an aft spherical thrust transmission pin and paired forward/aft arm mounts. Following burnout, the arms are severed pyrotechnically and internal gas-pressurised actuators provide separation velocity.

Scale model of the VLS launcher with four solid propellant strap-on booster rockets (Theo Parard) 0054363

VLS stage 1
Designation: S-43TM (similar to strap-on but nozzle reconfigured optimised for altitude operations)
Overall length: 8.86 m
Principal diameter: 1,006.6 mm
Stage mass: 8,720 kg (7,180 kg. propellant plus consumables)
Average thrust: 327 kN vacuum
Burn time: 59 s (62 s. action time)
Attitude control: nozzle flexure up to 3° provides pitch/yaw TVC; roll managed by solid thrusters

VLS stage 2
Designation: S-40TM
Overall length: 6.24 m
Principal diameter: 1,006.6 mm
Stage mass: 5,660 kg (4,450 kg propellant plus consumables)
Average thrust: 212.5 kN vacuum
Burn time: 56.4 s (62 s. action time)
Attitude control: nozzle flexure up to 3° provides pitch/yaw TVC; roll managed by liquid thrusters

VLS stage 3
Designation: S-44
Overall length: 1.75 m
Principal diameter: 1,006.6 mm
Stage mass: 1,025 kg (810 kg propellant plus consumables)
Average thrust: 33.9 kN vacuum
Burn time: 67.9 s (72 s. action time)
Attitude control: spin initiated by solids before separation of equipment bay
Heatshield/payload accommodation: The payload is mounted on its adaptor by a V-clamp and separated by springs following pyrotechnic initiation. The two heatshield aluminium halves provide an internal diameter of 1,180 mm (external 1,200 mm) and cylindrical payload section 1,180 mm high. Mass <150 kg. Pyrotechnic clamps and springs separate the fairing after stage 2 ignition.

China

Kaitouzhe Series

Current Status
A Kaitouzhe (KT) rocket is speculated to have carried the anti-satellite (ASAT) kill vehicle, on 11 January 2007, that destroyed an in-orbit Chinese Fengyun (FY-1C) meteorological satellite (see separate entry). As of early 2007, over 1,000 FY-1C debris pieces caused by the explosion have been catalogued. The weapons test caused widespread criticism, and confirmed suspicions by other space faring nations that the Chinese space programme is developing technologies that could be intended for aggressive purposes.

The KT family of small solid-fuel launchers has been under development since 2000. A launch was scheduled for 2005 but an exact date for this launch has not been confirmed. Four test launches reportedly took place previous to the 11 January ASAT launch from Xichang.

Background
China's Aerospace Solid Launch Vehicle Company, a subsidiary of China Aerospace Science and Industry Corporation (CASIC), was established in May 2000 to develop the Kaitouzhe (Pioneer) family of solid-fuel space launchers, based on the technology of the DF-21 ICBM. CASIC is China's leading missile developer and aerospace defence company.

The smallest Kaitouzhe variant, the mobile KT-1, has reportedly flown five times, but failed to reach orbit twice. Larger variants are planned using solid propellant strap-on boosters.

At the 2002 Zhuhai air show, CASIC showed models of the KT-2, based on the DF-31 ICBM, with a 300 kg payload, as well as the larger KT-2A with two strap-on boosters and an additional upper stage. The company has proposed the KT series as launch systems for small Earth-resources satellites, but there were also suspicions, now evidently confirmed, that the KT series is being developed for China's evolving ASAT programme.

The first KT-1 launch attempt took place in September 2002: it is believed that a failure involving the launch vehicle's second stage prevented the orbiting of the 50 kg OlympicSat payload (a 300 km polar orbit was intended). The second KT-1 launch took place a year later, and did not place a satellite into orbit, apparently due to another failure. The state-owned People's Daily reported in early 2004 that a launch was planned in that year and CASIC stated that the KT-2 and KT-2A were both expected to fly by the end of 2005.

Specifications
KT-1
Launch site(s) deployed: Taiyuan; Xichang
Maiden launch: 15 September 2002 (failed to reach orbit)
Orbital launch record: 5 launches, 2 failures to reach orbit
Latest launch: 11 January 2007
Number of stages: 4
Overall length: 13.6 m
Diameter: 1.4 m
Launch mass: 19.3 t
Payload capacity: 100 kg to 220 km LEO, 50 kg to 400 km SSO

KT-2
Launch site(s) deployed: Taiyuan?
Maiden launch: Planned for 2005 but did not take place
Number of stages: 4?
Diameter: 1.7 m (first stage), 1.4 m (upper stages). No performance data are available for the KT-2 vehicle.

KT-2A
Launch site(s) deployed: Taiyuan?
Maiden launch: Planned for 2005 or 2006
Number of stages: 2 strap-ons plus 4?
Diameter: 1.4 m (strap-ons) 1.7 m (first stage), 1.4 m (upper stages). No performance data are available for the KT-2A vehicle.

Long March (Chang Zheng)

Current Status
The Long March series of launch vehicles – also known as the Chang Zheng or CZ series – evolved from the Dong Feng ballistic missile series, developed in the 1960s, and like most ballistic missiles of that era they mainly use storable chemical propellants. The principal versions in use today are the human-rated CZ-2F, developed to launch the Shenzhou series of manned spacecraft; the CZ-2C and CZ-4B, mainly used to launch satellites into LEO and SSO; and the CZ-3A and CZ-3B vehicles, developed to launch large spacecraft into GTO. (Some sources, including some English-language documents of official Chinese origin, refer to these as the LM series). Development work, including rocket motor testing, has already started on an entirely new series of rockets to replace the current Long March series and provide China with an indigenous heavy-lift capability.

Background
China's space-launch programme is conducted by the China Aerospace Science & Technology Corporation (CASC), owned by the Chinese

A Long March 2C rocket takes off from JSLC on 6 October 1992 at 06:20 UT carrying the Chinese FSW-1 satellite and the Swedish Freja scientific satellite. Note the small fairing (CGWC) 1340462

government and headquartered in Beijing. Key subsidiaries include the China Academy of Launch Vehicle Technology, responsible for the design of the CZ-2 and CZ-3 series; the Shanghai Academy of Space Flight Technology, which developed the LEO CZ-2D and CZ-4; and the Academy of Space Chemical Propulsion Technology, responsible for liquid rocket engines. Another CASC subsidiary, the China Great Wall Industry Corporation, is responsible for marketing launch services to non-Chinese customers.

Rocket propulsion was included as one of the key technologies in the "Twelve Year Development Plan of Science and Technology" approved in 1958. This plan led to the launches of the first indigenously developed sounding rocket on 19 February 1960, and first successful ballistic missile on 29 June 1964 (failure of the first test on 21 March 1962 led to a major redesign). The initial CZ-1 space launcher, derived from the CSS-3 MRBM, utilised nitric acid/UDMH, but each subsequent operational and planned stage 1/2 has used N_2O_4 as oxidiser. Development of a LOX/LH_2 cryogenic upper stage for GEO missions began in 1977, leading to the heavyweight CZ-3 series, which are still the most powerful Chinese launchers.

By the end of October 2007, the Long March rocket series had been launched 103 times since their debut in April 1970. Between April 1970 and October 1997, there were seven failures in 49 launches. Since 1997, the success rate has been 100 per cent.

CZ-1/Long March 1

The CZ-1 rocket was derived from the CSS-3 missile (also called DF-4 – Dong Feng or "East Wind") with an extra solid fuel third stage. The first stage was 17.8 m long and weighed 64.1 tons, including 60 tons of nitric acid/UDMH propellant. It was powered by a YF-2A motor with four chambers that generated 1,100 kN at ground level and burned for 130 seconds (Isp = 241 s). The second stage was 5.4 m long and weighed 15.9 tons, including 13.2 tons of NTO and UDMH propellant. A YF-3 motor provided 294 kN of thrust in vacuum for 126 seconds (Isp = 287 s). The spin-stabilised third stage weighed 1.8 tons and used a GF-02 solid fuel motor to provide 29 kN of thrust. At launch, the CZ-1 was 29.45 m tall and weighed 81.8 tons. It could place a 300 kg payload into a 440 km orbit, 70° inclination. After a failure in November 1969, it successfully launched the first two Chinese satellites from Juiquan in April 1970 and March 1971.

CZ-2/Long March 2

The CZ-2 was the operational follow-on to the CZ-1, but it was flown only once. The CZ-2 debut launch in 1974 failed after a few seconds but the CZ-2C variant has since become the most-utilised Chinese launcher. The CZ-2C flew the first recoverable payload mission in November 1975, introducing China as only the third nation with such a capability and permitting reconnaissance and Earth resources imaging. The main current versions are the CZ-2C, CZ-2D and CZ-2F, which is man-rated and used to launch the Shenzhou spacecraft.

CZ-2 Launcher Family Record

First launch: 5 November 1974
Number launched: 27 (1-CZ-2A, 14-CZ-2C, 3-CZ-2D, 7-CZ-2E, 2-CZ-2F): 2 failures (CZ-2E)
Launch sites: Xichang, Jiuquan
Vehicle success rate: Overall 94.1 per cent to end 2007

CZ-2C Specifications

The CZ-2 was redesignated CZ-2C when it became operational. Its principal mission has been to launch recoverable satellites. Most of these have been defence reconnaissance spacecraft, but the platform has also been offered for microgravity research.

Another configuration, the CZ-2C/SD, was selected by Motorola in 1993 to launch its Iridium mobile communications satellites. It was fitted with a new steerable "Smart Dispenser" (SD) transfer stage, improved second stage fuel and oxidiser tanks and second stage engines with higher expansion ratio nozzles. The modified fairing, with a diameter of 3.35 m, could house two Iridium satellites. A successful CZ-2C/SD demonstration launch was conducted from Taiyuan on 1 September 1997. Six CZ-2C/SD launches in 1997–99 launched 12 Iridium satellites.

In May 2001, it was announced that another variant would be introduced to launch the TC (DoubleStar) and KOMPSAT-2 satellites. The CZ-2C with a CTS upper stage was a further development of the SD model, with a three-axis stabilised upper stage consisting of a spacecraft adapter and an orbital manoeuvre system. It was capable of putting 1,400 kg into a 900 km sun-synchronous orbit. Between November 1975 and November 2004, the CZ-2C, including the SD and CTS versions, flew 29 consecutive successful missions.

CZ-2C

Launch site(s) deployed: Jiuquan, Xichang, Taiyuan
Number of stages: 2
Overall length: 42 m
Diameter: 3.35 m
Launch mass: 233 t
Launch thrust: 2,961 kN sea level

CZ-2C stage 1

Designation: L-140
Engines: 4 × YF-20 liquid bipropellant single start engines (overall designator YF-6-2)
Overall length: 35.72 m
Diameter: 3.35 m
Oxidiser: Nitrogen tetroxide in forward tank
Fuel: UDMH in aft tank
Propellant mass: 162.7 t
Thrust: 2,961.6 kN sea level
Specific impulse: 2,556 N-s/kg sea level

CZ-2C stage 2

Designation: L-35
Engines: 1 × YF-20-1 liquid bipropellant single start main engine; 1 × YF-21-1 liquid bipropellant single start vernier engine (overall designator YF-24)
Overall length: 7.757 m
Diameter: 3.35 m
Oxidiser: Nitrogen tetroxide in forward tank
Fuel: UDMH in aft tank
Propellant mass: 54.7 t
Thrust: 741.4 kN main engine, 47.2 kN vernier engine (vacuum)
Specific impulse: 2,922 N-s/kg main engine, 2,834 N-s/kg vernier engine (vacuum)

CTS transfer stage

Diameter: 2.7 m
Length: 1.5 m
Propellant mass: 175 kg
Propellant: HTPB (main); hydrazine (RCS)
Thrust: 10.78 kN

CZ-2D

The CZ-2D was developed by the Shanghai Academy of Space Flight Technology in the late 1980s. It comprises the two lower stages of the CZ-4. Only eight vehicles have been flown since 1992, launching recoverable FSW satellites from the Jiuquan Satellite Launch Centre. It is offered with a choice of 2.9 m and 3.35 m fairings.
Launch site(s) deployed: Jiuquan
Maiden launch: 8 September 1992
Orbital launch record: 8 launches, 8 successful
Latest launch: 27 September 2004
Number of stages: 2
Overall length: 37.33 m
Diameter: 3.35 m
Launch mass: 232.7 t
Launch thrust: 2,961.6 kN sea level

A Long March 2D rocket launching from Jiuquan SLC. The small fairing houses a FSW recoverable reconnaissance satellite (CGWC) 1340463

The CZ-2F, which is rated for human spaceflight
0131955

CZ-2D stage 1
Engines: YF-21B liquid bipropellant single start engine
Diameter: 3.35 m
Oxidiser: Nitrogen tetroxide in forward tank
Fuel: UDMH in aft tank
Propellant mass: 182.07 t
Thrust: 2,961.6 kN sea level
Specific impulse: 2,550 N-s/kg sea level

ICZ-2D stage 2
Engines: 1 × YF-22B liquid bipropellant single start main engine; 1 × YF-23B liquid bipropellant single start vernier engine
Diameter: 3.35 m
Oxidiser: Nitrogen tetroxide in forward tank
Fuel: UDMH in aft tank
Propellant mass: 35.408 t
Thrust: 742 kN main engine, 46.1 kN vernier engine (vacuum)
Specific impulse: 2,910 N-s/kg main engine, 2,67 N-2/kg vernier engine (vacuum)

CZ-2E/F
The CZ-2E launcher had a relatively short and undistinguished career, but forms the basis of the CZ-2F manned launcher. The CZ-2E marked China's first venture into the international GTO launch market, and was produced by adding four liquid-fuel strap-on boosters to the core of the CZ-2C, together with a Hexi Chemistry & Machinery EPKM solid-rocket kick stage and a 4.2 m fairing. The Chinese government signed a contract with Hughes Communications Inc in November 1988 to start commercial operations in 1992. The new rocket was developed in only 18 months and made its first demonstration flight (with Pakistan's Badr 1 spacecraft as a ride-along) on 16 July 1990. It made six more flights by the end of 1995, two of them being failures. The last CZ-2E flight took place on 28 December 1995. China Great Wall Industrial Corporation has offered a version of the CZ-2E with a multi-spacecraft upper stage based on the CTS used with the CZ-2C, but no launches have taken place.

The CZ-2F is an upgraded version of the CZ-2E. The configuration, weight and performance are very similar, but the systems have been improved and made more redundant to increase overall safety, and the upper stages have been strengthened to handle the weight of the Shenzhou spacecraft, its fairing and escape tower. There is no upper stage. The CZ-2F made its first flight on 20 November 1999.

The total mass of the CZ-2F is about twice that of the CZ-2C, but it uses very similar components. The first-stage and second-stage propulsion and control systems are essentially the same, with four gimballed motors on the first stage and a propulsion motor and four-chamber vernier on the second stage. The fuel tanks have the same diameter, but are stretched to hold more fuel. The strap-on boosters comprise fuel and oxidiser tanks and a rocket engine basically similar to the second-stage engine. By the end of 2005, the CZ-2F had flown successfully six times.
Launch site(s) deployed: Xichang
Maiden launch: 16 July 1990
Orbital launch record: CZ-2E: 7 launches, 5 successful. CZ-2F; 6 launches, 6 successful
Number of stages: CZ-2E: 3 + 4 strap-on boosters; CZ-2F: 2 + 4 strap-on boosters
Overall length: CZ-2E: 49.686 m; CZ-2F: 58.34 m
Diameter: 3.35 m
Launch mass: CZ-2E: 460 t; CZ-2F: 479.7 t
Launch thrust: 5,923.2 kN sea level

CZ-2E/F strap-on booster
Number of strap-ons: 4
Designation: LB-40
Engines: 1 × YF-5-1 liquid bipropellant single start engine
Overall length: 15.326 m
Diameter: 2.25 m
Oxidiser: Nitrogen tetroxide in forward tank
Fuel: UDMH in aft tank
Propellant mass: 37.754 t
Dry mass: 3 t
Thrust: 740.4 kN sea level
Specific impulse: 2,556.2 N-s/kg sea level
Burn time: 127.26 s

CZ-2E/F stage 1 (core)
Designation: L-180
Engines: 4 × YF-20B liquid bipropellant single start engine (overall designator YF-21B)
Overall length: 28.465 m
Diameter: 3.35 m
Oxidiser: Nitrogen tetroxide in forward tank
Fuel: UDMH in aft tank
Propellant mass: 186.28 t
Dry mass: 12.55 t
Thrust: 2,961.6 kN sea level
Specific impulse: 2,556.2 N-s/kg sea level
Burn time: 160.43 s

CZ-2E/F stage 2
Designation: L-90
Engines: 1 × YF-22B liquid bipropellant single start main engine; 1 × YF-23B liquid bipropellant single start vernier engine (overall designator YF-24B)
Overall length: 14.223 m
Diameter: 3.35 m
Oxidiser: Nitrogen tetroxide in forward tank
Fuel: UDMH in aft tank
Propellant mass: 84.759 t
Dry mass: 4.955 t
Thrust: 738.4 kN main engine, 47.1 kN vernier engine (vacuum)
Specific impulse: 2,922.4 N-s/kg main engine, 2834.1 N-s/kg vernier engine (vacuum)
Burn time: 301.18 s main engine, 414.68 s vernier engine

CZ-2E stage 3
Designation: EPKM
Engines: EPKM
Overall length: 3.62 m
Diameter: 1.7 m
Propellant: solid
Propellant mass: 529 kg
Dry mass: 0.685 t
Specific impulse: 2,863.3 N-s/kg (vacuum)
Burn time: 70 s

CZ-2E payload shroud
Overall length: 10.5 m
Max diameter: 4.2 m
Mass: 1.9 t

CZ-3A/3B/3C
The CZ-3 family are three-stage vehicles with a LOX/LH_2 cryogenic third stage, designed for GTO/GEO missions. Development started in 1977, following feasibility studies in 1975 (the vehicle was then referred to as CZ-2B). The initial CZ-3 version was launched on 29 January 1984 and established China as only the third user of cryogenic propulsion, after the US and European Space Agency (ESA). On the first launch attempt the stage shut down shortly after re-ignition for GTO injection (which also happened on the eighth mission, December 1991), but the second launch three months later orbited the first Chinese GEO satellite. All CZ-3s have been launched from the Xichang launch centre, established specifically for GTO/GEO operations. The last basic CZ-3 was launched on 1 September 2000 and the type has now been retired.

The CZ-3A development programme started in the mid-1980s, with the goal of launching heavier and more effective communications satellites to support Chinese domestic services. Compared with the CZ-3, it has a stretched stage 1 and a considerably longer cryogenic third stage; it weighs some 20 per cent more than the CZ-3A at lift-off and has an almost-doubled GTO payload of 2,650 kg. Nine CZ-3A vehicles have flown without a failure since 1994, carrying Chinese GEO satellites.

The most powerful Chinese launcher today is the CZ-3B, developed to provide China with a launcher capable of lifting 5 t to GEO – almost twice the payload of the CZ-3A. Development started in 1986 and the first launch contract, with Intelsat, was signed in April 1992. The CZ-3B has four liquid-fuel strap-on boosters similar to those used on the CZ-2E/F and a considerably stretched second stage with 66 per cent more propellant, the core stage and the upper stage remaining unchanged. The first CZ-3B launch attempt – with Intelsat-7A on 15 February 1996 – was a failure due to a problem with an inertial measurement unit, but five subsequent flights have been successful. The most recent CZ-3B launch, in April 2005, was the first for the type in almost five years.

CASC has described and proposed a CZ-3C configuration. This would be identical to the CZ-3B apart from carrying two liquid strap-on boosters instead of four, and its GTO payload capacity of 3.8 t would fall directly between the CZ-3A and the CZ-3B. This configuration had not been flown by the end of 2007 and is not being marketed internationally.

Specifications
CZ-3A
Launch site(s) deployed: Xichang
Maiden launch: 8 February 1994
Orbital launch record: 15 launches, 15 successful
Number of stages: 3
Overall length: 52.52 m
Diameter: 3.35 m
Launch mass: 241 t
Launch thrust: 2,961.6 kN sea level
GTO payload: 2,650 kg

CZ-3A stage 1
Designation: L-180
Engines: 4 × YF-20B liquid bipropellant single start engines (overall designator YF-21B, also called DaFY6-2)
Overall length: 26.972 m
Diameter: 3.35 m
Oxidiser: Nitrogen tetroxide in forward tank
Fuel: UDMH in aft tank
Propellant mass: 171.775 t
Thrust: 2,961.6 kN sea level
Specific impulse: 2556.2 N-s/kg sea level
Burn time: 146 s

CZ-3A stage 2
Designation: L-35
Engines: 1 × DaYF20-1 liquid bipropellant single start main engine; 1 × YF-21-1B liquid bipropellant single start vernier engine
Overall length: 11.276 m
Diameter: 3.35 m
Oxidiser: Nitrogen tetroxide in forward tank
Fuel: UDMH in aft tank
Propellant mass: 30.752 t
Dry mass: 3.561 t
Thrust: 742 kN main engine, 47 kN vernier engine (vacuum)
Specific impulse: 2,922.4 N-s/kg main engine, 2,834 N-s/kg vernier engine (vacuum)
Burn time: 114 s main engine, 119 s vernier engine

CZ-3A stage 3
Designation: H-18
Engines: 2 × YF-75 liquid bipropellant multiple start engines
Overall length: 12.375 m
Diameter: 3 m
Oxidiser: Liquid oxygen in aft tank
Fuel: Liquid hydrogen in forward tank

A Long March 3A rocket launching from Xichang SLC. The CZ-3A is used to launch communications satellites (CGWC) 1340465

A Long March 3B rocket on the pad at Xichang SLC. The CZ-3B is used to launch communications satellites and is currently the most powerful commercial vehicle in the CZ series (CGWC) 1340466

Propellant mass: 18.193 t
Dry mass: 2.742 t
Thrust: 156.9 kN (vacuum)
Specific impulse: 4.312 N-s/kg (vacuum)
Burn time: 469 s

CZ-3A payload shroud
Overall length: 8.887 m
Max diameter: 3.35 m
Mass: 0.5 t

CZ-3B
Launch site(s) deployed: Xichang
Maiden launch: 14 February 1996
Orbital launch record: 9 launches, 8 successful, 1 first stage failure
Number of stages: 3 + 4 strap-on boosters
Overall length: 54.838 m
Diameter: 8.45 m
Launch mass: 426 t
Launch thrust: 5,923.2 kN sea level
GTO payload: 5,100 kg

CZ-3B strap-on booster
Number of strap-ons: 4
Designation: LB-40
Engines: 1 × DaFY5-1 liquid bipropellant single start engine
Overall length: 15.326 m
Diameter: 2.25 m
Oxidiser: Nitrogen tetroxide in forward tank
Fuel: UDMH in aft tank
Propellant mass: 37.75 t
Dry mass: 2.8 t
Thrust: 740.4 kN sea level
Specific impulse: 2,556.2 N-s/kg sea level
Burn time: 125 s

CZ-3B stage 1 (core)
Designation: L-180
Engines: 4 × YF-20B liquid bipropellant single start engines (overall designator YF-21B, also called DaFY6-2)
Overall length: 23.272 m

A Long March 4A rocket launching from Taiyuan SLC. Only two CZ-4A launches took place before the modified CZ-4B was introduced (CGWC) 1340464

Diameter: 3.35 m
Oxidiser: Nitrogen tetroxide in forward tank
Fuel: UDMH in aft tank
Propellant mass: 171.8 t
Dry mass: 12.12 t
Thrust: 2,961.6 kN sea level
Specific impulse: 2,556.2 N-s/kg sea level
Burn time: 146 s

CZ-3B stage 2
Designation: L-35
Engines: 1 × DaFY21-1 liquid bipropellant single start main engine; 1 × YF-23B liquid bipropellant single start vernier engine (overall designator is YF-24B)
Overall length: 9.943 m
Diameter: 3.35 m
Oxidiser: Nitrogen tetroxide in forward tank
Fuel: UDMH in aft tank
Propellant mass: 49.6 t
Dry mass: 3.848 t
Thrust: 742 kN main engine, 47 vernier engine (vacuum)
Specific impulse: 2,922.4 N-s/kg main engine, 2,834 N-s/kg vernier engine (vacuum)
Burn time: 178 s main engine, 184 s vernier engine

CZ-3B stage 3
Designation: H-18
Engines: 2 × YF-75 liquid bipropellant multiple start engines
Overall length: 12.375 m
Diameter: 3 m
Oxidiser: Liquid oxygen in aft tank
Fuel: Liquid hydrogen in forward tank
Propellant mass: 18.242 t
Dry mass: 3.062 t
Thrust: 156.9 kN (vacuum)
Specific impulse: 4.315 N-s/kg (vacuum)
Burn time: 478 s

CZ-3B payload shroud
Overall length: 9.561 m
Max diameter: 4 m or 4.2 m
Mass: 1.5 t

CZ-4/4B/4C
Developed by the Shanghai Academy of Spaceflight Technology, the CZ-4 family design started in 1982, and engineering development began in 1983. It was originally intended as a back-up to the China Academy's CZ-3, with a basically similar first and second stage. The principal difference is that the CZ-4 has a storable liquid-propellant upper stage rather than the LOX/LH_2 cryogenic stage used on the CZ-3. When the CZ-3 proved successful the CZ-4 was assigned to launch sun-synchronous meteorological and remote sensing satellites.

Only two CZ-4A launchers were flown, in 1988 and 1990. The programme was dormant for most of a decade, until flights were resumed with the slightly heavier and improved CZ-4B. 11 of these vehicles had flown successfully by the end of 2007, carrying weather and Earth-resources spacecraft. The CZ-4B has orbited three spacecraft in the China-Brazil Earth Resources Satellite (CBERS) programme.

The 2,700 kg Yaogan-1 (Remote Sensing Satellite-1) was launched by a modified design, known as CZ-4B Batch-02, on 27 April 2006. The satellite was housed in an enlarged fairing (diameter 3.8 m; length 10 m) on top of the three-stage CZ-4B. The launcher also featured a new third stage that is capable of reignition. This technology enabled the

A Long March 4B lifting off with the CBERS-2B satellite from Taiyuan space centre on 19 September 2007 (INPE) 1340618

Long March family (left to right): LM-3B, LM-2E, LM-3, LM-4A, LM-2D, LM-2C and LM-1D 0084624

launch vehicle to carry a heavier payload without increasing its fuel load.

Yet another version, known as the CZ-4C, was introduced on 12 November 2007. It was used to launch the 2,705 kg Yaogan 3 remote-sensing satellite into a 600 km sun-synchronous orbit. The CZ-4C also has an upgraded, restartable second-stage engine that increases the precision of delivering payloads to higher orbits. Instead of the open grid on the inter-stage sections of the CZ-4B, it has a protective cover that is ejected at liftoff. Structural rings have been added at the base of the first and second stages, opening up the possibility of launching heavier payloads or flying more stressful trajectories. Like other, recent CZ-4B missions, it carried a larger payload fairing. All CZ-4 launches have taken place from the Taiyuan Satellite Launch Centre.

CZ-4B
Launch site(s) deployed: Taiyuan
Maiden launch: 5 October 1999
Orbital launch record: 11 launches, 11 successful
Number of stages: 3
Overall length: 45.8 m
Diameter: 3.35 m
Launch mass: 249.2 t
Launch thrust: 2,962 kN sea level

CZ-4B stage 1
Designation: L-180
Engines: 4 × YF-20B liquid bipropellant single start engines (overall designator YF-21B)
Overall length: 24.65 m
Diameter: 3.35 m
Oxidiser: nitrogen tetroxide in forward tank
Fuel: UDMH in aft tank
Propellant mass: 182.07 t
Dry mass: 9.998 t
Thrust: 2,971 kN sea level
Specific impulse: 2,550 N-s/kg sea level
Burn time: 154 s

Preparation for launch of LM-4 at Xichang launch site 0084634

China Launch History 1975–2005

Launch Vehicle	Launch Date	Payload	Mission	Centre	Result
CZ-2C F01	26 November 1975	FSW-0	LEO	Jiuquan	Success
CZ-2C F02	7 December 1976	FSW-0	LEO	Jiuquan	Success
CZ-2C F03	26 January 1978	FSW-0	LEO	Jiuquan	Success
CZ-2C F04	9 September 1982	FSW-0	LEO	Jiuquan	Success
CZ-2C F05	19 August 1983	FSW-0	LEO	Jiuquan	Success
CZ-3 F01	29 January 1984	DFH-2	LEO	Xichang	Failure
CZ-3 F02	8 April 1984	DFH-2	LEO	Xichang	Success
CZ-2C F06	12 September 1984	FSW-0	LEO	Jiuquan	Success
CZ-2C F07	21 October 1985	FSW-0	LEO	Jiuquan	Success
CZ-3 F03	1 February 1986	DFH-2	GTO	Xichang	Success
CZ-2C F08	6 October 1986	FSW-0	LEO	Jiuquan	Success
CZ-2C F09	5 August 1987	FSW-0/ Piggyback	LEO	Jiuquan	Success
CZ-2C F10	9 September 1987	FSW-1	LEO	Jiuquan	Success
CZ-3 F04	7 March 1988	DFH-2A	GTO	Xichang	Success
CZ-2C F11	5 August 1988	FSW-1/ Piggyback	LEO	Jiuquan	Success
CZ-4 F01	7 September 1988	FY-1	SSO	Taiyuan	Success
CZ-3 F05	22 December 1988	DFH-2A	GTO	Xichang	Success
CZ-3 F06	4 February 1990	DFH-2A	GTO	Xichang	Success
CZ-3 F07	7 April 1990	AsiaSat-1	GTO	Xichang	Success
CZ-2E F01	16 July 1990	DP/BADR-A	LEO	Xichang	Success
CZ-4 F02	3 September 1990	FY-1B (plus 2 balloons)	SSO	Taiyuan	Success
CZ-2C F12	5 October 1990	FSW-1	LEO	Jiuquan	Success
CZ-3 F08	28 December 1991	DFH-2A	GTO	Xichang	Failure
CZ-2D F01	9 August 1992	FSW-2	LEO	Jiuquan	Success
CZ-2E F02	14 August 1992	Aussat-B1	GTO	Jiuquan	Success
CZ-2C F13	6 October 1992	Freja/FSW-1	LEO	Jiuquan	Success
CZ-2E F03	21 December 1992	Optus-B2	GTO	Xichang	Failure
CZ-2C F14	8 October 1993	FSW-1	LEO	Jiuquan	Success
CZ-3A F01	8 February 1994	SJ-4/DP2	GTO	Xichang	Success
CZ-2D F02	3 July 1994	FSW-1	LEO	Jiuquan	Success
CZ-3 F09	21 July 1994	Apstar-1	GTO	Xichang	Success
CZ-2E F04	28 August 1994	Optus-B3	GTO	Xichang	Success
CZ-3A F02	30 November 1994	DFH-3	GTO	Xichang	Success
CZ-2E F05	26 January 1995	Apstar-2	GTO	Xichang	Failure
CZ-2E F06	28 November 1995	AsiaSat-2	GTO	Xichang	Success
CZ-2E F07	28 December 1995	EchoStar-1	GTO	Xichang	Success
CZ-3B F01	15 February 1996	Intelsat-7A	GTO	Xichang	Failure
CZ-3 F10	3 July 1996	Apstar-1A	GTO	Xichang	Success
CZ-3 F11	18 August 1996	ChinaSat-7	GTO	Xichang	Failure
CZ-2D F03	20 October 1996	FHW-2/piggyback	LEO	Jiuquan	Success
CZ-3A F03	12 May 1997	DFH-3	GTO	Xichang	Success
CZ-3 F12	10 June 1997	FY-2	GTO	Xichang	Success
CZ-3B F02	20 August 1997	MabuhaySat	GTO	Xichang	Success
CZ-2C/SD F15	1 September 1997	Motorola MFS	LEO	Xichang	Success
CZ-3B F03	17 October 1997	Apstar-IIR	GTO	Xichang	Success
CZ-2C/SD F16	8 December 1997	Iridium 42, 44	LEO	Taiyuan	Success
CZ-2C/SDF17	26 March 1998	Iridium 51, 61	LEO	Taiyuan	Success
CZ-2C/SD F18	2 May 1998	Iridium 69, 71	LEO	Taiyuan	Success
CZ-3B F04	30 May 1998	ChinaStar-1	GTO	Xichang	Success
CZ-3B F05	18 July 1998	SinoSat-1	GTO	Xichang	Success
CZ-2C/SD F19	20 August 1998	Iridium 3, 76	LEO	Taiyuan	Success
CZ-2C/SD F20	19 December 1998	Iridium 11, 20	LEO	Taiyuan	Success
CZ-4B F01	10 May 1999	FY-1 / SJ-5	SSO	Taiyuan	Success
CZ-2C/SD F21	10 June 1999	Iridium 14A, 21A	LEO	Taiyuan	Success
CZ-4B F02	14 October 1999	CBERS-1 (ZY-1)/SACI-1	SSO	Taiyuan	Success
CZ-2F F01	20 November 1999	Shenzhou 1	LEO	Jiuquan	Success
CZ-3A F04	26 January 2000	ChinaSat-22	GTO	Xichang	Success
CZ-3 F13	25 June 2000	FY-2	GTO	Xichang	Success
CZ-4B F03	1 September 2000	ZY-2	SSO	Taiyuan	Success
CZ-3A F05	31 October 2000	BD 1-01	GTO	Xichang	Success
CZ-3A F06	21 December 2000	BD 1-02	GTO	Xichang	Success
CZ-2F F02	10 January 2001	Shenzhou 2	LEO	Jiuquan	Success
CZ-2F F03	25 March 2002	Shenzhou 3	LEO	Jiuquan	Success
CZ-4B F04	15 May 2002	FY-1D/Haiyang-1	SSO	Taiyuan	Success
CZ-4B F05	27 October 2002	ZY-2	SSO	Taiyuan	Success
CZ-2F F04	30 December 2002	Shenzhou 4	LEO	Jiuquan	Success
CZ-3A F07	25 May 2003	BD 1-03	GTO	Xichang	Success
CZ-2F FF05	15 October 2003	Shenzhou 5	LEO	Jiuquan	Success
CZ-4B F06	21 October 2003	CBERS-2 (ZY-1-2)/ SACI-1 Chuangxin-1	SSO	Taiyuan	Success
CZ-2D F04	3 November 2003	FSW-3	LEO	Jiuquan	Success
CZ-3A F08	15 November 2003	ChinaSat-20	GTO	Xichang	Success
CZ-2C/SM F22	30 December 2003	TC-1	LEO	Xichang	Success
CZ-2C/CTS F23	18 April 2004	Shiyan-1/Naxing-1	SSO	Xichang	Success
CZ-2C/SM F24	25 July 2004	TC-2	LEO	Taiyuan	Success
CZ-2C F25	29 August 2004	FSW-3	LEO	Jiuquan	Success
CZ-4B F07	8 September 2004	SJ-6A, SJ-6B	SSO	Taiyuan	Success
CZ-2D F05	27 September 2004	FSW-3	LEO	Jiuquan	Success
CZ-3A F09	19 October 2004	FY-2C	GTO	Xichang	Success
CZ-4B F08	6 November 2004	ZY-2C	SSO	Taiyuan	Success
CZ-2C F26	18 November 2004	Shiyan-2	LEO	Xichang	Success
CZ-3B F06	12 April 2005	Apstar-VI	GTO	Xichang	Success
CZ-2D F06	5 July 2005	SJ-7	LEO	Jiuquan	Success
CZ-2C F27	2 August 2005	FSW-3	LEO	Jiuquan	Success
CZ-2D F07	29 August 2005	FSW-3	LEO	Jiuquan	Success
CZ-2F F13	12 October 2005	Shenzhou 6	LEO	Jiuquan	Success
CZ-4B F09	26 April 2006	Yaogan 1 and a 1 kg picosat	LEO	Taiyuan	Success
CZ-2C F28	9 September 2006	SJ-8	LEO	Jiuquan	Success
CZ-3A F10	12 September 2006	Chinasat-22A	GTO	Xichang	Success
CZ-4B F10	23 October 2006	SJ-6C, SJ-6D	LEO	Taiyuan	Success
CZ-3B F07	28 October 2006	Sinosat 2	GTO	Xichang	Success
CZ-3A F11	8 December 2006	FY-2D	GTO	Xichang	Success
CZ-3A F12	3 February 2007	Beidou 2A	GTO	Xichang	Success

Lift-off of LM-2E from Xichang launch site (Theo Pirard) 0054375

Launch Vehicle	Launch Date	Payload	Mission	Centre	Result
CZ-2C F29	11 April 2007	Haiyang 1B	LEO	Taiyuan	Success
CZ-3A F13	13 April 2007	Beidou M1	MEO	Xichang	Success
CZ-3B F08	13 May 2007	Nigcomsat 1	GTO	Xichang	Success
CZ-2D F08	25 May 2007	Yaogan 2	LEO	Jiuquan	Success
CZ-3A F14	31 May 2007	Sinosat 3	GTO	Xichang	Success
CZ-3B F09	5 July 2007	Chinasat 6B	GTO	Xichang	Success
CZ-4B F11	19 September 2007	CBERS-2B	LEO	Xichang	Success
CZ-3A F15	24 October 2007	Chang'e-1	Escape	Xichang	Success
CZ-4C F1	11 November 2007	Yaogan 3	LEO	Taiyuan	Success

CZ-4B stage 2
Designation: L-35
Engines: 1 × YF-22 liquid bipropellant single start engine; 1 × YF-23F liquid bipropellant vernier single start engine
Overall length: 10.4 m
Diameter: 3.35 m
Oxidiser: nitrogen tetroxide in forward tank
Fuel: UDMH in aft tank
Propellant mass: 35.408 t
Dry mass: 2.932 t
Thrust: 742 kN main engine, 46.1 kN vernier engine (vacuum)
Specific impulse: 2,922.4 N-s/kg main engine, 2,834 N-s/kg vernier engine (vacuum)
Burn time: 126.8 s main engine, 136.8 s vernier engine

CZ-4B stage 3
Designation: L-14
Engines: 2 × YF-40 liquid bipropellant multiple start engine
Overall length: 4.932 m
Diameter: 2.9 m
Oxidiser: nitrogen tetroxide in forward tank
Fuel: UDMH in aft tank
Propellant mass: 14.3 t
Dry mass: 1.727 t
Thrust: 101.03 kN (vacuum)
Specific impulse: 2,971.4 N-s/kg (vacuum)
Burn time: 359.74 seconds

CZ-4B payload shroud
Overall length: 8.483 m
Max diameter: 2.9 m or 3.35 m
Mass: 1.35 t

CZ-5
The next generation of Long March rockets will be built at a new complex in Tianjin and launched from a new launch centre at Wenchang in the southern province of Hainan. Rocket stages will be delivered to the launch site by sea. The first project is the CZ- 5, which will have a core diameter of 5 m, height of 59.4 m, mass of 643 t and a lift-off thrust of 825 t. It will be powered by two new, cryogenic liquid hydrogen – liquid oxygen engines, one with a thrust of 120 t and the other with a thrust of 50 t. These have already undergone significant ignition testing. The largest version of the CZ-5 will have four 3.35 m diameter strap-on boosters and a 5 m diameter payload shroud. The rocket is likely to replace the Long March 2, 3 and 4. The CZ-5 project began in 2001, and the first flight is now planned for 2013, five years later than originally planned. It will be able to carry up to 25 t to near-Earth orbit and 14 t to geosynchronous orbits. Payloads will include large, multiple satellites and space station modules.

Long March Next Generation Launcher

Current Status
Late in 2005, China's state-owned Beijing Times reported that plans for a complete new family of launch vehicles were still waiting for an official launch decision, and that the first of the new rockets would enter service some six-and-a-half years after such a decision was taken. However, a full-scale technology development programme is under way, including the testing of prototype engines. The new launch vehicle is expected to use non-toxic fuels and the project is likely to include the construction of a new space launch centre on Hainan.

Background
China's government and industry have made great strides in the development of space launch systems with what by world standards is a very robust, simple approach. All of the Long March CZ series launchers, from the relatively small CZ-4B to the 5-tonne-to-GEO CZ-3B and the human-rated CZ-2F, use very similar first and second stage components, based on two motor designs. For example, the liquid strap-on boosters used on the heavier CZ models have the same motor as the standard upper stage.

The CZ programme, however, has its limitations. In terms of cost and reliability, it has yet to prove competitive on the international market, particularly at the upper end of its range: the heavyweight CZ-3B has flown only once since 1998, and competitors such as Proton M, Sea Launch and Ariane offer higher performance. While the US, France and the Ukraine have moved towards the use of LOX/kerosene and LOX/liquid hydrogen fuels, the CZ series still rely on toxic N204/UDMH fuels. China's space launch programme, not very large in volume, is divided among three launch sites, none of them on the coast. Each one specialises in different types of launches, influenced to some extent by their fly-out trajectories. All of them rely on rail or road transportation, which limits the diameter of booster components. For this and other reasons, the CZ series is at, or close to its maximum potential and there is little room for improvement in costs.

Design studies and preparatory work for the new rockets has been under way since the early 1990s. In 1992 a paper presented at the International Astronautical Federation Congress by Xiandong Bao discussed a possible new family of launch vehicles, built around the development of a few standard rocket stages which could be clustered in different ways. The rocket stages would use liquid oxygen/kerosene (for the first time in the Chinese space programme) lower stages and liquid oxygen/liquid hydrogen upper stages.

As part of the programme to develop LOX/kerosene engines, China attempted to purchase an RD-170/171 class engine from Russia, but the Russians refused to sell one of these 7,900 kN thrust engines. Instead the Chinese purchased three RD-120 class LOX/kerosene engines (thrust 835 kN), which are used on the second stage of the Zenit–2.

In June 2000 it was announced that a new LOX/kerosene engine had been fired at the Fengzhou Test Center. This was described as being "a major milestone in China's rocket engine development programme which paves the way in the new-generation, large thrust, non-pollution space launch vehicle". A paper presented at the International Astronautical Federation Congress in 2000 briefly mentioned the new LOX/kerosene launch vehicles which were under development, stating that: "A new expendable launch vehicle using non-poisonous propellants is being studied. It is a one-stage launch vehicle with four [strap-on] boosters, which is used to send satellites into LEO. When it is used to launch GTO satellites a second stage will be added". The most recent official information on the new project, however, was contained in two papers submitted to the International Astronautical Federation Congress, held in Bremen during September to October 2003.

Next Generation Concept
CASC and the China National Space Administration (CNSA) have been working on an ambitious plan to replace the CZ series with a new launcher family. It will be highly modular, covering payloads from 1.5 metric tons LEO to 14 metric tons GTO with a minimum number of basic components. The new launcher family will be based on LOX, LH2 and kerosene fuels. It is intended to be more reliable than the CZ series and will include human-rated versions.

The largest members of the family will have several times the lift performance of the CZ-3B, potentially supporting China's ambitions to develop a space station or send astronauts to the Moon. Included in the plan is a new launch center, probably on Hainan, which will be closer to the equator than current sites and accessible by sea, facilitating the delivery of large rocket components. The entire system is expected to cost 20–30 per cent less (in terms of cost per kg to orbit) making it more competitive on the international scene.

The following information is based on a paper delivered at the 2003 IAF Congress by Tangming Cheng, Xiaojun Wang and Dong Li of the Beijing Institute of Astronautical Systems Engineering. It identified the new system as "a new generation of launch vehicles in the Long March family" rather than as the CZ-5. An accompanying paper used a different designation system for the vehicles and modules than that described here, but the basic vehicle concept was the same.

The core of the new system is based on boost modules in three diameter sizes: 5 m, 3.35 m and 2.25 m. (In terms of cross-sectional area, each is roughly twice the size of the next smallest stage). The 5 m modules are LOX/LH2 fuelled and the two smaller classes burn LOX and kerosene. These modules use three new engine types: a 660 kN LOX/LH2 engine, two of which are fitted to the 5 m module; and a 1,200 kN LOX/kerosene engine, used singly on the 2.25 m module and in pairs on the 3.35 m module; and a 150 kN LOX/kerosene engine to be used on future medium rockets.

For GTO missions, the system also uses a LOX/LH2 upper stage which uses the proven YF-75 engine from the CZ-3B, but with a 5 m diameter structure.

The system can form a very large array of launcher configurations, because the two smaller

Specifications
New Generation Launch Vehicle Modules

Stage designator	**K2-1**	**K3-1**	**H5-1**	**H5-2**
Function	Core for small launch vehicle; strap-on for 335 and 500 series	Core for 335 series vehicle; strap-on for 500 series	Core for 500 series vehicle	Upper stage for 500 series
Oxidiser	LOX	LOX	LOX	LOX
Fuel	Kerosene	Kerosene	LH2	LH2
Number of engines	1	2	2	1
Total thrust, kN	1,200, sea level	2,400, sea level	1,320, sea level	160, vacuum
Specific impulse, sec (sl)	300	300	441	n.a.
Specific impulse, sec (vac)	335	335	426	438
Diameter, m	2.25	3.35	5.0	5.0
Length, m	25.0	26.3	31.0	
Propellant mass, tonnes	63.0	135	158	22.9
Total mass, tonnes	69	147	175	

module classes – 2.25 m and 3.35 m – can be used either as core or strap-on boosters. The 5 m cryogenic core, always flown with four strap-ons, can be fitted with three different combinations of boosters: two or four boosters of either size, or a combination of two 2.25 m and two 3.35 m units. The 5 m-based configurations take advantage of the high thrust possible with kerosene – required for lift-off and initial acceleration at maximum weight – and the higher specific impulse of LH2 for ascent and orbital injection. The LOX/kerosene engine is designed to operate at two thrust levels (65 per cent or 100 per cent) to optimise the ascent profile.

According to the paper published at the 2003 IAF congress, the first step in the programme will be the development of a series of heavy boosters, based on the 5 m module and designated as the 500 series, all of which provide greater payload than any of the CZ family. The second digit in the 500-series designation indicates the number of 2.25 m strap-ons, and the third corresponds to the number of 3.35 m modules, so that the smallest version is the 540, the middle version is the 522 and the largest is the 504. The suffix HO indicates a 2.5 stage GTO launcher with the new upper stage. The 540 configuration lifts 10 t to LEO and the 540/HO has a GTO capability of 6 t; the 504 can launch 25 t to LEO and the 504/HO has a GTO capability of up to 14 t.

The next step in the new-generation launcher programme will be to integrate the 3.35 m and 2.25 m modules into core boosters. The 335 series vehicle family covers the same range of launch capabilities as today's CZ series. It will have a 3.35 m, single-motor first stage, zero to four 2.25 m strap-ons, and a second stage with four new 150 kN LOX/kerosene engines. It will have a LEO capability of 3–10 t according to the number of strap-ons, and will have an updated version of the CZ-3B upper stage for a 1.5-6 t GTO capability.

Finally, a single 2.25 m module and a new second stage with a single 150 kN motor will be combined to form a new small vehicle with a 1.5 t LEO capability, slightly smaller than the CZ-4.

Design principles of the new components include an emphasis on cost rather than maximum performance-to-weight ratio, and the use of advanced fault-tolerant electronics. The goal is a launch reliability of 98 per cent for the entire family and 99 per cent (backed up by an escape tower) for manned missions. If the 335-series vehicle replaces the current man-rated CZ-2F, as seems likely, the core and strap-on systems should all have been developed and flown operationally on 500-series boosters before the 335 is tested.

A new "direct-to-pad" launch concept is planned for the new vehicle. The launch vehicle components will be delivered directly from the manufacturing site to the launch pad itself and assembled vertically with the help of a Mobile Service Tower (MST). Assembly, fuelling and final check-out all take place on the pad. The goal is a 20-day process from the arrival of the first-stage core to lift-off. Meanwhile, the payload will be integrated and encapsulated in a separate facility, and mated to the launcher on the pad three days before launch.

The People's Daily reported in October 2005 that a new site in Hainan was being considered for the launch of the new vehicles, because of its latitude, over-water fly-out zones and access to sea transport. In January 2006, the newspaper also reported that the LOX/LH2 engine for the core stage had completed a 200-second hot-firing test after five years of development.

500 series launch vehicles

The launch vehicles based on the 5 m diameter core stage represent the 500 family. All of the variants described by Chinese documents use four strap-on boosters which are based on the K2-1 and/or K3-1 modules. For missions to low Earth orbit the second stage H5-1 stage (ignited at launch) operates all of the way to orbital injection. When used for missions to geosynchronous orbit the H5-2 stage is added and the H5-1 stage is probably sub-orbital.

504
Strap-on boosters: 4 × K3-1
Second stage core: 1 × H5-1
Total length: 47.6 m
Launch mass: 779.5 t
Payload capacity: 25 t to LEO

504/HO
Strap-on boosters: 4 × K3-1
Second stage core: 1 × H5-1
Third stage: 1 × H5-2
Total length: 50 m
Launch mass: 789.5 t
Payload capacity: 14 t to GTO
10.6 t to trans-lunar trajectory

522
Strap-on boosters: 2 × K2-1 plus 2 × K3-1
Second stage core: 1 × H5-1
Total length: 47.6 m
Launch mass: 614.5 t
Payload capacity: 18 t to LEO

522/HO
Strap-on boosters: 2 × K2-1 plus 2 × K3-1
Second stage core: 1 × H5-1
Third stage: 1 × H5-2
Total length: 50 m
Launch mass: 636.6 t
Payload capacity: 10 t to GTO
8.1 t to trans-lunar trajectory

540
Strap-on boosters: 4 × K2-1
Second stage core: 1 × H5-1
Total length: 42.6 m
Launch mass: 442.5 t
Payload capacity: 10 t to LEO

540/HO
Strap-on boosters: 4 × K2-1
Second stage core: 1 × H5-1
Third stage: 1 × H5-2
Total length: 47 m
Launch mass: 462 t
Payload capacity: 6 t to GTO
4.4 t to trans-lunar trajectory

India

Geosynchronous Satellite Launch Vehicle (GSLV)

Current Status

By March 2008, GSLV had flown five times, with one launch failure. The indigenously developed Cryogenic Upper Stage was declared fully qualified after a 12 minute ground test on 15 November 2007. The first launch of the GSLV Mk. II with this stage is scheduled for 2008.

Background

GSLV was originally designed and developed by ISRO's Vikram Sarabhai Space Centre, Thiruvananthapuram, in order to create an Ariane 4-class Geosynchronous SLV with a 3.4 m diameter fairing, capable of handling 2.5 tonne Insat 2-class satellites. To develop the plans, some 500 configurations were reduced to four candidates during 1986–88, divided equally between solid/liquid cores and both with solid/liquid strap-ons. GSLV replaces PSLV's six solid strap-ons with four liquids powered by Vikas engines (although the initial version used PSLV strap-ons) and substitutes a cryogenic stage for the two upper stages. The vehicle is comparable to Ariane 44L except for the lower performance solid stage 1. The Vikas engine is derived from the SEP Viking 2 motor.

Ground test version of the Vikas engine used on the strap-on boosters for the GSLV; the original Vikas variant flew on the second stage of the PSLV (Liquid Propulsion Systems Centre) 0003431

Glavkosmos of Russia signed a RUR2,350 million agreement with ISRO in January 1991 to supply India with a cryogenic engine. This KVD-1 76 kN KB Khimmach engine was originally developed to power Russia's N-1 manned booster for the L-3M programme in the late 1960s. The contract called for two engines plus the technology transfer required for subsequent Indian production. However, the US government argued that this agreement was a violation of its Missile Technology Control Regime. Eventually, in late 1993, a compromise was reached which allowed Russia to supply seven engines to India without the transfer of critical technologies. The first engine was delivered in 1996 for the planned inaugural GSLV mission in late 1997 or early 1998. Test firings of lower stage GSLV motors were underway in 1994. ISRO noted that two years had been lost on the GSLV programme and forced a decision to develop an indigenous cryogenic engine. Meanwhile, the Indian government approved plans for a domestic water-cooled LOX/GH_2 subscale engine which ISRO tested successfully on 21 July 1989. Critics of international control regimes pointed out that the US action to stop the Russian sales actually created another source of such engines, namely India, and that a resentful India might not be restrained in efforts to sell its technology overseas.

The course of cryogenic engine development has not been easy for India. The contractor LPSC's first attempt at firing a 10 kN LOX/LH_2 engine ended with an explosion at ignition in early July 1993. LH_2 became available in the country only after a US company built a plant at LPSC's Mahendragiri site at the end of 1992.

The programme was restructured to include plans for a basic GSLV with 1,500 to 2,000 kg GTO compatibility, a GSLV Mk II with 2,500 kg GTO compatibility and a GSLV Mk III (C-20) with 4,000 kg GTO capability.

The 49 m tall GSLV Mk I is a three stage vehicle. The payload fairing is 3.4 m in diameter and 7.8 m in length. The first stage (GS1) comprises a S125 or S139 core motor with 129/138 tonnes of Hydroxyl Terminated Poly Butadiene (HTPB) based solid propellant and four L40/L40H strap-on motors, each carrying 40/42 tonnes of hypergolic liquid propellants. The first stage core motor, made up of five segments, is the same as the one used on PSLV and is amongst the largest solid propellant boosters in the world.

GSLV-F01 on one of the launch pads at Sriharikota (ISRO) 1343229

The strap-ons use indigenously developed Vikas L40 motors based on the Ariane Viking-2 engine of SEP. They are ignited on the ground to augment the first stage thrust. On GSLV-D1, each of the L40 motors carried 40 tonnes of UDMH and nitrogen tetroxide (N_2O_4), stored in two independent tanks. GSLV-D2 and later flights used an uprated (L40H) version of the Vikas engine, tested in December 2001, that develops higher chamber pressure of 58.5 bar against 52.5 bar. This uprated engine uses UH25 (a mixture of UDMH and hydrazine hydrate) as fuel and N_2O_4 as oxidiser, and has a silica-phenolic throat that allows extended burn time. The S125/S138 solid propellant core stage is ignited 4.6 seconds after confirming the normal operation of each of the L40/L40H stages.

The second stage (GS2) L37.5/L37.5H motor uses the indigenously-built Vikas engine based on the Viking-4A engine of SEP France and carries 37.5/39 tonnes of UDMH and N_2O_4. The vented inter-stage between the first and second stage enables the firing of the second stage 1.6 seconds before the first stage has completed firing.

The third stage (GS3) uses a Russian-built C12.5 cryogenic stage with 12.5 tonnes of liquid oxygen and liquid hydrogen. The 12KRB upper stage was developed and fabricated by the Khrunichev Space Centre for GSLV. Its sustainer was developed by the Isaev Chemical Machine Building Design Bureau. The restartable stage is powered by a 73.5 kN KB KhimMash KVD-1 RD-56M cryo engine (with two vernier engines) and burns for a duration of about 720 seconds. The engine has Indian avionics and software.

The GSLV utilises Insat 2's 119 kg (dry mass) unified bipropellant liquid propulsion system made up of a single 440 N LAM Liquid Apogee Motor and two redundant networks of 8 × 22 N RCS thrusters for final orbital placement. The qualification LAM was tested for a total 9,550 seconds, including a single burn of 3,000 seconds. RCS testing accumulated 30,500 seconds including a single continuous burn of 10,000 seconds. Pulsed firing mode logged >250,000 pulses, each component being tested over three life cycles. The engines, pressure regulator, check valve, fill, drain and vent valve, pyro valve and pressure transducer were developed indigenously; the tanks, gas bottles, filters and latch valves were procured from foreign suppliers. The system carries about 450 welded joints.

L40 strap-ons undergoing preparation (ISRO) 1343227

GSLV employs a flexible linear shaped charge separation system for the first stage, a pyrotechnically-actuated collet release mechanism for the second stage, and a Merman band bolt cutter separation mechanism for the third stage. Spacecraft separation is by spring thrusters mounted at the separation interface. The three-axis attitude stabilisation of GSLV is achieved by autonomous control systems provided in each stage. Single plane engine gimbal controls on the four strap-ons of the first stage are used for pitch, yaw and roll control. The second stage uses engine gimbal control for pitch and yaw and a hot gas reaction control system for roll control. Two swivelling vernier engines using LH_2 and LOX provide pitch, yaw and roll control for the third stage during thrust phases. A cold gas orientation system is used during third stage coast phases. The booster's inertial guidance system is located in the equipment bay above the third stage. The closed loop guidance scheme used by the on-board computer ensures the required accuracy in orbital injection conditions.

The maiden GSLV launch was delayed through technical difficulties in India and financial problems in Russia. The first launch of the GSLV (**GSLV-D1**) eventually took place from Sriharikota on 18 April 2001 with the launch of the 1,500 kg GSAT-1, a scaled down test satellite for future geosynchronous communications satellites. GSLV-D1's initial launch attempt on 18 March 2001 was aborted one second before lift off by the Automatic Launch Processing System when it detected that one of the L40 stage strap-ons was not providing the required thrust due to "defective plumbing in the oxidiser flow line". The problem was fixed by replacing it with a spare stage and the launch 18 days later was largely successful. However, there was a launch velocity shortfall of 0.6 per cent (mainly due to a cryogenic stage thrust shortfall of 4.1 seconds to 705.8 seconds instead of 709.9). An attempt was made to reach a usable orbit using the station-keeping motor of the GSAT satellite. After a series of burns, GSAT-1 ran out of propellant, leaving it in a 35,665 × 33,806 km, 0.99° orbit. The motor for the cryogenic upper stage had been purchased from Russia but the design had never flown in space before. In the end, the parameters of the drifting orbit (about 13°/day) were a period of 23 hours, apogee of 35,665 km, perigee of 33,806 km and inclination 0.99°. The fully functional transponders and transmitters on board were deactivated on instructions of the International Telecommunications Union.

The second flight of the GSLV (**GSLV-D2**) took place on 8 May 2003 carrying the experimental communications satellite GSAT-2, which weighed 1,825 kg at launch. About 17 minutes after lift off the satellite was placed in a 180.4 km × 36,000 km orbit at 19.2° inclination. The two test flights qualified the GSLV for launching 2 tonne satellites to GTO.

The first operational flight, **GSLV-F01**, successfully launched the 1,950 kg Edusat on 20 September 2004. Edusat was placed in a 180 × 35,985 km transfer orbit 1,014 seconds after lift off. It later reached its operational geostationary position at 74°E with the aid of its 440 N liquid apogee motor.

The fourth flight (**GSLV-F02**) on 10 July 2006 was unsuccessful. A defective propellant regulator on the fourth strap-on motor caused the vehicle to veer off course. The rocket and its Insat-4C payload had to be destroyed over the Bay of Bengal one minute after lift off from Sriharikota. The Failure Analysis Committee concluded that the design of GSLV is robust, but recommended implementation of stricter control on fabrication, inspection and acceptance procedures.

The fifth flight (**GSLV-F04**) successfully delivered a replacement satellite, Insat-4CR, to GTO on 2 September 2007. The 2,140 kg satellite was placed in orbit about 17 minutes after lift off.

GSLV-D3, scheduled for 2008, will be fully indigenous and carry the GSAT-4 experimental satellite. On 15 November 2007, ISRO announced

Launch Record

Vehicle/Flight	Launch Date	Launch Site	Payload	Outcome
GSLV-Mk I/GSLV-D1	18 April 2001	Sriharikota	GSAT-1 (1,540 kg)	Partial success. Cryogenic upper stage thrust shortfall left satellite in wrong orbit.
GSLV-Mk I/GSLV-D2	May 2003	Sriharikota	GSAT-2 (1,820 kg)	Successful delivery to GTO.
GSLV-Mk I/GSLV-F01	20 September 2004	Sriharikota	Edusat (1,950 kg)	Successful delivery to GTO.
GSLV-Mk I/GSLV-F02	10 July 2006	Sriharikota	Insat-4C (2,168 kg)	Failure – loss of thrust in one of four liquid propellant strap-on motors (S4).
GSLV-Mk I/GSLV-F04	2 September 2007	Sriharikota	Insat-4CR	Successful delivery to GTO.

GSLV-F04 Flight Profile

Event	Time (sec)	Altitude (km)	Velocity (km/sec)
GS1 burn out	149.9	70.4	2.8
Payload fairing separation	227.8	115.0	3.9
GS2 burn out	292.0	133.0	5.4
GS3 burn out	1,002.9	220.1	10.2
Satellite separation	13 seconds after GS3 burn out		

that it had successfully completed ground testing of its indigenously developed Cryogenic Upper Stage (CUS), to be used on the GSLV Mk. II. The test was conducted for the full GSLV flight duration of 720 seconds at ISRO's Liquid Propulsion test facility at Mahendragiri, Tamil Nadu. The flight stage will be used on the next mission of GSLV (GSLV-D3) in 2008, replacing the Russian stage used in the basic version. The CUS is powered by a regeneratively cooled cryogenic engine, which works on a staged combustion cycle developing a thrust of 69.5 kN in vacuum. Liquid Oxygen (LOX) and Liquid Hydrogen (LH_2) are fed by individual booster pumps to the main turbo-pump, which rotates at 39,000 rpm, to ensure a high flow rate of 16.5 kg/seconds of propellants into the combustion chamber. The main turbine is driven by the hot gas produced in a pre-burner. Thrust control and mixture ratio control are achieved by two independent regulators. LOX and Gaseous Hydrogen (GH_2) are ignited by pyrogen type igniters in the pre-burner as well as in the main and steering engines.

India hopes to launch its next-generation GSLV Mk. III booster by late 2009 or early 2010. The Mk. III will be able to lift a 4 tonne payload into GTO although there is the potential to increase this eventually to more than 5 tonnes. India plans to use the new vehicle to penetrate the commercial launch market. The Mk. III makes maximum use of technologies used by the baseline GSLV and its predecessor, the PSLV. This includes the two S200 solid strap-on boosters and storable propellant L110 core stage, as well as existing integration and launch facilities at Sriharikota. The C20 cryogenic upper stage is a higher power derivative of the CUS being developed for the GSLV Mk. II.

The GSLV Mark III will be able to launch 4 tonnes to GTO (Jane's/Patrick Allen) 1342961

Specifications
GSLV
First launch: 18 April 2001
Launch site: SHAR Centre (Sriharikota)
Principal uses: Medium-class GTO
Cost: USD45 million (in 1999 currency values)
Performance: 2,500 kg into GTO
Number of stages: 3 (1 solid/2 liquid) + 4 liquid strap-ons (6 PSLV solids initially)
Dimensions: Length 49 m, core diameter 2.8 m.
Launch mass: 402 tonnes
Guidance: Similar to PSLV

GSLV liquid strap-ons
Comment: PSLV's Vikas engine was adapted for GSLV's strap-ons. A successful 200 s run 24 July 1995 tested the new indigenously-developed silica-phenolic throat, designed to withstand the operational 1,300°C
Designation: L40/L40H
Overall length: 19.7 m
Principal diameter: 2.1 m
Dry Mass: 4 × 5,5 t
Total mass: 4 × 45.5 t
Propellant: 4 × 40 t UDMH/N_2O_4
Thrust: 4 × 735 kN vacuum average
Chamber pressure: 52.6 bar
Specific impulse: 260 s (vacuum)
Burn time: 158 s
Attitude control: Gimballed ±5° for 3-axis

GSLV core stage
Designation: S125/S139
Other specifications: Same as PSLV

GSLV stage 2
Designation: L37.5/L37.5H
Other specifications: Same as PSLV

GSLV stage 3
Designation: C12.5
Overall length: 8.72 m
Principal diameter: 2.80 m
Total mass: 14.60 t
Propellant: 12.50 t LOX/LH_2
Thrust: 76 kN vacuum average (450 s Isp), including two gimballed LOX/GH_2 verniers fed by main chamber pumps
Specific impulse: 76 s (vacuum)
Burn time: 720 s
Attitude control: Gimballed, auxiliary and cold gas
Payload fairing: 3.4 m and 7.8 m long, aluminium

LAM
First flight: July 1992 (Insat 2A)
Dry mass: 4 kg
Length: 570 mm
Maximum diameter: 276 mm
Mounting: Fixed
Engine cycle: Pressure-fed in regulated mode
Propellants: MMH/N_2O_4 from tanks at 16 atmosphere
Mixture ratio (O/F): 1.65
Thrust: 440 N vacuum nominal
Specific impulse: 310 s vacuum minimum
Expansion ratio: 160
Chamber pressure: 6.9 atmosphere
Injector: Coaxial titanium welded to chamber
Combustion chamber: Silicide-coated columbium alloy, radiatively cooled
Burn time: Qualified to 3,000 s continuous; minimum impulse bit not available

Insat 2 RCS Thruster
First flight: July 1992 (Insat 2A)
Dry mass: 850 g
Length: 249 mm
Maximum diameter: 60 mm
Mounting: Fixed
Engine cycle: Pressure-fed in blowdown mode
Propellants: MMH/NTO from tanks at 16 atmosphere
Mixture ratio (O/F): 1.65
Thrust: 22 N vacuum nominal
Specific impulse: 285 s vacuum minimum
Expansion ratio: 100
Chamber pressure: 6.9 atmosphere
Throat temperature: 1,200°C
Injector: Coaxial titanium welded to chamber
Combustion chamber: Silicide-coated columbium alloy radiatively cooled
Burn time: Qualified to 1,000 s continuous; minimum impulse bit not available

PSLV

Current Status
India has committed to moving its payloads from foreign launches to national launch vehicles and the PSLV is scheduled for an average of one flight a year for the foreseeable future. India has no plans to commercialise the PSLV for marketing.

Background
India's Polar Satellite Launch Vehicle (PSLV) represents the first stage in acquiring launcher autonomy for applications satellites. Sized for placing 1 tonne IRS Earth resources satellites in 817 km Sun-synchronous orbits from Sriharikota. The Indians have paid Rs 4,155 million, to develop the PSLV through its first two flights and the ground infrastructure. Each launcher itself costs about Rs 450 million. The government, in October 1994, approved six further tests known as PSLV C1-C6, (C = continuation) vehicles, after the first three development flights named PSLV D1-D3, (D = development). Some estimates state the total cost for all six flights is Rs 4,000 billion. The C version should improve performance to 1,300 kg. IRS-1D (1,200 kg) will use the PSLV.

PSLV employs an unusual combination of liquid and solid stages, using the first Indian liquid systems for stages 2 to 4 and clustering six ASLV strap-ons around the solid first stage. Stage 2's single Vikas engine is based on SEP's Viking from Ariane. In order to prepare for the eventual programme, the Indian Space Research Organisation began long-duration testing on the engine in January 1988 at LPSC's Mahendragiri facility.

In practice, Sriharikota has a launch constraint on azimuths of less than 140°, forcing PSLV to execute a 55° yaw manoeuvre after 100 seconds and stage 1 separation to reach a Sun-synchronous path. Without this constraint the PSLV could launch 1.6 tonne into Sun-synchronous orbit.

PSLV-D1 performed well but an uncorrected pitch problem after stage 2 separation left it at only 340 km altitude after stage 3's burn instead of the planned 414 km. Stage 4 had insufficient reserve to attain a useful orbit. Investigators later found

PSLV provides a 1 t Sun-synchronous capacity (ISRO) 0516904

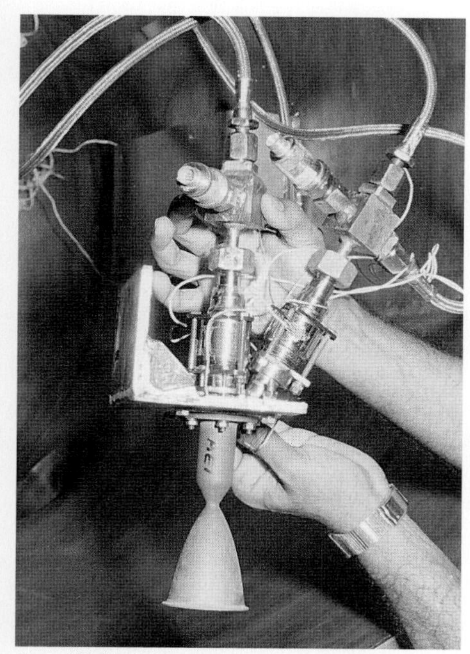

The 22 N RCS thruster carried on INSAT 2 (Liquid Propulsion Systems Centre) 0003444

stage 2 engine nulling occurring about 3.7 seconds before stage 3 ignition and as a consequence, the stages recontacted because two small solid retros failed, producing a pitch disturbance that exceeded the preset limit. PSLV could have removed it but a software error in the control loop prevented action. D2, with modified software, was successful, followed by D3 in March 1996. C1 launched in September 1997 placed the IRS satellite in an eccentric orbit instead of the planned circular orbit due to a fourth stage malfunction. C2 placed three satellites in Sun-synchronous orbit in May 1999. C3 also carried three satellites to orbit in October 2001. C4 was launched in September 2002 carrying Metsat to a GTO followed by C5 on 17 October 2003 carrying IRS-P6 (Resourcesat-1) to a sun-synchronous orbit.

Specifications
PSLV
First launch: 20 September 1993
Number launched: 8 through end of October 2003
Launch sites: SHAR Centre (Sriharikota)
Principal uses: 1 t-class remote sensing payloads into Sun-synchronous orbit, 3 t-class into LEO
Vehicle success rate: 87.5 per cent
Cost: Rs 450 million each D1-D3, Rs 650 million C1-C6
Availability: launches projected every year

The liquid propellant engine, two of which are carried on the fourth stage of the PSLV (Liquid Propulsion Systems Centre) 0003446

Performance: Sun-synchronous: 1,000 kg into 900 km, 99.1° from SHAR (1.6 t without range safety constraints)
GTO: 450 kg
LEO: 3,000 kg into 400 km
Number of stages: 4 (2 solid/2 liquid) plus 6 solid strap-ons
Overall length: 44.2 m
Fairing diameter: 3.2 m (5.1 m base circle with strap-ons)
Launch mass: 283 t
Guidance: Inertial Guidance System located in vehicle equipment bay surrounding stage 4 base. Redundant strapdown inertial navigation system (Resins, incorporating three dry-tuned gyros + four servo accelerometers) feeds the navigation processor, which produces navigation data every 500 ms for the guidance & control processor to issue steering commands at similar intervals. Open loop guidance is employed during stage 1 burn before switching to closed loop from stage 2 onwards. Target 3Σ injection accuracy for Sun-synchronous payloads ±35 km altitude, ±0.2 in c.
Launch sequence (typical IRS mission, min/s):

00:00	stage 1 ignite
00:00	2 solids ignite
00:30	4 solids ignite at 3 km
00:54	2 solids burnout at 17 km
01:13	2 solids sep at 24 km
01:19	4 solids burnout at 36 km
01:30	4 solids sep at 38 km
01:45	stage 2 ullage motors ignition, stage 1
–	sep + stage 1 retro motors ignition
01:45	stage 2 ignition at 48 km
02:32	fairing sep at 105 km
02:37	closed loop guidance begins
04:21	stage 2 shutdown, sep, retro
–	ignition, stage 3 ignition at 232 km
05:33	stage 3 burnout at 350 km
05:07	stage 3 sep at 405 km
10:17	stage 4 ignition at 700 km
17:06	stage 4 shutdown at 817 km

PSLV solid strap-ons
Designation: PSOM (PSLV strap-on motors).
Comment: The segmented motors are almost identical to the ASLV strap-ons. Two are ignited on the pad, followed 30 s later by the remaining four, with separation occurring at 73/90 s respectively. Nozzles are canted 9°. One ground-lit and one air-lit strap-on include TVC secondary injection to augment roll control.
Overall length: 11 m
Principal diameter: 1.0 m
Propellant: 8,628 kg of HTPB solid
Average thrust: about 440 kN vacuum each, 662 kN max
Burn time: 47 s
Attitude control: exhaust secondary injection
Separation: pyrotechnic nuts and springs

PSLV stage 1
Designation: PS1
Overall length: 20.3 m
Principal diameter: 2.8 m
Total mass: 156.0 t
Propellant: 129.0 t of HTPB solid in five maraging steel segments, each 3.4 m long and cast individually (centre three are interchangeable); segments held by 144 pins in tongue/groove joint.
Thrust: max 4,600 kN vacuum, 3,500 kN thrust at launch
Burn time: 103 s
Attitude control: pitch/yaw provided by secondary injection TVC of strontium perchlorate from two 70 cm diameter 1,650 litre steel tanks through quadrants of six valves from a total of 24 into stage 1 nozzle (35% of distance from throat to exit). Each quadrant can generate a 200 kN side force; valve flow rate is 12 litres maximum. Roll control provided by two continuous-burn swivelling 6.4 kN vacuum NTO/MMH thrusters below SITVC tanks.
Separation: flexible linear shaped charge and eight 48.5 kN mean thrust 1,414 × 209 mm diameter retros, each burning 26.8 kg propellant in 1.3 s

PSLV stage 2
Comment: The integrated system was first tested at Mahendragiri 21 March 1990. Final qualification test October 1992 for 150 s. Four 15.3 kN solid motors burn for 5.5 s at stage 1 separation to provide ullage for stage 2.
Designation: PS2/L37.5
Engine: Vikas open gas generator cycle engine (based on SEP Viking), gimballed for pitch/yaw control, single start, film + radiatively cooled, pumps driven by single 9,400 rpm shaft powered by gas generator.
Overall length: 11.5 m
Principal diameter: 2.8 m

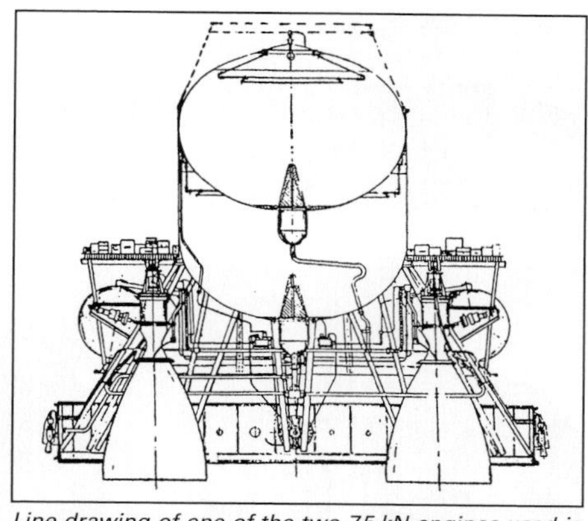

Line drawing of one of the two 7.5 kN engines used in PSLV stage 4 (ISRO) 0516906

Dry mass: 5.3 t
Propellant: 37.5 t of UDMH/NTO
Stage thrust: 725 kN vacuum
Burn time: 149 s
Attitude control: Vikas gimballing for pitch/yaw; roll control provided by two on/off 300 N nozzles drawing hot gas from Vikas gas generator
Separation: at 249 s/243 km, with four 22.6 kN 1.6 s retros providing separation from third stage

PSLV stage 3
Comment: Stage 3 is suspended inside a 2.8 m diameter shroud from the fourth stage skirt
Designation: PS3
Overall length: 354.1 cm
Principal diameter: 2.0 m
Propellant: 7,260 kg of HTPB in Kevlar casing
Thrust: max 386 kN vacuum
Burn time: 73.1 s (78.3 s action time)
Attitude control: pitch/yaw by ±2° nozzle flexure (qualified to ±3°) controlled by two actuators; roll control from the 50 N roll thrusters on stage 4.
Separation: ball release and springs

PSLV stage 4
Comment: The integrated stage was first tested at Mahendragiri on 27 September 1989, beginning a programme including two full duration trials and a demonstration of re-start capability.
Designation: PS4/L2
Overall length: 2.65 m
Principal diameter: 1.335 m
Dry mass: 920 kg
Propellant: 2 t MMH/NTO
Average thrust: 14 kN
Burn time: 425 s (D2: 397 s)
Attitude control: six 50 N thrusters in two blocks drawing propellants from stage 4's tanks provide 3-axis control during coast period; main engine gimballing employed during burn.
Separation: Merman clamp and springs

PSLV's stage 4 incorporates two 7.5 kN engines and, for coast attitude control, six 50 N thrusters. Surrounding the tank is PSLV's vehicle equipment bay. On top is IRS 1E (ISRO) 0517355

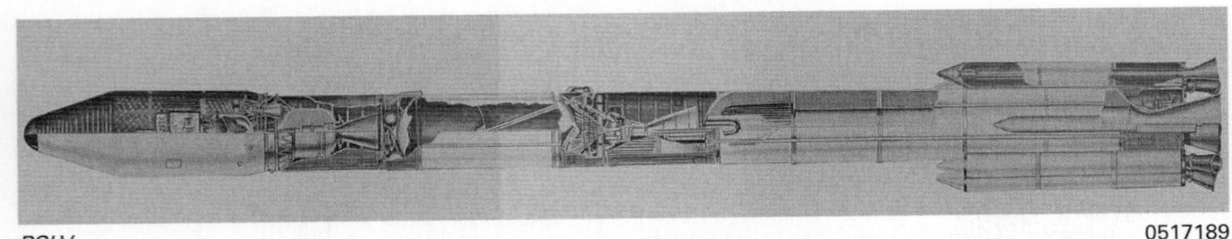

PSLV 0517189

PSLV stage 2: ISRO developed the Vikas engine from Ariane's Viking to power the first large Indian liquid stage. The toroidal water coolant tank is visible inside the base skirt, with solid ullage and retro motors on the exterior (ISRO) 0516905

PSLV stage 4 engines

Description: two ISRO 7.5 kN pressure-fed bipropellant engines, gimballed up to ±3° in orthogonal axes for 3-axis control, 8.37 atmosphere chamber pressure, 60:1 expansion ratio. The titanium alloy tanks are pressurised at 19.2 atmospheres by helium stored at 300 atmospheres in five titanium spheres. Tank incorporates surface tension-type propellant acquisition system to ensure supply under adverse acceleration conditions.
First flight: 20 September 1993
Dry mass: 28 kg
Length: 1.1 m
Maximum diameter: 0.63 m
Mounting: gimballed up to ±3° in two orthogonal planes by a closed loop servo control system to provide stage pitch, yaw and roll control during thrust phase.
Engine cycle: pressure-fed
Propellants: MMH/NTO from tanks at 19.2 atmosphere
Mixture ratio (O/F): 1.4
Thrust: each 7,500 N vacuum nominal
Specific impulse: 308 s vacuum minimum
Expansion ratio: 60
Chamber pressure: 8.37 atmosphere
Injector: AISI 304 stainless steel; 45 sets of triple elements
Combustion chamber: AISI 304 stainless steel film cooled along single helical grooving
Nozzle: silicide-coated columbium alloy radiatively cooled
Burn time: 425 s mission nominal
Qualification time: 530 s (continuous)

The 440 N LAM carried on INSAT (Liquid Propulsion Systems Centre) 0003445

PSLV L37.5 Vikas engine

Comment: India's first large liquid propellant rocket engine is derived from Viking technology acquired from SEP. The engine achieved its first full-duration 150 s firing in January 1988 on the principal test stand at the Liquid Propulsion Test Facility, Mahendragiri, and completed the first integrated stage test 21 March 1990 (qualification completed October 1992). A 686 kN version will be used for GSLV's strap-ons. A successful 200 s run 24 July 1995 tested the new indigenously-developed silica-phenolic throat, designed to withstand the operational 1,300°C.
First flown: 20 September 1993
Number flown: 4 to May 1999
Dry mass: 876 kg
Length: 3.509 m
Mounting: gimballed for pitch/yaw control
Engine cycle: gas generator (9400 rpm turbine)
Propellants: UDMH/NTO
Mixture ratio (O/F): 1.86
Thrust: 72 kN vacuum (686 kN for GSLV)
Specific impulse: 295 s vacuum
Time to full thrust: 2.4 s
Expansion ratio: 31 (13.88 C-SU strap-on)
Chamber pressure: 51.9 atmospheres
Injector: 216 like on like doublets set in six rows for each propellant in light alloy annular injector
Chamber cooling: UDMH film supplied through additional channels on the injector's lower section
Igniter: hypergolic
Nozzle materials: cobalt alloy with SEPHEN (phenolic resin/silica fibre) throat
Burn time: 160 s nominal mission

Payload fairing/accommodation

Hindustan Aeronautics Ltd. builds 3.2 m diameter aluminium fairing, 8.3 m long. The 4.5 m long isogrid central cylindrical section is topped by a nose with a 70 cm radius cap/20° sloping wall; the boat-tail narrows along a 14° slope to 2.8 m diameter. A Merman clamp, explosive bolt cutters and zip cord, effects separation. 937 mm payload adapter using 12.2 kN Merman clamp; 4 springs provide 80 cm/s max separation velocity.
Acceleration: 6.0 g max, during stage 3
Acoustic: payloads qualified to 146 dB overall for two min, acceptance testing to 142 dB for one min

PSLV stage 1 motor

PSLV's first stage is the most powerful model produced by ISRO; it was successfully static fired for the first time 21 October 1989. The second was successfully fired 23 March 1991.
Application: PSLV stage 1
First flight: 20 September 1993
Number flown: 6 to end 2001
Mass (kg): 145,000
Length (cm): 2,034.5
Diameter (cm): 280.4
Propellant:
type: HTPB, segment-cast with three central segments interchangeable
shape: forward segment deep 10-fin to burn 22 t loading in 20 s to generate high launch thrust; remainder tubular with average 120 cm diameter bore.
mass fraction: 0.890 (129 t loading)
Burn time (s): 100
Thrust (kN, vacuum): 3,463 mean, 4,600 max at 17 s
Specific impulse (s, vacuum): 265
Total impulse (kNs, vacuum): 326,000
Pressure (atmosphere): 58 max at 17 s
Nozzle:
throat diameter (cm): 80
length (cm): 2,034.8
exit diameter (cm): 227
expansion ratio: 8
materials: carbon phenolic throat insert, tape-wound carbon phenolic on divergent cone changing to silica phenolic below TVC inlets.
Casing materials: five segments of M250 maraging steel, 3.4 m long. Joints tongue/groove with two O-rings and 144 pins. Insulated by nitrile rubber, with silicon potting compound on joints.
Igniter type: pyrogen head-end 30 cm diameter, 125 cm long containing Pedpro 2661 HTPB and its own 2 kg-charge igniter

PSLV stage 3 motor

Comment: This is ISRO's most advanced solid propellant motor, designed for a 0.915 mass fraction and the first to adopt a submerged flex nozzle design for TVC. Off-loading is also possible. Test #5 made 29 July 1991, #6 17 October 1991. Tests 8, 9 and 10 were for qualification.
Application: PSLV stage 3 (PS3)
First flight: 20 September 1993
Number flown: 6 to end 2001
Mass (kg): 7,975
Length (cm): 244.20 (casing 208.35)
Diameter (cm): 198.8 (202.5 skirt)
Propellant:
type: HTPB casebound, 86% solid
shape: slotted
mass fraction: 0.910 (7,260 kg standard loading)
Burn time (s): 73.1; 78.1 action time
Thrust (kN, vacuum): 243.2 mean, 386 max
Specific impulse (s, vacuum): 293
Total impulse (kNs, vacuum): 20,850
Pressure (atmosphere): 57.4 max
Nozzle: the 19.6 per cent submerged contoured nozzle is provided with a ±3° (±2° flight) conical flex bearing. The elastomer, based on natural rubber of nominal 3 mm thickness is bonded with steel reinforcements of 4.5 mm thick run. Each seal consists of six reinforcements and seven elastomer pads. The seal positioned beyond the throat housing is driven by two 11.8 kN electromechanical actuators 90° apart; a thermal boot protects the bearing.
throat diameter (cm): 19.4
length (cm): 180.2
mass (kg): 280
exit diameter (cm): 141.4
expansion ratio: 52.3
materials: graphite throat insert, carbon and silica phenolic composites for nozzle.
Casing materials: Kevlar filament-wound, EPDM insulation and aluminium forgings for skirt and end-opening fittings; total mass 340 kg. Nozzle opening 750 mm diameter, 200 mm head end.
Igniter type: pyrogen head-end

PSLV strap-on motor

Comment: The same basic motor, derived from the SLV-3's stage 1, was employed for ASLV's stage 1 and strap-on, and adopted as PSLV's initial strap-on. The strap-ons are essentially identical to the ASLV stage 1 version, with the principal exception of canted nozzles. For PSLV, the secondary injection TVC systems are deleted on one ground-lit and three air-lit strap-ons. The specifications below are for ASLV's stage 1 model; differences for the strap-ons are noted where applicable.
Application: ASLV stage 1 + strap-ons, PSLV strap-ons
First flown: 24 March 1987, ASLV-D1 launch
Number flown: 24 PSLV strap-ons, 4 ASLV stage 1, 8 ASLV strap-ons to May 1999
Mass: 10.4 t
Length (cm): 998.99
Diameter (cm): 100
Propellant:
type: AP/18%Al/HTPB case-bonded
shape: star
mass fraction: 0.853 (8.9 t loading; 8.63 t for strap-on)
Burn time (s): 46 action time (49.95 action strap-on)
Ignition delay: about 200 ms strap-on
Thrust (kN, vacuum): 502.6 mean, 643 max (strap-on: 440 mean, 580 max)

Specific impulse (s, vacuum): 259 (252 strap-on)
Total impulse (kNs, vacuum): 22,600 (21,450 strap-on)
Pressure (atmosphere): 43.5 max (40.2 max strap-on)
Nozzle: secondary injection ports for TVC; 9°-canted nozzles
expansion ratio: 6.7
materials: graphite throat, carbon phenolic composite for convergent section, tape-wound carbon and then silica phenolic composites for divergent.
Casing materials: 3.5 mm thick low carbon steel with nitrile rubber insulation. Three-segment casing with tongue/groove joints.
Igniter type: pyrogen head-end

International

Ariane 5

Current Status

The first Ariane 5 ES ATV version was successfully launched on 9 March 2008. Six launches are scheduled for 2008.

Background

The Ariane programme has been wholly reliant on the all-new Ariane 5 rocket since the last Ariane 4 was launched on 15 February 2003. The project is supported by the European Space Agency and managed and marketed by Arianespace, a joint venture in which the largest shareholders are the French government, via its CNES research organisation, and European Aeronautical and Defense Systems (EADS).

The Ariane 5 programme was approved in 1987, to produce a vehicle offering 60 per cent additional GTO capacity compared with the Ariane 4, at significant cost reduction and able to carry a manned spaceplane – the long defunct European Hermes project. In service, its speciality has been double launches of large communications satellites into GTO.

Arianespace claims the world leadership in space launch. Since its creation in 1980, it has signed 267 launch contracts with 62 international operators. In early 2006, the company stated that it had a backlog of 37 satellite launches, many of then double launches on single vehicles – a unique capability among commercial launchers. In 2005, five successful Ariane launches carried eight satellites into orbit, and at the beginning of 2006 the Ariane 5 vehicle had notched up 21 successful launches (and two failures) and its success rate stood at eleven consecutive successful flights.

While the Ariane 2/3 launch vehicle variants were based upon the proven technology of Ariane 1, and Ariane 4 was similarly based upon the experience and hardware of the earlier versions, Ariane 5 marked the introduction of a totally new launch vehicle design and associated technology.

The current Ariane family members are all derived from the original Ariane 5 configuration, which is now known as the **Ariane 5 Generic or 5G**. It has a payload mass of six tonnes to GTO.

Ariane 5 is always flown in a single basic configuration, with a liquid-fuelled core flanked by two solid boosters. The core of the Ariane 5G is the cryogenic main stage, known as the Etage Principal Cryotechnique (EPC). It comprises pressure-stabilised tanks containing 133 tonnes of Liquid Oxygen (LOX) and 26 tonnes of Liquid Hydrogen (LH2), with structural skirts at each end of the tank assembly. The aft skirt transfers the thrust loads of the Snecma Vulcain engine to the booster, while the forward skirt transfers thrust loads to the upper stages and also carries the thrust loads from the two solid boosters. EADS Space Transportation is the prime contractor for the core stage. Cryospace provides the fuel tank structures, and the forward and aft skirts are produced by MAN and Dutch Space respectively.

The steel-cased, segmented solid boosters, known as the Etage d'Acceleration à Poudre (EAP) provide 92 per cent of the launcher's thrust on lift-off. The boosters each contain 238 tonnes of propellant, and the lower segments (containing most of the fuel) are loaded at the Arianespace spaceport in French Guiana. The booster nozzles can be vectored by up to 7.3° to control the launcher in roll during ascent. The solid boosters separate 132 seconds after launch at an altitude of 60 km, and can be fitted with a parachute kit for at-sea recovery.

The first Ariane 5 ES was launched on 9 March 2008, carrying the Jules Verne Automated Transfer Vehicle (ESA) 1343780

A cutaway view of the Ariane 5 ES which is designed to carry the 20 tonne Automated Transfer Vehicle to LEO. This is by far the heaviest payload ever carried by an Ariane 5 (CNES – ESA – D. Ducros) 1343787

The Ariane 5G upper stage is known as the EPS (Etage à Propergols Stockables, or storable propellant stage). It comprises pressurised tanks for Mono-Methyl-Hydrazine (MMH) fuel and Nitrogen Peroxide (N_2O_4) oxidiser, supplying the pressure-fed Aestus rocket engine. The engine can be shut down and reignited multiple times in a single flight and may operate for more than 1,000 seconds in a GTO mission. The nozzle is gimballed and driven by electrical actuators.

The EPS is carried in the centre of a ring-shaped Vehicle Equipment Bay (VEB). The VEB is the brains of the Ariane launcher and carries the flight control system, mission processors, communications equipment and inertial reference unit. It also incorporates a hydrazine-fuelled attitude control system which provides roll control after the separation of the solid boosters and

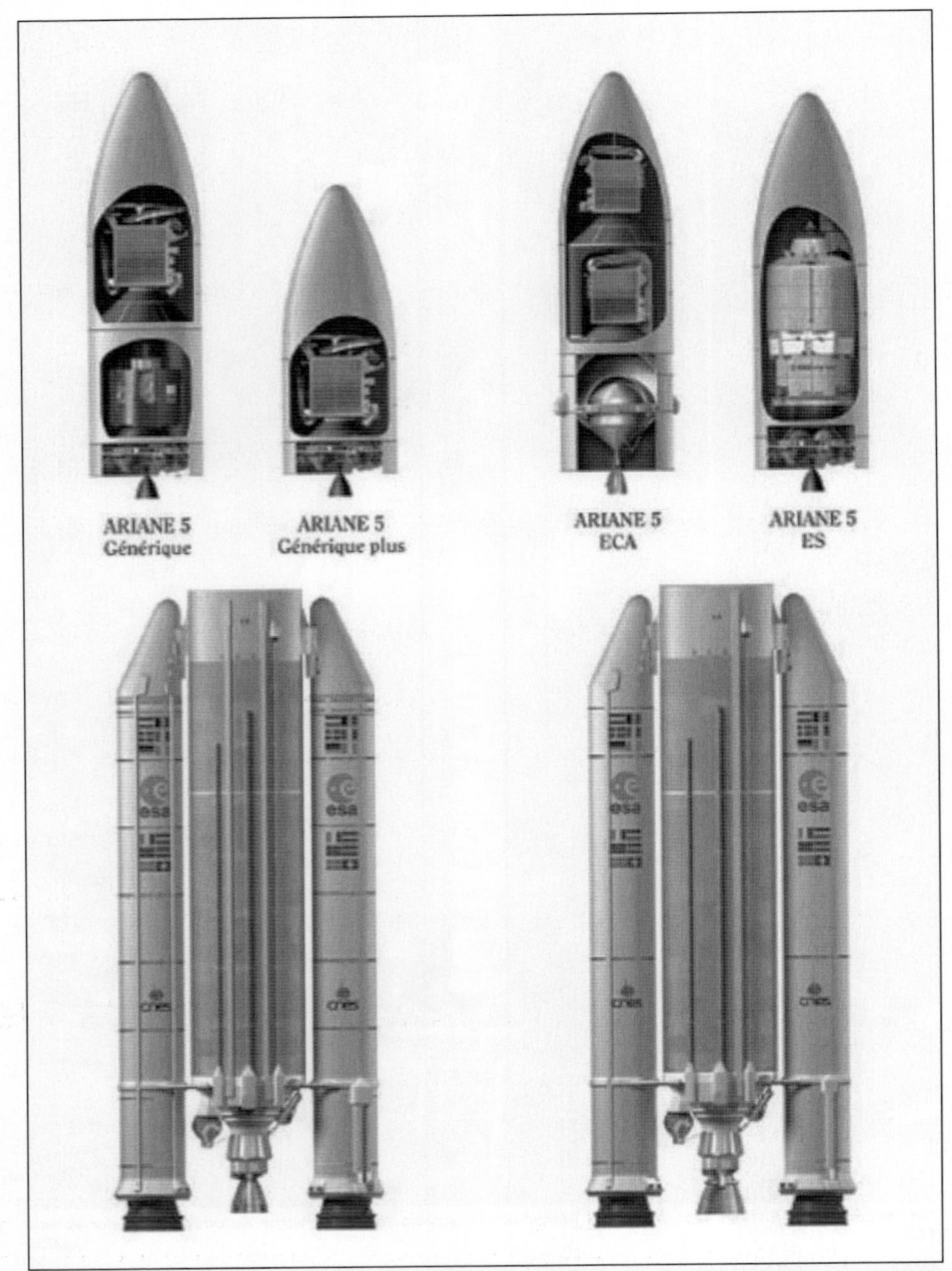

Earlier generic (5 G) versions of Ariane 5 were equipped with a storable propellant stage (EPS), replaced by the cryogenic upper stage (ESC-A) on Ariane 5 ECA. ESC-A burns liquid oxygen and hydrogen, whereas EPS uses other liquid propellants (mono-methyl-hydrazine and nitrogen peroxide) (ESA – D. Ducros) 1343821

three-axis control when the EPS is shut down. All the equipment in the VEB, including the attitude control system, is duplicated.

The upper stage carries one of two structures – the Speltra or the Sylda 5 – both of which are designed to carry two payloads and release them sequentially into their desired orbits. The Speltra has a larger volume than the Sylda 5 and is available in long or short versions, but the Sylda 5 is lighter than the Speltra. In each case, the system protects the second, lower satellite while the first one is released, and is separated by pyrotechnic actuators and springs once the first spacecraft has been placed in orbit. A triple-payload configuration is also available.

The Ariane 5G fairing is produced by Oerlikon-Contraves in Switzerland, and is produced in two half-shells with carbon-fibre skins over aluminium honeycomb cores. The fairing is split by pyrotechnic devices and jettisoned as soon as the launcher is clear of the atmosphere, at an altitude of 100 km, to reduce the weight being carried for the rest of the flight.

A set of improvements to the Ariane 5G was offered as the **5 Generic Plus (5G+)** model, first flown in 2004. This included the lighter P2001 nozzle on the EAP solid boosters; increased capacity in the upper-stage MMH tanks; and a composite structure for the VEB. Collectively, they increased payload by 150 kg.

A market study from the mid-1990s confirmed a demand for increased payload over the Ariane 5 Generic, which can lift some six tonnes to GTO. In 1999, Arianespace formed a new development division, jointly operated with CNES, to develop a higher-payload version of the Ariane 5, with a view to achieving a 10 tonne GTO payload by 2002. This effort led to the new **Ariane 5 ECA**.

The principal change in the Ariane 5 ECA is the new Snecma Vulcain 2 engine, with a 20 per cent thrust boost. It also operates at higher chamber pressures with a higher oxygen-to-fuel ratio, and the fuel load of the EPC is increased to 150 tonnes of Liquid Oxygen and 25 tonnes of LH_2. Ariane 5 ECA has redesigned EAP solid boosters with a 2.5 tonne increase in upper-segment propellant load, increasing take-off thrust from 600 to 650 tonnes. The ECA also uses a new cryogenic upper stage – Etage Supérieur Cryotechnique de type A, or ESC-A – based on the Snecma HM-7B LOX/LH_2 rocket formerly used on the Ariane 4 upper stage. The ESC-A also takes over the attitude control functions performed by the hydrazine rocket system on the VEB for the Ariane 5G, allowing the VEB to be simplified. On flight V176, 4 May 2007, the ECA lifted a record payload mass of 9,402 kg: the satellite masses totalled 8,605 kg, with payload adapters and dispensers making up the additional 797 kg.

The ECA designation originally denoted Enhanced Capability A, reflecting the fact that it was to be followed by a heavier-payload EC-B version. It was to use a new ESC-B upper stage with the Vinci cryogenic engine. However, this project has been deferred.

Some of the improvements in the ECA were included in the **Ariane 5GS** configuration, first launched in August 2005. The 5GS used the Vulcain 1B engine, improved solid boosters like those of the ECA, and ECA avionics, and offered a 700 kg boost in GTO payload.

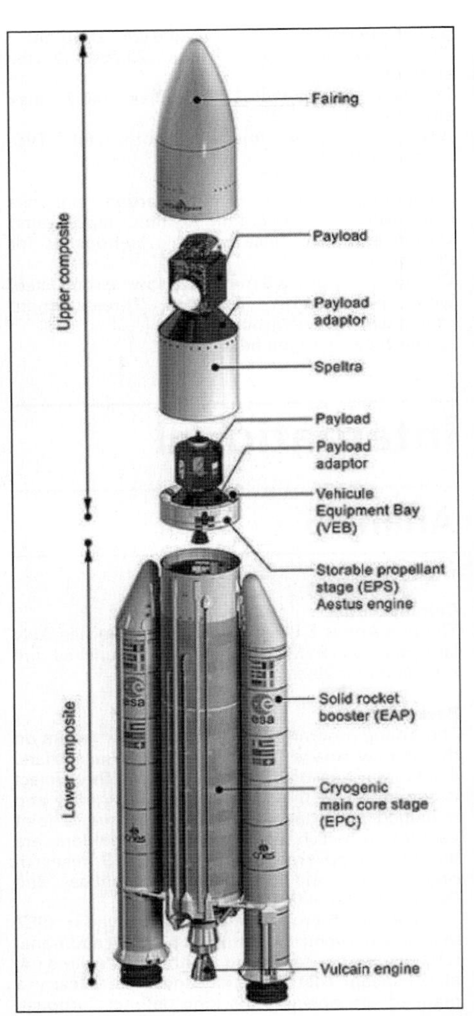

The various elements of the Ariane 5 g launch vehicle (ESA) 1343822

The final member of the current Ariane family is the **5 ES ATV** version, configured specifically to launch ESA's Automated Transfer Vehicle (ATV) on missions to support the International Space Station (ISS). It combines the Vulcain 2 engine and the more powerful EAP boosters of the ECA with the Ariane 5 Generic's hypergolic-fuelled EPS, which is required because of its reliable restart capability. The upper composite has a new, reinforced Vehicle Equipment Bay (VEB) to withstand ATV flight loads and a re-ignitable storable propulsion stage. For the ATV mission, first stage ignition takes place immediately following separation of the cryogenic main stage. The engine is then cut off and the VEB-storable propulsion stage-ATV composite commences a ballistic phase of about 45 minutes, at the end of which a second ignition occurs for a short duration before the ATV is separated and injected into the target LEO.

The first Ariane 5 ES launch, carrying the nearly 20 tonne Jules Verne ATV, was successfully launched on 9 March 2008. Jules Verne was more

Ariane 5's L9 EPS stage (ESA) 0517176

Solid Booster structures manufactured by SABCA of Belgium for Ariane 5 0084631

Rear skirt of Solid Booster for Ariane 5 0084632

than twice as heavy as the previous largest Ariane 5 payload. It was delivered to a 260 km circular orbit inclined at 51.6° to the equator. The unusual launch trajectory required the deployment of two new telemetry tracking stations, one on a ship in the Atlantic Ocean and one on the Azores Islands. The Ariane 5 upper stage performed an initial 8 minute burn over the Atlantic followed by a 45 minute coast phase over Europe and Asia. The EPS reignited for a 40-second circularisation burn over Australia. Separation of Jules Verne occurred at 06:09 CET and was monitored by a ground station in New Zealand.

The Ariane 5 programme has been marred by two launch failures. On 6 June 1996, the first Ariane 5 (Ariane 501, Volume 88) was launched from Kourou. About 40 seconds after the start of the flight sequence and at an altitude of 3.7 km the launch vehicle veered off its flight path, broke up and exploded with the total loss of the four Cluster scientific satellites.

The root cause of the accident was found to be the fact that the Ariane 5 used the same inertial reference system as the Ariane 4, with modified software. Part of the Ariane 4 software – used to re-align the inertial system in case of a late countdown hold – had been retained for commonality reasons although it had no function on Ariane 5 and operated for 40 seconds after lift-off. Part of this software is designed to shut the system down if it detects an excessive horizontal velocity, higher than the launcher itself is designed to attain, because this would indicate an error. However, Ariane 5 can actually attain much higher horizontal velocities immediately after lift-off than its predecessor. As the threshold values set for the Ariane 4 system were exceeded, the primary and back-up inertial systems both shut down, sending false information to the flight control system, which commanded the booster and core nozzles to make major corrections for a non-existent deviation. The launcher veered sharply off course and disintegrated.

The second loss of the Ariane programme was also the first flight of the ECA variant on 11 December 2002. This was attributed to fissures in the cooling tubes around the nozzle of the Vulcain 2 engine, which led to partial failure of the nozzle and a thrust imbalance that the control system could not overcome. Arianespace ordered more Ariane 5 Generic versions to ensure launch continuity while further changes were made to the Vulcain 2 design. The ECA version was flown successfully on 12 February 2005.

The failure of the first Ariane 5 ECA, together with the downturn in the worldwide launch market, triggered a new European assessment of the programme. It persuaded the sponsoring governments of ESA that while Ariane would not be a self-supporting commercial programme for many years, if at all, an autonomous European launch capability was politically desirable. As a result, during 2003, the European Union and ESA adopted a series of measures to stabilise the future of the programme. Under the European Guaranteed Access to Space (EGAS) programme, in force for 2005–2009, Ariane 5 became the chosen launch vehicle for European institutional requirements, and those institutions are guaranteed favourable launch prices and schedules. The goal was to establish a rate of six launches a year for the active period of EGAS. The European governments also agreed to continue supporting development of the ECA and to support a plan for future launch technology.

Ariane 501 on its roll-out from the Launcher Integration Building showing the enclosed aft area of the first stage (ESA) 0054380

A total of six Ariane 5 flights took place in 2007. These included a record-setting dual-passenger flight with ASTRA 1L and Galaxy 17 on 4 May which carried a total mass of just over 9,400 kg to GTO. Arianespace is in the process of accelerating its Ariane 5 launch rate to meet customer demand, switching to an annual rate of eight Ariane 5 missions by 2009. In February 2007, a declaration signed in Bremen, Germany, confirmed the increase in Ariane 5 production rate to meet growing demand in the launch services market place. With this agreement, the production of Ariane 5 ECAs – the most powerful version – will be raised to seven launch vehicles per year beginning in February 2008. It also provides the possibility of performing an additional mission each year – for instance to orbit the Automated Transfer Vehicle (ATV), which is to service the International Space Station. In June 2007, Arianespace signed an initial contract with prime contractor Astrium for the production of 35 more Ariane 5s.

All Ariane launches take place from Europe's Spaceport at Kourou in French Guiana. Launch facilities were completely rebuilt to handle Ariane 5 and include production plants for solid propellant, LH2 and LOX. A launch campaign takes five weeks

Hydraulic servo-activators and mechanical structures for Ariane 5 0084633

Ariane 5 Launch Record

	Launch	Type	Payload
V88*	4 June 1996	5G	Cluster
V101	30 October 1997	5G	Maqsat H, Teamsat, Maqsat B/EPS, YES
V112	21 October 1998	5G	Maqsat 3, ARD
V119	10 December 1999	5G	XMM-Newton
V128	21 March 2000	5G	AsiaStar, Insat 3B
V130	14 September 2000	5G	Astra 2G, GE7
V135	16 November 2000	5G	PAS-1R, Amsat P-3D, STRV 1C, STRV-1D
V138	20 December 2000	5G	Astra 2D, GE 8, LDREX
V140	8 March 2001	5G	Eurobird, BSat-2a
V142	12 July 2001	5G	Artemis, BSat-2b
V145	1 March 2002	5G	Envisat 1
V153	5 July 2002	5G	STELLAT 5, N-Star 3
V155	28 August 2002	5G	Atlantic Bird, MSG-1
V154*	11 December 2002	5 ECA	Hot Bird-7, Stentor
V160	9 April 2003	5G	Insat 3A/Galaxy 12
V161	11 June 2003	5G	Optus Defence C1, BSAT-2c
V162	27 September 2003	5G	Insat 3E, E-BIRD, SMART-1
V158	2 March 2004	5G+	Rosetta
V163	18 July 2004	5G+	Anik F2
V165	18 December 2004	5G+	Helios 2A, Nanosat 01, Essaim 1, Essaim 2, Essaim 3, Essaim 4, Parasol.
V164	12 February 2005	5 ECA	XTAR-EUR, Sloshsat, Maqsat-B2
V166	11 August 2005	5GS	Thaicom 4
V168	13 October 2005	5GS	Syracuse 3A, Galaxy 15
V167	16 November 2005	5 ECA	Spaceway 2, Telkom 2
V169	21 December 2005	5GS	Insat 4A, MSG-2
V170	11 March 2006	5 ECA	Spainsat, Hot Bird 7A
V171	27 May 2006	5 ECA	Satmex 6, Thaicom 5
V172	11 August 2006	5 ECA	JCSAT 10, Syracuse 3B
V173	13 October 2006	5 ECA	DirecTV 9S, Optus D1, LDREX 2
V174	8 December 2006	5 ECA	WildBlue 1, AMC 18
V175	11 March 2007	5 ECA	Insat 4B, Skynet 5A
V176	4 May 2007	5 ECA	Astra 1L, Galaxy 17
V177	14 August 2007	5 ECA	SPACEWAY 3, BSAT-3a
V178	5 October 2007	5 GS	Intelsat 11, Optus D2
V179	14 November 2007	5 ECA	Star One C1, Skynet 5B
V180	21 December 2007	5 GS	Horizons-2, RASCOM-QAF1
V181	9 March 2008	5 ES ATV	Jules Verne ATV

*indicates launcher failure: V88, control system failure; V154, nozzle failure on Vulcain 2 engine

from the arrival of the launcher components and payload in Kourou. The entire system is assembled vertically. The boosters, core stage, VEB and upper stage are mated in the launcher integration building (BIL) and are then moved on rails, on its mobile launch platform, to the final assembly building. This usually takes place about 12 days before launch. Here, the payloads and fairing are installed and (in the case of the 5G versions) the upper stage is fuelled with N204 and MMH.

The day before lift-off, the launcher and its platform are towed by tractor along the rail tracks to the launch pad itself. The cryogenic fuels are loaded and, finally, the automated launch sequence starts.

Snecma will continue work on the Vinci restartable upper stage engine for the Ariane 5 under a contract signed 22 December 2006. Designed to eventually replace the current upper stage, the Vinci would permit the Ariane 5 to carry a payload of 12,000 kg, an increase of about 2,000 kg. It would be capable of multiple ignitions, allowing Ariane 5 to place satellites in different orbits during a single mission. The cryogenic engine uses an expander cycle instead of a gas generator cycle, enabling a mid-flight engine restart. It will be the first European engine to use the expander cycle.

Specifications

Ariane 5 Generic (5G)

Launch site: Launch Complex ELA-3, Kourou, French Guiana
First launch: 4 June 1996
Number of launches: 24, including 16 5G models, three G+ models and five GS versions. (2 failed to deliver payloads to advertised orbit)
Success rate: 96%
Performance: 6,800 kg single or 6,000 kg double payload version into 7° GTO. 9,500 kg into 800 km, 98.6° sun-synchronous. 18 t into 70 × 300 km, 51.6° (Space Station transfer). 4,450 kg dedicated into lunar transfer orbit, shared GTO mission could deliver 400 kg lunar orbiter + 200 kg-payload lunar lander.
Number of stages: 2 + 2 strap-ons
Overall length: 45.71–51.37 m
Principal diameter: 5.40 m
Launch mass: 746 t
Guidance: Provided by the Vehicle Equipment Bay encircling stage 2. Accuracy (1 Σ): 26 km semi-major axis, 0.02° inclination GTO; 4 km, 0.04° 800 km 98.6° Sun-synchronous

Ariane 5's Vehicle Equipment Bay (MMS) 0517337

Ariane 5 solid boosters

Designation: EAP P238 (5G and 5G+). EAP P241 (5GS)
Contractor: EADS (stage integrator), Europropulsion (motors); SABCA (nozzles) Parachutostroenye (recovery parachutes)
Overall length: 27 m
Principal diameter: 3 m
Propellant type: PCA + HTPB + aluminium solid
Propellant mass: 237.7 t each
Empty mass: 37 t
Thrust: 5,500 kN sea level at launch
Burn time: 132 s

The boosters are recovered from the Atlantic 450 km downrange following separation at about 55 km altitude for inspection to check design/manufacturing margins. Six chutes (auxiliary, three drogues, secondary, main) for 27 m/s impact. First static test 16 February 1993; last two of the seven used flight-type motors (#6 first quarter was successful 10 March 1995; number 7 second quarter 21 July 1995). See the Europropulsion entry for motor details.

Ariane 5 G stage 1 (core)

The core stage delivered the upper composite into 50 × 1,300 km, 7° on GTO missions (which ARD employed for its re-entry mission).
Designation: EPC H158
Contractors: EADS (stage integrator), Snecma (engine), Dutch Space (thrust frame), MAN (forward skirt), Cryospace (tanks), SABCA (engine actuation system)
Engine: Single cryogenic gas generator cycle Snecma Vulcain 1. Began stage-level qualification firings 17 November 1994 using a battleship stage (concluded 27 January 1995); flight-type stage testing May 1995 – January 1996.
Overall length: 30.5 m
Principal diameter: 5.40 m
Oxidiser: 133 t max of liquid oxygen in 120 m^3 forward tank, pressurised to 3.5 atmosphere by helium (140 kg in 830 mm diameter, 400 atmosphere spheres)
Fuel: 25.60 t max of LH_2 in 390 m^3 aft tank, pressurised to 2.5 atmosphere by GH_2 from Vulcain

The Vulcain 2 engine is used on the Ariane 5 ECA core stage. Vulcain 2 is an upgraded version of the Vulcain cryogenic engine, providing a 20 per cent performance gain (CNES-ESA-Arianespace) 1343823

Dry mass: 12.6 t; launch mass 170 t
Average thrust: 1,350 kN in vacuum
Burn time: 580 s
Attitude control system: Vulcain gimballed ±6° for pitch/yaw

Stage 1 decays naturally from its low perigee. As the aluminium tanks are only 1.3 mm thick (for the hydrogen tank) and 4.7 mm thick (for the oxygen tank) they are pressure-stabilized when empty. During manufacture, the tank is first riveted to the forward skirt from which it is suspended. Then it is placed on the thrust frame. A 2 cm layer of expanded polyurethane insulates the tank. 18.5 cm diameter lines carry the propellants to the engine. On the launch pad the two boosters at the level of the forward skirt support the vehicle, and the EPC is suspended. The thrust from the boosters is transferred to the vehicle via the skirt.

Ariane 5 G stage 2 (upper stage)

Designation: EPS
Contractor: EADS (stage integrator, propulsion, structure); sub-contracts to Zeppelin (tanks).
Engine: Single EADS 29 kN Aestus storable propellant, open cycle, pressure-fed, reignitable, gimballed ± 16° for pitch/yaw control.
Overall length: 3.3 m
Principal diameter: 3.936 m
Propellants: 6.6 t N_2O_4 + 3.2 t MMH in four 1.410 m diameter tanks at 20 atmosphere, two 0.3 m^3 helium pressurant 830 mm diameter spheres (400 bar).
Dry mass: 1,275 kg
loaded mass: 11,000 kg
Thrust: 29 kN vacuum
Burn time: 1,100 s.

Ariane 5 VEB

EADS Astrium Vehicle Equipment Bay commands missions from its position encircling stage 2 and, unlike its predecessor Ariane 4 VEB, incorporates an attitude control system, providing roll control during stage ½ burns and three-axis control after burnout. The 5.4 m diameter, 1.56 m high, 1,300 kg unit carries 70 kg of hydrazine.

Payload fairing/carriers

Contraves provides three standard 5.40 m diameter fairings to accommodate 4.57 m diameter payloads: 12.70 m short, 13.8 m medium and 17 m long units. The ogive fairing jettisons at about 191 s/106 km during stage 1 burn when aerodynamic heating has reduced to 1,135 W/m^2. The Structure Porteuse Externe de Lancements Triples Ariane (Speltra) structure increases the length of the upper composite by 5.66 m and permits the launch of double payloads. Dornier (now part of EADS) was awarded the contract in early 1990; the first Speltra was delivered first quarter 1995. The long version (5.6 m high) weighs 820 kg and the short version (4.1 m high) weighs 704 kg. An alternative carrier system is the Sylda 5, which provides less volume but is lighter, weighing 440 kg . Standard payload adaptors are available with interface diameter 937; 1,194; 1,666; 2,624 mm.

The first two of four Cluster satellites attached to the Ariane 5 adapter 0084635

Ariane 5 ECA
Launch site: Launch Complex ELA-3, Kourou, French Guiana
First launch: 11 December 2002
First successful launch: 12 February 2005
Number of launches: 11 through March 2008 (1 failure)
Success rate: 91% through March 2008
Performance: 9.6 tonnes into GTO. 9.1 tonnes dual, 19–21 tonnes to ISS orbit
Number of stages: 2 + 2 strap-ons. EPC H173/Vulcain-2 (core), ESC-A H14.4/HM-7B (upper stage)
Overall length: 45.71–51.37 m
Principal diameter: 5.40 m
Launch mass: 780 t
Solid booster thrust: 6370 kN
Main engine: Snecma Vulcain 2, 1,342 kN

Ariane 5 ES ATV
Launch site: Launch Complex ELA-3, Kourou, French Guiana
First launch: 9 March 2008
Number of launches: 1 through March 2008
Success rate: 100% through March 2008
Performance: 19-21 tonnes to ISS orbit
Number of stages: 2 + 2 strap-ons. Reinforced Vehicle Equipment Bay (VEB).
Overall length: 53 m
Principal diameter: 5.40 m
Launch mass: 760 t
Solid booster thrust: 6,370 kN
Main engine: Snecma Vulcain 2, 1,342 kN

Dnepr-1

Current Status

On 26 July 2006 a Dnepr launched from Baikonur failed and brought down 18 small- and medium-sized satellites it was scheduled to carry into orbit, including Belarus' first. Science satellites constructed in the US, Norway, Korea, Italy and Russia were also aboard. The crash site was extensive and the accident has renewed Kazakh concerns over environmental damage in the Baikonur region. Kazakhstan has suspended Dnepr launches due to the incident.

Based on R-36M2 Intercontinental Ballistic Missiles (ICBMs) removed from service after the end of the Cold War, the Dnepr space launch system is managed by International Space Company (ISC) Kosmotras, a joint venture founded by the governments of Russia and the Ukraine. ISC Kosmotras plans to continue offering services until the stored missiles reach the end of their shelf lives, in 2016 to 2020.

Background

When NPO Yuzhnoye started to market this launch vehicle – based upon the R-36M2 missile – they took the NATO designator 'SS-18' and added a 'K' to indicate a space application. It remained the SS-18 K until 1996 when the name Dnepr started to be applied to the vehicle.

Russia and the Ukraine agreed to support the Dnepr programme jointly in May 1997 and subsequently established ISC Kosmotras. Partners in the effort include the Russian Aviation & Space Agency, which provides support facilities at Baikonur; the Russian Ministry of Defense, which maintains SS-18 launchers in storage; and the Ukraine's Yuzhnoye design bureau, which remains the design authority for the system.

Development of the silo-launched missile began in 1964 and it was first deployed in 1974. ISC Kosmotras states that the missile has achieved a 97 per cent success rate on 162 launch attempts since 1973, including five flights in Dnepr configuration. Production ended in 1991 under the first phase of the START arms control agreement. By 2003, 174 rounds had been destroyed despite START-2 not being ratified, and ISC Kosmotras states that 150 missiles are still available for space launch operations.

The Dnepr-1 is a three-stage vehicle with a "Space Head Module" (SHM) – named by analogy to the ICBM's warhead. The first two stages are taken from the ICBM without any modification. The third stage – formerly the warhead bus used to dispense multiple independently targeted re-entry vehicles (MIRVs) – has some changes to the flight control system and modifications to allow interface with the spacecraft, while the SHM replaces the standard warhead bus with a spacecraft interface. The R-36M2 warhead bus was designed with a high degree of manoeuvrability to defeat Anti-Ballistic Missile (ABM) systems, together with the ability to deliver more than 20 warheads and decoys, which translates in the flexibility to deliver large numbers of payloads.

The Dnepr-1 is fuelled by UDMH and N2O4. The first stage RD-274 engine has four motor chambers, and the second-stage RD-0255 has a single chamber and four attitude-control thrusters. Russian ICBMs are fuelled during production and leave the manufacturer's site ready for launch, and can be stored for years without any performance degradation. For instance, the first launch of Dnepr in April 1999 used a SS-18 missile that had been on duty for 20 years and 9 months. After this successful launch the guaranteed service life of this ICBM type was extended for up to 25 years.

The first and second stages of the Dnepr vehicle are stored in a Transport and Launch Container (TLC). Before launch, the TLC is trucked out to the launch silo, erected and lowered into the silo. The third stage and payload are integrated separately, the spacecraft is prepared and fuelled, and the upper components are driven to the silo in the type of transporter-erector originally designed for nuclear warheads. The third stage and SHM are mated with the booster in the silo. The vehicle is cold-launched from the silo with the aid of a black-powder gas generator and the first-stage motor ignites at a height of 20 m.

The first Dnepr was launched successfully from a silo at Baikonur on 21 April 1999, carrying the UK satellite UoSAT-12. The second Dnepr carried five satellites into orbit on 26 September 2000. These comprised Tinngsat 1, MegSat 1, SandiSat 1A, SandiSat 1B and UniSat. The third lifted six satellites to orbit on 20 December 2002. The fourth flight, on 29 June 2004, lifted a cluster of eight small satellites, including CNES' Demeter, three small communications satellites for Saudi Arabia's Space Research Institute, three US microsatellites and Italy's UniSat-3. The most recent mission on 24 August 2005 orbited the Japanese Aerospace Exploration Agency (JAXA) OICETS and INDEX spacecraft.

In February 2004, Kosmotras signed an agreement with SpaceTech GmbH to market Dnepr launches in Europe and the Far East. The company has also signed an agreement to launch a satellite developed by US-based space start-up Bigelow Aerospace, scheduled to fly during 2006. Other pending launches include an early-2007 flight carrying five satellites in the RapidEye earth imaging constellation, being developed by Surrey Satellite Technology Limited (SSTL), and launches in the CNES Myriade programme. A cluster mission has been booked by Belarus and EADS Astrium plans to launch its TerraSAR-X on the vehicle.

As part of Russia's strategy to reduce its dependence on Baikonur in Kazakhstan, a second launch site for the Dnepr is being commissioned at the base of the 13th Missile Division at Dombarovsky, close to Orenburg and the Kazakh border. ISC Kosmotras refers to this site by its civil name, Yasniy. A test firing in support of this programme took place in December 2004.

ISC Kosmotras is developing a higher-energy "space tug" to adapt the Dnepr to other missions, including GEO, high-elliptical, lunar and LaGrange point launches. The tug is offered in two versions – the ST-1 with an extra solid booster for GEO and other high orbits, and the liquid-only ST-3 for LEO missions. The ST-3 provides higher performance than the standard Dnepr for orbits higher than 500 km, while the ST-1 would allow the Dnepr system to place up to 200 kg in GTO.

For details of the latest updates to ***Jane's Space Systems and Industry*** online and to discover the additional information available exclusively to online subscribers please visit
jsd.janes.com

Specifications
Dnepr
First launch: 21 April 1999
Launch sites: Baikonur
Principal uses: delivery of medium-class payloads into LEO
Typical performance: 3,700 kg into 300 km circular at 50.5°; 2,300 kg into 300 km circular at 98°;
Injection accuracy: ±4.0 km altitude, ±3.0 s period, ±0.04° inclination.
Number of stages: three (liquid)
First-stage propellant: 147,900 kg
First-stage thrust: 461.2 t
Second-stage propellant: 36,740 kg
Second-stage thrust: 77.5 t
Third-stage propellant: 1,910 kg
Third-stage thrust: 1.9 t (full), 0.8 t (throttled)
Overall length: 34.30 m
Principal diameter: 3.00 m
Launch mass: 209.8 t
Guidance: computer-controlled inertial

KSLV (Korean Space Launch Vehicle)

Current Status
The maiden launch of the KSLV-1 is expected to take place in 2008.

Background
In the 1990s, South Korea developed and launched the solid-fuelled KSR-1 and KSR-2 sounding rockets. This was followed in 2000 by the start of construction of a new space centre on Naro Island in Goheung, 485 km south of Seoul, and by the launch of the 6,000 kg KSR-3 liquid-propellant sounding rocket on 28 November 2002. In the country's initial plans, announced in 2002, South Korea intended to develop a small satellite launch vehicle by 2005, based on technology flown on the KSR-3 test vehicle. The launcher would be entirely indigenous, based on the 122,500 kN thrust liquid oxygen/kerosene motor used in the KSR-3. It would comprise two KSR-3 strap-ons on either side of the core, which would be a single KSR-3. A solid propellant motor, based on the KSR-1 and KSR-2, would provide the third stage. The KSLV would be able to lift 100 kg to a 300 km, 38° orbit. Lift-off thrust would be 244,000 kN with a total mass of about 20,000 kg. Core diameter would have been about 3.9 m with an overall length of about 30 m.

South Korea's Ministry of Science and Technology estimated that the country could complete the development of the KSLV-1 by 2005, but problems with engine development led the Korea Aerospace Research Institute (KARI) to seek Russian technical assistance after the United States expressed concern over military applications and turned down similar requests. High level cooperation with Russia began in September 2004, when South Korean President No Mu-hyon signed a bilateral space cooperation agreement in Moscow.

A contract to design and build a small-lift launch vehicle was signed on 26 October 2004, during the visit of a Khrunichev SC delegation to South Korea. The contract covered a two stage launcher, with Russia developing the first stage and Korea the second stage. According to the contract, Russia would help South Korea to perform two launches. The Russian signatories included Alexander Medvedev, Director General of KhSC (the Prime Contractor responsible for the overall design of the vehicle), Boris Katorgin, Designer General of the Energomash Research and Production Association (designer and manufacturer of the 1st stage engines), and Gennady Biryukov, Director General of the Design Bureau for Transport Engineering (KBTM), the designer of the ground facility. On the Korean side, the small satellite launch vehicle development programme would involve numerous companies, universities and research institutes, under the supervision of the Space Launch Vehicle Division of KARI.

Joint development of the KSLV-1 has since run into some difficulties, causing delays to the programme. Although Russia and South Korea signed a bilateral Technology Safeguards Agreement (TSA) in October 2006 and South Korea's National Assembly ratified the agreement in December 2006, detailed negotiations on technology transfers continued. When South Korean Deputy Prime Minister and Minister of Science and Technology Kim U-sik visited Russia 28 March to 2 April 2007, one of his major objectives was to encourage the Russian Duma to ratify the agreement.

Under the National Space Development Plan, the joint Russian – South Korean KSLV-1 will make the first launch of a South Korean rocket from the new Naro space centre in 2008. The first orbital launcher to be developed by South Korea will be capable of launching a 100 kg scientific satellite into LEO. It has a kerosene – LOX first stage based on the Russian Angara and a solid-fuel second stage developed by Korea. 80 per cent of the launcher is developed by Russian industry. Lift off thrust is reported to be 1,700 kN and launch mass is approximately 140,000 kg. The overall length of the KSLV-1 is 33 m and the core diameter is 2.9 m.

South Korea intends to develop independent access to space. The original plan was to develop an uprated KSLV for flight in 2010. This intermediate rocket was cancelled in early 2005, and the programme was reorganised to concentrate on a more powerful vehicle that had been a longer-term objective. As currently envisaged, the KSLV-2 will be based on the KSLV-1, with a first stage powered by four 735 kN thrust engines and a second stage powered by a single engine with the same rating.

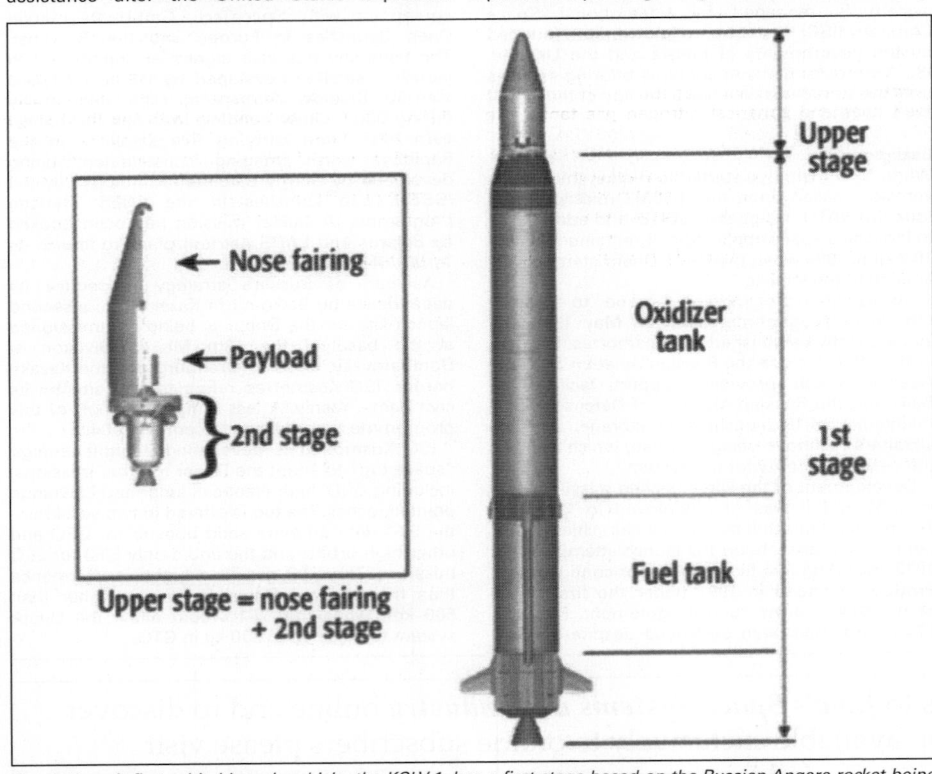

South Korea's first orbital launch vehicle, the KSLV-1, has a first stage based on the Russian Angara rocket being developed by Khrunichev. The solid fuel second stage is being developed in Korea (KARI) 1340467

The payload mass is about 1,500 kg. The launcher will be tested by 2017, two years later than previously reported. The Ministry of Science and Technology says test facilities for engines will be ready by 2014. KARI's objectives include independence in key areas of rocket technology by 2016.

Specifications
KSLV-1
First launch: 2008 ?
Engines: First stage: 1 RD-191 (Angara UM)
Second stage: 1 KSR-1
Overall length: 33 m
Diameter: 2.9 m
Payload capability: PSTSat-2: 100 kg to a 300 × 1,500 km, 80° orbit, period 103 min
First stage oxidiser: LOX
First stage fuel: Kerosene
Second stage propellant: Solid
Launch mass: Approx. 140, 000 kg
Lift off thrust: 1,700 kN

Rockot

Current Status
Rockot has so far flown 9 times with one failure. The most recent launch was on 28 July 2006, when South Korea's Kompsat-2 was placed in a 685 km sun-synchronous orbit. Eurockot's next mission will be the launch of GOCE for the European Space Agency in 2008. Rockot KM has also been accepted by ESA as a back-up to Vega.

Background
The Eurockot Launch Services company – a joint venture of EADS Space Transportation (51 per cent) and Khrunichev Space Centre (49 per cent) – markets the commercial Rockot launcher, derived from the SS-19 two-stage ICBM. The Khrunichev/DASA joint venture was announced 23 March 1995. Eurockot is responsible for all commercial activities, acting as the customer's prime contractor, but Khrunichev develops and produces all of the hardware. The first commercial launch was planned for the fourth quarter of 1998 from Plesetsk, using a modified Kosmos pad now designated exclusively for Rockot. Launches are from a tube, replicating a silo.

To prepare for commercial operations, Eurockot performed three demonstration launches from Baikonur. These took place with the Rockot K configuration and were launched with a small fairing from a silo at Baikonur Cosmodrome. The first two (20 November 1990 and 20 December 1991) were suborbital, demonstrating basic vehicle performance and KB Salyut's new Breeze third stage (inactive on the first ascent). After first and second stage burn-out, separation of the Breeze from the second stage was successfully performed and a sub-orbital controlled and stabilised flight of the upper stage, which carried scientific equipment, was undertaken (H max = 900 km, i = 65°). This included multiple restarts of the upper stage main engine.

The first orbital flight was 26 December 1994 from Baikonur, into 1,875 × 2,253 km, 64.8° carrying the 70 kg RS-15 Radio-Rosto amateur relay. Subsequent flights took place from complex LC133 at Plesetsk, using a Rockot KM fitted with a modified upper stage and a new, larger fairing developed by Khrunichev. The Commercial Demonstration Flight in May 2000 marked the operational debut of Eurockot. The mission simulated a typical LEO mission by carrying two identical satellite simulators (SIMSAT-1 and 2) into an 86.4° orbit at approximately 550 km. The third orbital flight from Plesetsk lifted two NASA/German Gravity Recovery And Climate Experiment (GRACE) satellites into low Earth orbit on 17 March 2002. The fourth flight conducted on 20 June 2003 carried two Iridium satellites. 30 June 2003 saw the Multiple Orbit Mission, when the 700 kg Monitor-E mass model was launched with eight small satellites weighing from 1 kg to 68 kg (Mimosa, DTUat, MOST, Cute-1, QuakeSat, AAU-Cubesat, Can X-1, Cubesat-XI), which were accommodated around the main payload. This launch also demonstrated Rockot's capability to inject satellites into different orbits with the same inclination. The 66 kg Czech Mimosa satellite was released into an elliptical orbit of 320 × 820 km. Following a re-ignition of its main engine, Breeze released the Canadian Space Agency's MOST satellite and six nano-satellites into a sun-synchronous orbit. At the end of the mission, the Breeze stage with the Russian Monitor mock-up satellite still attached was placed into a graveyard orbit. The sixth flight on 30 October 2003 lifted Japan's Servis-1 satellite into sun-synchronous orbit.

On 26 August 2005, a Rockot KM successfully placed the 750 kg Russian Monitor-E spacecraft into a 97.5° sun-synchronous orbit. The October 2005 launch of ESA's CryoSat failed when the flight control system in the Breeze upper stage did not execute the command to shut down the second stage engine. The combined stack of the upper two stages and the satellite fell into the ocean in the nominal drop zone north of Greenland. However, the Rockot successfully returned to flight in 2006 with the launch of South Korea's Kompsat-2.

Based on the first two stages of the SS-19 two-stage ICBM plus a re-ignitable Breeze upper stage, Rockot can provide precise orbit injections of small and medium-sized spacecraft weighing up to 2,000 kg into sun-synchronous, near polar and highly inclined orbits from its launch site at Plesetsk. The standard Rockot has a 1,850 kg payload capability to a circular 370 km orbit launched from Plesetsk to an orbital inclination of 63°. It can place 1,340 kg in an 800 km, 90° circular orbit or 1,530 kg in an 370 km, 90° orbit.

Development of the silo-launched missile began in 1965 and it was first deployed operationally in 1974. The latest Mod 3 entered service in 1979. By 1982, at the height of the Cold War, the Soviets had deployed a peak 360 SS-18s in silos. Some would be retired under the START 2 treaty. In 1991, Russia and other countries had 300 silos at four operational sites, Kozelsk (60 silos), Tatischevo (110), Khemilnitsky (90, Ukraine) and Pervomaysk (40, Ukraine). In December 1994, the operational stockpile had dwindled to 290 missiles and Khrunichev had 15 at its factory available for conversion as of October 1995.

The first stage contains N_2O_4 and UDMH tanks separated by a common bulkhead. Tank pressurisation is achieved by means of a hot gas system. The engines comprise four cardangimballed, closed-cycle, turbopump-fed engines with the designation RD-0233/RD-0234. The first stage also has four solid fuel retro-rockets for the first/second stage separation.

The second stage also contains a common bulkhead and a hot gas pressurisation system. It has a closed-cycle, turbopump-fed, RD-0235 main engine and verniers designated RD-0236 for directional control. The vernier engines are ignited just before the first and second stages separate. The exhaust gases are diverted by special hatches in the first stage. After separation, the first stage is braked by retro-rockets before the second stage engine is ignited.

The upper stage on the first three Rockot flights was the Breeze K. This has now been replaced by a modified version, known as the Breeze-KM. The structure of the Breeze-K equipment bay has been widened and flattened by redistributing the control equipment. This allows larger satellites to be accommodated and to reduces dynamic loads. The Breeze KM comprises three main compartments: the propulsion compartment, the hermetically sealed equipment compartment and the interstage compartment. The new equipment compartment can also double as a payload dispenser, accommodating multiple small satellites. The compartment has been stiffened by the insertion of walls to give adequate structural rigidity.

The Breeze-KM upper stage is no longer attached to the launcher at its base but is suspended within the extended transition compartment. This compartment is a load-bearing structure which provides a mechanical interface with the booster and accommodates the Breeze-KM separation system. Consequently, the fairing is now attached directly to the equipment compartment.

The Breeze-KM propulsion compartment includes a low pressure fuel tank (UDMH) and an oxidiser tank (N_2O_4) separated by a common bulkhead. The lower oxidiser tank surrounds the 20 kN main engine, which provides major the impulse to achieve the final orbit. Each tank contains equipment such as baffles, feed pipes and ullage control devices to facilitate main engine restarts during weightlessness. The main engine, attitude control and vernier engines are located at the base of the propulsion compartment, together with propellant feed lines and spherical nitrogen gas tanks. The 12 × 16N attitude control engines control the pitch, roll and yaw of the Breeze-KM. The 4 x 400 N verniers are for propellant settling and orbital manoeuvres.

A Rockot attached to the stationary mast on the pad at Plesetsk In the background is the mobile service tower where the upper stage with the payload is attached to the booster stages (Eurockot) 1340468

Breeze's restart capability, plus 7 hour flight life, allows payloads to be dispensed into different orbits. A single ignition is used for orbits up to 400 km. Breeze performs collision avoidance and can also de-orbit. During coast, it holds three-axis (1 to 10° accuracy) in any selected attitude within ±120° of Sun axis. It can provide 5 rev/minute about the longitudinal axis to the payload on release. Three burns are required for Sun-synchronous missions: (1) to attain 200 × 700 km, 93°; (2) circularise at 700 km, 63°; (3) plane change to 700 km, 98°.

No designators are available for the main or vernier engine systems used on the Breeze (Briz) third stage on Rockot. It has been indicated that the main engine has flown on the Phobos, VEGA, Venera and Mars spacecraft, suggesting that it is (or derived from) the Isayev bureau's (now Khimmash) KTDU-425. As a comparison with the third stage main engine data the KTDU-425A thrust is variable within the range 9.86 to 18.89 kN and the specific impulse within the range 293 to 315 seconds.

Eurockot has developed an 80 kg double launch system (DOLASY) to complement the Breeze upper stage. This allows two satellites to be launched together. The fairing is made from a three layer carbon fibre composite with an aluminium honeycomb core. It is mounted on top of the equipment bay of the Breeze third stage. The fairing separation and jettison take place by using a pyrodriver located in the nose of the fairing to release mechanical locks that hold the two half-shells together along the vertical split line. This pyrodriver has redundant firing circuits. Immediately after this, several pyrobolts on the fairing's horizontal split line are fired so that the half-shells are free to be driven apart by spring pushers. The half-shells rotate around hinges located at their base and are subsequently jettisoned. The design is based on that used for the commercial Proton fairing.

Payload integration and checkout is performed in a 180 m² class 100,000 facility at 18 to 25°C and 30 to 70 per cent humidity. After encapsulation in the adjacent 90 m² clean room, air conditioning maintains the payload at 10 to 25°C, 15 to 60 per cent humidity, class 100,000. Air conditioning is interrupted only during two periods, each of 1 hour max: mounting on the pad and final assembly of the 'silo' container. There is no access after payload encapsulation; the fairing is not radio transparent. Customer accommodation is provided in Mirny, some 42 km from the launch site, during campaigns.

Rockot Launch Record

Vehicle	Launch Date	Launch Site	Payload	Result
Rockot K	20 November 1990	Baikonur	Suborbital – geophysical experiments	Success
Rockot K	20 December 1991	Baikonur	Suborbital – geophysical experiments	Success
Rockot K	26 December 1994	Baikonur	RS-15 Radio-Rosto amateur relay satellite	Success
Rockot KM	16 May 2000	Plesetsk	Simsat-1 and Simsat-2	Success
Rockot KM	17 March 2002	Plesetsk	GRACE-1 and GRACE-2	Success
Rockot KM	20 June 2002	Plesetsk	Iridium SV97 and Iridium SV98	Success
Rockot KM	30 June 2003	Plesetsk	Monitor-E GVM, Mimosa, DTUat, MOST, Cute-1, QuakeSat, AAU-Cubesat, Can X-1, Cubesat-XI	Success
Rockot KM	30 October 2003	Plesetsk	SERVIS-1	Success
Rockot KM	26 August 2005	Plesetsk	Monitor-E	Success
Rockot KM	8 October 2005	Plesetsk	CryoSat	Failure
Rockot KM	28 July 2006	Plesetsk	Kompsat-2	Success

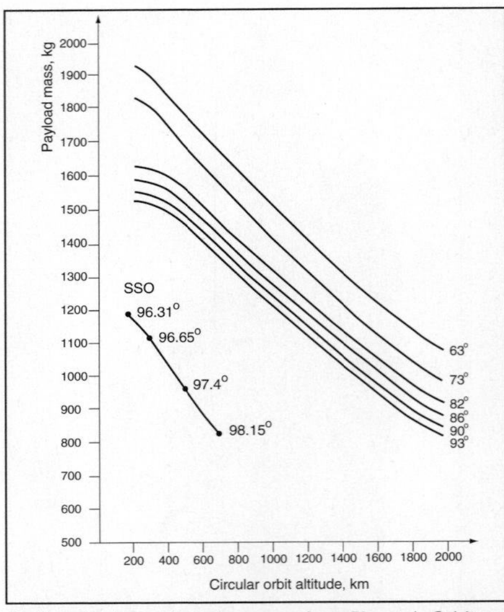

Rockot circular orbit performance from Plesetsk. Orbits up to 400 km are achieved by a single stage 3 burn; >400 km require Breeze re-ignition at apogee for circularisation. 63–93° inclinations are allowable without plane changes during launch (Khrunichev/DASA) 0519179

The airport is 35 km from the complex. A campaign requires 30 days: stage 1/2 assembly/testing L-30 to L-22; Breeze testing/fuelling L-21 to L-13; satellite fuelling L-13/12; upper composite integration L-11 to L-8; stage ½ transfer to pad L-7; transfer upper composite to pad L-6; launch preparations L-5 to L-1; fuelling/launch L-).

Specifications

Rockot

First orbital launch: 26 December 1994
Launch site: Plesetsk
Principal uses: Small-medium payloads (up to about 1,900 kg) into intermediate polar or sun-synchronous orbits.
Vehicle success rate: 88.9% through end 2007.
Availability: 5/yr (up to 12/yr) on 20-month cycle between contract signature and launch
Launch cost: Approx. USD12 – 15 million.
Number of stages: 3 (liquid)
Overall length: 29.15 m
Principal diameter: 2.5 m
Launch mass: 107 t
Guidance: Inertial, carried in compartment attached to front end of stage 3, provides three-axis control until payload separation commanded. Accuracy for 300 km 82° orbit ±3 km altitude, ±0.03° inclination; 700 km, 63° orbit ±14 km altitude, ±0.05° inc.
Launch sequence (mins) for 700 km, 63° (some values have been estimated)

0.00	Launch
0.52	Q_{max}, 605 m/s, 11 km altitude
2.01	Stage 1 shutdown & separation (stage 2 verniers provide hot separation)
2.27	Stage 2 main ignition
3.06	Fairing separation, 3.74 km/s, 119 km
4.44	Stage 2 main shutdown
5.04	Stage 2 verniers shutdown and separation
08.40	Breeze ignition 1,
–	Max altitude (257 km)
14.06	Breeze shutdown 1
–	Orbit 200 × 700 km
63.50	Breeze ignition 2 (circularisation)
64.26	Breeze shutdown 2
–	Orbit 700 × 700 km

Rockot stage 1

Bureau: Khrunichev/KB Salyut
Engines: RD-0233 and RD-0234
Overall length: 17.2 m
Diameter: 2.5 m
Oxidiser: Nitrogen tetroxide
Fuel: UDMH
Dry mass: not available
Propellant mass: 81 t
Total thrust: 2,030 kN (vacuum), 1,834 kN (sl)
Specific impulse: 310 s (vacuum), 285 s (sl)
Burn time: 121 s
Comment: Once the first stage engines have ignited they cannot be shut down: there are no brackets to hold the launch vehicle onto its launch table and no engine parameter verification before launch. The first stage carries four solid-propellant retro-rockets for separation from the second stage. The propellant tanks have a common bulkhead and tank pressurisation is achieved by means of hot gas from one of the main engines.

Rockot's new Breeze third stage was first displayed at 1995's Paris Air Show (Ted Hooton) 0517192

Rockot stage 2

Bureau: Khrunichev/KB Salyut
Engines: RD-0235 (main engine), RD-0236 (verniers)
Overall length: 3.9 m
Diameter: 2.5 m
Oxidiser: Nitrogen tetroxide
Fuel: UDMH
Dry mass: Not available
Propellant mass: 15 t
Total thrust: RD-0235 – 235 kN (vacuum), RD-0236 – 14.7 kN (vacuum)
Specific impulse: RD-0235 – 320 s (vacuum)
Burn time: RD-0235 – 158 s, RD-0236 – 183 s
Comment: Separation of the first and second stages is a 'hot' one since the RD-0236 vernier engines are ignited just before separation: the exhaust gases are diverted using special hatches within the first stage.

Rockot stage 3 (Breeze)

Bureau: Khrunichev/KB Salyut
Engines: KTDU-425? (main engine) ? (verniers)
Overall length: 3.38 m
Diameter: 2.28 m (maximum)
Oxidiser: Nitrogen tetroxide
Fuel: UDMH
Dry mass: = ~3.1 t
Propellant mass: 3.5 t
Total thrust: Main engine – 20 kN (vacuum), verniers – 3.9 kN (vacuum)
Specific impulse: Main engine – 325.5 s (vacuum), verniers – 275 s (vacuum)
Burn time: Main engine – 564 total (max. six burns)
In addition, the third stage carries 12 attitude-control thrusters: each has a thrust of 0.13 kN and a specific impulse of 270 s.

Payload processing/accommodation

The 8.5 m long upper composite includes the fairing, Breeze and interstage. The 7.9 m long, 2.5 m diameter cylindrical/biconic two-piece composite fairing, developed specifically for Rockot, provides a payload envelope of 4.757 m long, 2.260 m diameter. Bolted payload adaptor is 591 or 937 mm diameter. The separation system comprises pyrolocks (four for 937 mm, three for 591 mm), three springs, guide pins (eight for 937 mm, six for 591 mm) and two separation sensors. The upper ring of each adaptor remains bolted to the payload: 15 kg for 937 mm, 12 kg 591 mm. After the pyrolocks are cut by pyrotechnics, the springs impart a minimum 0.5 m/s. A Western standard system using a 937 mm Marman clamp band is also available. The adaptors carry up to eight 50-pin payload electrical connectors, relaying up to 20 A total 28 V DC during launch preparation. 100 lines allow payload checkout until launch. During launch, Rockot's batteries in the transition section between Breeze and the adapter can provide up to 5 A totalling 15 Ah over 7 hours. Up to 50 lines can be provided for Rockot's computer to issue payload commands. 32 payload telemetry channels are available.
Acceleration load: Longitudinal static 7.2 g at end of stage 1 burn, 3.0 g at end of stage 2 burn, 1.6 g max at end of Breeze burn (depends on payload mass); vibration 7.5 g rms long/lat.
Acoustic load: 142 dB max integrated level during lift-off and transonic.
Thermal load: <500 W/m² at any point of the fairing; <1,135 W/m² after fairing jettison.

Vega

Current Status

The Vega small launch vehicle is being supported by ESA as part of the European Guaranteed Access to Space (EGAS) policy, and a first flight from Kourou in French Guiana is now expected in late 2007. Most of the funding comes from the Agenzia Spaziale Italiana (ASI), the Italian space agency. Additional support comes from the French government, via CNES, which is using Vega to advance solid rocket technology in Europe.

Background

Vega's origins go back to 1988, when the Italian government accepted proposals from industrial concerns to upgrade the US Scout small launcher design for operations from the San Marco floating equatorial base, off the coast of Kenya. Among the various proposals was one from BPD Difesa (later acquired by Avio) to double the Scout's base performance.

BPD's offered its Zefiro (Zephyr) motor, developed from the company's Ariane experience, with thrust vector control. It conducted four successful static tests during 1991, but the first flight test ended by detonation from range safety 10 seconds into the planned 40 second burn on 18 March 1992 from the Sardinian military range. Ground qualification continued until final completion in mid-1995 and two qualification flights at the end of the year.

Italy's 1990–94 five-year plan approved development of the so-called San Marco Scout launcher for operations from Italy's equatorial platform off the Kenyan coast. The Italians expected to make their first launch in 1995. However, the University of Rome proposed a separate San Marco Scout programme closely based on the US Scout, prompting legal wrangling.

In the meantime, Italy's San Marco site became too expensive for the Italian government to maintain, leading to a decision that the new launcher – renamed Vega – would operate from Kourou and Vandenberg AFB, using transportable equipment to support campaigns of 7 to 15 days.

ESA took over the Vega programme in June 1998. In May 1999, ESA approved USD340 million (EUR317 million) for development, subject to successful completion of additional technical, financial and economic assessments in September 1999. In December 2000 European governments agreed to spend USD301.5 million on development of the Vega launcher and USD109.9 million on development of technology for the P80 first-stage solid rocket booster. Italy agreed to finance Vega development and part of the P80 effort. France agreed to support the development of the P80 motor by Europropulsion – an equally held JV between Snecma and Avio – because its solid rocket technology could be applicable to improved versions of Ariane.

In February 2003, European Launch Vehicle (ELV) SpA, a joint venture (JV) formed to be Vega's prime contractor, signed a contract with ESA for final development leading to manufacture and testing. ELV SpA is owned by Avio (70 per cent) and ASI (30 per cent). At the same time, Avio signed a contract with CNES to cover development and qualification of the P80. The value of the development contract for the launcher was EUR221 million, with EUR40 million for the P80. Under the contracts, ELV SpA is the integrating contractor for Vega, Avio leads development of the P80 first-stage booster and Italy's Vitrociset is prime contractor for the ground segment of the system. All launches will be based at Kourou, and the Ariane 1 launch pad (ELA-1) is being refurbished to support the new launcher.

Development of the launch vehicle and ground segment is being funded by Italy (65 per cent) and France (15 per cent) with minority contributions from Belgium, the Netherlands, Spain, Sweden and Switzerland. Customer management is supplied by a joint ESA/ASI/CNES team based in Frascati. France (65 per cent) is the largest contributor to the P80 motor project, followed by Belgium with 19 per cent, with CNES at Evry, France, as the managing agency.

Vega is designed to lift micro-, mini- and small satellites into low Earth orbits or sun-synchronous orbits, with a maximum payload of 2,500 kg and a benchmark target of 1,500 kg into a 700 km polar orbit. It is intended to launch either single or multiple satellites, and ESA estimates the potential market at three to five missions per year.

Vega comprises three solid-rocket boost stages with a liquid-fuel upper module. Avio is prime contractor for each stage and SABCA provides the thrust vectoring nozzles. On the first stage, Europropulsion is developing the P80FW motor; the second and third stages use Avio's Zefiro 23 and Zefiro 9 respectively.

The solid rocket stages use common technology, including a carbonfibre/epoxy filament-would case, protected by low-density microsphere thermal insulation; a composite nozzle featuring a carbon-carbon throat, made from a new low-cost material, flexible joint and carbon-phenolic cone; and an electrically actuated Thrust-Vector Control (TVC) system.

The upper stage, known as the Attitude and Vernier Upper Module (AVUM), incorporates a single-chamber pressure-fed engine developed by Avio and fuelled by UDMH (unsymmetrical dimethylhydrazine) and NTO (nitrogen tetroxide) under helium pressure, together with a cold-gas attitude control system. The motor can be restarted up to five times. The AVUM provides roll control throughout the flight, stabilises the payload after burn-out and provides the final boost to circularise the payload's orbit. It also carries the flight control, flight termination and communications systems. EADS CASA provides the AVUM structure and payload adapter, and Thales, Zodiac, Selex and SAAB provide the avionics. The fairing resembles a scaled-down Ariane design, with a vertical two-part split and a carbon-skinned aluminium honeycomb construction, and is provided by Contraves Space.

ESA conducted a system design review for Vega from May to July 2004. By the end of 2004, all but one of the main launcher development subcontracts had been finalised by ELV SpA, all subsystem preliminary design reviews had been held, and some items had reached the critical design review stage. In 2004 the manufacture and testing of the filament-wound cases for all of Vega's solid motors took place. The P80 Inert Loading Model verification took place at Kourou and work on the launch site started late in the year.

Another important test milestone was passed in December 2005 with the successful firing of the first Zefiro 9 third-stage motor at Salto de Quirra in Sardinia. The second test was due to take place in September 2006.

Specifications

Vega

First launch: End 2007
Launch sites: ELA-1 pad Kourou, French Guiana
Reference mission: 1,500 kg into 90° polar orbit
Performance: Up to 2,500 kg to LEO
Number of stages: 4 including circularisation/orientation AVUM stage
Overall length: 30 m
Maximum diameter: 3 m
Launch mass: 137 t
Guidance: Stage 1/2/3 TVC for pitch/yaw. ACS in fourth stage for roll control during stage 1/2/3 burn, and three-axis control during coast and AVUM burn

Vega stage 1
Motor: Europropulsion P80FW with TVC
Overall length: 10.5 m
Diameter (max): 3 m
Propellant mass: 88 t
Max thrust: 3,040 kN
Burn time: 107 s
Nozzle expansion ratio: 16

Vega stage 2
Motor: Avio Zefiro 23 with TVC
Overall length: 7.5 m
Diameter: 1.9 m
Propellant mass: 23.9 t
Max thrust: 1,200 kN
Burn time: 71.6 s
Nozzle expansion ratio: 25 s

Vega stage 3
Motor: Avio Zefiro 9 with TVC
Overall length: 3.85 m
Diameter: 1.9 m
Propellant mass: 10.1 t
Average thrust: 313 kN
Burn time: 117 s
Nozzle expansion ratio: 56

Vega stage 4
Motor: Avio UDMH/NTO storable liquid
Overall length: 1.74 m
Diameter: 1.9 m
Propellant mass: 550 kg
Max thrust: 2,450 N
Vacuum specific impulse: 315.2 s

Zenit-3SL (Sea Launch)

Current Status

On 22 August 2006, Sea Launch successfully launched Koreasat 5 aboard their Zenit-3SL launch vehicle. As of August 2006, the Zenit-3SL has enjoyed a consistent success rate, with only a single failure in 1999 out of 23 total launch attempts.

For further information on the Sea Launch system, its Odyssey Launch Complex and the Sea Launch joint venture, please see the World Space Centres, Launch Sites, Commercial section.

Background

The Sea Launch concept combines the Zenit-3SL three-stage rocket with a sea-mobile launch complex which transports the missile to a launch site in the Pacific Ocean, at zero° latitude and 154° West longitude. The system uses a minimum of hypergolic or toxic propellants and has a 6,000 kg payload to GTO.

The first two stages of the Zenit-3SL are produced by NPO Yuzhnoye and are based on the two-stage Zenit-2, the most modern Russian launch system. The Zenit-2 was designed to reconstitute military satellite constellations rapidly and the design emphasised robustness (Boeing engineers used to NASA-type vehicles were shocked to see Russians walking on the rocket casings in tennis shoes), ease of operation and fast reaction times. It was also intended to be man-rated, although this capability was never used. The third stage is the Energia Block DM, already proven in service on Proton. Boeing provides the payload fairing.

The main stages are among the most efficient in the world, Sea Launch claims, because of the high thrust/weight ratio of the advanced high-pressure engines and the absence of strap-on boosters. Without strap-ons, the first stage is simpler and more efficient, and reliability is improved. The streamlined configuration lends itself to robust control in all flight conditions.

Compared to the Zenit-2, the 3SL has modernised avionics and a stiffened first stage, to handle winds in the marine environment. The second-stage thrust has been increased and more propellant is carried in all three stages. Automated refuelling has been incorporated on the Block-DM upper stage. The Zenit-3SL propellants, kerosene and liquid oxygen, are chilled before loading to increase their density.

The NPO Energomash RD-171 first-stage engine has four thrust chambers, fed by a single turbopump, driven by two gas generators. The thrust chambers are differentially gimballed for directional control. The engine can be throttled down to 74 per cent of full thrust for trajectory control. The second stage is fitted with a single-nozzle RD-120 engine, together with an RD-8 engine, with a single turbopump feeding four gimballing thrusters, for control.

Energia's Block-DM upper stage includes its own flight control avionics. Its 11D58M thrust engine has a single gimballing nozzle for directional control in propulsive phases, backed up by two attitude control/ullage engines, fuelled by nitrogen tetroxide and UDMH, which provide control in coast. The main engine can be restarted five times.

In October 2003, Sea launch announced plans to offer commercial launches from Baikonur under the Land Launch brand name, using both two-stage and three-stage versions (Zenit-3SLB) of the Zenit launcher. The three-stage launcher differs from the Zenit-3SL primarily in using a Lavochkin fairing. Land Launch does not offer as great a GTO payload as Sea Launch – its maximum is 3,600 kg versus the proven 6,000 kg payload of Sea Launch – because of its more northerly location, but addresses the medium payload market at lower cost.

PAS-9 gets a boost into space from the Sea Launch Odyssey platform (Sea Launch) 0137129

Zenit-3SL Specifications

Stage 1
Length: 32.9 m
Diameter: 3.9 m
Gross weight: 354,582 kg
Kerosene: 88,768 kg
LOX: 233,512 kg
Engine/thrust: RD-171, 740 t (SL) 806 t (vacuum)

Stage 2
Length: 10.4 m
Diameter: 3.9 m
Gross weight: 90,757 kg
Kerosene: 22,832 kg
LOX: 58,908 kg
Main engine/thrust: RD-120, 101 t vacuum
Vernier engine/thrust: RD-8, 8.1 t

Upper stage (Block DM-SL)
Length: 5.6 m
Diameter: 3.7 m
Gross weight: 19,711 kg
Kerosene: 4,560 kg
LOX: 11,290 kg
Engine/thrust: 11D58M, 8.1 t vacuum

Fairing
Length: 11.39 m
Outer diameter: 4.15 m
Inner diameter: 3.9 m

Zenit-3SL flight record

Date (GMT)	Mission	Mass (kg)	Comments
27 March 1999	DemoSat	4,500	Inaugural demonstration launch
10 October 1999	DirecTV 1-R	3,447	
12 March 2000	ICO F-1	2,744	Second stage failure
28 July 2000	PanAmSat PAS-9	3,659	
21 October 2000	Thuraya-1	5,108	Heaviest GTO satellite to date
18 March 2001	XM-2 Rock	4,666	Radio broadcast satellite
8 May 2001	XM-1 Roll	4,672	
15 June 2002	Galaxy IIIC	4,850	
10 June 2003	Thuraya-2	5,177	Beat Thuraya-1 weight record
8 August 2003	EchoStar IX/Telstar 13	4,737	Dual-use
1 October 2003	Galaxy XIII/Horizons-1	4,090	
11 January 2004	Telstar 14/Estrela do Sul 1	4,694	
4 May 2004	DirecTV 7S	5,483	New weight record
29 June 2004	Telstar 18	4,640	Lower-than-planned apogee, satellite safely repositioned
1 March 2005	XM-3 Rhythm	4,703	
26 April 2005	Spaceway-1	6,080	New weight record
23 June 2005	Intelsat Americas-8	5,500	
8 November 2005	Inmarsat-4	5,958	
15 February 2006	EchoStar X	4,333	
12 April 2006	JCSAT-9	4,401	
18 June 2006	Galaxy 16	4,640	
22 August 2006	Koreasat 5	4,448	

Israel

Shavit

Background
Israel's development of a sovereign launch capability has been complicated by its northerly location and the fact that it is constrained to westward launchers, against the rotation of the Earth. The Shavit ('Comet') all-solid orbital launcher, built by Israel Aircraft Industries, is derived from the two-stage Jericho 2 missile, first fired into the Mediterranean in 1986. A full range firing of 1,400 km is believed to have taken place from South Africa's Overberg site near Cape Town in June 1989, followed on 14 September 1989 by a 1,300 km test into the Mediterranean. The Jericho 1 500 km-range missile was developed in the 1960s with French assistance, based upon the Dassault MD-600.

Launches took place from an area within the Palmachim Air Force Base south of Tel Aviv, Israel, near the town of Yavne, inaugurated with the launch of Ofeq 1 on 19 September 1988. A second Ofeq 2 was successfully launched on 3 April 1990 but a third launch on 15 September 1994 was unsuccessful.

Shavit 1 is an improved version with an extended first-stage motor. It successfully launched the Ofeq 3 reconnaissance satellite on 5 April 1995. A second Shavit 1 launch with Ofeq 4 failed on 22 January 1998. Ofeq 5 was launched by Shavit on 28 May 2002, but another failure caused the loss of Ofeq-6 on 6 September 2004.

Israel Aerospace Industries MLM division, the prime contractor for Shavit 1, states that it is developing an Advanced Shavit configuration with an increased payload, and is studying airborne launch systems. In 2002, the company announced that it was forming an international joint venture, known as Leolink, to market current and improved versions of Shavit, including two higher-performance versions known as LK-1 and LK-2. Around the same time, Israel was reported to be in discussions with Brazil regarding the use of the Alcantara launch site, but no agreement was reached.

Specifications
Shavit
First launch: 19 September 1988 (Ofeq 1 test satellite)
Number launched: 6 through end of 2002
Launch sites: Palmachim Air Force Base
Vehicle success rate: 60%
Principal uses: small payloads into LEO
Performance: 156 kg into 248 × 1,170 km, 142.9° orbit
Number of stages: three solid propellant
Overall length: 17.7 m
Principal diameter: 1.352 m
Launch mass: 22–23 t
Guidance: inertial, mounted on stage 2. Redundant; MIL-STD-1553B bus. Flight angle commanded as function of time (zeroed for stage 1 separation)

Shavit stage 1
Motor: TAAS Israel Industries Ltd. PBAN solid with filament wound graphite epoxy casing. SI 240 s sea level, expansion ratio 9
Overall length: 5.25 m
Principal diameter: 1.352 m
Stage mass: unavailable (9.1 t propellant)
Average thrust: 610 kN vacuum
Burn time: 43 s
Attitude control: four jet vanes (jettisoned after vertical phase) plus four air vanes
Staging: pyrotechnic separation of external V-clamp holding the two bulkheads with trapezoid teeth

Shavit stage 2
Motor: similar to stage 1 but with 23.4 expansion ratio for altitude performance
Overall length: 5.676 m
Principal diameter: 1.352 m
Stage mass: unavailable (9.1 t propellant)
Average thrust: 564 kN vacuum
Burn time: 52 s
Attitude control: pitch/yaw by LITVC modules. Four 0.8 kg modules each injects 0.35 kg/s (at 51.0 atmosphere) strontium perchlorate through a single orifice; max operating pressure 85.0 atmosphere)
Staging: as stage 1, hot separation

Shavit stage 3
Motor: Rafael AUS-51 'Marble'
Overall length: about 2.1 m

The Shavit satellite launch vehicle flanked on either side by missiles with propulsion systems developed by Israel Military Industries (Israel Military Industries) 0003371

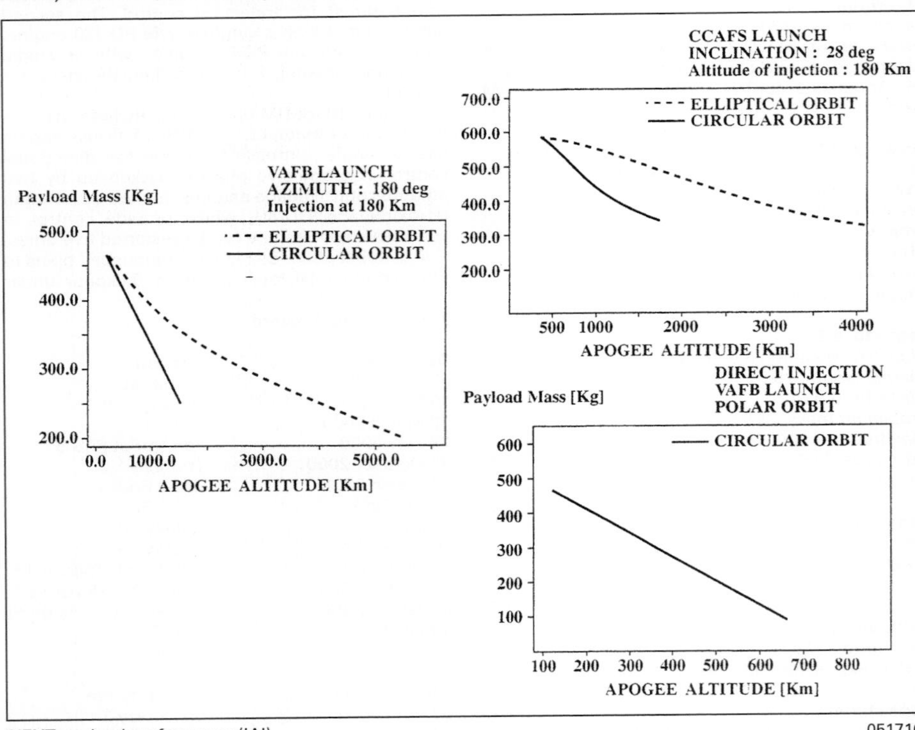

NEXT payload performance (IAI) 0517166

Principal diameter: 1.3 m
Stage mass: about 2 t (1,895 kg propellant)
Average thrust: 55
Burn time: 92.5 s
Attitude control: spin-stabilised, spun up/down by Rafael's ST-200N thrusters. NEXT's AUS-51 will employ LITVC. Rafael's ACT-25N thrusters provide attitude/nutation control

Japan

H-IIA

Current Status

Nine H-IIA rockets had been launched from Tanageshima by the end of February 2006, eight of them successfully. Five more launches – all for Japanese government projects – were scheduled in the 2006 and 2007 fiscal years. A larger-capacity version, the H-IIB, is under development and is expected to fly in 2008. One of the intended missions of the H-IIB is to launch the H-2 Transfer Vehicle to the International Space Station. An industry joint venture, the Rocket System Corporation, operates the H-IIA under contract to the Japanese Aerospace Exploration Agency (JAXA). Partners in RSC are Mitsubishi Heavy Industries, Kawasaki Heavy Industries, Ishikawajima-Harima Heavy Industries and NEC-Toshiba.

Background

Japan's Space Activities Commission gave the go-ahead, in February 1984, for the development of a large all-Japanese launcher, named H-II, capable of placing 2 tonnes in GEO and a small spaceplane in LEO. Prior to that time, Japanese industry had launched rockets based on McDonnell Douglas Delta stages, built under a 1969 licence that prohibited orbiting of third party payloads without US approval. Two demonstration missions flew 1986/87, with seven operational departures (five GEO) following. The interim H-I introduced Mitsubishi's cryogenic stage 2 and all-Japanese inertial guidance; H-II completed the transition in 1994 by adopting a large cryogenic first stage and solid strap-ons.

The H-II suffered some launch failures and proved very expensive. Consequently, in the mid-1990s, Japan's National Space Development Agency (NASDA) – which was merged into JAXA in 2003 – launched development of a simpler and more reliable rocket, the H-IIA, with the same performance characteristics and similar propulsion, but redesigned in many details based on experience with the H-II. The H-IIA made its first flight on 29 August 2001.

At least 80 per cent of lift-off thrust is provided by solid boosters which burn for 128 seconds. The core shuts down at 390 seconds, and the second stage performs LEO or GTO insertion.

The liquid hydrogen/LOX engines for the first and second stages of the H-IIA are the LE-7A and LE-5B, derived from the LE-7 and LE-5A engines of the H-II. They are simplified, with fewer welded joints – new high-precision casting technology is used instead – and fewer fuel-line joints. The same staged-combustion cycle (also used in the Space Shuttle Main Engine and the Energomash engines for Zenit and Atlas III and V) is retained: fuel-rich gas drives the turbopump and is then injected into the main combustion chamber. The H-IIA Solid Rocket Booster A (SRB-A) has a one-piece carbonfibre shell, replacing the four-segment structure of the previous SRB. The basic H-IIA configuration includes two SRB-A boosters, but its performance can be adjusted by adding either two or four smaller strap-on Solid Support Booster (SSB) units.

The H-IIA avionics are completely redesigned. Individual components use newer technology and are smaller, lighter and more reliable. The entire system, however, is changed to a databus-type architecture with digital connections among the different sensors and computers. This improved reliability, the effectiveness of diagnostics, and communications between the avionics and ground control systems. It also makes inter-stage connections simpler.

Structural changes include redesigned tank domes which are made in one piece rather than seven welded segments; a single integrated thrust structure for the LE-7A and SRB-As; and separate LH2 and LOX tanks on the second stage, rather than a single tank with a bulkhead. The new arrangement is heavier but easier to assemble. The interstage structure is changed from aluminium to carbonfibre, to save weight. As for systems, the H-IIA introduces electrical actuation for the gimballed rocket nozzles, eliminating hydraulic tubing. Overall, the number of parts in the rocket was reduced from 350,000 to 280,000 and its cost was halved to JPY9.3 billion.

The sixth H-IIA flight on 29 November 2003 ended in failure and led to a 15-month halt in flights. The direct cause of the accident as a burn-through of one of the SRB-A nozzles, which led to the failure of the booster to separate. However, the result was a comprehensive re-evaluation of the H-IIA design, which detected 786 potential problems with the design. The 77 most pressing issues would be fixed on the return-to-flight vehicle. The seventh and subsequent flights were entirely successful.

The basic H-IIA vehicle can be flown in four basic configurations, (please see table).

Basic configurations of H-IIA vehicle

Designation	Boosters	GTO payload
202	Two large SRB-A	3.7 t
2022	Two large SRB-A plus two SSB	4.2 t
2024	Two large SRB-A plus four SSB (burning consecutively in pairs)	4.6 t
204	Four large SRB-A	5.7 t

All except the 204 configuration had flown by early 2006. Six fairing designs are offered, including single- and double-spacecraft types.

Development of the 204 configuration was suspended after the November 2003 failure but it is expected to launch during 2006. The F7 flight tested an improved SRB design. The second flight after the failure, F8, tested some new features destined for the 204 configuration and other improvements to the system. One difference was the use of a longer regeneratively cooled nozzle on the LE-7B engine. The nozzle used on earlier flights was shorter than the original design, which had shown excessive lateral thrust in ground tests. The long nozzle boosts thrust from 1,074 kN to 1,098 kN and increases Isp from 429 to 440 seconds. The flight also tested improved insulation around the nozzle to protect it from exhaust heating in the four-SRB configuration.

Full development of a growth version of the rocket, the H-IIB, started in FY2006. It is designed to offer an 8 t payload to GTO. The first two units are expected to fly in FY2008, with the second flight carrying the first HTV service module to the ISS. The principal change in the H-IIB is a new first stage. The fuel tank diameter is increased from 4 m to 5.2 m and the stage now carries two LE-7A motors instead of one. Mitsubishi Heavy Industries is using Friction Stir Welding (FSW) to assemble the tank sections, to reduce cost, and is developing an in-house capability to produce spin-formed tank domes, which are currently manufactured by Zeppelin in Germany.

So far, no commercial or international launch contracts for H-IIA launches have been announced. Operations are restricted by an agreement with the fishing industry near Tanageshima, which limits launches to two 90-day windows each year, and this increases costs.

Specifications

H-IIA

First launch: 29 August 2001
Number launched: 9 to end February 2006
Industrial integrator: Rocket Systems Corporation
Launch sites: Yoshinobu Launch Complex, Tanegashima
Vehicle success rate: 88.9% for 9 launches
Performance:
202 configuration; LEO: 10,000 kg into 300 km circular 30° orbit
202 configuration; GTO: 3,700 kg
2022 configuration; GTO: 4,200 kg
2024 configuration; GTO: 4,600 kg
204 configuration; GTO: 5,700 kg
H-IIB GTO 8 t, LEO 16,500 t
Number of stages: 2+2 SRB-A boosters and up to four SSB boosters
Overall length: 53 m with 4S fairing
Principal diameter: 4.0 m
Launch mass:
202: 289 t
2022: 321 t
2024: 351 t
204: 445 t
H-IIB: 551 t
Guidance: Strapdown inertial using ring laser gyros

H-IIA orbital launch history

F1	29 August 2001	202	LRE, VEP-2
F2	4 February 2002	2024	MDS-1, DASH
F3	10 September 2002	2024	DRTS, USERS
F4	14 December 2002	202	ADEOS-II, FedSat, MicroLabSat
F5	28 March 2003	2024	IGS-1a
F6	29 November 2003	2024	IGS-2, failed
F7	26 February 2005	2022	FS-1300
F8	24 January 2006	2022	ALOS
F9	20 February 2006	2024	MTSAT-2

H-IIA SRB-A solid strap-ons
Motor contractor: IHI Aerospace
Length: 15.1 m
Principal diameter: 2.5 m
Mass at ignition: 77 t
Propellant mass: 66 t
Average thrust: 2,245 kN sea level
Specific impulse: 281 sec
Burn time: 120 s
Steering: TVC

H-IIA SSB solid strap-ons
Motor contractor: ATK
Length: 14.9 m
Principal diameter: 1 m
Mass at ignition: 15.5 t
Propellant mass: 13.1 t
Average thrust: 745 kN sea level
Specific impulse: 282 sec
Burn time: 60 s

H-IIA stage 1
Contractor: Mitsubishi Heavy Industries
Engine: single Mitsubishi LE-7A staged combustion cycle, single start, liquid oxygen/hydrogen, regenerative cooling, gimballed for pitch/yaw control.
Overall length: 37.2 m
Principal diameter: 4.0 m
Total stage mass: 114 t
Propellant mass: 101.1 t LOX and LH_2
Stage thrust: 1,098 kN vacuum
Specific impulse: 440 sec
Burn time: 390 s

H-IIA stage 2
Contractor: Mitsubishi Heavy Industries
Engine: single Mitsubishi LE-5B hydrogen bleed cycle, dual start, LOX/LH_2, regenerative cooling, 130:1 expansion ratio, 5.0:1 mixture ratio, gimballed 3.5° square for pitch/yaw control.
Overall length: 9.2 m
Principal diameter: 4.0 m
Total stage mass: 20 t
Propellant mass: 16.9 t LOX and LH2
Stage thrust: 137 kN vacuum,
Specific impulse: 448 sec
Burn time: 530 s max;.
Attitude control system: LE-5B gimballing provides pitch/yaw control during burn; two IHI hydrazine thruster modules (each 4×50 N $+ 2 \times 18$ N) provide roll control during burn and three-axis during coast

M-V

Current Status

The seventh and last M-V flew in September 2006.

Background

The M-V was the latest in a series of rockets and launchers developed by the Institute of Space and Astronautical Science (ISAS), now part of the Japanese Aerospace Exploration Agency (JAXA), to launch scientific satellites. ISAS has maintained its own stable of orbital and suborbital rockets for science missions since the late 1950s. All solid-fuelled, these have been integrated by ISAS with Nissan Motor – now IHI Aerospace – as prime industrial contractor. The largest sounding rocket, L-3H, was augmented by a kick stage to create the L-4S orbital vehicle as a technology demonstrator for the M-4S launcher, flying Japan's first satellite in 1970 after four failures.

Design work on a larger successor started in 1989. The goal was to develop a simple, all-solid-fuel launcher that would be compatible with launch facilities at Kagoshima Space Center and that would be inexpensive enough to permit annual flights. The Japanese government approved the M-V in 1989 and the first M-V launch took place on 12 February 1997.

The first two launches were successful, but M-V-4, launched on 10 February 2000, failed to deliver the Astro-E X-ray astronomy satellite into orbit. The failure was caused by imperfect combustion of the first stage rocket motor. The problem was traced to a structural problem in the nozzle throat of the first stage motor, which could not withstand the 3,000°C temperature. ISAS changed the material used in the nozzle throat of the first and third stages from graphite to 3D-woven carbon-carbon composite. It also fitted the second stage with a high-performance M-25 motor with high-pressure combustion and thrust vector control, an attitude-control unit using MN (Movable Nozzle) TVC in place of conventional LI (Liquid Injection) TVC to improve thrust performance. The rocket flew

An M-V rocket on the pad at Uchinoura Space Center, Kagoshima (JAXA – ISAS) 1340629

successfully on four more occasions before the programme ended in September 2006.

The M-V was a solid fuelled, three-stage rocket with an optional kick stage for lunar or planetary missions or for highly elliptical orbits. Each stage used a single motor. The first and second stage motor cases were made from HT-230M high-strength steel, and the third and kick stages had carbon fibre composite cases. An unusual feature, tested on an SY-735-2 vehicle in September 1994, was a "fire in the hole" separation system between the first and second stages. The second stage motor was ignited before the first stage burned out, eliminating a dedicated separation system and reducing second-stage inert mass. Flexible linear shaped charges blew away the connecting structure. The third and kick stages had extending nozzles to reduce the length of the interstage structures, and the guidance system used fibre optical gyro sensors.

The Japanese government has approved studies by JAXA to develop a new solid fuel launcher as the successor to the M-V. The plan aims to reduce costs and increase the efficiency of the new vehicle compared with the M-V. The launch vehicle will be developed by reusing and modifying technologies used in the H-2A and the M-V. The development costs of the new vehicle are estimated at about 20 billion Yen (USD175 million), with the first launch planned to take place in the Japanese fiscal year 2011.

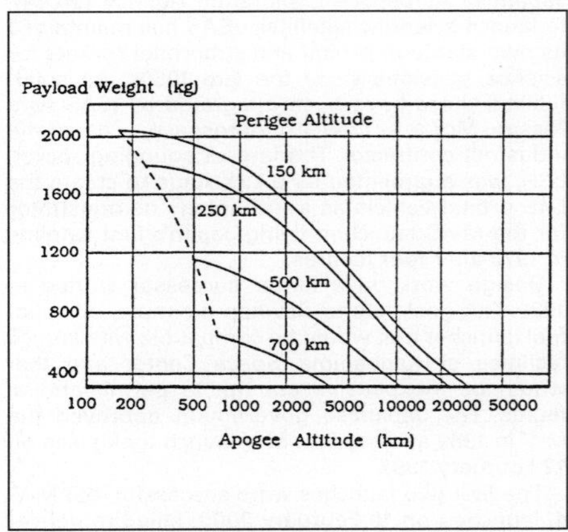

M-V performance from Kagoshima (ISAS) 0516917

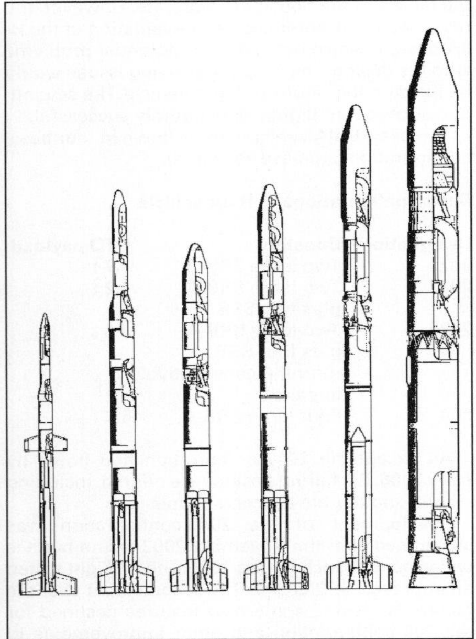

ISAS orbital launchers (from left): L-4S, M-4S, M-3C, M-3S, M-3SII, M-V 0517178

Specifications
M-V
Launch site: Kagoshima
Success rate: 85.7%
Performance: 250 km, 31°: 1,800 kg
Number of stages: 3 plus optional stage 4
Overall length: 30.7 m
Principal diameter: 2.5 m
Launch mass: 139 t (excluding payload)

M-V stage 1
Comment: Prototype stage 1 was successfully fired 21 June 1994. A second firing, using a thinner nozzle, was successful June 1995.
Designation: M-14
Contractor: IHI Aerospace
Overall length: 13.8 m
Principal diameter: 2.50 m
Mass at ignition: 83.5 t
Average thrust: 4,200 kN sea level
Specific impulse: 274 sec.
Burn time: 46 s (effective)

M-V stage 2
Designation: M-24/M-25
Contractor: IHI Aerospace
Overall length: 6.8 m
Principal diameter: 2.5 m
Mass at ignition: 34.5 t
Average thrust: 127 t vacuum
Specific impulse: 287 sec.
Burn time: 71 s (effective)

M-V stage 3
Designation: M-34
Contractor: IHI Aerospace
Overall length: 3.6 m
Principal diameter: 2.2 m
Mass at ignition: 11 t
Average thrust: 289 kN vacuum
Specific impulse: 302 sec.
Burn time: 103.5 s (effective)

M-V kick stage (optional)
Designation: KM-V1
Contractor: IHI Aerospace
Overall length: 1.57 m before extension; 1.97 m after extension.
Principal diameter: 1.2 m
Mass at ignition: 2.7 t
Average thrust: 51,900 kN vacuum
Specific impulse: 298 sec.
Burn time: 68 s

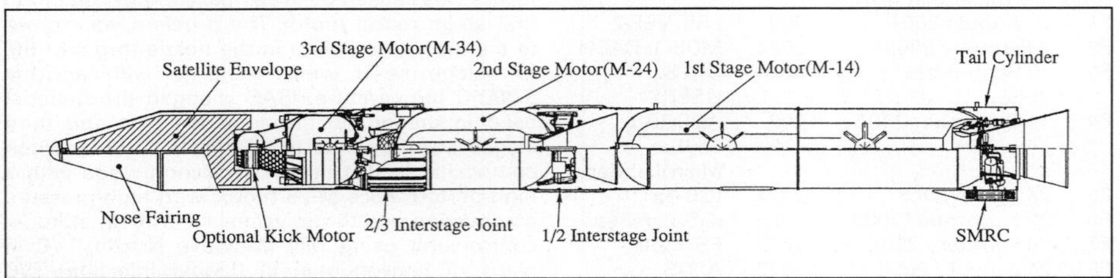

The M-V launcher (ISAS) 0516916

M-V Launch History

Vehicle/ flight	Date	Payload	Comments
M-V-1	12 February 1997	MUSES-B	Radio astronomy satellite renamed 'Halca' on orbit.
M-V-3	3 July 1998	Planet-B	Mars probe, renamed Nozomi after launch.
M-V-4	10 February 2000	Astro-E	Failure
M-V-5	9 May 2003	MUSES-C	Asteroid sample mission, renamed 'Hayabusa' after launch.
M-V-6	10 July 2005	Astro-E2	X-Ray astronomy, renamed 'Suzaku' on orbit
M-V-8	22 February 2006	Astro-F and Cute 1.7 student subsatellite.	Astro-F renamed 'Akari' (Light) on orbit.
M-V-7	22 September 2006	Solar-B, SSSAt solar sail experiment and Hitsat.	Solar-B renamed 'Hinode' on orbit.

Note: M-V-2, allocated to carry the Lunar A probe, was delayed due to spacecraft development problems and never flew.

Russian Federation

Angara

Current Status
The Russian Ministry of Defence awarded Khrunichev State Research and Production Space Centre a six-year, RUR16 billion (USD570 million) contract in May 2005 to continue development of the new Angara launch vehicle family. According to an August 2006 press release by Khrunichev, the first test launches are due to occur in 2010–2011, likely from Plesetsk Cosmodrome (see separate entry).

Background
The Angara project originated after the break-up of the Soviet Union. Its strategic goal was to free Russia from dependence on the Baikonur launch site, now located in independent Kazakhstan. This meant the development of a higher-performance rocket than the workhorse Proton, so that it could equal Proton's lift while being launched from Plesetsk in sovereign Russian territory. Another design goal is to reduce the use of toxic and environmentally damaging fuels of the kind used by Proton.

Khrunichev and RKK Energia competed to develop the new rocket, and the former was declared the winner of the competition in September 1994. However, lack of money has delayed the project considerably, and in the meantime the launcher has been redesigned substantially. The current design makes extensive use of existing technology and components to reduce development cost and risk.

Angara is based on the use of a Common Rocket Module (CRM), with one, three or five CRMs being combined to create small, medium and heavy-class launchers. Each CRM comprises fuel and oxidizer tanks and the Energomash RD-191 rocket engine. The single-chamber LOX/kerosene RD-191 shares many parts and design features with the two-chamber RD-180, used on Atlas V, and the RD-170 that NPO Energomash developed for Zenit. Thrust vectoring on the RD-191 provides pitch and yaw control. The engines used as upper stages can be throttled down to 30 per cent, and the centre stages are fitted with auxiliary off-axis thrust nozzles for roll control. The flight control system and power supply are housed in the intertank compartment.

All initial versions of the Angara system, except for the smallest configuration, use a common second-stage booster. It also uses LOX and kerosene, and is powered by the Energia RD-0124 four-chamber engine. The RD-0124 provides control in pitch, roll and yaw. A growth version of Angara A5 may use a new LOX/hydrogen second stage.

Specifications
Angara

	Angara 1.1	Angara 1.2	Angara A3	Angara A5
Lift-off mass, kg:	149,000	171,000	4,800,000	773,000
Payload to 200 km/63° orbit, kg:	2,000	3,700	14,000	24,500
Payload to GTO, kg:			2,400	6,600
CRMs:	1	1	3	5
Propellant mass, first stage, kg:	132.6	128.8	398	663
First stage sea-level thrust, t:	196	196	588	980
Second stage:	Breeze-M	RD-0124 LOX/ kerosene	RD-0124 LOX/ kerosene	RD-0124 LOX/ kerosene
Upper stage:	None	Breeze-M	Breeze-M	KVRB
Overall length, m:	34.9	41.5	45.8	55.4

The standard upper stage for the initial versions of Angara is the UDMH/NTO-fuelled Khrunichev Breeze-M, which can be used with or without its toroidal auxiliary propellant tank. Later and heavier versions will be fitted with the KVRB oxygen-hydrogen upper stage, being developed for both Angara and as an upgrade to Proton.

The Angara family, as currently defined, includes the following members:

The Angara 1.1 comprises a single CRM with a Breeze-M upper stage, without the auxiliary tank, and has a design payload of 2 tonnes. The Angara 1.2 has higher performance as a result of replacing the Breeze-M with the LOX/kerosene upper stage.

The Angara A3 is designed to lift 2.4 tonnes to GTO. It has three CRM stages, the RD-0124-powered second stage and a Breeze-M upper stage with auxiliary tank. It could also be flown with the KVRB cryogenic upper stage.

The Angara A5 heavy launcher has a 6.6 tonnes payload to GTO. It has five CRMs and is equipped with the KVRB upper stage as standard. One variant shown has a new LOX/hydrogen second stage.

Other versions of the Angara family have been discussed, but are not funded. In 2001, at the Paris air show, Khrunichev displayed a large-scale mock-up of a flyback booster for Angara, known as Baikal. It was fitted with an unswept wing which pivoted at its mid-point to lie flat against the fuel tank during ascent, tail surfaces, and a nose-mounted jet engine with a bifurcated exhaust. Reusable Baikal units would replace the booster CRMs in the A3 and A5 configurations.

Recent progress in the Angara project has included the completion of the first full-duration firings of the RD-191 engine, announced in October 2005, and the acceptance of the first launch pad for the rocket's base at Plesetsk. The Angara launch complex is being created by the completion and upgrading of a facility that was originally intended to support Zenit launches. Also, in December 2004, Russia and Kazakhstan reached an agreement on the establishment of a new complex at Baikonur, known as Baiterek, to support Angara operations. International Launch Services (ILS) plans to offer Angara launches on the international commercial market when the vehicle becomes operational.

Kosmos 3M

Current Status

Kosmos 3M is marketed commercially by Puskovie Uslugi Ltd., a Russian company created by governmental decree in 1998 to offer commercial satellite launch services. It is also marketed by Cosmos International, a German-registered subsidiary of the Fuchs Gruppe. The launcher was designed by NPO Yuzhnoye and was produced by the Polyot production company. Production was halted in 1994 but some vehicles remain available for operational launches.

Background

For many years the smallest vehicle in the operational inventory, Kosmos originated as the R-14 (11K65)/SS-5 Skean medium range missile, developed by the Yangel bureau (now NPO Yuzhnoye). In its first incarnation the Soviets called it Kosmos and then the current Kosmos 3M (11K65M). It is now the responsibility of AKO Polyot in Omsk. It remains as the only vehicle to be fired from all three Soviet sites – Baikonur, Kapustin Yar and Plesetsk, although almost all recent launches have been from Plesetsk Launch Complex 132. The first Kosmos 3M launch took place from Plesetsk on 15 May 1967, with Kosmos 158, a Tsiklon mass model as the payload. The last Kosmos 3M orbital launch from Kapustin Yar was on 28 April 1999. In the peak years of the programme, well over 20 launches were made each year, alongside an almost equal number of 11K65MP sub-orbital launchers which were used to test subscale models of spaceplanes and other re-entry vehicles. By the end of 2005, the Kosmos family had notched up 774 launches and a 96 per cent success rate. However, the rocket is described by foreign observers as being hampered by low injection accuracy. Payloads are launched into high inclination or sun-synchronous orbits.

Kosmos 3M released its first commercial payloads on 24 January 1995. They were Sweden's 28 kg Astrid 1 science satellite and Final Analysis Incorporated's 114 kg store/forward FAISAT 1. Since then, the programme has maintained a steady stream of international, commercial and military launches, including multiple small satellite launches.

One of the most important international contracts for Kosmos 3M launches is for the SAR-Lupe radar reconnaissance satellite developed by OHB System AG for the German armed forces. Five SAR-Lupe satellites are being built and will be launched from Plesetsk by Kosmos 3M. The spacecraft's large antenna requires a reshaped payload fairing, which was tested during the 24 January 2005 Kosmos 3M launch.

US marketing arrangements for Kosmos launches appear to be defunct, and mid-1990s plans for an improved version have not been proceeded with. Launch vehicle production was stopped in 1995, but production capability was retained. Stockpiled rockets were used to perform subsequent launches. In mid-2005, it was reported that a plan was being drawn up to resume production of the Kosmos-3M at the Polyot plant.

Kosmos 3M rocket being raised at the Plesetsk launch pad. Note the blue fairing used for SAR-Lupe (OHB-System)
1329930

Separation of the fairing from the Kosmos 3M rocket during launch of SAR-Lupe (OHB-System) 1329936

Recent Kosmos 3M Orbital Launches

Date	Payload	Site	Results
16 January 1996	Kosmos 2327	Plesetsk	
24 April 1996	Kosmos 2332	Plesetsk	
05 September 1996	Oscar 30	Plesetsk	
20 December 1996	Kosmos 2336	Plesetsk	
17 April 1997	Kosmos 2341	Plesetsk	
23 September 1997	FAISAT-2V	Plesetsk	
10 December 1998	Nadezhda 5	Plesetsk	
24 December 1998	Kosmos 2361	Plesetsk	
28 April 1999	ABRIXAS and Megsat-0	Kapustin Yar	
26 August 1999	Kosmos 2366	Plesetsk	
28 June 2000	Nadezda, Snap, Tsinhua	Plesetsk	
15 July 2000	CHAMP, MITA, BIRD-Rubin	Plesetsk	
21 November 2000	Quick Bird	Plesetsk	Failed
8 June 2001	Kosmos-2378	Plesetsk	
28 May 2002	Tsikada	Plesetsk	
8 July 2002	Kosmos 2390/2391	Plesetsk	Classified
26 Sept 2002	Nadezda-M	Plesetsk	
28 Nov 2002	AlSat/Mozhaets	Plesetsk	
4 June 2003	Kosmos 2398	Plesetsk	Classified
19 August 2003	Kosmos 2400/2401	Plesetsk	
27 September 2003	KAISTSAT-4, Bilsat-1, Nigeriasat-1, UK-DMC, Mozhaets-4, Larets	Plesetsk	
22 July 2004	Russian military	Plesetsk	Classified
23 September 2004	Kosmos 2408/2409	Plesetsk	
20 January 2005	Parus/Tatyana	Plesetsk	
27 October 2005	Mozhaets-5, Sinah-1, Beijing-1, TopSat, SSETI Express, Ncube-2, UWE-1, Rubin-V	Plesetsk	
21 December 2005	Gonets-1M, classified payload	Plesetsk	
19 December 2006	SAR-Lupe 1	Plesetsk	
2 July 2007	SAR-Lupe 2	Plesetsk	
11 September 2007	Kosmos 2429	Plesetsk	

Specifications
Kosmos 3M
First orbital launch: 18 August 1964 (Kosmos 38–40)
Number launched: 441 Kosmos 1 and Kosmos 3 or 3M vehicles attained orbit by end of September 2007.
Launch sites: Plesetsk, Kapustin Yar (KY currently suborbital only, but orbital available).
Principal uses: Small military store/dump communications, navigation and unknown minor missions at medium altitudes; Interkosmos science satellites; commercial (including German SAR-Lupe radar constellation).
Vehicle success rate: 94.8 per cent orbital attempts; 23 orbital failures: 23 October 1964, 16 November 1966, 26 June 1967, 27 September 1967 (pad explosion), 4 June and 15 June 1968, 27 December 1969, 27 June 1970, 23 December 1970, 22 July 1971, 17 October 1972, 25 May 1973, 3 June 1975, 19 December 1975, 29 November 1977, 20 December 1978 (Kosmos 1064 attained orbit but stage 2 failed to circularise), 4 March 1982, 18 June 1982 (Kosmos 1380 attained orbit but stage 2 failed to circularise), 30 August 1982, 24 November 1982, 25 January 1983, 23 October 1985, 25 June 1991, 6 October 1995 (stage 2 sticking fuel valve left Kosmos 2321 in low orbit, although still claimed as usable), 20 November 2000. Nine were killed 26 June 1973 at Plesetsk when Kosmos exploded during fuelling operations.
Cost: About USD10 million for complete capacity; USD6,500/kg for secondary payloads (Sweden was charged USD5,000/kg for Astrid).
Availability: Main payloads can be accommodated with three months' notice.
Performance: 1,400 kg into 180 km circular, but principally used for about 1 t payloads into 800-1,500 km circular orbits. Accuracy into 1,000 km ± 40 km perigee/apogee, + 0.04/–0.08° inc. 1,280 kg into 400 km circular at 66°; 1,210 kg into 400 km circular at 74°; 1,140 kg into 400 km circular at 83° and 1,070 kg into 200 × 1,500 km elliptical orbit at 51°.
Number of stages: 2
Overall length: 32.4 m
Principal diameter: 2.40 m
Launch mass: 109 t
Launch thrust: 1,726 kN
Guidance: Inertial, on stage 2 forward end, by Khartron

Kosmos 3M launcher with normal fairing (SSTL) 1329919

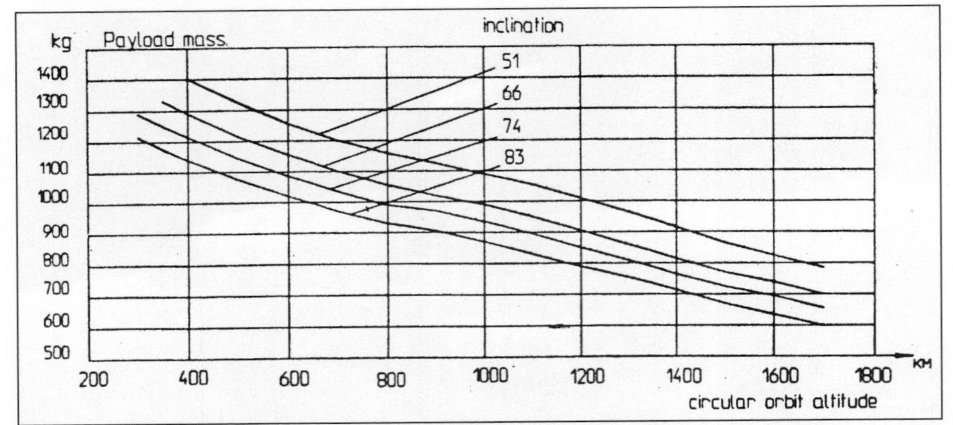

Kosmos 3M circular orbit performance from Plesetsk. The curves indicate different inclinations. Precision into a 1,000 km circular path is ±40 km; ±80 km for 1,600 km 0516873

Kosmos 3M Stage 1
Engines: RD-216 assembly (four fixed chambers) from NPO Energomash with storable propellants. Steering by graphite vane inserted into each exhaust (installed at end of pad preparations).
Overall length: 22.5 m
Diameter: 2.40 m
Thrust: 1,726 kN
Burn time: About 130 s
Oxidiser: N_2O_4 in NO_2
Fuel: UDMH (aft tank)
Propellant mass: 82.0 t; dry mass 5.30 t
Separation: Cold separation, assisted by solid braking rockets

Kosmos 3M Stage 2
Configuration: Stage 2 adds two side tanks for missions into 1,000–1,500 km circular orbits, using four smaller thrusters for the climb to the required altitude. The combined UDMH/N_2O_4 external tanks run the length of stage 2. They are not carried on the suborbital version, which also ignites stage 2 only once.
Engine: KB Khimmach 11D49: pump-fed fixed single chamber + four steering thrusters (used for steering, orbit trims, manoeuvres for multiple deployments, climb to high orbits).
Overall length: 4.205 m (6.585 m from nozzle to payload mount).
Diameter: 2.40 m (3.2 m across external tanks)
Thrust: 157.3 kN main chamber plus 4 × 1.4–1.8 kN steering/adjust thrusters (each with a pressure-fed 99 N thruster for three-axis coast control and ullage).
Propellants: As stage 1, carried in spherical tanks
Burn time: 325–335 s burn 1 plus 2–8 s second burn with 2 main engine
Propellant mass: 18.7 t; dry mass 1,434 kg. Fuel in aft tank.

Fairing/Payload Environment
The 2.40 m diameter, 5.720 m long 40° apex coni-cylindrical fairing provides 12.58 m³ payload volume, with four access doors in the 1.809 m high cylindrical section and seven in the conical. The vehicle is integrated horizontally and the payload attached in the same building. Access provides two 3 × 3 m class 100,000 clean rooms in the building for satellite integration/checkout, in addition to the building's own 28 m-long satellite area. Launcher preparation up to pad arrival requires 105 men over 34–36 h; pad activities 120–135 men for 8–10 h; launch is by a crew of 20–25. Fuelling is completed 4 h before launch and air conditioning to the payload fairing ends at T-1 h. Launches are made within –40/50°C and 20 m/s wind surface conditions. Kosmos provides 29.5 ± 2 V DC to the payload. The fairing halves separate mechanically at about 75 km. Payload is separated by four pyropushers within 20 s of shutdown; stage 2 ignites a small lateral solid 1.25 s after separation for collision avoidance. Payload angular disturbance at separation is ± 3.0°/s pitch/yaw + ± 0.5°/s roll for 1,000 kg.
Mechanical loads: 1.0–6.5 g longitudinal, <1.3 g lateral
Acoustic load: 140 dB rms max
Thermal: 280°C conical and 180°C cylindrical internal wall maximum temperatures

Proton

Current Status

The highly successful Proton family of heavy launch vehicles currently comprises two members: the long-serving Proton K and the revised and improved Proton M. The Proton M is now the only version on offer to commercial customers, and is available through International Launch Services (ILS), the joint venture between Proton's designer, the Khrunichev State Research and Production Space Centre, and Lockheed Martin. The Proton K is now used only for Russian government payloads, and will be replaced by Proton M.

Background

The Proton design originated as the Chelomei bureau's UR-500, developed as a "universal rocket" that could act as an ICBM as well as a launcher. However, the requirement for such large weapons disappeared before the UR-500 was tested, and the design was adopted exclusively as a launch vehicle. The first version was a two-stage rocket, designated 8K82. Four of these were flown, the first on 16 July 1965. These vehicles carried the Proton scientific satellite, and this name became attached to the launch vehicle.

This early Proton was supplanted from 1967 by a configuration with three main stages and an upper stage. The system used an upper stage that had been developed as the fifth stage of the N-1 lunar launch vehicle. The word for "stage" in Russian is the same as "block", and D is the fifth letter of the Russian alphabet, so this stage became known as Block D. In accordance with Soviet naming convention, the basic three-stage vehicle received a separate designation: 8K82K, UR-500K or Proton K. The four-stage configuration would be referred to as the Proton K Block D. In the early 1970s, the upper stage was upgraded to the Block DM, the principal difference being that it incorporates its own Guidance, Navigation and Control (GNC) avionics; in the Block D, those functions were performed by the payload. The Block D was still being flown on some Russian government missions in the early 2000s.

In the 1970s, Proton became the principal launch vehicle for Soviet interplanetary missions and also launched the Salyut space stations. Later in the decade, it also became the main launch vehicle for geostationary communications satellites, with launch rates accelerating from a half-a-dozen per year to 10–12 or more Proton flights per year.

Proton's potential was enough to arouse some market interest from outside the Soviet Union, even in the days of the Cold War. The Soviet Union made an early offer to ESA for launching the Marecs maritime satellite, followed by a more vigorous campaign in 1983 to launch the second-generation Inmarsat spacecraft. However, no customers were found, and discussions of international deals that would have opened up near-equatorial launching sites in Brazil, Australia or New Guinea also proved fruitless.

The first marketing success came in November 1992 when Inmarsat selected Proton and KB Salyut for a USD36 million contract to launch an Inmarsat 3; the contract was signed 27 April 1993. The success built on the Bush administration's June 1992 approval for one Proton launch of a US-built satellite.

After the break-up of the Soviet Union, KB Salyut (the design authority, then the name of what had been Chelomei's bureau) and Khrunichev (the builder) in recent years promoted Proton separately, disagreeing over rights to the vehicle. President Yeltsin issued a decree 7 June 1993 merging them as the Khrunichev State Research & Production Space Centre, which Lockheed joined as a minority shareholder to form Lockheed-Khrunichev-Energia (LKE) International in early 1993. ILS was formed in

Proton stage 1 production at Khrunichev's Moscow factory 0517365

1995, establishing the business arrangements that continue today. Lockheed Martin carried out a study of alternative launch sites for Proton, but Baikonur, in Kazakhstan, remains the only site today.

The Proton has flown more than 300 times. It has launched the Ekran, Raduga, and Gorizont series of geostationary communications satellites (which provided telephone, telegraph, and television service within Russia and between member states of the Intersputnik Organization), as well as the Zond, Luna, Venera, Mars, Vega, and Phobos interplanetary exploration spacecraft. All Russian unmanned lunar landing missions were flown by Proton. The Proton has launched the entire constellation of Glonass position location satellites and has carried the Salyut series space stations and the Mir space station modules. Proton launched the Zarya and Zvesda modules, which made up the first two elements of the International Space Station.

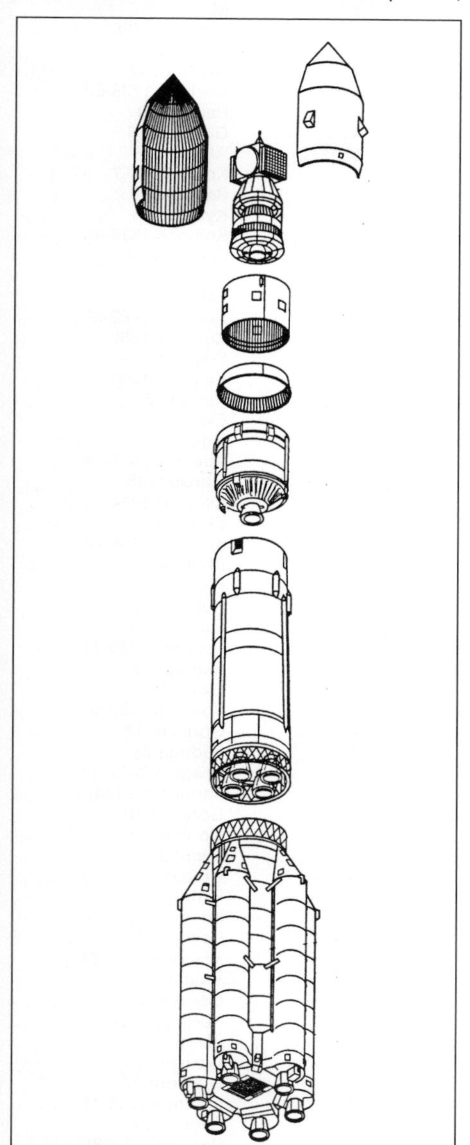

The Proton launcher (LKE International/C P Vick) 0516874

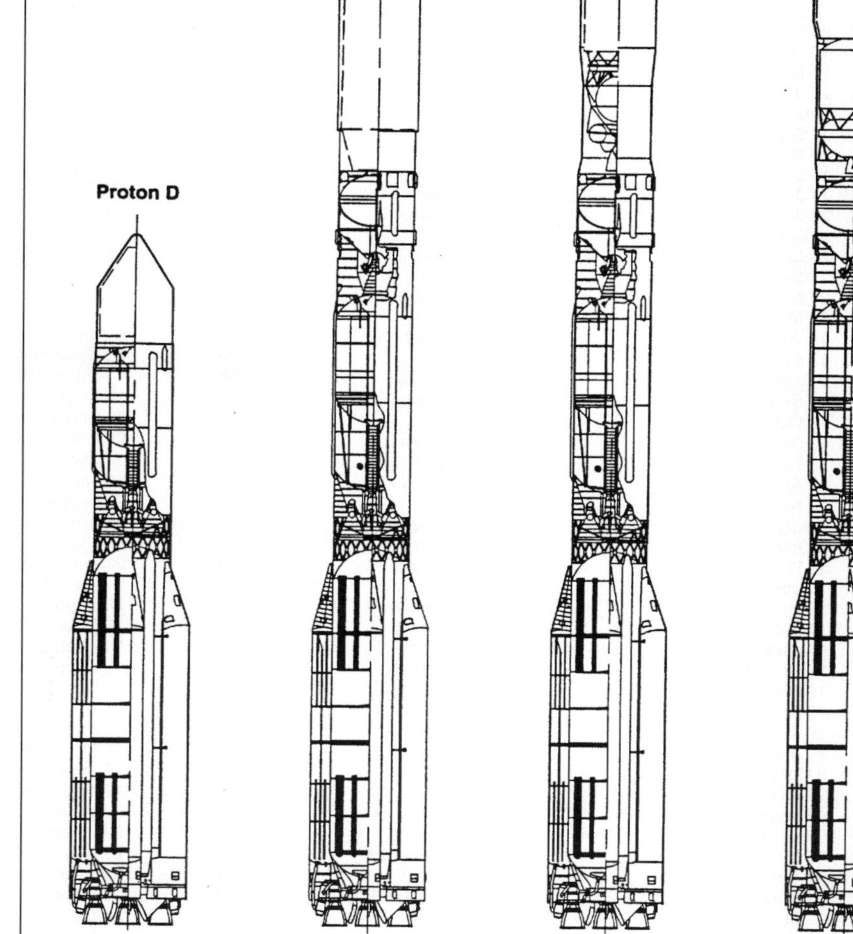

The Proton family (LKE International) 0516876

Proton launch history (all launches made from Baikonur)

Launch	Date	Stages	Payload
1	16 July 1965	2	Proton 1
2	2 November 1965	2	Proton 2
3*	24 March 1966	2	Proton
4	6 July 1966	2	Proton 3
5	10 March 1967	4	Kosmos 146/L1-P
6	8 April 1967	4	Kosmos 154/L1-P
7*	28 September 1967	4	Zond lunar
8*	22 November 1967	4	Zond lunar
9	2 March 1968	4	Zond 4 lunar
10*	23 April 1968	4	Zond lunar
11	14 September 1968	4	Zond 5 lunar
12	10 November 1968	4	Zond 6 lunar
13	16 November 1968	3	Proton 4
14*	20 January 1969	4	Zond lunar
15*	19 February 1969	4	Luna-1969A/Lunokhod
16*	27 March 1969	4	Mars-1969A
17*	2 April 1969	4	Mars-1969B
18*	14 June 1969	4	Luna-1969B
19	13 July 1969	4	Luna 15
20	7 August 1969	4	Zond 7 lunar
21*	23 September 1969	4	C300 (lunar?)
22*	22 October 1969	4	C305 (lunar?)
23*	16 November 1969	4	Kosmos/L1-E
24*	6 February 1970	4	Luna 1970A
25	12 September 1970	4	Luna 16
26	20 October 1970	4	Zond 8
27	10 November 1970	4	Luna 17
28	2 December 1970	4	Kosmos 382/L1-E
29	19 April 1971	3	Salyut 1
30*	10 May 1971	4	C419 (Mars)
31	19 May 1971	4	Mars 2
32	28 May 1971	4	Mars 3
33	2 September 1971	4	Luna 18
34	28 September 1971	4	Luna 19
35	14 February 1972	4	Luna 20
36*	29 July 1972	3	Salyut (DOS 2)
37	8 January 1973	4	Luna 21
38	3 April 1973	3	Salyut 2
39	11 May 1973	3	C557/Salyut
40	21 July 1973	4	Mars 4
41	25 July 1973	4	Mars 5
42	5 August 1973	4	Mars 6
43	9 August 1973	4	Mars 7
44	26 March 1974	4	Kosmos 637
45	29 May 1974	4	Luna 22
46	24 June 1974	3	Salyut 3
47	29 July 1974	4	Molniya-1S
48	28 October 1974	4	Luna 23
49	26 December 1974	3	Salyut 4
50	6 June 1975	4	Venera 9
51	14 June 1975	4	Venera 10
52	8 October 1975	4	Kosmos 775
53*	16 October 1975	4	Luna-1975A
54	22 December 1975	4	Raduga 1
55	22 June 1976	3	Salyut 5
56	9 August 1976	4	Luna 24
57	11 September 1976	4	Raduga 2
58	26 October 1976	4	Ekran 1
59	15 December 1976	3	Kosmos 881-882
60	17 July 1977	3	Kosmos 929
61	23 July 1977	4	Raduga 3
62*	4 August 1977	3	Dual Kosmos
63	20 September 1977	4	Ekran 2
64	29 September 1977	3	Salyut 6
65	30 March 1978	3	Kosmos 997-998
66*	27 May 1978	4	Ekran
67	18 July 1978	4	Raduga 4
68*	17 August 1978	4	Ekran
69	9 September 1978	4	Venera 11
70	14 September 1978	4	Venera 12
71*	17 October 1978	4	Ekran
72*	19 December 1978	4	Gorizont 1
73*	21 February 1979	4	Ekran 3
74	25 April 1979	4	Raduga 5
75	22 May 1979	3	Kosmos 1100-01
76	5 July 1979	4	Gorizont 2
77	3 October 1979	4	Ekran 4
78	28 December 1979	4	Gorizont 3
79	2 February 1980	4	Raduga 6
80	14 June 1980	4	Gorizont 4
81	15 July 1980	4	Ekran 5
82	5 October 1980	4	Raduga 7
83	26 December 1980	4	Ekran 6
84	18 March 1981	4	Raduga 8
85	25 April 1981	3	Kosmos 1267
86	26 June 1981	4	Ekran 7
87	30 July 1981	4	Raduga 9
88	9 October 1981	4	Raduga 10
89	30 October 1981	4	Venera 13
90	4 November 1981	4	Venera 14
91	5 February 1982	4	Ekran 8
92	15 March 1982	4	Gorizont 5
93	19 April 1982	3	Salyut 7
94	17 May 1982	4	Kosmos 1366
95*	23 July 1982	4	Ekran
96	16 September 1982	4	Ekran 9
97	12 October 1982	4	Kosmos 1413-15
98	20 October 1982	4	Gorizont 6
99	26 November 1982	4	Raduga 11
100*	24 December 1982	4	Raduga
101	2 March 1983	3	Kosmos 1443
102	12 March 1983	4	Ekran 10
103	23 March 1983	4	Astron 1
104	8 April 1983	4	Raduga 12
105	2 June 1983	4	Venera 15
106	6 June 1983	4	Venera 16
107	1 July 1983	4	Gorizont 7
108	10 August 1983	4	Kosmos 1490-92
109	25 August 1983	4	Raduga 13
110	29 September 1983	4	Ekran 11
111	30 November 1983	4	Gorizont 8
112	29 December 1983	4	Kosmos 1519-21
113	15 February 1984	4	Raduga 14
114	2 March 1984	4	Kosmos 1540
115	16 March 1984	4	Ekran 12
116	29 March 1984	4	Kosmos 1546
117	22 April 1984	4	Gorizont 9
118	19 May 1984	4	Kosmos 1554-56
119	22 June 1984	4	Raduga 15
120	1 August 1984	4	Gorizont 10
121	24 August 1984	4	Ekran 13
122	4 September 1984	4	Kosmos 1593-95
123	28 September 1984	4	C1603/Tselina 2
124	15 December 1984	4	Vega 1
125	21 December 1984	4	Vega 2
126	18 January 1985	4	Gorizont 11
127	21 February 1985	4	Kosmos 1629
128	22 March 1985	4	Ekran 14
129	17 May 1985	4	Kosmos 1650-52
130	30 May 1985	4	C1656/Tselina 2
131	8 August 1985	4	Raduga 16
132	27 September 1985	3	Kosmos 1686
133	25 October 1985	4	Kosmos 1700
134	15 November 1985	4	Raduga 17
135	24 December 1985	4	Kosmos 1710-12
136	17 January 1986	4	Raduga 18
137	19 February 1986	3	Mir
138	4 April 1986	4	Kosmos 1738
139	24 May 1986	4	Ekran 15
140	10 June 1986	4	Gorizont 12
141	16 September 1986	4	Kosmos 1778-80
142	25 October 1986	4	Raduga 19
143	18 November 1986	4	Gorizont 13
144*	29 December 1986	3	Almaz
145*	30 January 1987	4	Kosmos 1817
146	19 March 1987	4	Raduga 20
147	31 March 1987	3	Kvant 1
148*	24 April 1987	4	Kosmos 1838-40
149	11 May 1987	4	Gorizont 14
150	25 July 1987	3	Kosmos 1870
151	3 September 1987	4	Ekran 16
152	16 September 1987	4	Kosmos 1883-85
153	1 October 1987	4	Kosmos 1888
154	28 October 1987	4	Kosmos 1894
155	26 November 1987	4	Kosmos 1897
156	10 December 1987	4	Raduga 21
157	27 December 1987	4	Ekran 17
158*	18 January 1988	4	Gorizont
159*	17 February 1988	4	Kosmos 1917-19
160	31 March 1988	4	Gorizont 15
161	26 April 1988	4	Kosmos 1914
162	6 May 1988	4	Ekran 18
163	21 May 1988	4	Kosmos 1946-48
164	7 July 1988	4	Phobos 1
165	12 July 1988	4	Phobos 2
166	1 August 1988	4	Kosmos 1961
167	18 August 1988	4	Gorizont 16
168	16 September 1988	4	Kosmos 1970-72
169	20 October 1988	4	Raduga 22
170	10 December 1988	4	Ekran 19
171	10 January 1989	4	Kosmos 1987-89
172	26 January 1989	4	Gorizont 17
173	14 April 1989	4	Raduga 23
174	31 May 1989	4	Kosmos 2022-24
175	21 June 1989	4	Raduga-1 1 (24)
176	5 July 1989	4	Gorizont 18
177	28 September 1989	4	Gorizont 19
178	26 November 1989	3	Kvant 2
179	1 December 1989	4	Granat
180	15 December 1989	4	Raduga 25
181	27 December 1989	4	Kosmos 2054
182	15 February 1990	4	Raduga 26
183	19 May 1990	4	Kosmos 2079-81
184	31 May 1990	3	Kristall
185	20 June 1990	4	Gorizont 20
186	18 July 1990	4	Kosmos 2085
187*	9 August 1990	4	Ekran
188	3 November 1990	4	Gorizont 21
189	23 November 1990	4	Gorizont 22
190	8 December 1990	4	Kosmos 2109-11
191	20 December 1990	4	Raduga 27
192	27 December 1990	4	Raduga-1 2 (28)
193	14 February 1991	4	Kosmos 2133
194	28 February 1991	4	Raduga 29

Proton launch history (all launches made from Baikonur) — *Continued*

Launch	Date	Stages	Payload
195	31 March 1991	3	Almaz 1
196	4 April 1991	4	Kosmos 2139-41
197	1 July 1991	4	Gorizont 23
198	13 September 1991	4	Kosmos 2155
199	23 October 1991	4	Gorizont 24
200	22 November 1991	4	Kosmos 2172
201	19 December 1991	4	Raduga 30
202	29 January 1992	4	Kosmos 2177–79
203	2 April 1992	4	Gorizont 25
204	14 July 1992	4	Gorizont 26
205	30 July 1992	4	Kosmos 2204–06
206	10 September 1992	4	Kosmos 2209
207	30 October 1992	4	Ekran 20
208	27 November 1992	4	Gorizont 27
209	17 December 1992	4	Kosmos 2224
210	17 February 1993	4	Kosmos 2234–36
211	25 March 1993	4	Raduga 31
212*	27 May 1993	4	Gorizont
213	30 September 1993	4	Raduga 32
214	28 October 1993	4	Gorizont 28
215	18 November 1993	4	Gorizont 29
216	20 January 1994	4	Gals 1
217	5 February 1994	4	Raduga-1 3 (33)
218	18 February 1994	4	Raduga 34
219	11 April 1994	4	Kosmos 2275–77
220	20 May 1994	4	Gorizont 30
221	6 July 1994	4	Kosmos 2282
222	11 August 1994	4	Kosmos 2287–89
223	21 September 1994	4	Kosmos 2291
224	13 October 1994	4	Ekspress 1
225	31 October 1994	4	Elektro 1
226	20 November 1994	4	Kosmos 2294–96
227	16 December 1994	4	Luch 1
228	28 December 1994	4	Raduga 35
229	7 March 1995	4	Kosmos 2307–09
230	20 May 1995	3	Spektr
231	24 July 1995	4	Kosmos 2316–18
232	30 August 1995	4	Kosmos 2319
233	11 October 1995	4	Luch-1 1
234	17 November 1995	4	Gals 2
235	14 December 1995	4	Kosmos 2323–25
236	25 January 1996	4	Gorizont 31
237*	19 February 1996	4	Raduga 36
238	8 April 1996	4	Astra 1F
239	23 April 1996	3	Priroda
240	25 May 1996	4	Gorizont 32
241	6 September 1996	4	Inmarsat-3 2
242	26 September 1996	4	Ekspress 2
243	16 November 1996	4	Mars 8 (Mars-96)
244	24 May 1997	4	Telstar 5
245	6 June 1997	4	Kosmos 2344
246	18 June 1997	4	Iridium (×7)
247	14 August 1997	4	Kosmos 2345
248	28 August 1997	4	PAS-5
249	14 September 1997	4	Iridium (×7)
250	12 November 1997	4	Kupon 1
251	2 December 1997	4	Astra 1G
252	24 December 1997	4	AsiaSat 3
253	7 April 1998	4	Iridium (×7)
254	29 April 1998	4	Kosmos 2350
255	7 May 1998	4	EchoStar 4
256	30 August 1998	4	Astra 2A
257	4 November 1998	4	PAS-8
258	20 November 1998	3	ISS (Zarya)
259	30 December 1998	4	Kosmos 2362, 2363, 2364
260	15 February 1999	4	Telstar 6
261	28 February 1999	4	Raduga 1-4
262	21 March 1999	4	AsiaSat 3S
263	20 May 1999	4	Nimiq 1
264	18 June 1999	4	Astra 1H
265*	5 July 1999	4	Raduga
266	26 September 1999	4	LM-1
267*	27 October 1999	4	Ekspress-A1
268	12 February 2000	4	Garnda 1
269	6 June 2000	4	Gorizont
270	24 June 2000	4	Ekspress A-3
271	30 June 2000	4	Sirius 1
272	4 July 2000	4	Comsat
273	12 July 2000	3	ISS (Zvezda)
274	28 August 2000	4	Globus
275	5 September 2000	4	Sirius SR-2
276	1 October 2000	4	GEIA
277	13 October 2000	4	Glonass
278	21 October 2000	4	GE6
279	30 November 2000	4	Sirius SR-3
280	7 April 2001	4	Ekran-M
281	15 May 2001	4	PAS 10
282	16 June 2001	4	Astra 2C
283	24 August 2001	4	Kosmos 2379
284	6 October 2001	4	Raduga 1
285	1 December 2001	4	Glonass (×3)
286	30 March 2002	4	Intelsat 903
287	7 May 2002	4	DirecTV 5
288	10 June 2002	4	Express 4A
289	25 July 2002	4	Kosmos 2392
290	22 August 2002	4	Echostar 8
291	17 October 2002	4	Integral
292*	25 November 2002	4	Astra 1K
293	25 December 2002	4	Glonass (×3)
294	29 December 2002	4	Nimiq 2
295	24 April 2003	4	Kosmos 2397
296	6 June 2003	4	Americom 9
297	24 November 2003	4	Yamal-2000
298	10 December 2003	3	Glonass
299	29 December 2003	4	Express
300	16 March 2004	4	W3A
301	27 March 2004	4	Globus
302	27 April 2004	4	Express
303	17 June 2004	4	Intelsat 10-02
304	5 August 2004	4	Amazonas
305	14 October 2004	4	AMC-15
306	30 October 2004	4	Express
307	26 December 2004	4	Glonass
308	3 February 2005	4	AMC-12
309	30 March 2005	4	Express-AM2
310	23 May 2005	4	DirecTV 8
311	24 June 2005	4	Express AM3
312	9 September 2005	4	Anik F1R
313	25 December 2005	4	Glonass
315	29 December 2005	4	AMC-23
316	28 February 2006	4	Arabsat 4A (failed)

*indicates launcher failure. February 1996 stage 4 failed to reignite for GSO injection. The review board concluded that the triethyl aluminium ignition hypergol had not reached the gas generator or main chamber because of reduced pressure from a leaking joint in the pipe from the TEA bottle. The fitting nut had a broken lockwire, probably from poor installation, allowing it to back off during burn 1. Corrective actions include adding a second lockwire. May 1993 failure due to stage 2/3 engine burn-throughs caused by contaminated UDMH. 1990 stage 2 engine failure believed to be due to a worker leaving a rag in the vehicle. Stage 3 failure January 1988. Proton breakdown launch total for each type, followed by number of failures in ( ): two-stages 4(1), four-stages 235(30), three-stages 27(3): these give success rates of 75%, 87% and 89% respectively. Four stage failures include Block D missions where the failure was with the payload (for example Mars 8). Two-stages 4(1)/75%, four-stages 235(30)/87.3%, three-stages 27(3)/88.9%.

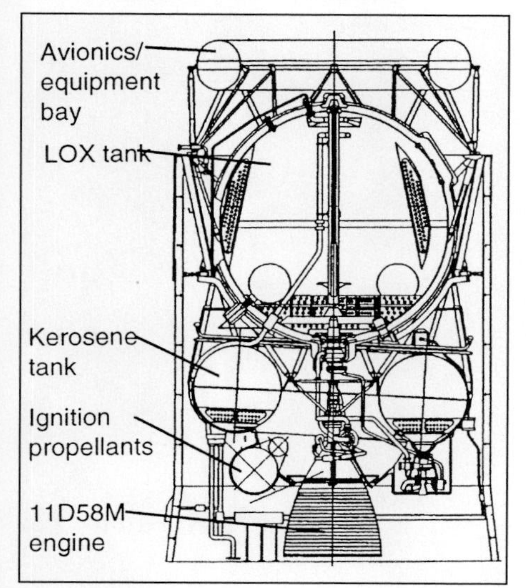

Block DM carries the vehicle's avionics in its forward torus (Boeing) 0517366

All Russian geostationary and interplanetary missions are launched on Proton.

Proton's first three stages are assembled by Khrunichev's factory close to Moscow. The factory can handle 16 per year and requires orders for six to remain a viable commercial entity. Only the propellant tanks are welded at Khrunichev; the vehicle itself is assembled at Baikonur over one and a half months as components arrive at the main 300 × 400 m, 20 m high assembly hall. The Block D/DM stage 4 is managed and built by RSC Energia for integration with the vehicle at Baikonur, where all Protons are launched.

Proton is fuelled by nitrogen tetroxide (N204) and unsymmetrical dimethylhydrazine (UDMH). The rocket's distinguishing feature is the "polyblock" design of the first stage. A central N204 tank is surrounded by six permanently attached modules, each comprising a UDMH tank and an Energomash RD-253 motor. The unique configuration was adopted to allow rail transport to Baikonur, where the outer tanks could be attached.

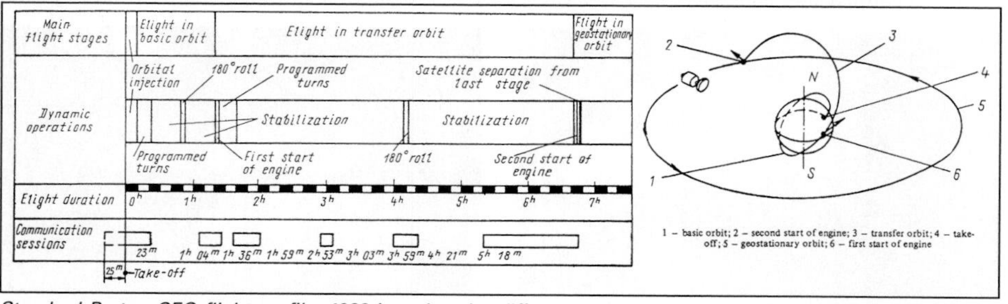

Standard Proton GEO flight profile. 1988 introduced a different technique in which the vehicle remained in its parking orbit for a further 6 hours 0516877

All six engines are gimballed to provide flight control.

The second stage has four independent, gimballed engines developed by the Khimavtomatika bureau. Three of them are of the RD-0210 and one is an RD-0211 – the difference is that the latter incorporates a system to pressurise the stage 2 fuel tanks. The third stage has an RD-0213 engine, a derivative of the second-stage engine, and a Khimavtomatika RD-0214 engine with four widely spaced gimballed chambers, which provides flight control.

Proton K launchers are all essentially identical, the actual flight configuration differing according to the upper stage. Launches using the Block D family employ the Block D to perform initial orbital injection, the first three stages of the vehicle being sub-orbital. The Block D then performs a second burn to launch the spacecraft out of a low parking orbit. The Block D family does not carry its own command and control system and it relies upon the payload to relay commands for engine firing, separation, and so on. These upper stages have been used exclusively for deep space missions. Different versions have appeared over the years: the original Block D (11S824), lunar and planetary missions during 1967-1983, the Block D-1 (11S824M) for Astron, Venera 11-16, VEGA 1-2 and Granat, and the Block D-2 (11S824F) used for Phobos 1-2 and Mars 8 (=Mars-96).

The Block DM family of upper stages have been used for launching GLONASS payloads. The Soviets used the original Block DM (11S86) for launches to geosynchronous orbit during 1974-1990 until its last flight for Gorizont 20. The Block DM-2 (11S861) put all GLONASS satellites into position, delivered Kosmos 1700 and the majority of GEO missions starting with Kosmos 1961 (in 1988), and put the Elint launches of Kosmos 1603/1656 into position. The Block DM-2M (11S861-01) has seen limited use in the launches of geosynchronous communications satellites. The commercial launch of Astra-1F introduced a new version to the Block DM family: the Block DM-3.

There have been two propellant versions used within the Block DM and Block D flight programme. Originally, the Soviets used a standard LOX/kerosene combination, but starting in the 1980s – possibly with the introduction of the Block DM-2 stage – the higher-performance LOX/syntin combination was introduced. Syntin is a synthetic hydrocarbon with a higher energy density than kerosene. In September 1996, the Russians halted production of syntin, but supplies are on hand to fuel any remaining Block D and Block DM upper stages.

The three-stage Proton-K places the fourth stage and payload assembly into a low parking orbit and then the fourth stage ignites for the first time in LEO to enter a transfer orbit and then a second time to circularise the orbit with an orbital inclination of around 47 to 48°. At the first pass through apogee the SOZ motors would fire to settle the remaining propellant in the main tanks and they would separate as the main engine ignited to circularise the orbit at geosynchronous altitude and reduce the inclination to 1.5° or less. The Block DM-3 performed the first burn to enter a geosynchronous transfer orbit but retaining the original 51.6° inclination: at the apogee pass the second burn changed the orbit to 11,970 to 35,940 km, but inclined at 6.95°.

Development of a thoroughly improved Proton launch system, including a new upper stage, started in the early 1990s. The resulting Proton M made its first flight on 7 April 2001.

According to Khrunichev, the most important new element of the Proton M is the replacement of the original GNC system, which dated back to the 1960s, used analog technology, and was not produced in Russia. Provided by NIIAP, the new digital system uses similar components to the Zenit GNC system. Its benefits include more efficient use of the fuel, improving performance and reducing the amount of fuel remaining when the first stage returns to Earth; better manoeuvrability for a wider range of orbit inclinations; improved control with narrower yaw limits, making it possible to increase the size of the payload fairing without loss of stability, an important issue in terms of keeping the launcher competitive; and easier programming and reprogramming.

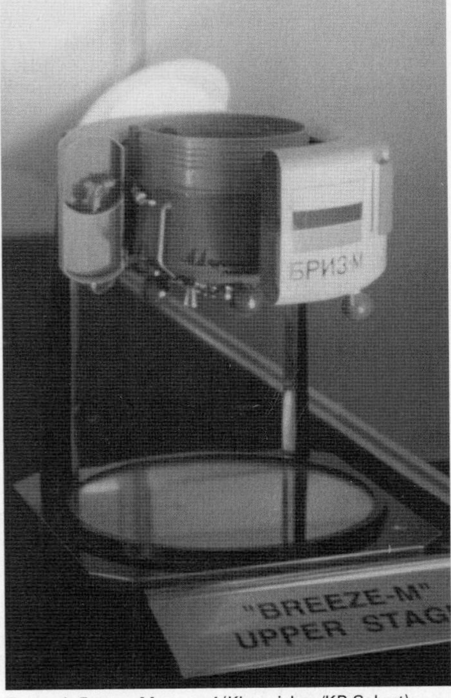

Proton's Breeze-M stage 4 (Khrunichev/KB Salyut) 0517381

Proton M addresses what has become a serious issue for the programme: the launch site and the impact zones for the first stage, with its toxic residual fuel, are in Kazakhstan and are leased by Russia, which is responsible for environmental clean-up after each launch. As noted above, the Proton M is designed to use fuel more efficiently. Parts of the fuel system have also been redesigned to reduce fuel remaining after shut-down by up to 50 per cent. The new launcher is also designed with better control over the first stage trajectory, so that the impact zone will be smaller.

The Proton M first stage is powered by six NPO Energomash RD-275 engines, uprated versions of the RD-253 fitted to the Proton K. Thrust is increased by 7 per cent by a modification to the propellant flow control valves. This modification was first carried out for the Proton launcher that launched the Mir space station core in 1986, to permit operations at 102 per cent of rated thrust, and the 107 per cent thrust rating has been flying since 1995. The second- and third-stage engines incorporate changes to improve reliability and combustion resistance. Proton M incorporates new aluminium alloys and composite materials to reduce weight, manufacturing improvements such as fully automated welding of propellant tanks, to reduce cost and improve quality.

Despite these major changes, Khrunichev says that 77 per cent of the parts, subassemblies and systems in Proton M are common to the Proton K, 18 per cent are upgraded and only 5 per cent are new.

The Proton M is mated with a new upper stage, Khrunichev's Breeze-M. It is derived from the UDMH/N204-fuelled Breeze and Breeze KM stages developed for Rockot, the principal change being the addition of a jettisonable, doughnut-shaped Auxiliary Propellant Tank (APT) which wraps around the core vehicle and almost quadruples the propellant capacity. This configuration makes the Breeze-M short and compact for its size,

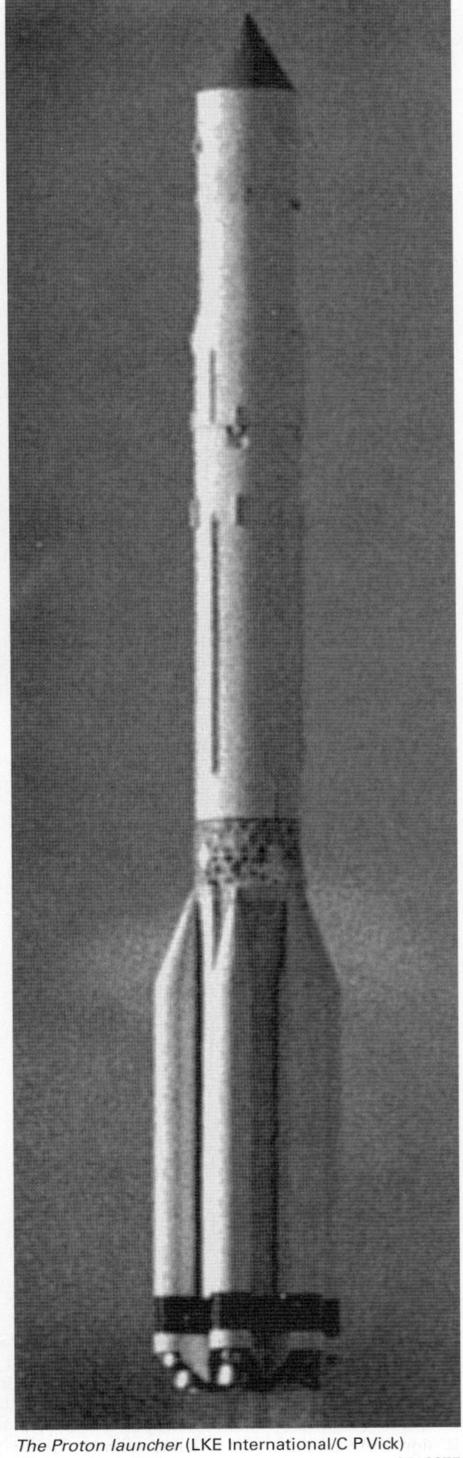

The Proton launcher (LKE International/C P Vick) 0516875

Proton with Block DM3 launches the Astra 1F satellite on 9 April 1996 (Theo Pirard) 0054381

Forward section of a completed three-stage Proton at the Khrunichev factory (Theo Pirard) 0054370

Elements of the Proton-launched vehicle in the Khrunisher production facility 0084618

reducing the vehicle's overall length and helping accommodate larger payloads. It incorporates separate thrust, vernier and attitude control motors, plus a gimballed motor on the thrust nozzle, for accurate attitude control. The system has an endurance of more than 24 h, and the propulsion system is tested and guaranteed for up to eight ignition events in a single flight. Its guidance system can use GPS or GLONASS signals for added accuracy.

Breeze-M can also be mated with the Proton K launcher and was deployed ahead of the Proton M. This combination was first launched on 5 July 1999, but the flight was unsuccessful due to a second-stage Proton failure. The first successful launch of Proton K/Breeze M was on 6 June 2000.

Ground tests of the Proton M and the Breeze-M were completed in mid-1999. Sub-scale and full-scale propellant tank structural and slosh dynamics tests were successfully conducted, along with full-scale propellant feed line flow tests, propellant conditioning, outflow, and pressurisation subsystem tests. The first integrated Proton M Breeze M was completed and shipped to the launch site in June 2000. Integration operations involving the newly renovated Hall 111 in Building 92A-50, the new Breeze M fuelling facility and the upgraded launch pad 24 took place from arrival until the successful first launch on 7 April 2001.

The first launch of Proton M/Breeze M for ILS took place on 30 December 2002. The next ILS launch, on 7 June 2003, was the final commercial flight of the Proton K.

Proton launcher components are delivered to Baikonur by rail. Spacecraft are typically delivered by air, arriving at Yubileiny Airport on the Baikonur site. The airport is connected to the rest of a site by a rail spur. Pre-launch operations are concentrated in Building 92A-50, which has two parallel main bays. One of these is used for receiving, storing and checking out spacecraft, and the other for the assembly of the complete launch vehicle. The spacecraft itself is fuelled before being moved into the launch vehicle assembly bay. Once the entire rocket is assembled, it is transferred to a rail-mobile transporter-erector unit.

After assembly, the rocket first moves to a covered fuelling facility located next to Building 92A-50, where the Breeze-M is fuelled and the spacecraft batteries are charged. Finally, the rocket moves to one of two launch pads for erection, check-out, fuelling of the launch vehicle itself, and launch.

Khrunichev and ILS have proposed further improvements of the Proton system. A package of modifications, including a further 5 per cent first-stage engine uprate, some material changes and a lighter payload fairing, are expected to raise GTO payload to 6,000 kg, and further changes allowing an increase to 6,200–6,300 kg are planned.

Khrunichev is developing a 5.1 m diameter, 16.37 m long payload fairing, giving the Proton an equivalent payload envelope to other current heavy launcher systems. The fairing can be used with the Breeze M upper stage, with mission design changes that retain the full 5,500 kg payload to GTO. Also under study is the use of the KVRB LOX/hydrogen upper stage, boosting GTO payload by as much as 20 per cent. It is based on experience with the 12KRB upper stage, developed for the Indian GLSV launch vehicle, but is larger. According to ILS, design work on the KVRB has been funded under a separate contract. Avionics and other equipment are based on those of the Breeze M. Dual-satellite launch systems have also been discussed.

Proton had seven launches in 2005, all successful – the highest flight rate of any launch system in that year. Four of the launches were for ILS and three for the Russian government. The type's record was, however, marred by a launch failure involving Arabsat 4A on 28 February 2006. ILS announced three new Proton contracts in the first quarter of 2006. In the long term, the Russian government expects to replace Proton with the Angara, eliminating toxic propellants and providing for large-payload launches from Plesetsk in Russian territory, but this will not happen in the present decade.

Restartable Breeze-M upper stage for Proton launcher
0084622

Comparison of Proton-K and Proton-M four-stage vehicles

	Proton-K	Proton-M
Payload fairing:		
Usable diameter	3.9 m	3.9 m
Usable length	7.3 m	10.0 m
Fairing mass	2,000 kg	2,300 kg
Payload capabilities:		
LEO (three-stage version)	20,100 kg	22,000 kg
GSO (four-stage version)	2,100 kg	2,920 kg
Length	60 m	58 m
Launch mass	688,000 kg	690,000 kg
Stage 1:		
Length	21.18 m	21.18 m
Diameter*	7.4 m	7.4 m
Dry mass	31,000 kg	30,600 kg
Propellant mass	419,410 kg	419,410 kg
Total thrust (vacuum)	10,000 kN	10,497 kN
Specific impulse	316 s	316 s
Stage 2:		
Length	17.05 m	17.05 m
Diameter	4.10 m	4.10 m
Dry mass	11,700 kg	11,400 kg
Propellant mass	156,113 kg	156,113 kg
Total thrust (vacuum)	2,325 kN	2,325 kN
Specific impulse	326.5 s	326.5 s
Stage 3:		
Length	4.11 m	4.11 m
Diameter	4.10 m	4.10 m
Dry mass	4,185 kg	3,700 kg
Propellant mass	46,562 kg	46,562 kg
Total thrust (vacuum)	613 kN	613 kN
Specific impulse	326.5 s	326.5 s
Stage 4:	Block DM-3	Briz-M
Length	6.28 m	2.61 m
Diameter	3.7 m	4.0 m
Dry mass**	3,130 kg	2,370 kg
Propellant mass	15,050 kg	19,800 kg
Total thrust (vacuum)	88 kN	19.6 kN
Specific impulse	361.0 s	325.5 s

*Diameter measured across a pair of external propellant tanks and the central core propellant tank.

**For the Block DM-3 this figure includes the fourth stage shroud discarded in LEO and the two ullage motors. The vehicles use nitrogen tetroxide throughout other than the Block DM-3 which uses LOX/SYNTIN. Some of the data quoted by ILS differs from those available from other Russian literature: for example it is known that some domestic Proton-K LEO payloads have been in excess of 20,600 kg mass and some GSO payloads are around 2,500 kg.

Proton launch vehicle on its pad at the Baikonur cosmodrome (ILS)
1047536

Specifications

Proton launcher

First launch: 16 July 1965 (two-stage version)
Number launched: 316 to end February 2006
Launch sites: Baikonur

Proton K (four stages)

First launch: 10 March 1967
Number launched: 302 to January 2005
Launch sites: Baikonur
Principal uses: geostationary, escape and Glonass navigation satellite missions
Vehicle success rate: 95 per cent
Availability: Retiring
Performance: GTO (48°): 4,800 kg
LEO: not used for LEO missions, but vehicle 20 t in 200 km parking orbit before GTO injection
GEO: 2,600 kg using Block DM-2M. 2,400 kg using Block DM and Syntin. 3,200 kg using Breeze-M.
Sun-synchronous: 2,800 kg (not yet flown)
Lunar delivery: 5,700 kg
Venus delivery: 5,300 kg
Mars delivery: 4,600 kg
Number of stages: 4
Overall length: SL-12 57.07 m, SL-13 57.76 m (42.340 m without stage 4/payload)

Elements of Proton launcher with strap-on boosters attached to core stage in dulame
0084620

Strap-on propulsion elements of Proton core stage
0084621

Breeze-M restartable upper stage adapter for Proton KM
0084623

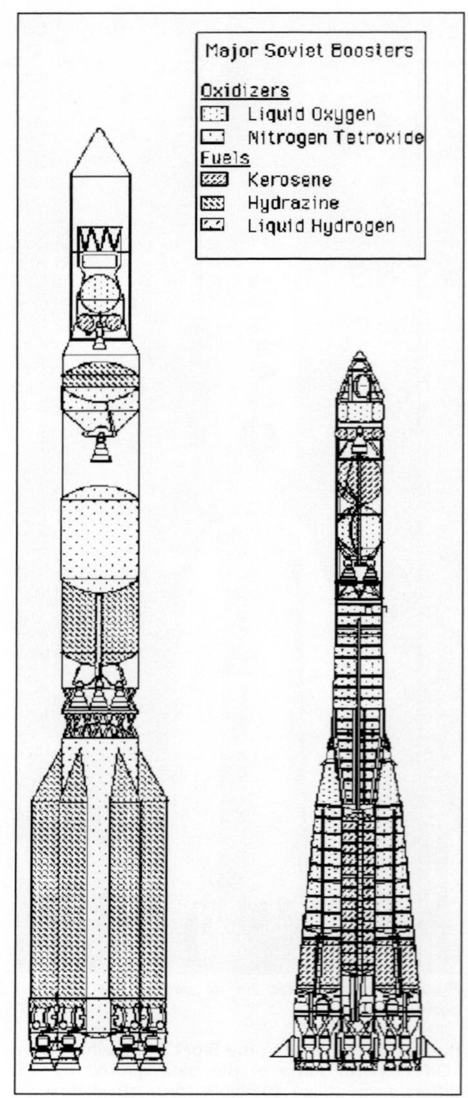

Comparative internal layouts for Soyuz and Proton launchers 0131960

Principal diameter: 7.400 m stage 1, 4.100 m upper stages
Launch mass: 690 t
Guidance: inertial, carried in stage 4, provides three-axis control by means unknown until payload separation commanded. GEO injection accuracy ±1°, ±20 min orbital period. Analogue control system carried by Block DM or by payload for Block D. GN&C systems provided by NPO AP.

Proton K stage 1
Bureau: Khrunichev
Engines: single storable liquid non-restartable NPO Energomash RD-253 carried in six cylindrical UDMH side modules around central oxidiser tank.
Overall length: 21.1 m

Proton prior to Granat launch 0131959

Principal diameter: 7.4 m overall, 1.6 m cylinders around 4.1 m core
Oxidiser: nitrogen tetroxide in core tank
Fuel: UDMH in side tanks
Propellant mass: 420 t; 31.0 t dry
Stage thrust: 8.8 MN at sea level, 9.8 MN at altitude
Burn time: 130 s

Proton K stage 2
Bureau: Khrunichev
Engines: gimballed liquid chambers, developed by KB Khimautomatiki. Three are RD-0210 and one is RD-0211 version (with gas generator).
Overall length: 14.56 m
Principal diameter: 4.15 m
Oxidiser: nitrogen tetroxide
Fuel: UDMH
Stage thrust: 2,376 kN vacuum
Burn time: up to 300 s
Propellant mass: 156 t, 11.7 t dry

Proton K stage 3
Comment: Stage 3 injects the payload and stage 4 into a 200 km circular, 51.6° parking orbit about 10 min after launch.
Bureau: Khrunichev
Engine: single fixed RD-0212 version of RD-0210 with four gimballed verniers for steering
Overall length: 6.52 m
Principal diameter: 4.15 m
Oxidiser: nitrogen tetroxide
Fuel: UDMH
Stage thrust: 593.6 kN main chamber plus 4 × 7.875 kN verniers
Burn time: about 250 s
Propellant mass: 46.6 t; 4,185 kg dry

Proton K stage 4 Block DM
Designation: Block D (11S824, first flown 1967), Block 'D-1' [?] (11S824M, 1978), Block D-2 (11S824F, 1988), Block DM (11S86, 1974), Block DM-2 (11S861, 1982), Block DM-2M (11S861-01, 1994), Block DM-3 (?, 1996): possibly a Block DM-4 in 1997.
Bureau: RSC Energia; control system by NPO AP
Engine: 58/58M restartable single chamber developed by Korolev bureau for D/DM.
Overall length: 5.366/6.218 m D/DM-2
Principal diameter: 3.70 m
Stage mass: 17.3 t D/18.46 t DM-2; dry mass 2.5 t D/3.37 t DM-2 (dry masses include the 800 kg casing + SOZ).
Oxidiser: 10,610 kg liquid oxygen in forward sphere
Fuel: 4,330 kg kerosene or syntin in aft torus
Stage thrust (vacuum): 85 kN kerosene; 83.5 kN Syntin.
Burn time: about 600 s total over two burns
Comment: Stage 4/Block D/DM-2 ignites about 80 min after stage 3 separation, making a maximum 450 s burn for injection into a 36,000 km, 48° GTO. About 400 min after launch, it ignites for a maximum of 230 s at first apogee to slot the payload directly into GEO – no satellite apogee boost motor is required. (DM injected Astra 1F into 12,100 × 36,000 km, 7.0° GTO). D/DM-2 carries two 110–120 kg (56 kg dry) 60 × 100 cm NTO/UDMH 'SOZ' thruster packages for three-axis control during coast and to settle the stage's propellants after coasting. Each houses five thrusters: 2 × 22 N pitch/roll, 1 × 44 N yaw and 2 × 11 N ullage.
Fairing: The spacecraft is protected until 351 s after launch at an altitude of 150 km. Class 100,000 air conditioning on the pad through a 45 cm^2 hatch provides 10–25°C (30°C possible for up to three days), 30–60% humidity inside the fairing until it is disconnected 2 h before launch. There is also a 4–5 h interruption while the vehicle is erected on the pad. Power and communications links are terminated 90 min before launch; telemetry contact is not available until stage 4 of the flight profile. Surface wind limit 15–18 m/s, depending on fairing type.
Acceleration load: about 4 g longitudinal ±1 g lateral at 5.7 Hz during ascent, 1.2 ±0.8 g longitudinal at 10–15 Hz/±3.0 g lateral at 5–7 Hz at lift-off; 1.25 g ±4.75 g longitudinal at 10–15 Hz/±1.5 g lateral at 5–7 Hz during separation.
Acoustic load: 144 dB rms
Temperature: 130°C sidewall max after 200 s

Proton M Breeze-M stage 4
Designation: Breeze-M
Bureau: Khrunichev/KB Salyut
Engine: single fixed restartable pump-fed engine from KB Khimmach, plus three-axis control provided by verniers

Proton KM cryogenic stage 4
Designation: KVRB Kislorodno-Vodourodny Razgonny Block (oxygen/hydrogen upper stage).
Bureau: Khrunichev (stage), KB Khimmach (engine).

Core element of the Proton launcher 0084619

Engine: single chamber cryogenic KVD-1M (with two verniers) capable of five burns and 7.5 h coast.
Overall length: 8.6 m
Principal diameter: 4.0 m
Oxidiser: liquid oxygen
Fuel: liquid hydrogen
Dry mass: 3.4 t
Propellant mass: up to 19 t
Stage thrust: 73.58 kN vacuum (69.66 kN main chamber plus 2 × 1.96 kN LOX/GH_2 two-axis verniers)
Burn time: 450 s total. First burn starts at 650 s.

Shtil

Current Status
The Shtil has so far completed two successful orbital launches. The German Tubsat satellites were carried into orbit on 7 July 1998. Kompass-2, a small scientific satellite, was delivered to LEO in May 2006.

Background
The Makeyev Design Bureau of Miass, Chelyabinsk, Russia, offers satellite launch services using its converted SLBMs. The bureau was founded by V P Makeyev (1925–1985) in 1955 as an offshoot of Korolev's bureau and developed the RSM-25 Zyb (SS-N-6 Serb), RSM-40 Vysota (SS-N-8 Sawfly), RSM-50 Volna (SS-N-18 Stingray), RSM-52 Rif (SS-N-20 Sturgeon) and RSM-54 Shtil (SS-N-23 Skiff) SLBMs. There were apparently three SLBM demonstrations 1991–93. A Shtil orbital demonstration launch was reportedly planned for 1996 carrying a 100 kg Izmiran satellite, but the first known launch did not take place until 7 July 1998.

Shtil (Calm) is a modified R-29RM (RSM-54/SS-N-23) three-stage Submarine-Launched Ballistic Missile (SLBM) used to orbit small satellites. It can lift up to 160 kg into a 200 km Earth orbit when launched from a submerged Delta IV class submarine in the Barents Sea. The satellite is arranged in a ring payload envelope after the warhead removal. To protect the payload during launch, it can be encapsulated, if required.

All three stages burn UDMH fuel with N_2O_4 oxidiser. The rocket is gas-ejected from a launch tube and the RD-0243 main engine ignites when the missile surfaces, producing about 82.2 tonnes of thrust at sea level. The engine comprises a main thrust chamber augmented by four smaller steering engines. The first stage burns for about 74 seconds. The second stage fires for about 94 seconds and the third stage for about 87 seconds. The third stage engine separates from the stage about one minute after shutdown. Small thrusters mounted on the exterior of the vehicle's nose perform a final orbital insertion or apogee burn. The spacecraft separation command is performed by the control system 5, 10 s after the last stage engine has been switched into low thrust mode. Capsule spacecraft separation is effected by spring pushers. Payloads are ejected from the rear of the stage.

NII Mashinostroyeniya developed the three-stage storable liquid-fueled R-29RM SLBM during the late 1970s and 1980s. Seven Delta IV submarines were deployed with 16 R-29RM missiles apiece. Each

The Shtil-1 is a converted, submarine-launched missile. The first commercial launch took place in 1998 with the launch to LEO of two German Tubsats (Makeyev) 1340459

missile carried four warheads, but were capable of handling up to ten warheads. In 1999, the Makeyev Design Bureau was ordered to resume production of the R-29RM missile. Unlike the missiles that are the basis of the 'Rockot' and 'Dnepr' launchers, the R-29RM is still being manufactured, so Shtil variants will remain available for many years to come.

During the mid-1990s, Makeyev began offering decommissioned R-29RM missiles, renamed "Shtil'", for commercial launch purposes. Payloads are simply located within empty warhead aeroshells attached to the missile's third stage. The first Shtil orbital launch, a success, occurred on 7 July 1998 when the German microsatellites TubSat-N and TubSat-N1 were launched from K-407 "Novomoskovsk" in the Barents Sea. This was the first known orbital launch from a submarine. The second Shtil orbital launch occurred on 26 May 2006, when the 80 kg Kompass-2 science satellite was orbited from submerged submarine "Ekaterinburg" in the Barents Sea east of Murmansk. Kompass-2 entered an approximate 400 × 500 km, 79° orbit.

The Shtil is regarded as a better option than the RSM-50 Volna for upgrading as a space launcher, since the missile has a larger lift capability. Development of the Shtil-2 version began in the late 1990s. It features a new payload module and nose fairing. This allows significantly heavier payloads to be the launched, even without any enhancement of the launcher's propulsion system. In comparison with the basic rocket, the Shtil-2 is slightly longer and heavier, but no changes are envisaged in the SLBM's main stages.

The first modification to be introduced is known as the Shtil-2.1. This has a shorter fairing with a simpler design than the Shtil-2 having a single-shell configuration instead of the Shtil-2's split-shell fairing. The satellite is encapsulated in a special Assembling Protection Unit (APU) to guard against the launcher environment. The shortened fairing allows the rocket to be launched from a submarine's launching tube fitted with a modified lid. The Shtil-2 with a longer nose fairing can not be accommodated in a submarine launching tube and is intended to be launched from a ground-based facility. The Shtil-2.1 and Shtil-2 have very similar payload capabilities and their launch profiles are also the same. The first and second stage drop (impact) zones are located in the Barents and Greenland Seas, while the drop zone for the third stage's jettisoned main engine is located in the Pacific Ocean. The third stage itself remains in orbit.

Since the flight trajectory mostly passes over remote ocean areas, there is a problem in receiving telemetry information. Makeyev literature suggests the following solution: During the initial stages of the flight, when radio visibility from the ground stations is not available, information from the computer or gyro integrator is stored on board. During the active stage of flight, the digital and analogue information is transmitted in real time to stations in Severodvinsk and Severomorsk, together with the stored data. During apogee leg and payload separation, data is stored on board and transmitted during the first orbit when the stage flies over Severodvinsk and Severomorsk.

Since the Shtil is based on stored stocks of the RSM-54 missiles, assembly of the launch vehicle is carried out at a specially allotted technical site in the North Navy Base which services the carrier submarines. Spacecraft payloads may be delivered to Makeyev, or directly to the Navy Base site. In the former case, the spacecraft is installed in an APU in the facility and then the APU is delivered to the technical site. In the latter case, an empty APU is delivered from the manufacturer and the assembly operation is carried out at the technical site.

The assembled and tested Space Head Module (SHM) is tilted into a horizontal position and loaded into a special isothermal transport vehicle which delivers it to the pier where the submarine is moored. The fuelled RSM-54 SLBM is already loaded in the submarine. The SHM is lifted into the launching tube by a crane and mated with the launcher's third stage. After completion of the integration and checks of the electric circuits which connect the spacecraft and launch vehicle, the tube's lid is closed and the submarine puts to sea. Final check-outs of the launcher and loading of the mission software are carried out on board the submarine during its journey to the launch location.

The first commercial launch of a Shtil-2.1 was planned for May-June 2007, carrying the 80 kg South African Sumbandila (ZA-002) mini-satellite into a 500 km sun-synchronous orbit, but this has yet to take place. The launch, brokered and managed by Commercial Space Technologies Ltd, is a joint venture involving the Makeyev company and the Russian Navy on behalf of the South African government, under the management of Stellenbosch University.

Makeyev is also considering a larger version, known as the Shtil-2R, which would have additional upper stage boosters and a larger lift capability. Two options are being studied. Option 1 would provide active attitude control of the orbital module (the space head module without a jettisoned nose fairing) during the final coast stage of the flight and a thrust impulse for the fixation of the required operational orbit. This option would enhance the accuracy of spacecraft insertion into a required orbit and improve the launcher's payload capability through the final thrust impulse. A liquid-propellant propulsion unit would have various configurations of propellant tanks, depending on specific missions. Payload capabilities of the 'Shtil-2R' equipped with the Option 1-type of booster will significantly enhance the launcher's payload capability (by 37.5 per cent for a 500 km orbit, for example).

Option 2 would have a liquid or solid propellant apogee propulsion system that would be ignited at apogee. The launcher would spin up the orbital module before injecting it into an elliptical orbit. Option 2 would provide roughly the same payload capabilities as Option 1 but would be cheaper to introduce since it involves a simplification of the booster design. However, the accuracy of satellite orbit insertion would be lower.

Another, larger, version, known as Shtil-3, has been proposed, with a new upper stage and a larger payload fairing with an internal volume of 3.6 cu m. The Shtil-3 would be able to place a 670 kg payload into a 200 km orbit, or a 410 kg payload into an 800 km orbit. This version would be land launched or deployed from an aircraft.

Specifications

Launch weight: Shtil-1: 39,300 kg. Shtil-2.1: 39,700 kg. Shtil-2R: 40,000 kg.
Engines: First stage: RD-0243 (1); second stage: ? (1), third stage: ? (1)
Overall length: Shtil-1: 14.8 m. Shtil-2.1: 16 m. *Shtil-2R:* 18.3.
Diameter: 1.9 m
Volume of payload (capsule) accommodation (m^3): Shtil-1: 0.195. Shtil-2.1: 0.25. Shtil-2R: 1.17.
Oxidiser: N_2O_4
Fuel: UDMH
Thrust: First stage: 82.2 tonnes at sea level.
Payload to LEO (orbit inclination i = 79°): Shtil-1: 130 kg (non-encapsulated) to 400 km; 70 kg (encapsulated) to 400 km.
Shtil-2.1 and Shtil-2: 295 kg to 200 km; 159 kg to 500 km; 50 kg to 740 km.
Payload mass: 200 km, 79° orbit – 160 kg.
600 km, 79° orbit – 80 kg.

Example Ascent Profile to 400 km, 79° Orbit from Barents Sea

Time	Event	Altitude (km)	Range (km)	Velocity (m/s)	Pitch Angle (°)
T+0 s	Lift off	0	0	0	90.00
T+75 s	Staging/stage 2 ignition	30	31	1,370	33.34
T+169 s	Staging/stage 3 ignition	169	258	4,370	10.91
T+256.4 s	Stage 3 cutoff	174	727	7,151	6.67
T+319.9 s	Stage 3 engine jettison	224	1,163	7,100	5.99
T+641 s	Apogee motor start	387	3,299	6,538	2.26
T+906 s	Apogee motor end	400	5,098	7,579	0.00
T+911 s	Spacecraft separation	400	5,134	7,581	0.00

Shtil space-related launches

Launch vehicle	Launch details	Launch date	Payload
Shtil-1	K-407 Novomoskovsk (Delta IV class submarine) in the Barents Sea.	7 July 1998	Tubsat N/Tubsat N1
Shtil-1	K-84 Ekaterinburg (Delta IV class submarine) in the Barents Sea.	26 May 2006	Kompass 2

Soyuz

Current Status

The Soyuz family of launchers is the world's oldest and most prolific, more than 1,700 vehicles having been launched since 1957. Vehicles of this type, derived from the R-7 ICBM, orbited the first satellite, Sputnik 1, and the first astronaut, Yuri Gagarin. It remains one of the world's most popular launchers and one of three human-rated launch systems in the world, and is essential to the support of the International Space Station.

Background

The Vostok, Soyuz and Molniya family of launch vehicles all derived from Sergei Korolev's R-7 (8K71) ICBM. Called SS-6 Sapwood in the West: the vehicle was first launched on 15 May 1957. In the 8K71PS version – little modified from the missile – it launched the first two Sputniks, while the further modified 8A91 was used for the Sputnik 3 launch. The Vostok variant launched all the early Soviet manned space missions, and the basic design was further adapted into the Voskhod, Molniya and Soyuz launchers. From the mid-1970s, the family of operational vehicles comprised the Soyuz and the specialised Molniya, used principally for high-elliptical orbital launches.

The current Soyuz enterprise is founded on two key developments in 1996. The Central Specialized Design Bureau (CSDB), which had been responsible for the technical development of the launcher systems since 1974, was merged with the Progress manufacturing plant, creating an organisation known as State Research and Production Space Rocket Centre (TsSKB) Progress. Also, a European/Russian joint venture named Starsem was established to market the Soyuz internationally. Success on the commercial market supported the continued production and improvement of the Soyuz family. This has included the incorporation of the Ikar and Fregat upper stages, permitting the Soyuz to address a full range of markets including GTO. Production has continued to run at 10–15 launchers per year, and roughly half of Soyuz-family flights in the past six years have been in support of Mir or ISS. In 2003, work started on preparing a second launch site for the Soyuz, at the European Space Agency's complex at Kourou in French Guiana.

By 2006, TsSKB Progress was well advanced with development of the Soyuz-2 family, which will replace all current Soyuz and Molniya models. A derivative of this family, the Soyuz-ST, is being developed for launches from Kourou. The 1,705th Soyuz-family launcher orbited GIOVE-E, the prototype of the European Galileo navigation satellite, on 28 December 2005.

Soyuz History

The launch vehicles that evolved from the R-7 missile are all built around the standard 'core plus four strap-ons' concept. Soviet era commentators designated the strap-ons as Block B, V, G and D while the core is Block A. Vostok added a stage designated Block E atop the core.

As the Soyuz family matured, the Soviets considered two variants to carry on the design. The first, the 8K72, incorporated the Kosberg

Soyuz launch of a manned Soyuz-TM spacecraft. Icing reveals liquid oxygen tank locations 0516869

RD-0105 engine into the Block E. The second, the 8K73 would have used Glushko's RD-0109 on that stage. The 8K72 eventually won out and the Soviets launched the initial Luna missions during 1958–1960 with it. Some Russian literature has called this the Vostok-L. Later, an improved Block E returned the RD-0109 engine to the first stage to give the 8K72K vehicle, used for the manned Vostok flights. Further improvements gave the Vostok-2 8A92, used within the photo reconnaissance programme and a final variant was called either Vostok-2M or Meteor (since it was used for launching Elint and Meteor satellites), the 8A92M.

The next variant to appear introduced two new upper stages: the Block I replaced the Block E and final Block L stage was added: that in turn spawned the Molniya launch vehicle. The Molniya carried the RD-0107 in its Block I, and this vehicle without the Block L was flown as the Voskhod launch vehicle. Western observers erroneously grouped the Voskhod launches with the Soyuz launch vehicles since they were very similar vehicles and flew the same types of missions. Although the Molniya-M launch vehicle had switched to an improved Block I engine compared with the original Molniya, there is no evidence that a similar switch took place with the Voskhod vehicle's Block I.

The true Soyuz launch vehicle, 11A511, came along in 1966, when the Soviet Union launched Kosmos 133 (the first Soyuz spacecraft launch). This was an improved version of the Voskhod vehicle, carrying the RD-0108 engine on the Block I. There were two minor variants of the Soyuz which flew a few missions, namely the Soyuz-L (three test flights of the lunar module and the Soyuz-M (eight flights of photo reconnaissance satellites).

RFAS/Soviet launchers are transported horizontally to the pad by rail and erected for firing. Manned shrouds carry four panels deployed on aborts to decelerate the escaping Soyuz craft 0516870

Soyuz U on pod with erector and propellant trucks 0131961

Soyuz-U

The Soyuz-U, 11A511U, replaced the older third-stage engine design with the RD-0110 Block I engine in May 1973 and put Kosmos 559 in orbit. In a design sense the U directly replaced and retired the Voskhod vehicle for good. The Voskhod saw its last mission in June 1976. Starting with launches in the ASTP programme and then the more general Soyuz programme which gave the world its first dedicated resupply vessel, the unmanned Soyuz 20, the Soyuz-U became the workhorse for manned and unmanned missions through to 1985. An upgraded version using higher-energy syntin (synthetic kerosene) fuel, the Soyuz-U2, was introduced in 1982 but was retired in 1996 when Russia ceased producing its fuel. This left the Soyuz-U as the backbone of the Russian manned missions until joined by the uprated Soyuz-FG in 2001.

The Soyuz-U is normally launched as a four-stage vehicle. The first stage comprises four liquid-fuelled boosters attached to the core stage. The tapered boosters contain LOX and kerosene and are each fitted with a single NPO Energomash RD-107 engine. The motors are fed by hydrogen-peroxide turbopumps. Flight control in the boost phase is provided by a combination of aerodynamic fins and two moveable vernier thrusters attached to each booster. The boosters are jettisoned 118 seconds after launch.

The second stage is the central core, which has a hammer-head shape to accommodate the tapered boosters. The core motor is an RD-108, which is similar to the RD-107 apart from incorporating four vernier thrusters for three-axis control. Both the core and the boosters are ignited at a low thrust setting, 20 sec before launch, allowing engine parameters to be monitored before a final decision to launch.

The third stage incorporates an RD-0110 four-chamber engine, also with four verniers for flight control. The three-stage configuration has mostly been used to launch Soyuz manned spacecraft missions and Progress unmanned resupply craft to Mir and ISS, and to place Yantar reconnaissance satellites in low earth orbit.

Support from Starsem allowed TsSKB Progress to offer the Ikar upper-stage module on the commercial market. Derived from the propulsion module of the Yantar reconnaissance satellite, Ikar can be restarted up to 50 times and allowed Starsem to orbit the GlobalStar constellation in six launches in 1999.

After one more Soyuz-U/Ikar flight in 2000, however, Starsem and Progress introduced the Fregat upper stage. Produced by NPO Lavochkin and based on components previously developed for the Phobos Mars-exploration spacecraft, the Fregat has much greater fuel capacity than the Ikar, opening up an even wider range of missions. Flight qualified in 2000, the Fregat upper stage is an autonomous and flexible upper stage that is designed to operate as an orbital vehicle. It extends the capability of the lower three stages of the Soyuz vehicle to provide access to a full range of orbits (MEO, SSO, GTO, escape). In order to provide the Fregat with high initial reliability and to speed up the development process, several flight-proven subsystems and components from previous spacecraft and rockets are incorporated into the upper stage. The stage is independent from the lower three stages, having its own guidance, navigation, control, tracking, and telemetry systems. The stage uses storable propellants (UDMH/NTO) and can be restarted up to 20 times in flight, thus enabling it to carry out complex mission profiles. It can provide the customer with 3-axis stabilisation or spin-up of their spacecraft.

Soyuz-U launches continued through late 2005, most of them being Soyuz or Progress launches from Baikonur. Another major user is the Yantar reconnaissance system, some of them being launched from Plesetsk. The type is due to be replaced by Soyuz-2, but no firm date has been set.

Soyuz-FG

Developed as an updated and uprated version of the Soyuz-U, the Soyuz-FG was first launched on 20 May 2001 and has since shared the burden of Progress and Soyuz launches in support of ISS. In particular, the FG has taken over manned Soyuz missions, its introduction coinciding with that of the automated three-passenger Soyuz-TMA spacecraft. The FG version has also been marketed by Starsem. On 13 August 2005, an FG with the Fregat upper stage became the first Soyuz-family booster to place a spacecraft in GTO. The system has a 1,660 kg payload to GTO.

The Soyuz-FG has uprated engines on the first three stages. The standard upper stage, when one is used, is NPO Lavochkin's Fregat, and a larger fairing (3.715 m in diameter and 7.7 m long) can be carried. The control system has been modernised, and the entire vehicle and propulsion system make less use of non-Russian components.

Molniya-M

Molniya is the most powerful of the R-7-based space launchers. It was employed until 1972 for planetary missions, but is now principally used for placing Molniya communications and early warning satellites into highly elliptical Earth orbital paths. These orbits, named Molniya after the first spacecraft that used them, are used to provide communications coverage over high Arctic regions. The Soviets and their Russian successors have not used it for a GEO mission, but the first Sun-synchronous mission came in December 1995.

The 8K78 Molniya first flew successfully in February 1961 carrying the Venera 1 Venus probe. The original 8K78 model employed the RD-0107 engine to power stage 2. This was replaced by the RD-0110 for the current 8K78M, first flown 19 February 1964 (failed to reach orbit).

There has only been one non-Russian-government Molniya mission – the December 1995 launch of India's IRS 1C, accompanied by Skipper, an experimental satellite for the US Ballistic Missile Defense Organization. Molniya's first Sun-synchronous mission uniquely used Block L for the orbital injection, the previous stages all being suborbital. This was also the last launch from Baikonur. Since then, all launches have involved Russian communications or early warning satellites launched from Plesetsk.

Molniya is due to be replaced by the Soyuz-2/Fregat launcher.

Soyuz-2

The Soyuz-2, which made its first flight from Plesetsk on 8 November 2004, is a thoroughly modernised version of the Soyuz family. It is designed to launch larger spacecraft than earlier members of the family, with a greater payload and a range of larger fairings. It will replace the older versions of the family for ISS support, for Russian military LEO launches and for Molniya high-elliptic launches, and will be flown from Kourou as well as from Baikonur and Plesetsk.

The Soyuz-2 programme is being completed with the aid of the European Space Agency (ESA). Early in its development, the modernisation

Soyuz Family Launches 2000–2005

Date	Variant	Payload	Site	Remarks
2 January 2000	Soyuz-U	Progress M1-1	Baikonur	
9 February 2000	Soyuz-U/Fregat	IRDT	Baikonur	
20 March 2000	Soyuz-U/Fregat	DUMSAT	Baikonur	
4 April 2000	Soyuz-U	Soyuz TM-30	Baikonur	
25 April 2000	Soyuz-U	Progress M1-2	Baikonur	
3 May 2000	Soyuz-U	Kosmos 2370	Baikonur	
16 July 2000	Soyuz-U/Fregat	Cluster II, FM6 & FM9	Baikonur	
6 August 2000	Soyuz-U	Progress M1-3	Baikonur	
9 August 2000	Soyuz-U/Fregat	Cluster II, FM5 & FM8	Baikonur	
29 September 2000	Soyuz-U	Kosmos 2373	Baikonur	
17 October 2000	Soyuz-U	Progress M43	Baikonur	
31 October 2000	Soyuz-U	Soyuz TM-31	Baikonur	
16 November 2000	Soyuz-U	Progress M1-4	Baikonur	
24 January 2001	Soyuz-U	Progress M1	Baikonur	
26 February 2001	Soyuz-U	Progress M-44	Baikonur	
28 April 2001	Soyuz-U	Soyuz-TM	Baikonur	
20 May 2001	Soyuz-FG	Progress M	Baikonur	
29 May 2001	Soyuz-U	Yantar-4K1	Plesetsk	
20 July 2001	Molniya-M	Molniya-3	Plesetsk	
21 August 2001	Soyuz-U	Progress M-45	Baikonur	
14 September 2001	Soyuz-U	Progress M-SO1	Baikonur	
21 October 2001	Soyuz-U	Soyuz-TM	Baikonur	
25 October 2001	Molniya-M	Molniya-3	Plesetsk	
26 November 2001	Soyuz-FG	Progress	Baikonur	
25 February 2002	Soyuz-U	Kosmos 2387	Plesetsk	
21 March 2002	Soyuz-U	Progress M1	Baikonur	
1 April 2002	Molniya-M	Oko	Plesetsk	Early warning
25 April 2002	Soyuz-U	Soyuz TM	Baikonur	
26 June 2002	Soyuz-U	Progress M-46	Baikonur	
25 September 2002	Soyuz-FG	Progress M1	Baikonur	
15 October 2002	Soyuz-U	Foton M	Plesetsk	Failure
30 October 2002	Soyuz-FG	Soyuz TMA	Baikonur	
24 December 2002	Molniya-M	Oko	Plesetsk	Early warning
2 February 2003	Soyuz-U	Progress M	Baikonur	
2 April 2003	Molniya-M	Molniya 1T	Plesetsk	
26 April 2003	Soyuz-FG	Soyuz TMA	Baikonur	
2 June 2003	Soyuz-FG/Fregat	Mars Express/Mars Orbiter/Beagle 2	Baikonur	
8 June 2003	Soyuz-U	Progress M	Baikonur	
19 June 2003	Molniya-M	Molniya-3	Plesetsk	
12 August 2003	Soyuz-U	Orlets-1	Baikonur	Kosmos 2399
29 August 2003	Soyuz-U	Progress M	Baikonur	
18 October 2003	Soyuz-FG	Soyuz TMA	Baikonur	
27 December 2003	Soyuz-FG/Fregat	Amos-2	Baikonur	
29 January 2004	Soyuz-U	Progress M	Baikonur	
18 February 2004	Molniya-M	Molniya-1T	Plesetsk	
19 April 2004	Soyuz-FG	Soyuz TMA	Baikonur	
25 May 2004	Soyuz-U	Progress M	Baikonur	
11 August 2004	Soyuz-U	Progress M	Baikonur	
24 September 2004	Soyuz-U	Yantar 4K1	Plesetsk	
14 October 2004	Soyuz-FG	Soyuz-TMA	Baikonur	
8 November 2004	Soyuz-2	Dummy mass	Plesetsk	Flight test
23 December 2004	Soyuz-U	Progress M	Baikonur	
28 February 2005	Soyuz-U	Progress M	Baikonur	
15 April 2005	Soyuz-FG	Soyuz TMA	Baikonur	
31 May 2005	Soyuz-U	Foton M-2	Baikonur	
16 June 2005	Soyuz-U	Progress M	Baikonur	
21 June 2005	Molniya-M	Molniya -2	Plesetsk	Failure
13 August 2005	Soyuz-FG/Fregat	Galaxy 14	Baikonur	First Soyuz GEO launch
2 September 2005	Soyuz-U	Yantar 1KFT	Baikonur	
8 September 2005	Soyuz-U	Progress M	Baikonur	
21 December 2005	Soyuz-U	Progress M	Baikonur	
1 October 2005	Soyuz-FG	Soyuz TMA	Baikonur	
9 November 2005	Soyuz-FG/Fregat	Venus Express	Baikonur	
28 December 2005	Soyuz-FG/Fregat	Giove-A	Baikonur	

programme was referred to as Rus, and was expected to be accomplished in one step. As a result of funding cutbacks, though, Soyuz-2 is being developed in two basic phases. The 2-1a phase introduces new avionics and controls and more powerful boost and core engines. Phase 2-1b introduces a new third stage, and Soyuz-ST is a configuration adapted for operations at Kourou. The system will be compatible with both Ikar and Fregat stages, the latter being used for launches into GTO. It will be available with four fairing sizes, the largest measuring 4.11 m in diameter and 11.4 m long. Even with the larger fairing, the combination of higher performance and an equatorial launch site almost doubles the payload to GTO, to 3,060 kg.

The Soyuz-2 features a digital avionics system, developed by NPO Avtomatiki of Ekaterinburg and located primarily in the third-stage equipment bay. (Earlier versions of Soyuz have separate guidance systems, one for the booster and core stages and one for the third stage.) It provides increased accuracy and, combined with a more efficient attitude control system, allows the vehicle to carry a larger payload fairing without compromising stability and control. The main engines are uprated with improved injection systems; these uprated engines were first flown on a standard Soyuz on 21 May 2001. The ST version, designed to carry large communications satellites, features the largest fairing. The rocket has the uprated booster and core engines of the Soyuz-FG.

The third stage of the 2-1b version has the RD-0124 engine. It is a staged combustion engine in which the turbopump is spun by an oxygen-rich exhaust from a gas generator. The combustion gases are mixed with additional kerosene from the engine cooling circuit. (On the earlier RD-0110 engine, the exhaust from the gas generator is fed to the attitude control verniers.) The result is a higher chamber pressure and a 34 second improvement in Isp, contributing to a substantial improvement in performance.

ESA formally decided to install a Soyuz launch pad at Kourou in February 2004. It forms part of ESA's European Guaranteed Access to Space (EGAS) policy, providing ESA with a medium-lift complement to Ariane and an alternative launch system if Ariane is unavailable. In April 2005, Arianespace and Russian space agency Roscosmos signed a contract covering the development of a Soyuz launch facility at Kourou. It covered the manufacture, assembly and installation of ground equipment at Kourou and the completion of development of the Soyuz 2-1b. Arianespace will be the prime contractor and manage the launch programme, while Starsem will act as the interface to the Russian-based members of the team.

The production contract for the first four Soyuz vehicles to be launched from Kourou was signed in February 2006. At that time, the first launch from Kourou was set for November 2008. Arianespace expects to launch two to three Soyuz missions per year from Kourou.

A Soyuz 2-1b with the ST-type fairing was planned for launch from Baikonur in mid-2006. This vehicle underwent dry-run testing at Baikonur in January 2006, and is scheduled to launch EuMetSat's MetOp 1 spacecraft. Another test of the Soyuz 2-1a, this time with the Fregat upper stage, was planned for a lift-off from Plesetsk in the third quarter of 2006. According to an Arianespace announcement in early 2005, the first Soyuz 2-1b flight would launch the CNES Corot astronomical satellite in mid-2006.

Specifications

Soyuz-U

Launch sites: Baikonur, Plesetsk
Principal uses: Soyuz manned spacecraft, Progress, reconnaissance, remote sensing, all LEO
Vehicle success rate: 98.3% claimed for all Soyuz vehicles as of February 2005
Commercial launches: 0
Performance: 6,855 kg into 220 km, 51.6°
Number of stages: 2 plus 4 strap-ons
Overall length (m): 42.5 unmanned (with 10.14 m-long fairing)
Principal diameter: 10.3 m across fins
Launch mass: 3,308 tonnes
Launch thrust (kN, vacuum): 4,146.5
Guidance: inertial. NPO AP is responsible for the GN&C systems

Soyuz-U Strap-ons

Number of strap-ons: 4
Engines: RD-107 from NPO Energomash, each with four fixed chambers and two gimballed verniers
Length: 19.6 m
Diameter: 2.68 m at base
Thrust: 8,385 kN SL; 102.1 kN vacuum
Burn time: about 118 s
Oxidiser: 27.8 t liquid oxygen
Fuel: 11.8 t kerosene
Propellant mass: 39.6 t each strap-on; dry mass 3.770 t

Soyuz-U Core Stage 1

The strap-ons separate at 120 s, but the hammerhead core continues for a further 190 s
Designation: Block A
Engines: single RD-108 from NPO Energomash, based on the RD-107 but with four gimballed verniers for steering
Overall length: 27.1 m
Diameter: 2.95 m maximum at strap-on attach points on LOX tank; 2.15 m lower kerosene tank
Thrust (vacuum): 990.2 kN
Burn time: 290 s
Oxidiser: liquid oxygen
Fuel: kerosene
Total mass: 99.5 t

Soyuz-U Stage 2

Designation: Block I
Engines: RD-0110 from KB Khimautomatiki, incorporating four fixed chambers and four gimballed verniers for steering
Overall length: 6.7 m
Diameter: 2.66 m
Oxidiser: liquid oxygen
Fuel: kerosene
Total mass: 25.2 t
Thrust: 298 kN vacuum
Burn time: about 24 s
Acceleration loads: 4.5 g longitudinal, 1.5 g lateral; payloads rated for 10 g longitudinal, 2 g lateral for 10 min
Acoustic: 144 dB for 60 s, payloads rated for 148 dB for 120 s
Vibration: 0.5 g 10–40 Hz actual
Thermal environment: fairing inner wall reaches maximum 57°C before separation at about 170 s

Fregat

Comment: NPO Lavochkin's stage has flown three times on Proton as the injection module for 1988's Phobos missions and for Mars-96
Engine: KB Khimash single chamber; up to 20 ignitions. Four clusters of 22.2 N hydrazine thrusters (SI 225 s) provides attitude control/ullage
Overall length: 1.500 m
Principal diameter: 3.350 m
Stage mass: up to 6,530 kg (up to 5,350 kg main engine propellant plus 85 kg hydrazine for RCS)
Oxidiser: N_2O_4 in two neo-spherical tanks
Fuel: UDMH in four neo-spherical tanks
Stage thrust (vacuum): 19.6 kN main (SI 327 s)
Operational time: Up to 48 hours

Start-1/Start

Current Status
On 25 April 2006, EROS-B1, the Israeli imaging satellite, was successfully launched aboard a Start-1 from Svobodny.

The Start-1 small launch vehicle is still advertised by the Puskovie Uslugi company of Moscow, formed by Russian space agency Roskosmos to market Soviet-era launch vehicles commercially. However, the last Start-1 to launch before the recent 2006 flight carried the ODIN satellite to orbit on 20 February 2001, and the SS-25 missile booster, on which the system is based, is approaching the end of its useful life.

Background
The first demonstration launch of the four-stage Start-1 was made from Plesetsk 25 March 1993, delivering a 260 kg communications payload into 695 × 966 km, 75.8°. The second Start-1 launch trailed along on 4 March 1997, inaugurating the Svobodny launch site. The third launch, on 5 December 2000, successfully launched EROS-A1, an Israeli Earth observation satellite.

Its sister version, the first five-stage Start, was lost 28 March 1995 on its only launch attempt. Stage 4 shut down 12 seconds early and stage 5 did not ignite. Israel's TechSat, Mexico's Unamsat and STC Complex's own test payload were lost in the failure.

Start-1 is a converted three-stage solid propellant road mobile SS-25 Sickle (RS-12MTopol) ICBM with an added solid stage 4.

Some elements from the related two-stage SS-20 IRBM are included. Both missiles came from the Nadiradze bureau and built at the Votkinsk plant, Udmurt. The SS-25 began development in 1971 and had its first test flights in 1982 (the first, 27 October 1982, was unsuccessful). Testing completed with the launch of 23 December 1987. It is capable of delivering a single 1 tonne 550 KT nuclear warhead over 10,500 km. It entered service in 1985. The missiles could be operational until 2015–2016.

All stages burn out fully (about 1 minute each). The 10 to 20 second coast after stage 1 burnout reduces dynamic pressure so that stage 2 is controllable after separation. The base aerodynamic fins and exhaust vanes provide control during stage 1 coast. N_2 jets in stage 5's aft section (stage 4 on Start-1) are used for roll control during stage 1 and three-axis during subsequent burns. There are no coast periods for the intermediate stages, but one of several 100 seconds before last stage firing, while the N_2 jets provide control.

Specifications
Start-1/Start
First launch: 25 March 1993/28 March 1995
Number of launches: 5/1, through September 2006
Success rate: 20%, through end of 2002
Number of stages: 4/5 (plus each can carry 70/80 kg post-boost stage)
Launch mass: 47/60 t
Length: 22.7/28.8 m
Maximum diameter: 1.8/1.8 m
Guidance: inertial; avionics in sealed compartment atop last stage
Performance: with post-boost stages 420/645 kg into 300 × 300 km, 90°; 300/530 kg into 500 × 500 km, 90°; 110/275 kg into 1,000 × 1,000 km, 90°. Accuracy 5 km apogee ±3' inclination with post-boost stage (solid propellant gas generator); 60–140 km apogee uncertainty without. Performance values increased by 70/80 kg without post-boost. Start-1's performance is reduced by 50 kg if larger fairing used.
Payload accommodation: Launcher is assembled horizontally in an 18 × 18.5 60 m class 100,000 building. Payload attachment requires 3 h, fairing/canister installation 5 h, tests 10 h and final launch operations 0.5 h. The campaign requires two to three days from when the payload is ready. The facility, canister and launch site maintain the payload at 10.5–26°C, 25–70% humidity. The payload can be accessed until 2–3 h before launch. The vehicle is transported horizontally to the 5 × 8 m launch box and raised. Start provides payload power and issues commands. Telemetry transmit: 203.27, 219.70, 75.67 MHz. On release, payload motion does not exceed 2°/s pitch/yaw + 1.5°/s roll with post-boost stage; 4°/s + 3.5°/s without.
Acceleration (g): 2.8/2.3 ejection, 0.15/3.3 stage 1, 6.5/4.4 stage 2, 6.5/6.1 stage 3, 9.0/6.3 stage 4, -/8.0 stage 5.
Acoustic load: ≤135 dB over 50–4,000 Hz

Strela

Current Status
The only launch of the Strela (Arrow) vehicle took place on 5 December 2003. However, NPO Mashinostroyenia is developing the Condor-E series of small remote sensing satellites for operation in LEO as future Strela payloads.

Background
The Strela was developed by NPO Mashinostroyenia from the SS-19 'Stiletto' (RS-18/UR-100N) ICBM, with very few changes apart from two fairing options and improved control system software. It is similar to Rockot, but only uses two stages of the SS-19. The missile is equipped with the APB (Agregatno-Priborny Otsek), a manoeuvrable platform designed to deliver multiple warheads to their targets. For the Strela project, the APB was reprogrammed to work as a third stage to deliver a payload into orbit. A special avionics section, diameter 2.4 m and length 0.8 m, contains guidance and flight control hardware and is fitted on top of the third stage. Although Strela is slightly inferior to Rockot in terms of payload mass capability, it has a lower orbital injection cost. The SS-19 ICBM has had over 150 flight tests and first entered service in 1975. ICBM reliability is said to be 98 per cent.

The Strela can be equipped with two upper stage sections (SHS-1 and SHS-2) with different payload fairings and internal volume. Each SHS consists of a measurement instruments section (MIC) with a payload adapter, an intermediate stage, a nose fairing and the spacecraft. The nose fairings consist of two halves made of metal sheet with longitudinal and lateral ribs. The jettison altitude of the nose fairing, 110 – 150 km, varies according to the requirements of the spacecraft's orbital altitude and possible thermal effects on the spacecraft. To reduce the heat flow affecting the spacecraft during the boost phase, the fairing's inner surface is covered by thermal insulating material.

The mechanisms and instruments section (MIS) on top of the second stage houses an autonomous control system with its own power source and propulsion system that can be used in the boost phase immediately after separation from the 2nd stage booster or during the boost trajectory at apogee. The MIS can perform the function of a single action booster unit. The MIS also houses the telemetry and trajectory measurement systems. Onboard power supply and a propulsion plant providing post-boost stabilisation in three channels are also mounted in this compartment.

During a baseline launch, first and second stage separation takes place after 126 sec at an altitude of 70 km and a velocity of 3.555 km/s. The nose fairing separates after 164.25 sec at an altitude of 114 km. The MIS and SHS sections separate from the second stage after 309 sec at a velocity of 7.727 km/s and an altitude of 240 km. The MIS propulsion plant is activated immediately. However, if a spacecraft is being delivered to a higher orbit, the MIS and SHS cluster continue in passive flight after separation from the second stage booster, with the MIC's propulsion system providing stabilisation. When the transfer orbit apogee is reached, the MIS propulsion plant is ignited and the spacecraft is accelerated up to the required circular orbital speed. The spacecraft is then separated, 29 minutes after lift off, at an altitude of about 1,000 km, and the MIS is taken away by the propulsion plant of the MIC.

The Strela can be launched from an upgraded silo at Svobodny (apparently not yet completed) or from silos in Baikonur. A post-boost module attached to the orbital payload can put satellites into orbits with an inclination of 63° from Baikonur and with inclinations of 51.8°–61° and 90°–104° (including sun-synchronous orbits) from Svobodny.

The first Strela demonstration launch took place from Baikonur in December 2003. (The press service of the Russian strategic missile forces mistakenly identified the mission as a sub-orbital training launch.) The rocket successfully delivered a 978 kg dummy payload into a 465 × 473 km orbit with an inclination of 67°. The altitude error was not more than 1 per cent (about 5 km) and the error in inclination was not more than 0.05°. The payload was reported to be a mockup of the Condor-E, a remote sensing and reconnaissance satellite being developed by NPO Mashinostroyeniya.

It was reported on 11 December 2002 that environmental concerns, related to the use of heptyl (UDMH) as a fuel, prompted residents in the city of Blagoveshchensk to protest about the possibility of future Strela launches from nearby Svobodny Cosmodrome. After public hearings were held, the Government of the Sakha (Yakutiya) Republic agreed to the use of certain areas as stage drop zones. The operational lifetime of the SS-19 is said to be about 25 years, after which the decommissioned boosters will no longer be available as launch vehicles. Launch price is approx. USD10 million.

Launch of an SS-19 missile (NPO Mashinostroyenia) 1329888

The second Strela launch may take place during the second half of 2008 carrying the Kondor remote sensing satellite for the Russian government.

Specifications
Designation: SS-19 'Stiletto' (RS-18/UR-100N)
Launch sites: Baikonur, Svobodny
Launch record: First launch: 06:00 UTC on 5 December 2003 from a silo at Site 132, Baikonur.
Engines: First stage: three single chamber RD-0233. Second stage: one single chamber RD-0234. 1 four-chamber RD-0236 steering engine.
Overall dimensions: Length 26.0 or 28.7 m (with SHS-2), body diameter 2.5 m
Launch weight: 104,000 kg
Propellant: N_2O_4/UDMH
Performance: 1,560 kg to LEO

Volna

Current Status
The Volna has not flown since 7 October 2005.

Background
The Volna (Wave) launcher was developed around the RSM-50 SLBM (NATO designation SS-N-18 Stingray) for suborbital and orbital launches. Approximately 350 missiles were built after it entered service in 1977. Also known as the R-29L, the submarine-launched ballistic missile was built by the Makeyev design bureau. The Volna is basically an RSM-50 with a space payload replacing the warheads plus modified mission software. This provides a very limited payload accommodation area of 1.3 m³.

The Volna can be launched from a surfaced or submerged submarine of the Kalmar (Delta III) type. The submarine cruises for 4 or 5 hours to its designated launch station, usually in the Barents Sea. The basic three-stage Volna is only capable of suborbital missions, carrying a recoverable vehicle of up to 720 kg or research equipment placed in a descent vehicle weighing up to 400 kg. With the addition of a small, liquid-fuelled fourth stage – known as an Orbital Booster Module (OBM) – developed by Babakin, the Volna-O version can launch small payloads to LEO. The basis of the OBM is the S5-89 liquid-propellant propulsion system with a displacer system for fuel feeding. Before separation the attitude of the module is controlled by the third stage of the LV. The standard Volna commercial microgravity orbit is about 2,200 × 200 km with an inclination of 76°.

The flight plan released prior to the Cosmos-1 launch show pressure being reduced inside the payload fairing after 3rd stage ignition and separation. 378 seconds into the flight, the stage would turn 180° to separate the spacecraft and the OBM. The OBM would then spin up and fire its apogee kick motor 1,153 seconds into the flight. This would be followed by separation of the kick motor and the protective fairing from the spacecraft, with orbital insertion at 835 km altitude after 1,168 seconds.

Makeyev developed the Volan re-entry vehicle for use in scientific experiments. It was designed to be

A Volna launch vehicle being lifted onto the Borisoglebsk Delta III class submarine prior to the test flight of the Planetary Society's Cosmos-1 solar sail in July 2001 (Makeyev Rocket Design Bureau, courtesy of the Planetary Society) 1329895

Space-related launches of the Volna

Launcher	Launch submarine	Launch date	Payload	Orbit/outcome
Volna	?	7 June 1995	Volan re-entry capsule with thermal-convectional Earth model developed by Bremen University.	Suborbital. Success.
Volna	K-496 Borisoglebsk (Barents Sea)	20 July 2001	Cosmos-1 test/IRDT	Suborbital. Failure.
Volna	K-44 Ryazan (Barents Sea)	12 July 2002	IRDT 2	Suborbital. Partial failure.
Volna-O	K-496 Borisoglebsk (Barents Sea)	21 June 2005	Cosmos-1	Orbital. Failure.
Volna	K-496 Borisoglebsk (Barents Sea)	7 October 2005	IRDT 2R	Suborbital. Launch success but spacecraft lost.

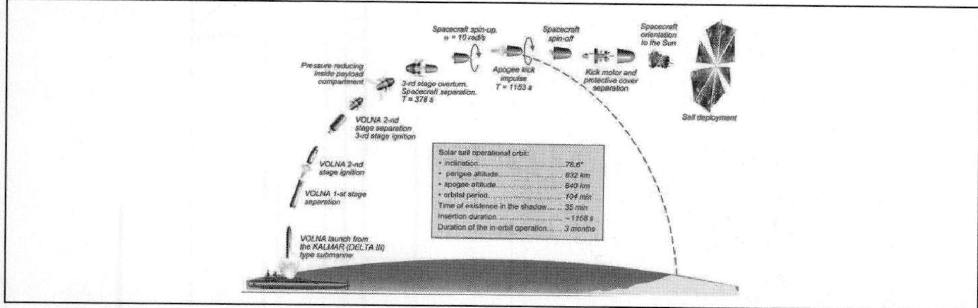

The planned launch profile of the Volna launch vehicle during the launch of the Planetary Society's Cosmos-1 solar sail (Babakin Design Bureau, courtesy of the Planetary Society) 1329896

A Volna launch vehicle being lifted by crane onto the launch submarine at Murmansk (Planetary Society) 1329898

injected into suborbital trajectories by the Volna to conduct experiments at microgravity levels of 10^{-5} to 10^{-6} g. The Volan could carry and return to Earth research equipment weighing up to 220 kg.

The first commercial launch took place in June 1995, when Makeyev, under contract with the German space agency DARA, used a Volna to launch a 120 kg experimental payload (electrically-driven thermal convection in a fluid shell between two concentric spheres) designed by the Centre for Applied Space Technologies and Microgravity at Bremen University. The suborbital microgravity flight lasted 20 minutes and recovery was 5,600 km downrange on Kamchatka.

Despite this initial success, the Volna has a poor record of reliability. Three of five commercial launches have been unsuccessful, with two of these failures due to a malfunction of the missile. Both of these failures were related to launches of the Cosmos solar sail spacecraft. In the first test flight, the spacecraft failed to separate from the launcher. The second failure was caused when the first stage engine failed 83 seconds into flight, and the first stage did not separate from the second stage. The failure was attributed to "critical degradation in operational capability of the engine turbo-pump". The flight ended 160 seconds after launch, with the LV probably reaching an altitude of about 200 km before falling back to Earth.

Other Volna missions have carried test versions of the IRDT inflatable re-entry device, designed to return cargo from orbit to Earth. These involved the three-stage Volna which was launched eastward towards Kamchatka. After Volna third stage separation, the IRDT fired a boost motor to increase its speed and then inflated the first stage of its heat shield. On 7 October 2005, a Volna carrying the Demonstrator D-2R inflatable braking device flew what appeared to be a normal flight toward the Kura impact range in the Kamchatka Peninsula. Efforts to locate the craft were unsuccessful. However, analysis of telemetry showed that it successfully separated from the rocket and inflated some 356 seconds after the launch at an altitude of 238 km.

Specifications

Launch weight: 35,400 kg
Engines: First stage: RD-0243 (1); second stage: ? (1), third stage: ? (1)
Overall length: 14.2 m
Diameter: 1.8 m
Oxidiser: N_2O_4
Fuel: UDMH
Payload to LEO (orbit inclination i = 79°): 100 kg to 800 km; 120 kg to 600 km; 150 kg to 400 km.
Volume of payload fairing: 1.3 m^3

Ukraine

Tsyklon-2

Current Status

Only two Tsyklon-2 launches have taken place since the end of 1999, both for the Russian Armed Forces. However, Yuzhnoye is offering an

For details of the latest updates to ***Jane's Space Systems and Industry*** online and to discover the additional information available exclusively to online subscribers please visit
jsd.janes.com

improved Tsyklon-2K variant for international and commercial customers.

Background

In the early 1960s, the Yangel design bureau (now Yuzhnoye) proposed a family of launch vehicles to be called Tsyklon. These would be based upon the bureau's R-16 and R-36 missiles and topped by a common C5M third stage for orbital missions. A government decree of 24 August 1965 ordered Yangel to develop the R-36 into the 11K67 to launch naval intelligence satellites and the Tsyklon 2 replaced the 11K67 from 1969. The R-16 had been approved for development as an ICBM in December 1956 and basic design was completed by November 1957. The first launch took place on 2 February 1961 following a launch pad disaster on 24 October, which killed almost 100 people. The Tsyklon-1 was the R-16 missile with the C5M added (launch mass 145.4 tonnes, payload to a polar 1,000 km orbit =~700 kg). The Tsyklon-2 would simply be a modified two-stage R-36 missile (launch mass 179 tonnes, 1.5 tonnes to a polar 200 km orbit). And the Tsyklon-3 would be the Tsyklon-2 with the S5M stage added (launch mass 185.5 tonnes, 2.5 tonnes to polar 1,000 km orbit). Of these three options, the Soviets backed away from the Tsyklon-1 and never developed it.

Launches of the dedicated space version of the R-36, the Tsyklon-2 vehicle, started in October 1967, but this variant had a short period of flights and was replaced by the launch vehicle referred to as both Tsyklon-2A and Tsyklon-M. For the now-retired RORSAT launches the payload used its own propulsion system to reach orbit, the two Tsyklon rocket stages being sub-orbital. While the spacecraft propulsion systems have failed either on their way to orbit or in orbit, the actual Tsyklon-2A launch vehicle has never failed in flight.

The Tsyklon-2K is a new version of the Tsyklon-2. The manufacturer claims low cost, low risk, improved performance and better accuracy for the new vehicle. It also offers, a high degree of automation and short pre-launch preparation, between 5 and 24 hours. Yuzhnoye says that the new variant will be able to launch 2,000 kg spacecraft into either a 1,000 km, 65° circular orbit or a 700 km sun-synchronous orbit from Baikonur, and that a launch could take place within 15 months of an order.

Tsyklon is based on Tsyklon M 0517328

Tsyklon is based on the SS-9 Scarp ICBM, with an unusual six-chamber first stage configuration 0516878

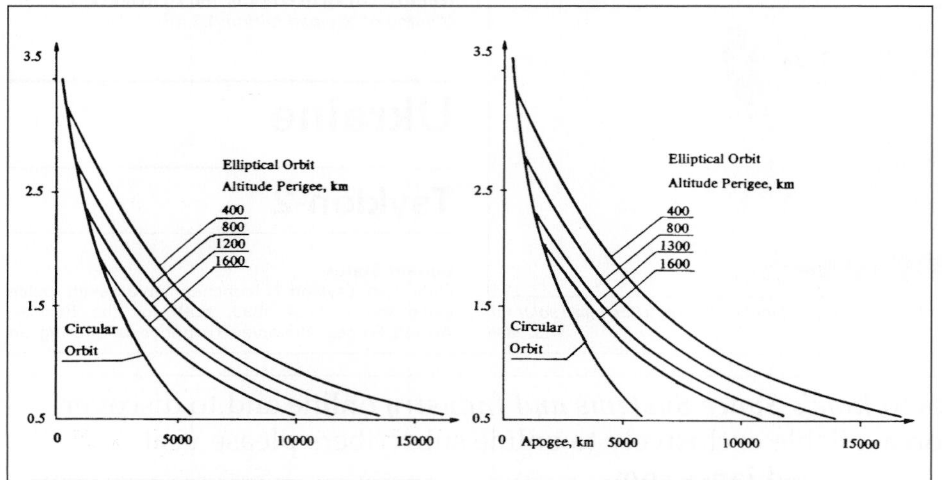

Tsyklon performance (t) into 82.5° (left) and 73.5° (right) 0517330

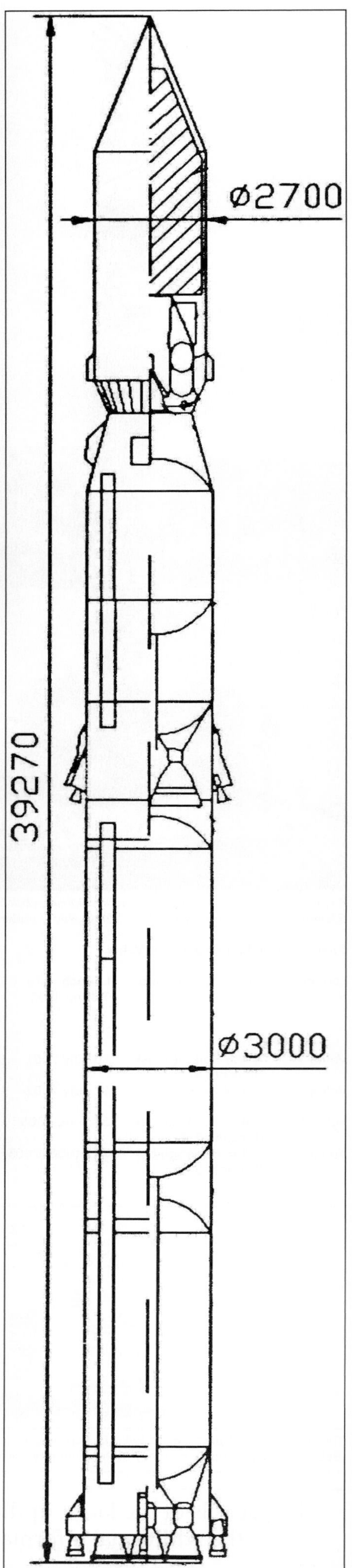

Tsyklon is based on Tsyklon M 0517329

Tsyklon-2 (M) launches

Launch	Date	Payload	Launch	Date	Payload
1	25 January 1969	Kosmos US-A	54	29 April 1982	Kosmos 1355
2	6 August 1969	Kosmos 291	55	14 May 1982	Kosmos 1365
3	23 December 1969	Kosmos 316	56	1 June 1982	Kosmos 1372
4	3 October 1970	Kosmos 367	57	18 June 1982	Kosmos 1379
5	20 October 1970	Kosmos 373	58	30 August 1982	Kosmos 1402
6	23 October 1970	Kosmos 374	59	4 September 1982	Kosmos 1405
7	30 October 1970	Kosmos 375	60	2 October 1982	Kosmos 1412
8	25 February 1971	Kosmos 397	61	7 May 1983	Kosmos 1461
9	1 April 1971	Kosmos 402	62	29 October 1983	Kosmos 1507
10	4 April 1971	Kosmos 404	63	30 May 1984	Kosmos 1567
11	3 December 1971	Kosmos 462	64	29 June 1984	Kosmos 1579
12	25 December 1971	Kosmos 469	65	7 August 1984	Kosmos 1588
13	21 August 1972	Kosmos 516	66	31 October 1984	Kosmos 1607
14	25 April 1973	Kosmos US-A	67	23 January 1985	Kosmos 1625
15	27 December 1973	Kosmos 626	68	18 April 1985	Kosmos 1646
16	15 May 1974	Kosmos 651	69	1 August 1985	Kosmos 1670
17	17 May 1974	Kosmos 654	70	23 August 1985	Kosmos 1677
18	24 December 1974	Kosmos 699	71	19 September 1985	Kosmos 1682
19	2 April 1975	Kosmos 723	72	27 February 1986	Kosmos 1735
20	7 April 1975	Kosmos 724	73	21 March 1986	Kosmos 1736
21	29 October 1975	Kosmos 777	74	25 March 1986	Kosmos 1737
22	12 December 1975	Kosmos 785	75	4 August 1986	Kosmos 1769
23	16 February 1976	Kosmos 804	76	20 August 1986	Kosmos 1771
24	13 April 1976	Kosmos 814	77	1 February 1987	Kosmos 1818
25	2 July 1976	Kosmos 838	78	8 April 1987	Kosmos 1834
26	21 July 1976	Kosmos 843	79	18 June 1987	Kosmos 1860
27	17 October 1976	Kosmos 860	80	10 July 1987	Kosmos 1867
28	21 October 1976	Kosmos 861	81	10 October 1987	Kosmos 1890
29	26 November 1976	Kosmos 868	82	12 December 1987	Kosmos 1900
30	27 December 1976	Kosmos 886	83	14 March 1988	Kosmos 1932
31	23 May 1977	Kosmos 910	84	28 May 1988	Kosmos 1949
32	17 June 1977	Kosmos 918	85	18 November 1988	Kosmos 1979
33	24 August 1977	Kosmos 937	86	24 April 1989	Kosmos 2033
34	16 September 1977	Kosmos 952	87	27 September 1989	Kosmos 2046
35	18 September 1977	Kosmos 954	88	24 November 1989	Kosmos 2051
36	26 October 1977	Kosmos 961	89	14 March 1990	Kosmos 2060
37	21 December 1977	Kosmos 970	90	23 August 1990	Kosmos 2096
38	19 May 1978	Kosmos 1009	91	14 November 1990	Kosmos 2103
39	18 April 1979	Kosmos 1094	92	4 December 1990	Kosmos 2107
40	25 April 1979	Kosmos 1096	93	18 January 1991	Kosmos 2122
41	14 March 1980	Kosmos 1167	94	30 March 1993	Kosmos 2238
42	18 April 1980	Kosmos 1174	95	28 April 1993	Kosmos 2244
43	29 April 1980	Kosmos 1176	96	7 July 1993	Kosmos 2258
44	4 November 1980	Kosmos 1220	97	17 September 1993	Kosmos 2264
45	2 February 1981	Kosmos 1243	98	2 November 1994	Kosmos 2293
46	5 March 1981	Kosmos 1249	99	8 June 1995	Kosmos 2313
47	14 March 1981	Kosmos 1258	100	20 December 1995	Kosmos 2326
48	20 March 1981	Kosmos 1260	101	11 December 1996	Kosmos 2335
49	21 April 1981	Kosmos 1266	102	9 December 1997	Kosmos 2347
50	4 August 1981	Kosmos 1286	103	26 December 1999	Kosmos 2367
51	24 August 1981	Kosmos 1299	104	21 December 2001	Kosmos 2383
52	14 September 1981	Kosmos 1306	105	28 May 2004	Kosmos 2407
53	11 February 1982	Kosmos 1337			

All launches from Baikonur; almost all either ocean surveillance satellites or ASAT tests. Failures 16 March 1966, 18 September 1966 (in orbit), 2 November 1966 (in orbit), 22 March 1967, 25 January 1969, 25 April 1973.

Specifications

Tsyklon 2

First launch: 25 January 1969
Number launched: 105 successful to end of 2005
Launch sites: Plesetsk
Number of stages: 3
Performance: 2,800 kg to 200 km orbit at 52.0°
Overall length: 39.7 m
Principal diameter: 3.00 m
Launch mass: 183,000 kg

Tsyklon 2K stage 1

Engines: Tsyklon 2 uses the RD-251 with six chambers and four gimballed NPO Yuzhnoye verniers.
Overall length: 18.9 m
Diameter: 3.00 m
Thrust: 2,797 kN vacuum; verniers 334.5 kN vacuum
Mass: 127,424 kg; 6,150 kg empty
Burn time: 120 s
Isp: 300 s

Tsyklon 2K stage 2

Engines: Tsyklon 2 stage 2 employs the RD-252 engine produced by NPO Energomash
Overall length: 9.4 m
Diameter: 3.00 m
Thrust: 955.2 kN vacuum; verniers 55 kN vacuum
Mass: 51,765 kg gross; 2,620 kg empty
Burn time: 160 s
Isp: 314 s

Tsyklon 2 stage 3

Engines: NPO Yuzhnoye RD-861 single restart embedded in toroidal tanks. Three-axis control during burns by eight fixed nozzles (6 × 20 N + 2 x 98 N) fed by main engine gas generator. Three-axis coast control by 10 thrusters of separate system.
Overall length: 2.5 m
Diameter: 2.0 m
Mass: 3,200 kg gross; 400 kg empty
Thrust: 78.0 kN
Burn time: 112 s
Isp: 317 s

Tsyklon 3/4

Current Status

Under an agreement signed with the Brazilian Space Agency (AEB) in 2004, the Ukrainian Space Agency is supporting development of Yuzhnoye's Tsyklon-4, a modernised version of the proven Tsyklon-3 light launch vehicle. It is planned for launch from Brazil's Alcantara Space Launch Center.

Background

Development of the Tsyklon 3 was authorised in 1970s as part of a plan to reduce the number of space vehicles. Based on the two stage 8K68 (SS-9 Mod 2), the Tsyklon 3 would adopt the S5M stage – a development of the R-36-0 deorbit stage – and be capable of lifting payloads up to 4 tonnes in mass to low earth orbit. Following the disintegration of the USSR Russia decided to cancel the programme and the Ukrainian government did not take over the production of this launcher. In 121 flights up to the last on 24 December 2004 the Tsyklon 3 had eight failures, a reliability rate of 93.3 per cent.

The three-stage variant of Tsyklon – referred to as either Tsyklon-3 or simply Tsyklon, was part of the original 'Tsyklon family' proposed in the early 1960s. It took over from the Meteor launch vehicle (a modification of the Vostok launcher) to accomplish the launches of Meteor and Elint satellites. Two Meteor launches carried paying piggyback payloads, Italy's Temisat (August 1993) and Germany's Tubsat B (January 1994). The UK's Surrey Satellite Technology Ltd used Tsyklon in August 1995 for Chile's 50 kg FASat-Alfa and was charged a fee of only USD400,000. Tsyklon production appears to have halted, but the Russians have hinted that some of Russia's R-36 strategic stockpile may be converted into Tsyklons.

Tsyklon is the three-stage version and was first offered as the commercial Tsyklon launcher in October 1987 at Sputnik's 30th anniversary forum in Moscow. It supports a range of medium military and civil missions but its single most numerous payloads are electronic intelligence spacecraft. It is launched only from Plesetsk.

Only four Tsyklon-3 vehicles have flown since the end of 1998. The most recent Tsyklon-3 launch carried the Sich-1M spacecraft along with the KS5MF2 micro-satellite. The vehicle is still on offer, but presumably from inventory on hand rather than production.

The Tsyklon-4 uses basically the same first and second stages as the Tsyklon-3. However, the third stage is redesigned and features a revised engine, the RD-861K, with improved specific impulse and provision for more restarts, together with 30 per cent greater fuel capacity. The entire vehicle can also be single-point refuelled on the launch pad. Tsyklon-4 also includes a new payload fairing with provision for separate clean-room assembly. It is expected to deliver 5,500 kg to a 500 km equatorial orbit and 1,700 kg to GTO, from the near-equatorial Alcantara site.

Discussions about collaboration between the Ukraine and Brazil started in early 2002, and it was announced in August 2002 that the two countries were considering the launch of an improved Tsyklon-4 from a new pad at Alcantara. A treaty covering the programme was signed in Brasilia in October 2003, and it was ratified in 2004. At that time, it was reported that the first Tsyklon-4 launch could take place as early as 2007 and that the team

Tsyklon 3 launches

Launch	Date	Payload
1	24 June 1977	C921
2	24 September 1977	C956
3	27 December 1977	C972
4	28 June 1978	C1025 electronic intelligence
5	26 October 1978	C1045/RS 1/2
6	12 February 1979	C1076 ocean
7	23 January 1980	C1151 ocean
8*	23 January 1981	GEO-IK
9	24 August 1981	C1300 ocean
10	21 September 1981	Aureole 3
11	30 September 1981	C1312 GEO-IK
12	3 December 1981	C1328 ocean
13	25 March 1982	Meteor 2-8
14	10 June 1982	C1378 ocean
15	16 September 1982	C1408 ocean
16	24 September 1982	C1410 GEO-IK
17	23 April 1983	C1455 electronic intelligence
18	23 June 1983	C1470 electronic intelligence
19	28 September 1983	C1500 ocean
20	24 November 1983	C1510 GEO-IK
21	15 December 1983	C1515 electronic intelligence
22	8 February 1984	C1536 electronic intelligence
23	15 March 1984	C1544 electronic intelligence
24	5 July 1984	Meteor 2-11
25	8 August 1984	C1589 GEO-IK
26	28 September 1984	C1602 ocean
27	18 October 1984	C1606 electronic intelligence
28*	27 November 1984	C1612/Meteor 3
29	15 January 1985	C1617-22
30	24 January 1985	C1626 electronic intelligence
31	6 February 1985	Meteor 2-12
32	5 March 1985	C1633 electronic intelligence
33	14 June 1985	C1660 GEO-IK
34	8 July 1985	C1666 electronic intelligence
35	8 August 1985	C1674 electronic intelligence
36	9 October 1985	C1690-95
37	24 October 1985	Meteor 3-1
38	22 November 1985	C1703 electronic intelligence
39	12 December 1985	C1707 electronic intelligence
40	26 December 1985	Meteor 2-13
41	17 January 1986	C1726 electronic intelligence
42	11 February 1986	C1732 GEO-IK
43	19 February 1986	C1733 electronic intelligence
44	15 May 1986	C1743 electronic intelligence
45	27 May 1986	Meteor 2-14
46	12 June 1986	C1758 electronic intelligence
47	28 July 1986	C1766 ocean
48	30 September 1986	C1782 electronic intelligence
49*	15 October 1986	Sextuplet communication satellites
50	2 December 1986	C1803 GEO-IK
51	10 December 1986	C1805 electronic intelligence
52	18 December 1986	C1809 science
53	5 January 1987	Meteor 2-15
54	14 January 1987	C1812 electronic intelligence
55	20 February 1987	C1823 GEO-IK
56	3 March 1987	C1825 electronic intelligence
57	13 March 1987	C1827-32
58	27 April 1987	C1842 electronic intelligence
59	1 July 1987	C1862 electronic intelligence
60	16 July 1987	C1869 ocean
61	18 August 1987	Meteor 2-16
62	7 September 1987	C1875-80
63	20 October 1987	C1892 electronic intelligence
64	6 January 1988	C1908 electronic intelligence
65	15 January 1988	C1909-14 communication satellites
66	30 January 1988	Meteor 2-17
67	15 March 1988	C1933 electronic intelligence
68	30 May 1988	C1950 GEO-IK
69	14 June 1988	C1953 electronic intelligence
70	5 July 1988	Okean 1 ocean
71	26 July 1988	Meteor 3-2
72	11 October 1988	C1975 electronic intelligence
73	23 December 1988	C1985
74	10 February 1989	C1994-99
75	28 February 1989	Meteor 2-18
76*	9 June 1989	Okean
77	28 August 1989	C2037 GEO-IK
78	14 September 1989	C2038-43
79	28 September 1989	Interkosmos 24
80	24 October 1989	Meteor 3-03
81	26 December 1989	C2053
82	30 January 1990	C2058 electronic intelligence
83	28 February 1990	Okean 2 ocean
84	27 June 1990	Meteor 2-19
85	30 July 1990	C2088 GEO-IK
86	8 August 1990	C2090-95
87	28 September 1990	Meteor 2-20
88	28 November 1990	C2106 electronic intelligence
89	22 December 1990	C2114-19
90	24 April 1991	Meteor 3-4
91	16 May 1991	C2143-48
92	4 June 1991	Okean 3
93	13 June 1991	C2151 electronic intelligence
94	15 August 1991	Meteor 3-5
95	28 September 1991	C2157-62
96	12 November 1991	C2165-70
97	18 December 1991	Interkosmos 3/Magion 3
98	13 July 1992	C2197-2202
99	20 October 1992	C2211-16
100	24 November 1992	C2221 electronic intelligence
101	22 December 1992	C2226 GEO-IK
102	25 December 1992	C2228 electronic intelligence
103	16 April 1993	C2242 electronic intelligence
104	11 May 1993	C2245-50
105	4 June 1993	C2252-57
106	31 August 1993	Meteor 2-21/Temisat
107	25 January 1994	Meteor 3-6/Tubsat
108	12 February 1994	C2268-2273
109	2 March 1994	Coronas-I
110*	25 May 1994	electronic intelligence?
111	11 October 1994	Okean 4
112	29 November 1994	GEO-IK 1
113	26 December 1994	C2299-2304
114	31 August 1995	Sich 1/FASat-Alfa
115	19 February 1996	C2328-30/Gonets
116	14 February 1997	C2337-9/Gonets
117	15 June 1998	C2352-7/Strela-3
118*	27 December 2000	Gonets/Strela-3
119	31 July 2001	Karonas-F
120	28 December 2001	Gonets/Strela-3
212	24 December 2004	Sich-1M

*indicates vehicle failure. All launches from Plesetsk.

would charge USD34 million for a LEO launch and USD38 million for a GTO operation.

Specifications

Tsyklon 3

First launch: 24 June 1977 (Kosmos 921)
Number launched: 121 to end of 2004 (116 attaining orbit)
Launch sites: Plesetsk
Principal uses: electronic intelligence, Meteor, oceanographic, geodetic, remote sensing
Vehicle success rate: 96.6% to December 2000
Number of stages: 3
Overall length: 39.270 m
Principal diameter: 3.000 m
Launch mass: 189 t
Launch thrust: 303.2 t L
Guidance: 600 km orbit ±15 km, 5 s, 3 arcmin

Tsyklon 3 stage 1

Engines: Tsyklon-M uses the six-chamber RD-251 engine from NPO Energomash and four gimballed NPO Yuzhnoye verniers: there are some reports that the three-stage Tsyklon uses a modification of this assembly which is designated RD-261.
Overall length: 19.38 m
Diameter: 3.000 m
Thrust: 2,364 kN SL (2,643 kN vacuum); verniers 296 kN SL (334 kN vacuum)
Mass: 127.0 t at launch; 8,300 kg dry
Burn time: 120 s (280 s quoted for combined stage 1/2 burn times)
Oxidiser: nitrogen tetroxide; 84,860 kg consumed. 60.15 m^3 tank pressurised to 2.5 atmospheres
Fuel: UDMH; 33,810 kg consumed. 44 m^3 tank pressurised to 2.1 atmospheres
Propellant mass: 120.7 t

Tsyklon 3 stage 2

Engines: Tsyklon-M uses the two-chamber RD-252 engine from NPO Energomash: there are reports that the three-stage Tsyklon uses a modification of this assembly which is designated RD-262
Overall length: 10.9 m
Diameter: 3.000 m
Thrust: 940.5 kN vacuum
Mass: 53.3 t at launch; 4,800 kg dry
Burn time: 160 s
Oxidiser: nitrogen tetroxide; 34,820 kg consumed. 25.00 m^3 tank pressurised to 3.7 atmospheres
Fuel: UDMH; 13,640 kg consumed. 18.14 m^3 tank pressurised to 2.2 atmospheres
Propellant mass: 49.8 t

Tsyklon 3 stage 3

Designation: S5M
Engines: NPO Yuzhnoye RD-861 single restart embedded in toroidal tanks. Three-axis control during burns by eight fixed nozzles (6 × 20 N + 2 × 98 N) fed by main engine gas generator. Three-axis coast control by 10 thrusters of separate system.
Overall length: 2.58 m
Principal diameter: 2.25 m
Thrust: 78.90 main kN vacuum
Mass: 4.63 t at launch; 1,407 kg dry
Burn time: 125 s in two burns (dual burn used for orbits >250 km)
Oxidiser: nitrogen tetroxide; 2,030 kg consumed. 1.6 m^3 tank pressurised to 7.5 atmospheres
Fuel: UDMH; 970 kg consumed. 1.5 m^3 tank pressurised to 4.2 atmospheres
Propellant mass: 3.2 t
Fairing: Length: 10.0 m.
Diameter: 2.700 m; payload volume 19 m^3 (15 m^3 in 2.4 m diameter cylindrical section).

The fairing can be installed up to 5 h before launch. Peak loads: 4.5/10 g stage ½ longitudinal, 1.5 g lateral

Vehicle/Payload Processing

Plesetsk vehicles are processed horizontally in a six-storey building, where stage 3 is fuelled, and not delivered to one of the two pads 40 km distant until aboutT-3 h by train. Processing takes 38 hours. Once at the pad, all operations are performed remotely. Propellant loading through the base requires about 1 hour from T-75 min. The complex can handle three launches monthly. A launch can be made 72 hours after request if Tsyklon is specially stored and the payload/s ready.

Zenit-2

Current Status

Zenit-2 was developed as a military satellite launcher for the Soviet Armed Forces. However, it was designed and produced by NPO Yuzhnoye in the Ukraine, so for the Russian Armed Forces it is

Zenit 2 roll out to one of the two Tyuratam pads 0517362

now an imported system. Only four flights have been carried out since 2000. However, its three-stage derivative, the Zenit-3SL, continues as the centrepiece of the Sea Launch programme (see separate entry) and the Zenit-2 is now on offer as part of the Sea Launch JV's new Land Launch programme. However, no Zenit-2SLB launches had been booked by early 2006.

Background

Zenit was planned as one of a versatile family of advanced launchers to be developed by the Yangel bureau, now NPO Yuzhnoye. The 11K55 and 11K66 vehicles would have been used for small satellite launches, the 11K77 (Zenit-2 itself) for medium satellites and the 11K37 was offered as a heavy launcher in the 3 to 6 tonne range.

Of these options only the 11K77 vehicle was actually built, and this was the first new Soviet launcher developed since the unsuccessful N-1 of 1969–1972. It provides a payload capability midway between those of Proton and Soyuz.

Zenit-2/3 and the similar Energia Zenit-1 strap-on were developed by NPO Yuzhnoye in Dniepropetrovsk, and its chief designer Dr Yuri A Smetanin. In the cavernous integration hall at Yuzhnoye's plant, workers can assemble about 20 first stages simultaneously. Titanium forging and tank production is performed on-site.

The Soviets claimed a manned capability (stage 1 is man-rated as Energia's strap-on), for the booster and stated a plan for carrying a two-man spaceplane called Uragan (Hurricane) to intercept US Shuttle missions out of Vandenberg. The project was cancelled in 1987 after west coast Shuttle launches were abandoned, in the wake of *Challenger's* loss. Zenit was also originally intended to replace Soyuz manned launches, and the launch facilities at Baikonur include a cosmonaut access tower.

The primary use of the Zenit-2 was to launch the Tselina-2 signals intelligence (SIGINT) spacecraft, although it has been used to launch other military LEO satellites, and this mission has continued in the post-Soviet era. In May 1995, Space Systems/Loral chose Zenith to orbit three clusters of 12 GlobalStar mobile communications satellites, but the first launch failed and the contract was switched to Soyuz and Delta. Planned launch pads at Plesetsk were never completed.

Various different versions of Zenit have been proposed, including a three-stage launcher with a cryogenic third stage, and a four-stage version with a 2.6 tonne NTO/UDMH apogee stage for 1.7-tonne GEO capacity from Baikonur. An air-launched version was described in 1994 for 1998 first test: released from an Antonov 225, it could handle 9-tonne LEO and 1-tonne GEO. Yuzhnoye began consideration of a Heavy Zenit in early 1989. Capable of delivering 25 tonnes to LEO, it was intended to supplant Proton by adopting the Proton approach of clustering Zenit stage 1's around a central tank. The Zenit's first stage appeared in parallel with Energia's strap-ons (designated Zenit-1).

The Zenit-2 version proposed for the Land Launch service is known as the Zenit-2SLB. It shares the same upgrades and improvements which have been applied to the Sea Launch Zenit-3SL.

Zenit-2 launches since 2000 (all Baikonur)

Date	Payload	Comments
3 February 2000	Tselina-2	SIGINT
25 September 2000	Yenisei-2	IMINT
10 December 2001	Meteor-3M	Plus 188 kg cluster of international payloads
10 June 2004	Tselina-2	SIGINT

Specifications

Zenit 2

First launch: 13 April 1985
Number launched: 36 to end of June 2004
Launch sites: Baikonur
Contractors: NPO Yuzhnoye (prime), NPO AP (GN&C), NPO Elektropribor (avionics), PO Komunar (inertial guidance package), NPO Energomash (main engines).
Performance: Zenit 2 13,740 kg into 200 km 51.6°, 11,380 kg into 200 km 99°, both from Baikonur; 15.7 t into 200 km 12° for equatorial site. Injection accuracy: ±3.5 km altitude, ±2 arcmin inclination, ±2.5 s period.
Number of stages: 2
Overall length: 57.4 m (61.40 m) including 13.65 m payload fairing
Principal diameter: 3.90 m
Launch mass: 445 t
Launch thrust: 740 t sea level
Guidance: inertial. 1 Mbit/s telemetry provides information on 1,000 parameters during launch. Receives navigation update just before launch via laser optical receiver package mounted on side; package is then blown free.
Comment: Zenit 2 stage 1 Stage 1 and Energia's strap-ons were developed in parallel. Both stages are aluminium with integrally machined stiffeners.
Engines: RD-171 from NPO Energomash, four gimballed chambers.
Overall length: 32.94 m
Diameter: 3.90 m
Thrust: 7,911 kN vacuum, 7,259 kN sea level
Burn time: about 144 s (146 s Resurs-O1)
Dry mass: 28,080 kg
Propellants: 318,800 kg of liquid oxygen (210.55 m³ forward tank) and kerosene (107.31 m³ aft)
Separation: four retro solids at aft end Zenit 2 stage 2
Engines: NPO Energomash RD-120 single fixed reignitable chamber with four NPO Yuzhnoye verniers
Overall length: 11.50 m
Diameter: 3.90 m
Thrust: 833.5 kN main + total 78.4 kN verniers, vacuum
Burn time: about 300 s total, verniers can burn for up to 1,100 s (Resurs-O1: 285 s main, 499 s verniers)
Dry mass: 8,300 kg
Propellants: 80,600 kg of liquid oxygen (53.43 m³ forward tank) and kerosene (27.87 m³ aft torus)
Separation: four retro solids at aft end

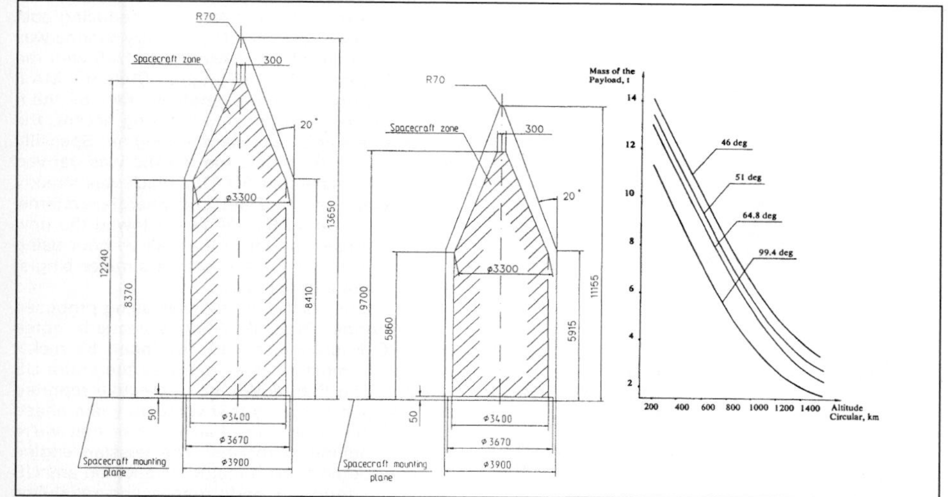

Zenit 2 fairings and performance 0517363

Vehicle/payload processing

Zenit is assembled horizontally, with the payload integrated on stage 2 before stage 1/2 mating. Assembly of the vehicle alone requires 80 h, increasing to 116 h with the payload. The vehicle can be stored for a year. Transfer to pad (45L/45R) is by rail; erection/launch processing is highly automated, requiring 21–80 h between initial integration and launch. Umbilicals retract 12 min before launch, returning to their protected positions. At 4 min, the train returns to the integration building. 4 m³/s of water is sprayed into the flame pit beginning at 15 s to reduce acoustic loads and temperature. The pad is automatically reconfigured within 5 h for the next launch. Zenit stages 1/2 separate at 150 s (Resurs-O1: 145 s) and the payload shroud during stage 2 burn between 160-1,240 s (Resurs-O1: 160 s). Payload volume is 90 m³ for the 13.65 m long shroud, reduced to 43–50 m³ for three-stage missions. An 11.155 m shroud is also available. Payload separation is achieved at 450–1,250 s, depending on the mission (Resurs-O1: 640 s). Acoustic load 140 dB.

United States

ARES

Current Status

The US Air Force plans to start development of an Affordable Responsive Spacelift (ARES) demonstrator in FY2007. Ultimately, the programme could lead to a new space launch vehicle, with a reusable, suborbital booster and an expendable upper stage, with a LEO payload of between 900 kg and 2,250 kg.

Background

The Military Spaceplane (MSP) and Space Operations Vehicle studies of the late 1990s envisaged the use of a suborbital, reusable rocket booster which could launch a variety of payloads at high mach and high altitude: the Common Aero Vehicle hypersonic glide bomb, an orbital Space Manoeuvre Vehicle or an expendable upper stage to launch satellites.

This concept was revived in 2005 by the Aerospace Corporation as part of USAF Space Command's Operationally Responsive Spacelift (ORS) Analysis of Alternatives (AoA). The Aerospace study found that what it called a Hybrid Launch Vehicle or HLV – hybrid indicating the combination of reusable and expendable components – could offer rapid turnaround but was economically attractive.

Compared with a fully reusable vehicle, the HLV requires the development of one-third as much reusable hardware; at the same time, it expends about one third as much mass as an expendable. The study found that the hybrid was the "best choice" in 85 per cent of the representative future scenarios considered in the AoA. Like the MSP, ARES could be used to launch satellites, CAVs or other payloads.

The first step in the ARES programme, according to USAF documents, is to define and build a small-scale demonstrator. This would be about one-quarter the size of a full-scale ARES and would weigh about 6,800 kg without booster fuel. Its main goals would be to demonstrate the integration of modern materials, avionics and propulsion technology in a low-maintenance, high-flight-rate vehicle – one specific target is to launch five times in ten days with a 15-person crew – and to demonstrate a flight profile in which the booster would release an upper stage at Mach 7 and return to its launch site. ARES is likely to use a small jet engine for flyback. The demonstration program is expected to cost USD200-250 million.

The USAF has not defined a choice of fuel or configuration for ARES, but LH2/LOX and RP/LOX are the most likely choices. The final size of the system is not defined: Aerospace briefings describe a larger vehicle than the FY2007 budget, using the Falcon upper stage for a 6,600 kg payload to LEO. It may also be significant that the USAF/Orbital Upper Stage Flight Experiment (USFE), which was initiated as part of the MSP effort, is still under way and could be launched by May 2007: the USFE is aimed at developing an upper stage fuelled by storable, non-toxic kerosene and hydrogen peroxide.

The FY2007 USAF research and development budget provides for a studies and analysis phase, continuing until the end of 2006, followed by development of the demonstrator. Preliminary design review would take place in the third quarter of 2007 with critical design review a year later. Flight demonstration would start in 2010.

The ARES subscale prototype would be followed by a full-scale demonstrator or Y-vehicle, which is expected to fly in 2018. Development of a production ARES spaceplane could start by 2019.

Atlas

Final Atlas assembly is now undertaken in Denver's Waterton facility 0516926

Current Status

The Atlas family has been consolidated around the the Atlas V version, developed to meet the USAF's Evolved Expendable Launch Vehicle (EELV) requirement and powered by the Energomash RD-180 propulsion system. All other Atlas versions have now been retired . The Atlas V is marketed worldwide by International Launch Services (ILS), a joint venture between Atlas developer Lockheed Martin and Russia's Khrunichev. As of late 2005, however, Lockheed Martin and Boeing were proposing to form another joint venture to support US military requirements.

Background

Atlas was developed by General Dynamics Convair Division as the first US ICBM, becoming operational in September 1959. A total of 159 were at one time deployed across the US before the last was withdrawn from service in 1965. As they were deactivated, they were refurbished as space launchers, serving alongside new-build rockets. The last of these refurbished launchers flew in 1989.

Since that time, more than 500 Atlas vehicles have been launched. They are responsible for orbiting the first US astronaut and dispatching the first successful probes to Mars, Venus, Mercury and Jupiter. Of the 140 Atlas E/F types originally stored at Norton AFB, the last departed in March 1995.

The most widely used version of the early Atlas was the Atlas Centaur combination, first launched in 1962. In January 1987, Atlas lost out to the McDonnell Douglas Delta as the USAF Medium Launch Vehicle for the GPS Navstar satellites. However, General Dynamics in 1987 committed USD100 million of company funding to 18 new commercial Atlas Centaur vehicles, providing a 2,335 kg GTO capability, and this became the foundation of the modern Atlas family. The company later increased the plan to USD400 million for 62 vehicles. Eutelsat was the first to select this Atlas I, for one of its second generation satellites in January 1990 (it actually flew on the first of the Atlas II family, December 1991). In October 1987, NOAA booked three GOES at a fixed price of about USD200 million for delivery into GTO and a free reflight in the event of a launcher failure. The last Atlas I was launched in 1997.

In May 1988, Atlas was selected as the USAF MLV-2 launch system. Fifteen of these launchers, uprated and modernised to launch the third-generation Defense Satellite Communications System (DSCS-3) spacecraft, were ordered by the USAF, and became the basis for the military/commercial Atlas II family.

Atlas II achieved the required DSCS 3 capability by stretching the booster and upper stage tanks by 2.75 m and 92 cm, respectively, uprating the Atlas booster powerplant from 1,668 kN to 1,841 kN and incorporating new lightweight and low cost avionics. Lockheed Martin's Integrated Apogee Boost Subsystem (IABS) provides insertion of DSCS into near GEO. This new Atlas II (the previous 18 Atlas Centaurs were retrospectively renamed Atlas 1) provides 2,812 kg GTO capability. The first Atlas II was launched under the MLV programme on 10 February 1992.

The MLV-2 success permitted GD to offer an Atlas IIA version (3,045 kg GTO) from 1992. The IIA was an enhanced version of the II model for commercial and government applications. The major improvement was the incorporation of Pratt & Whitney's RL10A-4 engines on the Centaur stage. A 508 mm long all-welded extension skirt, deployed following Centaur separation, provided a 6.5 second SI increase to 449.5 second vacuum (thrust 185 kN total vacuum). There was no thrust compartment length increase required and no additional mass. The avionics were further updated by replacing the sequence control unit and the servo inverter unit with a remote-control unit, providing 128 channels of solid state switching for vehicle/spacecraft sequencing instead of the previous electromechanical relays.

GD began offering a IIAS version in 1993. It was conceived to bid for the Intelsat 7 contract; the original intention was to propose the IIA, but Intelsat's mass growth dictated the addition of strap-on Solid Rocket Booster (SRB) units. The IIAS was uprated with the inclusion of four jettisonable Castor 4A boosters. Two ignited on the pad and were jettisoned at 88 seconds after their 54 second burn. The second pair ignited at 57.5 seconds for jettison at 114 seconds. The Atlas structure required strengthening to accommodate the boosters.

Intelsat awarded a USD9.5 million contract in 1991 to pay for enhancing Atlas 2AS capacity into GTO by 121 kg, the additional satellite propellant providing a 16 year life. The Block 1 upgrade became available in 1994 for IIAS/2A, able to improve GTO capacity by 8 per cent. In addition to software improvements, the two air-lit Castor 4As on the 2AS replace their 11° canted nozzles with 7°, the Centaur's RL10A-4 is upgraded to the 10A41 with thrust increase per engine of 6.67 to 99.19 kN. First flight was January 1995 for Intelsat 704.

In May 1994, GD sold its rocket business – based on the Atlas and Centaur – to Martin-Marietta. GD was in the process of selling off most of its businesses, and Martin-Marietta had enjoyed limited success with Commercial Titan. In August 1994, Martin Marietta and Lockheed announced their intention to merge, closing the deal in March 1995. At the Paris air show in June of that year, Lockheed Martin and Khrunichev announced the formation of International Launch Services (ILS) to market and support the Atlas and Proton launchers for commercial customers.

1995 was a peak year for Atlas launches, with eleven flights and eleven successes. The last Atlas I flew in April 1997, and the final basic Atlas II was launched in March 1998. The USAF's 15 MLV-2 vehicles were delivered as Atlas II, IIA, IIAS and IIIB configurations and were launched between February 1992 and February 2005. Eventually, eight of the MLV-2 rockets launched DSCS-III satellites and the other seven flew classified satellites under a series of Zodiac mission names. The secret payloads included three third-generation Naval Ocean Surveillance System (NOSS) doublets and Satellite Data System relay spacecraft. The last Atlas II was launched in March 1998 and the last IIA in December 2002. The last IIAS – and the last "classic" Atlas with Rocketdyne engines – was launched on 31 August 2004. It was the 30th Atlas IIAS, the 63rd Atlas II, and the 73rd consecutive successful Atlas flight.

Lockheed Martin, on 7 November 1995, announced development of the IIAR (R means re-engined) to replace all previous models of Atlas. The radical redesign was aimed at reducing cost by more than 20 per cent. The primary change was the replacement of the Rocketdyne MA-5 with the NPO Energomash RD-180 engine. While the MA-5 system comprised three engines, two of them being boosters and jettisoned during ascent, the RD-180 is a single two-chamber engine. Specially developed for the Atlas, the RD-180 was derived from the four-chamber RD-170, which was already flight-proven on the Zenit and Energia systems. The more powerful RD-180 also allowed the new Atlas to surpass the lift of the IIAS without using solid rocket boosters, and with a simpler single-engine Centaur upper stage.

Late in 1994, as the RD-180 was being proposed for the Atlas, Pratt & Whitney leaders noted that NPO Energomash had developed 53 rocket engines – almost twice as many as the entire US industry. One Pratt & Whitney executive reported that "the good news is that we're 20 years ahead in gas turbines, and the bad news is that we're 20 years behind in rockets." The Russian engine was considered to be a major advance on any US product in terms of performance and reliability, designed and tested for ten missions.

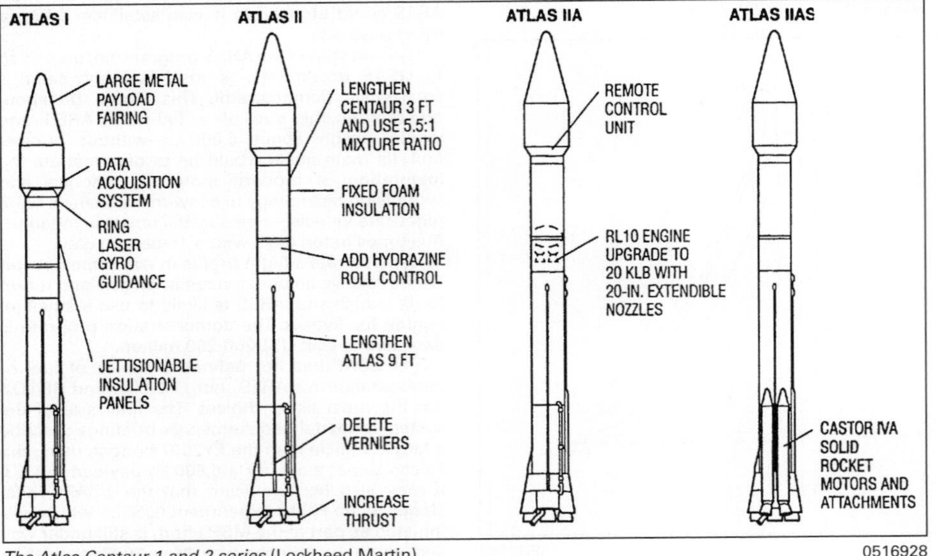

The Atlas Centaur 1 and 2 series (Lockheed Martin) 0516928

Lockheed Martin planned an initial series of 18 Atlas IIAR boosters, with Russian-built engines, after which manufacture of the RD-180 would be transferred to the US. On 8 April 1998, Lockheed Martin announced changes to the designation system whereby Atlas IIAR would become Atlas IIIA and a variant known as the IIARC would become the Atlas IIIB. The Atlas IIIB incorporated two RL-10A cryogenic engines to power a stretched version of Centaur and extends the lift capability to 4,500 kg to GTO.

Atlas IIIA made its first flight on 24 May 2000, followed by the first Atlas IIIB on 21 February 2002. However, with the slumping commercial market, competition from outside the US and the advent of the improved Atlas V, the Atlas III had a short career. Two IIIA launchers and four IIIBs were launched, all successfully. The last Atlas III, a IIIB, orbited a secret NRO payload on 3 February 2005.

In October 1998, Lockheed Martin was awarded contracts to complete development of a new family of launchers under the USAF Evolved Expendable Launch Vehicle (EELV) programme. The new launcher – officially named Atlas V in February 1999 – uses the same RD-180 motor as the Atlas III, and some of the same avionics, but is in many respects a new vehicle with a greatly increased payload capacity. The largest current Atlas V can lift more than twice the payload of the Atlas IIAS. The primary difference is that the signature feature of the original Atlas, the pressure-stabilised fuel tank system, is replaced by a structurally stable tank, 3.8 m in diameter. The tank and motor combination is known as a Common Core Booster (CCB) and has provisions for up to five solid boosters. Alternatively, three CCBs can be combined into a Heavy Launch Vehicle (HLV) configuration. All Atlas V vehicles use a common Centaur stage, equipped with either one or two RL-10 engines depending on the mission profile.

Two initial versions of the Atlas V have been developed: the 400 series, with a 4 m diameter payload fairing similar to that of the Atlas III, and the 500 series with a 5.4 m fairing, built by Contraves and available in short (20.4 m) and medium (23.5 m) lengths. Each Atlas V carries a three-digit designation: the first digit indicates the diameter of the fairing (4 or 5), the second the number of SRBs and the third the number of RL10s in the Centaur.

Advantages of the new design include a lower parts count in the new booster section, the ability to use new, composite-wound SRBs – developed by Aerojet – and a fault-tolerant inertial guidance system developed by Honeywell. The entire family features a high degree of commonality, with a single main engine, single SRB configuration and standard upper stage, reducing production and integration costs. Atlas V configurations (not including the HLV version) can launch payloads from 4,950 kg to 8,670 kg into Geostationary Transfer Orbit (GTO) and 12,500 kg to 20,520 kg into Low Earth Orbit (LEO). The top-end capability of this family is close to that of the Titan IVB.

The Atlas V is designed according to the principles of fault avoidance and fault tolerance. The essence of fault avoidance is creating a simpler design where fewer opportunities for failure exist, or through greater margins-moving away from the edge of the system's capability. An example of fault avoidance is reduction in the number of engines. Fault tolerance, on the other hand, is accomplished through redundancy, or by implementing processes like pre-launch check-out of the RD-180 engine. This capability allows verification that the engine has properly started and is up to 60 per cent power before the commitment to launch. If a problem is detected, the engine can be shut down and the problem corrected.

The RD-180 engine can be throttled between 47 per cent and 100 per cent of nominal thrust, making it possible to ease the vehicle's path through peak loads – transonic acceleration and maximum dynamic pressure – which provide high performance for ascent. A control system operated by RP-1 fuel under high pressure operates two servo-actuators on each thrust chamber and nozzle during ascent, providing pitch, yaw and roll control without the use of separate motors or thrusters.

The CCBs are identical, all featuring attachments for up to five SRBs. The SRBs are all identical and are all ignited on the ground, and have fixed nozzles: the thrust-vectoring and throttling capability of the RD-180 provides adequate control authority during ascent. The SRBs have graphite-epoxy composite cases and carbon-phenolic nozzles, and are delivered to the launch site in ready-to-fly condition.

All Atlas V configurations use a common Centaur hydrogen-oxygen upper stage, configured with either one or two Pratt & Whitney RL10A-4-2 rocket engines rated at 97.86 kN each. The upper stage also has a hydrazine-powered reaction control system to provide pitch and yaw control in coasting flight and (in the single engine version) roll control in powered flight.

Components for the Atlas V are produced at San Diego, California, and Harlingen, Texas, with the largest part of the manufacturing (including the CCBs, Centaur and final assembly) taking place in Denver. Major suppliers include AMROSS, for the main engines; Pratt & Whintey, for the Centaur engines; Aerojet, with the SRBs; and Contraves, which provides the 500-series payload fairing.

The first Atlas V, a 401 (the smallest configuration available) was launched successfully on 21 August 2002 from Launch Complex 41 at Cape Canaveral AFS. The vehicle is prepared in a vertical integration facility (VIF) and transported vertically to the launch pad, as little as 14 hours before launch. By the end of 2005, the launcher had logged six successful flights in six attempts. Five of these were for commercial customers and one for NASA (launching the Mars Reconnaissance Orbiter in August 2005.) The heaviest payload launched was Inmarsat's 4-F1, aboard an Atlas 431 in March 2005.

The Atlas V has been assigned 16 US Air Force launches under the EELV programme. The first of these is due in 2006. Under a September 2000 decision, Atlas launches would only be carried out from Cape Canaveral. In 2003, however, Lockheed Martin was given the go-ahead to prepare a West Coast launch site. "Booster on stand" tests, including stacking a complete booster, were carried out at Vandenberg AFB's Space Launch Complex 3 East (SLC-3E) in March 2005 and a propellant loading test was performed in June. The first launch will orbit a classified NRO payload. In December 2005, a contract for the launch of Defense Meteorological Satellite Program

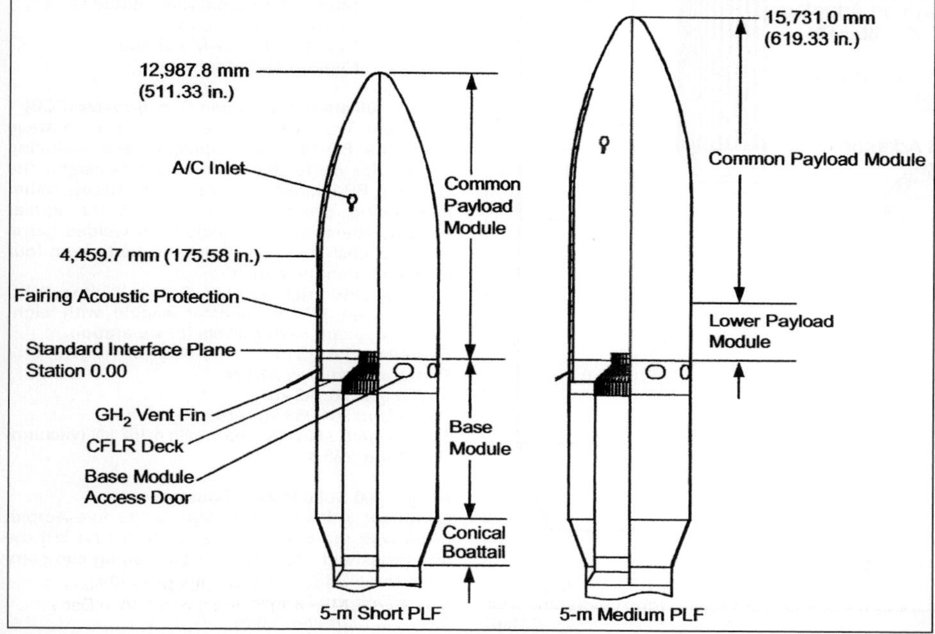

Atlas 5 m payload fairings 0131863

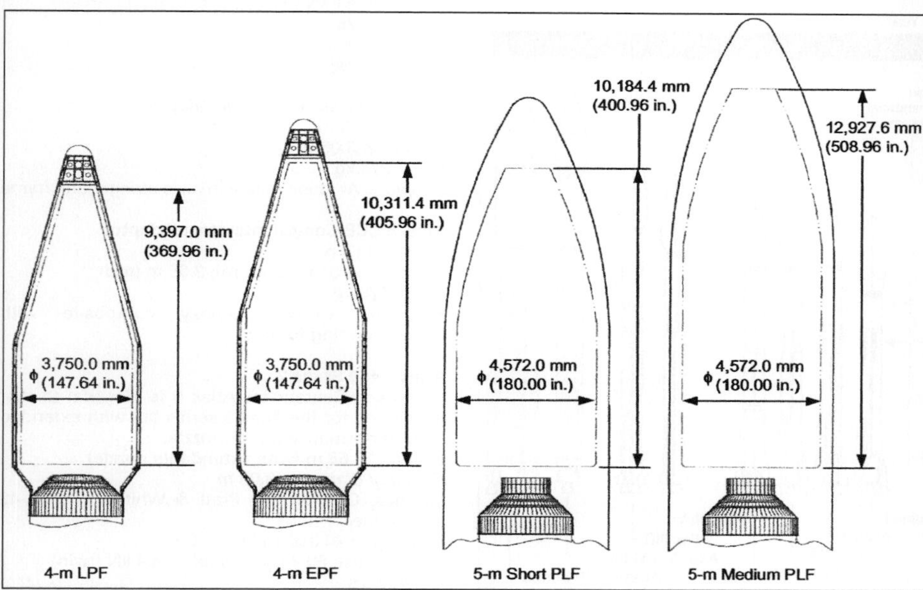

Atlas static payload envelope 0131862

NPO Energomash RD-180 engine adopted by Lockheed Martin for the 3 and V series (Theo Pirard) 0054362

(DMSP) satellites in late 2007 brought the number of EELV Atlas launches under firm contract to nine.

Test firings of an upgraded Block B SRB for the Atlas V started in April 2005 and continued with a second series in June 2005. The new rockets are expected to deliver improved performance and reliability. Other planned improvements include modified payload carriers for dual and secondary payloads (such as four microsats weighing as much as 200 kg each) and a modified payload truss for spacecraft weighing up to 20,400 kg. In the longer term, Lockheed Martin has studied a two-phase upgrade for Atlas V. Phase 1 would comprise a "wide-body" Centaur with the same diameter as the Atlas V-500 payload fairing, which could be fitted with as many as four RL10 engines. In Phase 2, the CCB would be expanded to the same diameter, and could be equipped with two RD-180s and up to six SRBs.

Unlike the Boeing Delta 4, the Atlas V depends critically on non-US technology. So far, all Atlas III and Atlas V vehicles have flown on RD-180 engines produced by NPO Energomash in Russia and delivered via RD Amross LLC, a joint venture between Energomash and Pratt & Whitney. When the EELV programme started, Pratt & Whitney planned to establish a US production line for the RD-180 by 2005, but this was delayed in 2001 because of the small numbers of engines required. Manufacturing data was delivered to RD Amross in April to August 2003 and some prototype components were built in 2004. In a 2004 report, the US Government Accountability Office noted that a US co-production facility was not due to open until 2008, and that US-built, mission-ready RD-180 engines would not be available until 2012.

Also uncertain is the status of the Atlas V Heavy Lift Vehicle (HLV) configuration. So far, the USAF has budgeted EELV funds to support the HLV configuration of the Delta IV, but to date has not funded a similar version of Atlas. Given budgetary constraints and the small numbers of these vehicles required, it is unlikely that the service will support such a development in the near future.

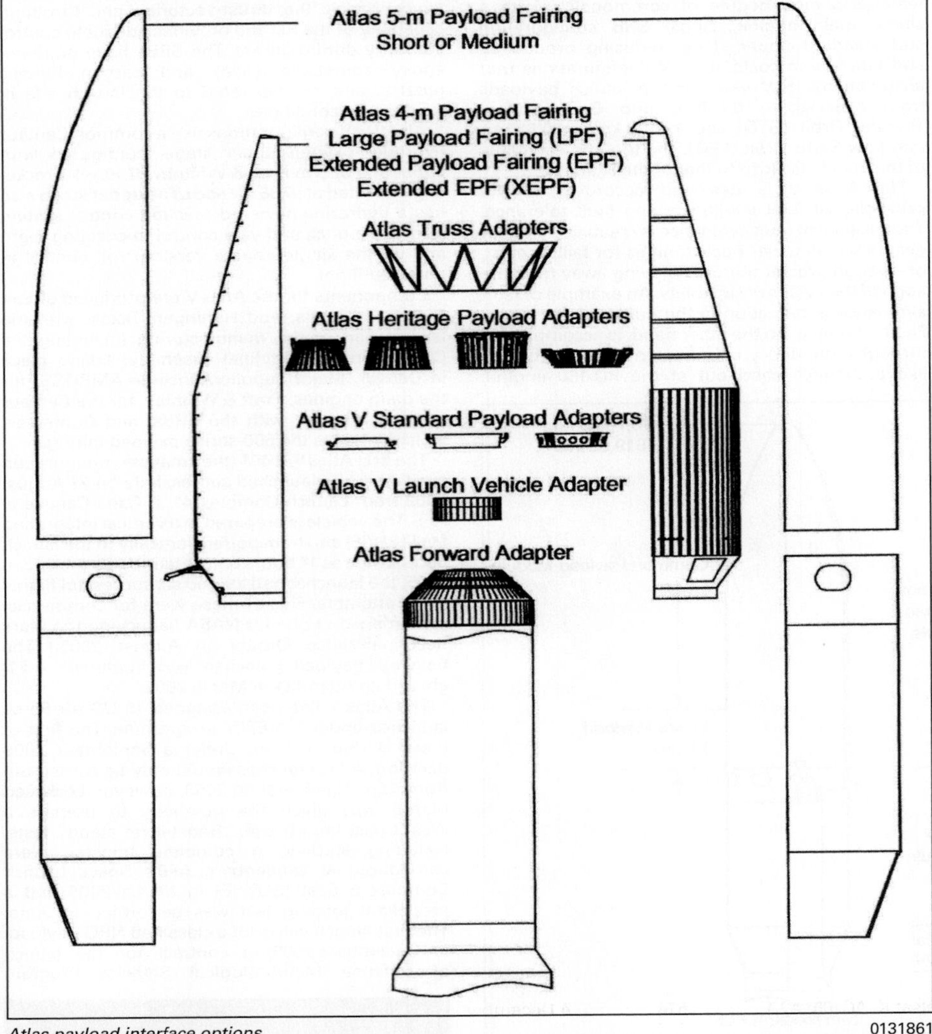

Atlas payload interface options 0131861

Atlas evolution from 1 CBM in the 1950s to the Atlas 5 series heavy satellite launcher 0131864

Specifications

Atlas V 400 series

First launch: 21 August 2002
Launch sites: CCAFS LC-41, Vandenberg AFB SLC-3E
Principal users: Heavy and medium class telecom and meteorological satellites into GTO; US military.
Performance:
LEO (185 × 185 km, 28.5°): 12,500 kg (no solids)
GTO (167 × 35,786 km, 27°): 4,950 kg, no solids; 5,950 kg one solid; 6,830 kg two solids; 7,640 kg three solids.
Number of stages: 2
Guidance: Honeywell Inertial Navigation Unit mounted on Centaur forward equipment module
Launch sequences into GTO 27° (sec):

00.0	Liftoff
236	Core engine cutoff
241	Core jettison
251	Centaur main engine start
259	Payload fairing jettison
937	Centaur main engine cutoff
1,162	Start turn to Centaur second start
1,502	Centaur second start
1,720	Centaur second cutoff
1,730	Start turn to separation attitude
1,945	Spacecraft separation
2,045	Start turn to CCAM attitude
4,070	Centaur end of mission

Atlas V 400 stage 1 Common Core Booster (CCB)

Comment: The CCB is a new, bigger, core stage using the RD-180 for propulsion and including options for up to five solid rocket boosters. The LOX and RP-1 tanks are structurally stable, rather than being pressure-stabilised like the earlier Atlas boosters, and are made from welded barrel sections. Each main tank barrel is made from four isogrid aluminium panels.
Engines: One dual-chamber Pratt & Whitney/NPO Energomash RD-180 booster engine with eight solid propellant retro-rockets for separation.
Overall length: 33 m
Principal diameter: 3.81 m
Inert mass: 20,743 kg
Propellant: 284,089 kg LOX + RP-1 (aft)
Stage thrust: 3,827 kN (sea level); 4,152 kN (vacuum)
Burn time: 236 s

Atlas V 400 Solid Rocket Boosters

Comment: Atlas 500 can carry up to five Aerojet SRBs with the 5 m diameter large fairing but the 400 series with the 4 m diameter fairing can carry up to three SRBs. The SRBs are ground lit.
Overall length: 20.4 m
Principal diameter: 1.5 m
Fuelled mass: 46,000 kg
Thrust: 1,361 kN (each)
Impulse: 275 s
Nozzle cant: 3°
Burn time: 99 s

Atlas 5 Centaur Interstage Adaptor

Length: 3.13 m
Diameter: 3.05 m
Mass: 374 kg
Structure: Aluminium lithium skin stringer and frame

Atlas 5 CCB Conical Interstage Adaptor

Length: 1.65 m
Diameter: 3.81 m (bottom); 3.05 m (top)
Mass: 420 kg
Structure: Graphite epoxy composite with aluminium ring frames

Atlas 5 400 Stage 2

Comment: Centaur for Atlas 5 is identical to that developed for the Atlas 3 series but with extended high-performance motor nozzle.
Length: 12.68 m (with extendable nozzle)
Principal diameter: 3.05 m
Engines: One or two Pratt & Whitney RL10A-42 restartable
Propellant: 20,830 kg LH_2 + LOX
Stage thrust: 99.2 kN (single); 198.4 kN (twin)
Isp: 450.5 s
Burn time: 904 s

Atlas launch history, from 1980

	Launch	Site	Payload
1980			
453	17 January	36A	Fltsatcom 3. AC 49
454	9 February	3E	GPS Navstar 5
455	3 March	3W	NOSS 3
456	26 April	3E	GPS Navstar 6
457*	29 May	3W	NOAAB
458	30 October	36A	Fltsatcom 4. AC 57
459	6 December	36B	Intelsat 502. AC 54
460*	8 December	3W	NOSS 4
1981			
461	21 February	36A	Comstar 1D. AC 42
462	23 May	36B	Intelsat 501. AC56
463	23 June	3W	NOAA 7
464	6 August	36A	Fltsatcom 5. AC59
465	15 December	36B	Intelsat 503. AC55
466*	18 December	3E	GPS Navstar 7
1982			
467	4 March	36A	Intelsat 504. AC 58
468	28 September	36B	Intelsat 505. AC 60
469	20 December	3W	DMSP F6
1983			
470	9 February	3E	NOSS 5
471	28 March	3W	NOAA 8
472	19 May	36A	Intelsat 506. AC 61
473	10 June	3E	NOSS 6
474	14 July	3W	GPS Navstar 8
475	17 November	3W	DMSP F7
1984			
476	5 February	3E	NOSS 7
477*	9 June	36B	Intelsat 509. AC 62
478	13 June	3W	GPS Navstar 9
479	8 September	3W	GPS Navstar 10
480	12 December	3W	NOAA 9
1985			
481	12 March	3W	Geosat
482	22 March	36B	Intelsat 510. AC 63
483	29 June	36B	Intelsat 511. AC 64
484	28 September	36B	Intelsat 512. AC 65
485	8 October	3W	GPS Navstar 11
1986			
486	9 February	3E	NOSS 8
487	17 September	3W	NOAA 10
488	4 December	36B	Fltsatcom 7. AC66
1987			
489*	26 March	36B	Fltsatcom 6. AC67
490	15 May	3E	NOSS 9
491	19 June	3W	DMSP F8
1988			
492	2 February	3W	DMSP F9
493	24 September	3W	NOAA 11
1989			
494	25 September	36B	Fltsatcom 8. AC 68
1990			
495	11 April	3W	Stacksat
496	25 July	36B	CRRES. AC69; A1
497	1 December	3W	DMSP F10
1991			
498*	18 April	36B	BS-3H. AC70; A1
499	14 May	3W	NOAA12
500	28 November	3W	DMSP F11
501	7 December	36B	Eutelsat 2 F3. AC102; A2
1992			
502	11 February	36B	DSCS 3B14. AC101; A2
503	14 March	36B	Galaxy 5. AC 72; A1
504	10 June	36B	Intelsat K. AC105; A2A
505	2 July	36A	DSCS 3B12. AC-103; A2
506*	22 August	36B	Galaxy 1R. AC71; A1
1993			
507*	25 March	36B	UFO 1. AC 74; A1
508	19 July	36A	DSCS 3B9. AC 104; A2
509	9 August	3W	NOAA 13
510	3 September	36B	UFO 2. AC75; A1
511	28 November	36A	DSCS 3B10. AC106; A2
512	16 December	36B	Telstar 401. AC108; A2AS
1994			
513	13 April	36B	GOES8. AC 73; A1
514	24 June	36B	UFO 3. AC 76; A1
515	3 August	36A	DBS 2. AC-107; A2A
516	29 August	3W	DMSP F12
517	6 October	36B	Intelsat 703. AC-111; A2AS
518	29 November	36A	Orion 1. AC 110; A2A
519	30 December	3W	NOAA 14
1995			
520	10 January	36B	Intelsat 704. AC113; A2AS
521	29 January	36A	UFO 4. AC112; A2
522	22 March	36B	Intelsat 705. AC 115; A2AS
523	24 March	3W	DMSP F13
524	7 April	36A	AMSC 1. AC 114; A2A
525	23 May	36B	GOES 9. AC77; A1
526	31 May	36A	UFO 5. AC 116; A2
527	31 July	36B	DSCS 3B-7. AC-118; A2A
528	29 August	36B	JCSat 3. AC 117; A2AS
529	22 October	36A	UFO 6. AC119; A2
530	2 December	36B	Soho.AC-121; A2AS
531	15 December	36A	Galaxy 3R. AC 120; A2A
1996			
532	1 February	36B	Palapa C1. AC126; A2AS
533	3 April	36A	Inmarsat 3F1. AC 122; A2A
534	30 April	36B	SAX. AC-78; A1
535	25 July	36B	UFO7. AC125; A2
536	8 September	36B	GE 1, AC123, Atlas 2A
537	21 November	36B	Hot Bird 2, AC 124, Atlas 2A
538	18 December	36A	INMARSAT-3 F3, AC-129, Atlas 2A
1997			
539	17 February	36B	JCSAT 4, AC-127, Atlas 2AS
540	8 March	36A	Tempo 2, AC 128, Atlas 2A
541	25 April	36B	GOES 10, AC-79, Atlas1 (final Atlas 1 launch)
542	27 July	36B	Superbird C, Atlas IIAS
543	4 September	36A	GE 3, Atlas IIAS
544	4 October	36B	EchoStar 3, Atlas IIAS
545	24 October	36A	DSCS-3. Atlas IIA
546	8 December	36B	Galaxy 8I, Atlas IIAS
1998			
547	29 January	36A	MLV-7, Atlas IIA
548	28 February	36B	Intelsat 806, Atlas IIAS
549	16 March	36A	UHF-FO 8, Atlas II
550	18 June	36A	Intelsat 805; Atlas IIAS
551	9 October	36B	Hot Bird 5, Atlas IIA
552	20 October	36A	UHF-FO 9, Atlas IIA
1999			
553	15 February	36A	JC-SAT 5, Atlas IIAS
554	12 April	36A	Eutelsat W3, Atlas IIAS
555	23 September	36A	Echostar 5, Atlas IIAS
556	23 November	36A	UF0-FO 10, Atlas IIA
557	18 December	SLC3E	Terra, Atlas IIAS
2000			
558	20 January	36A	DSCS-3, Atlas IIA
559	3 February	36B	Hispasat 1C, Atlas IIAS
560	3 May	36A	GOES L, Atlas IIA
561	24 May	36B	Eutelsat W4, Atlas IIIA; first Atlas III launch
562	30 June	SLC36	TDRS 8, Atlas IIA
563	14 July	36B	Echostar VI, Atlas IIAS
564	20 October	36A	DSCS B-11, Atlas IIA
565	6 December	36A	Satellite Data System-2, Atlas IIAS
2001			
566	19 June	36B	ICO-F2, Atlas IIAS
567	23 July	36A	GOES M, Atlas IIA
568	8 September	36B	NOSS-3, Atlas IIAS
569	11 October	36A	SDS-2, Atlas IIAS
2002			
570	21 February	SLC36	EchoStar 7, Atlas IIIB
571	8 March	36A	TDRS-I, Atlas IIA
572	21 August	SLC36	Hot Bird 6, Atlas V-401
573	18 September	SLC36	Hispasat 1D, Atlas IIAS
574	4 December	SLC36	TDRS-J, Atlas IIA
2003			
575	12 April	SLC36	Asiasat 4, Atlas IIIB
576	14 May	SLC41	HellasSat, Atlas V 401
577	17 July	SLC41	Rainbow 1, Atlas V 521
578	2 December	VAFB	NOSS-3, Atlas IIAS1
579	17 December	CCAS	UHF-FO, Atlas III
2004			
580	5 February	SLC-36A	AMC-10, Atlas IIAS
581	13 March	SLC-36A	MBSAT, Atlas III
582	15 April	SLC-36A	SuperBird 6, Atlas IIAS
583	19 May	SLC-36A	AMC-11, Atlas IIAS
584	31 August	SLC-36A	SDS-3, Atlas IIAS
2005			
585	3 February	SLC-36B	NOSS-3 Atlas IIIB
586	11 March	CCAS	Imarsat 4-F1, Atlas V-431
587	12 August	SLC-41	Mars Reconnaissance Orbiter, Atlas V-401

*denotes launch failure (June 1984 due to Centaur leak after abnormal separation shock; March 1987 due to lightning strike; April 1991 and August 1992 due to Centaur engine failures; March 1993 due to Atlas booster low thrust, payload attained inadequate low orbit). 36A/B = Canaveral launch; 3E/W = Vandenberg. AC is Atlas Centaur designator (Atlas 2 series began at #101). Launch number at left includes all Atlas vehicles flown. A1 = Atlas 1; A2 = Atlas 2; A2A = Atlas 2A; A2AS = Atlas 2AS; A2AR = Atlas 2AR

Atlas 5 500 series
First launch: 17 July 2003
Launch sites: CCAFS LC-41, Vandenberg SLC3W
Performance: LEO (185 × 185 km, 28.5°): 10,300 kg, no solids; 12,590 kg, one solid; 15,080 kg, two solid; 17,250 kg, three solid; 18,955 kg, four solid; 20,520 kg, five solid.
GTO (167 × 35,786 km, 27°): 3,970 kg, no solid; 5,270 kg, one solid; 6,285 kg, two solid; 7,200 kg, three solid; 7,980 kg, four solid; 8,670 kg, five solid.
GSO (35,786 km circ, 0°): 2,680 kg, two solid; 3,190 kg, three solid; 3,540 kg, four solid; 3,810 kg, five solid.

Atlas 5 HLV
Launch sites: CCAFS LC-41 Vandenberg SLC 3W
Principal users: Heavy and medium class telecom and meteorological satellites into GTO or GSO
Performance: LEO (185 × 185 km, 28.5°): 25,000 kg
GTO (167 × 37,786 km, 27°): 12,650 kg
GSO (37,786 km circ, 0°): 6,350 kg

Atlas 5 HLV Stage 1
Comment: CCB as developed for Atlas 3 and 5.

Atlas 5 Liquid Rocket Booster
Comment: Uniquely developed from the CCB differing only in structural attachments. All

parameters except length and mass are the same.
Overall length: 35.99 m
Inert mass: 21,902 kg

Atlas 5 Centaur Interstage Adapter
Length: 3.53 m
Diameter: 3.81 m
Mass: 1,096 kg

Atlas 5 Cylindrical Interstage Adapter
Length: 0.6 m
Diameter: 3.81 m
Mass: 278 kg

Atlas 5 HLV Stage 2
Comment: Centaur C3 as Atlas 3 and 5

Atlas early versions

Current Status
With the successful launches of the last Atlas IIAS in August 2004, and the last of only six RD-180-engined Atlas III boosters in February 2005, all Atlas versions prior to the Atlas V are now out of production and no longer offered to customers.

Specifications
Atlas I
First launch: 25 July 1990 (AC-69 CRRES)
Number launched: 11 to 1997 when the last in the series was launched
Launch sites: Canaveral pad 36B
Principal uses: Medium-class telecom and meteorological satellite payloads into GTO
Vehicle success rate: 82%
Performance: LEO (185 × 185 km, 28.5°): 5,900 kg medium fairing/5,700 kg large.
GTO (167 × 35,788 km, 27.0°): 2,375 kg medium fairing/2,255 kg large; 3Σ errors of 106 km apogee, 2.4 km perigee and 0.02° inc
Earth escape: 1,520 kg medium fairing, 1,400 kg large
Number of stages: 2½ (booster engines burn in parallel)
Overall length: 42.0 m with medium fairing, 43.9 m with large
Principal diameter: 3.05 m
Launch mass: 163,900 kg with medium fairing, 164,290 kg with large
Launch thrust: 1,953 kN sea level
Guidance: Honeywell's Inertial Navigation Unit mounted on Centaur's forward equipment module performs the inertial guidance and attitude control computations for both Atlas and Centaur. Some initial Atlas Is retained the existing Honeywell inertial unit and Teledyne flight control computer but subsequent vehicles incorporate a Honeywell ring laser gyro INU and Gulton digital data acquisition unit, saving 36 kg and enhancing reliability.

Atlas stage 1
Comment: Atlas uniquely incorporates two booster engines fired in parallel with the central sustainer until the base section is jettisoned by the release of pneumatically-actuated latches at about T+156 s/ 5.5 g longitudinal acceleration. The propellant tanks are thin-walled (0.361.04 mm), fully monocoque stainless steel separated by an ellipsoidal bulkhead. Structural integrity is maintained in flight by the pressurisation system and on the ground by either internal pressure (N_2 at 0.330.68 atmosphere) or application of mechanical stretch. The tank design is driven primarily by the 'max α Q condition', the angle of attack due to jet stream winds near the trajectory's maximum dynamic pressure. Atlas separation is ensured by eight solid propellant retros around the base firing angled at 40° to the vertical to prevent spacecraft contamination. An external Atlas pod houses range safety, propellant utilisation, pneumatics and instrumentation.
Engines: Rocketdyne MA-5 propulsion system of two booster, one sustainer and two vernier single start liquid bipropellant engines. Each booster yields 839.5 kN sea level, the sustainer 269 kN sea level, and the verniers 2,975 N each. The engines are gimballed hydraulically to provide three-axis control during Atlas burn.
Overall length: 22.16 m
Principal diameter: 3.05 m
Propellant: 137.53 t LOX (forward tank) + RP1
Burn time: 156 s boosters, 266 s sustainer

Interstage adaptor
Mass: 477 kg
Dimensions: 3.96 m long, 3.05 m diameter
Comment: The ISA supports Centaur until separation at about 268 s is effected by a flexible linear shaped charge around the forward circumference. Construction is an aluminium skin/stringer and frame.

Fairing/payload accommodation
Two fairing designs are available for spacecraft protection during ascent: 4.19 m diameter 12.22 m long 2,005 kg, or 3.30 m diameter 10.36 m long 1,375 kg mass. Usable diameters are 365.0 cm and 292.1 cm, respectively. Both employ an aluminium skin/stringer/frame structure and non-contaminating pyro separation bolts for jettison in halves at about 205 s before sustainer engine cut-off when the heating rate has reduced to 1,135 W/m². Four 762 mm² doors in the aft section provide payload access. The highest steady state longitudinal acceleration occurs at Atlas booster engine cut-off. It is typically 5.5 g but can be reduced with some loss of performance. Laterally, the highest steady state acceleration is 0.4 g caused by the vehicle reacting to winds. During the assembled payload/ fairing transport to the pad for vehicle mating, air conditioning can provide a 15 to 25°C environment around the spacecraft. On the pad, air with a 2°C dew point is used until 2 h before launch, when a switch to N_2 is made. Gas is provided to class 10,000 (5,000 available) with 1,029°C inlet temperature.
Acceleration load: 5.5 g maximum longitudinal, 0.4 g lateral
Acoustic: max 138.9 dB overall.
Atlas provides five basic payload adaptors compatible with the Ariane 937B (Atlas type A), 937A (A1), 1194A (B, B1) and 1666A (D, Hughes HS 601) interfaces, and two (C1, C2) providing bolted interfaces for satellite adapters or mission peculiar requirements. The numerical designators indicating diameter in millimetres. Separation is effected by springs following V-clamp release commanded by Centaur's guidance system. The upper stage's sequence control unit can also issue up to 10 control commands to the spacecraft.

Payload Processing
Cape Canaveral payload processing facilities, including those of Astrotech Space Operations in nearby Titusville, are described under the Boeing Delta entry. Each pad at complex 36 includes a 64 m Mobile Service Tower, 50 m Umbilical Tower and a Blockhouse for launch operations. The Atlas is erected on the pad typically 71 days before launch, followed by Centaur (69 days), its insulation panels and forward equipment module (45 days). Payload separation tests and terminal countdown demonstration are held 40 days and 29 days before launch, respectively. The spacecraft is encapsulated at one of the remote safe facilities and the assembly installed atop Centaur at 13-8 days. Atlas' RP1 fuel is loaded at 3 days. A countdown typically requires 910 h, including two 30 min and 5 min built-in holds. Centaur's LOX is loaded at T–75 min, followed by Atlas' LOX at 55 min and Centaur's LH_2 at 43 min. The booster and its upper stage switch to internal power at 4 min and 2 min, respectively.

Atlas II/IIA/IIAS
First launch: 7 December 1991
Number launched: 63, last in August 2004
Launch sites: Canaveral pads 36A/B, VAFB SLC 3E from 1998 for access to Sun-synchronous, polar and 63.4° orbits for military and Earth observation satellites
Principal uses: Delivery of DSCS- 3 satellites and commercial communications satellites; into GTO; high-inclination missions, including US navy reconnaissance satellites
Vehicle success rate: 100%
Performance: LEO (185 × 185 km, 27.0° Canaveral): 6,780 kg medium fairing, 6,580 1kg large
LEO (185 × 185 km, 90.0° VAFB): 5,510 kg large
GTO (160 × 35,788 km, 28.5°): 2,950 kg medium fairing, 2,810 kg large
Earth escape: 2,000 kg large fairing
Number of stages: Same as Atlas I
Overall length: 46.8 m with medium fairing; 47.4 m with large fairing
Principal diameter: 3.05 m
Launch mass: 187,170 kg with medium fairing, 187,560 kg with large
Launch thrust: 2,159 kN sea level
Guidance: see Atlas I

Atlas II stage 1
Comment: To complement the higher performance MA-5A engines, stage 1 tanks were stretched 1,702 mm (LOX) plus 1,041 mm (RP-1), adding 2,585 kNs total impulse.
Engines: Rocketdyne MA-5A single start liquid bipropellant consisting of two booster engines (1,841 kN total SL) and one sustainer (269 kN sea level). The sustainer remains unchanged from that of Atlas 1 but the booster portion now uses the Delta RS-27 engine for an SI increase from 259.1 s to 263.1 s sea level. The side mounted verniers have been deleted; hydrazine thruster modules on the interstage adaptor have replaced their roll control function.
Overall length: 24.9 m
Principal diameter: 3.05 m
Propellant: 156,260 kg of LOX/RP1
Stage thrust: 2,159 kN sea level
Burn time: 169 s boosters, 277 s sustainer

Interstage Assembly
The ISA is similar to that on Atlas I but two hydrazine thruster modules similar to those on Centaur provide roll control.

Atlas II stage 2
Comment: Centaur's LOX/LH_2 tanks are stretched 338/577 mm to increase propellant loading by 2,990 kg. Foam panels bonded to the tanks replace the four jettisonable insulation panels of Atlas 1.
Engines: as Atlas I but oxidiser/fuel ratio increased from 5.0 to 5.5
Overall length: 10.06 m
Principal diameter: 3.05 m
Propellant: 16,780 kg LOX (aft) + LH_2
Stage thrust: 146.8 kN vacuum
Burn time: typically as Atlas I

Fairing/payload accommodation
Same as Atlas 1.

Atlas IIA
First launch: 10 June 1992 (Intelsat K)
Number launched: 23 to end-2002
Vehicle success rate: 100%
Performance: LEO (185 × 185 km, 28.5° Canaveral): 7,316 kg large.
LEO (185 × 185 km, 90.0° VAFB): 6,190 kg large.
GTO (167 × 35,788 km, 27.0°): 3,040 kg medium, 2,900 kg large; Block 1 upgrade: 3,180 kg medium fairing, 3,066 kg large.
Earth escape: 2,160 kg large fairing

Atlas IIAS
First launch: 15 December 1993
Number launched: 30 to August 2004
Vehicle success rate: 100%
Performance: LEO (185 × 185 km, 28.5° Canaveral): 8,610 kg large fairing.
LEO (185 × 185 km, 90.0° VAFB): 7,210 kg large fairing.
GTO (167 × 35,788 km, 27.0°): 3,730 kg medium, 3.620 kg large; Block 1 enhancement: 3,810 kg medium, 3,700 kg large.
Earth escape: 2,680 kg large fairing

Castor 4A strap-ons
Motor contractor: ATK Thiokol, Huntsville division. TX7803 ground-ignited (quads 1 + 3), TX7804 air-ignited (quads 2 + 4) to minimise umbilical tower plume impingement
Length: 11.28 m
Diameter: 1,016 mm
Average thrust: each 433.6 kN sea level
Burn time: 53 s
Mass at ignition: 11,649 kg (ground), 11,743 kg (air)
Burnout mass: 1,452 kg (ground), 1,529 kg (air)
Propellant: 10,200 kg of HTPB solid

Atlas IIIA
First launch: 24 May 2000
Launch sites: Canaveral pad 36B
Principal uses: Medium-class telecom payloads into GTO.
Performance: LEO (185 × 185 km, 28.5°); 8,641 kg.
GTO (167 × 35,788 km, 27.0°): 3,810 kg (4,060 kg with two Castor 4Bs); 3Σ errors unavailable.
Availability: typically four launches/year per pad. Launch typically 24 to 30 months following contract
Cost: about USD45 million into GTO
Number of stages: 2
Guidance: Honeywell's Inertial Navigation Unit mounted on Centaur's forward equipment module performs the inertial guidance and attitude control computations for both Atlas/Centaur
Launch sequence (s) into GTO 27.0°:

00.0	stage 1 ignite (75 per cent thrust)
04.8	stage 1 throttle up to 85 per cent, 30 m
12.9	start pitch programme, 290 m
24.4	ramp to zero angle of attack
32.4	zero angle attack, 2.44 km
44.2	pre-max Q throttle down to 61 per cent
67.6	post-max Q throttle up to 85 per cent
147.9	max 5.0 g, begin throttle down
184.9	stage 1 cut-off/separation (49 per cent throttle)
199.9	Centaur ignition 1

209.9	jettison payload fairing
760.2	Centaur cut-off 1
1,449.9	Centaur ignition 2
1,625.3	Centaur cut-off 2

Atlas IIIA stage 1
Comment: The two-chamber throttleable RD-180 eliminates the booster section and provides a more benign environment. Max thrust 85%; 49% at cut-off. The LOX tank is extended by 2.60 m to accommodate the change in mixture ratio
Engines: NPO Energomash/P&W RD-180 two chamber engine throttleable 40–105%, nominal 3,827 kN SL/4,152 kN vacuum. The chambers are gimballed hydraulically to provide three-axis control during Atlas burn
Overall length: 28.91 m
Principal diameter: 3.05 m
Inert mass: 13,725 kg
Propellant: 183,200 kg LO_2 + RP1 (aft)
Stage thrust: 3,827 kN (sea level); 4,152 kN (vacuum)
Burn time: 184 s

Interstage adapter
Comment: The 4.42 m long, 3.05 m diameter ISA supports Centaur until separation at 185 s is effected by a flexible linear shaped charge around the forward circumference. Construction is an aluminium skin/stringer and frame

Atlas IIIA stage 2
Comment: Essentially the same Centaur stage as that used by the Atlas IIAS but with a single-engine RL10A-41
Overall length: 10.06 m
Principal diameter: 3.05 m
Propellant: 16,930 kg LH_2 + LOX (aft)
Stage thrust: 99.2 kN
Isp: 450.5 s
Burn time: ~ 651 s

Fairing/payload accommodation
Comment: IIIA used the Block 2 fairing upgrade, introduced on IIAS. Other specifications as for earlier models
Acceleration load: 5.0 g maximum longitudinal, 0.4 g lateral
Acoustic: unavailable
Max Q: 427 N/cm^2

Atlas IIIB
First launch: 21 February 2002
Launch site: Canaveral pad 36B
Performance: LEO (185 × 185 km, 28.5° Canaveral): 10,718 kg (extended fairing), 10,759 kg (large fairing)
GTO (167 × 35,786 km, 27.0°): 4,119 kg (single-engine Centaur), 4,400 kg (large fairing) or 4,477 kg (extended fairing) both with twin-engine Centaur
Number of stages: Two
Overall length: 54.5 m
Diameter: 3.05 m
Launch mass: 225,392 kg
Launch thrust: 3,827 kN sea level
Guidance: From upper stage

Atlas IIIB stage 1
Comment: The two-chamber throttleable RD-180 identified for the Atlas IIIA is the same as that used for the Atlas IIIB and all stage parameters are the same.

Atlas IIIB stage 2
Comment: The Centaur stage can be fitted with single- or double-engine configuration, each designated RL10A4-2. In either case the extended length Centaur is standard with an additional 180 kg inert mass for the double engine version
Engines: One or two Pratt & Whitney RL10A-42 restartable engines
Overall length: 11.74 m
Principal diameter: 3.05 m
Propellant: 20,830 kg LH_2 + LOX (aft)
Stage thrust: 99.2 kN (single), 198.4 kN (twin)
Burn time: ~ 907 s (single); 354 s (twin)
Isp: 450.5 s

Delta series

Current Status
Pinnacle of a development success story almost 45 years old, the latest generation of Delta launch vehicles successfully carried the Eutelsat W5 satellite into orbit when the first Delta IV lifted off from LC-37B on 20 November 2002. By early 2006,

Delta launches (from 1978)

No	Payload	Date	Pad	Type
138	IUE	26 January 1978	17A	2914
139	Landsat 3/Oscar	5 March 1978	2W	2910
140	BSE	7 April 1978	17B	2914
141	OTS 2	11 May 1978	17A	3914
142	GOES 3	16 June 1978	17B	2914
143	Geos 2	14 July 1978	17A	2914
144	ISEE3	12 August 1978	17B	2914
145	Nimbus 7	24 October 1978	2W	2910
146	NATO 3C	19 November 1978	17B	2914
147	Anik 4	16 December 1978	17A	3914
148	Scatha	30 January 1979	17B	2914
149	Westar 3	10 August 1979	17A	2914
150	RCA Satcom 3	7 December 1979	17A	3914
151	Solar Max	14 February 1980	17A	3910
152	GOES 4	9 September 1980	17A	3914
153	SBS 1	15 November 1980	17A	3910/PAM
154	GOES 5	22 May 1981	17A	3914
155	Dyn Exp 1/2	3 August 1981	2W	3913
156	SBS 2	24 September 1981	17A	3910/PAM
157	SME + Uosat 1	6 October 1981	2W	2310
158	RCA Satcom 3R	20 November 1981	17A	3910/PAM
159	RCA Satcom 4	16 January 1982	17A	3910/PAM
160	Westar 4	26 February 1982	17A	3910/PAM
161	Insat 1A	10 April 1982	17A	3910/PAM
162	Westar 5	9 June 1982	17A	3910/PAM
163	Landsat 4	16 July 1982	2W	3920
164	Anik D1	26 August 1982	17B	3920/PAM
165	RCA Satcom 5	28 October 1982	17B	3924
166	IRAS	26 January 1983	2W	3910
167	RCA Satcom 6	11 April 1983	17B	3924
168	GOES 6	28 April 1983	17A	3914
169	Exosat	26 May 1983	2W	3914
170	Galaxy 1	28 June 1983	17B	3920/PAM
171	Telstar 3A	28 July 1983	17A	3920/PAM
172	RCA Satcom 7	8 September 1983	17B	3924
173	Galaxy 2	22 September 1983	17A	3920/PAM
174	Landsat 5 + UoSat 2	1 March 1984	2W	3920
175	AMPTE	16 August 1984	17A	3924
176	Galaxy 3	21 September 1984	17B	3920/PAM
177	NATO 3D	13 November 1984	17A	3914
178*	GOES G	3 May 1986	17A	3914
180	DoD (SDI)	5 September 1986	17B	3920
179	GOES 7	26 February 1987	17A	3924
182	Palapa B2P	20 March 1987	17B	3920/PAM
181	DoD (SDI)	8 February 1988	17B	3910
184	Navstar 21	14 February 1989	17A	6925
183	Delta Star	24 March 1989	17B	3920
185	Navstar 22	10 June 1989	17A	6925
186	Navstar 23	18 August 1989	17A	6925
187	Marcopolo 1	27 August 1989	17B	4925
188	Navstar 24	21 October 1989	17A	6925
189	COBE	18 November 1989	2W	5920
190	Navstar 25	11 December 1989	17B	6925
191	Navstar 26	24 January 1990	17A	6925
192	RME/LACE	14 February 1990	17B	69208
193	Navstar 27	26 March 1990	17A	6925
194	Palapa B2R	13 April 1990	17B	69258
195	Rosat	1 June 1990	17A	692010
196	Insat 1D	12 June 1990	17B	4925
197	Navstar 2–8	2 August 1990	17A	6925
198	Marcopolo 2	18 August 1990	17B	6925
199	Navstar 29	1 October 1990	17A	6925
200	Inmarsat 2 F1	30 October 1990	17B	6925
201	Navstar 210	26 November 1990	17A	7925
202	NATO 4A	8 January 1991	17B	7925
203	Inmarsat 2 F2	8 March 1991	17B	6925
204	ASC 2	13 April 1991	17B	7925
205	Aurora 2	29 May 1991	17B	7925
206	Navstar 211+ Losat	4 July 1991	17A	7925
207	Navstar 2–12	23 February 1992	17B	7925
208	Navstar 2 13	10 April 1992	17B	7925
209	Palapa B4	14 May 1992	17B	7925 8
210	EUVE	7 June 1992	17A	6920-10
211	Navstar 2 14	7 July 1992	17B	7925
212	Geotail	24 July 1992	17A	6925
213	Satcom C4	31 August 1992	17B	7925
214	Navstar 2 15	9 September 1992	17A	7925
215	DFS 3	12 October 1992	17B	7925
216	Navstar 2–16	22 November 1992	17A	7925
217	Navstar 2 17	18 December 1992	17B	7925
218	Navstar 218	3 February 1993	17A	7925
219	Navstar 2–19/SEDS	30 March 1993	17A	7925
220	Navstar 220	12 May 1993	17A	7925
221	Navstar 2–21	26 June 1993	17A	7925
222	Navstar 222	30 August 1993	17B	7925
223	Navstar 2 23	26 October 1993	17B	7925
224	NATO 4B	8 December 1993	17A	7925
225	Galaxy 1RR	19 February 1994	17B	79258
226	Navstar 224/SEDS	10 March 1994	17A	7925
227	Wind	1 November 1994	17B	792510
228***	Koreasat 1	5 August 1995	17B	7925
229	Radarsat	4 November 1995	2W	792010
230	XTE	30 December 1995	17A	792010
231	Koreasat 2	14 January 1996	17B	7925
232	NEAR	17 February 1996	17	7925

No	Payload	Date	Pad	Type
233	Polar	24 February 1996	2W	7925 10
234	Navstar 225	28 March 1996	17	7925
235	MSX	24 April 1996	2W	7920 10
236	Galaxy 9	24 May 1996	17B	7925
237	Navstar 226	16 July 1996	17	7925
238	Navstar 227	12 September 1996	17	7925
239	Mars Global Surveyor	7 November 1996	17	7925
240	Mars Pathfinder	4 December 1996	17	7925
241*	Navstar 2R1	17 January 1997	17	7925
242	Iridium 4 8	5 May 1997	17	7920
243	Thor 2A	20 May 1997	17	7925
244	Iridium (×5)	9 July 1997	17	7920
245	Navstar 29	23 July 1997	17	7295
246	Iridium (×5)	21 August 1997	17	7920
247	ACE	25 August 1997	17	7920
248	Iridium (×5)	26 September 1997	17	7920
249	Navstar 38	6 November 1997	17	7925
250	Iridium (×5)	9 November 1997	17	7920
251	Iridium (×5)	20 December 1997	17	7920
252	Skynet 4D	10 January 1998	17	7925
253	Globalstar (×4)	14 February 1998	17	7420
254	Iridium (×5)	18 February 1998	17	7920
255	Iridium (×5)	30 March 1998	17	7920
256	Globalstar (×4)	24 April 1998	17	7420
257	Iridium (×5)	17 May 1998	17	7920
258	Thor 3	15 June 1998	17	7925
259*	Galaxy 10	27 August 1998	17	Delta 3
260	Iridium (×5)	8 September 1998	17	7920
261	Deep Space-1/SEDSAT	24 October 1998	17	7326
262	Iridium (×5)	6 November 1998	17	7920
263	Bonum 1	23 November 1998	17	7925
264	Mars Climate Orbiter	11 December 1998	17	7424
265	Mars Polar Lander	27 January 1998	17	7425
266	Stardust	7 February 1999	17	7426
267	Argos/Orsted/Sunsat	23 February 1999	17	7920
268	Landsat 7	15 April 1999	17	7920
269*	Orion 3	4 May 1999	17	8930
270	Globalstar 3	10 June 1999	17	7925
271	FUSE	24 June 1999	17	7925
272	Globalstar 4	10 July 1999	17	7925
273	Globalstar 5	25 July 1999	17	7925
274	Globalstar 6	17 August 1999	17	7925
275	GPS	7 October 1999	17	7925
276	Globalstar 7	8 February 2000	17B	7925
277	IMAGE	25 March 2000	SLC2W	7925
278	GPS	11 May 2000	17A	7925
279	GPS	16 July 2000	17A	7925
280	DMF3	23 August 2000	17B	8930
281	GPS	10 November 2000	17A	7925
282	F01	21 November 2000	SLC2W	7320
283	GPS	30 January 2001	17A	7925
284	Mars Odyssey	7 April 2001	17A	7925
285	Geolite	18 May 2001	17B	7925
286	MAP	30 June 2001	17B	7425
287	Genesis	8 August 2001	17A	7326
288	QuickBird	18 October 2001	SLC2W	7320
289	Jason 1/ Timed	7 December 2001	SLC2W	7920-10
290	Iridium (×5)	11 February 2002	SLC2W	7920
291	Aqua	4 May 2002	SLC2W	7920-10L
292	Contour	3 July 2002	17	7425
293	Eutelsat	20 November 2002	37B	4M
294	ICESat/CHIPSat	12 January 2003	2W	7320
295	Global Positioning System (GPS IIR-8) & XSS-10	29 January 2003	17	7925
296	DSCS III A3	10 March 2003	37	Delta IVM
297	Global Positioning System (GPS IIR-9)	31 March 2003	17	7925
298	Mars Exploration Rover-A	10 June 2003	17	7925
299	Mars Exploration Rover-B	7 July 2003	17	7925
300	NASA SIRTF	25 August 2003	17	7920
301	DSCS III B6	29 August 2003	37	Delta IVM
302	Global Positioning System (GPS IIR-10)	21 December 2003	17	7925
303	Global Positioning System (GPS IIR-11)	20 March 2004	17	7925
304	NASA Gravity Probe-B	20 April 2004	2W	7920
305	Global Positioning System (GPS IIR-12)	28 June 2004	17	7925
306	NASA Aura	15 July 2004	2W	7920
307	NASA Messenger	3 August 2004	17	7925
308	Global Positioning System (GPS IIR-13)	6 November 2004	17	7925
309	NASA Swift	20 November 2004	17	7320
310	Delta IV Heavy Demonstration Flight**	21 December 2004	37	Delta IVH
311	NASA Deep Impact	12 January 2005	37	7925
312	NASA NOAA-N	20 May 2005	2W	7320
313	Global Positioning System (GPS IIR-14 M)	25 September 2005	17	7925

*indicates launcher failure
**Partial success
***Koreasat 1's Delta did not achieve GTO because number 6 GEM failed to separate; the satellite's own station-keeping propellant was required to reach final GEO. The pad designator indicates launch site; 17A/B = Cape Canaveral; 2E/W = Vandenberg AFB.

however, there had only been three further Delta IV flights, one of them the first and partially successful flight of the Delta IV Heavy configuration, and in July 2003 Boeing announced that it would concentrate Delta IV efforts on the government market. At that time, Boeing had 24 contracted and awarded flight bookings on the Delta IV, but lost seven of those as a result of the Pentagon's investigation of alleged business misconduct in the programme.

In 2005, Boeing and Lockheed Martin announced plans to merge the Atlas V and Delta IV programs, both developed under the USAF's Evolved Expendable Launch Vehicle (EELV) project, into a new joint venture called the United Launch Alliance. This proposal was still awaiting approval in early 2006.

Boeing's Delta II rocket remains in production as the standard launcher for the Pentagon's Global Positioning Satellite (GPS) programme, until 2007, and as the most widely used launcher for NASA missions. A Delta II launch in September 2005 was the 42nd consecutive success for USAF Delta launches, a new record.

Over its 40-year life span, the Delta family of expendable launch vehicles has racked up what is perhaps the most successful flight record of any rocket currently in service. Of 309 flights, only 15 have been total failures, a success rate of 95.1 per cent.

Background

In 1959, NASA's Goddard Space Flight Center contracted the Douglas Aircraft Company to develop an interim space launcher from the company's Thor IRBM and Vanguard upper stages programmes. Douglas had been working on the Thor since 1954 for the USAF. NASA envisioned a vehicle capable of putting about 45 kg into GTO. As the space race matured, Delta's growth potential and the need for a medium launcher for communications, meteorological and science satellites paced a continuous set of improvements to the original vehicle. The first Delta launch took place on 13 May 1960 but a second stage malfunction prevented the vehicle from reaching orbit. On the second launch, on 12 August 1960, the Delta successfully launched NASA's Echo passive communications satellite. The basic Delta was used between 1960 and 1962 successfully launching Tiros weather satellites, scientific research satellites and the first UK satellite Ariel 1. In October 1962 the Delta A was introduced with the uprated MB-3 first stage motor and a payload capability of 320 kg with the Delta B incorporating increased tankage and a modified second stage motor increasing payload capacity to 375 kg.

Delta became a McDonnell Douglas product in 1967, when the St Louis company acquired Douglas.

Between 1962 and the 1980, the Delta was steadily evolved into a medium lift launch vehicle for both military and commercial missions. Strap-ons were first added in 1964, and in 1965 the booster propellant capacity was increased, a larger fairing was introduced and both upper stages were upgraded. The number of strap-ons was increased to six in 1970 and to nine in 1972, at the same time as Aerojet's Titan Transtage engine was adopted for stage 2. Further improvements were made in 1973 with the introduction of Rocketdyne's RS-27 first-stage engine and in 1975 with the introduction of Thiokol's Castor 4 strap-ons to replace the Castor 2s.

The development of the contemporary Delta began in January 1987, when the USAF selected McDonnell Douglas to develop the Medium Launch Vehicle (MLV), with the primary mission of launching the Global Positioning System (GPS) satellite constellation. The new Delta variant featured more powerful solid rocket strap-on boosters, a longer first-stage fuel tank and a larger payload fairing, and was identified as Delta II. The first Delta II was launched on 14 February 1989 and the last of the earlier Deltas was launched in June 1990. (Two interim vehicles – the Delta 4925 and 5920 – helped fill the gap between the 'classic' 3000-series Delta and the Delta II – but only two 4895s and a single 5920 were flown).

The first nine Delta IIs used metal-cased Thiokol Castor IV solid rockets, but subsequent vehicles – designated in the 7000 series – use graphite-epoxy-case GEM-40 motors from the same company.

In 1996, Boeing acquired Rockwell – including the Rocketdyne division, which produces the Delta's engine. (Boeing sold Rocketdyne to Pratt & Whitney in 2005.) In 1997, Boeing also acquired McDonnell Douglas, and Delta became a Boeing product, alongside the Boeing/Energia Sea Launch programme. Thiokol, producer of the GEM-40 solid rocket boosters, is now part of ATK.

McDonnell Douglas had started development of the Delta III or 8000 series in 1995, as a means of launching larger commercial satellites. Compared to the Delta II, it accommodated larger solid boosters and a new cryogenic upper stage. McDonnell Douglas had considered offering the Delta III for the USAF's EELV effort, which started in 1995, but it became clear that the vehicle did not have sufficient growth potential. Instead, the company was successful in proposing the almost entirely new Delta IV for the EELV requirement.

The first two Delta III launch attempts – on 26 August 1998 and 5 May 1999 – were both failures, and the third launch on 23 August 2000 was a partial success. Since Boeing was already active in the heavy-commercial market with Sea Launch, and Delta II and Delta IV were expected to meet US government requirements, Delta III was abandoned.

Delta II

Current Status

Delta II continues to serve customers until the Delta IV replaces it. Delta IV is expected to take over Air Force launches in 2007, but the rocket remains on offer for NASA. Although there were only three launches in 2005, there were seven in 2004 and seven in 2003. NASA has issued contracts to Boeing covering Delta II vehicles and launch services through 2010. The USAF is continuing to launch remaining GPS IIR satellites on Delta II, although the follow-on GPS IIF will fly on EELV. Reliability for Delta II has been 99.1 per cent since inception in 1989, or 100 per cent in the last 50 launches.

Background

NASA ordered Delta's Huntington Beach production line to close down in 1984 but Shuttle's failure in January 1986 resulted in its reactivation eight months later to produce three further vehicles (#182 to 184) under a USD130 million contract. Delta's subsequent emergence as a commercial vehicle resulted from its selection as USAF's Medium Launch Vehicle (MLV) in January 1987 under an initial USD316.5 million contract. The basic buy of seven eventually expanded to 20 when the air force exercised two options in February 1988 (seven vehicles) and December 1988 (six vehicles), totalling USD680 million, to launch the network of GPS Navstar satellites that were originally planned for Shuttle launch.

This version became known as Delta II. Like earlier rockets in the family, Delta II launchers are identified by a four-digit designation. The first digit indicates the type of first stage engine and solid rocket motors, the second digit indicates the number of solid rocket motors, the third digit indicates the type of second stage and the fourth digit indicates type of third stage. All current Delta IIs use the RS-27A first-stage engine and GEM-40 boosters, identified as the 7000 series, and the storable AJ10-118K second stage (identified by the digit 2).

Delta 7925 GPS launch from Canaveral's pad 17A 0517179

Delta became associated with civil science again when the USD140.6 million contract for the Geotail, Wind and Polar science satellites was signed with NASA in December 1990. It included VAFB pad modifications and pre-priced options for 12 further vehicles to mid-1998. The previous 3920/PAM models could handle 1,284 kg to GTO, but the Delta II 6925 raised that capacity to 1,447 kg, and the 7925 to 1,841 kg.

USAF Delta II requirements called for a switch to the more capable 7925 following the first nine launches to cope with the heavier GPS Navstar Block IIA models. The 7925 incorporates the same 2.9 m diameter fairing and 3.6 m longer first stage as the 6925, but adds graphite epoxy strap-ons and an uprated first-stage engine. 6925 made its last flight with Geotail in 1992.

On the 7925-series Delta, GPS performance is improved to 2,141 kg and GTO to 1,882 kg, principally by increasing the expansion ratios of the three air-lit strap-ons. It employs the triply redundant Redundant Inertial Flight Control Assembly (RIFCA) and the improved Flight Termination System (FTS). The fairing is modified so that the satellite, stage 3 and fairing can be transported to the pad as a single unit.

A Delta II Heavy configuration was first flown in July 2003. It uses larger-diameter GEM-46 solid boosters originally developed for the Delta III.

Although the majority of Delta IIs are USAF-type 7925 models, a number of lower-powered models have been flown. The 7420 vehicle, with four solids, was designed to deliver four GlobalStar satellites on a single launch for that company's low-earth-orbit communications network. Seven successful launches were carried out between 1998 and 2000.

The 7425 version, with four solids and a Star-48B solid-propellant kick stage, was developed to meet NASA requirements and was used for a number of scientific and planetary missions. The three-solid 7320 has been used for NASA scientific missions, two such vehicles having flown in 2005.

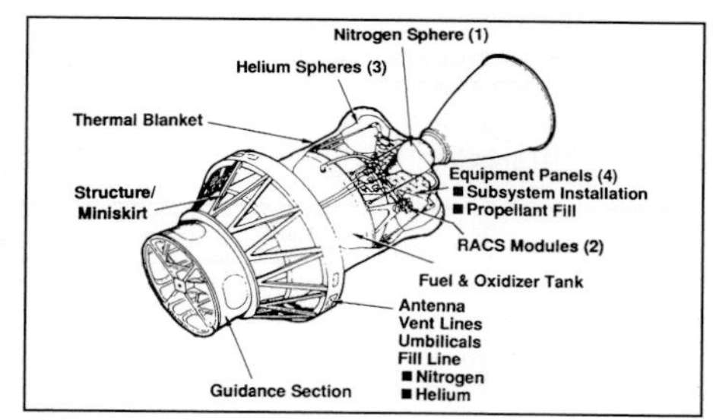

Delta second stage features (MDA) 0516932

From 2000, the bulk of Delta II production was moved to Decatur, alongside the new Delta IV. Delta II had problems with the launch of Koreasat 1: on this flight one of the GEM solid propellant boosters failed to separate. Orbit was reached, but it was lower than planned and the satellite had to use its own propellant to reach geosynchronous orbit, thus reducing its operational life. The first complete loss of a Delta II came on 17 January 1997 when a 7925 was destroyed seconds after launch, this was due to a malfunction of one of the strap-on boosters. The payload – the first Navstar IIR satellite – was destroyed.

Specifications

Delta II/Delta 7925/7920

First launch: 14 February 1989

Number launched: 120 to 1 Feburary 2006

Launch sites: Canaveral pads 17A/B, VAFB SLC2W.

Principal uses: launches of GPS satellite; medium class telecom satellites into GTO.

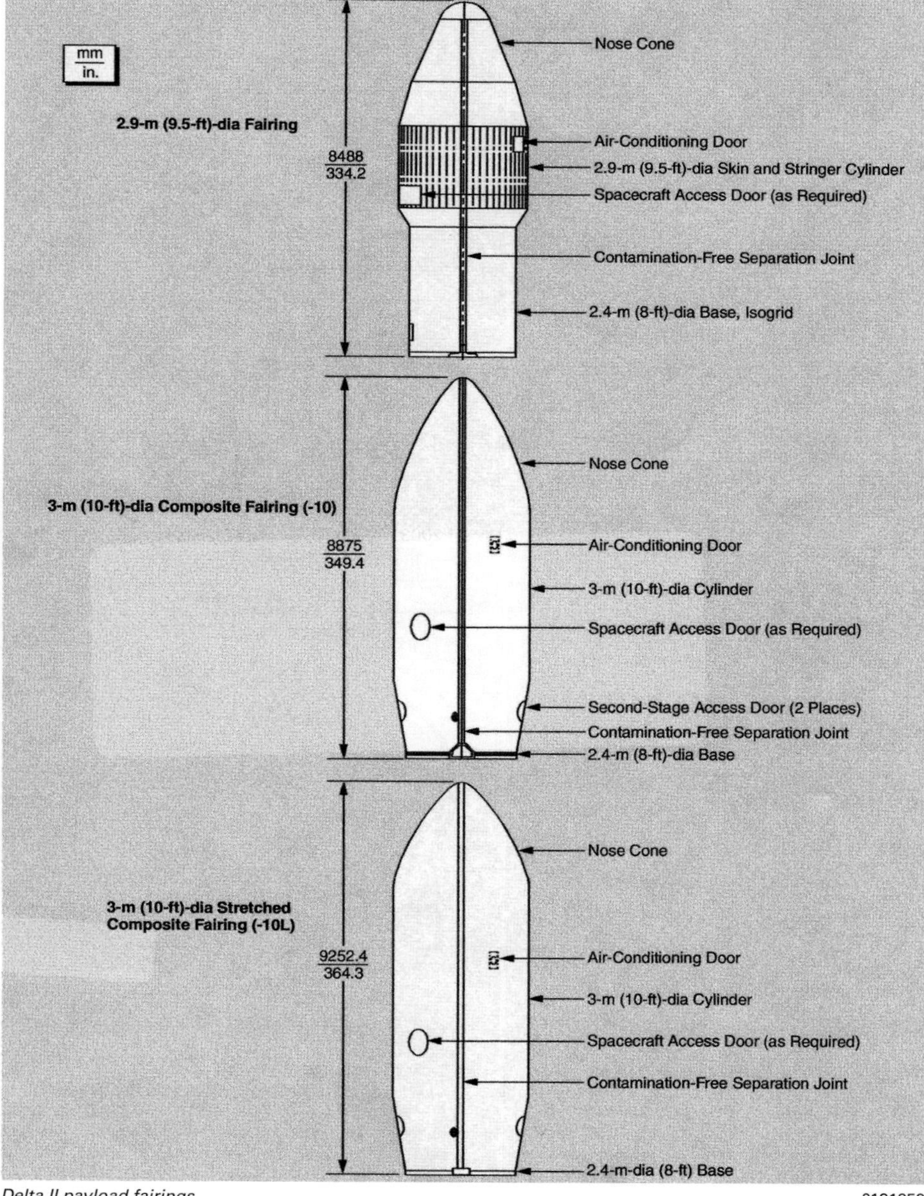

Delta II payload fairings 0131859

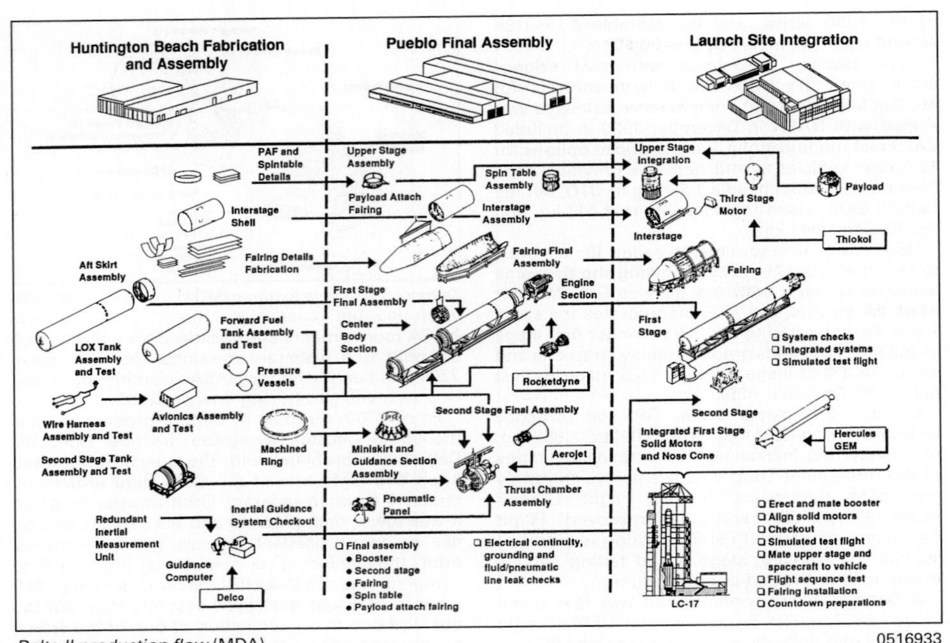

Delta II production flow (MDA) 0516933

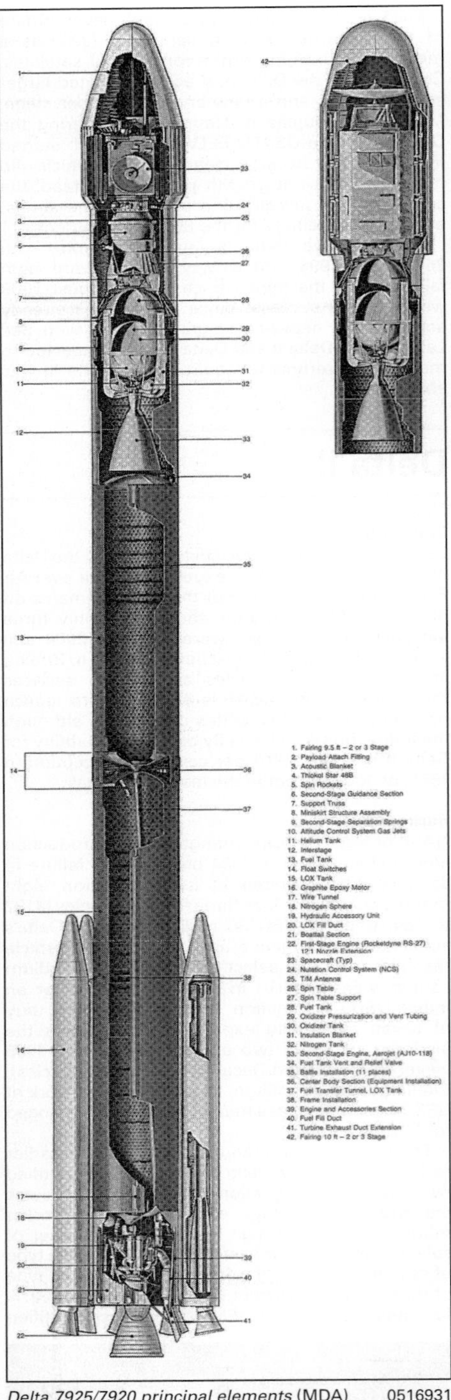

Delta 7925/7920 principal elements (MDA) 0516931

Delta II 2.9 m payload fairing 0131858

Vehicle success rate: 99.1%
Performance:
GTO: 1,869 kg (2064 kg with Heavy option)
LEO: 5,139 kg (28.7°); 5815 kg with Heavy option
Sun-synchronous: 3,175 kg, 830 km 7920 VAFB
Molniya-type (63.4°, 370 × 40,094 km): 1,275 kg
Lunar delivery: 1,240 kg
Number of stages: 3 plus 9 strap-ons (7920 excludes stage 3)

Overall length: 38.41 m
Principal diameter: 2.44 m
Launch mass: 231,870 kg
Launch thrust: 3,110 kN (six solids fire at launch)
Guidance: the digital inertial guidance system is mounted inside a cylinder at stage 2's forward end, controlling the vehicle during stage 1/2 and commanding spinup/separation of stage 3 in addition to triggering its fuze based sequencing system. The system is a strapped down all-inertial unit incorporating a Delco guidance computer and a Delta Redundant Inertial Measurement System (DRIMS) containing three gyros, four accelerometers and conditioning electronics. The computer also issues preprogrammed sequence commands and provides attitude control.

GTO perigee altitude error is typically <5.6 km + 0.6° inclination. Typical 3Σ errors for 2-stage missions into 1,851,850 km circular orbits are 2 km/0.05°. The triply redundant Redundant Inertial Flight Control Assembly (RIFCA) replaced DRIMS from the XTE launch of December 1995. The single-fault tolerant RIFCA comprises six RL 20 AlliedSignal ring laser gyros and six QA3000 Sundstrand accelerometers in a triply redundant modular architecture. Each lane contains two sensor channels, a 1750A processor and I/O interfaces. Two lanes have a 1553B bus, for vehicle data, and the third incorporates RS422 for GSE. There are two separate power sources. The data are continuously processed, updating the required course throughout ascent.

ATK GEM solid strap-ons

Comment: Hercules' Aerospace Group (now ATK) of Magna, Utah was contracted in February 1987 to provide 144 (16 flight sets) higher performance, lightweight Graphite Epoxy Motors for Delta 7925. To improve performance from 1995 onwards, the expansion ratio for the air-lit nozzles is increased

Hoisting Delta stage 2 at pad 17A (NASA) 0517180

from 10.6 to 16.3 by lengthening the nozzle by 30 cm and increasing diameter by 19.5 cm (debut was on XTE December 1995).
Length: 12.96 m
Diameter: 101.6 cm
Average thrust: each 446.02 kN SL; 499.18 kN vacuum
Burn time: 63.3 s
Mass at ignition: 13,232 kg each for six ignited on pad; 13,061 kg each for three air-lit
Propellant mass: 11,765 kg each
Propellant type: solid
Burn sequence: six GEMs ignite at launch, followed by the remaining three 2.5 s after the others burnout at 63 s. The original six are spring separated (under command of a sequencer between stage 1's tanks) in two symmetrical sets at 66/67 s; the final three burnout at 129 s and separate 3 s later.

Delta 7920/7925 stage 1
Comment: Incorporating uprated strap-ons, which provide the majority of the launch thrust, permitted increasing RS 27A main engine expansion ratio from 8:1 for improved altitude performance.
Engines: Rocketdyne 12:1 expansion ratio RS 27 single start liquid bipropellant of 890 kN sea level thrust (1,084.8 kN vacuum), hydraulically gimballed for pitch/yaw control; two Rocketdyne 4.45 kN LR101-NA11 verniers provide roll control.
Overall length: 26.047 m
Diameter: 2.44 m
Stage mass: 101,718 kg
Oxidiser: liquid oxygen
Fuel: RP-1
Propellant quantity (usable): 96,033 kg
Oxidiser tank length: 13.00 m
Fuel tank length: 8.38 m
Stage thrust: 890 kN SL + two 4.45 kN verniers
Burn time: 260.5 s
Comment: The two isogrid tanks are separated by a 77 cm centre body section that houses control electronics, ordnance sequencing and telemetry equipment. The package receives commands from RIFCA and drives servo amplifiers for engine gimbal; DC batteries in the centre section provide power. A rate gyro was added forward of this section to ensure adequate stability margins with the extended tanks and larger fairing.

Interstage
The 4.72 m-long isogrid interstage extends from the top of stage 1 to stage 2's miniskirt, encircling most of the upper stage until explosive bolt detonation 8 s after stage 1 burnout. Six spring driven separation rods at the forward end then separate the stages and stage 2 ignition occurs 5 s later.

Delta 7920/7925 stage 2
Engine: pressure-fed restartable 65:1 Aerojet AJ10 118K
Overall length: 5.89 m
Diameter: 1.70 m, suspended by 2.44 m diameter mini skirt and support truss.
Stage mass: 6,930 kg.
Oxidiser: nitrogen tetroxide.
Fuel: Aerozine-50.
Propellant mass (usable): 6,006 kg.
Thrust: 43.37 kN vacuum.
Burn time: 432 s (restartable).
Attitude control system: the nitrogen cold gas jet Redundant Attitude Control System (RACS) provides three-axis control during coast periods and roll control during powered flight; hydraulic gimballing of the main engine provides pitch/yaw control under the command of RIFCA at the forward end. Three helium spheres provide tank pressurisation. Four aft isogrid panels provide equipment attach points and tank fill ports; the gas spheres are filled through the miniskirt on pad.
Burn time: On a typical 7925 GTO mission, stage 2 will burn for 343 s to enter a low circular orbit, coast under control of RACS for 620 s and then burn near the equator for a final 70 s. For the two-stage 7920, the burns are 395/25 s, respectively. On some missions, the AJ10-118K has been restarted as many as seven times.

Delta 7925 stage 3
Motor: ATK Thiokol Star-48B or Star-37B solid
Stage length: (Star-48B) 2.03 m (Star-37B) 1.69 m
Diameter: (Star-48B) 1.24 m (Star 37B) 0.93 m
Propellant: 1,756–2,025 kg solid, depending on mission requirements.
Thrust: 66.4 kN vacuum
Burn time: 87.1 s
Mass: 2,141 kg before ignition; 132 kg after burnout.
Comment: Delta II's third stage is derived from components and concepts used on the Payload Assist Module-D (PAM-D) and the USAF SGS-2 upper stage. The Star 48B is a base supported by a spin table that mates to the top of the second stage guidance section. Spin rockets fire typically 50 seconds after stage 2 burnout to create a stabilising spin of 30,110 rpm for stage 3 firing. Four or eight spin rockets can be used in 334/823/934 N thrust versions, firing for 1 seconds. Stage 2's RIFCA triggers the stage 3 fuze based sequencing system, which produces spin-up, pyrotechnic/spring separation 2 seconds later and ignition after a further 38 seconds.

A Nutation Control System derived from SGS2 and mounted above the PAM suppresses coning during motor burn and post-burn phases; a single axis rate gyro senses the motion and fires a counteracting hydrazine 178 N thruster system (2.7 kg propellant loading). Stage 3 can also house a 3.6 kg C-band radar transponder tracking system to aid early orbit determination; if integrated with the telemetry unit it adds 2.3 kg. The latter returns spacecraft environment data such as vibration, acceleration transients, shock, velocity increment, indication of separation and temperatures.

Three stage missions adopt the standard 3712 Payload Attach Fitting at stage 3's forward end, capable of holding 3,084 kg. The spacecraft is secured by means of a two-piece V-block clamp assembly, held by two instrumented studs. Two ordnance cutters sever the studs for four springs to effect a 0.612.4 m/s separation rate. A yo-yo weight tumble system despins and imparts a coning motion to the expended motor 2 seconds after separation to modify its momentum vector and prevent recontact with the spacecraft. On Delta 6920 missions, a 6019 Payload Attach Fitting was used, designed originally for NASA's Multimission Modular Spacecraft and capable of holding 2,177 kg with three explosive nuts and a secondary latching system. Stage 2's nitrogen gas thrusters ensure a separation speed of 31 cm/s.

Delta Fairings
The Delta II launch vehicle offers the user a choice of three fairings: a 2.9 m diameter skin-and-stringer fairing and two sizes of 3 m diameter composite fairings with different lengths. Each of these fairings can be used on either two-stage or three-stage missions. The 2.9 m and standard-length 3.0 m fairings have been flight proven over many years. The new stretched-length 3.0 m (10 ft) composite fairing was developed to offer more payload volume. It has a reshaped nose cone and a cylindrical section 0.91 m longer than the standard 3 m version.

The 2.9 m fairing is an aluminum skin-and-stringer structure fabricated in two half-shells. These shells consist of a hemispherical nose cap, a biconic section, a cylindrical center section (the maximum diameter of the fairing), a 30° conical transition, and a cylindrical base section having the 2.4 m diameter of the core vehicle. The biconic section is a ring-stiffened monocoque structure; one-half of which is fibreglass covered with a removable aluminum foil lining to create an RF window. The cylindrical base section is an integrally stiffened isogrid structure, and the cylindrical center section has a skin-and-stringer construction. The half-shells are joined by a contamination-free linear piston/cylinder thrusting separation system.

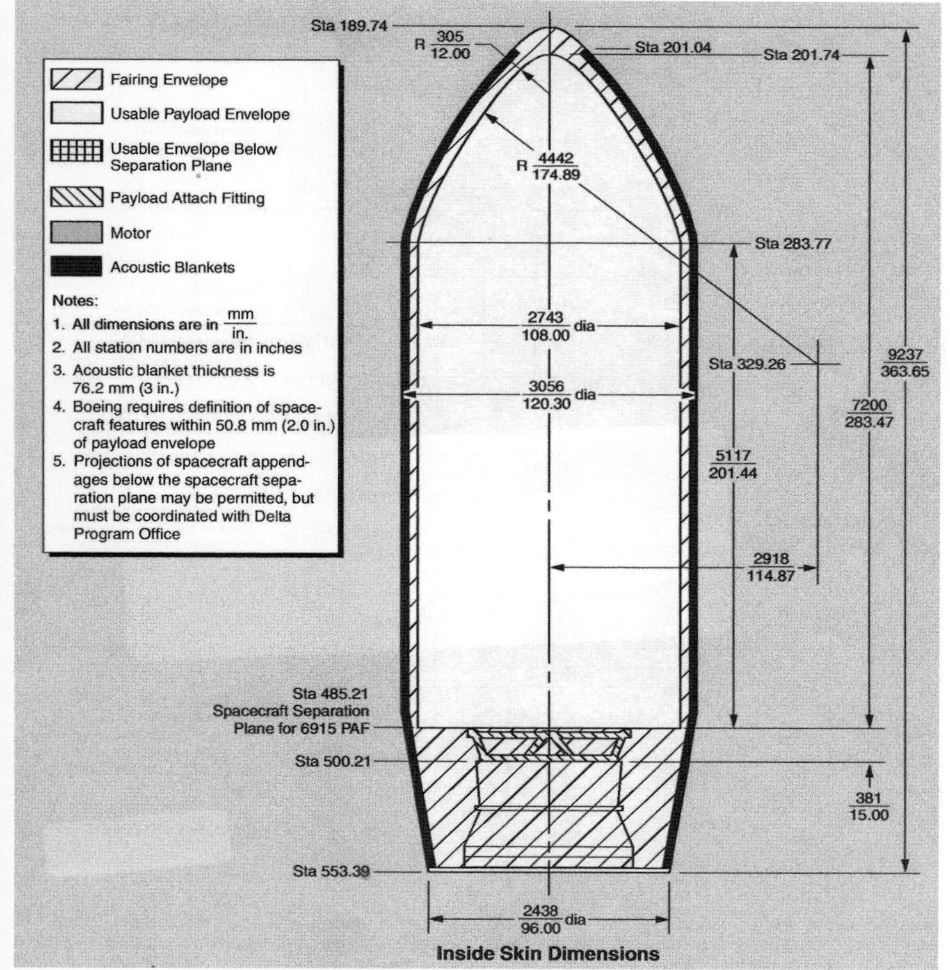

Delta II payload static envelope, 3 m diameter stretched composite fairing, two-stage configuration (6915 PAP) 0131857

The 3 m fairing is available for spacecraft requiring a larger envelope. It is a composite sandwich structure that separates into bisectors. Each bisector is constructed in a single co-cured layup, eliminating the need for module-to-module manufacturing joints and intermediate ring stiffeners. The resulting smooth inside skin enables the flexibility to install mission-unique access doors almost anywhere in the cylindrical portion of the fairing. The stretched 3 m (10 ft)-dia fairing, designated -10L, is available for payloads requiring a longer envelope.

Delta IV

Current Status

The first operational launch of a Delta IV Heavy took place on 10 November 2007, successfully placing the last of the Defense Support Program satellites into GEO.

Background

The Delta IV series of medium-weight and heavy-weight launchers are intended as the future of the Delta line. In a development cycle that began in 1995, the US Air Force began a programme to procure satellite launchers, designated as the Evolved Expendable Launch Vehicle (EELV). During the first phase, four competitors completed a 15-month contract to validate low-cost concepts and in December 1996 two contractors were selected to participate in the second phase known as the Pre-Engineering, Manufacturing and Development (Pre-EMD) phase under a fixed-price USD60 million contract for each company. In November 1997, the US Air Force announced that it would introduce competition into the EELV programme by sustaining a dual-source procurement strategy. On 16 October 1998, the US Air Force announced that it was to buy 19 Boeing Delta IV launchers for the EELV programme at a cost of USD1.38 billion.

After considering a version of the Delta III for the EELV programme, Boeing elected to develop an almost entirely new launch vehicle using different fuels and assembled at a brand-new site. Predicting a large and robust launch market, the design would be optimised for ease of production and assembly and for high-rate handling.

The Delta IV design is based on the Common Booster Core (CBC). This is an aluminium alloy isogrid structure containing liquid hydrogen (LH_2) and liquid oxygen (LOX) tanks and a Pratt & Whitney Rocketdyne RS-68 engine. Because of the fuel's high energy content, it was possible to accept higher weight and lower performance in the engine and structure in the interests of lower cost and reliability. For example, the 2,891 kN RS-68 engine shares some design features with the Space Shuttle Main Engine, but uses lower-cost materials and processes – including many cast rather than welded parts – and has a lower specific impulse and chamber pressure. The Delta IV second stage design, based on that of the Delta III, uses the proven Pratt & Whitney RL-10 engine, also fuelled by LH_2 and LOX.

The CBC is made of metal where it is in contact with cryogenic fuels, with composites used elsewhere to save weight. It represented one of the first applications of friction-stir welding in aerospace. To keep the fuel cool after it is loaded on the rocket, the CBC is covered with foam insulation. The high heat flux from the rocket exhausts scorches and in some cases ignites the insulation during launch, a phenomenon which is spectacular but apparently routine.

There are two basic Delta IV configurations: Delta IV Medium (IV-M) with a single CBC, and Delta IV Heavy (IV-H) with three CBCs. There are also three IV-M+ versions with ATK GEM-60 solid boosters and different upper stages and fairings. These are the 4,2 model with a 4 m upper stage, the same as that on the standard IV-M, and two GEM-60s; the 5,2 with two GEM-60s and a longer-burning 5 m upper stage and fairing; and the 5,4 with larger fairing and four GEM-60s.

There are four available fairings: the 4 m and 5 m composite fairings developed for the Delta IV-M, a longer 5 m composite fairing for the Delta IV-H, and a slightly larger aluminium fairing, developed by Boeing for the Titan IV, which can be adapted to the Delta IV-H.

The Delta IV Heavy launch system offers dual-manifest capability. This dual-manifest system provides payload autonomy similar to a dedicated launch, but at a significant cost reduction compared to a dedicated launch. The dual-manifest approach has the capability of launching two spacecraft totalling up to 10,700 kg to a standard 27° GTO orbit using a 5 m composite fairing that is 22.4 m long.

The 5 m diameter by 19.1 m long dual-manifest system consists of a 12.3 m long composite fairing and a 6.8 m long DPC. The 5 m diameter by 22.4 m long dual-manifest system consists of a 14.3 m long composite bisector fairing and an 8.1 m long DPC. Using standard PAFs, both payloads are mounted within two independent payload bays

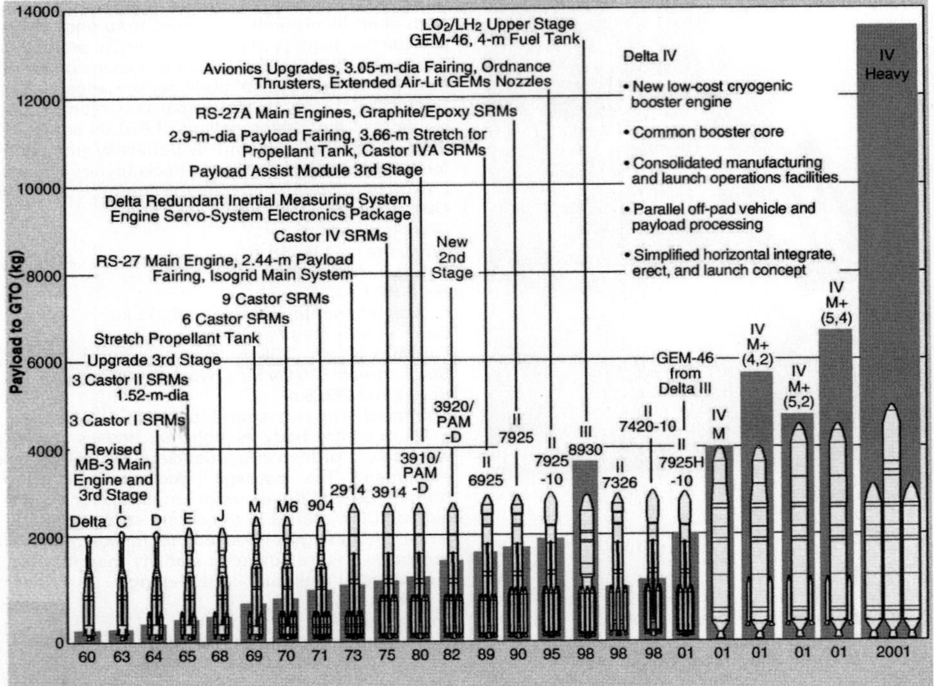

Delta family of launch vehicles

0131854

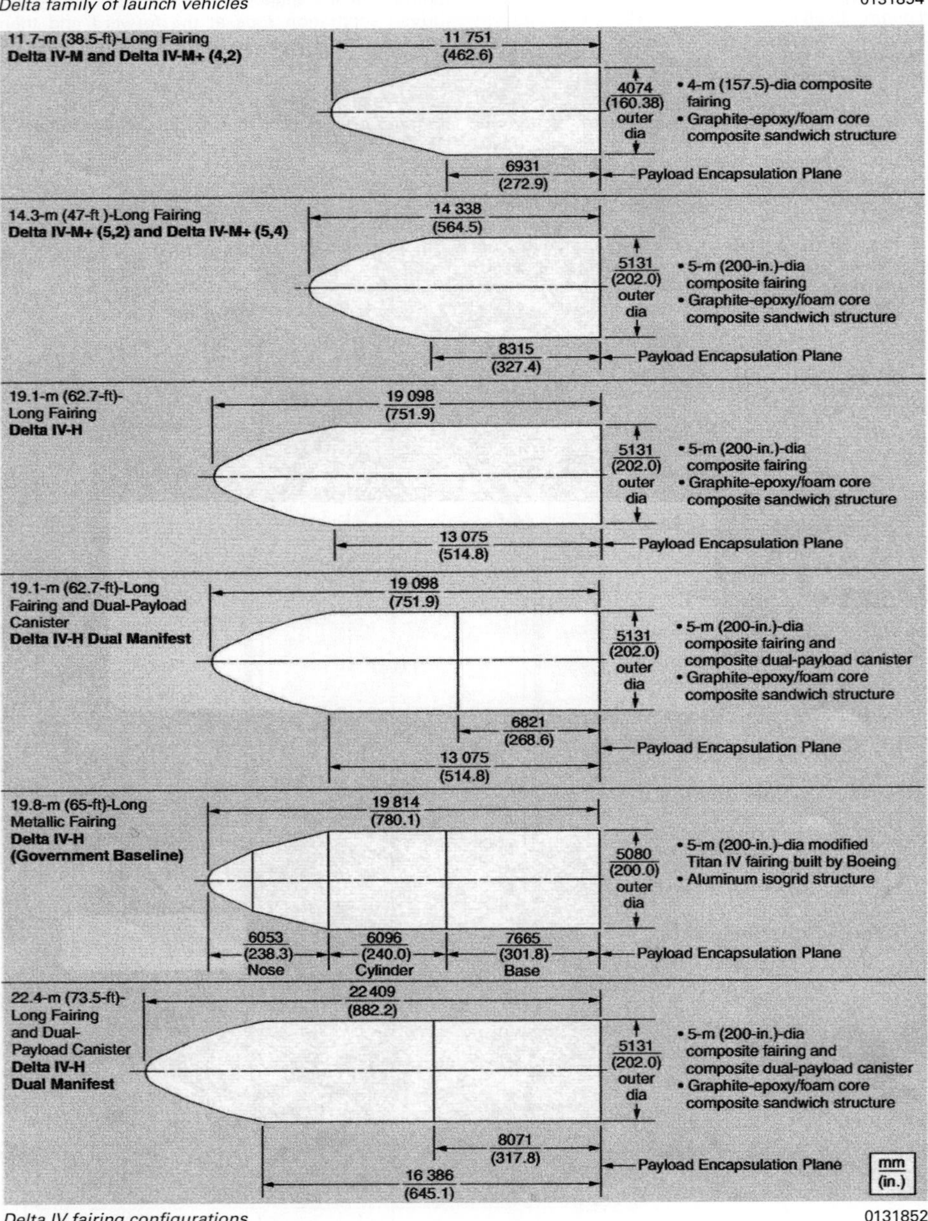

Delta IV fairing configurations

0131852

4-m Configuration (Delta IV-M, Delta IV-M+ (4,2))

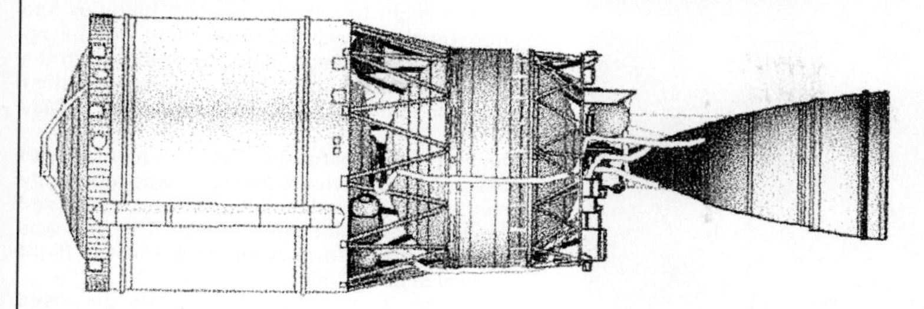

- **Modified Delta III second stage**
- **4-m-dia LO_2 tank**
- **Delta III Pratt & Whitney RL10B-2**

5-m Configuration (Delta IV-M+ (5,2), Delta IV-M+ (5,4), Delta IV-H)

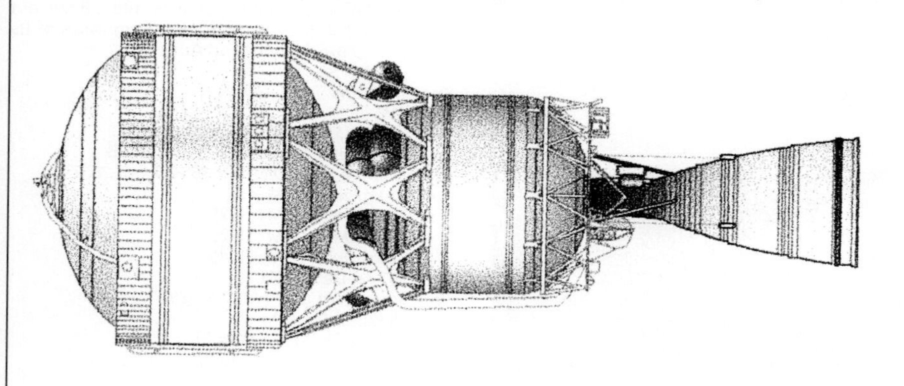

- **4-m-dia stretched LO_2 tank**
- **5-m-dia LH_2 tank**
- **Delta III Pratt & Whitney RL10B-2**

Delta IV second stage configurations 0131851

Configuration of the three main Delta IV variants (McDonnell Douglas) 1340422

that are similar in volume and vented separately. Separate fairing access doors of standard size are provided in the cylindrical section of each payload bay. As with the single-manifest mission, existing acoustic blankets are provided from the aft end of the fairing to the nose cone.

A contamination-free separation system that runs along the full length of the fairing is detonated to separate the fairing into halves, exposing the upper payload. For the lower bay, a circumferential Sure-Sep system (patented by Boeing) with spring actuators is used to deploy the DPC over the payload.

The Delta IV launch system uses a modified Delta III avionics system with a fully fault-tolerant avionics suite, including a Redundant Inertial Flight Control Assembly (RIFCA) and automated launch operations processing using an advanced launch control system.

The RIFCA, supplied by Honeywell, uses six RL20 ring laser gyros and six QA3000 accelerometers to provide redundant three-axis attitude and velocity data. In addition to RIFCA, both the first- and second-stage avionics include interface and control electronics to support vehicle control and sequencing, a power and control box to support power distribution, and an ordnance box to issue ordnance commands. A Pulse Code Modulation (PCM) telemetry (T/M) system delivers real-time launch vehicle data directly to ground stations or relays through the Tracking and Data Relay Satellite System (TDRSS). If ground coverage is not available, instrumented aircraft of TDRSS may be available, in co-ordination with NASA, to provide flexibility with telemetry coverage.

The flight software comprises a standard flight programme and a mission-constants database specifically designed to meet each customer's mission sequence requirements. Mission requirements are implemented by configuring the mission-constants database, which is designed to fly the mission trajectory and to separate the satellite at the proper attitude and time.

The RIFCA contains the control logic that processes rate and accelerometer data to form the proportional and discrete control output commands needed to drive the control actuators and hydrazine control thrusters.

Position and velocity data is explicitly computed to derive guidance steering commands. Early in flight, a load-relief mode turns the vehicle into the wind to reduce angle of attack, structural loads, and control effort. After dynamic pressure decay, the guidance system corrects trajectory dispersions caused by winds and vehicle performance variations, and directs the vehicle to the nominal end-of-stage orbit. Payload separation in the desired transfer orbit is accomplished by applying time adjustments to the nominal engine start/stop sequence, in addition to the required guidance steering commands.

Boeing constructed a new USD400 million, 140,000 m^2 plant in Decatur, Alabama, to produce up to 40 CBCs per year. Essentially, parts and raw materials were inducted into one end of the factory and emerged as complete, functionally tested CBCs at the other end. Boeing commissioned a 95 m roll-on, roll-off vessel, the Delta Mariner, to transport up to three CBCs at a time down the Tennessee/Tombigbee Waterway and thence to Cape Canaveral or Vandenberg. The use of the Delta Mariner removed length and diameter constraints on the CBC design that would have resulted from using air or ground transportation.

Unlike previous US launch vehicles, the Delta IV is designed to be assembled in a horizontal configuration rather than being "stacked" vertically on the launch pad or in a vertical assembly building. Boeing has built a new launch pad and Horizontal Integration Facility (HIF) on Cape Canaveral Air Force Station's Space Launch Complex 37 (SLC-37). The HIF has two processing bays, each large enough for a heavy vehicle, where the launch vehicles are assembled using laser alignment tools. A wheeled transporter rolls the vehicle to the pad, which incorporates a hydraulically operated erector.

Once the launcher is in position, a mobile service tower rolls up to and over the vehicle. This is where the payload, fairing and solid rockets are installed. Finally, the vehicle is fuelled. During the launch of a Delta IV Heavy, the central CBC throttles down to about 58 per cent thrust after 50 seconds to slow the rate of acceleration during maximum aerodynamic pressure. The two outer CBCs throttle down to about 58 per cent thrust at 3 minutes 55 seconds. At 4 minutes they shut down and separate at an altitude of about 93 km. The core engine then throttles up to 102 per cent power. Core shut down and separation occurs at about 5 minutes 39 seconds at 157 km altitude. The second stage is ignited three times to place the satellite in GEO, 6 hours 20 minutes after lift off.

At Vandenberg, Boeing and its subcontractor Raytheon have modified the long mothballed SLC-6 facility – originally built for the Manned Orbiting Laboratory and later designated for the Shuttle – to handle Delta IV. A HIF similar to that at CCAS has been built, along with a dock for Delta Mariner.

The first Delta IV was successfully launched from the converted ex-Saturn 1 launch complex 37B at CCAFS on 20 November 2002 carrying the Eutelsat W5 telecommunications satellite. Since then, there have been five launches of the Delta IV or Delta IV Medium, carrying both civil and military payloads. The fourth Delta IV flight, and the first launch of the Delta IV Heavy configuration, took place in December 2004 and was only partially successful, reaching a lower-than-expected orbit. This was caused by cavitation in a fuel feed line, which created a pocket of gaseous oxygen in the line. This caused an engine cut-off sensor to indicate a dry line, and the engine was shut off.

NASA's Geostationary Operational Environmental Satellite (GOES-N) was due to launch on a

Delta IV Launch Record

Launch Vehicle	Launch Date (local time)	Launch Site	Payload	Outcome
Delta IV-M+ (4,2)	20 November 2002	Cape Canaveral	Eutelsat W5	Success
Delta IV-M	10 March 2003	Cape Canaveral	DSCS III A3	Success
Delta IV-M	29 August 2003	Cape Canaveral	DSCS III B6	Success
Delta IV Heavy	21 December 2004	Cape Canaveral	Demosat, 3CS-1, 3CS-2	Partial failure (dummy payload placed in wrong orbit)
Delta IV-M+ (4,2)	24 May 2006	Cape Canaveral	GOES-13 (GOES N)	Success
Delta IV-M+	27 June 2006	Vandenberg	NROL-22	Success
Delta IV	4 November 2006	Vandenberg	DMSP F-17	Success
Delta IV Heavy	10 November 2007	Cape Canaveral	DSP-23	Success

A ULA – Boeing Delta IV Heavy launching the last of the DSP satellites (DSP-23) from Cape Canaveral on 10 November 2007 (ULA – Carleton Bailie) 1340420

A Boeing – ULA Delta IV M4+ (4, 2) launcher prior to launch of the GOES-N satellite from Cape Canaveral (NASA) 1340419

Delta IV-M+ in August 2005, but an attempt on 16 August was scrubbed because of a technical anomaly concerning battery voltage on the second stage. The flight termination system batteries had expired by that time and they could not be re-installed before the next opportunity to launch the vehicle. A strike by machinists at Decatur further delayed the launch, and it finally took place in May 2006.

The first Vandenberg AFB Delta IV – a IV-M+ with a classified NRO payload, NROL-22, was reportedly erected on the pad in late 2003 and was scheduled at the time for a late-2004 launch. The site was eventually dedicated in April 2005 and the flight took place in June 2006.

For future development, Boeing has proposed adding solid boosters to the Delta IV-Heavy; upgrading propulsion with a regenerative version of the RS-68, or an advanced engine known as RS-800; reducing dry weight with aluminium-lithium alloys; and, eventually, clustering as many as seven CBCs together.

The long-term future of the Delta IV lies with the future of the United Launch Alliance. Despite consolidating production of both the Delta II and Delta IV at Pueblo, Colorado, and Decatur – moving major manufacturing from Huntington Beach to Decatur – in June 2000, employment at Decatur has continued to decline and reached 550 by early 2005. There was some concern whether US Government work would be adequate to support both the Atlas V and Delta IV. However, in August 2007, NASA announced that it would phase out the Delta II within 3 years, transferring more of its payloads to the Delta IV and Atlas V.

Specifications
Delta IV
First launch: 20 November 2002
Launch sites: LC-37 CCAFS, SLC-6 VAFB
Performance to GTO (185 × 35,786 km, 28.5°):
4,231 kg Delta IV-M
5,941 kg Delta IV-M+ (4,2)
4,869 kg Delta IV-M+ (5,2)
6,822 kg Delta IV-M+ (5,4)
12,757 kg Delta IV-H
Performance to LEO (407 km, 28.7°):
9,106 kg, Delta IV-M
12,300 kg, Delta IV-M+ (4,2)
10,616 kg, Delta IV-M+ (5,2)
13,869 kg, Delta IV-M+ (5,4)
21,892 kg, Delta IV-H
Overall length:
Delta IV-M: 63.0 m
Delta IV-M+ (4,2): 66.0 m
Delta IV-M+ (5.2): 77.2 m
Delta IV-M+ (5.4): 77.2 m
Delta IV-H: 77.2 m
Core diameter: 5.1 m
Launch mass:
Delta IV-M: 249,500 kg
Delta IV-M+ (4,2): 292,732 kg
Delta IV-M+ (5,2): 292,732 kg
Delta IV-M+ (5,4): 404,600 kg
Delta IV-H: 733,400 kg
Lift-off thrust:
Delta IV-M: 2,893 kN
Delta IV-M+ (4,2): 4,182 kN
Delta IV-M+ (5,2): 4,148 kN
Delta IV-M+ (5,4): 5,865 kN
Delta IV-H: 8,666 kN
Number of stages:
Delta IV-M: 2
Delta IV-M+ (4,2): 2 + 2 SRM
Delta IV-M+ (5,2): 2 + 2 SRM
Delta IV-M+ (5,4): 2 + 4 SRM
Delta IV-H: 1 × triple CBC + 1

Delta IV stage 1 (CBC)
Comment: The first stage common to all Delta IV variants is the Common Core Booster (CBC) with a single Pratt & Whitney Rocketdyne RS-68 cryogenic motor throttleable between 60 and 100 per cent thrust. Stage elements include an engine section incorporating thrust mount, an aft LH2 tank, centrebody section and forward LOX tank. Delta IV-M and IV-M+ 4.2 variants have tapered interstage sections while the remainder have constant-diameter sections. The two CBC elements operating as liquid propellant boosters on the Delta IV-H carry nosecone elements. The RS-68 is gimballed for pitch and yaw control. Roll-control is provided by gas generator exhaust on the IV-M and by differential gimballing of the side CBCs on the IV-H.
Overall length: 40.8 m
Diameter: 5.1 m
Empty mass: 26,700 kg
Gross mass: 226,400 kg
Oxidiser: LOX

Fuel: LH_2
Isp: 420 s
Launch thrust: 3,313 kN
Burn time: 249 s

Delta IV solid strap-ons
Type: ATK GEM-60
Overall length: 12.95 m
Diameter: 1.55 m
Gross mass: 33,798 kg
Empty mass: 3,849 kg
Thrust (vac): 826.3 kN
Isp: 275 s
Burn time: 243 s

Delta IV stage 2
Overall length: 12 m
Diameter: 4 or 5.1 m
Gross mass: 24,170 kg or 30,710 kg
Empty mass: 2,850 kg or 3,490 kg
Thrust (vac): 110 kN
Isp: 462 s
Oxidiser: LOX
Fuel: LH_2
Motor: RL10B-2
Burn time: 850 or 1,125 s

Evolved Expendable Launch Vehicle

Current Status
In late 2005, Boeing and Lockheed Martin were still seeking approval from the US government, including the Department of Defense (DoD) and the Federal Trade Commission, to merge their Evolved Expendable Launch Vehicle (EELV) activities into a new joint venture, the United Launch Alliance.

Background
The Evolved Expendable Launch Vehicle (EELV) provides medium and heavy-lift expendable launch vehicles for the US Air Force, including launches on behalf of other services and the US intelligence community, and NASA. The programme is managed by the US Air Force Space & Missile Center at Los Angeles AFB, California, and covers two families of launch vehicles: Boeing's Delta 4 and Lockheed Martin's Atlas V. EELV launch sites are located at Cape Canaveral Air Force Station, adjacent to NASA's Kennedy Space Center, and Vandenberg AFB in California. Both EELV families are also available for commercial launches.

The EELV programme was launched under President Clinton's first administration. The loss of the Space Shuttle Challenger in January 1986 had put an end to the initial vision of the Shuttle as the sole and unique means of access to orbit for US government space programmes, so the DoD and NASA were forced to continue using their older launchers – Delta, Atlas and Titan – each of which had its own unique infrastructure and payload interfaces. As a result, the three rockets were becoming increasingly expensive to operate, and launch delays were mounting because there was no way to switch payloads from one launcher to another. The unique heavy-lift Titan 4 was notoriously prone to delays and backlogs. The older US launchers were also confronting competition from Europe, Russia – which was then entering the world launch market – and China.

After several unsuccessful attempts to develop a new launcher, including the Advanced Launch System study of 1987 to 1990 and the National Launch System project of 1991 to 1992, Congress directed the DoD to develop a new plan. In 1994, USAF Lieut Gen Thomas Moorman led a Space Launch Modernization Study, including participants from the intelligence community, NASA and commercial launch groups. In August 1994, the results of this study were embodied in President Clinton's National Space Transportation Policy, which gave NASA responsibility for the Shuttle and future reusable vehicles, while directing the DoD to produce an improved family of expendable vehicles.

In August 1995, the US Air Force issued USD30 million, 15-month risk-reduction contracts to four companies – Lockheed Martin, McDonnell Douglas, Boeing and Alliant TechSystems – covering their EELV proposals. Late in 1996, the USAF down-selected to Lockheed Martin and McDonnell Douglas, which each received USD60 million pre-engineering and manufacturing development (pre-EMD) contracts. At that time, the USAF expected to pick a single winner in late 1997.

However, a number of factors caused that plan to change. There was some concern that the plan placed too much reliance on a single rocket family. The loser in the competition would have a very difficult time remaining competitive. At the same time, the commercial satellite launch industry was forecast (incorrectly as it turned out) to grow rapidly. The result was a strategic policy shift announced by the Pentagon in November 1997. Both Lockheed Martin and Boeing – which had completed its acquisition of McDonnell Douglas in August 1997 – would build EELVs. However, rather than fully funding EMD, the Pentagon would contribute USD500 million to each company's development costs, and award each company a contract for an initial batch of launches. Industry was expected to pay the balance of the development cost and recoup it through future launch revenues, competing year-by-year for Pentagon as well as commercial business.

Despite their names, both EELV contenders were largely new systems. Development costs were high (USD2.5 billion in the case of the Delta 4) but Boeing and Lockheed Martin expected to recoup them through lower recurring costs, higher margins and a healthier continuing launch business. However, the commercial satellite launch business did not grow as expected. In 2000, the DoD found it necessary to renegotiate its contracts with Lockheed Martin and Boeing in order to ensure that neither company left the space launch market or – as was a possibility with Boeing, with its Sea Launch business – remained in the market in a manner that was dependent on non-US hardware and support.

The EELV programme has met many of its strategic goals. The two launcher families have replaced the three principal launch vehicles used earlier – a process which was largely completed when the last Titan 4 was launched on 19 October 2005. While the goal of a standard payload interface has not been fully met, the USAF has reported that the greater level of standardisation has helped reduce launch costs and schedule delays.

In July 2003, the DoD suspended Boeing from some EELV business when it was revealed that the company was under investigation for illegally obtaining information about its rival's EELV bids in 1996 to 1999. (In the first EELV launch competition, the Delta 4 was selected for 19 out of 28 missions.) Seven EELV launches were shifted to Lockheed Martin. The suspension remained in force until March 2005.

Facing continued cost increases and with a still-sluggish commercial market. Boeing and Lockheed Martin announced in May 2005 that they intended to merge their EELV activities in the form of a new joint venture company, named United Launch Alliance. The new company's headquarters and engineering centre will be located in Denver, Colorado, while major assembly and manufacturing of both the Atlas and Delta rockets would be located at Boeing's new Delta 4 facility in Decatur, Alabama. The two companies predict that the shared operation will save the customer up to USD100 million to USD150 million per year.

Falcon

Background
The Falcon programme name was originally an acronym for Force Application and Launch from CONUS (continental United States), but the full name is now downplayed because of Congressional sensitivity about the "weaponization" of space. This has also led to some changes and restrictions on the programme's activities.

Falcon is a joint venture between DARPA and the US Air Force (USAF), with support from NASA. It is a multi-faceted effort with several inter-related strands. In the near term, it could provide the USAF with a low-cost capability to orbit small spacecraft. Beyond that, it may lead to the development of a boost-glide vehicle capable of delivering weapons or other payloads to any point on Earth after a launch from the US. Its ultimate goal is to develop a Hypersonic Cruise Vehicle (HCV) which could either deliver boost-glide vehicles or act as the reusable first stage of a small-payload launch system.

Falcon has superseded the DARPA HyperSoar programme for a ramjet-powered hypersonic "skipper" vehicle, and the RASCAL (Responsive Access, Small Cargo, Affordable Launch) project for a turbojet-powered supersonic aircraft launching an expendable rocket booster stage.

Lockheed Martin's concept for a Hypersonic Cruise Vehicle (HCV) developed under the DARPA Falcon programme. Such a vehicle could be operational after 2020 (DARPA) 1127930

The Falcon programme started in June 2003, when DARPA solicited bids for two basic tasks. Task I is the development of the Small Launch Vehicle (SLV) and Task II covers the HCV.

Nine contractors were awarded Phase 1 SLV contracts in December 2003. In September 2004, four of them were selected to continue into Phase IIa: AirLaunch LLC of Reno, Nevada; Lockheed Martin; Microcosm Inc of El Segundo, California, and Space Exploration Technologies (SpaceX), also of El Segundo. Three of these are new companies, and the Lockheed Martin project is centered on the company's Michoud, Louisiana, operation rather than its established space-launch division.

The key goals for the SLV are as follows: launch within 24 hours of a request to orbit a spacecraft; a 450 kg launch capability into a 185 km circular Low Earth Orbit (LEO); a launch crew ideally comprising ten or fewer people; and a cost per launch of approximately USD5 million in current dollars.

Lockheed Martin's proposed SLV would be assembled and integrated with its payload in a horizontal position, mounted on a semi-trailer transporter. It uses hybrid propulsion with liquid oxygen (LOX) as the oxidizer and solid rubber-type fuel. The rocket would be fully assembled and then towed to a fuelling station where LOX would be taken on from tanker trucks. The vehicle would then be towed to the launch pad.

SpaceX's SLV is derived from its rocket design, also named Falcon, which is due to make its first flight in late 2005 carrying the USAF's experimental TacSat spacecraft. It is a LOX/kerosene system.

Microcosm's Eagle is intended to use pressure-fed LOX-kerosene engines, with mass-produced composite tankage to contain the pressure with reasonable weight and affordable cost. It would be assembled vertically on rolling pads.

AirLaunch LLC's QuickReach vehicle is designed to be air-launched from a Boeing C-17 cargo aircraft. Its pintle-type engines – claimed to have only one moving part per engine – are pressure-fed with LOX and propane. The 32-tonne vehicle is rolled from the rear cargo door of the transport aircraft, using a lanyard and parachute system to stabilize it in a launch attitude, and the motor fires when the launch aircraft is a safe distance away.

In August 2005, DARPA was in the process of deciding whether to fund one contractor through the next stage of SLV – Phase IIb, leading to flight test – or whether it could afford to fund two contenders. An announcement is expected late in September 2005, leading to a first flight in 2007 or 2008.

Falcon (SpaceX)

Current Status
The Falcon 1 launch vehicle was lost on its first flight attempt on 24 March 2006. The company lists a manifest of seven other launches, but the future schedule will depend on the results of the mishap investigation.

Background
Space Exploration Technologies (SpaceX) was founded in June 2002 by Elon Musk, the inventor of the PayPal online payment system, to develop a conventional but very low-cost family of space

Falcon family

	1	5	9	9-S5	9-S9
Lift-off mass:	27.2 t	154.5 t	290 t		
Length:	21.3 m	47 m	53 m	53 m	53 m
Diameter:	1.7 m	3.7 m	3.7 m	3.7 m	3.7 m
Payload to LEO:	570 kg	4.1 t	8.7 t*	16.5 t	24.75 t
Payload to GTO:		1.05 t	3.1 t*	6.4 t	9.65 t
Fairing diameter:	1.5 m	3.6 m	3.6/5.2 m	5.2 m	5.2 m
Strap-on engines:				10	18
Core engines:	1	5	9	9	9
Lift-off thrust:	342 kN	1,890 kN	3,400 kN	7,150 kN	10,200 kN
Second-stage engines:	1	1	1	1	1
Thrust:	31 kN	409 kN	409 kN	409 kN	409 kN
Total launch price:	USD6.7 m	USD18 m	USD27 m	USD51 m	USD78 m

*9.3 t LEO, 3.4 t GTO with 3.6 m fairing.

launch vehicles, together with a new family of rocket engines to power them. SpaceX plans to offer open, fixed pricing with discounts for multi-launch orders.

The family includes the Falcon 1, 5 and 9, the designations corresponding to the number of engines in the core vehicle. The company says that it has two launch orders for the Falcon 9, the balance being for the Falcon 1. One Falcon 9 order is for an unspecified US government customer. So far, the company has booked launches from Vandenberg AFB and Kwajalein.

The Falcon 1 is a small launch vehicle, designed to deliver a 570 kg payload into LEO. It is a two-stage rocket fuelled by LOX and rocket grade kerosene (RP). The first stage is constructed from conventional aluminium alloy, with a patented structural design which is pressure-stabilised in flight. According to SpaceX, this offers some of the weight advantages of pressure stabilisation but avoids the ground-handling problems of a structure that has to be pressurised when empty. The tanks are pressurised by helium in composite-wrapped Inconel tanks. The first stage is recovered by parachute to a water landing, and is intended to be recovered by a ship, restored and re-used.

The Merlin rocket engine for the Falcon 1 first stage uses a simple gas generator cycle: LOX and kerosene are burned in a separate combustion chamber and exhaust through a turbine that drives the dual-impeller turbopump which feeds the main motor chamber. High-pressure fuel from the turbopump is also used to drive pitch and yaw actuators for the motor's vectored nozzle. The turbopump exhaust is vectored to provide roll control.

The interstage structure is made from carbon-fibre honeycomb. The second stage is a structurally stable tank with skins machined from aluminium-lithium alloy. Automated welding is used for the major circumferential joints. The second-stage engine, the Kestrel, is pressure-fed, using stored compressed helium, augmented by a heat exchanger on the nozzle, which is made from a high-strength niobium alloy and is electrically vectored in pitch and yaw. The engine is fitted with dual torch igniters for multiple restarts.

The Falcon 5 and Falcon 9 designs use much the same technology as the Falcon, but with some unique features. The two larger rockets have exactly the same first and second stage tank structure and the same propulsion system – a single Merlin – on the second stage. The difference is the number of first stage engines (corresponding to the designation). The Falcon 5 is launched with the tanks only partially filled. This approach allows SpaceX to bring two different launch vehicles to market at the same time and reduces manufacturing and support costs. Both stages are intended to be reused, although SpaceX has not given details of how the second stage can be recovered from orbital injection speeds.

Another feature of the Falcon 5 and 9 is that the large number of independent engines, which gives them true engine-out reliability. The engines are Kevlar-jacketed to ensure that the failure of one engine does not take its neighbours out of action. The Falcon 5 has a 3.6 m payload fairing and the Falcon 9 – with a 3.4 t GTO payload – can be fitted with either the 3.6 m fairing or a 5.2 m structure. Before the loss of the first Falcon, SpaceX was predicting a first flight of Falcon 9 from Kwajalein in the second quarter of 2007.

Finally, SpaceX offers two Falcon 9 configurations – the S5 and S9 – with strap-on boosters. The S5 booster uses the same configuration as the Falcon 5, and the S9 booster has nine engines.

All versions of the Falcon have triple-redundant flight control systems with GPS for accuracy in orbit insertion.

SpaceX started tests of its upper-stage motor in August 2003. By that time, tests of the Merlin main engine had also been carried out and the company was predicting a first launch in early 2004. First orders were announced in November 2004, and the first flight was predicted for February 2005, but problems with the Merlin engine consumed more testing time. A flight date in November 2005 was missed due to a fuel leak before the failed March 2006 attempt.

In September 2005, SpaceX announced first orders for the Falcon 9 and said that the larger rocket would be brought into service ahead of Falcon 5, because of higher market demand. The company has launched legal action against Boeing and Lockheed Martin, to restrain them from joining their Evolved Expendable Launch Vehicle (EELV) programmes together under the United Launch Alliance joint venture. SpaceX asserts that the ULA will restrict competition and that the US Government could reap substantial savings by allowing the Falcon 9 to compete.

Specifications
Falcon 1
Number of stages: 2
Lift-off mass: 27.2 t
Length: 21.3 m
Diameter: 1.7 m
Payload to LEO: 570 kg
Propellant: LOX/RP
First-stage engine: SpaceX Merlin, gas generator cycle
Thrust: 342 kN sea level, 409 kN vacuum
Isp: 255 s sea level, 304 s vacuum
Second-stage engine: SpaceX Kestrel, pressure-fed
Thrust: 31 kN vacuum
Isp: 327 s vacuum

Minotaur

Current Status
The Minotaur family of rockets is used both to launch suborbital targets and test vehicles, and for orbital launches. It was developed for the USAF Space & Missile Command (SMC) and is offered only for US government applications. Since its first flight in 2000 the family has performed nine launches, four of them orbital, and all have been successful. At least three further vehicles were under contract to launch in early 2006. A more powerful version, Minotaur IV, is under development and is due to make its first flight in late 2008.

Background
The Minotaur was conceived as a smaller companion to Taurus, based on the use of first and second stages of now-retired Minuteman II ICBMs. The name is a contraction of Minuteman and Taurus. Development was initiated by the USAF Space & Missiles Center (SMC) under the Orbital/Suborbital Program (OSP), and the development contract was awarded to Orbital in September 1997. The first launch took place some 25 months later.

Minotaur I is a small launch vehicle, typically capable of lifting a 340 kg payload into a 740 km sun-synchronous orbit. This is roughly 50 per cent more than Pegasus. The two upper stages are equivalent to those of the Taurus and Pegasus XL, and the standard payload fairing is a modified version of the Pegasus fairing. A larger fairing is available, along with a multi-satellite adapter, developed by One Stop Satellite Systems, which allows the vehicle to launch one primary payload and up to four microsatellites.

The Minotaur family has expanded to include several versions. Minotaur II is essentially a demilitarised Minuteman II missile, using all three stages of the standard missile rather than just the first two, but equipped with the guidance, navigation and control avionics used on Pegasus, Minotaur I and Taurus. Also developed under the OSP programme, it is primarily a target launcher for missile defense programmes.

Minotaur III is also a suborbital vehicle, but uses components from the retired MGM-118A Peacekeeper ICBM rather than the Minuteman. Orbital provides a hydrazine upper stage to provide better accuracy and multiple targeting options. The first Minotaur III is under contract to launch the first Hypersonic Test Vehicle (HTV) for the Defense Advanced Research Agency's Falcon programme.

Minotaur IV is a larger orbital launcher, being developed as part of a ten-year OSP-2 contract. It combines the first three stages of Peacekeeper with a new Orbital-developed Guidance and Control Assembly (GCA). The GCA incorporates the same ATK Orion 38 motor as the Pegasus and Taurus third stage, along with Taurus/Minotaur avionics and a cold-gas attitude control system. Design payload to LEO is up to 1,720 kg. However, because the Peacekeeper stages are in the USAF inventory, the Minotaur IV costs only fractionally more than the Minotaur I. The USAF awarded Orbital a contract in early 2005 for the first Minotaur IV launch, which is scheduled to orbit the first Satellite Based Surveillance System (SBSS) spacecraft in December 2008.

Minotaur V is a proposed high-energy version of Minotaur IV, with an ATK Star-48GV-powered fourth stage inserted below the GCA. It would be able to lift a 600–700 kg class payload to GTO.

The Minotaur is intended to offer a variety of launch sites and includes a complete, transportable kit of Launch Support Equipment (LSE). Ground equipment includes a "sacrificial" launch tower that connects the upper stage and guidance equipment to the ground and is lost on launch; a 6 m high launch platform; and a movable gantry made from commercial scaffolding. Launch control equipment is housed in a van, and batteries and other equipment. The programme goal is to be able to operate from the USAF's primary launch sites at Cape Canaveral and Vandenberg, from NASA's Cape Canaveral and Wallops Island sites, and from Kodiak Island in Alaska.

Currently, the Minotaur family has an 18-month timeline from order to launch. However, the USAF and Orbital have studied ways to reduce the lead time. For example, it could be reduced to 12 months by acquiring a set of upper-stage components – the most critical being the solid rocket motor – to be held in reserve, rather than ordering these parts when the launch order is received. A three-month lead time would be possible with changes to range planning, and the USAF has looked at options – such as pads with transporter/erector vehicles and roll-away shelters – that would allow Minotaurs to be launched in 48 hours or less.

Specifications
Minotaur I
First launch: 26 January 2000
Number launched: 4 to March 2006
Launch sites: Vandenberg AFB
Principal applications: small LEO payloads
Vehicle success rate: 100 %
Performance: 580 kg to 185 km, 28.5° LEO; Taurus XL: 315 kg to 740 km, 28.5° LEO.
Number of stages: 4, optional 5 with HAPS
Overall length: 27.1 m
Principal diameter: 167 cm stage 1

Launch log

No	Date	Payload
1	20 January 2000	JawSat
2	19 July 2000	MightySat II.1
3	11 April 2005	XSS-11
4	22 September 2005	STP-R1

Pegasus

Current Status
The world's only operational air-launched satellite launch vehicle, Pegasus, had completed 37 space launches by the end of March 2006, with a clean record of success since 1997, and had also flown twice (with one failure) as the booster for NASA's X-43A hypersonic research vehicle.

Background
OSC was established in 1982 to develop Space Shuttle upper stages, resulting in the solid propellant Transfer Orbit Stage. In June 1988, OSC announced a partnership with motor manufacturer Hercules Aerospace (now ATK) to develop the Pegasus

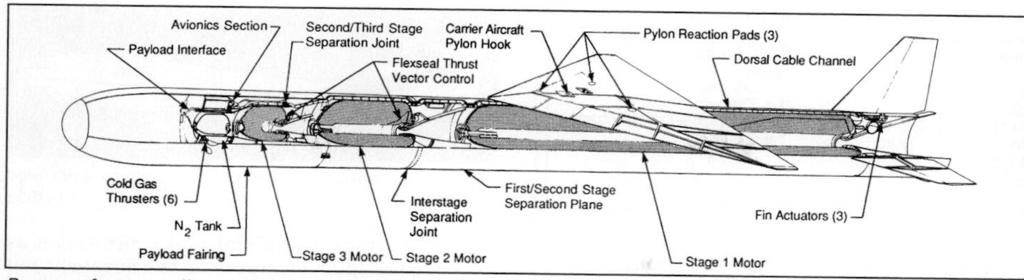

Pegasus features all-composite wing and solid rocket motor structures. The HAPS stage 4 replaces the N_2 tank shown here (OSC/Alliant)
0516935

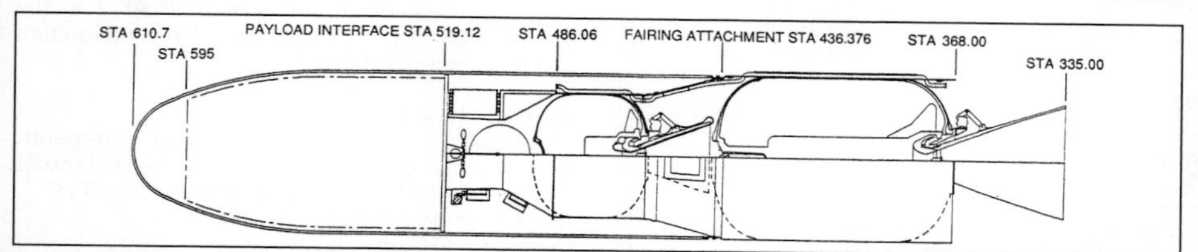

Pegasus payload accommodation. Station positions (standard Pegasus) are given in inches. Length is reduced 10 cm with HAPS (OSC/Alliant)
0516938

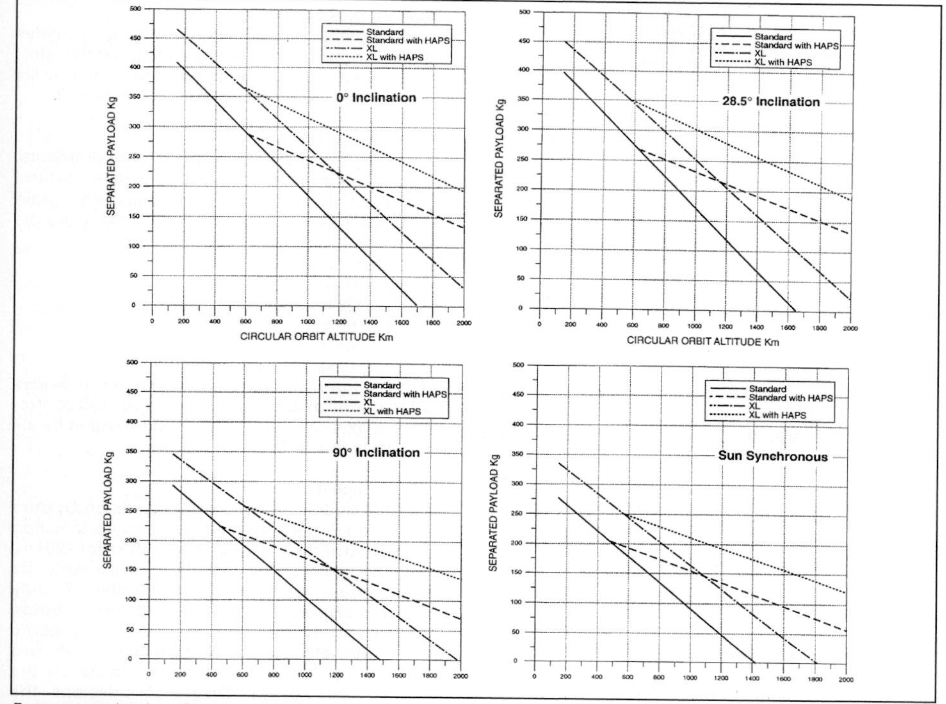

Pegasus payload performance (OSC)
0516936

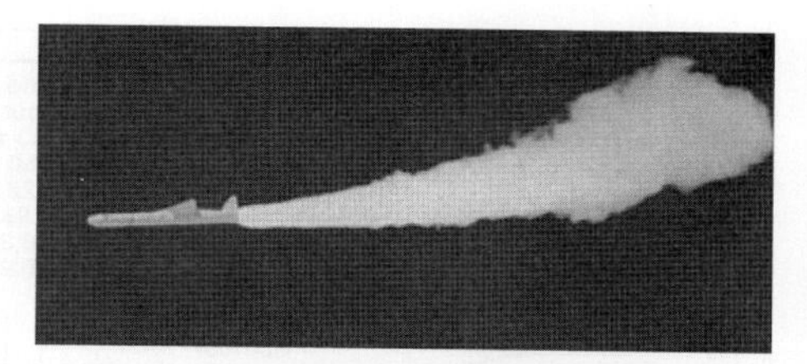

Pegasus stage 1 ignites 5 seconds after release (OSC)
0516934

air-launched small orbital delivery system. The 28-month development schedule was then already under way as a private venture, aiming for a first launch in July 1989 carrying a DARPA Glomar data relay and NASA Pegsat chemical release satellites.

Pegasus is capable of placing a 201 kg payload into a 460 km polar orbit (289 kg equatorial) following aircraft release at 11.9 km/Mach 0.8. Six launches (at 12.2 km) utilised the USAF-owned B-52 NASA-operated aircraft employed for X-15 hypersonic drops in the 1960s. A Lockheed L1011 Tristar, named 'Stargazer', was acquired by OSC from Air Canada in 1992 on a 10-year lease as the belly carrier beginning with STEP M1 in 1994. It is now leased from the owner, ITT Financial. Stargazer can handle 35.5 tonne vehicles and operate from most major runways.

The base vehicle consists of three graphite epoxy composite case ATK solid motors of conservative design, a fixed delta planform composite wing, an aft skirt assembly including three composite control fins, an avionics section forward of the stage 3 motor, and a two-piece composite payload fairing. Some 94 per cent of the structural mass is graphite epoxy, 5 per cent aluminium, 1 per cent titanium. Air-launching at 11.9 km altitude reduces ΔV requirements by 10 to 15 per cent.

The higher-performance Pegasus XL version was first launched in June 1994 and is now the only version on offer. Taking advantage of the extra lift capacity offered by the Stargazer over the B-52, it is 1.9 m longer and about 5,000 kg heavier than the original version. The last standard version was launched in October 2000. Pegasus XL is available with an optional Hydrazine Auxiliary Propulsion System (HAPS) stage 4, which provides higher performance and usefully improves accuracy.

NASA Goddard selected Pegasus in March 1991 to provide Small Expendable Launch Vehicle Services for its Small Explorer science satellite programme. The total contract, worth some USD80 million, covers seven firm and three optional launches from 1993. It was selected April 1991 for one firm (late 1992) and 39 optional launches under the AFSLV Air Force Small Launch Vehicle programme, bringing orders to 15 firm and about 45 optional. BMDO, in July 1992, awarded a USD14.7 million contract for one launch (plus nine options) for its MSTI 3 satellite. NASA selected Pegasus as its Ultralight launcher in November 1994, with two firm (USRA in 1997) and eight optional launches. Spain's INTA selected it in October 1994 for Minisat 01, which was the first west European orbital launch (from Spain's Torrejon AB).

Pegasus was used as the booster for the NASA X-43A hypersonic research vehicle, with a specially designed payload adapter to separate the vehicle at high speed. The first test on 2 June 2001 failed, partly because the effects of the lower-than-normal launch altitude on the booster had not been anticipated. The second and third tests, on 27 March 2004 and 16 November 2004, were successful. Pegasus launch rates peaked at six in 1998 and there have been only two flights since 2003.

Specifications
Pegasus
First launch: 5 April 1990
Number launched: 37 through end March 2006 (10 Pegasus; 27 Pegasus XL)
Launch sites: now hosted by L-1011 operating out of Vandenberg AFB, NASA Wallops and Cape Canaveral
Principal applications: small payloads into LEO
Vehicle success rate: 94.5 per cent
Performance: 3-stage typically 200 kg (XL: 279 kg) into 463 km polar, 288 kg (XL: 382 kg) into 463 km equatorial (payload fraction 2.2%). For 740 km polar orbit, accuracy is ±0.2° inclination, ±85 km 3Σ deviation from circular; with HAPS: ±0.1°/±19 km. 165 kg GTO capacity projected with XL + stage 4
Number of stages: 3
Overall length: 16.9 m
Principal diameter: 127.1 cm
PWing span: 6.7 m
Launch mass: 23,130 kg
Guidance: Digital distributed processor design centred on a 32-bit flight computer, using RS-422 digital links, Northrop Grumman LR-81 inertial measurement unit and GPS.

Pegasus stage 1
Stage 1 incorporates a Scaled Composites Inc graphite composite wing with 45° sweptback leading edge and 670.6 cm span. Subsonic L/D is 4.0. The airfoil is a double wedge with 2.5 cm radius leading edge. The wing thickness is truncated to 20.3 cm, with upper/lower parallel surfaces facilitating attachment to the motor case wing saddle. Three active aft fins provide control during stage 1 burn as part of an aluminium assembly attached to the motor's skirt extension. Control is augmented at end of stage 1 burn by firing small fin-mounted motors.
Motor: ATK Orion 50S XL, stretched by 137 cm for XL. All three motors use IM7/55A graphite epoxy composite cases with aramid filled EPDM rubber internal insulator and integral skirts. Each nozzle consists of a carbon phenolic exit cone with 3D carbon/carbon integral throat/entry. Propellant is HTPB with 88% solids, grain designed for low burn rates. The igniter is mounted on the forward dome. A flight termination charge is mounted on the aft dome of stage 1 and forward dome of stage 2 for range and aircraft safety requirements. Stage 1 employs a fixed nozzle.
Stage length: 10.28 m
Diameter: 1128 cm
Thrust: 726 kN max vacuum
Burn time: 68.6 s
Mass at ignition: 17,157 kg
Propellant mass: 15,014 kg

For details of the latest updates to ***Jane's Space Systems and Industry*** online and to discover the additional information available exclusively to online subscribers please visit
jsd.janes.com

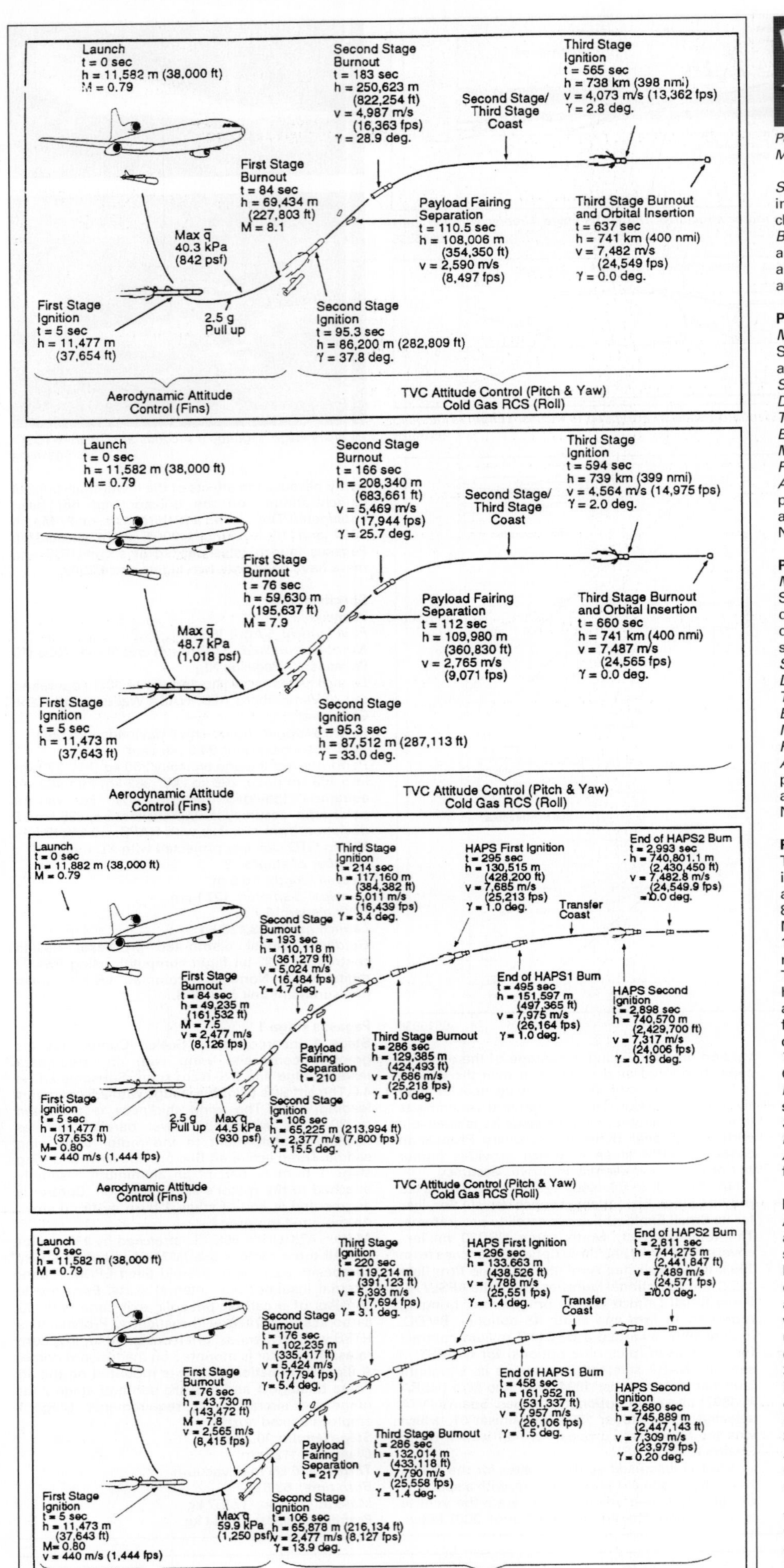

Pegasus mission profiles into 741 km circular polar. From top: standard Pegasus, 162 kg payload; Pegasus XL, 219 kg; standard + HAPS, 190 kg; XL + HAPS, 244 kg (OSC) 0516937

Pegasus expendable launcher attached to Lockheed Martin L1011 0131958

Separation: the forward skirt, which also serves as interstage adaptor, incorporates two linear-shaped charges.
Burn sequence: vehicle is released from L 1011, activating sequencer and autopilot. At 2 s, the arming sequence begins for stage 1 arming/ignition at 3/5 s, respectively.

Pegasus stage 2
Motor: ATK Orion 50 XL See stage 1 for comments. Stage 2 employs a silicon elastomer flexseal nozzle and Parker electromechanical actuators for TVC
Stage length: 383.7 cm; XL: 428.7 cm
Diameter: 128 cm
Thrust: 196 kN max vacuum.
Burn time: 69.4 s
Mass at ignition: 3,370 kg; XL: 4,314 kg
Propellant mass: 3,925 kg
Attitude control: motor nozzle flexing provides pitch/yaw control during powered flight; roll control and three-axis control during coast provided by six N_2 thrusters in two Parker modules on stage 3.

Pegasus stage 3
Motor: ATK Orion 38. See stage 1/2 for comments. Stage 3 also employs flexing for pitch/yaw control during burn. Igniter is toroidal. Head-end grain design maximises propellant density. XL uses the same stage 3 as the standard Pegasus.
Stage length: 208 cm
Diameter: 97 cm
Thrust: 36 kN max vacuum. .
Burn time: 68.5 s
Mass at ignition: 893 kg
Propellant mass: 770 kg
Attitude control: motor nozzle flexing provides pitch/yaw control during powered flight; roll control and three-axis control during coast provided by six N_2 thrusters in two Parker modules.

Pegasus stage 4
The HAPS Hydrazine Auxiliary Propulsion System is an optional stage carried to improve injection accuracy and performance by about 36 kg for 720 km, 82°. It was ordered by DARPA to cope with the Microsat mass growth (launch number 2, July 1991). It occupies the payload adaptor interior, replacing the N_2 tank of the 3-stage version. The attitude jets are supplied instead from two high-pressure (395 atmosphere) N_2 tanks on the adaptor's outer surface. On typical missions, the first burn establishes an elliptical orbit, followed by circularisation.
Thrusters: three fixed blowdown monopropellant General Dynamics MR-107s
Propellant: 72.6 kg of hydrazine stored in central sphere
Stage thrust: 3 × 222 N vacuum
Burn time: typically 2 burns of 131 and 110 s
Attitude control: nitrogen jets and differential firing of main thrusters

Fairing and Payload Accommodation
Alliant's 127 kg 2-piece composite fairing, which also covers stage 3 and the avionics packages (and stage 4 when carried), is jettisoned during stage 2 burn. The payload may remain attached to stage 3 or be separated by clamp, following pointing to ±2° accuracy or spin up. The payload can be provided with 60 V DC (up to 140 W) from the carrier aircraft until release, and up to 16 discrete commands can be issued by Pegasus' computer. Telemetry at 3 kbit/s from the vehicle's total 56 kbit/s downlink is available to the payload. On the ground, air conditioning maintains 21±5°C.
Payload volume: 1.17 m diameter, 2.13 m long for 3 stage; 1.76 or 1.79 m long when HAPS carried
Acceleration load: 8 g longitudinal max at end of stage 1 burn (lighter payloads can exceed 8 g during stage 3 burn).

Vehicle and Payload Processing
Vehicle and payload processing are performed in two 15 × 36 m bays of the OSC Vehicle Assembly Building at VAFB, controlled at 21±5°C and humidity of 40±10%. The stages and other subassemblies arrive typically 4 months before launch. Payload mating at T12 days is in a 3.66 × 7.32 m class 100,000 soft-walled clean room. The N_2 ACS tank is then filled and the fairing attached. All processing

Pegasus XL with the Celistis payload launched 21 April 1997 (Theo Pirard) 0054368

Pegasus launches

	Date	Type	Carrier/Base	Payloads
1	5 April 1990	std	B-52/DFRC	192 kg Pegsat + 68 kg SECS
2	17 July 1991	HAPS	B52/DFRC	7 × 21.8 kg Microsats
3	9 February 1993	std	B 52/KSC	115 kg SCD 1 + 14.5 kg CDS
4	25 April 1993	std	B52/DFRC	109 kg Alexis
5	19 May 1994	HAPS	B52/DFRC	180 kg STEP M2
6*	27 June 1994	XL	L 1011/VAFB	348 kg STEP M1
7	3 August 1994	std	B52/DFRC	261 kg APEX
8	3 April 1995	std	L1011/VAFB	2 × 40 kg Orbcomm 1/2 + 68 kg MicroLab 1
9*	22 June 1995	XL	L1011/VAFB	268 kg STEP M3
10	9 March 1996	XL	L 1011/VAFB	113 kg REX 2
11	17 May 1996	std	L1011/VAFB	211 kg MSTI 3
12	2 July 1996	XL	L-1011/VAFB	294 kg TOMSEP
13	21 August 1996	XL	L1011/VAFB	180 kg FAST
14*	4 November 1996	XL	L-1011/WI	2200 kg SAC B+NETE
15	21 April 1997	XL	L1011/C1	209 kg MINISAT 1/Celestis
16	1 August 1997	XL	L-1011/VAFB	309 kg Seastar
17	29 August 1997	XL	L-1011/VAFB	215 kg FORTE
18	22 October 1997	XL	L-1011/WI	395 kg STEP M4
19	23 December 1997	XL	L-1011/WI	8 × 43 kg Orbcomm
20	26 February 1998	XL	L-1011/VAFB	132 kg SNOE; 45 kg Teledisc T-1
21	2 April 1998	XL	L-1011/VAFB	250 kg TRACE
22	2 August 1998	HAPS	L-1011/WI	8 × 43 kg Orbcomm
23	23 September 1998	HAPS	L-1011/WI	8 × 43 kg Orbcomm
24	23 October 1998	Std	L-1011/K5C	540 kg STEX
25	6 December 1998	XL	L-1011/VAFB	283 kg SWAS
26	5 March 1999	XL	L-1011/VAFB	270 kg WIRE
27	18 May 1999	HAPS	L-1011/VAFB	123 kg TERRIERS; 45 kg MuBLcom
28	4 December 1999	XL	L-1011/VAFB	7 × 43 kg Orbcomm
29	7 June 2000	XL	L-1011/VAFB	TSX-5
30	9 October 2000	XL	L-1011/VAFB	HETE-2
31	5 February 2002	XL	L-1011/KSC	HESSI
32	25 January 2003	XL	L-1011/KSC	SORCE
33	28 April 2003	XL	L-1011/KSC	GALEX
34	26 June 2003	XL	L-1011/KSC	OrbView-3
35	12 August 2003	XL	L-1011/KSC	SCISAT-1
36	15 April 2005	XL	L-1011/VAFB	DART
35	22 March 2006	XL	L-1011/VAFB	STS5

*launcher failure. DFRC = Dryden Flight Research Center; KSC = Kennedy Space Center; VAFB=Vandenberg Air Force Base; WI = Wallops Island. Number 6 failure: the review board concluded the guidance software was unable to cope with unpredicted aerodynamic responses on the first XL vehicle. The design was then tested in wind tunnels to develop an aerodynamic model, rather than previous reliance only on computer modelling. Number 9 failure: stage 1/2 interstage remained attached during stage 2 burn (one of three separation skids improperly installed) satellite failed to separate from Pegasus final stage. Flight 14 from Wallops Island, 15 from Canary Islands.

is undertaken horizontally. Pegasus is attached to the carrier aircraft at T–1 day. Takeoff is 1 hour before release.

QuickReach

Current Status

In November 2005, the AirLaunch consortium, with its QuickReach booster, was selected by the Defense Advanced Research Projects Agency (DARPA) to carry out Phase 2B of the agency's Falcon responsive space launch programme. AirLaunch LLC was the only competitor selected for this phase, which leads to a Critical Design Review (CDR) in the fall of 2006. The contract includes an option for Phase 2C, which includes a flight test in 2008. The goal is a vehicle that can be launched for USD5 million or less.

Background

AirLaunch LLC was established in 2003 to compete in the Falcon programme. AirLaunch LLC is a consortium with a number of experienced partner companies including: Space Vector, which has integrated and launched suborbital target vehicles in support of missile defense programmes; Universal Space Lines LLC, based on the core team of the McDonnell Douglas Delta Clipper project; and propulsion supplier HMX of Reno.

The goal of the Falcon programme is to develop a small booster which can deliver a 450 kg payload to a 200 km orbit, that can be launched within 24 h of a decision to go and that can ultimately achieve a surge rate of 16 launches per day. AirLaunch is attempting to meet this goal with a two-stage launch vehicle, fuelled by LOX and liquid propane, which is launched from the cargo bay of a Boeing C-17 military transport.

The company ruled out sea launch because of the high cost of the ships of platforms that would be needed to meet surge launch requirements, and because of weather limitations. Ground launches cannot be responsive, AirLaunch argues, because launch is not possible if there is traffic in the range area. By using a standard, unmodified cargo aircraft, surge capacity is a matter of the number of aircraft that the Air Force can make available. If there is sea traffic in the first-stage impact zone, the launch point can be moved. The air-launch system is also largely immune to weather, and payload is increased by up to 30 per cent due to lower drag (the rocket does not have to accelerate through dense low-level air) and the use of a first-stage nozzle sized to near-vacuum conditions.

AirLaunch chose liquid propulsion because its higher Isp made it possible to meet performance goals with a two-stage vehicle, within the structural limits of a C-17 launch. Solid boosters are more expensive, rely on a small vendor base and have to be handled as munitions: the AirLaunch vehicle will be stored without any propellant. The company felt that hybrids were risky given the vehicle's size and timescale.

The rocket motors are pressure-fed with no turbopumps. No secondary pressurisation is required because of the high-altitude launch. An unusual feature is that the second-stage nozzle is submerged in the first-stage fuel tank, reducing overall length and weight and lifting constraints on the length of the second-stage nozzle.

The rocket is delivered in a Storage and Launch Container (SLC) and remains there until it is fired. The SLC provides thermal management when the rocket is fuelled and loaded on the aircraft. The rocket rests on rubber-tyred rollers (aircraft nosewheels, in fact) and is launched tail-first by gravity: the aircraft's rear ramp and the SLC open, the roller system extending to the lip of the ramp; the aircraft flies at a shallow deck angle and the rocket rolls out. A stabilising drogue attached to the rear of the rocket slows its pitch-up – the result of rolling off the ramp – and orients it into firing position.

AirLaunch released a dummy QuickReach booster from a C-17 in September 2005. Other milestones followed in January 2006 – with ground tests of the gas-pneumatic stage separation system – and in late February and early March, with the first hot-firings of the second-stage engine at a test stand in Mojave, California. Each test lasted 13 seconds, with two seconds of ignition and 11 seconds of full burn.

Specifications

Launch weight: Approximately 32 t to 34 t
Fuel: Propane
Oxidizer: LOX
Fuel feed: Vapor pressure
Launch: 25,000 ft, 100 m/sec
First stage burn: 110 sec
Staging: 161,000 ft altitude, 2,365 m/sec
Orbit insertion: T+366 sec, 1,150 km downrange
First stage impact: T+390 sec, 650 km downrange
Performance: 450 kg to 200 km, 28.5° orbit

Scorpius

Current Status

The Scorpius rocket family is being proposed by Microcosm Inc of El Segundo, California. As a low-cost spacelift system. The company was awarded a contract by the Defense Advanced Research Projects Agency (DARPA) for Phase 1A of the Falcon programme, but was not selected for Phase 2. However, hardware work continues to be supported by the Air Force Research Laboratory's (AFRL) Space Vehicles Directorate.

Background

Founded in 1984 as a diverse space technology company, Microcosm has been working with AFRL's Launch Vehicle Technology (LVT) programme since the mid-1990s on a modular rocket family based on clustered, scalable fuel tank and motor systems.

The root of the Scorpius philosophy is that in order to be responsive, a launch system must be inexpensive enough to be held in inventory. As a result, the design emphasises simplicity and commonality over high performance. Orbital systems have three stages to orbit: strap-on boosters, a first-stage core – all identical – and a third stage with the same tank design; 95 per cent of the vehicle weight is made from common components. The three-stage launch makes it possible to use high-margin, moderate-performance modules based on pressure-fed LOX and Jet-A fuel.

Scorpius modules are made from filament-wound carbon-fibre composite, using no-autoclave processes and automation to reduce cost. The goal is a tank cost of USD35 per litre of space, and a weight of under 1.8 kg per cubic foot at 4,130 kilopascals pressure, which is provided by gaseous helium. The motors are claimed to be 20 to 100 times less expensive than traditional pump-fed motors and have ablatively cooled nozzles.

The Scorpius system is designed to make efficient use of labour at the work site, avoiding the need for a standing army of people who are only fully occupied in the run-up to a launch. Major components of the modules – LOX and fuel tanks and motors – are delivered to the site and assembled into an inventory of vehicles on-site, so that the personnel are busy assembling vehicles between launches. The vehicle itself is "short and fat" and relatively light when it is not fuelled, so it requires no launch tower – a standard cherry-picker platform will suffice – and can be towed to the launch pad on a flatbed truck. Fuelling and pressurant loading take place at the pad.

Microcosm's near-term goal is the Scorpius Sprite small launch vehicle, with a 370 kg payload to a 185 km orbit. The first stage comprises six 107 cm diameter modules arranged around an identical common core. The company is proceeding towards this goal incrementally with AFRL support. A small suborbital demonstrator, the SR-S, was launched in January 1999. It had a 45 cm diameter airframe and a 22 kN engine. The SR-XM-1, with two similar engines and a 107 cm diameter, flew in March 2001. Both were launched from White Sands Missile Range.

Microcosm is under contract to launch a third demonstrator, SR-M, from White Sands in early 2007. Essentially, SR-M is a representative Sprite module with an 88 kN engine. Microcosm announced in May 2005 that it had successfully completed a series of tests on the 88 kN engine at Edwards AFB, using funding from the DARPA Falcon programme. The next intermediate step will be the SR-2 suborbital launcher, which essentially is a Sprite without the upper stage.

Microcosm also has plans for larger vehicles, starting with a light-medium lifter which represents a four-times scale-up of Sprite, with 135 cm diameter modules and 350 kN engines, and continuing to heavy-lift vehicles.

Specifications
Sprite LEO
Number of stages: 3
Lift-off mass: 36.3 t
Empty mass: 3.31 t
Length: 16.5 m
Diameter: 3.4 m
Payload to LEO: 260 kg to 400 km, 51.6° orbit
Propellant: LOX/RP
First stage propulsion: 6 × 8.8 kN Microcosm pressure-fed engines
Second stage propulsion: 8.8 kN Microcosm pressure-fed engine
Third-stage propulsion: 10.2 kN Microcosm pressure-fed engine

Taurus

Current Status
Taurus has successfully completed six out of seven launches, the most recent in May 2004.

Background
DARPA contracted OSC in 1989 to build the Taurus rapid response orbital launcher using the company's Pegasus as a baseline. OSC had already taken on Space Data Corporation of Tempe, Arizona in November 1988. SDC had launched a wide range of sounding rockets, including more than 600 suborbital boosters in 35 different configurations.

The original objective of the Taurus project was to create a self-contained road transportable system capable of rapid response from austere sites. The vehicle and all its support equipment was required to be transported up to 1,600 km before the launch site was established and the vehicle integrated/tested within five days. After this setup, the vehicle could wait for the payload to arrive, from when launch takes a further 72 hours. In practice, Taurus has been launched from fixed sites, but is still a simple vehicle to operate with a minimum of specialised infrastructure.

Taurus' three upper stages are derived from Pegasus. USAF launches use an ATK Peacekeeper stage 1. Commercial launches use an ATK Castor 120, which uses the same technology. An uprated option, flown in May 2004, is the Taurus XL, which uses the same uprated motors as the Pegasus XL and offers a 100 kg payload increase. A fourth stage (using an ATK Star 37 motor) has been offered but not flown, as has a configuration with strap-on boosters.

Taurus and Pegasus technology forms the basis of the Orbital Boost Vehicle (OBV) which Orbital produces for the US Missile Defense Agency's Ground-based Midcourse Defense (GMD) system.

Specifications
Taurus
First launch: 13 March 1994 from VAFB
Number launched: 7 to March 2006 (including 1 failure)
Launch sites: Vandenberg AFB
Principal applications: medium LEO payloads
Vehicle success rate: 85.7%
Performance: Standard Taurus: 1,250 kg to 400 km, 28.5° LEO; Taurus XL: 1,450 kg to 400 km, 28.5° LEO.
Number of stages: 4, optional 5
Overall length: 26.72 m launch 1; 27.56 m standard; 29.12 m XL/XLS
Principal diameter: 236.3 cm stage 1 (233.7 cm for launch 1)
Launch mass: 72,576 kg (68,040 kg for launch 1)
Guidance: autonomous autopilot hosted in Oettle & Reicher RCOM06OSC 68020-based flight computer contains the mission specific mission data load (MDL). Inertial/attitude data provided from Litton's LR-81 Inertial Measurement Unit (IMU) aided by position data from Trimble GPS receiver. The autopilot processor monitors GN&C performance and processor data are formatted within the flight computer for direct downlink via the vehicle telemetry system.

Taurus stage 0
Comment: The first vehicle (and any later USAF missions) utilised a Peacekeeper stage 1. Commercial launches adopt Thiokol's Castor 120 motor, details of which are listed below.
Motor: Thiokol Castor 120 solid. Employs Toray 100068 graphite epoxy wound composite case with an aramid-filled EPDM rubber internal insulator. External protection from aerodynamic heating also provided by EPDM rubber layer. The nozzle features a carbon phenolic ablative exit cone with a graphite epoxy structural over-wrap and a 3-D carbon/carbon integral throat/entry. Nozzle mounted in a laminated rubber/glass fibre flexseal for ±5° vectoring. The 88% solids HTPB propellant is machined for fine tailoring of thrust profile. Igniter mounted in the forward dome. Redundant linear shaped charge (LSC) in external raceway allows flight termination.
Stage length: 12.8 m
Diameter: 233.7 cm
Thrust: 1,615 kN average vacuum.
Burn time: 82.5 s
Propellant mass: 48,720 kg
Separation: aft skirt extension incorporates redundant MDF to sever connection with launch stand. Aluminium interstage uses linear-shaped charge to release stage 1. Interstage vent covers released when bolt cutters separate forward and aft clamp bands.
Flight sequence: stage 0 TVA enabled at T2 s. Internal count sequencer performs nozzle sweep then commands simultaneous release from stand and ignition. Flies preprogrammed pitch profile to limit aerodynamic loading; roll not controlled. Vent panels released just before burnout, relieving pressure for stage 1 ignition (keyed to predetermined thrust tail-off; staging occurs 0.9 s later).

Taurus stage 1
Motor: ATK Orion 50S (50XL stretched by 137 cm for XL/XLS). All three Alliant motors use IM7/55A graphite epoxy wound composite cases with aramid filled EPDM rubber internal insulator and integral skirts. Flight termination charges mounted on the aft domes of stage 1/2 motors for range safety requirements. Stage 1 employs silicon elastomer flexseal nozzle with AlliedSignal hydraulic cold gas blowdown TVAs. Changes from Pegasus include removal of the wing saddle, thickening of the aft skirt and skirt doublers, revision of the aft attach ring and aft skirt for a tapered unit that mates to the interstage, and a vectored nozzle (4°) to replace fins for vehicle control.

Orbital's Taurus rocket on the launch pad for GFO/ORBCOMM mission 9 February 1998 at VAFB 0024769

Stage length: 8.6 m
Diameter: 127.1 cm
Thrust: 471.511 kN (627.199 kN) average vacuum
Burn time: 72.4 s
Mass at ignition: 13,270 kg (16,230 kg)
Propellant mass: 12,143 kg (15,072 kg)
Separation: the forward skirt, which also serves as interstage adaptor, incorporates LSC
Flight sequence: stage 1 TVA enabled just before stage 0 burnout. Following release of the interstage vent panels, stage 1 ignites, followed immediately by the LSC firing in the 0/1 interstage. Stage 1 also follows a preprogrammed pitch profile. Nozzle vectoring provides pitch/yaw control during powered flight. Burnout occurs typically 154 s (151 s), followed by staging at 158.2 s (150 s). Springs push the stack away, providing clearance for stage 2 ignition.

Taurus stage 2
Motor: Alliant Techsystems Orion 50 (50XL stretched 45.0 cm for XL/XLS). Changes from Pegasus are reinforcement of the forward/aft skirts and attach rings. Stage 2 also employs silicon elastomer flexseal nozzle and Parker Bertea Aerospace electromechanical actuators for TVC. A graphite epoxy facesheet/syntatic foam core sandwich structure boat-tail transitions from the stage 2 diameter to the 160 cm diameter fairing. A self-contained non-contaminating frangible separation system on the boat-tail's forward end allows stage 2 to be jettisoned. Four springs provide separation for the upper stage/payload.
Stage length: (310.6 cm)
Diameter: 127.1 cm
Thrust: 115 kN (160.136 kN) average vacuum
Burn time: 75.1 s
Mass at ignition: 3,370 kg (4,314 kg)
Propellant mass: 3,024 kg
Separation: frangible joint at boat-tail forward end allows separation of avionics skirt/stage 3/payload
Flight sequence: stage 2 ignites at 165 s as the remainder of the 1/2 interstage is jettisoned. Payload fairing jettisoned at 168 s. Stage 2 burnout at 237 s. Roll control during burn and three-axis control during coast by six N_2 thrusters in two Parker Bertea Aerospace modules on avionics section.
Attitude control: as stage 1

Taurus stage 3
Motor: Alliant Techsystems Orion 38 (also for XL/XLS). Stage 3 also employs vectored nozzle for pitch/yaw control during burn. Head end grain with toroidal igniter. Unlike Pegasus, the motor is not a load-carrying member but is cantilevered from its forward attach ring. The annular avionics shelf encircles the motor to mount most of the vehicle's electronics in addition to the N_2 attitude control system. This arrangement allows replacement of

the motor with a spinning upper stage, such as Star 37, for perigee kick.
Stage length: 133.9 cm
Diameter: 96.5 cm
Thrust: 31.8 kN average vacuum.
Burn time: 68.5 s
Mass at ignition: 893 kg
Propellant mass: 771 kg
Flight sequence: stage 3 ignites at apogee for orbit circularisation, using nozzle flexing for pitch/yaw control and the N_2 jet system for roll control and three-axis control during coast. Following injection, stage 3 manoeuvres to prescribed payload deployment attitude and provides a separation command to the payload, typically 30 s after burnout.

Fairing and payload accommodation
Two bisector fairings are available: 1.6 and 2.3 m diameter external. Fairing shells are fabricated by Courtaulds Structural Composites, Inc (1.6 m) and R Cubed (2.3 m) from graphite epoxy facesheets and aluminium honeycomb core. An OSC proprietary frangible joint separation system secures the fairing halves at the base and along the longitudinal seams until jettison during stage 2 burn. A pneumatic thruster system ejects the fairing, which is guided by fall-away hinges. Class 10,000 is maintained inside the fairing at all times, beginning from encapsulation until launch. On the ground, air conditioning maintains 15.6+4/7°C and 2065% humidity. The payload can be provided with 28 V DC (up to 140 W) until T0. 90 passthrough circuits allow payload commanding and monitoring before launch. The Taurus flight computer can issue up to five discrete payload commands and can monitor five discrete talkbacks. RS-422/485 serial data interfaces can also issue commands and accept payload telemetry in the vehicle data stream. Four pyro commands can be issued. Data for 2.6 m fairing in ().
Payload volume: 1.37 m (4.06 m) diameter × 2.8 m (5.84 m) long
Acceleration load: 6.9 g longitudinal max at end of stage 1 burn
Acoustic load: 133.5 dB OASPL (XLS: 138 dB OASPL)

Vehicle and Payload Processing
Stages are pre-integrated and tested at a Stage Integration Facility. For early missions, this is the Missile Assembly Building at VAFB. Stage and launch support equipment are transported to the launch site as late as 8 days before launch. The LSV launch support van is positioned outside the caution/hazard corridor and connected to the pad via a fibre optic bundle. Portable generators can power the system if facility power is not available. Once the launch stand is bolted to the concrete pad, stage 0 is mounted and scaffolding erected up to its top. In parallel, the upper stages are processed in an adjacent integration tent. Jumper harnessing between the mated upper stages and the erected stage 0 allows end-to-end integrated system tests and mission simulations.

In parallel with (or before) vehicle integration, the payload is processed and encapsulated in an appropriate facility. OSC provides the payload cone, mounted on a GSE handling adaptor, for spacecraft integration/test. The fairing is then mated to the cone, encapsulating the cargo element. The environmental control system is connected and the cargo element moved to the launch. This can be as late as 72 h before launch. At the launch site, the cargo element is electrically and mechanically mated to the upper stages. This stack is then lifted and mated to stage 0. Final systems tests are completed, followed by alignment of the IMU. Vehicle closeout and final countdown starts at T7 h, controlled from the LSV. Dedicated consoles are provided for payload personnel, range safety personnel and the customer launch director as well as the OSC personnel controlling the launch.

Launch log

No	Date	Payload
1	13 March 1994	DARPASAT/STEP-0
2	10 February 1998	ORBCOMM G1/G2/ CELESTIS/GFO
3	3 October 1998	STEX/ATEX
4	21 December 1999	ACRIMSAT/ CELESTIS/KOMPSAT
5	12 March 2000	MTI
6	21 September 2001	QUICKTOMS/ ORBVIEW 4/ CELESTIS/SBD (Faliure)
7	20 May 2004	ROCSAT

SUB-ORBITAL

Canada

Black Brant

Current Status
NASA has been the most prolific user of Black Brant sounding rockets and has asked Magellan to evaluate a larger and more powerful variant. There have been some studies conducted by other manufacturers with a view to improving the lift capability.

Background
Bristol Aerospace is a member of the Magellan Aerospace group of companies. Named after a small, stocky goose with black head and neck, the Black Brant sounding rocket is manufactured by the Defence and Space business unit. Bristol Aerospace's solid propellant Black Brant launcher is offered in five basic versions (5, 8, 9, 10, and 12). Payloads of 70 to 850 kg can be delivered to 150 to 1,500 km from conventional boom rails or three to four fin towers, with the BB10 providing 18 minutes of μgravity conditions for 100 kg. >800 vehicles of all types have flown since 1962, of which some 280 (BB6/7) were meteorological rockets. The single stage BB5 forms the basis of all currently available versions, with the addition of upper stages and/or US Army surplus boost motors. Combination with Nike/Terrier boosters, respectively, creates the BB8/9 models. NASA continues to be the major user and the vehicle also flies under the CSA μgravity materials research programme. With the closure of the Churchill Research Range in 1984, the Canadian programme was interrupted but one or two missions are now conducted annually from US/European ranges. The BB9 also forms the basis for EER Systems Space Services Division's Starfire vehicle, first launched March 1989.

In addition to the supply of basic rocket vehicle hardware, Bristol provides a range of self-contained vehicle/payload support systems, including thrust termination, parachute recovery, payload separation, de-spin, telemetry systems and fairing deployment. Two launcher types are usually employed: a rail-type typically providing 6 to 8 m of guided rail travel and a 30 m travel tower. In cases where the normal dispersion is unacceptable, the Saab/SES S19 boost guidance system is available; Saab's Spinrac attitude control system provides attitude control for a spinning exoatmospheric stage.

A 16.0 kg and 22.6 cm high igniter housing module incorporating de-spin and payload separation systems caps the stage 1 motor. The forward end provides a separation interface, comprising a deploying manacle ring with associated pyrotechnically actuated release guns and springs to generate 0.5 to 5 m/s separation. A yo-yo system can accommodate the standard range and profiles of despin requirements. The housing also incorporates the deployment logic for each system, including battery power, sequencing timers and safety altitude switches. The payload can provide the latter functions.

A thrust termination system manufactured by Explosive Technology Inc is provided as a complete assembly with the exception of the range authority-supplied command receiver. Two 'paddles' housing the shaped pyrotechnic charge are attached to the motor's forward end, but all other components, including explosive transfer lines, safe/arm assembly, batteries and command receiver, are mounted in the igniter housing. If activated, two holes 180° apart are cut in the motor but the payload is unaffected and events continue as pre-programmed.

Black Brant has demonstrated reliability of over 98 per cent in more than 800 launches to date.

Specifications
Black Brant 5
Description: The single stage BB5, which forms the basis for all BB variants, is available in 3/4-fin versions, designated the 5B/5C, respectively.
First launch: June 1965.
Number operational launches: 139 to end-1995.
Success rate: 97% to end-1995.
Number of stages: 1 (solid).
Overall length: 5.29 m + payload bay/nosecone.
Principal diameter: 43.8 cm (151.4 cm fin circle diameter).
Launch mass: 1,263 kg + payload/nosecone.
Typical performance: 200 kg to 290 km.
Payload recovery system: two basic (forward/aft-mounted) designs are available. The forward version is housed in a 3:1 ogive stainless steel fairing, the forward section of which is deployed soon after burnout to expose the recovery section heatshield. The aft-mounted recovery system is housed in a cylindrical aluminium bay immediately forward of the payload separation module. Both are self-contained with the necessary deployment logic equipment. The system is automatically armed by the closure of timers and altitude switches on the upleg phase. The heatshield is deployed, beginning the parachute deployment sequence, at about 5 km during re-entry. Both systems are available with either an 8.5 m circular main canopy or a 14 m paraform type. Each recovery system is available in two versions. The forward type comes as a 44.1 kg/1.32 m long unit qualified to 340 kg/320 km apogee or a 50.9 kg/1.38 m long unit qualified to 450 kg/500 km. The aft cylindrical type comes as a 33.2 kg/40 cm long/43.8 cm diameter unit qualified to 340 kg/320 km or a 44.1 kg/52 cm long/44 cm diameter unit qualified to 450 kg/500 km. Two new recovery systems based on the existing forward-mounted hardware have been introduced. The HARS High-Altitude Recovery System extends the recoverable mass/apogee to 725 kg/1,000 km; it flew successfully on the first three Maxus flights, in May 1991 (150 km), November 1992 (718 kg/715 km) and November 1995 (505 kg/706 km). The second system is a water recovery version (WRSA) of the standard 450 kg/500 km system. It uses an integral parachute/flotation system and can accommodate both floating and sinking payloads. First operational flight was made successfully 9 December 1992. Two (successful) flights were made through 1995; no others are yet planned.

Black Brant 5 stage 1
Motor: Bristol Aerospace 26 KS 20000.
Overall length: 5.29 m.
Principal diameter: 43.8 cm.
Stage mass: 1,265 kg (propellant 1,000 kg).
Average thrust: 76.87 kN sea level.
Burn time: 33.0 s.
Payload environment/accommodation: A range of deployable and fixed nose fairings are available in conical (5:1) or ogival (3:1 or 4.25:1) fineness ratios. BB9/10 utilise lightweight 1.5 mm aluminium skins, the others 1.5 mm stainless steel. Deployable fairings are available in split clamshell design with a horizontal (for low spin rates) or vertical (high spin) hinge. Forward-ejecting fairings provide typical separation rates of 5 m/s and a lateral impulse can be provided to prevent vehicle re-contact or ensure a clear field of view for instrumentation.
Peak longitudinal acceleration: 13 g /181 kg payload.
Shock: instrument qualification to 100 g /11 ms.
Vibration: <10 g at any frequency.
Temperature: 50°C maximum

Black Brant 8
Description: BB8 was developed in 1974 as the first upgrade of the BB5, with the US Army surplus Nike booster providing a 40% apogee performance increase. Otherwise, vehicle configuration is identical.
First launch: December 1975.
Number operational launches: 104 to end-1995.
Success rate: 99% to end-1995.
Number of stages: 2 (solid).
Overall length: 8.995 m + payload/nosecone.
Principal diameter: 43.8 cm.
Launch mass: 1,850 kg + payload.
Typical performance: 200 kg to 400 km.
Payload recovery system: as BB5

Black Brant 8 stage 1
Motor: Nike M5E1.
Overall length: 370.4 cm.
Principal diameter: 41.9 cm.

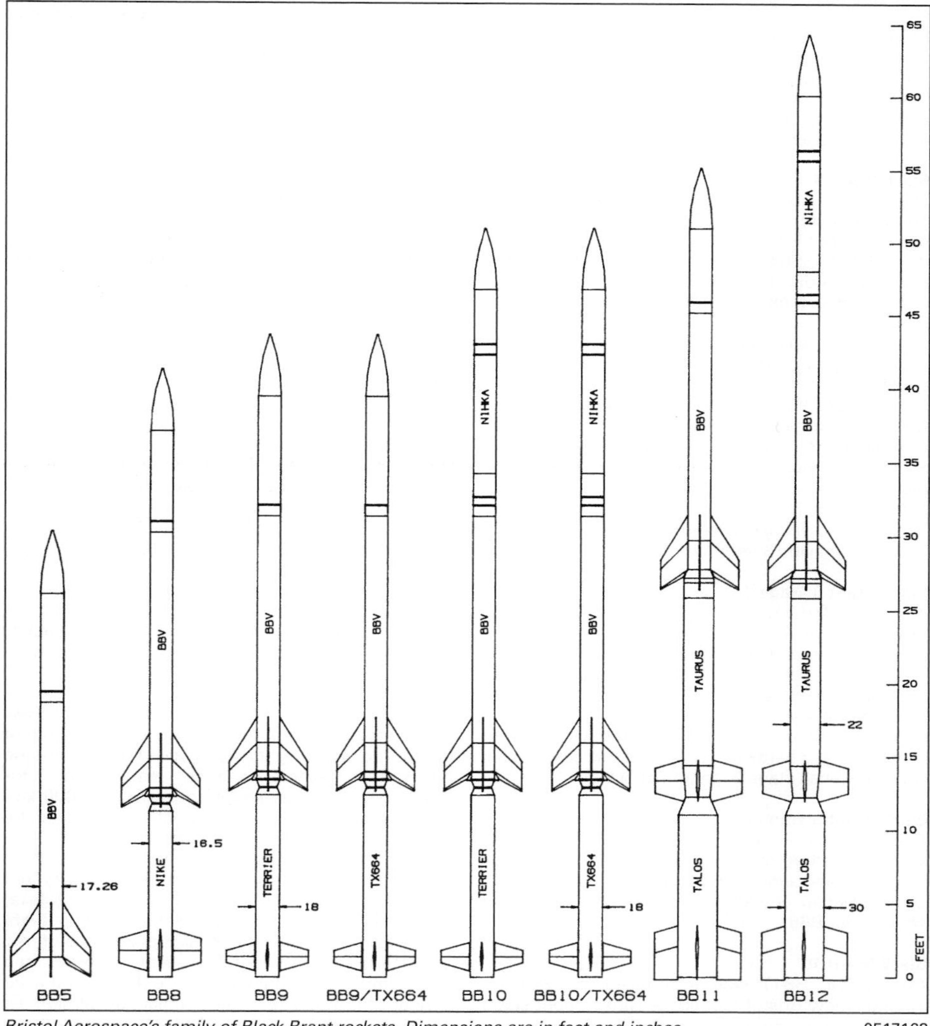

Bristol Aerospace's family of Black Brant rockets. Dimensions are in feet and inches 0517163

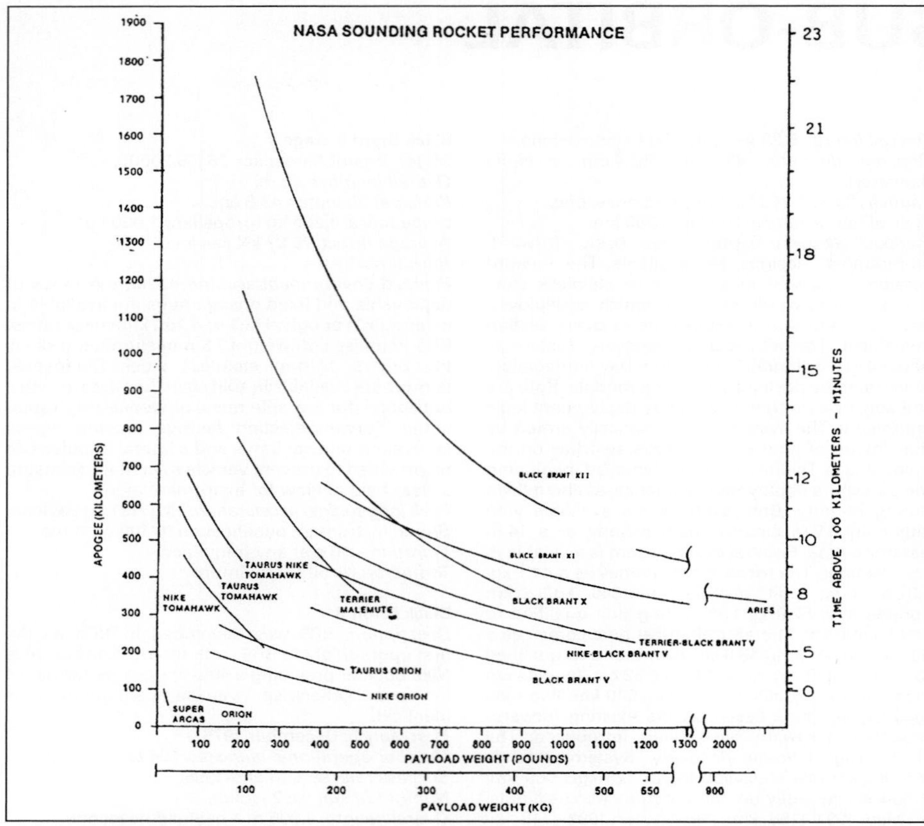

The Black Brant family provides NASA's sounding rocket programme, based at Wallops, with its highest performance vehicles (NASA) 0516864

Stage mass: 601 kg (propellant 361 kg).
Thrust: 195.6 kN average, sea level.
Burn time: 3.35 s.
Comment: At burnout, the two stages immediately separate by differential drag, augmented by a drag ring at the Nike's forward end. Stage 2 coasts for 5 s before ignition at 8.5 s/2.0 km.

Black Brant 8 stage 2
Configuration: Same as BB5, and the payload accommodation is identical to BB5's. Peak longitudinal acceleration is 10.3 g for 435 kg payload. Other payload environment specifications are the same as for BB5.

Black Brant 9
Description: The BB9 adds a US Army surplus Terrier booster to the BB5 to augment apogee performance by 75%. BB9/10 are also available with the Thiokol TX664 booster; 'mod 1' is added to the BB designation. It provides about 30% additional performance than the Terrier but is identical in geometry.
First launch: December 1981.
Number operational launches: 109 (+ 12 mod 1) to end-1995.
Success rate: 99% (+ 92% mod 1) to end-1995.
Number of stages: 2 (solid).
Overall length: 9.56 m + payload/nosecone.
Principal diameter: 43.8 cm (159.6 cm fin circle diameter).
Launch mass: 2,128 kg + payload.
Typical performance: 200 kg to 475 km.
Payload recovery system: as BB5

Black Brant 9 stage 1
Motor: US Army Terrier Mk 12 mod 1.
Overall length: 4.27 m.
Principal diameter: 45.7 cm.
Stage mass: 878.0 kg (propellant 534 kg).
Thrust: 257.7 kN average, sea level.
Burn time: 4.4 s.
Comment: Staging occurs at burnout using the same interface hardware as BB8. In this case, a 7.5 s coast precedes stage 2 ignition at 11.9 s.

Black Brant 9 stage 2
Configuration: Same as BB5's. Payload accommodation is the same as for BB5. Peak longitudinal acceleration is 13.4 g for 431 kg payload. Other payload environment specifications are the same as for BB5.

Black Brant 10
Description: BB10 was introduced in 1981 by adding a new Bristol Aerospace Nihka upper stage to the BB9 configuration. The flight sequence is as for BB9 until stage 2 burnout: the vehicle coasts for 33 s until the stage 2/3 separation mechanism is activated and the finless stage 3 ignites at 82 s/85 km. The payload fairing can be deployed before stage 3 ignition for a 4–8% apogee performance increase, typically at 60 s when dynamic pressure has fallen below 480 N/m². The Thiokol TX664 booster replaces the Terrier in the mod 1; see the BB9 for comments.
First launch: 14 August 1981.
Number operational launches: 30 (+ 4 mod 1) to end-1995.
Success rate: 93% (+ 100% mod 1) to end-1995.
Number of stages: 3 (solid).
Overall length: 11.88 m + payload/nosecone.
Principal diameter: 43.8 cm.
Launch mass: 2,560 kg + payload.
Typical performance: 200 kg to 750 km, 400 kg to 400 km.
Payload recovery system: high altitude version of the forward mount system

Black Brant 10 stage 1
Configuration: Same as BB9 stage 1

Black Brant 10 stage 2
Configuration: Same as for BB5

Black Brant 10 stage 3
Motor: solid propellant Bristol Aerospace Nihka 17 KS 12000, developed specifically for BB10.
Overall length: 1.92 m.
Principal diameter: 43.8 cm.
Stage mass: 402.1 kg (propellant 322 kg).
Thrust: 50.41 kN average, vacuum.
Burn time: 17.8 s.
Attitude control: none
Payload accommodation: Same as for BB5. Peak longitudinal acceleration 22 g for 136 kg payload.

Black Brant 12
Description: Bristol Aerospace and NASA Wallops completed studies in 1987 for a vehicle capable of delivering a 140 kg payload to 1,500 km altitude to satisfy plasma physics requirements for investigating ion acceleration over 900–2,000 km. A four-stage vehicle was selected, produced by replacing the BB10's Terrier booster with US Army surplus Talos and Taurus motors. Two demonstration flights were made from Wallops Flight Facility and the first operational flight from Poker Flat (Alaska) in March 1990. On the first demonstration flight in September 1988 the Nihka stage 4 suffered premature loss of thrust. The second vehicle, fired 5 December 1989, qualified the re-design and used a new spin motor to reduce vehicle dispersion. Of the six BB12s flown to end-1995, the one failure was due to Talos explosion at ignition. A BB11 3-stage version, without the Nihka stage 4, is available; its debut launch of 1 February 1990 carried a 547 kg plasma experiment to an altitude of 375 km from Poker Flat.
First launch: 30 September 1988.
Number launched: 2 BB11 + 7 BB12 to end-1995.
Success rate: 100% BB11 + 86% BB12 to end-1995.
Number of stages: 4 (solid).
Overall length: 16.11 m + payload.
Principal diameter: 43.8 cm (upper stages).
Launch mass: 5,253 kg + payload mass.
Typical performance: 140 kg to 1,500 km, 300 kg to 800 km.
Payload recovery system: as BB10

Black Brant 12 stage 1
Motor: US Army Talos.
Overall length: 3.51 m.
Principal diameter: 79.0 cm.
Stage mass: 2,053 kg (propellant 1,285 kg).
Thrust: 489.1 kN average, sea level.
Burn time: 5.42 s.
Separation: drag

Black Brant 12 stage 2
Motor: Taurus.
Overall length: 4.18 m.
Principal diameter: 58.1 cm.
Stage mass: 1361 kg (propellant 755 kg).
Thrust: 490.7 kN average, sea level.
Burn time: 3.30 s.
Separation: spring

Black Brant 12 stage 3/4
Configuration: Same as for BB10 stages 2/3, respectively.

India

Rohini Sounding Rockets (RSRs)

Background
India's space programme began in November 1963 with the launch of a two-stage US Nike-Apache sounding rocket from Thumba, near Trivandrum. Production of indigenous Rohini Sounding Rockets (RSRs) at VSSC created the core of expertise for an orbital launcher programme, beginning with the 75 mm diameter RH-75. Four Rohini models remain in service, launching at an average of one a week for meteorology (RH-200), middle atmosphere (RH-300) and ionosphere (RH-560) investigations from SHAR, Thumba and Balasore. The Indians favour the RH-200s, launching about 40 during 1987. These rockets carry radar chaff out of Balasore to test and calibrate Indian Meteorological Department equipment.

The RSR programme is sited at the Vikram Sarabhai Space Center, which includes a Rocket Propellant Plant and Rocket Fabrication Facility.

The Rohini 300 provides access to the middle atmosphere for Indian scientists (ISRO) 0516907

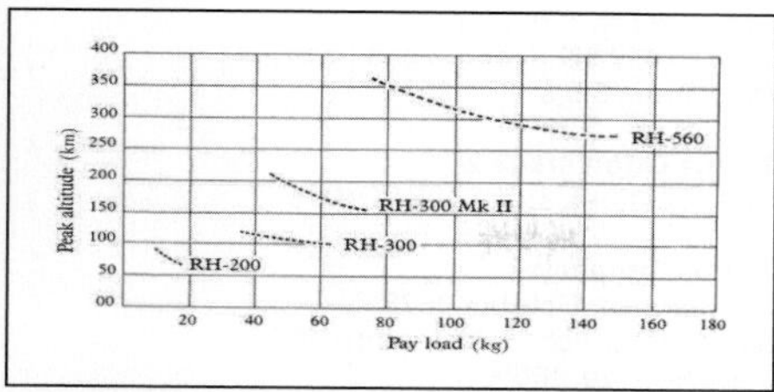

Rohini sounding rocket performance curves. The improved RH-560 will deliver 100 kg to 500 km
0516908

Several organisations in India have begun an extensive marketing campaign to offer Rohini Sounding Rockets in Europe and the United States.

Specifications

Rohini 200

Applications: meteorological sonde up to 80 km
Number launched: 0 in 1995, 18 in 1994, six in 1993
Number of stages: 2 (solid)
Overall length: 3.60 m
Principal diameter: 207 mm
Launch mass: 108 kg
Typical performance: 10 kg to 80 km

RH-200 stage 1

Motor: solid from VSSC
Overall length: 1.48 m
Principal diameter: 0.21 m
Average thrust: 16.89 kN sea level
Burn time: 5.7 s
Attitude control system: spin at max 270 rpm

RH-200 stage 2

Motor: RH-125 solid from VSSC
Overall length: 2.12 m
Principal diameter: 0.12 m
Average thrust: 10.49 kN sea level
Burn time: 2.9 s action time

Rohini 300

Applications: middle atmosphere probing above 100 km
Number launched: 8 to May 1999
Number of stages: 1 (solid)
Overall length: 4.14–6.84 m depending on payload
Principal diameter: 305 mm
Launch mass: 370 kg
Typical performance: 50 kg to 140 km

RH-300 stage 1

Motor: solid from VSSC
Overall length: 2.844 m
Principal diameter: 305 mm
Propellant mass: 240 kg
Average thrust: 37.94 kN sea level
Burn time: 14.5 s

Rohini 300 Mk II

Comment: The RH-300 Mk II was introduced during 1987 for investigations of the lower ionosphere, beyond the RH-300's capabilities. Major modifications included stretching motor length 1.2 m, reducing casing thickness by 0.4 mm from 2.0 mm, and adopting case-bonded HTPB propellant for high volumetric loading and burning characteristics resembling those of a two stage system. The resulting propellant mass increase was 100 kg.
Applications: upper middle atmosphere, lower ionosphere probing.
First launch: 8 June 1987
Number launched: 6 to end-1995
Number of stages: 1 (solid)
Overall length: 5.39–5.89 m depending on payload.
Principal diameter: 304.7 mm
Launch mass: 510 kg
Typical performance: 60 kg to 160 km
Spin rate: 6 rps

RH-300 Mk II stage 1

Motor: solid from VSSC
Overall length: 3.4 m
Principal diameter: 304.7 mm
Propellant mass: 330 kg of HTPB propellant
Average thrust: 39.14 kN sea level
Burn time: 20.2 s

Rohini 560

Comment: The RH-560 is India's largest sounding rocket, although a 560/300 Mk II is under development to deliver 100 kg to 500 km. A static test of an RH-560 with case-bonded HTPB was successful in 1995, and flew in 1995.
Applications: ionospheric and astronomical payloads
Number of stages: 2 (solid)
Number launched: 33 to end-1995 (1995:1; 1994: 0; 1993: 1)
Overall length: 8.376–9.176 m depending on payload
Principal diameter: 561 mm
Launch mass: 1,350 kg
Typical performance: 100 kg to 350 km
Spin rate: 6 rps

RH-560 stage 1

Motor: solid from VSSC
Overall length: 3.29 m
Principal diameter: 0.56 m
Average thrust: 75.97 kN sea level
Burn time: 19.3 s

RH-560 stage 2

Motor: solid from VSSC (based on RH-300)
Overall length: 3.78 m
Principal diameter: 0.30 m
Average thrust: 37.94 kN sea level
Burn time: 14.5 s action time

RH-560 Mk II

Overall length: 9.55 m
Number of stages: 2
Lift-off weight: 1,635 kg
Diameter: 56.1 cm
Payload: 100 kg
Spin rate: 6 rps

International

Maxus

Current Status

Astrium and the Swedish Space Corporation jointly developed the Maxus suborbital vehicle to extend the microgravity time provided by the Skylark-based Texus and the Terrier Black Brant-based Maser rockets. Thiokol's Castor solid propellant motor was selected following 1987–88 studies. Already qualified as a Delta strap-on booster, it can handle 420 kg useful payload to 850 km altitude, providing >14 min microgravity. Sweden's Guidance Control System ensures the rocket and payload descend within Esrange with <7 km 1 {gs} dispersion. Laboratories offering basic equipment are provided at the site to assist experimenters' final flight preparations. At least one flight was planned annually, but ESA budget limitations now restrict them to one every 1 ½–2 yr. A launch campaign typically requires 3 weeks in Apr or Nov/Dec, balancing daylight and the need for frozen lakes to avoid water recovery. User cost is about USD10,000/kg. DASA/ERNO has invested about DEM20 million (principally in 10 Castors) and SSC about SEK60 million.

The debut launch on 8 May 1991 attained only 200 km altitude instead of the 800 km planned. The flight was smooth to 37 s/21 km but then began pitching; at 46 s the all-ESA payload separated and was recovered. It was later determined that electrical cables of the nozzle's TVC system had burned through; thermal protection was added for future vehicles. ESA paid about DEM17 million for the flight and was guaranteed a reflight in the event of failure; insurance was 3 per cent. The second flight, Maxus 1B was fully successful 8 Nov 1992, providing 12 min 32 s μg time to the seven experiments. Five were controlled by telescience, including one via ESA's Olympus satellite. The third flight, Maxus 2 was successful 28 Nov 1995, again with an ESA payload (505 kg), attaining 706 km to provide min microgravity to eight experiments. The fourth flight, Maxus 3, took place on 24 November 1998, followed by the fifth flight, Maxus 4, on 29 April 2001. Maxus 5 is due for launch in 2004.
First launch: 8 May 1991
Number launched: 3
Success rate: 67%
Launch site: Esrange, Sweden
Number of stages: one (solid)
Overall length: 14.877 m min, 15.460 m max
Principal diameter: 1.018 m
Launch mass: 11,800 kg + 420 kg payload + 300 kg service systems
Typical performance: 720 kg total to 850 km, providing 14 min of useful microgravity time
Payload recovery system: flat spin deceleration, subsonic parachute descent (10 m/s landing), helicopter land retrieval

Maxus 1A launch preparations at Esrange (SSC)
0516856

Maxus stage 1

Motor: Thiokol Castor 4B (±6° nozzle TVC)
Overall length: 9.61 m
Principal diameter: 1.018 m
Thrust: 450 kN average
Burn time: 63 s

The motor is topped by an adapter, motor telemetry module and Saab Ericsson's Guidance Control System.

Payload Environment/Accommodation

Total length/mass of all payload modules must be <350 cm/420 kg; 6–8 can be carried. Modules are available on a rental basis, 640 mm dia, 5 mm aluminium wall (0.5 mm zirconium oxide thermal coating), 640 mm Radax mechanical joints. Access to experiments is possible up to 20 min before launch; the payload is returned to the launch site within an hour of landing. The module below the payload section provides telemetry (2.2–2.3 GHz S-band) + telecommand (450 MHz P-band) to all experiments via a standardised interface. If required, TV signals are transmitted at S-band in PAL B/G using pre-emphasis according to CCIR 405/625. After separation, the cold gas fibre optic gyro Attitude & Rate Control System in the service module above the payload section performs a despin manoeuvre before the microgravity phase begins at 90 s. If roll rate >1.5°/s or the cone angle >45° (for the RF downlink) during the microgravity phase, ARCS activates in low thrust mode to minimise experiment disturbance below 2×10^{-4}g. At T-891 s, a pyrotechnic valve opens to dump the residual gas through two roll nozzles to create a >60 rpm spin for thermal distribution during re-entry.

Japan

S-520 sounding rocket

Background

Nissan's S-520 is the largest of JAXA's range of sounding rockets, capable of delivering 250 kg to 400 km. In addition to space science missions, it is employed for research μgravity experiments. The single solid propellant motor, with steel casing, is supported by four aluminium honeycomb tail fins with niobium/titanium alloy leading edges. The SS-520 is derived from the S-520 but with a more powerful second-stage and was first launched in January 1998. The SS-520-2 was launched 4 December 2000 and reached an altitude of 1,040 km. In its former guise as the ISAS, JAXA continues to conduct analysis of improvements to the S-520 for national and internationally co-operative activities.

Specifications

S-520

First launch: 18 January 1980 (first recoverable flight 5 September 1981)
Number launched: 24 to end-April 2000 (20 Kagoshima, 4 Andøya)
Success rate: 100% to end-April 1999
Launch site: Kagoshima (Japan), Andøya (Norway)
Performance: 150 kg to 350 km, using 82° launch angle
Number of stages: 1 (solid)
Overall length: 931.0 cm

Principal diameter: 524 mm
Launch mass: 2,285 kg
Guidance: fin spin stabilisation, but de-spin system included.

S-520 stage 1
Contractor: Nissan Motor Co
Overall length: 6,053 mm
Principal diameter: 524 mm (179 cm across four fins)
Thrust: average 143.17 kN vacuum
Burn time: 28.7 s effective

Payload accommodation
GFRP clamshell ogive nose, providing 494 mm payload base diameter, narrowing to 156 mm after 1,252 mm length. The support section below houses telemetry, timing, ignition relay, radar transponder and power supply equipment. Immediately below are the attitude control and recovery sections.

United States

Orion and Orion derivatives (sounding rockets)

Current Status
Several proposed initiatives for development of an improved Orion sounding rocket matching reaction from the user community.

Background
The Orion is a single-stage unguided rocket, incorporating a surplus US Army motor, optimised for 39 to 68 kg payloads. An Orion version with 20 per cent increased total impulse was tested successfully 5 April 1994 as the second stage on a Terrier. The test included a GPS receiver, attitude/rate gyros and a new water recovery system. This design has been packaged into a much more powerful derivative with a variety of stage configurations.

Specifications
Orion
First launch: May 1974
Number launched: 33 to end-1996
Success rate: 100%
Number of stages: 1 (solid)
Overall length: 534 cm
Principal diameter: 35.6 cm
Launch mass: 422.3 kg plus payload
Typical performance: 39 kg/88 km, 68 kg/71 km
Payload recovery system: payload can be separated if required.
Payload environment: separable clamshell nose accommodates standard 35.6 cm diameter, 183–254 cm length.
Guidance: unguided, three fins provide 4 rev/s roll stabilisation at burnout.

Orion Derivative Terrier-(Improved) Orion
Comment: A new two-stage vehicle employing surplus military motors. Stage 2 employs an enhanced version of the basic Orion. Performance is similar to Taurus Orion, but the environment is more benign, allowing bulbous (42.8 cm diameter) payloads.
First launch: 5 April 1994
Number launched: 3 to end-1996
Success rate: 100%
Number of stages: 2 (solid)
Overall length: 10.6 m (6.6 m without payload)
Principal diameter: 45.7 cm
Launch mass: 1,316 kg plus payload
Typical performance: 290 kg to 118 km
Payload recovery system: IRMA Ignition and Recovery Modular Assembly
Payload environment: 35.6 cm or 42.8 cm diameter payloads with various deployable nosecones. 19 g peak for 290 kg payload.
Guidance: unguided, fin spin-stabilised. Spin motors. Burnout roll rate about 6.5 Hz.

Terrier (Improved) Orion stage 1
Motor: US Army Aerojet solid, surplus
Overall length: 268.1 cm
Principal diameter: 35.6 cm
Stage mass: 422.3 kg
Thrust: 7.74 kN average sea level
Burn time: 32 s

Terrier-(Improved) Orion stage 2
Motor: improved XM22E8 Orion
Overall length: 268 cm
Principal diameter: 35.56 cm
Stage mass: 438 kg (293 kg propellant)
Thrust: unavailable
Burn time: 25.4 s (T+15 s ignition typical).

Nike Orion
Comment: A boosted version of the basic Orion vehicle, extending payload performance to >200 kg
First launch: September 1975
Number launched: 103 to end-1996
Success rate: 97%
Number of stages: 2 (solid)
Overall length: 880.9 cm
Launch mass: 1,021.8 kg plus payload
Typical performance: 68 kg/190 km, 204 kg/90 km
Payload recovery system: Wallops' 35.56 cm (14 inch) recovery system
Payload environment: as for Orion
Guidance: unguided, fin spin stabilisation

Nike Orion stage 1
Specifications as for Nike Tomahawk.

Nike Orion stage 2
Specifications as for the Orion vehicle, but the motor ignites 9 s after lift-off and the four fins are canted to generate 4 rev/s at burnout.

Nike Tomahawk
First launch: March 1965
Number launched: 230 to end-1996
Success rate: 97%
Number of stages: 2 (solid)
Overall length: 7.3 m
Principal diameter: 22.9 cm
Launch mass: 885 kg
Typical performance: 45 kg/370 km, 113 kg/220 km
Payload recovery system: non-standard
Payload environment: normally 22.9 cm diameter under 3:1 tangent ogive nose; a 30 cm diameter version is available but requires the larger fins. Payload length 183–305 cm. Clamshell nose is separable; payload can be separated if required. De-spin module is standard.
Guidance: unguided, fin spin-stabilised

Nike Tomahawk stage 1
Motor: Nike M5-E1 solid (US Army Allegheny Ballistics Laboratory)
Overall length: 345 cm
Principal diameter: 41.9 cm
Stage mass: 597 kg (342 kg propellant)
Thrust: 190.29 kN average sea level, total impulse 651.8 kNs at sea level
Burn time: 3.2 s
Comment: An interstage adaptor is bolted to Nike's forward face, its conical portion slip fitting into stage 2's nozzle, allowing drag separation. Four fins, each 0.23 m^2, are canted to generate 2 rev/s at Nike burnout.

Nike Tomahawk stage 2
Motor: Thiokol TE-M-416 solid
Overall length: 361 cm
Principal diameter: 22.9 cm
Stage mass: motor 245 kg (175 kg propellant)
Thrust: 48.9 kN average sea level, total impulse 416.6 kNs sea level
Burn time: 9 s, ignited 12 s after launch
Comment: Fin and shroud assembly mass 22 kg. Four fins, each 0.13 m^2, are canted to generate 6 rev/s at burnout. 0.17 m^2 fins are available for increased aerodynamic stability.

Taurus Orion
First launch: July 1977
Number launched: 53 to end-1996
Success rate: 96%
Number of stages: 2 (solid)
Overall length: 986.8 cm
Principal diameter: 57.8 cm
Launch mass: 1,800 kg plus payload
Typical performance: 68 kg/260 km, 227 kg/140 km
Payload recovery system: Wallops' 35.56 cm (14 inch) recovery system
Payload environment: standard payload 35.6 cm diameter, length 183–380 cm. Separable clamshell nose; payload can be separated.
Guidance: unguided, four-stage 10.45 m^2 Ajax fins are canted to generate 2 rev/s spin at burnout

Taurus Orion stage 1
Motor: US Army surplus solid (Allegheny Ballistics Laboratory)
Overall length: 4.5 m
Principal diameter: 57.8 cm
Stage mass: 1,363 kg
Thrust: 457 kN average sea level
Burn time: 3.5 s

Taurus Orion stage 2
Specifications as for Nike Orion.

Taurus Tomahawk
First launch: October 1978
Number launched: 14 to end-1995
Success rate: 100%
Number of stages: 2 (solid)
Overall length: 821 cm
Principal diameter: 22.75 cm
Launch mass: 1,621 kg plus payload
Typical performance: 27 kg/590 km, 59 kg/490 km
Payload recovery system: non-standard
Payload environment: as Nike Tomahawk
Guidance: unguided, fin spin stabilisation

Taurus Tomahawk stage 1
Specifications as for Taurus Orion stage 1.

Taurus Tomahawk stage 2
Specifications as for Nike Tomahawk stage 2, but ignition is 18 s after launch.

Taurus Nike Tomahawk
A boosted version of the Nike Tomahawk. Stage 2 hardware includes an interstage locking device that holds the Nike/Tomahawk motors together during the coast before Nike ignition. Standard stage 3 hardware includes a Tomahawk Firing De-spin Module.
First launch: September 1983
Number launched: 15 to end-1996
Success rate: 100%
Number of stages: 3 (solid)
Overall length: 11.64 m
Principal diameter: 22.86 cm (Tomahawk)
Launch mass: 2,229.9 kg plus payload
Typical performance: 32 kg/700 km, 125 kg/400 km
Payload recovery system: non standard
Payload environment: typically 22.9 cm diameter, length 137 366 cm under 3:1 ogive cone. 30 cm diameter is possible but special analysis is required because of high dynamic load factors.
Guidance: unguided, fin spin stabilisation

Taurus stage 1
Specifications as for Taurus Orion stage 1 but fins canted for 1 rev/s spin at stage 1 burnout.

Nike stage 2
Specifications as for Nike Tomahawk stage 1 but ignition is made 11 s after launch.

Tomahawk stage 3
Specifications as for Nike Tomahawk stage 2 but ignition after 18 s and action time is 11.5 s.

Terrier-Malemute sounding rocket

Current Status
The first Terrier-Malemute flew on 11 November 1974. Since then it has been launched 46 times. The most recent flight was the EQUIS II ionosphere mission sponsored by NASA, which took place from Kwajalein in the Pacific on 15 August 2004. There were five failures, all in the 1970s.

Background
The Terrier-Malemute, developed by Sandia National Laboratories, Albuquerque, New Mexico, is a high performance, two-stage vehicle used for payloads weighing less than 180 kg. Performance drops appreciably as payload weight increases. Bulbous diameter payloads can be accommodated, although the high dynamic pressures result in high aerodynamic heating rates and high vehicle structural loads. The vehicle is usually rail launched and can be accommodated at most launch ranges. It is capable of lifting a 90 kg payload to an apogee of 700 km or a 227 kg payload to 400 km. The rocket has proven popular for plasma physics and it has also been selected for interactive plasma/magnetosphere experiments and measurements.

The first stage booster consists of a Terrier Mk 12 Mod 1 rocket motor with four fins arranged in a cruciform configuration. The Terrier was developed by Hercules Powder Company as the first stage of the Navy's Terrier anti-aircraft missile. For a payload weight of 90 kg, the longitudinal acceleration during the boost phase is 26 g. The second stage propulsion unit is a Thiokol Malemute TU-758 rocket motor, fitted with downswept triangular fins, which is designed specially for high altitude research rocket applications.

The Malemute was developed in 1974 in an interagency programme with NASA, Sandia Laboratories, and the Air Force Cambridge Research Laboratories as sponsoring agencies. Designed to be flown with either the Nike or the Terrier first stage, the Malemute began flight tests in 1974. It was named for the Alaskan Eskimo people by the contractor, Thiokol Corporation, in Thiokol's tradition of using Indian-related names.

After some early failures, the Malemute second stage motor was modified by adding insulation and using a propellant that produced less Al_2O_3 agglomerate in the chamber. This modification, designated Malemute II, reduced the sensitivity of the motor to the motor case burn-through induced by the roll rate and experienced on some earlier flights. Two flight tests, including a single stage Malemute II and a Terrier-Malemute II, were made by Sandia in the early 1980s to qualify this modification.

NASA's sounding rocket programme is based at the Wallops Flight Facility in Virginia, using vehicles from commercial sources or developed by the agency. Some 25 launches are made each year from a range of sites, including Wallops itself, Poker Flat and White Sands. NASA currently employs 10 sounding rocket types, including the Terrier-Malemute and the Terrier-Orion. Extensive use is made of military surplus motors and all are unguided.

Specifications

Terrier-Malemute

First launch: June 1975
Number launched: 46 to end 2007
Success rate: 89%
Number of stages: 2 (solid)
Overall length: 7.2 m
Principal diameter: 40.6 cm
Launch mass: 1,478 kg plus payload
Typical performance: 91 kg to 670 km
Payload recovery system: Custom built
Payload environment: 26 g peak acceleration for 90 kg payload, accommodated typically in 40.6 cm, 3:1 ogive nose
Guidance: Unguided, spin stabilised by four 0.22 m^2 fins

Terrier-Malemute stage 1

Motor: US Navy Terrier Mk12 mod 1 (Hercules Powder Co and Allegheny Ballistics Laboratory)
Overall length: 4.27 m
Principal diameter: 45.7 cm
Stage mass: 878 kg (534 kg propellant)
Thrust: 257.7 kN average sea level
Burn time: 4.4 s

Terrier-Malemute stage 2

Motor: Thiokol Malemute TU-758
Overall length: Motor 3.302 m
Principal diameter: 40.6 cm
Stage mass: 600 kg (499 kg propellant)
Thrust: 64 kN average SL (total impulse 1,223 kNs SL)
Burn time: 20 s

SPACECRAFT

Communications
Defence
Earth Observation
Human
Navigation
Scientific
Technology Demonstration

COMMUNICATIONS

Argentina

Nahuel series

Current Status
In July 2007, it was announced that more than 50 per cent of the thrusters on Nahuel 1 (formerly known as Nahuel 1A) had failed, making North-South station keeping impossible. As a result, the satellite entered an inclined orbit and services were increasingly disrupted.

Background
In 1991, Argentina issued a RfP for a domestic/regional satellite communication system in late 1991. Argentina wanted to operate two Nahuel ('Jaguar' or 'Tiger') satellites at one of two sets of locations 59° W, 71.8° W, 75.8° W, 80° W and 85° W. The country planned to pay about USD250-300 million for the capability and generate USD60-70 million revenue annually. Bolivia, Brazil, Chile and Uruguay all planned to benefit from the system. To establish a framework, the Argentine government organised the Comision Nacional de Telecomunicaciones on 16 December 1992 to negotiate with a consortium headed by DASA, and including Aerospatiale, Alcatel Espace and Alenia Spazio. The agency signed a contract 27 May 1993 for a term of 24 years and five orbital slots, to serve the southern part of the continent for the first time with high performance Ku-band links. NahuelSat SA created by the three European negotiation partners had USD100 million in equity to own/operate the system.

In 1993, NahuelSat began providing satellite services in Argentina, Chile and Uruguay through Paracom Satellites S.A., the owner of the Nahuel C1 and Nahuel C2 (formerly Anik C1 and C2) satellites. These made up the "Interim System" but were decommissioned and replaced by Nahuel 1A, launched by Ariane 4 on 30 January 1997. Nahuel 1A provided the equivalent of 36 TV programmes (180 with digital compression), 18,000 duplex telephone calls or 9,000 VSAT interactive connections. Connections required 1–2 m dishes. The Ku-band satellite had three different beams specifically designed for each region: one for Argentina, Chile, Paraguay and Uruguay; another exclusively for Brazil; and the third for all of Latin America, the Caribbean and southern United States.

In 1998, NahuelSat S.A. was allocated an orbital position of 81° W on C- and Ku-bands with full coverage in South, Central and North America. A second Nahuel satellite was planned to operate in this position, with launch in 2003. Nahuel 1B was originally to be launched on the maiden flight of the Chinese CZ-3C booster, but NahuelSat S.A. ran into financial difficulties and the project was cancelled in 2004. The satellite was then sold to GE Americom as GE 5. When GE Americom was taken over by SES, it was renamed AMC 5.

In 2006, the Argentine government authorised the formation of a new national enterprise, ArSat, to take over Nahuel 1 for the remainder of its life and to plan for follow-on satellites so that Argentina could keep its two active orbital slots. The satellite ran into severe technical problems in July 2007 when more than half of its thrusters malfunctioned and its orbit began to drift.

Specifications
Nahuel 1A
Launch: 30 January 1997 by Ariane 44 LP from Kourou, French Guiana
Location: 71.8° W geostationary until August 2007.
Design life: 12 years
Contractors: Aerospatiale (prime), Alcatel Espace (payload), DASA (AOCS, solar arrays)
Transponders: 18 55 W TWTA 14/12 GHz Ku-band South American beams, bandwidths 54 MHz, EIRP 50 dBW, orthogonal linear polarisation.
Principal applications: TV distribution, telephony, VSAT data and business services
Configuration: Spacebus 2000 platform, 3-axis control, 22.4 m span across solar panels, single dual reflector multifeed antenna. GEO location maintained within ±0.05°
Mass: 1,780 kg at launch; about 1,100 kg on-station BOL
Body dimensions: 1.64 × 1.46 × 2.2 m
AOCS: Provided by DASA's 22 × 10 N + 400 N MMH/MON bipropellant system
Power system: Two solar wings providing 2.9 kW equinox EOL.

Artist's impression of the Nahuel 1A communications satellite (Nahuelsat) 1340830

Brazil

Brasilsat/Star One series

Current Status
Brasilsat A1 was de-orbited in March 2002. Brasilsat A2 was removed from service in February 2004. At the end of 2007, all Brasilsat B satellites were still operational. Star One C1, the first of Embratel's third generation satellites, was launched in November 2007 and will replace Brasilsat B2 at the 65° W position. Star One C2 is expected to be launched in February 2008 and located at 70°W, replacing Brasilsat B1.

The Brasilsat B series satellites were based on the cylindrical Hughes 376W bus (Boeing) 1341068

Background
Embratel was a pioneer in providing satellite services in Brazil, with the inauguration in 1969 of the Tanguá Earth Station in Rio de Janeiro using the Intelsat-1 satellite. In August 1982, Canada's Spar Aerospace Ltd. was awarded a USD125 million contract to build, under licence, two satellites based on Hughes' HS-376 design. The cylindrical, spin-stabilised satellites comprised a single antenna on a despun platform, with hydrazine thrusters and body mounted solar cells. The first satellite, Brasilsat A1, was launched in 1985, followed by the A2 satellite in 1986. Brasilsat A1 was the first national communications satellite to operate in South America. After it completed its contract life, Embratel leased the satellite to PanAmSat Corp. of Wilton, Connecticut in November 1997. With its antenna re-aligned at North America, Brasilsat A1 continued to generate revenue for PanAmSat until it was taken out of service in March 2002, more than 17 years after it was launched. In 1995, Brasilsat A2 was moved to 92° W and later replaced by Brasilsat B4.

In 1990, Embratel purchased two Hughes (now Boeing) 376W models for the second generation Brasilsat B series. Like the 376 model, the 376W spacecraft are spin-stabilised satellites, each with two cylindrical, telescoping solar panels. The first of these, Brasilsat B1, was launched in 1994, followed less than one year later by Brasilsat B2. In December 1995, rapidly increasing demand led Embratel to exercise an option for a third spacecraft. A fourth was ordered in June 1998. Brasilsat B3 and B4 were launched in 1998 and 2000 respectively. As part of the contract, Hughes Space and Communications split the ground station engineering work with Promon Engenharia SA of São Paulo. The Brazilian Institute of Space Research (INPE) in Sao Jose dos Campos, was the site for final system testing of Brasilsat B1 and B2. Brasilsat B3 and B4 were tested at Hughes.

Like the first-generation satellites, the Brasilsat B spacecraft provide basic telecommunications services: telephone, television, facsimile and data transmission, and business networks. However, the satellites are larger and more powerful, with greater capacity. Each B-series spacecraft has 28 active C-band transponders. The B1 and B2 satellites have one X-band transponder for military use. B3 and B4 do not have X-band, but offer 20 per cent greater signal power in C-band. To provide this extra power, B3 and B4 use 24 SSPAs with 15.3 watts and four with 16.5 watts. The 2.44 m dual-polarized C-band antenna delivers nationwide communications coverage. There is also a regional beam for the densely populated east coast, with a high-gain area for business networking in four cities (Belo Horizonte, Curitiba, Rio de Janeiro, and São Paulo). The direct-radiating X-band antenna on B1 and B2 is much smaller, about 0.25 m in diameter, but it covers South America and the south Atlantic.

Embratel's satellite operating arm is now known as Star One. 80 per cent of the capital in Star One

Star One C1 is the first of Embratel's third generation comsats (Embratel – Star One) 1341069

belongs to Embratel and 20 per cent belongs to GE. Star One C1, the first satellite of the C Series third generation, was launched in November 2007 in order to guarantee the continuation of satellite telephone, television, radio, data transmission and Internet services in Brazil. Star One C1 was manufactured by Alcatel Space in France, and is based on SB 3000 – B3 platform. It has nearly twice as much power as Brasilsat B2, which it will replace at the 65° W position. Its coverage area will extend beyond Brazil into other parts of South and Central America, and Florida. It has 28 transponders in C-band, 16 in Ku-band and one in X-band. C- band provides voice, TV, radio and data signals, including the Internet. Ku-band offers video transmission services directly to users, plus broadband Internet and telephone services to remote areas. X-band is exclusively for military use. Star One C1 will be controlled by Embratel's Guaratiba Station in Rio de Janeiro. Star One C2 is expected to replace Brasilsat B1 at 70° W.

Specifications

Brasilsat A1 and A2

Launched: Brasilsat A1: 8 February 1985 by Ariane 3 V12 from Kourou, French Guiana.
Brasilsat A2: 28 March 1986 by Ariane 3 V17.
Location: Brasilsat A1: 65° W geostationary.
Brasilsat A2: 70°W geostationary.
Design life: Brasilsat A1: 9 years.
Brasilsat A2: 11 years.
Contractors: Spar Aerospace Ltd. (prime contractor). Hughes (bus).
Transponders: 24 C-band with 6 spares, 10 W TWTA, EIRP >34 dBW over most of Brazil.
Principal applications: Satellite telephone, television, facsimile and data transmission.
Configuration: Hughes (now Boeing) 376 platform. Spin stabilised. Star-30B solid fuel engine (A1), Star-30BP (A2).
Mass: 1,140 kg at launch. 671 kg in orbit.
Power system: Body mounted solar cells. 982 W BOL, 799 W EOL.

Brasilsat B1–B4

Launches: Brasilsat B1: 10 August 1994 by Ariane 44LP from Kourou, French Guiana.
Brasilsat B2: 28 March 1995 by Ariane 44LP.
Brasilsat B3: 4 February 1998 by Ariane 44LP.
Brasilsat B4: 17 August 2000 by Ariane 44LP.
Location: Brasilsat B1: 70° W geostationary.
Brasilsat B2: 65° W geostationary.
Brasilsat B3: 84° W geostationary.
Brasilsat B4: 92° W geostationary.
Design life: B1 and B2: 12 years. B3 and B4: 12.6 years.
Contractors: Hughes Space and Communications (now Boeing) – prime.
Transponders: 28 active C-band transponders powered by twenty-four 13 W and four 15.5 W solid-state power amplifiers (SSPAs), EIRP 33 – 41 dBW. B1 and B2 satellites have one X-band transponder powered by a 40 W TWTA, EIRP 33 dBW. B3 and B4 use 24 SSPAs with 15.3 W and four with 16.5 W.
Principal applications: Commercial and military communications. Satellite telephone, television, radio, data transmission and Internet services.
Configuration: Hughes (now Boeing) 376W platform. Spin stabilised. 2.44 m diameter dual-polarized C-band antenna. B1 and B2 only: 0.25 m diameter direct-radiating X-band antenna.
Mass: 1,757 kg at launch. 1,052 kg BOL.
Body dimensions: 8.3 × 3.65 m (in orbit)
Power system: Body mounted, large-area silicon solar cells. Up to 1800 W power. Hughes-designed 18-cell NiH battery.

Star One C1

Launch: 14 November 2007 by Ariane 5-ECA from Kourou, French Guiana.
Location: 65° W geostationary.
Design life: 15 years.
Contractors: Alcatel Space (now Thales Alenia Space) – prime.
Transponders: 28 C-band transponders, 16 Ku-band, 1 X-band.
Principal applications: Commercial and military communications. Satellite telephone, television, radio, data transmission and Internet services.
Configuration: Alcatel Spacebus 3000 – B3 platform.
Mass: 4,100 kg at launch. 1,750 kg dry.
Body dimensions: 4.0 × 3.2 × 2.4 m
Power system: 10.5 kW BOL. Two solar arrays, 22.4 m span.

SCD satellites

Current Status

Brazil's two Satélite de Coleta de Dados (SCD, data collection satellite) spacecraft were reported to be performing nominally in 2006.

Background

The SCD satellites were launched in the 1990s to relay data from remote environmental monitoring stations.

The Brazilian government launched SCD-1 on 9 February 1993 by Pegasus into 725 × 790 km, 24.97° orbit. Little equipment degradation was observed after one year of life. The 115 kg satellite provided real-time relay of data from environmental data collection platforms using the Argos system frequency (401.65 MHz) and 401.62 MHz, which alone will handle 200 platforms. 65 DCPs were in place at end-1995; another 250 were added during 1996, mainly for monitoring hydrographic basins and for flood control. The 1.70 m high octagonal prism carries University of Sao Paulo solar cells on the eight outer and top faces to generate 60 W average in sunlight, supported by an 8 Ah Ni/Cd battery. S-band/UHF antennas are mounted on the top and bottom.

SCD-1 monitors the condition of the Amazon River basin and the surrounding rain forests. It also monitors CO_2 and ozone levels, mainly produced from burning vegetation.

The Brazilians spun SCD at release to 120 rpm and thereafter maintained simple attitude control by a magnetorquer. The main requirement for satellite survivability is to hold the bottom face away from the Sun for heat dissipation. The satellite contains 160 telemetry channels.

SCD-1 was ready for flight in 1991, but had to wait until the Brazilian government could select a launcher. OSC's Pegasus won the USD11.5 million contract in October 1991 and it was signed 20 August 1992. The SCD-1 satellite cost USD20 million; 73 per cent of the funding was expended in Brazil.

SCD-2 was launched by Pegasus 23 October 1998. The 115 kg satellite was placed in a 742 × 768 km, 25°, orbit. Another vehicle in the series, SCD-2A, was launched on 2 November 1997 by Brazil's VLS rocket, but a failure during ascent destroyed the vehicle before it reached orbit.

Canada

Anik E series

Current Status

The Anik E satellites have been retired. Anik F1 was retired in January 2005 and E2 was retired in November of the same year.

Anik F series

Current Status

In April 2007, Telesat successfully launched its latest Anik F series satellite, the Anik F3. The Anik F3 provides increased broadcasting and telecommunications capacity throughout North America.

The satellite was equipped with 32 Ku-band transponders and 24 C-band transponders. The spacecraft had a launch mass of approximately 4,634 kg, a solar array span of 36 metres when deployed in orbit, and spacecraft power of 10 kW at end of life. The Anik F3 has an estimated life expectancy of about 15 years.

The F3 also carried a Ka-band payload to supplement services being carried out by the F2 satellite.

Background

On 27 November 1997, Telesat Canada announced that it had obtained government approval for two orbital slots to launch Anik F series satellites to replace current Anik E satellites from 2000. Telesat selected Hughes 702 series as the platform for its Anik F series, each of which would carry 84 active transponders to provide telecommunications

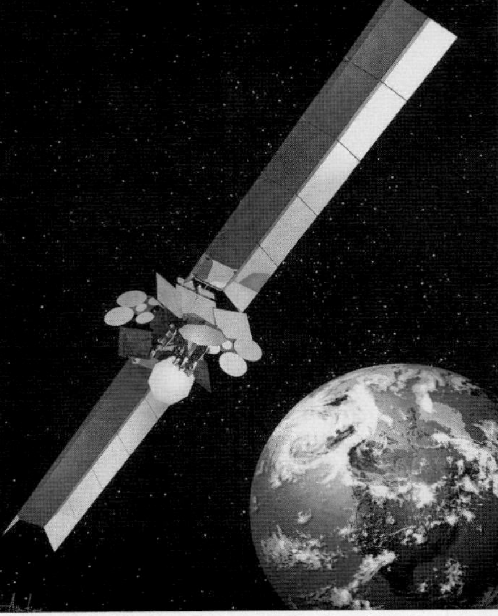

Anik F series are an on-going family of high-powered satellites, F2 being built by Boeing for communication services throughout North America (Boeing) 1047520

service for North and South America. The next generation Anik F series adopts the Hughes HS702 bus with a payload of 36 C-band and 48 Ku-band transponders. Anik F1 weighs 4,600 kg and was launched on 21 November 2000 by Ariane 44L. Anik F2 was ordered from Boeing Satellite Services in April 2000. It has 114 transponders consisting of 50 Ka-band, 40 Ku-band and 24 C-band. BSS-702 weighed 5,910 kg at launch on 18 July 2004 on an Ariane 5, and was designed for 15 kW EOL.

In September 2005, the Anik F1R satellite was launched in Kazakhstan. The F1R provides increased capacity for North American direct-to-home satellite television, along with a range of other telecommunications and broadcasting services.

The Anik F1R also had a navigation payload that enhanced Global Positioning Systems (GPS) for aviation use in Canada and the United States. The payloads were part of the Federal Aviation Administration's Wide Area Augmentation System geostationary communication and control segment initiative, operated by Lockheed Martin to provide precision guidance to aircraft.

The F1R took control of existing North American communications from the Anik F1 satellite, allowing the F1 to focus solely on services to South America. The F1R is equipped with 32 Ku-band transponders, 24 C-band transponders and two navigation transponders. The F1R had a launch mass of 4,471 kg (9,836 lb), a solar array span of 36 metres when deployed, and a spacecraft power of 10 kW at the end of its 15 year life expectancy.

Specifications
Anik F1
Launched: 21 November 2000 by Ariane 44LP
Location: 107.3°W
Design life: 15 years
Contractor: Boeing Satellite Systems
Transponders: 36 (8 redundant) 40 W TWTA C-band for North and South America via two radio antennas, 48 (10 redundant) 115 W TWTA Ku-band for North and South America. One west and one east antenna for North America Ku-band and C-band coverage, respectively
Configuration: Boeing 702 bus 3-axis stabilised. Two 2.2 m reflectors for South American Ku-band and C-band coverage. One west and one east reflector, each 2.4 m, for North American Ku-band and C-band coverage. One 433 N liquid propellant apogee motor, 10 × 43.3–26 N × IPS station-keeping thrusters. Stowed size 4 × 2.1 × 3.4 m; deployed size 40.4 m across solar arrays and 9 m across antennas.
Mass: 4,710 kg at launch, 3,015 on-orbit BOL
Power system: Two solar wings each carrying 5 panels of dual-junction GAaes solar cells, 17.5 kW BOL and 15 kW EOL with 56-cell NiH batteries.

MSAT

Current status
Telesat Mobile Inc (TMI) launched the MSAT-1 satellite in April 1996 to provide mobile communications across North America, in partnership with American Mobile Satellite Corporation (AMSC) and its AMSC-1 satellite. Both AMSC and TMI have now been absorbed into Mobile Satellite Ventures (MSV) and are covered in that entry.

Nimiq

Current Status
Four Nimiq satellites are on station in early 2007, two of them being leased from DirecTV, and two more are under contract for launch in 2008 and 2009.

Background
Telesat won Canadian government approval in April 1997 to develop and operate Canada's first direct-to-home (DTH) satellite TV service, marketed today as Bell ExpressVu. The first satellite was ordered from Lockheed Martin at the same time, and was named Nimiq (an Inuit word for a force that unites). It was launched successfully in May 1999, after an eight-month delay due to problems with its Travelling Wave Tube Amplifiers (TWTAs).

A second satellite, Nimiq 2, was ordered from Lockheed Martin in June 2001 and was launched on December 2002. However, 16 days after being placed in service on 4 February 2003, Nimiq 2 suffered a failure – believed to be caused by foreign object debris – which took one of its solar array wings out of action, shortening its life and power and forcing Telesat to use only 26 of its 32 transponders.

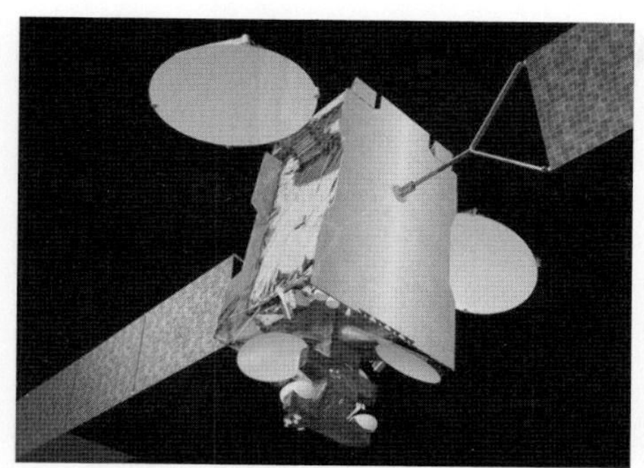

EADS Astrium is building the Nimiq 4 for Telesat (EADS Astrium) 1311353

Nimiq 1 prepared for launch. Canada's first DTH satellite is a Lockheed Martin A2100AX (Telesat) 1311354

In early 2004, Telesat announced an agreement with DirecTV under which an unused back-up satellite, DirecTV-3, would be leased to augment Nimiq 2. It started operations in August 2004 as Nimiq 3. Subsequently, Telesat also leased DirecTV-2, which started operations in early 2006 as Nimiq 4i, augmenting the capacity of Nimiq 1. (See DirecTV entry for specifications.)

Telesat has two more Nimiq satellites on order, replacing its current satellites in the company's two DTH orbital slots and supporting new broadcast services. Telesat selected EADS Astrium in January 2006 to build Nimiq 4, which will occupy the slot now used by Nimiq 1 and 4i. The E3000-class satellite is due to be launched in 2008. The all-Ku-band Nimiq 5 was ordered from Space Systems/Loral in January 2007 and is expected to launch in 2009, replacing Nimiq 1 and Nimiq 3 at 72.7°W.

Specifications
Nimiq 1
Launched: 20 May 1999 by Proton-K from Baikonur
Location: 91°W.
Design life: 12 years
Contractors: Lockheed Martin
Transponders: 32 Ku-band
Principal applications: DTH
Configuration: A2100AX bus
Mass: 3,600 kg

Nimiq 2
Launched: 29 December 2002 by Proton-M/Breeze-M from Baikonur
Location: 82° W.
Design life: 12 years
Contractors: Lockheed Martin
Transponders: 32 Ku-band (26 operational) and 2 Ka-band
Principal applications: DTH
Configuration: A2100AX bus
Mass: 3,529 kg

Nimiq 4
Launched: Scheduled for 2008 by Proton-M/Breeze-M from Baikonur
Location: 91° W.
Design life: 15 years
Contractors: EADS Astrium
Transponders: 32 Ku-band, 8 Ka-band
Principal applications: DTH, broadband internet
Configuration: E3000 bus
Mass: 4,800 kg
Power system: 39 m span solar array delivering 12 kW at end of life

China

Apstar series

Background
The Chinese government, which owns 75 per cent of The Asia Pacific Telecommunications Satellite Company launches and maintains the APstar satellites. APT plans to deliver a three satellite system for the Asia-Pacific region. The consortium signed a contact with Hughes Communications International Incorporated on May 25 1992 to build the APStar 1, and the Macau ground station for US$135 million. Hughes increased the price to US$200 million when APT included the hybrid HS-601 APStar 2 option, primarily for television broadcasting. Apstar also asked Hughes to extend the coverage region to Europe, India and Russia. Apstar 1 was launched 21 July 1994 and was operated until 1999 from 138° E until 1999. Apstar changed its plans when Apstar 2 suffered a launch failure on 25 January 1995 and instead substituted the APStar 1A HS-376 in March 1995. Apstar chose the smaller design because Hughes promised it could be ready in 11 months, rather than the HS-601's 18 months. Apstar 3 (1A) was launched by CZ-3 from Xichang launch site, a near identical twin to Apstar 1, on 3 July 1996.

The company originally placed APStar 1 at 131° E for operational service, but Rimsat & Japan (CS-3a) complained of signal interference and consequently APT moved Apstar 1 to 138° E. It presently resides at 138.04° E. APStar 4 (2R) was launched on 16 October 1997. APStar 5 will be launched in 2003 at 130°E as replacement for APStar 1. Based on a FS1300 bus it will have 38 60 W C-band transponders and 16 141 W Ku-band transponders. APStar 5 will be launched by LM-3B. In August 1999 Loral Space and Communications Ltd leased the satellite as a replacement for Orion 3.

APStar 1 was joined in 1996 by the near-identical 1A (Hughes) 0517202

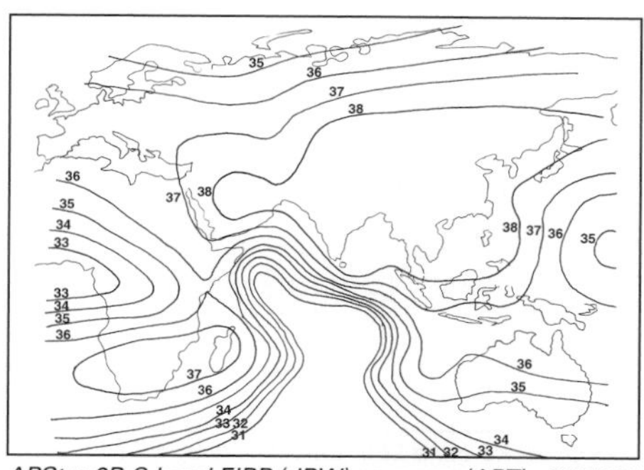

APStar 2R C-band EIRP (dBW) coverage (APT) 0517385

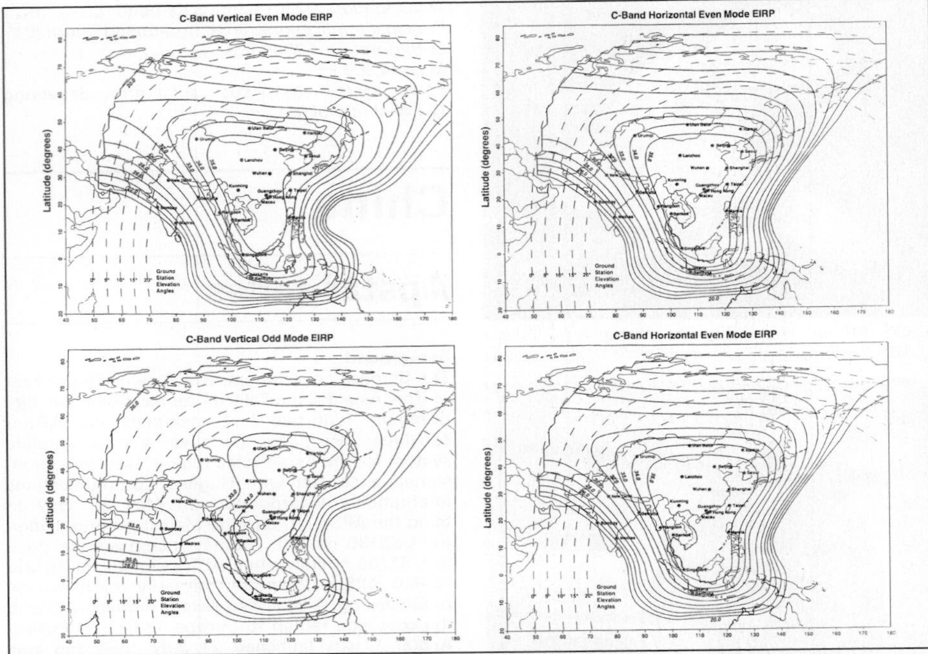

APStar 1A EIRP contours (dBW). Odd-numbered transponders are stretched for Indian coverage (APT) 0517383

Specifications

APStar 1

Launch: 21 July 1994 by CZ-3 from Xichang, China, People's Republic
Location: 138° E GEO. Initially located at 137° E for testing and was then to move to 131° E for operational service, but complaints from Rimsat & Japan (CS-3a) of signal interference with their nearby satellites resulted in positioning at 138° E, reportedly leased from Tonga
Design life: 12 years
Contractors: Hughes Space & Communications Co (contract with Hughes Communications International, Inc)
Transponders: 24 (plus six backup, 30-for-24) 16 W TWTA 5.850–6.425/3.625–4.200 GHz up/down C-band regional beam, 20 × 36 + 4 × 72 MHz bandwidth, EIRP (dBW) >35 China, 36.0 Hong Kong, 35.9 Philippines, 36.1 Thailand, 36.4 Korea & 35.1 Singapore, orthogonal polarisation
Principal applications: regional telecommunications; VSAT, data & video
Configuration: Hughes' HS-376 platform, cylindrical body 2.16 m diameter total height 6.59 m
Mass: 1,383 kg at launch, 557 kg dry
AOCS: spin-stabilised at 50 rpm by hydrazine thrusters. ±0.05° E/west and north/south station-keeping; antenna pointing accuracy 0.021° pitch & roll & 0.199° yaw
TT&C: through Macau station. Command uplink at 6,422 MHz; telemetry down at 4,198.125 and 4,199.625 MHz.
Power system: solar array mounted on cylindrical body and extension provide 1,070/970 W at solstice/equinox EOL.

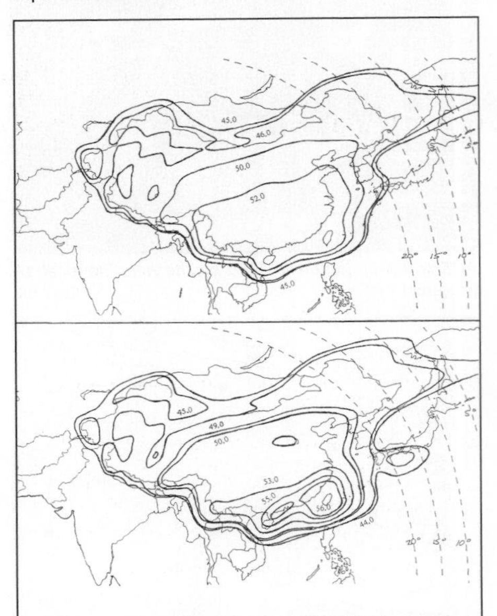
APStar 2R Ku-band EIRP (dBW) coverage; beam 1 top, beam 2 bottom (APT) 0517384

APStar 3

Launch: 3 July 1996 by CZ-3 from Xichang, China, People's Republic
Location: planned for 76.5° E GEO
Transponders: as APStar 1 but additional feedhorns extend coverage to all India (see EIRP contour diagram) Other specifications as for APStar 1.

APStar 4 (2R)

Launch: 16 October 1997 by CZ-3B from Xichang, China, People's Republic
Location: planned 76.5° E GEO
Design life: 15 years minimum
Contractors: Space Systems/Loral (prime)
Transponders: 28 (plus 12 backup, 36-for-28) 60 W TWTA 5.850–6.425/3.625–4.200 GHz up/down C-band regional beam, EIRP 38 dBW main region (see map), 27 × 36 MHz + 1 × 30 MHz bandwidth, linear orthogonal polarisation

16 (plus 8 backup, 24-for-16) 110 W TWTA 14.00–14.50/12.25–12.75 GHz up/down Ku-band China beams 1&2, EIRP see map, 15 × 54 MHz + 1 × 36 MHz bandwidth, linear orthogonal polarisation. Beam 1 has eight even-number channels; beam 2 eight odd number channels; the four horizontal polarisation beam 2 channels can switch to beam 1
Principal applications: China (Ku) and regional services
Configuration: 3-axis stabilised SS/L FS-1300 2.41 × 2.58 × 2.20 m box-shaped bus with twin solar wings on North/South sides, 27.3 m span
Mass: 3,700 kg at launch, 1,415 kg dry
AOCS: 3-axis bias momentum system; single Marquardt R-4D 490 N apogee thruster and two sets of six 22 N RCS thrusters, common NTO/MMH propellant system
Antenna pointing: 0.13°
TT&C: as APStar 1/1A
Power system: twin 4-panel GaAs wings providing 8,125 W BOL. Ni/H_2 battery provides 100 per cent eclipse protection.

AsiaSat series

Current Status

The Asiasat 3 satellite, a fully functional HS601HP, ended up in an incorrect orbit when the fourth stage of the Proton booster, carrying the satellite, failed. The final orbit, achieved after launch on 25 December 1997, resembled a GTO rather than a GEO orbit. On 10 April 1998 Hughes began a series of orbital manoeuvres resulting in a translunar fly-by in an attempt to reposition the satellite in its correct orbit.

The manoeuvre consisted of a series of satellite thruster firings by 7 May 1998 converting the elliptical orbit of 350 km by 36,000 km into a circumlunar trajectory. This is the first commercial use of the moon and the first lunar mission attempted by a non-government entity.

When owners Asia Satellite Telecommunications Co Ltd of Hong Kong filed an insurance claim of $215 million, Hughes Global Services, Inc obtained title for the salvage operation and renamed the satellite HGS-1. HGS would share any profits with etc insurers.

The first encounter on 13 May carried the satellite within 6,200 km of the lunar far side before looping back to Earth. Engineers adjusted the Earth encounter to effect a second fly-by of the Moon at a distance of 34,000 km on 6 June. Thrusters were fired at the second Earth fly-by to place HGS-1 in a 46 h orbit of 36,000 km by 82,000 km. Subsequent thruster firings settled it into a geostationary orbit at 36,000 km.

AsiaSat 3SA was launched by Proton K on 21 March 1999.

AsiaSat 4 was scheduled for launch in 2003 by Atlas to a position at 122° E as a replacement for AsiaSat 1. Satellite assembly at Boeing Satellite Systems officially began in September 2000. Based on an HS601HP bus, AsiaSat 4 will carry 20 transponders supporting Ku-band BSS services for DTH services to Hong Kong and 28 C-band transponders serving Asia, the Middle East, the CIS and Australasia in the 6/4 GHz band.

AsiaSat 1

AsiaSat 1 was manoeuvred off-station over 105.5° E and relocated to 122° E during June – August 1999. The satellite was manoeuvred away from this location at the end of January 2003, and was still drifting in mid-March 2003.

AsiaSat G

In order to fill a gap in its communications requirements, AsiaSat leased the Russian Gorizont 30 satellite which originally had been leased to RIMSAT. During December 1997 Gorizont 30 was relocated to 122° E where it was operated for nearly two years as 'AsiaSat G'. The satellite was relocated to 142° E in June 1999 when AsiaSat's agreement to use the satellite terminated.

AsiaSat 3 is planned for October 1997 (Hughes) 0517402

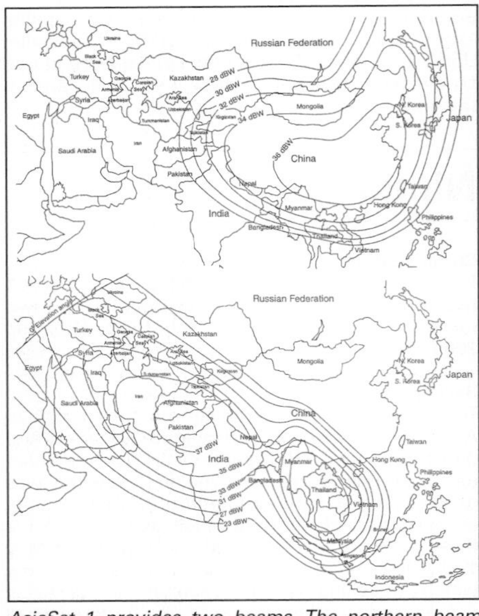

AsiaSat 1 provides two beams. The northern beam covers China, while the southern footprint is optimised for Thailand and Pakistan (AsiaSat) 0517203

With fully extended solar array panels, AsiaSat nears eclise (Boeing) 1047526

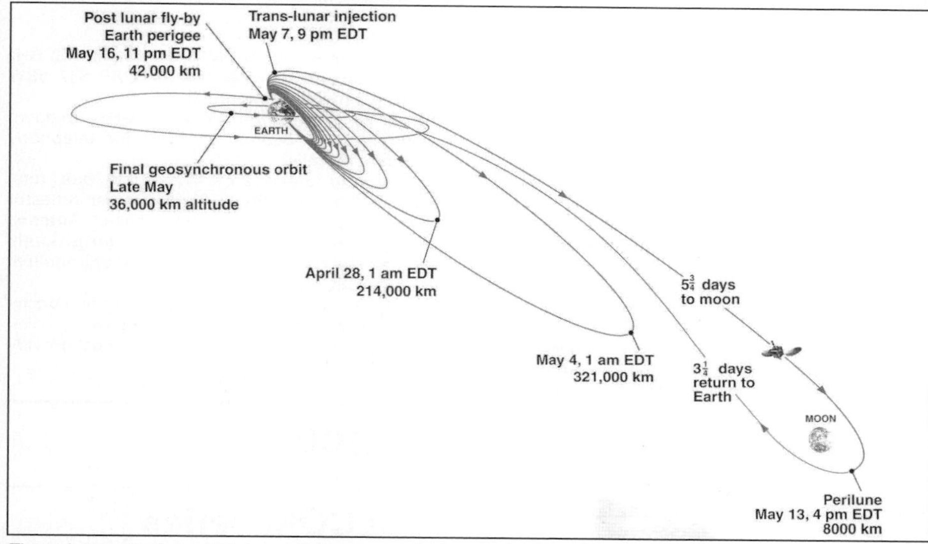

The trajectory for Asiasat 3's Lunar flyby rescue mission required nearly 45 days and used a large fraction of the satellite's station-keeping propellant. All dates shown on the diagram occurred in 1998 (Hughes Space and Communications) 0022575

Background

Lockheed Martin Astro Space, in competition with Hughes and Matra Marconi signed a contract for a hybrid AsiaSat 2 satellite on 26 October 1992. The contract had a potential worth of US$133 million if China exercised all of the options including a second satellite purchase. China paid COMSAT to give additional technical support.

The CZ-2E launch vehicle received the nod to put the satellites in space, but China withheld a launch decision until 1995 to allow time to determine the cause of the Telstar 402 (which used the same bus) and CZ- 2E failures to be identified. The first CZ-2E/EPKM launch vehicle combination built by the Chinese put AsiaSat 2 in space on 28 November 1995. It commenced commercial operations in January 1996.

With success assured, the consortium backing AsiaSat took out a US$220 million loan on 16 February 1996 to finance AsiaSat 3, and then signed a contract with Hughes Space & Communications International for the manufacture and on-orbit delivery of an HS-601 HP. On 22 August 1996, a US$250 million loan facility entered effect with 12 international banks called upon to refinance the existing loans for AsiaSat 1, AsiaSat 2 and AsiaSat 3 as well as to support future capital expenditure.

AsiaSat 3 carries 28 C-band and 16 Ku-band high-powered linearised transponders. The C-band footprint can cover Asia, the Middle East, the CIS and Australasia. There are also two separate footprints providing high-powered Ku-band coverage. In addition, a steerable Ku-band beam provides added market flexibility.

AsiaSat 4 has a launch delivery schedule of late 2001 when AsiaSat 1 reaches the end of its planned life. Asia Satellite Telecommunications selected Hughes to build a replacement for the AsiaSat 3 after its launch was unsuccessful on 25 December 1997. The AsiaSat 3S is an exact replica of AsiaSat 3 and was launched by Proton on 22 March 1999 toward its operating slot at 105.5° east. In November 1998, Hughes was contracted to build a backup, designated AsiaSat 3SB until its predecessor had been successfully launched subsequent to which the order has been converted to a future AsiaSat requirement.

AsiaSat's ultimate holding company went public in June 1996. Both the Hong Kong and New York Stock Exchange list it with 31 per cent of the company's shares held by the public. The remaining shares reside equally with Cable and Wireless PLC, China International Trust and Investment Corporation (CITIC) and Hutchinson Whampoa.

Specifications

AsiaSat 1

Launch: 7 April 1990 by CZ-3 from Xichang (launched initially as Westar 6 by STS-41B, 3 February 1984; retrieved by STS 51-A in November 1984)
Location: 105.5° E geostationary
Design life: 9–10 years
Contractor: Hughes Space & Communications Co
Transponders: 24 (plus six back-up) 8.2 W TWTA 5.925–6.425/3.700–4.200 GHz up/down C-band, 12 each in two beams (see map), 36 MHz bandwidth, China 32–36 dBW, Thailand 35–37 dBW, Pakistan 35–37 dBW, Korea 35 dBW, Japan 30–34 dBW, Singapore 32 dBW, linear polarisation (12 V+12 H)
Principal applications: TV, radio, teleconferencing and VSAT links. TV users include Star TV, China, Mongolia, Myanmar, Pakistan. Each transponder can handle 3,360 telephone channels via a 7 m station. TVRO diameter 2.6 m for 36 dBW
Configuration: Hughes' HS-376 platform, cylindrical body 2.16 m diameter, total height 6.59 m
Mass: 1,280 kg at launch, 654 kg on-orbit BOL
AOCS: spin-stabilised at 50 rpm by hydrazine thrusters
Power system: solar array mounted on cylindrical body and extension provide 900 W at BOL, about 700 W at EOL
TT&C: Stanley Earth Station, Hong Kong.

AsiaSat 2

Launch: 28 November 1995 by CZ-2E from Xichang
Location: 100.5° E GEO (arrived 14 December 1995, entered service mid-January 1996)
Design life: 13 years
Contractor: Lockheed Martin Astro Space
Transponders: 24 (plus 8 backup, 16-for-12) 55 W TWTA 5.845–6.425/3.620–4.200 GHz up/down C-band, 20 × 36 + 4 × 72 MHz bandwidth, 40 dBW EIRP max, linear orthogonal polarisation

9 (plus three backup, 12-for-9) 115 W TWTA 14.000–14.300/12.200–12.500 GHz up/down Ku-band, 54 MHz bandwidth, 53 dBW EIRP max, covering Hong Kong, China, Taiwan, Japan, Korea, linear orthogonal polarisation
Principal applications: TV and radio distribution, telecoms, VSAT
Configuration: series 7000 platform, 3-axis box-shaped bus with twin solar arrays spanning 23 m
Mass: 3,351 kg at launch
AOCS: nominal pointing accuracy ±0.1° pitch/roll, ±0.25° yaw. 1.8 kW hydrazine arcjets provide NSSK, China's EPKM perigee kick motor, Liquid Apogee Engine provides GEO insertion, Reaction Engine Assemblies for attitude control
Power system: twin 4-panel wings totalling 48.6 m², providing 4,780 W EOL, supported by Ni/H_2 battery
TT&C: Stanley Earth Station, Hong Kong.

AsiaSat 3

Launch: 25 December 1997 by Proton-K from Baikonur, Kazakhstan. Replacement AsiaSat 3S launched 21 March 1999.
Location: 105.5° E GEO (AsiaSat 35)
Design life: 15 years
Contractor: Hughes Space & Communications
Transponders: 28 (plus 6 back-up, 34-for-28) 55 W TWTA 5.845–6.425/3.620–4.200 GHz up/down C-band, 36 MHz bandwidth, 41 dBW EIRP maximum, coverage as AsiaSat 2, linear orthogonal polarisation

16 (plus 4 backup, 20-for-16) 138 W TWTA 14.000-14.500/12.25/12.75 GHz up/down Ku-band, 54 MHz bandwidth, 55 dBW EIRP max, three beams (North Asia, East Asia, steerable), linear orthogonal polarisation
Principal applications: TV distribution, telecoms, VSAT

After the failure of the geosynchronous placement motor on AsiaSat 3, manufacturer Hughes Space and Communications Company chose to salvage the satellite by repositioning it with a lunar gravity assist manoeuvre and onboard station-keeping propellant (Hughes Space and Communications Company) 0022572

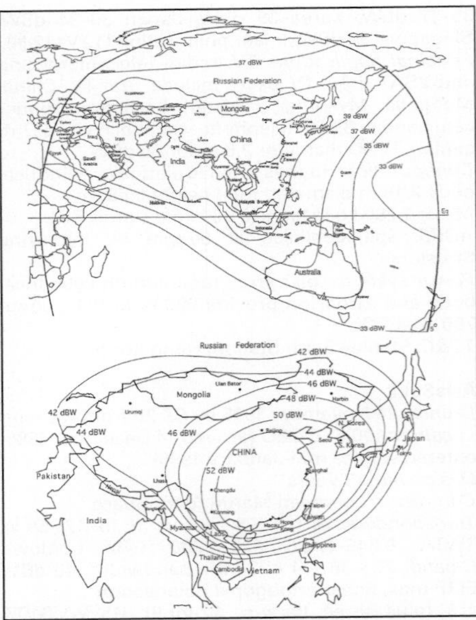

AsiaSat 2 dBW contours: C-band (top) and Ku-band (bottom) (AsiaSat) 0517204

Configuration: improved HS-601 HP (High Power) platform, featuring GaAs solar arrays, enlarged payload and lightweight contoured reflectors
Mass: 3,480 kg at launch, 2,560 kg on-station BOL
AOCS: Bipropellant system.
Power system: twin solar wings of four panels providing 8.7 kW BOL (7.8 kW total required); 350 Ah, Ni/H_2 battery provides 100% eclipse protection
TT&C: Stanley Earth Station, Hong Kong.

AsiaSat 3SA
Launch: 21 March 1999 by Proton K from Baikonur as replacement for AsiaSat 3 launched 25 December 1997
Design life: 15 years
Location: 105°E
Contractor: Hughes Space and Communications
Transponders: (as AsiaSat 3) plus two fixed 140 W ku-band
Mass: 3,465 kg at launch, 2,530 kg BOL, 1,770 kg dry

Feng Huo (Zhongxing 2X series)

Current Status
On 25 January 2000, a new communications satellite carrying the official designation *ZhongXing-22* (ChinaSat-22) was launched and located over 98° E by a CZ-3A rocket. Prelaunch reports had stated that the new satellite would be called *Feng Huo*.

The name *Feng Huo* means 'fire and smoke', and is taken from the ancient Chinese Army's practice of communicating an impending invasion by lighting a series of beacons along the Great Wall. This allowed warning to be transmitted in only a few days along the length of the Wall, while a human with a verbal or written warning would take far longer to cover the distance.

Based upon this derivation of the name and its military implications, it seems likely that *Feng Huo* 1 is being used primarily for military communications. It is believed that the satellite is being used for mobile communications work. Chinese literature has confirmed that the satellite, launched aboard a CZ-3A, uses a standard DFH-3 satellite platform.

The use of the *Zhongxing-22* (ChinaSat-22) location for the first *Feng Huo* satellite suggests that a constellation of five such satellites might be planned: the full *Zhongxing-2X* locations are: 21–103°E, 22–98°E, 23–75.5°E, 24–127.5°E and 25–117.5°E. However, no further Feng Huo launches had taken place by mid-2006.

STTW (Chinasat) series

Current Status
Chinasat 5 launched as Spacenet 1 by Ariane 1 on 23 May 1984 currently operated by China Telecom and Broadcasting Satellite. This was the first non-military satellite in the Chinasat series.

Background
MBB/CGWIC signed a DM51.2 million contract on 14 July 1987, effective from 8 March 1988, to develop China's communications and broadcast infrastructure and some satellite subsystems. The Chinese Space Agency CAST acts as prime contractor for both space/ground segments. MBB supported the system definition work and built the satellite antennas and solar array mechanical elements.

The result of this collaboration, the DFH-2A/Chinasat-1 flew successfully three times between 1988 and 1990. Each satellite carries four instruments (electron, proton and solar X-ray detectors and potentiometer) to monitor the GEO radiation environment, the satellite's surface charging and solar activity.

DASA and China Aerospace Corp signed a MoU 16 November 1993 to co-operate in building and launching the next generation satellites. These entities formed the EurasSpace joint venture 8 July 1994, with an initial staff of 12, to help produce China's domestic satellites and to bid for international contracts.

When it was completed, 80 per cent of China's first 3-axis GEO satellite came for indigenous sources. US sanctions did create delays. The satellites incorporate Teldix GmbH's SADA solar array drive assemblies. Officine Galileo provided the IRES IR Earth sensor to determine pitch/roll in GEO. M/A-COM Inc had contracted to supply a transponder but President Bush imposed a ban on export of US components for Chinese spacecraft in the summer of 1991. TIW Systems of the US received a contract in 1993 for the TT&C complex in Beijing.

Specifications
STTW-T1
Launch: 29 January 1984 by first CZ-3 from Xichang, initially into 451 × 309 km parking orbit then after 12 revs boosted to 359 × 6,479 km, 36°. First attempt at placing experimental DFH-1 telecom satellite at 125°E STW-1 but the stage 3 failed to re-ignite for GTO insertion. Satellite kick motor was fired as a demonstration.

STTW-T2
Launch: 8 April 1984 by CZ-3 from Xichang to 125° E GEO (STW-1 position). 420 kg on-station BOL, 915 kg at launch; design life 3 years. Provided two 8 W TWTA 6/4 GHz up/down 20 dBW global beams using horn antenna working to 10 m ground stations. The DFH-1 configuration is similar to STTW 2, but was 3.1 m high and employed a de-spun horn antenna. Operational life apparently ended 1987 when allowed to drift off-station.

STTW-1
Launched: 1 February 1986 by CZ-3 from Xichang to 103° E STW-2 position. DFH-2 design; specifications similar to STTW-2. Broadcast to 8 July 1989; corrections made only occasionally during 1990.

STTW-2/Chinasat-1A
Launched: 7 March 1988 by CZ-3 from Xichang
Location: 87.5° E geostationary (initially 88° E)
Design life: 4 years
Contractors: CAST Chinese Academy of Space Technology
Transponders: four 10 W TWTA 6.050–6.425/3.825–4.200 GHz up/down C-band regional beams using parabolic reflector providing 36 dBW centre, 32 dBW beam edge, 36 MHz bandwidth. Payload can handle 5 TV + 3,000 phone calls simultaneously
Configuration: 2.1 m diameter, 1.6 m height spin-stabilised cylindrical bus, including single de-spun parabolic antenna. Total height 3.68 m. 77.6 kg station-keeping propellant. Antenna pointing accuracy 0.6° N/South, 0.42° E/West
Mass: 441 kg on-station BOL, 1,024 kg at launch
TT&C: provided through XSCC Xian Satellite Control Centre
Power system: 351 W provided by 20,000-cell Si (11% efficiency) array on cylindrical exterior, supported by 15 Ah Ni/Cd batteries
Apogee kick motor: indigenous solid propellant

STTW-3/Chinasat-1B
Launched: 22 December 1988 by CZ-3 from Xichang
Location: 110.5° E geostationary (initially 111° E)
Other specifications as for STTW-2.

STTW-4/Chinasat-1C
Launched: 4 February 1990 by CZ-3 from Xichang
Location: 98° E geostationary
Other specifications are the same as for STTW-2.

STTW-5
Launched: 28 December 1991 by CZ-3 from Xichang but short stage 3 GTO burn left it in 219 × 2,451 km, 31.1°. Its kick motor was fired 29 December to attain 205 × 35,087 km, 31.6°. 4th/last flight of DFH-2A/Chinasat-I design. Possibly intended as STTW-2 replacement. Satellite was a ground spare upgraded to flight status.

Zhongxing series

Specifications
Zhongxing 5 /Spacenet 1
Launched: 23 May 1984 to 102° W by Ariane, was owned/operated by GTE at 120° W until it was sold in December 1992 to China and moved to 115.5° E by June 1993. CNPTAC China National Postal & Telecommunications Appliances Corp bought it for use by Chinasat for about US$30 million. See p326 1995–96 for specifications. It exhausted its propellant in second quarter 1996. Deorbited mid-December 1999. When DFH-3 1 failed to replace it, Chinasat 7 was ordered from Hughes.

Zhongxing 6/DFH-3/Chinasat-III
Launched: 29 November 1994, abandoned in geosynchronous drift orbit.

Zhongxing 7/ChinaStar 7
Launched: 18 August 1996, left stranded in a 27.2°, 200-17,230 km orbit when the second burn of the C2-3 third stage ended prematurely
Mass: 1,200 kg with propellant, Hughes HS376 satellite bus.

Zhongxing 8
Launched: 11 May 1997 using C2-3A from Xichang
Location: 115.3° E
Design life: >8 years
Transponders: six 16 W TWTA 6.2–6.4/4.0–4.2 GHz up/down C-band regional beams. EIRP >37 dBW (allowing 3 m TVRO antennas)
18 8 W SSPA 6.2–6.4/4.0–4.2 GHz up/down C-band regional beams. 35 dBW for telephony (8,000 duplex calls)
Configuration: 3-axis 2.2 × 1.72 × 2.0 m bus, total height 5.71 m with single 2 m diameter reflector deployed, 18.1 m solar array span. Antenna pointing 0.15° pitch/roll, 0.50° yaw. North South/East West station-keeping 0.10°. Liquid bipropellant motor for GEO insertion
Mass: 1,145 kg on-station BOL, 2,230 kg at launch
TT&C: provided through Beijing complex
Power system: two 3-panel solar wings provide 2,000 W BOL/1,700 W EOL.

France

TELECOM series

Current Status
Post-Telecom 2 series analysis has yet to reach decisions regarding successors. Probability of leasing from Inmarsat for satellite maritime services.

Background
France authorised France Telecom to build the Telecom satellite system in 1979 to link domestic and overseas territories with television, telephone, digital data and military links. To meet these needs, the quasi-public company immediately ordered the Telecom 1 satellite and then later ordered three

Telecom 2C in Matra Marconi Space's Toulouse test facilities 0517403

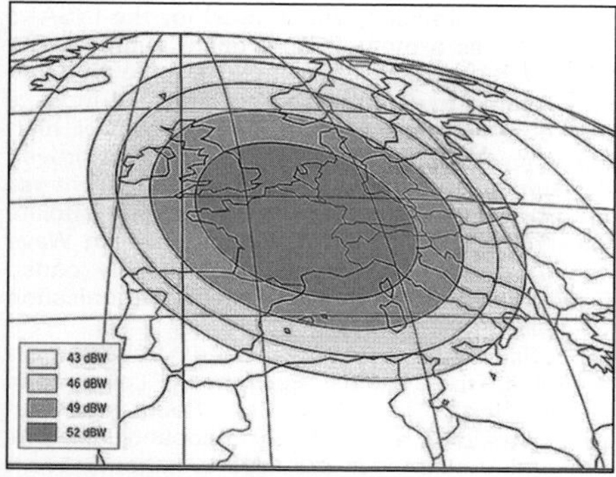

Telecom 2 Ku-band beam coverage 0517198

Telecom 2s in December 1987 from Matra/Alcatel. The Telecom 2 has three times the power of the Telecom 1. Each carries a Syracuse (Système de Radio Communications Utilisant on Satellite) package for secure Ministry of Defence links. The X-band transponders work through three Earth stations: in Brest (8 m dish), Paris (8 and 18 m dishes and the tracking, telemetry and command) and France-Sud (2, 18 m dishes, including back-up tracking, telemetry and command).

Syracuse 2 expands access to 40–90 cm antennas, permitting 75/200 bit/s telegraph; 2–4/16 kbit/s telephone; 75/2,400/16,000 bit/s data routeing and a secure backup in case of damage to the primary communication centre.

France Telecom may rely in future on leased Eutelsat coverage under privatisation. It is a member of the Satellite Aircom consortium providing aeronautical communications services through the Inmarsat system. Telecom 2A, 2B, 2C and 2D were still operating in February 2002.

Specifications

Telecom 1A

Launched: 4 August 1984, was raised above GEO in late 1992 after being replaced at 8° W by 2A. 1B, launched 8 May 1985, lost both attitude control systems 15 January 1988. 1C, launched 11 March 1988, retired from 3° E in February 1996 after being replaced by 2C.

Telecom 2A

Launched: 16 December 1991 by Ariane 44L from Kourou, French Guiana. In-orbit acceptance tests completed 15 March 1992 at 3° E; moved April to 8° W
Location: 8° W geostationary
Design life: 10.25 years minimum
Contractors: Telecom 2A: Matra (platform, integration), Alcatel Espace (payload), Sodern (Earth sensors), Galileo (Sun sensors), SEP/MMS (propellant tanks), Sagem (rate gyros), MBB (thrusters), Fokker (solar arrays), Eagle Picher (batteries), Teldix (reaction wheels), Thomson (TWTs)
Transponders: 10 (plus 4 back-up; 6-for-4 + 8-for-6) 8.5 W min SSPA 5.925–6.425/3.700–4.200 GHz up/down C-band Atlantic regional beams, 6 × 50 MHz + 4 × 92 MHz bandwidths, circular polarisation. Semi-global coverage at 32.5 dBW EIRP on channels C1-C4: mainland France, Reunion, Antilles/Guiana, St Pierre/Miquelon. Spot coverage: Antilles/Guiana (39.0 dBW; C5A, C6A, C7A), mainland France (42.4 dBW; C5B, C6B, C7B), St Pierre/Miquelon (39.0 dBW; C5B, C6A, C7A); channels C6A/C7A can be reconfigured onboard between Antilles/Guiana and St Pierre/Miquelon, C5B is divided

11 (plus 4 backup in 15-for-11) 55 W min TWTA 14.00–14.250/12.50–12.75 GHz up/down Ku-band European regional beams, 36 MHz bandwidth, 52.4 dBW min EIRP central zone, 49.4 dBW min EIRP surrounding zone, vertical polarisation. This payload also transmits a 12 GHz beacon signal for ground station pointing

3 (plus 3 backup; two groups, 3-for-2 + 3-for-1) 20 W min TWTA 7.900–8.395/7.250–7.745 GHz up/down X-band beams, circular polarisation, channel X1 60 MHz bandwidth central European beam EIRP 44.0 dBW, X4 40 MHz steerable spot EIRP 31.3 dBW, X5 80 MHz global 28.5 dBW

2 (plus 1 backup, 3-for-2) 40 W min TWTA 7.900–8.395/7.250–7.745 GHz up/down X-band beams, circular polarisation, channels X2/X3 40 MHz bandwidth global EIRP 31.3 dBW
Principal users: domestic/overseas telephony, data, radio, TV, military
Configuration: Matra/BAe Eurostar 2000 bus. Four military antennas: LHCP global horn transmitter, RHCP global horn receiver, 2.2 m diameter offset paraboloid 1.65 m fl for European coverage, and a steerable offset paraboloid spot reflector
Mass: 2,275 kg at launch, 1,380 kg on-station BOL, 1,124 kg dry mass (400 kg payload)
Pointing: antenna pointing accuracy typically 0.13°
AOCS: 3-axis, 12 × 10 N DASA hydrazine thrusters + Marquardt R-4D 490 N bipropellant apogee motor in unified bipropellant propulsion system
TT&C: via the Toulouse Space Centre working through antennas at Bercenay-en-Othe near Troyes and Aussaguel-Issus near Toulouse
Power system: twin 4-panel solar arrays, 2.050 m wide, spanning 22.02 m provide 3,770 W equinox EOL with full eclipse battery protection (78 Ah Ni/H_2). 2,600 W payload power requirement.

Telecom 2B

Launched: 15 April 1992 by Ariane 44L from Kourou, French Guiana
Location: 5° W from mid-1992 after commissioning at 3° E, now 24.98° EOther specifications as for Telecom 2A.

Telecom 2C

Launched: 6 December 1995 by Ariane 4 from Kourou, French Guiana. Replaced 1C February 1996 after testing at 1° E
Location: 5.05° E (replacing 1C)
Mass: 2,283 kg at launch, 1,360 kg on-station BOL, 1,120 kg dry mass (400 kg payload)
Other specifications as for Telecom 2A.

Telecom 2D

Launched: 8 August 1996 by Ariane 4 from Kourou, French Guiana
Location: 8° W (collocated with 2A)
The 11 Ku-band transponders use 14.25–14.50/11.45–11.70 GHz up/down.
Other specifications as for Telecom 2A.

TELEDIFFUSION DE FRANCE series

Current Status

TDF-1 launched 28 October 1988 has been retired and TDF-2, launched 24 July 1990, is still operating.

Background

TDF, created in 1975 as a limited company with the French government as majority shareholder (and a France Telecom subsidiary since 1988), operates the TDF television repeater satellites.

The Eurosatellite consortium was awarded the space segment contract in 1984, and the company built the original satellites basing TDFs design on Aerospatiale's Spacebus 300 platform. Aerospatiale acted as prime for TDF 1 and 2 and backup for TV-Sat 1 and 2. The French made the decision to build TDF 2 in December 1984. By 1986 the programme verged on cancellation because of developing competition from medium-power systems such as Astra, but by then the 1.5 billion francs of the 3.5 billion francs allocated to TDF had been spent. The French government ruled that it should proceed despite the financial difficulties in order to promote D2MAC as the standard for high definition television in the 1990s.

TDF 1 during final flight readiness review. The solar arrays have yet to be attached (Aerospatiale) 0517409

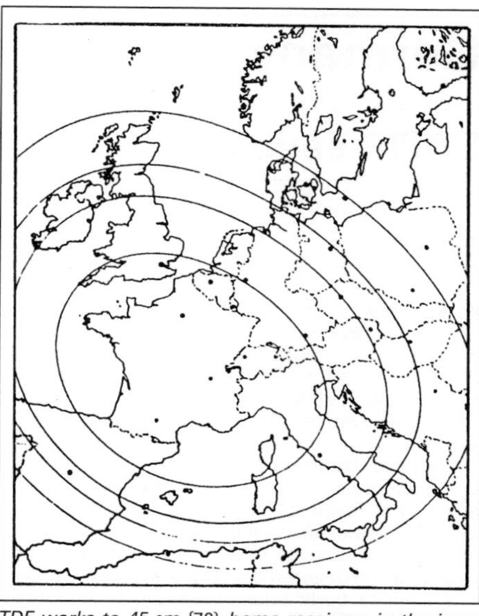
TDF works to 45 cm {70} home receivers in the inner France-centred ellipse 0516998

The TDF's under construction underwent minor modifications after TV-Sat 1 failed because one solar array remained stuck in November 1987 and thruster overheating occurred during orbital manoeuvres. At the same time, France decided there would be no TDF 3 and that the future system would lease Eutelsat space. By 1992, only about 35,000 locations had TDF dishes. France cut its annual transponder leasing cost from 80 million francs to 35 million francs. Both TDFs suffered from malfunctions that limited their usefulness. TDF 1's channel 1 transponder failed in August 1989 because of arcing promoted by leaking thruster propellant and TDF shut down channel 17 in September 1990 after 5 months of current variations. TDF 2's two Thomson TWTAs (channels 1/13) failed simultaneously in October 1990, when safety mechanisms irreversibly shut them down.

Specifications

TDF 1

Launched: 28 October 1988 by Ariane V26 from Kourou, French Guiana
Location: During August-September 1996 TDF 1 was relocated from 19° W to 21–22° W, and then on 4 October the satellite was boosted out of the geosynchronous orbit band, to retirement
Design life: 8 years (sufficient AOCS propellant after orbital positioning for 9 years operations)
Contractors: Eurosatellite (Aerospatiale/MBB prime, Alcatel Espace payload), AEG (three 260 W TWTAs), Thomson Tube Electroniques (three 230 W TWTAs)
Transponders: 5 (plus 1 backup only for channel 9) 230 W min TWTA 17.3-17.7/11.714-12.045 GHz up/down Ku-band elliptical beams providing 64 dBW min EIRP coverage of France, 27 MHz bandwidth, RHCP
Principal applications: direct TV/radio broadcast channels 1/5/9/13/17 and HDTV demonstrations
Configuration: Spacebus 300 platform, body 1.65 × 2.4 × 7.1 m box-shaped, solar array span 19.3 m, 2 m diameter circular receive antenna, 2.4 × 0.9 m elliptical transmission antenna (9-horn feed), both carbon fibre
Mass: 2,136 kg at launch, 1,318 kg on-orbit BOL
AOCS: 3-axis control by 14 × 10 N MBB thrusters of unified bipropellant system incorporating 400 N apogee kick motor
Antenna pointing: ±0.01° driven by RF sensor detecting beacon signal from control centre
TT&C: orbital positioning, station-keeping and performance monitoring by CNES from Toulouse; programme feeds allocated to TDF's Bercenay-sur-Othe station in the Champagne region
Power system: twin 4-panel solar arrays (43,000 cells) providing 4.3/3.3 kW BOL/EOL (8 years).

TDF 2

Launched: 24 July 1990 by Ariane V37 from Kourou, French Guiana
Location: 19° W geostationary
Mass: 2,096 kg at launch, 1,255 kg on-station BOL, 1,040 kg dry mass
Other specifications as for TDF 1.

Germany

SAFIR series

Current Status
SAFIR-2 has been in orbit since 10 July 1998.

Background
Germany uses the SAFIR (Satellite for Information Relay) as a global, two-way data collection/distribution system. OHB-System was responsible for developing all of the hardware and software of the SAFIR system, partly based on the BremSat satellite.

Germany launched the 38.5 kg SAFIR-R1 core demonstration package on 4 November 1994, attached to Resurs-O1-3, into 551 × 663 km, 98.05° orbit. The launch was delayed because the lifetime of the previous Resurs-O1 lasted longer than expected and deferred departure of its replacement. The satellite body housed the SAFIR electronics module, software and TM/TC box; Resurs provided the power. Two modified BremSat computers are used for SAFIR control and user data management. They are based on the INMOS T800 Transputer and are equipped with 5 Mbyte mass memories with error detection and correction and latch-up protection. OHB-System also operated the hub station in Bremen. SAFIR interrogated fleet stations and stored the data in its 10 Mbit solid-state memory for later transfer and/or multi-user access. In the operational system, 1.2 kg macro-stations provide 300 bit/s and 2.4 kbit/s of data transmission.

Co-funded by DARA and OHB-System, SAFIR-2 is an autonomous, cube-shaped, free-flying satellite based on SAFIR-1 and again using Bremsat technology. It carries an advanced Frequency Hopping Telemetry system (FHT) which allows 16 user stations to transmit within the same time slot in the same frequency channel. By means of this technique, the total throughput of the SAFIR 2 FHT has been increased by a factor of 16 without any loss in data security. SAFIR-2 was ejected from Russia's Resurs-O1-4 Earth observation satellite on 10 July 1998 for a DM200,000 fee. It was placed in a 815 × 818 km, 98.8°, 101 min. orbit after launch by Zenit-2. Transmission rates are 300, 600, 1,200, 2,400 and 4,800 Baud over two telemetry systems (TS1- 13.722 GHz, TS2-400 MHz).

Specifications
SAFIR-2
Launch: 10 July 1998 by Zenit-2 from Baikonur (ejected from Resurs-O1-4 satellite).
Orbit: 815 × 818 km, 98.8°, period 101 min.
Size: 45 × 45 × 50 cm cube; 60 kg
Power: 25 W orbit average, 50 W peak, from Si cells on five faces; 4 Ah Ni/Cd and 4 Ni/H_2 batteries
Stabilisation: ±5° nadir pointing accuracy using gravity gradient boom and magnetorquer, determination by magnetometer and Sun/star sensors
Data rate: 300 and 2,400 bit/s
Frequencies: Users, UHF up/down; TT and C VHF up/down
Configuration: Two modular compartments (lower 30 cm high; upper 20 cm high). The bottom plate carried launcher interface and two communications units. The middle plate housed the batteries, power supply unit, computer and gravity gradient boom interface. Controlled by modified Bremsat 1 OBC, based on Inmos T800 transputer with 10 Mbit mass memory. TT&C on 2 m band.

India

EduSat

Current Status
EduSat was successfully launched by India's Geosynchronous Satellite Launch Vehicle (GSLV) on 20 September 2004.

Background
The concept of beaming educational programmes via satellite was effectively demonstrated for the first time in India in 1975–76 through the Satellite Instructional Television Experiment (SITE) conducted using the American Application Technology Satellite (ATS-6). During this experiment, programmes related to health, hygiene and family planning were telecast directly to about 2,400 Indian villages spread over six states. Later, with the commissioning of the INSAT system in 1983, a variety of educational programmes began to be telecast. In the 1990s, the Jhabua Developmental Communications Project (JDCP) and the Training and Developmental Communication Channel (TDCC) further demonstrated the value of tele-education. With the success of the INSAT-based educational services, a need was felt to launch a satellite dedicated for educational service and ISRO conceived the EduSat Project in October 2002.

EduSat with one of its solar arrays deployed (ISRO) 1343500

EduSat is India's first satellite dedicated exclusively for educational services. It was expected to help solve two major problems of rural education – the lack of quality schools and shortage of qualified teachers – and make an impact on the high level of illiteracy. Its 12 transponders are used to support 'non-formal' education in hundreds of villages across the country, as well as for beaming programmes into schools and colleges. Through EduSat a teacher sitting in a television studio can simultaneously address hundreds of students in different schools and colleges all over the country as long as they have access to a receiving dish and a computer terminal.

Regarded as part of the INSAT system, it was developed by the ISRO Satellite Centre, Bangalore. The payloads were developed by the Space Applications Centre, Ahmedabad. The Master Control Facility at Hassan is responsible for all post launch operations of the satellite.

EduSat was launched into GTO on 20 September 2004 during the first operational flight of the GSLV from Sriharikota. It was co-located at 74°E with Kalpana-1 and INSAT-3C. After completion of the third apogee motor firing, the East side antenna of EduSat was deployed at 10:30 am (IST) on 24 September 2004. This antenna is used for transmitting and receiving Ku-band signals with multiple Spot beam coverage. Three-axis stabilisation of the satellite and its West side antenna deployment were carried out on 25 September 2004. The West side antenna is used for transmitting and receiving Extended C-band signals. In addition to the two deployable antennas, the satellite has one body-mounted antenna for transmitting and receiving Ku-band signals with the national coverage beam. In 3-axis stabilised mode, EduSat is locked onto Earth continuously through its optical sensors. Momentum wheels spinning at 4,500 revs per minute maintain 3-axis stabilisation.

Estimated to cost INR90 crore (USD22.5 million), the spacecraft is built around a standardised bus called I-2K, similar to those used for the INSAT-3 series. It has a multiple spot beam antenna with a 1.2 m reflector to direct precisely the Ku-band spot beams towards selected regions of India, a dual core bent heat pipe for thermal control, high efficiency multi-junction solar cells and an improved thruster configuration for optimised propellant use for orbit and orientation maintenance. The satellite uses radiatively cooled Ku-band Travelling Wave Tube Amplifiers (TWTAs) and dielectrically loaded C-band Demultiplexer for its communication payloads.

EduSat undergoing testing at Satish Dhawan Space Centre, Sriharikota (ISRO) 1343501

EduSat carries five Ku-band transponders providing spot beams. Each beam covers one region: North, North East, East, South and West while providing a 45 MBPS broadband link. One Ku-band transponder provides a national beam that covers all of mainland India. Six External C-band transponders also have national coverage beams. It joined the INSAT system which already had more than 130 transponders in C-band, Extended C-band and Ku-band providing a variety of telecommunication and television broadcasting services.

In the first of three EduSat development phases, educational pilot projects were taken up in Karnataka, Maharashtra and Madhya Pradesh ahead of its launch, using INSAT-3B. Once Edusat became fully functional in November 2004, there was a delay of a few months before the Visvesvaraya Technological University in Karnataka became its first user, linking 120 affiliated engineering colleges to reach students in the state with a one-way video and two-way audio channel. By March 2006, this university had successfully completed 1,000 sessions of lectures via EduSat. At the same time, about 2,700 schools were reported to have benefited from the programme and about 900 interactive networks for application of Higher and Technical education and teacher training were operational. A total of 33 interactive/Receive only Terminal networks were proposed, using the national beam and regional beams, of which 12 were operational.

By 2007, nine networks in the national beam and 22 using regional beams had been started and there were over 10,000 terminals in the EduSat network – most of them in the regional beams. In March 2008, the Orissa government announced that it would soon sign an agreement with ISRO to use EduSat for online education in technical institutions.

Specifications
Launch: 20 September 2004 by GSLV from Sriharikota, India.
Location: 74°E geostationary
Design life: 7 years
Transponders: Five lower Ku-band for spot beam coverage with 55 dBW Edge of Coverage-Effective Isotropic Radiated Power (EOC-EIRP). One lower Ku-band transponder for national coverage with 50 dBW EOC-EIRP. Six upper extended C-band transponders for national coverage with 37 dBW EOC-EIRP. One Ku-band beacon to help ground users with accurate antenna pointing and uplink power control.
Principal applications: One-way and two-way audio-visual educational services.
Configuration: 2.4 × 1.65 × 1.153 m box-shaped bus, 10.9 m span in orbit. 3-axis stabilised using Earth sensors, momentum/reaction wheels, magnetic torquers plus eight 10 N and eight 22 N bipropellant thrusters. 440 N liquid apogee motor using MMH and MON-3 for orbit raising.
Mass: 1,950 kg at launch. 850 kg dry.
TT&C: Through the Master Control Facility at Hassan.
Power system: Twin two-panel solar arrays, 2.04 kW (EOL). Two 24 Ah Ni-Cd batteries.

INSAT

Current Status
In March 2008, the operational INSAT system comprised INSAT-2E, INSAT-3A, INSAT-3B, INSAT-3C, INSAT-3E, INSAT-4A, INSAT-4B and INSAT-4CR. Kalpana-1, GSAT-2 and Edusat are also regarded as part of the system.

Background
The Indian National Satellite (INSAT) System provides geostationary platforms for simultaneous Indian domestic communications and Earth observation functions (Japan's 1999 MT-Sat adopts a similar approach).

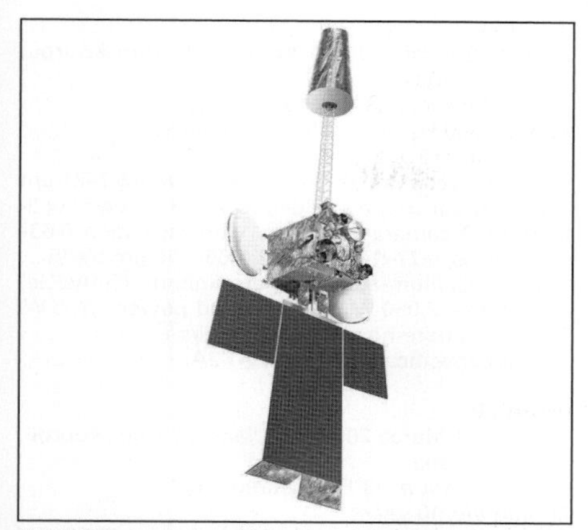

INSAT-3A was the third of the INSAT-3 series. The multipurpose satellite carried a single solar array on the south face, with a solar sail and boom on the north side. (ISRO) 1343330

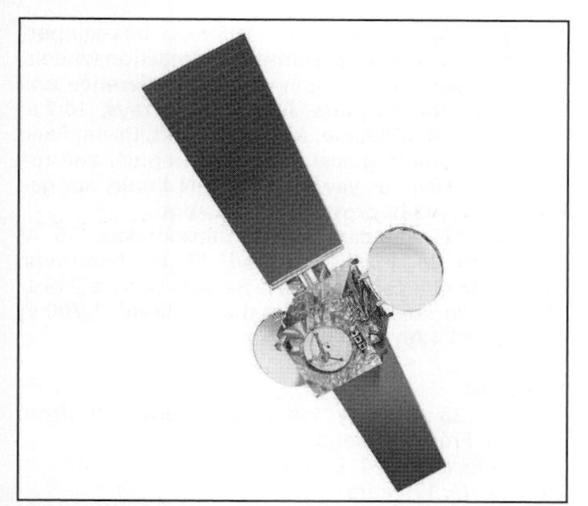

INSAT-3E was the fourth of the INSAT-3 series. It was devoted solely to telecommunications services (ISRO) 1343331

The INSAT system is a joint venture of the Department of Space (DOS), the India Meteorological Department (IMD), the Department of Telecommunications (DOT) and All India Radio. With ISRO Satellite Centre (ISAC), Bangalore, as lead centre, INSAT has been developed with major contributions from the Space Applications Centre (SAC), Ahmedabad, the Liquid Propulsion Systems Centre (LPSC) at Valiamala and Bangalore, the Vikram Sarabhai Space Centre (VSSC) and ISRO Inertial Systems Unit (IISU), Thiruvananthapuram. The INSAT office at Bangalore is responsible for controlling the system through the Master Control Facility (MCF) at Hassan in Karnataka. Meteorological imagery and data are received at the Delhi Earth Station and relayed by microwave to the key Meteorological Data Utilisation Centre (MDUC) at the India Meteorological Department (IMD) headquarters in New Delhi for processing and analysis. 11 VHRR images (each three visible and one infra-red) are acquired each day. The processed material is relayed to more than 20 Secondary Data Utilisation Centres (SDUCs), initially by land lines but converted by 1989 for retransmission through INSAT. Information from Data Collection Platforms can also be processed at the MDUC, through INSAT's Data Relay Transponder.

INSAT-1A was launched by a Delta on 10 April 1982 and positioned at 74°E, but its solar sail did not deploy and it was affected by a power shortage. The satellite lost the Earth lock because of an "unanticipated moon interference" on 4 September and was finally deactivated on 6 September 1982 following exhaustion of its attitude control fuel. **INSAT-1B** was launched by Shuttle in August 1983 but it was not until mid-September that Ford and ISRO engineers at India's Hassan Satellite Control Centre succeeded in deploying its solar array. Full operation was achieved in October 1983 and the satellite was retired in August 1993 after returning more than 36,000 images. **INSAT-1C** was launched by Ariane on 21 July 1988. A power system failure (due to a solar array isolation diode short) soon removed half of the communications capacity but the meteorological services, powered through either bus, remained unaffected from 93.5°E. However, the satellite lost Earth lock 22 November 1989 and was abandoned. INSAT 1C procurement cost was reported at INR800 million, with the Ariane launch cost adding INR740 million. The failed 1A/1C reportedly produced insurance payouts of about USD70 million each. **INSAT 1D** was damaged by a crane hook at the launch site and then while being repaired it suffered further damage. It eventually assumed the prime role in July 1990 from INSAT 1B.

The INSAT-1 satellites were all US-built. The more advanced INSAT-2 family was begun in 1983 in order to develop an indigenous, multi-purpose GEO spacecraft, although it still relied heavily on the previous Ford Aerospace design. The second generation of INSATs was all built by the ISRO Satellite Centre (ISAC), Bangalore. They were launched by Ariane 4 because India did not possess a launch vehicle that was powerful enough to put a 2,500 kg satellite into geostationary orbit. The five-satellite INSAT-2 system began in 1992 with **INSAT-2A**, which took over the prime imaging role from INSAT-1D on 6 October 1992. The almost identical **INSAT-2B** was retired in November 2000, followed by INSAT-2A in April 2002. **INSAT-2C/2D** replaced the imaging, DCP and DRT payloads with an enhanced communications element. INSAT-2D was abandoned on 4 October 1997, after four months in orbit, when it lost Earth lock because of a short circuit in one of its two power buses. **INSAT-2E** was more like 2A/B, but with expanded imaging capability.

INSAT-3B, the first of five successors to the INSAT-2 series, was launched 21 March 2000 by Ariane 5 and co-located with INSAT-2E at 83°E. Its design included the first use of an indigenously developed titanium propellant tank, with a special propellant management device to ensure bubble-free propellant supply. Its communications payload included 12 Extended C-band transponders and three Ku-band transponders that have coverage over the Indian region. INSAT-3B also incorporates a Mobile Satellite Services (MSS) payload with forward link between the hub and mobile station operating in CxS band and return link between the mobile station and the hub operating in SxC band. The launch was moved ahead of INSAT-3A to meet the immediate need for extended C-band capacity after the failure of INSAT-2D.

INSAT-3C was launched by Ariane 4 on 23 January 2002 and located at 74°E. Its payloads include 24 C-band transponders providing an EIRP of 37 dBW, six Extended C-band transponders with EIRP of 37 dBW, two S-band transponders to provide BSS services with 42 dBW EIRP and an MSS payload similar to that on INSAT-3B. All the transponders provide coverage over India.

Next came **INSAT-3A**, launched by Ariane 5 on 10 April 2003. The solar array and two antennas were deployed on 15 April and the satellite was placed in a three-axis stabilisation mode a day later followed by deployment of the solar sail/boom on the north side. The two meteorological cameras were tested over two days beginning 18 April. INSAT-3A reached its operating longitude at 93.5°E on 23 April.

Unlike its predecessors in the INSAT-3 series, 3A is a multi-purpose satellite. It can be used for communications, television broadcasting and meteorological services, including data relay and search and rescue services under the international COSPAR-SARSAT programme. It carries 25 transponders – 12 in the C-band frequency, six in the extended-C (XC) band, a frequency pioneered by ISRO, six in Ku-band and one in the S-band for search and rescue. In addition, there is an ultra-high frequency Data-Relay Transponder (DRT), which uses the XC-band for downlink.

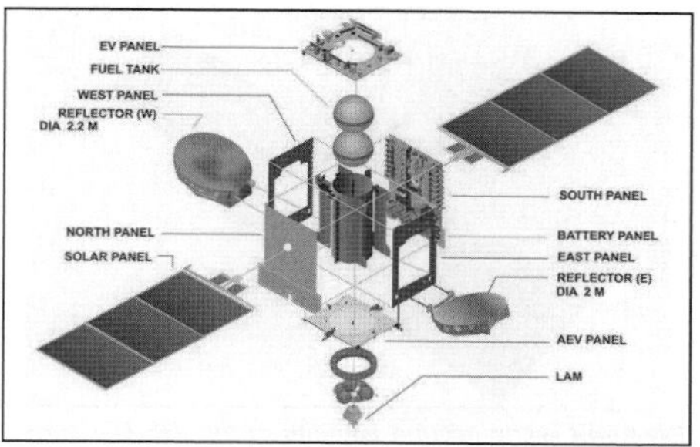

INSAT-4A exploded view (ISRO) 1343437

The meteorological payload aboard INSAT-3A has a Very High Resolution Radiometer (VHRR), which provides cloud images. The VHRR has a 2 km resolution in the visible band and an 8 km resolution in the infrared and water vapour bands. The CCD camera has a resolution of 1 km, operating in the visible, near-IR and short-wavelength IR bands.

Launched in September 2003, **INSAT-3E** is positioned at 55° E. It carries 24 C-band transponders that provide an edge of coverage EIRP of 37 dBW over India plus 12 Extended C-band transponders providing an edge of coverage EIRP of 38 dBW.

INSAT-3D is a meteorological satellite that is expected to be launched by GSLV during 2008–9. The satellite has many new technology elements including a star sensor, a micro stepping Solar Array Drive Assembly (SADA) to reduce disturbances to the spacecraft and a Bus Management Unit (BMU) for control, telecommunication and telemetry function. It also incorporates new features of bi-annual rotation and image and mirror motion compensations for improved performance of the meteorological payloads.

Launched in December 2005 by Ariane 5, **INSAT-4A** is positioned at 83°E, along with INSAT-2E and INSAT-3B. It carries 12 Ku-band, 36 MHz bandwidth transponders employing 140 W TWTAs to provide an EIRP of 52 dBW at the edge of coverage polygon with footprint covering the Indian mainland. 2 C-band, 36 MHz bandwidth transponders provide an EIRP of 39 dBW at the edge of coverage with expanded radiation patterns covering areas beyond India.

INSAT-4B was launched by Ariane 5 in March 2007. It carries identical payloads to its predecessor. Two Tx/Rx dual grid offset fed shaped beam reflectors of 2.2 m diameter for Ku-band and 2 m diameter for C-band are used. The satellite is intended to augment the high power transponder capacity over India in Ku-band and over a wider region in C-band.

INSAT-4C was lost during a failure of the GSLV in July 2006. An identical replacement, **INSAT-4CR**, was launched by GSLV in September 2007. While it normally took about 24 months to build an INSAT, INSAT-4CR was built in less than 12 months after the GSLV's failure. It carries 12 high-power Ku-band transponders designed to provide Direct-

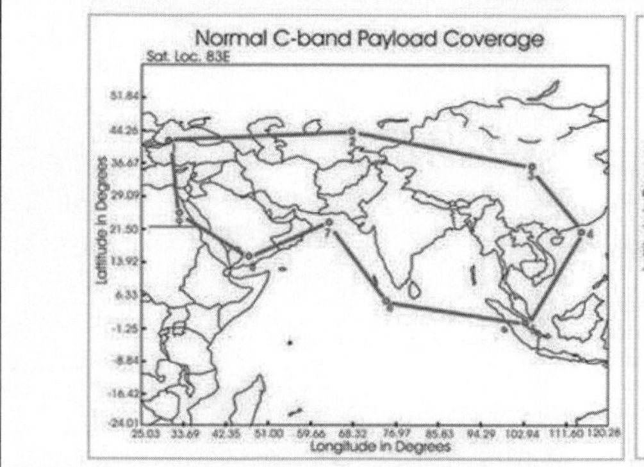

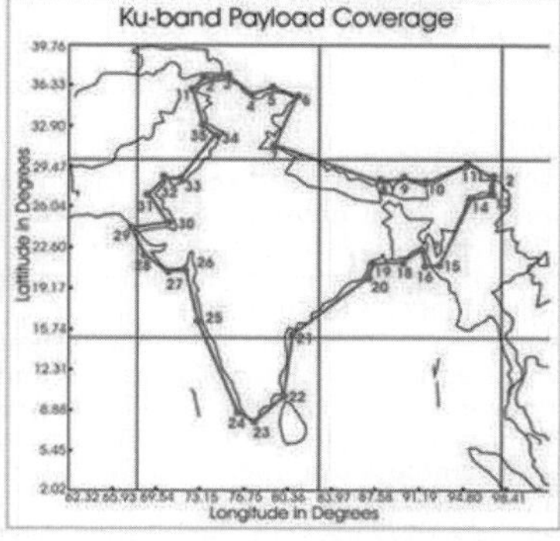

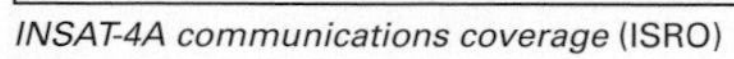

INSAT-4A communications coverage (ISRO) 1343436

For details of the latest updates to ***Jane's Space Systems and Industry*** online and to discover the additional information available exclusively to online subscribers please visit **jsd.janes.com**

INSAT-4B is the second satellite in the INSAT-4 series. Weighing 3,025 kg at lift-off, it is designed to augment Indian communication services in C-band and Ku-band (ISRO) 1343333

To-home (DTH) television services, Video Picture Transmission (VPT) and Digital Satellite News Gathering (DSNG). It is co-located with INSAT-3C, Kalpana-1 and Edusat at 74°E.

INSAT-4D will be configured as an exclusively C-band communication satellite. It will carry 12 C-band transponders and six Extended C-band transponders with wider uplink and downlink coverage over Asia, Africa and Eastern Europe as well as zonal coverage with a minimum of 35 dBW EIRP. It will be launched on board GSLV during 2008–09 and positioned at 82°E. The spacecraft antenna configuration includes two deployable shaped reflectors and one single shell shaped reflector.

INSAT-4E will carry digital multimedia broadcasting payload. The satellite will cover all of India through five S-band spot beams using SxC and CxS transponders. The C-band coverage for the feeder link will have India coverage. The satellite is planned to be launched during 2009–10 by GSLV.

INSAT-4F is proposed as a multi-band satellite carrying payloads in UHF, S-band, C-band and Ku-band. It is planned to be launched during 2010–11 by GSLV. **INSAT-4G** is proposed as a Ku-band satellite carrying 18 transponders, similar to INSAT-4A and INSAT-4B. It will also carry 2 BSS transponders and a GPS Aided Geo-Augmented Navigation (GAGAN) payload. The satellite is expected to be launched by Ariane 5 during 2009–10.

Specifications

INSAT-1D

Contractors: Ford Aerospace (prime)
Launched: 12 June 1990 by Delta 6925 from complex 17, Cape Canaveral. Entered service 17 July 1990
Orbital location: 74° E geostationary October 1999 to May 2002, then drifting between 68° and 72° E.
Spacecraft: 2.18 × 1.42 × 1.55 m box-shaped, 3-axis stabilised bus with asymmetrical 5-panel 11.5 m² solar wing providing 1,185 W BOL (930 W end) and creating total span of 19.4 m. The asymmetrical solar array configuration provides an unobstructed view into space for the radiative cooler of the VHRR and a solar sail is used for balance. A magnetorquer with current coil around the body provides fine control; a unified bipropellant system is used for orbit transfer, station-keeping and attitude maintenance.
Sensors: 2-channel Very High-Resolution Radiometer (VHRR) providing 0.55–0.75 µm visible + 10.5–12.5 µm infra-red images of full Earth disc every 30 minutes. Resolution: 2.75 km visible, 11 km infra-red. The Insats also relayed 402.75 MHz transmissions from up to 100 hydrological, meteorological and oceanographic data collection platforms (system dormant until Insat 2 appeared because of Insat 1 failures).

INSAT-2A

Launch: 9 July 1992 by Ariane 44L from Kourou, French Guiana
Design life: 7 years
Mass: 1,906 kg at launch. 1,162 kg BOL, 911 kg dry
Orbital location: 74° E geostationary July 1992 to January 2001, then 48° E from April 2001
Contractors: ISRO and Indian companies; ISRO Space Application Centre (VHRR)
Applications: Domestic long-distance telecommunications, meteorological Earth observation and data collection services, direct satellite TV broadcasting to community TV receivers in rural and remote areas, radio and TV, Satellite Aided Search and Rescue services.
Spacecraft: 1.92 × 1.77 × 2.37 m, box-shaped 3-axis controlled bus based around 930 mm diameter central cylinder with asymmetrical 5-panel solar wing providing 1,400 W BOL (1,180 W EOL) and creating 23 m span. 440 N integral liquid propellant (NTO/MMH) apogee boost motor and 16 associated 22 N attitude control thrusters. Two 60 Ah Ni-H_2 batteries.
Payload: Communication payload: 12 C- band transponders, seven with wide beam coverage and five zonal beams. 6 extended C-band transponders with zonal beam coverage. All channels provided Edge Of Coverage – Effective Isotopic Radiated Power (EOC-EIRP) of 36 dBW. 2 S-band transponders for BSS.

Meteorological payload: Very High Resolution Radiometer (VHRR) similar to Insat 1's but with 2 km resolution in visible band and 8 km resolution in water vapour band. Redundant Data Relay Transponder for Data Collection Platforms. VHRR detectors are redundant linear arrays of four Si photodiodes (visible) and redundant HgCdTe detectors at 105,000 (using passive radiative cooler) for thermal-IR. VHRR operates in 3 modes: full frame 20 × 20° full Earth scanned in 33 min.; normal mode 14°N-S 20°E-W scan 23 min.; sector scan 4.5°N-S 20°E-W scan 7 min. for rapid repetitive coverage during severe weather conditions such as cyclones.

CCD camera operating in visible, near infrared and shortwave infra-red band with 1 km resolution.

Single 402.75 MHz Data Collection System transponder.

Single 406 MHz Cospas/Sarsat search & rescue package.

INSAT-2B

Launch: 22 July 1993 by Ariane 44L from Kourou, French Guiana. Returned first image 21 July; declared operational 10 August 1993
Mass: 1,931 kg at launch
Orbital location: 93.5°E geostationary.
Other specifications as INSAT-2A.

INSAT-2C

Launch: 6 December 1995 by Ariane 44L from Kourou, French Guiana
Mass: 2,079 kg
Orbital location: 93°E geostationary
Payload: 12 C-band transponders (2 × 50 W TWTA, tx10WSSPA & 3 × 4 W SSPA) all-India beam. Six extended C-band (2 × 10 W & 4 × 4 W SSPAs), all-India beam. Three 20 W TWTA Ku-band all-India, 41 dBW EIRP Two (plus one backup) 50 W TWTA 5,858–5,930/2,550–2,630 MHz up/down S-band all-India BSS beam, 36 MHz bandwidth, 42 dBW EOL EIRP. One 50W TWTA 2,670–2,690/2,500 Mhz up/down S-band all-India mobile communications beam. 1 (plus one backup) 4W SSPA C-band all-India mobile communications feeder link.
Other specifications as INSAT-2A.

INSAT-2D

Launch: 3 June 1997 by Ariane 44L from Kourou, French Guiana
Mass: 2,079 kg at launch
Orbital location: 74.1°E geostationary, suffered electrical problems and declared dead 5 October 1997.
Other specifications as INSAT-2C.

INSAT-2E

Launch: 2 April 1999 by Ariane 42P from Kourou, French Guiana
Orbital location: 83°E geostationary
Mass: 2,550 kg at launch. 1,150 kg dry.
Design life: 12 years
Payload: 2E configured like 2A/2B, but with 5.7–7.1 µm water vapour channel added to VHRR and a new 3-band CCD camera providing 1 km resolution 0.63–0.69 µm vis, 0.77–0.86 µm NIR, 1.55–1.70 µm SWIR.
Power: Gallium-arsenide/germanium (GaAs/Ge) solar array, 2,050 W BOL. Payload power 1,755 W. Two 60 ampere-hour Ni-H_2 batteries.
Other specifications as INSAT-2A.

INSAT-3B

Launch: 21 March 2000 by Ariane 5G from Kourou, French Guiana.
Orbital location: 83°E geostationary
Design life: 10 years
Mass: 2,070 kg at launch. 970 kg dry. 1,100 kg of propellant (Mono-Methyl Hydrazine – MMH – and Mixed Oxides of Nitrogen – MON-3) for raising from GTO to final orbit and station-keeping/attitude control.
Configuration: 1.93 × 1.7 × 1.65 m, box-shaped, 3-axis stabilised using momentum/reaction wheels, Earth sensors, sun sensors, inertial reference unit and magnetic torquers. Twin solar arrays, 14.7 m span. Two deployable antennas and three fixed antennas, pointing accuracy ±0.2° in pitch and roll axes and ±0.4° in yaw axis. 440 N liquid apogee motor, unified bi-propellant thrusters.
Payload: 12 extended C-band transponders, 15 W, bandwidth 36 MHz. 3 Ku-band, 55 W, bandwidth 77/72 MHz. One S-band Mobile Satellite Service (MSS).
Power: Two solar arrays, total area 23 m², 1,700 W BOL. One 24 Ah Ni-Cd battery.

INSAT-3C

Launch: 23 January 2002 by Ariane 42L from Kourou, French Guiana.
Orbital location: 74° E geostationary
Design life: 12 years
Mass: 2,750 kg at launch. 1,210 dry.
Configuration: 2.0 × 1.77 × 2.8 m box-shaped. Span in orbit 15.445 m. 3-axis stabilised using Earth sensors, momentum and reaction wheels, sun sensors, inertial reference unit, magnetic torquers and eight 10 N plus eight 22 N bi-propellant thrusters. 440 N liquid apogee motor using MMH and MON-3. Two 2 m deployable transmit/receive antennas and three fixed antennas (0.7 m, 0.9 m and 1.1 m diameter).
Payload: 24 C-band transponders, uplink 5,930–6,410 MHz, downlink 3,705–4,185 MHz. EIRP 36 dBW. Receive G/T–5 dB/°k. Six Extended C-band,

INSAT-4B undergoing antenna tests in the Compact Test Facility at ISITE (ISRO) 1343337

Insat-4CR during a solar array deployment test (ISRO) 1343335

uplink 6,755–6,995 MHz, downlink 4,530–4,770 MHz. EIRP 36 dBW. Receive G/T–5 dB/°k. Two S-band, uplink 5,850–5,930 MHz, downlink 2,550–2,630 MHz. EIRP 42 dBW. Receive G/T–5 dB/°k. One Mobile Satellite Service transponder operating in S-band uplink and C-band downlink. MSS (forward link): uplink 6,450–6,470 MHz, downlink 2,500–2,520 MHz. EIRP 37 dBW. Receive G/T–5 dB/°k. MSS (return link): uplink 2,670–2,690 MHz, downlink 3,680–3,700 MHz. EIRP 30 dBW. Receive G/T–7.5 dB/°k.
Power: Twin three panel solar arrays generating 2.765 kW at equinox, 2.535 kW at summer solstice. (BOL). Two 60 Ah Ni-H_2 batteries.

INSAT-3A
Launch: 10 April 2003 by Ariane 5G from Kourou, French Guiana.
Orbital location: 93.5°E geostationary
Design life: 12 years
Mass: 2,950 kg at launch. 1,348 kg dry.
2.0 × 1.77 × 2.8 m box-shaped with solar array on south side, solar sail and boom on north side. Span in orbit 24.4 m (north-south), 8.5 m (east-west). 3-axis stabilised using Earth sensors, momentum and reaction wheels, solar flaps, magnetic torquers and eight 10 N plus eight 22 N bi-propellant thrusters. 440 N liquid apogee motor using MMH and MON-3.
Payload: Communication Payload. 12 C-band transponders, 9 with expanded coverage providing an Edge-of-Coverage (EOC) Effective Isotropic Radiated Power (EIRP) of 38 dBW and three with India beam coverage and EOC-EIRP of 37 dBW. 6 upper extended C-band transponders with India beam coverage providing EOC-EIRP of 37 dBW. 6 Ku-band transponders with India beam coverage and EOC-EIRP of 47.5 dBW. One Search & Rescue transponder.
Meteorological Payload. Very High Resolution Radiometer (VHRR) with 2 km resolution in visible band and 8 km resolution in infrared and water vapour bands. CCD camera operating in visible, near infrared and shortwave infrared bands with 1 km resolution.
One UHF Data Relay Transponder (DRT).
Power: Single 26.6 m² solar array generating 3.1 kW (BOL). Two 70 Ah Ni-H_2 batteries.

INSAT-3E
Launch: 21 September 2003 by Ariane 5G from Kourou, French Guiana.
Orbital location: 55°E geostationary
Design life: 12 years
Mass: 2,750 kg at launch. 1,181 kg dry.
Configuration: 2.0 m × 1.77 m × 2.8 m with two solar arrays on north and south sides. Span in orbit 15.445 m. 3-axis stabilised using Earth sensors, momentum and reaction wheels, magnetic torquers and eight 10 N plus eight 22 N bi-propellant thrusters. 440 N liquid apogee motor using MMH and MON-3.
Payload: 24 C-band transponders with India beam coverage, providing Edge of Coverage-Effective Isotropic Radiated Power (EOC-EIRP) of 37 dBW. 12 Extended C-band transponders with India beam coverage, EOC-EIRP of 38 dBW.
Power: Twin three panel solar arrays generating 2.4 kW (BOL). Two 70 Ah Ni-H_2 batteries.

INSAT-4A
Launch: 10 April 2003 by Ariane 5G from Kourou, French Guiana.
Orbital location: 83°E geostationary
Mass: 3,100 kg at launch. 1,987 kg dry.
Design life: 12 years
Configuration: 2.0 × 1.77 × 2.8 m box-shaped. 15.16 m span in orbit. 3-axis stabilised using Earth sensors, momentum and reaction wheels, magnetic torquers and eight 10 N plus eight 22 N bi-propellant thrusters. 440 N liquid apogee motor using MMH and MON-3.
2.2 m diameter Ku-band transmit/receive antenna, polarisation sensitive dual grid shaped beam deployable reflector with offset-fed feeds illumination. 2 m diameter C-band transmit/receive antenna, polarisation sensitive dual grid shaped beam deployable reflector with offset-fed feeds illumination.
Payload: 12 Ku-band 36 MHz bandwidth transponders employing 140 W TWTAs to provide an EIRP of 52 dBW at Edge of Coverage (EOC) polygon with footprint covering Indian mainland. 12 C-band 36 MHz bandwidth transponders employing 63 W TWTAs to provide an EIRP of 39 dBW at EOC with expanded coverage over India and the area beyond India in southeast and northwest regions.
Twin 3 panel solar arrays providing 5.5 kW (BOL). Three 70 Ah Ni-H_2 batteries.

INSAT-4B
Launch: 12 March 2007 by Ariane 5 ECA from Kourou, French Guiana.
Orbital location: 93.5°E geostationary
Mass: 3,025 kg at launch. 1,337 kg dry.
Payload: 12 Ku-band transponders, 9 with 36 MHz usable bandwidth and 3 with 27 MHz bandwidth, employing 140 W TWTAs to provide an EIRP of 52 dBW at Edge Of Coverage (EOC) polygon with footprint covering Indian mainland. 12 C-band transponders with 36 MHz usable bandwidth employing 63 W TWTAs to provide an EIRP of 39 dBW at EOC with expanded coverage encompassing India and the area beyond India in southeast and northwest regions.

INSAT-4CR
Launch: 2 September 2007 by GSLV from Sriharikota, India.
Orbital location: 74°E geostationary
Mass: 2,140 kg at launch. 941 kg dry.
Design life: 10 years
Configuration: 1.65 × 1.53 × 2.4 m box-shaped. 9.45 m span in orbit. 3-axis stabilised using Earth sensors, momentum and reaction wheels, magnetic torquers and eight 10 N plus eight 22 N bi-propellant thrusters. 440 N liquid apogee motor using MMH and MON-3. 2.2 m diameter Ku-band transmit antenna. 1.4 m diameter Ku-band receive antenna.
Payload: 12 Ku-band 36 MHz bandwidth transponders, 140 W TWTAs, EIRP 51.5 dBW at edge of coverage. Ku-band beacon to aid satellite tracking.
Power: Twin two panel solar arrays providing 2.87 kW (BOL). Two 70 A-h Ni-H_2 batteries.

Indonesia

Indostar (Cakrawarta) series

Current Status
Indostar 1 remains operational, as of March 2008, despite battery charger problems.

Background
PT Media Citra Indovision of Indonesia started operations with a five-channel C-Band Direct Broadcast Satellite (DBS) analogue service using the Palapa C-2 satellite. However, the company had ambitious plans to provide the world's first DBS dedicated to radio/TV for a single nation. Indostar 1 (renamed Cakrawarta 1 by the Indonesian President Suharto after launch) was to be the first of four planned satellites, providing five S-band transponders for digital TV, using 8:1 compression for 49 channels. The second was to have added L-band for CD-quality radio. The USD100 million contract for the first satellite and its launch was signed with CTA International (now Orbital Sciences Corp.) of the US on 8 December 1993. Indostar 1 was the first application of CTA's StarBus satellite platform. Funding was provided by 40 per cent private consortium, 30 per cent PT Amcol Graha Ltd. (a Jakarta electronics company) and 30 per cent PT Bimantra Citra. The Indostar 1 DBS system cost USD271 million. PT MediaCitra Indostar is the satellite company of the Indovision group. Indovision is a multichannel pay television company.

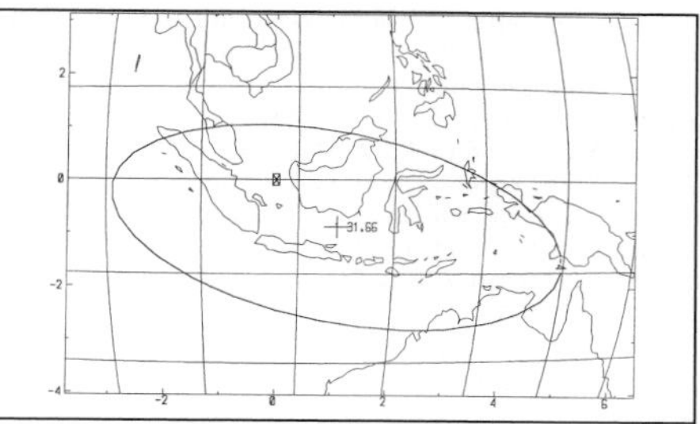

Indostar 1 TV coverage (CTA) 0517404

The commercial Indostar system is received via USD100 hand-held radios, USD100 analogue receivers and USD500 digital satellite decoders with satellite dishes 0.7 to 1 m in diameter. The signals are encrypted, which requires a compatible decoder and subscription authorisation. The use of S-Band frequencies, which are less vulnerable to rain and atmospheric interference than C-Band frequencies and are well suited to the tropical climate of Indonesia, is designed to guarantee quality reception within Indonesia.
Indostar 1 went into commercial service in June 1998. However, due to a failed power regulator in November 1997, only 80 per cent of the required power is available during the spring and autumn eclipse periods. At such times, two of the satellite's five transponders cannot be used. Insurers paid USD25 million in damages.
Indovision originally announced plans to locate Indostar 2 at 118.9°E, Indostar 3 at 107.5°E and Indostar 4 at 119.1°E. However, financial difficulties have caused these plans to be put on hold.

Specifications
Indostar 1
Launch: 12 November 1997 on Ariane 44L from Kourou, French Guiana
Location: 107.7°E geostationary
Design life: 10 years (12 years propellant)
Contractors: CTA International (now Orbital Sciences Corp.)
Transponders: 5 (plus 4 back-up, 5-for-4) 70 W TWTA 8.120–8.270/2.520–2.670 GHz up/down S-band domestic TV 3.6 × 8.35° beams (8:1 compression

Indostar 1 was the world's first geostationary DTH TV/radio lightsat (Orbital Sciences) 1166909

for 40 channels), 24 MHz bandwidth, EIRP 48–50 dBW min Indonesia, linear orthogonal polarisation
Principal applications: Direct to Home TV/radio
Configuration: CTA's Star lightsat, bus 165 × 185 × 150 cm. AKM Star 30C
Mass: 1,400 kg at launch; 430 kg on-station BOL
TT&C: Through master control station at Jakarta; backup control station at Surabaya
Power system: 1,700 W EOL; two SAFT 52 Ah Ni/H_2 batteries

Palapa series

Current Status
Palapa C2 was launched May 1996 and remains operational. None of the previous Palapa satellites is now operated by Indonesia. The first Palapa D satellite is expected to launch in 2009, eventually replacing Palapa C2.

Background
In the 1970s, Indonesia witnessed the commercially successful domestic satellite systems in Canada (Anik) and US (Westar) for large territorial communication and determined that its 6,064 inhabited islands arcing across 5,000 km along the equator would benefit from the same type of system. The country built 40 large Earth stations, some 200 5-metre telephony stations and more than 1,000 TVRO stations (television receive only). Indonesia decided to begin its satellite programme with the Palapa ('Fruit of labour'), but did not announce the plans until after negotiations with Hughes had been completed by the government-owned company Perumtel (later PT Telkom). The Palapa system became operational in August 1976 after the launch of **Palapa A1** by a Delta on 8 July 1976. Indonesia had a completely functioning network by summer 1977 after **Palapa A2** completed orbit tests.

The A series satellites ended their operational lives in June 1985 and January 1988 respectively, but in 1979 Indonesia had contracted with Hughes to build a follow-on system for two second-generation satellites at a cost of USD74.5 million. NASA threw in two Shuttle launches at a cut-rate price of USD18 million. The **Palapa B1** proved a success, operating continuously from its June 1983 Shuttle launch until retirement in 1990. PT Telkom retired the Palapa B1 at 118° E and sold it to PT Pasifik Satelit Nusantara (PSN) in 1991 to provide Pacific rim services. Still operated by PT Telkom, PSN's major shareholder, B1's inclination was allowed to drift in order to extend its life by three years.

The **Palapa B2** during the Space Shuttle 41-B mission in February 1984 but the PAM-D GTO insertion stage failed, stranding it in a low Earth orbit. Indonesia used the USD75 million insurance payment to purchase **Palapa B2P** (P means *pengganti* or 'replacement' in Indonesian) as a replacement for the stranded satellite. This was launched by Delta in March 1987 and located at 113° E. In 1996 it was sold by PT Pasifik Satelit Nusantara to PLDT for about USD3 million, becoming the first satellite of the Philippine **Mabuhay** system. It was sent to a graveyard orbit in January 1998.

Palapa C1 expanded coverage to southeast Asia, China and Australasia (Hughes Space & Communications) 0517405

Based on the powerful Hughes HS-601 bus, the Palapa C series was introduced in 1996 (Hughes Space & Communications) 0054340

The crew of Shuttle mission 51-A recovered B2 in November 1984 and the insurers sold it to Sattel Technologies Incorporated of Van Nuys, California for about USD18 million. Sattel planned to launch the refurbished satellite by Delta II in 1990 and then resell it to Indonesia once in-orbit. It was relaunched 13 April 1990 by Delta 6925 and sold back to Perumtel, when it became known as **Palapa B2R**. It was located at 108° E. Its replacement, **Telkom-1**, was launched in 1999 for PT Telkomunikasi Indonesia, and in January 2001 it was leased to US company **Newsat** and relocated at 42.5° E, in an inclined orbit. Its service life ended in June 2005, when it was moved to graveyard orbit. The last in the series, **Palapa B4**, was launched in May 1992 and remained operational until August 2005.

In January 1993, Indonesia set up a private company, PT Satelit Palapa Indonesia (PT Satelindo), to finance and operate the USD240 million Palapa C programme. The plan was to increase the number of telephones per 1,000 population from five to 200 by 2019, using follow-on systems to the Palapa. **Palapa C1**, the first of the third-generation series to be purchased from Hughes, was launched in January 1996 to replace Palapa B2P (which ended service in February 1996). It was able to expand coverage to China, India and Japan. However, a problem with the battery recharging system occurred on 24 November 1998, leaving the satellite without a power backup during the twice yearly eclipse periods. The satellite was declared unusable for its planned mission and the satellite was passed to the insurers. Hughes Global Services acquired the satellite and renamed it **HGS 3**. Hughes developed procedures that would enable it to maintain full operations except around the times of eclipse – a total of 88 days per year. The satellite systems had to be switched off for three hours per day on average. Overall availability of the spacecraft's transponders remained at 96 per cent. After an electronics failure, it was renamed **Anatolia 1** and operated at 50° E, before being sold in 2002 to the Pakistan government and renamed once again as **Paksat 1**, operating at 38° E.

Some of the second-generation Palapa satellites remained in service into the early 2000s (Hughes Space & Communications) 0517002

Palapa C2 was launched May 1996 and remains operational at 113° E.

In May 1995 the German company T-Mobil, a subsidiary of Deutsche Telekom MobilNet GmbH, purchased 25 per cent of PT Satelindo's shares. Satelindo had plans to develop the Palapa D series but it was delayed due to economic difficulties. Satelindo was merged with PT Indosat Tbk in November 2003, and on 29 June 2007 Indosat announced that it had signed a contract with Thales Alenia Space to build and launch the Palapa D satellite by the second half 2009. Based on the Thales Alenia Spacebus 4000B3 platform, the new satellite will eventually replace Palapa-C2 satellite, whose design life will end in 2011. It will have a larger capacity than Palapa C2, it will have 24 standard C-band, 11 extended C-Band and 5 Ku-band transponders, providing coverage of Indonesia, ASEAN countries, Asian countries, Middle East and Australia. Palapa D will have a launch mass of 4.1 t, payload power of 6 kW and a service life time of 15 years.

Specifications
Palapa B2/B2R
Launched: 6 February 1984 by Space Shuttle STS-41B, retired 12 November 1984. Refurbished and relaunched as Palapa B2R on 13 April 1990 by Delta II 6925 from Cape Canaveral.
Location: 108° E GEO. Replaced by C2 in 1996.
Mass: 1,240 kg at launch, 652 kg on-station BOL
Other specifications as for Palapa B2P.

Palapa B4
Launched: 14 May 1992 by Delta II 7925 from Cape Canaveral
Location: 118° E GEO. Taken out of service in August 2005.
Mass: 1,200 kg at launch. 692 kg on station BOL
Other specifications as for Palapa B2P.

Palapa B2P/Palapa Pacific 2/Agila 1/Mabuhay 0
Launched: 20 March 1987 by Delta from Cape Canaveral.
Location: 113°E. Replaced in 1996 by C1. Moved to 144°E GEO from August 1996. Taken out of service January 1998.
Design life: 8 years.

Contractors: Hughes Space & Communications Co.
Transponders: 24 (plus 6 back-up) 10 W TWTA 5.925–6.425/3.72–4.16 GHz up/down C-band regional beam, 36 MHz bandwidth, EIRP 34 dBW min Indonesia, 32 dBW min ASEAN countries (Philippines, Singapore, Thailand, Malaysia), linear orthogonal polarisation.
Principal applications: Telecoms, regional/domestic TV.
Configuration: Hughes' cylindrical HS-376 bus, 2.16 m diameter, 6.83 m deployed height, spin-stabilised, position/spin by hydrazine thrusters, GEO injection by solid AKM.
Mass: 1,200 kg at launch; 628 kg on-station BOL.
Pointing: 0.05° accuracy
TT&C: Through PT Telkom's Master Control Station at Cibinong, near Jakarta.
Power system: 1,062 W BOL/830 W EOL from K7 solar cells on cylindrical body and skirt.

Palapa C1/HGS 3/Anatolia 1/Paksat 1
Launched: 1 February 1996 by Atlas IIAS from Cape Canaveral.
Location: 113° E GEO (replacing B2P). Initially: in March 1996 manoeuvred off-station and re-located over 112° E; again manoeuvred off-station August 1996 and the following month re-located over 150° E. In 1999 moved to 50° E, and relocated at 38° E in 2002.
Lifetime: 12 years full performance; 14 years basic performance.
Transponders: 24 (plus 6 back-up) 21.5 W SSPA 5.925–6.425/3.700–4.200 GHz up/down C-band regional beam, 36 MHz bandwidth, EIRP 37 dBW min Indonesia, ASEAN countries (Philippines, Singapore, Thailand, Malaysia), Vietnam, Brunei, Papua New Guinea, New Zealand/Eastern Australia, linear orthogonal polarisation

6 (plus 2 back-up) 26 W SSPA 6.425–6.665/3.400–3.640 GHz up/down extended C-band beams for PT Pasifik Satelit Nusantara, 36 MHz bandwidth, EIRP 37 dBW min Pakistan, India, Burma, China, People's Republic, Korea, Taiwan, Japan, linear orthogonal polarisation

4 (plus 2 back-up, 6-for-4) 130 W TWTA 13.750–14.490/10.950–11.690 GHz up/down Ku-band regional beams, 72 MHz bandwidth, EIRP 50 dBW min Indonesia, Japan, Korea, Eastern China, Thailand, Vietnam, Singapore, linear orthogonal polarisation. A battery supply problem means this payload cannot operate while satellite is in eclipse.
Principal applications: Telecoms, regional/domestic TV.
Configuration: Hughes HS-601 platform, 3-axis, solar array span 21.0 m (satellite within 2.7 × 3.1 × 3.5 m envelope at launch). Payload mounted on North/South panels and Earth-facing floor; East/West faces carry hard points for two 216 cm diameter C-band antennas, fixed antenna mountings on Earth-facing wall for 178 cm diameter extended C-band + 152 cm diameter Ku-band; all antennas octagonal dual-gridded with single feed horns.
AOCS: 3-axis, unified 13 ARC 22 N and one Marquardt 490 N bipropellant thrusters incorporating onboard control processor, Sun/Earth sensors, and two 61 Nms 2-axis gimbaled momentum bias wheels. The NTO/MMH is contained in four spheres. Spin-stabilised in transfer orbit.
Mass: 2,989 kg at launch, 1,775 kg on-station BOL.
TT&C: As Palapa B.
Power system: Twin solar wings of three 2.16 × 2.54 m panels carrying K4 large area Si cells on Kevlar substrate providing 3,730 W; EOL 3,400 W. Ni/H_2 batteries provide 100% eclipse protection.

Palapa C2
Launch: 16 May 1996 by Ariane 44L from Kourou, French Guiana.
Location: 113° E GEO (replacing B2R).
Mass: 2,989 kg at launch, 1,803 kg on-station BOL, 1,669 kg dry
Other specifications as for Palapa C1.

Telkom series

Current Status
Telkom-1 developed problems with its south solar array motors during in-orbit testing, but its lifetime should not be affected and it remained operational in February 2008. Telkom-2 was launched on 16 November 2005.

Background
Lockheed Martin Commercial Space Systems built the Telkom-1 satellite for the state-owned PT Telkomunikasi (TELKOM) of Indonesia to link the country's thousands of islands. Lockheed Martin also provided launch support services, mission operation services, additional equipment to augment the existing ground station in Cibinong, Indonesia, and training of the Telkom-1 technical team.

Telkom-1 replaced the Palapa-B2R satellite which was relocated and leased to the US operator Newsat in 2000. The satellite supports a variety of telecommunication applications, including a rapid increase in Indonesia's demand for telephone services and high-speed digital traffic compatible with Very Small Aperture Terminal applications. Based on Lockheed's A2100A bus, the satellite's payload was also built at the Lockheed Martin Commercial Space Systems facilities in Newtown, Pennsylvania. Equipped with 24 C-band and 12 extended C-band transponders, it provides greater capacity than previous TELKOM-owned satellites, as well as a longer design and manoeuvring life. Positioned at 108° east, it provides coverage to all of Indonesia as well as parts of Southeast Asia and northern Australia.

Telkom-1 developed problems with its primary and back-up drive motors on the south solar array during in-orbit testing. These drives are used to manoeuvre the arrays in order to track the Sun. The north solar array was not affected. The problem seems to have been caused by an error during the assembly of Telkom-1. Occasional stoppages of the south solar array could be overcome by using the array motor in a high-torque mode, and TELKOM reported that the satellite is fully operational despite the problem. The solar arrays can be properly aligned and, if needed, are able to generate the maximum amount of electrical power.

Telkom-2 was built by Orbital Sciences Corporation for PT TELKOM. The contract included a firm order for one satellite and an optional order for another. Orbital also supplied the payload and supplemented PT TELKOM's existing ground segment with a new ground station and upgraded hardware and software. An Ariane 5 ECA successfully launched it into GTO about 34 minutes after lift-off. It was eventually located in geostationary orbit at 118°E. Telkom-2, which operates in C-band frequencies, replaced the on-orbit Palapa-B4 satellite, allowing Telkom to expand its satellite communications coverage area. The satellite's 84 spot beams and 3 shaped beams provide fixed satellite services to all of Indonesia, as well as areas of Southeast Asia, the Indian subcontinent, Australia and New Zealand.

Indonesia's Telkom-2 satellite was launched on 16 November 2005 (Orbital Sciences Corp.) 1342588

Specifications
Telkom-1
Launch: 12 August 1999 by Ariane 42P from Kourou.
Location: 108° east geostationary.

Telkom-2 in Orbital's Dulles, Virginia, satellite manufacturing facility (Orbital Sciences Corp.) 1342589

Design life: 15 years.
Contractors: Lockheed Martin Commercial Space Systems.
Transponders: 24 C-band, 12 extended C-band
Principal applications: Various telecommunication applications, including high-speed digital traffic compatible with Very Small Aperture Terminals.
Configuration: A2100A bus; 3-axis stabilisation. LEROS-1C liquid bi-propellant (hydrazine and nitrogen tetraoxide) transfer orbit system.
Mass: 2,763 kg at launch, 1,700 kg in GEO.
TT&C: Ground station in Cibinong, Indonesia.
Power system: Two solar arrays generating 4 kW of power.

Telkom-2
Launch: 16 November 2005 by Ariane 5 ECA from Kourou.
Location: 118° east geostationary.
Design life: 15 years.
Contractors: Orbital Sciences Corporation.
Transponders: 28 C-band. Two groups of 15-for-12 linearised TWTAs. Transponder power: 39 watts RF.
Principal applications: Telephony, image and data transmission services in Southeast Asia and the Indian subcontinent.
Configuration: Star-2 bus; 3-axis stabilisation. Two 2.0 m dual-gridded shaped-beam reflectors.
Dimensions: 2.4 × 3.3 × 1.9 m stowed; span in orbit 12.6 m
Mass: 1,975 kg at launch; 899 kg dry mass.
Propulsion: IHI 500 N liquid bi-propellant transfer orbit system; monopropellant hydrazine attitude control/orbit manoeuvre system
Power system: Two solar arrays, two panels per array, UJT gallium arsenide cells. 2.918 kW (EOL)

International

Amazonas series

Current Status
The first Amazonas satellite was launched on 5 August 2004. Its occupation rate exceeded 88 per cent at the end of 2006. In June 2007, EADS Astrium was awarded a contract to build a second satellite. This 5,400 kg, 12 kW, 64 transponder spacecraft will be launched in 2009 and located at 61°W.

Background
Amazonas-1 is owned by Hispamar Satellites, a joint venture involving Hispasat of Spain and Telemar of Brazil. The company, whose headquarters are located in Rio de Janeiro, is responsible for commercially exploiting the satellite and its communication services in Latin America. Marketing of the satellite, coupled with the extended capacity of the Hispasat satellites located at 30°W (see separate entry), allows both companies to provide a variety of telecommunication services throughout the Americas. Amazonas-1 doubled the space orbital capacity of Hispasat satellites and extended the company's commercial activities to the Americas. It is also provides trans-Atlantic capabilities which allow it to cover Europe and North Africa.

After a successful Proton launch, Amazonas was separated from the Breeze M upper stage 9 hours and 11 minutes into the flight. After three firings of the apogee motor, the satellite reached its GEO destination. Despite a sharp and unexplained drop in pressure in one of its oxidant tanks, opening up the possibility of a reduction in its 15-year operational life, the satellite became operational on 1 November 2004.

With a total mass exceeding 4.5 tonnes, the Amazonas satellite, based on the Astrium three-axis Eurostar 3000 stabilised platform, is equipped with 63 equivalent transponders, of which 36 operate in the Ku-band and 27 in the C-band. This capacity makes it the largest satellite serving Latin America, and the satellite with the largest number of transponders in that region. Of particular value is its ability to offer communication services in both the C- and Ku-bands. Until the launch of Amazonas-1, the Ku-band had mainly been used to offer Direct-To-Home television services (DTH) in Latin America, rather than business services and broadband.

Amazonas-1 incorporates advanced technologies in terms of antennas, repeaters and platform. Of particular note is **AmerHis**, an experimental processor designed to enhance the flexibility of two-way satellite communications. AmerHis incorporates a signal processor based on the DVB-S/DVB-RCS open standard. It was developed by a consortium led by Alcatel Espacio. The industrial team was made up of companies from Canada, France, Norway and Spain, co-ordinated by Spain's Centre for Industrial Technology Development and co-financed by ESA. The AmerHis system allows for the development of new, more flexible, higher quality and higher transmission speed broadband services, which do not require a double skip. By using a lattice network, the system allows for direct communication between small stations. Another important advance, is the interconnectivity between the coverage beams. With this advance, the system allows for the connection of one or more coverage zones using a single transmission, as well as the onboard merging of various signals into one, even if they come from different coverage zones. Both improvements can be applied to services such as video conferencing, corporate networks, etc. System fine-tuning was initiated through a pilot network in 2005, with the first demonstrations at the Le Bourget Fair.

The AmerHis system provides connectivity between user terminals located at any point within the Amazonas satellite coverage footprint. The main innovation introduced by the system is its capacity as a node to access, regenerate and switch broadband, as opposed to the conventional simple satellite signal re-transmitter function of most communication satellites. The system's nucleus is a digital processor, which manages four 36 MHz transponders and interconnects the four beams. Signals generated by users and service providers located within any one of the coverage zones, are processed and redirected to the destination channel, on board the satellite. The AmerHis system also respects the DVB-S/DVB-RCS open standards in order to guarantee compatibility with other systems and to reduce the costs of user terminals. This means that the user can send signals directly to the satellite without the need for teleports. One or more zones within its coverage footprint can also be connected using a single transmission, or various signals can be combined into one on board the satellite, even if they come from different coverage zones. In addition to the payload on board

Amazonas-1 during assembly and testing (Hispasat) 1166442

Amazonas-1 during assembly and testing (Hispasat) 1166441

Amazonas-1, AmerHis also consists of a Network Control Centre, which can reconfigure the useful load, assign capacities and efficiently manage user traffic. Several gateways and terminals have been developed and deployed to initiate and demonstrate the new services available.

Amazonas-2 will have a launch mass of about 5,400 kg and useful life cycle of 15 years. It will provide a wide range of communications services, including DTH, through 64 simultaneous transponders, of which 54 will operate in Ku-band and 10 in C-band. The satellite will be powered by a solar panel that will measure 39 m when deployed, providing over 14 kW at the end of its life cycle. Ku-band services will be available over all of the Americas, with C-band coverage over Latin America.

Specifications

Amazonas-1

Launch: 22:32 GMT, 4 August 2004 by Proton M – Breeze M from Baikonur.
Location: 61°W (GEO)
Design life: Minimum 15 years
Contractors: Prime contractor: EADS Astrium. Ku-band input multiplexers, filters and diplexers: Alcatel Espacio.
Transponders: 32 Ku-band and 19 C-band, equivalent to 36 MHz (36 Ku-band and 27 C-band). Polarization horizontal and vertical. Maximum EIRP: 52 dBW (Brazil). C-Band diaelectric resonator exit multiplexor. Linear channel amplifiers. Amplifier power: 50 W C-band, 100 W Ku-band.
Principal applications: Telecommunications services – TV, radio, telephone, broadband IP services, business networking.
Configuration: Astrium 3-axis Eurostar 3000 stabilised platform. 5 antennas. Solar array with 36.1 m span. Apogee motor with high specific impulse.
Body dimensions: 5.88 × 2.4 × 2.9 m
Mass: 4,605 kg at launch. Dry mass: 2,135 kg.
Power system: Gallium arsenide triple union solar panels with 7 kW output. Lithium ion batteries.

Arabsat 2 series

Current Status

Arabsat 2A was taken out of service in June 2006 and moved to a graveyard orbit. Arabsat 2B experienced a malfunction on 26 September 2007 which interrupted the service. It was restored within 2 days, but the satellite's orbit became inclined.

Background

Arabsat requested bids for a successor system to Arabsat 1 in early 1990, with a 15 October 1990 deadline. The Iraqi invasion of Kuwait forced a delay. The final Arabsat 2 RFP went to bidders in 1991 projecting a new launch date. In October 1992, Arabsat tentatively chose Hughes Space & Communications Company to build the new satellites on a USD258 million bid for two HS- 601 satellites. British Aerospace and Aerospatiale also submitted proposals. However, Hughes and BAe

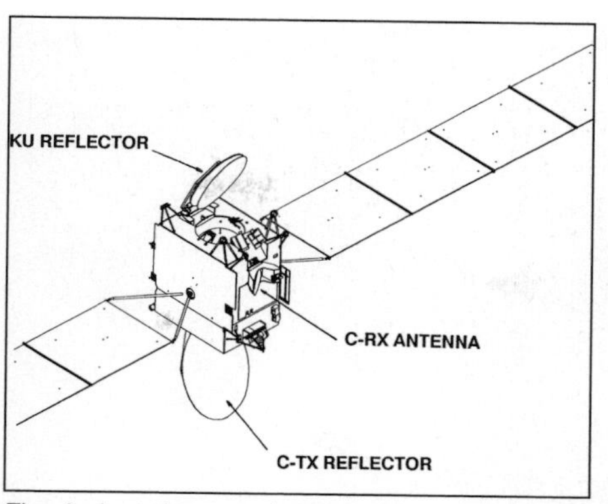

The Arabsat 2 generation was introduced in 1996 (Aerospatiale) 0517197

reportedly refused to agree to some of Arabsat's terms and the organisation instead placed the USD257.9 million order for Aerospatiale's Spacebus 3000 platform on 17 April 1993. The contract included Arabsat's first launch on Ariane 4, which took place in July 1996.

The Arabsat 2 satellites carried 22 C-Band and 12 Ku-band transponders that enabled Direct-to-Home Television services, as well as the possibility of establishing private networks for telephone and data transmission using very small aperture terminals. Arabsat 2A C-band had one channel each for Bahrain, Egypt, Mauritius, Oman, Sudan and the UAE, and three for Saudi Arabia. Ku-band provided three channels for Saudi Arabia and one each for Jordan, Kuwait, Qatar and Syria. To satisfy a growing demand and to guarantee services, Arabsat 2B was launched in November 1996. Previously intended as a ground spare, it was decided to launch it before Arabsat 1C was retired.

Specifications

Arabsat 2A

Launched: 9 July 1996 by Ariane 44L from Kourou, French Guiana.
Location: 26° E geostationary.
Design life: 12 years (16 years propellant).
Contractors: Aerospatiale (prime; structure, thermal control, solar arrays, integration/testing), Alcatel Espace (communication payload), Detecon (contract monitoring on behalf of Arabsat), DASA (propulsion/attitude control integration), Alenia (TT&C).
Transponders: 14 (plus 6 back-up: 10-for-7) 15 W SSPA 5.925–6.425/3.70–4.20 GHz up/down C-band regional beams, 12 × 36 + 2 × 54 MHz bandwidths, EIRP 35 dBW min, circular polarisation for telecom services.

8 (plus 4 back-up, 12-for-8) 57 W TWTA 5.925–6.425/3.70–4.20 GHz up/down C-band regional beams, 36 MHz bandwidth, EIRP 41 dBW min, circular polarisation for semi-DBS via 0.8–1.0 cm dishes

12 (plus 5 back-up, 17-for-12) 95 W TWTA 13.75–14.0/12.5–12.7 GHz up/down Ku-band regional beams, 8 × 36 + 4 × 30 MHz bandwidths, 47 dBW min EIRP, linear polarisation.
Principal applications: Regional TV, telephony, data, fax relay
Configuration: Aerospatiale/DASA Spacebus 3000 3-axis stabilised 1.8 × 2.6 × 2.3 m box-shaped bus (7 m deployed height) with 26.3 m-span twin solar wings. C-band: 2.1 × 2.2 m transmission west panel dish + 60 cm receive Earth face dish; Ku-band: single 1.6 m dual gridded dish on east panel. C-band repeater assembly on N panel; Ku-band on S panel.
Mass: 2,617 kg launch, 1,570 kg on-station BOL, 1,108 kg dry, 260 kg payload.
TT&C: As Arabsat 1.
Power system: 5,074 W EOL provided by twin Si solar wings totalling 46 m². Two 52 Ah Ni/H_2 batteries.

Arabsat 2B

Launch: 13 November 1996 by Ariane 44L from Kourou, French Guiana.
Location: 30.5° E geostationary.
All other specifications as Arabsat 2A.

Arabsat 3 series

Current Status

A major power failure occurred on Arabsat 3A (also known as **Badr 3**) in December 2001, less than three years after launch. The second in the series was never launched.

Background

Alcatel Space Industries was contracted to provide the first of two direct TV broadcast satellites based on the Alcatel Spacebus-3000B platform. Arabsat 3A was launched on 26 February 1999 for the Arab Satellite Communications Organisation based in Saudi Arabia. It was co-located with Arabsat 2A at 26°E.

The satellite had an expected design life of 13 years, but on 7 December 2001, eight of the 20 transponders on Arabsat 3A failed after the satellite lost about half its power. This was caused by a short circuit in the solar array drive mechanism which cut the electrical power input from one solar panel. Ground control was able to restore broadcasts to some of the transponders by frequency re-allocation and using alternate power supplies. The other solar array was reported to be functioning normally. Arabsat eventually filed a USD170 million claim for the total loss of the three year old satellite. As a result of the failure, many analogue channels and multiplexes were relocated to Arabsat 2A. Capacity was maintained by leasing **Panamsat 5** in May 2002 and co-locating it with Arabsat 3A at 26° E. Panamsat 5 (which was suffering from degradation of its nickel-hydrogen battery cells) was subsequently renamed **Arabsat 2C**, and is currently known as **Badr C**. This satellite is still operational. **Hot Bird 5**, later renamed **Eurobird 2**, was also leased to Arabsat, which renamed it **Arabsat 2D** and finally **Badr 2**. The solar arrays of this satellite were reported to be suffering from abnormal degradation which caused a loss in output power of more than 10 per cent.

Specifications

Arabsat 3A

Launched: 26 February 1999 by Ariane 44L from Kourou.
Location: 26° east, co-located with Arabsat 2A.
Planned design life: 13 years.
Transponders: 20 active Ku-band BSS, 8 back-up. Bandwidth 34 MHz. EIRP max. 48.5 dBW. G/T max. 0.2 dB/K. Polarization – linear H/V. Frequencies: uplink: 17.3–18.1 GHz; downlink: 11.7–12.5 GHz.
Configuration: Body dimensions: 3.83 × 3.35 × 2.26 m; solar array span 29 m; onboard power 6,400 W. DASA S400 liquid apogee engine.
Mass: 2,708 kg at launch; 1,646 kg BOL; 1,200 kg dry.

Arabsat 4 series

Current Status

Arabsat 4A was lost in a Proton-M/Breeze M failure in March 2006. It is due to be replaced by Arabsat 4AR (Badr 6) in 2008. Arabsat 4B (Badr 4) was successfully launched in November 2006 and remains operational.

Background

In October 2003, Arabsat signed a contract with EADS Astrium for the construction of Arabsat's fourth generation satellites, initially designated Arabsat 4A and Arabsat 4B. Scheduled to enter service in 2006, the satellites were to provide communications services over a wide area of the Middle East. EADS Astrium, as prime contractor, would design and build the spacecraft, supply the platforms, upgrade the Arabsat ground control centres and deliver the spacecraft in orbit. Alcatel Space was to supply the communications payloads.

Arabsat 4A was to carry 24 active channels in C-band and 16 active Ku-band channels, while Arabsat 4B would have 28 active Ku-band channels. Both satellites were based on the Eurostar-2000+ platform with a launch mass of around 3.3 tonnes and a 15-year design life. They were renamed **Badr 1** and **Badr 4** respectively in February 2006.

Arabsat 4A did not reach its planned orbit after a failure of the upper stage of the Proton-M Breeze-M launch vehicle. After evaluating a lunar fly-by manoeuvre to rescue the satellite, Arabsat 4A was deorbited over the Pacific Ocean on 24 March 2006. Three months later, a third satellite in this series (known as Arabsat 4AR or **Badr 6**) was ordered from Astrium as a replacement for the lost Arabsat.

Arabsat 4B/Badr 4 was successfully launched by Proton-M/Breeze-M on 8 November 2006. It was the first ILS-organised launch since Lockheed Martin sold its interest in the US-Russian joint venture. From the Arabsat orbital slot at 26° E it delivers TV, broadband and voice services to the Middle East, North Africa and parts of Europe. Astrium Satellites also developed the satellite control centres in Dirab (Kingdom of Saudi Arabia) and in Tunis (Tunisia).

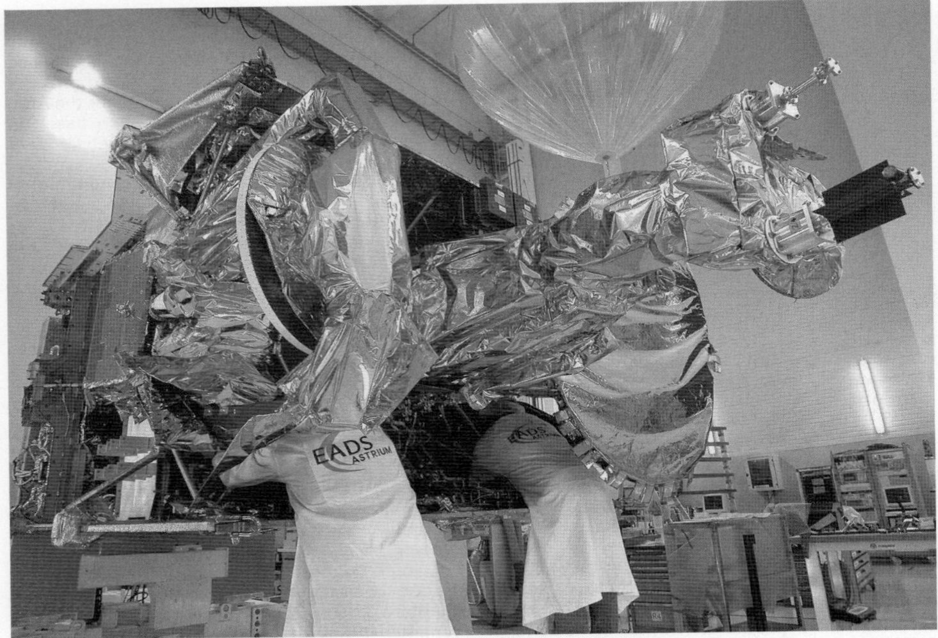

The Arabsat 4A communications satellite during assembly and testing (EADS Astrium) 1341042

At the time of launch, Badr 4 was offering services with the highest power level and the widest coverage of any comsat over the Arab countries.

Specifications
Arabsat 4B/Badr 4
Launch: 8 November 2006 by Proton-M Breeze-M from Baikonur, Kazakhstan.
Orbital location: 26° E geostationary.
Design life: 15 years.
Contractors: EADS Astrium – prime contractor. Alcatel Alenia Space – communications payloads.
Transponders: 32 Ku-band transponders, peak EIRP 51.5 dBW.
Principal applications: Television, broadband and voice services to the Middle East, N. Africa and parts of Europe.
Configuration: Astrium Satellites Eurostar E2000+ platform. Two 2.5m deployable antennas and one 1.35 m top floor antenna.
Mass: 3,280 kg at launch.
Power system: 25 m solar array span with 5 kW power output.

Arabsat 5 series

Current Status
On 16 June 2007, Arabsat signed a contract with EADS Astrium and Thales Alenia Space to build two 5th generation satellites, Arabsat 5A and Badr 5.

Background
Astrium, as the leading partner, will supply the platforms and integrate the satellites. Thales Alenia Space will design and build the communications payloads. The team will also upgrade the ground control segment for the extended Arabsat satellite fleet. Arabsat has chosen Arianespace and ILS/Proton to launch the satellites in 2009–2010.

Arabsat 5A, a Eurostar E3000 model, will have a launch mass of 4,800 kg and a spacecraft power of 11kW at the end of its 15-year service lifetime. Equipped with 16 active transponders in C-band and 24 in Ku-band, the satellite will take over from Arabsat 2B, which is reaching the end of its service life. It will provide additional capacity at 30.5° E for a wide range of satellite communications services, including television broadcasting, telephony, business communications, Internet trunking and the provision of VSAT and other interactive services over sub-Saharan Africa, North-Africa and the Middle East.

Badr 5 (Arabsat 5B), also a Eurostar E3000 model, will have a launch mass of 5,400 kg and a spacecraft power of 14kW at the end of its 15-year service lifetime. Equipped with 56 active transponders in Ku-band and Ka-band, it will primarily provide full in-orbit back-up capacity both for Arabsat 4B/Badr 4 and ArabSat 4AR/Badr 6. It will be co-located with the rest of Arabsat's Badr constellation of Direct-To-Home satellites at 26° E. Complementary uses include supporting the expected rapid increase in demand for HDTV and the development of sophisticated interactive services using its Ka-band capacity.

Artemis satellite

Current Status
Artemis (Advanced Relay and Technology Mission Satellite) remained operational at the end of 2007. In December 2006 it achieved a world first by establishing optical laser links with an aircraft.

Background
The ESA Council approved three new experimental payloads in June 1990 and folded them into the ARTEMIS research satellite. The new experiments ESA wants to investigate are the SILEX (Semiconductor Laser Intersatellite Link Experiment), SKDR data relay as a DRS precursor, and an L-band mobile channel to demonstrate services for European land vehicles in addition to backing up EMS. Technology experiments. Originally, ESA planned a launch aboard an Ariane 502 in April 1996 but technical problems delayed the launch date to late 1997.

Estimated to cost ECU665 million at 1992 rates, the cost of the Artemis satellite has swollen to about ECU920 million. The cost overrun prompted NASDA to offer a free H-2 launch in return for making Japan a full partner. Italy provided about 45 per cent of the original funding. ESA signed an ECU446 million contract with Alenia Spazio on 4 October 1993. Phase B2 began July 1989. The ultimate objective of the Artemis satellite is to demonstrate new communications techniques, principally related to data relay and mobile communications services. It acts as a relay satellite between low-Earth orbit satellites and ground stations and is to serve the planned EGNOS European navigation satellite system.

Artemis includes large reflectors to provide L-band mobile beams (left reflector) and S/Ka-band inter-satellite data relay (right reflector). Their feeder link antennas are centrally mounted: Ku-band for L-band (top right) and Ka-band (top left) for S/Ka/laser. The Silex laser system is the box unit at bottom right (Alenia) 0516976

A malfunction of the Ariane 5 upper stage Aestus engine prevented Artemis from achieving GTO. Artemis was injected into an orbit with a perigee of 590 km, an apogee 17,487 km and inclination of 2.94°, compared to expected values of 858 km perigee, 35,853 km apogee, and 2° inclination. During the recovery, the liquid apogee engine was fired during five perigee passes 18–20 July to raise the apogee to about 31,000 km. The engine then made three burns 22–24 July to produce a circular 31,000 km orbit. From 19 February 2002, the onboard ion propulsion system of four German xenon thrusters was used to slowly move it into position. The operation involved turning Artemis through 90° so that the ion thrusters could push in the orbital direction. About 20 per cent of the original onboard control software had to be rewritten. In all, the thrusters operated for about 6,000 hours. It was the first rescue operation utilising electric propulsion. The final orbital adjustments were made using the 10 N chemical thrusters, finally achieving the desired orbital location at 21.5° E on 31 January 2003. The remaining 40± 10 kg of MMH/N_2O_4 and 25 kg of Xe are sufficient to allow 7–10 years of operations.

The interval of several months prior to orbit-raising in February 2002 was used to commission the communications payload. All payloads met specifications. The most notable event was the world's first optical link between satellites, after initial commissioning of the SILEX payload via ESA's optical ground station on Tenerife. On 20 November 2001, in the course of four successive SPOT 4 orbits, the SILEX terminal on board Artemis activated its optical beacon to scan the area where SPOT was expected to be. When contact was made, SPOT 4 responded by sending its own laser beam to Artemis. On receiving the SPOT 4 beam, Artemis stopped scanning and the optical link was maintained for a pre-programmed period lasting from 4 to 20 minutes. During the period in which the two satellites were "communicating", test data were transmitted from SPOT 4 to the ground via Artemis at a rate of 50 Mbps.

Since March 2003, Artemis has provided L-band two-way links between fixed Earth stations and land mobiles in Europe, North Africa, and the Near East. From 1 April 2003, Artemis routinely provided the SILEX link for SPOT 4, as well as regular Ku-band links for ESA's Envisat. The latter involved using up to 14 links per day on two channels for ASAR and MERIS data, which Artemis transmits

Artemis and its SILEX payload (top) (Theo Pirard) 0054355

Artist's impression of the world's first laser data transmission in 2001, involving ESA's Artemis and SPOT 4 (ESA) 1340831

directly to the Envisat Processing Centre at ESRIN, Frascati. Artemis will also be used for S-band links from ESA's ATV, and S/Ku-band links from the Columbus laboratory at the ISS.

On 9 December 2005, the world's first bidirectional, optical inter-orbit communication took place, using a laser beam between JAXA's Optical Inter-orbit Communications Engineering Test Satellite "Kirari" (OICETS) and Artemis. In early December 2006, Artemis successfully relayed optical laser links from an aircraft. These airborne laser links were established over a distance of 40,000 km during two flights at altitudes of 6,000 and 10,000 m. The aircraft was a Mystère 20 equipped with the airborne laser optical link LOLA (Liaison Optique Laser Aéroportée).

Specifications

Launch: 12 July 2001 by Ariane-510 from Kourou, French Guiana.
Location: 21.4°E
Design life: 10 years minimum.
Mass: 3,105 kg at launch, (550 kg payload, 1559 kg propellant).
Contractors: Alenia Spazio (prime), Matra Espace (Silex prime), BPD/DASA (unified propulsion), DASA (S400 + S10MkII thrusters for AOCS), SAFT (Ni/H_2 batteries), CASA (structure), FIAR (power conditioning/distribution), Galileo (attitude sensors).
Transponders: Three (plus one back-up, 4-for-3) feeder link for Silex and SKDR, 27.5–30/18.1–20.2 GHz up/down, EIRP 43 dBW, G/T 0 dB/K, 234 MHz bandwidth, linear vertical polarisation Silex. Data rate 50 Mbit/s, bit error rate $<10^{-6}$. Two optical terminals are being built: the GEO element for Artemis and the LEO unit on Spot 4. Under an agreement with CNES, a pre-operational Silex service will be provided to Spot 4 after the initial experimentation phase. Each terminal employs a 25 cm diameter optical telescope mounted on a coarse pointing mechanism; total 140 kg; pointing accuracy 0.2 μrad (random), 0.8 μrad (static). Optical power source: 8,300 Å GaAlAs semiconductor laser diode with a peak output of 160 mW (60 mW continuous), and a beamwidth of 0.0004°.
Receiver: SI-APD silicon avalanche photodiode followed by a low noise trans-impedance amplifier; 1.5 nW useful received power. Sira's CCD acquisition/tracking sensors direct Teldix's fine pointing mechanism of orthogonal mirrors. Phase B concluded October 1990; C/D started mid-1991. A 1 m telescope at the Teide Observatory on Tenerife in the Canary Islands will be used for experiments with Artemis.

Astra series

Current Status

The successful launch of Astra 1KR on 20 April 2006 is the first of a new series of satellites for Astra. In early 2007, the company had four large satellites on contact to handle its core business of direct-to-home TV broadcast in Europe.

Background

The Grand Duchy of Luxembourg permitted SES to be incorporated as a private company in March 1985 for the purpose of establishing a medium-power satellite system for television distribution. SES now operates under a franchise extending to 2010 from the Duchy, which also holds a 20 per cent interest in the company. The company contracted with RCA-Astro Electronics in October 1986 for a single series 4000 satellite. The resulting Astra 1 launched on 11 December 1988 and became Europe's first private satellite when it began operational programming on 1 February 1989, after its orbital trials. It has now been retired, along with Astra 1B.

In December 1997, SES contracted with Alcatel Space and an international partner team to develop a massive satellite with 52 Ku-band channels and 2 Ka-band channels, to simultaneously replace multiple direct-broadcast satellites, expand coverage and provide new services such as direct data transmission. In November 2002 the Astra 1K satellite – then the largest commercial satellite in the world – was launched by a Proton rocket, but a failure of the Block DM upper stage left it stranded in low Earth orbit and it had to be de-orbited in December 2002.

In June 2003, Lockheed Martin was commissioned to build a less ambitious replacement satellite, Astra 1KR, and it was launched in April 2006, finally permitting SES Astra to retire the long-serving Astra 1A and 1B. SES also engaged Lockheed Martin to build Astra 1L, initially as a back-up to Astra 1KR.

Following the success of Astra 1KR, Astra 1L has been set for launch by Ariane 5 in early 2007. Like Astra 1KR, it will be based on a stretched A2100AX bus. It will carry 29 active Ku-band transponders as well as two Ka-band transponders. Its spacecraft's primary mission will be to replace Astra 1E at the company's prime Continental European DTH position of 19.2° East. The Ka-band payload will support interactive services like Astra BBI (Broadband Interactive) and Satmode, a low-cost satellite return channel for digital set top boxes.

Astra 1C-1H employ Hughes' HS-601 platform. Shown is Astra 1G/1H, featuring the newer lightweight contoured antenna and GaAs solar wings 0517201

Astra 2A is deployed at 28.2°E, using ion thrusters for NSSK (Hughes Space & Communications) 0517407

In January 2005, Lockheed Martin was selected to supply another A2100AX to SES, to be operated by SES-Sirius as Sirius 4, mainly serving Nordic markets. It is to be launched by a Proton-M/Breeze-M in 2007.

SES Astra announced in mid-2005 that EADS Astrium would build Astra 1M, to be launched in 2008. Based on a Eurostar E3000 platform, it will carry 36 transponders for the first five years (32 thereafter). The satellite will be specified for a minimum service life of 15 years. It will have a launch mass of 5,300 kg, a solar array span of 35 metres once deployed in orbit, and spacecraft power of 10 kW at end of life. In November 2006, EADS Astrium was awarded a follow-on contract for another E3000 spacecraft, Astra 3B, to be launched in late 2009.

Astra 1C

Launch: 12 May 1993 by Ariane 4 V56 from Kourou, French Guiana. Broadcasting began 1 July
Location: 19.2° E geostationary (above Zaire)
Design life: 15 years min
Contractors: Hughes Space & Communications Co
Transponders: 18 (plus six back-up) 63 W TWTA Ku-band European beams, 2 at 10.90–10.95 GHz +16 at 10.95–11.20 or (to replace Astra 1A) +16 at 11.20–11.45 GHz down, 26 MHz bandwidth, EIRP 50 dBW min (52 dBW in central region), orthogonal linear polarisation
Principal applications: as 1B
Configuration: HS-601 platform, 3-axis, solar array span 21 m (satellite within 229 × 254 × 254 cm envelope at launch). Payload mounted on North/

Wrapped ready for launch the Astra 1F satellite is shipped to Russia for a flight on a Proton launcher (Hughes Space & Communications) 0054346

Astra 3A during final tests at Boeing Satellite Systems 0132805

BSS technicians pack the Astra 2D HS376 HP for shipment prior to launch 0132808

South panels and Earth-facing floor; East/West faces carry hard-points for two dual-gridded antennas, fixed antenna mountings on Earth-facing wall
Mass: 2,790 kg at launch, 1,700 kg on-station BOL, 1,180 kg dry
AOCS: 3-axis, unified 12 ARC 22 N + one Marquardt 490 N bipropellant thrusters incorporating onboard control processor, Sun/Earth sensors + 2 × 61 Nms 2-axis gimbaled momentum bias wheels. 1,658 kg NTO/MMH is contained in four spheres. Spin-stabilised in transfer orbit.
TT&C: as Astra 1A
Power system: twin solar wings of three 2.16 × 2.54 m panels carrying K43/4large area Si cells on Kevlar substrate to satisfy 3.3 kW requirement. Ni/H_2 batteries provide 100 per cent eclipse protection.

Astra 1D
Launch: 1 November 1994 by Ariane 4 from Kourou, French Guiana. Broadcasting began 1 January 1995
Location: 23.5°E
Transponders: 18 (plus six back-up) 63 W TWTA Ku-band European beams, switchable as 16 at 10.70–10.95 or 10.95–11.20 (1C back-up) or 11.45–11.70 GHz (1B back-up), 26 MHz bandwidth or 18 at BSS 11.70–12.07 GHz (1E back-up) 33 MHz bandwidth, EIRP 50 dBW min (52 dBW in central region), orthogonal polarisation
Mass: 2,924 kg at launch, 1,700 kg on-station BOL, 1,250 kg dry
Other specifications as for Astra 1C.

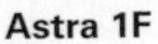

Astra 1E
Launch: 19 October 1995 by Ariane 4 from Kourou, French Guiana
Location: 19.2° E (above Zaire)
Transponders: 18 (plus six back-up) 85 W TWTA Ku-band European beams, switchable as 18 at BSS 11.70–12.10 GHz 33 MHz bandwidth, or 16 at 10.70–10.95 (Astra 1D back-up) or 10.95–11.20 (1C back-up) or 11.45–11.70 GHz (1B back-up), 26 MHz bandwidth, EIRP 50 dBW min (52 dBW in central region), orthogonal polarisation
Mass: 3,010 kg at launch, 1,803 kg on-station BOL, 1,343 kg dry
Power system: twin solar wings (spanning 26 m) of four 2.16 × 2.54 m panels carrying K43/4 large area Si cells on Kevlar substrate providing 4.7 kW.
Other specifications as for Astra 1C.

Astra 1F
Launch: 8 April 1996 on Proton-K from Tyuratam, Kazakhstan
Location: 19.2° E (above Zaire)
Transponders: 22 (plus two back-up) 82 W TWTA 12.10–12.50 GHz Ku-band European beams (switchable to 18 at 11.70–12.10 GHz as 1E back-up, or 16 at 11.2–11.45 GHz as 1A back-up), 33 MHz bandwidth, EIRP 50 dBW min (52 dBW in central region), orthogonal polarisation
Mass: 3,010 kg at launch, 1,900 kg BOL
Other specifications as for Astra 1E.

Astra 1G
Launch: 2 December 1997 on Proton-K from Baikonur, Kazakhstan
Location: 19.2° E (above Zaire)
Design life: 15 years min
Transponders: 32 100 W TWTA 12.50–12.75 GHz Ku-band European beams (switchable to 18 at 11.70–12.10 GHz as 1E back-up, or 22 at 12.10–12.50 GHz as 1F back-up), 26/33 MHz FSS/BSS bandwidths, EIRP 52 dBW min, orthogonal polarisation
Configuration: improved HS-601 HP (High Power) platform, featuring GaAs solar arrays, enlarged payload and lightweight contoured reflectors
Mass: 3,300 kg at launch, 2,485 kg on-station BOL
AOCS: Hughes' 68 kg XIPS xenon ion propulsion system will provide NSSK: two (+ two back-up) 18 mN 13 cm thrusters
Power system: twin solar wings of four panels providing 8 kW BOL (7,075 W total required) and 100% eclipse protection
Other specifications as for Astra 1E.

Astra 1H
Launch: 18 June 1999 on Proton from Baikonur
Location: 19.2°E (above Zaire)
Transponders: Same as Astra 1G, except two 29.50–30.00/18.80–19.30 GHz up/down Ka-band beams will introduce interactive services
Other specifications as for Astra 1G.

Astra 2A
Launch: 30 August 1998 on a Proton-K from Baikonur

Specifications
SES Astra/Sirius Fleet

Satellite	Type	Payload	Location	Launch date	Vehicle	Launch site
Astra 1C	Boeing 601	Ku-band	19.2°E	12 May 1993	Ariane 42L	Kourou
Astra 1D	Boeing 601	Ku-band	23.5°E	1 November 1994	Ariane 42P	Kourou
Astra 1E	Boeing 601	Ku-band	19.2°E	19 October 1995	Ariane 42L	Kourou
Astra 1F	Boeing 601	Ku-band	19.2°E	8 April 1996	Proton	Baikonur
Astra 1G	Boeing 601HP	Ku-band	19.2°E	2 December 1997	Proton	Baikonur
Astra 1H	Boeing 601HP	Ku-band	19.2°E	18 June 1999	Proton	Baikonur
Astra 2A	Boeing 601HP	Ku-band	19.2°E	30 August 1998	Proton	Baikonur
Astra 2B	Astrium	Ku-band	28.2°E	September 14 2000	Ariane 5	Kourou
Astra 2C	Boeing 601HP	Ku-band	19.2°E	16 June 2001	Proton	Baikonur
Astra 2D	Boeing 376HP	Ku-band	28.2°E	19 December 2000	Ariane 5	Kourou
Astra 3A	Boeing 376HP	Ku-band	23.5°E	29 March 2002	Ariane 44L	Kourou
Astra 1KR	Lockheed Martin A2100AX	Ku-band	19.2°E	20 April 2006	Atlas 5	Cape Canaveral
Astra 1L	Lockheed Martin A2100AX	Ku/Ka-band	19.2°E	2007 (Planned)	Ariane 5	Kourou
Astra 1M	Eurostar E3000	Ku-band	19.2°E	2008 (Planned)		
Astra 3B	Eurostar E3000	Ku/Ka-band	19.2°E	2009 (Planned)		
Sirius 2	Spacebus 3000	Ku-band	4.8°E	12 November 1997	Ariane 44L	Kourou
Sirius 3	Boeing 376	Ku-band	5°E	5 October 1998	Ariane	Kourou
Sirius 4	A2100	Ku-band		2007 (Planned)	Proton	Baikonur

A Boeing 601 HP, Astra 2C was launched by Proton from Baikonur on 16 June 2001 0132809

Astra 1KR, replacing the Astra 1K giant lost in 2002, is a stretched A2100AX platform (Lockheed Martin) 1310190

Location: 28.2°E
Design life: 15 years min
Other specifications as for Astra 1G.

Astra 2B
Launch: 14 September 2000 on an Ariane 44P.
Location: 28.2°E
Design life: 14 years min
Transponders: 30 108 W TWTA 11.70–12.75 GHz Ku-band European beams (16 switchable to steerable beam), EIRP 52 dBW min, orthogonal polarisation for first 5 years, 28 thereafter
Configuration: EAD Astrium Eurostar 2000+i
Mass: 3,300 kg at launch
Power system: twin solar wings providing >6.4 kW BOL; 100% eclipse protection.

Astra 2C
Launch: 16 June 2001 by Proton
Location: 28.2°E
Design life: 15 years
Transponders: 32 98.5 W TWTA Ku-band European beams (8 spare at BOL)
Configuration: Boeing 601 HP platform featuring GaAs solar arrays on two wings each with one panel dual-junction and three panels single-junction
Mass: 3,643 kg at launch: 2,494 kg BOL
AOCS: Hughes 490 N liquid apogee motor; 12 × 18 mN XIPS xenon ion propulsion system for NSSK
Power system: twin solar arrays of 4 panels providing 8.5 kW BOL, 6.7 kW EOL

The bulk of the SES Astra constellation still comprises Boeing 601 and 601HP spacecraft (Boeing) 1310191

Astra 1KR is launched successfully by an Atlas 5 in April 2006 (Lockheed Martin) 1310189

Dimensions: length (solar arrays) 26 m; width, 10 m. Stowed 5.5 m × 3.3 m × 3.3 m.

Astra 2D
Launch: 19 December 2000 by Ariane 5
Location: 28.2°E
Design life: 12 years
Transponders: 16 × 39 W TWTA for DTH
Configuration: Spin stabilised Boeing (HS) 376 HP with 18 TWTA, 2 spare, featuring two telescoping cylindrical panels
Mass: 1,445 kg at launch; 824 kg BOL
AOCS: Thiokol Star 30C solid apogee motor; 2 × 22.2 N N-S bipropellant; 2 × 22.2 N E-W bipropellant
Power system: Two telescoping cylindrical solar panels with large area GaAs solar cells and 16-cell, 141AhNiH batteries. Provides 1.6 kW BOL; 1.4 kW EOL
Antennas: 2.03 m diameter shaped reflector precision pointing antenna accuracy >0.10°. Comprises two shaped surfaces, one for vertical linear polarisation for communication and one for horizontal linear polarisation for communications and tracking
Dimensions: Stored height 3.15 m, width 2.17 m; deployed, height 7.97 m, width, 2.17 m

Astra 3A
Launch: 29 March 2002 by Ariane 44L
Location: 23.5°E
Design life: 10 years
Transponders: 20 × 30 W, 52 dBW. 24 Ku-band channel; 11.45–11.70 GHz on 12 channels, 12.50–12.75 GHz on 12 channels
Configuration: Spin stabilised Boeing (HS) 376 HP with 24 TWTA featuring two telescoping cylindrical solar panels
Mass: 1.5 kg BOL; 908 kG EOL
AOCS: Thiokol Star 30C solid apogee motor; 2 × 22.2 N N-S bipropellant; 2 × 22.2 N E-W bipropellant
Power system: as Astra 2D
Antennas: as Astra 2D
Dimensions: as Astra 2D.

Astra 1KR
Launch: 20 April 2006 by Atlas 5
Location: 19.2°E
Design life: 15 years
Transponders: 32 transponders in the first five years of life; 28 afterwards. 140W TWTA power.
Configuration: Stretched A2100AX platform
Mass: 4,331 kg at launch, 2,732 kg on orbit
Power system: 12 kW from 27 m span array

Eutelsat 2 series

Eutelsat itself manages the follow-on Eutelsat 2 programme that began with the 2-F1 satellite launched in August 1990. The cost of the six satellites plus launch and insurance ran about EUR1,094 million at 1994 rates. Each F2 offers 16 Ku-band transponders and improved on-board redundancy. When it built its economic model for the second generation, Eutelsat assumed all the traffic carried by the first generation would continue and included sufficient capacity for expansion of services.

The second generation offers 16 simultaneously-operational transponders. 2-F1 occupied the prime 13° E slot since 1990 to guarantee services without the need for users to re-direct their antennas. It was moved to 36° E, replacing TDF 2 in 1999 and then to 48° E until 2002 when it was moved to 70.5° E and, in January 2003, on to 76° E. Eutelsat contracted with Aerospatiale for three follow-on satellites in May 1986 at the price of EUR225 million. They took out an option on the fourth in June 1987, and planned to launch the new series at intervals of 6 months. Then due to increased demand Eutelsat added a fifth to the overall purchase in March 1989 at a cost of EUR52.8 million, and a sixth asking for delivery during 1990.

2-F6 features with higher power TWTAs in the expansion band frequencies to provide 49 dBW over Europe and to reach to the Urals. Collocation with 2-F1 at 13° E provides up to 39 TV channels from the one slot without receiver modification. Direct reception is possible with 70 cm antennas in all of central/western Europe, and with slightly larger antennas for cable and community reception as far east as Turkey and Ukraine. Commercial services began 29 April 1995. Built by Matra Marconi, the Eutelsat 2-F7/-8/-9/-10 series were launched between November 1996 and October 1998 providing digital and analogue television and multimedia series. The first in a series of widebeam telecommunications satellite built by Alcatel Eutelsat W-F2 was launched by Ariane 44L on 5 October 1998 to 16° east.

Logica built a new TCR system, for Eutelsat 2 including the Satellite Control Centre in Paris and TCR equipment centres in Rambouillet (France) and Sintra (Portugal). Demand exploded when Eastern Europe left the Soviet Union in 1990. Eutelsat met the demand for services to eastern Europe, by stretching 2-F4 and 2-F5's widebeams to the Urals and modifying 2-F6 (Hot Bird 1) by adding radically new and more powerful TWTAs.

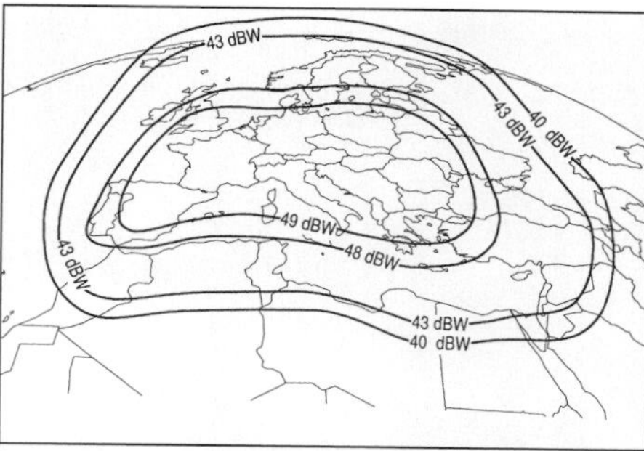

Instead of being divided into Widebeam/Superbeam coverage, 2-F6/Hot Bird 1 offers increased power European beams stretching to the Urals. Its new frequencies and collocation with 2-F1 at 13° east provide up to 39 TV channels from the one slot (Eutelsat) 0516982

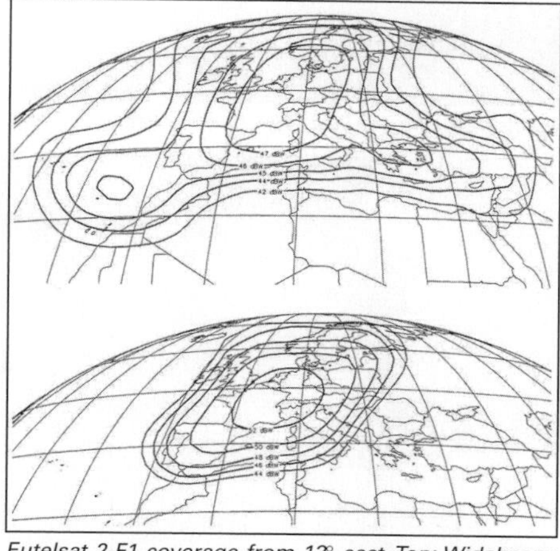
Eutelsat 2-F1 coverage from 13° east. Top: Widebeam; bottom: Superbeam 0516980

Specifications
Eutelsat 2-F1
Launch: 30 August 1990 by Ariane V38 44LP from Kourou, French Guiana
Location: 13°E geostationary; 36°E in 1999, 48°E in 1999, 70.5°E in 2002, 76°E in 2003
Design life: 9 years, reliability 0.74 for 16 channels and 7 years
Contractors: Aerospatiale (prime), Marconi Space Systems (payload), MBB (20% share: AOCS, solar arrays, TT&C equipment) Alcatel Espace, CASA, Aeritalia, ETCA, ERA
Transponders: 16 (plus 8 back-up; 12-for-8 ring redundancy) 50 W TWTA 14.0–14.5/10.95–11.20 + 11.45–11.7 + 12.50 12.75 GHz up/down Ku-band

Eutelsat 2-F4 integration at Aerospatiale's Cannes facility 0516979

Eutelsat satellite for East-West Communications (Theo Pirard) 0054354

A mock-up of the Eutelsat W satellite bus was test flown on the third Ariane 5G flight on 21 October 1998 (Arianespace) 1047472

regional Superbeam and Widebeam, bandwidths 72 MHz (seven transponders) and 36 MHz (nine), EIRP 44–52 dBW Superbeam over western Europe + 44–47 dBW pan-European Widebeam, linear orthogonal polarisation
Principal applications: TV/radio distribution to cable networks and domestic satellite systems
Configuration: derived from Spacebus 2000 platform, 3-axis control, 22.4 m span across solar wings, two 1.60 m diameter dual reflector multifeed antennas, one providing transmission/receive, the other transmission only, switchable between Widebeam/Superbeam as required. GEO location maintained within ±0.05°

The Eutelsat 2 series is based on Aerospatiale's Spacebus 2000 platform. Each carries 16 channels working through the East and West 1.6 m {70} reflectors 0516978

Mass: 1,878 kg at launch, propellant for 10 years operation; 1,123/866 kg on-station BOL/EOL
AOCS: provided by MBB's 22 × 10 N + 1 × 400 N MMH/MON bipropellant system
TT&C: initially through GSOC German Space Operations Centre at Oberpfaffenhofen, Germany, then via Eutelsat's TCR system, comprising the Satellite Control Centre in Paris and TCR equipment centres in Rambouillet (France) and Sintra (Portugal; also equipped as the remote SCC back-up facility)
Power system: twin solar wings 3,500/3,000 W BOL/EOL.

Eutelsat 2-F2
Launch: 15 January 1991 by Ariane V41 from Kourou, French Guiana
Location: 10°E geostationary; 21.5°W in 1999; 48° E in 2002
Principal applications: TV/radio distribution, domestic telephony, SMS business traffic
Other specifications as for Eutelsat 2-F1.

Eutelsat 2-F3
Launch: 7 December 1991 by first Atlas 2 (AC-102) from Cape Canaveral
Location: 16°E geostationary; 36°E in 1999; 21.5°E in 2000
Principal applications: TV/radio distribution, national/international telecom services, SMS business traffic
Other specifications as for Eutelsat 2-F1.

Eutelsat 2-F4
Launch: 9 July 1992 by Ariane 4 from Kourou, French Guiana
Location: 7° E geostationary; 10° E in 1999; 28.5° E in 2000; 25.5°E in 2002; 12.5° W in 2002
Other specifications as for Eutelsat 2-F1 but Widebeam modified for coverage as far as Urals, providing >42.5 dBW to Moscow and >39.0 dBW to Volgograd.

Eutelsat 2-F5
Lost in Ariane's 24 January 1994 failure. Planned for 36° E, with specifications as 2-F4 (Widebeam modified for coverage to Urals). Launch/satellite fully insured for ECU180 million.

Eutelsat 2-F6/Hot Bird 1
Launch: 28 March 1995 by Ariane 4 from Kourou, French Guiana
Location: 13°E geostationary (collocated with 2-F1)
Design life: 11 years
Transponders: 16 (plus 8 back-up; 12-for-8 ring redundancy) 70 W TWTA 12.895–13.25/11.20–11.55 GHz up/down Ku-band European beams, 36 MHz bandwidths, EIRP 40–49 dBW, linear orthogonal polarisation
Configuration: one transmit 1.60 m diameter dual reflector multifeed antenna and one elliptical receive antenna
Mass: 1,798 kg at launch, 1,078 kg on-station BOL, 840 kg dry
Other specifications as for Eutelsat 2-F1.

Eutelsat 2-F7/Hot Bird 2
Launch: 21 November 1996 by Atlas 2A from CCAFS
Location: 13° east geostationary (colocated with 2-F6)
Design life: 15 years
Transponders: 20 × 115 WTWTA in Ku-band (11.7–12.5 GHz)
Configuration: two 2.3 m transmit antennas for widebeam coverage from Kazakhstan to Persian

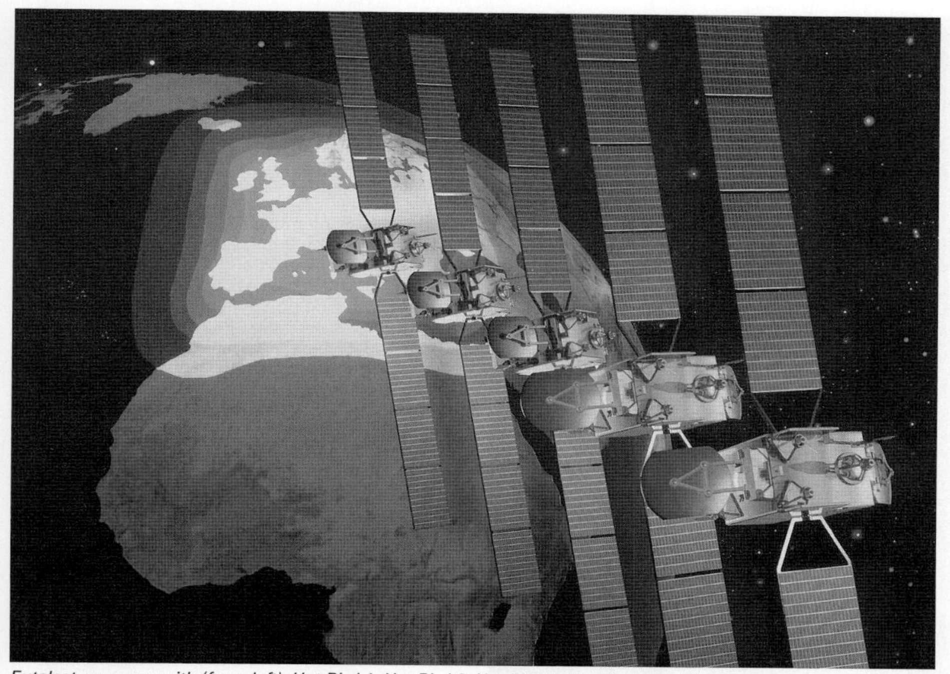
Eutelsat coverage with (from left): Hot Bird 1, Hot Bird 2, Hot Bird 3, Hot Bird 4 and Eutelsat 2F1 0084765

gulf and 'super beam' coverage for Europe and North Africa.
Mass: 2,923 kg at launch; approx 1,700 kg BOL
Bus: 3.65 × 2.30 × 3.99; span 27.9 m; onboard power 5,500 W (EOL).

Eutelsat 2-F8/Hot Bird 3
Launch: 2 September 1997 by Ariane 44LP
Location: 13° east geostationary
Lifetime: 12 years
Configuration: as Eutelsat 2-F7/Hot Bird 2
Mass: 2,915 kg at launch; 1,715 kg BOL; 1,310 kg dry.

Eutelsat 2-F9/Hot Bird 4
Launch: 27 February 1998 by Ariane 42P
Location: 13° east geostationary
Lifetime: 12 years
Mass: 2,885 kg at launch; 1,770 kg BOL; 1,310 kg dry.

Eutelsat W-F2
Launch: 5 October 1998 by Ariane 44L
Location: 16° east
Transponders: 24 × Ku-band (72 MHz and 36 MHz)
Configuration: 1 fixed widebeam over Europe, North Africa and the Middle East (40–49 dBW), 1 steerable beam positioned over Mauritius and Reunion Island (42–53 dBW)
Mass: 2,950 kg at launch, 1,810 kg BOL; 1,375 kg dry
Bus: 4.6 × 2.5 × 1.8 m; span 29 m; onboard power 5,900 W (EOL).

Eutelsat 2 F10/Hot Bird 5
Launch: 9 October 1998 by Atlas 2A
Location: 13° east

Eutelsat W-F3
Launch: 12 April 1999 by Atlas 2AS
Location: 7° east

Sesat
Launch: 18 April 2000 by Proton K
Location: 36° east

Eutelsat W-F4
Launch: 24 May 2000 by Atlas 3
Location: 36° east

Eutelsat W-F1
Launch: 6 June 2000 by Ariane 4
Location: 10° east

Eutelsat W1R
Launch: 6 September 2000 by Ariane 44P
Location: 10° east

Eurobird 1
Launch: 8 March 2001 by Ariane 5
Location: 28.5° east

Atlantic Bird 2
Launch: 25 September 2001 by Ariane 44P
Location: 8° west

Atlantic Bird 3
Launch: 5 July 2002
Location: 5° west

Hot Bird 6
Launch: 21 August 2002 by Atlas 5
Location: 13° east

Atlantic Bird 1
Launch: 28 August 2002 by Ariane 5
Location: 12.5° west

Eutelsat W5
Launch: 20 November 2002 by Delta 4
Location: 70.5° east

Eutelsats' purpose-built Paris Headquarters 0084764

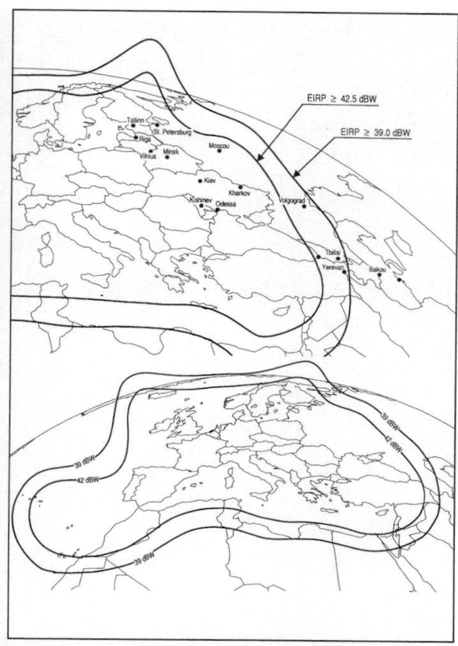

The Widebeam coverage of 2-F4/F5 was modified to reach the Urals (Eutelsat) 0516981

e-Bird
Launch: 27 September 2003
Location: 33° east

W3A
Launch: 16 March 2004
Location: 7° east

Horizons series

Current Status
Horizons-2 was launched on 21 December 2007 and will provide a variety of communications services to North America. Horizons-1 was launched in September 2003 and remains operational, providing communications services to North America and much of the Pacific region.

Background
In mid-2001, PanAmSat Corp. and JSAT International Inc. selected Boeing Satellite Systems to build a Boeing 601HP spacecraft for their Horizons joint venture, as well as PanAmSat's US domestic cable programme. Launched aboard a Sea Launch Zenit-3SL in September 2003, it was placed in an orbital slot at 127° W, between the Hawaiian Islands and the US West Coast, providing coverage over North America, Puerto Rico, Alaska, Hawaii and Mexico. It carries a total of 48 active transponders, 24 in Ku-band and 24 in C-band. The Ku-band transponders are each powered by 108-watt linear traveling wave tube amplifiers (LTWTAs). The Ku-band payload, known as Horizons-1, was constructed for the Horizons partnership and is jointly owned by PanAmSat and JSAT. It is used to offer a variety of digital video, Internet and data services. Using a Hawaii-based relay station, the Ku-band payload can also deliver content and services between the United States and Asia.

The satellite's C-band payload is known as **Galaxy XIII** and is operated separately as part of PanAmSat's Galaxy US cable service (see separate Galaxy entry). Galaxy XIII's 24 C-band transponders are each powered by 40-watt LTWTAs. The spacecraft's solar wings use dual-junction gallium arsenide solar cells manufactured by Spectrolab Inc., a Boeing subsidiary. These delivered 9.9 kW of power at the beginning of the satellite's 15-year design life.

Horizons-2 was built for Horizons 2 Satellite LLC, a joint venture of Intelsat and JSAT, by Orbital Sciences, based on its STAR satellite platform. The CONUS (Continental United States) and Caribbean beam features "hot spot" service to major North American cities transmitting a powerful linear Ku-band. The second footprint transmits a unique, boomerang-shaped beam covering the entire east coast of North America and the Caribbean. With coverage extending seaward over 550 km, it can support operations off the US eastern seaboard for homeland security, ship-to-shore communications and cargo tracking. The satellite

The Horizons-1/Galaxy XIII comsat built by Boeing Satellite Systems (Boeing) 1340827

Artist's impression of the Horizons-2 communications satellite, launched in December 2007 (Orbital Sciences) 1340828

acts as a multipurpose relay for digital video, high-definition television, IP-based content distribution networks, broadband Internet and satellite news services in the continental US, Caribbean and parts of Canada. It also supports a number of mobile communications applications for government customers.

Specifications
Horizons-1
Launch: 30 September 2003 by Zenit-3SL from the Odyssey platform in the Pacific Ocean.
Location: 127° W geostationary.
Design life: 15 years.
Contractors: Boeing Satellite Systems (prime).
Transponders: 24 Ku-band active (8 spare) 108 W LTWTAs. 24 C-band active (8 spare) 40 W LTWTAs.
Principal applications: Digital video, Internet and data services.
Configuration: Boeing 601HP platform. 3 axis stabilised. Two shaped Gregorian transmit/receive antennas (diameter 107 in) Two dual gridded, surface shaped transmit/receive antennas.
Liquid apogee motor: 110 lbf LAM. Station-keeping thrusters: N-S primary (xenon ion) 0.018 N; backup (bipropellant) 10N.
Mass: 4,060 kg at launch; 2,630 kg in orbit.
Body dimensions: 5.7 × 2.7 × 3.6 m
Power system: Twin solar arrays each with 4 panels of dual-junction gallium arsenide solar cells. 26.2 m span. Output 9.9 kW (BOL), 8.6 kW (EOL). Batteries: 30-cell NiH.

Specifications
Horizons-2
Launch: 21 December 2007 by Ariane 5 GS from Kourou.
Location: 74° W geostationary.
Design life: 15 years.
Contractors: Orbital Sciences Corporation (prime).
Transponders: 16 active Ku-band (36 MHz) transponders with 22-for-16 TWTAs at 85W, 4 active Ku-band (72 MHz) with 6-for-4 TWTAs at 150W. Polarization: Linear – Horizontal or Vertical. Downlink Frequency: 11.70 to 12.20 GHz. Typical Edge of Coverage EIRP: North America > 47.4 dBW for 85 W xpdr; > 50.0 dBW for 150 W xpdr; East Coast North America > 46.4 dBW for 85 W xpdr; > 49.0 dBW for 150 W xpdr. Uplink Frequency: 14.00 –14.50 GHz.
Principal applications: Digital video, Internet, mobile communications and data services.
Configuration: OSC STAR 2 platform. 3 axis stabilised. Two 2.3 m dual gridded shaped reflector antennas.
Mass: 2,304 kg at launch.
Body dimensions: 4.0 × 3.3 × 2.3 m
Power system: Twin solar arrays, span 18 m. Three panels per array of UJT gallium arsenide cells. 4.4 kW power output (EOL). Two 3,850 watt-hr capacity Li-Ion batteries.

Inmarsat 1 (Marecs) series

Current Status
The Marecs spacecraft which supported the first generation of Inmarsat services have been retired. Marecs-A, launched on 20 December 1981, was retired by Inmarsat in 1991 and deactivated by ESA in August 1996. Marecs-B2, launched on 10 November 1984, left Inmarsat service in December 1996 and was placed in a graveyard orbit in January 2002.

Inmarsat 2

Current Status
Inmarsat currently operates three of the original four Inmarsat 2 class satellites. They provide leased capacity and act as back-ups to Inmarsat 3 spacecraft.

Background
In 1975, the Inter-Government Maritime Consultative Organisation (IMCO) began discussing the need for a global satellite system serving shipping in the same way that Intelsat links countries. This body followed up with an agreement to establish Inmarsat, with headquarters in London, in 1979. Several countries contributed to the operating costs on inauguration day including the US 23.37

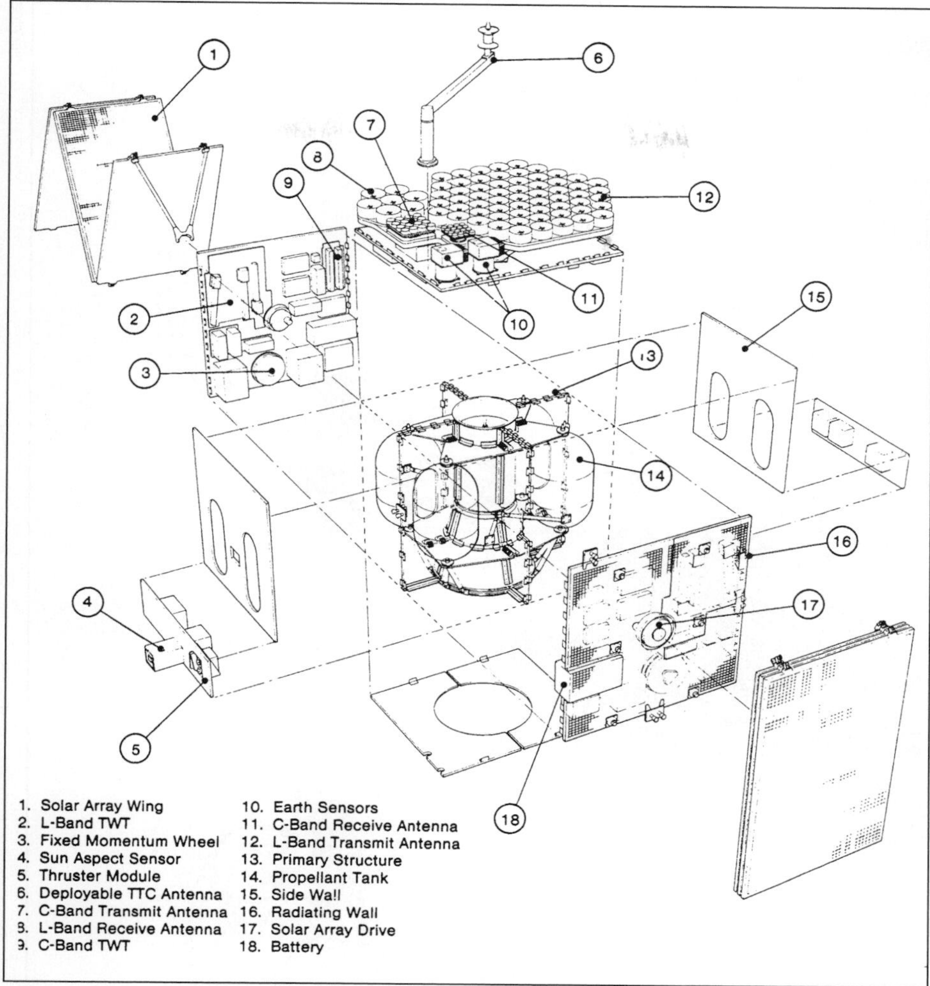

Inmarsat 2 principal features (Inmarsat) 0516984

per cent, USSR 14.09 per cent, UK 9.89 per cent, Norway 7.88 per cent and Japan 7.00 per cent.

Inmarsat took over responsibility for maritime links from the US-run Marisat on 1 February 1982. The organization started with a customer base of 900 ships and, by early 2007, reported that it was supporting more than 287,000 users on ships and aircraft with land-mobile terminals. Inmarsat was transformed from an intergovernmental organization (IGO) to a private company, London-based Intelsat plc, in April 1999.

Initially, Inmarsat leased Comsat's 10 telephone circuit capacity aboard each of the existing Marisats. But this still left spotty service. ESA built upon its experience with the existing ECS spacecraft to propose and develop three Marecs satellites, a maritime version of ECS, produced by the Mesh consortium. Two of these satellites were launched successfully (another was lost in an Ariane failure) and have now been retired.

Inmarsat requested bids for its first dedicated satellites in August 1983 and the consortium awarded a contract to British Aerospace in March 1985 for three satellites based on the Eurostar 1000 platform, with options for a further six. The Inmarsat board authorised a fourth satellite in March 1988. The overall programme cost about USD675 million, excluding cost of financing and TT&C capital/operating costs. Each satellite was designed to provide 125 voice channels forward, and 250 return, five times the previous capacity, and to have a ten-year lifetime.

Inmarsat finalised an agreement in October 1988 to sell and lease back its first three satellites from the North Sea Marine Leasing Co (a partnership of the leasing arms of four major UK banks: NatWest, Barclays, Lloyds, Midland (now HSBC)) for 10 years. The consortium insured each satellite for up to USD160 million (including an element of self insurance) to cover launch through the first year's operation.

These satellites marked the first time that Inmarsat managed its own satellites from the HQ's USD10 million Satellite Control Centre, working through ground stations in Pennant Point (Canada), Lake Cowichan (Canada), Fucino (Italy) and Beijing (China).

In early 2007, Inmarsat 2 F1 was located over the Pacific, providing lease capacity. Inmarsat 2 F2 was in orbit over the Western Atlantic, providing leased capacity and acting as back-up to Inmarsat 3 F4. Inmarsat 2 F4 was providing leased capacity and back-up for two other Inmarsat 3 satellites (F1 and F3) over the Indian Ocean. Inmarsat 2 F3 was placed in a graveyard orbit in early 2006.

Each Inmarsat 2 has a launch mass of 1,310 kg (Ariane) and 1,385 kg (Delta).

Specifications

Inmarsat 2-F1

Launch: 23.16 GMT 30 October 1990 by Delta 6925 from Cape Canaveral pad 17B. Three apogee burns 1–4 November provided GEO injection at 44° E; arrived 30° E 9 November, where payload activated 13 November for testing. The final move began 26 November, arriving/entering service 8 December, a week earlier than planned. It became a spare at the same location when 3-F1 became operational
Location: 64.4° E GEO over Indian Ocean, relocated to 179° E November 96. Currently deployed as a spare at 143° E.
Design life: 10 years minimum; estimated BOL sufficient hydrazine for 14 years
Contractors: BAe (prime; assembly, integration/test; thermal; power subsystems), Matra (AOCS), Fokker (solar arrays), Lockheed (propellant tanks), Marquardt (liquid apogee thruster), Hughes (payload, repeaters and, with Spar, antennas)
Transponders: four (plus two back-up) 45 W TWTA 1.621.64/1.53–1.54 GHz L-band global beams, 5+22.5 kHz bandwidths, 39 dBW min EIRP, circular polarisation, to provide 250 voice circuits shore to ship through the 43 cup-dipole radiating elements

One (plus one back-up) 10 W TWTA 6.42 6.44/3.60–3.62 GHz C-band global beams, 5+22.5 kHz bandwidths, 24 dBW min EIRP, circular polarisation, providing 125 voice circuits ship to shore through seven cup-dipole radiating elements. Each ring is separately excited for beam shaping, providing higher power at edge of coverage
Principal applications: maritime, air and land mobile
Configuration: based on box-shaped Eurostar 1000 bus, 2.557 m high, 1.586 m deep, 1.48 m wide, centred on 1.8 m high thrust cone. Payload and solar arrays mounted on North/South panels and Earth-facing floor. The payload uses separate receive/transmission antennas for both L/C band, all four cup-dipole phased arrays mounted on the Earth-facing wall

Inmarsat 2-F1 EMC test preparations 0516983

Mass: 1,385 kg, 824 kg on-station BOL; 130 kg payload mass, 624 kg dry
AOCS: unified redundant 3-axis control and apogee motor thrusters drawing on 110 kg MMH + 130 kg NTO; Kaiser Marquardt R- 4D 490 N AKM and two sets of six MBB 10 N thrusters. Roll/yaw control provided by solar sailing with solar array flaps. Single MW with redundancy
Antenna pointing: typically 0.1° half-cone
TT&C: through Inmarsat's London HQ on C-band (control assumed 10 November 1990 from CNES Toulouse facility)
Power systems: 1,200 W BOL/1,142 W EOL provided by twin Si wings, each comprising three 122 cm wide, 173 cm long panels and spanning 15.23 m. Regulated 42.5 V voltage. Eclipse protection provided by two Ni/Cd batteries.

Inmarsat 2-F2

Launch: 8 March 1991 by Delta 6925 from Cape Canaveral pad 17B. Entered service 13 April
Location: 15.5°W GEO over Atlantic Ocean until September 1996, located at 55°W from January 1997 until July 1999 when deployed as a spare at 98°W.

Other specifications as for F1.

Inmarsat 2-F4

Launch: 15 April 1992 by Ariane 4 from Kourou. Entered service 31 May 1992
Location: 54°W GEO over Atlantic Ocean (tested over 32°W); 17°W October 1997 to May 1999; 109°E from July 1999.

Other specifications as for F1.

Inmarsat 3

Current Status

Five satellites in this family remain operational alongside the new Inmarsat 4 heavyweights and the remaining Inmarsat 2s.

Background

Inmarsat issued a Request for a Proposal (RfP) on 2 October 1989 for its third generation of satellites. The consortium wanted to build a network of global, high-power multiple spot beams for mobile personal communications services, and L-band to L-band connections for direct mobile-to-mobile communications which it expected to be valuable in emergency and search and rescue operations. Responses to the RfP came in during February 1990 from British Aerospace leading a TRW/NEC/

Inmarsat 3 antenna feed assembly. At right is the navigation antenna (MMS) 0517222

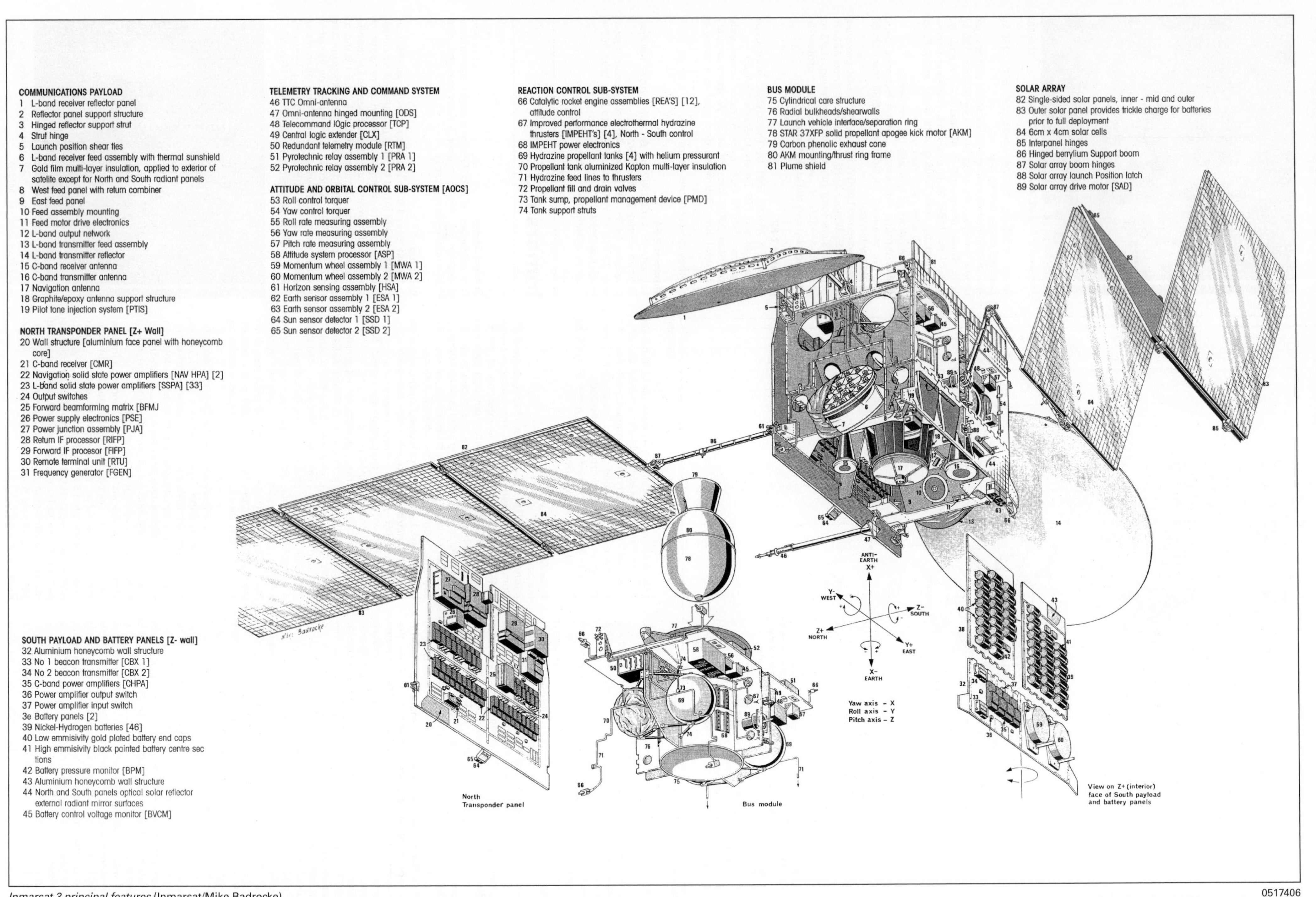

Inmarsat 3 principal features (Inmarsat/Mike Badrocke)

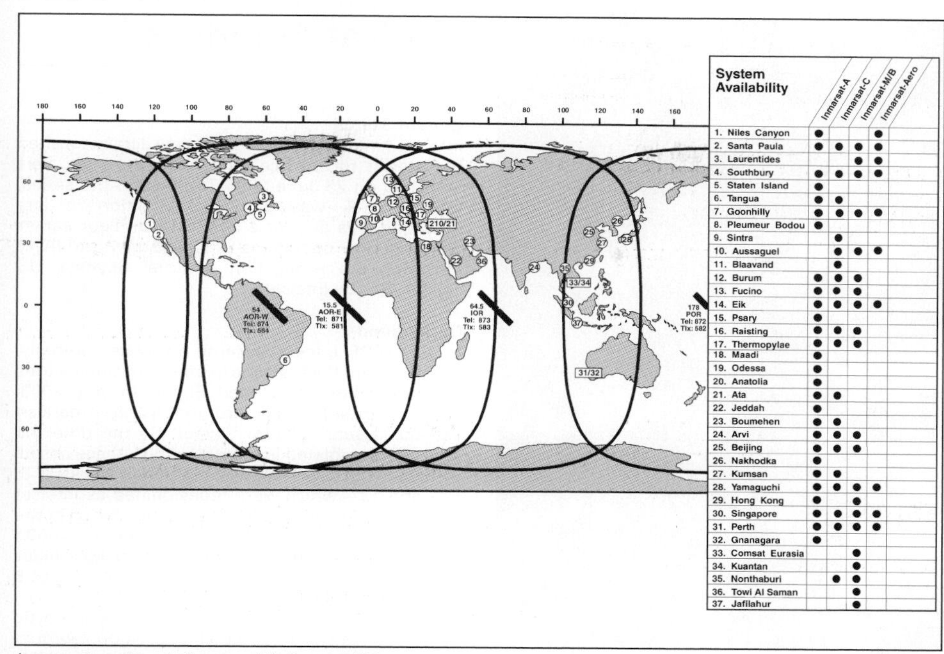

System Availability	Inmarsat-A	Inmarsat-C	Inmarsat-M/B	Inmarsat-Aero
1. Niles Canyon	●			●
2. Santa Paula	●	●	●	●
3. Laurentides			●	●
4. Southbury	●	●	●	●
5. Staten Island	●			
6. Tangua	●	●		
7. Goonhilly	●	●	●	●
8. Pleumeur Bodou	●			
9. Sintra		●		
10. Aussaguel		●	●	●
11. Blaavand		●		
12. Burum	●	●	●	
13. Fucino	●	●	●	
14. Eik	●	●	●	●
15. Psary	●			
16. Raisting	●	●	●	
17. Thermopylae	●	●	●	
18. Maadi	●			
19. Odessa	●			
20. Anatolia	●			
21. Ata	●	●		
22. Jeddah	●			
23. Boumehen	●	●		
24. Arvi	●	●	●	
25. Beijing	●	●	●	
26. Nakhodka	●			
27. Kumsan	●	●		
28. Yamaguchi	●	●	●	●
29. Hong Kong	●		●	
30. Singapore	●	●	●	●
31. Perth	●	●	●	●
32. Gnanagara	●			
33. Comsat Eurasia			●	
34. Kuantan			●	
35. Nonthaburi		●	●	
36. Towi Al Saman			●	
37. Jafilahur			●	

Inmarsat system stations and their capabilities 0517398

Matra team; Martin Marietta Astro Space/MMS UK; Aerospatiale/Alcatel with Ford Aerospace/ Mitsubishi; Hughes; and the Indian Space Research Organisation.

Inmarsat let the USD320 million contract for four satellites, with options for up to five more, to GE Technical Services Co on behalf of GE Astro Space/ Marconi Space on 1 February 1991. By the time the spacecraft were launched, GE had sold Astro Space to Martin Marietta, and Martin Marietta had merged with Lockheed. The responsible division is now Lockheed Martin Space Systems Company and Marconi Space is now part of EADS Astrium. The spacecraft were based on the Series 4000 satellite bus and were built at the Lockheed Martin facilities in East Windsor, New Jersey. Ultimately, five satellites of this type were launched. Inmarsat 3 F2 was one of the first non-Russian satellites to be launched from Baikonur by a Proton booster.

The satellites can reallocate power and bandwidths between the five spots and a single global beam to meet changing traffic requirements. Each satellite provides a capacity 10 times that of Inmarsat 2 and can transmit overlay signals almost identical to those of GPS/Glonass to provide three civil navigation functions: improvement of accuracy, compensation for coverage gaps; and continuously updated, independently monitored integrity monitoring. These functions identify faulty satellites, which is essential for air navigation safety.

The Inmarsat 3 satellites have supported the growth of ship-to-shore communications, but have also been the platform for Inmarsat's airborne communications services. These range from the low-rate, store-and-forward Aero C to the Swift64 service, delivered through the regional spot-beams of the Intelsat 3 and aimed at corporate and government users. First offered in 2002, Swift64 was the first service to deliver broadband-like Internet access – channels can be ganged together to provide 265kB/sec download speeds – to airborne platforms.

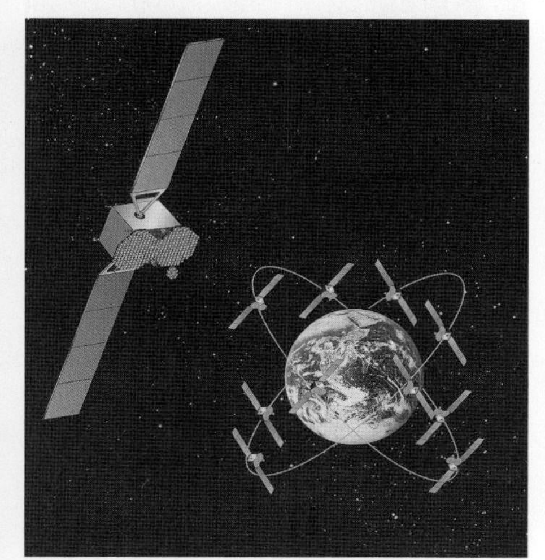

Mobile communication services provided by Inmarsat P satellites utilising the Hughes HS-601 bus (Hughes Space & Communications) 0054339

Specifications

Inmarsat 3-F1

Launch: 3 April 1996 by Atlas 2A from Cape Canaveral
Location: Initially 28°E for testing, moved to operational location of 64°E in May 1996
Design life: 13 years
Contractors: Lockheed Martin Space Systems Company (prime), EADS Astrium (communications payload), Olin Aerospace (hydrazine thrusters)
Transponders: 490 W SSPA matrix 1.63–1.66/1.53–1.56 GHz up/down L-band global + five spot beams, 34 MHz bandwidth, EIRP 48 dBW total (switchable between global/spot in any proportion of total EIRP), circular polarisation

Two (plus two back-up) 12 W SSPA 6.43 6.45/3.60–3.63 GHz up/down C-band global beams (feeder for L-band), 29 MHz bandwidth, EIRP 27 dBW in each of two circular polarisations

One (plus one back-up) 20 W SSPA 1,575.42 MHz GPS L1 L-band global beam, 2.2 MHz bandwidth, EIRP 27.5 dBW (translated from C-band uplink). GPS/Glonass overlay signals compensating for coverage gaps and enhancing their accuracy. Also, continuously updated information on these systems' integrity broadcast to identify faulty satellites. A C-to-C communication channel used by the uplink station for more precise signal tracking (less susceptible to ionosphere disturbances than L-band); channel can also be used for precise time and frequency applications
Principal applications: maritime, air and land mobile
Configuration: based on Astro Space boxed-shaped 4000 bus, body 2.1 × 1.8 × 1.7 m, 2.5 m high, 3.2 m radial envelope, 16.7 m span, centred on thrust cone. Payload and solar wings mounted on N/S panels. L band receive/transmission reflectors mounted on E/W panels and fed by cup-shaped radiating elements
Mass: 2,068 kg at launch, 1,100 kg on-station BOL, 190 kg payload, 860 kg dry
AOCS: 3-axis momentum bias stabilisation with MWs and magnetorquers; pivoted MW provides antennas pointing for up to 2.7° inc. Mono hydrazine thrusters, 283 kg capacity, 12 Reaction Engine Assemblies + four Improved Electrothermal Hydrazine Thrusters. GEO insertion by Thiokol Star 37XFP or 37FM
Power system: 2.8 kW EOL provided by twin Si wings (six panels total 30.5 m^2) spanning 16.7 m. Eclipse protection by 2 × 62 Ah Ni/H_2 batteries.

Inmarsat 3-F2

Launch: 6 September 1996 on Proton from Tyuratam
Location: 15.5° W Atlantic Ocean
Other specifications as for F1.

Inmarsat 3-F3

Launch: 18 December 1996 on Atlas-2A from Kourou
Location: Initially tested over 158° E, then relocated to operational 178° E 25 January 1997
Other specifications as for F1.

Inmarsat 3-F4

Launch: 3 June 1997 on Ariane-4 from Kourou
Location: 54°W in service from 26 July 1997. Currently 142°W.
Other specifications as for F1.

Inmarsat 3-F5

Launch: 4 February 1998 on Ariane 4 from Kourou
Location: 24.5°E and currently deployed as a spare
Other specifications as for F1.

Inmarsat 4

Current Status

Two of Inmarsat's heavyweight I-4 satellites were operational in early 2007, and Inmarsat announced in mid-2006 that it would launch the third spacecraft in the series in late 2007 to support global handheld satellite telephone users.

Background

In May 2000, just over a year after its transition from an International Governmental Organization (IGO) to a private company, Inmarsat awarded Astrium (now EADS Astrium) a USD700 million contract to develop and build three Inmarsat 4 satellites. Based on Astrium's Eurostar E3000 bus, the I-4s were among the largest commercial satellites ordered up to that time. They were designed to support Inmarsat's plan for a Broadband Global Area Network (BGAN), delivering high-rate data communications to mobile users on land, on ships and on aircraft.

The Inmarsat 4 spacecraft each offer more call capacity than all five of the previous Inmarsat 3 satellites combined, with 60 times more power than an Intelsat 3 and a 25-fold improvement in receiver sensitivity. The satellite is equipped with a single global beam, 19 wide-spot beams and 228 narrow spot-beams. The narrow beams support the delivery of data at up to 492 kbyte/sec into portable units the size of a laptop computer, or into antennas compatible with medium-sized business aircraft.

The spacecraft bus was produced in EADS Astrium's facility in Stevenage, England. The primary payload was developed by EADS Astrium in Portsmouth. Principal subcontractors included Northrop Grumman (formerly TRW) and Canada's EMS Technologies. Northrop Grumman has provided the 9 m diameter AstroMesh reflector for the transmit and receiver systems, together with its deployment system, and EMS provides the 120-element helical antenna feeds. Assembly and initial testing took place in Toulouse.

Inmarsat 4 fully assembled at Toulouse. The carbon-fibre rods that form the folded reflector are visible on the front side of the body (EADS Astrium) 1289733

The Inmarsat 4 satellite is distinguished by a 9 m diameter reflector supplied by Northrop Grumman (Inmarsat) 1289734

A key feature of the new spacecraft is the Astrium-developed Digital Signal Processor (DSP) system that feeds the antenna and handles signals on and off the satellite. One of the first applications of all-digital switching on a large spacecraft, the DSP allows the spacecraft to manage the many spot beams rapidly and reliably and controls the antennas, beam-forming and channel allocation. It can also be reconfigured to adapt to traffic demands or other changing conditions.

Other new technologies on the Inmarsat 4 include solar panels combining silicon and gallium arsenide (GaAs) cells for optimum efficiency, and combined chemical and plasma-ion propulsion. The Plasma Propulsion Subsystem (PPS) for station-keeping incorporates four Russian-designed stationary plasma thrusters. These flight-proven units are supplied by xenon gas and work by electrostatically accelerating plasma ions through a discharge chamber. Each unit contains one prime thruster and one back-up. A separate chemical propulsion system will be used to establish the I-4 in its final orbit, for some in-orbit manoeuvres, and to back up the PPS.

Each solar wing includes two GaAs panels and three silicon panels provided by Germany's RWE.

An Atlas V launches the first Inmarsat 4 from Cape Canaveral on 11 March 2005. This was the heaviest Atlas launch to that date (Lockheed Martin) 1289732

The former are more efficient but the latter are lighter and better proven, providing maximum assurance that the spacecraft will maintain power through its design lifetime.

Inmarsat selected International Launch Services in July 2001 to launch the first Inmarsat 4. The launch was carried out successfully on 11 March 2005 using an Atlas V-431 with a 4 m fairing, three solid boosters and a single-engine Centaur upper stage. The contract for the second launch was awarded to Sea Launch in March 2004, and the spacecraft was placed in orbit on 8 November 2005.

The first satellite covers Europe, Africa, the Middle East and Asia, as well as the Indian Ocean. The second brings broadband service to the Americas, the Atlantic Ocean and part of the Pacific Ocean. Together, according to Inmarsat, they cover 85 per cent of the world's land mass and 98 per cent of its population, and have been used to roll out the BGAN service – marketed to aviation users as SwiftBroadBand – and other mobile services. SwiftBroadBand is due to be available worldwide by mid-2007, following demonstration flights in a Gulfstream G550 in October 2006.

The third satellite was initially held in reserve in case of a problem with the first or second I-4s. However, in September 2006 Inmarsat announced a deal with Asia Cellular Satellite (ACeS), an Indonesia-based supplier of hand-held satellite phone services, to develop a global service based on an updated version of the R190 handset and the I-4 satellite. At the same time, Inmarsat announced that the third I-4 would be launched at the end of 2007. No launch contract had been announced by early 2007, but the satellite is compatible with Ariane 5 and Proton as well as with Atlas and Sea Launch/Zenit.

Specifications
Inmarsat 4 F1
Launched: 11 March 2005 by Atlas V-431 from Cape Canaveral LC-41
Location: 64°E
Design life: 18 years
Contractors: EADS Astrium: integrator, bus and payload; Northrop Grumman: reflector; EMS Technologies: antenna system
Transponders: 228 narrow-spot (1.1°) beams; 19 wide-spot beams; single global beam. Up to 630 200 kHz channels, dynamically allocated to beams.
Principal applications: Mobile communications
Configuration: Eurostar E3000 bus with 9 m deployable reflector. Body dimensions 7 m × 2.9 m × 2.3 m
Mass: 5,950 kg launch, 3,340 kg dry
Power system: 45 m-span hybrid (GaAs and Si) ten-panel solar array, delivering 13 kW

Inmarsat 4 F2
Launched: 8 November 2005 by Zenit 3SL from Sea Launch platform Odyssey from Cape Canaveral LC-41
Location: 53°W
Other specifications as 4 F1.

Intelsat Americas

Current Status
Intelsat Americas IA-8, the first spacecraft launched by Intelsat for service over North America, was launched on 23 June 2005. The next in the series, Intelsat IA-9, was under construction in late 2006 and was due for a 2007 launch. Four earlier satellites now operate as IA-5, IA-6, IA-7 and IA-13 following the acquisition of Loral Skynet's US Telstar fleet by Intelsat.

Background
In July 2001, Intelsat became a private company, enabling it for the first time to offer communications services within the United States. In July 2003, Intelsat agreed to acquire the North American satellite assets of Loral Skynet. By the time the deal was completed in February 2004, these assets included four on-orbit Telstar satellites – Telstars 5, 6, 7 and 13 – which were redesignated as Intelsat Americas (IA) spacecraft. (Telstar 4, included in the original sale, failed on orbit in September 2003.) The sale also included Telstar 8, which was under construction and was renamed Intelsat Americas 8 (IA-8) before launch.

The final satellite to be launched as a domestic Telstar was Telstar 13, placed into orbit on 7 August 2003 by a Zenit-3SL rocket from Sea Launch's Odyssey platform in the Pacific Ocean. It is unusual in that it is operated by two companies under two names. EchoStar Communications Corporation calls the spacecraft EchoStar IX, using its Ku-band transponders to support its DishNetwork direct-to-home satellite broadcasting system. It has Ka-band transponders to support direct-to-user satellite Internet access, but these are not currently in use. Intelsat identifies the satellite as IA-13 and uses its C-band transponders to serve cable TV providers. IA-13/EchoStar IX is based on the latest version of the SS/L 1300 series satellite bus.

IA-8 was launched on 23 June 2005, also by a Sea Launch Zenit-3SL. It represents a further advance in the SS/L 1300 family, with new technical features and greater power than any of its predecessors. IA-8 featured advanced thermal technologies not previously used on the 1300 bus, including deployable thermal radiators and loop heat pipes. It was also the second SS/L satellite to use Stationary Plasma Thrusters (SPTs) for orbital station-keeping manoeuvres – contributing to its 15-year design lifetime – and the first Intelsat spacecraft to carry Ka-band transponders, developed to support direct-to-home broadband Internet service. The IA-8 also has two high-powered zone beams specifically designed to provide coverage of Latin America.

IA-8 is designed to provide expanded coverage over the Americas, the Caribbean, Hawaii and Alaska, during a 15-year service life. IA-8 will host voice, video and data transmission and distribution services, as well as carry 28 C-band and 36 Ku-band transponders, along with the 24 Ka-band spot beams.

Intelsat Americas IA-8 flies into orbit aboard a Sea Launch Zenit-3SL on 23 June 2005 (Boeing) 1198514

Intelsat Americas IA-13 has a dual identity, also operating as EchoStar IX for the DishNetwork TV service (Space Systems/Loral) 1198515

Originally launched as Telstar 7, this 1300-bus spacecraft has been known as Intelsat Americas IA-7 since 2004 (Space Systems/Loral) 1198516

Space Systems/Loral was awarded a contract from Intelsat in November 2004 to design and produce Intelsat Americas 9 (IA-9). This is a two-band satellite with 24 C-band and 28 Ku-band transponders. Intelsat awarded a launch contract for IA-8 to Sea Launch in May 2006, with the flight scheduled for the last quarter of 2007. Intelsat plans to place the satellite at 263°E, serving data networking and video applications. Following the launch of IA-9, the IA-5 satellite is planned to be redeployed to a currently unserved location at 283°E.

Specifications

IA-5

Launch: 24 May 1997 by Proton K from Baikonur
Location: 263°E
Contractor: Space Systems/Loral
Design life: 12 years
Mass: 3,650 kg at launch; 1,400 kg dry
Specification: 24 × C-band 36 MHz at 20 W; 4 × Ku-band 54 MHz at 100W; 24 × Ku-band 27 MHz at 100W. Frequency band 4/6 GHz and 12/14 GHz.

IA-6

Launch: 15 February 1999 by Proton-K from Baikonur
Location: 267°E
Contractor: Space Systems/Loral
Mass: 3,650 kg at launch; 1,400 kg dry
Specification: 24 × C-band 36 MHz at 20 W; 4 × Ku-band 54 MHz at 100W; 24 × Ku-band 27 MHz at 100W. Frequency band 4/6 GHz and 12/14 GHz.

IA-7

Launch: 25 September 1999 by Ariane 44LP from Kourou, French Guiana
Location: 231°E
Contractor: Space Systems/Loral
Mass: 3,790 kg at launch; 1,537 kg dry
Specification: 24 × C-band 36 MHz at 37 W; 24 × Ku-band 36 MHz at 100 W. Frequency band 4/6 GHz and 12/14 GHz.

IA-13/EchoStar IX

Launch: 7 August 2003 by Zenit-3SL from Odyssey launch platform
Design life: 15 years
Location: 239°E
Contractor: Space Systems/Loral
Transponders: Intelsat: 24 × C-band at 36 MHz and 37 W; EchoStar: 32 Ku-band transponders at 110 W; two Ka-band at 120 W.
Mass: 4,740 kg
Power: 10 kW end-of-life

IA-8

Launched: 23 June 2005 by Zenit-3SL from Odyssey launch platform.
Location: 271° E
Design life: 15 years
Contractor: Space Systems/Loral
Transponders: 28 C-band, with 22 37 W and 6 100 W; 36 Ku-band rated at 130 W; 4 Ka-band at 105 W
Principal applications: Corporate networking; video; cable TV; direct-to-home internet
Configuration: Modified SS/L 1300 bus
Mass: 5,500 kg
Power system: 16 kW end of life

Intelsat 6 series

Current Status

Four of the original Intelsat 6 series satellites remained operational in late 2006, having greatly exceeded their original design life.

Background

Intelsat invited industry in 1980 to propose designs for a new communication satellite with three times the capacity of the Intelsat 5. Many international consortia bid on the request, led by Hughes and Ford. Intelsat closed the competition in April 1982 and awarded a USD700 million contract to Hughes for five satellites, with first launch in 1986. The contract called for an option on another six, which would bring the contract to USD1,300 million. At the time, Intelsat 6 was the largest unclassified satellite ever built, and the most expensive and ambitious single satellite programme.

Supported by a global team including British Aerospace (now BAE Systems) and the predecessor companies to EADS-Astrium, Hughes proposed a spin-stabilised hybrid that offered 33,000 2-way phone circuits and four TV channels. Originally Hughes designed the satellite for launch by Shuttle or Ariane 4; and the first (602) flew aboard an Ariane on 27 October 1989 and entered full service 7 April 1990.

In full service, the five satellites ordered provide the equivalent capacity of up to 120,000 simultaneous 2-way telephone circuits using digital modulation techniques and at least three TV channels utilising 6/4 and 14/11 GHz communication links. Intelsat became the first commercial satellite to employ Satellite Switched/Time Division Multiple Access (SS/TDMA), allowing beam interconnections according to traffic requirements. The satellites reuse the C-band frequencies six times through two hemispherical beams and four zone beams with dual circular polarisation. The K-band frequencies undergo reuse twice by spatially isolated east-west spots using linear orthogonal polarisation. All provide higher K-band power to promote the use of small Earth stations.

All five satellites under the initial contract were successful, but the options for follow-on satellites were never taken up. Intelsat launched the first (602) on 27 October 1989 on an Ariane and it entered service 7 April 1990. It was operating at 150.5°E in late 2006. Titan 3 stranded the second (603) in LEO following launch on 14 March 1990. Shuttle flight STS-49 carried a propulsion module to the stranded satellite and re-boosted it to a

Intelsat 6 communications platform (Hughes Space & Communications) 0516987

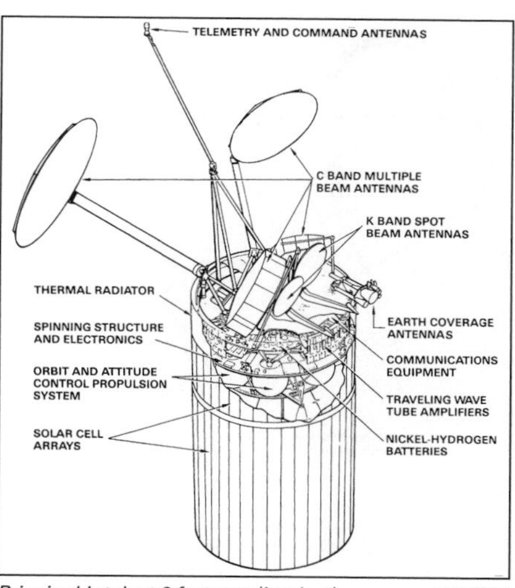

Principal Intelsat 6 features (Intelsat) 0516986

correct orbit May 1992, providing a 10 year life; still operating from 340°E. 604 went to space on a Titan 23 June 1990 but is no longer listed as operational. It was followed by 605 (14 August 1991 by Ariane), operating from 174°E; and 601 (29 October 1991 by Ariane) still operating from 64.25°E.

Prime contractor Hughes was absorbed by Boeing in 2000.

Specifications

Design life: 13 years min 602/3; 15 years 605
Contractors: Hughes Space & Communications Co (prime), British Aerospace (Ku/C-band reflectors, power electronics, Ariane adapter), Pilkington (solar cell covers), Alcatel Espace (Ku/C-band receivers, output multiplexer), Thomson-CSF (Ku-band TWTs), Alenia Spazio (spot antenna, telemetry transmitters/receivers, remote/central telemetry units), AEG (solar cells), MBB/ERNO (solar panels, contract worth about DM50 million), NEC (Ku-band receivers, SSPAs, master oscillators)
Transponders: 8 (plus 3 back-up) 20 W and 2 (plus 1 back-up) 40 W TWTA 14.0–14.5/10.95–11.2 + 11.45–11.70 GHz Ku-band steerable spots, 72 MHz bandwidth, EIRP 45 dBW min, dual polarisation
6 (plus 2 back-up) 16 W TWTA 5.85 6.42/3.62–4.20 GHz C-band hemispherical and zone beams, 72 MHz bandwidth, EIRP 30 dBW min, dual polarisation
32 (plus 10 back-up) 10W SSPA 5.85 6.42/3.62–4.20 GHz C-band hemispherical and zonal beams, 72 MHz bandwidth, EIRP 30 dBW min, dual polarisation
Principal applications: Intelsat services
Configuration: Hughes' HS-393 platform spin-stabilised, cylindrical body 3.63 m diameter, deployed height 11.84 m (height aboard Ariane 4 6.4 m, height with PKM 8.06 m). Each 3.2 m (30 kg carbon, Kevlar/boron fibre composite) + 2.0 m C-band antenna provides two fixed beams for hemispherical coverage and four isolated beams for zone coverage (these can be reconfigured in orbit to match the requirements of the three ocean regions). Steerable 1.1 m + 1.0 m diameter Ku-band reflectors provide spot coverage
Mass: 4,600 kg at launch (includes adapter), 2,546 kg on station BOL, 1,896 kg dry
Attitude control: 30 rpm spin-stabilised; thrusters fired automatically when attitude excursions exceed pre-set limits
Antenna pointing: ±0.05° maintained in beacon tracking mode, ±0.10° with Earth/Sun sensors
TT&C: through Intelsat's SCC HQ station
Power system: drum-mounted solar arrays provide 2.6 kW BOL/2.252 kW EOL; K7 cells on skirt, K4¾ on fixed body. Two Ni/H_2 batteries provide eclipse protection
Propulsion: perigee kick from Titan 3 or Shuttle provided by 8.76 t UTC/CSD Orbus 6S solid propellant stage. ACS unified with apogee kick: MMH/NTO supply four 22 N radial thrusters for E-W station-keeping + spinup/down, two 22 N axial for NSSK + attitude control, and two 490 N for apogee kick + reorientation manoeuvres.

Intelsat 7 series

Current Status

The Intelsat 7 series were built for Intelsat by Space Systems/Loral. Eight of the nine satellites built were launched successfully and seven were

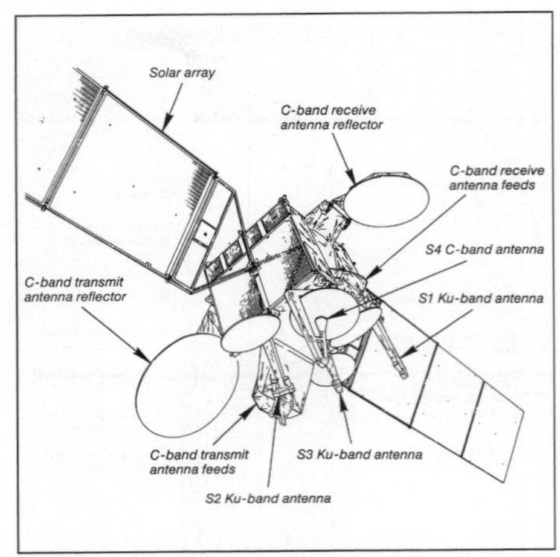

Intelsat 7 configuration (Space Systems/Loral) 0516988

operational in late 2006. NSS-703, now operated by SES New Skies, is expected to remain in service until March 2009.

Background

Intelsat requested proposals in October 1987 for fixed price bids for 2–3 satellites, with an additional bid for a fourth, for a new generation of Pacific region satellite. Intelsat expected to put the satellites in service in late 1992 or early 1993. Previous satellites had met Atlantic region requirements; which differed from Pacific service. Ford Space Systems received the contract in June 1988. Matra served as a back-up source. On 9 September 1988, Intelsat awarded a USD394.28 million contract for five Intelsat 7 satellites with the first launches in mid-1992 and early 1993.

The satellites, combining an Intelsat 5-type payload with Ford's FS-1300 bus, offer increased power and coverage for smaller ground stations and flexibility through three independently steerable Ku-band spots and a steerable C-band spot. Intelsat decided in December 1989 to increase the steerability and Ku-band power of F3 and its successors. The satellite operates inverted at 66° E/1° W, simplifying antenna design. General Dynamics' Atlas 2AS has launched three, charging Intelsat USD75 million each. ArianeSpace's Ariane 4 launched two at USD86.5 million each.

Ford sold its satellite operations to Loral in 1990 and it now operates as Space Systems/Loral.

Intelsat ordered two additional satellites of the same basic type in December 1990, for a total cost of USD202 million, to replace the Intelsat 5A satellites over the Atlantic. The so-called 7A satellites were based on the Intelsat 7s with the exception of higher Ku-band capacity, higher power transponders, and a stretched bus with improved power systems. A third satellite of this type was ordered in September 1992. Two of the three were launched successfully, but the third was lost when its CZ-3B Long March booster failed.

The first Intelsat 7 entered service in 1994 (Intelsat) 0516985

Specifications

Launch schedule: 701 22 October 1993 (Ariane 44LP; entered service 15 January 1994) operating from 180° E, 702 17 June 1994 (Ariane 44LP), operating from 55° E, 703 6 October 1994 (Atlas 2AS) operating as NSS-703 for SES New Skies at 57° E, 704 10 January 1995 (Atlas 2AS; entered service 25 February) now retired, 705 (Atlas 2AS) 22 March 1995 (Atlas 2AS) operating at 310° E since May 2002, 709 15 June 1996 (Ariane 44P) operating from 85° E (Ku-band only), 706 (Ariane) 17 May 1995 tested at 56° W prior to problems, now operating from 50.25° E; 707 (Ariane) 14 March 1996 now operating from 307° E; 708 (Long March) 14 February 1996, lost in launcher failure.
Design life: 65% transponder operability after 10.9 years; sufficient propellant for 15 years.
Contractors: Space Systems/Loral (prime), Alcatel Espace (payload), NEC (20/30 W SSPAs), MELCO (10/16 W SSPAs), AEG (TWTs).
Transponders (7 series): 26 SSPA (10, 16, 20, 30 W) in 7-for-5 rings 5.8–6.5/3.6–4.2 GHz up/down C-band global (26–29 dBW), hemispherical + zonal (33 dBW) and spot (33–36 dBW), 36/72 MHz bandwidth, dual polarisation.
Transponders (7A series): C-band global EIRP 29 dBW; C-band spot 36.1 dBW. Increased-power Ku-band transmitters and four more wideband transmitters. 44–47.5 dBW EIRP, 6 × 72 MHz + 4 × 112 MHz (allows 155 Mbit/s) bandwidth, dual polarisation.
Configuration: box-shaped SS/L FS-1300 bus with Intelsat 5 type communication platform with three steerable Ku-band dishes and one C band, in addition to fixed C-band receive (1.57 m diameter) + transmission (2.44 m diameter).
Mass (7 series): 1,450 kg dry mass, 3,650 kg GTO (for Ariane, with 2,160 kg propellant load), typically 1,800 kg on-station BOL.
Mass (7A series): 1,748 kg dry mass, 4180 kg GTO (for Ariane, with 2,820 kg propellant load).
AOCS: 3-axis (from immediately after launcher separation). 12 × 22 N N_2O_4/MMH, 490 N thruster permits Perigee Velocity Augmentation during Atlas orbital insertion, adding 2.1 years to life. Propellant loading 2,160 kg for 92% fill.
TT&C: through Intelsat's SCC HQ station.
Power system (7 series): twin 3-panel solar wings, spanning 21.843 m, providing 3.9 kW after 10.9 years. Supported by four Ni/H_2 batteries totalling 85.5 Ah.
Power system (7A series): twin 4-panel solar wings, providing 5.3 kW after 10.9 years. Supported by four Ni/H_2 batteries totalling 120 Ah.

Intelsat 8 series

Current Status

Six Lockheed Martin Intelsat 8 satellites were launched and five remained operational in late 2006. Three were listed as operational by Intelsat and two by SES New Skies, but the status of Intelsat 802 is uncertain following an on-orbit anomaly on 22 September 2006. Intelsat 804 failed on orbit on 14 January 2005.

Background

Intelsat issued a request for proposals in December 1991 for three new satellites, with options for a further four if required. The initial order covered only two new-model satellites (801 and 802), however, Intelsat were substituting a third Space Systems/Loral 7A-series satellite (Intelsat 709) for one of the new spacecraft. GE Astro Space won the contract to build the first two Intelsat 8 spacecraft based on its 7000 bus design. They featured a design to meet the growing Pacific region need, using 6-fold C-band and 2-fold expanded C-band frequency reuse.

GE Astro Space was acquired by Martin Marietta in 1993 and became part of Lockheed Martin following the 1995 merger of Lockheed and Martin.

Intelsat announced in December 1992 that it would use Ariane 4 boosters to launch the Intelsat 8 satellites. On September 1993, Intelsat exercised its options on the contract, ordering two satellites similar to the first pair (803 and 804) and a third satellite, Intelsat 805, equipped for land-mass coverage with added C-band power and steerable Ku-band antennas. A second satellite in this so-called 8A series, Intelsat 806, was ordered in March 1994. Both the 8A satellites were launched by Atlas IIAS rockets.

The first Intelsat 8 was launched in 1997 (Intelsat) 0005566

Specifications

Launch schedule: 801 (Ariane 44LP) 1 March 1997, damaged in on-orbit testing, currently at 328.5°E; 802 (Ariane 44LP) 25 June 1997 tested at 170°E, located at 33°E; 803 (Ariane 44LP) 23 September 1997, transferred to New Skies Satellites N.V. on 30 November 1998, moved from 27.5° W to 21.5° W in 1998, to 19.8° W in 2002 and to 183°E as NSS-53; 804 (Ariane 42L) 22 December 1997 operated from 44°E 1997 to 2002, then to 64.15°E in 2002 and 176°E in November 2002, failed on 14 January 2005; 805 (Atlas IIAS) 18 June 1998 operating at 304.5°E; 806 (Atlas IIAS) launched 28 February 1998, operating for SES New Skies at 319.5°E.
Design life: 10 years (13–19 years propellant).
Contractors: Lockheed Martin Astro Space.
Transponders (801–804): 18 27–38 W variable power SSPA C band, providing 6 global (EIRP 29.0 dBW) + 12 hemispherical (EIRP 34.5–36.0 dBW) bandwidths 34, 36, 41, 72, 77 MHz

20 10–20 W variable power SSPA C-band zonal beams, EIRP 34.5–36.0 dBW, bandwidths 34, 36, 41, 72, 77 MHz

6 43 W TWTA in two Ku-band steerable spots (2.8° circular + 1.9 × 4.3°), EIRP 44/47 dBW inner/outer, 602 MHz bandwidth (as 34, 72, 77, 112 MHz).
Principal applications: Intelsat services.
Configuration: 7000 series platform. 3-axis box-shaped 216 × 246 × 315 cm bus with central cylinder.
Mass: 3,245 kg at launch, 1,538 kg dry.
AOCS: spin-stabilised in transfer orbit; 3-axis on-station using roll/yaw magnetorquers in normal mode, hydrazine thrusters for station-keeping. Redundant 1.8 kW hydrazine arcjets provide NSSK. Pointing: ±0.1° pitch/roll, ±0.25° yaw accuracy; ±6.0° pitch + ±2.0° offset pointing control.
TT&C: through Intelsat's SCC HQ station.
Power system: twin 4-panel wings (totalling 48.6 m^2) providing 4,800 W BOL. 2 × 45 Ah Ni/H_2 batteries provide 100% eclipse protection.

Intelsat 9 series

Current Status

All seven of the Intelsat 9 satellites were in service with Intelsat in late 2006.

Background

In December 1996, Intelsat management chose to procure two follow-on spacecraft (FOS-11) from Space Systems/Loral, which it wanted to operate over the Indian Ocean as replacement for Intelsat 6 satellites. Re-named Intelsat 9, the programme expanded to include seven satellites located

Intelsat 907 is prepared for integration with its Ariane 44L launch vehicle before its February 2003 lift-off (Arianespace) 1198426

Spacecraft delivery and launch periods:

Spacecraft	Launch Vehicle	Launch	Orbit Location
Intelsat 901	Ariane 44L	9 June 2001	342° E
Intelsat 902	Ariane 44L	30 August 2001	62° E
Intelsat 903	Proton DM	30 March 2002	325.5° E
Intelsat 904	Ariane 44L	23 February 2002	60° E
Intelsat 905	Ariane 44L	5 June 2002	335.5° E
Intelsat 906	Ariane 44L	6 September 2002	64° E
Intelsat 907	Ariane 44L	15 February 2003	332.5° E

over the Atlantic, Europe and the Middle East, constituting the backbone of the Intelsat network.

The Intelsat 9 communications satellites began service in 2001, with Intelsat 907 launched in 2003. The Intelsat 9 spacecraft are multi-frequency communications satellites that provide voice, video, data transmission and distribution to virtually any location in the regions around the Atlantic and Indian Oceans. These spacecraft are among the world's largest and most powerful commercial communications satellites.

The Intelsat 9 is based on the same 1300-series bus as the Intelsat 7. It has a design lifetime of 13 years, and is stabilised on orbit with bipropellant propulsion and momentum-bias systems. It carries 22 Ku-band and up to 76 C-band transponders (in 36 MHz equivalents), and has solar arrays that generate more than 9.5 kW of power (beginning of life). Each Intelsat 9 spacecraft carries more high-power amplifiers and generates more solar array power than its predecessors with only a small increase in dry mass.

The Intelsat 9 provides greater power, coverage area and capacity than its predecessors, including high-power Ku-band spot beam coverage and more C-band capacity. Intelsat 9 is designed to provide an increase of up to 9 dB for downlink Effective Isotropic Radiated Power (EIRP) and up to 4.8 dB for G/T over the Intelsat 6 satellites. The later Intelsat 9s featured the ability to steer Ku-band beams based on demand.

Alcatel Alenia Space was Loral's major partner on the Intelsat 9, responsible for about 40 per cent of the satellite's construction. Alcatel Alenia Space's role on Intelsat 907 included the design, development, production and supply of the transponders, which were also integrated and tested at its facility in Toulouse. The satellite was assembled, integrated and checked out by a joint Space Systems/Loral-Alcatel Space team in the former's clean rooms at Palo Alto, California.

Intelsat 10-02

Current Status

The sole Intelsat 10 satellite, Intelsat 10-02, is on station and operational. Its 13-year design life should ensure operations until 2017.

Background

In early 2000, with development of the Intelsat 9 spacecraft well under way, Intelsat selected Matra Marconi Space to develop its next-generation comsat, the largest and most powerful spacecraft ever procured by the agency. Two spacecraft were ordered for operations over the Atlantic and Europe. Matra Marconi was to provide both the Eurostar E3000 payload bus and the communications payload.

Intelsat 10-02 was one of the world's largest comsats at the time of its launch in June 2004 (EADS Astrium) 1198513

Matra Marconi was merged with the space division of DaimlerChrysler Aerospace in 2000 to create Astrium. Preparations for the launch continued, and Intelsat contracted with Sea Launch in December 2001 to place the first satellite, 10-01, into space aboard a Zenit-3SL rocket. The first launch was set for 2003, but the programme fell behind schedule. In November 2002, Intelsat exercised its option to cancel the first of the two spacecraft, Intelsat 10-01.

Work continued on Intelsat 10-02 and a launch contract was signed with International Launch Services. In May 2004 the spacecraft was shipped to Baikonur for a launch by Proton. The launch was carried out successfully on 17 June 2004.

Intelsat 10-02 was one of the largest communications satellites ever built at the time of its launch. The payload features up to 70 C-band and 36 Ku-band 36 MHz equivalent unit transponders, with cross connectivity. The satellite's 150 W travelling wave tube amplifiers ensure the highest power available in the market, and the inclusion of Automatic Level Control enables the maintenance of constant power. Three steerable Ku-band spot beams, of which two are rotatable, allow for flexible coverage of target geographic markets.

Intelsat 10-02 was designed to support video, corporate networking, voice, government/military and Internet requirements. Most of EADS Astrium's facilities around Europe participated in the manufacturing process. The solar array, antennas and repeater equipment came from Germany, communications payload and structure came from the United Kingdom, some electronic equipment from Spain, and the avionics and final integration and test were completed in France.

Approximately 50 per cent of the satellite's Ku-band transponders are owned by the Norwegian company Telenor Broadcast Holdings AS, through its subsidiary Telenor Satellite Broadcasting, the leading satellite operator and satellite services provider in the Nordic Countries, and a long-standing customer of Intelsat.

Specifications

Launched: 17 June 2004 by Proton from Baikonur
Location: 359° E
Design life: 13 years
Contractors: EADS-Astrium (integrator and payload)
Transponders: Up to 70 C-band, up to 36 Ku-band
Principal applications: High speed Internet, corporate nets, telephony, broadcast content
Configuration: Eurostar E3000 bus
Mass: 5,600 kg
Power system: 45 m-span solar arrays, 11 kW end-of-life

Mobile Satellite Ventures (MSV) Series

Current Status

Mobile Satellite Ventures (MSV) announced in January 2006 that it had ordered three new large communications satellites from Boeing to deliver mobile satellite services across the Americas. The first two satellites are scheduled for launch in mid-2009 and 2010.

Background

MSV, in collaboration with Canada's Telesat Mobile Inc (TMI), launched the twin satellites, AMSC-1 and MSAT, in 1995 and 1996. These two spacecraft were still in service in early 2007, providing L-band telephone, data, fax and push-to-talk (PTT) two-way radio service across North America via MSV and its joint-venture partner, MSV Canada. They are expected to continue operating through 2009.

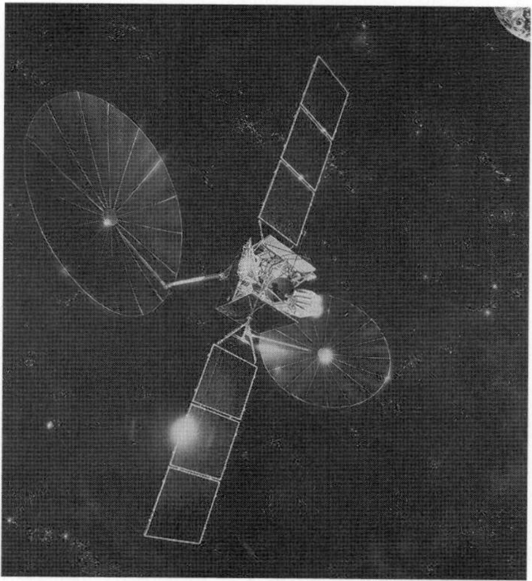

MSAT still provides mobile communications over North America (Spar Aerospace) 0517099

In January 2003, the Federal Communications Commission (FCC) ruled that spectrum allocated previously for Mobile Satellite Service (MSS) providers could also be used for terrestrial mobile services. Industry Canada made a similar ruling in May 2004. MSV plans to take advantage of these rulings with an integrated terrestrial and satellite wireless communications system, operating in the L-band, which will allow users to communicate anywhere in the US, with mobile phones and other devices which are similar in size, shape and cost to current mobile phones or mobile internet devices. The hybrid system handles dense user concentrations more efficiently and provides access for users in urban areas, where satellite signals are often blocked by buildings. Target markets include emergency and first-response services; fleet management services; "consumer telematics", similar to OnStar; coastal maritime users; and some consumers.

During 2005, MSV worked to consolidate technology for its system. Several patents were issued for its Ancillary Terrestrial Component (ATC) system. In February 2005, the Federal Communications Commission (FCC) released new rules for the use of ATC systems, welcomed by MSV because they contained minimal restrictions on ATC functions. MSV received FCC approval for its L-band satellite in May 2005, and announced in October 2005 that it had completed a proof-of-concept demonstration of a hybrid terrestrial-satellite system using their AMSC-1 and MSAT spacecraft. MSV also applied its ATC technology to the development of an S-band network named

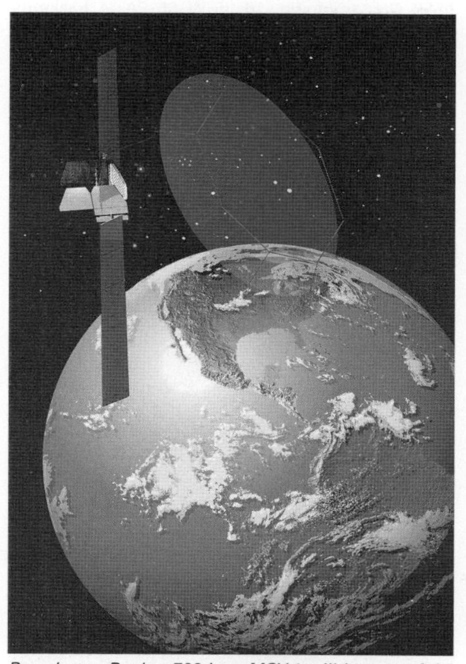

Based on a Boeing 702 bus, MSV-1 will be one of the largest commercial communications satellites when it flies in 2009 (Boeing) 1310361

MSAT 1's two 4.9×6.7 m graphite mesh antennas weigh only 20 kg each
(Communications Research Centre) 0517387

TerreStar, which was spun off into an independent company (see separate entries) in 2005.

MSV placed an order with Boeing in January 2006 for three 702-class satellites. The first pair, MSV-1 and MSV-2, will be launched in mid-2009 and early 2010 and will cover Canada; the United States, including Alaska, Hawaii, Puerto Rico; the Virgin Islands; Mexico and the Caribbean basin. A third satellite, MSV-SA, is to serve South America, but its deployment was deferred later in 2006. Hughes Network Systems will provide the Satellite Base Transceiver System which will link the MSV satellites to the terrestrial network.

The first large commercial satellites to be ordered from Boeing since 1997, the MSV spacecraft will generate 11 kW through five-panel solar array wings using triple-junction gallium arsenide cells. The 22 m L-band reflector will be one of the largest on any commercial satellite and will generate hundreds of simultaneous spot beams for mobile users, and a 1.5 m Ku-band reflector will communicate with the ATC system. The satellites are expected to handle tens of thousands of calls at the same time, along with broadband, IP-based messaging, audio and video.

For further MSV corporate information, please see the separate Mobile Satellite Ventures (MSV) entry.

Specifications

AMSC 1

Launch: 7 April 1995 by Atlas 2A from complex 36A, Cape Canaveral
Location: 101° W GEO (62/139° W also authorised)
Design life: 12 years
Contractors: Hughes Space & Communications Co HS601 (management/bus) and Spar Aerospace (payload)
Transponders: Forward (Ku/L) 13.023–13.232/1.530–1.559 GHz up/down, 16 35 W 55 PA (4 back-ups in 10-for-8 redundancy scheme) in 2×8 matrix feeding 6 elliptical spot beams (4 North America, a Caribbean beam for Puerto Rico, Virgin Islands and Mexico, and Alaska/Hawaii). Aggregate edge of coverage EIRP 56.5–57.3 dBW, circular polarisation. Bandwidth capacity for up to 2,000 5 kHz channels. Each '35 W SSPA' comprises four 20 W SSPAs in parallel but operated at 35 W for required 16 dB noise/power ratio
Reverse (L/Ku) 1.6315–1.6605/10.783–10.932 GHz up/down one (plus two back-up) 110 W TWTA 13.000–13.250/10.750–10.950 GHz up/down Ku-band 36 dBW feeder/return link in single North American beam
Principal applications: mobile radio, telephone, data (gateway to feeder 13.2405–13.2455/10.9405–10.9455 up/down Ku-Ku in single North American beam) aeronautical, maritime, safety and distress, position location, wide area paging. High quality voice, packet and circuit switched data
Configuration: Hughes' HS-601 3-axis bus, 2.3 m box-shaped body, 3.63 m stowed diameter, 4.44 m stowed height, twin solar wings spanning 20.96 m, 18.9 m across deployed graphite mesh antennas. Two Hughes 4.9×6.7 m 20 kg Springback mesh reflectors deployed from east/west faces provide separate L-band transmit/receive, respectively, illuminated by 23-cup dipoles. Single 76 cm Ku-band shaped reflector for North America coverage
Mass: 1,650 kg on-station BOL, 2,720 kg at launch (1,270 kg dry)
AOCS: onboard control processor for control of momentum wheel assembly. Unified 12×22 N + 1×490 N bipropellant thrusters (N_2O_4/MMH in four tanks)
Antenna pointing: 0.125° accuracy using Earth sensors
TT&C: Hughes Communications controls AMSC 1 under a five year contract from its OCC Operations Control Centre in El Segundo via its Spring Creek, New York ground station. AMSC's own Network Operations Centre in Reston, Virginia provides network control/channel access on demand
Power system: twin 9 m long 3-panel solar wings producing 3.6 kW BOL, 3 kW EOL from K4-¾ large area Si cells; 160 Ah Ni/H_2 battery provides 100 per cent eclipse protection
Thermal control: dissipation by north/south radiators, with heat transported by heat pipes.

MSAT

Launch: 20 April 1996 by Ariane 4 from Kourou, French Guiana
Location: 106.5° W GEO
Design life: 12 years
Contractors: Hughes Space & Communications Co HS601 (management/bus) and Spar Aerospace (payload)
Transponders: 16 (plus 4 backup in 10-for-8 open ring) 35 W SSPA 1.6315–1.6465/1.530–1.545 GHz up/down and 1.6465 1.6605/1.545–1.559 GHz up/down in 2×8 matrix feeding 6 elliptical spot beams (4 North America, a Caribbean beam for Puerto Rico, Virgin Islands and Mexico, and Alaska/Hawaii. Aggregate edge of coverage EIRP 56.5–57.3 dBW, circular polarisation. Bandwidth capacity for 2,000 5 kHz channels. Each '35 W SSPA' comprises four 20 W SSPAs in parallel but operated at 35 W for required 16 dB noise/power ratio 1 (plus 2 backup) 110 W TWTA 13.000 13.15 and 13.20–13.250/10.750–10.950 GHz up/down Ku-band 36 dBW feeder/return link in single North American beam
Principal applications: mobile radio, telephone, data, aeronautical, maritime, safety/distress, position location, wide area paging. High quality voice, packet and circuit switched data
Configuration: Hughes' HS-601 3-axis bus, 2.3 m box-shaped body, 3.63 m stowed diameter, 4.44 m stowed height, twin solar wings spanning 20.96 m, 18.9 m across deployed graphite mesh antennas. Two Hughes 4.9×6.7 m 20 kg Springback mesh reflectors deployed from east/west faces provide separate L-band transmit/receive, respectively, illuminated by 23-cup dipoles. Single 76 cm Ku-band shaped reflector for North America coverage
Mass: 1,710 kg on-station BOL, 2,855 kg at launch (1,270 kg dry)
AOCS: onboard control processor for control of momentum wheel assembly. Unified 12×22 N + 1×490 N bipropellant thrusters (N_2O_4/MMH in four tanks)
Antenna pointing: 0.125° accuracy using Earth sensors
Power system: twin 9 m long 3-panel solar wings producing 3.6 kW BOL, 3 kW EOL from K4-¾ large area Si cells; 160 Ah Ni/H_2 battery provides 100 per cent eclipse protection
Thermal control: dissipation by north/south radiators, with heat transported by heat pipes.

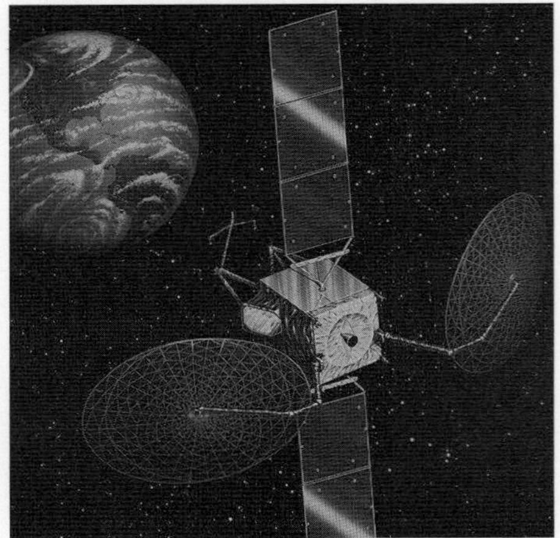

AMSC provides North American mobile services
(Hughes Space & Communications) 0517210

NATO series

Background

Before dedicated international geosynchronous satellites appeared, NATO paid the US government about USD60,000 annually to utilise IDSCS. However, beginning in the late 1960s, NATO began to assemble its own satellite communication network, generally referred to as the second-generation system. Philco-Ford built the NATO 2A and 2B, 243 kg GEO communications satellites, which operate about 5,950 km apart at 18° west and 26° west. These two initial satellites covered the northern hemisphere from Ankara (Turkey) to Virginia in the US, providing 57 voice and 100 telegraph point-to-point circuits through a dozen 12.8 m diameter fixed Earth stations. NATO retired 2A in May 1972 and 2B June 1976, after a total operational expenditure of US$15 million. NATO Communications and Information Systems Agency (NACISA) in Brussels manages NATO communication traffic and has had this responsibility since the inception of the second-generation system.

In February 1973, Ford Aerospace received a USD27.7 million contract to build three NATO 3 satellites and continue the programme into its third generation. Ford's new satellite weighed 720 kg and had a seven-year life. Each could handle hundreds of voice, telegraph, facsimile and wideband datalinks simultaneously on its three X-band channels. NATO manoeuvred NATO 3A into position, following its launch on 28 April 1976 by Delta from Canaveral, and stationed it between Africa and South America. The satellite retired December 1982. NATO 3B followed shortly after the NATO 3A launch on 28 January 1977 where it took up position off the US West Coast above the east Pacific. This satellite also filled in for the traffic that began overwhelming the US DSCS system after 1979. It retired from service in October 1986 and was boosted above GEO from 60° west in July 1993. NATO 3C joined 3A on 19 November 1978 in the slot midway between Africa and South America as an in-orbit spare.

Because NATO 3B continued to perform flawlessly beyond its expected lifetime, NATO 3C remained in orbital storage for seven and a half years, the longest time on record. In October 1986, 3C emerged from storage to begin communication on command from USAF's Sunnyvale facility, relayed through a New Hampshire link. It remained in service until retirement in October 1991, ending its life at 21° west. A small boost command lifted it above GEO in mid-1992. NATO 3D, launched 13 November 1984 by Delta from Canaveral completed the series. 3D cost USD80 million, including USD40 million for the satellite. It was placed initially at 21° west inclined before moving to its operational position at 18.5° west.

At the end of the NATO 3D lifetime, NATO began considering a more advanced, follow-on satellite, but deferred the plans because it normally needs only one satellite, and was owed two years' service from the US DSCS system in return for its use of NATO satellites. NATO 3D moved from 30° west to 21° west in mid-1993, where it remains as a back-up. At the beginning of 1996, the satellite had drifted into an inclination of 3.70°. NATO 3 operational control transferred on 11 July 1991 from the USAF CSTC Consolidated Space Test Center in Sunnyvale to the 3rd Satellite Control Squadron at the CSOC Consolidated Space Operations Center in Colorado, now referred to as the 3rd Space Operations Squadron of Space Command's 50th Space Wing. The control transferred again in 1994 to 5th SOPS.

NATO ordered two NATO 4 satellites, almost identical to the UK's Skynet 4 in February 1987 at a total cost of about £150 million. The total programme will ultimately cost NATO US$364 million at 1987 rates. 4A entered orbit from a Delta 7925 from Canaveral on 7 January 1991 for positioning at 18° west, where it remains as the principal satellite. Like Skynet, it has three 40 W SHF transponders to provide four channels of 60–135 MHz traffic and two UHF transponders which each carry a single 25 kHz channel on a full-Earth helix antenna. The payload includes signal processing and anti-jamming features that are normally associated with electronic intelligence satellites. 4B departed 8 December 1993, also aboard a Delta 7925 and NATO located it at 5° east, where it served as a hot standby for 4A until it was moved to 20° west. The United Kingdom manages the NATO 4 programme on behalf of NACISA.

British Aerospace and Marconi Space Systems act as prime and payload contractors respectively, while the operational control emanates from No 10001 Signals Unit at RAF Oakhanger in the UK. Since the Skynet 4 programme also originates at that facility, speculation is rife that the Zircon capability also exists on the NATO 4 satellites.

The NATO IV satellite employs the same basic configuration as the Skynet 4 and the NATO IV satellites for military communications (Matra-Marconi Space/BAe) 0019684

The Rascom-QAF1 communications satellite being lifted into position at Kourou prior to launch on 21 December 2007 (CNES/Arianespace) 1340829

Rascom

Current Status

Rascom-QAF1 was launched on 21 December 2007. A leak in the helium-pressurisation system forced a shutdown of its principal apogee motor while it was still in geostationary transfer orbit, and the satellite was placed in safe mode while an assessment of the problem took place.

Background

The Regional African Satellite Communications Organisation (Rascom) has been planning to launch its own communications satellite for a number of years, but funding proved to be a major obstacle. Rascom-QAF 1 was eventually built byThales Alenia Space as part of a turnkey business agreement with RascomStar-QAF of Abidjan, Ivory Coast. 26 per cent of the finance was provided by Rascom, 29 per cent by Libya's GPTC telecommincations service, 33 per cent by the Libyan African Investment Portfolio Bank and 12 per cent by Thales Alenia. In June 2007, Ariane 5 was selected as the launch vehicle instead of Long March, following a USD72 million loan from the African Development Bank and investment by Libya, which is thought to have preferred a European launch.

Rascom-QAF1, the first indigenous, pan-African fixed satellite service spacecraft, was launched on 21 June 2007. Thales Alenia Space was responsible for building the satellite and delivering it into orbit, along with the ground control station in Gharyan, Libya. Based on the Alenia Spacebus 4000 B3 platform, the satellite is fitted with 12 Ku-band and eight C-band transponders. Positioned at 2.85° E it is intended to serve the entire African continent, with telecommunications services to rural areas filling 65-70 per cent of capacity, as well as inter-city and international phone links, plus direct TV and Internet access service.

On 29 December 2007, Thales Alenia Space announced that Rascom-QAF1's Launch Early Operation Procedures had been stopped due to a leakage in the helium subsystem, which is used to pressurise the satellite's propulsion system. The leak occurred while it was still in geostationary transfer orbit. The satellite was immediately put in a stable configuration. The move to GEO was expected to be completed using back-up apogee motors, but, even if this is achieved, there are fears that the problem could shorten the operational life from 15 to 10 years. The satellite was insured for USD365 million.

A contract for a second Rascom satellite was expected to be signed in late 2008, once the Rascom-QAF1 was operartional.

Specifications

Rascom-QAF1

Launch: 21 December 2007 by Ariane 5 GS from Kourou, French Guiana.
Location: 2.85° E
Design life: 15 years.
Contractors: Thales Alenia Space (prime).
Transponders: 12 Ku-band and eight C-band. Ku-band transmissions, 10 active transponders. Two zones over Africa (Zone A and Zone B). C-C-band transmissions, 6 active transponders. One single zone over Africa (CONT zone). C-Ku and Ku-C-band transmissions, 4 active transponders. A capacity of 23 equivalent channels, each with 36 MHz bandwidth.
Principal applications: Fixed telecommunications services for the African continent.
Configuration: Alenia Spacebus 4000 B3 platform. 3 axis stabilised. Body dimensions 3.75 × 2.36 × 1.80 m. Span in orbit 31.8 m.
Mass: 3,160 kg at launch. 1,395 kg dry.
TT&C: Ground tracking, telemetry and control station in Gharyan, Libya. A second to be added in Abidjan, Ivory Coast.
Power system: Two solar arrays, 6.4 kW power EOL.

Sirius/Tele-X series

Current Status

Sirius 4 was launched on 17 November 2007 (GMT) by a Proton-M/Breeze M. Sirius 2 and 3 were also operational in November 2007. SES Sirius

For details of the latest updates to *Jane's Space Systems and Industry* online and to discover the additional information available exclusively to online subscribers please visit
jsd.janes.com

Artist's impression of Sirius 4, the largest and most recent of the SES Sirius fleet of communications satellites (SES Sirius) 1340418

communications satellites should not be confused with the Sirius digital radio satellites owned by Sirius Satellite Radio.

Background

SES Sirius AB owns and operates the Sirius system of communications satellites, which now comprises Sirius 2, 3 and 4. The company provides TV and radio broadcasts and broadband services in the Nordic countries, the Baltic States, Central and Eastern Europe. 75 per cent of SES Sirius AB is owned by SES Astra and 25 per cent by the state-owned Swedish Space Corporation (SSC).

The Nordic countries of Denmark, Finland, Iceland, Norway and Sweden began studying the feasibility of a joint regional satellite system in the mid-1970s. From the beginning, these countries sought television direct broadcasting via satellite for their far northern regions. In 1977, the World Administrative Radio Conference on broadcast satellite services (WARC) allocated eight channels to the Nordic group as an entity (excluding Iceland), plus another 17 to the individual countries (including five for Iceland). What became the Tele-X project assumed its current form in 1982, when the Nordic countries sought to provide both direct broadcast satellite and business communications services within Scandinavia, and also as a means of stimulating the space engineering capabilities of these countries. The 'X' represented Experimental, to acknowledge the developmental nature of this indigenous satellite.

Launched in 1998, Sirius 3 was co-located with Sirius 2, providing Ku-band services to the Nordic countries 0132813

By then, Denmark and Iceland had withdrawn from the project, though Denmark had earlier expressed interest in a beam covering south east Greenland. Sweden and Norway agreed to continue in a formal agreement signed in 1983, giving them 85 per cent and 15 per cent interest, respectively, in NSAB (Nordiska Satellite AB). 3 per cent of Sweden's share of cost/work in the project (but not in NSAB) was allocated to Finland later in a separate bilateral agreement.

Taking advantage of the Franco-German TDF/TV-Sat programme, initiated in 1980, the SSC ordered a similar satellite based on the Spacebus 300 platform. Nordiska Satellite AB (NSAB) planned to be proprietor of the system, while SSC planned to act as the procurement agent and executive agency. Shortly before launch, Norway withdrew from NSAB, trading its 15 per cent share for a lifetime lease of one channel. NSAB then formed a 50/50 partnership between SSC and Teracom, a major distributor of television/ radio programming. SSC became wholly responsible for operating the system.

The Sirius system eventually cost Sweden SEK1.25 billion at fixed 1982 rates, or USD280 million. With all three direct broadcast satellite television channels occupied, the Swedes sought other capacity. NSAB bought the UK's **Marcopolo 1** DBS satellite from National Transcommunications Ltd in 1993 for a reported GBP2.1 million to add five channels. NTL had acquired the satellite from BSkyB. It was renamed Sirius and moved from 3.1° W in January 1994 to begin services at the end of February from 5° E. At the beginning of 2000, it was relocated at 13°W and renamed **Sirius W**. It was taken out of service and sent to a graveyard orbit in June 2003.

On 4 July 1995, SSC signed a contract with Aerospatiale for a single **Sirius 2** satellite to replace Tele-X. Sirius 2 was initially operated on behalf of GE Americom, when it was re-designated GE-1E. At the time of its launch in November 1997, Sirius 2 was the largest telecom satellite ever built in Europe. The 32-transponder satellite transmits both analogue and digital signals and is used for direct-to-home TV transmissions as well as video and data communication services. It has two DTH beams, each with 13 transponders used for transmission of TV channels to homes equipped with parabolic antennas. One of the beams transmits to the Nordic area and the other to central and southern Europe. The Nordic beam has a signal power (EIRP) of 55 dBW in the primary reception area, exceeding 53 dBW for all Nordic and Baltic countries. The receiving antennas need to be no larger than 0.6 m. Although the European beam's footprint is larger than that of the Nordic beam, the signal power is still at least 50 dBW. Sirius 2 has a third beam for video and data communications. It consists of six 36 MHz transponders and covers northern and central Europe with a signal power of 46–48 dBW. The uplink band is 14.0–14.25 GHz and has high sensitivity, more than 9 dB/K in the central coverage area.

The Sirius 2 Aerospatiale contract included a duplicate payload for Sirius 3, but SSC subsequently changed to Hughes Space & Communications (now Boeing Satellite Systems). **Sirius 3**, a cylindrical HS 376 model, was placed under contract in May 1997 and launched in October 1998. Its signal power for the Nordic region is approximately the same as for Sirius 2. It has 15 active transponders and broadcasts TV signals in digital format and digital services based upon IP technology.

Sirius 4, designed and built by Lockheed Martin Commercial Space Systems for SES Sirius, was launched in November 2007 by a Proton-M/Breeze M. After the Proton-M reached a 942 × 164 km, 51.5° suborbital trajectory, there were three burns by the Breeze-M upper stage to place it in a 6,916 × 35,478 km, 17.4° geostationary transfer orbit. The move to its final slot in GEO was carried out by the satellite's Leros propulsion system. The largest spacecraft operated by SES Sirius, it will be co-located with Sirius 2 and 3. Based on the A2100AX platform, it will provide direct-to-home broadcast and interactive services across Scandinavia, Europe and Africa. In addition to 46 high-power, Ku-band transponders in the BSS and FSS frequency bands, it carries an active Ka-band transponder for interactive applications in the Nordic and Baltic regions. A beam to sub-Saharan Africa complements existing coverage of Africa, using 6 additional active Ku-band FSS (36 MHz) transponders. There is also an interconnect payload that links Africa and Europe with a second Ka-band transponder.

SES Sirius' main satellite ground station in Sweden is located at Kaknäs, Stockholm. The company also owns and operates a satellite station in Riga, Latvia, for services involving Sirius 3. The station is used for transmission of digital TV and radio and consists of rf-equipment, including a 3.7 m antenna, digital encoders, multiplexers and MPEG-2/DVB-S specific modulators. The uplink services from Riga station are monitored through Kaknäs. There is also a ground station in Kiev, Ukraine, operated by a Ukrainian company, providing digital TV services related to the Sirius 2 satellite.

Specifications

Tele-X

Launched: 17 November 2007 by Proton/Breeze M launch vehicle by Ariane V30 from Baikonur.

Location: 5° E GSO 1989–1998, 138.62° E September 2001.

Design life: 6 to 8 years.

Contractors: Aerospatiale/Eurosatellite were joint prime contractors. Aerospatiale also handled satellite AIT, thermal control, transmit reflectors/ antenna subsystems, and built the solar arrays. Scandinavian industry received 30 per cent of the programme contract value: Saab was associate prime, and was involved in structure, TT&C, antenna integration and OBC work. Ericsson was responsible for the communications payload, receive antennas & beam-tracking receiver. Other Eurosatellite members involved were AEG (solar cells, TWTs, power subsystems), ANT (TWTs), Alcatel Espace (communication module

Tele-X was based on the Spacebus 300 platform (Aerospatiale) 0517013

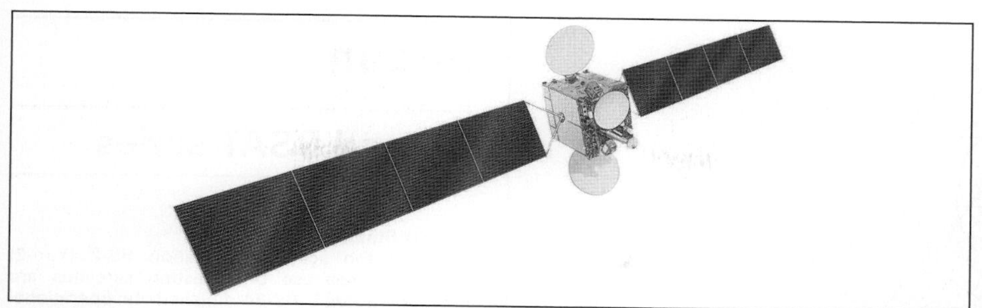

Artist's impression of the Sirius 2 communications satellite, operated by SES Sirius and formerly known as GE-1E (SES Sirius) 1340417

Amos 2 undergoing prelaunch testing (IAI) 1293442

integration, linear amplifiers, other components), ETCA (power system components) & MBB (AOCS). Ground segment contracts went to Elektrisk Bureau (Norway) and Nokia Research Centre, Teleste, Valmet, VTT Technology (Finland).
Transponders: Two (plus one back-up) 220 W (230 to 260 W planned) TWTA 17.7–18.1/11.7715–12.489 GHz up/down Ku-band Nordic TV DBS beams, 27 MHz bandwidth, EIRP 61 dBW max, LH circular polarised.

Two (plus one back-up) 220 W (230-260 W planned) TWTA 14–14.25/12.5–12.75 GHz up/down Ku-band Nordic data/video services, 40 & 86 MHz bandwidths, data rates of 0.64/2/34 Mbit/s, EIRP 61 dBW max at beam edge.
Configuration: Aerospatiale/MBB Spacebus 300 3-axis satellite with 2.4 × 1.65 × 2.4 m box-shaped bus. Unified liquid bipropellant propulsion system for GEO insertion + AOCS, in conjunction with MWs, IR Earth sensors. RF antenna pointing system. The same deployable receive/transmission antennas are used for each payload, with 0.165°/0.075° pointing accuracies, respectively. Body pointing accuracy is ±0.3°.
Mass: 2,130 kg at launch, 1,277 kg BOL on-station.
TT&C: Ku-band operational (S-band during transfer phase), via SSC Earth station at Kiruna, Sweden.
Power: 3,200 W EOL, generated by twin 4-panel solar wings with 19 m span. One 18 Ah Ni/Cd battery.

Sirius 1 (Marcopolo 1 or BSB-1)
Launched: 27 August 1989 by Delta 4925 from Cape Canaveral pad 17B.
Location: 5.2° E geostationary (1994 – April 2000). 13°W (May 2000 – June 2003).
Design life: 10 years. Taken out of service in June 2003.
Contractor: Hughes Space & Communications.
Transponders: Six paired 55 W TWTAs 17.385–17.992/11.7715–12.1054 GHz up/down Ku-band beam (BSS channels 4, 8, 12, 16, 20), 27 MHz bandwidth, 59 dBW min EIRP (63 dBW max), RHCP.
Principal applications: TV direct to home
Configuration: Hughes HS-376 spin-stabilised, 7.2 m high deployed (2.7 m stowed), 2.16 m diameter. Despun single reflector and communications shelf.
Mass: 1,250 kg at launch, about 660 kg on-station BOL.
Pointing: 0.05° accuracy by RF beacon.
Attitude control: Spin stabilised at 50 rpm by four 22.2 N hydrazine thrusters; GEO insertion provided by solid Thiokol Star 30B AKM.
Power system: K7 solar cells mounted on cylindrical exterior and deployed drum providing 1,024 W BOL (drum extended 25 cm over normal Hughes 376 configuration). Two Ni/Cd batteries provide 100 per cent eclipse protection.

Sirius 2 (GE-1E)
Launch: 12 November 1997 by Ariane 44L from Kourou, French Guiana.
Location: 5° E geostationary 1997–1999, 4.81° E September 2001.
Design life: >12 years (15 years propellant).
Contractors: Aerospatiale (prime), Space Systems/Loral (repeater), DASA (AOCS, solar arrays), Saab Ericsson Space (antennas, data handling), Alenia (TCR-RF).
Transponders: 26 BSS, 6 FSS. Frequency band: 11.7–12.5 GHz (BSS), 12.5–12.75 GHz (FSS). Transponder bandwidth: 33 MHz (BSS), 36 MHz (FSS).

13 57W TWTA 17.30–18.10/11.7–12.5 GHz up/down Ku-band Scandinavian BSS beams, 33 MHz bandwidth, EIRP 55.5 dBW edge, linear vertical polarised.

13 85W TWTA 17.30–18.10/11.70–12.50 GHz up/down Ku-band European BSS steerable beams, 33 MHz bandwidth, EIRP 49.7 dBW edge, linear horizontal polarised.

Six 85W TWTA 14.00–14.25/12.50–12.75 GHz up/down Ku-band Northern/Central European FSS beams, 36 MHz bandwidth, linear polarised.

Principal applications: DTH, TV distribution, and data services.
Configuration: Spacebus 3000B 3-axis with 1.8 × 2.3 × 2.86 m box-shaped bus. Two 1.8 m fixed dual gridded antennas for FSS and Scandinavian BSS coverage, plus steerable antenna for BSS European beam.
Mass: 2,920 kg at launch; 1,760 kg BOL; 1,250 kg dry.
Pointing: Standard body pointing accuracy 0.1–0.15° following in-orbit calibration. Fine pointing by RF sensors and antenna pointing mechanisms (0.05°).
AOCS: Unified bipropellant (NTO/MMH) system of single 400 N and redundant networks of 7 × 10 N thrusters provide GEO injection + 3-axis control. Two titanium propellant tanks, titanium/Kevlar pressurisation tanks. Two 2-axis IRES IR Earth Sensors, two Sun sensors (4 heads each, 60° FOV), two RIGA gyro packages, two 52–68 Nms fixed momentum pitch wheels controlled by digital programmable control electronics. Normal stabilisation is based on pitch by wheel, roll/yaw by Earth sensor using roll/yaw thrusters. Station-keeping by 10 N thrusters; yaw reference by Sun sensors or gyros.
TT&C: Via SSC Earth station at Kiruna, Sweden.
Power system: 5,845 W EOL from twin 4-panel wings spanning 26.3 m. Two 62 Ah Ni/H_2 batteries provide 100% eclipse protection.

Sirius 3
Launch: 5 October 1998 by Ariane 44L from Kourou, French Guiana.
Location: 28° E geostationary 1998–1999, 5° E September 2001.
Design life: 12 years.
Contractor: Hughes Space & Communications HS-376.
Transponders: 14 Ku-band 11.7–12.5 GHz.
Configuration: Body diameter 2.16 m, height extended from 3.32 m at launch to 7.76 m in orbit.
Mass: 1,465 kg at launch; 815 kg BOL; 630 kg dry.
Power: 1,400 watt.

Sirius 4
Launch: 22:39 GMT on 17 November 2007 by Proton-M/Breeze-M from Baikonur.
Location: 5° E geostationary.
Design life: >15 years.
Contractor: Lockheed Martin Commercial Space Systems.
Transponders: 53 active Ku-band transponders, including 46 Ku-band high-power transponders in the BSS (33 MHz) and FSS (36 MHz and 72 MHz) frequency bands. 2 active Ka-band transponders: one 250 MHz Interconnect and one 250 MHz Interactive transponder.
Configuration: Lockheed Martin A2100AX platform.
Mass: 4,385 kg at launch.
Power: 11.3 kW (EOL). Payload power: 8.1 kW.

Israel

Amos series

Current Status
Amos 1 and Amos 2 were both operational as of March 2008.

Background
Israel Aircraft Industries (IAI) (now Israel Aerospace Industries) began the Amos (Affordable Modular Optimised Satellite) programme in 1990, with the full go-ahead formally approved in January 1992, to develop, build, operate and market a medium-class GEO satellite, principally for sub-regional TV broadcasting. IAI retained ownership of Amos 1 and created a commercial entity named Space Communication Ltd. (Spacecom) in 1993 to sell capacity. Spacecom is in equal partnership with IAI, Gilat Communication Engineering, General Satellite Services Co. and Mer Services Group Ltd. Industry sources projected the cost through launch, plus operating costs over 10 years at USD250 million. Israeli banks provided USD100 million, guaranteed by the government and the government supported the programme at USD15 million per year.

The Israeli government initiated the project by guaranteeing it would use three transponders over the satellite's 10 year life, thereby replacing the three previously leased from Intelsat. Transponder cost was about USD3 million annually. Israel hoped that Arab users would be signed up before launch, since the majority of the 250 million people within the 700,000 km² Middle East beam are Arabic speakers. Amos 1 still feeds DBS service providers, cable TV companies and TV broadcasters, and HBO uses Amos 1 to deliver programmes to the Czech Republic, Hungary, Poland, Romania and Slovakia. It is also the "YES" digital DBS service provider to Israel.

IAI launched the larger Amos 2/CERES (Central European REgional Satellite) on 28 December 2003 aboard the Russian Soyuz launcher from Baikonur. It was the first Soyuz launch to GTO. The launch was originally planned for an Ariane 5 in late 2002, but IAI was late in completing the satellite due to component delivery problems and a decision to increase the power of the satellite's transponders. Spacecom invested some USD130 million in Amos 2, including USD70 million to build the satellite, USD35 million for the launch and USD25 million for insurance.

Amos 2 is based on the Amos 1 platform, but has some 50 per cent more capacity and transmission power than Amos 1. Co-located with Amos 1, the second satellite carries 11 active Ku-band, 72 MHz transponders (22 active 36 MHz segments). The satellite expanded the available bandwidth and coverage, offering a combined Hot Spot for European and Middle East customers. Amos 2 also has a third beam bridging the Atlantic from the US east coast to its other service areas, creating an important link for VSAT operators, government agencies and such. Services via this satellite include: direct distribution of TV and radio transmissions, TV and radio transmissions to communication centres, distribution of internet services and data transmissions to communication networks. At the time of launch, Spacecom had already signed contracts for about 70 per cent of the satellite's broadcast capacity. Early customers included the Israeli government, the Yes satellite television company, the Israel Broadcasting Authority, Gilat Satellite Networks, Germany's RTL television station and the American cable television station

Amos 2 serves clients in the Middle East, Europe and eastern coast of the US (IAI) 1166910

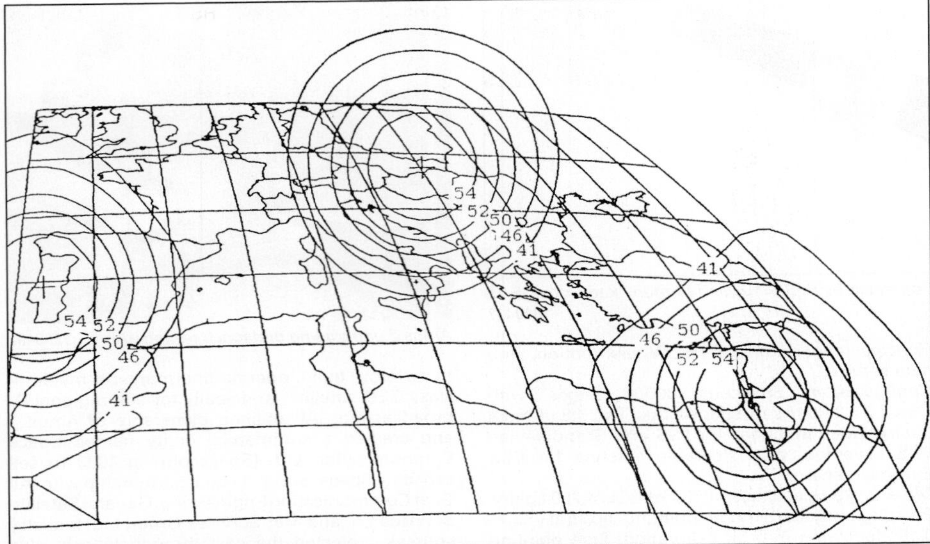

Amos 2 EIRP (dBW) footprints from 4°W. Boresight EIRP is 57 dBW (IAI/MBT) 0517196

HBO, which planned to broadcast to six countries in Europe via Amos 2.

The launch of Amos 3 is planned to coincide with Israel's 60th anniversary celebrations in 2008. The satellite will be decorated with the country's official anniversary logo and the Israeli flag, and renamed Amos-60 after launch. Amos 3 is currently scheduled for launch in late April 2008 aboard the inaugural Land Launch rocket — a Zenit-3SLB vehicle operated from Russia's Baikonur Cosmodrome in Kazakhstan. It will replace Amos-1, which will be retired. Beam coverage will include the east coast of the US, the Middle East and Europe. In November 2006, Spacecom announced that it had already sold more than 50 per cent of total capacity on Amos 3. Some existing and new customers on Amos 1 and Amos 2, including government users and two DTH platforms, YES and BOOM, will transition some of their capacity to Amos 3 once it becomes operational.

On 8 July 2007, IAI signed a contract to develop, manufacture and launch the Amos 4 communication satellite for Spacecom, a company in which IAI holds approximately 20.5 per cent of issued capital, with the Government of Israel. Scheduled for delivery in 2012, Amos 4 is to be operated from an undisclosed geostationary position between 64°E and 76°E for better coverage of Central and South Asia. The Israeli government is providing USD265 million to pay for the construction and launch. Spacecom is responsible for the remaining USD100 million but is not required to start making payments until 2010. The government will use most of the satellite's capacity and it is unclear how much capacity will be available for commercial sale.

Amos 4 will be larger than its predecessors, with a launch weight of approximately 3,400 kg. It will carry a large number of high power Ka- and Ku-band transponders. The satellite's lifetime is planned to be approximately 12 years.

The completed Amos 1 at IAI/MBT 0517395

Specifications

Amos 1

Launch: 16 May 1996 by Ariane 44L from Kourou, French Guiana

Location: 4°W (Israel also holds registrations at 39°E/1.5°E)

Design life: 12 years

Contractors: IAI (MBT; prime, systems engineering, mission life operations, structure, AOCS, TT&C, data handling, thermal), Alcatel (Ku-band payload), DASA/Dornier (power), DASA/MBB (thrusters), ANT (transponders), CDI (OBCs), Gates (Ni/H_2 battery, integrated by SS/L), Teldix (momentum wheels)

Transponders: 7 (plus 2 back-up in 7-for-2 ring) 36 W TWTA 14.0045–14.4955/10.9545–11.1935 and 11.4565–11.6955 GHz Ku-band up/down in two of three beams using single reflector: Israel beam can take 3–6 transponders, EIRP 55 dBW max, 14.0 dB/K max, −99 to −74 dBW/m²; European beam (centred on Hungary) of 1–4 transponders, one switchable to Portugal. 72 MHz bandwidth, orthogonal polarisation

Applications: Projected 70–80% TV distribution; DTH to 50–60 cm dishes; VSAT video, voice and data (including interactive learning) HDTV

Configuration: 1.21 × 1.670 × 1.930 m bus, solar array span 10.550 m. Single 1.73 m diameter deployed Ku-band antenna on west face. 3-axis stabilisation by MWs (no gyros); wheel unloading/station-keeping by redundant sets of 7 × 10 N thrusters. 400 N thruster accomplishes GEO insertion. Control by two 1,750A OBCs. Internal structure + East/West faces are composite; North/South are aluminium honeycomb. Ku-band horn for TT&C

Mass: 961 kg at launch, 479 kg dry, 580 kg on-station BOL

TT&C: From MBT's Yehud complex, Ku-band via two 9 m antennas. Same control room as Ofeq programme

Power system: 1.23 kW EOL from twin Si solar wings supported by 25-cell 50 Ah Ni/H_2 battery.

Amos 2

Launch: 27 December 2003 by Soyuz-Fregat from Baikonur, Kazakhstan

Location: 4° W

Design life: 12 years

Contractors: IAI (MBT; prime, systems engineering, mission life operations, structure, AOCS, TT&C, data handling, thermal).

Transponders: 11 active Ku-band, 3 back-up, bandwidth 72 MHz, power 75 W. Three beams: Middle East (ME), Europe (EU) and US. Ku-band *uplink:* 13.00 to 14.00 GHz. Ku-band downlink: 10.70 to 10.95 GHz and 11.45 to 11.70 GHz. EIRP at beam centre: ME – 57 dBW; EU – 57 dBW; US – 53 dBW. G/T at beam centre: ME – 15 dB/k; EU – 14.5 dB/k; US – 9 dB/k. Polarisation: linear.

Applications: DTH digital TV, internet, voice and data communications.

Configuration: 2.7 × 2.06 × 2.38 m bus, solar array span 11.03 m. 3-axis stabilisation. 400 N Liquid Apogee Boost Motor (ABM) and fourteen 10 N Reaction Control Thrusters for raising orbit from GTO to GEO and for attitude control.

Mass: 1,370 kg at launch. 646 kg dry.

TT&C: From MBT's Yehud complex, Ku-band via two 9 m antennas. Same control room as Ofeq programme

Power system: 1.9 kW power EOL from twin Si solar wings supported by 24 Ah Ni-Cd batteries.

Japan

BS and BSAT series

Current Status

None of the second generation BS-2 (Yuri-2) series of Japanese broadcasting satellites are now operational. BS-2A finished its operational programme in April 1989. BS-2B ended operations on 24 October 1991. BS-3A finished operation on 20 April 1998 and BS-3B ended operation on 1 December 1998. The satellites' remaining fuel was used to boost their orbits above the geostationary orbit.

BSAT-1A replaced BS-3A as prime in August 1997. BSAT-1B was launched in April 1998 and began commercial operations in August 1998. BSAT-2A was launched on 26 April 2001 followed by BSAT-2B on 12 July 2001. A launch vehicle failure prevented Ariane 5 placing it in the correct orbit and a contract for a replacement, BSAT-2C, was awarded to Orbital Sciences Corporation in October. BSAT-2C was launched on 11 June 2003, followed by BSAT-3A on 14 August 2007. The satellites are owned and operated by the Tokyo-based Broadcasting Satellite System Corporation (B-SAT), which was established in April 1993. They all operate from the 110°E orbital slot.

Background

Until 1989, Japan owned and operated all of its communications satellites through the Telecommunication Advancement Organisation of Japan (TAO). Organisationally the National Space Development Agency (NASDA) procured and launched (or obtained launch services for) the BS and CS satellites and then handed them over to TAO for operational use.

When it launched the second generation Broadcasting Satellites (BS), also known as Yuri, Japan became the first country to provide direct broadcast services to home receivers. Japan investigated the concept for two years with BS-1, launched 8 April 1978 by Delta from Cape Canaveral. The total programme cost the Japanese government 28.8 billion yen. Six years later, Japan launched the first operational satellite, BS-2A, on 23 January 1984 by N-2 from Tanegashima. Two of its three 100 W transponders failed within 3 months and Japanese DBS service then relied on the sole BS-2B. The Japanese removed the BS-2A from GEO in April 1989.

BS-2B brought the system to full capacity in 1986, but NHK purchased a 3000-class GE Astro Space satellite carrying three 200 W transponders, to maintain the system until the BS-3 series became operational in 1991. The total system cost the Japanese USD24.48 million for the satellite, about USD49 million for the launch and insurance.

BS-2X was lost in the Ariane 4 failure of 22 February 1990 and NHK initially indicated they

BSAT-1A began services, including high-definition TV broadcasting, in 1997 (Hughes Space & Communications International Inc) 0517209

BS-3N provided three active 120 W DBS transponders (Martin Marietta Astro Space) 0517195

would not seek a replacement craft because of BS-3's imminent appearance. Nonetheless, after a discussion with GE Astro Space, NHK purchased a second satellite for launch in January 1991. NHK had awarded GE the USD88 million contract (mostly paid for by the BS-2X insurance) to deliver the replacement BS-3H satellite in orbit. The replacement BS-3H was then itself lost in its April 1991 launch. BS-3A, launched 28 August 1990, replaced BS-2B but it operates on marginal power. 3B completed the system in 1991. Again faced with growing demand NHK procured BS-3N in late 1992 (launched July 1994) to guarantee services.

Japan carried on the tradition it pioneered with BS-2's DBS services, on the current generation BS-3 Yuri series. NHK announced in late 1991 that it intended to procure BS-3N for launch spring 1994. The proportion of indigenous systems and components on the BS-3s has reached 83 per cent. The programme cost Japan about JPY78.8 billion. As it prepared to launch the first in the series NHK reportedly self insured BS-3A. In-orbit testing revealed that one of the four solar array power segments failed, leaving the satellite with only marginal power supply.

After the creation of the Tokyo-based Broadcasting Satellite System Corporation (B-SAT) in April 1993, the company decided to purchase two Hughes HS-376 satellites (BSAT-1A and 1B) to replace the government's BS series. Commercial analogue operations with BSAT-1A began in July 1997. The next three satellites, BSAT-2A, 2B and 2C, were launched in 2001 and 2003 for digital broadcast services. They were based on the Orbital Sciences Star platform. A propulsion problem in the Ariane 5 upper stage stranded BSAT-2B in a much lower orbit than planned and since the satellite carried only a solid propellant apogee kick motor, it was unable to begin operations. The most recent addition, BSAT-3A, is based on the A2100A platform manufactured by Lockheed Martin Commercial Space Systems.

Specifications

BS 3A/Yuri 3A

Launch: 28 August 1990 by H1 from Tanegashima
Location: 110° E geostationary (arrived end-September 1990; AKM fired 30 August); collocated with BS-3b
Design life: 7 years (specified EOL reliability >0.8 bus, >0.9 broadcast transponder, >0.8 wideband transponder)
Contractors: NEC (prime, communications payload) with GE Astro Space (bus), Nissan (solid AKM)
Transponders: 3 (plus 3 back-up) 120 W TWTA 14.069–14.219/11.766–11.919 GHz up/down Ku-band domestic DBS beams, 27 MHz bandwidth, EIRP 55 dBW for mainland, circular polarisation. Channel allocations (up/down GHz, centre frequencies) number 3: 14.06584/11.76584; number 7 14.14256/11.84256; number 11 14.21928/11.91928

1 (no back-up) 20 W TWTA 14.37/12.64 GHz up/down Ku-band for 60 MHz wideband relay
Principal applications: DBS TV services: two channels (number 7 + 11) NHK (3rd channel, number 3, not used because of power limitation)
Configuration: GE 3000 series, 3-axis stabilised box-shaped 1.3 × 1.6 × 1.6 m bus with twin 3-panel 1.5 × 5 m solar wings and single 11 kg 80 × 170 cm offset parabolic elliptical reflector. In-orbit span 15 m, height 3.2 m
Mass: 1,115 kg at launch, 550 kg on-station BOL
AOCS: bias momentum 3-axis, antenna beam pointing accuracy ±0.1°, ±0.1° E/West + North/South. 4 × EHT + 12 × catalytic REA thrusters 0.39–0.91 N drawing on four hydrazine (170 kg total) surface tension tanks. Pitch control using MWs, roll/yaw using magnetorquers, active nutation damping. Solid AKM
TT&C: through TAO Kimitsu Satellite Control Centre; NHK programming uplinked through 8 m Tokyo antenna
Power system: 1,443/1,093 W BOL/EOL to be provided by twin solar wings (11,515 high voltage + 4,183 low voltage N-on-P 2 × 4 cm BSR) on North/South walls, but one of the four supply segments is not contributing. Supported by 2 × 17 Ah Ni/Cd batteries.

BS 3B/Yuri 3B

Launch: 25 August 1991 by H1 from Tanegashima
Location: 110° E GEO (collocated with BS-3a)
Design life: 7 years, as BS-3a, but insertion precision left 9 years propellant
Transponders: as BS-3a but using channels 5/9/15 (up/down GHz, centre frequencies): 14.1042/11.8042; 14.18092/11.88092; 14.2960/11.0660
Applications: one channel is providing JSB's DBS; one has been demonstrating HDTV 8 h daily since 25 November 1991 to some 400 sets; SDAB digital audio broadcasting is also a user

Other specifications as for BS-3a.

BS-3N/Yuri 3N

Launch: 8 July 1994 by Ariane 4 from Kourou, French Guiana
Location: 109.85° E geostationary, as 3a/b back-up
Transponders: 120 W TWTA, as BS-3a
Applications: as BS-3a/3b
Mass: 1,210 kg at launch, 699 kg on-station BOL, 575 kg dry
Configuration: as BS-3a except 50 cm circular antenna added for 17 GHz uplink
Power system: as BS-3a except Ni/H_2 batteries carried

Other specifications as for BS-3a.

BSAT-1A

Launched: 16 April 1997 by Ariane 4 from Kourou, French Guiana
Location: Initially over 121° E, but planned for 110° E (collocated with 1b)
Design life: 10 years minimum (12.5 years propellant)
Contractors: Hughes Space & Communications
Transponders: 4 (plus 4 back-up, 8-for-4) 106 W TWTAs 17.3–17.8/11.7–12.2 GHz up/down Ku-band domestic beams, 27 MHz bandwidth, >55 dBW EIRP mainland (as BS-3), RHCP
Principal applications: DTH, including high-definition TV
Configuration: Hughes' HS-376 spin-stabilised, 7.97 m high deployed (3.15 m stowed), 2.17 m diameter. Despun single reflector and communications shelf
Mass: 1,250 kg at launch, 720 kg on-station BOL
Pointing: <0.1° accuracy by RF beacon
AOCS: 50 rpm spin-stabilisation and station-keeping by four 22.2 N hydrazine thrusters. GEO insertion provided by solid Thiokol Star 30BP AKM
TT&C: by B-SAT
Power system: K4¾ solar cells mounted on cylindrical exterior and deployed drum providing 1,200 W BOL. Two Advanced Ni/Cd batteries provide 100 per cent eclipse protection.

BSAT-1B

Launch: 29 April 1998 by Ariane 44P from Kourou, French Guiana
Location: planned for 110° E (collocated with 1a)

Other specifications as for BSAT 1a.

BSAT-2A

Launch: 8 March 2001 by Ariane 5 from Kourou, French Guiana
Location: 110° E, geostationary
Design life: 11.5 years
Contractor: Orbital Sciences Corporation
Configuration: OSC Star™ platform, 3-axis stabilised box-shaped 3.8 × 2.8 m bus with twin solar arrays spanning 15.7 m
Mass: 1,317 kg at launch
Transponders: 4 of 8 × 130 W Ku-band conduction-cooled TWTAs. Ku-band 17.3–17.6 GHz/11.7–12.0 GHz receive/transmit
Propulsion: Solid propellant apogee motor, redundant liquid monopropellant systems for AOCS.
Application: DTH for Broadcasting Satellite System Corporation digital programming.

BSAT-2B

Launch: 12 July 2001 by Ariane 5 from Kourou, French Guiana
Location: Planned for 110° E, but a launch vehicle failure prevented the satellite reaching appropriate orbit.

BSAT-2C

Launch: 11 June 2003 by Ariane 5 from Kourou, French Guiana
Location:

BSAT-3A

Launch: 14 August 2007 by Ariane 5-ECA from Kourou, French Guiana
Location: 110°E, geostationary
Design life: 13 years
Contractor: Lockheed Martin Commercial Space Systems
Configuration: A1200A platform, 3-axis stabilised box-shaped 3.8 × 3.8 × 1.9 m bus with twin solar arrays spanning 14.65 m
Mass: 1,980 kg at launch. Dry mass 927 kg.
Transponders: 12 130-W Ku-band channels (eight operating at one time).
Propulsion: Atlantic Research Corporation's LEROS 1C Liquid Apogee Engine.
Application: DTH for Broadcasting Satellite System Corporation digital programming throughout Japan.

CS-3/JCSat series

Current Status

Successor to the ageing JCSat 2 launched on 1 January 1990, the JCSat-8 (JCSat 2A) satellite was carried to a GTO by Ariane 44L on 29 March 2002.

Background

Two Communications Satellite-3 (CS-3), also known as Sakura, superseded CS-2A/B and provide domestic Ka/C-band services, with both C-band area beams and seven of the 10 primary Ka-band spot beams dedicated to NTT operations. Japan estimates the programme cost 66.3 billion yen.

Japan created the JSB Japan Satellite Broadcasting Incorporated in 1984 by the Ministry of Posts & Telecommunications to commercially develop and launch communication satellites. The government ownership of JSB incited US pressure to open more of Japan's satellite purchases to foreign competition. Japan did relent and bids for the 'BS-4' generation resulted in the selection of Hughes Space & Communications as the prime contractor to build the satellites. Until then The Japan Broadcasting Corporation (NHK: Nippon Hoso

Kyokai) held a large stake in broadcast through BS-3 transponders.

Then in 1989, the Diet released all restrictions on commercial systems development and several projects resulted. Nippon Telegraph and Telephone provided communications services for the CS series through CS-3, but NTT turned to another manufacturer, Space Systems/Loral (SS/L), to build its own two NStars which it launched in 1995. Satellite Japan Corporation (Sajac) planned to launch its own Sajac Hughes HS-601 satellites in 1994 until financial problems forced it to merge with JCSat in 1993, creating Japan Satellite Systems (JSAT); owner and operator of the JCSat series of satellites. JCSat 3 began commercial services in November 1995 and JSAT ordered JCSat 4 in February 1996.

With no further Japanese government restriction on commercial venture, Broadcasting Satellite System Corporation (B-SAT) incorporated itself on 13 April 1993 as a consortium of NHK, WOWOW, five other broadcasters and eight banks to procure and operate the 'BS-4' generation of broadcasting satellites for launches which took place in 1997 and 1998.

About 65 per cent of the 64 transponders on the JCSat's satellites are now leased; and the company claims there are 55,000 subscribers for five television channels. In addition to network and cable television distribution, Japan Satellite Communications Network Corporation (JSNet) employs JCSat to provide VSAT business data services. Three transponders are used for the video services and two carry 12 stereo channels of high quality Digital Audio Broadcast programming. JCSat charges 300–650 million yen annually to lease a transponder.

Specifications

CS-3A/Sakura 3A

Launched: 19 February 1988 by H1 from Tanegashima Satellite operated over 132° E until late September 1996: it was then boosted to a retirement orbit

Design life: 7 years

Contractors: Mitsubishi Electric Corp (prime and antenna), NEC (transponders)

Transponders: 2 (plus 1 back-up) 4W min TWTA 5.925–6.425/3.7–4.2 GHz up/down C-band global beams, 180 MHz bandwidth, output power >=36.0 dBm, 25 dB min gain, circular polarisation

10 (plus 5 back-up) 4.2 W min TWTA 27.5–29.25/17.75–19.45 GHz up/down Ka-band spot beams, 100 MHz bandwidth, output power <=36.2 dBm, 33 dB min gain to main islands, 27 dB min to Okinawa islands, circular polarisation, plus 19.450 GHz beacon

Principal applications/users: NTT (7 Ka + 2 C; routine and emergency telephone, TV and datalinks for major and remote islands), National Police Agency (1 Ka), Construction Ministry (0.4 Ka), other users <0.4 Ka-band transponders

Configuration: 218.4 cm diameter, 242.9 cm high cylindrical bus, and overall height 356 cm including single despun antenna. Monocoque shell supporting two horizontal equipment platforms

Mass: 550 kg BOL on-station, 1,099 kg at launch

AOCS: spin-stabilised at 90 ± 9 rpm, 0.12° BOL spin axis adjustment, ±0.26° measuring accuracy of Earth width from GEO, four 20 N thrusters (2 radial, 2 axial; continuous mode or 90 ms pulses) drawing on 118.5 kg hydrazine in three spheres for attitude control and ±0.05° E/West-North/South station-keeping. GEO insertion by Thiokol Star 30B solid

Antenna pointing: <=0.2° half cone angle accuracy, ±0.03° stepping angle by despin assembly

TT&C: through TAO's Kimitsu Satellite Control Centre; 1–3 W DSN-compatible, near-isotropic S-band transponder up/down-converted to C-band; 125 bit/s 253-command command detector/decoder

Power system: 2-segment GaAs cylindrical solar array (36,621 P-on-N GaAs primary power; 396 N-on-P Si cells for battery charging), 833/750 W autumnal equinox/summer solstice EOL, two 35 Ah Ni/Cd batteries (55% discharge depth, 616 charge cycles in 7 years), main bus voltage 29.4 ±0.2 VDC

Thermal control: passive, augmented by heaters. Primary heat emitted through solar array and end shields.

CS-3B/Sakura 3B

Launched: 16 September 1988 by H1 from Tanegashima

Location: Initially over 136°E: during April 1997 relocated to 153–154°E

Other specifications as for CS-3a.

JCSat 1

Launched: 6 March 1989 by Ariane 44LP V29 from Kourou, French Guiana. It formally entered service 1 May 1989

Location: 150° E geostationary

Design life: 10 years (launch accuracy left sufficient propellant for 14 years operation)

Undergoing tests in the Hughes Space Environment Chamber, a JCSat Satellite has its extendable solar cell drum retracted (Hughes Space & Communications) 0054338

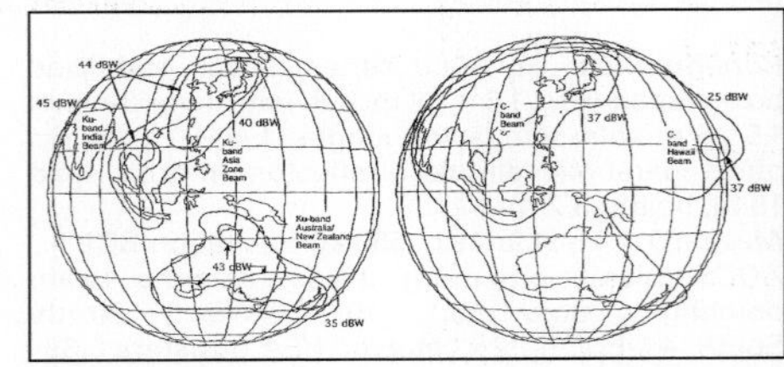

JCSat 3 Ku-band (left) and C-band EIRP footprints (JSAT) 0517217

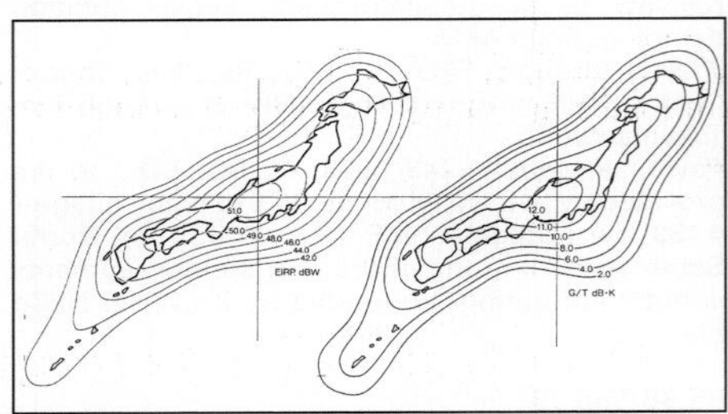

JCSat 1/2 EIRP (left) and G/T contours (JCSat) 0517003

Engineers check out the payload section of a Hughes JCSat (Hughes Space & Communications) 0054337

The hybrid JCSat 3 provides Ku/C-band services to the Pacific Rim and Asia (Hughes Space & Communications) 0517382

With nearly 55,740 sq metres of manufacturing space the Hughes Integrated Satellite Facility accommodates a 1,783 cu metre thermal vacuum chamber in which here seen is JCSat-6 (Hughes Space & Communications) 0054335

Transponders: 32 (plus 8 back-up, 5-for-4) 20 W TWTA 14.0–14.5/12.25–12.75 GHz up/down Ku-band domestic Japanese beams, 27 MHz bandwidth, EIRP 50 dBW min (51.8 dBW peak) over most densely populated regions, orthogonal linear polarisation
Principal applications: TV/cable distribution, telephone, business services, and satellite news gathering. Each transponder can carry one high-quality TV channel, a 45 Mbit/s data stream or 250 full duplex telephone circuits
Configuration: Hughes' HS-393 cylindrical platform, 3.66 m diameter, 3.6 m height stowed at launch, 10.0 m height deployed, payload and single 2.4 m diameter antenna, of two orthogonally offset grid reflectors, mounted on despun platform
Mass: 2,280 kg at launch, 1,376 kg on-station BOL, 1,097 kg dry
AOCS: station-keeping + 30 rpm spin-stabilisation maintained by six 22 N thrusters (two axial, four radial); GEO insertion and major manoeuvres performed by two Marquardt R-4D 489 N thrusters, both systems drawing on 1,200 kg NTO/MMH in four tanks each
Antenna pointing: 2-axis RF beacon tracking maintains <0.06° – accuracy pointing, backed up by Earth sensors
TT&C: though the company's Yokohama station with 11 m antenna for transfer orbit operations and two 5.5 m for on-station TT&C. A back-up facility is maintained in Gunma Prefecture
Power system: 2,240 W equinox BOL/1,808 W solstice EOL provided by Si solar array mounted on cylindrical body (K7 cells) and extension skirt (K4¾cells); 2 × 38.4 Ah Ni/H_2 batteries
Thermal control: quartz mirror radiator band encircling upper body provides primary heat rejection path.

JCSat 2
Launch: 1 January 1990 by Titan 3 from Canaveral
Location: 154°E geostationary
Principal applications: as JCSat 1 plus DTH TV + DAB
Other specifications as for JCSat 1, but Titan launch required perigee kick stage derived from Minuteman stage 3.

JCSat3
Launch: 29 August 1995 by Atlas 2AS from Cape Canaveral
Location: 128°E geostationary
Design life: 12 years
Transponders: 28 (plus 6 back-up, 17-for-14) 60 W TWTA 13.75–15.0/12.25–12.75 GHz up/down Ku-band beams: Japan (>55 dBW mainland, >51 dBW Okinawa, >46 dBW other), India (45 dBW), Asia zonal (>45 dBW), Australia/New Zealand

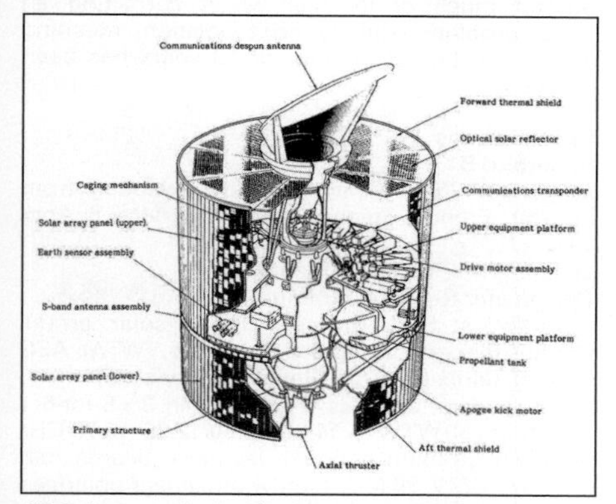

CS-3 has a single de-spun antenna for providing domestic communications links (NASDA) 0517007

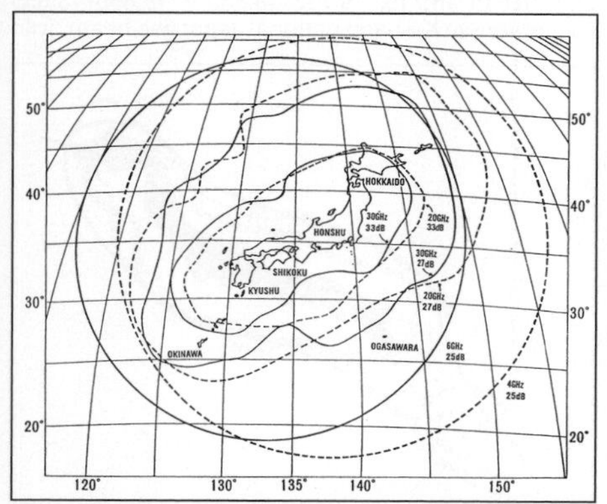

CS-3 coverage: C-band links the remoter communities, the higher-capacity Ka-band the main islands. The solid lines indicate CS-3a coverage from 132° east, the broken CS-3b from 136° east (NASDA) 0517008

A Hughes HS-393 derived from the Intelsat UI bus (Hughes Space & Communications) 0054336

(>46 dBW). 12 × 36 MHz + 16 × 27 MHz bandwidth (27 MHz switchable to 54 MHz), orthogonal linear polarisation
12 (plus 4 back-up, 16-for-12) 34 W SSPA 6.22–6.49/3.93–4.2 GHz up/down C-band zonal beam covering Russia, Korea, Japan, China, India, South Asia, Indonesia, Hawaii, Australia, New Zealand. 36 MHz bandwidth, orthogonal linear polarisation
Principal applications: as JCSat 1/2
Configuration: Hughes' HS-601 3-axis bus, 26.1 m solar array span
Mass: 3,100 kg at launch, 1,820 kg on-station BOL
Power system: twin solar wings providing 5.2 kW BOL, supported by 200 Ah Ni/H_2 battery.

JCSat 4
Launch: 17 February 1997 by Atlas 2AS from Cape Canaveral
Location: Initially over 150° E, but planned 124° E geostationary
Transponders: 16 60 W TWTA 13.75–14.5/12.25–12.75 GHz up/down Ku-band beams as JCSAT 3 except four 27 MHz transponders can be switched to 120 W/54 MHz
4 (plus 4 back-up, 8-for-4) 90 W TWTA 13.75–14.5/12.25–12.75 GHz up/down Ku-band beams: Japan (>55 dBW mainland, >53 dBW Okinawa), 36 MHz bandwidth, orthogonal linear polarisation. C-band as JCSat 3.
Mass: 3,100 kg at launch, 1,820 kg on-station BOL
Other specifications as for JCSat 3.

JCSat 5
Launch: 2 December 1997 using an Ariane-4 from Kourou
Location: 150°E
Contractors: Hughes Space & Communications Co
Bus: Hughes HS-601
Transponders: Ku-band, 60 W/90 W, 16 × 27 MHz and 16 × 36 MHz (4 switchable to 2 × 72 MHz)
Service area: Japan, Asia and Hawaii.

JCSat 6
Launch: 16 February 1999 aboard an Atlas 2 AS
Location: 124° E
Transponders: Ku-band, 70 W, 32 × 27 MHz
Service area: Japan.

JCSat 8 (JCSat 2A)
Launched: 29 March 2002 by Ariane 44L from Kouron

JCSAT 8 is checked out prior to delivery 0132824

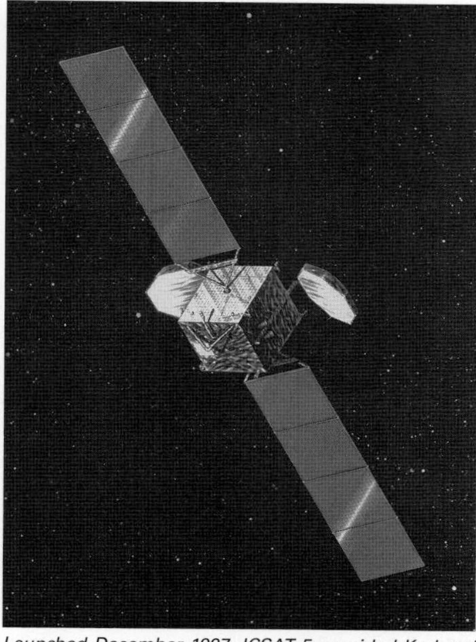

Launched December 1997 JCSAT 5 provided Ku-band services from 150° east 0132826

Location: 154°E
Contractor: Boeing Satellite Systems
Bus: HS-601
Transponder: Ku-band 16 × 120 W TWTA; C-band 16 × 34 W SSPA
Mass: 2,460 kg at launch, 1,500 kg BOL.

CS-4/NStar series

Current Status
NStar C has been operational since its launch on 5 July 2002. NStar B was operational until June 2007, when it was placed in a graveyard orbit. NStar A ended its service on 22 August 2006, after which it was sent to a graveyard orbit.

Background
The CS-4 series dates back to 1984 when two Super CS satellites were envisaged to replace CS-3 with an enlarged capacity of 60–90 multibeam transponders on a 2 tonne platform tailored around the H2 launcher. JPY2.57 billion was allocated to NASDA in FY90 to continue CS-4 studies. In 1989, MPT approved development of two satellites for launches in 1994/96. Transponders working at S/L-band would provide operational mobile services. The project would be closed to foreign bidders because it was designated as R&D. However, although NASDA was to provide 25 per cent of the funding, the domestic telephone monopoly NTT would fund the remainder in return for use of the operational transponders. US pressure resulted in early 1990 with CS-4 being transformed into two distinct projects: NASDA's 25 per cent was applied to the COMETS satellite and NTT acquired a satellite system on the open market.

NTT called for bids 20 June 1991, closing the competition 5 September 1991. Space Systems/Loral received the contract December 1991, against GE and Hughes competition, for two NStar satellites to be launched 1995 to replace CS-3. The on-orbit delivery contract was worth some USD600 million; the launch contract signing with Arianespace was announced 10 April 1992. Each of the SS/L satellites carried 26 active transponders: six Ka multibeam, five Ka-shaped beam, eight Ku-shaped beam, five C-shaped beam and one S plus its C feeder for mobile links. NStar A was damaged en route to the launch site and the launch was

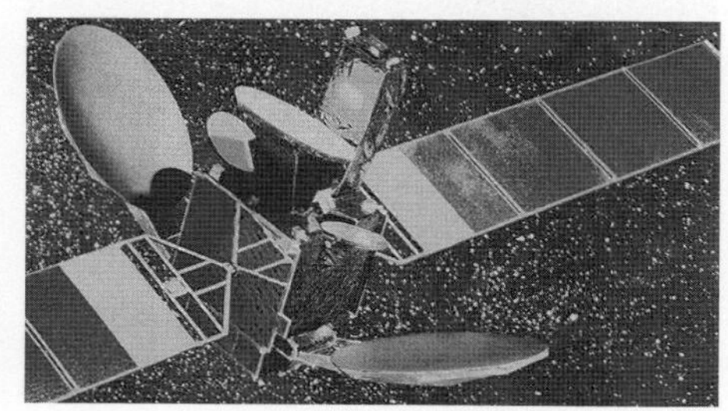

NStar A was launched in August 1995 and operated for 11 years 0517211

NStar C in Orbital's Dulles, Virginia, satellite manufacturing facilit (Orbital Sciences) 1342594

subsequently delayed. After launch in August 1995, it was delivered to the customer in October 1995. On 24 June 2002, JSAT Corporation announced an agreement with NTT East Corporation and NTT West Corporation to transfer the two NSat communications satellites and their operation to JSAT. Both satellites eventually operated successfully for about 11 years.

N-Star C was launched for NTT DoCoMo of Japan on 5 July 2002. The S-band satellite is designed for multiple applications including mobile telephony, data transfer and maritime communications. The communications payload was built by Lockheed Martin Commercial Space Systems, the prime contractor, which also participated in integration and test activities. The payload was integrated onto the Orbital STAR-2 bus in Orbital's Dulles, VA, facility. Final integration and test was performed at the Lockheed Martin Commercial Space Systems facilities in Newtown, Pennsylvania. N-Star C uses the Star 2 bus built by Orbital Sciences. It was the first flight of a new Orbital hydrazine/nitrogen tetroxide liquid apogee propulsion system with a 500N thrust apogee engine developed by Japan's IHI. Orbital provided the system design and the ground station, and was responsible for the assembly, integration and test of the spacecraft bus. Orbital also procured the launch services.

Specifications
CS-4A/ NStar A
Launch: 06:41 GMT on 29 August 1995 by Ariane 44P from Kourou, French Guiana.
Location: 132°E geostationary (replacing Sakura 3A) until 22 August 2006, then sent to a graveyard orbit.
Design life: 10 years minimum; sufficient propellant for 15 years.
Contractors: Space Systems/Loral (prime), MELCO, NEC, Thomson Tubes Electroniques (TH 3781 12 GHz; TH 3919 20 GHz TWTs), Hughes Electron Dynamics Div, Aerospatiale (reflectors).
Transponders: 5 (plus 2 back-up) 15 W min TWTA 27.5–29.25/17.75–19.45 GHz up/down Ka-band domestic beams, 100 MHz bandwidth, EIRP 44 dBW, circular polarisation.

6 (plus 3 back-up) 30W min TWTA 27.5–29.25/17.75–19.45 GHz up/down Ka-band multibeams, 200 MHz bandwidth, EIRP 51 dBW, circular polarisation, for 1.56 Mbit/s data transmission or broadband ISDN, with onboard switching capacity.

8 (plus 3 back-up) 55 W TWTA 13.84–14.56/12.19–12.91 GHz up/down Ku-band domestic beams, EIRP 53 dBW, 54 MHz bandwidth, linear polarised.

5 (plus 2 back-up) TWTA 5.925–6.425/3.7–4.2 GHz up/down C-band domestic beams, EIRP 38 dBW Tokyo, 72 MHz bandwidth, circular polarised; plus 6th TWTA as feeder link for S-band (see below).

1 (plus 1 back-up) 250W TWTA 2.655–2.690/2.500–2.535 GHz S-band domestic beam, EIRP 52 dBW, 15 MHz bandwidth; one C-band feeder link (as other C-band transponders; one of Ku-band payload serves as feeder backup). Principally for maritime links, replacing 200 terrestrial stations along Japan's coast.
Principal applications/users: Continued CS-3 services, mobile communications, and business services
Configuration: Box-shaped FS-1300 bus, two 2.6 × 3.25 m C/S + 2.6 × 4.5 m Ku/C deployable + one 2.1 m reflectors.
Mass: 3,410 kg launch, 1,617 kg dry, 2,057 kg on-station BOL.
AOCS: 3-axis stabilised (from immediately after launcher separation). 12 × 22 N N_2O_4/MMH; 490 N thruster permits perigee velocity augmentation. Propellant loading 2,160 kg for 92 per cent fill.
TT&C: Through TAO Kimitsu Satellite Control Centre; transponders controlled through NTT's Inuishi and Tokyo stations.
Power system: Twin 4-panel solar wings, spanning 27 m. 5,200 W BOL. Supported by Ni/H_2 batteries.

CS-4B/NStar B
Launch: 07:19 GMT on 5 February 1996 by Ariane 44P from Kourou.
Location: 136° E (replacing Sakura 3B) until June 2007, then placed in graveyard orbit.
Mass: 3,420 kg launch, 1,618 kg dry, 2,062 kg on-station BOL

Other specifications as for Nstar A.

CS-4C/Nstar C
Launch: 5 July 2002 by Ariane 5G from Kourou, French Guiana.
Location: 136°E geostationary.
Mass: 1,625 kg at launch, c. 800 kg dry.
Dimensions: 3.3 × 1.9 × 1.5 m (12.6 m span deployed)
Contractors: Lockheed Martin (prime, including communications payload). Orbital Sciences Corp. (bus).
Configuration: Star 2 bus, 3-axis stabilised. *Pointing:* <0.12 degrees circular error. 5.1 m unfurlable antenna.
Power: c. 2,600 W BOL. Multi-junction gallium arsenide cells
Payload: 1 C-band transponder (C-band feeder link), 20 S-band transponders. Repeater: S-band: 288 W multiport amplifier (three 5-for-4 groups of 24 W solid-state power amplifiers). C-band: 2-for-1 solid state power amplifiers (13 W). Frequencies: uplink 6.245–6.425 GHz; downlink 4.120–4.200 GHz.
Lifetime: 10–12 years.

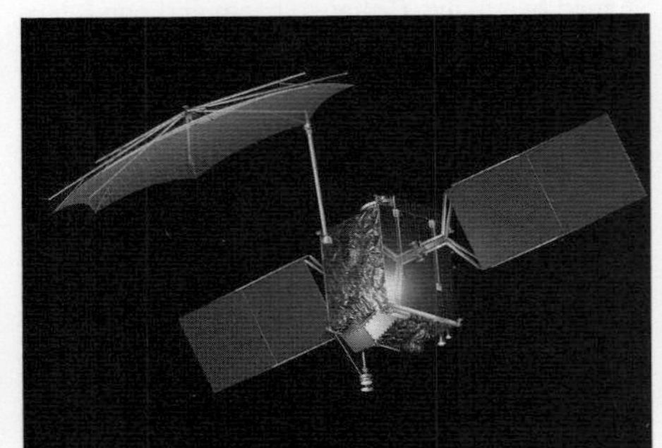

Japan's NStar C satellite was launched on 5 July 2002 (Orbital Sciences) 1342595

Superbird series

Current Status

In April 2006, it was announced that Arianespace had been chosen to launch the Superbird-7 telecom satellite for the Japanese Space Communications Corporation. Superbird-7 was the first commercial telecommunications satellite to be built in Japan as the country attempts to challenge the dominate US and Europe market.

The satellite was scheduled to be launched atop an Ariane 5 rocket in the first quarter of 2008 from Europe's Spaceport in Kourou, French Guiana, but this had not happened by mid-April 2008. The launch is now planned for later that year.

The spacecraft will weigh approximately 5,000 kg at launch and will occupy a geostationary orbit at 144°E. The Superbird-7 will be fitted with 28 Ku-band transponders to provide various communications services in the Asia-Pacific region, including video and audio broadcast to home, CATV and mobile terminals.

Background

A Japanese consortium planned to launch two Superbird satellites built by Space Systems/Loral and based on the FS-1300 platform. The Japanese stated the programme cost JPY70 billion (USD636 million), covering two satellites, launch services and insurance at premiums of about 20 per cent. An Ariane launched the Superbird A on 5 June 1989 to 158°E. Most of the satellite's station-keeping oxidiser bled off in December 1990 and commercial operations ended. Some customers transferred to the rival JCSat. Superbird claimed a reported USD170 million for the satellite. Superbird B, delayed from a December 1989 launch because of transponder problems, disappeared in the Ariane 4 accident of 22 February 1990. Insurance covered USD94.3 million. SS/L delivered the replacement Superbird B1 satellite within 24 months. Hughes received the order for Superbird C to expand coverage to the Asia Pacific region, including HDTV and 150 Mbit/s data. The satellite cost about 35 billion JPY (USD412 million) with launch insurance. On 6 April 1998 HSCI received an order for a fourth satellite, a HS60HP to be designated Superbird-4, launch was successfully carried out on 17 February 2000 by Ariane 4.

Superbird C provides business and TV services to Japan, south/east Asia and Hawaii. The steerable spot could extend coverage to Australia/New Zealand (Hughes Space & Communications) 0517223

In September 2001 Boeing Satellite Systems was awarded a contract from Space Communications Corp of Japan for a 601 series platform to fill the specification for Superbird 6. It was launched in 2003. Superbird 6 carried 23 active Ku-band and four active Ka-band transponders and occupies a location at 158°E longitude providing business telecommunications services using a Japan beam for both Ku-band and Ka-band services with a steerable spot beam for additional Ka-band services. Unfortunately the satellite burned much of its fuel whilst correcting an initial problem with its orbit position, meaning its original life expectancy of 13 years has been reduced.

Specifications
Superbird B1
Launched: 26 February 1992 by Ariane 4 from Kourou, French Guiana; entered service 6 April 1992
Location: 162°E geostationary
Design life: 10 years minimum
Contractors: SS/L (prime), DASA (solar array), Hughes Electron Dynamics Div (EPCs, TWTA), AEG (TWTs), Teldix (BAPTA), Gates (battery cells)
Transponders: 23 (plus 8 back-up, in 3 × 8-for-6 + 1 × 7-for-5) 50 W TWTA 14.00–14.50/12.25–12.75 GHz up/down Ku-band national Japanese beams, min EIRP 47.5 dBW, 36 MHz bandwidth, linear polarised (11 horizontal, 12 vertical)

3 (1 reserved as back-up) 29 W TWTA 28.4–28.8/18.615–18.715, 18.735–18.835 + 18.855–18.955 GHz up/down Ka-band national Japanese beam, min

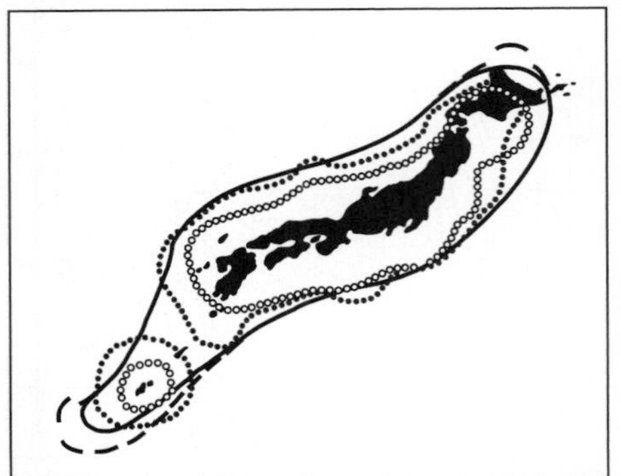

Superbird A/B coverage. Solid line: Ku 47.5 dBW EIRP; dashed: 4 dB/K. Dotted line: Ka 43 dBW; open dotted: 3 dB/K 0517006

From its location at 158° east, Superbird 6 will serve the business community in Japan 0132825

EIRP 43 dBW, 100 MHz bandwidth, LHCP up, RHCP down)
Principal applications: national TV and cable distribution, telephone, business services, and government
Configuration: 3-axis stabilised SS/L FS-1300 2.41 × 2.58 × 2.20 m box-shaped bus with twin solar wings on north/south sides, 20.3 m span; 2.1 m diameter Ku-band dual gridded reflector on West face, 2.0 m Ka-band reflector on East face
Mass: 2,560 kg at launch, 1,532 kg on-station BOL, 1,224 kg dry
AOCS: 3-axis bias momentum system; single Marquardt R-4D 490 N apogee thruster and two sets of 6 × 22 N AOCS thrusters, common NTO/MMH propellant system
Antenna pointing: 0.09° azimuth/0.08° elevation Ku-band; 0.09° azimuth/0.07° elevation Ka-band
TT&C: 14 GHz command frequency, 12 GHz telemetry frequency
Power system: twin 3-panel wings providing 3,984 W BOL summer solstice, 4,460 W BOL autumnal equinox, 3,479 W EOL summer solstice, 3,778 W EOL autumnal equinox. 42 ± 0.5 V bus (sunlight), 27.0–42.5 V bus (eclipse). Two 85.5 Ah Ni/H_2 batteries.

Superbird A1
Launch: 1 December 1992 by Ariane 4 from Kourou, French Guiana. Entered service 20 February 1993
Location: 158°E geostationary
Design life: 10 years (propellant for 13 years)
Mass: 2,780 kg at launch, 1,665 kg on-station BOL, 1,255 kg dry (propellant increase over B1 allowed by uprated Ariane stage 3)
Other specifications are the same as for Superbird B1.

Superbird C
Launch: 28 July 1997 on Atlas 2AS from Cape Canaveral
Design life: 13 years
Contractors: Hughes Space & Communications (prime), MELCO (steerable spot)
Transponders: 24 90 W TWTA Ku-band beams, orthogonal linear polarisation
Principal applications: as Superbird B + Asia-Pacific services
Configuration: Hughes' HS-601 3-axis bus, 26.1 m solar array span. Two 216 cm shaped-beam antennas plus

Superbird 4 (B2)
Launch: 17 February 2000 on Ariane 4 from Kourou French Guiana.
Design life: 13 years
Contractor: HSCI (now Boeing Satellite Systems)
Transponders: 23 82 W TWTA Ku-band beams, six 50 W TWTA Ku-band
Configuration: Hughes' 601 HP 3-axis bus, 26.1 m span, 7 m width across antennas (4 × 2.7 × 3.6 m stored). Three primary antennas: One 1.9 × 2.5 m ku-band dual-gridded, one 1 m Ku-band steerable
Mass: 4,060 kg launch 2,460 kg BOL
AOCS: 1 × 490 N liquid apogee motor; 4 × 10 N N-S bi-propellant and 4 × 22 N E-W bi-propellant station-keeping thrusters
Power system: Twin solar wings each with 3 panels of silicon cells and one panel GaAs + 31-cell NiH, 200 Amp batteries; 5.5 kW BOL.

Korea (South)

Koreasat (Mugunghua) series

Background
Korea Telekom, formerly owned by the South Korean government, asked satellite builders to supply a USD260 million Koreasat (Mugunghwa after the national flower) domestic satellite telecommunication system in March 1991. Four teams headed by GE Astro, Space Systems/Loral, Hughes and British Aerospace responded and KT issued a USD145 million contract for two satellites and the ground control system to GE Astro in December 1991, supported by Matra Marconi Space UK (payload), Goldstar Information & Communications of Korea, and Korean Air. Hughes Network Systems of the US provided a hub and network of 70–80 cm VSAT dishes.

At launch, Koreasat 1 had to expend its own onboard propellant to achieve orbit, because the Delta launch vehicle fired incorrectly for geostationary transfer orbit when a strap-on failed to jettison. Koreasat 1 used about half of its station-keeping propellant to achieve final geosynchronous orbit after the apogee kick motor fired. The satellite launch insurer withheld payment of USD104 million because the satellite remains operational although lifetime has been reduced from 10 to 4½ years. Beam now focused on Hungary.

Koreasat 5 based on a Spacebus-4000 from Alcatel Space will carry 36 transponders in Ku-band, C-band and SHF (military) band delivering advanced broadband multimedia and digital television transmission services.

Specifications
Mugunghua 1 (Koreasat 1)
Launched: 5 August 1995 by Delta 7925 from Cape Canaveral complex 17
Location: 116° E geostationary (with number 2) until July 2000, then at 45° E (Europestar B) and to 47.5°E in October 2000
Design life: 10 years (+2 years on-orbit storage), but Delta shortfall expected to limit to <5 years
Contractors: Lockheed Martin Astro Space (prime), MMS UK (payload), and Hughes Electron Dynamics Div (TWTAs)
Transponders: 12 (plus 4 back-up, 16-for-12) 14 W TWTA 14.00-14.50/12.25-12.75 GHz up/down Ku-band S Korean FSS 0.86° beam, 36 MHz bandwidth, EIRP peak/edge 52.5/50.4 dBW, orthogonal polarisation
3 (plus 3 back-up, 6-for-3) 120 W TWTA 14.50–14.80/11.70–12.00 GHz up/down Ku-band South Korean DBS 0.86 × 1.06° beams, 27 MHz bandwidth (4 channels per transponder), EIRP peak/edge 62/59.4 dBW, LHCP
Principal applications: domestic telecommunications, VSAT, data, video and DBS (HDTV from 1998)
Configuration: LMAS 3000 3-axis 163 × 142 ×174 cm box-shaped platform with 15.45 m solar array span and 1.52 × 1.83 m diameter reflector. Electronic units, batteries, propulsion and AOCS equipment are mounted on six honeycomb structural pallets or panels. Transponders and housekeeping components are mounted on four panels, two each on North/South sides. Additional housekeeping equipment is mounted on anti-Earth base panel
Mass: 1,459 kg at launch, 833 kg on-station BOL, 641 kg dry, 144 kg payload
AOCS: 3-axis by 6,000 rpm MW (pitch) and magnetorquing (roll/yaw) controlled by Earth Sensor Assembly error signals. Propulsion subsystem is a conventional blow-down monopropellant hydrazine type: four surface tension 50 cm diameter tanks (max capacity 50.3 kg each, initial 26.2 atmosphere; max blow-down 4:1) are manifolded to two redundant sets of eight Olin Aerospace thrusters (six catalytic 0.89 N MR-103C + two NSSK electro-thermal 0.4 N MR-501); 187.5 kg loaded. GEO insertion provided by Thiokol Star 30E solid AKM
TT&C: Yong-In primary, Taejon back-up; 11 m for TT&C, testing and manoeuvres, 6.4 m for on-station TT&C
Power system: twin Si arrays of three panels each provide 1,686/1,533 W BOL/EOL (10 years) supported by 2 × 42.5 Ah Ni/H_2 batteries.

Mugunghua 2 (Koreasat 2)
Launch: 14 January 1996 by Delta 7925
Location: 116° E GEO (with number 1), then to 113° E in June 2000 until present
Other specifications as for Mugunghua 1.

Mugunghua 3 (Koreasat 3)
Launch: 4 September 1999
Location: 116° E
Design life: 15 years
Contractor: Lockheed Martin A2100 platform and Spar Aerospace
Transponders: 6 × 120 W Ku-band DBS, 27 mHz bandwidth; 24 (plus 8 back-up) 45 W Ku-band FSS, 36 mHz bandwidth; 3 (plus 2 back-up) 70 W Ka-band, 200 mHZ bandwidth
Mass: 2,790 kg at launch; 1,322 kg dry

Malaysia

MEASAT/Africasat series

Current Status
MEASAT-3 was launched on 12 December 2006. MEASAT-2 is still operational at 148.0° E. In January 2008, MEASAT-1 was moved to the 46.0° E geostationary orbit location and renamed **Africasat-1**. Up to 5 years of life is expected from this satellite in the current orbital configuration. MEASAT-1R is scheduled for launch in the third quarter of 2008.

Background
In 1990, Malaysia, which had spent USD30 million annually on Intelsat and Palapa leases, made a national decision to take over responsibility for its own communications network. Binariang, a private company, was used to purchase satellites and launch services. Binariang signed a memorandum of understanding with Hughes in November 1991 to buy two HS-376 satellites to provide domestic and regional services. Considering Hughes production schedule at the time, Malaysia scheduled the first MEASAT (Malaysian-East Asian Satellite) launch for 1994 aboard Ariane. But Malaysia did not sign a final contract with Hughes Communications International Incorporated until 17 May 1994, following a four-month international competition. The agreement specified that Malaysia would launch one satellite by late 1995. The contract also kept an option open for a second satellite, which the Malaysian government authorised in January 1995, as well as the construction of a purpose-built satellite control facility located 915 m above sea level in Gunung Raya, Langkawi Island and training for the Malaysian operators.

MEASAT-1, launched in January 1996, was the first HS-376 model to carry GaAs solar cells, a lightweight shaped antenna (which improves gain and eliminates multiple feedhorns) and a bipropellant station-keeping propulsion system. The satellite provided C- and Ku-band telecommunications services over Malaysia and the rest of SE Asia. It pioneered the use of Ku-band services in the high rainfall region of South East Asia. The near-identical MEASAT-2 followed in November 1996.

On 21 March 2003, Binariang Satellite Systems Sdn. Bhd. and Boeing Satellite Systems announced a contract for the procurement of a high-power commercial communications satellite. The Boeing 601HP satellite, designated MEASAT-3, was to be co-located with MEASAT-1 at 91.5° E. It was to provide expansion and back-up capacity in meeting the increasing market demand for satellite services within the region. Boeing would also provide an upgrade to the MEASAT ground facilities on Pulau Langkawi, as well as training and launch support services.

MEASAT-3's launch occurred in December 2006, a year later than scheduled, due to delays at Boeing's El Segundo, California, manufacturing plant. Carrying 24 C-band and 24 Ku-band transponders, it was co-located with MEASAT-1 at 91.5° E. With switching capabilities between Malaysia, India and China and a global footprint covering Asia, Middle East, eastern Europe and eastern and central Africa, MEASAT-3 extended coverage to 70 per cent of the world's population through a single, high powered, C-band beam. Adding The three C-band beams on MEASAT-3 provide television and telecommunications broadcasts to 110 nations from Japan to Africa. Its three Ku-band beams provide direct-broadcast television services to South Asia and Southeast Asia, including Indonesia and Malaysia.

Hughes' HS 376 MEASAT-1, built for Binariang, was the first of its type to carry GaAs solar cells 0132811

In January 2008, MEASAT-1 was relocated at the company's 46.0° E orbital slot in order to help meet strong demand for telecom services. It was renamed **Africasat-1** to reflect its new African service area. The satellite is no longer operating at full capacity, but will offer 12 C-band and up to four Ku-band transponders for customers. The lifetime of Africasat-1 is expected to be up to 5 years in the current orbital configuration.

The next in the series, MEASAT-1R, is scheduled for launch by a Land Launch Zenit in the third quarter of 2008. Built by Orbital Sciences, the 3-axis stabilised satellite will have a launch mass of 2,417 kg and will carry 12 Ku-band and 12 C-band transponders with three antennas. MEASAT-1R will provide C-band communications services throughout Asia, the Middle East and Africa, and Ku-band direct-to-home television broadcasting to Malaysia and Indonesia. The satellite will be located in geostationary orbit at 91.5° E over Southeast Asia and will generate approximately 3.5 kW of payload power.

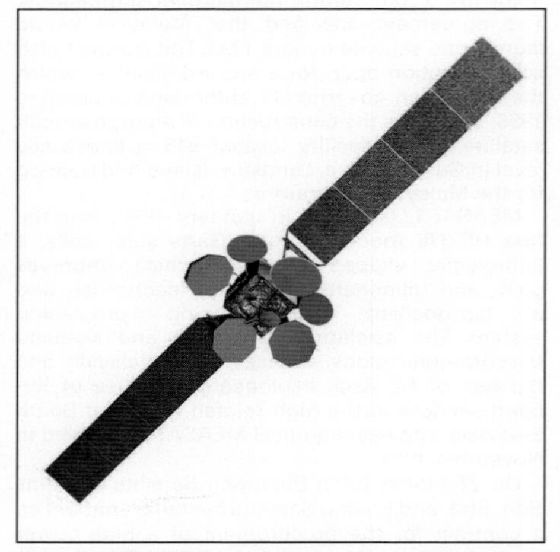

Boeing's MEASAT-3 was launched by an International Launch Services Proton-M on 12 December 2006 (MEASAT) 1342226

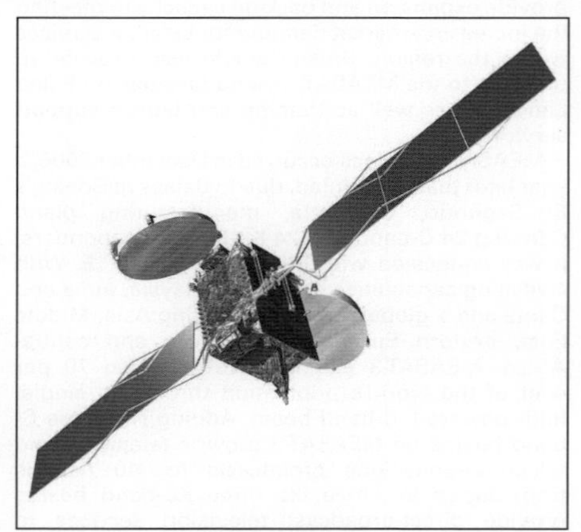

MEASAT-1R is scheduled for launch by a Land Launch Zenit in late 2008 (Orbital Sciences) 1342225

Specifications

MEASAT-1

Launched: 12 January 1996 by Ariane 44L from Kourou, French Guiana.
Location: Initially 91.5° E geostationary. Moved to 46.0° E (inclined) in January 2008 and renamed Africasat-1.
Design life: 12 years minimum.
Contractors: Hughes Space & Communications (prime).
Transponders: 12 (plus three back-up) 12 W SSPA 5.925–6.425/3.7–4.2 GHz up/down C-band regional beam, EIRP 35 dBW edge (40 dBW centre), 40 MHz bandwidth (36 MHz usable), orthogonal linear polarisation

5 (plus one back-up) 112 W TWTA 14.0–14.25 + 13.75–14.0/10.95–11.2 + 12.2–12.5 GHz up/down Ku-band domestic beam (India and Philippines selectable), EIRP 55 dBW edge (58 dBW centre), 54 MHz bandwidth, orthogonal linear polarisation.
Principal users: Ku-band for domestic direct to user services, including DTH TV via 50 cm dishes to Malaysia, Philippines and India; C-band for regional telecom services.
Configuration: Hughes' HS-376 spin-stabilised bus, 2.16 m diameter, and 6.6 m height with solar array drum extended (2.8 m stowed). Single-shaped antenna on despun platform.
Mass: 1,450 kg at launch, 839 kg on-station BOL, 626 kg dry.
AOCS: 50 rpm spin control maintained by four (two radial and two axial) 22.2 N NTO/MMH bipropellant thrusters. GEO insertion provided by Thiokol Star 30 solid.
TT&C: Satellite Control Facility on Langkawi Island.
Power system: GaAs cells on cylindrical surface provide 1,700 W (1,200 W BOL to payload); 16-cell 142 Ah Ni/H_2 battery.

MEASAT-2

Launch: 13 November 1996 on Ariane 44L from Kourou, French Guiana.
Location: 148° E geostationary, inclined.
Transponders: 3 (plus one back-up) 12 W SSPA 5.925–6.425/3.7–4.2 GHz up/down C-band regional beam, EIRP 35 dBW edge (40 dBW centre), 72 MHz bandwidth, orthogonal linear polarisation.

6 65 W TWTA Ku-band domestic, 54 MHz bandwidth, orthogonal linear polarisation.

6 100 W TWTA Ku-band domestic, 54 MHz bandwidth, orthogonal linear polarisation.Other specifications as for MEASAT 1.

MEASAT-3

Launch: 12 December 2006 by Proton-M/Breeze-M from Baikonur, Kazakhstan
Location: 91.5° E geostationary.
Design life: 15 years minimum.
Contractors: Boeing Satellite Systems (prime).
Transponders: 24 C-band, 36 mHz bandwidth each. 65 WTWTAs. Channel polarization: linear. EIRP (dBW): 40 (Global Beam)/44 (Asia Beam). G/T (dB/°K) + 0.5 (Global Beam)/+2.5 (Asia Beam).

24 Ku-band, 36 MHz. 120 W TWTAs. Channel polarization: linear. EIRP (dBW): 57 (max). G/T (dB/°K)+ 14 (max).
Principal users: Television and telecommunications broadcasts to 110 nations from Japan to Africa. Three Ku-band beams provide direct-broadcast television services to South Asia and Southeast Asia, including Indonesia and Malaysia.
Configuration: Boeing 601-HP spin-stabilised bus, 7.7 m diameter antennas.
Mass: 4,765 kg at launch. 3,220 kg in orbit (BOL)
Body dimensions: Diameter: 3.8 m. Height: 7.4 m.
AOCS: Liquid apogee motor 445 N. Twelve 10 N station-keeping thrusters (bipropellant).
TT&C: Satellite Control Facility on Langkawi Island.
Power system: Two solar arrays: 26.2 m span, each with 4 panels of triple-junction gallium arsenide solar cells. 10.8 kW (BOL), 9.8 kW (EOL). 32 cell NiH_2 batteries.

Mexico

Morelos/Solidaridad/Satmex series

Current Status

Satmex 6 was launched with Thaicom 5 on 27 May 2006. Satmex 5 (Morelos 3) has been operational since its launch by Ariane 42L on 6 December 1998.

Satmex 6 during launch preparation at Kourou. (CNES/ESA/Arianespace) 1342241

Satmex also operates Solidaridad-2, which has been functioning since its launch in 1994.

Background

Despite its participation in international communications bodies, Mexico decided in 1982 to develop an independent communications system. During that year, the Secretaria de Comunicaciones y Transportes (SCT) ordered two Hughes satellites to expand domestic television and other services available to the entire population of the country. Before that time, telecommunications only served the 18 million people living in and around Mexico City.

The Mexican government's Departamento Especial of the DGT Direccion General de Telecomunicaciones for Mexico's telecommunications systems gave authority to SCT to develop and service a communications network. Then, in 1990, the Mexican government created Telecomm, an independent non-stock company owned entirely by the federal authority, with SCT operating as a regulating body.

Mexico concluded a USD92 million contract with Hughes, which included the control centre at Iztapalapa. The Mexican government selected the name "Morelos" for its satellites to honour the revolutionary Jose Maria de Morelos y Pavon a hero of the Mexican revolution. The satellites were based on the Hughes cylindrical, spin-stabilised HS 376 design, weighing about 646 kg and generating 950 W of power. The Shuttle (STS 51-G) launched **Morelos 1** on 18 June 1985 to 113.5° west. A Thiokol Star-30B apogee kick motor was fired by command and placed the satellite into a circular synchronous orbit. When the satellite approached its operating position, the reflector antenna and the electronics shelf were despun with respect to Earth and achieved close pointing accuracy using a beacon tracking system. The remainder of the bus and solar panels continued to spin for stability. The satellite drifted into final orbit and was placed in operating position through the use of the onboard thrusters.

NASA allowed Mexican payload specialist Rodolfo Neri Vela to accompany **Morelos 2** when it was deployed from STS 61-B in November 1985. Morelos 2 was positioned at 116.8° west. It was later operated by GE Americom under the designation K2 and was moved to a graveyard orbit in June 2004.

When the Mexican government planned the implementation of a second-generation satellite system, it returned to Hughes Space and Communications Company (now Boeing Satellite Systems, Inc.) The new pair of spacecraft were called "Solidaridad", in recognition of the way satellite telecommunications were linking the urban and rural parts of the country, and the rest of the world. The Solidaridad contract was signed in May 1991. In selecting Hughes over two competitors, the Secretariat of Communications and Transportation cited technical excellence, lowest price, shortest delivery schedule, longest useful life, and best payload capacity. The spacecraft were operated by the government agency Telecomunicaciones de Mexico (Telecomm).

An HS-601 HP design, Satmex 5 relies on xenon propulsion for North/South station-keeping (Hughes Space & Communications) 0054329

Solidaridad was based on a body-stabilised Hughes HS601 bus. The high power and capacity spacecraft consists of a cube-shaped central section containing the electronic and propulsion systems, with a pair of three-panel solar arrays almost 21 m in length.

The first Solidaridad satellite was launched by Ariane 4 on 19 November 1993, followed by Solidaridad 2 on 7 October 1994. **Solidaridad 1** entered service in March 1994, when it replaced Morelos 1. On 28 April 1999, Solidaridad 1 developed a malfunction with its primary satellite control processors (SCP)s, which caused the satellite to spin. Most services were moved to backup capacity on Satmex 5 until all services were restored on 29 April. A further SCP problem caused another outage on 1 May, interrupting services for seven hours. On 27 August 2000, the remaining backup SCP malfunctioned, and a few days later the satellite was shut down.

Solidaridad 2 was located at 113° west until July 2006, when it was relocated at 114.9° west. Like Morelos, each Solidaridad spacecraft carried 18 active C-band transponders, but with much higher power that allowed reception by small terminals. The transponders were powered by Hughes-built solid-state power amplifiers (SSPAs) ranging from 10 to 16 W. There were also 16 active Ku-band transponders – four times the existing capacity – with 42.5 W travelling-wave tube amplifiers. In addition, Solidaridad had one L-band channel for mobile services or use in remote rural areas, using four 21 W SSPAs linked in parallel. All bands covered Mexico, while the C- and Ku-band coverage extended into the south western United States. Spot beams in Ku-band also reached major US cities from Los Angeles to Miami and New York. C-band coverage included the Caribbean, Central and South America. The existing satellite control center in Iztapalapa was upgraded for the more complex Solidaridad system.

As the replacement for Morelos 2, Satmex 5 carries 10 times the payload power and 48 C-band and Ku-band transponders (Hughes Space & Communications) 0054328

The specifications for a new Morelos satellite took more than a year to develop, but the bidding process started on 18 March 1996. Five companies responded initially, but only three presented proposals at the end of May. The contract with Hughes Space and Communications International, Inc. was signed in August 1996.

Satmex 5 (Morelos 3), initially called **Morelos 2-FOS**, was launched by Ariane 42L on 6 December 1998 to replace Morelos 2. Satmex 5 uses a Hughes 601HP ("high power") version of the body-stabilised 601 series bus. Its dual-junction, gallium arsenide solar cells produce more than 7 kW of payload power, at least 10 times the capacity of Morelos 2. Other innovations include radiation-cooled travelling-wave tube amplifiers (TWTAs), advanced battery technology and a xenon ion propulsion system (XIPS) for station-keeping. It carries 24 active C-band transponders and 24 active Ku-band transponders.

Satmex 6 was produced by Space Systems/Loral in Palo Alto, California. Originally scheduled for launch in 2003, it was grounded because Satelites Mexicanos S.A. de C.V. was unable to finance the final launch-related payments, in particular the insurance premium. The satellite was placed in storage at the Kourou launch site until SS/Loral and Satmex reached a settlement. In July 2005, it was agreed to ship Satmex 6 back to Loral's plant at Palo Alto for recertification. In return for performing launch and post-launch services, and for not pursuing its claims against Satmex, Loral secured Satmex's agreement to drop its claims against Loral regarding a late-delivery penalty for Satmex 6. In addition, Loral would continue to use three transponders on Satmex 5 and have exclusive use of four transponders – two C-band and two Ku-band – on Satmex 6. This capacity would be used by Loral to re-enter the transponder-lease market in North America.

Satmex 6 was eventually launched in May 2006. Located at 113° west, it provides Fixed Satellite Services (FSS) over the Americas, with hemispheric C- and Ku-band footprints, switchable coverage in both C-band and Ku-band, and high power coverage of South America and the Caribbean. 36 C-band transponders provide full coverage of the United States, Central and South America, while 24 Ku-band transponders serve the United States, Mexico, Central and South America, with concentrated coverage over major cities in South America.

Specifications

Morelos 2 (K2)

Launch: 27 November 1985 by Shuttle mission 61-B.
Location: 116.5°W geostationary. Moved to a graveyard orbit in June 2004.
Design life: 10 years, but operated for almost 20 years.
Contractors: Hughes Space & Communications Co.
Transponders: 12 (plus 4 back-up, 14-for-12) 7 W TWTA 5.925–6.425/3.70–4.20 GHz up/down C-band domestic beams, 36 MHz bandwidth, EIRP 36 dBW min, orthogonal polarisation.

6 (plus 2 back-up, 8-for-6) 10.5 W TWTA 5.925–6.425/3.70–4.20 GHz up/down C-band domestic beams, 72 MHz bandwidth, EIRP 39 dBW min, orthogonal polarisation.

4 (plus 2 back-up, 6-for-4) 20 W TWTA 14.00–14.50/11.70–12.20 GHz up/down Ku-band domestic beams, 108 MHz bandwidth, EIRP 44 dBW min, circular polarisation.
Principal applications: Domestic TV distribution, telephony and data services.
Configuration: Hughes HS-376 cylindrical spin-stabilised bus, 2.16 m diameter, 6.60 m high, single 1.8 m diameter dual reflector for C-band and Ku-band transmission, planar array for Ku-band receive (first Hughes civil satellite with planar array); AOCS provided by hydrazine thruster system, GEO insertion by Thiokol solid AKM, GTO insertion by McDonnell Douglas PAM-D.
Mass: 1,140 kg launch mass (including 133 kg hydrazine), 645 kg on-station BOL.
TT&C: Through the Iztapalapa Earth station 10 km South East of Mexico City. Full tracking antenna + two 11 m antennas.

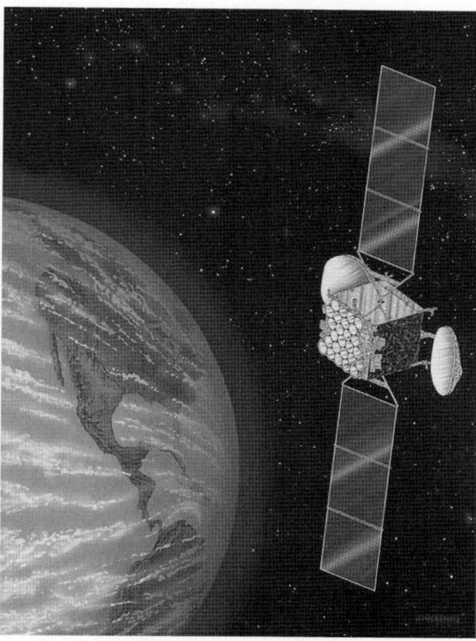

Two Solidaridad comsats were launched by Ariane 4 in 1993-94. (Hughes Space and Communications Co.) 1342286

Power system: Solar array of K-7 cells mounted on cylindrical body and skirt extension provides 950 W BOL/800 W EOL; two Ni/Cd batteries.

Solidaridad 1 (Satmex 3)
Launch date: 19 November 1993 by Ariane 44L from Kourou.
Location: 113.5° W geostationary.
Design life: 14 years. Actual lifetime < 7 years.
Mass: 1,641 kg BOL.
Configuration: Hughes 601, 3-axis stabilised.
Transponders: 18 active C-band. 16 active Ku-band, 42.5 W TWTAs. 1 L-band, four 21 W SSPAs.
Propulsion: Bipropellant system: one 490 N Marquardt liquid apogee motor, plus twelve 22 N thrusters for station-keeping.
Power system: Two 21 m, three-panel solar arrays provide 3.3 kW (BOL). 27 cell nickel-hydrogen battery.
Body dimensions: 2.7 × 3.5 × 3.1 m

Solidaridad 2 (Satmex 4)
Launch date: 8 October 1994 by Ariane 44L from Kourou.
Location: 113° W geostationary. Relocated to 114.9° W in July 2006.All other specifications for Solidaridad 2 are identical to Solidaridad 1.

Morelos 3 (Satmex 5)
Launch date: 6 December 1998 by Ariane 42L from Kourou.
Location: 116.8° W geostationary.
Design life: 15 years.
Mass: 4,144 kg at launch; 2,267 kg BOL
Configuration: Hughes 601HP, 3-axis stabilised.
Transponders: 24 C-band for continental coverage: bandwidth 36 MHz, TWT 36 W, EIRP 38.0 dBW. 24 Ku-band for regional and continental coverage: bandwidth 36 MHz, TWT 110 and 132.5 W, EIRP – Ku-1 49.0 dBW (EOC), Ku-2 – 46.0 dBW (EOC).
Propulsion: 490 N liquid apogee motor; station-keeping by 8 × 10 N bipropellant and 4 × 22 N bipropellant thrusters. N-S primary (xenon ion); N-S backup (bipropellant), E-W (bipropellant), Aft (bipropellant).
Power system: 2 × 4 panel wings of dual junction GaAs solar cells providing 10 kW BOL, 8.8 kW EOL. 32-cell NiH, 350 Ahr batteries.
Dimensions: Stowed: 5.5 x 3.5 m × 2.6 m. Deployed: length 26 m, width (antennas) 9.4 m.

Satmex 6
Launch date: 21:09 GMT on 27 May 2006 by Ariane 5 ECA from Kourou.
Location: 113° W geostationary.
Design life: 15 years.
Mass: 5,456 kg at launch. 2,310 kg in orbit.
Configuration: Space Systems/Loral LS-1300X 3-axis stabilised.
Power: 14.1 kW (BOL)
Transponders: 36 C-band (bandwidth 36 MHz): uplink: 6.1 to 6.4 GHz, downlink: 3.9–4.2 GHz. 24 Ku-band (bandwidth 36 MHz): uplink: 14.2–14.4 GHz, downlink: 11.9–12.1 GHz. Polarisation: orthogonal linear.
Power system: Two solar arrays. 12.69 kW (EOL)
Dimensions: 7.4 × 2.8 × 3.4 m. Span in orbit 31.4 m

Norway

Thor series

Current Status
Thor 2 is to be retired by early 2009 and replaced by Thor 5, which was launched on 11 February 2008. Thor 6, to be launched in 2009, will replace Thor 3.

Background
Thor 1 began life as the **Marcopolo 2** satellite, a Hughes HS-376 model that was part of the UK's British Satellite Broadcasting company. BSB merged into rival Sky Television in November 1990, in preparation to moving all programming to Sky's Astra carrier. In November 1995, Hughes Space & Communications received a USD300 million contract for the on-orbit delivery of **Thor 2**, a high-powered version of Hughes' spin-stabilised HS-376 model. Thor 2 was successfully launched in May 1997. Shortly after, Telenor announced the award of a follow-on contract for a second HS-376HP satellite. **Thor 3** was successfully launched in June 1998. Both contracts required Hughes to provide the spacecraft, launch services, Earth station upgrades at various sites, including the main site at Nittedal, Norway, and training. The two HS-376 satellites could provide five TV channels employing the D-MAC transmission standard. The contract represented Hughes' first sales of the HS-376 models configured for DBS, using paired 55 W TWTAs to achieve the required power levels. Norway bought the satellite for USD42 million.

The Thor antennas have shaped surface octagonal reflectors approximately 2 m in diameter, with single offset feeds. The antennas have three surfaces: one for horizontally polarised signals, one for vertically polarised signals, and one for on-station tracking and command. Both spacecraft use a bipropellant propulsion system for station-keeping and attitude control.

Thor 2 has 15 active Ku-band transponders (with three spares), powered by 40 W traveling-wave tube amplifiers (TWTAs), while Thor 3 has 14 active Ku-band transponders powered by 47 W TWTAs. The satellites deliver television and telephone/data services to Scandinavia and Northern Europe, with western offshore beams to the Faroes, Iceland and Greenland. Thor 2 and Thor 3 are co-located at Telenor's Nordic Hot-Bird position at 1°W. Thor 2 covers five zones stretching from Scandinavia across the north Atlantic to Greenland, with the primary zone comprising Norway, Denmark, Sweden, Finland, and the Baltic states. The Ku-band effective isotropic radiated power (EIRP) is 52 dBW in the primary zone. Thor 3 covers three zones stretching from Scandinavia across the north Atlantic to Greenland and into eastern Europe. The EIRP for Thor 3 is up to 50 dBW.

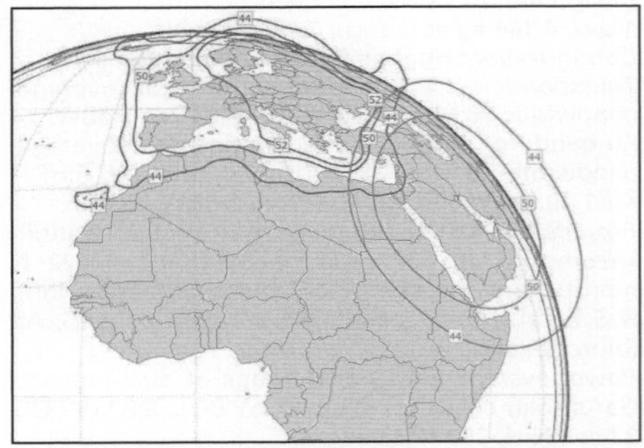

Thor 5 T2 (blue) and T3 (green) BSS coverage (Telenor) 1342672

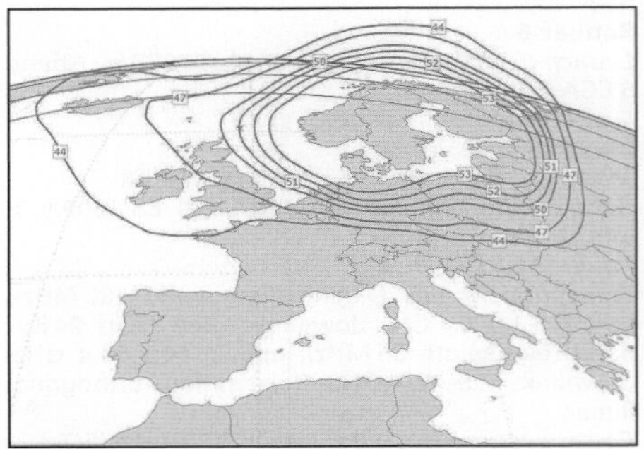

Thor 5 FSS coverage (Telenor) 1342673

Hughes' HS-376 bus was uprated to meet DTH power requirements for Thor 2 and 3 0517010

In September 2005, Telenor and Orbital Sciences Corporation announced that they had signed a contract for the delivery a new geosynchronous communications satellite, initially known as **Thor 2R**. The satellite, later named **Thor 5**, is based on Orbital's STAR-2 platform and carries 24 transponders with three times more payload power than Thor 2. The spacecraft features 15 active FSS transponders and nine active BSS transponders for fixed coverage regions. In addition, five more active transponders are routed through a steerable antenna that can be pointed toward any other region on Earth visible from 0.8°W. Thor 5 will carry all broadcasting services which currently reside on Thor 2 and provide additional capacity to allow growth in the Nordic region and expansion into Central and Eastern Europe.

Thor 3 is to be retired in 2010, after the launch of **Thor 6**. The new satellite will be based on the Thales Alenia Space Spacebus 4000B2 platform. Fitted with 36 active Ku-band transponders, THOR 6 will provide high power Direct-To-Home (DTH) television services for the Nordic countries, Central and Eastern Europe, from the 1° West orbital location. 16 transponders will point to the Nordic countries and 20 transponders will be positioned to serve the growing broadcasting demands within Central and Eastern Europe. The satellite will be launched on either an Ariane 5 or Soyuz from the Kourou Space Centre in French Guiana. Thor 6 is expected to start services in the summer of 2009 with an operational lifetime of 15 years. It was the first telecom satellite award for the manufacturer since it was taken over by Thales.

Specifications

Thor 1 (BSB 2)
Launched: 00.42 GMT on 18 August 1990 by Delta 6925 from Cape Canaveral. Satellite separated at 13.10 GMT, followed by AKM firing at 16.17 GMT 20 August for positioning at 50°W for a month of in-orbit testing. Ownership transferred to BSB in mid-December 1990 at 31°W collocated with Marcopolo. 1. Transferred over 15 September to 29 October 1992 to 1°W under Norwegian ownership
Location: 0.8°W geostationary until December 2002 when moved to 7.3° W and retired.
Design life: 10 years minimum.
Contractors: Hughes Space & Communications Co.
Transponders: Six paired 55 W TWTAs 17.385–17.992/11.7715–12.1054 GHz up/down Ku-band Nordic beam (BSB channels 4, 8, 12, 16, 20), 27 MHz bandwidth, 59 dBW min EIRP (63 dBW max), RHCP
Principal applications: DTH via 45 cm dishes
Configuration: Hughes' HS-376 spin-stabilised, 7.2 m high deployed (2.7 m stowed), 2.16 m diameter. Despun single reflector and communications shelf.
Mass: 1,250 kg at launch, 660 kg on-station BOL
Pointing: 0.05° accuracy by RF beacon
AOCS: Station-keeping and spin-stabilised at 50 rpm by four 22.2 N hydrazine thrusters; GEO insertion by solid Thiokol Star 30B AKM
TT&C: Via Nittedal station
Power system: K7 solar cells mounted on cylindrical exterior and deployed drum providing 1,024 W BOL (drum extended 25 cm over normal 376). Two Ni/Cd batteries provide 100% eclipse protection.

Thor 2 and 3 were high power versions of the Hughes spin-stabilised HS-376 model 0005568

Thor 2
Launch: 20 May 1997 by Delta II 7925 from Cape Canaveral.
Location: 0.8° W geostationary.
Design life: 10–12 years minimum.
Transponders: 15 40 W TWTAs 17/11.2–11.45 GHz up/down Ku-band Nordic beam, 26 MHz bandwidth, 52 dBW min EIRP
Principal applications: DTH via 45 cm dishes
Configuration: Hughes' HS-376HP spin-stabilised, 7.2 m high deployed (2.7 m stowed), 2.16 m diameter. Despun single reflector and communications shelf.
Mass: 1,467 kg at launch, 853 kg BOL
AOCS: Station-keeping and spin-stabilised at 50 rpm by four 22.2 N hydrazine thrusters; GEO insertion by solid Thiokol Star 30B AKM
TT&C: Via Nittedal station
Power system: GaAs solar cells mounted on cylindrical exterior and deployed drum providing 1,200 W BOL for payload. Batteries provide 100 per cent eclipse protection.

Thor 3
Launch: 11 June 1998 by Delta II 7925 from Cape Canaveral.
Location: ~1° W
Design life: >10–12 years.
Mass: 1,451 kg at launch, 853 kg BOL
Transponders: 14 Ku-band powered by 47 W TWTAs.

Other specifications same as Thor 2.

Thor 5
Launch: 11 February 2008 by Proton M/Breeze M from Baikonur.
Location: 0.8° W geostationary.
Design life: 15 years (fuelled for >18 years).
Transponders: 24 (15 FSS and 9 BSS) plus 5 active BSS available through steerable antenna. TWTA output power: 55 W (FSS) and 150 W (BSS). All transponders are linearised and eclipse protected. *Transponder bandwidth:* 27 MHz in FSS, 33 MHz in BSS.
Principal applications: Digital TV/radio to Nordic countries, Central and Eastern Europe.
Configuration: Orbital Star-2 bus, 3-axis stabilised. Two 2.3m dual gridded shaped deployable reflectors, one 0.75 m steerable antenna.
Mass: 2,012 kg at launch
AOCS: Station-keeping and spin-stabilisation using hydrazine thrusters.
TT&C: Via Nittedal station.
Power system: Twin solar arrays, four panels per array, UTJ GaAs solar cells. 3.6 kW of payload power.

Orbital Sciences Corporation built the Thor 5 communications satellite (Orbital Sciences) 1342637

Russian Federation

Ekran series

Background

Conceived in the late 1960s, Ekran (Screen) was the Soviet Union's first geosynchronous satellite and was originally to have been powered by a 5 kW nuclear reactor. Ekran 1, launched on 26 October 1976, brought direct broadcast services to the world for the first time. The original Ekran spacecraft were introduced in 1976 and upgraded to the Ekran-M model in the late 1980s. The more reliable variant resulted in a reduction in the launch rate. Ekran provides a unique direct television broadcasting service to the central federation region (Zone 3) of Siberia and the far north-east. All spacecraft in the series have been located at the Statsionar-T positions near 99°E over the Indian Ocean.

Kosmos 1817 in January 1987 introduced the improved Ekran-M, although a failure of the Proton fourth stage meant that it failed to reach GEO. A second spacecraft (Ekran 17) was launched in December 1987. Ekran 19 was placed in orbit in December 1988 and operated until early October 1996 when it performed an end-of-life manoeuvre and drifted off station. This left Ekran 20, launched October 1992, as the sole operating Ekran satellite. It continued in operation until 2001, when it was replaced by Ekran 18.

NPO PM has been controlling satellites itself instead of relying on the military since January 1994 through its Persei subsidiary at tracking, telemetry and command stations in Krasnoyarsk and Gauss Khroustalni. To service the system, the Soviet Union and its successor states built more than 5,000 receivers and now this network broadcasts television to some 20 million viewers in Siberia and the far north-east. Since no other signals interfere with the Ekran in these remote regions the satellite can broadcast directly at UHF (0.7 Ghz). The Ekran-M satellites each carry two 200 W transponders. They transmit directly to simple individual or communal receivers at 0.7 GHz. Their solar arrays have been augmented to provide 1.8 kW of power.

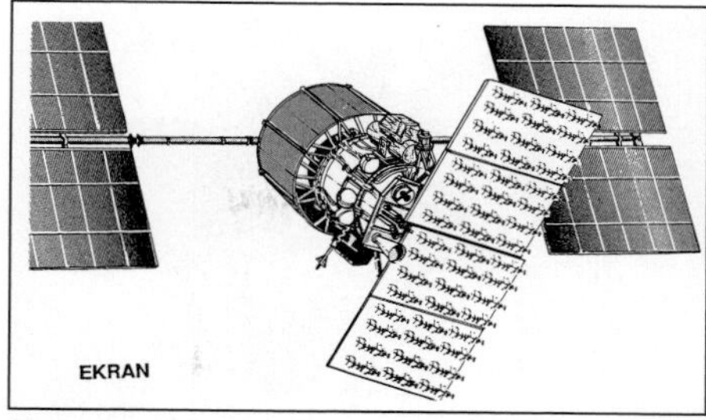

Ekran direct broadcast satellite, first launched 1976. Its 200 W transmitter provides TV coverage to remote Russian territory. Ekran-M has added a smaller panel to each of Ekran's four to boost power to 1.8 kW 0516991

At the time of the Ekran 18 launch in 2001, it was reported that this was the last flight of the Ekran M satellite bus, since the spacecraft carries only a single TV-relay channel, compared with modern communications satellites equipped with dozens of transponders. One more Ekran M spacecraft remained in inventory, but there were no launch plans because its storage life had expired. A modified version of the Ekran-M, called Ekran-D, was proposed to permit digital transmissions of a broader assortment of information, but has yet to appear.

Specifications

Launched: Total 27 between 1976 and 2001, made up of 21 launch attempts for Ekran and six of Ekran-M. The Ekran version was typically launched twice annually by 4-stage Proton-K from Baikonur. The first of the Ekran-M series were launched annually, but the most recent in the series was launched 7 April 2001 after a nine-year gap.
Locations: 99°E geostationary.
Design life: Ekran 1 year, Ekran-M >3 years.
Contractors: NPO PM (spacecraft), NPO Radio (payload)
Transponders: One (plus one back-up) C-band, 200 W covering Siberian region. Transponders bandwidth at –1dB, 24 MHz; Operating frequencies (transmission/reception), No 1 – 714/6,200 MHz, No 2 – 754/6,000 MHz; Radiation polarisation – circular; transmission – left-hand, reception – right-hand; EIRP at beam centre, 49.0 dBW; G/T at beam centre –12.0 dBW.
Principal users: Direct broadcast TV.
Configuration: KAUR-3 bus similar to Gorizont, but planar array carries 96 helical transmit aerials. Solar wings are configured to avoid antenna shadowing.
Antennas: one receiving antenna 3.5° × 3.5°, one transmitting antenna 3.0° × 8.5°
Mass: Ekran: 1,970 kg (320 kg payload).
Ekran-M: 2,100 kg
AOCS: 3-axis; east-west station-keeping to 0.5° by liquid thrusters
Power: Ekran-M's solar wings were enlarged to provide 1.8 kW.

Ekran-M launch record

Satellite	Launch date	Launch site	Launcher
Kosmos 1817 (Failure: Block DM-2 ignition failure, remained in LEO).	30 January 1987	Baikonur	Proton-K Block-DM-2
Ekran 17	27 December 1987	Baikonur	Proton-K Block-DM-2
Ekran 19	10 December 1988	Baikonur	Proton-K Block-DM-2
Ekran-M (14L). Failure: Third stage failure.	9 August 1990	Baikonur	Proton-K Block-DM-2
Ekran 20	30 October 1992	Baikonur	Proton-K Block-DM-2
Ekran 18	7 April 2001	Baikonur	Proton-M Briz-M

Ekran-M satellite being prepared for shipment (NPO-PM) 1330186

Express series

Current Status

The Russian Space Communications Company (RSCC) currently operates 11 satellites from 14°W to 145°E in geostationary orbit. These include three Express-A satellites and six Express-AM satellites.

Background

Gorizont's successor, called Express, appeared in 1994. InterKosmos operated and marketed the satellite service through its consortium of NPO PM, NIIKP, NPO Radio and the Vostok Bank (along with Gals). Most of the Express payload directly duplicated the Gorizont, although it deleted the L-band Volna service (transferred to Marathon). Express was the second Russian/Soviet satellite, after Gals, to provide NSSK. Intelsat agreed in March 1993 to lease up to three satellites beginning in mid-1994, but the agreement lapsed in 1994 because InformKosmos required payment in advance. Intelsat only agreed to begin leasing once each satellite was on-station.

Rimsat Ltd, an Asia-Pacific telecom company based in Fort Wayne, Indiana, planned to lease four Express satellites from InformKosmos for the Asia-Pacific region, with first launch in October 1994, but payments reportedly stopped in April 1994. The Express series spacecraft closely resembled the Gals spacecraft. It shared a similar spacecraft bus, though with slightly smaller dimensions of 3.6 m × 6.1 m, and identical solar arrays. It was expected to be deployed at 13 locations (40°, 53°, 80°, 90°, 96.5°, 99°, 103°, 140° and 145°E; 155°, 37.5°, 14° and 11°W) for domestic services and to support the Intersputnik Telecommunications Association. A typical Express payload included 10 C-band and two Ku-band transponders. **Express-1** was launched by Proton on 13 October 1994, and reached its checkout location of 70°E at the end of the month. It was moved to its operational Intersputnik position at 14°W in early 1995. The satellite experienced orientation difficulties on 12 June 2001 and was deactivated. Services were moved to Express-A3 at 11°W while Gorizont 26 was relocated to 14°W.

Express-2 was launched in September 1996 and located at 80°E. In July 2000 it was relocated at 103°E. On 11 June 2001 Express-2 suffered an emergency deactivation and services were transferred to Express A3. It was finally retired in May 2002.

The first Express-A satellite was destroyed in a Proton launch failure on 27 October 1999. **Express-A2** was launched on 12 March 2000 and located at 80.1°E. It was followed by **Express-A1R** on 24 June 2000 (40°E), built in only 15 months to replace the lost Express-A1, using insurance payment. **Express-A3** was launched 10 June 2002 by a Proton K/DM-2M but the parking orbit was off-nominal. This was

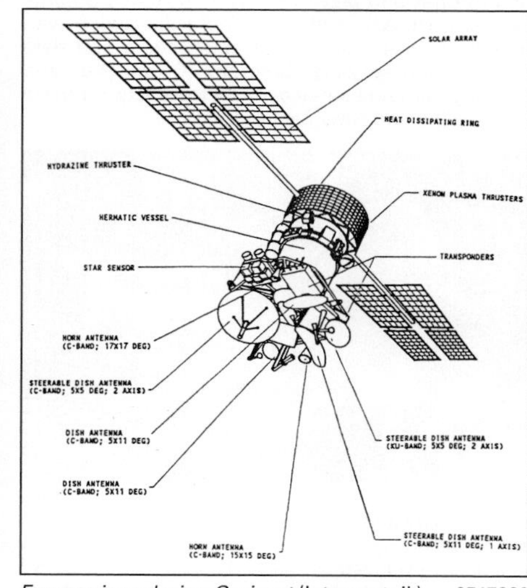

Express is replacing Gorizont (Intersputnik) 0517389

corrected by the second burn of the DM-21 upper stage and it was located at 11°W. The Express-A satellites are equipped with communications payloads produced by Alcatel Space (now Thales Alenia).

In 2001, the Russian Space Communications Company (RSCC) started a programme to renovate the civil communications satellite fleet as part of the 2001 to 2005 Russian Federal Space Programme. The purpose of the Express-AM programme was to secure the orbital and frequency resources of Russia, and to replace the obsolete Gorizont and Ekran-M series satellites in inclined orbits that had exceeded their lifetimes. The improved Express-AM series is based on the Express-M or 727M bus, first used on the Sesat satellite. It is designed to provide a wide range of Russian domestic communications services (digital TV, telephony, video conferencing, data transmission, the Internet access) and to apply VSAT technology. The spacecraft carry new antenna systems to provide high-quality communications and uniform coverage in C- and Ku- bands. The AM series communications payloads have been built either by NEC/Toshiba Space Systems or by Alcatel Space (now Thales Alenia).

Express-AM22, launched in December 2003, was the first of the third generation, built by Reshetnev (NPO PM) for Kosmicheskaya Svyaz. Located at 53°E, it became operational on 9 March 2004. It has 24 Ku-band transponders and four beams: European beam; wide European beam; plus steerable beams 1 and 2 covering North Africa, the Middle East and Central Asia. Intersputnik leases resources in the wide European beam to offer access to the Internet. A gyro failure in February 2008 threatened to curtail the spacecraft's life, but RSCC said it planned to upload a software patch to correct the problem. Eutelsat has a long-term lease on 12 transponders on AM22.

Express-AM11 was launched 26 April 2004 on behalf of RSSC and located at 96.5°E. The satellite suddenly failed at 23:41 GMT on 28 March 2006, as the result of a "sudden external impact", probably by space debris, according to a preliminary analysis by manufacturer NPO-PM. This impact caused the instantaneous decompression of the spacecraft's thermal control subsystem. The coolant leaked at high velocity, creating a torque that sent the satellite spinning. Although AM11 was brought back under control, it was suffering from rising temperatures after the loss of its cooling system. It was decided to put it into a graveyard orbit before high temperatures made it uncontrollable and caused its orbital slot to become unusable. All communication channels of state importance were transferred to Express-A2 at 103°E, Express-AM2 (80°E), and Express-AM3 (140°E).

Express-AM1 followed on 29 October 2004. Located at 40°E, it became operational on 1 February 2005. The satellite covers Russia, the CIS countries, Europe and the Middle East in C- and Ku-bands as well as South Asia in Ku-band. In Ku-band, Express-AM1 has two European beams and one high-power spot beam over India, and a single European beam in C-band. In the Intersputnik system, this satellite is used by mobile operators in VSAT networks in India, Afghanistan and Pakistan, as well as for e-learning. The wide European beam is mainly used for Internet traffic, telecommunications and broadcasting.

Express-AM2 was launched on 30 March 2005 and put into operation on 16 June 2005 at the 80°E slot. On 24 June 2005, **Express-AM3** was launched by Proton-K and placed at 140°E. It serves Siberia, the Far East, Asia and the Asia-Pacific region with digital television and radio, multimedia and data transmission services. Steerable C- and Ku-band antennas provide services from the eastern Russia to Australia and New Zealand.

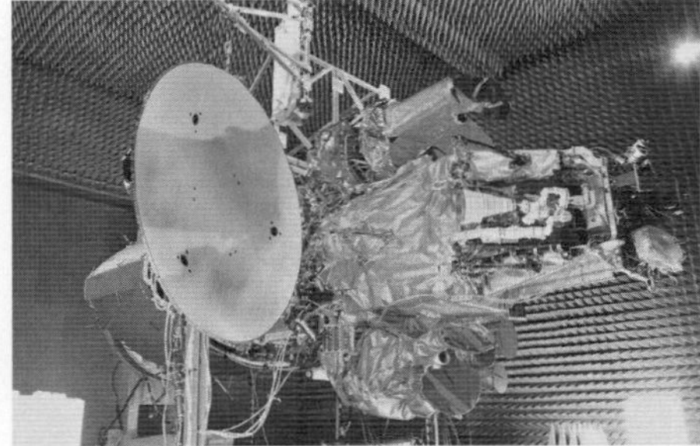

Express-AM33 is equipped with Ku-, C- and L-band transponders and differs from previous "Express-AM" spacecraft in having higher output power and new antennas developed by Thales Alenia Space (NPO-PM) 1343769

Express-AM22 with its antennas deployed during testing (RSSC) 1343766

In September 2004, Russia's Ministry of Information Technologies and Communications, the Federal Space Agency (Roscosmos), RSCC and NPO Prikladnoy Mechanicky (NPO-PM) signed a contract for the design, manufacture and launch of **Express-AM33** and **Express-AM44**. The same month, Alcatel signed a contract with RSCC to develop and deliver their payloads. The electronic equipment was manufactured in Alcatel Space's Toulouse plant and then integrated into the Express-AM platforms produced by NPO-PM in Krasnoyarsk, Russia.

Express-AM33 and AM44 have flexible configurations that permit RSCC to use them at any orbital position. Both satellites have C- and Ku-band transponders, and L-band capacity to provide mobile communications for the Russian president. The spacecraft are designed to provide digital TV and radio broadcasting, telephony, data transmission channels, videoconferencing services and Internet access. The satellites will also serve VSAT networks across Russia, the CIS countries, Europe, Asia and Africa.

Express-AM33 was launched in January 2008 and located at 96.5°E as part of the RSCC satellite constellation. Its high power characteristics reduce ground equipment costs. The steerable-beam antennas can meet varying market requirements. Intended to cover Russia and the Asia-Pacific region, Express-AM33 has a C-, Ku- and L-band payload, two fixed zone beams (C and Ku-band) and two Ku-band steerable spot beams.

Express-AM44 is set for launch in Q2 2008 to 11°W. It will replace the existing Express-A3 currently operating in this orbital position. It will have a C- and Ku-band payload, with a trans-Atlantic C-band zone beam, a C-band global beam and two Ku-band zone beams to cover Europe, the Middle East and North Africa.

The Federal Space Programme (FSP) for 2006–2015 foresees the creation of several high power communications satellites of the Express family. RSSC has imposed new requirements for these. The satellite mass shall be approximately 3200 kg; the lifetime shall not be less than 15 years. The satellite must ensure fixed, mobile president and government-level communications in C-, Ku- and L-bands, as well as TV broadcasting. The fixed service areas shall be enlarged, while there should be a possibility to have a number of steerable service areas. One of the main requirements is to a high reliability level and solely qualified flight proven equipment shall be installed on the satellite.

The first of these, and the largest Express satellite ever ordered, is **Express-AM4**, which will be placed in the 80°E orbital slot in 2010. On 17 March 2008, RSSC announced that it had selected Khrunichev Space Centre and Astrium to build the satellite. Express-AM4 will use Astrium's Eurostar E3000 satellite bus. It will carry 63 active transponders operating in L, C, Ku and Ka bands with payload power of 14 kW and a planned 15-year service life. Its 10 antennas will provide high performance coverage over the Russian Federation and the CIS countries and its steerable antennas will provide communication to any point within the satellite's visibility.

Meanwhile, on 6 December 2007, Thales Alenia Space (communications payload based on the Spacebus 4000 series) and NPO-PM (prime contractor) announced plans for a joint programme to develop a new multimission satellite platform, named Express-4000, which would be optimised for direct injection to GEO and also suitable for High-Elliptical Orbit (HEO) missions.

Express MD-1 and **MD-2** are due for launch in 2008. Alcatel Alenia Space Italia was contracted to provide their telecommunication payloads, while the bus is being built by Khrunichev. The 600 kg Express MD-1 is scheduled for a joint launch from Baikonur with Express-AM44. Smaller than the other Express satellites, they will each be equipped with eight C-band and one L-band transponders, plus a repeater panel and an antenna farm, to provide broadcasting and communications services across Russia and the CIS countries, as well as mobile presidential and governmental communications.

Express-AM22 carries a communications payload produced by Alcatel Space (now Thales Alenia) (RSSC) 1343767

Specifications

Express 1 and 2

Launch: Express 1: 13 October 1994 by Proton-K from Baikonur.

Express 2: 26 September 1996 by Proton-K from Baikonur.

Locations: Express 1: 70°E before moving to 14°W geostationary in February 1995.

Express 2: 80°E geostationary

Design life: 5–7 years

Contractors: NPO PM (spacecraft), NIIKP (payload)

Transponders: One (plus one back-up) 60–80 W TWTA 6.000/3.675 GHz up/down C-band beam, 40 MHz bandwidth, transponder number 6, switchable to A1 global (EIRP 35.6/32.6 dBW max/min) or A2 steerable (7.3° N, 5° S, 5.5° E/West) 4.6° spot (EIRP 46.1/43.0 dBW max min), RHCP, for single channel TV distribution to 2.5 m Moskva antenna

Nine 10.5 W TWTA 6.050–6.450/3.725 4.125 GHz up/down C-band global, hemispheric, zonal and spot beams, 36 MHz bandwidth, numbers 14/16 jointly switchable to A2 steerable (7.3° N, 5° S, 5.5° E/west) 4.6° spot (EIRP 38.0/35.0 dBW max/min) or A4 steerable 7.5° N-S 4.6 × 10.6° elliptical (EIRP 34.7/31.7 dBW max/min), numbers 8/10 jointly A4 steerable 7.5° N-S 4.6 × 10.6° elliptical (EIRP 39.2/36.2 dBW max/min) or A3 fixed 14.6° quasi-global (EIRP 33.4/30.4 dBW max/min), numbers 7/9/11 jointly A5 fixed 14.6° quasi-global (EIRP 28.9/25.9 dBW max/min) or A6 north-south steerable 4.6 × 10.6° elliptical (EIRP 34.7/31.7 dBW max/min), numbers 15/17 as 7/9/11, RHCP transmission.

Two 15 W TWTAs centred at 14.325/11.525 and 14.425/11.625 GHz up/down Ku-band beams, numbers 12/20 working through fully steerable (north-south and east-west) 4.6° spot antenna, 36 MHz bandwidth, EIRP 39.2/36.6 dBW maximum/minimum, orthogonal polarisation (vertical receive and horizontal transmission) for Luch service

Principal users: TV distribution via Moskva system for Russia, telecommunications services

Configuration: 3-axis stabilised, based on pressurised 2 m diameter central pressurised cylinder (maintained at 0–40°C). Seven antennas provided zonal, hemispheric and global beams. Power came from twin four-panel solar wings spanning 21.0 m, cylindrical base also carried solar cells; 2,400 W EOL. AOCS provided by liquid hydrazine thrusters (25.5 kg hydrazine); north-south/east-west station-keeping ±0.2° by eight OKB Fakel SPT-100 plasma thrusters (56 kg Xe loaded). ±0.1° pointing accuracy. Express could operate autonomously for 30 days.

Mass: About 2,500 kg at launch (payload 430 kg).

Express-A1

Launch: 27 October 1999 by Proton-K/Block DM2 from Baikonur. Launch failure due to early shut down in second-stage burn.

Express-A2 (Express-6A)

Launch: 12 March 2000 by Proton from Baikonur

Locations: Initially 96.5°E, 80°E (May 2000 – Oct 2005), 103°E (from Nov 2005)

Design life: 10 years

Contractors: NPO-PM (bus). Alcatel (communications payload)

Transponders: Eight 50 W TWTA 14.000–14.250/10.950 11.200 GHz up/down Ku-band 3 × 3° beams, 72 MHz bandwidth, EIRP 46.9 dBW, for 1,024 Mbit/s data services to 1.5–2.5 m dishes

12 (plus 4 back-up, 3 × 4-for-3 + 1 × 2-for-1) 32 W TWTA 14.340–14.500/11.540–11.700 GHz up/down

Ku-band 3 × 3° beams, 2 × 36 MHz bandwidth, EIRP 44.9 dBW, for TV distribution and telecom services
Principal uses: TV distribution, telecommunications services
Configuration: KAUR-4 MSO-2500 bus, 3-axis stabilised, based on pressurised 2 m diameter central cylinder. 4 transmission/receive antennas. Twin 6-panel solar wings spanning 26.6 m providing 6 kW (Ni/H_2 batteries) and the base also carries solar cells. AOCS by hydrazine thrusters; north-south/east-west station-keeping to ±0.1°, by Xe ion thrusters (120 kg Xe)
Mass: 2,570 kg at launch.

Express-A3
Launch: 24 June 2000 by Proton-K/DM-2M from Baikonur
Location: 11°W
Specification: As Express A2.

Express-A1R (A4)
Launch: 10 June 2002 by Proton-K/DM-2M from Baikonur
Location: 40°E (June 2002–Sept 2005), 14°W (from Oct 2005)
Specification: As Express A2.
Mass: 2,600 kg

Express-AM22
Launch: 29 December 2003 by Proton-K/DM-2M from Baikonur
Location: 53.1°E geostationary
Design life: 12 years
Mass: 2,600 kg at launch. Payload mass 590 kg
Contractors: NPO-PM (bus), Alcatel Space (communications payload)
Principal uses: TV & radio broadcasting, telephony, data transmission, Internet access, videoconferencing and other communications services.
Configuration: Express-M or 727M bus. 3-axis stabilisation. Station-keeping accuracy (longitude/incl.) −0.5°. Twin solar arrays.
Transponders: 24 Ku-band, 54 MHz bandwidth at −1dB. Output power, 103.5 W.
Power: Two solar arrays, 6 kW (EOL). Power for payload 4.2 kW.

Express-AM11
Launch: 26 April 2004 by Proton-K/DM-2M from Baikonur
Location: 96.5°E
Design life: 12 years
Mass: 2,543 kg at launch.
Contractors: NPO-PM (bus), Alcatel Space (communications payload)
Transponders: 4 Ku-band, bandwidth 54 MHz, 120 W EIRP 41/51 dBW. 26 C-band, bandwidth 36 and 40 MHz, power in one transponder 110 W, in 15 transponders 70 W, in 10 transponders 40 W. EIRP 37/46 dBW.
Power: 6.359 kW (EOL). Power for payload 4.4 kW.

Express-AM1
Launch: 29 October 2004 by Proton-K/DM-2M from Baikonur
Location: 40°E
Contractors: NPO-PM (bus), NEC/Toshiba Space Systems (communications payload)
Transponders: 18 Ku-band, 54 MHz bandwidth at −1dB, output power 95–100 W. 9 C-band, 40 MHz bandwidth at −1dB, output power 40 and 120 W. 1 L-band, 0.5 MHz bandwidth at −1dB, output power 30 W.
Mass: 2,542 kg at launch. Payload mass 570 kg.

Express-AM2
Launch: 29 March 2005 by Proton-K/DM-2M from Baikonur
Location: 80°E
Contractors: NPO-PM (bus), Alcatel Space (communications payload)
Transponders: 16 C-band, bandwidth 40 and 72 MHz, EIRP up to 47.5 dBW. 12 Ku-band, bandwidth 54 MHz, EIRP up to 53 dBW. 1 L-band, bandwidth 0.5 MHz.

Express-AM3
Launch: 24 June 2005 by Proton-K/DM-2M from Baikonur
Location: 140°E
Contractors: NPO-PM (bus), Alcatel Space (communications payload)
Transponders: 16 C-band, bandwidth 40 and 72 MHz, EIRP up to 47.5 dBW. 12 Ku-band, bandwidth 54 MHz, EIRP up to 53 dBW. 1 L-band, bandwidth 0.5 MHz.

Express-AM33
Launch: 0:18 GMT on 28 January 2008 by Proton-M/Breeze-M from Baikonur
Location: 96.5°E
Contractors: NPO-PM (bus, payload structure and L-band transponder), Alcatel Alenia Space France (communications payload)
Transponders: 16 Ku-band, 14/11 GHz, bandwidth 54 MHz, output power 150 W. Min. EIRP 49/53 dBW. Min. G/T −0.5/+6.0 dB/K. One 1.05 m antenna, one 0.65 m antenna.
10 C-band, 6/4 GHz, bandwidth 40 MHz, output power 100 W. Min. EIRP 36/47.5 dBW. Min. G/T −10/+3.5 dB/K. One 2.0 m antenna, two horns.
1 L-band, output power 80 W. EIRP (gateway-user) 36.5/40.5 dBW, (user-gateway) 23.0 dBW. G/T (gateway-user) −3.0/+3.5 dB/K, (user-gateway) −7.0/−3.5 dB/K. Two active phased array antennas.
Mass: 2,600 kg at launch. Payload mass 624 kg.
Power: 6.77 kW (EOL). Power for payload 4.41 kW.
Design life: 12 years

Gonets-D1 Constellation

Current Status
Gonets D1-1 to D1-3 were launched by Tsyklon 3 on 18 February 1996 followed by Gonets D1-4 to D1-6 by Tsyklon 3 on 14 February 1997. The system was never deployed to its planned constellation of 12 satellites in two orbital planes of six and commercial application failed through lack of funding. Only two sets of three satellites have been successfully launched to date.

Background
The civil Gonets system of small communications satellites come directly from the equivalent military Strela-3 communications satellite programme. Russia flew two Gonets-D demonstrator missions named Kosmos 2199 and Kosmos 2201 as part of the sextet launched on 13 July 1992. The two sextet missions launched on 19 February 1996 and 14 February 1997 each carried three of the military Strela-3 satellites and three Gonets-D1 test satellites known as Gonets D1-3 and D4-6, respectively.

The Gonets-D1 (Gonets means 'messenger') are cylindrical (covered in solar cells), 0.8 m diameter and 1.6 m long, plus an extended boom for gravity stabilisation and cone- antennae: the mass of each satellite is 225–230 kg. AKO Polyot builds the satellites for NPO PM. The Gonets-R satellite was to have been equipped with satellite intralinks with 45 satellites in five planes of nine each operating at L-band and Ku-band. Each satellite would have had the capacity of 15 ground-space and three space-ground channels with 8 Mb of on-board storage and a data transmission rate of 1.2–64 kbps.

The Smolsat consortium, including NPO PM, NPO Precision Instruments, Selkhoz Bank and Moscow's Soyuzmedinform Program Management (connected with the Ministry of Health), planned two systems. The first was to provide e-mail store/forward and real-time (within the 3,000 km diameter footprint) bulk data relay. The second, Gonets-R, was to offer mobile phone links.

Network Services International in New York claimed to be the western representative for Gonets and the Russian Space Agency provides state support. The first 36-satellite system was projected to be operational by end-1996, employing six satellites in each of six 82.5° planes at 1,300–1,500 km altitude for 4.8/9.4/64 kbit/s links. 312–315/387–390 MHz up/down 10 W link. Each 230 kg satellite would handle 1.85 Mbit daily, plus 250 Gbit for the 1.6/1.5 GHz up/down L-band, EIRP 10 dBW. This plan did not materialise and after a series of test launches to validate hardware the first trio of satellites was launched on 14 February 1997. A second trio was launched 27 December 2000 but failed to reach orbit. The third trio launched to date were placed in orbit on 28 December 2001.

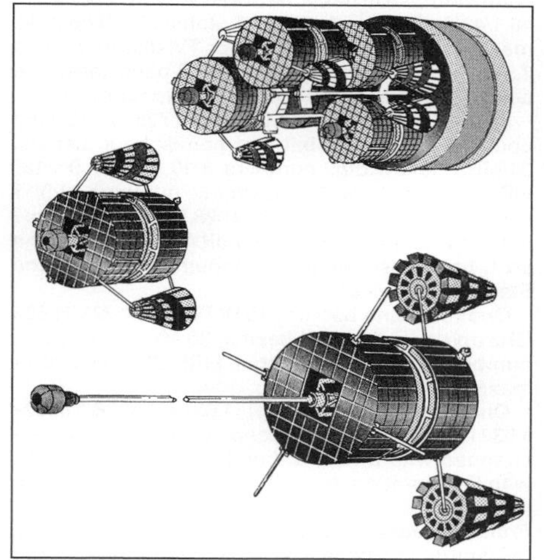

Derivatives of the military sextet store and forward satellites are offered commercially.
(Teledyne Brown Engineering) 0516993

Gorizont (Horizon) series

Current Status
The thirty-third and final Gorizont was launched on 6 June 2000.

Background
The Gorizont system was inspired by the 1980 Olympic Games held in Russia, under design article number 11F662 built by the MOM agency. The geostationary satellite series was later made a part of the YeSSS Unified Satellite Communication System. Gorizont 1 appeared 19 December 1978, although it achieved only 22,553 × 49,023 km, 11.3° due to a Proton Block DM malfunction. Despite its non-GEO path, it provided some service and was joined by three others for coverage of the Moscow Olympic Games.

The early Gorizont spacecraft had a launch mass of more than 2.1 metric tons and had an operational lifetime of nearly 10 years, although a 5-year service life was more common. The 3-axis stabilised satellite is approximately 2 m in diameter and 5 m long with two large solar arrays capable of generating 1.3 kW of power for the first three years. Seven separate transmission antennas permit a variety of reception patterns for both broad and localised service.

A typical Gorizont communications payload includes six general purpose (TV, audio, facsimile) 6/4 GHz transponders (five 12.5 W and one 60 W), one Luch 14/11 GHz transponder (15 W), and one Volna 1.6/1.5 GHz transponder (20 W). The Volna transponders are INMARSAT-compatible and are widely used by the Russian merchant marine fleet via the primary control centre in the Tomilino suburb of Moscow and the Odessa and Nakhodka ground stations. Gorizont was the primary GEO television rebroadcasting system, supporting all five federation time zones: Zone 1 from 140° E, Zone 2 from 90° E, Zone 3 from 80° E, Zone 4 from 53° E, and Zone 5 from 14° W. These transmissions are handled by the Orbita (12 m receiving antenna) and Moskva (2.5 m receiving antenna) ground stations in the 6/4 GHz band. The Moskva Globalnaya system was inaugurated in 1989 using 4 m receiving antennas and fed by Gorizonts at 96.5° E and 11° W.

A single 40 W transponder, similar to its Molniya 3 equivalent, relays TV to thousands of 2.5 m diameter dishes in the Moskva system. Gorizont took over the US-Russia hotline previously supplied by Molniya 3. The full Gorizont capacity was offered by Glavcosmos at about USD40 million, including launch. Intersputnik provides most of its services but Express is now replacing the Gorizonts.

Gorizont 22 appeared 23 November 1990, and was positioned at 40° E, the Statsionar-12 location. It was the first of three satellites for the Russian Federation's Ministry of Communications & Information Technology, which paid Rb100 million to the Ministry of Defence for launch services. The satellite provided TV services to 21,000 settlements in European Russia, while the second, Gorizont 23 launched 1 July 1991 to 103° E (Statsionar-21,) added 3,500 in the Urals and Siberia. The third appeared 2 April 1992 as Gorizont 25, also over 103° E (Gorizont 23 apparently failed mid-1992). Gorizont 20 carried a Mayak transmitter, possibly to demonstrate mobile L-band communications. What should have become the original Gorizont 28 was lost in the Proton launch failure of 27 May 1993. It was to have been the first occupant of the Statsionar-16 145° E slot, from where transponders numbers 9–11 were to have been marketed by SovCan Star and Vista Satellite.

Russia and the former Soviet Union launched a total of 33 named Gorizont satellites by end of 2000 into Statsionar and, later, Tonga slots. Russia said January 1996's Gorizont 31 would complete the series, but Gorizont 32 appeared in May 1996. Gorizont 29 was located at 130° E (a Tonga slot), and was leased from InformKosmos as Rimsat 1

Gorizont satellites

Launch Date	Satellite	Location	Location Name
19 December 1978	Gorizont 1	–	Failed to orbit
6 July 1979	Gorizont 2	14° E	Statsionar-4
28 December 1979	Gorizont 3	53° E	Statsionar-5
14 June 1980	Gorizont 4	14° W	Statsionar-4
15 March 1982	Gorizont 5	53° E, then 96.5° E	Statsionar-5, then Statsionar-14
20 October 1982	Gorizont 6	90° E, then 140° E	Statsionar-6, then Statsionar-7
1 July 1983	Gorizont 7	14° W, then 11° W	Statsionar-4, then Statsionar-11
30 November 1983	Gorizont 8	90° E	Statsionar-6
22 April 1984	Gorizont 9	53° E	Statsionar-5
2 August 1984	Gorizont 10	80° E	Statsionar-13
18 January 1985	Gorizont 11	140° E, then 53° E, then 11° W	Statsionar-7, then Statsionar-5, then Statsionar-11
10 June 1986	Gorizont 12	14° W, then 40° E	Statsionar-4, then Statsionar-12
18 November 1986	Gorizont 13	90° E	Statsionar-6
11 May 1987	Gorizont 14	140° E, then 103° E	Statsionar-7, then Statsionar-21
18 January 1988	Gorizont 25L	–	Failed to orbit
31 March 1988	Gorizont 15	14° W, then 11° W	Statsionar-4, then Statsionar-11
18 August 1988	Gorizont 16	80° E	Statsionar-13
26 January 1989	Gorizont 17	53° E, then 134° E, then 34° E	Statsionar-5, then Tonga, then Statsionar-2
6 July 1989	Gorizont 18	140° E	Statsionar-7
28 September 1989	Gorizont 19	97.5° E, then 34° E	Statsionar-14
20 June 1990	Gorizont 20	14° W, then 25° E, then 96° E	Statsionar-4, then Statsionar-14
3 November 1990	Gorizont 21	90° E, then 145° E	Statsionar-6, then Statsionar-16
23 November 1990	Gorizont 22	40° E, then 140° E	Statsionar-12, then Statsionar-7
2 July 1991	Gorizont 23	103° E	Statsionar-21
23 October 1991	Gorizont 24	80° E	Statsionar-13
2 April 1992	Gorizont 25	103° E	Statsionar-21
14 July 1992	Gorizont 26	349° E, then 14° W	Statsionar-11, then Statsionar-04
27 November 1992	Gorizont 27	53° E, then 96.3° E	Statsionar-05, then Statsionar-14
28 October 1993	Gorizont 28	90° E, then 96.5° E, then 103° E	Statsionar-6, then Statsionar-14, then Statsionar-21
18 November 1993	Gorizont 29 (Rimsat 1)	130° E, then 161° E, back to 130° E, then 3° W.	(Tonga)
20 May 1994	Gorizont 30 (Rimsat 2)	142.5° E	(Tonga)
25 January 1996	Gorizont 31	40.5° E, then 103° E, then 140° E	Statsionar-12, then Statsionar-21, then Statsionar-07
25 May 1996	Gorizont 32	53° E, then 14° W	Statsionar-5, then Statsionar-04
6 June 2000	Gorizont 33 (45L)	145° E	Statsionar-16

Three Gorizonts were located at positions which had not been registered to the RFAS: Gorizont 19 was close to the Statsionar-2 location (35° E) and Gorizont 20 close to Statsionar-19 (23° E). but there was no RFAS registration close to the Gorizont 29 location since it was in a Tonga slot, as was Gorizont 30.

by the US-Malaysian Rimsat company. Rimsat used the satellite for three years at a cost of USD12 million. In September 1995 the satellite was taken over by Intersputnik and in 1997 it was sold to the PASI group. Rimsat initially leased 'Tongasat 1' (Gorizont 17) at 134° E from July 1993, but when the satellite moved January 1995, the lease ended. Gorizont 30 (Rimsat 2) was launched 20 May 1994 towards 142.5° E.

Gorizont 25 was moved to a graveyard orbit in November 2006. Gorizont 26 was moved to a graveyard orbit in March 2007. Gorizont 27 was decommissioned in November 2003. Gorizont 29 was decommissioned in December 2003. Gorizont 31 was moved to a graveyard orbit in May 2006.

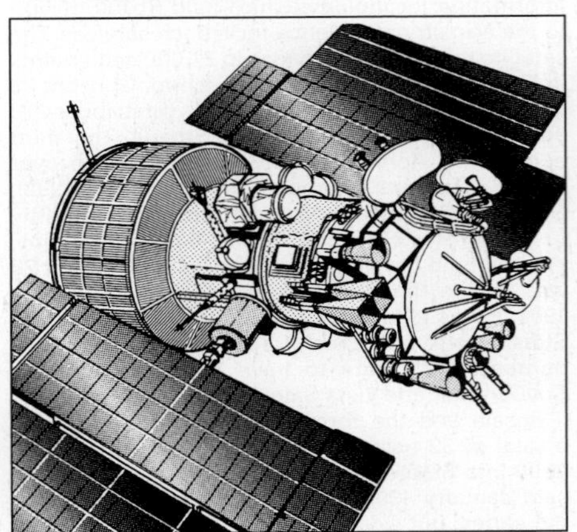

The Gorizont multipurpose communications satellites are being replaced by the Express series 0516992

Specifications

Launched: Until 1993 typically twice annually, by Proton-K from Baikonur.

Locations: 40/53/80/90/96.5/103/140° E and 14/11° W geosynchronous (plus Rimsat/Tonga slots), inclined 1.5° BOL

Design life: Minimum 3 years.

Contractors: NPO PM (spacecraft), NIIKP (payload).

Transponders: One (plus one back-up) 65 W TWTA 6.000/3.675 GHz up/down C-band spot or global beam for transponder number 6, 40 MHz bandwidth, switchable between A3 4.6° spot EIRP 46.1/43.0 dBW max/min or A1 global 35.6/32.6 dBW maximum/minimum, RHCP, for TV distribution to 2.5 m Moskva antenna. The six C-band (also see below) together comprise the Statsionar service.

Five 15 W TWTA 6.050–6.250/3.725–3.925 GHz up/down C-band global and hemispheric beams, 34 MHz bandwidth, numbers 8/10 on A4 9 × 18° EIRP 31.0/28.0 dBW max/min, numbers 7/9/11 switchable between A6 global 28.5/25.5 dBW max/min and A8 6 × 12° EIRP 34.0 dBW max; RHCP. The six C-band (also see above) together comprise the Statsionar service.

One (plus one backup) 15 W TWTA 14.325/11.525 GHz up/down Ku-band beams, 36 MHz bandwidth, number 12 on A9 4.6° spot EIRP 39.2/36.0 dBW max/min 4.6° Luch service.

One (plus one back-up) 1.6377–1.6386/1.5362–1.5371 GHz up/down L-band; transponder cross-strapped with a 15 W C-band transponder for link with Earth station, for Volna maritime service, EIRP 40 dBW.

Principal users: TV distribution via Moskva system, telecommunications services, maritime and aeronautical mobile.

Configuration: 3-axis stabilised, based on pressurised 2 m diameter central cylinder, about 5 m long. 11 antennas provide zonal, hemispheric and global beams. Twin solar wings, of two major/two minor panels each, spanning 9.460 m generate 1,280 W; cylindrical base also carries solar cells. Liquid bipropellant thrusters provide AOCS. East-West station-keeping to within 0.5° is exercised but north/south excursions are uncontrolled: 2° is possible after 3 years.

Mass: About 2,120 kg originally, later about 2,500 kg.

Luch/Altair/SDRN/SSRD series

Current Status

Luch-0 was taken out of service in October 1998. Luch-1 was taken out of service in 1999 after a payload failure.

Background

The Soviet Union gave notice in 1981 to the International Frequency Registration Board that it intended to establish a Satellite Data Relay Network (SDRN), similar to NASA's TDRSS. The mission requirements stemmed from a military specification but incorporated a strong civilian applications directive. Several changes in operational requirements were implemented in order to accommodate new requirements from the Ministry of Defence. Soviet officials planned to place satellites at 95°E, 16°W and 160°W between 1983 and 1985. The system would eliminate the Soviet problem of communicating with its spacecraft from the geographically limited Soviet territory. The system would feature an Eastern SDRN for communications with Salyuts and other low-orbiting spacecraft, and using frequencies of 10.82, 11.32, 13.7 and 13.52 GHz for downlink, and 14.62 and 15.05 GHz for uplink.

The first satellite in the series, **Kosmos 1700**, launched 25 October 1985, placed a Luch at 95°E (CSDRN) to relay communications with the new Mir space station (launched in 1986), and subsequently with Soyuz-TM. In October 1986, the satellite began drifting off station. A replacement, **Kosmos 1897**, appeared 26 November 1987. It moved to 12°E in July to August 1988 to support the Buran Shuttle mission, and then returned to 95°E in February 1989, but by the end of 1991 it had drifted to 77°E and was abandoned.

The WSDRN position at 16°W was occupied by Kosmos **2054** from mid-January 1990 (after launch on 27 December 1989). With both positions operational, the expanded Mir coverage reduced the requirements for Soviet communications ships. Together with Kosmos 1897, it enabled Mir to maintain contact with Mission Control in Moscow 70 per cent of the time.

The first named **Luch** (Beam) appeared 16 December 1994 and was positioned at 95°E, while Kosmos 2054 exceeded its planned operating lifetime. Its performance was said to almost treble the capacity of the channels and double the service lifetime compared to the Altair design. The second named spacecraft, **Luch-1**, was launched on 11 October 1995 and placed at 77°E. At the end of March 1997, Luch-1 left station and in early May was re-located close to the Kosmos 2054 longitude.

Excess Luch capacity was leased on a commercial basis through NPO PM's Mercury subsidiary, particularly for TV in the US and Argentina. The Luch name was given to both a service and a system. Some Gorizont and Raduga satellites carried Luch transponders, usually as only one of the spacecraft's payloads.

At the 2007 Paris Air Show, Thales Alenia Space announced that it had signed a contract with Russian satellite manufacturer NPO-PM to supply two sets of repeater equipment for the **Luch-5A** and **Luch-5B** communications satellites ordered by the Russian space agency (Roskosmos). The two geostationary relay satellites will be positioned at 16°W and 95°E, and will provide TV and data communications links between the International Space Station and the ground, as well as control of LEO spacecraft during all operational stages. The satellites are scheduled for launch in 2009 and 2010. The repeater equipment on Luch-5A will have six S- and Ku-band channels, while Luch-5B will carry four S- and Ku-band channels. The same repeater equipment will be integrated on platforms supplied by NPO-PM.

Specifications
Locations: 16°W, 77°E, 95°E GEO
Design life: 5 years
Contractors: NPO PM (spacecraft), NPO Radio (payload)
Transponders: TWTA 15.05/13.5 GHz up/down Ku-band 0.5° steerable, EIRP 56.2 dBW
TWTA 14.62/11 GHz up/down Ku-band 1.0° steerable, EIRP 42.3 dBW TV/34.2 dBW voice
TWTA 900/700 MHz up/down UHF 5.0° steerable, EIRP 40.7 dBW
Configuration: 16.0 m-span solar array provided 1.8 kW
Mass: 2,400 kg on-station
AOCS: 3-axis; east-west station-keeping to 0.5° by liquid thrusters.

Molniya (Lightning) constellation

Current Status
After initial attempts to upgrade the Molniya series, the last Molniya satellite was launched in June 2005. After the planned final launch in August 2007 of the Molniya 3K failed to take place, the programme closed and has now been superseded by the newer Meridian series of satellites.

Background
Sergei Korolev's design bureau first conceived of Molniya in 1962 as a method to cover the USSR's northern expanses from highly inclined orbits. Since boosters had not yet attained the ability to place a satellite in geosynchronous orbit, when the USSR launched the first Molniya it put it in a sub-geosynchronous orbit that became a unique signature of the satellite. With a usual perigee over the Southern Hemisphere, each satellite lingered around apogee over the northern territories for about 8 hours each day. Their unusual orbits also avoided the severe launch penalty imposed on geosynchronous satellites from latitudes at Tyuratam and Plesetsk. By coincidence, the 63.4° Molniya inclination also nulled orbital perturbations created by the oblateness of the Earth and so the satellites required much less station-keeping propellant than most other highly elliptical orbits. On the other hand, the orbit passed through the Van Allen radiation belt, which imposed a radiation dosage five to six times the amount experienced in geosynchronous orbit. Hardening to mitigate this radiation dose made the satellites about 20 per cent heavier than their non-hardened counterparts.

The USSR inaugurated communication service with the first Molniya 1 series satellite, launched 23 April 1965. This Molniya carried the first operational television and telephony links between Moscow and Vladivostok before decaying 16 August 1979. Molniya 1-3, launched almost exactly one year later, added Earth imaging to complement the Meteor series, and exchanged colour television signals with France. Three satellites in a constellation with the same orbital parameters provide 24-hour coverage. Generally however, the USSR placed eight satellites in planes 45° apart to ensure un-interruptible service. The original network of Molniya 1 satellites probably carried military/government links over one transponder. An improved Molniya-1T appeared in June 1970 equipped with the "Beta" transmitter and became operational in 1972. The satellite was part of the Korund system facilitating strategic communications and telephony with military units in Siberia and the Far East. The complete system included eight satellites on orbit continuously. The 16-constellation network were divided into two types and four separate groups. Eight Molniya-1T types were operated in two constellations of four with both consisting of four orbital planes spaced 90° apart with the ascending node of one displaced 90° to the other. Beginning with Molniya 1-75 all Molniya-1T types were launched from Plesetsk by Molniya launcher. The last Molniya-1T was launched in September 1998 followed by Molniya 1-92 in April 2003.

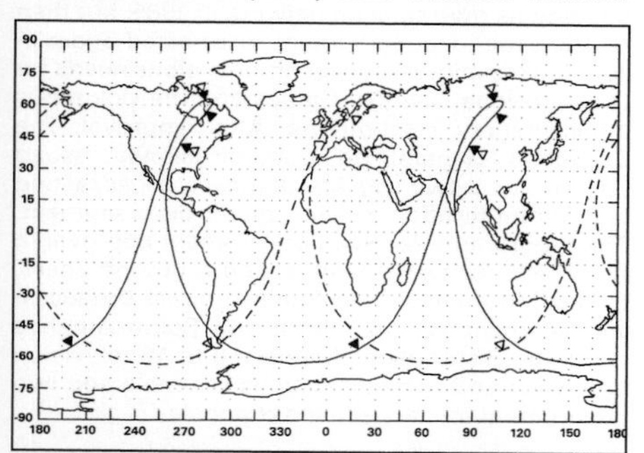

Molniya 1/3 satellites usually followed a common groundtrack taking them over the heart of the Russian landmass. Since 1983, some Molniya 3s had flown with groundtracks (dotted line) between the original constellations (Teledyne Brown Engineering) 0516996

The Molniya 1 appeared to be employed primarily for military and government links through at least a single 40 W X-band transponder. Some early models incorporated a camera to add synoptic Earth views to low-altitude Meteor imagery. Molniya 1-20, launched 4 April 1972 from Plesetsk carried France's piggyback research SRET 1. Molniya 1-30 similarly released SRET 2 on 5 June 1975. Molniya 1S, different from the standard series acted as a geosynchronous demonstration.

Beginning with Molniya 1-11 in 1969, the number above the horizontal at any one time was increased from two to three by reducing the equatorial spacing to 90°. Following the introduction of the Molniya 2 series in 1971 and Molniya 3 in 1974, they were positioned in groups of three employing different frequencies. As traffic increased, the Soviets had to decrease the spacing between Molniya 1s to 45°.

The Molniya constellation achieved great success at bringing communications to remote parts of the USSR. Starting with Molniya 1-11 in 1969, the USSR decreased the orbital stationing to 90° instead of 120° in order to keep at least two satellites above the horizon at all times. The Molniya 2 constellation followed in 1971, and the Molniya 3 constellation in 1974. But the newly increased traffic demands caused the USSR to rely more on geosynchronous satellites and the last Molniya 2 began its service in 1977.

The first Molniya 2, a more capable series employing higher frequencies orbited on 24 November 1971. The 17th and last satellite appeared on 11 February 1977.

Molniya 3 first appeared in November 1974 and was used to create the Orbita communications system for northern regions. It was derived from the Molniya 2M, development of which began in 1922. From early 1976, Molniya 3 provided one of the two independent satellite-based systems (the other via Intelsat) to maintain the Washington-Moscow hotline. Presidents Nixon and Brezhnev agreed to the hotline concept during the SALT negotiations of 1971. To ensure contact, the United States agreed to continually track the entire Molniya constellation from the US station at Fort Dietrick (Frederick, Maryland). Dietrick tested the circuits hourly. (Traffic from the hotline later shifted mostly to the Gorizont).

Until 1984, all Molniyas had a common ground track roughly bisecting Asia/North America. In late 1984, Molniya 3–21 manoeuvred to re-establish its ground track over the Pacific and the extreme eastern edge of the Atlantic. During 1985, Molniyas 3–25, 3–26 and 3–27 followed suit. These new satellites had a planar spacing of 90°, but the ascending nodes occurred mid-way between the other Molniya 3s. Western observers speculated

Each Molniya 1 carried at least one 40 W X-band transponder for government and military links 0516994

that this new constellation might be connected with maritime communications.

The Molniya 3, NPO-PM's newest class of satellites offered three C-band transponders for general telecommunication services, including television distribution over the 100 or more ground stations of the Orbita 2 network, 12 m ground stations in the CIS and affiliated countries such as Cuba and Mongolia.

157 named satellites were launched in the three series. Molniya 1-85 was at first described as a Molniya-1T, although the Russians subsequently indicated that Molniya-1T satellites had been operating for many years before this. In the same source the Russians referred to Molniya 1-88 as the 48th 1T, but they gave no indication of the difference with the standard design.

In 2003, eight Molniya 1 and four Molniya 3 satellites still operated in planes spaced at 45° intervals. Russia had formally ended the Molniya 2 programme. Russia continued to launch Molniyas, but not at the same rate as their original introduction in April 1965. Russia launched only four Molniya types between August 1996 and the end of 1999. No Molniya were launched in 2000 but two Molniya 3 series were launched in 2001 and one Molniya-1 was launched in April 2003. On 19 June 2003 another Molniya 3 was launched.

Specifications
Molniya 1
Launched: typically one a year by Molniya-M from Plesetsk (very rarely Tyuratam)
Orbit: typically 400 × 40,000 km, 63°; placed in eight orbital planes separated by 45°
Design life: 2 years
Contractors: NPO PM
Transponders: At least one 40 W 1.0/0.8 GHz up/down X-band; possibly total of three transponders
Principal users: military/government

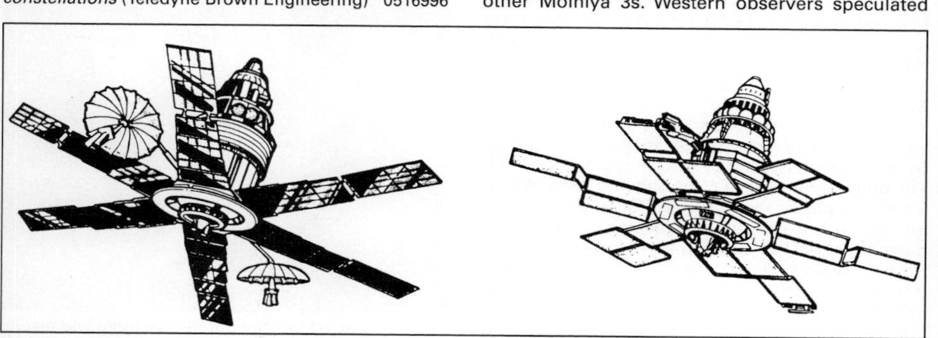
Molniya 3 (right) compared with Molniya 1 (Teledyne Brown Engineering) 0516995

A late Molniya 3 0132812

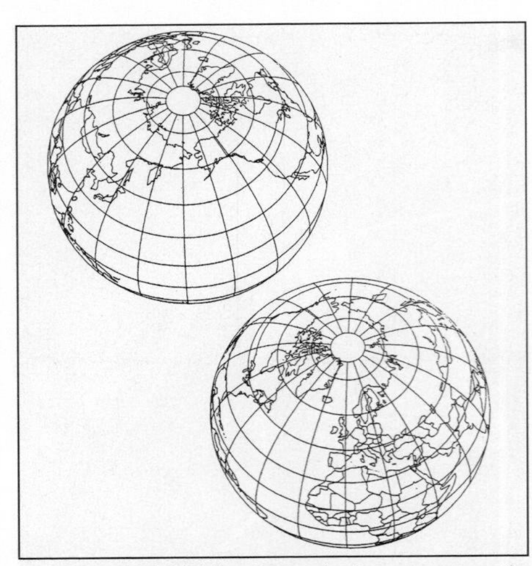

One group of Molniya 3s reached apogee over the Pacific and East Atlantic, possibly for maritime links. Shown is the vantage point on each of the two daily apogees (Teledyne Brown Engineering) 0516997

Configuration: pressurised cylinder with domed ends, 1.6 m diameter, 3.4 m long, accommodates electronics and provides mounting for two boom-mounted 90 cm diameter dishes. Six windmill solar wings generate 500–700 W. AOCS provided by liquid bipropellant engine assembly at other end; 3-axis control
Mass: about 1,600 kg.

Molniya 3
Launched: typically one or two per year on average
Orbit: typically 400 × 40,000 km, 63°; placed in eight orbital planes at 45° intervals
Design life: 4 years?
Transponders: 2 (plus 1 back-up) 30 W TWTA 5.9756.225/3.650–3.900 GHz up/down C-band global beams, 50 MHz bandwidth, 35 dBW edge EIRP, circular polarisation
Principal users: domestic TV/telecommunications
Configuration: pressurised cylinder with domed ends, 1.6 m diameter, 4.2 m long, accommodates electronics and provides mounting for six 3-panel windmill solar wings spanning 8.13 m and generating 1.2 kW (950 W payload requirement). AOCS provided by 1.96 kN liquid bipropellant engine assembly at other end; 3-axis control
Mass: 1,750 kg (payload 220 kg).

Potok/Geizer series

Current Status
The last launches in the Potok GEO military data-relay satellite programme took place in 1994, 1995 and 2000.

Background
The Potok military data relay system uses NPO PM's Geizer satellites launched within the Kosmos programme. Potok 1 resides at 346.5° E, Potok 2 at 80° E, and Potok 3 at 192° E, all potential useful orbits for civilian communication systems or data relay networks. The Russians have never occupied the third location. Each Potok employs C-band transponders at frequencies up and down of 4.40–4.68 and 3.95–4.00 GHz, using phased-array antennas designed by NPO ELAS.

The primary role of the Potok/Geizer system is to communicate with digital imaging satellites – that is, the Yantar 4KS1 and Araks satellites.

The first Potok (Kosmos 1366) was launched by Proton 8K82K on 18 May 1982 and stationed at 80° E and the second (Kosmos 1540) was launched by Proton on 2 March 1984 also to 80° E. Three more followed in 1986, 1987 and 1988 and a sixth was launched in 1990 with a seventh in 1991. Two more followed in 1994 and 1995 with the most recent in July 2000.

Raduga (Rainbow) series

Current Status
The first series of Raduga GEO satellites, providing communications to the Russian military, wound up with a last successful launch on 19 Feburary 1996 and a failed launch attempt on 5 July 1998. The follow-on Raduga-1, codenamed Globus, has replaced the older craft.

Background
Developed by NPO PM in Zheleznogorsk, Raduga satellites provide domestic telecommunications links and include an X-band military and government transponder for the Gals service dating to the 1970s. The L-band Volna maritime/aeronautical module and Ku-band Luch transponders also occupy spots on the Raduga satellites.

Raduga satellites have used the Statsionar geosynchronous slots as one of the three types of satellites to operate in these positions. Thirty-one Raduga satellites have orbited successfully, with the first orbit starting in December 1975. They carry the military codename Gran.

The second-generation Raduga-1 was first launched on 21 June 1989. Since then, six more have been launched including four between 28 February 1999 and 27 March 2004.

Raduga-1 was developed by NPO PM under the article number 17F15 and the codename Globus. It is the basis for the YeSSS-2 second generation Unified Satellite Communication System. The Raduga-1 is capable of communication with mobile platforms. The spacecraft has the T or C-band payload, and the same KAUR-3 spacecraft bus as the original Raduga.

Specifications
Launches: Less than one per year, using a four-stage Proton-K from Baikonur
Design life: two to three years, although operational lifetimes well in excess of this have been demonstrated
Contractors: NPO PM (spacecraft), NPO Radio (payload)
Transponders: 6 × 15 W TWTA, 5.75–6.25/3.42–3.92 GHz up/down C-band global, zonal and spot beams, 34 Mhz bandwidth, edge EIRP 26 dBW global, 35 dBW zonal, 45 dBW spot circular polarisation. Single 7.9–8.4/7.25–7.75 GHz up/down X-band beam.
Mass: 1,965 kg
AOCS: three-axis, east-west station-keeping to within 0.5° by liquid thrusters

Strela Series

Current Status
Strela-3 continues to be the prime Russian series for military store – dump satellites. These simple, low-rate devices serve mobile users and the intelligence community.

Background
Strela 1 Series
Soviet direct communications between ships, aircraft and bases used to be passed over a constellation of eight 61 kg 100 cm diameter Strela 1 (Arrow) satellites put in orbit by a single SL-8 Kosmos launched from Plesetsk into 1,500 km circular orbits at 74° with 115-minute periods. Strela 1s offered medium-range links between forces in the field, and possibly at sea, over VHF/UHF bands. The system was developed in the 1960s with 26 flight tests of the Strela 1 from 18 August 1964 to 18 September 1965.

At any one time, the Soviets had to keep about 24 satellites operating in order to ensure global coverage. Each had an operational life of about two years, but as their orbital life is about 10,000 years, with a 20,000-year life for the rocket stages, the orbits had to accommodate more than 350 satellites and stages. These orbits are now largely avoided because of the debris left behind from the Strela-1.

The Strela 1M was used with Strela 2M for Soviet military and intelligence communications. The first was launched on 25 April 1970 and 360 were launched before the system was phased out with the launch of Mozhaets on 3 June 1992.

Strela 2 Series
The Soviets conducted longer-range military communications on a constellation of three 875 kg Strela 2 satellites, built by AKO Polyot for NPO PM. The practice remains in place with the RFAS. An SL-8 Kosmos places them in 786–810 km orbits at 74° inclination and separated in spacing on the orbital plane by 120°. They apparently receive information from low-power spy and other clandestine transmitters around the world, and store the information until a suitable pass enables it to be dumped to RFAS receiving stations. Their ground tracks repeat approximately every 85 revolutions. Two or more satellites have been operated in recent years in each plane of the system, which began with Kosmos 103 on 28 December 1965, the first of five launched by 27 August 1968. The first operational Strela 2M flew on 27 June 1970, the last of 52 on 20 December 1994.

Strela 3 Series
The third-generation Strela 3 started development in 1973 and has now replaced the Strela 1 and 2. Heavier than the Strela-1M, it was designed to be launched in sixtuplets on Tsiklon-3 boosters. Twelve spacecraft constituted the operational constellation. The 225 kg spacecraft was stabilised by passive means, using a gravity-gradient boom. The system has 12 Mbytes of onboard storage (that is, about 1000th as much as a first-generation IPod) with a transmission rate of 2.4 kbyte/sec. Beause the Russian armed forces ceased using the Ukrainian-developed Tsiklon, Strela is now launched in pairs by the smaller Kosmos-3M. Some reports suggest that the most recent such launch, on 21 December 2005, involved a new-generation Strela replacement known as Rodnik.

Spain

Hispasat series

Current Status
The increase in demand on Hispasat's capacity due to digital television and the liberalisation of satellite services led Hispasat SA to purchase a third and fourth satellite, Hispasat 1C and ID respectively, before the first generation satellites required replacement. Hispasat 1C was launched on 3 February 2000. Hispasat 1D, also built around an Alcatel Space Industries Spacebus 3000 platform, followed in September 2002. The most recent addition to the Hispasat fleet was the Amazonas satellite, which is operated by Hispasat's Brazilian affiliate HISPAMAR (see separate entry). Hispasat 1A was placed in a graveyard orbit and taken out of service in July 2003, followed by Hispasat 1B in June 2006. As of September 2007, Hispasat satellites were located in three orbital positions: a Trans-Atlantic position at 30°W, where Hispasat 1C, 1D and Spainsat were located; an American position at 61°W, where the Amazonas satellite was located; and an oriental position at 29°E, where Xtar-Eur was located (see separate entry).

Background
The Spanish government originally considered building its own satellite network in 1985, but then rejected the idea. Parliament resurrected again in July 1988 to address the communications needs for the Barcelona Olympic Games and the Columbus quincentenary celebrations. After approval, The Secretaria General de Comunicaciones issued tenders for a two-satellite domestic broadcasting and communications system, plus ground segment. The programme received government approval in April 1989 for two satellites, and a ground spare, to provide three to five Spanish direct broadcast service channels, two North and South America television distribution channels, 8 to 16 television broadcast and communications channels and two government/military X-band channels.

The Hispasat 1D satellite during assembly and testing (Hispasat) 1328224

Spain gave the satellite construction contract to the Satcom International partnership of Matra/British Aerospace in June 1989 for Pts19,898 million. The design reflected closely Matra's Telecom 2 because Spain sought to have the satellite in space by 1992. Spain chose an Ariane 4 instead of an Atlas 2 to launch the satellite despite the lower US bid because of the presumed greater stability of the peseta against the franc.

Hispasat 1A became operational in January 1993 despite the fact the 2.2 m direct broadcast satellite antenna pointed slightly too far north, probably because of thermal distortion. The X-band payload failed on 18 December 1997 and it was taken out of service in July 2003. The addition of 1B created a total of five operational DBS channels. The Spanish military used the Hispasat for the first time in March 1993 when Loral Western Development Labs installed an SCT-10 anchor terminal in Madrid to provide 12 voice/ data circuits to a transportable unit in Bosnia.

Hispasat 1A and 1B reached an occupancy of 100 per cent of their capacity in 1998, with the entry into service of Vía Digital, the Spanish digital satellite TV operator. 1A was finally deactivated and sent into a graveyard orbit on 25 July 2003. The satellite's Ku-band services were transferred to Hispasat 1D. In September 2003 Hispasat 1B was placed in an inclined orbit to guarantee continuity of the X-band transmissions for the Spanish Ministry of Defence until the launch of SpainSat.

Hispasat 1C has 24 Ku-band transponders with the capability to provide flexible up-links and down-links in Europe and America. This capacity allows the satellite to provide direct, analogue or digital radio transmission services in Europe, the distribution of analogue or digital television channels in Europe and America and other telecommunication services such as corporate networks, VSAT networks, SCADA, as well as modern broadband services.

Hispasat 1D was launched to maintain the Ku-band coverage offered through 1A and 1B. Based on the Alcatel 3-axis stabilised SPACEBUS 3000B platform, it has three antennas, 28 transponders and connective flexibility in the Ku-band frequencies. Hispasat 1D carries an additional four transponders with American and trans-Atlantic connectivity in order to meet an expected growth in demand. A beam over the Middle East gives access to Asian satellites for American and European customers. Operational lifetime is 15 years

Specifications

Hispasat 1A

Launch: 10 September 1992 by Ariane 4 from Kourou, French Guiana

Location: 30° W geostationary

Design life: 10 years min. Re-orbited and taken out of service on 25 July 2003.

Contractors: Satcom International (MMS France prime), MMS UK (payload), Thomson Tubes Electroniques (TWTs), ANT (EPCs)

Transponders: Three (plus two back-ups shared with TVA) 110 W TWTA 17.314–17.648/12.136–12.470 GHz up/down Ku-band supporting two DBS beams over Spain and the Canary Islands working to 30–60 cm dishes, 27 MHz bandwidth, 55–58 dBW Spain (54-56 dBW Canary Islands) EIRP

One (plus two back-ups shared with DBS) 110 W TWTA 14.249/12.078 GHz (12.015 GHz on 1B) Ku-band supporting TVA Spanish-speaking TV distribution service to North/South America through 1.5–2.5 m stations, 36 MHz bandwidth, 41–47 dBW EIRP

Eight (plus four back-up) 55 W TWTA 14.00–14.50/11.450–11.700 + 12.50–12.750 GHz up/down Ku-band FSS European/Canary Islands beams supporting TV broadcast + telecom services, 4 × 36, 2 × 54, 2 × 72 MHz bandwidth, 49.5–52 dBW

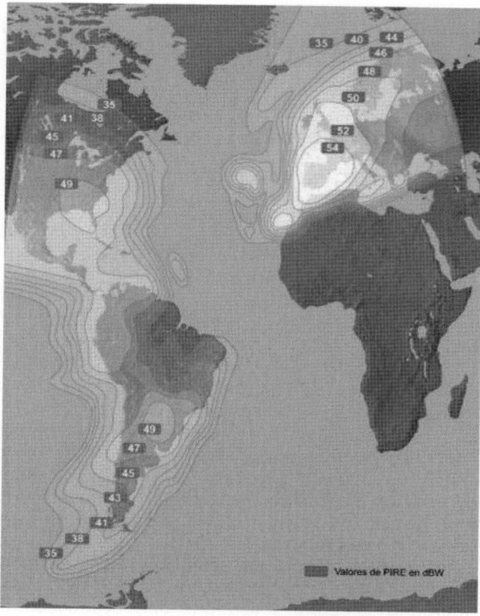

The PIRE footprints of Hispasat 1D (dBW) (Hispasat)
1328223

Three (plus one back-up) 40 W TWTA 7–8 GHz X-band Spanish beams for government/military applications, 20–40 MHz bandwidth, EIRP 42 dBW

Principal applications/users: DBS, TV distribution, datalinks, point-to-point applications, rural telephony, and defence. FSS controlled by Telefonica, Retevision and Spanish Post Office; military/government by INTA

Configuration: Matra/BAe Eurostar 2000 bus

Mass: 2,194 kg at launch, 1,325 kg on-station BOL, 1,013 kg dry, 280 kg communications payload

Attitude control: 3-axis employing inertial/Earth sensors in conjunction with hydrazine thrusters

TT&C: Identical to Telecom 2 Satellite Control Centre Three dedicated Ku-band + single S-band antennas at the Arganda facility, near Madrid

Power system: Twin 4-panel solar wings, 2.05 m wide, spanning 22 m, providing 3,237 W at equinox BOL and full eclipse protection with Ni/H_2 batteries. Communications payload requirement 2.50 kW

AOCS: Unified 3-axis control and apogee GEO insertion motor using NTO/MMH; two sets of 10 N thrusters and single 490 N motor.

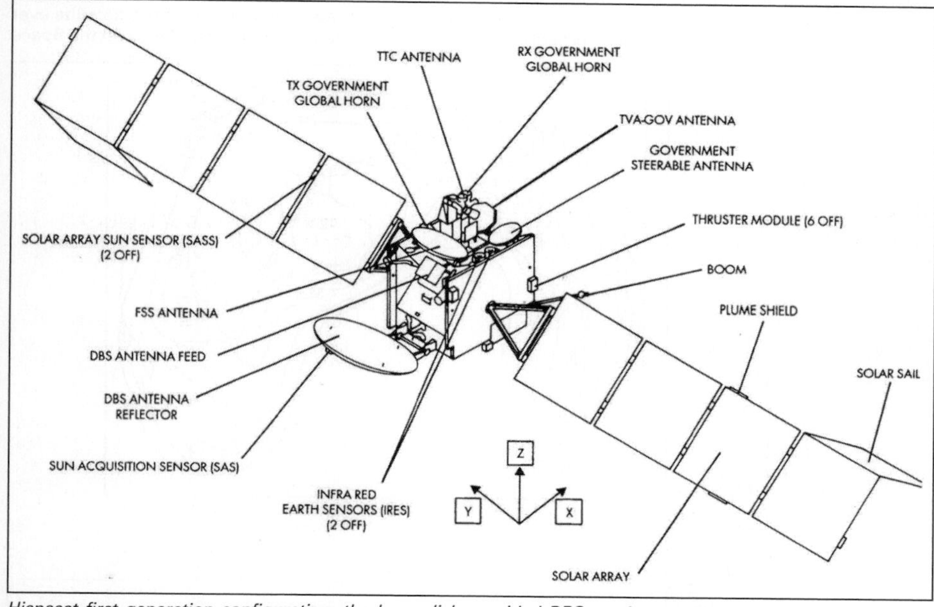

Hispasat first generation configuration: the large dish provided DBS services to 60 cm receivers within Spain
0517012

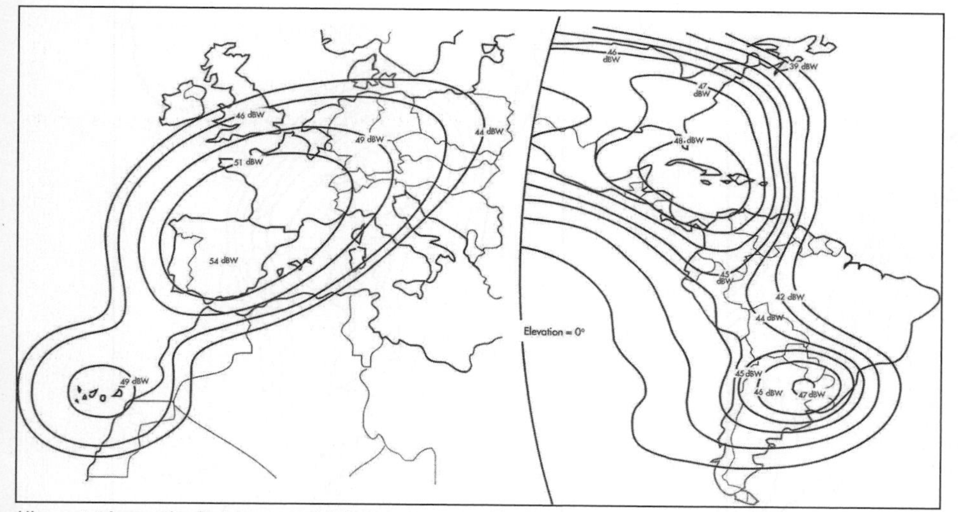

Hispasat 1A was the first commercial European communications satellite to address American countries. Left: FSS (DBS is similar but 5–6 dBW higher); right: broadcast to America. Not shown is the classified footprint of the military payload
0517392

Hispasat 1B

Launch: 22 July 1993 by Ariane 4 V58 from Kourou, French Guiana

Location: 30° W (co-located with Hispasat 1A). Re-orbited and taken out of service on 6 June 2006.

Transponders: Same as 1A but two FSS channels (1 × 54 + 1 × 72 MHz) have switchable uplink to relay TV (TVR) from the Americas to Spain

Mass: 2,210 kg at launch; 1,330 kg on-station BOL; 1,052 kg dry

Other specifications are the same as for Hispasat 1A.

Hispasat 1C

Launch: 3 February 2000 by Atlas 2 AS from CCAFS, Florida

Location: 30° W

Transponders: 24 Ku-band. Number of channels: Europe 12–20 (Tx), 12–24 (Rx); America 4–12. Power level 110 W. PIRE: Europe 54/45 dBW, America 47/41 dBW. G/T: Europe 8.5/3.5 dB/K, America 1.5/4.5 dB/K.

Mass: 3113 kg at launch; 1,304 kg dry

Power system: Total power available: 5,848 W. Power consumption 5,471.7 W.

Configuration: Alcatel Space Industries Spacebus 3000B2

Design life: 15 years

Contractors: Alcatel Espacio, CASA Espacio, G.M.V., Indra Espacio, Rymsa y Sener

Hispasat 1D

Launch: 18 September 2002 by Atlas 2 AS from Cape Canaveral AFB, Florida

Location: 30° W co-located with Hispasat 1C

Transponders: 28 Ku-band with optimised bandwidths between 36 MHz and 72 MHz. Number of channels: Europe 16–23; America 5–12. Power level 103.5 W. PIRE: Europe 53.5/48 dBW, America 47/41 dBW. G/T: Europe 9/3 dB/K, America 1/-5 dB/K

Mass: 3,288 kg at launch. Dry mass 1,382.3 kg.

Power system: Power available 6,257 W. Power consumption 5834.7 W.

Other specifications are the same as for Hispasat 1C.

Thailand

Thaicom series

Current Status
Thaicom 3 was deorbited on 2 October 2006. Thaicom 1A, 2, 4 (Ipstar) and 5 remain operational, as of late February 2008.

Background
Thailand requested that satellite offerers provide them with quotes for a new satellite broadcasting system in 1987 to reach its entire population of 55 million persons. The announcement predated Thailand's eventual entry into Inmarsat in January 1995. The country expected to offer a 30 year concession to lease transponders to government and private sector users, with excess capacity offered to neighbouring countries.

Shinawatra Computer and Communications Group (SC&C), the largest group of Thai companies operating telecommunications and broadcasting service concessions, won an eight year monopoly in 1991 to supply satellite capacity to domestic users. The founder and major shareholder of Shinawatra was Dr Taksin Shinawatra. SCC began formal negotiations June 1991 with Hughes, and signed a contract on 8 October 1991 for about USD100 million to buy two satellites, ground equipment and training.

As a lighter version of Hughes' HS-376 bus, Thaicom was the second to come within Ariane's SDS (Spelda Dedicated Satellite) 1 tonne launch limit. Shinawatra signed a letter of intent on 1 November 1993 to negotiate with Hughes for a third satellite. The total cost for Thaicom 3 on station was estimated at USD240 million.

Thailand inaugurated its first national satellite communications network 1993–1994 with the launches of **Thaicom 1** and **Thaicom 2**. Thaicom 1 was handed over by Hughes to Nonthaburi (Bangkok) station on 31 December 1993 and entered full service in February 1994. The Hughes HS-376L spacecraft were operated by the Shinawatra Satellite Company of Bangkok under a lease arrangement with the Thai government. Both satellites were stationed at 78.5°E and carried 10 C-band and two Ku-band transponders. C-band footprint coverage included all of South-East Asia as well as the Philippines, Korea, Japan and the East Coast of China with a nominal EIRP of 36 dBW. They also provided a high-powered Ku-band spot beam for Thailand with a nominal EIRP of 50 dBW which was ideally suited to direct broadcasting applications. Thaicom 1 was renamed **Thaicom 1A** after it was moved to 120°E in August 1997, following the launch of Thaicom 3.

Thaicom 3, launched in 1997, was based on Aerospatiale's SpaceBus 3000A platform. It provided communications services and direct to home television programming during its nine year lifetime. The satellite functioned in four different coverage modes from its orbital position of 78.5°E. Using its 25 C-band transponders, the satellite provided coverage for Asia, Australia, Europe, and Africa, as well as regional coverage for South and South-East Asia, including India and Thailand. The 14 Ku-band transponders were used mostly for domestic coverage in Thailand.

Thaicom 3 experienced an anomaly in its power system in 2003, leading to some outages in customer services. The cause of the failure was reported to be a short circuit in the solar array drive mechanism. These problems led to the early retirement of the spacecraft in 2006 and the order of Thaicom 5 as a replacement.

Thaicom 4 (Ipstar 1), launched in August 2005, was conceived by Shin Satellite plc as a new generation of Internet Protocol (IP) satellite that would meet the future demand for high-speed broadband

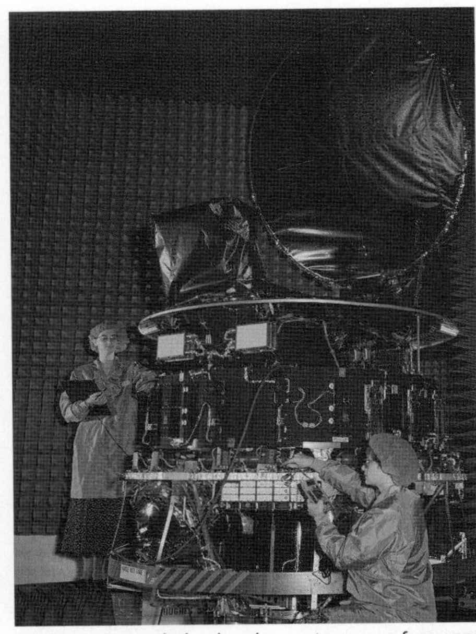

Thaicom 2 anechoic chamber antenna performance tests (Hughes Space & Communications) 0517207

Internet access. SSA developed Ipstar technology to increase system capacity and efficiency while reducing considerably the cost of service. Ipstar 1 was envisaged as the first of a new generation of broadband satellites that will act both as an Internet backbone connection to fibre optic cables for ISPs and as a last-mile broadband Internet service to consumers, competing with cable modem and ADSL. With a launch mass of more than 6,500 kg, it was the heaviest telecommunications satellite ever placed into geostationary orbit. Based on the Space

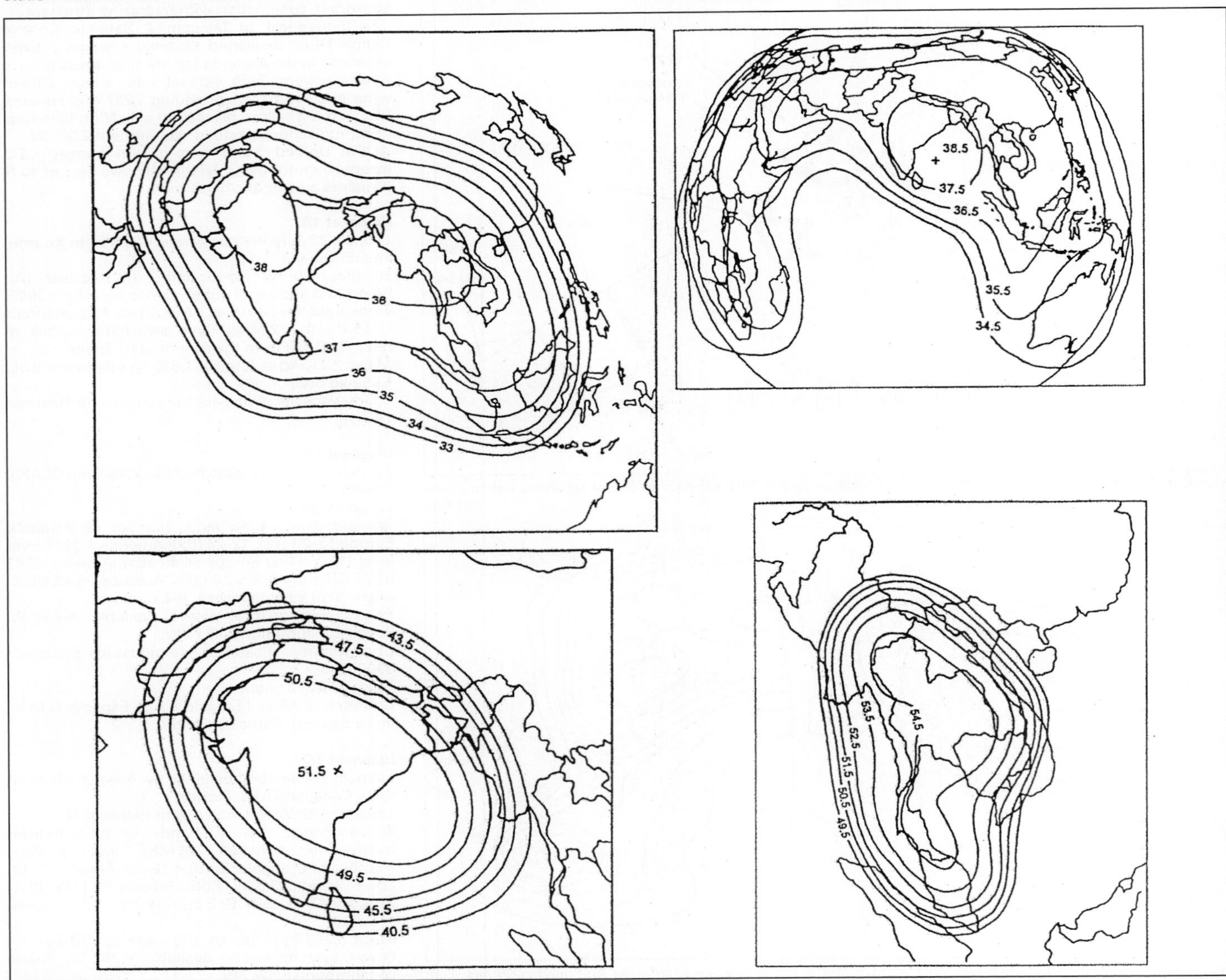

Thaicom 3 footprints (dBW). Top left: C-band India/Indochina; top right: C-band hemispherical; bottom left: Ku-band steerable (India); bottom right: Ku-band Thailand (Shinawatra Satellite Public Co Ltd) 0517386

Systems/Loral LS-1300 SX platform, the satellite is located at 119.5°E in order to provide broadband services to 14 countries in the Asia-Pacific region.

It provides users with data speeds of up to four Mbps on the forward link and two Mbps on the return link. 84 Ku-band spot beams can be aimed toward population centres, while 10 shaped and regional beams provide more general coverage to rural markets. The spot beams offer 20 times more bandwidth than traditional Ku-band systems. The Ka-band payload operates 18 feeder beams and uses gateways to connect to external networks such as the Internet backbone and telephone lines. The total digital bandwidth capacity aboard iPSTAR equals that offered by over 1,000 transponders using conventional coding, approximately 45 gigabytes per second.

Thaicom 5 was launched in May 2006 to the Shin Satellite geostationary slot at 78.5°E. Like Thaicom 3, it is based on the SpaceBus 3000A platform, provided by Alcatel Alenia (now Thales Alenia). Global beam coverage on Thaicom 5 spans Asia, Europe, Australia and Africa. Its high-powered Ku-Band transponders, with both spot and steerable beams, are ideally suited to Digital DTH services for Thailand and other countries in South-East Asia.

Specifications

Thaicom 1

Launch: 18 December 1993 by Ariane 44L from Kourou, French Guiana.
Location: 78.5°E GEO (co-located with Thaicom 2); moved to 120° E in August 1997 after the launch of Thaicom 3.
Design life: 13.5 years minimum, 15 years goal.
Contractors: Hughes Space & Communications (prime), Telespace Ltd. (technical support).
Transponders: 10 (plus 2 back-up, 12-for-10) 11 W SSPA 5.925–6.425/3.700–4.200 GHz up/down C-band Thailand, Indochina, North Pacific rim beam, EIRP 35 dBW edge (37.5 dBW over Thailand), 36 MHz bandwidth, orthogonal linear polarisation

2 (plus 1 back-up, 3-for-2) 47 W TWTA 14.3159–14.4951/12.5679–12.7471 GHz up/down Ku-band Thailand beam, EIRP 51 dBW edge (54 dBW centre), 54 MHz bandwidth, orthogonal linear polarisation.
Principal users: Domestic phone, TV, cable TV, voice, video, data, VSAT services
Configuration: Lightweight version of Hughes' HS-376 spin-stabilised bus, 2.16 m diameter, 6.7 m height with solar array drum extended (2.6 m stowed). Single antenna comprising two offset grid reflectors with horizontal/vertical feed system on despun platform.
Mass: 1,080 kg at launch, 629 kg on-station BOL, 436 kg dry
AOCS: 55 rpm spin control/position maintained by four Hughes 22.2 N hydrazine thrusters. GEO insertion provided by Thiokol Star 30BP solid.

Thaicom 1, ANHS-376L gets an antenna beam checkout in an anechoic chamber
(Hughes Space and Communications) 0054345

Thaicom 4 (Ipstar) was the largest commercial satellite ever launched into geostationary orbit
(ESA – CNES – Arianespace) 1166895

TT&C: Through the Nonthaburi station, 15 km north of Bangkok.
Power system: Si cells on cylindrical surface provide 801/671 W BOL/EOL (summer solstice) in conjunction with 51.6 Ah Ni/H_2 battery.

Thaicom 2

Launch: 8 October 1994 by Ariane 44L from Kourou, French Guiana.
Location: 78.5°E GEO (co-located with Thaicom 1).
Other specifications as for Thaicom 1.

Thaicom 3

Launch: 16 April 1997 on Ariane 44LP from Kourou, French Guiana.
Location: 78.5°E GEO (co-located with Thaicom 2).
Design life: 14 years (16 years propellant).
Contractors: Aerospatiale (prime; structure, thermal control, solar arrays, integration, testing), Alcatel Espace (communication payload), DASA (AOCS integration).
Transponders: 18 (plus 6 back-up; 24-for-18) 19 W SSPA 5.925–6.725/3.405–3.700 GHz up/down C-band India/Indochina regional beams, 36 MHz bandwidths, EIRP 37 dBW min, orthogonal linear polarisation

7 (plus 3 back-up; 9-for-6) 42 W SSPA 6.425–6.725/3.405–3.700 GHz up/down C-band hemispheric beams, 36 MHz bandwidths, EIRP 34.5 dBW min (38.5 maximum), orthogonal linear polarisation

Technician checks the feedhorn array for Thaicom 1
(Hughes Space & Communications) 0054344

Thaicom 5 during launch preparation at Kourou
(ESA – CNES – Arianespace) 1166896

14 (plus 4 back-up, 18-for-14) 97 W TWTA 14.000–14.500/12.200–12.750 GHz up/down Ku-band domestic DTH beams and one steerable beam (max 7 36 MHz transponders), 2 × 54 MHz + 12 × 36 MHz bandwidth, EIRP 54 dBW min Thailand and 45 dBW India (51.5 maximum), orthogonal linear polarisation
Principal applications: Ku-band principally for DTH.
Configuration: Aerospatiale/DASA Spacebus 3000 3-axis stabilised 1.8 × 2.6 × 2.3 m box-shaped bus (7 m deployed height) with 25 m-span twin solar wings.
Mass: 2,650 kg launch, 2,450 kg on-station BOL, 1,160 kg dry
TT&C: As Thaicom 1/2
Power system: 5,300 W EOL provided by twin Si solar wings totalling 46 m². Two 57 Ah Ni/H_2 batteries.

Thaicom 4 (Ipstar)

Launch: 11 August 2005 by Ariane 5GS from Kourou, French Guiana.
Location: 119.5°E GEO.
Design life: 16 years.
Contractors: Space Systems/Loral (prime).
Transponders: 87 Ku-band, 10 Ka-band. 84 spot beams, 3 shaped beams, 7 regional broadcast beams. 45 Gbps. Equivalent to 1,000 + 36 MHz transponders of conventional coding and modulation.
Principal applications: Broadband Ku-band/Ka-band data relay and digital telecoms in the Asia-Pacific region.
Configuration: LS-1300 SX platform. 3-axis stabilised. In orbit span approx. 25.9 m.
Mass: 6,505 kg at launch. 3,400 kg dry.
Power system: Twin solar arrays. Min. 14.4 kW EOL.

Thaicom 5

Launch: 27 May 2006 by Ariane 5 ECA from Kourou, French Guiana.
Location: 78.5°E GEO.
Design life: 14 years (16 years propellant).
Contractors: Alcatel Alenia Space (prime).
Transponders: 25 C-band (36 MHz), 14 Ku-band (two 54 MHz and twelve 36 MHz).
Principal applications: Global and regional broadcasting, telecommunication and broadband services.
Configuration: Spacebus-3000A platform. 3-axis stabilised.
Mass: 2,766 kg at launch. 1,600 kg BOL.
Power system: Twin solar arrays. Min. 5,300 W EOL.

For details of the latest updates to *Jane's Space Systems and Industry* online and to discover the additional information available exclusively to online subscribers please visit
jsd.janes.com

Turkey

Turksat series

Current Status
Turksat 1C and Turksat 2A have been in service since 1996 and 2001 respectively. Turksat 3A will replace Turksat 1C in 2008. Turksat 1B was removed to a graveyard orbit in November 2006.

Background
Turkey chose to build a completely turnkey satellite operation in order to reach Turkish speaking people in the Central Asian land mass. The government released a request for proposals in June 1989 and received offers from Aerospatiale, Hughes/Selenia Spazio, BAe-Matra/ANT and Aerospatiale/MBB by the deadline of December 1989. Aerospatiale signed the USD315 million prime contract on 21 December 1990 for turnkey delivery of two satellites, associated ground stations, insurance, financing and training. It was the first complete system ever delivered on a turnkey basis by a European team. Each satellite carried 16 active Ku-band transponders providing television/radio distribution (also serving the large Turkish populations in Germany/Austria and CIS), data and voice services.

Turkey's first satellite, **Turksat 1A**, was lost in the Ariane 44LP launch failure of 24 January 1994. In the contract, Turkey required Aerospatiale to provide on-orbit delivery, and the company had taken full insurance, so it provided a replacement, **Turksat 1C**, in 1996. **Turksat 1B** was successfully launched on 10 August and was positioned at 42° E. It entered service on 10 October 1994. Turksat 1C was successfully launched in July 1996, and initially placed at 31.3° E. On completing its orbital tests, it was moved to 42° E and took over the broadcast traffic of Turksat 1B. The latter spacecraft was then transferred to 31.3° E. Both satellites were based on the Aerospatiale Spacebus-2000 platform with an on-orbit mass of a little over 1,000 kg. Their communications payloads consisted of 16 Ku-band transponders.

Turkish Telecom and Aerospatiale formed the Eurasiasat joint venture in August 1996 before ordering **Turksat 2A**, which became known as **Eurasiasat 1** following launch on 10 January 2001. Turksat 2A was co-located with Turksat 1C at 42° E and entered operation on 1 February 2001. Turksat 2A carries 32 high-powered Ku-band transponders with 20 × 33 mHz fixed beam in BSS band and 12 × 36 mHz spot transponders with two steerable beams in FSS. It was designed to provide communications links between Western Europe, the Middle East, Central Asia and the Far East. The fixed beams cover Europe, Turkey and Central Asia, while the steerable beams can be pointed where the market requires – including the Middle East, Russia, India and South Africa. It has switching capability between the fixed and steerable beams.

On 26 February 2006, Alcatel Alenia Space (now Thales Alenia) announced that it had signed a turnkey contract with Turksat AS to build and deliver into orbit **Turksat 3A**. The fifth Turksat contract awarded to Alcatel continued the relationship between the two companies that stretched back more than 10 years. As prime contractor, Alcatel Alenia Space wil be responsible for the satellite design, construction and test activities, in-orbit delivery, and modernisation of the ground segment. Alcatel Alenia Space will also provide training assistance to Turksat engineers. Turksat 3A will enable Turksat to offer telecommunication services as well as direct TV broadcasting over Turkey, Southern Europe and Central Asia. Based on the Spacebus 4000B2 platform, it will be fitted with 24 Ku-band transponders and will offer beginning of life power of about 8kW. Positioned at 42° E, Turksat 3A is due to start services in 2008, replacing Turksat 1C.

An Assembly, Integration and Test Centre is to be built at the Turksat campus to manufacture telecommunication satellites for Turkey and other countries around the globe. The Turksat satellite design centre will provide concurrent designs for future satellites such as the **Turksat 4A**, with a possible launch date of 2010.

Specifications
Turksat 1B
Launch: 10 August 1994 by Ariane 44LP from Kourou, French Guiana.
Location: 42° E GEO initially, relocated to 31° E during October 1996.
Design life: 10–13 years (10 years contracted).
Contractors: Aerospatiale (prime), Alcatel Espace (payload), DASA/MBB (AOCS, solar arrays), Thomson Tubes Electroniques (TWTAs), Eagle Picher (battery), SEP (BAPTA), Contraves (structure), Sextant Avionique (power control, distribution unit), ETCA (power conditioning), Teletas (TV receive antennas).
Transponders: 16 (plus 8 back-up; 12-for-8 ring redundancy) 55 W TWTA 14.000–14.500/10.950–11.700 GHz up/down Ku-band Central European, Turkey, Central Asia, bandwidths 6 × 72 MHz + 10 × 36 MHz, EIRP 51 dBW Turkish, 48 dBW Central European, 45 dBW Central Asia, linear polarisation. Up to four channels can be switched to Central Europe and up to two to Asia.
Principal applications: TV/radio direct (12 TV + 20 radio to 50 cm antennas), telephony, VSAT data and business services, mobile links.
Configuration: Spacebus 2000 platform, 3-axis control, 22.4 m span across solar panels, single dual reflector multifeed antenna. GEO location maintained within ±0.05°.
Mass: 1,783 kg at launch; 1,078 kg on-station BOL; dry 827 kg
AOCS: Provided by DASA/MBB's 22 × 10 N + 400 N MMH/MON bipropellant system
TT&C: Ankara (prime). Istanbul (back-up).
Power system: Twin solar wings providing 2.9 kW equinox EOL.

Turksat 1C
Launch: 9 July 1996 by Ariane 44L from Kourou, French Guiana.
Location: Initially over 31° E for testing July–September 1996, then 42° E geostationary.
Mass: 1,747 kg at launch; 1,062 kg on-station BOL; dry 789 kg.
Other specifications are the same as Turksat 1B.

Turksat 3A with its solar arrays deployed during ground tests (Thales Alenia)  1166891

Turksat 2A (Eurasiasat) showing the 4 beam dishes. One of the solar panels is visible stowed across the top of the satellite (Eurasiasat) 1166892

Turksat 2A (Eurasiasat 1)
Launch: 10 January 2001 by Ariane 44P from Kourou, French Guiana.
Location: 42° E geostationary.
Design life: 15 years.
Contractors: Alcatel Space Industries (prime).
Transponders: 32 high-powered Ku-band transponders with 20 × 33 mHz fixed beam in BSS band and 12 × 36 mHz transponders with two steerable beams in FSS. EIRP in excess of 50 dBW.
Principal applications: Direct to Home TV (analog and digital), radio services and data transmission (Internet, IBS, VSAT etc.)
Configuration: Spacebus 3000B3 platform, 3-axis control. Body dimensions: 5.0 x 2.5 x 3.5 m. 37 m span in orbit.
Mass: 3,535 kg at launch
Power system: Twin solar wings providing 8 kW EOL.

United Kingdom

Skynet Series

Current Status
The British Ministry of Defence's (MoD's) Skynet 5B was launched from Kourou on 14 November 2007. Skynet 5A was successfully launched on 11 March 2007, along with the ISRO's Insat 4B, aboard an Ariane 5 from Kourou. Paradigm Secure Communications, a wholly-owned EADS Astrium subsidiary through Astrium Services, expects that Skynet 5C will follow, also from Kourou, in 2008. Skynet 5C will act as an in-orbit backup for the first two. The Skynet 5A launch slipped by about a year; it was originally expected to take flight in 2006. The contract for the three-satellite constellation, a partial-build ground spare, satellite ground segment upgrades and communications services is costing the MoD GBP3.6 billion; Skynet 5A and 5B are so far the most powerful X-band satellites in orbit.

In 2002, the MoD decided in favour of the Paradigm team for development and production of the Skynet 5 system, in the largest UK government Private Finance Initiative (PFI) of its kind. Paradigm Secure Communications supports the prime system manufacturer, Astrium Satellites. Under the terms of the contract, signed in 2003 and restructured in 2005, Paradigm is responsible for MoD secure beyond-line-of-sight communications services until 2020. Paradigm owns and operates the Skynet 4 and 5 satellite generations; the company also leases satellite bandwidth to provide secure military and government communications to NATO and other countries, in addition to the UK.

As of early 2007, Skynets 4C - 4F, as well as 5A, remained operational. The Skynet 5 satellites are based on Astrium's Eurostar E3000 satellite platform. They have improved anti-jamming capability via a nulling antenna, multiple steerable spot beams, and about five times the capacity of the Skynet 4 generation, increasing the systems' ability to quickly relay secure data. The system's communications capabilities are enhanced by the commercial Iridium, Intelsat and Inmarsat satellite constellations (see other entries). Ground facilities remain at RAF Oakhanger for the Skynet 5 generation, augmented by a new control centre at RAF Colerne, and a new system operation centre at DCSA Corsham.

Background
The full Skynet military communications coverage is provided by multiple spacecraft in GEO at 34° W, 17.8° W, 1° W, 6° E and 53° E, offering X-band and UHF coverage from approximately 100° W to 120° E. The UK began planning to build a network of communication satellites for its far-flung outposts even before the country had begun to withdraw from Asia and the Middle East. Britain needed secure voice, telegraph and fax links for strategic and tactical communications between UK military headquarters and ships and bases around the globe. However, the first attempts to obtain this capability proved unsatisfactory and the UK began to lose interest.

Skynet 1 was launched on a Delta rocket in November 1969 from Cape Canaveral, almost as soon as geosynchronous orbits became accessible. But Skynet 1's Travelling Wave Tube Amplifiers (TWTA) failed within a year. Skynet 1B followed shortly, but it failed to reach the correct orbit and remained stranded in a highly elliptical orbit. The first three satellites in this series offered two UHF global channels for submarine and other mobile users, four SHF channels ranging from global to 3° spot beam coverage, and an experimental EHF uplink channel for propagation studies.

For its second-generation satellites the UK sought to build an indigenous capability. Ironically, the UK, with the most sophisticated electronics industry in the world, contracted with Philco-Ford to build the first Skynet. With help from Philco, the UK's Marconi Space & Defence Systems built the first communications satellite produced outside the US or USSR. A launch failure set the Skynet 2A tumbling through space until it plunged back into the atmosphere six days after being launched. Skynet 2B succeeded and broke the string of failures for the UK. In 1974, it linked the UK directly with Western Australia for the first time in history.

The original Skynet system included 17 ground stations, with a 12.8 m master terminal at RAF Oakhanger, air-transportable terminals for use on land, and shipborne terminals. By the time the system became operational, successive defence cuts and British withdrawals from Asia and the Near East made it seem like an unnecessary luxury. The British quietly cancelled the planned Skynet 3 series in 1975 and filled UK military satellite communications requirements by leasing capacity from US/NATO satellites.

In the decade before the Falklands conflict of 1982, the UK's military leased channels from the US, and the plan worked well for all UK defence needs. Nonetheless, the Falklands conflict itself proved the difficulty of routing military communications through the good offices of a foreign provider, in this case the US DSCS circuits. The UK revived attempts shortly thereafter to establish an independent and secure communications network for the Royal Navy and other UK forces. The idea passed in Parliament and the budget included funds for the Skynet 4 programme.

After several delays, the UK finally scheduled Skynet 4A and 4B for launches in June and December 1986, both by Space Shuttle despite strong French pressure to use the Ariane from Kourou. The UK finally agreed to the Shuttle launches because of NASA's standing offer to let a British astronaut go along and the hope that the US would help the UK make Skynet 4 the standard NATO system. By early 1988, Skynets 4B and 4C had berths on the Ariane 4 and in late 1988 and early 1990, and 4A had a Shuttle booking.

The continuing Shuttle problems and delays following the Challenger accident in 1986 forced Skynet 4A to move to a Titan 3 launch in August 1989, at a reported cost of USD55 million. With the launch vehicles chosen, the UK government ordered Skynet 4C from BAE in May 1985.

The Skynet programme, popular in Parliament, seemed to have assured backing, but in January 1987, a journalist reported that Skynet actually cloaked an electronics intelligence (ELINT) programme known as Zircon. In a secret intelligence sharing deal with the United States, the UK would provide all the message traffic that the Skynet satellites snatched from their Asian orbital slots, in return for US information. The United States benefited from this by averting suspicion from ELINT satellites that it needed to have in the region through the mechanism of the false flag. The satellite eavesdropped on other electronic communications at geosynchronous orbit. In the published article, the author cast suspicion on the location registered for the satellite, which was far from any conceivable orbit for communications. The MoD eventually admitted its role in Project Zircon and confessed to spending GBP70 million on the programme.

In operation it would have resembled the US Rhyolites and the later Magnum signal intelligence (SIGINT) satellites. Reports indicated that the UK would have shared intelligence with the US via the UK government's monitoring centre GCHQ Cheltenham. Zircon was abandoned following disclosure amid concerns over its ageing technology. By the time Zircon was ready to go though, the ageing satellites, delayed after years of waiting for a Shuttle ride, were simply obsolete.

The Skynet programme also suffered from US-related funding problems. Originally, the British military projected Zircon would cost about GBP400–500 million, including the construction, launch and operation of two to three satellites providing ELINT, reportedly from GEO at 53° E. The UK Parliamentary Space Committee received a report in June 1990 that the entire Skynet series had cost almost GBP1 billion, even before Skynet 4D-4F were developed.

The Skynet 4 satellite generation assured continuity of UK tactical and strategic defence communications into the 21st century. The fourth-generation Skynet added a considerable capability to the communication system with steerable antennas for SHF spot beams, increased power, improved anti-jamming and a fully tunable UHF system for increased flexibility. The more recent Skynet 4 spacecraft are expected to continue operations until at least 2012. Their capacity is now being augmented by the introduction of three Skynet 5 satellites.

Manufacture, transport and assembly of the Skynet 5 satellite structure are facilitated by having two independent platforms around a central tube on which the payload equipment is installed. The lower, Service Module (SM) is largely derived from the Eurostar 3000 platform and carries most of the Attitude Determination and Control System and the power units. Above this is the Communication Module, which carries most of the communication equipment. A Liquid propellant Apogee Engine (LAE) is used for orbit circularisation. Four identical tanks (two MMH and two NTO) are secured symmetrically round the central tube and supply the LAE mounted on the central cone, and the seven modules housing the main, redundant nozzles. The helium pressurisation tank is installed inside the central tube. The Service Module equipment is mounted on the North and South panels for the thermal dissipation units. The Chemical Propulsion System (CPS) is mounted on the bottom of the SM. The batteries, which require efficient temperature control, are mounted on the side walls of the SM. Electrical power is delivered by two solar panels, folded during launch and deployed after 3-axis stabilisation of the satellite, with the outer panel of each wing deployed one hour after separation from the launcher. Energy is supplied by a Li-Ion cell battery in transfer orbit and during eclipses. Low-thrust (10 N) thrusters are used in the pulse or continuous mode for housekeeping and attitude control operations.

Skynet 5B during launch preparation at Kourou (EADS Astrium) 1330385

The Communication Module (CM) carries two helicoidal UHF antennas. These are deployed when the satellite reaches geostationary orbit. The central part of the CM also carries an S-band telemetry and telecommand deployable mast antenna. The North and South panels of the Communication Module carry the travelling wave tubes and other high thermal dissipation equipment, and, like the Service Module panels, also provide a heat-sink function.

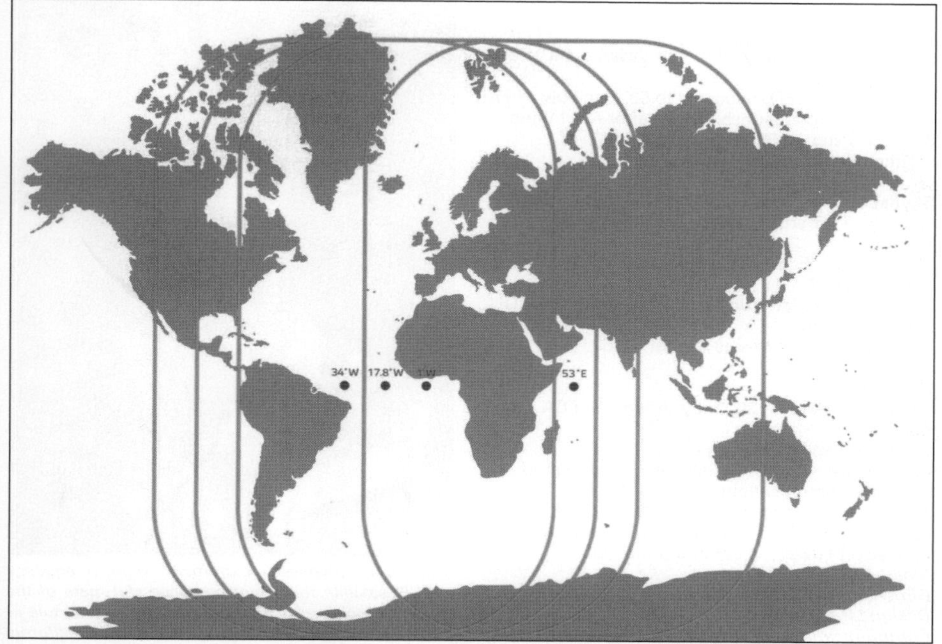

Coverage provided by the Skynet 4 and 5 military communications satellites (Paradigm) 1330389

Specifications

Skynet 1
Launched: 22 November 1969 from Cape Canaveral by Delta
Launch mass: 243 kg, 129 kg on-station. Payload of two 3 W X-band TWTAs. Cylindrical bus 137 × 81 cm high (157 cm with solid AKM), 90 rpm spin-stabilised with a despun antenna. 7,236 solar cells supported by two 16 Ah Ni/Cd batteries. Prime contractor Philco-Ford; same design as NATO 2A/B. Positioned in GEO over the Indian Ocean but its TWTAs failed within a year

Skynet 1B
Launched: 19 August 1970 by Delta from Cape Canaveral. Abandoned in 270 × 36,058 km GTO after AKM failure. Used for meteorology and communications tests. Other details as Skynet 1

Skynet 2A
Launched: 19 January 1974 by Delta from Cape Canaveral. Launch mass 435 kg; 235 kg on-station. Built by the UK's Marconi Space & Defence Systems (with Philco-Ford providing technical direction) as the first communications satellite produced outside US or USSR. Payload of two 16 W X-band TWTAs provided redundant 2 + 20 MHz channels. Cylindrical bus 191 × 208 cm high (with solid AKM),

90 rpm spin-stabilised with despun antenna. Launch failure left it tumbling in 120 × 1,857 km orbit; re-entered after six days

Skynet 2B
Launched: 23 November 1974 by Delta from Cape Canaveral, placed in GEO over Indian Ocean. With a design life of at least three years, it provided links from the UK to W Australia. Other details as 2A. It began drifting in late 1992, indicating its propellant was exhausted. Tests in 1994 showed that its two 16 W TWTAs still met specifications. It remained usable until that time, although it was inclined at >13° and slowly drifting.

Skynet 4A
Launched: 1 January 1990 by Titan 3 from Cape Canaveral
Location: 6° E (1990); 29° E (1991); 65° E (1991); 34°W (1992 to 1999)
Other specifications as for Skynet 4B, except McDonnell Douglas PAM-D2 provided perigee kick

Skynet 4B
Launched: 11 December 1988 by Ariane 44LP V27 from Kourou, French Guiana, declared operational 22 February 1989
Location: 1° W (1989 to 1990); 53° E (1990 to 1998)
Design life: 7 years. Taken out of service and deorbited June 1998.
Contractors: BAE (prime), Marconi Space Systems (payload), and CAL Ltd. (UHF antenna)
Transponders: 3 × 7.250–8.400 GHz X-Band 40 W TWTAs providing four SHF channels made up of channel 1, 31 dBW EIRP 17.0 dB gain global 135 MHz bandwidth; channel 2, 34 dBW EIRP 24.7 dB gain narrow (Europe) 85 MHz beam; channel 3, 35 dBW EIRP 21.7 dB gain wide (hemispherical) 60 MHz beam; channel 4, 39 dBW EIRP 29.5 dB gain 3° spot (central Europe) 60 MHz beam
2 × UHF 305–315/250–260 MHz up/down global 40 W transponders, 25 kHz bandwidth, 26 dBW EIRP, working through single helical antenna. With EMP hardened anti-jamming and ECM interference features
Principal applications: UK Ministry of Defence. Submarines, aircraft, maritime patrol, tankers, AWACs, transportable Earth terminals, 1.7 m mobile terminals, manpacks
Configuration: Based on MMS ECS/OTS bus, 2.1 m high, 1.9 m deep, 1.4 m wide; 16.0 m span with solar wings deployed. The Communications Module is a U-shaped structure sitting on the Service Module; major structural elements are predominantly aluminium alloy honeycomb with aluminium or CFRP skins. Payload components are mounted on both sides of the upper floor and on the sidewalls. The central 2.2 m diameter thrust cone is CFRP honeycomb with aluminium support struts
Mass: 1,433 kg at launch, 790 kg after AKM firing, 770 kg on-station BOL, 655 kg dry
AOCS: 3-axis control with ±0.1°E-W station-keeping incorporating IR Earth sensors (Officine Galileo, accuracy ±0.04°) and Sun acquisition sensors (TNO/TDP, accuracy ±2° coarse) maintained by two 25 Nms and one 16 Nms Teldix MWs and 20 catalytic hydrazine thrusters. (0.5 N roll/yaw, 2 N east-west station-keeping, 5 N pitch, 20 N active nutation damping). The thrusters draw on 70 kg propellant in two N_2-pressurised tanks. Pointing accuracy 0.07° roll/pitch, 0.35° yaw
TT&C: Through RAF Oakhanger in Hampshire, UK, which also acts as the primary communications node. A dedicated Skynet in-orbit checkout facility is also maintained at Oakhanger. The TT&C subsystem, working through the SHF receivers with S-band utilised during launch and as back-up, can handle 512 8-bit words about every 16 s; a total of 750 telecommands is available
Power system: Twin 228 × 730 cm 3-segment (each 128 × 205 cm) Si solar wings and auxiliary body panels, providing 1,200 W at regulated 42 V in sunlight; eclipse power provided by two banks of 14-cell 35 Ah Ni/Cd batteries with 30–37 V

Skynet 4: several in the series of UK milcom satellites will continue beyond 2007 (MMS) 0517317

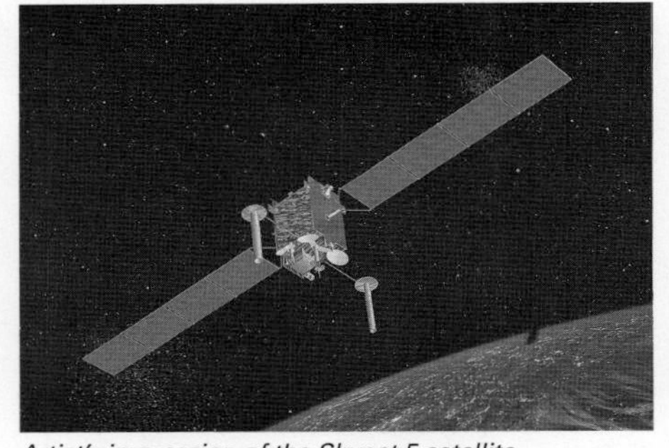
Artist's impression of the Skynet 5 satellite (Paradigm Secure Communications) 0096865

unregulated output. Eclipse power requirement 940 W, maximum requirement 1,083 W (equinox). Degradation of AEG back surface reflector cells 9% over 7 years
Thermal control: Heaters plus passive radiators, including Secondary Surface Mirrors and 10–15 layers
Apogee kick motor: Thiokol Star 30E solid propellant, burn time 49 s, specific impulse 290 s

Skynet 4C
Launch: 30 August 1990 by Ariane 44LP V38 from Kourou, French Guiana. AKM fired 2 September, 69 h after launch; solar wings deployed 3 September; UHF antenna deployed 4 September
Location: 1° W GEO until April 2007 (moved from storage position at 53° E in 1992, where it was ideally placed to provide links during the 1990–91 Gulf conflict); 6° W since June 2007, inclination 1.1°

Skynet 4D
Launch: 10 January 1998 by Delta II 7925 from Cape Canaveral
Location: 5° E GEO, relocated to 53° E (June 1998 to July 1999), 34.5° W (October 1999 – October 2002), 39° E (February 2003–December 2003), 34.5° W since February 2004.
Life: 7 years
Platform: As 4A-C but reworked TT&C (including increased autonomy); improved power system; improved reaction control system
Transponders: 3 × 7.250–7.725 GHz X-Band 50 W TWTAs providing four SHF channels: channel 1, more than 31 dBW EIRP 17.0 dB gain global 125 MHz bandwidth; channel 2, >34 dBW EIRP 26 dB gain narrow (Europe) 75 MHz beam; channel 3, >36 dBW EIRP 21.5 dB gain wide (hemispherical) 75 MHz beam; channel 4, >40 dBW EIRP 30 dB gain 3° steerable spot (central Europe) 60 MHz beam; two UHF 295–318/254–258 MHz up/down global 50 W transponders, 25 kHz bandwidth, >25 dBW EIRP, working through single helical antenna (up/down frequency selection independently tunable). With EMP-hardened anti-jamming and ECM interference features
Mass: 1,510 kg at launch, 868 kg after AKM firing, 851 kg on-station BOL, 256 kg payload
Power system: As Skynet 4A-C, except 2 × 29 Ah Ni/Cd batteries.
Other specifications as Skynet 4A-C

Skynet 4E
Launch: 26 February 1999 by Ariane 44L from Kourou
Location: 5° E GEO, relocated to 53° E in July 1999.
Mass: 1,510 kg at launch, 873 kg after AKM firing, 851 kg on-station BOL, 256 kg payload
Other specifications as Skynet 4A-C

Skynet 4F
Launch: 7 February 2001 by Ariane 44L from Kourou
Location: 5° E GEO, relocated over 34° W on 1 August 2007.
Mass: 1,510 kg at launch, 851 kg on-station BOL, 256 kg payload
Other specifications as Skynet 4A-C.

Skynet 5A
Launch: 11 March 2007 by Ariane 5 ECA from Kourou
Location: 1° W GEO
Contractors: Astrium Satellites and Paradigm Secure Communications
Spacecraft Platform: EADS Astrium Eurostar E3000S; 3 axis stabilised; 34 m solar array span
Spacecraft Power: >6 kW minimum EOL
Mass: 4,635 kg
Dimensions: 4.5 × 3.7 × 2.9 m
Design Life: 15 years
Communications Payload: 15 SHF transponders, ranging in bandwidth from 20 MHz to 40 MHz. 9 UHF channels. High power 160 W TWTAs on all transponders, giving 56 dBW peak EIRP in each transmit spot beam and 41 dBW EIRP in each global beam per transponder. Multiple, fully steerable transmit spot beams. High performance, active receive antenna capable of generating multiple shaped uplink beams. Highly flexible switching, allowing connectivity between any uplink beam and at least two different downlink beams.
Payload Power: >4.5 kW EOL

Skynet 5B
Launch: 14 November 2007 by Ariane 5 ECA from Kourou
Location: 53° E GEO
Other specifications as Skynet 5A.

United States

AEHF (Advanced Extremely High Frequency)

Current Status
The Advanced Extremely High Frequency (AEHF) satellite is a new global, secure, radiation-hardened and jam-resistant military communications system for both mixed and mobile users. It will replace the 1990's-technology Milstar series, providing 10 times greater total capacity and channel data rates six times higher than the Milstar II satellites. Lockheed Martin is the prime contractor and Northrop Grumman is developing the payload. Lockheed Martin is under contract to provide three Advanced EHF satellites and the command control system to its customer, the Military Satellite Communications Systems Wing, located at the Space and Missile Systems Center, Los Angeles Air Force Base, California. The first launch is due in late 2008.

Background
The AEHF programme was approved by the Pentagon in October 2001. In November 2001, a Systems Development and Demonstration (SDD) contract, including the delivery of the first two satellites, was awarded on a sole source basis to Lockheed Martin and TRW (acquired by Northrop Grumman in 2002). The SDD contract was valued at USD2.7 billion. In July 2002, BAE Systems was awarded a USD55 million contract from TRW to provide radiation-hardened electronics for AEHF. The Information & Electronic Warfare Systems (IEWS) Space Systems and Electronics business area would provide its RAD750 single-board computers for the AEHF system's digital payload as

The first Advanced Extremely High Frequency (AEHF) satellite following the successful mate of the propulsion core structure and the payload module at Lockheed Martin's facilities in Sunnyvale, California (Lockheed Martin) 1192626

Artist's impression of an AEHF military strategic and tactical relay satellite (Lockheed Martin) 1192625

well as a variety of radiation-hardened custom logic and memory components built at IEWS' facility in Manassas, Virginia. In January 2006, Lockheed Martin Space Systems received a USD491.2 million cost-plus-award-fee, cost-plus-fixed-fee, firm-fixed-price contract modification to build a third satellite, which is expected to be launched in 2010.

Among the largest geostationary satellites ever built, the AEHF craft is based on Lockheed Martin's A2100 spacecraft bus, with a launch mass of about 6,600 kg.

Low gain antennas deliver communications anywhere within the satellite's footprint. The phased array antenna provides super high-gain coverage for all users, including small portable terminals and submarines. Six Medium Resolution Coverage Antennas (MRCA) are provided by dwelling spot beam antennas and up to 24 time-shared MRCA coverages are provided by phased array antennas. For global communications, the AEHF system uses inter-satellite crosslinks, eliminating the need to route messages via terrestrial systems.

Major advances over Milstar include maximum data rates increased from 1.544 Mbps to 8.192 Mbps per user, and a six-fold increase in crosslink speed, from 10 Mbps to 60 Mbps. The AEHF will be backward-compatible with the existing Milstar constellation and Milstar terminals – offering the existing low data rate and medium data rate capabilities of the Milstar – but will add a new eXtended Data Rate (XDR) option. Uplinks and crosslinks operate in the Extremely High Frequency (EHF) range, at 44 GHz, and downlinks in the Super High Frequency (SHF) 20 GHz range.

Time-sharing increases the number of spot beams from two to 24, and the AEHF satellite is fitted with two high-resolution antennas which permit communications even in the presence of in-beam jamming. Frequency-hopping is used to defeat jamming and interception attempts. Like Milstar, the system is intended to operate in all forms of combat, including general nuclear war.

AEHF will be introduced alongside new terminals, including the airborne Family of Advanced Beyond Line-of-Sight Terminals (FAB-T), under development by Boeing and Rockwell Collins and used by all of the US services and the international partners (Canada, Netherlands and UK) who are involved in the AEHF programme. Maritime, vehicle-mounted and man-pack AEHF terminals will also be developed.

In 2005, Lockheed Martin delivered the flight structure and payload support assembly for the first AEHF. On 18 June 2007, the company announced that it had integrated the spacecraft propulsion core structure and the payload module for this satellite. The core structure contains the integrated propulsion system as well as panels and other components that serve as the structural foundation of the satellite. The payload module consists of spacecraft electronics as well as the complete set of payload processing, routing and control hardware and software that perform the satellite's communications function. The spacecraft's Baseline Integrated System Test (BIST) was conducted from 27 July to 23 October 2007 at Lockheed Martin's Space Systems facilities in Sunnyvale. This test characterised the performance of the integrated satellite and established a performance baseline prior to environmental testing. After launch, the satellite will undergo some 15 months of on-orbit testing before it becomes operational. The second AEHF is due to launch in 2009. The system will reach Initial Operational Capability in 2010 when the second satellite becomes operational.

A fourth satellite was considered, but the Pentagon decided in 2004 that its place will be taken by the first TSAT, which will initially join the AEHF constellation to provide worldwide XDR coverage.

Specifications

Launches: First launch late 2008.
Orbit: Geosynchronous
Mass: Approx. 6,590 kg at launch. Approx. 4,090 kg on orbit.
Telecommunications: Data rates from 75 bps to 8.192 Mbps.
Antennas: 2 SHF Downlink Phased Arrays, 2 Crosslinks, 2 Uplink/Downlink Nulling Antennas, 1 Uplink EHF Phased Array, 6 Uplink/Downlink Gimbaled Dish Antenna, 1 each uplink/downlink Earth coverage horns.
Antenna coverages: 1 Earth coverage beam; 4 agile beams; 24 time-shared spot beams; 2 nulling spot beams; 6 dwelling spot beams.
Crosslinks: 2 per satellite, each bi-directional. Compatible with Milstar and AEHF requirements at 60 Mbps.
Dimensions: Length –9.76 m (across payload axis); solar arrays span 23.1 m
Payload: Onboard signal processing, crossbanded EHF/SHF communications.
Design life: 14 years.

High Resolution Coverage Antenna (2)
Medium Resolution Coverage Antenna (6)
Crosslink Antenna (2)
Transmit Phased Array Antenna (2)
Receive Phased Array Antenna
Low Gain Earth Coverage Antenna

The AEHF antenna suite. Low gain antennas deliver communications anywhere within the satellite's footprint. The phased array antenna provides super high-gain coverage for all users, including small portable terminals and submarines. Six medium resolution coverage antennas are provided by dwelling spot beam antennas and up to 24 time-shared MRCS coverages are provided by phased array antennas (Northrop Grumman) 1329315

AMSC

Background

American Mobile Satellite Corporation (AMSC) launched the AMSC-1 satellite in April 1995 to provide mobile communications across North America, in partnership with Telesat Mobile Inc (TMI) of Canada and its MSAT satellite. Both AMSC and TMI have now been absorbed into Mobile Satellite Ventures (MSV) and are covered in that entry.

Comstar/SBS series

Current Status

SBS-1 was retired in June 1990, SBS-2 was retired in September 1996 and was de-orbited, SBS-3 was retired in June 1995 and de-orbited, SBS-4 is still operating, SBS-5 was retired in March 2000 and was replaced by Galaxy 10R, SBS-6 is still operating as of late February 2004. It is now used primarily for occasional access for coverage of special events and for breaking news.

Background

Comsat established the Comsat General Corporation subsidiary in 1973 to act as the satellite operator of its growing inventory of orbiting hardware. The subsidiary now manages the five satellites of the Marisat, Comstar and SBS 2 networks. Another subsidiary, Comsat Video Enterprises owns and operates a satellite-based network distributing entertainment and video conference services to the US lodging industry.

Comsat General's 4-satellite Comstar system was the first to be integrated into the US national telephone network, providing voice, data and TV services for the whole of the US, Alaska, Hawaii and Puerto Rico. Hughes received COMSAT's USD65 million contract in September 1973. Comsat immediately leased all of the capacity to AT&T for the 7-year satellite operational lives at USD1.3 million monthly for each. Comstar D1, launched 13 May 1976, took its slot at 76° W next to D2 as their capacity declined. D3 replaced it 29 June 1978. Each satellite provided 18,000 duplex voice channels. COMSAT announced that it planned to move D4 over China in 1993; AsiaSat had an option on D4 but it expired 31 July 1993. D2, launched 22 July 1976, drifting January 1994.

Comsat also established a new Satellite Business Systems (SBS) subsidiary in December 1975 with ownership held at 41.3 per cent. IBM holds an equal 41 3 per cent share and Aetna Life & Casualty 17.4 per cent. In April 1984, after the successful launches of three satellites, Comsat sold its interest to its partners. The Federal Communications Commission approved of transfer of SBS to MCI Communications Corporation in November 1985. Comsat then announced the purchase of SBS 1 and 2 from MCI in March 1987.

Then, after Comsat successfully launched three SBS satellites through April 1984, Comsat sold its interest to its partners. Part of the partnership arrangement of commercial owners entailed a transfer of SBS to MCI Communications Corporation, approved by the FCC in November 1985. But SBS 4 and subsequent satellites remained majority partner IBM's wholly-owned Satellite Transponder Leasing Corporation (STLC) subsidiary. SBS 4 began operations in October 1984 to be joined in late 1988 by the last of the series' HS-376-class satellites, SBS-5. The more powerful HS-393-class SBS 6 had a Shuttle launch date of September 1988 when the *Challenger's* loss occurred in January 1986. It eventually departed October 1990 aboard Ariane.

STLC services provided by SBS 4 and 5 included network TV, wideband TDMA (multiple voice and high speed data) and corporate network VSATs for data networks and video broadcasting. HCI reached agreement with IBM for purchase of the SBS system in July 1989.

SBS 6 is the only HS-393 satellite of the SBS series, offering triple the capacity of its predecessors. Galaxy 4 replaced it at 99° W in November 1993. HCI moved SBS 6 to 95° W in March 1994 until Galaxy 3R's December 1995 launch. With the Galaxy in place HCI then moved the SBS satellite to 74° W. Galaxy 7 replaced SBS 4 at 91° W in January 1993. SBS 4 then moved in February 1993 to 77° W pending a decision on its future. NBC bought its entire capacity in September 1993, and began using it on a full-time basis by September 1995. It is now operating in an inclined orbit to extend its life beyond 2000.

Specifications
Comstar D4
Launched: 21 February 1981 by Atlas Centaur from Cape Canaveral complex 36
Location: 76° W geostationary, inclined 8.6°
Design life: 10 years
Contractors: Hughes Space & Communications
Transponders: 24.5/5.5 W TWTA 5.925–6.425/3.70–4.20 GHz up/down C-band CONUS/Hawaii/Alaska/Puerto Rico beams, edge EIRP 33 dBW, 34 MHz bandwidth, orthogonal polarisation. 19.0/28.6 GHz beacons for propagation research
Principal users: national voice, data, TV
Configuration: similar to Intelsat 4A. Spin-stabilised (56 rpm) cylindrical satellite with despun communication platform incorporating two offset parabolic 127 × 178 cm reflectors each handling 12 channels. Stationkeeping provided by hydrazine thrusters
Mass: 1,516 kg at launch, 911 kg on-station BOL
TT&C: Comsat's Washington facility working through Southbury, CT and Santa Paula, CA stations
Power: 570 W BOL from 17,000 silicon cells on cylindrical surface; two Ni/Cd batteries
Apogee kick motor: Thiokol solid propellant.

SBS 2
Launched: 24 September 1981 by Delta/PAM from Cape Canaveral
Location: 71° W GEO, inclination 6.7°; 97° W May 1992 to October 1993; 71° W June 1994 to August 1996
Design life: 7 years
Transponders: 10 (plus 4 back-up) 20 W min TWTA 14.025–14.466/11.725–12.166 GHz up/down Ku-band CONUS beams, 43 MHz bandwidth, EIRP 43.7 dBW edge CONUS
Principal users: domestic business voice, video, data, and VSAT services
Configuration: based on Hughes' HS 376 spin-stabilised bus, 2.16 m diameter, 6.6 m height with solar array drum extended (2.8 m stowed). Single antenna comprising two offset grid reflectors with horizontal/vertical feed system on despun platform
Mass: 546 kg on-station BOL
Attitude control: position + 50 rpm spin control maintained by four Hughes 22.2 N hydrazine thrusters
TT&C: through Comsat's Clarkburg, MD facility
Power system: >16,000 Si body cells provide 900 W EOL in conjunction with two Ni/Cd batteries
Apogee kick motor: Thiokol Star 30 solid.

SBS 4
Launched: 30 August 1984 by Shuttle STS-41D/PAM-D from Cape Canaveral
Location: 77° W GEO, inclination 2.4°; 91° W May to October 1992; 77° W from October 1993
Design life: 10 years minimum GEO
Mass: 1,117 kg at launch, 571 kg on-station BOL
Other specifications as for SBS 5 except that SBS 4 cannot double-up its transponders for 40 W output and does not incorporate the four 110 MHz units.

SBS 5
Launched: 8 September 1988 by Ariane V25 from Kourou, French Guiana
Location: 123° W geostationary (to be replaced by Galaxy 10)
Design life: 10 years minimum
Transponders: 10 (plus 4 back-up) 43 MHz and 4 (plus 2 back-up) 110 MHz bandwidth 20 W min TWTA 14.0–14.46/11.72–12.16 GHz up/down Ku-band CONUS plus Alaska and Hawaii spot beams, EIRP 39 dBW edge CONUS (45–46 dBW E/W concentrations), orthogonal polarisation. Four of the 43 MHz and all four of the 110 MHz transponders can be paralleled to provide 40 W output, principally for VSAT services
Principal users: domestic business voice, video, data, and VSAT services
Configuration: based on Hughes' HS-376 spin-stabilised bus, 2.16 m diameter, 6.6 m height with solar array drum extended (2.8 m stowed). Single antenna comprising two offset grid reflectors with horizontal/vertical feed system on despun platform
Mass: 1,241 kg at launch, 725 kg on-station BOL
AOCS: 50 rpm spin control and station-keeping maintained by four Hughes 22.2 N hydrazine thrusters
TT&C: through the Castle Rock, Colorado station
Power system: >16,000 Si cells on cylindrical surface provide 1,200/1,150 W BOL/EOL in conjunction with two Ni/Cd batteries
Apogee kick motor: Thiokol Star 30 solid.

SBS 6
Launched: 12 October 1990 by Ariane 4 from Kourou, French Guiana. Four apogee firings 13/14 October for GEO insertion
Location: 77° W geostationary; May 1992 to October 1995 at 95° W; from November 1995 at 74° W
Design life: 10 years (hydrazine for 15.6 years remaining after Ariane launch)
Transponders: 19 (plus 11 back-up; 30-for-19 ring redundancy) 41 W min TWTA 14.00–14.49/11.700–12.19 GHz up/down Ku-band CONUS plus switchable Hawaii/Alaska spot beams, 43 MHz bandwidth, CONUS EIRP 49–53 dBW, orthogonal polarisation
Configuration: Hughes HS-393 spin-stabilised bus 3.66 m diameter, 3.6 m stowed height/10.0 m deployed, payload and single 2.4 m diameter antenna mounted on despun platform
Mass: 2,478 kg at launch, 1,514 kg on-station BOL, 1,138 kg dry
AOCS: spin-stabilisation maintained by six 22 N thrusters, GEO insertion and major manoeuvres performed by two Marquardt R-4D 489 N thrusters, both systems drawing on 1,200 kg of NTO/MMH in four tanks each
Antenna pointing: 2-axis RF beacon tracking maintains 0.06°-accuracy pointing, backed up by Earth sensors
TT&C: Same as SBS 5
Power system: 2,300 W equinox BOL/2,200 W solstice EOL provided by Si solar array on cylindrical body (K7 cells) and extension skirt (K4¾ cells) in conjunction with two 48 Ah Ni/H_2 batteries
Thermal control: quartz mirror radiator band encircling upper body provides primary heat rejection path

DirecTV Series

Current Status
DirecTV is one of two leading suppliers of direct-to-home (DTH) entertainment services in the US market. It is continuing to build up its satellite network to support the bandwidth demands of high definition TV services.

Background
Hughes spun off DirecTV in 1992 after Sky Cable collapsed to provide a domestic US DTH service using Ku-band transmissions from the three DBS satellites launched in 1993, 1994 and 1995, renamed DirecTV-1, -2 and -3, respectively. DirecTV-IR was launched in 1999 as an on-orbit spare. DirecTV 4S was launched on 27 November 2001. DirecTV-5 was based on a FS-1300 platform from Space Systems/Loral. It was originally planned for launch on a Proton in October 2000, then moved to February 2001, 21 May 2001 and 19 October 2001 as the launch shifted to an Atlas 2AS before finally being launched on 7 May 2002 on a Proton.

Since 1999, DirecTV has mainly ordered satellites from Boeing. Most of these were designed from the outset for DTH services. However, two of them were originally commissioned as part of the

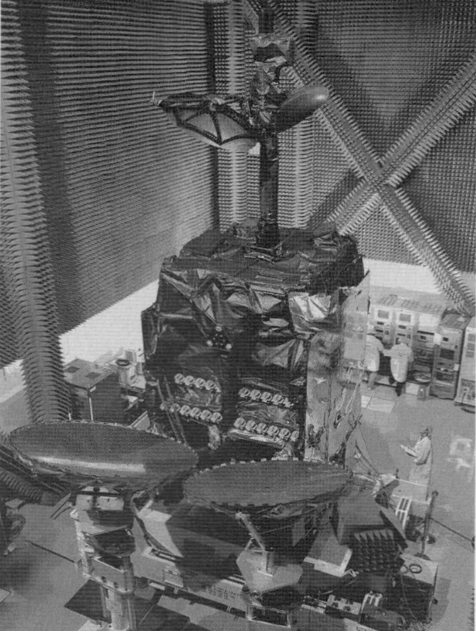

DirecTV-4S, a Boeing (HS) 601HP prior to final shipment for launch in November 2001 0132823

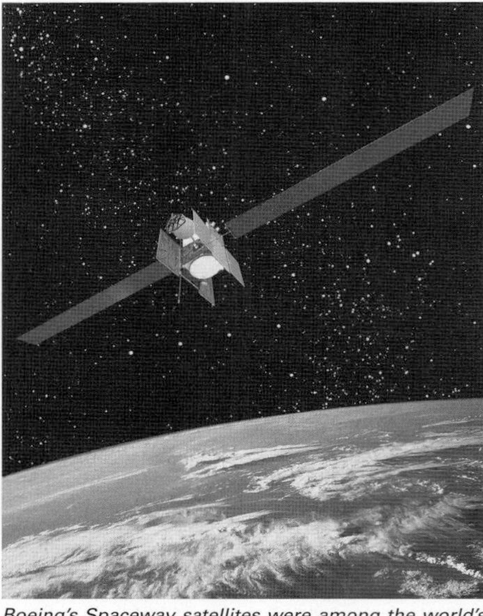

Boeing's Spaceway satellites were among the world's most advanced at the time of their launch (Boeing) 1311343

Hughes Spaceway global communications system. After News Corporation took over Hughes' DTH business, and renamed it DirecTV, the Spaceway satellites were retrofitted for DTH HDTV services.

DirecTV ordered two satellites from Space Systems/Loral in October 2003, as part of a deal in which the then-troubled satellite company agreed to complete the long-delayed DirecTV-7S within a certain period.

In 2003, DirecTV-3 was leased to Canada's Telesat to back up the partially failed Nimiq-2. It is known to Telesat as Nimiq-3 and has been moved into the 82°W position. Later, in 2006, DirectTV leased DirecTV-2 to Telesat as Nimiq-4i. It is now operational in the 91°W position pending the arrival of the new Nimiq-4.

Two more Boeing DirecTV satellites were due for launch in 2007. Both are 702-series spacecraft. DirecTV 10 is due to launch about mid-year on a Proton from Baikonur, and DirecTV-12 was due to be launched on Sea Launch later in 2007. The 30 January 2007 explosion of a Sea Launch rocket has placed that schedule in doubt.

Specifications
DirecTV-1
Launched: 18 December 1993 by Ariane 4 from Kourou, French Guiana. HSC handed over control of number 1 to HCI 6 January 1994; broadcasting began 17 June
Location: 101.2°W geostationary
Design life: 12–15 years
Contractors: Hughes Space & Communications Co, Hughes Aircraft Company (prime)
Transponders: 16 (plus eight back-up; 12-for-8) 120 W TWTAs (can be paired into 8 × 240 W) 17.3–17.84/12.2–12.74 GHz up/down Ku-band CONUS beams, 27 MHz bandwidth, 48–53 dBW EIRP, 40 Mbit/s, R/LHCP for DBS 1, L/RHCP DBS 2. BSS channels 1–32. Each shaped reflector is fed by a single feed horn; the shape concentrates power on areas such as the SE affected by rain fade
Principal applications: direct broadcast to 45 cm diameter antennas
Configuration: Hughes' HS-601 3-axis bus, 26.2 m solar array span, two 2.14 m transmission reflectors, 102 cm receive reflector, 36 cm beam tracking antenna. Each antenna array is mounted at only one point to minimise thermal distortion to shaped profile

DirecTV-8 awaits its May 2005 launch by a Proton-M from Baikonur (Space Systems/Loral) 1311341

DirecTV-9S, a Space Systems/Loral satellite, is checked out before its launch on Ariane 5 on 13 October 2006 (Arianespace) 1311342

Mass: 2,860 kg at launch, 1,270 kg dry mass, 1,725 kg on-station BOL
TT&C: from HCI's Operations Control Center in El Segundo, CA, via Castle Rock, CO + Spring Creek, NY
Power system: twin solar arrays providing 7,000 W BOL. 100 per cent eclipse protection by Ni/H_2 battery.

DirecTV-2 (Nimiq-4i)
Launched: 3 August 1994 on Atlas 2A from Cape Canaveral
Location: 91° W GEO
Other specifications as for DBS 1.

DirecTV-3 (Nimiq-3)
Launched: 10 June 1995 by Ariane 4 from Kourou, French Guiana
Location: 82° W GEO
Mass: 2,934 kg at launch, 1,259 kg dry mass, 1,707 kg on-station BOL
Other specifications as for DBS 1.

DirecTV-IR
Launched: 9 October 1999 by Zenit 3SL from SeaLaunch
Location: 101° W GSO
Contractor: Boeing Satellite Systems
16 (+4 spare) × 200 W MPA (dual 100 W TWTA) Ku-band (12.2–12.7 GHz) with 50–55 dBW radiated power. Two 272 cm shaped Gregorian reflectors for transmit and one 137 cm shaped single-surface reflector
Configuration: BSS (Hughes) 601HP 3-axis bus, 26 m solar array span, 7 m width across antennas.
Stowed size: H 4 m, W 2.7 × 3.6 m
Mass: 3,446 kg at launch; 2,304 kg at BOL
Power system: Twin solar arrays 8.7 kW BOL, 7.7 kW EOL
AOCS: 1 × 490 N liquid apogee motor; 1 × 17 mN N-S ion propulsion system; 4 × 10N E-W bipropellant thruster; 4 × 22N N-S bipropellant backup thrusters.

DirecTV-4S
Launched: 27 November 2001 by Ariane 4 from Kouron Fr Guiana
Location: 101° W
Contractor: Boeing Satellite Systems

Spaceway F1 is launched by a Sea Launch Zenit-3SLB on 26 April 2005 (Sea Launch) 1311340

Communications: spot beam payload: 38 × 30 W, 45 W, 65 W, 80 W TWTA to varying combinations. National beam payload: 2 × 280 W TWTA. Supplemental beam payload: 2 × 240 W TWTA; 6 × 120 W TWTA.
Configuration: as DirecTV-IR
Mass: 4,260 kg at launch; 2,800 kg at BOL
Power system: as DirecTV-IR
AOCS: as DirecTV-IR.

DirecTV-5
Launched: 7 May 2002 on a Proton 8K82K/11S861-01 from Baikonur
Location: 129° W
Contractor: Space Systems/Loral
Configuration: Space Systems/Loral FS-1300 with Tempo 1 payload
Mass: 3,640 kg at launch

DirecTV-7S
Launched: 4 May 2004 on a Sea Launch Zenit 3-SLB from launch platform Odyssey
Location: 119° W
Contractor: Space Systems/Loral
Configuration: Space Systems/Loral FS-1300
Payload: 54 Ku-band transponders
Mass: 5,483 kg at launch

DirecTV-8
Launched: 22 May 2005 on a Proton M/Breeze M from Baikonur
Location: 101° W
Contractor: Space Systems/Loral
Configuration: Space Systems/Loral FS-1300
Payload: 16 Ku-band transponders plus Ka-band transponders
Mass: 5,500 kg at launch

DirecTV-9S
Launched: 13 October 2006 on Ariane 5 ECA from Kourou, French Guiana
Location: 101°W
Contractor: Space Systems/Loral
Configuration: Space Systems/Loral FS-1300
Payload: 52 Ku-band channels and 2 Ka-band channels
Power: 13.9 kW start-of-life
Mass: 5,505 kg at launch

Spaceway F1/F2
Launched: F1: 26 April 2005 on a Sea Launch Zenit 3-SLB from launch platform Odyssey. F2: 16 November 2005 on Ariane 5 ECA from Kourou, French Guiana
Location: F1: 103° W; F2: 99° W
Contractor: Boeing Satellite Systems
Configuration: Boeing 702
Payload: Ka-band channels
Power: 15.9 kW start-of-life, 13.9 kW end-of-life from 40.9 m span solar arrays
Mass: 6,116 kg at launch

DSCS III

Current Status
Defense Communications Satellite System Phase III (DSCS-III) spacecraft provide high priority, secure command communications for the US armed forces. The last DSCS-III was launched in August 2003. The Wideband Gapfiller Satellite, which is due for its first launch in 2007, is intended to replace DSCS.

Background
DSCS III satellites have a ten-year design life, but many of the satellites on orbit have surpassed it by a large margin. The programme started in 1975 and the first satellite, A1, was launched by a Titan 34D in October 1982. The contract was originally awarded to RCA Astro Space, which subsequently went through three acquisitions and mergers, becoming first General Electric, then Martin-Marietta and finally Lockheed Martin.

Two double DSCS-III launches – one by the Shuttle Atlantis on 3 October 1985 and a second on a Titan 34D on 4 September 1989 – established full global DSCS coverage.

The spin-stabilised satellites were along the first military satellites to operate in the Super High Frequency (SHF) band for increased data rates and jamming resistance. Their multiple-beam antennas employ advanced encryption to protect their data and 1,300 duplex voice links for both tactical and strategic users. In addition to nuclear hardening,

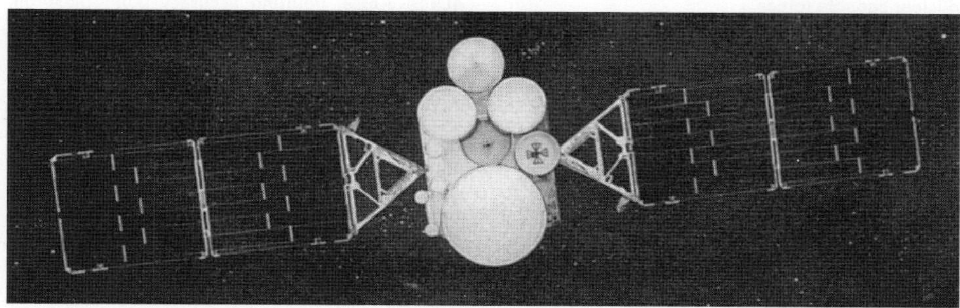

DSCS Phase III: these models are the primary user of the Atlas 2 launcher. At bottom is the large 61-beam receive array; the 19-beam transmit antennas are at top, below the single steerable spot. The two UHF antennas are central. At left are the global horns (Lockheed Martin Astro Space) 0516831

Lockheed Martin produces the third-generation Defense Satellite Communications System (DSCS III). It features high-data rate and around the globe, secure communications. Each satellite has six independent SHF transponders with a 500 MHz bandwidth (Lockheed Martin) 0022567

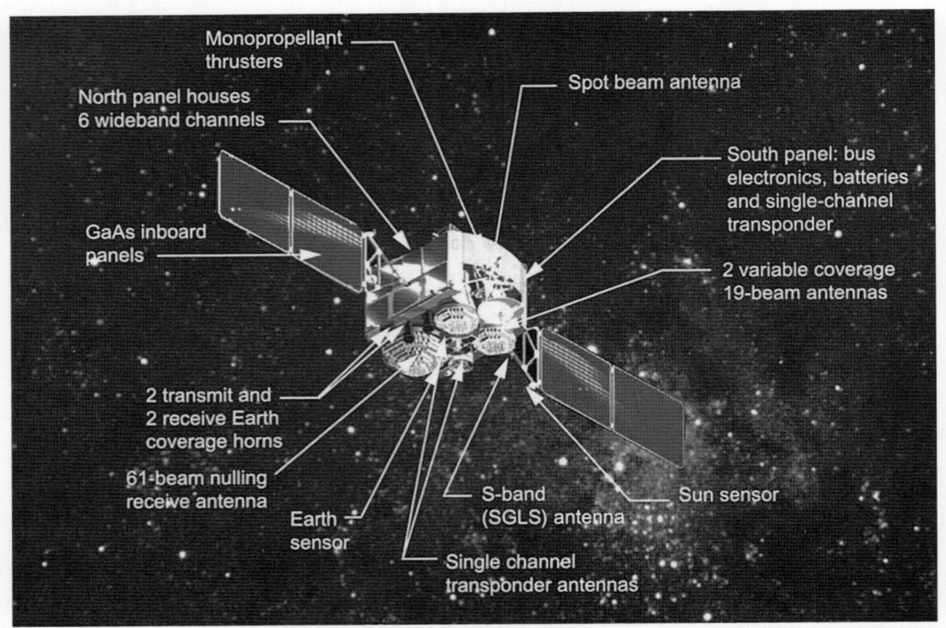

All of the major features of the DSCS III are shown in this drawing 0022568

sensors can detect a jamming attempt, report it and wait for ground control to plot the location of the jammer and dispatch instructions on how to use its steering capacity to null the jammer. The nuclear hardening cannot protect against direct attack, but enables it to withstand low-level flash effects, X-rays, gamma rays and ElectroMagnetic Pulses (EMPs).

The DSCS communications payload includes six independent SHF transponder channels that cover a 500 MHz bandwidth. Three receive- and five transmit-antennas provide selectable options for Earth coverage, area coverage and/or spot beam coverage. A special purpose single-channel transponder is also on board, used for disseminating emergency action and force direction messages to nuclear capable forces.

The primary on-orbit DSCS constellation comprises five satellites in geosynchronous orbit. However, most of the 14 DSCS satellites launched since 1982 were still operational late in 2005, and all were being used in various capacities, from operational communications in Southwest Asia to research and development of ground-based support capabilities.

The USAF launched a DSCS Service Life Extension Program (SLEP) in the mid-1990s to extend the active life of the overall system. The final four satellites were upgraded to provide a 200-per cent increase in communication capacity, with new 50W Travelling-Wave-Tube (TWT) amplifiers. Tactical users with small terminals can gain a 700 per cent increase in capacity in certain scenarios. New solar cells provided more than 1,700W compared with 1,330W for earlier versions. Additionally, the new satellite offered greater flexibility for mixing high- and low-power users. The first DSCS III SLEP satellite was launched on 21 January 2000. The remaining three spacecraft flew on 20 October 2000, 11 March 2003 and 29 August 2003.

The 50th Space Wing's 3rd Space Operations Squadron at Schriever AFB, Colorado, provides satellite bus command and control for all DSCS satellites. Ground terminals range from fixed stations with 18.3 m antennas, to transportable stations carried on large transport aircraft, to new generation systems offered by Raytheon, ITT and L-3 which are fitted to a Humvee. Harris provides the USAF and Marine Corps with systems that are transported in standard transit cases.

Specifications
Contractors: Lockheed Martin Astro Space
Transponders: 6 × SHF 7.900–8.400/7.250–7.750 GHz up/down channels: 1 40 W 50 MHz; 2 40 W 75 MHz; 3/4 10 W 85 MHz; 5 10 W 60 MHz; 6 10 W 50 MHz. Two (plus two back-up) 40 W TWTAs + four (plus two back-up in 1-for-2) 10 W SSPAs.

Working through: One high gain (30.2 dB min) steerable 3.0° spot dish for channel 4 (EIRP 37.5 dBW) + channels 1/2 (EIRP 44 dBW).

Two 19-beam waveguide lens Multi-Beam Antennas (MBAs) for ch1-4 with beam-forming networks to produce selected antenna patterns accommodating the ground receiver network. EIRP: Earth coverage channels 1/2 29.0 dBW, channels 3/4 23.0 dBW; 1° narrow coverage channel 1/2 40.0 dBW channels 3/4 34.0 dBW. Receive through 61-beam waveguide lens MBA, with associated beam-forming network to provide selective coverage and jamming protection.

Two global horns receive and two transmit for channels 3–6 (EIRP 25.0 dBW)

UHF bow tie 300–400 MHz receive and cross dipole 225–260 MHz SSPA transmit for Single Channel Transponder to support AFSATCOM. AFSATCOM II signals handled at either UHF or SHF (channels 1)
Configuration: 206 cm long body (279 cm across antennas), 193 cm wide, and 196 cm deep to antenna tips. Transponder amplifiers and associated components are mounted on the north panel. Modular structure, 3-axis stabilised
Mass: about 1,170 kg at launch
AOCS: 3-axis, pointing accuracy 0.09° pitch/roll, 1.0° yaw, 16 × 4.45 N thrusters, 276 kg capacity hydrazine system utilising four spheres. Inserted into near-GEO by IUS stage 2 (III on Shuttle) or IABS (IIIB)
TT&C: via S-band (1,807.764 and 1,823.779 up/2,257.5 and 2,277.5 MHz down) and SHF (channels 1/5 up/7,600.0 and 7,604.70588 MHz down). The SHF link's primary function is to control channel and antenna configuration in real time through the defence Information Systems Agency's SCCE Satellite Configuration Control Elements
Power system: 2-panel Si dual solar wings spanning 11.62 m, totalling 11.72 m², providing 1,240 W BOL, 980 W after 10 years, supported by 3 × 35 Ah Ni/Cd batteries. 190 W provided by stowed array during spin transfer phase.

EchoStar series

Current Status
EchoStar Communications has launched a fleet of ten satellites and retails DTH television services under the DishNetwork brand. An 11th satellite in the series is due to fly in 2007 on a Sea Launch rocket.

Background
When EchoStar Satellite Corporation announced its intentions to enter the direct-to-home broadcast market it became the second largest corporation to do so, setting up a competitive market for TV services. EchoStar signed a contract with Martin Marietta Astro Space in October 1992 to build one 7000-series satellite, plus options on six others.

Since that time, EchoStar has expanded its fleet with Lockheed Martin A2100 and A2100AX and Space Systems/Loral 1300-series spacecraft. One of these, EchoStar IX, is jointly operated with Intelsat as Galaxy 23, which uses its C-band transponders. (When it was launched, the C-band segment was known as Telstar 13 and operated by Loral Skynet). In January 2005, EchoStar acquired the on-orbit Rainbow 1 satellite from Cablevision Systems and renamed it EchoStar XII.

EchoStar, like its US rival DirecTV, is expanding its fleet in the race with hybrid satellite-cable systems to provide high-definition TV services in the US market. EchoStar XI, a high-power (20 kW through-life) all-Ku band 1300-series, is set for launch in 2007 under a May 2006 contract with Sea Launch. In June 2006, Space Systems/Loral announced another satellite order from EchoStar. Most likely, this spacecraft will be used by EchoStar's Hong Kong-based affiliate, China Satellite Mobile Broadcast Limited, to provide S-band mobile video service in time for the 2008 Summer Olympics in Beijing.

Specifications
EchoStar I
Launch: 28 December 1995 by CZ-2E from Xichang; entered service 3 March 1996
Location: 148°W geostationary
Design life: 12 years' propellant carried
Contractors: Lockheed Martin Astro Space
Transponders: 16 130 W TWTAs 14/12.2–12.7 GHz up/down Ku-band CONUS beams (BSS channels, odd numbers 1–21), 24 MHz bandwidth, EIRP CONUS 51 dBW edge, circular polarisation
Principal applications: DTH TV
Configuration: Lockheed Martin Astro Space 7000 series 3-axis platform
Mass: 3,288 kg at launch
AOCS: station-keeping ±0.5° lat/long. NSSK by 1.8 kW hydrazine arcjets

EchoStar II
Launch: 11 September 1996 by Ariane 4 from Kourou, French Guiana
Location: 148°W geostationary
Contractors: Lockheed Martin Astro Space
Transponders: 16 transponders
Principal applications: DTH TV
Configuration: Lockheed Martin Astro Space 7000 series 3-axis platform
Mass: 2,865 kg at launch
Other specifications as for EchoStar 1.

EchoStar III
Launch: 4 October 1997 by Atlas 2AS from Cape Canaveral
Location: 61.5°W geostationary
Contractors: Lockheed Martin Space Systems
Transponders: 16 transponders
Principal applications: DTH TV
Configuration: Lockheed Martin A2100X series 3-axis platform
Mass: 2,865 kg at launch

EchoStar IV
Launch: 7 May 1998 on Proton-K from Baikonur, Kazakhstan
Location: 119°W geostationary
Contractors: Lockheed Martin Space Systems
Transponders: 32 transponders
Principal applications: DTH TV
Configuration: Lockheed Martin A2100AX
Mass: 3,478 kg at launch

EchoStar V
Launch: 23 September 1999 by Atlas IIAS from Cape Canaveral
Location: 119°W
Contractors: Space Systems/Loral
Transponders: 32 transponders
Principal applications: DTH TV
Configuration: SS/L FS-1300
Mass: 3,602 kg at launch

EchoStar I provides US direct to home TV broadcasts (Lockheed Martin Astro Space) 0517019

EchoStar X launches into orbit aboard a Sea Launch Zenit-3SL on 15 February 2006 (Sea Launch) 1311346

EchoStar VI
Launch: 14 July 2000 by Atlas II2AS
Location: 110°W
Contractors: Space Systems/Loral
Transponders: 32 transponders
Principal applications: DTH TV
Configuration: SS/L FS-1300
Mass: 3,700 kg at launch

EchoStar VII
Launch: 21 February 2002 by Atlas IIIB
Location: 119°W
Contractors: Lockheed Martin Space Systems
Transponders: 32 transponders plus spot beams
Principal applications: DTH TV
Configuration: Lockheed Martin A2100AX
Mass: 4,027 kg at launch

EchoStar VIII
Launch: 22 August 2002 by Proton-K from Baikonur Cosmodrome, Kazakhstan
Location: 110°W
Contractors: Space Systems/Loral
Transponders: 32 transponders plus spot beams
Principal applications: DTH TV

High-gain antennas for spot beams distinguish the Loral-buyild EchoStar VIII (Space Systems/Loral) 1311344

Echostar IX prepared for launch by Space Systems/ Loral. DBS spacecraft are among the largest comsats in service (Space Systems/Loral) 1311345

Configuration: SS/L FS-1300
Mass: 4,660 kg at launch

EchoStar IX
Launch: 7 August 2003 by Sea Launch Zenit-3SLB from launch platform Odyssey
Location: 121°W
Contractors: Space Systems/Loral
Transponders: Ku-band, Ka-band spot-beam and 24 C-band. C-band used by Intelsat as Galaxy 23.
Principal applications: DTH TV, cable TV programming
Configuration: SS/L FS-1300
Mass: 4,737 kg at launch

EchoStar X
Launch: 15 February 2006 by Sea Launch Zenit-3SLB from launch platform Odyssey
Location: 110°W
Contractors: Lockheed Martin Commercial Space Systems
Transponders: Ku-band transponders
Principal applications: DTH TV
Configuration: A2100AX
Mass: 4,333 kg at launch

EchoStar XII
Launch: 17 July 2003 by Atlas V from Cape Canveral, Florida (as Rainbow 1)
Location: 61.5°W
Contractors: Lockheed Martin Commercial Space Systems
Transponders: Ku-band and C-band transponders
Principal applications: DTH TV
Configuration: A2100
Mass: 4,328 kg at launch

Galaxy

Current Status

Now operated by Intelsat, the Galaxy fleet continues to grow with the launch of Galaxy 16 in June 2006 and the launches of Galaxy 17 and 18, scheduled for the first half of 2007.

Background

Hughes Electronics Corporation established Hughes Communications Incorporated (HCI) as a wholly owned subsidiary to operate the world's largest fleet of privately-owned commercial communications satellites, known as the Galaxy series. HCI began satellite operations with the launches of its three initial Galaxy satellites during 1983–84. In 1993, HCI also established DirecTV Inc to operate a domestic US direct broadcast system. It became an independent Hughes Electronics subsidiary in November 1993 and was acquired by Rupert Murdoch's News Corporation in December 2003.

Galaxy Satellite Services was sold by Hughes to PanAmSat in May 1997. In turn, Intelsat acquired PanAmSat in July 2006.

The current Galaxy fleet comprises ten satellites including two on-orbit spares. The oldest of the primary satellites was launched in December 1999, reflecting the robust development of the cable TV market and the growing demand for high-definition TV service and video-on-demand. The network's primary role is to deliver TV programming to 'head-end' receiver stations across North America, which then feed local cable networks. For example, a single 'head-end' with one multi-beam antenna can receive dozens of channels simultaneously from three Galaxy satellites – 10R, 13 and 14. The fleet includes one jointly-operated satellite: Intelsat operates the C-band payload on Galaxy 13 and partner JSAT uses the Ku-band payload as Horizons-1.

Galaxy 3C on its way to its launch in June 2002 0132822

Galaxy 4R, Boeing 601HP satellite, is prepared for its April 2000 launch (Boeing) 1198722

Recent Galaxy retirements and failures have included Galaxy 1R, launched in February 1994 and replaced by Galaxy 15 in early 2006; Galaxy 3R, retired in early 2006; and Galaxy 8I, which was launched in December 1997 but was retired in 2004 after its Xenon-Ion Propulsion System (XIPS) failed. Galaxy 11 – like another Intelsat satellite, PAS-1R – faces a shortened lifetime because of a class problem with early Boeing 702 satellites, which have a novel solar array design that loses power more quickly than expected.

Given its history, it is no surprise that many of the Galaxy satellites are Boeing (formerly Hughes) designs, and Galaxy 11 was the first satellite based on the Boeing 702 bus. However, since the system was divested by Hughes, new satellites have been ordered from Orbital, Alenia Alcatel Space and Space Systems/Loral. The Orbital satellites are all-C-band, dedicated to cable TV service and much smaller than the Boeing spacecraft.

Specifications

Galaxy 3C
Launched: 15 June 2002 by Zenit (Sea Launch)
Location: 95° W longitude GSO
Design life: 15 years
Configuration: BSS-702 bus; two solar cell wings each with six panels of improved triple-junction GaAs cells producing 15 kW EOL; 56-cell NiH battery with 328-Amp/hr life; single 490 N liquid apogee motor; 4 × 0.165 N XIPS; 4 × 0.165 N transfer orbit; 4 × 0.08 N attitude control.
Mass: 4,860 kg launch; 2,873 at BOL
Size: length 48 m across extended solar arrays; width 9.4 m across antennas; height 7.3 m to tip of antennas; stowed 7.3 m × 3.2 m × 3.3 m
Transponders: 24 (6 spare) × 36 MHz (34 W SSPA) C-band; 8 (2 spare) × 54 MHz (240 W each configurable to 16 active/4 spare 120 W TWTA), 13 (3 spare) × 27 MHz (127 W TWTA), and 24 (4 spare) × 24 MHz (120 W LTWTA) in Ku-band.
Antennas: One 216 cm, one 140 cm, one 114 cm dual gridded shaped reflectors and two 284.5 cm single surface reflectors.
Principal application: North American TV and corporate distribution and inventory management networks.

Galaxy 4R
Launched: 18 April 2000 by Ariane 4 from Kourou, French Guiana
Location: 99° W
Design life: 15 years
Contractor: Boeing Satellite Systems
Coverage: North America
Transponders: C-band: 24 × 36 MHz 40 W transponders; Ku-band: 24 × 36 MHz 108 W transponders.

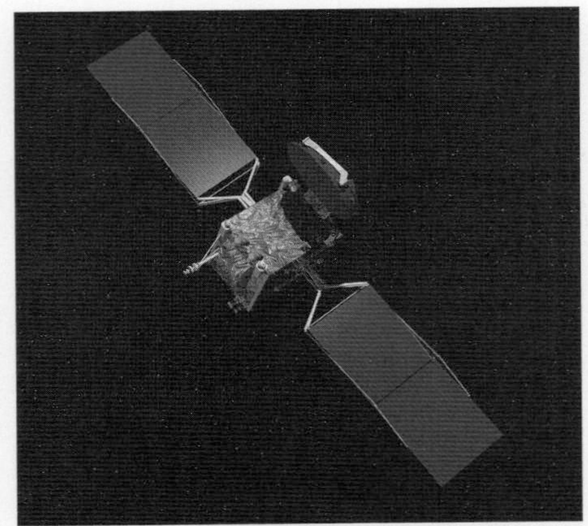

Galaxy 12, launched in April 2003, is the first of a series of smaller all C-band satellites built for the Galaxy constellation by Orbital Sciences (Orbital Sciences) 1198726

Antennas: One 216 cm, one 140 cm, one 114 cm dual gridded shaped reflectors and two 284.5 cm single surface reflectors.
Power: Two solar array wings spanning 26 m in total, each with four panels of GaAs cells; 8.8 kW beginning of life, 8 kW end of life
Launch mass: 3,668 kg
Principal application: North American TV distribution; replaced 2006 by Galaxy 16.

Galaxy 9
Launch: 24 May 1996 by Delta 7925 from Cape Canaveral
Location: 269° E
Design life: predicted end of life 2008
Transponders: 24 (plus 6 back-up; 30-for-24 ring redundancy) 16 W TWTA 5.925–6.425/3.700–4.200 GHz up/down C-band CONUS/Alaska/Hawaii beams, 24/36 MHz bandwidth, EIRP 40/37.5 dBW peak/min over CONUS, orthogonal polarisation
Configuration: stretched HS-376, 3.10/7.49 m height stowed/deployed
Mass: 1,397 kg at launch
Principal application: On-orbit spare

Galaxy 10R
Launch: 24 January 2000 by Ariane 4 from Kourou, French Guiana. Replaced Galaxy 10, destroyed on launch 27 August 1998.
Location: 123° W
Design life: 15 years
Transponders: C-band: 24 × 36 MHz 40 W transponders; Ku-band: 24 × 36 MHz 108 W transponders.
Power: Two solar array wings spanning 26 m in total, each with four panels of GaAs cells; 8.8 kW beginning of life, 8 kW end of life
Launch mass: 3,475 kg
Configuration: Boeing 601HP
Coverage: North America; cable TV distribution, voice and data.

Galaxy 11
Launch: 22 December 1999 on Ariane 44L (V125) from Kourou, French Guiana.

Galaxy 14 is prepared for its August 2005 launch. The C-band satellite was built by Orbital Sciences (Orbital Sciences) 1198725

Location: 91° W.
Transponders: 24 × 20 W C-band (6 spare); 24 × 75 W Ku-band (6 spare); 16 × 140 W Ku-band (4 spare).
Configuration: Boeing 702 platform. GaAs solar arrays span 31 m provide 10.4 kW EOL; 42 cell Ni/H_2 batteries for eclipse protection. Liquid apogee motor (490 W); AOCS by bi-propellent thrusters (4 × 22 N; 4 × 10 W); Hughes Xenon is a propulsion subsystem for NSSK (4 × 0.08 W for transfer orbit; 4 × 0.08 N for altitude control).
Antennas: 2 × 2.4 m/2 × 1.8 m dual-gridded shaped reflections.
Size: Stowed: h 6.2 m × w 3.8 m by 3.3 m; in orbit h 31 m, w 9.0 m
Launch mass: 4,488 kg
Coverage: North America; Brazil; cable TV and Internet

Galaxy 12
Launch: 8 April 2003 on Ariane 5 from Kourou, French Guiana
Location: 125° W geostationary
Contractor: Orbital Sciences
Design life: 15 years
Configuration: Orbital Star 2 platform; power 2.64 kW EOL; tri-axis, zero momentum stabilisation
Transponders: All C-band: two groups of 16-for-12 redundant LTWTA (37 W)
Power: GaAs cells delivering 4.7 kW
Dimensions: 1.5 m × 1.9 m × 3.3 m stowed
Launch mass: 1,730 kg
Coverage: North America – on-orbit spare to support cable TV distribution.

Galaxy 13/Horizons 1
Launch: 30 September 2003 on Zenit from Sea Launch platform
Location: 127° W geostationary
Design life: 15 years
Transponders: Ku-band (Horizons 1) 24 × 108 W (+8 spare); C-band (Galaxy 13) 24 × 40 W (+8 spare)
Configuration: BSS-601HP platform; 2 × four-panel GaAs dual junction solar cell arrays 9.9 kW BOL, 8.6 kW EOL plus 30-cell NiH batteries; one 490 N liquid apogee motor; AOSC by 4 × 13 cm (0.018 N) North-South primary (xenon backup) thrusters; 12 × 10 N thrusters (4 × north-south backup, 4 × east-west, 4 × aft).
Antennas: two shaped Gregorian transmit/receive each 42 cm antennas; two dual gridded surface shaped transmit/receive antennas each 20 cm.
Size: length 26.2 m across solar arrays; width 7 m across antennas; stowed 5.7 m × 2.7 m
Mass: 4,060 kg at launch; 2,630 kg BOL
Coverage: US and Asia

Galaxy 14
Launch: 14 August 2005 on Soyuz-Fregat from Baikonur, Kazakhstan
Location: 125° W geostationary
Contractor: Orbital Sciences
Design life: 15 years
Configuration: Orbital Star 2 platform; power 2.64 kW EOL; tri-axis, zero momentum stabilisation
Transponders: All C-band: two groups of 16-for-12 redundant LTWTA (37 W)
Power: GaAs cells delivering 4.7 kW
Dimensions: 1.5 m × 1.9 m × 3.3 m stowed
Launch mass: 1,730 kg
Coverage: North America – dedicated cable satellite

Galaxy 15
Launch: 13 October 2005 on Ariane 5 from Kourou, French Guiana
Location: 133° W geostationary
Design life: 15 years
Contractor: Orbital Sciences
Configuration: Orbital Star 2 platform; power 2.64 kW EOL; tri-axis, zero momentum stabilisation
Transponders: C-band: two groups of 16-for-12 redundant LTWTA (37 W); L-band, specialized Geostationary Communications and Control Segment (GCCS) broadcasting GPS data as part of experimental precision air navigation system.
Power: GaAs cells delivering 4.7 kW
Dimensions: 1.5 m × 1.9 m × 3.3 m stowed
Launch mass: 1,892 kg
Coverage: North America – dedicated cable satellite plus support for FAA navigation system.

Galaxy 16
Launch: 17 June 2006 on Sea Launch Zenit-3SL from launch platform Odyssey
Location: 99° W geostationary
Design life: 15 years
Contractor: Space Systems/Loral

Ariane 5 launches Galaxy 15 in October 2005 (Arianespace) 1198724

Configuration: Loral 1300 Omega 2
Transponders: C-band: 24 channels at 36 MHz; Ku-band: 24 channels at 36 MHz
Power: 10 kW end-of-life
Launch mass: 4,640 kg
Coverage: North America – dedicated cable satellite for HDTV and data

Galaxy 17
Launch: Ariane 5 from Kourou, set for February 2007
Location: 99° W or 91° W geostationary
Design life: 15 years
Contractor: Alcatel Alenia Space
Configuration: Spacebus 3000 B3
Transponders: C-band: 24 channels at 36 MHz; Ku-band: 24 channels at 36 MHz
Power: 8.6 kW end-of-life
Launch mass: 4,000 kg
Coverage: North America – dedicated cable satellite for HDTV and data. Replaces Galaxy 11.

Galaxy 18
Launch: Sea Launch Zenit-3SL from launch platform Odyssey, due by second quarter of 2007.
Location: 123° W geostationary
Design life: 15 years
Contractor: Space Systems/Loral
Configuration: Loral 1300 Omega 2
Transponders: C-band: 24 channels at 36 MHz; Ku-band: 24 channels at 36 MHz
Power: 10 kW end-of-life
Launch mass: 4,640 kg
Coverage: North America – dedicated cable satellite for HDTV and data

Globalstar Constellation

Current Status
After a seven-year gap, launches of Globalstar satellites resumed on 30 May 2007, when a Starsem Soyuz FG vehicle orbited four satellites which had been in storage as ground spares since 2002. The satellites were delivered into a 920 km orbit, and then used their onboard thrusters to reach the operational orbit. Four more satellites are expected to follow on 29 July 2007. Globalstar has selected Alcatel Alenia Space (now Thales Alenia Space) to design and build a new generation of 48 satellites, with deliveries scheduled to begin in the summer of 2009. The satellites are being designed to provide service until at least 2025.

Background
Globalstar achieved commercial operating capability with its 48-satellite network in February 2000, following the 22 November 1999 launch of four satellites by a Soyuz-Ikar from Baikonur. On 8 February 2000, the full constellation of 52 satellites was achieved with the launch of four spare satellites on a Delta II from CCAFS. In response to satellite failures and anomalies, Globalstar

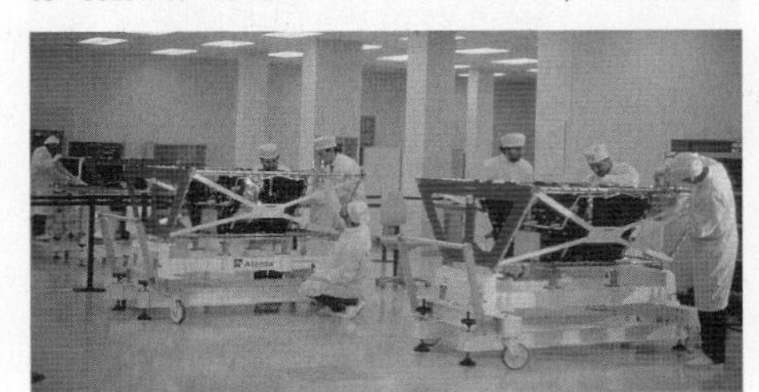

Engineers at Alenia Aerospazio begin assembly of a Globalstar satellite bus 0084770

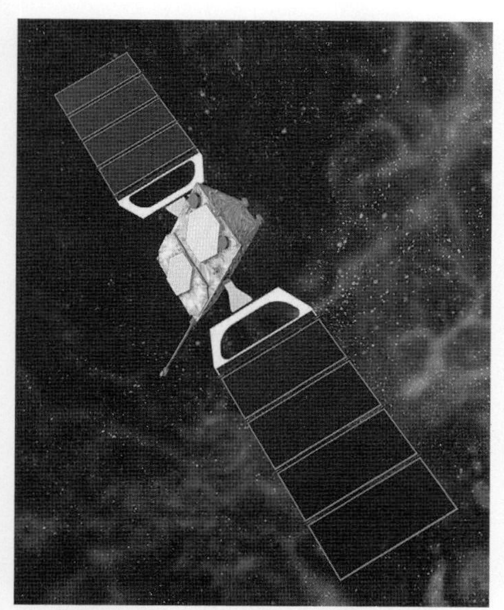

The first Globalstar launch took place in 1998 aboard a Delta II (Globalstar) 0005569

reconfigured the system in mid-2003 from a 48-satellite constellation to a 40-satellite constellation with in-orbit spares. Eight orbital planes have been maintained, but with five service satellites per plane rather than six. The satellites are available to relay communications over more than 80 per cent of the Earth's surface.

The Globalstar system is based on the direct-repeater or "bent-pipe" concept. The user's call is received by a satellite, which relays it directly to a ground station or gateway in its field of view. The ground station then passes the call to its destination via terrestrial or normal geostationary satellite links. Globalstar service is therefore dependent on the availability of a Globalstar gateway. At the end of 2006 there were 25 gateways, each of which served an area of approximately 700,000 to 1,000,000 miles^2.

The satellite's trapezoidal body is fabricated from a rigid aluminium honeycomb. The trapezoidal shape was selected to conserve volume and to allow the mounting of multiple satellites under the launch vehicle's payload fairing. The satellite operates in a body-stabilized, three-axis attitude control mode and uses sun sensors, Earth sensors and a magnetic sensor to help maintain attitude. The satellite contains momentum wheels and magnetic torquers to minimize propellant consumption for attitude control, and utilizes five thrusters for orbit-raising, station-keeping manoeuvres and attitude control. The spacecraft's mono-propellant hydrazine thrusters are fed from a single on-board propellant tank, sufficient to keep the satellite in its proper orbit for the full lifetime.

Solar arrays provide the primary source of power for the Globalstar spacecraft, while batteries are used during eclipses and peak traffic periods. The solar panels provide 1,100 W of power, and they automatically track the sun as the satellite orbits the Earth, providing the maximum possible exposure to sunlight. The communications systems are mounted on the Earth deck, which is the larger of the two rectangular faces on the satellite's body. There are C-band antennas for communications with Globalstar gateways, and L- and S-band antennas for communications with user terminals. These antennas are of a phased array design that projects a pattern of 16 spot beams on the Earth's surface, covering a service area, or 'footprint', several thousand kilometres in diameter.

Final assembly of the Globalstar satellites (Theo Pirard) 0054357

The original Globalstar satellites had a design life of 7.5 years. Nine of the original 52 satellites had failed on orbit by December 2006, according to a Globalstar document filed with US regulators. Relocation of these satellites to disposal orbits as high as 2,000 km began in late 2005. Eight of the failures were attributed to common anomalies in the satellite communications subsystem S-band antenna. The ninth satellite failure was attributed to an anomaly with the satellite command receiver. The company announced in February 2007 that the performance of the solid state amplifiers that feed the S-band antennas and permit two-way communications links is eroding, probably because of the high daily radiation dose from the Van Allen Belts at the constellation's orbital altitude of 1,414 km. The degradation does not affect adversely the one-way 'Simplex' data transmission services, which utilise only the L-band uplink from a subscriber's 'Simplex' terminal to the satellites.

In October 2005, Globalstar and Starsem signed a contract covering one firm Soyuz launch from Baikonur with four satellites, with an option on a second. Eight ground-spare satellites were in storage at a Space Systems/Loral facility in Palo Alto, California, in early 2006, and Globalstar received government approval in August 2006 to export the satellites to Europe for launch preparation. By late 2006 all eight satellites had been delivered to Rome, and both Starsem launches were on schedule for the first half of 2007. According to a company statement, the launch of these satellites should limit any gaps in coverage caused by degradation of the constellation's S-band voice and data service.

In April 2006, Globalstar awarded Alcatel Alenia Space (now Thales Alenia Space) a contract to design a second-generation LEO satellite to replace the original spacecraft. This was followed by a EUR661 million contract, signed on 4 December 2006, for the design, development, manufacture and launch support of a constellation of 48 new satellites. Launches will start in 2009 or 2010 and the satellites will have twice the design life of their predecessors, sustaining service until 2025.

The new satellites will be heavier than the first-generation Globalstars, weighing some 700 kg at launch. They will reuse the basic bus of the first generation satellites, but with higher gain antennas and other payload improvements, many derived from Alcatel's Spacebus 4000 and the Galileo satellite navigation programmes. They will have an end-of-life power of 1.7 kW and carry 32 transponders operating in C-band, S-band and L-band. The introduction of data compression and other techniques should allow data rates of 250–350 Kbps, or more. The new constellation is also intended to be capable of supporting a wide variety of potential handset services such as push-to-talk and multi-cast networking. They will be launched in groups of 6 or 8. The satellites will be assembled and integrated in Thales Alenia Space's facility in Rome. The payloads will be provided by the company's facility in Toulouse (France); the structures as well as the thermal subsystems being provided by its facility in Cannes (France).

For further Globalstar corporate details, please see the separate Globalstar Inc. entry.

Specifications

Launch: 1–4 on Delta 7420 (14 February 1998); 5–8 on Delta 7420, 24 April 1998; 9–12 on Soyuz-Ikar, 9 February 1999; 13–16 on Soyuz-Ikar, 15 March 1999; 17–20 on Soyuz-Ikar, 15 April 1999; 21–24 on Delta 7420, 10 June 1999; 25–28 on Delta 7420, 9 July 1999; 29–32 on Delta 7420, 25 July 1999; 33-36 on Delta 7420, 17 August 1999; 37–40 on Soyuz-Ikar, 22 September 1999; 41–44 on Soyuz-Ikar, 18 October 1999; 45-48 on Soyuz-Ikar, 22 November 1999; 49–52 on Delta II, 8 February 2000; 53-56 on Soyuz-Fregat, 30 May 2007.

Orbits: 1,414 km, 52°; 5 active satellites in 8 planes (plus 1 spare in each plane)

Design life: 7½ years (EOL de-orbit)

Contractors: Space Systems/Loral (prime: satellite and ground/space operations), Loral (operations control centre, TT&C stations), Alcatel Espace (payload modules), Alenia Spazio (payload/platform integration), DASA (AOCS, solar array), Qualcomm (gateways, operations control centres, handsets)

An engineer makes final adjustments to a Globalstar satellite (Theo Pirard) 0054356

Principal applications: 4.8 kbit/s voice/fax, 9.6 kbit/s data, and position location (300 m)

Transponders: 1.610–1.6265/2.4835–2.5000 GHz user up/down, 5.091–5.250/6.700–7.075 GHz gateway up/down. Overall EIRP 39 dBW

Configuration: 3-axis 0.60 × 1.50 × 1.60 m box-shaped bus. Attitude/orbit determination by onboard GPS receivers using IBM RISC 6000 processors, reducing ground operations and eliminating Earth sensors. 1,100 W from twin solar wings spanning 12 m. AOCS by hydrazine thrusters

Mass: c. 450 kg at launch, 400 kg dry

TT&C: Two Satellite Operation Control Centres (SOCC).

Iridium Constellation

Current Status

The Iridium constellation consists of 66 operational satellites and 14 spares operating from six polar planes, each plane hosting 11 mission satellites at an altitude of 780 km. Late in 2006, the company announced plans to develop a replacement constellation of satellites.

Background

Motorola submitted its Iridium proposal to the Federal Communications Commission (FCC) on 3 December 1990 and the Commission granted an experimental licence in August 1992 for five satellites. A full licence followed on 31 January 1995. Lockheed Space Systems Division (now Lockheed Martin) received a USD700 million contract in August 1993 for 120 Iridium spacecraft, to be built through the year 2003. Lockheed built the satellite bus in Nashua, New Hampshire, at Lockheed Martin Sanders (now part of BAE Systems) and then integrated it with Motorola's communications module in Motorola's final assembly facility in Chandler, Arizona. Chandler's production line was set up to handle five satellites simultaneously. Please see the separate Iridium LLC entry for further Iridium corporate details.

Lockheed drew on its Agena upper stage experience to build the satellite's power and control systems. Volume production allowed the use of then-new GaAs solar arrays. Launch contracts were signed with McDonnell Douglas (later Boeing) for Delta II, Khrunichev/ILC and China.

The principle of the Iridium network is that users communicate with satellites in L-band, and

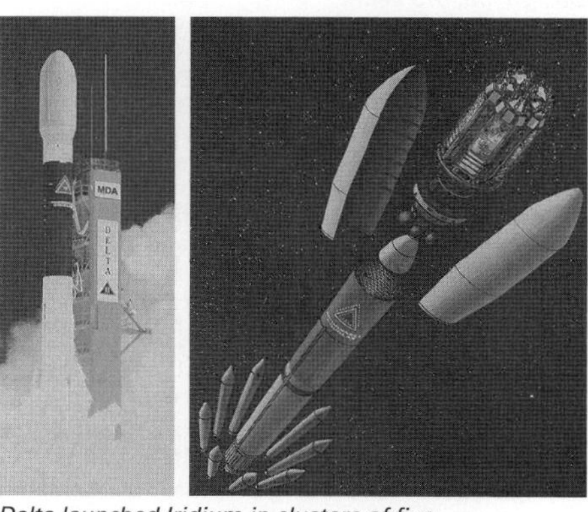

Delta launched Iridium in clusters of five (McDonnell Douglas Aerospace) 0517225

The final spare Iridium spacecraft were launched in 2002 and the company is now looking at a replacement constellation (Iridium LLC)
0022577

that calls are then routed across space and to the system's terrestrial gateways via Ka-band links. Individual satellites broadcast to 48 L-band cells 670 km across using FDMA/TDMA. The gateways, meanwhile, connect to the rest of the worldwide public telephone system, so an Iridium user can be reached from any mobile or fixed phone worldwide. Because signals are passed from satellite to satellite before being downlinked to a gateway, the system is truly global as long as a satellite is in sight: the principal limitation is that Iridium relies on line-of-sight transmission, so it does not work within buildings.

In December 1998, a CZ-2C/5D launch from Taiyuan completed the 72 satellite (66 prime, 6 spares) Iridium constellation. The service was launched immediately.

On 11 February 2002, a Boeing Delta II launch vehicle from Vandenberg AFB launched five more Iridium satellites to add to the seven on-orbit spares then available. A final pair of spacecraft were launched from Plesetsk by a Rockot in June 2002.

The current satellite constellation is considered able to continue operating until 2010. Iridium says that it is working with the satellite industry to develop and deploy a new constellation which will support services until 2030. It will be IP-based and will be able to support portable and fixed broadband data access and wide-area broadcast services.

Specifications

Launches: 5 × Delta 2, 5 May 1997; 7 × Proton K4, 18 June 1997; 5 × Delta 2, 9 July 1997; 5 × Delta 2, 21 August 1997; 7 × Proton K4, 14 September 1997; 5 × Delta 2, 26 September 1997, 5 × Delta 2, 9 November 1997; 5 × Delta 2, 20 December 1997; 5 × Delta 2, 18 February 1998; 2 × CZ-2C/SD, 25 March 1998; 5 × Delta 2, 30 March 1998; 7 × Proton K4, 7 April 1998; 2 × CZ-2C/SD, 2 May 1998; 5 × Delta 2, 17 May 1998; 2 × CZ-2C/SD, 19 August 1998; 5 × Delta 2, 8 September 1998; 5 × Delta 2, 6 November 1998; 2 × CZ-2C/SD, 19 December 1998; 2 × CZ-2C/SD, 11 June 1999; 5 × Delta 2, 11 February 2002; 1 × Rockot, 20 June 2002
Orbits: 780 km circular, 86.4°; 11 active satellites in 6 planes (plus 1 spare in each plane)
Design life: 5–8 years
Contractors: Motorola (payload, integration), Lockheed Space Systems Division (bus), Raytheon (phased array antennas), COM DEV (inter-satellite and feeder link antennas), Olin Aerospace (thrusters), Barnes Engineering Division (miniature infra-red Earth horizon sensors), Tecstar Inc (solar cells), Ithaco, Johnson Controls, Schaeffer Magnetics, Pressure Systems Inc (hydrazine tanks)
Principal applications: voice, fax and data global handheld services
Transponders: three phased array antennas each provide 16 user up/down L/S-band beams from 106 elements at 1.610–1.6265/2.4835–2.500 GHz

Single phased array antenna provides inter-satellite up/down Ka-band beam

Single dish antenna provides feeder link up/down Ka-band beam
Configuration: 3-axis triangular bus, 1 m diameter, accommodating twin solar wings, body-mounted Ka-band inter-satellite antenna, and lower Earth-pointing section with three angled 86 × 188 cm 30 kg L/S-band antennas, each with 106 radiating elements. 115 kg hydrazine (sufficient for >10 years) in single 53.5 cm diameter 71.1 cm height titanium tank
Mass: 690 kg at launch
TT&C: 10 worldwide System Control Segment terminals, controlled by master site at Landsdowne, Virginia (back-up at Nuova Telespazio, Rome, Italy).
Power: two deployed 5-panel wings of GaAs cells; 50 Ah battery.

Leasat/Syncom 4 series

Current Status

Only Leasat 5, launched 9 January 1990, is currently operational, several years beyond its design life.

Background

Also known as Syncom IV, Leasats took full advantage of the Shuttle's wide payload bay, the first satellites to require the full capacity of the Shuttle. A Leasat launch uses the unique 'Frisbee' or rollout method. The satellite has a built-in upper stage that remains in place on station. Each satellite sits 6.1 m high with deployed antennas and has a 4.2 m diameter. Two large helical UHF antennas provide, receive and transmit signals at 240 to 400 MHz. The principal Fleet Broadcast function includes an SHF uplink and SHF/UHF downlinks.

Developed, owned and operated by Hughes Communications Inc, the system was designed around five HS-381 satellites (which included one ground spare) which DoD leased for worldwide UHF links between ships, aircraft and fixed facilities. DoD and the navy specified that USD83.75 million would be paid annually for each satellite once it was operational in its GEO position – but not before. This clause served the US Navy well when Leasat 3 became stranded in the wrong orbit for eight months and Leasat 4 failed completely after attaining GEO.

TRW technicians integrate the LDR payload on to DFS 2 in Lockheed Martin's facilities in Sunnyvale. Prominent is the 60 GHz inter-satellite antenna at top (TRW)
0517308

One Leasat continues in military use. Leasat 5, having reached the end of its planned lifetime, was relocated by satellite service provider G2 Satellite Services, which specialises in the recovery of old satellites. G2 Satellite Services was acquired in 2006 by Intelsat General Corporation, a subsidiary of Intelsat which provides services to US and allied governments. Leasat 5 has been placed at 100° East and provides services to the US and Australian armed forces.

Milstar series

Current Status

The final Milstar spacecraft was placed in orbit on 8 April 2003, bringing the number of active satellites to five. The system is due to serve until the Advanced EHF and TSAT replace it, starting around 2010.

Background

The concept of a survivable military communications system operating in the Extremely High Frequency (EHF) band originated in the 1970s. The USAF formally started the Milstar programme in 1983 to become operational in the second half of that decade. The 4.67-tonne satellites offer tactical communications via the 44.5 GHz uplink and 20.7 GHz downlink frequencies. The EHF band ensures that communications continue during nuclear conflicts and for the first time provides interservice links. Frequency-hopping techniques eliminate jamming to some degree and EHF inter-satellite transmissions permit users to communicate around the globe without routing via intermediate

Leasat Launch History

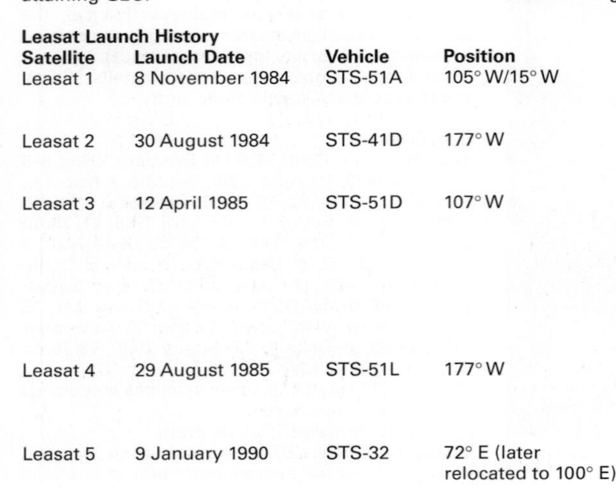

Satellite	Launch Date	Vehicle	Position	Comments
Leasat 1	8 November 1984	STS-51A	105° W/15° W	In October 1985, the only fully operational Leasat. Moved to 33,093 × 39,532 km early 1993.
Leasat 2	30 August 1984	STS-41D	177° W	Operational until September 1985, when its wideband channel began to malfunction
Leasat 3	12 April 1985	STS-51D	107° W	Automatic timing sequence failed to activate after deployment. Hughes claimed USD85 million insurance and paid NASA USD8.5 million for a repair operation. NASA left the satellite in low orbit until 27 October to warm up the solid motor.
Leasat 4	29 August 1985	STS-51L	177° W	The primary communications system failed eight days after deployment. Hughes claimed the USD85 million insurance.
Leasat 5	9 January 1990	STS-32	72° E (later relocated to 100° E)	

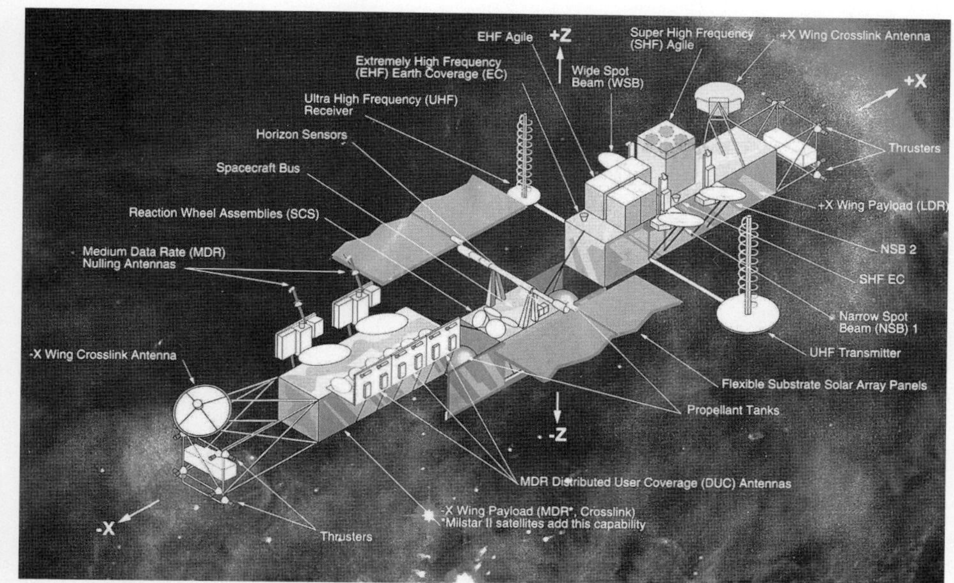

A drawing showing the major features of the Milstar satellite (Lockheed Martin) 0022566

Milstar 1 LDR payload final test/integration at TRW during 1989. The helical antennas provide US Navy UHF links (larger for receive, smaller transmit) and the two dishes provide EHF/SHF spot beams. The smaller arrays provide agile EHF links; missing at centre is the SHF agile unit. Milstar's other payload wing, originally planned for the Science Prime package to relay KH11 imagery, will carry the MDR package from number 3 0516834

The USAF established the Milstar programme in 1983 to provide military communication before and after a massive nuclear exchange. Each satellite carries 13 assorted antennas. The US Air Force paid a reported US$1.5 billion for the first Milstar. A total constellation requires at least four satellites (TRW) 0022565

ground stations, which were considered vulnerable and politically sensitive.

Milstar was the first satellite system to allow all branches of the US armed forces to communicate with one another on the same secure network; the first to operate at extremely high frequencies; and the first to provide satellite-to-satellite communications. Milstar's "switchboard-in-the sky" operational concept was considered revolutionary because the satellites handle all processing and traffic management chores without ground station relays, greatly enhancing data security and jam resistance.

Milstar underwent a long and difficult development. In the early 1990s, the USAF attempted to cancel the project, on the grounds that the cost of the high degree of jamming resistance and radiation-hardening provided was no longer justified in the post-Cold War era. Resistance to jamming and electromagnetic pulse effects had also taken precedence over performance, and the system did not offer high data rates. However, by that time other service users were committed to Milstar and had integrated it into their communications networks. Instead, the programme was restructured, with the final three spacecraft in the series being completed as Milstar IIs, with reduced survivability features and the addition of a new Medium Data Rate (MDR) payloads.

At one time, provisions on Milstar allowed it to relay encrypted KH-11 imagery, but that apparently has been dropped. Milstar's space testing phase began with the launch of FltSatCom 7 in December 1986 which carried its highly jam-resistant 44 GHz EHF transmitter. The US headquarters in Riyadh, Saudi Arabia used a prototype Milstar terminal during 1991's Desert Storm to provide a demonstration link with the Pentagon.

The first of six Milstar satellites left for orbit on 7 February 1994 aboard the first Titan 4 Centaur. At first DFS 1 (Development Flight Satellite) took up a position over the 90° west test slot until the air force moved it about September 1995 to 120° west. DFS2 was launched 6 November 1995 for testing at 90° west, before the air force put it in the planned operational slot at 4° east, inclined at 4°.

Milstar is the first system to allow all three armed services to communicate with each other on the same network. The extensive onboard signal processing and resource management allows it to automatically set up, maintain and reconfigure voice/data networks in real time. It is the first payload to operate at EHF using unique frequency hopping to avoid interception and interference. Crosslinking between satellites eliminates the need for costly, politically sensitive and vulnerable ground stations.

Each satellite has a 0.5 Mbit/s total capacity, which is limited because of the nuclear countermeasures, spread over 192 channels at 752,400 bit/s, with onboard processing and five agile antennas covering 185 service areas. The first two satellites, Milstar 1A, carry only this LDR Low Data Rate payload.

The first Milstar II was launched by Titan 4B/Centaur on 30 April 1999 from Cape Canaveral. Designated DFS 3 (or USA 143 in the official launch announcement) Milstar II has an added medium data rate data transmission system. A failure of the Centaur upper stage to fire properly left the first Milstar II satellite stranded in a useless orbit of 1,097 × 5,149 km, 20.2° inclination orbit instead of the geostationary orbit planned. The fault was traced to an incorrect mathematical constant that steered the Centaur off course shortly after the first of three planned burns.

The 4th Space Operations Squadron of Space Command's 50th Space Wing, based at Schriever AFB in Colorado, operates Milstar.

Milstar is the first communications satellite designed to serve the needs of all three services. Land units use Rockwell Collins Scamp 18 kg terminals with an 0.5 m dish capable of supporting four simultaneous 2,400-bps voice or data channels, and larger Raytheon Smart-T terminals mounted on Humvees and capable of higher data rates. Milstar EHF capability is built into some Navy shipboard terminals such as the USC-38. However, very few aircraft are fitted for Milstar. Some USAF aircraft – including Air Force One and the E-4B Airborne Command Post – are fitted with Milstar terminals, along with the Navy's E-6B TACAMO aircraft. The B-2 bomber was intended to communicate via Milstar, but will receive this capability around 2006 with the addition of an EMS Technologies satcom antenna.

Specifications

EHF/SHF Services: EHF 44.5 GHz (2 GHz bandwidth) up/ SHF 20.7 GHz (1 GHz bandwidth) down
Data rates: LDR 75–2,400 bit/s; MDR 4.8–1,544 kbit/s
Channels: LDR 192 (100 at 2,400 bit/s); MDR 32
Antenna coverage: LDR 1 up/down (Earth), 5 up/1 down agile, 2 up/down narrow spot, 1 up/down wide spot; MDR 2 up nulling spots (allowing operation in presence of in-beam jammers), 2 coincidental down spots, 6 up/down spots (distributed user coverage)
UHF Services: 4 transmit/rx 75 bit/s AFSATCOM IIR channels (Earth coverage) and 1 transmit-only 1.2 kbit/s Fleet Broadcast channel (Earth coverage)
Crosslinks: (60 GHz) 2 per satellite (1 each direction), compatible with LDR/MDR

MUOS

Current Status

A team led by Lockheed Martin is developing the Mobile User Objective System (MUOS) satellite under contract to the US Navy Space and Naval Warfare Systems Command (SPAWAR). Five MUOS satellites are due to replace the current UHF

Follow-On (UFO) satellites and provide narrow-band mobile communications to military users. The first satellite is due for launch in 2010.

Background

Along with the Advanced EHF and Wideband Gapfiller Satellite systems, MUOS is one of a troika of new US military space systems that address different communications requirements. AEHF is a secure, hardened system; WGS uses commercial technology to provide very high data rates; and MUOS provides narrow-band communications to a very large number of mobile users.

MUOS emerged from a Joint Mobile User Study started in November 1996. By 2000, the Navy – which is leading MUOS because it takes over the role of UFO and UFO's predecessor, FltSatCom – had arrived at a plan to sustain and replace its mobile satellite system. This involved procuring a gapfiller satellite – UFO F11 – to extend the life of the UFO constellation, using commercial services (such as Iridium) where possible and embarking on development of a new satellite.

The decision was taken to retain the UFH band for MUOS, for two reasons: compatibility with existing terminals and the fact that UHF suffers less attenuation from foliage than higher frequencies do. UHF also works well for penetrating buildings. However, the UHF frequency is also crowded, requiring some advanced processing technology to maximise capacity and performance.

The Navy issued a request for proposals covering a Component Advanced Development (CAD) phase of MUOS in early 2002, awarding CAD contracts in September 2002 to teams led by Lockheed Martin and Raytheon. At the end of this process, in February 2004, the two teams submitted their proposals for a risk-reduction, design development, acquisition and support contract. Lockheed Martin was selected as the winner in September 2004, and was awarded a USD2.1 billion contract including the first two spacecraft, plus options on the remaining three spacecraft.

Lockheed Martin Space Systems in Sunnyvale, California, is the prime contractor for MUOS, which is based on the A2100 spacecraft bus. General Dynamics leads development of the ground segments of the system, and Boeing, which produced the original UFO satellites, provides the UHF payload. Harris has since been selected to produce the 5 m and 12 m unfurlable mesh reflector antennas for the spacecraft.

MUOS will use commercial third generation (3G) mobile phone technology to increase the system's capacity to serve multiple users. By using a Wideband Code Division Multiple Access waveform (WCDMA) – compared to a roomful of people all speaking different languages, allowing one conversation to be filtered from another – MUOS is expected to provide even hand-held terminals with a high-speed "always-on" connection including voice, data and video. The goal is a system that can support more than 16,000 simultaneous accesses worldwide, providing 15 times the capacity of the current UFO system.

The first of five MUOS satellites – a four-satellite constellation with one on-orbit spare – is due to be in service by mid FY2010, with the remaining four satellites arriving once per year after that. Full operational capability will be reached in mid-2014.

The MUOS team announced in late October 2005 that the system had successfully completed its Preliminary Design Review (PDR), clearing the way to the purchase of long lead items for manufacture. Production of the first two satellites is due to start at the end of 2006.

Orbcomm workers prepare a stack of eight LEO satellites for launch by an OSC Pegasus in 1998 (Orbcomm Global) 0022576

Orbcomm Constellation

Current Status

In June 2006, Orbcomm announced its first new satellite orders since the early 1990s. Six new satellites will be delivered for launch in 2007. Orbcomm also anticipates the launch of 18 further satellites beginning in 2008.

Background

After launching one unsuccessful experimental satellite (Orbcomm-X) by Ariane on 17 July 1991, Orbcomm launched a total of 36 satellites using Orbital Sciences Corporation's (OSC) own Taurus and Pegasus boosters between 1995 and 1999.

The Orbcomm constellation now comprises 30 operational satellites and five on-orbit spares. 28 of the operational satellites reside in four 45°, 800 km altitude orbital planes, and two in high Polar orbits. They use wheel-augmented gravity-gradient technology for stabilisation and cold gas and differential drag control for station-keeping, minimizing fuel use, and so far have substantially exceeded their design life: the first satellites need to be replaced in 2008.

The Orbcomm system is designed to handle data messages (not voice) of around 200 bytes. The system operates in near-real-time mode if the satellite is in view of an Orbcomm ground gateway when it receives a subscriber message. Otherwise, the system automatically switches to store-and-forward mode. The satellites communicate in VHF and UHF with 12 Gateway Earth Stations (GES), connected in turn to six Gateway Control Centres (GCC), two of which are located at Orbcomm's headquarters and its Network Control Centre in Dulles, Virginia.

In January 2006, Orbcomm was able to announce a new round of equity financing to start the replenishment of its constellation. Under contracts announced in June 2006, an initial batch of six new satellites is being jointly produced by OHB-System AG, as prime contractor and integrator, with OSC, Orbcomm's former parent company, producing the payloads. The satellites are planned for launch in early 2007 by a Cosmos-3M from Plesetsk. Larger than the current Orbcomm craft – weighing 80 kg with 300 W onboard power – the new satellites will feature an additional set of receivers, and will incorporate the Automatic Identification System (AIS), an internationally adopted system for reporting the location and identity of ships.

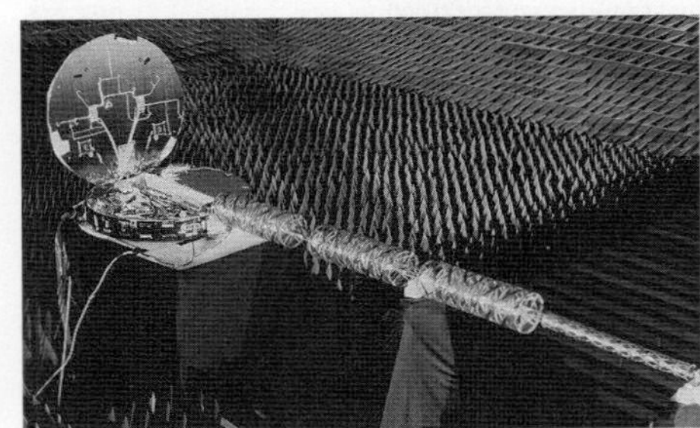

Orbcomm anechoic chamber testing (OSC) 0517399

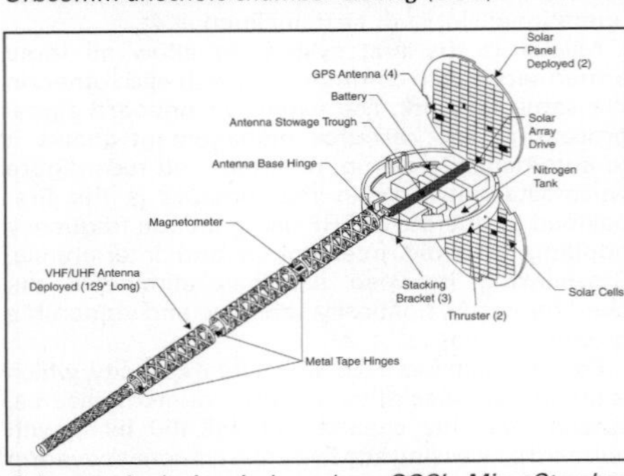

Orbcomm's design is based on OSC's MicroStar bus 0517027

Orbcomm also announced in 2006 that it was evaluating suppliers for its next-generation satellite, with at least 18 new and more capable spacecraft to be launched starting in 2008. The company made an initial public offering in October 2006, and announced at the end of the year that its base of billable subscriber communicators had almost doubled in 2006, reaching 225,000 active devices on its system.

For further Orbcomm corporate information, please see the separate Orbcomm Inc entry.

Specifications

Orbcomm First Generation

Launch: 3 April 1995 (numbers 1–2 Pegasus); 8×Pegasus-XL, 23 December 1997; 2×Taurus, 10 February 1998; 8×Pegasus-XL+HAPS, 2 August 1998; 8×Pegasus-XL+HAPS, 23 September 1998; 7 × Pegasus XL, 4 December 1999

Orbits: 28 operational satellites in four orbital planes, A-D, at 800 km and 45°, with three spares; one satellite and one spare in each of two 70° planes, at 710 and 820 km.

Design life: 4–6 years; actual life up to 11 years.

Transponders: 4,800/9,600 bit/s downlink uses 320 kHz in the 137.0–138.0 MHz VHF band (10 channels) and 50 kHz at 400.1 MHz UHF (1 channel, for GPS time data only). EIRP 16.5 dBW at 137 MHz and 19.5 dBW at 400 MHz. For 2,400 bit/s uplinking, a Dynamic Channel Activity Assignment System directs terminal users to a block of 24 out of a possible 760 channels at 148.00–149.90 MHz depending on instantaneous usage by other systems

Principal applications: two-way packet data communications

Mass: 42 kg at launch

Configuration: MicroStar design, allowing eight to be stacked on single Pegasus. Stowed size 104 cm diameter × 16.5 cm. Deployed length 432 cm, width 224 cm (across solar arrays). ±5° nadir pointed by active magnetorquers and gravity gradient; attitude sensing by Earth sensors + magnetometer. 328 cm VHF/UHF antenna deployed by four hinges. Orbital spacing provided by N_2 thrusters

TT&C: via Dulles Satellite Control Centre; 57.6 kbit/s VHF

Power system: twin circular Sun-tracking solar arrays hinge out from main body providing 230 W BOL; 14 V system; 5V regulated bus for electronics. Supported by Ni/H_2 battery.

Orbcomm CDS

Launch: Due early 2007 on Cosmos 3M from Plesetsk

Orbits: 83° and 1000 km

Principal applications: Two-way packet data communications and AIS

Mass: 80 kg at launch

Configuration: Modular bus by OHB; 23 kg payload from Orbital Sciences

Orion (Orion Atlantic LP) series

Current Status
Orion was acquired by Loral Skynet in 1999 and the Orion 1 and 2 satellites are operated as Telstar 11 and 12.

PanAmSat series

Current Status
The PanAmSat series of communications satellites, providing worldwide links over the Pacific, Atlantic and Indian Oceans, are now operated by Intelsat. PanAmSat became a wholly owned subsidiary of Intelsat on 1 July 2006. The latest satellite in the series, PAS-11, was under construction in late 2006 and was due to be launched in the first half of 2007.

Background
PanAmSat was created to operate the first global geosynchronous private sector telecommunications system, with satellites over the Pacific, Atlantic and Indian Ocean regions. The company selected Hughes in 1991 to provide four of its HS-601 models for PAS 2-4. PanAmSat paid Hughes about USD300 million for all of the satellites. The fourth satellite remained on the ground as a ground spare. Each bus carries cross-strapped C-/Ku-band 54 MHz transponders optimised for TV programmers and corporate telecommunications. Ariane received the contract to launch the entire series in June 1992, in return for which Credit Lyonnais arranged French export loans covering 85 per cent of the USD240 million launch costs. Then Televisa SA of Mexico stepped up to purchase 50 per cent of PanAmSat's equity, to finance the USD700 million expansion in December 1992, for USD200 million in cash and USD440 million raised from a bond sale.

The 2,985 kg PAS 3, intended for 43° W to provide Direct To Home (DTH) services to Latin America disappeared in Ariane's 1 December 1994 failure. Insurance providers paid an estimated USD214 million at 17.5 per cent. PanAmSat outfitted its ground spare which became PAS 3R, and began the Latin America service in early 1996.

The company announced plans for further satellites and ordered PAS 5 from Hughes Space & Communications in March 1995 and PAS 6 from Space Systems/Loral in August 1994 to expand the DTH Latin America capacity to more than 120 digital channels per customer. PanAmSat raised USD262 million in April 1995 in a preferred stock offering to finance PAS 5/6. SS/L took an order for PAS 7/8 in April 1996.

In May 1997 PanAmSat Corporation and the Galaxy Satellite Services business element of Hughes Communications Inc, merged under the PanAmSat Corporation name, and the new company operated the PanAmSat series, the

Another utilisation of the HS-601HP, this PanAmSat has 24 high-power transponders in C-band and Ku-band
(Hughes Space & Communications) 0054341

A 601HP, PAS 10 serves Asia, Africa, the Middle East and Europe 0132819

Galaxy series – concentrated over the Americas – and other Hughes (now Boeing) satellites. Seven new PanAmSats were launched between August 1997 and May 2001, but a long gap followed before the scheduled launch of PanAmSat 11 as the company built up the domestic Galaxy fleet for the US market.

At the end of 2006, the PanAmSat line-up comprised ten active satellites. PAS-6 was retired following power losses in March and April 2004 and has been replaced by PAS-6B. PAS-5 experienced battery problems soon after its launch in 1997 and its capacity has been leased to Arabsat, which operates it as Badr-C. The other satellites in the constellation provide video transmission services to broadcasters, and also support DTH services in different regions. In July 2005, PanAmSat acquired Europe*Star and renamed the Europe*Star 1 satellite as PAS-12.

The on-orbit PanAmSats are a mix of Boeing 601, Boeing 702 and Space Systems/Loral 1300 satellites. PAS-1R, launched on 15 November 2000, was a Boeing 702 and one of the largest communications spacecraft launched at the time, and the first to be launched by the Ariane 5. PAS-1R – like another Intelsat satellite, Galaxy 11 – faces a shortened lifetime because of a class problem with early Boeing 702 satellites, which have a novel solar array design that loses power more quickly than expected.

PAS-11 is the first PanAmSat to be supplied by Orbital Sciences Corporation, following three OSC-built Galaxy spacecraft. It is also due to be the first satellite placed into orbit by the Land Launch service offered by Sea Launch, using a Zenit rocket launched from Baikonur. PAS-11 will serve TV broadcasters in Latin America as well as offering C-band coverage in the Atlantic region.

A PAS-14 satellite is under study but no firm contract had been let by late 2006. The US Strategic Command has funded Cisco to look at the possibility of adding intersatellite links and an IP router to PAS-14, using laser or radio frequency technology.

Intelsat also continues to operate the all-Ku-band SBS-6, one of the oldest functioning communications satellites in orbit. It is largely used by US TV networks to augment breaking news coverage.

Specifications
PAS-1R
Launch: 15 November 2000 by Ariane 5 from Kourou, French Guiana
Location: 315° E
Design life: 15 years minimum
Contractor: Boeing Space Systems (prime)
Transponders: 12 (plus 3 spare) 34 W SSPA, C-band; 24 (plus 6 spare) 55 W TWTA C-band; 24 (plus 6 spare) 125 W TWTA ku-band; 12 (plus 4 spare) 140 W TWTA ku-band
Configuration: Boeing Space Systems 702 bus, 3-axis, colour array span 40.8 m, antenna width 8.2 m. Stowed dimensions 6.2 m × 3.8 m × 3.4 m.
Antennas: 2 × 37 cm and 2 × 27.5 cm dual-gridded shaped reflectors
Mass: 4,792 kg at launch; 3,059 kg bol
AOCS: 3-axis, 1 × 490N liquid apogee motor; 4 × 0.165 N transfer orbit XIPS; 4 × 0.08 N attitude control XIPS
Power system: twin solar arrays each with 5 panels of dual-junction gallium – arsenide cells; 54-cell NiH, 328-Ahr, battery for 100% eclipse protection. System provides 14.3 kW EOL
Coverage: Americas, Caribbean, Europe, Africa

PAS-2
Launch: 8 July 1994 by Ariane 4 from Kourou, French Guiana
Location: 169° E geostationary
Design life: 15 years
Contractors: Hughes Space & Communications, now Boeing (prime)
Transponders: 16 (plus 4 back-up) 30 W SSPA two C-band 5.925–6.425/3.70–4.20 GHz up/down Pacific beams (Australasia, Northeast Asia, China; Western US/Hawaii), 54 + 64 MHz bandwidth, EIRP 33–38 dBW for >=2.4m Earth stations, orthogonal linear polarisation. 8 transponders can be cross-strapped C-Ku

16 (plus 4 back-up in 20-for-16) 63 W TWTA three 14.00–14.50/12.250–12.750 GHz up/down Ku-band Pacific beams (Australia/New Zealand; Northeast Asia/China; China/Japan), 54 + 64 MHz bandwidth, EIRP 44–51 dBW for >=90 cm Earth stations, orthogonal linear polarisation. 8 transponders can be cross-strapped Ku-C.
Principal applications: TV distribution and business communications
Configuration: HS-601 platform, 3-axis, solar array span 26.2 m (satellite within 2.7 × 3.8 × 4 m envelope at launch). Payload mounted on north/south panels and Earth-facing floor; east/west faces carry two dual-gridded rectangular antennas, Earth-facing wall carries two dual-gridded dishes
Mass: 2,920 kg at launch, 1,802 kg on-station BOL, 1,291 kg dry
AOCS: 3-axis, unified 12 ARC 22 N and one Kaiser Marquardt 490 N bipropellant thrusters incorporating onboard control processor, Sun and Barnes Earth sensors and two 61 Nms 2-axis gimbaled momentum bias wheels. NTO/MMH in four spheres. Spin stabilised in transfer orbit
Power system: twin solar wings (spanning 26.2 m) of four 2.16 × 2.54 m panels carrying K43/4 large area Si cells on Kevlar substrate provide 4.225 kW EOL. 100% eclipse protection by Ni/H_2.
Coverage: Asia-Pacific

PAS-3
Launch: 12 January 1996 by Ariane 4 from Kourou, French Guiana
Location: 317° E geostationary
Transponders: Same as PAS-2 but two C-band beams cover Americas and Africa/Europe and five Ku-band US, Canada, Europe and South America
Mass: 2,918 kg at launch, 1,753 kg on-station BOL, 1,318 kg dry.

Other specifications as for PAS 2
Coverage: Americas, Caribbean, Europe, Africa.

PAS 4
Launch: 3 August 1995 by Ariane 4 from Kourou, French Guiana
Location: 72° E
Design life: 15 years
Transponders: C-band as PAS 2/3 but three beams cover Africa; Middle East/India; Asia
24 (plus 4 back-up) 60 W TWTA four 14.00–14.50/12.250–12.750 GHz up/down Ku-band Pacific beams (southern Africa; Northeast Asia; Europe/

Final alignments in the Mass Properties Laboratories for PAS 10, a 601HP series satellite 0132821

PAS-7 9 in an anechoic chamber for antenna tests 0132816

Eastern Russia; Middle East/India), 16 × 27 + 8 × 54/64 MHz bandwidth, EIRP 44–53 dBW for >=90 cm Earth stations, orthogonal linear polarisation. 8 transponders can be cross-strapped Ku-C
Mass: 3,043 kg at launch, 1,868 kg on-station BOL
Other specifications as for PAS 2
Coverage: Asia, Africa, Middle East, Europe.

PAS-6
Launch: 8 August 1997 on Ariane 44LP from Kourou, French Guiana
Location: 317° E
Design life: 15 years minimum
Contractors: Space Systems/Loral (prime)
Transponders: 36 (plus 12 back-up, in 12-for-9) 100 W TWTA Ku-band South American beams, 36 MHz bandwidth, linear orthogonal polarisation
Principal applications: DTH Latin America
Configuration: 3-axis stabilised SS/L FS-1300 2.41 × 2.58 × 2.20 m box-shaped bus with twin solar wings on North/South sides, 27.3 m span; 2.4 m diameter Ku-band reflectors on East/West faces plus 1.2 m receive on Earth face
Mass: about 3,420 kg at launch, 1,285 kg dry
AOCS: 3-axis bias momentum system; single Marquardt R-4D 490 N apogee thruster and two sets of six 22 N RCS thrusters, common NTO/MMH propellant system
Antenna pointing: 0.09° azimuth/0.08° elevation Ku-band
TT&C: under development
Power system: twin 4-panel GaAs wings providing about 8 kW BOL. Ni/H_2 battery provides 100 per cent eclipse protection
Coverage: South America.

PAS- 6B
Launch: 21 December 1998 by Ariane 4 from Kourou, French Guiana
Location: 317° E GEO
Design life: 15 years
Contractor: Boeing Space Systems
Transponders: All Ku-band, 32 × 36 MHz, 16 at 105 W and 16 at 140 W
Configuration: Boeing 601 HP with XIPS xenon-ion propulsion and bipropellant control motors
Launch mass: 3,470 kg
Power: 26 m span solar arrays; 10 kW beginning of life, 8.7 kW end of life
Coverage: South America, dedicated to Direct-To-Home (DTH) television services.

With a total of 72 transponders, PAS-IR launched in 1997 was the second of the powerful 702 series platforms 0132817

PAS-7
Launch: 16 September 1998 on Ariane 44LP from Kourou, French Guiana
Location: 68.5° E
Contractor: Space Systems Loral
Design life: 15 years minimum
Transponders: 14 (plus 4 back-up in 9-for-7) 50 W TWTA 6.425–6.725/3.40–3.70 GHz up/down C-band beams, 36 MHz bandwidth, maximum EIRP 39 dBW, orthogonal linear polarisation
30 (plus 10 back-up in 40-for-30) 100 W TWTA 13.75-14.25/10.95-11.70 GHz up/down Ku-band, 36 MHz bandwidth, max EIRP 53 dBW
Principal applications: telecom + DTH
Configuration: 3-axis stabilised SS/L FS-1300 2.41 × 2.58 × 2.20 m box-shaped bus with twin solar wings on N/south sides, 27.3 m span; 1.8 m diameter C/Ku-band reflectors on east/W faces plus 1.2 m receive on Earth face
Mass: about 3,838 kg at launch, 2,118 kg BOL, 1,466 kg dry
AOCS: 3-axis bias momentum system; single Marquardt R-4D 490 N apogee thruster and two sets of six 22 N RCS thrusters, common NTO/MMH propellant system
Antenna pointing: 0.09° azimuth/0.08° elevation Ku-band
TT&C: location to be determined
Power system: twin 4-panel GaAs wings providing about 9 kW BOL. Ni/H_2 battery provides 100 per cent eclipse protection
Coverage: Asia, Africa, Middle East, Europe.

PAS 8
Launch: 4 November 1998 on Proton-K from Baikonur, Kazakhstan
Location: 166° E geostationary
Design life: 15 years minimum
Transponders: 24 (plus 4 back-up in 28-for-24) 50 W TWTA 5.925–6.425/3.70–4.20 GHz up/down C-band beams, 36 MHz bandwidth, orthogonal linear polarisation
24 (plus 8 back-up in 32-for-24) 100 W TWTA 13.75–14.00/12.25–12.75 GHz up/down Ku-band, 36 MHz bandwidth
Principal applications: telecom + DTH
Configuration: 3-axis stabilised SS/L FS-1300 2.41 × 2.58 × 2.20 m box-shaped bus with twin solar wings on N/south sides, 27.3 m span; 2.4 m diameter C/Ku-band reflectors on East/West faces plus 1.2 m receive on Earth face
Mass: about 3,800 kg at launch, 2,100 kg BOL; 1,466 kg dry
AOCS: 3-axis bias momentum system; single Marquardt R-4D 490 N apogee thruster and two sets of six 22 N RCS thrusters, common NTO/MMH propellant system
Antenna pointing: 0.09° azimuth/0.08° elevation Ku-band
TT&C: location to be determined
Power system: twin 4-panel GaAs wings providing about 9 kW BOL. Ni/H_2 battery provides 100 per cent eclipse protection
Coverage: Asia-Pacific with connectivity to United States.

PAS-9
Launch: 28 July 2000 by Sea Launch
Location: 302° E
Design life: 15 years
Contractor: Boeing Space Systems
Transponders: 24 × 55 W C-band TWTA; 24 × 108 W TWTA ku-band
Principal applications: C-band and Ku-band services for the Americas, the Caribbean and Western Europe plus DTH series for Mexico in Ku-band
Configuration: 3-axis stabilised 601 HP bus with twin solar arrays spanning 26 m. Antennas span 7 m.
Stowed size: L 4 × 2.7 × 3.6 m
Mass: 3,659 kg at launch; 2,389 kg at BOL
AOCS: 1 × 445 N liquid apogee engine; 1 × 13 cm, 0.017 N N-S primary xenon in thruster; 4 × 10 N N-S backup bi-propellant thrusters; 4 × 10 N E-W bi-propellant thrusters; 4 × 10 N aft bi-propellant thruster
Power system: twin 4-panel dual-junction gallium arsenide solar cell arrays, 30-cell NiH, 350 Ahr, battery.
Total power levels: 9.9 kW BOL, 8.9 kW EOL
Coverage: Americas, Caribbean, Europe.

PAS-10
Launch: 14 May 2001 by Proton
Location: 68.5° E
Design life: 15 years
Contractor: Boeing Space Systems
Transponders: 24 (+8 spare) × 55 W C-band TWTA; 24 (+8 spare) × 98 W Ku-band.
Principal applications: DTH, internet, video and data broadcasting services

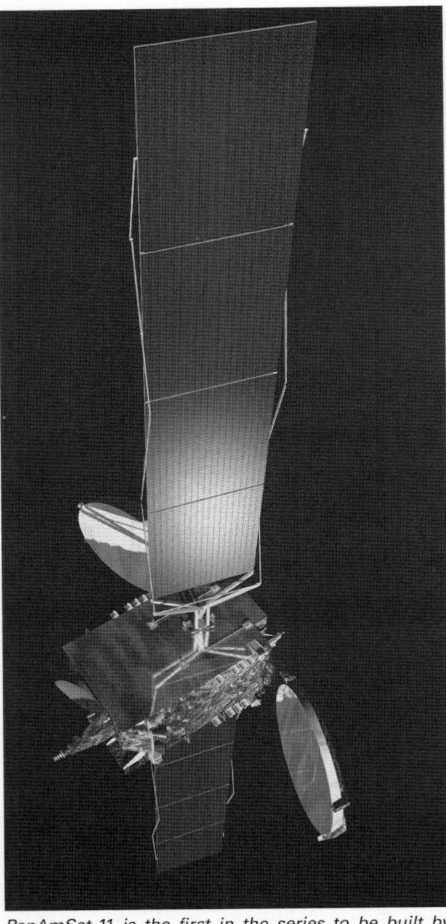

PanAmSat-11 is the first in the series to be built by Orbital Sciences, on a smaller bus than other craft in the constellation (Orbital Sciences) 1198723

Configuration: 3-axis stabilised 601 HP with twin solar arrays spanning 26 m. Antennas span 7 m
Stowed size: H 4 m, W 2.7 × 3.6 m
Mass: 3,739 kg launch, 2,510 kg BOL
AOCS: 1 × 490 N liquid apogee motor; 1 × 13 cm, 0.017 N N-S primary xenon ion thrusters; 4 × 10 N N-S lockup bipropellant thrusters; 4 × 10 N E-W bipropellant thrusters; 4 × 10 N aft bipropellant thrusters
Power system: twin 4 panel dual-junction GaAs solar cell arrays, 30 cell NiH 35O A/h battery, 9.6 kW BOL, 8.5 kW EOL
Coverage: Asia, Africa, Middle East, Europe.

PAS-12
Launch: 29 October 2000 by Ariane 4, as EuropeStar 1
Location: 45° E
Design life: 15 years
Contractor: Alcatel Space (prime) Space Systems Loral (bus)
Transponders: 30 Ku-band at 36 MHz
Principal applications: DTH, high-definition TV, on-demand satellite services and digital satellite news-gathering.
Configuration: 1300 series bus
Solar array: Span 24.68 m, 9.6 kW
Mass: 4,150 kg launch
Coverage: Asia, Africa, Middle East, Europe.

PAS-11
Launch: Scheduled for second quarter of 2007, by Land Launch Zenit-3 from Baikonur
Location: Atlantic
Design life: 15 years
Contractor: Orbital Sciences Corporation
Transponders: C-band: 16-for-12 redundant transponders at 55 W; Ku-band: Two groups of 12-for-9 redundant transponders at 110 W
Principal applications: Broadcast and corporate links; DirecTV
Configuration: Orbital STAR design
Solar array: Two arrays with four panels each, using GaAs cells
Mass: 2,500 kg launch
Coverage: C-band: Americas and Western Europe; Ku-band: South America.

SBS-6
Launch: 12 October 1990 by Ariane from Kourou, Fr Guiana
Location: 74° W longitude.
Design life: 10 years

Configuration: BSS-393 platform; solar cells on telescoping cylindrical sections telescoped for launch and deployed on orbit delivering ~2 kW and with Ni-H batteries for eclipse coverage; 2 × 22.2 N axial and 4 × 22 N radial AOCS thrusters; 2 × 490 N thrusters for apogee manoeuvres and longitudinal transfer.
Transponders: 30 × 43 MHz (41 W) TWTA (inc 11 spares) Ku-band.
Mass: 2,445 kg at launch, 1,481 kg BOL.
Principal applications: Occasional coverage of US breaking news.

SDS series

Current Status
Four of the most recent generation of Satellite Data System (SDS) spacecraft were launched between 1998 and 2004 and all are believed to be operational.

Background
The Satellite Data System spacecraft operate in Highly Elliptical Orbit (HEO) similar to those of the Russian Molniya. Their primary mission is to relay imagery from Improved Crystal reconnaissance satellites to ground stations in the United States. They also provide communications links with US aircraft, nuclear and other forces in the polar regions that have limited coverage from GEO satellites. They relay data to a ground station at Fort Belvoir, Virginia.

There have been three generations of SDS, all apparently based on standard Hughes (now Boeing) communications satellite buses. The seven satellites of the first generation were apparently based on the Intelsat IV bus (also used for some SIGINT satellites) and were launched between 1976 and 1987 – essentially in parallel with the first generation of KH-11 satellites. Some reports suggest that they carried a 12-channel UHF payload, similar to the Navy's Fltsatcom satellites.

Four second-generation satellites were launched between 1989 and 1996. The first three were launched by the Shuttle. The imaging satellites that they supported, the Improved Crystal series, were also intended to fly on the Shuttle but were adapted after the NRO abandoned the Shuttle in 1986–88. They were based on the commercial HS-376 bus. The SDS-2 design adds a K-band antenna downlink, and reportedly features two 5 m diameter antennas.

The SDS-3 satellite series parallels the last generation of heavy reconnaissance satellites, launched between 1998 and 2004. They are believed to be based on the HS-601 bus.

SES AMERICOM

Current Status
SES Americom continues to update its satellite fleet, with major concentration on direct-to-home broadcast services and – via subsidiary Americom Government Services – to the US civilian government and the Department of Defense. Current plans call for the launch of AMC-14 in 2007 and AMC-21 in 2008.

Background
The backbone of the SES Americom fleet remains the Lockheed Martin A2100 mid-sized satellite. Both SES Americom and Lockheed Martin's commercial satellite unit were formerly units of General Electric, and before that were part of RCA; the A2100 was originally conceived as a co-development between GE's space operating and manufacturing units. Of SES Americom's 16 primary satellites, 12 are A2100 models; the rest are Alcatel Alenia Space products. Orbital is under contract to deliver a single satellite bus for SES Americom, AMC-21. The oldest spacecraft in the primary fleet was launched in September 1996.

Satcom C-3 and C-4, launched in August (C-4) and September (C-3) 1992, remain in service to provide C-band backup. Spacenet 4 (originally launched by GTE Spacenet) was replaced by AMC-23 in 2006.

The first group of A2100 satellites (AMC-1 to AMC-4 and AMC-6) were launched as hybrid C-band and Ku-band platforms, with the C-band transponders supporting cable TV services and Ku-band aimed at direct-to-home broadcasting. They were followed by a series of C-band A2100s. The first Alcatel Alenia satellite in the network was the Ku-band AMC-5, based on a Spacebus 2000 platform and orbited in 1998.

SES Americom announced launch contracts for four new satellites in June 2002. AMC-10 and AMC-11 were C-band A2100s, intended to support the North American cable market. AMC-13 was a C-band Spacebus 4000 to serve the Pacific market; and AMC-15 was the first of a pair of Ku-band and Ka-band A2100s for direct-to-home TV and Internet services, the second being AMC-16. In late 2006, AMC-18 became the third all C-band A2100 for the North American market, joining AMC-10 and AMC-11, which were already near their maximum capacity.

Uncertainties and variations in the market have led to some changes in SES Americom plans. In January 2004, SES Americom announced the formation of a new subsidiary, SES Worldsat, which was linked to the Connexion by Boeing project for delivering high-speed Internet access to commercial aircraft. The AAP-1 Ku-band A2100, launched to serve Asia in 2000, was redesignated as Worldsat 1, the AMC-12 as Worldsat 2 and AMC-13 as Worldsat 3. With the demise of Connexion by Boeing, however, Worldsat 1 and Worldsat 2 reverted to their previous designations and Worldsat 3 became AMC-23. It is based on a heavyweight Spacebus 4100 platform.

AMC-14 is a Broadcast Satellite System (BSS) satellite based on an A2100 bus. Its entire capacity is covered by a contract from EchoStar to provide North American services under the DishNetwork brand. However, the launch has been delayed until 2008 because of problems with both the satellite and the Proton launcher. AMC-21, meanwhile, is a smaller Ku-band satellite being developed by Alcatel Alenia Space on an Orbital Star-2 bus, and is not expected to fly before 2008. In mid-2006, SES Global contracted with Sea Launch to orbit AMC-21 in mid-2008, using a Land Launch Zenit-3SLB from Baikonur.

Lockheed Martin A2100 satellites
Design life: 15 years
Contractors: Lockheed Martin Space Systems Company (prime)
Transponders (C/Ku-band satellites): 24 (plus 8 back-up; 16-for-12) 12–18 W variable SSPA 5.925–6.425/3.7–4.2 GHz up/down C-band CONUS/Alaska/Hawaii beams, 36 MHz bandwidth, edge EIRP 37 dBW CONUS, 29 dBW Alaska, 32 dBW Hawaii, 33 dBW Caribbean, linear orthogonal polarisation

24 (plus 12 back-up; 18-for-12) 60 W TWTA 14.0–14.47/11.73–12.17 GHz up/down Ku-band CONUS + Hawaii beams, 36 MHz bandwidth, edge EIRP 45 dBW CONUS, 42 dBW Hawaii beam, orthogonal polarisation
Transponders (C-band satellites): 24 (plus 8 back-up; 16-for-12) 20 W C-band beams
Transponders (Ku/Ka-band satellites): 28 × 36 MHz, 120 W TWTA Ku-band beams; amplifier redundancy 11 for 8, receiver redundancy 6 for 4. Frequencies: 14–14.5 GHz uplink, 12.25–12.75 GHz downlink.

12 × 125 MHz Ka-band spot beams with 3 × 39 MHz subchannels per beam, 75W LTWTAs.
Principal users: Direct-to-home, cable TV, broadcast, education, government and business
Configuration: A2100A platform, 3-axis stabilisation
Mass: 2,585 kg at launch, 1,575 kg on-station BOL,
AOCS: 3-axis by MW, magnetorquing, arcjets and hydrazine thrusters. Arcjets and thrusters divided into two redundant sets

An Atlas IIAS lofts AMC-11 (Lockheed Martin) 1310186

The A2100 was one of the first spacecraft to feature a modular design which makes it easier to adapt it to different payloads (Lockheed Martin) 1310185

Specifications
AMC Inventory

Satellite	Type	Payload	Location	Launch date	Vehicle	Launch site
AMC-1	A2100A	C/Ku-band	105°W	8 September 1996	Atlas IIA	Cape Canaveral
AMC-2	A2100A	C/Ku-band	85°W	30 January 1997	Ariane 44L	Kourou
AMC-3	A2100A	C/Ku-band	87°W	4 September 1997	Atlas IIA	Cape Canaveral
AMC-4	A2100A	C/Ku-band	101°W	13 November 1999	Ariane 44LP	Kourou
AMC-5	Spacebus 2000	Ku-band	79°W	28 October 1998	Ariane 44L	Kourou
AMC-6	A2100AX	C/Ku-band	72°W	22 October 2000	Proton K	Baikonur
AMC-7	A2100A	C-band	137°W	14 September 2000	Ariane 5G	Kourou
AMC-8	A2100A	C-band	139°W	19 December 2000	Ariane 5G	Kourou
AMC-9	Spacebus 3000B3	C/Ku-band	83°W	7 June 2003	Proton K	Baikonur
AMC-10	A2100A	C-band	135°W	5 February 2004	Atlas IIAS	Cape Canaveral
AMC-11	A2100A	C-band	131°W	19 May 2004	Atlas IIAS	Cape Canaveral
AMC-12	Spacebus 4000C3	C-band	37.5°W	3 February 2005	Proton-M	Baikonur
AMC-14	A2100	Ku-band		2007–2008 (Planned)		
AMC-15	A2100AX	Ku/Ka-band	105°W	15 October 2004	Proton-M	Baikonur
AMC-16	A2100AX	Ku/Ka-band	118.7°W	17 December 2004	Atlas V	Cape Canaveral
AMC-18	A2100AX	C-band	105°W	8 December 2006	Ariane 5	Kourou
AMC-21	Star-2	Ku-band		2008 (Planned)		
AMC-23	Spacebus 4100	C/Ku-band	172°E	29 December 2005	Proton-M	Baikonur
AAP-1	A2100AX	Ku-band	108.2°E	2 October 2000	Proton-K	Baikonur

Most of SES Americom's satellites are Lockheed Martin's A2100 type. Here, AMC-11 is prepared for launch (Lockheed Martin) 1310187

AMC-12, an Alcatel Alenia Spacebus satellite, is launched by Proton M (ILS) 1310184

Power system: twin 4-panel solar wings providing 5,000 W EOL; 2 × 100 Ah Ni/H_2 batteries.

AMC-23

Contractors: Alcatel Alenia Space
Design life: 16 years
Transponders: C-band: 12 × 36 MHz, 6 × 72 MHz with 80W TWTAs. Receiver redundancy 4 for w, tranponder redundancy 22 for 18.
Ku-band: 20 channels, switchable among 27, 36 and 72 MHz bandwidths. 138 W TWTA, Transponder redundancy 26 for 20.
Principal applications: TV broadcast, Internet service, education, private networks. Ku-band payload for space-to-aircraft links.
Configuration: Alcatel Spacebus 4100
Mass: 4,967 kg at launch

Satcom C-4

Launch: 31 August 1992 by Delta 7925 from Cape Canaveral
Location: 135°W geostationary
Transponders: 24 (plus 8 back-up; 8-for-6) 17.5 W TWTA 5.925–6.425/3.7–4.2 GHz up/down C-band CONUS/Alaska/Hawaii beams, 36 MHz bandwidth, edge EIRP 39 dBW CONUS (40 dBW peak), 27 dBW Alaska, 31 dBW Hawaii, linear orthogonal polarisation (evens horizontal down)
Principal users: cable programming
Mass: 1,402 kg at launch, 791 kg on-station BOL
Power: 1,950 W BOL; 2 × 50 Ah Ni/Cd batteries

Satcom C-3

Launch: 10 September 1992 by Ariane 4 from Kourou, French Guiana
Location: 131°W geostationary
Mass: 1,375 kg at launch, 789 kg on-station BOL
Other specifications as for Satcom C-4.

Spaceway series

Current Status

The first two Spaceway satellites were originally built by Boeing for broadband internet access via HughesNet, but they were subsequently taken over by News Corporation/DirecTV and retrofitted to deliver HD local TV channels to US customers. They were both launched in 2005 (see separate entry). The launch contract for Spaceway 3 was signed with Arianespace in February 2007, with a successful launch in August 2007. Spaceway 3 is the first satellite wholly owned and operated by Hughes Network Systems, (a subsidiary of Hughes Electronics Corporation). This new satellite broadband platform is the basis of the next generation of the HughesNet service. The HNS SPACEWAY commercial service in North America is expected to begin in early 2008.

Background

Spaceway 3 was launched with BSAT-3a on board an Ariane 5 ECA launcher. Spaceway 3 was released first, approximately 27 minutes into the flight. One of the largest telecommunications satellites ever built, it features innovative, on-board digital processors, packet switching and spot beam technology. Dynamic multi-beam switching will deliver bandwidth-on-demand and direct site-to-site mesh networking. The spot beam technology will enable the satellite to provide services to small terminals, while on-board routers will enable mesh connectivity. Users will be able to directly communicate with any other user of the system without requiring connection through a central hub.

The spacecraft was produced in El Segundo, California by Boeing Satellite Systems, Inc., and will be operated by Hughes Network Systems, LLC of Germantown, Maryland for satellite-delivered broadband services to enterprise, government and consumer users throughout North America. Operating in the globally-assigned Ka-band spectrum, SPACEWAY 3 will deliver a wide range of new high-speed communications services for IP data and multimedia applications to North American enterprise, consumer, and government customers.

Specifications

Spaceway 3

Launch: 7:44 pm. EDT on 14 August 2007 on Ariane 5 ECA from Kourou, French Guiana
Location: 95°W
Launch mass: 6,075 kg
Body dimensions: 5.1 × 3.4 × 3.2 m
Design life: 12 years
Contractor: Boeing Satellite Systems
Payload: Ka-band regenerative processing payload with hopping downlink spot beams, capacity of 10 GBps per spacecraft. Receive antennas – multihorn shaped beam with dual offset fed Cassegrain reflectors. Transmit antennas – 1,500 element phased array, 2 m diameter, forming multiple-hopping spot beams.

Spaceway 3 is one of the first commercial satellites to employ onboard traffic switching and routing capability (Boeing) 1311343

Principal applications: High-speed, two-way broadband communications for internet, data, voice, video and multimedia applications.
Configuration: Boeing 702
Power system: End of life power 12.8 kW. 2 solar arrays each with 5 panels of triple-junction gallium arsenide solar cells. Batteries: 45 cell NiH_2
Propulsion: Liquid apogee engine 100 lbf (445 N). Station keeping thrusters – XIPS: 4 × 25 cm 0.018 lb (79 mN) (low power); 0.038 lb (165 mN) (high power). Bi-propellant axial: Bi-propellant E-W 4.5 lbf (20 N) 4.2 lbf (18.7 N)

TDRS Constellation

Current Status

The first in a series of three Advanced TDRS satellites (H, I and J) was successfully launched by Atlas 2A on 30 June 2000. The second followed on 8 March 2002 but a loss of pressure in one of four onboard fuel tanks four days later prevented it from reaching GSO and it remained in GTO while contingency operations to raise it to geosynchronous orbit were developed. On 30 June 2000, Boeing reported that five of the 18 communication services on TDRS-H were performing below specification due to a specific material used in the fabrication of the phased array antenna, an anomaly corrected for TDRS I and J. Based on the Boeing Space Systems, (formerly HASC) HS 601, TDRS H uses dual telemetry and command antennas, advanced RF electronics and, uniquely in the TDRS series, on-board beam forming for the multiple access system with Ku- and Ka-band services.

Pre-Phase A studies of an advanced TDRS satellite, were conducted 1981–82 and followed by Phase A preliminary analysis between September 1987 and April 1989. Phase B definition began in August 1990 and were completed in July 1991 but a programme redirection in October 1992 preceded a contract award in February 1995 to (then) Hughes Space and Communications. Under a USD481.6 million contract BSS will provide three Advanced TDRS derived from the HS 601 bus augmenting existing S-band and Ku-band frequencies with Ka-band capability.

A Preliminary Design Review, in July 1996, preceded the Critical Design Review in June 1997. TDRS-H (TDRS 8 on orbit) was followed by TDRS-I on 8 March 2002 and TDRS-J, the last in the series, on 4 December 2002.

Background

TDRSS provides NASA with communications coverage of satellites and manned vehicles below 3,000 km, permitting links during 85 per cent of each orbit and allowing NASA to close much of its extensive ground system. NASA planned for TDRSS to be the backbone of the Space Shuttle communication system, and provide 80 per cent of the voice datalinks instead of the 15 per cent coverage previously available via ground stations.

When it first proposed the system in December 1979, NASA projected a cost of US$796 million, but by late 1985 costs had exploded to about USD3 billion. Part of the cost overrun resulted form NASA's realisation in 1978, when the satellites were already designed and partially constructed, that they might suffer radio frequency interference from Soviet radars in Eastern Europe. NASA immediately made USD70 million available to correct the problems and harden the satellites. Then NASA entertained a USAF proposal in 1981 to add a Batson 11 security system to protect against Soviet electronic interference. The air force offered the system to NASA for USD500 million for the six satellites. In the end, the air force dropped the plan, but the bureaucratic discussion

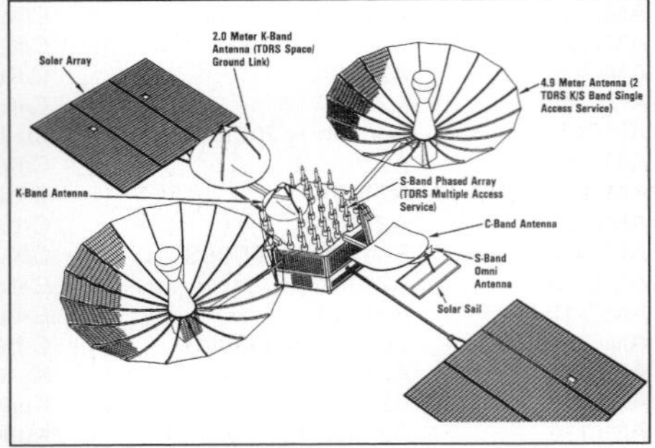

NASA's TDRS system provides communications links for LEO spacecraft over more than 85 per cent of each orbit. The K-band antenna is not used; that and the C-band payload are deleted on TDRS 7 (TRW) 0517025

Launched by Atlas 2A on 8 March 2002, TDRS-I is the second of three data relay satellites built by BSS for NASA
0132807

delayed the programme for several months and further added to the costs.

In the first independent satellite launch from the Shuttle, the Boeing IUS failed and the satellite limped to its correct orbit six months later after 30 separate burns of its control thrusters. At that point the system was two years behind schedule and Spacelab 1, originally designed to use both TDRS 1 and 2 for full exploitation, went into space with only the replacement TDRS 1 available. Because of the delays, NASA had to reprogramme the use of data relay for the Landsat Thematic Mapper images and ironically denied the air force the use of the system until full service could be established.

Up and running, TDRS 1 operations led ground controllers to conclude a fundamental flaw in the satellite design caused several timing-oriented problems, which would have been compounded by multiple satellite operations.

Besides technical problems, TDRS suffered from financing problems as well. NASA attempted to spread the cost over 10 years of operation by leasing the satellite to itself instead of paying the purchase price. It entered into its first large-scale joint venture in the form of a leasing contract with Western Union's SpaceCom. Funding problems, including interest charges totalling USD600 million, created another string of delays. TRW originally contracted to build six satellites; two for lease by NASA, one for SpaceCom for commercial Westar users, the fourth as a joint on-orbit spare for NASA/SpaceCom, and two as flight-ready spares. Two months before the TDRS 1 launch, NASA/SpaceCom agreed on cancellation of the commercial service, making it a 100 per cent NASA system, though still owned by SpaceCom. Though increasing NASA's costs by USD216 million, this decision extended TDRSS' life in the relay role to 15 years, allowing numerous ground stations to be closed.

When the Shuttle resumed operations after the Challenger accident, NASA allotted two out of the first three missions to TDRS. TDRS 1 continued operations despite an additional loss of capacity following an antenna failure on 28 November 1986. NASA planned to replace TDRS 2, lost in the Challenger accident, at a cost of about USD250 million compared with the original cost of USD100 million. In the meantime, NASA kept six ground stations open at Guam, Hawaii, Santiago, Ascension Island, Dakar and Yarragadee, which were to have been closed after TDRS 2 became operational. These stations cost NASA USD35 million per month.

With much less demand for satellite communications during the Shuttle stand down, an international team used TDRS-1 in July and August 1986 for radio astronomy observations. In the first use of a synchronous satellite in such a role, the Jet Propulsion Laboratory linked TDRS-1 with ground radio telescopes in Australia and Japan in a Very Long Baseline Interferometry (VLBI) technique to accurately locate quasar sources.

Despite deployment difficulties with their 4.9 m antennas, TDRS 3 and 4 brought the system to full operational status in September 1989, when NASA formally accepted TDRS 4. One of TDRS 3's two 4.9 m steerable antennas become locked in LHCP (Left Hand Circular Polarisation) in January 1990, preventing RHCP reception from the Shuttle's RHCP datalinks. NASA decided to shunt TDRS 3 to 174° west, augmenting it with TDRS 1 at 171° west to cover for the lost Ku-band capacity.

NASA stopped using the Shuttle as a launch vehicle, when it decided in 1991 to transfer such payloads to expendable vehicles, rather than using the costly and fragile Shuttle.

The appearance of TDRS 6 in 1993 created for the first time a full system of two operational satellites and a stored spare. The military pays some USD100 million annually for use of the system. NASA's operating costs are about USD65 million annually. The satellites work through the White Sands Complex in New Mexico, reporting to Goddard's Network Control Center (NCC). The station can handle up to 2,400 satellite passes each day at data rates up to 300 Mbit/s.

Congress halted funding for an Advanced TDRS successor in 1993. But NASA still awarded a USD481.6 million fixed price contract to Hughes for three modestly enhanced satellites. Congress has yet to approve the funding.

TDRS 7 was launched in July 1995 as a spare, the last of the first-generation type built by TRW plans.

In February 1995, NASA selected Hughes Space and Communications to build three Tracking and Data Relay Satellites (TDRS-H,-I and J) for launch from 2000. They were each to be functionally equivalent to number 7 but adding Ka-band, to help avoid interference from commercial satellites and to offer higher data rates. Based on HSC's HS-601 bus, each will have two 4.58 m Springback steerable single access antennas. These upgrade satellites will provide simultaneous S-/Ku-band or S-/Ka-band operations: 6/0.30 Mbit/s receive/transmit S-band, 300/25 Mbit/s receive/transmission Ku-/Ka-band. An S-band phased-array will receive from five spacecraft.

Specifications

TDRS 1

Launched: 4 April 1983 by Shuttle STS-6/IUS
Location: 49° W GEO (inclination 8.7° January 1996). Replaced as TDRS-E by TDRS 4 in 1989 and moved to 79° W as system spare; in May 1990 transferred to 171° W as cover for lost TDRS 3 Ku-band capacity. Arrived 85° E 7 February 1994 (after number 6 appeared) to provide Compton Observatory link through Australia's Tidbinbilla for relay to White Sands via Intelsat; moved off-station May 1995, stopped 139° W August to December 1995 for Antarctica science operations, before arriving at 49° W June 1996
Design life: 10 years minimum, still operating in September 2000
Contractors: TRW Space & Communications (spacecraft), Harris Electronics Systems Group (ground/space antennas)
Transponders: 2 (plus 4 back-up) 26 W TWTA 2.030–2.113/2.219–2.290 GHz forward/return S-band 2.08° steerable ±22.5° E-west/±31° N-S beams working via 4.9 m antenna, 20/12 MHz forward/return bandwidth, <=0.300/12 Mbit/s forward/return, EIRP >=43.6 dBW min, LH/RHCP selectable

2 (plus 2 back-up) 1.5 W TWTA 13.775/15.0034 GHz forward/return Ku-band 0.28° steerable ±22.5° E-W/±31° N-S beams working via 4.9 m antenna, 50/225 MHz forward/return bandwidth, <=25/12 Mbit/s forward/return, EIRP >=46.5 dBW min, LH/RHCP selectable

30 3.5 W SSPA phased-array for accessing up to 20 satellites simultaneously in 13° beam. 30 elements receive: 2.2875 GHz, 5 MHz bandwidth, LHCP, <=50 kbit/s. 8 elements (plus 4 back-up) also transmission: 2.1064 GHz, 6 MHz bandwidth, LHCP, <=10 kbit/s, 4° beam, EIRP >=34.0 dBW

2 (plus 4 back-up) 25 W TWTA 14.6–15.23/13.4–14.05 GHz up/down Ku-band 0.70° beams linking White Sands with TDRS, 650 MHz bandwidth, <=300 Mbit/s, EIRP >=50.8 dBW min, orthogonal linear polarisation

12 5 W TWTA 5.925–6.425/3.700–4.200 GHz C-band up/down beams, 36 MHz bandwidth, EIRP 32 dBW, linear polarisation
Principal users: NASA (including orbiting spacecraft), US government agencies and commercial C-band leasing
Configuration: hexagonal bus/payload modules, 17.4 m span across deployed solar wings
Antennas: two 24.3 kg steerable ±22.5° E-W/±31° N-S 4.9 m S/Ku-band single access molybdenum wire mesh dishes (that is, each accesses one target spacecraft at a time), one 2 m 15.0 kg Ku-band ground link antenna, a 30-element S-band receive phased array for accessing up to 20 satellites and 12 of those elements for S-band transmit, 1 m 5.0 kg K-band dish (not used), 1.2 × 1.6 m 13.6 kg C-band, S-band omni TT&C antenna
Mass: 2,200 kg at launch, 2,120 kg on-station BOL
AOCS: 3-axis, maintained by Earth sensors, RWs + 24 × 4.45 N thrusters (385 kg hydrazine)
TT&C: NASA's White Sands Complex
Power system: two 380 × 380 cm solar wings provide 2.4 kW BOL, 1.85 kW after 10 years (1.7 kW required), supported by 3 × 40Ah Ni/Cd batteries.

TDRS 3

Launched: 29 September 1988 by Shuttle STS-26/IUS
Location: 85° E geostationary and still operating in September 2000

Other specifications as for TDRS 1.

TDRS 4

Launch: 13 March 1989 by Shuttle STS-29/IUS
Location: 41° W GEO TDRS-E (replaced number 1)

Other specifications as for TDRS 1.
Comments: In addition to the NASA mission TDRS 4 – as well as TDRS 5 and TDRS 6 – are providing commercial C-band services.

TDRS 5

Launch: 2 August 1991 by Shuttle STS-43/IUS
Location: 174° W GEO TDRS-W (replaced number 3) and still operating in September 2000

Other specifications as for TDRS 4.

TDRS 6

Launch: 13 January 1993 by Shuttle STS-54/IUS
Location: 46° W geostationary (from November 1993) and still operating in September 2000

Other specifications as for TDRS 4.

TDRS 7

Launch: 13 July 1995 by Shuttle STS-70/IUS
Location: 150° W as reserve until May 1996: the satellite was then re-located to 172° W at the end of June 1996 and was still operating in September 2000

Advanced TDRS satellite seen following assembly and prior to pre-delivery tests (Boeing) 1047465

Mass: 2,225 kg at launch
Comments: The K-band antenna and C-band packages were deleted (the C-band antenna was replaced by a larger solar sail); SSPAs replaced the S/Ku-band TWTAs of the single-access system.
Other specifications as for TDRS 1.

TDRS 8
Launch: 30 June 2000 by Atlas 2A
Location: 150° W GEO
Mass: 1,600 kg (dry)
Comment: First TDRS built by Hughes, consists of HS-601 bus features S-band phased array antenna and two Ku-. Ka-band reflectors 4.6 m in diameter
Design life: 11 years
Configuration: HS-601 bus with 2 × 4.5 m diameter flexible graphite mesh antenna reflectors folded for launch and deployed on orbit for simultaneous transmit/receive at either S-band or combined Ku-band and Ka-band. Receive data at 300 mbits/s at Ku-band or Ka-band and 6 mbits/s at S-band. Transmit rates at 25 mbits/s for Ku-band and Ka-band or 300 kbits/s. Stored size 3.4 × 8.4 m; solar arrays deployed 21 m length, antennas 13 m across.
Mass: Launch 3,180 kg; BOL 1,781 kg; propellant 1,671 kg
Power system: 2 × solar arrays, 2.3 kW at BOL.

TDRS 9
Launch: 8 March 2002 by Atlas 2A
Location: (planned)
Specification: as TDRS 8

TDRS 10
Launch: 4 December 2002 by Atlas 2A

Telstar

Current Status

Loral Skynet announced in February 2006 that construction of Telstar 11N had started, for service entry in 2008. It will be based on sister company Space Systems/Loral's 1300 series bus and will serve North and Central America, Europe and Africa. It will bring Loral Skynet's active fleet to six spacecraft.

Background

Telstar's history dates back to the world's first commercial satellites, Telstar 1 and 2, built by AT&T's Bell Laboratories in 1962 and 1963. AT&T revived the designation two decades later to commemorate the achievement when the FCC authorised it to proceed with the Telstar 300 series to replace the Comstar satellites. AT&T Skynet was acquired by Loral Space & Communications in 1997.

In 2003, Loral encountered financial difficulties and, as part of its recovery plan under Chapter 11 of US Federal bankruptcy law, sold its spacecraft serving the US domestic market to Intelsat, which operates them under the Intelsat Americas title.

In late 2006, Telstars 10, 11, 12, 14 (also known as Estrela do Sul) and 18 were in operation under the Telstar name. Telstar 10 is Loral Skynet's name for Apstar IIR. Loral and APT Satellite Holdings of Hong Kong agreed in October 1999 that Loral would lease all the transponders on Apstar IIR for the rest of the spacecraft's life. It is based on a Space Systems/Loral 1300-series bus. Telstar 11 was originally built by Matra Marconi Space (now EADS Astrium) on a Eurostar-2000 bus as Orion 1 for Orion Network Systems, acquired by Loral in 1998. Telstar 12, on an SS/L-1300 bus, was originally Orion 2.

Telstar 14/Estrela do Sul was built by SS/L on a 1300 bus. It carries 41 Ku-band transponders and was designed to serve markets such as broadcast video and cable programming, Internet backbone connectivity, VSAT data and other telecommunications services. More than 50 per cent of the satellite's power was to be dedicated to Brazil, providing Ku-band solutions for the Brazilian marketplace. It was also intended to support the Connexion by Boeing Internet-to-aircraft service. Telstar 14/Estrela do Sul was launched successfully, but only part of its solar array deployed correctly and it entered service in March 2004 with only 17 transponders.

Telstar 18, another SS/L 1300-series satellite, was launched successfully on 28 June 2004 and is expected to achieve its full 13-year design life. It is also known as Apstar V. In consideration for funding a portion of the satellite project's cost, APT Satellite Company Limited, Hong Kong, initially acquired use of 68.5 per cent of Telstar 18's capacity for Apstar V services.

The start of construction on Telstar 11N was announced in February 2006 and it is expected to enter service in the second quarter of 2008. It will replace Telstar 11, which will by then have reached its 13-year design life, and like Telstar 11 is an all-Ku-band spacecraft.

Specifications

Telstar 10/Apstar-IIR
Launch: 16 October 1997
Location: 76.5° E
Mass: 3,747 kg at launch; 1,415 kg dry

Telstar 11
Launch: 29 November 1994 aboard Atlas 2A from Cape Canaveral
Location: 37.5° W geostationary (arrived 25 December 1994, MMS handed over control 16.00 GMT 22 January 1995 after completing checkout)
Design life: 13 years
Contractors: BAe Space Systems (prime, now part of Matra Marconi Space and now absorbed into EADS Astrium), Martin Marietta (launch and insurance services), Matra (platform electronics), NEC (communications payload), Lockheed (propellant tanks), Kaiser Marquardt (apogee motor), MBB (10 N thrusters), Fokker (solar arrays), Spectrolab (solar cells), Eagle Picher (Ni/H_2 batteries), COM DEV (output multiplexers) and Telesat Canada (LEOP services)
Transponders: 34 (plus 8 back-up: 5-for-4 54 MHz and 7-for-6 36 MHz) 15 W SSPAs 14.0–14.5/11.45–12.2 and 12.5–12.75 GHz up/down Ku-band Europe and North America spots & regional beams, bandwidth 28 × 54 and 6 × 36 MHz, EIRP 50 dBW (54 dBW max) spots and 45 dBW (48 dBW max) regional. Beam connectivity via switch matrix provides not only trans-Atlantic services but also regional and domestic coverage for Europe and US. 17 transponders allocated to North America (11.7–12.2 GHz down) and 17 to Europe (11.45–11.7 and 12.5–12.75 GHz down). Spots: 3 Europe (North, South and Central) and 4 North America (Northeast, North, South and Central). Regional: 1 North America and 1 Europe
Principal applications: optimised for VSAT applications (1.2 m diameter 2-way; 85 cm diameter receive only). Flexible interconnectivity between high powered spot beams provide for both trans-Atlantic and regional/domestic networking
Mass: 2,361 kg at launch, 1,140 kg dry, payload 330 kg
Configuration: MMS Eurostar 2000 bus. Deployed size 2.1 × 1.7 × 1.9 m; span 22.0 m. Two 2.3 m antennas on east/west faces
AOCS: 490 N R-4D for apogee raising, 12 × 10N MBB for station-keeping and wheel unloading; NTO/MMH in 4 tanks
TT&C: 15 m fully steerable and two 9 m antennas at Mount Jackson, Virginia
Power system: Twin 4-segment Si solar wings providing 3.5 kW EOL, supported by two 78 Ah Ni/H_2 batteries. Payload power is 2,200 W.

Telstar 12
Launch: 19 October 1999 by Ariane 4 from Kourou
Location: 345° E
Design life: 13 years
Contractors: Space Systems/Loral
Transponders: 38 Ku-band at 54 MHz. Coverage of North America as far West as Cleveland, Ohio; the

Telstar 11 began providing trans-Atlantic services as Orion 1 in January 1995 (Orion Atlantic LP) 0517199

majority of South America; Europe as far east as the United Arab Emirates and South Africa. Telstar 12 also has the capability to provide intercontinental connectivity including trans-Atlantic to the Mid-East.
Mass: 3,814 kg at launch
Configuration: SS/L 1300-series bus

Telstar 14/Estrela do Sul
Launch: 10 January 2004, Sea Launch Zenit 3SL
Location: 297° E geostationary
Design life: 13 years
Contractors: Space Systems/Loral
Transponders: 17 Ku-band transponders available
Principal applications: Ku-band satellite with coverage in North and South America and the North Atlantic Ocean Region. The satellite offers focused coverage in South America covering Brazil, the Andean region including Colombia and Panama, and the Mercosul region including Argentina and Chile
Mass: 4,694 kg at launch
Configuration: SS/L 1300 bus

Telstar 18/Apstar V
Launch: 28 June 2004, Sea Launch Zenit 3SL
Location: 138° E geostationary
Design life: 13 years
Contractors: Space Systems/Loral
Transponders: 38 C-band transponders; standard at 36 MHz, extended at 54 MHz; extended at 36 MHz. 16 Ku-band transponders.
Principal applications: Data and broadcast services in China, Hawaii, Hong Kong, the Indian subcontinent, Korea, New Zealand, the South Pacific Islands and Taiwan
Mass: 4,640 kg at launch
Configuration: SS/L 1300 bus

TerreStar Series

Current Status
TerreStar Networks ordered a heavyweight communications satellite from Space Systems/Loral in 2005 and plans to offer space-based mobile communications services in the North American market before the end of 2008.

Background
TerreStar announced in April 2005 that construction of the TerreStar-1 satellite was under way at Space Systems/Loral. Based on the SS/L 1300 bus design, it is claimed to be one of the largest commercial satellites under construction. An antenna more than 18 m in diameter will generate some 500 simultaneous spot beams to communicate with mobile users, and Ku-band feeder links will connect the satellite to ground stations. A second satellite was ordered in August 2006: the availability of a spare satellite is a requirement imposed by the US Federal Communications Commission (FCC) before a permanent licence to operate the network is granted. TerreStar-2 is due to be delivered in 2009. TerreStar's satellite will operate in the S-band (2–2.2 GHz) range, providing North American mobile voice and data communications services to cell-phone-sized user devices; it was designed to have a service life of at least 15 years.

In November 2006, TerreStar announced that it had selected Arianespace to launch the first satellite. Lift-off is scheduled for November 2007. TerreStar-1 will be located in an orbital slot allocated to Telesat Mobile Inc (TMI), one of the company's shareholders. For further corporate details on TerreStar, please see the separate TerreStar Networks entry.

TSAT

Current Status
Industry teams led by Boeing and Lockheed Martin are competing to develop the spacecraft segment of the Transformational Satellite Communications (TSAT) programme. TSAT is a highly ambitious project, with a price tag well above USD12 billion, intended to provide Internet-type communications to deployed and mobile military users worldwide. It is also designed to provide protected communications (low probability of detection, low probability of intercept and jam resistance) to mobile users with small antennas. The Pentagon plans to select a team to build the TSAT satellites at the end of 2007.

Background
The Pentagon's vision of Network-Centric Warfare (NCW) envisages replacing the user-to-user communications, which constitute the current model of military communications, with an Internet-type system in which information and imagery resident with any user or platform can be shared with any other authorised user. TSAT is intended to extend this model – the Global Information Grid (GIG) – to mobile users worldwide, retaining a high degree of protection and security.

TSAT and the GIG are the planned solution to the shortfall of communications capability – particularly for the rapid transmission of large digital files, such as reconnaissance imagery – which has been apparent throughout recent conflicts.

TSAT evolved from DoD studies of future communications needs that started in 1996 and led to new systems such as MUOS, WGS and AEHF. Along with these was another new project known as the Advanced Wideband Satellite (AWS), a predecessor of TSAT. This project had not been completely defined before 2001, when the Pentagon developed a new "transformational" communications architecture aimed at much greater speeds and Internet-like access. This depended on even more secure satcoms capability than AWS could deliver, and led to the definition of TSAT as an essential element of the new architecture.

TSAT is intended to enable real-time and persistent worldwide connectivity to airborne and spaceborne intelligence, surveillance and reconnaissance assets: any user could access the imagery being gathered by a UAV, like an Internet user accessing a webcam. The USAF describes TSAT as combining the data rates associated with DSCS and WGS with the security of Milstar and AEHF.

The TSAT constellation, comprising a ring of five GTO satellites cross-linked by laser communications, is expected to deliver a total worldwide capacity of 28.5 Gbps. Another goal is for a ground-mobile user to connect to the GIG via TSAT at 1.5 Mbps – equivalent to a high-speed landline – using a 30 cm antenna, and with full security and jamming resistance. The system will operate in the EHF band with additional receive-only K-band, plus lasers for high-rate, secure cross-linking. The use of laser communications and cross-linking will reduce the need for ground stations outside the US.

Dynamic Bandwidth Resource Allocation (DBRA) will allow users to log on to a satellite and dynamically obtain the needed bandwidth based on the information demand, jamming or adverse weather conditions. Dynamic bandwidth resource allocation allows the system to take full advantage of the available margin based on current conditions, providing significant increases in system capacity.

Network-centric interoperability is an essential element of TSAT. The network aspects of TSAT are managed through the TSAT Mission Operations System (TMOS). TMOS includes both the operations management element that provides the long-term policy and operational planning functions, and the network operations element that provides real-time management of the operation and configuration of the TSAT network.

In January 2004, teams led by Boeing and Lockheed Martin with Northrop Grumman were each awarded USD514 million contracts to carry out Phase B – risk reduction and design development – of the TSAT space segment.

Boeing is teamed with Raytheon, Ball Aerospace, General Dynamics, Cisco Systems, IBM, L-3 Communications, BBN Technologies, Hughes Network Systems, Lucent Technologies, Harris, EMS Technologies, and Alpha Informatics. The Lockheed Martin/Northrop Grumman team includes ViaSat, Rockwell Collins, General Dynamics Advanced Information Systems, L-3 Communications, Stratogis and Caspian Networks. Both space segment teams completed their Interim Space Segment Design Review (ISSDR) milestones in July 2005.

The USAF originally planned to select a single space-segment contractor in 2006, but restructured the programme significantly in late 2005. In 2006 and 2007, the teams' competing approaches will be evaluated in detail, including subsystem hardware testing, at MIT's Lincoln Laboratory. The goal is to perform an extensive and independent investigation of the maturity of key new technologies. This move delays the selection of a single space segment contractor team until the first quarter of FY2008 (the end of calendar year 2007). Under the new schedule, the first launch is expected in the last quarter of FY2014 and full operational capability will be attained in 2019.

For details of the latest updates to ***Jane's Space Systems and Industry*** online and to discover the additional information available exclusively to online subscribers please visit **jsd.janes.com**

In another risk-reduction move, the DoD divided TSAT deliveries into two blocks, relaxing specifications for the most difficult new technologies – laser communications and the Internet-type router – on the first two satellites, which will constitute Block 1. The remaining three satellites (Block 2) will be designed to higher performance standards, so that the overall network will meet Key Performance Parameter (KPP) requirements. The first TSAT, when launched, will act as the fourth satellite in the AEHF constellation.

Competitive research and development contracts for the TMOS ground segment were awarded to Raytheon, Lockheed Martin and Northrop Grumman in November 2003. In January 2006, Lockheed Martin's team, including Northrop Grumman, Telcordia Technologies, SAIC and Verizon, was awarded a USD2 billion contract to develop TMOS.

UHF Follow-On (UFO)

Current Status

Ten satellites in the Navy-sponsored UHF Follow-On (UFO) series remain on orbit. The constellation is planned to be phased out of use between 2011 and 2020 as the Mobile Objective User System (MUOS) replaces it.

Background

UFO was developed to replace the 1970s-vintage FltSatCom and Leasat.

The initial contract for development and production of the UFO series was awarded to Hughes (acquired by Boeing in 2000) in July 1988. The contract covered one satellite with options on nine more. Eight of the options were exercised before the first satellite was launched: two in May 1990, three in November 1990 and three more in November 1991.

The first UFO was launched by an Atlas I in March 1993, but a launcher problem left it in an unusable low orbit. UFO F2 departed successfully in September 1993 and the USN formally accepted command authority on 2 December. All subsequent launches have been successful. In January 1994, the Navy ordered a tenth satellite to replace the lost UFO F1.

Based on the HS-601 standard bus, the first ten UFO satellites were delivered in three blocks, differing in size and payload. F1, F2 and F3 constituted Block I. They weighed 1,180 kg, had an 18.3 m solar array delivering 2,500 W and carried UHF and SHF communications payloads. They have 21 narrowband (5 kHz), 17 relay (25 kHz) and one fleet broadcast (25 kHz) channel working through 11 solid-state phased array antennas, using time-division multiple access. UFO features enhanced anti-jamming and EMP hardening. It can function for up to 30 days without ground contact of any kind. The UFO satellites were designed to last for 15 years and are expected to serve for 18–20 years.

Four Block II UFOs – F4 to F7- added an EHF capability for improved protection against jamming and electromagnetic pulse. The first three of this group provided 11 EHF channels in two beams – a theatre-coverage beam and a spot beam. The EHF payload is compatible with Milstar terminals. F7 introduced an enhancement to the EHF package that boosted its capacity to 20 channels. They were larger than their predecessors, with a 1,360 kg weight on orbit and a 2,800W power supply.

Three substantially upgunned UFOs – F8, F9 and F10 – made up Block III. The key change, under a March 1996 contract modification, was the incorporation of a Global Broadcasting System (GBS) payload. GBS is a high-power, high-speed system similar to a commercial broadcasting system, and is used for many functions, from the dissemination of intelligence imagery to entertainment programming. GBS uses high-power satellite transponders to provide high-speed wideband, simplex broadcast signals. The GBS payload utilises four 130 W, 24 Mbit/s military Ka-band transponders operating in the 30/20 GHz frequency. The Block IIIs discarded the SHF capability and retained the EHF that had been introduced in Block II.

To integrate the GBS payload required several spacecraft modifications. The payload's high-power amplifiers were integrated on to a larger south radiator panel with heat pipes added to accommodate the increased thermal dissipation. The GBS fixed receive antenna, together with the forward SGLS omni antenna, was mounted on the structure that supported the SHF antennas on earlier UFOs. The GBS steerable receive antenna

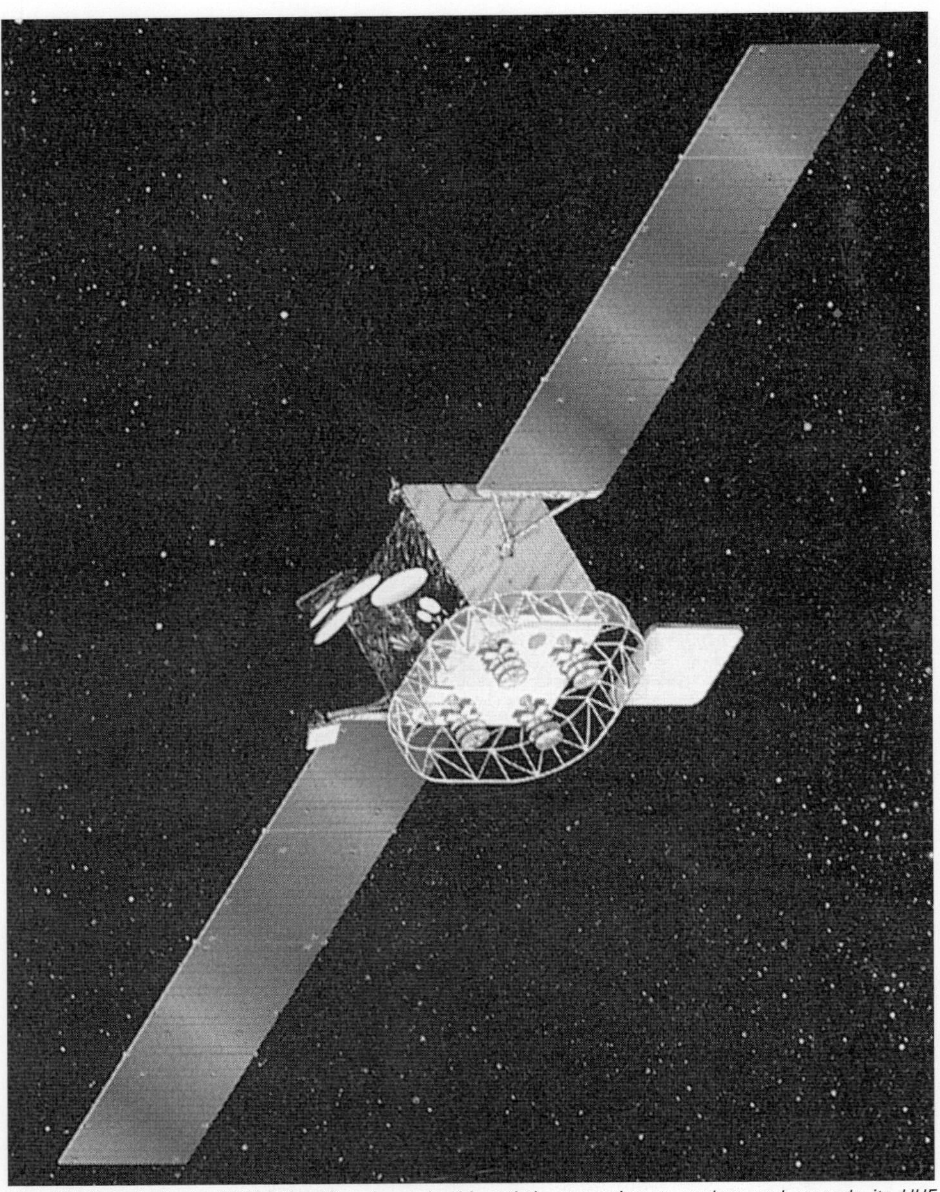

The US Navy built the UHF Follow-On, shown in this artist's conception, to replace and upgrade its UHF (Ultra High Frequency) *satellite communications network. Navy communications needs have prompted it to order 11 satellites in the series* (Hughes Space and Communications Division) 0022569

Spacecraft	Blk	Weight (lbs)*	Power (w)**	L (ft)	W (ft)	Payload
F1-F3	I	2,600	2,500	60	23	UHF/SHF
F4-F7	II	3,000	2,800	60	23	UHF/SHF/EHF
F8-F10	III	3,400	3,800	86	22	UHF/EHF/GBS

*Begining of life **End of life

Height stowed
11 ft
Width stowed
10.5 x 11.1 ft
Block I
UHF
F1 - F3
SHF
Block II
UHF
F4 - F7
SHF
EHF
Block III
UHF
F8 - F10
GBS
EHF
Block I
Block II
Block III

Major features of the UHF Follow-On satellite are shown in this diagram (Hughes Space and Communications Division) 0022570

UFO Satellite Launch History

Satellite	Launch Date	Position	Comments
UFO 1	March 1993		In an unusable low orbit because of an Atlas 1 failure. Hughes was required to pay the navy a USD137 million penalty.
UFO 2	3 September 1993	174° west/71.5° east	Navy and Hughes originally tested it at 174° west. Hughes formally handed it over on 2 December 1993.
UFO 3	24 June 1994	15° west	
UFO 4	29 January 1995	161° west/177° west	
UFO 5	31 May 1995	17.1° east/71.5° east	
UFO 6	22 October 1995	106° west	Inclination of 4 to 5°.
UFO 7	25 July 1996	337° east	
UFO 8	16 March 1998	188° east	
UFO 9	20 October 1998	186° east	
UFO 10	23 November 1999	170° west	
UFO 11	17 December 2003	172° east	Atlas III

was mounted on a deployable boom. A new pallet structure was added to integrate the three GBS transmit spot beam antennas. The Block III developed 3,800W from 23 m solar wings and weighed 1,720 kg.

Launched between March 1998 and November 1999, the three Block III UFOs are instrumental in providing GBS service to all the US forces.

The final UFO satellite was F11, intended to extend the life of the UFO constellation until its replacement is ready. Boeing was authorised to begin production of F11 in January 2001. Identified as Block IV, it is similar to Block II, but is more modern (with a new UHF receiver) and, like the Block III, dispenses with the SHF payload. It was launched successfully on 17 December 2003.

The F11 launch was the first under the auspices of the Naval Satellite Operations Center (NAVSOC). Previous launches had been supervised by the 3rd Space Operations Squadron of the USAF's 50th Space Wing, but control of the UFO constellation was transferred to the Navy in 1999–2000. NAVSOC, based in Point Mugu, California, is a component of the Naval Network and Space Operations Command in Dahlgren, Virginia.

Wideband Global Satcom

Current Status

The first Wideband Global Satcom was launched on 10 October 2007 and is expected to begin service in the first quarter of 2008. Five more satellites are to be launched by 2012.

Background

The Wideband Global Satcom (formerly Wideband Gapfiller Satellite) constellation is based on commercial satellite communications technology, and is intended to boost the Pentagon's in-house satcoms capacity. WGS brings a massive increase of capacity relative to the early-1990s DSCS design. Each WGS provides more capacity than the entire DSCS constellation today. Boeing is the prime contractor and the US MILSATCOM Joint Program Office (MJPO) at the USAF Space and Missile Systems Center (SMC) is the management office.

By the late 1990s it was becoming clear that the US military's rapidly expanding demand for satellite capacity was outstripping the relatively limited bandwidth available from existing military communications satellites, and that the military could not rely on access to commercial satellites to meet its needs. Furthermore, existing satellites such as DSCS III, Milstar and UHF Follow-On were expected to start reaching the end of their useful lives in the early 2000s. Protected, hardened replacement satellites designed for military use could not offer enough bandwidth in time to meet military needs.

The chosen solution was to develop a "gapfiller" satellite using commercial technology, but tailored to military needs and operated exclusively by the military. An Operational Requirements Document (ORD) for the Wideband Gapfiller was approved in May 2000. Boeing was awarded the contract in January 2001, including production of the first three satellites, plus the associated ground-based command and control elements. Boeing would also provide logistics, training and engineering support. A follow-on contract for up to three Block II satellites, worth a potential USD1.067 billion, was received by Boeing Integrated Defense Systems of St Louis in October 2006.

The WGS is based on a Boeing 702 satellite bus. The initial constellation will provide global coverage with three Block I satellites, but three Block II spacecraft are planned to augment coverage in key areas. The orbital positions for the first three satellites are over the Pacific, Indian and Atlantic Oceans. The follow-ons will include new features such as support for high-bandwidth airborne intelligence, surveillance and reconnaissance (ISR) missions, and providing unmanned air vehicles such as Global Hawk with dedicated communications channels. WGS will augment and eventually replace the X-band communications now provided by the DSCS and the one-way Ka-band service provided by the Global Broadcast Service (GBS). WGS will also provide a new two-way Ka-band service. The Block II satellites will be similar to Block I, but with an additional radio-frequency (RF) bypass capability to support airborne intelligence, surveillance and reconnaissance platforms that need ultra-high bandwidth for data rates used by unmanned aerial vehicles. The RF bypass supports data rates of up to 311 Megabits per second.

The spacecraft carries eight transmit/receive X-band phased-array antennas, ten Ka-band gimballed dish antennas and one broad-area X-band antenna. Phased array technology allows the eight X-band beams to be steered and shaped to apply gain and power exactly where it is required. The WGS payload can filter and route 4.875 GHz of instantaneous bandwidth. Depending on the combination of ground terminals, data rates and modulation schemes employed, each satellite can support data transmission rates ranging from 2.1 Gbps to more than 3.6 Gbps (8–14 times the rate offered by DSCS III satellite). Using a digital channeliser-router, the WGS can serve 39 channels. The main channels can be subdivided into as many as 1,900 subchannels, and connectivity features allow any two users in the downlink and uplink areas to communicate with each other, even if they are using different frequency bands.

The WGS communications payloads will be controlled by four Army Wideband Satellite Operations Centers (WSOCs), using equipment supplied by Boeing, Raytheon and ITT Industries. The platforms will be controlled by the USAF's 3rd Space Operations Squadron at Schriever AFB.

Launch of the first satellite was originally scheduled for 2006 but was delayed until October 2007 because of quality control problems. The second satellite is due to be launched by Atlas V in June 2008, followed by SV3 on a Delta IV Medium in late 2008.

In February 2006, Boeing was awarded a USD148 million contract to begin work on the fourth satellite. It is due to be launched in FY2011, with the fifth WGS following in FY2012. On 21 December 2007, Boeing announced that the US Air Force had exercised an option for a sixth WGS satellite. The Australian government is funding the procurement as part of a co-operative agreement with the US government. A memorandum of understanding signed by both governments on 14 November 2007 agreed that the Australian Defence Force will have access to WGS services worldwide in exchange for funding the sixth satellite. The satellite is expected to launch in the fourth quarter of 2012 or early 2013. The six WGS satellites are valued at USD1.8 billion, which includes associated ground-based payload command and control systems, mission unique software and databases, satellite simulators, logistics support and operator training. Boeing also performs final satellite processing and preparations for launch, as well as initial orbital operations and on-orbit testing.

Artist's impression of a Block I Wideband Global Satcom satellite (USAF) 0577400

Specifications

Wideband Global Satcom

Launches: WGS-SV1: 10 October 2007 by Atlas V 421 from SLC-41 at Cape Canaveral.

Location: Block I: three geostationary satellites operating over Pacific (WGS-SV1), Indian and Atlantic regions.

Design life: 12 years.

Contractors: Boeing Integrated Defense Systems.

Payload: Transponded, cross-banded X- and Ka-band communications suite. Antennas: 8 beam,

transmit and receive X-band phased arrays and 10 Ka-band gimbaled dish antennas, 1 X-band Earth coverage. Thirty nine 125 MHz channels via digital channeliser/router.
Principal applications: High-capacity military communications.
Configuration: Boeing 702 bus with xenon-ion propulsion system (XIPS). Four 25 cm thrusters used to remove orbit eccentricity during transfer orbit operations and for orbit maintenance. Deployable radiators with flexible heat pipes.
Mass: 5,769 kg at launch. 3,455 kg in orbit.
TT&C: Communications payloads controlled by four Army Wideband Satellite Operations Centers (WSOCs). Spacecraft controlled by the USAF's 3rd Space Operations Squadron at Schriever AFB.
Power system: Triple-junction gallium arsenide solar cells. 13 W of power BOL, >11 kW EOL.

United Arab Emirates

Thuraya series

Current Status

Thuraya-3 was launched on 15 January 2008. Thuraya-2 is also operational. Thuraya-1 has a problem with its solar arrays.

Background

The Thuraya family is owned and operated by Thuraya Satellite Telecommunications Co. of Abu Dhabi, United Arab Emirates (UAE). The satellites provide regional mobile telecommunications services to 100 countries in the Middle East, Europe, India, Africa and central Asia. The USD960 million contract with Hughes was signed in 1997. It included the manufacture of two high-power Hughes (now Boeing Satellite Systems) satellites, the launch of the first spacecraft, insurance, ground facilities, and user handsets. The second satellite was to be a ground spare, with an option for a third.

The first operational satellite, Thuraya-1, was successfully launched on 21 October 2000 and full commercial service was gradually introduced from April 2001. The 5,108 kg Thuraya-1 was the first of the Boeing GEM (Geosynchronous-Mobile) spacecraft, based on the 702 design, and the heaviest commercial payload ever launched at that time. It initiated the first mobile communications service in the Middle East. Thuraya-1 transmitted and received calls through a single 12.25 m aperture reflector with an L-band feed array. The reflector weighed 78 kg and was compressed into a 1.3 m × 3.8 m package for launch.

Originally positioned in geosynchronous orbit at 42° E, inclined at 6.3°, it was replaced by Thuraya-2 after a problem with defective solar arrays was later found to be associated with the reflectors attached to the satellite's solar arrays. This caused Thuraya Satellite Telecommunications Co. to reduce the number of spot beams used and perform other energy conservation manoeuvres in order to extend Thuraya-1's operational life. It was moved to 51.5° E in September 2003, then to 97.5° E in August 2004. Thuraya-1's solar array defect began to have a severe effect on satellite power output in 2005, but company officials hoped to continue operations in Asia by shutting down non-essential satellite functions. In 2002, capacity on the satellite was leased to Inmarsat to provide mobile broadband service at 144 kbps.

Thuraya-2 and 3 were Boeing GEM satellites fitted with traditional solar arrays. The first of these, Thuraya-2, was launched on 10 June 2003 and located in an inclined geostationary orbit at 44° E. Thuraya-3, launched in January 2008, was added in order to extend Thuraya's mobile communications coverage into East Asia. Its launch was delayed from February 2007 by a January pad explosion during launch of a Sea Launch Zenit-3SL and by poor sea conditions in late 2007.

The satellites use onboard digital signal processing to create more than 200 spot beams that can be redirected on-orbit. This facilitates interconnectivity between the common feeder link coverage and the spot beams, in order to make effective use of the feeder link band and to enable mobile to mobile links between any spot beams. Digital beam-forming capability allows Thuraya to reconfigure beams in the coverage area, to enlarge beams and to activate new beams. It also allows the system to maximise coverage of "hot spots" or areas where excess capacity is required. There is also a possibility to allocate 20 per cent of the total power to any spot beam or reuse the spectrum up to 30 times, increasing efficient use of the spectrum. Calls are routed directly between handheld units or to a terrestrial network, and the system has the capacity for 13,750 simultaneous voice circuits.

The Thuraya ground segment includes terrestrial gateways plus a co-located network operations centre and satellite control facility in the UAE.

Specifications

Thuraya-2
Launched: 10 June 2003 by Zenit-3SL from the Odyssey platform in the Pacific Ocean.
Location: 44° E geostationary

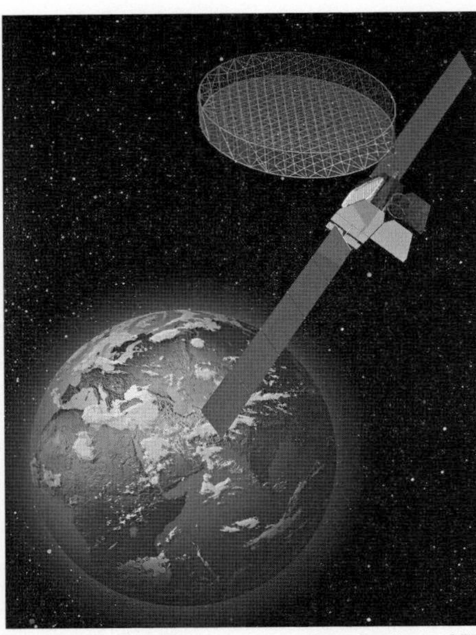

Thuraya-2 and 3 are identical Boeing GEM communications satellites, two of the largest and most powerful yet launched. (Boeing) 1342062

Design life: 12 years.
Contractors: Boeing Satellite Systems (prime).
Communications: 12.25 × 16 m mesh transmit-receive reflector. L-band: 128 active elements, dipole feed array, 17 W SSPAs.
C-band: two active (2 spare) feeder link 125 W TWTAs. 1.27 m round, dual-polarised shaped reflector for C-band communications.

FDMA carrier channel BW 27.7 kHz. Channel bit rate: 46.8kbps.
Principal applications: Mobile voice and data communications in the Middle East, Europe, India, Africa, central and eastern Asia.
Configuration: Boeing GEM (702) body-stabilised bus. 12.25 m diameter deployable antenna. R-4D liquid apogee motor (436 N). Station-keeping thrusters: two 10N and eight 22N.
Mass: 5,177 kg at launch. Approx. 3,200 kg in orbit.
Body dimensions: 7.6 × 3.75 × 3.75 m
Power system: Two solar arrays of 5 panels each with dual-junction gallium arsenide cells. 13 kW of power BOL. 11 kW EOL. Batteries 328 A-hr cells.

Thuraya-3
Launch: 15 January 2008 by Zenit-3SL from Odyssey platform in Pacific.
Location: 98.5° E GEO
Thuraya-3 is identical to Thuraya-2.

DEFENCE

France

Essaim

Current Status
France's Essaim (meaning 'swarm' in English) SIGINT constellation is continuing with its operational demonstration, and the satellites are expected to continue operations until late 2007.

Background
Sponsored by the DGA's Centre for Armament and Electronics (centre d'électronique de l'armement, Celar), the Essaim constellation comprises four 120 kg microsatellites, built by EADS-Astrium on the government-developed Myriade platform design and equipped with a Thales payload. They were launched as an Ariane 5 hitch-hiker payload in December 2004, along with a Helios imaging satellites.

The Essaim system is intended to locate electronic emitters through the Time Difference Of Arrival (TDOA) technique: the signal reaches each satellite at a slightly different time, and the difference yields an accurate, instantaneous three-dimensional location for the emitter.

The Essaim satellites are on a 700 km sun-synchronous orbit that passes over each point on the globe four times in 24 hours, with 10 minutes observation of any given point per pass. Three satellites are needed to locate an emitter using TDOA techniques, and the fourth is a spare. The Essaim craft are tasked as they pass within range of the ground control station in Toulouse, record signals as they pass over the target and download them to Celar's station at Bruz, near Rennes.

The French government is reportedly advocating the development of a European SIGINT capability based on Essaim technology.

Another French military microsat programme is Spirale (Système Préparatoire Infra-Rouge pour l'Alerte), an experimental infra-red missile-warning system using an upgraded version of the Myriade platform. Alcatel was awarded a EUR124 million contract to develop Spirale in January 2004 and the satellites will be launched as hitch-hikers on an Ariane 5 in 2008.

Helios series

Current Status
Two out of three Helios optical reconnaissance satellites are currently operational, with the launch of a fourth in the series, Helios 2B, foreseen for the first quarter of 2009. The most recent addition was the Helios 2A optical reconnaissance satellite, launched on 18 December 2004. Helios 1A, launched in July 1995, continues to operate after 12 years. Helios 1B was launched on 3 December 1999 but its batteries failed in October 2004, resulting in it being shifted to a graveyard orbit.

The Helios 1 satellites can acquire high resolution images of any point on the globe, with daily revisit capability. The French Ministry of Defence procurement agency DGA runs the programme, retains direct control over the management of the ground segment, and delegates responsibility for the space segment to the French space agency CNES. In March 1998, Germany announced it was withdrawing from the Helios 2 partnership and France began to re-evaluate its continuation of the programme.

Background
France joined the rank of nations capable of performing autonomous military reconnaissance from space when it launched the Helios 1 surveillance satellite in July 1995 aboard an Ariane. On a platform based on the civil Spot Earth resources observation satellite, Helios provides the fourth independent military surveillance capability after those of the US, Russian Federation and China. During the 1987 parliamentary season, France announced it had paid FFr7.6 billion for the programme, and had saved FFr300 million by joint development with Spot 4. France claimed the satellite allowed it autonomy from the US decision-making process in the western alliance, by giving it an independent source of corroboration for US Intelligence claims. France justified the decision at the time by saying it needed to upgrade its data gathering ability to prepare for the expected improvements in Soviet strike capability in response to the US Strategic defence Initiative. Germany, for a while, agreed to participate in the programme but then later backed out, citing a desire to build a radar imaging system rather than a visible spectrum satellite. After this, CNES proposed that the Helios and Spot 4 production programmes be merged to reduce costs. The world recognised France's sophistication in this new field when it first published Spot photographs of 10 m resolution, showing Soviet naval, aircraft and nuclear weapon storage facilities at Murmansk and Severemorsk in 1987. The US and the CIA immediately challenged France's right to publish such data at the time. The White House Senior Interagency Group for Intelligence had to file a report on the national security aspects of civil remote-sensing programmes.

The Italians and Spanish share part of the financial burden and the image products with France. Helios is controlled from the Centre de Maintien Poste (CMP) at Toulouse, and has a back-up control facility at Francazal Air Base. The Fresnel building at CNES Toulouse handles both Helios and Spot control and schedule uplinking. The images are downlinked to three Centre de Reception d'Images (CRI): Colmar in northeast France, Lecce in southeast Italy and Maspalomas in Spain. Each of the three capitals has an individual processing centre. Controllers can select onboard encryption so that imagery is restricted to one or more of the three principal nations or is more widespread. Initial reports declared Helios 1A imagery to be excellent. Helios 1A initially flew in a 673 × 676 km orbit, but the French adjusted it 14 July 1995 to a 680 × 682 km, 98° inclination, Sun-synchronous orbit with a spot revisit time of 48 hours for one satellite.

Helios 1B was also based on the Matra Marconi Space Spot 4 bus. It was described as an advanced version of Helios 1A, with more computer memory and new batteries. Like Helios 1A, it was placed in a 680 km, sun-synchronous polar orbit, but with a 180° separation to halve the time between overflights of a given region. Helios 1A has so far sent back some 100,000 pictures, shared according to the partners' financial stakes – 79 per cent for France, 14 per cent for Italy and 7 per cent for Spain.

In mid-October 2004, Helios 1B was taken out of service, its orbit was lowered from 679 km × 681 km, inclination 98.2°, to 637 km × 640 km, inclination 98.2°, taking it out of the path of Helios 1A and future successors.

Helios 1A platform testing at Intespace. See France's National entry for an artist's impression of the satellite in orbit (Ministère de la Défense) 0517311

The Helios 2A optical – infrared reconnaissance satellite during testing (EADS Astrium) 1191257

Meanwhile, Aerospatiale was contracted to develop the Instrument Haute Resolution (IHR) for the second generation optical reconnaissance system. This gives the Helios 2 series an improved spatial resolution of 50 cm and an additional infra-red capability for night imaging. The series comprises two satellites, Helios 2A and Helios 2B, in sun-synchronous orbit. Images may be obtained at short notice by reserving 'priority' viewing windows. The Helios 2 is based on a nearly identical platform built by EADS Astrium for the Spot 5 civil-commercial optical observation satellite launched in May 2002. The main difference is in the more precise attitude and orbital control system used on Helios 2.

Belgium and Spain have each purchased a 2.5 per cent stake in the Helios 2 programme, and Greece has also agreed to join with a small ownership share. The Helios 2 programme is budgeted at about USD2.7 billion, including the satellites' construction, launch and ground installations. The DGA has been designated as programme manager by the Belgian, French and Spanish defence ministries, while CNES is the system prime contractor and EADS Astrium is prime contractor for the two satellites.

Specifications
Helios 1
Launch: Helios 1A: 7 July 1995, by Ariane 40 from Kourou, French Guiana. **Helios 1B:** 3 December 1999 by Ariane 40 from Kourou.
Lifetime: 5 years
Orbit: **Helios 1A:** released into 673 × 676 km, adjusted by 14 July 1995 to 680 × 682 km, inclination 98.07°, period: 98.4 min. **Helios 1B:** 679 × 681 km, inclination 98.1°, period: 98.4 min. Both sun-synchronous.
Resolution: 1 m.
Mass: **Helios 1A:** 2,537 kg. **Helios 1B:** 2,544 kg.
Comment: Italy provided 14.1 per cent funding and Spain 7 per cent. Based on the extensive experience derived from the civil Spot, Helios provides the 4th independent military surveillance capability after those of the USA, RFAS/USSR and China. Programme cost about FFr10 billion. Helios 1B was launched to back-up and complement 1A operations; orbits were phased to provide 24-hour revisit periods.
Contributors: Délégation Générale pour l'Armement/Direction des Missiles et de l'Espace (DGA/DME, executive programme management under AGEX Agence Exécutive), CNES (system/spacecraft architect), Aerospatiale (system industrial architect, EPV imaging system, GSR-3 solar array), Matra Marconi Space France (satellite, processing centre), SEP (processing centre in collaboration with Matra), Alenia/Inisel (ground stations), Alcatel Espace (payload electronics, command/communications systems), Sodern (DTA 04 cameras).
Comment: Spacecraft: based on Matra's Spot 4 bus 3-axis control: Spot 4's 0.05° RMS pointing

accuracy improved to 0.005° RMS by addition of star tracker. 5-panel GSR-3 solar wing.
Sensors: EPV (Ensemble de Prises de Vues) imaging system using DTA 04 multispectral CCD cameras, and electronic intelligence equipment.

Helios 2
Launch: **Helios 2A**: 16:26 GMT on 18 December 2004 by an Ariane 5G+, serial number 520, on Arianespace flight V165 from Kourou.
Orbit: **Helios 2A**: 688 × 690 km, inclination 98.08°, period 98.39 min. Sun-synchronous.
Resolution: 50 cm
Mass: 4,200 kg
Body dimensions: 6 × 3.7 × 3.4 m
Payload: High resolution and wide-field instruments operating in the visible and infrared.
Communications: S-band (2 GHz) for housekeeping, telecommand and telemetry. X-band for image data.
Power: Solar array generating 2,900 W.

Germany

SAR-Lupe

Current Status
The first of these was launched on 19 December 2006, followed by SAR-Lupe 2 on 3 July 2007 and SAR-Lupe 3 on 1 November 2007. The remaining two satellites will be launched in intervals of four or five months, with the entire system to be fully operational in 2008.

Background
SAR-Lupe (Synthetic Aperture Radar) is Germany's first satellite-based radar reconnaissance system, which will eventually comprise a constellation of five identical satellites. (In German, the word 'Lupe' means 'magnifying glass', a reference to high-resolution imagery.) The SAR-Lupe concept is based on five identical radar satellites in three offset polar orbits at a height of approximately 500 km. Due to the physical mode of operation of a SAR, each of the five satellites can operate in swath and spotlight imagery modes. The side-looking radars can take images either side of their orbit. The constellation of five satellites ensures that there are no gaps between the areas under surveillance. The system also includes a ground segment for controlling the satellites and receiving, processing and evaluating image data. The all-weather system works day and night and supplies real-time, high-resolution radar images of virtually any area on Earth. The satellites have several imaging modes, depending on whether programmers want to study a small area or scan larger swaths of territory at lower resolutions. The constellation is equipped with cross-links communications technology, permitting the satellites to relay commands among themselves to speed response time.

OHB-System developed the system on behalf of the German government, in particular the German Ministry of Defence (BMVg) and the Federal Office of Defence Technology and Procurement, referred to as BWB (Bundesamt für Wehrtechnik und Beschaffung), Koblenz, Germany. BWB manages the procurement of the ground and space segments.

Following extensive testing at the IABG satellite test centre in Ottobrunn, the first satellite was delivered to Plesetsk Cosmodrome in mid-December 2006. Shortly after launch, the satellite started transmitting via the Kerguelen ground station in the southern Indian Ocean. The first

The first SAR-Lupe satellite being installed into the Kosmos 3M launch vehicle (OHB-System) 1184861

Launch rehearsal at Plesetsk in January 2005 involving a Kosmos 3M rocket fitted with an enlarged payload fairing for the SAR-Lupe satellites (OHB-System) 1191259

contact between the control centre and the satellite was established 64 minutes after launch on 19 December 2006. Initial control of the satellite was handled by the German Space Agency DLR (DLR-GSOC) in Oberpfaffenhofen. This was followed by comprehensive testing of the individual satellite subsystems, including the deployment of the antenna boom. The ground station of the German Armed Forces in Gelsdorf also tracked the satellite and assumed responsibility on 8 January 2007, when activation of the radar system and operations began. The satellite was reported to be supplying superb, high resolution images and to be operating very successfully.

A similar sequence of events took place after the launch of SAR-Lupe 2 on 2 July 2007. The Gelsdorf centre took over responsibility for it at the end of July. After the commissioning of the second satellite by late 2007, the German Federal Armed Forces had access to an operational reconnaissance system.

It was necessary to modify the fairing of the Kosmos 3M rocket to accommodate the broad parabolic antenna fitted to the SAR-Lupe radar satellites. Polyot developed a special payload fairing, demonstrating its functionality during a launch rehearsal in January 2005.

To demonstrate the capabilities of the satellite before the first launch, various inverse tests were conducted. The satellites on the ground were aligned to obtain high-resolution radar images of the International Space Station. The tests were performed successfully several times at the new satellite integration hall on the Bremen site of OHB-System. The hall was designed as a clean room and has a 'Radome' radar dome especially for such tests, which protects the satellite from external soiling during imaging. The co-ordinates of the ISS and the imaging command were transferred to the satellite, which performed all steering functions required for recording images on its own. The image data was temporarily stored in the satellite and was encrypted, downloaded and processed. After that, a radar image was produced, the quality of which far exceeded expectations for this test. To allow a meaningful comparison, the Forschungsgesellschaft für Angewandte Naturwissenschaften (Research Institute for Applied Natural Sciences, FGAN), provided radar images that were taken at the same time.

SAR-Lupe will be part of the European Reconnaissance System. At Franco-German summit meetings held in Mainz, Germany, on 9 June 2000, France and Germany announced plans to develop an independent European satellite reconnaissance network. Germany would contribute its SAR-Lupe all-weather radar satellite system and France its complementary optical satellite system HELIOS II. The ministers of defense of France and Germany signed a first treaty on the execution of co-ordinated studies about the SAR-Lupe/HELIOS II system network in Schwerin, Germany, on 30 July 2002. Under the ESGA project (German acronym for "Europeanisation of Satellite-Based Reconnaissance"), which is co-financed by Germany and France, OHB-System is creating the technical basis allowing France to use the SAR-Lupe radar system. Various modifications to the German SAR-Lupe ground segment are necessary to implement this project. Interfaces are to be installed to allow France to request and receive images from SAR-Lupe. In addition, the system will be enlarged to permit the addition of further partners in the future.

Specifications
Launches: SAR-Lupe 1, 14:00:19 UT on 19 December 2006.
SAR-Lupe 2 at 20:38:41 UT on 2 July 2007. All launches by Kosmos 3M from Plesetsk.

The first SAR-Lupe satellite in the clean room during assembly and testing. The spacecraft is dominated by the large, fixed parabolic antenna (OHB-System) 1191264

SAR-Lupe 3 at 00:51 UT on 1 November 2007.
Orbit: SAR-Lupe 1: 468 x 505 km, 98.2°, period 94.3 min.
SAR-Lupe 2: 469 x 507 km, 98.2°
SAR-Lupe 3: 465 x 493 km, 98.2°
3 orbital planes. Orbital plane 1 contains 2 spacecraft; orbital plane 2 contains 1 spacecraft, orbital plane 3 contains 2 spacecraft. The angle between orbital plane 1 and 2 is 64°; the angle between orbital planes 2 and 3 is 65.6°. The phase angles of the spacecraft are: orbital plane 1 = 0° and 69°; orbital plane 2 = 34.5°; orbital plane 3 = 0° and 69°.
Dimensions: Approximately 4 m × 3 m × 2 m
Design life: 10 years
Contractors: OHB-System (prime contractor). *Other participants:* Alcatel Alenia Space, Toulouse; Saab Ericsson of Göteborg, Sweden (SAR antennas); Carlo Gavazzi Space, Milan, Italy; TESAT-Spacecom GmbH, Backnang, Germany (development of high-performance amplifier); RST Radar Systemtechnik GmbH, Salem, THALES of Ulm, Germany, EADS (ground segment), DLR (Launch and Early Orbit Phase and operational backup function).
Power: Average power consumption approximately 250 W. NiH2 batteries for storage.
Principal applications: Radar military reconnaissance. Other uses include: a) elevation models from multipass interferometric data, b) multipass stereo products, c) change detection products, d) products with enhanced radiometric resolution.
Configuration: The satellite comprises a satellite bus and the radar payload. The satellite can be aligned with great precision to a specific location on the ground using 3-axis stabilisation. Attitude is sensed by sun sensors and star sensors; attitude control is provided with reaction wheels and magnetic torquers. The synthetic aperture radar image data are collected via a 3.3 m × 2.7 m fixed parabolic SAR/communication dual-use antenna attached to the satellite bus. Propulsion system uses hydrazine thrusters.
Mass: Approximately 725 kg
Communications: Data are transmitted via X-band. Encrypted S-band is used for telemetry and command transmissions direct to ground stations or via SAR-Lupe intersatellite links. XSAR (X-band SAR instrument) observes in X-band (centre frequency of 9.65 GHz corresponding to a wavelength of 3.1 cm). An onboard image storage capability of 128 Gbit (EOL).
Imaging capability: Spatial resolution of SAR data less than 1 m in spotlight mode for a scene of about 5.5 km × 5.5 km; a stripmap scene has a size of 60 km × 8 km.

Israel

Ofeq series

Current Status
The latest military reconnaissance satellite, Ofeq 7, was launched on 10 June 2007. Ofeq 6 failed to reach orbit in 2004. Ofeq 5, launched in 2002, has outperformed its intended lifespan. Ofeq 7 is reportedly the first of Ofeq's third generation satellites; design improvements include increased optical resolution, possibly around the 0.5 m range. Ofeq 7 imagery may be marketed, which would categorise the satellite as dual use.

The Ofeq (also Ofek and Offek) series has elevated Israel's satellite reconnaissance capabilities to the level of countries that were previously more advanced in that area. Israel likely observes its neighbours, especially Iran, Syria and Iraq, via Ofeq 5 and 7, for national security purposes. The Israel Aerospace Industries IAI/MBT Division operates Ofeq satellites on behalf of the Israel Space Agency and MoD.

The Ofeq series is closely related to ImageSat's commercial Earth Remote Observation Satellite (EROS) series (see separate entry). El-Op developed and manufactured the remote sensing equipment for both satellite series, and IAI manufactures the spacecraft bus for both series. It is therefore reasonable to assume that the Ofeq series has very similar specifications to the EROS series, although much information regarding the satellites is classified.

Background

While most launch vehicles fly eastward to pick up energy from the Earth's rotation, Israel launches Ofeq (and other national satelllites) to the west, into a 142° retrograde orbit, to avoid dropping stages on its neighbours. The first Ofeq (Horizon) satellite was a technology demonstrator developed by Israel Aerospace Industries IAI/MBT Division for the Israel Space Agency (ISA), and like subsequent satellites in the series, was controlled from MBT's ground station. The satellite itself resembled an irregular octagonal prism of conventional aluminium construction, measuring 2.3 m high, and 1.2 m at the base, shrinking to a top diameter of 0.7 m. It weighed 156 kg, broken down by: structure, 33 kg; power system, 58 kg; computer, 7 kg; and communications package, 12 kg. The thermal control weighed 5 kg; the wiring 9 kg; and the instrumentation 32 kg. To maintain control, the Israelis spin-stabilised the satellite at the rate of 400°/s. Sixteen solar panels provided 246 W through a 7 Ah Ni/Cd battery. The satellite consumed 53 W on average. It transmitted 2.5 kbit/s in S-band from a 128 kbyte memory.

The Israelis refused to disclose the full payload, but did say it included magnetometers and housekeeping monitoring. Attitude sensing results from the output of a three-axis rate gyro, magnetometer and Sun sensor. During its short lifetime, the only malfunction was in telemetry memory when the computer switched to back-up. Originally, the Israelis expected the satellite to last only a few weeks, but it did not re-enter until 14 January 1989 because of a higher than expected orbit.

Almost identical to Ofeq 1, Ofeq 2 carried an improved computer, telemetry system back-up, thermal protection, gyros, and magnetometer. It could also execute some commands, whereas Ofeq 1 followed a pre-set program. The second satellite had a mass budget of: structure, 34 kg; power system, 59 kg; computer, 7 kg; communications system, 14 kg; thermal control, 5 kg; wiring, 10 kg; and instrumentation/balance masses, 31 kg. The orbit differed from Ofeq 1's to provide new perspective on radiation belts and to test Shavit's injection precision. It re-entered 9 July 1990 with an orbital life 40 days longer than expected because of reduced solar activity.

Some observers have speculated that the Israelis attempted another launch shortly after that of Ofeq 2, but that it failed.

By the time the Israelis were ready to proceed to their second-generation system, they had built a prototype 36 kg satellite. They described it in 1994 as a demonstration satellite with no science or other payload, but post-launch reports suggested a 2 m resolution Earth imager, supported by the orbit providing suitable lighting conditions over the region. El-Op (see separate entry), in 1995, confirmed there was a camera similar to its Earth Resources Monitoring System (ERMS) aboard.

Ofeq 3, launched in April 1995, was the first of the second-generation satellites. Power is provided by

Ofeq 3 was almost identical to South Africa's cancelled Greensat. Rafael's hydrazine propulsion module is visible at top (IAI) 0516733

Launch of the Ofeq 7 surveillance satellite on board a Shavit SLV, 11 June 2007 (IAI) 1165946

twin, three-panel solar wings totalling 3.6 m². The Sun sensors were developed by the Netherlands' TNO Institute of Applied Physics. Rafael provided the blowdown Hydrazine Propulsion Module for de-spin from 65 rpm, final orbit positioning, station-keeping and MW unloading. The thrusters are mounted on the top platform supported by eight composite struts, above the single 420 mm diameter positive expulsion sphere carrying 30 kg of hydrazine pressurised at 22 atmospheres. There are two independent branches of six LT-5N 5 N plus one HT-25N 25 N thrusters (HT-25N is principally for final positioning and orbit control). This propulsion system, with two tanks, appears to have been planned for South Africa's Greensat. Photographs show them to be identical. Payload mass was 36 kg.

Ofeq 5 is believed to be capable of delivering both panchromatic and colour images, at resolutions substantially exceeding those provided by Israeli commercial satellite services (0.8 m). Built by IAI/MBT, it is a three-axis stabilised, lightweight satellite platform.

Ofeq 7 was launched in a retrograde orbit by a Shavit rocket on 10 June 2007. The first images were broadcast to the ground station on 14 June. It was reported to be transmitting high-quality images with a resolution of better than 0.7 m from digital cameras produced by Elbit Systems Electro-Optics Elop Ltd. Other Israeli hi-tech companies involved included IMI, Rafael, Tadiran-Spectralink and Elisra. Ofeq 7 is described as an advanced satellite carrying newer surveillance equipment than its larger predecessor, Ofeq 5, and has been in operation for five years. The second-generation system is intended to fill the gap created by the failed launch of Ofeq 6 in September 2004.

Ofeq's east-to-west orbit at 36° inclination is phased to give optimal daylight coverage of the Middle East, with six or more daylight passes per day over Israel and the surrounding countries. Although this optimal coverage degrades after several months, the satellites maintain very good coverage of the Middle East.

At the June 2007 Paris Air Show, IAI/MBT Space Division unveiled its third generation, low weight, very high performance optical observation satellite, the OptSat 3000. This was described as presenting a breakthrough in terms of weight-to-performance ratio, achieving high agility, very high quantities of widely spread images in each satellite pass, and high levels of autonomy. OptSat 3000 is based on a new generic small platform which can accommodate various types of payloads. It contains panchromatic and multi-spectral imaging cameras, sharing a common optical assembly and capable of simultaneous operation and the creation of pan-sharpened images. The satellite can be controlled by a single ground control atation and it may serve multiple users. The satellite is designed for a mission life of more than six years.

Specifications

Ofeq 1

Launched: 09:32 19 September 1988 by Shavit from Palmachim Air Force Base south of Tel Aviv
Orbit: 248 × 1,170 km, 142.86°, 98.79' period, retrograde.
Mass: 156 kg

Ofeq 2

Launched: 12:02 on 3 April 1990 by Shavit from Palmachim Air Force Base
Orbit: 208 × 1,584 km, 143.23°
Mass: 160 kg

Ofeq 3

Launch: 11:16 on 5 April 1995 by Shavit from Palmachim Air Force Base
Orbit: 247 × 732 km, 143.4° and raised itself at first apogee to 369 × 730 km, 143.4°. The goal was about 500 km circular but no further changes were made.
Mass: 225 kg at launch
Design life: 2 years
Imaging Resolution: about 2 m

Ofeq 3's 1995 launch aboard the Shavit, a launch vehicle derived from Israel's Jericho 2 missile (IAI) 0517283

AOCS: 3-axis control system providing 0.1° astronomical-class pointing accuracy
Data rate: 15 kbit/s with a command rate of 5 kbit/s.
Power consumption: 180 W

Ofeq 4

Launch: Failed in launch attempt from Palmachim on 22 January 1998 due to a problem with the second stage of the Shavit launcher. Failed to reach orbit.

Ofeq 5

Launch: 14:55 on 28 May 2002; Military reconnaissance satellite launched retrograde from Palmachim
Orbit: 370 × 600 km; 143°
Mass: 300 kg at launch
Dimensions: 2.3 × 1.2 m
Imaging Resolution: Likely between 1 – 0.7 m
Design Life: Four years

Ofeq 6

Launch: Third stage launch failure from Palmachim on 6 September 2004; Ofeq 6 fell into the Mediterranean Sea.

Ofeq 7

Launch: 23:40 UT on 10 June 2007 by a Shavit rocket from Palmachim.
Orbit: 142° retrograde, 340 × 575 km
Mass: ~300 kg
Imaging Resolution: Likely between 0.7 – 0.5 m
Design Life: Five years

TecSAR/Polaris

Current Status

TecSAR (also known as **Polaris-1**) was launched by an Indian PSLV on 21 January 2008.

Background

Israel's first Synthetic Aperture Radar (SAR) Technology Demonstrator Observation Satellite, TecSAR, was funded by the MoD Defence Research & Development Directorate (DRDD). The satellite was designed, built and integrated by Israel Aerospace Industries (IAI), with the company's MBT Space division acting as prime contractor and Elta Systems Ltd., an IAI subsidiary, provided the EL/M-2070 SAR payload. The project also included contributions from other IAI divisions, as well as from other Israeli companies such as Rafael, Rokar and Tadiran Spectralink.

The small, low-cost spacecraft features a lightweight design and a configuration (shape of bus with low moments of inertia about the roll axis) that provides maximum pointing agility. The highly agile bus design, combined with the body-pointing parabolic antenna, permits greatly increased viewing capabilities from the spacecraft. The spacecraft is three-axis stabilised, using a newly developed bus, with a launch mass of about 300 kg, including a payload mass of 100 kg. The design keeps the bus and payload well separated, allowing plenty of scope for future changes in either element. The Attitude and Orbit Control

A model of TecSAR at the 2007 Paris Air Show (Patrick Allen) 1326191

Subsystem (AOCS) provides a high degree of pointing accuracy.

The SAR instrument consists of five major subsystems: the Radar Signalling and Control system; the Multi-Tube Transmitter; a deployable paraboloid mesh with electronic beam steering; the Onboard Data Recorder (ODR); and the Data-Link Transmission Unit (DLTU). The payload modules are located close to the antenna and separated from the bus so that only cables and wires connect the two. The OBR and DLTU components are located on the bus modules. The XSAR's radiating element is a deployable parabolic antenna with a rigid Carbon Fibre Reinforced Plastic (CFRP) central dish and knitted mesh gores stretched by skeleton ribs. The mass of the reflector mesh is <0.5 kg. The electronic beam steering capability is achieved by using an antenna feed array in the focal plane.

TecSAR has multiple modes of operation, including high resolution spot (1 m resolution), strip (3 m), mosaic (1.8 m using electronic steering) and wide area coverage (8 m resolution). Image enhancement for better target discrimination is also supported through the use of multipolarisation. Combining high manoeuvrability with electronic antenna beam steering, TecSAR can collect a large number of high resolution images per orbit or hundreds of medium resolution swaths. Using electronic beam scanning, TecSAR can also provide very wide coverage at lower resolutions. The multimode, X-band radar payload can provide rapid response, near-real-time, day-night and all-weather imaging capability to meet the immediate needs of front line forces as well as those of the wider intelligence community.

The TecSAR spacecraft during integration at IAI's MBT Space Division (IAI) 1341969

TecSAR was scheduled for launch in September 2007, under a commercial contract signed in November 2005 between Israel Aerospace Industries (IAI) and India's ANTRIX Corporation, but it was delayed until October by a sub-assembly problem on the PSLV launch vehicle. Cyclones associated with the monsoon forced a further delay, followed by more launcher issues. It was decided to use the Indian PSLV to launch the satellite since it is not possible to reach a polar orbit using Israel's Shavit launcher from Palmachim AB. The launch of TecSAR finally took place on 21 January 2008. It was the first time an Israeli satellite had been launched by an Indian rocket. In-orbit tests were scheduled to continue for several weeks, with receipt of the first images 14 days after launch.

In April 2007, Northrop Grumman Corporation and Israeli Aerospace Industries (IAI) signed a teaming arrangement to build and launch operational surveillance satellites in order to provide US government users with all-weather, day-night imaging capabilities. The US government was seeking operationally responsive capabilities to meet the challenge posed by the proliferation of anti-satellite capabilities, particularly in China, and it was hoped that this programme would be included in the US government's Fiscal 2009 budget. Such satellites could provide an early, though relatively basic, capability to the Pentagon long before the introduction of the **Space Radar** satellite system. Northrop Grumman was planning to have the first satellite, to be known as "Trinidad", ready for operational use within 28 months of receiving authorisation to proceed. The company would enlarge and modify the Israeli platform to carry additional equipment required by the US government, including mission assurance systems and secure communications. Northrop Grumman would also conduct final integration and testing at its facilities in Redondo Beach, California. The satellites could be individually launched by a low-cost Minotaur or Falcon 1, or in a group of four or more on an EELV-class launcher.

Specifications

Launch: 03:45 GMT on 21 January 2008 by PSLV-C10 from Satish Dhawan Space Centre, Sriharikota, India.
Orbit: 450 × 580 km, 41° inclination, sun-synchronous. 15.22 orbits/day, repeat cycle of 36 days.
Design life: 5 years minimum.
Contractors: MBT Space – IAI (prime). Elta Systems Ltd. – IAI (SAR payload).
Communications: High-rate X-band downlink.
Principal applications: Military SAR radar surveillance/technology demonstration.
Configuration: Improved Multi-Purpose Satellite bus, 3-axis stabilised. Parabolic mesh SAR antenna. Onboard memory 240 Gbit.
Mass: Approximately 300 kg (260 kg dry mass, including a payload mass of 100 kg).
Power: 750 W (EOL) provided by two solar arrays.

Russian Federation

Oko Constellation

Current Status

The long-serving Oko (Eye) early warning constellation continues in service, with a launch as recently as June 2006.

Background

The system of three Oko satellites, known as US-KS and developed under the military designation 73D6, roughly corresponds to the US DSP system. The Soviets intended for the system to produce an early warning of a ballistic missile attack. Work on the system began in 1967 with TsNII Kometa as prime contractor on a 15 to 30 minute warning system. It took 15 difficult years for the Soviets to place the system and get it fully operational in 1987. US military observers claim that 'it is capable of providing about 30 minutes warning of a US attack and of determining the general area from which it originated'. Okos probably also monitor nuclear tests, as do the US systems.

The Soviets began a four-year test programme for the first-generation system in 1972 with launch of Kosmos 520 on 19 September 1972, and operationally deployed the system starting in 1976. A Molniya LV, launched from Plesetsk, places each Oko in a highly elliptical (about 600 × 40,000 km) semi-synchronous orbit at 40° intervals. A full system requires nine satellites operating 160 minutes apart. With apogees over Western Europe and the Pacific and overlapping coverage, each views ICBM sites in the US for 5 to 6 h/revolution (12 hours daily) and transmits data to a command centre at Serpukhov-15. Oko also monitor routine launches from the US and other countries.

When Kosmos 1849 reached orbit on 4 June 1987, the Soviets filled all their requested operational slots for the first time. The RFAS keeps one or two older satellites on a more westerly ground track, possibly for better viewing of the western US. Despite its age and the lessening of tension with the United States, Russia shows no sign of abandoning the system and, during the last days of the Soviet Union, Oko established a new orbital plane, in late 1990, by Kosmos 2105 (20 November 1990).

Since the late 1990s, a decline in launches has caused the constellation to fall below its full population. In the 1980s, the Soviet Union launched four to six Okos each year. In 1990, there were six launches to replenish the system, then none in 1991, four in 1992, three in 1993, one in 1994, one in 1995 and none in 1996. The Russians launched a new Oko satellite in April 1997 with the new Kosmos 2340. Kosmos 2222 apparently died in the second half of 1996.

Interfax News Agency initially stated that the planned launch of a Kosmos satellite aboard a Molniya-M on 18 December 1996 had been delayed without giving a specific reason. After the Russians rescheduled the launch attempt for 10 January 1997, it was again postponed 'for technical reasons'. After a launch on 9 April 1997, the Russians named the

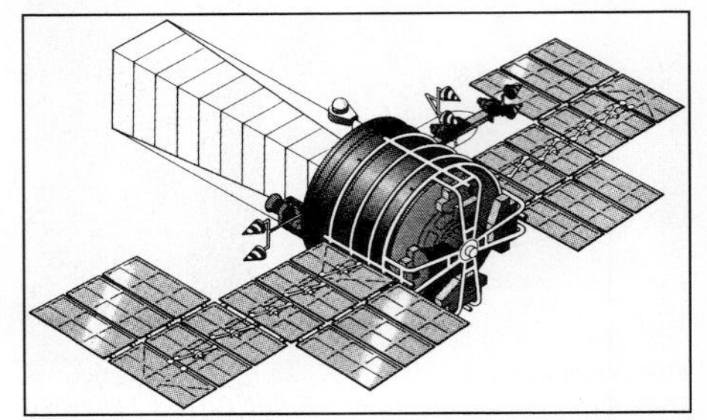

The original Oko early warning satellite (Europe and Asia in Space, 1991–92; USAF Phillips Lab) 0517154

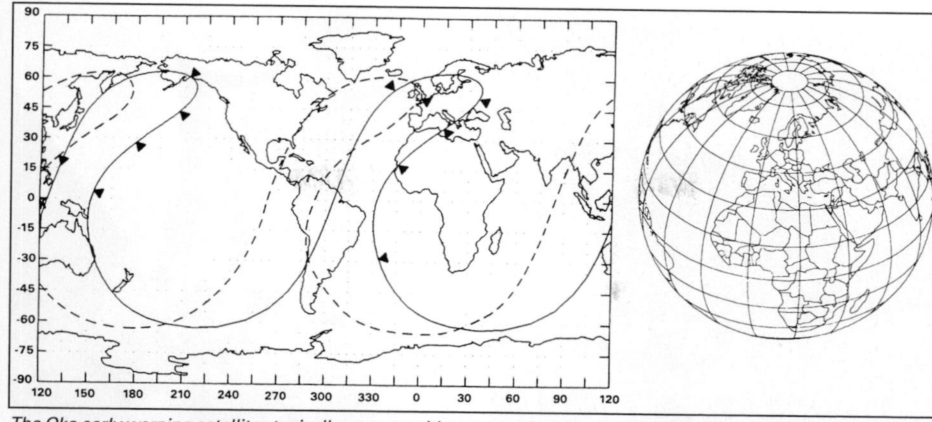

The Oko early warning satellites typically operate with apogees that provide surveillance of the US while remaining in contact with the RFAS/USSR. The globe illustrates the view from one apogee. The solid groundtrack depicts the daily path of the primary satellite, while one or two satellites are usually maintained in the more westerly dotted track (Teledyne Brown Engineering)
0516818

satellite Kosmos 2340 once in orbit. It was followed within a month by Kosmos 2342, with further launches of one each in 1998 and 1999.

Recent Oko launches have taken place on 2 April 2002, 24 December 2002 and 21 July 2006. The latest launch appears to have brought the system up to three operational satellites.

Russian drawings have depicted the Oko satellites, showing a cylindrical drum about 2 m diameter and 2 m long, plus two sets of solar panels and a telescopic housing for the instruments. They appear to weigh about 1,250 kg. Oko is the responsibility of TsNPO Kometa, with NPO Lavochkin responsible for the satellite bus.

Russian media reported in mid-2006 that a new GEO warning satellite named Faza was being developed for the RFAS. It would be capable of detecting strategic and tactical aircraft, and presumably missile launches as well.

Prognoz series

Current Status

A single operational Prognoz (Forecast) GEO strategic missile warning satellite remained on orbit and operational in mid-2006.

Background

Prognoz was developed as a GEO supplement and potential replacement for Oko, but never reached its full constellation strength. The system is also known as US-KMO.

As a follow-on plan to Oko, the Soviets requested seven GEO positions in 1981 and named the satellites Prognoz. The Soviets claimed the Prognoz would study the Earth's oceans and natural resources beginning in 1982. The Soviets requested slots at: 12° east, 35° east, 80° east, 130° east. 166° east, 159° west, and 24° west. The first Prognoz satellite took up its position 26 April 1988 named Kosmos 1940. The spacecraft was formally identified as an early warning system in 1993. A Proton-K carried it into orbit, but the flight plan left it in GTO for several weeks prior to transfer.

Kosmos 2133 (14 February 1991) established the series as the second vehicle launched under the Prognoz programme. It occupied two slots first at 80° east and Western observers claimed it handled Potok communications at the time. Then it manoeuvred first to 35°, 12°, 336° and 80° east – each of the Prognoz locations. In August 1995, it drifted off its assigned longitude and appears to be no longer operational.

Prognoz 2 was joined at 336° east by Kosmos 2209, launched 10 September 1992 then Kosmos 2224 17 December 1992 at 12° east, and later moved to 336° east. It returned to 12° east in April 1994. Kosmos 2282 was launched on 6 July 1994 and settled at 335° east; but began drifting in October 1995. Other satellites in the series were launched on 14 August 1997 and 29 April 1998.

The satellite is made up of a 2 m diameter main instrument section, two large solar wings and a large telescope tube housing a beryllium mirror. The PbS linear CCD can apparently detect aircraft on afterburners.

According to some sources, a new version of Prognoz was launched in 24 August 2001. It has improved sensors and can detect missile launches against the Earth background. The most recent launch was on April 24 2003. It was reported in mid-2006 that this is the only operational satellite in the series. Both Prognoz and Oko may eventually be replaced by the new Faza satellite.

Tselina

Current Status

The Soviet armed forces applied the programme name Tselina to SIGnals INTelligence (SIGINT) satellites deployed to support land and air warfare and intelligence gathering. Launches of the current operational version, Tselina-2, still take place at extended intervals.

Background

Tselina-2 was designed in the early 1970s as a new SIGINT spacecraft that would combine the attributes of the different small craft that had been flown until that time under the Tselina-O and Tselina-D programmes. In particular, it would be designed with both high electronic and directional sensitivity, and with a greater ability to identify emitters on the ground. TsNIRTI Minradioprom was the prime contractor for the SIGINT payload and KB Yuzhnoye provided the spacecraft bus.

The result was a spacecraft that was too heavy to fly on the Tsyklon-3 booster which had been used for earlier SIGINT missions. The programme was changed in 1979 to use the new Zenit-2 rocket. However, this not only delayed the development of the spacecraft but resulted in further delays when development of the Zenit ran late. More than 20 Tselina 2 satellites have been launched. The Zenit launcher places the 3.2 tonne spacecraft into an 839 km circular orbit.

The new satellite appeared as Kosmos 1603, launched 28 September 1984, and Kosmos 1656, launched 30 May 1985. Both were launched by Protons from Baikonur because of Zenit teething troubles. Both performed very extensive manoeuvres. Starting from an initial 198 by 186 km orbit in an inclination of 51.6°; the satellite and its upper stage remain in this parking orbit for only about 30 minutes until it uses the booster to raise the apogee to 860 km. Some 50 minutes later, another manoeuvre raises the perigee to 850 km and changes the inclination to 66.6°.

On 22 October 1985, the Soviets launched the third satellite in the series, Kosmos 1697. It left no debris in the 51.6° inclination and 66.6° inclination orbits it used as it adopted its final position. It was the first Tselina to be launched by the Zenit-2 launchers, which has flown all subsequent satellites in the system.

Flights resumed in 1987 with Kosmos 1833 (18 March) and Kosmos 1844 (13 May), and two more in 1988: Kosmos 1943 (15 May) and Kosmos 1980 (23 November). Kosmos 2082 appeared next on 22 May 1990. The Soviet attempt to complete a constellation of four satellites in orbital planes spaced 45° apart failed 4 October 1990 when the Zenit exploded within seconds of launch destroying both the payload and pad. The Russians experienced another major setback 30 August 1991 when the next Zenit exploded after launch from a new pad. The launch authority issued a statement saying: it was a 'military-technical satellite to verify the fulfilment of disarmament treaty commitments.' A third Zenit failure followed on 5 February 1992, but the Zenit of 17 November 1992 carried Kosmos 2219 into an orbit almost coplanar with 1988's Kosmos 1943. The second Zenit in 6 weeks delivered Kosmos 2227 on 25 December 1992, into an orbital plane 90° separated from Kosmos 2219's. Kosmos 2237 appeared 26 March 1993, apparently replacing Kosmos 1980. Kosmos 2263 appeared 16 September 1993. These most recent additions form a 3-satellite system with planes 120° degrees apart. Kosmos 2278 appeared 23 April 1994 with its plane 40° west of Kosmos 2263's and Kosmos 2297 appeared 24 November 1994 with its plane 40° west of Kosmos 2278's. Kosmos 2322 appeared 31 October 1995. A

The most recent Tselina-2 launches took place on 28 July 1998, 3 February 2000 and 10 June 2004. Another launch was expected to take place in late 2006.

US-P

Current Status

Russia continues to launch US-P maritime signals intelligence (SIGINT) spacecraft of the US-P class, the last vestiges of an ambitious Cold War project.

Background

The US-PU is a modernised version of the US-P (17F17) spacecraft, which were launched between 1974 and 1991. Developed by KB Arsenal, the US-P was a passive maritime signals intelligence (SIGINT) spacecraft intended to locate and identify Western fleets by detecting their radar and radio emissions, and formed part of a maritime surveillance and targeting system along with the nuclear-powered US-A radar ocean reconnaissance satellite (RORSAT). The overall system was intended to detect and track ships and allow Russian ships, submarines and bombers to attack them with cruise missiles from beyond the horizon. The US-P had a greater detection range than US-A, and was simpler to build, but could be negated by the target using emission control (EMCON) procedures. The programme operates under the military codename Legenda.

After the RORSAT programme ended in 1988, due to technical difficulties and environmental objections, the Soviet Union switched to the improved US-PU (17F20). Originally designed for launch by the Zenit rocket, it was modified to be launched by the Tsyklon-2 after the break-up of the Soviet Union, which left the Zenit programme in the Ukraine.

The US-PU is a 3,150 kg low-Earth-orbit (400 km) spacecraft with two large solar arrays, parallel to the body. Photographs depict a cross-shaped receiver array, carrying a number of antennas. Unlike the US Navy Ocean Surveillance System (NOSS) satellites, they do not operate in close formation, but carry substantial manoeuvring fuel to sustain consistent orbits. The system is believed to be accurate to 2 km.

The most recent US-PU launches took place on 21 December 2001, 18 May 2004 and 25 June 2006, using Tsyklon-2 launchers from Baikonur. The May 2004 satellite re-entered in May 2006. According to some sources, the June 2006 launch was not successful because the solar array did not fully deploy; it was also reported that the spacecraft had been in storage for as much as 15 years before it was launched.

United States

AIRSS

Current Status

The Alternative Infrared Satellite System (AIRSS), a new programme started in the FY2007 Pentagon budget, is intended to substitute for the GEO satellite segment of the troubled SBIRS High programme and produce a replacement for the US Defense Support Program (DSP) missile-warning satellites.

Background

In 2005, against a background of continued schedule slippage and budget overruns in the SBIRS programme, and under pressure from Pentagon and Congressional critics, the USAF planned an alternative system. The USAF is estimating a budget of USD1.9 billion through

FY2011 for the development of AIRSS, which is a potential replacement for the third SBIRS-GEO satellite and the planned substitute for the fourth and fifth in the series.

The most important goal of the programme is to ensure that a satellite with DSP-like warning capability can be launched in 2015, when the current constellation will start to lose its effectiveness. The current plan is to start parallel technology development and system definition phases in October 2006, at the beginning of FY2007, and to award a single systems design and development in the third quarter of FY2008.

Defense Support Program (DSP)

Current Status

DSP-23, the last of the series, was launched on 11 November 2007 from Cape Canaveral by a Delta-IV Heavy launch vehicle. The series of geosynchronous satellites has provided early warning of missile launches since 1970. They will eventually be replaced by the Space Based Infra-Red System (SBIRS) or, if that programme is unsuccessful, the Alternative Infra-Red Satellite System (AIRSS).

Background

The Defense Support Program (DSP) has been the US Air Force umbrella term since 1966 for satellites intended to provide early warning of ballistic missile launches. The entire programme was classified until November 1991, when DSP-16 was launched by the Space Shuttle Atlantis – the only spacecraft in the series, until DSP-23, that has not been mated to a Titan. The number of operational satellites in the constellation is still classified, although the Congressional Research Service has reported that four satellites are needed for worldwide coverage and that six are usually operational. The cost of each satellite is about USD400 million.

The second-generation DSP was first launched and entered orbit in 1975; launches continued until 1987 (TRW) 0516836

The first generation of DSP spacecraft, described as Phase I, comprised four satellites launched in 1970–73, the first launch having taken place on 6 November 1970. Three Phase II satellites were launched between 1975 and 1977. They were followed by four Multi-Orbit Satellite Performance Improvement Modification (MOS/PIM) versions, launched in 1979–84, and two Phase II Upgrade vehicles with improved sensors. These spacecraft were all quite similar in size and weight.

The current configuration, known as DSP-I (although individual satellites in this family started with DSP-14, in 1989) is a larger vehicle with more power and more effective radiation hardening. These enhanced spacecraft incorporate more sensitive IR and Vela-type nuclear detectors for improved discrimination. They also employ two IR wavelengths to combat laser jamming. Recent technological improvements in sensor design include above-the-horizon capability for full hemispheric coverage and improved resolution. Increased on-board signal-processing capability improves clutter rejection. DSP-I-type spacecraft can indicate missile type, launch point and azimuth, and with the help of improved ground signal processing can predict the weapon's likely impact point. Each third-generation DSP has a seven- to nine-year lifetime and senses targets at two IR wavelengths (with HgCdTe and PbS detectors) to avoid laser jamming and improve discrimination. The air force claims this new two-colour system has the sensitivity to detect aircraft operating on afterburners. Scientists have also used DSP infra-red instruments to detect volcanic eruptions and wildfires.

DSP-16 was released by Shuttle Atlantis (STS-44) on 24 November 1991. Clustered below the main barrel are optical sensors looking for the characteristic double flashes of terrestrial nuclear explosions (NASA) 0516835

The Defense Support Program (DSP-22) spacecraft during a spin balance test at Northrop Grumman's Space Park facility (Northrop Grumman) 1340461

DSP spacecraft also carry nuclear detonation sensors. DSP-21 carried the Compact Environmental Anomaly SEnsor (CEASE), designed to measure ionising radiation hazards. DSP-23 also carried a 24.5 kg payload to detect low yield nuclear blasts in space.

The original DSP satellite weighed 900 kg and had 400 W of power, 2,000 detectors and a design life of 1.25 years. Throughout the life of the programme, the satellite has undergone numerous improvements to enhance reliability and capability. The weight grew to 2,380 kg, the power to 1,275 W, the number of detectors increased three-fold to 6,000 and the design life increased to a goal of five years (though most satellites have survived for at least 15 years).

The 1st Space Operations Squadron of the 50th Space Wing at Schriever AFB is responsible for spacecraft control, including launch, early orbit and tracking. DSP signals and warnings are handled by the 2nd Space Warning Squadron of the 460th Space Wing, based at Buckley AFB, Colorado. A new Mission Control Station (MCS) at Buckley AFB, declared operational in 2002, represents a major change in DSP operations. Originally, the DSP system used three control sites – Buckley, Nurrungar in Australia and Kapuan AS in Germany – which received raw signals and generated warnings. With the MCS, developed as part of the SBIRS programme by Lockheed Martin and Northrop Grumman, operations are consolidated at Buckley and the DSP system operates with simpler relay stations.

The DSP system's role has expanded to include tactical missile defense. The satellites monitored 166 exchanges of Soviet SS-1 Scud medium-range missiles between Iraq and Iran during the missile exchange known as the 'War of the Cities'. During 1991's Gulf War, the last two second-generation DSPs – the first to carry the 6,000 element detectors – moved their orbits from their ordinary positions

The Air Force Space Command-operated Defense Support Program (DSP) satellites are a key part of North America's early warning systems. In their 35,970 km geosynchronous orbits, DSP satellites are designed to detect missile launches, space launches and nuclear detonations (Radarsat) 0022563

to give stereo coverage. The 69° E slot provided an ideal position to view the launch field of the H2 airfield in Iraq. Reports from field commanders indicate the satellites provided launch information within two minutes of the launch and therefore could give the Patriot batteries information for targeting well before the end of the Scud's seven-minute flight. Target field identification was said to be less than 6 km of uncertainty. In 1995, DOD added the Attack and Launch Early Reporting to Theater (ALERT) system to DSP satellites to augment their theater missile warning capabilities.

Prompted by experience in Desert Storm, two programmes developed ground stations for tactical commanders to receive DSP data directly. A prototype of the Army/Aerojet Tactical Surveillance Demonstration station began operating in Germany in 1993 and was transferred to South Korea in 1994 as a precaution against North Korean missiles. The station merges data from two or more satellites. The Army's Space and Missile Defense Command (SMDC) proceeded to develop the operational Joint Tactical Ground Station (JTAGS), a transportable station based in mobile shelters, designed to cue Patriot interceptor missiles. During the invasion of Iraq, JTAGS provided the estimated launching point and time of enemy Theater Ballistic Missiles (TBM), their projected impact point and time, and their trajectory. This information was available in-theater, linked directly to the missile early warning architecture, and then disseminated within minutes to both theater and worldwide users by data and voice. JTAGS, in conjunction with missile early warning capabilities belonging to the Navy and Air Force, helped ensure 100 per cent detection of all 20 Iraqi TBM launches and the effective dissemination of early warning alerts to targeted forces.

A more modern replacement, the MultiMission Mobile Processor (M3P), is under development by Lockheed Martin. Hardware and software integration was completed at the end of 2003, paving the way for field testing. M3P is designed eventually to support the SBIRS constellation as well as DSP.

The DSP-I spacecraft has a unique configuration. The spinning bus, its exterior covered with solar cells, carries a 3.63 m long Schmidt telescope with a 92 cm aperture, to collect IR radiation and permit target discrimination by ground controllers within 1.20 seconds. The satellite spins at 5.7 rpm to maintain orientation. The telescope's optical axis points at 7.5° off the spin axis to generate a conical scanning pattern as the vehicle rotates. From GEO, each of the IR array's 6,000 IR detector cells views a small region on Earth.

Early in the programme, engineers used a so-called "zero momentum" approach that allows for spacecraft attitude control with a minimum of fuel expenditure. The DSP satellite is spun about its Earth-pointing axis to provide a scanning motion for the infrared sensor. To reduce the satellite spin momentum to a nominal value of zero, engineers introduced a reaction wheel that achieved an equal and opposite momentum. To maintain the satellite's pointing accuracy throughout its lifetime, the DSP has an Inertial Properties Adjustment Device (IPAD) which balances the changes in moments of inertia as fuel and other consumables are depleted.

The assembly and launch of DSP satellites was slowed down to match the long lifetimes achieved by the spacecraft, which meant that replacements were not needed as soon as expected. It has also been important to extend the DSP system's life in view of the long delays to the SBIRS replacement programme. TRW – the prime contractor for all DSPs, taken over by Northrop Grumman in 2002 – received a USD743 million contract from the USAF's Space and Missile Systems Center in mid-1987 to build a second batch of DSP-I-type satellites, numbers 18–22, to follow on from 14–17. The USAF ordered numbers 23 to 25 in June 1993 under an eight-year, USD724 million contract. However, DSP-24 and DSP-25 were later cancelled. GenCorp Aerojet's Electronic Systems unit (also taken over by Northrop Grumman, in 2001) delivered the infra-red sensor for DSP-23 to TRW in 1998, but its launch was postponed for nine years. Northrop Grumman eventually delivered DSP-23 to the USAF's Cape Canaveral Air Station in May 2005 for an autumn launch as the first operational satellite to be launched by a Delta 4 Heavy. Rescheduled for 1 April 2007, the launch was further delayed by the discovery of two structural cracks in the metallic launch table.

Specifications
DSP-I
Launch mass: 2,386 kg
Orbital altitude: 35,970 km geosynchronous
Dimensions: Height: 10 m on orbit (with solar panels deployed); 8.5 m at launch. Diameter: 6.7 m on orbit; 4.2 m at launch.
Power generation: 1,485 W

Phase I
Phase II
MOS/PIM (Multi-Orbit Satellite/ Performance Improvement)
Phase II Upgrade – SED (Sensor Evolutionary Development)

The changing design of the DSP since the introduction of Phase I in the early 1970s. Five versions of DSP have flown since DSP-1 in 1970 0022562

NOSS

Current Status
The US Navy continues to rely on passive signals intelligence (SIGINT) spacecraft, operating in formations in LEO, to detect and track ships at sea. The latest generation of Naval Ocean Surveillance System (NOSS) comprises three groups of two satellites each – rather than the triplets used in earlier generations – and was launched between 2001 and 2005.

Background
The use of spaceborne SIGINT to track ships is traced back to 1962, when the Naval Research Laboratory proposed a spacecraft which became known as Poppy. Primarily intended to intercept radar signals, seven Poppy satellites were launched between December 1962 and August 1977. The existence of Poppy was declassified in September 2005.

Poppy was directly succeeded by the first-generation NOSS spacecraft, which were launched in triplets by Atlas H rockets from Vandenberg AFB. According to a Russian paper, the NOSS satellites were preceded by an experimental triplet launched in 1971. Eight groups of the operational first-generation NOSS were launched, between 1977 and 1987. The system has been known by the codenames White Cloud and Parcae. (In Greek mythology, the Parcae were the three daughters of Zeus and Themida, who together spun, cut and measured the thread of life for mortals).

The NOSS-1 and NOSS-2 systems are believed to use the Time Difference Of Arrival (TDOA) principle to locate targets. Any Radio-Frequency (RF) signal is received by all three satellites at a minutely different time that is proportional to the difference in range from the source to each of the three satellites. Three signals and two differences yield a unique two-dimensional location for any object on the sea surface, even if the signal is instantaneous.

The Russian paper estimated that the NOSS satellites could receive signals from a zone with a radius of 3,500 km. With four satellite groups, most regions could be monitored more than 30 times a day.

The original system was operated by the Naval Security Group and included some half-a-dozen ground stations located at Naval Security Group

Activity (NSGA) centres in the US, Guam, Diego Garcia, Scotland and other areas.

The second-generation NOSS satellites were presumably larger than their predecessors because they required a larger launch vehicle. Since the requirements for the first-generation spacecraft had been written, the Navy had acquired more over-the-horizon weapons and the Soviet Navy had deployed new classes of ships with different emitters. Moreover, the adversary had presumably identified the NOSS system and would have taken more serious steps to implement emissions control (EMCON) in the fleet.

The system comprised three spacecraft attached to a large dispenser that would manoeuvre each satellite into its designated orbit. Developed by the Naval Research Laboratory and Fairchild (now Orbital), it was originally known as the Shuttle Launch Dispenser (SLD), but with the Defense Department's withdrawal from the Shuttle programme it became the Titan Launch Dispenser (TLD).

Four NOSS-2 triplets were produced. Launches on 8 June 1990 and 8 November 1991 were successful, but the third launch on 2 August 1993 failed. A replacement launch was scheduled and carried out on 12 May 1996.

In the following year, the NSGAs at Adak Naval Air Station in Alaska and RAF Edzell in Scotland were closed. This suggests that the NOSS-2 used a different communications system from its predecessor, probably using space-based relays and more effective onboard processing, making the ground stations unnecessary. It also suggests that the NOSS-1 was retired when the NOSS-2 constellation reached full strength.

The third generation of NOSS followed its predecessor by a decade, with launches on 8 September 2001, 2 December 2003 and 3 February 2005. The first two missions were nicknamed Gemini and Libra, indicating that the spacecraft were twins, rather than triplets, and all three launches used medium-lift Atlas rockets rather than Titans. (The first two twins flew on Atlas IIIAS and the third flew on an Atlas IIIB).

The technique used by NOSS-3 to locate targets with two receiving points is not known. However, it may be related to the ability to locate the satellites more rapidly and accurately with GPS, or the use of newer SIGINT techniques such as long-baseline interferometry.

SBIRS

Current Status

The first highly elliptical SBIRS payload, HEO-1, was launched on a classified satellite on 27 June 2006 and has been providing data since at least November 2006.

Background

SBIRS has evolved from efforts to develop a replacement for the Defense Support Program (DSP) early warning satellites, combined with the Pentagon's desire for a system that would track missiles rather than just detect a launch. After the failure of several programmes in the 1980s – attributed to immature technology and high cost – the Pentagon in 1994 transferred a Strategic Defense Initiative programme called **Brilliant Eyes** to the USAF, and gave that service the task of developing an integrated IR surveillance and warning system operating in several different orbits.

The new system was named the Space Based Infra-Red System (SBIRS). Brilliant Eyes, which relied on small LEO satellites, was renamed SBIRS Low. The name SBIRS High was given to a DSP replacement programme, combining new GEO satellites and payloads riding on classified HEO satellites. The Air Force initially planned to buy five geosynchronous orbiting satellites for the SBIRS programme, plus two sensors to be hosted on classified satellites in highly elliptical orbits. The budget allocation for the space infrastructure and related ground equipment was about USD2 billion when Lockheed Martin was awarded the contract in 1996. At that time, the first of the geosynchronous orbiting satellites was expected to be launched in 2002.

In 2001, SBIRS Low was sent back to the Ballistic Missile Defense Office (now the Missile Defense Agency) which later renamed it as the **Space Tracking and Surveillance System (STSS)**.

A Lockheed Martin/TRW team (TRW became part of Northrop Grumman in 2002) was selected to build SBIRS-High in 1996. However, the programme dropped behind schedule, with the launch of the first GEO satellite not expected until 2006, and

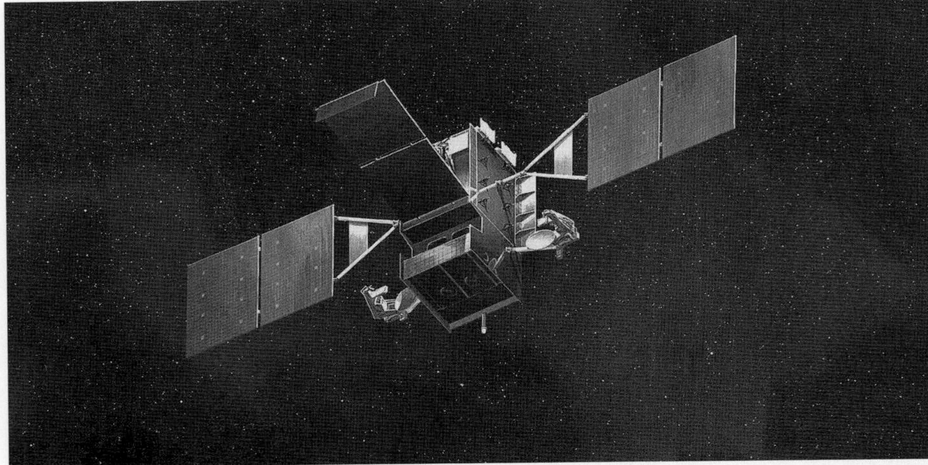

Artist's impression of the SBIRS spacecraft (Lockheed Martin)

0590541

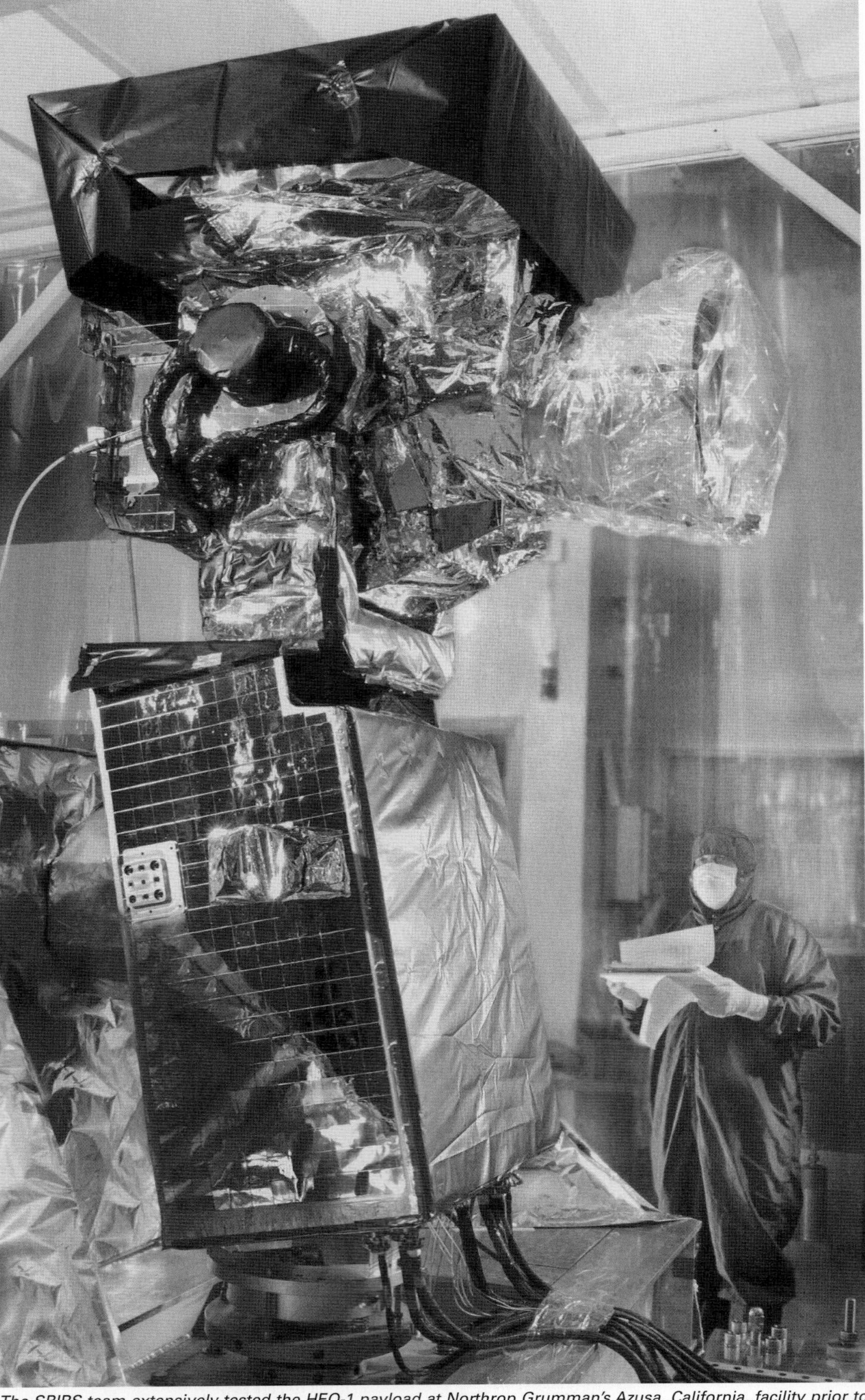

The SBIRS team extensively tested the HEO-1 payload at Northrop Grumman's Azusa, California, facility prior to shipment for spacecraft integration. The HEO payload will scan a larger region than the current system for ballistic missile launches. (Russ Underwood/Lockheed Martin) 1328332

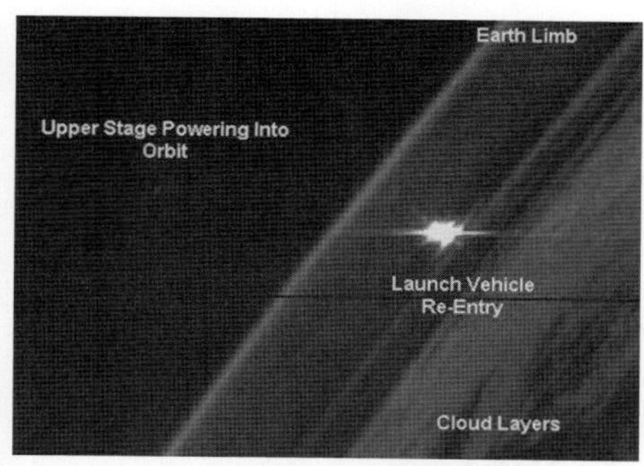

An infra-red image taken by the SBIRS HEO-1 payload in highly elliptical orbit showing the heat signature of an unidentified satellite as the upper stage motor boosts it into orbit. The infra-red signature of the launch vehicle re-entering the atmosphere is also shown. The sensor also detects cloud layers above Earth's horizon. The horizontal black line shows a drop in the sensor data. The resolution of the image was degraded for public release and other features that might identify its location were also removed (USAF) 1341643

Technicians at Lockheed Martin's facilities in Sunnyvale, Calif., prepare for the "jitter" test of the first Space-Based Infrared System (SBIRS) geosynchronous orbit (GEO) satellite. (Lockheed Martin) 1328333

costs increased. In 2002, the programme was restructured, so that the first and second of the five GEO spacecraft would be paid for with research and development funds and the subsequent three would be purchased later with procurement dollars.

A Government Accountability Office (GAO) report in October 2003 cited numerous problems with software and hardware development and concluded that the programme was at risk of more delays and cost increases. In the following two years, this proved to be the case. Delivery of the first HEO sensor was delayed by a year, from summer 2003 to August 2004, because of electromagnetic interference between the sensor and other spacecraft equipment, and the first GEO launch had slipped by two years (from 2006 to 2008) by the second half of 2005. However, the goal of the programme remained to produce five GEO satellites (an orbital constellation of four spacecraft, plus one spare to be stored on the ground in case of launch failure or early-on orbit failure), two HEO payloads and the SBIRS ground segment.

By the end of 2005, the first GEO launch had slipped into 2009, with operations starting in late 2010, and the cost had increased to more than USD10 billion. Moreover, senior Pentagon leaders were becoming increasingly concerned that the system – already under development for 10 years – was based on outdated technology, including the 1980s-era Ada programming language. As a result, the Pentagon decided to limit the SBIRS programme to two HEO payloads and two GEO satellites, with the establishment of a back-up programme. The FY2007 budget funded development of the Alternative Infra-Red Satellite System (AIRSS) to replace at least two and possibly three of the remaining GEO spacecraft in the SBIRS programme. AIRSS was to be considered as a potential alternative to the third SBIRS GEO spacecraft and an intended replacement for the fourth and fifth vehicles.

The primary mission of SBIRS High (now simply known as SBIRS) is to warn of a ballistic missile attack, and it is intended to provide faster warning and better tracking of any such attack, whether from an ICBM, a sea-launched missile or a Tactical Ballistic Missile (TBM). It is reported to use both scanning and staring sensors. The scanning sensor detects a launch and cues the staring sensor to track the object and provide more detailed information.

Compared with DSP, SBIRS operates in both the mid-wave and short-wave IR bands (DSP is short-wave) and also offers "see-to-the-ground" bands to detect targets despite atmospheric attenuation. The spacecraft is based on Lockheed Martin's A2100 bus design.

The HEO payloads operate in a highly inclined elliptical orbit and are intended to detect ballistic missile launches from northern polar regions. The first of a new generation of SBIRS sensors, this payload has improved sensitivity to detect dim theatre missiles and can be commanded to scan off-pole areas of military interest. The first HEO payload was manifested on the first Boeing Delta IV launch from Vandenberg AFB, attached to National Reconnaissance Office primary payload which was being launched into HEO. This launch, NROL-22, was scheduled for 2005 but slipped first into April and then into October before being scrubbed due to a fuel-sloshing issue. HEO-1 was finally launched on 27 June 2006. It is expected to be declared operational by mid-2008. The second HEO payload was delivered in September 2005 and HEO operations were due to start by the end of FY07. The launch is now anticipated for late 2008.

The programme appeared to be getting back on track during 2007, with the successful operation of the HEO-1 payload and the mating of the spacecraft bus and the payload for the first geosynchronous orbit (GEO-1) satellite. However, in September 2007, USAF Secretary Michael Wynne notified the Pentagon of a potential USD1 billion overrun after a problem was identified in the safe-hold software system of the GEO satellites, a stand-by system which engages if the spacecraft encounters in-orbit anomalies. The concerns arose after safe-hold system problems were encountered on a classified satellite launched during 2007, causing the termination of the mission only 7 seconds after its systems were switched on. The launch of GEO-1 is now unlikely to occur before 2009.

SBSS

Current Status

The USAF plans to launch a constellation of four Space Based Space Surveillance (SBSS) satellites to detect and track small objects in space. The goal is to improve military space situational awareness, tracking potentially hostile objects and also protecting spacecraft from orbital debris.

Background

Since 1996, the USAF has made use of a Space-Based Visible (SBV) sensor on the Midcourse Space Experiment (MSX) satellite to track space objects. In the early 2000s, the service initiated a follow-on programme to improve its tracking of space objects, which is still largely reliant on ground-based telescopes. The SBSS is expected to provide timelier and more accurate tracking and vastly improve the system's ability to track multiple objects.

Northrop Grumman Mission Systems (a former TRW unit) manages the SBSS programme on behalf of the USAF's Space and Missile Center (SMC). In March 2004, Northrop Grumman selected a Boeing team, including Ball Aerospace and Harris Technologies, to develop, build and launch the SBSS Block 10 satellite, also known as

Pathfinder. Boeing is the prime contractor and Ball is producing the satellite and its electro-optical payload. The satellite is due to be launched at the end of 2008 and the contractor team will operate it for up to a year after launch. Initially, it will replace the SBV/MSX capability.

The SBSS Pathfinder is to be followed by a constellation of four operational SSBS Block 20 spacecraft. A new competition for the Block 20 contract is due to start in 2006, leading to the start of development in 2008. Fabrication of the Block 20 spacecraft would start in 2011.

SIGINT

Current Status

Signals intelligence (SIGINT) satellites in the US are developed and launched jointly by the USAF and the National Reconnaissance Office, with the support of the National Security Agency (NSA). The current family is believed to focus on two large satellites based on the Lockheed Martin Milstar bus, supported by smaller Mercury satellites which also exist in data relay versions.

Background

The first US SIGINT satellites were launched in the late 1960s and were based on the same Lockheed Agena bus as the early KH photo-reconnaissance satellites. The USAF sponsored the Canyon series, which had the primary objective of establishing an Electronic Order of Battle (EOB) for the Soviet Union. Rhyolite, developed by the CIA's Science & Technology Directorate, focused on TELINT (telemetry intelligence). The CIA's main goal was to determine the performance parameters of Soviet ICBM and space launch boosters.

The slightly later Hughes Jumpseat, first launched in 1971, was based on the same bus as the Intelsat 4 commercial GTO communications satellite. However, Jumpseat was placed in a Highly Elliptical Orbit (HEO) in order to provide maximum observation time over the northern Soviet Union.

The success of Jumpseat led to the development of Mercury. It has also been called Chalet and Vortex, but Mercury is the only name known to be accurate. Building on the success of Jumpseat, Mercury was developed by the Air Force for the NSA, primarily for COMINT, and first launched in 1978. Mercury has a similar drum-shaped, spin-stabilised bus that was also used for the Intelsat VI commercial satellites and NRO's SDS data relay satellites.

In 1975, TRW's ambitious 3-axis-stabilised Argus TELINT satellite programme was cancelled, and its role was absorbed by Mercury. Beginning with the Mercury, launched in October 1979, a TELINT capability was included. TELINT data from Mercury was apparently fed to the CIA, and Mercury gradually took over the role of Rhyolite. The CIA's Pine Gap ground control station, which had controlled Rhyolite, underwent an expansion from 1983–1985, presumably to allow Mercury satellites to be controlled from that location as well as from the USAF/NSA stations in Europe. That would allow the satellites to be positioned over that part of the world to monitor Soviet and Chinese ICBM tests.

The second and third TELINT-capable Mercury satellites were launched in 1981 and 1984. There is a widespread belief that the satellite known as Magnum, launched by the space shuttle in 1985, was a new CIA-developed replacement for Rhyolite. The only evidence for this appears to be the new launch mode and the contemporaneous expansion of Pine Gap. There is no reason to assume that Magnum was a new type of satellite. On the other hand, it very well may have been the first Mercury to be placed over the Pacific and controlled by Pine Gap.

A total of five Mercurys were launched by Space Shuttles. Of these, the two SIGINTs used the Boeing Inertial Upper Stage (IUS), and the three lighter data-relay versions used Orbus 6 or Orbus 7 solid fuel Perigee Kick Motors (PKMs). Due to extended delays in the availability of the Titan 4, Centaur needed to launch the Milstar-based Trumpet satellite described below, three Mercury satellites were configured as data relay platforms and launched by the shuttle. These launches were in August 1989, November 1990 and December 1992. A fourth data relay, Mercury, was launched in July 1996 on a Titan 4. In all, 15 Mercury satellites were launched. Six were launched by Titan 3/34/Transtage, two by Shuttle/IUS, three by Shuttle/Orbus, three by Titan 4/Centaur, and one by Titan 4/Orbus.

The pinnacle of US SIGINT satellite development was the adaptation of Lockheed's Milstar bus to the SIGINT role, creating a 4,500 kg intelligence collection platform capable of surviving a full-scale nuclear war. It exists in two versions. The HEO version, with Boeing as prime integrator, is called Trumpet. Three Trumpets were built, and they were launched in 1994, 1995, and 1997. Boeing's Trumpet payload was probably developed in parallel with Milstar in 1983–91. In addition to its SIGINT role, Trumpet also serves as the polar component of the Milstar communications system.

The GEO version is generally reported to be called Orion. Orions were launched in 1995 and 1998, and there were probably one or two remaining to be launched in early 2006. Launches of Trumpets and Orions were distinguished by the 26 m Titan 4 payload shroud versus the 23 m shroud for Mercury.

NRO is currently working on an Integrated Overhead SIGINT Architecture (IOSA) designed specifically to integrate the large Trumpet and Orion satellites and the smaller Lockheed Martin Intruder, which is based on the Advanced EHF satellite and uses the A2100 bus. The follow-on IOSA-2 architecture would have added LEO SIGINT satellites using Time Difference of Arrival (TDOA) geolocation techniques. However, the IOSA-2 programme was cancelled in September 2000 due to projected cost.

STSS

Current Status

Conceived as Brilliant Eyes in the 1980s, renamed as SBIRS-Low in the 1990s, this programme was restructured by the Missile Defense Agency (MDA) in 2002 and renamed as the Space Tracking and Surveillance System (STSS). Two research and development satellites are due to be launched in 2007.

Background

The STSS is designed as a constellation of LEO satellites equipped with infra-red and visible-light sensors, capable of acquiring missiles and tracking missiles through both the boost and coast phases of their flights, and cueing interception systems on to their targets. They are also expected to discriminate between decoys and missiles. It was originally conceived by the Strategic Defense Initiative Organization as Brilliant Eyes, then merged into the SBIRS programme as SBIRS-Low in 1996.

Two industry teams were chosen in 1999 for Programme Definition and Risk Reduction (PDRR): Spectrum Astro/ Northrop Grumman, and TRW/ Raytheon. The Pentagon was expected to select one of the teams for the next phase in mid-2002 and the satellites were to have been launched between 2006 and 2010.

However, in 2002, the programme was restructured. The two teams were merged, with TRW – by then part of Northrop Grumman – as prime contractor and Spectrum Astro as a major subcontractor, producing the satellites. A competition between Raytheon and TRW to produce the sensor was ended in favour of Raytheon. Since that time, General Dynamics acquired Spectrum Astro.

The payload comprises a wide-angle scanning refractive telescope fitted with a shortwave infra-red array, and an agile narrow-field-of-view tracking sensor. The high speed processor uses only 175 W of power – one tenth as much as previous processors – and can detect and track more than 100 objects in real time.

Under the current plan, two research and development satellites will be launched in early 2007 by single Delta II launch vehicle. Referred to as Block 2006, they are refurbished versions of experimental SBIRS-Low spacecraft that were partially built but not completed in 1999. A ground segment will be delivered ahead of the Block 2006 launch. The goal is to track representative targets and pass those tracks to missile defense interceptors quickly and accurately enough for a successful engagement. Targets will include short and long-range missiles and airborne vehicles. The Block 2006 vehicles are expected to operate for two to four years after launch.

Block 2008 is apparently an improvement to the STSS ground segment, and Block 2010 is a classified programme, possibly involving another pair of demonstration satellites. The MDA plans to start fielding operational spacecraft, known as Block 2012, in 2012. In September 2005, the MDA said that it expected to award a prime contract for development of the Block 2012 in late 2006 or early 2007. They will use the same sensors as the Block 2006 satellites but will have improvements in producibility and longevity and will include better processing and communications systems.

The eventual size of the constellation has not been settled, but a minimum of 9–12 vehicles may be needed to ensure that satellites can communicate with each other. A constellation of 18–20 vehicles would provide better coverage of key threat regions, and a 25–30-strong constellation would offer global coverage.

EARTH OBSERVATION

Canada

RADARSAT

Current Status
RADARSAT-1 remains operational. RADARSAT-2 was launched on 14 December 2007.

Background
After a successful launch aboard a McDonnell Douglas Delta II from Vandenberg Launch Facility in California on 4 November 1995, RADARSAT began full operations 28 March 1996.

RADARSAT-1 was built by Spar Aerospace of Canada and developed by a government/industry collaboration led by the Canadian Space Agency (CSA). It was the world's most powerful commercial radar remote sensing satellite totally dedicated to operational applications. During its design life of five years, RADARSAT's C-Band Synthetic Aperture Radar (SAR) was intended to gather data for numerous global applications including ice surveillance, crop monitoring, disaster assessment, forest applications including clearcut monitoring, mapping, geological exploration, and coastal zone and ocean monitoring. RADARSAT's unique and state-of-the-art characteristics enable RADARSAT International, the Canadian company tasked with international marketing and distribution of RADARSAT data, to customise products to meet specific client requirements. Seven beam modes of 10 to 100 m resolution and 50 to 500 km swath widths combined with 25 beam positions ranging from 10 to 60° incidence angles mean a wide range of products are available for use in environmental monitoring and natural resource management applications worldwide.

In the ScanSAR wide beam mode, the whole of Canada can be covered every 72 hours with daily coverage of the Arctic. Imagery can be provided to high priority users, such as the Canadian Ice Centre to navigate ships through the high Arctic, and within 4 hours of data acquisition. The RADARSAT system is designed to operate with no backlog, so that all new imagery can be processed and distributed (or stored) within 24 hours. For most of its life, RADARSAT has been flying with the SAR antenna 'looking' to the right, providing coverage over the Arctic region. However, the spacecraft has been yawed 180° several times to switch the beam to the left for Antarctic mapping, during maximum and then minimum sea ice cover. Satellite images obtained in 1997 for the RADARSAT Antarctic Mapping Project (RAMP) culminated in the world's first high-resolution map of Antarctica and completed the mapping of the Earth. The RADARSAT mission control centre is in Saint-Hubert, Quebec, Canada (near Montreal), and the principal data processing centre, operated by RADARSAT International, is in Gatineau, Quebec.

Total cost of the RADARSAT-1 programme reached CAD642 million, with the Federal government contributing CAD520 million, which included the CAD75 million launch, the provinces CAD59 million, and the private sector, namely RADARSAT International, CAD63 million. NASA provided the launch in exchange for US research access to 15 per cent of RADARSAT SAR on-time and US private sector participation in distribution. The projected operations costs of RADARSAT over five years was expected to be recovered from RADARSAT International by the Canadian Space Agency in the form of 20 per cent royalty payments on the estimated CAD250 million revenue. In September 1999 the primary pitch momentum wheel suffered from excess friction and control was shifted to a back-up wheel. This suffered similar problems and was taken off-line on 27 November 2002 leaving the satellite safe but tumbling. Service was restored in December.

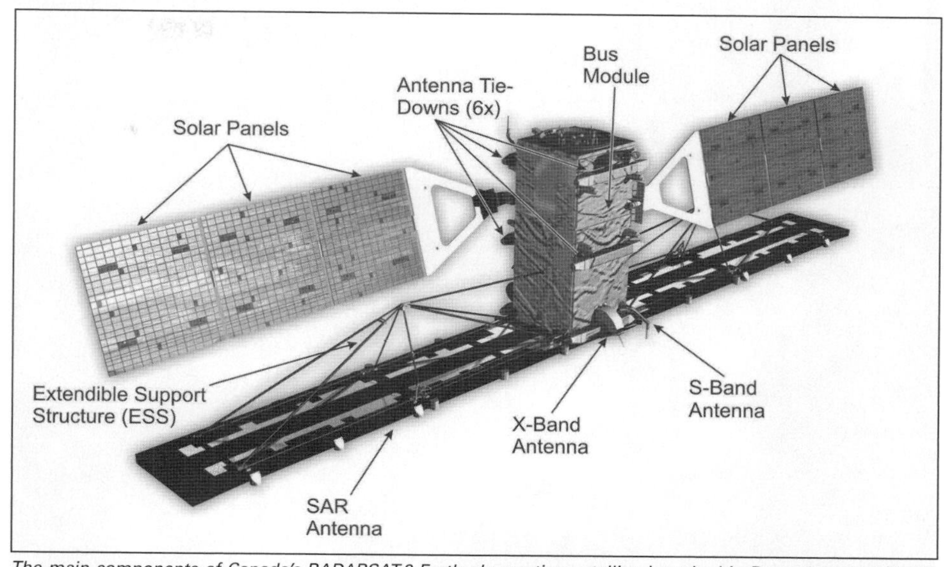

The main components of Canada's RADARSAT-2 Earth observation satellite, launched in December 2007 (MDA) 1341070

US government regulations related to the export of satellite technology and the distribution of high-resolution imagery, caused CSA to cancel an existing satellite bus contract with Orbital Sciences Corporation and to award a new contract to Alenia Spazio of Italy in January 2000. The next satellite in the series, RADARSAT-2, was contracted to McDonald Dettwiler and was initially scheduled for launch late in 2005 on a Delta II. It was eventually launched by a Starsem Soyuz on 14 December 2007. Development of RADARSAT-2 was undertaken by a government/industry collaboration involving the Canadian Space Agency and MacDonald, Dettwiler and Associates Ltd. (MDA). MDA owns and operates the satellite and its ground segment, while the CSA contributed about 75 per cent of the funding for RADARSAT-2's construction and launch in exchange for providing imagery to government agencies. ORBIMAGE, an affiliate of Orbital Sciences Corporation (OSC), Dulles, Virginia, is the exclusive distributor of RADARSAT-2 imagery to US customers.

RADARSAT-2 ensures continuity of data. Like its predecessor, RADARSAT-2 will provide commercial C-band synthetic aperture radar imagery in such areas as marine surveillance, ice monitoring, disaster management, environmental monitoring, resource management and mapping in Canada and around the world. Technical advancements include 3 m high-resolution imaging, flexibility in selection of its scanning polarisation, left and right-looking imaging options, superior data storage and more precise measurements of spacecraft position and attitude. Yaw steering (unlike RADARSAT-1), combined with improved attitude control will simplify image processing and improve image quality. RADARSAT-2 will be able to switch from right- to left-looking mode within about 10 minutes. About 75 percent of imaging will be performed in right-looking mode. The left-looking mode will be used for Antarctic mapping, emergency situations and to improve market access. Several imaging modes are available based upon incidence angles, resolution and polarization. A single image 'exposure' may vary from a minimum of five seconds to a maximum of 28 minutes.

Specifications
RADARSAT-1
Launch: 4 November 1995 by Delta II 7920-10 from Vandenberg, California.
Contractors: Spar Aerospace (prime), Ball Aerospace Systems Division (bus), CAL Corp. (slotted waveguide panels), Telesat, COM DEV (low-power electronics), SED (satellite operation, special test and ground equipment), MacDonald Dettwiler (mission control and ground processor), Odetics (two digital tape recorders), Dornier/AEG (TWT), SAFT (batteries), Astro Aerospace Corp. (SAR antenna deployment mechanism).
Resolution: Variable over 7.6–100 m
Mass: 2,749 kg (dry bus, solar array 1,315 kg, payload 1,366 kg, hydrazine 68 kg), plus 55 kg adapter and 352 kg margin.
Orbit: 783 × 787 km at 98.6° sun-synchronous, 18.00 ascending node, 24 day repeat cycle to provide optimum revisit opportunities (Arctic regions daily; 49–70° N every 3 days). 14 revisits daily (28 min SAR on-time/orbit), 100.7 min/orbit).
Spacecraft: 3-axis controlled to 0.1° in all axes by reaction wheels, magnetometers, magnetorquers, Earth and Sun sensors. Hydrazine thrusters to maintain orbit altitude within at least 10 km to ensure track repeat. Downlinking on two X-band (8.2 15–8.400 GHz) channels (one for real time, other recorder dumping) at 85–100 Mbit/s to Prince Albert (Saskatchewan) and Gatineau (Quebec) stations in Canada, plus NASA's Fairbanks station in Alaska. Additional international network stations currently receiving RADARSAT data include Tromso Satellite Station (Norway), West Freugh (UK), and Singapore. Rigid Spar array of two 5-panel wings provide 3 kW, supported by three 48 Ah SAFT Ni/Cd batteries. Two tape recorders store 10 min of data each at 85 Mbit/s SAR antenna 1.5 × 15 m.

Artist's impression of RADARSAT-1 (CSA) 1341071

Beam modes

Mode	Incidence angle / Resolution	Swath	Beam positions
Fine	36–48° incidence angle, 8 m Resolution	*50 km swath width, 1 × 1 look	5 beam positions
Standard	20–49° incidence angle, 25 m Resolution	100 km swath width, 1 × 4 looks	7 beam positions
Wide	20–49° incidence angle, 30 m Resolution	150 km swath width, 1 × 4 looks	3 beam positions
ScanSAR Narrow	20–46° incidence angle, 50 m Resolution	300 km swath width, 2 × 2 looks	2 beam positions
ScanSAR Wide	20–49° incidence angle, 100 m Resolution	500 km swath width, 2 × 4 looks	1 beam position
Extended Low	10–23° incidence angle, 35 m Resolution	170 km swath width, 1 × 4 looks	1 beam position
Extended High	49–59° incidence angle, 25 m Resolution	75 km swath width, 1 × 4 looks	6 beam positions

*Tape recorded images may vary in size.

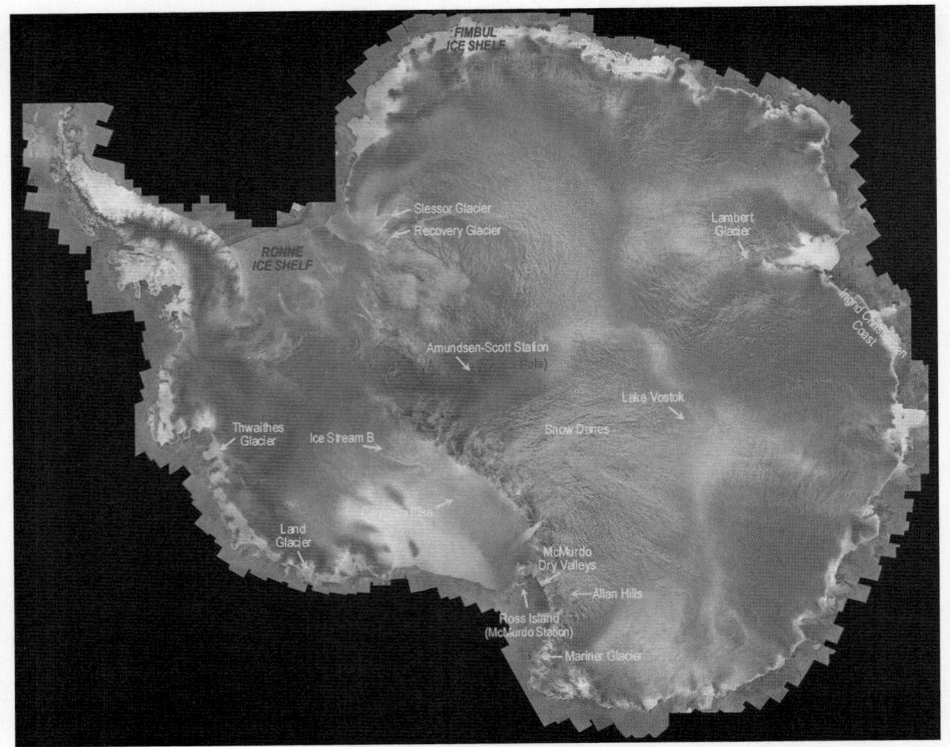

Satellite images obtained in 1997 for the RADARSAT Antarctic Mapping Project (RAMP) produced the world's first high resolution map of Antarctica (CSA) 1341072

Sensors: 5.3 GHz C-band SAR, HH polarisation, 50–500 km selectable swath width, 10–60° selectable incidence angle. The 300 W (average power; 5 kW peak) TWT is slightly modified from ERS-1. RF bandwidth 11.6, 17.3 or 30.0 MHz, transmit pulse length 42.0 μs, pulse repetition frequency 1,270–1,390 Hz.
Processing Levels: Path Image, Path Image Plus, Map Image, Precision Map Image, Signal Data, Single Look Complex
Derived products: Mosaics, Digital Elevation Models.

RADARSAT-2
Launch: 14 December 2007 by Soyuz-Fregat from Baikonur, Kazakhstan.
Design lifetime: 7.25 years.
Contractors: MacDonald, Dettwiler and Associates Ltd. (prime). Thales Alenia Space (bus). EMS Technologies (payload). Astrium (SAR sensor electronics). Alliant Techsystems (extendable support structure). Galileo Avionica S.p.A.(star trackers). LAGRANGE of Laben SpA (GPS receiver).
Resolution: Variable over 3-100 m.
Mass: 2,200 kg at launch.
Orbit: 798 km sun-synchronous orbit, 98.6° inclination, 100.7 min. Ascending node 18:00 hrs ±15 min (dawn-dusk orbit), repeat cycle 24 days. Same orbit, ground track and repeat cycle as RADARSAT-1, separated by 30 minutes.
Spacecraft: Based on the Alenia PRIMA bus, dimensions 3.7 m × 1.36 m. 3-axis stabilised. ACS uses two star trackers for precision pointing. Attitude knowledge ±0.02°, attitude control ±0.05° (3σ in each axis). GPS receiver provides real-time position knowledge of ±60 m. Two solar panels each 3.73 m × 1.8 m, provide 4.8 kW at EOL. NiH battery with 89 Ah. Six 1 N monopropellant hydrazine thrusters for orbit manoeuvring and maintenance (better than ± 5km, with a goal of ± 1km). Two Solid-State Recorders, each with 150 Gbit BOL capacity. Payload downlink on two parallel X-band channels at 105 Mbit/s. S-band for TT&C communications: downlink data rates 15, 128, 512 kbit/s (2230.00 MHz), uplink data rate 4 kbit/s (2053.458 MHz).
Sensors: SAR antenna 15 m × 1.5 m, mass > 700 kg. C-Band (5.405 GHz), channel bandwidth 100 MHz, channel polarization HH, HV, VH, VV. Maximum orbit average power consumption 745 W EOL. The phased array SAR antenna has 8192 radiating elements fed by 512 T/R modules. Two subapertures give a limited Ground Moving Target Indication capability based on along-track interferometry.
Imaging modes: All RADARSAT-1 modes plus polarimetric imagery (retrieval of full vectorial polarization information), ultra-fine (3 m resolution) beams. Left or right look direction. Single, dual and quad-polarisation imaging (HH, HV, VV, VH).

China

Fengyun Series

Current Status
On 11 January 2007, the Chinese destroyed their own ageing Fengyun 1C (FY-1C) with an anti-satellite (ASAT) weapon believed to have been launched aboard a Kaitouzhe (KT) rocket (see separate entry). The explosion created more than 1,000 pieces of orbiting debris that could conceivably damage other in-flight or launching spacecraft. The weapons test attracted international criticism, and suspicions that China has a well developed ASAT programme seem to have been confirmed.

Background
China launched its first Fengyun (FY-1) polar orbiting meteorological satellite in 1988 aboard the first CZ-4 from the Taiyuan polar site. The Chinese began developing the satellite in the late 1970s. The country then planned to build the FY-2 family and introduce it in April 1994 for geosynchronous observations.

The Shanghai Institute of Satellite Engineering, an arm of the Shanghai Academy of Spaceflight Technology (SAST), built the Fengyun 1 (wind and cloud) and Fengyun 2 series satellites with funds and specification requirements from the China Meteorology Administration (CMA). FY-1A, with a one year design life, operated for only 39 days but yielded high-quality imagery. Its far infra-red (IR) channel, however, failed to produce useful information because water vapour contaminated the detectors and radiation cooler, a fault rectified on 1B. The heavier FY-1B appeared a year later, accompanied by two balloons for upper atmosphere density measurements. It lost attitude control in February 1991 because of radiation upsets, but was recovered after a 50 day effort. The recovery came in time for it to monitor severe floods and help forecast the weather for the Asian Games. However, radiation damage broke control again. Chinese government officials noted in January 1994 that contact was maintained but data were unusable.

The Chinese used the FY-1A and 1B as test satellites, preceding the operational FY-1C and 1D satellites, launched in 1999 and 2002, respectively. To ensure that the satellites could survive the radiation hazards of space, the Chinese tested sub-assemblies of equipment at their Lop Nor nuclear site. FY-1C was the first satellite to be launched from the FY-1, batch 2 series of operational satellites. It was launched along with the Shijian-5 satellite on the maiden flight of the CZ-4B launch vehicle. FY-1C was destroyed by the Chinese in an 11 January 2007 ASAT technology demonstration. FY-1D, the second batch 2 satellite in the programme, was launched along with the Haiyang-1 satellite. As of early 2007, FY-1D was operational.

FY-2A and FY-2B, China's first geosynchronous orbiting meteorological satellites, were launched in 1997 and 2000, from Xichang and are regarded as test satellites. These two successful launches followed a terrible FY-2 launch failure in 1994. The first FY-2 satellite was destroyed in a fire and explosion during final checkout in test hall number 2 at Xichang on 2 April. One worker died and the fire injured 31 more.

FY-2C and FY-2D, the first of the FY-2, batch 2 operational satellites, were launched on 19 October 2004 and 8 December 2006, aboard CZ-3As from Xichang. A further FY-2, batch 2 satellite, FY-2E, is planned to launch by 2010. FY-2B was replaced by FY-2C. The main difference between the FY-2 batch 1 and batch 2 satellites is that the latter carry radiometers operating in five, rather than three, channels. Various sources estimated in 1994 that the FY-2 programme cost CNY300 million (USD35 million).

SAST has been developing the Fengyun-3 satellite series to replace the existing FY-1 Polar orbiting satellites; the last of these, FY-1D, remains operational, however, it has already far exceeded its intended lifespan. The first FY-3 satellite, a second-generation polar satellite following the FY-1 series, will carry a 10-channel scanning radiometer, IR and microwave radiometers, a space environment monitor and an Earth radiation budget scanning radiometer.

The Fengyun-3 series was originally planned for introduction in 2005, however Chinese comments in March 2004 indicated that the maiden launch was scheduled for 2006. The FY-3 satellite series is now projected for launch beginning in 2007, after several years delay. Xinhua reported that the first FY-3 satellite would lift-off during the second half of the year. As of March 2007, none of the series had yet flown.

The Fengyun-4 satellite series has not been described in any detail, but the Chinese plan to eventually replace the existing FY-2 series meteorological satellites with the FY-4 series. The final FY-2 series satellites will likely fly prior to any FY-4 series launches.

Fengyun ground stations are located in Beijing, Guangzhou and Urumqi, in Western China. CMA announced in 2007 that China would fly 22 meteorological satellites by 2020, however that number seems optimistic given the Fengyun launch rates over past years.

Specifications
According to the Chinese designations, odd numbered Fengyun series satellites are polar orbiting and even numbered series satellites are geosynchronous orbiting craft.

Fengyun-1 Series (Polar Orbit)
FY-1A
Launched: 6 September 1988 by CZ-4A from Taiyuan; design life one year but failed after 39 days
Mass: 757 kg at launch
Orbit: 881 × 904 km, 99.13° (afternoon orbit)
Power: 774 W generated BOL.

FY-1B
Launched: 3 September 1990 by CZ-4A from Taiyuan; nominal lifetime 2 years
Mass: 881 kg at launch
Contractors: Shanghai Institute of Satellite Engineering (platform), Shanghai Institute of Technical Physics (radiometers)
Resolution: 1.1 km visible high, 4 km infra-red
Orbit: 886 × 900 km, 98.9° (morning orbit), providing seven passes per day over Chinese territory and real-time image coverage of 55–145° E/0-65° N in conjunction with three stations able to receive data within 2,600 km
Spacecraft: body 1.4 × 1.4 × 1.4 m, height 1.76 m. Solar array span 8.6 m. 750 W 42.5 VDC BOL (design goal) power provided by 14,000 2 × 4 cm + 2 × 6 cm Si cells covering 6.8 m², supported by two 48 Ah Ni/Cd batteries. 3-axis attitude control by N_2 gas thrusters, RWs (magnetorquer unloading), gyros, infra-red horizon sensors. High-resolution image transmissions made at 6.654 kbit/s in real time on 1.6955/1.7045 GHz, APT on 137.795/137.035 MHz; housekeeping telemetry on 180.0 MHz. Global data stored on tape recorder
Sensors: Two 95 kg 360 rpm scanning 5-channel Very High-Resolution Scanning Radiometers (VHRSR) operating at (in μm): 0.48–0.53 (ocean colour, land), 0.53–0.58 (ocean colour, land), 0.58–0.68 (daytime cloud, land), 0.725–1.1 (daytime

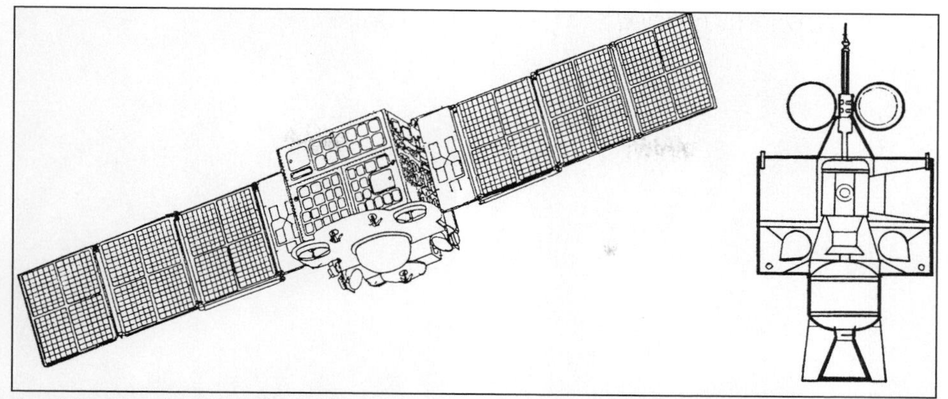

Four FY-1 meteorological satellites (left) have been launched into polar orbit. The geostationary FY-2 is on the right. Scales are comparable
0517038

cloud, water, snow, ice, vegetation), 10.5–12.5 (diurnal cloud, land/seaT). 4–24 MeV/n solar/cosmic ray composition monitor.

FY-1C
Launch: 10 May 1999 using a CZ-4B from Taiyuan
Orbit: Polar: 98.8°, 849–868 km
Lifetime: nominal mission 2 years; destroyed on 11 January 2007
Contractor: Shanghai Academy of Spaceflight Technology
Configuration: hexahedral body, 1.42 × 1.42 × 1.2 m plus two vanes of solar panels, giving a total satellite span of 10.556 m
Mass: 958 kg
Instruments 10-channel radiometer, operating in 0.433–0.453, 0.49–0.51, 0.54–0.56, 0.58–0.68, 0.745–0.785, 0.725–1.1, 1.58–1.64, 3.55–3.95, 10.3–11.3 and 11.5–12.5 μm

FY-1D
Launch: 15 May 2002 using a CZ-4B from Taiyuan
Lifetime: nominal mission 2 years; operational as of early 2007
Orbit: Polar: 98.8°, 851–873 km
Other specifications as FY-1C.

Fengyun-2 Series (Geosynchronous orbit)
FY-2A
Launch: 10 June 1997 by CZ-3 from Xichang; retired in 2000
Mass: 1,250 kg at launch, 600 kg on-station
Contractor: Shanghai Institute of Satellite Engineering
Resolution: 1.25–1.41 km visible, 5–5.76 km infra-red, 5–5.76 km water vapour
Orbit: geosynchronous over 105° E. China stated that a de-spin problem developed on 3 March 2000 and the satellite ceased operations. On 19 April the satellite was manoeuvred off-station over 105° E and five days later the longitude was restabilised over 85–86° E (this was not announced by China).
Spacecraft: similar configuration to Meteosat/GMS. Spin-stabilised with despun antenna platform. Control system derived from DFH 2. 2.1 m diameter, 4.5 m height, 280 W generated by drum-mounted solar cells. GEO insertion by solid AKM (separated after insertion)
Sensors: 0.55–0.75 visible, 10.50–12.50 IR, 5.70–7.10 μm water vapour scanning radiometer. Payload includes data collection and image relay systems
TT&C: China's three S/VHF-band stations at Beijing (116°16′ east/40°03′ north), Guangzhou (113°20′ east/23°09′ north) and Urumqi (87°34′ east/43°52′ north) receive real-time high resolution and APT imagery from NOAA and FY polar satellites for relay to the data processing/archiving facility at Beijing's National Satellite Meteorological Centre, part of the State Meteorological Administration. The local station utilises a microwave link to the Centre but the two remote sites communicate via GEO communications satellites through ground stations 10 km distant. The stations receive data when satellites are within 2,600 km, the spread providing pan-China coverage of 55–145° E/0-65° N. Data processing is accomplished by three IBM 4381 model 3s in conjunction with two IBM 7350 image processors and two IBM 5080 graphic processors.

FY-2B
Launch: 25 June 2000 using a CZ-3 from Xichang
Mass: all specifications as FY-2A
Contractors: Shanghai Institute of Satellite Engineering
Lifetime: nominal mission 2 years; replaced by FY-2C

FY-2C
Launch: 19 October 2004 using a CZ-3A from Xichang; intended to replace FY-2B
Mass: 1,380 kg
Contractors: Shanghai Institute of Satellite Engineering
Lifetime: nominal mission 3–4 years; operational as of early 2007

FY-2D
Launch: 8 December 2006 using a CZ-3A from Xichang
Mass: 1,380 kg
Contractors: Shanghai Institute of Satellite Engineering
Lifetime: nominal mission 2 years

Fengyun-3 Series (Polar Orbit)
FY-3A
Launch: Late 2007 at the earliest, using a CZ-4B from Taiyuan
Orbit: Polar: 98.728°, 836.4 km
Lifetime: nominal mission 3-4 years
Contractor: Shanghai Academy of Spaceflight Technology
Configuration: 4.4 × 2 × 2 m hexahedron, with solar panel vanes and instrumentation deployed in orbit – 4.46 × 10 (solar panel span) × 3.79 m
Mass: 2,400 kg
Power: average at EOL – 2.48 kW
Instruments: medium-resolution imaging spectrograph, infra-red spectrometer, scan radiometer, microwave radiometer, microwave imager, ultra-violet ozone sounder, Earth-radiation budget instrument and space environment monitor for global, all-weather, multi-spectrum, three-dimension quantitative detection
Data transmission: real time will use L-band BPSK modulation, data rate 4.2 Mbps, and X-band QPSK modulation, data rate 18.2 Mbps: stored data will use X-band QPSK modulation, data rate 93 Mbps.

Haiyang Series

Current Status
China first anticipated the launch of Haiyang 1B in 2005, however, this has been delayed. A launch date may be assigned for 2007, however as of early 2007, no further Haiyang satellites have been flown.

Background
The Haiyang (Ocean) oceanographic satellite uses the CAST-968 satellite platform which was first test flown as the Shijian 5 (see separate entry) satellite in 1999. Haiyang 1 (HY-1) was launched piggy-back with Fengyun 1D (FY-1D). China has planned a series of five Haiyang satellites to be flown at two-year intervals.

HY-1 observes China's surrounding seas and collects maritime data such as sea surface temperature, chlorophyll density, and seawater contamination.

Specifications
HY-1
Launch: 15 May 2002 using a CZ-4B from Taiyuan
Orbit: Initial orbit was 98.8, 851–872 km; during the first two weeks in orbit the satellite manoeuvred to a Sun-synchronous 98.8°, 792–795 km orbit
Lifetime: Nominal mission 2 years; still operating in early 2007
Contractor: China Academy of Spaceflight Technology (CAST)
Configuration: Box-shaped body, approximately 1 m each side, plus two vanes of solar panels, giving a total span of approximately 4 m
Mass: about 360 kg
Instruments: 10-channel infra-red ocean scanner and four-band CCD camera.

Jianbing Satellites

Background
China's original photo reconnaissance satellite programme was the Fanhui Shi Weixing (FSW) series: within Chinese Government circles the programme was called Jianbing, which means 'pathfinder'. The designator Jianbing was only applied to the specific FSW satellites that were launched on photo reconnaissance missions: other FSW satellites which flew on nominally civil remote sensing and microgravity missions were *not* given Jianbing designators, although such designators appear to have been applied throughout the FSW programme by Western intelligence circles.

From the history of the FSW programme, it would be reasonable if the FSW-0 series of satellite comprised the Jianbing-1 programme and that the specific photo reconnaissance missions in the FSW-1 programme were given the Jianbing-2 designator. It is possible that the latter group might be further divided into Jianbing -2A and -2B series, although such a division might be a figment of western intelligence designator systems. Open literature does not allow a definitive decision concerning which FSW satellites had which Jianbing designators.

The recoverable FSW-0 and -1 satellites were designated Jianbing-1, while the FSW-2 series was an improved Jianbing-1. The designator Jianbing-2 refers to an as-yet unflown satellite design (and not the FSW-2 series as some western sources have speculated).

The China-Brazilian Earth Resources Satellite (CBERS) programme is discussed elsewhere: the Chinese name for the programme is Ziyuan (ZY) which means 'resource', with the CBERS satellites specifically comprising the ZY-1 series.

With no prior announcement, China launched the first Ziyuan-2 satellite on 1 September 2000. While the first ZY-1/CBERS satellite would operate in a 774 to 775 km orbit, ZY-2 1 manoeuvred to a much lower 484 to 500 km orbit. The Sun-synchronous orbit was co-planar with that of the first ZY-1/CBERS and the orbital periods were in the ratio (ZY-2:ZY-1) 17:16.

Unlike the CBERS satellite system, China has released no information concerning the design of the ZY-2 satellite or the imaging system which it is carrying. Officially, the satellite is a Chinese domestic equivalent of the ZY-1/CBERS satellite, but western observers believe that it is a photo reconnaissance satellite in the Jianbing-3 series, returning images through a digital link.

At the same time, the Chinese government has continued to sponsor the development of recoverable spacecraft. An improved version of the FSW recoverable satellite series made its debut in November 2003: western reports indicated that the satellite was the first in the FSW-3 series and that it bore the designator Jianbing-4, but the Chinese have not confirmed this speculation. The satellite was launched on 2 November 2003 and recovered on 21 November 2003.

A pair of FSW satellites was launched and recovered in 2004. The first operated was launched on 29 August and recovered on 24 September. The second was launched on 27 September and recovered on October 15, landing off course in an apartment building. Another pair was launched on 2 August and 29 August 2005.

Specifications
Ziyuan-2 1 (Jianbing-3)
Launch 1 September 2000 using a CZ-4B from Taiyuan
Design life: 2–4 years?
Contractor: China Academy of Space Technology
Orbit: initially 97.42°, 475 to 493 km, initial operational orbit 485 to 499 km, standard operational orbit 490 to 495 km
Configuration: no illustrations of the ZY-2 satellite have been released, but it is believed to use the same satellite bus as the ZY-1/CBERS satellite. This implies a 1.8 × 2.0 × 2.2 m body plus a single vane of solar panels 2.6 × 6.3 m: 3-axis control by RWs and hydrazine thrusters. Orbit-raising manoeuvres

take place at ~2 week intervals (with a large amount of variation in this interval)
Mass: ~1,500 kg at launch
Power system: solar panels provide ~1.1 kW BOL

Ziyuan-2 2 (Jianbing-3)
Launch: 27 October 2002 using a CZ-4B from Taiyuan
Orbit: initially 97.40°, 471 to 483 km, initial operational orbit 478 to 506 km.
Other specifications as ZY-2 1

***Ziyuan-2* 3 (*Jianbing-3*)**
Fanhui Shi Weixing-3 1 (Jianbing-4 ?)
Launch: 3 November 2003 using an uprated CZ-2D from Jiuquan
Contractor: China Academy of Space Technology
Orbit: 62.99°, 194–324 km
Configuration: Three module spacecraft, comprising conical instrument/propulsion module plus retrorocket module plus dome-shaped recoverable module, maximum (base) diameter 3.35 metres, length ~5.5 metres. Recoverable module is approximately 1.75 metres base diameter and 1.5 metres high.
Mass: ~3,500 kg at launch ?

The Chinese referred to the satellite as being "scientific;" although scientific experiments could have been carried, the primary mission is believed to be military photo-reconnaissance, possibly including some "civil" remote sensing work as well. Satellite manoeuvred six times in orbit before the descent module was recovered on 21 November in Sichuan Province. The instrument/propulsion module continued in orbit until 18 December. The launch was the first from the new launch pad close to the one used for the piloted *Shenzhou* launches and it was the first to use a stretched version of the CZ-2D vehicle.

France

Pleiades

Current Status
France is developing the Pleiades-HR electro-optical surveillance satellites under the Orfeo (Optical and Radar Federated Earth Observation) project, jointly sponsored by France and Italy. The two spacecraft are due to be launched in 2008 and 2010.

Background
France and Italy are looking to small satellites to sustain their optical imaging capability. Under an agreement signed in January 2001 and ratified in 2004, the two nations are jointly leading development of the Orfeo (Optical and Radar Federated Earth Observation) project. Austria, Belgium, Spain and Sweden are minority partners. Within Orfeo, France is developing the Pleiades-HR small satellites – under 1000 kg mass – to replace the current Helios. (Italy is responsible for the Cosmo-Skymed radar satellite.) The first of these is due to be launched at the end of 2008 and the second in March 2010, using Russian Soyuz launchers from Europe's spaceport in French Guiana. The goal is to deliver 0.7 m resolution from a 694 km orbit, and to deliver any imagery within 24 hours of a request. Under an October 2003 contract, EADS-Astrium is the prime contractor for the Pleiades satellites, and Alcatel is providing the payload. France's CNES space agency is the primary design authority. It will have a 12.9 m-focal length optical instrument based on the Korsch folded optical design, with a 650 mm primary mirror.

Spot

Current Status
The Spot 4 earth observation satellite was successfully launched by an Ariane 40 launcher from the European spaceport in Kourou, French Guiana, on 24 March 1998 at 2.46 am French time. Spot 4's HRVIR instruments possess an additional spectral band in the Short-Wave Infra-Red (SWIR). The SWIR band offers better discrimination between different types of crops and plant cover. With Spot 1 and Spot 2 still in service, the enhanced capability of Spot 4, provides commercial stereographic images at the 5 m resolution level. Spot 5 was launched on 4 May 2002 by Ariane 42P. Today, Spot Image holds a 60 per cent share of the market,

Spot 3 launch preparation (Arianespace) 0517121

supported by a worldwide commercial network of three subsidiaries (United States, Australia and Singapore), nearly 90 distributors and 23 receiving stations.

Currently Spot 2 and Spot 4 are operational with Spot 1 in reserve.

Background
CNES began the Spot programme in 1978 and subsequently established Spot Image to market the data. Spot Image had a start-up capitalisation of US$41.1 million, allocated by public institutions and private companies in Belgium, France and Sweden. Initially the company had several partners. The original percentages were: (current values shown in parentheses), CNES 39.0 (35.32), Institute Geographique National 10.0 (11.27), Matra 8.8 (28.01), Aeropart (5.40), Bureau de Recherches Geologiques et Minieres 7.4 (1.00), Alcatel Espace (3.33), Crédit Lyonnais (1.19), Banque Nationale de Paris (1.19), Société Générale (1.19), Paribas (0.60), Swedish Space Corp 6.0 (5.94), Nuova Telespazio (2.32) and Belgium 4.0 (3.27). Spot Image has three subsidiaries, the Spot Image Corp in the US, Spot Imaging Services in Australia and Spot Asia in Singapore.

As the satellite operator, CNES is at the centre of the organisational structure, directly responsible from Toulouse for orbit maintenance, payload programming, reception and pre-processing. Data downlinks to two SRIS (Station de Reception des Images Spatiales) stations at Aussaguel near Toulouse (SRIS-T) and one at Esrange, near Kiruna, Sweden. These receive both real-time and recorded data as Spot passes over the north polar region, Europe and North Africa within 2,500 km range, in addition to the stored images from other regions. Together, they have a reception capacity of 500,000 images annually. Each SRIS is associated with a pre-processing/archiving centre or CAP (Centre d'Archivage et de Prétraitement), with the equivalent of 700 scenes archived every 24 hours at Toulouse.

Spot Image has responsibility for the Direct Receiving Stations (SDRS) around the world, which receive real-time imagery only. SDRS are located at Prince Albert and Gatineau (Canada), Maspalomas, Spain, station managed by ESA), Cuiaba (Brazil), Lad Krabang (Thailand), Hatoyama (Japan), Islamabad (Pakistan, commissioned 1989), Hartebeesthoek (South Africa, 1989), Riyadh (Saudi Arabia, 1989), Alice Springs (Australia, 1990), Tel Aviv (Israel, 1 February 1991), Cotopaxi (Ecuador, July 1992), Taipeh (Chung-Li University, Taiwan, July 1993), Pare-Pare (Indonesia, October 1993), Fucino (Italy, July 1995) and Singapore (September 1995); the USAF Eagle Vision transportable system began operating in June 1994 (another 20 will be ordered). South Korea was added in 1996, and Russia's Resurs-O station at Obninsk will be upgraded for Spot 2 imagery during the two warmest months.

CNES categorises processing operations into several levels of sophistication. Level 1 provides basic radiometric and geometric corrections and it does not involve ground control points or satellite attitude restitution data. Level 1A downlinks essentially raw data, apart from the normalisation of CCD detector response in each spectral band. It is useful for stereoplotting and basic radiometric studies. Level 1B gives full radiometric (desmearing) and limited geometric corrections with the pre-processing level

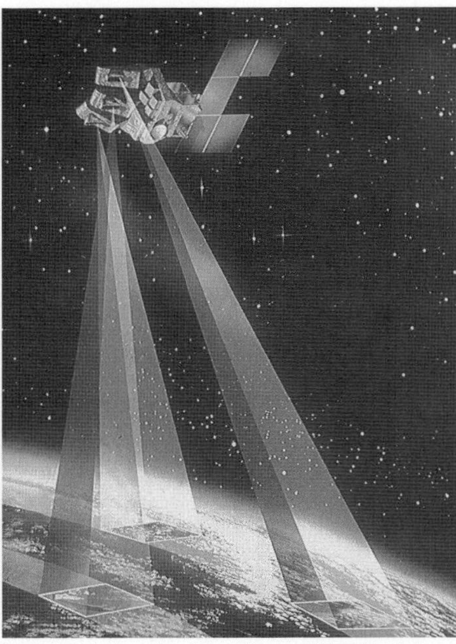
Spot 5 is being developed in parallel with Helios 2 (CNES/D Ducros) 0517120

for photo interpretation and thematic analysis. Users may also attain stereoscopic pairs at this level. Level 2 gives advanced rectifications according to a given cartographic projection and Level 2A corresponds to Level 2 precision processing but can be implemented without use of map ground control points. The Spot introduced Level 1AP in 1990, which is optimised for photogrammetric applications using analytical stereoplotters. The commerical suppliers provide products in standard Computer-Compatible Tapes, photographic form and CD-ROM.

Spot's 4 millionth scene was archived 19 July 1995. 1995's revenue from data/product sales and receiving station fees totalled FFr207.5 million (US$41.5 million); 1994 220; 1993 178, profit 0.031; 1992 218, profit 1.38; 1991 204; 1990 165; 1989 123; 1988 90; 1987 55 and 1986 15. 1993's decrease reflected lost sales because of Spot 2 recorder problems before Spot 3 became available. Asia/Pacific (largely Japan) accounted for 25.3 per cent of 1995's revenue; Europe 33 per cent, North America 25 per cent, Middle East 6.7 per cent, Latin America 3.8 per cent, Africa 6.3 per cent. Accumulated sales exceeded FFr1.4 billion (US$273 million) by end-1995 (70 per cent from outside Europe); the government had invested FFr8 billion. Europe provided 32 per cent revenue in the first 10 years, followed by North America 22 per cent and Asia/Pacific 20 per cent. The enhanced Spot 4 design was approved in 1989 for launch about 1994, but actually launched in 1998.

Spot 1
Spot 1 was placed in orbital storage 15 January 1991, but with its health monitored at least every 6 revolutions so that it could resume operational service and meet all image quality specifications within two weeks. It became operational again in March 1992 to help Spot 2 meet increasing demand through to late October 1992 during the northern hemisphere's growing season. It provided real-time transmissions (recorders are inoperative; some 60,000 to 80,000 images were received) for Europe/Middle East, while Spot 2 worked in recording mode

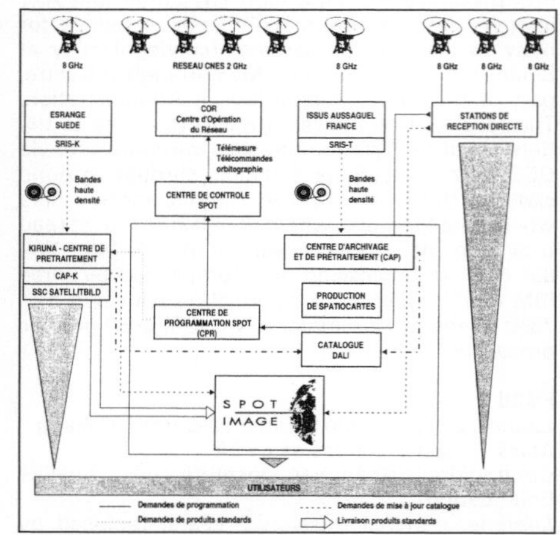

Spot control and data reception network 0517413

Spot 5 will provide routine 5 m imaging (CNES/Spot Image/ISTAR) 0517232

(and normally for the rest of the world). The two were phased to optimise stereopair acquisition above 41° latitude. It was revived similarly for April–July 1993, at a total cost of FFr7.7 million, requiring a third control centre, as the second is now used for Spot 3. The European Commission this time contributed €350,000, in addition to paying for each scene used for its agricultural monitoring work. Spot 1 was again retired after Spot 3 entered service in late 1993, but can still be called upon at a few days' notice. Spot 1 and 2 carry two identical push-broom CCD High Resolution Visible (HRV) imagers, pointed for a total swath width of 117 km with a 3 km overlap. Site revisit period is 26 days but up to 27° off-nadir viewing (creating total swath width of 950 km) via a tilting mirror permits a single area to be imaged on seven successive passes at equatorial latitudes and 11 passes at 45°. Image width is 60 km for nadir viewing, stretching to 80 km at maximum mirror tilt. The capability also permits stereo imaging. Through 1995, Spot 1's cameras totalled 2,927 h, returning 1.8 million images.

Spot 1 was returned to active duty on 13 January 1997 to fill a gap caused by the loss of Spot 3. Although the recorders have failed it can operate when in contact with a ground station.

Spot 2

The satellite returned its first multispectral image, from HRV 2, at 10.18 GMT 23 January 1990, 32 hours after attaining orbit. It arrived in the desired orbit 180° from Spot 1 on 30 January, having expended 12 kg hydrazine. The image quality review of 5 March concluded it to be of commercial quality, and Spot 2 was declared ready for operational exploitation from 21 March 1990. 1.9 million images were returned by end-1995. One tape recorder failed after one year. A second tape recorder in February 1993 began suffering from recording-head wear, leaving spots on imagery. The prime image telemetry channel has failed but the spacecraft is otherwise in perfect order. Spot 3 replaced it as prime satellite November 1993. Spot 2 requires FFr12 million in annual sales to cover operations costs; contracts from the European Commission allowed simultaneous Spot 2/3 operations through 1994–95.

Spot 3

The first image was transmitted within 33 hours, showing the Strait of Bonifacio separating Corsica/Sardinia. Spot 3 was declared operational 29 November 1993. It expended only 17 kg hydrazine attaining its final orbit, 42 and 208° ahead of Spot 1/2, respectively. It is exhibiting slightly better gain control and dark current characteristics than its predecessors. 686,028 images had been returned by end-1995. Spot 3 cost FFr1.4 billion, including FFr650 million for launch. Spot 3 unexpectedly ceased operating on 14 November 1996 when altitude control was lost after gyroscope problems.

Spot 4 was fast-tracked to launch and Spot 1 was reactivated.

Spot 4

The French government approved Phase C/D in July 1989 for the expanded Spot 4. It employs the same augmented multimission bus as the Helios 1 reconnaissance satellite, adds a mid-infra-red band to the HRV (now HRVIR) by replacing Sodern's DTA 01 detection unit with the DTA 03 model, incorporating Thomson's 7811 CCDs. Sodern's DTA 05 detection unit gives images at 1 km resolution for monitoring crops, natural vegetation and geosphere/biosphere interactions. The bands are equivalent to HRVIR plus a blue channel for atmospheric corrections. The European Union in January 1994 agreed to contribute half of VGT development costs; Sweden, Belgium (10 per cent) and Italy are also contributing separately. The total VGT programme is expected to cost ECU109.7 million.

Spot 4's Vegetation is an independent payload, having its own recording, transmission and management systems, working to its own ground segment for payload programming, image reception/processing and product distribution. Local users can acquire 1 km resolution imagery broadcast in real time at 1.704 GHz L-band 1 Mbit/s on their existing metsat receivers. The global data, recorded on the 2.25 Gbit SSR, are downlinked at 3.4 Mbit/s at 8.153 GHz X-band using Spot 4's main antenna. The CPV Centre de Programmation de VGT in Toulouse draws up the daily plan for uplinking by the main Spot 4 control centre. The SRIV Station Reception des Images VGT at Esrange takes 4/5 data dumps daily totalling about 8 Gbit. The CTIV Centre de Traitement des Images VGT in Mol (Belgium) archives/processes the data for a daily synthesis, increasing progressively to a 10-day global synthesis. It can deliver dedicated products within three days if ordered in advance. The QIV Centre Qualité Image VGT in Toulouse is in charge of instrument calibration and image quality.

Spot 5

Definition of Spot 4's successor began in 1993, following an extensive user market survey in 1991–92. 5 m resolution is necessary to address new markets such as civil engineering and an along-track 5 m stereo capability was to generate simultaneous stereo pairs, whereas its predecessors are limited to separate passes. Resolution is legally barred from being less than 5 m. The goal is to provide simple generation of 1/50,000–1/100,000 standard maps (1/25,000 in some cases) and digital elevation models. The government approved the FFr6.3 billion programme 4 October 1994 for the first of two launches in 2002 (but available late 1999), employing a bus jointly developed for Helios 2. Sweden and Belgium (4 per cent) are again contributing; Belgium is separately paying 10 per cent towards VGT. The programme was reduced in early 1996 to FFr4.4 billion, cutting Spot 5B and losing the simultaneous stereo imaging by dropping one camera and nadir pointing the remaining two. Spot 5 was launched in May 2002.

Specifications

Spot 1

Launched: 22 February 1986 by Ariane 1 from Kourou, French Guiana (2 year design life; initially retired 31 December 1990)
Mass: 1,830 kg at launch

In this view of the Spot 4 satellite undergoing testing in the Interspace anechoic chamber, the vegetation camera appears as a cylinder in the horizon position, under the twin imaging cameras (Matra Marconi) 0038492

Contractors: Matra (prime, imaging system), Aerospatiale (structure, thermal control, solar array), Sodern (optical sensors)
Resolution: 20 m multispectral, 10 m panchromatic
Orbit: 824 × 829 km (822 km after 3 years) in 98.7° Sun-synchronous, crossing equator at about 10.30 local time, with 26 day repeat cycle (369 revs). Altitude corrections executed every 2 months and inclination adjustments annually.
Spacecraft: 3.5 m high, 2 m², consisting of Multimission Platform (PFM) and forward Spot payload module. 2 year design life. 15.6 m solar wing provided 1,382/1,326 W BOL/EOL (equinox values; 36 V) with 3 × 24 Ah Ni/Cd batteries. Panel degradation after 3 year and 18,000 thermal cycles was 4 per cent instead of the projected 21 per cent. 3-axis platform pointing accuracy of 0.1° (0.15° specified) maintained by three MWs unloaded by two magnetic coils. System includes digital Earth sensors, 2 × 2° FOV. orbit control by 8 × 15.6 N hydrazine thrusters (two 75 kg tanks, 130 kg remained after 3 years). Each of the two HRV imagers returns data at 25 Mbit/s on 8.025–8.400 GHz real-time using 20 W Thomson-CSF TWTA
TT&C: Spot Mission Centre is located at Toulouse. 8.307 GHz beacon multiplexed with the transmitted

Spot 5 will provide routine 5 m imaging (Theo Pirard) 0054347

For details of the latest updates to ***Jane's Space Systems and Industry*** online and to discover the additional information available exclusively to online subscribers please visit **jsd.janes.com**

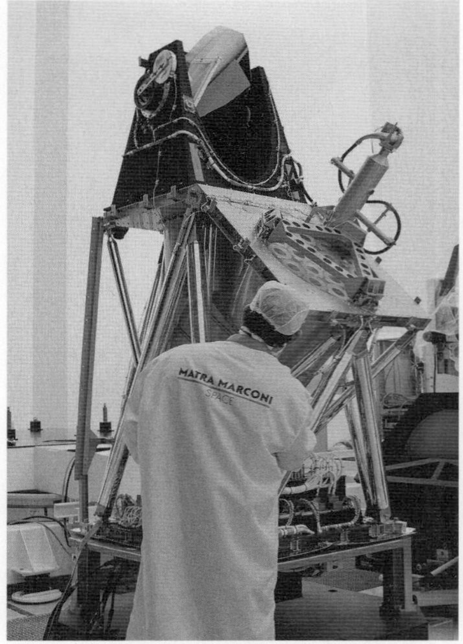

The Spot 4 IR instrument receives a final vibration checkout before launch (Matra Marconi) 0038493

signal to facilitate ground antenna pointing. Communications subsystem mass 240 kg; power consumption 170 W for direct tx, 270 W for recorder playback. The 70 kg Odetics recorder could hold 22 min of data. Recorder number 1 (a qualification model flown to save money) failed in Sepember 1986, number 2 was shut down after problems appeared July 1990, and the prime telemetry TWT failed August 1989. 3.7 kg 2.025–2.120 GHz S-band transponder provides housekeeping telemetry (2,048 bit/s) +TC (2 kbit/s)
Sensors: High Resolution Visible (HRV) Spot was the first satellite to incorporate push-broom sensors, avoiding the need for a mechanical scanning mirror. On Spot 1, each spectral band utilises Fairchild's 122 DC detectors; Spot 2/3 employ four Thomson-CSFTH 7801A 1,728-element CCD arrays, combined in each HRV's Sodern DTA 01 detection unit. The spectral ranges (in µm) of the bands are: Band 1 0.50–0.59, Band 2 0.61–0.68, Band 3 0.79–0.89; panchromatic 0.51–0.73
Passenger: radar transponder for calibration of Kourou's ground radar and operator training.

Spot 2

Launch: 01.35.27 GMT 22 January 1990 by Ariane 40 from Kourou, French Guiana; declared operational 21 March 1990
Mass: 1,870 kg at launch
Resolution: same as Spot 1
Orbit: 819 × 822 km, 98.7°, initially 180° from Spot 1, now 166° behind Spot 1
Spacecraft: same as Spot 1 but with 18.5 kg 20 W DORIS (Doppler Orbitography and Radiopositioning Integrated by Satellite) unit for accurate orbit determination in preparation for 1992's joint Franco-US Topex mission and hydrazine loading 160 kg
Sensors: Same as Spot 1.

Spot 3

Resolution: same as Spot 2
Launch: 26 September 1993 by Ariane V59 from Kourou, French Guiana
Mass: 1,907 kg launch; dry 1,749 kg; 158 kg hydrazine
Orbit: 816 × 818 km, 98.68°
Spacecraft: as Spot 2, but TWTs and recorders modified slightly for longer lives
Sensors: Same as Spot 2, plus 25 kg POAM II Polar Ozone & Aerosol Measurement instrument carried for USAF Space Test Program.

Spot 4

Launch: 24 March 1998 by Ariane 40 from Kourou, French Guiana
Mass: 2,755 kg launch; dry 2,600 kg: payload 1,400 kg
Orbit: 832 km, 98.7°, 101 min period
Contractors: MMS France (prime, assembly, integration, test, onboard data management, AOCS, inertial units, Sun/star sensors), Aerospatiale (VGT, mechanical subsystem, thermal control, solar array), Alenia (structural elements, wheels/magnetorquer drive electronics), Alcatel Espace (transponder), CASA (cabling harness), CRISA (software, EGSE), Sextant (housekeeping unit, bus couplers, decoding/reconfiguration unit), ETCA (electrical power supply), Laben (wheels/magnetorquers drive electronics), Saab (central communications unit), SAFT (batteries), SAGEM (gyroscopes), SEP (propulsion), Sodern (Earth sensors, star sensors, optical head), Schlumberger (tape recorders)

Spot 4 immediately prior to encapsulation on top of the Ariane launcher at Kourou 0084763

Resolution: HRVIR 20 m multispectral/10 m monospectral; VGT 1 km/4 km
Spacecraft: based on Spot 1–3 design but with additional compartment between platform and payload to accommodate the new tape recorders and passenger experiments. Design life is 5 years (0.8 reliability after four years). New rigid 74 kg 5-panel 1.90 × 2.60 m Aerospatiale GSR-3 solar array will provide 2.5 kW BOL/equinox and 2.2 kW EOL (5 years)/Summer solstice, feeding four SAFT 40 Ah Ni/Cd batteries. 3-axis control to <0.03°, employing three RWs and two magnetorquers, Sodern STD-16 Earth horizon infra-red sensors, Sun sensors and Sagem's Regys 10 gyro orbit and yaw adjustment by four pairs of 15.6 N thrusters; roll/pitch by four pairs of 3.5 N thrusters supplied from four 75 kg hydrazine tanks. Control system based on radiation hardened microprocessor: Central Communication Unit (that is, main computer) incorporates a Fairchild 9450 microprocessor, providing 128,000 words redundant RAM. Telemetry rate at 4,096 bit/s; up 2.025–2.120, down 2.200–2.290 GHz. Two 121 kg 50 Mbit/s 120 Gbit Schlumberger redundant tape recorders store up to 40 min imagery each (175 W required in record mode, 260 W replay, 12 W standby; 70,000 on/off cycle specification). Imagery can be encrypted. VGT has its own recorder and telemetry system: 2.25 Gbit SSR (3 Mbit/s replay at 8.153 GHz); real-time at 1.704 GHz. Spot 4/Helios 1 controlled from the new Fresnel building at Toulouse
Sensors: HRVs paired as in Spot 1-3 are upgraded to HRVIRs (High-Resolution Visible Infra-Red) by the addition of a mid-infra-red band for vegetation water registration; independent and simultaneous observing schedules possible. The new wide-angle VGT is optimised for monitoring of vegetation, crops and geosphere/biosphere interactions

HRVIR Paired instruments with spectral ranges (in µm): Band B1 0.50–0.59; Band B2 0.61–0.68; Band B3 0.79–0.89; SWIR 1.580–1.750; yielding 20 m resolution in spectral mode. In monospectral mode, covering 0.61-0.68, resolution is 10 m (the previous panchromatic band is replaced by sampling the B2 band at 10 m or 20 m resolution). Data rate 25 Mbit/s. Viewing characteristics as Spot 1–3. See Sodern's entry in the Space Industry GN&C section for detailed imager information

VGT bands as HRVIR B2, B3, SWIR but with B1 replaced by B0 0.43-0.47 µm for atmospheric corrections. Swath 2,200 km (101° FOV), allowing daily global coverage
Passenger payloads: DORIS (Doppler Orbitography and Radiopositioning Integrated by Satellite), improved for precise (1 m) onboard real-time autonomous orbit determination and distribution of timing pulse; 10 kg, 14 W

Vega Radar Transponder for calibration of Kourou's ground radars and operator training.

PASTEC (Passager Technologique), a group of experiments for studying the orbital environment.

PASTEL (Passager Spot de Telecommunication Laser), demonstration of optical 50 Mbit/s datalinks via GEO satellites as part of ESA's Artemis SILEX programme. ESA's Artemis satellite will be used to relay HRVIR images.

POAM III Polar Ozone & Aerosol Measurement instrument carried for USAF Space Test Program.

Spot 5

Contractors: MMS France (prime, assembly, integration, test, onboard data management, AOCS, inertial units, Sun/star sensors), Aerospatiale (VGT, mechanical subsystem, thermal control, solar array), ETCA (electrical power supply), Saab (central communications unit), Alcatel Espace/IBM (solid-state recorders)
Resolution: HRG 5 m PAN, 10 m XS, 20 m SWIR; VGT 1 km/4 km
Launch: 4 May 2002 by Ariane 5 from Kourou, French Guiana
Mass: 2,600 kg launch
Orbit: 825 × 826 km, 98.8°, 101.4 min period
Spacecraft: based on Spot 4/Helios 1 design. Design life 5–7 years. 3-axis control. Two 100 Gbit SSRs. 150 Mbit/s real time/storage
Sensors: HRG three 270 kg High-Resolution Geometry cameras operating independently and simultaneously were planned, but there will now be two nadir-pointing (losing the simultaneous stereo pairs). The PAN band is as Spot 1-3 but at 2.5–5 m resolution; the visible/NIR B1/B2/B3 are as Spot 4, but at 10 m resolution; the 20 m SWIR band for vegetation monitoring is as Spot 4. For three cameras, they would be sighted at nadir, 19.2° fore, 19.2° aft. 60 km nadir swath + ±27° cross-track tilting (as on Spot 1–4) provides swath widths up to 160 km for two cameras operating in conjunction, and 240 km for three. Three operating modes: multi-spectral, 60–240 km swath, 10 m B1/B2/B3 + 20 m SWIR; high resolution, 60–240 km swath, 5 m PAN + 10 m B1/B2/B3; stereo, 60–80 km swath, 5 m PAN + 10 m B3 VGT as Spot 4
Passenger payloads: DORIS as Spot 4, PASTEC as Spot 4.

Germany

TerraSAR-X

Current Status

The TerraSAR-X radar imaging satellite was launched from Baikonur on 15 June 2007.

Background

TerraSAR-X is high-resolution radar satellite operating in the X-band (9.65 GHz). The mission's objectives are the provision of high-quality, multi-mode, Synthetic Aperture Radar (SAR) data for scientific research and applications, as well as the establishment of a commercial EO market for SAR products.

It is the first German satellite to be developed as the result of a public-private partnership. The

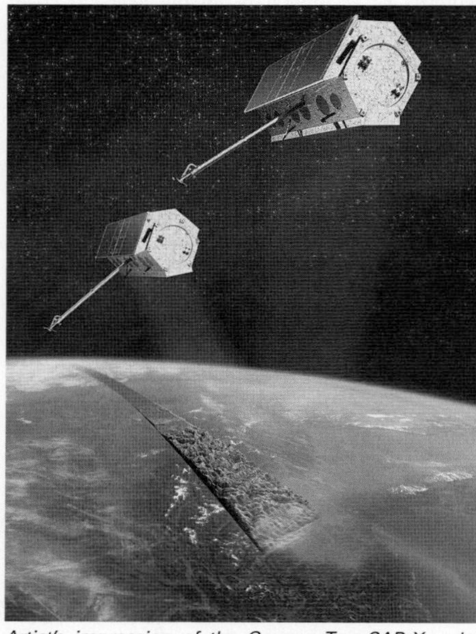

Artist's impression of the German TerraSAR-X radar imaging satellite, launched 15 June 2007 (DLR) 1340718

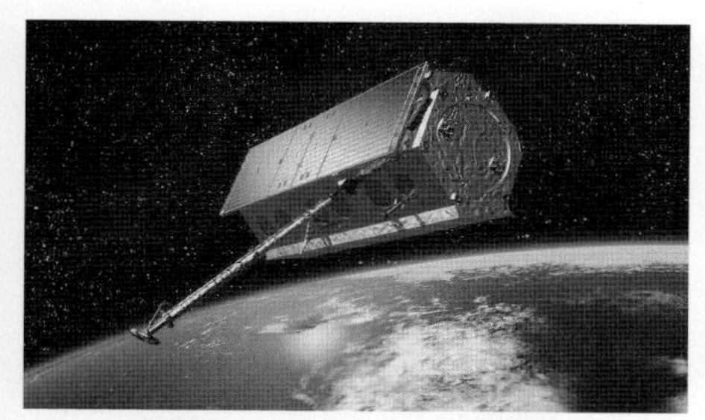

Artist's impression of the German TerraSAR-X and TanDEM-X radar imaging satellites flying in close formation. TerraSAR-X was launched 15 June 2007 and Tandem-X is due to be launched in 2009 (EADS Astrium) 1340719

partners are the German Ministry of Education and Science (BMBF), the German Aerospace Centre (DLR) and EADS Astrium GmbH. DLR is responsible for the satellite control system and the payload ground segment for receiving, processing, archiving and distribution of the SAR data. DLR is also responsible for instrument calibration, operations and scientific use of the TerraSAR-X data. Astrium developed and built the satellite under contract to DLR. Astrium also set up a distribution system for commercial use of the TerraSAR-X data and products. Distribution and commercial operations are the responsibility of Infoterra GmbH.

TerraSAR-X carries a high frequency SAR sensor that can be operated in different modes (resolutions) and polarisation. The radar sensing system emits multiple short bursts of microwave signals. The echoes from the Earth's surface are received by the satellite's antenna and stored on board. After complex digital processing, radar images with a resolution of up to one metre can be generated. Mission life is a minimum of 5 years.

The satellite follows a circular, sun-synchronous, dawn-dusk orbit with a local time of ascending node at the equator of 18:00 (± 15 min). Nominal revisit period is 11 days and ground track repeatability is within ± 500 m per repeat cycle. TerraSAR-X can image anywhere on Earth within 4.5 days max., and 90% of the surface within 2 days.

The satellite is based on the successful X-SAR/SIR-C (1994) and SRTM (2000) radar missions. The spacecraft design, based on an AstroSat-1000 bus, is mainly driven by the SAR instrument, the body mounted solar array, and the size of the Dnepr-1 launcher fairing. The bus features a central hexagonal CFRP structure. The sun-facing side carries the solar array, while the SAR antenna is mounted on one nadir-facing side. The other nadir looking side carries the S-band TT&C antenna, a SAR data downlink antenna – deployed on a 3.3 m boom to avoid RF interferences during simultaneous radar imaging and data transmission to ground – and a Laser Retro Reflector to support precise orbit determination. The space-facing side is used for the Laser Communication Terminal and as a thermal radiator.

In flight attitude, the SAR antenna points 33.8° off nadir. The sensor has three different operation modes: the "Spotlight" mode with 10 x 10 km ;es at a resolution of 1 – 2 m; the "Stripmap" mode with 30 km wide strips at a resolution of 3 – 6 m; and the "ScanSAR" mode with 100 km wide strips at a resolution of 16 m. TerraSAR-X also supports the reception of interferometric radar data for the generation of digital elevation models.

Science users are provided with their data directly from the DLR Ground Segment, while Infoterra GmbH deals with the commercial customers. Most of the TerraSAR-X data are transmitted to DLR's central receiving station in Neustrelitz. Data can also be downloaded to ground stations around the world that are operated by Direct Access Partners (DAP) and Direct Access Customers (DAC) of Infoterra.

On 30 August 2006, DLR and EADS Astrium GmbH signed a contract to build the TanDEM-X satellite. TanDEM-X ('TerraSAR-X add-on for Digital Elevation Measurement') is almost identical to TerraSAR-X. The pair will orbit in close formation, completing a survey of the entire land surface within three years. One goal will be the production of a global Digital Elevation Model of unprecedented accuracy, with elevation data accurate up to 2 m. Launch of TanDEM-X is due in 2009.

Specifications

Launch: 02:14 GMT on 15 June 2007 by a Dnepr-1 from Baikonur LC109.
Orbit: 499 x 512 km, 97.5°
Launch mass: 1,346 kg
Body dimensions: 5 x 2.4 m
Command and Data Handling: Data storage 256 Gbit. Data transmission 300 Mbit/s. X-Band (9.65 GHz) downlink. Integrated Control and Data System Electronics core consists of two redundant 32 bit processor modules, implementing the ATMEL ERC32SC (Embedded Real-time computing Core – 32 bit Single Chip) processor, with a processing performance of more than 18 MIPS.
Telecommunications: S-band TT&C antenna. The German Space Operations Center (GSOC) of DLR in Oberpfaffenhofen controls the satellite from DLR's ground station in Weilheim.
Power: Single solar array 5.25 m^2, triple-junction GaAs solar cells, average orbital power 800 W EOL.
Attitude control: Reaction wheels for fine-pointing, with magnetorquers for desaturation. Propulsion system also capable of attitude control in order to achieve rapid rate damping during initial acquisition. Attitude measurement is performed with a GPS/Star Tracker system (MosaicAODS) during nominal operation and a CESS (Coarse Earth and Sun Sensor) in safe mode. An Inertial Measurement Unit and a magnetometer support rate measurements. In fine pointing mode, a pointing accuracy of 65 arcsec is achieved.
Propulsion: Monopropellant hydrazine blow-down mode propulsion system for orbit maintenance and safe mode attitude control. Propellant mass 78 kg.
Payload: Mass 394 kg.

SAR Antenna 4.8 m × 0.7 m × 0.15 m utilising 384 transmit/receive modules. Resolution 1 m for a 5 × 10 km scene.

Carrier frequency – 9.65 GHz; wavelength – 3.1 cm, PRF between 3.0 and 6.5 kHz; chirp bandwidth range 5–300 MHz, 150 MHz nominal, 300 MHz experimental; ADC sampling rates 330 MHz, 165 MHz, 110 MHz; raw data characteristics 8-bit quantisation, BAQ compression to 4,3 or 2 bits or bypass; transmit duty cycle 13 – 20%; radiated peak output power 2260 W, system noise figure 5 dB; phased array antenna with azimuth steering capability; look direction right-looking nominal, left-looking experimental, mechanical offset angle (off nadir) – 33.8°.

Tracking, Occultation and Ranging Instrument provided by the Geoforschungszentrum Potsdam (GFZ) and the Center of Space Research of the University of Texas.

Laser Communication Terminal (LCT) provided by DLR and manufactured by TESAT in Germany.

India

IRS (Indian Remote Sensing) series

Current Status

In March 2008, the operational IRS system included IRS-1D, Oceansat-1, Resourcesat-1, Cartosat-1 and Cartosat-2. The Technology Experiment Satellite (TES) is also regarded as part of the system.

Background

The Indian Remote Sensing (IRS) satellite system, India's first domestic satellite programme dedicated to Earth resources studies, is an element of the National Natural Resource Management System. Between 1988 and 2007 India launched 12 polar orbiting satellites in the IRS series, including the Technology Experiment Satellite (TES). These satellites put India in a position to satisfy its own applications needs, but also to dominate a significant portion of the global commercial market. The spacecraft are designed and built at ISAC (ISRO Satellite Centre), Bangalore, India, with contributions from SAC (Space Applications Centre), Ahmedabad.

EOSAT and ISRO signed an agreement 21 October 1993 for the US company to distribute IRS data globally. EOSAT began receiving IRS imagery at its Norman, Oklahoma station on a daily basis in June 1994. EOSAT and the Indian government signed an agreement 2 February 1995 for exclusive worldwide marketing rights for IRS data. EOSAT anticipated that, within five years, the majority of its new imagery would be from IRS.

The country launched the first in the series, the Rs 650 million **IRS-1A** by a Soviet Vostok booster in March 1988, the first commercial satellite launched by the organisation Glavkosmos. All three imaging systems remained operational from 7 April 1988 until ISRO retired 1A from routine service 17 March 1995. ISRO signed a INR220 million contract for the

IRS-1C undergoes dynamic balancing. The large drum-shaped PAN camera is at top right; WiFS and MOS at top left (IRS) 0517417

identical **IRS-1B** in November 1988 for an August 1991 launch.

IRS-1C 's December 1995 launch followed a contract signed on 18 January 1991. IRS-1C was the first Russian launch into a retrograde orbit using the Molniya M four stage launcher. **IRS-1D,** launched 29 September 1997, was to have replaced IRS-1C but it was left in a 300 × 823 km orbit by the PSLV launcher and remained in an elliptical sun-synchronous orbit.

1C/1D offered improved spatial and spectral resolution, onboard recording, stereo viewing capability and more frequent revisits, making them the world's most advanced civil remote sensing satellites. The VNIR resolution was about 20 m, the SWIR about 70 m and the PAN <10 m. IRS-1C was the first commercial satellite to provide imagery with sub-10 m resolution and regular repeat coverage. SWIR provided data on water stress and pest infestation. The WiFS Wide Field Sensor with 180 m resolution monitored vegetation. All three cameras returned their first images on 5 January 1996. Outside observers estimated the launch cost India INR500 million.

The 846 kg **IRS-1E**, 1A's refurbished engineering model, was lost on PSLV-D1 on 20 September 1993. It was carrying Germany's MEOSS Monocular Electro-Optical Stereo Scanner (as flown on the failed SROSS 2) and India's LISS 1 camera and CO_2-band Earth radiance monitor.

The demonstration **IRS-P2** , using the 1C/1D bus, was launched on PSLV number 2 in October 1994. It was declared operational on 7 November 1994. It carried a LISS 2 camera and during the first year, it returned more than 60,000 images of India, in addition to EOSAT imagery received at two US stations.

IRS-P3, launched by PSLV in March 1996, carried a German Modular Opto-electronic Scanner (MOS) and an improved wide-field imager. IRS-P3 covered four major applications: ocean chlorophyll analysis, vegetation assessment, snow studies and geological mapping for identifying prospective mineral sites.

Oceansat-1 (**IRS-P4**), launched by PSLV-C2 in May 1999, was designed to study many physical and biological aspects of the oceans. It was the first time an Indian launch vehicle had carried more than one payload. Oceansat carries an Ocean Colour Monitor (OCM) and a Multi-frequency Scanning Microwave Radiometer (MSMR). The OCM has 360 m spatial resolution and covers a 1,420 km swath. With a repeat cycle of two days, Oceansat provided global oceanographic and meteorological data on sea surface temperature, sea surface height, wind speed and such.

The MSMR instrument was shut down in 2003. The OCM was still intermittently operational in 2007, but only 30 per cent of the nominal power was available for spacecraft and payload operations after eight solar array string failures since launch. The spacecraft operations team introduced new procedures to generate maximum power by orienting the spacecraft towards the sun in inertial mode and returning it to an Earth pointing orientation for periodic observations. From this mission onwards, ISRO replaced the IRS-P designations in favour of specific names.

Resourcesat-1 (IRS-P6) was launched by PSLV-C5 in October 2003. It was intended to provide continuity to the services provided by IRS-1C and IRS-1D, which had far outlived their design lives, as well as providing imagery with better spatial resolution and additional spectral bands. It carries three cameras. The Linear Imaging Self Scanner (LISS-4) operates in three spectral bands in the Visible and Near Infrared Region (VNIR) with 5.8 m spatial resolution. The medium resolution LISS-3 operates in three spectral bands in VNIR and one in Short Wave Infrared (SWIR) with 23.5 m spatial resolution. The Advanced Wide Field Sensor (AWiFS) operates in three spectral bands in VNIR and one band in SWIR with 56 m spatial resolution. A combined swath of 730 km is achieved through two AWiFS cameras. Resourcesat-1 has a Solid State Recorder to store the images.

Cartosat-1 (**IRS-P5**), launched by PSLV in May 2005, carries two panchromatic cameras of PAN heritage, similar to those flown on IRS-1C/D. The cameras are mounted with a tilt of +26° and –5° along the track with respect to nadir to provide stereo pairs of images that can be used to generate Digital Terrain Model (DTM)/Digital Elevation Models (DEM). Spatial resolution is 2.5 m and swath width 30 km. The orbital revisit cycle is 116 days, but a revisit capability of five days is provided by body-pointing the spacecraft about its roll axis by ±26°. The data are used for cartography, land use and GIS applications.

Cartosat-2, launched in January 2007, carries a panchromatic camera that takes black and white pictures in the visible spectrum. The swath is 9.6 km and spatial resolution is less than one metre. The satellite can be steered up to 45° along, as well as across, its track and is capable of providing scene-specific spot imagery. New technologies developed for the mission include: two mirror-on-axis camera using Carbon Fibre Reinforced Plastic (CFRP) based electro-optic structure with 12,000 CCD, large size lightweight zerodur mirrors, indigenous JPEG-like data compression with rate regulation, advanced solid state recorder, bus management unit housing telemetry, tracking and command, attitude and orbit control system, sensor processing, thermal management, power monitoring, high torque reaction wheels, high performance DTG based IRU and high performance star sensors.

The Cartosat-2 data are used for detailed mapping and other cartographic applications, urban and rural infrastructure development and management, as well as applications in land information systems and Geographical information systems. The first images were received on 12 January 2007 and the satellite was declared operational at the end of July 2007. The spacecraft is operated by ISTRAC (ISRO Telemetry, Tracking and Command Network) of Bangalore, using its network of stations at Bangalore, Lucknow, Mauritius, Bearslake in Russia and Biak in Indonesia. The National Remote Sensing Agency (NRSA) at Hyderabad is the processing centre for CartoSat data.

Cartosat-2A has been described as India's first dedicated military satellite – reports refuted by ISRO officials. It was scheduled to launch with the Israeli Polaris/TecSAR satellite on a PSLV in August 2007, but the launch took place without Cartosat-2A due to "technical issues". The satellite is reported to be identical to Cartosat 2 with a panchromatic camera that provides black and white imagery with <1 m spatial resolution.

Oceansat-2 is expected to replace Oceansat-1. It will carry an Ocean Colour Monitor (OCM) and a Ku-band pencil beam scatterometer, plus a Radio Occultation Sounder for Atmospheric Studies (ROSA), developed by the Italian Space Agency (ASI). The satellite will be used for identification of potential fishing zones, sea state forecasting, coastal zone studies and provide inputs for weather forecasting and climatic studies. It is scheduled for launch by PSLV into a near polar, sun-synchronous orbit of 720 km. This orbit, combined with the wide swath of both payloads, will provide an observational repetitivity of two days.

The Oceansat-2 OCM is an eight-band multi-spectral camera operating in the visible – near infra-red spectral range. This camera provides an instantaneous geometric field of view of 360 m covering a swath of 1,420 km. There is provision to tilt OCM by + 20° in the along-track direction. The Ku-band pencil beam scatterometer is an active microwave radar that operates at 13.515 GHz, providing a ground resolution cell of 50 km × 50 km. It consists of a parabolic dish antenna of 1 m diameter which is offset mounted with respect to the yaw axis (Earth-viewing axis). This antenna is continuously rotated using a scan mechanism with the scan axis along the positive Yaw axis. The back scattered power in each beam from the ocean surface is measured and used to derive the wind vector.

The 1,750 kg **Radar Imaging Satellite (RISAT)** is a planned microwave remote sensing mission with Synthetic Aperture Radar (SAR) operating in C-band and having a 6 × 2 m planar active array antenna based on trans-receiver module architecture. The SAR will have a spatial resolution of 3–50 m and a swath of 10–240 km. With all-weather remote sensing capability, RISAT will enhance remote

Resourcesat-1 was the tenth in the Indian Remote Sensing satellite (IRS) series. It was the most advanced remote sensing satellite yet built by ISRO (ISRO) 1168183

A three-dimensional view of Jaipur, India, generated from Cartosat-1 data (ISRO) 1343659

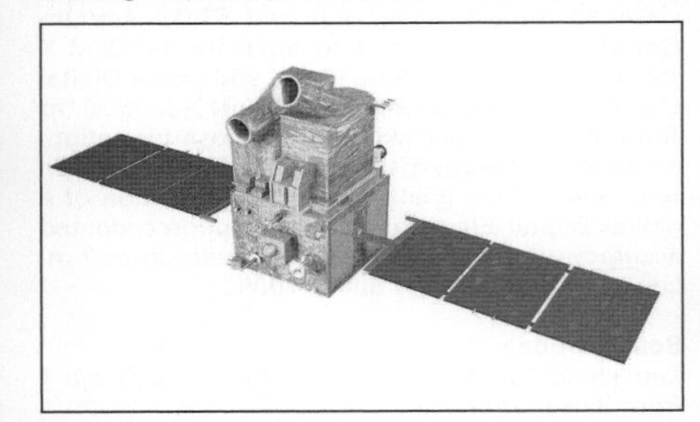

Artist's impression of Cartosat-1. PSLV-C6 launched Cartosat-1 and HAMSAT on 5 May 2005 (ISRO) 1168182

(2) The Panchromatic camera (PAN) on Cartosat-2 provides images with a spatial resolution of better than one metre (ISRO) 1343660

sensing applications in the areas of agriculture and disaster management.

Resourcesat-2 will have imaging sensors similar to Resourcesat-1. Payload electronics are being miniaturised to reduce the overall weight. The PSLV launch is planned for 2009–10.

Specifications

IRS-1A

Launch: 17 March 1988 by Russian SL-3 Vostok from Baikonur
Mass: 975 kg in orbit at beginning of 3 year life
Contractors: ISRO Satellite Centre (prime), Hindustan Aeronautics (structure), and ISRO Space Applications Centre (imaging system)
Resolution: 72.5 m LISS 1, 36.25 m LISS 2
Orbit: 867 × 913 km (aiming for 904 km circular), 99.03° sun-synchronous, crossing equator on descending node at 10.25 local time (allowed to drift to 10.10 after 2 years), with 22 day repeat cycle (307 revs) and 2,872 km ground track separation at equator.
Spacecraft: Box-shaped 1.6 × 1.56 × 1.1 m bus with two Sun-tracking solar wings totalling 8.58 m², providing 709 W EOL. Two Ni/Cd 40 Ah batteries provide eclipse power. 3-axis control with a 0.3° pitch/roll + 0.5° yaw pointing accuracy provided by zero-momentum RW system utilising Earth/Sun/star sensors + gyros; 1 N hydrazine thrusters (80 kg loaded) for AOCS + momentum dumping. Real time LISS 2A/B data downlinked at 10.4 Mbit/s each to the 10 m dish at Shadnager on 20 W X-band + LISS 1 data on 5 W S-band at 5.2 Mbit/s. No onboard recorder. Satellite control exercised from Bangalore, with Lucknow station providing backup
Sensors: Linear Imaging Self-Scanning (LISS). Three push-broom CCD units operate in four bands compatible with Landsat TM and Spot HRV. Band ranges (μm) and applications: Band 1 0.45–0.52 (coastal environment, soil/vegetation); Band 2 0.52–0.59 (vegetation vigour, rock/soil discrimination, turbidity, bathymetry); Band 3 0.62–0.68 (chlorophyll absorption, plant species); Band 4 0.77–0.86 (delineation of water features, land forms).

LISS 1. Four 2,048-element linear CCD imagers with spectral filters; total unit mass 38.5 kg. Focal length 162.2 cm, FOV 9.4°, generating resolution of 72.5 m over 148 km swath, framing LISS 2 image pairs. Data transmitted on S-band at 5.2 Mbit/s

LISS 2A/B. Eight 2,048 element linear CCD imagers with spectral filters. Focal length 324.4 mm, FOV 4.7° each, 36.25 m resolution over 74 km swath. The two 80.5 kg units are positioned either side of LISS 1 and view either side of the ground track with a 3 km lateral overlap; two pairs thus cover a single LISS 1 frame. Data transmitted on X-band at 10.4 Mbit/s each.

IRS-1B

Launch: 29 August 1991 by Russian SL-3 Vostok from Baikonur. Declared operational 16 September 1991
Orbit: 857 × 919 km, 99.25° sun-synchronous.

Other specifications as for IRS-1A.

IRS-1C

Launch: 28 December 1995 by Russian SL-6 Molniya from Baikonur
Orbit: 817 km 98.6° sun-synchronous, crossing equator on descending node at 10.30 local time, with 24-day repeat cycle (341 revs) for LISS 3 and 5 days for PAN/WiFS; 2,820 km ground track separation at equator
Mass: 1,250 kg in orbit at beginning of 3 year life
Contractors: ISRO Satellite Centre (prime), Hindustan Aeronautics (structure), ISRO Space Applications Centre (imaging system), Thomson Tubes Electroniques (40 W X-band TWTA), ISRO Inertial Systems Unit for momentum/reaction wheels
Resolution: 23.5/70.5 m LISS 3, 5.8 m PAN, 188 m WiFS
Configuration: Box-shaped 1.6 × 1.56 × 1.1 m bus with two solar wings. 3-axis control with a 0.15° pitch/roll + 0.2° yaw pointing accuracy provided by a zero-momentum RW system (four 5 Nm wheels) utilising Earth/Sun/star sensors and gyros. Propulsion system employs 1/11 N hydrazine thrusters for orbit control, attitude manoeuvres and momentum dumping. Lockheed 62 Gbit tape recorder
Communications: Satellite control exercised from Bangalore, with Lucknow station providing back-up. Recorded data downlinked on the 40 W 8 GHz X-band link
Sensors: Three cameras all utilise push-broom CCD units, continuing and expanding the IRS 1A/B imagers.

LISS 3. Similar to LISS 1/2 but replaces one visible band with Short Wave IR (SWIR). 85 W power operating. Three 6,000-element linear visible CCD imagers with spectral filters for bands 2,3 and 4. 23.5 m resolution, 142 km swath width. Data rate 35.70 Mbit/s. One 2,100-linear SWIR InGaAs CCD imager cooled to –10°C for band 5 (1.55–1.70 μm). 70 m resolution, 148 km swath width. Data rate 2.02 Mbit/s.

PAN. Single-band panchromatic (0.50–0.75 μm), 5.8 m resolution, three linear CCDs with 23.9 km swath widths combined for 70 km width. Three off-axis mirrors with focal plane splitting by isosceles prism. Swath steering of ±398 km by Payload Steering Mechanism with ±26° steerability (entire camera rotates, yielding better S/N + MTF). 0.2°PSM step yields 2.57 km nadir. 55 W power operating. Data rate 84.903 Mbit/s

WiFS. Wide Field Sensor similar to LISS 1 of IRS 1A/B. Four 2,048-element linear CCDs, dual band (0.62–0.68, 0.77–0.86 μm), focal length 56.420 mm, 188 m resolution, 774 km swath width (yielding 5-day repeat cycle), FOV ±27° (±13.5° each imager), data rate 40.43 Mbit/s.
Power: Two sun-tracking solar arrays (three 1.1 × 1.46 m panels) totalling 9.6 m², providing 830W EOL. Two Ni/Cd 21 Ah batteries provide eclipse power.

IRS-1D

Launch: 28 September 1997 by PSLV from Sriharikota, India. Suffered fourth stage malfunctions that left 1D in 308 × 822 km orbit rather than the planned circular orbit at 820 km.

Other specifications as for IRS-1C.

IRS-1E (P1)

Launch: 20 September 1993 by PSLV from Sriharikota but destroyed in ascent failure.

IRS-P2

Launched: 15 October 1994 on PSLV-D2, from Sriharikota, India
Mass: 804 kg at launch
Orbit: 817 km, 98.7° sun-synchronous, crossing equator on descending node at 10.40 am, with 24-day repeat cycle (341 revs). The first IRS with a frozen perigee to minimise scale variations.
Configuration: As 1C, with 6.42 m² 4-panel wing providing 510 W (with two 21 Ah Ni/Cd batteries). Data rate 2 × 10.4 Mbit/s
Payload: LISS 2. Single camera of 1A/1B design except each band carries staggered CCDs for one unit to provide same coverage as two LISS 2 previously. 131 km swath width.

IRS-P3

Launch: 21 March 1996 by PSLV from Sriharikota, India
Mass: 922 kg at launch

Orbit: 819 × 821 km, 98.8° sun-synchronous, crossing equator on descending node at 10.30 am local time, with 22 day repeat cycle.
Sensors: WiFS. Same as 1C but with added 1.5–1.7 µm SWIR for determination of vegetation dynamics. In conjunction with 1C, provided 2–3 day coverage of India.

MOS. German (DLR) Modular Opto-electronic Scanner optimised for oceanography. 18-channel VNIR, 248 km swath. MOS-A: 0.7567, 0.7606, 0.7635, 0.7664 µm, resolution 2,520 × 2,520 m; MOS-B: 0.408, 0.443, 0.445 0.485, 0.520, 0.615, 0.650, 0.685, 0.750, 0.815, 0.870, 1.010 µm, resolution 720 × 580 m; MOS-C: 1.600 µm, resolution 720 × 1,000 m.

X-ray astronomy payload (XAP) for time variability and spectra of sources and detection of X-ray transients. Three Pointed-mode Proportional Counters 2–20 keV. XSM X-ray Sky Monitor 3–6 keV pinhole camera (PSPC). The Earth imagers had no onboard storage; real-time imagery was transmitted to one Indian and two German stations. XSM operated continuously, but PPC required P3 to operate in sky-pointing mode, precluding Earth observations.

Oceansat-1 (IRS-P4)
Launch: 26 May 1999 on PSLV-C1 from Sriharikota, India
Orbit: 720 km circular; 98.28° inclination; period 99.31 min., repeat cycle 2 days, equator crossing time at 12:00 hours on descending node.
Payload: Ocean Colour Monitor (OCM) operates in 402–422, 433–453, 480–500, 500–520, 545–565, 660–689, 745–785 and 845–885 nm bands with 360 metre spatial resolution and 1420 km swath. IFOV 250 m VNIR, 500 m SWIR. +20/0/–20° along track steering, 17.35 Mbit/s, 134 W, 78 kg.

MSMR (Multi-frequency Scanning Microwave Radiometer) 6.6/10.6/18/21 GHz, 3 dB beam widths 4.2/2.6/1.6/1.4°, <1.0 K T resolution, 120/75/45/40 km resolution over 1,500 km swath, 86 cm diameter antenna. 5 kbit/s, 76 W, 65 kg
Configuration: IRS-1C/1D, and IRS-P3 heritage box-shaped bus, aluminium/aluminium honeycomb structure with CFRP elements. 1.8 m × 1.98 m × 2.57 m; length 11.67 m fully deployed. 3-axis stabilised using Earth sensors, sun sensors, magnetometer, reaction wheels, magnetic torquers and hydrazine thrusters (1N & 11N). RF system X-band (8,350 MHz), data handling 20.8 Mbit/s. Onboard data recorder.
Mass: 1,050 kg at launch
Power: Twin 9.6 m² sun-tracking solar arrays generating 800 W (EOL). Two 21 Ah Ni-Cd batteries.
Design life: 5 years.

Resourcesat-1 (IRS-P6)
Launch: 17 October 2003 on PSLV C5 from Sriharikota, India
Orbit: 817 km, 98.7° sun-synchronous, period 101.35 min. crossing equator on descending node at 10.30 am local time, with 24 day repeat cycle (LISS-3) and 5 day revisit period.
Mass: 1,360 kg at launch.
Configuration: Box-shaped bus of IRS-1C/1D and -P3 heritage, 2 m × 2.1 m. Two-tier payload module: payload module deck accommodates LISS-3, AWIFS-A and AWIFS-B camera modules; LISS-4 camera mounted on a rotating deck which is attached to a PSM (Payload Steering Motor) which can rotate by ±26°. Payload module attached to CFRP monocoque cylinder, which is attached to main cylinder of bus through a strut assembly. 3-axis stabilised using reaction wheels, magnetic torquers and hydrazine thrusters. RCS (Reaction Control System) with 12 nozzles, eight with 1 N thrusters, four with 11 N thrusters. Solid state data recorder. TT&C communications in S-band with downlink data rates of 1-16 kbit/s, uplink frequency 2071.875 MHz, downlink frequency: 2250 MHz. Payload data transmitted in X-band at rate of 105 Mbit/s. LISS-4 data transmitted on carrier-1 at 8,125 MHz and LISS-3 + AWiFS data on carrier-2 at 8,300 MHz.
Payload: Linear-Imaging Self-Scanner (LISS-4) operating in three spectral bands in the visible and near infrared region (VNIR) with 5.8 m spatial resolution. Steerable up to ±26° across track to obtain stereoscopic imagery and achieve five-day revisit capability.

Medium resolution LISS-3 operating in three spectral bands in VNIR and one in Short Wave Infrared (SWIR) band with 23.5 m spatial resolution and 142 km swath.

Advanced Wide Field Sensor (AWiFS) operating in three spectral bands in VNIR and one band in SWIR with 56 m spatial resolution.
Power: Twin solar arrays generating 1.25 kW (EOL). Two 24 Ah Ni-Cd batteries.
Design life: 5 years

Cartosat-1 (IRS-P5)
Launch: 5 May 2005 by PSLV-C6 from Sriharikota, India
Orbit: 618 km, 98.87°, sun-synchronous, period 97 min. crossing equator on ascending node at 10.30 am local time, with 126 day repeat cycle and 5 day revisit period.
Mass: 1,560 kg at launch.
Configuration: Based on IRS-1C/1D and P3 bus. 2.4 m × 2.7 m (body dimensions). 3-axis stabilised using star sensors in loop, magnetic bearing reaction wheels in tetrahedral configuration, 16 nozzles with 1 N thrusters, 4 nozzles with 11 N thrusters. Pointing accuracy ±0.05° in all axes. GPS receiver for orbit determination. Solid-state data recorder, 120 Gbit capacity. Payload data transmission: 2 X-band carriers, one for each camera, QPSK modulated, single polarised, at 105 Mbit/s.
Payload: Two Pan cameras. PAN-F (Panchromatic Forward-pointing Camera), fixed forward tilt of 26°. PAN-A (Panchromatic Aft-pointing Camera), fixed aft tilt of –5°. Each has spectral range of 0.5–0.85 µm, spatial resolution 2.5 m, swath width 30 km. Each 1.5 m × 8.5 m × 1 m, mass 200 kg. Linear CCD detector array of 12,288 pixels. Three-mirror off-axis reflective telescope with concave hyperboloidal primary mirror and concave ellipsoidal tertiary mirror made from Zerodur glass.
Power: Twin 15 m² solar arrays generating 1.1 kW (EOL). Two 24 Ah Ni-Cd batteries.
Design life: 5 years

Cartosat-2
Launch: 10 January 2007 by PSLV from Sriharikota, India
Orbit: 635 km, sun-synchronous, 97.92°, period 97.4 min. crossing equator on descending node at 9.30 am local time, with 310 day repeat cycle and 4 day revisit period.
Mass: 680 kg at launch
Configuration: Hexagonal, lightweight, IRS generic bus. 3-axis stabilised using high-torque reaction wheels, magnetic torquers and hydrazine thrusters. Pointing accuracy ±0.05° in all axes, attitude knowledge 0.01°, stability (attitude drift) is 0.0003°/s, and the ground location accuracy is 100 m. The satellite is very agile providing a body-pointing capability in along-track and cross-track of up to ±45°. Two solar arrays. Dual gimballed antenna. Improved IRU (Inertial Reference Unit), high performance star sensor. Eight-channel GPS receiver. Image data downlinked in X-band (8,125 MHz), 105 Mbit/s. TT&C data link in S-band (2,067 MHz). Solid state data recorder, 64 GB capacity.
Payload: PAN panchromatic camera, <1 m resolution, swath width 9.6 km, spectral band 0.5–0.85 µm. CFRP-based mirrors, on-axis telescope.
Power: Twin solar arrays with advanced triple junction solar cells generating 900 W (EOL). Two 18 Ah Ni-Cd batteries.
Design life: 5 years

Kalpana (MetSat)

Current Status
Kalpana-1 was launched by PSLV from Sriharikota on 12 September 2002.

Background
Kalpana-1 (originally known as MetSat-1) is ISRO's first dedicated GEO weather satellite. Previous meteorological services had been combined with telecommunication and television services on the INSAT satellite series. Kalpana-1 is a precursor to the future INSAT system that will have separate satellites for meteorology and telecommunication and broadcasting services. This will enable larger capacity to be built into INSAT satellites, both in terms of transponders and radiated power, without the design constraints imposed by meteorological instruments.

In a commemorative ceremony on 5 February 2003, the MetSat series was renamed Kalpana by Indian Prime Minister Atal Bihari Vajpayee in

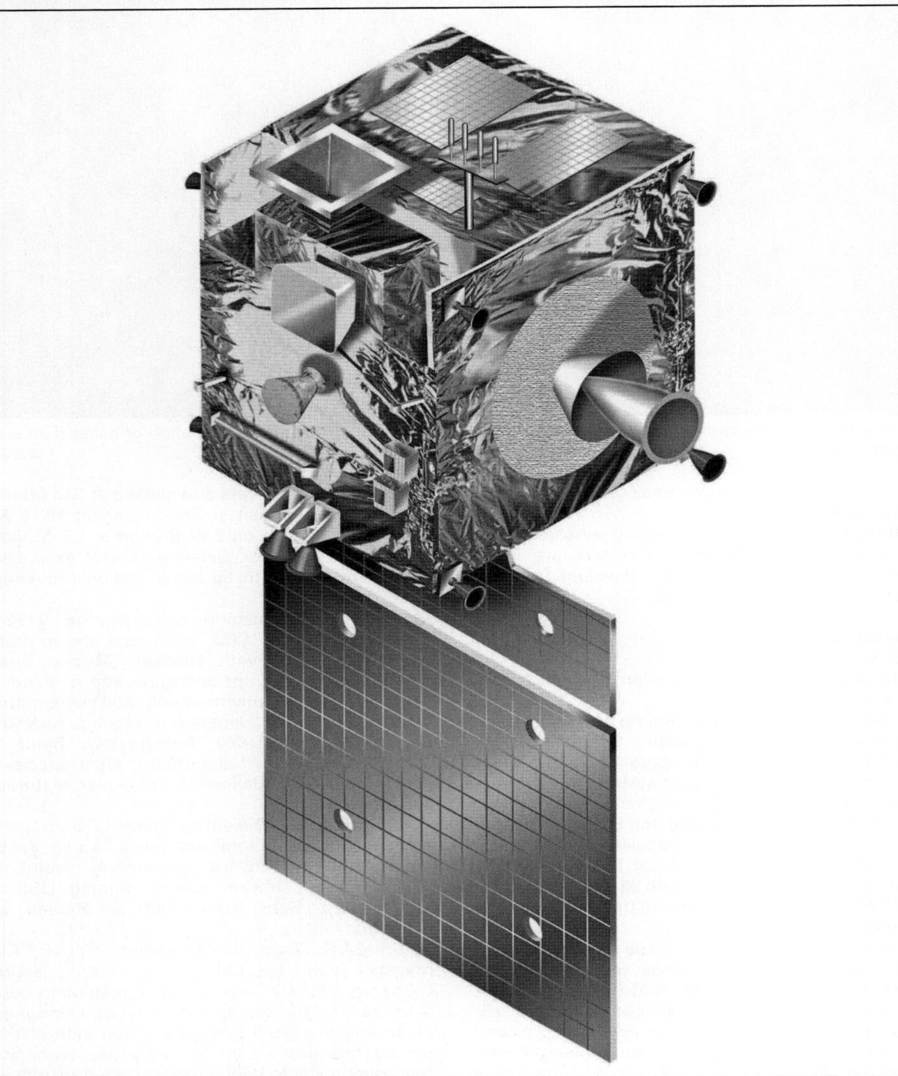

Kalpana-1 was India's first dedicated weather satellite (ISRO) 1328530

honour of Kalpana Chawla, born in Karnal, India, who died as a NASA astronaut on 1 February 2003 when Space Shuttle Columbia was destroyed during re-entry.

With a project cost of project cost of INR75 crores (~USD17 million), Kalpana-1 was launched by PSLV from Sriharikota on 12 September 2002. It was the first time that the PSLV had been used to take a satellite to GTO. Its solar array was automatically deployed immediately after injection into GTO. The deployment of the array and the general health of the satellite were monitored by a ground station of the ISRO Telemetry, Tracking and Command network (ISTRAC) located on the Indonesian island of Biak. The Master Control Facility (MCF) at Hassan in Karnataka took over control for all its post launch operations. Ground stations at Lake Cowichan (Canada), Fucino (Italy) and Beijing (China) supported the MCF in monitoring the health of the satellite and its orbit raising operations. Kalpana's orbit was raised to the final geostationary orbit by firing the satellite's Liquid Apogee Motor. The first orbit raising operation took place on 13 September, by command from MCF.

On 19 September 2002 the Very High Resolution Radiometer (VHRR) was successfully switched on and sent back its first full-frame Earth image in the visible spectral band. Images in the water vapour and infra-red spectral bands were returned on 20 September. The satellite spacecraft reached its designated orbital slot of 74°E on 23 September when detailed in-orbit testing of the payload began. Operational meteorological services commenced in the first week of October 2002.

Kalpana-1 was developed by ISRO Satellite Centre, Bangalore. The meteorological payloads were developed by the Space Applications Centre, Ahmedabad. The box-shaped satellite was designed using a new I-1000 spacecraft bus that used lightweight structural elements such as Carbon Fibre Reinforced Plastic (CFRP). It has a new, lightweight, planar array antenna to transmit data from the Very High Resolution Radiometer (VHRR) and Data Relay Transponder (DRT). Two high capacity magnetic torquers deal with all disturbances, including solar radiation pressure effects experienced in GEO.

Kalpana carries a VHRR that images the Earth in the visible, thermal infra-red and water vapour bands with spatial resolution of 2 km in VIS band and 8 km for the others. VHRR/2 is a modified version of VHRR instruments flown on INSAT-2A, 2B, and 2E. The instrument operates in three scanning modes: full frame mode (20° × 20°) covering the entire Earth disk in about 33 min; normal frame mode (14° North-South × 20° East-West) in about 23 min; sector frame mode (4.5° North-South × 20° East-West), which is particularly suited for rapid, repetitive coverage during severe weather conditions like a cyclone. It also carries a Data Relay Transponder (DRT) for collecting data from unattended meteorological platforms. Data sent by these platforms is relayed to the Meteorological Data Utilisation Centre at New Delhi.

Specifications

Launch: 12 September 2002 by PSLV from Sriharikota, India
Location: 74°E geostationary
Design life: 7 years
Mass: 1,055 kg at launch, including about 560 kg of propellant. 495 kg dry.
Configuration: 1.5 × 1.5 m box-shaped bus. 3-axis stabilised using gyros, Earth sensors, digital sun sensor, coarse analog sun sensors, solar panel sun sensor, magnetic torquers. One single, side-mounted solar array. Eight 22 N bipropellant thrusters for orbit and attitude control. One 440 N liquid apogee motor for orbit raising using MMH and MON-3.
Telecommunications: RF communication: C-band downlink frequency 4503.05 MHz at data rates of up to 526.5 kbit/s. The EIRP 18 dBW.
Power: One single panel 2.15 m × 1.85 m rotatable solar array of GaAs solar cells. 550 W of power (BOL). One 18 Ah Ni-Cd battery.
Payload: Very High Resolution Radiometer (VHRR/2) for imaging Earth in the visible, thermal infra-red and water vapour bands. VIS spectral band 0.55–0.75 µm; spatial resolution 2 × 2 km. TIR spectral band 10.5–12.5 µm; spatial resolution 8 × 8 km. MWIR (water vapour) spectral band 5.7–7.1 µm; spatial resolution 8 × 8 km.

Data Relay Transponder (DRT) for collecting data from unattended meteorological platforms. Uplink frequency 402.75 MHz; downlink 4,506.05 MHz; bandwidth 100 kHz; EIRP 21 dBW (min).

International

China-Brazil Earth Resources Satellite (CBERS)/Ziyuan series

Current Status

The third China-Brazil Earth Resources Satellite, CBERS-2B, was launched by Long March 4B on 19 September 2007. CBERS-1 ceased operations in August 2003, but CBERS-2 remained operational in September 2007. CBERS-3 and 4 will follow, in 2009 and 2011, respectively. Note that the name Ziyuan is also used by a series of Chinese Earth imaging satellites (see separate entry under Jianbing).

Background

An agreement calling for the development and launch of two remote sensing satellites was signed in November 1993, between China and Brazil's National Institute for Space Research (INPE). The project was 70 per cent funded by China and 30 per cent by Brazil. The Chinese name for the satellites is Ziyuan, meaning 'Resource'. The joint programme is called the China-Brazil Earth Resources Satellite (CBERS). The satellites are primarily used for monitoring changes in land use and natural resources, e.g. estimation of forest reserves, crop land, damage due to floods and earthquakes; and environmental pollution.

The CBERS satellite consists of a payload module and a platform. The main body is box-shaped, with a single solar array on one side of the satellite. Thermal control is mainly passive, such as thermal coatings, multilayer insulation blankets, and heat pipes. Only in special circumstances is an electric heater employed. The power supply subsystem includes a solar array, a NiCd battery, regulators and converters. The satellite is designed to operate in a sun-synchronous orbit and the local time at the descending node is 10:30 a.m. The repeat cycle is 26 days and the satellite can provide global imaging coverage.

CBERS-1 and CBERS-2 carry three imaging sensors with different spatial resolutions and data-collection frequencies: the Wide-Field Imager (WFI), the CCD Camera (CCD) and the Infra-red Multi-Spectral Scanner (IR-MSS). The WFI is used to obtain low-resolution, wide-field image information in two visible spectral bands: 0.63 to 0.69 µm and 0.77 to 0.89 µm. The ground resolution is 250 m and the swath width is about 900 km. The repeat cycle is reduced to five days due to the wide swath width.

The CCD camera acquires high resolution images in five visible and near-infrared spectral bands: 0.45 to 0.52 µm, 0.52 to 0.59 µm, 0.63 to 0.69 µm, 0.77 to 0.89 µm, and 0.51 to 0.73 µm. The ground resolution is 20 m and the swath width is 120 km. Earth images in the five spectral bands can be obtained simultaneously. The CCD camera, using scanning pushbroom techniques, mainly takes images of the areas around nadir, but also has side-looking capabilities to cover any specified area within a three-day period.

The IR-MSS obtains medium resolution panchromatic images in four spectral bands: 0.5 to 1.1 µm, 1.55 to 1.75 µm, 2.08 to 2.35 µm and 10.4 to 12.5 µm. The ground resolution is 80 m for the first three bands and 160 m for the fourth band. The instrument only takes the images around nadir. Its imaging mode can be performed by ground command in real time or deferred by time tagging.

In 2002, an agreement was signed for the continuity of CBERS programme, with the construction of two new satellites – CBERS-3 and -4, with new payloads. Under this agreement, Brazil and China would each contribute 50% fof the funding. However, since CBERS-3 could not be launched before 2009, by which time CBERS-2 might have reached the end of its operational life, it was decided in 2004 to build CBERS-2B as an interim solution. CBERS-2B is very similar to CBERS-1 and 2, but IR-MSS is replaced by a High-Resolution Panchromatic Camera (HRC). HRC operates in a single spectral band which covers visible and near-infrared wavelengths. It generates images of 27 km width and resolution 2.7 m. Other improvements are the installation of GPS (Global Positioning System) and a star sensor, which assist orbit determination and attitude control.

In addition to the imaging payload, the CBERS-1 satellite carried a Data Collection System (DCS) for environmental monitoring; a Space Environment Monitor (SEM) for detecting high-energy radiation in space; and an experimental High Density Tape Recorder (HDTR) to record imagery onboard.

The CBERS primary imaging system was developed by the Xi'an Institute of Radio Technology. It has characteristics similar to the French SPOT, using linear CCDs. Brazil provided the wide field imager. INPE was responsible for the development of the data collection system, structure, power supply, TT&C, electrical ground support equipment, integration and testing (30 per cent) and management (30 per cent). The first satellite was manufactured by the China Academy of Space Technology (CAST), whereas INPE was responsible for the manufacture, integration and testing of the second and third satellites.

In the CBERS-2 control schedule, the first 9 months were the responsibility of China, including

Artist's impression of the CBERS-2/Ziyuan-1B satellite, launched in 2003 (INPE) 1340520

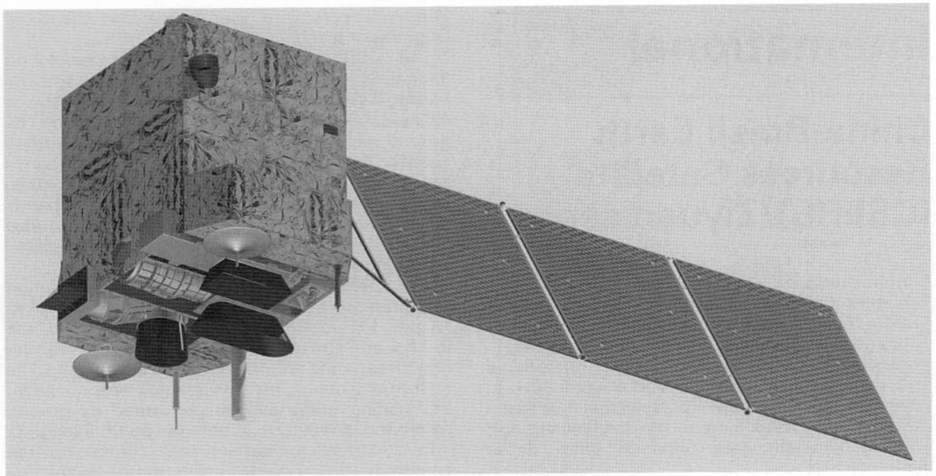

Artist's impression of the CBERS-3 satellite, to be launched in 2009 (INPE) 1340521

the 2 months of in-orbit acceptance tests. After that, Brazil was to take over control of the satellite for 8 months, then China for the following 7 months. For the third year of operations, INPE would have responsibility for the first 6 months and China for the second 6-month period. Control of the satellite will then be shared by each country, in 3 monthly periods, starting with Brazil.

CBERS-1 was operational for almost 4 years. The WFI instrument experienced a malfunction in May 2000, after seven months of service. On 13 August 2003, CBERS-1 experienced an X-band malfunction causing an end to all image data transmissions. CBERS-1 surpassed its design life of two years by almost two years.

CBERS-3 and CBERS-4 will carry improved CCD cameras which will permit a ground resolution of 5 m. The four cameras are: PanMux Camera (PANMUX), Multi-spectral Camera (MUXCAM) – an upgrade of the 20m-High Resolution CCD Camera of CBERS-1 and 2, Scanning Medium Resolution Scanner (IRSCAM) – an upgrade of the Infra-red Multispectral Scanner (IR-MSS) of CBERS-1 and 2, and Wide Field Imaging Camera (WFICAM) – an upgrade of the Wide Field Imager (WFI) on CBERS-1 and 2. The orbits of the two satellites will be the same as for CBERS-1 and 2.

China and Brazil have also undertaken studies for the launch of satellites based upon the CBERS platform to geosynchronous orbit for meteorological, remote sensing or communications missions. No further details are available concerning these studies.

Specifications

CBERS-1/Ziyuan-1

Launch: 14 October 1999 by a CZ-4B from Taiyuan.
Orbit: Initially 98.54°, 733–746 km, standard operational orbit Sun-synchronous 773-773 km; repeated ground track every 373 circuits (26 days): repeat cycle also permits a 3-day repeat with +32° side-looking permitted.
Resolution: 20–260 m visible, 80 m IR (160 m in TIR).
Mass: 1,450 kg at launch.
Lifetime: Nominal lifetime was 2 years, but actually operated until 13 August 2003; on this date it was raised to a 774–782 km retirement orbit.
Spacecraft: 1.8 × 2.0 × 2.2 m body, 6.3 × 2.6 m solar panel providing 1,100 W. 3-axis control provided by RWs plus 1 N thrusters. 20 N thrusters used for orbit maintenance (a small orbit-raising manoeuvre was completed at least once every 6–8 weeks).

53 Mbit/s experimental high-density tape recorder. Image data transmission was at 2 × 53 Mbit/s (each using 20 W TWTA) plus 6.13 MBit/s (8 W SSPA) on 8.103–8.321 GHz.

Argos-compatible data collection system can handle up to 500 platforms: 401.65 + 401.62 MHz up and 462.6 + 2,267 MHz down.

Sensors: CCD: 5-channel linear CCD providing visible/PAN coverage: 0.51–0.73, 0.45–0.52, 0.52–0.59, 0.63–0.69 and 0.77–0.89 µm. 120 km swath.

IR-MSS: 2 × 53 Mbit/s IR/thermal XS scanner covers 0.50–1.10, 1.55–1.75, 2.08–2.35 and 10.40–12.50 µm, swath 120 km. 6.13 Mbit/s

WFI: CCD wide field imager covers two visible channels: 0.63–0.69 and 0.77–0.89 µm, swath 900 km. 1.1 Mbit/s.

Space Environment Monitor also carried.

CBERS-2/Ziyuan-1B

Launch: 21 October 2003 by CZ-4B from Taiyuan.
Orbit: Initially 98.54°, 731–750 km, standard operational orbit sun-synchronous 773–773 km.
Mass: 1,550 kg
Lifetime: Nominal ~2 years. Still operational in September 2007.

Other specifications as CBERS-1/Ziyuan-1.

Although it takes the same name, Ziyuan-2 is a reconnaissance satellite, and is discussed in the Jianbing entry. It is not part of the CBERS programme.

CBERS-2B/Ziyuan-2B

Launch: 19 September 2007 by CZ-4B from Taiyuan.
Orbit: Initially 98.54°, 731–750 km, standard operational orbit sun-synchronous 773–773 km.
Mass: 1,550 kg
Lifetime: Nominal ~2 years. Still operational in September 2007.

Other specifications as CBERS-1/Ziyuan-1.

Sensors: In addition to WFI and CCD, it carries HRC: spectral band 0,50–0,80 µm (panchromatic); field of view 2,1°; spatial resolution 2.7 × 2.7 m; imaged strip width 27 km (nadir); temporal resolution 130 days; image data rate 432 Mbit/s (before compression).

Earth Remote Observation Satellite (EROS) Series

Current Status

EROS B, the second satellite in the commercial imaging satellite series, was successfully launched on 25 April 2006. EROS A was also operational as of 2007. EROS C is under development and will likely launch around 2009. The EROS B launch reportedly cost USD30 million.

Background

The Earth Remote Observation Satellite (EROS) series is owned and operated by ImageSat International NV, an international company based in the Netherlands Antilles, with offices in Tel Aviv, Israel and Limassol, Cyprus. Israel Aircraft Industries (IAI) and El-Op (an Elbit Systems subsidiary), hold significant interests in ImageSat. IAI is the company's primary investor, and El-Op owns 14 per cent; Core Software Technology, a US company, and other investors, also hold stakes in the company. ImageSat has expressed aspirations to issue an Initial Public Offering (IPO) in the future. EROS ground control facilities are located at IAI/MBT's site in Israel.

EROS capabilities are marketed primarily toward the defence community, although the imagery could have civil applications. ImageSat customers have the ability to temporarily control the EROS satellite camera, pointing it toward the desired target, while the spacecraft are over the customer's area of interest; customers can also downlink satellite imagery in real time, to a private ground station, eliminating ImageSat from their intelligence loop. ImageSat manages and controls the satellite, ending its link to the client, once the client's area of interest has passed from EROS' view, and renewing the link at the satellite's next revisit time.

The EROS series is closely related to Israel's Ofeq reconnaissance satellite series (see separate entry). El-Op developed and manufactured the remote sensing equipment for both satellite series, and IAI manufactures the spacecraft bus for both series. Additionally, both companies hold interests in the ImageSat International joint venture. It is therefore reasonable to assume that the Ofeq series has very similar specifications to the EROS series.

Specifications

EROS A and EROS B use the same IAI/MBT satellite bus and are structurally identical. EROS B upgrades include faster downlink speed, a star sensor and a 50 cm Cassegrain telescope, improved over EROS A's 30 cm aperture size. EROS B's optical imaging resolution was greatly enhanced over EROS A's capabilities, and EROS C is expected to further build upon the series' success, offering even higher downlink speeds and potentially higher resolution images. EROS A's Panchromatic Imaging Camera (PIC) and EROS B's high resolution PIC-2, both of the Charge Coupled Device (CCD) type, were designed and built by El-Op, in Israel. PIC-2 adds Time Delay Integration (TDI).

EROS A

Launch: 5 December 2000 aboard a Start-1 from Svobodny
Orbit: ~500 km, circular, sun-synchronous; 97.4° inclination; 94.7 minute period
Mass: 250 kg
Dimensions: 2.3 × 4 m
Design Life: Up to 10 years
Ground Sampling Distance (GSD)/PIC Imaging *Resolution:* 1.9 m; 14 km swath
Satellite Revisit Time: At 10°, 2.5–10 days

EROS B

Launch: 25 April 2006 aboard a Start-1 from Svobodny
Orbit: ~500 km, circular, sun-synchronous; 97.4° inclination; 94.7 minute period
Mass: 300 kg
Dimensions: 2.3 × 4 m
Design Life: Up to 10 years
Ground Sampling Distance (GSD)/PIC-2 Imaging *Resolution:* 0.7 m; 7 km swath
Satellite Revisit Time: At 10°, 2.5–10 days

EROS C

Launch: 2008–2009 aboard a Start-1
Mass: 350 kg
Design Life: Up to 10 years
Imaging Resolution: At least 0.7 m

ERS

Current Status

ERS-1 mission operations were concluded 18 March 2000. The operational phase was divided into seven phases: orbit acquisition, commissioning phase (5.5 months), first-ice phase (28 December 1991 to 31 March 1992), experimental roll-tilt-mode campaign (4–14 April 1992), multidisciplinary phase (14 April 1992–6 December 1993), second ice phase (1 January-7 April 1994), geodetic phase (April 1994–March 1995).

Background

ESA built the ERS for global measurements of sea wind/waves, ocean and ice monitoring, coastal studies and a small amount of land imagery. Following ERS 1 approval in October 1981, West Germany, as 24 per cent contributor, provided the instrumentation. The Phase C/D contract, worth about DM840 million, was signed with the Dornier group 29 October 1986 for development, construction, launch preparation and the associated ground station. Under a January 1986 ESA/NASA agreement, direct readout of ERS 1 SAR data at the Fairbanks, Alaska, station is permitted, and NASA scatterometer and radar imagery are exchanged for other ERS data of interest. The exchange enhances NASA/ESA-supported polar ice research, and complements NASA experimental activities with the Topex/Poseidon (see International) and Shuttle Imaging Radar-C. By permitting NASA direct data readout from ERS, ESA reciprocates earlier NASA provisions allowing European data readout from Seasat and Nimbus 7.

The European Remote Sensing satellite's three primary all-weather instruments provides systematic, repetitive global coverage of ocean, coastal zones and polar ice caps, monitoring wave height/wavelengths, wind speed/direction, precise altitude, ice parameters, sea surface T, cloud top T, cloud cover and atmospheric water vapour content. The Active Microwave Instrument (AMI) can operate as a wind scatterometer or Synthetic Aperture Radar (SAR, with wave scatterometer mode), the Along-Track Scanning Radiometer & Microwave Sounder (ATSR-M) provides the most

ERS 2 final preparations. Prominent is the ATSR (curved apertures) (DASA) 0517414

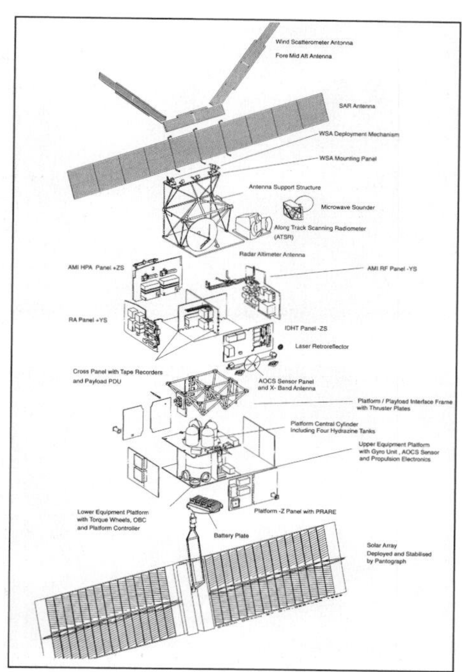

ERS: ESA's ocean and ice monitoring platform 0517028

ERS 1 SAR image of flooding in Northern Europe in January 1995 (ESA) 0517243

accurate sea surface T data to date, and the Radar Altimeter (RA) measures large-scale ocean/ice topography and wave heights. The follow-on ERS 2 was approved in June 1990 to provide continuity until Envisat 1 appears in 1999 (see the POEM entry at end). ERS 1/2 operated simultaneously 16 August 1995 to mid-May 1996, the first time that two identical civil SARs worked in tandem. The orbits were carefully phased to provide 1- day revisits, allowing the collection of interferometric SAR image pairs and improving temporal sampling. Although still working perfectly, ERS 1 funding required it to be put on standby from the end of May 1996. ERS-1 was formally retired in 1999 when it ran out of propellant. ERS-2 is expected to remain operational for several more years. Extending ERS 2 operations to 2000 is under discussion.

ESA's Council announced in June 1990 that subscriptions were sufficient to start the ERS-2 programme. Cost is 371 MAU (1989 rate). Dornier was awarded the DM480 million prime contract (DM360 million went to non-German subcontractors). Similar instruments and mission to ERS 1 provided coverage until Envisat 1 became available. ERS 2's first SAR image, covering Italy's Campania region, was acquired 2 May 1995. The first ATSR 2 image was acquired 5 May 1995, covering Britain/Ireland. AMI's wind scatterometer mode did not begin operations until 22 November 1995 because of an anomaly. ATSR stopped normal operations 22 December 1995 because of a problem with its scan mirror. The SAR, RA and Microwave Sounder operated nominally since September 1995; ATSR, PRARE and GOME since January 1996.

Matra of France (18.31 per cent) supplied a Spot-type bus. The UK (13.34 per cent) provided the AMI lead contractor (Marconi). Other contributors to the 13-nation project include Italy (10.61 per cent), Canada (9.1 per cent), Netherlands (5 per cent), Sweden (3.9 per cent), Belgium (3.72 per cent), Spain (2 per cent), Denmark (1.99 per cent), Switzerland (1.7 per cent), Norway (1.5 per cent). ERS 1 development phase was estimated at 584 MAU at 1984 rates; ESA projected cost to completion was 728 MAU at 1989 rates.

The ERS 1 payload was delivered to Matra at the end of February 1990 for integration with the platform. Following launch, the solar array began deployment within 1 minute of separation, followed by the SAR antenna after 75 minutes and the wind scatterometer three-part antenna after 4 hours. All the instruments were activated in the first two weeks and the first SAR image (of Spitzbergen) was returned to Kiruna 27 July 1991. The operational phase began January 1992. Apart from PRARE's failure, the only significant in-flight anomalies have been some interruption to AMI operations during 1994 caused by recoverable problems in the transmitter, a sudden gain loss in the SAR telemetry downlink requiring a switch to the redundant unit, and failure of ATSR's 3.7 μm channel. All instrument and satellite performances meet or exceed expectations. >100,000 SAR images were returned in the first year.

Specifications
ERS 1
Launch: 17 July 1991 by Ariane 4 from Kourou, French Guiana
Mass: 2,384 kg in-orbit BOL (1,100 kg payload, 317.6 kg hydrazine)
Contractors: Dornier (prime), Matra (bus), Marconi Space Systems (AMI), Alenia Spazio (RA), Laben (payload data handling), Contraves (payload module structure), Fokker (payload module thermal control, harness, integration), Aerospatiale (solar array, laser reflectors), British Aerospace (ATSR), Insitut für Navigation Universität Stuttgart (PRARE)
Resolution: 30 m SAR (100 km swath), 1 km/0.5 K ATSR-M, 3 cm Radar Altimeter, 2 m/s wind scatterometer, 5–10 cm PRARE satellite range (failed)
Orbit: controlled within 765–825 km, 98.5° Sun-synchronous for required track repeat period, crossing equator southbound at 10.30 am local time. 3-day (43 rev) repeat cycle for 3-month commissioning phase, then 35-day cycle alternating with 3-day cycle periods in early 1992 and 1994. From April 1994, a 168 day repeat 'geodetic' orbit has been flown
Spacecraft: 11.8 m high, 11.7 m span on-orbit. Based on Matra's 3-axis Spot bus. Payload electronics housed in compartment mounted on platform. Two 2.4 × 5.8 m solar wings (total of 22,260 cells) to provide 2.2 kW after 2 years, supported by 4 × 24 Ah SAFT Ni/Cd batteries. ESOC Mission Management & Control Centre in Darmstadt controls via Kiruna
Sensors: Active Microwave Instrument (AMI) operates in the SAR mode to produce C-band Earth imagery and separately in the AMI-wind mode as a 3-beam C-band scatterometer. Synthetic Aperture Radar: spatial resolution 30 m; frequency 5.3 GHz (C-band), VV polarisation, pulse length 37.1 μs, data rate 105 Mbit/s; swath width 100 km, with 23° incident angle at mid-swath (up to 35° possible via experimental'roll-tilt' attitude control system mode). Wave Scatterometer: the programmable wave mode operates every 200–300 km providing a 6 × 6 km image for extraction of information on wave length and direction. Wave length accuracy of ±25% over 100–1,000 m, direction accuracy ±20°. AMI Wind Scatterometer: three antennas (mid 0.35 × 2.3 m, fore/aft 0.25× 3.6 m) providing fore/mid/aft beams sweep a 500 km swath in 50 km cells at 5.3 GHz, allowing the radar returns to be analysed for surface wind vectors: 4–24 m/s + 0–360° with ±20° accuracy. Incident angle varies over 27–58°. Spatial resolution: 50 km.

Radar Altimeter (RA). The nadir- viewing 13.8 GHz Ku-band 1.3° beamwidth 1.2 m dia antenna altimeter measures, in Ocean Mode (330 MHz bandwidth), wind speed (2 m/s accuracy from measuring

ERS orbital configuration (ESA) 0517029

Alcatel Bell is developing the Ka-band on-board communicators system for the Envisat platform 0084762

backscattering coefficient to 0.5 dB), wave heights of 1–20 m with 0.5 m accuracy over 1.6–2.0 km footprints, and determine altitude to about 5 cm (<1 cm resolution). Ice Mode (82.5 MHz bandwidth) operates with a coarser resolution to determine ice sheet topography, ice type and sea/ice boundaries.

Along-Track Scanning Radiometer & Microwave Sounder (ATSR-M). An experimental passive instrument comprising an advanced four-channel infra-red radiometer (UK-Australia) and a two-channel nadir-viewing microwave sounder (France) for T and water vapour measurement, respectively. ATSR-M IR Radiometer: res 0.5K over 50 × 50 km square, 1 km spatial res; wavelengths 1.6/3.7/11/12μm; swath width 500 km. Microwave Sounder: passive nadir-viewing radiometer operating at 23.8/36.5 GHz measuring the vertical column water vapour content within a 22 km footprint, providing corrective data for ATSR sea surface T + RA measurements.

Precise Range And Range Rate Experiment (PRARE) for orbit determination (ranging accuracy 3–7 cm) using 8.489 GHz X-band 1 W signals transmitted to a network of 60 cm 2 W mobile ground transponders, with 2.248 GHz 1 W S-band transmissions permitting ionospheric corrections. PRARE failed after 3 weeks in orbit, concluded to be destructive proton-induced memory latch-up over the South Atlantic Anomaly. Russia's Meteor 3–06 launched January 1994 is successfully operating a PRARE provided by Kayser-Threde under DARA contract.

Laser Retroreflector permits precise range/orbit determination, although less frequently, and RA calibration.

ERS 2

Launch: 21 April 1995 by Ariane 4 from Kourou, French Guiana
Mass: 2,516 kg at launch, 2,516 kg on-station BOL, 2,110 kg dry

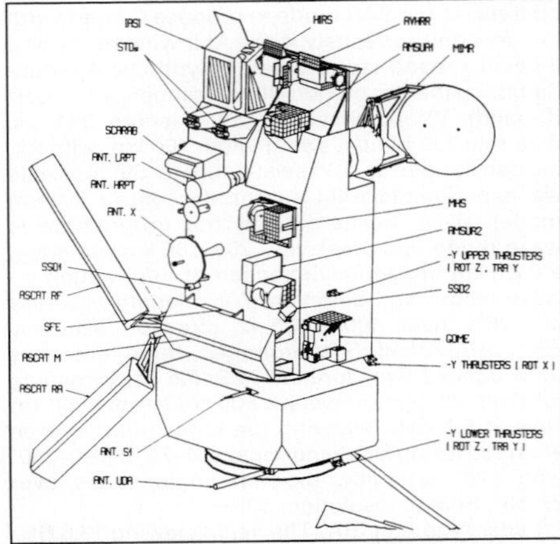

Metop will provide operational meteorology and climate monitoring. Scarab and MIMR have been deleted (ESA) 0517228

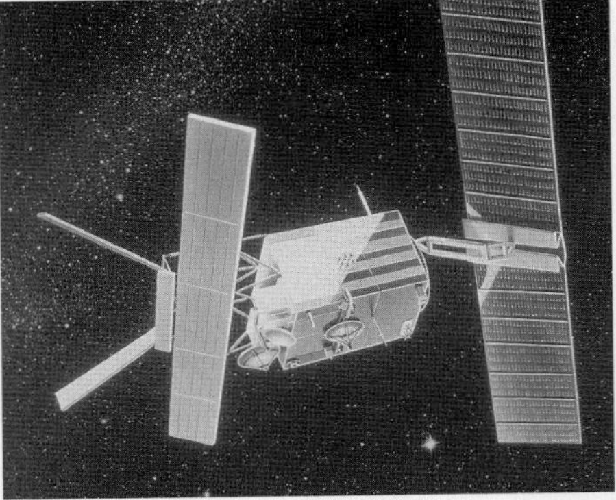

The ERS-1 carriesan ESA SAR, and a NASA scatterometer 0038495

Contractors: as ERS 1 except Officine Galileo (GOME), Schrack (microwave radiometer)
Orbit: similar to ERS 1, phased for 1 or 8 day revisits in turn. 771 × 797 km 98.55° at injection
Spacecraft: Saab developed a new computer memory due to ERS 1 design obsolescence
Sensors: Active Microwave Instrument as ERS 1

Radar Altimeter as ERS 1

Global Ozone Monitoring Experiment (GOME) scanning near-UV/visible spectrometer measuring backscattered Earth radiance in 3,500 channels over four bands, 240–295, 290–405, 400–605, 590–790 nm (2–4 Å res), to determine ozone/trace gases in tropo/stratosphere. Scan angle and integration variable within ±2–31°/0.1–3,000 s. Internal, Sun, Moon calibration. 52 kg, 30 W operating; 50 × 60 × 70 cm. Each band focused on 1,024-pixel photodiode array cooled to –40±1°C by Peltier coolers. Essentially a scaled-down version of SCIAMACHY

ATSR augmented by 0.55, 0.67, 0.78 μm visible channels to improve monitoring of land applications (vegetation moisture, state, species). Also improvements to mechanical and electronics design.

PRARE redesign of control processor and memory with radiation-tolerant components plus redundant unit, to avoid ERS 1-type loss
Ground segment: ERS are controlled from ESOC at Darmstadt, Germany with ESA ground receiving stations at Salmijärvi, near Kiruna (Sweden, the primary station with 15 m antenna; also used for TT&C), Fucino (Italy), Gatineau (Canada), Maspalomas (Spain), plus national stations at Fairbanks (Alaska), Prince Albert (Canada), West Freugh (UK), Alice Springs (Australia). SAR's 105 Mbit/s image data are returned in real time only, available only when the wave/wind modes are inactive (other data are recorded onboard, thus providing global coverage). ESA's ESRIN ERS Central Facility (EECF) facility at Frascati, Italy is the data management centre and prepares the mission operation plan for ESOC, with processing/archiving facilities at Brest (France), Farnborough (UK), DLR Oberpfaffenhofen (Germany) and Matera (Italy). Some products, such as from the wind scatterometer, are available within 3 hours of observation. Dornier was awarded a DM5 million contract in April 1989 for 20 to 30 mobile PRARE ground stations. Some were deployed in 1994 in support of Meteor 3 and the remainder in 1995.

POEM series

To solidify its civil remote sensing plans ESA developed the Polar Orbit Earth Observation Mission Programme, envisaged as two series (Envisat/Metop) of polar platforms providing data to a wide user spectrum. POEM's bus derives from the ESA designed modular platform built as part of the Columbus programme until it was unified in 1994 with Envisat 1. POEM-1 Phase 1 was approved at 1991's Munich Ministerial Council meeting and the 1,174.5 MAU Ph 2 at Spain in 1992. Envisat 1, the largest and most expensive satellite ever built by Europe, will be launched in 1999 for environmental research, providing continuity for ERS. 40 MAU preliminary work on Metop 1 was approved; the full 760 MAU (1994 rates) was reduced in early 1996 by 45 per cent when it became clear it could not be met. The cost was cutting the programme from 15 years to 10 years and deleting the microwave imager. Developed in co-operation with EUMETSAT for 2001 launch, its climatic research instruments will be accompanied by operational meteorological units. The PPF prime contract with MMS is worth ECU502 million (1988 rate; ECU675 million 1995). Ph C/D contract was signed 24 July 1995 with MMS, including Envisat 1 payload integration. DASA has 30 per cent.

ESA plans to gather environmental data with the Envisat using eight major instruments to perform atmospheric ozone and turbulence monitoring 0038494

France provides 24 per cent of the funding, UK 19 per cent, Germany 18 per cent, Italy 11 per cent; 5 per cent remains uncovered. The Phase C/D contract, worth ECU675 million (1995 rates; ECU502 million 1988) was signed with MMS 24 July 1995 for Polar Platform development and Envisat 1 integration. MMS Bristol is responsible for platform development and Envisat integration, and MMS Toulouse for the service module. ASAR, RA-2, MERIS, MIPAS, GOMOS, MWR and LRR will be provided by ESA; AATSR (UK/Australia), SCIAMACHY (Germany/Netherlands) and DORIS (France) will be funded nationally. The ECU810 million contract for payload development and system engineering was signed 17 July 1996 with Dornier.

Metop 1 is being jointly developed by ESA and EUMETSAT in the frame of a single space segment as the prototype of a series of operational satellites addressing operational meteorology and climate monitoring missions of EMETSAT. The single space segment, comprises Metop 1, 2 and 3 satellites and instruments like ASCAT, GOME and GRAS. The procurement will be done by a single space segment team led by an ESA project manager assisted by an EUMETSAT project manager. The other instruments will be provided by EUMETSAT, NOAA and the 'Centre National d'Etudes Spatiales' (CNES). EUMETSAT is responsible for launcher procurement, ground segment, satellite operations and data distribution. In addition to complementing the meteorological payload, the climate instruments will establish a global ocean, ice and ozone monitoring system. Phase A started October 1993 for completion 1994; Phase B began July 1996, EUMETSAT's cost is projected at ECU 1.5 billion (1994 rate) for two satellite launches, and 14 years of operations: ESA contributes financially, partially to Metop-1, the development of ASCAT, GOME and GRAS. Metop- 2 as planned for 2007 but it is to be available for launch within 18 months of the launch of Metop-1. Metop-3 is envisaged for 2011.

Envisat 1

Launch: 1 March 2002 by Ariane 5 from Kourou, French Guiana
Design life: 5 years
Mass: about 8 t; 1,900 kg payload
Contractors: Dornier (mission prime, MIPAS), MMS (platform, service module, integration, GOMOS, ASAR, AATSR), Alenia Spazio (RA, MWR), Fokker Space (solar array, SCIAMACHY), CASA (service/payload module structures, ASAR panels), Aerospatiale (MERIS)
Orbit: planned 800 km 98.55° morning orbit, crossing equator 10.00 am, 35 day repeat cycle
Spacecraft: 3 axis 10 m long, 3 m wide, comprising the payload module + service module. Single 5 × 14 m 14-panel solar array generates 6.7 kW after 4 years (1.9/4.1 kW average/peak to payload); 8 × 40 Ah Ni/Cd batteries. Data handling: one 100 Mbit/s channel (for ASAR) + 10 × 32 Mbit/s for others. 4 × 30 Gbit tape recorders; record 5 Mbit/s, playback 50 Mbit/s. FOCC Flight Operations Control Centre at ESOC (via Kiruna S-band or DRS); PDCC

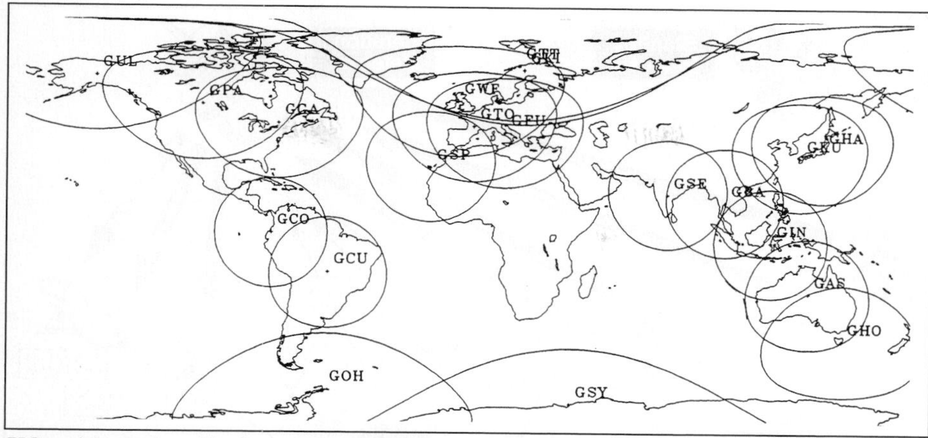

ERS receiving stations. Alice Springs (GAS, Australia), Aussaguel (GTO, France), Bangkok (GBA, Thailand), Cotopaxi (GCO, Ecuador), Cuiaba (GCU, Brazil), Fairbanks (GUL, US), Fucino (GFU, Italy), Gatineau (GCA, Canada), Hatoyama (GHA, Japan), Hobart (GHO, Australia), Hyderabad (GSE, India), Kiruna (GKI, Sweden), Kumamoto (GKU, Japan), Maspalomas (GSP, Spain), O'Higgins (GOH, Antarctica/Germany), Pare-Pare (GIN, Indonesia), Prince Albert (GPA, Canada), Syowa (GSY, Antarctica/Japan), Tromso (GTT, Norway) and West Freugh (GWF, UK) (Eurimage) 0517030

Payload Data Control Centre at ESRIN. X-band real time and recorded data direct and Ka-band via DRS (steerable 90 cm antenna on 2 m mast); each two channels 50/100 Mbit/s. Hydrazine thrusters (300 kg loading) for attitude control; RWs + magnetorquers for fine control (0.1° 3)

Sensors: Advanced Synthetic Aperture Radar (ASAR) improved version of ERS SAR. 5.331 GHz, 1.3 × 10 m array of 20 66.4 × 99.5 cm radiating panels (each 16 rows of 24 microstrip patches). Imaging mode HH or VV, 29 × 30 m/2.5 dB resolution, 7 swaths 100–56 km at 20–45°, 96.3 Mbit/s, 1,200 W DC power consumption; alternating polarisation mode HH + VV, 29 × 30 m/3.5 dB, 7 swaths 100–56 km at 20–45°, 96.3 Mbit/s, 1,200 W; wide swath mode HH or VV, 150 × 150 m/2.5 dB, >400 km 5 subswaths at 17–42°, 96.8 Mbit/s, 1,200 W; global monitoring mode HH or VV, 1,000 × 1,000 m/1.5 dB, >400 km 5 subswaths at 17–42°, 0.9 Mbit/s (allowing onboard storage), 750 W; wave mode HH or VV, 30 m/2.0 dB, two vignettes of 5 × 5 km every 100 km in any swath at 20–45°, 0.9 Mbit/s, 520 W. Up to 30 min of high resolution imagery each orbit.

Radar Altimeter (RA-2) fully redundant nadir pointing pulse limited radar using single 1.2 m dish at 13.575 & 3.3 GHz. Derived from ERS RA but improved performance without significant increase in mass and power consumption. 3.3 GHz channel added to correct for ionosphere propagation effects. Fixed pulse repetition frequencies of 1,800/450 Hz are respectively used by the two channels. RA-2 onboard autonomous selection of transmitted bandwidth (CW/20/30/320 MHz) avoids dedicated operational mode commanded from ground and makes possible continuous operation over ocean, ice, land and their boundaries. Altitude accuracy after ionospheric correction improved to <4.5 cm for Significant Wave Height up to 8 m. Operation supported by MWR/LRR/DORIS measurements.

Medium Resolution Imaging Spectrometer (MERIS) is the first programmable imaging spectrometer. VNIR (400–1,050 nm), 250 m/12.5 nm resolution (adjustable as required), swath width 1,130 km at 800 km altitude. Aimed at water quality measurements, such as phytoplankton content, depth and bottom-type classification and monitoring of extended pollution. Secondary goals: atmospheric monitoring and land surfaces processes.

Michelson Interferometer for Passive Atmospheric Sounding (MIPAS) is a Fourier transform spectrometer observing mid-IR 4.15–14.6 μm limb emissions with high spectral resolution (<0.03 cm^{-1}), allowing day/night measurement of trace gases (including the complete nitrogen- oxygen family and several CFCs) in stratosphere and cloud-free troposphere. Global coverage, including poles.

Global Ozone Monitor by Occultation of Stars (GOMOS). Two spectrometers observe setting stars over UV to near-infra-red for 50 m vertical resolution and 0.1% annual variation sensitivity of ozone/related gases. Occultation method is self-calibrating and avoids the long term instrumental drift problems of current sensors. Spectrometer A: 2,500–6,750 Å, res 0.3 nm/pixel; B: 9,260–9,520 Å (H_2O) + 7,560–7,730 Å (O_2), res 0.05 nm/pixel. The Fast Photometer Detection Module provides 1 kHz 2-band scintillation monitoring of the star image.

Advanced Along-Track Scanning Radiometer (AATSR). Performance almost identical to ATSR 2 on ERS 2. Provided by UK Department of Environment + Australia.

MicroWave Radiometer (MWR) is a nadir-viewing 23.8/36.5 GHz Dicke radiometer with 3 dB IF bandwidth 600 MHz. ERS's MWR design modified mainly in the mechanical layout and antenna configuration. Radiometric stability <0.5 K over 1 year. Periodic onboard calibration by switching receiver input between two references: a horn pointing at the cold sky and a hot radiator at ambient. MWR determines tropospheric columnary water vapour content by measuring the radiation received from Earth's surface to correct RA-2 altitude measurements.

Scanning Imaging Absorption Spectrometer for Atmospheric Chartography (SCIAMACHY) 2,400–23,800 Å grating spectrometer (limb/nadir viewing) for detrimental trace gas measurement in troposphere/stratosphere. resolution 2.4 Å UV and 2.2–14.8 Å vis/infra-red. Swath 1,000 km in nadir mode.

DORIS as Spot 3/4 + Topex/Poseidon for precise (1 m) onboard real time autonomous orbit determination and distribution of timing pulse; 10 kg, 14 W.

LRR Laser Retro Reflector for precise orbit determination

Ground segment: Comprises two major elements: FOS Flight Operations Segment and PDS Payload Data Segment. FOS employs the FOCC Flight Operations Control Centre at ESOC to provide mission and operations planning, and command/control via the primary S-band TT&C station at Kiruna-Salmijärvi. The PDS is composed of the PDCC Payload Data Control Centre at ESRIN (controlling and monitoring all services offered by PDS facilities); two Payload Data Handling Stations at Kiruna-Salmijärvi (X-band) and at ESRIN (via DRS) for acquiring global and regional data and providing near-realtime and FD services; the PAC Processing and Archiving Centres (offline services). The PDS will offer a comprehensive set of user services, with online access to catalogue, browse and ordering services. ESA's products will range from raw reformatted data (level 0) to geophysical products (level 2) and images (from browse to high resolution). PDS development has been awarded to a consortium led by Thomson-CSF.

Metop 1

Launch: planned for 2005 by Ariane 5 from Kourou, French Guiana
Design life: 5 years
Mass: 4,244 kg, payload 975 kg
Contractors: Matra Marconi Space/Dornier
Orbit: 830 km Sun-synchronous, 09.30 descending node
Spacecraft: reduced Envisat platform, PPF/Spot avionics, 10- panel solar array feeding 5 × 40 Ah Ni/Cd batteries. Eight paired 15 N hydrazine thrusters drawing on four tanks, fine pointing (±0.15°) by three 40 Nms RWs
Sensors: Advanced Visible/Infra-Red High-Resolution Radiometer (AVHRR/3). Day/night imaging in 6 bands over 0.68–12.50 μm, sea surface T, ice, snow and vegetation cover. 31.3 kg, 29 W, 822.0 kbit/s.

High-Resolution Infra-red Temperature Sounder (HIRS/3). 20 bands over 0.69–14.95 μmm, vertical T profile. Moisture content, cloud height, surface albedo. 33.1 kg, 24 W, 2.88 kbit/s.

Advanced Microwave Sounder Unit (AMSU-A1/A2). Two modules covering 23.8 MHz–89 GHz: A1 13 channels + A2 2 channels for T/humidity atmospheric profiles. A1: 53.3 kg, 88.3 W, 2.080 kbit/s; A2: 47.4 kg, 37.25 W, 1,120 bit/s.

Microwave Humidity Sounder (MHS). 90–190 GHz 5-channel self-calibrating for humidity profiling. 66 kg, 100 W, 3.95 kbit/s.

Argos. DCS 2 data collection/location. 47 kg (inc antennas), 27.5 W, 2.56 kbit/s.

IR Atmospheric Sounding Interferometer (IASI). 3.5–15.5 μm Fourier Transform spectrometer. 148 kg, 196 W, 1.5 Mbit/s.

Advanced Wind Scatterometer (ASCAT). 6.255 GHz C-band radar for ocean surface wind vectors, 213 kg, 250 W, 45 kbit/s.

Global Ozone Monitoring Expt (GOME). As ERS 2. 59 kg, 44 W, 50.0 kbit/s for Metop 1, 2.

OMI-Ims considered for Metop-3.

Global Navigation Satellite System Receiver for Atmospheric Sounding (GRAS) Spaceborne GPS/GLONASS receiver for Earth science applications with dual frequency measurement from GPS and GLONASS carrier phase measurements to mm-level precision good quality (<1 m) code phase measurements.

Meteosat series

Current Status

Meteosat 5, located at 63°E, has ended its service over the Indian Ocean. It sent back its last operational image on 16 April 2007 and was retired from service on 26 April, when it was boosted into a higher graveyard orbit. Meteosat 7 is the 'new' Indian Ocean Data Coverage (IODC) primary satellite after its move from 0° to 57°E during 2006. Meteosat 6, which had been providing the 10 minute Rapid Scan Service (RSS, disseminated via EUMETCast) from 10°E, was moved to the Indian Ocean region as back-up satellite for Meteosat 7 IODC in early 2007. From 1 June to 31 August 2007, Meteosat 8 (MSG-1) will operate in Rapid Scan mode for the Convective and Orographically-induced Precipitation Study (COPS). It was replaced as the operational satellite at 0° longitude by Meteosat 9 (MSG-2) in April 2007 and is now the back-up for Meteosat 9. Three more satellites are planned. Launch of MSG-3 is projected for early 2011, followed by MSG-4 no sooner than 2013.

Background

Meteosat constitutes ESA's contribution to the international World Weather Watch of the Global Atmospheric Research Programme (GARP) which began with Meteosat 1, launched 23 November 1977 by Delta from Cape Canaveral. Although fully operational for only two years of its projected three year planned life, it gathered data until it exhausted its supply of hydrazine propellant in October 1985. During this lifetime, the satellite played a key role in the Global Weather Experiment, a major exercise involving almost all of the 147 member nations of the World Meteorological Organisation. GWE had almost ended when, a day after its second anniversary, Meteosat 1 suffered an onboard radiometer failure. Despite efforts spread over some months, its imaging capacity never reached its full programme again, although its data collection function remained unimpaired. By then, however, ESA had gathered more than 40,000 images and submitted an outline proposal for the five-satellite Meteosat programme for 1984–1994.

Engineers modified Meteosat 2 after M1's radiometer problems and repaired damage during ground tests that were designed to assess its resistance to the additional vibration involved in

Meteosat integration at Aerospatiale 0517031

A three-channel Meteosat image is returned every 30 min: visible (EUMETSAT) 0008086

A three-channel Meteosat image is returned every 30 min: infra-red (EUMETSAT) 0008087

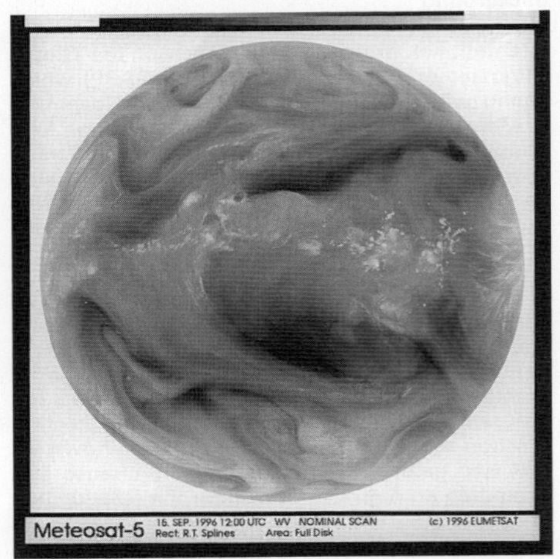

A three-channel Meteosat image is returned every 30 min: water vapour (EUMETSAT) 0008088

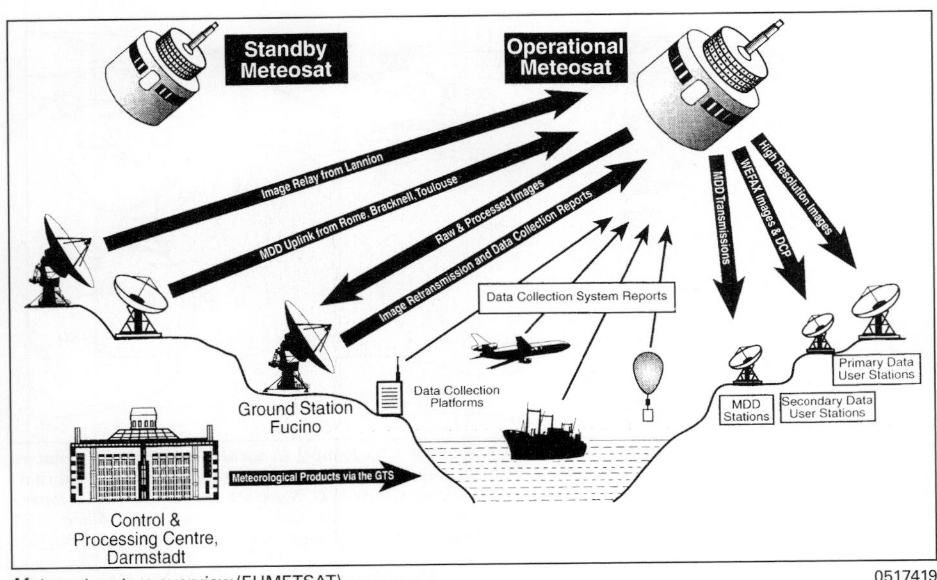

Meteosat system overview (EUMETSAT) 0517419

an Ariane launch instead of the Delta. The launch occurred on 19 June 1981. Its primary imaging mission successfully started, but ESA ground controllers could not activate the DCP system. Despite the problems the two satellites together provided a full service until the end of 1984, when M1 began to malfunction and its hydrazine approached exhaustion. NOAA then loaned ESA the partially operational GOES 4, moving it to 10° W by June 1985, as a return for M2's help following a GOES failure over the Atlantic. Data collection by GOES 4 continued from 43°W into 1988, with M2 providing the images, until Meteosat 3 (P2) assumed the relay role. Although M2 had a three year design life, the quality of its imagery continued to be excellent into 1991. As part of the effort to extend its life into 1989 (when Meteosat 4 became available), ESA performed a small inclination manoeuvre at end of 1986 to hibernate it at 10°W. ESA planned to raise it above GEO, using the 2 kg hydrazine remaining, when Meteosat 5 was commissioned in 1991. The satellite returned a commemorative photo on its 10th anniversary image 19 June 1991. In all, the satellite returned some 284,000 images during its career. Its retirement move began 2 December 1991, when its thrusters fired for eight minutes to raise altitude by 140 km. During the third burn sequence two days later, the hydrazine became exhausted and the orbit only increased by 334 km instead of the planned 700 km. The transmitter was shut down 6 December 1991.

ESA built M3/P2 as a qualification model for the pre-operational programme and launched it 15 June 1988 as a back-up to the ageing M2 until the delayed MOP series could begin. Financial constraints dictated that the 12 year old spacecraft could not be upgraded to full flight standards, but its primary role was only to provide the DCP service unavailable from M2. Although its electronically despun antenna experienced 3 dB fluctuations, it operated as prime satellite from August 1988 until M4 appeared. ESA considered loaning M3 to NOAA at 50°W to compensate for the GOES 6 failure of January 1989. Both organisations approved the move June 1989 and completed the transfer from 3°W to 50°W on 4 November 1989. It returned to assume the primary role in January 1990 after the appearance of M4's imaging problems. M4 again assumed the prime role in April 1990 after M3 lost dissemination channel 1. M3 moved to 3°W as standby; after Meteosat 5's appearance it began the move again 29 April 1991 to 50°W, arriving 19 July, again to substitute for GOES. Service began 1 August 1991.

Limited visibility to Earth stations limited a move further west for Meteosat. After a new relay station came online February 1993 at Wallops Island (Virginia), this problem disappeared and Meteosat began to move again 27 January 1993 to 75°W, arriving 24 February, to provide coverage of the whole continental US. The Wallops station was operated remotely by ESOC. All imagery received there returned to Germany via PanAmSat, then after processing returned to Wallops, where it was uplinked to Meteosat for broadcast to users. M3 moved to 70°W by 14 March 1995, from where it was boosted 21 November 1995 into 36,725 × 36,762 km, 2.82°, using the last 5.1 kg hydrazine.

MOP-1/M4 was launched 6 March 1989 and began service as the prime satellite 19 June 1989. But the satellite began to generate erroneous grey levels in October 1989 and image rectification became impossible. A switch to the second Synchronisation and Imaging Chain (SIC-2) in November 1989 resumed nominal imaging, but it was replaced as prime at 0° by M3 24 January 1990 to allow detailed failure analysis and an intensive test programme. The cause was found to be a malfunction of the DC/DC converters supplying the SICs. All future satellites were modified. Procedures were initiated to switch periodically from one SIC to the other and the image processing software was improved to detect/correct the erroneous grey levels (if not too numerous). Following the failure of an M3 dissemination channel, M4 returned as prime in April 1990. It was then replaced as prime by M5 in February 1994 and held at 8°W until 10 May 1995, when it moved to 8°E before being boosted 9 November 1995 into 36,619 × 36,777 km, 1.50°.

M5 has remained the operational satellite, but the on-board fuel reserves for inclination control have been consumed and thus inclination is increasing at about 0.8°/year. Due to the high inclination and the impact upon user station reception, the operational mission was switched to M6 in February 1997: M6 was relocated from 10°W to 0° in early 1997.

When the EUMETSAT Convention came into force on 19 June 1986, the organisation assumed overall and financial control of the MOP Meteosat Operational Programme beginning on 12 January 1987. A special Council controls EUMETSAT Earth observation programmes and it represents the National Meteorological Services of 17 European states. Each European country makes a contribution to the programmes based on GNP (per cent): Austria 2.23, Belgium 2.70, Denmark 1.76, Finland 1.84, France 16.78, Germany 22.29, Greece 0.96, Ireland 0.54, Italy 15.46, Netherlands 4.03, Norway 1.47, Portugal 0.86, Spain 6.96, Sweden 3.20, Switzerland 3.33, Turkey 1.50 and UK 14.09.

In order to make a smooth transition from a research and technology programme to a fully operational system, ESA planned to develop its Earth imaging resources in three phases: the Meteosat Operational Program (MOP), the Meteosat Transition Program (MTP) and the Meteosat Second Generation Program (MSG). Each phase uses distinct satellites and ground procedures. Meteosat 5 served as the geostationary system's primary satellite, with Meteosat 6 as a standby. Two older satellites, Meteosat 3/4, predated this pair but EUMETSAT de-orbited both in 1995. Meteosat 7, the MTP Meteosat Transition Programme satellite, joined the others in geosynchronous orbit in the second half of 1997 to maintain a continuous service until the MSG (Meteosat Second Generation) appears in 2000. Encryption of high resolution image data began 1 September 1995. On 1 December 1995, Eumetsat inaugurated a new facility dedicated to control the satellites at its MTP HQ of satellite launch and operations Primary Ground Station in Fucino, Italy. Before then, ESA provided these functions.

The Meteosat Operational Program (MOP) began 23 November 1983 under ESA's auspices and ended 30 November 1995. ESA transferred overall responsibility to EUMETSAT in January 1987 but in-orbit control remained with ESA. The Meteosat Operational Program received funds from the EUMETSAT member states in contributions of (per cent): Belgium 4.4, Denmark 0.58, Finland 0.35, France 25.60, Germany 26.39, Greece 0.30, Ireland 0.11, Italy 12.00, Netherlands 3.00, Norway 0.50, Portugal 0.30, Spain 5.24, Sweden 0.93, Switzerland 3.03, Turkey 0.50, UK 16.76 (0.01 per cent not covered). Various sources estimated the total cost for the three satellite series in 1982 terms at ECU378 million (revised to ECU721 million, or USD901 million, in 1995 terms).

Aerospatiale received an ECU139.1 million contract in May 1984 for three flight model MOP satellites and one spare. MOP had three primary missions: Earth imaging, dissemination of image and other meteorological data, and data collection/distribution, with two secondary objectives of meteorological processing and data archiving/retrieval. These MOP/Meteosat satellites transmit real-time image data (1,686.833 MHz) with up to 66 channels of DCP data (1,675.181–1675.381 MHz) at 333 kbit/s to EUMETSAT's Primary Ground Station in Fucino for relay to the Mission Control Centre in Darmstadt.

EUMETSAT expected to put MOP 2 in service as the prime satellite in Summer 1991, replacing Meteosat 4. It received the first image on 3 April during commissioning at 4°W. However, an imaging anomaly required EUMETSAT to make

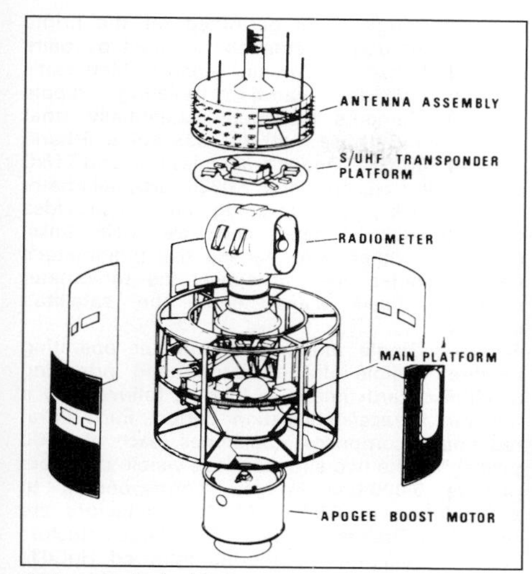

Meteosat principal features 0517032

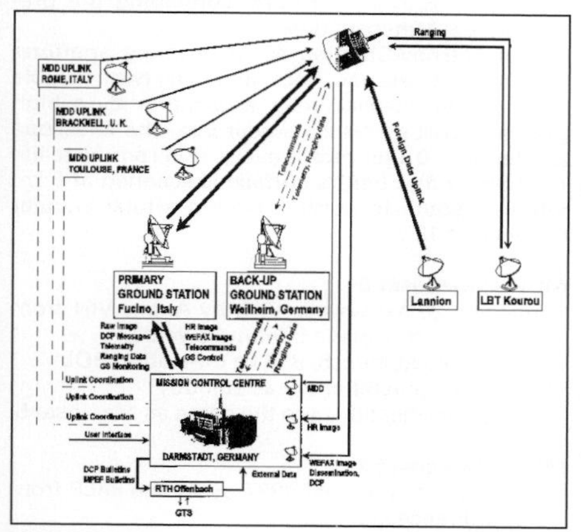

EUMETSAT ground segment from December 1995 (EUMETSAT) 0517033

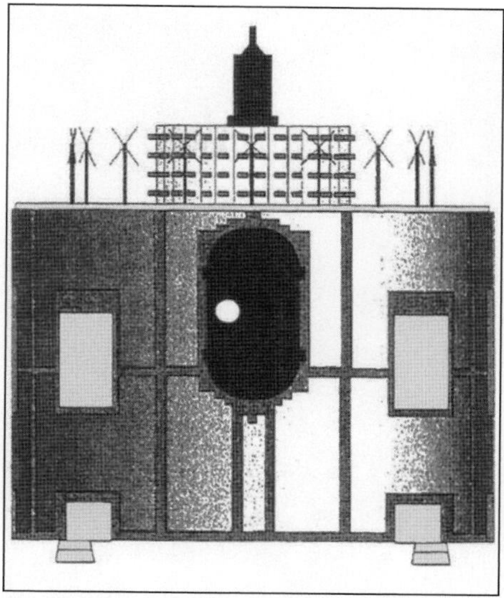

Meteosat Second Generation retains the spin-stabilisation of the smaller first generation satellites 0517418

The first satellite in the second generation of Meteosats, (MSG-1), is shown here in Aerospatiale's facilities in Cannes during instrument integration (Aerospatiale) 0038496

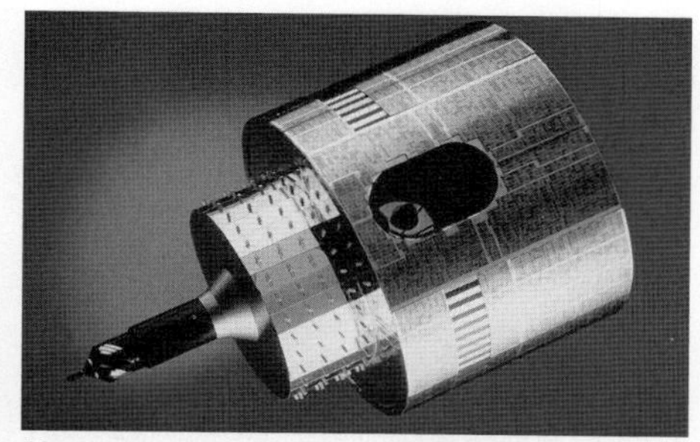

Meteosat contributes ESA's contributors to the World Weather Watch 0084760

modifications to ground processing software. Then the satellite entered full service as prime in February 1994. ESA formally transferred the satellite to EUMETSAT 14 January 1992.

Then MOP 3 followed MOP 2 as part of the programme after it transmitted its first image (visible) on 29 November 1993. An onboard anomaly in the infra-red/water vapour imagery was detected in further testing, caused by an apparent movement of the cold optics within the radiometer. Additional ground processing software corrects the anomaly sufficiently to allow normal extraction of image products.

Darmstadt processes MOP images into a range of formats covering Europe only, or covering most of the full Earth disc with up to 24 frames of data. Final processed images include lat/long grids and coastlines which are added by transmission up to Meteosat (channel 1 2,101.5 MHz; channel 2 2,105.0 MHz) and relayed to users on two channels. Channel 1 (1,691.0 MHz) operates on conventional analogue WEFAX, and is compatible with other GEO metsats and NOAA's Automatic Picture Transmission service. This information is available to simpler Secondary Data User Stations. Channel 2 (1,694.5 MHz) provides high-resolution digital transmissions to Primary Data User Stations. Some of NOAA's GOES images are also available via Meteosat using the Lannion, northwest France relay station. Three other channels (1,695.725, 1,695.756, 1,695.787 MHz) primarily relay digitised facsimile and selected conventional met observations to Africa, as the Meteorological Data Dissemination (MDD) service.

The MARF Meteorological Archive and Retrieval Facility holds, as the single repository, all Meteosat image data and derived products acquired since 1978. Live data continuously enters the archive and EUMETSAT copies historic data for later retrieval. All of the data are recorded on 6.6 Gbyte optical disks, with online access through the MARF's catalogue. During MOP, ESOC gathered some 40,000 tapes' worth of data, dating back to Meteosat 1.

The Meteosat Transition Programme began in 1991 when EUMETSAT planned to orbit one Meteosat follow-on satellite, identical to MOP, to maintain a continuous service until MSG appeared. The organisation signed FFr630 million contract in June 1993 with Aerospatiale to cover one satellite plus components for a second. ESA managed satellite procurement on EUMETSAT's behalf, but EUMETSAT had responsibility for the ground segment, launch and operations.

A new ground segment was developed by EUMETSAT to take full control of all satellites after 1 December 1995. It includes the MCC Mission Control Centre in Darmstadt and Primary/Secondary Ground Stations, replacing the system previously operated from ESOC. The PGS Primary Ground Station is at Fucino, Italy and includes facilities for operational support of two satellites. A high-speed link connecting PGS/MCC transmits data, telemetry and satellite commands. The back-up station is at Weilheim, Germany.

The first generation Meteosat series (Meteosat 1 to 7) of EUMETSAT is gradually being replaced by a second generation series (MSG). The first satellite, MSG-1 (later renamed Meteosat 8), was launched on 28 August 2002, followed by MSG-2 (now Meteosat 9) on 22 December 2005. This programme will ensure operational continuity in GEO for at least 16 years. The 2005 EUMETSAT budget included a total expenditure of EUR291.67 million. 44 per cent of this (EUR128.4 million) was accounted for by MSG. This expenditure mainly covered the development of the satellites, launchers, ground segments and consultancy support.

MSG-1 commissioning got under way on 25 September 2002. On 17 October 2002, SSPA-C (Solid State Power Amplifier-C) failed just before switching on the SEVIRI instrument. SSPA-C was to be used to re-broadcast the data that had been processed by the EUMETSAT control centre. The failure led to an automatic payload switch-off. All attempts to restart the SSPA failed and commissioning was suspended. A new satellite configuration was applied, and commissioning activities resumed on 26 November 2002. MSG-1 became Meteosat-8 on 29 January 2004, when the mission was declared operational at the end of the commissioning phase. MSG-2 became operational in July 2006.

The Meteosat Second Generation (MSG) programme definition started in 1993, and Phase C/D plan began in July 1995. EUMETSAT planned to launch its first satellite under this programme in the fourth quarter of 2000. This did not actually take place until August 2002. Eventually the organisation plans to launch three satellites, the second after 18 months and the third as needed. Arianespace signed an Ariane 5 launch contract for all three, worth more than ECU200 million on 27 July 1995. The programme is funded by ESA and EUMETSAT. ESA is responsible for designing and developing the first of the four satellites in the MSG programme (the fourth spacecraft of the MSG series was approved in 2003). EUMETSAT has overall responsibility for defining the end-user requirements, developing the ground segment and procuring the launchers. EUMETSAT is also the operator of the MSG system. EUMETSAT is contributing about 30 per cent of the development costs for MSG-1 and fully financing the three subsequent flight units (MSG-2, MSG-3, and MSG-4).

MSG's mission is to provide basic multispectral imagery. Improved spatial, temporal and spectral resolution, for weather forecasting. It must also provide high resolution imagery. With AVHRR-type spatial resolution in visible band, for mesoscale convective monitoring over Europe. The information provided includes air mass analysis for water/CO_2 absorption and meteorological product extraction for improved wind and temperature data. EUMETSAT emphasises support of climate and environment monitoring, continuity of data collection and dissemination, the relay of search and rescue distress signals, and observation of the global Earth radiation budget.

The MSG satellites also carry Global Earth Radiation Budget (GERB), a small instrument flying as a science 'passenger'. GERB-1 made the first ever geostationary measurements of the Earth's energy balance. It was provided by a UK-led consortium: Natural Environment Research Council; Rutherford Appleton Lab (technical management, data processing, archiving, distribution); Imperial College of Science, Technology and Medicine (PI Prof John Harries, science, calibration); Royal Meteorological Institute of Belgium, (Advanced Mechanical and Optical Systems – AMOS) and Italy (Officine Galileo). The University of Leicester provided the detector arrays that ultimately sense the radiation from the Earth. Subsequent GERB instruments are being funded by EUMETSAT, through contracts with the same consortium members and same structure as for the first instrument.

Hemispherical data from Meteosat are now combined with higher resolution data from Europe's first polar orbiting meteorological satellite, Metop-A, which was launched on 19 October 2006. Metop, the Meteorology Operational Polar satellite, provides the European component of a joint NOAA/EUMETSAT polar system that aims to provide continuous meteorological global observation from Sun-synchronous LEO. NOAA will launch the morning and afternoon (local time) orbits and EUMETSAT will eventually become responsible for the morning element.

EUMETSAT implemented a technical means of controlling access to Meteosat HRI High-Resolution Image data on 4 September 1995. Now the satellite encrypts the data normally received by PDUSs, while the analogue (WEFAX) transmissions remain unencrypted. Over the past decade, a rapid increase in value-added commercial activity by the private sector has developed, based on meteorological forecasts and data provided by meteorological services. Simultaneously, government satellite organisations have brought increasing pressure on space resources to recover their costs by charging for services. The encryption of data began an effort to register all users of the high-resolution Meteosat data. Regular test transmission of encrypted data began in 1994. Decryption requires an MKU Meteosat Key Unit from EUMETSAT and some adaptation to the PDUS for MKU interfacing. Some exceptions exist. For example, HRI images at 00/06/12/18 GMT remain un-encrypted, and the three hourly encrypted data are free to all National Meteorological Services. In general, the NMS of countries with a GNP per capita less than USD2,000 have free access to hourly and half-hourly data for internal use; wealthier countries pay according to their GNPs. Educational and science programmes have free access to all data.

One of the first images returned from the Meteosat 7 weather satellite 0084761

One of the early images from MGS-1 declared operational as Meteosat 8 in January 2004 0572904

Specifications

MOP-2/Meteosat 5

Launched: 2 March 1991 by Ariane V42 from Kourou, French Guiana (lifetime five years)
Mass: 681 kg at launch, 316 kg after firing of Mage apogee kick motor (including 39 kg hydrazine thruster propellant, with 2 kg conserved for removal from GEO at EOL)
Contractors: Aerospatiale (prime), MMS (radiometer, AOCS), BoschTelecom (transponders), Alenia (TT&C), ETCA (power supply), MBB (structure, thermal)
Resolution: 2.5 km visible, 5 km water vapour/TIR
Orbit: 0° geostationary
Spacecraft: 2.1 m diameter, 3.195 m high stepped cylindrical body with solar cells on six main body panels providing 300 W at beginning of five year life (200 W EOL). Spin-stabilised at 100 rpm around main axis aligned almost parallel to Earth's axis, with spin regulated by two small hydrazine thrusters. Two pairs of larger thrusters provide spin axis precession/inclination control and east-west station-keeping, respectively. East-west drifting of ±1° about 0° longitude is permitted for the prime satellite. Attitude information provided by pairs of Earth horizon + Sun-slit sensors. Meteosat's smaller cylinder carries radiating dipole antenna elements activated sequentially (that is, electronically despun) for 333 kbit/s S-band (1.670–2.300 GHz) image transmissions and TT&C operations (telecommand 2.0980 GHz; telemetry 1.675928 GHz); the protruding antenna provides toroidal-pattern S-band and low UHF links. Meteosat's other pole houses the radiometer's passive cooler open to space; the radiometer telescope itself scans along the satellite's equator
Sensors: Single Imaging Radiometer operating in three visible-infra-red bands and providing a full-disc Earth image in 25 min, followed by a five min retrace/stabilisation period. Each infra-red image comprises 2,500 lines (each of 2,500 pixels) but the two simultaneous visible detectors provide 5,000-line images, corresponding to a resolution of 2.5 km. All four detectors are redundant. The two visible Si photodiode detectors cover 0.4–0.9 μm, the thermal-infra-red HgCdTe detector 10.5–12.5 μm, and the water-vapour/infra-red HgCdTe 5.7–7.1 μm, continuing the pre-operational Meteosat data
Optics: Ritchey-Chrétien 40 cm primary aperture, 365 cm fl telescope mounted on two flexible plate pivots attached to a step motor by a high precision jack screw via a gear box. The telescope is stepped 0.125 mrad every 100 rpm satellite rotation so that Earth's surface is scanned at 5 km intervals south to north, covering a total angular distance of 18°.

MOP-3/Meteosat 6

Launched: 20 November 1993 by Ariane V61 from Kourou, French Guiana (lifetime 5 years)
Mass: 704 kg at launch, 316 kg on-station BOL
Orbit: 10°W geostationary as standby
Other specifications are the same as Meteosat 5.

MOP-4/Meteosat 7

Launched: 2 September 1997 by Ariane 44LP from Kourou (French Guiana)
Mass: 703 kg at launch, 321 kg onstation BOL
Orbit: 349°E initially, operated at 0°
Other specifications are the same as Meteosat 5.

Meteosat Second Generation

Launch: Meteosat 8 (MSG-1) 28 August 2002 by Ariane 5 from Kourou, French Guiana. Meteosat 9 (MSG-2) by Ariane 5G from Kourou on 22 December 2005.
Mass: 2,010 kg
Orbit: 0° geostationary (operational). 3.4°W (back-up).
Contractors: The satellite is developed by an industrial consortium led by Aerospatiale, with MMS responsible for the SEVIRI instrument
Resolution: 1 km visible, 3 km infra-red
Spacecraft: Height 3.7 m, diameter 3.2 m. The satellites continue 100 rpm spin stabilisation, though considerably larger: 3.2 m diameter, 3.7 m high. Unified bipropellant system. 750 W BOL from Si panels. Seven year life. The communications package's three channels will downlink 3.2 Mbit/s raw images, housekeeping and relayed DCP data to the primary ground station, and retransmit processed imagery and other data to users in high-/low-rate data schemes (up to 1 Mbit/s without compression). MSG's data volume will be an order of a magnitude greater than the current system. MSG-1 was the first European geostationary satellite to carry a COSPAR-SARSAT Search and Rescue Transponder.
Sensors: SEVIRI Spinning Enhanced Visible/IR Imager, scanning Earth's disc every 15 min in 12 channels. The 0.50–0.90 μm HRV High-Resolution Visible channel provides 1 km resolution. The other 11 channels return images of max 3,750 lines of 3,750 samples, corresponding to 3 km resolution visible 0.56–0.71, 0.74–0.88, 1.50–1.78; window 3.40–4.20, 8.3–9.1, 9.8–11.8, 11.0–13.00; water vapour 5.35–7.15, 6.85–7.85; ozone 9.38–9.94; CO_2 12.40–14.40 μm. Typical radiometric resolution 0.25 K window channels, 0.75 k water vapour and 1.50 k ozone/1.80 k CO_2.

GERB Global Earth Radiation Budget. Together with SEVIRI, GERB enables study of water vapour and cloud forcing feedback, two of the most important (and poorly understood) feedback processes in climate prediction. It measures Earth's long-wave and short-wave radiation every 15 minutes to an absolute accuracy of about 1 W/m². Bands: 0.35–4.0, 0.35–30 μm (4.0–30 μm by subtraction), 48 km nadir pixel, absolute accuracy 1 per cent SW/0.5 per cent LW, S/N 1,500 SW/LW, dynamic range 0–450/0–130 Wm² SW/LW, <30 kg, <70 kbit/s, <35/40 W avg/peak.

Small Multimission Spacecraft (SMMS)

Current Status

The Small Multimission Spacecraft (SMMS) launch has been delayed by several years. First anticipated for launch in 2004, lift-off will occur, at the earliest, sometime in 2007.

China, Iran, Thailand, the Republic of Korea, Mongolia and Pakistan are co-operating in an international regional effort through the Asia Pacific Multilateral Co-operation in Space Technology and Applications (AP-MCSTA) organisation, to produce a Small Multimission Spacecraft (SMMS) capable of providing an autonomous civil remote sensing and imaging capability. China views the cooperative venture as a platform on which further international space ventures can be built and sees this programme as an investment, creating new Asian markets for its launch services. Pakistan is cooperating with China on regional remote sensing activities and the SMMS is intended to accommodate the needs of that country also. The partners will likely use a CAST-968 platform to form the basis of the satellite. China wants the satellite bus to provide a common platform for other users in the region.

SMMS will form part of the China Environment and Disaster Monitoring Small Satellite constellation. The original project was proposed by China, Pakistan and Thailand in November 1992.

Specifications

Launch: earliest 2007, using a CZ-4B from Taiyuan
Orbit: Sun-synchronous at 650 km
Lifetime: nominal mission about 3 years
Contractor: Chinese Academy of Space Technology (CAST)
Mass: 350–400 kg
Instruments: Hyper-Spectrum Imager (HSI) and Wide-Swath CCD camera (WSCCD)

Topex/Poseidon

Current Status

The Topex/Poseidon mission officially ended on 5 January 2006, after nearly 62,000 orbits of Earth. The spacecraft lost its ability to manoeuvre, bringing to a close a successful 13-year mission. On 9 October 2005, the satellite's pitch reaction wheel, which helped to keep the spacecraft in its proper orbital orientation, ceased to function. Operating well beyond its five year design lifetime, Topex/Poseidon became history's longest Earth-orbiting radar mission. "It provided, on average, more than 98 per cent of the science data it was designed to collect in every 10-day measurement cycle, a remarkable achievement," said Project Scientist Dr. Lee-Lueng Fu of NASA's Jet Propulsion Laboratory. The satellite remains in orbit 1,336 km above the Earth.

Background

NASA/JPL's combined its Topex (The Ocean Topography Experiment) with CNES' similar Poseidon payload in FY87 to provide long-term observation of global ocean circulation and surface topography. The satellite carries two Ku-band altimeters to determine sea height to within 2 to 5 cm. NASA's dual frequency (C/Ku) altimeter built on experience from NASA Goddard and the Applied Physics Lab at The Johns Hopkins University of 1978's Seasat instrument. The CNES Ku- band device is a new design employing solid-state technology.

Topex was originally planned as a USD270 million NASA mission, targeted for 1989 launch, as a successor to Seasat. In 1985, NASA merged Topex with Poseidon. In April 1984, JPL awarded an eight month, USD1 million satellite definition contract to Fairchild, RCA Astro-Electronics and Rockwell based on their respective Solar Max, Tiros/DMSP and Navstar platforms. But Congress refused to grant new money in FY86 for the FY87 budget. NASA then issued requests for a proposal to the same three companies, in July 1986 with the intention of selecting a single satellite contractor in December. Fairchild won in January 1987, basing their design on the company's successful Solar Max multimission. Topex became the first complete spacecraft built by Fairchild. The total satellite cost ended 42 per cent over the USD121 million initial budget, largely due to NASA imposing more stringent requirements.

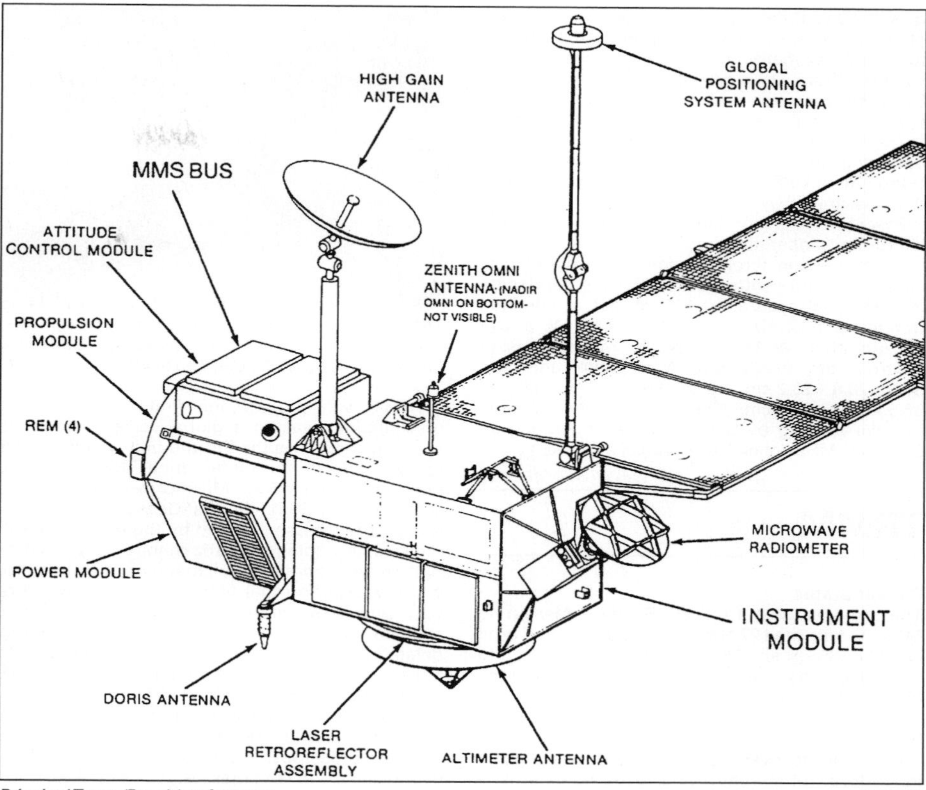

Principal Topex/Poseidon features 0517034

NASA projected total costs including launch and tracking at USD485 million in 1985 dollars and USD125 million for CNES. France's independently estimated its costs in 1991 to be about FFr900 million.

Topex/Poseidon was part of the World Ocean Circulation Experiment (WOCE), a major oceanographic field programme under the auspices of the World Climate Research Program (WRCP). WCRP combined satellite-based data with traditional observations to generate global 3-D ocean current structure model. The operational orbit was phased to overfly a NASA calibration site on the Harvest oil platform near Pt Conception (California) and a CNES calibration site on Lampione Rock near Lampedusa in the Mediterranean (the sites provide independent measures of satellite height and sea level). A three year extended mission began in August 1995 after the USD30 million annual operation costs were reduced to less than USD20 million. In late 1998, Topex/Poseidon was used to experiment with an autonomous orbit control mode. A software program was uploaded to allow the satellite to control its orbit. In May 1999 the satellite began to operate with the backup altimeter.

NASA anticipated at first that inaccuracies in orbital determination would reduce certainty relative to the geoid to 13.4 cm. But data from the DORIS system meant that orbital determination was no longer the major source of error and sea surface height accuracy proved to be less than 5 cm rms.

Topex/Poseidon's major science and application achievements include:

- the first decade-long global descriptions of seasonal and yearly ocean current changes;
- refined scientists' estimates of rising global sea level during the past decade;
- a new understanding of the role tides play in mixing the deep ocean;
- development of the most accurate ever global ocean tide models;
- provided the first global data set to test ocean general circulation model performance;
- demonstrated that global positioning system measurements in space could determine spacecraft positions with unprecedented accuracy, enabling rapid delivery of data.

Specifications

Topex/Poseidon

Launch: 10 August 1992 by Ariane 42P from Kourou. Operational mission began late February 1993
Mass: 2,380 kg in-orbit BOL
Contractors: JPL (project management, Topex payload), Fairchild Space Co (bus), Alcatel Espace (Poseidon altimeter), Honeywell Satellite Systems Div (antenna pointing system, RW assemblies), Dassault Electronique (DORIS receiver)
Altitude precision: 2–3 cm Topex, 2–5 cm Poseidon
Orbit: 1,331 × 1,332 km, 66.05°, repeating within 1 km every 9.9 days (127 revs). This is a 'frozen' orbit, holding a fixed eccentricity and argument of perigee to minimise manoeuvres to maintain the ground track. The selected altitude was a trade-off between gravity anomalies and atmosphere introducing unacceptable perturbations versus increased altimeter power for higher resolution in a higher orbit
Spacecraft: Based on NASA/Fairchild standard 3-axis Multi-Mission Spacecraft (MMS) bus designed for 3 to 5 year lifetime and 3.38 kW BOL solar power (2.14 kW after 5 years). 5.5 m length, 11.5 m span, 6.6 m high. Communications via TDRS. Power supply supported by 3 × 50 Ah batteries. Attitude control (nadir pointing) by RWs + magnetorquers; attitude determination by Earth/Sun sensors, star trackers, magnetometers, 3-axis gyros. Orbit control by 1/22 N hydrazine thrusters providing total ΔV 176 m/s (sufficient for 12 years)
Sensors: Topex/Poseidon carries two altimetry systems (see below), a microwave radiometer to correct for atmospheric effects, and three tracking systems:

Topex (US) Radar altimeter: operating with a prime channel at 13.6 GHz (Ku-band) and a secondary at 5.3 GHz (C-band), the two frequencies permitting corrections for ionospheric effects. The US/French altimeters share one 1.5 m dia dish, providing a 3 dB beamwidth of 2.7/1.1° at C/Ku. US radiated

Topex/Poseidon: built by NASA and launched by France, carrying sensors from both 0517231

power 2/9 W C/Ku. Topex mass 206 kg, consumed power 237 W. Microwave radiometer: 18/21/37 GHz to correct altimetry data for tropospheric water vapour effects. Orbit determination: operationally by TDRS, precision (science) by laser reflectors (13 cm radial accuracy) and experimental NASA/Motorola GPS Demonstration Receiver operating at 1.2276/1.5754 GHz, using a new technique of GPS differential tracking for accuracy of a few cm.

Poseidon (France) Radar altimeter: solid-state instrument operating on a time-share basis with US version through same 1.5 m dia dish at single Ku-band centred on 13.65 GHz, bandwidth 330 MHz, pulse duration 100 μs, pulse repetition frequency 1,700 Hz. Radiated power 5 W, power consumption 49 W. An 80C86 microprocessor controls the instrument. Microwave unit size 36.5 × 28.8 × 23.2 cm; 13 kg. Processing unit size 35.5 × 25.5 × 20.3 cm; 12 kg. Orbit determination: Doppler ranging by DORIS payload operating on 2,036/401 MHz uplink (10 cm radial accuracy).

TRMM

Current Status

The co-operative Tropical Rainfall Measuring Mission, launched 27 November 1997, entered orbit at 367 × 385 km and 35° inclination. Descending to 350 km, on board propulsion was used to raise the orbit to 402 km in August 2001.

Background

TRMM plans to investigate the interactions of water in all three of its physical phases to determine the contribution it makes to the Earth's process. Water substantially alters incoming and outgoing radiation and affects global air motions and heat fluxes through condensation and evaporation. Moreover, the latent heat released by tropical rain provides 75 per cent of the energy the atmosphere receives, thus playing a key role in driving global atmospheric circulation. Yet scientists know the amount of tropical rainfall to within only a factor of two over the oceans and little better, if any, over the arid continents and jungles. As a result, five of the leading climate models differ widely in their estimates of rainfall under identical boundary conditions – they differ by factors of two to three for the mean annual rain in the equatorial belt and even more widely for rainfall on a regional scale.

TRMM will provide the first comprehensive precipitation data on a global scale. It is the largest spacecraft ever built in-house by NASA Goddard (NASDA) 0517035

TRMM will provide the first comprehensive precipitation data on a global scale. The rainfall package comprises the PR (the first weather radar to fly in space), the multichannel TMI microwave radiometer and VIRS. TMI improves rain remote sensing of its SSM/I Special Sensing Microwave/Imager predecessor, carried by the military DMSP metsats, by adding a 10 GHz channel. TRMM's data system will operate in collaboration with EOSDIS, which will perform its first collection/archiving of very large space/ground validation data with this mission. CERES and LIS, funded by EOS (LIS was deleted from EOS in 1992), were added later to improve investigation of global change. CERES will measure upwelling cloud radiation, helping to identify how clouds warm/cool the planet. LIS will clarify the relation between cloud electrification and rain processes, and why it apparently differs over oceans and land masses.

TRMM will provide monitoring of tropical rainfall over three years, yielding monthly averages over 5 × 5° (500 × 500 km) cells. The US provided the spacecraft and four instruments (project management at NASA Goddard as an Earth Probe) and Japan launched the satellite aboard an H-2 rocket from Tanegashima. In addition, Japan added one instrument to the payload. NASA approved TRMM in 1990 following a 1986-88 Phase A evaluation programme. NASA put forward FY96 funding of US$23.1 million. (Previous years: FY95 48.1; FY94 63.1; FY93 51.5). NASDA for its part put up FY96 ¥3,630 million. (Previous years: FY95 4,964; FY94 2,537.5; FY93 919.1; FY92 874.) Total costs for the project are estimated at approximately US$650 million. NASA has proposed a similar follow-on mission is being studied, flying at 55° to cover all the continental US.

CERES will help to identify how clouds affect Earth's temperature (TRW) 0517416

Specifications

TRMM Satellite

Launch: 27 November, 1997 by H2 from Tanegashima, Japan

Mass: 3,820 kg (725 kg hydrazine to maintain altitude to ±1.25 km)

Orbit: planned 350 km circular, 35°

Spacecraft: 3-axis (0.2° knowledge) by Earth sensor (primary attitude reference), coarse/digital Sun sensors, gyros, four Ithaco 75 Nms RWs, magnetorquers, hydrazine thrusters, 1.1 kW min from twin GaAs solar wings + 2 × 50 Ah Ni/Cd batteries, S-band via TDRSS: data 32 kbit/s real-time; 2.048 Mbit/s playback. Instrument data rate 166.4 kbit/s. Two 264 Mbyte SSRs.

Sensors: PR Precipitation Radar (NASDA, Japan; Toshiba prime). 0.7 mm/h measurable rainfall. 13.796/13.802 GHz, pulse width 1.67 μs, pulse repetition frequency 2.776 Gz, peak power 578 W, beam width 0.71°, cross track scan angle ±17°, 93.5 kbit/s, 128 slotted waveguide antennas (using SSPAs) in 2.3 × 2.3 m planar array, mass 470 kg, power 250 W.

VIRS Visible Infra-Red Scanner (NASA). Channels (μm): 1 0.63, 2 1.6, 3 3.75, 4 10.8, 5 12. IFOV ±0.210° at 350 km, ±45° cross track. 2 km res at 10 kbit/s/channel. Mass 49 kg, power 36 W.

TMI TRMM Microwave Imager (NASA). 9 channels (GHz): 10.65, 19.35, 21.3, 37, 85.5, mass 53 kg, power 43 W, data rate 8.5 kbit/s

CERES Cloud/Earth Radiant Energy System (NASA EOS; TRW). Channels (μm): 1 0.3–3.5, 2 8.0–12.0,

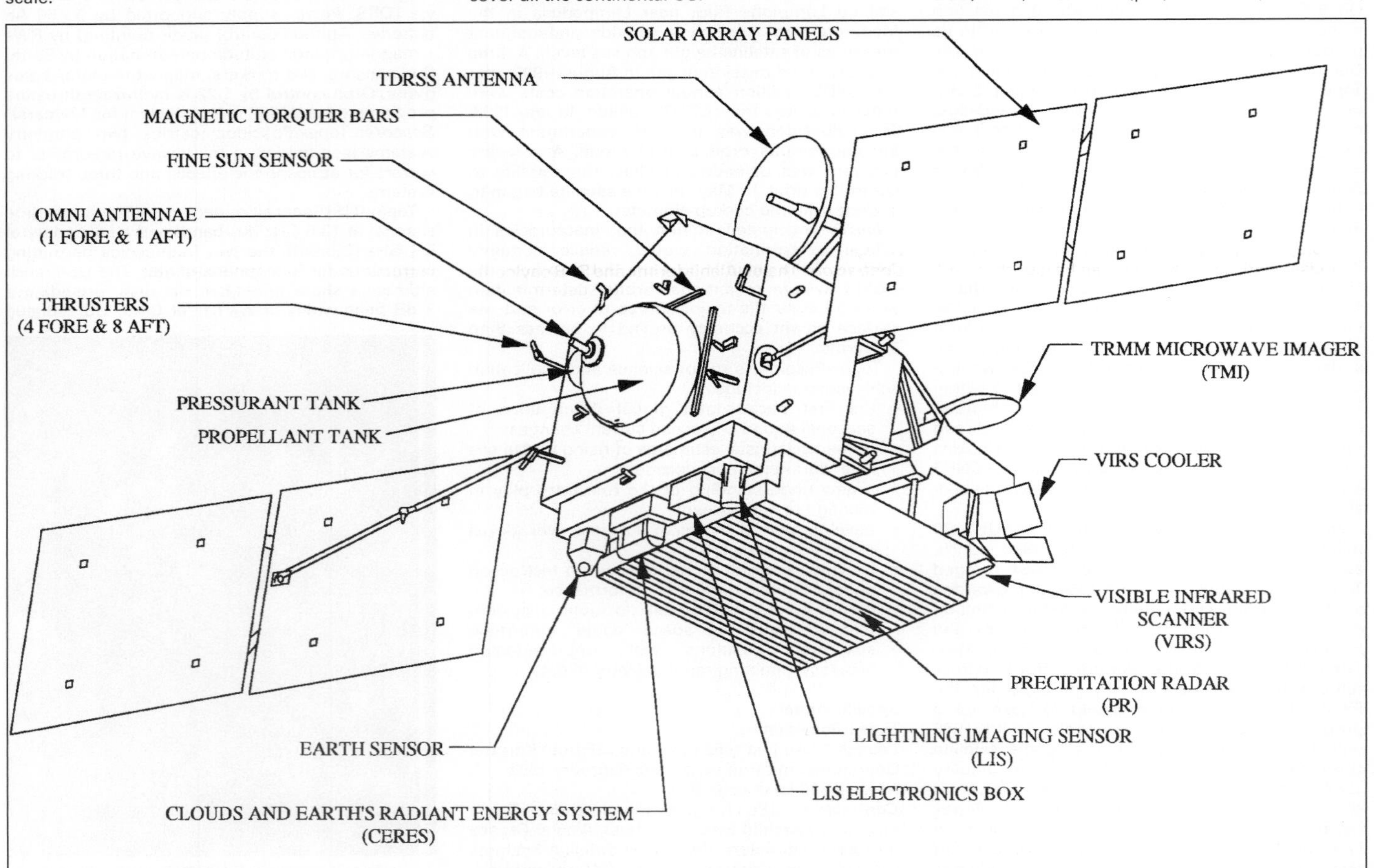

TRMM principal features (NASA) 0517230

3 0.3–>50, scan angle ±42.5° in 6.6 s cycle, 8 kbit/s, IFOV 25 km nadir, mass 45 kg, power 45 W.

LIS Lightning Imaging Sensor (NASA MSFC in-house). 0.7774 μm, scan angle ±42.5° cross track, res 5 km nadir, 0.5 kbit/s (6 kbit/s at full background), mass 20 kg, power 24 W.

Italy

COSMO-SkyMed

Current Status

COSMO-Skymed 1, the first in a series of radar imaging satellites developed for the Italian Space Agency (ASI) and the Italian Ministry of Defence, was launched on 8 June 2007. The second was launched on 10 December 2007, and the third has been scheduled for the second half of 2008.

Background

COSMO stands for 'Constellation of Satellites for Mediterranean Basin Observation'. In 1996 the Italian government provided initial funding for a national Earth observation programme. The 1998–2002 Italian National Space Plan included the COSMO-SkyMed dual-use Earth observation programme of ASI. In the summer of 2001, the Italian Ministry of Defence became a partner with the Italian Ministry of Research in the COSMO-SkyMed programme. However, the dual-use nature of COSMO-SkyMed, i.e. civil (research and commercial) and military use of its data, resulted in a virtually classified programme.

In 2000, ASI signed a cooperative agreement with CONAE (Comisión Nacional de Actividades Espaciales) of Argentina. Under this agreement, referred to as SIASGE (Sistema Italo Argentino de Satélites para la Gestión de Emergencias or Italian-Argentinian satellite system for emergency management), two L-band SAR satellites in the Argentine SAOCom (SAR Observation & Communications Satellite) system will operate jointly with the Italian COSMO-SkyMed constellation of X-band SAR satellites to twice-daily coverage capability for emergency management. Launch of SAOCom-1A is expected by 2008.

On 29 January 2001, an intergovernmental memorandum of understanding, known as ORFEO (Optical and Radar Federated Earth Observation) was signed by Italy and France. The objective of this agreement was bilateral co-operation on a "dual high-resolution Earth observation system" comprising two Pleiades optical satellites, developed by France, and the COSMO-Skymed radar constellation.

The COSMO-SkyMed SAR constellation is composed of four identical satellites, having each an operational in-orbit service of at least 5 years. The deployment of the constellation is performed in a step-like progression. After one satellite is commissioned, a new satellite can be launched and deployed. Eventually there will be four satellites with phased orbits 90° apart in the same orbital plane.

They are based on the Thales-Alenia Prima bus, which was developed for small Earth observation satellites and communications satellites such as Globalstar. Each satellite is three-axis stabilised and comprises the main body (bus), two deployable solar arrays, and a SAR antenna. A lower service module carries all vital subsystems, including the propulsion module and the payload support hardware; the payload module at the top carries the Payload Data Handing and Transmission subsystem, and the AOCS (Attitude and Orbit Control Subsystem). The bus structure is made of carbon fibre reinforced plastic, while the service module and payload module consist of aluminium alloys.

The SAR-2000 (Synthetic Aperture Radar-2000) payload is a multi-mode instrument. The programmable system can be modified in terms of swath size, spatial resolution, and polarization. The SAR transmitter/receiver system has an electrically steerable, multi-beam antenna which concentrates the transmitted energy into narrow beams in the cross-track direction. The characteristics of the transmitted pulses and the reflected signal determine the spatial resolution and coverage. The swath width is 10–200 km, depending on support mode.

Each satellite follow a circular sun-synchronous orbit with a ground track that repeats after approx. 237 orbits or 16 days, although each satellite has a near revisit time of 5 days. The overall system requirements include: full global observation coverage with all weather, day/night acquisition capability; collection capability over large areas during a single pass, with along-track stereo imaging; high image quality – spatial and spectral resolution – that allows accurate image interpretation with different degrees of detail; ground track repeatability of better than 1 km; fast response times for user requests.

The constellation can be operated in the nominal orbital configuration, when the key requirement is ground track repeatability, or in the interferometric orbital configuration, which permits 3-D SAR imagery by combining two radar measurements of the same target obtained from slightly different incidence angles.

Although the first constellation satellite SAR instruments (SAR-2000) observe in X-band (9.6 GHz with a wavelength of 3.1 cm), multi-mode scenarios (X-, C- L- and P-band) are planned for the future. Once the constellation is fully operational, ASI intends to develop and operate a bistatic, interferometric mission, known as SABRINA. This would use a passive satellite, named BISSAT (Bistatic and Interferometric SAR Satellite), which co-orbits in close formation with one of the COSMO-SkyMed spacecraft.

Specifications

Launches: COSMO-Skymed 1, 02:34 UT on 8 June 2007 by Delta II 7420 from SLC-2 at Vandenberg AFB, California.

COSMO-Skymed 2, 02:31 UT on 9 December 2007 by Delta II 7420 from SLC-2 at Vandenberg AFB, California.

Orbit: 621 × 625 km, 97.9° orbit. 4 equi-phased orbits, sun-synchronous, dawn-dusk frozen. Orbit cycle (1 satellite) 16 days. Local Time of Ascending Node at 6:00 a.m.

Launch mass: 1,870 kg

Design life: 5 years (nominal)

Contractors: Thales-Alenia (prime and SAR payload), Galileo Avionica (solar arrays), COM DEV (Li-ion battery), Laben SpA (Lagrange GPS receiver).

Payload: SAR-2000 (Synthetic Aperture Radar-2000) instrument. X-band SAR with a 5.75 × 1.5 m active antenna.

Principal applications: Full global observation coverage with all weather, day/night acquisition capability.

Configuration: Thales-Alenia Prima bus. Onboard payload recording with a 300 Gbit solid-state memory. Payload data are compressed, encrypted and downlinked in X-band at 300 Mbit/s.

AOCS: Star trackers and gyros plus a GPS receiver. AOCS provides an antenna steering capability of ±2° in yaw and a re-pointing capability to the left side of the ground track.

TT&C: S-band.

Power system: Two solar arrays (total area 18.3 m^2) provide a minimum of 3.6 kW at EOL, using triple-junction solar cells. Li-ion battery.

Italy is developing the COSMO-Skymed series of 4 synthetic aperture radar satellites (Thales Alenia) 1340630

Russian Federation

Meteor

Current Status

Meteor 3M-1 stopped functioning on 5 April 2006 after a malfunction of its power supply system.

Background

Similar to the US NOAA series, more than 50 Meteor polar orbiting spacecraft have been launched over four generations. The series was preceded by a Kosmos series of 10 experimental meteorological satellites, starting with Kosmos-44, which was launched on 25 August 1964. After three years of testing under the Kosmos label, the first Meteor 1 appeared in 1969, followed by the more capable Meteor 2 in July 1975 and Meteor 3 in November 1984. The latest version, known as Meteor 3M, was introduced in 2001.

The All-Russian Research Institute for Electromechanics (VNII Elektromekaniki), designed and built the Meteor/Resurs-O satellites. The Soviet Union retired the last of the 31 Meteor 1s in June 1984. These included four Meteor-Prirodas, based on the Meteor 1 bus, that were placed in sun-synchronous orbits 1977–81 as test beds for the Resurs-O programme. In 1982, the SL-14 Tsyklon began to replace the Vostok SL-3 as the launch vehicle for putting the Meteors into orbit. Meteor 1 and the first Meteor 2 satellites typically flew at 800 to 900 km, but when the Soviet Union switched to the Tsyklon, the altitude was increased to 950 km for Meteor 2 and to 1,200 km for Meteor 3. This provided a more comprehensive coverage. The new generation 3M-1 flew at 1,012 km. Apart from 3M-1, all Meteors have launched from Plesetsk. Meteor 3–05 broke new ground in 1991 when it carried NASA's TOMS ozone monitoring instrument.

In conjunction with the normal launch some Meteors have released third party satellites. Italy's Temisat entered space on a Meteor 2 and Germany's Tubsat B on Meteor 3–6, which also included France's Scarab Earth radiation budget instrument and DARA's PRARE unit.

Meteors have provided a daily weather review for more than two-thirds of the globe, on clouds, ice cover, atmospheric radiation, weather fronts and jet streams. Kosmos 122, launched 25 June 1966 from Baikonur, 589 × 643 km, 65°, later became the first dedicated meteorological satellite identified by the Soviets. President de Gaulle of France witnessed the launch, the first Westerner to visit a Soviet launch site. By 1978, according to Moscow Radio, Meteors had transmitted one million images. These satellites cut the sailing time of Soviet ships because the satellites indicated the best courses to avoid storms, winds and ice. This maritime contribution alone was officially estimated to be worth more than R1 million. Meteors also monitored clear-air turbulence so that the airliner crews could be warned of potential hazards. In 1980, Meteor satellites made their first hail warnings, a prediction of particular value in Georgia, with its 600,000 hectares of vineyards.

Meteor 1

Meteor 1 used a cylindrical bus made up of two sealed compartments. The upper section contained the power supply, attitude control, stabilisation systems and radio equipment. The lower section housed two TV cameras, an infra-red scanner and a radiation budget sensor. After the launch of Meteor 1 on 26 March 1969 from Plesetsk (644 × 713 km, 81°), launches followed at the rate of three to four annually so that the system was always operational

Meteor Launch Record

Name	Launch date	Orbit	Payload
Meteor [failed]	1 February 1969		
Meteor 1-1	26 March 1969	644 × 713 km, 81.2°	TV, IR, AC
Meteor 1-2	6 October 1969	630 × 690 km, 81.2°	TV, IR, AC
Meteor 1-3	17 March 1970	537 × 635 km, 81.2°	TV, IR, AC
Meteor 1-4	28 April 1970	637 × 736 km, 81.2°	TV, IR, AC
Meteor 1-5	23 June 1970	863 × 906 km, 81.2°	TV, IR, AC
Meteor 1-6	15 October 1970	633 × 674 km, 81.2°	TV, IR, AC
Meteor 1-7	20 January 1971	630 × 679 km, 81.2°	TV, IR, AC
Meteor 1-8	17 April 1971	620 × 646 km, 81.2°	TV, IR, AC
Meteor 1-9	16 July 1971	618 × 650 km, 81.2°	TV, IR, AC
Meteor 1-10	29 December 1971	880 × 905 km, 81.2°	TV, IR, AC
Meteor 1-11	30 March 1972	878 × 903 km, 81.2°	TV, IR, AC
Meteor 1-12	30 June 1972	897 × 929 km, 81.2°	TV, IR, AC
Meteor 1-13	26 October 1972	893 × 904 km, 81.2°	TV, IR, AC
Meteor 1-14	20 March 1973	882 × 903 km, 81.2°	TV, IR, AC
Meteor 1-15	29 May 1973	867 × 909 km, 81.2°	TV, IR, AC
Meteor 1-16	5 March 1974	853 × 906 km, 81.2°	TV, IR, AC
Meteor 1-17	24 April 1974	877 × 907 km, 81.2°	TV, IR, AC
Meteor 1-18 (Meteor-Priroda 1)	9 July 1974	877 × 905 km, 81.2°	MSU-M, MSU-S, TV, IR, SHF
Meteor 1-19	28 October 1974	855 × 917 km, 81.2°	TV, IR, AC
Meteor 1-20	17 December 1974	861 × 910 km, 81.2°	TV, IR, AC
Meteor 1-21	1 April 1975	877 × 906 km, 81.2°	TV, IR, AC
Meteor 1-22	18 September 1975	867 × 918 km, 81.2°	TV, IR, AC
Meteor 1-23	25 December 1975	857 × 918 km, 81.3°	TV, IR, AC
Meteor 1-24	7 April 1976	863 km × 906 km, 81.2°	TV, IR, AC
Meteor 1-25 (Meteor – Priroda 2)	15 May 1976	866 × 908 km, 81.3°	TV, IR, AC
Meteor 1-26	15 October 1976	871 × 904 km, 81.3°	TV, IR, AC
Meteor 1-27	5 April 1977	869 × 909 km, 81.2°	TV, IR, AC
Meteor 1-28 (Meteor – Priroda 3)	29 June 1977	602 × 685 km, 97.9°	TV, IR, AC
Meteor 1-29 (Meteor – Priroda 4)	25 January 1979	628 × 656 km, 98.0°	MSU-M, MSU-S, SI-GDR, R10-M
Meteor 1-30 (Meteor – Priroda 5)	18 June 1980	589 × 678 km, 98.0°	MSU-M, MSU-S, Fragment, R10-M,
Meteor 1-31 (Meteor – Priroda 6)	10 July 1981	574 × 612 km, 97.9°	MSU-SK, MSU-E, MSU-M, MSU-S, R10-M, SHF
Meteor-2-1	11 July 1975	872 × 906 km, 81.3°	TV, IR, SM, RMK-2
Meteor 2-2	6 January 1977	893 × 932 km, 81.3°	TV, IR, SM, RMK-2
Meteor 2-3	14 December 1977	872 × 906 km, 81.2°	TV, IR, SM, RMK-2
Meteor 2-4	1 March 1979	857 × 908 km, 81.2°	TV, IR, SM, RMK-2
Meteor 2-5	31 October 1979	877 × 904 km, 81.2°	TV, IR, SM, RMK-2
Meteor 2-6	9 September 1980	868 × 906 km, 81.2°	TV, IR, SM, RMK-2
Meteor 2-7	14 May 1981	868 × 904 km, 81.3°	TV, IR, SM, RMK-2
Meteor 2-8	25 March 1982	954 × 976 km, 82.5°	TV, IR, SM, RMK-2
Meteor 2-9	14 December 1982	836 × 904 km, 81.2°	TV, IR, SM, RMK-2
Meteor 2-10	28 October 1983	780 × 901 km, 81.2°	TV, IR, SM, RMK-2
Meteor 2-11	5 July 1984	954 × 974 km, 82.5°	TV, IR, SM, RMK-2
Meteor 2-12	6 February 1985	950 × 975 km, 82.5°	TV, IR, SM, RMK-2
Meteor 2-13	26 December 1985	952 × 975 km, 82.5°	TV, IR, SM, RMK-2
Meteor 2-14	27 May 1986	953 × 974 km, 82.5°	TV, IR, SM, RMK-2
Meteor 2-15	5 January 1987	950 × 973 km, 82.5°	TV, IR, SM, RMK-2
Meteor 2-16	18 August 1987	954 × 974 km, 82.6°	TV, IR, SM, RMK-2
Meteor 2-17	30 January 1988	952 × 975 km, 82.5°	TV, IR, SM, RMK-2
Meteor 2-18	28 February 1989	947 × 973 km, 82.5°	TV, IR, SM, RMK-2
Meteor 2-19	29 June 1990	951 × 974 km, 82.5°	TV, IR, SM, RMK-2
Meteor 2-20	28 September 1990	950 × 974 km, 82.5°	TV, IR, SM, RMK-2
Meteor 2-21	31 August 1993	953 × 975 km, 82.6°	TV, IR, SM, RMK-2
Kosmos 1612 (failure)	27 November 1984		
Meteor 3-1	24 October 1985	1,235 × 1,263 km, 82.6°	MR-2000M, MR-900B, IR, SM, RMK-2, Ozon-M
Meteor 3-2	26 July 1988	1,198 × 1,221 km, 82.5°	MR-2000M, MR-900B, IR, SM, RMK-2, Ozon-M
Meteor 3-3	24 October 1989	1,191 × 1,228 km, 82.6°	MR-2000M, MR-900B, Klimat, SM, RMK-2, IR, Ozon-M
Meteor-3-4	24 April 1991	1,195 × 1,230 km, 82.6°	MR-2000M, MR-900B, Klimat, SM, RMK-2, Ozon-M
Meteor-3-5	15 August 1991	1,195 × 1,230 km, 82.5°	MR-2000M, MR-900B, Klimat, SM, RMK-2, IR, TOMS
Meteor-3-6	25 January 1994	1,186 × 1,207 km, 82.5°	MR-2000M, MR-900B, Klimat, SM, RMK-2, IR, ScaRaB, PRARE
Meteor-3M-1	10 December 2001	996 × 1,015 km, 99.7°	MR-2000M1, Klimat, MIVZA, MTVZA, MSU-E, SFM-2, KGI-4C, MSGI-5EI, RRA, SAGE-III

and provided a continuous survey of atmospheric conditions from pole to pole in a swath up to 1,500 km wide. The information was dumped to ground stations in passes lasting only a few minutes. Three sites received these transmissions, the USSR Hydrometeorological Service in Moscow, Novosibirsk (Siberia) and Khabarovsk (Pacific Coast). It took about 1 hour 30 minutes for the Soviets to process the data and transmit it to users. From Meteor 1–10, launched on 30 December 1971, the Meteor 1s used an 81° orbit at about 890 km. This satellite carried the first Soviet meteorological automatic picture transmission compatible with Western receivers. The bus had a small Fakel plasma thruster for orbital adjustments.

From Meteor 1-28 (also known as Meteor-Priroda 3), launched 29 June 1977 (re-entered 28 August 1993), orbits were retrograde, sun-synchronous and circular, at 630 km, 98°. This was the first of four Meteor 1/Prirodas flown as test beds for Resurs-O, which became operational in 1988 with Kosmos 1939. The last Meteor 1 was 1–31, launched 10 July 1981.

Meteor 1–31 also known as Priroda ('Nature') went into space 10 July 1981 aboard a SL-3 from Baikonur into 630 km, 98° sun-synchronous orbit.

The Meteor 1 had a main sensor consisting of a 32-channel scanning radiometer and 33-channel microwave radiometer for surface observations. The Bulgaria 1300 programme provided some of

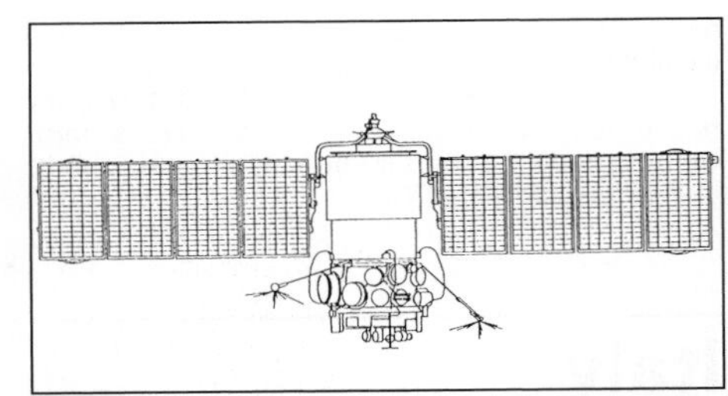

Meteor 2 second-generation weather satellite 0517041

the equipment. At the end of 1983, this equipment was being used to determine snow depths on all Soviet agricultural regions to assist in planning the spring work season and forecasting the 1984 harvest.

Meteor 2

Meteor 2 was a smaller, lighter spacecraft than Meteor 1. Introduced experimentally, while the Meteor 1 programme continued, Meteor 2-1 (11 July 1975), established the new operational orbits of 900 km, 82.2°. The satellites had a longer life, and included scanning radiometers. Various combinations of orbital planes were tried, usually with two to three satellites giving passes at 6 to 8 h intervals. From 1981, the Soviets maintained two or three in operation at any one time, although the Russians let the two remaining satellites fall silent in summer 1992. Automatic picture transmissions for visible imagery were switched off when the satellite was in the Earth's shadow.

Meteor-2-22 (launched 31 August 1993) carried an additional payload, a Fizeau RetroReflector Array for Satellite Laser Ranging (SLR) applications. The retroreflector array consisted of three corner cubes, with the two outer corner cubes pointing at 45° angles relative to the central cube. SLR tracking was used for precise orbit determination until October 1998.

Meteor 3

The main design change over the Meteor 2 was the addition of a payload truss which facilitated the addition of new instruments. The first official Meteor 3 launch occurred on 24 October 1985 from Plesetsk using SL-14, after a launch failure involving the prototype spacecraft (Kosmos 1612) the previous year. Although employing the same 82.6° inclination as the SL-14 Meteor 2s, Meteor 3-1 was inserted into a much higher 1,227 × 1,251 km. In retrospect, the launch of Kosmos 1612 on 27 November 1984 (135 × 1,230 km) appears to have been a precursor Meteor 3. This launch failed when Tsyklon's upper stage did not re-ignite. Meteor 3-1 incorporated plasma thrusters, which were used to lower its orbit to a mean altitude of 1,200 km in November-December 1985. In late June 1986, the Soviets lowered the orbit by another 4 km. Meteor 3-2 appeared 26 July 1988, followed by Meteor 3-3 on 24 October 1989. Both satellites were replaced in 1991 by Meteors 3-4 (24 April) and 3-5 (15 August). 3-5 carried NASA Goddard's Total Ozone Mapping Spectrometer to extend Nimbus 7 observations, which discovered Antarctica's ozone hole in 1987. The instrument failed December 1994, well beyond its two year design life. Data went via NASA's Wallops Flight Facility to the Russian Obninsk station.

Meteor 3-6 carried France's ScaRaB Earth radiation balance instrument and a German PRARE (Precision Range and Range Rate Experiment) geodetic instrument. The ScaRaB (Scanner for Radiation Budget) project began in 1987, made up of three 40 kg units built for FFr65 million. Data from Scarab went first to Obninsk and Medvezhi Ozera for relay to NPO Planeta in Moscow and then to CNES Toulouse. The instrument monitored four bands: visible 0.5–0.7, solar 0.2–4, total 0.2–50, infra-red 10.5–12.5 μm. The US NOAA and ERBS satellites correlated the data. The first instrument operated for one year (March 1994 – March 1995). However, the satellite was in a non sun-synchronous polar orbit, with a long precessional cycle (about 200 days), so that diurnal coverage was poor.

Meteor 3M

The Russian Federation planned to continue the successful series during the mid-1990s, but funding problems hampered progress. The new generation Meteor 3M-1, eventually launched in December 2001, was based as much as possible on the old platform. The satellite operated in a circular, sun-synchronous orbit inclined at 99.6° with a 09:15 a.m. ascending node. The payload included the MIVZA and MTVZA radiometers, which had limited capabilities due to technical problems related to their scanning mode. Due to a non-functioning 466 MHz transmitter, the satellite had limited capabilities for MR-2000M and KLIMAT direct data broadcast.

In 2002, the original Meteor 3M satellite design was revised considerably. It is planned to develop two Meteor 3M satellites on the basis of a "Resurs" type of unified heavy platform. The satellites will operate in a sun-synchronised orbit and provide operational hydro-meteorological and helio-geophysical information on the atmosphere, ground surface and oceans. The first of these is expected to be launched on a Soyuz-2.1b-Fregat in summer 2008.

Meteor 3M-1 carried NASA's SAGE III Stratospheric Aerosol and Gas Experiment (NASA) 1340670

Meteor 3M-1 was the first of Russia's third generation of polar orbiting weather satellites (NASA) 1340671

Specifications
Meteor

Launched: **Meteor 1**: Vostok from Plesetsk. **Meteor 2 and 3**: typically annual by Tsyklon from Plesetsk; 3M on 10 December 2001 by Zenit 2 from Baikonur.

Mass: **Meteor 1**: 2,200 kg. **Meteor 2**: 1,300 kg on-orbit; **Meteor 3**: 2,215 kg on-orbit, payload 500–700 kg; **Meteor 3M**: 2,500 kg on-orbit, payload 900 kg.

Orbit: **Meteor 1**: approx. 650 km for the first nine satellites, then approx. 900 km, 82.5° **Meteor 2**, approx. 900 km for the first 10 satellites, then approx. 950 km, 82.5° **Meteor 3**: approx. 1,200 km, 82.5° **Meteor 3M-1**: 1,020 +/–20 km, 99.64° sun-synchronous, period: 105.5 min. 09.15 ascending node equatorial crossing

Spacecraft: 3-axis stabilisation (0.5° accuracy; better for 3M), 4.2 m long, 1.4 m diameter cylindrical body with twin 10 m span, 2 m-wide solar wings providing 500 W to the payload (3M: 1 kW). Design life 2 years (3M, 3 years). Meteor 3 adjusts orbit by Fakel SPT plasma thrusters. 3M uses GPS/Glonass for precise orbit determination.

Transmissions directed at three primary sites (Moscow, Novosibirsk in Siberia and Khabarovsk on the Pacific coast; Tashkent was lost with independence; new sites were planned for Murmansk and Petropavlovsk-Kamchatski) in conjunction with >80 smaller ground stations.

Internationally-compatible APT transmissions are made on 137–138 MHz to 15,000 CIS terminals; there is also a 10 W analogue store/forward mode at 466.5±0.125 MHz 3M used 1.69–1.71 GHz digital). Data are processed within 90 minutes by the Hydrometeorological Centre.

Sensors: **Meteor 1.** TV (also known as MR-600A) vidicon camera TV system, spectral range: 0.4–0.8 μm, swath width 1000 km; resolution 1.25–3 km.

IR (Lastocha) infra-red radiometer, spectral range 8–12 μm; swath width 1000 km; resolution 15 km.

AC radiation budget sensor, spectral range 0.3–30 μm, swath width 2,500 km, resolution 40 km × 50 km.

Meteor 2. TV (MR-600A) vidicon camera TV system, spectral range 0.5–0.7 μm, swath width 2,100 km; medium resolution (2 km), 0.5–0.7 μm, swath width 2,400 km; high resolution (1 km), 8–12 μm. Two satellites covered 80 per cent of the Earth's surface in 6 h.

IR global scanning infra-red radiometer, spectral range 8–12 μm, swath width 2,600 km, resolution 8 km.

SM multichannel spectrometer (Device 174-K), spectral range 9.65–18.7 μm, swath width 1,000 km, resolution 42 km.

RMK-2 radiation measurement complex measured flux densities of protons in the 5–90 MeV energy range and electrons in the 0.15–3.0 MeV energy range.

Meteor 3. MR-2000M TV camera system, 0.7–1.4 km resolution; spectral range 0.5–0.8 μm, swath width 3,100 km.

MR-900B TV camera system, spectral range 10.5–12.5 μm, swath width 2,600 km, resolution 1–2 km.

Klimat infrared radiometer, spectral range 10.5–12.5 μm, swath width 3,100 km, spatial resolution 3 × 3 km.

SM 10-channel infra-red spectrometer, swath width 1,000 km, 42 km resolution, spectral range 9.4–19.68 μm.

8-channel ultraviolet backscatter spectrometer, swath width 200 km, 3–5 km altitude resolution.

Ozon-M 4-channel UV ozone monitor, spectral range 0.25–0.29, 0.37–0.39, 0.6–0.64, 0.99–1.03 μm, 2 km altitude resolution.

RMK-2 electron/proton detectors, energy range 0.15–3.1 and 1–600 MeV.

Meteor 3-5 also carried NASA's TOMS. Meteor 3-6 carried France's Scarab Earth radiation balance instrument, DARA's PRARE unit (see ERS-1 for background).

Meteor 3M-1: US SAGE-III instrument.

MR-2000-M1 TV Camera System (0.5–0.8 μm) at a local solar angle not less than 5°, spatial resolution 0.7–1.4 km.

Klimat infrared radiometer 10.5 to 12.5 μm.

MIVZA microwave humidity sounder, channel scanning radiometer within the 20–90 GHz.

MTVZA microwave imaging/sounding radiometer (19, 33, 36.5, 42, 48, and 91.65 GHz plus oxygen absorption lines at 52–57 GHz and water vapour at 22.235 and 183.31 GHz).

A model of the Meteor 3 weather satellite first launched in 1985 0054352

MSU-E high-resolution multispectral pushbroom imager (0.5–0.6, 0.6–0.7, 0.8–0.9 μm), spatial resolution 38 m, swath width 76 km.

SFM-2 (Spectrophotometer-2) estimates vertical ozone distribution using 4 channels, spectral range 0.25–0.6 μm.

KGI-4C radiation monitoring system.

MSGI-MKA spectrometer for geoactive measurements.

KGI-4C radiation monitoring system.

RRA retroreflector array.

Resurs-DK

Current Status

Resurs-DK1 was launched on 15 June 2006 from Baikonur cosmodrome. The PAMELA experiment was switched on for the first time on 21 June and it has operated continuously since 11 July. The satellite commissioning phase was completed on 15 September and it remains operational as of late January 2008.

Background

The Resurs-DK1 is an experimental remote sensing satellite manufactured by the Russian space company TsSKB-Progress. It is based on the Terilen/Neman class reconnaissance satellite in use since the 1980s, using the Yantar satellite bus. Originally developed in 1996, the first Resurs-DK was supposed to be launched in 1999 or 2000. However, its launch was postponed several times due to financial constraints. As a result, a new, simpler version evolved with a smaller number of spectral bands, and in 1998 it was designated Resurs-DK1.

It is Russia's first civil Earth observation satellite designed to perform high resolution, multi-spectral remote sensing of the Earth's surface and to acquire high-quality images in near real-time. Unlike earlier Resurs missions, which stayed in orbit for a few months, captured images on film and released the canisters over Russia, this satellite transmits data via a high-speed radio link to Russian ground stations.

The status of natural resources, natural disasters, sea ice conditions and polar weather are available within a few hours to national and international organisations, as well as to private commercial customers. The imaging system has a wide swath width (28.3km) and high resolution (1 m panchromatic and 2–3 m in narrow spectral bands). The Resurs-DK1 can image 450,000 to 750,000 km^2 of the terrestrial surface per day with a resolution of one metre. The satellite relies on Russia's constellation of Glonass satellites for navigation.

In addition to the main payload, Resurs-DK1 has plenty of space and power available for scientific and applied research projects. The PAMELA and Arina scientific instruments are "piggybacking" on board the satellite. **PAMELA** is a part of the Russian-Italian Mission (RIM) research programme, which began with experiments on the Mir space station in the mid-1990s. The full name of the first experiment is RIM-PAMELA (Russian-Italian Mission – Payload for Antimatter Matter Exploration and Light-Nuclei Astrophysics). A joint Italian-Russian project which also involves Sweden, the United States and Germany, it is the first project specifically designed to investigate dark matter and dark energy from Earth orbit. In its first few months of operation, PAMELA recorded more of these particles than all previous observations combined.

PAMELA is mounted in a dedicated pressurised container attached to the side of the satellite. During launch and orbital manoeuvres, the container is secured against the body of the satellite. During data-taking it is swung up to give Pamela a clear view into space. PAMELA uses a permanent magnet spectrometer with a variety of specialised detectors that can measure with unprecedented precision and sensitivity the abundance and energy spectra of cosmic rays electrons, positrons, antiprotons and light nuclei over a very large energy range, from 50 MeV to hundreds of GeV, depending on the species. Additional objectives are: long-term monitoring of the solar modulation of cosmic rays; measurements of energetic particles from the Sun; high-energy particles in the Earth's magnetosphere and Jovian electrons.

ARINA is a small (8 kg) Russian automatic scintillation spectrometer designed to record charged particle bursts with the objective of refining a new technique for using space observations to predict earthquakes. The intention is to detect energetic particles that appear several hours before a tremor and are thought to be short-lived precursors of an earthquake.

The main Resurs-DK1 ground station is located at the Research Centre for Earth Operative Monitoring (NTsOMZ) in Moscow, with a back-up downlink station is located at Khanty-Mansiysk West Siberia. There are two to three downlink sessions per day. The reception antenna at NTsOMZ is a parabolic reflector 7 m in diameter, equipped with an azimuth-elevation rotation mechanism and two frequency multiplexed radio channels. The average volume of Pamela data transmitted is 15 Gigabytes/day. Raw data and preliminary processed data are sent by a normal internet line to the main storage centre at MEPhI in Moscow. From here, GRID infrastructure is used to move data to CNAF (Bologna, Italy), a specialized computing centre of INFN.

Resurs-DK1 will be replaced by the new-generation Resurs-II spacecraft, which will image Earth's surface with a resolution of 0.5 – 2 m in the visible and infra-red bands. The first of these may be launched in 2009–2010.

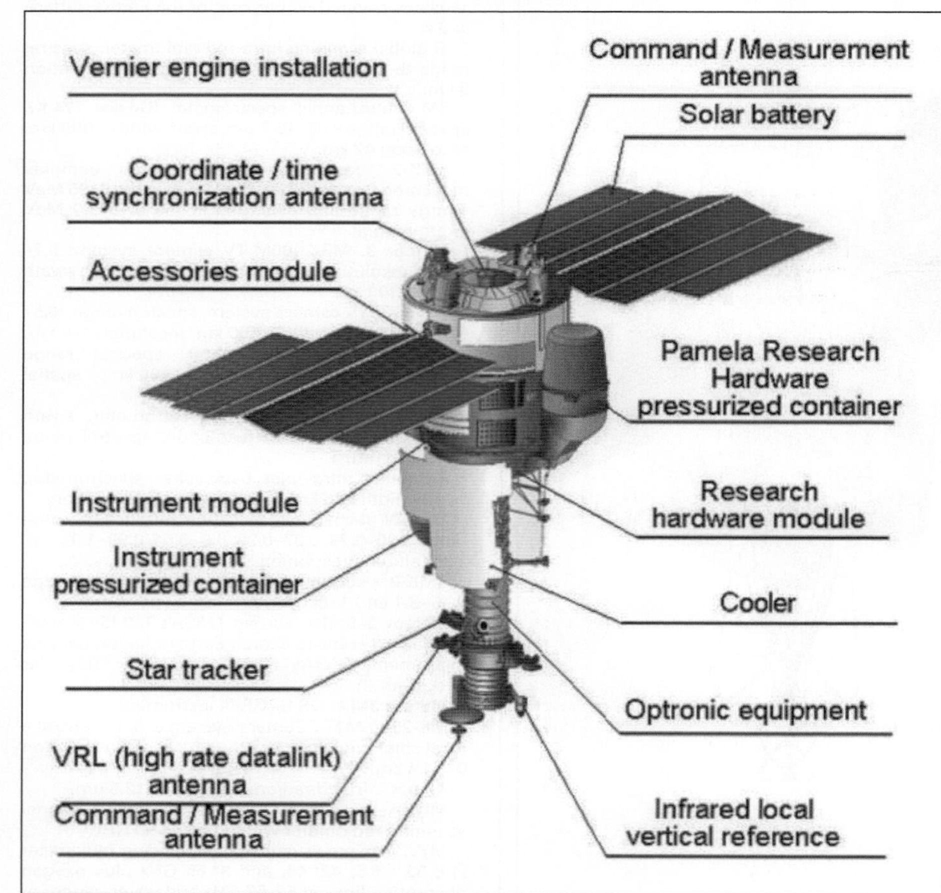

The main features of the Resurs-DK1 spacecraft (INFI) 1342012

Specifications

Launch: 08:00 GMT on 15 June 2006 by Soyuz-U from Baikonur.

Orbit: 356 × 585 km, 69.9° inclination, period 94 min.

Mass: 6,570 kg at launch. Payload mass 1,200 kg.

Dimensions: Height: 7.4 m

Design life: 3 years

Configuration: Yantar bus, modular design: assembly compartment with propulsion module and two solar arrays; instrumentation bay and the purpose equipment bay. 3-axis stabilised. Axis orientation accuracy is 0.2 arcmin; angular velocity stabilisation accuracy 0.005°/s. Onboard data storage capacity of 768 Gbit.

Communications: Payload data link X-band at 8.2–8.4 GHz; downlink data rate of up to 300 Mbit/s.

TT&C: Primary ground station: Research Centre for Earth Operative Monitoring (NTsOMZ) in Moscow. Back-up downlink station at Khanty-Mansiysk, West Siberia.

Power: Solar array span: approx. 14 m. Area 36 m^2. Average power 2 kW (PAMELA requirement 360 W).

Payload: Geoton-1 opto-electronic pushbroom imaging instrument. Panchromatic and multispectral imagery in 4 bands. Survey observation mode provides image scenes of up to 2,100 km along-track. Spacecraft body-pointing (± 30°) in cross-track direction gives wider field of view.

RIM-PAMELA (Payload for Antimatter Matter Exploration and Light Nuclei Astrophysics): mass 470 kg, size: 1.3 × 0.7 × 0.7 m. Comprises: magnetic spectrometer, calorimeter, shower tail catcher scintillator, neutron detector.

ARINA earthquake precursor energetic particle detector, mass 8 kg. Identify and determine the energy of electrons in the energy range 3–30 MeV, and protons in energy range 30–100 MeV.

Ukraine

Okean-O1/Okean-O/Sich

Okean Programme Launches[1]

	Launch	Orbit	End Ops
E1/Kosmos-1076[2]	12 February 1979	593 × 620 km, 82.5°	31 March 1980
E2/Kosmos-1151[2]	23 January 1980	613 × 640 km, 82.5°	13 October 1981
OE1/Kosmos-1500	28 September 1983	643 × 679 km, 82.6°	16 July 1986
OE2/Kosmos-1602	28 September 1984	629 × 664 km, 82.5°	5 December 1986
O1/Kosmos-1766	28 July 1986	631 × 662 km, 82.5°	24 October 1988
O2/Kosmos-1869[3]	16 July 1987	634 × 667 km, 82.5°	3 May 1989
O3/Okean 1	5 July 1988	635 × 666 km, 82.5°	14 June 1990
O4[4]	9 September 1989	–	–
O5/Okean 2	28 February 1990	639 × 666 km, 82.5°	18 July 1991
O6/Okean 3[5]	4 June 1991	634 × 666 km, 82.5°	4 January 1994
O7/Okean 4	11 October 1994	632 × 666 km, 82.6°	n/a
O8/Sich 1[6]	31 August 1995	632 × 669 km, 82.5°	n/a
O9/Okean 2-1	17 July 1999	633 × 661 km, 98.1°	December 1999
Sich 1M	24 December 2004	280 × 650 km, 82.56°. (Third stage apogee burn too short, raising perigee from 78 km to 280 km, rather than the planned 650 km.)	8 August 2005

Notes: [1] All launches made by SL-14 Tsyklon 3 from Plesetsk
[2] Experimental, did not carry SLR
[3] Radar antenna failed to deploy but other sensors operational
[4] Stage 3 failure
[5] Plane 90° from number 2's
[6] Chile's 50 kg FASat-Alfa failed to separate.

Current Status

Sich 1M re-entered the atmosphere on 15 April 2006, after being launched into the wrong orbit on 24 December 2004. Communication was lost with the first Okean-O2 series satellite in December 1999.

Background

The Okean-O1 (now known as Sich), is a civilian remote sensing spacecraft with radar imaging capability. It closely resembles the top-secret Tselina D and Tselina R satellites. The Okean all-weather radar, ice and oceanographic satellite system became operational with the launch of Okean-O1-1 by Tsyklon 3 on 5 July 1988. Kosmos-1076 and -1151 flew as precursor satellites, but 1983's Kosmos-1500 prototype was the first to incorporate 1.5 km resolution side-looking radar utilising a 12 m deployable antenna. Data return was through the three primary Meteor ground stations. Ice forecasts were uplinked four times weekly to vessels via the Ekran geostationary satellites. In December 1983, Kosmos-1500 helped free 70 Soviet ships trapped in heavy Arctic ice near Wrangel Island.

Each Okean-O1 spacecraft has a mass of about 1,950 kg, with a payload capacity of about 550 kg. They are launched into sun-synchronous orbits from Plesetsk Cosmodrome by the Tsyklon-3 vehicle. The spacecraft bus is a three-segmented, vertical cylinder, 3 m with a base diameter of 1.4 m and an upper diameter of 0.8 m. Okean's primary structure is pressurised and maintained at normal temperatures to protect the support system and internal payload electronics. There are two small, rotatable solar arrays (1.6 m wide and 2.0 m tall) and stabilisation is partially provided by a gravity-gradient boom extending from the top of the satellite. At the bottom, four panels (1.0 m wide and 2.9 m long), attached at 90° intervals, support the payload receivers and transmitters. An arrow, 11 m long radar antenna is fixed to the base of one panel. Okean-O1 spacecraft transmit data in real time on 137.4 MHz using APT formats similar to that employed by Meteor satellites with a scan rate of 4 lines per second. Data is also stored and retransmitted on 466.5 MHz to the three principal data reception and processing centersat Moscow, Novosibirsk, and Khabarovsk.

The major Okean-O1 payload is the side-looking RLS-BO radar, operating at a vertically polarised 9.5 GHz frequency. This instrument, developed by the Radio Engineering and Electronics Institute (IRE) in Kharkov, determines surface characteristics of land, sea, and ice but also near-surface winds.

Okean-O1 satellites also act as the central node in the Condor system, which collects environmental data from small, remote surface stations. Designated Condor-1, these stations are interrogated by Okean satellites at 460.03 MHz and then transmit their data at 1553.4 MHz during a 4–12 sec. contact. The Okean satellite, also known as the Condor-2 node, then relays the data to special Condor-3 processing stations at 460.03 MHz. Okean satellites can interrogate Condor-1 stations within 800 km and can store up to 64 kbits of data for subsequent relay to a Condor-3 site.

NPO Yuzhnoye in Dniepropetrovsk, Ukraine, built and designed the Okean, and the VKS Military Space Forces operated the system for NPO Planeta. The last of the 1.9 tonne Okean-O1 designs was launched in August 1995 on a Tsyklon as a joint Russian/Ukrainian project, designated **Sich 1**. The Ukraine took over full control from 11 October 1995, using Yevpatoria.

The first **Okean-O** series satellite (also called Okean-O1-9) was launched by Zenit-2 on 17 July 1999 as a joint venture between the RSA and the

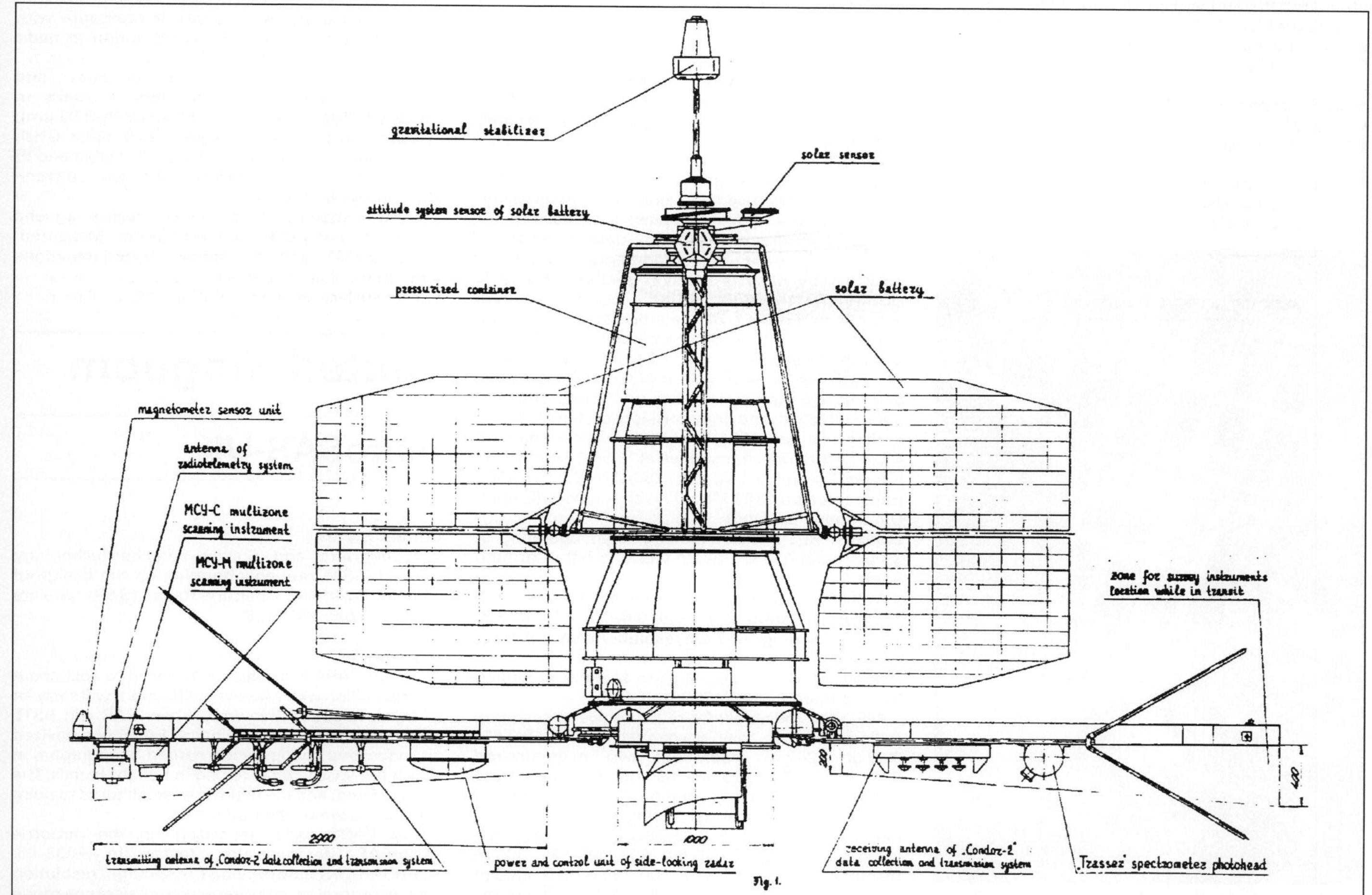

Okean-O1 configuration; the large base radar antenna is not visible

0517044

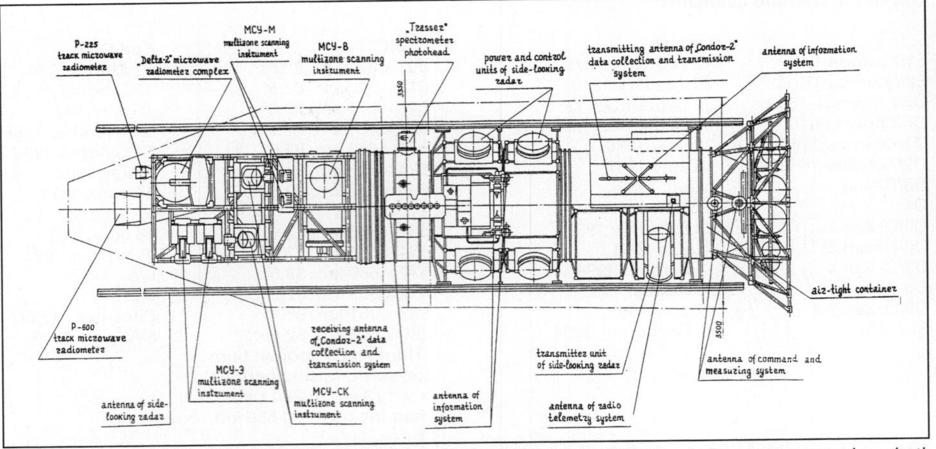

Earth-pointing face of 6.3 tonne Okean-O with potential sensor payload. The two SLR antennas run along both sides; the propulsion system is at far right. The dashed line indicates the Zenit payload fairing 0517045

Ukrainian National Space Agency. With a launch mass of 6,150 kg and a payload mass of 1,520 kg, the spacecraft was of a completely different design to the Okean-O1 series. The satellite bus comprised a pressurised cylinder about 6.6 m long, with a single articulated solar array deployed above the spacecraft. Two side-looking radars (RLS-BO) with right- and left-viewing capability provided full coverage of the Earth's surface from sun-synchronous orbit. Major failures plagued operations and it was silent between December 1999 and May 2000.

Okean-O1 was renamed Sich and has become the Ukraine space agency's main satellite programme. However, there was a major setback in December 2004 when the Tsyklon 3 placed the modified **Sich 1M** into the wrong orbit, although Microsatellite-1 (MS-1-TK) of Ukraine Space Agency was successfully separated from the main spacecraft on 25 December. Like its predecessors, the satellite used a gravity boom for stabilisation and used the same platform. The spacecraft orientation accuracy was 1° in roll, 2° in yaw, and 3° in the pitch axis. It carried a GPS receiver and had an average power consumption of 380 W. The first remote sensing images and data were returned during tests of onboard equipment on 27 January 2005. Operations ended on 8 August 2005, and Sich-1M re-entered the atmosphere in April 2006.

In June 2007, the Ukraine Space Agency announced that Sich 2 would be launched by a Dnepr in 2008.

Specifications
Okean-O1
Resolution: 1.5 km radar, 0.35–1.5 km visible/infra-red, 6–15 km radiometer.
Design life: 6 months.
Mass: 1,950 kg (550–650 kg payload)
Orbit: Typically 630–660 km, 82.5°

The first 6.3 tonne Okean-O was launched on 17 July 1999 (Theo Pirard) 0517415

Spacecraft: Stepped cylindrical pressurised bus 3 m high, maximum 1.4 m diameter, 3-axis stabilisation (base nadir pointing, accuracy 0.75–1.0°, determination to 2 arc/min), aided by gravity gradient boom. Twin solar arrays spanning 4.82 m (each an irregular 2 m high, 1.6 m wide); power consumption 1.1 kW operating, 110–270 W daily average. Sensors and single 11.8 m radar antenna mounted on four 2.9 m panels. APT transmissions on 137.4 MHz. Condor 2 digital data collection from small automatic Condor 1 stations at 500 bit/s on 1533.4 MHz, relay at 2 kbit/s on 460.03 MHz). No orbit or yaw control.
Sensors: RLS-BO side-looking radar operating at 3.15 cm and 450 km swath width to left of ground track. Power limits operations to 10 min sessions.

MSU-S visible/near-infra-red scanning radiometer (channels 0.5–0.7, 0.7–0.9 μm) providing 350 m resolution over 1,380 km swath width, nadir pointing.

MSU-M multispectral visible/near-infra-red scanner, 0.5–0.6, 0.6–0.7, 0.7–0.8, 0.8–1.1 μm, 1.5 km resolution over 1,930 km swath, nadir pointing.

RM-08 8 mm-wavelength scanning radiometer, 550 km swath, to left of track. C1500/1602/1766 carried the MSU-K circular scanning radiometer, 0.5–0.9 μm single band, 900 km swath, 500 m resolution, 6.5 kg.

Okean-O
Resolution: 1.5 km radar, 25–200 m visible, 100-600 m infra-red, 16–130 km radiometer.
Design life: 3 years (1 year Okean-O number 1).
Mass: Up to 6.5 t (up to 2 t payload); Okean-0 number 1 6.3/1.55 t
Orbit: 660 km, 98°
Spacecraft: Total length 10.6 m, 1.8 m diameter, 6.6 m long pressurised bus, aluminium-magnesium alloy structure. 3-axis stabilisation (base nadir pointing, accuracy 5 arcmin, determination to 2 arc/min). orbit adjust/momentum dumping provided by base propulsion system: 10 × 30 N UDMH/NTO thrusters. Attitude control by RWs and magnetorquers. Single articulated solar wing; power consumption 5 kW operating, 1.5 W daily average (Okean-O number 1 3.5/0.80 kW). Sensors and radar antenna mounted on sides/base of bus and forward instrument platform. Condor 2 digital data collection and transmission system (receive at 500 bit/s on 1.5334 GHz; transmit at 2 kbit/s on 460.03 MHz). Data transmitted at 15.36–122.88 Mbit/s (Okean-O number 1 up to 64 kbit/s) on 8.2 GHz; APT transmitted on 137.4 MHz. 19 Gbit memory (Okean-O number 1 5.5 Gbit)
Sensors: RLS-BO side-looking radar operating at 3.15 cm and 450 km swath width to left of ground track

MSU-M multispectral visible/near-infra-red scanner operating at 0.5–0.6, 0.6–0.7, 0.7–0.8, 0.8–1.1 μm providing 1–1.7 km resolution over 1,930 km swath.

MSU-A 3-channel 0.5–0.9 μm scanner providing 30 m resolution over ±300 km swath.

MSU-SK visible/infra-red conical scanning radiometer. Two visible channels 0.53–0.59/0.61–0.69 μm 200 m resolution; two near-infra-red 0.7–0.8/0.9–1.0 μm 200 m resolution; one infra-red 10.4– 12.6 μm 600 m resolution. 600 km swath.

MSU-V conical scanner. Eight channels 0.45–0.52 (50 m resolution), 0.52–0.62 (50 m), 0.62–0.74 (50 m), 0.76-0.9 (50 m), 0.9–1.1 (50 m), 1.55–1.75 (100 m resolution), 2.1–2.35 (250 m), 10.0–12.0 (100 m resolution) μm. 180 km swath, 1.28–5.12 Mbit/s. Ice/snow imaging, land use.

The Okean-O1 radar satellite was first displayed at 1985's Paris Salon. The SLR antenna is seen diagonally in the bottom left corner (Graham Turnill) 0517043

Trasser-0 spectroradiometer. 62 channels over 0.430–0.800 μm, <1 n mile resolution.

Land Delta 2 scanning radiometer, 7.0/13.0/22.5/36.5 GHz, 2,600 km swath.

R600 microwave radiometer. 6 cm/4.9 GHz, two polarisations, 130 km resolution, 1 kbit/s.

R225 microwave radiometer. 2.25 cm/13.3 GHz, two polarisations, 130 km resolution, 2 kbit/s.

Sich 1M
Resolution: 10/50 m SAR, 1.5 km SLR, 30 m visible
Launch: 24 December 2004 by Tsyklon 3 from Plesetsk.
Design life: 3 years.
Mass: 2,223 kg at launch, payload 650 kg
Orbit: Planned 670 km, 98°
Spacecraft: Used structural platform of Sich-1 and Okean-O1 satellites.
Sensors: RLBO double side-looking radar (with scatterometer mode), 12 m antenna, 3.2 cm wavelength (9.7 GHz, X- band), 700 km swath each side ground track. Data resolution: 1.7 – 2.8 km in flight direction, 1.3 – 0.7 km in cross track direction. Built by Kharkov IRE, Ukraine.

MSU 3-channel visible/near-infra-red scanner, 0.5–1.0 μm, 48 km scan with 700 km for 30 m resolution.

RM-08 Passive Microwave Scanning Radiometer built by Kharkov IRE, Ukraine. Wavelength/frequency: 0.8 cm/36.6 GHz; swath 550 km. Measure cloud, sea ice and sea surface temperature with an accuracy of 1–2 K. Spatial resolution at nadir 25 km × 25 km.

MTVZA-OK module for temperature and humidity sounding of atmosphere: 4 bands in visible (0.37–0.78 μm), 1 band in IR (3.55–3.93 μm), 11 bands in microwave regions (6.9–183.3 GHz); swath width – 2000 km; resolution in visible and IR regions – 1.1–1.1 km, in microwave range – between 112–260 and 8–19 km.

Variant instrument to detect electromagnetic waves as precursors to earthquakes. Measured: electric current density, magnetic field fluctuations and electric field vector.

United Kingdom

AstroSAR-UK

Current Status
EADS-Astrium and Surrey Satellite Technology Limited (SSTL) are collaborating on the design of a small Synthetic Aperture Radar (SAR) satellite known as AstroSAR-UK.

Background
The AstroSAR-UK satellite is expected to cost about GBP40 million with a five-year lifetime. It will carry an EADS-developed active-array X-band SAR, with SSTL providing the platform avionics. EADS has devised an innovative configuration named Snapdragon, in which the spacecraft is folded in half for launch. The array is fixed, and the entire spacecraft turns to point the radar towards the target area.

The EADS radar is based on the airborne MicroSAR demonstrator, tested in 2003–05. AstroSAR-UK could provide 1 m spotlight resolution and incorporates a maritime reconnaissance mode covering a 900 km swath at 20 m resolution.

TopSat

Current Status

The UK's first military surveillance satellite, TopSat (Tactical Optical Satellite), was launched on 27 October 2005. It is seen as a precursor of a UK national surveillance programme based on a constellation of small spacecraft.

Background

TopSat has been developed by QinetiQ and Surrey Satellite Technology Limited (SSTL), and it was providing operational information to military users by mid-2006.

UK satellite studies started with a detailed review of what users required in terms of resolution, coverage, revisit rates and data dissemination. Some important conclusions emerged from this process. One was that most of the users' resolution requirements, for an optical sensor, resided between 1 m and 1.5 m. The study also showed that the quality of an Intelligence, Surveillance, Target Acquisition and Reconnaissance (ISTAR) constellation should be measured not only in spatial resolution but also in temporal resolution – measured in revisit rates.

The result was a vision of an ISTAR constellation including optical, Synthetic Aperture Radar (SAR) and signals intelligence (SIGINT) craft, with the SAR satellites cueing the optical craft and the SIGINT satellites cueing both SAR and optical sensors. TopSat is a first step towards fielding such a system.

Because of budget limitations, TopSat was designed to achieve only half the resolution of an operational Electro-Optical (EO) micro-satellite – around 2.5 m from a 600 km sun-synchronous orbit. (The actual resolution is lower because the launch placed the satellite in a 686 km orbit). It carries a fixed camera developed by the Rutherford Appleton Laboratories and uses a novel imaging technique: in order to increase exposure time, and get an acceptable image with a smaller aperture and camera, the entire spacecraft rotates as it passes over the target, keeping the camera pointing at the objective without blurring.

The 120 kg TopSat is based on SSTL's enhanced microsatellite family. The satellite is assembled from a 'layer-cake' stack of machined aluminium trays carrying its main electronic components – the structure also provides some protection against radiation – with solar arrays forming a box around them. Momentum wheels provide attitude control and resistojets – using electrically heated gas – provide orbital adjustment.

Although TopSat is limited in some ways – memory and datalink capacity, for example, permit only five images per day, and it has a design lifetime of a year – it has proved very useful since it was launched on 27 October 2005. The TopSat project includes a mobile ground station, carried on a Land Rover with a trailer-mounted antenna. The satellite can be tasked and controlled in-theater from the mobile station. Operational users have included the special-operations community.

An operational system would be slightly larger – weighing some 150 kg. It would have a 1–1.5 m resolution and a greater roll angle – up to 45° either side of its track – to cover a wider area, and a capacity for 30–60 images per day and two images per pass.

SSTL – reflecting the thinking of space advocates in the MoD – sees a satellite constellation comprising ten EO satellites, providing multiple morning and evening passes, plus four SAR satellites and two swarms of four SIGINT spacecraft.

United States

Defense Meteorology Satellite Program (DMSP) series

Current Status

The long-running DMSP programme remains crucially important to the US Department of Defense and other users. Originally launched to avoid wasting irreplaceable spy-satellite film on cloud decks, the DMSP family now provides information for all military operations, including combat and routine flight planning, in conjunction with the civilian Polar Orbiting Environmental Satellite (POES) programme. Launches are scheduled through 2012 and the system is eventually to be replaced by the National Polar Orbiting Environmental Satellite System (NPOESS).

Background

DMSP is the US services' only assured source of global weather data, providing visible and infra-red cloud cover imagery (600 m constant resolution) and other meteorological, oceanographic, land surface, and space environmental data. At least two satellites (one in each of two orbit planes) are required in sun-synchronous, 830 km polar-orbit at all times.

DMSP has evolved through several generations and 43 spacecraft launches since the first satellites were launched in 1965. In early 2006, the two primary operational spacecraft were DMSP F-15, launched on 12 December 1999, and DMSP F-16, launched on 18 October 2003. F-17 was scheduled to launch in 2006. Four older DMSP satellites (F-11, F-12, F-13 and F-14) remain available as on-orbit spares.

F-15 was the first Block 5D-3 version of the DMSP series, developed under a 1989 contract to General Electric Astro Space (now Lockheed Martin). Compared with earlier generations, the Block 5D-3 series accommodates larger sensor payloads than earlier generations. They also feature a larger capability power subsystem; a more powerful on-board computer with increased memory — allowing greater spacecraft autonomy — and increased battery capacity that extends the mean mission duration. They have a more advanced attitude control system for precision pointing and a higher rate command link for shorter ground contact times. F-15 was a hybrid; prime contractor Lockheed martin refers to F-16 as the first Block 5D-3, but the customer identifies F-15 as a Block 5D-3 with legacy sensors. DMSP F-16 and later satellites carry new series of highly capable microwave and ultraviolet sensors.

The primary sensors aboard the current DMSP satellites include the Operational Linescan System (OLS), an electro-optical telescope system with visible, low-light and infra-red sensors; the SSM/I passive microwave imager, gathering data as diverse as ice formation, soil moisture and ocean surface wind speed; and instruments to gather vertical profiles of atmospheric moisture and to measure the Earth's magnetic field.

Control of the DMSP satellites was passed from the USAF to the Department of Commerce in June 1998, as part of the Clinton Administration policy that also created the joint USAF/DoC NPOESS replacement programme. The National Oceanic & Atmospheric Administration (NOAA) Office of Satellite Operations is the DoC's office responsible for satellite control. Two operational POES satellites and two operational DMSP satellites are positioned so that they can observe the Earth in early morning, mid morning, and early afternoon polar orbits. Together, they ensure that, for any region of the Earth, the data provided to users are generally no more than six hours old.

Tracking stations at New Boston Air Force Station, in New Hampshire, Thule Air Base, Greenland, Fairbanks, Alaska, and Kaena Point, Hawaii, receive DMSP data. Under a shared processing agreement among four satellite data processing centres-NOAA's National Environmental Satellite Data and Information Service (NESDIS), the Air Force Weather Agency (AFWA) at Offutt AFB, Nebraska, the Navy's Fleet Numerical Meteorology and Oceanography Center, and the Naval Oceanographic Office-different centres are responsible for producing and distributing, via a shared network, different environmental data sets, specialised weather and oceanographic products, and weather prediction model outputs.

Each of the four processing centres is also responsible for distributing the data to its respective users. For the DoD centers, the users include regional meteorology and oceanography centers, as well as meteorology and oceanography staff on military bases. NESDIS forwards the data to NOAA's National Weather Service for distribution and use by government and commercial forecasters. The processing centers use the Internet to distribute data to the general public. NESDIS is responsible for the long-term archiving of data and derived products from POES and DMSP.

Field terminals that are within a direct line of sight of the satellites can receive real-time data directly from the polar-orbiting satellites. In 2003 there were an estimated 150 such field terminals operated by US and foreign governments and academia. Field terminals can be taken into areas with little or no data communications infrastructure-such as on a battlefield or a ship-and enable the receipt of weather data directly from the polar-orbiting satellites. These terminals have their own software and processing capability to decode and display a subset of the satellite data to the user. Early terminals were large and complex, but more modern devices are not only smaller but can access DMSP, POES and non-US weather satellites.

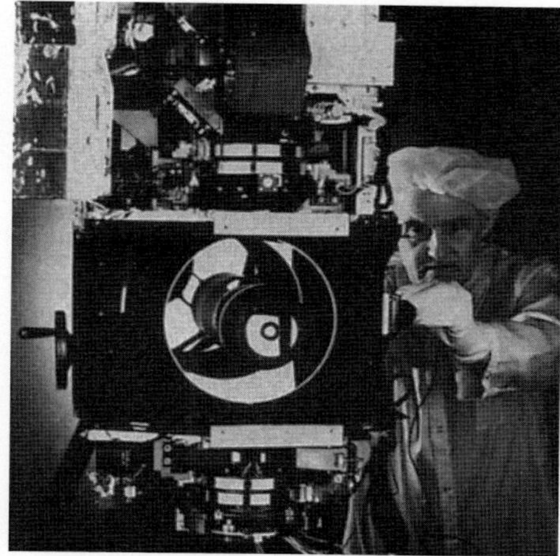

DMSP's primary imager is the Operational Linescan System (Northrop Grumman)
(Westinghouse Electric Corporation) 0516840

In early 2006, DMSP launches were expected to continue into 2012. DMSP F-15, launched in December 1999, has experienced premature failures in some of its attitude-determination gyros, exposing a fleet-wide, potentially life-limiting problem with all subsequent Block 5D-3 satellites. To remedy this problem, F-17 and subsequent satellites carry a redundant Miniature Inertial Measurement Unit (MIMU) as an insurance against gyro failures; F-16 had to be launched for operational reasons before the MIMU could be fitted. Also, on-orbit testing of F-16 revealed a number of sensor problems that have been addressed before the launch of F-17.

DMSP F-17 is expected to be launched in the last quarter of FY2006. F-16 was the last satellite to be launched by a Titan 2 rocket and F-17 will be flown on a Delta 4 Medium, the first in the series to fly on the Evolved Expendable Launch Vehicle (EELV). F-18 is due to be launched by an Atlas 5 in the third quarter of 2008, with F-19 and F-20 following in 2010 and 2012 respectively. The programme office is looking at life-extension options in the event of further delays with NPOESS.

Specifications

DMSP 2-10 (S15)

Launched: 12 December 1999 by Titan 23G from Vandenberg into a 837 × 851 km, 98.9° orbit.

Contractors: Lockheed Martin (prime), Northrop Grumman (Operational Linescan imaging system), Raytheon (microwave imager), Aerojet (microwave temperature sounder), Barnes Engineering (infra-red temperature/moisture sounder)

Resolution: typically 0.5 km

Design life: 48 months

Mass: 823 kg at launch, including 180 kg sensor payload

Orbit: typically 820–860 km Sun-synchronous at 98.8°, providing repeated passes at either 6 am or 10.30 am local time. Scan width 2,960 km, creating continuous coverage at equator and global coverage in 12 h

Spacecraft: based on Astro Space's NOAA Tiros bus, 1.22 m wide bus, 6.4 m long with single solar wing (1 kW from 8 panels of 12,500 Si cells) deployed. 127 kg structure provides 5.4 m² of

DMSP Block 5D3 incorporates the first Special Sensor Microwave Imager Sounder instrument 0516841

payload mounting area. Zero momentum 3-axis control by MWs plus magnetorquers; 4 hydrazine (15.9 kg propellant) plus 8 N_2 (2.3 kg propellant) thrusters. Earth pointing accuracy of 0.01° provided by strapdown star sensor and gyros; 0.12° accuracy backup system of static Earth sensor, Sun sensor, gyros. Final orbit insertion provided by Thiokol Star 37S solid. TT&C assigned to AF Space Command's 6th Space Operations Squadron (part of 50th Space Wing, Falcon AFB, which can also provide control) at Offutt AFB's Satellite Operations Center. Data downlink on 2,207.5, 2,252.5, 2,267.5 MHz; telemetry downlink 2,237.5 MHz; command uplink 1,792 MHz at 2 kbit/s (minimum of 300 stored commands); data stored on Odetics DDS-5000 tape recorder. Passive thermal control provided by finishes and radiators; active control by electrically-controlled louvres and heaters (thermal subsystem mass 24 kg)
Sensors: primary meteorological imaging is provided by the visible/infra-red OLS Operational Linescan System. The system includes an 85.5 GHz Microwave Imaging Sensor (SSM/I, an enhanced version rides on NASA's TRMM) measures ocean surface wind speed, ice coverage/age, areas/ intensity of precipitation, cloud water content and land surface moisture, the microwave temperature sounder (SSM/T-1) returns cloud temperature profile data, the SSM/T-2 microwave water vapour sounder will be added later for atmospheric humidity monitoring to improve global forecasting. The precipitating electron spectrometer (SSJ/4) monitors electron plus proton densities at different energy levels, and the electrostatic ionosonde locates auroral boundaries to aid radar and long-range radio operations. Gamma/X-ray Spectrometer (SSB/X2) and SSM Magnetometer also carried.

Future Imagery Architecture (FIA)

Current Status

Future Imagery Architecture (FIA) is a largely classified programme that was originally planned to replace the National Reconnaissance Office's current optical reconnaissance satellites, known as Advanced Crystal or Enhanced Imaging System (EIS) and descended from the pioneering KH-11 system, and also the NRO's Lacrosse radar. After severe delays and problems, the radar version of FIA was cut back in early 2005 and Boeing was replaced by Lockheed Martin as the supplier of the electro-optical satellites later in the year. NRO L-21, an experimental satellite that is believed to have been designed to test features of the FIA optical reconnaissance spacecraft, was launched on 14 December 2006, but was declared a total failure on reaching orbit.

Background

FIA arose in the mid-1990s from the need to replace the spacecraft from the KH-11 and Lacrosse families with a more modern system that would cost less to maintain and would respond to the needs of the post-Soviet world. Commanders in the first Gulf War had complained that the NRO's large spacecraft could not respond to their needs for timeliness and wide-area coverage. One proposed solution was to replace these satellites with smaller spacecraft, built and launched in larger numbers and replaced at higher rates.

The existence of FIA is not officially classified, but it has been very little mentioned in unclassified documents since 2003. Budget numbers and schedules are secret.

In 1995, Director of Central Intelligence John Deutch commissioned a Small Satellite Review Panel. It considered two inputs from the NRO. The office's imaging intelligence (IMINT) directorate, responsible for in-service systems, advocated a 50 per cent reduction in satellite size and cost. The NRO Office of Special Applications went much further, suggesting that future satellites could be as little as 20 to 25 per cent of the mass of earlier NRO IMINT spacecraft – that is, with a launch weight around 3 to 3.5 t.

The small-satellite panel endorsed the smaller "one-quarter-size" option, not because it would be cheaper but because it would be more robust and more flexible. Keeping replacement satellites available, whether on orbit or on the ground, would be more affordable. The panel envisaged separating the functions of today's satellites: for example, putting medium-resolution, high-resolution and radar sensors on different satellites. This would mean that medium-resolution revisit rates could be increased in a crisis without also investing in high-resolution intelligence capability.

The NRO, in turn, backed the panel's conclusions, stating in July that it "fully supported the concept that smaller is better... The question facing the intelligence community and the NRO is not whether to move towards smaller satellites but, rather, how small our satellites should be".

By 1996, the NRO was proceeding with definition of FIA, and conducted a competition to select a prime contractor for the programme. In September 1999, the NRO announced that Boeing had won the contract, toppling Lockheed Martin from a position that it had held unchallenged since 1960. Raytheon, Harris, Kodak and Marconi Federal (now part of Ericsson) have been reported as members of the FIA team.

Details of the programme are classified, but at the start of the project it was estimated to involve a constellation of 12 to 24 satellites, each about one-third the size of the 17 t EIS/Crystal. The electro-optical satellites would feature infra-red sensors for increased night-time surveillance capability, and other satellites within the constellation would carry Synthetic Aperture Radar (SAR) payloads. The overall contract was expected to be worth some USD25 billion to the Boeing team. The contract was due to be completed by 2010. In 2005, Boeing Space & Intelligence Systems – focused on FIA – employed 11,500 people.

Early concerns over the ability of the Tasking, Processing, Exploitation, and Dissemination (TPED) system to cope with the larger number of spacecraft caused some delays and an increase in the budget, as part of the FY2001 programme. Boeing CEO Phil Condit stated in April 2002 that, in 2001, the company "finished the preliminary design phase of the FIA programme on cost and on schedule".

By 2002, the first launch, originally scheduled for 2003, had slipped to 2006, and a review panel headed by former NRO director Martin Faga warned that there was a potential for a further 18 to 30 month slip. Moreover, the NRO was informing users that existing satellites should not be used when commercial imaging spacecraft would be adequate, indicating some concern over remaining lifetime on the older spacecraft.

In August 2002, a task force formed by the Defense Science Board and the USAF Scientific Advisory Board began a study of space acquisition programmes. In November 2002, in its oral report, the task force concluded that the FIA programme as it stood was not executable.

Although it was unable to reveal specific information about FIA, the task force noted that outsiders had ousted incumbents in many space acquisition competitions, and that non-incumbents could be "more optimistic" in their assessments of likely risk.

From the task force's final report, it was clear that the team considered that Boeing had underestimated the cost and difficulty of the programme and was now trapped between inflexible requirements, an inflexible launch plan (restricting the spacecraft's weight) and an inflexible schedule – presumably defined by the lifetimes of existing spacecraft. The result was increased risk. "Multiple small increments of accumulated risk can result in an unacceptably high cumulative probability of mission failure," the report stated. "The task force believes that the FIA programme under contract in the fourth quarter of calendar year 2002 fits this scenario".

Tension between Congressional intelligence committees, the NRO and the USAF continued in 2004. By 2005, the NRO was reported to have spent USD10 billion on the programme, USD4 billion to USD5 billion of that being accounted for by overruns, and the first launch had reportedly slipped into 2009. Early in the year, the Space Based Radar (SBR) programme became the SR programme and was identified as the sole national space radar project, indicating that the radar segment of FIA would at best be a stop-gap.

In September 2005 it was announced that much of the FIA contract was transferred from Boeing to Lockheed Martin, including the completion of the electro-optical imaging satellites. However, the new programme has not gone well. NRO L-21, an experimental satellite that is believed to have been designed to test features of the new Lockheed Martin spacecraft, was launched on 14 December 2006, but was declared a total failure on reaching orbit.

It seems likely that Boeing's "better and lighter" FIA approach relied on new mirror technology. Under the Advanced Mirror System Demonstrator (AMSD) programme, a four-year partnership between NASA, the NRO and the USAF, Kodak produced a lightweight experimental glass mirror and Ball Aerospace produced a beryllium mirror. The latter was selected for the 6.5 m segmented mirror on NASA's new James Webb space telescope. Silicon carbide has also been used as a substitute for the special Ultra-Low-Expansion (ULE) glass used for earlier imaging satellites.

The fact that a test satellite could fly little more than a year after the contract was moved to Lockheed Martin suggests that much of the technology from the Boeing proposal has been carried over to the follow-on programme. The NRO has now apparently decided not to repeat the experiment – given the age of the on-orbit constellation, there is probably no time to do so – and will go directly to the operational system.

This leaves Boeing with a small number of unique radar FIA satellites, which are affected by two issues. It seems most likely that the early Lacrosse satellites used a version of the Hughes (later Raytheon) Advanced Synthetic Aperture Radar System (ASARS-2) developed for the U-2, and that later vehicles used a radar related to the improved, electronically scanned ASARS-2. By this logic the radar version of FIA would be closely related to the Multi-Platform Radar Technology Insertion Program (MP-RTIP) radar, also led by Raytheon. However, the demise of the Boeing E-10A aircraft, which was intended as the sole platform for the full-size MP-RTIP sensor, leaves this programme substantially cut back, with only the small version for the RQ-4B Block 40 Global Hawk surviving.

The other problem for the radar FIA is that it is now seen as a stop-gap until the definitive SR is available. For the time being, the radar FIA's primary mission is a medium-term, very costly replacement for Lacrosse.

Geosat Follow-On (GFO)

Current Status

In April 2008, Geosat Follow-On (GFO) entered its eleventh year of operation.

Background

The US Navy's Space & Naval Warfare Systems Command awarded a USD46 million contract to Ball Corp in August 1992 to build the Geosat Follow-On. Ball's Aerospace Systems Division built the spacecraft bus, procured the payload – consisting of an altimeter, radiometer, doppler beacon and global positioning system (GPS) receiver – and supplied the system software and hardware for the two ground support stations. The altimeter was supplied by Raytheon. A microwave radiometer supplied by AIL systems Inc is also carried. The spacecraft is made of lightweight graphite composite, is approximately 3 m long with a fixed, side-mounted solar array that generates up to 126 W average power, and has 96 Mbytes of on-board data storage. The spacecraft cost was USD46 million.

The original **Geosat** was a larger research satellite launched on an Atlas in 1985. It provided sea surface height information (and thus information on the shape of the Earth's gravity field) using a radio altimeter. The GFO radar altimeter satellite was built to provide accurate measurement of ocean topography for the Naval Meteorology & Oceanographic Command in Bay St Louis, Missouri. By measuring sea height, the satellite provides the USN with information on sea floor topography and water density, surface roughness and by association, an index of wind speed. GFO data, when properly interpreted, form a vital link to the scientific understanding and accuracy of global ocean circulation models. The data also identify

Geosat Follow-On provides accurate sea height measurements. It directly support snaval operations such as ship routing, anti-submarine warfare and amphibious operations (Ball Aerospace) 0517319

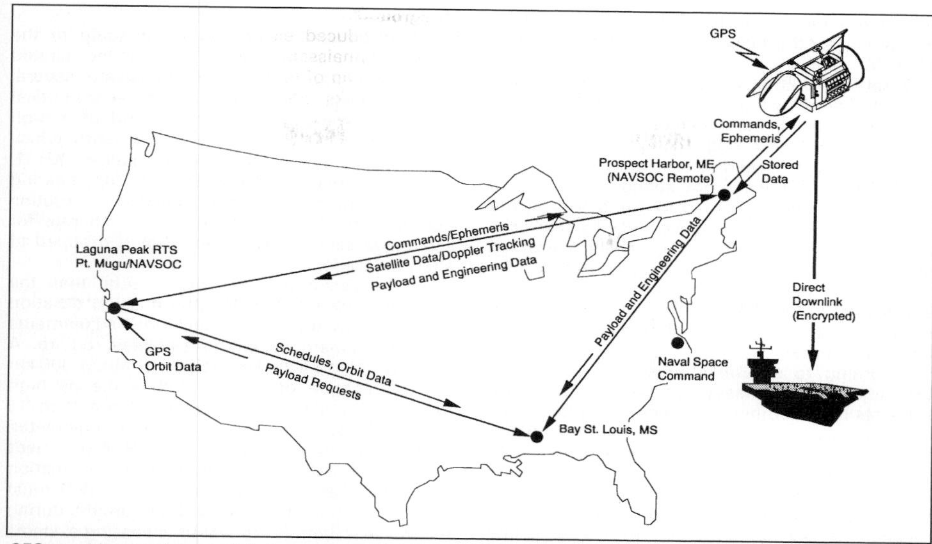

GFO provides continuous global oceanographic data to ships and shore facilities (Ball Aerospace) 0517159

warm and cool water masses, which is important for anti-submarine warfare. For the first time, the USN planned to use the satellite for real-time direct encrypted readout to 65 ship/shore terminals.

Although primarily a Navy tactical satellite, the data collected by GFO are important for weather research. NOAA is responsible for distribution to NASA and the civil and scientific communities. GFO data are used to improve long-term weather predictions, such as forecasting El Niño events, mapping variations in the position of major ocean currents and predicting drought and flood patterns resulting from oceanic changes. GFO can also be used for fisheries management and routing of shipping, to help search-and-rescue efforts, and to monitor pollution and oil spills.

A Taurus rocket launched from Vandenberg successfully deployed GFO on 9 February 1998. The satellite was released from the Taurus's fourth stage into a 777 × 875 km orbit. Early in the mission, momentum wheel directions were found to be reversed, although that was rectified by software fixes. The GPS receivers, supplied by Allen Osborne Associates (AOA) and based on JPL designs, had trouble acquiring a locational fix, and the onboard computer sometimes resets unexpectedly. The GPS receivers failed despite four unit redundancy. With the loss of the primary orbit determination system and the loss of precision time tagging, orbit determination now relies on satellite laser ranging and doppler. The satellite was accepted by the Navy on 29 November 2000.

During its mission life, the satellite stays in the Geosat Exact Repeat Mission (ERM) orbit (800 km altitude, 108° inclination, 0.001 eccentricity, and 100 min. period). This 17-day Exact Repeat Orbit (ERO) retraces the ERM ground track to +/–1 km. The satellite uses a radar altimeter and a water vapour radiometer to measure differences in sea surface height associated with ocean currents and eddies to an accuracy of 3.5 cm. Geosat performance-stable oscillators and doppler beacons allow orbits to be determined with 1.9 cm rms radial orbit error for mesoscale oceanography after tilt and bias removal along a 3000 km arc-filter length. The lightweight (47 kg total) payloads feature complete redundancy and low power consumption (121 W total).

All GFO payload data are provided on an encrypted, continuously operating tactical downlink to AN/SMQ-11 Tactical Terminal-equipped Navy ships and facilities. The Naval Meteorology and Oceanography Command's Naval Oceanographic Office at the Stennis Space Center near Bay St Louis, Mississippi, processes GFO data at their Payload Operations Center.

The GFO satellite completed 154 exact repeat cycles in February 2008 and began its 11th year of operation in April 2008, still providing continuous worldwide oceanographic data for ships at sea and the Navy's shore-based facilities. Currently the radar altimeter is being power cycled on/off during eclipse periods to maintain the satellite's electrical power system.

Specifications

Launch: 13.20 GMT on 9 February 1998 by Taurus from VAFB, California.
Orbit: 785 × 788 km, 108°, 17 day repeat
Design life: 8 years minimum
Mass: 369 kg at launch.
Contractors: Ball Aerospace (prime). Raytheon (radar altimeter). AIL systems Inc. (microwave radiometer). Allen Osborne Associates (GPS receivers).
Payloads: Pulse radar altimeter, single frequency (13.5 GHz) with 3.5 cm sea surface height precision.

Passive water vapour radiometer to correct altimetry data for tropospheric water vapour effects – dual frequency (22 and 37 GHz) nadir-looking with a path correction accuracy of 1.9 cm rms.
AOCS: 3-axis control. GPS provides precise orbit determination (altitude precision <5 cm). Eight hydrazine thrusters drawing on two tanks
TT&C: Telemetry downlink 400.033 MHz. S-band data downlink at 16 kbit/s (direct) and 512 kbit/s in dump mode. All datalinks to/from GFO are encrypted. Uses NAVSOC Naval Satellite Operations Center at Pt Mugu, California via ground stations at Prospect Harbor, Maine, and Laguna Peak, California.

Improved Crystal

Current Status

The last Titan IV launch on 19 October 2005 is generally believed to have carried the last in a 30-year series of electro-optical imaging spacecraft produced by Lockheed Martin for the National Reconnaissance Office (NRO).

Background

Military reconnaissance was one of the first applications of space technology in the US. Under the cover of a scientific project named Discoverer, the CIA and USAF launched their first experimental spy satellite in February 1959. It failed: but the Discoverer 14 mission in August 1960 was a success. A new agency – the National Reconnaissance Office – was formed in secrecy in September 1961 to operate the spy satellite programme, and thereafter all such missions were classified. The existence of the NRO was officially secret until 1992 and it was not until 1996 that NRO launches were acknowledged.

The NRO adopted the codename Keyhole for its satellites and issued designations in the KH series, retroactively applying the designation KH-1 to Discoverer. These satellites comprised a space vehicle containing blunt-cone re-entry capsules which housed the photographic film. Once the film was used, the capsule was ejected, re-entered the atmosphere and descended by parachute to be captured in mid-air by a modified transport aircraft.

In the 1960s, the NRO rapidly improved the quality and capability of the reconnaissance spacecraft, fielding the KH-4 Corona, KH-9 Argon, KH-6 Lanyard and KH-7/8 Gambit. The last of the film-return satellites was the KH-9 Hexagon, nicknamed Big Bird, with four film capsules: it soldiered on for many years, the last being lost in a launch failure in 1986.

The next step beyond KH-9 was a subject of intense debate. The USAF, in the 1960s, backed the KH-10 Dorian – identified in public as the Manned Orbiting Laboratory – while some in the CIA advocated the development of a high-speed rocket-powered transatmospheric vehicle codenamed Isinglass. Both projects were terminated before 1970s in favour of a new reconnaissance satellite system known as the KH-11 Kennan or Crystal.

KH-11 introduced electro-optical imaging to the photo-reconnaissance inventory of the United States. Instead of returning a film canister several days to weeks after a launch, an electro-optical imaging system takes the equivalent of a high resolution television picture of the Earth which is then transmitted to a ground station. KH-11, despite a lower resolution than its film bearing predecessors, became more practical for routine use, and the 1970s semi-annual launch rate for espionage satellites with film aboard dropped to a few a decade.

KH-11s also occupied a higher orbit than the Big Bird because their lengthy mission duration did not allow for the frequent orbit adjustments that atmosphere skimming satellites required. A usual KH-11 orbit had a perigee of about 300 km and an apogee of 500 km with some as high as 1,000 km. Each KH-11 weighed more than 13 tonnes and was about 19.5 m long and 3 m in diameter.

An SDS satellite relays the data to Fort Belvoir in Virginia as the KH-11 takes each picture. As configured by the USAF, the KH-11 operational system consisted of a constellation of two satellites in planes 48.7° apart so that one covers a target in mid-morning, while the second follows mid-afternoon. Each satellite repeated its track every four days.

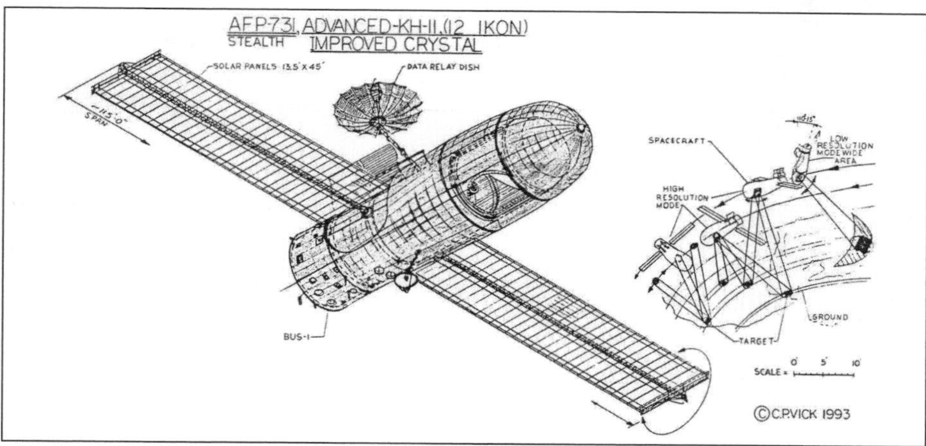

Advanced KH-11/Improved Crystal conceptual study. Dimensions are in feet (C P Vick) 0003438

For details of the latest updates to ***Jane's Space Systems and Industry*** online and to discover the additional information available exclusively to online subscribers please visit
jsd.janes.com

The last of nine initial-series KH-11s (including one lost on launch on 28 August 1985) was launched on 6 November 1988.

Subsequent satellites remain classified and their designations are unknown. From the early 1970s, though, the NRO had planned a successor to the KH-11, to be launched – and possibly refuelled and replenished on orbit – by Space Shuttle missions from Vandenberg AFB's Space Launch Complex 6. The first West Coast Shuttle mission was scheduled for October 1986, but was put on hold because of the loss of Challenger in January. Months later, in August 1986, the USAF modified its Titan IV Complementary Expendable Launch Vehicle (CELV) programme to include a two-stage version suitable for launching KH-class satellites, and in May 1988, the USAF formally abandoned the Shuttle programme and mothballed SLC-6.

It is a reasonable hypothesis that the NRO replaced the Shuttle-carried spacecraft with a second-generation electro-optical reconnaissance satellite that incorporated some of the same technology. It may be heavier than the original KH-11, because the Titan IV was designed with a greater LEO payload. According to researcher Jeffrey Richelson, this generation of spacecraft featured higher resolution, an infra-red capability developed under the codename Dragon, and improved area-imaging capability. The extent of the redesign, and the disruption caused by the abandonement of the Shuttle, may be reflected in the four-year gap between the launch of the last classic KH-11 and the first Improved Crystal, three times longer than the average interval up to that point.

The new satellite has been given many names, but is most often referred to as Improved Crystal. The Shuttle-based version may have been KH-12, but the KH series designation was so thoroughly compromised by 1986 that it had lost much of its usefulness.

The high orbit would require improved optics to maintain the resolution of earlier KH-11s, let alone achieve an improvement, and some circumstantial evidence exists for such a change. As part of its promotion for using the 'Bus I' service module for the International Space Station in 1993, Lockheed claimed that the bus was operational on classified missions. Improved Crystal is a prime candidate. Its 4.04 m diameter and 2.67 m length accommodates a larger mirror than KH-11's. Bus 1 is also reported to accommodate as much as 5,300 kg of propellant, accounting for the weight increase associated with Titan IV.

Observers recognise three launches for the second-generation Improved Crystal: 28 November 1992 (USA 86); 5 December 1995 (USA 116); and 20 December 1996 (USA 129).

Operations in Desert Storm showed that the NRO's spacecraft were not suited to the needs of tactical planners. This led to a debate over the future of space reconnaissance and – by 1994–95 – to the emergence of ideas for smaller spacecraft that culminated in the Future Imagery Architecture (FIA). Even then, FIA was not expected to be operational before 2005, so replacements for the early 1990s Improved Crystals would be needed before that time.

According to several sources, the latest and last generation of heavy optical reconnaissance satellites include other changes to meet the needs of tactical commanders, and are known as 8X or the Enhanced Imaging System. Two launches – USA 161 on 5 October 2001 and USA 186 (the last Titan IVB launch) on 19 October 2005 – are widely agreed to be Crystal/EIS launches, while observers differ on whether USA 144, on 22 May 1999, was the first of the EIS flights or a Misty stealth reconnaissance spacecraft. Currently, the NRO's concern will be to extend the lives of these satellites as long as possible, to avoid a gap in coverage before the long delayed FIA enters service.

KH-11 Kennan/Crystal series

Current Status

No KH-11 Kennan/Crystal is operational today and the lack of a shuttle launch facility for polar orbit flights undoubtedly had a part to play in the retirement of this type. Vandenbeng AFB, Calif, was to have been the base from which shuttle missions would have been available to extend the life of the satellite. It was the concept of in-orbit-servicing for military satellites that inspired the mission design for the Hubble Space Telescope.

Background

KH-11 introduced electro-optical imaging to the photo-reconnaissance inventory of the United States. Instead of returning a film canister several days to weeks after a launch an electro-optical imaging system takes the equivalent of a high resolution television picture of the earth which is then transmitted to a ground station. KH-11, despite a lower resolution than its film bearing predecessors, became more practical for routine use, and the 1970s semi-annual launch rate for espionage satellites with film aboard dropped to a few a decade.

KH-11s also occupied a higher orbit than the Big Bird because their lengthy mission duration did not allow for the frequent orbit adjustments that atmosphere skimming satellites require. A usual KH-11 orbit had a perigee of about 300 km and an apogee of 500 km with some as high as 1,000 km. Each KH-11 weighs more than 13 tonnes, is about 19.5 m long and 3 m in diameter. TRW builds the satellites for the USAF who uses the Titan 3D/34D to place them in 97° inclination orbits. Even at the higher altitude the KH-11 must raise its altitude about every three months during a three-year lifetime. Its real-time sensing systems (8–12 images per minute) and high resolution CCD cameras enable it to distinguish military from civilian personnel on the ground. The infra-red and multi-spectral sensing devices of Advanced KH-11 can locate mobile missiles, trains and launchers by day or night, and distinguish camouflage and artificial vegetation from real plants and trees. Each satellite has a side-looking capability, doubling the number of daily passes over a site from two to four.

An SDS helper satellite relays the data to Fort Belvoir in Virginia as the KH-11 takes each picture. As configured by the USAF, the KH-11 operational system consists of a constellation of two satellites in planes 48.7° apart so that one covers a target in mid-morning while the second follows mid-afternoon. Each satellite repeats its track every four days.

There had been no public reference to the KH-11 designation on the part of the CIA until a CIA agent sold an operating manual to a KGB agent in Greece in 1978. The KH-11 scored its largest intelligence coup by monitoring a shed in early 1980 at the Severodvinsk naval yard on the White Sea.

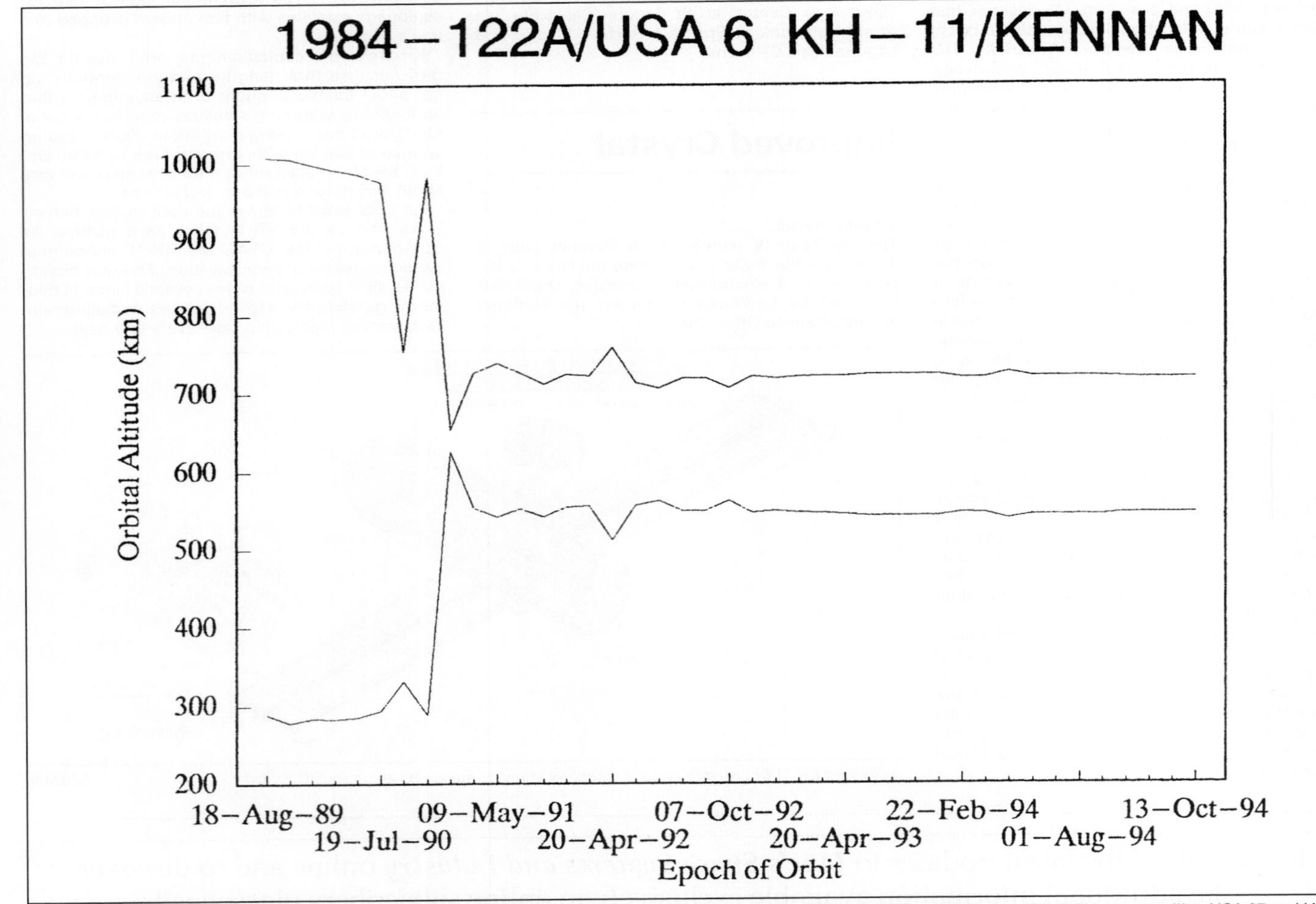

Graphs showing the orbital altitude regimes for photo-reconnaissance missions operating during the 1980s to 1990s. USA 6 was a KH-11/Kennan satellite, USA 27 and USA 33 were KH-11/Crystal satellites and USA 53, USA 86, USA 116 and USA 129 were Improved Crystal satellites. These orbits are based upon amateur observations of the satellites, since the actual operating orbits are classified

0003400

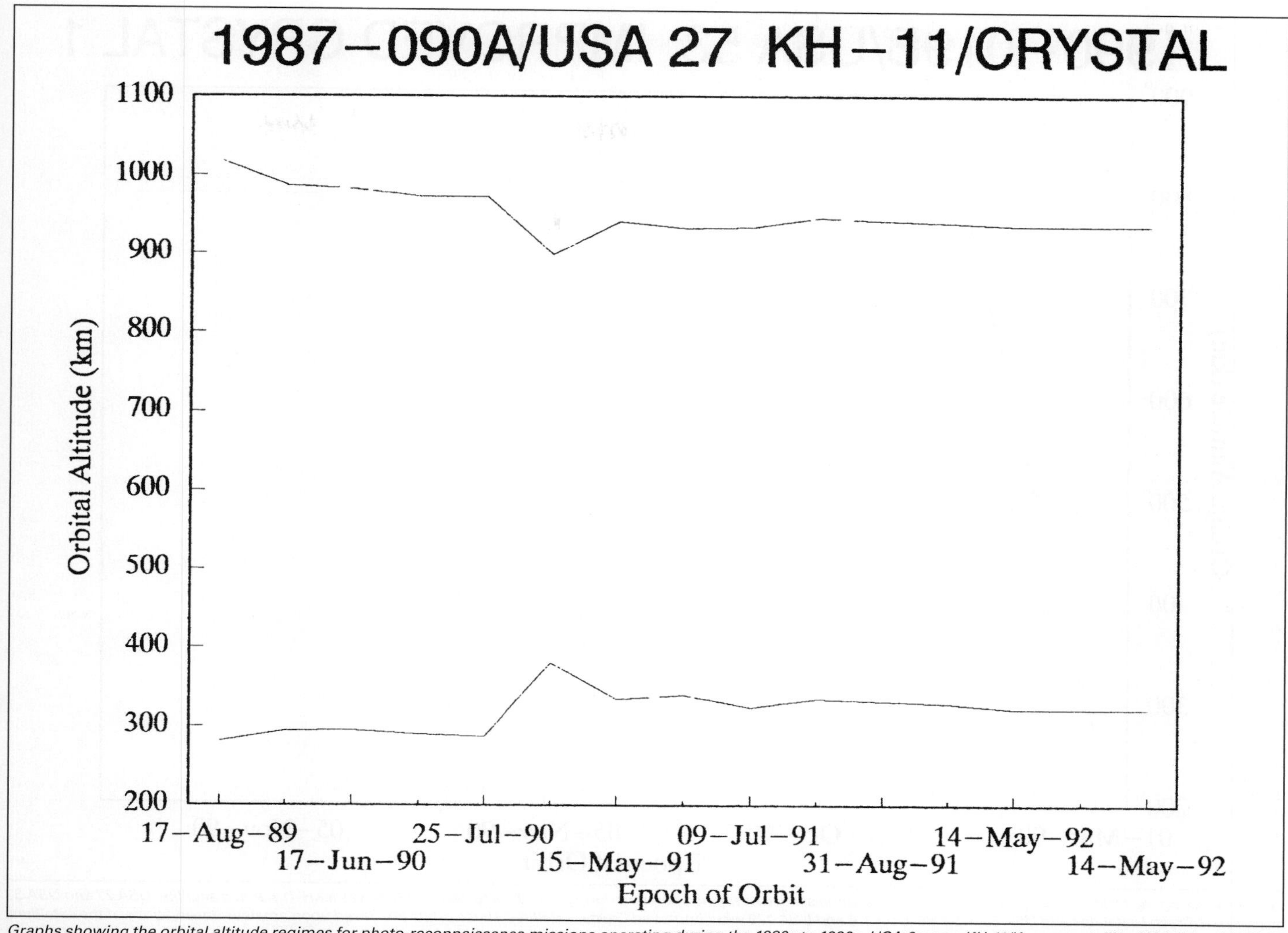

Graphs showing the orbital altitude regimes for photo-reconnaissance missions operating during the 1980s to 1990s. USA 6 was a KH-11/Kennan satellite, USA 27 and USA 33 were KH-11/Crystal satellites and USA 53, USA 86, USA 116 and USA 129 were Improved Crystal satellites. These orbits are based upon amateur observations of the satellites, since the actual operating orbits are classified
0003401

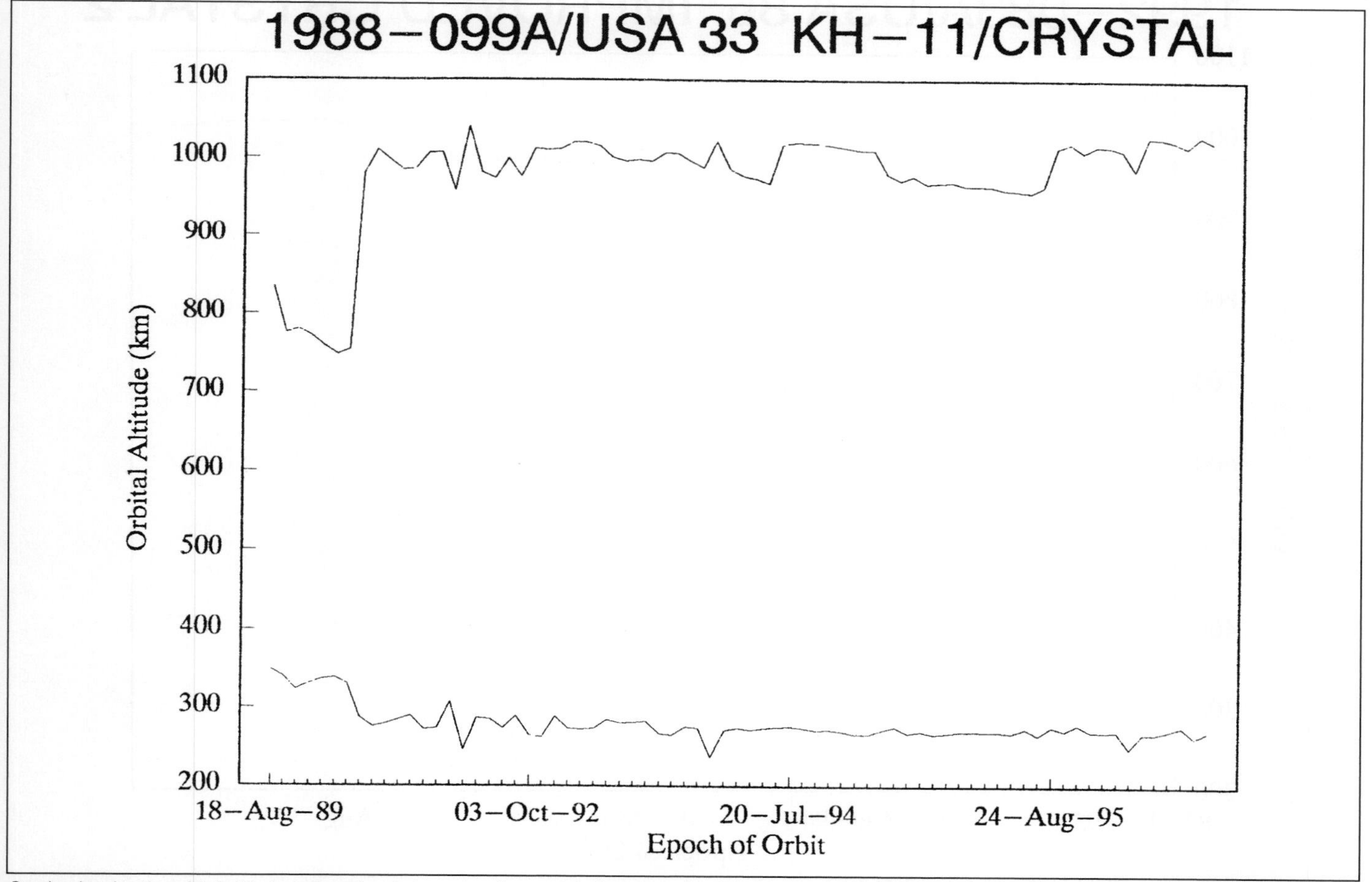

Graphs showing the orbital altitude regimes for photo-reconnaissance missions operating during the 1980s to 1990s. USA 6 was a KH-11/Kennan satellite, USA 27 and USA 33 were KH-11/Crystal satellites and USA 53, USA 86, USA 116 and USA 129 were Improved Crystal satellites. These orbits are based upon amateur observations of the satellites, since the actual operating orbits are classified
0003439

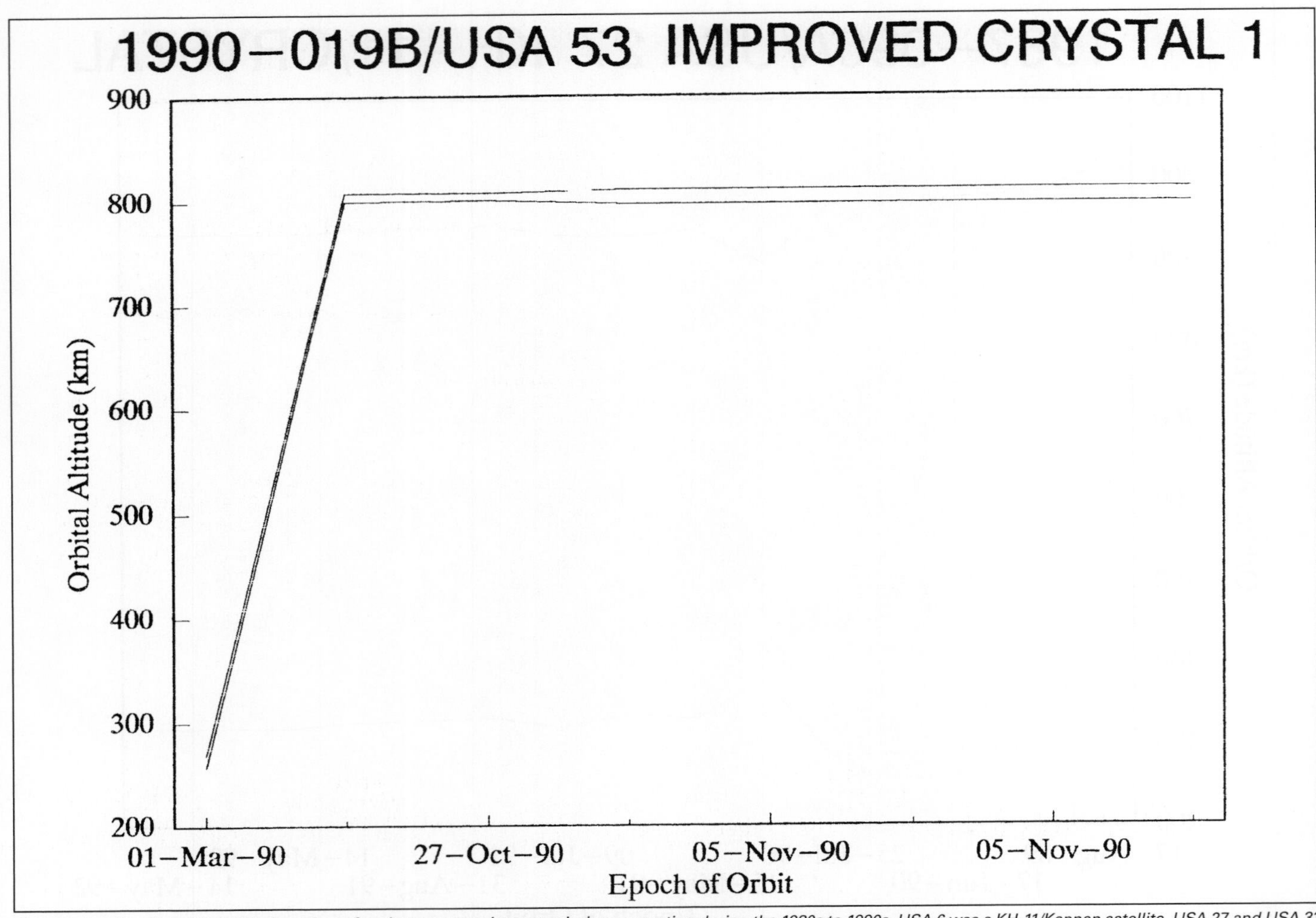

Graphs showing the orbital altitude regimes for photo-reconnaissance missions operating during the 1980s to 1990s. USA 6 was a KH-11/Kennan satellite, USA 27 and USA 33 were KH-11/Crystal satellites and USA 53, USA 86, USA 116 and USA 129 were Improved Crystal satellites. These orbits are based upon amateur observations of the satellites, since the actual operating orbits are classified
0003440

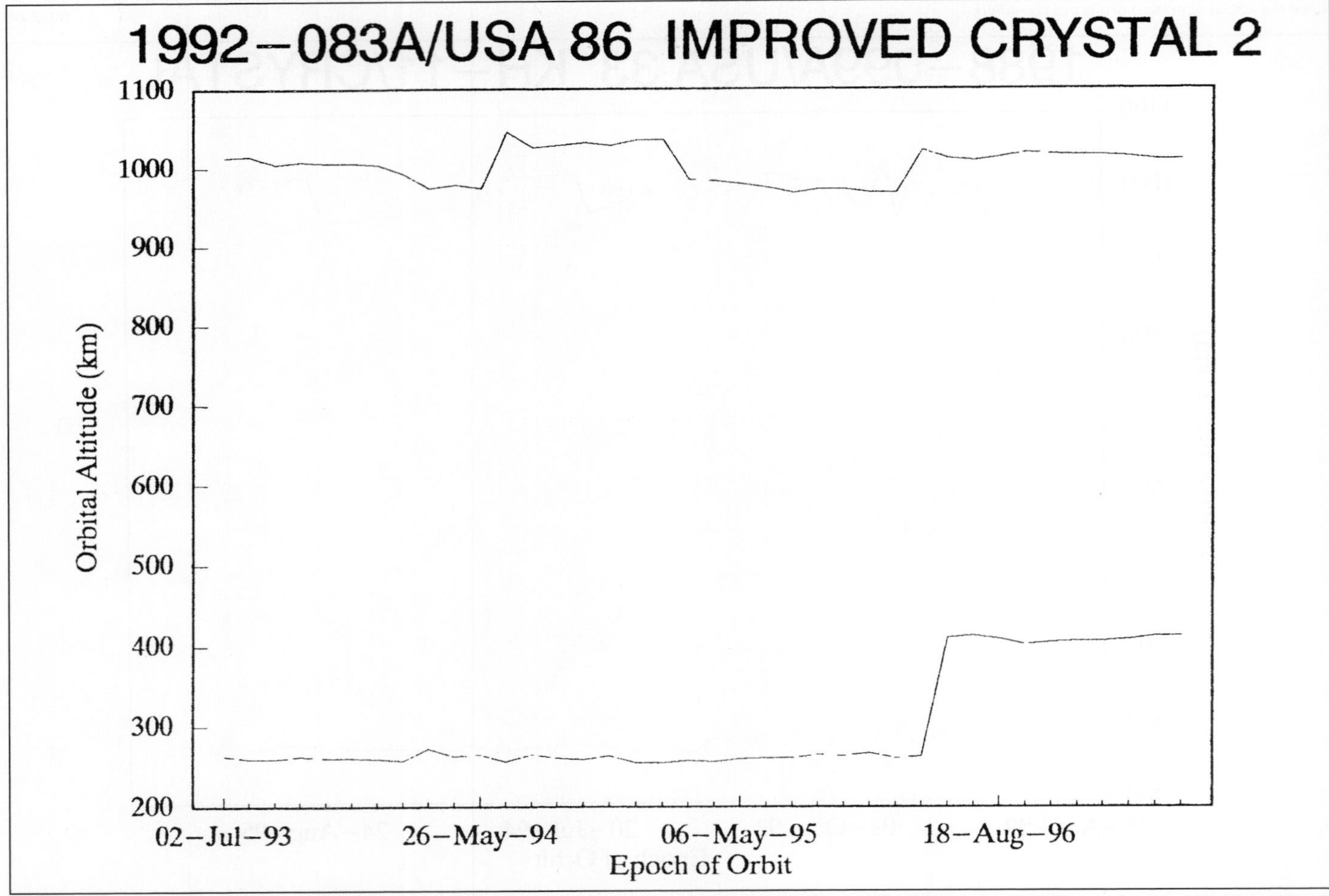

Graphs showing the orbital altitude regimes for photo-reconnaissance missions operating during the 1980s to 1990s. USA 6 was a KH-11/Kennan satellite, USA 27 and USA 33 were KH-11/Crystal satellites and USA 53, USA 86, USA 116 and USA 129 were Improved Crystal satellites. These orbits are based upon amateur observations of the satellites, since the actual operating orbits are classified
0003441

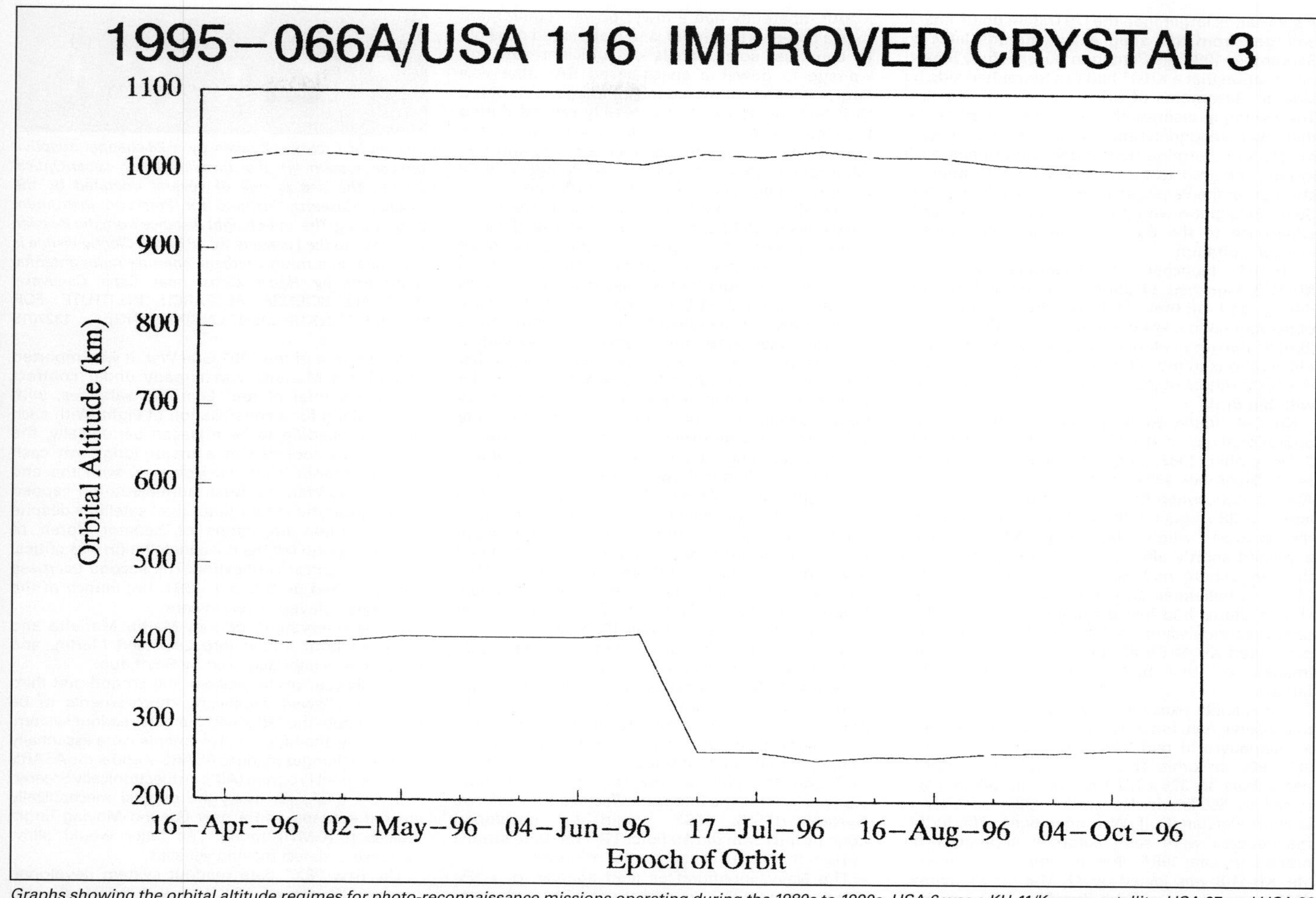

Graphs showing the orbital altitude regimes for photo-reconnaissance missions operating during the 1980s to 1990s. USA 6 was a KH-11/Kennan satellite, USA 27 and USA 33 were KH-11/Crystal satellites and USA 53, USA 86, USA 116 and USA 129 were Improved Crystal satellites. These orbits are based upon amateur observations of the satellites, since the actual operating orbits are classified
0003442

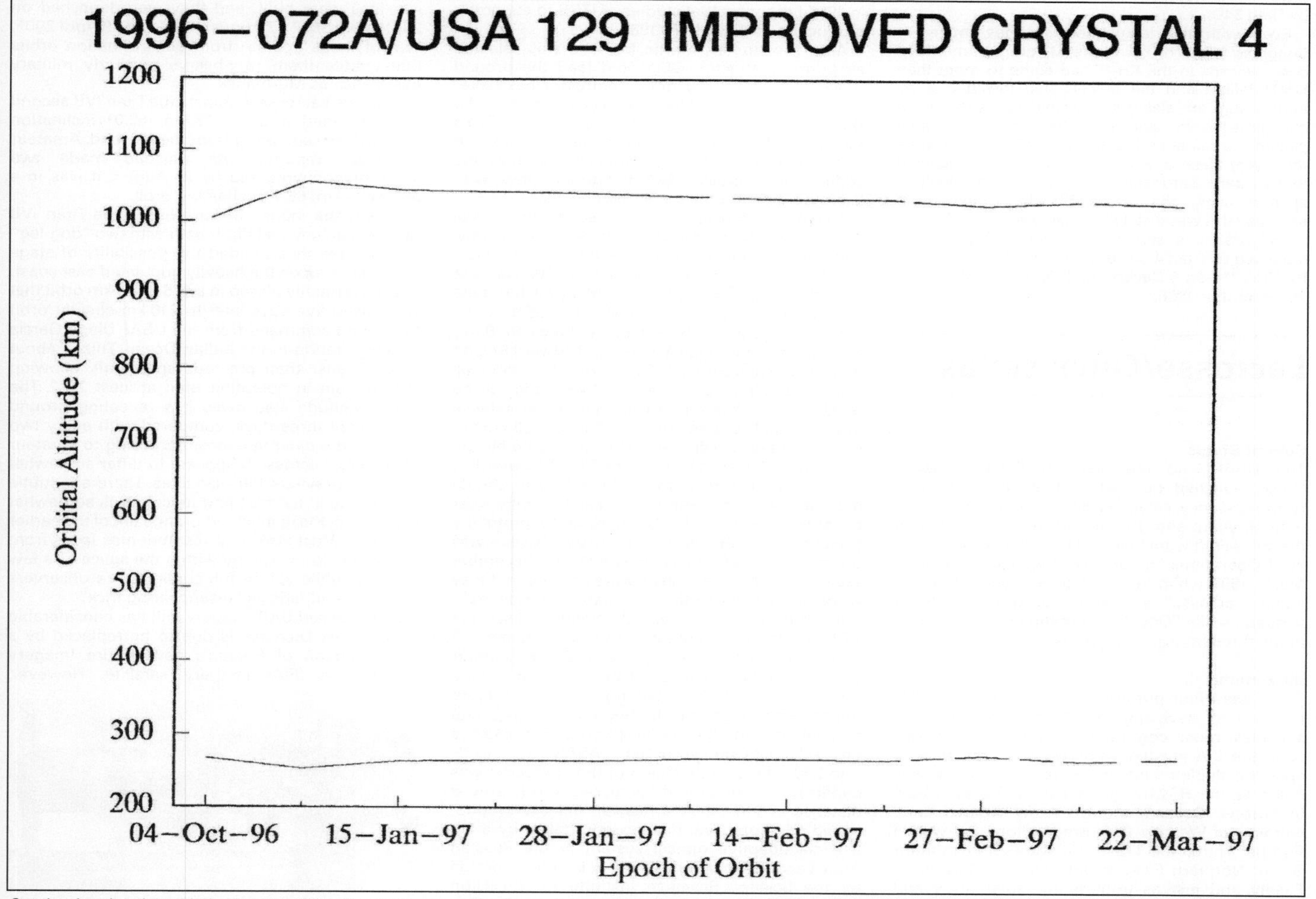

Graphs showing the orbital altitude regimes for photo-reconnaissance missions operating during the 1980s to 1990s. USA 6 was a KH-11/Kennan satellite, USA 27 and USA 33 were KH-11/Crystal satellites and USA 53, USA 86, USA 116 and USA 129 was an Improved Crystal satellite. These orbits are based upon amateur observations of the satellites, since the actual operating orbits are classified
0003443

A submarine larger than the US Trident boats finally emerged from this shed a few months later. In September 1980 the Pentagon purposefully leaked information that a KH-11 had photographed side by side emplacements of SS-20 and SS-16 missiles. The Pentagon claimed the Soviets planned to use their own reconnaissance satellites as surrogate KH-11s to determine how much camouflage each missile required to be mistaken for the other. In short order the Pentagon cited this as evidence that Soviet deception would not allow the US to verify adherence to the Salt 2 agreement on ballistic missile limitation.

KH-11-1, launched 19 December 1976, and KH-11-2, launched 14 June 1978, worked in pairs during part of their lifetimes. Both made their observational passes in the morning. After KH 11-3 joined them on 7 February 1980, the pairs alternated observations in mornings and afternoons to watch the early stages of the Iran-Iraq war in conjunction with Big Bird.

KH-11-4, launched 3 September 1981, set an operational record at 1,177 days, KH-11-5, launched 7 November 1982, and later joined by KH-11-6 on 4 December 1984, worked as a pair until 13 August 1985, when KH-11-5 re-entered. Two weeks later, on 28 August 1985, its KH-11-7 replacement disappeared when the Titan 34D launcher exploded shortly after lift-off, seriously damaging the Vandenberg pad. Because of the loss of the Shuttle Challenger and the Titan explosion, the United States had lost a significant portion of its photo-reconnaissance ability and heroic measures prolonged KH 11-6's life by using it to only relay images of the highest priority intelligence targets.

The satellite executed few manoeuvres in order to conserve fuel, but amateur observations revealed a manoeuvre in mid-1989 into a still unexplained 350 × 900 km orbit. The CIA allowed the orbit to decay back to 336 × 732 km when its perigee was raised to 559 km in November 1990, possibly to support Persian Gulf War operations. No further manoeuvres were seen before it apparently de-orbited in late 1994 after an impressive 10-year life. KH-11-8 was joined by KH-11-6 on 26 October 1987 in an afternoon orbit. The Titan 34D launch of 6 November 1988 was the last launch of a KH-11, arriving at the morning slot in unison with KH-11-8. It was initially placed in a 156 × 1,012 km, 97.9° orbit by Titan 34D.

Even with the lengthened lifetimes, many in Congress talked of killing the KH programme. Total cost overruns in the KH-11 had come to more than USD1 billion and the Soviets had thwarted it by introducing an elaborate camouflage system. To overcome the objection of cost the air force at one time planned to refuel each satellite in orbit to stretch its three-year lifetime. Some observers have suggested that military astronauts made a practice refuelling attempt using the Space Shuttle. Others have indicated this was just a plan that was never tried.

The first in a series of Improved Crystal was launched by Titan 4 on 28 November 1992, followed by USA-116 on 5 December 1995 and USA-129 on 20 December 1996.

Lacrosse/Onyx series

Current Status

The USAF and the National Reconnaissance Office launched the last of five Lacrosse radar reconnaissance satellites on 29 April 2005, more than 16 years after the first of them flew. Four of the spacecraft were believed to remain in service in 2007. Operations for Lacrosse-1 were completed in March 1997, when the satellite was deorbited. The name "Lacrosse" is generally used to refer to all variants, while "Onyx" is sometimes used to refer to the three improved versions.

Background

There were four primary drivers in the late 1970s that led to development of the radar imaging satellites most commonly known as Lacrosse. First, the CIA needed an all-weather capability to spot the deployment of theater nuclear missiles (such as the RT-21M, known to NATO as SS-20) in Europe. Second, the US Army wanted early warning of Warsaw Pact preparations for war in Europe. Third, the US Navy wanted to monitor Soviet Northern Fleet activity around Murmansk. Finally and just as importantly, solar array and Synthetic Aperture Radar (SAR) technologies both had improved significantly by the late 1970s.

SAR inherently had a lower power requirement than a real-aperture radar, and it appeared possible to construct solar panels of sufficient size and wattage to power a space-based SAR that was large enough to produce useful imagery. Airborne SAR technology had been steadily refined during the Vietnam War, with systems typically carried in pods by tactical reconnaissance aircraft, and had been refined to a point where SAR imagery could be transmitted electronically to the ground.

The Air Force, however, had little interest in a space-based SAR because it was clear from the start that the imagery from such a system would not be of sufficient quality for targeting. All three of the major requirements for Lacrosse disappeared with the fall of the Soviet Union, causing the Lacrosse programme to slow down but not stop completely.

Soviet radar ocean surveillance satellites known as Rorsats were orbited beginning in December 1967. Rorsats used an X-band phased array radar powered by a small nuclear reactor. The US Navy began development of a similar system in the late 1970s under the programme name Clipper Bow.

Three contractor teams performed competitive studies for Clipper Bow in 1978–79: Lockheed/Westinghouse, Martin Marietta/Hughes, and McDonnell Douglas/General Electric. Westinghouse, Hughes, and GE were of course the three primary makers of airborne SARs. Significantly, Lockheed and Hughes began work in 1977 on the TR-1 (U-2 derivative) and its ASARS-2 SAR payload, respectively. Plans called for two prototype Clipper Bow satellites to be orbited in 1982–83.

As part of the FY1980 budget deliberations in 1979 however, Defense Secretary Harold Brown directed the Navy to cancel Clipper Bow in favour of a restructured programme that could meet CIA and Army needs and take advantage of emerging technologies. An OSD official explained before a Senate committee that Clipper Bow's cost was too high and its utility too low. The new programme was designated the Integrated Tactical Surveillance System (ITSS). ITSS would be developed concurrently with the Air Force TR-1 reconnaissance system that actually had similar missions.

The Navy continued as lead agency for ITSS. A new competition was opened, with Grumman, Hughes and TRW joining the bidding as primes. In the end however, only the same three contractor teams from Clipper Bow were given new contracts in 1980 for conceptual studies of ITSS to support a production decision in FY1982.

After the studies, Martin Marietta was selected as prime contractor. GE would lead the ground processing effort. The radar contractor has never been revealed, but it is generally believed to be Hughes, using a design scaled up from the TR-1's ASARS-2. ITSS then went black and disappeared from public view. It has since been known by the code names Lacrosse, Vega, Indigo and Onyx, with Lacrosse entering the public vocabulary.

The TR-1 first flew in 1981, and it became operational in 1983. ASARS-2, however, matured more slowly, deploying operationally for the first time in 1985. The Lacrosse production decision in 1982 was CIA Director William Casey's to make. Many felt he would balk at the cost, but in the end he approved it.

That triggered a Senate battle between Barry Goldwater supporting Lacrosse and Edward Boland opposing it. Since ASARS-2 was not yet operational, Goldwater cited the success of SAR pods carried by USAFE F-4 fighters in Germany in his defense of Lacrosse. Lacrosse survived. The overall cost of each Lacrosse mission was almost USD1.5 billion, including USD411 million for the Titan IVB launch.

In the mid 1980s, unclassified Space Shuttle missions deployed both a huge experimental solar panel believed to be for Lacrosse and a prototype space-based SAR. The first Lacrosse satellite was orbited by the Shuttle Atlantis after a 2 December 1988 launch. Its solar panels were so large that they were visible from Earth with amateur telescopes.

In 1998, the NRO released photos of Lacrosse without its radar antenna and solar panels. It appears to use the same Lockheed "Bus 1" Support Systems Module developed for KH-11, and there does not seem to be anything remarkable about the satellite other than its huge solar wings and its synthetic aperture radar antenna. The latter is probably quite similar to that of ASARS-2.

In 2007, the overall design of the spacecraft was confirmed in images of Lacrosse-2 taken from a distance of 816 km by a Russian military ground-based telescope. The images were taken by a 28-channel adaptive optical system at the Russian Altay Laser/Optical Centre, one of several operated by the Science Research Institute for Precision Instrument Engineering. Although the Russian images are blurred, they show the huge mesh dish antenna at different angles.

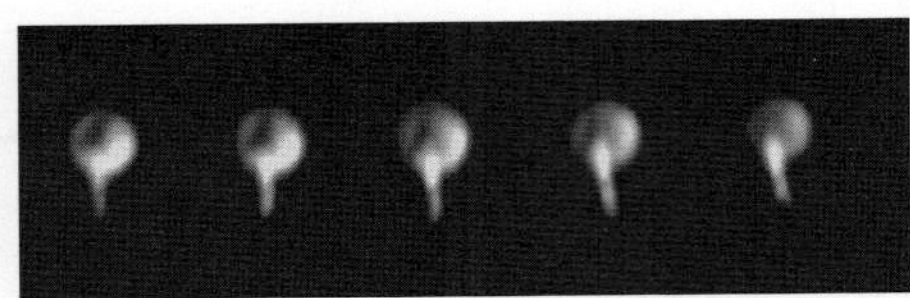

Images of Lacrosse-2 taken by a 28-channel adaptive optical system at the Russian Altay Laser/Optical Center. The site is one of several operated by the Science Research Institute for Precision Instrument Engineering. The line-of-sight distance from the Russian telescope to the Lacrosse was 816 km. Clearly visible is the secret 8 m mesh synthetic aperture radar antenna, developed by Harris Corp. near Cape Canaveral (RUSSIAN SCIENCE RESEARCH INSTITUTE FOR PRECISION INSTRUMENT ENGINEERING) 1327015

At the time of the 1991 Gulf War, it was reported that Martin Marietta was already under contract to build a total of four Lacrosse satellites, with plans calling for a constellation of eight. With each of those needing to be replaced periodically, the programme looked like a major long-term cash cow for Martin Marietta. However, with the end of the Cold War, the Bush administration capped the programme at four (later five) satellites despite the vehement objections of Senator Boren of Oklahoma who felt the full constellation was critical for arms control verification. The second Lacrosse was launched on 8 March 1991, but launch of the third was delayed several years.

In the intervening period, Martin Marietta and GE were both folded into Lockheed Martin, and Hughes was later acquired by Raytheon.

The six-year delay between the second and third launches allowed significant improvements to be made to both the radar and the data readout system. Most likely, the radar improvements were essentially the same changes made to ASARS-2 under the ASARS Improvement Program (AIP): an Electronically Steered Antenna (ESA) replacing the original mechanically steered antenna and a new Ground Moving Target Indication (GMTI) mode. The latter would allow Lacrosse to detect moving vehicles.

The new "8X" data readout system developed for the EIS version of KH-11 was reportedly also incorporated in the improved Lacrosse. This would allow eight times as much imagery per pass to be downloaded. A total of three Improved Lacrosse satellites were built, and they were launched on 23 October 1997, 17 August 2000 and 29 April 2005. Two of these operate from 68° inclination orbits that enable them to observe northerly military bases such as Murmansk.

After the Lacrosse-4 launch, the Titan IVB second stage reached a 572 × 675 km, 68.0° inclination orbit before separating from the payload. Amateur observers reported the payload made two small manoeuvres and by 23 August it was in a 681 × 695 km, 68.1° inclination orbit.

During the launch of Lacrosse-5, the Titan IVB followed an unusual flight path with two "dog-leg" manoeuvres that avoided the possibility of stage debris landing on the heavily populated east coast. It was eventually placed in a 475 × 713 km orbit that was raised five days later to 716 km circular orbit by ground command from the USAF Diego Garcia tracking station in the Indian Ocean. This is about 30 km higher than previous spacecraft, allowing it to remain in operation until at least 2012. The higher altitude also gives it a repeating ground track every three days, compared with every two days for the older versions. According to amateur observers, Lacrosse-5 appears to differ somewhat from the previous four satellites. There are subtle differences in its orbit and its colour is somewhat whiter than the distinct red-orange tint of the earlier satellites. Most strikingly, it sometimes fades from brightness to invisibility within the space of a few seconds while still in full sunlight. The observers call this rapid fade its "disappearing trick".

Space-based SAR imagery still has considerable value, since Lacrosse is due to be replaced by a radar version of Boeing's new Future Imagery Architecture (FIA) imagery satellite. However,

An early version of the US Lacrosse radar reconnaissance satellite during integration by Lockheed Martin (NRO) 1327002

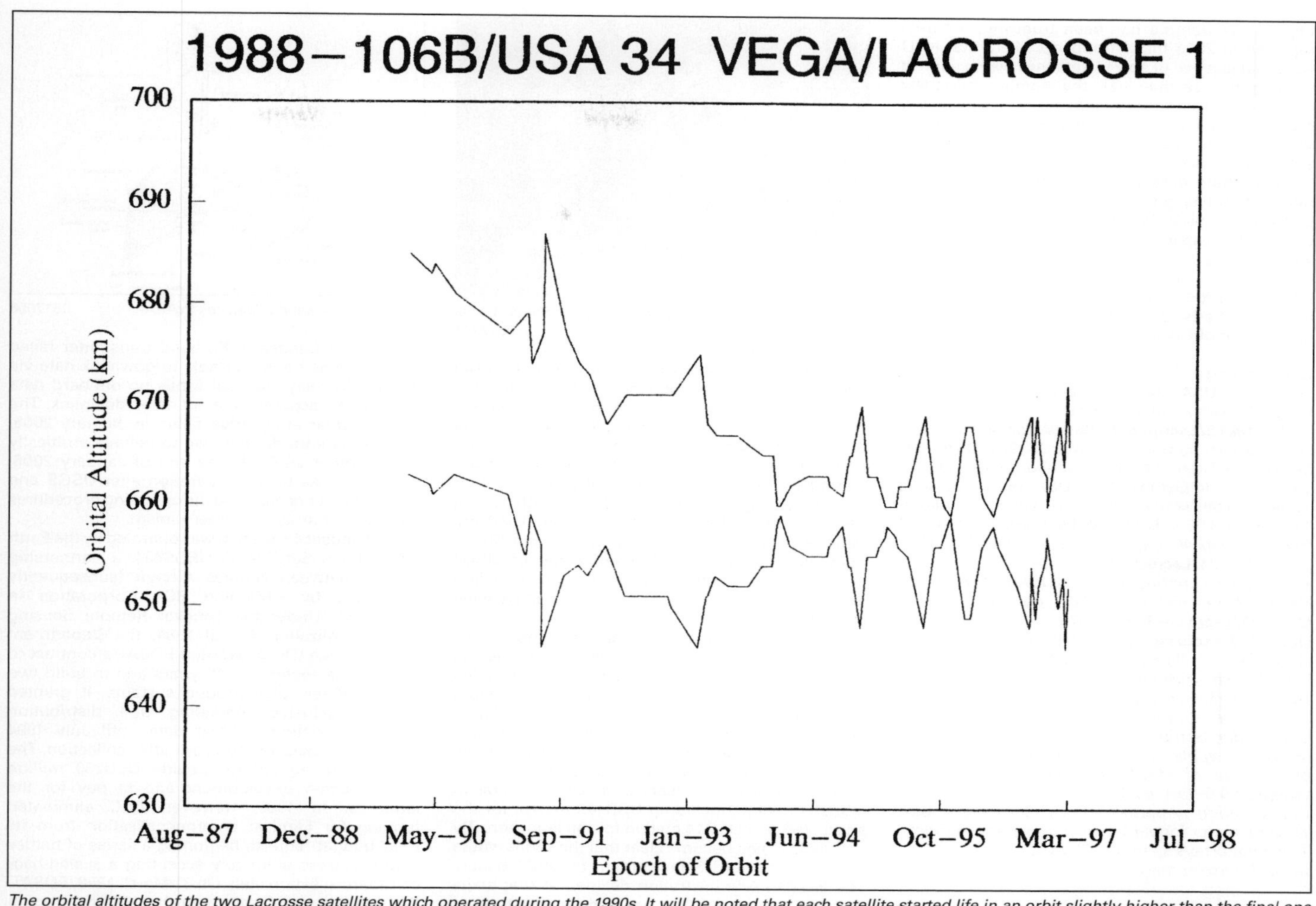

The orbital altitudes of the two Lacrosse satellites which operated during the 1990s. It will be noted that each satellite started life in an orbit slightly higher than the final one and was allowed to slowly drift down until the altitude was stabilised at around 660 to 670 km. USA 34/Lacrosse 1 disappeared from orbit in March 1997.These orbits are based upon amateur observations of the satellites, since the actual operating orbits are classified
0003436

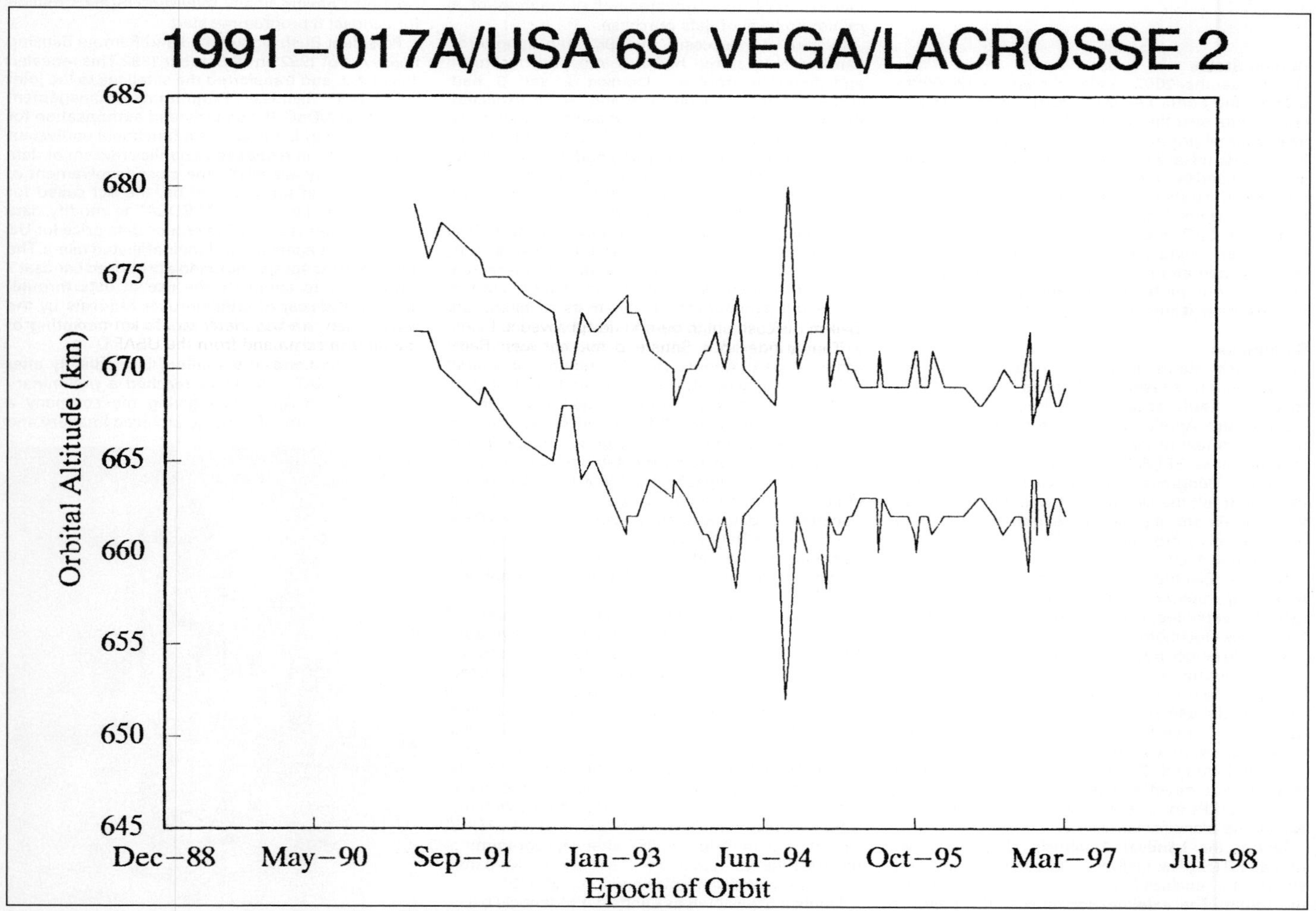

The orbital altitudes of the two Lacrosse satellites which operated during the 1990s. It will be noted that each satellite started life in an orbit slightly higher than the final one and was allowed to slowly drift down until the altitude was stabilised at around 660 to 670 km. USA 34/Lacrosse 1 disappeared from orbit in March 1997.These orbits are based upon amateur observations of the satellites, since the actual operating orbits are classified
0003437

the FIA programme has been truncated. *Reuters* reported in 2003 that "it appears likely that NRO will build just five or six FIA satellites over a shorter timeframe, less than half the number originally envisioned". This implies that no more than two radar FIAs will be built. Moreover, those satellites will likely be the last dedicated SAR imaging satellites as such.

The primary mission of the satellites developed under the DoD's new Space Radar programme will be Ground and Air Moving Target Indication (GMTI and AMTI, respectively), with SAR imaging as a secondary capability. In January 2005, Secretary of Defense Donald Rumsfeld and Director of Intelligence Porter Goss signed a memo designating the Space Radar as the replacement for the radar component of FIA.

Specifications

Launches: **USA 34/Lacrosse-1:** Shuttle Atlantis (STS-27) from Kennedy Space Center, 2 December 1988. **USA 69/Lacrosse-2:** Titan IV from Vandenberg space launch complex 4, 8 March 1991. **USA 133/Lacrosse-3:** Titan IV from Vandenberg space launch complex 4, 24 October 1997. **USA 152/Lacrosse-4:** Titan IVB from Vandenberg space launch complex 4, 17 August 2000. **USA 182/Lacrosse-5:** Titan IVB from complex 40, Cape Canaveral, 29 April 2005.
Orbit: **USA 34/Lacrosse-1:** 437 × 447 km. *Inclination:* 57°, period: 93.4 min. **USA 69/Lacrosse-2:** 420 × 662 km, inclination 68°, period: 95.5 min. **USA 133/Lacrosse-3:** 666 × 679 km, inclination 57°. **USA 152/Lacrosse-4:** 689 × 695 km, inclination: 68°, period: 98.53 min. **USA 182/Lacrosse-5:** 718 × 712 km, inclination 57°.
Design life: Seven years
Contractors: Prime contractor: Lockheed Martin. *Mesh radar:* Harris Corp., Florida.
Principal applications: All-weather, 24 hour observation of stationary and moving military targets at 0.6–1 m resolution.
Configuration: Approximately 15 m main bus. 7.5 m diameter forward dish antenna (mesh radar). Two solar arrays spanning 33 m.
Mass: Approximately 14,500 kg

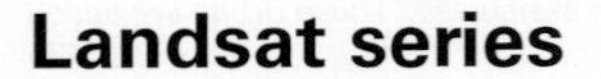

Landsat series

Current Status

On 1 September 2007, Landsat 5 made its 125,000th orbit of the Earth. Landsat 5 is still being used to collect data over the United States and for selected international ground stations. The expected end of life of the Landsat 5 mission, based on fuel reserves, is 2010. Landsat 7 has been returning degraded imagery since its main instrument malfunctioned in 2003. Landsat 5 imaging was suspended on 6 October 2007 due to a loss of a cell within one of two remaining batteries. However, Landsat 5 resumed operations in January 2008, and is now operating on one battery. Its end of life, based on fuel reserves, is projected as no later than 2010.

Background

The United States floated the first proposals for a civil remote sensing satellite in the late 1960s following Earth observations from the manned Gemini and Apollo. Serious study began when the Department of Interior (DOI) in 1968 requested funding for an EROS Earth Resources Observation Satellite. Congress denied DOI the funding, because it felt the Department was not authorised to undertake space projects. Nevertheless DOI built a co-operative programme with NASA and funds appeared in 1969.

General Electric built Landsat 1-3. They were enhanced versions of the Nimbus weather and research satellites, originally known as the Earth Resources Technology Satellites (ERTS). NASA planned to launch two in consecutive years, 1972–73, as an experiment in systematically surveying the Earth's surface to study the health of its crops and the potential use and development of its land and oceans. 300 investigators in the 50 participating countries made use of the dramatic and vivid Landsat 1 images. Nations quickly realised these images were directly relevant to the management of the world's food, energy and environment and NASA laid plans for Landsat 2.

Before the **Landsat 1** launch, NASA decreed that all ERTS data, including 9,000 weekly images, should be unclassified and made available to the public. The satellite downlinked the pictures through NASA ground stations to the US Department of Interior's EROS Data Center in Sioux Falls, South Dakota where they could be bought for prices starting at USD1.25. Unexpectedly, news agencies and private individuals studied detailed pictures of Soviet- and Chinese-launch centres and missile sites. The immediate success of the project caused the processing and distribution systems to be swamped with requests. As a result, a number of countries subsequently installed their own receiving stations, including South Africa, Canada, Brazil, Italy, Sweden, Japan, Thailand, India, Australia, Argentina, Zaire, Iran and China. They initially paid construction costs of USD4 million to 7 million, USD1 million to 2 million annual operational costs and an annual service fee of USD200,000.

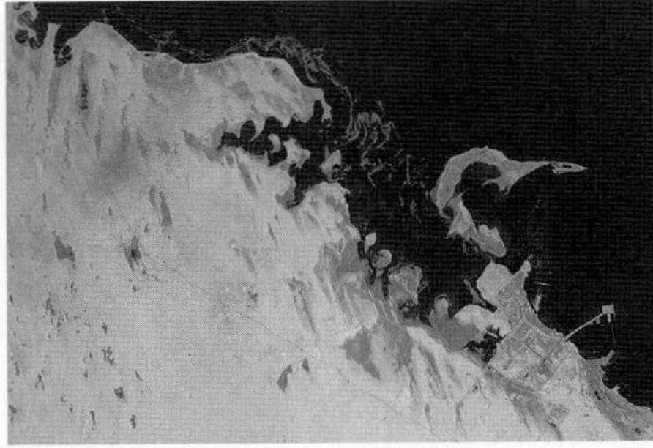

Landsat 5 image including an oil slick off Saudi Arabia (EOSAT) 0517241

The first three Landsats carried combinations of Return Beam Vidicon (RBV) and Multi-Spectral Scanner (MSS) imagers. Technical problems resulted in data from the RBVs of Landsat being seldom used. For **Landsat 3**, even though RBV resolution was improved from 80 m to 40 m, the spectrally and radiometrically superior MSS was still preferred. RBVs were deleted from later spacecraft.

The first three Landsats cost NASA a total of USD251 million, including USD149 million for the spacecraft and USD14 million for the launchers. By July 1981, it had become clear that the series would represent a total US investment of USD1 billion. The Reagan Administration decided at that stage that the remote sensing programme must either end, or be absorbed by the private sector. COMSAT, Hughes and several other companies expressed interest, subject to government guarantees of a minimum level of data purchase.

Despite its success, by 1982 the long-term Landsat programme had run into both technical and financial troubles. Landsat 2 and 3 had surpassed their design lives and NASA expected them to fail at any time. GE, whose prime contract for Landsat D/D-Prime (as Landsat 4 and its back-up were known before launch) had trouble selling its USD505 million contract in Congress. Hughes also seemed to be behind in the development of the advanced Thematic Mapper.

Finally, NASA managed to solve the technical and bureaucratic problems and **Landsat 4** was launched into space on 16 July 1982. The spacecraft suffered numerous malfunctions which limited spacecraft functionality early in its mission. On 27 July 1982, the high gain Ku-band antenna failed to deploy. After many attempts, success was finally achieved on 15 August. Then the redundant central unit in the spacecraft Command and Control Unit (CCU) failed. It also suffered a malfunction in its solar panels in March 1983, when power cabling from the solar panel broke up under the day/night thermal cycling, leaving the satellite with only 50 per cent power capacity. The TM X-band link failed after transmitting 6,000 images. As a result, the downlink of data was not possible until the TDRS (Tracking and Data Relay Satellite) system became operational. Landsat 4 was then able to use its Ku-band transmitter to send data to ground stations via TDRS.

This situation prompted the early launch of Landsat 5 to guarantee continued coverage. NOAA, faced with a monthly revenue loss of USD600,000, because of these failures, requested the USD20.4 million launch of the back-up Landsat D-prime, undertaken in March 1984. NASA studied a Landsat rescue mission to retrieve the USD60 million satellite in June 1986, using the second Vandenberg Shuttle mission. The end to Shuttle preparations at Vandenberg shelved the idea. Instead, EOSAT maintained full operations with the MSS and resumed TM transmissions via the TDRS satellite link in October 1987 after reprogramming the satellite to reactivate the back-up Ku-band. Landsat 4 was decommissioned in mid-2001.

Landsat 5 was built as Landsat 4's identical back-up (then modified to prevent similar failures) and launched when the prime vehicle suffered failure of its redundant power and transmitter systems.

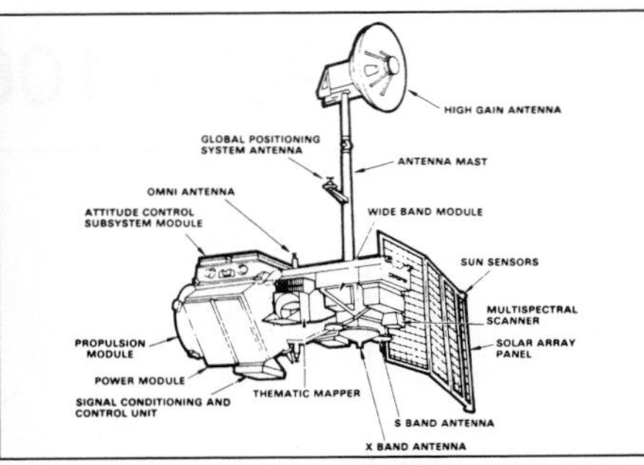

Principal Landsat 4/5 features (NASA) 0517056

In 1987, the Landsat 5 Ku-band transmitter failed so that it was no longer able to downlink data via the TDRSS relay. Landsat 5 has no onboard data recorder to acquire data for later downlink. The primary solar array drive failed in January 2005, and the back-up drive began to behave erratically in November 2005. By the end of January 2006, Landsat 5 was back in operation after USGS and NASA engineers modified the operating procedures for the solar array drive mechanism.

The Landsat 4/5 system was operated by the Earth Observation Satellite Co (EOSAT), a partnership formed between Hughes Aircraft (subsequently purchased by GM) and RCA Corporation in May 1984. Under the Landsat Remote Sensing Commercialization Act of 1984, the Department of Commerce (DoC) awarded EOSAT a contract to operate the system for 10 years and to build two new satellites, plus ground systems. It granted EOSAT exclusive marketing and distribution rights to existing Landsat data until July 1994 and to new data for 10 years after collection. The government agreed to provide USD250 million for spacecraft development and to pay for the launches. In 1986, however, DoC eliminated funding for Landsat commercialisation from its FY87 budget request, beginning a series of battles with Congress ultimately accepting a scaled-back commercialisation plan. On 31 March 1988, EOSAT/DoC signed modifications to the 1985 contract to bring it into line with the compromise reached between Congress and the Reagan Administration, Congress having finally released USD62.5 million for Landsat 6 programme start.

President Bush signed the Land Remote Sensing Policy Act of 1992 on 28 October 1992. This repealed 1984's Act, and transferred the satellites to the joint NASA-DoD Landsat Programme Management from NOAA/DoC. It also provided authorisation for the funding of Landsat 7 as a functional equivalent of Landsat 6. In response to public criticism of data release policy especially the close involvement of the military in the programme, the Act called for negotiations between LPM/EOSAT to modify data policy for Landsat 4 to 6 to reduce data price for US government agencies and their affiliated users. The 1992 Landsat Act also outlined a proposed Landsat 7 data policy to stimulate the use of data through sales at the cost of fulfilling user requests by the government and/or the competitive marketing of unenhanced data.

Following **Landsat 6**'s failure immediately after launch, EOSAT and NASA reached a preliminary agreement in April 1994 giving the company a further five years of rights to archived imagery and

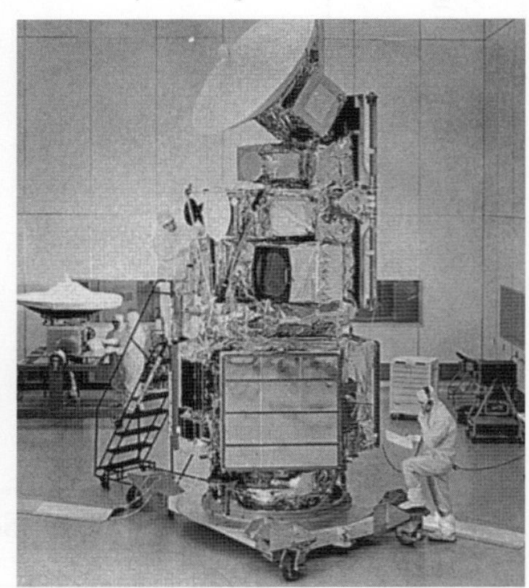

Landsat 5 checkout; Fairchild's MMS section is at bottom, with the instrument section and folded TDRS antenna above (EOSAT) 0517057

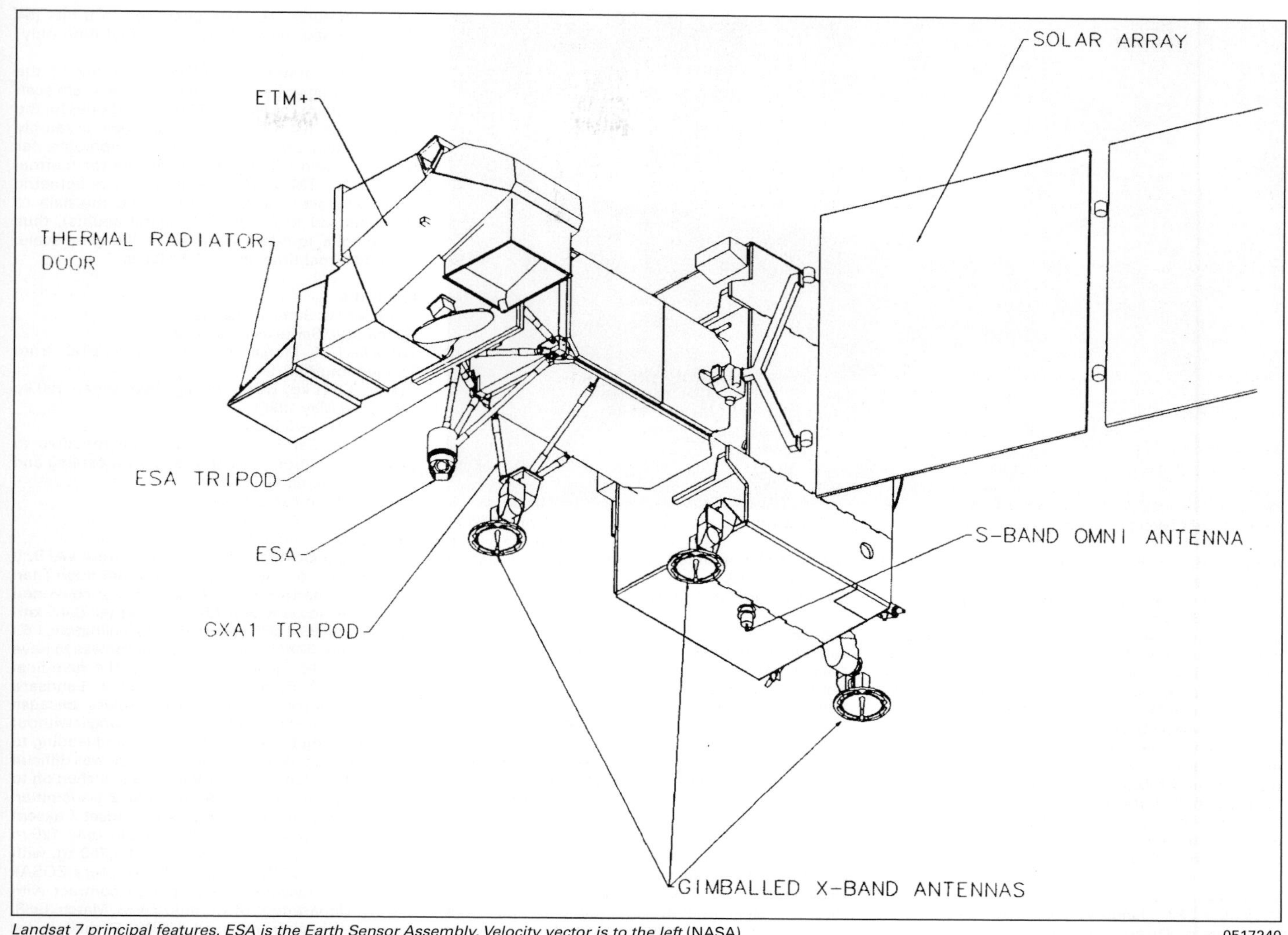

Landsat 7 principal features. ESA is the Earth Sensor Assembly. Velocity vector is to the left (NASA)

0517240

exclusive rights to new Landsat 4/5 scenes (and for five years after the satellites' deaths). In return, EOSAT would charge government users reduced rates: USD3,500 per scene for the rest of 1994, and USD2,500 from 1995, instead of the commercial USD4,400 rate. Once bought, these scenes could be copied for other government users at no extra cost. However, the DoC, in July 1994, decided to open the contract for bidding; EOSAT's contract was extended through 30 June 1995 while the matter was pursued; EOSAT was successful.

In November 1996, Space Imaging and Lockheed Martin announced that they had reached an agreement concerning the purchase of EOSAT, a Lockheed Martin company, by Space Imaging. Under the terms of the agreement, Space Imaging would acquire all of the assets of EOSAT through a new subsidiary – Space Imaging, the Earth Observation Satellite Co. (SII/EOSAT) – which would maintain the existing distribution agreements between EOSAT and its global satellite imagery suppliers. These included agreements to distribute Landsat imagery.

The National Space Council established a joint DoD/NASA **Landsat 7** programme in early 1992. The NSC designated the Air Force for the development of the spacecraft, instruments and launch services, and NASA for the ground system. Because of severe budgetary pressures, coupled with Landsat 6's loss, the National Science and Technology Council reassessed the purpose of Landsat 7. Congress appropriated USD84 million for DoD in FY93 for the procurement of Landsat 7, with launch planned for 1997. In October 1992, the USAF awarded the USD398 million contract to Martin Marietta Astro Space (now Lockheed Martin). The DoD agreed to build Landsat 7 in the expectation that it would receive 5 m imagery from the new USD207 million High-Resolution Multispectral Stereo Imager, but NASA continued to resist funding the high costs for the required ground system. The Clinton Administration decided, in February 1994, that NASA should develop Landsat 7 for a 1998 launch.

This produced Presidential Decision Directive/NSTC-3, dated 5 May 1994, changing it to a joint NASA, NOAA and US Geological Survey programme, dropping the military component. The Landsat 7 programme aims to maintain data continuity and minimise any gap as well as make Landsat 7 data available to all users at the cost of fulfilling requests.

NASA operated Landsats 5 and 7 until 1 October 2000, when control was turned over to the U.S. Geological Survey. USGS also processes and distributes the data, continuing the operation and maintenance of the National Satellite Land Remote Sensing Data Archive for the long-term preservation of Landsat and other satellite land remote sensing data. In support of the International Polar Year (2007–2008), the USGS, British Antarctic Survey and NASA, with funding from the National Science Foundation, released the Landsat Image Mosaic of Antarctica (LIMA) in November 2007. The mosaic was compiled from 1,065 hand-selected Landsat scenes taken with the ETM+ sensor on board Landsat 7. LIMA is the most geometrically accurate and highest spatial resolution mosaic ever made of Antarctica.

The new Newcastle ground station site at Norman, Oklahoma was developed in late 1991 for integrated TT&C and data reception of Landsat 6-era operations, adding 158 m^2 to an existing 279 m^2 building. The mid-continent location allows real-time reception of data covering the contiguous 48 states. The station was formally dedicated 8 May 1992, discontinuing the datalink through NASA Goddard. The data are then transferred to EOSAT's Lanham headquarters for processing and distribution as CCTs (6,250/1,600 bpi), 8 mm Exabyte tapes and photo products. Swedish Space Corporation's Kiruna site provides back-up TT&C. Control is maintained from HQ's LSOC Landsat Spacecraft Operations Center via Newcastle or Kiruna.

After a bid to commercialise the collection and delivery of Landsat data fell through in 2003, the White House supported an inter-agency arrangement to use the future National Polar-orbiting Operational Environmental Satellite System (NPOESS) satellites to carry an imager that would continue the Landsat observations. Rising costs associated with development of this sensor, combined with severe budgetary and technical problems that threatened lengthy delays in the deployment of NPOESS, led to a policy reversal. On 23 December 2005, the White House directed NASA to pursue development of a free-flying Landsat replacement, later dubbed the **Landsat Data Continuity Mission**.

In July 2007, Ball Aerospace and Technologies Corp. of Boulder, Colorado was selected to develop the Operational Land Imager instrument for the Landsat Data Continuity Mission. The instrument will provide 15 m panchromatic and 30 m multi-spectral spatial resolution capability, with a 185 km swath that will enable the entire globe to be imaged every 16 days.

On 3 October 2007, NASA announced a USD124 million contract with Lockheed Martin Commercial Launch Services for the launch of the Landsat Data Continuity Mission. The firm-fixed-price contract includes launch services for an Atlas V 401 rocket, payload processing, launch vehicle integration, and tracking, data and telemetry support. The spacecraft is scheduled to be placed into a 689 km high polar sun synchronous orbit in July 2011, lifting off from Vandenberg Air Force Base, California. The Landsat Data Continuity Mission will extend the three decade record of land surface observations from previous Landsats. NASA's Goddard Space Flight Center in Greenbelt, Maryland, manages procurement and acquisitions for the Landsat Data Continuity Mission in partnership with the USGS. The USGS will manage the satellite after launch and in-orbit checkout. As of February 2008, a decision on the manufacturer of the satellite had yet to made.

Specifications

Landsat 1

Launched 23 July 1972 by Delta from Vandenberg. Launch mass was 891 kg and the orbit was 901 × 920 km, 99° inclination. Operations ended 6 January 1978. By 30 March 1973, when faults occurred in the tape recorder, the North American continent had been covered 10 times; the total of >34,000 images included all the world's major landmasses at least once. The recorder fault meant that images could no longer be stored for transmission, though when the satellite was over one of NASA's three ground receiving stations in California, Alaska and Maryland it was still possible to transmit real-time imagery of the entire North American continent. By the time Landsat 1 had to be retired in 1978, it had returned 300,000 images and demonstrated the potential of remote sensing.

Landsat 7 has been operational since its launch in 1999 (Lockheed Martin) 1342671

Sensors: Return Beam Vidicon (RBV). Landsat 1/2 each carried three 80 m instruments sensitive to Band 1 0.475 to 0.575 μm (blue-green), Band 2 0.580 to 0.680 μm (yellow-red); and Band 3 0.690 to 0.830 μm (red-near-infra-red). Simultaneous views were provided of 185 km² areas. Landsat 3 carried two single-band 40 m RBVs providing adjacent 98 km² images, covering 0.505 to 0.750 μm. TM replaced the RBV on later satellites.

Multi-Spectral Scanner (MSS). All three included 80 m resolution four-band sensors, with L3 adding a 5th, thermal 120 m band. Band 4 0.50–0.60 (green), Band 5 0.60 to 0.70 (red), Band 6 0.70 to 0.80 (red-near infra-red), Band 7 0.80 to 1.10 μm (near-infra-red). L3 only: Band 8 10.40 to 12.50 (thermal-infra-red). Coverage was identical to RBVs and subsequent Landsat MSS.

Landsat 2

Launched 22 January 1975 by Delta from Vandenberg. 816 kg; 907 × 918 km, 99°. Operations ended 25 February 1982. This Landsat incorporated a 14 kg digital computer with 4,096-word memory able to handle 55 separate commands from ground stations and carry out routine operations for up to 24 h – a vital function as it was out of tracking station range for 80 per cent of the time. Landsat 2 was intended to supplement and later replace Landsat 1, and the two were able to work together until 1978, providing repeat coverage every 9 instead of 18 days. Retired on its fifth birthday in 1980 because of attitude control problems, L2 was revived six months later by Goddard engineers, who developed a magnetic compensation system to provide stabilisation. Though its tape recorders had failed, it still provided real-time imagery in 1981, supplementing the limited activities of Landsat 3.
Sensors: Same as Landsat 1.

Landsat 3

Launched 5 March 1978 by Delta from Vandenberg. 960 kg; 900 × 918 km, 99°. Operations switched to standby 31 March 1983. Further sensor refinements included the addition of a thermal band to the MSS, able to detect temperature differences in vegetation, bodies of water and urban areas by day/night; and an improved RBV offering a 50 per cent resolution increase and enabling areas as small as half an acre to be identified. The three year Large Area Crop Inventory Experiment (LACIE), concluded in November 1978, showed that Landsat's multispectral scanners revealed the Soviet Union's wheat crop to be 91.4 million tonnes, less than 1 per cent below the official Soviet figure. There was less success with predictions of wheat yields in Canada and the US, where the long, narrow fields can easily be confused with adjoining spring-planted crops. Improvements in Landsat 4's instruments were intended to remedy this, making it impossible for countries such as the Soviet Union to repeat past successes in buying up world wheat surpluses cheaply when anticipating a poor crop of its own. As wheat is the world's most important grain crop, Australia, China, Brazil and Argentina were also included in LACIE. As a follow-on, it was decided to monitor wheat, corn, rice, barley, sorghum, soya bean and sunflower crops in the Soviet Union, China, Brazil, Mexico, Australia, Canada and US. In January 1981, after two years of operations, Landsat 3's MSS malfunctioned but the RBV continued to complement Landsat 2's operational MSS.
Sensors: Same as Landsat 1.

Landsat 4

Launch: 16 July 1982 by Delta 3920 from VAFB.
Mass: 1,941 kg BOL
Contractors: General Electric (later Lockheed Martin Astro Space) prime, Fairchild for Multimission Spacecraft, Hughes Santa Barbara Research Center for sensors.
Resolution: 30 m Thematic Mapper (120 m thermal infra-red), 80 m Multi-Spectral Scanner.
Orbit: 705 km circular, 98.2° sun-synchronous, crossing equator at 09.45 local time, with 16 day repeat, 2,760 km ground-track separation at equator between revolutions, 172 km between tracks in completed coverage.
Spacecraft: About 4 m long and is based upon the NASA/Fairchild Multimission Modular Spacecraft (MMS), as used initially by Solar Max, attached to the instrument/solar panel/antenna forward section. The MMS is essentially a triangular polyhedron with a propulsion module on one end and three 1.2 × 1.2 × 0.3 m box-like side modules handling power, AOCS and data/communication links. 3-axis control with pointing accuracy of 0.01° provided by MWs in conjunction with an inertial reference system using attitude updates from two star trackers. The aft propulsion module houses 11 N + 0.5 N hydrazine thrusters for maintaining the 16-day repeat cycle and MW unloading. Real-time TM data transmitted at 84.9 Mbit/s on X-band and MSS data at 15 Mbit/s on S-band; an S/Ku-band dish on a 4 m mast accesses the TDRS network. No data recorders. The single 4-panel solar wing was designed to provide 990/814 W BOL/EOL; this failing power system prompted the launch of Landsat 5.
Sensors: Thematic Mapper (TM). Landsat 4 was the first to incorporate the TM, with its greater number of spectral bands and enhanced resolution. The 245 kg 1.1 × 0.7 × 2.0 m TM draws 345 W. The spectral ranges (μm) and physical characteristics measured by the bands are: Band 1 0.45–0.52 (chlorophyll and carotenoid for soil/vegetation and deciduous/coniferous differentiation), Band 2 0.52–0.60 (green reflectance by healthy vegetation), Band 3 0.63–0.69 (chlorophyll absorption for plant species differentiation), Band 4 0.76–0.90 (near-infra-red reflectance of healthy vegetation for biomass surveys), Band 5 1.55–1.75 (vegetation moisture and snow/cloud reflectance differences), Band 6 10.40–12.50 (thermal mapping), Band 7 2.08–2.35 (vegetation moisture and hydroxyl ions in soils). Band 6 exhibits 120 m resolution; the remainder are 30 m. Each uses an oscillating mirror to scan the Earth's surface in the cross-track direction as Landsat's motion provides along-track coverage. The 170 km long 185 km wide images have a 14 per cent side overlap at the equator, increasing with latitude.

Multi-Spectral Scanner. The Landsat 4/5 MSS are similar to those flown on Landsat 1-3, with the four bands offering 80 m resolution and the same coverage as the TM. The 58 kg 35 × 40 × 90 cm MSS draws 81 W. Spectral ranges (μm): Band 1 0.50–0.60 (green); Band 2 0.60–0.70 (red); Band 3 0.70–0.80 (near-infra-red); Band 4 0.80–1.10 (near-infra-red). Bands 1-3 in the Ritchey-Chrétien telescope system utilise six photomultiplier tubes; band 4 uses four silicon photodiodes. Data quantised to 6 bits (64 digital counts); transmission rate (real time only) 15 Mbit/s.
Optics: The prime focal plane assembly of the Ritchey-Chrétien telescope with its 40 × 53 cm scan mirror contains four sets of 16 Si photodiodes for the VNIR bands 1–4. The cooled focal plane assembly contains two arrays of 16 InSb photodiodes for bands 5/7, and four HgCdTe detectors for thermal band 6. The TM also offers improved radiometric sensitivity over the MSS, permitting the data to be quantised to 8 bits (256 digital counts), thus providing a four-fold increase in the grey scale. Data rate (real-time only) 84.9 Mbit/s.

Landsat 5

Contractors: Same as Landsat 4
Resolution: Same as Landsat 4
Launched: 1 March 1984 by Delta 3910 from Vandenberg AFB
Mass: 1,941 kg BOL (249 kg hydrazine, 150 kg remained May 1995)
Orbit: Same as Landsat 4
Spacecraft: Same as Landsat 4, but modified to prevent repetition of solar array power cabling and X-band transmitter failures
Sensors: Same as Landsat 4.

Landsat 6

Launched 5 October 1993 by Titan 2 from VAFB. It appears not to have attained orbit, although Titan apparently performed as expected. The converted ICBM released it with a 724.7 km (target 724.5 km) apogee and 98.0107° (target 98.006) inclination. L6's Thiokol Star 37XFP solid apogee motor was to have ignited at 819 seconds to establish the near-final 704 km, 98.2° circular orbit. However, Landsat's attitude control thrusters were useless because their hydrazine tank had ruptured, leaving it without control during the kick motor's burn and leading to it burning up over the Pacific. The cause was difficult to identify as Landsat telemetry was switched off to avoid interference with Titan's stage 2 transmitter. Landsat 6 was almost identical to Landsat 7, except that ETM's thermal-IR band provided only 120 m resolution. Launch mass was about 2,750 kg, with 1,740 kg on-orbit BOL. Design life five years. EOSAT signed L6's rescoped development contract with the US Department of Commerce 31 March 1988, Congress having released USD62.5 million for programme start. Total cost was USD268.9 million, including USD36.5 million for launch, USD220 million for the space/ground segments, with USD10.8 million contributed by EOSAT. Landsat 6 was to continue its predecessors' TM products, while adding a co-registered 15 m PAN band to the USD92 million 390 kg ETM.

Landsat 7

Contractors: Lockheed Martin Astro Space (satellite, flight systems integration), Hughes Santa Barbara Research Center (ETM+).
Resolution: ETM+ 30 m (6 bands), 15 m panchromatic, 60 m thermal-infra-red
Launch: 15 April 1999 by Delta II 7920 from VAFB.
Mass: About 2,170 kg at launch, about 2,020 kg on-orbit BOL
Orbit: 705 km circular in 98.2° Sun-synchronous, descending node at 10.00±00.15 local time, with 16-day repeat cycle (233 revolutions) and 172 km ground track separation at equator.
Spacecraft: About 4.3 m long and based on Tiros/DMSP meteorological satellite bus but with extensive subsystem modifications for Landsat-unique requirements and to improve mission reliability. Sensor mounted on the forward module. Three steerable high-gain X-band antennas operating at 8.0825, 8.2125 or 8.2434 GHz. Each provides two 75 Mbit/s streams (I/Q) of ETM+ data, with bands 1–6 on one channel and 7–8 on other. 375 Gbit SSR can store about 100 full ETM+ scenes. Stored and real-time data acquired at USGS EROS Data Center in Sioux Falls. Real-time data available to international stations on X-band. Command uplink 2.1064 GHz; telemetry downlink 2.2875 GHz. 1,550 W generated by a 4-panel single solar wing driven at orbital rate to follow the Sun. Eclipse power by 2 × 50 Ah Ni/H_2 batteries. 3-axis attitude control by RWs and magnetorquers, accuracy 180 arcsec; attitude information from star sensors, gyros, Earth sensors and magnetometer. Ground track held within 5 km by 12 × 4.45 N hydrazine thrusters on lower module.
Sensors: Enhanced Thematic Mapper + ETM+ is a nadir-viewing cross-track scanning 8-band multispectral radiometer, developed by Hughes SBRC. It provides continuity of TM data with Landsat 4/5 and adds a 15 m PAN band, improved resolution (60 m) for thermal, and improved radiometric accuracy (5 per cent). All other bands

remain at 30 m. ETM+ can image >525 scenes daily, with a 185 km swath width and 170 km along-track length. Spectral ranges (μm): Band 1 0.45–0.52, Band 2 0.52–0.60; Band 3 0.63–0.69; Band 4 0.76–0.90; Band 5 1.55–1.75; Band 6 10.4–12.5 (thermal mapping); Band 7 2.08–2.35; Band 8 (PAN) 0.50–0.90. Imagery can be collected in low- or high-gain modes; high gain doubles the sensitivity.

Misty

Current Status

Misty is the codename for a highly classified imaging reconnaissance satellite incorporating low-observable or stealth features. Originally developed to bypass Soviet efforts to track and monitor US imaging spacecraft, the first Misty was launched in 1990 and has now been retired. A second craft was launched in 1999 and debate in late 2004 suggested that a third satellite was being planned.

Background

From the earliest days of satellite reconnaissance, both the Soviet Union and US devoted efforts to tracking each other's imaging satellites and taking elementary precautions to protect secret activities. For instance, the USAF's classified flight-test center at Groom Lake in Nevada has long featured shelters close to the runway, known as "scoot-and-hides", where an aircraft can be concealed, and flight tests are planned around known satellite passes.

When the first KH-11 reconnaissance satellite was launched in 1976, replacing ejectable film canisters with an electronic datalink, the National Reconnaissance Office (NRO) practiced strategic deception. After initial operations, the KH-11 was placed in an orbit that mimicked that of a KH-9 which had used all its film canisters. The Soviets took the bait and ceased taking active measures when the satellite passed over. The result, according to some analysts, was that KH-11 imagery differed from pictures acquired by other satellites, revealing a more sophisticated and extensive pattern of deception on the part of the Soviet Union than anyone had suspected. Within a matter of months, though, CIA employee William Kampiles sold a KH-11 manual to Soviet intelligence agents, and the ruse was over.

The Reagan administration decided in 1983 to regain that advantage by developing a stealth satellite under the auspices of the CIA's Office of Development & Engineering (OD&E). It was assigned the codename Misty, under a special classification level (known as Zirconic) within the Byeman code system.

A stealth satellite is not the same as a stealth aircraft, which is primarily designed to evade radar. Any satellite operates at an extreme range from radar and fairly simple measures to reduce its Radar Cross-Section (RCS) will make detection and tracking difficult. It is more important to reduce the visual signature by either reducing or controlling sun glint from the spacecraft.

Glint can be reduced to some extent by the use of light-absorbing coatings, and can also be controlled in a manner analogous to the faceted shape of the F-117 stealth fighter: that is, by designing a shape that glints only in a finite number of directions and then orienting the spacecraft (with reference to the sun and likely observing locations) so that observers will not see it.

The launch of the satellite could not be concealed, using any known launch vehicle, so the Misty concept apparently combined stealth with deception, using visible decoys that would emulate either a failed satellite or an operational non-stealthy spacecraft.

The first Misty satellite was launched on 1 March 1990 by the Shuttle Atlantis. At the time, sources reported that it had malfunctioned and would re-enter the atmosphere. Civilian observers, however, spotted the spacecraft itself (officially identified as AFP-731) and were able to track it until November 1990. Observers believe that it was de-orbited some time after 1997.

The second Misty spacecraft was launched on 22 May 1999. In this case, the spacecraft was apparently accompanied by a decoy that observers believed for some time was the satellite itself, but as civilian observers acquired and used more sophisticated equipment – and applied new processing techniques to old data – it became clear that the spacecraft was somewhere else, if it was still in orbit.

In December 2004, a dispute within the US Senate's normally secret Intelligence Committee became public, with Senator John Rockefeller (Democrat from West Virginia), objecting openly but without specifics to what he described as a highly expensive but unnecessary intelligence programme. Reports described it as a stealthy reconnaissance satellite, either a third Misty spacecraft or a follow-on design, intended for launch some time before 2009.

NOAA series (TIROS/TOS/ITOS/NOAA/GOES)

Current Status

NOAA currently operates 17 meteorological satellites in three separate constellations: GOES, POES and DMSP. The future NPOESS constellation will merge the two polar orbiting series into a single programme. NOAA launched NOAA-K on 13 May 1998.

The GOES-West/East satellites, at 135° and 75°W GEO, provide synoptic visible/infra-red imaging and infra-red/thermal sounding for atmospheric temperature profiles. GOES 7 remains in orbit, left as the last representative of the earlier technology, after a bulb failed in GOES 6's scanning encoding system January 1989. The replacement new-generation GOES 8 did not occupy the 75°W slot until February 1995; GOES 9 took over at 135° W in January 1996.

Background

The United States has launched 35 civil Tiros/north polar meteorological satellites (and suffered three launch failures) in three distinct programmes since the beginning of the space age. NASA named the first series, launched during 1960–65, Tiros. Then it followed the Tiros programme with nine more satellites over the 1966–1969 period. It designated these the Tiros Operational Satellite (TOS) and sometimes called them the ESSA (Environmental Science Services Administration) satellites after the organisation that managed the programme during that period. NOAA introduced the second-generation Improved TOS (ITOS) in 1970. In order to enhance the ITOS satellites, NOAA built a prototype Synchronous Meteorological Satellites (SMS) in 1974, to provide synaptic coverage of short-term and other weather features from GEO. These geosynchronous satellites set the stage for a permanent weather station in geosynchronous and acted as the testbed and forerunners of GOES, which now provides continuous coverage of North and South America and its oceans. NOAA polar spacecraft distribute their data to Earth stations in 120 nations as they pass overhead, and thousands of schools, private individuals and others receive NOAA imagery.

TIROS

The Television & Infra-red Observation Satellite programme began as a joint NASA/Department of Defence project to develop meteorological satellites. Tiros 1 was launched in April 1960, and orbited at 692 × 740 km. The satellite demonstrated the value of an orbital viewpoint for meteorological applications. During its 78 day battery life, it returned 22,952 cloud cover images. The original Tiros satellites were basically cylindrical 18-sided prisms, 1.07 m in diameter and 0.55 m high. Solar cells covered the sides and top, with apertures for two TV cameras on opposite sides of the spacecraft body. Each camera captured 16 images a revolution at 128 second intervals. The image system passed its information to two tape recorders capable of storing up to 48 scenes when ground stations were out of range. Tiros 1's 119 kg increased to 138 kg over the course of design changes to Tiros 9/10.

GOES 9 began operations at 135° west in January 1996
(Space Systems/Loral) 0517424

NOAA 14 became the prime satellite in 1995
(Lockheed Martin Astro Space) 0517246

By the time Tiros 10 had been launched, NASA had already begun to phase in the more advanced Nimbus and ESSA satellites. The first eight Tiros', all operating in similar orbits, transmitted several hundred thousand images, together with information on the Earth's heat budget. Tiros 9, marked the first time that NASA attempted to reach polar orbit from Cape Canaveral. It used a series of three Delta dogleg manoeuvres to reach the 82° inclination angle, but a stage 2 overburn meant that the orbit, instead of being 644 km circular, was 700 × 2,578 km.

Tiros 10 transmitted the first photomosaic of the entire world's cloud cover in 450 high-quality images. By the time Tiros 10 shut down on 3 July 1967, 500,000 cloud-cover pictures had been returned.

TOS/ESSA

TOS/ESSA satellites, similar but more advanced, carried two Automatic Picture Transmission (APT) cameras able to image a 3,000 km wide strip with 3 km resolution at the centre of the picture. The ESSA satellites transmitted Images every 352 seconds allowing a typical APT station to receive 8 to10 daily. They began operating on 3 February 1966 from a 702 × 845 km orbit. The series ended with ESSA 9 launched 26 February 1969 into a 1,427 × 1,508 km orbit. By that time, various countries had set up 400 receiving stations and the weather services of 45 countries, as well as 26 universities, and up to 30 US TV stations as well as an unknown number of private citizens relying on home-built receivers, used the imagery routinely. In 1969, an ESSA 7 scene made history by revealing that the snow cover over America's Midwest, in Minnesota and the Dakotas, was three times thicker than normal. Measurements showed that it was equivalent to 15 to 25 cm of water covering thousands of square miles. A disaster area was declared before the event, and when the floods came much had been done to control the situation.

ITOS

The Improved Tiros Operational System (ITOS) added IR sensors to permit night observations. NOAA superseded ESSA during the period that the agency launched the satellites and so they were designated ITOS before launch and once in orbit given a new series of NOAA designations. Thus NOAA 1, the second-generation prototype, and launched 11 December 1970 from Vandenberg into a 1,429 × 1,472 km, 101° inclination orbit began the series. ITOS-B since it failed to achieve orbit never received a NOAA designation. And the NOAA designation resumed with NOAA 3/4 launched in November 1973 and 1974 respectively.

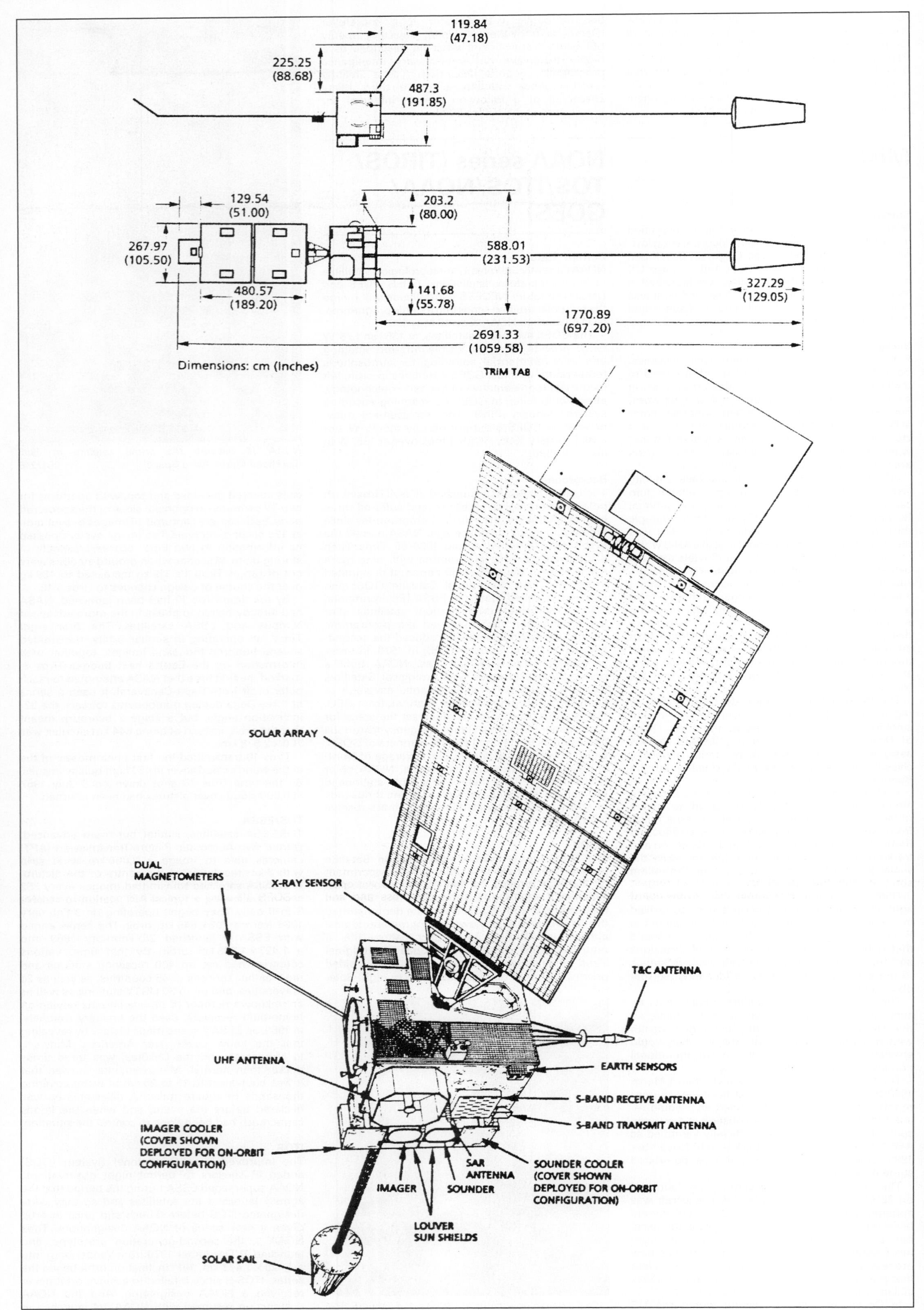

GOES 8–12 principal features (NOAA) 0517058

NOAA

The US National Oceanic and Atmospheric Administration (NOAA) operates the polar-orbiting NOAA and geostationary GOES environmental systems of two satellites each, procured for NOAA by NASA through its Goddard operations centre. Launched into near-polar Sun-synchronous orbits, each NOAA can view almost all of Earth's surface twice every 24 hours. In addition to returning weather imagery, they monitor atmospheric temperature and humidity, sea surface temperature to 1.6°C accuracy, snow/ice cover, and total ozone content near-Earth proton/electron fluxes atmospheric aerosols and radiation fluxes, and relay Argos surface platform environmental data and Cospas/Sarsat distress signals. NOAA designs its satellites for a nominal two year lifetime.

NOAA satellites carry two basic sensors (AVHRR/TOVS), and afternoon satellites such as NOAA 11/14 include the Solar Backscatter UV Spectral Radiometer, Mod 2 (SBUV/2, non-scanning 1,600-4,000 Å) for determining ozone levels to 1 per cent absolute accuracy. It also carries a Cospas/Sarsat distress relay, Argos Data Collection & Location System transponder, and a Space Environment Monitor.

NOAA 8 was the first of RCA's Advanced Tiros-N (ATN) satellites operating as part of the international Cospas-Sarsat search and rescue system. In June 1984 it began tumbling following failure of the master timing system, and the refusal of the back-up system to activate. All attempts to save the USD30 million satellite failed, and it had to be abandoned, presumed dead, leaving NASA with no search/rescue capability.

NOAA 9 introduced an ozone mapper being flown on subsequent satellites with near-noon equator crossings. NOAA launched the morning orbit satellite counterpart NOAA 10 in September 1986 as NOAA-G, which it replaced, by NOAA-D/12 in 1991. NOAA 9, after a series of weather and technical delays, restored both weather and rescue services from its afternoon orbit. It also carried NASA's Earth Radiation Budget Experiment scanner and non-scanner instruments. The morning satellite NOAA-10 replaced the degraded NOAA 6, which was brought back into use following the failure of NOAA 8. NOAA 10's Sarsat equipment restored the ability of the US to contribute to the international Cospas/Sarsat search/rescue system. Within six days of becoming operational, it had picked up its first distress signal, which led to the rescue of four Canadians whose aircraft had crashed in a remote area of Ontario.

One of the three attitude control gyros failed on NOAA 11 in August 1989 and the single back-up unit was brought on line. Software for operating with only two gyros was uplinked 8 August 1990 a month before the expected second failure. Software has been developed for operations with no gyros. AVHRR failed 13 September 1994 after operating since 8 November 1988. Placed on standby 1995 after being replaced by NOAA 14. The satellite cost USD53.5 million.

NOAA intended to use NOAA 13 as the prime afternoon satellite. However, failure of the power link from the solar panel lost contact 21 August during checkout. The September 1994 review board report identified a short-circuit in the battery charge assembly that prevented the solar arrays from charging the batteries. The short was probably caused by a 31 mm screw penetrating insulation and contacting a radiator plate. Inspection of NOAA-J (NOAA 14) showed 10 of the 12 screws to be in danger of causing the same fault.

NOAA plans to continue launches, on average, at 16-month intervals for the foreseeable future. It launched NOAA 13 (I) 9 August 1993 to replace NOAA 11 but this new satellite suffered a power failure within days. As a replacement NOAA converted NOAA-J, originally planned to replace NOAA 12, into an afternoon role and launched it in December 1994. In the coming decade NOAA plans to drop its morning operational satellites and turn meteorological monitoring into a co-operative effort with Europe's Metop.

President Clinton, in May 1994, approved a long awaited plan to combine NOAA's polar system with the military DMSP, halving the number of operational satellites to two. NOAA had planned a new series of Tiros-Next satellites (designated 0-0) for 2005, 2008 and 2011. The first joint military/civilian satellites may appear in 2007, controlled by NOAA from its Suitland, Maryland centre. To prepare, Suitland will take over DMSP control beginning in 1998, and DoD will phase out its own station retaining only one as back-up. The plan projects a USD300 million saving through 1999, by eliminating duplications and then at least USD1 billion over the next 15 years.

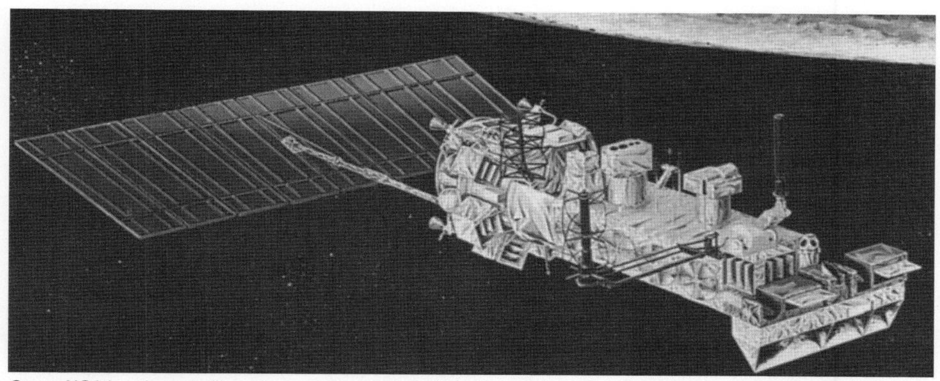

Some NOAA polar satellites also operate as part of the Cospas/Sarsat international search and rescue system, relaying distress calls from ships and aircraft (Lockheed Martin Astro Space) 0517060

NASA placed NOAA-11 on standby 1995 after being replaced by NOAA 14, crossing the equator in the afternoon. The Centre de Meteorologie Spatiale in Lannion, France can relay stored/real-time data. Afternoon satellites such as NOAA 11/14 include the Solar Backscatter UV Spectral Radiometer, Mod 2 (SBUV/2, non-scanning 1,600–4,000 A) for determining ozone levels to 1 per cent absolute accuracy.

NOAA planned to place NOAA 14 in the morning orbit as a replacement for NOAA 12, but then the loss of NOAA 13 forced the agency to reconfigure the satellite for NOAA 11's afternoon role. This required replacing the Naval Research Lab's RAIDS Remote Atmosphere Ionospheric Detection System with SBUV. RAIDS is a precursor to two instruments planned for DMSP 5D3 that will improve prediction of ionospheric disruptions to communications and Over-the-Horizon radar systems. It probably will never fly since the next generation of satellites cannot carry it.

NASA built a new fourth-generation prototype for NOAA and named it the NOAA-N, and launched it into an 850 × 866 km orbit at 102° inclination. NOAA 6 joined it on 27 June 1979. Developed, like the earlier generations, by RCA, these were the start of a series based on the Block 5D bus developed for the USAF DMSP. NOAA-B failed to achieve the correct orbit, but by the time NOAA 7 was in orbit (launched 23 June 1981), NASA claimed another success for the programme. Fishermen in California, Oregon, Washington and Alaska noted improved catches as a result of following the NOAA sea surface temperature charts. One 60-vessel towing and transportation company announced fuel savings of 20 to 40 per cent as a result of choosing routes with the help of NOAA's stream and loop current information. NOAA 7 cost USD15 million plus USD7.5 million launch. NOAA-D was the last Tiros-N, launched out of sequence in 1991 as NOAA 12.

Beginning with NOAA-M, the US will fly only afternoon satellites; and Europe will be responsible for the morning orbit making NOAA-K the last US morning satellite. This satellite includes a channel for humidity profiles, delineating sea ice and open water, and monitoring snow thickness and soil moisture.

SMS/GOES

While the NOAA satellites added a significant improvement to the weather gathering capabilities of meteorologists, NOAA sought a continuous scene generator, which required a move to geosynchronous orbit. NASA built two prototype Synchronous Meteorological Satellites (SMS) to answer this need which it later released to NOAA as part of the Geostationary Operational Environmental Satellite system. NASA remains responsible for spacecraft procurement. SMS/GEOS satellites have a Visible/Infra-red Spin Scan Radiometer (VISSR) to differentiate between water land and clouds. By GOES 4 the more advanced VAS added data on atmospheric temperature and water vapour content. NASA stationed SMS 1 initially over the east Atlantic, and then moved it to 75° west over Bogota, to give the first day/night storm watch and alert capability, with full disc pictures of the Western hemisphere every 30 minutes.

As part of the Global Atmospheric Research Program (GARP), SMS-1 provided the first continuous coverage of a major hurricane, designated Carmen, in September 1974. SMS 2 joined the first satellites when NASA stationed it over 115° west, just east of Hawaii, completing total coverage for the western hemisphere. NOAA assigned it the tasks of watching California's forest areas to give warning within 90 minutes of fire outbreaks.

During the Global Weather Experiment of 1977–78, NOAA positioned GOES 1 over the Indian Ocean at 60° east, with ESA's ground station in Spain processing data. The satellite later moved to 116° west and 135° west.

GOES 2 performed as expected until a VISSR primary encoder failed 26 January 1979 and the back-up encoder failed 18 December 1978, ending this satellite's imaging function after returning 20,591 full disc images.

On GOES 3, the VISSR problem continued to plague the GOES system and this satellite the replacement for GOES 1 operated in a degraded mode from 14 September 1979 until the encoder finally failed 6 May 1981 after returning 36,190 full disc images. After its failure, NASA moved the satellite to 176° west from 130° west in June 1990 for use by Pacific island nations as a communication satellites through the Peacesat Pan-Pacific Educational and Cultural Experiments. But perhaps the most successful of all of the meteorological satellites, in 764 operational days, the GOES 4 provided warning of many Pacific storms, transmitting 28,500 scenes.

NASA intended to use the GOES 6 as the east coast satellite, but following GOES 5 failure, the Agency moved it to 98° west to cover both US coasts; then to 108° west to monitor winter storms in the Pacific off the US northwest coast. For almost three years following the failure of GOES 5, GOES 6 had to do the work of two satellites by being moved with the seasons between 98 and 108° west.

At the beginning of 1991, NOAA planned that it would follow the mid-1992 launch of GOES-I, eight months later by GOES-J and by the 1996 debut of GOES-K. However, instrument technical problems surfaced during the year, including the use of incorrect wiring, prompted the delay of GOES-I to at least December 1993. At one time during the correction process, NOAA considered launching at end of 1992 with below specification instruments. Congress set aside an additional USD110 million to ensure the programme continued, if required. The Department of Commerce decided instead to press for a full-performance GOES-I while buying time with Europe's Meteosat.

GOES 7 began at the GOES position, but later moved west following the GOES 6 failure, to 108° west for the winter season, then to 98° west in the Spring for observing hurricanes over the Atlantic and returning further west in the autumn. This GOES carries the Space Environment Monitor, a Data Collection System. To improve station-keeping lifetime, NASA included a small conical sail on a 17 m boom, on GOES 8 to balance solar radiation torque and reduce station-keeping propellant use; a trim tab on solar wing provides fine control.

NOAA 7 transmitted this image of Hurricane Lili over the north Atlantic on 20 December 1984, only the third recorded hurricane in that region (the others were in 1887 and 1954) 0517061

GOES-L/M will take the series into the next century, with launches tentatively set for 2000/2004. NOAA began work on the series beyond GOES-M during 1988, holding meetings in early 1989 to define requirements. These were passed on to NASA in June 1989 for Goddard Space Flight Center to undertake Phase A, completed during 1990.

NOAA revealed in February 1995, in its FY96 budget request, that it intends to order three more similar satellites, plus an option on a fourth, before beginning Phase A of the GOES-R next generation in 1998 for first launch around 2008. FY96's request includes funding to begin procurement of GOES-N/O. The contract will be awarded in late 1997 (for 2002 first delivery) after the competition is opened by end-1996. It is expected that sounder resolution will be improved. Adding star trackers will improve location identification to 1 km from the current 4 to 6 km, improving forecasting.

Studies may show that this payload should be divided between several smaller carriers. Similar to the previous polar orbiters and built by Lockheed Martin Astro Space, this new generation carries different instruments. Three were initially ordered at USD160 million (excluding instruments and launches); N/N-prime were added in 1993 (NASA made the USD151 million award January 1995) to allow more time to prepare the next series. Beginning with NOAA-M, the US will fly only afternoon satellites; Europe will be responsible for the morning orbit. DSB has been eliminated, although sounder data are available through HRPT broadcasts. An eight-channel ocean colour instrument for VNIR sensing was considered.

Specifications

Tiros 1
Launched: 1 April 1960
Orbit: 692 × 740 km

TOS
Launched: as ESSA 1 on 3 February 1966 (702 × 845 km) and ended with ESSA 9 (launched 26 February 1969, 1,427 × 1,508 km)

NOAA 1
Launched: 11 December 1970 from Vandenberg
Orbit: 1,429 × 1,472 km, 101° 114.8 min
Mass: 307 kg

NOAA 2 (ITOS-B)
Launched: 21 October 1971
Comment: This launch was never allocated an NOAA designation because it failed to achieve a satisfactory orbit

NOAA 3
Launched: 6 November 1973
Orbit: 1,499 × 1,508 km, 102.2°, 116.1 min
Mass: 340 kg

NOAA 4
Launched: 15 November 1975
Orbit: 1,442 × 1,482 km, 102°, 114.9 min

NOAA 5
Mass: 340 kg
Launched: 29 July 1976.
Orbit: 1,504 × 1,520 km, 102.2°, 116.3 min

Tiros N/NOAA-A
Launched: 13 October 1978 (first of the series known as Tiros N – the NASA prototype)
Mass: 734 kg
Orbit: 850 × 866 km 102°
Comment: The complementary NOAA 6 joined the satellite on 27 June 1979. Developed, like the earlier generations, by RCA, these were the start of a series based on the Block 5D bus developed for the USAF DMSP.

NOAA 6
Launched: 27 June 1979
Orbit: 785 × 800 km, 98.7°, 100.7 min
Out of service: 31 March 1987

NOAA 7
Launched: 23 June 1981
Orbit: 828 × 847 km, 98.9°, 101.7 min
Out of service: June 1986 due to power failure

NOAA-B
Launched: 29 May 1980
Orbit: 264 × 1,445 km, 92.2°, 102.1 min
Comment: Failed to achieve the correct orbit, but by the time NOAA 7 was in orbit (launched 23 June 1981) it was possible to claim that fishermen in California, Oregon, Washington and Alaska were improving their catches by using the sea surface temperature charts. One 60-vessel towing and transportation company announced fuel savings of 20–40 per cent as a result of choosing routes with the help of NOAA's stream and loop current information. Decayed 3 May 1981.

NOAA 8/NOAA-E
Launched: 28 March 1983 by Atlas from Vandenberg
Mass: 1,712 kg
Orbit: 833 km circular at 98.3°
Comment: First of RCA's Advanced Tiros-N (ATN) operating as part of the international Cospas-Sarsat search/rescue system. High-resolution instruments measured both surface and vertical temperature; a UK stratospheric sounding unit monitored upper atmosphere temperature and the French Argos system relayed data from balloons, buoys and remote weather stations. Attitude control problems began once in orbit. In June 1984, the satellite began tumbling following failure of the master timing system, and the refusal of the back-up system to activate. All attempts to save the USD30 million satellite failed, and it had to be abandoned, presumed dead, leaving NASA with no search/rescue capability. NOAA 6, placed on standby following a malfunction of its AVHRR, was reactivated to fulfil NOAA 8's other functions. Then, in June 1985, NOAA 8's back-up attitude control system unexpectedly came back online and its use was fully recovered. However, an overcharged battery burst in December 1985, some contact was restored, but NNOAA finally shut it down on 8 January 1986.

NOAA 9
Launched: 12 December 1984 by Atlas from Vandenberg
Comment: After a series of weather and technical delays, restored both weather and rescue services from its afternoon orbit. Carried NASA's Earth Radiation Budget Experiment scanner and non-scanner instruments (scanner failed 20 January 1987, non-scanner remains operational); see the ERBS satellite entry in the main national US section. Also the first to carry the SBUV/2 ozone mapping instrument. Satellite cost was USD43.5 million plus USD11.4 million launch. Placed on standby following the launch of NOAA 11. Its MSU channel 2 failed 8 March 1987, number 3 7 May 1987, prompting launch of NOAA 11. Collection of SBUV and ERBE non-scanner data continues. S&R function decommissioned March 1995. AVHRR remains usable, although 3.7 µm channel out of specification. The power system is marginal due to array shunt failures and battery problems.

NOAA 10
Launched: 17 September 1986 by Atlas E from Vandenberg after 16 delays (planned for August 1985)
Orbit: Morning 808 × 826 km, 98.8°
Comment: This satellite replaced the degraded NOAA 6, which was brought back into use following the failure of NOAA 8. NOAA 10's Sarsat equipment restored the ability of the US to contribute to the international Cospas/Sarsat search/rescue system. Within six days of becoming operational, it had picked up its first distress signal, which led to the rescue of four Canadians whose aircraft had crashed in a remote area of Ontario
Sensors: Same as NOAA 11 but does not carry SBUV/2 and incorporated Earth Radiation Budget Experiment scanner and non-scanner for NASA Langley. ERBE scanner failed 22 May 1989 (non-scanner solar monitor shutter stuck open, but continues to provide usable data). The Sarsat processor receiver failed 8 September 1988. Replaced by NOAA 12 in May 1991. NOAA 10 remains on standby, but has some degraded performance in the AVHRR, SEM and power systems.

NOAA 11
Launched: 24 September 1988 by Atlas E from Vandenberg AFB; two year design lifetime
Mass: about 1,700 kg BOL
Cost: USD53.5 million
Contractors: GE Astro Space (prime), ITT (HIRS, AVHRR), NASA/JPL (microwave sounding unit)
Resolution: 1.1 km AVHRR, 20 km HIRS/2, 147 km SSU, 105 km MSU
Orbit: 849 × 865 km, 98.9° Sun-synchronous, crossing equator in afternoon (originally 13.40; drifted to 16.02 by May 1993) on ascending node with daily repeat cycle
Spacecraft: Astro Space's Advanced Tiros-N platform, 1.80 m wide, 4.18 m long with single 2.37 × 4.91 m 8-panel 11.6 m² solar wing deployed, providing 1,500/1,400 W BOL/EOL; 3 × 26.5 Ah Ni/Cd batteries. Injection provided by Thiokol Star 37S solid. ±0.2° 3-axis control by RWs using magnetic unloading; 8 GN_2 thrusters (4.23 kg supply) provide control during solid burn and until wheels take over (also auxiliary wheel unloading); orbit adjust by 4 hydrazine (28.4 kg supply). Payload 386 kg. Real-time transmissions made for Automatic Picture Transmission (APT, 4 km resolution, 137.5 or 137.62 MHz), High Resolution Picture Transmission (HRPT, full resolution, 1.698 or 1.707 GHz) services and DBS Direct Broadcast Sounder (full resolution sounder data on 1.698, 1.7025 or 1.707 GHz). Spacecraft command/control at 148.56 MHz provided by NOAA's CDA stations in Wallops Island, Virginia and Fairbanks, Alaska; Satellite Operations Control Center is in Suitland, Maryland. The Centre de Meteoroloogie Spatiale in Lannion, France can relay stored/real-time data
Sensors: Advanced Very High-Resolution Radiometer (AVHRR). The instrument operates in five channels (µm): 0.58–0.68 (Si), 0.725–1.0 (Si), 3.55–3.93 (InSb), 10.3–11.3 (HgCdTe) and 11.4–12.4 (HgCdTe), primarily for assisting weather forecasting, snow/ice monitoring and sea surface temperature, but also applied to marine oil pollution mapping, volcanic eruption monitoring, assessment of vegetation vigour on international scales and estimating atmospheric aerosols for climate monitoring. 1.3 mrad IFOV. 20 cm diameter Cassegrain telescope, beryllium scan mirror, ±55° scan angle.

Tiros Operational Vertical Sounder (TOVS) is a three-instrument system consisting of a four-channel Microwave Sounding Unit (MSU) for tropospheric temperature soundings in cloudy regions, a three-channel Stratospheric Sounding Unit (SSU) for 25–50 km stratospheric temperature probing, and the 20-channel High-Resolution Infra-red Sounder (HIRS/2) for vertical temperature profile, water vapour and total ozone content to 40 km. JPL's MSU, 50.30, 53.74, 54.96, 57.95 GHz, 220 MHz bandwidth, ±47.4° scan width from nadir, two scanning reflector antennas (9.5° steps through 360°), 0–350 K dynamic range, 0.3 K NEδT. Matra Marconi Space's SSU, pressure modulated CO_2 cell in each optical path for 14.926, 14.934, 14.940 µm, 10° IFOV, 0.25 NEδT at 273 K, no collecting optics (5 cm aperture), ±40° scan width from nadir. ITT's HIRS/2 channels 1–12 6.72–14.95 µm HgCdTe, numbers 13–19 3.76–4.57 µm InSb, number 20 0.69 µm Si. 24 mrad IFOV, ±49.5° scan width from nadir, 15 cm diameter Cassegrain telescope, 1.8° step scanner covers 56 steps then retraces.

Space Environment Monitor records radiation levels to determine the energy deposited by solar particles in the upper atmosphere and to provide a solar storm warnings. TED cylindrical electrostatic analyzer and spiraltron: 0.3–20 keV protons/electrons in 11 bands. MEPED solid state telescopes and omnidetectors: 30–20 keV protons, 11 bands; >30–>300 keV electrons, 3 bands; >6 MeV ions; >16, >36, >80 MeV omniprotons.

NOAA 12 (NOAA-D)
Launched: 15.52 GMT 14 May 1991 by Atlas 50E from Vandenberg AFB
Orbit: 807 × 826 km, 98.7° Sun-synchronous in morning orbit. Replaced NOAA 10 as the prime morning satellite
Sensors: Same as NOAA 11 but does not carry SBUV/2, the S&R payloads or the SSU instruments. MSU has degraded because of gain changes. Other specifications as for NOAA 11
Comment: The last Tiros-N. First known as NOAA-D, but launched out of sequence in 1991 as NOAA 12.

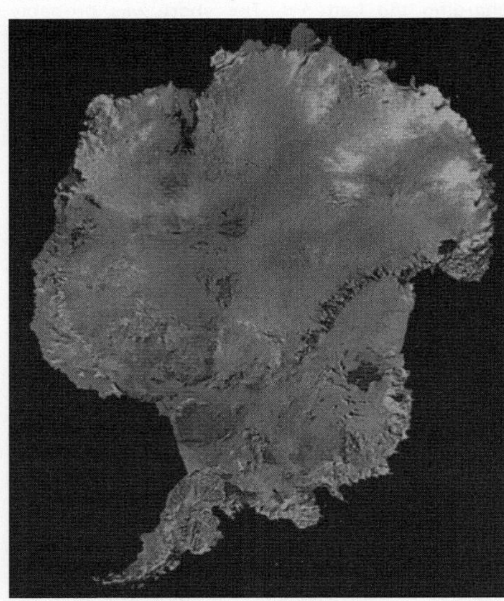

Mosaic of 28 NOAA AVHRR images covering Antarctica produced in 1988 in a joint venture between NOAA and the UK's National Remote Sensing Centre (NRSC)
0517062

GOES 8's first visible engineering test image, 9 May 1994 (NOAA) 0517247

NOAA 13
Launched: 9 August 1993 by Atlas 34E from Vandenberg AFB to replace NOAA 11
Orbit: Afternoon satellite
Comment: Failure of the power link from the solar panel lost contact 21 August during checkout. The September 1994 review board report identified a short-circuit in the battery charge assembly that prevented the solar arrays charging the batteries. A 31 mm screw penetrating insulation and contacting a radiator plate probably caused the short. Inspection of NOAA-J (NOAA 14) showed 10 of the 12 screws to be in danger of causing the same fault. The same design had flown on 16 NOAA and DMSP satellites
Sensors: Same as NOAA 11 plus the experimental MAXIE and EHIC sensors, sponsored by the Office of Naval Research. The Magnetospheric Atmospheric X-ray Imaging Experiment, provided by Lockheed, assisted by the Aerospace Corp and Norway's University of Bergen, was to map the intensities and energy spectra of X- rays generated by electrons in the upper atmosphere, and the associated auroral and substorm imaging. Good data were returned until the power failure. The Energetic Heavy Ion Composition Experiment, provided by the University of Chicago and Canada's NAC HIA, was to measure the chemical and isotopic composition of energetic particles between hydrogen and nickel over 0.5–200 MeV/nucleon.

NOAA 14 (NOAA-J)
Launch: 30 December 1994 by Atlas 11E from Vandenberg AFB. Control transferred from NASA to NOAA 3 January 1995
Mass: 1,712 kg at launch, 1,030 kg BOL on-orbit
Orbit: 848 × 863 km, 98.9° afternoon Sun-synchronous to replace NOAA 11.
SARP failed, SBUV/2 failed, DTR 4A/B inoperable February 1995.

NOAA 15 (NOAA-K, POES 1)
Launch: 13 May 1998 by Titan 2, first of five POES satellites with improved imaging and sounding capabilities. Imaging instrument failed July 2000
Orbit: 833 km, 98.7°, 101 min
Mass: 2,234 kg at launch, 1,454 kg BOL on-orbit
Power system: as NOAA 11 but 11-panel 17 m^2 solar wing
Sensors: AVHRR/3. Same as NOAA 11 but 6th time-shared channel (3A: 1.58–1.64 μm InGaAs) improves snow/cloud discrimination and changes in several channels improve calculation of worldwide vegetation levels and estimation solar and terrestrial radiation levels previously monitored by the two ERBEs. 33 kg, 28.5 W, ITT Aerospace/Communications Division.
HIRS/2 20-channel High-Resolution IR Sounder for vertical T profile, water vapour and total ozone content to 40 km. Channels 1–12 6.52–14.95 μm HgCdTe, numbers 13–19 3.76–4.57 μm InSb, number 20 0.69 μm Si. 24 mrad IFOV, ±49.5° scan width from nadir, 15 cm diameter Cassegrain telescope, 1.8° step scanner covers 56 steps then retrace. 34 kg, 24 W, ITT Aerospace/Communications Division.
AMSU-A 15-channel advanced microwave sounding unit for atmospheric T soundings from surface to 45 km at 23.8, 31.4, 50.3–57.3 and 89 GHz. 50 km nadir resolution, 2,200 km swath, 2.2 kbit/s.
AMSU-B 5-channel on NOAA-K/L/M for humidity profiles, delineating sea ice and open water, and monitoring snow thickness and soil moisture. 89, 157 and 183 GHz, 15 km nadir resolution, 2,200 km swath, 4 kbit/s. UK Met Office instrument; Matra Marconi Space was awarded a £9.5 million contract in March 1989.
MHS 5-channel microwave humidity sounder on NOAA-N/N-prime for humidity profiles up to 42 km. 89, 150 and 183 GHz, 15.4 km nadir resolution, 2,348 km swath, 3.95 kbit/s, 66 kg, 100 W. EUMETSAT (also on Metop). Argos system capacity is quadrupled, able to handle eight messages simultaneously at 2,560 bit/s over 80 kHz (currently 24 kHz) bandwidth
Cospas-Sarsat S&R. Same as NOAA 11
SEM. Same as NOAA 11
SBUV/2. Same as NOAA 11; not on NOAA-L.

NOAA 16 (NOAA-L)
Launched: 21 September 2000 by Titan 2
Orbit: 870 km, 98.7°, 102 min. APT system failed in September 2001

NOAA 17 (NOAA-M)
Launched: 24 June 2002 by Titan 2
Orbit: 833 km, 98.7°, 101.3 min

SMS 1
Launched: 17 May 1974 by Delta from Cape Canaveral
Mass: 627 kg
Orbit: Initially over the East Atlantic, and then at 75° W over Bogota, 132° W in 1980
Comment: This satellite provided the first day/night stormwatch, with full disc pictures of the western hemisphere every 30 min. As part of the Global Atmospheric Research Programme (GARP), it provided the first continuous coverage of a major hurricane, designated Carmen, in September 1974. Out of service 24 January 1981.

SMS 2
Launched: 6 February 1975
Orbit: 125 ° W, east of Hawaii; *75* ° W in 1980
Comment: Positioned so that it and SMS 1 could cover the western hemisphere. One of its tasks was to keep watch on California's forest areas to give warning within 90 min of fire outbreaks. Out of service 5 August 1982.

GOES 1
Launched: 16 October 1975 by Delta from Cape Canaveral. 293 kg
Mass: 293 kg
Orbit: During the Global Weather Experiment of 1977–78, it was positioned over the Indian Ocean at 60° E, with ESA's ground station in Spain processing data. Later moved to 116° W and 135° W
Comment: Following GOES 4 failure in November 1982, it was functioning only in the visible band, but it was reactivated to transmit images via GOES 4. Retired 7 March 1985.

GOES 2
Launched: 16 June 1977
Orbit: 105° W. At beginning of 1992, positioned over 60° W (±1.0°), inclined at 9.4° for relay of Mode AAA (Stretched VAS) from GOES 7, relay of East WEFAX and for East Data Collection Platform Interrogation (time-code). These functions were principally for western European and African users, who could not see GOES 7. It was moved in 1992 to 135° W to replace GOES 6 supporting all non-imaging functions of the west spacecraft (WEFAX, DCS, and SEM)
Comment: VISSR primary encoder failed 26 January 1979 (redundant encoder failed 18 December 1978), ending its imaging function after returning 20,591 full disc images and 6,838 partial. Its functions were replaced by GOES 7 in January 1995 and it retired out of service in May 2001.

GOES 3
Launched: 16 June 1978, replacing GOES 1
Orbit: Moved to 176° W from 130° W in June 1990. It was held at 176° W, inclined at 10.6°, but it has been drifting since 1995 although still usable
Comment: The move to a new location occurred under an agreement with NASA and DoC's National Telecommunications & Information Administration for use by Pacific island nations through the Peacesat Pan-Pacific Educational & Cultural Experiments by Satellite project to re-establish the communications service inaugurated in 1971 by ATS 6. This provides 2-way voice/data services to small Earth stations at more than100 sites in 22 countries. Peacesat is controlled for NOAA by NASA from the Kokee Park Geophysical Observatory on Kauai, Hawaii; NTIA co-ordinates user operations.

GOES 4
Launched: 9 September 1980
Orbit: 135° W
Comment: In 764 operational days provided warning of many Pacific storms, transmitting 28,500 scenes. Imaging system failed in November 1982, after which it was used as a standby transponder relaying data. In May 1985, it was loaned to ESA as temporary substitute for Meteosat 1 and was moved 4° daily to place it over the Atlantic at 43° west by June. This was in return for Meteosat 2's help to NOAA in 1984 following the GOES 5 failure. As GOES 4 was incompatible with ESA's ground equipment, NOAA operated GOES 4 for ESA from Suitland, Maryland and Wallops, Virginia. It was to be boosted out of GEO at the end of the loan. In August 1985 however, unidentified UHF interference at 401.9 MHz rendered its data collection capability useless as it was being moved. It was therefore returned to 43° W. It was boosted above GEO on 22 November 1988 and deactivated.

GOES 5
Launched: 22 May 1981
Orbit: 75° W. Hydrazine was depleted November 1989 and it began drifting westwards. It will drift to about 145° W, then return east towards its original starting point of 60° W. This pattern will continue, with outer limits decreasing, until it settles around 105° W
Comment: Intended to monitor US east coast and tropical storm formations in the west Atlantic Ocean. VAS failed 29 July 1984 due to two tungsten filament lamp failures; then used to relay data from GOES 6 and then GOES 7. Relay of GOES 7 imagery to Europe was transferred to GOES 2 in May 1990. On 18 July 1990 all other functions ended and it was deactivated.

GOES 6
Launched: 28 April 1983
Orbit: 135° W, moved to 98° W to cover both US coasts; then to 108° W to monitor winter storms in the Pacific off US northwest coast. Following the failure of GOES 5, GOES 6 had to do the work of two satellites by being moved with the seasons between 98/108° W. Into 1992, it was at 135.6° W (inclined 3.6°) out of service 21 January 1989 when VAS imager failed.
Comment: Intended to cover western US. With one of its four encoder lamps burned out, there was concern that the other three might fail; otherwise there was sufficient propellant to keep it operational until late 1988 or early 1989. Meanwhile, it could provide coverage for the main US continent, but could not provide hurricane coverage for some Atlantic islands, the Caribbean and areas west or south of Hawaii. The last lamp failed 21 January 1989 and GOES 6 was reduced to acting as a data relay, supporting all non-imaging functions of the west spacecraft (WEFAX, DCS, and SEM). It was low on hydrazine and each station-keeping manoeuvre could have depleted the supply. This happened in 1992 and it began slowly drifting, following a pattern similar to GOES 5's. GOES 2 took over its relay role, but return of SEM data continues.

GOES 7
Launched: 26 February 1987 by Delta
Mass: 835 kg at launch, 399 kg on-station BOL
Orbit: Initially GOES-East satellite, nominally at 75° W GEO. Following GOES 6 failure, it was moved to 108° W for the Winter season, then to 98° W in the spring for observing hurricanes over the Atlantic, returning further west in the Autumn. At 112° W from 1992. Arrived 135° W 27 January 1995 after GOES 8 began operations; replaced 11 January 1996 at 135° W by GOES 9. Started moving 26 January to 96° W storage position (arriving 10 July). The inclination is being allowed to increase (3.0° by April 1996, increasing at 0.9°/year)
Contractor: Hughes Aircraft Co (prime)
Resolution: 900 m visible, 6.9/13.8 km infra-red
Spacecraft: 2.16 m diameter 3.53 m high cylindrical spin-stabilised satellite; body-mounted solar array provides 450/330 W BOL/EOL. Two Ni/Cd batteries provide TT&C eclipse protection. Injected into GEO by solid AKM
Sensors: Visible Infra-red Spin-Scan Radiometric Atmospheric Sounder (VAS). Scans the full Earth disc in 1,820 successive steps by combination of spacecraft spin and stepping optics. VAS carries single visible channel (eight photomultiplier tube detectors in push-broom array) providing 900 m resolution, and six infra-red channels: two 6.7–14.7 μm with 6.9 km resolution primarily for imaging, two 3.7–7.3 μm and two 7.7–14.7 μm. The satellite also carries a Space Environment Monitor, a Data Collection System and was the first GEO satellite equipped for Cospas/Sarsat trials (see GOES-I to -M). DCS signals are relayed to a Command & Data Acquisition station, which performs error checks and routes them to NOAA offices at the World Weather Building in Camp Springs, Maryland for distribution to users. WEFAX broadcasts are made in the 10 minute intervals between VAS readouts.

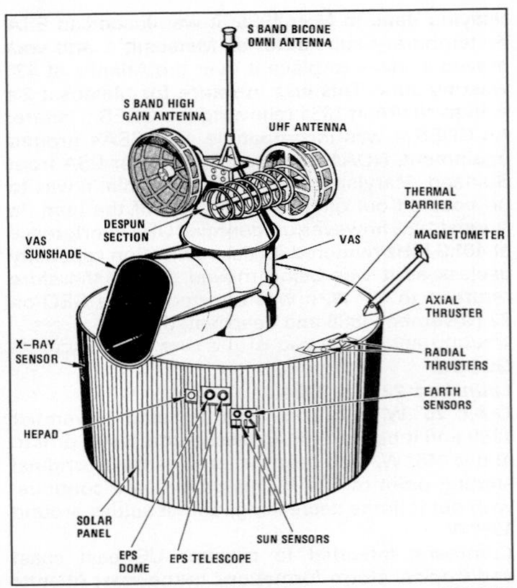

Principal GOES 7 features (Hughes) 0517059

GOES 8

Launched: 13 April 1994 by Atlas 1 from Cape Canaveral; lifetime five years (propellant sufficient for 8 years). Formally transferred from NASA to NOAA 26 October 1994; declared fully operational 9 June 1995. One of two momentum wheels lost on 9 January 1997. Now functioning on remaining momentum wheel and reaction wheel for altitude control

Mass: 2,105 kg at launch, 1,140 kg on-station BOL

Orbit: arrived 75° W GEO (GOES-East) 27 February 1995 after testing at 90° W (departed 20 January 1995)

Contractor: Space Systems/Loral (prime), ITT Aerospace/Communications Division (sensors)

Resolution: 1 km visible, 4/8 km infra-red

Spacecraft: Space Systems/Loral's 3-axis stabilisation box-shaped platform, with single two-panel Si solar wing on south side and associated solar sail on north, allowing passive north-facing Imager/Sounder coolers to view cold space. Conical sail on 17 m boom balances solar radiation torque; a trim tab on solar wing provides fine control. Total span 26.91 m. Attitude control by two 51 Nms MWs, one 2.1 Nms yaw RW + two magnetorquers; during orbital manoeuvres by two sets of 6 × 22 N MMH/NTO thrusters. GEO insertion provided by three firings of 490 N Kaiser Marquardt R-4D. 431 kg MMH + 694 kg NTO each in single 103 cm diameter titanium sphere, pressurised by helium. Payload pointing accuracy: ±9.1 μrad roll; ±9.4 μrad pitch; ±73.3 μrad yaw (in 90 min). 268 × 481 cm solar wing can provide 1,057 W EOL Summer solstice; supported by 2 × 12 Ah Ni/Cd batteries; 42.0 V DC (30.0 V DC min in eclipse). Command/control provided via NOAA's Command & Data Acquisition station on Wallops Island, Virginia; the Satellite Operations Control Center is in Suitland, Maryland. Command transmitted 2.03420 GHz. Data are processed at the CDA and uplinked for broadcast

Sensors/payload: Imager. Imaging radiometer sweeping 8 km high swath along east-west/west-east path at 20°/s via servo-driven 2- axis gimballed mirror and 31.1 cm diameter aperture Cassegrain telescope. 120 kg, 119 W daily average. Channel 1: 0.55–0.75 μm, 8 Si detectors, nadir instantaneous FOV 1 km each, cloud cover; 2: 3.80–4.00 μm, 4 InSb detectors, nadir IFOV 4 km each, night cloud cover; 3: 6.50–7.00 μm, 2 HgCdTe detectors, nadir IFOV 8 km each, water vapour; 4: 10.20–11.20 μm, 4 HgCdTe detectors, nadir IFOV 4 km each, surface temperature; 5: 11.50–12.50 μm, 4 HgCdTe detectors, nadir IFOV 4 km each, sea surface temperature and water vapour. Full Earth scan in <26 minutes; 3,000 × 3,000 km in 3 minutes; 1,000 × 1,000 km in 41 seconds. Data rate 2.621 kbit/s. Imaging stability 42 μrad for noon ±8 h, 70 μrad for midnight ±4 h. Passive cooler maintains infra-red detectors at 94/101K Winter/Summer. Full Disk Mode repeatedly scans the full disc every 30 minutes, except once every 6 hours when it performs an extended northern hemisphere scan to allow 10 minutes for housekeeping work. Routine Mode is a repeated 3 h sequence. It begins with a 30 minute full disc scan, followed by a 30 minute sequence (performed five times) of a northern hemisphere scan, Conus scan and southern hemisphere scan. The last southern scan is omitted every 6 hours for 10 minutes housekeeping. Star looks (for precise attitude registration) and blackbody calibrations occur every 30 minutes. Rapid Scan Mode is similar but it emphasises Conus scans.

Sounder. 19-channel radiometer for atmospheric vertical temperature and moisture profiles, surface/cloud top temperature, and ozone distribution. 127 kg, 106 W daily average, 31.1 cm aperture. Optics and scan patterns similar to Imager. Wheel carries 18 filters in three concentric rings, one for each infra-red detector group. Longwave channels: 1–7, 14.71–12.02 μm, 4 HgCdTe detectors, and circular IFOV 242 μrad. Midwave channels: 8–12, 11.03–6.51 μm, 4 HgCdTe detectors, circular IFOV 242 μrad. Shortwave channels 13–18, 4.57–3.74 μm, 4 InSb detectors, circular IFOV 242 μrad; visible channel 19, 0.70 μm, 4 Si detectors, circular IFOV 242 μrad. Data rate 40 kbit/s. Scans in 10 km steps east-west/west-east, moving 40 km north-south at end of scan line. Each step 100/200/400 m/s, controlled by filter wheel.

Space Environment Monitor. Of four instruments for GEO particle radiation and solar X-ray output. EPS Energetic Particles Sensor: 0.8–500 MeV protons, 3.2–400 MeV α, 0.55–4.0 MeV electrons. HEPAD High Energy Proton & Alpha Detector: 370–970 MeV protons, 640–>850 MeV α. XRS x-ray Sensor: 0.5–3.0 and 1.0–8.0 A real-time solar flux; mounted on solar wing yoke. Magnetometers: redundant 3-axis fluxgate magnetometers on 3 m boom, ±1,000 nT (±4 nT accuracy). GOES-M and successors will include a Solar X-ray Imager to aid prediction of geomagnetic storms. SEM data downlinked on 1.694000 GHz to Environmental Research Lab, Boulder, Colorado.

Cospas/Sarsat Distress Relay (406 MHz beacon relay only, no location service). Downlink at 1.544500 GHz.

WEFAX transponder. Processed imagery is returned to GOES at 2.03300 GHz for broadcast to users at 1.6910 GHz at 11 W to users

Data Collection System: Up to 233 100 bit/s signals received from DCP Data Collection Platforms on 401.9 or 402.2 MHz. Downlinked at 1.69450 or 1.69480 GHz to CDA and users. DCP interrogated by request uplinked to GOES from CDA on 2034.9000 or 2034.9125 MHz; DCP interrogated by GOES at 468.8250 or 468.8375 MHz. >12,000 DCPs active in 1995 (increases 15% yearly).

GOES 9

Launched: 23 May 1995 by Atlas 1 from Cape Canaveral. NOAA assumed control from NASA 31 October 1995. Scheduled for 155°W in mid-2003 to bolster Japanese meteorology support after GMS-5 end of life. Monthly rental USD127,500

Orbit: GEO at 135° W (GOES-West) from 11 January 1996, replacing GOES 7 after testing at 90° W. First (visible) image received 12 June 1995.

Other specifications are the same as for GOES 8.

GOES 10

Launched: 25 April 1997 from Cape Canaveral using an Atlas-1 (final Atlas-1 launch). Solar array rotation problem. Satellite inverted for optimum solar energy

Orbit: GEO over 106° W (early June 1997)

Other specifications as for GOES 8.

GOES 11

Launched: 3 May 2000 by Atlas 2A from LC-36A, CCAFS

Orbit: GEO over 106° W

GOES 12

Launched: 23 July 2001 by Atlas 2A

Orbit: GEO at 75° W from where enters service April 2003 replacing GOES 8

NPOESS

Current Status

The National Polar-orbiting Operational Environmental Satellite System (NPOESS) is jointly funded by the Department of Defense and Department of Commerce, and is planned as a six-satellite constellation to replace in-service civilian and military weather satellites. In late 2005, the Government Accountability Office reported that NPOESS was 17 months late, a probable USD3.2 billion over budget and the programme was facing a major review in early 2006.

Background

In May 1994, the Clinton Administration issued a directive that required the Pentagon and the Commerce Department's National Oceanic and Atmospheric Administration (NOAA) to merge their polar-orbiting weather satellite programmes into a single effort.

To manage this programme, the DoD, NOAA, and the National Aeronautics and Space Administration (NASA) formed a tri-agency integrated programme office, located within NOAA. Within the programme office, each agency has the lead on certain activities. NOAA has programme management responsibility for the converged system and for satellite operations; the DoD has the lead on the acquisition; and NASA has primary responsibility for facilitating the development and incorporation of new technologies into the converged system. NOAA and DoD share the costs of funding NPOESS, while NASA funds specific technology projects and studies.

Each of the six satellites incorporates 13 instruments or sensors plus an imaging sensor to take over the functions of the Landsat programme. Four of the new sensors are to be flown aboard a test satellite, the NPOESS Preparatory Project (NPP), before the NPOESS spacecraft itself is launched.

The NPOESS is expected to bring about a great improvement in the quality and timeliness of environmental data gathered from space, allowing military users to plan operations with a more accurate weather picture, improving weather tracking and forecasting and providing a wealth of data for Earth sciences. However, the programme has run behind schedule, in part because of the complexity of its onboard sensors.

The NPOESS project started with a concept and technology development phase, which ran from 1995 to early 1997. This was followed by a Programme Definition and Risk Reduction (PDRR) phase, from early 1997 to August 2002. The NPOESS programme office awarded a development and production contract to a Northrop Grumman/Raytheon team in August 2002. At that point, the NPP was due to be launched by May 2006. The launch schedule for the NPOESS itself was dependent on whether the final NOAA and Defense Meteorological Satellite Program launches were successful. The first and second NPOESS spacecraft were accordingly planned to be available to back up the final NOAA launch in March 2008 and the last DMSP in October 2009.

However, because of a delay in the launch of a DMSP satellite, the DoD "bumped" the replacement NPOESS launch date into 2010 and reduced its annual spend on the programme. Because the programme is funded 50:50 by the DoD and NOAA, this caused NOAA to reduce its own funding and slowed the programme down. Further problems have occurred with a key sensor, the Visible/Infra-red Imager Radiometer Suite (VIIRS). The result is that the NPOESS has been delayed to the point where a coverage gap of 33 months could occur if the final NOAA launch is not successful.

The NPOESS programme encompasses a ground control system and data dissemination network in addition to the six spacecraft. Based on a stretched version of the satellite bus developed for NASA's Aqua and Aura Earth-observing satellites, the six NPOESS spacecraft will carry different combinations of the 13 sensors types and will operate in different orbital planes.

Sensors carried on NPOESS include infra-red and microwave sounding instruments to measure temperatures, humidity and pressure in the atmosphere at different levels. Other new sensors will collect data to measure the effect of space weather – electromagnetic effects and radiation levels – on the satellite and its radio links. The VIIRS will collect unique data on ocean and land surfaces.

NPOESS is one of the first spacecraft to use an onboard data network based on the IEEE 1394 FireWire standard. The fault-tolerant, radiation-hardened FireWire circuits used on NPOESS for communications between the sensors and spacecraft will operate at speeds as high as 100 Mb/second, 100 times faster than the hardware on most spacecraft.

The primary dissemination system for NPOESS is named SafetyNet. The satellites will downlink mission data at 150 Mb/second to 15 receive-only Ka-band terminals located around the world and connected to commercial fibre-optic networks. The satellites will be within sight of the ground terminals more than half the time. The result is a major increase in speed compared with today's systems: weather networks should receive most of the raw data gathered by the spacecraft within 15 minutes, compared with the 2–3 hour delay encountered with existing systems. NPOESS will also transmit weather data directly to mobile and fixed field terminals operated by civil and military users, using X-band and L-band datalinks.

Ball Aerospace is providing the NPP test satellite, and in March 2006 it was due to be launched in the April 2008. The NPOESS programme completed its Preliminary Design Review (PDR) in June 2005, with Critical Design Review (CDR) set for the second quarter of FY2007. The launch of the first of two spacecraft under the research and development contract is expected in September 2010, and the second in December 2011.

According to FY2007 budget documents, however, the entire programme is undergoing a review because of cost overruns. Possible changes would include cancellation of the NPP and further delays to the launches of the first two spacecraft. The review was due to be completed in May 2006.

OrbView series

Current Status

OrbView-1 (originally known as MicroLab-1) was launched in 1995 and operated for five years. Orbimage launched OrbView-2 (SeaStar) on 1 August 1997 from Vandenberg AFB, California using the Pegasus XL booster. OrbView-3 was also launched by a Pegasus XL from Vandenberg on 26 June 2003. OrbView-4, launched 21 September 2001, was lost when the Taurus 2110 vehicle encountered problems with its T6 second stage and final velocity at cutoff was far too low to reach a sustainable orbit after all succeeding stages fired correctly. The satellite burned up in the Earth's atmosphere, northeast of Madagascar, before reaching orbit. Note that in January 2006, the commercial imaging company GeoEye was established, made up of the former Orbimage of Dulles, Virginia and Space Imaging of Thornton, Colorado. As a result, OrbView-5 was renamed GeoEye-1.

Background

OrbView-1 (MicroLab-1)

Orbital Sciences Corp's Eyeglass International, was formed in 1994, to build and operate the Eyeglass satellite system and to market imagery products and services. In 1995, Orbimage decided instead to pursue a smaller, lower cost satellite, derived from its MicroLab design. The cost of deploying one satellite and establishing the ground system was estimated at more than USD100 million. OSC's MacDonald Dettwiler Associates managed the ground segment requirements. Saudi Arabia's EIRAD Co. signed June 1994 to become a major investor (operating a receive station). The United States' Department of Commerce approved the project 5 June 1995. EIRAD decided to take a 20 per cent stake in Orbimage since US regulations imposed a 25 per cent limit.

The OrbView-1 satellite included two government-sponsored sensor systems: the Optical Transient Detector (OTD), a lightning imaging system, and GPS/MET (GPS/Meteorology), an atmospheric measurement system that used radio occultation techniques. In August 1993, Orbital Sciences Corporation (OSC)/NASA Marshall signed an all-inclusive contract for the provision of atmospheric research data from NASA's payload aboard MicroLab-1. The Optical Transient Detector studied the spatial/temporal distribution of global lightning to better understand and predict major atmospheric storm systems and climate changes. It was a precursor to the LIS Lightning Imaging Sensor on the 1997 NASA/Japanese TRMM mission. 500 images were recorded and the real-time event processor handled 10 million pixels to extract lightning signals from the bright background. Microlab data was downlinked twice daily to OSC's tracking station in Fairmont, West Virginia. The data was analysed in NASA Marshall's Global Hydrology & Climate Center and archived/distributed by the EOSDIS Earth Observing System Data Information System. The OTD was the world's first space-based sensor capable of detecting and locating lightning events in the daytime, as well as during the night-time, with high detection efficiency.

OSC's OrbView-2 (SeaStar) provides Earth colour data
0517063

One useful finding of the satellite resulted from observations of a 17 April 1995 storm in Oklahoma. Results suggested that increased intracloud lightning presages tornadoes, a possible method to perform tornado predictions and provide population alerts from space.

MicroLab's GPS meteorological experiment was a proof-of-concept demonstration instrument of UCAR/JPL. It used an onboard eight-channel TurboStar GPS receiver (donated by Allen Osborne Associates), daily recording 500 occultations of GPS signals to map atmospheric temperature and water vapour content. The experiment was managed by the UCAR (University Corporation for Atmospheric Research) for the National Science Foundation, which paid some USD500,000 to OSC. The total price paid to OSC for the two year mission was more than USD7 million. MicroLab was based on OSC's MicroStar small standard platform, developed for the Orbcomm satellite messaging system.

OSC retained ownership of Microlab-1 with NASA renting space for experiments and their operations. When OSC's affiliate, Orbital Imaging Corporation (Orbimage), renamed its satellites in 1997, MicroLab-1 became known as OrbView-1. The satellite successfully completed its mission in April 2000, after five years of operations.

OrbView-2 (SeaStar)

NASA Goddard selected OSC in March 1991 for a USD43.5 million contract to deliver multispectral ocean colour data to NASA investigators for five years as part of the agency's Earth Probes programme. This is the first time that the US government has purchased global environmental data from a privately-designed and operated remote sensing satellite. SeaStar, based on OSC's PegaStar bus, carries the SeaWiFS Sea-viewing Wide Field Sensor of Hughes' Santa Barbara Research Center. The instrument, originally planned for Landsat 6, is a next-generation design of Nimbus 7's Coastal Zone Colour Scanner.

SeaWiFS measures ocean surface-level productivity of phytoplankton and chlorophyll for ocean dynamics and marine life research. It will contribute to understanding the global carbon cycle and its effect on global warming. Orbimage retains exclusive worldwide rights to imagery for commercial and operational uses. It expects to find markets in ocean fishing, coastal monitoring, land management and the military. SeaStar was originally designed for ocean colour but SeaWiFS gains were changed during development to add higher land radiance dynamic range. It will provide a more environmentally stable Vegetation Index than NOAA's AVHRR, which is inaccurate under hazy atmospheric conditions because of its single visible and near-infra-red channels. Orbimage will sell to both end-users and value added resellers at a price point of US$500 per image set.

A delay to mid-1994 was announced August 1993 for instrument modifications to cure a problem with stray light. The onboard computer then required redesign (the original computer from OSC's APEX was found to have insufficient throughput to process SeaWiFS data) and 1994's Pegasus XL failure resulted in further delays. SeaStar was eventually launched by a Pegasus XL in August 1997.

OrbView-3

OrbView-3 was one of the first commercial satellites to provide high-resolution imagery from space. The spacecraft carries one imaging camera, the OrbView High Resolution Imaging System, capable of collecting one metre resolution panchromatic and four metre resolution multispectral imagery. One-metre imagery enables the viewing of houses, automobiles and aircraft, and makes it possible to create highly precise digital maps and three-dimensional fly-through scenes. Four-metre multispectral imagery provides colour and infra-red information to further characterise cities, rural areas and undeveloped land from space. This satellite collects up to 210,000 km² of imagery each day.

The OrbView-3 bus serves as the platform for the high resolution imaging instrument. The bus design originated with the Space Test Experiment Programme (STEP) in 1991 and consists of 12-sided horizontal honeycomb equipment decks and aluminium plate or honeycomb side panels that stiffen the structure and provide additional radiation protection for the equipment mounted inside. The bus structure is divided into three modules (propulsion, core and payload), each of which is constructed so components can easily be attached to each other for final assembly and check out.

OrbView-3's imaging instrument provides both one-meter panchromatic imagery and four-meter multispectral imagery with a swath width of 8 km. The satellite revisits each location on Earth in less than three days with an ability to turn from side-to-side up to 45 degrees. OrbView-3 imagery can be downlinked in real-time to ground stations located around the world or stored on-board the spacecraft and downlinked to ORBIMAGE's master US ground stations. Orbimage's control centre in Dulles, Virginia, is equipped to provide full telemetry, tracking and command of the satellite, as well as the archiving of imagery data and the distribution of the imagery products to customers. OrbView-3 provides imagery for a variety of applications such as telecommunications and utilities, oil and gas, mapping and surveying, agriculture and forestry and national security.

OrbView-4

OrbView-4 was scheduled for launch by a Taurus vehicle on 21 September 2001 but the satellite was lost due to launcher failure.

OrbView-5 (GeoEye-1)

The fifth OrbView satellite was renamed GeoEye-1 after Orbimage completed its acquisition of Space Imaging in January 2006 and changed its name to GeoEye. The rebranded company plans to launch its next-generation Earth imaging satellite, GeoEye-1, from Vandenberg Air Force Base, California in 2007. GeoEye-1 will be the world's highest resolution and most accurate commercial imaging satellite with a ground resolution of 0.41 m.

Specifications

OrbView-1 (MicroLab-1)

Launch: 3 April 1995 by Pegasus XL launched from a L-1011 aircraft flying out of Vandenberg AFB, California.
Resolution: ½ m PAN (½ km swath), 4 m XS (8 km swath)
Lifetime: 5 years
Mass: 74 kg
Orbit: Circular, non-sun-synchronous, altitude 734–747 km, inclination 70°, period 100 min
Spacecraft: 1.04 m diameter and 38 cm in height. Two circular solar arrays 96.5 cm in diameter. Some onboard storage but most imagery downlinked in real time
Sensors: OTD. Composed of six major subsystems: an imaging system, a focal plane assembly (including a CCD array detector of 128 × 128 pixels, preamplifiers and multiplexers), a Real-Time Event Processor (RTEP) and background remover, an event processor and formatter, power supply and interface electronics. The imaging system was a simple telescope consisting of a beam expander, an interference filter, and reimaging optics. The filter was narrow band (8.4Å) and centred about a prominent neutral oxygen emission triplet in the lightning spectrum to optimise SNR in the presence of a brightly illuminated cloud top. OTD parameters: view direction: nadir; FOV = 100° × 100°; spatial resolution 10 km; temporal resolution 2 m; wavelength 777.4 nm; sensor mass 2 kg; power 3 W. The two modules with electronics units had a mass of about 18 kg. The combination of the wide field-of-view lens and the altitude of the orbit allowed OTD to observe simultaneously an area of the Earth equivalent to 1,300 km × 1,300 km. 'Flashes' were determined by comparing the luminance of adjoining frames of OTD optical data. If the difference was more than a specified threshold value, an 'event' was recorded.

OrbView-2 (SeaStar)

Launch: 1 August 1997 by Pegasus XL
Mass: 309 kg on-orbit (SeaWiFS 50 kg)
Orbit: Altitude 705 km, inclination 98.2°, equator crossing at noon, period 99 min (16 day cycle)
Resolution: 1.13 km Local Area Coverage and 4.5 km Global Area Coverage. L-band transmits encrypted continuous LAC data when SeaStar is visible to standard HRPT stations equipped with OSC's licensed decoder. GAC data are stored and downlinked every 12 h on S-band to central location
Spacecraft: 213 cm high, 112 cm diameter OSC PegaStar bus. 5-year design life; 10 year goal. Redundant 3-axis momentum biased ACS (0.5°; 1.23 mrad knowledge) of momentum wheels, magnetorquers, horizon sensors, Sun sensors, and 2-axis magnetometers. Redundant GPS receiver provides precise real-time position data to within 100 m. 200 W BOL orbit average from four 55.9 × 152.4 cm rigid solar panels supported by 10 Ah Ni/H_2 battery. TT&C and 0.665/2.0 Mbit/s

data via L/S-band downlinks; 1.25 Gbit recorder. Four 22.2 N hydrazine thrusters to raise/maintain mission orbit
Sensors: SeaWiFS. The main telescope rotates 360° about a pivot axis to scan each scene, thereby avoiding the use of a scan mirror and its associated polarisation effects. Specular Sun reflection is avoided by tilting the telescope in the plane perpendicular to the scan plane to one of three positions: +20°, 0° or –20°. The continuous 360° scan permits reference sources to be viewed during the non-scene portion, in addition to including a deep space scan for zero referencing before the scan begins. The spread of observing bands addresses both chlorophyll and pigment absorption values, as well as water optical properties and suspended sediment measurements. Spectral band centres, (in μm, with bandwidth and application in brackets): Band 1: 0.412 (0.020, gelbstoffe), Band 2: 0.443 (0.20, chlorophyll), Band 3: 0.490 (0.020, pigment), Band 4: 0.510 (0.020, chlorophyll), Band 5: 0.555 (0.20, suspended sediments), Band 6: 0.670 (0.020, atmospheric aerosols), Band 7: 0.765 (0.40; atmospheric aerosols), Band 8: 0.865 (0.044, atmospheric aerosols). The bands for atmospheric aerosols provide atmospheric corrections.

OrbView-3
Launch: 26 June 2003 by Pegasus XL from Vandenberg.
Launch mass: 304 kg (including ca. 50 kg of hydrazine for orbit raising)
Orbit: Perigee: 452 km. Apogee: 455 km. Inclination: 97.3°. Period: 93.7 min. Nodal crossing 10:30 a.m. Revisit time less than 3 days.
Resolution: 1 m panchromatic and 4 m multispectral. Swath width 8 km.
Spacecraft: Dimensions: 1.9 m height; 1.2 m diameter. Solar Arrays: 5 GaAs panels. Power: 625W. Geolocation Accuracy: ±12 m/10 m Uplink/Downlink: UHF/X-band. Maximum Data Rate: 150 Mbps. Expected mission life: 7 years
Sensors: Type: Body-scanning whiskbroom PAN/MS imager Weight: 66 kg. PAN spectral range (one channel) 450–900 nm. Multispectral range (4 channels) 450–520 nm (blue), 520–600 nm (green), 625–695 nm (red) and 760–900 nm (near-IR). Dynamic Range: 11 bits per pixel. The satellite revisits each location in less than three days with an ability to turn up to 50° either side of the ground track.

Space Radar

Current Status
The Space Radar (SR) is a joint programme operated by the USAF Space & Missile Center, the National Reconnaissance Office and the National Geospatial Intelligence Agency, together with the US Army and Navy. Its goal is to develop and launch a constellation of spacecraft capable of detecting and imaging moving ground targets. The first launch is planned for no earlier than 2015.

Background
Space Radar evolved as a response to the absence of any other system that could provide all-weather, 24-hour detection and surveillance of targets and locations inside hostile territory, with particular reference to the difficulty of detecting tactical ballistic missile launchers with sufficient timeliness and uncertainty to destroy them. This was different from the task performed by Lacrosse/Onyx, which has a relatively narrow field of view from its sensor and is unable to track moving targets.

The project originated as a joint initiative by the USAF, NRO and the Defense Advanced Research Projects Agency. It was named Discoverer II, in a deliberate echo of the first US reconnaissance satellite programme, and was started in 1998. The basic goal of Discoverer II was to exploit Active Electronically Scanned Array (AESA) technology and advanced processing to combine Moving Target Indication (MTI), Synthetic Aperture Radar (SAR) and three-dimensional terrain imaging in a single, affordable spaceborne radar. MTI would be used to detect moving targets; SAR, combined with high-range-resolution MTI, would be used to classify and identify targets; and terrain imaging would be used to eliminate some targets, to identify potential places of concealment and to locate targets accurately enough for weapon guidance.

The aim of Discoverer II was to launch and test a demonstration satellite by 2008. Study contracts were issued to Lockheed Martin, Spectrum Astro and TRW, but the programme was cancelled at the preliminary design review stage in 2000 because of cost and schedule uncertainties.

The goals of Discoverer II were transferred in 2001 to a new Air Force program named Space Based Radar (SBR). In July 2003, however, an independent cost assessment team estimated that SBR would cost no less than USD28.6 billion through its lifetime, for a constellation of nine spacecraft, and Congress reduced the programme's budget.

A June 2004 Defense Science Board report on SBR stated that the programme was – at that time – due to begin deployment in 2010 to 2012.

In January 2005, the Pentagon restructured the programme to stress its joint nature and reduce risk. It was renamed Space Radar, and the project office, which had been located within the USAF Space & Missile Center at Los Angeles AFB, was replaced by a new JPO based in Chantilly, Virginia, closer to its intelligence-community customers. First launch, which had been originally set for 2010 and later for 2012, has moved back to 2015.

The Space Radar has now been designated as the sole radar-based space sensor for "national" – that is, intelligence – and military users. As such it will be the replacement for today's Onyx satellites, and also replaces the radar element of the under-performing and over-budget Future Imagery Architecture (FIA) programme.

The new programme also places more stress on innovation and affordability, and on a system-of-systems approach in which SR will work with other assets to provide global coverage rather than attempting to provide global coverage on its own. It is also linked to a DARPA demonstration programme aimed at developing an extremely large antenna which could image targets from Medium Earth Orbit (MEO).

Industry work on Space Radar continued under contracts awarded to Lockheed Martin and Northrop Grumman in April 2004. These contracts were originally due to continue into April 2006, and called for the teams to develop system and software architectures, conduct basic studies to balance performance, affordability, risk and schedule; and validate some basic technologies. As part of the restructuring, both teams will remain on contract until FY2009. In early FY2009, at what is known as Key Decision Point B (KDP-B), the JPO intends to select one contractor team to continue design development. This would lead to the launch of the first of a constellation – still estimated at eight-to-ten satellites – in 2015.

According to the DSB report, the system as defined in 2004, comprised a combination of satellites in Medium Earth Orbit (MEO) at an altitude of 10,000 km and LEO satellites at 1,000 km. The LEO satellites would provide MTI, with the ability to track a truck-sized 10 dB radar target at 2,800 km slant range. The report described three LEO options with different-sized spacecraft – available launchers could deliver one, two or three vehicles. An all-LEO system would require 21 satellites to provide persistent global access; adding MEO satellites would reduce timeliness (because SAR images would take longer to process) but would reduce the number of spacecraft required.

The JPO has been directed to look at whether a space-based technology demonstrator should be launched before the final decision to start development. So far, no decision in this area has been made public, but the USAF has discussed launching a quarter-scale demonstrator by 2008 or 2009. A demonstrator programme might affect the timing of KDP-B.

Northrop Grumman's team includes Boeing, responsible for the design, development and production of the SR space segment, including launch vehicle integration; Raytheon, which will provide global mission and system management, and support to other ground-segment elements; and General Dynamics, providing mission and ground-system analysis and mission data processing elements. BAE Systems will supply large-scale information storage and retrieval technologies. Lockheed Martin is teamed with Harris, Honeywell and Cisco Systems.

Space Radar is heavily dependent on the Transformational Satellite Communications (TSAT) programme, which is the only means in sight by which SR imagery can be handled fast enough to hit moving targets or to provide operationally useful data to the warfighter.

A closely related initiative is the Innovative Space-Based Radar Antenna Technology (ISAT) programme, being conducted by DARPA's Special Projects Office and the USAF Research Laboratory's Sensors Directorate. Because radar performance is affected by both power and antenna size, ISAT is looking at the development of a deployable AESA measuring 300 m in length, using advanced materials such as shape-memory polymers and rigidized inflatable structures. According to DARPA, antennas of this size are enabling technologies for a constellation that can provide continuous coverage with 10 to 12 satellites in MEO, 9,200 km from the Earth; less capable satellites would have to operate in LEO to provide the same detection capability, and more of them would be needed because each spacecraft would have a smaller field of view.

Two teams – Lockheed Martin teamed with Harris, and Boeing with Raytheon, each corresponding to the SR team – are competing to produce a one-third-size 100 m LEO demonstrator for the ISAT programme, due to be launched in 2010. The teams are currently funded to continue until the Critical Design Review (CDR) phase. After that, in June 2006, DARPA and the USAF expect to down-select to a single contractor.

TOMS-EP satellite

Current Status
In December 1998 satellite was hit by a high-energy particle which disrupted onboard computer and put satellite in safe mode. Stabilisation process exhumed fuel and TRW developed a control method using magnetic torque.

Background
The Total Ozone Mapping Spectrometer Earth Probe (TOMS-EP) extends the work of Nimbus 7's similar instrument in monitoring atmospheric ozone levels. It helps establish the continuity in the gathered data so a long-term picture of the Earth's ozone layer can be formed. NASA originally selected TOMS in 1989 as the fourth Small Explorer but then dropped its funding before reviving it as the first Earth Probe in NASA's Mission to Planet Earth.

NASA Goddard released a request for spacecraft proposals in late 1990; and the TRW's Eagle bus won out over its competitors for the June 1991 contract of US$29.3 million (now US$57 million). Mission cost, including launch but excluding operations has amounted to about US$80 million

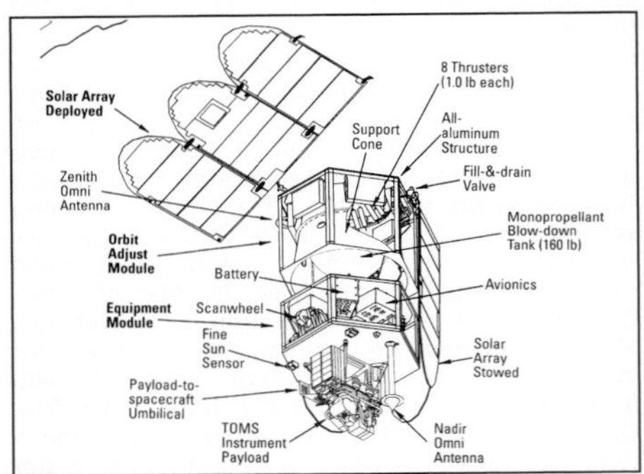

TOMS-EP principal features (TRW) 0517227

TOMS-EP is mapping global ozone levels (TRW) 0517226

for NASA. TRW built the spacecraft and integrated the instrument.

In operation, the satellite stores up to 24 h of data stored on 16 Mbyte solid-state recorder and downlinks it via S-band transponder at 202.5, 50.6 or 1 kbit/s. The Total Ozone Mapping Spectrometer is based on 3,086–3,600 A Fastie-Ebert monochrometer, for 1 per cent/decade ozone trend determination in six wavelength bands.

Specifications

Launch: 2 July 1996 by Pegasus XL from Lockheed L1011 out of Vandenberg AFB
Mass: 294 kg (including 54 kg hydrazine)
Orbit: planned 500 km circular 97.4° Sun-synchronous at 11.00–11.30 ascending node
Contractors: TRW (bus), Orbital Sciences Corp (TOMS instrument)
Resolution: 50 × 50 km
Lifetime: minimum 2 years, goal 3 years
Spacecraft: TRW Eagle platform, two compartments, hexagonal prism of three equipment decks. Bus 99 cm diameter, 168 cm long without TOMS; stowed/deployed 112/388 cm max width, 178/243 cm long with TOMS. Two 3-panel dual-sided solar wings of composite material over honeycomb core providing 6.25 m^2 total cell area, mounted at 45° cant to body, for 233 W EOL (β=20°). 200 μm Si cells protected by 150 μm cover glass. 9 Ah Super Ni/Cd battery provides eclipse power. 3-axis control to 0.5° pitch/roll (0.25° knowledge, 3Σ) pitch/roll, <1.0°/0.25° 3Σ yaw by momentum bias RW system utilising Earth/Sun sensors + gyros. 4 N dual-seat hydrazine thrusters for orbit insertion, trim and station-keeping. Up to 24 h TOMS data stored on 16 Mbyte solid-state recorder and downlinked via 2 W S-band transponder to DSN & Wallops stations at 202.5, 50.6 or 1 kbit/s. Satellite designed for 24 h autonomous operations. Combined Mission Operations Center and Science Operations Center at NASA Goddard
Sensors: Total Ozone Mapping Spectrometer. Based on 3,086-3,600Å Fastie-Ebert monochrometer, for 1%/decade ozone trend determination in six wavelength bands. 3.0 × 3.0° FOV crosstrack scanning capability allows daily high spatial resolution (50 × 50 km) global O_3 maps. SO_2 monitoring also undertaken. Sensor mass is 32 kg, 1,525 cm^3, 25 W average power consumption (including heaters); onboard solar and electronic calibration capabilities, plus reflectance calibration mode not previously flown. Microprocessor-based electronics provide flexibility in control and data formatting; data rate up to 736 bit/s.

For details of the latest updates to ***Jane's Space Systems and Industry*** online and to discover the additional information available exclusively to online subscribers please visit **jsd.janes.com**

HUMAN

China

Shenzhou series and Project 921

Current Status

Shenzhou 6, launched in October 2005, was the most recent Shenzhou mission to fly; it carried Fei Junlong and Nie Haisheng into orbit. The Shenzhou 6 orbital module remains aloft and continues to make small orbital adjustments. Shenzhou 7 is expected to fly in 2008, with reports that it will carry at least two crew and include China's first EVA.

China became the third nation to launch its own manned space mission in October 2003. Yang Liwei, a People's Liberation Army Air Force (PLAAF) pilot was the country's first 'taikonaut'.

According to comments made by the China National Space Administration's (CNSA) Administrator in 2006, Shenzhous 1 to 5 cost roughly USD2.4 billion, and Shenzhou 6 cost under USD125 million.

Background

China's human spaceflight programme designator is 'Project 921'. Project 921 came to fruition almost three decades after the cancellation of China's first attempts at a manned space programme. In 1966 a programme called Shuguang (Dawn) was initiated. There has been Western speculation that the Shuguang spacecraft would have been a manned variant of the Fanhui Shi Weixing/Jianbing (FSW/JB) photo reconnaissance satellite, but this was not confirmed from reliable Chinese sources. If correct, then the spacecraft would have been similar to the United States Mercury 'man in a can' concept, allowing a single person to fly in orbit for a maximum of one to two days. The original FSW satellite had a mass of 1.8 tonnes and was launched aboard the CZ-2A launcher.

In 1971, 19 trainee astronauts were selected for the initial Shuguang missions, which were then expected to begin in 1973. Before test flights could take place, politics and budgets led to the cancellation of the programme in 1972.

Project 921, China's second attempt at a manned space programme, is an offshoot of Project 863 (merging numbers that signify the third major government project in 1986 yields the designator 863), which was intended to improve China's science and technology capabilities and standards. Project 921 is a direct derivative of Project 863, and it is acknowledged by China to be the largest and most expensive project ever undertaken by the country.

Seven Elements of Project 921

When authorised in 1992, Project 921 comprised seven elements which were designated Project 921-1 to 921-7, although in the public domain it is not known which program element refers to which designator. The seven elements are:

- Launch site system
- Astronaut system
- Measurement and control system
- Manned delivery rocket system
- Manned spacecraft system
- Landing system
- Spacecraft application system

By the end of the second Shenzhou test flight, China had demonstrated all of these elements.

Jiuquan Launch Facility

The launch of SZ-1 introduced the new launch pad, constructed at Jiuquan for operational launches. The Jiuquan pad is associated with a nearby assembly building, akin to the Vehicle Assembly Building at the Kennedy Space Center (KSC) which was constructed for the stacking of Saturn-5 and Apollo vehicles and is now used for stacking space shuttles prior to roll-out to the launch pad. The Chinese assembly building is large enough to allow the simultaneous assembly and integration of two, possibly three, launch vehicles and spacecraft.

The new assembly building and launch pad brought a new philosophy to China's preparations for launches. Previously, all launch vehicles had been transported to the launch pads stage-by-stage and then stacked on the launch pad. For the manned programme the launch vehicles would be stacked and integrated in the assembly building, and then rolled out to the launch pad a few days before the planned launch.

This change makes sense when one considers that at present there is only one launch pad available for the manned programme. If the Chinese want to launch rendezvous or docking missions – like the Soviet-era Soyuz 4 and Soyuz 5 missions, launched on consecutive days but using two pads – then they need to have a minimum on-the-pad time for their launch vehicles. Also, this approach allows for a simplified launch pad design and construction – so that if the pad is damaged or destroyed during a launch mishap it should not be a major job to repair or rebuild it.

In 2002 it was reported that the Chinese were working on a second launch pad close to the one used for the CZ-2F launches. The new pad is smaller than the one used for Shenzhou and it has no rail links to the assembly building. Completed by early 2003, the new pad saw its first launch in November 2003 when the first CZ-2D carried the first of a new series of recoverable Jianbing reconnaissance satellites into orbit.

The Hangtianyuan Team

Independently, Chen Lan and Chiew Lee Yih coined the term 'taikonaut' for a Chinese astronaut, and this term appears to have come into popular use, even though it is unofficial. Official Chinese sources have used the words 'yuhang yuan' , 'taikongren' and 'hangtianyuan' – the latter apparently being the most "official" of the various terms.

The first Chinese hangtianyuans to be named were Li Qinglong and Wu Jie. The Russians announced in late 1996 that these two men would be undertaking a year's cosmonaut training at the Zvezdny Gorodok (Star City) centre, and they were met and photographed by the British Neil Da Costa and Rex Hall when they visited the training centre in April 1997. Following the successful SZ-4 mission, the name 'Chen Long' circulated as a possible candidate for China's first piloted space mission, although in reality there has never been a Chinese trainee with this name. Some sources indicate that Chen Long might have been an incorrect transliteration for 'Li Qinglong'.

The Russian announcement referenced above noted that the two Chinese trainees had been selected by the Chinese as "national astronauts" in March 1995 and had been given the designation "instructor astronauts" by the Chinese. They were to complete a one-year long general space training course, which would include three months learning Russian, physical training and familiarisation with the technical details of piloted space missions. The Russians said that the two trainees would return to China "where the launch of the first national manned spacecraft carrying a crew of two is scheduled for 1999". Of course, this target was not met, and it is possible that the Russians misunderstood Chinese plans which called for the maiden unmanned launch of a Shenzhou during the 1998–1999 period.

The training at Zvezdny Gorodok was conducted on a purely commercial basis, with the Chinese paying "several hundred thousand dollars" for the two men to complete the training course. Privately, the Russians said that they believed that some of the "trainers" who accompanied Li Qinglong and Wu Jie were also astronaut candidates, although the Chinese said nothing to confirm this.

The Chinese had announced that they had 14 hangtianyuans in training but did not immediately release the names of the candidates. It was not until April 2003 that the authoritative *Novosti Kosmonavtiki* website listed 14 hangtianyuan names which had been quoted in the German publication *Fliegerrevue*, issue 5, 2003:

- Chen Quan
- Deng Qingming
- Fei Junlong
- Jing Haipen
- Li Qinglong
- Liu Buoming
- Liu Wang
- Nie Haisheng
- Pan Zhanchun
- Wu Jie
- Yang Liwei
- Zhai Zhigang
- Zhao Chuandong
- Zhang Xiaoguan

The Chinese have described the basic requirements for the members of the hangtianyuan team. These are:

- Age, approximately 30 years old
- Height, approximately 1.7 metres
- Weight, approximately 50 kg (65 kg is quoted in some sources)

Additionally, all of the trainees were said to hold a master's degree in "mathematics, physics, engineering and biology, etc". They had a strict training regime of ten hours per day in the "aerospace city" centre, located in the suburbs of Beijing. The instruction included "specific" training relating to the aerospace environment and spaceflight simulation – general knowledge of "astrology" (sic – astronomy was probably meant), geography, medicine, flight dynamics, and rocket and spacecraft technology.

We now know that Yang Liwei became China's first hangtianyuan on 15 October 2003 and that his back-ups for the flight were Zhai Zhigang and Nie Haisheng. All three names had appeared in the *Fliegerrevue* listing earlier in 2003.

In March 2004 the Chinese press began reporting about the future selection of women hangtianyuans. Like their male counterparts, the women would be drawn from pilots within the People's Liberation Army Air Force (PLAAF).

A second team of hangtianyuans is said to be in the process of being recruited and trained.

Yuanwang Tracking Ships

There are four Yuanwang ships operated by the Chinese for the tracking of satellite and missile launches, tracking newly-launched payloads into orbit and assisting in the tracking and recovery of missile re-entry vehicles. The first two Yuanwang ships were operating in 1980 (possibly accepted for operations in 1979). Their first operations appear to have been in connection with the DF-5 missile test on May 18, 1980 which fired a re-entry vehicle from Jiuquan into the Pacific Ocean, near the Solomon Islands. The Yuanwang 3 was delivered in March 1995 and Yuanwang 4 in July 1999.

The original three ships underwent a major systems upgrading which was announced in January 1999. The work had been carried out in Shanghai and had taken 16 months to complete. The ships were now carrying more advanced equipment on board than they had previously used, and this ensured that they were capable of global tracking and control. The data transfer speed had been increased by a factor of 400.

At the times of the Shenzhou flights the ships are deployed as follows:-

Yuanwang 1: North Pacific (off Chinese coast)
Yuanwang 2: South Pacific (east of Australia)
Yuanwang 3: South Atlantic (west of Angola)
Yuanwang 4: Indian Ocean (west of Australia)

Launch Vehicle and Spacecraft Systems

The CZ-2F launch vehicle used for the Shenzhou missions is a human-rated version of the CZ-2E vehicle which has so far flown six times. The 2F has improved propulsion systems, improved control and computer systems and carries a payload shroud with an escape tower at the top for use in emergencies.

The manned spacecraft system is Shenzhou itself. The landing system, of course, is incorporated into the Shenzhou design, but the Chinese probably include the ground recovery vehicles and associated systems as part of the landing system.

The spacecraft applications system relates to the equipment carried aboard the Shenzhou orbital module which functions in orbit as an independent satellite once the descent module has returned to Earth.

Shenzhou Design

The name Shenzhou is said to have been chosen personally by President Jiang Zemin, and can be translated into English in different ways: "shen" can translate as "God", "heavenly" or "divine", while "zhou" can translate as "ship" or "vessel".

The prime contractor for the Shenzhou spacecraft is the China Academy of Space Technology (CAST), and the spacecraft's chief designer is Qi Faren.

Whenever the Chinese talked about possible manned space missions during the 1990s, the goal was always said to be the establishment of an orbiting laboratory. As a result, the Chinese have based the design of Shenzhou on the World's most-used space station ferry, the Soviet Soyuz, which first flew a space station mission in April 1971. The Chinese took the basic Soviet three-module concept and modified it for their programme.

This has sometimes been either misunderstood or completely misrepresented in the West, with even the official US Cox Report claiming that when China started its manned space programme it would be using no more than surplus Soyuz spacecraft, purchased from Russia. Overall, the design of Shenzhou is very much like Soyuz, but physically it is slightly larger and about half a tonne more massive. In the detail there are major design differences, the most obvious being that the forward orbital module on Shenzhou carries its own set of solar panels and its own orbit manoeuvring system. On the SZ 2 to 6 missions the orbital module operated as an independent satellite for six months or more after the descent module returned to Earth. The Soyuz orbital module was never designed for this added function.

However, the Chinese have clearly benefited from Russian experience. They purchased single items of Russian technology to see how the Russians solved specific problems and then incorporated those solutions into the Shenzhou design. Examples of the technology purchases include a Soyuz life support system; the shell of a Soyuz descent module, with most of the interior instrumentation and equipment stripped out; an androgynous docking unit; and a Soyuz pressure suit of the type worn at launch and landing – not an EVA suit.

Physically, the Chinese spacecraft is slightly larger than the Soyuz; Russia's Soyuz-TM was the variant designed to take crews to and from the Mir complex. The improved Soyuz-TMA is used as the "lifeboat" spacecraft for the International Space Station (ISS). A comparison with the Shenzhou is given in the accompanying table.

Soyuz-TM – Shenzhou Specification Comparison

Spacecraft	**Soyuz-TM**	**Shenzhou**
Total spacecraft mass, kg	7,250	7,84
Total length, metres	7.48	9.25
Service module:		
Mass, kg	2,950	3,000
Propellant mass, kg	900	1,000
Length, metres	2.60	2.94
Diameter (cylindrical), metres	2.17	2.50
Diameter (base), metres	2.72	2.80
Area of solar panels, m²	10	24
Descent module:		
Mass, kg	3,000	3,240
Length, metres	1.90	2.06
Diameter, metres	2.17	2.52
Orbital module:		
Mass, kg	1,300	1,500
Propellant mass, kg	0	40
Length, metres	2.98	2.80
Diameter, metres	2.26	2.25
Area of solar panels, m²	0	12

The return to Earth of the Shenzhou spacecraft follows the philosophy of the Russian Soyuz closely, and was described in general terms for the flight of Shenzhou 1. The orbital module separates in orbit and approximately ten minutes later the combined descent and instrument modules are oriented as they are passing over the Yuanwang 3 ship off the coast of Angola; the main engine on the instrument module fires for a last time to bring the assembly out of orbit. Atmospheric re-entry is considered to begin at an altitude of 80 km. Once the craft has descended to an altitude of about 40 km, a communications blackout starts. Once the blackout has ended, the descent is followed by radar with helicopters converging on the predicted landing area. At a distance of 30 km from the landing site, the spacecraft's electrical power is turned on, followed by the separation of the drogue parachute and the deployment of the main parachute with an area of 1,200 m². As the landing draws close, the descent module's base falls away, exposing the four solid-propellant motors which ignite at an altitude of 1.5 metres to cushion the landing.

Future Manned Missions

Even before the first manned Shenzhou was launched, China made it clear that its goal was the establishment of a small space laboratory, to be followed by something similar to the Russian Mir modular space station by around 2010. Wang Yongzhi, named as one of the people behind China's manned space programme, has described a three-step programme for Chinese manned operations:

- Launch unmanned and manned spacecraft and conduct suitable Earth observations and scientific experiments
- Conduct EVAs; rendezvous and docking activities; and launch a space laboratory
- Build a large, permanent manned space station

2010 was an optimistic target, and it is very doubtful that the Chinese will have a manned space station in orbit by that time. They could, however, conduct docking exercises before then. Since the Shenzhou spacecraft was designed from the outset for docking operations, an initial docking mission will come relatively early in the flight programme. SZ 6 flew in 2005 and carried two men on a flight that lasted five days. The Chinese specifically said that no docking was planned for that mission, but it was originally thought that the SZ 6 orbital module could be used as a docking target for SZ 7.

Several announcements were made concerning the Chinese manned space programme following SZ 6 's successful touch down. The SZ 7 launch is expected to take place in 2008. The Chinese have announced that the mission will include an EVA. SZ 8 and 9 will follow in quick succession; both will be unmanned. SZ 8, at least, will be a different craft to the previous seven SZ launched. SZ 8 and 9 will perform unmanned rendezvous and docking manoeuvres as soon as 2009, however the mission date will likely slip, as SZ 7 before. SZ 10 may again be a manned mission, possibly involving further docking exercises, however that mission will not launch until 2012 at the earliest.

An initial space station module can be launched using the CZ-2F vehicle, and this has been depicted in Chinese animations released following the SZ 5 mission. Without the need for the heavy payload shroud and escape tower, the CZ-2F should be capable of orbiting 9.5–10 tonnes. This squares with the animations of a Shenzhou docking with a two-cylinder space laboratory which is slightly larger than the piloted spacecraft. Such a mission could begin around 2009.

It is possible that the orbital module left by one Shenzhou mission could be used as an expansion module for the crew on the next Shenzhou, and a 'train' of orbital modules could be assembled over a period of time. One problem with this scenario is that there is a docking unit at only one end of the module and it would therefore be impossible to build up a 'train' of such modules as an interim space station.

Once the initial piloted test flights of Shenzhou are complete, the logical goal will be the docking of two spacecraft in orbit, much as the Soviets did with Soyuz 4 and Soyuz 5 in January 1969. Unlike the Soyuz of that era, Shenzhou has an internal transfer system built into it, so the 'taikonauts' will be able to move from one craft to another with ease. If one of the Shenzhou spacecraft for such a mission were to be launched unmanned the two or three 'taikonauts' on the other spacecraft could fly a longer mission with the same consumables that could be undertaken if two crews were launched.

A larger space laboratory could be launched as the development of the CZ next generation launch vehicles proceeds. The 504 vehicle has a payload capability of 25 tons to LEO, enough for a laboratory significantly larger than either the Shenzhou orbital module or a complete unmanned Shenzhou spacecraft. The launcher could conceivably orbit space station modules similar to those of the Soviet Salyut programme, which later comprised the Mir complex.

A manned lunar programme is in the early stages, with the Chang'e I robotic orbiter to launch in 2007. The Chinese aspire to land humans on the Moon around 2020. This goal may be optimistic, as the lunar and launcher programmes will require continuous funding to meet this target and the human spaceflight programme has seen delays.

One orbit scenario presents itself for Shenzhou loops around the Moon, akin to the Soviet Zond/L-1 missions launched during 1967–1970. The CZ-5/504 variant with the planned LOX/hydrogen third stage could easily launch a complete Shenzhou spacecraft on a circumlunar mission, although without a larger launch vehicle for follow-on lunar orbit and lunar landing missions such a project would be a dead-end.

Another launch scenario possibility is that the manned lunar assembly could be launched on two vehicles, with Earth orbit rendezvous before trans-lunar injection. A further possibility is that a single 70 ton rocket stage could be launched, followed by the manned, upgraded Shenzhou and the lunar lander on separate launches of CZ-2F class launchers.

It is too early for even the Chinese to have decided upon the exact mission profile and thus launch vehicle requirements for a manned lunar landing programme, and any related analysis in this entry can only be guesswork at present. However, it is likely the Chinese will eventually meet their manned lunar mission goals, even if they don't meet the initial 2020 target date. They have also indicated that their plan will escalate to include a permanently manned base on the Moon and Mars exploration.

Specifications

Shenzhou Spacecraft

Prime contractor: China Academy of Space Technology (CAST) and Shanghai Academy of Spaceflight Technology (SAST)
Launch vehicle/site: CZ-2F from Jiuquan
Dimensions:
Overall: 2.8 × 9.25 m
Service module: 2.5 × 2.94 m
Descent module: 2.52 × 2.06 m
Orbital module: 2.25 × 2.8 m
Power: Instrument/propulsion module: two vanes of solar panels with a total surface area of 24 m²
Orbital module: Two vanes of solar panels with a total surface area of 12 m²

Shenzhou 1 (SZ 1)

Launch: 20 November 1999, 22.30 GMT
Descent module recovery: 21 November 1999, 19.41 GMT
Orbital module decay: 1 December 1999

Western observers were ready for the expected maiden launch of what would become China's manned spacecraft in mid-November 1999, since the four Yuanwang tracking ships were deployed around the World for the first time. China announced nothing until the descent module had returned to Earth, although Western observers had been following the mission using orbital data issued through NASA's Goddard Space Flight Center. The first orbital data for what would be announced as Shenzhou 1 showed the spacecraft in a 42.6°, 197–325 km orbit. No orbital manoeuvres took place until retrofire brought the spacecraft out of orbit after less than a day in flight.

SZ 1 was a basic test flight, with the objectives being the successful launch and recovery of the spacecraft. No life support system was carried and the orbital module appears to have carried only 'dummy' solar panels which remained folded throughout the flight. SZ 1 is the only flight where the orbital module apparently had no orbital activities after the descent module's return to Earth; it simply decayed from orbit.

Shenzhou 2 (SZ 2)

Launch: 9 January 2001, 17.00 GMT
Descent module recovery: 16 January 2001, 11.22 GMT
Orbital module decay: 24 August 2001

The SZ 2 launch was originally planned for October 2000, but nothing was heard until December when a 5 January 2001 launch date was announced. The day before the planned launch, a delay was announced, with the launch subsequently planned for 10 January (Beijing time). When the launch took place, it was the first launch of the new Millennium, which, of course started on 1 January 2001; the final successful launch of the old Millennium had been China's Beidou-2 navigation satellite (see separate entry).

SZ 2 was launched into a 42.6°, 197–336 km orbit at about 17:00 GMT on 9 January; the spacecraft made the same geographical orbital pass as had SZ 1 at the time of retrofire. SZ 2's propulsion system then fired to circularise the orbit at 329–334 km. This orbit was an almost perfect 31-circuit repeater, the kind of orbit which will be useful when rendezvous and docking missions begin. A further manoeuvre on 12 January raised the orbit to 331–340 km and a final one on 15 January resulted in a 330–346 km orbit. The orbital module separated from the combined instrument/propulsion and descent modules on 16 January at 10.23 GMT and retrofire took place ten minutes later. The descent module landed in Inner Mongolia.

Ever since the SZ 2 descent module's return to Earth, there have been persistent reports that something had gone wrong with the landing. Unlike the SZ 1 recovery, there were no photographs or television coverage of the SZ 2 descent module at the landing site or after its return to Beijing. Although the Chinese have denied that there was a serious problem, reliable sources suggest that one of the parachute attachments failed during the descent, and as a result the descent module made a hard landing.

Various animals, plants and seeds were carried on board SZ 2, and the Chinese said that these were recovered intact. According to a *Reuters* report after the landing, the descent module had carried a monkey, a dog and a rabbit, along with an unspecified number of snails.

Even though the descent module was back on Earth, the SZ 2 mission was not over. A fully functional orbital module was carried on this flight.

On 17 January, the module used its thrusters to raise its orbit from 330–345 km to 388–404 km. Two further manoeuvres took place: on 20 February to raise the orbit from 375–391 km to 389–403 km and on 15 March to raise the orbit from 382–390 km to 394–405 km. On 11 July, the Chinese announced that the module was still working after 170 days of operations. The module finally decayed from orbit on 24 August.

SZ 2 was still not a human-rated spacecraft but the flight had tested far more of its systems than had the SZ 1 mission. The spacecraft had remained in orbit for a week and had performed all of the manoeuvres necessary to fly a docking mission. A total of 64 scientific experiments were carried aboard the spacecraft: 15 were inside the descent module, 12 inside the orbital module and 37 in what was called the "attached segment" – possibly the assembly on the orbital module in place of the docking system.

While details of the orbital module's operations have not been disclosed in detail, it is believed that an Earth imaging system was carried, possibly for military reconnaissance observations.

Shenzhou 3 (SZ 3)
Launch: 25 March 2002, 14.15 GMT
Mass: 7,800 kg (orbital module 2,000 kg)
Descent module recovery: 1 April 2002, 08.51 GMT
Orbital module decay: 12 November 2002

The original plan was for China to launch SZ 3 in August or September 2001, and indeed the launch vehicle was seen by Western imaging satellites as it travelled to and from the launch pad during this time period. The Chinese admitted that the launch vehicle had to be returned to Beijing for modifications. In 2002 it was said that some modifications to the spacecraft were not found to be improvements during 2001, and the launch was delayed for some real improvements to be completed.

SZ 3 was not planned as a simple re-run of SZ 2. For the first time, SZ 3 would carry a fully functioning life-support system, and 'dummy' 'taikonaut' would be carried to test these systems. When previewing the flight it was explained why a simulated man was being carried rather than a live animal: if a monkey were to be launched then there was the danger of it breaking loose and it would be "sure to make some troubles, skipping and fumbling about when loosened" (sic). One might wonder whether this statement might be based upon past experiences aboard SZ 2.

The SZ 3 launch was next rumoured for mid-January 2002, but it did not come until 25 March. The initial orbit was not catalogued by the US Air Force, but they tracked the CZ-2F final stage in a 42.4°, 198 to 326 km orbit. The slight shift in orbital inclination appeared to be linked with the new inclination, allowing better rendezvous and repeating orbits to be maintained.

SZ 1 had performed its retrofire 20 hour 19 min after launch, and SZ 2 had performed its orbit-raising manoeuvre 20 hour 23 min after launch. However, SZ 3 performed its orbit-raising manoeuvre far earlier in the flight, only 7 hours after launch. Calculations showed that this would give a perfect rendezvous orbit for a hypothetical second spacecraft launched about two days later. After the first manoeuvre, the SZ 3 orbit was 332 to 338 km. As with SZ 2, two further manoeuvres followed: on 29 March to raise the orbit from 331–336 km to 331–338 km, and on 31 March to raise the orbit from 330–337 km to 330–340 km. After flying for nearly seven days, like the SZ 2, the SZ 3 descent module returned to Earth on 1 April. This time the Chinese media immediately released photographs and television pictures of the scene at the landing site.

On SZ 2 the first orbital module manoeuvre had waited for just over a day after the descent module recovery, but on SZ 3 the first orbital module manoeuvre came only a few hours after the descent module's return. On the afternoon of 1 April the orbit was raised from 330–340 km to 353–358 km. There were four further manoeuvres of the orbital module with the following post-manoeuvre orbits: 25 April – 382 to 388 km, 13 – 14 June – 375 to 385 km, 16 July – 373 to 378 km and 17 July 377 to 382 km. After this, the orbital module was allowed to decay naturally from orbit.

It is noted that the SZ 3 orbital module orbits are lower than those seen on SZ 2. It is possible that this was because either a different imaging system was being carried or the same imaging system was being carried but the lower altitude was allowing higher resolution images.

Shenzhou 4 (SZ 4)
Launch: 29 December 2002, 16.40 GMT
Mass: 7,794 kg (orbital module 2,000 kg)
Descent module recovery: 5 January 2003, 11.16 GMT
Orbital module decay: 9 September 2003

In an interview conducted before the launch of SZ 4, but published once the spacecraft was in orbit, Yuan Jiajun, a General Director of the manned space programme, previewed the forthcoming flight. For the first time, the spacecraft would be identical to the version which would carry crews into orbit. It carried both automatic and manual controls, as well as all of the emergency systems and equipment which a crew might need. The descent module's instrument panel now used 'sound prompts' for data to be input or to alert crew members to tasks. The spacecraft carried a fully functioning life support system, sleeping bag, food, medicine, a fire extinguisher and "hygiene articles and other necessities".

Prior to the spacecraft's launch, 'taikonauts' completed training exercises using the spacecraft on the launch pad, simulating emergency escape from the spacecraft. Should inclement weather have prevented a landing at the primary site in Inner Mongolia, the descent module was capable of landing at Jiuquan.

SZ 4, the final unmanned Shenzhou test flight, was launched on 29 December 2002. Like SZ 3, the orbit-raising manoeuvre took place about seven hours after launch. The second stage rocket body was tracked in a 42.4°, 198 to 332 km orbit, while SZ 4 itself was in a 42.4°, 331 to 337 km orbit. On this flight, smaller orbital manoeuvres took place, probably because solar activity was not affecting the Earth's atmosphere, and thus the spacecraft's decay rate, as much as on the SZ 2 and SZ 3 missions. Small manoeuvres were noted on 2, 4 and 5 January, and the descent module was recovered after another flight lasting seven days.

Again, like SZ 3, the SZ 4 orbital module performed its first orbital manoeuvre only a few hours after the descent module's return to Earth: on 5 January, the orbit was raised from 331–339 km to 354–366 km. Subsequent manoeuvres were as follows: 9 February to 359–366 km, 2 March to 359–373 km, 17 April to 345–366 km, 22 April to 346–382 km and 27 April to 345–381 km. The orbital module was then allowed to decay from orbit.

Shenzhou 5 (SZ 5)
Launch: 15 October 2003, 01.00 GMT
Mass: 7,790 kg (orbital module 2,000 kg)
Descent module recovery: 15 October 2003, 22.23 GMT
Orbital module decay: 30 May 2004

The official goal for China's first manned space launch was the end of 2005, when that era's Five Year Plan came to an end, but it was clear as 2002 drew to a close that – barring major failures- the first manned flight would come during 2003. When SZ 4 was launched it was clearly stated that this would be the final test flight, and that SZ 5 would carry China's first crew around October 2003.

The number of 'taikonauts' to be carried on the first manned flight was initially unclear. It was known that Shenzhou was capable of carrying three people in the descent module, but it seemed unlikely that the full compliment would fly on the maiden manned mission. On the other hand, it was thought that China might want to launch more than one person on SZ 5, simply because both the United States and Soviet Union launched a single crewman on their debut missions. Therefore, two crewmen was considered to be a distinct possibility. The Chinese themselves were ambiguous concerning the number of people for the first flight, and even in summer 2003 the official People's Daily was saying that it was still to be decided whether to fly one or two people on the maiden flight.

Previous to the flight, the duration of the SZ 5 mission was also uncertain to Western observers. Since the second, third and fourth Shenzhou test flights had all lasted for seven days, and the spacecraft on the fourth mission was said to be identical in all respects to the future manned spacecraft, there was the possibility that SZ 5 might fly for about seven days.

If China had flown two men on a flight lasting for nearly seven days on the first manned mission, then that flight would have virtually accomplished the equivalent of the United States August 1965 Gemini 5 flight which saw two men in a single spacecraft for eight days.

The Chinese decided to err on the side of caution, and in September 2003 it was confirmed that the first flight would carry a single man and last for approximately one day. The launch period was expected to be during the 10 days or so starting 10 October, and it was announced that State Television would be carrying the launch "live".

In early October it was announced that the target launch date would be 15 October and the launch time would be approximately 9.00 am Beijing Time. The four unmanned tests had all been launched during local darkness and recovered during local darkness, the rationale behind this being that it would ease visual tracking of both the launch and re-entry phases. For SZ 5 this rationale was reversed – the launch and landing times were chosen so that both events would come during daylight, although landing would be made very close to local dawn: in the case of an emergency it would be easier to find and recover the descent module during daylight.

As the announced launch date approached, somewhere within the Chinese bureaucracy, nerves were lost. This can be understood when the various parties involved with the Shenzhou programme are considered. Being a major government programme, Shenzhou is run by the People's Liberation Army (PLA). There is also the Chinese government, as well as the scientists and technicians who construct the spacecraft and launch vehicle and plan the missions, and finally, the Chinese media. One can imagine that the PLA was horrified with the thought of live launch coverage when something could go seriously wrong. A few days before the launch it was announced that the launch would be shown on Chinese television shortly after it took place, but not live.

There were also Chinese press reports that the flight would only last for about 90 minutes. These could be quickly discounted because it would have meant a landing far to the west of the normal landing site. The reports, quickly corrected by official Chinese media, were possibly a misunderstanding of the spacecraft entering an orbit which would take 90 minutes to circle the Earth.

Also, the Chinese started to hedge their bets as far as the launch date and time were concerned. Rather than being on 15 October the launch became sometime during 15–17 October, and even the 9.00 am launch time seemed to disappear from media reports. However, the Chinese media continued to build up as if the launch would be taking place as previously announced.

The spacecraft and launch vehicle were delivered to the vehicle assembly building at Jiuquan from Beijing on 8 October, they were integrated and checked, and the roll-out to the main launch pad took place on 11 October.

The complete hangtianyuan group arrived at Jiuquan during the days leading up to launch and all of the members got a chance to practice inside the spacecraft while it stood on the launch pad. Although there were fourteen men in the group, it was known that five unnamed men had been selected for the final SZ 5 training. On the day of launch this had already been reduced to three men: Nie Haisheng, Yang Liwei and Zhai Zhigang. Finally, on the morning of launch, Lt-Col Yang Liwei was introduced to the Chinese media as China's first hangtianyuan, with Nie and Zhai named and identified as his two back-ups.

The Chinese press witnessed the launch, as did local people and sightseers, even though they were kept well outside the launch base. Launch was rapidly reported to the western media about four minutes after the event. The Chinese media quickly showed pictures of the launch, showing the vehicle climbing away with the four main engines on the first stage core and the four strap-on booster engines (one per strap-on) clearly firing.

A major event for the mission was the circularisation of the SZ 5 orbit which was completed 6 h 57 m after launch. China said that post-manoeuvre, the spacecraft was in a 343 km circular orbit, while finally USSSN started to issue orbital data for the spacecraft – showing it to be in a 331–338 km orbit.

The rationale behind this manoeuvre is the planned mission of the SZ 5 orbital module. This carries 40 kg of propellant, allowing manoeuvres totalling 60 m/s after the manned phase of the flight is over. On the SZ 2 to 4 flights the module was left in the 335 km circular orbit by the returning spacecraft. If SZ 5 had not manoeuvred to the same circular orbit then the orbital module would have had to use 60–70 per cent of its propellant to reach the circular orbit, and this would mean that there was insufficient propellant left for the orbital module to remain in orbit for the planned six to eight months.

While in orbit, Yang had the traditional conversations with Chinese political leaders as well as his family, he had two periods of rest and he ate traditional Chinese food, especially prepared for his flight. He also found time to take excellent-quality video images of the Earth from his spacecraft, showing one of the solar panels at the right of the image. It was not clear whether the video was showing one of the orbital module solar

Artist's impression of Shenzhou 1 (CGWC) 1340626

panels ahead of the descent module or one of the instrument module panels behind the module.

As with the previous four unmanned Shenzhou flights, the Yuanwang 3 space tracking ship was located off the west coast of Africa, close to the equator: it would have the honour – as always – of transmitting the command to bring the spacecraft back to Earth. The commands were transmitted to the ship on 16 October at 5.04 am Beijing Time. As Shenzhou 5 rose in the sky as seen from the ship, the orbital module was first separated and then two minutes later the command for retrofire was sent to the spacecraft and executed. As the orbital module continued in solo flight, the manned part of SZ 5 started its return to Earth.

As with the Russian Soyuz descent capsule, after atmospheric entry and separation from the instrument module, a drogue parachute opens from the descent module, is discarded and then a main parachute opens and continues to slow the descent module. The base plate of the heat shield is discarded, exposing four solid-propellant retro-rockets. At an altitude of 1.5 m these fire to cushion the final landing on terra firma. Landing came in the Dorbod Xi area of Inner Mongolia which is located at 41.3°N, 111.4°E: the descent module had landed a mere 4.8 km from the planned touchdown point – remarkably accurate for the first piloted recovery.

Chinese television was showing programmes devoted to the flight, and it was interesting to note that the television transmissions from the recovery crew as shown by CCTV4 had a delay of about 6–7 minutes. This continued until the camera team had reached the spacecraft and the transmissions suddenly switched to "live" once the hatch was opened and Yang was seen sitting in the descent module.

The landing must have been more than a little rough on Yang. After the retrorockets fired the parachute must have toppled the descent module onto one side. The result was that Yang was initially hanging upside-down inside his descent module until he was able to carefully release himself and prepare for the arrival of the rescue team. The flight of SZ 5 appears to have been an unqualified success.

The orbital module was equipped with a photoreconnaissance camera carrying a reported ground resolution of 1.6 m and an electronic intelligence-gathering package. There was also a retinue of scientific instruments on board the module. The module performed its first orbit-raising manoeuvre close to midnight (GMT) of 21–22 October, manoeuvring from a 330–332 km orbit to 339–346 km. Subsequent manoeuvres resulted in the following post-manoeuvre orbits: 3 November – 343–354 km, 12 November – 357–367 km, 24 December – 343–358 km and 31 December – 341–387 km.

Shenzhou 6 (SZ 6)
Launch: 12 October 2005, 01:00 GMT
Mass: 8,040 kg
Descent module recovery: 16 October 2005, 20:33 GMT
Orbital module decay: As of mid-2007 the module was still in orbit. Two small orbital corrections recorded on 28 January and 1 February 2007 maintained the module in a 31-circuit repeating orbit, followed by a return to the standard operating orbit of 346 × 356 km, inclination 42.42°, period 91.56 min on 7 February.

Fei Junlong and Nie Haisheng became the second and third Chinese 'taikonauts' to view Earth from orbit, when they took flight aboard Shenzhou 6 in October 2005, spending nearly five days in space. Initially, the Chinese did not report what activities the 'taikonauts' would be engaged in during their five-day flight, and only vague reports surfaced. Later, it was said that the two performed routine mobility and systems exercises, removing and donning spacesuits, moving between the descent and orbital capsules, and testing the sanitary facilities, in addition to conducting biomedical and material sciences experiments. However, there is speculation that Fei and Nie, as well as Yang, engaged in military Intelligence, Surveillance and Reconnaissance (ISR) and ELINT-related activities and experiments during the mission. Like SZ 5, SZ 6 is thought to have carried a 1.6 m resolution camera in its orbiting module. However, unlike the SZ 5 mission, Fei and Nie moved between the descent and orbital modules, whereas Yang remained in the descent capsule. This fueled speculation that the 'taikonauts' were engaged in some sort of Human-In-The-Loop (HITL).

SZ 6 reportedly underwent a number of improvements and upgrades based upon SZ 5's performance and possibly Yang's feedback. A more modern computer system was included, as well as several comfort enhancements for the 'taikonauts'. Fei and Nie slept in turns in the descent module, in a single sleeping bag.

Thirteen landing sites were selected: the primary site in Inner Mongolia and 12 secondary sites. Further, at least nine ground stations supported the mission, in addition to the four existing Yuanwang ships. The orbital module remained aloft and continued to make small orbital manoeuvres 18 months after launch.

Initially, it was a surprise when the Chinese stated that the second piloted mission would not come until approximately two years after the first manned Shenzhou flight. One may speculate that this is because the Chinese Five-year Plan at the time only called for two piloted flights, and the test programme was so successful that the initial piloted flight came earlier than expected.

The Chinese discussed various scenarios for the SZ 6 flight, considering a flight of up to three hangtianyuan lasting for more than one day. By the end of 2003 these reports had settled to there being two people on the second piloted mission, with the flight lasting about a week.

Shenzhou 7 (SZ 7)
Launch: 2008 (delayed from 2007)
Mass: TBA
Descent module recovery: TBA
Orbital module decay: TBA

SZ 7 will be launched in late 2008, a delay of more than one year beyond the originally expected launch date. The delay was blamed on the time needed to perfect a pressure suit to be used in the EVA. The craft's flight will likely take place around the time Beijing hosts the 2008 Summer Olympics.

Three 'taikonauts' are expected to man SZ 7, and at least one, one-man EVA is scheduled. Yang Liwei, now deputy director of the China Astronaut Research and Training Centre, said the astronauts would perform work outside the capsule such as installing equipment and "tightening screws". Rendezvous and docking manoeuvres are not expected on this mission.

International

Automated Transfer Vehicle (ATV)

Current Status
The first ATV, named Jules Verne, was launched on 9 March 2008 and docked with the International Space Station on 3 April.

Background
The European Space Agency's contributions to the International Space Station were decided in October 1995 by the ESA Council at Ministerial Level. One of the main elements was the Automated Transfer Vehicle, an unmanned vehicle that would be launched by Europe's Ariane 5 and used to perform regular reboost, refuelling and payload supply/removal missions to the Station. In the original plan, the ATV would be operational from 2003, flying up to eight servicing missions to the ISS until 2013 or beyond. The ATV would be a payment in kind for Europe's share of the ISS common operations costs.

The first ATV mission (using a vehicle named 'Jules Verne') was the first rendezvous and docking ever by a European spacecraft, the first automated rendezvous with a space station using optical sensors, and the first ever where the target's attitude motion was compensated for by the chasing vehicle. It was also the first use of the Ariane 5 ES launcher with a re-ignitable upper stage.

In a typical mission, Ariane 5 ES will inject ATV into a 51.6° orbit at an altitude between 200 km and 300 km, 50–150 km below the ISS, depending on the Station altitude at the time. On separation from Ariane, the ATV subsystems are automatically activated, then ATV stabilises its attitude, establishes radio communication with the its Control Centre (ATV-CC) in Toulouse via the NASA TDRSS and ESA Artemis data relay network, and deploys its solar arrays. A sequence of orbital manoeuvres brings ATV to an orbit 10–15 km below the Station, where the natural drift due to orbital dynamics corrects the relative phase angle. When ATV is 200 km behind the ISS, trim manoeuvres position it in exactly the same orbital plane as the Station, still at a slightly lower altitude, so that the relative drift brings them ever closer. When the separation distance is 30 km (point S0), direct ATV-ISS radio communications are established, allowing ATV to initiate relative navigation to the Station using GPS data from both ATV and ISS receivers, processed by ATV.

At point S1, about 20 km behind the Station, the ATV-CC commands ATV to begin its homing manoeuvre, a two-boost transfer lasting 45 minutes. This brings ATV to point S2 on the same orbit and about 3.5 km behind the ISS, still outside the ISS 'Approach Ellipsoid' (an ellipsoid of 4 × 2 × 2 km centred on the ISS centre of mass, with the long axis along the velocity vector, or V-bar). All ATV operations inside the Approach Ellipsoid are 'combined operations' and involve the mission control authorities in Houston, Moscow and Toulouse. S2 is the first stable hold point of ATV with respect to the ISS; ATV stays there for up to 90 min, in preparation for final approach.

Once the Mission Control Centre in Houston gives the go-ahead for approach Initiation, the ATV-CC commands the vehicle to enter the Approach Ellipsoid. The closing manoeuvre brings it to point S3, 250 m behind the ISS. It must not enter the ISS 'Keep Out Zone' – a safety sphere of radius 200 m, centred around the ISS centre-of-mass. If the braking burn fails at S3, then ATV naturally returns to S2 in another half-orbit.

Between S3 and docking, ATV approaches on a 'forced' translation along the ISS velocity vector to the docking port at the aft end of the Russian Zvezda module on the ISS. In addition to using the 220 N thrusters to move towards the ISS, perpendicular thrusters are fired towards Earth to stop the orbit rising. The final automated approach from S3 is performed in several steps from one hold point to the next, each initiated by ATV-CC command.

From S3 to S4, 20 m from docking, ATV stays on the V-bar and controls its motion relative to the ISS based on range and line-of-sight information from the Videometer. Before reaching S4, the sensor mode changes to relative attitude measurement. From then on, ATV's docking port always aligns with that of the ISS Zvezda module, thereby tracking any movements of the ISS. On reaching S41, about 12 m out, ATV performs a last hold for the crew and ground to confirm that all is ready for docking. It resumes the approach on command from the ATV-CC. The crew closely monitors this last part of the approach using information independent of ATV's onboard systems: the Zvezda video camera visual image, and range rate readings from the Russian Kurs rendezvous system on Zvezda.

Jules Verne, the first European Automated Transfer Vehicle, undergoing electromagnetic tests at ESTEC (EADS Astrium) 1343781

The Equipped External Bay of Jules Verne's Integrated Cargo Carrier is rotated to vertical ahead of mating with the spacecraft's Propulsion Module (ESA/CNES/Arianespace/Photo optique video du CSG) 1166930

To ensure proper capture and acceptable docking loads, ATV's docking probe has to meet Zvezda's docking cone within a radius of 10 cm and a lateral velocity of less than 2 cm/s. To meet these conditions, the relative navigation during final approach is based on ATV's optical sensors, with corresponding passive target patterns close to Zvezda's port. While measurements from the Videometer primary sensor are used in the active Guidance, Navigation and Control (GNC) loops to control ATV's motion, information from the Telegoniometer secondary sensor is also provided to the Flight Control Monitoring (FCM) system.

The Videometer delivers range and line-of-sight angles to the GNC system, and, from within 30 m, relative attitude angles, based on triangulation. A diverging laser beam emitted by diodes on the ATV front cone towards Zvezda is returned by a pattern of reflectors, imaged by a CCD. The pattern size provides the range, its position on the CCD yields the line-of-sight angles, and its apparent shape gives information about the relative attitude angles.

The Telegoniometer delivers range and line-of-sight angles to the FCM, based on time-of-flight measurement. Collimated laser pulses from a diode scan the ISS vicinity and are returned by three reflectors on Zvezda. The light-pulse travel time provides the distance, and the positions of two beam steering mirrors give the two line-of-sight angles.

With a total launch mass of up to 20.5 t, ATV is designed to deliver up to 9 t of supplies, payloads, crew items and propellant. During its stay of six months, ATV also provides reboost and attitude control for the Station. On departure, ATV will take up to 6,300 kg of waste (including up to 840 kg of liquid wastes), for destructive re-entry into the atmosphere.

The ATV project involves dozens of companies from 10 European countries under the prime contractorship of EADS SPACE Transportation (France). The European companies include Thales Alenia Space, EADS-Astrium, Oerlikon Space, Dutch Space, Snecma, Alcatel Espacio, Crisa and OHB-System. Eight companies from Russia were also involved under the main contractor, RSC Energia, which built the ATV docking mechanism, the refuelling system and the associated electronics.

The ATV is an aluminium cylinder, 10.3 m long and up to 4.5 m in diameter. The external structure is covered with a foil insulation layer and meteorite protection panels. Four independent, rotatable solar arrays in an X arrangement, 90° apart, extend from the spacecraft. When deployed 100 min after lift-off, the arrays span 22.3 m. They are able to produce an average of 4.8 KW and provide power to rechargeable batteries used when the spacecraft is in Earth's shadow.

It consists of two main modules: the Integrated Cargo Carrier (ICC) and the Service Module. The ICC takes up 60 per cent of the total ATV volume. The Integrated Cargo Carrier itself is designed in two parts: the pressurised section accounts for roughly 90 per cent of the volume. Its front end docks and connects, through an open hatch, to the ISS. At the rear of the ICC is a non-pressurised section, called the Equipped External Bay.

The 48 m³ pressurised section of the ICC has room for up to eight standard racks which are designed with modular aluminium elements to store equipment and transfer bags. It carries the entire re-supply payload to the ISS with a maximum upload capacity of 7,667 kg. The dry cargo, including hardware and personal items, is stored in the pressurised part of the ICC where it is accessible to the ISS crew. The fluid cargo, including propellant, water and gas (air or oxygen or nitrogen) for the Station, is stored in the Equipped External Bay, which contains 22 spherical tanks of different sizes and colours. Fluids are transferred through dedicated pipes to the Station's own plumbing or through manually operated hoses. Air carried in the ATV is released manually from the Cargo Carrier into the Station cabin through the hatch.

The front cone of the ICC accommodates the 235 kg Russian docking system with a 80 cm-diameter hatch that provides the ISS crew access to the ICC, an alignment mechanism and a one metre long extendible probe. During rendezvous with ISS, ATV is the active spacecraft. The docking system enables physical, electrical and propellant connections with the Station. The associated Russian-made electronics are installed on the side of the front racks in the pressurised module.

The 'nose' of the ICC also houses various kinds of rendezvous sensors and manoeuvring thrusters. These include two telegoniometers, which continuously calculate distance and direction from ATV to ISS; two videometers, an image processing system able to compute distance and orientation of the ISS; two star trackers, which are able to recognise constellations in the sky; two visual video targets, used by the ISS crew for visual monitoring of the ATV's final approach; and eight thrusters for attitude control.

Since the ATV is the first European vehicle to perform an automated docking with a crewed space station, very tight safety constraints have been imposed. The approach and docking phase is controlled by the ATV's computers under close monitoring by the teams of ESA, CNES and EADS Astrium (the prime contractor) at the ATV Control Centre at CNES Toulouse, France, as well as the ISS crew inside the Zvezda module. In case of an anomaly, both ends can trigger pre-programmed manoeuvres to hold position, retreat to the previous reference point or escape to a safe distance. The ATV's behaviour is also checked by its own independent Monitoring and Safing Unit (MSU), which uses a separate set of sensors and computers to check that the approach manoeuvre is conducted safely. In case of major anomaly, the MSU can take over command and order a Collision Avoidance Manoeuvre (CAM) through dedicated avionics chains and thrusters.

The ATV Jules Verne was launched by Ariane 5 ES from Kourou, French Guiana, on 9 March. Three days later, it successfully demonstrated its autonomous CAM capability and was cleared for ISS proximity operations. It moved to a parking orbit 2,000 km ahead of the ISS during the visit of space shuttle Endeavour to the ISS. On 29 March, Jules Verne demonstrated its ability to navigate safely from a point 39 km behind the ISS to a stand-off point just 3.5 km away using relative GPS navigation. On 31 March, it conducted a second docking rehearsal, approaching to 11 m from the Station. This included the first demonstration of the critical optical navigation system, using the European-developed Videometer technology.

Loading dry cargo into Jules Verne's pressurised module (ESA /CNES/Arianespace/Photo optique video du CSG) 1166931

Docking was successfully completed on 3 April. The 19-ton ATV manoeuvred from a holding position 39 km behind the 275-ton space outpost and conducted a 4-hour staged approach with several stops at reference points for checks. It autonomously computed its own position through relative GPS (comparison between data collected by GPS receivers both on the ATV and the ISS) and in close range it used videometers pointed at laser retroreflectors on Russia's Zvezda module to determine its distance and orientation relative to its target. Final approach was at a relative velocity of 7 cm/s, with a positional accuracy of less than 10 cm, while both the ATV and the ISS were orbiting at about 28,000 km/h, some 340 km above the Eastern Mediterranean. The ATV's docking probe was captured by the docking cone at the aft end of Zvezda at 16:45 CEST (14:45 GMT). Docking was completed when the hooks closed at 16:52 CEST (14:52 GMT).

The ISS crew opened the hatch to Jules Verne on 4 April 2008 and briefly entered to deliver an air filtering device that would be operated for 8 hours to remove any unwanted gases and small particles of floating debris. On 5 April the crew opened the hatch once more to turn on the internal lights and dismount the air scrubber. They installed portable breathing apparatus, a fire extinguisher and handrails, together with a flexible hose for additional ventilation.

Jules Verne will become an additional module of the ISS for about four months. The astronauts will enter its pressurised cargo module and retrieve 1,150 kg of dry cargo, including food, clothes and equipment as well as two original manuscripts handwritten by Jules Verne and a 19th century illustrated edition of his novel "From the Earth to the Moon". In addition, they will pump 856 kg of propellant, 270 kg of drinking water and 21 kg of oxygen into Zvezda's tanks. The ATV can carry about three times as much payload as Russia's Progress freighters but on this mission, most of it is propellant to be used by the ATV's own propulsion system for periodic manoeuvres to increase the altitude of the ISS in order to compensate for the natural decay caused by atmospheric drag. If required, the ATV will also be able to provide redundant attitude control to the ISS or perform evasive manoeuvres to move the Station out of the way of potentially dangerous space debris. The first of Jules Verne's reboost manoeuvres was scheduled for 21 April.

Specifications

Jules Verne (ATV-1)

Launch: 04:03 GMT on 9 March 2008 by Ariane 5 ES from Kourou, French Guiana.

Orbit: 300 × 300 km, 51.6°, initial transfer orbit

Principal applications: Orbit reboost, refuelling and payload supply/removal missions to ISS.

Programme cost: 2 billion euro (USD2.93 billion)

Contractors: EADS Astrium (prime). RSC Energia (ATV docking mechanism, refuelling system and associated electronics). Thales Alenia (Integrated Cargo Carrier, passive thermal protection control, power control unit, TTC communication equipment). MT Aerospace AG (thruster module bearing structure, fuel tanks, high-pressure gas tanks and water tanks). Oerlikon Space (ATV propulsion module structure, payload racks, upper stage separation mechanism). Dutch Space (solar arrays). Snecma (200-N bipropellant engine). Aerojet (490 N engines). EADS Sodern (videometers, star trackers).

Artist's impression of an ATV closing on the International Space Station, prior to docking with the Zvezda module (ESA – D. Ducros) 1121434

Flight longevity: Up to six months.
Configuration: Pressure shell: Al – 2219. Micrometeoroid and Debris Protection System: primary bumper: Al-6061-T6; secondary bumper: Nextel/Kevlar blankets. Internal structure (racks): Al-6061-T6. Thermal insulation: goldised Kapton multi-layer insulation blanket and aluminised beta cloth.
Payload: Eight payload racks with 2 × 0.314 m³ and 2 × 0.414 m³: each 1.146 m³ in front of four of these eight racks.
Nominal cargo mass: dry cargo: 1,500–5,500 kg; water: 0–840 kg; gas (nitrogen, oxygen, air, two gases/flight): 0–100 kg; ISS refueling propellant: 0–860 kg (306 kg of fuel, 554 kg of oxidiser); ISS re-boost and attitude control propellant: 0–4,700 kg. Total cargo upload capacity: 7,667 kg.
Actual cargo mass (Jules Verne): 1,150 kg of dry cargo, 856 kg of propellant, 270 kg of drinking water, 21 kg of oxygen.
Dimensions: Length: 10.269 m (probe deployed), 9.794 m (probe retracted), max. diameter: 4.48 m. In-orbit span 22.281 m.
Mass: 19,357 kg at launch. Dry mass: 10,470 kg. Vehicle consumables: 2,613 kg
Propulsion: 4 × 490 N thrusters (main propulsion); 28 × 220 N thrusters for attitude control – both pressurized liquid bi-propellant system using MMH and nitrogen tetroxide; helium pressurant at 31 MPa.
Communications: To ground: S-band via TDRS; ATV to ISS: S-band via proximity link.
Power: Four independent, rotatable solar arrays of four panels each, silicium solar cells on 4 Carbon Fibre Reinforced Plastic Sandwich panels. Total power output 4,800 W after six months. 40 Ah rechargeable batteries. Required power: <400 W dormant mode; supplied by ISS <900 W active mode.
Habitable volume: 48 m³

International Space Station (ISS)

Current Status

As of April 2007, Expedition 15 crewed the ISS, replacing Expedition 14 that same month. Expedition 15 launched on 7 April 2007; Charles Simonyi, the fifth Space Tourist, accompanied Expedition 15 aboard the Soyuz flight to the ISS, and returned with the Expedition 14 crew on 21 April.

The Space Shuttle Atlantis flight STS-117, also ISS assembly flight 13A, was originally scheduled to launch in March 2007, however it was delayed due to the spacecraft's exposure, on the launch pad, to a hailstorm, and the damage that was caused to its external fuel tank as a result. Atlantis must undergo repairs before it can launch. The damage has caused further Shuttle launch delays, and consequently, slips in the ISS assembly schedule. Additionally, Sunita Williams, an Expedition 14 crew member, must extend her original tour length due to the shuttle delay as well. She is expected to return in June 2007, in keeping with the revised Atlantis STS-117 mission schedule. Six ISS assembly missions have been subject to launch delay due to the storm and hail damage caused to Atlantis: STS-117; STS-118; STS-120; STS-122; STS-123; STS-124. The delays also affect ESA and JAXA schedules, as ESA's Columbus Laboratory Module and JAXA's Kibo Module components are scheduled to fly on these missions (see the ISS assembly sequence, below).

The largest space station yet built, the ISS has been a technological and co-operative triumph as well as a budgetary burden for the participating nations, especially the US. The ISS suffered repeated delays and cost overruns, long before the 2003 Columbia disaster, which, due to Shuttle groundings, directly contributed to further ISS assembly postponements. Following the Columbia accident, and previous to the Shuttle's return to flight mission in 2005, the ISS was only manned by a crew of two (instead of three) and was serviced only by Russian Soyuz and Progress flights.

As of April 2007, previous to the STS-117 flight, the ISS weighs 213,843.4 kg and is about 50 per cent complete. The participating space agencies have set a target completion date of 2010. This target could be optimistic if past and present delays are any indication of the pace to be realistically expected for the next three-plus years. In total, more than 80 flights will be required to assemble, outfit and service the ISS to its completion date.

NASA's 2007 ISS Independent Safety Task Force Report expressed grave concern regarding damage that may be caused to the station by orbiting debris. The ISS can, of course, be moved to avoid debris that is identified in advance as a possible threat. However, the generation of further fragments, such as the debris which the anti-satellite (ASAT) test, conducted by China in January 2007, created is considered most unwelcome by space faring nations. China has expressed interest in becoming an ISS participating nation, however it has so far been unable to overcome the political obstacles that prevent it from doing so. The Chinese announced some years ago that they intend to build their own orbiting laboratory.

Background

Following years of delay and postponements, ISS assembly began on 20 November 1998 with the launch of the Functional Cargo Block (FGB) Zarya by a Proton launch vehicle from Baikonur. Known as Flight 1A/R, it put the first element of the ISS into space. It came almost 15 years after President Reagan announced the goal of operating a permanently manned space station, 12 years after the launch of Russia's Mir station and nine years after the Russians began permanent habitation of their laboratory in orbit. The second ISS flight, 2A, was launched on 3 December 1998 when the Shuttle Endeavour (STS-88) carried the Node 1 Unity module to a docking with Zarya.

The third element, Russia's Service Module (SM), was modelled after the Mir core module; the basic structure comprised DOS-8, which would have been the core module of Mir 2, which was never built. Known as Zvezda, the ISS Service Module was launched by Proton from Baikonur on 12 July 2000. It docked to the aft end of the FGB, which serves as a connecting and propellant module between Unity and the Service Module. Unity is the connecting link to US, Japanese and European elements while the Service Module is the strongback for Russian elements.

On 31 October 2000, the Soyuz TM-31 spacecraft launched to the ISS from Baikonur carried the Expedition 1 crew, comprising Russian cosmonauts Yuri Gidzenko and Sergei Krikalev, and NASA astronaut William Shepherd. TM-31 docked to the Zvezda port on 2 November, a day after Progress M1-3 vacated that position and de-orbited over the Pacific Ocean, replaced with Progress M1-4 launched 16 November and docked at Zvezda two days later.

The STS-98/5A assembly flight was launched on 7 February 2001. On 10 February, the Shuttle Remote Manipulator System (RMS) was used to unberth PMA-2 from Unity and attach it to the Z1 truss at the start of the first mission EVA. The Destiny US laboratory module was removed from the cargo bay and docked to Unity during this EVA, followed by the re-location of PMA-2 to the Destiny end port. Two more EVA's were conducted before Atlantis returned to Earth on 20 February. On 24 February, the ISS Expedition 1 crew entered Soyuz TM-31, undocked from Zvezda and re-docked to the -Z port on Zarya. Two days later Progress M44 was launched and docked to the vacated -Y port on Zvezda on 28 February 2001.

Assembly flight STS-102/5A.1 was launched 8 March 2001 with the Leonardo logistics module and docked to the USS PMA-2 port two days later. Known as the Expedition 2 crew, cosmonaut Yuri Usachev and NASA astronauts James Voss and Susan Helms, replaced the Expedition 1 crew. The Leonardo module was berthed at Unity's nadir port on 12 March and returned to the shuttle before Discovery landed back at KSC on 21 March 2001. Two EVAs had been conducted.

For further information on ISS assembly missions, please see the Assembly Sequence table below. Please also see the Human Spaceflight and Space Shuttle tables in the Space Logs section for further information on specific human spaceflight missions.

Programme History

The evolution, final configuration and Russian involvement planning processes in the US International Space Station all came to a conclusion in 1994. NASA had named the assembly Alpha until then, but now simply calls it the International Space Station (ISS). 'Alpha' normally meaning first, followed by a decade the permanent manned presence in space that the Soviet Union pioneered.

President Reagan directed NASA to begin development of a permanent Space Station, crewed by six to eight astronauts, in his January 1984 State of the Union Address. Echoing Kennedy's May 1961 Moon landing pronouncement, he proposed that it should be in orbit 'within a decade' and should constitute a bold and imaginative programme to maintain US space leadership into the 21st century. The ISS began as an eight-year, USD8 billion project requiring possibly seven Shuttle flights for initial assembly. NASA administrator James Beggs was requested to discuss participation with European nations, Canada and Japan.

Almost immediately, NASA's various centres suggested methods of building the Station. Johnson Space Center's (JSC) Space Station Office advocated an evolutionary facility with 'modular add-ons'. Tasks would include servicing of satellites in low orbits and refuelling of Orbital Transfer Vehicles for GEO satellites. JSC also advocated design of the station as a stepping stone towards a lunar base for mining, military surveillance and communications, and perhaps also for hazardous industrial activities. NASA emissaries seeking international support in Europe in 1983 noted that both polar and LEO stations would be required by 2000, by which time returns from the stations would be worth USD2.4 billion annually.

In its first incarnation, the Station would be placed in a 500 km, 28.5° orbit, with five modules,

The ISS as seen from Shuttle Discovery flight STS-116, also ISS assembly flight 12A.1, in December 2006 (NASA) 1133889

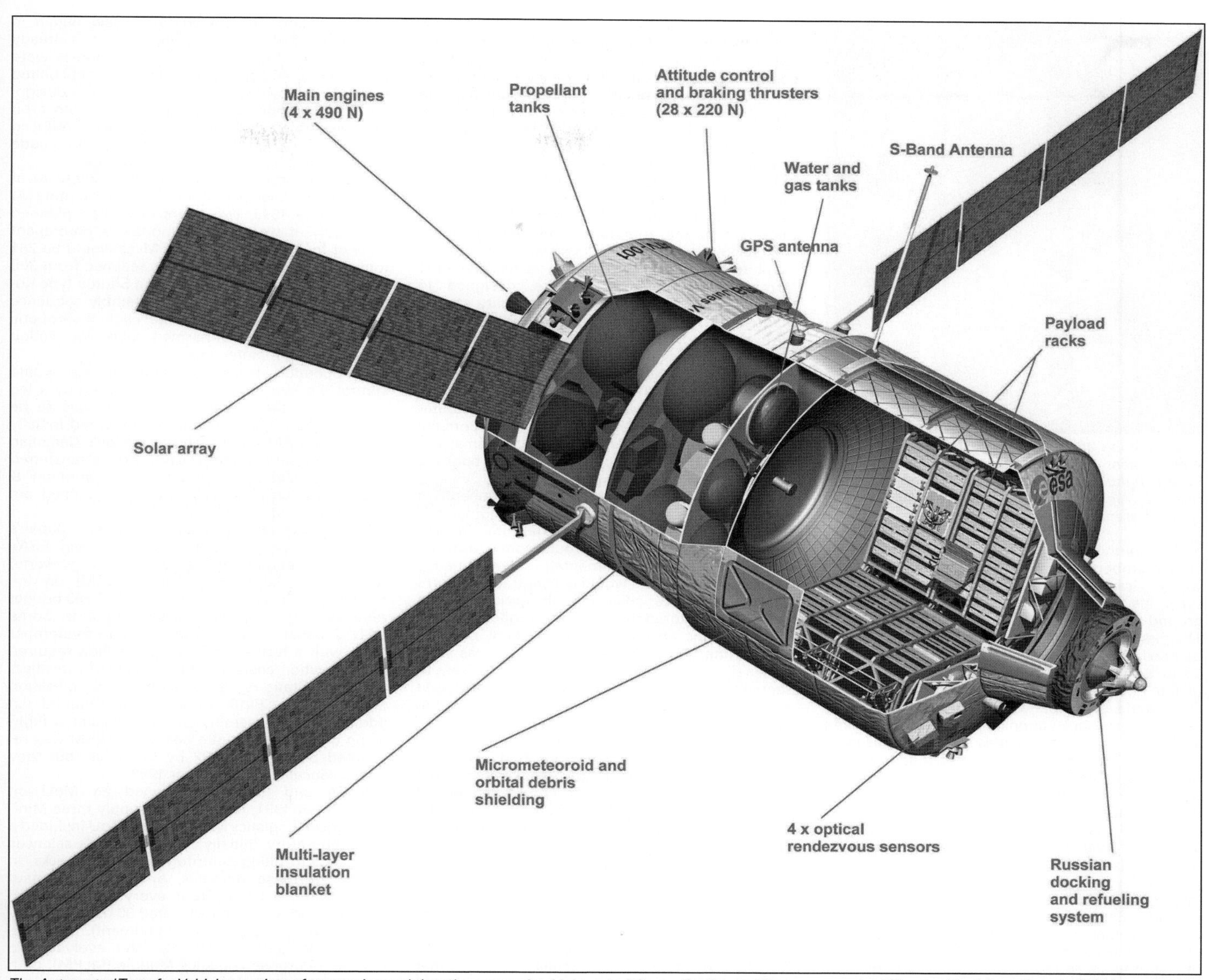

The Automated Transfer Vehicle consists of two main modules: the pressurised Integrated Cargo Carrier and the unpressurised Service Module at the rear (ESA) 1343785

The ATV makes an automated rendezvous and docking with the Russian Zvezda module on the ISS by analysing images of its emitted laser beams returned by passive retroreflectors (ESA) 1343782

Space Station after 2002 completion of the original Alpha plan. Russia's elements are clustered to the right; the only point of contact with the rest of Alpha is the node's docking adaptor (John Frassanito & Associates/NASA) 0517129

each requiring three to four hours work per day by crew members. A two-year trial by two crewmen was envisaged in the 1990s, to validate closed-cycle life support systems, ultimately enabling ground supply requirements to be much reduced. The plan specified that another 10 modules would be needed by 1996, with station mass rising from 36 to 94 tonnes by 2000. Included in the USD8 billion price tag were two free-flying platforms. One would accompany the Station, undertaking astrophysics observations, materials processing and other research activities which, if based in the main Station, could be disturbed by crew movements. The second, Shuttle-serviced in a 98° Sun-synchronous polar orbit, would perform Earth observation.

Grumman, under a contractor study, advocated a Power Tower concept, which was eventually accepted by NASA because of the advantages provided by assembling the Station around a 120 m-long central tower. By keeping one end nadir pointing, Earth viewing occurred simultaneously with celestial-viewing. Grumman placed the power generation system between the two sets of instruments, so that it did not obscure the view of either. A massive cross-arm supported either eight solar panels on gimballed joints so that they faced the Sun for about 60 per cent of each orbit; or alternatively, in the solar dynamic concept, four dish reflectors. Further down, above the modules, a pair of large thermal radiators constantly orientated the system to avoid solar illumination as they discharged waste heat. Tether experiments could be simply accommodated, because both Tower and tether orientation stay approximately constant in relation to Earth.

Furthermost from the power system, the Tower offered a large clear area to accommodate the approach and departure of Shuttle Orbiters or Orbital Manoeuvring Vehicles (OMV) and their payloads. OMVs flew in and out of the lower end for removal and servicing by the Remote Manipulator System (RMS). The tower's 2.4 m^2 truss structure provides a convenient track for RMS, as well as supplying a versatile storage area. Finally, the Tower configuration was attractive because it was not sensitive to mass changes from additional modules. By using gravity gradient stabilisation, it inherently accounted for mass changes and therefore did not place tight constraints on the locations of payloads or upon the direction of growth. Attitude control gyros could easily compensate for any stability loss.

Then NASA abandoned the Power Tower approach in October 1985 in favour of the Dual Keel concept. Largely based on Lockheed and McDonnell Douglas designs, it provided the users with a better microgravity environment (10^{-5} g for all modules) by housing the modules centrally. The plan also increased usable area on the structure for attaching external payloads, allowed better pointing accuracy due to the stiffer structure, and reduced traffic through the lab modules. This radically changed plan, announced on 14 May 1986 (and modified again, four months later), was the result of criticisms of the Power Tower expressed in both an internal report and by a group of external engineers and retired astronauts. According to them, the Power Tower design paid insufficient attention to the needs of science and commercial users; in addition, the arrangement of the crew modules, mounted end-to-end in a racetrack pattern, offered poor safety. Individual modules could not be sealed off in emergency – for instance, if punctured by debris. The new design had the modules placed in parallel, and connected with external airlock hubs, enabling any one module to be sealed off without affecting the others. It also provided more room inside the modules.

The design was based on two vertical keels 110 m long, connected by upper and lower horizontal booms 44.5 m long. The central transverse boom upon which the modules would be mounted was 153.3 m long and the truss would consist of graphite/epoxy tubes 5.4 cm in outer diameter, forming a 5 m^2 open structure, designed to provide maximum stiffness for stability during manoeuvres. Astronauts using assisting fixtures during EVAs would erect it. The upper/lower booms were designed for mounting astronomical instruments at the top, and Earth sensors at the bottom, with the central transverse boom supporting the pressurised modules. The outer ends of the transverse boom would support the power generation and heat rejection systems, and Canada's mobile RMS would traverse the boom to reach required points of operation.

The new station needed two power generation systems. One was the well-proven Si photovoltaic system, providing 37.5 kW of the total 87.5 kW required. The other 50 kW would come from the newer, and still experimental, solar dynamic system. Large mirrors would concentrate the Sun's energy at a receiver adjacent to the generator, with energy being stored thermally in a molten salt. The solar dynamic collectors must be kept pointing towards the Sun within about 0.1°, whereas solar panels do not have to track the Sun so precisely. In September 1986, the surrounding truss was stripped away to cut costs, leaving the central beam and the configuration in basically the current form.

Before the project was even three years old, NASA began to experience its first real cost overruns. That year NASA spokesmen told Congress the Station cost to completion would be USD21.5 billion, with USD14.5 billion paid by the US, excluding launch costs. By 1989, Station development and initial operations costs by one estimation would be USD24.7 billion at 1989 rates; with launch costs added, the total was around USD30 billion. NASA projected total US cost through FY99 in 1990 at USD38.3 billion. NASA dropped this figure to USD30 billion for its restructured programme unveiled in March 1991. The US Government Accountability Office (at that time the US General Accounting Office) estimated costs in April 1991, at USD40 billion, followed by USD78 billion up to 2027.

A major programme review held in the Summer of 1989, under threats from Congress of budget cuts, stated the first element would be in place by March 1995. The Man Tended Configuration (MTC) milestone slipped by five months to April 1996, Permanent Manned Configuration (PMC) by seven months to July 1997 and assembly completion after 29 Shuttle launches by one and a half years to August 1999. NASA issued a new baseline assembly sequence in November 1989, requiring 387 elements to be delivered over four and a half years.

System modifications made as a result of 1989's review included changing from a hydrogen/oxygen propulsion system, involving 4 kW electrolysis of Shuttle waste water, to a modular hydrazine approach. The hybrid AC/DC system was switched to an all-DC network. Development of hard high pressure EVA suits was postponed indefinitely in favour of current Shuttle models. It was decided that various subsystems, such as the closed loop environmental control system, data management, communications and tracking would be incrementally brought to full capability as assembly progressed. It was expected that only 37.5 kW would be available by the time PMC was attained in July 1997. The active thermal control system of fluid/gas heat pipes was replaced by a passive system of body-mounted radiators.

The mass ceiling set for the November 1989 assembly sequence called for a 232 tonne station (including 87 ton modules, 58 tonne framework, 39.5 ton power system), with the basic power requirement of 45 kW leaving 30 kW for experimenters. The mass projection peaked at 298 tons in June 1990 (and power at 61 kW), before changing the storage tanks' stainless steel to Inconel alloy yielded 283 tons the following month. Other measures reduced the power level to 56 kW.

In Autumn 1990, Congress ordered a restructuring of the programme that would reduce expenditure by USD5.7 billion over FY92–96 and peak annual spending from USD4 billion to USD2.6 billion. NASA's 90-day review began in November 1990, encompassing reductions already under consideration. The main US modules were shortened by 40 per cent, and the hexagonal bolted aluminium alloy truss by 50 per cent. In this design, the Shuttle would carry the ISS pieces into orbit fully assembled. Assembly elements were reduced from 122 to 17 – an 18th was later added for a node carrying the 2.5 m life sciences centrifuge.

The goal was to reduce in-orbit activity to at most one EVA weekly and preferably monthly. At PMC in late 1999, two years later than planned previously, it would accommodate a permanent crew of four, instead of eight. Mass would be 281 tons. Data system capacity was reduced from 300 to 50 Mbit/s and now employing a Shuttle-type Ku-band antenna. The simplified assembly sequence resulted in cancellation of the Flight Telerobotic Servicer (FTS), a multi-armed robot for station assembly and maintenance.

An operator viewing via four cameras would control the two manipulators, supported by a leg clamped on the nearest structure. It was to be available from the first mission, mounted initially on Shuttle's RMS and then the station's Canadian Mobile Servicing System. The project was transferred to NASA's Office of Aeronautics, Exploration & Technology and a late 1993 Shuttle test planned, but it was cancelled in the FY92 budget.

Some modules, such as the US lab, Japan's Japanese Experiment Module (JEM) and ESA's Columbus, required the use of the more powerful Advanced Solid Rocket Motor (ASRM) on the Shuttle. It was cancelled from NASA's FY93 budget request in January 1992 on cost grounds. Some USD1.2 billion was allocated through September 1992, with a further USD300-400 million required in termination costs. If it had remained cancelled, then the Station's 26 assembly and utilisation flights before PMC would have required an additional two assembly and one utilisation flight using the current Shuttle boosters. ASRM was re-instated September 1992 by Congress, but they finally cancelled it in October 1993.

NASA and Italy's ASI signed an MoU on 6 December 1991, for Alenia to supply three Mini-Pressurised Logistics Modules. The MoU included a Mini Laboratory, initially dedicated to life sciences with a 2.5 m tilting centrifuge and three racks. In return for these modules, an Italian astronaut would serve a 90-day tour every six years and one user rack would be allocated 90 days annually (in addition to Italy's ESA entitlement). The Mini-Pressurised Logistics Module later evolved into the Multi-Purpose Logistics Module (MLPM), both using the acronym MPLM.

FY93's budget allocation USD150 million shortfall in September 1992 prompted further schedule slips: March 1996 first launch (four-month delay), June 1997 MTC (six months) and June 2000 PMC (nine months). The programme and its management were overhauled and by the end of 1993, Russia had agreed to become a fully-fledged international partner. The arrival of a new administration and concern over costs resulted in President Clinton in February 1993 ordering a restructuring to cut cost to completion to USD9 billion (USD9 billion has already been expended), capping annual spending at USD2.1 billion. After years of struggling to survive annual budget procedures, the possibility of cancellation could still not be ruled out. The restructuring aimed to reduce life from 30 years to 10 years (extendable to 15 years) and permanent occupancy was no longer guaranteed. Much simpler schemes, possibly employing Shuttle- and Spacelab-based stations or cooperating on Russia's Mir, were also examined. The 45-member Redesign Team began work 10 March 1993 and submitted its final report 7 June 1993.

The team proposed three options: *Option A* using existing flight proven hardware and cost-effective *Freedom* systems, *Option B* making maximum use of existing *Freedom* designs, *Option C* using a Shuttle-derived vehicle placing a relatively simple station in orbit on a single launch. The White House announced on 17 June 1993, the President's selection of Option A. The plan based on Option A was submitted 7 September 1993 as the 'Alpha Station'. Cost and schedule projections were added 20 September (the USD9 billion goal could not be achieved, but the administration stuck with the USD17.4 billion projection NASA offered for work from 1994 to 2002, excluding launch costs.

The heads of all the involved space agencies met in Montreal 7 November 1993 to review the outcome of Station's redesign and discuss Russia's participation. The international partners on 7 December formally invited Russia to join the project. Russia's funding problems prompted suggestions in 1995 that Mir should, particularly with the new Spektr and Priroda modules, form the basis for the ISS. NASA declined this late suggestion, insisting that any changes must not alter ISS's cost or schedule. Instead, Russia kept Mir operational until 2001, as a much-needed source of revenue, and NASA added two 1998 Shuttle missions for Mir supplies. The Shuttle was also to deliver most of Russia's Science Power Platform (SPP) in 1999, saving three Zenit flights, however this element was later cancelled, with all participants using a universal power system.

From 1985 to 1993, NASA spent USD10.1 billion on the station. The USD17.4 billion comprises: USD9.1 billion development, USD4.6 billion operations, USD1.0 billion utilisation plus USD2.7 billion for payloads and Mir support. Operations costs 2003–2012 are projected at USD13.0 billion. The United States plans to pay Russia more than USD1 billion for its contributions, including the USD400 million already agreed on for FY94-97, largely encompassing the Shuttle-Mir missions. Upon completion, NASA will own the FGB under a USD190 million agreement signed by Boeing Defense & Space Group with Khrunichev on 15 August 1995 for in-orbit delivery.

Boeing Integrated Defense Systems (Defense and Space Group) received the prime contract for the revised station on 17 August 1993, with the award being USD5.63 billion. Boeing is responsible for delivering the full-up vehicle and for coordinating and integrating the US portion with international elements, as well as the design, development, physical and analytical integration, test, delivery and launch of the vehicle. Under the arrangement, Boeing also handles the first one year of sustaining engineering following launch of each package, including spares.

As prime contractor, Boeing oversees three Product Group subcontractors: first, the McDonnell Douglas, truss, external thermal control, command/data handling, communications and tracking, node/cupola integration, GN&C, pressurised mating adaptor components; second, the Rocketdyne, solar arrays, Ni/H_2 batteries, power management/distribution systems; and lastly, Boeing's own, US lab module, node/cupola primary structures, lab integration, life support systems, international thermal control, internal audio/video, secondary power subsystem.

Assembly was projected to require about 36 Shuttle and nine Russian launches. From the beginning, NASA has had a troublesome time managing the disparate elements of the programme. The FGB was supposed to be on-orbit in November 1997 with the initial Shuttle assembly flight following the next month. However, the Service Module, scheduled to follow FGB by about five to six months, was not finished by then. NASA paid Russia for the manufacturing of the Service Module, but those funds had not been reaching the Khrunichev factory home of the Service Module construction. Because of the lack of funds the work on the Service Module fell behind schedule and it became clear that it would be impossible to meet the scheduled launch in April 1998. As a result it was decided to delay all of the launches for ISS.

Finances and launch delays were not the only problems that ISS suffered during 1996. The first resident crew to the ISS was named as A Solovyov and S Krikalev from Russia and W Shepherd from the United States – the latter being backed-up by D Thomas. Both Russian cosmonauts had gained a great deal of experience on multiple visits to the Mir space station, while only Thomas has any space station flight experience. While Solovyov was not questioned as the Soyuz-TM commander, the United States insisted that Shepherd should become the commander once the crew was on board the fledgling ISS. The Russians became reluctant to cede the commander's duties to anybody without space station experience, a matter eventually taken up by the Russian Duma.

Legal precedents were sought out to back up the Russian claim that Solovyov and Krikalev should not be under the command of a foreigner. Of course, the US is the major financial partner in the ISS programme, but at the time of the flight there would be more Russian units comprising ISS than US ones.

Under pressure from NASA, the Russian FSA removed Solovyov and replaced him with the less experienced Y Gidzenko, while Gidzenko's place on the Soyuz-TM 26 mission to Mir was taken by Solovyov.

In December 1996, ITAR-TASS announced that there were nine Russians in the training group for flights to the ISS. From the Russian Air Force, there were Y Gidzenko, Y Malenchenko, Y Onufrienko, V Dezhurov, V Korzun and S Sharipov. The Energia bureau had supplied N Kuzhelnaya, S Krikalev and M Tyurin. At the time of the report, eight of the nine people were in training at Zvezdny Gorodok (Star City) for ISS missions. Korzun would join the group at a later date since at that time he was in orbit aboard Mir. Despite having all of these people in training, the Russians had not, at the time, scheduled any visits beyond Soyuz-TM 29.

A NASA flight director in Houston controlled ISS, even when only Soyuz had access in the early phase. The physical control in that Russian-only phase was from Kaliningrad, but under direction from Houston. NASA assumed full control when the US Lab was attached. English is the working language.

Initially, three Shuttle Orbiters were modified for station operations. The RMS was recertified for loads up to 265 tons and a remotely-operated power/data connector was carried on one side of the cargo bay for early flights until the pressurised docking system could be used. The existing internal airlock was to have been refurbished and mounted outside, attached to the cabin by a short flexible tunnel. On top would be an ASTP-type androgynous docking unit, with integral power/datalinks, and crew hatch. Three docking units have been built and carried when required.

International Space Station Specifications Upon Completion

Mass: 419,600 kg

Crew size: Six, including four science officers. Crews serve six-month tours and are typically comprised of a Commander and Flight Engineers. Commanders alternate between Russian and US nationals. Spaceflight Participants, or Space Tourists, have also boarded the ISS for much shorter 1–2 week periods

Total pressurised volume: 935 m^3

Orbit: 370–460 km circular at 51.6° inclination; assembled at 390 km, allowing each Shuttle to carry about 1.8 tons more cargo to the ISS. The inclination overflies 85 per cent of Earth's landmass and 95 per cent of population

International Standard Payload Racks (ISPR): 34 total 1.054 m wide × 2.019 m height × 1.50 m^3 usable volume (1.97 m^3 total) ISPR racks. US Lab Destiny: 13; Japanese Experiment Module (JEM): 11; Columbus Module: 10. Both JEM and Columbus share 5 racks each with the US

Microgravity experimentation time per year: Between 180 and 266 days quiet microgravity annually. The Active Rack Isolation System (ARIS) protects payloads from vibration

Configuration: Pressurised modules in T-arrangement with transverse 3.7 × 4.9 m hexagonal truss carrying four end-mounted solar arrays for total 110 m width, 74 m length and 2,500 m^2 solar array span. The US provided two modules, Node 1 (Unity), and the US Lab (Destiny) outfitted as separate habitation and laboratory facilities, and two connecting nodes (one for storage, other for power control); node has six 2 m diameter attachment ports. Node 1 has an attached EVA airlock and a 1.4 m high × 2.2 m diameter cupola with seven windows. The 14.5 ton (24 tons with full payload), 8.5 m long, 4.3 m diameter US Lab has a 1.3 m^2 hatch, 8 latches, and a central 51 cm diameter window 60 mm thick. Russia has provided two major elements: the 13 m long × 4.1 m diameter Zarya Functional Cargo Block (FGB) and the Zvezda Service Module (SM). The 20 ton SM includes galley (oven, fridge/freezer, trash compactor, hand washer, water supply and wardroom (table, 2 × 51 cm windows) at one end and bathroom plus shower in mid section. US modules are principally 2219 aluminium alloy with skin panels featuring external waffle grid.

Canada's Remote Manipulator System (RMS) moves along truss on mobile transporter carrying 17 m long seven-jointed manipulator with 265 ton load capacity, tactile feedback, a machine vision system and, if required, Special Purpose Dexterous Manipulator (SPDM or Dextre) with two 2 m arms. RMS applications include station assembly and servicing, Orbiter docking and undocking, payload manipulation and EVA assistance. Russia provided 19.3 ton FGB, 21 ton Mir Service Module; two modified Soyuz-TM as interim Assured Crew Return Vehicles; science modules; Progress M and further FGBs to deliver propellant. FGB will provide initial AOCS, power and control; after spacecraft docks, its primary function is as a propellant depot/distribution. It also forms a link in the ISS overall power and data distribution system. The Japanese Experiment Module (JEM), or Kibo, is a 10-rack, 11.2 m long 4.4 m diameter internal diameter pressurised module with attached remote manipulator and exposed platform; a 4 m long pressurised Experiment Logistics Module (ELM) will be delivered once annually.

ESA's 6.8 m long × 4.5 m diameter Columbus lab will initially carry a 2.5 ton payload in 10 racks, and has a 9 ton maximum payload capacity for materials sciences, fluid physics and life science research. As its contribution to operations costs, ESA will provide servicing and reboost using its Automated Transfer Vehicle (ATV) launched on Ariane 5; it receives a percentage of ISS resources in return. Italy provided three (Leonardo, Raffaelo and Donatello) 16-rack (including up to five powered refrigerators/freezers) Shuttle-launched Multi-Purpose Logistics Module (MLPM): 9 ton payload, 4 ton empty, 6.4 m long, 4.5 m diameter.

Electrical Power System (EPS): 110 kW, including 30 kW average for payload. Single US array

Zarya and Unity modules viewed from STS-88
0062157

Space Station configuration after the US Lab module and Canada's RMS are attached (at right). Although occupation was possible from May 1998, the main solar wing was not due to be attached to the US node until September 1998 (NASA)
0517130

International Space Station Assembly Sequence

Date	Flight	Launch vehicle	Element(s)
20 November 1998	1A/R	Russian Proton	Zarya Control Module/Functional Cargo Block (FGB)
4 December 1998	2A	US Orbiter STS-88	Unity, Node 1 (1 Stowage Rack) 2 Pressurized Mating Adaptors attached to Unity
27 May 1999	2A.1	US Orbiter STS-96	Spacehab – Logistics Flight
19 May 2000	2A.2a	US Orbiter STS-101	Spacehab – Maintenance Flight
12 July 2000	1R	Russian Proton	Zvezda Service Module (SM)
8 September 2000	2A.2b	US Orbiter STS-106	Spacehab – Logistics Flight
11 October 2000	3A	US Orbiter STS-92	Integrated Truss Structure (ITS) Z1 Pressurized Mating Adapter-3 Ku-band Communications System Control Moment Gyros (CMGs)
31 October 2000	2R	Russian Soyuz	Soyuz **Expedition 1 Crew**
30 November 2000	4A	US Orbiter STS-97	Integrated Truss Structure P6 Photovoltaic Module Radiators
7 February 2001	5A	US Orbiter STS-98	US Destiny Laboratory Module
8 March 2001	5A.1	US Orbiter STS-102	Logistics and Resupply; Lab Outfitting Multi-Purpose Logistics Module (MPLM) carries equipment racks
19 April 2001	6A	US Orbiter STS-100	(Lab outfitting) Ultra High Frequency (UHF) antenna Space Station Remote Manipulator System (SSRMS)
12 July 2001	7A	US Orbiter STS-104	Joint Airlock High Pressure Gas Assembly
10 August 2001	7A.1	US Orbiter STS-105	MPLM
14 September 2001	4R	Russian Soyuz	Docking Compartment 1 (DC-1) Strela Boom Delayed from February 2001
5 December 2001	UF-1	US Orbiter STS-109	MPLM Photovoltaic Module batteries Spares Pallet (spares warehouse)
8 April 2002	8A	US Orbiter STS-110	Central Truss Segment (ITS SO) Mobile Transporter (MT)
5 June 2002	UF-2	US Orbiter STS-111	MPLM with payload racks Mobile Base System (MBS)
7 October 2002	9A	US Orbiter STS-112	First starboard-side truss segment (ITS S1) with radiators Crew & Equipment Translation Aid (CETA) Cart A
23 November 2002	11A	US Orbiter STS-113	First port-side truss segment (ITS P1) Crew & Equipment Translation Aid (CETA) Cart B
1 February 2003	N/A	US Orbiter STS-107	**Shuttle Columbia re-entry accident** Shuttle grounding contributes to a further 2.5 year ISS assembly delay
26 July 2005	LF1	US Orbiter STS-114	Space Shuttle return to flight Supply delivery External Stowage Platform installation Control Moment Gyroscope replacement
4 July 2006	ULF1.1	US Orbiter STS-121	MLPM Mobile transporter maintenance Return to 3 crewmember capacity
9 September 2006	12A	US Orbiter STS-115	Second port-side truss segment (ITS P3/P4) Solar array and batteries
9 December 2006	12A.1	US Orbiter STS-116	Third port-side truss segment (ITS P5) Spacehab supply delivery
TBD 2007	13A	US Orbiter STS-117	Second starboard-side truss segment (ITS S3/S4) Solar array set and batteries (Photovoltaic Module)
TBD 2007	13A.1	US Orbiter STS-118	Spacehab Single Cargo Module Third starboard truss segment (ITS S5) External Stowage Platform 3 (ESP3)
TBD 2007	10A	US Orbiter STS-120	**US Core complete upon conclusion of this mission** US Node 2 Sidewall – Power and Data Grapple Fixture (PDGF)
TBD 2007 or 2008	ATV1	Ariane 5	European Automated Transfer Vehicle (ATV) Supply vehicle first launch
TBD 2007	1E	US Orbiter STS-122	Columbus European Laboratory Module Multi-Purpose Experiment Support Structure – Non-Deployable (MPESS-ND)
TBD 2008	1J/A	US Orbiter STS-123	Kibo Japanese Experiment Logistics Module – Pressurized Section (ELM-PS) Spacelab Pallet – Deployable 1 (SLP-D1) with Canadian Special Purpose Dexterous Manipulator, Dextre
TBD 2008	1J	US Orbiter STS-124	Kibo Japanese Experiment Module Pressurized Module (JEM-PM) Japanese Remote Manipulator System (JEM RMS)
TBD 2008	15A	US Orbiter STS-119	Fourth starboard truss segment (ITS S6) Fourth set of solar arrays and batteries
TBD 2008	ULF2	US Orbiter STS-126	MLPM
TBD 2008	3R	Russian Proton	Multipurpose Laboratory Module with European Robotic Arm (ERA)
TBD 2009	2J/A	US Orbiter STS-127	Kibo Japanese Experiment Module Exposed Facility (JEM EF) Kibo Japanese Experiment Logistics Module – Exposed Section (ELM-ES) Spacelab Pallet – Deployable 2 (SLP-D2)
TBD 2009	HTV-1	Japanese H-IIB	Japanese H-II Transfer Vehicle (HTV) Supply vehicle first launch
TBD 2009	17A	US Orbiter STS-128	**ISS six crew capability upon mission completion** MPLM Lightweight Multi-Purpose Experiment Support Structure Carrier (LMC) Three crew quarters, galley, second treadmill (TVIS2), Crew Health Care System 2 (CHeCS 2)
TBD 2009	ULF3	US Orbiter STS-129	EXPRESS Logistics Carrier 1 (ELC1) EXPRESS Logistics Carrier 2 (ELC2)
TBD 2009	19A	US Orbiter STS-130	MPLM Lightweight Multi-Purpose Experiment Support Structure Carrier (LMC)
TBD 2010	ULF4	US Orbiter STS-131	EXPRESS Logistics Carrier 3 (ELC3) EXPRESS Logistics Carrier 4 (ELC4)
TBD 2010	20A	US Orbiter STS-132	Node 3 with Cupola
TBD 2010	ULF5	US Orbiter STS-133	**ISS Assembly Complete upon mission conclusion** EXPRESS Logistics Carrier 5 (ELC5) EXPRESS Logistics Carrier 1 (ELC1)
TBD 2010	9R	Russian Proton	Research module

Notes: Automated resupply and crew transfer missions are not listed.
Following ISS completion the Space Shuttle will be retired and the ISS will be supported by the Orion Crew Exploration Vehicle, and Commercial Orbital Transportation Services (COTS) as well as other spacecraft.
The ISS assembly sequence is subject to frequent change.

provided 23 kW during initial assembly, in addition to the small Russian arrays. Four photovoltaic Si solar arrays with a total 262,400 solar cells. Each thermal cover 9.8 × 29.3 m. Supported by six Ni/H_2 batteries on each array module. All DC system: 160 V primary, 124 V secondary.
AOCS: Initially by FGB then by Service Module (SM), plus ATV reboost. FGB stores 6.14 tons UDMH/NTO in eight external cylinders, refuelled via SM. Attitude determination by US, principally by GPS receivers.
Communications: Data, audio and video via Tracking and Data Relay Satellites (TDRS); 150 Mb/s Ku-band simultaneous same-rate uplink and downlink; 72 kb/s S-band uplink and 192 kb/s S-band downlink. UHF and Ham Radio capabilities also
Thermal: 42 1.8 m × 3 m liquid ammonia radiator panels totalling 312 m^2
Command and Control: From NASA's Johnson Space Center, Houston; 20-console Flight Control Room 1, upgraded in 2006; Blue Flight Control Room and Shuttle Flight Control Room formerly used (1998–2006). Working language is English.

Russian Federation

Kliper (Clipper)

Current Status
RSC Energia updated its Kliper design and rollout phases in 2006. Similar to NASA's Orion (see separate entry), Kliper was meant to replace the Soyuz as the ISS's Expedition crew transfer vehicle, after the ISS is complete. ISS assembly completion is targeted for 2010. However the Russian Federal Space Agency (FSA) cancelled its 2006 tender, did not selected the Kliper concept (or any competing concept), and the spacecraft programme has not been funded.

FSA, or Roscosmos, and ESA confirmed at the 2006 Farnborough Air Show, that the two agencies would work together to develop future space transportation systems. The agencies will partner on the Advanced Crew Transportation System (ACTS); this project decision comes after ESA initially turned down financing and participation on the Kliper project. ACTS may provide ESA with an opportunity to contribute more toward a transportation programme and perhaps later allow for a reassessment of its participation in an evolved Kliper programme. Phase one of the ACTS programme involves substantial Soyuz upgrades in preparation for European-Russian human-rated lunar exploration missions.

Energia has not yet given up on the Kliper concept. It may resurface if funding is secured, or it may evolve as a late-phase spacecraft within the ESA-FSA partnership framework.

Background
The reusable Kliper spacecraft was RSC Energia's and FSA's parallel proposal to NASA's Orion Crew Exploration Vehicle (see separate entry). Originally intended to fly piloted missions to the International Space Station (ISS – see separate entry) by about 2010, Kliper was to replace the Soyuz (see separate entry) series, the most travelled craft type in the history of human space flight. FSA projected that, in the future, Kliper could journey to the Moon, and perhaps also be included in missions to Mars.

RSC Energia (see separate entry), Russia's largest spaceflight concern, has managed Kliper (also known as Clipper) development since 2000. Kliper design has undergone several iterations, including the latest proposal for the winged lifting body spaceplane design, capable of carrying up to six crew members and passengers. Sukhoi contributed to the spaceplane's development. A model of the winged design appeared at the 2005 Paris Air Show in Le Bourget as well as the 2005 Moscow Air Show.

Although Kliper has struggled to secure funding, FSA initially included the craft in Russia's space programme. Projected cost for the vehicle was RUR10 billion up to 2015. Russia had previously proposed a partnership with the European Space Agency (ESA – see separate entry), with the additional possibility of the use of European launch facilities. The inability of the US Space Shuttle (see separate entry) to ferry personnel and equipment

The Kliper mock-up on display at the 2005 Paris Air Show, Le Bourget (Jane's/Patrick Allen) 1144581

to the ISS between 2003 and 2005 perhaps made the proposal attractive to ESA. However after consideration, in December 2005, ESA member states rejected European funding participation for Kliper trials. ESA's 2005 decision does not, however, bar the agency from future Kliper programme participation.

Russia has also sought partnership with Japan. Although the Japanese Aerospace Exploration Agency (JAXA) has yet to make a final Kliper funding decision, after ESA's 2005 pullout, JAXA may be reluctant to pursue any involvement in the Kliper programme.

The 2005 Kliper design dictates that it be launched from a fixed facility using a heavy-lift launch vehicle such as the proposed Soyuz-2-3 or Angara-3. FSA's RUR10 billion estimate did not include any launch vehicle development costs that will be necessary for launch vehicle upgrade or modification programmes. Neither the Soyuz-2-3 nor the Angara-3 will be developed to the point where either could be used to launch the Kliper near the 2010 ISS completion date.

The 2005 Kliper was modular in concept, consisting of separately launched segments. The Kliper reusable crew re-entry vehicle was designed to be launched along with an expendable service module designed for ISS linkage. Additionally, the Parom, an unmanned module acting as a space tug, could be launched earlier, in anticipation of a Kliper docking and subsequent travel to the ISS. Still planned, the Parom could provide additional habitational and cargo facilities as well as further manoeuvrability capabilities. Parom is also a re-usable vehicle and would remain with the ISS until required for additional missions.

Kliper may be instrumental in Russia's efforts to develop its space tourism industry. The craft was designed to carry three more passengers than Soyuz and has added cargo capabilities.

Progress Series

Current Status
The Progress M1 has continued to reliably service the International Space Station (ISS) since 2000. Typically taking two days after launch to reach the station, Progress vehicles can remain docked for up to six months.

M-47, the first Progress cargo tanker launched after the loss of Shuttle Columbia on 1 February 2003, lifted off on 2 February, on a planned logistics flight; it docked with the ISS on 4 February. Between February 2003 and the Space Shuttle's return to flight in July 2005, the Russian Progress and Soyuz were the only spacecraft to service the ISS.

Background
The first Progress (11F615A15, the A15 mod version of the 11F615 Soyuz class vehicle) was launched in 1978, followed by the upgraded 11F615A55 Progress-M version vehicle in 1989. Vehicle designation for the basic Progress was 7K-T9 and for the Progress-M, 7K-T9M. The latest version, the Progress-M1, is officially designated 7K-TGM1, and was developed for use with the ISS.

The constant demand by the Salyut 6 and 7 stations and their Mir successor for consumables sparked the Soviet Union to build an automatic resupply vessel. RSC Energia designed and constructed the Progress vessel based on its manned Soyuz vehicle. The Soviets replaced the Soyuz manned descent module with a compartment for transporting up to 940 kg of propellants, other liquids and compressed gases. The Progress design features a fully autonomous docking port with pipes to carry the consumables into the station.

The orbital module accommodates up to 1.8 tons of dry, removable cargo. It has been used to carry experimental equipment, food, film and air regeneration cylinders. For the return trip, the crew fills the module with ISS waste materials. A Progress may also raise the host station's orbit to combat atmospheric drag. For re-entry, ground controllers command the ferry to undock and retrofire and then target it to re-enter the atmosphere over Russia for decay along a Pacific corridor. Cargo accounts for about 35 per cent of a Progress launch mass, the high proportion made possible by the absence of escape systems and re-entry thermal protection.

Progress delivered a total of 24.97 tonnes to Salyut 7 by 12 named flights. Kosmos 1669 added 2,254 kg to this total. Mir's first 17 deliveries totalled more than 40 tons, or double the station's original mass. The first Progress design lacked solar panels, which limited it to an independent flight time of at least three days, although the capsule itself remained docked to Mir for two or three months. The vehicle stretches to a slightly greater length than Soyuz because RSC Energia added an additional pressurised instrument section to the rear unit.

Progress-M incorporates Soyuz-TM's Kurs rendezvous and docking system, computer, propulsion unit and solar arrays. Kurs allows forward port docking, whereas the previous Progress could only approach a space station through the aft docking unit.

The last original Progress variant flew between 5 and 27 May 1990. M-7 was the first to suffer docking failure, in March 1991, because of a broken antenna on Mir. Consequently, it became the first to take propellant from a station for orbit raising. M-4 carried an open cargo carrier in place of the refuelling mid-section for the first time.

Russian space technology has made great strides in docking compared to other space faring nations. Manakov Poleshchuk demonstrated this when he undocked Progress M-16 on 26 March 1993, took it out to 70 m and brought it back in for docking. In September 1994, TM-19's Malenchenko brought in the M-24 (with its malfunctioning Kurs) manually from 150 m using TV signals from the unmanned craft. Remote-control docking from the ground may be added, with a cosmonaut in the Flight Control Center controlling Progress using Soyuz-type controls and a TV monitor.

Some Progress's delivered a coni-cylindrical re-entry capsule (VBK: ballistic return capsule) among their cargo. Loaded with 80–150 kg of film and material samples, the capsule, mounted in Progress's hatch by the crew, releases itself after retrofire at about 120 km for recovery by parachute (initially in Kazakhstan's normal manned landing area but beginning with M-18 within Russia to avoid political problems). M5 carried the first of these configurations and landed with at least some crystals among its 113 kg payload. Unfortunately, the M-7's Raduga and 94 kg payload was lost.

The first Progress M1 enhanced cargo ferry was launched to Mir from Baikonur on 1 February 2000. It was deorbited 26 April 2000 and was followed by M1-2, launched 25 April 2000. Progress M1-3 became the first Progress vehicle to service the ISS.

Having re-supplied and replenished three Russian space stations and begun the process of sustaining the ISS, more than 23 years after it was first used, a Progress, M1-5, was used to de-orbit the Mir station on 23 March 2001. Having dispatched Mir, Progress was turned fully and exclusively to support for the International Space Station. Progress M1-6 launched 20 May 2001, was launched by the first Soyuz FG as ISS supply flight 4P, carrying 2.5 tons of food, fuel, water, life-support equipment and spares including redundant computer equipment for the Destiny Module. An adapted version of

The International Space Station photographed on 17 September 2006, by the departing STS-115 crew; Progress M-56 is docked with the Zvezda Service Module, at the top of the image (European Space Agency) 1133890

Progress to ISS Launch and Decay Schedule

Progress Mission:	Launch Date to ISS:	End of Mission Decay Date:
M1-3	6 August 2000	1 November 2000
M1-4	16 November 2000	8 February 2001
M-44	26 February 2001	16 April 2001
M1-6	20 May 2001	22 August 2001
M-45	21 August 2001	22 November 2001
M-S01	14 September 2001	27 September 2001 (engine and instrument module only)
M1-7	26 November 2001	20 March 2002
M1-8	21 March 2002	25 June 2002
M-46	26 June 2002	14 October 2002
M1-9	25 September 2002	1 February 2003
M-47	2 February 2003	28 August 2003
M1-10	8 June 2003	3 October 2003
M-48	29 August 2003	28 January 2004
M1-11	29 January 2004	3 June 2004
M-49	25 May 2004	30 July 2004
M-50	11 August 2004	22 December 2004
M-51	23 December 2004	09 March 2005
M-52	28 February 2005	15 June 2005
M-53	16 June 2005	7 September 2005
M-54	8 September 2005	3 March 2006
M-55	21 December 2005	19 June 2006
M-56	24 April 2006	19 September 2006
M-57	24 June 2006	17 January 2007
M-58	23 October 2006	28 March 2007
M-59	17 January 2007	Expected 2007

For Salyut and Mir dates, please see *Jane's Space Directory* editions 2006–2007 and older, or the online archive.

Progress, the SO1 vehicle, carried the Pirs docking and airlock module instead of standard cargo and fuel sections; it launched on 14 September 2001 and docked to the Zvezda nadir port three days later. The engine and instrument module separated and deorbited on 27 September 2001.

Specifications
Progress
First launch, all variants: 20 January 1978; 19 July 1985 for original Mir design; 23 August 1989 for Progress M (27 September 1990 for M with return capsule); 1 February 2000 for Progress M1 variant
Launch site: Baikonur
Launch vehicle: Soyuz-U and Soyuz-FG
Principal applications: Unmanned LEO cargo ferry to the ISS
Performance: M1 variant can carry up to 3,200 kg total in 6.6 m^3 of cargo space. Up to 1,800 kg dry cargo and up to 1,700 kg fuel, plus another 250 kg of Progress surplus fuel can be transferred to the ISS. Progress-M versions can also transfer their own excess propellant. Independent flight time is 30 days or 3 days without solar arrays, attached to the ISS, a maximum stay of 180 days can be expected
Dimensions: Progress M and M1: 7.23 m length; 2.2 m diameter
Launch mass: M1 variant 7,250 kg
Propulsion: derived from ODU unified propulsion system of Soyuz-T/TM. Hypergolic UDMH/NTO is employed with the 3.1 kN main engine and a network of 14 × 98 N (four back-up) plus 12 × 9.8 N (four back-up) thrusters.
Power: TM-type solar arrays generating 1.3 kW from 10 m^2; batteries carried in pressurised instrument compartment.
Docking: To the ISS Zvezda service module or Pirs docking bay

Soyuz Series

Current Status

Soyuz TMA-10 took off from Baikonur, the only Soyuz launch site, on 8 April 2007 carrying Expedition 15 toward the International Space Station (ISS). Launched along with the latest Expedition crew was Charles Simonyi, the billionaire US entrepreneur and fifth space tourist. Simonyi reportedly paid USD25 million to join the Expedition crew and board the ISS.

Soyuz will be undergoing further upgrades associated with future ISS servicing and also ESA's Aurora exploration programme. The Russian Federal Space Agency and ESA have teamed to send manned missions to the Moon and then to Mars. To do this, the space agencies will design and build the Advanced Crew Transportation System (ACTS). ESA had failed to add funding to the Kliper project (see separate entry), in December 2005, however agreed in 2006 to involvement and funding of the ACTS programme. A joint ACTS concept study is underway, and Russian and European officials are expected to announce a preferred design in mid-2007. It is likely that the design will adhere to the traditional, three-module Soyuz model and be large enough to carry at least 5 or 6 crew, including paying space tourists.

Background

Soyuz is the most-utilised manned spacecraft and will soon enter its fifth decade in flight. There have also been dozens of unmanned orbital tests and it probably has a version that houses a photo-reconnaissance satellite. RSC Energia in Kaliningrad is responsible for design and production. Russian planners have long floated plans for an upgraded and upsized version of the Soyuz but, in 1991, Energia halted design work on the 13 tonne Zarya super-Soyuz, capable of carrying five to six crew, and to be launched from a Zenit booster, because of funding difficulties. The overall form of the descent module was similar to that of Soyuz, but with a 3.7 m base diameter. The manned version would have included 1.5 tonnes of cargo and the unmanned 3.8 tonnes.

An improved Soyuz TMA was selected 1 June 1996 as the initial Assured Crew Return Vehicle (ACRV) for the International Space Station. NASA signed an agreement with RSC Energia in December 1992, when the Phase A study ended, affirming that Soyuz was technically suitable. Plans originally called for two docked three-man Soyuz-TMs to be used as the emergency escape vehicle in much the same role as the spacecraft was used aboard the Mir for American astronauts. The vehicle also had to be able to land within 20 km of a designated landing spot. Crews do not wear pressure suits. The Soyuz TMA ('anthropometric' design) Descent Module interior was modified to accommodate the larger torso size ranges of US astronauts. In 1995, Scott Parazynski stood down from the May 1997 Mir visit because the Soyuz TM could not accommodate his torso length. Although he was to fly up and down on Shuttle, he had to be able to return using Soyuz in an emergency. TMA is also equipped with improved parachutes and two new Amber display units. Because the ISS crew size is limited to a maximum of 3 people until the Station is complete, only one Soyuz TMA is needed on the station at any one time. Between the loss of Columbia on 1 February 2003, and the Space Shuttle's return to flight in July 2005, the Soyuz TM series spacecraft were the only means of maintaining an ISS caretaker crew.

Programme History

Originally, Chief Designer Sergei Korolev intended to build several versions of Soyuz. His ideas called for a full three-module craft for Earth-orbit operations, a stripped-down two-module Zond for demonstrating lunar mission techniques, and a lunar-landing orbital version. As flown, Soyuz has had three distinct mission categories. Soyuz 1–9 (1967–69) carried three-man LEO solo missions with durations of up to 2½ weeks. Then when it became evident to the Soviets that they had lost the moon race, design work began on the Salyut space station in late 1969 and Soyuz 10 and 11 functioned as crew ferries with integral transfer tunnels. After the Soyuz 11 failed, killing its crew during descent, the Soviets initiated a thorough redesign and Soyuz became a two-man station ferry capable of only 2½ days' independent flight. At this time, the Soviet space program evolved towards a more automated program than its American counterpart and as a result, the unmanned Progress (see separate entry) cargo version of the Soyuz capsule came about.

The post-Soyuz 11 version of the craft travelled 24 times in its manned configuration when Soyuz 40 retired the model in May 1981. After Soyuz 40, the Soviet Union introduced the Soyuz-T with its updated systems and three-man crews. This capsule took over for Salyut 7 operations until a further reworking produced the TM Mir ferry. The TM designates all Russian craft as modified transports. Soyuz had thus been transformed from a spacecraft with lunar missions to a dedicated space station ferry, through several generations of space exploration objectives.

The Soyuz Orbital Module (OM) provides space for cargo, food, waste management services, orbital experimentation on solo missions and EVA operations through the side hatch which is employed primarily for crew boarding on the pad. The forward hatch houses the docking apparatus of a central cone and eight metal blades inside a 1.3 m diameter docking collar that has eight mechanical latches (Soyuz 16/19/TM16 carried androgynous docking units). Two crew use sleeping bags in the OM, while the third sleeps in the Descent Module (DM) for safety (a rule not always adhered to). The toilet swings out from behind a panel, using pressurised airflow as a gravity substitute.

When delivering payloads to space, the Soyuz uses built- in connections for electrical, communications and hydraulics links. The original Soyuz design included four OM waist portholes, reduced to two on Soyuz-T and one on TM. Since 1991 and possibly before, TM has carried at least one front-viewing cupola in the OM, with controls for flyaround operations using direct vision. The OM is jettisoned by 12 pyrotechnic charges after retrofire. Until the TM5 failure of September 1988, cosmonauts ejected the OM before retrofire. This sequence saved about 10 per cent of the propellant. Failure investigations proved that crew safety required the post-retrofire jettisoning operation.

The Soyuz Descent Module (DM) accommodates a crew of two or three, with the Commander occupying the central couch, the flight engineer the port side and the cosmonaut-researcher the starboard side. Alternatively, a 100 kg cargo pod can replace one crew member. These Kazbek-Y couches are supported on shock-absorbing struts and can be lowered to provide increased space. Ten minutes before landing, compressed N_2 raises the couch 10 cm to help absorb the shock. Each seat liner is customised for the occupant and when cosmonauts return in different craft from their delivery craft, then the liners must be changed.

The commander provides manual thruster control via translational and rotational handles to his left and right, respectively, performing rendezvous and docking operations using the range and range-rate radar data and the 15° Field of View (FoV) VSK-4 periscope below the centre of the main control panel. From the introduction of Soyuz-T, the onboard computer has been capable of performing a fully automatic docking. The DM's separate 67 N H_2O_2 thruster network is activated only for re-entry: two nozzles near the apex some 60° apart provide roll control, with pitch control from a pair located between, and two at the base for yaw.

The Service Module (SM) completes the vehicle, supporting the DM on its forward face. The SM carries the propulsion unit. In the old Soyuz, the propulsion system employed a KTDU-35 assembly at the rear of the SM. Pumps fed propellant and oxidisers to a single main engine, which fired for

TM16 displays the forward-looking cupola (arrowed) and androgynous docking system 0517094

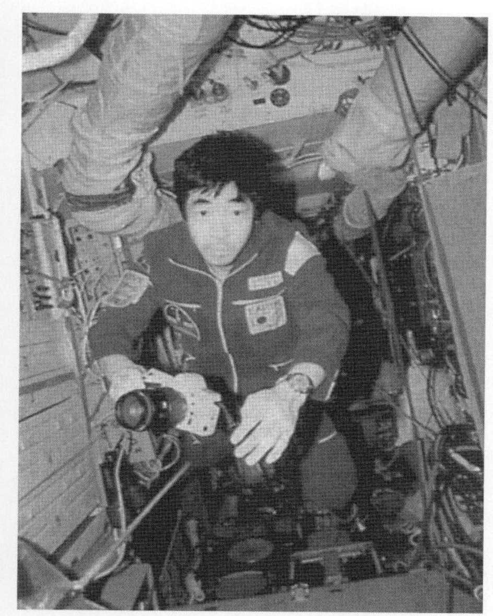

Tokyo Broadcasting Service paid JPY12 million for its journalist, Toyohiro Akiyama, to travel to Mir aboard Soyuz-TM 11 0516708

2½ minutes to initiate re-entry. A single start twin-nozzle engine served as a back-up unit. AOCS thrusters distributed around the SM's forward and aft faces provided guidance and control of the vehicle. Soyuz-T introduced the ODU unified propulsion system in which all engines were supplied from the same tank. Soyuz T's smaller AOCS provided a back-up re-entry capability.

As a safety measure, the propulsion system tanks now use metallic membranes to avoid pressurising N_2 which would have leaked through the previously elastic barrier. Some 150 kg of the propellants are reserved for rendezvous and docking and the retrofire operation consumes an additional 200 kg (400 kg allocated) for Earth return.

In the event of an emergency, an escape tower can fire the capsule away from the immediate area. From the top of the launcher on the pad, the capsule would reach an altitude of 1 km and land some 2½ km away, using the 27 m diameter reserve parachute. On a normal mission, descent is initiated by a 3 to 4 minute ~155 m/s retroburn by the main engine over the south Atlantic, followed by OM and then SM ejection. Re-entry occurs over Africa at an angle of attack of about 30°; and a lift to drag ratio between 0.25–0.30. The Soyuz TM can land cross-range of ±65 km, with a landing accuracy of 20 × 30 km ellipse. During re-entry, the external temperature reaches 3,000°C, but internal temperature maintains a 25–30°C constraint.

When landing, a pressure switch, activated at 9 to 11 km altitude and 850 km/h, unfurls two sequential stabilising drogues releasing the 4.25 m braking parachute from the port compartment. The primary canopy is deployed reefed at 8 km some 40 to 45 minutes after retrofire. It is freed to its full 35.5 m diameter at a descent rate of around 35 m/s to reduce sink rate to 8 m/s. Soyuz-TM retained the same size parachutes but their lighter material accounted for 110 kg of the craft's increased lift capacity. The 90 kg reserve system yields 10 m/s descent rate with its 25 m parachute, activated at 6 km. The heatshield is dropped at about 3 km some 5 minutes before landing to clear the base retromotors for a soft landing.

In the final burn, four solids initiate upon command from a radar altimeter about 2 m above the ground to cushion the impact. Soyuz-TM carried an improved system that reduced contact speed from 3 to 4 m/s to 1 to 2 m/s. Following the Soviet design, the Soyuz generally touches down on land but the craft is equipped for water landings. Coping with splashdown is a routine element of cosmonaut training.

Each capsule carries standard Granat-6 survival kits of Forel nylon flotation suit, Neva self-inflating life preserver, TZK cold weather suit, dried food, medical supplies, flares, radio, machete and canteen. The equipment is designed to sustain the crew for up to 3 days in the capsule in severe conditions and up to 12 hours in 2°C water with −10°C ambient air. A recovery beacon automatically activates on touchdown from behind a small ejectable cover to the left of the parachute compartment. In the event of a water landing, a balloon inflates from the same location. A radioactive source also acts as a beacon, something the recovery crew has to avoid after landing.

Specifications

Soyuz

First manned launch: 23 April 1967 Soyuz; 5 June 1980 Soyuz-T; 5 February 1987 Soyuz-TM; 30 October 2002 Soyuz TMA
Principal applications: Space station ferry, LEO solo operations
Availability: Typically 2–4 launched annually.
Performance: TMA can accommodate a crew of three and 100 kg cargo for ISS-type operations, returning a full crew and 50 kg to Earth.
Flight longevity: Independent 3.2 days and attached to ISS 200 days.
Cargo size: 45 × 60 × 100 cm up and 30 × 40 × 50 cm down.
Length: TMA: 6.98 m
Mass: TMA 7,220 kg at launch
Habitable volume: About 10 m³ (T/TM), previously ~9 m³
Landing mass: About 2,900 kg
Power: Solar cells with a span of 10.6 m. Soyuz locks on to the Sun with a sensor and slowly cartwheels around the Sun axis. DM includes a small battery supply for power after SM separation.
Thermal protection: 8 m² of SM radiators with the rest of the surface area covered by green thermal blanketing. On the descent module, eight blankets held by apex and base rings released when the other modules separate.
Life support systems: Pressure: Hermetically sealed on the pad under sea level conditions at 184–263 mbar partial O_2 pressure.
Regeneration: Potassium superoxide and LiOH cylinders.
Suits: Sokol-KV2 (Falcon) pressure suits. Wearer has to be minimum of 1.64 m tall.

NPO Zvezda is responsible for all Russian life support systems and suits. The bulk of the system is housed in the OM but the DM carries a smaller, independent unit under the couches. O_2 bottles provide emergency protection. Waste management and eating is handled entirely in OM; DM offers air sufficient for 48 hours and only food/water packs for landing emergencies.

Avionics/Control: Docking Type: S-band Kurs ('course') docking system. Allows approach to the ISS from any angle without the entire station having to rotate, as under the early Mir Igla ('needle') system.
Sensors: 15° FoV VSK-4 periscope for final approach. Two external 625-line 25 pictures TV cameras. The OM's waist carries VHF TV, radio and telemetry/command aerials, T-shaped telemetry/command antennas encircle its base.
Communications: Direct voice communications on 121.75 MHz. Data relay performed by the Rassvet system working via Mir through GEO satellites providing simultaneous dual-voice channels.
Landing/escape system: Type: 1,971 kg solid propellant tower, 6.680 m long, 1.415 m diameter.
Operation: Pyrotechnics separate below the DM and halfway down the shroud for the main 441 to 715.4 kN solid (1,607 kN total impulse) to pull the DM/OM combination free.
Peak Acceleration: about 14 g. The motor attains 70 per cent thrust in less than 0.07 seconds. The main escape motor has two components, a larger motor with four nozzles at its apex, plus an upper smaller motor with four smaller nozzles at its base. Both fire for aborts up to 20 seconds, but from 20 seconds only the larger one operates. The firing sequence lasts for about 5 seconds and then four shroud panels pivot outwards to slow the ascent. Throughout the escape, a small apex unit with four nozzles curves the vehicle onto its recovery path.
Engines/Motors: 150 kg, 98 to 171.5 kN (98 kN total impulse) 12 nozzle tower separation motor.
Soyuz Launch Vehicle: Tractor escape motor system providing protection for the first 115 s; the fairing's own separation rockets can then separate with Soyuz attached up to 165 s.

Soyuz TMA Orbital Module

Mass: 1.2 tons
Size: 2.2 m diameter spheroidal
Volume: 6 m³
Hatches: 65 cm diameter side hatch, 80 cm forward hatch

Soyuz TMA Descent Module

Size: 2.2 m base diameter, 2.1 m high
Mass: 2.9 tons
Usable volume: ~4 m³
Cosmonaut size limitations: 1.9 m standing maximum, 1.5 m standing minimum; 99 cm when seated. Some cosmonauts exceeded the TM's 94 cm height limit while still within the seat range. Cosmonaut chest circumference is unlimited, although they each must weigh between 50–95 kg, with a maximum foot length of 29.5 cm.
Hatches: A single 70 cm diameter overhead hatch, opened from either side and two 70 cm diameter side hatches for the primary and reserve parachute systems. The hatch lips extending from the 7° slope wall add to the DM's lift during re-entry and help to flip the hatches away in the airstream. The DM provides occupants with two 20 cm diameter portholes, one either side of the control panels; the discoloured outer pane is ejected after entry to restore vision.

Soyuz TMA Service Module

Size: 2.3 m long, 2.2 m diameter (flaring to an aft 2.72 m)
Power systems: Twin Si solar wings (spanning 10.6 m).
Thermal Control: ~8 m² of thermal radiators.

United States

Kistler K-1

Current Status

In February 2006 Kistler Aerospace was acquired by George French, the majority owner of Oklahoma-based Rocketplane. Rocketplane plans to resume development of the long-dormant K-1 vehicle and offer it as a supply craft for the International Space Station.

Background

Kistler Aerospace was formed in 1994 to develop a two-stage, fully-reusable space launch system with a 5.7 t LEO payload. The company hired a strong team of technical leaders, and completed its preliminary design in 1995. Between then and 1999, Kistler's investors spent USD600 million on subcontracts to major suppliers, including Lockheed Martin, Northrop Grumman, GenCorp Aerojet and Honeywell. In January 1997, Space Systems/Loral signed a contract with Kistler for ten launches in 1999–2002, and the K-1's first test flight was expected in the second half of 1998.

The Kistler K-1 is a two-stage LOX/RP rocket, launched vertically and recovered with the aid of parachutes and airbags. The second stage carries a payload compartment in its nose, with a movable cap that also acts as a heat shield on re-entry. A restartable, expendable K-1 Active Dispenser upper stage has also been designed for GTO and other high orbits.

The K-1 is powered by Russian-built Samara NK-33 and NK-43 engines, originally built for the N-1 Moon rocket in the 1960s. GenCorp Aerojet acquired the remaining engines from Russia, fitted modern avionics and other components, and redesignated them as the AJ-26 series. Kistler negotiated a right of first refusal on the entire supply of engines, considered adequate for more than 150 flights. Kistler secured a lease on a dedicated launch site in Woomera, Australia, and was discussing a second site in Nevada, United States.

However, the collapse of the "big LEO" market caused Kistler's funding to dry up at a point where it was claimed that 75 per cent of the first vehicle's components were complete. The company filed for Chapter 11 bankruptcy protection in July 2003. Kistler managed to exit Chapter 11 in March 2005, but in November 2005 the company's main investor withdrew its support. Rocketplane has acquired rights to the system, including parts in storage at subcontractor facilities.

Orion (Crew Exploration Vehicle)

Current Status

NASA officially named its Crew Exploration Vehicle (CEV) Orion in August 2006, just prior to selecting Lockheed Martin to build the spacecraft. The initial contract was valued at USD3.9 billion. In April 2007, NASA extended Lockheed's contract to 2013, two years beyond the original agreement. The extension adds USD385 million to the contract, placing its value at about USD4.3 billion. An International Space Station cargo carrier design that was included in the original contract has been deleted, and two further launch abort tests were added to the contract.

The Orion concept remains the same as the original CEV concept that gave the world a ballistic capsule design similar to the Apollo spacecraft, amid widespread criticism. Larger than the Apollo capsule, but similar in shape, Orion will have about 2.5 times the volume of Apollo and will be capable of carrying six astronauts to the International Space Station (ISS) or four to the Moon. It is comprised of two modules: the Crew Module and the Service Module. Unlike the Space Shuttle, the original design also includes a launch escape system, as with Apollo, Mercury and Soyuz.

In April 2007, NASA announced that it had purchased abort test boosters in preparation for Orion crew abort and escape trials. Orbital Sciences Corporation will provide launch services for tests, which will be conducted beginning in 2008 through to 2011. An ongoing US Air Force (USAF) sounding rocket programme enabled NASA to team with the USAF in order to procure the boosters and services. The 3rd Space Test Squadron, a USAF Space Development and Test Wing (with which NASA has entered into the booster agreement) unit, will manage the programme for NASA from Kirtland AFB and the White Sands Missile Range in New Mexico, where the trials will be conducted. The abort system will include small rocket boosters that will propel the crew to safety in the event of a launch failure or accident. Four (out of a total of six) tests requiring the boosters will be conducted; the USAF contract for the trials, which incorporates integration and launch services, is valued between USD35 million and USD57 million and includes two boosters, with an option for two more.

Orion belongs to NASA's Constellation programme, along with Ares I, formerly the Crew Launch Vehicle (CLV), named in June 2006. Ares V will be the launch vehicle's heavy lift version; it will launch cargo and rendezvous with Orion in LEO; the spacecraft would then proceed to the Moon or Mars.

Orion's contractor team consists of Lockheed Martin, United Space Alliance, Honeywell, Hamilton Sundstrand, and Orbital Sciences. Orion final assembly and checkout will take place at the Operations and Checkout building at Kennedy Space Center (KSC).

Background

First known as the CEV, Orion was conceived as the intended replacement for NASA's Space Shuttle and the first element of a new space exploration system, ultimately intended to take astronauts to the Moon and Mars.

The CEV programme was officially launched by President George Bush on 14 January 2004, in a speech announcing the administration's Vision for Space Exploration, following the loss of the Shuttle Columbia on 1 February 2003. Bush set NASA the goals of completing the ISS and retiring the Shuttle by 2010, while testing the CEV in 2008 and performing a manned mission by 2014. The CEV would support human flights to the Moon as early as 2015, leading to progressively longer stays on the lunar surface. NASA now projects that it will return astronauts to the Moon before 2020, however the agency remains committed to launching Orion on human missions by 2014. An Orion cargo version will likely be tested and launched on ISS logistics missions before that time.

NASA formally solicited bids on the CEV in March 2005, but NASA Administrator Michael Griffin changed some of the goals of the programme shortly after he was confirmed in the position in April 2005, replacing Sean O'Keefe. He directed that the CEV should be brought into use as early as possible, to minimize the gap between the last Shuttle flight and the first crewed CEV flight, and that the CEV should be available to support the ISS.

This change was reflected in the Phase 1 CEV development contracts awarded by NASA on 13 June 2005, to two competitors: a Lockheed Martin led team and a team comprising Northrop Grumman and Boeing. The first phase called for industry "to mature their crewed vehicle designs and demonstrate their ability to manage the cost, schedule, and risk of human-rated spacecraft development". It was originally due to continue until 2008, with the first human-rated CEV to be delivered in 2014.

After NASA issued its Exploration Systems Architectural Study (ESAS), also initiated by Griffin, the agency issued a "call for improvements" to the CEV competitors, outlining final requirements for the CEV. This was the basis for their Phase 2 proposals. The Phase 2 award was pushed forward to 2006; the Lockheed Martin led team was selected on 31 August 2006 to build the spacecraft.

Parts of the Orion Crew Module will be reusable, however the heat shields will be replaced for each flight. Specific details on the level of reusability of the craft have not yet come to light. Safety throughout flight is a paramount requirement, as is reliability and the use of affordable, dependable technology. Crew abort features have been integrated into the design, so astronauts could abort at any stage of the mission. Orion will be land recoverable, similar to Soyuz and Shenzhou.

One of Lockheed Martin's previous CEV designs showed details of a winged core vehicle which would have glided in to the Earth's atmosphere before being recovered by parachute on land. It would have been mated to a mission module containing supplies and fuel. This module would have also carried solar panels and radiators for long missions. According to Lockheed Martin, this concept provided safe means of abort during ascent, had enough cross-range to recover to land within the required time from any orbit, and provided acceptable g-loads during re-entry from the Moon or Mars. Many of the design features and concepts, aside from the winged re-entry module, may survive until Orion sees its first launch.

Boeing and Northrop Grumman, in 2005 and 2006, revealed fewer details of their failed CEV proposal; their team-members included Alenia Spazio, Ares Corporation, Draper Laboratory and United Space Alliance.

Specifications
Orion
Mass: Dry Mass: 14,045 kg
Launch Mass, including propellant: 23,395
Landing Weight: 7,337 kg
Diameter: 5 m
Crew Capacity: Six astronauts on LEO ISS missions; four astronauts on Lunar missions; further modifications required for future Martian missions
Volume: Pressurised: 20 m^3
Habitable: 11 m^3
Service Module Propulsion: Velocity: 1,738 m/s
Thrust: 33,362 N
Lunar Return Payload Capacity: 100 kg

Rocketplane XP

Current Status

The Rocketplane company plans to offer suborbital passenger spaceflights with the XP, a rocket-boosted aircraft which uses the cabin structure of a Learjet 25 business jet. The first XP is under construction for a planned first flight in January 2007.

Background

Rocketplane was founded in 2001, with a view to capturing the Ansari X-Prize, but did not open its first office until May 2004, by which time the Rutan SpaceShipOne was well on the way to victory. However, the company continued with plans for a vehicle that could transport at least four people from a runway to a 100 km altitude and return to the same runway.

Rocketplane has acquired the intellectual property of the Pioneer Rocketplane company, which grew out of the Air Force's Black Horse studies of the mid-1990s. Pioneer Rocketplane had envisaged a vehicle that would take off on turbojet power and be refuelled with rocket fuel – kerosene and oxidizer – in flight. Rocketplane simplified this concept for the suborbital mission, eliminating the inflight refuelling and focusing on a vehicle that would climb to medium altitude on turbojet power, use a rocket for ascent to 100 km, perform a gliding descent and relight the turbojets for return to the launch site and landing. (The vehicle is also designed for glide landing.) The goal is a vehicle that can perform two flights to mission altitude per week.

The initial Learjet-based design passed its preliminary design review in May 2005 and evolved considerably in the subsequent months. The resulting configuration retains the fuselage structure and rear-mounted General Electric CJ610 engines of the original aircraft but has a new delta wing with leading-edge root extension and a V-tail. The delta wing is being manufactured by Spirit Wing, an Oklahoma-based company that develops Learjet modifications.

Much of the fuselage accommodates propellant for the rocket engine. The operational vehicle is still planned to be powered by a new Polaris Propulsion AR-36 engine fitted with a Barber-Nichols hydrogen peroxide/RP turbopump, but first flights will use an off-the-shelf Pratt & Whitney Rocketdyne RS-88 engine. The engine was originally developed as a pad-abort emergency propulsion system for the now-cancelled Orbital Spaceplane, and has been loaned to Rocketplane by NASA.

Most of the XP airframe will be covered by a Thermal Protection System (TPS) comprising a low-density nanoparticle ceramic coating capable of maintaining its emissive characteristics at temperatures as high as 1,650° C. The coating emits 93 per cent of the energy that it receives. The nose and leading edges will be constructed from titanium. The XP will have a complete fly-by-wire system and a fault-tolerant cold nitrogen Reaction Control System (RCS) produced by Space Vector Corporation. Operations will take place from Spaceport Oklahoma, the former Clinton-Sherman in southwestern Oklahoma.

In early 2006, Rocketplane was conducting high-speed wind-tunnel tests and continuing fabrication of the first vehicle. The company was aiming for first flight in early 2007, leading to a 25-sortie test programme and passenger-carrying flights before the end of the year.

Specifications
Mission: Suborbital tourism and microgravity experiments
Length: 13.1 m
Diameter: 1.5 m
Wing span: 7.74 m
Take-off weight: 8.85 t
Empty weight: 3.85 t
Maximum speed: Mach 3.6
Maximum altitude: 100 km
Take-off propulsion: Two GE CJ610, 12.7 kN each
Ascent propulsion: Polaris Propulsion AR-36, 160 kN

SpaceShipOne

Current Status

After winning the USD10 million Ansari X-Prize for the first demonstration of a privately funded, reusable, sub-orbital passenger-carrying vehicle in October 2004, the SpaceShipOne rocket craft was permanently retired to the National Air & Space Museum in Washington, DC in September 2005. The White Knight carrier vehicle has been retained by Scaled Composites as a test asset for other programmes. Scaled Composites has teamed with the UK's Virgin Group to develop a commercial follow-on vehicle, designated SpaceShipTwo.

Background

The X-Prize was announced by US entrepreneur Peter Diamandis in 1996, as a way to stimulate the development of commercial passenger spaceflight. The USD10 million prize was offered to the first team to fly a piloted vehicle, capable of carrying three people, to an altitude of 100 km, and then repeat the flight within seven days. At least 90 per cent of the vehicle had to be reusable.

Burt Rutan, founder and President of the Scaled Composites company, almost immediately launched a study of how to win the X-Prize with a vehicle that could form the basis of a commercial suborbital craft. Inspired by the success of the North American X-15 in the 1960s, Rutan focused initially on an air-launched vehicle, but sought a solution that would be inherently safer than the X-15 and that would not be dependent on a sophisticated flight control system. Early designs were based on the Scaled Composites Proteus high-altitude research aircraft, carrying a simple wingless rocket with a shuttlecock-like deployable aerodynamic device to stabilise the craft on re-entry, the most critical phase of the flight. It would then descend under a parachute and be retrieved in mid-air by

A view from SpaceShipOne showing the Los Angeles Basin covered in fog. The photo was taken during the 12 June 2004 test flight (Scaled Composites) 1047182

SpaceShipOne being carried aloft by the White Knight aircraft (Scaled Composites) 1340611

a helicopter. However, Rutan considered parachute recovery to be risky.

By 2000, Rutan had evolved a unique design based on similar principles. The upper-stage vehicle would be a winged rocket-powered craft, with a tail section that could pivot 65° upwards during re-entry to perform the same functions as the shuttlecock device – it was internally dubbed the "feather". This would also orientate the vehicle's body and wing at right angles to its trajectory for maximum aerodynamic braking effect. Once the vehicle was back in the atmosphere, the tail would return to normal flight position and the craft would land as a glider.

In March 2001, Microsoft co-founder Paul Allen agreed to fund Rutan's X-Prize attempt. Allen formed a new company, Mojave Aerospace, that would own rights to the new technology in the vehicle.

Scaled Composites completed the SpaceShipOne system, also known as Tier One, in two years for a total cost of USD20 million, and it was unveiled at Mojave, California, on 18 April 2003. The system comprised the spacecraft, named SpaceShipOne, and the White Knight carrier vehicle. SpaceShipOne was powered by a hybrid rocket, using nitrous oxide (N_2O) as an oxidizer and rubber as the fuel. The N_2O is contained in a pressurised tank, produced by ATK, behind the cabin. Rutan selected two companies, SpaceDev and Environmental Aerospace, to develop competing Case, Throat and Nozzle (CTN) designs for the motor. The CTN, which couples to the N_2O tank, contains the rubber-based fuel and is the only expendable part of the system.

The SpaceShipOne vehicle was constructed from carbonfibre/epoxy composite material with a thermal protective layer on surfaces exposed to re-entry heat. Because of its high operating altitude, the cabin featured small, dual-paned circular windows. The vehicle had a GPS-inertial navigation system driving specially developed flight instruments, with symbology designed to guide the pilot in a near-vertical climb out of the atmosphere. It was controlled by a combination of electrical trim surface, used in supersonic ascent; manually operated cold-gas thrusters for exo-atmospheric control; and conventional manual controls for gliding flight. The wheeled main landing gears and nose skid were extended by gravity and springs.

The White Knight carrier vehicle, with its twin-body layout and inverse-gull-wing profile, was designed specifically to carry the SpaceShipOne. However, it was also designed with an identical cabin section, avionics, environmental control system and other hardware, and with oversized speedbrakes to allow a steep descent, so that it could be used as a flying system testbed and pilot trainer for SpaceShipOne. It is powered by two afterburning turbojets, selected because they were available, inexpensive and delivered high thrust at high altitude. The name White Knight resulted from the resemblance of the small cabin windows to an armoured knight's vizor and from the lance-like body extensions that carried the twin nosewheels.

The project also included the design and construction of a flight simulator, a rocket test stand and a mobile N_2O refuelling and storage system.

SpaceShipOne during flight trials (Scaled Composites) 1047183

The White Knight started its flight tests without publicity in August 2002. The first captive flight test of SpaceShipOne took place on 20 May 2003. After a manned captive flight on 29 July 2003 Mike Melvill performed the first glide test on 7 August. A total of seven glides were performed before Brian Binnie flew SpaceShipOne on its first powered flight on 17 December, the 100th anniversary of the Wright brothers' first flight. This was also the first supersonic flight by a vehicle developed without government sponsorship.

Two more powered flights were performed before the first suborbital flight attempt on 21 June 2004. Mike Melvill reached a maximum speed of Mach 2.9 after a 76 second rocket burn. The burn ended at 180,000 feet and the vehicle coasted to 328,491 feet – just above the 100 km mark – despite a loss of trim control. The vehicle re-entered as designed and returned to Mojave.

The programme was completed with the two X-Prize flights, flown by Brian Binnie on 29 September 2004, and Mike Melvill on 4 October. The project demonstrated the practicality of an air-launched system, the hybrid motor concept and the "care-free re-entry" made possible by the "feather". These features are expected to be carried over into the commercial SpaceShipTwo/White Knight Two system.

Specifications
SpaceShipOne
Designation: Scaled Composites Model 316
Engine: SpaceDev hybrid rocket
Overall length: 8.5 m
Span over tails: 8.17 m
Oxidiser: N_2O
Fuel: HTPB
Propellant mass: 2,600 kg
Dry mass: 1,200 kg
Thrust: 74 kN
Specific impulse: 240–250 s
Burn time: 83 s

White Knight
Designation: Scaled Composites Model 316
Engine: Two General Electric J85-GE-5 turbojets, 17.1 kN with afterburning
Wingspan: 25 m
Length overall: 15.7 m
Maximum take-off weight: 8,150–8,500 kg

SpaceShipTwo

Current Status
On 27 July 2005, Scaled Composites President Burt Rutan and Virgin Group founder Sir Richard Branson announced the formation of The Spaceship Company to design, certificate and manufacture a commercial sub-orbital craft based on Scaled Composites' SpaceShipOne design.

The new company will licence the key technologies used in SpaceShipOne, including the motor and aerodynamic design, from Mojave Aerospace, the company owned by SpaceShipOne sponsor Paul Allen. It will market the operational system – comprising the SpaceShipTwo (SS2) suborbital vehicle and White Knight Two (WK2) carrier aircraft – to commercial spaceflight operators. Under the agreement establishing The Spaceship Company, Branson's Virgin Galactic company has placed orders for the first five SS2s and the first two WK2s, securing exclusive access to the design for the first 18 months of passenger operations.

Late in 2005, Virgin Galactic officials were quoted as saying that tests of the new system would start in 2007, with the first commercial flights in the following year.

Details of the SS2/WK2 system have not been revealed. However, Rutan stated at the unveiling of SpaceShipOne that an operational suborbital craft would carry between six and ten passengers, for economic reasons. It would also fly faster and higher than the prototype, to give the passengers a longer period in weightlessness, and would have a large enough cabin to allow passengers to float in zero gravity. More recently, Virgin Galactic President Will Whitehorn has characterised the SS2 cabin as being similar in size to that of a Gulfstream G550 business jet, seating two pilots and seven passengers. The WK2 is expected to be about as large as a Boeing 737 (suggesting a take-off weight of around 70 tonnes).

SpaceShipThree would be an orbital transport system. So far, it has not been defined in detail, although Rutan has said that the White Knight could be scaled up into an eight-engined aircraft, capable of releasing a 270 metric tonne vehicle at high altitude.

Space Shuttle (Space Transportation System)

Current Status
External fuel tank damage caused by a hail storm to the Space Shuttle Atlantis, while awaiting mission STS-117 on the pad, has again delayed Shuttle flights. The Atlantis launch, originally scheduled for March 2007, has been pushed back to June. Subsequently, NASA has reassessed and reworked the Shuttle launch manifest through April 2008, with the final launch adjustment date being applied to STS-124. The move also delays STS-118; STS-120; STS-122 and STS-123. All six Shuttle flights affected are ISS assembly flights, and therefore the launch delays will negatively affect the ISS assembly schedule. Shuttle delays will also impact ESA's Automated Transfer Vehicle (ATV) launch, scheduled to fly from Kourou in late 2007, as well as ESA's Columbus Laboratory module, and JAXA's Kibo module components, formerly scheduled to launch aboard the Shuttle in 2007 and 2008.

In January 2004, President George W Bush announced the Vision for Space Exploration, which is aimed at returning US astronauts to the moon before 2020, and then ultimately to Mars. The plan will use a new launch architecture, including Orion, also known as the Crew Exploration Vehicle (CEV), to replace the Shuttle in its human transport mission. At the same time, President Bush assigned NASA the goals of completing the ISS and retiring the Shuttle by 2010.

Flying since 1981, Space Shuttle technologies, while advanced at the programme's initiation, have been getting on in years. The Vision for Space Exploration followed the Columbia disaster by one year. Neither the Bush administration, nor NASA could afford another catastrophic Shuttle and

Typical Space Shuttle launch. Note the 'white room' (centre bottom), which provided crew access and the only pad escape route (NASA) 0517287

Columbia approaches touchdown at the conclusion of STS-73 0517288

A completed External Tank leaves NASA's Michoud plant 0517266

crew loss after Columbia in 2003 and Challenger in 1986; modernisation was in order. Although Orion and Ares I, its launch vehicle, will integrate more up-to-date technologies, the designs are hardly revolutionary. Ares I will use Shuttle and Saturn rocket engine technologies (as will the cargo launch vehicle, Ares V) and Orion is much like the conical/ballistic Apollo in its design; both project concepts have brought criticism to NASA's Constellation programme, to which Orion and Ares both belong. It seems that after the Shuttle is retired, the spaceplane concept will be retired also, at least for some years.

The Space Shuttle will support ISS logistics missions and transport ISS components until the Station's 2010 completion. This target may be ambitious, as Shuttle flights have once again encountered delays. It may be possible that the Shuttle will fly on some extended missions past 2010, or an alternative means of transporting ISS components and supplies will have to be employed if the ISS is not completed by that time. Orion is scheduled to take over ISS flights in 2014, leaving a four-year void when no NASA transportation spacecraft will service the ISS. NASA's Commercial Orbital Transportation Services (COTS) programme could help fill the ISS logistics gap, as could ESA's ATV and JAXA's H-II Transfer Vehicle (HTV). Russia's Soyuz will continue to fly human rated missions but will also undergo upgrades under a Russian-European programme.

The first return-to-flight mission, incorporating changes recommended by the Columbia Accident Investigation Board (CAIB), was STS-114, flown by Shuttle Discovery, and launched on 26 July 2005. On 9 August 2005, Discovery, NASA's third orbiter, and its crew, landed safely at Edwards Air Force Base in California. Crew members successfully completed three spacewalks, external repairs to the Shuttle, and supply deliveries to the ISS during the mission. However, problems encountered during STS-114, most notably, the separation and loss of insulating foam upon Discovery's launch, continued to call into question Space Shuttle programme viability.

Following the launch of STS-114, NASA suspended future flights because the large piece of foam was seen to separate from the external tank during the ascent. Foam debris damage to Columbia's wing edge, occurring at STS-107's launch, caused that spacecraft's catastrophic failure upon its re-entry in 2003. The agency formed two engineering teams – a Marshall Action Team tasked with examining possible causes of the incident, and a Tiger Team to review all engineering issues associated with the ET – and announced that the Shuttle would not be flown again until the foam issue was resolved. Immediately affected was the second return-to-flight mission, STS-121, to be flown by the Shuttle Atlantis. The 11-day mission to the ISS was initially planned for a September 2005 launch. STS-121 finally took flight on 4 July 2006, nearly a year later. Two further Shuttle missions flew that year: STS-115 in September and STS-116, in December 2006.

For a schedule detailing each Space Shuttle flight, please see Table 3: US Space Shuttle Flights, in the Space Logs section.

Background

When conceived in the mid-1960s, the Space Transportation System (STS) was to be fully recoverable, but the concept was abandoned because of its high development cost, in favour of the present system. This involves the Solid Rocket Boosters (SRB) parachuting into the sea for recovery and the External Tank (ET) being irrecoverably jettisoned. This compromise should have enabled the system to reduce overall programme costs by – compared with the expendable launch vehicles available at the time, which were to be phased out. However, the cost target was not achieved because manpower requirements and turnaround times were not reduced to the expected levels.

Shuttle manifest changes

A significant shift in the Shuttle flight manifest was forced upon NASA by a presidential decision following the loss of Orbiter Vehicle (OV) Challenger in January 1986, not to fly commercial satellites on future missions.

The recovery programme cost the United States an estimated USD2.4 billion, with an additional USD1.8 billion spent on the replacement vehicle Endeavour. Rockwell delivered the Orbiter in 1991 for its 1992 debut. At the time, NASA aimed to maintain an annual flight rate of seven Shuttle launches, spending no more than 88 days processing each Orbiter after return from orbit. In this 88 day objective, NASA budgeted 23 days on the pad.

In order to make operations run more smoothly, NASA removed all commercial satellites from Shuttle manifests, leaving only a backlog of science and defence missions and beginning in November 1998, International Space Station assembly and utilisation launches. STS-53 in December 1992 entered history as the last dedicated military mission. DoD dramatically reduced its Shuttle requirements and transferred most of its launches to Atlas, Titan and Delta launchers. One of the last DoD entries on the Shuttle mission docket was the September 1999 Shuttle Radar Topography Mission, added in July 1996 after struggling for years to find a slot.

While the Shuttle was able to maintain most of its target launch dates (apart from minor delays generally due to the weather), NASA announced repeated delays in the first missions dedicated to the International Space Station. NASA tried to blame these delays on Russian contributions to the Station, but later NASA's own contractors admitted they were behind schedule as well, and over budget. Initially, NASA delayed Shuttle missions starting with STS-88 by about eight months. After the May 1998 release of a study that showed that NASA had provided unrealistic estimates of the Station's cost to the American public, all Shuttle missions related to the Space Station's construction were rescheduled.

The Shuttle launch manifest has endured repeated changes during its history. This trend shows no signs of reversing, and schedule slippage has come to be a programme norm.

Shuttle components

Orbiter Vehicles/Space Shuttles

The Orbiters are about the size of a DC-9 airliner. Long-term initial planning was based on production of five, each capable of at least 100 flights. Before being named, what was intended to be the structural test article Orbiter Vehicle (OV) was designated STA-099 and the five flight-standard vehicles OV-101 to OV-105. Shortly before the Approach & Landing Tests (ALT) began in 1977, NASA intended to name OV-101 Constitution, but a letter campaign to President Ford produced a change to Enterprise after the starship in the 'Star Trek' TV series. However, it was decided shortly afterwards to upgrade STA-099 as the second flight vehicle, replacing it with Enterprise as the ground test vehicle after the ALT programme. The first four operational Orbiters were then named after pioneer sailing ships. Columbia (OV-102) is named after one of the first navy frigates to circumnavigate the world (it was also the name of the Apollo 11 Command Module). Challenger (OV-099) is named after the ship that explored the Atlantic and Pacific in 1872–76 (also the name of the Apollo 17 Lunar Module). Discovery (OV-103) was named after two ships, the vessel that discovered Hudson's Bay in 1610–11 and the one in which Captain Cook discovered the Hawaiian Islands. Atlantis (OV-104) obtained its name after the ketch that performed oceanographic research from 1930–66. Challenger's OV-105 replacement was named in May 1989 as Endeavour, after Cook's first command; it was also used by Apollo 15's Command Module.

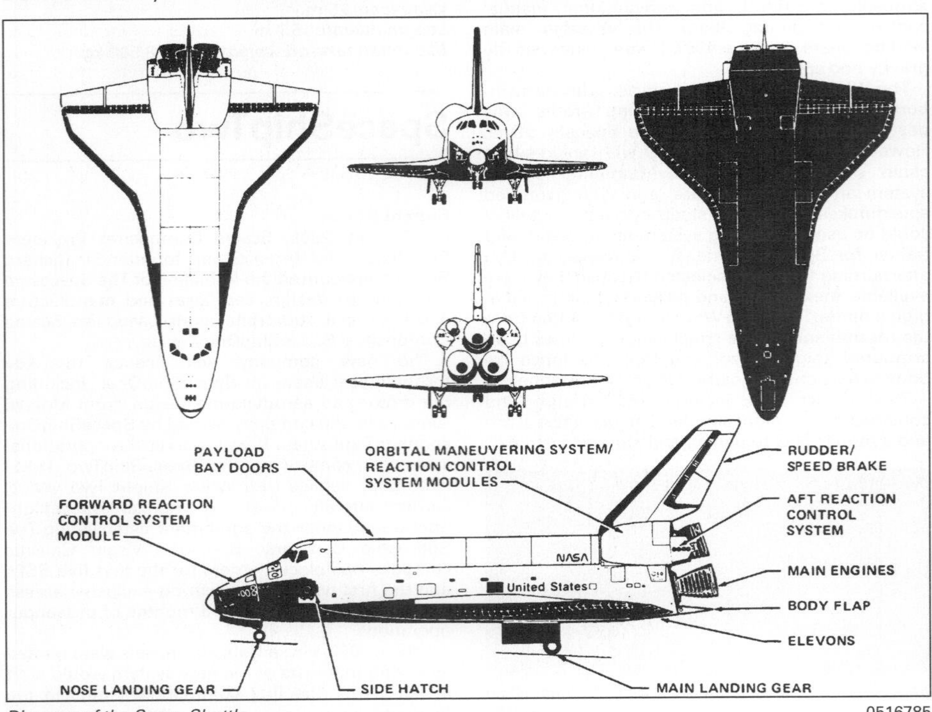

Diagram of the Space Shuttle 0516785

STS-57's Wisoff on the RMS watches David Low working on Eureca's antenna. Spacehab is in the foreground (NASA) 0516776

About 210 significant hardware changes were made to each Orbiter between missions 51L and STS-26, in addition to 100 software and 35 Space Shuttle Main Engine (SSME) changes. The system now contains some 800 Criticality 1 items – that is, equipment without back-up and creating significant mission impact should they fail. Improvements included RCS rewiring so that thrusters cannot fail into the 'on' position, replacement of the limited-life APUs (the retrofitted models offer 75 h lives), and rerouteing of fuel cell waste water outlets so that freezing cannot disable the whole power production system. The main axles of the landing gear were thickened and the capacity for nosewheel steering, introduced on Columbia for mission 61C, was added to the other two vehicles. Higher temperature thermal protection was added in the wing elevon cove areas, along with RCC protection between the nose cap and landing gear door. Columbia's low-T thermal protection along its mid-fuselage, payload doors and tail was replaced by Advanced Flexible Reusable Surface Insulation.

OV-103/104 now have strengthened wings after analyses showed that the weight reduction programme during construction left them vulnerable to flight loads higher than predicted. Closure of the flapper valves on the 43 cm External Tank propellant lines during SSME burn was a major concern before Challenger's loss because of the fatal consequences, but latching mechanisms were added to ensure they cannot close while liquid is passing through to the engines.

Enterprise (OV-101)

Used for the Approach & Landing Tests (ALT) of 1977–78, Enterprise subsequently served as a ground test vehicle, ending at Vandenberg in 1985. It was used there for preflight test matings on the SLC-6 complex, later closed. It now belongs to the Smithsonian's National Air and Space Museum Steven F Udvar-Hazy Center near Washington DC's Dulles Airport. In June 1987, NASA borrowed OV-101 for arrester net tests at Dulles. The need for such nets at emergency landing sites was under consideration before the Challenger accident and particularly at Dakar (Senegal) and the Pacific Ocean runways on Easter Island and Hoa. The net is based on those already in use at military airfields.

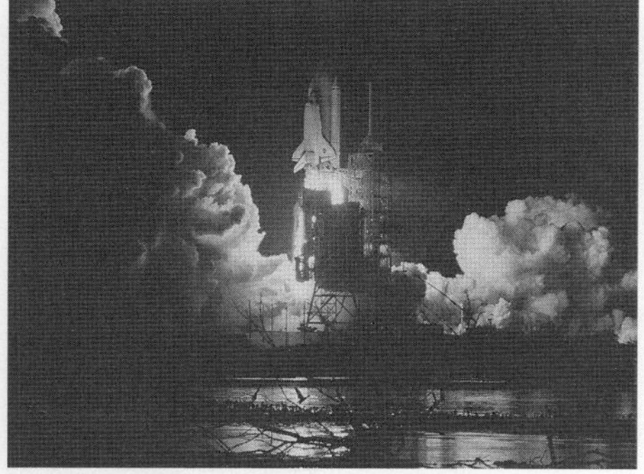

Endeavour lifts off into a night sky for the start of mission STS-88, 4 December 1998 0054451

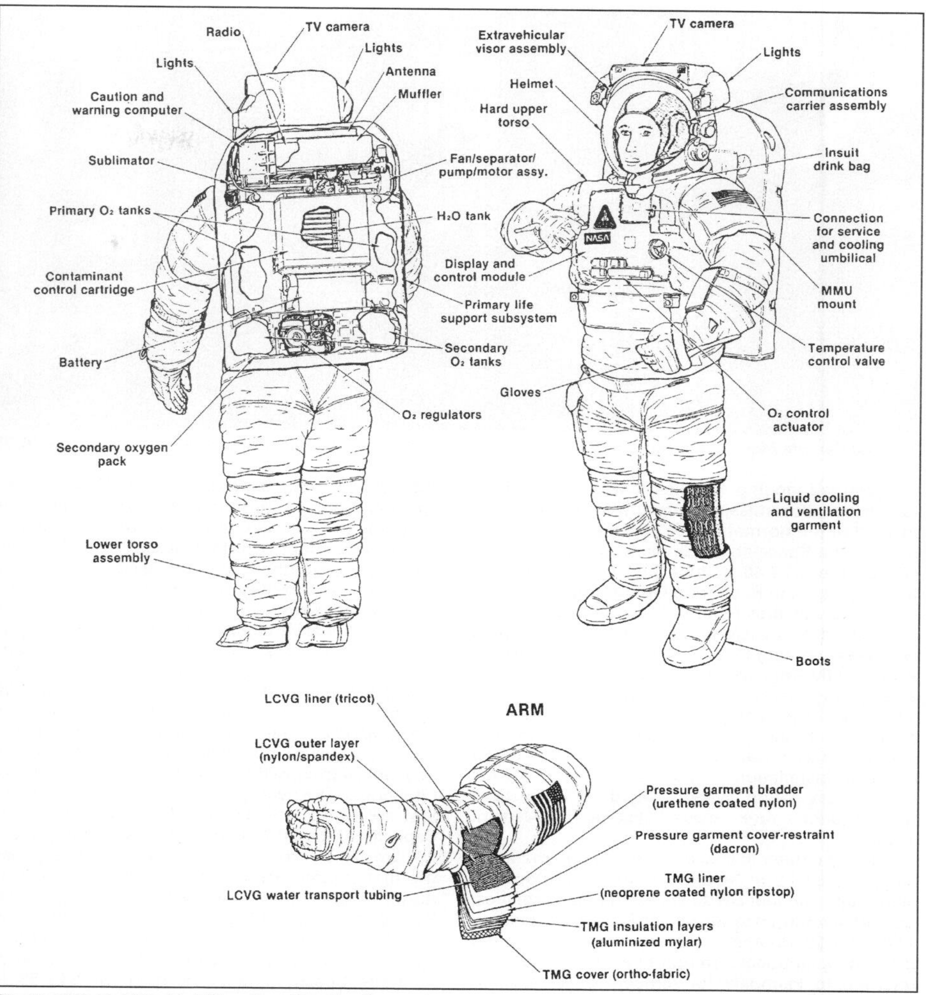

Shuttle EVA Mobility Unit (Hamilton Standard) 0516769

Columbia (OV-102)

After 27 successful flights into space, on 1 February 2003, NASA's first Orbiter specifically built from the outset for operational space missions was destroyed on re-entry when its thermal protection systems failed to defend its aluminium structure from the searing heat of kinetic friction during atmospheric descent. Carrying a crew of seven returning from a 16 day solo science mission that began with a 16 January 2003 launch, Columbia was conducting its 28th orbital flight in a series that began with the first launch of a Shuttle on 12 April 1981. For two years OV-102 was the only Shuttle flying and had conducted five flights before Challenger (OV-099) joined it in April 1983. Challenger had been converted to operational flight status from its original role as a structural test article and was originally built as a ground vehicle only. Columbia completed the Shuttle flight test programme with its fourth flight in June 1982, conducted the first in-orbit satellite launch in November 1982 and carried the first European built Spacelab science laboratory into space during November 1983. After a period of modifications conducted after Challenger began flight operations in 1983, Columbia had a large number of items removed or changed, including the ejection seats, fully adapting the orbiter from a test and development role to a fully fledged operational spaceship. In 1984 Discovery joined the fleet, followed by Atlantis in 1985, completing the approved inventory of four flight orbiters with funds for a fifth flight vehicle not approved by Congress.

On returning to flight in early January 1986, Columbia performed the last Shuttle mission before Challenger was destroyed during launch at the end of that month. Columbia returned to space in August 1989 when it launched a DoD mission and in 1990 recovered the Long Duration Exposure Facility from orbit. Columbia continued to support science missions and Spacelab flights, launching the Chandra X-ray observatory in July 1999 before stand-down for extensive modifications and refurbishment. As the first flight orbiter, Columbia carried a lot of electrical wiring and data lines unnecessary for operational flights. Although much had been removed during its earlier refurbishment, much wiring built in for test and operational monitoring equipment used only on the first few flights remained in the vehicle. Columbia returned to flight in March 2002 supporting the fourth Hubble Space Telescope servicing mission (Servicing Mission 3B) before preparing for its science mission, STS-107, during which it was destroyed. Before the catastrophic STS-107 flight, Columbia had completed the equivalent of 284 days 19 hour 19 minutes 1 second in flight, more than 40 days more than Discovery, 65 days more than Atlantis and 78 days more than Endeavour; Challenger had totalled more than 62 days when it was destroyed.

The Columbia Accident Investigation Board (CAIB) chaired by Admiral Harold Gehman (retired), published its report in August 2003. The CAIB concluded that the Shuttle could be made acceptably safe, given engineering changes, but that NASA's flawed safety culture was also to blame for the accident. The CAIB also recommended that the Shuttle should be replaced as soon as possible as NASA's means of carrying humans into orbit. At that time, however, there was no successor to the Shuttle on NASA's plans, following the failure of successive attempts to develop a new-technology launch system – including the National Aerospace Plane (NASP), the McDonnell Douglas Delta Clipper, the NASA/Lockheed Martin X-33 Delta Clipper and the early-2000s Space Launch Initiative.

As the heaviest flying Orbiter, Columbia's utilisation potential was enhanced by the increase

For details of the latest updates to *Jane's Space Systems and Industry* online and to discover the additional information available exclusively to online subscribers please visit **jsd.janes.com**

The first Remote Manipulator System (RMS), being assembled in Toronto (Spar Aerospace) 0516770

NASA astronaut Newman during EVA at Unity module (NASA) 0062167

in permitted landing weights. It was modified for longer duration missions of nine to twelve days, instead of the normal five to seven days, and flew most of the Spacelab missions. It was taken out of service after STS-40, in June 1991 for about a year, for conversion in Rockwell's Orbiter final assembly facility as the first Extended Duration Orbiter to accommodate missions of up to 16 days (plus two days' contingency). Dry mass was 82,267 kg. Additionally, improved APUs, GPCs, carbon brakes and a drag parachute were fitted and a major structural inspection recertified the airframe for a further three years. Obsolete development instrumentation was removed and provision made for later installation of the new Collins' TACAN landing navigation system. Its third inspection/recertification took place October 1994–April 1995.

The first Orbiter in space, Columbia had made six flights by STS-9 in November 1983, had flown 18 astronauts and launched two commercial satellites. Columbia returned to service for STS-61C in January 1986 after 18 months' major modifications to bring it to an operational configuration. It should have returned to Palmdale for removal of its ejection seats and overhead escape windows after STS-5 in late 1982, but DoD insisted that two Orbiters should be available in case of loss, so the work was delayed until after STS-9/Spacelab 1. When it was returned to Palmdale, the planned 24 modifications became more than 240. Because of the greater dry mass, monitoring instrumentation and extended mission facilities such as additional cryogen storage, Columbia's maximum payload capacity for a 204 km, 28.5° orbit was 3,800 kg less than that of OV-103 and 104. Work included structural strengthening, to enable it to withstand higher aerodynamic loading, wiring changes and avionics updating and a reinforced carbon-carbon nose cap, incorporating pressure sensor inlets as part of the programme to collect data on dynamic pressure during re-entries. A pod was also installed on the vertical stabiliser to obtain detailed infra-red (IR) imagery of upper surface heating of the port wing and fuselage during re-entry. A plan to take Columbia to Vandenberg for Shuttle handling tests before the facilities were shut down, was abandoned in October 1986. NASA estimated that each of Columbia's first six flights cost USD300 million.

Challenger (OV-099)
Challenger was the second Orbiter to fly, in April 1983. So much was learned during construction that it was upgraded from the Structural Test Article (STA) and replaced by Enterprise in that role. By November 1985, it had made nine flights, flown 45 astronauts and launched five commercial satellites. Its milestones included the first 7/8-member crews, the first EVA (including the first untethered), and the first capture, repair and redeployment of a disabled satellite. Following STS-41G in 1984, 13,984 thermal tiles had to be replaced in an emergency operation involving 200 Rockwell tile specialists being sent to KSC, after 2,500 new tiles had been manufactured at Palmdale. Challenger and its crew, including two civilian astronauts, were lost in the launch explosion of 28 January 1986; the debris is sealed in two Minuteman silos at Cape Canaveral. The average cost of each of its eight flights was estimated at USD290 million. OV-099 performed 10 flights totalling 62 days, 7 hours, 56 minutes and 22 seconds.

Discovery (OV-103)
The third Orbiter to fly, OV-103, undertook STS-26, the first mission of the new post-Challenger series. It debuted in August 1984 and by November 1985, it had made six flights, flown 35 astronauts and launched 13 commercial satellites. Its firsts included the first capture/return to Earth of two disabled satellites, the first orbital test of a laser tracking system, the first trial of a 32 m solar array and the first classified military flight. Before the Challenger accident, it was to have been based permanently at Vandenberg from its seventh flight, for polar operations, but its delivery from Kennedy Space Center was delayed by problems with the new lightweight SRBs, followed by delays with completion of the Vandenberg facilities. Between the accident and the cancellation of Shuttle/Centaur, Discovery was modified, at a cost of USD5 million, to enable it to carry the Centaur payloads in place of Challenger. The average cost of each of Discovery's first six flights was estimated at USD170 million. Capacity of the 78,700 kg inert-mass Orbiter into a 204 km, 28.5° orbit was 24,950 kg. Discovery was the first to use carbon-carbon brakes, on STS-31 during 1990. It was removed from service after STS-42 in January 1992 for extensive modification (the 78 major changes included addition of a drag' chute and nosewheel steering) and inspection at Kennedy Space Center, returning for STS-53 in December 1992. It returned to Palmdale following the July 1995 STS-70 mission for its regular three year major service, this lasting from September 1995 to June 1996. During the servicing a complete structural inspection was made of the orbiter, which found no major problems.

More than 750 kg of performance-enhancement weight savings were achieved to reach the requirements for International Space Station missions, the first of which was then planned to be STS-94 in November 1998; this mission was cancelled. AFT ballast was added, as well as a fifth cryogenic tank which set provisions for additional on-orbit capabilities, additional nitrogen supply, installation of lightweight seats and a reinforcement of the crew module's floor under seats 1, 2 and 5 to meet the requirements for a 20 g crash overload. Modifications in support of the International Space Station and Shuttle-Mir missions included air-cooling provisions for mid-deck locker payloads, installation of a UHF space station communications system, orbiter provisions for ISS assembly power conversion units, removal of the internal airlock and the installation of a new airlock which provided a Mir docking capability. Design improvements included a redesign of the main propulsion system gaseous hydrogen flow control valves and a wing doubler modification. Four next-generation toughened single-piece fibrous insulation tiles were installed on Discovery's side hatch. Discovery returned to flight for the STS-82 second Hubble Space Telescope servicing mission in February 1997.

The 1995–1996 Discovery servicing marked the first use of a new vehicle automated power-on checkout system at the Palmdale facility. This demonstrated programme savings which reduced the number of data storage tapes from 1,060 to 10, cutting the number of checkout support staff from 21 to 3 and cutting the start-up time from four hours to five minutes.

Atlantis (OV-104)
OV-104 performed STS-27, the second mission of the post-Challenger series, reportedly deploying the new Lacrosse imaging radar reconnaissance satellite. Its first mission, the classified STS-51J, for DoD in September 1985, was claimed to be the most trouble-free of all Shuttle flights until then. It had been used for only one more, STS-61B in November 1985, before the 2.5 year shutdown following the Challenger accident. The scheduled December 1984 delivery to Kennedy Space Center was delayed until 13 April 1985 for modifications to enable it to carry the Centaur upper stage in the payload bay. These modifications included extra plumbing to load/vent cryogenic propellants and controls in the aft flight station for loading, sensing and monitoring the cryogenics. When delivered, Atlantis' inert weight was 76,658 kg – slightly above Discovery's, which had not undergone wing spar modifications. Atlantis was protected with 21,801 high-temperature reusable insulation tiles and 1,977 thermal protection tiles. Final assembly began at Palmdale in 1983, with experience of earlier Orbiters enabling it to be completed with a 49.5 per cent reduction in man-hours. In October 1986, it was rolled out to Pad 39B and fully stacked with ET/SRBs for tests of a new weather protection system, emergency evacuation and slide wire tests. Following the post-Challenger modifications, its dry mass was 78,255 kg, with a 24,950 kg

Endeavour makes the 16th Shuttle landing in Florida, to conclude STS-57 (NASA) 0516777

The upgraded Multifunction Electronic Display Subsystem (MEDS) is shown in the cockpit of Orbiter Atlantis 0054448

capacity into a 204 km, 28.5° orbit; dry mass afterward was 78,372 kg. Atlantis was the first to fly the upgraded GPCs, on STS-37 in 1991. It was returned to Palmdale following return from STS-46 in August 1992 for conversion as the second EDO and modifications to allow it to carry an RSC Energia docking unit for Mir flights. It returned to KSC 29 May 1994.

Endeavour (OV-105)

Work began 1 August 1987, at Rockwell's Palmdale and Downey facilities, on construction of the Challenger replacement. Final assembly was performed in Palmdale Building 150. NASA awarded a USD1.3 billion contract to Rockwell (total cost about USD1.8 billion), which at peak production employed 2,000. OV-105 was rolled out, on schedule, 25 April 1991 and formally handed over to NASA. It arrived at Kennedy Space Center 6 May, minus its main engines, forward RCS module and OMS pods. The 20 seconds on-pad Flight Readiness Firing (FRF) standard for all new Orbiters was made 6 April 1992. While the FRF was considered successful, all three of the new engines were replaced because of minor irregularities. Rockwell completed mating of the major structural elements, apart from the OMS pods, in January 1991. Onboard systems were powered up for the first time 6 July 1990. Endeavour made its debut as an Extended Duration Orbiter in March 1995 for STS-67.

While duplicating the original OV-103 and 104 configurations in most respects, post-Challenger modifications included the new GPCs with higher-density memories. They also included quick disconnect valves from ET to Orbiter; carbon brakes to resolve the overheating problems experienced on most landings; modified RCS valves to prevent them sticking open and operating continuously or dumping fuel overboard; and a crew escape hatch. A manually deployed 12.2 m diameter nylon/Kevlar drag parachute on a 26 m line is carried at the tail base to shorten rollout by 760 m and cut nosewheel contact speed by 37 km/h to 260 km/h, reducing tyre/brake wear. It also improved handling characteristics, particularly in crosswinds and on wet runways. The system was retrofitted to the other Orbiters. Eight parachute trials were conducted July to October 1990 at the Dryden Flight Research Center by NASA's B-52, employed the previous April to launch OSC's orbital Pegasus vehicle. OV-105 tested the parachute on its six initial missions under a USD33 million Rockwell contract.

OV-105 carried the same thermal tile patterns as 103 and 104 but employed a new type of gap filler: Ceramic Ames Gap Filler (CAGF), developed by Rockwell. The older rubber-impregnated quartz fabric leaves a whitish residue on the tiles as it burns away. CAGF avoids that build-up and can undertake at least 30 missions. It gradually appeared on the other Orbiters as attrition occurred.

Five assemblies (mid-fuselage, body flap, wing gloves, vertical stabiliser and main landing gear doors) were already available, having been built as operational spares under a 1983 NASA contract. Structural assemblies remaining to be built included wings, forward and aft fuselage sections, crew module, payload bay doors and the forward RCS module. Endeavour's waste management system is based on the Space Station design. Previous Orbiters carried nine computer-controlled 23 × 56 cm vent doors along each fuselage side to manage pressure variations during launch and landing, but assemblies 4 and 7 were eliminated on OV-105 because mission experience showed the original design to be conservative. Following the STS-77 mission in May 1996, Endeavour was returned to Palmdale for its first servicing since being introduced to flights in 1992. OV-105 returned to flight with its STS-89 mission to the Mir space station in January 1998 and flew the first US International Space Station mission (STS-88) in December 1998.

OV-106

In mid-1989, NASA was planning to include, in its FY91 budget request, funding for two further Orbiters, partly to support demands on the fleet from Space Station assembly and operations; the funding was ultimately not requested. It was then considered for the FY92 request, aiming for 1997 delivery, but again was not included. The Augustine committee report of December 1990 was against a sixth Orbiter, preferring the funding for a heavy-lift unmanned vehicle. President George HW Bush signed a domestic launch policy in 1991 that rejected any new Orbiters.

As a follow-on to Endeavour's August 1987 contract, NASA and Rockwell began discussions on replacing the spares used for 105. Although for use primarily as spares to replace possible damage on existing Orbiters, they could also be used for construction of a new Orbiter. Rockwell was awarded a 4.5-year, USD375 million contract in late 1989. The contract included USD35 million for FY90, covering fabrication of major structural spares: aft fuselage, crew compartment, lower/upper forward fuselage, mid-fuselage, wings (including elevons), vertical stabiliser (including rudder and speed brake), body flap, forward RCS system and one set of OMS pods. The spares were to be of the same configuration as Endeavour. However, NASA decided, in January 1994, to cancel the work on cost grounds.

External Tank (ET)

The largest Shuttle component (46.88 m high, 8.40 m diameter), the ET dominates the cluster and carries typically 730 tonnes of propellant for the Orbiter's three main engines. It remains the only part of the stack not recovered. As its 29,930 kg dry mass equates with that of the Shuttle's original theoretical maximum payload, there were continuing proposals for it to be taken into orbit. It is jettisoned shortly before orbital speed to ensure a controlled re-entry. Releasing it at the most economic moment would leave it tumbling in low orbit, to re-enter within days.

The ET provides the cluster's backbone at lift-off, absorbing the SRB/SSME thrust. It incorporates three units, the LOX tank forward, containing 625,850 kg at −183°C, the intertank, with instrumentation, and aft, the LH_2 tank, containing 104,308 kg at −253°C. Total loaded mass, which varies for each mission, is about 760 tonnes. External insulation is provided by a 2.5 cm multilayered polyurethane foam coating to withstand the extreme thermal differential. At launch, the propellants are pressure-fed at a combined rate of 3,900 litres/s (1,407 kg/s) through 43.2 cm diameter feedlines to the three SSMEs. The tanks are self-pressurised: 3.0 atmosphere LH_2; 1.43 atmosphere LOX. The ET is jettisoned after 8.5 minutes, at about 114 km altitude. Venting residual LOX was used to begin a slow tumble, preventing skipping off the atmosphere and promoting break-up over a remote ocean area. The valve has been omitted since STS-65 as the tank tumbled anyway when venting was deliberately not used on STS-31. A 'beanie cap' on the launch tower was added, to siphon off vented oxygen from the apex during countdown, to prevent ice formation and possible launch damage to the thermal tiles. It was first fully implemented for STS-2. The ET carries typically 770 kg of residual propellants at separation, intended as a safeguard against premature depletion before main engine cut-off. Early exhaustion of LH_2 would be particularly serious as running the SSMEs on an oxygen-rich mixture would cause erosion.

NASA decided, in 1994, to proceed with development of an Super Light Weight Tank version (SLWT), awarding Lockheed Martin Manned Space Systems USD172.5 million for the work. Replacing 2219 aluminium with the Weldalite 2195 aluminium-lithium alloy, changing to an orthogrid design and refining the foam insulation spraying technique will save 3,600 kg. The alloy is 30 per cent stronger and 5 per cent less dense than the current material and comprises 1 per cent Li, 4 per cent Cu, 0.4 per cent Ag, 0.4 per cent Mg and 94.2 per cent Al. The saving trades directly to payload capacity, necessary for Space Station missions. A 12.2 m long full-diameter version, essentially a segment of the hydrogen tank with a LOX tank dome at one end, was delivered to NASA Marshall 1 February 1996 for 6 months' dynamic and pressure testing to verify the structural design. A one-piece graphite phenolic nose cap that does not require the spray-on foam replaced the former version in 1997. Studies showed it was possible for the insulation to break off and impact an Orbiter window, with fatal consequences. The new design also eliminated 961 fasteners, requiring only about 50. It was used for the first time on the STS-91 flight launched in June 1998.

Shuttle Discovery lifts off on 27 May 1999 to deliver supplies and cargo to the ISS 0054452

The major post-Challenger modification involved the installation of a mechanical latch on the Orbiter side of the 43 cm feed lines to prevent the fatal closing of either flapper valve before separation. These 43 cm disconnect mechanisms were replaced by simplified 35.6 cm versions in 1993, with flapper valves folded against the walls (instead of opening sideways). This cleaner design, using 51 per cent fewer parts, permitted the same propellant flow; the rest of the 43 cm plumbing remains. The projected cost of an ET rose from USD1.8 million in 1971 to USD10.1 million 10 years later. A September 1983 contract for 26 ETs plus material for another 21 (following the existing contract for 15 ETs) was worth USD505 million. ETs are manufactured by Lockheed Martin Space Systems at NASA's Michoud Assembly Facility in New Orleans for NASA's Marshall Space Flight Center and shipped by ocean-going barge to Kennedy Space Center. After the 1986 Challenger accident, production of the 59 contracted tanks was halted, with seven available, with a 1985 contract for a further 60 held in abeyance. The USD1,797 million contract was finally awarded in August 1989.

Solid Rocket Boosters (SRB)

The SRBs produce their thrust rapidly upon ignition and burnout in about 130 seconds, having provided the main thrust to lift the Shuttle cluster up to about 45 km. Thiokol is the motor prime contractor, selected in 1974. Other contractors included McDonnell Douglas (now a part of Boeing) for the structures and USBI (now a part of United Space Alliance) for SRB checkout, assembly, launch and refurbishment. Each booster generates a sea level thrust of 14,678 kN at launch and together they provide 71.4 per cent of the thrust at lift-off and during first stage ascent. After separation, they coast upwards to about 65.9 km. They then fall in a ballistic trajectory for four minutes; at 4,700 m, a barometric switch causes the 3.5 m drogue first and then three 41.4 m parachutes to deploy; the SRBs splashdown at about 95 km/h for recovery in the Atlantic about 227 km downrange. Each SRB is 45.46 m long, 3.7 m diameter and (from

The ISS Unity connecting module is moved into the payload bay of Orbiter Endeavour prior to STS-88 0054449

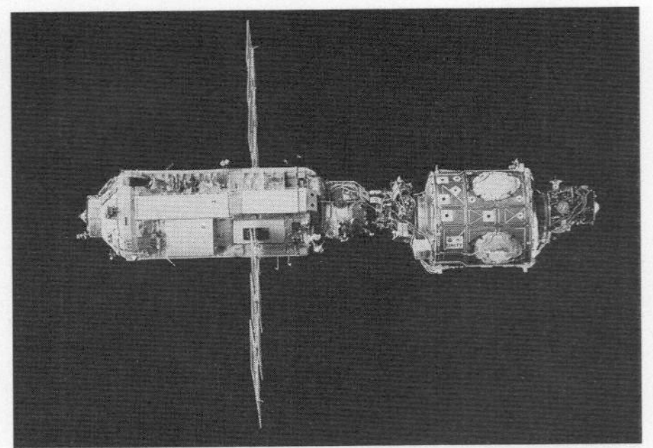

Zarya and Unity modules viewed from Endeavour (STS-88) (NASA) 0062168

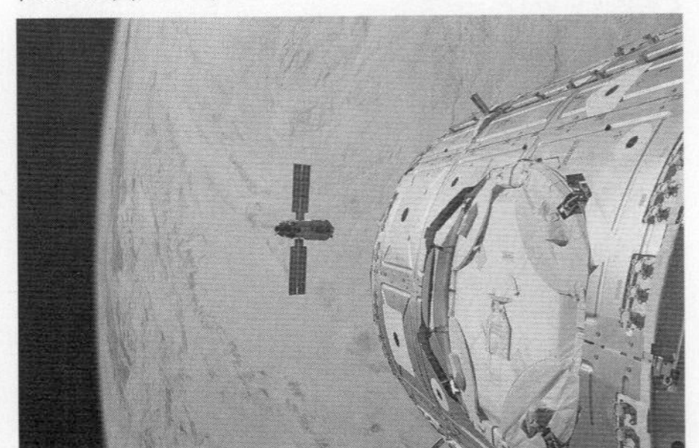

The Russian-built Zarya Functional Cargo Block (FGB) approaches Shuttle Endeavour and the Unity module (NASA) 0062166

STS-6) weighs about 589,670 kg, of which about 503,950 kg is propellant – a mixture of ammonium perchlorate (oxidiser, 69.6 per cent by mass), aluminium (fuel, 16 per cent), iron oxide (a catalyst, 0.4 per cent), a polymer binder and an epoxy curing agent (1.96 per cent). It is configured to provide high thrust at ignition, reducing by about one-third after 50 seconds to avoid overstressing the vehicle during maximum dynamic pressure. The SRBs are interchangeable matched pairs, each comprising four motor segments. SRB ignition can occur only when the ground crew manually removes a lock pin from the Safe & Arm device. The computer launch sequence starts the three SSMEs in sequence (each within 1/4 second), at –6.6 second; only when they have all attained 90 per cent, rising to 104 per cent, will the four onboard computers command SRB ignition. Two Orbiter-type APUs provide hydraulic power from T-28 seconds until separation for TVC. Each APU is fuelled by 10 kg of hydrazine.

On 1 February 1987, a year after the Challenger accident, NASA reported that the PDR of the Redesigned Solid Rocket Motor (called the Reusable SRM since 1995) was complete; CDR followed in October 1987. The field joints now incorporate a tang capture feature to reduce joint deflection, a third O-ring, changes to the design of the joint insulation, sealing without putty and external heaters. The heaters maintain seal temperature at greater than or equal to 24°C in all weather conditions and a weather band prevents water penetration. The nozzle/casing joint, which showed evidence of erosion on some flights, was similarly redesigned. Five full-scale RSRM firings were made, culminating in the Product Verification Motor (PVM-1) test at Thiokol's Brigham City facility on 18 August 1988. At the time of Challenger's loss, 11 motors remained to be used. Thiokol fired these over November 1988 to late 1991 to empty them for conversion to the current design; the firings also generated further engineering data. NASA annnounced July 1991 a USD2.6 billion Buy 3 contract extension to Thiokol for 67 further flight pairs to 1999/STS-105, plus eight ground test motors.

Proposals by contractors for a second-generation SRM design led to the award, in August 1987, by NASA's Marshall facility, of study contracts to five companies for an Advanced SRM, providing the Shuttle with payload increases of up to 5,440 kg. This almost made the Shuttle deliver as much payload as NASA originally promised, being close to the original 29.5 tonnes). Capacity was to be 27.67 tonnes into a 280 km orbit and raised Space Station 28.5° orbit capability from 20.9 tonnes to 26.3 tonnes, essential for station assembly. Thrust profile was designed to greatly reduce Orbiter main engine throttling during maximum dynamic pressure, removing some 175 SSME and APU critical failure modes associated with the potential inability to throttle up. Lockheed Missile Systems Division, teamed with Aerojet Space Booster Co, was selected April 1989 to develop ASRM. An associated 3.5-year USD550 million contract required construction of a NASA production facility at Yellow Creek in Iuka, Mississippi capable of handling 32 motors annually. Programme cost was estimated at USD3.9 billion, but Congress cancelled it in October 1993 after about USD2 billion had been expended. The cost of cancellation was estimated at USD240 million, and Lockheed/Aerojet handed over the Yellow Creek facility to NASA in December 1994. Thiokol then planned to transfer its RSRM nozzle manufacturing there, but was directed 2 May 1995, by NASA, to discontinue work.

Studies were also conducted into liquid boosters for future Shuttle and other launch vehicle use. They would have increased Shuttle payload capability by 5.5 to 9 tonnes and added an abort capability during early ascent, but would have required USD5 billion and eight years to develop, plus USD500 million more for pad modifications. NASA's initial studies ended in Spring 1989, with General Dynamics and Martin Marietta completing their studies in December 1989.

The intention was to order five SRB pairs, at about USD50 million per pair, for each Orbiter and reuse them 20 times. However, one pair was lost on STS-4 as a result of excessive water impact speeds, leading to enlargement of the main parachutes from the original 35 m. A second pair was lost with Challenger on STS-51L. After recovery, each is broken down into 14 major components for refurbishing, with the sections being reused separately. Total mass was steadily reduced – for example, reducing wall thickness in some areas produced a 1.8 tonne saving. A proposal was also made to use the SRBs as sounding rockets on each mission, since each could carry 90 kg of instruments without loss of performance. A USD46 million programme was planned (but cancelled when Orbiter weight reduction proved cheaper) to reduce the weight of each SRB by 2,720 kg, adding 540 kg capacity into the planned 51.6° Space Station orbit. A further 1,025 kg capacity would be created if the parachutes and various recovery structural items were omitted, saving 5,125 kg per booster.

Space Shuttle Main Engines (SSME)

The SSMEs underwent upgrades beginning in 1983 under a 10-year USD1 billion improvement programme. Improved turbopumps were introduced for STS-26. In the 18 months before that mission, SSME firing trials accumulated the equivalent of more than 40 missions. On 15 November 1991, total testing and flight time passed 500,000 seconds, the equivalent of 320 missions; it exceeded 600,000 seconds in December 1993. Pratt & Whitney high-pressure turbopumps, allowing 7.5 hr of firing (55 mission equivalents) between major overhauls replaced Rocketdyne's versions. The older pumps required refurbishment after 33 minutes use (three flights). Problems increased development cost to USD504 million, so NASA halted fuel pump work from December 1991 in order to focus on the more critical oxygen unit for the Block 1 upgrade.

It restarted work May 1994 for a 1997 debut as part of the Block 2 upgrade once the LOX pump was on schedule for its July 1995 flight debut. The Block 1 upgrade included a single coil heat exchanger (eliminating seven weld joints) and a two-duct hot gas manifold, which allowed the fuel pump to run at lower temperatures and pressures. A single Block 1 upgrade engine (number 2036) flew successfully on STS-70 in July 1995. All three engines were upgraded for STS-77 in May 1996. Testing of the Block 2 upgrade began at NASA's Stennis Center on number 0521, once Block 1 certification was completed 26 May 1995. Block 2 included P&W's fuel pump and a 10 per cent diameter larger throat combustion chamber that reduced wear, reducing pump discharge pressure by 4.5 to 6 per cent and pump operating temperature by 7 to 12 per cent. All three engines on STS-86 in September 1997 introduced the Block 2. STS-85 was the last mission to fly with an old engine.

Computer and automated control systems

Each Orbiter's five IBM AP-101 General Purpose Computers (GPCs) of 1972 vintage were replaced by IBM AP-101S models designed in 1984, offering 2.5 times the capacity and three times the speed for half the size and mass. At peak, they are used at more than 50 per cent of their capacities, contrasted with 80 per cent for the older models. The upgraded GPCs first flew aboard Atlantis on 1991's STS-37. New Block 2 Honeywell SSME controllers were initially installed on OV-105 and then retrofitted to the others, providing four times the capacity with the first space-qualified 68000 series microprocessors. They also eliminated a separate 13.6 kg vibration monitoring box and added 20 engine parameter monitoring channels to the previous 120. The last of 30 ordered was delivered in April 1993. The KT-70 Inertial Measurement Units (IMUs) were augmented and eventually replaced by the High Accuracy Inertial Navigation System, of similar configuration but with increased accuracy and reliability; the first one flew on Atlantis during STS-44 in 1991. A new TACAN system was also due. Endeavour was delivered with the new items installed.

Remote Manipulator System (RMS)

Spar Aerospace developed the RMS as Canada's principal Shuttle contribution. It is 38 cm in diameter, 15.3 m long, has a mass of 408 kg, and is capable of handling a 29.5 tonne payload. Shoulder, elbow and wrist joints are driven by DC electric motors and can be controlled in five different modes, ranging from manual hand controllers for the crew, using direct observation and TV cameras on the elbow/wrist, to full computer control. Stowed on the bay's port side and latched into three cradle pedestals, it can be restowed by EVA if it jams in use or, in the last resort, jettisoned by pyrotechnics. Each RMS is designed for a 10 year, 100 mission life. The first, the number 201 DDT&E arm and associated ground equipment, was delivered April 1981 at a cost to Canada of CAD100 million. The operational 301/302/303 models were delivered May 1982, 14 December 1983 and 29 March 1985, respectively. 302 were lost on Challenger, but were not replaced until the refurbished 202 qualification model debuted on STS-66 in November 1994. RMS' highly successful use as a mobile crew work platform, in addition to its primary role as a manipulator, established it as an indispensable tool. A servicing centre based on RMS derivatives is Canada's primary Space Station contribution. Spar was contracted in 1991 to upgrade the arms to handle up to 265 tonnes for berthing the Orbiter to Space Station. This includes a CAD30 million redesign of arm control electronics, followed by a refurbishment and overhaul of various critical components, completed in 1996.

Extravehicular Mobility Unit (Space Suit)

ILC Dover has produced NASA's space suits since the Apollo programme and works as a sub-contractor under Hamilton Sundstrand, NASA's prime contractor for the Extravehicular Mobility Unit (EMU).

EVAs have played a major role in Shuttle activities. Through February 1996, 35 different astronauts had conducted 60 EVAs totalling 378.2 hours, had repaired and redeployed four satellites and brought back to Earth two more. Crew members must be able to leave the Orbiter for routine work and emergencies. Three of the 14, 117 kg, suits are normally carried on each EVA mission (four were carried for the first time on STS-49 in May 1992 and again STS-61 December 1993) and two for emergencies on non-EVA missions. They are stowed in the airlock, where they are donned and doffed, their size adjusted to fit the crewmembers trained for EVA on a particular mission. The crew uses 30.5 m tethers when working in the payload bay, unless they are working in the MMU or on the RMS foot restraint.

Cheaper and more flexible than Apollo's lunar suits, donning the suit begins with a zip-on one-piece Spandex liquid cooling/ventilation garment. Weighing 3 kg and with a projected life of 15 years, it includes urine collection facilities for up to 950 ml and a drink bag containing 0.6 litres of water.

Shuttle Endeavour's remote manipulator arm re-positions the Unity module prior to docking with the Russian Zarya module during STS-88 0085781

EVA astronauts Newman and Ross work on the Zarya and Unity module during STS-88 0085780

A Snoopy cap with headphones and microphone fits over the head and chin, and also provides caution and warning tones. Over this fits the two-piece spacesuit itself, comprising: hard upper torso, made in five sizes with hard waist ring. The lower torso comes in various sizes with hard waist ring; the helmet (one size only), and the visor assembly, which protects against micrometeoroids and the Sun's UV/IR radiation fits over the helmet. A more dextrous glove was tested during STS-37 in 1991. Expected life is 15 years for the hard sections and six years for the softer portions. There are no zippers on the suit; the components connect with hard snap-rings. The suit is composed of several bonded layers, beginning with a polyurethane-on-nylon pressure bladder, followed by Kevlar layers with folded/tucked joints (for mobility) and ending with a Kevlar, Teflon and Dacron anti-abrasion layer. The hard upper torso has a glass fibre shell. The materials are designed to prevent fungal or bacterial growth; unlike the individual Apollo suits, they are used by different astronauts and must be cleaned and dried after each use. Each suit can be used up to 25 times during a mission.

The Portable Life-Support System (PLSS), attached to the back of the upper torso, provides sufficient oxygen and water for 7 h inside the suit, including 6 h for EVA and a 30-minute reserve. There is also a secondary pack to supply O_2 and maintain suit pressure for 30 minutes if the primary system fails. Astronauts can plug in to the Orbiter airlock for fresh supplies of O_2 and water. A liquid crystal display/control module on the front of the upper torso indicates remaining EVA capacity, and warns of excessive consumption or problems with the water pressure and T in the cooling garment. A suit vents 0.7 kg of water/O_2 hourly, a potential source of contamination. Hamilton Standard (now Hamilton Sundstrand, a United Technologies Corporation subsidiary) was awarded a USD97.3 million contract in 1991 for two further PLSS (adding to the 15 built) and continued engineering support through September 1997, bringing contract total value to USD212 million. Astronaut training and suit maintenance costs NASA some USD100 million annually.

NASA planned to provide suitless crew members with a Personal Rescue Enclosure (PRE), an 86 cm diameter inflatable sphere made of suit material. Sitting inside with a portable oxygen system, the occupant would have had a communications system and viewing port, and sufficient pressurising gas for a 1 hr rescue. If an Orbiter became disabled, suited crew members could carry the others across to a rescue Orbiter; alternative methods included use of the RMS and passing them along a line as in a ship-to-ship rescue. The PREs were not produced.

The initial disadvantage of the Shuttle's suit is that, as it is pressurised to only 0.29 atmosphere (4.3 psi) of pure O_2, the wearer must prebreathe, in-suit, pure O_2 for 3.5 hr prior to an EVA to prevent N_2 bubbles forming in the joints, a condition known as 'the bends' among divers. Previously on the Shuttle, prebreathing began with an hour, during which cabin pressure is reduced from 14.7 psi to 10.2 psi. After a 12 hr 'campout', the EMU was donned and, during checkout, another 40 minutes was spent breathing pure O_2 before depressurizing the airlock. Prebreathing has been reduced to a 2.5 hour process after studies showed that exercise improved denitrogenisation; new hose and mask prebreathing equipment was also introduced. The process includes one hour of in-suit prebreathing.

To overcome the prebreathing necessity, which might delay an EVA to meet an emergency, NASA, ILC Dover and Hamilton Sundstrand began work on a zero prebreathe, 8.3 psi high-pressure hard suit called Mark III. NASA decided later, because of the cost (estimated in 1995 at USD700 million), to use Shuttle-type suits on the Space Station. The ISS uses '1 atmosphere' (equal to Earth sea level), requiring the full 3.5 h pre-breathe. Hamilton also proposed a USD10 million minor upgrade that would operate the suit at 0.4 atmosphere, the same as Russia's Orlan suit, which requires only 30 minutes prebreathing. Thermal modifications were made as Station EVA work is away from Shuttle's warm cargo bay. STS-69 in September 1995 tested fingertip heaters and the body cooling loop can be shut off.

Manned Manoeuvring Unit (MMU)

The 153 kg Manned Manoeuvring Unit, which snapped on to the back of the PLSS, enabled an astronaut to operate free of tethers. Two hand controllers, operating 24 × 7.6 N N_2 thrusters, provided 6 DoF and a total 6,190 Ns impulse. Two were carried in the forward cargo bay when required. Martin Marietta built only three at USD10 million each. On missions 10/11/14 they were used nine times, with an accumulated flight time of 10 hr 22 minutes. NASA began a USD600,000 study in mid-1990 to determine the cost of restoring them to flight readiness, possibly for a flight in late 1991 carrying an IMAX camera and other detectors to observe Orbiter thruster firings and water dumps. A total of about USD5 million was projected. However, NASA management ordered them returned to storage in 1991 as they were developed for demonstration purposes and are too limited for effective deployment.

The 37.6 kg Simplified Aid for EVA Rescue (SAFER) was tested September 1994 during STS-64 to show its value in an emergency, such as when an astronaut comes adrift from Space Station. Astronauts must work in pairs, carefully tethered, and with the Orbiter ready to provide rescue. A drifting astronaut would activate SAFER, to stop any tumbling, using its automatic attitude hold mode and then jet back to safety. SAFER is attached below the PLSS, providing a 3 m/s total impulse from a 1.4 kg pressurised N_2 supply via 24 3.5 N thrusters. It was first used operationally during March 1996's STS-76 EVA at Mir, as it would have taken Atlantis 20 minutes to undock for a rescue.

Shuttle support equipment, maintenance and reliability

Shuttle Carrier Aircraft (SCA)

A Boeing 747-123 N-9668, purchased by NASA from American Airlines in June 1974 after logging 8,999 hr, was modified (and registered as N-905NA) to ferry Orbiters between Shuttle facilities and to carry Orbiter *Enterprise* for the original ALT programme. For ferry flights, a tail cone fairing is installed on the Orbiter to reduce aerodynamic drag and buffet, and aerosurface control locks are added to the Orbiter's elevons. The Orbiter is carried unmanned with its systems inert. A bailout system is carried for the 747 crew. With use of the Kennedy runway limited, the aged lone 747 was a vital element of the Shuttle system. Had it been damaged or lost, the Shuttle fleet would have been grounded. In September 1987, NASA signed a letter of agreement with Boeing for a second aircraft. An ex-Japan Air Lines 747-SR was delivered to Boeing's Wichita plant 17 April 1989 under a USD55 million contract to undergo conversion; NASA accepted it in November 1990. Designed to carry 109 tonnes (22 tonnes more than the older aircraft), N-910NA has a 1,700 km range and was designed for 265 ferry flights over 15 years.

The USAF's 433rd Military Airlift Wing at Kelly AFB, Texas, operates two Lockheed C5As modified by the company under an October 1985 USD133 million contract for carrying a loaded Shuttle cargo container and its towing tractor. These C5A SCM (Shuttle Cargo Modification) aircraft, delivered November 1988/Oct 1989, can handle a full cargo bay load such as the Hubble Space Telescope.

Shuttle operations

Processing/turnaround

NASA initially planned an annual flight rate of seven for a full four-Orbiter fleet, through the Space Station era. The processing goal for each mission was 88 days, covering 60 days in the Orbiter Processing Facility, 5 days in the VAB and 23 days on the pad. Endeavour set a 52-day OPF record preparing for STS-57 in 1993, beating a previous record, of 1993's STS-45, by 3 days. Platforms and equipment from the mothballed Vandenberg site were installed in KSC's Orbiter Maintenance & Refurbishment Facility, creating a third OPF position from late 1991, and allowing simultaneous processing of three Orbiters.

In September 1983, it was announced that Lockheed Space Operations Co (now Lockheed Martin Space Operations), a consortium including Grumman, Thiokol and Pan American World Services, had won the NASA/USAF Shuttle processing contract, effective for three years from 1 October 1983, with four three-year options. NASA exercised the USD1.60 billion second option in September 1989, extending the contract to 30 September 1992. The third was exercised in 1992, worth USD1.830 billion through September 1995, bringing the total for 1983–95 to USD6.3 billion. A one year USD638 million extension began 1 October 1995, in preparation for the new approach of awarding Shuttle operations to a single contractor. LMSOC assumed all the work from October 1995. Before then, Lockheed Martin's share as consortium leader was 75 per cent. Grumman, responsible for launch processing, had a 13 per cent stake. Thiokol's task of processing the ET and SRB (including recovery after launch) was worth 11 per cent, and Johnson Controls World Services (who bought out Pan American), responsible for operations analysis, process planning/control and logistics engineering, received 1 per cent.

Orbiter pilot training

Pilot training includes at least 30 flights on Shuttle Training Aircraft (STA), with about 900 practice landing approaches. Some 10 training sorties are made weekly at the White Sands Space Harbour. The aircraft are housed in the NASA hangar at El Paso International Airport, Texas. STAs are Gulfstream II executive jets, with Rolls-Royce engines and extensive structural, wiring and cockpit changes and with installation of direct lift control systems and thrust reversers to simulate Orbiter flight characteristics. In early 1985, NASA ordered a third STA from Grumman, the modifications costing USD14 million. A fourth was being considered before the Challenger accident to support a launch rate of about 10 per year. It became operational in April 1991. Honeywell's glass cockpit displays were fitted from 1997 as they are installed on the Orbiters. The two pilots during STS-58 in 1993 introduced a new training aid designed to refresh landing skills on long missions: the Pilot In-flight Landing Operations Training (PILOT) unit is a portable computer with an Orbiter-type hand controller. It is mounted on the console in front of the pilot's flight deck seat and the controller attached to the actual Orbiter controller.

Shuttle maintenance and reliability

All of the Orbiters return to Palmdale every three years for a major inspection, recertification, and maintenance. Realistic analyses of the chance of a Shuttle loss resulting from a 1988 NASA analysis, state the chance of loss at a 1 in 78 median probability of loss during ascent. In 1989, NASA recalculated the values at 1 in 94. Then in 1995, NASA again upgraded safety projections for any Shuttle mission as 1 in 145 (estimates ranged from

1 in 76 to 1 in 230), made up in part of a factor of 1 in 248 for ascent only. The main engines are the major contributors with a mean probability of failure of 1 in 410. With the upgraded engines, the chance will level off at less than 1 in 200 for the whole mission and less than 1 in 300 for ascent. The first estimates that NASA provided to the American public in 1975 indicated that the loss rate would be less than 1 in 1,000,000.

Orbiter systems modifications

Space Shuttle systems have undergone hundreds of modifications and improvements over the 25-plus years since the spacecraft entered service. Following the Columbia accident, a number of improvements were made to the External Tank (ET) and Thermal Protection System (TPS). Notably, the foam around the ET's bipod fitting of the forward attach point, which was at risk of falling from the ET during launch, was eliminated and improved foam spraying processes were applied to areas where insulation foam remains part of the ET design. Additionally, Inspection, repair and replacement procedures for the Shuttle's TPS Reinforced Carbon-Carbon (RCC) wing edge insulation tiles were made more rigorous. NASA also examined inspection, repair and replacement of parts to the Shuttles' Rudder Speed Brake (RSB) and foreign object debris processes, and made component and process improvements to these as well. Ground and launch segment equipment was also upgraded and augmented, including the Shuttle pad crawler transporter and launch and Orbiter camera equipment.

Through the first several years of launch operations, Shuttle in-orbit missions usually lasted less than 10 days. Columbia and Atlantis underwent major modifications to provide additional fuel cell cryogenic supplies and improved life support systems for 16-day missions and, as a result, were referred to as Extended Duration Orbiters (EDO). The Columbia retrofit was completed in 1992. The 13-day US Microgravity Lab STS-50, in June 1992, used the EDO for the first time. Atlantis underwent the same conversion following return from STS-46 in August 1992 and was also equipped for docking with Mir in June 1995. Endeavour was built with the required plumbing and electrical connections and flew its first EDO mission as STS-67 in March 1995.

In order to make the modifications, Rockwell added a cryogenic pallet in the aft cargo bay to extend fuel cell capacity and enlarged the storage space in the cabin for additional food and clothing. In addition, they installed an improved waste collection system (first flown operationally STS-65 July 1994) and a reusable CO_2 scrubber to replace the previous lithium hydroxide canisters. Atlantis and Endeavour are capable of carrying EDO kits allowing up to 28-day missions, although none is yet planned.

During the preliminary planning for the Shuttle, NASA announced that the spaceplane would have a mission duration of up to 30 days. When it actually flew, NASA could not promise an on-orbit duration of more than eight days and returned to Congress to request additional funds for modification. In November 1987, NASA provided Congress with a report on methods of extending the in-orbit capability of at least one Orbiter to 16 days. This extended the time from the pre-Challenger limitation by fuel-cell life of 9–10 days. In May 1983, Rockwell proposed a two-stage plan to extend stay-time to 15 to 18 days by 1985 and to 45 days by 1988–89, but NASA did not proceed with it. The first stage, comparable to the current improvements, was projected to cost USD200 million. The second stage, costing USD200-250 million, would have required an orbiting solar array and battery system with which the Shuttle could dock. This required relocation of the airlock and mid-deck modifications and gained some DoD support as a cheaper alternative to space station development. NASA agreed to Congressional suggestions in early 1989 of providing OV-105 with a 28 day mission capability at a FY90 cost of USD52 million.

For EDO, the 2,926 kg cryo pallet plumbs four tank sets to the standard five, each capable of supporting two days' activities. The pallet could carry four further sets on its aft side but Columbia is too heavy to be converted to a 28 day Orbiter. Hamilton Standard's (now Hamilton Sundstrand) 147 kg Regenerative CO_2 Removal System, forward of the mid-deck lockers, operates two beds in a 15-minute cycle, one absorbing the gas while the other is exposed to vacuum. A similar system was flown on the Skylab space station. Some N_2 is also recovered for addition to the two extra tanks carried under the payload bay against the aft bulkhead. STS-50, the first EDO flight, still carried 50 LiOH canisters as back-up (astronauts had to repair the RCRS in flight). Although RCRS is now fully qualified, EDO missions continue to carry sufficient canisters for four days.

The EDO 22 kg trash compactor was demonstrated aboard 1990's STS-35, carried in place of a mid-deck locker. Rubbish is loaded in a 0.014 m^3 plastic bag, equivalent to one person's daily waste and compacted by pistons at 41 N/cm^2 into quarter the volume. The bag is then stored under the mid-deck floor. Some items, such as the rehydratable food package, have been redesigned to allow crushing. STS-35 also demonstrated the new Waste Containment System. The old toilet has to be removed/cleaned after each flight, but Hamilton Standard's newer design collects the waste in bags, which are then compacted into storage tubes with odour/bacterial filters. Using removable bags means the toilet has unlimited capacity. Development versions were carried by Endeavour on STS-54/57 in 1993, allowing the then record-breaking STS-65 July 1994 to be the first operational mission.

NASA has studied the Longer Duration Orbiter (LDO) modification for longer missions. The preliminary results indicated that 30 days were possible and 60 days appeared feasible if the Shuttle remains attached to the Space Station for mutual support. Power (fuel cells) is the limiting factor beyond 30 days. A solar array module was proposed, which could also make it independent of Space Station. LDO was deleted from FY94's budget to save USD43 million, particularly as long missions were available aboard Mir, and now, the ISS.

Beginning in 1994, NASA and its contractors modified Atlantis to fly seven Mir docking missions between 1995 and 1997 and Discovery for two missions. Engineers believed the Orbiter structures would be good for more than 100 missions, allowing flights until around 2030. The thrust structures are the principal limiting factor on a Shuttle lifetime. Discovery, Atlantis and Endeavour are modified for Space Station operations. Orbiter Atlantis stood down for 10 months while engineers and technicians replaced electromechanical cockpit instruments with 11 colour flat-panel screens, improving pilot efficiency and reducing costs of eliminating the need to maintain outdated equipment. Atlantis arrived at Boeing's orbiter processing facility at Palmdale, California, on 14 November 1997 and returned with more than 130 major modifications on 27 September 1998.

Space Shuttle specifications

Shuttle payload capacity

Structural analyses and reviews of landing stresses following Challenger's loss showed that landing weight could be increased by 8,619 kg to 104,328 kg. Abort landings, which constrain launch mass, are limited to 108,860 kg. Discovery, Atlantis and Endeavour can deliver 24,990 kg into 204 km orbits; they are referred to as 55,000 Orbiters because of their 55,100 lb capacities into NASA's standard 120 n mile (220 km) orbit. Originally NASA promised the Shuttle would deliver 65,000 to the standard orbit. Columbia could handle 3,152–3,840 kg less because of its greater dry mass. As a result, it was not planned for Station assembly use.

Shuttle costs

NASA's own figures for the first 20 flights made by Columbia, Challenger and Discovery showed an average cost of USD257 million. The improved Discovery, new Atlantis and modified Columbia were expected to reduce the average, but not to the USD74 million per 'dedicated flight' (use of the whole payload capacity) which NASA intended to charge after 1988. NASA quoted a cost per mission of USD286 million in 1990, but charges only USD130 million at 1988 rates. True amortised rates suggest that the real cost of Shuttle missions is probably in excess of about USD350 million (1986 rates). The true cost is debatable, depending on what costs are included. NASA in 1992 estimated the *marginal* cost (that is, the saving by reducing the flight rate from eight to seven) at only about USD40 million. But NASA's Shuttle operations budget produces a figure of USD450 million for FY95. Adding other programme costs such as TDRS, ASRM and civil service salaries puts the figure at more than USD500 million in 1992 terms.

NASA estimates the Shuttle's original development at USD5,150 million (FY71 rate). The actual cost was USD10.1 billion (real year cost), or USD6,744 million (FY71 rate). The 1971 estimate of USD250 million for each Orbiter rose to about USD2 billion in 1983 dollars, though this included, for the first time, engines and equipment, Remote Manipulator System, galley and the closed-circuit TV.

Major Shuttle Contracts in 1982

Contract:	Value in USD millions:
Rockwell (Shuttle Orbiter)	3,560
Grumman (wings)	45
McDonnell Douglas (OMS/RCS pods)	85
McDonnell Douglas (Support)	52
Rocketdyne (Main Engines)	1,546
Thiokol (Solid Rocket Boosters)	206
USBI (SRB assembly/retrieval)	89
Martin Marietta (External Tank)	529

Following the Challenger accident, NASA submitted, in February 1987, a series of proposed actions in response to the Rogers Commission recommendations, with a total estimated cost of USD1,379 million. The estimated total expenditures to meet this goal were actually USD2.4 billion at the time of STS-26.

Other costs may be estimated from subsequent actual mission costs. As an example, President Reagan's 15 point policy, issued in March 1988, was designed to create opportunities for US commercial space interests. The proposals emphasised the Shuttle-launched Industrial Space Facility as a man-tended microgravity platform and the use of Spacehab inside the payload bay. It was projected that the former price of USD74 million for the use of the whole cargo bay could go as high as USD255 million by the early 1990s. As the 1990s began, NASA cited a new price of around USD180 million, which represented full cost recovery if there were 14 flights annually. NASA reached a *quid pro quo* agreement with the DoD in October 1987 citing USD115 million (FY89 rate) for a full cargo bay but, in 1996, NASA quoted USD200 million to USD250 million for DoD.

President Reagan's August 1986 edict, prohibiting NASA from using the Shuttle for commercial satellite launches, made his earlier Shuttle pricing policy. The intention was to charge a minimum of USD74 million (1982 USD) for a full payload bay and to offer three such missions by auction two years in advance for commercial and foreign use. At less than half the cost of a Shuttle mission, NASA was hoping that demand would increase the price.

Aborts

Four basic abort modes are available to Shuttle controllers, the precise scheduling of each varying with the mission ascent profile. Return To Launch Site (RTLS) would be used in the event of a main engine failure in the first 260 seconds. The remaining engines and RCS thrusters would achieve a pitch-around, enabling the Orbiter to jettison the ET 45 km from the coastline and glide back to the KSC runway. RTLS is considered to be a very risky manoeuvre.

The Transatlantic Landing (TAL) abort is required when an SSME fails after the last opportunity for

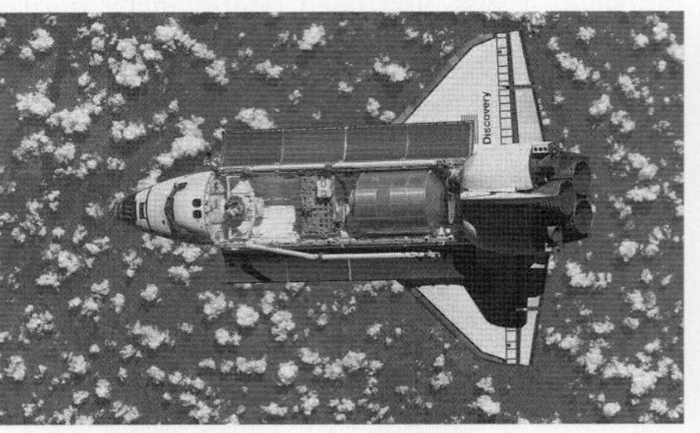

Shuttle Discovery, during STS-114, as taken from aboard the ISS; the Raffaello MPLM is visible in the cargo bay (NASA) 1133867

Space Shuttle Discovery, during STS-114, docked to the ISS's Destiny laboratory (NASA) 1133868

The new cockpit upgrade was flown for the first time on STS-101 0085851

RTLS and before an Abort to Orbit is possible. There are four TAL sites where an Orbiter would land about 25 minutes after launch: Moron Air Base (Spain), Yundum International Airport (Banjul, Gambia), Zaragoza Air Base (Spain, prime site for high-inclination flights) and Ben Guerir Airfield (Morocco). All sites provide Shuttle visual and navigational aids (principally TACAN and a Microwave Landing System); two or three are activated for each ascent and one (usually Ben Guerir) remains active throughout in case of emergency de-orbit. There are also two ACLS (Augmented Contingency Landing Site), at Hickham AFB (Honolulu) and Andersen AFB (Guam). They are similar to TAL sites in facilities and equipment but used instead for contingency de-orbits and not ascent aborts.

Abort Once Around (AOA) would be used from about 2 minutes after SRB separation, again in the event of a main engine failure. The procedure relies on two OMS firings after ET jettison to place the Orbiter in a sub-orbital coast and free return orbit for re-entry and glide back to the runway at KSC, Edwards or White Sands. Abort To Orbit (ATO) is available in the event of a main engine failure after passing the AOA point. Again the procedure relies upon two OMS firings, one for orbit insertion and the other to circularise the path. An alternative mission might then be possible, depending on mission rules. An ATO occurred on STS-51F in July 1985, but most of that mission was completed. Each mission has its emergency landing sites, which vary with the launch profile, but they carry no Shuttle equipment or aids. In the case of STS-51F, a landing could have been made at Zaragoza, although it would have taken at least 5 weeks to recover Challenger. The AOA site was White Sands Space Harbour, New Mexico.

One abort site safety device is a 13.7 m high arresting net attached to an energy-absorbing tape drum, capable of stopping an Orbiter travelling at 185 km/h. Tests of this Shuttle Orbiter Arresting System (SOAS) were made with Enterprise in June 1987 at Washington DC Dulles airport. Nets were installed at the four TAL sites and at Hickham AFB. Similar abort procedures were devised for Vandenberg launches and arrangements were made for emergency landings on several Pacific Islands, including Chile's Easter Island.

Crew escape system

Following Challenger's loss, each Orbiter was retrofitted with a telescoping pole escape system to permit evacuation while the craft is in a controlled glide descent. As such, it provides very limited cover and would primarily be called upon in the event of a water ditching, which would probably be non-survivable for a crew remaining with the craft. The system comprises a 109 kg, 7.5 cm diameter aluminium/steel pole that would be extended 2.67 m through the jettisoned hatch to guide evacuating crew 45° down and 15° back to avoid the port wing. The abort decision must be made below 18,300 m, with the Cdr and GPCs achieving a controlled glide descent before the designated jumpmaster, occupying the mid-deck seat closest to the door, manually depressurises the cabin and blows the hatch pyrotechnically at 6,710 m. Bale-out begins at 6,100 m, with each of up to eight crew members attaching the suit lanyard to a hook to slide down the pole. The main parachute opens automatically within 6 seconds. All crew members now wear 32 kg high-altitude pressure suits carrying emergency oxygen, liferaft and other survival equipment.

The transcript of onboard crew conversation, plus the fact that at least two of the air-breathing packs were activated before Challenger fell into the Atlantic, convinced the astronauts that even though it was a 'worst case' accident, at least some crew members could have survived if there had been an escape system.

Tspace CXV

Current Status

Transformational Space Corporation (t/Space) is a consortium of private-enterprise space companies, formed to offer manned and unmanned space launch systems. The company's most visible project is the Crew Transfer Vehicle (CXV), proposed to NASA as a low-cost, low-risk means of ferrying crew to the International Space Station after the Shuttle is retired.

Background

Partners in t/Space include AirLaunch LLC, which is developing the QuickReach air-launched rocket for the Defense Advanced Research Projects Agency's Falcon programme; Scaled Composites, developer of the SpaceShipOne manned suborbital technology demonstrator and the SpaceShipTwo suborbital commercial transport; and Universal Space Lines.

The CXV concept combines technology from QuickReach and SpaceShipOne with proven features from other programmes. The CXV vehicle measures 27.5 m long, and the booster is a scaled-up version of the QuickReach launcher. Like QuickReach, it uses propane fuel and LOX. A small amount of fuel and oxidizer is heated around the rocket nozzle and the vapor pressure is used to pressurise the fuel tanks and feed the motor. As on QuickReach, the upper-stage nozzle is submerged in the lower-stage propane tank.

The release system differs from QuickReach because the rocket is carried externally. It is similar in that the rocket is allowed to rotate tail-down, with the rotation being braked by a parachute attached to the tail until a near-vertical position is reached. On QuickReach, the rotation takes place as the rocket tips over the launch aircraft's ramp; on the CXV, a lanyard restrains the rocket's nose as it falls away from the carrier.

The 4–6 passenger CXV capsule is based on the truncated-cone aeroshell shape used by US reconnaissance satellites to return film to Earth before the development of all-electronic imaging systems. t/Space says that the shape is naturally stable on re-entry and less sensitive to re-entry angle than the conical Apollo capsule. The capsule will descend by parachute into water, like NASA capsules. Airborne launch obviates the need for an escape tower: in the event of a problem on launch, the capsule separates from the rocket and descends by parachute.

Scaled Composites would produce the capsule structure, and is also developing a Very Large Aircraft (VLA) to carry the launch vehicle. In concept, it would resemble Scaled's White Knight carrier vehicle, but would be much larger and fitted with at least four engines of the type fitted to large commercial aircraft.

In early 2006, the CXV programme was waiting for a strong expression of interest from NASA. The agency appeared to be leaning at that time towards fielding the in-house Crew Exploration Vehicle as soon as possible, as a human-rated Shuttle successor.

NAVIGATION

China

Beidou series

Current Status
Beidou M1, the first Chinese navigation satellite to be placed in a medium Earth orbit, was launched on 13 April 2007. The fourth satellite to be placed in GEO, Beidou 2A, was reportedly the first of a new Beidou generation. It was launched from Xichang aboard a CZ-3A on 2 February 2007. It experienced difficulty in deploying its solar arrays after an apparent explosion of the liquid propellant main engine during the transfer to geostationary orbit.

Background
China developed a two-satellite navigation satellite technology in the late 1980s, and it was tested in 1989 on two communications satellites. The Beidou (BD) system was given the go-ahead in 1993. The name Beidou, meaning 'north dipper', comes from the seven stars which are called The Plough in the UK and The Big Dipper in the US. The Chinese also give the name "Compass" to the Beidou constellation.

The Beidou satellite navigation experimental system, based on satellites in GEO, covers the region between longitude 70–140°E and latitude 5 – 55°N. The system will provide two navigation services. The service available to commercial customers will provide positioning accuracy within 10 m, speed accuracy within 0.2 m/s and timing accuracy within 50 nanoseconds. In addition to the "open" service, the system will also offer a restricted positioning, velocity and timing communications service to government and military authorities. The BD system is fundamentally different from GPS or Glonass, being based on two-way signals and a central control station. The satellites transmit at 2491.75+/–4.08 MHz and the ground receiver transmits back to the satellite on 1615.68MHz. The main control station sends inquiry signals to the users via two satellites. When the user terminal receives a signal from one satellite, it transmits a response to both satellites. The central station receives the user's responding signals from the two satellites and the time difference of arrival between the two signals is then compared with a terrain database to provide a position accurate to 100 m. The system can handle 540,000 requests per hour (GPS and Glonass handle unlimited users) and user equipment is bulkier and requires much greater power than GPS. The ground systems include a main control centre, three ground tracking stations (at Jamushi, Kashi and Zhanjiang) for orbit determination and ground correction stations.

In the early 2000s China registered four geosynchronous orbit locations named Compass with the International Telecommunications Union. The four Compass locations were simply identified by their longitudes. In June 2006 it was reported that Compass was the designation of a planned all-new Chinese satellite navigation system similar to GPS or Galileo. The Compass Navigation Satellite System will eventually include at least 35 satellites, made up of five in Geostationary Earth Orbit (GEO) and 30 in Medium Earth Orbits (MEO).

The Beidou 1 series of four experimental satellites was developed by the China Academy of Space Technology and based on the DFH-3 communications satellite bus. The box-shaped satellite weighs approx. 2,300 kg, and carries 1,200 kg propellant for its main engine. Design life is approximately 5 years.

The first Beidou satellite was launched into GEO on 30 October 2000. At the time, it was not clear whether this was BD-1 or the first in the BD-1 series. This led to the first four satellites in the series being given the designations 1A – 1D, although western sources also named them as Beidous 1 – 4. Since the fourth satellite is reported to have been the first of a new generation, it is now generally designated as Beidou 2A.

The location of BD 1A over 140°E was registered for ChinaSat-32, and therefore it was expected that the next BD satellites would eventually use the ChinaSat –31, –32 and –33 locations. This was verified with the launches of two more BD satellites by 2003, and the navigation and positioning services were opened up to civilian users in April 2004. Beidou 1C was described as a back-up to the first two satellites in GEO.

Early reports indicated that Beidou 2A failed to achieve geostationary orbit after an explosion at the time of the first planned apogee burn by the spacecraft's liquid propellant engine on 2 February 2007. The US Space Surveillance Network reported 70–100 pieces of debris and large numbers of smaller pieces of debris. However, the malfunction seems not to have been catastrophic, and the satellite eventually reached GEO. On 12 April, Chinese official media reported that the satellite had experienced trouble deploying its solar panels, but this problem had now been overcome. It was initially to serve as a backup satellite.

The first satellite to be placed in MEO was Beidou M1, launched by a Long March 3A from Xichang at 20:11 UT on 13 April 2007. It has been speculated that this constellation may eventually comprise four operational satellites plus one spare in six orbital planes.

Specifications
Beidou 1A
Launch: 30 October 2000 using a CZ-3A from Xichang.
Location: 140°E (not announced by China: this is derived from US tracking data): Chinasat-32 location.
Design life: >8 years?
Contractor: China Academy of Space Technology (CAST).
Configuration: Uses the DFH-3 satellite bus: 3-axis stabilised, 2.2 × 1.72 × 2 m box plus two vanes of solar panels on communications satellites. The solar panels span 18.1 m.
Mass: ~2,300 kg at launch, ~1,150 kg on-station at the beginning of operations.
Power system: Solar panels provide ~2 kW BOL, 1.7 kW at the EOL.

Beidou 1B
Launch: 20 December 2000 using a CZ-3A from Xichang.
Location: 80°E (not announced by China: this is derived from US tracking data): Chinasat-31 location. Other specifications as BD-1A.

Beidou 1C
Launch: 24 May 2003 using a CZ-3A from Xichang.
Location: 110.5°E (not announced by China: this is derived from US tracking data): ChinaSat-33 location. Other specifications as BD-1A.

Beidou 1D/2A
Launch: 16:28 UT on 2 February 2007 from Xichang by CZ-3A; launch delayed by some three years.
Location: 144°E (not announced by China: this is derived from US tracking data).

Beidou M1
Launch: 20:11 UT on 13 April 2007 from Xichang by CZ-3A.
Orbit: The first to be placed in MEO. Initial orbital parameters were apogee 21,545 km, perigee 21,519 km, inclination 55.3°, period 773.4 min.

International

Galileo

Current Status
Galileo is the civil-operated navigation satellite system being developed in Europe both to complement the US GPS and Russian Glonass and to assure a continued capability for European users in the event of any future denial of service by these military-owned systems.

To be controlled ultimately by the European Union, the 30-satellite Galileo, comprising ten spacecraft in each of three orbital planes, is due to enter full operation from 2010. It is designed to work with GPS and Glonass to deliver real-time positioning accuracy down to the metre range, which is unprecedented for a publicly available system. But it will differ from the military systems in offering service availability guarantees to civil users under all but the most extreme circumstances and will be capable of informing users within seconds of a failure of any satellite. This will make it particularly suitable for safety-related applications such as aircraft approach and landing support.

Background
Initial development work is being carried out by the European Commission (EC)/European Space Agency (ESA) Galileo Joint Undertaking (GJU). Among the GJU's tasks are the procurement and launch of satellites for the In-Orbit Validation (IOV) phase of the programme, management of supporting research and development work, integration of Galileo into Europe's EGNOS satellite navigation enhancement system, and establishment of the operating company that will manage the constellation and offer service under a Public-Private Partnership (PPP) structure.

The operating company is being formed by two formerly competing consortia: Eurely (AENA, Alcatel, Finmeccanica, Hispasat) and iNavSat (EADS, Inmarsat, Thales). In June 2005 the two groups announced that they would merge and work with the GJU to agree a final contract covering Galileo operation and commercialisation.

Once the contract has been agreed, the members of the operating company will continue their early work to define and eventually implement a portfolio of revenue-generating services aimed particularly at the telecoms, transport and security sectors, as well as taking over and carrying forward the IOV programme and the subsequent full deployment of the system.

The first stage of IOV, based on the Galileo In-Orbit Validation Element (GIOVE) A and B satellites, is already under way. Development of GIOVE-A by Surrey Satellite Technology Ltd (SSTL) of the UK began in July 2003 and the spacecraft was launched at the end of 2005.

The mission of the GIOVE satellites is to prove new technologies for operational use; demonstrate the feasibility of broadcasting the high-accuracy, near-real-time orbit-determination and time-synchronisation data envisaged for the full Galileo system; work in the frequencies provisionally allocated to Galileo by the ITU and so secure them for operational use; and characterise the radiation environment in the medium Earth orbit that the operational satellites will occupy.

GIOVE-A started transmitting test navigation signals in January 2006. In March ESA submitted evidence to the ITU that the spacecraft had transmitted successfully on all the required frequencies.

GIOVE-A is due to be joined in September by GIOVE-B, built by Galileo Industries (a consortium comprising Alcatel Alenia Space, Astrium, Galileo Sistemas y Servicios and Thales) and currently undergoing final integration and testing at Alcatel Alenia's Rome facilities. Larger than GIOVE-A, the second spacecraft will be used primarily to test the extremely accurate Passive Hydrogen Maser (PHM)-based atomic clocks that will be the primary time source aboard the operational Galileo satellites.

The second stage of IOV will see four representative Galileo spacecraft – ordered from Galileo Industries at a cost of USD1.15 billion in January 2006 – launched into operational orbits from 2008. They will be used to validate the space and ground segments of the full system before the remaining 26 satellites –23 operational and three spares – are launched and full operational capability is achieved by around 2010.

The ground segment will comprise two European-located Galileo control centres (GCCs), each capable of controlling the satellites and managing the navigation/positioning service. The GCCs will receive data via a dedicated network from 20 Galileo Sensor Stations (GSSs) around the world and use it to update integrity information and synchronise the satellite and ground-station clocks. Data will be exchanged between the GCCs and the satellites via five S-band uplink stations and 10 C-band uplink stations located around the world.

Once the full Galileo constellation is in place, services will be available worldwide at latitudes up to 75° degrees north. Five types of service will be offered.

The Open Service (OS) will be free to all and based on signals broadcast in two bands, 1164–1214MHz and 1563–1591MHz. Receivers using both bands will achieve an accuracy of 4 m horizontally and 8 m vertically; single-band accuracy will be 15 m horizontally and 35 m vertically, comparable with the current GPS C/A service.

The encrypted Commercial Service (CS) will be available for a fee and will offer an accuracy of better than 1 m. It could also be complemented by ground stations to increase accuracy to better than 10 cm. The signal will be broadcast in three bands, the two used for OS plus 1260–1300MHz. The encrypted Public Regulated Service (PRS) and Safety of Life Service (SoL) will both provide accuracy comparable to that of OS. They will be characterised by robustness against jamming and the quick and reliable detection and reporting to users of system problems affecting integrity and accuracy. Their respective target markets are security authorities (police and military), and safety-critical transport applications as air traffic control and aircraft approach and landing.

The Galileo satellites will also carry a transponder to detect and report signals from the established Cospas-Sarsat search-and-rescue system, based on aircraft and ship beacons operating in the 406.0–406.1 MHz band. Galileo will thus contribute to the Global Maritime Distress and Safety System (GMDSS).

Specifications
GIOVE-A
Launch: 28 December 2005 by Soyuz-Fregat from Baikonur
Location: Near-circular orbit at 56° inclination and 23,258 km altitude
Design Life: Two years
Contractors: Surrey Satellite Technology Ltd (SSTL)
Payload: Two rubidium atomic clocks, two navigation signal-generation units operating in the 1.1–12 GHz and 1.5 GHz Radio Navigation Satellite Service (RNSS) bands, L-band phased-array transmit antenna, two radiation sensors; laser retro-reflectors for accurate obit determination
Principle Applications: Test navigation signal broadcasts to secure the use of the frequencies allocated by the International Telecommunications Union (ITU) for the Galileo system; demonstrate technologies for the navigation payload of operational Galileo satellites; characterise the radiation environment of the operational orbit
Configuration: Cuboid three-axis-stabilised bus with a pair of rectangular solar panels spanning 7 m from tip to tip when deployed
Mass: 660 kg (at launch)
AOCS: Momentum wheels and butane thrusters
Power: Two solar panels; 2 hr-duration batteries

GIOVE-B
Launch: Scheduled for September 2006 by Soyuz-Fregat from Baikonur
Location: Near-circular orbit at 56° inclination and 23,258 km altitude
Design Life: Two years
Contractors: Galileo Industries
Payload: Similar to that of GIOVE-A, but with the addition of more accurate passive hydrogen maser (PHM)-based atomic clocks and a third navigation signal-generation unit.
Principle Applications: Similar to GIOVE-A, with the addition of proving of the PHM clocks to be used in the operational spacecraft and transmissions of a third signal format
Configuration: Similar to GIOVE-A
Mass: Similar to GIOVE-A
AOCS: Similar to GIOVE-A
Power: Similar to GIOVE-A

Galileo
Launch: Launch of the first of a total of 30 operational and spare Galileo satellites is scheduled for 2008
Location: Near-circular orbit at 23,222 km altitude. The final constellation will comprise three orbital planes, each inclined at 56° to the equator and containing nine operational spacecraft and a spare. The three planes will be evenly separated at 120° intervals around the equator, and within each plane each satellite will be 40° from its neighbours
Design Life: At least 12 years
Contractors: Galileo Industries
Payload: Primary payload is the Navigation Signal Generation Unit (NSGU) and associated systems such as the mission receiver (MISREC) used for reception of data from the ground in relation to functions such as system integrity monitoring, and the security units, used for encryption/decryption of communications links with the ground to deny unauthorised access. Antennas include the L-band signal broadcast unit, C-band MISREC unit and S-band transmit/receive antenna for satellite command and control of the satellites
Principle Applications: A range of high-accuracy navigation and positioning services for both civil and government users, available worldwide up to and beyond 75° latitude.
Configuration: 2.7 m × 1.2 m × 1.1 m cuboid three-axis-stabilised bus with a pair of rectangular solar panels spanning 18.7 m from tip to tip when deployed
Mass: 675 kg
Power: Solar arrays yielding 1,500W at end of life

Russian Federation

Glonass Series

Current Status
The Russian government is continuing with a plan to re-establish its Glonass navigation satellite system alongside the US GPS and Europe's new Galileo. A new triplet of spacecraft has been launched each year since 2000, the improved Uragan-M satellite has been deployed and the new, more modern, Uragan-K is under development.

Background
The GPS equivalent Global Navigation Satellite System, Globalnaya Navigatsionnaya Sputnikovaya Sistema (Glonass) system has been under deployment since 1982 and became fully operational in December 1995. Glonass follows in a long tradition of Soviet navigation satellites going back to 1967's Kosmos 192 'Tsyklon'. The Soviet Union informed the International Telecommunications Union it would inaugurate the Glonass service in 1982. On 12 October 1982, a Proton launched the first three satellites on a single booster from Baikonur and identified them as Kosmos 1413–1415. In order to achieve the constellation spacing the satellites first entered a low altitude parking orbit at 51–60 kms. Then the upper stage performed two manoeuvres to place the three payloads into their required orbits. The Soviets repeated this pattern on their next five launches during the next three years. Curiously, each trio had only one or two Glonass pre-operational satellites and the remaining seven were dummies, although still called Kosmos and given Kosmos numbers. The 10 operating satellites weighed 1,260 kg and had one-year design lives. But they achieved an average of 14 months. During this pre-operational testing phase the Soviets only had six satellites simultaneously operating at any one time. The last launch of the pre-operational phase included the first Block 2a, transmitter with improved clocks. Six Block 2a satellites eventually appeared and each had an average life of 17 months.

With the pre-operational phase complete, the Soviets entered into a new phase and launched the satellites on a more direct profile, to distinguish it from the earlier phase. Each Proton carried the new satellites directly into an initial inclination of almost 65°. Again, two manoeuvres by the upper stage spaced the satellites properly, but no orbital plane change occurred during this manoeuvre. All subsequent missions have used this more direct route. The operational satellites carry the name Uragan.

With September 1986's launch, the Soviets put up the first payload in which all three satellites were live. But a launch failure marred the debut of the Block 2v which the Soviets claimed had a 24-month design life. The Soviets launched 12 Block 2Bs between 1987 and 1988, but only half reached orbit. These satellites achieved a 22-month average life. The 1,415 kg Block 2v first appeared in September 1988. The new satellite has a reported three-year lifespan and is the basis of the current Glonass system.

Two 1990 missions raised the number of operational craft to 11 and 1992's first launch maintained the active number at 11. There were 13 active satellites by September 1992. Lifetimes began to improve significantly following the launch on April 1994, which had a radiation, hardened satellite. This launch also filled every slot in planes one and three at some time. Then August 1994's launch began the process of filling plane two.

Glonass planners wanted to have a Phase 1 system of 10 to 12 operational satellites ready by the end of 1990, but didn't realise that goal until 11 March 1992. From this original deployment the Russians planned to grow the system to a full operational capability of 21 craft, plus three in-orbit spares by the end of 1995. Still, by end of 1993 the Russians had 13 active satellites (the oldest from 1990) plus one reserve. The system continued to grow in 1994, building to 16 satellites before the end of that year. When the Russians announced a full deployment of 24 satellites after the launch of three in December 1995 they had thus made up for schedule slippage and achieved a considerable success. At least part of the urgency centred on keeping pace with the west in having a commercial navigational system in place to compete with GPS. However, Glonass was not adopted on the same massive scale as GPS.

Russia's own Institute of Radionavigation & Time estimated in 1993 that a scant 200 to 500 civil Glonass receivers had made it into widespread use compared with the hundreds of thousands of contemporary GPS receivers. As the Soviets originally conceived the system, maritime and aeronautical users would be the principal consumers of signals allowing positional accuracy within 100 m surface location and 150 m altitude and 15 cm/s velocity readings.

The system has an operating philosophy that closely mirrors GPS. One difference arises from the geoid model upon which the two systems base their calculations. GPS uses the WGS-84 geodetic systems. Glonass uses SGS-85 (Soviet Geocentric System), adding an error of a few metres whenever a receiver averages between signals from the two systems. Satellites transmit in narrow bands centred around 1.250 and 1.6035 GHz. The higher frequency band creates complaints from radio astronomers that cite its interference with the hydroxyl line at 1.612 GHz. In response the Russians began changing the frequencies of some satellites beginning in September 1993.

In terms of constellation the Glonass does differ from the US system, which uses circular semi-synchronous orbits near 20,000 km (718-minute period). Glonass satellites fly at a slightly lower altitude of 19,100 km (676 minute period) at inclinations of 64.8° which gives a precise retrace of the ground track every 17 revolutions (eight days). The VKS Military Space Forces launch and operate Glonass and monitors it through five principal monitoring and command stations: Moscow, St. Petersburg, Eniseyisk, Komsomolsk-on-Amur and Balkhash. The VKS performs control functions from a single location at Golitsyno-2 in the Moscow region, the core of VKS's command system.

NPO PM has responsibility to produce the satellites which are manufactured at AKO Polyot, using the NIIKP Institute Space Device Engineering's payloads. NIIKP has responsibility for manufacturing both the civil and military receivers. The Institute of Radionavigation and Time maintains the ground time/frequency standards.

After December 1995, there was a three-year gap before the next triplet was launched, and two more years passed before another launch in 2000. The result was that the constellation dropped below the 18 needed to maintain global real-time coverage, placing Glonass on the path to extinction. Efforts to promote Glonass as a European substitute for GPS failed. In 2001, however, the Russian government decided to invest in the system and rebuild it, planning to reach 18 satellites in 2007 and to reach 24 (21 plus three on-orbit spares) by 2010–11.

The 10 December 2003 launch introduced the Uragan-M spacecraft. The 1,570 kg Uragan-M improves stability of the three cesium clocks from 5×10^{-13} daily to 1×10^{-13} and extends the series lifetimes to five to seven years per satellite. The new satellite is associated with other improvements, including a civil reference signal on

Glonass satellites	Launch		Plane	Slot	Block
1	12 October 1982	Kosmos 1414	1	1	1
2	10 August 1983	Kosmos 1490	1	3	1
3		Kosmos 1491	1	2	1
4	29 December 1983	Kosmos 1519	3	18	1
5		Kosmos 1520	3	17	1
6	19 May 1984	Kosmos 1554	3	19	1
7		Kosmos 1555	3	18	1
8	4 September 1984	Kosmos 1593	1	2	1
9		Kosmos 1595	1	3	1
10	17 May 1985	Kosmos 1650	1	1	1
11		Kosmos 1651	1	2	2a
12	24 December 1985	Kosmos 1710	3	18	2a
13		Kosmos 1711	3	17	2a
14	16 September 1986	Kosmos 1778	1	2	2a
15		Kosmos 1779	1	3	2a
16		Kosmos 1780	1	8	2a
17*	24 April 1987	Kosmos 1838	3		2b
18*		Kosmos 1839	3		2b
19*		Kosmos 1840	3		2b
20	16 September 1987	Kosmos 1883	3	18	2b
21		Kosmos 1884	3	17	2b
22		Kosmos 1885	3	24	2b
23*	17 February 1988	Kosmos 1917	1		2b
24*		Kosmos 1918	1		2b
25*		Kosmos 1919	1		2b
26	21 May 1988	Kosmos 1946	1	8	2b
27		Kosmos 1947	1	7	2b
28		Kosmos 1948	1	1	2b
29	16 September 1988	Kosmos 1970	3	24	2v
30		Kosmos 1971	3	18	2v
31		Kosmos 1972	3	19	2v
32	10 January 1989	Kosmos 1987	1	2	2v
33		Kosmos 1988	1	3	2v
34	31 May 1989	Kosmos 2022	3	19	2v
35		Kosmos 2033	3	24	2v
36	19 May 1990	Kosmos 2079	3	17	2v
37		Kosmos 2080	3	19	2v
38		Kosmos 2081	3	20	2v
39	8 December 1990	Kosmos 2109	1	4	2v
40		Kosmos 2110	1	7	2v
41		Kosmos 2111	1	5	2v
42	4 April 1991	Kosmos 2139	3	21	2v
43		Kosmos 2140	3	22	2v
44		Kosmos 2141	3	24	2v
45	29 January 1992	Kosmos 2177	1	3	2v
46		Kosmos 2178	1	8	2v
47		Kosmos 2179	1	1	2v
48	30 July 1992	Kosmos 2204	3	24	2v
49		Kosmos 2205	3	18	2v
50		Kosmos 2206	3	21	2v
51	17 February 1993	Kosmos 2234	1	3	2v
52		Kosmos 2235	1	2	2v
53		Kosmos 2236	1	6	2v
54	11 April 1994	Kosmos 2275	3	17	2v
55		Kosmos 2276	3	23	2v
56		Kosmos 2277	3	18	2v
57	11 August 1994	Kosmos 2287	2	12	2v
58		Kosmos 2288	2	16	2v
59		Kosmos 2289	2	14	2v
60	20 November 1994	Kosmos 2294	1	3	2v
61		Kosmos 2295	1	6	2v
62		Kosmos 2296	1	4	2v
63	7 March 1995	Kosmos 2307	3	20	2v
64		Kosmos 2308	3	22	2v
65		Kosmos 2309	3	19	2v
66	24 July 1995	Kosmos 2316	2	15	2v
67		Kosmos 2317	2	10	2v
68		Kosmos 2318	2	11	2v
69	14 December 1995	Kosmos 2323	2	9	2v
70		Kosmos 2324	2	–	2v
71		Kosmos 2325	2	13	2v
72	30 December 1998	Kosmos 2362	1	–	2v
73		Kosmos 2363	1	–	2v
74		Kosmos 2364	1	–	2v
75	13 October 2000	Kosmos 2376	1	–	2v
76		Kosmos 2374	1	–	2v
77		Kosmos 2375	1	–	2v
78	1 December 2001	Kosmos 2380	2	–	2v
79		Kosmos 2382	2	–	2v
80		Kosmos 2381	2	–	2v
81	25 December 2002	Kosmos 2394	3	–	2v
82		Kosmos 2395	3	–	2v
83		Kosmos 2396	3	–	2v
84	10 December 2003	Kosmos 2402	Unk	–	2v
85		Kosmos 2403	Unk	–	2v
86		Kosmos 2404	Unk	–	2v
87	26 December 2004	Uragan-M	Unk	–	2v
88		Uragan	Unk	–	2v
89		Uragan	Unk	–	2v
90	25 December 2005	Uragan-M	Unk	–	2v
91		Uragan-M	Unk	–	2v
92		Uragan	Unk	–	2v

*Launch failure. The slot listing is the initial position for each; some later transferred to other slots.
Note: 1: spare.

L2 frequency to substantially increase the accuracy of navigation relaying on civil signals. The new civilian band includes an ionospheric correction transmission, which allows for compensation based on the comparison of effects on signals of two different frequencies. Passage through the ionosphere generally creates the largest error in a receiver's position. Some Uragan-Ms have been launched on mixed manifests with older Uragans.

A third-generation spacecraft, Uragan-K, is under development. It is built on the Express platform, designed by NPO PM, weighs under half as much as the Uragan-M and has a longer lifetime, estimated at 12–15 years. It will also have a third civil reference signal within the new frequency band of 1164–1215 MHz to enable civil airlines to use satellite-aided positioning as the primary navigation facility. The proposed new signal will also supply users with system integrity data and deliver high-precision differential corrections to ensure user real-time positioning in absolute space accurate to 12 cm or less. Through co-ordination with GPS and Galileo, Glonass-K will also improve the efficiency of emergency search and rescue operations. Uragan-K satellites can be launched either in groups of six from a Proton-M from Baikonur, or in pairs by a Soyuz-2/Fregat from Plesetsk.

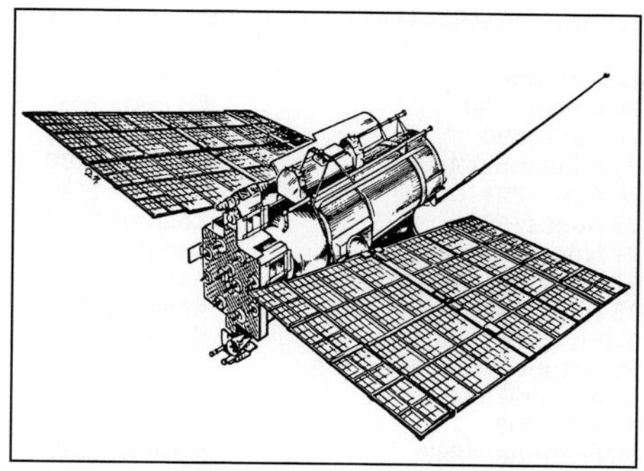

Glonass navigation satellites are launched by Proton, typically in triplets. Height is 7.840 m with magnetometer boom deployed; 7.230 m span across solar wings (Nicholas L Johnson/Kaman Sciences) 0517156

Parus

Current Status

The latest Parus satellite was launched as Kosmos 2414 by a Kosmos 11K65M launch vehicle from Plesetsk on 20 January 2005.

Background

Developed by NPO PM, the Parus series of satellites was launched primarily to provide navigation services to the Soviet Union's warships and submarines. The first trace of this system appeared on an SL-8 Kosmos launch in 1967. Following a two-phase test programme at 74° inclination between 1967 and 1972, the Soviets declared the system operational following the launch of Kosmos 514, which completed a three-satellite constellation at 83° inclination. Eventually the Soviets phased this system out in 1978 (the last launch, Kosmos 1027, occurred on 27 July 1978). In its place they established the current constellation of six Parus 'sail' satellites in orbital planes spaced 30° apart and inaugurated by Kosmos 700 (26 December 1974).

Operationally, the Parus spacecraft form part of a system known as Tsiklon-B, providing navigational data and store-dump communications to naval units. The constellation continues to be maintained by Kosmos 11K65M launches from Plesetsk. In late 2006, one launch was expected before the end of the year.

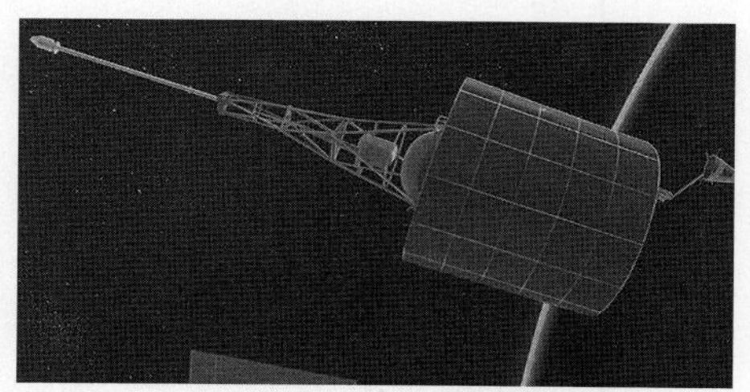

The Parus military navsats are probably almost identical to this Nadezhda civil version. See the Cospas-Sarsat entry for an illustration showing Nadezhda under construction 0517157

Launches of Parus Navigation satellites 1990–2005

Launch Date	Satellite	Node	Plane °E
20 March 1990	Kosmos 2061	90	3
20 April 1990	Kosmos 2074	60	2
14 September 1990	Kosmos 2100	120	4
16 April 1991	Kosmos 2142	150	5
22 August 1991	Kosmos 2154	90	3
27 November 1991	Kosmos 2173	120	4
17 February 1992	Kosmos 2180	180	6
15 April 1992	Kosmos 2184	60	2
1 July 1992	Kosmos 2195	30	1
29 October 1992	Kosmos 2218	90	3
9 February 1993	Kosmos 2233	150	5
1 April 1993	Kosmos 2239	120	4
2 November 1993	Kosmos 2266	30	1
26 April 1994	Kosmos 2279	180	6
22 March 1995	Kosmos 2310	60	2
6 October 1995	Kosmos 2321	[30]	[1]
16 January 1996	Kosmos 2327	30	1
5 September 1996	Kosmos 2334	30	1
20 December 1996	Kosmos 2336	120	4
17 April 1997	Kosmos 2341	60	2
23 September 1997	Kosmos 2346	30	1
24 December 1998	Kosmos 2361	–	–
26 August 1999	Kosmos 2366	90	3
8 June 2001	Kosmos 2378	–	4
28 May 2002	Kosmos 2389	–	1
4 June 2003	Kosmos 2398	–	N/A
22 July 2004	Kosmos 2409	–	N/A
20 January 2005	Kosmos 2414	–	1

The nodes are referred to an arbitrary zero point. Kosmos 2321 was left in its transfer orbit due to a malfunction with the second stage of the Kosmos-3M launch vehicle.

Tsikada/Nadezhda series

Current Status

The Nadezhda navigation satellite constellation remains in use for civil and military missions. The COSPAS payload on Nadezhda 6 was decommissioned on 6 August 2007, leaving no SARSAT search and rescue satellite in operation.

Background

The Tsikada/Nadezhda system provides location information, as well as warning and distress signal identification under the COSPAS-SARSAT system. Tsikada was a civilian system that complemented the Parus military naval navigation satellite system for the Soviet merchant navy and the Academy of Sciences. Development began after the first flight tests of Parus began in 1974. The Tsikada system began with the launch of Kosmos 883 on 15 December 1976, and deployment of the operational Tsikada system began after the launch of Kosmos 1000 in 1978. The first named Tsikada appeared in 1995, probably replacing Kosmos 2123. A total of 20 Tsikadas were launched, including the last (Kosmos 2315) on 5 July 1995. The operational system consisted of a minimum of four operational satellites, each deployed in one of four planes, spaced 45° apart. The Tsikada planes were offset from those of the military Parus satellites to minimise the delay between sightings. Typical mission lifetime was 18 to 24 months.

The first COSPAS search and rescue payload was carried on Kosmos 1383, launched on 29 June 1982. This version of theTsikada was known as Nadezhda (Hope). Since 1995, a modified version, known as Nadezhda-M, has carried the Kurs rescue location system for improved positional information of aircraft and ships in distress, with transmission of digital data to the Kurs Centre. In addition to search and rescue information, the spacecraft are used for location of vessels in the Soviet merchant marine and fishing fleet. Flying below 1,000 km altitude, they provide a positional accuracy of 3 km for 406 MHz emergency beacon receivers and 20 km for 121.5 MHz beacons. Nadezhda satellites carry COSPAS transponders with a 1544.50 MHz downlink. This downlink is only active when the satellite is receiving a distress beacon on 121.5 or 406 MHz. The International COSPAS-SARSAT system will cease satellite processing of 121.5 MHz beacon signals from 1 February 2009. The system has been credited with saving thousands of lives.

The Russian Nadezhda navigation satellite uses a gravity gradient boom for attitude control. At the bottom are the payload and telemetry antennas (Teledyne Brown Engineering) 1330193

Nadezhda launch mass is 830 kg. Spacecraft diameter is 2 m, height 3 m. An internally pressurised compartment houses the primary payload (approx. 0.86 × 0.55 m, and mass 200 kg). Power output from the body mounted solar cells is 140–200 W. Attitude control is achieved using a 10 m gravity gradient boom extending from the top of the spacecraft. Payload and telemetry antennas are attached to the lower, Earth-facing section. Navigational information is derived from Doppler-shifted VHF transmissions of satellite position and orbital data at 150 and 400 MHz. The navigational payload was developed with the help of the Institute of Space Device Engineering. The satellites were originally provided by NPO-PM, but they were later built by PO Polyot in Omsk. They were launched singly from Plesetsk by Kosmos 3M rocket.

Two Nadezhdas have carried small secondary payloads. Nadezhda 5 was launched with a Swedish scientific satellite named Astrid-2. A proposed Tubsat launch on Nadezhda 5 was transferred to a different vehicle prior to launch. Nadezhda 6 carried the SNAP-1 nanosatellite and the 50 kg Tsinghua-1 microsatellite, both designed and built in the UK by Surrey Satellite Technology Ltd. (the latter in collaboration with China's Tsinghua University). Images of the Nadezhda were acquired by SNAP-1 seconds after the satellite was deployed into orbit by four micro-miniature single-chip video cameras. Nadezhda 6 was the first sun-synchronous launch from Plesetsk. Nadezhda 7 was also called Nadezhda-M. The Nadezhda COSPAS role is expected to be taken over by the new Sterkh satellite being developed by PO Polyot, with first launch tentatively set for 2008.

Launches of Tsikada and Nadezhda 1990–2007

Launch date	Satellite	Orbit	Plane
27 February 1990	Nadezhda 2 (COSPAS 5)	956 × 1,021 km, 83.0°	14
5 February 1991	Kosmos 2123 (Tsikada 17)	963 × 1,005 km, 82.9°	13
12 March 1991	Nadezhda 3 (COSPAS 6)	958 × 1,018 km, 82.9°	12
9 March 1992	Kosmos 2181 (Tsikada 18)	973 × 1,014 km, 83.0°	11
12 January 1993	Kosmos 2230 (Tsikada 19)	973 × 1,007 km, 82.9°	11
14 July 1994	Nadezhda 4 (COSPAS 7)	951 × 1,007 km, 82.9°	14
24 January 1995	Tsikada 1 (Tsikada 20)	964 × 1,020 km, 82.9°	13
5 July 1995	Kosmos 2315 (Tsikada-M1)	970 × 1,013 km, 82.9°	11
10 December 1998	Nadezhda 5 (-M) (COSPAS 8)	976 × 1,013 km, 83.0°	14
28 June 2000	Nadezhda 6 (-M) (COSPAS 9)	686 × 712 km, 98.14°	N/A
26 September 2002	Nadezhda 7 (-M) (COSPAS 10)	987 × 1,022 km, 82.93°	12

Nadezhda 5 satellite with Astrid-2 satellite prior to launch from Plesetsk, 10 December 1998. Note red tags on the gravity gradient boom (Swedish Space Corp.) 1330188

United States

GPS constellation

Current Status

The widely used GPS satellite constellation comprises 27 spacecraft – 24 in each of six orbital planes and three on-orbit spares – including Block II, Block IIR and Block IIR-M variants. The USAF plans to launch 12 of the next version, Block IIF, before placing the first Block III in service around 2013.

Background

The Navstar Global Positioning System (GPS) is a space-based radio positioning, navigation and time transfer system operating at L-band frequencies, providing 16 m, 0.1 m/s and 0.1 μs accuracy to military users. The operational space segment consists of 24 Navstar (Navigation System using Timing & Ranging) satellites, including three hot in-orbit spares, in six orbital planes (A-F). Each carries four atomic clocks, two caesium with a stability of 1 second in 300,000 years and two rubidium with about half that stability (Block 1 satellites carry only rubidium clocks). The satellites fly in circular 20,182 km, 55° inclination, 11.967 hour (half a sidereal day) orbits, spaced so that a minimum of four appear to a receiver worldwide. They transmit information on L1 (1,575.42 MHz) and L2 (1,227.60 MHz), phase modulated by the data they carry and by spread spectrum pseudorandom number codes.

The GPS programme began in 1973 as a US Navy initiative to replace the 200 m accuracy of the Transit system. Successful as the Transit had been, it could not match or support the accuracy of newer submarine-launched missiles. The navy awarded Rockwell the prime contract in 1974 to build the first 12 Navstars, but then delayed the programme by deciding to harden the satellites and to re-designate the 11 already launched by Atlas as prototypes. The 12th satellite became a ground qualification unit to prepare for the Block II production series. The navy added funding in 1979 to develop an onboard sensor able to report whether the satellite had been illuminated by laser energy or interfered with by another spacecraft. The design review completed in mid-1982 resulted in the Block II being one-third larger, with lifetime increased from 5 years to 7 years 6 months, larger solar panels increasing power supply from 400 to 700 W, and the adoption of improved atomic clocks. All Block I satellites are retired.

An L1 carrier provides the C/A coarse/acquisition code at 1.023 MHz and both Ll/L2 are modulated by the P precise code at 10.23 MHz. These codes contain a message at 50 bit/s giving details about the satellite ephemeris, atmospheric propagation correction and clock bias information. By measuring the time interval between transmission and reception, the receiver calculates the distance to a satellite. Using the distance to at least three satellites in an algorithm provides the position fix. A minimum of three satellites signals provides latitude/longitude; the fourth adds altitude.

The system was originally designed to deliver an accuracy of about 100 m to civilian users, using the C/A code Standard Positioning Service (SPS), and an accuracy of about 16 m to military users with the encrypted P-code PPS Precise Positioning Service. However, even the earliest commercial C/A SPS receivers originally provided 20 to 40 m accuracy. Until May 2000, the US government degraded the C/A code by Selective Availability (SA) to 100 m by broadcasting slightly erroneous or "dithered" clock and ephemeris information. Differential GPS (DGPS) overcomes SA and provides increased overall accuracy. With DGPS, a transmitter is placed at a known location and its position information is used to calculate corrections which are then communicated to other receivers in the area. With DGPS, civilian users get, 3–5 m possible for post-processing calculations in static positions.

Differential GPS is now widely available through government service such as the Wide Area Augmentation System (WAAS), developed by the US government to aid air navigation, and the US Coast Guard's own DGPS system. Specialised DGPS systems have been deployed by the US to aid in weapon guidance.

GPS has an additional A-S (anti-spoofing) signal to prevent friendly forces being deceived by an enemy deliberately transmitting inaccurate signals. A-S encrypts the P-code to become the Y-code, which is accessed by an auxiliary output chip in the receiver.

The USAF 50th Space Wing's 2nd Space Operations Squadron at Schriever AFB, Colorado, is responsible for monitoring and controlling the GPS satellite constellation. The GPS Operational Control Segment (OCS) consists of a Master Control Station (MCS) at Schriever AFB, Colorado, connected to a global network of six monitor stations (Schriever, Cape Canaveral, Hawaii, Kwajalein Atoll, Diego Garcia and Ascension Island) and uplink antennas at Kwajalein, Diego Garcia, Ascension Island and Cape Canaveral.

Originally designed as an adjunct to strategic weapon systems, GPS has become literally ubiquitous for two reasons: a GPS receiver is entirely passive, so the number of users is unlimited, and the computer technology has made the processing required for GPS inexpensive and compact. Commercial hand-held GPS receivers were used by the US military in the first Gulf War. In the early 1990s, the USAF started development of its first mass-produced weapon using GPS guidance, the Joint Direct Attack Munition (JDAM). It was first used in Operation Allied Force in 1999 and more than 140,000 guidance kits had been produced by early 2006.

GPS: the first Navstar Block 2 satellite (Navstar 14) undergoes solar panel deployment testing at Rockwell's Seal Beach facility. It was delivered to the US Air Force 27 April 1987 as a Pathfinder in the Navstar Processing Facility at Cape Canaveral 0517316

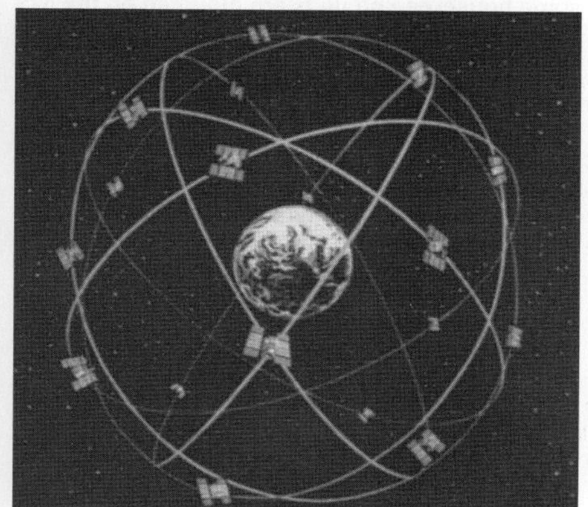

The operational GPS network comprises 24 satellites in six orbital planes. The system can fix the positions of military users to within 16 m and offers a downgraded civil facility (Rockwell) 0516842

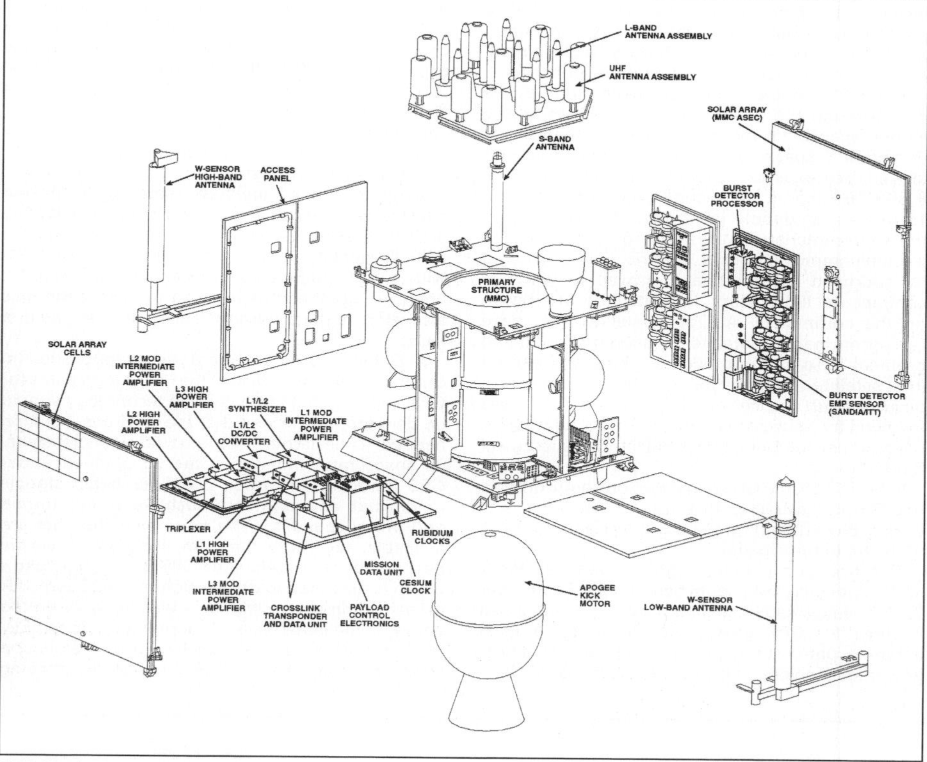

GPS Block 2R principal features (ITT) 0517310

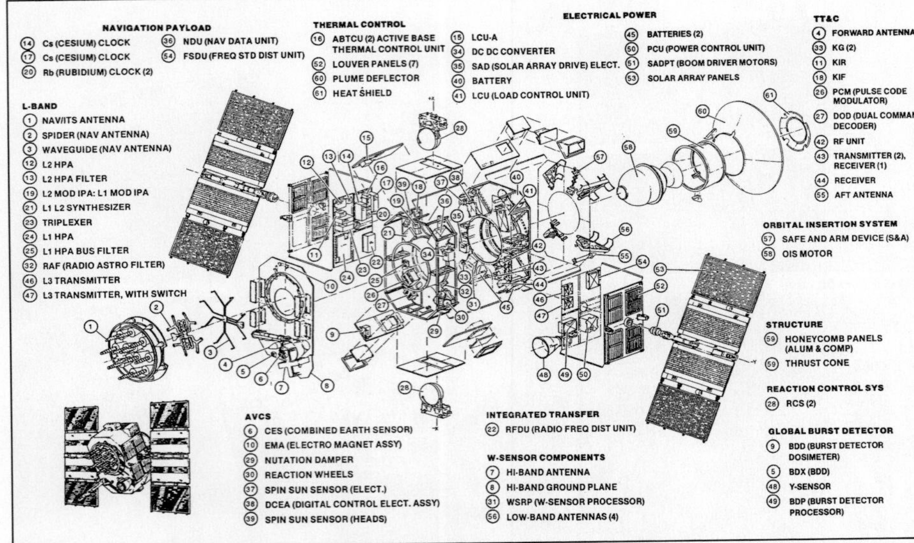

Navstar Block 2A satellite configuration; Block 2 was very similar (Rockwell) 0516843

GPS receivers are now almost universally installed on commercial and private aircraft and are fitted to many automobiles, and are widely used for recreation such as boating or backpacking. The demise of S/A led to the invention of the new sport of "geocaching" in which players hunt for hidden objects using GPS co-ordinates.

The spacecraft themselves have evolved through multiple generations. Rockwell received the USD1.2 billion production contract in 1983 for 28 Block II satellites. The navy planned to deliver them to orbit aboard the Shuttle (10 planned for 1987 alone) in conjunction with McDonnell Douglas' PAM-D2 solid upper stage. These plans had to be dropped after the Challenger accident and the satellites had to be retrofitted for the Delta II expendable launcher.

Each Block II incorporated the Integrated Operational Nuclear Detection System (IONDS) devices to detect nuclear explosions for assessment of nuclear attacks. Detonations can be pinpointed with 100 m accuracy, with every point on the globe within view of at least four satellites. The Dept of Energy's Sandia National Labs provide the Earth-facing 'bhangmeter' optical sensor that detects the characteristic double flash of an Earthbound nuclear explosion, while the Los Alamos National Lab's X-ray/ neutron detectors are spread around each satellite to monitor for space events. This sensor is also referred to as the Nuclear Detonation Detection System (NDS) or NUDET.

In practice the GPS Navstar system has been operating at full strength (24 Block satellites, including three hot spares) since March 1994. One ground spare remains in the military's inventory to be launched as an emergency replacement. Another can be built with two months' notice.

The GPS constellation has been kept up-to-date through the launch of new satellites as the earlier spacecraft reach the end of their lives. (Many GPS spacecraft have remained in service longer than expected: in October 2005, the Block II SVN-15 became the second Block II to pass the 15-year mark, double its design life). The first of the replacements are designated GPS Block IIR (replenishment). General Electric (later acquired by Lockheed Martin) beat Rockwell to win this contract in 1989. Block IIR satellites are based on the commercial AS-2000 satellite bus, have reprogrammable satellite processors enabling problem fixes and upgrades in flight, and feature inter-satellite ranging for greater accuracy. An unsuccessful launch on 17 January 1997 was followed by a successful mission on 23 July 1997. A 6 November 2004 launch orbited the 12th successful Block IIR.

In April 2005, the GPS constellation compromised one Block II satellite; 15 Block IIA satellites; and twelve Block IIRs. Eight Block IIR-M and 19 Block IIFs were in the pipeline.

The first of a planned eight modernised Block IIR-M satellites was launched on 25 September 2005. The new satellites offer a variety of enhanced features for GPS users, such as a modernised antenna panel that provides increased signal power to receivers on the ground, two new M-Code signals providing improved accuracy, enhanced encryption and anti-jamming capabilities to the military, and a second civil signal (L2C) that will provide users with an open access signal on a different frequency.

The next planned generation of GPS satellites is the Boeing-developed Block IIF. Intended improvements include an extended design life of 12–15 years, larger solar cells with greater power, faster processors with more memory, and a new civil signal on a third frequency. The Block IIF will also be designed for launch on the Evolved Expendable Launch Vehicle (EELV) family.

In 1996 Boeing was selected for a contract with options for up to 33 Block IIF satellites. However, the programme has proceeded slowly and the first IIF is not expected to be launched before 2008 and only six spacecraft were on firm order at the end of 2004. In January 2005, the USAF exercised options on three more Block IIFs, bringing the total order to nine, plus long-lead hardware for three further craft. Fortunately, a key to the original IIF design was to build in growth flexibility to accommodate new or evolving requirements, as well as other upgrades, and increased capability has been added to them during the programme. However, together with the cut in production numbers this has increased their unit cost from the planned USD30 million to USD70-80 million.

Boeing and Lockheed Martin/General Dynamics are competing to develop the GPS Block III spacecraft, due for its first launch in FY2013. The two companies received Phase A concept exploration and risk-reduction contracts in January 2004 and the USAF expects to select a winning contractor later in FY2006. Improvements and new capabilities on GPS III will include a new signal (L1C) which will be compatible with Europe's Galileo, and high-data-rate crosslinks. A high power spot beam ("theater"-size) is intended to meet new anti-jam requirements for +20 dB signal strength improvement by providing additional power directly from the satellites, rather than making substantial changes to user equipment antennas and processing technology. First launch for the GPS III satellite was originally planned for FY09 with a fully populated constellation by 2016/17; however, go-ahead delays and funding shortages elsewhere in the programme have delayed the first launch.

A Defense Science Board report published in October 2005 was critical of the Pentagon's plans for GPS, arguing that anti-jamming technology should be delivered faster and that the GPS constellation needed to be larger – at least 30 and preferably 36 spacecraft, rather than the 30 planned under GPS Block III – in order to deliver better signals in difficult environments such as rough terrain and urban canyons. (The more satellites that are available, the more they are likely to be visible despite obstacles). Also, the GPS Block III payload – which is designed to deliver tighter spot beams for greater jamming resistance – was likely to be too heavy, in the DSB's view, to permit a medium EELV to launch two spacecraft – making the constellation much more costly. The DSB suggested removing other payloads from the spacecraft in order to stay within the two-spacecraft-per-launch limit.

Specifications

Navstar Block II

Navstars 13–21 weigh about 840 kg in orbit and each unit cost USD38.8 million in FY87 dollars. Block II and IIA launched 1989 (5), 1990 (5), 1991 (1), 1992 (6), 1993 (6), 1994 (1), 1996 (3), 1997 (1). Newer Block IIs were still active in 2005.

Navstar Block IIA

Navstars 22–40 have enhanced survivability in the event of a nuclear attack and weigh about 1,881 kg at launch. Navstar 40, the last of the contracted 28 satellites, was delivered in 1992 and launched by Delta 7925. The first (Navstar 22 or GPS 23) appeared in November 1990, increasing the operational system to 16 satellites. However, a problem developed in the circuit board of one solar array's control system 12 December just two days after launch. This glitch turned out to be a design fault and delayed the launch of Navstar 23 (GPS 24) from February to July 1991 while it was corrected. Then number 24 (GPS 25) experienced a further delay from August 1991 to February 1992 after a reaction wheel failed in a ground test because its extended storage had affected lubrication. Navstar 40 was launched 12 September 1996.

Navstar Block II/IIA

Signal spread: –160 dBW C/A &–163.0 dBW P on L1; –166.0 dBW C/A or P on L2, –165.2 dBW C/A on L3
Atomic clocks: two rubidium and two caesium clocks
Mass: 843/930 kg on-orbit BOL (1,667/1,881 kg launch)
Span: 5.34 m across deployed solar panels
Design life: 7.5 years
Structure: aluminium honeycomb panels, 1.95 m^2 radiating array
AOCS: spin-stabilised, 22 hydrazine thrusters in two modules with 60 kg supply sufficient for >10 years of normal operations. Orbital injection by 47.1 kN (2,097 kN total impulse) Thiokol Star 37XFP
Power: 710 W EOL from 7.26 m^2 array of two Si cell panels; 3 × 35 Ah Ni/Cd batteries.

Navstar Block IIR

Navstar 41–60 became the first round of replenishment satellites to maintain the fully occupied constellation. Each has been improved to provide six months of autonomous satellite operations without ground control corrections, increased service life and improved survivability. The atomic clocks, normally updated daily, offer five times the stability of their predecessors. First delivery was planned for October 1995 for launch March 1996, although payload development problems delayed first launch to 23 January 1997. First Block 2R satellite was launched aboard a Delta-2 (7925) 17 January 1997, but following a fault with one of the GEM strap-on boosters the launch vehicle self-destructed 21 s after launch, resulting in the loss of the satellite. The satellite had been intended to fill the vacant F5 slot. Twelve satellites in this series had been launched successfully by the end of November 2004. The first of eight modernised Block IIR-M spacecraft was launched on 26 September 2005.
Mass: 2,032 kg at launch; 1,075 kg BOL on-station
Deployed dimensions: 152 × 193 × 191 cm body, 19.3 m span across solar arrays
Design life: 10 years minimum
Structure: equipment is mounted on six aluminium honeycomb structural panels mounted on central aluminium core. Communications equipment and some housekeeping units are mounted on the N/S faces; antennas and Earth sensors are on the Earth-facing panel
Power: twin 2-panel solar wings (totalling 13.4 m^2) using N-on-P Si cells providing 1,136 W EOL (970 W requirement); Ni/H_2 batteries provide 100 per cent eclipse protection
Pointing: not available
TT&C: at 2,227.5 MHz S-band
AOCS: 3-axis by momentum wheel, magnetorquing and 16 blowdown monopropellant hydrazine thrusters. The thrusters are divided into two redundant sets of eight: six 0.89 N catalytic Reaction Engine Assemblies (REAs) accompanied by two 22.2 N Electrothermal Hydrazine Thrusters (EHT) for increased performance NSSK, drawing on four tanks (94.8 kg total hydrazine). Insertion by Thiokol Star 37FM solid
Thermal systems: passive and heaters.

SCIENTIFIC

Argentina

Satélite de Aplicaciones Científicas (SAC) series

Current Status

Argentina's Satélite de Aplicaciones Científicas (SAC), Scientific Applications Satellite programme, active for more than 25 years, remains operative. The SAC programme orbits Earth observation satellites primarily for environmental science applications and relies heavily upon international co-operation and participation for its success. SAC-C has already outlived its nominal lifespan but continues to return data. Argentina has several further satellites in development.

Background

Argentina's space programme emphasises space science. Until May 1991, the air force executed all space research projects. In that year, the CNIE National Commission for Space Research was transformed into the civil National Commission for Space Activities (CONAE – see separate entry), under direct supervision of the country's President. CONAE and its predecessor have worked on the SAC series since 1981.

SAC-A, built by INVAP SE, and a component of the SAC-C mission, was primarily a technology demonstration mission that aimed to assist Argentina to develop and validate more sophisticated SAC series technology. The satellite and its instruments were all developed and manufactured by Argentina. NASA launched the small satellite from the Space Shuttle Endeavour in December 1998 using a Hitchhiker canister and ejector, the development for which NASA was responsible. SAC-A carried three primary instruments: a Differential Global Positioning System (DGPS); a panchromatic camera; and a magnetometer, as well as Argentine built solar panels, and data communication, transmission and processing systems, which included a whale tracking experiment. CONAE controlled the spacecraft from the Córdoba Ground Station in Argentina, and also provided tracking and telemetry from the same location. Spacecraft power problems occurred at the mission's start, but were corrected, and the successful mission ended in August 1998.

SAC-B, also built by INVAP, was Argentina's first science satellite. It was launched by NASA in November 1996, two years prior to SAC-A. SAC-B was to have been placed in a 550 km, 38° orbit, however a power failure in the Pegasus launch vehicle's third stage prevented both SAC-B and the High Energy Transient Experiment (HETE), an accompanying small satellite, from ejecting out of the launcher. Ground control and tracking and telemetry were able to make contact with the craft, but it would not function properly while still attached to the Pegasus third stage as well as to HETE, which also had not separated from the third stage. Ground control was successful, however, in commanding the spacecraft to deploy its solar panels. SAC-B was a solar physics mission intended to monitor solar output and its terrestrial effects during a solar maximum. Its instruments included an Argentine Hard X-Ray Spectrometer (HXRS); the NASA-provided Goddard X-Ray Experiment (GXRE) and Cosmic Unresolved X-Ray Background Instrument using CCDs (CUBIC); and an Imaging Particle Spectrometer for Energetic Neutral Atoms (ISENA) provided by the Italian Space Agency (ASI – see separate entry). ASI also provided the solar arrays for the craft. Argentina's space segment cost was estimated at USD9 million and NASA's overall cost at USD13.4 million, including USD10.8 million for the Pegasus launch. The satellites and faulty vehicle finally decayed in April 2002.

SAC-C is Argentina's first Earth observation satellite and is part of the AM satellite constellation that also includes Landsat 7, EO-1 and Terra; its mission primarily lies with the atmospheric sciences. SAC-C supplies data on the Earth's atmospheric temperature and water vapor content as well as Earth magnetic and gravitational fields; its objective is to assist with the study of the structure and dynamics of the atmosphere and ionosphere. SAC-C is an international co-operative mission between CONAE, NASA, the French Centre National d'Études Spatiales (CNES), the Brazilian Space Agency (AEB), the Danish Space Research Institute (DSRI) and the Italian Space Agency (ASI). Each agency designed and built the three instruments and six cameras aboard the spacecraft, with CONAE sponsoring the Multispectral Medium Resolution Scanner (MMRS); the High Resolution Technological Camera (HRTC); the High Sensitivity Technological Camera (HSTC); and the Data Collection System (DCS). NASA has provided the GPS Occultation and Passive Reflection Experiment (GOLPE) and in partnership with the DSRI, the Magnetic Mapping Payload (MMP). CNES provided the Influence of Space Radiation on Advanced Components (ICARE); and ASI provided both the Italian Navigation Experiment (INES) and the Italian Star Tracker (IST). SAC-C was also built by INVAP. Although it was designed to achieve a four-year lifetime, as of 2006, SAC-C was still operational.

SAC-D, also known as Aquarius, is scheduled for launch in 2008 and will carry at least eight cameras and instruments; it is also a joint effort led by CONAE and NASA, with CNES and ASI also taking part. SAC-D is primarily a climatic Earth observation satellite. The mission aims to track ocean circulation, and measure sea salinity, precipitation and wind speed in order to learn more about the Earth's water cycle and atmosphere. SAC-D will also be able to detect fire hot spots.

SAC-E, also known as the Satélite Argentino-Brasileño de Información en Alimento, Agua y Ambiente (SABIA), or the Argentine-Brazilian Satellite for Information on Food, Water and Environment is still in development. The satellite will monitor food and water production in Argentina and Brazil as well as in Paraguay, Uruguay, and Venezuela: the Mercosur area. Along with the satellite, ground stations will also be built in order to receive and distribute SAC-E's data. The satellite's projected launch date is sometime in 2009.

SAC-F, or ALSAT-2, also remains under development. It is a joint effort between CONAE and the Algerian Space Agency. SAC-F may be launched as early as 2007 and will carry out an Earth observation mission primarily focused upon disaster management, collecting data on environmental issues such as desertification.

Specifications

SAC-A

Launch: 14 December 1998 by Space Shuttle on STS-88; after fulfilling its short mission, SAC-A re-entered Earth's atmosphere on 25 October 1999
Orbit: altitude: 389 km; inclination 51.6°
Mass: 68 kg

SAC-B

Launch: 4 November 1996 by Pegasus-XL from NASA's Wallops Island; SAC-B never fully deployed from its launch vehicle and the mission failed
Orbit: intended altitude: 550 km; inclination: 38°
Mass: 191 kg

SAC-C

Launch: 21 November 2000 by Delta II from Vandenberg; as of 2006, SAC-C was operational
Orbit: altitude: 705 km, sun syncronous; inclination: 98.2°
Mass: 485 kg

SAC-D/Aquarius

Launch: scheduled for September 2008 from Vandenberg, aboard a Delta II
Orbit: planned altitude: 657 km; planned inclination 98.2°
Mass: 1,600 kg

SAC-E/SABIA

Launch: 2009
Orbit: planned altitude: 705 km; planned inclination 98.2°
Mass: not yet announced

SAC-F/ALSAT-2

Launch: 2007
Orbit: planned altitude: 705 km; planned inclination 98.2°
Mass: 700–800 kg

Chile

FASat Series

Current Status

As of 2006, only two FASat series satellites have been launched. The civilian Agencia Chilena Del Espacio, the Chilean Space Agency, founded in 2001, has satellite development on its agenda, however no further concrete satellite development or launch plans have been released.

Background

On 31 August 1995, the MOD's Fuerza Aerea de Chile (FACH), the Chilean Air Force, Space Division launched the 50 kg FASat-Alfa into polar orbit aboard a Ukrainian Tsyklon (see separate entry) from Plesetsk as a secondary payload. FASat-Alpha was Chile's first satellite. Chile purchased from the UK's Surrey Satellite Technology Ltd (SSTL), in May 1994, the satellite, ground station and on-the-job training for eight FACH engineers; launch cost was additional. A Chilean team built the actual satellite at SSTL, working alongside SSTL staff. The programme helped students to gain Masters' Degrees on University of Surrey's satellite engineering course. The satellite carried 5 g of Chilean soil on board and was to have transmitted regular information on UV radiation and the ozone layer over the Antarctic. Unfortunately, several hours after launch, a pyrotechnic problem on a locking system built by the UK left FASat-Alfa attached to the primary payload, the Ukranian SICH-1 satellite. Moscow ground control was unable to separate the two craft, and the mission failed. FASat-Alpha's Chilean ground station was located at Los Cerrillos Air Base, in Santiago.

After FASat-Alpha's failure, FACH signed a GBP1.3 million contract with SSTL to build the identical FASat-Bravo. FASat-Bravo was launched as a menage relay for the Chilean Air Force by a Zenit 2 from Baikonur on 10 July 1998. The satellite was retired from service when its batteries ran out in June 2001.

Both FASat projects were intended to help develop Chile's space professionals and technology. Additionally, educational activities using FASat-Bravo were designed to help create space technology awareness in the country. Chile planned to launch two other satellites to form a constellation providing LEO communications services and the development of the FASat-Gamma mini-satellite for remote sensing began in 2003, however Chile's satellite programme has now been inactive for some time.

FASat-Alfa Flight Model; Bravo was identical (FACH)
0517272

Specifications
FASat-Alfa and FASat-Bravo
Launch: Alfa 31 August 1995 piggyback on Sich 1 by Tsyklon from Plesetsk; Bravo 10 July 1998 piggyback on Zenit 2 from Baikonur
Orbit: Alfa 632 × 669 km, 82.5° attached to Sich 1; Bravo 817 × 818 km, 98.8°
Mass: 50 kg on-orbit; 53 kg including launcher fittings

China

Chang'e series

Current Status
Chang'e-1 was launched on 24 October 2007 and arrived in its operational orbit around the Moon on 7 November.

Background
Although some theoretical studies had been carried out for many years, China did not begin serious preliminary studies for lunar exploration until the early 1990s. The Advancement of Selenology, completed in 1977 by a team led by Ouyang Ziyuan at the CAS Institute of Geochemistry in Guiyang, is probably the most important work on the subject published in China. Under the direction of the Project 863 Experts Committee, a team of scientists led by Ouyang Ziyuan and Zhu Guibo of China Aerospace Industry Corporation began in 1993 to study the feasibility and necessity of Chinese lunar exploration. Based on a comprehensive survey of the nation's space technology and infrastructures, the feasibility study completed in 1995 believed it was possible to orbit a lunar satellite by 2000. In April 1997, CAS members Yang Jiachi, Wang Daheng and Chen Fangyun issued the "Proposal for Development of Our Nation's Lunar Exploration Technology" as part of the Project 863. The research and development of robotic Rovers for lunar exploration began the following year. In May 2000 and January 2001, Tsinghua University organised two symposia on lunar exploration technology. The third lunar conference was held in March 2001 at Beijing University of Aeronautics and Astronautics to discuss China's lunar exploration and human spaceflight in the 21st century. A feasibility study was unveiled at the conference for the first time.

China's three step lunar exploration programme was first announced in 2003. Chang'e-1, China's first deep-space mission, is named after a Chinese goddess who, in a popular fairy tale, lives on the Moon. The spacecraft is based on the Dongfanghong 3 telecommunication satellite platform and carries a 127 kg science payload of eight instruments. These will image and map the lunar surface in three dimensions, analyse the abundance and distribution of elements on lunar surface, investigate the depth and other characteristics of the regolith, and study the particle environment in translunar space and around the Moon. Also on board are some 30 songs, among them Chinese folk songs and "The East is Red" – China's national anthem. The cost of Chang'e-1 was said to be CNY1.4 billion, equivalent to USD175.5 million.

The onboard control system includes gyros, accelerometers, a Sun sensor, Moon UV sensor, star sensor, momentum wheel control and unloading, panel driving mechanism and thruster control. The spacecraft guidance, navigation and control system has self diagnosis and certain fault treatment functions. The computer has a redundant design, with autonomous control of the multiple-heater system and emergency back-up systems. Thermal control includes reflective coatings, heat pipes, cushions, sheets and louvers. The spacecraft has a high-gain directional antenna and medium-gain, omni-directional antennae. The omni-directional antenna was used in the Earth-Moon transfer orbit phase and the directional antenna is used in lunar orbit.

A CCD stereo camera will produce three-dimensional images of the lunar surface by obtaining three separate views of each target area. The interferometer-spectrometer-imager will overlay optical measurements with spectra to depict the regional distribution of surface minerals. A laser altimeter takes precise elevation measurements of the lunar surface, while gamma ray and X-ray spectrometers detect up to 14 elements – among them iron, potassium, uranium and titanium. A microwave/millimetre-wave radiometer operates on four different frequencies that penetrate to different depths of the lunar regolith, while a high-energy solar particle detector and low-energy ion instrument measure the solar wind environment.

In the baseline flight plan, Chang'e-1 would have a perigee of 600 km and gradually increase its velocity at perigee in order to raise its apogee altitude. Three engine firings would raise the apogee to about 51,000 km, 71,000 km and 120,000 km respectively, with the corresponding orbital periods of 16 h, 24 h and 48 h. Three more deceleration burns would be required to reach the operational circular polar around the Moon. These would change the orbital period from 12 h, then 3.5 h, and finally 127 min.

Chang'e-1 was launched on 24 October 2007 and separated from the upper stage 24 minutes after lift off. It entered a 16 hour orbit of 205 × 50,930 km, inclination 31°. The high energy solar particle detector was activated on 25 October. The first orbital manoeuvre took place on 25 October, raising the perigee to 600 km. On 26 October, a second burn moved it to a 24 hour orbit with an apogee of 70,000 km. Instructions for the third orbital transfer burn on 29 October were transmitted by the Yuanwang-3 space tracking ship in the south Pacific, transferring the satellite to a 48-hour orbit with its apogee raised to 120,000 km. The transfer to the Moon began on 31 October with a 13 minute main engine burn which placed it in an orbit with an apogee of about 380,000 km. This increased its speed to 10.916 km/s. The orbit was modified on 2 November during an eight minute burn by two small engines on Chang'e-1.

More than 10 days after lift off, Chang'e-1 successfully completed its first braking manoeuvre at perilune on 5 November. The 22 minute burn slowed it from 2.4 km/s to 2.06 km/s, ensuring that it was captured by lunar gravity and entered a 12 hour, 210 × 8,660 km orbit at about 03:37 GMT. This was followed by a second braking manoeuvre at perilune on 6 November. This reduced its speed to 1.8 km/s and placed it in a 3.5 hour orbit of 200 × 1,700 km.

After a journey of 1,800,000 km, Chang'e-1 completed its third braking manoeuvre on 7 November (the eighth orbital manoeuvre since launch), and entered its final operational orbit around the Moon. This slowed the spacecraft's speed to 1.59 km/s. All the instruments were expected to be checked out and operational by late November. The stereo camera will image the entire lunar surface once a month, and the microwave detector twice per month. By mid-January 2008, all the instruments were expected to have scanned the entire surface at least once. The orbiter is expected to operate in a 127 minute, 200 km circular, polar orbit for at least one year.

ESA's 15 m European Space Tracking (ESTRACK) network stations at Maspalomas (Spain) and Kourou (French Guiana) and the 35 m DS1 station at New Norcia, Australia, were used in rotation to supplement Chinese tracking stations during Chang'e-1's cruise to the Moon and orbit insertion manoeuvres.

The Chang'e-1 mission is the first part of China's three-phase plan for lunar exploration, which includes a Lander and Rover mission in 2012, and a sample return flight in 2017. Since the more powerful Long March 5 is unlikely to be ready in time, Chang'e-2 is expected to be launched by a Long March 3B.

Specifications
Launch: 10:05 UT on 24 October 2007 by Long March 3A from No. 3 launch pad at Xichang.
Launch mass: 2,350 kg
Telecommunications: Chang'e-1 uses the S-band Telemetry, Tracking and Command network of 12 m diameter antennae designed for China's manned space programme. To provide accurate navigation for the probe during its Earth-Moon flight and initial lunar orbits, China's VLBI system designed for astronomical observations will be used in addition to the ranging and range rate measurement capabilities of the S-band TT&C network. This will provide 100 m accuracy in position determination during lunar orbit.
Payload: Eight instruments with mass 127 kg:

- CCD stereo camera with a resolution of 160 m.
- Interferometer spectrometer imager from 0.48 μm to 0.96 μm wavelength.
- Laser altimeter with 1064 nm, 150 μJ laser and a resolution of 1 m.
- Gamma/X-ray spectrometer with an energy range from 0.5 eV to 50 keV for X-rays and 300 keV to 9 MeV for gamma rays.
- Microwave radiometer operating at 3, 7.8, 19.35 and 37 GHz, with penetration depth of up to 30, 20, 10 and 1 m respectively, and a thermal resolution of 0.5 K.
- High-energy solar particle detector and low-energy ion detector capable of detecting electrons and heavy ions up to 730 MeV.

Shijian series

Current Status
Shijian 8 (SJ-8) was launched on 9 September 2006 from Jiuquan. A recoverable satellite, SJ-8 had a microgravity and life science mission; it carried 215 kg of vegetable and fruit seeds aloft for 15 days, with the recoverable section returning to Earth in Sichuan Province on 24 September. An orbital segment of the satellite will continue to operate until its batteries fail. Shijian 8 was China's 23rd recoverable satellite launch, and the 90th launch to use the Long March series rocket. Shijian 9 may be launched in late September 2007.

Shijian 6, Group 2 was launched on 24 October 2006. The pair are designated by China as a science mission, however some suspicion exists that the satellites may be engaged in an intelligence mission.

Background
The Shijian (Practice), or SJ series, is China's scientific research and technological experiment satellite series. They have a variety of configurations. Historically, the majority of Chinese satellite launches in the domestic programme were dedicated to practical uses, primarily for remote sensing, including meteorological satellites, or communications missions. Science missions were few and far between. The Chinese space programme now has goals, however, intended to strengthen its space science sector. China plans to develop further capabilities in space physics and astronomy, as well as in microgravity and space life sciences. SJ series launches have taken place at the rate of one to three per year since 2004.

Shijian 1
Launched 3 March 1971 as the second satellite launched by China and placed in a 268 × 1,830 km, 69.9°, 106.1 min orbit. Apparently, the Chinese used the same bus for most of their mission applications and the Shijian 1 duplicated the Dongfanghong 1 (DFH-1) communications satellite configuration. Similar in shape to Telstar-1, it carried the first true Chinese science payload – a magnetometer, and cosmic ray and solar X-ray detectors. It flew about a year after the successful launch of China's first satellite. The Chinese learned from their earlier experience. This time the satellite carried solar cells for power and the temperature of the platform was controlled by using louvres.

Shijian 2
Shijian 2 was originally planned to be a single payload, but it was later decided to take advantage of spare launch capacity by adding two small satellites. An attempt to launch China's first triple payload failed on 28 July 1979. The mission was reflown as Shijian 2, with the triple payload made up of SJ-2, 2A and 2B. SJ-2 was the back-up from the previous attempt. Launched on 19 September 1981, the satellites were deployed into slightly differing orbits close to 234 × 1,600 km. 2A was bell-shaped and larger than 2B, with two cone antennas. It studied the ionosphere through transmissions of radio signals to the ground at 40.5 and 162 MHz. 2B was a metal sphere and balloon linked by a thin wire. It studied atmospheric density and operated for six days before its orbit decayed.

Shijian 3
Shijian 3 is a designator used for an abandoned 1980s meteorological satellite programme, and was never used for an orbital mission.

Shijian 4
Shijian 4 was launched on 8 February 1994 from Xichang aboard the maiden flight of the CZ-3A, mounted above a second satellite identified as Kua Fu 1, named in honour of a character in Chinese mythology who chases the Sun. Kua Fu 1 was a drum-shaped DFH-3 test satellite, with an approximately 2.5 m base diameter, 2.2 m height and a mass of 2,200 kg. SJ-4's mission was devoted to solar research.

Shijian 5
Shijian 5 was launched piggy-back on the maiden flight of the CZ-4B, the primary payload being the

third FY-1 meteorological satellite, FY-1C. SJ- 5 was a test flight of the new China Academy of Space Technology (CAST) 968 platform. Its main goal was to develop a modular minisatellite platform to be used for various applications. The Centre for Space Science & Applied Research (CSSAR) was responsible for the development of the payload and the ground system. The overall objectives of the mission included: (1) Demonstration of advanced space technology experiments including the operation of multiple-mode attitude control, an integrated satellite electronic system, a unified S-band TT&C system, S-band data transmitter, onboard data management system and large-capacity solid state storage. (2) Detection of single event upsets and investigation of suitable countermeasures. (3) Dose measurements of high-energy particles in near-Earth space. (4) Demonstration of a fluid physics experiment in microgravity.

Shijian 6
The first Shijian 6 group comprised a pair of satellites, SJ-6A and SJ-6B, which were launched on 8 September 2004. Shijian 6, Group 2 was launched on 24 October 2006 out of Taiyuan aboard a Long March 4B and also consisted of two satellites designated SJ-6-2A and SJ-6-2B. The second group replaced the first in orbit. China declared that the satellites are conducting scientific measurements of space radiation, but some sources suspect the SJ-6 satellites have a signals intelligence (SIGINT) or Earth observation role.

Shijian 7
Shijian 7 was launched from Jiuquan on 5 July 2005 aboard a Long March 2D. China stated that this was also a space environment monitoring satellite designed to carry out space science and technical experiments over a three-year lifetime.

Shijian 8
SJ-8 was a recoverable microgravity seed growing mission designed to study the effects of radiation and microgravity upon 2,000 varieties of fruit and vegetable seeds weighing a total of 215 kg. China hopes that further study and growth of the exposed seeds will later deliver higher yield crops. It was launched from Jiuquan on 9 September 2006 at 07:00 UTC by a Long March 2C. The recoverable capsule, based on the FSW, landed in Sichuan province at 02:43 UTC on 24 September 2006. Another section of the satellite remained in orbit until its orbit decayed on 1 November 2006.

Specifications
Shijian 2
Launch: 20 September 1981 using Feng Bao 1 from Jiuquan
Orbit: 240 × 1,610 km; 59.5° inclination, 103.4 min period.
Operational lifetime: SJ-2: 332 days. **SJ-2A**: 382 days. **SJ-2B**: 6 days.
Contractor: Beijing Institute of Spacecraft Systems Engineering.
Configuration: **SJ-2**: 8-sided prism, 1.23 m diameter, 1.1 m tall. Hydrazine thrusters to rotate spacecraft at 15–20 rpm and maintain solar-pointing attitude. Louvres for thermal control. Tape recorder to store 520,000 bits of data. Single, ultra-short wave transmission system.
Mass: **SJ-2**: 257 kg. **SJ-2A**: 483 kg. **SJ-2B**: 28 kg.
Instruments: **SJ-2** was devoted to space physics studies. Payload included: magnetometer; semi-conductor proton directional probe; semi-conductor proton semi-omni-directional probe; semi-conductor electronic directional probe; scintillation counter; long wave infrared radiometer; short wave infrared radiometer; Earth atmosphere ultraviolet radiometer; solar ultraviolet radiometer; solar X-ray probe; thermoelectric ionisation barometer. **SJ-2A** transmitted radio signals at 40.5 and 162 MHz to study the ionosphere. **SJ-2B** used a balloon to study the density of the upper atmosphere.
Power system: 4 small solar panels, each 1.14 × 0.56 m (area 2.55 m^2) and made up of 5,188 solar cells generating 140 W.
Secondary satellites (SJ-2A and 2B): **SJ-2A**: 483 kg launch mass; orbit 235 × 1,615 km, 59.5° inclination, 103.5 min period. **SJ-2B**: 28 kg launch mass; orbit 232 × 1,597 km, 59.5° inclination, 103.2 min period.

Shijian 4
Launch: 8 February 1994 using a CZ-3A from Xichang
Orbit: 118 × 236 km; 28.5° inclination
Contractor: Shanghai Institute of Satellite Engineering
Configuration: Drum, approximately 1.6 m diameter, 2.18 m tall
Mass: 396 kg
Instruments: SJ-4 was devoted to solar research. It carried: 5-channel, 0.5–4 MeV high-energy electron spectrometer, electrostatic analyser (0.1–40 KeV protons/electrons), high-energy proton (3–400 MeV) and heavy ion (4–25 MeV) detector, charging meter registering –6.4/+1 kV at the top and bottom of the satellite and –70/300 V around the waist of the satellite and stationary and dynamic single-event monitors
Power system: Solar panels made up of 11,000 2 cm × 2 cm solar cells covered the waist of the satellite.

Shijian 5
Launch: 10 May 1999 using a CZ-4B from Taiyuan
Orbit: 849 × 868 km; 98.8° inclination
Contractor: CAST; Shanghai Institute of Satellite Engineering
Configuration: Box-shaped body, approximately 1 m along each side plus two three-segment vanes of solar panels
Mass: 298 kg
Instruments: SJ-5 carried instruments to monitor the particle precipitation and the Earth's radiation belt particles over the South Atlantic Anomaly, including a Single-Event Upset (SEU) monitor, single-event shield effect tester, single-event upset tester, single-event synthetic tester, single-event latch-up tester, high-energy proton and heavy ion detector, high-energy electron detector and radiation dosimeter
Lifetime: Three month mission completed; satellite operation continued past the mission end date.

Shijian 6
Launch: Shijian 6, Group 1: 8 September 2004; Shijian 6, Group 2: 24 October 2006, both launches from Taiyuan aboard a Long March 4B
Orbit: Group 1: 590 × 602 km, 97.7° inclination; Group 2: 592 × 600 km, 97.1° inclination
Contractor: Groups 1 and 2: Shanghai Academy of Spaceflight Technology (SAST); Dongfanghong (The East is Red) Satellite Company
Mass: Group 2: estimated 1,000+ kg
Lifetime: Groups 1 and 2: two years plus

Shijian 7
Launch: 22:40 UT on 5 July 2005 aboard a Long March 2D from Jiuquan
Orbit: 550 × 569 km; 97.6° inclination, 95.9 min period.
Contractor: Unknown
Mass: Unknown
Lifetime: 3 years

Shijian 8
Launch: 07:00 UTC (approx) on 9 September 2006 from Jiuquan aboard a Long March 2C
Orbit: 177 × 428 km; 63° inclination; period 90.6 min
Contractor: China Aerospace Science and Technology Corp (CAST)
Mass: Payload: 215 kg
Lifetime: Recoverable mission lasted 15 days; orbital mission ended on 1 November 2006.

Space-based Multiband Variable Objects Monitor (SVOM)

Current Status
As of 2006, the Space-based Multiband Variable Objects Monitor (SVOM) micro-satellite is in development. The Chinese space programme has kept a space-based observatory on its space science priority list, however. The Observatory's pre-phase A development was funded by the National Natural Science Foundation of China (NSFC) from 2003–2005. The project is similar to that of the CNES/ESA COnvection, ROtation and planetary Transits (COROT) programme, and the Chinese team consulted with French/ESA COROT team during SVOM's development. As of 2006, the project is evolving into a joint Chinese/French endeavour, merging the former French ECLAIRS project into the SVOM programme. The National Astronomical Observatories of the Chinese Academy of Sciences NAOC/CAS and Tsinghua University are leading the SVOM programme in China.

Background
The SVOM satellite was described early on in a Chinese paper at the 2002 International Astronautical Federation (IAF) Congress. Although the SVOM project has evolved, the primary research areas remain the same and include the detection and measurements of stellar vibrations, exoplanet detection, and zero-delay X-ray and optical synchronous observation of rapidly varying objects such as Gamma-Ray Bursts (GRB). The satellite will possibly be launched piggy-back with another science or remote sensing satellite. SVOM and the Hard X-ray Modulation Telescope (HXMT) satellite will be the first in a series of Chinese space-based observatories, with the expectation that Chinese science satellite sophistication will grow until a large observatory can be developed and launched around 2020.

Specifications
Launch: earliest 2011, likely from Taiyuan
Orbit: sun-synchronous; 98.6°; 800 km
Lifetime: three to five years
Developers: National Astronomical Observatories, CAS (NAOC); Tsinghua University, Beijing; Institute of High-energy Physics, CAS; Astronomical Department, Nanjing University; Department of Physics, Hong Kong University
Configuration: box-shaped body, 0.8 × 0.8 × 0.9 m, plus four solar panel vanes
Payload Mass: 150 kg, including an additional, unspecified payload of 50 kg
Instruments: a wide field-of-view optical photometric telescope; a wide-field photometric telescope; high-resolution wide field-of-view soft and hard X-ray telescopes; an optical patrol telescope
Communications: 9.6 kps uplink, 2 Mbps downlink
Power: four folding solar panels with silicon cells: Ni Cd battery packs will deliver 144 W towards the end of operations

France

Starlette and Stella

Current Status
Both Starlette and Stella continue to be operational, and are tracked from time to time by the SLR network. Due to their passive nature and construction, both could function for hundreds of years.

Background
Starlette (Satellite de Taille Adaptée avec Réflecteurs Laser pour les Etudes de la Terre), launched in 1975, was the world's first passive laser satellite for solid Earth research. It is covered with 60 laser retroreflectors for geodetic ranging. SLR tracking data is used to study the long wavelength gravity field and its temporal changes. It is also used to determine the tidal responses of the solid Earth and to improve knowledge of the long wavelengths of the ocean tides. Stella, launched 18 years later, is almost identical and has the same purpose.

In appearance, Stella is a 35.74 kg icosahedron core of depleted U_{-238} alloy, surrounded by 20 Al-Mg alloy spherical segments each carrying three reflectors of 3.29 cm diameter entry pupil. Stella was placed on top of Ariane's 3rd stage with a special device that included a spring to give the required velocity and a spin of 5 to 8 rpm on the axis perpendicular to the ecliptic. It was launched into polar orbit to complement Starlette's Earth coverage. The Centre National d'Etudes Spatiales

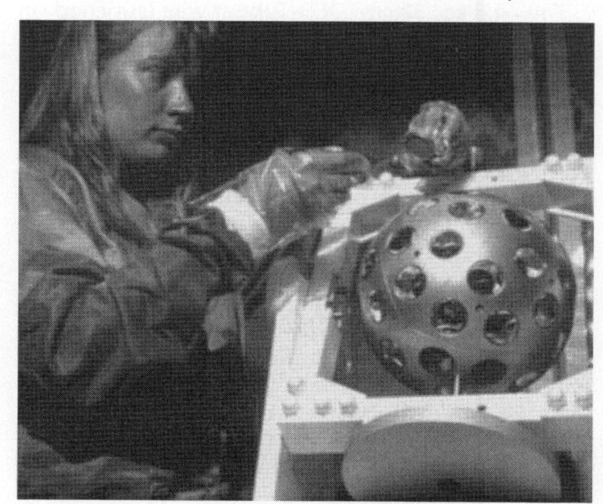

Stella joined the network of laser geodetic satellites in 1993 (CNES) 0516731

(CNES) built Stella as an in-house effort. CEA (the atomic energy authority) at Bruyères-le-Châtel provided the structure, and Aerospatiale the reflectors. Both spacecraft are designed to be tracked by laser with 1 cm accuracy. They continue to be used by international agencies in many countries, in addition to the geodetic institutes and university groups in France.

Specifications
Starlette
Launch: 6 February 1975 by Diamant BP4 from Kourou
Orbit: 804 × 1,138 km, 49.83°, 104.2 min
Mass: 48 kg
Size: 24 cm sphere carrying 60 laser reflectors for geodetic studies

Stella
Launch: 26 September 1993 by Ariane 40 from Kourou ELA-2, with Spot-3
Orbit: 798 × 805 km, 98.68°, 100.8 min
Mass: 47.978 kg
Size: 24 cm sphere carrying 60 laser reflectors for geodetic studies.

Germany

TubSat Series

Current Status
TubSat is a series of science and Earth observation technology demonstrator satellites built by the Technical University of Berlin's (TUB), Institut für Luft- und Raumfahrttechnik (ILR), or Institute for Air and Space Technology. All are microsatellites of various mass and application. Some remain operational after many years in orbit, despite the shorter projected lifespans typical of small, lower cost satellites. TUB has partnered with other nations' organisations to both launch and build its satellites, and will likely continue this trend as its microsatellite expertise grows, attracting as customers, developing countries such as Morocco and Indonesia, that wish to orbit less costly, small satellites.

Background
TUB's first satellite, TubSat A, was released into polar orbit from ERS 1's Ariane launch on 17 July 1991. The 35 kg, 38 cm^3 carried an L-band store/forward payload to demonstrate electronic mail applications, GaAs test cells and a 3-axis star sensor. It is orbiting at 780 km with an inclination of 98° and remains operational.

Following the successful first launch, TUB put the 40 kg TubSat B technology demonstrator into orbit, 25 January 1994, on Russia's Meteor 3-06 launched by Tsyklon 3. It was placed into a 1,250 km 82° inclination orbit. Ground controllers lost contact after 5 March and the mission failed.

The TubSat-N mobile communications satellite was launched on 7 July 1998 by Shtil-1 and decayed 22 April 2002. Partnered with TubSat-N, TubSat-N1 decayed 21 October 2000. While active, the satellites functioned in a 700 × 400 km orbit. TubSat-N and N1 were nanosatellites and had masses of only 8.5 kg and 3 kg, respectively. The satellites demonstrated store and forward communications, and attitude control technologies. The total cost for both satellites, including launch, was less than USD700,000.

The 44.8 kg, 32 cm^3 DLR-TubSat was launched on 25 May 1999 by India's Polar Satellite Launch Vehicle, PSLV-C2. It remains in operation and is in a 720 km orbit with 98° inclination. It is on an Earth observation mission and carries two fore field cameras and a 1 m high resolution telescope with the ability to return Earth observation data at 6 m ground resolution. The mission is a co-operative project between TUB, which provided the TubSat-C satellite bus, and the Deutsches Zentrum für Luft- und Raumfahrt (DLR) – the German Aerospace Centre.

Maroc-TubSat, a joint effort with Morocco's Centre Royal de Télédétection Spatiale (CRTS), the Royal Center for Remote Sensing, was launched on 10 December 2001 aboard a Zenit 2 from Baikonur. It was placed in a 1,000 km orbit with 98° inclination. CRTS built and supplied the satellite's instruments, while TUB only provided the TubSat-C satellite bus. The satellite also follows a cubic design of 32 × 34 × 36 cm, and has a mass of 47 kg. Maroc-TubSat has communications, Earth observation and attitude control missions; it carries a monochromatic camera with a 200 m ground resolution capability. Maroc-TubSat remains operational.

Scheduled for launch by the end of 2006 or beginning of 2007 from Satish Dhawan Space Centre, LAPAN-TubSat is a joint project between TUB and Indonesia's Lembaga Penerbangan dan Antariksa Nasional (LAPAN), or National Institute of Aeronautics and Space. An Indian PSLV will be used to launch the 56 kg satellite into orbit along with its primary payload, Cartosat-2, and a recoverable satellite. LAPAN-TubSat is a 45 × 45 × 27 cm cube. It carries as its payload, two color video cameras with 6 m and 200 m ground resolution capabilities for a remote sensing, surveillance and attitude control demonstration mission.

International

Cassini-Huygens

Current Status
Scientists announced in 2006 that evidence has been found of a new class of small moonlets residing within Saturn's rings. It is thought that there could be as many as 10 million of these objects within just one of Saturn's rings. Astronomers are researching how the moonlets might help to explain how Saturn's rings were formed. Scientists have also found that small moons are located within some of Saturn's newly discovered rings and have put forth the hypothesis that moon collisions and debris may be creating Saturn's rings.

Although the Huygens probe completed its mission in 2005, Cassini will continue to orbit Saturn, returning streams of data, at least until mid-2008. Cassini's mission reached its midpoint in June 2006.

Background
Cassini was launched at 08:43 UT on 15 October 1997, aboard a Titan 4-B/Centaur. Cassini performed a fly-by of Jupiter on 30 December 2000, at a distance of 9.7 million km/h gaining 2 km/s. Using Jupiter to accelerate the spacecraft on to Saturn, Cassini entered the Jovian magnetosphere on 9 January 2001. It reached Saturn and entered orbit 1 July 2004 UT. ESA's Huygens probe was released and landed on Titan's surface on 14 January 2005.

NASA chose Titan as the target for Cassini, partly because the organic processes taking place on Saturn's largest moon provide the only planetary-scale laboratory for studies of a pre-life terrestrial atmosphere. Titan has a dense, predominantly nitrogen atmosphere and the mission will permit intensive scrutiny. Then, suffering from budget constraints due to the Shuttle problems, NASA asked ESA to participate in the programme as well. Following a joint NASA/ESA study, the agencies agreed in June 1985 that NASA would supply a 1,540 kg (dry mass) Saturn orbiter and ESA a 190 kg Titan atmospheric probe.

After launch, Cassini entered into a Venus/Venus/Earth/Jupiter gravity assist multi-planet swing-by path, thereby reducing launch energy from 80 km^2/s^2 to 22 km^2/s^2. Cassini first encountered Venus on 26 April 1998 when it flew by the planet at a distance of 284 km. The 3 December 1998 450 m/s deep space manoeuvre at aphelion set up the second Venus gravity assist for 24 June 1999 at 600 km. Then a 1,171 km Earth fly-by of 18 August 1999 provided the final energy to reach Jupiter. A 115–140 R_J (10 million km) Jupiter swing-by on 30 December 2000 set up Saturn arrival for 1 July 2004.

Cassini traversed the asteroid belt December 1999 to April 2000, but did not gather data because the science group committee did not turn on their instruments again until two years before Cassini's Saturn arrival (apart from a 40-day gravitational wave experiment about December 2001). On 23 January 2000 Cassini's cameras took images of the asteroid 2685 Masursky from a distance of 1.6 million km. Too small to image from earth, 2685 Masursky was found to be roughly 15 × 20 km in size.

After getting into orbit about Saturn, Cassini also performed a 2,000 km Phoebe fly-by on 11 July 2004, where a 96-minute 621 m/s braking burn established an initial 147-day, 1.3 × 176 R_S 17° inclination orbit around Saturn. This was the closest approach to Saturn during the entire mission and the insertion burn precedes its optimal point (centred on periapsis) by 48 minutes for science observations. The orbiter will fly by Titan on 44 of the 74 planned orbits over its four-year mission about Saturn , sometimes approaching to within 950 km and eventually raising the orbital inclination to 85° (Voyager photographed the planet from an inclination of less than 30°). Fly-bys of Iapetus, Enceladus, Dione, Rhea, Tethys, Mimas and Hyperion have also been completed or planned before the mission formally ends in 2008.

During the January 2005 Titan fly-by, Europe's Huygens took 2.5 h to descend through the Moon's atmosphere. The probe entered at 6.15 km/s over Titan's day side at about 18° N/200–220° longitude (where 0° is defined as the sub-Saturn point; Hubble observations confirmed that Titan is tidally locked to Saturn). ESA used the probe to determine the atmosphere's chemical composition and measure winds, temperature and pressure profiles from 170 km down to the surface. Huygens continued returning data about an hour after touchdown and as of 2006, scientists are still interpreting the data.

Cassini Specifications
Launch: 15 October 1997 by Titan 4-B/Centaur from Cape Canaveral
Mass: 2,150 kg dry orbiter, 3,132 kg propellants, 373 kg Huygens (335 kg probe), 165 kg launch adapter. Total injected mass 5,820 kg
Design life: 11 years
Communications: 4 m diameter HGA and GHz X-band 20 W TWTA will return 20 bit/s – 169 kbit/s. Data stored on two 1.8 Gbit SSRs, providing up to 403.2 kbit/s record/playback
Power: Three RTGs provide 815 W BOL (638 W EOL; no batteries)
AOCS: 3-axis attitude control (2 mrad pointing accuracy, 6.98 mrad/s max slewing) maintained via SRU, four gyros, RWs and four clusters of four 1.0–0.6 N hydrazine blowdown thrusters on propulsion module. (primary: two redundant 445 N MMH/NTO bipropellant R-4D thrusters will be used for up to 200 manoeuvres totalling 2,360 m/s)

Cassini Payload
Cassini Plasma Spectrometer Developed by the Southwest Research Institute, the Cassini Plasma Spectrometer (CAPS) measures the flux of ions as a function of mass per charge and the flux of ions and electrons as a function of energy per charge and angle of arrival relative to CAPS. Its science objectives are: to measure the composition of ionised molecules originating from Saturn's ionosphere and Titan; to investigate the sources and sinks of ionospheric plasma (ion inflow/outflow, particle precipitation); to study the effect of magnetospheric/ionospheric interaction on ionospheric flows; to investigate auroral phenomena and Saturn Kilometric Radiation (SKR) generation; to determine the configuration of Saturn's magnetic field; to investigate the plasma domains and internal boundaries; to investigate the interaction of Saturn's magnetosphere with the solar wind and solar-wind driven dynamics within

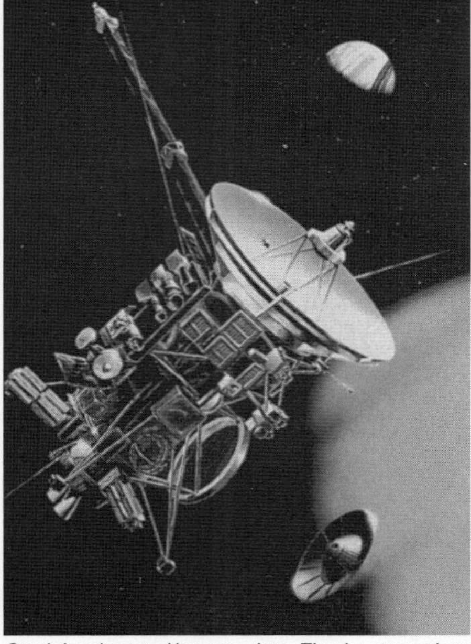

Cassini releases Huygens into Titan's atmosphere
0516789

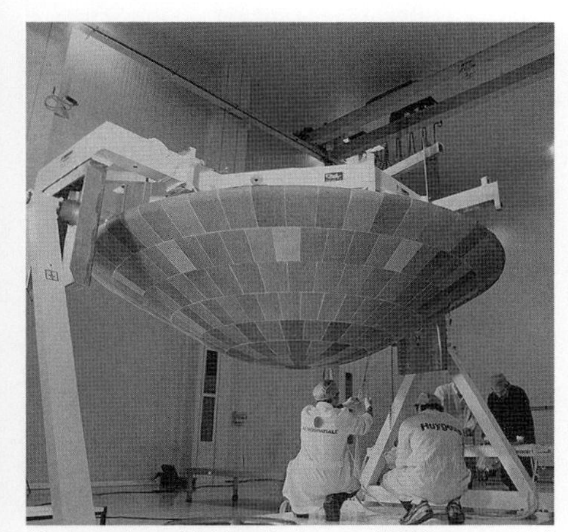

Huygens heat shield and aft cover (ESA/Aerospatiale) 0517147

the magnetosphere; to study the microphysics of the bow shock and magnetosheath; to investigate rotationally driven dynamics, plasma input from the satellites and the rings and radial transport and angular momentum of the magnetospheric plasma; to investigate magnetotail dynamics and substorm activity; to study reconnection signatures in the magnetopause and tail; to characterise the plasma input to the magnetosphere from the rings; to characterise the role of ring/magnetosphere interaction in ring particle dynamics and erosion; to study dust-plasma interactions and evaluate the role of the magnetosphere in species transport between Saturn's atmosphere and rings; to investigate auroral phenomena and SKR generation; to study the interaction of the magnetosphere with Titan's upper atmosphere and ionosphere; to evaluate particle precipitation as a source of Titan's ionosphere; to characterise plasma input to magnetosphere from the icy satellites and; to study the effects of satellite interaction on magnetospheric particle dynamics inside and around the satellite flux tube. CAPS incorporates an ion mass spectrometer, ion beam spectrometer, electron spectrometer, data processing unit and a scan motor.

VIMS Visual Infra-red Mapping Spectrometer (320 channels; 40.0 kg, facility instrument). The VIMS comprises a pair of imaging grating spectrometers designed to measure reflected and emitted radiation from atmospheres, rings, and surfaces over wavelengths from 0.35–5.1 micrometers to determine their composition, temperatures and structures. Scientific objectives are: to map the temporal behaviour of winds, eddies and other features on Saturn and Titan; to study the composition and distribution of atmospheric and cloud species on Saturn and Titan; to determine the composition and distribution of the icy satellite surface materials; to determine temperatures, internal structure and rotation of Saturn's deep atmosphere; to study the structure and composition of Saturn's rings; to search for lightning on Saturn and Titan and for active volcanism on Titan; to observe Titan's surface.

ISS Imaging (65.1 kg, 250 mm fl wide angle, 2,000 mm narrow angle, 1,024 × 1,024 element CCDs; facility instrument). The Imaging Science Subsystem is a remote sensing instrument that captures images in visible light and some in infrared and ultraviolet light and has a camera that can take broad, wide angle, pictures and a camera that can record small areas in fine detail and the ISS is expected to return hundreds of thousands of images of Saturn and its rings and moons. The ISS consists of a wide angle camera with angular resolution of 60 microradians per pixel and a narrow angle camera with angular resolution of 6.0 microradians per pixel, the sensors being 1,024 × 1,024 CCD arrays. Science objectives are: to map the three-dimensional structure and motions within the Saturn/Titan atmosphere; to study the composition, distribution and physical properties of clouds and aerosols; to investigate scattering, absorption and solar heating within the Saturn/Titan atmospheres; to search for evidence of lightning, aurorae, airglow and planetary oscillations; to study the gravitational interactions between the rings and Saturn's satellites; to determine the rate and nature of energy and momentum transfer within the rings; to determine ring thickness and sizes, composition and physical nature of ring particles; to map the surfaces of the satellites (including Titan) to study their geological histories; to determine the nature and composition of the icy satellite surface materials; and to determine the rotation of the icy satellites. The ISS sensing instruments include a wide angle camera (WAC), incorporating a 20 cm f/3.5 refractor (380–1100 nm) with 18 filters and a 3.5° × 3.5° fov, and a narrow angle camera (NAC), incorporating a 2 Metre f/10.5 reflector (200–1100 nm) with 24 filters and a 0.35 × 0.35° fov.

RADAR (58.0 kg, 13.8 GHz, 120 W raw power, employs HGA; facility instrument). The Radio Detection and Ranging Instrument is a remote active and remote passive sensing instrument that is designed to produce maps of Titan's surface and measures the height of surface objects. The RADAR facilitates imaging, altimetry and radiometry. Each mode allows for the collection of different types of data from straightforward imaging and 3-D modelling to passive collection of information such as recording energy emanating from the surface; an important aspect is the opportunity it affords to observe and map the synchrotron emissions at a new frequency (13.8 GHz) not possible from earth-based telescopes, this opportunity stemming from the Cassini radiometer's ability to separate atmospheric thermal emission from synchrotron emission. The Cassini RADAR experiment uses the five-beam Ku-band antenna feed assembly associated with the spacecraft high gain antenna to direct radar transmissions toward targets and to capture blackbody radiation and reflected radar signals from targets. Scientific objectives aim: to determine whether oceans exist on Titan and if so to determine their distribution; to investigate the geologic features and topography of the solid surface of Titan; to acquire data on the rings and icy satellites as conditions permit.

Ion Neutral Mass Spectrometer The Ion and Neutral Mass Spectrometer (INMS) is intended to measure positive ion and neutral species composition and structure in the upper atmosphere of Titan and magnetosphere of Saturn and to measure positive ion and neutral environments of Saturn's icy satellites and rings, study the interaction of Titan's upper atmosphere with the magnetosphere and the solar wind and to measure ion and neutral species composition during ring plane crossings. The sensing instruments include a closed source (neutrals only), an open source (ion and neutrals), a lens system and a quadrupole mass analyser and detector, the INMS having a mass of 9.25 kg with average operating power of 27.7 W and an average data rate of 1.5 kbps.

Cosmic Dust Analyser (15.1 kg, Max Planck Institute für Kernphysik, Germany). The Cosmic Dust Analyser is designed to provide direct observations of particulate matter in the Saturnian system, to investigate the physical, chemical and dynamical properties of these particles and to study their interactions with the rings, icy satellites and the magnetosphere of Saturn. Specific science objectives being to: extend studies of interplanetary dust (sizes and orbits) to the orbit of Saturn; define dust and meteoroid distribution (size, orbits, composition) near the rings; map the size distribution of ring material in and near the known rings; analyse the chemical composition of ring particles; study processes (erosional and electromagnetic) responsible for the E ring structure; search for ring particles beyond the known E ring; study the effect of Titan on the Saturn dust complex; study the chemical composition of icy satellites from studies of ejecta particles; determine the role of icy satellites as a source for ring particles; determine the role of icy satellites as a source for ring particles; determine the role that dust plays as a magnetospheric charged particle source/sink.

Radio Plasma Wave Spectrometer (37.4 kg, University of Iowa). The major functions of the Radio and Plasma Wave Science (RPWS) instrument are to measure the electric and magnetic fields and electron density and temperature in the interplanetary medium and planetary magnetospheres. The scientific objectives of the mission are to: study the configuration of Saturn's magnetic field and its relationship to Saturn kilometric radiation (SKR); to monitor and map the sources of the SKR; to study daily variations in Saturn's ionosphere and search for outflowing plasma in the magnetic cusp region; to study radio signals from lightning in Saturn's atmosphere; to investigate Saturn's electrical discharges; to determine the current systems in Saturn's magnetosphere and study the composition, sources and sinks of magnetospheric plasma; to investigate the dynamics of the magnetosphere with the solar wind, satellites and rings; to study the rings as a source of magnetospheric plasma; to look for plasma waves associated with ring spoke phenomena; to determine the dust and meteoroid distributions throughout the Saturnian system and interplanetary space; to study waves and turbulence generated by the interaction of charged dust grains with the magnetospheric plasma; to investigate the interactions of the icy satellites and the ring systems; to measure electron density and temperature in the vicinity of Titan; to study the ionisation of Titan's upper atmosphere and ionosphere and the interactions of the atmosphere and exosphere with the surrounding plasma; to investigate the production, transport and loss of plasma from Titan's upper atmosphere and ionosphere; to search for radio signals from lightning in Titan's atmosphere, a possible source for atmospheric chemistry; to study the interaction of Titan with the solar wind and magnetospheric plasma; to study Titan's vast hydrogen torus as a source of magnetospheric plasma; to study Titan's induced magnetosphere.

Ultra Violet Imaging Spectrograph (14.9 kg, University of Colorado). The Ultra Violet Imaging Spectrograph (UVIS) is a set of detectors designed to measure ultraviolet light reflected or emitted from atmosphere, rings and surfaces over wavelengths from 55.8 to 190 nanometres to determine their compositions, distribution, aerosol content and temperatures. Science objectives aiming to: map the vertical/horizontal composition of Titan's and Saturn's upper atmosphere; determine the atmospheric chemistry occurring in Titan's and Saturn's atmospheres; map the distribution and

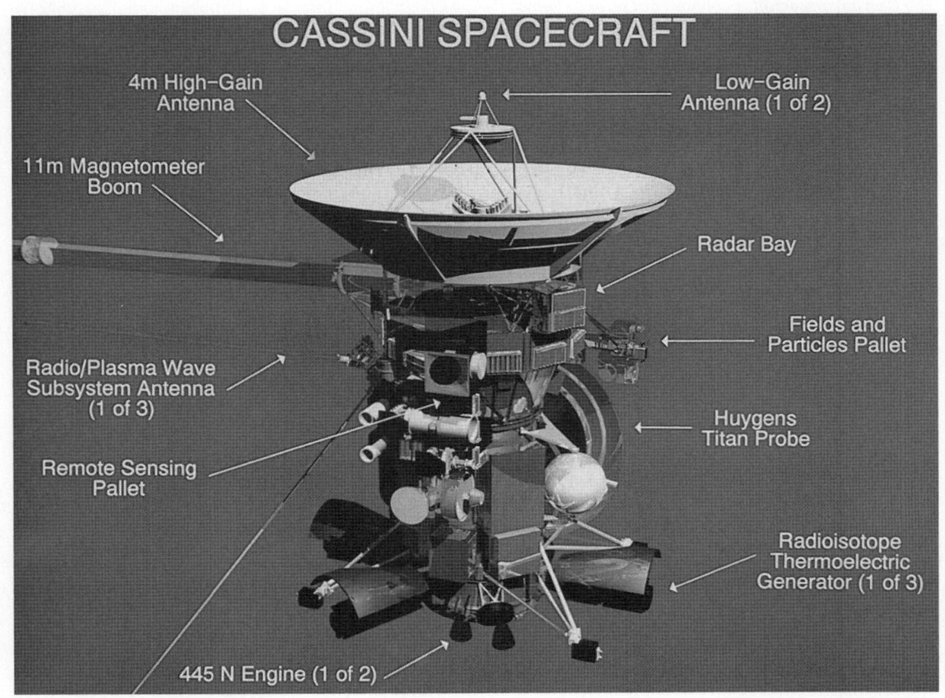

Cassini spacecraft (an image) 0085829

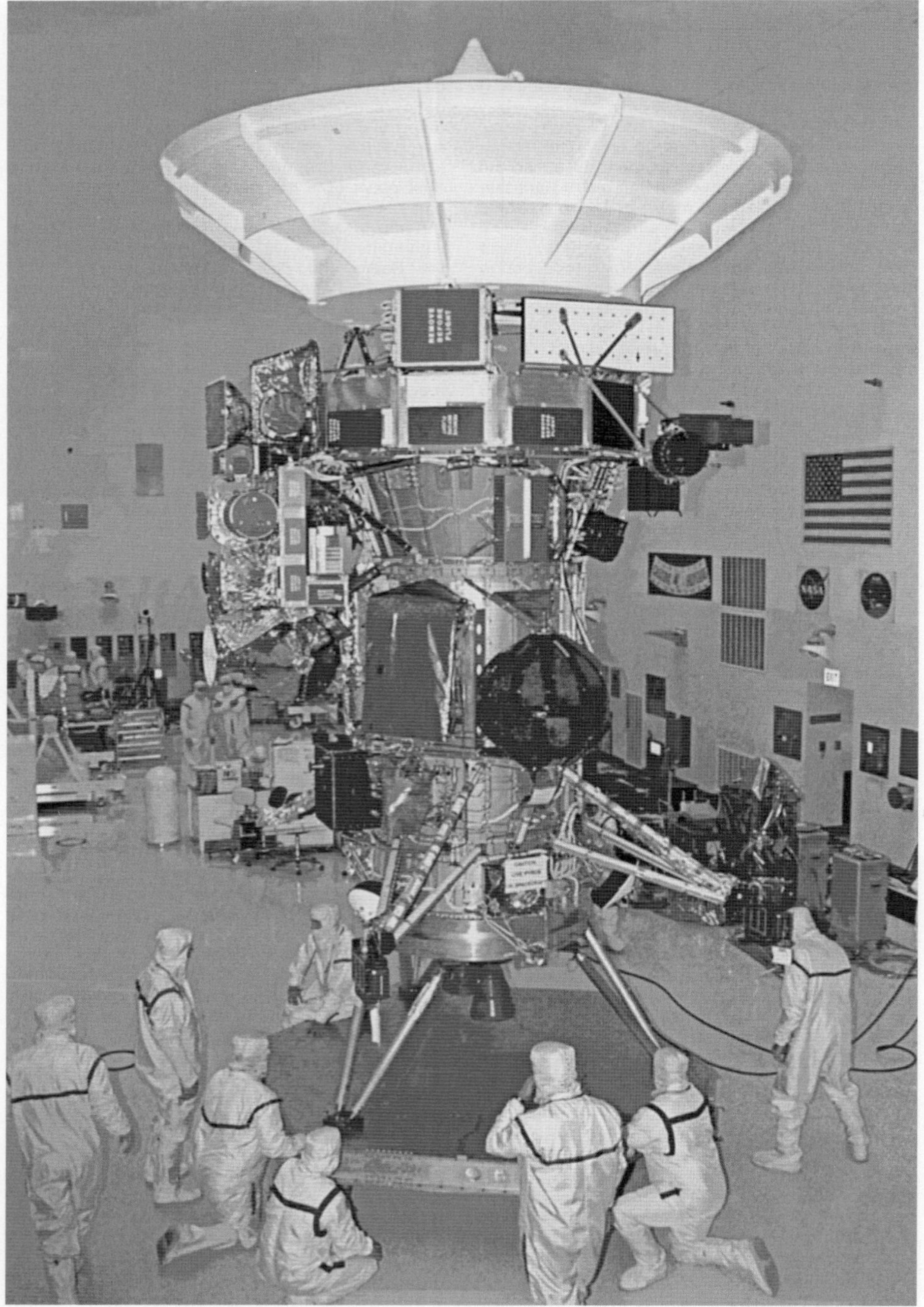

The Cassini spacecraft is checked out prior to launch (NASA) 0572498

properties of aerosols in Titan's and Saturn's atmospheres; infer the nature and characteristics of circulation in Titan's and Saturn's atmospheres; map the distribution of neutrals and ions within Saturn's magnetosphere; study the radial structure of Saturn's rings by means of stellar occultations; study surface ices and tenuous atmospheres associated with icy satellites. The UVIS contains four separate instruments:

Magnetospheric Imaging Instrument (24.9 kg, The Johns Hopkins University). The Magnetospheric Imaging Instrument (MIMI) is designed to: measure the composition, charge state and energy distribution of energetic ions and electrons; and to conduct remote imaging of the Saturn's magnetosphere, information which will be used to study the overall configuration and dynamics of the magnetosphere and its interactions with the solar wind, Saturn's atmosphere, Titan, rings and icy satellites. The MIMI is designed to accomplish the following scientific objectives: to determine the global configuration and dynamics of hot plasma in the magnetosphere of Saturn; to monitor and model magnetospheric substorm-like activity and correlate this with Saturn kilometric radiation (SKR) observations; to study magnetosphere/ionosphere coupling through remote sensing of aurora and measurements of energetic ions and electrons; to investigate plasma energisation and circulation processes in the magnetotail of Saturn; to determine through imaging and composition studies the magnetosphere/satellite interactions at Saturn and understood the formation of clouds of neutral hydrogen, nitrogen and water products; to measure electron losses due to interactions with whistler waves; to study the global structure and temporal variability of Titan's atmosphere; monitor the loss rate and composition of particles lost from Titan's atmosphere due to ionisation and pickup; to study Titan's interaction with the magnetosphere of Saturn and the solar wind; to determine the importance of Titan's exosphere as a source for the atomic hydrogen torus in Saturn's outer magnetosphere; to investigate the absorption of energetic ions and electrons by Saturn's rings and icy satellites; to analyse Dione's atmosphere.

Dual Technique Magnetometer (8.7 kg, Imperial College, UK). The primary objective of the Dual Technique Magnetometer (MAG) is to determine the planetary magnetic fields and the dynamic interactions of the planetary environment. Its science objectives being: to determine the internal magnetic field of Saturn; to develop a three-dimensional model of Saturn's magnetosphere; to determine the magnetic state of Titan and its atmosphere; to derive an empirical model of the Titan electromagnetic environment; to investigate the interactions of Titan with the magnetosphere, magnetosheath and the solar wind; to survey the ring and dust interactions with the electromagnetic environment; to study the interactions of the icy satellites with the magnetosphere of Saturn; to investigate the structure of the magnetotail and the dynamic processes therein.

Radio Science Subsystem (13.2 kg). The Radio Science Subsystem (RSS) uses the spacecraft X-band communication link as well as S-band downlink and Ks-band uplink and downlink to study compositions, pressures and temperatures of atmospheres and ionospheres, radial structure and particle size distribution within rings, body and system masses and gravitational waves. The RSS science objectives are: to search for and characterise gravitational waves coming from beyond the solar system; to study the solar corona and general relativity when Cassini passes behind the Sun; to improve estimates of the masses and ephemerides of Saturn and its satellites; to study the radial structure and particle size distribution within Saturn's rings; to determine temperature and composition profiles within Saturn's/Titan's atmospheres; to determine temperatures and electron densities within Saturn's/Titan's ionospheres.

Composite Infra-Red Spectrometer (36.2 kg, NASA Goddard) The Composite Infra-Red Spectrometer (CIRS) evolved from the Voyager Infrared Interferometer Spectrometer (IRIS) exhibiting significant improvements with spectral resolution 10 times higher than the Voyager IRIS, and a larger wavelength with more closely spaced data points greatly increasing the detail seen in the spectrum. The CIRS consists of dual interferometers that measure infrared emission from atmospheres, rings and surfaces with wavelengths from 7 to 1000 micrometers (1400 to 10-cm-1) to determine their composition and temperatures. CIRS science objectives are: to map the global temperature structure within Titan's and Saturn's atmospheres; to map the global composition within Titan's and Saturn's atmospheres; to map global information on hazes and clouds within Titan's and Saturn's atmospheres; to collect information on energetic processes within Titan's and Saturn's atmospheres; to search for new molecular species within Titan's and Saturn's atmospheres; to map the global surface temperatures at Titan's surface; to map the composition and thermal characteristics of Saturn's rings and icy satellites.

Huygens Specifications

Mass: 335 kg probe containing an 87 kg 2.7 m diameter decelerator. Science instrumentation: 48 kg.
Payload: 15.4/15.8 GHz phased array radar to show altitude and surface reflectivity data from 10 km to 150 m.

Huygens Instruments

Huygens Atmospheric Structure Instrument (HASI): 6.7 kg. Science objectives: to collect data on the temperature and pressure of Titan's atmosphere, winds and turbulence, atmosphere electricity. (University of Rome)

Gas Chromatograph Neutral Mass Spectrometer (GCMS): 19.5 kg. Science objectives: to collect data on atmosphere composition. (NASA Goddard)

Aerosol Collector/Pyrolyser (ACP): 6.7 kg. Science objectives: to collect data on atmosphere aerosol composition. (France)

Descent Imager/Spectral Radiometer (DISR): 8.5 kg. Science objectives: to collect data on image clouds/surface and UV-NIR spectroscopy. (University of Arizona)

Surface Science Package (SSP): 4.2 kg. Science objectives: to collect data on surface state after touchdown. (University of Kent, UK)

Doppler Wind Experiment (DWE): 2.1 kg. Science objectives: to collect data on high accuracy zonal wind characteristics. (University of Bonn, Germany)

Cluster Series

Current Status

Cluster II remains operational six years after launch and continues to enable scientists to make discoveries about Earth's magnetosphere. ESA closely manages the satellites' propellant consumption and expects Cluster II to remain on its extended mission until 2009.

Background

The Cluster II group of four satellites fly in close formation and take field and plasma measurements of the Earth's magnetosphere. Cluster maps the three dimensional extent and dynamic behaviour of small scale structures (from a few hundred km to a few thousand km) in the Earth's plasma. The satellites observe how solar particles interact with Earth's magnetic field and study the interactions of solar-terrestrial magnetic and electrical fields by direct measurement. Cluster is part of the Inter-Agency Solar Terrestrial Physics (IASTP) programme.

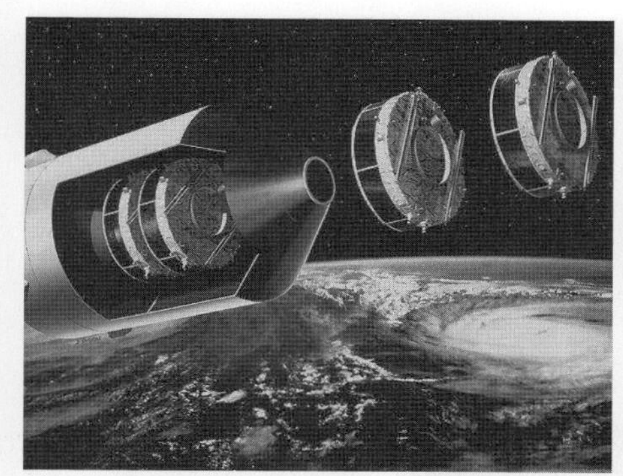

The four Cluster craft were launched in two double stacks (ESA) 0517302

The Cluster I group had four identical spacecraft (FM1-FM4) scheduled to launch on board the first Ariane 5 on 4 June 1996. The rocket exploded and destroyed all four satellites. Their loss on Ariane 5's debut was a major scientific blow. Only a month later at the ESA Science Programme Committee (SPC) meeting held in London on 2–3 July 1996, it was agreed that a Cluster-II/Phoenix mission should be readied as soon as possible, with a final decision coming at the November 1996 meeting.

Studies to determine Cluster I's replacements included launching the ground spare which was shelved because the Cluster concept required a group of satellites in orbit, not just one. The next most plausible solution called for ESA to build three new satellites to accompany the ground spare, thus duplicating the planned 1996 mission, or developing a set of four new smaller satellites. At the November 1996 meeting, the ESA Board chose the option of flying the existing ground spare with three new identical satellites, on condition that the cost could be capped at 210 Million Accounting Units (MAU).

In April 1997, ESA announced that it had chosen to launch the Cluster-II mission using two Russian Soyuz-U launch vehicles, each carrying two Cluster-II satellites. Satellites FM6 and FM7 were launched by Soyuz, with a Fregat upper stage on each, on 16 July 2000 followed by FM5 and FM8 by a similar launcher on 9 August 2000.

Specifications

Launch: Cluster I: 4 June 1996 by Ariane 501 from Kourou; Cluster II: 16 July 2000 and 9 August 2000 by Soyuz from Baikonur

Mass: FM 1 1,183/531 kg launch/dry; FM 2 1,169/518 kg; FM 3 1,171/520 kg; FM 4 1,184/533 kg 650 kg propellant, 72 kg payload; FM5-8 identical

Orbit: Elliptical Polar orbit; initial 10° GTO, rising to planned 25,513 × 140,318 km, 90°, 66 h in five manoeuvres Tuned so that the four satellites would fly in tetrahedral formation, passing in and out of Earth's magnetic field, crossing cusp, magnetopause, bow shock, magnetotail and other features. Separations adjusted between 200 km in cusp and 18,000 km in plasma sheet. FM5–8 were placed in a 19,000 × 119,000 km, 65°, 57 h orbit.

Lifetime: 5 years nominal; extended mission to 2009

Contractors: Dornier (prime), Alcatel (TT&C), BAE (AOCS, RCS), Contraves (structure), FIAR (power), Fokker (thermal control), Laben (OBDH), DASA/MBB (solar array, thrusters), Saab (harness, separation system), Sener (booms), Odetics (tape recorder), Cubic (HPA)

Spacecraft: 2.9 m diameter 1.3 m high cylinders, with conductive surfaces, spin stabilised at 15 rpm. Each spacecraft mass is 1,200 kg, 650 kg of which is propellant, and 71 kg of which is scientific instrumentation. Magnetic field experiments mounted on two 5 m radial booms; two pairs of 100 m tip to tip wire antennas for electric field measurements. Solar array to provide 224 W, including 47 W for the payload. Science data rate 16.9 kbit/s normal operation and 105 kbit/s burst mode; payload data outside of ground coverage stored in 1 Gbit tape recorder or 2.25 Gbit SSR for subsequent dumping. Telemetry downlink 2–262 kbit/s. Data from four satellites synchronised via highly stable onboard clock and time stamping at ground stations. Operated from European Space Operations Centre (ESOC), Germany. Science operations co-ordinated through the Joint

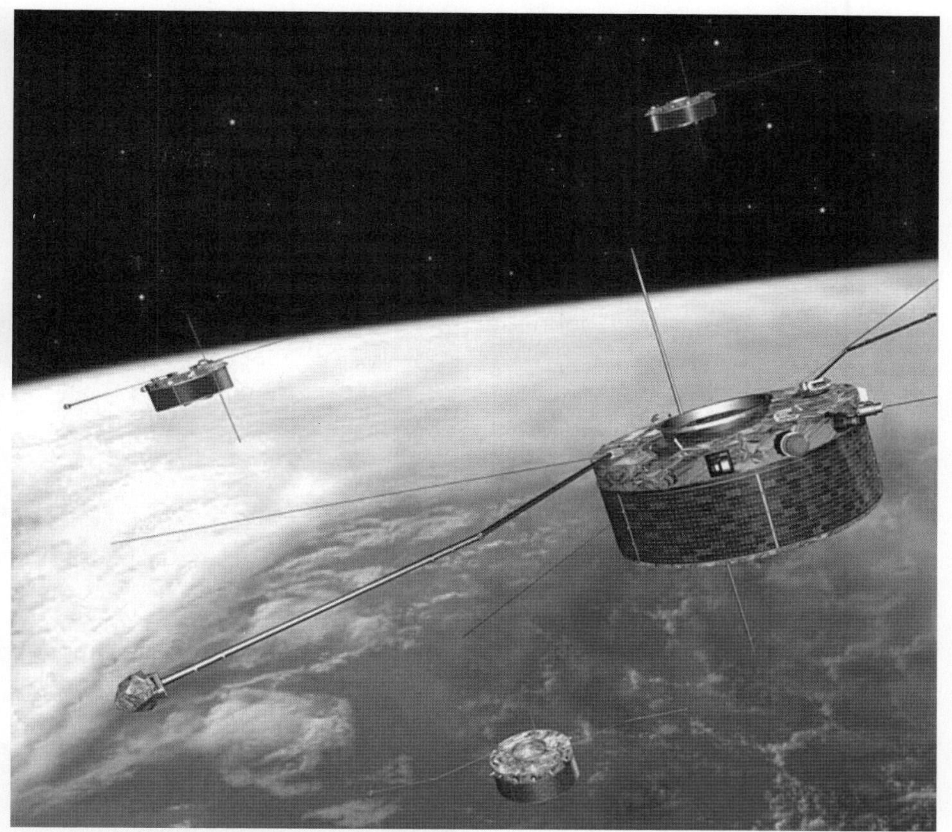

An artist's view of the Cluster constellation in earth orbit (ESA) 0572507

Cluster satellites in a test cell (ESA) 0572506

For details of the latest updates to *Jane's Space Systems and Industry* online and to discover the additional information available exclusively to online subscribers please visit
jsd.janes.com

Science Operations Centre, at Rutherford Appleton Laboratory, UK; ground stations in Villafranca and Maspalomas, Spain; data distributed via Cluster Science Data System, using ESANET to transfer it to centres in Austria, China, France, Germany, Hungary, Scandinavia, UK and US
Payload: STAFF Spatio-Temporal Analysis of Field Fluctuations (AC magnetometer; France)
EFW Electric Fields & Waves (utilising long wire antennas; Sweden)
WHISPER Waves of High Frequency and Sounder for Probing of Density by Relaxation (active density measurement of solar plasma; France)
WBD Wide Band Data (high frequency electric fields of several 100 kHz; US)
DWP Digital Wave Processor (controls STAFF, EFW, WHISPER, WBD wave consortium experiments; UK)
FGM Fluxgate Magnetometer (DC magnetometer; UK)
EDI Electron Drift Instrument (measurement of electric field by firing electron beam in circular path for many tens of km around satellite, detector on other side picks up return beam; FRG)
CIS Cluster Ion Spectrometry (composition/dynamics of slowest ions; France)
PEACE Plasma Electron/Current Analyser (distribution, direction, flow and energy distribution of low/medium-energy electrons; UK)
RAPID Research with Active Particle Imaging Detectors (essentially a pinhole camera to measure high energy electrons/ions; FRG)
ASPOC Active Spacecraft Potential Control (removal of satellite excessive charge by emitting indium ions; Austria)

TC-2, the second of two Chinese Double Star (Tan Ce) spacecraft, undergoing testing prior to launch in July 2004 (ESA) 1168198

Double Star (Tan Ce)

Current Status
TC-1 was decommissioned and re-entered the atmosphere on 14 October 2007. TC-2 remained operational as of November 2007.

Background
The Double Star programme involves the launches of two satellites – the first already placed in orbit in June 2004 and a second was scheduled for June 2004 – each carrying experiments provided by European and Chinese institutes. In China, the spacecraft are known as Tan Ce, which means Explorer. Their mission was to study the Earth's magnetic field and particle environment, operating simultaneously with ESA's four Cluster spacecraft from much lower elliptical orbits. TC-1 was the "equatorial" spacecraft, travelling into the solar wind and down the magnetotail, while TC-2 was the "polar" spacecraft that monitored the energy input from the solar wind into the polar ionosphere. The orbits were originally synchronised so that all six spacecraft could study simultaneously the same region of near-Earth space. The unique six-point study would provide new insights into the processes that trigger magnetic storms and auroral displays.

In 1992, the Chinese approached ESA with a proposal to establish a Data and Research Centre in Beijing for Europe's Cluster mission. China's request to participate in an international space programme for the first time culminated in an official co-operation agreement that was signed on 25 November 1993. As a result of this agreement, a number of Chinese scientists and engineers were hosted by ESA and the institutes of Cluster Principal Investigators (PIs), while five of their compatriots became co-investigators on Cluster. In 1997, Professor Liu Zhenxing of the Chinese Centre for Space Science and Applied Research (CSSAR) gave a presentation on the proposed Double Star programme at the Cluster Science Working Team meeting. Six Cluster PIs responded to the invitation to participate by offering to provide flight spare models of their experiments. In September 2000, Double Star was approved by the Chinese government. On 9 July 2001, the ESA Director General, Antonio Rodotà, and the CNSA Administrator, Luan Enjie, signed the agreement to develop the joint mission.

ESA agreed to contribute EUR8 million to the Double Star programme. This funding would be used for refurbishment and pre-integration of the European instruments, acquisition of data for four hours per day and co-ordination of scientific operations. China's contribution would include the two spacecraft buses, eight scientific experiments, launch and operations.

The hardware test phase began in late 2002 with a successful compatibility test of European and Chinese equipment at Imperial College, London. In parallel, the assembly of the spacecraft structural-thermal model in China culminated with the successful completion of the environmental test programme in February 2003. Despite delays caused by a SARS outbreak in China, the payload and the CSSAR subsystems were delivered to Beijing by 10 July for integration into the TC-1 spacecraft. Chinese communications equipment was installed at ESA's Villafranca ground station.

Artist's impression of the two Chinese Double Star (Tan-Ce) spacecraft (CNSA/CSSAR) 1330085

The launch campaign for the first Double Star satellite began on 25 November 2003, and lasted five weeks. TC-1 was launched on 29 December 2003, when the first and second stages of the Long March 2C/SM rocket fired for seven minutes, and the solid fuel upper stage injected the spacecraft into its operational orbit. Although the apogee was about 12,000 km higher than expected, due to over-performance of the upper stage engine, this had a minimal effect on the scientific results. The higher orbit meant that the spacecraft was able to observe the Earth's bow shock, which was not in the original science plan. The reduced number of conjunctions with Cluster was compensated by longer co-ordinated conjunctions.

Check-out of the TC-1 spacecraft started in early January 2004 with the successful deployment of the solid boom carrying the FGM magnetometer. A second boom that carried the search coil for the STAFF/DWP experiment failed to deploy. Analysis showed that the stability of the spacecraft was unaffected and its spin axis was, as expected, close to the north ecliptic pole.

TC-1 was placed in an elliptical, low inclination orbit and was expected to operate for 18 months. TC-2 was launched in July 2004 into a highly elliptical

TC-1, the first of two Chinese Double Star (Tan Ce) spacecraft, undergoing mechanical testing prior to launch in December 2003 (ESA) 1330082

polar orbit with a one year prime mission. Enhanced geomagnetic activity during the early stages of the mission caused both satellites to lose their attitude computers, so attitude information was subsequently provided by the magnetometers. Both satellites survived the 2006 eclipse season and the Double Star mission was extended twice, first to January 2007 and then to 30 September 2007. During 2007, the orbital planes of Cluster and TC-1 were separated by around 60° in azimuth or 'local time' to allow different types of observations. TC-1 was decommissioned and re-entered the atmosphere on 14 October 2007.

The Chinese DFH Satellite Co. Ltd. was responsible for building the satellite buses and payload integration. Data collection, mission management and communications, together with the development of some instruments, was the responsibility of the Centre for Space Science and Applied Research (CSSAR). The drum-shaped satellites are spin-stabilised, rotating at 15 rpm. They carry two 2.5 m experimental rigid booms and two axial telecommunication antenna booms. About eight hours after launch the two solid booms holding the magnetometers were deployed. ESA contributed eight scientific instruments, seven of which were based on hardware provided for Cluster. The exception is NUADU. These are the first ever European experiments to fly on a Chinese satellite. The other eight experiments were provided by China. Spacecraft tracking and communication was provided by ground stations at Villafranca, Spain (average 3 h per day), Miyun in Beijing and Sheshan in Shanghai.

Specifications

Launch: TC-1: 19:06 UT on 29 December 2003 by Long March-2C/SM from Xichang.

TC-2: 07:05 UT on 25 July 2004 by Long March-2C/SM from Taiyuan.

Mass: 335 kg

Body dimensions: 2.1 × 1.4 m

Orbit: TC-1: 570 × 78,970 km, 28.5°, period 27.5 h.

TC-2: 690 × 38,230 km, 90.1°, period 11.58 h.

Telecommunications: Two axial telecommunication antenna booms.

Power: 6.33 m² solar cells generating 280 W (BOL); nickel-cadmium batteries.

Attitude control: Spin stabilised, rotating at 15 rpm.

Payload: Mass 30 kg.

- Active Spacecraft Potential Control (ASPOC) DS-1 only
- Fluxgate Magnetometer (FGM) DS-1 and DS-2
- Plasma Electron and Current Experiment (PEACE) DS-1 and DS-2
- Neutral Atom Imager (NUADU) DS-2 only
- Hot Ion Analyzer (HIA) DS-1 only
- Low Energy Ion Detector (LEID) DS-2 only
- Spatio-Temporal Analysis of Field (STAFF)/ Digital Wave Processor (DWP) DS-2 only
- Low Frequency Electromagnetic Wave (LFEW) DS-2 only
- High Energy Electron Detector (HEED) DS-1 and DS-2
- High Energy Proton Detector (HEPD) DS-1 and DS-2
- Heavy Ion Detector (HID) DS-1 and DS-2.

Geotail

Current Status

The Geotail satellite is operational as of 2006, however, NASA has not funded the project beyond FY2008. Geotail scored lower than other solar science programmes, such as Cluster and SOHO, on a 2005 Sun-Solar System Connection (SSSC) mission operations and analysis review examining the scientific contributions of active solar satellite missions. Since NASA's science funding has, with the inception of the Exploration Program, been stretched, NASA Headquarters and the Science Mission Directorate must take care in prioritising missions. Geotail may be a science programme casualty in the future if its programme team cannot make a strong case for the satellite's operation beyond 2008. The review panel also recommended termination of the Polar, FAST, Ulysses and TRACE programmes by 2008, with the Polar mission ending in 2006. Geotail may survive if its mission is adjusted to support THEMIS, a five-satellite swarm due to launch by early 2007. If the Geotail team succeeds in refocusing the mission to support THEMIS, Geotail could remain operational beyond 2010.

Background

Geotail was the first Inter-Agency Solar-Terrestrial Physics (IASTP) programme spacecraft to fly. Other craft in the international series include Polar, Wind and SOHO. Geotail investigates the energy mechanisms in the magnetosphere tail and the physical processes in other important regions, such as the dayside magnetopause and bow shock. Scientists are interested in the geomagnetic field lines on the dayside, in a hemisphere of some 10 Earth radii, because they are compressed by the solar wind. The nightside lines are correspondingly stretched into a tail extending beyond 2,000R_E. A substantial fraction of the solar wind energy imparted to the magnetosphere is stored in those stretched lines, and enhanced activity arises in the inner magnetosphere and polar ionosphere when this energy is released.

Geotail is a joint Japanese – US programme. NEC developed the Geotail under JAXA (formerly the Institute of Space and Astronautical Science – ISAS) direction. EPIC, CPI, the Multichannel Analyser and the Inboard Magnetometer on the Magnetic Field Measurement Investigation are funded by NASA Goddard.

Specifications

Launch: 24 July 1992 by Delta II 6925 from Cape Canaveral

Mass: 1,009 kg (360 kg propellant; 105 kg science payload)

Orbit: for distant tail Sun-synchronous Moon orbit plane double lunar swing-by, max apogee 200R_E; for near tail 7.5° ecliptic inclination, 8 × 30R_E. Orbit 41,363 × 58,542 km, 22.4°, 4,750.6 min. Time spent in Earth's shadow was 1.7 h max, including penumbra. In the first part of the mission, lasting 1.75 years, its apogees were held in the magnetotail (on Earth's night side) by repeated paired lunar swing-bys. After 4½ Earth orbits, the first lunar

Geotail seen atop its kick stage motor on the Delta launch pad at Cape Canaveral (NASA) 1047490

Japan's Geotail was the first IASTP craft (ISAS) 0516798

swing-by was 13,657 km at 15.34 GMT 8 September 1992, producing a $136R_E$ apogee in the magnetotail 18 days later. Farthest apogee was $199R_E$ on 25 April 1994. Four δV manoeuvres totalling 154.2 m/s in November 1994 reduced apogee to $50R_E$ and then six more totalling 161.5 m/s in February 1995 reduced it to $30R_E$ to study the neutral sheet current. Geotail is presently at 9 R_E × 30 R_E with an inclination of −7° to the ecliptic plane. No further δV manoeuvres are planned; it will remain in that orbit to support other IASTP missions

Life: 1.75 years in the distant tail and 1.45 years in the near tail, the nominal mission was completed in July 1996; extended mission is expected to continue into 2008, perhaps longer

Spacecraft: 20 rpm spin-stabilised cylinder 2.2 m diameter 1.6 m height with mechanically despun antennas. 340 W EOL provided by Si N-on-P BSFR cells + 3 × 19 Ah Ni/Cd batteries. Data is recorded at 16.384, 65.536 or 131.072 kbit/s real-time or playback (two 450 Mbit Odetics tape recorders) to NASA DSN, Usuda, Kagoshima (Japan). Operation/control from SOCC Sagamihara Operation Control Centre

Payload: Electric Field (EFD). Comprising the Spherical Probe, Wire Antenna and Electron Boomerang.

Magnetic Field (MGF). Fluxgate and Search Coil.

Plasma (LEP). Made up of Ion/Electron Energy Analyser, Solar Wind Ion Analyzer and Ion Mass Spectrometer.

Plasma (CPI). Consisting of the Hot Plasma Analyser, Solar Wind Ion Analyser and Ion Composition Analyser.

Energetic Particles (HEP). Constituted by the Low Energy Detector, Burst Detector and Isotope Telescopes.

Energetic Particles (EPIC). Composed of the Supra-Thermal Ion-Composition Spectrometer and the Ion Composition Subsystem.

The Plasma Waves (PWI). Made up of the Multichannel Analyser, the Sweep Frequency Analyser and the Wave Form Capture (both E/B fields).

Inter-Agency Solar Terrestrial Physics Series (IASTP)

Current Status

IASTP spacecraft that remain fully operational are: the Cluster II group; Geotail, Polar, SOHO and Wind. IMP 8 is acting as an adjunct to the IASTP programme, as well as providing data to the Voyager and Ulysses programmes, but is no longer an independent programme. Please see each spacecraft's separate entry for more detailed information.

Background

IASTP, also known as the International Solar Terrestrial Physics (ISTP) programme, is a project involving a constellation of satellites developed and operated by NASA, ESA, JAXA (formerly ISAS) and Russia's Institute of Space Research (IKI). The programme also has ground-based and theoretical elements. IASTP plans and executes detailed investigations of the Sun, Earth's space environment and Sun-Earth interaction. The origins of the Solar-Terrestrial science programme trace back to a decision to investigate new subjects made by the Inter-Agency Consultative Group (IACG) following the highly successful Comet Halley collaboration. IASTP began in 1977 and originally the IACG planned the programme's completion in 2001; at its peak, it involved more than 100 universities, research labs and major contractors in 16 countries. The IASTP programme continues, as several spacecraft remain operational.

The following is a list of IASTP spacecraft: Cluster II; Geotail; Polar; SOHO; and Wind. IASTP has also collaborated with other spacecraft programmes, collecting data from some on-board instruments of each, that are, or were, used to augment IASTP spacecraft data; some spacecraft programmes that have engaged in data sharing are: IMP 8; GOES; LANL; SAMPEX; FAST; MSX; Equator-S; Interball-Aurora; Interball-Tail; and ACE.

James Webb Space Telescope (JWST)

Current Status

On 18 June 2007 ESA Director General Jean-Jacques Dordain and NASA Administrator Michael Griffin signed the official agreement that defines the terms of the co-operation on the JWST. Under the agreement, NASA – responsible for the overall management and operations of the JWST mission – builds the spacecraft, the telescope and the platform that will host the instruments. ESA will provide the launch with an Ariane 5 ECA rocket. Launch is currently anticipated for 2013.

The final segment of the flight primary mirror for NASA's James Webb Space Telescope (JWST) was completed delivered for grinding and polishing in early February 2007. The beryllium mirror segment, one of 18 segments that comprise the telescope's primary mirror, was delivered from Axsys Technologies, Inc. in Cullman, Alabama, to L-3 Communications SSG-Tinsley in Richmond, California. The prime contractor is Northrop Grumman Space Technologies. The Space Telescope Science Institute will operate JWST after launch. The mission's estimated cost is USD4.5 billion, including spacecraft development, launch and operations.

Background

The James Webb Space Telescope was formerly known as the Next Generation Space Telescope (NGST). JWST is a mission to investigate the origin and evolution of galaxies, stars and planetary systems, involving international co-operation between NASA, ESA and the Canadian Space Agency (CSA). NASA and ESA, joined by the CSA, have collaborated on JWST since 1997. Although optimised to operate over a different range of wavelengths, JWST is considered to be the successor of the Hubble Space Telescope. It is due for launch in 2013 and will operate for at least five years. JWST will operate at the 'second Lagrangian point' (L2), located 1.5 million km from Earth in the direction opposite to the Sun.

At the heart of the JWST observatory is a telescope with a 6.5 m diameter primary mirror (compared to 2.4 m for Hubble), providing a relatively large field of view. JWST will carry four instruments. NASA will provide the Near-Infrared Camera (NIRCam), to be built by the University of Arizona. ESA will provide the Near-Infrared spectrograph (NIRSpec) operating over similar wavelengths, with the detectors and slit selector device provided by NASA. The Mid-Infrared Instrument (MIRI), is being built by a consortium of nationally-funded European institutions (responsible for the MIRI optical assembly) and NASA, with co-ordination via ESA. The fourth instrument, the Fine Guidance Sensor/Tuneable Filter Imager (FGS/TFI), used for precision pointing, will be provided by the CSA.

Advanced technologies being developed for the JWST include lightweight cryogenic mirrors, microshutters to block unwanted light from the telescope's infrared detectors; materials for the sunshade that will keep the detectors cool at the L2 point; and the complex wavefront sensing and control algorithms that will keep the telescope's multi-element 6.5 m reflecting mirror in focus.

The microshutters are designed, tested and built by NASA-Goddard. They will work in conjunction with the Near Infrared Spectrograph, allowing the large-format detector to measure optimally the infrared spectra of distant, faint sources. The shutters allow spectroscopy on up to 100 targets simultaneously. Each of the 62,000 shutters measures 100 by 200 microns. They are arranged in four identical grids that have a layout of 171 rows by 365 columns. These shutter grids are in front of an eight million pixel infrared detector that records the light passing through the open shutters. Each shutter grid array is etched from a single piece of silicon, leaving a sculpture of cavities and doorframes with microscopic hinges and moving doors. The tiny shutters are laced with magnetic cobalt-iron strips. A passing magnet will open all the doors, pulling them down into the cavity. While the doors are opened, engineers can apply a combination of voltages to keep the selected microshutters open. The remainder close when the magnet moves away. The microshutters must perform at a temperature of 40 K (−233°C), which is the temperature of the Near Infrared Spectrograph.

The primary mirror comprises 18 hexagonal-shaped segments each a little more than 1.3 m across, and weighing approximately 20 kg after light-weighting. The completed JWST optics will have more than nine times the effective light-collecting area of the Hubble Space Telescope's optics, yet the JWST primary mirror will weigh only about half as much as Hubble's. Ball Aerospace is the principal optical subcontractor for the JWST programme, led by prime contractor Northrop Grumman Space Technology, under a contract from the NASA Goddard Space Flight Center. A secondary and tertiary mirror, plus flight spares, will be delivered to Ball Aerospace from its mirror manufacturing team that includes Brush Wellman, in addition to Axsys and L-3 Communications. As each mirror completes grinding and polishing and is delivered to Ball Aerospace during the next four years, it will be mounted onto a lightweight, actuated strong-back assembly and undergo functional and environmental testing.

Mars Express

Current Status

Mars Express is reported to be in a healthy condition. During 2006, the spacecraft successfully survived a period characterised by significant power limitations (eclipses longer than 70 minutes combined with near-aphelion conditions). As a consequence, science operations ceased from 21 August 2006. In the last week of September 2006, the spacecraft was reconfigured to the nominal configuration and prepared for the conjunction season in October 2006, when the only possible scientific observations were performed by Radio Science for solar corona measurements. Normal science operations for all experiments resumed on 6 November 2006. Mars Express participated in joint observations with Rosetta during the Mars flyby on 25 February 2007. The mission has been extended until May 2009.

Background

Mars Express was described as the most cost-effective Mars mission ever launched. It was the first of the so-called 'Flexible Missions' in ESA's 'Horizons 2000' programme. This class of mission had a cost cap of EUR175 million, including all industrial, launch and operation costs. The maximum reuse of existing hardware and software, together with the standardisation of all interfaces, including those of the scientific payloads, enabled a streamlined, no-risk development approach. It also enabled a streamlined development with only one year from engineering concept to start of development. The spacecraft made maximum use of existing hardware, including technology developed for Rosetta and instruments developed for Mars-96.

The box-shaped orbiter carries two solar arrays for electricity generation, with three newly developed, rechargeable lithium-ion batteries which supply

Artist's impression of Mars Express in Earth orbit prior to ignition of the Fregat upper stage (ESA) 1047485

Mars Express above Mars, artist's impression 1020782

power for periods of up to 90 minutes when the spacecraft is in shadow.

In 2003, the Earth and Mars were closer than they had been for thousands of years, so the spacecraft was able to take advantage of the planets' relative proximity by travelling the shortest possible route. Mars Express was inserted into an Earth parking orbit by a Starsem Soyuz-Fregat rocket. An interim orbit around the Earth was reached following a first firing of the Fregat upper stage. One hour and thirty-two minutes later the Fregat upper stage fired again to inject the spacecraft into an interplanetary orbit. The 400 million km journey to Mars took just over six months.

On 5 June 2003, the clamps that held the Beagle 2 lander securely in place during launch were released. A separate mechanism held Beagle 2 in place during the trip to the planet. However, if the launch clamps had not released, this second mechanism would have failed. The release of the launch clamps started at 10.10 CEST and lasted about 30 minutes. The unusual release mechanism consisted of a sleeve over a clamp bolt. An electric current heated the sleeve to about 100°C, causing the sleeve to expand and break the bolt. The three bolts broke in sequence.

Mars Express was reactivated at the end of November 2003. Since Beagle 2 had no propulsion system, the spacecraft's trajectory had to be adjusted on 16 December to ensure that the lander would enter the atmosphere of Mars on target. Beagle 2 was released on 19 December 2003, six days before arrival at Mars. At 8:31 GMT, the pyrotechnic device was fired to slowly release a loaded spring. The spin up and eject mechanism pushed Beagle 2 away at a speed of about 0.5 m per second. An image from the onboard Visual Monitoring Camera (VMC) showed the lander drifting away.

After ejection of Beagle 2, Mars Express was on a collision course with Mars. On 22 December, the main engine fired to alter its trajectory and reduce its speed by 1.3 km/s, providing the deceleration needed to acquire a highly elliptical transition orbit. The main engine fired at 02:47 GMT on 25 December for a 37-minute burn that inserted Mars Express into a transitional Mars orbit. At 09:00 GMT on 30 December, the main engine was fired for four minutes to move the spacecraft from an equatorial orbit into a polar orbit with an apocentre of 188,000 km. All commands were transmitted via ESA's new Deep Space Station in New Norcia, Australia. On 4 January 2004, at 13:13 GMT, a five-minute main engine burn changed the orbit to 40,000 km by about 250 km. Two more scheduled main engine burns enabled it to reach the final operational orbit of about 11,000 km by 300 km, inclination 86.6°, on 28 January 2004. The primary mission was to observe the planet for an entire Martian year (687 Earth Days). During this time, the low point of its orbit would be allowed to change so that the scientific instruments could observe the entire surface at high spatial resolution from different viewing angles.

The switch-on of instruments began on 5 January 2004, with the first public release of images and data on 23 January. On 6 February 2004, Mars Express relayed commands between Earth and NASA's Spirit rover, the first in-orbit communication between ESA and NASA spacecraft. Mars Express payload operations were stopped on 27 April 2005 in order to perform an avoidance manoeuvre with Mars Odyssey and to start the MARSIS deployment phase. The Planetary Fourier Spectrometer (PFS) malfunctioned early in the science operations phase, but recovery was possible by taking advantage of internal instrument redundancy. After switching to the back-up motor (more powerful than the primary motor), the PFS returned to full operation in early November 2005.

Concerns about deployment of the MARSIS experiment's long, whip-like antennas delayed the start of radar sounding for nearly two years. The three radar booms of MARSIS, consisting of a pair of 20 m hollow fibreglass cylinders, each 2.5 cm in diameter, and a 7 m boom, were initially to have been deployed towards the end of the instrument commissioning phase in April 2004. No satisfactory ground test of deployment in flight conditions was possible, so that verification of the booms' performance was based on computer simulation. Just prior to their scheduled release, improved computer simulations carried out by the manufacturer, Astro Aerospace (California), revealed the possibility of a whiplash effect and possible impact with the spacecraft before they locked in their final outstretched positions. Following advice from NASA-JPL, which contributed the boom system to the Italian-led MARSIS radar instrument, and the Mars Express science team, ESA put a hold on deployment. After a review board gave the go-ahead, deployment of the first boom started on 4 May 2005. Although 12 of the 13 boom segments of Boom 1 were correctly locked into position, one of the final segments, possibly No. 10, had deployed but was not positively locked into position. On 10 May, the 680 kg spacecraft was slewed so that the Sun would heat the cold side of the boom. After an hour, Mars Express was pointed back to Earth, and contact re-established at 04:50 CET on 11 May. A detailed analysis of the data received showed that all segments had successfully locked and Boom 1 was fully deployed. The second 20 m antenna was successfully deployed on 16 June, followed by the 7 m monopole boom on 17 June. MARSIS started its science operations on 4 July 2005, obtaining the first ever soundings of the planet's surface and ionosphere.

Beagle 2

Beagle 2 was largely developed and built in the UK. The prime contractor was Astrium UK and the lead institute was the Open University in Milton Keynes. It was designed to search for chemical evidence of life (past or present) on Mars. Packed inside a protective aeroshell less than 1 m across, the Beagle 2 probe consisted of the lander itself, which weighed 34 kg (including 11 kg of scientific instruments) and an entry, descent and landing system. Its clam-like design would enable the lander to open out, revealing four panels covered with solar cells. The limited room or mass available meant that miniaturisation of the instruments and tools was essential. For example, the Position Adjustable Workbench (PAW), a multi-functional suite of instruments attached to the lander's arm, weighed 2.5 kg. Mounted on this unit were tools for the collection of soil samples as well as stereo cameras, a microscope, chemical and microbiological spectrometers and various environmental sensors. Samples would be obtained with a mole that could burrow under the surface, and a corer/grinder that could drill into nearby rocks.

Protected by a heatshield, Beagle 2 entered the Martian atmosphere at more than 19,000 km/h on 25 December 2003, around the same time as the arrival of Mars Express. When its speed had reduced to 1,600 km/h, two parachutes would deploy to provide further deceleration. With the redundant heatshield now jettisoned, three large, gas-filled bags would inflate to protect the spacecraft when it impacted the surface at around 65 km/h. This crucial phase in the mission was to last 10 minutes. After Beagle 2 stopped bouncing, the gas bags were to separate and roll away, leaving the lander to drop the final 1 m to the surface. The petal-like solar panels would then open out, ensuring that the lander was sitting upright, and the cameras would start to image the landing site. From these images, scientists in the lander operations planning centre would recreate the local terrain before deciding on a strategy for collecting samples. Communications with Beagle 2 were to be handled by the lander operations control centre at the National Space Centre in Leicester, UK. After a couple of days the detailed analyses of rock and soil were to begin, using the instruments mounted on the PAW. Samples delivered to the gas analysis package inside the lander would be analysed for evidence of life.

Beagle 2 was targeted to land in Isidis Planitia, a large lowland basin a little to the north of the Martian equator (latitude 11.6°N, longitude 90.5°E). The exact location of the landing site depended on factors such as the angle of descent and wind speed. Isidis Planitia (up to 3 km below the average Martian 'sea level') was considered to be suitable from both the engineering and science points of view, with night time temperatures that are not too extreme (the southern hemisphere of Mars was going into winter when Beagle 2 arrived). The target ellipse covered what was thought to be an ancient impact crater which was later filled with sedimentary deposits, possibly even mud flows. Although the region slopes from the SW to NE, there are vast areas where the variation in height is not more than 250 m with slopes of 1° or less. The small ridges and hillocks are thought to be volcanic cones, which could indicate there have been local thermal hot spots associated with the release of volatiles, including water. These may be good places to search for life. Measurements of the albedo (reflectivity) and thermal inertia suggest a region of hard pan covered with 5–15 per cent rock fragments and a thin veneer of dust, but very few sand dunes.

The landing was anticipated at about 02:52 GMT. However, the first attempt to communicate with Beagle 2 via NASA's Mars Odyssey orbiter, three hours after landing, did not establish radio contact. Efforts to contact the lander via ground-based radio telescopes and orbiters, including Mars Express, continued for several weeks, but without success. Attempts to image the lander have also failed. A six month internal investigation by the Beagle 2 Project Team conclude that it is not possible to identify an exact failure mode, although Beagle 2 probably failed during the entry, descent and landing phase. Apart from possible mechanical failures in the deployment of systems such as the parachutes or the gags, it was suggested that a lower-than-expected atmospheric density 20–40 km above the surface might have been a contributory factor. However, an ESA-UK committee of enquiry decided that the following factors contributed to increased mission risk:

- Treating Beagle 2 as a scientific instrument rather than as a complex innovative spacecraft;
- Lack of guaranteed funding during the early phases of development;
- The withdrawal of Martin Baker Aircraft Co., the designer of the Entry, Descent and Landing System;
- Lack of an adequate management organisation with the relevant experience;
- Lack of adequate funding margins to manage and mitigate risks.

Specifications

Mars Express

Launch: 2 June 2003 at 23.45 local time (17:45 GMT) by Soyuz-Fregat from Baïkonur, Kazakhstan.

Launch mass: 1,120 kg

Bus dimensions: 1.5 × 1.8 × 1.4 m

Command and Data Handling: Data transfer rate: 500–4,300 Mbits.

Telecommunications: 1.6 m dish high gain antenna. 40 cm long low gain antenna.

Power: Two solar panels, area 11.42 m². Output at Mars: 660 W. Three lithium-ion batteries: 22.5 Ah each. Power consumption: 270 W-445 W.

Attitude control: Eight 10 N thrusters on each corner of the bus

Propulsion: 400 N

Payload: Orbiter scientific payload mass: 116 kg. Six instruments. **High Resolution High Resolution Stereo Colour Imager (HRSC).** Lead institutes: DLR, Berlin; FU Berlin, Germany. The stereoscopic camera provides global coverage in full colour at a spatial resolution of 10 m, with resolution of 2 m/pixel over selected areas. The images are used to produce a geological map showing the location of different minerals and rock types, and providing evidence of temporal changes in volcanism, tectonics. HRSC is a modified flight model of a camera originally developed for the Russian-led Mars 96 mission, with the addition of a Super Resolution Channel.

Subsurface Sounding Radar/Altimeter (MARSIS). Lead institutes: University of Rome, Italy; NASA-JPL. A low frequency radar sounder and altimeter,

Mars Express in final assembly and checkout by Astrium SAS in Toulouse, France 0572491

Artist's impression of the Beagle 2 lander on Mars. Note the PAW and robot arm (left) and the clam-like solar arrays which were to be deployed after landing (ESA) 0572500

the first to be used at Mars. Uses synthetic aperture techniques. Two 20 m booms and a secondary receiving antenna. The primary objective is to map the distribution of water and ice by detecting reflections of radio waves in the upper 3 – 5 km of the Martian crust. Secondary objectives include study of subsurface geology, surface roughness, topographic measurement and ionospheric sounding.

Analyser of Space Plasma and EneRgetic Atoms (ASPERA). Lead institute: RFI, Kiruna, Sweden. An energetic neutral atom analyser. Studies the interaction between the charged particles of the solar wind and the Martian atmosphere, and characterises the plasma and neutral gas environment above Mars.

Infrared Mineralogical Mapping Spectrometer (OMEGA). Lead institute: IAS, Orsay, France. Derived from the spare model built for the Mars-96 mission. A visible and near-infrared spectrometer operating at 0.38–5.1 microns. Global coverage at 1–5 km resolution, down to a few hundred metres in selected areas. Designed to characterise the composition of surface materials; study temporal and spatial changes in distribution of carbon dioxide, carbon monoxide, water vapour; identify aerosols and dust particles in atmosphere; monitor processes of transporting dust.

Planetary Fourier Spectrometer (PFS). Lead institute: CNR, Frascati, Italy. An infrared spectrometer optimised for atmospheric studies. Derived from a similar instrument developed for Mars-96. Covers wavelength ranges 1.2–5 microns and 5–45 microns with a spectral resolution of 0.2 cm and a spatial resolution of 10–20 km. Main objectives are 3-D measurement of temperature in lower atmosphere, variations in water vapour and carbon monoxide, studies of size, chemical composition and optical properties of clouds, dust and aerosols. Also measures thermal inertia of planet's surface.

UV and IR Atmospheric Spectrometer (SPICAM). Lead institute: CNRS, Verrieres, France. Studies the atmosphere's photochemistry, density-temperature structure, upper atmosphere – ionosphere escape processes and interaction with the solar wind. Uses limb viewing and stellar occultation, plus ability to perform nadir measurements.

Beagle 2

Mass: Total 68.8 kg. Lander 33.2 kg (including 11.4 kg of scientific instruments).

Dimensions:

Diameter: 0.64 m. Depth: 0.23 m (prior to deployment).

Telecommunications:

UHF frequency: forward 437 MHz, return 410 MHz. Data rates: forward 2.8 kbps, return 2–128 kbps.

Power:

Solar panels: 30 W. Battery: 160 Whrs

Payload:

Six instruments and two tools plus sampling arm. Attached to the end of the arm was a Payload Adjustable Workbench (PAW) where most of the experiments were located.

Gas Analysis Package (GAP). GAP comprised eight miniature ovens linked to a mass spectrometer. Samples of soil or rock would be heated at different temperatures, each increment being supplied with freshly generated oxygen. Any carbon compound present would burn to give off carbon dioxide. The gas generated at each temperature step would be delivered to a mass spectrometer. This would distinguish between carbon-12 to carbon-13, the two stable isotopes of the carbon and quantify their ratio. Other gases to be analysed included methane, nitrogen and noble gases. The latter would allow rock ages to be determined. Lead institute: Open University.

Rosetta and Mars Express Power Management Unit 1020795

Environmental Sensor Suite (ESS). 11 sensors on the lander platform and PAW. Designed to: detect surface UV flux at 200–400 nm; measure the total long-term dose and dose rate of solar protons and high-energy cosmic rays; detect air temperature variations to within ±0.05°C; measure air pressure to an accuracy of 0.1 mb; monitor wind speed (to 0.1 m/s) and direction (to 5°) in two orthogonal directions; register the momentum, direction and rate of accumulation of dust; record information about the upper atmosphere and the spacecraft during the landing sequence. Lead institutes: Leicester University, Open University.

Stereo Camera System (SCS). Two identical CCD cameras and integrated filter wheels, to be used to construct a Digital Elevation Model of the landing site. This would be sued to position the PAW. Landing site investigations to include panoramic, 360°, multi-spectral imaging to determine mineralogy and close-range imaging to infer texture of rocks and soil. Assessment of atmospheric properties through observations of the day and night sky. Lead institute: MSSL.

Microscope (MIC). Investigate the nature of rocks, soils and debris on the particulate scale (a few microns). Able to provide direct evidence of microfossils, microtextures and mineral inclusions of biogenic origin. The acquisition of complete sets of images allows 3-D reconstruction of sample surfaces in the visible and ultraviolet parts of the spectrum. Lead institute: MSSL.

Mössbauer Spectrometer (MBS). Designed to provide information about the iron and mineralogical content of rocks and soil. An X-ray detector would provide elemental compositions. A primary goal is measurement of the potassium content for age-dating purposes. Measurements of elements Mg, Al, Si, S, Ca, Ti, Cr, Mn and Fe enable identification of rock types. Lead institute: MPI für Chemie, Mainz, Germany.

X-ray Spectrometer (XRS). Used X-ray fluorescence to determine the elemental constituents of rocks. Able to detect Mg, Al, Si, Ca, Ti, Cr, Mn and Fe. Able to perform radiometric dating of rocks *in situ* Lead institute: Leicester University.

Mole. Tube-shaped subsurface sampling device which can retrieve samples from a depth of about 1.5 m and from beneath a large boulder. Also makes *in situ* temperature measurements and allows soil mechanical properties and layering to be analysed. Lead institutes: DLR, University of Hong Kong.

Corer/grinder. Allows removal of weathered rinds. Provides a flat, fresh surface suitable for spectrometer measurements. Lead institutes: DLR, University of Hong Kong.

Orbiting Satellites Carrying Amateur Radio (OSCAR)

Current Status

As of November 2006, at least 22 fully or partially functional OSCAR satellites were in orbit. The most recent OSCAR launch occurred on 22 September 2006 from Uchinoura Space Center, with an M-V carrying HITSat-OSCAR 59 into space. AMSAT-OSCAR 7 is the oldest of the series still in semi-functional orbit; it was launched on 15 November 1974 out of Vandenberg along with ITOS-G (NOAA 4) and INTASAT. More than 85 OSCAR launches have taken place since 1961.

Background

The Orbiting Satellites Carrying Amateur Radio (OSCAR) series of small satellites was initially created to enable amateur radio operators (Hams) located world-wide to experience satellite tracking and participate in radio propagation experiments. The World Administrative Radio Conference has allocated numerous frequencies for the Amateur Satellite Service. OSCARs have become increasingly sophisticated over the 45 years that the series has operated. Designed and built by volunteers, OSCARs have delivered services for amateurs in many capacities, including providing emergency communications for disaster relief efforts; acting as technology demonstrators; serving academic groups; and transmitting Earth imagery. The UK's UoSAT series alone has involved hundreds of schools and thousands of groups worldwide in activities using simple antennas, receivers and personal computers. OSCAR work includes telemetry transmission; amateur radio communications; and advanced experiments such as testing indium phosphide solar cells and transputers and monitoring radiation effects on electronics, representing satellite engineering for a fraction of the time and cost than the more advanced and conventional large satellites.

With a mass of only 4.5 kg, OSCAR 1, the first amateur satellite, was launched on 12 December 1961 from Vandenberg. More than 570 amateurs in 28 countries forwarded observations to the Project OSCAR data reduction centre before the satellite burned up in the atmosphere 22 days later. OSCAR 1 was also the first secondary payload to accompany a primary payload into space aboard the same launch vehicle. OSCAR 1 led to the creation of the Radio Amateur Satellite Corporation (AMSAT) in 1969; the membership organisation is responsible for OSCAR co-ordination and information dissemination to the present day.

The terminated US military Transit programme satellites also carry OSCAR designations; the last of these satellites was launched in 1988 and the programme was replaced by Navstar.

Polar

Current Status

NASA will not fund the Polar programme past a minimum mission extending through March 2007. Polar was one of NASA's lowest priority solar science missions, according to the 2005 Sun-Solar System Connection (SSSC) mission operations and analysis review, which examined the scientific contributions and merits of NASA's active solar satellite missions. In any event, Polar's propellant was fully depleted by October 2006 and several of its instruments are no longer in working order. The spacecraft will remain in a position, until March 2007, that will enable it to continue to provide useful data to NASA, especially in connection with other solar programmes, after which the programme will officially end.

Background

Polar is one of the Inter-Agency Solar-Terrestrial Physics (IASTP) programme satellites; it carries instruments designed and built by US universities and aerospace organisations. Polar monitors the ionosphere's role in substorm phenomena and in overall magnetospheric energy balance. It measures plasma energy input through the dayside cusp, and determines the characteristics of ionospheric plasma outflow and energised plasma inflow to the atmosphere. Like the Wind satellite, it also studies characteristics of the auroral plasma acceleration regions, and provides multispectral auroral images of the footprint of the magnetospheric energy disposition into the ionosphere/upper atmosphere.

Lockheed Martin Astro Space built the Polar satellite. It carries 11 instruments all with high data compression capabilities including experiments utilising the telemetry transmission at 2265.0 MHz (realtime 55.6 kbps, playlock 512 kbps) and command at 2085.688 MHz.

Specifications

Launch: 24 February 1996 by Delta II 7925-10 from Vandenberg

Mass: 1,300 kg (301 kg propellant, 226 kg science payload, 999 kg dry)

Orbit: highly elliptical; 5,141 × 50,605 km, 85.9°

Lockheed Martin Astro Space was both Polar and Wind prime contractor 0516799

Life: 3 years nominal; extended mission continued into 2006; minimum mission extension through March 2007
Spacecraft: 2.7 m diameter 2.5 m height cylinder (stowed envelope) with conductive surfaces, 10 rpm spin-stabilised with a 180° precession manoeuvre every 6 months. Despun platform for imagers inertially stabilised to 0.2°. Two deployable 6 m lanyard booms for magnetic field instruments, two deployable 5.5 m Z-axis booms and four radial wire antenna (two 50 m, two 80 m) for electric field measurements. 14.9 m^2 solar array provides 511 W BOL, including 186 W for payload; 3 × 26.5 Ah Ni/Cd batteries. Four Olin MR-111 2.2 N spin, four MR-111 4.4 N spin axis precession + four MR-50 22 N δV thrusters. Telemetry received by DSN over four 45 min nominal daily contacts. Science data 55.5 kbit/s real-time and up to 512 kbit/s playback. Payload data outside of ground coverage stored on 1.3 Gbit digital tape recorder. Mission operations/ data processing from GSFC
Payload: MFE Magnetic Fields Experiment (University of California at LA).
EFI Electric Fields Instrument (University of California, Berkeley).
PWI Plasma Wave Instrument (University of Iowa).
TIDE/PSI Thermal Ion Dynamics Experiment/ Plasma Source Instrument (NASA Marshall).
HYDRA Fast Plasma Analyser Experiment (University of Iowa).
TIMAS Toroidal Imaging Mass-Angle Spectrograph (Lockheed Palo Alto Research Lab, California).
CEPPAD/SEPS Comprehensive Energetic Particle Pitch Angle Distribution/Source Loss Cone Energetic Particle Spectrometer (Aerospace Corp, California).
CAMMICE Charge & Mass Magnetospheric Ion Composition Experiment (Boston University).
PIXIE Polar Ionospheric X-ray Imaging Experiment (Lockheed Palo Alto Research Lab, California).
VIS Visible Imaging System (University of Iowa).
UVI Ultraviolet Imager (University of Washington).

Rosetta

Current Status

Rosetta came out of hibernation mode on 26 July 2006; its Mars gravity assist is set for 25 February 2007, and the swingby phase began on 28 July 2006. Although the Standard Radiation Monitor (SREM) has so far been the only constantly active instrument, all the spacecraft's instruments appear to be healthy more than two years after launch. From 4 to 9 July 2006, Rosetta conducted plasma measurements in the Comet Honda's tail and the spacecraft's magnometer was determined to be operating properly, as it detected the comet's tail. Future observations of this kind will be carried out as Rosetta crosses paths with other comets on its way to Comet 67 P.

As of October 2006, Rosetta was 285.8 million km, or 1.91 Astronomical Units (AU) from Earth, and 150.4 million km, or 1 AU, from the Sun. The spacecraft's communications signals took 15 minutes and 17 seconds to reach Earth.

During 11 to 12 March 2006, observation of the asteroid Steins took place by use of the OSIRIS camera on board the Rosetta spacecraft. From the end of May through to the end of July 2006, the spacecraft was placed into a 'Near Sun Hibernation Mode' (NSHM).

Background

The Rosetta mission was approved in November 1993 by the ESA Science Programme Committee as the Planetary Cornerstone Mission in ESA's long-term science programme. Comet 46 P/Wirtanen was initially the comet with which the spacecraft was planned to rendezvous; however, after a delayed launch, a new target was set – Comet 67 P/Churyumov-Gerasimenko. The mission was named Rosetta after the Rosetta Stone, which provided the key to deciphering Egyptian hieroglyphics.

Rosetta was launched on 2 March 2004, at 07:17 UT on board an Ariane 5 G+ from Kourou, French Guiana. In order to enter orbit about comet 67P, the mission trajectory requires three Earth and one Mars gravity-assist manoeuvre. The spacecraft will then remain with the increasingly active comet during its journey towards the Sun. Rosetta will carry out observations on the comet's nucleus, and a lander named Philae will attempt to execute the first ever-controlled landing on a comet.

In order to achieve the scientific objectives of the mission, the spacecraft instruments will observe and monitor the following key aspects:

- Characteristics of the comet's nucleus, dynamic properties, surface morphology and composition
- Chemical, mineralogical and isotopic compositions of volatiles and refractories within the nucleus
- Development of cometary activity during its approach to the Sun, the processes within the surface layer of the nucleus and the inner coma
- Characterisation of asteroids: dynamic properties, surface morphology and composition.

Spacecraft operations take place at the European Space Operations Centre (ESOC) based in Darmstadt, Germany; early commissioning and near-Earth phases took place at the ESA 15 m ground station in Kourou. Spacecraft control is also provided by ESA's 35 m ground station at New Narcia, Australia.

Rosetta Mission Schedule

EVENT	NOMINAL DATE
Launch	2 March 2004–07: 17 UT
First Earth gravity assist	March 2005
Mars gravity assist	February 2007
Second Earth gravity assist	November 2007
Third Earth gravity assist	November 2009
Rendezvous manoeuvre	May 2014
Global Mapping	August 2014
Lander delivery	November 2014
Perihelion Passage	August 2015
Mission End	December 2015

Specifications

The Rosetta spacecraft design, within which all subsystems and payload equipment are mounted, comprises a cubic central structure, measuring 2.8 × 2.1 × 2.0 m. The solar panels, which extend out to 14 m, cover an area of 64 m^2. The total mass of the spacecraft equals 2,900 kg, with the propellant mass making up 1,578 kg and the lander 90 kg.

Rosetta Orbiter Instruments

ALICE: This Ultraviolet Imaging Spectrometer will be used to analyse and study the composition of the comet's nucleus and coma. The wavelength range within which the spectrometer operates is between 70 and 205 nm, with observations being carried out in both the extreme and far ultraviolet regions. Furthermore, ALICE will be used to determine the production rate and structure of H_2O, CO and CO_2 gas surround the nucleus, the far-UV properties of solid grains in the coma, and noble gas abundances. The instrument has a mass of 2.7 kg and an average power consumption of 5.6 W.
CONSERT: The Comet Nucleus Sounding Experiment by Radio wave Transmission will be used to perform tomography of the comet's nucleus. The instrument consists of two components: one component will be used to pass radio signals to the other on the comet surface, and then back to the source. The propagation delay of the radio waves will be used to understand the dielectric properties of the nuclear material. The instrument itself has a mass of 3.1 kg and operates on a centre frequency of 90 MHz, and a bandwidth of 10 MHz, with a power consumption of 2.5 W.
COSIMA: The Cometary Secondary Ion Mass Analyser is a secondary ion mass spectrometer, which will study dust from the near comet environment, via the use of a dust collector, primary ion gun and optical microscope target characterisation. The instrument has a mass of 19.1 kg and consumes power at a rate of 20.6 W. The primary ion source produces a 10 keV monoisotopic beam of 115 In ions and 2000 pulses per second maximum repetition rate.
GIADA: The Grain Impact Analyser and Dust Accumulator uses an optical grain detection system and mechanical grain impact sensor to measure the scalar velocity, size and momentum of dust particles within the coma of the comet. The amount of dust collected will be measured by five microbalances. The mass of the instrument is 6.35 kg and the power consumption occurs at a rate of 20.7 W. The GIADA instruments works with the measurable grain size of 10 micrometers (diameter), a grain momentum sensitivity of 7 × 10–11 kg m/s, and has a scalar velocity sensitivity of 1 m/s.
MIDAS: Micro Imaging Dust Analysis System will conduct a micro textural and statistical analysis of the cometary dust particles, down to a spatial resolution of 4 nm. The instrument has a mass of 8 kg and an average power consumption rate of 16 W.
MIRO: Microwave Instrument for the Rosetta Orbiter uses a millimetre wave mixer receiver and

Rosetta's fairing, prior to launch
(European Space Agency) 1133878

a submillimetre heterodyne receiver to measure the near surface temperature of the comet, thus allowing scientists to approximately calculate the thermal and electrical properties of the comet surface. The spectrometer portion of MIRO will also take measurements of water, carbon monoxide, ammonia and methanol within the comet coma. The instrument has a mass of 20.4 kg and average power consumption rate of 18.3 to 70.7 W, depending on the operating mode. MIRO's millimetre wave receiver operates at a frequency of 188 GHz and the submillimetre wave receiver operates at a frequency of 562 GHz.
OSIRIS: The Optical Spectroscopic and Infra-red Remote Imaging System makes use of a dual camera imaging system operating within the visible, near infra-red and near ultraviolet wavelength ranges. The two independent camera systems use a narrow angle camera system to produce high spatial resolution images of the target comet's nucleus; and a wide-angle camera, which also has a high stray light rejection, in order to image the dust and gas directly above the surface of the comet's nucleus. Each camera uses filter wheels, thus allowing the selection of imaging wavelengths for various purposes. The instrument mass is 34.4 kg and it has a maximum power consumption of 57.2 W. The wide-angle camera has a two mirror off axis optical design with a focal length of 140 mm; the narrow angle camera has a three mirror off axis optical design with a focal length of 700 mm.
ROSINA: The Rosetta Orbiter Spectrometer for Ion and Neutral Analysis uses two mass spectrometers in order to resolve and accurately analyse the required range of molecular masses. Two pressure gauges also provide density and velocity data for the cometary gas. The instrument has a mass of 36 kg and a maximum power consumption (depending on the operating mode) of 53 W.
RPC: The Rosetta Plasma Consortium uses a suite of five instruments to make complementary measurements of the plasma environment. All five instruments share a common electrical and data interface with the Rosetta orbiter: the PIU. RPC's instruments are:

- Ion Composition Analyser (ICA) – measures the 3-D distribution of the velocity and mass of positive ions.
- Ion and Electron Sensor (IES) – the flux of the electrons and ions within the plasma surrounding the comet are measured simultaneous via the use of this device.
- Langmuir Probe (LAP) – measures the density, temperature and flow density of the cometary plasma.
- Fluxgate Magnetometer (MAG) – measures the magnetic field of the region where the solar wind plasma and comet interact.
- Mutual Impedance Probe (MIP) – studies the electron gas density, temperature and drive velocity of the inner coma.
- Plasma Interface Unit (PIU) – provides one path for all five instruments to relay commands and scientific communications to and from ground control

RSI: Radio Science Investigation uses the communication system to study the nondispersive frequency shifts and dispersive frequency shifts (due to the ionised propagation medium), the signal power, and the polarisation of the radio

An artist's drawing of Rosetta and its instruments, including the Philae lander (European Space Agency) 1133879

carrier waves. Variations within the aforementioned parameters will provide data on the motion of the spacecraft, and the disturbing forces acting upon the spacecraft and the propagation medium.
VIRTIS: Visible and Infra-red Thermal Imaging Spectrometer contains three data channels. Two of these channels perform spectral mapping and the third is used for spectroscopy. The spectral range for the visible channel is 0.25–1.0 μm, 0.95–5 μm for the infra-red channel and 2.03–5.03 for the high-resolution spectrometer. The spectral resolutions are 100–380, 70–360 and 1300–3000, for the visible, infra-red and high-resolution spectrometer, respectively.

Philae Instruments
The cube-shaped Rosetta lander, Philae, carries nine instruments and uses a base plate, instrument platform and polygonal sandwich construction (all made of carbon fibre) in order to complete its structure. Data is transmitted to Earth via the Orbiter. Before arrival at Comet 67P/Churyumov-Gerasimenko, the lander is carried on the side of the orbiter. Once the correct alignment is attained, Philae will be sent the command to eject itself from the main spacecraft and prepare to land on the comet. The lander follows a ballistic descent, at the end of which it unfolds its three legs in preparation for a gentle touch down. Immediately after landing, Philae will be anchored to the comet by a harpoon, in order to prevent it escaping from the comet's weak gravity.

Philae's instruments will investigate the composition and structure of the comet 67/P by measuring the elemental, molecular, mineralogical and isotopic composition of the comet's surface and subsurface material. The near-surface strength, density, texture, porosity, ice phases and thermal properties will also be measured by the nine instruments onboard the lander. A Sampling Drilling and Distribution device (SD2) is also carried by Philae in order to collect samples for inspection.
APXS: The alpha-p-X-ray spectrometer will determine the chemical composition of the landing site and its potential alteration during the comet's approach to the Sun.
CIVA: Uses six identical micro-cameras to take panoramic pictures of the surface. The spectrometer studies the composition, texture and albedo of samples collected from the surface.
CONSERT: A complex experiment, which will reveal the internal structure of a comet's nucleus for the first time. Components of the instrument are found on both the orbiter and lander.
COSAC: Specialises in the detection and identification of complex organic molecules.
MUPUS: Multi Purpose Sensors for Surface and Sub surface Science will be used to understand the properties and layering of the near-surface matter as the planet approaches the Sun; to understand the energy balance and the surface and its variation with time and depth; to understand the mass balance at the surface and it's evolution with time; to provide support for the other instruments on the Lander.
Ptolemy: Devised to tackle the analytical challenge of making isotopic measurements of solar system bodies.
ROLIS: The Rosetta Lander Imaging System will be used to deliver the first close-ups of the environment of the landing site. It will also make high resolution investigations of the structure and mineralogy of the surface.
ROMAP: The Rosetta Lander Magnetometer and Plasma Monitor is a multi-sensor experiment, measuring the magnetic field, local pressure, and ions and electrons.
SD2: The Sampling, Drilling and Distribution device provides microscopes, advanced gas analysers with samples collected at different depths below the surface of the comet. SD2 can bore up to 250 mm into the surface of the comet.
SESAME: The Surface Electrical, Seismic and Acoustic Monitoring Experiments will help to further understand how comets have formed and thus, how the solar system (including Earth) formed.

Satellite per Astronomia raggi-X (SAX)/BeppoSAX

Current Status
On 29 April 2003, the Satellite per Astronomia raggi-X, (SAX) re-entered Earth's atmosphere and decayed over the Pacific Ocean, West of Columbia. SAX was also known as **BeppoSAX**, in honour of physicist Giuseppe Occhialini, one of the Italian founders of the European Space Research Organisation.

Background
The Satellite Astronomia raggi-X (X-Ray Astronomy Satellite), was an Italian-Dutch bilateral project. The main scientific characteristic of the BeppoSAX mission was the wide spectral coverage, ranging from 0.1 to over 200 keV. X-ray observations were obtained at moderate angular resolution (1 arcmin) with emphasis on spectral/time variability measurements of X-ray binaries, pulsars, transients, supernovae remnants, stellar coronae, active galactic nuclei and normal galaxies. 80 per cent of its observing time was devoted to a set core programme; the remainder was available for opportunity and guest observations.

In December 1992, the Italian Space Agency (ASI) , quoted SAX's budget as 408 billion lire; the cost was reported in January 1994 at 614 billion lire, including launch. The Science Data Centre was sited at ESRIN in Frascati, Italy. Data were returned via a Kenyan station (required because of the low inclination), with Kourou as back-up. The Netherlands Agency for Aerospace Programmes (NIVR) was responsible for Dutch involvement in the project. The Space Research Organisation Netherlands (SRON) contributed two X-ray Wide Field Cameras (WFC) and Fokker Space provided the AOCS and solar arrays. The bilateral MoU, worth 65 million guilders to Dutch companies, was signed 29 June 1990.

Italian instruments comprised a concentrator/spectrometer, 0.1–10 keV 1 arcmin resolution (three concentrators constituted the 1–10 keV Medium Energy Concentrator Spectrometer, a 0.10–10 keV Low Energy Concentrator Spectrometer (provided by ESA's Space Science Department), a High Pressure Gas Scintillation Proportional Counter, 3–120 keV, and a Phoswich Detector System 15–300 keV (also operating as 100–600 keV gamma-burst monitor).

Two SRON Wide Field Cameras viewing in opposite directions (each 42 kg, 14.5 W 1.8–30 keV, 5 arcmin resolution over 20 × 20'1 FOV; coded mask 70 cm from 25 × 25 cm multiwire proportional counter filled with 2.1 atmosphere Xe). As the Narrow Field Instrument (NFI) group studies a source, the 3-axis SAX rotates about its pointing axis for the perpendicular WFCs to scan the sky for simultaneous observations.

By December 2001 all three primary gyros and the three spares had failed, after which SAX used the three star trackers and was orientated through the reaction wheels. Battery 1 also began to deteriorate, with the loss of four of its 32 cells during January-February 2002. As planned, on 30 April 2002 at 13:38 GMT the BeppoSAX spacecraft was permanently switched off, terminating all in-orbit operations. During its six years of active life BeppoSAX made 30,720 contacts with the Malindi ground station and performed nearly 1,500 observations of most types of cosmic sources, discovering over 50 gamma ray bursts.

Specifications
Launch: 30 April 1996 by Atlas 1 from Cape Canaveral.
Contractors: Alenia Spazio (satellite, system, assembly, integration/test; structure, thermal control, harness subsystems), Alenia Spazio/Fokker (AOCS, TT&C), Fokker (solar arrays), FIAR (power distribution), Laben (OBDH, NFI Narrow Field Instruments), BPD (RCS), Teldix (four RWs), Nuova Telespazio (ground segment).
Mass: 1,400 kg at launch (490 kg payload)
Orbit: 583 × 603 km, 3.96°, 97 min
Design life: 26 month minimum; up to 4 years projected.
Spacecraft: 3.623 m high, 2.718 m diameter at launch,; 3-axis stabilised, 1.5/0.5 arcmin accuracy/stability. Power (2.42 kW EOL) from two fixed deployable wings of three 2.58 × 1.15 m panels (spanning 8.97 m) each plus a single body-mounted panel; supported by 2 × 30 Ah Ni/Cd batteries

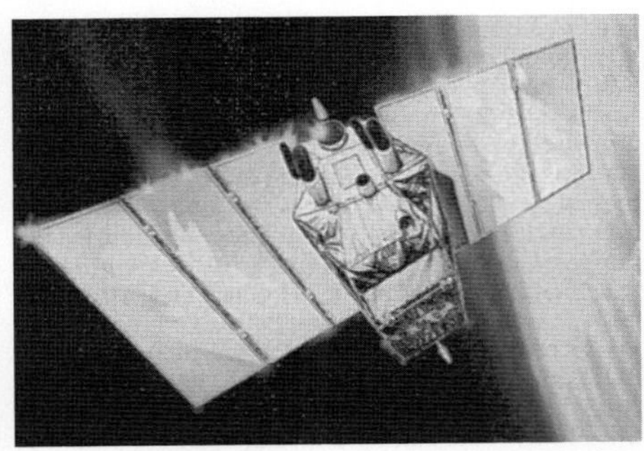

The Italian-Dutch SAX made X-ray observations 0516806

(650 W mean over eclipses of up to 37 min). Data production 50 kbit/s mean, 100 kbit/s max, 300 Mbit per orbit. Downlinked at 2.2455 GHz S-band from 450 Mbit redundant tape recorder memory at 917 kbit/s (high) or 16 kbit/s (low) or 131 kbit/s real-time over Malindi (Kenya) during 5–8 min contact period each pass. 2.0675 GHz uplink at 2 kbit/s. RCS by 12 × 10 N thrusters. Attitude control by four RWs, three magnetorquers, gyro package, magnetometer, three star trackers, four Sun acquisition sensors (requirement 1.5 arcmin error for 10^5 s observation).

Small Missions for Advanced Research and Technology (SMART-1)

Current Status
On 3 September 2006, after three years in space, the SMART-1 probe impacted in the Moon's Lacus Excellentiae region. Observers conducted a composition analysis on the plume that was sent up upon the spacecraft's lunar impact, which could be viewed by Earth-based telescopes. Images taken by SMART-1, of the Shackleton crater, at the Moon's South Pole, have assisted scientists with further investigation into the possibility of whether water ice is present on the Moon.

SMART-1's operational mission has ended, although analysis of the spacecraft's observations and impact plume continue. In addition to its regular mission, SMART-1 was also used to calibrate ground instruments in preparation for future Lunar missions that will be launched by the Chinese and Indian Space Agencies.

Background
ESA's Small Missions for Advanced Research and Technology (SMART-1) spacecraft was launched from Kourou aboard an Ariane 5 rocket, alongside three other satellites, on 27 September 2003. It was Europe's first Lunar mission. The mission's goal was to investigate the origin of the Moon and search for ice within the craters of the Moon's south pole.

The mission executed a 16-month orbit transfer from the Earth to the Moon. The spacecraft itself entered orbit around the Moon on 15 November 2004. The instruments on board the spacecraft investigated and contributed clues to questions regarding the origin and evolution of the Moon, its crust composition, cold traps at the lunar poles and potential lunar resources.

Specifications
For its journey from the Earth to the Moon, the SMART-1 spacecraft used electric propulsion, allowing in-flight spacecraft checks and cruise science to be completed prior to arrival. Upon arrival, the spacecraft began a nominal six-month mapping phase of the Moon's surface. The spacecraft's orbit was, at this time in the mission, highly elliptical, with its lowest point occurring between 300 km and 1,000 km from the Moon's South Pole, and at its farthest point, approximately 10,000 km.

Built by the Swedish Space Corporation, which led a group of 20 European organisations, the spacecraft itself had a height and width of 1 m, a mass of 350 kg, and an aluminium primary cube structure. The propulsion system used on the SMART-1 spacecraft was the PPS-1350 developed by SNECMA. This stationary plasma thruster (SPT) had a diameter of 100 mm and a maximum discharge power of 1.5 kW. At maximum power, the qualified lifetime of the thruster was 7,000 hours. The spacecraft carried a number of instruments.

SMART-1 Instrumentation
OBAN – Onboard Autonomous Navigation: An experiment investigating onboard autonomous navigation for future inter-planetary missions. Using the bearings of stars recorded by SMART-1's star trackers, and the Earth and Moon as seen by the AMIE camera, OBAM was able to evaluate a computer technique for onboard autonomous navigation.
EPDP – Electric Propulsion Diagnostic Package: Monitored the functionality of the electric propulsion thruster: the ion energy and current density was measured, a Langmuir Probe was used and measured the plasma potential and electron density and temperature; a solar cell and a Quartz-crystal micro-balance was used to provide data on the extent of contamination.
SPEDE – Spacecraft Potential Electron and Dust Experiment: Consisting of two electric sensors on the ends of 60 cm booms, these sensors were also used to monitor the state of the electric propulsion system. Furthermore, SPEDE was also used to conduct some space plasma physics.
KaTE – X/Ka-band Telemetry and Telecommand Experiment: Used to validate new communication digital technology; demonstrated for the first time the performance of a new higher range of communication frequencies in the X-band and Ka-band; tested new data encoding techniques; and validated corresponding ground-based infrastructure.
RSIS – Radio Science for Smart-1: Studied the feasibility and accuracy of determining the rotational state of a planet, or satellite from orbit.
Laser Link: ESA has laser links with telecom satellites from an optical ground station in Spain's Canary Islands. Although Smart-1 was far away and moved away rapidly, the onboard camera AMIE was still be able to view Tenerife illuminated with laser light.
SIR – Smart-1 Infrared Spectrometer: Was used to perform reflectance infrared spectroscopy in the 900 to 2400 nm range, providing detailed analysis of the surface composition.
D-CXIS – Demonstration of a Compact Imaging X-ray Spectrometer: Was used to provide the first high resolution global map of the lunar surface composition with relative abundances of the different elements in the lunar surface.
XSM – X-ray solar monitor: Used to measure the X-ray spectrum of the Sun.
AMIE – Asteroid Moon Micro-Imager Experiment: Used to survey the terrain using visible and near-infra-red light.

Solar Heliospheric Observatory (SOHO)

Current Status
According to the 2005 Sun-Solar System Connection (SSSC) mission operations and analysis review, which examined the scientific contributions and merits of NASA's active solar satellite missions, the Solar Heliospheric Observatory (SOHO) mission has been approved to operate at least through 2010. SOHO was ranked near the top, for science contributions, of 13 solar science programmes reviewed by the panel.

On 29 March 2006, SOHO data supported rare total solar eclipse observations made by numerous organisations. The eclipse's total phase lasted four minutes, best seen from Libya; NASA, assisted by The Exploratorium and the University of California at Berkeley, aired eclipse images observed from ground-based telescopes in Turkey, in real time, on the internet.

Background
During June 1998, ground controllers at the NASA Goddard facility lost contact with SOHO. Using the Deep Space Network, Goddard engineers managed to locate the satellite one month later. Apparently operator fault turned the satellite antenna away from ground stations. SOHO's instruments are in good health after more than 10 years in space, although some are beginning to show their age and the craft has been operating in gyro-less mode since 1999. SOHO's life was extended from an initially projected 6 years to a possible 20 years as a result of highly accurate launch vehicle insertion parameters. The programme is more expensive to fund than other solar satellites, given that it is a complex observatory. However, NASA will continue to fund SOHO at least through 2010, downscaling the funding as the years progress.

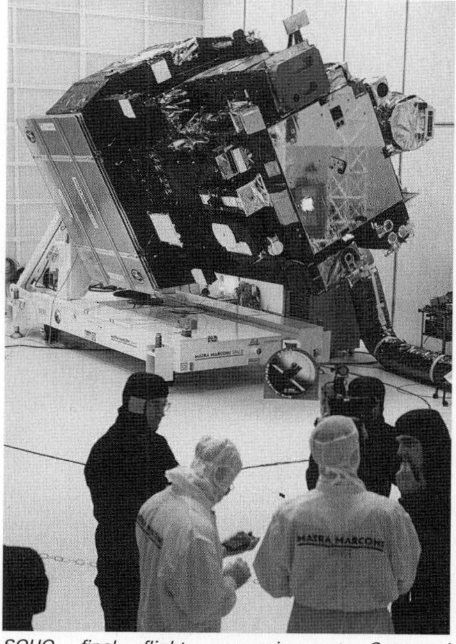

SOHO final flight processing at Canaveral (Theo Pirard/Space Information Centre) 0517301

In 2010, a 'Bogart' mission will begin, and only one or two (LASCO and MDI) of SOHO's instruments will remain in operation.

SOHO makes continuous observations of the solar surface, corona and wind, from a vantage point in the L1 Lagrangian libration point, about 1.5 million km out. The observatory is a joint NASA-ESA Inter-Agency Solar-Terrestrial Physics (IASTP) series satellite. ESA and the international scientific community seek the data to investigate physical processes that form and heat the corona; maintain it and give rise to the expanding solar wind; and to investigate internal structure by helioseismology and solar irradiance variations.

Matra Marconi Space France was the prime contractor, with help from BAE (AOCS), MMS UK (payload module, propulsion subsystem), Alenia (structure, harness), CASA (thermal control), Saab Ericsson (communications, data handling, software), FIAR (power), SAFT (batteries), Odetics (tape recorder), and Cubic (HPA).

Specifications
Launch: 2 December 1995 by Atlas-IIAS from Cape Canaveral. First solar image recorded 19 December 1995 and commissioning formally completed 16 April 1996
Mass: 1,850 kg at launch (610 kg payload)
Orbit: Circling L1 libration point since 14 February 1996, 1.5 million km out on Earth-Sun line
Lifetime: 2 years nominal; consumables sized for 6 years, however an accurate launch left 20 years worth of propellants
Spacecraft: 4.3 × 2.7 × 3.65 m, with solar array deployed width is 9.5 m; 3-axis stabilised Sun-pointing with 10 arcsec accuracy and a pointing stability of 1 arcsec per 15 min (10 arcsec per 6 months). Instruments accommodated in payload module, with a high degree of decoupling from the service module. Solar panels generate about 1,400 W (payload requires 440 W). S-band telemetry received by NASA's Deep Space Network over 3 × 1.3 h + 1 × 8 h periods each day. Continuous science data stream 40 kbit/s, increased to 200 kbit/s when MDI solar oscillation imaging instrument is operating in high rate mode; 1 Gbit tape recorder + 2 Gbit SSR. The Experiment Operations Facility (EOF) at NASA Goddard co-ordinate and plan science operations. Its main tasks are to organise the payload's real-time operation and to control the imaging/spectrometric instruments during the daily 8 h contacts. MDI's solar oscillations imaging data requires a specialised data facility at Stanford University, Palo Alto
Payload: SUMER Solar Ultraviolet Emitted Radiation (Max Planck Institute for Solar System Research, Germany)
CDS Coronal Diagnostic Spectrometer (Rutherford Appleton Laboratory, UK)
EIT Extreme UV Imaging Telescope (NASA-Goddard, US)
UVCS UV Chronograph Spectrometer (Harvard-Smithsonian Center for Astrophysics, US)
LASCO White Light/Spectrometric Chronograph (Naval Research Laboratory, US; Max Planck Institute for Solar System Research, Germany); C1 has not functioned since SOHO's temporary loss in 1998; C2 and C3 remain in working order
SWAN Solar Wind Anisotropies (FMI, Finland; Service d'Aeronomie, France)
CELIAS Charge, Element/Isotope Analysis (Universität Bern, Switzerland)
COSTEP Suprathermal/Energetic Particle Analyser (Universität zu Kiel, Germany)
ERNE Energetic Particle Analyser (University of Turku, Finland); COSTEP/ERNE are combined in the single CEPAC instrument
GOLF Global Oscillations at Low Frequencies (Institut d'Astrophysique Spatiale, France)
VIRGO Variability of Solar Irradiance (Institut d'Astrophysique Spatiale, France)
MDI/SOI Michelson Doppler Imager/Solar Oscillations Investigation (Stanford University, US).

Ulysses

Current Status
Ulysses has been returning solar science data for over 17 years. On 14 January 2008, almost a year after visiting the south solar pole, the spacecraft reached the highest point of its orbit over the Sun's northern polar cap, completing its third rapid south-to-north transit. The mission has been extended until March 2009.

Background
Until 1981, the International Solar Polar Mission (ISPM) was a two-craft, NASA-ESA project to be launched by a single Space Shuttle and three-stage Inertial Upper Stage (IUS) in February 1983. ESA had the responsibility to provide one spacecraft and about half of the 16 experiments, while NASA promised to provide the other craft, the rest of the experiments, RTG and the Shuttle launch. The two agencies signed the Memorandum of Understanding (MoU) on 29 March 1978.

The original plan called for the two spacecraft to fly for 15 months to Jupiter, one passing above and the other below the planet at distances of 450,000 km, causing both to enter high inclination, but opposite (posigrade and retrograde solar) orbits. Four years after launch, they would have passed over opposite solar poles at 250 million km, through the ecliptic plane and then over the other poles. The five-year mission would have been the first to investigate the Sun in three dimensions, providing an accurate assessment of the solar magnetic field and solar wind and studying solar flares, coronal holes and loss of mass/energy.

NASA's decision in 1980 not to proceed with the three-stage IUS raised doubts about the feasibility of the dual launch plan. The project switched to the cryogenic liquid Centaur upper stage, but US budget cuts in February 1981 forced NASA to save USD250 million to USD300 million on ISPM by cancelling its craft. Shuttle delays and other problems had already delayed the launch to 1985. It was finally agreed that NASA would continue to provide launch/tracking services and an RTG for the remaining ESA spacecraft. With ESA's increased prominence as the majority partner in the project, it was renamed Ulysses (from ISPM) at ESA's request to honour both Homer's mythological hero, and a reference in Dante's *Inferno* of Ulysses' urge to explore "an uninhabited world behind the Sun".

NASA programme delays and problems, including the impact of the Challenger disaster, continued to plague the project, and it was not until 13 years after ESA's Science Programme Committee approved the mission that NASA launched Ulysses from Shuttle Discovery during the 5-23 October 1990 Jupiter window. The IUS and Payload Assisted Module (PAM-S) motor accelerated Ulysses to a record 15.4 km/s, exceeding the velocities of NASA's Pioneers and Voyagers and making it the fastest man-made object ever launched.

Ulysses' outbound path would have taken it to 17 AU, almost to Uranus' orbit, if Jupiter had been avoided. A 0.29 m/s adjustment on 8 July 1991 set up the Jupiter gravity assist encounter 16 months after launch. After travelling 993 million km and already collecting important information on the interplanetary medium, Ulysses began the Jupiter encounter by detecting the bow shock crossing at 17:33 GMT on 2 February 1992, some 113 R_J out. This was more distant than expected: Pioneer's was 100 R_J and Voyager's 60 R_J. It approached through the late morning region of the magnetosphere at about 30° N, came within 378,400 km of the cloud tops (5.3 R_J) at 12:02 GMT on 8 February and then

Ulysses was launched on an Inertial Upper Stage/PAM deployed from the payload bay of Space Shuttle Discovery on 6 October 1990 (NASA-ESA) 1342189

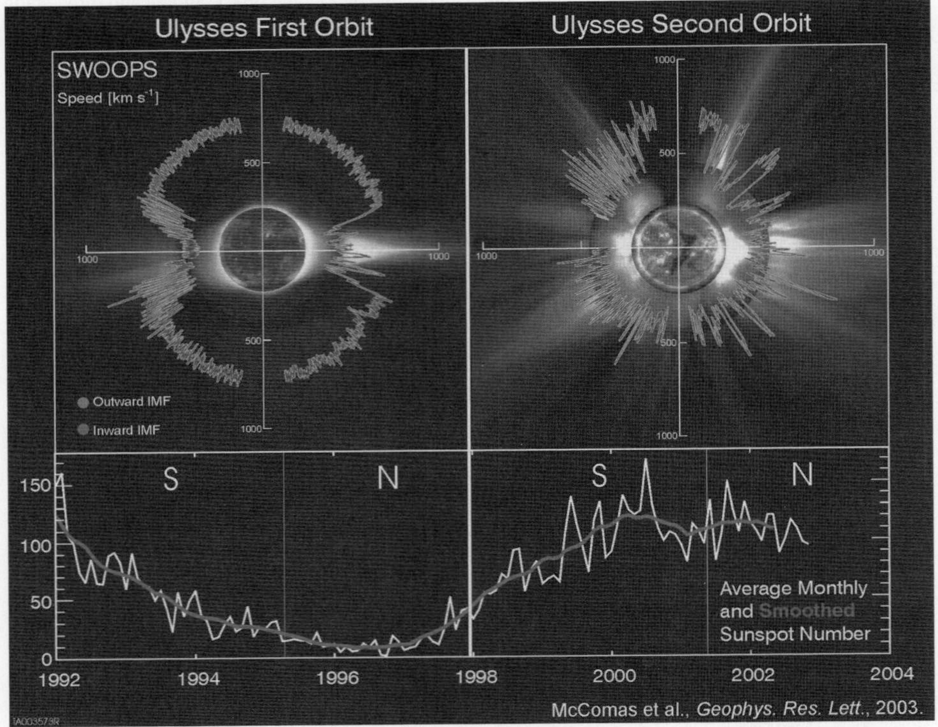

Polar plots of solar wind speed as a function of latitude for Ulysses' first two orbits (top). The solar wind speed data was obtained by the SWOOPS instrument (Solar Wind Observations Over the Poles of the Sun). The bottom panel shows the sunspot number over the period 1992-2003. The first orbit occurred through the solar cycle declining phase and minimum while the second orbit spanned solar maximum. Both solar wind plots are plotted over solar images characteristic of solar minimum (17 August 1996) and maximum (7 December 2000) (ESA/NASA/ D. McComas) 1342191

exited unscathed on the previously unexplored evening side, at high southern latitudes, having spent more than a week in the magnetosphere.

The close 13.5 km/s flyby was necessary to produce the spacecraft's high ecliptic inclination – IUS/PAM could have reached only 23° without the Jupiter gravity assist. Ulysses found that Jupiter's intense radiation belts reach only 40° latitude, whereas Earth's extend to 70°. Ulysses then passed directly through the Io plasma torus (only Voyager 1 preceded it), which plays a key role in refuelling the magnetosphere with plasma. The overall density was about as expected, implying injection from Io has not changed dramatically since Voyager, but significant azimuthal asymmetries were found. Other important Ulysses observations included the first direct detection of ionised O, N and He (and neutral He) atoms arriving from interstellar space and the measurement of μm-sized dust grains from interstellar space.

Ulysses' Polar Passes

Polar Pass	Start date	Maximum solar latitude	End date
1st South	26 June 1994	13 September 1994, – 80.2°	5 November 1994
1st North	19 June 1995	31 July 1995, 80.2°	29 September 1995
2nd South	6 September 2000	27 November 2000, – 80.2°	16 January 2001
2nd North	31 August 2001	13 October 2001, 80.2°	10 December 2001
3rd South	17 November 2006	7 February 2007, – 79.7°	3 April 2007
3rd North	30 November 2007	14 January 2008, 79.8°	15 March 2008

Following the flyby of Jupiter, the spacecraft entered an elliptical, heliocentric orbit inclined at 80.2° to the solar equator, with an orbital period of 6.2 years. Aphelion was at ~5.4 AU from the Sun and perihelion distance was ~1.34 AU. As a result, it became the first probe to undertake detailed investigations of the heliosphere's third dimension. On 9 June 1993, it became the first spacecraft to exceed 32° solar latitude, surpassing Voyager 1's record. During its first solar pass, the trajectory brought Ulysses over the Sun's South Pole during 1994, with 132 days spent above 70° solar latitude between 26 June and 6 November. A maximum 80.2° was reached on 13 September 1994, at a distance of 2.3 AU from the Sun. The Sun's equator was crossed on 5 March 1995. Passage over the North Pole came during 19 June to 30 September 1995; 80.2° was attained on 31 July.

In high inclination heliocentric orbit, Ulysses passed over the solar South Pole and conducted the first long-term in situ observations of high speed solar wind flowing from the large coronal hole that covers that polar regions at solar minimum. Ulysses' measurements of ionised interstellar neutralised gas particles, so-called interstellar pick-up ions, have led to a major advance in the understanding of the processes affecting this component of our heliospheric particle population.

Unexpectedly, the magnetometers detected a wide variety of fluctuations at many spatial scales in the Sun's high latitude field. They are believed to represent relatively unevolved turbulence originating at the southern polar coronal hole. Another important finding was that the magnetic field radial component varies relatively little with latitude. This is contrary to the expectation that the imprint of the underlying dipole-like magnetic field at the Sun's surface, showing an increase in the radial field at the poles, would be detected at Ulysses' position. In simple terms, a south magnetic pole was absent at high latitudes. Undoubtedly related to the presence of the large-scale fluctuations in the polar magnetic field was the detection of an unexpectedly small increase in the measured influx of cosmic ray particles over the poles compared with model predictions.

Ulysses' first mission extension and second solar orbit began in October 1995. During its second orbit, which coincided with solar maximum, Ulysses found an entirely different heliosphere compared with that observed near solar minimum. At solar minimum, the heliosphere was dominated by the fast wind from the southern and northern polar coronal holes. During solar maximum, the large polar coronal holes had disappeared, and the heliosphere appeared much more symmetric. The solar wind flows showed no systematic dependence on latitude, while the wind itself was generally slower and much more variable than at solar minimum. However, when Ulysses reached high northern latitudes in late 2001, it observed the formation and growth of a new polar coronal hole. The solar wind recorded at Ulysses then became faster and more uniform, resembling the flows seen over the poles at solar minimum.

Although the solar magnetic field, corona and solar wind were highly variable, the magnetic field at Ulysses (~1.5–2.5 AU from the Sun) maintained a surprisingly simple, dipole-like structure. In contrast to the situation at solar minimum, the equivalent magnetic poles were located at low latitudes rather than in the polar caps. It was found that the complex process of magnetic polarity reversal at the end of one soalr cycle and the beginning of the next takes place over a period of several months.

Due to the mission's scientific success, both ESA and NASA have continued to fund the programme. In June 2000, ESA's Science Programme Committee agreed to continue operating the spacecraft from the end of 2001 to 30 September 2004, a decision that was subsequently supported by NASA. Ulysses flew within 0.8 AU of Jupiter – its second Jovian encounter – on 4 February 2004, when it was able to explore previously unexplored regions of the planet's magnetosphere. It is now in its third six-year orbit around the Sun. The third South Polar pass took place November 2006 to February 2007, followed by the third North Polar pass November 2007 to January 2008. This coincided with the onset of the new solar cycle.

On 15 November 2007, ESA's Science Programme Committee unanimously approved to continue the operations of the highly successful Ulysses spacecraft until March 2009. This latest extension, for a period of 12 months, is the fourth in the history of the mission. Since the power output from its Radioisotope Thermoelectric Generators (RTGs) is falling, the spacecraft operations team has devised a new operational strategy that will

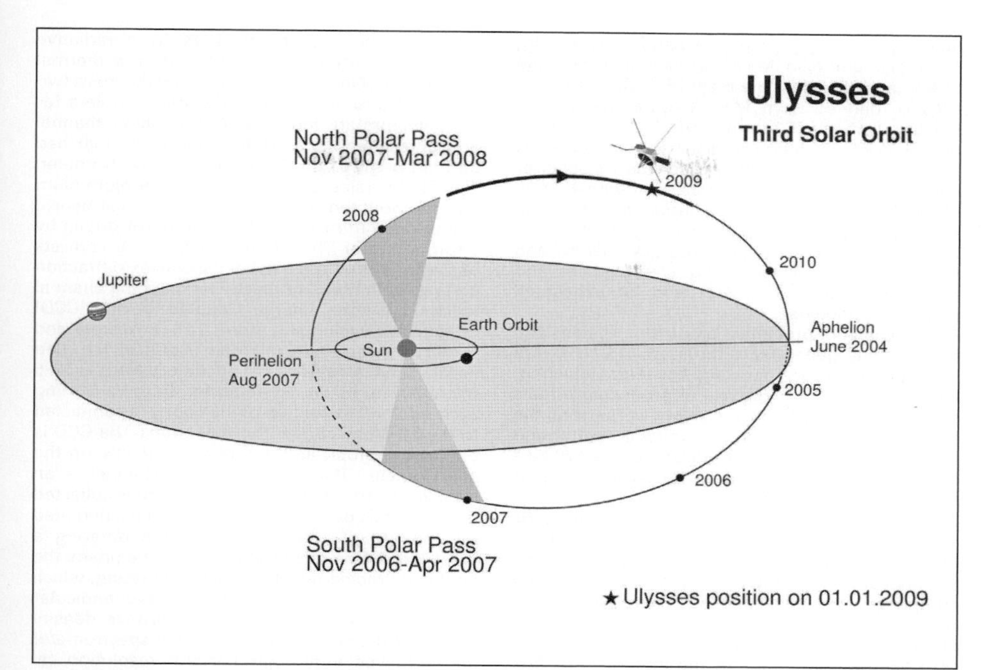

Ulysses is now on its third orbit of the Sun (ESA) 1342188

Preparing to insert the RTG's fuel elements on the Shuttle pad (NASA) 0517304

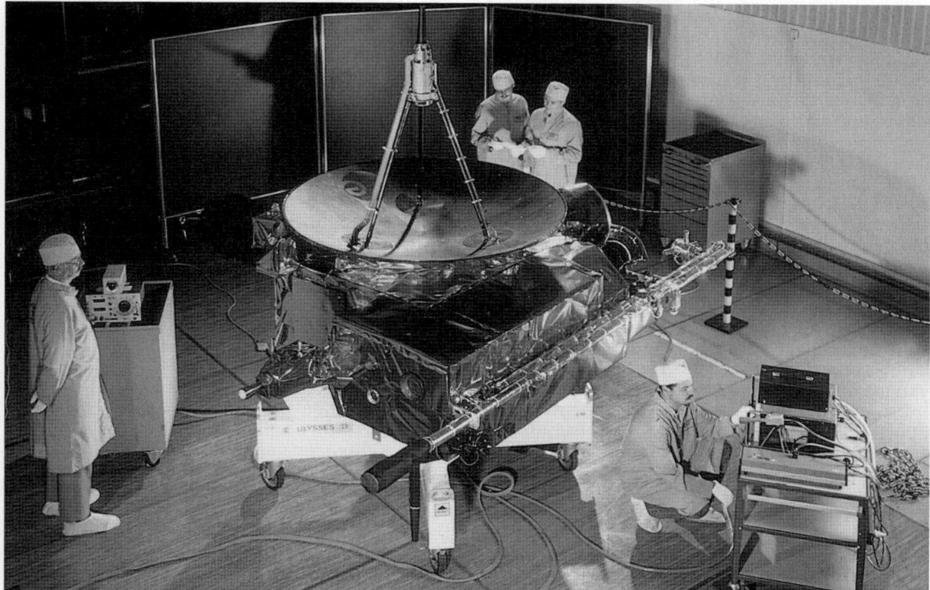

Ulysses in final assembly 0085789

allow the majority of the scientific instruments to operate throughout the fourth extension without much power sharing.

Ulysses is tracked eight hours per day throughout the mission by the 34 m dishes of NASA's Deep Space Network (DSN). A joint NASA/ESA team at the Jet Propulsion Laboratory's Mission Control and Computing Center conducts operations. In the normal operating mode, the scientific data from Ulysses' instruments are stored on an onboard tape recorder for approximately 16 hours and downlinked to the DSN once a day, together with the real time data, during the nominal 8-hour tracking pass.

Specifications

Launch: 6 October 1990 aboard Shuttle Discovery, STS-41 from Kennedy Space Center. Three-stage IUS-PAM upper stage.
Mission: Fields/particles and cosmic dust investigations of the heliosphere up to high solar latitudes.
Contractors: Dornier (prime), BAE (AOCS, HGA), Fokker (thermal, nutation damper), FIAR (power), Officine Galileo (Sun sensors), Laben (data handling), Thomson-CSF (telecommand), MBB (thrusters).
Mass: 371 kg at launch; 366.6 kg at PAM separation. Science payload 55 kg.
Dimensions: 3.2 × 3.3 × 2.1 m
Orbit: 6.2 year elliptical heliocentric orbit, 80° inclination; aphelion at ~5.4 AU and perihelion at ~1.34 AU.
Description: Spin-stabilised at 4.99 rpm with the 1.65 m diameter HGA pointed continuously towards Earth. Powered by a single DoE RTG providing 280 W at BOM and 220 W by December 2001 (4.5 kW heat). Double-hinged 5.55 m radial boom carries four sensors for HED, STO and HUS. 7.41 m axial boom as monopole for STO. Two wire booms 72.5 m tip-to-tip as dipole for STO.
Communications: X-band (20 W, 8,408.209 MHz, 2° 3 dB beamwidth); 5 W 2 GHz S-band 2,111.607/2,293.148 MHz up/down is used for dual frequency radio science investigations and early manoeuvres. During contact periods, real-time data at 1,024 kbit/s are interleaved with playback of stored data. Two redundant tape recorders can each hold 45.8 Mbit, stored at 128/256/512 bit/s.
Propulsion: 33.5 kg high purity hydrazine in single bladder tank with dual manifold feeds two clusters of four 2 N catalytic thrusters (total delta-V 185 m/s)
Payload: **COSPIN**: Cosmic Ray and Solar Particle Investigation; University of New Hampshire, US.
DUST: Cosmic Dust Experiment; Max-Planck-Institut für Sonnensystemforschung, Germany.
EPAC: Energetic Particle Investigation; Max-Planck-Institut für Aeronomie, Germany.
GAS: Interstellar Neutral Gas Experiment; Max-Planck-Institut für Aeronomie, Germany.
GRB: Gamma-Ray Burst Experiment; University of California, Berkeley Space Science Laboratory, US.
GWE: Gravitational Wave Experiment; Universita di Pavia, Italy.
HI-SCALE: Heliosphere Instrument for Spectra, Composition & Anisotropy at Low Energies; Bell Laboratories, Lucent Technologies, US.
SCE: Coronal Sounding Experiment; Universität Bonn, Germany.
SWICS: Solar Wind Ion Composition Experiment; University of Maryland, US/International Space Science Institute, Switzerland.
SWOOPS: Solar Wind Plasma Investigation; Southwest Research Institute, US.
URAP: Unified Radio & Plasma Wave Investigation; NASA Goddard.
VHM/FGM: Vector Helium Magnetometer/Flux Gate Magnetometer; The Blackett Laboratory, Imperial College of Science and Technology, UK.

Venus Express

Current Status

Venus Express has been in orbit around Venus since 11 April 2006. The mission is expected to last for at least two Venus sidereal days (486 Earth days), ending in October 2007. A mission extension was approved in February 2007, meaning Venus Express operations will now continue until early May 2009, by which time observations will have been performed for more than four Venus sidereal days.

The first Venusian day of the mission was devoted to global latitude, longitude, and local

Venus Express in orbit around the planet 1179612

time coverage of the planet. The second day provided opportunities for studying in more detail objectives selected on the basis of the first day's observations, and for investigating the time variability of previously observed phenomena.

Background

Venus Express is the first ESA mission to visit the second planet from the Sun. Since the spacecraft is a virtual twin of Mars Express, it was developed very quickly and at low cost under the leadership of EADS Astrium. Most subsystems have not been modified from Mars Express. The major differences are in thermal control with larger, more efficient radiators and 23 layers of multi-layer insulation which are gold instead of black. The two lateral solar arrays were also redesigned. They are about half the size of those on Mars Express and use 'triple junction' gallium arsenide technology suitable for a high temperature environment. The spacecraft bus is a honeycomb aluminium box. The payload consists of seven instruments. Five of them are re-used from the Mars Express and Rosetta projects with small modifications. The scientific instruments are located on three sides of the spacecraft. There are two high-gain antennas – a 1.3 m dish and a 0.3 m antenna – mounted on different faces. The main engine is on the bottom, while minor manoeuvres involve four pairs of thrusters at the four bottom corners.

Venus Express was successfully launched into Earth parking orbit by Soyuz-Fregat. After the second firing of the Fregat upper stage, the spacecraft was successfully placed into an escape trajectory. It arrived at Venus on 11 April 2006, after a five-month journey. The capture orbit was an ellipse ranging from 330,000 km to less than 400 km, with a period of 9 days. A series of manoeuvres gradually reduced the apocentre and the pericentre altitudes. The main engine was fired twice (on April 20 and 23) and the thrusters were ignited five times (15, 26 and 30 April, 3 and 6 May). On 7 May 2006, after 16 orbits around the planet, it reached the nominal science operations orbit. On 3 June at 13:42 UT, after 207 days of flight, 43 orbits around Venus and many test activities, Venus Express completed its commissioning phase and entered the routine science phase. Pericentre is located at 80°N and each orbit lasts 24 hours. Joint observations were undertaken with NASA's Messenger during its Venus flyby on 6 June 2007.

High rate communications are performed 8 hours per day in X-band, in order to transmit to Earth all science data stored in the Solid State Mass Memory. An average of 2 Gbits of science data is downlinked every day to the new ESA ground station of Cebreros, Spain.

Specifications

Launch: 03:33 UTC, 9 November 2005 from Baikonur by Soyuz-Fregat.
Launch Mass: 1,270 kg
Body Dimensions: 1.65 × 1.7 × 1.4 m
Operational Venus Orbit: 250 × 66,000 km, c. 90° inclination, period 24 hours. The only major perturbation affecting the orbit is the effect of solar gravity, which raises the periapse. Over one Venus sidereal day (243 Earth days) the periapse altitude increases by approximately 170 km. To counteract this effect, the periapse is lowered as necessary by using the thrusters.
Power: Two symmetrical gallium arsenide solar arrays with a total area of 5.7 m^2. They generate about 1,400 W in Venus orbit. During eclipse or when spacecraft power demand exceeds solar array capacity, power is supplied by three 24 Ah Lithium-Ion batteries that combine low mass with high-energy efficiency.
Communications: A transponder capable of transmitting and receiving in S-Band as well as X-Band, plus four different antennas: two low-gain S-band antennae, one dual band high-gain 1.3 m antenna, and one high-gain X-band antenna with 0.3 m diameter dish. Maximum downlink data rate at Venus is 228 kbps. The main High Gain Antenna (HGA1), derived from Mars Express but with a smaller diameter (1.3 m instead of 1.6 m), is used during ¾ of the mission, when the spacecraft is most distant from Earth, either side of superior conjunction. The second High Gain Antenna (HGA2), derived from Rosetta, is used during the remainder of the mission, when Venus is closer to Earth.
Attitude Control: Attitude measurements and control are performed using star trackers and gyros. Reaction wheels are used for almost all attitude manoeuvres.
Propulsion: A bi-propellant propulsion system. The main engine generates a thrust of 415 N, and is used for the Venus capture and apocentre lowering. Eight thrusters each capable of producing thrusts of 10 N are located in the lower corners of the spacecraft. They are used for attitude control and orbit maintenance. Total propellant load is 570 kg – higher than on Mars Express because of higher Delta V requirements for the capture burn.
Payload: **Analyser of Space Plasma and Energetic Atoms (ASPERA)** characterises the plasma environment around Venus and studies escape processes in the outer atmosphere. ASPERA-4 comprises four sensors: two Energetic Neutral Atom (ENA) sensors, plus electron and ion spectrometers. The Neutral Particle Imager (NPI) provides measurements of the integral ENA flux (0.1–60 keV) with no mass and energy resolution but relatively high angular resolution of 4° × 11°. The Neutral Particle Detector (NPD) measures the ENA flux, resolving velocity (0.1–10 keV) and mass with a coarse angular resolution of 4.5° × 30°. The ELectron Spectrometer (ELS) is a top-hat electrostatic analyser with a very compact design to measure electron fluxes in the energy range 1 eV–20 keV with 8 per cent energy resolution. These three sensors are located on a scanning platform. The Ion Mass Analyser (IMA) is a separate unit eectrically connected with the main unit. It provides ion measurements in the energy range 0.01–36 keV. This is the most comprehensive plasma particle package ever sent to Venus.

Planetary Fourier Spectrometer (PFS) is a high-resolution infrared spectrometer optimised for atmospheric studies, in particular the structure and composition of the Venus middle atmosphere. It has a short wave channel that covers the spectral range 0.9–5.5 microns and a long wave channel that covers 5.5–45 microns. The spectral resolution is 4 nm at 5.5 microns. FOV is about 1.6° for the short wavelength channel and 2.8° for the long wavelength channel. This corresponds to a spatial resolution of 7 and 12 km from an orbital altitude of 250 km.

Ultraviolet and Infrared Atmospheric Spectrometer (SPICAV/SOIR) is a versatile suite of three spectrometers that studies the structure and composition of the upper atmosphere during solar/stellar occultation and nadir observations. The Acousto-Optic Tunable Filter (AOTF) spectrometer analyses near-infrared (0.7–1.65 microns, resolution about 1,500) thermal emission coming from the deep atmosphere on the night side. This reveals the composition of the lower atmosphere and water content, as well as detecting hot spots on the surface. The UV spectrometer (110–320 nm, resolution 1.5 nm) studies natural emissions from the upper atmosphere and probes the vertical distribution of haze layers and SO2 in order to determine the nature of a mysterious UV absorber. The Solar Occultation in the InfraRed (SOIR) instrument is a new type of high-resolution spectrometer to measure constituents above the clouds and their vertical distribution. It consists of a high dispersion grating spectrometer (2.2–4 microns) working in high diffraction orders, combined with an AOTF. The detector is a 2D Sofradir HgCdTe matrix, cooled to 100 K.

Venus Express Magnetometer (MAG) measures the magnetic field "frozen" in the circumplanetary plasma, supporting the ASPERA observations and the solar wind interaction with the atmosphere. It has two fluxgate sensors to measure magnetic field and direction. It consists of two sensors, an electronics box, and a 1 m long boom made from carbon fibre. One sensor is on the tip of the boom and the other is on the spacecraft body.

Ultraviolet/visible/near-infrared Mapping Spectrometer (VIRTIS) is a visible and near-infrared spectro-imager and high resolution spectrometer to investigate the structure, composition, and dynamics of the middle and lower atmosphere. It combines three observing channels in one instrument. Two of the channels are devoted to spectral mapping (mapper optical subsystem), while the third channel is devoted to spectroscopy (high resolution optical subsystem). The optical subsystems are housed inside a common structure – the cold box – cooled to 130K by a radiative surface supported on a truss having low thermal conductivity. On the pallet supporting the truss, two sets of electronics and two cryogenic coolers for the detectors are mounted. The mapping channel optical system is a Shafer telescope matched through a slit to an Offner grating spectrometer. The Shafer telescope consists of five aluminium mirrors mounted on an aluminium optical bench. The primary mirror is a scanning mirror driven by a torque motor. The Offner spectrometer consists of a relay mirror and a spherical convex diffraction grating, both made of glass. The mapping channel utilizes a silicon Charge Coupled Device (CCD) to detect wavelengths from 0.25–1 micron and a mercury cadmium telluride (HgCdTe) InfraRed Focal Plane Array (IRFPA) to detect from 0.95–5 microns. The IRFPA is cooled to 70K by a Stirling cycle cooler. The cold tip of the cooler is connected to the IRFPA by copper thermal straps. The CCD is operated at 155K and is mounted directly on the spectrometer. The high resolution channel is an echelle spectrometer. The incident light is collected by an off-axis parabolic mirror and then collimated by another off-axis parabola before entering a cross-dispersion prism. After exiting the prism, the light is diffracted by a flat reflection grating, which disperses the light in a direction perpendicular to the prism dispersion. The low groove density grating is the echelle element of the spectrometer and achieves very high spectral resolution by separating orders seven through sixteen across a two-dimensional detector array. The high-resolution channel employs a HgCdTe IRFPA to perform detection from 2–5 microns. The detector is cooled to 70K by a Stirling cycle cooler.

Venus Monitoring Camera (VMC) studies the cloud morphology, atmospheric dynamics, and water distribution in the middle and lower atmosphere. It also provides thermal imaging of the surface and searches for volcanic activity. The VMC has four channels that share a single CCD. These channels have narrow band filters chosen for particular scientific objectives. The UV channel is centred on 365 nm and is used to track cloud features. There are two near-infrared filters. One of these is centred at 1,010 nm and studies night emission from the hot surface. The other, centred at 935 nm, studies global distribution of water vapour at the cloud tops and in the lower atmosphere. The fourth filter (513 nm) observed the oxygen nightglow in the lower thermosphere.

Venus Radio Science Experiment (VeRa) uses the radio system of the spacecraft to sound the structure of the middle and lower atmosphere and ionosphere during Earth occultation. It also carries out bi-static radar sounding of selected targets on the surface and sounding of the solar corona. It uses radio signals at X- and S-band (3.5 cm and 13 cm wavelengths) to probe the surface, neutral atmosphere and ionosphere, the Venus gravity field and the interplanetary medium.

Wind

Current Status

As of 2006, after 12 years in orbit, Wind is operational. NASA's 2005 Sun-Solar System Connection (SSSC) mission operations and analysis review, which examined the scientific contributions and merits of the agency's active solar satellite missions, reviewed the Wind programme and determined that the spacecraft can continue to be used effectively to track energetic particles and observe the solar wind, especially in conjunction with SOHO and RHESSI (see separate entries). Although ACE has assumed some of Wind's particle measurement responsibilities, Wind is still viewed as a valuable solar science resource by NASA, and will likely function and be funded for years to come.

Background

Wind is one of the Inter-Agency Solar-Terrestrial Physics (IASTP) programme satellites. The craft was built to investigate sources, acceleration mechanisms and propagation processes of energetic particles and the solar wind. It also provides complete plasma, energetic particle and magnetic field input data for magnetospheric and ionospheric studies; determines the magnetospheric energy output to interplanetary space in the upstream region; provides baseline ecliptic plane observations for heliospheric studies; and provides gamma ray burst measurements for astrophysics research.

Lockheed Martin Astro Space built the satellite. Wind's initial programme cost was USD173 million.

Wind is installed on Delta (NASA) 0517299

Specifications

Launch: 1 November 1994 by Delta II 7925-10 from Cape Canaveral

Mass: 1,250 kg (300 kg propellant, 198 kg science payload, 752 kg dry)

Orbit: lunar swing-by (first 27 December 1994) to 250 R_E during its first 2 years; L1 halo (3×10^5 km) from November 1996

Life: 3 years nominal; extended mission plans approved through at least 2009

Spacecraft: 2.6 m diameter 2.5 m height cylinder with conductive surfaces, spin-stabilised at 20 rpm. Two deployable 12 m lanyard booms for magnetic field instruments, two deployable 5.28 m Z-axis booms plus four wire antennas (two 7.5 m, two 50 m) for electric field measurements. 11.5 m^2 solar array provides 472 W BOL, including 144 W for payload; 3 × 26.5 Ah Ni/Cd batteries. Four Olin MR-111 2.2 N spin plus eight MR-50 22 N δV thrusters. Telemetry received by DSN over 2 h nominal daily contact. Science data rates 5.65/11.3 kbit/s real-time and up to 128 kbit/s playback. Payload data outside of ground coverage stored on 1.3 Gbit digital tape recorder. Mission operations/data processing from NASA Goddard. Telemetry 2275.5 MHz (realtime 5.56 kbps, playlock 32/64 kbps). Command 2094.9 MHz (realtime 250 kbps).

Payload: MFI Magnetic Fields Investigation (NASA Goddard).

WAVES Radio & Plasma Wave Experiment (Observatoire de Meudon, France).

SWE Solar Wind Experiment (NASA Goddard).

EPACT Energetic Particles Acceleration, Composition, Transport (NASA Goddard); EPACT is only partially operational.

TGRS Transient Gamma Ray Spectrometer (NASA Goddard); TGRS failed early in the mission.

SMS (SWICS/MASS/STICS) Solar Wind and Suprathermal Ion Composition Studies – Solar Wind Ion Composition Spectrometer (SWICS); High Mass Resolution Spectrometer (MASS); and Suprathermal Ion Composition Spectrometer (STICS) (University of Maryland); SMS is partially operational.

3D PLASMA Energetic Particles & Three Dimensional Plasma Analyser (University of California, Berkeley).

KONUS Gamma Ray Burst Investigation (Ioffe Physical Technical Institute, St Petersburg).

Israel

TechSat Gurwin Series

Current Status

As of late 2006, TechSat II, also known as TechSat-1b, was still functioning; it continues to return data long after its mission began in 1998. All but three of the spacecraft's instruments are fully operational. As of 2004, TechSat II was enlisted as a resource for the amateur radio community; it is known in that capacity as Gurwin-OSCAR 32 (please see the separate OSCAR entry for more details on that satellite series).

Background

The Technion Israel Institute of Technology, in the early 1990s, proposed building three or four small satellites for continuous early warning coverage. The Gurwin I (named in honor of Joseph and Rosalind Gurwin, whose interest and support of Israeli space research allowed for TechSat's development) TechSat civil satellite may have been its first attempt to provide technology demonstrations of future imaging systems. The satellite came about as a result of a collaboration between the Faculty of Aerospace Engineering at Technion Institute (under the direction of Professor Chaim Eshed), Israel Space Agency (ISA), industry and Amateur satellite-IL to develop a simple, low-cost, low-power platform. The flight model was completed October 1994, but the first TechSat, also known as TechSat-1a, was lost when its Start launch vehicle failed.

The Asher Space Research Institute (ASRI) at Technion manages and conducts TechSat II operations from the university-based station. TechSat II carries instrumentation to perform a range of experiments, including an Earth imaging device. The satellite's total cost, including launch services was estimated at USD5 million; it took 30 months to build and test.

Specifications

TechSat I

Launch: 28 March 1995 aboard a Russian Start from Plesetsk; planned orbit was 670 km, 75.4°

Mass: 52 kg

Configuration: box-shaped 460 mm height, 430 mm across (six Al plates, 2.5 mm thick). Earth-pointed face carried antennas, horizon sensor, and camera. 80C186EC-based CPU with two 512 kbyte independent RAM memories. Controlled from Technion station; command reception at 145.910/1,269.85 MHz

Power: solar cells on four faces (1,000 2 × 4 cm Si, 17 W average, 9 W required for housekeeping, Ni/Cd battery)

Attitude control: 3-axis Earth-pointing control (5° coarse, 0.1–0.5° fine) by 2.5 kg 3,600 rpm 4 Nms MW 3 × 1 Am^2 magnetorquers, 1 kg 1 W static horizon sensor (0.05° accuracy), 0.22 kg 0.8 W ±0.6 gauss 3-axis magnetometer

Payload: Elisra Earth-observing 350 × 570 pixel CCD camera, compressed image data broadcast using millimetre-wave link experiment (also on UHF).

Technion UV CsTe spectroradiometer to monitor Earth's albedo at 0.24–0.30 μm (0.01 μm/2 mrad resolution); 15° FOV. The spectral profile is a function of the vertical ozone profile. Goals are investigating the long-term stratospheric impact of the Kuwaiti oil fires and the extent of the region's ozone depletion.

Technion x-ray detectors (up to 200 keV). Principal purpose is to prove that CdTe crystals can be used for x-ray imaging.

TechSat II

Launch: 10 July 1998 by Zenit 2 from Baikonur, along with four other small satellites

Mass: 48 kg microsatellite

Orbit: sun-synchronous; 820 km altitude; 98.8° inclination

Configuration: box-shaped: 44.5 cm^3; 1-year design life

Power: thin-film photovoltaic solar cells on four faces; Ni/Cd battery for eclipse operations

Attitude control: 3-axis Earth-pointing control; actuators: momentum wheel and three magnetorquers; attitude sensor: 3-axis magnetometer; about 2°–2.5° nadir-pointing accuracy; total 3 W power consumption

Payload: **SOREQ**: Single Event Monitor for Detecting Protons and Heavy Ions in Space; measures solar protons and heavy particles.

ERIP: Earth Remote-Sensing Imaging Package; panchromatic CCD video camera; 52 × 60 m resolution, 25 × 31 km images; the instrument has performed below expectations as images are overexposed.

SLRRE: Satellite Laser Ranging Retroreflector Experiment; passive laser retroreflector array for ground-based laser range measurement and orbit determination.

XDEX: X-Ray Detector Experiment; in-flight instrument calibration failed after launch, as did the experiment.

SUPEX: Superconductivity Experiment; cooler degradation compelled the experiment to end in 2000; power generation technology demonstrator.

OM-2: Ozone Meter-2; failed 10 months after launch.

Global amateur radio community software: equipped with a digital store and forward multi-user system; this application came about only after significant software development work aimed at correcting TechSat II's bulletin board system problems was completed in 2004.

ADCS & EPS: Attitude Control and Electric Power subsystems.

Japan

Akari (Astro-F)

Current Status

Akari's on-board supply of liquid helium ran out at 08:33 (UT) on 26 August 2007, signalling the completion of observations at far-infrared and mid-infrared wavelengths, including the All Sky Survey. The non-cryogenically-cooled instruments continue to send back near-infrared data and are expected to operate for at least one year.

Background

Akari ("a light"), previously known as ASTRO-F, is Japan's 21st Scientific Satellite and the second infrared astronomy mission of the Japanese Institute of Space and Astronautical Science (ISAS). It was Japan's first astronomical satellite to perform an all-sky survey at infrared wavelengths. Its instruments have much greater sensitivity and higher resolution than those carried by the Infrared Astronomy Satellite (IRAS). It uses a Ritchey-Chretien telescope with a 68.5 cm diameter mirror and focal length of 4.2 m. The primary mirror is made of silicon carbide (SiC) and hollowed out in order to make it lighter. The actual weight of the 71 cm diameter (effective diameter 67 cm) primary mirror is 11 kg. This is the first time a SiC mirror has flown in space. The primary mirror is gold-coated and the trusses which support the secondary mirror are made of beryllium. It observes in the wavelength range from 1.7 to 180 microns (near-infrared to far-infrared). The operational lifetime of Akari's two main science instruments was based on the supply of liquid helium stored in a cryostat that cooled the telescope to 5.8 K (–267.4°C).

During launch in February 2006, the M-V launcher was set to a vertical angle of 81.5° and the flight azimuth was 143.0°. After the satellite separated from the M-V-8 third stage, it was inserted into 301 × 718 km, 98.2° orbit. Its onboard propulsion system was used to reach the final operational orbit above the day/night twilight boundary of the Earth. The nominal mission lifetime was 550 days, after which it was expected to run out of cryogenic coolant.

The three-axis stabilized satellite has an octagonal bus topped by a cryostat and a sunshade. A telescope and scientific instruments are stored in the cryostat and cooled by 170 litres of liquid helium and two sets of Stirling-cycle mechanical coolers. The addition of the mechanical coolers extended the helium life and reduced the quantity of helium to be carried into space. The bus module includes attitude control, data handling and communications. The key structure of the bus module is a cylindrical thrust tube (1 m high and 1.2 m diameter) made of carbon-fibre reinforced plastic. The propellant tanks of the reaction control system are stored inside this thrust tube. Subsystems of the bus module are installed on eight instrument panels, and the panels are integrated around the thrust tube. The lower end of the thrust tube is connected to the top of the M-V rocket third stage. The cryostat and the bus module have independent structures so as to decrease heat inflow into the cryostat. The 5.5 m span solar arrays are extended in orbit. The cryostat has an aperture lid on the ground so as to maintain a tight vacuum and prevent stray light inflow. This lid is ejected in space after the satellite reaches the correct attitude. The telescope is directed away from Earth to shield it from the planet's heat. An early problem with a Sun sensor and the cooling system of the star trackers caused attitude control to be handed over to the Earth sensor and gyroscopes.

Two instruments, the Far Infrared Surveyor (FIS) and Infrared Camera (IRC) observed the entire sky in six wavebands from the near-infrared to the far-infrared. These instruments also performed detailed photometric and spectroscopic observation of

Artist's impression of Japan's Akari (Astro-F) infrared space observatory (JAXA) 1330380

selected astronomical targets over the 2–180 micron range in 13 bands. Two detectors of the FIS are photoconductors which use semiconductor crystal Ge:Ga, germanium doped with gallium. Stressed Ge:Ga chips are sensitive to far-infrared light of longer wavelength than normal ones. Each detector is used with filters, so the FIS effectively has four observation bands. The FIS was also used for pointing observations to detect faint objects or to perform spectroscopy using a Fourier transform spectrometer.

The IRC is composed of three independent camera systems. The NIR camera is assigned to near-infrared wavelengths in the 1.7–5.5 micron range. The MIR-S camera is assigned to shorter mid-infrared wavelengths over the 5.8–14.1 micron range and the MIR-L camera is assigned to longer mid-infrared wavelengths of 12.4–26.5 micron. One of the advantages of the IRC is that it can observe 10 sq. arcmin at a time because of large format detector arrays (512 × 412 for NIR, 256 × 256 for MIR). Each camera can select a specific waveband to observe by using filters. In addition, IRC is equipped with prisms and grisms to perform spectroscopic observations. The IRC observations are generally made in pointing mode, but two of the wide band filters of the IRC (9 microns and 18 microns) were used for the All Sky Survey.

The liquid helium ran out on 26 August 2007, ending observations at far-infrared and mid-infrared wavelengths. The availability of mechanical coolers allows limited observations with the near-infrared camera for up to five years. The all-sky survey observations began on 8 May 2006 and the first survey was completed in November 2006, covering about 80 per cent of the sky. The second mission phase was dedicated to pointed observations as well as gap-filling observations for the All-Sky Survey.

Before its supply of liquid helium ran out, Akari completed the most detailed far-infrared All Sky Survey, covering about 94 per cent of the sky. It also carried out a mid-infrared survey and more than 5,000 individual pointed observations. The mission continues with warm phase observations using the near-infrared instrument that can still operate by using the on-board mechanical coolers.

The Japan Aerospace Exploration Agency (JAXA) mission is supported by Nagoya University, the University of Tokyo, the National Astronomical Observatory Japan, the European Space Agency (ESA), Imperial College London, the University of Sussex, the Open University (UK), the University of Groningen/SRON (The Netherlands) and Seoul National University (Korea). The far-infrared detectors were developed in collaboration with The National Institute of Information and Communications Technology. ESA provides ground station coverage (for data transfer between the telescope and the ground) and pointing reconstruction (determining where on the sky the telescope is pointing) for the All Sky Survey. In return, ESA is given 10 per cent of the pointing opportunities available when the telescope interrupts its all sky survey to make detailed studies of particular objects, and during the non-survey and post-helium-depletion phases of the mission.

Specifications
Launch: 21:28 UT on 21 February 2006 (22 February, 06:28 local time) on M-V-8 rocket from Uchinoura.
Orbit: 695 km × 710 km, inclination 98.2°, Sun-synchronous.
Mass: 952 kg
Dimensions: 2.0 × 3.7 m (launch config.). 3.3 × 5.5 m in observation mode with solar panels deployed.
Data and communications: Downlink rate 4 Mbps for scientific data. Data generation rate approximately 2 GBytes per day. Data recorder capacity 2 GBytes.
Payload: Far-Infrared Surveyor (FIS) and InfraRed Camera (IRC)

EGS (AJISAI) satellite

Current Status
In March 2008, a research team led by Paolo Villoresi and Cesare Barbieri from Padova University, Italy, reported that they had successfully fired individual photons directly at Ajisai. The researchers claimed to have proved that the photons received back at the Matera ground station in southern Italy were the same as those originally emitted.

Background
NASDA and Kawasaki Heavy Industries developed the Experimental Geodetic Satellite (EGS), also known as Ajisai, which means Hydrangea. EGS is a simple 685 kg, 2.15 m diameter, spin stabilised sphere carrying 318 mirrors and 120 laser reflector assemblies (1,436 corner cube reflectors) for precise tracking to provide Earth crustal movement and other data.

The H-I rocket also carried NAL's 295 kg MABES Magnetic Bearing Flywheel Experimental Satellite (although it was not designed for release) to test a two-axis actively controlled flywheel. A 2-D CCD star sensor and a linear CCD Sun sensor measure the attitude. The battery provided a three day life for the satellite.

EGS, carrying 120 laser reflector assemblies and 318 mirrors, is a passive geodetic satellite. (NASDA) 0517136

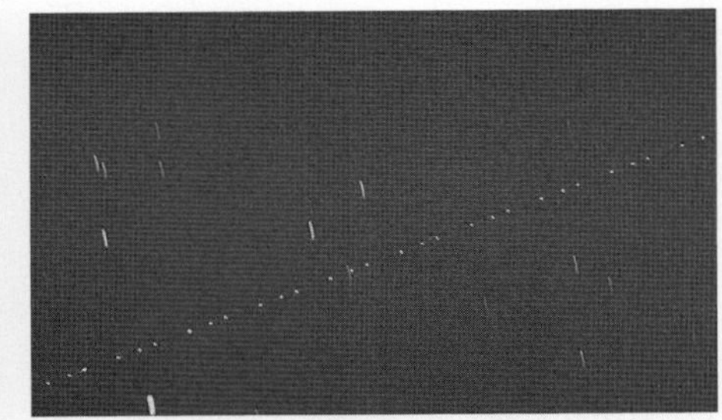

This time-lapse exposure showing EGS was obtained at Tsukuba Space Centre soon after launch (NASDA) 0517135

The primary short-term objective for Ajisai was as a test payload for NASDA's 2-stage H-I launch vehicle. The satellite applications included a survey aimed at rectifying Japan's domestic geodetic triangular net – determining the exact position of many isolated Japanese islands and establishing Japan's geodetic point of origin. The survey was conducted by the Geographical Survey Institute of the Ministry of Construction and the Hydrography Department of the Maritime Safety Agency, Ministry of Transport.

Specifications
Launched: 13 August 1986 on inaugural flight of H-I from Osaki launch complex, Tanegashima, Japan
Orbit: 1,483 × 1,497 km, 49.98°, 115.7 min
Mass: 685 kg
Diameter: 2.15 m

Hayabusa (MUSES-C)

Current Status
In late November 2005, Hayabusa suffered a chemical engine fuel leak following its second touch-down on the surface of near-Earth asteroid, Itokawa. Since the chemical engines were no longer operational, a strong attitude disturbance was experienced, causing loss of communications on 8 December 2005. Although radio communications were resumed in January 2006, Hayabusa was significantly crippled. JAXA decided upon an alternative flight plan and trajectory that would result in its return to Earth in June 2010, a three-year delay in its original schedule. The return to Earth began on 25 April 2007, with only one of its four ion engines working properly.

Background
The main purpose of the Hayabusa (Falcon) mission was to test advanced technologies, such as ion propulsion and autonomous navigation, that could be used in future sample return missions. In order to do this, the spacecraft would rendezvous with an Earth-approaching asteroid, attempt one or more touchdowns on its surface and return a sample of material to Earth for further analysis. If successful, it would be the first opportunity to analyse material from an asteroid, with the potential to provide valuable clues about conditions in the early Solar System. Prior to launch it was known as MU Space Engineering Spacecraft C (MUSES-C).

The original MUSES-C mission was to explore and sample the near-Earth asteroid Nereus. Launch by a Japanese M-V launch vehicle would take place in July 2002, with arrival at the asteroid in 2003 and return to Earth in June 2006. NASA offered to provide the 1 kg **MUSES-CN** nanorover as part of its contribution to the mission. Data from the rover's camera, near-infrared spectrometer and alpha/X-ray spectrometer were to have been transmitted back to Earth via the Japanese lander. In August 2000, the target was changed to Itokawa (1998 F36). The new mission profile involved a launch in November or December 2002, with arrival at the asteroid in September 2005. On 3 November 2000, NASA ordered the Jet Propulsion Laboratory in California to stop work on the rover. The primary reasons for the cancellation were given as rising costs and weight.

Hayabusa was built by Japan's Institute of Space and Astronautical Sciences (ISAS). A key technology on board Hayabusa is its electric propulsion system. After the launch vehicle placed the spacecraft into a transfer orbit, subsequent manoeuvres were carried out by the spacecraft's four ion engines. These engines use a microwave

Artist's impression of Hayabusa attempting a sample return at asteroid Itakawa. The sample collection horn is pointing towards the surface. In the foreground is the spherical target marker. In the background is the Minerva lander (JAXA) 1329686

discharge to ionize xenon gas and then use high-voltage electrodes to accelerate the plasma and eject it at high speed through four thruster heads. Other new technologies used on Hayabusa include: a bi-propellant small thrust reaction control system; X-band up-down communications; complete CCSDS packet telemetry; duty guaranteed heater control electronics assuring heater power constraint; wheel unloading via ion engines; PN-code ranging; lithium ion re-chargeable batteries and multi-junction solar cells.

After arriving in the vicinity of the asteroid, the spacecraft was to hover at about 10 km altitude to investigate the asteroid's size, shape, topography, rotation axis and period, as well as surface composition and structure. The Optical Navigation Camera (ONC), the Light Detection and Ranging (LIDAR), the Laser Range Finder (LRF) and the Fan Beam Sensors (FBS) would be used to gather topographic and range information. Other instruments on board the spacecraft included the Telescopic Narrow Angle Camera (AMICA) for multi-wavelength surface imaging, the Near Infra-Red Spectrometer (NIRS) for mineralogical observation in longer wavelengths, and the X-ray Fluorescent Spectrometer (XRS) for atomic abundance measurement. Autonomous 'intelligent' navigation and guidance technologies would be used to land on the surface. A small lander, named **Minerva**, would be released onto the surface, then hop over the surface using the reaction of two torquers inside it. Minerva would take stereo images and record surface temperatures.

After obtaining enough data on the asteroid for a decision to be made on a landing point, the spacecraft starts its descent, measuring the relative position and attitude to the surface with ONC and LIDAR. The spacecraft autonomously adjusts the propulsion engine and also transmits images to Earth. Although the control station on Earth cannot manoeuvre the spacecraft with a two-way radio delay of about 20 minutes, it may send an order to stop the descent when assessing a danger. At about 100 m altitude, the spacecraft releases a 10 cm wide target marker onto the asteroid. This reflects the flash beam from the spacecraft, by which the ONC navigates the spacecraft. From this point onward, LRF is used in place of LIDAR, measuring the distance between the spacecraft and the surface as well as the gradient of the surface with multiple radial beams. FBS are also used to detect a dangerous contact with the surface.

Before the final touchdown, the spacecraft stops using its propulsion system and enters into a free fall descent. This is to prevent the jets from contaminating the asteroid surface. Sampling starts as soon as the landing of a sampler horn on the asteroid surface is detected. Immediately after sampling, the spacecraft starts its engine, lifts off, resumes its hovering position at 100 m altitude and waits for the next command from Earth.

The sampling device on the base panel comprises a funnel-shaped horn 1 m long and 20 cm in diameter. Immediately after sensing contact between the bottom of the horn and the asteroid's surface, the device fires a 5 g bullet made of tantalum into the surface at a speed of 300 m/s. Fragments broken away from the surface are captured by the sampler horn and funnelled into a sample container at the top of the horn. This system was expected to perform several sample extractions from different locations. The time for each contact with the asteroid surface was estimated to be about 1 second.

Hayabusa was launched towards Itokawa by the M-V-5 rocket on 9 May 2003. It was supposed to arrive at Itokawa in mid-summer of 2005, but the arrival was delayed after a solar flare in late 2003 slightly degraded the solar panels which supply the ion engines with electrical power. Hayabusa lost one reaction wheel (X-axis) on 31 July and entered a flight mode using two wheels for attitude stabilisation. Orbital manoeuvring was handed over from the ion engines to the bi-propellant thrusters on 28 August 2005, and the spacecraft reached its planned location, 12 km from the asteroid, on 12 September. On 3 October, the Y-axis reaction wheel was found not to be functioning. This was compensated by using the RCS chemical engines.

Hayabusa remained close to the asteroid, observing it for approximately two months prior to rendezvous, hovering between 3 and 20 km above it. Images showed that Itokawa measures only 540 m × 310 m × 250 m, and its deformed potato-shaped surface is mostly covered with sizeable boulders, separated by the occasional smooth "sea". After constructing a global map of the asteroid, the project team selected a suitable landing location.

On 4 November 2005, Hayabusa performed a practice touchdown procedure, during which the Minerva miniature lander was to be released. Mission controllers detected an anomalous signal at the critical Go – No Go time and aborted both the release of the target marker and Minerva within 700 m of the landing. On 9 November, controllers manoeuvred Hayabusa to within 70 m of Itokawa to take another look at the proposed touch down and sampling site. During another rehearsal on 12 November, the Minerva deployment command arrived when Hayabusa was drifting away from the surface at an altitude of about 200 m. As a result, the 600 g lander was sent on a flight path that took it away from the asteroid and contact was lost.

The Hayabusa team commanded the spacecraft to make its first touch down on Itokawa on 20 November. At an altitude of 54 m, a wire that connected the target marker to the bottom of the Hayabusa was cut. The spacecraft landed within 30 m of its designated target point, but a false FBS reading caused a Touch Down Sensor malfunction, which resulted in the craft lifting and landing several times before controllers discovered the errors. The touch down was aborted. On 26 November, the team tried once more to land Hayabusa and collect samples, using the first target marker. Several further malfunctions occurred and the team could not determine whether the craft had landed. It was later found that Hayabusa had probably been unable to fire its sampling projectile. Hayabusa was again ordered away from Itokawa and put into safe mode. When the RCS thrusters were switched from subsystem A to subsystem B, one of the jets began to leak hydrazine. Controllers began an emergency operation to stop the leak by shutting off the valves. A few days later, when ground control was attempting to bring Hayabusa out of safe mode, they discovered that the thrust of each of the engine on subsystem A was very small, indicating a complete failure of the RCS. Communications were disrupted for several weeks and controllers were obliged to turn to the ion thrusters. At the same time, four of the 11 cells in the lithium battery became useless leaving the craft with less than two thirds of its intended storage capacity. The schedule for the return to Earth was put back three years.

Communications were eventually restored in March 2006. Hayabusa started its "D" ion engine – the only ion engine still functioning normally – on 25 April 2007, and began its journey back to Earth. The "D" engine had fired for 11,000 hours and will need approximately 9,000 hours more to complete the two orbits of the Sun required to return to Earth. Both the "A" and "C" engines are unstable for continuous firing. Engineers are concerned that a voltage buildup in the "B" ion engine that is close to 50 V will lead to a breakdown, although there is still a possibility it will be useful again when spacecraft temperatures drop as Hayabusa gets farther from the Sun. Controllers are also trying to correct the instability of the "C" engine, which is due to its inability to turn itself on because of a malfunctioning safety switch. Meanwhile, the last of the spacecraft's three reaction control wheels is experiencing longer-than-expected spin cycles. To compensate, its spin rate has been lowered from 3,500 rpm to 1,800 rpm.

The insulated and cushioned re-entry capsule is 40 cm in diameter and 25 cm high, with a mass of about 20 kg. Attached to the spacecraft near the sample collection horn, it has a convex nose covered with a 3 cm thick ablative heat shield to protect the samples from the high velocity (~13 km/s) re-entry. The capsule will be detached from the main spacecraft about 350,000 km from Earth and then make a ballistic re-entry into the atmosphere. Peak deceleration will be about 25 G with heating rates about 30 times greater than those experienced by the Apollo spacecraft. The final descent by parachute is planned for central Australia, near Woomera.

Specifications

Launch: 04:29:25 UTC on 9 May 2003 by M-V-5 from Uchinoura.
Launch mass: 530 kg including 50 kg of chemical propellant and 65 kg of xenon gas.
Body dimensions: 1.5 × 1.5 × 1.05 m
Telecommunications: One 1.5 m diameter high-gain (X-band) parabolic antenna mounted on a two-axis gimbal with a transmitted power of 20 W. At high bit rate communication of 8 kbps, the spacecraft orientates the HGA towards Earth. X- and S-band low gain antennas.
Power: Two fixed gallium arsenide solar arrays, total area of 12 m^2, producing 2.6 kW at 1 AU from Sun. One 15 A-hr rechargeable nickel-metal hydride (Ni-MH) battery.
Attitude control: Three reaction wheels.
Propulsion: Nitrogen tetroxide/hydrazine thrusters with peak thrust of 22 N. Four xenon ion engines with peak thrust of 20 mN using 1 kW power.
Payload: Asteroid Multi-band Imaging Camera (AMICA) – CCD imaging camera with eight filters and polarizers; Optical Navigation Camera; Light Detection and Ranging Instrument (LIDAR) using a YAG pulse laser emitter and a Si-APD receiver; Near Infrared Spectrometer (NIRS) – a InGaAs 64 pixel linear array detector covering 0.85–2.10 micron; X-Ray Fluorescence Spectrometer (XRS) with energy resolution 100 eV at 1.5 keV; Laser Range Finder (LRF); Fan Beam Sensors (FBS); Minerva lander equipped with three miniature cameras to obtain stereoscopic images and different focal length images of the asteroid surface plus six platinum thermometers; two target markers.

Kaguya (SELENE)

Current Status

Kaguya has been in lunar orbit since 4 October 2007. All of the main orbiter's systems and the two subsatellites are operating normally.

Background

Originally known as SELENE (SELenological and ENgineering Explorer), Japan's lunar orbiter mission was subsequently renamed after a popular poll of the Japanese public produced the nickname "Kaguya", after the Moon-born heroine of a Japanese folk tale. The largest spacecraft sent to the Moon since the Apollo missions, it consists of the main orbiter and two 53 kg (117 lb) subsatellites: Rsat (Relay satellite) and VRAD (VLBI Radio satellite). The Relay satellite and the VRAD satellite are respectively nicknamed "OKINA" meaning an "honourable elderly man" and "OUNA" meaning an "honourable elderly woman."

The orbiter's main bus is box-shaped, with an upper module which contains most of the scientific instruments and a lower propulsion module. A 21.6 m^2 solar array is mounted on one side of the spacecraft. A 1.3 m high-gain antenna is mounted 90° from the solar panel. A 12 m magnetometer boom projects from the top of the spacecraft and four 15 m radar sounder antennas protrude from the top and bottom corners of the upper module.

Kaguya carries an instrument suite which was designed to gather data aimed at answering the basic questions of lunar science: How was the Moon formed? How has it evolved? What does it tell us about the history of the Solar System? In addition to the radio sources for the gravity-field measurements, which should help plan future robotic and human missions to the Moon, the instruments will collect data on surface chemistry and mineralogy; surface and subsurface structure to a depth of 5 km; the remnants of the Moon's magnetic field, which is only 1/10,000 as strong as Earth's; and charged and neutral particles in the Moon's environment. The mission will also use the lunar vantage point to investigate Earth's plasmasphere and auroras.

Kaguya carries an X-Ray Spectrometer (XRS) that will map the surface elemental composition using X-ray fluorescence spectrometry via bombardment of the surface with solar X-rays. The XRS instrument has a resolution of 20 km and an energy range of 0.7–8 keV. Also measuring surface composition will be the Gamma Ray Spectrometer (GRS), which will map the distribution of such key elements as hydrogen, potassium, uranium and thorium. The GRS is built around a 250 cc germanium crystal to

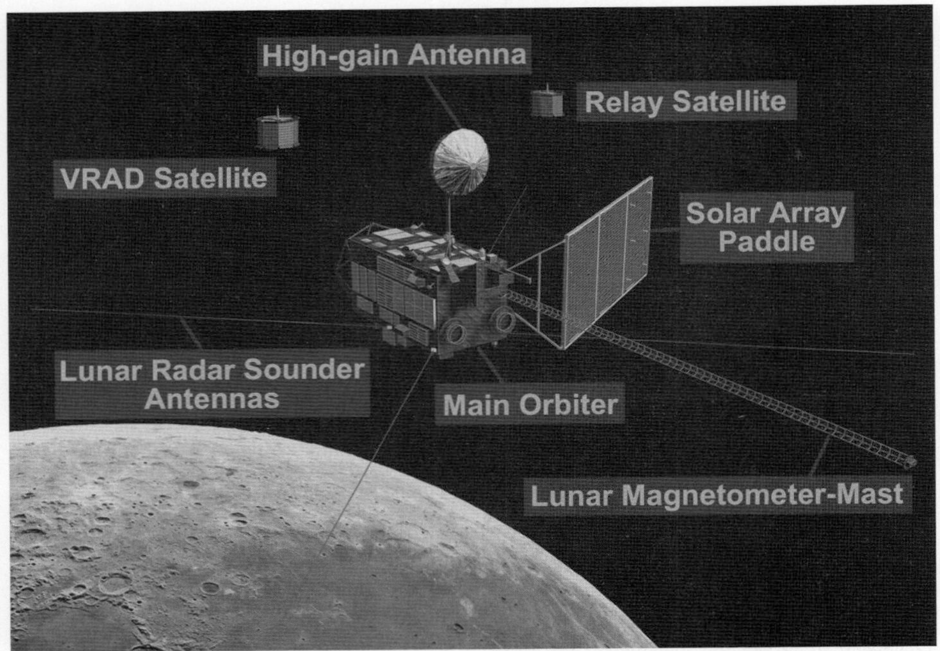

The main features of Japan's Kaguya (SELENE) lunar orbiter (JAXA) 1330407

measure reflected gamma rays in the 0.1–10 MeV range with a resolution of 160 km.

Mineral distribution will be mapped with a Multiband Imager (MI) that collects visible and near-infrared images of the lunar surface in nine spectral bands. The MI has a spatial resolution of 20–60 m. Also mapping mineral distribution will be the spectral profiler, which provides a continuous spectral profile in visible and near-infrared with a spatial resolution of 500 m.

Three instruments will be used to map the surface and subsurface structure of the Moon. The terrain camera will produce stereo images of the surface with a resolution of 10 m, while the Lunar Radar Sounder (LRS) will use active 5 MHz radar sounding to map subsurface structure to a depth of 5 km, with a resolution of 75 m. Terrain altitude data will be provided by the laser altimeter which uses a 100 m J YAG laser with a 1 Hz pulse rate to map terrain with a 5 m height resolution and a spatial resolution of 1,600 m.

Kaguya has five instruments to study the lunar environment, as well as Earth's magnetosphere and ionosphere. The lunar magnetosphere will be measured with a flux-gate magnetometer deployed on a 12 m boom. A charged particle spectrometer will measure high-energy alpha radiation from the lunar surface and the abundance of cosmic rays. The plasma energy angle and composition experiment will provide energy, angle and composition measurements of low-energy electrons and ions around the Moon. A radio science experiment will use small deviations in the S- and X-band carriers from the VRAD microsatellite to detect the Moon's tenuous ionosphere, while the upper atmosphere and plasma imager takes images of Earth's magnetosphere and ionosphere to study the behaviour of the plasma.

To encourage public interest in space exploration, Kaguya carries a High Definition Television camera that has returned the first HDTV video of the Earth and Moon seen from orbit. Through a collaboration with Japanese Broadcasting Corp. (NHK), the first HDTV "Earthset" video was taken on 7 November, showing the Earth appearing to set above the Moon's horizon close to the South Pole.

Kaguya's science data is returned in X-band at a rate of 10 Mbps. It is expected to deliver some 4.3 terabytes of raw data and 25.6 terabytes of high-level data over the one year nominal mission. Data is sent to the JAXA Usuda Deep Space Centre. Mission control and data analysis are based in a new facility near the Sagamihara Space Operations Centre.

The two spin-stabilised microsatellites are designed to map lunar gravitation by using Doppler ranging to measure changes in Kaguya's orbit caused by variations in the density of the terrain below. Together the three spacecraft will produce the first global gravity map of the Moon with a positional error of less than 1 m. The Rsat relays the Doppler ranging signal between the orbiter and the ground station for the first direct measurements of the gravity field on the lunar far side. The octagonal subsatellites are spin-stabilised at 10 rpm and have no propulsion. Power is provided by a 70 W silicon solar cell array covering their sides. Both carry one X-band and three S-band radio sources, enabling differential VLBI observations from the ground. On 6 November, a direct observation test of the gravity field on the far side of the Moon was carried out using Kaguya and Rsat. It was the world's first direct, four-way Doppler observation of the gravity field of the Moon's far side. The radio signal from the VRAD is also used to search for and study a lunar ionosphere.

The launch of Kaguya was delayed by four weeks when a ground test of the Wideband Internetworking Engineering Test and Demonstration Satellite (WINDS) showed that the onboard condenser had been installed with its polarity reversed. JAXA checked the Kaguya and found the same problem in the condensers of the two subsatellites (one condenser on each satellite.) The condensers were replaced.

Kaguya was placed in a highly elliptical Earth orbit after launch. After three orbit change manoeuvres, which increased the apogee, it was inserted into a lunar orbit of 11,741 × 101 km, period 16 h 42 min at 06:20 on 4 October 2007 (JST). The Relay satellite was released into a 100 km × 2,400 km polar orbit at 09:36 on October 9 (JST), followed by the VRAD, which was released into a 100 × 800 km orbit on 12 October (JST). The main spacecraft reached its operational circular, polar orbit at an altitude of about 100 km on 18 October 2007 (JST). Normal scientific operations were due to begin in mid-December. Kaguya will operate for at least one year, using correction burns roughly every two months to maintain the orbit. The orbit may then be lowered to 40–70 km.

Specifications

Launch: 10:31 on 14 September 2007 (JST) by H-IIA 2022 from Tanegashima.
Launch mass: 2,885 kg
Body dimensions: 2.1 × 2.1 × 4.8 m
Payload:

- High Definition Television System (HDTV).
- Monitor Camera. A 3.2 megapixel (656 × 488 = 320,128) CCD camera to verify the deployment of the high-gain antenna, solar array paddle, and UPI (plasma imager) as well as the separation of the two subsatellites.
- X-Ray Spectrometer (XRS). Global mapping of Al, Si, Mg, Fe distribution using 100 cm² CCD, spatial resolution 20 km, energy range 0.7–8 keV, 5μm Be film, solar X-ray monitor.
- Gamma-Ray Spectrometer (GRS). Global mapping of U, Th, K, major element distribution using a large pure Ge crystal of 250 cm³, spatial resolution 160 km, energy range 0.1–10 MeV.
- Multi-band Imager (MI). UV-VIS-NIR imager, spectral ranges from 0.4 to 1.6 microns, 9 band filters, spectral resolution 20–30 nm, spatial resolution 20–60 m.
- Spectral Profiler (SP). Continuous spectral profile ranging from 0.5 to 2.6 microns, spectral resolution 6–8 nm, spatial resolution 500 m.
- Terrain Camera (TC). High resolution stereo camera, spatial resolution 10 m.
- Laser Altimeter (LALT). Nd: YAG laser altimeter, 100 mJ output power, height resolution 5 m, spatial resolution 1.6 km with pulse rate 1 Hz, Beam divergence 3 mrad.
- Lunar Radar Sounder (LRS). Mapping of subsurface structure using active sounding, frequency 5 MHz, echo observation range 5 km, resolution 75 m. Also detection of radio waves (10 kHz–30 MHz) from the Sun, the Earth, Jupiter, and other planets.
- Charged Particle Spectrometer (CPS). Measurement of high-energy particles, 1–14 MeV (LPD), 2–240 MeV (HID), alpha particle detector 4–6.5 MeV.
- Plasma energy Angle and Composition Experiment (PACE). Charged particle energy and composition measurement, 5 eV/q–28 keV/q.
- Upper atmosphere and Plasma Imager (UPI). Observation of terrestrial plasmasphere from lunar orbit, XUV(304A) to VIS.
- Lunar Magnetometer (LMAG). Magnetic field measurement using flux-gate magnetometer, accuracy 0.5 nT.
- Rsat (Relay satellite). Far-side gravimetry using 4-way range rate measurement from ground station to orbiter via relay satellite.Doppler accuracy 1 mm/s.

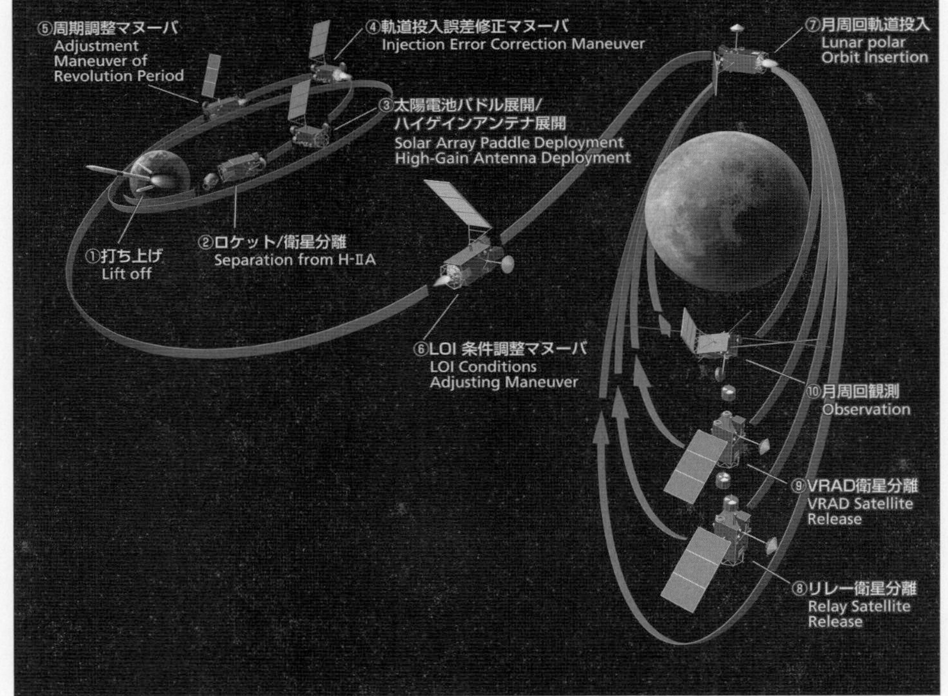

Japan's Kaguya (SELENE) lunar orbiter spiralled outwards from Earth for 20 days before arriving in lunar orbit on 4 October 2007. 14 days later, the orbit was circularised at an altitude of 100 km (JAXA) 1330408

The 53 kg Rsat or Relay satellite (Okina) and VRAD or VLBI Radio Satellite (Ouna) – two subsatellites carried to the Moon by Japan's Kaguya (SELENE) spacecraft and deployed into lunar orbit on 9 and 12 October 2007 respectively (JAXA)
1330409

- VRAD (VLBI Radio satellite). Differential VLBI observation from ground stations, selenodesy and gravimetry, onboard two sub-satellites, 3 S-bands and 1 X-band.

Lunar-A

Current Status

Lunar-A, Japan's first lunar mission, has suffered numerous launch delays since the late 1990's, primarily due to financial constraints. It is likely that the orbiter and its associated probes will not be launched before late 2007 or 2008; Lunar-A has not been integrated into the launch schedule. Still a JAXA-approved mission, the mission plan remains very similar to the original. The project's cost estimate is about USD135 million.

Background

The Institute of Space and Aeronautical Science (ISAS – now integrated into the Japan Aerospace Exploration Agency – JAXA – see separate entry) first selected a lunar orbiter/penetrator mission in 1990 for its launch vehicle M-V's second mission. It then chose NEC to act as system integrator. As it laid out the programme in the 1990's, ISAS planned for an orbiter to fire three 13.6 kg, 13.8 cm diameter, 71.4 cm-long (1.44 m with motor attached) penetrators carrying seismometers and heat flow probes 1–3 m into the surface.

JAXA now plans to fire two penetrators, both of which should remain operational for at least one year. The probes will penetrate both the far side and the near side of the moon. Japan has designed the seismic data experiments to clarify long-standing questions on the far side crustal and deep mantle structure and existence of the lunar core. The orbiter will act to relay data to Earth, fleshing out the Apollo programmes near side measurements. In order to test its plans, ISAS set up an N_2-powered gas gun at its Noshiro facility to fire penetrator designs into a simulated lunar surface. Lower speed drop tests began in August 1986. A one-stage S-520 sounding rocket, launched on 17 September 1995, also tested penetrator separation and attitude control plans, but during the test the ground station lost telemetry after 25 seconds.

Physically, Lunar-A is a 540 kg spin-stabilised unit, 2.2 m in diameter and 2 m high vehicle. Launched from Uchinoura Space Center, the three-stage M-V plus two solid kick stages will inject it into a Geotail-type highly eccentric Earth orbit with an apogee of about 500,000 km. N_2O_4/N_2H_4 thrusters account for 189 kg of the weight budget and the penetrator modules about 124 kg. Six 20 N and four 1 N/N_2H_4 thrusters augment the main 500 N bipropellant engines.

In order to reach the final trajectory, Lunar-A will fly-by the moon to obtain an apogee of 1,185,000 km and reduce the delta-V to 550 m/s for the main engine to establish a 200 × 200 km, 20° orbit during the second encounter.

For each penetrator release, the orbiter will descend to 45 × 200 km before returning to its higher orbit to use its Lunar Imaging Camera (with an image compression chip provided by CNES). The orbiter will spin up to 120 rpm for release from its normal 30 rpm. A 30 cm diameter spherical solid motor (20 second burn, 5 kN max thrust) will brake the penetrator out of orbit. Its N_2 attitude control system will ensure a head-on impact at 8° and a velocity of 250–300 m/s, a deceleration of 8,000 g, burrowing 1–3 m into the surface. One penetrator is planned for the near side, close to the Apollo 12 and 14 region, and one is planned for the far side. The orbiter will then rise to a 300 × 300 km orbit for data relay. Data transfer from penetrator to the orbiter is at 256 bps plus and from the spacecraft to Earth at 8 kbits/sec.

Nozomi (Planet-B)

Current Status

Nozomi (Hope) encountered several problems during its five-year mission. Ground control finally abandoned the mission on 9 December 2003, after the spacecraft suffered an irreparable thruster malfunction. Nozomi failed to reach Mars orbit and on 14 December passed about 1,000 km above the planet's surface after ground control commands, relayed on 9 December, changed the craft's trajectory to prevent it from impacting with Mars. The incapacitated Nozomi continues to fly in heliocentric orbit.

Background

Japan launched the Planet-B spacecraft on 3 July 1998 aboard an M-V rocket, built by ISAS (now integrated into the Japan Aerospace Exploration Agency, or JAXA – see separate entry) and launched from Kagoshima. After lift-off, ISAS renamed it Nozomi. The cost of the spacecraft was approximately JPY11 billion (USD80 million). At Nozomi's launch, Japan was only the third country to launch a planetary mission.

The spacecraft platform is 0.58 m high, and takes the form of 1.6 m square prism with truncated corners. Extending from two opposite sides are solar panels containing silicon solar cells which provide power directly to the spacecraft or via Ni-MH (nickel metal hydride) batteries. On the upper surface is a high-gain dish antenna, while a propulsion unit protrudes from the bottom. A 5 m deployable mast and a 1 m boom extend from the sides, together with two pairs of thin wire antennas which measure 50 m tip to tip. Other instruments are also arranged along the sides of the spacecraft. Spacecraft communications were via X-band at 8,410.93 MHz and S-band at 2,293.89 MHz. The launch mass of Nozomi was 540 kg, including 282 kg of propellant.

Nozomi carries 14 instruments from five nations: an imaging camera, neutral mass spectrometer, dust counter, thermal plasma analyzer, magnetometer, electron and ion spectrum analysers, ion mass spectrograph, high energy particles experiment, UV

Japan's Nozomi (Planet-B) spacecraft failed to enter orbit around Mars in December 2003 (ISAS/JAXA)
1340460

imaging spectrometer, sounder and plasma wave detector, LF wave analyser, electron temperature probe, and a UV scanner. The overall mass of the science instruments is 33 kg. Radio science experiments were also possible using the radio system and an ultrastable oscillator. Its main objectives were to study the planet's atmosphere and ionosphere, including examination of the Martian upper atmosphere's interaction with the solar wind and the escape of oxygen molecules from the thin atmosphere. The spacecraft was also to send back images of the planet's surface. Although no Mars science results were returned, Nozomi's Ultraviolet Imaging Spectrometer was able to measure hydrogen Lyman-alpha light in interplanetary space and detect helium ions leaving the Earth's ionosphere.

After launch on the M-V-3 launch vehicle, Nozomi was placed in an elliptical parking orbit with a perigee of 340 km and an apogee of 400,000 km. The spacecraft performed two lunar flybys on 24 September and 18 December (2,809 km) to increase the apogee of its orbit. It swung by Earth on 20 December at a perigee of 1,003 km. The gravitational assist from the swingby, coupled with a seven minute burn of the bipropellant engine, put Nozomi in an escape trajectory towards Mars. However, a malfunction of a thruster valve during the perigee burn provided insufficient velocity and a course correction burn on 21 December used more propellant than planned, leaving the spacecraft short of fuel. Instead of reaching Mars on 11 October 1999 and entering a 300 × 47,500 km orbit, mission controllers decided on a longer trajectory that would add four years to the journey. Nozomi conducted a second Earth swing-by in January 2002, when it was directed into an out-of-ecliptic trajectory that led to a third Earth fly-by in June 2003.

On 21 April 2002, the spacecraft was bombarded by extremely energetic particles from a solar flare over a period of more than six hours. The solar activity caused one of the electrical power converters to "latch up", knocking out the main power, and causing a breakdown in communications and a failure of a heater unit. This resulted in the freezing of the hydrazine fuel as the spacecraft neared its aphelion of 1.4 AU – just inside the orbit of Mars. By September, Nozomi had warmed as it moved to 1.25 AU from the Sun, so enough of the fuel melted to allow critical course-correction burns. Communications were restored, enabling the spacecraft to change its attitude during a second Earth flyby at a distance of about 11,000 km on 19 June 2003. En route to Mars, engineers were unable to reactivate the power control system. Attempts to revive the spacecraft were abandoned on 9 December, when ISAS sent a command to alter Nozomi's trajectory so that it would pass about 1,000 km above the Martian surface on 14 December. It then continued into a heliocentric orbit.

Pakistan

Badr Series

Current Status

As of November 2006, Badr-B, also known as Badr-2, is operational after five years in orbit, and has surpassed its two year design life.

The UK's cloud camera flies on Badr-B
(Rutherford Appleton Lab) 0517293

Background
Pakistan is developing an independent remote-sensing, science and communications space presence. The Badr series science goals include the development of indigenous space-based and ground capabilities and technologies that would enable Pakistan to maintain an independent microsatellite programme. To pursue its space science goals, Pakistan built the Badr-A, or Badr-1, in partnership with the UK and China, which provided launch services. Badr-A passed over the country about five times a day to give ground controllers experience in handling spacecraft operations and preparing for an eventual store and forward communications capability.

After a successful launch on 16 July 1990 by CZ-2E from Xichang, Badr-A entered into a 205 × 983 km, 28.49°, 96.7 min orbit. The demonstrator satellite, built by Pakistan Space and Upper Atmosphere Research Commission (SUPARCO) engineers trained at the University of Surrey/Surrey Satellite Technology Ltd (SSTL), weighed 52 kg, and was a 48.3 cm diameter 26-sided polyhedron. It carried a UHF/VHF store and forward payload. Chinese officials quoted the launch cost to Pakistan as USD300,000. The orbit provided four or five usable passes daily for its controllers, each lasting as much as 20 minutes, instead of the two pairs separated by at least 10 hours available from a polar orbit. The SUPARCO Karachi campus provided TT&C, supported by the Sparc station at Lahore. The sites also operated two low-cost tracking and receiving stations. Unfortunately, contact with Badr-A was lost after 35 days, seriously affecting the communications experiments. Badr-A was taken out of service on 21 August 1990 and re-entered the Earth's atmosphere 8 December 1990 after 145 days.

Built in partnership with Satellite International Ltd (SIL), the Badr-B microsatellite was launched by Zenit-2, on 10 December 2001, from Baikonur into a 986 ×1,015 sun-synchronous orbit at 99.7° with a period of 105 min. The satellite weighs 68.5 kg, is a 51 × 51 × 46.5 cm polyhedron shaped craft and has a design life of two years plus. Gravity gradient stabilisation using 4 kg tip mass on a 6 m boom will maintain the transponder orientation. Badr-B carries an Earth Imaging Camera (EIC) Charge Couple Device (CCD) camera for cloud monitoring, provided by the UK's Rutherford Appleton Laboratory (RAL). Other experiments are the Store & Forward Experiment (S&FE), provided by SUPARCO; the radiation Compact Dosimeter Experiment (CDE); and Battery End of Charge Detector, both provided by SIL/ESA. S-band and UHF provide 128 and 9.6 kbit/s data rates, respectively. RAL's Chilton location acts as Lahore's TT&C back-up. SIL also provided the entire attitude control system, consisting of one MFM-3L fluxgate magnetometer, two MTR-25 magnetorquers, two DSS-256 Digital Sun Sensors (each a pair) and attitude control electronics. It also provided the S-band transceiver, 4 Ah Ni/Cd battery, OBDH and power conditioning system. The UK's EEV Ltd provided the gallium arsenide (GaAs) solar panels. The CCD camera's specifications are: 0.840–0.890 μm, 2.5 kg, 7 W, 17° FOV, 72 mm fl, f/6, EEV 576 × 770-pixel CCD for cloud monitoring with 310 m resolution (from 1,000 km). The camera is carried free of charge for RAL in return for SUPARCO using it for land imaging.

The Badr (Full Moon) series was named for the Battle of Badr, which took place near Medina, in present day Saudi Arabia, during Muhammad's lifetime.

ArabSat operates its own, independent communications satellite series; these are also known as the Badr, as well as the ArabSat series, and should not be confused with the Pakistani Badr technology demonstrator microsatellites described above.

Russian Federation

Etalon

Current Status
The two Etalon satellites continued to be used in 2006 as targets for geodetic measurements. According to the International Laser Ranging Service, they have an expected life of "hundreds of years".

Background
Etalon 1, launched 10 January 1989 as Kosmos 1989 and Etalon 2 launched 31 May 1989 as Kosmos 2024 are 1,415 kg, 1,294 mm diameter geodetic satellites covered with 306 antenna arrays each with 14 corner cubes for laser reflection. Etalon reflects laser light in a passive role. The first was placed in 19,102 × 19,149 km at 64.9° along with two Glonass satellites. Etalon carries 306 laser retro-reflector assemblies for international geodetic studies; six germanium reflectors may be used for future IR interferometric measurements. Etalons supplement Glonass information, aiming to improve the navigation system's accuracy. Three laser stations at Ternopol, Yevpatoria and Maydanak provide 25 cm ranging accuracy, and were later joined by 10 foreign sites.

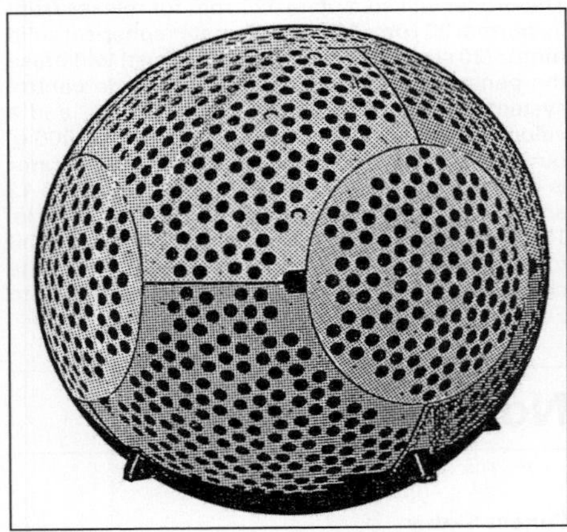

The Etalon laser-reflecting geodetic satellite complements the Lageos, Starlette and EGP spheres
(Teledyne Brown Engineering) 0516693

Foton Series

Current Status
Foton-M3 was originally scheduled to launch in October 2006; its launch has been delayed to September 2007. As of October 2006, the European Space Agency's (ESA) total payload mass for the flight was 292 kg. The Russian Federal Space Agency (FSA), Roscosmos, launched Foton M-2 from Baikonur on 31 May 2005. Its recoverable descent module successfully returned to Earth on 16 June 2005 and was retrieved in Kazakhstan. M2 carried 39 experiments in the materials science, exobiology and fluid physics fields. Foton-M1 was destroyed on 15 October 2002, due to a Soyuz-U launch failure, and subsequent explosion, that killed one person.

Foton M series illustration clearly showing all three spacecraft modules (European Space Agency) 1133881

Foton integration with its carrier 0517248

ESA, in 2003, signed an agreement with Roscosmos for two Foton missions, M2 and M3, to carry a combined science payload of 660 kg. The agreement binds FSA, ESA and two Russian partner companies, KBOM and TsKB-Progress, through at least 2006; since the M3 launch has been delayed into 2007, it is likely that this agreement will be extended. Foton-M2 largely re-flew experiments that were lost in both the M1 and Columbia accidents.

An upgraded vehicle, the Foton M series has a mass of 6,425 kg; it carries science experiments, primarily for ESA. As with all past Foton flights, with the exception of Kosmos/Fotons-1-4, Russia markets Foton capacity to users, primarily university departments and laboratories working through ESA, who wish to conduct longer-term microgravity (μg) experiments than a sounding rocket launch or other short-term means can provide.

Background
The Soviets began marketing space facilities and services to the international μg community in 1987, notably at that year's Paris Salon. The USSR first offered temporary positions aboard the Mir-Kvant orbital complex, with equipment delivered by the Soyuz manned or Progress unmanned ferries, and accompanied by a specialist if required. The Russians offer the Foton capsule, for dedicated bookings on 12 to 18 day missions; the spacecraft's lifespan is a maximum of 30 days.

The first three Foton flights were not designated at launch as such by the USSR. It was not until Foton-4, launched in 1988, that the Soviets began

The Foton-12 descent module with FluidPac experiment; the Foton M series descent module improves upon the original design by allowing a larger payload and higher experiment power consumption
(European Space Agency) 1133880

Foton Microgravity Missions

Spacecraft	Launch	Orbit	Days in orbit
Foton-1			
Kosmos 1645	16 April 1985, Plesetsk	215 × 378 km, 62.8°	12.6
Foton-2			
Kosmos 1744	21 May 1986, Plesetsk	218 × 371 km, 62.8°	13.6
Foton-3			
Kosmos 1841*	24 April 1987, Plesetsk	218 × 381 km, 62.8°	13.6
Foton-4			
Foton-1*	14 April 1988, Plesetsk	217 × 375 km, 62.8°	13.6
Foton-5			
Foton-2*	26 April 1989, Plesetsk	216 × 380 km, 62.8°	14.6
Foton-6			
Foton-3	11 April 1990, Plesetsk	225 × 398 km, 62.8°	15.6
Foton-7			
Foton-4	4 October 1991, Plesetsk	216 × 396 km, 62.8°	15.6
Foton-8			
Foton-5	8 October 1992, Plesetsk	221 × 359 km, 62.8°	15.6
Foton-9			
Foton-6	14 June 1994, Plesetsk	221 × 364 km, 62.8°	17.6
Foton-10			
Foton-7**	16 February 1995, Plesetsk	220 × 369 km, 62.8°	14.6
Foton-11			
Foton-8	9 October 1997, Plesetsk	218 × 375 km, 90.45°	13.6
Foton-12			
Foton-9	7 September 1999, Plesetsk	217 × 384 km, 62.8°	14.6
Foton-M1	15 October 2002, Plesetsk	Launch failure	0
Foton-M2	31 May 2005, Baikonur	258 × 291 km, 63°	15.6
Foton-M3	September 2007, Baikonur (Planned); Delayed from October 2006	260 × 305 km, 63° (typical)	13 (planned)

*reused descent capsule; same capsule used on Kosmos 1841 and following two flights in 1988 and 1989
**capsule dropped by the recovery helicopter after safe landing; most of the experiments were destroyed or unusable due to this accident; the following 1997 mission re-flew the destroyed experiments

designating the spacecraft publicly as Foton. Consequently, there are two pre-2000 mission designation sequence number sets for the Foton: the original Soviet (and subsequently Russian) progression, and a sequence that designates the first launch as Foton-1, which is used by some sources, including ESA. Both sets end when Russia adapted the M series and the Foton-M1 launch attempt was conducted in 2002. Both mission designation sequence sets are represented in the Foton Microgravity Missions table, in the left-hand column.

Foton is based on the well-proven Vostok spacecraft introduced in 1960 for the Soviet manned programme; the same craft that carried Yuri Gagarin into orbit and brought him safely to Earth. The spacecraft are built by the Central Specialized Design Bureau of Russia's State Research and Production Space Rocket Centre, TsSKB-Progress in Samara. Both the original Foton and M series spacecraft consist of a spherical descent module, a service module with solid propellant deorbit motor, and a separate battery module, and are launched aboard a three-stage Soyuz-U. The spacecraft provide a $\leq 10^{-5}$ g gravity level environment in which researchers can conduct a wide range of microgravity experiments. In addition to the interior experiment capacity, Foton can carry exterior experiments that need to be exposed to the space environment and/or the re-entry environment; this experiment programme is called "Stone" by ESA. ESA also offers Biopan and Biobox, among other options, to carry biological experiments; Biopan is mounted on Foton's exterior.

ESA Foton users monitor their experiments from two locations: the Payload Control Centre at TsSKB Samara and the Telescience Centre at the Esrange launch facility in Kiruna, Sweden (see separate entry). Mission control is executed from the TsUP Korolev mission control centre near Moscow.

Specifications

Foton

Launch: Soyuz-U from Plesetsk
Availability: 1–1.5 years on signature of contract
Spacecraft mass: 6,200 kg
Payload mass: 400 kg aggregate maximum
Descent Module: recoverable module; 2.3 m diameter sphere
Service Module: 3.2 × 2.5 m polyhedron; included retrorocket, attitude control and TT&C equipment
Battery Module: 1.8 m diameter, 47 cm deep cylindrical battery pack; provided 400 W of power daily or 700 W max for 90 min/day

Foton M

Launch: Soyuz-U from Baikonur
Availability: approximately 1–2 years after experiment approval
Spacecraft mass: 6,425 kg
Payload mass: 600 kg aggregate maximum
Descent Module: recoverable module; 2.2 m diameter sphere; 2,500 kg mass; 4.6 m^3 total capacity; 1.6 m^3 experiment capacity; maximum allowable single experiment dimensions 26.5 × 15.7 × 15.7 cm, or use of a 17.8 × 12.7 × 11.6 ESA Type III experiment container
Service Module: 3.2 × 2.5 m polyhedron; includes retrorocket, attitude control and TT&C equipment
Battery Module: 1.8 m diameter, 47 cm deep cylindrical battery pack uses lithium cells and AgZn batteries; provides 800 W of power daily

Spektr Series

Current Status

None of the satellites in the Spektr (Spectrum) series have yet achieved launch or orbit. Due to financial difficulties that resulted in space science programme cut-backs, the Russians have delayed the observatory series' development and launch time again over the last decade. In an attempt to breathe new life into Russia's space science programme, as of 2006, three Spektr series satellites have been funded and integrated into Russia's Federal Space Programme though 2015. Spectr-R, also known as RadioAstron, now has a potential launch date of late 2007 or 2008, after development and launch delays stretching back to the 1990s.

Spektr-UF, an ultraviolet observatory, will be launched around 2010. Roscosmos perceives Spektr-UF, also dubbed the World Space Observatory, as the Hubble's successor, as well as an improvement upon the aging observatory.

Spektr-X-gamma, or Spektr-RG, was originally the first Spektr satellite to be conceived of; initially, it was supposed to have been launched around 1996. The original project was cancelled around 2001, but has been resurrected as a joint European Space Agency (ESA) project called Spektr-RG, or Spectrum-RG; it will be equipped with an 'extended ROentgen Survey with an Imaging Telescope Array' (eROSITA), and the wide field X-ray monitor "Lobster" instrument; it is to be launched around the 2011 timeframe, according to The Russian Federal Space Agency (FSA), or Roscosmos.

Also in the design and test stages is the new NPO Lavochkin-designed satellite platform, called Navigator, that will succeed the heavy Spektr design. Spektr-R, weighing in at about 5,000 kg, was originally designed to be carried into orbit aboard a Proton launch vehicle. This has proved very expensive and has contributed to Spektr-R's launch delays over the years. About one third the mass of the 2,500 kg Spektr bus, Navigator's design allows it to be launched aboard a more cost-effective Zenit or Soyuz. The first Navigator craft to fly, the Electro-L remote sensing mission, is scheduled to liftoff in May 2007. The Millimetron observatory, with a projected launch year of 2016, will also use the Navigator design.

Background

In the late 1980s, Soviet scientists introduced a new class of astronomical observatory platform design to supersede the Venera-based Astron/Granat type. Sources initially indicated that the Spektr would appear in late 1995, but FSA financing problems have delayed the series and the first launch date has still not arrived. The Spektr class bus provides 128 Mbit/s data rates, 30–40 arcsec pointing stability and 500 W power for a 2.7 tonne payload. While these features seemed impressive in the early 1990s, in fact, it became apparent in 1993 that Lavochkin's bus was already in service for the Prognoz GEO early warning satellites (see separate entry), the first of which flew as Kosmos 1940 in April 1988. Considerable effort was expended on garnering international support for Spektr, to spread resources and funding among several investing research groups.

Spektr-R/RadioAstron

The Lebedev Physics Institute's Astro Space Centre of the Russian Academy of Sciences (RAS) was the RadioAstron project initiator. NPO Lavochkin is responsible for the spacecraft's design and manufacture. With its 1,500 kg, 10 m Space Radio Telescope (SRT), RadioAstron will undertake, in conjunction with ground stations worldwide, high resolution Very Long Baseline Interferometry (VLBI) radio mapping of active galaxies, black holes, neutron stars and other exotic objects. Spektr-R is an international effort, with space science organisations participating at the global level, providing ground support and instruments.

As of 2006, the launch plan for Spektr-R dictates that it be orbited from Baikonur aboard a Zenit-2SLB (see separate entry) with a Fregat upper stage, instead of a Proton launch vehicle, as originally planned; the joint venture Land Launch will provide launch services. This may lower launch costs and potentially avoid further launch delays. Spektr-R will fly in a highly elliptical orbit reaching above Earth's radiation belts, which permits protracted observations of X-ray sources and monitoring for gamma-burst events.

In 2005, NASA informed the Astro Space Centre (ASC) and RadioAstron participants that it would no longer provide support to the project. ASC would prefer that NASA honour its previous commitments. If NASA exits the programme, RadioAstron VLBI and tracking capabilities will be reduced. NASA Science programme funding has been tight, and it is likely that the Agency prefers to fund and support US projects.

Spektr-R/RadioAstron Specifications

Launch: projected for late 2007–2008 from Baikonur
Payload: 10 m parabolic SRT with 27 panels around 3 m central disc; surface accuracy rms 0.7 mm; operating frequencies: 0.324, 1.66, 4.83 and 18.4–25.1 GHz.
Participating organisations: Lebedev Physics Institute's Astro Space Centre: lead science organisation.
NPO Lavotchkin: Spektr/Navigator spacecraft bus; SRT.
FSA: control station, Bear Lake, Russia; Tracking station, Puschino, Russia.
ESA: Vacuum chamber panel test; Neuchâtel Observatory, Switzerland; Hydrogen and rubidium on-board frequency standards.
NASA, JPL: DSN Tracking stations, Tidbinbilla, Australia; Very Long Baseline Array (VLBA) recorders; near-real time correlator; ground radio telescopes; (NASA may not provide these services).
Canadian Space Agency (CSA) and Dominion Radio Astronomical Observatory (DRAO): S2-correlator; ground radio telescope.
Ukrainian Space Agency and Ukrainian Institute of Radio Astronomy (IRA), Kharkov: On-board attitude control system; ground radio telescope; control station, Evpatoria.
National Radio Astronomy Observatory (NRAO), US: Tracking station, Green Bank; VLBA correlator; low-noise band 18–25 GHz amplifier.
Helsinki University of Technology, Finland: 22 GHz on-board receiver; ground radio telescope.
Tata Institute for Fundamental Research (TIFR), India: 92 cm on-board receiver; ground radio telescope.
Australia Telescope National Facility (ATNF): 18 cm on-board receiver; ground radio telescopes.
Mass: payload: 2,500 kg; total: 5,000 kg
Orbit: highly elliptical; 10,000–70,000 × 310,000–390,000 km; 51.6° inclination; 7–10 day period
Lifetime: minimum five years; possible additional five year mission extension

Spektr-UF/Spectrum-UV/World Space Observatory (WSO/UV)

Spektr-UF was originally planned for launch around 1997, and then again in 2002; a prototype of the spacecraft was completed as soon as the early 1990's. As its name implies, Spektr-UF, or Spectrum-UV, is an ultraviolet observatory. The Russians see it as a follow-on and improvement to the United States' Hubble Space Telescope. Spektr-UF is an international collaborative project and is also known as the World Space Observatory (WSO or WSO/UV), or International Space Observatory. Spektr-UF will observe black holes and star formation, and will enable astrophysicists to study the evolution of the universe and the nature of dark matter.

FSA hopes to complete development of the Navigator bus, and launch Spektr-UF previous to Hubble's eventual demise around 2013, after a 2008 Space Shuttle servicing mission to extend its lifespan. Roscosmos has included the spacecraft in its 2006–2015 federal space programme and has projected its launch date to occur at the end of 2010. Spektr-UF is an international project, with several institutions around the world contributing instruments and ground support. Additionally, China has proposed to manage launch services for the spacecraft. Spektr-UF will be built around a primary T-170 mirror with a 1.7 m diameter.

Spektr-UF Specifications

Launch: projected for 2010 from China aboard a Long March CZ-3B
Payload: 1.7 m mirror UV telescope
Participating organisations: FSA: lead organisation; ground control.
NPO Lavochkin: Navigator spacecraft bus.
Crimean Astrophysical Observatory (CrAO) and Lytkarino Optical Glass Plant (LZOS), Ukraine: improved T-170M telescope optics – 1,715 mm diameter main mirror and 400 mm diameter secondary mirror; possible ground segment support.
Italian Space Agency (ASI): two UV and one optical Field Camera Units (FCU).
Institut für Astronomie und Astrophysik, Eberhard Karls Universität Tübingen (IAAT) and Deutsches Zentrum für Luft- und Raumfahrt (DLR), Germany: High Resolution Double Echelle Spectrograph (HIRDES).
Instituto de Astronomía y Geodesia (IAG)/ Universidad Complutense de Madrid (UCM)/ Consejo Superior de Investigaciones Científicas (CSIC), Spain: ground segment.
China National Astronomical Observatories (NAO) and China National Space Administration (CNSA): launch services (cost to be shared among international partners); tracking, ground stations and ground support; Long Slit Spectrograph (LSS); Science Operations Centre
Particle Physics and Astronomy Research Council (PPARC), UK: has expressed interest but has not yet committed funding for an instrument or ground support services; a contribution may materialise in the form of data processing services or an LSS.
Indian Institute of Astrophysics (IIA) and Indian Department of Science and Technology (DST): India has expressed interest in the project but has not yet committed funds; support could occur in the form of calibrations and testing services or software design and services; other possibilities include optical, electrical, mechanical and thermal systems design and manufacture.
Tel Aviv University and Israel Space Agency (ISA), Israel: Israel had previously expressed interest, but is likely to only provide nominal funding and support for the project due to possible lack of funds and political will caused by budget cycles and a change in ISA leadership; limited design participation only.
Mass: spacecraft: 1,147 kg; payload: 1,600 kg; total: 2,747 kg
Orbit: highly elliptical, 500 × 300,000 km, L2 Lagrangian point; 28.5°; 7 day period
Lifetime: at least five years

Spektr-RG/Spectrum-RG/Spektr-X-Gamma

Spektr-RG will investigate black holes, galaxy clusters, gamma ray bursts, various star types, and the nature of dark matter and dark energy; it will be equipped with a 0.2–12 keV band X-ray telescope and gamma ray burst detector to accomplish these goals. As of 2006, the spacecraft was projected to launch from Kourou aboard a Soyuz-2, using a Fregat upper stage/Payload Assist Module (PAM). Spektr-RG's eROSITA and Lobster instruments were originally designed to fly on the International Space Station (ISS) as well as the failed ABRIXAS mission.

The Spektr-RG concept has been around since the 1980s. The programme requirements were outlined in the early 1990s and evolved through the years until the programme was shelved due to lack of funds. The project has been picked up again, with ESA as a major partner. The instrumentation requirements have changed since the 1990s and none of the experiments that were slated to fly on the 1996 version will be launched on the 2011 version.

Spektr-RG is still in the planning stages; the spacecraft is included in the Russian space programme through 2015, however it remains to be seen whether the programme will run according to schedule during this decade.

Spektr-RG Specifications

Launch: projected for 2011 from Kourou, or Baikonur as a back-up
Payload: X-ray telescopes and monitor; gamma ray burst detector
Participating organisations: NPO Lavochkin or RSC Energia and FSA: Navigator (932 kg), or Yamal (687 kg), spacecraft bus; ground control.
ESA: launch site; ground support and telemetry.
Italian Space Agency (ASI): ground station, Malindi, Kenya.
Max Planck Institut für extraterrestrische Physik (MPE), Germany: extended ROentgen Survey with an Imaging Telescope Array (eROSITA) Wolter-telescopes/seven 35 cm diameter X-ray mirror telescopes.
Leicester University, UK: wide field X-ray monitor – Lobster.
Space Research Institute (IKI), Russia: Astronomical Roentgen Telescopes (ART) – X-ray concentrator (ART-XC) or coded-mask telescopes with silicon detector (ART-XS); or Hard X-ray coded-mask telescope with gamma-camera (ART-HX); gamma ray burst detector (GRB).
Mass: spacecraft: 687–932 kg; payload: 1,250 kg; mass requirement total: ≤2,180 kg
Orbit: planned circular 580–600 km equatorial orbit with ≤5° inclination (≤30° if launched from Baikonur); 96 min period
Lifetime: at least five years, possibly up to 10 years

This Spektr series should not be confused with the Mir Space Station's Spektr remote sensing module, launched aboard a Proton in May 1995 and disabled in a Progress collision in 1997.

United States

2001 Mars Odyssey

Current Status

During its first five years imaging Mars, 2001 Mars Odyssey has confirmed that there is a large amount of water on the planet, has helped solve the mystery of black deposits, or spots, at Mars' poles, and has mapped the largest canyon in the solar system. Odyssey has also relayed to Earth approximately 85 per cent of the data collected by the Mars Rovers, Spirit and Opportunity.

Mars Odyssey was launched by Delta II on 7 April 2001 and arrived at Mars on 24 October 2001 at 02:30 UT. The primary science mission lasted from February 2002 through August 2004; Odyssey is currently on its second extended mission. In addition to its main mapping mission, the spacecraft supports other Mars missions and serves as a relay for the two Mars Exploration Rover spacecraft launched in mid-2003 (see separate entry). Odyssey assisted with the landing site selection for the Mars Rover Opportunity, and is also supporting the site selection process for the Mars Phoenix, due to launch in 2007. Odyssey is scheduled to remain operational, working alongside NASA's younger Mars Reconnaissance Orbiter (see separate entry), through at least the next two fiscal years; NASA will likely continue to fund Odyssey until the craft malfunctions, loses power, or is rendered predominantly obsolete by the Mars Reconnaissance Orbiter and other US and international Mars exploration programmes. Odyssey has enough fuel to survive well beyond 2010; all but one instrument on the spacecraft are in good health, as are the power, navigation and communications systems.

Background

Originally called the Mars Surveyor 2001 Orbiter, 2001 Mars Odyssey is an orbiter carrying three packages of science experiments designed to make global observations of Mars to improve our understanding of the planet's climate and geologic history, including the search for liquid and ice water and evidence of possible past life. The primary mission extended across a full Martian year, or 29 earth months. The Jet Propulsion Laboratory (JPL) manages the 2001 Mars Odyssey programme, named in honour of the science fiction writer Arthur C Clarke. Lockheed Martin Space Systems built the spacecraft bus. The primary mission's cost was USD297 million; the first extended mission, approved from August 2004 through September 2006, had a budget of USD35 million.

The chief scientific goal of the Mars Odyssey programme is to map the chemicals and minerals that make up the Martian surface. As on Earth, the geology and elements that form Mars chronicle its history, and while neither elements nor minerals can convey the entire story of a planet's evolution, both contribute significant pieces to the puzzle. These factors have profound implications for understanding the evolution of Mars' climate and the role of water on the planet, the potential origin and evidence of life, and the possibilities that may exist for future human exploration.

Other major Odyssey mission goals are to determine the abundance of hydrogen, most likely in the form of water ice, in the shallow subsurface; globally map the elements that make up the surface; acquire high-resolution thermal infra-red images of surface minerals; provide information about the structure of the Martian surface; and record the radiation environment in low Mars orbit as it relates to radiation-related risk to human exploration.

Odyssey's interplanetary cruise phase lasted about 200 days. It began with the first contact by the Deep Space Network (DSN) after launch and extended until seven days before Mars arrival. Primary activities during the cruise included check out of the spacecraft in its cruise configuration, check out and monitoring of the spacecraft and the science instruments and navigation activities necessary to determine and correct Odyssey's flight path to Mars.

The science mission began about 45 days after the spacecraft was captured into orbit about Mars. The primary science mission lasted 917 Earth days. Odyssey's orbit inclination is 93.1°, nearly Sun-synchronous, with a two hour period.

During the orbiter's science operations, the Thermal Emission Imaging System (THEMIS) has been taking multispectral thermal-infra-red images, making a global map of Martian surface geology and minerals; THEMIS also acquires visible images at a resolution of about 18 m. The orbiter images the Martian landscape in successive ground tracks, which are separated in longitude by approximately 29.5°. Odyssey repeats the entire ground track every two Martian days, or sols. The Gamma Ray Spectrometer (GRS) takes global gamma ray measurements during all Martian seasons. The Martian Radiation Environment Experiment (MARIE) operated during part of the primary science mission to collect data on the planet's radiation environment. Early data collected by this instrument revealed that low Mars orbit radiation levels are at least double those found in LEO. MARIE ceased operation in October 2003, when it was adversely affected by a strong solar flare.

Specifications

Spacecraft: Built by Lockheed Martin Space Systems, the Mars Odyssey orbiter is a three-axis stabilised spacecraft consisting of a bus 2.2 × 1.7 × 2.6 m in size composed of a frame fabricated from aluminium and titanium. Odyssey's systems are fully redundant, the exception being the memory card that collects imaging data from THEMIS. The spacecraft is comprised of two modules: the propulsion module containing the tanks, thrusters and associated mechanisms, and an equipment module, with separate equipment and science decks; the decks support engineering components and the science instruments, MARIE, THEMIS amd GRS (see below) as well as Odyssey's High-Energy Neutron Detector (HEND), the neutron spectrometer and star cameras. The structures subsystem weighs 81.7 kg.
Mass: 725 kg at launch; 331.8 kg dry plus 348.7 kg of propellant and 44.5 kg of science instruments.
Command and Data Handling: RAD6000 32-bit processor with Mars Pathfinder heritage in the Orbiter Command and Data Handling subsystem, provides a central processing capability for all spacecraft systems including payload elements. It can be switched between clock speeds 5, 10 or 20 MHz and includes 128 Mbyte of RAM of which approximately 20 per cent is used for running spacecraft programmes. The rest of the memory is

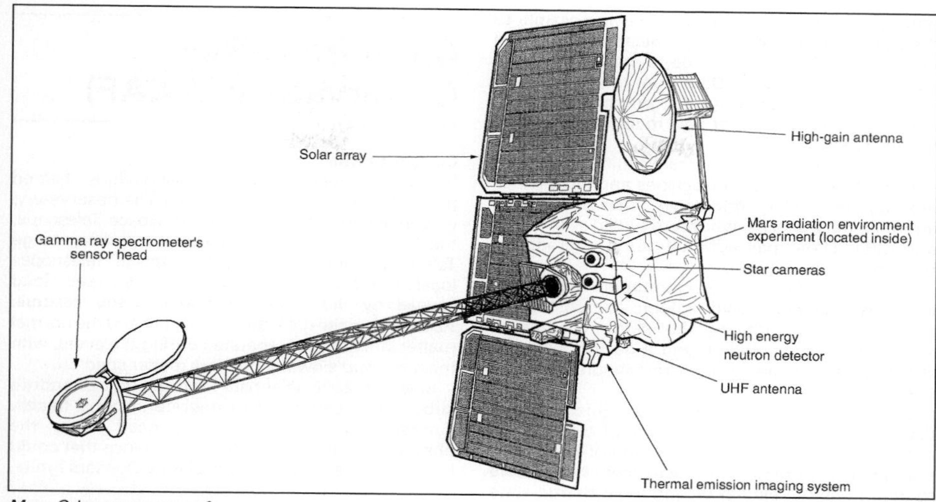

Mars Odyssey spacecraft

0105939

used for science and for storing data later relayed to Earth. An interface card handles communication with Odyssey's attitude sensors and science instruments. The spacecraft is equipped with several more interface cards that perform various functions such as electrical power and telecommunications subsystems control. Odyssey also has a 1Gbyte mass memory card used to store imaging data; it is non-redundant. The command and data handling subsystem weighs 11.1 kg.

Telecommunications: Comprises a radio system operating in the X-band microwave frequency range for direct communications with Earth and a system that operates in the UHF range for communication with the Mars Rovers and any other possible future landers. The system weighs 23.9 kg.

Power: The two-axis, articulated, gallium arsenide, three-panel, solar array (7.4 m^2 cell area) provides electrical power and served as the most significant drag brake during aerobraking. Nickel-hydrogen (NiH_2) 16 Ah common pressure vessel batteries provide power during eclipses peak power needs.

Thermal Control: The TC system uses passive methods and louvres to control the temperature of the batteries and the solid-state power amplifiers. Passive coatings as well as multilayer insulation blankets are used to control temperatures along with thermostatically controlled and computer controlled heater circuits. Where needed, radiators are used to remove excess heat from spacecraft components to maintain proper operating temperatures. The TC system weighs 20.3 kg.

Attitude control: The orbiter is three-axis stabilised in all mission phases except following separation from the launch vehicle. Primary attitude determination is via star camera derived from the Clementine spacecraft star camera and an inertial measurement unit with analogue star sensors as back-up. Three reaction wheels (and one further back-up) provide primary attitude control, desaturated via RCS thrusters. The inertial measurement units (IMU) are switched off during significant portions of cruise and mapping to prolong the vehicle's lifetime, and the spacecraft is operated in all-stellar mode except during orbit insertion, aerobraking drag passes, orbital trim manoeuvres and safe mode. Mars Odyssey has four 0.98 N thrusters for roll control and four 22.56 N thrusters for pitch and yaw control. Using three redundant pairs of sensors, the guidance, navigation and control subsystem determines spacecraft orientation. A sun sensor is used to detect the position of the sun as a backup to the star camera, used primarily to locate star fields. Between star camera updates the IMU collects information. The guidance, navigation and control subsystem weighs 23.4 kg.

Propulsion: Derived from the propulsion system for Mars Global Surveyor, the Mars Odyssey spacecraft has a single LEROS 640 N main engine operating on hydrazine and nitrogen tetroxide bipropellants for Mars orbit insertion. A single gaseous helium tank pressurises the fuel and oxidiser tanks. The propulsion subsystem weighs 49.7 kg.

Science Payload: **THEMIS**: The Thermal Emission Imaging System (THEMIS) is responsible for determining Mars' surface mineralogy. The instrument can see in both visible, at 18 m resolution, and infra-red, thus collecting imaging data that has been previously invisible to the scientists In the infra-red spectrum. THEMIS weighs 11.2 kg, is 54.5 cm long, 34.9 cm tall and 28.6 cm wide; it runs on 19 W of electrical power.

GRS: The Gamma Ray Spectrometer (GRS) plays a lead role in determining the elemental makeup of the Martian surface. Using a GRS and two neutron detectors, the experiment detects and studies gamma rays and neutrons emitted from the planet's surface. GRS data is collected at 300 km resolution. GRS consists of two main componets: the sensor head and the central electronics assembly. The sensor head is separated from the rest of the Odyssey spacecraft by a 6 m boom, which was extended after Odyssey entered its mapping orbit at Mars. This was done to minimize interference from any gamma rays coming from the spacecraft itself. The two neutron detectors, the neutron spectrometer and the high-energy neutron detector are mounted on the main spacecraft structure. GRS weighs 30.3 kg and uses 32 W of power. Including its cooler, the GRS measures 46.8 cm long, 53.4 cm tall and 60.4 cm wide. The neutron spectrometer is 17.3 cm long, 14.4 cm tall and 31.4 cm wide. The high-energy neutron detector measures 30.3 cm long, 24.8 cm tall and 24.2 cm wide.

MARIE: The Martian Radiation Environment Experiment (MARIE) characterised aspects of the radiation environment both on the way to Mars and while in Martian orbit. Space radiation presents an extreme hazard to crews of interplanetary missions; the experiment attempted to predict anticipated radiation doses that would be experienced by future astronauts, and help determine the possible effects of Martian radiation on human beings. MARIE was supposed to have operated throughout the entire primary science mission, however it was permanently damaged during a particularly strong solar flare in October 2003, and has ceased to function since that time. MARIE weighs 3.3 kg and used 7 W of power. It measures 29.4 cm long, 23.2 cm tall and 10.8 cm wide.

Advanced Composition Explorer (ACE)

Current Status

The Advanced Composition Explorer (ACE) programme was rated highly on science contributions and the contribution toward NASA's solar science goals in the 2005 NASA solar missions senior review. ACE will continue to be funded by NASA through at least 2010 due to its scientific merit and relevance.

11 years after launch, ACE continues to operate well and return important solar science data. It remains in Earth-sun libration point (L1) orbit. Scientists continue to submit papers analysing the data received from the spacecraft. In 2004–2005 alone, ACE scientists generated at least 138 scientific papers.

NASA expects that if ACE remains in good health, it will function until 2025, at which point its fuel and power will be depleted beyond operational capabilities.

Background

Formally proposed to NASA in 1986, ACE is part of the Explorer Concept Study Program. The Johns Hopkins University Applied Physics Laboratory (APL) designed and built the ACE spacecraft and was involved in planning for mission definition phase. Spacecraft construction began in 1994 and the launch took place on 25 August 1997 on board a Delta II rocket. Goddard Space Flight Centre (GSFC) Explorer Projects Office of the Flights Directorate managed mission development.

From its orbit about the Sun-Earth libration point, about 1.5 million km sunward of Earth, ACE carries six high-resolution spectrometers, which measure the elemental, isotopic and ionic charge state composition of nuclei from hydrogen to nickel (from solar wind energies to galactic cosmic ray energies) as part of the spacecraft. ACE has an elliptical orbit, with a semi-major axis of 200,000 km, thus providing the spacecraft with an excellent view of the Sun and beyond. In addition, it is worthwhile to note that the spacecraft is carrying enough propellant to maintain its orbit at the Sun-Earth libration point until approximately 2022.

A further three instruments are also carried by the ACE spacecraft; these monitor the state of the interplanetary medium. By studying these accelerated particles, which arrive from the interstellar and galactic sources and the Sun, scientists will be able to further understand the evolution of the solar system. In addition, ACE provides real-time solar wind measurements to the National Oceanic and Atmospheric Administration (NOAA) from its Real-Time Solar Wind (RTSW) data. The data is used for forecasting space weather. This allows ACE to provide an advance warning of approximately one hour, of geomagnetic storms, which can cause such hazards as overloading power grids and disruption of communications on Earth.

ACE is a spinning spacecraft, which rotates at a rate of 5 rpm. The science payload contributes a mass of 156 kg to the spacecraft, which generally points towards the Sun; this instrumentation can be found mounted on both the flat and sun-facing sides of the cylindrically shaped spacecraft. ACE operates at the L1 point, thus allowing scientists to study the incident particles during both solar quiet and solar active periods: all phases of the solar cycle.

Study of the accelerated particles from the Sun, interstellar and galactic sources, have allowed scientists to research problems in four major areas: determination of the elemental and isotopic composition of matter from which nuclei are accelerated; the origin of the elements and their subsequent evolution; study of the formation of the solar corona and how the subsequent acceleration of solar wind occurs; and understanding the nature of particle acceleration.

In 1994, NASA and NOAA completed an inter-agency agreement. This agreement explained that NOAA would use the real-time, continuous broadcast of solar wind parameter measurements in order to forecast space weather. The data, provided via RTSW, covers measurements such as: density, magnetic field vectors from the MAG instrument, solar wind velocity, temperature measurements from the SWEPAM instrument and energetic particle fluxes (provided by EPAM and SIS). Data is transmitted via ground link to NOAA, during the 3 hours a day that ACE is in contact with the Deep Space Network (DSN). The space weather data generated by ACE's RTSW is available to the public.

The spacecraft operates in a spin-stabilised mode at a rate of 5 rpm. The body, cylindrical in shape, is 1.6 m in diameter and 1 m in length; it consists of octagonal aluminium honeycomb decks joined by eight flat side panels. With the solar panels and magnetometer boom extended, ACE has a wingspan of approximately 8.3 m. All components, except the propulsion system, are mounted on the external surfaces of the body; this simplifies access to the instruments and spacecraft subsystems. As well as retaining a clear field-of-view for the many instruments, a practical balance of weight around the spacecraft is also kept.

The two sensors belonging to MAG, are mounted on booms in order to reduce the magnetic effects from the spacecraft. In addition, in order to correct

For details of the latest updates to *Jane's Space Systems and Industry* online and to discover the additional information available exclusively to online subscribers please visit
jsd.janes.com

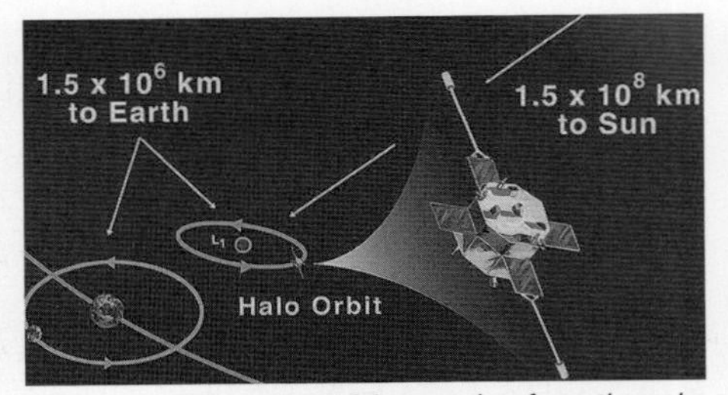

ACE studies charged particles ranging from the solar wind to galactic matter (NASA) 0517090

launch vehicle dispersion errors, make trajectory corrections, introduce the spacecraft into the L1 orbit, adjust orbit and spin axis pointing and maintain the 5 rpm spin rate – ACE needs a propulsion subsystem. This was provided for via the use of a monopropellant hydrazine blow down design, which uses nitrogen gas as a pressurant.

Four axial thrusters and six radial thrusters provide the spacecraft with the appropriate velocity control along the spin axis and spin plane rate control, respectively. The power system on board the ACE spacecraft is regarded as quite simple. After launch, an array of four silicon solar panels unfolded. The quartz covers provide appropriate radiation shielding, allowing the power system to provide 443 Watts after launch and many years into service. Power levels decrease after years of service and exposure to the space environment.

Attitude control is determined via the use of a star scanner and redundant Sun sensors. In addition, gyroscopic stability of the spinning spacecraft, two fluid filled nutation dampers and ten thrusters of the propulsion subsystem provide for attitude control.

Two identical and independent communications subsystems can be found on board the ACE spacecraft: a single high-gain, dual polarized, parabolic reflector antenna and two sets of broad-beam antennas. The RF system operates at S-band. The mission is designed to communicate with the DSN once a day, for a duration of approximately 3 hours. It was also designed in order to miss one pass without losing any data.

ACE carries nine scientific instruments. The instrumentation on board the spacecraft was designed in order to provide several key capabilities. It was important that contributions from various particle populations over a broad energy range could be isolated, and so the instruments had to be designed in order to provide co-ordinated measurements of charge and mass of the elements from hydrogen to nickel. This would then allow scientists to relate the observed composition patterns back to the appropriate acceleration process and sample of source material. The instruments are described in the specifications.

Specifications
Launch: 25 August 1997 from Cape Canaveral aboard a Delta II
Orbit: elliptical, at the L1 point
Mass: 785 kg at launch
Lifespan: projected up to 28 years
Payload: Cosmic Ray Isotope Spectrometer (CRIS) – measures the elemental and isotopic composition of galactic cosmic rays over an energy range of approximately 100 to 600 MeV/nucleon. Two CCD cameras, using a scintillating optical-fibre trajectory system, track the paths of the energetic nuclei, which stop in four co-aligned stacks of large-area silicon detectors.

Solar Isotope Spectrometer (SIS) – using two identical stacks of silicon detectors, the elemental and isotopic composition of energetic nuclei within an energy range of 10 to 100 MeV/nucleon, are measured. This energy range includes transient fluxes of energetic nuclei accelerated in large solar particle events, anomalous cosmic rays and low-energy galactic cosmic rays.

Ultra Low Energy Isotope Spectrometer (ULEIS) – by covering an energy range that includes solar energetic particles, particles accelerated by interplanetary shocks and low-energy anomalous cosmic rays, the mass and kinetic energies of nuclei can be measured. ULEIS has suffered some degradation but remains operational.

Solar Energetic Particle Ionic Charge Analyser (SEPICA) – measured the charge state, kinetic energy and nuclear charge of energetic ions between 0.2 and 3 MeV/nucleon. Once the particles entered SEPICA's multi-slit collimator, they were electrostatically deflected between six high-voltage carrying electrode plates. By measuring these charge states, it was possible to then acquire information on the temperature of the source plasma and charge-to-mass dependant acceleration processes. Data from this instrument is no longer available due to component failure. The ACE team is looking into methods to extract further SEPICA data.

Electron, Proton and Alpha Monitor (EPAM) – studies the behaviour of electrons and ions, which are accelerated by impulsive solar flares and interplanetary shocks, which are linked with CMEs and CIRs. EPAM has suffered some degradation but remains operational.

Solar Wind Ion Mass Spectrometer (SWIMS) – measures the solar wind isotopic composition in all solar wind conditions. The instrument has an excellent mass resolution, and a design, which is based on that of similar instruments on Wind and SOHO.

Solar Wind Ion Composition Spectrometer (SWICS) – by using a combination of electrostatic deflection, post acceleration, time-of-flight and energy measurements, this instrument will be able to determine the elemental and ionic charge state of all major solar wind ions from hydrogen (H) to iron (Fe).

Solar Wind Electron, Proton, and Alpha Monitor (SWEPAM) – measures the three-dimensional characteristics of the solar wind and suprathermal electrons and ions. SWEPAM has suffered some degradation but remains operational.

Magnetometer (MAG) – a flight spare of the magnetometer instrument flown on the 'Wind' spacecraft, MAG will measure the dynamic behaviour of the vector magnetic field, which will include measurements of interplanetary shocks, waves and other features governing the acceleration and transport of energetic particles.

Chandra X-Ray Observatory (AXAF)

Current Status
In August 2006, Chandra observations helped prove the existence of dark matter. The observatory, in conjunction with the Hubble Space Telescope, the European Southern Observatory's Very Large Telescope, and the Magellan optical telescopes located a one-hundred-million degree gas cloud caused by the collision of two galaxy clusters. According to the data, the dark matter and the normal matter, the hot gas, separated during the event, with the gas cloud slowing; the dark matter sped on.

In March 2006 scientists reported that Chandra's subsystems continued to operate in good health. Furthermore, scientists announced that the observatory might have found evidence that could help to answer the question of how Quasars ignite.

Background
The Smithsonian Astrophysical Observatory in Cambridge, Massachusetts operates Chandra for NASA, Marshall. The spacecraft communicates through NASA's Deep Space Network (DSN). The Chandra Science Center is located on the Harvard University campus. The observatory enables astrophysicists to study black holes, quasars, galaxy clusters, dark matter and the evolution and nature of the universe. At launch, Chandra was the most powerful X-ray telescope ever orbited.

NASA renamed the Advanced X-ray Astrophysics Facility (AXAF) after the renowned astrophysicist and Nobel-Laureate Subrahmanyan Chandrasekhar,

The mirror for the Chandra telescope is subjected to environmental tests (NASA) 1047481

The Chandra X-ray observatory satellite during final assembly and checkout (TRW) 0572504

giving it the name Chandra in late 1998. Chandra was launched by the Space Shuttle STS-93 mission on 23 July 1999. Following a series of five engine firings, Chandra reached its desired orbit on 7 August 1999. The official first light image was received from the Advanced CCD Imaging Spectrometer, a picture of Cassiopiea A, on 19 August 1999.

Chandra constitutes the third in NASA's Great Observatories Series, following the Hubble Space Telescope and the Compton Observatory. Chandra resumed the work started by Uhuru in 1970 and continued by HEAO 2 until re-entry in 1981. Its four pairs of nested mirrors for focusing 0.1–10 keV high-energy x-rays offer about 10 times the spatial resolution and 100 times the sensitivity of HEAO 2. With 1.2 m outer diameter, the 10 m focal length mirrors will provide 1,100 cm^2 collecting area (instead of the 1,700 cm^2 of the original six pairs).

Preliminary plans called for the USD1.4 billion, 5,200 kg AXAF-I (imaging), later renamed Chandra, to be launched by the Shuttle with an unspecified upper stage in September 1998 into a 10,000 × 140,000 km, 28.5° orbit. The USD320 million, 2,700 kg AXAF-S (spectroscopy), which later became JAXA's failed ASTRO-E mission, would follow in December 1999 aboard Delta into polar orbit. NASA wanted both to last at least five years in orbit. Chandra consists of a four-mirror pair-shell grazing incidence telescope (instead of the original six parabola/hyperbola pairs), two gratings and two focal plane instruments. Boeing agreed to provide an IUS upper stage in May 1994 under a USD25 million contract to establish the final orbit. AXAF-S changed the replicated optics telescope with the supercooled x-ray spectrometer in the focal plane.

NASA adopted a phased programme to avoid Hubble-type cost overruns. Hughes Danbury Optical Systems, now BF Goodrich Optical Systems, began grinding the largest parabola/hyperbola pair (P1/H1) of the four paired nested grazing incidence mirrors in December 1989 under TRW's USD158.1 million contract. Congress, as a measure to check NASA's repeated history of cost escalation, required the agency to make a semi-annual progress report through the completion of testing on 1 October 1991. Testing on P1/H1 was completed that September in MSFC's new USD17 million X-Ray Calibration Facility (XRCF).

All eight flight mirrors were completed in early 1995, followed by coating at Optical Coating Laboratories, Inc (H1 was the last, in January 1996) and delivered to Eastman Kodak. Kodak provided the optical bench and assembled the eight mirror elements (HRMA High Resolution Mirror Assembly) and the telescope. Ball Aerospace Electro-Optics/Cryogenics Div supplied a visible wavelength aspect camera for attitude reference. Ball also provided the Science Instrument Module (SIM), which supports the focal plane instruments, provides focus control and translates the instrument in/out of the focal position as required.

Specifications

Launch: 23 July 1999 by shuttle mission STS-93 from Kennedy Space Center using IUS as boost stage
Contractors: TRW Space and Electronics Group
Mass: 5,864 kg at launch
Orbit: 10,000 × 140,000 km, 64 hr period
Spacecraft: 13.8 m long, 19.5 m span across deployed solar panels. Mission life >5 years. Power provided by 3-panel, silicon solar arrays (2.35 KW) and 3 × 40 Ah metal/hydrogen batteries for eclipse power. Data recording from solid-state recorder with 1.8 giga-bytes (16.8 hr) capability
Design life: 5 years plus
Payload: AXAF CCD Imaging Spectrometer, sub-arcsec angular resolution (ACIS, Pennsylvania State University/MIT).

High-Resolution Camera, microchannel plate imager with less than 0.5 arcsec angular resolution (HRC, Smithsonian Astrophysical Observatory).

High-Energy Transmission Grating Spectrometer, 0.4–8.0 keV, resolution 800 at 1.0 keV.

ACIS, images spectral dispersion of grating (HETGS, MIT).

Low-Energy Transmission Grating Spectrometer, 0.1–3.0 keV, resolution 750 at low energies, HRC images spectral dispersion of grating (LETGS, Utrecht Lab for Space Research).

Payload weight 954 kg. Focal length 10 m, outer diameter 1.2 m

Dawn

Current Status

The once cancelled Dawn mission has been reinstated after a mission review. The mission will now aim to launch on 20 June 2007. The mission launch was originally planned to take place on 27 May 2006, with arrival at Vesta in 2010 and Ceres in 2014. The asteroid arrival dates have been revised to 2011 and 2015 respectively.

In 2006, the International Astronomical Union (IAU) reclassified Ceres (formerly and asteroid) as a "Dwarf Planet" along with Pluto, formerly classified as a Planet, and Eris, a body that was found to be larger than Pluto. The ruling has proved to be controversial among astronomers, and it remains to be seen whether there will be further reclassifications of known solar system bodies in the coming years. Dawn is a mission within NASA's Discovery programme, along with Lunar Prospector, Mars Pathfinder and Deep Impact, among others.

Background

The Dawn mission will study the two largest asteroids in the solar system: Ceres and Vesta. Both Ceres and Vesta have remained intact since their formation millions of years ago. Dawn's investigation will allow scientists to analyse the conditions and processes of the solar system during its earliest era. Both asteroids, thought to have formed from protoplanets, can be found within the asteroid belt between Mars and Jupiter. Scientists know that Ceres and Vesta have experienced varied evolutionary paths due to the diversity of processes operating during the first few million years of solar system evolution. Ceres, on the one hand, is viewed as primitive and wet; whereas, Vesta is evolved and dry. It is believed that Ceres may have active hydrological processes, which lead to seasonal polar caps of water frost, thus changing the understanding of the interior of these bodies. In addition, it has been suggested that Ceres could have a thin, permanent atmosphere, thus separating it from the other minor planets. Vesta, on the other hand, may have rocks that are more strongly magnetised than those on Mars. Ceres and Vesta are the largest asteroids.

The three key scientific objectives for the Dawn mission are as follows:

- To capture the earliest moments during the origin of the solar system in order to assist scientists in understanding the circumstances under which these planets formed.
- To compliment understanding of the formation of these planets, it is also necessary to determine the nature of the building blocks from which the terrestrial planets formed.
- To compare and contrast the varied paths of formation and evolution for both planets, thus helping scientists to understand what controls evolution.

The Dawn spacecraft has a fully redundant electrical system. Additionally, Orbital's spacecraft bus has been used on a number of missions, including SORCE and Deep Space 1. The solar-electric/ion propulsion technology that will be used to propel Dawn to its multiple destinations was also used on Deep Space 1; the propulsion system was developed, built and tested at NASA's Jet Propulsion Laboratory (JPL). The spacecraft contains a 100 W travelling wave tube amplifier, tracking, telemetry and command system with fixed 1.5 m high-gain antenna, medium-gain fan beam, omni antennas and small deep-space transponders. The attitude control system and flight software is the same as that on the Orbview imaging mission; the structure is made from aluminium and the command and data handling uses off-the-shelf components, again, the same as that on Orbview. NASA's Deep Space Network (DSN) will track Dawn.

In order to achieve the additional velocity needed to reach Vesta, the Dawn spacecraft will use ion propulsion, an efficient way to use the onboard xenon fuel. The spacecraft's Gallium Arsenide solar arrays, spanning about 19.7 m, store the Sun's energy, then transfer that energy to the ion engines to ionize the xenon; this in turn, powers Dawn. Ion propulsion will be used to drop the spacecraft down to lower altitudes whilst at Vesta. Once this segment is complete, ion propulsion will be used to cruise towards Ceres. It is interesting to note that while Dawn's engines have a specific impulse of 3,100 s and a thrust of 90 mN, a chemical rocket on a spacecraft might have a thrust of up to 500 Newtons; Dawn's smaller engines achieve an equivalent trajectory change by firing for a longer period of time.

Specifications

Launch: 20 June 2007, from Cape Canaveral aboard a Delta II
Mission Participants and Payload: University of California, Los Angeles (UCLA): lead science organisation; magnetometer; data analysis and processing.
Orbital Sciences Corporation: spacecraft bus.
NASA JPL: mission control; navigation; ion propulsion; radio science.
Max Planck Institute for Solar System Research (MPL)/Deutsche Zentrum für Luft- und Raumfahrt (DLR), Germany: Framing Camera (FC).
Istituto Nazionale di Astrofisica (IFSI-INAF)/Italian Space Agency (ASI), Italy: Visual and Infrared Imaging Spectrometer (VIR) mapping spectrometer – virtually the same instrument as Rosetta's VIRTIS.
Los Alamos National Laboratory (LANL), US: Gamma Ray and Neutron Detector (GRaND).
NASA, Goddard: laser altimeter.
Mass: 1,210 kg at launch
Orbit: 25 × 4,500 km about each asteroid
Lifespan: 10 years

Deep Impact

Current Status

On 3 July 2005, the Deep Impact spacecraft rendezvoused with the comet Tempel 1. As planned, the impactor hit the comet on its sunlit side, resulting in a rapid release of dust. The brightness and amount of released debris suggested that microscopic dust, water, carbon dioxide and hydrocarbons could be present underneath the surface. The impact resulted in a crater which was 100 m in diameter and 25 m deep. Following the impact, on 9 August 2005 the spacecraft was put into a deep sleep, and delivery of the data to the Planetary Data System began during January 2006. The Spitzer Space Telescope observed Tempel 1 before and after the impact; it noted a strong difference in Tempel's thermal radiation emissions subsequent to the rendezvous. Scientists continue to analyse Deep Impact's data.

Deep Impact's team considered sending the spacecraft back to Tempel 1 after the impact, to observe the comet and collect post-impact data. Although a feasible trajectory existed for the craft's return to the comet, with arrival taking place in January 2011, due to the time investment, costs, and toll it would take on Deep Impact's instruments, the team decided not to take this option. Instead, Deep Impact will likely visit the Comet Boethin on an extended mission called Deep Impact eXtended Investigation (DIXI), a proposal submitted by the project's existing University of Maryland team. The Earth gravity assist manoeuvre for this trip was initiated on 20 July 2005 and Earth flyby will occur on 31 December 2007. Deep Impact will reach Boethin in December 2008. The spacecraft will use its existing instruments, which remain in good health, to compare Boethin data with that already collected on Tempel 1. A second proposal, called Extrasolar Planet Observations and Characterization (EPOCh), submitted by NASA-Goddard, proposes that Deep Impact be used to search for Earth-sized planets outside our solar system; these observations would take place previous to the Boethin visit. Neither proposal has yet received full NASA funding, however, approval of DIXI at least, looks hopeful, since it would be a cost-effective way to further observe and compare new and existing comet data.

Background

On 12 January 2005 the Deep Impact spacecraft was launched from Cape Canaveral, Florida, on board a Delta II rocket. Deep Impact is part of NASA's Discovery Series and was designed to study the composition of the interior of the comet Tempel 1; planning and mission design for the Deep Impact spacecraft took place between November 1999 and May 2001. In the past, missions have travelled to comets, for example the Giotto and Stardust missions. However, Deep Impact was the first mission to eject material from a comet's surface. The spacecraft contained a larger flyby spacecraft, which carried an impactor spacecraft; the impactor was then released into the comet's path in order for the collision to take place. NASA's Jet Propulsion Laboratory (JPL), the University of Maryland, College Park, and Ball Aerospace & Technologies Corporation were the partners on the initial Tempel 1 mission; the same institutions will work on DIXI.

Following launch, the spacecraft's journey took 174 days and approximately 429 million km at a cruising speed of 28.6 km/s, to reach its target, comet Tempel 1. On arrival, the spacecraft separated into two portions, the impactor and the flyby probe. The impactor, using its thrusters, then moved into the path of the comet, closest approach being at a distance of 500 km from the comet, with impact occurring 24 hours after release, at a relative speed of 10.3 km/s.

The impactor spacecraft was battery powered and able to operate independently of the flyby probe for 24 hours; the impactor was able to control its own navigation and manoeuvres into the path of the comet. A camera was also placed on the impactor, allowing images of the comet's nucleus, just before collision, to be captured and relayed back to the flyby spacecraft. Although the impactor collided with the comet, the impact was not great enough to make a substantial change in the comet's orbital path.

In addition to the camera placed on the impactor, the flyby spacecraft also observed and recorded the impact. The flyby probe was able to observe the ejected material which was blasted from the crater and subsequently the structure and composition of the crater's interior. The flyby probe's shields were used to protect the spacecraft from the passing comet's dust tail. Once this was complete, it was then possible for the spacecraft to turn and look at the comet once again, allowing it to record additional data, for example, changes in the comet's activity. Resulting photographs of the impact indicated the comet to be dustier and less icy than expected.

The Deep Impact mission sought to answer seven key scientific questions:

- What are the basic properties of the comet's nucleus (such as density, size, visual features)?
- How has the comet changed over its lifetime?
- What kinds of ice remain unchanged from the comet's early days?
- Is it possible for the heat of the Sun to eventually drive all the ice out of a comet so that it becomes extinct?
- When smaller comets collide, do larger comets form as a result of this collision?
- Are there impact craters on comets, similar to what we find on moons and asteroids?
- Is it possible to alter the trajectory of a comet?

Specifications

The flight system of the Deep Impact spacecraft comprises two parts: the flyby probe and the impactor, both of which were able to receive and transmit data.

The flyby spacecraft, with its high throughput RAD750 CPU, 1553 data bus-based avionics architecture and high stability pointing control system, also carries two instruments which make use of optical imaging and infra-red spectral mapping. These two instruments were used to monitor and observe the impact, crater and subsequent debris. The optical navigation and ground-based navigation assist the flyby spacecraft with manoeuvring itself into as close a position as possible, so as to observe the collision with Tempel 1.

An X-band radio antenna communicated near-real-time images of the impact back to Earth with an approximate transmission of 8 GHz, whilst also listening to the impactor on the S-band frequency. NASA's Deep Space Network (DSN) was used for communication for the majority of the mission; however for the short period of time from encounter to impact, due to the increase in volume of data, overlapping antennas from around the world were also used.

After release of the impactor, the flyby spacecraft not only observed and transferred data itself, however, it also received data from the impactor, which was then transferred to the DSN ground receivers. The two primary instruments, which are used for imaging, infrared spectroscopy and optical navigation and carried by the flyby spacecraft are: the High Resolution Instrument (HRI) and the Medium Resolution Instrument (MRI).

The three axis stabilised flyby spacecraft has a mass of 650 kg and makes use of a fixed solar array and small NiH_2 battery for the power system. The aluminium structure is used within a honeycomb construction, and blankets, surface radiators and heaters passively control the temperature of the spacecraft. A simple blow down hydrazine design that provides 190 m/s of delta 5 provides the propulsion for the flyby probe.

Debris shielding was used in order to avoid damage caused by small particles as the flyby spacecraft passed through the inner coma of the comet. By rotating the spacecraft prior to its journey through the inner coma, the debris shielding provided complete protection and prevention of damage to the control, imaging and communication systems.

Discovery Series

Current Status

As of 2006, ten Discovery missions have been selected, and eight have been launched. The Dawn and Kepler missions will launch in 2007 and 2008 respectively. Please see the individual entries on each recent mission for a more detailed description of the spacecraft and the mission's science objectives.

NASA has also selected two Discovery Missions of Opportunity, mission extensions to already orbiting craft and/or contributions to another nation's mission. These are: the Analyzer of Space Plasma and Energetic Atoms (ASPERA-3), and the Moon Mineralogy Mapper, or M-Cube (M3). The ASPERA-3 instrument is aboard the European Space Agency's (ESA) Mars Express. NASA will fund further study of the interaction of Mars' atmosphere with the solar wind. M3 is an instrument that will fly on Chandrayaan-1, India's first Lunar mission, scheduled to launch in 2008. M3 is designed to map the Moon's mineral composition.

MESSENGER reached Venus on 24 October 2006, and is scheduled to carry out a second flyby on 5 June 2007. Deep Impact also remains in orbit after accomplishing its primary mission. NASA will likely approve two further Deep Impact Missions of Opportunity, extending the scientific usefulness of that spacecraft until the end of 2008.

Discovery Series Missions

Spacecraft:	Launch Date:
Near Earth Asteroid Rendezvous (NEAR)	17 February 1996
Mars Pathfinder	4 December 1996
Lunar Prospector	6 January 1998
Stardust	7 February 1999
Genesis	8 August 2001
Comet Nucleus Tour (CONTOUR)	3 July 2002
Mercury Surface Space Environment Geochemistry and Ranging (MESSENGER)	3 August 2004
Deep Impact	12 January 2005
Dawn	20 June 2007 (scheduled)
Kepler	October 2008 (planned)

Discovery Series Missions of Opportunity

Spacecraft, Mission or Instrument:	Mission Date:
Deep Impact Extrasolar Planet Observations and Characterization (EPOCh)	2007 (Proposed)
Mars Express Analyzer of Space Plasma and Energetic Atoms (ASPERA-3)	2007 (Planned)
Chandrayaan-1 Moon Mineralogy Mapper, M-Cube (M3)	February 2008 (Planned)
Deep Impact eXtended Investigation (DIXI)	December 2008 (Proposed)

Background

Discovery missions are science missions that concentrate efforts and funding on exploring the bodies and phenomena within our solar system. NASA's 1991 strategic plan included the Discovery programme of near-Earth missions for the first time. Discovery missions are rapid turnaround missions, each costing less than a total of USD425 million. This price cap must include everything: spacecraft design, development and assembly; launch costs; mission operations; and data processing. NASA expects to receive for its investment, a focused science return from each Discovery mission; mature instruments and spacecraft technology should provide simplicity and reliability. Only flight proven launchers, no larger than Delta II, are employed. Each proposal requires a consortium that could include a NASA centre or other federally funded research centre and industry and university partners. Each mission must take no longer than 36 months to plan, from project beginning to launch. NASA aims to launch a Discovery mission every 12 to 24 months.

The first two Discovery missions were NEAR in February 1996, and Mars Pathfinder in December 1996. These projects were first approved and funded by Congress in 1994.

Explorer Series

Current Status

At the end of May 2007, 16 NASA Explorer Series spacecraft were in flight and operational or partially operational. At least two more missions are planned to launch through 2009. NASA is also funding four Missions of Opportunity through the Explorer programme, with at least three more (including the second TWINS instrument) planned to launch. For a more in-depth description of each mission and its science objectives, please see each separate entry.

Current Explorer Series Missions

Mission:	Mission Number and Class:	Launch Date:
Interplanetary Monitoring Platform 8 (IMP 8/IMP J)	Explorer 50; Explorer	26 October 1973
Solar Anomalous and Magnetospheric Particle Explorer (SAMPEX)	Explorer 68; SMEX	3 July 1992
Rossi X-ray Timing Explorer (RXTE)	Explorer 69; Explorer	30 December 1995
Fast Auroral SnapshoT (FAST)	Explorer 70; SMEX	21 August 1996
Advanced Composition Explorer (ACE)	Explorer 71; Explorer	25 August 1997
Transition Region and Coronal Explorer (TRACE)	Explorer 73; SMEX	2 April 1998
Submillimeter Wave Astronomy Satellite (SWAS)	Explorer 74; SMEX	5 December 1998
Far Ultraviolet Spectroscopic Explorer (FUSE)	Explorer 77; MIDEX	24 June 1999
Imager for Magnetopause-to-Aura Global Exploration (IMAGE)	Explorer 78; MIDEX	25 March 2000
High Energy Transient Explorer 2 (HETE-2)	Explorer 79; MO	9 October 2000
Wilkinson Microwave Anisotropy Probe (WMAP/MAP)	Explorer 80; MIDEX	30 June 2001
Reuven Ramaty High Energy Solar Spectroscopic Imager (RHESSI/HESSI).	Explorer 81; SMEX	5 February 2002
INTErnational Gamma-Ray Astrophysics Laboratory (Integral)	MO	17 October 2002
Cosmic Hot Interstellar Plasma Spectrometer (CHIPS)	Explorer 82; UNEX	12 January 2003
Galaxy Evolution Explorer (GALEX)	Explorer 83; SMEX	28 April 2003
Swift	Explorer 84 MIDEX	20 November 2004
Suzaku (Astro-E2)	MO	10 July 2005
Two Wide-angle Imaging Neutral-atom Spectrometers (TWINS)	MO	TWINS-A 28 June 2006 (aboard NROL-22) and TWINS-B 2007 (Planned; both launches delayed by several years)
Time History of Events and Macroscale Interactions during Substorms (THEMIS)	Explorer 85-89; MIDEX	17 February 2007
Aeronomy of Ice in the Mesosphere (AIM)	Explorer 90; SMEX	25 April 2007
Coupled Ion-Neutral Dynamics Investigation (CINDI)	MO	June 2008 to launch aboard DoD C/NOFS (Planned; delayed for years)
Interstellar Boundary Explorer (IBEX)	Explorer —; SMEX	June 2008 (Planned)
Wide Field Infrared Explorer (WISE)	Explorer —; MIDEX	2009 (Planned)
Spectrum-X-Gamma (SXG)	MO	2011 (Planned)

Ground Control lost communications with IMAGE in December 2005, however the team is hopeful that it will be able to restore communications in October 2007, during the eclipse season. Although chances of revival are not great, if communications and operations are restored, the mission would be able to continue as it did before communications loss. SWAS has largely been in hibernation since July 2004.

Background

The venerable Explorer Series is the oldest US satellite series, and will continue to fly science payloads for the foreseeable future. As of the end of May 2007, 90 Explorer missions had been launched since Explorer 1 became the first US satellite orbited in February 1958.

There are five Explorer spacecraft mission classes: Explorer Class; Medium-class Explorers (MIDEX); Small Explorers (SMEX); University-class Explorers (UNEX), or Student Explorer Demonstration Initiative (STEDI) craft; and Explorer Missions of Opportunity (MO). MIDEX spacecraft are smaller and cheaper than the more numerous Explorer line. The SMEX programme is providing a return to lower cost, rapid turnaround LEO missions emphasising space physics and astrophysics. Explorer MOs provide a means for NASA to fund instruments or mission segments for craft that were not originally launched within any NASA-funded programme, including international projects run by the European and Russian space agencies, such as Suzaku and Integral. Each Explorer class has a project cost ceiling; MIDEX craft must cost less than USD180 million, SMEX less than USD120 million, and MOs less than USD35 million.

NASA also plans the UNEX series, giving the Principal Investigator (PI) authority and management responsibility for less complex missions than SMEX. It is managed under a fixed cost ceiling of USD15 million, with NASA retaining oversight authority sufficient only to discharge its responsibility for handling public money.

The Explorer programme is headed by NASA's Goddard Space Flight Center (GSFC) Explorer Program Office, within the NASA Office of Space Science.

SWAS has investigated the chemistry of interstellar clouds (Ball Aerospace) 0517128

Far Ultraviolet Spectroscopic Explorer (FUSE)

Current Status

The FUSE mission was terminated on 18 October 2007.

Background

NASA initially studied FUSE as a 1 m telescope for very high-resolution spectroscopy over 900–1,200 Å, expanding the preliminary investigations made by OAO 3 and complementing HST and various X-ray missions. FUSE was planned for launch by Delta in 2000 in a total USD254 million programme, but in 1994, Congress reduced its total budget. NASA reduced the budget again in 1998. As with the Hubble Space Telescope, NASA decided to save money by lowering the apogee of the orbit at the expense of the scientific mission. With FUSE, NASA lowered the original 71,000 km apogee to LEO, reducing the number of observations but not affecting the basic science.

FUSE was one of the Explorer Series satellites (see separate entry). Instead of NASA-Goddard managing the programme and building the spacecraft for the Principal Investigator's (PI) payload, PI Warren Moos of Johns Hopkins University's (JHU) Center for Astrophysical Sciences had full responsibility for the mission, and selected Orbital Sciences Corporation in August 1995 for the USD37 million bus contract. The mission's other partners were The University of California, Berkeley; the University of Colorado, Boulder, and France's Centre National d'Études Spatiales (CNES).

One FUSE goal was to study trace species in interstellar and intergalactic gas using absorption spectroscopy of distant faint sources. Scientists believe the primordial deuterium/hydrogen ratio is a critical parameter in understanding the first seconds of the Big Bang. Most of the spacecraft was taken up by the single instrument, the FUSE spectrograph mounted on top of the spacecraft, which contained four co-aligned telescopes fitted with curved mirrors. The telescopes gathered spectra on an individual focal plane assembly carrying four spectrograph gratings: 1.5 × 20 arcsec (highest resolution), 4 × 20 (extended objects), 30 × 30, and 0.2 diameter pinhole (bright objects that would overwhelm the detector). Each telescope illuminated a holographic diffraction grating and the output from all four was detected by two 10 × 179 mm microchannel plates.

Artist's impression of the FUSE spacecraft (NASA/Johns Hopkins University Applied Physics Laboratory) 1330125

The FUSE spacecraft during assembly and testing (Orbital Sciences Corp.) 1330126

FUSE used four reaction wheels – one in each of the three body axes plus one backup – to manoeuvre and maintain attitude. Two of the wheels, along the x and y axes, showed friction anomalies early in the mission, causing occasional erratic behavior and resulting in several short duration (less than one day) autonomous shut downs of science operations. On 25 November 2001, the x-axis reaction wheel stopped. Science operations continued using the three remaining operable wheels. Then, on 10 December, the y-axis wheel also stopped, causing a cessation of science operations.

Teams from the Johns Hopkins University, NASA's Goddard Space Flight Center, Orbital Sciences Corporation and Honeywell Technology Solutions, Inc. tried to work around the problem. They developed new procedures and software that made it possible to use the Earth's magnetic field to help point the satellite. Controllers were able to generate local magnetic fields by running electric current through FUSE's three magnetic torquer bars. The polarity of these fields could be altered by changing the direction of current flow. As a result, full operations were restored in March 2002.

There was also concern over the spacecraft's gyros. One gyroscope failed in May 2001, and the five remaining gyroscopes were all showing signs of wear. FUSE had two packages of three ring-laser gyroscopes. One operating gyroscope on each of the three axes was needed to conduct normal science operations. In mid-April 2003, new software was uploaded to all three spacecraft computers: the Attitude Control System, the Instrument Data System, and the processor on the Fine Error Sensor guide camera, provided by the Canadian Space Agency. This would enable it to continue operations, if necessary, without any gyros.

FUSE's third reaction wheel was lost in December 2004. After the control system was modified again to use magnetic control on two axes, observations resumed on 1 November 2005, with full operations by February 2006. On 8 May 2007, the last reaction wheel malfunctioned, probably due to excess friction. It was briefly restarted 12 June to 12 July, but then failed completely.

FUSE scientific data is available in a data archive at the Space Telescope Science Institute. Its successes included:

- Measuring abundances of molecular hydrogen, showing that a large amount of water has escaped from Mars, possibly enough to form a global ocean 30 m deep.
- Observing a debris disk that is surprisingly rich in carbon gas orbiting the young star Beta Pictoris.
- Discovering far more deuterium in the Milky Way galaxy than astronomers had expected.
- Detecting an atmosphere of very hot gas around the Milky Way.
- Detecting highly ionised oxygen atoms, showing that about 10 per cent of matter in the local universe consists of million-degree gas between the galaxies.

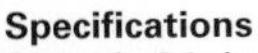

Specifications

Launch: 24 June 1999 by Delta II 7320 from Cape Canaveral

Lifetime: Three year nominal mission. End of mission 18 October 2007.

Mass: 1,360 kg (bus 580 kg, payload 780 kg, no propellant)

Body dimensions: 0.9 × 0.9 × 1.3 m

Orbit: 760 km; 25°; 100 minute period

Spacecraft and Payload: 3-axis 0.9 × 0.9 × 1.3 m box-shaped bus with simple mechanical/thermal links to payload. 0.5 arcsec pointing (1σ) with FES (2° without) using four MWs (16.6 Nms at 5,100 rpm), 3 × 100 Am^2 torque rods, Sun sensors and magnetometers. No propulsion. 1 Mbit/s on 5 W S-band downlink (2 kbit/s command link) from 240 Mbyte memory. Twin GaAs 3.5 m^2 solar wings provide 500 W EOL (payload requires 340 W); 40 Ah Ni/Cd battery.

Fast Auroral Snapshot Explorer (FAST)

Current Status
More than ten years after its launch, and nine years after its primary mission ended, the FAST spacecraft continues its extended mission in good health.

Background
FAST is the second in NASA's Small Explorer (SMEX) series (see separate entry) developed by NASA-Goddard. The Fast Auroral SnapshoT Explorer (FAST) is investigating plasma phenomena in auroral processes discovered in previous satellite and sounding rocket missions. Its orbit takes it through the source region for much of the energy that appears as auroral light. A transportable 8 m antenna is positioned in Alaska to collect real-time data as FAST passes through the northern aurora. These observations are being complemented by data from higher altitude satellites, placing FAST's data in a global context.

FAST had difficulty making it into space because August 1994 and 1995's six-week launch windows were both missed due to 1994–1995 Pegasus failures. FAST was placed in storage waiting its mid-1996 launch in order to make it synchronous with peak northern auroral activity.

Specifications
Contractors: GSFC provided the spacecraft for the UCB University of California at Berkeley's payload
Launch: 21 August 1996 by Pegasus XL/Lockheed L-1011 from VAFB
Lifetime: 1 year minimum
Mass: 191 kg total; 65 kg science payload
Orbit: highly elliptical, 350 × 4,175 km; 83°; 133 minute period
Spacecraft: bus irregular octagonal prism. 50 W orbit average provided by GaAs body panels (5.6 m^2 total) + one 9 Ah Ni/Cd battery (science instruments require 15 W orbit average, 31 W operating, so frequently turned off). 1 Gbit solid-state memory, 0.90/1.5/2.25 Mbit/s rate via NASA standard S-band 5 W transponder. 2 kbit/s commands up, 4 kbit/s housekeeping down. 12 rpm spin stabilisation (normal to orbit) by two magnetorquers, spin Sun sensor, horizon crossing indicator, magnetometer. Attitude post-knowledge 1°. 86 cm diameter × 1.5 m length launch envelope. Four 30 m radial wire booms, 2 axial booms, 2 magnetometer booms
Payload: EESA. Quadrispherical Electrostatic Electron Analyser (UCB, 11.2 kg, on AMPTE, Giotto, Mars Observer, Wind, Cluster).
TEAMS Time-of-Flight Energy Angle Mass Spectrograph (University of New Hampshire + Lockheed Palo Alto Research Lab, 6.4 kg, on Cluster).
EFPE Electric Field Plasma Experiment (UCB, 13.6 kg, on ISEE 1, CRRES, Polar, Cluster).
Magnetometer (UCLA, 3.7 kg, on Pioneer, ISEE 1/2, Polar, Cluster).

FAST is investigating how electrical and magnetic fields accelerate charged particles in the auroral regions. It is shown during spin balancing at Goddard (NASA) 0517113

Fast On-orbit Recording of Transient Events (FORTE)

Current Status
Los Alamos National Laboratory's (LANL) Fast On-orbit Recording of Transient Events (FORTE), or USAF Space Test Program (STP) P94-1, satellite has detected many thousands more radio bursts from lightning strikes and other phenomena than previously reported. Los Alamos scientists are using another major FORTE instrument, an event classifier, to distinguish the inherent structures of Radio Frequency (RF) signals in the 30 to 300 MHz range, which includes commercial television, FM radio, aircraft navigation and communication bands. Los Alamos claims a library of these sources will assist it in sorting out suspected nuclear detonations from ordinary events, making the satellite essential as a monitoring tool for nuclear test ban treaty compliance purposes. Instruments developed through the FORTE programme will be incorporated into the GPS III satellites. FORTE remains in orbit and operational as of late 2006.

Background
LANL and Sandia National Laboratories (SNL) launched FORTE on a Pegasus XL on 29 August 1997 into an 800 km, 70° circular orbit to test the detection of nuclear detonations by RF pulses. The mission is sponsored by the US Department of Energy (DoE) and jointly executed by the Los Alamos and Sandia National Laboratories. The purpose of the mission is to detect and distinguish between electromagnetic pulses caused by events such as low-technology nuclear detonations and lightning-strikes. This characterisation will be aided by optical sensors in order to correlate the optical signals with the RF signals. The satellite is also providing a proof of concept for an advanced composite bus made of graphite epoxy and cut into shape with a water jet. The advanced structure allows the satellite to carry an additional 23 kg of payload. The satellite is three axis stabilised, and most of the exterior panels are covered with solar cells to provide 160 W of power.

The satellite had an expected lifetime of at least one year, with a three-year goal; FORTE has been functioning, as of 2006, for nine years. SNL supplied an optical monitor to eliminate lightning flashes. The satellite serves as a prototype for the GPS 2F-satellite generation, which the USAF intends to use to look for the pulses from low technology explosions. The 10.7 m antenna, built by Astro Aerospace and comprising two log-periodic orthogonal arrays, is held pointing Earthwards by magnetorquers. Sandia is contributing the data to a database of lightning's global distribution. The optical monitor uses a 10 km^2 resolution wide field of view imager to locate lightning and a 50 kHz sampling rate to record separate flashes and trigger RF recording. Forte tests whether it can sort out lightning automatically.

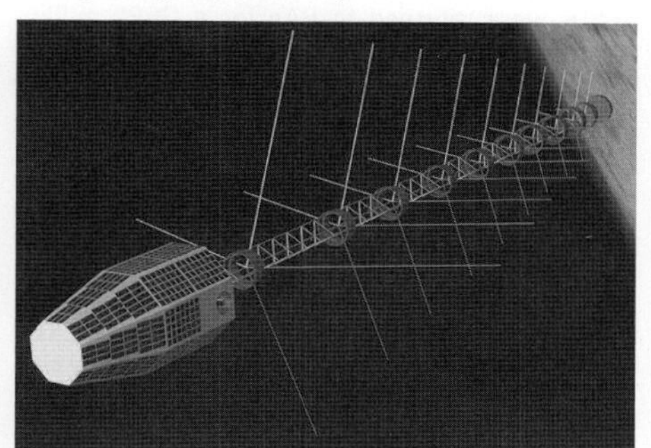

Forte studies RF pulses from nuclear detonations (Los Alamos National Laboratory) 0517153

Gravity Probe B

Current Status
At of the end of 2006, Gravity Probe B continued to operate in good health. The spacecraft points towards the guide star, HR 8703, or IM Pegasi. The spacecraft's mission is complete, however physicists have not yet completed their data analysis. The science team expects to complete

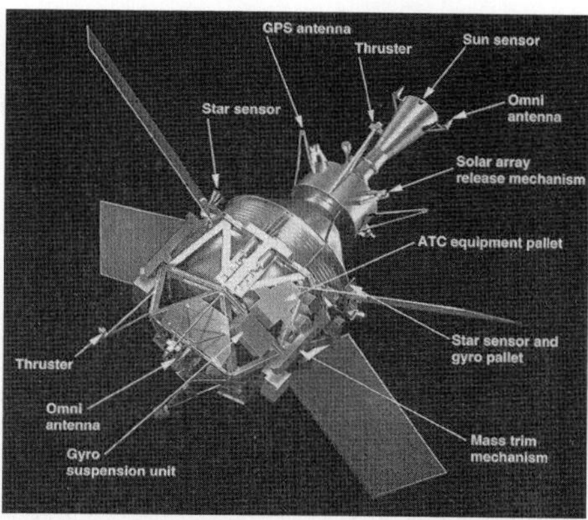

Gravity Probe B tests Einstein's theory of general relativity by measuring the precession of gyroscopes in Earth orbit (NASA-Marshall) 0517106

the last of its three-phase data analysis in February 2007. NASA and Stanford expect to announce the Gravity Probe B experiment results around April 2007. With that announcement will come a possible validation of Einstein's theory of General Relativity.

Sometime in 2007, the United States Air Force Academy (USAFA) will take total control of Gravity Probe B and use it as a spacecraft control training satellite. USAFA is currently using the craft part-time, along with Stanford personnel.

Background
Gravity Probe B was considerably delayed beyond its originally planned launch date in 2001, finally taking flight on 20 April 2004.

Gravity Probe B compares the observed precession of gyroscopes in Earth orbit with that predicted by Einstein's theory of General Relativity. Stanford University is leading the experiment for NASA-Marshall. Stanford put angular inertial measuring instruments on a free-flying 642 km polar satellite to look for the expected 6.6 arcsec/year precession perpendicular to Earth's rotational axis and the 0.042 arcsec/year parallel to it.

In order to distinguish between fine grained effects, the 3,100 kg symmetric spacecraft slowly rolls about the line of sight to the guide star. The spin axis is measured to 0.1 arcsec by superconducting readout of the gyro's magnetic 'London moment' and is compared to the line of sight observations from the reference telescope. Maintaining the proper conditions requires near absolute zero temperatures. All four gyros are cooled to 1.8 K by 2,441 litres of superfluid helium, which should be sufficient to keep them cooled for at least 18 months.

Each gyro is electrically suspended and spins at more than 100 Hz and operates in a 10^{-14}atm vacuum and a less than 10^{-7} gauss magnetic field. To reduce translational contamination, a drag compensation system reduces the mean gyro acceleration to less than 10^{-10} g employing thrusters utilising the Helium boil-off from the dewar. Up to 600 W of power comes from the GaAs wings, spanning 8.8 m and supported by two 35 Ah batteries.

Stanford researchers built a ground-based prototype Science Instrument Assembly (SIA) with two gyros operating at 1.8 K. The second configuration began tests in February 1995, using a flight prototype SIA with four gyros supported by the flight back-up hardware and the engineering development dewar. The third configuration in 1996 tested a protoflight SIA with the flight back-up support hardware and the flight dewar. Ground test of the full flight science payload in 1998 verified the payload for flight. A Shuttle Test Unit was planned to establish the characteristics of the suspension system in μg, but it was cancelled in early 1994 in a major programme reorganisation.

Gravity Probe-A started the series in 1976, when a Scout suborbital package launched it to 10,000 km altitude, on a 116 minute mission over the Atlantic on 18 June 1976, carrying a hydrogen maser clock for telemetric comparison with a duplicate terrestrial version to test the equivalence principle of general relativity.

Hubble Space Telescope

Current Status
On 31 October 2006, NASA announced the approval of the fifth and final HST servicing mission. In June 2007, the launch date was scheduled for

Hubble's first servicing mission (NASA) 0517143

10 September 2008. During the 11-day flight, termed Servicing Mission 4 (SM4), the seven-astronaut crew of Space Shuttle Atlantis will repair and improve the observatory's capabilities.

Although there are no plans for further servicing missions after next year's visit, the crew of STS-125/HST-SM4 will attach a next-generation docking device to the telescope that could be used by automated or piloted vehicles after the Shuttles are permanently grounded in 2010. The crew will attach the passive half of the Low Impact Docking System (LIDS) that NASA intends to use for the Orion crew exploration vehicle and other human spacecraft in the post-Shuttle era. Plans call for a future robotic vehicle to use the LIDS ring and a retro-reflector homing target that will be installed with it to attach a deorbit motor to the telescope for a controlled re-entry over the Pacific Ocean. The telescope's natural orbital decay will bring it back into the atmosphere between 2022 and 2028, provided the final servicing mission is a success.

On 27 January 2007, Hubble's Advanced Camera for Surveys (ACS) ceased to function and the telescope switched to safe mode; HST was recovered from safe mode less than two days later, however engineers were unable to restore ACS operations. Much of HST's astronomy observation work was performed through the ACS. Subsequent investigation indicated that HST's side B power connections had failed. The ACS UV Solar Blind Channel (SBC) was restored to working order on 20 February. However, the SBC represents only a fraction of ACS capabilities. The ACS Wide Field Channel (WFC) and High Resolution Channel (HRC) remained inoperable. ACS was installed during the 2002 Shuttle-Hubble servicing mission 3B, and had reached its 5-year design life limit in 2007. Hubble's other instruments continue to function in good health.

First launched in 1990, the Hubble Space Telescope will remain operational through at least 2013. SM4 will extend Hubble's life to 2013, at which time the James Webb Telescope, due to launch no earlier than that year, will take over observations. The last Hubble servicing mission by a shuttle crew occurred in March 2002.

SM4 astronauts will provide general servicing to extend Hubble's life, such as replacing batteries and all six gyros. The Shuttle crew will also install two new instruments: the Cosmic Origins Spectrograph (COS) and the Wide Field Camera 3 (WFC3). COS will replace COSTAR, which has been obsolete since 2002. WFC3 will replace the second WFC, which has been in operation since 1993. Due to technological advancements over the last decade, the Hubble team expects that WFC3 will provide even more outstanding images over the remainder of Hubble's lifespan than the previous WF cameras collected. It is also planned to restore the Space Telescope Imaging Spectrograph to operational status. In May 2007 it was reported that engineers need "a few more months" to decide if an ACS repair will be possible during the 2008 servicing mission.

On 31 August 2005, the Hubble team shut down one of Hubble's four gyros. One of its four had already ceased to function. The observatory is now, for the first time, operating with only two gyros. It will continue operating in this way until 2008, when the gyros are replaced by the servicing crew.

Background

Major changes to the Shuttle programme, resulting from the loss of Columbia on 1 February 2003, as well as the Bush Administration's Vision for Space Exploration, resulted in a decision announced 21 April 2004 not to fund any further Hubble servicing missions. A major campaign recruiting support from the scientific community and from the general public for a final servicing mission was partly successful in forcing re-evaluation of this decision. During a Congressional hearing in April 2004, NASA Administrator Sean O'Keefe confirmed that the agency was re-evaluating options in the wake of its decision not to fly the Shuttle to a location from where a safe haven could not be secured. He testified that NASA was examining the possibility of using either NASA's Robonaut, developed at the Johnson Space Centre, or the University of Maryland's Ranger robot. Robonaut is a human-like android designed by the Robot Systems Technology Branch at JSC and is a collaborative effort with the Defense Advanced Research Agency. It was developed as a demonstrator system that can perform similar duties to a space-walking astronaut. Ranger has a series of dextrous manipulators and has already been tested against Hubble servicing tasks. It was even mooted that one or both of these systems might be able to replace science instruments in a payload performance upgrade. An October 2006 announcement by Administrator Michael Griffin nullified the need for robotic assistance to upgrade Hubble.

History

Named the Edwin P Hubble Space Telescope, after the eminent American astronomer in October 1983, this much-postponed project was described by NASA as the most important scientific instrument ever flown. NASA expected the Hubble to resolve objects 50 times fainter and seven times more distant than those visible from Earth-based telescopes. In the most distant views, astronomers would be observing the universe as it was around 14 billion years ago.

ESA agreed to make a 15 per cent contribution in October 1976. In 1976 terms, the full ESA cost was projected to be USD595 million, but delays and technical problems caused a rise to more than USD1 billion. NASA promised a return of 15 per cent of observation time for European astronomers. In fact, observing allocation continues to be 20 per cent in open competition.

NASA awarded the USD69.4 million contract for the mirrors and fine guidance sensors to Perkin-Elmer in 1977 following a competition with Kodak Federal System Division of Rochester, New York. Corning Glass, New York cast the 900 kg primary mirror in October 1977. Perkin-Elmer began nine months of fine polishing in August 1980, and aluminised the completed mirror in December 1981. Kodak produced a back-up mirror, although never fully polished it. The Kodak mirror remains in storage.

The main elements came together during 1985 for the first verification programme to begin, although NASA never asked for an end-to-end optical test. BAe (now BAE Systems) supplied its two solar array wings in 1985 under a 1979 GBP13 million contract for the 1986 launch. Following the Challenger accident in January 1986 however, the arrays were returned to BAe for a GBP2 million modification programme. BAe added features to facilitate EVA exchange with new arrays, and the new arrays arrived at Lockheed in April 1989 for installation.

Following the 1986 Shuttle accident, NASA delayed Hubble's launch from October 1986 until a possible window in 1988 or 1989. Many astronomers doubted whether the Hubble would have been ready for the 1986 launch anyway. One principal stumbling block was the data transmission of collected images from Hubble to the ground. The Hubble, like its KH-9 counterparts had to use relay satellites or foreign ground stations to receive information. KH-8 satellites used ground stations at remote locations and relays in common orbits with the spy satellites. The TDRS constellation was designed to replace these expensive links. Unfortunately, the air force decided, after the TDRS programme started, that the links needed to be hardened against radio frequency and laser attack from the Soviet Union. The air force argued that the Soviet Union could blind all US intelligence gathering capability with such attacks. In order to make TDRS attack-hardened, the transponders had to be retrofitted at a large cost over-run to the programme and about a two year delay in deployment of the full network. This similarly delayed the relays for the Hubble and without the national security argument behind it, it was unclear how Hubble could use a ground-based data relay network for its image production, let alone do it in a timely fashion. Hence, the Telescope would have remained under-used in orbit while the ground system was brought to full operational capability.

As NASA grappled with this problem, the guide star catalogue for steering the Shuttle had to be finished as well. ST ScI completed the initial Guide Star Catalogue in 1989 after four years of work. When finished, it constituted the largest-ever sky survey. A total of 1,477, 6.4 × 6.4° plates exposed by the US Palomar Schmidt and UK Australian Schmidt telescopes had to be digitised to generate position and magnitude data on 18,819,291 objects for Hubble's three guidance sensors. Some 15 million objects in the catalogue were stars, most of the remainder were galaxies.

Hubble arrived at Kennedy Space Center on 11 October 1989 aboard a USAF C-5 aircraft for its eventual Shuttle launch. Ground technicians powered it up for the first time on 28 October by remote command from Sunnyvale. Then Hubble underwent an 11 day GST-8 ground systems test that concluded on 8 December 1989, including simulation of early orbit operations and safemode recovery.

After its 24 April 1990 launch on the Shuttle Discovery, Hubble entered orbit at 20.39 GMT 25 April 1990 619 km over the east Pacific. NASA planned that Orbital Verification (OV) would take three months, followed by five months' Science Verification (SV) in which the instruments would be checked out and calibrated, while allowing some of the Guaranteed Time Observer programmes. Instead the OV took over 19 months.

Once in its lower and less energetic non-sun synchronous orbit, Hubble passed out of the Earth's shadow. Differential heating caused the telescope to ring like a bell for nearly half of the orbit. This ringing prevented precise tracking of the stars that is required for gathering a high resolution image. Lockheed attempted several software 'patches' to correct the problem, but all of them failed to fully alleviate the problem. Later the air force admitted it had known about the problem in advance of the launch, because of similar problems experienced on its KH-9 spy satellites. The air force had refused to alert NASA to the problem, however, for national security reasons. This decision cost the Hubble more than half of its available data gathering time.

The air force nearly cost the United States the use of Hubble at another point as well. After release from the Shuttle, Hubble entered its deployment phase in which the solar panels are unfurled. Many problems had developed as the panels unfurled, but the crew, which could have assisted, had moved on to deliver a classified payload to a different orbit.

Cycle 1 of the full observational schedule was to begin towards the end of 1990. The first image, of Carina's NGC 3532 open star cluster, chosen not for scientific reasons, but so that NASA could release a publicity photo, was returned by WF/PC on 20 May 1990. Resolution was 0.7 arcsec, as against the expected 1.5 arcsec, but an improvement by a factor of seven was still expected. This first light observation showed light, which appeared to ground-based telescopes as a single star, was, in fact, composed of two stars. Astronomers had long suspected the fact, but let NASA use the information anyway, for a press release.

After first light was achieved on 6 June, it soon became apparent that a large flaw existed in the optics. The first images showed that Hubble suffered from classical spherical aberration, caused by a 2 μm grinding error around the primary's edge. The mirror's 24 actuators could not compensate and would complicate the 'pure' aberration (the final compromise focus was not decided on until November 1991). Specifications required 70 per cent of the incoming light to fall within a 0.2 arcsec area, and measurements showed 1.4 arcsec. Some 15 per cent covered 0.1 arcsec, with the remainder spread across 1 arcsec (equivalent to ground-based performance). Computer processing for objects seen against a dark background could achieve the target 0.1 arcsec resolution. Low sky background at shorter wavelengths meant that ultra-violet observations were largely unaffected – the visible work suffered most. WF/PC was the most serious loss: it was to have taken 40 per cent of the observing time, but occupied only about 10 per cent.

As the data gathering entered its full programme, astronomers became disenchanted with the programme as the consequences of working with NASA became apparent. At first, NASA promised to allow scientists unfettered access to the Telescope through the TDRS data linking network. In fact, NASA had an agreement with the US Air Force to allow the military to have priority linkage through the data relays for nuclear attack planning.

Hubble deployment by STS-82. Photographed by an aft-mounted wide-angle IMAX camera (NASA) 0006567

NASA also promised the Hubble could be repaired on orbit at least every 30 months. The solar arrays were to be retracted, first every five years, then every three years, and the whole vehicle brought to Earth for refurbishment and relaunch about a year later. By 1987, NASA announced it would avoid Earth return if at all possible and make in-orbit servicing more infrequent.

Hubble was launched with several principal scientific objectives. One key objective was the measurement of the Universe's expansion rate (Hubble constant) by observing the brightness of magnitude 24 and fainter stars (Cepheid variables) in distant galaxies with WF/PC (see Specifications, below). With the flaws in the optics, definitive observations had to wait for WF/PC-II, later. ESA reduced its FOC work by about half after visible observations offered no improvement over ground systems.

The independent six-man Hubble Optical Systems Board of Investigation established 2 July under JPL Director Dr Lew Allen reported in late November 1990 that the primary mirror had been well figured but to the wrong shape. Perkin-Elmer's Reflective Null Corrector laser interferometer, used to check the figure around 1981, apparently had a lens 1.3 mm out of position. Perkin-Elmer had some indication of the problem, but disregarded the test results. Kodak's bid for the 1977 contract included USD10 million for an end-to-end test, which would have immediately identified the problem.

The Board concluded that the defect could not be solved by a global repair in-orbit but individual instruments could be corrected optically. NASA planned anyway, that WF/PC would be replaced by a second-generation version on SM1. FOC, HRS and FOS are corrected by replacing HSP (see Specifications, below) with the 290 kg 220 cm long 87.6 cm^2 Correcting Optics Space Telescope Axial Replacement (COSTAR), carrying a boom to insert 10 mirrors into the light paths (one picks off the light for a second to provide the correction) before they reach the instruments. The coatings call for 56 per cent throughput of 1,216 Å UV. HSP was lost but it was the least-used instrument. Ball Aerospace began work on COSTAR under a February 1991 letter contract, finalised October 1991 at USD30.4 million. ESA considered an on-orbit FOC internal modification and flying a new instrument, although at an estimated USD180 million it proved too costly.

SM1, the first telescope maintenance mission solved a second major problem in December 1993. It corrected the pointing flaws due to the thermal ringing. The ringing produced two modes: a 0.1 Hz end-to-end movement lasting up to 10 minutes after terminator crossing, and a lesser 0.6 Hz transverse action during daylight. The solar array tips flexed up to 1 m under worst conditions. The wobble prevented long exposures. Previous software fixes took up too much of Hubble's antiquated computer memory. SM-1 put in place replacement solar arrays (STSA-2), and an aluminium layer supported by 900 plastic discs in an accordion-like structure covered each modified boom. This reduced thermal gradients by a factor of 20. Then BAe also built a system for countering expansion at the array tips with frictionless springs and immobilised the storage drums by an electrical brake. They were found to reduce jittering by a factor of 10. These arrays were sufficient until replaced in 2002.

SM1 also replaced two gyro pair packages. The number 6 rate gyro failed 3 December 1990, but one of two back-ups assumed its duties. Then number 4 failed in July 1991, possibly from radiation damage of the hybrid circuits generated by high solar activity. Number 5, too, began showing signs of distress, possibly because of dirt in its rotor mechanism. Number 3 failed in November 1992, probably because of a random electronic part failure in the control unit. If they had all failed simultaneously before replacement, Hubble would have entered a magnetic torque mode to hold safe attitude. SM1 achieved all objectives in December 1993 during five EVAs totalling more than 35 hours by four crew members.

The second Hubble servicing mission began with the launch of Shuttle Discovery on 11 February 1997 (STS-82). When the Shuttle had captured Hubble and the astronauts were able to look in detail at the telescope's thermal insulation cover, it was found that it had degraded more than expected during the three and a half years since its first servicing mission. As a result, an extra EVA was scheduled so that some running repairs could be done. Discovery was used to re-boost the altitude of Hubble: before being captured it had been in a 590–599 km orbit and when released it was in a slightly higher 599–620 km orbit.

During SM2, the Space Telescope Imaging Spectrograph (STIS) was installed in Axial Science Instruments Bay 1. The crew also installed the Near Infrared Camera and Multi-Object Spectrometer (NICMOS), which was built by Ball Aerospace. NICMOS was cooled by a dewar containing a 105 kg block of nitrogen ice. The dewar successfully cooled the detectors for about two years, but ran out of coolant prematurely.

Degradation of several key systems and the failure of two crucial gyroscopes, plus intermittent performance from a fourth, stimulated the Hubble team to request an interim servicing mission before a planned, but delayed, visit in late 2000. The third STS servicing mission, dubbed Servicing Mission 3A (SM3A), was assigned to Shuttle Discovery (OV-103). STS-103 launched 19 December 1999 and the mission lasted almost eight days. After rendezvous and capture, the Hubble was placed in the payload bay. During the period of servicing five EVA operations were carried out.

SM3A replaced all six gyroscopes, as well as one of Hubble's three Fine Guidance Sensors and a transmitter. The crew also installed an advanced central computer, a digital data recorder, an electronics enhancement kit, battery improvement kits, and new outer layers of thermal protection. Hubble was released back into orbit on 25 December.

Servicing mission 3B (STS-109/SM3B), took place 1–12 March 2002. Following launch, rendezvous and capture, the crew of Space Shuttle Columbia conducted five scheduled EVAs. The crew removed the Faint Object Camera (FOC) and replaced it with the Advanced Camera for Surveys (ACS), which doubled Hubble's field of view and collected data 10 times faster than the Wide Field and Planetary Camera 2. ACS was the first new instrument to be installed in Hubble since 1997. In addition, the four, flexible solar panels, which had been on board for eight years, were replaced with smaller, rigid ones, built by Lockheed Martin Missiles and Space, that produce 30 per cent more power.

Astronauts also replaced the outdated Power Control Unit, which distributes electricity from the solar arrays and batteries to other parts of the telescope. Replacing the original unit required the telescope to be completely powered down for the first time since its launch in 1990. During the last spacewalk, astronauts installed a new mechanical "cryocooler" for the Near Infrared Camera and Multi-Object Spectrometer (NICMOS), which became inactive in 1999. One of the four Reaction Wheel Assemblies that make up Hubble's Pointing Control System was also replaced.

The Space Telescope Imaging Spectrograph (STIS) failed and went into safe mode on 3 August 2004, after performing on-orbit operations for approximately 65,000 hours (seven years, 171 days). The instrument accounted for about 30 per cent of all Hubble scientific observation programmes. The repair of STIS during servicing mission 4 will restore Hubble's full spectroscopic capability. This will be accomplished by replacing a low voltage power supply board that contains the failed component. The STIS repair will be performed with STIS remaining inside Hubble. The astronauts will use existing EVA tools to gain access to the instrument. A new main electronics box cover and cooling system will fully restore STIS.

Ball Aerospace built the Cosmic Origins Spectrograph (COS) and the Wide Field Camera 3 (WFC3), both of which will be installed by shuttle astronauts during the 2008 servicing mission. Both COS and WFC3 contain advanced technology sensors, which far surpass what has been available on Hubble to date, and improvement factors of 10X-70X are expected in certain key performance areas. COS will be over 30 times more sensitive in the far-ultraviolet than earlier Hubble ultraviolet spectrographs, and will be able to observe distant quasars too faint for detection by previous spectrographs.

WFC3 will be sensitive to wavelengths from the near-ultraviolet to the near-infrared. Using a 4,000 × 4,000 pixel charged couple device detector with a large field of view, WFC3 provides images with less 'background noise' than previous instruments. WFC3 will take over the role of NICMOS for most near–infrared imaging, and will complement ACS by providing Hubble's first wide field, high sensitivity imaging capability at near–ultraviolet wavelengths. The WFC3 provides some backup to ACS for visible light imaging, but is not as sensitive as ACS at the red wavelengths of interest for studies of galaxy evolution.

Total cost to the end of 15 years of operations was projected in 1989 at USD2,800 million, inclusive of USD1,450 million for development and activities up to launch. In 1990, NASA added USD600 million to the estimate. SM1 was priced at USD674 million, including USD378 million for the Shuttle's element. Cost of repairs and correcting for the mirror defects was included at USD86 million.

As launch was originally planned for 1983 and full operations were not possible until 1994, advances in the technology of ground-based astronomy had removed some of Hubble's anticipated unique advantages. It does maintain superiority, however, with its 0.01 arcsec angular resolution (ground telescopes can achieve 0.3 arcsec and then only briefly) and its ability to work with the ultra-violet wavelengths blocked by the atmosphere.

In its first 10 years of operation, the Hubble studied 13,670 objects, made 271,000 individual observations, returned 3.5 Tbytes of data and stimulated 2,651 scientific papers.

Discoveries

Early observations by the FOC showed that the specified 0.1 arcsec performance was still possible in some circumstances. Hubble then gathered images of Supernova 87A's ring, Pluto/Charon and of the R Aquarii red giant/white dwarf double star nova. Scientists then made a major step towards determination of the Hubble constant by accurately measuring the distance to the Large Magellanic Cloud. Hubble/IUE combined to measure the diameter of SN87A's ring at 1.37 light years, translating to an LMC distance of 169,000 light years ±5 per cent (replacing the previous ±30 per cent spread).

As the Telescope settled into its observation schedule, in 1991 it used the HRS to measure the Universe's deuterium content and found the amount supports the Big Bang and ever-expanding cosmos concepts. HRS observed evidence of planetary formation around star Beta Pictoris and also discovered huge hydrogen clouds that, theoretically, should have evolved into galaxies long ago. The WF/PC instrument found evidence of a black hole 2.6 billion times the Sun's mass in a region 1–10 times the size of the solar system at the centre of galaxy M87.

In 1992, the Hubble uncovered strong evidence that discs of planet-forming material surround many stars. WF/PC found 15 in the Orion Nebula; only four were known before Hubble. Work continued on refining the Hubble constant, but it really required improved optics.

By 1994, NASA had corrected Hubble's defective optics and observations with the improved optics provided seemingly conclusive evidence for the massive black hole. WF/PC also began a long-term programme of Mars observations to characterise its climatic processes and surface changes at a resolution of 50 km. 1994 observations showed that Mars' average temperature had dropped 20°C since Viking, but the climate now appears more stable.

In 1994, WF/PC-II observations of 20 Cepheid variables in M100 placed the galaxy at 56 ± 6 million light-years, yielding a Hubble constant equivalent to an age of 8 to 12 billion years for the Universe, instead of the generally accepted 15 billion years. If confirmed, it would require a major reworking of cosmology. However, many more galaxies must be measured because those around the Virgo cluster are mutually perturbed by their concentration. The FOC in 1994 provided the first image of star G1623b, the red dwarf companion in a double star system in Hercules and one of the smallest stars known in our Galaxy. It has only 10 per cent of the Sun's mass and is 60,000 times fainter – it would appear only eight times brighter than the

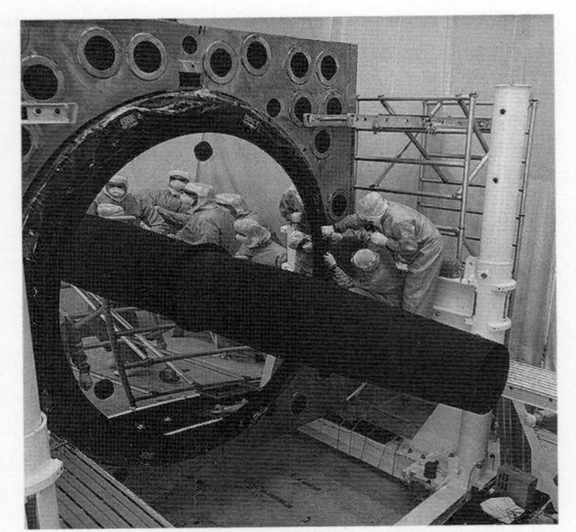

HST's primary 2.4 m {70}, 830 kg mirror with central light baffle installed. The surface is so precise that if it were expanded to cover the entire US its largest bumps would be only a few cm high (Perkin-Elmer) 0517144

full Moon from 1 AU. Its presence was known from astrometric measurements of its parent but ground-based telescopes were unable to resolve it. Red dwarfs were believed to be the most abundant stars but Hubble's observations showed them to be surprisingly rare. They cannot therefore account for much of the Universe's missing mass.

From a study of very young galaxies, Hubble provided the surprising result that elliptical galaxies formed very early on, but spirals took much longer and most were torn apart. FOC observations of a distant quasar revealed the type and quantity of helium predicted by the Big Bang theory. FOC and WF/PC II observations of 525 km asteroid 4 Vesta in November/December 1994 showed the solar system's oldest terrain at 80 km resolution. The gouged surface revealed details below the crust, lava flows and impact basins. The once-molten interior shows that Vesta should be regarded more as a mini-planet, rather than an inert lump of rock.

WF/PC-II observations in 1995 confirmed the existence of the postulated Kuiper Belt of comets 40–500 AU out, beginning just beyond Neptune and the source of comets that orbit the Sun in periods of greater than 200 years. The observations indicate the belt hosts at least 200 million comets that have remained essentially unchanged since the solar system formed. Other 1995 observations strongly suggest that volcanoes are still erupting on Jupiter's moon Io and that Ganymede has a tenuous oxygen atmosphere, created from the bombardment of water ice by Jupiter's charged particles. The first unambiguous brown dwarf was confirmed: Gliese 229B is the faintest object seen orbiting another star. At 20–50 Jupiter masses, it is too large to be classed as a planet but too small and cool (720°C) to shine as a star.

In 1999 it completed its objective to measure the rate at which the universe is expanding. In 2004 Hubble looked into the universe farther than ever before, imaging the earliest formed galaxies using its ACS and NICMOS instruments during the Hubble Ultra Deep Field (HUDF) project. Hubble watched as Deep Impact collided with comet Tempel on 4 July 2005. In 2006, Hubble found evidence that dark matter has existed from the Universe's infancy, encouraging expansion, at least nine billion years ago. The finding supports Einstein's contention that a repelling gravitational force in space exists.

Specifications

Launched: 24 April 1990 by Shuttle STS-31 from Kennedy Space Center, released 20.39 GMT 25 April.

Mass: initially 10,843 kg on-station; 10,960 kg after SM1

Orbit: released at 613 × 620 km, 28.5° (altitude to minimise atmospheric drag), 586 × 596 km by time of SM1 capture, raised to 592 × 601 km for SM1 release

Lifetime: Originally 15 years, with 3 to 5 yearly revisits by Shuttle for maintenance, replacement of instruments and reboost

Currently planned to remain operational until 2013

Configuration: Hubble was originally 13.1 m long (without aperture door), 4.27 to 4.7 m in diameter carrying a 2.40 m diameter primary mirror and optical system held rigidly in an Optical Telescope Assembly (OTA) relaying light to five (four after SM1) aft-mounted focal plane instruments. The Support Systems Module (SSM) encircling the OTA's base and protecting the instruments provides housekeeping functions. 426 kg OTA is required to maintain the 5 m mirror-to-mirror spacing to within 3 μm/0.003 arcsec over 24 h using the 5.08 m long graphite epoxy metering truss. The primary is held in a main support ring inside a large square tooling fixture. Inside the truss is a 3 m diameter aluminium light baffle. Externally, the OTA is covered forward with the SSM light shield and end-aperture door. The twin 33 m² solar arrays and primary communications antenna are mounted on the SSM forward shell. The SSM equipment section encircles the metering truss base, accommodating power, control, communications, RWs, rate gyros and other subsystems in its 10 bays, accessible externally for servicing

AOCS: Hubble avoids thruster exhaust contamination by employing four 3,000 rpm RWs for attitude control (turn rate of 90° in 20 min), with angular momentum offloading by magnetorquers. Periodic reboosting during Shuttle servicing missions combats orbital decay. Six rate gyros (four prime plus two back-up; two pairs replaced by SM1) and three fixed head star trackers in the SSM equipment section provide overall attitude information; three Sun sensors mounted on the aft face prevent pointing within 50° of the Sun (aperture door available for protection). Two magnetometers at the front end sense attitude relative to Earth's magnetic field. Pointing stability requirement is 0.007 arcsec for up to 10 hours, using three interferometric Fine Guidance Sensors on the field of view's periphery. Any two are sufficient for guidance; magnitude +13 was detection requirement, but +15.5 appeared feasible pre-mission. FGS are also operated as a science instrument (see Payload section). The Guidance Star Catalogue, providing position and magnitude data on 18,819,291 objects, was completed in 1989 but continues to be expanded, including proper motion and colour information

Power: Originally, two ESA/British Aerospace solar arrays provided 5.0 kW BOL, 4.5 kW after two years and 4.3 kW after five years at 34 V; 5 per cent degradation measured after one year. Each 150 kg 2.83 × 11.8 m wing carried 24,380 Si cells, rolled into a 20 cm diameter drum for launch. Unfurling was achieved by Bi-stem actuator units at either end unwinding and interleaving two spring steel elements to form 2.2 cm diameter booms either side of the thermal covers. The arrays can be restowed or jettisoned, and all functions can be achieved manually. The original arrays were replaced during SM1 because incorrect design allowed excessive flexing during day/night cycling. The second set of arrays reduced jittering by ×10. Each component was qualified for 30,000 thermal cycles of −100/100°C over a five year life. Payload requirement was 2.4 kW. The six 90 Ah Eagle Picher Ni/H_2 batteries, in three modules, can be replaced on each servicing mission

Solar Array 3 (SA3) installation was performed during SM3B in 2002. Hubble's latest solar arrays are rigid and stationary. SA3 provides better performance at a smaller size than former arrays, yielding 30 per cent more power.

TT&C: 1 Mbit/s data via NASA's TDRS satellite system to White Sands for relay to the ST Operations Control Center at Goddard and to the ST Science Institute at Johns Hopkins. Goddard exercises vehicle control, Space Telescope Science Institute (STScI) manages the observations. Hubble is controlled by a triply redundant 50 kg Rockwell Autonetics DF-224 computer (48,000 word total memory; two of six memories have failed, three required for full operation; a co-processor was added by SM1 to increase memory/speed) and Lockheed's 37.7 kg Data Management Unit. All incoming commands are decoded/relayed by the DMU and it receives the data from the Scientific Instrument Control & Data Handling computer, which relays DMU's commands and formats the science data

Thermal: Heaters/coverings maintain mirror at 21°C. Elements such as batteries requiring greater heat dissipation are positioned on the anti-solar side

Payload: **The Ritchey-Chretien f/24 Cassegrain Optical System** comprises an 829 kg, 2.40 m diameter primary mirror of ultra-low expansion titanium silicate glass and a 30 cm diameter Zerodur secondary 5 m distant. Effective focal length is 57.6 m. The primary is constructed of 2.5 cm-thick face and rear glass plates separated by a honeycomb layer, and provides {1/20} rms at 6,330 Å. Both hyperboloid mirrors are polished to less than 0.05 μm, and coated with a 0.064 μm layer of aluminium and a 0.025 μm layer of protective magnesium fluoride. That outer layer is required for adequate ultra-violet reflectance, although it reduces the aluminium's visible light reflectivity to 85 per cent. Reflection at 1,216 Å is 70 per cent; spectral range is 1,150 Å-1 mm. The secondary directs the light cone through the primary's 60 cm diameter central aperture to a focus 1.5 m behind the face plate for dispersion to the science instruments and FGS.

Wide Field/Planetary Camera (WF/PC) the WF/PC was designed/built by JPL at a cost of USD69 million. It was used for general astronomical far-ultraviolet to near-infrared applications over 1,200–11,000 Å, including astrometric searches for planetary bodies about other stars (which was prevented by the mirror defect), solar system and galactic/extra-galactic observations. WF/PC I was replaced during SM1 in 1993 (at a cost of USD101 million) by the 278 kg WF/PC-II, which included integral optics to correct Hubble's defect, a revised filter set (including far-ultra-violet), updated CCDs reaching into far-ultra-violet, articulation of the pick-off mirror and 3 of the 4 fold mirrors to ensure accurate pointing required by optical correction. A 1991 decision reduced detectors from eight to four (one PC plus three WF) on cost grounds.

Advanced Camera for Surveys (ACS) is Hubble's latest camera, installed in 2002 during SM3B. ACS doubled Hubble's field of view and was able to collect data 10 times faster than the Wide Field and Planetary Camera 2. ACS was optimised for wide-field imagery in the visible, and although it can detect UV light the field of view is small. It can also make limited observations in the near infrared. It comprises three separate instruments: the high resolution camera, the solar blind camera, and the wide field camera. These provide imaging coverage from 120–1,000 nanometers. The **Wide Field Camera (WFC)** is a high throughput, wide field, optical and near-infrared (I band) camera. This channel will be optimised for surveys in the near-infrared to search for galaxies and clusters of galaxies in the early universe. Its characteristics are: 350–1,050 nanometer spectral response; 202″ × 202″ field of view; 0.049″ pixel size; 2 butted 2,048 × 4,096, 15 μm/pixel CCD detectors; 45 per cent throughput at 700 nanometers, (including the HST optical telescope assembly) half critically sampled at 500 nanometers; three mirror optical design (overcoated silver). The **High Resolution Camera (HRC)** is designed for high angular resolution imaging and coronography. The HRC optical path includes a coronagraph which improves the HST contrast near bright objects by about a factor of 10. This channel is used for diffraction-limited studies of the light in the centres of galaxies with massive black holes, as well as ordinary galaxies, star clusters, and gaseous nebulae. Its characteristics are: 200–1,050 nanometer spectral response; 29.1″ × 26.1″ field of view; 0.028″ × 0.025″ pixel size; 1024 × 1024, 21 μm/pixel, near UV-enhanced CCD detector; 25 per cent throughput at 600 nanometers; critically sampled at 500 nanometers; three mirror optical design (MgF_2 on aluminium). The **Solar Blind Camera** is a far-ultraviolet camera that has a relatively high throughput and is intended primarily for faint object and extended object imaging. This channel will be used to find hot stars, quasars, and to study aurora on Jupiter. Its characteristics are: 115–180 nanometer spectral response; 34.59″ × 30.8″ field of view; 0.033″ × 0.030″ pixel size; 1,024 × 1,024, CsI 25 μm/pixel MAMA detector (STIS spare); 6 per cent throughput at 121.6 nanometers; two mirror optical design (MgF_2 on aluminium). In January 2007, ACS experienced an electrical short that put the WFC and HRC out of commission.

Near Infrared Camera and Multi-Object Spectrometer (NICMOS) NICMOS is an infrared instrument that is capable of identifying objects obscured by interstellar gas and dust. It was installed in 1997 during SM2. It provides imaging capabilities in broad-, medium- and narrow-band filters, broad-band imaging polarimetry, coronographic imaging, and slitless grism spectroscopy, in the wavelength range 0.8–2.5 microns. NICMOS has three adjacent but not contiguous cameras, designed to operate independently, each with a dedicated array at a different magnification scale. NICMOS employs three low-noise, high QE, 256 × 256 pixel HgCdTe arrays. In 1999, NICMOS encountered cooling problems, crippling the device soon afterwards. The experimental NICMOS Cryogenic Cooler and Cooling System Radiator (NCS) were installed in 2002. Active cooling by NCS keeps the detectors at a temperature between 75 K and 86 K.

Space Telescope Imaging Spectrograph (STIS) STIS replaced the Goddard High Resolution Spectrograph. Developed jointly by NASA-Goddard and Ball Aerospace, it was installed during SM2 in 1997. It views objects from far-ultraviolet (FUV) to near-infrared wavelengths. It is similar to a prism, in that it refracts light into component wavelengths. STIS gives Hubble its

2-D spectroscopy capabilities. It has three large-format (1,024 × 1,024 pixel) detectors: **CCD**: Scientific Image Technologies (SITe) CCD with ~0.05 arcsecond square pixels, covering a nominal 52 × 52 arcsecond square field of view, operating from ~2,000 to 10,300 Å. **NUV-MAMA**: Cs2Te Multi-Anode Microchannel Array (MAMA) detector with ~0.024 arcsecond square pixels, and a nominal 25 × 25 arcsecond square field of view, operating in the near ultraviolet from 1,600 to 3,100 Å. **FUV-MAMA**: Solar-blind CsI MAMA with ~0.024 arcsec-pixels, and a nominal 25 × 25 arcsecond square FOV, operating in the far ultraviolet from 1,150 to 1,700 Å.STIS stopped functioning in 2004 due to a power supply failure, and is currently in 'safe mode'. It is scheduled for repair during SM4.

Wide Field Camera 3 (WFC3). WFC3 is being constructed mostly at Goddard Space Flight Center in Maryland and Ball Aerospace in Colorado. Some parts are also being provided by other contractors in the US and the UK. WFC3 will study a diverse range of objects and phenomena, from young and extremely distant galaxies, to much more nearby stellar systems, to objects in our own solar system. Its key feature is its ability to span the electromagnetic spectrum from the ultraviolet to the near infrared. WFC3 is configured as a two-channel instrument using two detector technologies. The incoming light beam from the Hubble telescope is directed into WFC3 using a pick-off mirror, and then directed to either the Ultraviolet-Visible (UVIS) channel or the Near-Infrared (NIR) channel. The light-sensing detectors in both channels are solid-state devices. For the UVIS channel a large format Charge Coupled Device (CCD) is used. In the NIR detector the crystalline photosensitive surface is composed of mercury, cadmium and tellurium (HgCdTe). The high sensitivity to light of the 16 megapixel UVIS CCD, combined with a wide field of view (160 × 160 arcseconds), yields about a 35 times improvement in discovery power compared with the ACS High Resolution Channel. The NIR channel's HgCdTe detector is a highly advanced and larger (one megapixel) version of the 65,000 pixel detectors in NICMOS. The combination of field-of-view, sensitivity, and low detector noise results in a 15-20x enhancement in capability for WFC3 over NICMOS. An important design innovation for the WFC3 NIR channel results from tailoring its detector to reject infrared light longer in wavelength than 1,700 nm. The detector is chilled with an electrical device called a Thermo-Electric Cooler (TEC). This greatly simplifies the design and should give WFC3 a longer operational life.

Cosmic Origins Spectrograph (COS). COS is the most sensitive ultraviolet spectrograph ever flown on Hubble. It has two channels, the Far Ultraviolet (FUV) channel covering wavelengths from 115 to 177 nm, and the Near Ultraviolet (NUV) channel, covering 175–300 nm. The FUV channel is particularly sensitive because it contains only one selectable optical element that performs triple duty: correcting for the spherical aberration of Hubble's primary mirror, dispersing the light into its component wavelengths, and focusing the light beam onto a light-sensing detector. While the NUV channel must employ four optical elements to do its job, it still provides unprecedented sensitivity in its wavelength range. The light-sensing detectors of both channels are designed around microchannel plates. These are thin plates comprising thousands of tiny curved glass tubes, all aligned in the same direction. Incoming photons of light induce showers of electrons to be emitted from the walls of these tubes. The electron showers are accelerated, captured, and counted in electronic circuitry immediately behind the microchannel plates. The FUV channel is supported by a cross-delay line detector supplied by the University of California at Berkeley. The detector has dual curved microchannel plates with a total active area of 170 mm by 10 mm. This large active area is essential to the COS design concept as it allows the spectrograph to cover a wide band of wavelengths at one time. The NUV channel uses a spare microchannel-plate-based detector of a different configuration, which was built and spaceflight-qualified originally for the STIS instrument. The prime contractor for the design and manufacture of COS was Ball Aerospace & Technologies Corporation.

Faint Object Camera (FOC) provided by ESA (Dornier prime), it used Hubble's full resolution capabilities for faint objects/structures (magnitude 24–29) using cumulative exposures. FOC was replaced by ACS in 2002.

Faint Object Spectrograph (FOS) FOS was built by Martin Marietta for University of California. It was replaced by STIS and NICMOS in 1997.

Goddard High Resolution Spectrograph (GHRS) GHRS was built by Ball Aerospace for NASA Goddard. It was similar to FOS but optimised for ultra-violet (1,100–3,200 Å). It was replaced by STIS and NICMOS in 1997.

High-Speed Photometer (HSP) the HSP was designed and built by the University of Wisconsin. Four photon-counting image dissector tubes, a photomultiplier and >50 focal plane/aperture combinations could detect temporal variations down to 16 μs. It was sacrificed to install the COSTAR optics correction module in 1993.

Correcting Optics Space Telescope Axial Replacement (COSTAR) COSTAR originally corrected Hubble's vision, which was blurred by the spherical aberration discovered soon after operations began in 1990. COSTAR was rendered obsolete during SM3B, in 2002, due to the installation of ACS, which is equipped with built-in optics correcting equipment, however the instrument has not yet been removed from Hubble. It is scheduled for disassembly during SM4.

Fine Guidance Sensors (FGS) the FGS are primarily used for Hubble's pointing system. Only two are required for a 0.007 arcsec pointing goal, but a third was added for astrometric observations to 0.002 arcsec. Astrometry could not begin until the end of 1991 because of Hubble's problems.

International Cometary Explorer (ICE)/ International Sun-Earth Explorer 3 (ISEE 3)

Current Status

ICE is expected to remain active until at least 2014, when it returns to the Earth-Moon vicinity.

Background

The International Sun-Earth Explorers, part of the International Magnetosphere Study, followed up the IMP Explorer series. NASA built ISEE 1 and 3 and ESA provided 2. ISEE 1 and 2 were launched jointly on 22 October 1977. ISEE 1 and 2 decayed 26 September 1987. ISEE 3 followed on 12 August 1978 by Delta from Cape Canaveral. ISEE 3 had the distinction of being the first spacecraft to attain a halo orbit around the L1 libration point, 1.5 million km out on the sunward side of Earth.

ISEE 3, also known as Explorer 59, had two 3 m booms for the magnetometer and plasma wave sensors, and four 49 m wire antennas for radio and plasma wave studies. It provided 1 hour warning of solar radiation before events reached ISEE 1 and 2. In late 1982, NASA diverted ISEE 3, due to remain as a solar flare sentinel for 10 years, towards Earth so that it could be directed into a series of lunar swing-bys (the last, in December 1983, within 100 km) for the first ever comet encounter. Its speed was increased from 4,679 km/h to 8,278 km/h by sweeping 120 km behind the Moon on 22 December 1983, surviving an eclipse period of 28 minutes without solar power (its battery failed in 1981). A total of 15 manoeuvres and five lunar flybys were needed to complete the transfer from a halo orbit to an escape trajectory on a heliocentric orbit. After the final lunar swing-by, ISEE 3 became the International Cometary Explorer, or ICE.

On 11 September 1985 it passed 7,860 km behind Giacobini-Zinner's 2.4 km diameter nucleus at a closing speed of 75,300 km/h. This was the first ever flyby of a comet. Goddard controllers commanding the vehicle had feared cometary dust could coat its solar cells and damage its 91 m-span antennas. Throughout the 20 minute coma passage its 5 W transmitter and small antenna, designed to provide data from no further than 1.6 million km, transmitted 70.8 million km to Earth. NASA's Deep Space Network (DSN) had six large dishes trained on it (Goldstone, Madrid and Canberra), with additional facilities in Japan and Arecibo, Puerto Rico. Though lacking an imaging system, it returned data on plasma densities, flow speeds, temperature and heavy ions in the tail. Initial analysis revealed water ice to be the primary constituent, with cold slow-moving plasma, and water and carbon monoxide ions inside the tail, supporting Whipple's dirty snowball model.

Unexpectedly, ICE began encountering atmospheric ions one million km out, and the 25,000 km-wide tail was much wider than expected. The first instrument to detect the comet was a European energetic proton instrument. ICE also passed within 40.2 million km of Halley's Comet on 28 March 1986. ICE then became the first spacecraft to directly

NASA's ISEE-3/ICE spacecraft during assembly and testing (NASA) 1330122

Artist's impression of NASA's ISEE-3/ICE spacecraft flying past comet Giacobini-Zinner in September 1985 (NASA) 1330121

investigate two comets. In January 1990, it was in a 355 day heliocentric orbit with an aphelion of 1.03 AU, perihelion of 0.93 AU and inclination of 0.1°.

NASA ended ICE operations on 5 May 1997 after extending the mission in 1991. Its telemetry transponders were turned off on 19 December 2005 for JPL to begin a radio science mission. Low bit rates mean that communications cannot be commanded until about 2010. Hydrazine tank 1 is empty due to a leak and number 2 holds 135 kg. The solar array is 6 A. On 10 August 2014, it will return for capture by the Earth-Moon system, when it could be retrieved for analysis of its cometary dust coating. Ownership has already been formally transferred to the Smithsonian National Air & Space Museum, where the craft may be displayed if successfully recaptured.

Specifications

Manufacturer: Fairchild

Launch: 15:12 UT on 12 August 1978 by Delta 2914 from Cape Canaveral pad 17B.

Mass: 478 kg at launch. Hydrazine 89 kg. Science experiments 104 kg.

Body dimensions: 1.7 × 1.6 m

Telecommunications: 5 W transmitter. Transmission rate was nominally 2048 bps during the early part of the mission, and 1024 bps during the flyby of comet Giacobini-Zinner. The rate then dropped to 512 bps (12 September 1985), 256 bps (1 May 1987), 128 bps (24 January 1989) and 64 bps (27 December 1991).

Power: 160 W from solar cells mounted on spacecraft's "waist".

Attitude control: Spin stabilised at 20 rpm.

Propulsion: Twelve 18 N thrusters.

Payload:

- Solar Wind Plasma Experiment
- Magnetometer
- X-Ray and Low Energy Cosmic Ray Experiment
- Interplanetary and Solar Electrons Experiment
- Plasma Wave Experiment
- Plasma Composition Experiment
- Radio Wave Experiment
- Energetic Particle Anisotropy Spectrometer
- Gamma Ray Bursts Experiment (0.05–6.5 MeV)
- X-Ray and Gamma-Ray Bursts Experiment (5–228 keV)
- Medium Energy Cosmic Rays and Electrons Experiment
- High Energy Cosmic Rays Experiment
- Low Energy Cosmic Ray Experiment
- Cosmic Ray Isotope Spectrometer

Interplanetary Monitoring Platform 8 (IMP 8)

Current Status

For several years, IMP 8 (Explorer 50) has been used as an adjunct spacecraft for the Inter-Agency Solar-Terrestrial Physics (IASTP) programme; it continues to monitor the solar wind, and provides in-ecliptic, one AU baseline data for the Voyager and Ulysses missions. The Solar Plasma Faraday Cup instrument continues operation, providing data to the Massachusetts Institute of Technology (MIT). Data returned from the satellite continues to assist scientists in understanding long term solar phenomena. The Canberra, Australia station is responsible for telemetry at about 30–50 per cent coverage. IMP 8 has been in orbit and returning data for 33 years.

Background

Launched from Cape Canaveral on 26 October 1973, IMP 8, also known as Explorer 50 and IMP J, was retired as an independent programme on 28 October 2001, when the last commands were sent to the spacecraft. Following the failure of the magnetometer in 2000, a review panel advised NASA's office of space science that funds should be diverted to more productive missions.

Ten Interplanetary Monitoring Platforms (IMPs), each a spacecraft within NASA's long-lived Explorers Program, monitored Earth's radiation environment during a complete 11-year cycle of solar activity, defining the nature and extent of the magnetosphere; IMP 8 was the last of these. IMPs provided continuous warning of solar flare radiation during the Apollo and Skylab missions, and provided baselines for Pioneer 10 and 11 and Mariner 10. In December 1992, IMP 8 provided key data as NASA's Galileo passed through Earth's magnetotail.

IMP 8 is a 135.6 × 157.4 cm cylindrical craft, operating in a near-circular, 35 Earth Radii, 12-day orbit. It carries instruments to measure cosmic rays, energetic solar particles, plasma, and electric and magnetic fields.

Kepler

Current Status

Having experienced some fiscal problems, the launch of the Kepler spacecraft has been delayed at least twice. The planned launch is now set to occur some time during October 2008. Additionally, the design of the spacecraft itself has been slightly changed in order to reduce the risk, complexity and cost of the mission. The gimballed high-gain antenna has been replaced with a fixed antenna which requires the body to downlink the acquired scientific data.

Background

Kepler was selected as NASA's tenth Discovery mission. Designed to determine the frequency of Earth-size and smaller planets within and close to the Habitable Zone (HZ), the objective of the mission is to explore the structure and diversity of planetary systems. This will be done via the observation of a large array of stars, thus helping scientists to determine the number of terrestrial and larger planets, which are present either within or near to the HZ (over a wide range of spectral types of stars); the different types of sizes and shapes of the orbits of these planets; the number of planets that are present in multiple-star systems; the properties of the stars that harbour planetary systems; and the range of orbit size, brightness, mass and density of short-period giant planets.

The mission will also test a number of hypotheses including the theory that most stars, which are similar to the Sun, should have terrestrial planets in or near to the HZ. The Kepler mission also supports the objectives of the Space Interferometry Mission (SIM) and Terrestrial Planet Finder (TPF), both of which are NASA Origins theme missions. Kepler will allow scientists to identify common characteristics of host stars for future planet searches, thus allowing SIM to target a volume of space known to have terrestrial planets.

In order for a transit to take place; a planet must pass in front of its star as viewed by the observer. When a terrestrial planet performs such a transit, there will result a small change in the star's brightness, lasting between two and sixteen hours. Subsequently, if the transit is caused by a planet, then the change must be period, therefore allowing scientists to calculate the orbit size and planet size and the amount of luminosity decrease during the transit. It is important to note here, if all transits are produced by the same planet, then the change in brightness and the amount of time taken for this to occur, would be the same. Furthermore, once the orbit size is calculated, it is possible to then compute the temperature of the planet and therefore, answer the question of whether or not the planet is habitable.

PHASE	TIME
Concept Study	March 2001 to July 2001
Discovery selection	21 December 2001
Phase B	February 2002 to October 2004
Phase C/D	November 2004 to July 2008
Launch	October 2008
Commissioning	Launch +30 days
Phase E	
Flight operations	For 4 years from end of commissioning
Data analysis	For 5 years from end of commissioning
Optional extended mission	Additional 2 years of operations and data analysis

Specifications

Kepler's orbit will be an Earth-trailing heliocentric orbit with a period of 372.5 days. Such an orbit provides the advantage of a very-low disturbing torque, thus leading to a very stable pointing attitude. Within this orbit, the spacecraft will slowly drift away from the Earth, the worst-case scenario resulting in the spacecraft being at a distance of 0.5 AU at the end of the four-year mission. The largest external torque to note will be caused by solar pressure; therefore, although the orbit avoids high radiation dosage that would be associated with an Earth orbit, it will be subjected from time to time to solar flares.

The key instrument to be used on the Kepler spacecraft, a 0.95 m photometer or light meter, will observe the brightness of a range of stars (more than 100,000) for the entire four years of the mission. The design of the differential photometer itself is based on a modified Schmidt telescope design. The instrument itself has a wide Field Of View (FOV) of 100 square degrees, allowing the simultaneous and continuous observation of the brightness of 100,000 main sequence stars. It has enough precision in order for it to detect the transits of Earth-size planets orbiting G2 dwarfs. The corrector has an aperture of 0.95 m and a 1.4 m diameter F/1 primary, which reduces the Poisson noise level sufficiently enough to obtain a 4σ detection for a single transit from an Earth-size planet transiting a 12th magnitude G2 dwarf with a 6.5 hour transit. The focal plane is composed of forty-two 1,024 × 2,200 backside-illuminated CCDs with 27 μm pixels.

It is required that the FOV of the photometer is out of the ecliptic plane so that the Sun or the Moon does not block it periodically. Therefore, a star field, which meets these viewing constraints and has a high star density, has been selected within the Cygnus-Lyra region.

The spacecraft bus surrounds the base of the photometer and also supports the arrays and communication, navigation and power equipment. Heat pipes cool the detector focal plane, which is at prime focus, by carrying the heat out to a radiator, which can be found in the shadow of the spacecraft. Four guidance sensors, mounted within the photometer focal plane, provide the spacecraft with stable pointing and thrusters are used for the de-saturation of the momentum wheels. If it is decided that the mission is to be extended, the spacecraft carries a sufficient amount of expendables so that the mission life can be extended to six years.

The scientific operations, such as photometer development, mission and operations and scientific analysis, will take place at the NASA Ames Centre. Data management will be taken care of by the Space Telescope Science Institute. Routine contact will be made every 4 to 5 days, and the Deep Space Network (DSN) will be used for telemetry. The spacecraft will be launched into its Earth-trailing heliocentric orbit via a D2025-10L (Delta II) rocket. The photometer and spacecraft will be built by Ball Aerospace and Technology Corporation (BATC) in Boulder, Colorado and JPL will manage the spacecraft and mission development.

The current best estimates for mass and power of the spacecraft (made in July 2005) are 955 kg and 527 W respectively.

Laser Geodynamics Satellite (LAGEOS) Series

Current Status

LAGEOS 1 and 2 remain in use for laser ranging tasks. The two satellites are expected to have an operational lifetime of many decades, due to

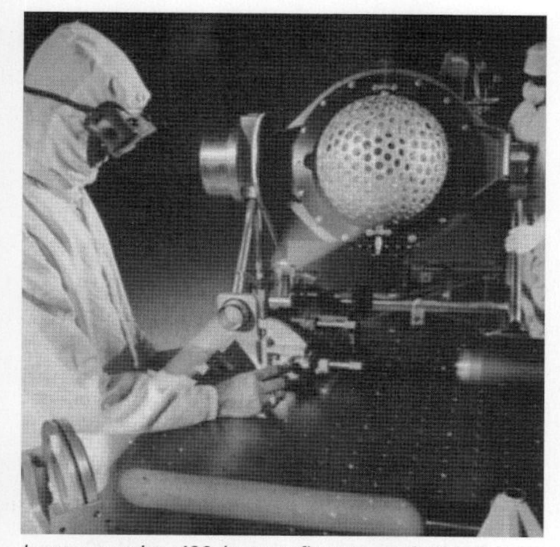

Lageos carries 426 laser reflectors embedded in its aluminium shell (NASA) 0516749

their relatively simple spherical structure and passive nature. Even after operational capacity has disappeared, the satellites are not expected to decay for another 8 million years.

Background

NASA designed the Laser Geodynamics Satellite as a tool to aid geodetic research, including earthquake prediction. It was NASA's first effort devoted wholly to laser ranging. With a high orbit and resulting long life, LAGEOS 1 acts as a permanent reference point to measure distances between ground stations. From this information scientists derive knowledge of the Earth's crustal motion, Earth's rotational variations, solid Earth and ocean tides and movement of the polar axis. By paying particular attention to earthquake areas such as the San Andreas Fault in California, LAGEOS may ultimately provide information on impending earthquakes. Station co-ordinates are now determined to within 1 to 3 cm and relative motions of 5 to 10 mm per year can be measured. NASA Goddard, through its Laboratory for Terrestrial Physics manages the programme.

LAGEOS allows researchers to determine position by bouncing LASER light off of its 426 3.8 cm diameter corner cube reflectors. The four germanium and 422 fused silica reflectors return laser beams to their source, regardless of the incident angle. To be effective, the satellite had to accommodate a large number of reflectors but at the same time be small enough to minimise solar radiation pressure and atmospheric drag effects. By the same arguments, NASA built the satellite from a combination of aluminium and brass. A pure aluminium construction would have been too light and brass too heavy, but their combination yielded a sufficiently high mass to surface area ratio to meet the objectives. NASA had to consider the interaction between the satellite and Earth's magnetic field in its selection of construction materials as well.

The LAGEOS carries a symbolic mission in addition to its scientific ones. Each end of the bolt connecting the hemispheres carries a copy of a message prepared by Dr Carl Sagan. It includes three maps of Earth's surface. One shows the period 225 million years ago when, it is believed, the landmasses were one super-continent (sometimes called 'Pangaea'). The second shows the position of the continents as they are now, and the third their estimated position 8.4 million years from now when the satellite, so solid it will survive re-entry, falls back to Earth.

LAGEOS 1's first four years were devoted to determining its precise orbit and establishing a global network of 14 permanent Satellite Laser Ranging (SLR) stations supported by mobile truck-mounted SLR systems. By accurately measuring the round trip time for a laser pulse, the position of the laser system could be determined to within about 10 cm. LAGEOS 2 improved this accuracy by a factor of 2.

As part of its Geodynamics Program, NASA in 1979 initiated the international Crustal Dynamics Project (CDP) to measure contemporary tectonic plate velocities, regional crustal deformation and various Earth orientation and rotation parameters. CDP was superseded in 1992 by the Dynamics Of Solid Earth (DOSE) programme. During the 1980s, 36 permanent SLR stations were operated by 18 countries (US, Mexico, France, UK, Germany, Italy, Netherlands, Poland, Czechoslovakia, Austria, Australia, USSR, Egypt, Israel, China, Japan, Chile, Peru), making repeated measurements with LAGEOS and providing their data to the CDP. NASA currently has 10 stations. Three sites (Texas, Hawaii and France) also range to reflectors on the Moon placed by Apollo 11/14/15 and the unmanned Lunokhod 1/2.

Typical velocities between tectonic plates vary over 1 to 20 cm per year, and the contemporary plate motions measured by space geodetic techniques are in good general agreement with the long-term average motion indicated by the geological record. Models derived from LAGEOS provided the first evidence of gravity field temporal variations, believed to result from continuing crustal relaxation following the last ice age.

Originally due for Shuttle launch, with the IUS in 1987, Italy's identical LAGEOS 2 was released by STS-52 on 23 October 1992. The altitude is the same as LAGEOS 1's but the 52° inclination improves coverage of the seismically active Mediterranean Basin and should enable investigators to identify the source of irregularities noted in LAGEOS 1's orbit. Some 16 mobile laser ranging facilities supplement the 10 fixed stations around the world that track both satellites.

Specifications

LAGEOS 1

Launched: 4 May 1976 by Delta 2913 from Vandenberg
Orbit: 5,858 × 5,958 km 109.8°, 225.4 min
Mass: 405 kg
Lifetime: Operational life of at least 50 years; orbital life of 8 million years
Size: 60 cm diameter, 407 kg
Description: Two aluminium hemispherical shells, totalling 117 kg, surrounding a 31.76 cm diameter 27.5 cm long cylindrical brass core
Payload: 426 3.8 cm diameter laser reflectors for ground stations to make precise distance measurements

LAGEOS 2

Released: from the Shuttle STS-52 on 23 October 1992
Orbit: 5,610 × 5,950 km, 52.84°; 222.5 minutes
Other specifications: identical to LAGEOS 1

Mars Exploration Program

Current Status

NASA's Mars Exploration Program is an integral part of the 2004 Vision for Space Exploration. As one of NASA's goals is to send a manned mission to Mars sometime after a successful human Moon mission by 2020, the current and planned Mars robotic and orbiter missions are of great importance to NASA. As of the end of 2006, Mars Global Surveyor, Mars Odyssey, the Mars Rovers, and Mars Reconnaissance Orbiter were all operational. It is possible, however, that the Global Surveyor mission has ended; Opportunity did not receive communications from the craft on 22 November 2006, and ground stations last heard from Global Surveyor on 2 November 2006. The Mars Reconnaissance Orbiter has also attempted to image Global Surveyor, without luck. Scientists will try further to raise communications with the spacecraft for a short time, however, after Global Surveyor's 10 years in space, far past its nominal mission, they are not overly hopeful that the mission will continue.

Please see each current mission's separate entry for more specific information.

Background

NASA's strategic plan for the robotic exploration of Mars evolved as a result of recommendations from the Solar System Exploration Committee (SSEC) in 1983. It sought to resume the exploration of Mars after completion of the Viking programme, which had put two spacecraft in Mars orbit and two on the surface in 1975. The strategy was substantially modified after the loss of Mars Observer in August 1993, and was changed again in response to the dual loss of the Mars Climate Orbiter in September 1999 and the Mars Polar Lander in December 1999. The first launch in this revised strategy was Mars Odyssey, launched in 2001.

In 1983, the SSEC recommended a series of low-cost Planetary Observer missions beginning with Mars Geoscience/Climatology Orbiter (subsequently renamed Mars Observer), but budget restrictions delayed launch from 1990 to 1992. Based on tried and tested hardware, the spacecraft bus was built up from the RCA Satcom K design using electronics derived from the company's Tiros/DMSP meteorological satellite programme. Originally designed to be carried into low Earth orbit by Shuttle and boosted on its way to Mars by an Orbital Science Corporation TOS upper stage it (and TOS) was assigned a Titan 3 launcher, following the Challenger disaster of January 1986, and sent on its way in September 1992.

Prior to the loss of the 2,565 kg Mars Observer in August 1993, when it was believed an explosion occurred in orbit-insertion propellant lines, NASA had proposed a follow-on programme of small lightweight landers under what it called the Mars Environmental Survey (MESUR) programme. MESUR would place 12–16 100 kg landers on the surface with proof-of-concept performed by a precursor Pathfinder mission funded under NASA's Discovery programme. With the loss of Mars Observer, NASA cancelled the MESUR programme and re-examined exploration strategy along more fiscally stringent lines with Pathfinder and its Sojourner rover retained as a spring-board to a new generation of low-cost orbiters and landers. As part of the revised Mars exploration strategy NASA began a series of Surveyor orbiters and landers, placing the newly defined Mars Global Surveyor (MGS) alongside Mars Pathfinder for the 1996 launch opportunity. MGS would refly some instruments previously carried on the failed Mars Observer spacecraft and utilise the same launch window as Russia's Mars 96 mission. This would involve a single 6,000 kg orbiter, two small 50 kg lander stations and two 65 kg penetrators. The landers would be released during the final orbiter approach phase and the penetrators from Mars orbit.

For future extended mission operations NASA decided on a strategy shaped by a series of scientific goals based on studying Mars in three primary areas: evidence of past or present life; climate (weather, processes and history); and resources (environment and utilisation). The Mars Science Working Group laid out a strawman strategy for fitting science goals into a set of missions which evolved throughout 1995 and would begin with Mars Pathfinder and MGS launched in late 1996 in the same window as Russia's Mars 96 mission. NASA decided to launch in 1998 Surveyor orbiter and lander spacecraft, the former reflying another instrument from Mars Orbiter and the latter searching for life near the south pole of Mars. The final element lost with Mars Observer would fly on a Surveyor orbiter launched in 2001, along with a lander which would analyse rocks and surface geology. Original plans were to launch the lander on a Delta and the orbiter on a Russian Molniya.

For 2003, NASA wanted to join forces with ESA and add three surface landers to a European Mars orbiter launched on a single Ariane 5 in a

Mars Exploration Program Spacecraft

Mission:	Launch Date:
Mars Observer	25 September 1992; failure; lost on route
Mars Global Surveyor	7 November 1996
Mars Pathfinder (Carl Sagan Memorial Station) & Sojourner	4 December 1996
Mars Climate Orbiter	11 December 1998; failure; lost on arrival
Mars Polar Lander & Deep Space 2	3 January 1999; failure; lost on arrival
Mars Odyssey	7 April 2001
Mars Express (joint ESA-NASA mission)	2 June 2003 (Beagle 2 Lander lost on descent 12/2003)
Mars Exploration Rovers: Spirit and Opportunity	10 June and 7 July, 2003
Mars Reconnaissance Orbiter	12 August 2005
Phoenix (first Mars Scout mission)	August 2007 (planned)
Mars Science Laboratory	Late 2009 (planned)
Mars Telesat Orbiter	Late 2009 (planned)
Mars Scout; Second Mars Scout mission	Earliest 2011 (projected)
Mars Sample Return	Earliest 2011 (projected)
Astrobiology Field Laboratory	Earliest 2011 (projected)
Mars Deep Drill Lander	Earliest 2011 (projected)

joint endeavour called InterMarsNet. The 415 kg landers would explore the interior of the planet using seismology to detect 'Marsquakes' and study geochemistry at three sites. In addition, NASA contemplated sending a network of small and complementary weather stations around the planet. The earliest opportunity to retrieve samples from the surface of Mars and return them to Earth would come, said NASA, with a possible launch in 2005 and, because it would contravene the stringent low-cost standards of the Surveyor missions, involvement with either Russia or Europe. Beyond 2005 (a decade after the 1995 planning date) NASA wanted to fill the gap between this published sequence of missions and possible human expeditions, which it believed it could mount from 2018, with precursor robotic missions paving the way for the first scientific research base.

Of the three Mars 96 missions, only the two NASA spacecraft (Mars Global Surveyor and Mars Pathfinder/Sojourner) were successful, with Russia's Mars 96 being destroyed in a launch failure. MGS reached Mars orbit in September 1997 but experienced aerobraking difficulties due to a technical problem with the –Y solar cell array. A modified orbit construction sequence delayed the start of the mapping mission until 9 March 1999. Meanwhile, Pathfinder and its Sojourner rover landed on the surface in July 1997 and carried out a spectacular survey of the landing site. During 1998, and in direct response to the successful return to Mars, NASA reconsolidated its revised Mars exploration strategy, reaffirming its intention to send Mars Climate Orbiter and Mars Polar Lander at the 1998 window. It also reaffirmed its decision to launch identical sample collecting rovers in the 2001 and 2003 windows complemented by respective orbiters. Extant plans for the 2001 and 2003 missions included surface sample collection by rovers that could cover sufficient ground to ensure a diverse suite of rock and soil samples delivered to the prospective landing site for the 2005 surface sample return mission. The two Rovers, identical to save costs, were also to have carried an Athena payload designed to locate, select, collect and store multiple fresh samples. The 2005 sample collection mission envisaged Mars-orbit-rendezvous from the surface instead of direct ascent and anticipated delivery of those samples to Earth in 2008.

Following the loss of Mars Climate Orbiter in September 1999 and Mars Polar Lander in December 1999, NASA conducted several major internal reviews of its 'faster-better-cheaper' strategy which underpinned plans developed after the SSEC review in 1983. Criticism from scientific institutions and determined investigations by Congressional subcommittees responsible for science and budgetary oversight claimed NASA had cut corners in an attempt to do too much with too few resources. As a result, a refined strategy slowed the pace of the robotic exploration of Mars and plans to move from the robotic to the human exploration of Mars were deferred indefinitely. The existing plan for Mars 2001 was cancelled and the lander was eliminated, but the orbiter Odyssey was allowed to continue.

Phoenix, scheduled to fly in 2007, will carry upgraded instruments similar to those that were on the failed Polar Lander mission as well as a Lander that was intended for the cancelled Mars Surveyor lander 2001 mission.

After the success of Mars Odyssey and the Rovers Spirit and Opportunity, NASA proposed to develop and launch a next-generation 'mobile surface laboratory' with potentially long-range roving capabilities, greater than 10 km, and more than a year of surface operational lifetime as a pivotal step toward a future Mars sample return mission. By providing a major leap forward in surface measurement capabilities and surface access, the Mars Science Laboratory (MSL) mission will also demonstrate the technology needed for accurate landing and surface hazard avoidance in order to allow access to potentially compelling, but difficult to reach, landing sites. Its suite of scientific instruments could include new devices that will sample and probe the Martian subsurface of organic materials. Launch is planned for 2009 with arrival at Mars in 2010.

Mars Telesat Orbiter is a multiband (X, Ka, UHF) telecommunications relay satellite that will be launched into an optimal orbit to maximise coverage of orbital, sub-orbital and surface assets. Set to launch in 2009, it will support the 2009 MSL and provide a significant increase in the quantity of data relayed from the surface of Mars to the Deep Space Network (DSN) on Earth for this and successor missions.

NASA plans to launch the first in a series of human precursor missions to Mars, robotic testbeds which will demonstrate technologies such as improved aerodynamic entry, Mars orbit rendezvous and docking, high precision landing and resource extraction and utilisation. These missions will also obtain critical data for future human missions on chemical hazards, resource locations and research sites and may prepare resources and sites in anticipation of human landings.

NASA is studying additional scientific orbiters, rovers and landers, as well as approaches for returning the most promising samples of Martian materials (rocks, soils, ices and atmospheric gases/dust) back to Earth. While schedules initially called for the first of several sample return missions to be launched in 2014, with a second mission in 2016, options that could move the date up to 2011 are under examination. Technology development is underway for advanced capabilities including a new generation of miniaturised surface instruments such as mass spectrometers and electron microscopes, as well as deep drilling instruments to bore up to 20 m or more.

The following lists past, present and future Mars Exploration Program spacecraft and launches.

Mars Exploration Rovers: Spirit and Opportunity

Current Status

Spirit and Opportunity have both surpassed their intended lifetimes by ten times. Their primary mission ended in April 2004; as of 2006's end, both Rovers continued to travel the Red Planet, returning valuable data. Spirit and Opportunity have provided over 70,000 and 58,000 Martian images respectively. Although Spirit is crippled, with only five of its six wheels working, it continues to creep along collecting data, and has done so for nearly 10 km over the years. Opportunity is in good health for its age and has travelled about the same distance. The Mars Reconnaissance Orbiter has taken several high resolution images of the Rovers and their landing sites, showing the vehicles, their tracks, the landers, heat shield pieces, and parachutes. The Rovers will receive NASA funding at least up to 1 October 2007.

Background

Spirit and Opportunity won their names in a student competition in which over 10,000 essays and names were submitted. The Rovers are also known as Columbia (Spirit) and Challenger (Opportunity) Memorial Stations. Launched separately in June and July 2003, the twin Rovers, able at first to travel almost as far in one Martian day as Sojourner did over its entire lifetime, landed at separate sites and set out to determine the history of climate and water on the planet, where conditions may once have been favourable for life. By means of sophisticated sets of instruments and access tools, the twin Rovers evaluate the composition, texture and morphology of rocks and soils at a broad variety of scales, extending from those accessible to the human eye to microscopic levels. The Rover science team selects targets of interest such as rocks and soils on the basis of images and infra-red spectra sent back to Earth. Two different Martian sites were chosen on the basis of an intensive examination of information collected by the Mars Global Surveyor and Mars Odyssey orbiters as well as previous missions.

Spirit descended to Mars on 4 January 2004, landing in the Gusev Crater, and has been a NASA workhorse ever since. The same is true of Opportunity, which landed on the Meridiani Planum on 25 January 2004. Spirit touched down near the area the Viking 1 craft explored 30 years before it. Opportunity is on the other side of the planet. The Rovers have found geological evidence that water existed on Mars many years ago; the first discovery was made almost immediately by Opportunity in early 2004. After landing, Spirit encountered both lander deployment problems as well as communications problems, occurring on 21 January. NASA teams successfully solved both problems, allowing the Rover to roll along for the past three years. There is a 20-minute communications delay between Mars and Earth.

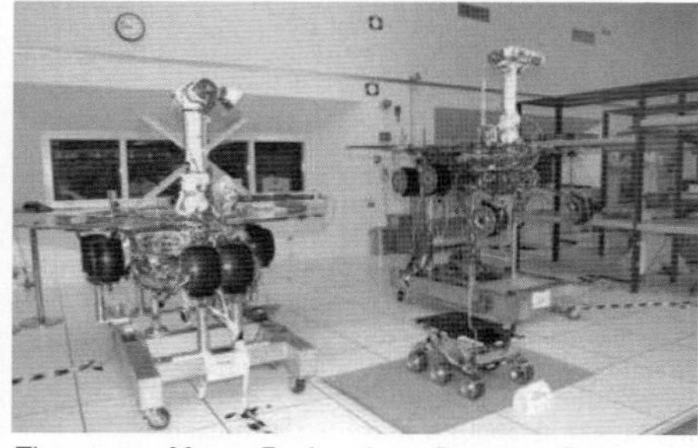

The two Mars Exploration Rovers, Spirit and Opportunity, with Sojourner's flight spare from the Pathfinder mission in the foreground (NASA) 1158895

An artist's rendering of an Exploration Rover on Mars (NASA) 1158894

Cameras are mounted on the identical Rovers' 1.5 m masts to give the science team a panoramic view of the area; scientists choose geological features and destinations to study, based on these images. The Rovers can travel at a top speed of 5 cm per second, however each usually averages about 1 cm/s due to hazard avoidance procedures and geologic investigations. The vehicle can tilt up to 45° without overturning, however its software prevents it from tipping more than 30°.

Due to the impressive amount of data returned by the Rovers, it is likely that NASA will run both until they irreparably malfunction and/or their batteries degrade to the point where they can no longer hold a charge and the Rovers permanently lose power. Given their survival nearly three years past their original mission end date, the Rovers have furnished NASA with outstanding value for money.

Specification

Rover Specifications

Launch: Spirit: 10 June 2003; Opportunity: 7 July 2003; both from Cape Canaveral aboard Delta II launch vehicles

Rover Spacecraft: Cruise stages, landers and aeroshells; NASA Jet Propulsion Laboratory (JPL).

Rover Body/Warm Electronics Box (WEB), including the Rover Equipment Deck (RED) and Rover Electronics Module (REM), multi-panel solar array (provides 140 W for ≤4 hr/day), communications, rechargeable batteries, computers and temperature controls; JPL.

Rover Wheels; also act as trench diggers; front and rear pairs swivel 360° to turn the vehicle; each cleated wheel is equipped with an independent motor; JPL.

Robotic Arm/Instrument Deployment Device (IDD); Alliance Spacesystems Inc, Pasadena, California; carries four instruments (see below).

Navigation Cameras (Navcams); two located on the Rovers' mast; JPL.

Hazard-Avoidance Cameras (Hazcams); four located on the lower front and rear WEB sections; JPL.

Science Payload: Panoramic Camera (Pancam); JPL; CCD stereo pair mounted on a 1.5 m mast.

Miniature Thermal Emission Spectrometer (Mini-TES), Arizona State University, Tempe; infra-red spectrometer for minerology; located on the mast.

Mössbauer Spectrometer (MB), Johannes Gutenberg University, Mainz, Germany; for iron analysis; located on the Robotic Arm.

Alpha Particle X-ray Spectrometer (APXS), Max Planck Institute for Chemistry, Mainz, Germany; for elemental composition analysis; located on the Robotic Arm.

Microscopic Imager (MI), JPL; microscopic view CCD camera located on the Robotic Arm.

Rock Abrasion Tool (RAT), Honeybee Robotics, New York, New York; for grinding geologic samples; located on the Robotic Arm.

Magnetic Targets/Magnet Arrays, Niels Bohr Institute, Copenhagen, Denmark; 3 sets to collect and examine magnetic dust samples mounted on the Robotic Arm and WEB.

Mass and Dimensions: 1,063 at launch, including cruise stage spacecraft, 2.65 × 1.6 m aeroshell, parachute, airbags and 31 kg hydrazine propellant. Each Mars Rover is 174 kg; 1.6 × 1.5 m.
Mission Length: nominal, three months; total extended missions to 3 years as of January 2007.

Mars Global Surveyor

Current Status

After 10 years in Martian orbit, returning valuable images and supporting other Mars missions, the Mars Global Surveyor (MGS) mission came to a sudden end in November 2006. The Rover Opportunity was unable to communicate with MGS on 22 November 2006 and ground stations last heard from the spacecraft on 2 November 2006. The Mars Reconnaissance Orbiter, launched in 2005, also attempted to image MGS, to no avail. A preliminary report by an internal review board, released on 13 April 2007, noted that orbiter appeared to have succumbed to battery failure caused by a complex sequence of events involving the onboard computer memory and ground commands.

On 2 November 2006, after the spacecraft was ordered to perform a routine adjustment of its solar panels, MGS reported a series of alarms, but then indicated that it had stabilised. That was its final transmission. Subsequently, the spacecraft reorientated to an angle that exposed one of two batteries carried on the spacecraft to direct sunlight. This caused the battery to overheat and ultimately led to the depletion of both batteries. Incorrect antenna pointing prevented the orbiter from telling controllers its status, and its programmed safety response did not include making sure the spacecraft orientation was thermally safe.

Mars Global Surveyor operated longer at Mars than any other spacecraft in history, and for more than four times longer than the prime mission originally planned. MGS pioneered the use of aerobraking at Mars, using careful dips into the atmosphere so that friction would modify a long elliptical orbit into a nearly circular one. After the primary mapping phase began in April 1999, the spacecraft returned detailed information that changed our understanding of the Martian surface. Major findings included evidence that water still flows in short bursts down hillside gullies and identification of deposits of water-related minerals leading to selection of a Mars Rover landing site. Images of the same locations taken in 1999 and 2006 revealed two fresh deposits, each several hundred metres long, in gullies on crater walls, together with 20 fresh impact craters. Altogether, the Mars Orbiter camera returned more than 240,000 images.

MGS also supported the landings of the Mars Exploration Rovers, Spirit and Opportunity, which reached their landing sites in January 2004. A mission aimed primarily at providing global visible and topographic and thermal emission spectroscopy of Mars was transformed into a flight support operation backing up the science assignments of NASA's Mars Odyssey and ESA's Mars Express, and when called upon it also served as a relay station. MGS imaged more than 3 per cent of the planet's surface, representing nearly 5 million km^2. After an unsuccessful attempt to locate the Mars Polar Lander, contact with which was lost on 3 December 1999, MGS continued to provide high-resolution images, stimulating renewed conjecture about the origin and evolution of the Red Planet's surface features. The mission was formally extended on 1 February 2001. Mars Odyssey arrived at Mars in October 2001 and a second year of MGS Extended Mission operations began in February 2002. In all, NASA extended its mission four times, with the final extension beginning on 1 October 2006. The orbit of MGS is not expected to degrade significantly for another 40 years.

Background

NASA began a new initiative, the Mars Surveyor programme in the wake of the Mars Observer loss in 1993. In effect the administration intended to demonstrate a new commitment towards cheaper, faster and more frequent missions. Rather than flying a duplicate mission of the lost Mars Observer, NASA shook up its approach by including USD78.4 million in FY95's request (receiving USD59.4 million) to begin the Mars Surveyor series of small orbiters and landers. Congress capped the programme to USD100 million annually, and contrary to NASA's track record the budget cap seemed to hold.

The 1 tonne first orbiter, Mars Global Surveyor, carried five of Mars Observer's seven experiments. The Mars Pathfinder carried one of MO's two remaining instruments and acted as a relay for the separate lander.

Lockheed Martin Astronautics built MGS, employing Mars Observer's spare thrusters and HGA but with a graphite composite bus instead of aluminium, because of mass constraints. Major elements drew on MO spares: most major electronic assemblies were MO spares, retrofitted to eliminate MO-identified discrepancies. The dual-mode propulsion system drew on Cassini designs.

NASA launched MGS at 17:00:50 GMT on 7 November 1996 and initially the Delta-II launch vehicle placed the satellite and PAM-D third stage into a near-circular orbit at 28.5°, 185 km. At 17.41, the second stage of the Delta-2 re-ignited to place the assembly into an orbit reaching out to 4,700 km and then separated from the combined PAM-D/MGS. PAM-D ignition at 17:44:30 put the spacecraft into a heliocentric, trans-Mars orbit.

At about 17:52, MGS spread out its solar array, but telemetry indicated that one of the solar panels (-Y) did not fully open, tilting instead about 20° from the expected angle relative to the spacecraft's main bus. Both solar panels were receiving sunlight, while still fully charging the batteries. The orientation of the (-Y) solar panel had little effect during the cruise phase to Mars.

On 10 November 1996, the spacecraft used its star sensor to establish its orientation in space and later that day it was put into an 'array normal spin' orientation which was used for the rest of the trans-Mars coast. On 21 November, MGS fired its main engine and performed the first of its four planned flight path correction manoeuvres. The 44 s burn achieved a change in spacecraft velocity of about 27 m/s. By performing the burn, the third stage of the Delta-II launch vehicle was steered away from a Mars collision.

On 22 January 1997, the spacecraft's flight computer activated a 53 W heater in the Mars Orbiter Camera to begin 'baking' the instrument and removing residual moisture. Without this 14-day bakeout period, the moisture in the camera's tube-like structure would leak into space at a slow rate and cause a gradual shift in the camera's focus. The bakeout removed all of the moisture at once and stabilised the focus of the camera. Flight controllers also performed a series of very slight manoeuvres on 22, 23 and 24 January to attempt to manipulate the MGS solar array and characterise the exact condition of the debris that prevented the panel from fully deploying.

A 26 second burn on 20 March 1997, refined MGS's flight path to Mars and achieved a change in spacecraft velocity of about 3.87 m/s. The burn was performed in two stages, in which flight controllers first commanded the spacecraft to fire its small thrusters for 20 seconds, then to fire its main engine for another 6 seconds. During April the spacecraft was in a quiet phase and conducted a search campaign to detect gravity waves.

MGS arrived at the Red Planet on 11 September 1997, and made a 973.03 m/s delta-V burn of 22 minutes 39 seconds to put it into an initial 250 × 54,000 km, 48 hour orbit. Aerobraking over a five-month period was to have changed the initial, elliptical path to a near circular 378 km orbit with an inclination of 92.9°. This would have provided an orbital period of 117.65 minutes and a 2 pm equatorial sun-synchronous repeat cycle of 7 days (88 revolutions) so that full mapping could begin in March 1998. On 1 October 1997, when the spacecraft descended through periapsis of 110 km, engineers observed movement in the –Y solar cell array that had given trouble shortly after launch. This generated concerns about possible damage caused by successive aerobraking passes through the tenuous outer layers of the Mars atmosphere and the periapsis altitude was temporarily raised back up to 121 km while an engineering analysis was performed. On 11 October, it was decided to raise periapsis to 171 km and on 27 October a management meeting was held to discuss options; at this date apoapsis was about 45,135 km and the orbital period 35.4 h. Periapsis reduction recommenced 7 November in a burn that put the low point of the orbit down at 134.8 km but aerobraking would be performed at a much slower rate than originally planned, enabling engineers to carefully monitor any stress on the –Y panel. Engineers worked out that during initial deployment, a damper arm had fractured resulting in structural damage to one end of the solar cell frame.

The revised aerobraking plan reduced the periapsis pressure exerted on the solar panels to 0.2 N/m^2, one-third the level assumed in the original plan. Global Surveyor would reach a circular mapping orbit in March 1999 after a

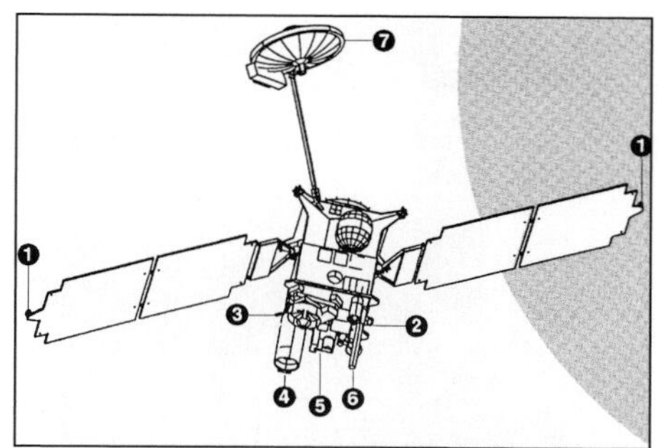

Mars Global Surveyor. 1) magnetometer, 2) electron reflectometer, 3) laser altimeter, 4) camera, 5) thermal spectrometer, 6) Mars Relay, 7) HGA (NASA/JPL) 0517127

total of more than 900 orbits. To achieve that, aerobraking continued until 23 March 1998 when it was suspended for four months during the period covered by solar conjunction. Four days later, the science instruments were turned on for a period of observation and imaging. On 23 September, Global Surveyor performed the first of three propulsive manoeuvres, completed two days later, totalling 11.62 m/sec to step back into the atmosphere, lower periapsis to 121 km and settling it into a pressure regime of 0.14 N/m^2. On 26 and 27 September, further manoeuvres of 0.18 m/sec lowered periapsis to 116 km but engineers recorded peak aerobraking pressures of 0.25 N/m^2. As the spacecraft continued to reduce its orbital period it experienced peak pressures of 0.28 N/m^2. At 12:40 UT on 12 December it passed within 300 km of the moon Phobos.

On 29 January 1999, with apoapsis altitude down to 1,000 km and periapsis of 103 km, the spacecraft conducted the first of four tiny 'walk-out' manoeuvres, each less than 1 m/sec, allowing it to step out of the atmosphere and stabilise in the desired orbit. Having slowed the spacecraft by more than 1,200 m/sec, aerobraking was completed at 20:11 UT on 4 February when a 61.9 m/sec burn of the main propulsion engine raised periapsis and put the spacecraft in a 1.97 hour orbit with a local solar mean time of 2:04 pm. At 22:00 UT on 19 February, a further burn marginally lowered the newly raised periapsis from 405 to 367 km establishing a fully sun-synchronous orbit. On 9 March the mapping mission began and on 28 March the high-gain antenna was successfully deployed for optimised data return to Earth. On April 15, the spacecraft put itself into an automatic contingency mode when the azimuth hinge on the high-gain antenna failed to point the antenna toward Earth. The entire spacecraft was turned to align the antenna with Earth and normal mapping operations resumed 5 May. Having completed an 88-orbit repeat pattern, a propulsive manoeuvre of 3.5 m/sec on 7 May started the spacecraft 'walking' round the planet by 58 km on each successive revolution for the next mapping phase. The primary objective of initial mapping operations during June and July was the Mass Polar Lander sector near the South Pole. On 4 November the MGS control team reported success with a test of the Deep Space 2 microprobes relay system, which was to relay telemetry from the probes when they reached the surface of Mars on 3 December 1999. When contact with Mars Polar Lander was lost, the MGS spacecraft supported attempts to communicate with the spacecraft via UHF. On 22 December, MGS completed 16 of 19 planned imaging targets in the hope of finding the spacecraft on the surface but this was unsuccessful. MGS returned excellent data and high-quality imagery and performed science mapping operations during its second extended mission phase, after which it served as a communications relay during the entry, descent and landing phase of the 2003 Mars Rovers when they reached the planet in January 2004. MGS imagery also contributed to the selection of landing sites for the Mars Phoenix and Mars Science Laboratory missions.

Specifications

Mass: 1,062.1 kg at launch (388.3 kg propellant, 75.8 kg science, 598 kg bus); body approx 1.5 m square by 3 m high.
Attitude Control: 3-axis pointing to 10 mrad (3 mrad knowledge, stability 1 mrad for 0.5 s and 3 mrad for 12 s) using 4 RWs, Sun sensors, IRU, star sensor, Mars Horizon sensor. 596 N bipropellant thruster for large delta-Vs; 12 × 4.45 N for momentum unloading, small delta-Vs and TVC.
Communications: 2 × 1.5 Gbit SSRs. 25 W X-band transmitter provides max 85.3 kbit/s via 1.5 m

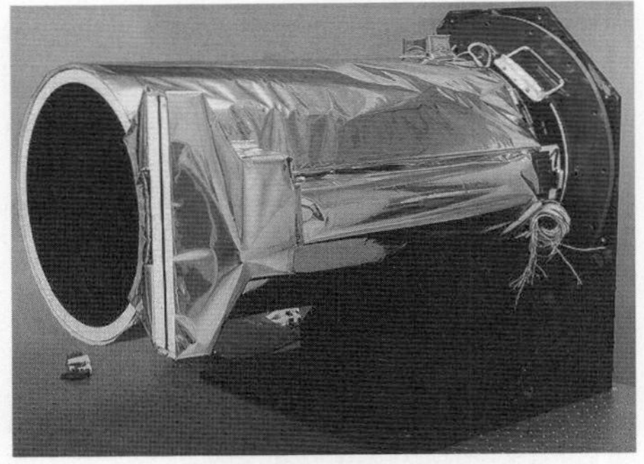

The completed MGS camera. The narrow-angle camera views through the main 35 cm aperture; the two wide-angle cameras (red/blue filters) view through the narrow slit (Malin Space Science Systems) 0517276

parabolic articulated boom-mounted HGA. Command uplink 7.8–500 bit/s.

Power: Twin 2-panel solar wings (12 m top-to-top extended) provide 940/660 W aphelion/perihelion (GaAs inner panel, Si outer); 2 × 20 Ah Ni/H_2 batteries

Payload: (all surface instruments are nadir body-mounted)

MOC Mars Orbital Camera. CCD line-scan camera for wide angle (f/6.5, 11.3 mm focal length, 140° FOV, 3,456 CCD array length, resolution 7.5 km/240 m) and narrow angle (f/10, 350 cm fl, 2,048 CCD array length, resolution 1.4 m) imaging to monitor surface and atmospheric changes. 20.5 kg, 6 W. Imaging systems include 96 Mbit solid-state memory. (California Institute of Technology)

MOLA Mars Orbital Laser Altimeter. 10 Hz pulses, surface topography within 30 m (1–10 m relative) and surface reflectance at 1.06 μm. 160 m diameter footprint. NASA Goddard

TES Thermal Emission Spectrometer. Mapping of rock mineral composition, frost coverage and cloud composition at up to 3 km resolution across 9 km FOV using 6.25–50 μm Michelson interferometer

MAG/ER Magnetometer/Electron Reflectometer. Two NASA Goddard triaxial fluxgate magnetometers and French CNES ER to map magnetic field and monitor solar wind-Mars plasma interactions; resolution ±0.004 to ±16 nT

RS Radio Science. Uses radio system's X-band transponder + ultrastable oscillator to measure vertical profile of atmospheric refractive index (indicating pressure, temperature, density) by monitoring Earth-received signal after atmospheric passage; map gravity field by spacecraft Doppler tracking. JPL

MR Mars Relay. Relays data from NASA Mars landers. Data stored in MOC memory. Downlink beacon activates data uplink.

Mars Reconnaissance Orbiter

Current Status

After successful entry into Martian orbit and six months of precise aerobraking in order to position the spacecraft into an appropriate orbit for optimal data return, Mars Reconnaissance Orbiter (MRO) began returning science data in November 2006. The spacecraft initially entered into an elliptical 35-hour orbit. After this, the orbiter briefly dropped into Mars' atmosphere during its orbital passes, using its solar panels and high-gain antenna to create resistance and therefore, slow the spacecraft and move it into a more circular orbit. After aerobraking, the spacecraft was left with an orbit period of approximately two hours.

The HiRISE instrument (see below) began imaging the Martian landscape when MRO was first captured into Martian orbit, and returned its first test images of the Red Planet to the NASA ground crew on 24 March 2006. HiRISE has returned thousands of images of the Martian landscape, including photos that show the Mars Rovers Spirit and Opportunity (see separate entry) and their heat shields and parachutes, shed after landing. Other instruments have imaged water and carbon dioxide frost at the planet's polar regions.

Background

Built by Lockheed Martin Space Systems and managed for NASA by the Jet Propulsion Laboratory (JPL), MRO was launched on 12 August 2005. Its goal is to search for evidence supporting the theory of water persisting on Mars during the course of the planet's history. Whilst previous missions have already shown that water once flowed across the Martian surface, it remains to be proven whether the water was actually around long enough to provide a sufficient habitat for life. The orbiter will endeavour to search for clues and answers to four key scientific points: whether life ever arose on Mars; the climate of Mars and its physical mechanisms of seasonal change; the geology of Mars – studying the complex terrain and identification of water-related landforms; the most promising areas for scientific study; and also identification of future promising landing spots for human missions. Data will also be returned to the spacecraft from future Mars landers and rovers during a relay phase.

Mars Reconnaissance Orbiter Mission Schedule

Date	Activity
2002 – August 2005	Mission preparation, delivery to Cape Canaveral
12 August 2005	Launch aboard an Atlas V rocket
August 2005 to March 2006	Seven month cruise from the Earth to Mars
March 2006	The spacecraft is captured into orbit around Mars
March 2006 – November 2006	Six months of aerobraking in order to settle into a lower, circular orbit for science data collection
November 2006 – November 2008	Primary science mission
November 2008 – December 2010	Orbiter will be used to communicate with future landers

Specifications

Mars Reconnaissance Orbiter's primary science mission will last two years, during which, its instruments will carry out investigations into three key areas: global mapping, regional surveying, and high-resolution targeting of specific surface areas. MRO carries a total of 11 instruments, including six science instruments, three engineering instruments and two science facility instruments.

MRO had a launch mass of approximately 2,180 kg. Over half of the launch mass was for the fuel required for the 20 onboard thrusters. Its dry mass is about 1,031 kg. It is 6.5 m in height. When the solar panels are deployed, the width from one end of the solar panels to the other is 13.6 m. In order to store and process large amounts of data, the spacecraft's solid state recorder has 160 Gbits of memory, provided by 700 microchips, as well as a processor operating at up to 46 million instructions per second. Finally, the spacecraft is topped by a 3 m radio antenna dish. Most MRO systems are redundant, adding confidence by providing backup should a system fail.

MRO Instrumentation

HiRISE – High Resolution Imaging Science Experiment: HiRISE is used to photograph areas of Mars' surface in extraordinary detail. It measures thousands of Martian landscapes at 1 m resolution and is able to reveal small-scale objects and details such as the geologic structure of canyons, craters and layered deposits. The high-resolution images allow scientists to view 1 m sized objects on Mars. HiRISE works in both the visible and infra-red wavelengths. Observations within the infra-red wavelength will provide information on the mineral groups that are present. HiRISE is managed by the University of Arizona Lunar and Planetary Laboratory.

CTX – Context Camera: The CTX collects grayscale images of Mars, and is used simultaneously with HiRISE and CRISM (see below). The resolution of CTX is not as sharp as HiRISE; it is used to provide a broader image of the planet's surface to put HiRISE and CRISM high resolution images in context. The camera has a resolution of 6 m per pixel and images 30 km wide swaths of the planet. CTX is managed by Malin Space Science Systems.

NASA's Mars Reconnaissance Orbiter, launched in August 2005, is conducting the most detailed reconnaissance of the planet Mars since missions to the Red Planet began over 40 years ago (NASA) 1047538

MARCI – Mars Colour Imager: MARCI is used to produce a global map to study the daily, seasonal and yearly variations in Mars' climate. The camera's five visible wavelengths and two ultraviolet wavelengths are used to detect the presence and variations of the planet's atmospheric ozone, water vapour, dust and carbon dioxide. MARCI is managed by Malin Space Science Systems.

CRISM – Compact Reconnaissance Imaging Spectrometer for Mars: CRISM uses its infra-red and visible spectrometers at about 38 m resolution to search for mineral residues within hot springs, thermal vents or lakes that may have been present on Mars far back in its history. It can cover swaths between 5.8 and 11.9 km, depending on MRO's altitude. In targeted mode, it is capable of taking high resolution images at 18 m resolution in 10 × 10 km swaths. CRISM is managed by the Johns Hopkins University Applied Physics Laboratory.

MCS – Mars Climate Sounder: The MCS radiometer is used to observe the temperature, humidity and dust content of the Martian atmosphere, further helping scientists to understand the Martian climate. It creates a weather map of the planet by collecting atmospheric data in 3 km vertical swaths. MCS is managed by JPL.

SHARAD – Shallow Radar: SHARAD uses radar to probe up to 1 km beneath the Martian surface to investigate whether liquid or ice water still exists in the planet's crust. SHARAD operates its radar at a frequency of 15–25 MHz in order to achieve its high depth resolution; horizontally it is capable of resolving between 0.3 and 3 km, and vertically 15 m in free space and 10 m in the planet's crust. SHARAD is managed by the Italian Space Agency (ASI), with Washington University, JPL and the Smithsonian National Aeronautics and Space Museum participating.

The three engineering experiments carried by the orbiter assist with spacecraft navigation and communications. Initially, the Electra communications package will be used for spacecraft communications. Towards the end of MRO's primary science mission, other Mars missions will begin to arrive, and so the Electra package will also be used to provide UHF coverage to the Mars landers and rovers on the surface, replacing 2001 Mars Odyssey (see separate entry) in this capacity. An optical navigation camera is also being tested during this mission. If successful, the camera will be used to improve navigation capabilities in future missions. Tests are also being carried out on the Ka-band frequency in order to demonstrate its potential for greater performance whilst using less power; current deep space communication operates on the X-band frequency.

Two science facility experiments are being carried out via the use of the engineering data. By studying the Doppler shift within the radio communications signal, the Gravity Field Investigation Package can use the resulting data to determine the subsurface structure of Mars on the scale of several hundred kilometres. The second science facility experiment – Atmospheric Structure Investigation Accelerometers – determines the density of the atmosphere present at the altitude of the spacecraft, thus providing the science team with information on coupling between the lower and upper atmosphere, variations in seasonal winds and the effects of dust storms on atmospheric density.

Mercury Surface, Space Environment, Geochemistry and Ranging (MESSENGER)

Current Status

MESSENGER completed its first flyby of Mercury on 14 January 2008. All systems were nominal.

Background
The Mercury Surface, Space Environment, Geochemistry and Ranging (MESSENGER) is only the second spacecraft to visit Mercury, following on from Mariner 10 in the 1970s. The seventh mission in NASA's Discovery Series, MESSENGER will be the first spacecraft to orbit the nearest planet to the Sun. The data received will help scientists to answer questions about the composition and structure of Mercury's crust; how the high density of Mercury came about; the nature and dynamics of its thin and Earth-like magnetosphere; the possible presence of water ice in shaded polar craters; and how its magnetic field is generated. The orbital mission will be preceded by three Mercury flybys in order to achieve a preliminary study of high priority targets and investigation of the atmosphere and magnetosphere, and to slow the spacecraft sufficiently to enable it to enter orbit around the planet.

MESSENGER was launched from Cape Canaveral Air Force Station in Florida on 3 August 2004. It was placed into a solar orbit 57 minutes after launch. Once in orbit, MESSENGER automatically deployed its two solar panels. The mission operations team at APL acquired the spacecraft's radio signals through tracking stations in Hawaii and California. Its 7.9 billion km voyage includes 15 orbits of the Sun, with one Earth flyby, two flybys of Venus and three flybys of Mercury before it eventually enters orbit around the planet in March 2011. These flybys provide gravity assists that slow the spacecraft and help it to match the planet's speed and location prior to the orbit insertion manoeuvre. They also allow the spacecraft to gather data critical to planning the year-long orbit phase.

The spacecraft's Earth flyby took place on 2 August 2005, with closest approach approximately 2,347 km over central Mongolia at 20:13 GMT. Distant approach shots of Earth and the Moon were obtained with the narrow angle camera on 11 May, from a distance of 29.6 million km. During the flyby sequence, MESSENGER's camera took several approach images of Earth and the Moon, as well as a series of colour images that documented one full Earth rotation. The atmospheric and surface composition spectrometer also made several scans of the Moon in conjunction with the camera observations. In addition, the particle and magnetic field instruments spent several hours measuring Earth's magnetosphere.

Prior to the Earth flyby, MESSENGER's Mercury Laser Altimeter (MLA) successfully accomplished a two-way laser link with Earth and received laser pulses from the NASA Goddard Geophysical and Astronomical Observatory (GGAO). In three observing opportunities, the MLA laser was fired for 5-hour periods at a distance of nearly 24 million km. Forty MLA downlink observations and 90 uplink observations were obtained during observing sessions on 27 and 31 May 2005. Precise standard ground timing allowed a solution for spacecraft range, range rate, and acceleration, as well as clock bias. This experiment established a new distance record for laser detection.

After five trajectory correction manoeuvres using the small thrusters, the first Deep Space Manoeuvre (DSM 1) took place on 12 December 2005, using the large bipropellant thruster. The 524 second burn changed MESSENGER's velocity by about 316 m/s and aligned the spacecraft for a 3,140 km altitude flyby of Venus. The 10th trajectory correction manoeuvre (TCM-10) was carried out on 22 February 2006. This short duration manoeuvre lasted just over two minutes, allowing the spacecraft to adjust its velocity by approximately 1.4 m/sec. The first Venus flyby took place on 24 October 2006, with closest approach at 08:34 GMT. About 18 minutes after the approach, an anticipated solar eclipse cut off communication between Earth and the spacecraft. Contact was re-established at 14:15 GMT. No scientific observations were made during the flyby since the spacecraft entered superior conjunction, on the far side of the Sun from the Earth, shortly before the Venus flyby, making communications very difficult.

TCM-13 took place on 2 December 2006 to prepare MESSENGER for its second Venus flyby. It flew within 338 km of Venus on 5 June 2007, over 12° S, 107° E, in the uplands of Ovda Regio. 614 images were taken by the Mercury Dual Imaging System (MDIS) during the Venus 2 encounter. The night side was imaged in near-infra-red bands, while colour and higher-resolution monochrome mosaics were made of both the approaching and departing hemispheres. The Ultraviolet and Visible Spectrometer on the Mercury Atmospheric and Surface Composition Spectrometer (MASCS) instrument made profiles of atmospheric species on the day and night sides as well as observations of the exospheric tail on departure. The Visible and Infra-red Spectrograph on MASCS was used to sense cloud chemical properties and near-infra-red returns from the lower atmosphere and surface. The laser altimeter served as a passive 1064-nm radiometer and measured the range to one or more cloud decks for several minutes near closest approach. The Energetic Particle and Plasma Spectrometer (EPPS) observed charged particle acceleration at the Venus bow shock and elsewhere. The Magnetometer provided measurements of the interplanetary magnetic field (IMF), bow shock signatures, and pickup ion waves as a reference for EPPS and Venus Express observations. The MESSENGER observations were complementary to those being obtained simultaneously by ESA's Venus Express orbiter, and were of particular value for characterisation of the particle and field environment at Venus. This enabled two-point measurements of IMF penetration into the Venus ionosphere, primary plasma boundaries, and the near-tail region.

The first of three Mercury flybys took place at 19:04 GMT on 14 January 2008, at an altitude of 200 km. The closest approach occurred when the spacecraft was in eclipse and on battery power, as well as out of contact with Earth. It missed its targeted aim point by only 8.25 km. The following day, MESSENGER turned toward Earth and began downloading the 500 megabytes of data that had been stored on the solid-state recorder during the encounter, including 1,213 images from the MDIS cameras. The data were received by the Science Operations Center at JHU-APL. The other Mercury flybys will occur in October 2008 and September 2009. Together, they will allow the spacecraft to map nearly the entire planet in colour, study the composition of the surface, atmosphere and magnetosphere and image most of the areas that were unseen by Mariner 10.

MESSENGER Mission Schedule

DATE	MISSION ACTIVITY
Launch:	07:15 GMT on 3 August 2004 by a Delta II 7925-H rocket from LP 17-B, Cape Canaveral.
2 August 2005	Earth flyby
24 October 2006	Venus flyby
5 June 2007	Venus flyby
14 January 2008	Mercury flyby
6 October 2008	Mercury flyby
29 September 2009	Mercury flyby
18 March 2011	Year long science orbit of Mercury begins

MESSENGER's 12-month primary mission covers two Mercury solar days or just over four Mercury years.(One Mercury solar day is equal to 176 Earth days.) The first solar day will be focused on obtaining global map products from the different instruments and the second will focus on targeted science investigations.

Science Instruments
MESSENGER carries seven scientific instruments:
- Mercury Dual Imaging System (MDIS) – these wide and narrow angle CCD cameras will allow scientists to view and map landforms, track variations in surface spectra and gather topographic information.
- Gamma-Ray and Neutron Spectrometer (GRNS) – will be used to detect gamma rays and neutrons that are emitted by radioactive elements on Mercury's surface. This instrument will give scientists a good idea of the relative abundances of various elements and will also help to determine if ice is present at Mercury's poles.
- X-Ray Spectrometer (XRS) – the emitted X-rays from Mercury's surface will allow scientists to measure the abundance of different elements in the materials of Mercury's crust.
- Magnetometer (MAG) – this instrument has been placed at the end of a 3.6 m boom in order to search for magnetised rocks within Mercury's crust and also to investigate Mercury's magnetic field.
- Mercury Laser Altimeter (MLA) – by monitoring the amount of time it takes to send laser light to the planet's surface and a sensor to gather the light after its reflection from the surface, highly accurate measurements of Mercury's topography will be obtained.
- Mercury Atmospheric and Surface Composition Spectrometer (MASCS) – this instrument will be used to measure the abundance of atmospheric gases and to detect minerals on the surface.
- Energetic Particle and Plasma Spectrometer (EPPS) – will be used to study Mercury's magnetosphere and its composition, distribution and energy of charged particles (electrons and various ions).
- There is also a Radio Science (RS) experiment, which will use the Doppler effect to measure the smallest changes in the spacecraft's velocity during its orbit around Mercury, scientists will be able to study and evaluate Mercury's mass distribution as well as variations within the thickness of its crust.

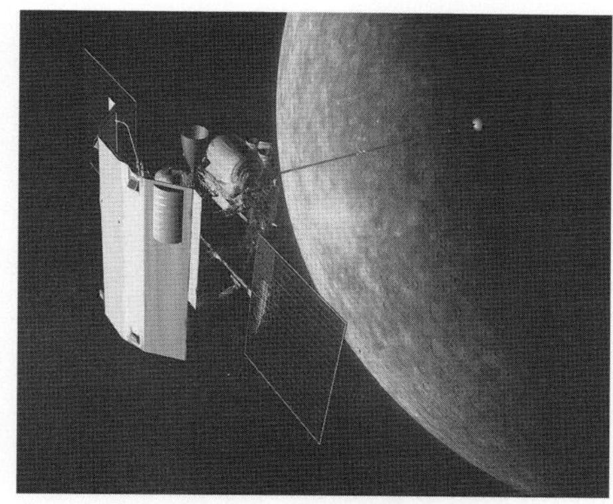
An artist's concept of MESSENGER as it approaches Mercury (NASA/Johns Hopkins University Applied Physics Laboratory/Carnegie Institution of Washington)
1133887

The MESSENGER spacecraft was designed and built by the Johns Hopkins University Applied Physics Laboratory (APL) in Laurel, Maryland. JHU-APL also manages the MESSENGER programme on behalf of NASA. GenCorp Aerojet, Sacramento, California, provided MESSENGER's propulsion system and Composite Optics Inc., San Diego, provided the composite structure. KinetX Inc., Simi Valley, California, leads the navigation team. MESSENGER's instruments were built by NASA's Goddard Space Flight Center (GSFC), APL, the University of Colorado, and the University of Michigan. NASA's Deep Space Network (DSN), operated by the Jet Propulsion Laboratory (JPL), is the primary communication link with MESSENGER. The cost of the mission is USD446 million (including spacecraft and instrument development, launch vehicle, mission operations and data analysis).

The spacecraft makes economical use of its mass via the integration of the dual-mode, liquid chemical propulsion system into the spacecraft's structure. The structure is primarily made up of a graphite epoxy material, while power is provided by two large solar panels, which are further enhanced by a nickel-hydrogen battery. The spacecraft structure also encloses redundant integrated electronics modules, both of which house two processors each.

Attitude determination is achieved via the use of star-tracking cameras and an Inertial Measurement Unit (IMU) containing four gyroscopes and four accelerometers, with six Digital Solar Sensors as backup. The four reaction wheels retain control of the spacecraft most of the time, with the small thrusters being used only when necessary. Commands and data are sent primarily through the circularly polarised X-band phased-array antennas.

The key factor that scientists had to consider when designing MESSENGER was the intense heat that will be experienced by the spacecraft while at Mercury. At Mercury, the Sun is up to 11 times brighter than on Earth and surface temperatures can reach up to 450°C. In order for the spacecraft to operate nominally at these temperatures, MESSENGER was provided with a heat-resistant ceramic cloth, which acts as a sunshade for the spacecraft.

As MESSENGER uses the gravitational attraction of the planets to adjust its orbit about the Sun, the fuel requirements for the spacecraft are greatly reduced. Due to the launch date and operational rules for the spacecraft, for example, prevention of spacecraft from overheating, the path from Earth to Mercury requires the least amount of fuel.

MESSENGER's orbit around Mercury will be highly elliptical, with the spacecraft's lowest point occurring at approximately 200 km above Mercury's surface and the highest point occurring at 15,193 km. The low point of the orbit will be attained at a latitude of 60°N, thus allowing MESSENGER to carry out a detailed investigation of the geology and composition of Mercury's giant Caloris impact basin – the planet's largest known surface feature.

The Mercury orbit insertion will use approximately 33 per cent of the spacecraft's propellant. The thrusters will have to slow the spacecraft as it approaches Mercury, while the largest thrusters will be used to point the spacecraft in the forward velocity direction. This manoeuvre will last around 15 minutes and place the spacecraft into a stable orbit.

Once in its primary science orbit, MESSENGER will experience small forces such as solar radiation pressure, which will slowly change the spacecraft's orbit. Although these forces have only a small effect on the 12-hour orbit period of the spacecraft, there will be an ultimate increase in the spacecraft's minimum altitude. Therefore, propulsive manoeuvres will be carried out in pairs, once every 88 days, which is once every Mercury year, in order to retain the spacecraft's minimum altitude below 500 km. In order to protect the spacecraft from direct sunlight, the propulsive manoeuvres are limited to a few days and thus carried out when Mercury is near the same point in its orbit as it was at MESSENGER's Mercury orbit insertion.

Specifications
MESSENGER
Orbit: Heliocentric, then near-polar elliptical around Mercury: 200 × 15,193 km, 80° inclination to Mercury's equator. Periapsis at 60°N. Two orbits of Mercury every 24 hours.
Design life: 7 years.
Contractors: JHU-APL (prime). GenCorp Aerojet, Sacramento, California (propulsion system). Composite Optics Inc., San Diego, (composite structure). KinetX, Inc., Simi Valley, Calif., leads the navigation team.
Body dimensions: 1.42 × 1.85 × 1.27 m. Front-mounted ceramic-fabric sunshade 2.5 × 2 m.
Propulsion: Dual-mode system with one bipropellant (hydrazine and nitrogen tetroxide) thruster for large manoeuvres; 16 hydrazine monopropellant thrusters for small trajectory adjustments and attitude control.
Mass: 1,107 kg at launch. 599.4 kg of propellant and 507.9 kg dry.
TT&C: NASA Deep Space Network (DSN).
Power system: Two body-mounted gallium arsenide solar arrays; 390 W near Earth, 640 W in Mercury orbit. Nickel-hydrogen battery.

Near Earth Asteroid Rendezvous (NEAR) Shoemaker

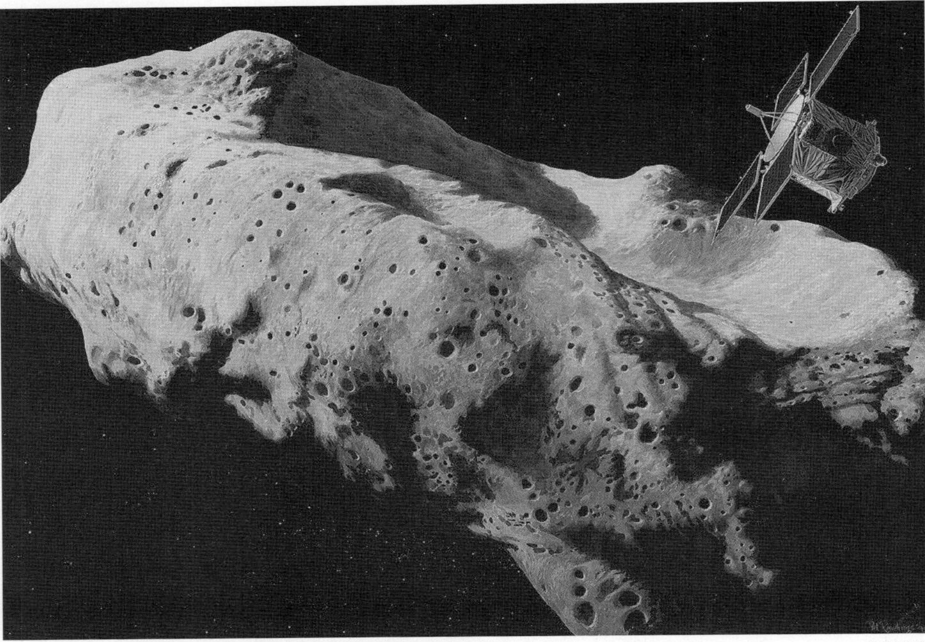
Artist's impression of NEAR in orbit around Eros (JHU-APL) 0517274

Current Status
The Near Earth Asteroid Rendezvous (NEAR) Shoemaker spacecraft reached the end of its primary mission and the spacecraft was brought down to the surface of the asteroid, Eros, at 20:01:52 UT on 12 February 2001, at a speed of 1.9 m/s. It continued to return a signal from the surface, but NASA shut down its communications at midnight UT, 28 February 2001. On 10 December 2002, NASA attempted to regain communications with the craft, but was unsuccessful.

Background
The Near-Earth Asteroid Rendezvous (NEAR) mission was the first in NASA's Discovery programme. The spacecraft was completed for less than the USD150 million budget cap. NEAR would be the first solar-powered spacecraft to fly beyond the orbit of Mars and the first to enter orbit around an asteroid in order to perform a lengthy, close-proximity study. It later became the first spacecraft ever to land on an asteroid.

NEAR's primary target was the stony (S-type) asteroid catalogued as 433 Eros, the second largest of the near-Earth asteroids (only three are larger than 10 km). NEAR's objective was to determine the asteroid's physical and geological properties as well as its chemical and mineralogical composition. As the first asteroid orbiter, NEAR was also designed to evaluate the environment for orbital missions during its lifetime. NEAR promised to throw light on the nature of planetesimals (the building blocks for the terrestrial planets), the origin of objects impacting Earth, and the relationship between asteroids, comets and meteorites. The data would significantly enhance interpretation of data from the flybys of the main belt S-type asteroids, Gaspra and Ida, by NASA's Galileo spacecraft in 1991 and 1993.

The Applied Physics Laboratory of The Johns Hopkins University (JHU-APL) was selected in 1992 to build/control NEAR, with the design reflecting simplicity and reliability, and the payload addressing the highest priority science issues. It was the first NASA planetary mission to be conducted by a non-NASA centre. APL conducted mission and science operations from its campus.

Simplicity and low cost were the main drivers in developing the spacecraft. This meant that the instruments, solar panels and high-gain antenna were all fixed and body-mounted, despite the fact that this increased the complexity of spacecraft operations. Mounted on the outside of the forward deck were the X-band high-gain antenna, four solar panels and the X-ray solar monitor system. The solar arrays were produced by Spectrolab Inc. of Sylmar, California.

NEAR was designed to be highly fault tolerant. Fully redundant subsystems included the complete telecommunication system (except the high-gain and medium-gain antennas), as well as the solid-state recorders, command and telemetry processors, data buses, attitude interface unit and flight computers for guidance and control, and power subsystem electronics. Additional fault tolerance was provided by use of redundant components: NEAR had two inertial measurement units (one operational, one backup), five sun sensors, and 11 small thrusters. The spacecraft structure was an octagonal box made of 1.7 m^2 aluminium honeycomb panels connected to forward and aft aluminium honeycomb decks.

The NEAR propulsion subsystem, supplied by Gencorp Aerojet of Sacramento, California, comprised the fuel and oxidiser tanks, 11 small monopropellant thrusters, a large bipropellant thruster, and a helium pressurisation system. The monopropellant system was composed of four 21 N fine velocity control thrusters and seven 3.5 N fine velocity control thrusters, all fuelled by pure hydrazine. The specific impulses of the monopropellant thrusters ranged from 206 to 234 seconds. They were arranged in six thruster modules mounted to the forward and aft decks and were located so that the loss of any one thruster did not affect performance. The large thrusters, which pointed in the same direction as the main thruster, were used for thrust vector control during the bipropellant burns. The small thrusters were used for momentum dumping and orbit maintenance around the asteroid. A minimum change-in-velocity increment of 10 mm/s was achievable in all directions.

The bipropellant thruster, or large velocity adjustment thruster, burned a mixture of hydrazine and nitrogen tetroxide (NTO) to produce a maximum 450 N of thrust, with a specific impulse of 313 sec. The large thruster was used for major velocity changes during the mission. The propulsion system carried 209 kg of hydrazine in three fuel tanks and 109 kg of NTO oxidiser in two oxidiser tanks. The 55.1-litre oxidiser tanks were located along the launch vehicle spin axis equidistant from the spacecraft centre of mass. The 91-litre fuel tanks were arranged 120° apart in the main thruster plane.

The science payload consisted of five instruments. Several of the instruments were derived from designs developed by JHU-APL for US DOD spacecraft. A Multispectral Imager was fitted with a CCD imaging detector capable of imaging details on Eros' surface as small as 3 m in diameter. A Near-Infra-red Spectrometer provided the main evidence for the distribution and abundance of surface minerals such as olivine and pyroxine. An X-Ray/Gamma-Ray Spectrometer measured and mapped abundances of several dozen elements at and near the surface of Eros using X-ray fluorescence. A Laser Range-finder determined the distance from the spacecraft to the asteroid by precisely measuring the delay time between the firing of a laser pulse and its return reflection from the surface. The ranging data were used to construct a global shape model and a global topographic map of Eros with horizontal resolution of about 300 m, plus some detailed topographic profiles of surface features with a best spatial resolution of about 4 m. A Magnetometer was included to measure the strength of Eros' magnetic field. A radio science experiment was also conducted to study the asteroid's mass distribution and gravity field.

Despite the lower cost and rapid development schedule, the instrument designs incorporated many technical innovations:

- First space flight of a laser incorporating an inflight calibration system (Laser Range-finder)
- First space flight using a near-infra-red system with a radiometric calibration target and an indium-gallium-arsenide focal plane array that does not require cooling with liquid nitrogen (Near-Infra-red Spectrometer).
- First space flight of a silicon solid-state detector viewing the sun and measuring the solar input X-ray spectrum at high resolution (X-Ray Spectrometer).
- First space flight of a bismuth germanate anticoincidence shielded gamma-ray detector (Gamma-Ray Spectrometer).

After launch on 17 February 1996, NEAR made a flyby of the main belt asteroid 253 Mathilde on 27 June 1997, coming within 1,212 km of its surface and taking a series of 534 images. Data showed that the heavily cratered carbonaceous (C-type) asteroid is very dark, with an average geometric albedo of only 4.7 per cent, making it twice as dark as a lump of charcoal. Although the asteroid measured 66 × 48 × 46 km, NEAR's Multispectral Imager, one of six instruments on the spacecraft, found at least five craters larger than 20 km in diameter on the asteroid's sunlit side alone. Mathilde showed no color or albedo variations over the 60 per cent of its surface that was visible to the spacecraft. The volume derived from the images and the mass of the asteroid determined from the spacecraft tracking data yielded a bulk density for Mathilde of 1.3 grams per cubic centimetre, only about half that of CM chondrite meteorites. This suggests that asteroid Mathilde may have a very porous interior structure.

NEAR prior to launch (JHU-APL) 0517273

NEAR performed an Earth flyby on 23 January 1998, passing 540 km above south-western Iran. An attempted engine burn on 20 December 1998 to align the trajectory for the encounter with Eros failed and the spacecraft was placed in 'safe' mode. In a revised mission plan, NEAR performed a flyby of Eros on 23 December 1998 coming to within 3,830 km and taking 222 pictures. Potato-shaped Eros was found to measure 13 × 14 × 35 km.

A successful engine burn on 3 January 1999 ensured that NEAR returned to Eros for the planned 57 sec (10 m/sec) orbit insertion burn on 14 February 2000, about a year later than originally planned. Just over an hour after entering orbit around Eros, NEAR took its first picture of the asteroid from a range of 330 km above the surface. Orbital parameters for NEAR were 327 × 450 km. A series of manoeuvres placed Near in a 204 × 200 km orbit by 3 March until transferred to a 200 × 100 km orbit on 2 April. Also during March, the spacecraft was renamed NEAR Shoemaker after the legendary astrogeologist, Dr Eugene M Shoemaker, who had died in 1997 after a car accident.

Through a series of orbit changes, NEAR Shoemaker entered a circular 50 km orbit of Eros on 30 April 2000 and a circular 35 km orbit by 13 July. This was changed to a 35 × 50 km on 24 July and to a 50 km orbit on 31 July. NEAR Shoemaker executed a 16° orbit inclination change, moving from 90 to 106°, on 8 August and shifted to a 50 × 100 km orbit on 26 August, circularised at 100 km on 5 September. A further series of manoeuvres placed the spacecraft in a 200 km circular orbit on 3 November, followed by further manoeuvres which placed it in a circular 35 km orbit on 13 December. During January 2001, several separate manoeuvres placed NEAR Shoemaker in several different elliptical orbits around this maximum 35 km trajectory, dipping to 19 km, placing it at times just 2.7 km above the surface (all orbit data are referred to the centre of the asteroid).

More than 100,000 images of the asteroid were returned during the year in orbit. They showed thousands of impact craters and major depression named "the saddle". The number of small craters was much lower than expected. There were also some smooth patches and a relatively homogeneous interior. The asteroid was covered with regolith – a loose layer of rocks, gravel and dust – that was embedded with numerous large boulders. The spacecraft also found places where the regolith apparently had slumped, or flowed downhill, exposing fresh surface underneath. At the end of its mission, it was decided to attempt a landing on the surface of Eros. Touchdown occurred at 20:01:52 UT on 12 February 2001 at a speed of 1.9 m/s. It continued to return gamma ray and magnetometer readings from the surface until NASA shut down communications at midnight UT, 28 February 2001.

NEAR Science Payload

Instrument	Mass (kg)	Power (W)
Multispectral Imager System	10	7
X-ray/Gamma-Ray Spectrometer	26	31
Near-IR Spectrograph	18	9
Magnetometer	1	1
Laser Rangefinder	5	22

Specifications

Launched: 17 February 1996 by Delta II 7925 from Cape Canaveral.

Gravity assists and flybys: Mathilde flyby 27 June 1997.

Earth flyby 23 January 1998.

Eros flyby 23 December 1998.

Lifetime: 4 years.

Configuration: Three-axis stabilised. Dual-mode propulsion system. Octagonal bus, 1.7 m² aluminium honeycomb panels connected to forward and aft aluminium honeycomb decks. Instruments, solar panels and high-gain antenna all fixed and body-mounted. X-band high gain antenna. Two solid-state recorders: 1.6 gigabit capacity.

Mass: 805 kg. at launch. Propellants: 325 kg. *Experiments:* 56 kg.

Power system: Four fixed 1.8 × 1.2 m gallium arsenide solar arrays, providing 400 W at 2.2 AU (327 million km) and 1,800 W at 1 AU (150 million km). Super nickel cadmium (NiCad) battery.

New Horizons

Current Status

The New Horizons mission to Pluto and the Kuiper belt was launched on 19 January 2006 on a Jupiter gravity assist trajectory. Jupiter closest approach occurred on 28 February 2007. It will cross Saturn's orbit in mid-2008. The flyby of Pluto and Charon will take place on 14 July 2015.

Background

New Horizons is the first of NASA's New Frontiers missions. New Horizons will be the first spacecraft to visit Pluto and its moon Charon, as well as the first to fly past one or more Kuiper belt objects.

After the scrapping of NASA's Pluto Express proposal, New Horizons was competitively selected by NASA for development on 29 November 2001. NASA then shelved the mission after no funds were proposed in the 2003 White House budget, but Congress approved the necessary funds for the mission.

The Johns Hopkins University Applied Physics Laboratory (JHU-APL) manages the mission and operates the spacecraft for the NASA Science Mission Directorate. The Southwest Research Institute (SwRI) leads the science investigations. A number of other commercial, governmental and institutional partners are participating in the project, Including Stanford University, Ball Aerospace and NASA's Goddard Space Flight Center.

New Horizons' primary structure comprises an aluminium central cylinder that supports honeycomb panels and served as the payload adapter fitting that connected the spacecraft to the launch vehicle, supports the interface between the spacecraft and its power source, and houses the propellant tank. The propulsion system includes 16 hydrazine thrusters mounted in eight locations. Four 4.4 N thrusters are used mostly for course corrections. Twelve 0.8 N thrusters are used for attiude control and spin stabilisation. Eight of the 16 thrusters are considered the primary set; the others are retained as a backup. At launch, the spacecraft carried 77 kg of hydrazine, stored in a lightweight titanium tank. Helium gas is used to pressurise the system. The spacecraft is covered in lightweight, gold-coloured, multilayered thermal insulation blankets.

The command and data handling system uses a radiation-hardened 12-megahertz Mongoose V processor that distributes operating commands to each subsystem, collects and processes instrument data, and sequences information sent back to Earth. It also runs the advanced "autonomy" algorithms that allow the spacecraft to check the status of each system and, if necessary, correct any problems, switch to backup systems or contact Earth for help. New Horizons has two low-power solid-state recorders (one backup) that can each hold up to 8 gigabytes (64 gigabits) of data.

Electrical power is provided by a single Radioisotope Thermoelectric Generator (RTG), supplied by the US Department of Energy (DoE). The RTG contains 11 kg of plutonium dioxide. At the start of the mission, the RTG provided 240 W at 30 volts. Due to the decay of the plutonium, the power output decreases about 3.5 W per year. By July 2015, the RTG will be producing 200 W. The science instruments combined draw less than 28 W. New Horizons uses an RTG because solar power cannot be employed during this mission, and therefore the RTG is suitable due to its proven reliable power supply and operation under severe environmental conditions for a number of years.

In August 2006, Pluto was reclassified by the International Astronomical Union (IAU) as a 'dwarf planet', and lost its status as the ninth planet in our Solar System. Astronomers project that there may be hundreds of dwarf planets in solar orbit, at least some of them larger than Pluto. The controversy as to whether Pluto is a planet has continued among astronomers for many years, and this decision is not universally accepted. Pluto was demoted because, using the current definition of a planet, it "has not cleared the neighbourhood around its orbit, and is not a satellite."

On 19 January 2006, at 19:00 UT, the New Horizons spacecraft was launched from Launch Complex 41 at Cape Canaveral Air Force Station in Florida on board a Lockheed-Martin Atlas V-551 fitted with a Centaur upper stage and a Boeing STAR-48B solid-propellant rocket third stage. This combination enabled New Horizons to leave Earth at a speed of 57,935 km/h, making it the fastest spacecraft ever launched. It reached lunar orbit distance (about 384,000 km from Earth) approximately nine hours after launch.

The first phase of the mission, which lasted approximately 13 months, was called the 'early cruise' phase. During this time, spacecraft and instrument checkouts, instrument calibrations, trajectory correction manoeuvres and rehearsals for the Jupiter encounter took place. On 13 June 2006, it flew about 102,000 km from asteroid 2002 JF56 and determined its colour, shape and size. The first images and other Jupiter data were obtained on 4 September 2006.

The second phase, the Jupiter gravity assist, occurred on 27 February 2007 (05:45 UTC on 28 February) at a distance of 2.305 million km (less than 32 Jupiter radii), just outside the orbit of Callisto. This was more than four times closer to Jupiter than Cassini in 2000. The flyby took place at a speed of 21 km/s relative to Jupiter: As a result of the gravity assist, New Horizons' speed away from the Sun increased by nearly 4 km/s and its trajectory was raised 2.34° above the ecliptic plane.

During the flyby, all of the spacecraft's instruments were used. Observations included studies of Jovian meteorology, the Great and Little Red Spots, the rings and auroras; sampling of the magnetosphere and dust environment; and ultraviolet mapping of the gas ring (torus) around Jupiter's moon, Io. Surface mapping, compositional mapping and atmospheric studies of Jupiter's largest moons were also conducted during about 100 image sequences. It became the first spacecraft ever to fly down the Jovian magnetotail.

The third phase, the 'Interplanetary Cruise,' will last approximately eight years, with the spacecraft in hibernation mode for much of the time. In this mode, most of its electronics will be turned off and the spacecraft will rotate at 5 rpm with its high gain antenna pointing toward Earth. The onboard flight computer will monitor system health and broadcast a weekly beacon tone through the medium-gain antenna. New Horizons will transmit a "green" coded tone if all is nominal, or one of seven coded "red" tones if it detects a problem and requires help from the operations team. The second cruise phase will also include annual spacecraft and instrument checkouts, trajectory corrections, instrument calibrations and Pluto encounter rehearsals in preparation for the final phase of the mission.

The cameras and spectrometers on New Horizons will start taking data on the Pluto system five months before the spacecraft arrives at Pluto and Charon. About three months prior to closest approach, about 100 million km from Pluto and Charon, the first visible and near-infrared maps will be produced. From then on, there will be continuous imaging and spectral measurements of Pluto, Charon, and the small moons Nix and Hydra as each rotates. For the last four Pluto days before encounter (26 Earth days), the team will compile maps and obtain spectral measurements of Pluto, Charon, Nix and Hydra every half Pluto day in a search for short-term changes. Pluto and Charon each rotate once every 6.4 Earth days. The spacecraft will also study ultraviolet emissions from Pluto's atmosphere.

In current mission plans, the spacecraft will approach as close as 12,500 km to Pluto and approximately 27,360 km from Charon, 14 minutes

NASA's New Horizons spacecraft will make the first ever flyby of Pluto and Charon on 14 July 2015. The RTG is just visible to the right (JHU-APL)
1341681

after Pluto closest approach. This will enable close-up pictures in both visible and near-infra-red wavelengths with a spatial resolution of about 100 m. New Horizons will approach Pluto's sunlit southern hemisphere at a solar phase angle of 15°. Pluto will then be more than 4.9 billion km from Earth, so the one-way time delay for a radio signal to reach New Horizons from Earth will be 4 hours and 25 minutes. As it moves away from the system, the spacecraft will observe the mostly dark side of Pluto and Charon to look for haze in the atmosphere, possible rings and surface topography. Flying through the shadows cast by Pluto and Charon it will detect radio waves from Earth and use the occultation measurements to reveal the composition, structure and thermal profile of Pluto's atmosphere. It may then be targeted towards a Kuiper belt object.

Science Instruments
The New Horizons spacecraft carries seven instruments: three optical instruments, two plasma instruments, a dust sensor and a radio science receiver/radiometer. They are:

- **Ralph** Using two separate channels, this imaging instrument will obtain high-resolution colour maps and surface composition maps of both Pluto and Charon. The Multispectral Visible Imaging Camera (MVIC) and Linear Etalon Imaging Spectral Array (LEISA) make up the two channels, and a single 6 cm aperture telescope is used to focus the light which is used in both channels. MVIC operates at visible wavelengths, using four different filters to produce the colour maps. One of these filters will measure the distribution of methane frost over the surface, while the other filters will cover blue, red and near-infra-red colours. LEISA operates at infra-red wavelengths and uses its etalon to analyse each wavelength separately. LEISA will also be used to measure the distribution of methane frost, molecular nitrogen, carbon monoxide and water over the surface of Pluto and the water frost distribution over Charon (SwRI).
- **Alice** This spectrometer will use ultraviolet imaging to provide information on the atmospheric composition of Pluto. The instrument uses two modes of operation: an 'airglow mode' and an 'occultation mode'. The airglow mode allows the measurement of emissions from atmospheric constituents; whereas the occultation mode is used when the Sun or a bright star is viewed through the atmosphere, thus producing absorption by the atmospheric constituents. The latter mode will be used when the New Horizons spacecraft passes behind Pluto and looks back at the Sun through the planet's atmosphere (SwRI).
- **REX (Radio Experiment)** is an essential part of the integrated New Horizons radio telecommunications system as all communication, including the transfer of science data, occurs through this radio package, thus playing a critical part in the mission's success. REX will also be used to probe Pluto's atmosphere. The powerful radio transmitters of the Deep Space Network will be used to send radio signals to the spacecraft. As the spacecraft goes past Pluto, REX will record these detected radio waves and finally send them back to Earth for analysis. In addition to the above, REX's radiometry mode will also measure the weak radiation emission from Pluto when looking back at the planet after the flyby, this will then allow scientists to calculate a temperature value for the night side of Pluto (Stanford University).
- **LORRI (Long Range Reconnaissance Imager)** – consists of a telescope with a 20.8 cm aperture, which focuses visible light onto a Charged Couple Device (CCD), resulting in the provision of the highest spatial resolution on New Horizons. During closest approach, LORRI will capture images of Pluto's surface with a resolution of approximately 100 m across (JHU-APL).
- **SWAP (Solar Wind Analyser around Pluto)** will be used to measure the charged particles from the solar wind close to Pluto, in order to determine whether Pluto has a magnetosphere and the rate at which the atmosphere of Pluto is escaping (SwRI).
- **PEPSSI (Pluto Energetic Particle Spectrometer Investigation)** will be used to detect neutral atoms escaping from Pluto's atmosphere and subsequently becoming charged due to their interaction with the solar wind (JHU-APL).
- **Student Dust Counter** is an education and public outreach project, will be used to count and measure the number and sizes of dust particles along the entire trajectory of the New Horizon's spacecraft (University of Colorado, Boulder).

Specifications
New Horizons
Launch: 19:00 UT on 19 January 2006 by Atlas V-551 from LC-41 Cape Canaveral.
Mass: 478 kg at launch; 77 kg hydrazine fuel; 30 kg science instruments.
Body dimensions: 0.7 × 2.1 × 2.7 m
Communications: X-band. 2.1 m diameter high-gain antenna. High-gain beam is 0.3° wide. Data rate: 38 kilobits per second at Jupiter; 0.6 to 1.2 kilobits per second at Pluto. 30 cm diameter medium-gain dish antenna with 14° beam. Two broad-beam, low-gain antennas on opposite sides of the spacecraft for near-Earth communications.
Power: One RTG containing 11 kilograms (24 pounds) of plutonium dioxide. 240 Watts BOL. 200 W by July 2015.
Propulsion: 16 hydrazine thrusters for trajectory adjustments and attitude control. Eight primary and eight back-up. Four 4.4 N thrusters and twelve 0.8 N thrusters.

New Millennium series

Current Status
NASA's New Millennium programme has supported five missions, one of which, Deep Space 2, failed. Earth Observing 1 remains operational as of February 2008. Two further missions are planned for launch by 2010. Earth Observing 3, which was scheduled for launch in 2006, was cancelled, primarily due to lack of sufficient Navy funding for launch services.

Background
NASA's Office of Space Science (OSS) and Office of Earth Science jointly established the New Millennium Programme (NMP) in 1995. 23 partners from industry and universities – representing all sectors of the US technological community, were selected to work alongside NASA. They were to participate in four of the five Integrated Product Development Teams (IPDTs) that were to lead in the development and delivery of selected advanced technologies in four primary spaceflight development areas: autonomy, communications, microelectronics and modular architectures and multifunctional systems. A fifth IPDT was selected to focus on science instruments and microsystems, electronic systems and mechanical systems.

The low cost programme covers the identification, development and flight validation of key, innovative technologies for incorporation into 21st century science missions without the risks inherent in their first use. NASA achieves this by conducting frequent and affordable science missions with highly focused objectives: micro-spacecraft carrying advanced miniaturised instruments, returning a continuous information flow. New Millennium demonstrates revolutionary technologies and architectures. The New Millennium flights also provide opportunities for meaningful science, although the programme was refocused in 2000 so that the emphasis is now on technology development and science is no longer a requirement driver.

The programme encompasses small (USD50 million – USD100 million) projects with shared launches (e.g. Space Technology 7) to medium USD100 million – USD150 million) class projects (e.g. Deep Space 1) involving system-level technologies for flight validation. It also identifies subsystem technologies (such as those on Space Technology 6 and Space Technology 8) costing USD25 million – USD40 million. Both classes of technology experiments help enable NASA's Discovery and Explorer Programme (medium-class, small, and university-class) space science missions. Some NMP projects (Deep Space 1, Earth Observing 1, Space Technology 5 and Space Technology 7) incorporate several spacecraft systems and instrument technologies on a single flight validation mission. Other projects (Deep Space 2, Earth Observing 3, and Space Technology 6) fly new instrument concepts for technological advances as secondary payloads (experiments) on spacecraft that have other primary objectives.

The programme is now funded and managed solely under NASA's Science Mission Directorate (SMD). The Jet Propulsion Laboratory (JPL) manages the New Millennium programme, and the advanced technologies for flight validation are identified and developed by integrated product development teams of representatives from NASA, industry, academia and other government centres.

Deep Space 1
DS1, the first New Millennium mission, was a comet/asteroid flyby using solar electric propulsion. The industrial partner for mission definition and implementation was Spectrum Astro, Inc. DS1 weighed 365 kg at launch, including 125 kg advanced technologies, 23 kg hydrazine, and 58 kg Xe. Its design incorporated a fixed 1.5 m X/Ka-band HGA, two LGAs, two star trackers and hydrazine ACS into an aluminium bus, 96 × 112 × 79 cm with a hexagonal spaceframe. Autonomous remote software planned and executed activities such as autonomous optical navigation. The optical system imaged asteroids and comets against star background for position and trajectory information as well as imaging target for flyby corrections. Two 152 × 533 cm solar concentrator arrays generated 2.6 kW.

DS1 carried 12 instrument demonstrators, including the SEPS, a solar concentrator array, a miniature integrated camera/spectrometer, a small deep space transponder, an autonomous optical

JPL and NASA Lewis developed this 30 cm {70} 92 mN Xe thruster under the NASA SEP Technology Application Readiness (NSTAR) project to validate low-power ion propulsion technology. It was carried on DS1. The thruster is shown before a 2,000 h test at LeRC; a 7,000 h test at JPL was also conducted in 1996 (NASA) 0517282

navigation, an integrated space physics package, a beacon mode operations, a composite HGA, and a 3D stack processor. DS1 was launched by Delta II from Cape Canaveral at 13:08 UT on 24 October 1998. On 29 July 1999 DS1 flew to within 15 km of the asteroid 9969 Braille. The primary mission ended on 18 September 1999 but an extended mission was approved and took DS1 to comet Borrelly for an encounter in September 2001. DS1 was retired on 18 December 2001.

Deep Space 2

DS2 flew two microprobes that piggybacked on NASA's Mars Polar Lander to demonstrate key technologies that would have enabled network science as well as *in situ* and sub-surface science data acquisition analysis. Within their non-ablative aeroshell, each had a micro-telecommunications subsystem with programmable transceiver, power microelectronics with mixed digital and analogue ASIC, an ultra-low temperature lithium battery, a microcontroller, flexible interconnects for system cabling, a meteorological high-g pressure sensor, a sample/H_2O experiment for soil conductivity, and a high-g temperature sensor. Each probe's mass was 6.5 kg (including the interface structure and aeroshell). Microinstruments in the lower spacecraft (forebody) were to collect a sample of soil and analyse it for water content. Data from the forebody would be sent through a flexible cable to the upper spacecraft (aftbody) at the surface. A telecommunications system on the aftbody would relay data to the Mars Global Surveyor spacecraft in orbit around the planet. The DS2 probes failed to provide information to ground controllers in association with the loss of the carrier, Mars Polar Lander, in December 1999.

Earth Observing 1

EO-1 was developed, built and integrated by Swales Aerospace of Beltsville, Maryland, as prime contractor. It was launched by Delta 7320 on 21 November 2000 and inserted into an orbit flying in formation with Landsat 7 for comparison of 'paired' images. EO-1 weighs 370 kg without its payload and weighed 529 kg at launch; it was sent to a 98° sun-synchronous, 705 km low Earth orbit in a position which follows Landsat 7 by one minute, or approximately 450 km.

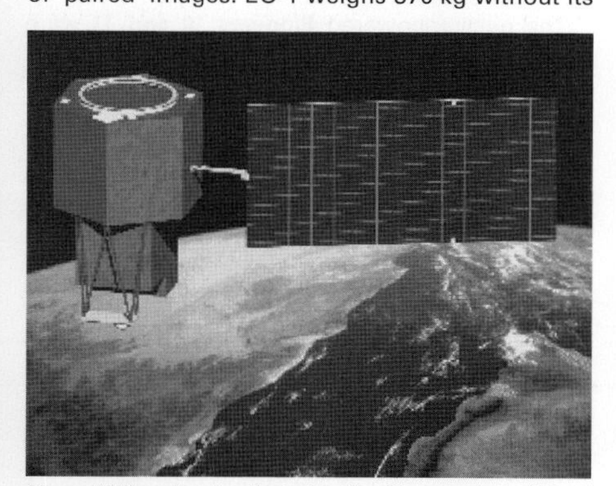

New Millennium's first Earth Observation (EO-1) mission carries an advanced land imager 0517296

As of February 2008, EO-1 was operational and returning data. It is on an extended mission; its primary validation mission was completed. Managed by NASA's Goddard Space flight Center, EO-1 is the responsibility of NASA's Earth Science office and ensures the continuity of Landsat data. Three land imaging instruments collect multispectral and hyperspectral data in co-ordination with the Enhanced Therratic Mapper (ETM+) on Landsat 7. EO-1 provides on-orbit demonstration of spacecraft technologies to enable future spacecraft of this type to be a magnitude smaller and lighter than current versions.

Beginning in 2004, onboard decision making artificial intelligence software, called Sciencecraft (see Space Technology 6, below), was installed on EO-1 via datalink. Sciencecraft notices and points the spacecraft toward unexpected geological, topographical or weather changes upon Earth; the software then decides whether the information is valid and meaningful, and then records the changes without being commanded to do so.

Earth Observing 3

EO-3 was a joint NASA-Office of Naval Research mission. The technology demonstration was to carry a Geostationary Imaging Fourier Transform Spectrometer (GIFTS) and the Indian Ocean METOC Imager (IOMI) on a Geostationary Meteorological Satellite (GMS) satellite. Originally set for launch during the 2005–2006 time frame; EO-3 was supposed to have launched as a primary payload aboard a Delta II DoD flight from Cape Canaveral. The US Air Force announced in 2003 that EO-3's launch date would be delayed, since the US Navy encountered a budget shortfall; the launch was at first delayed to 2008 or beyond and now has been cancelled completely. It would have tested advanced technologies for measuring temperature, water vapour, wind and chemical composition at high resolution. EO-3 would have expanded observations of the atmosphere and its oceans from a few spectral bands to several hundred.

Space Technology 5

The three ST5 microsats were launched on 22 March 2006 aboard a single Pegasus XL from Vandenberg; the mission was successfully completed 30 June 2006. ST5 operated in a 'string of pearls' constellation from polar orbit, the primary mission being to take multi-point measurements of Earth's magnetic field.

Also historically known as the Nanosat Constellation Trailblazer mission, ST5 comprised three very small satellites, or microsats, each octogon shaped and measuring 53 × 48 cm. The full service microsats each weighed about 25 kg when fully fuelled. The purpose of ST5 was to validate key technologies for future microsat constellations. Technologies included a Cold Gas Micro-Thruster (CGMT); an X-Band Transponder Communication System; Variable Emittance Coatings for Thermal Control; and CMOS Ultra-Low Power Radiation Tolerant (CULPRiT) Logic.

Space Technology 6

The ST6 mission validated two technologies: the Autonomous Sciencecraft Experiment (Sciencecraft), and the Inertial Stellar Compass (Compass). Although a part of the ST6 programme, Sciencecraft was flown aboard EO-1 (see above). Sciencecraft demonstrated onboard decision making artificial intelligence capabilities and was installed aboard EO-1 in 2004 using the spacecraft's datalink. Sciencecraft was developed in partnership between NASA, academia, the military, and several commercial companies.

Compass was launched aboard a Minotaur I rocket from the Wallops Flight Facility Mid-Atlantic Regional Spaceport (MARS) on 16 December 2006, as a payload on the TacSat-2 (see separate entry in the Technology Demonstration section) microsatellite. TacSat-2 is a joint US Air Force, Navy and DoD project. Compass, designed by Draper Laboratory, is a 3 kg, low-power instrument that enables the satellite to autonomously reposition itself using a miniaturised star camera and gyro system. Compass also allows the microsat to adjust its attitude and orientation following power loss or spacecraft malfunction. TacSat-2 ceased operations on 21 December 2007.

Space Technology 7

In March 2002, NASA announced that it had selected the Disturbance Reduction System (DRS) for the ST7 project. The DRS will be flown on the European Space Agency's Laser Interferometer Space Antenna 'LISA' Pathfinder mission in late 2009 or 2010. DRS incorporates advanced thruster and sensor technologies to enable precision control of spacecraft. It consists of clusters of micronewton thrusters and controlling software on a dedicated computer. DRS is based on the concept of a freely floating mass, shielded from non-gravitational forces, on board a host spacecraft. This technology validation is important for future interferometer missions that require precise spacecraft position control, as well as missions to measure gravitational forces and test Einstein's theory of relativity. One such mission is the joint ESA and NASA Laser Interferometer Space Antenna (LISA).

Space Technology 8

ST8 is scheduled for launch on 28 February 2009, into a sun-synchronous, elliptical 300 × 1,300 km orbit at 98.6° inclination. Its mission will last seven months.

Much like Deep Space 1, ST8 will validate four technology demonstration instruments aboard a single, hexagonal spacecraft: the Dependable Multiprocessor; UltraFlex 175; the Scalable Architecture for the Investigation of the Load Managing Attributes of a SlenderTruss (SAILMAST) Ultra Lightweight Boom; and the Thermal Loop. The Dependable Multiprocessor uses Commercial Off The Shelf (COTS) software, the aim of which is to validate cost-effective, existing technologies adapted for use in the space environment. UltraFlex 175 is a lightweight, circular, fanned, deployable, Gallium Arsenide (GaAs) solar array and will provide the spacecraft's solar power. SAILMAST is a deployable mast that will be used to support large lightweight solar sails, solar arrays and other large lightweight technologies. The Thermal Loop will manage and maintain the spacecraft's temperature using structures and methods appropriate for microsatellite-sized spacecraft.

Space Technology 9

ST9 is a future mission, still in the planning stages. NASA has not yet specified the technologies that will be validated on this mission, but the demonstrations will be selected from specific concept areas. These include: precision formation flying; descent and terminal guidance technology for pinpoint landing and hazard avoidance; solar sail flight system technology; system technology for large space telescopes; and aerocapture system technologies for planetary missions.

Phoenix

Current Status

The Phoenix lander, the first NASA Mars Scout mission, was successfully launched on 4 August 2007. The launch was the longest ever to involve a Delta II, lasting nearly 88 minutes from lift-off to separation, after completion of nearly one orbit around the Earth. It was also the first time that the inner wall of the rocket's fairing was baked to destroy any organisms that could contaminate the scientific instruments on Phoenix. The launch was so accurate that an extra 10 kg of manoeuvring propellant will be available during the descent and landing on Mars. The cruise stage deployed its solar arrays within three minutes of separation from the Star 48 solid propellant upper stage. After the first and largest of six course corrections planned during the flight to Mars was successfully completed on 10 August, Phoenix was travelling at 33,180 m/s in relation to the sun. The spacecraft fired its four mid-size thrusters for three minutes and 17 seconds, changing its velocity by about 18.5 m/s.

For details of the latest updates to *Jane's Space Systems and Industry* online and to discover the additional information available exclusively to online subscribers please visit
jsd.janes.com

Phoenix is targeted to touch down within a landing ellipse near Mars' North Pole on 25 May 2008. The favoured landing site is 68°N, 233°E, in Vastitas Borealis. Part of the US Mars Exploration Programme, Phoenix was built and tested by Lockheed Martin Space Systems. NASA's Jet Propulsion Laboratory (JPL) is managing the mission, along with the University of Arizona's Lunar and Planetary Laboratory, where the Principal Investigator (PI) is located.

Phoenix is a lander programme; once it touches down on Mars, it will collect its site samples in a stationary state; it will not roam the planet like the Mars Exploration Rovers (see separate entry). Phoenix will help scientists answer the following scientific questions: can the Martian arctic support life; what is the history of water at Mars' North Pole; and how is the Martian climate affected by polar dynamics?

Phoenix was named after the legendary bird that resurrects itself from its own ashes, since the lander's instruments have been resurrected from past cancelled and failed NASA Mars programmes.

Background

NASA has created a new line of small 'Scout' missions that are competitively selected from proposals submitted by the scientific and aerospace community. Exciting concepts could be developed into solid science contributions by means of this innovative approach, either through observations made from UAVs, networks of small surface landers, or from highly focused orbital laboratories. NASA aims to compete Scout missions as often as possible, potentially every four years, depending on resource availability. The second Scout mission is scheduled to launch in 2011.

Phoenix's total cost is USD420 million, including launch. The Scout programme initially instituted a cost cap of USD325 million, when the first projects were competed in 2002. Selected as the first Scout programme mission in August 2003, Phoenix beat 24 other proposals, including three other finalists: Sample Collection for Investigation of Mars (SCIM); Aerial Regional-scale Environmental Survey (ARES); and Mars Volcanic Emission and Life Scout (MARVEL), each of which received USD500,000 in NASA funding to conduct feasibility studies before final programme selection.

Phoenix carries upgraded duplicates of instruments that were intended to fly on the failed Mars Polar Lander mission, as well as on the cancelled lander segment of the 2001 Mars Odyssey/Surveyor mission, to an ice-rich region close to Mars' North Pole. It will scoop up soil to analyse at the landing site and send back its results about the history of Martian water and the possibility of current or past life via the Mars Reconnaissance Orbiter (see separate entry). Imaging data returned from the Mars Reconnaissance Orbiter will also help the Phoenix team identify the most appropriate landing site for the spacecraft.

Phoenix is scheduled to reach its landing site on 25 May 2008. The lander will conduct its primary science investigations over a three month period, until the sun completely sets over the Pole, and Mars enters its winter season, probably coating the lander with a layer of ice. A stereo colour camera and a weather station will study the surrounding environment while other instruments examine excavated soil samples for water, organic chemicals and conditions that could indicate whether the arctic site was ever hospitable to life.

Although developed, managed and funded primarily by NASA and other US commercial and academic interests, the Phoenix programme is an international endeavour, including scientific participants from the Canadian Space Agency, engaging in its first Mars mission; the Max Planck Institute in Germany; the University of Neuchatel in Switzerland; the universities of Copenhagen and Aarhus, Denmark; and the Finnish Meteorological Institute.

Specifications

Launch: 4 August 2007 at 10:26 GMT from Cape Canaveral Pad 17A aboard a Delta II 7925 launch vehicle

Spacecraft: Cruise vehicle dimensions: diameter 2.64 m, height 1.74 m; span of cruise solar arrays 3.6 m. Lander dimensions: height to top of meteorology mast 2.2 m, or slightly less depending on legs absorbing impact; span of deployed solar arrays 5.52 m; deck diameter 1.5 m; length of robotic arm 2.35 m. Launch mass: 670 kg, consisting of 350 kg lander plus cruise stage, aeroshell, parachute and propellant.

Built by Lockheed Martin Space Systems and based upon cancelled 2001 Mars Odyssey/Surveyor Lander technology; beginning in 1999, the lander was stored in a clean room, until the Phoenix programme resurrected the original, unfinished Surveyor vehicle in 2003. The spacecraft's upgraded design includes an aeroshell, inside which the lander will cruise to Mars, as well as the standard propulsion, command, data handling, telecommunications, navigation, thermal and power subsystems.

Science Payload: **Robotic Arm (RA)**: built by JPL, the 2.35 m long RA will dig into the Martian crust, up to 0.5 m to collect soil and ice samples for analysis by Phoenix's other instruments. The end of the RA is equipped with a scoop, tines and serrated blades to delve into the Martian surface. A similar robotic arm was first developed for the Mars Polar Lander.

Robotic Arm Camera (RAC): a joint project of the University of Arizona and the Max Planck Institute for Solar System Research, RAC was originally built for the Mars Surveyor lander. The RA mounted RAC will image in colour as the RA digs trenches and collects samples, and will give the science team a view of the RA's operations and soil samples.

Surface Stereoscopic Imager (SSI): SSI will provide high resolution, panoramic images of the Martian Arctic landscape about the lander; SSI is a stereoscopic camera, set on a mast about 2 m in height, and acts much like a pair of eyes. It will provide range mapping as well as taking atmospheric dust and cloud measurements. SSI can look down at the sample collections as well as looking up and out at the landscape, supporting RAC and adding multispectral image capabilities to the soil analysis process. SSI was developed by the University of Arizona.

Thermal and Evolved Gas Analyzer (TEGA): the RA will deliver soil and ice samples to TEGA's eight single-use ovens, which will heat the eight unique samples in temperatures up to 1,000°C, so that TEGA's mass spectrometer can analyse the vaporised samples' chemical composition and characteristics, including identification of possible organic molecules. TEGA was built by the University of Arizona and the University of Texas, Dallas.

Microscopy, Electrochemistry, and Conductivity Analyzer (MECA): MECA will examine soil samples delivered to it by the RA to determine soil and ice chemical properties, mineralogy, and water ice content. MECA is a "mini-lab" composed of four scientific instrument processes. It includes a wet chemistry lab, with four sample beakers filled with soaking solution; 69 wheel-mounted substrates, designed to attract and adhere to specific sample types; optical and atomic force microscopes to examine the wet samples; and a thermal and electrical conductivity probe that will insert its prongs into the trench dug by the RA. MECA was built jointly by JPL, the University of Arizona, and the University of Neuchatel, and was originally designed for the Mars Surveyor mission.

Mars Descent Imager (MARDI): MARDI will colour image the Martian landscape in wide angle as Phoenix descends into the Arctic, attached to its parachute. MARDI will begin its image series about 8 km above Mars, after Phoenix's aeroshell is jettisoned. Malin Space Science Systems built MARDI, an instrument originally designed for use on both the Mars Polar Lander and Mars Surveyor missions.

Meteorological Station (MET): Project managed by CSA and built by MD Robotics and Optech Inc, MET will monitor landing site weather. MET equipment includes a LIght Detection And Ranging (LIDAR) instrument, and atmospheric temperature and pressure sensors. MET will help the Phoenix team better understand the solid and gas water cycles in the Martian arctic.

Pioneer series

Current Status

NASA finally lost contact with Pioneer 10 after 31 years. As of late January 2008, Pioneer 10 was about 95 AU from the Sun and 94.3 AU (14.1 billion km) from Earth. Its speed relative to the Sun was 12.122 km/s. Contact with Pioneer 11 ceased on 30 September 1995 when its power levels dropped too low to operate its instruments and transmit data. In late January 2008, Pioneer 11 was 75.4 AU from the Sun. Its speed relative to the Sun was 11.507 km/s. The spacecraft are heading towards interplanetary space in opposite directions. Pioneer 6 was last contacted on 8 December 2000, on its 35th launch anniversary. At the time of contact, Pioneer 6 was the oldest partially functional NASA spacecraft still in orbit. Contact may also be possible with Pioneers 7 and 8.

The flight of Pioneer 10. Its firsts include: main asteroid belt crossing January 1973; Jupiter flyby December 1974; crossing of Saturn's orbit February 1976; crossing of Uranus' orbit August 1979; crossing of Neptune's orbit June 1983 (NASA) 0516751

Background

The name "Pioneer" was first used for a series of launches towards the Moon, initiated by the US Advanced Research Projects Agency in 1958 and developed by the Air Force and Army. The first DOD launch was a failure, and was named **Pioneer 0**. **Pioneer 1**, launched on 11 October 1958, was nominally the first NASA deep space probe, having been transferred to NASA by the USAF. Pioneers 1–4 failed to attain their lunar target but probed Earth's radiation belts and magnetosphere. **Pioneer 4**, launched in March 1959, passed the Moon at a distance of about 60,000 km and went into solar orbit. The 43 kg **Pioneer 5**, launched on 11 March 1960, also went into solar orbit and sent back data from deep space. It monitored solar particle radiation and provided up to 15 days' warning of solar flares until 26 June 1960.

The relatively simple Pioneer craft paved the way for more sophisticated missions to study the Sun, Jupiter, Saturn and Venus. In the late 1950s, Ames officials proposed a series of spacecraft that would fly in heliocentric orbit near the Earth, collecting information about the space environment. NASA Headquarters gave its approval for Ames to begin work on the Pioneer project in 1962. **Pioneer 6** was launched on 16 December 1965, followed by **Pioneer 7** in August 1966, **Pioneer 8** in December 1967 and **Pioneer 9** in November 1968. A fifth spacecraft (**Pioneer E**) was lost due to a Delta launch failure.

The virtually identical Pioneers 6 to 9 were cylindrical and spin stabilised, with experiment booms designed to collect data on the interplanetary environment and the magnetosphere, including the solar wind, solar cosmic rays and solar flares. They were the first spacecraft to detect the Earth's magnetotail, measuring some 5.6 million km on the anti-solar side. Originally designed for six months' operation, they provided coverage from widely separated points during an entire solar cycle of 11 years and were all still predicting solar storms at the end of 1981. Pioneer 9 was the first to fail, in 1983. Pioneer 6 may still be re-activated after more than four decades in orbit.

The nuclear-powered **Pioneers 10 and 11** were designed to be the first spacecraft to explore the outer Solar System. Pioneer 10 was the first to leave Earth, in March 1972. With an initial speed of 51,800 km/h, it was the fastest man-made object ever launched. Pioneer 11 followed in April 1973. These flights introduced a new generation of planetary probes designed as precursors to more sophisticated vehicles, such as Voyager. Pioneer 10 became the first man-made object to fly beyond Mars, the first to enter the asteroid belt and the first to reach Jupiter. It proved that the asteroid belt contains fewer dispersed particles than at one time believed and showed that traversing vehicles faced little hazard from collisions.

Pioneer 10 crossed Neptune's orbit in May 1983, and for many years was the most remote man-made object in the solar system. (On 17 February 1998, it was 'overtaken' by Voyager 1.) It became the first spacecraft to reach 50 AU from the Sun, on 22 September 1990. After a gravity assist from Jupiter in December 1974, Pioneer 11 looped across the Solar System and made the first flyby of Saturn, passing 20,920 km above its cloud tops on 1 September 1979. It was the first spacecraft to detect Saturn's magnetic field and radiation belts. It also discovered the F ring and a small satellite.

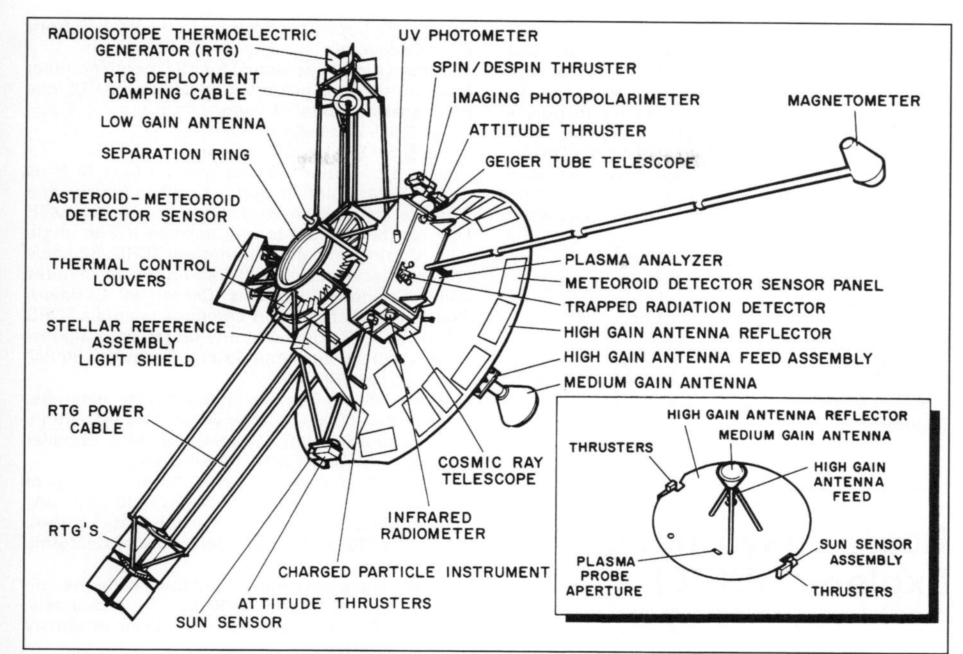

Pioneer 10 and 11 principal features (NASA) 0516752

Pioneers 10 and 11 have shown the heliospheric boundary with the interstellar medium to be significantly more distant than expected. Measurements of the solar wind indicate the boundary of the heliosphere may lie some 90–120 AU from the Sun. Inference on the heliosphere's extent is also provided by monitoring the solar modulation of cosmic ray intensities and from UV observations of changing atomic interactions with sunlight that result from incursion of extra-solar particles into the heliosphere.

In October 1990, Pioneer 11's primary receiver developed a problem that required very high power commands and prevented measurement of spin axis orientation. The radio science search for extra sources of gravitation ended with that failure.

Pioneers 10 and 11 were built to last for seven years of data transmission. Out of their total mass of 258 kg, they carried 30 kg of science instruments and 27 kg of propellant. The only drain on this supply was an insignificant use during periodic precession manoeuvres.

Six 4N thrusters each provided control and mid-course corrections and using the original propellant supply produced a total delta-V of 670 km/h. The Pioneers retained their attitude with respect to the fixed stars by two Sun and one star sensors mounted on a spin-stabilised platform having a full-circle scan rate of about four times a minute. The spin rate was found to be decreasing slowly; the cause is unknown but has been theorised as a build-up of electrostatic charge interacting with the interplanetary magnetic field. Four nuclear radio-isotope generators (RTGs), located on 2.7 m booms to avoid interference with the science experiments, provided power.

The heart of the communications system was the fixed 2.7 m diameter dish antenna and redundant 8 W S-band transmitter, 2,110/2,292 MHz up/down through NASA's Deep Space Network (DSN).

Pioneer 6 was the first spacecraft to operate for more than a quarter century. This is the similar Pioneer 8 undergoing final balance checkout at Cape Canaveral 0516750

Pioneer 10 and 11's instruments enjoyed a remarkable and unpredicted longevity and continued returning data on the solar wind from the outer reaches of the Solar System for an extraordinary period of time. Rather than instrument malfunction, operations failed due to decreasing power, which diminished by 81 m annually. Only those instruments with the lowest power consumption were operated toward the end, and even then not all simultaneously. On 31 March 1997 – more than 25 years after launch and nearly 10 billion km from Earth, Pioneer 10 fell silent. Only a few weeks' of data had been lost due to spacecraft problems during the entire mission.

Pioneer 10 was last able to send telemetry data on 27 April 2002. NASA received its last, faint signal on 23 January 2003, and the last, unsuccessful, attempt to communicate with the spacecraft was on 3–5 March 2006. Pioneer 10 continues to move outward at about 2.7 AU annually in a direction generally away from the Galactic centre and nearly opposite the direction of the basic Solar System motion with respect to the nearby stars. It is heading towards the bright star Aldebaran, in the constellation of Taurus, a journey that will take millions of years. Pioneer 11 continues to depart at about 2.48 AU per year, towards the Galactic centre in the general direction of Sagittarius. It will fly near the star Lambda Aquila about four million years in the future.

One of the most intriguing puzzles involving the Pioneer 10 and 11 spacecraft is the so-called "Pioneer Anomaly", first noticed as a navigation discrepancy while bouncing microwaves off each Pioneer probe. The data showed a tiny, but unexpected drift in each probe's Doppler frequency. The discrepancy found that Pioneer 10 and 11 were each about 400,000 km closer to the Sun than they should be according to Newtonian physics. Some force seemed to be pushing against the outbound spacecraft. Whether that force originated from the probes themselves, from some unusual source such as dark matter, or some new facet of physics or gravity, remains under study. However, analysis and modelling of how the Pioneer 10 spacecraft emits heat from various sources, including its RTG, found that they account for between 55 and 75 per cent of the Pioneer Anomaly. It has been suggested that heat from the Pioneers' electronics or RTGs could be emitting infra-red photons that are reflected from the main antenna, causing a recoil effect similar to a solar sail.

The last Pioneer spacecraft were **Pioneer Venus 1 and 2**, otherwise known as the **Pioneer Venus Orbiter** and **Pioneer Venus Multiprobe**, respectively. Pioneer Venus 1 was designed to study the planet's thick clouds and upper atmosphere, and make the first map of the hidden surface. Six months after launch, on 4 December 1978, it was injected into a highly elliptical orbit around Venus. Circling the planet every 23 hours 11 minutes, it passed within 150 km of the surface. Radar mapping took place around the close approaches. Although the radar instrument malfunctioned between 18 December 1978 and 20 January 1979, it mapped most of the surface between 73°N and 63°S at a resolution of 75 km. The map revealed broad plains, two large

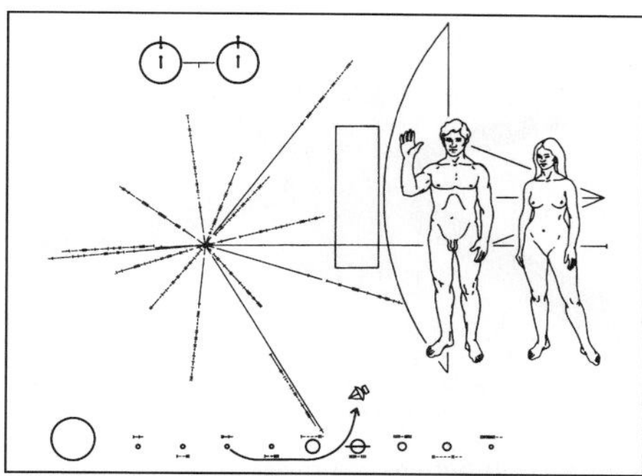

The Pioneer 10 and 11 craft both carry a 15×23 cm plaque indicating its origin should it be recovered by an intelligent alien species. The key is the hydrogen atom (top left), the most common in the Universe. 14 lines from Sun indicate distance to pulsars. Human figures, with the man's hand raised in goodwill gesture, are compared with the spacecraft to give scale. Bottom, plan of solar system, showing Pioneer originating from Earth 0516753

"continents" and many volcanic features. There were also global observations of the clouds and atmosphere, studies of the solar wind – ionosphere interaction, and observations of several comets, including Halley, with the ultraviolet spectrometer.

For the first 19 months, the periapsis of Pioneer Venus 1 remained at about 150 km. As propellant began to run low, the manoeuvres were discontinued, and the orbit's low point rose to about 2,300 km. From 1986, the orbit was lowered again, enabling further measurements within the ionosphere. The spacecraft was a flat cylinder. The scientific instruments and electronic subsystems were in the upper end. Below these were 15 louvers that controlled heat radiation from the spacecraft. There was a 1.1 m diameter, despun, high gain dish antenna which could be turned to continuously face the Earth. A solid-propellant rocket motor was used to brake into orbit. On 8 October 1992, with its fuel supply exhausted, the spacecraft burnt up in the planet's atmosphere.

Pioneer Venus 2 consisted of a Bus which carried four atmospheric probes. Their mission was to make a ballistic entry into the planet's dense, hot atmosphere and send back data to Earth during their descent. The drum-shaped Bus was similar to the Pioneer Venus Orbiter, but without a high gain antenna and orbit insertion motor. Although it had no heat shield, onboard instruments were able to return data before it burnt up in the upper atmosphere. Solar cells on the Bus provided power to all probes during the flight to Venus. The Large Probe was about 1.5 m in diameter. It carried seven science experiments, contained within a sealed spherical pressure vessel which was encased in a nose cone and aft protective cover. After deceleration from atmospheric entry at about 11.5 km/s near the equator on the Venus night side, a parachute was deployed at 47 km altitude. The three identical Small Probes were 0.8 m in diameter. They also consisted of spherical pressure vessels surrounded by an aeroshell, but they had no parachutes and the aeroshells did not separate during descent.

After a three month journey to Venus, the Large Probe was released from the Bus on 16 November 1978, followed by the small probes on 20 November. The probes and the Bus entered the planet's atmosphere on 9 December. The radio signals from all four probes were used to characterise the winds, turbulence, and other characteristics of the atmosphere. The small probes were each targeted at different parts of the planet. The North Probe entered the atmosphere at 59.3°N, 4.8°E on the day side. The Night Probe entered over the night side at 28.7°S, 56.7°E. The Day Probe entered at 31.3°S, 317°E, well into the day side, and was the only one to send radio signals back after impact, surviving for over an hour.

Reuven Ramaty High Energy Solar Spectroscopic Imager (RHESSI)

Current Status

In February 2006, NASA's Sun-Solar System Connection Senior Review endorsed the continuation of the RHESSI mission, stating that

Artist's impression of RHESSI (NASA Goddard Space Flight Center/Spectrum Astro Inc) 1133888

'future contributions promise to be compelling'. RHESSI was highly rated by the panel, achieving top marks for NASA programme relevance and scientific value.

Background

The primary objective of the Reuven Ramaty High Energy Solar Spectroscopic Imager (RHESSI) is to explore the basic physics of particle acceleration and explosive energy release in flares. The RHESSI spacecraft was launched on board the Pegasus XL vehicle on 5 February 2002 at 3:58 PM (EDT). The Principal Investigator oversees the programme from the University of California in Berkeley, where the primary ground station is also located. RHESSI is part of NASA's Explorer's programme and is managed by NASA's Goddard Space Flight Center (see separate entry). Spectrum Astro, Inc, acquired in 2004 by General Dynamics, built the spacecraft bus. A number of academic institutions are also involved in the RHESSI project.

When a flare occurs, it is believed that much of the energy is used for the acceleration of electrons, protons and other ions to very high energies. The electrons which are accelerated are those that primarily emit X-rays, and the ions emit gamma-rays. The RHESSI mission uses a combination of high-resolution imaging in hard X-rays and high-resolution spectroscopy with gamma-rays to obtain a detailed energy spectrum, thus allowing researchers to further understand the fundamental high-energy processes that occur at the core of the solar flare problem. Due to the proximity of the Sun, it is easier to resolve the intense high-energy emissions both spatially and temporally.

The RHESSI mission aims to investigate and understand the impulsive energy release and the acceleration of particles that occur within the magnetised plasmas of the solar atmosphere. These high-energy processes are found all over the universe, from magnetospheres to active galaxies. Therefore, researchers hope that via the use of RHESSI's scientific instruments, further knowledge will be gained within the fields of space physics and astrophysics.

Specifications

The single spin-stabilised RHESSI spacecraft rotates at approximately 5 rpm and is inclined 38 degrees to the Earth's equator within a low altitude orbit. As the spacecraft points close to the Sun's centre, this allows researchers to achieve the required rotation in order for RHESSI's imaging technique to be successful. The low altitude equatorial orbit was chosen in order to prevent the charged particles within the Earth's radiation belts from damaging the germanium detectors.

The imaging spectrometer on board the spacecraft allows scientists to receive movies (in colour) of the solar flares within X-rays and gamma rays. This imaging capability is achieved via the use of two complementary technologies: fine grids made of tungsten and/or molybdenum modulate the solar radiation and germanium detectors are used to measure the energy of each photon. The use of the aforementioned complementary technologies means that up to 20 detailed images are obtained each second, thus allowing scientists to track the acceleration of electrons from the solar corona (the acceleration site) to the lower solar atmosphere. In addition, ground based optical and radio telescopes will provide complementary data on the regions where X-ray and gamma-ray emissions are generated.

The spacecraft bus carries a honeycomb structure equipment platform, the power system (i.e. battery, solar panels, and control electronics), the attitude control system, thermal control, Command and Data Handling (C&DH) and telecommunications components. The Imaging Telescope Assembly, constructed and tested at the Paul Scherrer Institut in Switzerland, comprises the telescope tube, grid trays, the Solar Aspect System (SAS) and the Roll Angle System (RAS).

The germanium detectors contained within the spectrometer are cryogenically cooled and maintained at a temperature of –198°C; a high voltage is then put across them, in order to convert the incoming X-rays and gamma-rays to pulses of electric current. The germanium crystals were artificially grown and manufactured by the ORTEC division of PerkinElmer. Attenuator discs are used in order to protect the germanium detectors from intense solar flares and prevent saturation effects.

Height	2.16 m
Width (before solar panel deployment)	5.76 m
Width (after solar panel deployment)	1.1 m
Total Weight	293 kg
Total Power Consumption	414 W
Onboard Memory	4 Gbytes
Telemetry	Up to 11 Gbits/day (3.5 Mbits/sec)

Rossi X-ray Timing Explorer (RXTE)

Current Status

The Rossi X-ray Timing Explorer (RXTE) remains operational, 11 years after its launch, and continues to process and relay data on black holes and numerous stars. The spacecraft is healthy, however sporadic problems with the HEXTE instrument portion, notably a Cluster A rocking anomaly where the on- and off-source modulation stopped, have been reported; team members have delivered software to RXTE that will alleviate data collection problems caused by this anomaly.

Background

RXTE carries three instruments for intensive studies of x-ray source luminosity variations, for the first time ranging from μs to months, to provide information on processes and structures of white dwarfs, black holes, neutron stars and other exotic objects. The large effective area (0.8 m^2) and broadband sensitivity (2–200 keV) are valuable for timing intensity variations and determination of spectra from high energy sources.

The Proportional Counter Array (PCA) and High Energy X-ray Timing Experiment (HEXTE) form a single telescope, viewing a single source in their common 1° FOV. The large areas and low backgrounds provide high sensitivity to weak sources. The All-Sky Monitor (ASM) scans 70 per cent of the sky every 1.5 hours to monitor about 75 of the sky's brightest sources. It thus measures long-term variations and identifies changes for PCA/HEXTE to study at short notice. The microprocessor Experiment Data System (EDS) provides onboard processing/compression of PCA/ASM data from 500,000 bit/s to the telemetry rate.

Originally simply the X-ray Timing Explorer, RXTE was renamed in February 1996 after x-ray astronomy pioneer Professor Bruno B Rossi. Satellite cost, including ground equipment, was USD195 million; annual operations cost is about USD11 million.

Specifications

Contractors: NASA Goddard (project management; development/integration of observatory, development of PCA), MIT (ASM, Experiment Data System), UCSD (HEXTE)

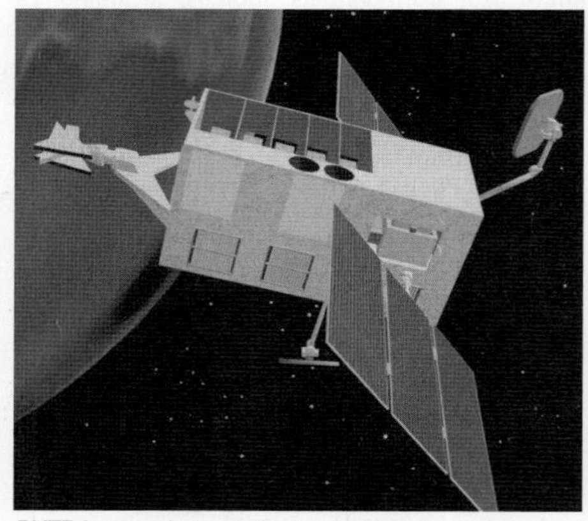

RXTE is aimed at monitoring luminosity variations of X-ray sources (NASA) 0517297

Launch: 30 December 1995 by Delta 7920-10 from Cape Canaveral
Lifetime: Originally planned for a 24 month nominal mission; up to 5 years was desirable; RXTE has been operational for 11 years
Mass: 3,045 kg at launch
Orbit: 565 × 583 km, 23°, 96.1 min
Spacecraft: >6°/min manoeuvres via four 75 Nms Ithaco RWs, pointing accuracy <0.1° (knowledge from star trackers + gyros to 60 arcsec). PCA/HEXTE FOV can be pointed to any position if Sun angle >30°. Command/datalinks through TDRS: 32 kbit/s (26 kbit/s science) plus 256 kbit/s for 30 minutes daily. Science Operations Center at Goddard. Two pointable high gain antennas maintain TDRS link while XTE pointed at any target. Twin 3-panel rotating solar wings (total 17.86 m^2) provides 800 W; 2 × 50 Ah batteries
Payload: PCA. Five Xe proportional counters, 6,250 cm^2, 2–60 keV, 1 × 1° FWHM, 1 μs resolution, 0.1 mCrab sensitivity (10 min), 18 kbit/s. Smaller version flown on HEAO 1. NASA Goddard.

HEXTE. Two rocking (every 15 s) clusters of four NaI/CsI detectors, 1,600 cm^2, 20–200 keV, 1 × 1° FWHM (rocking), 10 μs resolution, 1 mCrab sensitivity (10^5 s), 5 kbit/s. University of California at San Diego.

ASM. Three scanning shadow cameras on rotating boom, 90 cm^2, 2–10 keV in 3 channels, 0.2 × 1° FOV, 1.5 h resolution, 30 mCrab sensitivity (1.5 h), 3 kbit/s. MIT.

Solar Anomalous and Magnetospheric Particle Explorer (SAMPEX)

Current Status

The Solar Anomalous and Magnetospheric Particle Explorer (SAMPEX) was the first in the Small Explorer (SMEX) satellite series. Launched in 1992, SAMPEX was designed to operate for up to three years; as of early 2007, the satellite was still capable of returning data. In 2004, NASA discontinued SAMPEX funding, although the satellite was, after 12 years of service, functional. Since 1997, Mission support was run from the University of Maryland's Flight Dynamics and Control Laboratory in co-operation with NASA Goddard. Following project cancellation, Goddard was working with the Aerospace Corporation, for one year, on continued data return and the possibility of an extended mission. A new Data Center was constructed to handle SAMPEX data, as the satellite's data was not initially compatible with the World Wide Web and was not adequately accessible to the public.

Background

SAMPEX studies questions in space physics, astrophysics and upper atmosphere science. It is detecting solar/interplanetary charged particles and galactic cosmic rays of energies from 0.4 MeV to hundreds of MeV while over the poles, and observing magnetospheric electrons before they interact with the middle atmosphere. Of the instruments on board, LEICA and HILT were built for GAS (Getaway Special) Shuttle missions but not flown because of delays. Another instrument,

SAMPEX was the first in the Small Explorer series 0516741

MAST/PET, was partially fabricated to fly on the US probe half of Ulysses.

SAMPEX science operations began 10 July 1992. The LEICA initially had power problems, which later resolved themselves. MAST confirmed a belt of anomalous cosmic rays (particles resulting from the Sun's interaction with the interstellar medium) at 600 km within the inner van Allen belt.

Specifications

Contractors: NASA Goddard provided the satellite for the University of Maryland's payload
Launch: 3 July 1992 by Scout from Vandenberg AFB
Lifetime: 3 years nominal; operations far surpassed this goal
Mass: 158 kg (41 kg science)
Orbit: 550 × 675 km, 81.67°, 96 min
Spacecraft: 100 W orbit average provided by four GaAs solar panels (3.4 m^2 total) + one 9 Ah Ni/Cd battery (science instruments require 22 W). 32 Mbyte memory, telemetry by NASA standard 5 W S-band transponder: data down at 900 kbit/s for 10 minutes twice daily. Attitude control by MW, three magnetorquers, two-axis fine Sun sensor, four coarse Sun sensors, 3-axis magnetometer. 86 cm diameter × 1.5 m length launch envelope
Payload: LEICA Low-Energy Ion Composition Analyser (8.0 kg, 0.16–25 MeV, University of Maryland).

HILT Heavy Ion Large Telescope (21.0 kg, 3.9–177 MeV, Max Planck Institute Germany).

MAST Mass Spectrometer Telescope (7.5 kg, 7–470 MeV, California Institute of Technology).

PET Proton Electron Telescope (part of MAST, 1–500 MeV).

Solar Radiation and Climate Experiment (SORCE)

Current Status

The Solar Radiation and Climate Experiment (SORCE) spacecraft continues to operate and send data on incoming X-ray, ultraviolet, visible, near-infra-red, and total solar radiation. SORCE is operated and managed by the University of Colorado Laboratory for Atmospheric and Space Physics (LASP) for NASA's Earth Science Enterprise. Goddard Space Flight Center (GSFC) oversees the programme for NASA. Orbital Sciences provided the LEOStar-2™ spacecraft bus.

Background

The SORCE spacecraft, launched on board a Pegasus XL vehicle on 25 January 2003, provides detailed measurements of solar radiation. The aim of the mission is to study the effects of the Sun on the Earth and its subsequent effect on humankind. Specifically, the measurements of incoming X-ray, ultraviolet, visible, near infra-red and total solar radiation, particularly attend to topics such as long-term climate change, natural variability and enhanced climate prediction, and atmospheric zone and UV-B radiation. Also, this knowledge could be used to estimate future solar behaviour and climate response.

This is done by making precise measurements of the Total Solar Irradiance (TSI); the solar ultraviolet irradiance (120 to 300 nm, spectral resolution = 1 nm); visible solar irradiance for future climate studies (between 0.3 and 2 μm, spectral resolution = 1/30); and further understanding the variability occurring at the Sun and its subsequent effect on our atmosphere and climate.

Specifications

Spacecraft design: 3-axis inertially stabilised, zero momentum design (capable of precision pointing and attitude knowledge during science observations)
Satellite Mass: 290 Kg
Power: 248 W (six solar array wings generate power)
Communications: Redundant S-band Transceivers (compatible with NASA's ground network) Uplink: 2 kbps Downlink: 1.5 Mbps Data Storage: 1024 Mbit
Inertial Pointing: Slew Rate >1 degree per second Knowledge <60 arcsec
Solar Arrays: Fixed GaAs
Redundancy: Nearly fully redundant
Orbit: 645 km, 40 degree inclination
Mission life: 5 years (with a 6-year goal)
Dimensions: 62 in height, 45.5 in diameter
Launch Vehicle: Pegasus XL
Attitude control: Reaction wheels, torque rods, star trackers, sun sensors and magnetometers
Science data storage: More than 24 hours of spacecraft data
LEO-T ground stations: Wallops Island, Virginia and Santiago, Chile, two passes per day

The primary goal of the SORCE mission is to obtain precise measurements of the solar spectral irradiance at the ultraviolet, visible, and near-infra-red wavelengths. Four instruments, three spectrometers and one photometer, are used in order to achieve this goal.

SORCE Instrumentation

Total Irradiance Monitor (TIM): Provides measurements of the Total Solar Irradiance (TSI). The instrument has a mass of 7.9 kg, 14 W power usage and 12.8 ° wide field of view
Spectral Irradiance Monitor (SIM): Provides the first long-duration solar spectral irradiance measurements in the visible and near infra-red regions. The wavelength range varies between 300 and 2000 nm; wavelength resolution = 0.25 to 33 nm; mass = 22 kg; peak power = 35.4 W at 25 V.
The Solar Stellar Irradiance Comparison Experiment (SOLSTICE): Performs daily measurements of the solar ultraviolet irradiance within the range 115 to 320 nm. The instrument has an absolute accuracy of 5%; a total mass of 36 kg; power usage of 33.2 W (total); and a resolution of 0.1 nm (Sun) and 1.1–2.2 nm (stars)
XUV Photometer System (XPS): Measures solar XUV irradiance. The instrument has a wavelength range of 1–34 nm and 121–122 nm; a wavelength resolution of 1–10 nm; an absolute accuracy of 20%; its mass (with GCI) is 3.6 kg; and average orbit power of 9 W

Solar Terrestrial Relations Observatory (STEREO)

Current Status

On 26 October 2006 UT, both STEREO spacecraft were successfully launched into highly eccentric Earth orbit from Cape Canaveral aboard a Delta II 7925-10L. Both STEREO observatories remained in Earth orbit until December, when STEREO A was transferred into Solar orbit following its Lunar swingby; STEREO B will remain in Earth orbit until the end of January 2007, at which point it will follow STEREO A into heliocentric orbit. Primary science operations will begin once the two spacecraft achieve their respective, opposite heliocentric orbits.

Background

The goal of STEREO's two-year mission is to use two nearly identical space-based observatories to provide the first ever three dimensional images of the Sun. These images will allow scientists to study Coronal Mass Ejections (CMEs) in further detail. The third mission in NASA's Solar Terrestrial Probes Program, the twin observatories will fly as mirror images of one another in order to provide 'stereo' views of the Sun.

STEREO will assist in providing alerts for Earth-directed solar ejections. Although coronal mass ejections are powerful drivers of the Sun-Earth connection, their origin and evolution is still not fully understood. The study of CMEs will allow scientists to further understand their structure and role in interplanetary science.

Specifications

Each observatory carries two instruments and two instrument suites, resulting in 16 instruments on each observatory. Both observatories launched onboard a single rocket; they will use lunar swingbys to place them in their respective heliocentric orbits. The "Ahead" observatory has been directed to a position ahead of the Earth, and the "Behind" observatory will be directed to a position trailing behind Earth in its orbit. STEREO A has completed this manoeuvre; STEREO B has completed its first Lunar flyby, but it requires two gravity assist manoeuvres in order to place it into heliocentric orbit opposite STEREO A. It will complete its second swingby in January 2007.

The Johns Hopkins University Applied Physics Laboratory (JHU/APL) designed and built both spacecraft; mission management is carried out by NASA Goddard. The Naval Research Laboratory leads SECCHI investigations; The University of California at Berkeley, the University of New Hampshire, and the University of Minnesota manage IMPACT, PLASTIC and SWAVES respectively.

STEREOs A and B Instrumentation

Sun Earth Connection Coronal and Heliospheric Investigation (SECCHI):

Four instruments are used on SECCHI: an extreme ultraviolet imager, two white-light coronagraphs and a heliospheric imager. These instruments aid the study of the evolution of CME's, from their creation at the Sun's surface, through to their eventual impact on Earth.

STEREO/WAVES (SWAVES): This interplanetary radio burst tracker will study the generation and evolution of radio disturbances from the Sun to the Earth.

In-situ Measurements of Particles and CME Transients (IMPACT): The 3-D distribution sample will allow scientists to study the plasma characteristics of solar energetic particles and the vector field.

PLAsma and SupraThermal Ion Composition (PLASTIC): Provides plasma characteristics of protons, alpha particles and heavy ions.

Stardust

Current Status

In October 2006, the proposed Stardust NExT mission was selected by NASA to complete a USD250,000 concept study. If selected as a fully funded USD35 million NASA mission of opportunity, Stardust NExT will make its way to the Comet Tempel 1, last visited by Deep Impact (see separate entry) in 2005, to investigate whether the comet's surface has changed after its closest pass to the Sun.

In 2006, the Stardust Preliminary Examination Team (PET) discovered methylamine and ethylamine, fixed nitrogen rich molecules, in the Comet Wild 2 samples Stardust sent to Earth. These organic molecules strengthen the hypothesis that comets could have provided to Earth some of the building blocks of life.

On 30 January 2006, the Stardust spacecraft was placed into hibernation mode to maintain it for possible future use. The Stardust mission itself

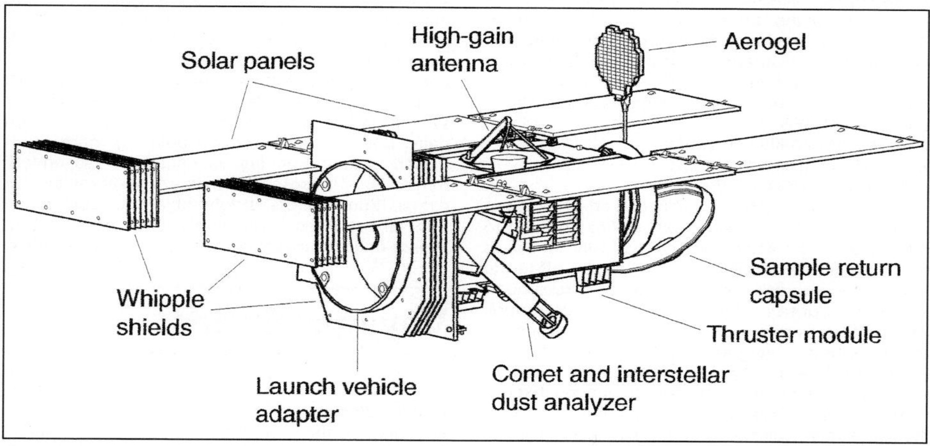

Stardust experiment and sample return capsule location 0089027

experienced a successful ending on 15 January 2006 when the sample return casing containing cometary and interstellar particles was successfully returned to Earth at 22:10 UT. Data analysis continues at NASA's Johnson Space Center.

Background
Stardust, managed by NASA's Jet Propulsion Laboratory (JPL), was one of NASA's Discovery Programme missions sponsoring low-cost solar system exploration projects with highly focused scientific goals. Stardust was selected in 1995 as a mission designed to obtain samples from the cloud of dust that surrounds the nucleus of comet Wild 2 and bring it back to Earth for analysis. Comets are thought to hold many of the original ingredients of the mixture of materials that formed the solar system almost 5 billion years ago, rich in organic material which provided the Earth with many of the molecules that gave rise to life. Due to their unique cocktail of organic and inorganic molecules, samples obtained directly from a comet will contain important chemical and physical information about the earliest history of the solar system.

Encounter with Wild 2 occurred on 2 January 2004 at a radius of 1.86 AU with Stardust travelling at a velocity of 6.1 km/sec relative to the comet. The velocity had been selected to optimise particle capture by the aerogel collector and passage through the most intense region lasted about eight minutes. Just before closest approach the spacecraft deployed the second side of the aerogel collector facing the direction of incoming particles. When Stardust completed its passage through the coma, the aerogel collector was stowed for the third and final time, sealing the interstellar and cometary particle collection inside the re-entry capsule.

Sub-millimeter Wave Astronomy Explorer Satellite (SWAS)

Current Status
The Sub-millimeter Wave Astronomy Explorer Satellite (SWAS) was in 'hibernation mode' in 2004, after a mission lasting more than five years, when it was called upon to support NASA's Deep Impact mission in 2005. SWAS was active for three months and continued to monitor water emissions from the comet Tempel 1 upon Deep Impact's (see separate entry) collision on 4 July 2005; it continued to support the mission through August 2005. The spacecraft returned new data until mid-2006. Afterwards, SWAS was sent back into hibernation mode.

Background
During its initial mission, SWAS observed dense galactic molecular clouds, particularly at water, molecular oxygen, neutral carbon and carbon monoxide wavelengths for information on chemistry, energy balance and structure. Ball Aerospace's Electro-Optics/Cryogenics Division received a USD14.5 million NASA contract in the autumn of 1989 for the primary SWAS instrument. For the first year of operation SWAS concentrated on exploratory observations of more than 70 targets in the galactic plane. It is one of NASA's Small Explorer (SMEX) missions.

Specifications
Contractors: NASA Goddard provided the spacecraft for the Harvard-Smithsonian Center for Astrophysics payload; Ball Aerospace (antenna, star tracker, instrument integration); Millitech Corp (sub-mm heterodyne receiver)
Launch: 9 December 1998 by Pegasus XL/Lockheed L-1011 from VAFB
Lifetime: nominally two years
Mass: 283 kg (science 102 kg)
Orbit: planned 637 × 653 km, 70°
Spacecraft: 97 cm diameter. Payload attaches to top of bus as single module. 3-axis stellar pointing to 38 arcsec (<19 arcsec jitter), typically three to five targets per orbit. Nods up to 3° off-target in 15 s increments every 40 s for background reading. Control by three magnetorquers, digital Sun sensor, five coarse Sun sensors, three MWs, magnetometer, three inertial gyros, high accuracy CCD star tracker. 250 W orbit average provided by four deployed fixed + one body panel (3.4 m^2 total GaAs) plus one 21 Ah Ni/Cd battery (science instruments require 50 W orbit average). 100 Mbyte bulk memory, 1.8 Mbit/s rate via NASA standard S-band 5 W transponder. 2 kbit/s commands up Telemetry: 2,215 MHz (realtime 18.75 kbps, playlock 1.8 mbps). 5 W transmitter. Command 2,039.65 MHz (2kbps).
Payload: 55 × 71 cm diameter off-axis f/4.8 Cassegrain antenna feeding two sub-mm receivers (cooled to 150K) for University of Cologne's acousto-optical spectrometer (1.4 GHz bandwidth, 1,400 channels, 2.2 GHz centre frequency).

Transition Region And Coronal Explorer (TRACE)

Current Status
The Transition Region And Coronal Explorer (TRACE) is in good health and remains operational, far surpassing its original one-year nominal mission; it continues to collect and return data. TRACE received adequate scientific value and programme relevancy scores from the 2005 Sun-Solar System Connection Review Panel; consequently, NASA will continue to fund TRACE for the foreseeable future. Spacecraft power degradation due to orbit decay, causing longer eclipse durations and multiple eclipse periods, will become an issue toward the end of 2008.

Background
TRACE is one of NASA's Small Explorer (SMEX) Missions, along with FAST, SWAS and WIRE. TRACE was the third in the series.

TRACE studies the connections between fine scale magnetic fields in the solar surface and features in the photosphere, chromosphere, transition region and corona. Scientists want to use it to explore the relation between diffusion of the surface magnetic fields and the changes in heating and structure throughout the solar transition region and corona. The simultaneous ultra-violet films of the 6,000–10,000,000,000 regions at 1 arcsec spatial and 1 second time resolutions will reveal the rate of change of magnetic topology. The launch coincides with the onset of a new solar cycle, so the emerging magnetic flux will be observed in a relatively uncomplicated atmosphere.

TRACE was built at the Lockheed Martin Solar and Astrophysics Laboratory (LMSAL), Palo Alto, California. TRACE's payload consists of a 30 cm diameter ultra-violet/extreme ultra-violet telescope with image motion compensation. Four quadrant normal-incidence coatings on the primary/secondary mirrors form identically-sized and perfectly co-aligned images. A 1,024 × 1,024 CCD collects images over 8.5 × 8.5 arcmin FOV. TRACE's image processor can continuously monitor the data stream and adapt the observing programme and instrument pointing. Guide telescope is integral part of satellite/instrument pointing system.

Specifications
Contractors: NASA Goddard provided the spacecraft for the Lockheed payload
Launch: 2 April 1998 by Pegasus XL/Lockheed L-1011 from VAFB
Lifetime: 1 year minimum
Mass: 250 kg (59 kg science)
Orbit: 599 × 641 km, Sun synchronous 97.8°, 97 min
Spacecraft: bus irregular octagonal prism, 86 cm diameter × 1.5 m length launch envelope. 220 W orbit average provided by four deployed GaAs wings + one 9 Ah Super Ni/Cd battery (science instruments require 83 W). 80C86 instrument processor, 300 Mbyte solid-state memory (66 used by error detection/correction), 2 kbit/s uplink, 0.0234/1.125/2.25 Mbit/s rate via NASA standard S-band 5 W transponder. Four to six daily passes at Wallops/Poker Flat; 320–480 Mbyte science down daily. 3-axis solar 20 arcsec pointing momentum biased around Sun line (stabilised by secondary mirror to <1 arcsec), 3-axis magnetometer, six coarse Sun sensors, bright object sensor, digital Sun sensor, three gyros, guide telescope (fine Sun sensor on instrument), three torque rods, four RWs.

Voyager Series

Current Status
Voyager 1 is the most distant human-made object from Earth. In August 2006, it reached 100 AU, about 15 billion km, from the Sun; Voyager 2 was at about 80.5 AU. The spacecraft travel about 1.6 million km per day. Voyager 1 reached the termination shock in December 2004 when it was 94 AU from the Sun; it will take about 10 more years for the spacecraft to pass completely from the Sun's influence, and enter interstellar space. Voyager 2 will reach the termination shock between 2008 and 2010. Voyager 1 has entered the Heliosheath and Voyager 2 will reach the Heliosheath after it passes the termination shock; the Voyagers will continue their journey on to the Heliopause and bow shock point. No human-made spacecraft has ever journeyed this far into space. NASA expects Voyager 1 to reach the Heliopause by about 2015, at which point it will enter interstellar space.

The Twin Voyagers have been on the Voyager Interstellar Mission (VIM) since 1989, after the completion of the 12-year Voyager primary mission. Both Voyager 1 and 2 are in good health, returning data to NASA; the spacecraft are tracked by NASA's Deep Space Network (DSN). The Voyagers will continue to collect and return data on the solar wind, and will be able to detect where the Sun's influence ends and interstellar space begins.

Although Voyagers 1 and 2 were the third and fourth spacecraft to exit the solar system, following Pioneers 10 and 11 (see separate entry), the Voyagers have now sped past the Pioneers and are the most distant human-made objects.

Background
Dr Gary Flandro of the University of Utah, first noticed that a rare alignment of the planets late in the 1970s allowed a single spacecraft to visit all of the major outer planets of the solar system. Working with NASA's Jet Propulsion Laboratory (JPL), he and JPL engineers conceived of the 'Grand Tour' in the 1960s. In the mission profile, gravitational forces of the outer planets served to swing one spacecraft in turn from Jupiter to Saturn, Uranus and Neptune and another from Jupiter to Saturn and Pluto. Financial cuts at NASA required to pay for the burgeoning budget of the Shuttle pared the project to a mere extension of the Mariner programme, which had explored the inner planets. Work began in January 1972 but the project underwent many changes. For example, NASA lacked the funds for Jupiter orbiters and postponed a visit until Galileo. The craft were designed, built and operated by JPL.

As the flight date neared, JPL prepared the craft for launch. But following the four-day road journey from JPL to Cape Canaveral, engineers discovered faults in the attitude and articulation control and flight-data subsystems of Voyager 2, due to depart first because its trajectory included the Uranus option. The two Voyagers thus had to be interchanged, with consequent swapping of equipment.

The overall project cost about USD320 million not including launchers and flight support activities, and NASA sources indicated the entire programme cost after 1986's Uranus encounter was USD600 million. Voyager 1 launched on a Titan-IIIE/Centaur on 5 September 1977 and came within 278,000 km of Jupiter on 5 March 1979, and 124,000 km of Saturn 12 November 1980. Voyager 2 launched on 20 August 1977, and came within 650,000 km of Jupiter on 9 July 1979, 101,300 km of Saturn on 25 August 1981, 71,000 km of Uranus on 24 January 1986 and 5,016 km of Neptune on 25 August 1989.

Voyager 2's Neptune fly-by formally ended 2 October 1989, the sixth and last planetary encounter of the two-craft programme. When this happened, NASA formally renamed the project Voyager Interstellar Mission (VIM) and estimated the programme had cost USD863 million. Under its new designation, the VIM cost was USD70 million through 1995, exclusive of tracking. Voyager's two Jupiter, two Saturn and the first Uranus and Neptune fly-bys sustained NASA's planetary programme during an 11-year dearth of launches. Voyager 1 captured the last of some 67,000 images in February 1990 to create a mosaic portrait of the solar system.

The Voyagers and Pioneers 10 and 11 (see separate entry) are exiting the Sun's system in different directions, heading towards the hypothetical heliopause, where the solar wind is turned back by the interstellar wind. Voyager 1 has been heading 35° N of the ecliptic since its November 1980 Saturn encounter; Voyager 2 has been on a 48° S trajectory since Neptune. Instruments no longer collecting data and non-essential heaters on Voyager 2's scan platform were turned off in November 1998 as a power conservation move. Five instruments continue to provide data. The Voyager 1 spacecraft scan platform was turned off in mid-2000 to

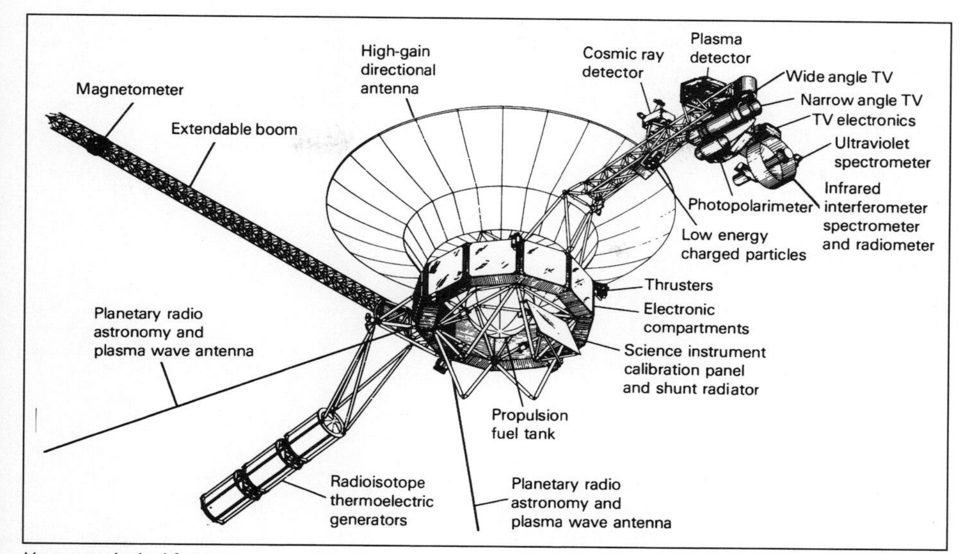

Voyager principal features 0516784

conserve power by saving 43.9 W. Both Voyager spacecraft confirmed the earlier Pioneer finding that the heliopause exists further than an earlier estimate of its position. The satellites scan the path ahead for low energy cosmic rays penetrating the outer reaches of the solar system from nearby supernovae remnants. The plasma wave receivers have been recording intense 2–3 kHz emissions since August 1992, believed to originate at the heliopause where the solar wind interacts with the cold interstellar plasma.

JPL's DSN continues to receive daily ultra-violet and fields/particles data. The Laboratory's scientists turned off the imaging cameras, photopolarimeter and IR spectrometer on both craft after Voyager 2's Neptune fly-by as they are of no further scientific use. The remaining seven instruments: the UV spectrometer; cosmic ray telescope; low energy charged particle experiment; magnetometer; plasma and plasma wave subsystems; and the planetary radio astronomy instrument continue to return data at a combined rate of 160 bit/s. Their data return and communications could be maintained to around 2017, when power levels will reduce to the minimum 230 W required for critical life functions. Voyager 1 and 2 will then be about 138 and 113 AU out. Electrical power outputs by April 2000 were 318 and 220 W (480 W at launch). With sufficient power, contact would be lost with their 20 W transmitters around 2033 because of the distances (29,000 million km for Voyager 1). Attitude control propellant is not a limiting consumable: the 32.05 and 34.07 kg of hydrazine remaining in each craft at April 2000 (104 kg at launch) is sufficient for operations well beyond 2020.

Specifications

Launch date: Voyager 1: 5 September 1977 on a Titan IIIE/Centaur from Cape Canaveral; Voyager 2: 20 August 1977 of a Titan IIIE/Centaur from Cape Canaveral.
Mass: 2,016 kg launch mass; Voyager 792 kg.
Communications: 3.7 m antenna dish, normally Earth-pointing (largest flown on a planetary mission). Transmitter power is 20 W.
Power: 3 RTGs, each 39 kg, encased in beryllium on a deployed boom. (To avoid their radiation, the most sensitive science instruments are mounted on an opposite 2.3 m boom).
Computation: 536 Mbit tape recorder memory (allowed the storage of the equivalent of 100 full-resolution images.) Six computers (three back-up), total 32 kbit capacity, reprogrammable to meet the needs of each encounter.
Propulsion: Spherical propellant tank at the centre of the basic 10-sided aluminium bus supplies hydrazine (104 kg loaded) for the 12 attitude control and four trajectory correction thrusters.
Payload: TV: 200 mm f/3 and 1,500 mm f/8.5 cameras.

IR spectrometer: 51 cm diameter gold-plated telescope covering 2,000 wavelengths over 4–50 μm ultra-violet spectrometer: 128 wavelengths over 500–1,700 Å

Photopolarimeter: 15 cm Cassegrain telescope

Cosmic ray telescope: covering 0.5–500 MeV

Low energy charged particle experiment: 0.01–11 MeV electrons and 0.015–150 MeV protons/ions in the planets' magnetospheres

Planetary radio astronomy: radio emissions

Plasma particles: 10–6,000 eV charged particles in the solar wind and the planet/moon magnetospheres

Plasma waves: 0.01-56 kHz oscillations in surrounding plasma to determine temperature and density

Magnetometers: magnetic fields around planets/moons and in interplanetary space; $20^{-2} \times 10^{-8}$ gauss

Each Voyager carries a 30 cm gold-plated copper 'Golden Record' containing greetings in 55 languages, sounds of birds, whales and other animals, samples of music, and 115 analog images, in case the spacecraft are ever retrieved by intelligent beings. The record's aluminium jackets carry playing instructions using the cartridge and needle provided along with our Sun's location, a hydrogen atom diagram, and a uranium-238 "clock". Dr Carl Sagan chaired the committee that assembled the recordings and images.

TECHNOLOGY DEMONSTRATION

France

Cerise

Current Status
Cerise collided with a piece of debris from an Ariane third stage, which orbited Spot 1 in 1986, making it the first proven casualty of space debris. The collision with the debris 1986-019RF occurred at 09.48 GMT on 21 July 1996 breaking off the upper portion of the Cerise gravity gradient boom. French ground controllers regained control over the satellite following the collision and it remains operational.

Background
The 50 kg Cerise (Caractérisation de l'Environnement Radioélectrique par un Instrument Spatial Embarqué) was carried into orbit in order to characterise the Earth's radio environment in research that could lead to a 'Zenon' Elint satellite during the next decade. It came with Ariane 40 on 7 July 1995. Alcatel Espace built the satellites for DGA, using the University of Surrey's/Surrey Satellite Technology Ltd (SSTL) UoSAT bus. The French Ministry of Defence initially expected to use the satellite for about 2.5 years. In 1993, the MoD requested industry proposals for 1998's Clementine Cerise follow-on to investigate environmental interaction at different radio frequencies.

The Jason-1 Proteus bus undergoing integration at Alcatel Space's facility in Cannes (Alcatel Space/JL. Bazine) 1342240

Cerise is contributing to the development of an electronic intelligence satellite (SSTL) 0517268

Proteus bus

Current Status
The first mission to use Proteus was the Jason-1 ocean altimetry satellite, launched in December 2001. Jason-1 continues to operate and performance has exceeded specifications. The Proteus bus has also been used for the Calipso and COROT missions, both launched in 2006. Future Proteus-based missions are SMOS and Jason-2, both scheduled for launch in 2008. (See separate spacecraft entries).

Background
The generic Proteus bus, developed in partnership by CNES and Alcatel Space Industries, was intended as a low cost, minisatellite bus that would be adaptable and quick to develop for scientific or applications missions, using one or two Proteus buses according to needs. The bus systems are grouped according to function and easily tailored to specific mission requirements, such as the type of orbit and pointing, mass, and payload power consumption. The interfaces between the bus and payload are standardised to further facilitate mission adaptations. Integration equipment and facilities are in place to sustain the series, and the frequent recurrence in testing a similar platform guarantees a streamlined satellite validation process.

Proteus is designed for use in sun-synchronous, polar and near-equatorial orbits at altitudes from 500 to 1,500 km, with a minimum orbital lifetime of 5 years. It offers precision attitude control (0.02–0.05°) with the main instrument package aligned with the centre of the Earth, or precision inertial stabilisation and attitude control. Attitude can also be programmed to follow a customer-specified pattern. The platform provides 300 W of power, a 2-Gbyte mass memory and a 690 kbit/s S-band telemetry transmitter. The Proteus service module is compatible with all launch vehicles with a payload capacity between 500 and 1,000 kg. The precision AOCS (attitude and orbit control system) features a fully redundant configuration of rate gyros, star trackers, GPS receivers, reaction wheels and magnetic torquers. Delivery is advertised as within 27 months of an order.

CNES planned the Topex-Poseidon Follow-On mission (later named Jason-1) for launch in 1999 as the first of the 200 to 500 kg Proteus (Platforme Reconfigurable pour l'Observation, les Télécommunications et les Usages Scientifiques) multipurpose LEO platforms. Jason-1 was eventually placed in orbit by Delta II 7925 on 7 December 2001.

France showed its determination to make the TPFO/Jason-1 programme cost effective by establishing a capital cap on the programme of FFr100 million for the platform and payload accommodation. Contractors studied the programme goals in the Phase A investigation, which ran from November 1994 to April 1995. Then CNES selected Aerospatiale as the Phase B prime contractor in May 1996. A final contract called for one satellite plus one option.

After the launch of Jason-1, CNES ordered five buses in 2002, to be used for the Calipso, COROT, SMOS and Jason-2 missions. All of these missions carry larger payloads than Jason-1, necessitating a number of changes. Many of these were qualified by Calipso, which was launched on 28 April 2006, followed by COROT on 27 December 2006. In the case of COROT, the telescope and two cameras are mounted on the Proteus platform. Jason-2 is scheduled for launch in mid-2008, with SMOS scheduled in autumn 2008. Megha-Tropiques, to be launched on a PSLV in 2009, was originally intended as a Proteus spacecraft, but responsibility for the spacecraft was later switched to ISRO.

A Proteus Generic Ground System (PGGS) has also been developed in Toulouse to operate a Proteus spacecraft. It includes an Earth terminal (TTC-ET), a command and control centre (CCC) and a data communication network (DCN). It was used to perform satellite operations during the early phase of the Jason-1 mission (launch, station acquisition, in-orbit acceptance). Since the Project Operations Control Centre (POCC) took over routine satellite operations, it has performed spacecraft analysis and navigation functions.

India

GSAT

Current Status
GSAT-1 was successfully launched in April 2001, but placed in the wrong orbit. This adversely affected the planned in-orbit operations and the communications payloads have been shut down. GSAT-2 was launched in May 2003 and remains operational.

Background
India's GSAT satellites were intended to demonstrate indigenously developed satellite communications technologies, such as digital audio, data and video broadcasting.

GSAT-1 was designed with two S-band, a high power C-band and two indigenous C-band transponders, similar to the INSAT 2 satellites. It was launched on the first developmental flight of India's Geo-synchronous Satellite Launch Vehicle, GSLV-D1. The experimental communication satellite was successfully placed in GTO as planned. It was the first time that validation of an Indian geosynchronous launch had included acquisition of signals at satellite injection and during orbit raising manoeuvres through a network of international stations, coordinated and controlled by INSAT Master Control Facility at Hassan.

On 18 April 2001, GSAT-1 was placed in a transfer orbit of 181 km × 32,051 km, inclination 19.2°. Its velocity was 99.4 per cent of the intended 10.2 km/sec, but this resulted in reduction of its orbital apogee. This was probably due to a shortfall in the thrust of the GSLV's cryogenic third stage motor provided by Russia. Six subsequent orbital manoeuvres 19–23 April 2001, using the satellite's Liquid Apogee Motor, raised the orbit close to geosynchronous height with an apogee of 35,665 km, perigee 33,806 km, inclination 0.997°. The satellite used two different propellant tanks, built in Germany and India, which resulted in an unequal flow of fuel, causing the spacecraft to tilt. Recovery required the use of more propellant than planned, so GSAT-1 had no propellant left to complete its final circularisation manoeuvre. The antenna reflector, solar array and solar sail were successfully deployed, and the satellite was put in 3-axis stabilisation mode using momentum wheels. However, it was not possible to fulfil the original purpose of demonstrating digital TV and audio broadcasts, and Internet services.

To counteract the torque generated by the sun's radiation falling on the solar array on the south side, a solar sail mounted on a boom was located on the north side. Several new communication satellite technologies were evaluated using GSAT-1. These included: (i) Fast Recovery Star Sensor (FRSS) which provides enhanced accuracies of measuring satellite orientation and for quick Earth-lock recovery in case of loss of lock; (ii) a new Earth sensor using pyro electric detectors; (iii) a new technique of using 22 N thrusters in a combination of four as an alternate strategy for orbit raising; (iv) thermal control of satellites using axially grooved, aluminium ammonia heat pipes developed by the ISRO Satellite Centre, (ISAC) Bangalore; (v) a new power management technique.

GSAT-1 also carried two C-band transponders employing 10 W Solid State Power Amplifiers, one C-band transponder using 50 W Travelling Wave Tube Amplifier (TWTA) and two S-band transponders using 70 W TWTA. The Antenna Positioner Mechanism (APM) which switched the antenna beam between two locations was tested and a few experiments such as Digital Video Transmission, Digital Audio Transmission and Internet broadcasting were also carried out. Since the satellite was drifting in the wrong orbit, its fully functional transponders and transmitters were deactivated on instructions of the International Telecommunications Union.

GSAT-2 is an Indian experimental communication satellite launched by GSLV on 8 May 2003. It is regarded as part of the INSAT system. About 17 minutes after lift off, GSAT-2 was successfully placed in an orbit of 180.04 km × 36,000 km with an orbital inclination of 19.2°. The first signals from GSAT-2 were received at the ground station at Biak, Indonesia. GSAT-2 was tracked, monitored and controlled from ISRO's Master Control Facility (MCF) in Hassan. During the initial phase of operation, MCF also used INMARSAT's ground stations at Beijing (China), Fucino (Italy) and Lake Cowichan (Canada). The satellite's orbit was precisely determined by continuous ranging from the participating Telemetry Tracking and Command (TTC) ground stations.

Subsequent orbit-raising manoeuvres were carried out from MCF by firing the 440 N Liquid Apogee Motor to place it in geostationary orbit. At the end of these manoeuvres on 11 May, the satellite was allowed to drift towards its designated orbital slot at 48°E. In this drift orbit, deployment of the antenna and solar arrays was carried out on 12 May. Station acquisition manoeuvres were conducted 14–19 May by firing the 10 N Reaction Control Thrusters.

GSAT-2 carries four C-band transponders, two Ku-band transponders and a Mobile Satellite Service (MSS) payload operating in S-band forward link and C-band return link.

GSAT-2 also carries four piggyback experimental payloads:

- Total Radiation Dose Monitor (TRDM) to compare the estimated radiation doses inside the satellite with the directly measured radiation doses using a Radiation Sensitive Field Effect Transistor (RADFET).
- Surface Charge Monitor (SCM) to indicate the state of the charging environment in the vicinity of the spacecraft.
- Solar X-ray Spectrometer (SOXS) to study the solar flare emission in 4 KeV-10 MeV energy range using semiconductor devices and a Phoswich Scintillation Detector. During initial tests, a series of small solar flares was detected by SOXS on 8 June 2003.
- Coherent Radio Beacon Experiment (CRABEX) to investigate the spatial structure, dynamic and temporal variations of Ionosphere and several aspects of equatorial electrodynamics.

GSAT-3 (Edusat) was launched by GSLV in September 2004, and is covered in a separate entry.

GSAT-4 is envisaged as a technology demonstrator carrying a multi-beam Ka-band bent pipe and regenerative transponder and navigation payload in C, L1 and L5 bands. GSAT-4 will also carry a scientific payload, TAUVEX, comprising three UV telescopes developed by Tel Aviv University and the Israel space agency (ELOP) for surveying a large part of the sky in the 1400–3200 A wavelengths. GSAT-4 will employ several new technologies like electric propulsion with four stationary plasma thrusters, Bus Management Unit (BMU), miniaturised dynamically tuned gyros, 36 AH Lithium ion battery, 70 V bus for Ka-band TWTAs, on-board structural dynamic experiment, thermal control coating experiment and vibration beam accelerometer. GSAT-4 is planned for launch by GSLV during 2008.

GSAT-5/INSAT-4D will be an exclusively C-band communication satellite, with 12 normal C-band transponders and six extended C-band transponders with wider coverage in uplink and downlink coverage over Asia, Africa and Eastern Europe as well as zonal coverage with minimum of 35 dBW EIRP. It will be launched on board GSLV during 2009 and positioned at 82°E deg E longitude. During the year, GSAT-5/INSAT-4D payload configuration has been finalised. Spacecraft antenna configuration has been finalised to include two deployable shaped reflectors and one single shell shaped reflector.

GSAT-6/INSAT-4E will carry a digital multimedia broadcasting payload. The satellite will cover all of India through five S-band spot beams using SxC and CxS transponders. The satellite is planned to be launched during 2009-10 by GSLV.

GSAT-7/INSAT-4F is proposed as a multi-band satellite carrying payloads in UHF, S-band, C-band and Ku-band. It is planned to be launched during 2010–11 by GSLV.

GSAT-8/INSAT-4G is proposed as a Ku-band satellite carrying 18 transponders similar to that of INSAT-4A and INSAT-4B. It will also carry 2 BSS transponders and a GPS Aided Geo Augmented Navigation (GAGAN) payload. The satellite is expected to be launched during 2009–10.

Specifications

GSAT-1

Launch: 18 April 2001 by GSLV-D1 from Satish Dhawan Space Centre, Sriharikota, India

Orbit: 35,665 km × 33,806 km, 0.997°. (failed to reach geostationary orbit)

Mass: 1,540 kg at launch

Configuration: Box-shaped, 1.6 m × 1.7 m × 1.9 m. One asymmetric, accordion type solar panel offset on north side of the bus by an extendible solar sail.

Power: Single Asymmetric, accordion type solar array, 1.38 kW (BOL). Two 24 Ah Ni-Cd batteries.

Propulsion: 440 N Liquid Apogee Motor (LAM) and sixteen 22 N reaction control thrusters.

Payload: 2 C-band transponders, 10 W SSPAs; one C-band transponder, 50 W TWTA; two S-band transponders, 70W TWTA.

GSAT-2

Launch: 8 May 2003 by GSLV-D2 from Satish Dhawan Space Centre, Sriharikota, India

Orbit: 48°E geostationary

Mass: 1,800 kg at launch. 840 kg of propellant (Mono Methyl Hydrazine and MON-3).

Configuration: Box-shaped, 2.406 m × 1.651 m × 1.530 m with solar arrays on north and south side. 9.55 m span in orbit. 3-axis body stabilised using sun and Earth Sensors, momentum and reaction wheels, magnetic torquers and bi-propellant thrusters.

Power: Twin solar arrays, 1380 W, two 24 Ah Ni-Cd batteries.

Propulsion: 440 N Liquid Apogee Motor (LAM) and sixteen 22 N reaction control thrusters.

Payload: Communication Payloads: 4 C-band transponders, 2 Ku-band transponders, 1 MSS (18 MHz) payload consisting of S-band forward link and C-band return link.

Experimental Payloads:

Total Radiation Dose Monitor (TRDM) to compare the estimated radiation doses inside the satellite with the directly measured radiation doses using a Radiation Sensitive Field Effect Transistor (RADFET).

Surface Charge Monitor (SCM) to indicate state of charging environment around the satellite.

Solar X-ray Spectrometer (SOXS) to study solar flare emission in 4 KeV-10 MeV energy range using state of the art semiconductor devices and Phoswich Scintillation Detector.

Coherent Radio Beacon Experiment (CRABEX). To investigate the spatial structure, dynamic and temporal variations of Ionosphere and several aspects of equatorial electrodynamics

Space Recovery Experiment (SRE)

Current Status

The Space capsule Recovery Experiment (SRE-1) was launched by PSLV on 10 January 2007 and successfully recovered on 22 January 2007.

Background

India's first Space capsule Recovery Experiment (SRE-1) was intended to demonstrate the capability of conducting automated experiments in microgravity and recovering an orbiting space capsule. It was intended to test reusable thermal protection system, navigation, guidance and control, hypersonic aero-thermodynamics, management of communication blackout, deceleration and floatation system and sea recovery operations. It is also seen as a first step towards an Indian human spaceflight programme.

The INR20-25 crores (~USD5 million to USD6 million) SRE experiment comprised an aero-thermo structure, spacecraft platform, deceleration and flotation system and two microgravity payloads. It had a sphere-cone-flare configuration with a spherical nose of about 0.5 m radius, base diameter of 2 m and 1.6 m height. The capsule was made of mild steel and the heat shield comprised carbon phenolic ablative material and silica tiles developed by the Vikram Sarabhai Space Centre at Thiruvananthapuram. The parachute, pyro devices, avionics packages of triggering unit and sequencer, telemetry and tracking system and sensors for measurement of system performance parameters were inside the capsule. Pre-launch preparations included helicopter drop tests on land at Agra, in the Pulicat lake, off Sriharikota island and in the Bay of Bengal.

SRE-1 was launched as a co-passenger of Cartosat-2 on board PSLV on 10 January 2007. It was placed in a circular, sun-synchronous near-polar orbit at an altitude of 637 km. Over the next 12 days, the two experiments on board were successfully conducted under microgravity conditions. One of the experiments was related to the study of metal melting and crystallisation. Designed by the Indian Institute of Science, Bangalore, and Vikram Sarabhai Space Centre, Thiruvananthapuram, it was performed in an Isothermal Heating Furnace. The second experiment, designed by the National Metallurgical Laboratory, Jamshedpur, studied the synthesis of nano-crystals in order to design better biomaterials.

On 19 January 2007, in preparation for its re-entry, SRE-1 was commanded by the Spacecraft Control Centre (SCC) of ISTRAC at Bangalore to move into a 485 × 639 km orbit. The de-boost operations were executed from SCC, supported by a network of ground stations at Bangalore, Lucknow, Mauritius, Sriharikota, Biak in Indonesia, Saskatoon in Canada and Svalbard in Norway, as well as shipborne and airborne stations.

Re-orientation of the SRE-1 capsule for de-boost operations commenced at 08:42 am (IST) on 22 January 2007. The de-boost started at 09:00 am as it flew southward over Mexico with the firing of onboard rocket motors and the operations were completed at 09:10 am. At 09:17 am, the capsule was reorientated and re-entry began at 09:37 am over Lucknow, India, at an altitude of 100 km with a velocity of 8 km/s. A minute later, due to severe heating and formation of plasma, the capsule entered the RF blackout phase. During re-entry, the capsule was protected from the intense heat by carbon phenolic ablative material and silica tiles on its outer surface.

SRE-1 being dropped from an Indian Air Force helicopter near Sriharikota, 19 August 2004 (ISRO)
1343773

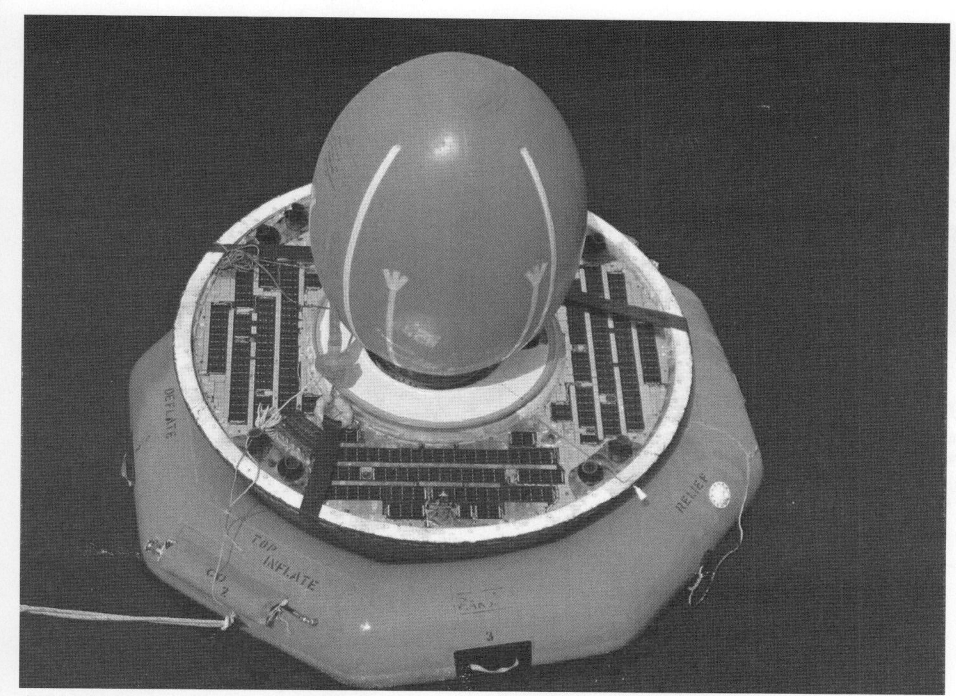

SRE-1 after splashdown in the Bay of Bengal, with its orange "balloon" and floatation collar (ISRO) 1343776

When it reached an altitude of 5 km, aerodynamic braking had reduced its velocity to 101 m/s. Deployment of the pilot and drogue parachutes further reduced its velocity to 47 m/s. The main parachute was deployed at about 2 km altitude and SRE-1 splashed down in the Bay of Bengal, about 140 km East of Sriharikota, with a velocity of 12 m/s at 09:46 am. The flotation system was immediately triggered, enabling the successful recovery of the capsule. Recovery operations were carried out by the Indian Coast Guard and Navy using ships, aircraft and helicopters.

Soon after the Coast Guard was informed that the satellite had splashed down, a helicopter took off from the Coast Guard vessel *Sarang*, which was anchored many kilometres away. The *Sarang* carried 16 ISRO personnel and divers. About 30 minutes after the SRE-1 landed, the helicopter spotted it bobbing in the water. The capsule was located with the aid of a UHF beacon that was triggered on hitting the water; and a greenish chemical released in the water. The divers were dropped into the sea and took underwater photographs to see whether the satellite was safe. It was later towed and put on board the *Sarang*. ISRO had earmarked a 30 × 15 km splashdown area.

A more complex version, SRE-2, has been approved and could take place in 2009.

Specifications
SRE-1
Launch: 10 January 2007 by PSLV-C7 from Satish Dhawan Space Centre, Sriharikota, India.
Orbit: Initial 637 km, 97.92°, sun-synchronous. 19 January 2007, changed to 485 × 639 km.
Recovery: 22 January 2007, Bay of Bengal about 140 km east of Sriharikota.
Design life: 30 days
Principal applications: Demonstration of low cost platform for microgravity science and technology experiments in space and return of specimens from space.
Contractors: Vikram Sarabhai Space Centre, ISRO Satellite Centre.
Configuration: Cone-flare configuration with spherical nose about 0.5 m radius, base diameter 2 m, height 1.6 m. Mild steel capsule, heat shield of carbon phenolic ablative material and silica tiles. Navigation, Guidance and Control
Processing: Mission Management Unit (MMU), Inertial Measurement Unit (IMU), sun sensors and magnetometer sensors. Actuators: magnetic torquer/22 N thrusters. Propulsion: eight 22 N bi-propellant thrusters. TT&C: S-Band with Belt Array antenna. Deceleration: three stage parachute system.
Mass: 555 kg at launch
Payload: Isothermal Heating Furnace (IHF) – heating of metallurgical samples. Growth of Ico Ga-Mg-Zn quasi crystal.

Biomineralisation of Inorganic Materials (Biomimetic) – synthesis of calcium hydroxyapite nano-crystals.
Power: Battery and body mounted solar panels.

Japan

ETS-VIII/Kiku-8

Current Status

Engineering Test Satellite (ETS-VIII/Kiku-8) was launched successfully in December 2006. In January 2008 two of its four ion thrusters malfunctioned, but the satellite remains operational.

Background

ETS-VIII is the eighth in Japan's series of Engineering Test Satellites. Its main purpose is to demonstrate mobile satellite communication system technology that will enable audio/data communications with hand-held digital terminals, such as mobile phones.

The ETS-VIII measures 40 m × 37 m after deployment of two solar arrays and two large deployable antennas. The satellite has two sets of two ion engines – System A and System B – on its northern and southern sides to perform orbital manoeuvres. The satellite is to verify the following four technologies in geostationary orbit:

1) An advanced, 3 ton spacecraft bus, comprising the satellite body, power system, command and data handling, thermal control and attitude control system. Major features of the ETS-VIII include a lightweight body structure to improve the payload to bus system ratio to 40 per cent, an increase in bus power supply voltage to 100 V, and use of CCSDS-compliant packet transmission and MIL-STD-1553B data bus. In addition, the satellite's heat pipe, which connects the north and south panels, expands the effective radiation surface while the attitude control system is equipped with fault tolerant functions and in-orbit programming capability.

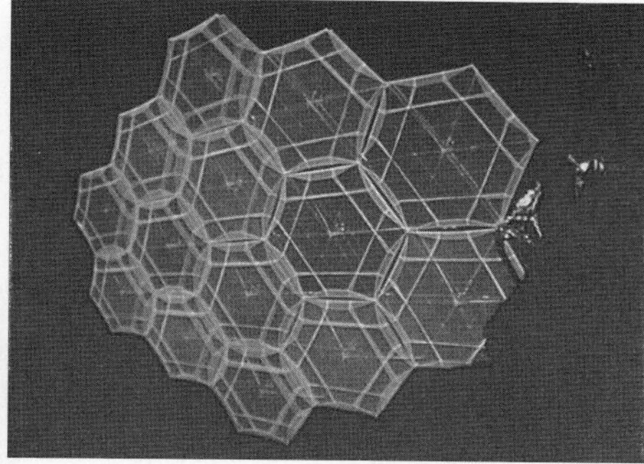
Deployment of the Large Deployable Receiving antenna on ETS-VIII was imaged by onboard cameras on 25 December 2006 (JAXA) 1343957

2) Two Large Deployable Reflector (LDRs) developed by the Japan Aerospace Exploration Agency (JAXA) to establish communication links with a cellular phone sized ground terminals. ETS-VIII is equipped with two LDR modules, the largest deployable antennas ever placed in orbit, one for data transmission and the other for reception. The LDR uses a modular structure to meet the requirements of reflector surface precision (2.4 mm rms surface precision) and antenna diameter expandability. During launch the LDR system is packed in a cylindrical container 1 m diameter by 4 m in length. Each LDR is deployed on orbit by expanding simultaneously 14 hexagonal umbrella-like modules which are connected to each other by cables. Once deployed, the reflector forms a parabola surface 19.2 m × 16.7 m across with a 13 m aperture.

The phased array feeder systems are combined with the LDRs to create multiple steerable Mobile Satellite communication Systems (MSS) and to support mobile Broadcasting Satellite System (BSS) experiments. A 31 element active phased array feeder of 400 W gross output and beam forming networks have been developed to synthesise signals into several beams to cover the entire country. The transmission side feeder consists of 31 solid state power amplifiers, and the receiving side feeder of 31 low noise amplifiers. The onboard beam-forming network controls the amplitude and phase of the signals to/from the feeder elements to create multiple steerable beams and to form the desired beam patterns.

3) High-Power Transponder (HPT), developed by Advanced Space Communications Research Laboratory (ASC). ETS-VIII is designed to conduct orbital experiments on mobile satellite communications and high-speed

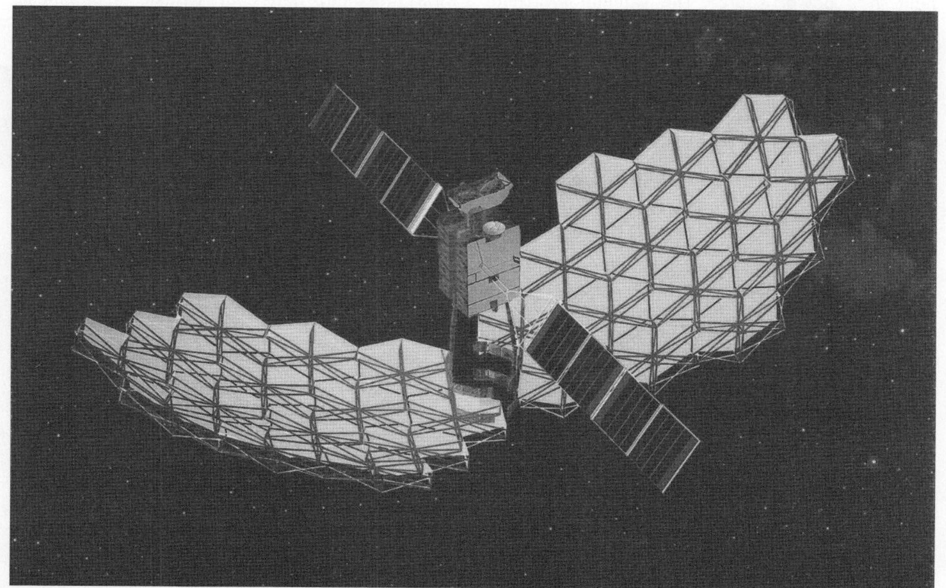
ETS-VIII is Japan's eighth engineering test satellite. It is designed to demonstrate mobile satellite communication technology that will enable audio/data communications with hand-held terminals. Its two deployable antennas form a tennis-court-size parabola (JAXA) 1343956

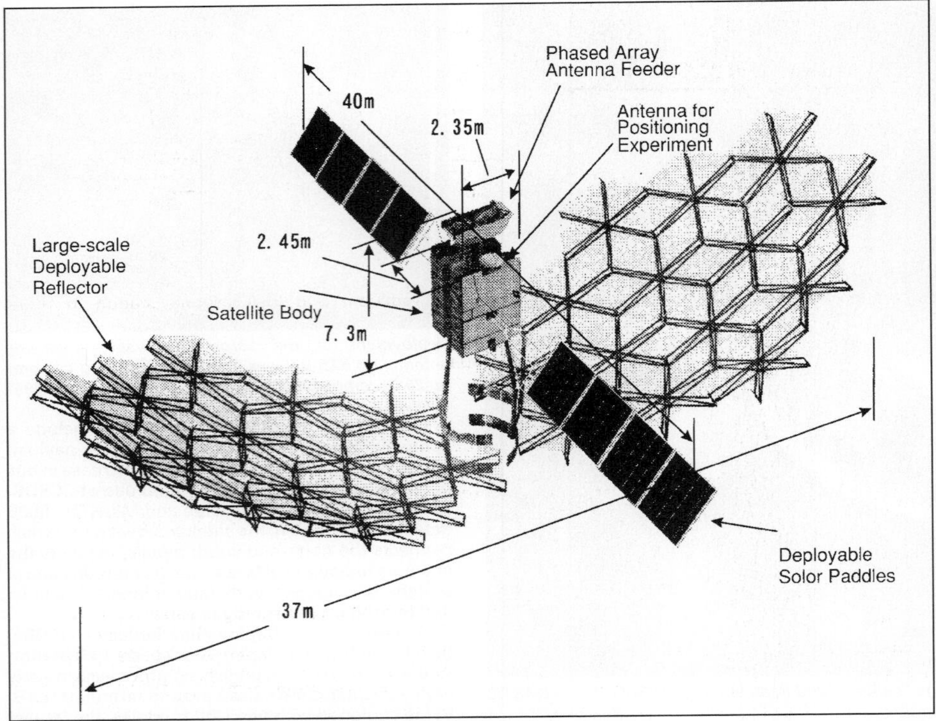

ETS-VIII is one of the largest satellites ever placed in geostationary orbit (JAXA) 1343961

packet communications, providing voice/data communications with hand-held terminals in the S-band frequency.

4) On-Board Processor (OBP), provided by National Institute of Information and Communications Technology (NICT). The processor switches links of cellular phones and high-speed packets, making it possible to establish a single-hop communication link with the ETS-VIII without ground switchboards along the path. Two sets of onboard baseband switching equipment are installed for signal routing. The OBP is designed to perform voice channel switching among mobile users or between mobile users and ground network users and has 500 channel capacity. The onboard packet switch (OPS) is used for high-speed packet data communications.

ETS-VIII also carries a high precision atomic clock and time transfer equipment for satellite positioning experiments, developed by JAXA and NICT. Combining the extremely accurate time signals with GPS data, ETS-VIII conducts experiments on basic satellite positioning system technology.

ETS-VIII was launched by H-IIA in December 2006. Successful deployment of the LDRs took place on 25 and 26 December. This was confirmed by telemetry and images taken by two onboard cameras.

In January 2008, problems were reported with two of its four ion engine thrusters. The southern thruster on System A failed 15 January due to a fault in a power source. The satellite switched to System B on 23 January, but controllers discovered the southern thruster on System B was unstable for continuous firing. JAXA stated that ETS-VIII was carrying sufficient propellant to substitute chemical thrusters for the duration of its mission.

Specifications

ETS-VIII/Kiku-8

Launch: 18 December 2006 by H-IIA F11 from Tanegashima Space Centre, Japan.

Location: 146° E geostationary.

Design life: Three years (mission); bus life 10 years.

Contractors: Mitsubishi Electric (prime). Advanced Space Communications Research Laboratory (High-Power Transponder). National Institute of Information and Communications Technology (On-Board Processor).

Transponders: Feeder link Ka-band, EIRP 44 dBW at NPR 20 dB, G/T 15 dB/K. Mobile Link (service link), S-band, EIRP 65 dBW at NPR 16 dB, G/T 15 dB/K, beam scanning range ±2°.

Principal applications: Demonstration of mobile satellite communication technology involving audio/data communications with hand-held digital terminals.

Configuration: Rectangular box with two deployable solar arrays. 3-axis stabilised using Earth sensor assembly, a sun sensor and integrated rate gyro assembly for attitude sensing. Attitude accuracy: roll/pitch axis ±0.05° max, yaw axis ±0.15° max. Four reaction wheels, each 50 Nms, in a skewed arrangement. 13 accelerometers to monitor dynamic behaviour of deployed structure. MIL-STD-1553B data bus for onboard data handling. Autonomous Fault Detection, Isolation and Reconfiguration (FDIR) functions.

Bi-propellant 500 N apogee engine, 22 N thrusters for attitude control and E-W station-keeping. Two xenon ion engines, thrust 25 mN each, for N-S station-keeping.

Mass: 5,800 kg at launch, 2,900 kg on orbit. Payload mass 1,100 kg.

TT&C: S-band. All data transmission with CCSDS protocols.

Power system: 7.5 kW (EOL at summer solstice) with regulated 100 V of bus voltage. NiH_2 batteries provide 100 Ah for eclipse operations.

Kirari (OICETS)

Current Status

Operations of Kirari, officially known as the Optical Inter-orbit Communications Engineering Test Satellite (OICETS), were completed on 16 October 2006.

Background

Kirari (meaning "glitter" or "twinkle" in Japanese) was launched by a Dnepr from Baikonur Cosmodrome, Kazakhstan, in August 2005.

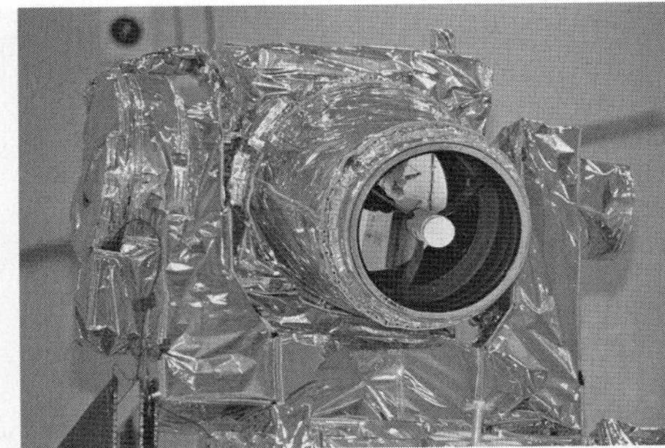
A close-up view of the Kirari laser utilising communications equipment (LUCE) (JAXA) 1343959

The Dnepr was chosen after a lengthy delay caused by the grounding of Japan's H-IIA launcher. It was originally intended to launch Kirari on a Japanese J-1 rocket during 2000.

Research and development for the Kirari/OICETS mission started in 1993, preliminary design was completed in 1994 and the confirmation tests for the onboard laser communications equipment were completed in 2000. This was a follow on from the flight demonstration of the Laser Communication Experiment (LCE) on Engineering Test Satellite VI (ETS-VI).

Kirari carried a reflecting telescope and other optical communication subsystems designed to enable optical inter-orbit communications tests with ESA's ARTEMIS satellite over distances of tens of thousands of kilometres, as well as tests involving ground stations. Such an optical inter-orbit communications system would offer various advantages: more stable communications with less interference; lighter, more compact communications equipment; and higher data transmission rates. The tests were expected to lead to new technologies that would support the development and utilisation of space, including global data reception from Earth observation satellites and continuous communication links with a crewed space station.

New technologies required for the tests included high-gain optical antennas, a high-power semiconductor laser, and highly sensitive signal detectors. This made it possible to test such operations as beam acquisition, reception of an incoming laser beam; beam tracking, detection and control of the angle of the incoming laser; and beam pointing so that a laser beam could be transmitted accurately in the right direction, taking into consideration the relative motion between the two satellites.

Other experiments included evaluation of methods of acquisition, tracking, and pointing (tracking better than 1 microrad, and pointing better than 2 microrad), measurement of micro-vibration of the satellite and precise orbit determination by laser ranging. By operating inter-satellite optical communication equipment on the ground, tracking performance could be confirmed and data obtained on the effects of atmospheric damping for optical communications with the low orbit satellite.

The prime contractor for Kirari was NEC Toshiba Space Systems. The main payload on the box-shaped Kirari was the movable Laser Utilising

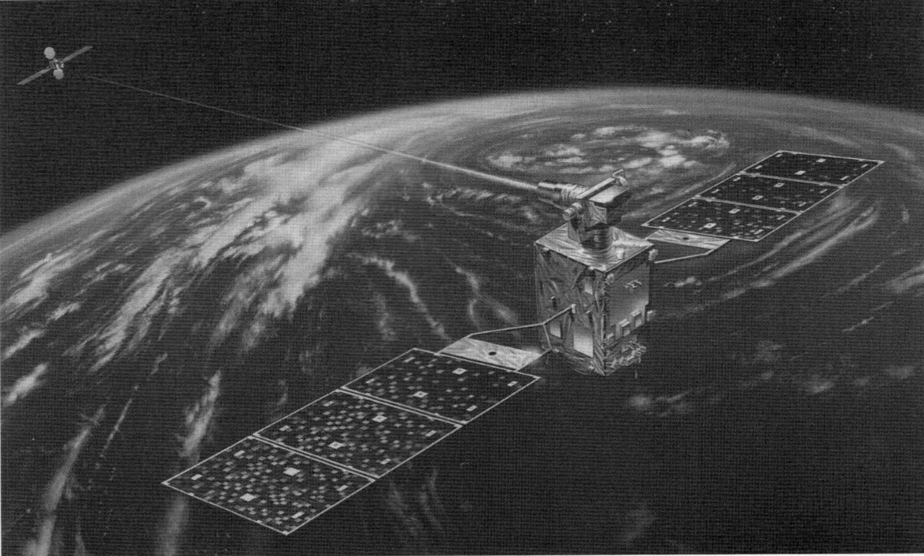
The Kirari (OICETS) spacecraft was designed to demonstrate optical inter-orbit communications (JAXA) 1343962

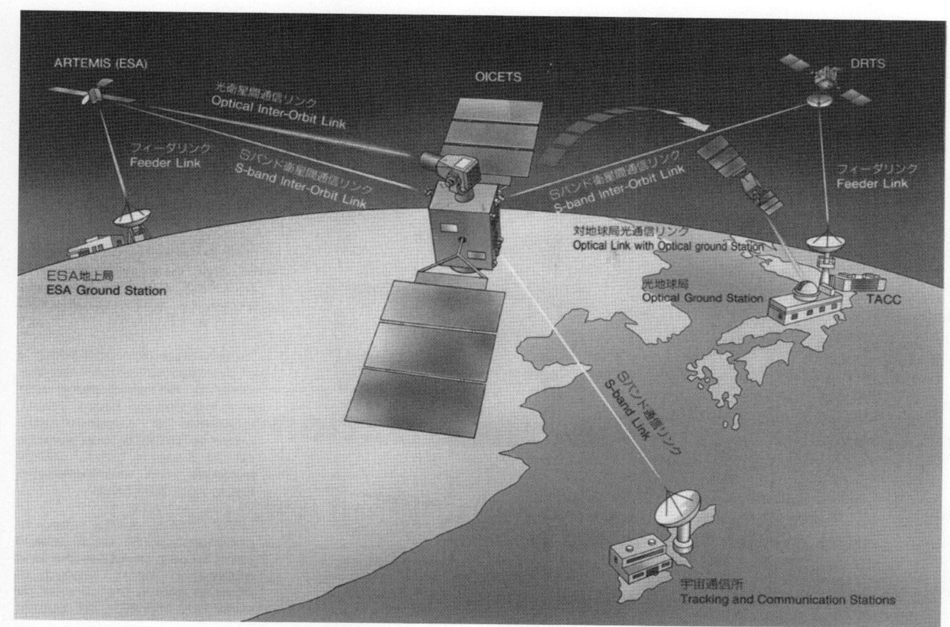

The Kirari (OICETS) operational concept. (JAXA)

1343960

Communications Equipment (LUCE) installed on the anti-Earth panel. The 140 kg LUCE was made by NEC. The 26 cm diameter telescope was made of ceramic for thermal stability, with Canon reflective optics. Power was provided by two solar arrays.

After launch in December 2005, initial functional tests were interrupted by solar flares 9–20 September. At 12:13 JST on 24 November 2005, JAXA found that one of the four reaction wheels was disconnected from the onboard attitude control system of Kirari. The satellite continued to operate normally using three wheels to control its attitude. JAXA believed that the safety function of the attitude control system disconnected the reaction wheel because an abnormal value for the number of revolutions was indicated. JAXA started to send a command from the Masuda Tracking and Communication Station to reactivate the disconnected reaction wheel at 01:44 on 20 April (JST). No problems were detected and at 02:13 on 21 April the reaction wheel was reintegrated into the attitude control system by a command from the Perth ground station, returning the satellite to an attitude control mode with four operational attitude control reaction wheels.

The historic first bi-directional optical link between Kirari and ESA's Advanced Relay and Technology Mission (ARTEMIS) was made on 9 December 2005 at 02:05 GMT. Kirari was in LEO at an altitude of about 600 km while ARTEMIS was in geostationary orbit. As with the previous ARTEMIS/SPOT-4 links, the data between Kirari and the ground were transmitted at 50 Mbps in the return direction and 2 Mbps in the forward direction. The transmissions through ARTEMIS were linked to the ESA ground station at Redu, Belgium, which was connected via data lines to the Kirari control centre at Tsukuba, Japan.

In order to precisely target a laser beam between the two satellites, they first exchanged positioning information which enabled each satellite to calculate the other's position and speed. This enabled them to aim their laser communications equipment according to the other satellite's motion. ARTEMIS then transmitted a beacon beam at the estimated position of Kirari. After receiving the beam, Kirari sent a laser beam back toward ARTEMIS. The entire procedure had to take place within a window of opportunity of no more than 5 minutes.

Further experiments involving ARTEMIS, lasting a total of 15 hours, were successfully carried out during 2006, as well as optical communication experiments with the ground optical stations of the National Institute of Information and Communication Technology (NICT) in March, May, and September 2006. The KODEN experiment was the world's first in-orbit laser communication demonstration between a low Earth orbiting satellite and an optical ground station (at Koganei, Tokyo). This involved rotating the satellite 180° so that its optical communications equipment was facing in the direction of the ground station. For this experiment, Kirari's laser beam had to point within an area about 5 m wide. During 10 trials in March and May, acquisition and tracking of the satellite were successful on seven of the assigned test days. The tests were unsuccessful during three out of the four days when it was cloudy or raining, but they were successful on all six of the days when partly clear skies were predominant. For the uplink, the fluctuation of the received signal power was successfully minimised by using multi-beam laser transmissions. The bit error ratio (BER) on the downlink was measured to be only 10'–5'. Optical links could always be established even under atmospheric turbulence when partly clear skies were predominant.

At 10:13 CET on 7 June 2006, Kirari successfully carried out an optical communication experiment with the optical ground station "OGS-OP" (Optical Ground Station Oberpfaffenhofen) of the German Aerospace Centre (DLR) in Wessling, Bayern. The optical communication downlink from Kirari was performed for 3 minutes.

After the successful completion of the Kirari mission, operations were shut down on 16 October 2006.

Specifications
Kirari (OICETS)
Launch: 24 August 2005 by Dnepr from Baikonur, Kazakhstan.
Orbit: 595 km, 97.8°, sun-synchronous
Design life: 1 year
Payload: Laser Utilising Communications Equipment (LUCE): weight 140 kg, communication power 220 W, optical antenna 26 cm Cassegrain telescope, max. laser power 100 mW, divergence angle (FHWM) approx. 5µrad, wavelength 848 nm, optical receiver APD detector, coarse pointing tracking accuracy ±0.01°, optical sensor CCD FOV ±0.2°, fine pointing tracking accuracy ±1 µrad, optical sensor quadrant detector, FOV 200 µrad.
Forward link 2.048 Mbps, beacon frequency 819 nm (ARTEMIS to OICETS); return link 49.372 Mbps, 847nm (OICETS to ARTEMIS). Also S-band links to ARTEMIS, DRTS and ground stations.
Principal application: Inter-orbit satellite and orbit-ground optical communications demonstrations.
Configuration: Box-shaped, 0.78 m × 1.1 m × 1.5 m, with LUCE on anti-Earth side. Span in orbit 9.36 m. *RRA diameter:* 16 cm. 3-axis stabilised. Laser ranging reflectors: 6 corner cubes. Two three-panel solar arrays.
Mass: 581.5 kg at launch
TT&C: S-band inter-orbit link with DRTS or direct links with ground stations in Japan (main centre at Tsukuba) and ESA ground station in Redu, Belgium.
Power: Twin solar arrays, 1.4 kW (max)

Kizuna (WINDS)

Current Status
Kizuna was launched by H-IIA on 23 February 2008. It reached its operational position in geostationary orbit on 14 March 2008.

Background
Kizuna ("ties between people") was originally known as the Wideband InterNetworking engineering test and Demonstration Satellite (WINDS). The super high speed Internet satellite was jointly developed by the Japan Aerospace Exploration Agency (JAXA) and the National Institute of Information and Communications Technology (NICT), as part of the e-Japan Priority Policy Programme of the Japanese government's IT strategy headquarters. The USD484 million five-year mission is the first step in a government-led demonstration of technologies that could revolutionise Internet-based communications. JAXA provided most of the mission's funding, with the National Institute of Information and Communications Technology providing the remaining USD56 million.

Kizuna has two 2.4 m Ka-band dish reflectors attached to a multi-beam antenna. It is intended to establish the world's most advanced information and telecommunications network, with a much higher speed and capacity than anything achieved previously. Kizuna aims for a maximum speed of 155 Mbps (receiving) and 6 Mbps (transmitting) for users with 45 cm aperture antennas, and ultra-high speed 1.2 Gbps communication for users with 5 m antennas. It will establish ultra-high speed Internet access for Japan and other Asian Pacific countries.

One key objective is to enable smaller ground antennas to attain ultra-high data rate communication previously reserved for large and expensive dishes. This is particularly useful in remote rural and mountainous regions, or where the communications infrastructure is not well developed. The system will also link up with small mobile stations in demonstrations of its usefulness during disasters and of multicasting between teachers and students in a distance

Kizuna (WINDS) is a Japanese communications satellite that enables super high-speed data communications of up to 1.2 Gbps (JAXA)

1342634

Kizuna has been developed by JAXA and the National Institute of Information and Communications Technology as part of the e-Japan Priority Policy Programme of the Japanese government's IT strategy headquarters (JAXA) 1342635

learning programme. 53 other experiments have been selected from private proposals during its mission.

Kizuna is equipped with three new technologies:

1) An ultra-high speed antenna which consists of a Multi-Beam Antenna (MBA) and a Multi-Port Amplifier (MPA). The MBA has two 2.4 m diameter high precision reflectors. One dish will focus on nine regions in Japan, plus parts of Korea, Beijing and Shanghai. The other provides communications to 19 areas in Japan and the Asia-Pacific region. The MPA is a high output power amplifier with eight ports. The MBA can allocate its power capacity to regions based on both weather conditions (Ka-band frequencies are typically hindered by rain) and real-time demand. This provides a level of flexibility unmatched by most current communications satellites.

2) An active phased array antenna with an arrangement of transmitters and receivers capable of rapidly changing coverage zones in response to demand over an area stretching from the central Pacific to India.

3) An onboard high-speed base band switching router developed by NICT which can instantly route uplinked messages to their intended destinations without intervention from the ground. This can conduct packet cell based switching up to 155 Mbps × 3 channels between I/O ports.

Kizuna was launched by H-IIA in February 2008. On 24 February, JAXA confirmed that its solar arrays were deployed through telemetry and image data received at the Santiago Station, Chile, and Maspalomas Station, Canary Islands. Over the next three weeks, the apogee engine was fired four times and the 20 N thruster once to modify its drift orbit. At 4:00 pm on 2 March 2008 (JST), JAXA found an anomaly in the signal processing of the main system of the sun sensor. JAXA began an investigation of the cause and revised the operational procedures. Meanwhile, stable attitude control was achieved with the Earth sensor and gyro.

On 14 March 2008, Kizuna reached its operational position at 143°E geostationary. After functional verifications for onboard equipment in co-operation with the National Institute of Information and Communications Technology (NICT), it will start normal operations in late June.

Specifications

Launch: 23 February 2008 by H-IIA F14 from Tanegashima Space Centre, Japan.
Location: 143°E geostationary
Design life: 5 years
Contractors: JAXA, NEC Toshiba Space (prime), National Institute of Information and Communications Technology (high-speed base band switching router)
Payload: Ka-band (18 GHz) Multi-Port Amplifier (MPA), 10 TWTAs (including two backups), maximum output power >280 W. Multi-Beam Antenna (MBA) up to 1.2 Gbps. Ka-band 128 element Active Phased Array Antenna (APAA) based on Monolithic Microwave IC. High-speed base band switching router: packet cell based switching up to 155 Mbps × 3 channels between I/O ports.
Principal applications: Demonstration of ultra-high speed Internet and telecommunications technology.
Configuration: 8 m × 3 m × 2 m bus, span 21.5 m. 3-axis stabilised. Two 2.4 m Ka-band dish reflectors attached to MBA.
Mass: 4,850 kg at launch. 2,750 kg on orbit.

Sweden

Viking/Astrid-Freja Engineering Test Series

Background

The Swedish government funded spaceborne scientific investigations primarily to gain experience in system development/management. It intended to follow up Sweden's sounding rocket studies of the interaction of the solar wind and Earth's magnetosphere. These tests would expand the investigations to 2 Earth Radii (RE) to gain insight into the behaviour of the Aurora Borealis. Since Sweden did not plan to develop an indigenous launcher capability, the government sought a satellite that could ride aboard existing launch vehicles dedicated to other missions. After an initial design phase, the Viking vehicle resulted from the Swedish effort.

Sweden funded its effort through the Swedish Space Corporation. SSC awarded a prime contract to Saab-Scania in September 1980 to begin work on a small precursor satellite. The resulting small satellite platform became known as Mesa. It was based on the USAF Small Scientific Satellites programme and NASA's Atmospheric Explorers. Given the space available on the Ariane launcher at the time of the first design review, the satellite met the objective of fitting below the main payload on Ariane launches. Swedish technical organisations and universities provided the first payload consisting of electric/magnetic field experiments, particle and magnetospheric wave experiments, and two Canadian UV imaging cameras.

As the satellite neared its launch date of 22 February 1986, the Swedish government officially named it the Viking. All of the experiments on the satellite performed well, until it ceased operations as planned 12 May 1987 after 444 days (goal: 8 months). The only reason the satellite finally shut down was due to a degrading power supply. It cost the Swedes SKr238 million (1991 rate) for the satellite and its launch.

By September 1986, the Canadian imaging cameras had returned more than 20,000 images showing the complete auroral oval and previously unobtainable sequences of the full cycle of auroral activity. The cameras were recording sequential shots at 20 second intervals and global views every 80 seconds, for real-time transmission to Kiruna. They were then relayed by tape or telemetry to Canada's Centre for Space Science and the University of Calgary.

SSC followed the programme with the Freja-C ('Compact') platform, which is a scaled-down version of the original Freja. Astrid became the first application for the Freja bus. In keeping with a tradition started by the British with their satellite names drawn from Shakespeare, the Swedes named the satellite after author Astrid Lindgren and the instruments after her characters. SSC states the first Astrid launch cost SKr9 million, including launch. Three instruments successfully monitored magnetosphere neutral particles and electrons and imaged aurora in ultra-violet, but their power supply failed 1 March 1995. Ground controllers had the last Astrid contact on 27 September 1995.

The Freja version, which SSC now offers to clients, is 45 × 45 × 39 cm and weighs 20 kg. The bus can carry 10 kg payload and it has 0.3 mm aluminium facesheets glued to aluminium honeycomb core. SSC promises attitude determination within 2° by analogue Sun sensors, coupled with two ±60 μT two-axis fluxgate magnetometers. A 12 Am^2 magnetorquer gives precession control, along with a 7 Am^2 for spin. In operation, the bus spins up and down by 10–20 Ns HTPB solids space-qualified motors. Power comes from deployable Si solar panels that generate 80 W BOL. Users communicate with the satellite by a 5 W downlink, 512 kbit/s, 2.20–2.30 GHz S-band antenna.

Specifications
Viking
Launched: 22 February 1986 piggyback with Spot 1 by Ariane V16 from Kourou
Mass: 286 kg (535 kg launch)
Orbit: 811 × 13,536 km, 98.8°, 261.6 min
Size: 1.9 m diameter, height 50 cm

Astrid 1
Launched: 24 January 1995 by Kosmos 3 m from Plesetsk
Orbit: 965 × 1,026 km, 82.93°, 105 min
Comment: The first application of the Freja-C generic micro platform
Mass: 28 kg
Dimensions: 0.45 × 0.45 × 0.45 m
Power: 4 solar arrays; 28 V DC
Stabilisation: spin-stabilised
Design life: 3 months

Astrid 2
Contractors: prime Swedish Space Corporation on behalf of Swedish National Space Board
Mission life: 1 year
Mass: 30 kg
Orbit: planned 979 × 1,1013 km, 105.04°, 105 min
Cost: budgeted at SKr10 million, including development of S-band ground station
Mission: electric and magnetic field polar phenomena investigations, electron density studies and a continuation of Astrid 1's neutral/charged particle imaging (secondary)
Spacecraft: Astrid 1 mechanical design slightly modified for increased solar cell area to provide more than 55 W peak and deployable booms. Dimensions with stowed solar panels: 40 × 45 × 95 cm. Payload power 15 W continuously. Sun-pointing spin axis is under automatic attitude control. A star imager will probably be test flown giving accurate attitude knowledge. 128 kbit/s 5 W S-band downlink to a 1.8 m ground station antenna.

United Kingdom

UoSat series

Current Status
A new SSTL 100kg enhanced microsatellite based on the heritage of the 50kg microsatellite has been launched to demonstrate new technologies for Enhanced Small Satellite users.

Background
Satellites built by University of Surrey (Surrey Satellite Technology Ltd). UoSAT-1 launched 6 October 1981 to a 372× 374 km, 97.6° orbit, retired October 1989. UoSAT-2 launched 1 March 1984 by Thor-Delta from VAFB. UoSAT-3 launched 22 January 1990 by Ariane, retired late 1996. UoSAT-4 launched 22 January 1990 but retired 23 January 1990. UoSAT-5 was launched 17 July 1991 by Ariane.

UoSAT-12 was launched by Dnepr on 21 April 1999 from Baikonur. Surrey University built UoSat-12 as a modular multipurpose bus which will

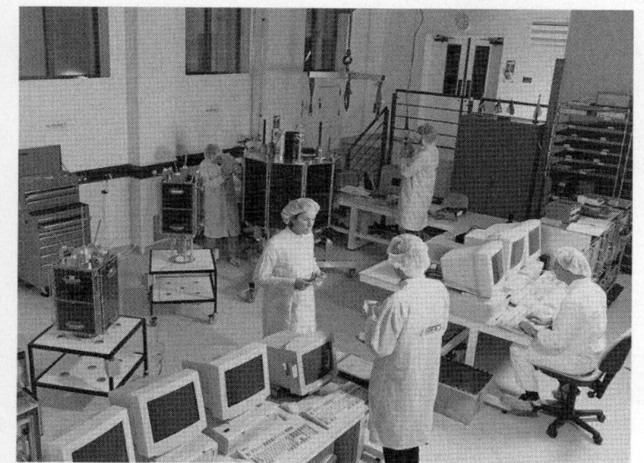

Surrey Space Centre's new mini satellite UoSat-12 was launched on 21 April 1999 on a Dnepr (SSC) 0054359

Surrey University's series of successful 50 kg microsatellites will be joined in 1997 by the 300 kg UoSAT 12 minisat
(Theo Pirard/Space Information Centre) 0517291

SSTI built UoSat-12 to demonstrate advanced leo communications and Earth observation payloads
(SSTL) 0054360

provide S-band communications and enhanced OBDH for 180 kg payloads. The first payload carries RWs, a 40 m resolution Earth camera and propulsion system for orbit control. The bus is known as Bus Alpha and it weighs 23 kg and measures 600 × 600 × 500 mm, accommodates four body-mounted GaAs solar panels providing 38W average power. A full 3-axis altitude control system allows payload pointing accuracy better than 0.2°.

Bristol University emphasised simplification by making the satellite transmit-only and stripping the bus of attitude or orbit control. The 2 kg single main experiment can be varied for each satellite. On the first satellite a new type of battery cell will be tested. The other experiments examine GaAs solar cell degradation by exposing 14 experimental cells on the X dimensional face panel, radiation environment measurement, and monitoring the satellite's nutation and spin behaviour. The SSTL enhanced satellite can carry 5 CCD remote-sensing cameras providing 3-band multispectral imaging with 50 metres ground resolution and panchromatic imaging with 20 metres ground resolution.

United States

ARGOS (P91-1, Advanced Research and Global Observation Satellite)

Current Status
Argos (P91-1) was launched on 23 February 1999 by Delta 7925 from Vandenberg AFB pad SLC2W to a 831 × 847 km orbit at 98.7° inclination. Co-orbited by the same launch vehicle were the magnetophonic mapping satellite Orsted and the South African student satellite Sunsat.

Background
The ARGOS mission is to fly and operate joint service payloads which include two technology demonstrators and seven experiment payloads for global and celestial observation. It is categorised as USAF Space Test Programme P91-1 built by Boeing at its Seal Beach facility. ARGOS weighs approximately 2,700 kg and has a goal of three years of on-orbit operations. ARGOS incorporates an accurate attitude determination and control system with an embedded Global Positioning System (GPS) receiver for position determination and time reference and a solid-state recorder, providing 2.6 Gbits of onboard data storage. ARGOS also includes an automated mission planning system, developed by the University of Colorado's Laboratory of Atmospheric and Space Physics, to optimise ground contacts as well as onboard power and data storage requirements. The automated mission planning tool adds tremendous capability to manage the ARGOS downlink of 9.6 Gbits each day. On board experiments are:

The High Temperature Superconductivity Space Experiment, or HTSSE 2, developed by the Naval Research Laboratory (NRL), Naval Center for Space Technology. HTSSE 2 will space qualify super-conducting digital and RF subsystems to demonstrate an operational space capability. Expected performance of super-conducting components include factors of 100–1,000 in power reduction, more than 10 times higher speed and similar weight reduction than today's silicon or gallium arsenide (GaAs) based electronics.

The Extreme Ultra-Violet Imaging Photometer (EUVIP) payload is sponsored by the US Army Space and Strategic Defense Command and built by the University of California at Berkeley Extreme Ultra-Violet Astrophysics Laboratory. EUVIP will establish the behaviour of the upper atmosphere and plasmasphere as needed for Army RF secure communication systems design, prediction of magnetic storms, and characterisation of the aurora.

The Naval Research Laboratory, Space Science Division, sponsors the Unconventional Stellar Aspect (USA) payload. ARGOS will be one of the first research satellites to fly an embedded Global Positioning System (GPS) receiver, while characterising astronomical x-ray sources for potential use as autonomous position, attitude and time-keeping references for military space systems. The augmentation of a proven x-ray detector design with advanced computers will help to demonstrate the feasibility of autonomous satellite navigation using x-ray pulsars in place of GPS timing and navigation signals.

Electric Propulsion Space Experiment or ESEX, developed at TRW's Space and Electronics Group facilities will demonstrate reliable high powered arcjet thruster operations in space. ESEX will also

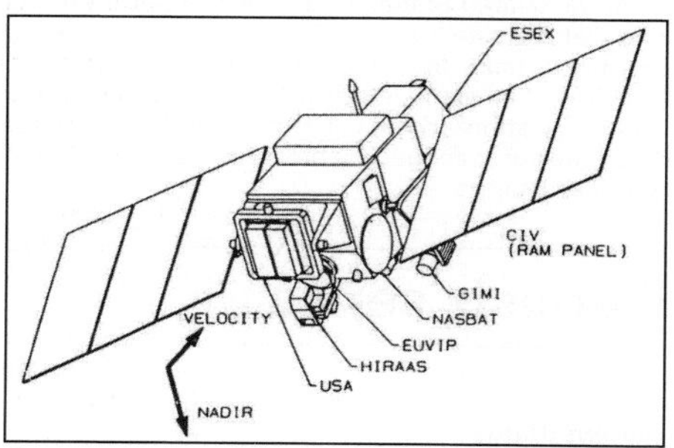

Argos: the artist's impression highlights the operating ammonia arcjet (TRW) 0516847

Argos: the artist's impression highlights the operating ammonia arcjet (TRW) 0516848

execute orbit transfers and verify compatibility with spacecraft subsystem controls. Electric propulsion is projected to double the payload-to-orbit capability of current space propulsion systems. The managing organisation for this payload is the Air Force Phillips Laboratory, Electric Propulsion Laboratory.

The Space Dust Experiment of SPADUS will provide a 3-D survey map of the present dust distribution in low Earth polar orbit and allow prediction of orbital debris 'showers' which could affect spacecraft such as the Space Shuttle. SPADUS will also obtain early flight experience for sensors and electronics, which are planned for International Space Station and the Cassini mission to Saturn in 1998. This payload is sponsored by the Office of Naval Research, Electronics Division and built by the University of Chicago Enrico Fermi Institute, Laboratory for Astrophysics and Space Research.

The Critical Ionisation Velocity experiment of CIV will release xenon and carbon dioxide to study ionisation caused by plasma and collision processes in the upper atmosphere. The results will be used to help identify plumes and atmospheric wakes of launch and orbital vehicles. CIV is developed and sponsored by the Air Force Phillips Laboratory, Satellite Assessment Division.

The High Resolution Airglow Aura Spectroscope (HIRAAS) payload, developed and sponsored by the Naval Research Laboratory, Space Science Division, consists of three high spectral resolution ultra-violet spectrographs designed to measure the naturally occurring thermospheric and ionospheric airglow. Measurements of the ionosphere can improve a number of DoD systems that depend on radio and microwave propagation through the upper atmosphere and ionosphere. Neutral density measurements also support NORAD operational requirements for satellite drag forecasting and the ability to predict orbital life and re-entry impact locations.

Global Imaging Monitor of the Ionosphere or GIMI will demonstrate operational charged coupled device (CCD) sensor technology for environmental monitoring of upper atmospheric perturbation due to meteors, rocket exhausts, and aurora. Use of wide-field sensors in three separate wavelengths will enable GIMI to continuously image 560 m^2 of the Earth's limb. GIMI is developed and sponsored by the Naval Research Laboratory, Space Science Division.

The Coherent Radio Topography Experiment Payload (CERTO) consists of a three-frequency radio beacon and radiating antenna mounted on the space vehicle. Receivers on the ground use the differential phase techniques to derive the integrated electron density. Characterising CERTO data plays a critical role in assessing impacts on navigational accuracy, communication systems, and remote sensing by radars. CERTO is managed by the Naval Research Laboratory Plasma Physics Division.

Co-manifested on the Delta-2 launch vehicle are two secondary payloads from Denmark and South Africa. Assembled is a coalition of international test teams to support space test programmes from the United States, Denmark and South Africa. The Danish satellite ORSTED will map the Earth's magnetic field and the South African satellite SUNSAT will map the Earth's gravitational fields. The integration and development for these micro-satellites are managed by the NASA Orbital Launch Services Project.

Genesis series

Current Status

As of late February 2008, both Genesis I and Genesis II remain operational.

Background

Bigelow Aerospace is working to validate expandable structures and provide a low-cost, Low Earth Orbit (LEO), human-rated space complex that is suitable for commercial use. The pair of Genesis pathfinder modules is the first stage in Bigelow Aerospace's vision to provide an affordable and flexible space complex architecture that can be adapted for any manned or unmanned mission requiring a large pressurised volume.

The inflatable technology under development by Bigelow is derived from the Transhab concept studied at NASA Johnson Space Center during the 1990s. Transhab had been under consideration for use as the NASA Habitation Module for the International Space Station. The Transhab was composed primarily of multi-layered fabric, foam and Kevlar. A shell of carbon composite ribs maintained the shape of the module. Its design included floors, much like a multistorey house. In October 2000, funding for Transhab was blocked by Congress, and in February 2001 the project was put on indefinite hold. This was followed by several licensing agreements between NASA and Bigelow Aerospace in which some of the Transhab technology, including debris shielding, was transferred from JSC to the company. JSC engineers continued to provide technical assistance.

By 2004, Bigelow had begun development of 25 per cent and 45 per cent scale inflatable test modules for a series of unmanned orbital flights 2005–2007. This included two flights of the Genesis modules in 2005 and 2006 to test inflation technology, pressure integrity and debris shield deployment. These would be followed by two "Guardian" flights, also to be launched on the Russian Dnepr converted ballistic missile, in 2007. These would test life support systems before the launch of a human-tended commercial "Nautilus" module 2008–2010. This module, which would be capable of docking with Russian Soyuz and space station modules, would differ from Transhab by having an open internal layout without a large central core; a strap-matrix air bladder design rather than constant weave; and lower cost materials, including carbon fibre composite in the debris shielding. This plan was subsequently modified, with the Genesis modules to be followed by increasingly large modules known as Galaxy and Sundancer. The full-scale manned version was dubbed the BA-330.

Genesis I was launched in July 2006, under a launch agreement with ISC Kosmotras of the Ukraine. The Dnepr vehicle, a modified SS-18 ballistic missile, was launched from an underground silo at the Yasny spaceport in southern Russia. On reaching orbit, the module was successfully inflated using the onboard tank of nitrogen gas, expanding from its launch width of 1.9 m to a flight diameter of 2.54 m, a length of 4.4 m and an internal volume of 11.5 m^3. Power was provided by deployment of eight solar arrays – four at each end. The initial internal temperature was recorded as 26°C. There were 13 cameras on board Genesis I, as well as a "Fly Your Stuff" prototype, employee photos and colour test cards.

Genesis II, launched in June 2007 after a five-month delay caused by a Dnepr launch failure investigation, was externally similar to Genesis I. However, it carried a large billboard sign on its exterior reading "Blair", after the granddaughter of CEO Robert T. Bigelow – ostensibly as a test subject for the onboard imaging system. It also carried new sensors and avionics to monitor and control the spacecraft's internal pressure, temperature, vehicle attitude control and radiation levels. Unlike its predecessor, Genesis II also had a multi-tank system to inflate the module with compressed air. This was to provide redundancy in the inflation process and allow better control of the craft's gas supplies.

In December 2007, Bigelow reported that Genesis I's original orbit of 557 km had degraded to 547 km, suggesting that it will remain in orbit for well over 10 years before re-entering the Earth's atmosphere. The battery life span was anticipated to be at least 7 years. The external envelope remained perfectly intact, while leak rates were lower than achieved during ground tests. Bigelow reported some problems with a computer that controlled several of the cameras, but all of the interior lights, fans and other internal systems remained in perfect working condition. The vehicle was passively thermal controlled, so the interior air temperature varied with the use of electronic systems and amount of Sun exposure, ranging from about 4°C to about 32°C. Some communications problems that required use of backup systems occurred during and after a very severe solar radiation event on 14 December 2006. This problem was remedied with a reset of the primary system.

Genesis II carried 22 cameras on both the interior and outside of the spacecraft. The internal cameras were used to image personal items provided by paying customers under the firm's "Fly Your Stuff" campaign. The vision system included articulated cameras with dual FireWire and Ethernet interfaces, as well as a wireless boom camera for exterior views. Space-to-ground communications were provided by UHF, VHF and S-band antennas. Magnetic torque rods, GPS and sun sensors and a reaction-wheel system provided attitude control and stabilisation.

Another payload on Genesis II was a Space Bingo game that was mainly intended as entertainment, with no actual wagering involved. It was expected to begin operations a few months after launch.

An artist's conception showing one of the Genesis series of pathfinder modules developed by Bigelow Aerospace. Genesis I was launched on 12 July 2006, followed by Genesis II on 28 June 2007. Both were orbited by a Dnepr rocket – a converted Cold War SS-18 Intercontinental Ballistic Missile – launched from a silo at the Yasny Launch Base in Siberia (Bigelow Aerospace) 1329380

An image taken with an external camera on Genesis I in late January 2007. Visible in the background is the big island of Hawaii. Two solar arrays are visible at the far end. (Bigelow Aerospace) 1329382

The Bingo Box would use fans and levers to randomly mix and select bingo balls during games presented on the company's website.

A third payload was the Biobox, a three-chamber, pressurised vessel with compartments for biological specimens to be observed by onboard cameras. The Biobox contain red harvester ants, a group of South African flat rock scorpions, and Madagascar hissing cockroaches, the same species that flew aboard Genesis 1. Images of these specimens were to be made available on the Bigelow Aerospace website during the mission.

In December 2007, Genesis II was reported to be in a nearly circular orbit (eccentricity 0.028) with a minimum altitude of 550 km – a reduction of 5 km since launch. Genesis II should remain in orbit for more than 12 years. The Attitude Control System (ACS) successfully damped out post-launch tumbling within 48 hours and oriented the antennae toward Earth. Initial pressure following the expansion of the outer shell was approximately 6 psia. After initial evaluation, inflation valves were commanded open from the ground, and pressure in the vehicle was increased. The internal pressure of Genesis II then held steady in a range of 10.1 to 10.5 psia. The pressure inside the spacecraft varies due to thermal effects as the vehicle passes in and out of sunlight.

Bigelow planned to launch **Galaxy** – another pathfinder module that built on the Genesis vehicles – in 2008. However, on 13 August 2007, the company announced that the flight had now been cancelled, blaming inflation, rising Russian launch costs and the falling value of the US dollar. To conduct another subscale demonstrator mission would cost two to three times what it had in the past. Bigelow still intended to construct and test on the ground the Galaxy spacecraft and/or various parts of it in order to gain familiarity and experience with critical subsystems. This policy would help the company to expedite the Sundancer programme, with a possible launch in 2011. Galaxy was to have 45 per cent more habitable space than the Genesis craft, with a pressurised volume of about 16.7 m^3.

The first crew-rated spacecraft, the three-person, 180 m^3 volume **Sundancer** is expected to be launched around 2010. The module will have life support systems, three windows, measure 8.7 m × 6.3 m and weigh 8,618 kg. It will later be fitted with a propulsion system that can be used for orbital manoeuvring and deorbiting, a Russian docking system at one end and a NASA lightweight Low Impact Docking System on the other end.

Successful tests with Sundancer will lead to the launch of the BA 330 module. The larger BA 330, capable of housing up to six occupants, is expected to have a 330 m^3 habitable volume when fully inflated. Bigelow has outlined plans to launch three inflatable complexes by 2015. Each would comprise two BA-330 modules docked to a Sundancer module attached to a node-propulsion bus.

On 5 February 2008, Bigelow Aerospace and Lockheed Martin Commercial Launch Services announced that they were engaged in discussions to supply Atlas V launch vehicles to provide crew and cargo transportation services to a Bigelow-built space complex. The Atlas V 401 configuration has been selected by Bigelow as its desired launcher. This will require "modest" system upgrades that will augment existing safety features prior to flying the first passengers. During the operational phase, which is currently planned to begin in 2012, up to 12 missions per year are envisioned, increasing as demand dictates.

Bigelow Aerospace operates four ground stations that serve as communications relays between the company's spacecraft and the BA Mission Control Centre in Las Vegas. They are in North Pole, Alaska; the big island of Hawaii; Limestone, Maine; and North Las Vegas. There are communication links with both Genesis I and II several times a day.

Specifications

Genesis I

Launch: 14:53 GMT on 12 July 2006 by Dnepr from Yasny (Dombarovsky military missile base), Russia.
Orbit: 557 km, 64°
Mass: Approx. 1,364 kg at launch.
Design life: 7–10 years, based on battery life.
Contractors: Bigelow Aerospace (prime).
Dimensions: 4.4 × 1.6 m at launch. 4.4 × 2.54 m inflated.
Principal applications: Validation of expandable structures for low-cost, low Earth orbit (LEO), human-rated commercial use.
Configuration: Inflatable cylinder, multilayer skin 15.25 cm thick. 11.5 m^3 of usable volume. One communications antenna. One exterior window.
TT&C: Bigelow Aerospace Mission Control Center in North Las Vegas. Ground stations in North Pole, Alaska; the big island of Hawaii; Limestone, Maine; and North Las Vegas.
Power system: eight deployable solar arrays, four at either end arranged at 90° intervals. Battery.

Genesis II

Launch: 15:02 GMT on 28 June 2007 by Dnepr from Yasny (Dombarovsky military missile base), Russia.

Other specifications are similar to Genesis I.

JAWSAT (P98-1, Joint Academy, Weber State University Satellite)

Background

STP continued to prepare for the JAWSAT (P98-1), Joint USAF Academy/Weber State University Satellite. JAWSAT, along with the Air Force Academy's FalconSat, was due to be launched on the first Orbital/Suborbital Program (OSP) launch vehicle in September 1999 but the flight was delayed to 27 January 2000 by Minotaur from Vandenberg AFB, pod CLF (Commercial Launch Facility) on a leased site. The US Air Force refers to Minotaur on the Orbital/Sub-orbital Program Space Launch Vehicle (OSPSLV), it comprises the Minuterran and Taurus stages in a configuration put together by Orbital Services. JAWSAT was placed in a 746 × 810 km, 100.2°, polar orbit. JAWSAT carries three experiments: a Thermospheric Temperature and Nitric Oxide Spectrograph, an Ionospheric occultation Experiment, and the Coherent Electromagnetic Radio Tomography Probe (CERTO) series experiment. Success with this flight confirmed the JAWSAT concept for future science platform applications, a secondary objective of the P98 effort.

MiTex

Current Status

A Boeing Delta II booster launched the MiTEX (Microsatellite Technology Experiment) on 21 June 2006. It is aimed at developing technology for small satellites in Geostationary Earth Orbit (GEO).

Background

MiTex combined two 225 kg satellites – one produced by Lockheed Martin and one by Orbital Sciences – with an upper stage built by the Naval Research Laboratory. The exact goals of MiTEX are not clear and very little information was released concerning the launch, apart from the statement that the craft were placed in geostationary orbit. Budget documents state that the Microsatellite Demonstration Science and Technology Experiment Program (MiDSTEP) experiment, of which MiTEX is a part, is aimed at developing technology for high-performance microsatellites "across the continuum from Low Earth Orbit (LEO) to deep space geosynchronous orbit (GEO) environment," including high-power solar-thermal structures, light weight and stable payload isolation. Neither DARPA nor the NRL has released details of the spacecraft.

The Center for Defense Information (CDI) – a Washington watchdog group – noted in a paper at the end of July 2006 that the technologies in MiDSTEP and MiTEX "are consistent with proposals for antisatellite weapons". In particular, the CDI noted that the NRL upper stage seemed to constitute overkill if its job was simply to place the satellites into GEO – and that it might be intended to demonstrate the ability to support the craft for an extended period and manoeuvre at will, delivering satellites into a wide range of orbits as commanded from the ground. That could place satellites in position to monitor or negate (through jamming, for example) a GEO satellite. The CDI also pointed out that the MiTEX satellites are effectively stealth vehicles, since the US military operates the only system that is capable of detecting such small vehicles in GEO.

MTI (P97-3, Multispectral Thermal Imager)

Background

The Multispectral Thermal Imager is a Department of Energy (Sandia National Laboratory, Los Alamos National Laboratory, and Savannah River Technology Center) and STP sponsored mission. MTI was placed into a 574 × 609 km circular, 97.5° inclination Sun-synchronous orbit on 12 March 2000 by a Taurus launch vehicle from VAFB. The 1,610 kg spacecraft is 3-axis stabilised, provides an average of 540 W of power, and is designed for a 14-month mission life with a three-year goal. The objective of the MTI mission is to demonstrate and evaluate multispectral and thermal imaging technology of nuclear proliferation monitoring and for other military and civilian applications. The MTI sensor will provide multispectral data with high ground resolution and radiometric accuracy. Ball Aerospace is building the MTI spacecraft bus and Hughes Santa Barbara Research Center, Hughes Danbury Optical Systems, TRW, and Sandia National Laboratory are building the sensor.

Near Field Infra-red Experiment (NFIRE)

Current Status

NFIRE was launched on 24 April 2007 and began operations in May 2007. As of April 2008 it was still operational.

Background

The Near Field Infra-red Experiment (NFIRE) satellite is the key component of a Kinetic Energy Boost Phase research programme which began with a study and mission system engineering of approaches for boost phase intercept of ICBMs. Sponsored by the Missile Defense Agency (MDA), NFIRE is designed to gather near field, high resolution phenomenology data that will assist in development of boost phase intercept systems. USD44 million was appropriated for NFIRE in FY2004, with another USD68 million requested by MDA for FY2005. MDA is requesting USD9 million for NFIRE for FY2009, the last planned request.

The primary mission of NFIRE is to collect high and low resolution images of a boosting rocket to improve understanding of missile exhaust plume observations and plume-to-rocket body discrimination during three plume signature types: targets of opportunity, dedicated missile flybys, and ground observations.

NFIRE is also assessing the viability of a laser communications system for missile defence applications, such as use of laser cross-links between satellites to transmit missile tracking information faster, particularly over long distances. The development of the laser terminal was sponsored by the German Ministry of Economy and Technology (Bundesministerium für Wirtschaft und Technologie (BMWi)) via the German Space Agency (DLR-Raumfahrtagentur); the Ministry of Defence (Bundesministerium der Verteidigung (BMVg)) managed the co-operation with the US DoD for the in-orbit demonstration of the system.

The German experiment was added to NFIRE in 2005 after a controversial payload known as the "Generation 2 kill vehicle (KV)" was cancelled by the MDA. The data it was to collect would "support KEI kill vehicle hardware and algorithm development and boost phase plume models and simulation". Once NFIRE was in orbit, this payload, part of the Kinetic Energy Interceptor programme, was to separate from the main spacecraft to take a closer look at a specially launched missile target. A small, non-manoeuvrable, sensor platform mounted on a missile defense kill vehicle would be deployed and attempt to get as close as possible to the missile in order to collect imagery from very close (sub-metre) range. DoD indicated that the KKV might move so close that impacted the missile. Air Force Undersecretary Peter Teets commented: "It is true that the kind of capability that NFIRE will have could, with a different concept of operations, be used as a space-based weapon capability."

The NFIRE programme began in August 2002, when the initial contract for NFIRE was awarded to prime contractor Spectrum Astro (now General Dynamics). NFIRE was originally scheduled to launch in summer 2004, but anomalies in the sensor payload delayed its delivery and put back the launch until late 2005. The time it took the DoD to finalise the agreement with Germany for the laser payload and DoD's desire for more capable software in the spacecraft contributed to a launch delay from late 2006 into early 2007.

In 2004, the KEI programme office signed a memorandum of agreement that transferred day-to-day management and execution of NFIRE to the Space Tracking and Surveillance System (STSS) programme. NFIRE supports the design and development of space-based sensors like those to be used in the STSS. In FY2006, NFIRE was transferred from the Kinetic Energy Interceptor programme to the Advanced Technology programme. In FY2007, MDA requested that NFIRE and all associated funding be moved into the BMD System Space Program. In September 2006 it was announced that prime contractor General Dynamics C4 Systems Inc of Gilbert, Arizona, received a cost-plus-award fixed-fee contract modification of USD11.1 million to equitably adjust the contract for the Government-caused NFIRE launch delay.

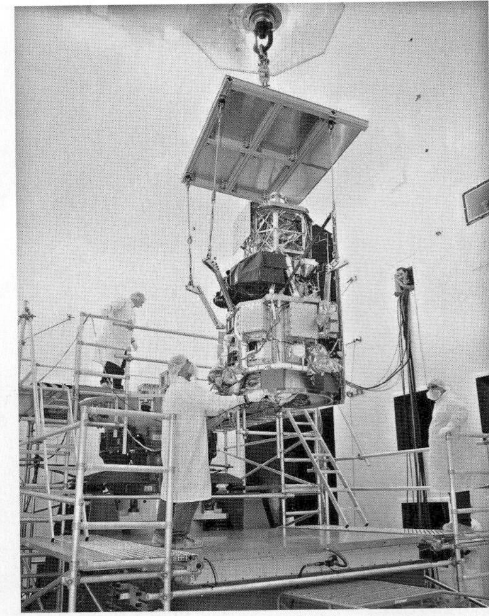

The Missile Defense Agency's NFIRE during integration by General Dynamics C4 (formerly Spectrum Astro)
(General Dynamics) 1166939

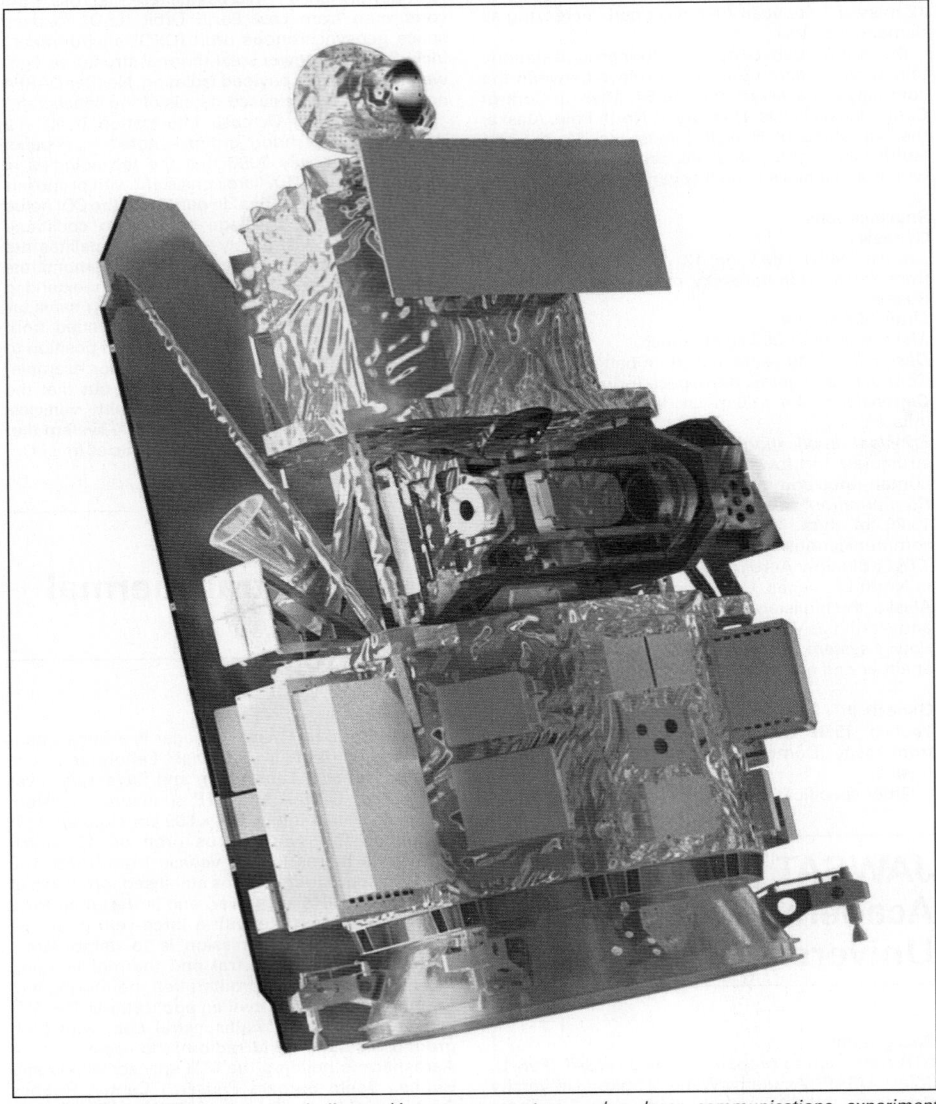

The NFIRE satellite carries a missile tracking sensor system and a laser communications experiment
(General Dynamics) 1166938

NFIRE comprises an agile, low-Earth orbiting satellite with an onboard Track Sensor Payload provided by the Air Force Research Laboratory (AFRL) and SAIC-San Diego, plus a Laser Communication Terminal built by Tesat-Spacecom of Germany. There are two ground-based Mission Operations Centres (MOCs) operated by the US MDA. General Dynamics was the system integrator, and also responsible for the design and manufacture of the spacecraft, payload integration, full satellite system testing including EMI/EMC and environmental, configuration of the MOCs, and a year of on-orbit operations support. General Dynamics also led the Mission Assurance and Systems Engineering Integrated Product Teams.

A Minotaur I launched the experimental missile warning satellite from Wallops Island, Virginia, into a 255 × 465 km, 48.2° orbit on 24 April 2007. 3–18 May the orbit was raised to 489 × 497 km. On 9 August the orbit was lowered again to 243 × 487 km and by 23 August it was 219 × 450 km. On that day, NFIRE successfully completed its first experiment for MDA when it successfully tracked a modified Minuteman II TLV-7 (named NFIRE-2a) that was launched from LF06 silo, Vandenberg Air Force Base (AFB), California at 08:30 UTC. This version of the Minuteman used a larger third stage than normal and followed a precisely-controlled target trajectory. The missile flew within 3.5 km of the satellite while its third stage motor was burning.

The exercise provided an opportunity for NFIRE to collect high and low resolution images of a boosting rocket in order to improve understanding of missile exhaust plume observations and plume-to-rocket body discrimination. Data from the NFIRE satellite were downlinked to the Missile Defense Space Experimentation Center (MDSEC) at the Missile Defense Integration & Operations Center (MDIOC) at Schriever AFB, Colorado. Data collected during this experiment will be used to help with the development of future missile defence technology efforts. The second and final flight test (NFIRE-2b) was scheduled for the second quarter of FY2008 with a target flyby at a range of less than 10 km.

In addition to two missile data collection experiments, NFIRE has been used to conduct laser crosslink satellite-to-satellite and satellite-to-ground communication experiments with its secondary payload. On 12 March 2008, Tesat-Spacecom announced that the two German Laser Terminals on NFIRE and the German TerraSAR-X satellite had been successfully tested in space since November 2007, achieving a data rate of 5.5 Gbit/s in both directions – hailed as a new milestone in space-borne communications. NFIRE and TerraSAR-X both operate in low Earth orbit with several encounters per day (depending on orbital alignment), enabling laser links to be established for up to 20 minutes at a range of about 5,000 km. Optical inter-satellite links for broadband data transmission were established and operated flawlessly.

A German laser communications terminal is carried on the NFIRE and TerraSAR-X satellites (Tesat-Spacecom) 1166932

Specifications
NFIRE
Launch: 06:48 GMT on 24 April 2007 by Minotaur I from Mid-Atlantic Regional Spaceport, Wallops Island, Virginia.
Orbit: 255 × 465 km, 48.2°
Design life: 2 years.
Contractors: Spectrum Astro (now General Dynamics Advanced Information Systems) (prime). Air Force Research Laboratory and SAIC-San Diego (Track Sensor Payload). Tesat-Spacecom (Laser Communication Terminal)
Principal applications: Near field, high resolution, missile phenomenology data collection; test of laser communications system for missile defence applications.
Configuration: Dimensions: 2.65 m × 1.31 m stowed. Aluminium primary bus structure, externally mounted components for easy access, riveted aluminium frame with honeycomb panels. Passive cold-biased thermal system, thermostatically-controlled heaters. 40 Gbit Solid State Recorder for payload data storage.
Mass: 494 kg at launch. 248 kg dry. Propellant: 114 kg blowdown hydrazine. Modular Architecture with cPCI backplane
AOCS: 3-axis stabilised, Zero Momentum Biased (ZMB) on-board attitude determination with star trackers, IRUs, and GPS. Pointing accuracy (3s): 360 arcsec. Pointing knowledge (3s): 14.8 arcsec. Ephemeris accuracy (3s): 19 m.
Communications: SGLS narrowband DL: 40.96 kbps SOH. SGLS narrowband UL: 2 kbps commands, AFSCN. X-band wideband PL Data DL: 51.2 Mbps, USN. SGLS wideband PL Data DL: 1.024 Mbps, AFSCN, TCP/IP. Auxiliary SGLS wideband UL receiver: 1.024 Mbps, TCP/IP.
Power system: Triple junction GaAs solar array, body mounted. 503 W (EOL). 16 amp-hr NiH_2 CPV battery.
Payload: Track Sensor: visible spectrum, Si CCD; long wave infra-red, HgCdTe (MCT) hybrid; medium and short wave infra-red, InSb.

TESAT Laser Communications Terminal: 12.7 cm aperture full hemisphere pointing. Mass: 30 kg. Power consumption: 130 W. Optical data rate: >5 Gbps. Modulation: BPSK. Detection: coherent, homodyne.

Orbital Express (ASTRO/NextSat)

Current Status

The Orbital Express mission was terminated 21–22 July 2007 after a successful series of technology demonstrations.

Background

Orbital Express (OE) involved the simultaneous launch of two satellites – the Autonomous Space Transfer and Robotic Orbiter (ASTRO), which was the servicing satellite, and the Next Generation Satellite/Commodity Spacecraft (NextSat/CSC), which functioned as a supply depot for ASTRO and as the satellite to be serviced by ASTRO. Once on-orbit, eight scenarios were to be conducted over a mission lifetime of one year.

Boeing Phantom Works, Huntington Beach, California, was the prime integrator for the Orbital Express (OE) programme and builder of the ASTRO spacecraft. Ball Aerospace constructed NextSat. Other members of the OE team included NASA; Northrop Grumman Space Technology; MacDonald, Dettwiler and Associates; Charles Stark Draper Laboratory; and Starsys Research. The USD300 million programme was sponsored by the Defense Advanced Research Projects Agency (DARPA). NASA contributed about USD25 million to the project in 2001 and engineers at the space agency's Marshall Space Flight Center helped develop the spacecraft's automated guidance system.

Technology firsts to be demonstrated were:

- First fully autonomous capture and servicing of a satellite without client assistance.
- First fully autonomous onboard navigation and guidance to approach and station keep within 10 cm of client using passive systems.
- First on-orbit use of IEEE 1394 (Firewire) standard allowing ASTRO to remove and replace a flight computer.
- First fully autonomous "soft" capture of a satellite while station keeping at 10–18 cm or berthing from 1.5 m.
- First fully autonomous transfer of a component from one vehicle to another using advanced robotics.
- First fully autonomous transfer of propellant from one satellite to another with U.S. technology.
- First fully autonomous capture of a free-flying vehicle and component transfer using a robotic arm with closed-loop servo vision system and autonomous fault recovery.

ASTRO had an octagonal bus with two solar arrays. On its exterior were a replaceable battery unit and a 3 m long, hinged robotic manipulator arm. At the front were various range-finders and sensors, the fluid coupler system and the capture system which consisted of three claws that extend and grapple the fixture on NextSat. The rendezvous braking thrusters were aligned 30° away from the docking axis to avoid contamination of NextSat.

NextSat/CSC employed architecture adapted from the Deep Impact Impactor, including software, command and data handling, and power control, together with elements from Ball Aerospace Commercial Platform (BCP) 2000 spacecraft, such as the narrow-band telecom architecture from CloudSat. It had an octagonal bus with a grapple fixture, an Orbital Replacement Unit and a single solar array. At the front were fluid couplers and a petal-shaped capture fixture.

The rendezvous and proximity operations subsystem provided navigation and guidance for all unmated conditions, ranges, rates, angles, lighting (day/night) and Earth background. NASA

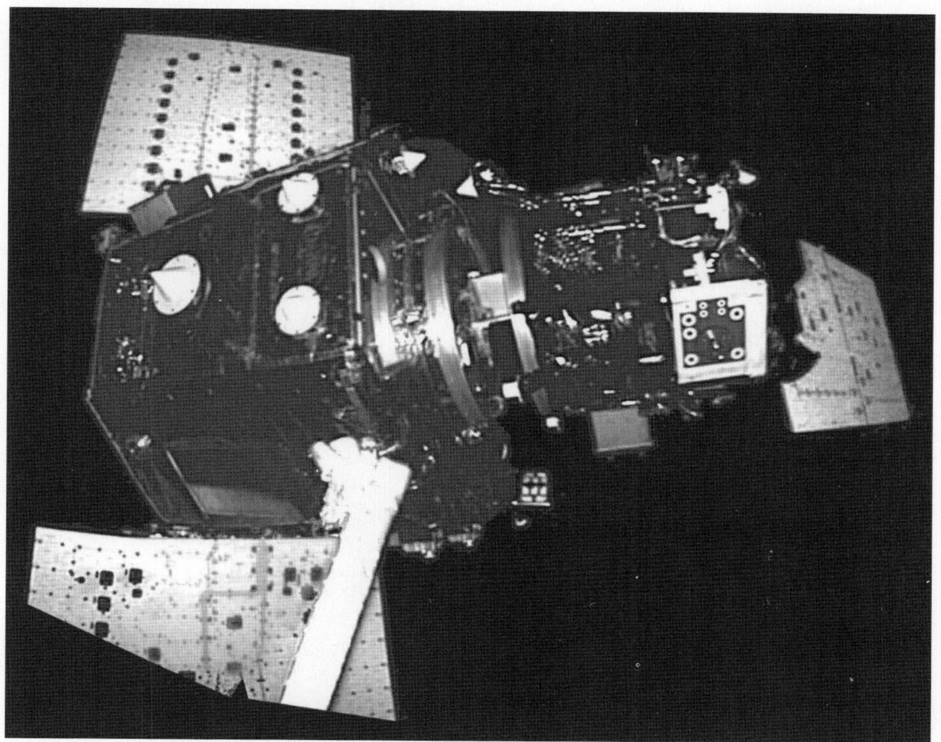

An in-orbit photo of the combined ASTRO (left) and Nextsat Orbital Express spacecraft taken with the robotic arm on ASTRO on 22 June 2007 (DARPA/Boeing) 1166940

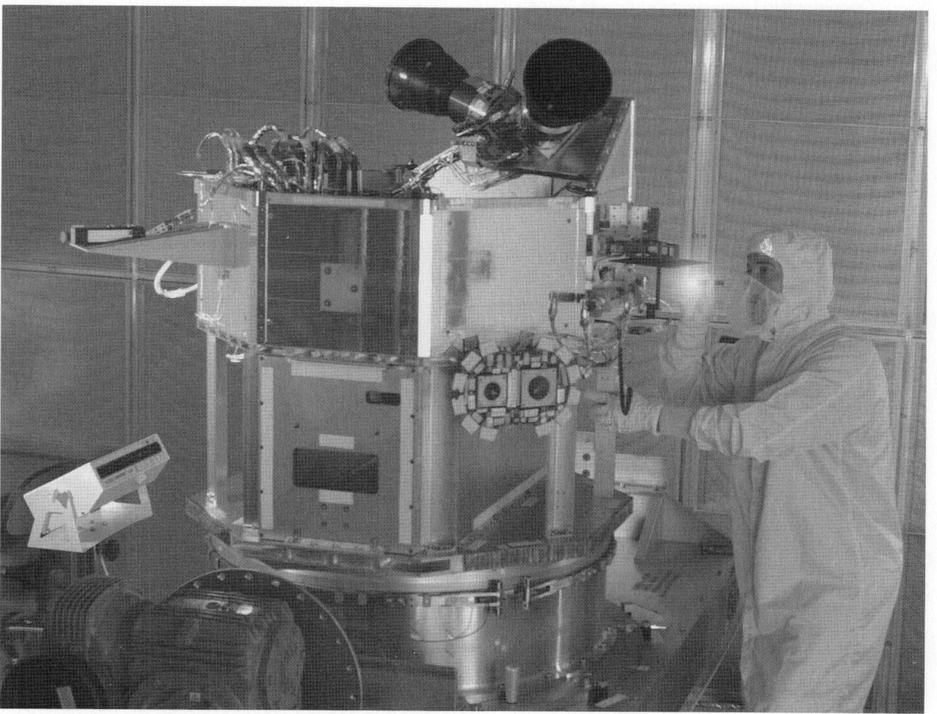

Ball's NextSat during pre-launch checkout (Ball Aerospace) 1169300

Boeing technicians perform a post-integration inspection of the DARPA Orbital Express spacecraft at Astrotech Space Operations in Titusville, Florida (Ball Aerospace) 1166934

The Orbital Express mission involved autonomous rendezvous and docking manoeuvres by ASTRO (right) and NextSat (Boeing) 1169299

AVGS and Boeing sensors operated concurrently to maximise sensor validation within rendezvous allocations. Vis-STAR software passively tracked NextSat and determined its position and orientation. AutoNav software used data from all sensors to compute precise and self-correcting positions and orientations of both spacecraft. AutoGuide software computed and commanded ASTRO translation manoeuvres and pointing to fly a desired trajectory without ground assistance. Mission manager software monitored the status of ASTRO and NextSat subsystems and commanded aborts if conditions warrant. Abort commanding from the ground was available if required.

OE used two methods of fully autonomous "soft" capture of the client satellite: "direct capture" was performed while stationkeeping within the mechanism capture envelope at 10–18 cm; "berthing" required the advanced robotic arm to grapple NextSat from a distance of 1.5 m and position it within the capture envelope.

Fluid transfer involved the two-way transfer of hydrazine propellant between the satellites using either the fluid transfer pump or ullage recompression. ASTRO controlled all fluid transfer operations, including transfer from NextSat to ASTRO. This was undertaken using a wide range of flow rates, pump-fed and pressure-fed transfers, multiple gauging methods, and range of fill fractions.

The two spacecraft were launched on 8 March 2007 (local time) in a mated configuration with a special launch ring that diverted potentially damaging structural loads away from the ASTRO spacecraft's capture mechanism. Following launch, ASTRO experienced an anomaly with its guidance and control systems because its attitude control wheels had been installed backward, forcing major software changes. To allow time to correct the problem, control of the mated pair was shifted from ASTRO to NextSat and the OE team used NextSat's guidance system to successfully point the mated stack towards the sun. Modified software was successfully uploaded and validated, enabling ASTRO to resume control of the OE stack in the planned nominal mode.

The first scheduled sequence of tests took place while they remained mated. Called Scenario 0, the 20 day series included a robotic video survey of the vehicles, successful demonstrations of autonomous refuelling with the Fluid Transfer System, and robotic transfer of a 24 kg battery Orbital Replacement Unit (ORU) from ASTRO to NextSat. This was the first time hardware had been autonomously transferred between unmanned spacecraft on orbit.

On 18 April ASTRO's robotic arm was used to successfully separate them for the first time. The launch ring between them was safely ejected at 1:54 pm EDT after ASTRO's robotic arm moved NextSat to a safe position out of the ring's projected departure path. A wide-field optical camera and an infra-red camera on ASTRO observed the ring slowly tumble away, remaining visible in both cameras' field of view for a number of minutes. Following the ring ejection, the robotic arm placed the NextSat close to ASTRO. NextSat was then grasped by the arm and the two spacecraft returned to a mated configuration for characterisation and calibration of the ASTRO's rendezvous and capture sensors.

The first free-flying undocking and re-docking demonstration took place on 5 May. The satellites undocked, moved about 10 m apart and flew separately for a full orbit around the Earth before re-docking.

Scenario 3 was to include backing ASTRO about 30 m from NextSat, followed by separation of 60 m, several circumnavigations, and finally a rendezvous from a distance of 7 km. However, a major computer problem occurred on 11 May. The two craft were attempting to fly in formation approximately 30 m apart when ASTRO suffered a main processor failure in its sensor flight computer, triggering an automatic abort due to the loss of critical navigation and guidance data. The AC-2 computer was used to collect and process data from ASTRO's rendezvous instruments, including visible and infra-red sensors and a laser range-finder, before sending it to the satellite's navigation system. ASTRO's control software was switched to free drift mode, and the two satellites moved about 6 km apart. The backup sensor computer, called AC-3, which was one of the mission's ORUs, was powered up successfully and began to process data nominally. After completing several thruster burns to reposition itself, ASTRO was able to close within 3 km of NextSat and re-establish reliable tracking data using its infra-red camera and laser range-finder. On 17 May Boeing software engineers

This image of NextSat was captured by the ASTRO servicing vehicle on 17 April. The battery Orbital Replacement Unit is the large box on the left side of the image (DARPA) 1165803

uploaded an autonomous guidance profile that was executed by ASTRO the following day. ASTRO's Starsys mechanism reported capturing NextSat at 23:00 ET. On 19 May and a few minutes later the vehicles were remated.

The spacecraft completed their second final rendezvous and capture manoeuvre on 29 June when NextSat was grappled by ASTRO. The USAF then planned to deactivate the satellites 5–8 July, after they separated to a distance of 1,000 km, but this was later postponed. The Pentagon extended the mission by two weeks to enable one final rendezvous manoeuvre to demonstrate ASTRO's ability to locate and rendezvous with NextSat from long range. The demonstration entailed moving the satellites 400 km apart so that ASTRO's sensor suite lost track of NextSat. With input from the ground-based Space Surveillance Network, it was able to relocate NextSat and close in. ASTRO then switched over to its on-board sensors to complete the rendezvous.

On 21 July, flight controllers turned NextSat's solar arrays away from the sun and turned off its on-board computer. The next day, ASTRO was decommissioned by turning off its computer and dumping the remaining onboard fuel when the two satellites were about 1,000 km apart. The spacecraft had approximately 88 per cent of their original propellant left. In order to decommission NextSat, its remaining hydrazine had to be depleted. Since NextSat had no independent means to do this (no thrusters or overboard vents), the first requirement was to transfer as much propellant as possible from the NextSat fluid transfer tank to the ASTRO propulsion tank. Any remaining propellant and pressurant (helium) was then vented overboard.

ASTRO and NextSat were left in trajectories with no possibility of re-contact, with both of their orbits naturally decaying within 25 years. A total of 8 mission scenarios was completed under various lighting conditions and approach conditions.

Specifications

ASTRO (Autonomous Space Transfer and Robotic Orbiter)

Launch: 03:10 GMT on 9 March 2007 by Atlas V 401 from SLC-41, Cape Canaveral AFS, Florida

Initial orbit: 490 × 498 km, 46.0°

Design life: Min 3 months, max 1 year

Contractors: Boeing Phantom Works (prime). Starsys Inc (docking capture system, spacecraft separation system). MacDonald Dettwiler (ASTRO robot arm, associated fixtures and software). Northrop Grumman (fluid transfer and propulsion system). ATK (solar arrays) NASA MSFC/Orbital Sciences (AVGS). Draper Laboratory (mission manager software).

Principal applications: On-orbit demonstration of an autonomous guidance, navigation, and control system; autonomous rendezvous, proximity operations, and capture; fluid transfer between satellites; component transfer.

Configuration: Body dimensions 1.78 m × 1.75 m. Span in orbit 5.59 m. 3 m, 71 kg manipulator arm.

Mass: 953 kg dry. Propellant mass 136 kg.

AOCS: 16 3.5N hydrazine thrusters.

TT&C: OE command and control functions executed at the Air Force's Research, Development, Test and Evaluation Support Complex (RSC) at Kirkland Air Force Base, Albuquerque, New Mexico. The RSC is owned by the Space and Missile System Center (SMC), Space Development and Test Wing (SDTW). RSC was the primary interface between the OE team and the AFSCN/TDRSS for telemetry, commanding, orbital determination and scheduling.

Power system: Twin solar arrays, 1,560 W.

NextSat (Next Generation Satellite)

Contractors: Ball Aerospace (prime). ATK (solar arrays). Northrop Grumman (fluid transfer and propulsion system).

Configuration: Octagonal bus, body measurements 1 m × 0.99 m. Span in orbit 2.1 m. One solar array.

Mass: 226.9 kg at launch.

AOCS: Momentum wheels and magnetic torquer rods.

Power system: Single solar array, 500 W.

Other specifications as ASTRO.

PICOSat (P97-1) 7/8

Background

Surrey Satellite Technology Ltd (SSTL) is building PICOSat. It will provide a test of an off-the-shelf micro-satellite to determine if cost effective spaceflight of DoD experiments can be achieved in this manner. The four experiments to be flown on the satellite are: the Polymer Battery Experiment, the Ionospheric Occultation Experiment (IOX) to demonstrate the use of occultation of GPS signals to characterise the ionosphere, Coherent Electromagnetic Radio Tomography Ionospheric (CERTO), and Optical Precision Platform Experiment (OPPEX). P13EX is an all plastic battery that has virtually limitless charge/discharge cycles and other space-favourable characteristics. IOX and CERTO are experiments measuring electron content and densities in the ionosphere. OPPEX is an experimental platform that demonstrates active and passive vibration isolation for future precision instrument applications. The carrier satellite MightySat 2 was launched 19 July 2000 to a 547 × 581 km, 97.8°, orbit. PICOSat 7/8 were launched from MightySat 2 on 7 September 2001 into a 511 × 539 km, orbit.

RADCAL (P92-1 Radar Calibration Satellite)

Background

A Scout booster launched the 95 kg RADCAL 25 June 1993 from Vandenberg into 759 × 888 km, 89.5° orbit. It provides radar calibration via two-C band transponders for more than 13 agencies. The Doppler transmitter permits use of ground processed ephemeris (by Naval Surface Warfare Center) as a standard reference for the radars. Two onboard GPS receivers using 4 patch antennas demonstrate a new technique for generating ephemeredes using ground processing. The USAF Phillips Laboratory Small Satellite Power System Regulator (SSPSR) experiment is also testing improved methods of battery charging. The RADCAL experiments are housed on a modified X-Sat bus. Defense Systems Inc (then CTA Space Systems, now Orbital Sciences Corporation) was prime contractor for the 87 kg spacecraft at a total cost of less than USD4 million. Both satellite bus and payload integration were used as a potential template for future missions of this type.

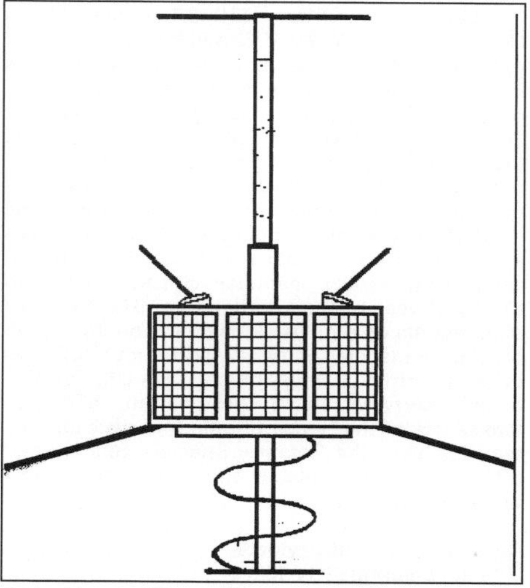

Radcal is providing calibration for >70 C-band radars 0516849

Future Payloads

POGS 2 (S92-1, Polar Orbiting Geomagnetic Survey 2)

The experiment uses a Special Sensor Magnetometer to collect data, which would be used to update the geomagnetic maps of the Earth. POGS 2 was due to be carried onboard defence Meteorological Satellite Program spacecraft S-15, which had a predicted launch need date of 1998. The magnetometer was to be placed on an extendible 5 m boom in order to reduce the effects of electromagnetic noise from the spacecraft, planned to fly on DMSP's S15-S20. First was planned for the second quarter of 1997.

CERTO PLUS (S97-2, Coherent Electromagnetic Radio Tomography/Profiling the Limb with Ultra-violet Sensors)

CERTO PLUS will provide measurements of the integrated electron density of the ionosphere in the satellite orbit plane. It will also provide a database for global models of the ionosphere and will test tomographic algorithms for reconstruction of ionospheric densities and irregularities. CERTO PLUS is scheduled to fly on STRV ID, a spin stabilised, 100 kg class spacecraft to be launched into a geosynchronous transfer orbit by an Ariane 5 launch vehicle.

REX II (P94-2) satellite

Background

STP's follow-on mission to 1991's REX (PS9-1) for research into the physics of electron density irregularities that cause disruptive scintillation effects on radio signals. The payloads include a Rome Laboratory Communications experiment and an OPS Attitude Determination and Control System (ADACS) experiment to perform on-orbit attitude determination and direct feedback to the control system using the Global Positioning System. REX I was launched by Scout G on 29 June 1991 to a 766 × 869 km, 89.6° orbit.

The air force successfully launched the spacecraft by a hybrid Pegasus on 9 March 1996 from Vandenberg into a 799 × 835 km, 90°, orbit. Built by CTA Space Systems, the 113 kg, gravity gradient stabilised with three-axis magnetotorque coils and a momentum wheel for pitch momentum bias, with four deployable solar arrays and gravity gradient boom. A failure occurred in the REX payload in July 1996, which prevents normal operations, but efforts continue to identify and compensate for the cause of the failure. The spacecraft and ADACS payload continue to operate nominally. The air force paid less than USD14 million for REX, including USD9 million for Pegasus.

STEP series

Current Status

STEP M5 was launched on 7 June 2000, by Pegasus XL to a 405 × 1,689 km, 69°, orbit.

Background

TRW signed a USD5.48 million contract 17 April 1990 for STP's Space Test Experiments Platform (STEP). The STEP designs and flies satellites below 450 kg, with lifetimes of one to three years, which communicate with small ground stations. The bus upon which the air force places its experiments provides up to 300 W power and is 3-axis, or gravity gradient stabilised. The TRW contract allows for options on a further 11 satellites. OSC's Pegasus became the prime launch vehicle after selection in April 1991 for one firm (APEX) and 39 optional launches under the AFSLV Air Force Small Launch Vehicle Program.

STEP MO (P90-5)

STEP MO carried a payload of ten TAOS Technology for Autonomous Operational Survivability experiments, including seven computers. It weighed 503 kg at launch on 13 March 1994 into 539 to 560 km, 105° on Taurus with the classified DARPAsat. The air force declared it fully operational on 31 March 1994. The satellite package cost USD56 million, including USD16 million for

TAOS is the first STEP mission (TRW) 0516846

launch. It demonstrated Phillips' Lab project for satellite autonomy from ground controllers using a 6-channel GPS receiver, horizon scanner and high speed computer for 400 m and 0.03° accuracy position determination. It had a design life of 18 months and two articulated solar panels, generating 195 W (100 W available to payload).

MO suffered a serious blow on 20 July 1994 when the IMU failed and full capability was not restored with new software circumventing the problem until 28 February 1995. Responsibility for operating the MO spacecraft has now been transferred to Air Force Space Command.

STEP M1 (P90-1)
The air force originally planned for a 500 km 90° orbit on a Pegasus-XL to investigate HF radio propagation in the ionosphere below 250 km, atmospheric drag over 195–400 km, and plasma instabilities and electric fields. The satellite weighed 348 kg, had 3-axis control and 105 W of power from deployed solar panels. It carried payloads of Ducted HF Propagation; Satellite Electrostatic Triaxial Accelerometer (SETA); Absolute Density Mass Spectrometer (ADMS); Composition and Density Sensor (CADS); Coordinated Heating and Modification Process in the Ionosphere (CHAMPION), Plasma Environment Analyser (PEA). The satellite was lost in the 27 June 1994 Pegasus launch failure.

STEP M2 (P91-2)
The air force launched the 180 kg SIDEX (Signal Identification Experiment) on 19 May 1994 for the USAF Rome Laboratory Air Development Center into 603–821 km, 81.96° orbit (830 km circular planned) aboard Pegasus. The low orbit degraded data collection. The STEP M2 spacecraft failed December 1995 and investigators never determined the cause. Full operations of the spacecraft did not begin until October 1994 because of several anomalies, and then suffered power problems because of overheating. The satellite payload consisted of a narrow beam parabolic gimbaled antenna; wide beam fixed helix antenna, and a GPS antenna. The experiments were designed to evaluate communications in dense signal environments. Controllers flipped the M2 180° in yaw 22 December 1994 to position the battery on the anti-Sun side to resume normal operations after the over-heating.

STEP's first four missions: M0 TAOS (bottom), M1 (centre), M2 SIDEX (top left) and M3 (top right) (TRW) 0517161

STEP M3 (P92-2)
The fourth STEP was lost 22 June 1995 due to failure of the Pegasus-XL launch vehicle. It carried five experiments: ACTEX advanced controls technology experiment (BMDO), EDMM erasable disk mass memory, SAMMES space active modular materials experiment (BMDO), SAWAFE strategic attack warning and assessment flight experiment ('smart skin', BMDO), SQUOD space qualified optical disk. The mission cost the air force about USD40 million. M3 was completed in August 1994 but then had to be stored because of Pegasus-XL's 1994 failure.

STEP M4 (P95-1)
The fifth and final STEP Mission is a 402 kg spacecraft launched on 22 October 1997 aboard a Pegasus XI, into a 430 × 511 km parking orbit at 45°. Ground radar confirms that the solar arrays are not deployed, leaving the satellite in a deep under-voltage condition. STEP M4 hosts three experiments. They are an Orbiting Ozone and Aerosol Measurement (OOAM) experiment, which measures global ozone depletion and water vapour in the middle atmosphere to help gauge infra-red sensor performance, the Electromagnetic Propagation Experiment (EMPE), which characterises ionospheric effects and the Digital Ion Driftmeter (DIDM), which measures the flow of ionospheric ions. Mission requirements demand that the attitude control system operate in both spin and 3-axis modes. In July 1996, CTA Space Systems delivered the spacecraft to TRW S&EG in Chantilly, Virginia for final integration and test. The mission cost the air force an estimated USD80 million.

STP (Space Test Programme) payloads

Current Status
Since inception in 1966, STP has flown more than 420 experiments on over 130 missions supporting scientific research and operational satellite programs.

Background
The Air Force Space and Missile Systems Center, created in May 1965, runs a programme of science and engineering test payloads known as the Space Test Program. The Department of Defense relies on the STP (formerly called STEP) programme to test new concepts and gather information about the space environment. STP put its first payload in space in 1967 and since that time had either launched or participated in the launches of 415 different experiments. DoD provided USD38.3 million for STP operations and experiments this year. In past years the budget has been FY97 USD43.1 million, FY96 USD43.8 million, FY95 USD53.l million, FY94 USD73.1 million, FY93 USD77.9 million, FY92 USD69 million, FY91 USD62.4 million. STP participates in space experiments in three ways: free-flyers, Shuttle pallets and piggyback flights on other experimental packages. For experiments with unique orbital requirements that can best be met by free-flying spacecraft, STP contracts for spacecraft development, experiment integration, and launch service. A recent example is TRW's Space Test Experiments Platform. These experiments usually have a 'P' designation and carry a numerical identifier closely related to the fiscal year of their completion. STP also flies experiments as piggyback payloads on spacecraft of various agencies of the US (NASA and DoD) and other countries (CNES-France, DRA-UK, Russia), usually carrying an 'S' designation. STP also makes use of the Space Shuttle mid-deck lockers, the Hitchhiker, the Getaway Specials, Spartan and other low cost available space with the NASA programme.

Current and recent payloads
STS-83, Columbia (OV-102)
STS-83 launched on 4 April 1997 carried the Cryogenic Flexible Diode (CRYOFD), a DoD Cargo Bay payload. CRYOFD is designed to determine the behaviour of Cryogenic 2-phase thermal control components in microgravity, demonstrate oxygen and methane heat pipe startups from a super-critical condition, demonstrate operations, verify analytical performance models, and establish the correlation between one g and microgravity thermal performance. A secondary objective is to validate the performance of an American Loop Heat Pipe with Ammonia (ALPHA).

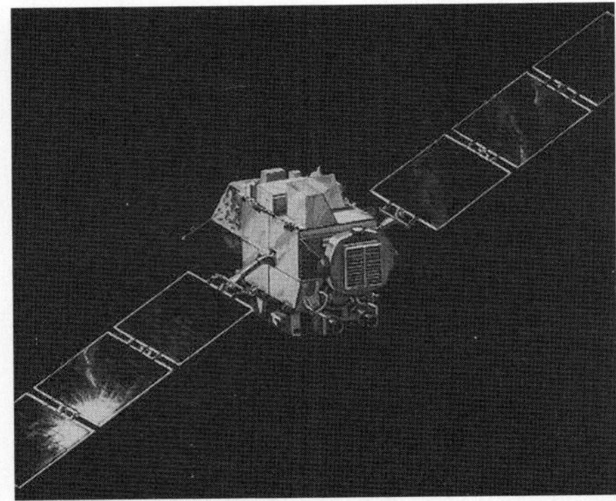

Argos was STP's next major satellite, planned for 1997 (Naval Research Laboratory) 0517160

STS-84, Atlantis (OV-104)
STS-84 launched on 15 May 1997 as the sixth Shuttle/Mir docking mission. Atlantis carried two DoD payloads, which were the first DoD sponsored experiments to operate inside the Mir Space Station. The Radiation Monitoring Equipment – RME III – attempted the first ever direct measurement of the East-West effect, a theory that predicts higher radiation levels (by a factor of 2 to 4) on the West facing side of the Mir. Data was collected throughout the docked phase of the mission and also from inside the Spacehab on the way back to Earth. The Cosmic Radiation Effects and Activation Monitor is composed of passive and active radiation detectors that work together to better understand the effects of cosmic radiation on both the crew and electronics. CREAM recorded over 151 hours of data.

STS-94, Columbia (OV-102)
STS-94, launched on 1 July 1997, again carried a Cryogenic Flexible Diode (CRYOFD). It performed successfully.

STS-85, Discovery (OV-103)
STS-85 was launched from KSC on 7 August 1997 carrying the COOLLAR Flight Experiment activated nine hours into the mission aboard the TAS-01 bridge. CFE completed all of its planned activities and was able to run for an additional 18 hours to gather secondary data.

STS-86, Atlantis (OV-104)
STS-86 launched on 25 September 1997 carried three DoD STP payloads, the Cell Culture Module (CCM), Cosmic Radiation Effects and Activation Monitor, and Shuttle Ionospheric Modification with Pulsed Localised Exhaust (SIMPLEX). CCM performed very well throughout the mission. CREAM was deployed and activated inside the Mir Space Station on 28 September. The experiment ran flawlessly until its deactivation and partial stowage in the Shuttle on 2 October. The remainder of CREAM's components (all passive) remained deployed in the Mir module, Kvant-II module, and Kristall module until STS-89 (January 98). NASA supported SIMPLEX with a pass over the Jicamarca, Peru ground station at approximately 15:30 CDT on 4 October 1997. The pass consisted of a 10-second OMS burn producing an exhaust plume northward. The orbiter aft cameras focused on the starboard OMS engine throughout the burn which produced excellent video.

STS-87, Columbia (OV-102)
STS-87 lifted off from KSC on 19 November 1997 with the Midcourse Space Experiment (MSX) and Shuttle Ionospheric Modification with Pulsed Localised Exhaust (SIMPLEX). A primary objective of this mission was the deployment and retrieval of a Spartan free-flyer. Due to problems with the Spartan deploy and retrieve operations, propellant was at a premium for this mission causing SIMPLEX to 'stand down' halfway through the mission. MSX was able to observe a 10-second, 2-engine, RAM burn (from the Orbiter's rear engines). The ram burn was visible using the LTV narrowband sensor, and the spectrographic imagers. The result was excellent data, making the experiment a success.

STS-89, Endeavour (OV-105)
STS-89 launched on 22 January 98 with the Cosmic Radiation Effects and Activation Monitor (CREAM) active monitors on board the Spacelab Payload Processing Facility (SPPF). Some modifications took place in order to extend the data collection capacity during the Mir phase of the flight. CREAM was successfully deployed and activated inside the Mir Space Station on 25 January. CREAM's active and passive components were deployed in the Mir, Kvant-II module, and Kristall module. All of the CREAM components from STS-86 aboard the Mir were returned to Endeavour at the same time that the STS-89 components were deployed.

CHARGECON-GEO (SP90-3)
The Satellite Charge Control Experiment at GEO flies on the DSCS 3B B-7 communications satellite. The satellite tests the build-up of an electrostatic charge on the spacecraft and methods to then dissipate the charge by releasing an ionised xenon plasma. The experiment started gathering data in August 1995 and through July 1996 the experiment had completed 20 automatic charge dissipation operations. The experiment continues to run.

SWIM (S91-4)
The Solar Wind Interplanetary Measurements instrument on NASA's Wind satellite, launched 1 November 1994.

POAM 2 (S88-1)
NRL's 25 kg POAM Polar Ozone and Aerosol Measurement package orbits attached to France's SPOT 3 launched 26 September 1993. ThermoTrex of La Jolla, California built the instrumentation. Nine UV-IR channels observe sunsets/rises over the poles to provide vertical profiles of polar ozone, aerosols, water vapour and nitrogen tetroxide, and stratosphere and upper troposphere density and temperature. POAM 2 characterises and clarifies the intense turbulence in the middle atmosphere that adaptive optics in systems such as laser communications would have to accommodate. Two of POAM's channels are at important laser wavelengths: 0.353 and 1.059 µm. The other frequencies are 447.4, 452.5, 550.0, 600.0, 763.4, 780.0 and 945.0 µm. When NASA's TOMS instrument aboard Russia's Meteor-3 failed in December 1994, POAM became the only operating US ozone sensor, returning data in unprecedented detail on the ozone hole. The French announced the end of the SPOT 3 mission in December 1996 after an unexplained failure rendered the spacecraft inoperative.

POAM 3 (S95-A)
The Naval Research Laboratory Polar Ozone and Aerosol Measurement 3 is a congressionally mandated follow-on to the POAM 2 project in support of an integrated NRL effort to globally monitor environmental parameters. NRL contracted with the Centre National d'Etudes Spatial (CNES) to integrate, test and launch POAM 3 on the SPOT 4 spacecraft. An Ariane 5 launch vehicle boosted POAM 3/SPOT 4 to an 833 km Sun-synchronous orbit from Kourou, French Guiana.

BINRAD (S93-1, Beryllum-7 Induced Radiation Experiment)
STP conducted a third reflight of the experiment from 14 March 1996 to 15 June 1996. The experiment first flew with Kosmos 2331, launched on a Soyuz-U launch vehicle. The analysis of the foil showed a Be-7 level only 1 per cent of the expected level and analysts are still struggling to explain these results. The experiment followed-up on NASA's LDEF to help explain how LDEF accumulated Be-7 at a rate several thousand times greater than modelled for that altitude. Some researchers have theorised a heretofore unknown transport mechanism from a lower altitude exists. Two previous reflight attempts in 1992 and 1993 on Resurs-F spacecraft failed when the aluminium foil remained unexposed because of improper release from the experiment containment system.

MPTB (S96-1)
The Naval Research Laboratory Microelectronics & Photonics Test Bed is a piggyback experiment that was launched into a high radiation orbit in Autumn 1997. MPTB is presently returning data vital to predicting the performance of advanced microelectronic and photonic devices. Once proven, these off-the-shelf devices will be used in military SPO procurements. The experiment contains 24 individual experiment boards and a charged particle telescope. The decline of the Cold War means that a full complement of radiation hardened parts is no longer available for space systems, leaving commercial unhardened devices to be employed.

TacSat series

Current Status
TacSat-2 was launched on 16 December 2006 and ceased operations on 21 December 2007. TacSat-1 was cancelled in August 2007.

Background
The Tactical Microsatellite (TacSat) programme is currently planned as a series of four launches of experimental small satellites. In April 2003, Brig. Gen. Pete Worden, then director for development and transformation within the USAF's Space and Missile Command, attended the roll-out of the SpaceShipOne suborbital craft at Mojave airport in California. In a side discussion at the event, Worden described an emerging vision of an alternative space architecture based on smaller, low-cost, spacecraft launched by inexpensive rockets, and capable of being generated or replenished rapidly as needed.

The concept of "responsive space" – as opposed to the traditional space architecture, in which launches are programmed years in advance and users take whatever space services are made available to them – was adopted by the Pentagon's Office of Force Transformation (OFT), then headed by the late Vice Adm. Arthur Cebrowski. In May 2003, the OFT reached an agreement with the Naval Research Laboratory (NRL) to develop the TacSat 1 demonstrator. It was delivered 12 months later for a cost of about USD9.3 million. Three more TacSat demonstrators were subsequently contracted, with the four vehicles being due for launch at approximate six-month intervals between the late summer of 2006 and April 2008.

One goal of the TacSat programme is to demonstrate the technology for small satellites that carry payloads which are useful to the warfighter. The new small satellites should also be inexpensive enough to be built and held in readiness for a contingency and should be able to be put into service very quickly after launch. The project is also aimed at demonstrating reduced development time and cost, so that the system can meet an emerging need to put a new payload in orbit in a time-span of six months to two years, rather than ten years as is the case with current spacecraft. Another principle is that the payload should be controlled, and data accessed, through the military's Secret Internet Protocol Router Network (SIPRNET).

A June 2004 briefing from the OFT defined the tactical satellite as complementary to both Unmanned Air Systems (UAS) and traditional satellites. Like the UAS, the TacSat offers launch on demand (by smaller, lower-cost launch vehicles), with coverage and payloads tailored to the mission, and local tactical control. Like a traditional spacecraft, though, the TacSat offers access to denied areas, global access, long-duration missions and the ability to operate contemporaneously in several theatres.

The TacSat, according to USAF studies, lends itself to tactical applications that are different from standard satellites. One example is the so-called Magic Orbit, an asymmetrical 500 × 8,000 km elliptical path that provides a combination of low perigee (for Earth observation) and high perigee (for persistence and area coverage) which provides consistent access to the same area at constant lighting conditions. A traditional satellite would avoid such an orbit because it passes through zones of high radiation. For a TacSat, with a typical lifetime of six months to a year, this is less important.

Another inherent advantage of the TacSat concept is that satellite launches and orbits can be arranged according to the desired region of interest, arranging tracks, altitude and timing, for best coverage. This is likely to be more efficient than long-lasting satellites, where the coverage has to be arranged with many potential areas of interest in mind.

TacSat-1 was developed by the NRL but never flew. The spacecraft was completed in less than one year, from go-ahead to the end of system-level testing, at a cost of less than USD10 million. The 132 kg minisatellite was based on the Orbital Sciences MicroStar bus, including avionics and two circular solar arrays. It carried two low-resolution imaging sensors – an uncooled thermal infra-red camera (non-cryogenically cooled microbolometer FPA) and a Hitachi HanVision HVDUO-F7 industrial camera that used new technology to simultaneously image red, green, and blue on each pixel. Both were to be qualified for use in space. One of the primary goals was to demonstrate that the sensors

Preparing TacSat-2 for thermal vacuum testing at the Space Vehicles Directorate (USAF) 1341968

For details of the latest updates to *Jane's Space Systems and Industry* online and to discover the additional information available exclusively to online subscribers please visit **jsd.janes.com**

The USAF TacSat-2 microsatellite was launched on 16 December 2006 and ceased operations on 21 December 2007 (USAF) 1341967

could be tasked by the SIPRNET and deliver data the same way. Also a Radio-Frequency (RF) emitter identification payload based on the Copperfield-2, an existing sensor system developed for flight on the Global Hawk Unmanned Aerial Vehicle (UAV). The NRL Copperfield-2 payload detects, tracks, and identifies pulsed radio frequency signals.

TacSat-1 was available for launch by 2004 but had to wait for the availability of its contracted launch vehicle, the private-enterprise SpaceX Falcon-1. TacSat-1 was originally intended to be launched on the first flight of Falcon, which was due to take place from Vandenberg AFB, but SpaceX was forced to move the first flight to Kwajalein, with a different payload, because of conflicts with other uses at the California site. Following the loss of the first Falcon in March 2006 and a second failure to reach orbit in March 2007, the TacSat-1 mission was cancelled by DoD in August 2007. Military officials currently see no reason to launch the mission, unless a payload upgrade of TacSat-1 may be of value, since TacSat-2 has already flown and demonstrated many of the capabilities TacSat-1 was intended to test.

TacSat-2, also known as RoadRunner, was launched by Minotaur from Wallops Island on 16 December 2006, the first of two TacSat launches contracted to Orbital Sciences Corporation (OSC) by the USAF in May 2006. The launch was postponed from 11 December after last-minute concerns with TacSat 2's flight software. It was the first orbital launch from Wallops in seven years. Communications failed 77 minutes after the launch, and were only restored after a faulty ground-segment configuration was discovered and corrected on 18 December. USAF ended operations on 21 December 2007.

TacSat-2, was primarily developed by the USAF Research Laboratory at Kirtland AFB in New Mexico. Other participants were the Space Vehicles Directorate, the Space and Missile Systems Center's Space Development and Test Wing, the Naval Research Laboratory, the Army Space Program Office, Air Force Space Command, US Strategic Command, and the Office of Under Secretary of Defense for Acquisition, Technology and Logistics. Weighing some 300 kg, it was based on a spacecraft bus originally produced by MicroSat Systems for the TechSat 21 project, which was cancelled in early 2003.

TacSat-2 mission milestones included demonstrating the rapid development and deployment of a spacecraft suitable for tactical use; accomplishing web-based, time critical payload tasking and information distribution from in-theatre systems employed for aircraft and UAV communications; collecting tactically relevant imagery and signals intelligence data; performing real-time signal geo-location and identification of emitters using satellite and aircraft-based collection platforms.

TacSat-2 carried 11 experiments. Its primary payload was an Earth Surface Imager (ESI) developed by AFRL and SAIC (Science Applications International Corporation), with a resolution of around 1 m. The 51 cm optical telescope was designed to produce photos of targets soon after receiving orders from military commanders. Imagery was downlinked to a ground site at the US Naval Air Warfare Center facility at China Lake, California. Officials hoped this experiment would lead to reduced response times for image requests from deployed troops by directly linking to any common datalink compatible ground station across the globe.

Other experiments included an ion thruster, a global positioning system occultation receiver and an experimental thin film solar array. The spacecraft also featured software for autonomous operations and a new technology instrument sponsored by the NASA New Millennium Programme as part of the Space Technology 6 project. The instrument, called the Inertial Stellar Compass (ISC) was developed by Draper Laboratory in Cambridge, Massachusetts. The ISC comprised a miniature star camera and MEMS gyro suite designed to allow future small spacecraft to autonomously find their orientation in space. It also carried an X-band, two-way Common Data Link (CDL) communications package.

TacSat-3 has been developed by a team that includes NRL, Air Force Research Laboratory (AFRL), Defense Advanced Research Projects Agency (DARPA), the Army Space and Missile Defense Center, and the USAF Space and Missile Systems Center. Programme cost was USD75 million. It is the first of the series to be developed after formal consultation with US field commanders. It will also be the first to demonstrate features such as a standard bus-to-payload interface, and will be first to be based on what is intended to be a standard, multipurpose bus designed for the ORS mission by Swales Aerospace (now ATK). The 160 kg bus is designed to provide an average power of 440 W to the payload.

TacSat-3 will carry a HyperSpectral Imaging (HSI) sensor developed by Raytheon and named ARTEMIS (Advanced Responsive Tactically Effective Militarily Imaging Spectrometer), combining a panchromatic imager and a mercury-cadmium-telluride focal plane array. An onboard digital processor provided by SE Corporation develops a "cube" of information – a two-dimensional picture plus waveband response data – and transmits it to the ground via a space-protected version of the L-3 Communications Common DataLink (CDL), used by USAF aircraft and UAS. Once calibrated, the 160 kg HSI is expected to detect vehicle-sized targets under foliage or camouflage. The primary payload contract was awarded in the summer of 2005 and it completed its Critical Design Review (CDR) in May 2006.

A secondary payload has been developed under contract to the Office of Naval Research: part of the Ocean Data Telemetry Microsatellite Link (ODTML), developed by Praxis Inc. of Alexandria, Virginia. The ODTML collects and disseminates environmental data from a variety of open ocean sensors mounted on buoys, moorings, *in situ* floats, and Autonomous Underwater Vehicles (AUV). Central to the entire concept is the Space Communication Payload (SCP) that would fly on host microsatellites and act as a "router in the sky" by communicating with buoys as nodes in an Internet-like network.

The third payload, the AFRL-designed Space Avionics Experiment (SAE),will validate plug-and-play avionics capability, which involves the use of reprogrammable components to integrate the SAE experiment and the spacecraft structure.

The ORS Modular Bus (ORSMB) was developed by Swales Aerospace (now Alliant Techsystems). It was delivered to the US Air Force Research Laboratory (AFRL) for final integration and test on 18 September 2007. Launch has been delayed several times, and is currently expected in September 2008. The delays are reportedly due to thermal issues with the optical link between the spacecraft's sensor processor and communications system, and the need to upgrade mirror mounts on TacSat-3's ARTEMIS hyperspectral imaging payload to minimise the effects of launch vibration.

TacSat-4, currently due for launch from Vandenberg by Minotaur IV in October 2008, will carry a UHF mobile communications payload and a Blue Force Tracking sensor, which will help troops on the ground to identify and locate friendly forces. The communications payload is similar to that of the Mobile User Objective System (MUOS) comsat, but TacSat-4 will have far fewer transponders on board. Such a system would enable ground troops and vehicles to communicate with other forces located beyond their line of sight. This would be done via low-power transmitters sending their identification and position to small satellites in orbit, which would bundle that data and pass it both to GEO satellites for wider use and to receivers in an air operations centre. Tactical aircrews and other operators would therefore be able to identify any friendly forces on the ground. A third mission for the satellite is to relay data collected by buoys floating in the ocean.

TacSat-4, also known also known as Com-X, is led by NRL and funded by a joint partnership between the Office of Naval Research (ONR) and the Office of Force Transformation (OFT). There is also some involvement from the National Reconnaissance Office (NRO). It will be placed into a 400 × 12,000 km elliptical orbit that will keep the satellite over target areas for hours rather than minutes, maximising the utility of its communications payload.

Various proposals are under consideration for TacSat-5.

Specifications

TacSat-1

Launch: Planned for Falcon-1 from Vandenberg. Cancelled in August 2007.
Orbit: Planned 500 km near-circular, 64°.
Design life: 1 year
Mass: 132 kg at launch
Contractors: Naval Research Laboratory
Payload: Hitachi KP-F100C colour CCD camera. *Spectral response:* 400–660 nm (with IR cutoff filter). 1,030 × 1,300 pixels, 6.7 micron square 67 m GSD, 8.0 × 10.0° FOV, 69.0 × 87.1 km image area from 500 km. 10 bit data (Bayer filter, 30 bits after interpolation) 4.4 cm wide × 4.4 cm high × 11 cm diameter plus lens, 200 g plus lens, 3.6 W max.

Indigo Omega Uncooled Microbolometer Camera. Spectral response: 7.5 to 13 microns. Noise equivalent temperature: 85 mK. 120 × 160 pixels, 51 micron square 850 m GSD, 11.5 × 15.2° FOV, 102 × 136 km image area from 500 km. 14 bit data (8 bits used for TacSat-1). 3.43 cm wide × 3.68 cm high × 4.83 cm diameter plus lens. 120 g plus lens. 1.5 W typical.

Copperfield-2 Radio-Frequency (RF) emitter identification payload with sensor system and high-speed interface that provides conversion from the TCP/IP payload communication protocol to OX.25-based communications of MicroStar/Orbcomm bus. Communication via Ethernet network.
Configuration: Orbital Sciences MicroStar bus, 1.05 m diameter, 0.5 m height/width). 3-axis stabilised. Two circular solar arrays.

TacSat-2 (Joint Warfighting Space Demonstrator 1)

Launch: 16 December 2006 by Minotaur I from NASA's Wallops Island Flight Facility, Virginia.
Orbit: 410 km circular, 40°
Design life: 1 year
Mass: 415 kg at launch
Contractors: MicroSat Systems (bus).
Payload: ESI (Earth Surface Imager) developed by AFRL and SAIC (Science Applications International Corporation). Pushbroom imager, <1 m GSD (Ground Sample Distance), Pan band and 3 multispectral bands in the visible range across a swath of 6,144 pixels (5 km swath width). COTS telescope, 50 cm aperture, from RC Optical Systems.

TIE (Target Indicator Experiment) to locate targets based on RF signatures and in conjunction with P3 aircraft. Uses 11 antennas in total. Palyoad power 7 W (70 W on TacSat-1).

CDL (Common DataLink). AFRL experimental communication system for use with ground-based CDL terminals. Downlink data rates up to 274 Mbit/s.

ROPE (Roadrunner On-orbit Processing Experiment). Array of FPGAs (Field Programmable Gate Array) to process imagery into standard military imaging formats. Supports event processing to identify likely targets, and JPEG data compression for real-time data transmission to ground segment.

HET (Hall Effect Thruster), provided by AFRL. Demonstration ion engine based on the Busek Tandem Hall Thruster BHT-200-X3 (Busek Co. Inc., Natik, MA). Isp up to 1,600 s, variable thrust levels, variable power usage. Nominal 200 W thruster, operating envelope 50 to 300 W. Thrust range 4–17 mN, nominally 12.4 mN. Instrument mass <1 kg, 10 cm diameter, 10.5 cm length. Main use to maintain orbit against drag.

IGOR (Integrated GPS On-orbit Receiver), developed by Broad Reach Engineering, Tempe, AZ. IGOR is of BlackJack heritage from JPL. To conduct ionospheric reflection and transmission experiments plus precise navigation in support of high-precision imaging.

ADS (Atmospheric Density Specifications). Originally two experiments to characterise the upper atmosphere. ACME (Anemometer Cross-track Measurement Experiment) measures cross-track component of rarefied atmosphere at 350 km; developed by University of Texas at Dallas. Removed to save weight. ADMS (Absolute Density Mass Spectrometer) measures atomic mass of the gases in the range 1 to 50. Developed at AFRL, Hanscom AFB, MA.

MVIS (Miniaturised Vibration Isolation System) to demonstrate damping of spacecraft jitter. Consists of series of actuators placed at the imager telescope assembly center of gravity.

ESA (Experimental Solar Array), developed by MSI. A flexible, thin film photovoltaic array designed to demonstrate two different cell technologies and two different deployment mechanisms. FITS (Folded Integrated Thin film Stiffeners) provide improved stiffness. These arrays offer the potential to drastically reduce the storage volume needed for high-power microsatellites.

AE (Autonomy Experiment), developed by ICS (Interface & Control Systems). Two subsystems: OOCE (On-Orbit Checkout Experiment) has enabling technology for autonomous commissioning of the spacecraft during its first day in orbit. ATE (Autonomous Tasking Experiment) permits non-expert warfighters to send data requests to spacecraft and receive a direct response (depending on current spacecraft orbital location. Also serves as long-term and short-term experiment scheduler for all payloads.

Inertial Stellar Compass (ISC). NASA New Millennium payload to continuously determine spacecraft attitude and recover orientation after temporary malfunction or power loss.

Configuration: MicroSat Systems bus with two solar arrays. Irregular octagon with "payload" and "separation" deck. 3-axis stabilised, pointing accuracy of <0.15°. The ADCS (Attitude Determination and Control Subsystem) uses 3 reaction wheels (Dynacon MW-1000) and 3 torque rods as actuators, IMU (Inertial Measurement Unit), analog sun sensors, and magnetometer. A star camera provides high-precision attitude measurements.

Power: Twin three-panel solar arrays, 550 W. Li-Ion 30 Ah battery.

TacSat-3 (Joint Warfighting Space Demonstrator 2)
Launch: September 2008 (?) by Minotaur I from Mid-Atlantic Regional Spaceport, Wallops Island, Virginia
Orbit: 320 km
Design life: 1 year
Mass: Approx. 400 kg
Contractors: Swales Aerospace (now ATK) bus.
Payload: ARTEMIS (Advanced Responsive Tactically Effective Military Imaging Spectrometer) hyperspectral imager. Spectral range 0.4–2.5 µm (VNIR and SWIR) Pan imagery in the visible region only. HSI data in the VNIR and SWIR region. Spectral sampling 5 nm bands. HgCdTe detector, 16 GB storage. Instrument mass 170 kg.

ODTML (Ocean Data Telemetry Microsatellite Link). Uses COTS (Commercial-off-the-Shelf) technologies to improve data collection and dissemination involving sensor platforms, such as free-floating buoys or UGS (Unattended Ground Sensors). Includes Spacecraft Communications Payload (SCP) router; Internet communications.

SAE (Space Avionics Experiment) based on plug-and-play avionics architecture.

Configuration: Swales Aerospace (now ATK) ORSMB (Operationally Responsive Space Modular Bus). Hexagonal service/payload module with three solar arrays at 120° intervals. Cylindrical Ritchey-Chrétien telescope.

TSX-5 (P95-2) Tri-Service Experiments Mission – Mission 5

Background
The US Air Force (USAF) launched TSX-5 on a Pegasus XL launch vehicle on 7 June 2000. It was placed in a 405 × 1,689 km, 106.1 min orbit at a 69° inclination controlled by the Research and Development, Test and Evaluation Support Complex (RSC) at Kirtland Air Force Base, New Mexico. TSX-5 carried the Space Technology Research Vehicle II (STRV 2) experiment package, sponsored by the former Ballistic Missile Defense Organization (now the Missile Defense Agency) and the USAF Research Laboratory's Compact Environment Anomaly Sensor (CEASE). The experiments were designed to remain operational for up to a year after launch.

TSX-6 (P97-x) Tri-Service Experiments Mission – Mission 6

Background
Tri-Service Experiment Mission 6 is expected to carry the Optical Reflection Experiment (OPREX) Spacecraft Evaluation Model (OSEM) and the Polymer Battery Experiment (PBEX) into an 835 km orbit at an inclination of 82 to 98°. Mission requirements were set in 1997 and initial launch capability was expected in November 1999.

XSS (Experimental Satellite System) series

Current Status
XSS-10 completed its brief mission on 29–30 January 2003. XSS-11 was launched in April 2005.

Background
The XSS (Experimental Satellite System) series comprises small, low cost satellites designed to test and demonstrate new technologies that may eventually be used for space applications including servicing, repair, and resupply, as well as, potentially, anti-satellite operations.

Boeing was awarded the contract for **XSS-10** in 1997 under a project funded by the US Air Force Research Laboratory (AFRL). Boeing's Space and Intelligence Systems and Rocketdyne Propulsion and Power, both business units of Boeing Integrated Defense Systems, designed, developed and built the USD40 million satellite.

Its mission was to demonstrate the complex interactions of line-of-sight guidance with basic inertial manoeuvring. The spacecraft consisted of a primary interface platform, known as SPP (Sconce Payload Platform), developed by Swales Aerospace. The SPP supported the satellite and its ejection system and was integrated with the second stage guidance section of the Delta II. A camera was mounted on the SPP to image the ejection of the satellite. The other primary interface platform, known as SEP (Sconce Electronics Platform), functioned as a support platform for the microsatellite prior to its ejection. It supported the satellite Support Electronics Unit (EIU) and an antenna that was mechanically interfaced with the Delta II guidance section at the opposite side of the SPP. The ACS aimed the vehicle in the direction of the Delta II second stage while a digital camera sent an image of the target to the Digital Signal Processor. Spacecraft power was provided by a lithium ion polymer battery consisting of two series-connected battery module subassemblies – the first use of this technology in space. Each subassembly had a total capacity of 12Ah and contained four series-connected cell elements operating from 3.0 to 4.1 V. Communications used an S-band carrier and a miniature SGLS (Space Ground Link Subsystem) system developed at NRL. SGLS consisted of a receiver/demodulator transmitter.

The 28 kg micro-satellite was launched on 29 January 2003 as a secondary payload attached to a Delta II second stage. The primary payload was a Global Positioning Satellite (GPS). Once the second stage separated from the GPS satellite, 16 hours after launch, the XSS-10 was successfully ejected from the secondary launch carrier (SCONCE). Swales Aerospace was selected by the AFRL to design, develop, fabricate, assemble, test and integrate the micro-satellite carrier system with the XSS-10.

After ejection, the satellite successfully commenced an autonomous inspection sequence around the Delta II, and live video was transmitted to ground stations. XSS-10 spent approximately 8 hours flying a series of station keeping manoeuvres with the help of a small TV camera. The micro-satellite flew within 100 m of the Delta II second stage and transmitted images to the ground from its low Earth orbit. XSS-10 then backed away from the stage before repeating the process twice more. Telemetry was briefly lost between the micro-satellite and the ground tracking station at the end of the primary mission sequence but picked up again through a different ground station within 10 minutes. The team then successfully put the micro-satellite into sleep mode and reactivated it as planned. The integrated visual camera, propulsion system, and guidance and control software were all reported to have performed extremely well. The demonstration mission lasted a total of 12 hours.

The 31 kg XSS-10 micro-satellite, covered in gold-coloured thermal blankets, attached to the Delta II launch vehicle (Boeing) 1344024

Key technologies being tested were: lightweight propulsion system, guidance, navigation & control (GNC), miniaturised communications system; primary lithium polymer batteries, integrated camera and star sensor. Octant Technologies, San Jose, California, provided guidance, navigation and flight code design, processor-in-the-loop testing, mission training simulators and command and telemetry ground systems.

XSS-11 was developed by the US AFRL, Space Vehicles Directorate, Kirtland Air Force Base (AFB), New Mexico. Prime contractor was Lockheed Martin Space Systems, with a contract signed in 2001. The project represented a partnership between DOD and industry. Jackson and Tull, Albuquerque, New Mexico, provided technical support to the programme office in design, build, integration, test, and launch activities as well as continued support for ongoing mission operations. Lockheed Martin Astronautics was the spacecraft's contractor for structure, propulsion, and system support.

The XSS-11 micro-satellite was designed to perform several rendezvous and proximity operations with its launch vehicle upper stage (USAF) 1344027

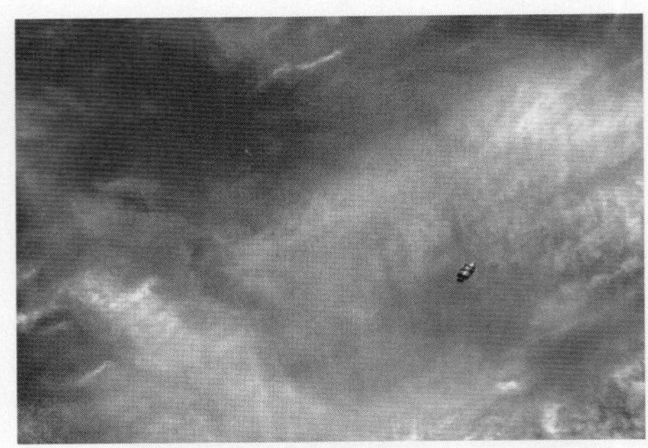

XSS-11 imaged the upper stage of the Minotaur I launch vehicle at a distance of 0.5 km (USAF) 1344025

Other private sector companies participating in the programme included Broad Reach Engineering, Tempe, Arizona (Integrated Avionics Unit); Octant Technologies, San Jose, California (ground segment command and control software); Draper Laboratory, Cambridge, Mass (autonomous mission management software); and SAIC Space & Defense Group, San Diego, California. The Air Force's Space Test Program administered by Space and Missile Systems Center's Detachment 12, located at Kirtland AFB, provided launch, on-orbit command and control oversight.

The lightweight XSS-11 had a box structure containing a propellant tank and a number of subsystems. A radiation-hardened Power PC 750 processor served as the master avionics OBC, enabling onboard autonomous operations and mission planning. An advanced propulsion subsystem using hot gas hydrazine for the main propulsive burns enabled a high degree of translational manoeuvrability.

XSS-11 was extremely mobile and used a Light Detection and Ranging (LIDAR) and autonomous navigation system to locate its target. During its 18 month flight, the spacecraft was to conduct rendezvous and proximity manoeuvres with several US-owned dead or inactive space objects near its orbit, as well as demonstrating increased autonomy. Although never mentioned in any of the official descriptions of the XSS-11's mission, the satellite was able to do everything necessary to intercept and collide with an enemy satellite.

The Rendezvous Laser Vision payload, developed by Optech Inc of Toronto and MD Robotics of Brampton, Ontario, Canada, was designed to demonstrate accurate detection and tracking of on-orbit targets for rendezvous and docking operations. XSS-11 also carried the Naval Research Laboratory's Miniature Space Ground Link System, a small transponder to communicate with the spacecraft. Launched in April 2005 from Vandenberg AFB, California, XSS-11 had rendezvoused three times with the upper stage of the Minotaur I launch vehicle at distances between 1.5 km and 500 m by the end of October 2005. It had also completed more than 75 "natural motion" circumnavigations of the expended Minotaur rocket body. The projected mission cost of XSS-11 was USD82 million.

NASA's Gravity Recovery And Interior Laboratory (GRAIL) mission to create a global lunar gravity map will use two identical spacecraft based on the flight-proven XSS-11.

Specifications

XSS-10

Launch: 18:06 GMT on 29 January 2003 by Delta II 7925-9.5 from Cape Canaveral AFS, Florida.
Orbit: 515 × 805 km, 39.7°, period 98 min.
Design life: 24 hours
Contractors: Boeing IDS/NASA Systems (primes). Swales Aerospace (secondary launch carrier). Octant Technologies (flight code, simulators, test equipment).
Principal applications: Technology demonstration of a miniature communications system, a lightweight propulsion system and advanced lithium polymer batteries involving autonomous navigation test and visual inspection of Delta II second stage.
Configuration: 28 kg microsatellite
Power system: Lithium ion polymer battery. Two series-connected battery module subassemblies, each with total capacity 12 Ah.

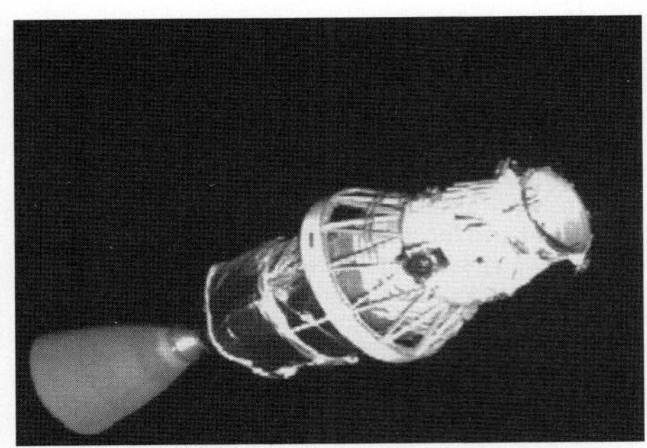

The US Air Force Research Laboratory XSS-10 microsatellite used its camera system to view the second stage of the Boeing Delta II rocket during mission operations on 30 January 2003 (US AFRL/Boeing) 1344023

XSS-11 (USA 165)

Launch: 13:35 GMT on 11 April 2005 by Minotaur I from SLC-8, Vandenberg AFB, California.
Orbit: 839 × 875 km, 98.8°, 102.1 min.
Design life: 18 months
Contractors: Lockheed Martin (prime – structure, propulsion, and system support). Optech (lidar sensor). MD Robotics (Rendezvous Laser Vision payload). Scientific Systems Company, Inc (autonomous navigation); Broad Reach Engineering (Integrated Avionics Unit). Broad Reach Engineering (Integrated Avionics Unit). Octant Technologies (ground segment command and control software). Draper Laboratory (autonomous mission management software). Naval Research Laboratory (Miniature Space Ground Link System). Lithion Inc (Lithium-ion battery)
Principal applications: Technology demonstration of autonomous space operations for the purpose of servicing and inspecting other satellites.
Configuration: Approximately 145 kg at launch. Box structure, 3-axis stabilised. Advanced hydrazine propulsion subsystem. Two solar arrays.
Power system: Twin solar arrays, triple-junction solar cells. High energy density Li-ion battery 28 V, 30 Ah.

CONTRACTORS

CONTRACTORS

3S Navigation
545 East Glendale Avenue, Orange, California, 92865, United States
Tel: (+1 972) 99 55 21 35
Fax: (+1 972) 99 58 78 69
e-mail: ahass@inter.net.il

30th Space Wing
Customer Support Office
806 13th Street, Suite 3A, Vandenberg, AFB, California, 93437-5244, United States
Fax: (+1 805) 606 79 79

A

AB LM Ericsson Finans
EFS
Telefonvägen 30, SE-12625 Stockholm, Sweden
Tel: (+46 8) 719 00 00
Fax: (+46 8) 719 90 50

Academician M F Reshetnev's Nauchno-Proizvodstvennoe Obiedinenie Priklandnoi Mechaniki (NPO PM)
Nauchno-Proizvodstvennoe Obiedinenie Priklandnoi Mechaniki, Academician M.F. Reshetnev's 52 Lenin Street, Zheleznogozsk, 662972 Krasnoyarsk, Russian Federation
Tel: (+7 391) 972 80 08
Fax: (+7 391) 975 61 46
e-mail: npopm@npopm.ru
Web: www.npopm.ru
■ Design of communications systems, TV broadcasting, navigation and geodite satellites. Manufacture of ground antennas for satellite communications. Projects include Express-AK, Express-2000, Express-K2 (Troika project) and SESAT communications satellites and Gonets low earth orbit communications satellite.
Representing: Aerospatiale, France
Alcatel Espace, France
European Telecommunications Satellite Organization, France
Russian Satellite Communications Company, Russia
Russian Space Agency, Russia
SRI Precise Instruments, Russia

Academie Internationale d'Astronautique (IAA)
International Academy of Astronautics
PO Box 1268-16, F-75116 Paris, Cedex 16, France
6 rue Galilée, F-75766 Paris, France
Tel: (+33 1) 47 23 82 15
Fax: (+33 1) 47 23 82 16
e-mail: sgeneral@iaanet.org
Web: www.iaanet.org
Publications:
Acta Astronautica (monthly), IAA Newsletter, Proceedings of Symposia.

Académie Nationale de l'Air et de l'Espace (ANAE)
1 Avenue Camille-Flammarion, Ancien Observatoire de Joliment, F-31500 Toulouse, France
Tel: (+33 5) 34 25 03 80
Fax: (+33 5) 61 26 37 56
e-mail: anae@wanadoo.fr
Web: assoc.wanadoo.fr/anae
■ Development of scientific, technical and cultural activities in areas of air and space.

Academy of Sciences
Institute of Astronomy
76–A Khoroshevskoye Shosse,
123007 Moskva, Russian Federation
Tel: (+7 095) 195 23 63
(+7 095) 195 15 00
(+7 095) 195 63 35
Fax: (+7 095) 195 22 53
(+7 095) 195 15 00

ACC I&M
Engineering and Systems
32 rue du Pré la Reine, F-63017 Clermont Ferrand, Cedex 2, France
Tel: (+33 4) 73 98 38 38
Fax: (+33 4) 73 98 38 20
Web: www.accim.com
■ Ground segment, positioners, software and control units for antenna measurement. Servo-controlled pedestals for flightpath analysis. Earth stations, microwave relays, naval stabilisers for antenna.

Access eV
Access Materials and Processes
Materials and Processes
Intzestrasse 5, D-52072 Aachen, Germany
Tel: (+49 241) 809 80 00
Fax: (+49 241) 385 78
e-mail: welcome@access-technology.de
Web: www.access-technology.de
■ Development and construction of microgravity facilities. Casting processes, process simulation, multicomponent materials, microstructural modelling and metallic solidification.
Accreditations: ISO 9000ff

ACREO AB
Swedish Institute of Microelectronics
Electrum 236, SE-164 40 Kista, Sweden
Tel: (+46 8) 632 77 00
Fax: (+46 8) 750 54 30
e-mail: info@acreo.se
Web: www.acreo.se
■ Facilities of interest for space projects. Projects connected to signal reception from satellites; contract with ESA. Focal plane arrays at 10 μm. Bipolar IC on SIMOY 20 GHz. Applied research, development and consulting services in electronics.

Actel Corporation
Corporate Headquarters`
2061 Stierlin Court, Mountain View, California, 94043 United States
Tel: (+1 650) 318 42 00
Web: www.actel.com

Ad Astra Rocket Company (AARC)
141 West Bay Area Boulevard, Webster, Texas, 77598 United States
Tel: (+1 281) 526 05 00
Fax: (+1 281) 526 05 99
e-mail: aarcinfo@adastrarocket.com
Web: www.adastrarocket.com/home.html

Adaptive Broadband Corporation
China Office
California Microwave Inc
N° 8 Jianguomenbei Avenue, China Resources Building–Room 901, 100005 Beijing, China
Tel: (+86 10) 85 19 13 88
Fax: (+86 10) 85 19 13 87

Advanced Products Co
33 Defco Park Road, PO Box 296 North Haven, Connecticut, 06473, United States
Tel: (+1 203) 239 33 41
Freephone: (+1 800) 243 60 39
Fax: (+1 203) 234 72 33
Cable: ADVPRODCO
e-mail: sales@advpro.com
Web: www.advpro.com
■ Static metal gasket: C/E/O/V-rings, wire rings. Spring-energized PTFE seals (dynamic and static).
Subsidiaries
Advanced Products France
Advanced Products NV, Belgium
Advanced Products Dichtungen, Germany
Advanced Products (Seals & Gaskets) Ltd, UK

Advanced Products (Seals & Gaskets) Ltd
Unit 25c, Number One Industrial Estate, Consett, County Durham, DH8 6SR United Kingdom
Tel: (+44 1207) 50 03 17
Fax: (+44 1207) 50 12 10
■ Manufacture of seals and gaskets.

Advanced Products NV (APNV)
a subsidiary of Advanced Products Co, US
Industrieterrein Krekelenberg, Rupelweg 9
B-2850 Boom, Belgium
Tel: (+32 3) 880 81 50
Fax: (+32 3) 888 48 62
e-mail: sales@advpro.be
■ Manufacture of seals for extreme environmental conditions. Markets: aircraft and aerospace, automotive, chemical process and equipment manufacturing industries.

Advent Communications Ltd
Nashleigh Hill, Chesham, Buckinghamshire, HP5 3HE, United Kingdom
Tel: (+44 1494) 77 44 00
Fax: (+44 1494) 79 11 27
e-mail: sales@advent-comm.co.uk
Web: www.adventcomms.com
■ Design and manufacture of satellite earth stations for fixed transportable and flyaway applications in commercial, government and military markets.

AEA Technology
Aberdeen Office
Exploration House, Offshore Technology Park, Bridge of Don, Aberdeen, AB23 8GX, United Kingdom
Tel: (+44 01224) 25 42 00
Fax: (+44 01224) 25 42 01
e-mail: energy@aeat.co.uk

AEA Technology
Technical Products
Culham, Abingdon, Oxfordshire, OX14 3ED, United Kingdom
Tel: (+44 1235) 52 18 40
Fax: (+44 1235) 43 29 16
e-mail: enquiry@aeat.co.uk
■ Specialises in technology support to clients in sectors outside the nuclear industry. Activities cover advanced materials engineering for aerospace, ordnance and armour including metal matrix, polymer and ceramic composites and depleted uranium, ion inplantation, non-destructive inspection equipment, advanced batteries, sensors, surface engineering, radio frequency systems, radar cross section modelling, partical beam technology, instrumentation, dosimeters, EM, ETC gun technology, plasma technology, advanced power generation, space hardware design and manufacture, plasma modelling and testing, space environmental testing, space tribology design, lubrication and test. Space technology, antennae and black bodies, neural networks, computational modelling and codes, advanced computing, NBC technology, biotechnology and biomedical services, thin layer activation, combustion. Develops and provides advanced robotic systems and 3D video camera systems.

AEA Technology
Thurso Office
Dounreay, Thurso, Caithness, KW14 7TZ, United Kingdom
Tel: (+44 01847) 80 40 80
Fax: (+44 01847) 80 28 18
e-mail: enquiry@aeat.co.uk

AEA Technology Plc
Harwell, Didcot, Oxfordshire, OX11 0QJ, United Kingdom
Tel: (+44 870) 190 19 00
(+44 870) 190 29 21
(+44 870) 112 10 24
Fax: (+44 870) 190 29 20
(+44 870) 190 82 61
(+44 1235) 43 29 16

AEDC Group
877 Avenue E, Arnold Air Force Base, Tennessee, 37389-5051, United States
Tel: (+1 931) 454 70 63
Fax: (+1 931) 454 61 88

Aero Astro
Corporate Headquarters
20145 Ashbrook Place, Ashburn, Virginia, 20147, United States
Tel: (+1 703) 723 98 00
Fax: (+1 703) 723 98 50
Web: www.aeroastro.com

Aero Sekur SpA
Irvin Aerospace SpA
Irvin Manufatture Industriall SpA
a subsidiary of Irvin Group Ltd, UK
Via delle Valli 46, I-04011 Aprilia, Latina, Italy
Tel: (+39 06) 92 01 61
Fax: (+39 06) 92 72 71 65
e-mail: commerciale@sekur.it
Web: www.aerosekur.it
■ Parachutes for personnel, cargo, weapon and aircraft retardation. Recovery systems for missiles/space vehicles and cargo delivery systems. Safety harnesses and air drop sequence control devices. Inflatable survival products and flexible fuel tanks. Nuclear, biological and chemical equipment. Camouflage, lifevests, aerial delivery platforms and field hospital equipment (air transportable). Flexible fuel tanks for aircraft and military vehicles.

AeroAstro Corp
20145 Ashbrook Place, Ashburn, Virginia, 20147, United States
Tel: (+1 703) 723 98 00
Fax: (+1 703) 723 98 50
e-mail: info@aeroastro.com
Web: www.aeroastro.com
■ Design and manufacture of small, low cost space systems and components for small satellites.

Aerodata AG
Aerodate Systems GmbH
Herman Blenk Strasse 34-36, Braunschweig, D-38108, Germany
Tel: (+49 531) 235 90
Fax: (+49 531) 235 91 58
e-mail: mail@aerodata.de
Web: www.aerodata.de

Aeroflex Incorporated
35 South Service Road, Plainview, New York, 11803-0622, United States
Tel: (+1 516) 694 67 00
Freephone: (+1 800) 843 15 53
Fax: (+1 516) 694 06 58
Web: ams.aeroflex.com
www.aeroflex.com
Subsidiaries
Aeroflex Laboratories Inc, US
Aeroflex International Inc, Puerto Rico
Aeroflex Systems Corp, US
Vibration Mountings & Controls Inc, US

Aerojet
Alabama Office
1500 Perimeter Parkway, Suite 315, Huntsville, Alabama, 35806, United States
Tel: (+1 256) 837 33 40
Fax: (+1 256) 837 38 69

Aerojet
Heavy Metals
1367 Old State Route 34, Jonesborough, Tennessee, 37659, United States
Tel: (+1 615) 753 12 00
Fax: (+1 615) 753 86 45

Aerojet
Munition Loading and Packing
604 Spring Street, Socorro, New Mexico, 87801, United States
Tel: (+1 505) 835 20 70
Fax: (+1 505) 835 49 11

Aerojet
New Jersey Office
100 Stierli Court, Suite 103, Mount Arlington, New Jersey, 07856, United States
Tel: (+1 201) 770 32 40
Fax: (+1 201) 770 32 87

Aerojet
Propulsion Systems
PO Box 13222, Sacramento, California, 95813, United States
Highway 50 and Aerojet Road, Rancho Cordova, California, 95670, United States
Tel: (+1 916) 355 10 00
Fax: (+1 916) 351 86 67

Aerojet
Redmond Operations
General Dynamics Space Propulsion Systems
Olin Aerospace Company
Primex Aerospace Company
Rocket Research Co
11411 139th Place, North East, Redmond, Washington, 98052, United States
Tel: (+1 425) 882 57 84
Web: www.aerojet.com

Aerojet
Tennessee Operations
Old State Route 34, Jonesborough, Tennessee, 37659, United States
Tel: (+1 615) 753 12 00

Aeropribor-Voskhod JS
Aeropribor-Voskhod Joint Stock Company
19 Tkatskaya Street, 105318, Moscow, Russian Federation
Tel: (+7 495) 363 23 01
Fax: (+7 495) 363 23 43
e-mail: aerovoskhod@sovintel.ru
■ Manufacture of information systems, air data measurement sets for military and civil aircraft, back-up height/velocity measuring instruments, pitot/static tubes, flight emergency mode warning systems, automatic chute deployment devices, instrumentation for light and super-light aircraft, automatic devices for aerial/space vehicle life support systems, wide scope of precision pressure sensors and sensor-based test equipment, electronic barometric altimeters with the function of FL deviation alert system. Digital air pressure meters.

Aerospace Industries Association of America Inc (AIA)
1000 Wilson Boulevard, Suite 1700, Arlington, Virginia, 22209, United States
Tel: (+1 703) 358 10 02
(+1 703) 358 10 00
Fax: (+1 703) 358 10 11
(+1 703) 358 10 12
(+1 703) 358 10 13
e-mail: aia@aia-aerospace.org
Web: www.aia-aerospace.org
■ Trade association representing US manufacturers of commercial, military and business aircraft, helicopters, aircraft engines, missiles, spacecraft and related components and equipment.
Associate member companies
Abelconn, LLC
ADI American Distributors, Inc
Advanced Technical Products, Inc
Aeroflex Inc
Air Industries Machining Corp
Allen Aircraft Products, Inc
Ambel Precision Manufacturing Corp
American Data & Computer Products
AMI Metals, Inc
Arkwin Industries, Inc
Arrow Gear Company
Arrow/Zeus Electronics
AUSCO, Inc
Auto-Valve, Inc
AV Chem, Inc
Avexus, Inc
Avionics Specialities, Inc
Avnet Electronics Marketing
Banneker Industries, Inc
Berkshire Industries, Inc
Bill-Jay Machine Tool Corporation
Brek Manufacturing Company
BTC Electronic Components, Inc
Burton Industries Aerospace Heat Treating
California Screw Products
Chandler/May, Inc
Cherokee Nation Distributors
Cincinnati Machine
Circle Seal Controls, Inc
Cohesia Corporation
Compass Aerospace Corporation
Composites Atlantic Limited
Consolidated-PAC Foundries
CPI Aerostructures, Inc
Cristek Interconnects, Inc
Cytec Engineered Materials
Dy 4 Systems, Inc
Dynabil Industries, Inc
Eaton Aerospace
Electro-Methods, Inc
Electronic/Fasteners, Inc
EMS Technologies, Inc
ENSCO, Inc
Ensign-Bickford Air & Defense Co.
Envision, LLC
F A G Bearings Limited
Faber Enterprises, Inc
Fastener & Hose Technology, Inc
Fortner Aerospace Manufacturing
Frontier Electroni Systems Corp
G S Precision, Inc
Genmech Aerospace Corporation
GEAR Software
Greene, Tweed & Company
Hagemeyer North America
Hamby Corporation
Hangsterfer's Laboratories, Inc
Hartwell Corporation
Hi-Tech Aero Spares
Hi-Temp Insulation Inc
Hobart Machined Products, Inc
Holaday Circuits, Inc
H & S Swanson's Tool Company
Hughes Bros. Aircrafters, Inc
Hughes-Treitler Manufacturing Corp
Industrial Precision, Inc
Integrated Aerospace
ISPA, Inc
JAMCO Aerospace
JCM Engineering Corp
John Hassall, Inc
Kennebec Tool & Die Co, Inc
Kulite Semiconductor Products, Inc
Lefiell Manufacturing Company
Lewis & Saunders
Lilly Software Associates, Inc
LMI Aerospace, Inc
M/A-Com, Inc
Magnetico, Inc
Manufacturers' Services Control US Operations
Manugistics
Manzi Metals, Inc
Marotta Controls, Inc
McCann Aerospace Machining Corp
Meyer Tool Inc
Micro-Coax, Inc
Millitech, LLC
Morris Machine Company, Inc
MPC Products Corporation
Natel Engineering Co. Inc
National Machine Group (National Aviation Products, National Machine Company)
Navigant Consulting
Neuvant Aerospace
Norfil Manufacturing
Northwest Composites, Inc
Omni International
Onboard Sofware, Inc
Paravant, Inc
Park Engineering & Manufacturing Co, Inc
PC Guardian
Perkinelmer Fluid Sciences
Plymouth Extruded Shapes
Precision Machine & Manufacturing Co
Precision Machine Works Inc
Precision Plymouth Tube Bending
Precision Tube Bending
Primus International
PRMS Incorporated
Pro Fab
Production Engineering Corp.
PTI Technologies, Inc
Quick-Wright Associates, Inc
Radant Technologies, Inc
Ram Manufacturing Inc
REMEC Microwave, Inc
Ryder Systems, Inc
Safe Flight Instrument Corporation
Sechan Electronics, Inc
Service Steel Aerospace
Servotronics, Inc
Sparton Corporation
Spectra Lux Corporation
Spirit Electronics, Inc
Spring Board technology Corp
STADCO
Sunshine Metals
Sypris Electronics, LLC
Texas Composites, Inc
Thayer Aerospace
The Deutsch Company
The Ferco Group
Therm, Inc
Thermal Solutions, Inc
Thomas James International
TMX Aerospace
Transtar Metals, Inc
Tru Circle Aerospace Corp
Trylon Machine Company
Tyco Electronics
Tyco Printed Circuit Group LP
UFC Aerospace Corp
Uni-Tek, LLC
Vaupell Industrial Plastics
Viking Metallurical (Firth Rixson)
Welding Metallurgy, Inc
Wems Electronics
West Cobb Eng & Tool Co Inc
Williams International
Windings, Inc
Xelus
Member companies
AAI Corporation
The Aerostructures Corporation
Alliant Techsystems Inc
American Pacific Corporation
Analytical Graphics Inc
Argo-Tech Corporation

Astro Vision International, Inc
Aviall, Inc
BAE Systems, North America
Ball Aerospace & Technologies Corp
Barnes Aerospace
B.H. Aircraft Company, Inc
The Boeing Company
Celestica Corporation
Computer Sciences Corporation
Cordiem, LLC
Crane Aerospace
Cubic Corporation
Curtiss-Wright Corporation
Curtiss-Wright Flight Systems, Inc Metal Improvement Company
Dassault Falcon Jet Corporation
DRS Technologies, Inc
Ducommun Incorporated
Dupont Company
EDO Corporation
EFW Inc.
Embraer Aircraft Holding Inc
ESIS, Inc
Esterline Technologies
Exostar LLC
Federation Inc
General Atomics Aeronautical Systems, Inc
General Dynamics Corporation
General Electric Company
GKN Aerospace Services
Goodrich Corporation, Airframe Systems, Electronic Systems, Engine Systems
Groen Brothers Aviation, Inc
Harris Corporation
HEICO Corporation
Hexcel Corporation
Hitco Carbon Composites, Inc
Honeywell
ITT Industries Defense and Electronics
Jedco Inc
Kaman Aerospace Corporation
Kistler Aerospace Corporation
L-3 Communications Holdings, Inc
Lockheed Martin Corporation
Martin-Baker America Inc
Matrixone, Inc
MD Helicopters, Inc
MOOG Inc
Northrop Grumman Corporation, Mission Systems, Space Technology
Omega Air, Inc
Orbital Sciences Corporation, Advanced Systems Division
Parker Aerospace
The Purdy Corporation
Raytheon Company
Remmele Engineering, Inc
Rockwell Collins, Inc
Rolls-Royce North America Inc
RTI International Metals, Inc
SiliconGraphics, Inc
Smiths Aerospace Actuation Systems Los Angeles
Spectrum Astro, Inc
Stellex Aerostructures, Inc
Teleflex Inc
Textron Inc
Titan Coporation, Advanced Systems Division
Triumph Group, Inc
United Defense
United Technologies Corporation
(Hamilton Sundstrand, Pratt & Whitney, Sikorsky)
Vought Aircraft Industries, Inc
Woodward Governor Company

Aerospace Medical Association (ASMA)

320 South Henry Street, Alexandria, Virginia, 22314-3579, United States
Tel: (+1 703) 739 22 40
Fax: (+1 703) 739 96 52
Web: www.asma.org

■ Non-profit association of physicians, physiologists, nurses and other professionals specialising in maintaining the health and efficiency of man in atmospheric and space flight.

Publications: Aviation, Space & Environmental Medicine

Aerospace Research Center of the Indonesian National Institute of Aeronautics and Space

a subsidiary of the Indonesian National Institute of Aeronautics and Space
Pusat Riset Dirgantara, Lapan, Jalan Drive Djundjunan 135, PO Box 26, 40001 Bandung, Indonesia
Tel: (+62 22) 61 26 02
Telex: 28229

■ Research and development in aerospace science through exploration and exploitation of the earth and atmosphere.

Aeroterra SA

Avenida E Madeiro 1020, PB 1106 Buenos Aires, Argentina
Tel: (+54 1) 225 40 30
Fax: (+54 1) 311 85 91
Web: www.aeroterra.com

Aeroterra SA

Gorostiaga 2465, 1426 Buenos Aires, Argentina
Tel: (+54 1) 771 58 81
(+54 1) 775 62 05
Fax: (+54 1) 774 61 83

Aetheric Engineering Ltd

Katana House, Fort Fareham Trading Estate, Fareham, PO14 1AH, United Kingdom
Tel: (+44 1329) 82 35 83
Fax: (+44 1329) 28 86 75
e-mail: sales@aetheric.co.uk
Web: www.aetheric.co.uk

■ Consultancy in all aspects of satellite communication systems, including system planning and operation, earth station implementation and satellite procurement. Services include feasibility and design studies; preparation of specifications and procurement documentation; evaluation of proposals and tenders; project management; supervision and witnessing of in-factory activities, including manufacture, test and integration; supervision and witnessing of on-site activities, including installation, test and commissioning; system operation; and frequency co-ordination and access planning. Provision of support to the establishment of effective communication regulatory organisations.

Afro-Asian Satellite Communications Ltd (ASC)

135 Dr Annie Besant Road,
Worli, Mumbai (Bombay), 400018, India
Tel: (+91 22) 493 96 85
Fax: (+91 22) 493 51 88

Agência Espacial Brasileira (AEB)

Centro Tecnico Aeroespacial
SPO Área 5 Quadra 3, Bloco A CEP 70.610, Brasília, 200-DF, Brazil
Tel: (+55 61) 34 11 55 72
(+55 12) 39 47 32 10
(+55 12) 39 47 33 10
(+55 12) 39 47 33 11
Fax: (+55 61) 34 11 56 88
Web: www.aeb.gov.br

■ Responsible for education and R&D programs formulated to meet the objectives of the National Aerospace Policy

Comprising following institutes
ITA Instituto Technologico de Aeronautica (Aeronautical Technological Institute)
IAE Instituto de Atividades Espaciais (Space Activities Institute)
IPD Instituto de Pesquisos e Desenvolvimento (Research & Development Institute)
IFI Instituto de Fomento e Coordenação Ind (Industrial Coordination & Fostering Institute)
IEAv Instituto de Estudos Avançados (Advanced Studies Institute)

AGM Container Controls Inc

PO Box 40020 Tucson, Arizona, 85717–0020, United States
3526 East Fort Lowell Road, Tucson, Arizona, 85716, United States
Tel: (+1 520) 881 21 30
Freephone: (+1 800) 995 55 90
Fax: (+1 520) 881 49 83
e-mail: sales@agmcontainer.com
Web: www.agmcontainer.com

■ Manufacture breather valves, desiccators, desiccant, humidity indicators, records holders, access ports, shock overload indicators, tie down straps and shelving for the environmental control of moisture and shock sensitive electronics, optics, and weapons in and out of sealed re-usable shipping and storage containers, mobile shelters and shipboard stowage areas.

Overseas agents/representatives
Adtech Inc., Japan
Airsec SA, France
Brownell LTD, UK
Diatero, France
Karsel LTD, Turkey
Sued-Chemie Espana, Spain
Sued-Chemie AG, Germany
Yail-Noa Ltd Agencies, Israel

Ahmed Satellite Earth Station

Dehra Dun
P O Lachhiwala, Doiwala, Dehra Dun, 248140, India
Tel: (+91 135) 69 51 03
Fax: (+91 135) 336 48 45

Air Force Space Command (AFSPC)

21st Space Operations Squadron
Air Force Satellite Test Center
a unit of the 50th Space Wing
1080 Lockheed Martin Way, Box 129, Sunnyvale, California, 94089-1235, United States
Tel: (+1 408) 752 40 35
Fax: (+1 408) 751 60 82
Web: www.schriever.af.mil/fact_sheets/21sops

Air Force Space Command (AFSPC)

21st Space Wing
Peterson AFB, Colorado Springs, Colorado, 80914-1294, United States
Tel: (+1 719) 556 78 25
Fax: (+1 719) 556 78 48

Command & Control unit at
Cheyenne Mountain AS

Space warning units at
Beale AFB
Cape Cod AS
Cavalier AS
Clear ASN
Thule AB (Greenland)
Woomera AS (Australia)

Surveillance units at
Diego Garcia
Eglin AFB
Maui
Misawa AB (Japan)
Peterson AFB
RAF Feltwell (UK)

Air Force Space Command (AFSPC)

30th Space Wing
747 Nebraska Avenue, Vandenberg AFB, California, 93437-6267, United States
Tel: (+1 805) 606 35 95
(+1 805) 606 58 14
(+1 805) 606 58 15
Fax: (+1 805) 606 83 03
Web: www.vandenberg.af.mil

■ The unit's mission is to spacelift satellites to polar orbit, operate and maintain the western range, support ICBM test launches, provide host base support and support commercial space launch.

Launch unit at
Vandenberg AFB

Tracking stations at
Kaena Point
Pillar Point

Air Force Space Command (AFSPC)

45th Space Wing
Patrick AFB, Florida, 32925-6655, United States
Tel: (+1 407) 494 59 33
(+1 407) 494 11 10
Fax: (+1 407) 494 73 02

Launch unit at
Cape Canaveral AS

Tracking stations at
Antigua AS
Ascension Aux Field (St Helena)
Cape Canaveral AS
Patrick AFB

Air Force Space Command (AFSPC)

50th Space Wing
Schriever AFB, Colorado, 80912-5000, United States
Tel: (+1 719) 567 50 40
Fax: (+1 719) 567 53 06
Web: www.schriever.af.mil

Satellite operations units at
Cape Canaveral AS
Onizuka AS
Schriever AFB

Satellite remote tracking stations at
Andersen AFB
Diego Garcia
Kaena Point
New Boston AS
RAF Oakhanger (UK)
Schriever AFB
Thule AB (Greenland)
Vandenberg AFB

Air Force Space Command (AFSPC)

90th Space Wing
5305 Randall Avenue, FE Warren Air Force Base, Wyoming, 82005-2266, United States
Tel: (+1 307) 773 20 74
Fax: (+1 307) 773 33 81
Web: www.warren.af.mil

Air Force Space Command (AFSPC)
91st Space Wing
201 Summit Drive, Minot Air Force Base, 58705-2000, United States
Tel: (+1 701) 723 62 12
Fax: (+1 701) 723 65 34
Web: www.minot.af.mil

Air Force Space Command (AFSPC)
341st Space Wing
2177th Street North, Room 150A, Malmstrom Air Force Base, 59402-7538, United States
Tel: (+1 406) 731 40 50
Fax: (+1 406) 731 40 48
e-mail: 3415wpa@nalmstrom.af.mil
Web: www.malmstrom.af.mil

Air Force Space Command (AFSPC)
721st Support Group
Cheyenne Mountain Air Force Station, Colorado, United States
Tel: (+1 719) 554 73 21

Air Force Space Command (AFSPC)
Headquarters
150 Vandenberg Suite 1105, Peterson AFB, Colorado Springs, Colorado, 80914-4500, United States
Tel: (+1 719) 554 37 31
Fax: (+1 719) 554 60 13
e-mail: afspc/pa@peterson.af.mil
Web: www.peterson.af.mil/hqafspc
■ The primary missions of Air Force Space Command are: space and intercontinental ballistic missile operations, space warfare centre, including ballistic missile warning, space control spacelift and satellite operations. To integrate space capabilities and techniques into daily AF operations.
Major units include
Fourteenth Air Force (14AF)
–21st Space Wing–Ballistic Missile Warning Space Surveillance, Peterson AFB
–30th Space Wing–DoD, Civil and Commercial Spacelift, Vandenberg AFB
–45th Space Wing–DoD, Civil and Commercial Spacelift, Patrick AFB and Cape Canaveral AS
–50th Space Wing–Satellite Operations, Schriever AFB
–460th Air Base Wing–Missile Warning, Buckley AFB
Twentieth Air Force (20AF)
–90th Space Wing–Peacekeeper and Minuteman III, FE Warren AFB
–91st Space Wing–Minuteman III, Minot AFB
–341st Space Wing–Minuteman III, Malmstrom AFB
–721st Space Group–Mission Support Cheyenne Mountain AS

Air Force Space Command (AFSPC)
North Dakota Office
Grand Forks AFB, North Dakota, United States
Tel: (+1 701) 747 30 00
■ Air Combat Command Base.

Air Force Space Command (AFSPC)
Public Affairs Office
150 Vandenberg Street, Suite 1105, Peterson Air Force Base, Colorado, 80914-4500, United States

Air Launch LLC
5555 Lakeview Drive, Suite 201, Kirkland, Washington, 98033, United States
Tel: (+1 425) 285 15 55
e-mail: info@airlaunchllc.com
Web: www.airlaunchllc.com

Air Liquide
Advanced Technology Division
2 rue Clémenciére, F-38360, Sassenage, France
Tel: (+33 4) 76 43 60 30
Fax: (+33 4) 76 43 60 98

Air Liquide
Headquarters
75 quai d'Orsay, F-75007 Paris, France
Tel: (+33 1) 40 62 55 55

Air Products plc
Hersham Place, Molesey Road, Walton-on-Thames, Surrey, KT12 4RZ United Kingdom
Tel: (+44 1932) 24 92 00
Fax: (+44 1932) 24 95 65
Web: www.airproducts.com
■ Liquid hydrogen and oxygen suppliers, industrial gases plant designers, builders and suppliers of all other gases.

Airborne Air and Fuel Products
711 Taylor Street, Elyria, Ohio, 44035, United States
Tel: (+1 440) 937 13 15
Freephone: (+1 800) 382 84 22
Fax: (+1 440) 937 54 09
e-mail: techhelp@parker.com
Web: www.parker.com/airborne

Airborne Systems
Para-Flite Inc
a subsidiary of Wardle Storeys Ltd, UK
part of Airborne Systems NA (North America), US
5800 Magnolia Avenue, Pennsauken, New Jersey, 08109-1399, United States
Tel: (+1 856) 663 81 20
Fax: (+1 856) 663 81 59
e-mail: marketing@paraflite.com
sales@paraflite.com
Web: www.paraflite.com
■ Tactical gliding parachute systems, automatic aerial cargo delivery systems.

Aircraft Tanks Ltd
a subsidiary of MSM Group of companies
Spring Vale Works, Middleton, Manchester, M24 2HS, United Kingdom
Tel: (+44 161) 643 24 62
Fax: (+44 161) 643 34 90
e-mail: info@msmgroup.org
Web: www.msmgroup.org
■ Maintenance, repair and overhaul of aircraft and aircraft powerplants. Manufacture of fuel tanks for fighting vehicles and aircraft and aircrew seats. Sheet metal work and component manufacture. Subcontract work on missile programmes and Ariane space launcher.

AIT Recorders Ltd
Walton Radar Systems Ltd
24 The Green, Twickenham, Middlesex, TW2 5AB, United Kingdom
Tel: (+44 208) 858 00 50
Fax: (+44 208) 858 00 60
e-mail: enquiries@ait_recorders.com
Web: www.ait-recorders.com
■ Recording systems for surveillance, archival and training use, datalogging and real-time data processing systems and graphics. Design, development and production of specialist electronic systems.

Alaska Aerospace Development Corporation (AADC)
4300 B Street, Suite 101, Anchorage, Alaska, 99503, United States
Tel: (+1 907) 561 33 38
Fax: (+1 907) 561 33 39
■ Development of aerospace related economic, technical and educational opportunities for the State of Alaska.

Alaska Aerospace Development Corporation (AADC)
Kodiak Launch Complex
Kodiak Administrative Office, 323 Carolyn Street, Suite 102, Kodiak, Alaska, 99615, United States
Tel: (+1 907) 561 33 38
Fax: (+1 907) 561 33 39
Web: akaerospace.com

Alcatel Espace
Headquarters
a part of the Alcatel Alsthom Group
5 rue Noël Pons, F-92737, Nanterre, Cedex, France
Tel: (+33 1) 46 52 62 00
Fax: (+33 1) 46 52 62 50
■ Design and development, production and integration of complete systems, satellites and ground segments for telecommunications, earth observation and science for civilian and military applications. Communication payloads for worldwide satellite manufacturers. Equipment, instruments and space sub systems.

Alcatel Espacio SA
Alcatel Space Business Line
a subsidiary of Alcatel Space Industries
c/o Einstein 7,
PTM Tres Cantos, E-28760, Madrid, Spain
Tel: (+34 91) 807 79 00
Fax: (+34 91) 807 79 99
e-mail: comunicacion.espacio@alcatel.es
Web: www.alcatel.es/espacio
■ Design, development and manufacture of satellite communications equipment including on board equipment and subsystems for space vehicle applications. Products include transponders and transceivers, TTC subsystems, amplifiers, transmitters, modulators, multiplexers, filters, diplexers and other passive devices. Digital processing units and on-board data handling and signal processing equipment for advanced communication payload.

Alcatel Space Denmark AS
Alcatel Kirk AS
a subsidiary of Alcatel Space Industries, France
Toldbodgade 12, DK-1253, Ballerup, Denmark
Tel: (+45) 44 80 75 00
Fax: (+45) 44 80 75 75
e-mail: info@alcatel.dk
Web: www.alcatel.dk
■ Design and manufacture of control electronics, power distribution units, DC/DC power converters and transformers and coils for space vehicle applications. Satellite programmes.

Alcatel Space Norway A/S
a subsidiary of Alcatel Space Industries, France
Knudsrødveien 7, PO Box 138, N-3191, Horten, Norway
Tel: (+47 33) 03 27 00
Fax: (+47 33) 03 28 00
e-mail: asn@alcatel.no
Web: www.asn.no
■ Manufacture of on-board space electronics for satellites and launchers.

Alcoa Corporate Center
Alcoa Industrial Components
201 Isabella Street, Pittsburgh, Pennsylvania, 15212-5858, United States
Tel: (+1 412) 553 45 45
Fax: (+1 412) 553 44 98
Web: www.alcoa.com

Alcoa Industrial Components
15 West South Temple, Suite 1600, Salt Lake City, Utah, 84101-1532, United States
Tel: (+1 801) 933 41 93

Alcoa Manufacturing (GB) Ltd
a subsidiary of ALCOA, US
Waunarlwydd Works, PO Box 68, Swansea, SA1 1XH, United Kingdom
Web: www.alcoa.com

Alenia Aeronautica SpA
a wholly owned subsidiary of Finmeccanica, Italy
Via Campania 45, I-00187, Roma, Italy
Tel: (+39 6) 42 08 81
Fax: (+39 6) 42 82 45 28
e-mail: communication@alenia-aeronautica.it
international-sales@aeronautica.it
Web: www.alenia-aeronautica.it
■ Design, build, manufacture and certification of military and commercial aircraft and mission systems including the unmanned air vehicles (UAVs). Aircrafts include: Eurofighter Typhoon, C-27J, ATR42MP, ATR72 ASW, A380 and Boeing 787.

Allen Osbourne Associates Inc
756 Lakefield Road, Westlake Village, California, 91361-2624, United States
Tel: (+1 805) 495 84 20
Fax: (+1 805) 373 60 67
e-mail: aoa@aoa-gps.com

Alliance for Microgravity Materials Science and Applications (AMMSA)
NASA/Marshall Space Flight Center, Building 4481, Alabama, 35812, United States
e-mail: ammsa@space.hsv.usra.edu

Alliance Spacesystems Inc
1250 Lincoln Avenue, Suite 100, Pasadena, California, 91103-2466, United States
Tel: (+1 626) 296 13 73
Fax: (+1 626) 296 00 48
Web: www.asi-space.com

Alliant Aerospace Propulsion Company
a part of Alliant Aerospace business segment
Bacchus Works, 8400 West 5000 South, Magna, Utah, 84044, United States
■ Supply of solid propulsion systems for commercial space launchers and for strategic missiles.

Alliant Missile Products Company
a part of Alliant Defense Systems business segment
Allegany Ballistics Laboratory, 210 State Route 956, Rocket Center, West Virginia, 26726, United States
Tel: (+1 304) 726 50 00
Fax: (+1 304) 726 47 30
■ Suppliers of propulsion systems, warheads and structures for tactical missiles. Manufacture of fibre-placed composite structures.

All-Power Manufacturing Co
13141 Molette Street, Santa Fé Springs, California, 90670, United States
Tel: (+1 562) 802 26 40
Fax: (+1 562) 921 99 33
e-mail: sales@allpowermfg.com
Web: www.allpowermfg.com
■ Aerospace bushings, spacers and precision hardware.

Ambitech International
Axiom Operations,
Axiom Electronics, Inc
14924 North West Greenbrier Parkway, Beaverton, Oregon, 97006-5733, United States
Tel: (+1 503) 350 49 06
Freephone: (+1 800) 643 66 01
Fax: (+1 503) 350 43 66
e-mail: doug.robertson@axiom.ambi-intl.com
Web: www.axiomsmt.com
■ Services include complete procurement service, surface mount assembly, mixed technology assembly, precision rework of existing products and complete final assembly and unit integration services.
Accreditations: ISO 9002, IPC-A-610, J-STD-001, NASA STD-8739.3, UL, Boeing QMS, FDA, ITAR

American Aerospace Controls Inc (AAC)
570 Smith Street, Farmingdale, New York, 11735-1115, United States
Tel: (+1 631) 694 51 00
Freephone: (+1 888) 873 85 59
Fax: (+1 631) 694 67 39
e-mail: aac@rcn.com
Web: www.a-a-c.com

American Astronautical Society (AAS)
6352 Rolling Mill Place #102, Springfield, Virginia, 22152-2354, United States
Tel: (+1 703) 866 00 20
Fax: (+1 703) 866 35 26
e-mail: aas@astronautical.org
Web: www.astronautical.org
■ America's network of space professionals, technical and non-technical, dedicated to advancing all space activities.
Publications: AAS Directory (every 2 years)
AAS Magazine (bi-monthly),
Journal of the Astronautical Sciences (quarterly).
Science & Technology series, Advances in the Astronautical Sciences, History series (bound books)

American Fuel Cells and Coated Fabrics Co (AMFUEL)
601 Firestone Drive, PO Box 887, Magnolia, Arkansas, 71753, United States
Tel: (+1 870) 235 72 39
(+1 870) 234 33 81
Fax: (+1 870) 235 72 70
Web: www.amfuel.com
■ Provide design and fabricating of composite material products for aviation, aerospace, military ground forces, and commercial transport industries.
Accreditations: ISO 9001:2000

American Institute of Aeronautics & Astronautics (AIAA)
1801 Alexander Bell Drive, Suite 500, Reston, Virginia, 20191-4344, United States
Tel: (+1 703) 264 75 00
Freephone: (+1 800) 639 24 22
Fax: (+1 703) 264 75 51
Web: www.aiaa.org
■ Main purpose is to advance the arts, sciences and technology of aeronautics and astronautics. US representative on the International Astronautical Federation and the International Council on the Aeronautical Sciences.
Publications: Aerospace America

American Meteorological Society (AMS)
45 Beacon Street, Boston, Massachusetts, 02108, United States
Tel: (+1 617) 227 24 25
Fax: (+1 617) 742 87 18
e-mail: amspubs@ametsoc.org
Web: www.ametsoc.org/ams
■ Committees on Aviation Meteorology, Satellite and Aerospace Systems.

Americom Government Services (AGS)
wholly-owned subsidiary of SES Americom, US
2 Research Way, Princeton, New Jersey, 08540, United States
Tel: (+1 609) 987 45 00
Fax: (+1 609) 987 44 11
e-mail: ags-info.americom@americom-gs.com
Web: www.americom-gs.com

Ametek Aircontrol Technologies
Howden Aircontrol Ltd
111 Windmill Road, Sunbury onThames, Middlesex, TW16 7EF, United Kingdom
Tel: (+44 1932) 76 58 22
Fax: (+44 1932) 76 10 98
e-mail: mail@aircontroltechnologies.co.uk
Web: www.aircontroltechnologies.co.uk
■ Design and manufacture of military and aerospace air and liquid conditioning systems including vapour cycle or air cycle cooling, NBC filtration equipment, engine cooling systems, ground support equipment and environmental control systems.

Ampex Great Britain Ltd
a subsidiary of Ampex Corp, US
Ampex House, Beechwood, Chineham Business Centre, Chineham, Basingstoke, Hampshire, RG24 8WA, United Kingdom
Tel: (+44 1256) 81 44 10
Fax: (+44 1256) 81 44 74
e-mail: info@ampexdata.com
Web: www.ampexdata.com
■ Rugged airborne and commercial laboratory DSRs series of recorders using solid state and raid disc arrays to store the data at up to 2.5 GBits per sec. with storage capacity up to 1 Tera Byte. Applications include airborne image recording, radar recording, and acoustic sonar recording. Interfaces include 8bit parallel, fibre channel, ethernet and many custom interfaces to meet user's requirements. Small modular multiplexers and solid state recorders for the capture of flight test, elint/comint, telemetry, cockpit video data.

Amptek Inc
14 De Angelo Drive, Bedford, Massachusetts, 01730, United States
Tel: (+1 781) 275 22 42
Fax: (+1 781) 275 34 70
e-mail: sales@amptek.com
Web: www.amptek.com
■ Development and manufacture of low-power, space qualified hybrid electronic circuits. Products include radiation hardenedhybrid amplifiers and preamplifiers which have been used on Galileo, NEAR and Mars Pathfinder; custom hybrid circuits.

Amsat Deutschland eV
Ernst-Giller Strasse 20, D-35039 Marburg, Germany
Tel: (+49 6421) 68 41 12
Fax: (+49 6421) 282 56 65
e-mail: p3e-project@amsat-de.org
■ Non-profit-organisation. Satellites for communication and science.

Analex Corporation
2001 Aerospace Parkway, Brook Park, Ohio, 44142, United States
Tel: (+1 216) 925 11 71
Fax: (+1 216) 925 10 40
Web: www.analex.com

Analytical Graphics
Dag Lane, Stoke Goldington, Buckinghamshire, MK16 8NY, United Kingdom
Tel: (+44 1908) 55 11 05
Fax: (+44 1908) 55 10 63

Analytical Graphics Inc (AGI)
220 Valley Creek Boulevard, Exton, Pennsylvania, 19341, United States
Tel: (+1 610) 981 80 00
Freephone: (+1 800) 220 47 85
Fax: (+1 610) 981 80 01
e-mail: info@agi.com
info@stk.com
Web: www.stk.com
■ Developers of the Satellite Tool Kit (STK) family of products, an interactive, graphical software package that helps analysts, developers, operators and users of satellite systems access, manage, display and manipulate their data. Users can analyse aerospace systems both graphically and numerically. Operates on UNIX platforms, Windows 95/98, Windows 2000 and Windows NT. Programmer's Library and 3D visualisation option available. The satellite orbit analysis software addresses market segments including: design, operations, communications, navigation, remote sensing, science and defence.

Andøya Rocket Range
PO Box 54, N-8483 Andenes, Norway
Tel: (+47) 76 14 44 00
Fax: (+47) 76 14 44 01
e-mail: info@rocketrange.no
Web: www.rocketrange.no
■ Provides products, services and infrastructure with regard to space science and environmentally related science and surveillance. Launch of sounding rockets and balloons for investigation of upper atmosphere at high latitudes. Recovery of payloads. Users from more than 70 institutes and universities in Europe, USA, Canada and Japan. Owns and operates 'SvalRak' a non-permanent launching facility at Svalbard.

Andrew Canada Inc
Andrew Antenna Co Ltd
a subsidiary of Andrew Corporation, US
606 Beech Street West, PO Box 177, Whitby, Ontario, L1N 5S2, Canada
Tel: (+1 905) 668 33 48
(+1 905) 665 43 00
Fax: (+1 905) 668 85 90
(+1 905) 665 43 16
(+1 905) 430 39 64
Web: www.andrew.com
■ Microwave terrestrial antennas, HELIAX coaxial cable and elliptical waveguide, military earth station antennas, radar and navigation aid systems, HF antennas, direction finding tactical antennas, weather radar.

Andrew Corporation
Corporate Offices
10500 West 153rd Street, Orland Park, Illinois, 60462-3099, United States
Tel: (+1 708) 349 33 00
■ Microwave antennas and antenna systems, telemetry and missile tracking antennas, scatter antennas and reflectors. Rigid and flexible transmission lines and waveguides. Antennas (airborne, microwave, telemetry, broadcast earth station; transmission lines). Defence electronics.
Branches in
Argentina, Australia, Austria, Brazil, Canada, China, France, Germany, India, Indonesia, Italy, Japan, Malaysia, Mexico, Netherlands, Philippines, Russia, South Africa, Spain, Sweden, Switzerland, Taiwan, Thailand

Andrew Ltd
a subsidiary of Andrew Corporation, US
The Avenue, Lochgelly, Fife, Scotland, KY5 9HG, United Kingdom
Tel: (+44 1592) 78 05 61
Fax: (+44 1592) 78 23 80
Web: www.andrew.com
■ Antenna and electro-optical communications systems equipment.
Representing:
Andrew Corp, US

Anite Systems GmbH
Robert-Bosch Strasse 7, D-64293 Darmstadt, Germany
Tel: (+49 6151) 872 51 00
Fax: (+49 6151) 872 51 51
Web: www.anitesystems.de

Antenna Technology
Eastern Office
289 Atlas Street, Simpson, Pennsylvania, 18407, United States
Tel: (+1 717) 282 35 90
Fax: (+1 717) 282 32 58

Antenna Technology Communications Inc (ATCI)
450 North McKemy Chandler, Chandler, Arizona, 852210, United States
Tel: (+1 480) 844 85 01
Fax: (+1 480) 898 76 67
Web: www.atci.net

■ Design, manufacture, sales and installation of wireless broadband services and ground based commercial satellite communications systems. Products for domestic and international; corporate, broadcast and cable television; government and educational institutions.

Antrix Corp Ltd
Antariksh Complex, Near New Bel Road, Bangalore, 560094, India
Tel: (+91 80) 341 62 73
(+91 80) 341 62 74
Fax: (+91 80) 341 89 81
e-mail: krs@antrix.org
Web: www.antrix.org
■ Markets space products and services from Indian Space Research Organisation including satellite systems and subsystems (telecommunication, remote sensing and scientific), TTC and mission support, launch services and value added services based on satellite applications.

Aon Group Ltd
Aviation Information Resources
Leslie & Godwin Aviation Holdings Ltd
Nicholson Leslie Aviation Ltd
a member of the AON Corporation, US
8 Devonshire Square, London, EC2M 4PL, United Kingdom
Tel: (+44 20) 623 55 00
Fax: (+44 20) 216 38 60
Telex: 929464
e-mail: aviation.aerospace.aon.co.uk
Web: www.aon.com
■ Specialise in risk management and insurance brokerage services for all satellite related risks, as well as major aerospace and aviation companies.

APCO Technologies SA
Avenue de Corsier 1, CH-1800 Vevey, Switzerland
Tel: (+41 21) 922 11 22
Fax: (+41 21) 922 11 26
e-mail: vevey@apco-technologies.ch
Web: www.apco-technologies.com
■ Ground support equipment for satellites and launchers, flight structures for satellites, scientific instruments, delocated services.

Applied Dynamics International (ADI)
3800 Stone School Road, Ann Arbor, Michigan, 48108-2499, United States
Tel: (+1 734) 973 13 00
Fax: (+1 734) 668 00 12
e-mail: info@adi.com
Web: www.adi.com
■ Manufacture of digital computer systems for simulating models of dynamic, continuous systems. The SIM integrated system, provides an environment for real-time simulation of technologically advanced equipment such as aircraft, turbines, missiles and satellites. Beacon code generator for embedded microprocessors for safety critical applications such as Fadec engine control systems. Beacon offers graphical controller definition with C, Fortran, or Ada outputs and also generates unit test vectors.
Overseas Representatives:
Anacon Technology, Sweden
Anawell Corporation, Korea
Applied Dynamics International, UK
China National Aero-Technology International Supply Corp, China
Hinditron Infosystems (Pvt) Ltd, India
Kyokuto Boeki Kaisha Ltd, Japan
Otopos, Italy

Arab Satellite Communications Organisation (ARABSAT)
PO Box 1038, 11431 Riyadh, Saudi Arabia
Tel: (+966 1) 482 00 00
Fax: (+966 1) 488 79 99
Web: www.arabsat.com
■ Provide satellite communications services to the Arab League and neighbouring countries.

Arch Chemicals Inc
Olin Chemicals
PO Box 5204, Norwalk, Connecticut, 06856-5204, United States
Tel: (+1 203) 229 29 00
Fax: (+1 203) 229 26 13
Web: www.archchemicals.com

Arde Inc
Head Office and Plant
500 Walnut Street, Norwood, New Jersey, 07648, United States
Tel: (+1 201) 784 98 80
Fax: (+1 201) 784 97 10
e-mail: admin@ardeinc.com
Web: www.ardeinc.com
■ Research, development and manufacturing of lightweight metal and metal and fibre composite pressurant and propellant tanks, with and without propellant management system and ASME code.

Arianespace SA
Headquarters
Boulevard de l'Europe, PO Box 177, F-91006 Evry, Courcouronnes Cedex, France
Tel: (+33 1) 60 87 60 00
Web: www.arianespace.com
■ Manufacture, sales and launch operations of satellites.
Subsidiaries and offices:
Arianespace Inc, US
Arianespace, Japan
Arianespace, Singapore

Arianespace SA
Japan Liaison Office, Tokyo
Kasumigaseki Building, 31st Floor, 3-2-5 Kasumigaseki, Chiyoda-Ku, Tokyo, 100-6031, Japan
Tel: (+81 3) 35 92 27 66
Fax: (+81 3) 35 92 27 68
Representing:
Arianespace SA, France

Arianespace SA
Kourou Facility
BP 809, F-97388 Kourou, French Guiana
Tel: (+594) 33 67 07
Fax: (+594) 33 69 13

Arianespace SA
Singapore Liaison Office
Shenton House #25-06, 3 Shenton Way, Singapore, 68805, Singapore
Tel: (+65) 62 23 64 26
Fax: (+65) 62 23 42 68
Representing:
Arianespace SA, France

Armadillo Aerospace
18601 LBJ FWY, Suite 460, Mesquite, Texas, 75150, United States
Fax: (+1 972) 279 74 70
e-mail: pr@armadilloaerospace.com
Web: www.armadilloaerospace.com

Array Systems Computing Inc
1120 Finch Avenue West, 7th Floor, Toronto, Ontario, M3J 3H7, Canada
Tel: (+1 416) 736 09 00
Fax: (+1 416) 736 47 15
e-mail: marketing@array.ca
Web: www.array.ca
■ Software and systems integration in sonar, radar, synthetic aperture radar (SAR) and remote sensing.

Arrowhead Products
4411 Katella Avenue, Los Alamitos, California, 90720, United States
Tel: (+1 714) 828 77 70
Web: www.arrowheadproducts.net

Arsenal Machine-Building Plant
Arsenal Design Bureau named after MV Frunze
Federal State Unitary Enterprise
KB Arsenal
Arsenal Mashinostroitelniy Zavod
1-3 Komsonmola Street, 195009 St Petersburg, Russian Federation
Tel: (+7 812) 542 28 46
(+7 812) 542 79 00
Fax: (+7 812) 542 71 27
(+7 812) 542 31 56
Telex: 122774 grot
e-mail: info@mza.spb.ru
■ Research, development and creation of space platforms, various spacecraft and space systems.

Artisys sro
Stursova 71, CZ-616 00 Brno, Czech Republic
Tel: (+420 541) 22 48 36
Fax: (+420 541) 22 48 70
e-mail: info@artisys.aero
Web: www.artisys.aero
■ Software house and system integrator focused on real-time systems and related technologies in the aerospace and space fields. Production of real time information and control systems based on mission critical key technologies. Products and services include: ATC systems and simulators, airport systems, time servers, data engineering, computer technologies consulting, MMI development, Lynxos and Kinesix/Sammi authorised distributor, Linux consulting.
Representing:
Kinesix Corporation, US
Lynx Real-Time Systems Inc, US

Arzamas Instrument Factory
Ulitsa 50 let VLKSM, Nizhny Novgorod, 607220 Arzamas, Russian Federation
Tel: (+7 83147) 991 21
(+7 83147) 991 20
(+7 83147) 991 22
Fax: (+7 83147) 419 26
(+7 83147) 446 68
e-mail: gazapz@nts.nnov.ru
■ Manufacture of avionics, control instruments of technical processes and medical equipment.

Asher Space Research Institute, Technion
Satellite Integration, Remote Sensing
Technion City, IL-32000 Haifa, Israel
Tel: (+972 4) 829 30 20
Fax: (+972 4) 829 56 43
e-mail: aerglmn@aerodyne.technion.ac.il
Web: www.technion.ac.il
■ Satellite development.

Ashtech
Park Place Moscow, 113 Leninski Prospekt, 117198 Moskva, Russian Federation
Tel: (+7 502) 956 54 00
Fax: (+7 502) 956 53 60
e-mail: kupff@ashtech.msk.ru

Ashtech Inc
Manufacturer's Representative
62 Kuhl Avenue, Hicksville, New York, NY 11801, United States
Tel: (+1 516) 937 28 00
Fax: (+1 516) 937 76 68

Asia Satellite Telecommunications Co Ltd (ASIASAT)
17/F, The Lee Gardens, 33 Hysan Avenue, Causeway Bay, Hong Kong
Tel: (+852) 25 00 08 88
(+852) 25 00 08 80
Fax: (+852) 25 00 08 95
e-mail: as-mkt@asiasat.com
Web: www.asiasat.com
■ Operates AsiaSat regional satellite systems.

Asian Broadcasting and Communications Network Co Ltd
SM Tower, 979/66-70 Phaholyothin Road, Phayathai, Bangkok, 10400, Thailand
Tel: (+66 2) 298 00 44
Fax: (+66 2) 271 30 77
(+66 2) 271 30 78

Associate Administrator for Commercial Space Transportation
Federal Aviation Administrator
800 Independance Avenue, SW, Room 331, Washington, District of Columbia, 20591, United States
Tel: (+1 202) 267 77 93

Association for the Advancement of Space Science & Technology
IKI Building, 84/32 Profsoyuznaya Street, 117810 Moskva, Russian Federation
Tel: (+7 095) 333 24 45
Fax: (+7 095) 330 12 00
■ Independent organisation offering space expertise to its members.

Association in Scotland to Research into Astronautics Ltd (ASTRA)
ASTRA
Flat 65, Dalriada House, 56 Blythswood Court, Glasgow, G2 7PE, United Kingdom
Tel: (+44 141) 221 76 58
Web: www.astra.org.uk
■ Lectures, exhibitions, amateur astronomy and rocketry, observatory management and waverider re-entry vehicle research and development. Meetings in Airdrie (weekly) and Glasgow (monthly). Operates Airdrie public observatory for North Lanarkshire District Council.
Publications:
Asgard
International Spacereport
Spacereport
The Night Sky

Association of Andean Community Telecommunications Enterprises (ASETA)
La Pradera 510 y San Salvador, Quito, Ecuador
Tel: (+593 2) 256 38 12
(+593 2) 250 98 21
(+593 2) 250 98 22
Fax: (+593 2) 256 24 99
e-mail: info@aseta.org
Web: www.aseta.org
■ Association of telecommunications service operating companies within the Andean community, supporting development of telecommunications, to contribute to their countries' process of integration.

Astech Engineered Products Inc
Astech Manufacturing Inc
a GKN Aerospace Chem-tronics Company
3030 Red Hill Avenue, Santa Ana, California, 92705-5866, United States
Tel: (+1 949) 250 10 00
Fax: (+1 949) 250 49 82
e-mail: sales@astechmfg.com
Web: www.astechmfg.com
■ Engineering, design, development, manufacture and test of aerostructures. Acoustic panels.

Astelit
Building 6, Khohlovskyi per 10, 101028 Moskva, Russian Federation
Tel: (+7 095) 50 59 16 99 11
e-mail: astcom@astelit.ru
■ Provide digital communications technology.

Astelit
Moscow Office
Khlebny per. 19B, 121069 Moskva,
Russian Federation
Tel: (+7 095) 916 99 11
(+7 505) 916 99 11
Fax: (+7 095) 916 99 66
(+7 505) 916 99 66
e-mail: astcom@astelit.ru
Web: www.astelit.ru

Astelit
St Petersburg Office
Europe-House, Artilleriiskaya strasse apartments 1, 191104, St Petersburg, Russian Federation
Tel: (+7 812) 118 81 21
Fax: (+7 812) 279 66 15
e-mail: astptr@astelit.ru

Astrium NJRS
PO Box 38, NL-5340 AA Oss, Netherlands
Tel: (+31 24) 641 99 29
e-mail: gkspace@zonnet.nl
Web: www.njrs.nl
■ Space magazine published in Dutch.

Astrium SAS
Headquarters
6 rue Laurent Pichat, F-75216 Paris, Cedex 16, France
Tel: (+33 1) 77 75 80 37
(+33 1) 40 69 20 14
Fax: (+33 1) 77 75 80 08
Web: www.astrium.eads.net
■ Specialists in the production of vehicles for space tourism.

Astrolab
185 rue de Solignac, F-87000 Limoges, France
Tel: (+33 5) 55 31 20 59
e-mail: buryph@voila.fr
■ Magazine for the introduction, studies, documentation on astronautics, astronomy and space. Booklets on astronomy for beginners.

Astron Rechen-Institut
Monchhofstrasse 12-14, D-69120 Heidelberg, Germany
Tel: (+49 6221) 40 50
Fax: (+49 6221) 40 52 97

Astronautics Corporation of America
4115 North Teutonia Avenue, PO Box 523, Milwaukee, Wisconsin, 53209, United States
Tel: (+1 414) 449 40 00
Fax: (+1 414) 447 82 31
Web: www.astronautics.com
■ Advanced colour and monochrome (AMLCD/CRT) displays and aircraft instruments including electronic and electromechanical EFI, HSI, ADI, BDHI and EICAS displays. Air data computers, airborne server units, multi-functions displays, mission and display processors, EFIS/FMS cockpit integration, flap control systems and autopilots. Mechanical and ring laser gyros, spaceborne, shipborne, airborne and ground-based inertial navigation systems. Synchros, resolvers, brushless motors, actuators and cryogenic refrigeration.
Joint venture:
Astronautics Kearfott Electroautomatica (AKE), Russia
Subsidiaries:
Astronautics C A Ltd
Bnei-Brak, Israel
Kearfott Guidance & Navigation Corp Wayne, N J

Astronomical Institute AV CR
Ondrejov Office
Fricova 1, CZ-251 65 Ondrejov, Czech Republic
Tel: (+420 2) 04 64 92 01
(+420 2) 04 64 92 12
Fax: (+420 2) 04 62 01 10
Telex: 121579 astro
e-mail: dpivova@asu.cas.cz
Web: www.asu.cas.cz
■ Research into solar and meteor physics and stellar and galactic astronomy.

Astronomical Institute AV CR
Prague Office
Bocni II/1401a, CZ-141 31 Praha, Czech Republic
Tel: (+420 2) 67 10 31 11
e-mail: zuzana@ig.cas.cz
Representing:
Research into solar and meteor physics and stellar and galactic astronomy.

Astronomical Observatory
Box 515, SE-751 20 Uppsala, Sweden
Tel: (+46 18) 51 35 22
Fax: (+46 18) 52 75 83

Astrophysikalisches Institut Potsdam (AIP)
An der Sternwarte 16, D-14482 Potsdam, Germany
Tel: (+49 331) 74 93 81
Fax: (+49 331) 74 92 57
e-mail: gharinger@aip.de
deliebacher@aip.de

Astropower, Inc
Solar Park, Newark, Delaware, 19716-2000, United States
Tel: (+1 302) 366 04 00
Fax: (+1 302) 368 64 74
e-mail: sales@astropower.com
Web: www.astropower.com
■ High performance silicon solar cells, high performance multi-junction solar cells, high performance light-emitting diodes and detectors.

Astrotech
Vandenberg Office
Vandenberg Air Force Base, PO Box 5097, Vandenberg, 934387, United States
Tel: (+1 805) 734 11 02
Fax: (+1 805) 734 25 51

Astrotech Space Operations
Headquarters
1515 Chaffee Drive, Titusville, Florida, 32780, United States
Tel: (+1 407) 268 38 30
Fax: (+1 407) 268 38 34
■ Pre-launch payload processing at Cape Canaveral and Kennedy Space Center; aerospace engineering services.

AT&T Alascom
210 East Bluff Drive, Anchorage, Alaska, 99501, United States
Tel: (+1 907) 264 70 00
Fax: (+1 907) 276 59 37
■ Provides long-distance telecommunications within and outside Alaska. Maintains three major toll centers and provides dedicated line service, WATS, computer transmission and access, telegram and mailgram service, marine radio and live television and radio signals downlinked from the Alascom-owned satellite AURORA-1.
■ Produces the Army Transportable Satellite Communications Terminal.

Atea Rocketlab
Head Office
24 Balfour Road, Parnell, Auckland, New Zealand
Tel: (+64 9) 373 27 21
Fax: (+64 9) 373 27 21
Web: www.rocketlab.co.nz

Atec Inc
12600 Executive Drive, Stafford, Texas, 77477-3604, United States
Tel: (+1 281) 276 27 00
Freephone: (+1 866) 753 23 84
Fax: (+1 281) 240 26 82
e-mail: sales@atec.com
Web: www.atec.com
■ Diagnostic age and instrumentation systems for jet engine and aircraft systems testing at flight line, intermediate and depot maintenance levels.

Athena Global
5 Victoria, Suite C, Knowlton, Lac Brome, Québec, J0E 1V0, Canada
Tel: (+1 450) 243 42 78
Fax: (+1 450) 243 42 78
e-mail: athena@athenaglobal.com
Web: www.athenaglobal.com
■ Strategic management consultancy and international independent think-tank, applying space-based solutions to environmental problems, sustainable development, security, disasters and oceans. Main service areas include independent business intelligence and event organisation.
Publications: Athena Global Earth Observation Guide (annual)
Associated based in:
Ankara, Turkey
Hannover, Germany
Montreal, Canada
Ottawa, Canada
Prague, Czech Republic
San Francisco, US
St John's, Canada
Victoria, Canada

ATK Launch Systems
Launch Systems Group
ATK Thiokol Propulsion
Thiokol Propulsion
PO Box 707, Brigham City, Utah, 84302-0707, United States
Tel: (+1 801) 251 28 19
Fax: (+1 435) 863 23 91
e-mail: businessdevelopment@atk.com
Web: www.atk.com
■ Manufacture, delivery, stacking, retrieval and refurbishment of space shuttle solid rocket booster propulsion.

ATK International Products
Alliant Aerospace Composite Structures Company
a part of Alliant Aerospace business segment
1215 South Clark Street, Crystal Gateway 3, Suite 1510, Arlington, Virginia, 22202, United States
Tel: (+1 703) 412 32 16
(+1 801) 775 10 82
Fax: (+1 801) 775 12 07
e-mail: international@atk.com
Web: www.atk.com
■ Supplier of composite structures for space launch vehicles and military and commercial aircraft. Full service design and manufacturing capability which includes fibre-placed filament-wound, hand lay-up and precision bonded assemblies.

ATK-Able Engineering Company Inc
AEC-Able Engineering Co Inc
600 Pine Avenue, Goleta, California, 93117, United States
Tel: (+1 805) 685 22 62
Fax: (+1 805) 685 13 69
Web: www.aec-able.com

Atlantic Positioning Systems (APS)
Xybion/Sensor Positioning Systems
a division of Cobham Defense Electronics Systems
a wholly owned subsidiary of Cobham plc, UK
6950 112th Circle, Largo, Florida, 33773, United States
Tel: (+1 727) 299 01 50
Fax: (+1 727) 541 64 03
(+1 727) 299 08 04
e-mail: custserv@atlanticpositioners.com
Web: www.atlanticpositioners.com
■ Design, manufacture and supply of high-precision positioners and controls for electro-optics, electromagnetic and laser designation and directed energy systems.

Atmel
2325 Orchard Parkway, San Jose, California, 95131, United States
Tel: (+1 408) 441 03 11
Web: www.atmel.com

Austernetics Pty Ltd
a division of Austernetics Pty Ltd, Australia
PO Box 61, St Albans, Hertfordshire, AL1 2XA, United Kingdom
Tel: (+44 1727) 85 90 00
Fax: (+44 1727) 83 17 43
Telex: 299840 aaiuk
e-mail: aaiuk@aol.com
■ Consultancy services, design services and project management for the aerospace, space and defence industries.

Australian Space Industry Chamber of Commerce (ASICC)
GPO Box 7048, Sydney, New South Wales, 2001 Australia
Tel: (+61 2) 92 93 38 72
Fax: (+61 2) 99 88 02 62
■ Industry association promoting growth of Australian space industry.

Australian Space Research Institute (ASRI)
PO Box 3890, Manuka, Australian Capital Territory, 2603, Australia
e-mail: asri@asri.org.au
Web: www.asri.org.au

Austrian Aerospace GmbH
Schrack Aerospace/ORS
a subsidiary of Saab Space AB, Sweden
Stachegasse 16, A-1120 Wien, Austria
Web: www.space.at
■ Manufacture of electrical, mechanical equipment and thermal hardware, for communications, earth observation, navigation, scientific satellite missions and electronic satellite check out equipment. Also provides thermal protection, multi-layer insulation, cryogenic and insulation.

Austrian Research Promotion Agency/ Aeronautics and Space Agency (FFG)
Austrian Space Agency
Die Österreichische ForschungsfÖrderungsgesellschaft mbH (FFG)
Sensengasse 1, A-1090 Wien, Austria
Tel: (+43 5) 77 55 60 10
Fax: (+43 5) 775 59 60 10
e-mail: office@ffg.at
Web: www.ffg.at
■ Provide management services and promotion of research, technology, development and innovation for the benefit of Austria in fields of aeronautics and astronautics. Also provide domestic aerospace policy and operational representation of Austria in international bodies.

Aut Inc
300, rue Cormier, Sorel-Tracy, Québec, J3R 1V2, Canada
Tel: (+1 450) 743 31 31
e-mail: aut@qc.aira.com
Web: www.aut.qc.ca

Auxiliar de Recursos y Energia SA (Aurensa)
San Francisco de Sales 38, E-28003 Madrid, Spain
Tel: (+34 1) 554 38 65
Fax: (+34 1) 554 47 80
e-mail: aurensa@aurensa.es

Auxitrol Co
a subsidiary of Auxitrol Technologies SA, France
c/o Fluid Regulators Division,
313 Gillett Street, Painesville, Ohio, 44077, United States
Tel: (+1 440) 352 61 82
Fax: (+1 440) 354 29 12
Web: www.auxitrol.com

Auxitrol Iberico SA
a subsidiary of Auxitrol SA, France
Carretera Caucho 18, Poligono Industrial, Apartado 30, E-28850 Torrejón de Ardoz, Madrid, Spain
Tel: (+34 91) 675 23 50
Fax: (+34 91) 656 62 48
e-mail: aeroespacial@auxitrol.es
■ Level gauging systems and switches. Temperature sensors for aerospace (cryogenic and high temperature).

Auxitrol SA
Centre of Research and Production
5 Allée Charles Pathé, F-18941 Bourges, Cedex 9, France
Tel: (+33 2) 48 66 78 78
Fax: (+33 2) 48 66 78 77
e-mail: cep.dir@auxitrol.com

Avalon Electronics Inc
a subsidiary of Avalon Electronics Ltd, UK
100 Bartow Municpal Airport, Bartow, Florida, 33830, United States
Tel: (+1 863) 519 09 05
Fax: (+1 863) 519 07 63
e-mail: us-info@avalon-electronics.com
Web: www.avalon-electronics.com
■ Manufacture single and multi-channel S-VHS based analogue and video cassette recorders with 12 MHz bandwidth for airborne data acquisition, aerospace, satellite and high speed communications applications. Ruggedised disk-based analogue and digital recorders, single and multi-channel, recording up to 75 MHz, 1 Gigabits. Recording systems, SIGINT.

Avibras Fibras Oticas E Telecomunicacoes SA
Rua Ricardo Hausen 100, PO Box 229 12227-820 São José dos Campos, São Paulo, Brazil
Tel: (+55 12) 355 60 00
Fax: (+55 12) 351 62 77

Avibras Indústria Aeroespacial SA
Rodovia dos Tamoios km 14, Jacareí, PO Box 278 12300-000 São Paulo, Brazil
Tel: (+55 12) 39 55 60 00
(+55 12) 39 51 66 44
Fax: (+55 12) 39 51 62 77
e-mail: gspd@avibras.com.br
Web: www.avibras.com.br
■ Rocket research and sounding systems and equipment. Meteorological sounding rockets, command post, radio vehicles, gendarmerie and utility vehicles. Ground-based weapon fire control systems. Data processing systems. Missiles and missile systems.
■ Products include: AV-BAFG, AV-BI, AV-BP 250, BFA-230/2, BADI, BLG-120, BLG-252 bombs; gun pods; SBAT-37, SBAT-70, SBAT-127, Skyfire AV-SF-70, LM-70/19 rockets.

Avica
a subsidiary of Meggitt plc, UK
Boundary Way, Hemel Hempstead, Hertfordshire, HP2 7SL, United Kingdom
Tel: (+44 1442) 26 47 11
Fax: (+44 1442) 23 00 35
Web: www.avicauk.com
■ Design, development and manufacture of a range of products and systems for aircraft, space vehicles, nuclear, cryogenic and marine applications. These include metallic ducting systems, gimbals, flexible joints, flexible metallic hoses, bellows, flanges, clamps and total ducting design service.

Avio SpA
Colleferro (Roma) Plant
Corso Garibaldi 22, I-00034 Colleferro, Roma, Italy
Tel: (+39 06) 97 28 51 11
Fax: (+39 06) 97 28 52 01
■ Space products: solid propellant apogee motors. Ariane 4 and 5 boosters and separation motors. Liquid propellant and cold gas propulsion systems. Magneto plasma dynamic electric propulsion. Small launcher (VEGA). Defence products: powders and propellants for missile motors.

Avio SpA
Headquarters
Fiat Avio SpA
Via I° Maggio, 99, I-10040 Torino, Rivalta, Italy
Tel: (+39 11) 685 81 11
(+39 11) 008 21 11
Fax: (+39 11) 008 20 00
(+39 11) 685 98 32
Web: www.aviogroup.com
■ Design, development and production of aero engine components for military and civil aircraft, helicopter transmission, helicopter engines, jet-derivative turbines for marine applications, energy generation, maintenance, repair and overhaul and spare parts. Also provide planning, development and construction of solid propellant motors for the ariane launchers, separation motors, liquid, gas and arc jet fuel orbital propulsion systems, liquid oxygen turbopump for the Vulcan and Vinci motors, mechanical components of launcher nozzles, tactical propulsion, research and development of hybrid propulsion and in the automation sector includes; design and development of control systems and marine automation.

AXA Space
International Technology Underwriters
4800 Montgomery Lane, 11th Floor, Bethesda, Maryland, 20814, United States
Tel: (+1 301) 654 85 85
Fax: (+1 301) 654 75 69

Axel Neumann Versicherungsmakler GmbH (ANV)
Hauptstrasse 19, D-72124 Pliezhausen, (OT Rübgarten), Germany
Tel: (+49 7127) 975 40
Fax: (+49 7127) 97 54 44
e-mail: info@axelneumann.de
Web: www.axelneumann.de
■ National and international aviation and space insurance.

Aydin Displays Inc
1 Riga Lane, Birdsboro, Pennsylvania, 19508, United States
Tel: (+1 610) 404 53 53
Fax: (+1 610) 404 81 90
e-mail: sales@aydindisplays.com
Web: www.aydindisplays.com
■ Design and manufacture command and control equipment and systems, air traffic control systems and air defence consoles.

Azerbaijan National Aerospace Agency (ANASA)
National Aerospace Agency, Azerbaijan 159 Azadlyq Prospect, 1106 Baku, Azerbaijan
Tel: (+994 12) 62 93 87
Fax: (+994 12) 62 17 38
■ State policy and co-ordination of national aerospace and space programmes focusing on two areas: aerospace environmental monitoring; thematic mapping, developing GIS, and space applications. Remote sensing of the earth's surface using airborne and satellite systems
Managing the following:
Ecology Institute
Experimental Industry Work
Institute of Aerospace Informatics
Institute for Space Research of Natural Resources
Special Design Unit

B

Ball Aerospace & Technologies Corp (BATC)
a subordinate unit of Ball Corporation, US
1600 Commerce Street, PO Box 1062 Boulder, Colorado, 80301, United States
Tel: (+1 303) 939 61 00
(+1 303) 533 60 59
Fax: (+1 303) 939 51 00
(+1 303) 939 61 04
Telex: 45605
e-mail: info@ball.com
Web: www.ballaerospace.com
■ Provide systems engineering services. Design and manufacture of complete spacecraft and space systems, space and scientific sensors, cryogenic subsystems, antenna systems and video products for commercial and government customers.

Ball Aerospace & Technologies Corp (BATC)
Washington DC Office
2200 Clarendon Boulevard, Suite 1202, Arlington, Virginia, 22201, United States
Tel: (+1 703) 284 54 00
Fax: (+1 703) 284 54 49

Barr Associates Ltd
a subsidiary of Barr Associates Inc., US
3 and 4 Home Farm Business Units, Yattendon, Newbury, Berkshire, RG18 0XT, United Kingdom
Tel: (+44 1635) 20 13 17 **Fax:** (+44 1635) 20 20 30
e-mail: info@barr-associates-uk.com
Web: www.barr-associates-uk.com
■ Advanced optical filters. Infra-red coating services. Consultancy on most aspects of infra-red optical design. Filters have been supplied for the IBSS as part of the SDI project.

Barrios Technology Inc
2525 Bay Area Boulevard, Suite 300, Houston, Texas, 77058-1556, United States
Tel: (+1 281) 280 19 00
Fax: (+1 281) 280 19 01
e-mail: info@activeiweb.com
Web: www.barrios.com
www.activeiweb.com
■ Products include Active iWeb, a solution to information challenges that provides the tools necessary to publish information to

the web in a secure environment and facilitate communications throughout an organisation. It allows administrators to post content without the need for extensive web development knowledge or training. It has an interface that is customisable allowing different user groups to see only the information they need.

Bayern-Chemie Gesellschaft für Flugchemische Antriebe mbH

Bayern-Chemie GmbH
a subsidiary of LFK (100%)
PO Box 1131 D-84544 Aschau, am Inn, Germany
Tel: (+49 8638) 60 10
Fax: (+49 8638) 60 13 99
e-mail: info@bc.eads.de
Web: www.bayernchemie-protac.com

■ Development and manufacture of propulsions with solid propellant in steel or lightweight structures for: artillery rockets, guided missiles, multiple pulse motors, hypersonic missiles, side thrusters, mini rockets for model aeronautics and rockets for rescue systems. Programmes: Alarm, Crotale, DWS, Dar, Kormoran, Kriss, Mica, Magic, Otomat, Pars, Patriot, Martel, Super 530, Shahine. Throttleable and non-throttleable air breathing propulsions (ram rocket motors) with high energy propellant (boron). Applications on medium and long range missiles. Programmes for A3M, Aramis and Meteor. Components and propellants: Helical wound motor structures and flow turned technology. Non-asbestos thermal insulations. Solid propellant grains of extruded double base and cast composite propellants. Gas generators and pyrotechnics: gas generators with cool burning propellant for applications including ejection of sub-munitions, power supply for guided missiles, starter cartridges for turbines, inflation, deflation and fire extinguishing. Pyrotechnical devices and ignition mixtures for applications including valves, actuators, infra-red radiators and igniters. Chemical and physical analysis, environmental testing and simulation, x-ray of large objects, component testing and quality management. Consulting and sale of machinery for propellants. Environmental testing, laboratory analysis and modifications of production machines. Design and adaptation of special equipment. Research and development on hypervelocity propulsion (Mach 5), multiple pulse motors, ram rockets, throttleable and non-throttleable. Nozzleless booster (propellant technology).

BBN Technologies

10 Moulton Street, Cambridge, Massachusetts, 02138, United States
Tel: (+1 617) 873 80 00
Fax: (+1 617) 547 89 18
(+1 617) 873 37 76
Web: www.bbn.com

■ Design, develop and deploy innovative systems for government and industry including: distributedinteractivesimulationsystems,combat development simulations, network monitoring systems, packet communication systems, sensor and surveillance systems and C3 systems. Provide contract research and development in physical sciences, communications, computer and information sciences.

Beijing Astronomical Observatory

W Suburb, 100080 Beijing, China
Tel: (+86 1) 255 19 68
(+86 1) 255 12 61
Fax: (+86 1) 256 10 85

Beijing University of Aeronautics & Astronautics (BUAA)

a subsidiary of Aviation Industries of China
37 Xue Yuan Road, Haidian District, 100083 Beijing, China
Tel: (+86 10) 62 01 72 51
(+86 10) 62 01 75 61
Fax: (+86 10) 62 02 83 56

■ Research and development whose areas include advanced metallics, super conductive materials, composite materials, fine casting, telemetry, radio and satellite navigation, microwave technology and antennas. Further interests include automatic controls, gyro and inertial navigational sensors, hydraulic transmissions, and control, aeroengines, space propulsion, fluid mechanics, aerodynamics, solid mechanics, astronautical human factor engineering, environmental engineering, aircraft design, computer software engineering, artificial intelligence, airships and ultralight aircraft.

Belgacom

Boulevard du Roi Albert II, 27, B-1030 Bruxelles, Belgium
Tel: (+32 2) 202 41 11
(+32 2) 205 40 00
Fax: (+32 2) 203 65 93
(+32 2) 205 40 40
e-mail: baudhuin.pringiers@belgacom.be
ann.roegies@belgacom.be
Web: www.belgacom.be

■ National Eutelsat, Intelsat and Inmarsat communications carrier.

Belgian Federal Science Policy Office

Office for Science, Technology and Cultural Affairs
Rue de la Science 8 Wetenschapsstraat, B-1000 Bruxelles, Belgium
Tel: (+32 2) 238 34 11
Fax: (+32 2) 230 59 12
Telex: 24501 prosci
Web: www.belspo.be

■ Management of Belgium's space programme.

Belgian Institute of Space Aeronomy

Institut D'Aeronomie Spatiale De Belgique
3 Avenue Circulaire, B-1180 Bruxelles, Belgium
Tel: (+32 2) 373 04 04
(+32 2) 373 04 13
Fax: (+32 2) 375 93 36

■ Research into space physics and aeronomy.

Belgian Inter-Trade Association for Space Activities (BELGOSPACE)

80 Building A. Reyers, B-1030 Bruxelles, Belgium
Tel: (+32 2) 706 81 48
Fax: (+32 2) 706 79 52
e-mail: belgospace@agoria.be
Web: www.agoria.be/belgospace

■ Association of Belgian companies working in space technology.

Association Members
Alcatel Bell Space, Alcatel-ETCA, Cegelec, Gillam Fei, Newtec Cy, Nexan HARNESSES, SABCA, SONACA, Space Applications Services, Spacebel, Techspace Aero, Verhaert Design & Development

BES Engineering

1880 South Dairy Ashford, Suite 606, Houston, Texas, 77077, United States
Tel: (+1 281) 496 09 27
Fax: (+1 281) 496 47 05
Web: www.bergaila.com

Beyond-Earth Enterprises

520 E.Challenger, Roswell, New Mexico, 88203, United States
Tel: (+1 505) 347 98 08
Web: www.beyond-earth.com

Bezeq, The Israel Telecommunication Corp Ltd (Bezeq)

Bezeq House, 2 Azrielli Centre, IL-61620 Tel-Aviv, Israel
Tel: (+972 3) 626 26 00
Fax: (+972 3) 626 26 09
e-mail: bzqspk@attmail.com
Web: www.bezeq.co.il

■ Telecommunications services and systems. Digital local, long distance and international telephone services, cellular communications, data communications, satellite services, leased lines and corporate networks.

Bharat Electronics Ltd (BEL)

Bangalore Unit
Jalahali Post, Bangalore, Karnataka, 590 013, India
Tel: (+91 80) 28 38 26 26

■ Defence and satellite communications, sound and vision broadcasting, radars, simulators, systems and components.

Bharat Electronics Ltd (BEL)

Corporate and Registered Office
Outer Ring Road, Bangalore, Nagavara, 560 045, India
Tel: (+91 80) 25 03 93 22
(+91 80) 25 03 93 00
(+91 80) 25 03 92 48
(+91 80) 25 03 92 87
(+91 80) 25 03 92 88
Fax: (+91 80) 25 03 93 05
(+91 80) 25 03 92 86
e-mail: imd@bel.co.ind
Web: www.bel-india.com

■ Manufacture of products in the areas of defence communications, radars and sonars; telecommunications, sound and vision broadcasting, opto and tank electronics, solar systems and components. Products include: Secondary Surveillance Radar (SSR), Monopulse Secondary Surveillance Radar (MSSR), Airport Surveillance Radar and Automatic Visual Range Assessor (AVRA).

Accreditations: ISO 9001, ISO 9002

Bharat Electronics Ltd (BEL)

Ghaziabad Unit
Site IV, Sahibabad Industrial Area, Bharat Nagar Post, Ghaziabad, Utter Pradesh, 201 010, India
Tel: (+91 120) 261 95 00
(+91 120) 261 97 86
Fax: (+91 120) 277 67 30

■ Production of radars, antennas and and microwave components for defence.

Bigelow Aerospace

4640 S Eastern Avenue, Las Vegas, Nevada, 89119, United States
Tel: (+1 702) 688 66 00
(+1 702) 688 60 16
e-mail: contact_ba@bigelowaerospace.com
Web: www.bigelowaerospace.com

Binariang Sdn Bhd

Maxis
Ground Floor, Block B, Wisma Semantan, 12 Jalan Gelenggang, Bukit Damansara, 50490 Kuala Lumpur, Malaysia
Tel: (+60 3) 252 20 00
Fax: (+60 3) 252 32 99
e-mail: binaria@jaring.po.my

Bioserve Space Technologies

Department of Aerospace Engineering Science, University of Colorado, Campus Box 429 Boulder, Colorado, 80309, United States
Tel: (+1 303) 492 10 05
Fax: (+1 303) 492 88 83

■ Developers of life sciences experiments and enabling hardware for space operations, ranging from KC135 research aircraft through sounding rockets, space shuttle, COMET and space station applications.

Birkhäuser Verlag AG

Viaduktstrasse 40-44, CH-4051 Basel, Switzerland
Tel: (+41 61) 205 07 07
Fax: (+41 61) 205 07 99
e-mail: sales@birkhauser.ch
Web: www.birkhauser.ch

■ Publications on astronomy, space flight and aeronautics in English and German.

Black Holes BV

Witte Singel 85, NL-2311 BP, Leiden, Netherlands

Blackhawk Management Corporation

1335 Regents Park Drive, Suite 130, Houston, Texas, 77058, United States
Tel: (+1 281) 286 57 51
Fax: (+1 281) 286 57 52
Web: www.blackhawkmgmt.com

Böhler Edelstahl GmbH & Co KG

a subsidiary of Böhler Uddeholm AG, Austria
Mariazeller Strasse 25,
PO Box 96 A-8605 Kapfenberg, Austria
Tel: (+43 3862) 20 61 70
(+43 3862) 200
Fax: (+43 3862) 20 75 70
(+43 3862) 20 75 25
e-mail: office@bstg.buag.co.at
Web: www.bohler-forging.com
www.bohler-schmiedetechnik.com
www.bohler-edelstahl.at

■ Precision forged and machined parts for engines, compressors and aerospace applications.

Booz Allen Hamilton

Arlington Office
1725 Jefferson Davis Highway, Suite 1100, Arlington, Virginia, 22202, United States
Tel: (+1 703) 769 77 00
Fax: (+1 703) 685 65 08
Web: www.boozallen.com

■ Provide consultancy services of strategy, operations, organisation, change information technology and systems integration.

Booz Allen Hamilton Inc

Aberdeen Office
4692 Millennium Drive, Suite 200, Belcamp, Maryland, 21017, United States
Tel: (+1 410) 297 25 00
Fax: (+1 410) 297 25 50

■ Provide consultancy services of strategy, operations, organisation, change information technology and systems integration.

Booz Allen Hamilton Inc
Colorado Springs Office
121 South Tejon Street, Suite 900, Colorado Springs, Colorado, 80903, United States
Tel: (+1 719) 570 31 00
(+1 719) 637 68 80
Fax: (+1 719) 597 81 31
■ Provide consultancy services of strategy, operations, organisation, change information technology and systems integration.

Booz Allen Hamilton Inc
Dayton Office
1900 Founders Drive, Suite 300, Dayton, Ohio, 45420, United States
Tel: (+1 937) 781 28 00
Fax: (+1 937) 781 28 08
■ Provide consultancy services of strategy, operations, organisation, change information technology and systems integration.

Booz Allen Hamilton Inc
Headquarters
8283 Greensboro Drive, Mclean, Virginia, 22102, United States
Tel: (+1 703) 902 50 00
Fax: (+1 703) 902 33 33
Web: www.boozallen.com
■ Provide consultancy services of strategy, operations, organisation, information technology and systems integration.

Booz Allen Hamilton Inc
Honolulu Office
Pacific Guardian Tower – Makai Tower, Suite 3000, 733 Bishop Street, Honolulu, Hawaii, 96813, United States
Tel: (+1 808) 545 68 00
Fax: (+1 808) 545 68 08
■ Provide consultancy services of strategy, operations, organisation, change information technology and systems integration.

Booz Allen Hamilton Inc
Huntsville Office
6703 Odyssey Drive, Suite 200, Huntsville, Alabama, 35806, United States
Tel: (+1 256) 922 27 60
Fax: (+1 256) 922 27 69
■ Provide consultancy services of strategy, operations, organisation, change information technology and systems integration.

Booz Allen Hamilton Inc
Lexington Park Office
46950 Bradley Boulevard, Lexington Park, Maryland, 20653, United States
Tel: (+1 301) 862 31 10
Fax: (+1 301) 866 69 10
■ Provide consultancy services of strategy, operations, organisation, change information technology and systems integration.

Booz Allen Hamilton Inc
New York Office
101 Park Avenue, New York, New York, 10178, United States
Tel: (+1 212) 697 19 00
Fax: (+1 212) 551 67 32
Telex: 620196
Cable: BOOZ ALLEN
■ Provide consultancy services of strategy, operations, organisation, change information technology and systems integration.

Booz Allen Hamilton Inc
Norfolk Office
5800 Lake Wright Drive, Suite 400 Twin Oaks 11, Norfolk, Virginia, 23502, United States
Tel: (+1 757) 893 61 00
Fax: (+1 757) 893 61 50
■ Provide consultancy services of strategy, operations, organisation, change information technology and systems integration.

Booz Allen Hamilton Inc
Omaha Office
1299 Farnam Street, Suite 1230, Omaha, Nebraska, 68102, United States
Tel: (+1 402) 522 28 00
Fax: (+1 402) 522 28 08
■ Provide consultancy services of strategy, operations, organisation, change information technology and systems integration.

Booz Allen Hamilton Inc
San Antonio Office
700 North Saint Mary's Street, Suite 700, San Antonio, Texas, 78205, United States
Tel: (+1 210) 244 42 00
Fax: (+1 210) 244 42 06
■ Provide consultancy services of strategy, operations, organisation, change information technology and systems integration.

Booz Allen Hamilton Inc
San Diego Office
1615 Murray Canyon Road, Suite 140, San Diego, California, 92108, United States
Tel: (+1 619) 725 65 00
Fax: (+1 619) 725 66 99
■ Provide consultancy services of strategy, operations, organisation, change information technology and systems integration.

Bradford Engineering BV
PO Box 323, NL-4600 AH Bergen op Zoom, Netherlands
De Wijper 26, NL-4726 TG Heerle (NB), Netherlands
Tel: (+31 165) 30 51 00
Fax: (+31 165) 30 44 22
e-mail: info@bradford-space.com
Web: www.bradford-space.com
■ The company is divided into two sections, the Microgravity Division and the Space System and Components Division. The first group handles the design and manufacture of space science equipment for NASA and the European Space Agency whilst the second group comprises propulsion and thermal systems.

Brandt Fijnmechanische Industrie BV
P O Box 1385, NL-1300 BJ Almere, Netherlands
De Stubbenweg 15, NL-1327 GB Almere, Netherlands
Tel: (+31 036) 523 13 98
Fax: (+31 036) 533 26 33
e-mail: info@brandtfmi.nl
Web: www.brandtfmi.nl
■ Manufacture of precision components, tools and assemblies. Repair and overhaul of mechanical components.
Representing:
Scheer & Co GmbH, Germany

Brasilsat Harald SA (Brasilsat)
Rua Guilherme Weigert, 1955, 82720-000 Curitiba, Paraná, Brazil
Tel: (+55 41) 2103 03 00
Fax: (+55 41) 2103 05 55
e-mail: brasilsat@brasilsat.com.br
Web: www.brasilsat.com.br
■ Ground antennas for satellite communications. Communication towers, SHF microwave antennas, UHF and VHF antennas, radar antennas, UHF and HF military multicouplers, RF coaxial connectors, radios and repeaters.

Bristol Aerospace Ltd (BAL)
a Magellan Aerospace Company
PO Box 874, Winnipeg, Manitoba, R3C 2S4, Canada
660 Berry Street, Winnipeg, Manitoba, R3C 2S4, Canada
Tel: (+1 204) 775 83 31
Fax: (+1 204) 775 74 94
Web: www.bristol.ca
■ CRV-7, 70 mm (2.75 inch) air to ground rocket weapon system, Black Brant family of sub orbital research rocket vehicles. Wire Strike Protection Systems (WSPS) for helicopters, large airframe structures, composites and gas turbine engines. Repair, overhaul and modification of high performance fixed wing fighter aircraft.

Bristol Industrial & Research Associates Ltd (BIRAL)
PO Box 2, Portishead, Bristol, BS20 7JB, United Kingdom
Unit 8, Harbour Road Trading Estate, Portishead, Bristol, BS20 7BL, United Kingdom
Tel: (+44 1275) 84 77 87
Fax: (+44 1275) 84 73 03
e-mail: enquiries@biral.com
info@biral.com
biodetection@biral.com
Web: www.biral.com
■ Meteorological systems and sensors.

British Sarozal Ltd
Handrail House, Maygrove Road, London, NW6 2EG, United Kingdom
Tel: (+44 20) 73 28 21 11
Fax: (+44 20) 76 24 51 44
e-mail: sales@bslexpress.com
Web: www.bslexpress.com
■ Solid state electronic valves replacing thermionic valves for communications and radar installations. Supply of spare parts for British and American civil and military electronic equipment. Installation and maintenance of electronic systems for ATC. Manufacture of image intensifiers, synchros tacko-generators, servo motors, avionic tubes, display units (radar), avionic tubes, military tubes, transformers and cable assemblies. Pressure gauges, power supply units, integrated circuits, military tubes, image intensifiers, synchros, tacko generators and avionic tubes.

Britte SA
27 rue de Cheratte, B-4683 Vivegnis, Belgium
Tel: (+32 4) 256 90 69
Fax: (+32 4) 264 08 63
e-mail: info@britte.be
Web: www.britte.be
■ Machining of high precision mechanical parts for defence, space and aerospace applications.

Broadcasting Satellite System Corporation
Parkside-Yamamotokan, 3rd Floor, 1-16-4 Tomigaya, Shibuya-ku, Tokyo, 151-0063, Japan
Tel: (+81 3) 54 53 57 07
(+81 3) 54 53 65 21
Fax: (+81 3) 54 53 65 24
■ Broadcast satellite procurement, transfer and leasing of broadcast satellite transponders, control and management of satellites, commissioned broadcasting operations.

Brüel & Kjaer North America Inc
Headquarters
2815 Colonnades Court, Norcross, Georgia, 30071-1588, United States
Tel: (+1 770) 209 69 07
Fax: (+1 770) 448 32 46
e-mail: bkinfo@bksv.com
Web: www.bkhome.com
■ Instrumentation for sound, vibration, illumination, thermal environment and signal analysis.
Representing:
Brüel & Kjaer, Denmark

Brush Wellman Inc
Electrofusion Products
44036 South Grimmer Boulevard, Fremont, California, 94538-6346, United States
Tel: (+1 510) 623 15 00
Fax: (+1 510) 623 76 00
■ Specialised metal fabrication; beryllium, tungsten, tantalum, moly. Beryllium X-Ray window, IF-1 beryllium foil

Brush Wellman Inc
Engineered Materials
14710 West Portage River South Road, Elmore, Ohio, 43416, United States
Tel: (+1 419) 862 27 45
Fax: (+1 419) 862 41 74
e-mail: beproducts@brushwellman.com
Web: www.brushwellman.com
■ Beryllium based materials including: beryllium metal matrix composites, aluminium beryllium alloys, beryllium oxide ceramics and beryllium copper. These materials are utilized in aerospace and defense applications where low density, high rigidity, high thermal conductivity and low co-efficient of thermal expansion are useful in design.

Bundesverband der Deutschen Luft- und Raumfahrt-Industrie eV (BDLI)
German Aerospace Industries Association
ATRIUM Friedrichstrasse 60, D-10117 Berlin, Germany
Tel: (+49 30) 206 14 00
Fax: (+49 30) 20 61 40 90
e-mail: info@bdli.de
Web: www.bdli.de
■ Association representing 120 companies organised into four manufacturer groups: aerospace systems, engines, equipment, and materials technology and components. Organisation of Berlin-Brandenburg International Aerospace Exhibition.
Ordinary Members:
AeroTech Peissenberg GmbH
Aerodata-Flugmesstechnik GmbH
Airsys Navigation Systems GmbH
AOA Apparatebau Gauting GmbH
ASG Luftfahrtechnik und Sensorik GmbH
Astrium GmbH
Autoflug GmbH & Co
Base Ten Systems Electronics GmbH
Bayern-Chemie GmbH

Becker Flugfunkwerk GmbH
Behr Industrietechnik GmbH & Co
Bernd Stephan Stahlhandel GmbH
Bodenseewerk Geraetetechnik GmbH
Böhler Uddeholm Deutschland GmbH
Bosch SatCom GmbH
CAE Electronics GmbH
Cargolifter AG
Carl Zeiss
Corus Aluminium Walzproduct GmbH
Deutsche Titan GmbH
Diehl Avionik Systeme GmbH
Diehl Luftfahrt Elektronik GmbH
Diehl Munitionssysteme GmbH & Co KG
Doncasters Precision Castings – Bochum
Dornier GmbH
Dornier Flugzeugwerft GmbH
Draeger Aerospace GmbH
EADS Airbus GmbH
EADS Deutschland GmbH
EATON Fluid Power GmbH
E I S Electronics GmbH
Elan GmbH
Elektro-Metall Export GmbH
Elettronica GmbH
Entrak GmbH & Co KG
ESG Elektroniksystem- und Logistic GmbH
ESW – Extel Systems Wedel Gesellschaft für Ausrüstung
Eurocopter Deutschland GmbH
Europa Fasteners GmbH
EVAC GmbH Kanalisationssysteme
Fairchild Dornier GmbH
Gebr Hoever GmbH & Co
GKN Aerospace GmbH
Goodrich Aerospace Europe GmbH
Goodrich Rosemount Aerospace GmbH
Grässlin Präzisionstechnik GmbH
Hella Aerospace GmbH
Holmburg GmbH & Co KG
Honeywell Aerospace GmbH
Honeywell Regelsysteme GmbH
Honsel AG
IABG Industrieanlagen-Betriebsgesellschaft mbH
Jena-Optronik GmbH
Jenoptik Laser, Optik & Systeme GmbH
Jeppesen GmbH
Kayser-Threde GmbH
KID-Systeme GmbH
Klaus Stegmann GmbH & Co KG
Krupp VDM GmbH
Labinal Aero & Defence Systems GmbH
Leistritz Turbomaschinen Technik GmbH
Liebherr-Aerospace Lindenberg GmbH
LITEF GmbH
Lockheed Martin GmbH
MAN Technologie AG
Mannesmann Rexroth AG
MST Aerospace GmbH
MTU Aero Engines
Nord-Micro AG & Co OHG
OHB – Orbital- und Hydrotechnologie Bremen System GmbH
Otto Fuchs Metallwerke
Pace Aerospace Engineering and Information Technology GmbH
Panavia Aircraft GmbH
Parker Hannifin GmbH
Pfalz-Flugzeugwerke GmbH
Pierburg Luftfahrtgerate Union GmbH
Praezisionsmechanik Gilchinp GmbH
Recaro Aircraft Seating GmbH & Co KG
Rhode & Schwarz GmbH & Co KG
Roeder Praezision GmbH
Rolls-Royce Deutschland GmbH
SFIM Industries Deutschland GmbH
Sitec Aerospace GmbH
STN Atlas Elektronik GmbH
TDW Gesellschaft für verteidigungstechnische Wirksystems mbH
Telair International GmbH
Teldix GmbH
Test Fuchs GmbH
Thales Electron Devices GmbH
Thyssen Umformtechnik Turbinen Komponenten GmbH
Tital – Titan-Aluminium Feinguss GmbH
Turbomeca GmbH
VCS Nachrichtentechnik GmbH
VIDAIR Avionics Aktiengesellschaft
Von Hoerner and Sulger GmbH
Wittenstein Motion Control GmbH
W L Gore & Associates GmbH
ZARM Technik GmbH
ZF Luftfahrttechnik GmbH
Z11 Imaging GmbH

Sponsoring Members:
Aeronaval Ingenieurtechnik GmbH & Co KG
AirTruck GmbH
Allianz Versicherungs-AG
Arianespace
BAE Systems Deutschland GmbH
Bayerische Landesbank Girozentrale
Bayerische Vereinsbank AG
BTG Messe – Spedition GmbH
Deutsche Bank AG
DLR Deutsches Zentrum für Luft- und Raumfahrt
Dresdner Bank AG
Gebrueder Krose
Edgar Hausmann GmbH
IBCOL Technical Services GmbH
Intospace GmbH
Ing Horst Kegler GmbH
Lufthansa Consulting GmbH
Schenker-Eurocargo AG
Vereinigte Motor-Verlage GmbH & Co KG
Walter Kostelezky GmbH & Co KG

Busek Co Inc
11 Tech Circle, Natick, Massachusetts, 01760, United States
Tel: (+1 508) 655 55 65
Fax: (+1 508) 655 28 27
■ Development of advanced electric propulsion thrusters for use on military, government and commercial satellites.

C

Cable & Wireless Optus Pty Ltd
Belrose Office
Optus Communications Pty Ltd
PO Box 235, Frenchs Forest, New South Wales, 2086, Australia
2 Challenger Drive, Belrose, New South Wales, 2085, Australia
Tel: (+61 2) 93 42 31 00
Fax: (+61 2) 93 42 31 11

CAESAR Consultancy
35 Millington Road, Cambridge, CB3 9HW, United Kingdom
Tel: (+44 1223) 35 38 39
Fax: (+44 1223) 30 38 39
Web: www.caesarconsultancy.com
www.mjrycroft.com
■ Provide advice on scientific, technical and commercial topics on atmospheric, environmental and space activities and research.

Calian
United States Office
5175 Parkstone Drive, Suite 250, Chantilly, Virginia, 2015, United States
Tel: (+1 703) 378 82 28
Fax: (+1 703) 378 82 55
e-mail: info@calian.com

California Space Authority Inc (CSA)
Corporate Office
3201 Airpark Drive, Suite 204, Santa Maria, California, 93455, United States
Capital Office, 1107 Ninth Street, Suite 1005, Sacramento, California, 95814, United States
Tel: (+1 805) 349 26 33
(+1 916) 551 15 43
Fax: (+1 805) 349 26 35
(+1 916) 551 15 79
Web: www.californiaspaceauthority.org

California Space Grant Consortium (CaSGC)
University of California at San Diego, 9500 Gilman Drive, Department 0411, La Jolla, California, 92093-0411, United States
Tel: (+1 858) 822 15 97
Fax: (+1 858) 534 78 40
e-mail: spacegrant@ucsd.edu
Web: casgc.ucsd.edu

California Space Institute
CalSpace 0524, University of California at San Diego, La Jolla, California, 92093-0411, United States
Tel: (+1 619) 534 29 08
Fax: (+1 619) 534 74 80

California Space Institute
CalSpace Center of Excellence
University of California, Institution for Computational Earth System Science, Santa Barbara, California, 93106-4060, United States

CAM Systems GmbH
CAM Computer Anwendung fur Management GmbH
Beta-Strasse 10 c, D-85774 Unterföhring, Germany
Tel: (+49 089) 189 08 00
Fax: (+49 089) 189 08 01 11
Web: www.cam-comp.de
www.cam-systems.de
■ Control centre software.

Canadian Space Agency
6767 Route de l'Aéroport, St Hubert, Québec, J3Y 8Y9, Canada
Tel: (+1 450) 926 48 00
Fax: (+1 450) 926 43 52

CAP Gemini Ernst & Young Norway AS
Beddingen 10, Trondheim, N-7485, Norway
Tel: (+47) 241 12 80 00
Fax: (+47) 73 84 60 01
■ Computer and software development and associated services, from consultancy to turnkey projects. Software development and systems integration, installation, commissioning and related services during warranty and maintenance periods.

Carl Zeiss Optronics GmbH
Zeiss Optronik GmbH
Zeiss-Eltro Optronic GmbH
Carl Zeiss-Strasse 22, D-73447 Oberkochen, Germany
Tel: (+49 7364) 20 65 30
Fax: (+49 7364) 20 36 97
e-mail: optronics@optronics.zeiss.com
Web: www.zeiss.com/optronics
■ Provide manufacture, innovation, operational and logistical needs of optronic instruments, sensor systems, submarine optronics and aerospace optics for land, sea, air forces and national security organizations.
Business units:
Sensor Systems
Surveillance and Targeting Systems
Submarine Mast Systems
Airborne Sensor Systems

Carlo Gavazzi Space SpA (CGS)
Via Gallarate 150, I-20151 Milan, Italy
Tel: (+39 02) 38 04 81
Fax: (+39 02) 308 64 58
e-mail: cgs@cgspace.it
Web: www.cgspace.it
■ Satellites: turn-key systems for scientific and commercial missions, scientific payloads, unmanned space vehicles, earth observation, ground segment and applications.

Caval Tool Division Chromalloy Gas Turbine Corp
a division of Chromalloy Gas Turbine Corp., Texas
275 Richard Street, PO Box 310158, Newington, Connecticut, 06131-0158, United States
Tel: (+1 860) 667 21 34
Fax: (+1 860) 667 00 57
■ Manufacture of engine parts, hubs, discs, spacers, satellite hardware and rocket motor cases. Overhaul and repair of gas turbine engine hardware.

CBA International
Naka, PO Box 12, Yokohama, 231-0057, Japan
Tel: (+81 45) 664 24 84
Cable: CBAI
■ Non-profit research organisation, technical investigation for radar/visual UFO reports by civilian and military aviation pilots.
Publications:
Aerospace UFO News (Quarterly)
Divisions
Aerospace UFO Research Center
HQ International UFO Observer Corps

CD Adapco Group
1516 East Palm Valley Boulevard, Suite B-1, Round Rock, Texas, 78664, United States
Tel: (+1 631) 549 23 00
e-mail: starinfo@adapco.com
Web: www.cd-adapco.com

Celab Ltd
a subsidiary of Celab Power Management Ltd
Woolmer Way, Bordon, Hampshire, GU35 9QE, United Kingdom
Tel: (+44 1420) 47 70 11
Fax: (+44 1420) 47 80 43
e-mail: celab@murata-ps.com
Web: www.celab.co.uk
www.murata-ps.com/celab
■ Design and manufacture of high density, switching regulator military power supplies for rigorous applications in airborne, ship and ground environments with AC and DC inputs. Multiple rail modular systems with full BITE facilities. Comprehensive sub-contract manufacturing facility including post design.

Center for Advanced Space Studies (CASS)
3600 Bay Area Boulevard, Houston, Texas, 77058-1113, United States
Tel: (+1 281) 486 21 39
(+1 281) 244 20 00
Fax: (+1 281) 486 21 62
(+1 281) 244 20 06
e-mail: info@lpi.usra.edu
info@dsls.usra.edu
hammond@sop.usra.edu
cardenas@sop.usra.edu
■ Research and education in lunar and planetary matters.

Center for Astrophysics
60 Garden St, Cambridge, Massachusetts, 02138, United States
Tel: (+1 617) 495 72 44
Fax: (+1 617) 495 72 31

Center for Cell Research
117 Research Office Building, The Pennsylvania State University, University Park, Pennsylvania, 16802, United States
Tel: (+1 814) 865 24 07
Fax: (+1 814) 865 24 13

Center for Commercial Space Communications
Virginia Tech, Blacksburg, Virginia, 24061-0111, United States
Tel: (+1 703) 231 44 61
Fax: (+1 703) 231 33 55
Telex: 3331861 vpi bks
e-mail: bowen@vtvm1

Center for Macromolecular Crystallography
University of Alabama at Birmingham, UAB Station, Box 79 THT, Birmingham, Alabama, 35294-0005, United States
Tel: (+1 205) 934 53 29
Fax: (+1 205) 934 04 80

Center for Satellite & Hybrid Communication Networks
Institute for Systems Research, University of Maryland, A V Williams Building 115, College Park, Maryland, 20742, United States
University of Maryland, Engineering Annex Building 093, College Park, Maryland, 20742, United States
Tel: (+1 301) 405 79 00
Fax: (+1 301) 314 85 86
Web: www.cshcn.com
■ Conducts research and development in the field of satellite and terrestrial communications technologies.

Center for Space and Planetary Sciences
Old Museum Building Muse 202, University of Arkansas, Fayetteville, Arkansas, 72701, United States
Fax: (+1 479) 575 77 78
e-mail: csaps@uark.edu
Web: www.spacecenter.uark.edu

Center for Space Power (CSP)
Texas Engineering Experiment Station, The Texas A&M University, Wisenbaker Building, College Station, Texas, 77843-3118, United States
Tel: (+1 979) 845 87 68
■ Design and development of space power related technology to support commercial efforts and NASA mission needs. Projects combine university researchers with industry partners. Current work includes photovoltaics, batteries, thermal management hardware and electronic materials and devices.

Central Research Institute for Machine Building
TSNIIMASH
4 Pionerskaya Street, 141070 Korolev, Moscow Region, Russian Federation
Tel: (+7 095) 513 50 00
(+7 095) 187 03 22
e-mail: corp@tsniimash.ru
Web: www.tsniimash.ru

Central Scientific and Research Institute of Machine Building
TsNIIMASH
Pionerskaya Street 4, 141070 Korolev, Moscow Region, Russian Federation
Tel: (+7 095) 516 59 10
Fax: (+7 095) 187 15 88
e-mail: corp@tse.ru
Web: www.korolev.ru/english/e_tsniimash.html
www.tse.ru
■ Development of rocket and space technology. Development and production of long range ballistic missiles, air defence missiles and their propulsion units.

Centre Commun de Recherche des Communautes Européennes (JRC)
Directorate For Science Strategy
SDME 10/53, B-1049 Bruxelles, Belgium
Tel: (+32 2) 295 80 43
Fax: (+32 2) 295 01 46
Telex: 21877 comeu b
Web: www.jrc.cec.eu.int
■ Joint research centre acting as scientific and technical body for the CEE in the fields of remote sensing applications, environment, safety technology, prospective technological studies, advanced materials, reference materials and measurements, transuranic elements, systems engineering and informatics, health and consumer protection.

Centre d'Étude des Environnements Terrestre et Planetaires (CETP)
Centre Universitaire de Vélizy, 10-12 Avenue de l'Europe, F-78140 Vélizy, France
Tel: (+33 1) 39 25 49 04
Fax: (+33 1) 39 25 49 22
e-mail: trangbuiqoc@cetp.ipsl.fr
■ Laboratory research into surface physics, atmosphere, ionosphere and magnetosphere of the Earth and other bodies in the solar system. Activities include experimental work (radars, electromagnetic waves and plasma analysis) theory and modelisation. Numerical simulation.

Centre de Lancements de Ballons
F-40800 Aire sur L'Adour, France
Tel: (+33 5) 58 71 61 82
Fax: (+33 5) 58 71 67 44
Telex: 560951

Centre de Météorologie Spatiale (CMS)
BP 147 F-22302 Lannion, Cedex, France
Tel: (+33 2) 96 05 67 00
Fax: (+33 2) 96 05 67 37
Web: www.meteo.fr
■ This centre belongs to the French Weather Service. Operational reception and processing of meteorological satellite data. Archiving and distribution. Research activities.

Centre for Applied Special Technology (CAST)
40 Harvard Mills Square, Suite 3, Wakefield, Massachusetts, 01880-3233, United States
Tel: (+1 781) 245 22 12
Fax: (+1 781) 245 52 12
e-mail: cast@cast.org
Web: www.cast.org
■ Innovative, technology-based educational resources and strategies based on the principles of Universal Design for Learning (UDL).

Centre for Research in Earth and Space Technology (CRESTECH)
Corporate Office
4850 Keele Street, Toronto, Ontario, M3J 3K1, Canada
Tel: (+1 416) 665 33 11
Fax: (+1 416) 665 20 32
Web: www.crestech.ca

Centre for the Development of Industrial Technology (CDTI)
4 Cid Street, E-28001 Madrid, Spain
Tel: (+34 91) 581 55 00
Fax: (+34 91) 581 55 94
e-mail: info@cdti.es
Web: www.cdti.es

Centre National d'Études Spatiales (CNES)
Centre de Lancements de Ballons
F-40800 Aire sur L'Adour, France
Tel: (+33 5) 58 71 40 40
Fax: (+33 5) 58 71 40 45
Telex: 560951 f

Centre National d'Études Spatiales (CNES)
Centre Spatial Guyanais
PO Box 726, F-97387 Kourou, Cedex, French Guiana
Tel: (+594) 33 51 11
Fax: (+594) 33 47 19
■ A launch range and test facility of CNES (the French Space Agency) made available to the European Space Agency (ESA) and its member states for the purpose of their programmes under the terms of an agreement with the French Government. Incorporates Ariane launching pads for putting satellites into orbit and all relevant electronic and optical telemetry monitoring equipment.

Centre National d'Études Spatiales (CNES)
Établissement d'Evry
Rond-Point de l'Espace, F-91023 Evry, Cedex, France
Tel: (+33 1) 60 87 71 11
Fax: (+33 1) 60 87 70 47
(+33 1) 60 87 70 48
Telex: 604701

Centre National d'Études Spatiales (CNES)
Headquarters
2 Place Maurice Quentin, F-75039 Paris, Cedex 1, France
Tel: (+33 1) 44 76 75 00
Fax: (+33 1) 44 76 76 76
Web: www.cnes.fr
■ French Space Agency. Analysis of long term issues and future course of space activities, and submission of proposals to the French Government. Major development programmes undertaken nationally as part of European Space Agency activities (Ariane).

Publications:
CNES Magazine (in French and in English, 4 per year)

CNES subsidiary and participations:
Arianespace
Cerfacs
CLS
DERSI
GDTA
IFRTP
Intespace
MEDES-IMPS
Medias France
Novespace
OST
Prospace
Renater
Satel Conseil
Scot
Semeccel
Simko
Skybridge
Spot Image
Telespace Participation

Centre of Space Science and Applied Research (CSSAR)
Chinese Academy of Sciences
PO Box 8701, 100080 Beijing, China
Tel: (+86 10) 62 55 99 44
Fax: (+86 10) 62 57 69 21

Centre Royal de Télédétection Spatiale du Maroc (CRTS)
16 bis avenue de France, Agdal, Rabat, Morocco
Tel: (+212 7) 77 63 05
Fax: (+212 7) 77 63 00

Centro de Estudios Espaciales (CEE)
Universidad de Chile, A. Prat 1171, Casilla 411-3, Santiago de Chile, Chile
Tel: (+56 2) 555 34 00
Tel: (+56 2) 555 33 71
Fax: (+56 2) 556 83 82
e-mail: infocee@cec.uchile.cl
Web: www.cee.uchile.cl
■ Satellite tracking station and ground support services. Offers remote sensing, a geographic information system, geomatics, and telecommunications.

Centro de Investigaciones Aplicades
Avenida Fuerza Aérea Argentina km 5 1/2, 5103 Córdoba, Argentina
Tel: (+54 51) 69 06 66
Fax: (+54 51) 69 06 31
Telex: 51965 ancorar

Centronic Ltd
Centronic House, King Henry's Drive, New Addington, Croydon, Surrey, CR9 0BG, United Kingdom
Tel: (+44 1689) 80 80 00
Fax: (+44 1689) 84 51 17
Telex: 896474
Cable: CENTRONIC London
e-mail: eosales@centronic.co.uk
Web: www.centronic.co.uk
■ Manufacture of UV, visible and near infra-red detectors for military target acquisition, guidance, laser marker warning and training systems. Sensors for satellite star mapping, attitude detection and radiometry.

Chang Woo Inc
Yeoeuido, PO Box 747, Seoul, 150-607, Korea, South
Keum Young Building, 12th Floor, 15-11 Yeoeuido-dong Youngdeungpo-gu, Seoul, 150-872, Korea, South
Tel: (+82 2) 782 90 56
Fax: (+82 2) 782 90 58
e-mail: salescwinc@hanafos.com
Web: www.changwooinc.com

Cheyenne Mountain Public Affairs
1 Norad Road, Suite 101-213, Cheyenne, Mountain, Colorado, 80914-6066, United States
Tel: (+1 719) 474 22 38

China Academy of Launch Vehicle Technology (CALT)
No.1 Nan Da Hong Men Road, Fengtai District, 100076 Beijing, China
Tel: (+86 10) 68 38 10 35
(+86 10) 68 75 75 98
(+86 10) 68 38 13 83
e-mail: caltinfo@calt.com.cn
■ Launch vehicle research base. Products include liquid fuel strategic rockets and long march launch vehicles, spacecraft satellite communications systems, precision measuring and testing instrumentation; applied computer technology and automatic control technology.

China Aerospace Corporation
China National Space Administration
9 Fucheng Road, Haidian District, PO Box 949, Beijing, 100830, China
Tel: (+86 10) 68 37 00 43
(+86 10) 68 37 06 99
Fax: (+86 10) 68 37 00 80
Web: www.spacechina.com
■ Management of national civil space programme.

China Atomic Energy Authority (CAEA)
2 Guanganmen Nanjie, 100053 Xuanwuqu, Beijing, China
Tel: (+86 10) 83 98 33 81
(+86 10) 83 98 33 83
Fax: (+86 10) 83 98 35 16
■ Governmental agency dealing with international affairs in the field of peaceful use of nuclear energy.

China Great Wall Industry Corp (CGWIC)
67 Beisihuan Xilu, Haidian District, 100080 Beijing, China
Tel: (+86 10) 88 10 21 88
(+86 10) 88 10 20 00
Fax: (+86 10) 88 10 21 07
e-mail: cgwic@cgwic.com
Web: www.cgwic.com
■ Trading company acting as the main foreign trade and marketing channel for China Aerospace Corporation (CASC) and is the sole commercial organisation authorised by the government to provide commercial satellite launch services and space technology co-operation to the world market. Development of international activities and aerospace technology collaboration specialising in space technology products. Import and export of equipment, precision machinery, instruments and meters. Technology consulting, project contracting, employment services, technology transfer, joint production, equity, co-operative joint ventures, investment co-ordination, export capital construction, storage and transportation, international exhibition, tourism, satellite communications, property management and domestic trade.

Locations:
Jiuquan Satellite Launch Center
Taiyuan Satellite Launch Center
Xian Satellite Control Center
Xichang Satellite Launch Center

China Great Wall Industry Corp (CGWIC)
California Office
21515 Hawthorne Boulevard, Suite 1065, Torrance, California, 90503, United States
Tel: (+1 310) 540 77 06
Fax: (+1 310) 540 34 75
e-mail: gwahuang@aol.com
Representing: China Great Wall Industry Corp., China

China Great Wall Industry Corp (CGWIC)
Division
22 Fu Cheng Road, PO Box 129, 100036 Beijing, China
Tel: (+86 10) 68 37 23 63
Fax: (+86 10) 68 42 91 12

China Jiangnan Space Company Group (061 Base)
36 Beijing Road, 563003 Zunyi, Guizhou, China
Tel: (+86 852) 861 11 43
e-mail: info@cjspace.com.cn
Web: www.cjspace.com.cn
■ Production, research and development of military space systems. Products include tele- communications facilities, batteries and power sources, electrical motors, metal wire drawing facilities, production machinery, engineering machines, moulds, dies and tools.
Representing:
38th Research Institute
302nd Research Institute (General Institute of Military Products)
303rd Research Institute
Chaohui Electromechanical Factory
Guizhou Gaoyuan Machinery Factory – SAM Launchers
Honggang Electromechanical Factory
Jiangnan Electromechanical Design Institute
Meiling Factory
Nanfeng Factory
Qunjian Machinery Factory
Wujiang Machinery Factory
Xinfeng Instrument Manufacturing Corporation – Tracking and Control Systems

China National Electronics Import & Export Corp (CEIEC)
Electronics Building A23, Fuxing Road, PO Box 140, 100036 Beijing, China
Tel: (+86 1) 821 95 32
Fax: (+86 1) 821 23 52
(+86 10) 68 21 23 61
e-mail: ceiec@ceiec.com.cn
Web: www.ceiec.com.cn
■ Products include air, naval and army radios and radars, air defence systems, navigation systems, optical systems, cryptographic equipment, mine detection equipment, fibre and laser optics, command, control and communications systems, electronic warfare systems, simulators, components and spare parts. Services include the modification, overhaul and upgrading of radar systems, military communication systems, fire control systems and air defence networks, operates a spare parts supply service and export of military electronics systems.

China National Machinery Import & Export Corp
Wuhan Office
Wuchang, 430070 Wuhan, China
Tel: (+86 27) 71 52 63
Fax: (+86 27) 71 66 64
■ Manufacture of machine tools, nuclear reactors, mechanical appliances, machinery, boilers and parts; photographic, cinematographic, optical, measuring and precision equipment; ball and roller bearings.

China National Precision Machinery Import & Export Corporation (CNPMIEC)
30 Haidian Nanlu, Haidian Qu, 100080 Beijing, China
Tel: (+86 10) 68 74 88 77
(+86 10) 68 74 88 91
Fax: (+86 10) 68 74 88 44
Web: www.cpmiec.com.cn
■ Import, export and production of tactical defensive missiles and related equipment, including: anti-ship missiles, air defence missile systems from hand held to vehicle mounted and multilaunch rocket systems. Other products include space equipment, precision machinery, electronic products, space hardware and optical instruments, missile components such as gimbals and gyroscopes. Research and development facilities have wind tunnel, plasma etching technology, IC photolithograph capabilities as well as microwave and infra-red homing guidance simulation laboratories. Deals with joint ventures, compensation trade, contracts for the design and construction of industrial and civil projects.

China National Space Administration (CNSA)
8A, Fucheng Road, 100037 Haidian, Beijing, China
Tel: (+86 10) 88 58 13 77
(+86 10) 88 58 13 79
(+86 10) 88 58 13 80
(+86 10) 88 58 13 86
Fax: (+86 10) 88 58 15 15
e-mail: cnsa@cnsa.gov.cn
Web: www.cnsa.gov.cn

China Sanjiang Space Group
Alloy First Qianjin Road, Jiangham District, 430022 Hankou, Wuhan, Hubei, China
Tel: (+86 27) 586 40 37
Fax: (+86 27) 586 40 37

China Space Civil & Building Engineering Design & Research Academy (CSCBA)
93 Fengtai Road, PO Box 2964, 100071 Beijing, China
Tel: (+86 10) 874 97 68
Fax: (+86 10) 837 08 65

Chinese Academy of Space Technology (CAST)
Beijing Institute of Control Engineering
31 Bai Shi Qiao Lu, PO Box 2417, 100081 Beijing, China
Tel: (+86 10) 68 37 82 30
(+86 10) 68 37 94 39
Fax: (+86 10) 68 37 82 37
(+86 10) 68 37 91 63
Telex: 22473 ccsc
■ Production of recoverable spacecraft.

Chinese Society of Aeronautics & Astronautics (CSAA)
5 Liangguochang Road, Dongchen District, 100010 Beijing, China
Tel: (+86 10) 64 02 14 16
Fax: (+86 10) 64 02 14 13
e-mail: mail@csaa.org.cn
Web: www.csaa.org.cn
■ Consists of 24 technical committees aerodynamics, flight dynamics and testing, structural design and analysis, propulsion systems, materials technology, manufacturing technology, aviation medicine and human engineering, maintenance engineering, management science, composites, instrumentation and measurements, weaponry and fire control, helicopters, light flying vehicles, avionics, automatics, information, reliability) and organises academic conferences, meetings, forums, technical exhibitions and information exchanges. Over 40,000 members.
Publications:
Acta Aeronautica et Astronautica
Sinica (with English abstracts)
Aerospace Knowledge Magazine (in Chinese)
Model Airplane (in Chinese with English text)
Regional societies:
Beijing Aerospace Society and the Aeronautics and Astronautics Societies of Guangdong, Guangxi, Ghuizou, Heilongjiang, Hubei, Hunan, Jiangsu, Jiangxi, Jilin, Liaoning , Shaanxi, Shanghai, Sichuan, Zhejian, Fujian, Henan

Chinese Society of Astronautics (CSA)
No.2, Yue Tan Beixiao Jie, Beijing, China
Tel: (+86 10) 68 76 80 85
Fax: (+86 10) 68 05 10 70
e-mail: iaf@public.bta.net.cn

For details of the latest updates to *Jane's Space Systems and Industry* online and to discover the additional information available exclusively to online subscribers please visit
jsd.janes.com

■ An academic organisation which carries out academic exchanges on space science and technology at home and abroad, sponsors various kinds of national and international academic symposia and education of the public, particularly the young, about space, science and technology.

CHL Netherlands BV

Christiaan Huygens Laboratorium BV
PO Box 3072, NL-2220 CB Katwijk, Netherlands
Lageweg 16, NL-2222 AG Netherlands
Tel: (+31 71) 402 55 14
Fax: (+31 71) 402 50 78
e-mail: marketing@chl.nl
Web: www.chl.nl

■ Design and production of complete radar antenna systems. Production of Artemis microwave range-bearing position fixing system, radar based traffic detection systems. Consultancy on radar and microwave based systems, vessel traffic counter, a radar-based system, pulse doppler sensor, sail in-sail out detection system, doppler radar system for car traffic speed measurement equipment.

CIC International Ltd

Five Marine View Plaza, Hoboken, New Jersey, 07030, United States
Tel: (+1 201) 792 18 00
Fax: (+1 201) 792 57 55
e-mail: 103227.250@compuserve.com
cicinternational@compuserve.com
Web: www.cic-international.com

■ Supplier of defence products.

Cicoil Corporation

24960 Avenue Tibbitts, Valencia, California, 91355, United States
Tel: (+1 661) 295 12 95
Fax: (+1 661) 295 08 13
e-mail: sales@cicoil.com
Web: www.cicoil.com

■ Flat, flexible, silicone encapsulated cables. Applications include: aerospace and weapons, space and satellite, medical, robotics and automation.

Cimarron

1115 Gemini, Houston, Texas, 77058-3584, United States
Tel: (+1 281) 226 51 00
Fax: (+1 281) 226 51 90
e-mail: cimarron@cimarroninc.com
sales@cimarroninc.com
Web: www.cimarroninc.com

Circle Seal Controls Inc

a division of Circor International Inc, US
2301 Wardlow Circle, Corona, California, 92880-2881, United States
Tel: (+1 951) 270 62 00
Fax: (+1 951) 270 62 01
Web: www.circle-seal.com

■ Valves and controls for industrial and aerospace and military applications. Products range from discrete components to integrated assemblies to complete fluid control systems. Products include check, relief, shutoff, solenoid, cryogenic, fuse, and motor operated valves, pressure regulators, tyre fill valve gauges, restrictors and manifold systems.

CISET International SA

350 rue Saint Jacques, B-5500 Dinant, Belgium
Tel: (+32 82) 22 29 38
Fax: (+32 82) 22 43 16
e-mail: marketing@ciset-int.com

■ Operation and maintenance of Redu Station. Design, integration and realisation of turnkey systems such as fixed, mobile and fly-away earth stations for satellite broadcast applications.

CLS

Collecte Localisation Satellites
a subsidiary of Centre National d'Études Spatiales (CNES), France
8-10 rue Hermès, Parc Technologique du Canal, F-31520 Ramonville Saint Agne, France
Tel: (+33 5) 61 39 47 00
Fax: (+33 5) 61 75 10 14
Telex: 531752
e-mail: info@cls.fr
Web: www.cls.fr

■ Turnkey services, subsurface mooring monitoring, moored buoys monitoring. Location and data collection by satellite.

CLS America Inc

1441 McCormick Drive, Suite 1050, Largo, Maryland, 20774, United States
Tel: (+1 301) 925 44 11
Fax: (+1 301) 924 89 95
Web: www.clsamerica.com

■ Technical expertise in the data processing chain from data collection through to development of technologies and custom applications.

CMC Electronics Cincinnati

a subsidiary of CMC Electronics Inc, Canada
7500 Innovation Way, Mason, Ohio, 45040-9699, United States
Tel: (+1 513) 573 62 82
Fax: (+1 513) 573 67 67
e-mail: sales@cinele.com

Codan Limited

Head Office
81 Graves Street, Newton, South Australia, 5074, Australia
Tel: (+61 8) 83 05 03 11
Fax: (+61 8) 83 05 04 11
e-mail: info@codan.com.auussales@codan.com.au
Web: www.codan.com.au

Codan Pty Ltd

Perth Office
Suite 11A, 2 Hardy Street, Perth, Western Australia, 6151, Australia
Tel: (+61 8) 93 68 52 82
Fax: (+61 8) 93 68 52 83

Codan Pty Ltd

Queensland Office
532 Seventeen Mile Rocks Road, Sinnamon Park, Queensland, 4073, Australia
Tel: (+61 7) 32 91 63 33
Fax: (+61 7) 32 91 63 50

Codan (UK) Ltd

Gostrey House, Union Road, Farnham, Surrey, GU9 7PT, United Kingdom
Tel: (+44 1252) 71 72 72
Fax: (+44 1252) 71 73 37
e-mail: uksales@codan.com.au
Web: www.codan.com.au

■ Communications equipment: HF radio, microwave systems and satellite communications equipment.

Codan US Inc

8430 Kao Circle, Manassas, Virginia, 20110, United States
Tel: (+1 703) 361 27 21
Fax: (+1 703) 361 38 12

Coherent Optics (Europe) Ltd

Vinten Electro-Optics Ltd
a subsidiary of Coherent Inc, US
Unit 28-35 Ashville Way, Whetstone, Leicestershire, LE8 6NU, United Kingdom
Tel: (+44 116) 284 62 00
Fax: (+44 116) 275 16 73
Web: www.coherent.com

■ Precision optical manufacture and thin film coating. Areas of specialist expertise are optical components, modules, sub-assemblies and assemblies for infra-red imaging and tactical sighting equipment. Design and manufacture is offered throughout the visible and far infra-red spectral regions.

COM DEV Europe Ltd (CDE)

a subsidiary of Com Dev International Space Group, Canada
Triangle Business Park, Stoke Mandeville, Aylesbury, Buckinghamshire, HP22 5SX, United Kingdom
Tel: (+44 1296) 61 64 00
Fax: (+44 1296) 61 65 00
e-mail: info@comdev.co.uk
Web: www.comdev.co.uk

■ Manufacture and design of microwave and electronic equipment for communications and earth observation, remote sensing and defence satellites. Products include: ferrite switches, isolators and circulators, input and output multiplexers, filters, couplers and waveguides, detector and preamplification electronics, ASICs design.
Accreditations: ISO 9001

COM DEV International Ltd

Corporate Headquarters and R & D Centre
155 Sheldon Drive, Cambridge, Ontario, N1R 7H6, Canada
Tel: (+1 519) 622 23 00
Fax: (+1 519) 622 16 91
(+1 519) 622 21 58
Web: www.comdev.ca

■ Strategic microwave, mm-wave and signal processing components and subsystems for worldwide space and defence markets.

COM DEV Xian

Manufacturing Plant
East High Technology, Development Zone, 4 Building, 2nd Floor, 1 Huoju Road, 710043 Xian, China
Tel: (+86 29) 222 80 94
(+86 29) 222 81 27
Fax: (+86 29) 223 06 54

Comision Nacional De Telecomunicaciones

15012 Tegucigalpa, Honduras
Tel: (+504 234) 86 00
Fax: (+504 234) 86 11

Command and Control Technologies Corporation (CCT)

1425 Chaffee Drive, Suite 1, Titusville, Florida, 32780, United States
1311 North Highway US-1, Suite 129, Titusville, Florida, 32796, United States
Tel: (+1 321) 264 11 93
Fax: (+1 321) 383 50 96
e-mail: info@cctcorp.com

Commercial Space Technologies Ltd (CST)

67 Shakespeare Road, Hanwell, W7 1LU, United Kingdom
Tel: (+44 20) 88 40 10 82
Fax: (+44 20) 88 40 77 76
e-mail: cst@commercialspace.co.uk
Web: www.commercialspace.co.uk

■ Assessment studies of space activities (both technical and commercial) and assistance with the instigation or development of client's business activities in the space sector. Contacts with organisations in the Russian Federation and associated states.

Commission d'Astronautique de l'Academie de Roumanie

125 Calea Victoriei, R-71102 Bucharest, 1, Romania
Tel: (+40 1) 650 76 80
Fax: (+40 1) 312 02 09
Telex: 11907 acad

■ Main commission of the Romanian Academy, joining approximately 90 scientist-members, devoted to the promotion of non-governmental original scientific works in space related fields.

Commission of Science, Technology and Industry for National Defense (COSTIND)

2A Guanganmen, Xuanwu District, 100053 Beijing, China
Tel: (+86 010) 63 57 13 97
Fax: (+86 010) 63 57 13 98

■ Deliberating and drawing up guidelines, policies and regulations for defence industries and military conversion and managing the implementation of technology policies and development planning for nuclear, aviation, aerospace, shipbuilding and weaponry.

Committee for Aviation & Space Industry Development (CASID)

Suite 1712, 333 Keelung Road, Sec.1, Taipei, Taiwan
Tel: (+886 2) 757 61 57
Fax: (+886 2) 757 60 43
e-mail: casid@ms2.hinet.net
Web: www.casid.org.tw

Committee on Space Research (COSPAR)

Comité pour La Recherche Spatiale
c/o CNES, 2 place Maurice Quentin, F-75039 Paris, Cedex 01, France
Tel: (+33 1) 44 76 75 10
Fax: (+33 1) 44 76 74 37
e-mail: cospar@cosparhq.cnes.fr
Web: cosparhq.cnes.fr

■ Promote internationally scientific research through the organisation of scientific assemblies, publications and other activities. Members include 44 National Scientific Institutions engaged in space research, and 13 International Scientific Unions adhering to the International Council for Science (ICSU). Acts as a forum with contributions from most countries engaged in space research, for the presentation of the lastest scientific results, for the exchange of knowledge and also for the discussion of space research problems. Acts as a scientific committee advising,

as required, the UN and other intergovernmental organisations on space research matters or on the assessment of scientific issues in which space can play a role. Acts as a panel for the preparation of scientific and technical standards related to space research. Acts as an entity promoting, on an international level, research in space, much of which has grown into large international collaborative programs in the mainstream of scientific research. Strives to promote the use of space science for the benefit of mankind and for its adoption by developing countries and new space-faring nations. Objectives include: promoting on an international level, scientific research in space, with emphasis on the exchange of results, information and opinions, and to provide a forum, open to all scientists, for the discussion of problems that may affect scientific space research.
Publications:
Advances in Space Research, 24 times a year
Space Research Today, 3 times a year
COSPAR Colloquia Series, published in advances in space research

Communications & Power Industries (CPI)
Satcom Division
811 Hansen Way, M/s -400, PO Box 51625, Palo Alto, California, 94303, United States
Tel: (+1 650) 846 37 00
Fax: (+1 650) 424 17 44
■ Low to high power amplifiers for satellite communications.

Communications & Power Industries
Varian Associates Inc
PO Box 51110, Palo Alto, California, 91303-1110, United States
Tel: (+1 650) 846 29 00
(+1 650) 846 28 00
(+1 650) 846 37 00
Fax: (+1 650) 846 32 76
(+1 650) 424 17 44
Web: www.cpii.com

Communications Center
2723 Green Valley Road, Clarksburg, Maryland, 20871-8599, United States
Tel: (+1 301) 831 67 00
Fax: (+1 301) 865 55 77
■ Technical, marketing and financial expertise in the areas of telecommunications capacity leasing, sale and leaseback, due diligence studies, market assessment and forecasting, competitive analyses, economic analyses, regulatory requirements, satellite and fibre optics. A private research laboratory is available.

Communications Research Laboratory
Strategic Planning Division
4-2-1 Nukui-Kitamachi, Koganei, Tokyo, 184-8795, Japan
Tel: (+81 42) 327 53 92
Fax: (+81 42) 327 75 87

Computadoras, Redes e Ingeniería SA (CRISA)
a subsidiary of EADS Astrium, France
Torres Quevedo, 9 (PTM), E-28760 Tres Cantos, Madrid, Spain
Tel: (+34 91) 806 86 00
(+34 91) 806 02 35
Fax: (+34 91) 806 02 35
e-mail: info@crisa.es
Web: www.crisa.es
■ Development and manufacture of electronics and software products for aerospace, defence and telecommunications sectors.
Accreditations: ISO 9001: 2000; AQAP 110; ESA PSS-01-XXX

Computer Sciences Corp (CSC)
Corporate Headquarters
2100 East Grand Avenue, El Segundo, California, 90245, United States
Tel: (+1 310) 615 03 11
Fax: (+1 310) 322 36 82
e-mail: generalinformation@csc.com
Web: www.csc.com
■ Design and development of system software and application programs, system engineering, facilities management, systems integration, remote computing services and communication systems.

Computer Sciences Corp (CSC)
Federal Sector
3170 Fairview Park Drive, Falls Church, Virginia, 22042, United States
Tel: (+1 703) 876 10 00
(+1 703) 641 22 22
Fax: (+1 703) 849 10 00
(+1 703) 204 83 55
Web: www.csc.com

Comsol Inc
8 New England Executive Park, Suite 310, Burlington, Massachusetts, 01803, United States
1 New England Executive Park, Suite 350, Burlington, Massachusetts, 01803, United States
Tel: (+1 781) 273 33 22
Fax: (+1 781) 273 66 03
Web: www.comsol.com

Comtech Antenna Systems Inc
Comtech Telecommunications
a subsidiary of Comtech Telecommunications Corp.
3100 Communications Road, St Cloud, Florida, 34769, United States
Tel: (+1 407) 892 61 11
Fax: (+1 407) 892 09 94
e-mail: dcreasy@comtechantenna.com
sales@comtechantenna.com
customerservice@comtechantenna.com
technical@comtechantenna.com
info@comtechantenna.com
Web: www.comtechantenna.com
■ Manufacture of satellite antenna systems for the broadcast and commercial markets. Sizes ranges from 0.9 to 7.3 meters and antennas are produced as fixed, motorised and flyaway.

Conduant Corporation
1501 South Sunset Street, Suite C, Longmont, Colorado, 80501, United States

Consiglio Nazionale delle Ricerche (CNR)
Istituto di Radioastronomia
Viap P Gobetti 101, I-40129 Bologna, Italy
Tel: (+39 051) 639 93 85
Fax: (+39 051) 639 94 31

Consortium for Materials Development in Space
University of Alabama in Huntsville
Research Institute, Room M-65, 301 Sparkman Drive, Huntsville, Alabama, 35899, United States
Tel: (+1 205) 890 66 20
(+1 205) 890 64 14
Fax: (+1 205) 895 67 91
e-mail: lundquist@email.uah.edu

Constellation Services International Inc
2313 Via Puerta, Suite B, Laguna Woods, California, 92637, United States
Tel: (+1 818) 710 38 77
Web: www.constellationservices.com

Control Techniques Dynamics Ltd
Moore Reed & Co Ltd
a division of Emerson Electric Co, US
South Way, Walworth Industrial Estate, Andover, Hampshire, SP10 5AB, United Kingdom
Tel: (+44 1264) 38 76 00
Fax: (+44 1264) 35 65 61
e-mail: sales@ctdynamics.com
Web: www.ctdynamics.com
■ Components, synchros, resolvers, AC servo motors, servo motor generators, DC motors and generators, magnetic pick-offs, contact, optical and solid state encoders, stepper motors, packaged servo systems, absolute and incremental digital displays. Torque motors and brushless DC motors.

Co-operative Research Centre for Satellite Systems (CRCSS)
GPO Box 1483, Canberra, Australian Capital Territory, 2601, Australia
Building 2, Wilf Crane Crescent, Yarralumla, 2602, Australia
Tel: (+61 2) 62 81 85 20
(+61 2) 62 16 72 70
(+61 2) 62 81 85 21
Fax: (+61 2) 62 16 72 72
e-mail: satsys@crcss.csiro.au
Web: www.crcss.csiro.au
■ Development and launch of a small scientific satellite FedSat. Research and development in areas of space science, communications, remote sensing and satellite systems.
Core partners:
Auspace Limited, CSIRO, Queensland University of Technology, University of Newcastle, University of South Australia, University of Technology, Sydney, VIPAC Engineers and Scientists Ltd.

Copernicus Astronomical Centre
Ul. Bartycka 18, PL-00716 Warszawa, Poland
Tel: (+48 22) 41 00 41
(+48 22) 41 10 86
Fax: (+48 22) 41 08 28
Telex: 813978 cbk

Cosmos International Satellitenstart GmbH
Univeritätsalle 29, D-28359 Bremen, Germany
Tel: (+49) 42 12 02 08
Fax: (+49) 42 12 02 07 00
e-mail: cosmos@fuchs-gruppe.com
Web: www.fuchs-gruppe.com/cosmos

Council of Defense & Space Industry Associations (CODSIA)
2111 Wilson Boulevard, Suite 400, Arlington, Virginia, 22201, United States
Tel: (+1 703) 247 94 90
Fax: (+1 703) 243 85 39
e-mail: codsia@ndia.org
Web: www.codsia.org
■ Provides a central channel of communications for improving industry-wide consideration of many policies, regulations, implementation problems, procedures and questions involved in federal procurement actions.
Member associations:
Aerospace Industries Association
American Electronics Association
American Shipbuilding Association
Contract Services Association
Electronic Industries Alliance
Manufacturers Alliance/MAPI
National Defense Industrial Association
Professional Services Council

Cranfield Aerospace Ltd
UK Office
CAe Ltd
Hangar 2, Cranfield, Bedfordshire, MK43 0AL, United Kingdom
Tel: (+44 1234) 75 40 46
Fax: (+44 1234) 75 11 81
e-mail: cae@cranfieldaerospace.com
Web: www.cranfieldaerospace.com

Cranfield University
Cranfield Campus, Cranfield, Bedfordshire, MK43 0AL, United Kingdom
Tel: (+44 1234) 75 01 11
Fax: (+44 1234) 75 08 75
e-mail: info@cranfield.ac.uk
Web: www.cranfield.ac.uk
■ Institute of applied research including aeronautical engineering and research, defence systems, propulsion, electronics, communications, data processing, weapon systems and explosives, materials and manufacturing science, management, chemical systems.

Cryospace
PO Box 2, F-78133 Les Mureaux, Cedex, France
Tel: (+33 1) 39 06 11 73
Fax: (+33 1) 39 06 10 52
■ Development and construction of cryotechnical reservoirs.

CSC-DynSpace
a subsidiary of Dyncorp, US
3190 Fairview Park Drive, VTC-A, Suite 700, Falls Church, Virginia, 22042, United States
Tel: (+1 703) 261 47 46
Fax: (+1 703) 261 54 20
■ High technology company specialising in the launching of commercial satellites into low earth orbits.

CSIR
Satellite Applications Centre
PO Box 395, 0001 Pretoria, South Africa
Meiring Naude Road, Brummeria, 0001 Pretoria, South Africa

Tel: (+27 12) 334 50 00
Fax: (+27 12) 334 50 01
Web: www.sac.co.za
■ Provision of spacecraft tracking, telemetry and command services to launchers, operators and owners. Remote sensing and information services.

Cubic-i Ltd
Bluebell Building 7F, 2-15-9 Nishi-Gotanda, Shinagawa-ku, Tokyo, 141-0031, Japan
Tel: (+81 3) 37 79 55 06
Fax: (+81 3) 37 79 57 83
e-mail: argos@cubic-i.co.jp
cubici@kt.rim

Custom Electronics Inc
Browne Street, Oneonta, New York, 13820, United States
Tel: (+1 607) 432 38 80
Freephone: (+1 877) 735 92 34
Fax: (+1 607) 432 39 13
e-mail: ceisales@customelec.com
■ Manufacture of high voltage reconstituted mica paper capacitors for use in radar power supplies, ECM power supplies, high-voltage transmitters for missile applications, TWT power supplies for satellites, detonation systems, ignition systems and many other high voltage applications.

Cynetics Corporation
2603 South Highway 79, Rapid City, South Dakota, 57701, United States
Tel: (+1 605) 394 64 30
Fax: (+1 605) 343 72 40
e-mail: cynetics@rapidnet.com
Web: www.rapidnet.com/~tepco
■ Spacecraft and ground TT and C, data communications links for lightsats, microwave transmitters and receivers. MPEG2 DBS receivers and equipment.

Czech Academy
Astronomical Institute
Bocni Ii 1401, CZ-141 31 Praha, 4, Czech Republic
Tel: (+420 2) 267 10 30 61
Fax: (+420 2) 76 90 12

D

Daco Scientific Ltd
Vulcan House, Calleva Industrial Park, Aldermaston, Berkshire, RG7 8PB, United Kingdom
Tel: (+44 118) 981 73 11
Fax: (+44 118) 981 99 63
Web: www.daco.co.uk
■ Design and manufacture of trackballs, control grips and joysticks for defence and aerospace related applications.
Accreditations: ISO 9001

Dage Corporation
1011 High Ridge Road, Stamford, Connecticut, 06905, United States
Tel: (+1 203) 461 90 00
Fax: (+1 203) 461 97 95
e-mail: emailus@dage.com
■ Full service export management. Supplies custom designed, catalogue and spare parts components and subsystems for the commercial and military markets: communications of all types, cellular, line-of-sight, mobile earth stations, satellites. Radars: ground-based, fixed and mobile, shipborne, missile and space. Electronic warfare system components. TV, radio transmitter and receiver parts. Linear and ring accelerator components. Industrial heating products.
Representing:
Antelope Valley Microwave
Atlantic Microwave Corporation (Antennas)
Custom Microwave
DB Products
EMF Systems
EMR Corporation
Mac Technology
Mega Industries
Microwave Development Co. (MDC)
Microwave Development Labs. (MDL)
Microwave Device Technology
Microwave Resources Inc. (MRI)
Microwave Resources Corporation (MRC)
Omniyig
Pendulum Electromagnetics
Reactel
Resin Systems
Res-Net Microwave
Sonoma Scientific
Telgaas Inc.

Daicel Chemical Industries Ltd (DCI)
Osaka Head Office
1 Teppo-cho, Sakai, Osaka, 590-8501, Japan
Tel: (+81 722) 27 31 11
Fax: (+81 722) 27 30 00
Telex: 5374213 daicel

Dallas Remote Imaging Group (DRIG)
4209 Meadowdale Drive, Carrollton, Texas, 75010, United States
Tel: (+1 972) 898 35 63
e-mail: consulting@drig.com
Web: www.drig.com
■ Provides consulting and educational services on weathersatelliteimagery, satellitecommunications and tracking, and the use of image acquisition and processing in education. Provides remote imaging and space related content on a wide variety of platforms.

Danish National Space Center (DNSC)
Danmarks Rumcenter
Danish Space Research Centre
30 Juliane Maries Vej, DK-2100 København, Denmark
Tel: (+45 35) 32 57 00
(+45 35) 32 57 01
(+45 35) 32 57 21
Fax: (+45 35) 36 24 75
e-mail: office@spacecenter.dk
Web: www.spacecenter.dk
■ Government agency concerned with space research, x-ray astronomy, plasma physics and instrument development.

DARA/DLR Bureau Paris
17 avenue de Saxe, F-75007 Paris, France
Tel: (+33 1) 42 19 94 26
Fax: (+33 1) 42 19 96 29

Darchem Engineering Ltd
a subsidiary of Esterline Corporation, USA
Stillington, Stockton-on-Tees, Cleveland, TS21 1LB, United Kingdom
Tel: (+44 1740) 63 04 61
Fax: (+44 1740) 63 05 29
e-mail: sales@darchem.co.uk
Web: www.darchem.co.uk
■ High temperature insulation materials and systems. Precision fabrications and honeycomb core structures in stainless steel, other alloys and components for aerospace and other industries. Marine gas turbine intake and exhaust systems. Sheet metal profiling. Specialist forming techniques. Heat shields.

Darchem Engineering Ltd
Gloucester Office
Insumat Ltd
Eastbrook Road, Eastern Avenue, Gloucester, GL4 3DB, United Kingdom
Tel: (+44 1452) 37 72 00
Fax: (+44 1452) 37 72 01
■ Thermal insulation and fire protection. Sheet metal fabrication in stainless steel, titanium, nimonic and composites. Air intake and exhaust silencers and exhaust systems for naval vessels. Gun barrel insulation for MBT's.

Dassault Aviation
Head Office
jointly owned by Groupe Industriel Marcel Dassault, France (50.55 per cent), EADS France (46.30 per cent), Public (3.15 per cent)
78 Quai Marcel Dassault, F-92214 Saint Cloud, Cédex 300, France
Tel: (+33 1) 47 11 40 00
Fax: (+33 1) 47 11 56 60
Web: www.dassault-aviation.com
■ Scientific research and studies. Manufacture of civil and military aircraft and flight control systems. Maintenance and repair of aeronautical material and space systems.
Corporate structure:
Aero Precision Inc, US (50%)
Corse Composites Aeronautiques, France (25%)
Dassault Aircraft Services, US (100%)
Dassault, France (100%)
Dassault Falcon Jet, US (100%)
Dassault Falcon Jet Wilmington (100%)
Dassault Falcon Service, France (100%)
Dassault International, France (100%)
Dassault Procurement Services, US (100%)
Dassault Réassurance, France (100%)
Embraer, Brazil (2.1%)
Eurotradia International, France (16%)
Falcon Training Center, France (50%)
Midway, US (100%)
Secbat, France (36%)
Sogitec Industries, France (99.7.%)
Sofema, France (6.7%)
Sofresa, France (6%)

Dassault Aviation
Saint Cloud Factory
78 Quai Marcel Dassault, F-92552 Saint Cloud, Cedex 300, France
Tel: (+33 1) 47 11 40 00
Fax: (+33 1) 47 11 49 01

Datamat Rome SpA
Datamat Ingegneria dei Sistemi SpA
a Finmeccanica company
Via Laurentina 760, I-00143 Roma, Italy
Tel: (+39 6) 502 71
Fax: (+39 6) 50 51 14 07
e-mail: info@elsagdatamat.com
info@datamat.it
Web: www.datamat.it
■ C3I systems, training simulators, satellite ground stations, software technologies, system integration, logistic support.

Dateno
ZAC Es-Passants, BP 90145, F-35801 Dinard, Cedex, France
ZAC. Es Passants II, Rue Amiral Bérenger, F-35801 Dinard, Cedex, France
Tel: (+33 2) 99 46 24 75
Fax: (+33 2) 99 46 47 27
e-mail: dateno@dateno.fr
■ Manufacture of power equipment for satellite earth stations.

Datum GmbH
Fichtenstrasse 25, D-85649 Hofolding, Germany
Tel: (+49 8104) 66 24 29
Fax: (+49 8104) 66 24 28
e-mail: sales.gmbh@datum.com

Debis Systemhaus Solutions for Research
a joint venture of Daimler Chrysler Services and German Aerospace Center.
Bunsenstrasse 10, D-37073 Gottingen, Germany
e-mail: dir-infomaster@dlr.de

Defence Communications Research Centre (DCRC)
Institute for Telecommunications Research, University of South Australia, The Levels, South Australia, 5095, Australia
Tel: (+61 8) 83 02 38 97
Fax: (+61 8) 83 02 38 73

Defence Geographic and Imagery Intelligence Agency Headquarters (DGIA)
an agency of the Ministry of Defence, UK
Room 17, Watson Building, Elmwood Avenue, Feltham, Middlesex, TW13 7AH, United Kingdom
Tel: (+44 20) 88 18 24 22
Fax: (+44 20) 88 18 22 46
■ Provide imagery intelligence and geographic support to defence policy, planning, operations and training.

Defense Advanced Research Projects Agency (DARPA)
3701 North Fairfax Drive, Arlington, Virginia, 22203-1714, United States
Tel: (+1 703) 696 24 00
(+1 703) 696 22 37
(+1 703) 696 24 11
(+1 703) 696 22 28
(+1 703) 526 41 70
Fax: (+1 703) 696 22 07
(+1 571) 218 43 56
Web: www.darpa.mil
■ Develop imaginative, innovative and often high-risk research ideas offering a significant technological impact that will go well beyond the normal evolutionary development, and to pursue these ideas from the demonstration of technical feasibility through the development of prototype systems.

Defense Information Systems Agency (DISA)
Defense Communications Agency
DISA Headquarters Building #12, 701 S Courthouse Road, Arlington, Virginia, 22204-2199, United States
Tel: (+1 703) 692 00 18
(+1 703) 607 69 00
(+1 703) 607 62 65
Web: www.disa.mil
■ Provider of information systems support.

Dendron Resource Surveys Inc
880 Lady Ellen Place, Suite 206, Ottawa, Ontario, K1Z 5L9, Canada
Tel: (+1 613) 725 29 71
Fax: (+1 613) 725 17 16
Web: www.dendron.com

Department di Astronomia
Via dell Osservatorio 5, I-35122 Padova, Italy
Tel: (+39 49) 829 34 36
Fax: (+39 49) 875 98 40

Department of Commerce
Office of Space Commercialization
Room 7060, Washington, District of Columbia, 20230, United States
Tel: (+1 202) 482 61 25
Fax: (+1 202) 482 25 68
Telex: 892536 usdocwashdc
e-mail: space.commerce@noaa.gov
Web: www.nesdis.noaa.gov/space

Department of Energy (DoE)
Headquarters
1000 Independence Avenue South West, Washington, District of Columbia, 20585, United States
Tel: (+1 202) 586 50 00
Freephone: (+1 800) DIAL DOE
Fax: (+1 202) 586 44 06
(+1 202) 586 84 03
Web: www.energy.gov

Department of Energy (DoE)
Office of Nuclear Energy Science and Technology
MA 75, 19901 Germantown Road, Germantown, MD 20874, United States
Tel: (+1 301) 903 55 59

Department of Geophysics & Planetary Sciences
Tel-Aviv University, IL-69978 Tel-Aviv, Israel
Tel: (+972 3) 640 69 72
Fax: (+972 3) 640 92 82

Department of Industry, Science and Resources
Space and Aerospace Industries Section
Australian Space Office
20 Allara Street, GPO Box 9839, Canberra, Australian Capital Territory, 2601, Australia
Tel: (+61 2) 62 13 72 24
Fax: (+61 2) 62 13 72 49
Web: www.isr.gov.au
■ Advisory services to the Australian government and industry on issues of space and aerospace policy. Promotion of Australian space and aerospace capabilities on the international market. Support for and regulation of the Space Activities Act 1998. Negotiation for intergovernmental agreements with the Russian Federation and NASA. Development of the commercial spaceport facilities of Kistler Aerospace and Spacelift Australia at Woomera, South Australia; Asia-Pacific Space Centre on Christmas Island; and, United Launch Systems near Gladstone, Queensland. Study reports for the Australian space industry, remote sensing sector and the prospects and impact of commercial space launch activities.

Department of Industry, Tourism and Resources
GPO Box 9839, Canberra, Australian Capital Territory, 2601, Australia
4/40 Allara Street, Canberra, Australian Capital Territory, 2600, Australia
Tel: (+61 2) 62 13 60 00
Fax: (+61 2) 62 13 70 00
Web: www.isr.gov.au
www.minister.industry.gov.au
www.ausindustry.gov.au

Department of Land Information (DLI)
PO Box 2222, Midland, Western Australia, 6936, Australia
1 Midland Square, Midland, Western Australia, 6056, Australia
Tel: (+61 8) 92 73 73 73
Fax: (+61 8) 92 73 76 66
e-mail: mailroom@dli.wa.gov.au
sales@dli.wa.gov.au
landsales@dli.wa.gov.au
Web: www.dli.wa.gov.au

Department of Space
Indian National Satellite System
Antariksha Bhavan, New BEL Road, Bangalore, 560094, India
Tel: (+91 80) 333 44 74
Fax: (+91 80) 333 42 29
Telex: 8452499

Department of Space
Regional Remote Sensing Services Centre
Natural Resources Management System
Antariksh Bhavan, New BEL Road, Bangalore, 560094, India
Tel: (+91 80) 33 26 77
Fax: (+91 80) 33 42 29
Telex: 8458867

Department of Trade and Industry (DTI)
British National Space Centre
BNSC
British National Space Centre
151 Buckingham Palace Road, London, SW1W 9SS, United Kingdom
Tel: (+44 20) 72 15 50 00
(+44 20) 72 15 08 06
(+44 20) 72 15 08 07
Fax: (+44 20) 821 53 87
(+44 20) 72 15 09 36
e-mail: bnscinfo@dti.gsi.gov.uk
Web: www.bnsc.gov.uk
www.open.gov.uk/bnsc/bnschome.htm
■ Co-ordinates the UK's space activities. Technology Centres at DERA Farnborough and Rutherford Appleton Laboratory, Oxford.
Affiliated to
European Space Agency

DETECON GmbH
Oberkasseler Strasse 2, D-53227 Bonn, Germany
Telex: 886384 dtc
e-mail: info@detecon.com

Detica
Surrey Research Park, Guildford, Surrey, GU2 5YP, United Kingdom
Tel: (+44 1483) 81 60 00
Fax: (+44 1483) 81 61 44
e-mail: info@smithgroup.co.uk
■ Construction of intelligence systems to aggregate and analyse data to generate useful and relevant intelligence.

Deutsche Agentur für Raumfahrtangelegenheiten (DARA) GmbH
Königswinterer Strasse 522-524, Postfach 300364, D-53183 Bonn, Oberkassel, Germany
Tel: (+49 228) 44 70
Fax: (+49 228) 44 77 00

Development Consultants Pvt Ltd (DCL)
24B Park Street, Kolkata (Calcutta), 700016, India
Tel: (+91 33) 249 76 01
Fax: (+91 33) 249 28 97
e-mail: dcl@dclgroup.com
Web: dcpl.net.in
■ Design and fabrication consultancy for Polar Satellite Launch Vehicle (PSLV) service structure.

Digital Diecutting Inc
1405 16th Street #4002, PO Box 1824, Racine, 53403, United States
Tel: (+1 800) 331 96 63
Fax: (+1 888) 703 03 83
e-mail: info@digitaldiecutting.com
Web: www.digitaldiecutting.com

DigitalGlobe
1601 Dry Creek Drive, Suite 260, Longmont, Colorado, 80503, United States
Tel: (+1 303) 684 40 00
Freephone: (+1 800) 496 12 25
Fax: (+1 303) 684 40 48
e-mail: info@digitalglobe.com
Web: www.digitalglobe.com
■ Provide services of earth imagery from company owned satellites and other sources such as aerial survey.

DIRECTV
a subsidiary of Hughes Electronics Corp., USA
PO Box 92600, Los Angeles, California, 90009, United States
2230 E Imperial Highway, El Segundo, California, 90245, United States
Tel: (+1 310) 535 50 00
Fax: (+1 310) 535 52 22
Web: www.directv.com

Drägerwerk AG
Moislinger Allee 53-55, PO Box 1339, D-23542 Lübeck, Germany
Tel: (+49 451) 88 20
Fax: (+49 451) 882 20 80
e-mail: info@draeger.com
Web: www.draeger.com
■ Development, production and product support of breathing equipment, gas protective equipment, compressed air systems, gas and particle filter systems, gas detecting and monitoring systems and medical rescue equipment. Aviation oxygen systems, portable oxygen units, air supply systems, liquid oxygen system and solid oxygen unit.

Draeger Safety UK Limited
Draeger Ltd
a subsidiary of Draegerwerk AG, Germany
Ullswater Close, Kitty Brewster Industrial Estate, Blyth, Northumberland, NE24 4RG, United Kingdom
Ullswater Close, Blyth Riverside Business Parl, Blyth, Northumberland, NE24 4RG, United Kingdom
Tel: (+44 1670) 35 28 91
Fax: (+44 1670) 35 62 66
e-mail: marketing.uk@draeger.com
Web: www.draeger.com
■ Manufacture and supply of breathing apparatus and gas detection systems for professional firefighters and industry worldwide.

Dress Making Factory
Ulitsa Kolkhoznaya, 2a, Kulebaki, 607010 Nizhnyi Novgorod, oblast, Russian Federation
Tel: (+7 8317) 65 07 22
(+7 8317) 65 05 83
Fax: (+7 8317) 65 58 47
■ Production and sale of special and working clothes for chemical, oil, gas, wood, mining, metallurgy, food, medical, the Ministry of Defence and Ministry of Internal Affairs industries.

DSIR Division of Physical Sciences
Information Technology Group
PO Box 31311, Lower Hutt, New Zealand
Tel: (+64 4) 66 69 19
Fax: (+64 4) 69 00 67

Dspace Inc
28700 Cabot Drive, Suite 1100, Novi, Michigan, 48377, United States
Tel: (+1 248) 567 13 00
Fax: (+1 248) 567 01 30
Web: www.dspaceinc.com

Dunmore Europe GmbH
Hausener Weg 1, D-79111 Freiburg, Germany
Tel: (+49 761) 490 46 21
Fax: (+49 761) 490 46 79
Web: www.dunmore.de
■ Custom manufacturing of coated, laminated and metallised films for spacecraft and MLI multi layer insulation/ thermal protection.

Dutch Space BV (DS)
Fokker Space BV
PO Box 32070, NL-2303 DB Leiden, Netherlands
Newtonweg 1, 2333 CP, NL-2333 CP Leiden, Netherlands
Tel: (+31 71) 524 50 00
Fax: (+31 71) 524 59 99
e-mail: info@dutchspace.nl
Web: www.dutchspace.nl

Dynamic Instruments Inc (DI)
3860 Calle Fortunada, San Diego, California, 92123, United States
Tel: (+1 858) 278 49 00
Freephone: (+1 800) 793 33 58
Fax: (+1 858) 278 97 00
e-mail: pwhitte@dynamicinst.com

E

E2V Technologies
EEV Ltd
English Electric Valve Co Ltd
Marconi Applied Technologies
a subsidiary of Marconi plc, UK
106 Waterhouse Lane, Chelmsford, Essex, CM1 2QU, United Kingdom
Tel: (+44 1245) 49 34 93
Fax: (+44 1245) 49 24 92
e-mail: enquiries@ev2.com
Web: www.ev2.com
■ Electron tubes for radar, communications and EW, IR devices, night vision, TV camera tubes, image intensifiers, display devices including LCDs, hydrogen thyratrons, microwave tubes, travelling wave tubes, frequency agile magnetrons, receiver protection. CCD sensors and cameras for industrial, scientific, security purposes. Slapper detonators for electronic safety and arming circuitry. UV detectors. Caesium arc lamps.

EADS Astrium
Astrium Satellites – Friedrichshafen
Claude-Dornier-Straße, D-88090, Immenstaad, Germany
Tel: (+49) 754 58 91 23
Web: www.astrium.eads.net

EADS Astrium
Astrium Services, Satellites, Space transportation
Dornier Satellitensysteme GmbH
Robert-Koch Strasse 1, D-82024 Taufkirchen, Ottobrunn, Germany
Tel: (+49 89) 60 70
Web: cs.space.eads.net/sp
■ Production and development of Earth observation and science satellites, microwave instruments, thermal components and power electronics.

EADS Astrium
Astrium Space Transporataion & Satellites
EADS Construcciones Aeronáuticas SA
part of European Aeronautics Defence and Space Comapny NV (EADS), Netherlands
Avenida de Aragon 404, E-28022 Madrid, Spain
Tel: (+34 915) 86 37 41
e-mail: communications@casa.eads.net
sales@casa.eads.net
Web: www.astrium.eads.net
■ Design, produce, market and provide product support for its own products (C-212, C-101, CN-235 and C-295), as well as a number of international joint operations. The main international joint operations in which CASA participates are European: Airbus, Eurofighter, Airbus Military and Arianespace. In each case, a joint company has been established to direct overall programme management. The company also participates in other aircraft manufacturing programmes and is responsible for the development and standard production of components, including the MD-11 horizontal stabilizer, APU doors and other elements, in addition to the SAAB 2000 wing. The company is a qualified supplier of aerodynamic surfaces of the Boeing 737, 757 and 777 aircraft and of various components of the McAir's F-18 fighters. Design, produce, inspect and repair various aeroestructures made in CFC.

EADS Astrium
German Headquarters
D-81663 München, Germany
Tel: (+49 89) 60 70
Fax: (+49 89) 60 72 64 81
Web: www.astrium.eads.net
■ Production and development of LEO communications satellite platforms, navigation systems, avionics and electronics; solar arrays and antennas and provision of telecommunications operations and services.

EADS Astrium
Portsmouth Division
Anchorage Road, Portsmouth, Hampshire, PO3 5PU, United Kingdom
Tel: (+44 23) 92 70 57 05
Fax: (+44 23) 92 70 57 06
e-mail: jeremy.close@eads.net

EADS Astrium
Stevenage Office
Astrium Ltd
Matra Marconi Space UK Ltd
Gunnels Wood Road, Stevenage, Hertfordshire, SG1 2AS, United Kingdom
Tel: (+44 1438) 31 34 56
Fax: (+44 1438) 77 36 37
Web: www.astrium.eads.net
Web: www.space.eads.net
■ Production of end to end space systems, space craft construction, launch and control, satellite platforms, communications and surveillance paylods, land, sea and air communications terminals, satellite and network control stations, surveillance systems and services.

EADS Astrium SAS
Astrium Satellites
31 rue des Cosmonautes, Z I du Palays, F-31402 Toulouse, Cedex 4, France
Tel: (+33 5) 62 19 62 19
Fax: (+33 5) 62 54 57 10
■ Manufacture of satellite subsystems for telecommunication, navigation, earth observation and sciences. Special interest in satellite level prime contracting, assembly integration and tests for telecommunications, sciences and avionics, opticals instruments, onboard software, ground systems and GMES applications at subsystem level.

EADS CASA
Espacio
Avenida de Aragón 404, E-28022 Madrid, Spain
Tel: (+34 91) 586 37 00
Fax: (+34 91) 747 47 99
Telex: 48540 casa e
e-mail: espacio@casa-de.es
■ Design, development, production and testing of space frames for satellites and launchers.

EaglePicher Technologies LLC (EPT)
Eagle-Picher Industries
a division of Eagle-Picher Industries Inc, US
C and Porter Street, PO Box 47, Joplin, Missouri, 64802-0047, United States
Tel: (+1 417) 623 80 00
Fax: (+1 417) 781 19 10
e-mail: inquiry.power@eaglepicher.com
Web: www.eaglepicher.com
■ Special purpose aerospace batteries, solar cell components, metal fabrication/machining, electro-explosive devices, electronics and test equipment.

Earth Data Analysis Center
University of New Mexico
Bandelier West, Room 111, Albuquerque, New Mexico, 87131-6031, United States
Tel: (+1 505) 277 36 22
Fax: (+1 505) 277 36 14
e-mail: edac@spock.unm.edu
Web: rgis.unm.edu

Earth Remote Sensing Data Analysis Center (ERSDAC)
Earth Resources Satellite Data Analysis Center
Forefront Tower, 3-12-1 Kachidoki, Chuo-Ku, Tokyo, 104-0054, Japan
Tel: (+81 3) 35 33 93 10
Fax: (+81 3) 35 33 93 83
e-mail: ersdesk@ersdac.or.jp
Web: www.ersdac.or.jp
www.gds.aster.ersdac.or.jp
■ Research and development of image processing and analysis techniques of satellite remote sensing to non-renewable resources, development, environment protection and global monitoring. Dissemination of remote sensing technology. Research co-operation and information exchange with foreign organisations on remote sensing technology.

Earth Resource Mapping
Blenheim House, Crabtree Office Village, Eversley Way, Egham, Surrey, TW20 8RY, United Kingdom
Tel: (+44 1784) 43 06 91
Fax: (+44 1784) 43 06 92
Web: www.ermapper.com
www.earthetc.com
■ Image processing and internet distribution software development.

Earth Resource Mapping Pty Ltd
4370 La Jolla Village Drive, Suite 900, San Diego, California, 92122-1253, United States
Tel: (+1 619) 558 47 09
Fax: (+1 619) 558 26 57
Web: www.ermapper.com
www.earthetc.com

Earth Resource Mapping Pty Ltd
Level 2, 87 Colin Street, Perth, Western Australia, 6005, Australia
Tel: (+61 8) 93 88 29 00
Fax: (+61 9) 93 88 29 01
e-mail: ianc@erm.oz.au
Web: www.ermapper.com
Web: www.earthetc.com
■ Manufacture and distribution of image processing and integrated mapping software.

Earth Resources Observation Systems (EROS)
US Geological Survey
EROS Data Center
Mundt Federal Building, Sioux Falls, South Dakota, 57198, United States
Tel: (+1 605) 594 65 11
Fax: (+1 605) 594 61 54
■ Reproduces and distributes digital and photographic data products of Earth acquired by aircraft and spacecraft.
■ Development of geographic information systems and digital spatial databases.
■ Respository of earth observation data. More than 10,000,000 images.

Earth Satellite Corporation
6011 Executive Boulevard, Suite 400, Rockville, Maryland, 20852, United States
Tel: (+1 301) 231 06 60
Fax: (+1 301) 231 50 20
Telex: 1248618 esco ut
Web: www.earthsat.com
www.geocover.com
■ Consulting and professional services firm specialising in the application and development of remote sensing and GIS for the exploration, development monitoring and management of the earth's resources. Image processing, interpretation services, environmental, geologic and GIS consulting. Distributor of Spot, Landsat, ERSI, AVARR, GOES and other satellite data.

Eaton Aerospace
Eaton Aerospace Controls
a division of Eaton Corp, US
3 Park Plaza, Suite 1200, Irvine, California, 92614, United States
Tel: (+1 601) 987 52 12
Fax: (+1 949) 253 21 11
Web: www.aerospace.eaton.com
■ Manufacture and design of fluid control valves, switches, relays, control panels, actuators, regulators, hydraulic fluid power, systems monitoring, used in aircraft, space, missile and marine applications.

Ecliptic Enterprises Corp
398 West Washington Boulevard, #100, Pasadena, California, 91103-2000, United States
Tel: (+1 626) 798 24 36
Fax: (+1 626) 798 22 51
e-mail: info@eclipticenterprises.com
Web: www.eclipticenterprises.com
■ Produce data-transport systems and onboard imaging systems for use with rockets, spacecraft and other remote platforms.

École Nationale SupÉrieure de l'AÉronautique et de l'Espace
Supaero
10 Avenue Édouard Belin, PO Box 4032, F-31055 Toulouse, Cedex 4, France
Tel: (+33 5) 62 17 80 80
Fax: (+33 5) 62 17 83 30
Web: www.supaero.fr
■ Higher education establishment open to students of all nationalities in the fields of aeronautics and space.

École Nationale SupÉrieure de MÉcanique et d'Aerotechnique (ENSMA)
Site du Futuroscope, BP 109, F-86960 Futuroscope, France
Tel: (+33 5) 49 49 80 80
Fax: (+33 5) 49 49 80 00
e-mail: armanet@ensma.fr
Web: www.ensma.fr

- Teaching and research in the mechanical, metallurgical, thermal, aeronautical, space and nuclear industries.

Editions Cépaduès
111 rue Vauquelin, F-31100 Toulouse, France
Tel: (+33 5) 61 40 57 36
Fax: (+33 5) 61 41 79 89
e-mail: cepadues@cepadues.com
Web: www.cepadues.com
- Publisher specialising in aviation and space (SFACT, CNES, SUP'AERO, INP).

Eidetic Digital Imaging Ltd
1210 Marin Park Drive, Brentwood Bay, British Columbia, V8M 1G7, Canada
Tel: (+1 250) 652 93 26
Fax: (+1 250) 652 52 69
e-mail: eidetic@eidetic.bc.ca
- Development of digital image processing systems for the analysis of remotely sensed images of the Earth's surface.

Eidsvoll Electronics AS (Eidel)
Nedre Vilberg vei 8, N-2080 Eidsvoll, Norway
Tel: (+47) 63 95 97 21
(+47) 63 95 97 00
Fax: (+47) 63 95 97 10
e-mail: in@eidel.no
eidel@eidel.no
Web: www.eidel.no
- Analogue and digital designs with telemetry applications for aerospace and defence. Military control systems.

EKA
Offices 503, 504, 2B Tchouïkova Street, 420094 Kazan, Tatarstan, Russian Federation
Tel: (+7 8432) 19 55 74
Fax: (+7 8432) 19 55 75
e-mail: ekalira@mi.ru

Elbit Systems Ltd
Elbit Ltd
owned by the Federman's Group
Advanced Technology Center, PO Box 539, IL-31053 Haifa, Israel
Tel: (+972 4) 831 53 15
Fax: (+972 4) 855 00 02
(+972 4) 855 16 23
e-mail: marcom@elbit.co.il
elbit-systems@elbit.co.il
Web: www.elbitsystems.com
- Elbit Systems Ltd is an international defence electronics company engaged in defence-related programs throughout the world. The Company, which includes Elbit Systems and its subsidiaries, operates in the areas of aerospace, land and naval systems, command, control, communications, computers, intelligence surveillance and reconnaissance (C4ISR), unmanned air vehicle (UAV) systems, advanced electro-optics, electro-optic space systems, EW suites, airborne warning systems, ELINT systems, data links and military communications systems and radios. The company also focuses on the upgrade of existing military platforms and development of new technologies for defense, homeland security and commercial aviation applications.

Subsidiaries and shareholdings:
Cyclone Aviation Products Ltd, Israel
EFW Inc, US
Elbit Systems Electro-Optics Elop Ltd, Israel
Elisra, Israel
Elsec Homeland Defense Limited, Israel
Kinetics Ltd, Israel
Kollsman Inc, US
Semi-Conductor Devices, Israel
Silver Arrow LP, Israel
Tadiran Communications, Israel

Electro Optic Systems Pty Ltd (EOS)
111 Canberra Avenue, Griffith, Australian Capital Territory, 2603, Australia
Tel: (+61 2) 62 22 79 00
(+61 2) 62 99 24 70
Fax: (+61 2) 62 99 76 87
e-mail: fcs@firecontrolsystems.com.au
Web: www.firecontrolsystems.com.au
- Aerospace lasers and electronics. Turnkey systems or sub-systems in multi-disciplinary project areas. Spaceborne or airborne systems. Military rangefinders.

Accreditations: ISO 9001:2000

Electron Tubes Ltd
Thorn EMI Electron Tubes
a subsidiary of Electron Technologies Ltd
Bury Street, Ruislip, Middlesex, HA4 7TA, United Kingdom
Tel: (+44 1895) 63 07 73
Fax: (+44 1895) 63 17 74
e-mail: info@electron-tubes.co.uk
Web: www.electrontubes.com
- High technology, low light level detectors. Photomultipliers and related accessories. Design, develop and manufacture of light and x-ray detector packages to customer specifications.

Accreditations: BS EN ISO 9001

Electronics Corporation of India Ltd (ECIL)
Corporate Business Development
ECIL Post Office, Hyderabad, Andhra Pradesh, 500 062, India
Tel: (+91 40) 27 12 01 31
e-mail: ecilweb@ecil.co.in
cbdg@ecil.co.in
Web: www.ecil.co.in
- Air traffic management systems, battlefield C3I systems, radar data processing and multi-sensor tracking; switching, transmission and access products, V-Sat networks, radio communication and EW products.

Electronics Corporation of India Ltd (ECIL)
East Zone
4th Floor, Apeejay House, 15 Park Street, Kolkata (Calcutta), 700 016, India
Tel: (+91 33) 249 55 23
Fax: (+91 33) 249 55 23
e-mail: eczmcal@ecil.sprintrpg.sprint.com

Electronics Corporation of India Ltd (ECIL)
North Zone
B-7, DDA Shopping Centre, Naraina, New Delhi, 110 028, India
Tel: (+91 11) 589 50 41
(+91 11) 589 50 42
Fax: (+91 11) 589 50 48
e-mail: zmnorth@ecil.siril.in

Electronics Corporation of India Ltd (ECIL)
South East Zone
Ground Floor, Panagal Building, 1 Jeenis Road, Saidapet, Chennai (Madras), 600015, India
Tel: (+91 44) 434 90 85
Fax: (+91 44) 434 01 30
e-mail: eczm.mas@sprintrpg.ems.vsnl

Electronics Corporation of India Ltd (ECIL)
South Zone
Leeman's Complex 30/1, Cunningham Road, Bangalore, 560052, India
Tel: (+91 80) 226 70 82
Fax: (+91 80) 225 06 49
e-mail: ec.dzm.bng@sprintrpg.ems.vsnl.net.in

Electronics Corporation of India Ltd (ECIL)
West Zone
1207 Veer Savarkar Marg, Dadar, Prabhavdevi, Mumbai (Bombay), 400028, India
Tel: (+91 22) 422 34 43
Fax: (+91 22) 430 21 05
e-mail: eczm.bom@ecil.sprintrpg.ems.vsnl.in

Elettronica SpA
Via Tiburtina Valeria km 13.700, Settecamini, PO Box n.56, I-00131 Roma, Italy
Tel: (+39 6) 415 41
Fax: (+39 6) 415 46 91
(+39 6) 415 49 24
(+39 6) 415 49 23
e-mail: info@elt.it
Web: www.elt-roma.com
- Manufacture of RWR, ESM, ECM, ELINT and EWOS equipment for airborne, naval and ground applications.

Elisra
Corporate Headquarters
Elisra Electronic Systems Ltd
part owned by Elbit Systems Ltd (70 per cent) and IAI ELTA Electronics Industries Ltd (30 per cent)
48 Mitzva Kadesh Street, IL-51203 Bene Baraq, Israel
Tel: (+972 3) 617 51 11
(+972 3) 617 55 22
Fax: (+972 3) 617 58 50
e-mail: marketing@elisra.com
Web: www.elisra.com
- Provide design, develop, manufacture, integrate and supports advanced system solutions for Information Warfare including the activities of Electronic Warfare, Intelligence, Command and Control and Communications for air, sea and land deployed in over 25 countries. Products include: New Generation Airborne Self-Protection Suites including Passive IR Missile Warning for Fighters, Helicopters, Transport Aircraft and Commercial Aircraft; Homeland Security solutions; COMINT/COMJAM/ ECM/ESM/ELINT; C3IRST Solutions–Battlefield Command Chains, Via IP Technology; Battlefield Management and Control; Theater Missile Defense including Test Bed; Artillery C4I; Advanced Search and Rescue Systems; Data Links for Unmanned Platforms in the Air, Land and Sea, and for Guided Weapons.

Representing:
Elisra Electronic Systems Ltd
Tadiran Electronic Systems Ltd
Tadiran Spectralink Ltd

ELOX
Leninsky Prospekt 32A, 117993 Moskva, Russian Federation
Tel: (+7 095) 938 61 10
Fax: (+7 095) 938 61 12
e-mail: elox@online.ru

Elsevier Science
United States & Canada Regional Sales Office
Customer Support Department, PO Box 945, New York, New York, 10159-0945, United States
Tel: (+1 212) 633 37 30
Fax: (+1 212) 633 36 80
e-mail: usinfo-f@elservier.com
Web: www.elsevier.com
- Publisher of scientific, technical and medical information.

ELTA
CEIS TM
Compagnie pour l'Electronique, l'Informatique et les Systèmes
4 Avenue Didier Daurat, PO Box 48, F-31700 Blagnac, France
Tel: (+33 5) 61 16 32 00
Fax: (+33 5) 61 16 32 01
(+33 5) 61 16 32 02
(+33 5) 61 16 32 31
Telex: 531280
e-mail: commercial@elta.fr
Web: www.elta.fr
- Electronic components, qualified power supplies. AC/DC and DC/DC convertors. Special applications in airborne and land based electronics and data processing. Process control equipment. Electronics engineering. Emergency Locator Transmitter (ELT).

ELTA Systems Ltd (IAI/ELTA)
Elta Electronics Industries Ltd
a subsidiary of Israel Aircraft Industries Ltd, Israel
100 Yitzhak, Hanasi Boulevard, PO Box 330, IL-77102 Ashdod, Israel
Tel: (+972 8) 857 21 55
(+972 8) 857 23 12
Fax: (+972 8) 856 18 72
(+972 8) 856 45 68
e-mail: market@elta.co.il
Web: www.iai.co.il
www.elta-iai.com
- Provision of airborne, naval and ground based military electronics systems including intelligence, surveillance and reconnaissance; early warning and control; homeland security; defence and self protection; target acquisition and fire control; advanced technologies–analog and digital subsystem.

Emerson & Cuming Microwave Products Sarl
a subsidiary of Emerson & Cuming Microwave Products NV, Belgium
9 rue du Colonel Moraine, F-92360 Meudon, France
Tel: (+33 1) 46 01 58 20
Fax: (+33 1) 46 01 58 25
e-mail: contact@emerson-cuming-microwave.fr
Web: www.eccosorb.com
- Microwave absorbers, dielectric and shielding materials.

EMS SATCOM
Headquarters
CAL Corporation
400 Maple Grove Road, Ottawa, Ontario, K2V 1B8, Canada
Tel: (+1 613) 727 17 71
Freephone: (+1 800) 600 97 59
Fax: (+1 613) 727 12 00
e-mail: info@emssatcom.com
Web: www.emssatcom.com
■ Specialise in the design and development of satellite-based terminals and antennas for the aeronautical, land-mobile, maritime, and search and rescue markets. Development of steerable antenna systems to provide live television to commercial aircraft by a wireless link to a direct broadcast satellite.

EMS Technologies Inc
Space & Technology Group, Canada
Satellite and Communications Systems Group
Sovcan Star Satellite Communications Inc
Spar Space Systems
21025 Trans Canada Highway, St Anne-de-Bellevue, Québec, H9X 3R2, Canada
Tel: (+1 514) 457 21 50
Fax: (+1 514) 457 27 24
Telex: 05822792
e-mail: marketing@ems-t.ca
Web: www.ems-t.com
■ Design and manufacture of products and systems for space-based communications, remote sensing and manned-space missions. Technological achievements in satellite equipment include: the Alouette payload in 1961, various payload products and systems for the ANIK series of satellites, the MSAT mobile communications satellite, the Radarsat 1 satellite and the radar antenna and antenna control electronics for the Radarsat 2 programme. Also provides products and systems from the following product lines: antennas, microwave and power products, digital command and control products, optical products and broadband products.

EMS Technologies Inc – Corporate Headquarters
Defense and Space Systems Division
660 Engineering Drive, Technology Park, Atlanta, PO Box 770, Norcross, Georgia, 30092, United States
Tel: (+1 770) 263 92 00
Fax: (+1 770) 446 57 39
e-mail: marketing@ems-t.com
Web: www.emsdss.com
■ Design and development of satellite-based terminals and antennas for the aeronautical, land mobile and search and rescue markets. Developed a steerable antenna system to provide live television to commercial aircraft by a wireless link to a direct broadcast satellite.

Energia Deutschland GmbH
Wolfratshauser Strasse 48, D-81379 München, Germany
Tel: (+49 89) 72 49 50
Fax: (+49 89) 72 49 52 91
e-mail: info@kayser-threde.com
Web: www.kayser-threde.com
■ Marketing and distribution of Earth observation data.

Energia Ltd
631 South Washington Street, Alexandria, Virginia, 22314, United States
Tel: (+1 703) 836 19 99
Fax: (+1 703) 836 19 95
e-mail: energia@energialtd.com
Web: www.energialtd.com
■ American marketing organization for SC Energia, Russia's oldest and largest space Corporation. SC Energia is the Russian prime contractor for space station Alpha.

Enertec SA
Schlumberger Industries DAR Division
Z A de Courtaboeuf 2, 3 avenue du Canada, Immeuble Parnasse, LP 880, BP 316, F-91966 Courtaboeuf, Cedex, France
Tel: (+33 1) 64 86 34 00
Fax: (+33 1) 64 86 34 12
e-mail: sales_enertec@zodiac.com
Web: www.enertecgroup

■ Design and manufacture of high-end data and video acquisition, storage and processing systems for aerospace, defence, security and related markets.

Engineered Arresting Systems Corporation (ESCO)
Military Products
Engineered Systems Co
2550 Market Street, Aston, Pennsylvania, 19014, United States
Tel: (+1 610) 494 80 00
Fax: (+1 610) 494 89 89
Web: www.esco.zodiac.com
www.zodiac.com
■ Manufacture of aircraft arresting systems and special material handling systems.

Engineering Consultants for Environmental Analysis & Remote Sensing BV (EARS)
EARS BV
PO Box 449, NL-2600 AK Delft, Netherlands
Kanaalweg 1, NL-2628 EB Delft, Netherlands
Tel: (+31 15) 256 24 04
Fax: (+31 15) 262 38 57
e-mail: ears@ears.nl
Web: www.ears.nl
■ Remote sensing, terrestrial thermography, NDT. Feasibility studies of satellite remote sensing techniques.

Entran Devices Inc
10 Washington Avenue, Fairfield, New Jersey, 07004, United States
Tel: (+1 973) 227 10 02
Freephone: (+1 800) 635 06 50
(+1 888) 8 ENTRAN
Fax: (+1 973) 227 68 65
e-mail: sales@entran.com
Web: www.entran.com
■ Semiconductor accelerometers, pressure transducers, pressure transmitters, load cells, strain gauges, instrumentation modules and digital pressure meters.

Environmental & Remote Sensing Services Center
5 Al Ahram Street, PO Box 1041, Cairo, Egypt
Tel: (+20 2) 291 83 30
(+20 2) 419 07 18
Fax: (+20 2) 290 69 16
e-mail: etss@titsec1.com.eg
Web: www.etss.com
■ Remote sensing, image processing, geographic information systems, mapping and environmental studies.

Environmental Analysis and Remote Sensing BV (EARS)
PO Box 449, NL-2600 AK Delft, Netherlands
Kanaalweg 1, NL-2628 EB Delft, Netherlands
Tel: (+31 15) 56 24 04
Fax: (+31 15) 62 38 57
Telex: 26401
e-mail: ears@ears.nl

EOSAT
4300 Forbes Boulevard, Lanham, Maryland, 20706-9954, United States
Freephone: (+1 800) 344 99 33

Epsori Space Systems
PO Box 711685, San Diego, California, 92171, United States
Tel: (+1 858) 492 80 82
Web: www.epsori.com

ERA Technology Ltd
is a part of Cobham Plc, UK
Cleeve Road, Leatherhead, Surrey, KT22 7SA, United Kingdom
Tel: (+44 1372) 36 70 00
Fax: (+44 1372) 36 70 99
e-mail: info@era.co.uk
Web: www.era.co.uk
■ Design, development, testing, assessment and consultancy for communications, aerospace, defence, information technology, manufacturing, transport, electronics and energy.

ERG Materials and Aerospace Corp
ERG Inc
900 Stanford Avenue, Oakland, California, 94608, United States
Tel: (+1 510) 658 97 85
Fax: (+1 510) 658 74 28
e-mail: sales@ergaerospace.com
Web: www.ergaerospace.com
■ Manufacture of Duocel, an open celled, reticulated, rigid foam material of aluminium, carbon, ceramics and composites. Design and manufacture of high performance components such as heat exchangers, lightweight structures, energy absorbers, baffles, diffusers, optics and filters for the aerospace, semiconductor, energy, nuclear, military, optical, defence, biomedical and other high-tech industries.

Ericsson Anslutningssystem AB (EZS)
Svedjevägen 12, SE-93136 Skellefteå, Sweden
Tel: (+46 910) 845 00
Fax: (+46 910) 846 00

Ericsson Treasury Services AB (TSS)
Telephonvägen 30, SE-12625 Stockholm, Sweden
Tel: (+46 8) 719 00 00
Fax: (+46 8) 681 22 90

ESA/ESRIN
Information Retrieval Service
Via Galileo Galilei, PO Box 64, I-00044 Frascati, Roma, Italy
Tel: (+39 06) 94 18 01
Fax: (+39 06) 94 18 02 80
e-mail: esaweb@esrin.esa.it
Web: www.esa.int
■ Computerised scientific and technical information.
Publications: News & Views

ESG Elektroniksystem und Logistik GmbH
Livry-Gargan-Strasse 6, D-82256, Fürstenfeldbruck, Germany
Tel: (+49 89) 92 16 27 45
(+49 89) 921 60
(+49 89) 92 16 26 16
(+49 89) 92 16 29 14
Fax: (+49 89) 92 16 29 14
(+49 89) 92 16 26 31
(+49 89) 92 16 26 32
(+49 89) 92 16 27 36
Telex: 522594
e-mail: info@esg.de
Web: www.esg.de
■ Development of complex electronic and information systems, command and control systems. Cost-reducing logistics procedures.

ESO
Casilla, 19001 Santiago, 19, Chile
Tel: (+56 2) 698 87 57
Fax: (+56 2) 695 42 63

ESO Astrophysics
Karl Schwarzschildstrasse 2, D-85748 Garching, Munchen, Germany
Tel: (+49 89) 32 00 62 76
Fax: (+49 89) 320 23 62

Essex Corporation
9150 Guildford Road, Columbia, Maryland, 21046-2306, United States
Tel: (+1 301) 939 70 00
Fax: (+1 301) 953 78 80
Web: www.essexcorp.com
■ Optical and communications engineering.

Essex Industries Inc
Corporate Headquarters
7700 Gravois Avenue, St Louis, Missouri, 63123, United States
Tel: (+1 314) 832 45 00
Fax: (+1 314) 832 16 33
e-mail: corp@essexind.com
Web: www.essexind.com
■ Design, development and manufacture of life-saving, high technology components, systems and custom-design applications for the military, aerospace, medical, automotive and general industry fields.

Subsidiary Companies:
Alar Parmeko Inc
Alar Products Inc
Dolphin Systems Corp
Essex Cryogenics of Missouri
Essex Fluid Controls
Essex Manufacturing Division
Essex Medical Products
Essex Portable Breathing & Rescue Products
Essex Precision Controls
Essex Screw Products
Propellex Corp

Esys plc
Unit 3, Stirling House, Guildford, Surrey, GU2 7RF, United Kingdom
Unit 3, Stirling House, Surrey Research Park, Guildford, Surrey, GU2 7RF, United Kingdom
Tel: (+44 1483) 30 45 45
Fax: (+44 1483) 30 38 78
e-mail: info@esys.co.uk
■ Divided between four consulting practices serving space, telecoms, defence and education, and research. Each of these practices offers a portfolio of consulting services, namely strategy and policy, business assessment, project management, technology evaluations and system engineering.

E-Systems Inc
Headquarters
11 Penn Plaza, 5th Floor, New York, New York, 10001-2606, United States
Tel: (+1 212) 946 26 79
Fax: (+1 800) 613 74 56
e-mail: info@esystemsinc.com
Web: www.esystemsinc.com

ETS – Lindgren LP
1301 Arrow Point Drive, Cedar Park, Texas, 78613, United States
■ Specialists in RFI/EMC/NEMP and Tempest protection systems offering shielded enclosures, anechoic chambers, filters, fibre optics and suppression equipment. system design, manufacture, installation and testing are available on a worldwide basis.

ETS-Lindgren Ray Proof Ltd
Lindgren-Rayproof
Ray Proof Ltd
Unit 4, Eastman Way, Pin Green Industrial Area, Stevenage, Hertfordshire, SG1 4UH, United Kingdom
Tel: (+44 1438) 73 07 00
Fax: (+44 1438) 73 07 50
e-mail: info@ets-lingren.co.uk
Web: www.ets-lindgren.com
■ RFI/EMC/NEMP and Tempest protection systems offering shielded enclosures, anechoic chambers, filters, fibre optics and suppression equipment. System design, manufacture, installation and testing are available on a worldwide basis.

Eumetstat GmbH
Am Kavalleriesand 31, D-64295 Darmstadt, Germany
Tel: (+49 6151) 80 77
Fax: (+49 6151) 80 75 55
Web: www.eumetsat.de

Eurimage SpA
Via E D'Onofrio 212, I-00155 Roma, Italy
Tel: (+39 6) 40 69 41
Fax: (+39 6) 40 69 42 32
e-mail: info@eurimage.com
Web: www.eurimage.com
■ Provides global satellite data and imagery.
Shareholders:
Telespazio, Italy
Astrium, Germany/France/UK

Euro Space Center Belgium
1 rue Devant les Hêtres, B-6890 Transinne, Belgium
Tel: (+32 61) 65 64 65
Fax: (+32 61) 65 64 61
e-mail: escinfo@skynet.be
■ Educational facility consisting of a permanent space exhibit and a space camp for youth, available for seminars and special events about new technologies.

Eurockot Launch Services GmbH
Airport Centre, Flughafenallee 26, D-28199 Bremen, Germany
Tel: (+49 421) 539 65 19
(+49 421) 539 65 01
Fax: (+49 421) 539 65 00
e-mail: eurockot@space.eads.net
Web: www.eurockot.com
■ Commercial launch provider for Leo Satellite Industry.

Eurockot Launch Services GmbH
Rockot
Hünefeldstraße 1-5, Postfach 10 59 46 Bremen, Germany
Tel: (+49 421) 539 65 01
(+49 421) 539 65 10
Fax: (+49 421) 539 65 00
■ Launch service provider.

Euroconsult
World Space Industry Survey
71 Boulevard Richard Lenoir, F-75011 Paris, France
Tel: (+33 1) 49 23 75 30
Fax: (+33 1) 43 38 12 40
e-mail: marketing@euroconsult-ec.com
Web: www.euroconsult-ec.com
■ Strategic consulting in the satellite industry and satellite service providers. Multi-client studies and publications on the space industry.
Fields of expertise: telecommunications, television, multimedia, navigation, launch services, ground equipment and services.
Services provided: corporate strategy, business plans, risk assessment, valuations of satellite systems; demand modelling, market forecasting and systems definition.
Publications:
Government Space Markets, Worldwide Survey (the age of co-operation)
World Satellite Communications and Broadcasting Markets Survey, Prospects to 2009
Launch Services Market Survey, Worldwide Prospects 2009
Space Business in Europe 1999 edition
Satellite Communications Ground Stations Market Survey Worldwide Prospects 1997–2007
Asia-Pacific Space Programs and Industry Prospects to 2005
Asia-Pacific Satellite Communications and Broadcasting Market, Opportunities, Prospects to 2005

European Aeronautic Defence and Space Company NV (EADS NV)
Le Carré, Beechavenue 130–132, NL-1119 PR Schiphol, Rijk, Netherlands
Tel: (+31 20) 655 48 00
Fax: (+31 20) 655 48 01
e-mail: press@eads.com
Web: www.eads.net
■ Manufacture of military and civil airborne, spaceborne and land-based platforms, systems, equipment and components.
Operations and activities of the five business units:-
Airbus: manufacture of Airbus A300-600F, A300-600ST, A310, A318, A319, A319CJ, A320, A321, A330, A340, A380 and A380F.
Military Transport Aircraft Division: manufacture of C-212, C-295, CN-235 and A400M. Manufacture and supply of special mission aircraft for duties such as maritime surveillance and anti-submarine warfare.
Aeronautics Division: manufacture of military aircraft (Eurofighter Typhoon, Mako, C-101 Aviojet), helicopters (Eurocopter NH 90, Tiger, EC 120, EC 135, EC 155), regional aircraft (ATR 42-500, ATR 72-500), light aircraft (Socata TBGT, TBM 700, Epsilon) and aircraft conversion and maintenance (Sogerma, Elbe Flugzeugwerke GmbH). Technical and logistic support for F/A-18 Hornet, MiG-29, Tornado, F-4 Phantom, Mirage F-1, C-130 Hercules, Transall C-160, P-3 Orion and E-3A AWACS.
EADS Astrium Division: design, development and production of satellites (broadcast satellites for Intelsat, Eutelsat and Inmarsat), earth observation and scientific satellite systems for civil and military applications (Envisat, Metop, Spot 5, Helios, XMM), the European navigation systems (Galileo), orbital infrastructure (space laboratory Columbus and the ATV for the international Space Station) and launchers (Arianespace heavy-lift launchers, Starsem medium-lift launchers, Eurockot small-lift launchers). Optronics and laser technologies through the subsidiaries of Sodern and Cilas.
■ Defence and Civil Systems Division: manufacture of missiles and missile systems, defence electronics, telecommunications and services. The Missiles and Missile Systems sector comprises MBDA. Products include: Meteor, Aster, Exocet, Kormoran, Roland, Milan, HOT, Mistral, ASRAAM, Mica, Seawolf, RAM, Patriot, Stinger, Taurus, Trigat MR and Trigat LR, Polyphem and Scalp EG/Storm Shadow. The Defence Electronics sector produces C3I systems, reconnaissance and surveillance systems, airborne multi-mode radars and electronic warfare units. The Civil Communications sector supplies internet protocol networks, high-speed and long-distance networks, switching products, local loop equipment and fibre-optics and cellular telecommunications networks.
Corporate companies:
Airbus
ATR
EADS Astrium
EADS CASA
EADS EFW Elbe Flugzeugwerke
EADS Socata
Eurocopter
Sogerma
Shareholder structure:
DaimlerChrysler: 22.47%
SOGEADE (Lagardère, together with French financial institutions and the French state holding company Sogepa): 29.95%
SEPI (Sociedad Estatal de Participaciones Industriales), Spain: 5.5%
Free Float: 42.10% (including EADS employees and about 3 percent held directly by DaimlerChrysler and the French state)

European Association of Remote Sensing Laboratories (EARSEL)
Association Européenne de Laboratoires de Télédétection
2 Avenue Rapp, F-75340 Paris, Cedex 7, France
Tel: (+33 1) 45 56 73 60
Fax: (+33 1) 45 56 73 61
e-mail: earsel@meteo.fr
■ Co-ordination of European research in earth observations, from aircraft and satellites through special interest groups. Remote sensing from satellites and aircraft. Aerial photography.
Publications:
EARSeL Newsletter (quarterly on the web)
Proceedings GA (annual)
Proceedings Symposium (annual)
Workshop Reports

European Organisation for the Exploitation of Meteorological Satellites (EUMETSAT)
European Organisation Exploitation Meteo
Am Kavalleriesand 31, D-64295 Darmstadt, Germany
Tel: (+49 6151) 80 73 45
(+49 6151) 80 77
Fax: (+49 6151) 80 75 55
e-mail: press@eumetsat.int
Web: www.eumetsat.de
www.eumetsat.int
■ European intergovernmental meteorological satellite organisation, representing the national interests of 18 member states and 6 co-operating states. Primary objective is to establish, maintain and exploit European systems of operational meteorological satellites and to contribute to the operational monitoring of the climate, the detection of global climatic changes and associating in environmental monitoring.

European Space Agency (ESA)
Brussels Office
Avenue de Cortenbergh 52, B-1000 Bruxelles, Belgium
Tel: (+32 2) 743 30 70
Fax: (+32 2) 743 30 71

European Space Agency (ESA)
ESRIN Earthnet Program Office
Via Galileo Galilei, PO Box 64, I-00044 Roma, Italy
Tel: (+39 06) 940 11
(+39 06) 94 18 01
(+39 06) 94 18 03 50
(+39 06) 94 18 09 51
Fax: (+39 06) 940 13 61
(+39 06) 94 18 02 80
■ Remote sensing satellite data acquisition preprocessing and distribution.
Publications: Earthnet Review

European Space Agency (ESA)
European Astronauts Centre
Linder Höhe, D-51147 Köln, Germany
Tel: (+49 2203) 600 10
Fax: (+49 2203) 60 01 66
Web: www.estec.esa.int/spaceflight/astronaut
■ ESA centre for astronaut selection and training.

European Space Agency (ESA)
European Space Research Institute
Via Galileo Galilei, PO Box 64, I-00044 Frascati, Roma, Italy
Tel: (+39 06) 94 18 01
(+39 06) 94 18 03 50
(+39 06) 94 18 09 51
Fax: (+39 06) 94 18 02 80
Web: www.esa.int

European Space Agency (ESA)
European Space Research & Technology Centre
PO Box 299, NL-2200 AG Noordwijk, Netherlands
Keplerlaan 1, NL-2201 AZ Noordwijk, Netherlands
Tel: (+31 7) 15 65 65 65
Fax: (+31 7) 15 65 60 40
Web: www.estec.esa.nl
■ Technical centre responsible for the design and execution of scientific and application satellite programmes plus the human space flight programme. Also holds a large test centre for space craft.

European Space Agency (ESA)
Headquarters
Agence Spatiale Européenne
Europaische Weltraumorganisation
8-10 rue Mario Nikis, F-75738 Paris, Cedex 15, France
Tel: (+33 1) 53 69 76 54
(+33 1) 53 69 71 55
Fax: (+33 1) 53 69 75 60
(+33 1) 53 69 75 61
(+33 1) 53 69 75 62
(+33 1) 53 69 76 90
Web: www.esa.int
www.esoc.esa.de
■ Space research and application. Provides and promotes, for exclusively peaceful purposes, co-operation among European states in space research and technology. Also promotes their space applications for scientific purposes and for operational space applications systems.
Member states:
Austria
Belgium
Denmark
Finland
France
Germany
Greece
Ireland
Italy
Luxemburg
Netherlands
Norway
Portugal
Spain
Sweden
Switzerland
UK

European Space Agency (ESA)
Kourou Office
F-97388 Kourou, French Guiana
Tel: (+594 33) 71 59
Fax: (+594 33) 47 54
■ Space activities.

European Space Agency (ESA)
Moscow Office
Sretensky Boulevard 6/1–122, 101000 Moskva, Russian Federation
Tel: (+7 095) 928 75 29
Fax: (+7 095) 928 53 52
e-mail: emodesk@esa.int

European Space Agency (ESA)
Washington Office
955 L'Enfant Plaza South West, Suite 7800, Washington, District of Columbia, 20024, United States
Tel: (+1 202) 488 41 58
Fax: (+1 202) 488 49 30
Web: www.esa.int

European Union Satellite Centre
Western European Union
Union de l'Europe Occidentale Satellite Centre Satellitaire
Apartado de Correos N° 511, E-28850 Torrejón de Ardoz, Madrid, Spain
Tel: (+34 91) 678 60 00
Fax: (+34 91) 678 60 06
e-mail: info@weusc.es
Web: www.weu.int

EUTELSAT SA
European Telecommunications Satellite Organisation
Organisation Europeenne de Telecommunications par Satellite
70 rue Balard, F-75502 Paris, Cedex 15, France
Tel: (+33 1) 53 98 47 47
Web: www.eutelsat.com
■ Intergovernmental organisation enabling TV and radio broadcasters, internet service providers, telecommunications companies and corporations to transmit and receive content in Europe, Africa, Asia and the Americas.

Evans Consoles Corporation
Head Office
1616 27th Avenue North East, Calgary, Alberta, T2E 8W4, Canada
Tel: (+1 403) 291 44 44
Fax: (+1 403) 250 65 49
e-mail: info@evansonline.com
Web: www.evansonline.com
■ Design and manufacture of technical furniture that includes consoles, open desks and rear projection display walls for control centres.

Explorocean Technology Ltd (ETL)
Unit B4, Thorpe Industrial Estate, Crabtree Road, Thorpe, Surrey, TW20 8RW, United Kingdom
Tel: (+44 1784) 47 21 30
Fax: (+44 1784) 47 30 32
e-mail: sales@explorocean.com
Web: www.explorocean.com
■ Oceanographic, hydrographic, geophysical and meteorological equipment sale and hire.

F

Fairchild Communications Data Processing
900 Circle 75 Pky NW, Atlanta, Georgia, GA 30339-3035, United States
Tel: (+1 770) 980 60 00

Fairchild Imaging
Lockheed Martin Fairchild Systems
Loral Fairchild Imaging Sensors
1801 McCarthy Boulevard, Milpitas, California, 95035, United States
Tel: (+1 408) 433 25 03
Freephone: (+1 800) 325 69 75
Fax: (+1 408) 435 73 52
e-mail: sales@fcimg.com
■ Design, manufacture and test of high performance, high-end charge coupled devices (CCD)s and cameras for both military and commercial applications.

Fakel Experimental Design Bureau
OKB Fakel
181 Moscovskiy Prospekt, 236001 Kaliningrad, Russian Federation
Tel: (+7 112) 46 19 64
Fax: (+7 112) 46 17 62
Web: users.gazinter.net/fakel/index_eng.html
■ Development and manufacture of electric thrust units and plasma sources for different applications. Attitude and orbit control systems of different spacecraft, full-scale production of stationary plasma thrusters (SPT) and SPT systems.

Farran Technology Ltd (FTL)
Ballincollig, County Cork, Ireland
Tel: (+353 21) 487 28 14
Fax: (+353 21) 487 38 92
e-mail: sales@farran.com
Web: www.farran.com
■ Waveguide mixers, frequency multipliers for commercial and defence applications. MM and SUBMM components (mixers, detectors, frequency sources, diodes.) Low noise and medium power amplifiers. Custom MMIC-based subsystems in frequency range from 10 GHz to 100 GHz for use in radar, wireless communications and EW applications. Amplifiers, transceivers, passive receivers, up-converters, down-converters, multipliers and mixers.

Farsound Engineering Ltd (FEL)
a subsidiary of Farsound Group plc
Unit 3, Highams Park Industrial Estate, Jubilee Avenue, London, E4 9JD, United Kingdom
Tel: (+44 20) 84 98 38 88
Fax: (+44 20) 84 98 38 87
e-mail: sales@farsound.co.uk
info@farsound.co.uk
Web: www.farsound.co.uk
■ Design, manufacture and distribution of a range of components for the aerospace, transport, marine, medical and petrochemical industries, utilising a range of engineering skills including: sheet metal fabrication, presswork, machining, welding and assembly, waterjet and laser cutting, overhaul and kitting solutions and NDT approved inspection and testing facilities.

Federal Space Agency
Russian Aviation and Space Agency
Russian Space Agency
42 Schepkin Street, 107996 Moskva, Russian Federation
Tel: (+7 095) 631 94 44
Fax: (+7 095) 288 90 63
(+7 095) 975 44 67
e-mail: cosmpress@roscosmos.ru
■ Responsible for the implementation of state policy in the field of exploration and use of outer space for peaceful purposes and for the general management of the development of new aeronautics and space technologies and products.

Federal State Unitary Enterprise 'Russian Institute of Space Device Engineering' (RISDE)
53 Aviamotornaya Street, Moskva, Russian Federation
Tel: (+7 095) 509 12 01
(+7 095) 509 12 02
Fax: (+7 095) 509 12 00
Web: www.rniikp.ru
■ Development and production of ground stations for satellite data reception, satellite control and orientation systems; remote sensing and data transmission satellite payloads; and passive microwave units.

Ferriere Cattaneo SA
Via Ferriere 12, Casella Postale, CH-6512 Giubiasco, Ticino, Switzerland
Tel: (+41 91) 850 91 91
Fax: (+41 91) 850 91 92
e-mail: fcsa@ferrierecattaneo.ch
Web: www.ferrierecattaneo.ch
■ Capabilities include: freight wagons, workshop-machining, frameworks, engineering, research and development.

Ferrite Domen Co (FDC)
SPA Ferrite Domain
Ferrite Domain Co
8 Chernigovskaya Street, 196084 St Petersburg, Russian Federation
Tel: (+7 812) 294 71 48
Fax: (+7 812) 298 37 91
e-mail: info@domen.ru
domen@domen.ru
Web: www.domen.ru
■ Microwave ferrite materials and components, microwave and millimetre wave equipment for radar and space communications; soft (poly- and mono-crystalline) and hard ferrites; molybdenum permalloy powder cores (MPP-cores).

Finmeccanica SpA
partly state-owned, the Italian Treasury owns 32.8 per cent of the capital.
Piazza Monte Grappa 4, PO Box 101, I-00195 Roma, Italy
Tel: (+39 06) 320 86 21
(+39 6) 32 47 33 13
(+39 3) 31 81 38 08
Fax: (+39 3) 31 86 96 65
e-mail: ufficiostampa@finmeccanica.com
news@finmeccanica.com
investor_relations@finmeccanica.it
Web: www.finmeccanica.it
www.finmeccanica.com

■ Design and manufacture of helicopters, multi-role defence aircraft, trainer and tactical transport aircraft, defence electronics, energy generation, aerostructures, defence systems, air traffic management and control systems, space infrastructure and satellite services. Also provide operational services of rail transport and systems.

Companies Represented:
Agusta
Alenia Aeronautica SpA
Avio SpA
Elettronica SpA
Finmeccanica
Galileo Avionica SpA
Aeronavali Venezia SpA
Selex Communications SpA
Selex Sistemi Integrati

Finnish Astronautical Society
Sallskapet for Astronautisk Forskning
Suomen Avaruustutkimusseura
PO Box 507, FI-00101 Helsinki, 10, Finland
Tel: (+358 10) 521 58 56
Fax: (+358 10) 521 59 10
Publications:
Avaruusluotain (a quarterly journal, in Finnish) relating to interests in space.

Finnish Geodetic Institute
Geodeettinen Laitos
Ministry of Agriculture and Forestry, Geodeetinrinne 2, PO Box 15, FI-02431 Masala, Finland
Tel: (+358 9) 295 55 50
Fax: (+358 9) 29 55 52 00
Web: www.fgi.fi
■ Research in geodesy, photogrammetry, remote sensing, cartography and space geodesy. Activities include satellite tracking, GPS network operation, development of digital satellite image interpretation methods and GIS.

Finnish Meteorological Institute (FMI)
Geophysical Research Division
15A Vuorikatu, PO Box 503, FI-00101 Helsinki, Finland
Tel: (+358 9) 192 91
Fax: (+358 9) 19 29 46 03
Web: www.fmi.fi
■ Scientific and technical research in space-physics, geomagnetic, climatological and air quality. Relevant computer software and instrumentation.

Finnish Space Committee
c/o TEKES, PO Box 69, FI-00101 Helsinki, Finland
Tel: (+358 10) 521 58 56
Web: www.tekes.fi/space
■ An interdepartmental advisory body responsible for planning the national space strategy.

Firetrench
Integrated Secure Information Consortium Limited
Monks Farm, St James Road, All Saints South Elmham, Halesworth, Suffolk, IP19 0HG, United Kingdom
Tel: (+44 1986) 78 25 47
Fax: (+44 1986) 78 25 25
e-mail: team@firetrench.com
sales@firetrench.com
Web: www.firetrench.com
■ Integrated, electronic data processing and communications systems for standard and secure environments together with associated consultancy and training services and avionics simulation systems.

Firstmark Aerospace Corporation
1176 Telecom Drive, Creedmoor, North Carolina, 27522, United States
Tel: (+1 919) 956 42 00
Freephone: (+1 888) 832 77 22
Fax: (+1 919) 682 37 86
e-mail: info@firstmarkaerospace.com
repair@firstmarkaerospace.com
Web: www.firstmarkaerospace.com
■ Design, manufacture and repair of precision electronic, electromagnetic, and mechanical components and systems for aerospace and defence industries.

Fischer Advanced Composite Components GesmbH (FACC)
a joint venture between of Fischer Ges.mbH and Austrian Salinen AG
Fischerstrasse 9, PO Box 192, A-4910 Ried, Austria
Tel: (+43 7752) 61 60
Fax: (+43 7752) 61 63 50
e-mail: info@facc.co.at
■ Composite components. Aircraft interiors, furnishings, aircraft access doors, fairings, structural aircraft assemblies, galley equipment, landing gear doors.

Fisher Space Pen Co
711 Yucca Street, Boulder City, Nevada, 89005, United States
Tel: (+1 702) 293 30 11
Freephone: (+1 800) 634 34 94
Fax: (+1 702) 293 66 16
e-mail: fisher@spacepen.com
Web: www.spacepen.com
■ Manufacture of pens that write in extreme temperatures and at any angle. Can be used over grease, under water, through dirt, oil and in other harsh conditions. The full line includes a variety of colours and styles. Used by NASA and all manned space flights.

Fleet Industries Ltd
a subsidiary of Magellan Aerospace Corp, Canada
1011 Gilmore Road, PO Box 400, Fort Erie, Ontario, L2A 5N3, Canada
Tel: (+1 416) 871 21 00
Fax: (+1 416) 871 27 22
■ Manufacture of complex aerospace structural assemblies, specialising in flight control components with advanced metal to metal bonding techniques.

Fleximage
Headquarters
113 avenue Aristide Briand, F-94117 Cedex, France
Tel: (+33 1) 49 08 76 00
Fax: (+33 1) 49 08 76 02
e-mail: info@fleximage.fr
Web: www.fleximage.fr

Fleximage
Toulouse Agency
Voie l'Occitane no 6, BP 171, F-31676 Labège, Cedex, France
Tel: (+33 5) 61 00 36 36
Fax: (+33 5) 61 00 35 35

FLIR Systems Ltd
European Operations
AGEMA Infrared Systems Ltd
Broadcast & Surveillance Systems Ltd
Inframetrics Inc
a subsidiary of Flir Systems Inc, US
2 Kings Hill Avenue, West Malling, Kent, ME19 4AQ, United Kingdom
Tel: (+44 1732) 22 00 11
Fax: (+44 1732) 22 00 14
e-mail: marketing@flir.uk.com
Web: www.flir.com
■ Design and manufacture of thermal imaging systems for air, land, sea and security applications for the military, law enforcement and civilian sectors.
Accreditations: MoD

Flometrics
5900 Sea Lion Place, Suite 150, Carlsbad, California, 92010, United States
Tel: (+1 760) 476 27 70
Fax: (+1 760) 476 27 63
Web: www.flometrics.com

Florida Space Research Institute (FSRI)
Building M6-306, Room 9030, Kennedy Space Centre, Florida, 32899, United States
Tel: (+1 321) 452 26 53
Fax: (+1 321) 456 99 61
e-mail: marketing@fsri.org
■ Centre for space research, leveraging academic and economic development resources with those of the industry, NASA and the military to expand and diversify Florida's space-related enterprise.

FLS Aerospace (DK) A/S
Copenhagen Airport South, Hangar 243, DK-2791 Dragør, Denmark
Tel: (+45) 32 45 00 45
Fax: (+45) 32 45 00 12
Web: www.flsaerospace.com
■ Undertakes light maintenance of civil aircraft.

Fluent
10 Cavendish Court, Centerra Resource Park, Lebanon, New Hampshire, 03766-1442, United States
Tel: (+1 603) 643 26 00
Fax: (+1 603) 643 39 67
Web: www.fluent.com

Fluid Regulators Company
a division of Auxitrol SA, France, an Esterline Group Company
313 Gillett Street, Painesville, Ohio, 44077, United States
Tel: (+1 440) 352 61 82
Fax: (+1 440) 354 29 12
e-mail: rdb@fluidreg.com
■ Fluid control and measurement components. Custom designed aerospace hydraulic and pneumatic control valves. Specialists in integrating sensing and fluid controls in modular sub-systems.

Fluor Corporation
Headquarters
One Enterprise Drive, Aliso Viejo, California, 92656-2606, United States
Tel: (+1 949) 349 20 00
Fax: (+1 949) 349 25 85
Web: www.fluordaniel.com
■ Provides worldwide engineering, construction, procurement, project management, operations and maintenance services to defence, aerospace, and government agencies. Services include project management, feasibility and engineering studies, planning, engineering studies, planning engineering, design, reliability, availability and maintainability (RAM) analysis, environmental studies and design, construction and construction management, procurement, telecommunications design, studies, and project management. Operations and maintenance.

Fokker Elmo BV
PO Box 75, NL-4630 AB Hoogerheide, Netherlands
Tel: (+31 164) 61 76 44
Fax: (+31 164) 61 76 37
e-mail: jn11175@elmo.nl
Web: www.elmo.nl

Fortech – Tecphy (Group H.T.M.)
6 rue Condorcet, F-63063 Clermont Ferrand, Cedex 1, France
Tel: (+33 4) 73 28 75 13
Fax: (+33 4) 73 28 75 47
■ Production of parts in specialised steel, aluminium alloy, titanium, super alloy (nickel/cobalt) and in power metallurgy (parts and coatings). Superstructure frames, pylons, thrust reversers, engine nacelles, landing gears and wheels, engine discs, shafts and gear boxes, helicopter rotors and hubs rings and hemispheres for satellites and launchers.

Forum Weltraumforschung RWTH Aachen (RWTH)
Templerrgraben 55, D-52062 Aachen, Germany
Tel: (+49 241) 80 35 13
Fax: (+49 241) 40 44 72
■ Research and technology for the peaceful utilization of space.

Fourteenth Air Force
747 Nebraska Avenue, Vandenberg AFB, California, 93437-6268, United States
Tel: (+1 805) 606 38 88
Fax: (+1 805) 605 88 56
■ Operators of the Air Force Space Command's space forces.

Framatome Connectors France
145 rue Yves le Coz, F-78035 Versailles, France
■ Manufacture of aircraft components.

Francis & Lewis International Ltd (F&L)
Waterwells Drive, Waterwells Business Park, Gloucester, GL2 4AA, United Kingdom
Tel: (+44 1452) 72 22 00
Fax: (+44 1452) 72 22 44
e-mail: postmaster@fli.co.uk
Web: www.fli.co.uk
■ Design and manufacture of towers, masts and poles for telecommunications, radar and meteorological applications and the supply of associated accessories. Design and installation of helical piled foundations in addition to braces, trusses and bridges.

Franke Holding AG

Franke-Strasse 2, CH-4663 Aarburg, Switzerland
Tel: (+41 62) 787 34 43
Fax: (+41 62) 787 30 37
e-mail: info@franke.com
Web: www.franke.ch

■ Manufacture of powerplant components for combustion chambers and afterburners. Sheet-metal components, including mechanical processing. Machining of highly heat-resistant special alloys. Deep drawing expansion forming, inert-atmosphere welding and laser drilling.

French Aeronautics and Space Industry Group (GIFAS)

8 rue Galilée, F-75116 Paris, France
Tel: (+33 1) 44 43 17 00
Fax: (+33 1) 40 70 91 41
e-mail: com@gifas.asso.fr
Web: www.gifas.asso.fr

■ Co-ordination of the industrial and economic activities of its 240 members engaged in the development, manufacture, maintenance and sales of aircraft, helicopters, engines, missiles, space vehicles and their associated components, equipment, accessories, products and operating facilities. Develops and defends interests of the profession in dealings with the Government. Co-ordinates the industrial activity and economic action of its members and organises lectures and exhibitions to promote the image of the French aerospace industry. Organisers of the Paris Air Show – Le Bourget.

Publications:
GIFAS Annual Report
GIFAS Directory
GIFAS Info News Letter

Friemann & Wolf Batterietechnik GmbH (FRIWO)

Exide Verwaltungs GmbH, Büdingen
Industriestraße 22, D-63654 Büdingen, Germany
Tel: (+49) 60 42
(+49) 95 40
Fax: (+49) 60 42
(+49) 95 41 90
e-mail: info@friwo-batterien.de
Web: www.friwo-batterien.de

■ Manufacture of batteries such as traditional nickel-cadium storage, lithium, silver-zinc, special batteries and battery systems with a heavy-duty capability.

Frontier Astronautics

PO Box 127, Chugwater, Wyoming, 82210, United States
Tel: (+1 307) 331 30 43
e-mail: info@frontierastronautics.com
Web: frontierastronautics.com/

Fuji Heavy Industries Ltd (FHI)

Head Office
Fuji Jukogyo KK
1-7-2 Nishishinjuku, Shinjuku-ku, Tokyo, 160-8316, Japan
Tel: (+81 3) 33 47 21 11
(+81 3) 33 47 25 25
Fax: (+81 3) 33 47 23 38
Web: www.fhi.co.jp

Fujitsu Ltd

Kawasaki Main Office
4-1-1 Kamikodanaka, Nakahara-ku, Kawasaki, Kanagawa, 211-8588, Japan
Tel: (+81 44) 777 11 11
Web: www.fujitsu.co.jp, www.fujitsu.com

■ Spaceborne electronics including remote sensing radiometers, telemetry systems, microwave amplifiers, small earth stations, data processing systems and satellite networks.

Fujitsu Ltd

Shiodome City Center, 1-5-2 Higashi-Shimbashi, Minato-Ku, Tokyo, Japan
Tel: (+81 3) 625 22 20
Web: www.jp.fujitsu.com

G

G2P

consortium formed by Snecma Moteurs, France and SNPE, France
Parc Kennedy, Bât. E., 3 Avenue Henri Becquerel, F-33700 Mérignac, France
Tel: (+33 5) 57 92 26 60
Fax: (+33 5) 56 34 33 72

■ Co-ordinate activities in design, development and manufacture of large solid-propellant motors.

Representing:
Snecma, France
SNPE Materiaux Energetiques, France

GAF Company for Remote Sensing and Information Systems

Arnulfstrasse 197, D-80634 München, Germany
Tel: (+49 89) 121 52 80
Fax: (+49 89) 12 15 28 79
e-mail: info@gaf.de
Web: www.gaf.de

■ Consultancy, project planning and monitoring, market studies. Digital image processing, standard image products, thematic image enhancement. Laser film writing (CIRRUS, LC 3000). Interpretation and analysis. Training and development of special image processing software, cadastral applications.

Galileo Avionica SpA

Rome Office
Selex Sensors and Airborne Systems SpA
Selex Galileo
Alenia Difesa
Via Giulio Vincenzo Bona 85, I-00156 Roma, Italy
Tel: (+39 0641) 88 31

■ Ground-based and naval optical fire control systems for anti-aircraft guns. Laser fire control systems for main battle tanks and light armoured fighting vehicles. Field artillery fire control computers and systems. Thermal imaging devices for ground, naval and airborne use (aiming and surveillance). Electro-optic missile guidance systems. Electro-hydraulic and electrical servos. Periscopes for day/night aiming and observation, panoramic stabilised sight for battle tanks. Optronic equipment for military applications. Electro-optic attitude sensors for satellites. Observation cameras and spare robotics. Electro-optic instruments for astrophysical research and remote surveying in space. Surface Military Systems Division: electro-optic fire control systems for battle tanks, automation systems for field artillery, infra-red night vision systems. Avionic activities: IR equipment for fixed- and rotary-wing aircraft, missile guidance systems and their components, electro-optic seekers. Space Division: electro-optic attitude sensors for satellites, observation cameras and space robotics, electro-optical instruments for astrophysical research and remote surveying in space.

Garmin (Europe) Ltd

European Office
Unit 5, The Quadrangle, Abbey Park Industrial Estate, Romsey, Hampshire, SO51 9AQ, United Kingdom
Tel: (+44 1794) 51 99 44
Fax: (+44 1794) 51 92 22
Web: www.garmin.com

■ Design, manufacture and marketing of navigation and communication electronics equipment, particularly GPS products.

Garmin International Inc

1200 East 151 Street, Olathe, Kansas, 66062, United States
Tel: (+1 913) 397 82 00
Freephone: (+1 800) 800 10 20
Fax: (+1 913) 397 82 82
Web: www.garmin.com

■ Design, manufacture and market navigation and communications electronics worldwide for consumer, business and military applications.

Garvey Spacecraft Corporation

PO Box 15706, Long Beach, California, 90815, United States
Tel: (+1 562) 498 29 84
(+1 562) 431 42 86
Fax: (+1 562) 498 29 84
(+1 562) 431 42 86
e-mail: info@garvspace.com
Web: garvspace.com

Gascom JSC

Lenin Street 4a, 141070 Kaliningrad, Moscow Region, Russian Federation
Tel: (+7 095) 513 69 72
Fax: (+7 095) 513 69 70
(+7 095) 513 69 02
Web: www.gascom.ru

■ Develops satellite communication systems.

GE Aviation

Corporate Office
Smiths Aerospace
a subsidiary of Smiths Industries Aerospace, UK
Suite 1013, 1601 North Kent Street, Arlington, Virginia, 22209, United States
Tel: (+1 703) 351 58 90
Fax: (+1 703) 351 07 64

■ Focus on government relations and sales.

GE Aviation

Mechanical, Wolverhampton
Dowty Aerospace Wolverhampton
Smiths Aerospace
a member of Smiths Group plc
Wobaston Road, Wolverhampton, West Midlands, WV9 5EW, United Kingdom
Tel: (+44 1902) 39 42 56
Fax: (+44 1902) 39 43 94
Telex: 338220 dowty

■ Design, development and manufacture of primary and secondary flight control actuation systems, airframe accessory drive gearboxes and servo-valves.

GenCorp Inc

Headquarters
Tire & General Rubber Company
PO Box 537012, Sacramento, California, 95853-7012, United States
Tel: (+1 916) 355 40 00
e-mail: comments@gencorp.com
Web: www.gencorp.com

■ Technology-based manufacturing company within the automotive, aerospace and defence industries. Products include liquid and solid propulsion systems for strategic and tactical defence systems and manned and unmanned space launch missions.

Business units:
Aerojet
Aerojet Fine Chemicals
GDX Automotive

General Atomics Aeronautical Systems Inc (GA)

Reconnaisance Systems Group
PO Box 85608, San Diego, California, 92186-5608, United States
13322 Evening Creek Drive North, San Diego, California, 92128, United States
Tel: (+1 858) 964 67 00
(+1 858) 455 30 00
Fax: (+1 858) 312 29 01
(+1 858) 455 36 21
(+1 858) 964 69 52
Telex: 695065
e-mail: pr-asi@ga-asi.com
Web: www.uav.com
www.ga.com

■ Research and development of electromagnetic systems, advanced materials, robotic systems, chemical weapon demilitarisation, signature control, automated non-destructive inspection, repair technologies, hazardous waste elimination, unmanned air vehicles, synthetic aperture radar, electronics, superconducting magnets and materials, nuclear spacepower and advanced computing.

General Dynamics C4 Systems

Headquarters
General Dynamics, Decision Systems
Motorola Integrated Information Systems Group
8201 East McDowell Road, Mail Drop H3184, Scottsdale, Arizona, 85257, United States
Tel: (+1 480) 441 28 85
Freephone: (+1 877) 449 06 00
(+1 877) 466 94 67
(+1 877) 449 05 99
Fax: (+1 480) 441 50 67
e-mail: info@gdds.com
Web: www.gdc4s.com

■ Prime systems integrator of technologies, products and systems for government, defence and industrial customers worldwide, specialising in command and control, information assurance and interoperable communication solutions.

Accreditations:
ISO 9001; CMMI Level 5; OSHAS 14001/18001

General Dynamics C4 Systems

SATCOM Technologies
General Dynamics Decision C4 Systems
1500 Prodelin Drive, PO Box 850, Newton, North Carolina, 28658, United States
Tel: (+1 828) 464 41 41
Fax: (+1 828) 464 41 47
e-mail: info@gdsatcom.com
Web: www.gdsatcom.com

General Dynamics Corp
Headquarters
2941 Fairview Park Drive, Suite 100, Falls Church, Virginia, 22042-4513, United States
Tel: (+1 703) 876 31 99
Fax: (+1 703) 876 31 86
Web: www.gd.com
www.generaldynamics.com
- The company is involved in mission-critical information systems and technologies, land and amphibious combat systems, shipbuilding and marine systems and business aviation.

Gentex Corporation
Carbondale Office
PO Box 315, Carbondale, Pennsylvania, 18407-0315, United States
324 Main Street, Simpson, Pennsylvania, 18407, United States
Tel: (+1 570) 282 85 18
e-mail: info@gentex.com
- Manufacture of military life support systems, oxygen masks, air, ground and space shuttle helmets. Manufacture microphones, optical and opthalmic products and laser protective optics. Aluminised, engineered, ballistic chemical and biological defence textiles and garments, chemical and biological casualty care systems. Also produce plastic tecnologies, filters and laserwelding.

GeoEye
12076 Grant Street, Thornton, Colorado, 80241, United States
Tel: (+1 303) 254 20 00
Fax: (+1 303) 254 22 11
(+1 301) 552 37 62
(+1 303) 254 22 15
e-mail: info@spaceimaging.com
custserv@spaceimaging.com
Web: www.spaceimaging.com
- Supply of visual information products and services derived from space imagery and aerial photography.

Geofizika-Cosmos NPP
Geokos Special Design Bureau (SKB)
Scientific Production Enterprise Geofizika-Cosmos
11/17 Irkutskaya Street, 107497 Moskva, Russian Federation
Tel: (+7 495) 462 03 43
Fax: (+7 495) 462 13 14
e-mail: info@geofizika-cosmos.ru
Web: www.geofizika-cosmos.ru
- Design and manufacture of opto-electronics including automatic opto-electronic instruments for altitude measurement using the sun, stars and planets, optical visualisation instruments for use by astronauts during manual manoeuvering in daylight and night conditions, instruments for studying the earth's natural resources and atmosphere surrounding space vehicles (radiometers, photometers, spectroradiometers, spectrophotometers, actiometers, lightmeters and laser measuring devices), ground-based instruments for orientating heliostats towards the sun, star and earth simulators, radiometers for star simulator calibration, simulators for the visualisation of external effects during training of astronauts, high precision optical angle-code encoders. Test facilities include precise opto-mechanical stands equipped with sun, star and planet simulators, test centre for vibration, shock, vacuum, acoustic and climatic environment, graduation complex for calibration of spectrometric instrumentation.

Geoimage Pty Ltd
Darwin Office
GPO Box 1569, Darwin, Northern Territory, 0801, Australia
Suite 4, First Floor, 59 Smith Street, Darwin, Northern Territory, 0800, Australia
Tel: (+61 8) 89 41 36 77
Fax: (+61 8) 89 41 36 99
e-mail: darwin@geoimage.com.au

Geoimage Pty Ltd
Head Office
PO Box 789, Indooroopilly, Queensland, 4068, Australia
Tel: (+61 7) 38 71 00 88
Fax: (+61 7) 38 71 00 42
e-mail: geoimage@geoimage.com.au
Web: www.geoimage.com.au

Geo-Informatics and Space Technology Development Agency (GISTDA)
Thailand Remote Sensing Centre
196 Phahonyothin Road, Chatuchak, Bangkok, 10900, Thailand
Tel: (+66 2) 579 03 45
(+66 2) 940 64 20
(+66 2) 940 64 21
(+66 2) 940 64 22
(+66 2) 940 64 23
(+66 2) 940 64 24
(+66 2) 940 64 25
(+66 2) 940 64 26
(+66 2) 940 64 27
(+66 2) 940 64 28
(+66 2) 940 64 29
(+66 2) 940 63 45
Fax: (+66 2) 561 30 35
(+66 2) 562 04 29
(+66 2) 326 42 91
(+66 2) 326 42 83
(+66 2) 561 30 35
(+66 2) 561 24 72
Web: www.gistda.or.th
- Development of space technology and geo-informatics applications, application of satellite imagery, and development of satellite database.

Geoinformation Laboratory & Consultancy
Bartycka 18a St., PL-00-716 Warszawa, Poland
Tel: (+48 22) 851 11 66
e-mail: office@geosystems.com.pl
Web: www.geosystems.com.pl

Geomat International Inc
75 Queen, Suite 3200, Montréal, Québec, H3C 2N6, Canada
Tel: (+1 514) 369 50 00
Fax: (+1 514) 369 50 59
e-mail: geomat@geomat-intl.com
Web: www.groupealta.com

GEOSPACE Beckel Satellitenbilddaten GmbH
Jakob-Haringer-Strasse 1, A-5020 Salzburg, Austria
Tel: (+43 662) 45 81 15
Fax: (+43 662) 458 11 54
e-mail: office@geospace.co.at
Web: geospace.co.at

Geosys
20 Impasse René Couzinet, BP 65815, F-31505 Toulouse, Cedex 5, France
Tel: (+33 5) 62 47 80 80
Fax: (+33 5) 62 47 80 70
e-mail: info@geosys.com
- Provision of decision makers in agriculture, agri-business and rural land management.

Geosys Data
20 Impasse René Couzinet, Parc d'activité de la Plaine, BP 5815, F-31505 Toulouse, Cedex 5, France
Tel: (+33 5) 62 47 80 77
Fax: (+33 5) 62 47 80 70
e-mail: data@geosys.fr

German Aerospace Center
Aerodynamics and Flow Technology
Lilienthalplatz 7, D-38108 Braunschweig, Germany
Tel: (+49 531) 295 24 00
Fax: (+49 531) 295 23 20
Web: www.dlr.de

German Aerospace Center
Berlin-Adlershof Office
Deutsche Zentrum für Luft- und Raumfahrt
Rutherfordstrasse 2, D-12489 Berlin, Adlershof, Germany
Tel: (+49 30) 67 05 50
Fax: (+49 30) 67 05 51 20
Web: www.dlr.de

German Aerospace Center
Brussels Office
Deutsche Forschungsanstalt für Luft- und Raumfahrt
Rue du Trône 98, B-1050 Bruxelles, Belgium
Tel: (+32 2) 500 08 41
Fax: (+32 2) 500 08 40
e-mail: internet@dlv.de
brussels@dir.de
Web: www.dlr.de

German Aerospace Center
Deutsches Zentrum für Luft- und Raumfahrt
DLR–Lampoldshausen, D-74239 Hardthausen, Germany
Tel: (+49 6298) 280
Fax: (+49 6298) 281 12
Web: www.dlr.de
- European research and testing centre for rocket propulsion systems.

German Aerospace Center
Executive Headquarters
Deutsches Zentrum für Luft- und Raumfahrt eV
Linder Höhe, D-51147 Köln, Germany
Tel: (+49 2203) 60 10
Fax: (+49 2203) 673 10
Telex: 8810-0 dlr
Web: www.dlr.de

German Aerospace Center
Gottingen Site
Institute of Aerodynamics and Fluid Mechanics
Institute of Aeroelasticity
Deutsche Zentrum für Luft-und Raumfahrt eV
Bunsenstrasse 10, D-37073 Gottingen, Germany
Tel: (+49 551) 70 90
Fax: (+49 551) 709 21 01
Web: www.dlr.de

German Aerospace Center
Köln-Porz Office
Deutsches Zentrum für Luft und Raumfahrt eV
Linder Höhe, D-51147 Köln-Porz, Germany
Tel: (+49 2203) 601 21 16
(+49 2203) 60 10
Fax: (+49 2203) 673 10
(+49 2203) 601 35 02
(+49 2203) 601 32 49
e-mail: pressestelle@dlr.de
Web: www.dlr.de

German Aerospace Center
Oberpfaffenhofen Office
Arbeitsgruppe Lidar
Flight Department, Oberpfaffenhofen
Institut fur Hochfrequentecchnik und Radarsysteme
Münchner Strasse 20, PO Box 1116, D-82234 Weßling, Germany
Tel: (+49 8153) 28 29 81
(+49 8153) 28 29 86
Fax: (+49 8153) 28 13 47
- Research in robotics, systems dynamics, radio frequency technology, atmospheric physics, communications technology, opto-electronics, flight operations and remote sensing.

German Aerospace Center
Press Office and Public Relations
Deutsches Zentrum für Luft- und Raumfahrt
D-51170 Köln, Germany
Tel: (+49 2203) 60 10
Fax: (+49 2203) 673 10
e-mail: pressestelle@dlr.de
Web: www.dlr.de

German Aerospace Center
Stuttgart Office
Institut für Verbrennungstechnik
Institute for Combustion Technology
Institute for structures and Design
Deutsche Forschungsanstalt für Luft- und Raumfahrt
38-40 Pfaffenwaldring, D-70569 Stuttgart, Germany
Tel: (+49 711) 686 20
Fax: (+49 711) 96 86 23 49

German Aerospace Center
Washington Office
Deutsche Forschungsanstalt für Luft- und Raumfahrt
1627 Eye Street, #540 N.W, Washington, 20006-4020, United States
Tel: (+1 202) 785 44 11
Fax: (+1 202) 785 44 10
e-mail: wash.office@dlr.org

German Remote Sensing Data Center (DFD)
DLR Oberpfaffenhofen
D-82234 Weßling, Germany
Tel: (+49 8153) 28 28 02
Fax: (+49 8153) 28 14 46
e-mail: helpdesk@dfd.dlr.de
Web: www.dfd.dlr.de

Gidroagregrat

Ulitsa Kommunisticheskaya, 78, Pavlovo, 606100 Nizhnyi Novgorod, Russian Federation
Tel: (+7 83171) 615 16
Fax: (+7 83171) 342 60
■ Manufacture of high precision hydro-mechanical and electric-hydraulic units of aircraft and helicopter control systems, air propellers with solid metal blades, salvage equipment, plunger-type pumps, industrial pipe line fittings and locks.

Gilat Satellite Networks Ltd

Beijing Representative Office
a subsidiary of Gilat Satellite Networks Ltd, Israel
1725, Tower 2, Bright China ChangAn Building, No. 7, Jianguomeng Nei Avenue, 100005 Beijing, China
Tel: (+86 10) 65 10 28 38
Fax: (+86 10) 65 10 28 39
Web: www.gilat.com.cn

Gilat Satellite Networks Ltd

Headquarters
21 Yegia Kapayim Street, Kiryat Arye, IL-49130 Petah Tikvah, Israel
Tel: (+972 3) 925 22 01
Fax: (+972 3) 925 22 52
Web: www.gilat.com

Gisat

Charkovská 7, CZ-10100 Praha, 10, Czech Republic
Tel: (+420 2) 71 74 19 35
Fax: (+420 2) 71 74 19 36
Web: www.gisat.cz
■ Offer satellite data (in photographic and digital form), satellite data processing and evaluation, space topographic maps, thematic maps for geology, geomorphology, hydrology, water resource management, agriculture, forestry, regional planning and environmental monitoring. Digital terrain models, geographical information systems. Satellite and air photo digital image processing, geographic information systems and GIS creation. Consultancy and advisory services.

GKN Aerospace FPT Industries

FPT Industries
GKN Westland Aerospace Ltd
Special Products (Portsmouth)
Airport Service Road, Portsmouth, Hampshire, PO3 5PE, United Kingdom
Tel: (+44 23) 67 52 00
Fax: (+44 23) 67 08 99
Web: www.fptind.co.uk
■ Flexible, crash-resistant and self-sealed fuel cells and assemblies. Emergency flotation equipment for aircraft, helicopters and vehicles, flotation collars and salvage bags. Hyclad flexible gaskets, rubber proofed fabrics, rubber mouldings, extrusions, adhesives and lacquers. Rigid plastic mouldings for fuel and hydraulic reservoirs, explosion suppressants and air portable fuel containers.

GKN Aerospace Transparency Systems Kings Norton

Pilkington Aerospace Ltd
Triplex Aircraft & Special Products Ltd
a division of GKN Aerospace
Triplex House, Eckersall Road, Kings Norton, Birmingham, B38 8SR, United Kingdom
Tel: (+44 121) 606 41 00
(+44 121) 606 41 44
Fax: (+44 121) 433 35 41
(+44 121) 606 41 91
(+44 121) 606 41 61
e-mail: sales@gknaerospace.com
Web: www.gkntransparencysystems.com
■ Design and manufacture of heated/unheated, curved/flat, framed/unframed impact resistant transparencies for railway and transit industries.

Global Communications Co

a division of Global Defense Products Inc., USA
20 Waterside Plaza 33A, New York, 10010, United States
Tel: (+1 212) 689 51 50
(+1 212) 689 96 44
Fax: (+1 212) 889 73 59
Telex: 237593
Cable: GLOBCOM
e-mail: 103227.250@compuserve.com
■ Design, assembly and supply of military electronics including C3 systems design and development.

Globalstar

461 So Milpitas Boulevard, Building 5, Suite 1 and 2, San Jose, California, 95035, United States
Tel: (+1 403) 933 40 00
Fax: (+1 403) 933 41 00
e-mail: pr.group@globalstar.com
Web: www.globalstar.com

Glocom Inc

20010 Century Boulevard, Germantown, Maryland, 20874, United States
Tel: (+1 301) 916 21 00
Fax: (+1 301) 916 94 38
e-mail: info@glocom-us.com
Web: www.glocom-us.com

GMV S A

Defence & Security
Grupo de Mecanica del Vuelo SA
Isaac Newton 11, Parque Technologico de Madrid, Tres Cantos, E-28760 Madrid, Spain
Tel: (+34 91) 807 21 00
Fax: (+34 91) 807 21 99
e-mail: info@gmv.es
Web: www.gmv.com
■ Provide consulting and engineering services, software development, and turnkey systems integration to the aerospace and defence markets. Activities include: engineering and mission analysis, satellite control centres, satellite navigation systems, back-end information systems and defence systems.

Goodrich Actuation Systems

Cergy-Pontoise
Lucas Aerospace
Lucas Aerospace FCS Paris
a part of Goodrich Airframe Systems business segment
13 Avenue de l'Eguillette, Saint Ouen l'Aumone, BP 7186, F-95056 Cergy-Pontoise, Cedex, France
Tel: (+33 1) 34 32 63 00
Fax: (+33 1) 34 32 63 10
Web: www.goodrich.com

Goodrich Actuation Systems

France Headquarters
TRW Aeronautical Systems, Lucas Aerospace
a part of Goodrich Airframe Systems business segment
106 rue Fourny, F-78530 Buc, France
Tel: (+33 1) 39 20 52 00
e-mail: marketing.buc@goodrich.com
Web: www.goodrich.com
■ Provision of a range of actuation systems for airborne and land based platforms. Products include primary and secondary flight controls; helicopter main and tail rotor, engine and nacelle, utility, precision weapon and land vehicle actuators.
■ Provision of complete systems including hydraulic, fly-by-wire and power-by-wire controls using technologies such as hydraulic, rotary geared, electro-hydrostatic and electro-mechanical actuators.

Goodrich Corporation

Corporate Headquarters
BFGoodrich Aerospace
Four Coliseum Centre, 2730 West Tyvola Road, Charlotte, North Carolina, 28217-4578, United States
Tel: (+1 704) 423 70 00
e-mail: corporate.communications@goodrich.com
Web: www.goodrich.com
Organised into three business segments:
–Actuation and Landing Systems
–Goodrich Actuation Systems
–Goodrich Engine Components
–Goodrich Turbine Fuel Technologies
–Goodrich Turbomachinery Products
–Goodrich Power Transmission
–Goodrich Aviation Technical Services
–Goodrich Landing Gear
–Goodrich Wheels & Brakes
–Electronic Systems
–Goodrich Sensors and Integrated Systems
–Goodrich Propulsion Systems
–Goodrich De-Icing & Speciality Systems
–Goodrich Fuel & Utility Systems
–Goodrich Sensor Systems
–Goodrich Hoist and Winch
–Goodrich Missile Actuation
–Goodrich Engine Control and Electrical Power Systems
–Goodrich Engine Control Systems
–Goodrich Power Systems
–Goodrich ISR Systems
–Goodrich Nacelles and Interior Systems
–Goodrich Aerostructures
–Goodrich Engineered Polymer Products
–Goodrich Interiors
–Goodrich Aircraft Interior Products
–Goodrich Lighting Systems
–Goodrich Cargo Systems
–Goorich Customer Services

Goodrich ISR Systems

Surveillance & Reconnaissance Systems, Malvern
Goodrich Intelligence Surveillance and Reconnaissance Systems
Goodrich Optical & Space Systems
a part of Goodrich ISR Systems business segment
Malvern Hills Science Park, Geraldine Road, Malvern, Worcester, WR14 3SZ, United Kingdom
Tel: (+44 1684) 58 52 25
Fax: (+44 1684) 58 53 67
Web: www.goodrich.com
■ Development and production of high performance cameras and ground stations for capturing and interpreting reconnaissance and surveillance data. Produce imagery products for intelligence, surveillance and reconnaissance (ISR) missions, sensors for day, night, and multi spectral image capture, along with the ground exploitation and data handling systems.

Grinaker Telecom

PO Box 8492, 0046 Hennopsmeer, South Africa
86 Oak Avenue, Highveld Technopark, Centurion, South Africa
Tel: (+27 12) 665 00 34
Fax: (+27 12) 665 00 81

Gromov Flight Research Institute (Gromov FRI)

High Computer Technologies, Computer Based Training (CBT) and Training Methodology Laboratory
Zhukovsky, 140182 Moskva, Russian Federation
Tel: (+7 095) 556 56 07
Fax: (+7 095) 556 53 34
■ Research and production of Computer Based Training (CBT) systems.

Groupe SNPE

Head Office
SME GROUP SNPE
SNPE Explosives and Propellants
12 Quai Henri IV, F-75004 Paris, France
Tel: (+33 1) 48 04 68 45
(+33 1) 48 04 66 66
Fax: (+33 1) 48 04 68 87
(+33 1) 49 96 74 01
e-mail: dircom@snpe.com
Web: www.snpe.com
www.sme-propulsion.com
www.energetic-materials.com
www.materialspyrotecs.com
■ Design, manufacture, research and development of energetic materials including propellant charges, industrial explosives, chemicals and speciality chemicals for military and aerospace and industrial applications. Industrial safety services, environmental protection and pyrotechnic pollution solutions. Services include dismantling old units, pollution clearing of land, re-establishing industrial operations and restructuring of sites.
Subsidiaries:
SNPE Materiaux Energetiques
Nobel Explosives France
ISOCHEM
Bergerac NC

GTE Spacenet Corp

Headquarters
1750 Old Meadow Road, Mclean, Virginia, 22102, United States
Tel: (+1 703) 848 10 00
Fax: (+1 703) 848 10 10

H

Hamilton Sundstrand

Flight and Undersea Systems
a United Technologies Company, US
4747 Harrison Avenue, PO Box 7002, Rockford, Illinois, 61125-7002, United States
Tel: (+1 860) 654 60 00
Fax: (+1 815) 226 72 88

■ Aircraft pumps, electrical generating systems, gearboxes, actuation systems and components, generators, electric motors, starting systems, pneumatic conditioning and control systems, fans, refrigeration systems, missile power units and emergency turbine systems.

Hamilton Sundstrand

Headquarters
Hamilton Standard
a division of United Technologies Corp, US
One Hamilton Road, Windsor Locks, Connecticut, 06096-1010, United States
Tel: (+1 860) 654 60 00
e-mail: hs.general@hsd.utc.com
Web: www.hamiltonsundstrand.com
■ Technological products and systems for the aerospace business market. Electric power generation systems, engine controls, environmental control systems, propeller systems, flight control systems, actuation systems and auxiliary power units.
International subsidiaries:-
Hamilton Standard Marston Aerospace, Ltd, UK
Hamilton Standard-Nauka, Russia
Microtecnica, Italy
Nord-Micro, Germany
Ratier-Figeac, France

Hamilton Sundstrand

Power Systems
4400 Ruffin Road, PO Box 85757, San Diego, California, 92186 5757, United States
Tel: (+1 858) 627 65 27
Fax: (+1 860) 660 41 85
e-mail: businessdev@hs.utc.com
Web: www.hs-powersystems.com
■ Small gas turbine engines, generator sets and APU's for commercial and military aircraft and vehicles. Fans and vapour cycle systems. Propulsion engines for RPVs.

Harris Corporation

Government Communications Systems Division
2400 Palm Bay Road, North East, Melbourne, Florida, 32905, United States
Tel: (+1 321) 727 91 00
Fax: (+1 321) 727 45 00
e-mail: govt@harris.com
Web: www.harris.com
■ Design, development and integration of assured communications and information systems design for US government customers including aerospace, defence and intelligence.

Harris Corporation

Headquarters
1025 West NASA Boulevard, Melbourne, Florida, 32919-0001, United States
Tel: (+1 321) 727 92 07
(+1 321) 727 92 07
Freephone: (+1 800) 4 HARRIS
Fax: (+1 321) 727 96 46
Web: www.harris.com

Harris Corporation

RF Communications Division
1680 University Avenue, Rochester, New York, 14610, United States
Tel: (+1 585) 244 58 30
Fax: (+1 585) 242 47 55
Web: www.rfcomm.harris.com
■ Secure wireless radio communication products, systems and services for defence, government and law enforcement markets. Multiband (HF-VHF-UHF) radios are provided in manpack and vehicular applications.

Hawker de Havilland (HdH)

a wholly owned subsidiary of The Boeing Company
361 Milperra Road, PO Box 30, Bankstown, New South Wales, 2200, Australia
Tel: (+61 2) 97 72 81 11
Fax: (+61 2) 97 73 08 98
■ Airframe and aircraft component design and manufacture, defence aircraft production.
Subsidiaries
Hawker de Havilland Aerospace Support Pty Ltd
Hawker de Havilland Components Pty Ltd

HBL Power Systems Limited

Aviation Division
HBL Limited
Hyderabad Batteries Ltd
8-2-601 Road No 10, Banjara Hills, Hyderabad, 500034, India
Tel: (+91 8418) 24 46 42
Fax: (+91 40) 23 35 31 89
(+91 40) 23 35 50 48
e-mail: contact@hbl.in
Web: www.hbl.in
■ Manufacture of silver, zinc and nickel-cadmium batteries for aircraft, torpedoes and UPS applications. Lithium, thermal, valve regulated lead acid, tubular and fuse batteries. Supplier of sealed maintenance-free lead acid generator start batteries and maintenance-free, rechargeable VRLA monoblock batteries for main battle tanks. Supplier for internal and export markets. Battery monitoring system for submarines.

Helsinki University of Technology (HUT)

Space Technology Laboratory
PO Box 1000, FI-02015 Finland
Otakaari 1, FI-02150 Espoo, Finland
Tel: (+358 9) 45 11
(+358) 451 23 71
Fax: (+358) 46 02 24
(+358) 451 29 98
Web: www.tkk.fi
■ Education and research in science and engineering.

Hernandez Engineering Inc (HEI)

16055 Space Center Boulevard, Suite 725, Houston, Texas, 77062, United States
Tel: (+1 281) 280 51 59
Fax: (+1 281) 480 75 25
Web: www.hernandezengineering.com
■ Capabilities include: engineering analysis, operations, and science support; design and system engineering; system safety, reliability and quality engineering, product assurance, and independent assessment; documentation and information management; industrial and occupational safety; and multi-media computer-based training.

High-Reliability Components Corp

BIF, Shiba Masuda Building, 14-4 Shiba 1-chome, Minato-ku, Tokyo, 105-0014, Japan
Tel: (+81 3) 54 44 76 91
Fax: (+81 3) 54 44 76 95
Web: www.hirec.co.jp
■ Quality assurance and distribution of high reliability components for space applications.

Hindustan Aeronautics Ltd (HAL)

Aerospace Division
New Thippasandra, PO Box 7502, Bangalore, 560075, India
Tel: (+91 80) 524 14 03
Fax: (+91 80) 524 16 51
■ Light-alloy structures for space launch vehicles.

Hindustan Aeronautics Ltd (HAL)

European Liason Office
Room No 602, India House, High Commission of India, Aldwych, London, WC2B 4NA, United Kingdom
Tel: (+44 20) 74 97 23 60
Fax: (+44 20) 74 97 83 98
e-mail: marketing@hal-india.com
Web: www.hal-india.com
■ Liaison work for design, development, manufacture and servicing of fixed and rotary-wing aircraft, engines, systems, accessories, equipment and avionics. Light alloy structures for space and ground handling equipment.
Representing: Hindustan Aeronautics Ltd, India

Hitachi Ltd

Transportation Systems Division
18-13, Soto-Kanda 1-chome, Chiyoda-Ku, Tokyo, 101-8608, Japan
Tel: (+81 3) 45 64 55 40
(+81 3) 52 95 55 40
e-mail: chiaki.yamada.ty@hitachi.com
Web: www.hitachi-rail.com
■ Communications equipment, computers, software, electronic components and computer circuits.

Hitco Carbon Composites, Inc

1600 West 135th Street, PO Box 1097, Gardena, California, 90249, United States
Freephone: (+1 800) 421 54 44
Fax: (+1 310) 516 57 14
e-mail: refrasil@hitco.com
Web: www.hitco.com
■ Supply of composite structures, rocket nozzles and high temperature materials to aerospace and industrial markets. Design and manufacture of specialty insulation components fabricated from a range of high temperature fabrics, non-wovens and metal foils. The design of insulation products can be tailored to customer specifications. Markets served include: aircraft, chemical processing, personnel safety, foundry and forging, heavy industrial, power generating, automotive and basic metal processing.
Accreditations: ISO 9001

Holscot Fluoroplastics

Holscot Industrial Linings Limited
Alma Park Road, Alma Park Industrial Estate, Grantham, Lincolnshire, NG31 9SE, United Kingdom
Tel: (+44 1476) 57 47 71
(+44 845) 456 05 74
Fax: (+44 1476) 56 35 42
(+44 845) 456 05 76
e-mail: sales@holscot.com
Web: www.holscot.com
■ Production of fluoroplastic items for a wide range of industries and applications, including melt processible fluoroplastics, FEP, PFA, PVDF or ECTFE.

Honeywell Aerospace

Aerospace Headquarters
AlliedSignal Aerospace
AlliedSignal Inc Aerospace
1944 East Sky Harbor Circle, Phoenix, Arizona, 85034, United States
Tel: (+1 800) 365 21 80
(+1 800) 601 30 99
Freephone: (+1 800) 421 21 33
Web: www.honeywell.com
■ Manufacture of aircraft guidance, navigation, communication and identification systems. Flight controls and cockpit instrumentation that include weather radar, fuel controls and ignition systems. Commercial and military aircraft wheels, brakes and struts. Gas turbine engines, turboprop and turbofan engines. Environmental systems, cabin pressurisation systems, systems and controls for aircraft, missiles and undersea applications. Provision of technical services and spacecraft and satellite tracking support for NASA.

Honeywell Aerospace

Yeovil Facility
Honeywell
Honeywell Normalair-Garrett Ltd
Normalair-Garrett Ltd
a subsidiary of Honeywell Inc., US
Bunford Lane, Yeovil, Somerset, BA20 2YD, United Kingdom
Tel: (+44 1935) 47 51 81
Fax: (+44 1935) 42 76 00
e-mail: yeovil.sales@honeywell.com
Web: www.honeywell.com
■ Manufacture and sale of electromechanical, electronic and environmental control systems and hydraulic equipment. Oxygen systems, data recorders, investment castings, heat transfer, weapon launch and actuation systems. Environmental and air systems: cooling and pneumatic equipment for aerospace applications. Aerospace breathing systems: life support anti-G protection and oxygen generating systems. Aerospace and defence hydraulics and components. Actuation and weapon control systems. Aerospace and industrial electronics, severe environment data acquisition and recording systems, ground replay equipment, analogue and digital recorders.

Honeywell Aerospace, Aerospace Electronic Systems

Glendale
19019 North 59th Avenue, Glendale, Arizona, 85308, United States
Tel: (+1 602) 822 31 58
Fax: (+1 602) 822 36 04

Honeywell Aerospace, Engines, Systems & Services

Airframe Systems
AlliedSignal Aerospace, Engines & Systems
2525 West 190th Street, Torrance, California, 90504-6099, United States
Tel: (+1 310) 323 95 00

Honeywell Aerospace, Engines, Systems & Services

Manufacturing Center
AlliedSignal Aerospace, Engines & Systems
Honeywell Aerospace, Engines & Systems
PO Box 52187, Phoenix, Arizona, 85034, United States
402 South 36th Street, Phoenix, Arizona, 85072-2187, United States
Tel: (+1 602) 231 10 00

Honeywell AG
a subsidiary of Honeywell Inc, US
Kaiserleistrasse 39, D-63067 Offenbach, am Main, Germany
Tel: (+49 69) 806 40
Fax: (+49 69) 81 86 20
Web: www.honeywell.de
■ Space and aviation control systems.
Representing: Honeywell Inc, US

Honeywell Data Control
Commercial Electronic Systems
AlliedSignal Data Control
Sundstrand Data Control
a unit of Honeywell Aerospace, USA
15001 North East 36th Street, PO Box 97001, Redmond, Washington, 98073-9701, United States
Tel: (+1 425) 885 37 11
Fax: (+1 425) 885 20 61
Web: www.egpws.com
■ Components and subsystems such as avionics systems for flight safety, accelerometers, thermal products, navigation and flight management equipment, telecommunication and satellite communication programmes, airborne communications systems and data management equipment.

Honeywell Inc (HON)
Headquarters
101 Columbia Road, Morristown, New Jersey, 07962, United States
Tel: (+1 973) 455 52 81
Freephone: (+1 800) 328 51 11
Fax: (+1 973) 455 48 07
Telex: TWX 910-576-2692
Web: www.honeywell.com
■ Manufacturer of automotion and control systems for homes, buildings, industry, space and aviation.

Honeywell Inc
Space and Aviation Control Headquarters
21111 North 19th Avenue, Phoenix, Arizona, 85027, United States
■ Manufacture products and systems for large commercial airliners, corporate aircraft and twin-turbine helicopters. Integrated systems provider for aircraft landing, airport ground vehicle monitoring and airfield/obstruction lighting.

Honeywell Technical Services Corporation
Headquarters
AlliedSignal Technical Services Corporation
One Bendix Road, Columbia, Maryland, 21045-1897, United States
Tel: (+1 410) 964 70 00
Fax: (+1 410) 730 67 75
Telex: 087860 benfld colb
■ Integrated management and field engineering services, systems engineering, logistics planning and support, data acquisition and presentation, computer programming, systems analysis and design, RFI/ECM measurement analysis, repair and overhaul of electronic and electromechanical equipment. Operation, maintenance and support of traffic control facilities and equipment, including operation and maintenance of aircraft navigational and communication equipment. Design and modification of airport surveillance radar systems, planetary and interplanetary space mission tracking, laser satellite tracking.

HR Textron
Textron Motion Control
a subsidiary of Textron Inc, US
25200 West Rye Canyon Road, Santa Clarita, California, 91355-1265, United States
Tel: (+1 661) 294 60 00
Fax: (+1 661) 259 96 22
Web: www.hrtextron.com
www.systems.textron.com
■ Flight control systems and components for high-performance aircraft, space launch vehicles, helicopters, missiles and turbine engines; servovalves, fuel and pneumatic systems components, automatic test equipment and product support. Land vehicle turret control and stabilisation systems.

Hughes Network Systems (HNS)
Corporate Headquarters
11717 Exploration Lane, Germantown, Maryland, 20876, United States
Tel: (+1 301) 428 55 00
Fax: (+1 301) 428 18 68
(+1 301) 428 28 30

Hungarian Space Office (HSO)
Dob u. 75-81, H-1077 Budapest, VII, Hungary
Tel: (+36 1) 461 36 39
Fax: (+36 1) 351 03 53
e-mail: hso@hso.hu
hso.mui@omfb.x400gw.itb.hu
Web: www.hso.hu
■ Co-ordination of the activities of the Hungarian institutes dealing with space research and the international activities of the Hungarian space research community.

Hydro Fitting Manufacturing Corp
733 East Edna Place,
PO Box 1558, Covina, California, 91722, United States
Tel: (+1 626) 967 51 51
Fax: (+1 626) 339 44 54
■ Fittings, AN, MS, NAS, NSA approved Airbus. Valves, air charging; MIL-SPEC to order; 100% CNC shop. Covers, engine magnesium inspection plates.

Hymec BV
Hymec is part of the Neways Electronics International group of companies.
Dr Nolenslaan 107b, PO Box 336, NL-6130 AH Sittard, Netherlands
Tel: (+31 46) 420 19 99
Fax: (+31 46) 452 68 28
e-mail: sales@hymec.nl
Web: www.hymec.nl
■ Design and production of thick and thin film hybrid circuits. High-reliability screening to MIL-STD 883. Micro electronics hybrid assembly.
Accreditations: ISO 9001, space qualified according ESA. ESA-PSS-01-608

I

IFI Srl
Impresa Forniture Industriali
Circonvallazione Nomentana 180, I-00162 Roma, Italy
Tel: (+39 06) 862 23 41
Fax: (+39 06) 86 39 88 89
e-mail: info@ifisrl.it
Web: www.ifisrl.it

IGG Component Technology Ltd
Waterside House, Waterside Gardens, Fareham, Hampshire, PO16 8RR, United Kingdom
Tel: (+44 1239) 82 93 11
Fax: (+44 1239) 82 93 12
e-mail: sales@igg.co.uk
Web: www.igg.co.uk
■ Parts engineering of electronic components and centralised parts procurement. Procurement and testing of components for complete satellites. Also provides engineering and consultancy, obsolescence management, customer training, upscreening, destructive physical analysis, failure analysis, evaluation of new technologies and materials, and electronic component kitting.

IHI Aerospace Co Ltd
Head Office
1-1, Toyosu 3-Chome, Koto-Ku, Tokyo, 135-0061, Japan
Tel: (+81 3) 62 04 80 00
Fax: (+81 3) 62 04 88 10
Web: www.ihi.co.jp
■ Design, development, production and sales of space equipment systems, defence rocket systems and other aerospace related products.

IHI Aerospace Co Ltd
Tomioka Plant
Nissan Motor Co Ltd
a division of IHI Aerosapce Co Ltd, Japan
900 Fujiki, Tomioka, Gunma-Ken, 370-2398, Japan
Tel: (+81 274) 62 41 23
(+81 81) 62 76 11
Fax: (+81 274) 62 77 11
Web: www.ihi.co.jp
■ Design, development, test and manufacture of launch vehicles, space equipment systems, Multiple Rocket Systems (MRS), mine clearing vehicle and mine breaching rocket reload system.

ILC Dover Inc
a subsidiary of ILC Industries, Inc US
One Moonwalker Road, Frederica, Delaware, 19946-2080, United States
Tel: (+1 302) 335 39 11
Freephone: (+1 800) 631 95 67
Fax: (+1 302) 335 07 62
Web: www.ilcdover.com
■ Space suits and accessories, aircrew equipment, aerostats, airships and balloons. NBC masks, weapon decelerators, impact attenuation airbags, pressure suits and inflatable space structures.

IMEC
Interuniversity Microelectronics Centre
Kapeldreef 75, B-3001 Leuven, Belgium
Tel: (+32 16) 28 12 11
Fax: (+32 16) 22 94 00
e-mail: info@imec.be
Web: www.imec.be
■ Research and development in heterogeneous design methodology for systems-on-chip; deep sub-micron CMOS processing technologies, with special focus on optical lithography. Microsystems and solar cells. Training.

Impuls
Ulitsa Volodarskogo, 83,
Arzamas, 607220 Nizhnyi Novgorod, Russian Federation
Tel: (+7 83147) 416 53
Fax: (+7 83147) 416 53
■ Manufacture of aircraft instrumentation and provision of testing services.

Indian Institute of Remote Sensing
No. 4 Kalidas Road, Dehra Dun, 248001, India
Tel: (+91 135) 245 83
Telex: 595224

Indian Space Research Organisation (ISRO)
an agency of the Department of Space
Antariksh Bhavan, New BEL Road, Bangalore, 560094, India
Tel: (+91 80) 234 15 275
(+91 80) 234 15 357
Fax: (+91 80) 234 12 253
(+91 80) 234 15 328
Telex: 8452326
8452499
Cable: ISRO
e-mail: info@isro.org
kitta@isro.org
Web: www.isro.org
■ Development and operation of satellites for telecommunication, television broadcasting, meteorology, disaster management, resources survey and management. Design and development of launch vehicles including the Polar Satellite Launch Vehicle (PSLV) for launching IRS class of remote sensing satellites and Geo-synchrous Satellite Launch Vehicle (GSLV) to launch INSAT class of communications satellites. Conducts space science research. Products are available to international customers through Antrix Corporation.

Indian Space Research Organisation (ISRO)
Development and Educational Communication Unit
Ambavadi 4 Vistar Post Office, Jodhpur Tekra, Ahmedabad, 380015, India
Tel: (+91 79) 677 31 24
(+91 79) 677 31 22
(+91 79) 677 25 35
(+91 79) 677 31 28
(+91 79) 26 92 97 94
Fax: (+91 79) 26 76 85 56
(+91 79) 674 67 15
(+91 79) 26 91 58 14
■ Development of educational and social development software. Conducts research into space applications.

Indian Space Research Organisation (ISRO)
Inertial Systems Unit
Vattiyoorkavu Post, Thiruvananthapuram (Trivandrum), 695013, India
Tel: (+91 471) 36 18 97
(+91 471) 36 30 92
(+91 471) 36 32 88
(+91 471) 36 32 89
(+91 471) 36 32 82
(+91 471) 36 32 94
(+91 471) 236 10 13
Fax: (+91 471) 236 19 73
(+91 471) 36 18 13
■ Design and development of inertial systems for satellites and launch vehicles.

Indian Space Research Organisation (ISRO)
Liquid Propulsion Systems Centre
Valiamala P. O.,Thiruvananthapuram (Trivandrum), 695547, India
Bangalore Unit Address, 80 Feet Road, Hal II Stage HPO, Bangalore, 560008, India
Tel: (+91 471) 56 77 15
(+91 471) 56 72 64
(+91 471) 56 72 95
(+91 80) 527 22 40
(+91 471) 56 77 36
(+91 471) 256 75 25
Fax: (+91 471) 56 72 96
(+91 80) 526 20 85
(+91 471) 56 74 92
(+91 80) 526 04 15
(+91 471) 46 26 86
(+91 80) 526 20 47
(+91 471) 256 72 42
(+91 471) 256 78 99
(+91 471) 256 72 80
Telex: 8452469
Cable: PROPULSION
■ Design and development of liquid propulsion systems for satellites and launch vehicles.

Indian Space Research Organisation (ISRO)
Satellite Centre
Airport Road, Vimanapura Post, Bangalore, 560017, India
Tel: (+91 80) 508 21 02
(+91 80) 508 21 08
(+91 80) 526 62 51
(+91 80) 527 45 16
(+91 80) 252 683 61
(+91 80) 508 21 29
(+91 80) 252 659 89
Fax: (+91 80) 250 821 10
(+91 80) 508 21 99
(+91 80) 526 94 90
(+91 80) 526 64 28
(+91 80) 526 36 22
(+91 80) 526 85 44
(+91 80) 526 90 21
(+91 80) 508 21 98
(+91 80) 508 23 21
(+91 80) 252 69 40
(+91 80) 252 654 07
Cable: UPAGRAM
■ Design and development of communication, remote sensing and scientific satellites.

Indian Space Research Organisation (ISRO)
Satish Dhawan Space Centre
Sriharikota Range, Nellore District, Andhra Pradesh, 524124, India
Tel: (+91 8623) 450 55
(+91 8623) 450 43
(+91 8623) 24 50 60
(+91 4425) 36 44 50
Fax: (+91 8623) 451 54
(+91 8623) 451 59
(+91 8623) 450 67
(+91 8623) 24 51 2
(+91 8623) 24 51 59
(+91 8623) 24 50 02
(+91 8623) 451 53
Telex: 416723
Cable: SPACE
■ Launch station providing infrastructure and facilities for testing of large solid propellant motors.

Indian Space Research Organisation (ISRO)
Space Applications Centre
Ambavadi Vistar Post Office, Jodhpur Tekra, Ahmedabad, 380015, India
Tel: (+91 79) 674 20 10
(+91 79) 674 88 07
(+91 79) 676 93 91
(+91 79) 676 34 42
(+91 79) 676 55 32
(+91 79) 26 91 20 00
(+91 79) 674 19 39
(+91 79) 26 91 50 00
Fax: (+91 79) 26 76 80 73
(+91 79) 676 85 56
(+91 79) 26 76 17 60
(+91 79) 676 54 10
Cable: SPACE
ANTARIX
■ Space applications activities including development of satellite payloads for communications and remote sensing.

Indian Space Research Organisation (ISRO)
Technical Liaison Unit
Embassy of India, 8 rue Halevy, F-75009 Paris, France
Tel: (+33 1) 42 66 33 62
Fax: (+33 1) 42 66 32 72
(+33 1) 40 50 09 96
e-mail: itlu@wanadoo.fr
Web: www.isro.org
■ Promoting development and application of space technology.

Indian Space Research Organisation (ISRO)
Telemetry, Tracking & Command Network
A1-6 Peenya Industrial Estate, Peenya, Bangalore, 560058, India
Tel: (+91 80) 839 42 61
(+91 80) 839 42 30
(+91 80) 837 63 16
(+91 80) 837 63 11
(+91 80) 28 09 40 00
(+91 80) 28 39 51 31
Fax: (+91 80) 28 39 45 15
(+91 80) 28 09 40 03
(+91 80) 28 09 40 21
(+91 80) 839 82 57
Telex: 8455010 trac
■ Provision of telemetry, tracking and command network support for satellite and launch vehicle missions.

Indian Space Research Organisation (ISRO)
Vikram Sarabhai Space Centre
ISRO P.O, Thumba, Thiruvananthapuram (Trivandrum), 695022, India
Tel: (+91 471) 41 58 20
(+91 471) 56 41 23
(+91 471) 256 24 44
(+91 471) 256 25 55
Fax: (+91 471) 41 51 76
(+91 471) 41 53 58
(+91 471) 270 61 36
(+91 471) 256 41 05
Telex: 435201
435202
Cable: SPACE
■ Design and development of launch vehicles and related technologies.

Indonesian National Institute of Aeronautics & Space (LAPAN)
Ground Segment and Space Mission Centre
Rancabungur, Bogor, Jawa Barat, Indonesia
Tel: (+62 251) 62 16 67
Fax: (+62 251) 62 30 10

Indonesian National Institute of Aeronautics & Space (LAPAN)
Jalan Pemuda Persil No. 1, Rasamangun, Jakarta, 13220, Indonesia
Tel: (+62 21) 489 50 40
(+62 21) 489 28 02
(+62 21) 489 49 89
Fax: (+62 21) 489 48 15
e-mail: humaslapan@yahoo.com
humas@lapan.go.id
■ Government agency responsible for space programme.

Indonesian National Institute of Aeronautics & Space (LAPAN-LRB)
Space Communication Technology Division
Semplak, PO Box 13, 16310 Bogor, Indonesia
Tel: (+62 251) 62 16 67
Fax: (+62 251) 62 30 10

Indonesian Satellite Corp (INDOSAT)
Indosat Building, Jalan Medan Merdeka Barat 21, 10110 Jakarta, Indonesia
Tel: (+62 212) 815 23 45
■ Provides international telecommunications services via satellite.

Indra
Corporate Headquarters
Indra Sistemas, SA
Ceselsa and Inisel
Arroyd de la Vega, Avenida de Bruselas 35, Alcobendas, E-28108 Madrid, Spain
Tel: (+34 91) 480 50 00
Fax: (+34 91) 480 50 57
e-mail: indra@indra.es
Web: www.indra.es
■ Radars, automatic test systems, air defence, simulation, avionics, air traffic management, radar display consoles, electronic warfare, command and control, fire control systems, satellite communications, earth observation systems, maintenance and logistic support.
Member companies:
Indra ATM
Indra EMAC
Indra Espacio SA
Indra Sistemas de Seguridad

Industrie Cometto SpA
Via Cuneo No. 20, I-12011 Borgo San Dalmazzo CN, (Cuneo), Italy
Tel: (+39 0171) 26 33 00
Fax: (+39 0171) 26 63 35
e-mail: cometto@comettoind.com
Web: www.comettoind.com
■ Low-bed units for the transportation of heavy equipment; tanks and special loads in terms of weight and dimensions. Self-propelled units for various purposes including a NASA space shuttle SPU transporter and shipyard elevating transporters.

Industrieanlagen-Betriebs GmbH (IABG)
Einsteinstrasse 20, D-85521 Ottobrunn, Munich, Germany
Tel: (+49 89) 60 88 27 84
(+49 89) 896 08 80
Fax: (+49 89) 60 88 13 27 84
(+49 89) 60 88 40 00
e-mail: info@iabg.de
Web: www.iabg.de
■ Scientific and technical services including systems engineering, operations research studies, logistics, aircraft and spacecraft technology. Qualification and acceptance testing, fatigue and static strength tests, environmental tests, realistic field tests and defence weapons systems simulation.

Infoterra Ltd
Barwell Office
National Remote Sensing Centre Ltd
Arthur Street, LE9 8GZ Barwell, Leicestershire, United Kingdom
Tel: (+44 1455) 84 92 29
(+44 1455) 84 92 03
Fax: (+44 1455) 84 17 85
e-mail: info@infoterra-global.com
Web: www.infoterra-global.com
■ Deliver advanced geo-information solutions to DGIA, European defence mapping agencies, JARIC, ESA, ordnance survey and the UK Armed services.

Infoterra Ltd
National Remote Sensing Centre Ltd
a wholly owned subsidiary of Astrium
Atlas House, 41 Wembley Road, Leicester, LE3 1UT, United Kingdom
Tel: (+44 116) 273 23 00
Fax: (+44 116) 273 24 00
(+44 1455) 84 17 85
e-mail: info@infoterra-global.com
Web: www.infoterra-global.com
■ Leading supplier of data and information products and services from Earth observation satellites, aircraft and ground observation.

Inmarsat Ltd
99 City Road, London, EC1Y 1AX, United Kingdom
Tel: (+44 20) 77 28 17 77
Fax: (+44 20) 77 28 11 42
e-mail: customer_care@inmarsat.com
Web: www.inmarsat.com
■ Provision of global mobile satellite communications, delivering voice and high-speed data solutions for use in the air, at sea and on land to governments and enterprises.

IN-SNEC
Les Ulis Office
5 Avenue des Andes, BP 101 F-91943, Les Ulis, Cedex A, France

Tel: (+33 1) 69 82 78 00
Fax: (+33 1) 69 07 39 50
■ Vehicle borne TM kits or launchers, missile, aircraft. PCM telemetry pre-processing for the ground counterpart. TT and C base band station.

IN-SNEC
a subsidiary of Zodiac, France
5, Avenue des Andes – ZA de Courtaboeuf, BP 101
F-91943 Les Ulis, Cedex A, France
Tel: (+33 1) 69 82 78 00
(+33 2) 31 29 49 01
(+33 2) 31 29 49 49
Fax: (+33 2) 31 80 65 49
(+33 1) 69 07 39 50
Web: www.in-snec.com
■ Design, engineering and manufacture of complete ground station (telemetry and tracking system) CE-GPS. Mobile satellite communications terminals.

Inspace Group of Companies
Informcosmosconsult LLC
9 Krasnoproletarskaya Street, 103030 Moskva, Russian Federation
Tel: (+7 095) 978 85 54
(+7 095) 978 85 65
(+7 095) 978 85 69
Fax: (+7 095) 978 85 38
e-mail: group@inspace.ru
Web: www.inspace.ru
■ Commercial ventures in satellite manufacture, launch and operation, formation of strategic partnerships and legal consultancy. Official council to the Russian Aviation & Space Agency.
Offices in
Hong Kong
Russia
USA

Inspec Foams Inc
IMI-Tech Corp
101 East Park Boulevard, Suite 201, Plano, Texas, 75074, United States
Corporate Center One, 800 North Watters Road, Suite 190, Allen, Texas, 75013, United States
Tel: (+1 972) 516 07 02
Freephone: (+1 800) 666 76 10
Fax: (+1 972) 516 06 24
e-mail: moreinfo@solimide.com
Web: www.inspecfoams.com
■ Manufacture of Solimide polyimide flexible foams used for fire protection, thermal and acoustic insulation in marine and aircraft applications. Cushioning for aircraft and space applications.

Institut d'Astrophysique
98bis Boulevard Arago, F-75014 Paris, France
Tel: (+33 81) 34 67 11 11

Institut für Astrophysik und Extraterrestrische Forschung (IAEF)
Auf dem Hügel 71, D-53121 Bonn, Germany
Tel: (+49 228) 73 36 76
(+49 228) 73 36 76
Fax: (+49 228) 73 36 72
e-mail: user@astro.uni-bonn.de
■ Space science and astronomy

Institute for Astronautics Information (IAI)
a subsidiary of China Aerospace Corp.
1 Binhe Lu, Hepingli, PO Box 1408, 100013 Beijing, China
Tel: (+86 010) 68 37 34 00
Web: www.space.cetin.net.cn
■ Provides project consultancy, translation and interpretation services, networked information dissemination, advertising, database building, patent services, audio/video services and space policy analysis. Co-ordinates STI services within CASC, manages 16 specialised information networks and sponsors 38 aerospace related periodicals.

Institute for Biomedical Problems (RFSSC-IMBP, RAS)
State Scientific Center of Russian Federation–Institute of Biomedical Problems of Russian Academy of Sciences
Institut Mediko-Biologicheskikh Problemi, Khoroshovskoye Shosse 76a, 123007 Moskva, Russian Federation
Tel: (+7) 195 23 63
(+7) 195 15 00
e-mail: info@imbp.ru
Web: www.imbp.ru
■ Research in space biomedicine. Specific areas include the effect of stress on personnel under certain conditions–naval personnel on board atomic-powered submarines, on remote sites and in space.

Institute for Computer Applications in Science and Engineering (ICASE)
6 North Dryden Street, Building 1298, Room 103, Hampton, Virginia, 23681-0001, United States
Tel: (+1 757) 864 21 74
Fax: (+1 757) 864 61 34
e-mail: info@icase.edu

Institute for Geodesy, Cartography & Remote Sensing (FÖMI)
Remote Sensing Centre
H-1149 Budapest, Hungary
Tel: (+36 1) 163 66 69
Fax: (+36 1) 252 82 82
e-mail: h3874win@ella.hu

Institute for Geodesy, Cartography & Remote Sensing
Bosnyák tér 5, H-1149 Budapest, Hungary
Tel: (+36 1) 222 51 01
Fax: (+36 1) 222 51 12
Web: www.fomi.hu

Institute for Global Change Research and Education (IGCRE)
National Space Science and Technology Center
Global Hydrology and Climate Center
320 Sparkman Drive, Huntsville, Alabama, 35805, United States
Tel: (+1 256) 961 79 77
Fax: (+1 256) 961 77 23
e-mail: igcre@space.hsv.usra.edu

Institute for Space and Nuclear Power Studies (ISNPS)
University for New Mexico, Farris Engineering Center, Room 239, Albuquerque, New Mexico, 87131-1392, United States
Tel: (+1 505) 277 04 46
Fax: (+1 505) 277 28 14
e-mail: isnps@unm.edu
■ Research and development organization with a focus on space power and propulsion technologies and related fields.

Institute for Space Physics, Astrophysics and Education (ISPAE)
NASA, Marshall Space Flight Center, Building 4481, Alabama, 35812, United States
e-mail: ispae@space.hsv.usra.edu

Institute for Telecommunications Research (ITR)
University of South Australia, The Levels, South Australia, 5095, Australia
Tel: (+61 8) 83 02 33 36
Fax: (+61 8) 83 02 38 73
Web: www.itr.unisa.edu.au
■ Satellite communications systems and telecommunication services.

Institute for the Social Science Study of Space (ISSSS)
Fairmont State University, 1201 Locust Avenue, Fairmont, West Virginia, 26554, United States
Tel: (+1 304) 367 46 74
Fax: (+1 304) 367 47 85
■ Co-ordinating institution for common interest among professionals in sixteen social science disciplines in space use and exploration.

Institute for Unmanned Space Experiment Free Flyer (USEF)
Shinko Building, 2-12, Kanda-Ogawamachi, Chiyoda-Ku, Tokyo, 101-0052, Japan
Tel: (+81 3) 32 94 48 34
Fax: (+81 3) 32 94 71 63
Web: www.usef.or.jp
■ Promote research and development of unmanned space experiment systems.

Institute of Geodesy and Cartography
instytut Geodezji I Kartografii
Jasna 2/4, PL-00-950 Warszawa, Poland
Tel: (+48 22) 827 03 28
e-mail: igik@igik.edu.pl
stan@igik.edu.pl

Institute of Meteorology and Water Management Maritime Branch
Gdynia Office
ul. Waszyngtona 42, PL-81-342 Gdynia, Poland
Tel: (+48 2) 052 21
Telex: 016 41
■ Monitoring of radioactive pollution and participation in experiments on oceanographic and intercalibration exercises.

Institute of Sound and Vibration Research (ISVR)
University of Southampton, Highfield, Southampton, SO17 1BJ, United Kingdom
Tel: (+44 23) 80 59 22 94
Fax: (+44 23) 80 59 31 90
e-mail: mzs@isvr.soton.ac.uk
Web: www.isvr.soton.ac.uk
■ Teaching, research and consulting centre for the study of all aspects of sound and vibration. Defence and aerospace activities include work on spacecraft, satellites, aircraft, data processing recording systems, active control, automotive engineering, subjective acoustics and environmental noise.

Institute of Space and Astronautical Science (ISAS)
3-1-1 Yoshinodai, agamihara, Kanagwa, 229-8510, Japan
Tel: (+81 42) 751 39 11
Fax: (+81 42) 759 42 51
Web: www.isas.ac.jp
■ Government agency responsible for scientific research of space.

Instituto Astronomia
UNAM
Apt 70-264, DF 04510 Mexico
Tel: (+52 5) 622 39 06
Fax: (+52 5) 616 06 53

Instituto de Aeronáutica e Espaco (IAE)
Prac Mal.Eduardo Gomes, 12228-904 São José dos Campos, São Paulo, Brazil
Tel: (+55 12) 347 65 55
Fax: (+55 12) 341 25 22
Web: www.iae.cta.br

Instituto de Astronomia y Fisica del Espacio (IAFE)
Ciudad Universitaria, 1428 Buenos Aires, Argentina
Tel: (+54 1) 783 26 42
(+54 1) 781 67 55
(+54 11) 47 83 26 42
Fax: (+54 1) 786 81 14
e-mail: difusion@iafe.uba.ar
Web: www.iafe.uba.ar

Instituto Iberoamericano de Derecho Aeronautico y del Espacio y de la Aviacion Comercial
Plaza del Cardenal Cisneros 3, E-28040 Madrid, Spain
Tel: (+34 91) 336 63 74
Fax: (+34 91) 543 98 59
e-mail: en201067@encomix.net
■ Congress and publications on aeronautical and space law.

Instituto Nacional de Pesquisas Espacias (INPE)
Avenida dos Astronautas 1758, Jardim de Granja, CP 515 12227-010, São José dos Campos, São Paulo, Brazil
Tel: (+55 12) 345 69 82
(+55 12) 345 60 00
(+55 12) 345 69 84
(+55 12) 345 60 29
Fax: (+55 12) 345 69 80
(+55 12) 341 20 77
Telex: 123 2530

e-mail: escada@dir.inpe.br
makayano@dir.inpe.br
Web: www.inpe.br
■ Responsible for the ground and space segments of Brazilian applications satellite programmes.

Instituto Nacional de Tecnica Aeroespacial (INTA)
Carretera de Ajalvir, Km 4, E-28850 Torrejón de Ardoz, Madrid, Spain
Tel: (+34 91) 520 21 81
(+34 91) 520 17 74
Fax: (+34 91) 520 15 86
(+34 91) 520 19 39
Web: www.inta.es
■ Ministry of Defence autonomous agency acting as aerospace, scientific, technological and experimental centre.

Instrumentation Technology Associates Inc (ITA)
Administration Office
110 Pickering Way, Suite 100, Exton, Pennsylvania, 19341, United States
Tel: (+1 610) 363 83 43
Fax: (+1 610) 363 85 69
e-mail: itaincusa@aol.com
Web: www.itaspace.com

Integral Systems Inc
5000 Philadelphia Way, Lanham, Maryland, 20706-4417, United States
Tel: (+1 301) 731 42 33
Fax: (+1 301) 731 96 06
Web: www.integ.com

Intel Corp
Chandler
5000 West Chandler Boulevard, Chandler, Arizona, 85226, United States
Tel: (+1 602) 554 80 80
Web: www.intel.com
■ Manufacture of wireless communication systems.

Intelsat (Bermuda) Ltd
Wellesley House North, 2nd Floor, 90 Pitts Bay Road, Pembroke, HM 08, Bermuda
Tel: (+1 441) 294 16 50
Fax: (+1 441 292) 292 83 00
Web: www.intelsat.com
■ Provides telephony corporate network, video and internet solutions around the globe.

Intelsat Brasil Ltd
Sala 1601 Centro Empresarial Internacional, 20090-003 Rio de Janeiro, Brazil
Tel: (+55 21) 213 89 00
Fax: (+55 21) 213 89 01
Web: www.intelsat.com
Representing: Intelsat Global Service Corporation, US

Intelsat China (Hong Kong) Ltd
Room 2502A, 25/F, Tower 1, Admiralty Centre, 18 Harcourt Road, Admiralty, Hong Kong
Tel: (+852) 25 20 20 32
Fax: (+852) 28 66 20 22
Web: www.intelsat.com
Representing: Intelsat Global Service Corporation, US

Intelsat France SAS
72, rue du Faubourg, Saint Honore, F-75008 Paris, France
Tel: (+33 1) 40 07 81 13
Fax: (+33 1) 40 07 80 20
Web: www.intelsat.com
Representing: Intelsat Global Service Corporation, US

Intelsat Germany GmbH
Schierholzstrasse 27, D-30655 Hannover, Germany
Tel: (+49 511) 956 99 58
Fax: (+49 511) 956 99 33
Web: www.intelsat.com
Representing: Intelsat Global Service Corporation, US

Intelsat Global Sales & Marketing Ltd
566 Chiswick High Road, Building 3, Chiswick Park, London, W4 5YA, United Kingdom
Tel: (+44 20) 89 99 60 35
Fax: (+44 20) 88 99 61 94
Web: www.intelsat.com

Intelsat Global Service Corporation
Headquarters
3400 International Drive, North West, Washington, District of Columbia, 20008, United States
Tel: (+1 202) 944 68 00
(+1 202) 944 75 00
(+1 202) 944 74 06
(+1 202) 944 85 35
Fax: (+1 202) 944 78 98
e-mail: media.relations@intelsat.com
corporate.communications@intelsat.com
investor.relations@intelsat.com
Web: www.intelsat.com
■ Provide telephony corporate network, video and internet solutions around the globe via capacity on 26 geosynchronous satellites in prime orbital locations.

Intelsat Global Services Corporation
Beijing Representative Office
Suite 1916, China World Trade Tower 1, No 1 Jian Guo Men Wei Avenue, 10004 Beijing, China
Tel: (+86 10) 84 86 18 22
Fax: (+86 10) 84 86 18 21
Web: www.intelsat.com
Representing: Intelsat Global Service Corporation, US

Intelsat Marketing India Private Ltd
International Trade Tower, Nehru Place, New Delhi, 110 019, India
Tel: (+91 11) 51 60 85 73
Fax: (+91 11) 51 60 85 74
Web: www.intelsat.com
Representing: Intelsat Global Service Corporation, US

Intelsat Singapore Pte Ltd
80 Robinson Road, #14-02, 68898, Singapore
Tel: (+65) 62 27 30 24
Fax: (+65) 62 24 59 88
Web: www.intelsat.com
Representing: Intelsat Global Service Corporation, US

Intelsat South Africa (pty) Ltd
138 West Street, Second Floor, 2146 Sandon, South Africa
Tel: (+27 11) 535 47 00
Fax: (+27 11) 535 47 69
(+27 11) 884 72 69
Web: www.intelsat.com
Representing: Intelsat Global Service Corporation, US

Intelsat United Arab Emirates
Office No. 411, Building 11, 4th Floor, Dubai Internet City, Dubai, United Arab Emirates
Tel: (+971 4) 390 15 15
Fax: (+971 4) 390 86 76
Web: www.intelsat.com
Representing: Intelsat Global Service Corporation, US

Interdisciplinary Centre for Technological Analysis & Forecasting (ICTAF)
Space and Remote Sensing Division
Tel-Aviv University, IL-69978 Tel-Aviv, Israel
Tel: (+972 3) 640 75 75
Fax: (+972 3) 641 01 93
e-mail: ictaf@post.tau.ac.il
Web: www.ictaf.tau.ac.il
■ Interdisciplinary research centre carrying out policy related projects, as well as technology surveys in space related fields. Promotion of remote sensing applications in Israel. Applications especially in thermal imagery analysis, in geology, forestry and urban environmental issues. ICTAF is the Israeli distributor for all types of satellite imagery for Space Imaging Eurasia, SPOT Image and Eurimage as well as representative for the Israeli Space Agency's receiving station and international distribution companies.

Intermap Technologies Corp
Corporate Office
#1000, 736-8th Avenue, Calgary, Alberta, T2P 1H4, Canada
Tel: (+1 403) 266 09 00
Fax: (+1 403) 265 04 99
e-mail: info@intermaptechnologies.com

Intermap Technologies Corp
Ottawa Office
2 Gurdwara Road, Suite 200, Nepean, Ontario, K2E 1A2, Canada
Tel: (+1 613) 226 54 42
Fax: (+1 613) 226 55 29

International Astronautical Federation (IAF)
Fédération Internationale d'Astronautique, 3-5 rue Mario Nikis, F-75015 Paris, France
Tel: (+33 1) 45 67 42 60
Fax: (+33 1) 42 73 21 20
Web: www.iafastro.com

International Institute of Air and Space Law
PO Box 9520, NL-2300 RA Leiden, Netherlands
Faculty of Law, Leiden University,
Steenschuur 25, NL-2311 ES Leiden, Netherlands
Tel: (+31 71) 527 77 24
Fax: (+31 71) 527 76 00
Web: www.iiasl.leiden.edy
■ Education, research and advisory activities on all legal and policy issues related to international space activities, including satellite communications, satellite navigation, launching and institutional issues, at all levels.

International Institute of Space Law (IISL)
Institut International de Droit Spatial
3-5 rue Mario Nikis, F-75015 Paris, France
Tel: (+33 1) 45 67 42 60
Fax: (+33 1) 42 73 21 20

International Launch Services (ILS)
Headquarters
1660 International Drive, Suite 800, McLean, Virginia, 22102-3900, United States
Tel: (+1 571) 633 74 00
Fax: (+1 571) 633 75 00
Web: www.ilslaunch.com
■ Provide launch services for satellites.

International Organization of Space Communications (Intersputnik)
Headquarters
2 Smolensky per, 1/4, 121099 Moskva, Russian Federation
Tel: (+7 495) 244 03 33
Fax: (+7 495) 253 99 06
e-mail: dir@intersputnik.com
Web: www.intersputnik.com
■ Direct access global organisation which can be used by licensed users of satellite services. Offers all satellite-based communication services (voice, data, TV and audio broadcasting) through LMI-1 and Express-A, and Express-AM geostationary satellites together with full range of project engineering services.

International Science and Technology Center (ISTC)
Luganskaya Ulitsa 9, 115516 Moskva, Russian Federation
Tel: (+7 501) 797 60 10
Fax: (+7 501) 797 60 47
e-mail: istcinfo@istc.ru
Web: www.istc.ru
■ Provide peaceful research opportunities to weapons scientists in CIS countries.

International Space Brokers (ISB)
1300 Wilson Boulevard, Suite 990, Rosslyn, Virginia, 22209, United States
Tel: (+1 703) 276 49 00
Fax: (+1 703) 276 49 01
Web: www.isbworld.com

International Space University (ISU)
North America Office
409 3rd Street, Suite 206, Washington, District of Columbia, 20024, United States
Tel: (+1 703) 522 67 31
Fax: (+1 703) 533 94 42
Web: www.isunet.edu

International Space University (ISU)
Parc d'Innovation, 1 Rue Jean-Dominique Cassini, F-67400 Illkirch, France
Tel: (+33 3) 88 65 54 30
Fax: (+33 3) 88 65 54 47
e-mail: info@isu.isunet.edu
Web: www.isunet.edu

International University Bremen ((IUB))
PO Box 750561, D-28725 Bremen, Germany
International University Bremen (IUB), Campus Ring 1, D-28759 Bremen, Germany
Tel: (+49 0421) 200 44 55
Fax: (+49 0421) 200 44 53
e-mail: iub@iu-bremen.de
Web: www.iu.bremen.de

Interorbital Systems
PO Box 662, Mojave, California, 93502-0662, United States
Tel: (+1 661) 965 07 71
e-mail: ios@interorbital.com

Interpoint
Integrated Circuits Incorporated
a subsidiary of Crane Co., US
10201 Willows Road, PO Box 97005, Redmond, Washington, 98073-9705, United States
Tel: (+1 425) 882 31 00
Freephone: (+1 800) 822 87 82
Fax: (+1 425) 882 19 90
e-mail: power@intp.com
Web: www.interpoint.com
■ Manufacture of high-density, high-reliability, microelectronics and power conversion products for the aerospace, space, military, medical, industrial and commercial markets. Products qualified to MIL-PRF-38534 Class H or K and space products available to radiation level R. Provides turn-key, co-design and build-to-print options specialising in light weight, high-reliability microelectronic assembly and test. Capabilities include BGA, flip-chip, COB, COF and hermetically sealed devices.

Intertechnique SA
Headquarters
61 rue Pierre Curie, BP 1, F-78373 Plaisir, Cedex, France
Tel: (+33 1) 30 54 82 00
Fax: (+33 1) 30 55 71 61
■ Fuel systems, fuel gauging, flowmeters, fuel management systems, fuel pumps and flow proportioners, valves, couplings, in-flight refuelling equipment. Air temperature and pressure control systems, oxygen and pneumatic systems, military oxygen regulators, anti-g valves, LOX converters, quick-donning masks, emergency oxygen equipment. Seekers and triggering devices for missiles and munitions, infra-red line-scan camera for surveillance and observation. Ground-based and airborne PCM telemetry.
Subsidiaries:
ECE, France
EROS (GIE Intertechnique – SFIM), France
IN-LHC, France
IIN-FLEX, France
IN-Aerospace Ltd, UK

Intespace
Ingenerie Test and Environmental Spatial
a joint venture between Telespace Participation, Astrium and Alcatel Space Industries, France
18 avenue Edouard Belin, BP 4356, F-31029 Toulouse, Cedex 4, France
Tel: (+33 5) 61 28 11 11
Fax: (+33 5) 61 28 11 12
e-mail: marketing@intespace.fr
Web: www.dynaworks.com
■ Environmental test including mechanical, thermal and electrical, for all sectors of industry. On-site assistance for European customers for operating and maintaining test facilities. Engineering and consulting services in environmental technology. Test centre engineering, principally for the space industry.

Irvin Aerospace Canada Ltd
Irvin Industries Canada Ltd
a subsidiary of Airborne Systems UK
35 Wilson Avenue, Belleville, Ontario, K8P 1R7, Canada
Tel: (+1 613) 967 80 69
Fax: (+1 613) 967 04 69
e-mail: marketing@irvincanada.com
Web: www.airbornesystems-na.com/main.html
■ Parachutes for personnel, cargo, weapon and aircraft retardation. Recovery systems for missiles and space vehicles, cargo delivery systems. Safety harnesses and air drop sequence control devices. Inflatable survival products and systems, missile recovery parachutes, flare chutes, jump tower chutes and safety harnesses.
Accreditations: ISO 9001:2000

ISC Kosmotras
PO Box 7, 123022 Moskva, Russian Federation
Tel: (+7 495) 745 72 58
Fax: (+7 495) 956 16 59
(+7 495) 232 34 85
e-mail: info@kosmotras.ru
Web: www.kosmotras.ru

Iskra Science and Production Association
Academician Vedeneyev St 28, 614038 Perm, Russian Federation
Tel: (+7 3422) 72 80 00
Fax: (+7 3422) 72 58 98
(+7 3422) 72 67 54
■ Development of gas generators and solid-propellant rocket engines.

Israel Aerospace Industries Ltd (IAI)
Headquarters
Bedek Aviation
Israel Aircraft Industries Ltd
Ben Gurion International Airport, IL-70100 Tel-Aviv, Israel
Tel: (+972 3) 935 31 11
(+972 3) 935 33 43
(+972 3) 935 81 11
(+972 3) 935 85 09
(+972 3) 935 85 14
(+972 3) 935 85 41
Fax: (+972 3) 935 33 96
(+972 3) 935 85 16
(+972 3) 935 82 78
e-mail: hpaz@iai.co.il
istern@iai.co.il
Web: www.iai.co.il
■ Research, design, development, integration, testing, manufacture, upgrading, support and service of land, sea, air systems equipment and space technologies. Military and civil aircraft, UAVs, missiles, hydraulic, electro-mechanical, electronic and optronic systems, ordnance and navigation equipment, patrol boats, mine clearance devices, C3I systems, NCW Network Centric Warfare, airborne and ground based radars.
Accreditations: ISO 9001
Subsidiaries:
Israel Aircraft Industries-Europe, France
Israel Aircraft Industries International Inc, US

Israel Aerospace Industries Ltd (IAI)
MBT Missile Systems Division
a part of Israel Aerospace Industries (IAI), Systems Missile and Space Group
Yehud Industrial Zone, PO Box 105, IL-56000 Yehud, Israel
Tel: (+972 3) 531 40 01
(+972 3) 531 40 09
Fax: (+972 3) 531 41 10
e-mail: marketing@mbt.iai.co.il
istern@iai.co.il
Web: www.iai.co.il
■ Design and development of satellites, full satellite integration and testing, launching and in-orbit operations. Also provides mission control centres, tracking and multi-satellite remote sensing stations. Design and manufacture of smart bombs, fire control systems, offensive and defensive precision and tactical battlefield missiles. Products include Nimrod, Gabriel, Harpy air to surface missiles; Griffin and Guillotine bombs. Design and manufacture of precision optical and navigation systems.

Israel Aerospace Industries Ltd (IAI)
MLM Division
Israel Aircraft Industries Ltd
a part of Israel Aerospace Industries (IAI), Systems Missile and Space Group
PO Box 45. IL-70350 Beer Yaakov, Israel
Tel: (+972 8) 972 24 25
(+972 8) 927 30 26
Fax: (+972 8) 927 28 90
e-mail: mlm_marketing@mlm.iai.co.il
Web: www.iai.co.il
■ Prime activites include: Anti-ballistic missile defence systems such as the Arrow, launchers and space systems including the Shavit small satellite launchers; advanced avionics, training and simulation systems including the EHUD autonomous ACMI for fighter aircraft and similar tailored systems for trainers, helicopters and situational awareness systems (SAW). Information Technology (IT) and command, control, communication and intelligence (C3I) systems. Communication, telemetry and test/launch range equipment and systems.

Israel Aerospace Industries Ltd (IAI)
Tamam Division
Israel Aircraft Industries Ltd
Tamam Precision Instrument Industries
a part of Israel Aerospace Industries (IAI), Systems Missile and Space Group
Yehud Industrial Zone, PO Box 75, IL-56100 Yehud, Israel
Tel: (+972 3) 531 53 34
(+972 3) 531 50 03
Fax: (+972 3) 531 51 40
e-mail: infotamam@tamam.iai.co.il
Web: www.iai.co.il
■ Design and manufacture of inertial measurement, stabilisation, navigation and optronic systems and components for land, sea and air, manned and unmanned platforms, specialised civil space components.

Israel Aerospace Medicine Institute (IAMI)
PO Box 4572, IL-91044 Jerusalem, Israel
Tel: (+972 2) 641 43 33

Israel Military Industries (IMI)
TAAS Israel Industries Ltd
64 Bialik Avenue, PO Box 1044, IL-47100 Ramat Hasharon, Israel
Tel: (+972 3) 548 57 45
(+972 3) 548 63 89
Fax: (+972) 548 53 65
e-mail: had-marketing@imi-israel.com
Web: www.imi-israel.com
■ Manufacture of small arms, armoured fighting vehicles, ammunition, electronics systems and equipment, command and control equipment, airborne systems, countermeasures, rockets, unmanned aerial vehicles, personal protection equipment and security systems. Ammunition produced includes small arms: 5.56×45 mm M193, 7.62 × 39 mm M1943, 7.62 × 51 mm, .30-06, .50 BMG. Mortar: 51 mm, 52 mm, 81 mm and 120 mm. Rifle grenades. Tank: 60 mm APFSDS, 105 mm APFSDS, TPFSDS-T, I-HEAT-T M152/3, APAM, LAHAT and 120 mm APFSDS, LAHAT. Artillery: 130 mm HE, 152 mm ICM, 155 mm HE M107, Smk WP M110AS, Smk HC M116A1, Illg M485A2, 155 mm HE DPICM M395, 175 mm Cargo M366. Rockets 160 mm, 240 mm, 350 mm.
■ Air launched weapons include Delilah, STAR-1 ITALD, MAPATS and MSOV air to surface missiles; ATAP-300/-500/-1000, MSOV, Condib 70/120, M-2000, PB-500A1, IBDU-33, Mk 81, 82, 83, 84 GP and RAM bombs; Negev 5.56 mm LMG guns.

Israel Space Agency (ISA)
26a Chaim Levanon Street, PO Box 17185, IL-61171 Tel-Aviv, Israel
Tel: (+972 3) 642 22 97
Fax: (+972 3) 642 22 98
e-mail: aby@most.gov.il
Web: www.most.gov.il
■ Sub-committees: research and technology, space infrastructure and industrial applications, external relations.

Ist di Radioastronomia
Via Gobetti 101, I-40129 Bologna, Italy
Tel: (+39 51) 639 03 84
Fax: (+39 51) 639 94 31

ISTAR
2600 Route des Crétes, BP 282 F-06905 Sophia-Antipolis, Cedex, France
Tel: (+33 4) 97 23 23 23
Fax: (+33 4) 93 95 83 29
Web: www.istar.com
■ Cartographic databases. RS satellite image processing, specialising in digital terrain models.

Italian Small and Medium Enterprise Aerospace Association
Via del Tempio 1, Roma, Italy
Tel: (+39 06) 686 92 22
Fax: (+39 06) 686 92 22
Web: www.aipas.it

Italian Space Agency (ASI)
Centro di Geodesia Spaziale
Agencia Spaziale Italiana, Localtia Terecchia, I-75100 Matera, Italy
Tel: (+39 0835) 37 79
Fax: (+39 0835) 33 90 05
Telex: 812535
Web: www.asi.it

Italian Space Agency (ASI)
Headquarters
Agencia Spaziale Italiana, 26 Viale Liegi, I-00198 Roma, Italy
Tel: (+39 06) 856 71
Fax: (+39 06) 440 41 86
e-mail: graziani@asi.it
Web: www.asi.it
■ Government agency in charge of the co-ordination and management of the Italian space programme.

Italian Space Agency (ASI)
Stratospheric Balloons Launch Site
Base di Lancio, SS 113 N174 Contrada Milo, I- 91100 Trapani, Italy
Tel: (+39 0923) 53 99 28
(+39 0923) 53 90 36
Fax: (+39 0923) 53 84 93
Telex: 911253-910263

Iteris
Odetics
1515 S Manchester Avenue, Anaheim, California, 92802, United States
Tel: (+1 714) 774 50 00
(+1 714) 780 72 81
Fax: (+1 714) 780 78 57
(+1 714) 780 72 46
e-mail: lss2@iteris.com
Web: www.odetics.com
■ Manufacture of equipment for security systems, telecommunications and transportation industries.

Itochu Corporation
Tokyo Head Office
C Itoh & Co Ltd
2-5-1 Kita-Aoyama, Minato-ku, Tokyo, 107-8077, Japan
Tel: (+81 3) 34 97 21 21
Fax: (+81 3) 349 79 91
Affiliates
JAMCO Corp., Japan
ITOCHU Aviation Inc., USA

ITT Corporation
Space Systems Division
1447 Saint Paul Street, Rochester, New York, 14653, United States
Tel: (+1 585) 269 50 60

ITT Corporation A/CD
Aerospace/Communications
ITT Aerospace/Communications
ITT Industries
an ITT Industries Company, US
756 Lakefield Road, Building J, Westlake Village, California, 91361-2624 United States
Tel: (+1 219) 451 60 00
(+1 260) 451 63 84
(+1 260) 451 60 42
Fax: (+1 260) 451 50 92
(+1 260) 451 61 26
Web: www.acd.itt.com
■ Design, development and production of forward area battlefield, tactical communications, signal processing equipment and systems. Voice and data communications receiving and transmitting systems; key management systems, surveillance and reconnaissance receiver systems; voice and data secure communications equipment and meteorological and navigation spaceborne payloads. Qualified for ground, airborne and space applications. Featuring microprocessor control and standard digital interface.

ITT Corporation Headquarters
Corporate Offices
International Telephone & Telegraph
ITT Industries Inc
4 West Red Oak Lane, White Plains, New York, 10604, United States
Tel: (+1 914) 641 20 00
Fax: (+1 914) 696 29 50
Web: www.itt.com
■ Engineering and manufacturing company supplying pumps, systems and services to move, control and treat water and other fluids. Also supplies military defence systems and provides advanced technical and operational services to a broad range of government agencies. Produces electrical connectors for telecommunications, computing and aerospace as well as industrial applications. Industrial components for the transportation, construction and aerospace markets.

ITT Defence Ltd
ITT Defense International UK
Jays Close, Viables Estate, Basingstoke, Hampshire, RG22 4BA, United Kingdom
Tel: (+44 1256) 31 16 00
Fax: (+44 1256) 84 05 56
e-mail: ittdefence@uk.itt.com
Web: www.ittdefence.co.uk
■ Systems integration company with major interest in tactical communications systems. Supports the UK interests of ITT Defence in the areas of airborne electronic warfare, night vision image intensification devices, ATC radar and displays, tactical and strategic radar, communications systems, VHF radios and space-based sensors.
Representing: ITT Defense, US

ITT Defense
ITT Defense & Electronics
1650 Tysons Boulevard, Suite 1700, Mclean, Virginia, 22102, United States
Tel: (+1 703) 790 63 00
Fax: (+1 703) 790 63 60
Web: defense.itt.com
■ Supply of advanced technology command, control and communications, electronic defence, electro optical and surveillance systems, satellite imaging systems and information services, navigation and meteorological space payloads and related operations and management services. the company's primary business markets include avionics and data communications for military, space and transportation customers.
Group Companies:-
ITT Industries, Advanced Engineering and Sciences
ITT Industries, Aerospace/Communications
ITT Industries, Avionics
ITT Industries, Defence International
ITT Industries, Night Vision
ITT Industries, Space Systems
ITT Industries, Systems
ITT Industries, Gilfillan

ITT Defense Electronics and Services
EDO Corporation
60 East 42nd Street, 42nd Floor, New York, New York, 10165, United States
Tel: (+1 212) 716 20 00
(+1 212) 716 20 71
(+1 212) 716 20 38
Fax: (+1 212) 716 20 50
e-mail: ir@edocorp.com
Web: www.edocorp.com
■ Provide military and commercial products and professional services, with core competencies in a wide range of critical defence areas, including: Aircraft armament, defence electronics, undersea warfare, Command, Control, Communications, Computers and Intelligence (C4I); and integrated composite structures.
Companies represented:
Antenna Products & Technologies, Communications & Networking Systems, Communications and Countermeasures Systems, Defense Systems, Reconnaissance and Surveillance Systems, EDO-EVI/NexGen, EDO-IST, EDO (UK) Ltd (MBM Technology Ltd), Fiber Innovations, Fiber Science, MTech, Artisan, Specialty Plastics, Acquisition and Logistics Management Operations, EDO-CAS, Technical Services Operations, Technical Services Operations, Electro-Ceramic Products, Naval Command & Sonar Systems

Izmiran
Troitsk, 142092 Moskva, Russian Federation
Tel: (+7 095) 334 01 20
Fax: (+7 095) 334 01 24
Telex: 412623 scstp

J

J C Aviation International Pty Ltd (JCAI)
5b Jubilee Avenue, Warriewood, New South Wales, 2102, Australia
Tel: (+61 2) 99 79 17 77
Fax: (+61 2) 99 79 17 88
e-mail: jc@jc-aviation.com
Web: www.jc-aviation.com
■ Supplier of Ni-Cad batteries, borescopes, diesel and electric GPU's, frequency changers, rechargeable hi-intensity flashlights, ground support equipment, aircraft spares and tooling, air-conditioning units and aircraft tugs.

J S Chinn Engineering Co Ltd
J S Chinn Engineering
Chinn Engineering Co Ltd, J S
a subsidiary of J S Chinn Holdings Ltd, UK
Faraday Road, Harrowbrook Industrial Estate, Hinckley, Leicestershire, LE10 3DE, United Kingdom
Tel: (+44 1455) 63 33 53
(+44 1455) 23 83 33
Fax: (+44 1455) 25 12 83
(+44 1455) 89 05 85
■ Precision aerospace fabrications to 20, assembly and handling jigs, ground equipment and test rigs, satellite packaging and handling equipment. Special purpose test equipment for aerospace applications.

J S Chinn Holdings Ltd
Chinn Holdings Ltd,
J S, Coventry Road, Exhall, Coventry, West Midlands, CV7 9FT, United Kingdom
Tel: (+44 24) 76 36 94 00
■ Sub-contractors in precision fabrication and machining of components and complete assemblies for aero engines, airframe, satellites, missiles and underwater weapons and a wide range of aerospace applications.
Subsidiaries:
A O Henton Engineering Co Ltd
Colledge & Morley (Gears) Ltd
J S Chinn & Co Ltd
J S Chinn Engineering Co Ltd
J S Chinn Project Engineering Ltd

J. Allen Hynek Center for UFO Studies (CUFOS)
2457 W Peterson Avenue, Suite 6, Chicago, Illinois, 60659, United States
Tel: (+1 773) 271 36 11
e-mail: infocenter@cufos.org
Web: www.cufos.org/index.html
■ Articles and news on all aspects of the UFO phenomenon.
Publications:
International UFO Reporter (quarterly journal)
Journal of UFO Studies, (annual scholarly journal)

Jackson and Tull
2705 Blandensburg Road North East, Washington, District of Columbia, 20018, United States
Tel: (+1 202) 333 91 00
Fax: (+1 202) 526 28 76
e-mail: infogen@jny.com
Web: www.jacksonandtull.com
■ Provision of engineering and technical services.

Jacobs Technology
Jacobs Sverdrup
600 William Northern Boulevard, Tullahoma, Tennessee, 37388, United States
Tel: (+1 931) 455 64 00

For details of the latest updates to *Jane's Space Systems and Industry* online and to discover the additional information available exclusively to online subscribers please visit **jsd.janes.com**

e-mail: webmaster@jacobssverdup.com
Web: www.jacobstechnology.com
■ Complete planning, design, construction and programme management services for defence/aerospace applications (including military bases, systems and facilities); specialised launch, test and manufacturing facilities; advanced technology and research facilities and operation/maintenance of major test facilities.
Subsidiaries:
Sverdrup Civil Inc
Sverdrup Environmental Inc
Sverdrup Facilities Inc
Sverdrup Investments Inc
Sverdrup Technology Inc

Jacobs Technology Inc (JS)
Jacobs Sverdrup Inc
a subsidiary of Jacobs Engineering Group, Inc., US
600 William Northern Boulevard, PO Box 884, Tullahoma, Tennessee, 37388, United States
Tel: (+1 931) 455 64 00
Fax: (+1 931) 393 63 89
e-mail: webmaster@jacobssverdup.com
Web: www.jacobssverdup.com
■ Provide scientific and engineering services in support of the design, construction, operations, maintenance, testing and evaluation of advanced facilities and systems for customers including DoD, NASA and a diversity of commercial clients.

Jane's Information Group
Asia
a branch of Jane's Information Group, UK
78 Shenton Way, #10-02, UIC Building, Singapore, 079120, Singapore
Tel: (+65) 63 25 08 66
Fax: (+65) 62 26 11 85
e-mail: asiapacific@janes.com
Web: www.janes.com
Representing: Jane's Information Group, UK

Jane's Information Group
Australia
a branch of Jane's Information Group, UK
PO Box 3502, Rozelle, Delivery Centre, New South Wales, 2039, Australia
Tel: (+61 2) 85 87 79 00
Fax: (+61 2) 85 87 79 01
e-mail: oceania@janes.com
Web: www.janes.com
Representing: Jane's Information Group, UK

Jane's Information Group
Europe, Middle East and Africa
Sentinel House, 163 Brighton Road, Coulsdon, Surrey, CR5 2YH, United Kingdom
Tel: (+44 20) 87 00 37 00
Fax: (+44 20) 87 63 10 06
e-mail: info.uk@janes.com
Web: www.janes.com
■ Jane's Information Group is the leading supplier of professional information to the defence and aerospace communities and a major information provider to selected areas of transportation and law enforcement worldwide. All Jane's publications, including magazines, are available on CD-ROM for personal computers and many UNIX platforms. All publications are now available ON-LINE (http://www.janes.com), updated a minimum of every three months. Jane's publishes Special Reports and provides consultancy services, data services and an extensive image library.
Data Services:
Defence Forecasts -
Military Aircraft Programmes
Combat Vehicle Programmes
Military Vessel Programmes
Defence Equipment Library:
Aero-Engines
Air-Launched Weapons
Aircraft Upgrades
All the World's Aircraft
Ammunition Handbook
Armour and Artillery
Armour and Artillery Upgrades
Avionics
C4I Systems
Electro-Optic Systems
Explosive Ordnance Disposal
Fighting Ships
Infantry Weapons
Land-Based Air Defence
Military Communications
Military Vehicles and Logistics
Mines and Mine Clearance
Naval Weapon Systems
Nuclear, Biological and Chemical Defence
Radar and Electronic Warfare Systems
Strategic Weapon Systems
Underwater Warfare Systems
Unmanned Aerial Vehicles and Targets
Defence Magazines Library:
Defence Industry
Defence Weekly
Foreign Report
Intelligence Digest
Intelligence Review
International Defence Review
Islamic Affairs Analyst
Missiles and Rockets
Navy International
Terrorism and Security Monitor
Market Intelligence Library:
Aircraft Component Manufacturers
All the World's Aircraft
Defence Industry
Defence Weekly
Electronic Mission Aircraft
Fighting Ships
Helicopter Markets and Systems
International ABC Aerospace Directory
International Defence Directory
Marine Propulsion
Naval Construction and Retrofit Markets
Police and Homeland Security Equipment
Simulation and Training Systems
Space Systems and Industry
Underwater Security Systems and Technology
World Armies
World Defence Industry
Security Library:
Amphibious and Special Forces
Chemical-Biological Defense Guidebook
Facility Security
Fighting Ships
Intelligence Digest
Intelligence Review
Intelligence Watch Report
Islamic Affairs Analyst
Police and Homeland Security Equipment
Police Review
Terrorism & Security Monitor
Terrorism Watch Report
World Air Forces
World Armies
World Insurgency and Terrorism
Sentinel Library:
Central Africa
Central America and the Caribbean
Central Europe and the Baltic States
China and Northeast Asia
Eastern Mediterranean
North Africa
North America
Oceania
Russia and the CIS
South America
South Asia
Southeast Asia
Southern Africa
The Balkans
The Gulf States
West Africa
Western Europe
Transport Library:
Aero-Engines
Air Traffic Control
Aircraft Component Manufacturers
Aircraft Upgrades
Airport Review
Airports and Handling Agents –
Central and Latin America (inc. the Caribbean)
Europe
Far East, Asia and Australasia
Middle East and Africa
United States and Canada
Airports, Equipment and Services
All the World's Aircraft
Avionics
High-Speed Marine Transportation
Marine Propulsion
Merchant Ships
Naval Construction and Retrofit Markets
Simulation and Training Systems
Transport Finance
Urban Transport Systems
World Airlines
World Railways

Jane's Information Group
Japan
a branch of Jane's Information Group, UK
CERAS 51 Building, 1-21-8 Ebisu, Shibuya-ku, Tokyo, 150-0013, Japan
Tel: (+81 3) 57 91 96 63
Fax: (+81 3) 54 20 64 02
e-mail: japan@janes.com
Web: www.janes.com
www.janes.jp
Representing: Jane's Information Group, UK

Jane's Information Group
London
1st Floor, The Quadrangle, 180 Wardour Street, London, W1A 4YG, United Kingdom
Tel: (+44 20) 87 00 37 00
Fax: (+44 20) 72 87 47 65
(+44 20) 72 87 77 65

Jane's Information Group
Middle East
a branch of Jane's Information Group, UK
PO Box 502138, Dubai,
United Arab Emirates
Tel: (+971 4) 390 23 35
(+971 4) 390 23 36
Fax: (+971 4) 390 88 48
e-mail: mideast@janes.com
Web: www.janes.com
Representing: Jane's Information Group, UK

Jane's Information Group
North/Central/South America
a branch of Jane's Information Group, UK
110 North Royal Street, Suite 200, Alexandria, Virginia, 22314, United States
Tel: (+1 703) 683 37 00
(+1 703) 236 24 10
Freephone: (+1 800) 824 07 68
Fax: (+1 703) 836 02 97
(+1 800) 836 02 97
e-mail: info.us@janes.com
Web: www.janes.com
■ Visit Jane's on the net at http://www.janes.com.

Japan Aerospace Exploration Agency (JAXA)
Bangkok Office
National Space Development Agency of Japan
B B Building, Floor 13, Room 1305, 54 Asoke Road, Sukhumvit 21, Bangkok, 10110, Thailand
Tel: (+66 2) 260 70 26
Fax: (+66 2) 260 70 27

Japan Aerospace Exploration Industry (JAXA)
Bonn Office
National Space Development Agency of Japan
H 1204 Bonn Centre, 2-10 Bundeskanzlerplatz, D-53113 Bonn, Germany
Tel: (+49 228) 91 43 50
Fax: (+49 228) 911 21 50

Japan Aerospace Exploration Agency (JAXA)
Earth Observation Center
National Space Development Agency of Japan
1401 Aza-Numanoue, Oaza-Oohashi, Hiki-gun, Saitama, 350-0302, Japan
Tel: (+81 492) 98 12 00
Fax: (+81 492) 96 02 17
■ Satellite data acquisition and earth observation.

Japan Aerospace Exploration Agency (JAXA)
Houston Office
National Space Development Agency of Japan
Cyberonics Building 16511, Space Center Boulevard, Suite 201, Houston, Texas, 77058, United States
Tel: (+1 281) 280 02 22
Fax: (+1 281) 486 10 24
(+1 281) 228 04 89

Japan Aerospace Exploration Agency (JAXA)
Katsuura Tracking and Communication Station
1-14 Hanatateyama, Haga, Katsuura-shi, Chiba-ken, 299-5213, Japan
Tel: (+81 470) 73 06 54
Fax: (+81 470) 70 70 01
■ Tracking and control.

Japan Aerospace Exploration Agency (JAXA)
Masuda Tracking and Communication Station
National Space Development Agency of Japan
1897, Masuda, Nakatane-machi, Kumage-gun, Kagoshima, 891-3603, Japan
Tel: (+81 9972) 719 90
Fax: (+81 9972) 420 00
■ Tracking and control.

Japan Aerospace Exploration Agency (JAXA)
Nagoya Liaison Office
National Space Development Agency of Japan
Kanayama Sougo Building 10F, 1-12-14, Kanayama, Naka-ku, Nagoya-shi, Aichi, 460-0022, Japan
Tel: (+81 52) 332 32 51
Fax: (+81 52) 339 12 80

Japan Aerospace Exploration Agency (JAXA)
National Aerospace Laboratory
7-44-1 Jindaiji Higashi-machi, Chofu-shi, Tokyo, 182- 8522, Japan
Tel: (+81 4) 22 40 30 00
(+81 3) 62 66 60 61
(+81 3) 62 66 61 14
(+81 3) 62 66 61 49
(+81 3) 62 66 62 10
(+81 3) 62 66 62 49
(+81 3) 62 66 64 00
(+81 3) 62 66 64 94
Web: www.jaxa.jp
■ Aerospace research and development agency.

Japan Aerospace Exploration Agency (JAXA)
National Space Development Agency of Japan
Triton Square X-23F, 1-8-10 Harumi, Chuo-ku,Tokyo, 104-6023, Japan
Tel: (+81 3) 62 21 90 00
Fax: (+81 3) 62 21 91 91
■ Earth observation, research and development and study services.

Japan Aerospace Exploration Agency (JAXA)
Ogasawara Downrange Station
National Space Development Agency of Japan
Kuwanokiyama Chichijima, Ogasawara-mura,Tokyo, 100-2101, Japan
Tel: (+81 4998) 225 22
Fax: (+81 4998) 223 60
■ Tracking and control.

Japan Aerospace Exploration Agency (JAXA)
Okinawa Tracking and Communication Station
National Space Development Agency of Japan
1712, Kinrabaru, Aza Afuso, Onna-son, Kunigami-gun, Okinawa-Ken, 904-0402, Japan
Tel: (+81 989) 67 82 11
Fax: (+81 989) 83 30 01
■ Tracking and control.

Japan Aerospace Exploration Agency (JAXA)
Paris Office
National Space Development Agency of Japan
3 Avenue Hoche, F-75008 Paris, France
Tel: (+33 1) 46 22 49 83
Fax: (+33 1) 46 22 49 32

Japan Aerospace Exploration Industry (JAXA)
Tanegashima Space Center
Aza Mazu, Minamitane-machi, Kumage-gun, Kagoshima-Ken, 891-3703, Japan
Tel: (+81 9972) 621 11
Fax: (+81 9972) 440 04
■ Rockets, launching service and space vehicles.

Japan Aerospace Exploration Agency (JAXA)
Tokyo Office
National Space Development Agency of Japan
Marunouchi Kitaguchi Building, 1-6-5 Marunouchi, Chiyoda-ku,Tokyo, 100-8260, Japan
Tel: (+81 3) 62 66 64 00
Fax: (+81 3) 62 66 69 10
e-mail: proffice@jaxa.jp
Web: www.jaxa.jp
■ Established as a corporate entity charged with developing launchers and applications satellites. Undertakes development and launch of vehicles development and operation of earth-observation satellites; tracking and control of satellites. Launching services, research and development and rockets. Space science, aircraft, earth observation and satellite data aquisition.
Overseas offices
Bangkok, Bonn, Houston, Kennedy Space Center, Los Angeles, Paris, Washington DC,

Japan Aerospace Exploration Agency (JAXA)
Tsukuba Space Center
National Space Development Agency of Japan
2-1-1 Sengen, Tsukuba-shi, Ibaraki-ken, 305-8505, Japan
Tel: (+81 29) 868 50 00
Fax: (+81 29) 868 59 88
■ Research and development, study services, satellites, tracking and control, satellite data acquisition and spacecraft.

Japan Aerospace Exploration Agency (JAXA)
Washington Office
National Space Development Exploration Agency of Japan
2020 K Street NW, Suite 325, Washington, District of Columbia, 20006, United States
Tel: (+1 202) 333 68 44
Fax: (+1 202) 333 68 45

Japan Aviation Electronics Industry Ltd (JAE)
21-2 Dogenzaka, 1-chome, Shibuya-ku, Tokyo, 150-0043, Japan
Tel: (+81 3) 37 80 29 32
(+81 3) 37 80 27 52
Fax: (+81 3) 37 80 29 45
Web: www.jae.co.jp
■ Inertial navigation and guidance systems, automatic flight control systems, avionics. Connectors, switches and relays, fibre-optic products and electronic equipment for aerospace and related applications.

Japan Manned Space Systems Corp (JAMSS)
Otemachi Building, 1-6-1 Otemachi, Chiyoda-ku, Tokyo, 100-0004, Japan
Tel: (+81 3) 32 11 20 02
Fax: (+81 3) 32 11 20 04
Web: www.jamss.co.jp

Japan Radio Co Ltd (JRC)
Nittochi Nishi-Shinjuku Building, 10–1 Nishi-Shinjuku 6 Chome, Shinjuku-Ku, Tokyo, 160-8328, Japan
Tel: (+81 3) 33 48 01 51
Fax: (+81 3) 33 48 39 37
e-mail: pr_jrc@m1.jrc.co.jp
Web: www.jrc.co.jp
■ Manufacture of communications systems and equipment. Radio equipment, navigation systems and simulators.

Jehier
Jehier Ingénierie Informatique
Route de Saint Lézin, BP 29, F-49120 Chemillé, France
Tel: (+33 2) 41 64 54 00
Fax: (+33 2) 41 64 54 01
e-mail: info@jehier.fr
Web: www.jehier.fr
■ Thermal and acoustic insulation for performance applications in aerospace, marine, military vehicles and industry. Manufacture of materials and equipment with metallic or non-metallic cover.
Representing: Inspec, US

Jena-Optronik GmbH (JOP)
Subsidiary of Jenoptik AG (100%)
Prüssingstrasse 41, D-07745, Jena, Germany
Tel: (+49 3641) 20 01 10
Fax: (+49 3641) 20 02 20
(+49 3641) 20 02 22
e-mail: info@jena-optronik.de
Web: www.jena-optronik.de
■ Development and production of components and systems for aerospace and security. Development, manufacture, assembly and test of opto-electronic sensors and instruments; optical, electrical and mechanical ground support equipment (hardware and software); software for satellite on-board data handling and instrument data processing.

Jet Propulsion Laboratory NASA (JPL)
NASA, Jet Propulsion Laboratory, 4800 Oak Grove Drive, Pasadena, California, 91109, United States
Tel: (+1 818) 354 43 21
Fax: (+1 818) 393 18 99
Web: www.jpl.nasa.gov
■ Operated by the California Institute of Technology for NASA, with planetary exploration as its primary mission. The laboratory also carries out research for the US Department of Defence and other federal agencies. Working on the Mars Pathfinder mission. In addition, JPL manages the worldwide Deep Space Network, which communicates with spacecraft and conducts scientific investigations from its complexes in California's Mojave Desert near Goldstone, near Madrid, Spain and near Canberra, Australia.

Jodrell Bank Observatory (JBO)
Nuffield Radio Astronomy Laboratory
University of Manchester, Macclesfield, Cheshire, SK11 9DL, United Kingdom
Tel: (+44 1477) 57 13 21
Fax: (+44 1477) 57 16 18
Web: www.jb.man.ac.uk
www.jodrellbank.manchester.ac.uk
■ Radio astronomy teaching and research.

Johns Hopkins University
Applied Physics Laboratory
11100 Johns Hopkins Road, Laurel, Maryland, 20723-6099, United States
Tel: (+1 240) 228 83 49
e-mail: paul.biege@jhuapel.edu
Web: www.jhuapl.edu
■ Deals with a broad spectrum of academic and applied disciplines.

Johns Hopkins University
Chemical Propulsion Information Agency
10630 Little Patuxent Parkway, Suite 202, Columbia, Maryland, 21044-3204, United States
Tel: (+1 410) 992 73 01
Fax: (+1 410) 730 49 69

Johns Hopkins University
Department of Physics and Astronomy
Baltimore, Maryland, MD 21218, United States
Tel: (+1 410) 516 72 17
Fax: (+1 410) 516 50 96

Johns-Manville
Schuller International
PO Box 5108, Denver, Colorado, 80217-5108, United States
Tel: (+1 303) 978 20 00
Freephone: (+1 800) 654 31 03
Web: www.jm.com
■ Acoustical and thermal insulation for aircraft, Q-Fiber component of space shuttle tiles and blankets.

Joint Stock Company Research and Production Association Molniya (JSK NPO Molniya)
Molniya Scientific and Industrial Enterprise
OAO Nauchno-Proizvodstvennoye Obedinenie Molniya, NPO Molniya JSC, 6 Novoposelkovaya Street, 123459, Moskva, Russian Federation
Tel: (+7 095) 493 50 53
(+7 095) 497 47 60
Fax: (+7 095) 497 47 23
(+7 095) 497 59 30
e-mail: molniya@dol.ru
Web: www.buran.ru
■ Design and manufacture of re-usable orbital vehicle Buran. Multipurpose aerospace systems, passenger, business and cargo aircraft (triplane) with payload from 500 to 450,000 kg.

Jotron Electronics AS
Kirkestien 1, PO Box 54, N-3280Tjodalyng, Norway
Tel: (+47) 33 13 97 00
Fax: (+47) 33 12 67 80
e-mail: sales@jotron.com
Web: www.jotron.com
■ Design and manufacture of ground-to-air radio communications equipment. VHF 117-137 MHz AM base stations, UHF 225-400 MHz AM/FM base stations, remote control systems, hand-held transceivers, data transceivers, base station change-over units.

Joyce Telectronics Corp
2049 Range Road, Clearwater, Florida, 33765-2124, United States
Tel: (+1 727) 461 35 25
Fax: (+1 727) 461 61 72
Web: www.joycetelectronics.com
■ Aircraft and shipboard headset microphones, amplifiers, telephone handsets, headset adapters, ground communication cables and switches used in airlines, helicopters, military and space applications.Telephone, telegraph and aircraft equipment.

JRA Aerospace Ltd
JRA Aerospace & Technology Ltd
JRA House, Taylors Close, Marlow, Buckinghamshire, SL7 1PR, United Kingdom
Tel: (+44 1628) 89 11 05
Fax: (+44 1628) 81 65 81
Web: www.jra-aero.com
■ Safety management systems.

JSAT Corporation

Japan Satellite Systems Inc
Pacific Century Place Marunouchi 17-18F, 1-11-1 Marunouchi, Chiyoda-Ku, Tokyo, 100-6216, Japan
Tel: (+81 3) 52 19 77 78
Fax: (+81 3) 52 19 78 78
e-mail: info@jsat.net
Web: www.jsat.net
■ Provide telecommunication and broadcast satellite services.

K

K&K Associates

10141 Nelson Street, Westminster, Colorado, 80021, United States
Tel: (+1 303) 702 12 86
Fax: (+1 877) 873 91 10
Web: www.tak2000.com
■ Develop and market thermal analysis software. Primary end use orbiting scientific packages and satellites. Recent applications include GEO and COSTAR, NCMOS, STIS for the Hubble Space Telescope.

Kagoshima Space Center (KSC)

Uchinoura-cho, Kimotsuki-gun, Kagoshima, 893-1402, Japan
Tel: (+81 994) 67 22 11
Fax: (+81 994) 67 38 11
■ Facilities for launching rockets, telemetry, tracking and command stations for rockets and satellites, and optical observation posts.

Kal-Gard UK Ltd

KG Coating Ltd
Scanwel Scientific Ltd
Canalwood Industrial Estate, Chirk, Clwyd, LL14 5RL, United Kingdom
Tel: (+44 1691) 77 20 70
(+44 1691) 77 80 70
Fax: (+44 1691) 77 83 03
e-mail: tory@kal-gard.co.uk
general@kgcoating.co.uk
Web: www.kal-gard.co.uk
■ Sub-contract vacuum coating, IVD aluminium and cadmium, anti-corrosion, low friction and specialised coating service. BAe, CAA and Lucas approved.

Kaman Aerospace Corp (KAC)

Bloomfield Office
a subsidiary of Kaman Corp, US
Old Windsor Road, PO Box 2, Bloomfield, Connecticut, 06002-0002, United States
Tel: (+1 860) 243 73 36
Fax: (+1 860) 243 70 43
e-mail: frenchm-kac@kacman.com
Web: www.kamanaero.com
■ Helicopters (SH-2, K-Max), aircraft components, remotely piloted vehicles, electro-optical systems, precision measuring equipment, memory systems and fusing and safety devices.

Kamatics Corporation

PO Box 3, Bloomfield,
Connecticut, 06002-0003
United States
1330 Blue Hills Avenue, Bloomfield, Connecticut, 06002-0003, United States
Tel: (+1 860) 243 97 04
Fax: (+1 860) 243 79 93
e-mail: kamnrkt-kam@kaman.com

Kaveri Telecoms Limited

Kaveri Industrial Complex, Plot No. 31 to 36, 1st Floor, 1st Main, 2nd Stage, Arakere Mico Layout, Bannerghatta Road, Bangalore, 560076, India
Tel: (+91 80) 658 05 18
(+91 80) 658 08 63
(+91 80) 658 13 01
(+91 80) 658 30 90
(+91 80) 658 75 43
Fax: (+91 80) 658 55 01
e-mail: kaveritl@vsnl.net
mktg@kaveritelecoms.com
Web: www.kaveritelecoms.com
■ Manufacture of antennas and RF products catering to civilian, military and satellite applications up to 18 GHz frequency range.
Accreditations: ISO 9001

Kawasaki Heavy Industries Ltd (KHI)

Aerospace Company
World Trade Centre Building, 4-1 Hamamatsu-cho, 2-chome, Minato-ku, Tokyo, 105-6116, Japan
Tel: (+81 3) 34 35 21 11
Fax: (+81 3) 35 78 15 73
e-mail: webtekkou@khi.co.jp
Web: www.khi.co.jp
■ Manufacture, repair and overhaul of equipment for commercial aircraft, anti-submarine aircraft, fighters, trainers, helicopters, missiles and space systems.

Kawasaki Heavy Industries Ltd (KHI)

Gifu Works
1 Kawasaki-cho, Kakamigahara, Gifu, 504-8710, Japan
Tel: (+81 583) 82 57 12
(+81 583) 82 57 22
(+81 583) 82 52 74
Fax: (+81 583) 82 29 81
(+81 583) 82 51 30
(+81 583) 82 42 43
(+81 583) 82 45 61
■ Aircraft and space systems. Manufacture, repair and overhaul of equipment for commercial aircraft, anti-submarine aircraft, fighters, trainers, helicopters, missiles and space systems: P-3C ASW patrol aircraft, F-15J fighter, Boeing B767 and B777 passenger aircraft, T-4 medium trainer, T-4 aerobatic aircraft, low-noise STOL experimental aircraft, C-1 medium transport aircraft, T-33A trainer, repair of E-2C early warning aircraft and C-130H transport. Manufacture of Kawasaki BK117 helicopter, CH-47J helicopter, Kawasaki Vertol 107-IIA helicopter, Kawasaki Hughes 369D helicopter. Type 64 anti-tank missile (ATM), type 79 anti-landing craft and anti-tank missile (H-ATM), type 87 anti-tank missile (M-ATM), H-II rocket launch, H-II rocket fairing, geodetic satellite (GS), machinery control and surveillance system for destroyer.

Kawasaki Heavy Industries Ltd (KHI)

Nagoya Works 1
3-11 Oaza Kusunoki, Yatomi-cho, Ama-gun, Aichi, 498-0066, Japan
Tel: (+81 567) 68 51 17
Fax: (+81 567) 68 50 90
Web: www.khi.co.jp
■ Manufacture and assembly of aircraft and equipment for space exploration. Fuselage for Boeing B777.

Kawasaki Heavy Industries Ltd (KHI)

Tokyo Head Office
Kawasaki Jukogyo
World Trade Centre Building, 4-1 Hamamatsu-cho 2-chome, Minato-ku, Tokyo, 105-6116, Japan
Tel: (+81 3) 34 35 21 11
(+81 3) 67 16 31 11
Fax: (+81 3) 34 36 30 37
Web: www.khi.co.jp
■ Manufacture full range of products for land, sea, air and outer space. Products include: aircraft and jet engines, spacecraft and equipment, ships, marine engineering, rolling stock, plant and factory automation systems, construction machinery, small engines, motorcycle and jet-ski personal watercraft. Aerospace products: aircraft; fuselage of Boeing B777 and B767, fuselage panel of Airbus A321, wing rib of B737, outboard flap for B747, flap hinge fairing of MD80 and wing components ERJ-170. T-4 medium jet trainer and T-4 aerobatics trainer (Blue Impulse), wing, tail and after fuselage of F-15, fuselage of F-2. Rotary-Wing: BK117 (joint development with MBB (ECD), Germany), OH-1 Observation Helicopter, CH-47J, OH-6.
■ Space-related activities: payload fairing for H-2 rocket, geological experiment satellite, engineering test satellite VII, H-2 orbiting plane (in design) and automatic landing flight experiment/hypersonic flight experiment, Japanese experiment module (part of space station Alpha), astronaut training facility.
■ Engines: T53 and T55-K-712 turboshaft engines, V2500, RB211/Trent, PW4000 and auxiliary power units (international joint programmes), transmission for OH-1, BK117 and MD900, Air-Turbo-Ramjet engine (under development).
■ Other products: construction of airport facilities for domestic and international airports, maintenance hangars, docks, passenger bridges, aircraft cleaning systems, engine run-up noise suppressors, air cargo terminals, cargo storage and distribution systems. Hang-glider, motorcycle and missile simulators.

Overseas offices:
Bangkok, Thailand
Beijing, China
Jakarta, Indonesia
Kuala Lumpur, Malaysia
Manila, Philippines
Seoul, South Korea
Shanghai, China
Sydney, Australia
Taipei, Taiwan

Subsidiaries:
Canadian Kawasaki Motors Inc, US
Glory Kawasaki Motors Co Ltd, Thailand
Hainan Sundiro-Kawasaki Engine Co Ltd, China
Kawasaki Construction Machinery Corp of America, US
Kawasaki do Brasil Indústria e Comércio Ltda, Brazil
Kawasaki Gas Turbine Europe GmbH, Germany
Kawasaki Heavy Industries (Europe) BV, Netherlands
Kawasaki Heavy Industries (HK) Ltd, Hong Kong
Kawasaki Heavy Industries (Singapore) Pte Ltd, Singapore
Kawasaki Heavy Industries (UK) Ltd, UK
Kawasaki Heavy Industries (USA) Inc, US
Kawasaki Heavy Industries GmbH, Germany
Kawasaki Machine Systems Korea Ltd, Korea, South
Kawasaki Motoren GmbH, Germany
Kawasaki Motors Corp USA, US
Kawasaki Motors Enterprise (Thailand) Co Ltd, Thailand
Kawasaki Motors Holding (Malaysia) Sdn Bhd, Malaysia
Kawasaki Motors Manufacturing Corp USA, US
Kawasaki Motors Netherlands NV, Netherlands
Kawasaki Motors Pty Ltd, Australia
Kawasaki Motors (Phils) Corp, Philippines
Kawasaki Motors (UK) Ltd, UK
Kawasaki Precision Machinery (UK) Ltd, UK
Kawasaki Rail Car Inc, US
Kawasaki Robotics GmbH, Germany
Kawasaki Robotics (UK) Ltd, UK
Kawasaki Robotics (USA) Inc, US
KHI Design & Technical Service Inc, Philippines
KHI (Dalian) Computer Technology Co Ltd, China
Nantong COSCO KHI Ship Engineering Co Ltd,
PT Kawasaki Motor Indonesia, Indonesia
Shanghai Cosco Kawsaki Heavy Industries Steel Structure Co Ltd, China
Tiesse Robot SpA, Italy
Wuhan Kawasaki Marine Machinery Co Ltd, China

Kayser-Threde Moscow

Technical Center 'Splav', ulitsa Baikalskaja, 9, 107 497 Moskva, Russian Federation
Tel: (+7 95) 462 46 88
Fax: (+7 95) 462 40 35
e-mail: ktmoskau@glasnet.ru
ktmoskau@online.ru

Kayser-Threde NA Inc

8469 South Saginaw Street, Grand Blanc, Michigan, 48439, United States
Tel: (+1 810) 695 69 10
Fax: (+1 810) 695 61 30
e-mail: support@kt-na.com
support@kayser-threde-na.com
Web: www.kayser-threde-na.com

Kearfott Guidance & Navigation Corp (KGN)

a subsidiary of Astronautics Corp of America (ACA), US
1150 McBride Avenue, Little Falls, New Jersey, 07424, United States
Tel: (+1 973) 785 60 00
(+1 973) 785 61 83
Fax: (+1 973) 785 60 25
(+1 973) 785 59 05
Telex: 133440
e-mail: marketing@kearfott.com
Web: www.kearfott.com
■ Advanced inertial and stellar-inertial navigation and guidance using ring laser and tuned rotor gyroscopes. Design, manufacture and integration of systems for navigation, guidance and control of aircraft, space, land and marine vehicles, inertial components including gyroscopes, platforms, accelerometers, servo mechanisms, pilots displays and control panels, GSE, hydraulic and electro mechanical controls, motors, synchros, resolvers and actuators.

KEC Limited

Kern Electrical Components Ltd
Orpheus House, Calleva Park, Aldermaston, Reading, Berkshire, RG7 8TA, United Kingdom
Tel: (+44 118) 981 15 71
Fax: (+44 118) 981 15 70
e-mail: sales@kec.co.uk
Web: www.kec.co.uk
■ Electrical connector accessories for EMC applications. Through bulkhead fittings for EMC applications. Cable and enclosure systems specialising in EMI/RFI/EMP and Tempest areas and conduit system.
Representing:
Fastener Specialty Inc, US
Inotec GmbH, Germany
Uponor GmbH, Germany

Keiser Engineering Inc
2046 Carrhill Road, Vienna, Virginia, 22181-2917, United States
Tel: (+1 703) 281 95 82
Fax: (+1 703) 281 95 82
■ Consulting in satellite communications systems engineering, feasibility studies, interface problems, applications and seminars.

Keldysh Institute of Applied Mathematics
Acad Sciences
Miusskaja Place, 125047 Moskva, Russian Federation
Tel: (+7 499) 978 13 14
Fax: (+7 499) 972 07 37
e-mail: info@keldysh.ru
Web: www.keldysh.ru

Keldysh Research Centre (KeRC)
8 Onezhskaya Street, 125438 Moskva, Russian Federation
Tel: (+7 495) 456 46 08
Fax: (+7 495) 456 82 28
e-mail: kerc@elnet.msk.ru
Web: www.kerc.msk.ru
■ Development, manufacture and testing of rocket engines, space propulsion systems, high energy beam generators, particle accelerators, and lasers. Energy saving developments in the interests of ecology include: purification of natural and sewage water and water desalination, power supply on the basis of multi-stage processing of solid propellants and wastes, elimination of balneological infectious wastes, ozone-friendly refrigerant C1-freon substitute, elimination of oil spots by laser beam at the water surface, metallic filters, and vibrodampers.

Kelly Space & Technology
294 South Leland Norton Way, Suite 3, San Bernardino, California, 92408, United States
Tel: (+1 909) 382 20 10
Fax: (+1 909) 382 20 12
e-mail: katadmin@kellyspace.com
Web: www.kellyspace.com

KFKI Research Institute for Particle and Nuclear Physics (RIPNP)
PO Box 49, H-1525 Budapest, Hungary
Konkoly Thege Miklós út 29-33, H-1121 Budapest, Hungary
Tel: (+36 1) 392 22 22
Fax: (+36 1) 395 91 51
(+36 1) 392 25 98
Web: www.rmki.kfki.hu
■ Research activities in: experimental particle and nuclear physics, theoretical physics, bio-physics, materials science, plasma physics and space physics.

Kharthik Engineering
36, J.C. Road, Bangalore, Karnataka, India
Tel: (+91 80) 22 22 48 25
Fax: (+91 80) 22 21 37 03
Web: www.karthikengineering.com

Khrunichev State Research and Production Space Centre
Gosudarstvennyi Kosmicheskii Naucho-Proizvodstvennyi Tsentr Imeni M V Krunicheva
18 Novozavodskaya Street, 121087 Moskva, Russian Federation
Tel: (+7 495) 145 92 10
Fax: (+7 495) 142 59 00
e-mail: proton@khrunichev.com
Web: www.khrunichev.com
■ Production of the launch-vehicle Proton, Angara and Rockot. Also produces multi-purpose space stations and orbital modules.

Kinesix Corporation
7700 San Felipe, Suite 200, Houston, Texas, 77063, United States
Tel: (+1 713) 953 83 00
Freephone: (+1 800) 953 53 30
Fax: (+1 713) 953 83 06
e-mail: kx_info@kinesix.com
Web: www.kinesix.com

Knight Electronics
784 Park Place, Uniondale, New York, 11553, United States
Tel: (+1 516) 833 72 34
Fax: (+1 516) 481 16 75
■ Broadline distributor in the electronics' procurement market servicing manufacturers, research and development facilities and repair shops. Distributor of 1N, 2N & 4N; JAN, JANTX, JANTXV and JANS level semiconductor products for commercial, industrial, military, and space level applications. Also house integrated circuits including memory products, connectors and electromechanical, relays, switches, passive components, computer processors, computer storage (HDD/SCSI Drives), development tools and compilers.
Accreditations: EOS/ESD compliant
3M, US
AMD, US
American Power Devices, US
AMP, US
Amphenol Corporation, US
Analog Devices, US
API, US
Arcotronics Italia SpA, Italy
Areva, France
ATC, US
Augat, US
AVX, US
BC Comp, US
Berg, France
Beyschlag, US
BKC, US
Bourns, US
Burndy, France
Burr-Brown, US
Cal-Chip, US
Central Semiconductor, US
Centralab, US
Chemicon, Japan
Cherry Semiconductor, US
Cirrus Logic, US
Compensated Devices, US
Conexant, US
Cornel-Dubilier, US
Cypress Semiconductor, US
Dale, US
Draloric, US
Epcos, Germany
Erie Technology, US
ERNI, Germany
ETC, US
Evox Rifa, Finland
Fairchild Semiconductor, US
FCI, France
Film Microelectronics, US
Germanium Power Devices, US
Glenair Inc, US
Hirose, Japan
Hitachi, Japan
Honeywell Bendix/King, US
IDT, US
Integrated Device Technology, US
Intel, US
International Rectifier, US
Intersil, US
Isocom Components, US
ITT Cannon, US
Johnson Components, US
JST, US
Kemet, US
Kings, US
Knox Semiconductor, US
KOA Speer, US
Kyocera, Japan
Lattice Semiconductor, US
Linfinity Microelectronics, US
Mallory Sonalert, US
Maxim, US
Microdot, US
Micron, US
Micropac Industries Inc, US
Microsemi, US
Mikro Elektronika, Serbia and Montenegro
Mill-Max, US
Molex, US
Motorola, US
Murata, Japan
National Semiconductor, US
NIC, US
Nichicon, Japan
Nippon Chemicon, Japan
Novacap, US
Omnirel LLC, US
On Semiconductor, US
Paccom, US
Panasonic, Japan
Panduit Corp, US
Pericom, US
Philips Semiconductor, Netherlands
Powerex, US
Presidio, US
Protek, US
Raychem Corporation, US
Renesas, Japan
Roederstein, US
Rohm, Japan
Rubycon, Japan
Samsung, South Korea
Sanyo, US
Saraonix, US
Semicoa, US
Semicon Inc, US
Semiconductor Technology, US
Semtech, US
Sensitron, US
Sharp, US
Siemens AG, Germany
Signetics Inc, US
Silicon General, US
Siliconix, US
Solitron Devices, US
Sourio, US
Spectrum Control, US
Sprague, US
Syfer, UK
Taiyo Yuden, Germany
Tansistor, US
TDK, Japan
Telefunken, US
Texas Instruments Inc, US
Toshiba, US
Tyco Electronics, US
United Chemicon, Japan
Vishay, US
Vitramon, US
Xilinx, US
Yageo, China
Zetex, US
Zicor, US
Zilog, US

Kometa Central Science and Research Institute
TsNPO Kometa
Velozavodskaya Street 5, 109280 Moskva, Russian Federation
Tel: (+7 095) 275 15 33
Fax: (+7 095) 274 08 70
■ Development of early warning and anti-satellite (ASAT) systems.

Kompozit Scientific Production Association
NPO Kompozit, Pionerskaya Street 4, 141070 Korolev, Moscow Region, Russian Federation
Tel: (+7 095) 516 81 72
(+7 095) 516 57 80
Fax: (+7 095) 516 61 12
Telex: 411204 nex
■ Research, development and manufacture of metallic and non-metalic materials for space applications. Manufacture of solid rocket motor castings, beryllium alloys, ceramics, fibreglass, polymetals, polymers and aluminium lithium alloys, carbon-carbon composites. Materials for spacecraft systems and heat protective equipment.

Komsomolsk-on-Amur Aircraft Production Association
1 Sovetskaya Street, 681018 Komsomolsk, on Amur, Russian Federation
Tel: (+7 4217) 52 62 00
(+7 4217) 22 85 25
Fax: (+7 4217) 52 64 51
(+7 4217) 22 98 51
e-mail: info@knaapo.com
knaapo@kmscom.ru
■ Military aviation and equipment, ultralight and space technology.

Kongsberg Defence & Aerospace AS (KDA)
Headquarters
Kongsberg Aerospace
PO Box 1003, N-3601 Kongsberg, Norway
Kirkegårdsveien 45, N-3601 Kongsberg, Norway
Tel: (+47 32) 28 82 00
Fax: (+47 32) 28 86 20
e-mail: kda.office@kongsberg.com
office.defence-aerospace@kongsberg.com
Web: www.kongsberg.com
■ Activities are focused on anti-ship missiles that can be launched from ships, fighter aircraft and helicopters. Also development and manufacture of military products based on related propulsion navigation and construction technologies.

Kongsberg Spacetec A/S
a subsidiary of Kongsberg Defence & Aerospace A/S, Norway
PO Box 6244, N-9292 Tromsø, Norway
Tel: (+47) 77 66 08 00
Fax: (+47) 77 65 58 59
e-mail: marketing@spacetec.no
Web: www.spacetec.no
■ Satellite data receiving station equipment.

Konstruktorskoye Byuro Mashinostroyeniya (KBM)
42 Oksky Avenue, 140402 Kolomna, Moscow Region, Russian Federation

Tel: (+7 4966) 12 19 44
(+7 4966) 16 33 44
Fax: (+7 9661) 330 64
(+7 4966) 12 19 44
(+7 4966) 15 50 04
e-mail: kbm@kolomna.ru
Web: www.kbm.ru
■ Manufacture short range guided surface-to-air missile systems, anti-tank missile systems, theatre tactical missile systems and surface-surface active protection systems for armoured vehicles against explosive weapons.

Korea Aerospace Industries Ltd (KAI)

Head Office
Hyundai Space and Aircraft
Samsung Aerospace Industries Ltd
135 Seosomun-dong, Jung-gu, Seoul, 100-737, Korea, South
Tel: (+82 2) 20 01 31 14
Fax: (+82 2) 20 01 37 77
Web: www.koreaaero.com
■ Production of satellites; fixed-wing aircraft include Korea Fighter Program KF-16, KT-1, T-50 and UAVs; rotary-wing aircraft include SB427, SOKOL, Korean Multipurpose Helicopter(KMH) and upgrade of ROKN Lynx helicopters; aerostructures including aircraft fuselages, wing assemblies and helicopter dynamic components. Development of composites and defence electronics including F-5E and M109 self propelled howitzer simulators.

Korea Aerospace Industries Ltd (KAI)

Seoul Office
135 Seosomun-dong, Jung-gu, Seoul, 100-737, Korea, South
Tel: (+82 2) 20 01 31 14
(+82 2) 20 01 30 46
Fax: (+82 2) 20 01 30 11
Web: www.koreaaero.com
■ Manufacture of aerostructures, systems integration of KTX-1 basic trainer, design and development of UAVs and manufacture of satellite platforms. Prime contractor for Korean Basic Trainer (KTX-1), Korean Light Helicopter (KLH) and major partner in Korean Fighter Programme (KFP) supplying components including F-16 centre fuselages, side panels and ventral fins. Supplier of Boeing B-747 wings and fuselage, construction of Fairchild Dornier 328 fuselage shells and supply of complete wings for BAE Hawk and Lockheed Martin P-3B/C outer wings. Agreement with PZL-Swidnik (Poland) for worldwide marketing of W-3A Sokol helicopter, marketing alliance with Asiana Airline. Research and development activities include UAV program for reconnaissance and surveillance, co-axial rotating unmanned helicopter programme for agriculture and other services. Participant member of Korean Commercial Aircraft Development Consortium (KCDC) for mid-sized commercial aircraft.

Korea Aerospace Industries, Ltd (KAI)

Sacheon Plant
802 Yucheon-ri, Sanam-Myeon, Sacheon, Gyeongsangnam-do, 664-710, Korea, South
Tel: (+82 55) 851 10 00
Fax: (+82 55) 851 10 04

Korea Aerospace Industries, Ltd (KAI)

Sacheon Plant 2
Yongdang-ri,Sacheon-Eup,Sacheon,Gyeongsangnam-do, 664-802, Korea, South
Tel: (+82 55) 851 25 14
Fax: (+82 55) 854 86 38

Korea Aerospace Industries, Ltd (KAI)

Space Development & Research Center
48-6, Moonpyung-Dong, Daeduck-Ku, Daejun, 306-220, Korea, South
Tel: (+82 42) 939 35 19
Fax: (+82 42) 939 35 00

Korea Telecom Corporation

206 Jungja-Dong, Bundang-Gu, Sungnam-Shi Kyonggi-Do, 463-000, Korea, South
Tel: (+82 342) 727 31 14
■ Provision of telecommunication services.

Korolev Rocket & Space Corporation Energia (RSC Energia)

Energiya Scientific Production Association
4a Lenina Street, 141070 Korolev, Russian Federation
Tel: (+7 095) 513 77 03
Fax: (+7 095) 187 98 77
(+7 095) 513 86 20
e-mail: mail@rsce.ru
Web: www.energia.ru
■ Development and operation of orbital stations, spacecraft, communication satellites and launchers. Main programs are launch services for the International Space Station, Sea Launch services, the Universal Space Launch Platform, Yamal and a future manned mission to Mars.

KPMG Peat Marwick

Commercial Space Group
2001 M Street North West, Washington, District of Columbia, 20036-3310, United States
Tel: (+1 202) 533 30 00
Web: www.us.kpmg.com
■ Business advisory services on space transportation, materials processsing, remote sensing, communications, life sciences and industrial services.

Krasnoyarsk Machine Building Plant

29 Prospekt Gazety, 660123 Krasnoyarsk, Russian Federation
Tel: (+7 3912) 64 66 01
(+7 3912) 64 65 37
(+7 3912) 64 48 95
Fax: (+7 3912) 64 48 91
e-mail: root@krasm.krasnojarsk.su
■ The company has six branches. with defence activities including production facilities for rockets and rocket stages, cryogenic rocket engines blocks, missile fuelling and rocket engine testing facilities.

Krauss – Maffei Wegmann GmbH & Co KG

MAN Mobile Bridges GmbH
MAN Technologie AG
Military Mobile Bridges GMBH
PO Box 3151, D-55021 Mainz, Germany
Wilhelm-Theodor-Romheld Strasse 24, D-55130 Mainz, Germany
Tel: (+49 6131) 215 52 88
Fax: (+49 6131) 215 53 83
e-mail: info@kmweg.de
Web: www.kmweg.com
■ Ground systems and facilities used by Ariane, the European launcher; ground facilities used in satellite-based communication systems and for telescopes conducting extra-terrestrial research. Production of bridging systems including: Leguan 26 m wheeled vehicle comprising a Man 8 × 8 launcher vehicle with shifting frame capable of laying a 26 m long MLC 70 bridge in seven minutes, the system can be automatically launched from such tank chassis as M1A1/A2 Wolverine heavy assault bridge, Leopard 1 and M60; can be fully automated, controlled by remote control or from inside the vehicle. The Leguan bridge can be provided with pontoons, featuring an integrated pump jet drive, and hydraulically adjusted ramps to enable it to float and be used as a ferry. Man 22 m medium assault bridge and the Leguan modular bridge system (PSB2) on Leopard 2 containing 3 bridge modules each 9.7 m which allow several combinations of bridge to be laid.

Kyokuto Boeki Kaisha Ltd (KBK)

Head Office
7th Floor New Otemachi Building, 2-1 Otemachi, 2-Chome, Chiyoda-Ku, Tokyo, 100-0004, Japan
Tel: (+81 3) 32 44 37 00
Fax: (+81 3) 32 46 27 65
Web: www.kbk.co.jp
■ Aircraft and aerospace related products including engines, communications, navigation, ground support equipment, satellite equipment, launch related equipment. Global positioning systems, electronics, measuring and control technologies.
Representing:
L-3 Communications Corp, US
Rockwell Collins Inc, US

L

L-3 Communication Aviation Recorders

Lockheed Martin Advanced Recorders
Loral Data Systems
a subsidiary of L-3 Communications Corp., US
6000 Fruitville Road, PO Box 3041, Sarasota, Florida, 34232, United States
Tel: (+1 813) 371 08 11
Fax: (+1 813) 377 55 98
(+1 914) 377 55 91
Web: www.l-3ar.com
■ Design and manufacture of Fairchild Aviation recorders.

L-3 Communications ESSCO (L-3 ESSCO)

Electronic Space Systems Corp
48 Old Powder Mill Road, Concord, Massachusetts, 01742-4697, United States
Tel: (+1 978) 369 72 00
Fax: (+1 978) 369 76 41
e-mail: info.essco@l-3com.com
Web: www.l-3com.com/essco
■ Design and manufacture of ground-based, airborne, shipboard radomes (metal space frame, sandwich, solid laminate, and RF composite structures) for the military, air traffic control, weather radar, secure communications, SATCOM and remote sensing market. Design and manufacture of high-end millimetre wave antenna systems (aircraft, communications, radar) for specialised RF applications, including satellite communications, radio astronomy, precision pointing and tracking applications. Design studies, structural and electromagnetic analysis, manufacture, installation and field support of all company products.

L-3 Communications ESSCO Collins Ltd (L3Comm/ECL)

a subsidiary of L3 Communications Corp, US
Kilkishen, County Clare, Ireland
Tel: (+353 61) 36 72 44
(+353 61) 26 72 44
Fax: (+353 61) 31 10 44
e-mail: info.essco@l-3com.com
Web: www.esscoradomes.com
www.l-3com.com/essco
■ Design and manufacture of ground-based, shipboard radomes (metal space frames, sandwich, solid laminate, and RF composite structures) for the military, air traffic control, weather radar, secure communications, satcom and remote sensing market. Design and manufacture of high-end millimetre wave antenna systems (aircraft, communications, radar) for specialised RF application, including satellite communications, radio astronomy, precision pointing and tracking applications. Design studies, structural and electromagnetic analysis, manufacture, installation and field support of all company products.

L-3 Communications Interstate Electronics Corporation (IEC)

Interstate Electronics
Interstate Electronics Corp
602 East Vermont Avenue, PO Box 3117, Anaheim, California, 92803, United States
602 East Vermont, Anaheim, 92805, United States
Tel: (+1 714) 758 05 00
Freephone: (+1 800) 854 69 79
Fax: (+1 714) 758 42 22
Web: www.iechome.com
■ Design, manufacture and support of a range of test instrumentation and missile tracking systems, ruggedised displays for military and industrial applications and secure communications equipment. Applies GPS technology to 'smart' munitions, missiles and precision-guided bombs. Supply of test instrumentation for the US Navy's fleet ballistic missile systems (including Trident submarines).

L-3 Communications Telemetry-East

Aydin Corp
L-3 Communications Aydin Corp
1515 Grundy's Lane, Bristol, Pennsylvania, 19007, United States
Tel: (+1 267) 545 70 00
Fax: (+1 267) 545 01 00
e-mail: sales/mktg.te@l-3com.com
Web: www.l-3com.com
■ Design, manufacture and support of advanced telemetry and other electronic products for military, space and other defence-related and commercial applications.

L-3 Communications Titan Corporation

Corporate Headquarters
The Titan Corp
Titan Corp, The, 3033 Science Park Road, San Diego, California, 92121, United States
Tel: (+1 858) 552 95 55
Fax: (+1 858) 552 96 68
■ Provision of satellite communications systems, information technology solutions, sterilisation systems and services for commercial and governmental customers worldwide.

L-3 Communications Titan Group

Linkabit Division
Titan Wireless Inc
Linkabit
The Titan Corp
Titan Systems Corporation
3033 Science Park Road, San Diego, California, 92121, United States
Tel: (+1 858) 552 95 00
(+1 858) 552 99 08
Fax: (+1 858) 552 99 09
Web: www.titan.com

L-3 TRL Technology Ltd

Sigma Close, Shannon Way, Tewkesbury, Gloucestershire, GL20 8ND, United Kingdom
Tel: (+44 1684) 85 28 14
e-mail: info@trltech.co.uk
Web: www.trltech.co.uk
■ Design, development and integration of hardware and software products with a particular interest in radio and satellite communications systems for government and defence applications. Digital processing and radio frequency engineering, real-time and off-line analysis software tools, radio electronic warfare systems, radio receivers, jammers, signal analysers and robust data links. Radio surveillance systems and radio direction finding systems. Bespoke DSP hardware and software design, real time/time critical software systems, and long-range optical surveillance equipment. Design and implementation of secure private networks. Key products include the Tarax Lightweight EW system – a versatile EW tool that can be deployed in a wide cross-section of scenarios and applications, and a portfolio of satellite monitoring solutions which enable the identification of transmissions of interest, accurate and near real-time location of Inmarsat terminals under scrutiny, and traffic analysis.

Laben SpA

a subsidiary of Finmeccanica, Italy
SS Padana Superiore 290, I-20090 Vimodrone, Milano, Italy
Tel: (+39 02) 25 07 51
Fax: (+39 02) 250 55 15
e-mail: info@laben.it
■ Development of scientific and remote sensing satellites, deep-space probes, commercial launchers and telecommunications satellites, as well as microgravity experiments.

Laboratoire d'Astrophysique de Marseille

Spatiale du CNRS
Traverse du Siphon, PO Box 8, F-13376 Marseille, Cedex 12, France
Tel: (+33 4) 91 05 59 00
Fax: (+33 4) 91 66 18 55
Cable: ASTROSPA
■ Manufacture of scientific optical equipment carried in ground based observatories, rockets and satellites; astronomical studies in space, essentially in the ultra-violet and infra-red range.

Laboratoire de Recherche ONERA-Ecole de l'Air

Salon-de-Provence Lab
Base Aérienne 701, F-13661 Salon de Provence Air, Cedex, France
Tel: (+33 4) 90 17 01 10
Fax: (+33 4) 90 17 01 09

Laboratoire d'Hyperfréquences

Université Catholique de Louvain, Bâtiment Maxwell, 3 Place du Levant, B-1348 Louvain la Neuve, Belgium
Tel: (+32 10) 47 80 95
Fax: (+32 10) 47 87 05
e-mail: secretariat@emic.ucl.ac.be
Web: www.emic.ucl.ac.be
■ Development of hardware and software products for industrial partners in the field of satellite communications.

Labtech Ltd

a subsidiary of Intelek plc, UK
Broadaxe Industrial Park, Presteigne, Powys, LD8 2UH, United Kingdom
Tel: (+44 544) 26 00 93
Fax: (+44 544) 26 03 10
e-mail: ptfe@labtech.ltd.uk
Web: www.labtechcircuits.com
■ Manufacture of soft-board and metal backed PTFE circuits in prototype and medium volumes. Applications include radar, microwave, links and satellite communications systems. Special Projects Division. Design of broadband microwave components using high accuracy thin film ceramic circuit manufacturing facility.
Accreditations: European Space Agency approval

Lacroix PyroTechnologies

BP 213, F-31607 Muret, France
Route de Gaudiès, F-09270 Mazeres-sur-le-Salat, France
Tel: (+33 5) 61 56 65 00
Fax: (+33 5) 61 56 65 74
Web: www.etienne-lacroix.com
■ Aeronautic and space programmes include signalling, pyromechanisms, actuation devices, aircrew protection and airfield bird scare cartridges. Other programmes cover transportation signalling, avalanche initiation, intrusion protection, fire suppression, law enforcement, meteorological control and industrial smokes.

Lagardère

Head Office
part own EADS NV, Netherlands (15.07 per cent)
4 rue de Presbourg, F-75116 Paris, Cedex 16, France
Tel: (+33 1) 40 69 16 00
Fax: (+33 1) 40 69 18 35
e-mail: presse@lagardere.fr
Web: www.lagardere.com

Lansdowne Technologies Inc

a subsidiary of Canadian Shipbuilding & Engineering Limited, Canada
275 Slater Street, Suite 1001, Ottawa, Ontario, K1P 5H9, Canada
Tel: (+1 613) 236 33 33
Fax: (+1 613) 236 44 40
e-mail: info@lansdowne.com
Web: www.lansdowne.com
■ Project management, planning and control, proposal and logistic management for defence and space industries. Non-tactical software development services.

Larsen & Toubro Limited

Heavy Engineering Division
EPC Block (5th Floor/B-Wing), Saki-Vihar Road, Powai Works, Mumbai (Bombay), 400072, India
Tel: (+91 22) 28 58 10 91
(+91 22) 67 05 39 61
(+91 22) 28 58 26 11
Fax: (+91 22) 67 05 16 52
(+91 22) 28 58 10 76
e-mail: ranga9022@hed.ltindia.com
Web: www.lnthed.com
■ Defence equipment including rocket launchers, torpedo tubes. Marine systems including shafting, steering gear and stabilisation systems. Also manufacture warships.

LAS

Traverse Du Siphon
Les Trois Lucs, F-13012 Marseille, France
Tel: (+33 4) 91 05 59 32
Fax: (+33 4) 91 66 18 55

Latécoère Société Industrielle

Head Office and Works
135 rue de Périole, F-31079 Toulouse, Cedex, France
Tel: (+33 5) 61 58 77 00
Fax: (+33 5) 61 58 76 17
Telex: 531714
Cable: LATECOERE
Web: www.latecoere.fr
■ Aircraft, missiles. Centrifuge devices. Mechanical constructions. Testing and environmental equipment. Space technologies electronics.

Lavochkin Scientific Production Association

Leningradskaya Street 24, Moscow Region, 141400 Khimki, 2, Russian Federation
Tel: (+7 095) 251 67 44
(+7 095) 573 56 75
Fax: (+7 095) 573 35 95
Web: www.laspace.ru
■ Design, development, testing and manufacturing of satellite systems for scientific research and space-borne telescopes as well as small multi-purpose satellites.

Le Cinq Chemins

Bordeaux – Le Haillan Plant
Les Cinq Chemins, F-33187 Le Haillan, Cedex, France
Tel: (+33 5) 56 55 30 00
Fax: (+33 5) 56 55 30 01
Telex: 560678

Le Fiell Manufacturing Co

13700 Firestone Boulevard, Santa Fé Springs, California, 90670, United States
Tel: (+1 800) 451 59 71
(+1 562) 921 34 11
Fax: (+1 800) 373 33 61
(+1 562) 921 54 80
e-mail: sales@lefiell.com
Web: www.lefiell.com
■ Precision tubular metal parts and assemblies, swaging and expanding, flaring, upsetting, extrusion, forming, tapering, re-drawing, seamless drawn tubing custom made, control rods, missile cases, turbine engine shafts.

Leach International UK Ltd

LRE Relays & Electronics Ltd
a sales office of Leach International Europe SA, France
Green Acres, The Patches, Ruardean Woodside, Ruardean, Gloucestershire, GL17 9XZ, United Kingdom
Tel: (+44 777) 84 78 29
Fax: (+44 1594) 54 25 64
Web: www.leachintl.com
■ Supply of components, modules and equipment, central warning systems, land light control boxes and lightweight 200 contractors. Also, thin integrated keyboard systems, advisory and central warning panels, illuminated push button switches, anti icing controllers and electronics. Relays, contactors, time delay relays, solid state intelligent switching devices, knobs and fasteners.

Leica Geosystems Inc

Navigation and Positioning Division
23868 Hawthorne Boulevard, Torrance, California, 90505, United States
Tel: (+1 310) 791 53 00
Fax: (+1 310) 791 61 08

Lewicki Microelectronic GmbH

a subsidiary of Silicon Sensor International AG
Allee 35, D-89610 Oberdischingen, Germany
Tel: (+49 7305) 960 20
Fax: (+49 7305) 96 02 50
e-mail: info@lewicki-gmbh.de
Web: www.lewicki-gmbh.de

L'Garde Inc

15181 Woodlawn, Tustin, California, 92680-6487, United States
Tel: (+1 714) 259 07 71
Fax: (+1 714) 259 78 22
Web: www.lgarde.com
■ Research and development of inflatable space structures (including decoys, targets, antennas, solar concentrators, solar array supports), military data recorders and specialized elastomeric material for harsh environments.

Liquidmetal Technologies

Head Office
100 North Tampa Street, Suite 3150, Tampa, Florida, 33602, United States
Tel: (+1 813) 314 02 80
Fax: (+1 813) 314 02 70
e-mail: information@liquidmetal.com
defense@liquidmetal.com
space@liquidmetal.com
Web: www.liquidmetal.com
■ Development of alloy composites for use in environmentally safe Kinetic Energy Penetrators (KEP) and space borne applications.

Liquidmetal Technologies

Lake Forest
25800 Commercentre Drive, Suite 100, Lake Forest, California, 92630, United States
Tel: (+1 949) 206 80 00
Fax: (+1 949) 206 80 08

Liskovsky Electrotechnical Plant

Lyskovsky Electromechanical Plant
7-aya Zavodskaya, 1a, Liskovo, 606211 Nizhnyi Novgorod, Russian Federation
Tel: (+7 8314) 92 02 05
Fax: (+7 8314) 92 07 81
■ Manufacture of motor and tractor electrical equipment products and generators for the defence industry

Lista International Corp

a subsidiary of Lista AG, Switzerland
106 Lowland Street, Holliston, Massachusetts, 01746-2094, United States
Tel: (+1 508) 429 13 50
Freephone: (+1 800) 722 30 20
Fax: (+1 508) 626 03 53
e-mail: sales@listaintl.com
Web: www.listaintl.com

■ Design and manufacture of modular storage and workspace systems-drawer storage cabinets, industrial and technical electronic workbenches. Lista International sells in the United States, Canada, Mexico and Latin America.

Litef GmbH

a subsidiary of Northrop Grumman Electronic Systems, Navigation Systems Division, US
Loerracherstrasse 18, D-79115 Freiburg, Germany
Tel: (+49 761) 490 10
Fax: (+49 761) 490 14 80
e-mail: info@litef.de
Web: www.litef.de
www.litef.com
■ Development, production and support of inertial navigation systems, strapdown heading and attitude reference systems (AHRS), gyros and accelerometers, mobile digital high performance computers. Development of special electronic units and systems, application of specific integrated circuits (ASICs) for airborne and space applications, on-board ships and land vehicles.

Lithion

division of Yardney Technical Products, Inc
82 Mechanic Street, Pawcatuck, Connecticut, 06379-2154, United States
Tel: (+1 860) 599 11 00
Fax: (+1 860) 599 39 03
e-mail: vyevoli@lithion.com
Web: www.lithion.com
■ Manufacture of lithium-ion batteries.

Lobo Systems Ltd

26 The Parker Centre, Mansfield Road, Derby, DE21 4SZ, United Kingdom
Tel: (+44 1332) 36 56 66
Freephone: (+1 800) 640 54 92
Fax: (+44 1332) 36 56 61
e-mail: sales@lobosystems.com
Web: www.lobosystems.com
■ Manufacture of fast assembly access systems, work platforms and scaffolding.

Lockheed Martin Aeronautics Company (LM Aero)

Palmdale Operations
Lockheed Martin Skunk Works
1001 Lockheed Way MS 0819, Palmdale, California, 93599-3740, United States
Tel: (+1 661) 572 41 53
■ Advanced development projects. Programmes have included F-117 stealth fighter, U-2 high altitude reconnaissance aircraft, C-130 special mission aircraft, rapid vehicle and system prototyping and classified programmes.

Lockheed Martin Corporation

Lockheed Corporation
Martin Marietta Corporation
6801 Rockledge Drive, Bethesda, Maryland, 20817 , United States
Tel: (+1 301) 897 60 00
Fax: (+1 301) 897 62 52
Web: www.lockheedmartin.com
■ A global enterprise engaged in the research, design, development, manufacture and integration of advanced-technology systems, products and services for government and commercial customers. The core businesses span systems integration, aeronautics, space technology services and global telecommunications.
Operating Units:
–Aeronautics Business Area
–Aeronautics
–Electronic Systems Business Area
–Lockheed Martin Canada
–Maritime Systems & Sensors
–Missiles & Fire Control
–Simulation, Training & Support
–Systems Integration – Owego
–Transportation & Security Solutions
–Information and Technology Services Business Area
–Aircraft & Logistics Centers
–Information Technology
–KAPL Inc (Knolls Atomic Power Laboratory)
–Sandia Corporation (Sandia National Laboratory)
–Space Operations
–Systems Management
–Technical Operations
–Technology Ventures
–Integrated Systems & Solutions Business Area
–Integrated Systems & Solutions
–Lockheed Martin Orincon
–Space Systems Business Area
–Space Systems

Lockheed Martin Management & Data Systems (M&DS)

Washington Metro Office
1725 Jefferson Davis Highway, Suite 403, Arlington, Virginia, 22202-4127, United States
Tel: (+1 703) 416 62 00

Lockheed Martin Missiles and Fire Control

Dallas Office
Lockheed Martin Vought Systems
PO Box 650003 M/S PT-42, Dallas, Texas, 75265-0003, United States
1701 West Marshall, PT-42, Grand Prairie, Texas, 75051, United States
Tel: (+1 972) 603 14 01
(+1 972) 603 16 15
(+1 972) 603 10 00
Fax: (+1 972) 603 10 09
Web: www.missilesandfirecontrol.com
■ Development, manufacture and support of advanced combat, missile, rocket and space systems.

Lockheed Martin Space & Strategic Missiles (LMSS)

Lockheed Martin Space Systems Company
Mail Stop DC 1020, PO Box 179, Denver, Colorado, 80201, United States
Fax: (+1 303) 971 49 02
■ Design, development, testing and manufacture of a variety of advanced technology systems for the space and defence sectors. Chief projects include Titan II and Titan IV; Atlas, Centaur, Atlas V, crew return vehicle for use by International Space Station astronauts, Athenai I and II launch vehicles; Multi-Service Launch Systems (MSLS) and Target Delivery Vehicle (TDV).

Lockheed Martin Space Mission Systems & Services (LESC)

Loral Aerosys
7375 Executive Place, Suite 301, Seabrook, Maryland, 20706, United States
Tel: (+1 301) 352 26 01

Lockheed Martin Space Operations

Lockheed Martin Space Mission Systems and Services
2625 Bay Area Boulevard, Houston, Texas, 77058-8999, United States
Tel: (+1 281) 283 44 00
Fax: (+1 281) 283 46 60
■ Engineering and scientific support services; management of government technical facilities; spacecraft avionics; test range instrumentation; computer sciences support and management.

Lockheed Martin Space Systems

Lockheed Martin Missiles & Space
1111 Lockheed Martin Way, PO Box 3504, Sunnyvale, California, 94089, United States
Tel: (+1 408) 742 43 21
(+1 408) 742 71 51
(+1 408) 742 66 88
Fax: (+1 408) 756 86 54
(+1 408) 742 84 84
e-mail: lmms.communications@lmco.com
■ Design, development, test, manufacture and operation of advanced technology systems for military, civil, and commercial customers. Products include: remote sensing and communications satellites for commercial and government customers, space observatories and interplanetry spacecraft, fleet ballistic missiles and missile defence systems.

Lockheed Martin Space Systems Company

Headquarters
12257 South Wadsworth Boulevard, Littleton, Colorado, 80125-8500, United States
Tel: (+1 303) 977 30 00
■ Provide space launch services, production and services relating to commercial satellites, government satellites and strategic missiles.

Lockheed Martin UK (LMUK)

Ampthill
Insys Limited
Lockheed Martin UK Insys Limited
Reddings Wood, Ampthill, Bedfordshire, MK45 2HD, United Kingdom
Tel: (+44 1525) 84 10 00
Fax: (+44 1525) 40 58 61
e-mail: marketing@insys-ltd.co.uk
Web: www.lockheedmartin.co.uk
■ Development and large scale production (as prime or sub contractor) of weapons and communication systems for HM Government and approved overseas customers.
Sister companies:
Euro-Shelter Limited, France
Insys Limited, STL Pershore
Insys Limited, CBDE Porton Down

Loral Skynet

500 Hills Drive, PO Box 7018, Bedminster, New Jersey, 07921, United States
Tel: (+1 908) 470 23 00

Loral Space & Communications

Washington Operations
1755 Jefferson Davis Highway, Suite 1007, Arlington, Virginia, 22202, United States
Tel: (+1 703) 414 10 40
Fax: (+1 703) 414 10 75

Loral Space & Communications

600 3rd Avenue, New York, New York, 10016, United States
Tel: (+1 212) 697 11 05
Fax: (+1 212) 338 56 62
Web: www.loral.com
■ Satellite manufacturing and satellite-based services including: broadcast transponder leasing and value added services, domestic and international corporate data networks, global wireless telephony, broadband data transmission and content services. Internet services and international direct-to-home satellite services.
Business units:
Cyberstar, L.P.
Globalstar Limited Partnership
Loral Orion
Loral Skynet
Space Systems/Loral

Lord Industrial Ltda.

Via Anhanguera, km 63.5, Distrito Industrial, 13200 Jundiai, São Paulo, Brazil
Tel: (+55 11) 73 92 77 55
Fax: (+55 11) 73 92 35 81
Telex: 36339 liki br
■ Rubber parts for tank volute suspension, rocket and missile components.

Louisiana State University

Department of Physics/Astronomy
Baton Rouge, Louisiana, LA 70803 4001, United States
Tel: (+1 225) 388 11 60
Fax: (+1 225) 334 10 98

Lowell Observatory

1400 W Mars Hill Road, Flagstaff, AZ 86001, United States
Tel: (+1 520) 774 33 58
Fax: (+1 520) 774 62 96

Luch Scientific Production Association

NPO Luch
Zheleznodorozhnaya Street 24, 142100 Podolsk, Russian Federation
Tel: (+7 095) 137 92 58
(+7 095) 137 93 39
Fax: (+7 095) 137 93 84
e-mail: postmaster@npouch.msk.su
■ Research and production of metals for nuclear industry. Satellite nuclear reactors.

Lunar Rocket & Rover Co Inc

3597 Sausalito Street, Los Alamitos, California, 90720, United States
Tel: (+1 562) 596 78 47
Fax: (+1 562) 596 29 50
Web: www.lunar-rocket.com

Lund Observatory

Box 43, SE-221 00 Lund, Sweden
Tel: (+46 222) 72 97
Fax: (+46 222) 46 14

M

Météo-France

Meteorologie Nationale
1 quai Branly, F-75340 Paris, Cedex 7, France
Tel: (+33 1) 45 56 71 71
Fax: (+33 1) 45 56 70 05
(+33 1) 45 56 71 11
Web: www.meteo.fr
■ Meteorological services

Mackay Communications Inc
Headquarters
3691 Trust Drive, Raleigh, North Carolina, 27616-2955, United States
Tel: (+1 919) 850 30 00
Fax: (+1 919) 954 17 07
e-mail: info@mackaycomm.com
Web: www.mackaycomm.com
■ Offer a range of mobile satellite communications products and services, with products that have been designed to withstand rugged environments. Installation and service of shipboard and terrestrial-based Inmarsat satellite systems. Inmarsat A, B, C, M, Mini-M and Fleet F products and airtime provide reliable voice, fax, 64/128kbps digital data, and video conferencing with ISDN connectivity. Also offer network and data application engineering services for new or existing applications, in addition to a full line of marine navigation and below-deck electronic equipment from manufacturers.

Magellan Aerospace Corporation
Bristol Division
Avenida Jose Miguel Carrera, Paradero 36 1/2, 11087 El Bosque, Chile
Tel: (+1 204) 775 83 31
Fax: (+1 204) 775 74 94
■ Manufacture of rocket weapon systems, warheads, rocket launchers and motors.

Major Tool & Machine Inc
1458 East 19th Street, Indianapolis, Indiana, 46218, United States
Tel: (+1 317) 636 64 33
Fax: (+1 317) 634 94 20
Web: www.majortool.com

Malin Space Science Systems
PO Box 910148, San Diego, California, 92191-0148, United States
Tel: (+1 858) 552 26 50
Fax: (+1 858) 458 05 03
e-mail: info@msss.com
Web: www.msss.com

MAN Technologie AG (MT)
PO Box 10 22 35, D-86012, Augsburg, Germany
Franz-Josef-Strauss Strasse 5, D-86153 Augsburg, Germany
Tel: (+49 82) 15 05 01
Fax: (+49 82) 15 05 10 00
e-mail: spacetransport@mt.man.de
Web: www.man-technologie.de
■ Development and manufacture of lightweight structures, tanks and propulsion components for space transportation systems (e.g. the European ARIANE, the Japanese H2A launcher), structural components and thermal protection systems for re-usable space vehicles, high-pressure vessels, fuel tanks and structural components for satellites.

MAPAERO – Aerospace Coatings
a subsidiary of Peintures Maestria, France
10 Avenue de la Rijole, Zone Industrielle, F-09100 Pamiers, France
Tel: (+33 5) 34 01 34 01
Fax: (+33 5) 61 60 23 30
e-mail: sales.mapaero@mapaero.com
Web: www.mapaero.com
■ Development, manufacture and sale of paints for aircraft, space launchers and satellites. Products include aircraft exterior paint; water based primers for aluminium structures; fire retardant water based paints for cabin interiors and special paints for high temperature, anti-skid, anti-static and anti-erosion applications.
Accreditations: ISO 9001, ISO 2000, EN 9100

MarathonNorco Aerospace Inc
Marathon Power Technologies
PO Box 8233, Waco, Texas, 76714-8233, United States
8301 Imperial Drive, Waco, Texas, 76712-6588, United States
Tel: (+1 254) 776 06 50
(+1 254) 741 54 00
Fax: (+1 254) 776 65 58
e-mail: marathon@mptc.com
Web: www.mnaerospace.com
■ Sintered nickel-cadmium batteries, battery chargers, airborne static inverters, and custom electronics for military aviation, communications and standby power applications. Hold open rods and struts.

Marconi Mobile Ltd
Marine Office
Marconi House, New Street, Chelmsford, Essex, CM1 1PL, United Kingdom
Tel: (+44 1245) 27 55 88
Fax: (+44 1245) 27 56 89
e-mail: marime-sales@marconi.com
Web: www.marconi-marine.com
■ Sale of marine electronics, servicing, repairs, installation and ship survey.

Marlin Yug Ltd
2 Kapitanskaya Strasse, 99000 Sevastopol, Ukraine
Tel: (+380 692) 54 04 50
Fax: (+380 692) 54 04 50
e-mail: marlin@stel.sebastopol.ua
Web: marlin.stel.sebastopol.ua
■ Manufacture of the WOCE SVP-B Drifter deployable autonomous buoy and the Diving Drifter surface buoy. Also manufacture the Platform Transmitter Terminal (PTT) and testing device for the data platform.

Marsh Space Projects Ltd
Sedgwick Space Services
No.1, The Marsh Centre, London, E1 8DX, United Kingdom
Tel: (+44 20) 73 57 52 74
(+44 20) 73 57 10 00
Fax: (+44 20) 73 57 52 78
(+44 20) 79 29 27 05
Web: www.marsh.com
global.marsh.com
■ Spacecraft risk management – manufacture, prelaunch, launch, in-orbit and Third Party Liability insurance.

Marshall of Cambridge Aerospace Ltd
Marshall Aerospace
Marshall of Cambridge (Engineering) Ltd
a division of the Marshall Group
The Airport, Newmarket Road, Cambridge, CB5 8RX, United Kingdom
Tel: (+44 1223) 37 37 37
Fax: (+44 1223) 37 33 73
Web: www.marshallaerospace.com
■ OEM capabilities. Activities include conversion, modification and maintenance of military and civil aircraft; international engineering support services undertaken at the Cambridge facility or on site with the customer. UK design authority for the RAF C130 and sister design authority to Boeing on the E3D AWACS aircraft. Design authority for the L1011 Tristar Tanker responsible for the design and conversion of the aircraft. L1011 freighter conversions; B747 and MD11 interior reconfigurations; B767 maintenance. Approved Citation service centre.
Accreditations: ISO 9001, UK MoD PE design organisation approval, UK CAA ANO approval for design and manufacturing, UK CAA ANO approval for maintenance, JAR-145 approval CAA 00031, FAA FAR-145 approval.

Mashinostroenia NPO
33 Gagarina Street, Reutov, 143952 Moskva, Russian Federation
Tel: (+7 095) 307 91 94
Fax: (+7 095) 302 20 01

Masten Space Systems
PO Box N, Mojave, California, 93502, United States
e-mail: info@mastern-space.com
Web: mastern-space.com

Matra Marconi Space UK Ltd
Ground Systems Directorate
Gunnels Wood Road, Stevenage, Hertfordshire, SG1 2AS, United Kingdom
Tel: (+44 1438) 77 36 98
Fax: (+44 1438) 77 30 69

Matrix Systems Ltd
161 Alden Road, Markham, Ontario, L3R 3W7, Canada
Tel: (+1 905) 447 44 42
Fax: (+1 905) 477 36 06
■ Manufacture of space, defence, communications, information and electronic ATC systems.

MB Aerospace Ltd
a subsidiary of Motherwell Bridge Holdings, UK
Logons Road, PO Box 4, Motherwell, ML1 3NP United Kingdom
Tel: (+44 1698) 26 22 77
(+44 1698) 26 61 11
Fax: (+44 1698) 26 97 74
(+44 1698) 27 54 87
e-mail: sales@mbaerospace.com
Web: www.motherwellbridge.com
■ Highly specialised sub-contract machine company, supplying critical quality components to the aerospace and defence industries.

MBDA
Headquarters
British Aerospace (Dynamics) Ltd
Matra BAE Dynamics UK Ltd
a joint venture between BAE Systems (37.5 per cent), EADS (37.5 per cent) and Finmeccanica (25 per cent)
11 Strand, London, WC2N 5RJ, United Kingdom
Tel: (+44 207) 451 60 00
Fax: (+44 207) 451 60 01
Web: www.mbda.co.uk
www.mbda-systems.com
■ Design, development, manufacture and supply of missile systems and countermeasures. Missiles for air, land and sea operations.

McCormick Selph Inc (MSI)
Teledyne McCormick Selph
3601 Union Road, PO Box 6, Hollister, California, 95024-0006, United States
Tel: (+1 831) 637 37 31
Fax: (+1 831) 637 14 50
Web: www.mcselph.com
■ Design, development and manufacture of controlled explosive products for the aerospace, automotive and petroleum industries. Products include linear explosives, ordnance systems and components.

MCL Inc
501 South Woodcreek, Bolingbrook, Illinois, 60440-4999, United States
Tel: (+1 630) 759 95 00
(+1 312) 461 45 36
Freephone: (+1 800) 743 46 25
Fax: (+1 630) 759 50 18
e-mail: sales@mcl.com
Web: www.mcl.com
■ Manufacture of high power microwave oscillators, amplifiers and severe environment ECM power supplies. Amplifiers for satellites and communications ground stations.

MD Robotics
MacDonald Dettwiler Space and Advanced Robotics Ltd
Spar Aerospace Limited
a wholly owned subsidiary of MacDonald, Dettwiler and Associates Ltd, Canada
9445 Airport Road, Brampton, Ontario, L6S 4J3, Canada
Tel: (+1 905) 790 28 00
Fax: (+1 905) 790 44 00
e-mail: info@mdrobotics.ca
Web: www.mdrobotics.ca
■ Development and manufacture of robotics for use in space. Products include the Canadarm robot arm used in the US space shuttles and a robotic servicing system to be used in assembly, maintenance and servicing of the International Space Station. On-orbit servicing.

Mecachrome SA
Avenue Eugéne Casella, PO Box 9, F-18700 Aubigney-Sur-Nère, France
Tel: (+33 2) 48 81 22 00
Fax: (+33 2) 48 58 20 84
e-mail: mecachrome@groupe-mecachrome.com
■ Mechanical machining and assembly of structural parts and main assemblies; engine parts and components, landing gear.

MEDA Inc
22611 Markey Court, Suite 114, Dulles, Virginia, 20166, United States
Tel: (+1 703) 471 14 45
Fax: (+1 703) 471 91 30
e-mail: sales@meda.com
Web: www.meda.com
■ Manufacture of TAM-1 and TAM-2 space qualified magnetometers for satellite attitude determination and control applications.

Medes-Imps
Institut de Médecine et de Physiologie Spatiales an affiliate of Centre National d'Études Spatiales (CNES), France
1 avenue Jean Poulhès – BP 4404, F-31405 Toulouse, Cedex 4, France
Tel: (+33 5) 62 17 49 50
Fax: (+33 5) 62 17 49 51
e-mail: contact@medes.fr
Web: www.medes.fr
■ Research into medicine and physiology in space. Specialise in life sciences and biomedical research.

Metal Mart Inc
12225 Coast Drive, Whittier, California, 90601, United States
Tel: (+1 562) 692 90 81
Fax: (+1 562) 699 68 68
Web: www.metal-mart.com

Meteorological Satellite Center (MSC)
Japan Meteorological Agency
3-235 Nakakiyoto, Kiyose-shi, Tokyo, 204-0012, Japan
Tel: (+81 4) 24 93 49 70
Fax: (+81 4) 24 92 24 33
e-mail: syskan@msc.kishou.go.jp

Metria
Satellus
PO Box 820, SE-981 28 Kiruna, Sweden
Rymdhuset 1, Osterleden 15, Kiruna, Sweden
Tel: (+46 980) 670 00
Fax: (+46 980) 670 67
e-mail: lantmateriet@lm.se
satellite@lm.se
Web: www.lantmateriet.se
■ Space information retrieval services.

MF Tekhnoinvest
Aerospace Air Filters Division
Mantulinskaya Ulitsa 2, 123100 Moskva, Russian Federation
Tel: (+7 095) 205 76 82
Fax: (+7 095) 205 76 81
e-mail: spacecraftair@cabonfilter.net
Web: www.carbonfilter.net
■ Design and production of recirculation and regeneration active carbon air filters for space craft living compartments. Products used in Space Station Mir and are currently installed in the Russian module of the International Space Station.

Michigan Aerospace Corporation
1777 Highland Drive, Suite B, Ann Arbor, Michigan, 48108-2285, United States
Tel: (+1 734) 975 87 77
Fax: (+1 734) 975 02 39
Web: www.michiganaero.com

Microcosm Inc
4940 West 147th Street, Hawthorne, California, 90250-6708, United States
Tel: (+1 310) 219 27 00
Web: www.smad.com
■ Activities include space mission architecting, mission and systems engineering and related orbit and attitude analysis services.

Microfiltrex
Fairey Microfiltrex Ltd
Fareham Industrial Park, Fareham, Hampshire, PO16 8XG, United Kingdom
Tel: (+44 1329) 28 56 16
Fax: (+44 1329) 82 24 42
e-mail: info@porvairfiltration.com
■ Design and manufacture of specialist filtration equipment for fuel, hydraulic, lubrication and air systems. Supply filters that protect vital subsystems for flight controls, fuel management, thrust reversers, breaking and steering, power generation and air intakes in aircraft, helicopters, military vehicles and space.
Accreditations:
AS9100
Defstan 05-91
EASA Part 145
EASA Part 21 subpart G
ISO 9001
ISO 2000

Microspace Communications Corporation
3100 Highwoods Boulevard, Raleigh, North Carolina, 27604, United States
Tel: (+1 919) 850 45 47
Fax: (+1 919) 850 45 18
Web: www.microspace.com

Microtecnica-Torino Srl
a subsidiary of Hamilton Sundstrand, US
Piazza Arturo Graf, PO Box 1210, I-10126 Turin, Italy
Tel: (+39 011) 693 21
Fax: (+39 011) 693 22 42
■ Design, development and servicing of navigation and guidance systems for aircraft, helicopters and missiles. Primary and secondary flight controls, drive, control and actuation systems, air conditioning, pressurisation and heat transfer systems. Power generation and control. On-board power units for missiles, test stands.

Mid-Atlantic Regional Spaceport (MARS)
Virginia Space Flight Center
NASA/GSFC/Wallops Flight Facility, Building N-134, Wallops Island, Virginia, 23337-5009, United States
Tel: (+1 757) 824 23 35
Fax: (+1 757) 824 23 32
e-mail: info@marsspaceport.com
Web: www.marsspaceport.com
■ Provide space launch facilities and services for commercial, government and scientific/academic users, both foreign and domestic.

Miltope Corp
Corporate Headquarters
3800 Richardson Road South, Hope Hull, Alabama, 36043, United States
Tel: (+1 334) 284 86 65
Freephone: (+1 800) MILTOPE
Fax: (+1 334) 613 63 02
(+1 203) 250 92 85
(+1 334) 613 65 91
e-mail: info@miltope.com
custserv@miltope.com
Web: www.miltope.com

Ministry for Posts and Telecommunications
7 Gorky Street, 1033 Moskva, Russian Federation
Tel: (+7 215) 986 40 11
Fax: (+7 215) 98 66 31 70

Ministry of Foreign Affairs (MFA)
Security Affairs and Disarmament Division
32/34 Smolenksaya-Sennaya, 121200 Moskva, Russian Federation
Tel: (+7 95) 244 22 30
Fax: (+7 95) 253 90 82

Ministry of Science, Technology and Environment
196 Paholyotin Road, Chatuchak, Bangkok, Thailand
Tel: (+66 2) 579 13 70
(+66 2) 579 13 79
Fax: (+66 2) 561 30 35
(+66 2) 561 30 49
(+66 2) 326 91 50

MirCorp
Media and Press Contact
The Info West Group, 8027 Leesburg Pike, Suite 303, Vienna, Virginia, 22182, United States
Tel: (+1 703) 448 56 69
Fax: (+1 703) 448 61 30

MirCorp
North American Office
661 South Washington Street, Alexandria, Virginia, 22314, United States
Tel: (+1 703) 836 19 99

Missiles & Space Batteries Ltd (MSB)
MSA (Britain) Limited, Catalyst Division
Hagmill Road, East Shawhead, Coatbridge, Scotland, ML5 4UZ, United Kingdom
Tel: (+44 1236) 43 77 75
Fax: (+44 1236) 43 66 50
■ Design and manufacture of thermal batteries.

Missionspace Corporation
4519 Castle Road, La Canada, California, 91011, United States
Tel: (+1 818) 542 35 80
(+1 818) 542 35 81
Fax: (+1 818) 541 00 67
Web: www.missionspace.com

Miteq Inc
100 Davids Drive, Hauppauge, New York, 11788-2034, United States
Tel: (+1 631) 436 74 00
Fax: (+1 631) 436 74 31
(+1 631) 436 74 30
Web: www.miteq.com
■ Manufacture of high performance components and systems for the microwave electronics industry. Products include satellite communications earth station equipment and space qualified equipment such as amplifiers, mixers, synthesizers and super components. Recent space platforms include P-97, TOPEX, SPINSAT, SEAWINDS, GEOSAT, SEASAT, SSMIS and AMSV-B.

Mitsubishi Corporation
Aerospace Division
2-6-3 Marunouchi, Chiyoda-ku, Tokyo, 100-8086, Japan
Tel: (+81 3) 32 10 46 04
(+81 3) 32 10 47 11
(+81 3) 32 10 21 21
Fax: (+81 3) 32 10 47 83
Web: www.mitsubishi.co.jp
■ Commercial and military helicopters, military aircraft, aircraft engines, guided missiles and defence electronics. Satellite and space systems and aircraft interior equipment including galley, lavatory fittings and seats. Aircraft financing and leasing.
Representing:
ADI Limited, Australia
Aerojet Gen Corp, US
Arch Chemicals Inc, US
Atlantic Research Corp, US
Aurora Flight Sciences Corp, US
Control Flow Inc, US
Del Mar Avionics, US
EMS Technologies Canada Ltd, Canada
Ferrostaal AG, Germany
GE-P&W Engine Alliance LLC, US
Honeywell Inc, US
Indal Technologies Inc, Canada
International Aero Engines AG, US
Jena-Optronik, Germany
L-3 Communications Corp, US
Lockheed Martin Corporation, US
Longbow International, US
Macdonald Dettwiler, Canada
NASAM Inc, US
Northrop Grumman Corporation, US
Pratt & Whitney Canada, Canada
Pratt & Whitney, US
Raytheon Company, US
Saab Ericsson Space AB, Sweden
Saft, France
Science Applications International Corp, US
Sermatech International Inc, US
Sikorsky Aircraft Corporation, US
Space Bridge Network Corp, US
Spacehab Inc, US
Stephen Ward, Australia
Strachan & Henshaw Ltd, UK
Thales Avionics In-Flight Systems LLC, US
The Aerospace Corp, US
The Boeing Company, US
Vertical Circuit Inc, US

Mitsubishi Electric Corporation (MELCO)
Head Office
Tokyo Building, 2-7-3 Marunouchi, Chiyoda-Ku, Tokyo, 100-8310, Japan
Tel: (+81 3) 32 18 34 30
Fax: (+81 3) 32 18 28 95
(+81 3) 32 18 90 48
Web: www.mitsubishielectric.com
■ Missile systems, communication equipment, combat direction systems, EW equipment, radars, fire control systems, signal processing equipment, power equipment, computers, semiconductors and satellites.

Mitsubishi Heavy Industries Ltd (MHI)
Head Office
Mitsubishi Jukogyo Kabushiki Kaisha, 16-5 Konan, 2-Chome, Minato-ku, Tokyo, 108-8215, Japan
Tel: (+81 3) 67 16 31 11
(+81 3) 67 16 30 22

Fax: (+81 3) 67 16 58 65
(+81 3) 67 16 58 00
(+81 3) 67 16 58 23
Web: www.mhi.co.jp
www.mhi-ir.jp
■ Provide Fuselage panels and doors for Boeing 777/767, wings for Bombardier Global Express and parts for the PW4000 turbofan engine. Flaps for the Boeing 737/747 and cargo doors for the Airbus A380. Shipbuilding; steel structures; power systems; nuclear energy; machinery; general machinery and special vehicles for defence applications.

Mitsubishi Heavy Industries Ltd (MHI)
Nagoya Guidance and Propulsion Systems Works
1200 Higashi-Tanaka, Komaki-shi, Aichi Ken, 485-8561, Japan
Tel: (+81 568) 79 21 11
Fax: (+81 568) 78 25 52
e-mail: h.ikuno@daiya-pr.co.jp
Web: www.mhi.co.jp
■ Manufacture, repair, testing and calibration of missiles and aerospace engines and control equipment. Type 93 air-to-ship guided missiles, Type 90 air-to-air guided missiles, Patriot surface-to-air missile systems. Rocket engines LE-5B and LE-7A.

Mitsubishi Heavy Industries Ltd (MHI)
Technical Centre
MitsubishijukoYokohama Building, 3-1 Minatomirai 3-chome, Nishi-ku, Yokohama, 220-8401, Japan
Tel: (+81 45) 224 98 88
Fax: (+81 45) 224 99 01

Mitsubishi Precision Co Ltd (MPC)
Head Office
Mita 43 MT, Building F, 3-13-16 Mita, Minato-ku, Tokyo, 108-0073, Japan
Tel: (+81 3) 34 53 64 21
Fax: (+81 3) 34 53 64 34
Web: www.mpcnet.co.jp
■ Design, manufacture, repair, maintenance and sale of space, aviation and vehicular equipment and systems. Simulator, trainer and visual systems, information processing systems, parking management systems and traffic control systems.

MLM System Engineering and Integration
PO Box 45 IL-70350 Beer Yaakov, Israel
Tel: (+972 8) 927 30 26
Fax: (+972 8) 927 28 90/26 60
e-mail: mlm_marketing@mim.iai.co.il

Mobile Telesystems Inc
2 Metropolitan Court, Suite 6, Gaithersburg, Maryland, 20878, United States
Tel: (+1 301) 963 59 70
Fax: (+1 301) 948 46 74
Web: www.mti-usa.com
■ Develop, manufacture and distribution of mobile Inmarsat satellite communication systems and portable, collapsible Ku-Band antennas for commercial and government military users.

Modular Devices Inc (MDI)
One Roned Road, Shirley, New York, 11967, United States
Tel: (+1 516) 345 31 00
Fax: (+1 516) 345 31 06
e-mail: sales@mdipower.com
Web: www.mdipower.com

Moog Controls Ltd
a subsidiary of Moog Inc, US
Ashchurch, Tewkesbury, Gloucestershire, GL20 8NA, United Kingdom
Tel: (+44 1684) 29 66 00
Fax: (+44 1684) 29 67 60
Web: www.moog.com
■ Electro-hydraulic/pneumaticandelectro-mechanical servo control systems and components, valves, actuators, motors and gassing systems and fuel control valves.

Moog Inc
Chatsworth Operations
Schaeffer Magnetics Inc
21339 Nordhoff Street, Chatsworth, California, 91311, United States
Tel: (+1 818) 341 51 56
Fax: (+1 818) 341 38 84
■ Supplier of spaceflight quality motors and actuators.

Moog Inc
Corporate Headquarters
Jamison Road, PO Box 18, East Aurora, New York, 14052-0018, United States
Tel: (+1 716) 652 20 00
Fax: (+1 716) 687 44 57
Telex: 415239
e-mail: info.usa@moog.com
Web: www.moog.com
■ Design and manufacture of electrohydraulic, electromechanical and electropneumatic controls, components and systems to the defence and aerospace industries.

Moog Italiana Srl
a subsidiary of Moog GmbH, Germany
Via G. Pastore 4, I-21046 Malnate, Italy
Tel: (+39 0332) 42 11 11
Fax: (+39 0332) 42 92 33
e-mail: info.italy@moog.com
Web: www.moog.com
■ Flow servovalves, brushless servomotors and electronics. Propellant valves, redundant valves and servoactuators.
Representing: Moog Inc, US

Moog Japan Ltd
Moog Japan Ltd is a wholly owned subsidiary of Moog Inc
1-8-37 Nishi Shindo, Hiratsuka, Kanagwa, 254-0019, Japan
1532 Shindo, Hiratsuka, Kanagwa, 254-0017, Japan
Tel: (+81 463) 55 36 15
Fax: (+81 463) 54 47 09
e-mail: sales@moog.co.jp
Web: www.moog.com
■ Manufacture of precision control components and systems, used in a wide range of high performance aircraft, strategic and tactical missiles, space vehicles and military ground vehicles.
Representing: Moog Inc, US

Morcom International Inc
PO Box 220824, Chantilly, Virginia, 20153-0824, United States
14210-B Sullyfield Circle, Chantilly, Virginia, 20151, United States
Tel: (+1 703) 263 93 05
Fax: (+1 703) 263 93 08
Web: www.morcom.com
■ Supply of communications and weather systems.

Moreton Hall Associates (MHA)
Morar House, Altwood Close, Maidenhead, Berkshire, SL6 4PP, United Kingdom
Tel: (+44 1628) 78 34 55
Fax: (+44 1628) 63 75 86
■ Consultancy, information services and support for projects in space, computer applications and risk management. Fields include product assurance, systems engineering, operations, software, technology and insurance.
Publications: Monthly bulletin: Projected Future Launches
Annual bulletin: Space Launches Pocket Book

Moscow Scientific Research Institute of Instrument Engineering
NPO Vega-M MNIIP
Scientific & Research Institute of Instrumental Engineering
34 Kutuzovsky Prospect, 121170 Moskva, Russian Federation
Tel: (+7 95) 249 76 10
(+7 95) 247 07 04
Fax: (+7 95) 148 79 96
Telex: 412268 vega
■ Research, develop and manufacture radar early warning systems, surface surveillance radar and control systems.

Moscow State Aviation Institute (Technical University) (MAI)
Moscow Aviation Institute
4 Volokolamskoe Shousse, 125872 Moskva, Russian Federation
Tel: (+7 095) 158 04 65
Fax: (+7 095) 158 29 77
Telex: 411746 sokol
e-mail: intdep@mai.ru
rus@mai.ru
■ Institute for aerospace education.

Motorola Inc
Corporate Office
1303 East Algonquin Road, Schaumburg, Illinois, 60196, United States
Tel: (+1 847) 538 46 88
(+1 847) 576 50 00
Freephone: (+1 800) 247 23 46
Fax: (+1 847) 576 76 53
Web: www.motorola.com

Motorola Ltd
Automotive & Industrial Electronics Group
a subsidiary of Motorola Inc, US
Integrated Electronics Systems Sector, Taylors Road, Stotfold, Hitchin, Hertfordshire, SG4 5AY, United Kingdom
Tel: (+44 1462) 83 11 11
Fax: (+44 1462) 83 58 79
Web: www.motorola.com
■ Communication systems and equipment.

MPB Electronics Test Centre (MPB ETC)
Eastern Canada Division
302 Legget Drive, #100, Kanata, Ontario, K2K 1Y5, Canada
Tel: (+1 613) 599 68 00
Fax: (+1 613) 599 76 14
e-mail: emc@mpb-technologies.ca

MPB Electronics Test Centre (MPB ETC)
Western Canada Division
27 East Lake Hill, Airdrie, Alberta, T4B 2B7, Canada
Tel: (+1 403) 912 00 37
Fax: (+1 403) 912 00 83
e-mail: inquire@etc-mpbtech.com

MPB Lasertech
9924-45 Avenue, Edmonton, Alberta, T9E 5J1, Canada
Tel: (+1 780) 436 97 50
Fax: (+1 780) 437 12 40
e-mail: lasertec@oanet.com

MPB Technologies Inc (MPBT)
Experimental Fusion Facility
Centre Canadien de Fusion Magnétique, 1804 Montee Suite Julie, Varennes, Québec, J3X 1S1, Canada
Tel: (+1 514) 652 87 01
Fax: (+1 514) 652 86 25

MPB Technologies Inc (MPBT)
Space and Photonics
151 Hymus Boulevard, Pointe Claire, Québec, H9R 1E9, Canada
Tel: (+1 514) 694 87 51
e-mail: info@mpbc.ca
Web: www.mpb-technologies.ca
■ Specialises in advanced technology products and systems, research, development and measurement services. Activities include: communications; electromagnetics (radar, defence electronics, microwave systems); electronic systems; fusion technology; lasers and electro-optics; space technology. Inter-satellite links, ASIC design and qualification, MMIC design, fast processing, laser flourescence spectroscopy, high speed communications, multiplexing, atmospheric laser communication system for surveillance.

MPE Ltd
Dublier Limited
Wego Condenser Co Ltd
PO Box 11, Liverpool, L33 7UL, United Kingdom
Hammond Road, Knowsley Industrial Park, Liverpool, L33 7UL, United Kingdom
Tel: (+44 151) 632 91 00
Fax: (+44 151) 632 91 12
e-mail: sales@mpe.co.uk
Web: www.mpe.co.uk
■ Provision of EMC/RFI electrical filters, specialist plastic film and ceramic filters. EMC/RFI system analysis, test and solutions. Specialise in NEMP/HEMP/EMPP/TEMPEST filter solutions, MilStud 188 125 filter solutions, shelter/rapid response equipment filter solutions, fighting vehicle filter solutions and telecommunications filter solutions.

MTS Systems Corp
14000 Technology Drive, Eden Prairie, Minnesota, 55344, United States
Tel: (+1 612) 937 40 00
Telex: 290521 mts system enpe
e-mail: info@mts.com
Web: www.mts.com
■ Structural fatigue and analysis test systems, data acquisition, landing gear test, materials test, component fatigue and analysis test.

Muirhead Aerospace Limited
Muirhead Vactric Components Ltd
an Esterline Group company
Oakfield Road, Penge, London, SE20 8EW, United Kingdom
Tel: (+44 20) 86 59 90 90
Fax: (+44 20) 86 59 99 06
e-mail: sales@muirheadaerospace.com
■ Design and manufacture of high performance motion technology. Products include: servo components and control systems, synchros/resolvers, AC and DC motors, engine control components, torque motors, shaft encoders, gearheads, stepping motors, hybrids, brakes, flight controls, thrust reversal and fuel metering. Repair and overhaul of avionics and accessories, civil/military aircraft, fixed- and rotary-wing, including support chain management, ILS.
Accreditations: Lloyd's Register Quality Assurance Ltd to ISO 9001 Certificate No. 912120, CAA, DAI/4376/54, JAR145 repair station, CAA 00477, FAA, FAR145 repair station, M2UY050N, MoD approved No. IMRM06, Honeywell authorised warranty repair station.

Multispectral Solutions Inc (MSSI)
20300 Century Boulevard, Germantown, Maryland, 20874, United States
Tel: (+1 301) 528 17 45
Fax: (+1 301) 528 17 49
e-mail: info@multispectral.com
Web: www.multispectral.com

Myasishchev Design Bureau (MDB)
Eksperimentalnyi Mashinostroitelnyi Zavod Imeni V M Myasishcheva
Zhukovsky 5, 140185 Moskva, Region, Russian Federation
Tel: (+7 095) 728 41 35
(+7 095) 556 78 29
Fax: (+7 095) 728 41 30
(+7 095) 556 55 83
(+7 095) 556 76 94
e-mail: mdb@mail.sitek.net
Web: www.corbina.net
■ Aircraft design. Activities include defence and civil aircraft, manufacture of aerospace equipment and space technology. Company divided into four divisions: design bureau, experimental manufacturing base, flight test base and aviation transportation squad.

N

Nahuelsat SA
Lavelle 472, No. 1 Piso, 1047 Vapital Federal, C1047AAJ Buenos Aires, Argentina
Tel: (+54 11) 58 11 26 00
Fax: (+54 11) 58 11 26 88
e-mail: info@nahuelsat.com.ar
Web: www.nahuelsat.com.ar

Nallatech
Boolean House, One Napier Park, Cumbernauld, Glasgow, G68 0BH, United Kingdom
Tel: (+44 1236) 78 95 51
Fax: (+44 1236) 78 95 99
Web: www.nallatech.com

Nammo AS
Head Office
PO Box 142 N-2831 Raufoss, Norway
Tel: (+47 6115) 36 00
Fax: (+47 6115) 36 20
e-mail: info@nammo.com
nammo@nammo.com
Web: www.nammo.com
■ Manufacture of small arms ammunition including medium calibre 12.7 mm to 40 mm, anti armour weapons, ammunition artillery, ground based, APFSDS-T (30 × 173 mm), mortar bombs, pyrotechnical products, 155 mm shell bodies, 40 mm brass cases, specialised aluminium components for the defence industry, rockets, rocket motors, warheads, space products, composite materials and pressure vessels, aerospace products include boosters and components, missile components and propulsion.

Nammo Raufoss AS
Head Office
Raufoss ASA
PO Box 162 N-2831 Raufoss, Norway
Tel: (+47) 61 15 36 50
Fax: (+47) 61 15 36 60
Web: www.nammo.com
■ Manufacture of Small Arms Ammunition; Cannon 20 × 102 mm M39, 20 × 110RB, 25 × 137 mm KBA, 25 mm Aden, 27 × 145 mm Mauser, 30 mm Aden, 30 mm DEFA; Mortar 81 mm; Air defence 40 mm L/60 MPT, 40 mm L/70 MPT; Naval 76 mm ASP; 105 mm ASP, 127 mm ASP; Field 155 mm HE NM28. Research, development and production in the space and automotive sectors. Ammunition and aerospace defence products.
Associated companies/subsidiaries:
Nammo Raufoss AS
Raufoss Service AS
Raufoss Metall GmbH, Germany
Raufoss France SARL, France
Raufoss do Brasil Ltda, Brasil
Kongsberg-Raufoss AS

Nanjing Astronomical
Instruments Research Centre
210042 Nanjing, China
Tel: (+86 25) 541 17 76
Fax: (+86 25) 541 18 72

Nanjing University of Aeronautics & Astronautics (NUAA)
Office of International Co-operation and Exchange
29 Yudao Street, PO Box 213 210016 Nanjing, Jiangsu, China
Tel: (+86 25) 84 89 24 40
Fax: (+86 25) 84 49 80 69
e-mail: icedao@nuaa.edu.cn
Web: www.nuaa.edu.cn
■ Development of super-light pilotless fixed- and rotary-wing aircraft.

NASA European Representative
c/o American Embassy, 2 Avenue Gabriel, F-75382 Paris, Cedex 8, France
Tel: (+33 1) 43 12 21 00
Fax: (+33 1) 42 65 87 68

NASA Institute for Advanced Concepts (NIAC)
75 5th Street, North West, Suite 318, Atlanta, Georgia, 30308, United States
Tel: (+1 404) 347 96 33
Fax: (+1 404) 347 96 38
Web: www.niac.usra.edu

NASA Japan Representative
US Embassy, Unit 45004, Box 235, Tokyo, 96337-5004, Japan
Tel: (+81 3) 32 24 58 27
Fax: (+81 3) 32 24 52 29
e-mail: gkirkham@hq.nasa.gov

NASA Specialized Center of Research
Strahlenzentrum der Universitaet Giessen, Leihgesterner Weg 217, D-35392 Giessen, Germany
Tel: (+49 641) 991 53 00
Fax: (+49 641) 991 50 09
■ Radiation health in space.

National Aeronautics & Space Administration (NASA)
Ames Research
NASA Ames Research
Moffett Field, California, 94035, United States
Tel: (+1 650) 604 50 00
(+1 650) 604 39 37
(+1 650) 604 90 00
Freephone: (+1 877) NSSC 123
e-mail: nssc-contactcenter@nasa.gov
Web: www.nasa.gov/centers/ames/home/index.html
■ Provision of research services to support NASA's space exploration programmes.

National Aeronautics & Space Administration (NASA)
Dryden Flight Research Center
PO Box 273, Edwards, California, 93523-0273, United States
Tel: (+1 661) 276 34 49
Fax: (+1 661) 276 35 66
Web: www.dfrc.nasa.gov

National Aeronautics & Space Administration (NASA)
Flight Simulation Laboratories
Ames Research Center, Mail Stop 243-1, Moffett Field, California, 94035-1000, United States
Tel: (+1 650) 604 32 71
Fax: (+1 650) 604 39 52
e-mail: sims@artemis.arc.nasa.gov
Web: www.simlabs.arc.nasa.gov

National Aeronautics & Space Administration (NASA)
Glenn Research Center
21000 Brookpark Road, Cleveland, Ohio, 44135-3191, United States
Tel: (+1 216) 433 40 00
Fax: (+1 216) 433 81 43
Web: www.grc.nasa.gov

National Aeronautics & Space Administration (NASA)
Goddard Institute for Space Studies
2880 Broadway, New York, New York, 10025, United States
Tel: (+1 212) 678 55 00
Web: www.giss.nasa.gov

National Aeronautics & Space Administration (NASA)
Goddard Space Flight Center
Greenbelt, Maryland, 20771, United States
Tel: (+1 301) 286 20 00
Web: www.gsfc.nasa.gov
■ A national resource for the pursuit of exploration of the Earth and space. Provides customer centered leadership to implement the goals of NASA and its six enterprises with principal responsibilities in support of the Earth Science Enterprise and the Space Science Enterprise and secondary roles in support of the other four NASA enterprises.

National Aeronautics & Space Administration (NASA)
John C Stennis Space Center
NASA Public Affairs, PA00, Mississippi, 39529-6000, United States
Tel: (+1 228) 688 18 80
Web: www.ssc.nasa.gov

National Aeronautics & Space Administration (NASA)
John F Kennedy Space Center
Kennedy Space Center, Florida, 32899-0001, United States
Tel: (+1 321) 867 50 00
(+1 321) 867 44 44
Web: www.ksc.nasa.gov

National Aeronautics & Space Administration (NASA)
Johnson Space Center
Automation, Robotics and Simulation Division, Mail Code ER, Houston, Texas, 77058, United States
Tel: (+1 281) 483 15 55
Web: www.vesuvius.jsc.nasa.gov

National Aeronautics & Space Administration (NASA)
Langley Research Center
100 NASA Road, Hampton, Virginia, 23681-0001, United States
Tel: (+1 757) 854 85 54
Web: www.larc.nasa.gov
■ Research in atmospheric flight supports programs in earth and space science, space access and exploration, aerospace vehicle systems technology, airspace systems, and aviation safety and security. Specialising in aerodynamics, aerothermodynamics, acoustics, aerospace systems concepts and analysis, airborne systems, atmospheric sciences, structures and materials, and systems engineering.

National Aeronautics & Space Administration (NASA)
Lyndon B Johnson Space Center
Room 476, Building 1, Mail Code KT, Houston, Texas, 77058, United States
Tel: (+1 281) 483 01 23
(+1 281) 483 86 93

National Aeronautics & Space Administration (NASA)
Marshall Space Flight Center
Huntsville, Alabama, 35812, United States
Web: www.msfc.nasa.gov

National Aeronautics & Space Administration (NASA)
Michoud Assembly Facility
PO Box 29300, New Orleans, Louisiana, 70189, United States
Tel: (+1 504) 255 33 11

National Aeronautics & Space Administration (NASA)
NASA Headquarters
Suite 5K39, Washington, District of Columbia, 20546-0001 United States
Tel: (+1 202) 358 00 00
Fax: (+1 202) 358 32 51
e-mail: info-center@hq.nasa.gov

National Aeronautics & Space Administration (NASA)
NASA Inspector General
NASA HQ, 300 E Street, Washington, District of Columbia, 20546, United States
Tel: (+1 202) 358 12 20

National Aeronautics & Space Administration (NASA)
Public Affairs Office
National Space Technology (NSTL)
Stennis Space Center, Mississippi, 39529, United States
Tel: (+1 228) 688 33 33
Fax: (+1 228) 688 10 94
e-mail: public-inquiries@ssc.nasa.go
Web: www.nasa.gov

National Aeronautics & Space Administration (NASA)
Space Telescope Science Institute
John Hopkins Homewood Campus, 3700 San Martin Drive, Baltimore, Maryland, 21218, United States
Tel: (+1 410) 338 47 00
Fax: (+1 410) 338 47 67

National Aeronautics & Space Administration (NASA)
White Sands Test Facility
PO Drawer NM, Las Cruces, New Mexico, 88004-0020, United States
Tel: (+1 505) 524 57 71
Web: www.wstf.nasa.gov

National Aeronautics & Space Administration (NASA)
Building J-17, Wallops Island, Virginia, 23337-5099, United States
Tel: (+1 757) 824 22 98
(+1 757) 824 22 97
(+1 757) 824 13 44
(+1 757) 824 10 00
(+1 757) 824 15 79
(+1 757) 824 14 79
Fax: (+1 757) 824 17 76
(+1 757) 824 19 71
e-mail: bbland@pop100.gsfc.nasa.gov
troutman@pop100.gspc.nasa.gov
Web: www.wff.nasa.gov
■ Launch site managing the space shuttle small payloads projects and university class projects for NASA earth and space science enterprises; conducts observational earth science studies, provides aircraft flight services for scientific investigations and operates an orbital tracking station and test range.

National Aeronautics & Space Center (NASA)
KSC VLS Resident Office
PO Box 425, Lompoc, California, 93438, United States
Tel: (+1 805) 866 58 59

National Aerospace Laboratory of the Netherlands (NLR)
Nationaal Luchten Ruimtevaartlaboratorium
PO Box 90502 NL-1006 BM Amsterdam, Netherlands
Anthony Fokkerweg 2, NL-1059 CM Amsterdam, Netherlands
Tel: (+31 20) 511 31 13
Fax: (+31 20) 511 32 10
e-mail: info@nlr.nl
Web: www.nlr.nl
■ Independent non-profit research institute conducting contract research for national and international customers. The NLR owns and operates several research facilities. Seven main divisions of research: Fluid Dynamics, Structures and Materials, Space, Aircraft, Air Transport, Information and Communication Technology, Avionics. Research and test facilities wind tunnels, ATC simulators, research aircraft, supercomputers and environmental test facilities. Participation in DNW and ETW. Member of AEREA (Association of European Research Establishments) in Aeronautics.
Accreditations: ISO 9001

National Aerospace Laboratory of the Netherlands
Library
PO Box 905 02 NL-1006 BM Amsterdam, Netherlands
Tel: (+31 527) 24 84 44
(+31 527) 24 83 17
Fax: (+31 527) 24 82 10

National Aerospace Laboratory of the Netherlands
Noordoostpolder Laboratory
PO Box 153, NL-8300 AD Emmeloord, Netherlands
Tel: (+31 527) 24 84 44
Fax: (+31 527) 24 82 10
■ Aeronautical research and development in computational and experimental fluid dynamics, aeroacoustics, structures and materials, flight dynamics and operations research. Space research in the fields of fluid dynamics, stability and control, robotics environmental simulation, materials and structures, remote sensing, micro-gravity. Extensive facilities and additional equipment for experimental research and development in all fields.

National Agency for Aerospace Programs (NIVR)
PO Box 35, NL-2600 AA Delft, Netherlands
Kluyverweg 1, NL-2629 HS Delft, Netherlands
Tel: (+31 15) 278 80 25
Fax: (+31 15) 262 30 96
e-mail: info@nivr.nl
Web: www.nivr.nl
■ Government agency for the promotion of aerospace activities in the Netherlands; management of national space projects. The NIVR acts as a management agency for government sponsored aerospace research and design. The agency does not execute such research itself, but monitors the definition and execution of research and design activities conducted by Netherlands industry and laboratories, both in national projects and in international collaborative projects.

National Air and Space Museum (NASM)
Smithsonian Institution
6th and Independence Avenue South West, Washington, District of Columbia, 20560, United States
Tel: (+1 202) 357 27 00
Fax: (+1 202) 633 81 74
Web: www.nasm.si.edu
■ Research, education, exhibits, collections management, public service, library and photographic collection and publications.
Publications: Air & Space/Smithsonian

National Astronomical Observation
Astrometry Cel Mech Division
Osawa Mitaka, Tokyo, 181, Japan
Tel: (+81 422) 34 36 13
Fax: (+81 422) 34 37 93

National Astronomical Observatories Chinese Academy of Sciences (NAOC)
20A Datun Road, 100012 Chaoyang District, Beijing, China
Tel: (+86 10) 64 88 87 12
Tel: (+86 10) 64 88 87 08
Web: www.bao.ac.cn
■ Conduct astronomical observations, theoretical astronomical work and fundamental astronomy orientated high technology research.

National Astronomical Observatory (NAOJ)
Kokuritsu Tenmondai
Osawa Mitaka-shi, Tokyo, 181-8588, Japan
Tel: (+81 422) 34 36 00
Fax: (+81 422) 34 36 90
Telex: 2822307 taomtk
Web: www.nao.ac.jp

National Center for Microgravity Research (NCMR)
Case Western Reserve University, Glennan 312, 10900 Euclid Avenue, Cleveland, Ohio, 44106-7074, United States
Tel: (+1 216) 368 07 48
Fax: (+1 216) 368 07 18
■ Perform critical-path microgravity research on fluids and combustion, supporting NASA missions, pursue knowledge transfer, outreach and education. Work to enhance the pool of microgravity research scientists, develop concepts leading to experiments on the International Space Station, and assist NASA in developing strategies for future microgravity research efforts.

National Central University
Centre for Space and Remote Sensing Research
320 Chung-Li, Taiwan
Tel: (+886 3) 425 72 32
(+886 3) 422 71 51
Fax: (+886 3) 425 55 35
(+886 3) 425 55 35
(+886 3) 425 49 08
e-mail: service@csrsr.ncu.edu.tw
Web: www.csrsr.ncu.edu.tw
■ Remote sensing utilisation and data provider.

National Committee for Space Research
Israel Academy of Sciences and Humanities, PO Box 39040 IL-69978, Tel-Aviv, Israel
Tel: (+972 3) 640 86 20
Fax: (+972 3) 640 92 82

National Environmental Satellite Data & Information Service (NESDIS)
1335 East-West Highway, SSMC1, Room 7216, Silver Spring, Maryland, 20910, United States
Tel: (+1 301) 713 35 78
Fax: (+1 301) 713 12 49
Web: www.nesdis.noaa.gov
■ Manages polar and geo-stationary operational satellite systems and gathers, archives and undertakes applied research in support of NOAA programs.
Affiliated organisations:
Space Environment Laboratory

National Geospatial Intelligence Agency (NGA)
Public Affairs Office
National Imagery and Mapping Agency
4600 Sangamore Road, MS D-54, Bethesda, Maryland, 20816-5003, United States
Fax: (+1 301) 227 39 20
e-mail: queries@nga.mil
Web: www.nga.mil
■ NGA is a Department of Defence combat support agency and a member of the National Intelligence Community. Provision of timely, relevant and accurate geospatial intelligence in support of national security. Geospatial intelleligence is the exploitation and analysis of imagery and geospatial information to describe, assess and visually depict physical features and geographically referenced activities on the Earth.

National Institute of Telecommunications
Szachowa Strasse 1, PL-04894 Warszawa, Poland
Tel: (+48 22) 512 81 00
Fax: (+48 22) 512 86 25
e-mail: info@itl.waw.pl
Web: www.nit.eu

National Museum of the US Air Force (NMUSAF)
National Museum of the United States Air Force
1100 Spaatz Street, Wright-Patterson AFB, Dayton, Ohio, 45433-7102, United States
Tel: (+1 937) 255 32 86
(+1 937) 253 IMAX
Fax: (+1 937) 255 05 23
Web: www.wpafb.af.mil/museum
www.nationalmuseum.af.mil
■ Display over 300 aircraft and missiles plus many artifacts from the Wright Brothers to the present.

National Oceanic and Atmospheric Administration (NOAA)
14th Street & Constitution Avenue, NW, Room 6217, Washington, District of Columbia, 20230, United States
Tel: (+1 202) 482 60 90
Fax: (+1 202) 482 31 54
Web: www.noaa.gov

National Oceanic and Atmospheric Administration (NOAA)
World Weather Building, Room 601 E/RA 21, Washington, 20233, United States
Tel: (+1 301) 763 82 51

National Reconnaissance Office
14675 Lee Road, Chantilly, Virginia, 20151-1715, United States
Tel: (+1 703) 808 11 98
Fax: (+1 703) 808 11 71
Web: www.nro.gov

National Remote Sensing Agency (NRSA)

Balanagar, Hyderabad, 500037, India
Tel: (+91 40) 387 87 88
(+91 40) 387 96 77
(+91 40) 238 795 72 76
Fax: (+91 40) 238 772 10
(+91 40) 387 88 65
Telex: 4256522
Cable: REMOSEN
e-mail: sales@nrsa.gov.in
Web: www.nrsa.gov.in
■ Remote sensing satellite data reception, processing and dissemination.

National Research Council (NRC)

Associateship Programs
500 Fifth Street North West, Washington, District of Columbia, 20001, United States
Tel: (+1 202) 334 27 60
Fax: (+1 202) 334 27 59
e-mail: rap@nas.edu
Web: www.national-academies.org/rap

National Snow & Ice Data Center (NSIDC)

University of Colorado, Boulder, Colorado, 80309-0449, United States
Tel: (+1 303) 492 6199
Fax: (+1 303) 492 24 68
e-mail: nsidc@nsidc.org
Web: www.nsidc.org
■ Data management for snow and ice and atmospheric research.

National Space Agency of Ukraine (NSAU)

NKAU
Bozhenka 11 Street, 03022 Kyiv, Ukraine
Tel: (+380 044) 226 25 55
Fax: (+380 044) 529 50 58
Telex: 131406 sky
e-mail: yd@nkau.gov.ua
Web: www.nkau.gov.ua
■ Development of state policy concepts, in the sphere of research and peaceful uses of space, as well as in the interests of national security.

National Space Development Agency of Japan (JAXA-KSC)

KSC Office
O&C Building, Room 1014, Code: NASDA-KSD, John F Kennedy Space Center, Florida, 32899, United States
Tel: (+1 321) 867 38 79
(+1 321) 867 32 95
Fax: (+1 321) 452 96 62

National Space Development Agency of Japan (NASDA)

Los Angeles Office
633 West 5th Street, Suite 5870, Los Angeles, California, 90071, United States
Tel: (+1 213) 688 77 58
(+1 213) 688 11 71
Fax: (+1 213) 688 08 52
■ Governmental space agency in charge of Japanese space development programmes.

National Space Program Office (NSPO)

8th Floor, 9 Prosperity 1st Road, Science-Based Industrial Park, 30077 Hsin-Chu, Taiwan
Tel: (+886 3) 578 42 08
Fax: (+886 3) 578 42 34
(+886 3) 578 42 10
e-mail: service@nspo.gov.tw
Web: www.nspo.gov.tw
■ Development and operation of the national space programme including space technology, Earth observation and telecommunications.

National Space Society of Australia (NSSA)

GPO Box 7048, Sydney, New South Wales, 2001, Australia
Tel: (+61 2) 99 88 02 52
Fax: (+61 2) 99 88 02 62
e-mail: nssa@nssa.com.au
Web: www.nssa.com.au
■ Promotion of space development, education and industry liaison.

National University of Singapore

Block SOC-1 Level 2, Faculty of Science, Lower Kent Ridge Road, Singapore, 119260, Singapore
Tel: (+65) 68 74 63 96
Fax: (+65) 67 75 77 17
e-mail: crisp@nus.edu.sg

NAV Portugal

Communication and Image Office
Navegacao Aérea de Portugal – NAV Portugal EPE Rua D – Edificio 121, Aeroporto de Lisboa, P-1700 – 008 Lisboa, Portugal
Tel: (+351 218) 55 31 43
(+351 218) 55 31 00
(+351 218) 55 35 07
(+351 218) 49 50 87
(+351 296) 88 62 08
Fax: (+351 218) 55 31 47
(+351 218) 55 36 00
Telex: 12936 cctal
lpppyoyc
lpppzqzx
lppozqzx
132.3mhz
e-mail: gabcim@nav.pt
desica@nav.pt
acclis@nav.pt
dinis.resendes@nav.pt
Web: www.nav.pt
■ Aeronautical information services. Responsibility for the management of air traffic in the Lisbon and Santa Maria flight information regions, under the jurisdiction of the Portugese State, in full compliance with the provisions of international agreements and those of the international civil aviation organisations to which Portugal subscribes or of which Portugal is a member state. En-route services.

Naval Center for Space Technology

1515 South Manchester Avenue, Anaheim, California, 92802-2907, United States
Tel: (+1 714) 780 78 13
Fax: (+1 714) 780 78 57
■ Conception and development of space and aerospace systems to meet navy and DoD needs.

Naval Research Laboratory (NRL)

Code 5712 Division
4555 Overlook Avenue South West, Washington, District of Columbia, 20375, United States
Tel: (+1 202) 767 25 41
Fax: (+1 202) 767 69 91
Web: www.nrl.navy.mil
■ Programme of scientific research and technological development targeted at maritime applications of new and improved materials and techniques.

Navigation Maritime Bulgare Ltd

1 Primorski Boulevard, 9000 Varna, Bulgaria
Tel: (+359 52) 63 29 14
(+359 52) 68 32 35
(+359 52) 68 32 36
Fax: (+359 52) 22 24 91
(+359 52) 60 03 60
(+359 52) 63 29 47
(+359 52) 68 32 36
Telex: 77351

Navitrak International Corporation

1660 Hollis Street, Suite 904, Halifax, Nova Scotia, L6J 6G6, Canada
Tel: (+1 902) 429 14 38
Fax: (+1 902) 429 15 82
e-mail: sales@navitrak.com
Web: www.navitrak.com

NEC

Seoul Liaison Office
15th Floor, Star Tower, 737 Yeoksam-Dong, Kangnam-Ku, Seoul, 135-984, Korea, South
Tel: (+82 2) 21 12 27 80
Fax: (+82 2) 21 12 27 90
e-mail: seouloffice@necseoul.com

NEC

Tehran Office
House No 114, Shahid Khaled Eslambuli Avenue, Tehran, Iran
Tel: (+98 21) 879 52 84
Fax: (+98 21) 879 53 74
■ Provides support activities such as supply of technology and product information.

NEC Australia Pty Ltd (NECA)

National Head Office
a subsidiary of NEC Corp, Japan
649-555 Springvale Road, Mulgrave, Victoria, 3170, Australia
Tel: (+61 3) 13 16 32
Fax: (+61 3) 92 62 13 33
(+61 3) 92 64 31 07
e-mail: contactus@mail.neca.nec.com.au
Web: www.nec.com.au
■ Marketing of telecommunications products including: television and radio broadcasting equipment, portable transceivers, fibre-optic communications, facsimile, telephone video systems, semiconductors, electronic components, satellite and mobile earth stations.
Affiliate:
NEC Business Solutions

NEC Corporation

Headquarters
7-1 Shiba 5-chome, Minato-ku, Tokyo, 108-8001, Japan
Tel: (+81 3) 34 54 11 11
Fax: (+81 3) 37 98 15 10
(+81 3) 37 98 15 11
(+81 3) 37 98 15 12
e-mail: webmaster@nec.co.jp
Web: www.nec.co.jp
■ Solutions: supercomputers, computers, PCs, printers. Networks: network systems and equipment, mobile communications and related software and services. Electron Devices: compound semiconductors, display modules, rechargeable batteries, capacitors.
Group Divisions:
NEC Electron Devices
NEC Networks
NEC Solutions

NEC Malaysia Sdn Bhd

33rd Floor, Menara TA One 22, Jalan P Ramalee, 50250 Kuala Lumpur, Malaysia
Tel: (+60 3) 21 64 11 99
Fax: (+60 3) 21 64 14 68

NEC Semiconductors (Malaysia) Sdn Bhd

Telok Panglima Garang
Free Industrial Zone KM 15 Jalan, Banting Kuala Langat, Selanor Darul Ehsan, Malaysia
Tel: (+60 3) 352 62 01
Fax: (+60 3) 352 89 01

Neptec

Suite 105, 16850 Saturn Lane, Houston, Texas, 77058, United States
Tel: (+1 281) 280 85 55
Fax: (+1 281) 280 86 16
Web: www.neptec.com

Nerac Inc

One Technology Drive, Tolland, Connecticut, 06084, United States
Tel: (+1 860) 872 70 00
Fax: (+1 860) 875 60 62
e-mail: info@merac.com
Web: www.nerac.com
■ Bibliographic reviews of aerospace technology and companies, technology tracking services and document retrieval. Information on aerospace engineering, agriculture and plant biology, automotives, biochemistry, biomedical and medical devices, business combined with management and marketing, coatings and adhesives, computers and software, electrical engineering, energy and power, environmental biology, good chemistry, inorganic chemistry, mechanical engineering, metallurgy, military and govenment, molecular and cell biology, optics combined with imaging and lasers, organic chemistry-synthesis, packaging, patent law, pharmaceuticals, polymer chemistry and textiles combined with paper and pigments.

Netherlands Aerospace Group (NAG)

Boerhaavelaan 40, PO Box 190, NL-2700 AD Zoetermeer, Netherlands
Tel: (+31 79) 353 13 56
Fax: (+31 79) 353 13 65
e-mail: nag@fme.nl
Web: www.nag.aero
■ National association of aerospace companies and organisations engaged in education, research and development, engineering, manufacturing, maintenance repair and the overhaul of civil and military systems and equipment. The association facilitates the setting up of national pavilions at airshows and conferences, organises international trade missions, courses, seminars and meetings on specific areas of interest.

Netherlands Agency for Aerospace Programs
Nederlands Instituut voor Vliegtuigontwikkeling en Ruimtevaart (NIVR)
PO Box 35, NL-2600 AA Delft, Netherlands
Kluyverweg 1, NL-2629 HS Delft, Netherlands
Tel: (+31 15) 278 80 25
Fax: (+31 15) 262 30 96
e-mail: info@nivr.nl
Web: www.nivr.nl
■ Aims to promote the scientific, industrial and user support activities in the Netherlands, in the field of aeronautics and space. Advise the Dutch government on matters concerning aerospace policy in national and international framework. Acts as an intermediary between research institutes, trade, industry, users and government agencies regarding aircraft development and use, and the various aspects of space technology and development. Manages projects and programmes on behalf of the relevant ministries. Core activities include: aircraft development – stimulating basic research, technology innovation and improvements in industrial production in civil and military aircraft programmes; utilisation of aircraft–advancing sophisticated technology and concepts to improve air traffic management and ariport operations with an eye to safety and the environment; space technology – contributing to solutions concerning scientific, social and economic issues by participating in important space programmes. Managing national technology schemes and multilateral space projects and assist ESA in technology programmes for Dutch companies.

Netherlands Astronautical Society
Nederlandse Vereniging v. Ruimtevaart
Zonnenburg 2, NL-3512 NL Utrecht, Netherlands
Tel: (+31 30) 231 13 60
Fax: (+31 30) 234 28 52
e-mail: info@ruimtevaart-nvr.nl
Web: www.ruimtevaart-nvr.nl
■ The NVR, member of the International Astronautical Federation (IAF), was established with the purpose to inform individuals interested in space research and space technology and to bring them together, by organising lectures, excursions, symposia and film-shows.
Publications: Ruimtevaart

Netherlands Industrial Space Organisation (NISO)
PO Box 32070 NL-2303 DB Leiden, Netherlands
Tel: (+31 71) 525 51 24
Fax: (+31 71) 524 51 25
e-mail: info@pipistrel.si
Web: www.niso.nl
■ Comprised of Dutch industries, research organisations and users in the space field with the aims of: defining common interests and strategies; promoting industrial involvement; stimulating the use/application of space facilities and products by industrial enterprises and institutes; broadening the social base of space; and promoting technology spin-offs. Powered hand-gliders and composites.

New Mexico Spaceport Authority
1100 St. Francis Drive, Santa Fé, New Mexico, 87505, United States
Tel: (+1 505) 476 37 47

New Zealand Space Flight Association Ltd
Wellesley Street, PO Box 5829, Auckland, New Zealand
Tel: (+64 9) 480 79 00
Fax: (+64 9) 480 79 01
■ Promotion, education and liaison on space matters.

Newtec America Inc
1250 Summer Street, Suite 305, Stamford, Connecticut, 6905, United States
Tel: (+1 203) 323 00 42
Fax: (+1 203) 323 84 06
e-mail: general@newtecamerica.com
sales@newtecamerica.com
support@newtecamerica.com
hrm@newtecamerica.com

Newtec Cy
Laarstraat 5, B-9100 Sint-Niklaas, Belgium
Tel: (+32 3) 780 65 00
Fax: (+32 3) 780 65 49
e-mail: general@newtec.be
sales@newtec.be
techsupport@newtec.be
webmaster@newtec.be
Web: www.newtec.be
■ Communications equipment and services.

Nexans Harnesses NV
Alcatel Fabrisys NV
Heideveld 1, B-1654 Huizingen, Belgium
Tel: (+32 2) 363 17 11
Fax: (+32 2) 363 17 87
e-mail: mkt.sales@nexans.com
Web: www.nexans.com
■ Provision of electric and electro optic connection systems and harnesses, feasibility studies, design, manufacturing and installation of industrial vehicles, satellites and aircrafts.
Accreditations: ISO 9001: 2000, EN 9100, AQAP 110

Niagara University, Space Settlement Studies Project (3SP)
Department of Sociology
Space Settlement Studies Project (3SP), Niagara University
Niagara University, New York, 14109, United States
Tel: (+1 716) 286 80 94
Fax: (+1 716) 286 85 81
■ Research and study of social and cultural aspects of long-term space habitation.

Nigel Press Associates Ltd (NPA)
Crockham Park, Edenbridge, Kent, TN8 6SR, United Kingdom
Tel: (+44 1732) 86 50 23
Fax: (+44 1732) 86 65 21
e-mail: info@npagroup.com
Web: www.npagroup.com
■ Provide services of acquire, process and distribute both optical and radar satellite image data including landsat, SPOT, RADARSAT, ERS, QUICKBIRD and IKONOS. Also provide remote sensing applications including interpretation services and the generation of Digital Elevation Models (DEMs).

Nigeria Telecommunications Ltd (NITEL)
PO Box 12550 Lagos, Nigeria 3-5 Tafewa Balewa Square, Lagos, Nigeria
Tel: (+234 1) 260 07 71
(+234 1) 261 35 42
Fax: (+234 1) 262 28 45
Telex: 11001 netad
28112 cirtel
■ National signatory to Intelsat and Inmarsat.

NII Parachutostroeniya
2 Irkutskaya Ulitsa, 107421 Moskva, Russian Federation
Tel: (+7 095) 462 13 19
Fax: (+7 095) 462 52 33
■ Development, design, testing and manufacture of parachute systems for different purposes and other relevant technologies products.

NII Radio (NIIR)
16 Kazakova Street, 105064 Moskva, Russian Federation
Tel: (+7 095) 261 36 94
Fax: (+7 095) 261 00 90
e-mail: info@niir.ru
niir@ccs.ru
■ Research centre of the Ministry of Communications.

Ningbo Donglian Mechanical Seal Co Ltd
Jiangshan Science and Technology Zone, Ningbo, Zhejiang, China
Tel: (+86 574) 88 45 23 48
(+86 574) 88 45 23 54
(+86 574) 88 45 83 54
Fax: (+86 574) 88 45 74 35
e-mail: marketing2@dyseals.com
Web: www.dyseals.com
■ Production of ceramic, tungsten carbide and carbon graphite mechanical seals.
Accreditations: ISO 9002, ISO 9001-2000

Nippi Corporation
Japan Aircraft Manufacturing Co Ltd
3175 Showa-machi, Kanazawa-ku, Yokohama, 236-8540, Japan
Tel: (+81 45) 773 51 18
(+81 45) 773 51 00
Fax: (+81 45) 771 12 53
(+81 45) 773 51 01
e-mail: webchief@mail.nippi.co.jp
Web: www.nippi.co.jp
■ Manufacture of aircraft parts, rocket parts and space equipment. Maintenance and modification of aircraft. Manufacture of non-destructive inspection systems, industrial fans and marine equipment.

Nippon Hoso Kyokai (NHK)
Japan Broadcasting Corporation
2-2-1 Jinnan, Shibuya-ku, Tokyo, 150-8001, Japan
Tel: (+81 3) 34 65 11 11
Fax: (+81 3) 34 69 81 10
e-mail: intl@pr.nhk.or.jp
Web: www.nhk.or.jp
■ Operates two terrestrial TV channels, three satellite TV channels including one HDTV channel (all satellite channels are transmitted in digital and analog), three radio channels and three world-wide services (two TV and one radio). Broadcasts three channels through BSAT-1a and BSAT-1b and will in future be using BSAT-2a and BSAT-2b for its broadcasting.

Nissho Iwai Corporation
Aircraft & Advanced Technology Department
3-1 Daiba 2-Chome, Minato-ku, 135-8655, Japan
Tel: (+81 3) 55 20 32 43
Fax: (+81 3) 55 20 32 46
■ Marketing consultant for military and aerospace companies.

NMB Minebea UK Ltd
Lincoln Plant
Rose Bearings Ltd
a subsidiary of Minebea Co Ltd, Japan
Doddington Road, Lincoln, LN6 3RA, United Kingdom
Tel: (+44 1522) 50 09 33
Fax: (+44 1522) 50 09 95
Cable: ROSE
Web: www.minebea.jp
■ Rod end and spherical bearings to aerospace, railway, automotive and power generation industries. All types of military standard to customer bespoke. Specialist in coupling bearings for articulated PSV.

Norsk Elektro Optikk A/S (NEO)
PO Box 384 N-1471 Loerenskog, Norway
Tel: (+47 67) 97 47 00
Fax: (+47 67) 97 49 00
e-mail: neo@neo.no
sales@neo.no
support@neo.no
Web: www.neo.no

Nortel Limited
Radio Infrastructure
Brixham Road, Paignton, Devon, TQ4 7BE, United Kingdom
Tel: (+44 1803) 66 20 00
Fax: (+44 1803) 66 28 01
e-mail: benroome@nortel.com
■ Electronic test equipment (GPS simulators).

Northern Telecom Inc
200 Athens Way, Nashville, Tennessee, TN 37228-1803, United States
Tel: (+1 615) 734 40 00
Fax: (+1 615) 734 51 89

Northrop Grumman Corporation
Corporate Headquarters
1840 Century Park East, Los Angeles, California, 90067-2199, United States
Tel: (+1 310) 553 62 62
Fax: (+1 310) 201 30 23
(+1 310) 553 20 76
(+1 310) 555 45 61
e-mail: onewebmaster@ngc.com
Web: www.northropgrumman.com
■ Design, manufacture and systems integration: military surveillance, combat aircraft, defence electronics and systems, airspace management and information systems, marine systems, precision weapons, space systems and commercial and military aerostructures.
Northrop Grumman Corporation is split into eight sectors:
Electronic Systems
Information Technology
Integrated Systems
Mission Systems
Newport News
Ship Systems
Space Technology
Technical Services

Northrop Grumman Corporation
Electronic Systems
Northrop Grumman Electronic Systems
Litton Industries Inc
Mail Stop A255, PO Box 17319, Baltimore, Maryland, 21203-7319, United States
1580-A West Nursery Road, Linthicum, Maryland, 21090, United States
Tel: (+1 410) 765 10 00
(+1 410) 993 68 48
Freephone: (+1 800) 443 92 19
Fax: (+1 410) 981 48 03
e-mail: mccalde@northropgrumman.com
es_communications@ngc.com
Web: www.es.northropgrumman.com
■ Production of radars and electronics for combat aircraft and battlespace management systems including the F-22 and F-16 fighters, B1-B bomber, Apache/Longbow attack helicopter, C-130 transport, E-3 AWACS and E-8 Joint stars as well as military space and undersea systems, electronic countermeasures, precision weapons, space systems, logistics systems and automatic standard intervention systems.

Northrop Grumman Corporation
Electronics and Systems Integration Division
1111 Stewart Avenue, Bethpage, New York, 11714-3580, United States
Tel: (+1 516) 575 51 19
Fax: (+1 516) 575 36 91

Northrop Grumman Electronic Systems
Defensive Systems Division
600 Hicks Road, Rolling Meadows, Illinois, 60009-1098, United States
Tel: (+1 847) 259 96 00
Fax: (+1 847) 870 57 05
Web: www.dsd.es.northropgrumman.com

Northrop Grumman Electronic Systems
Navigation Systems Division
Litton Advanced Systems
Litton Industries Inc
21240 Burbank Boulevard, Woodland Hills, California, 91367-6675, United States
Tel: (+1 818) 715 40 40
(+1 818) 715 20 00
(+1 818) 715 24 70
Freephone: (+1 866) NGNAVSYS
e-mail: customerservice.nsd@northropgrumman.com
Web: www.nsw.es.northropgrumman.com
■ Provides situational awareness electronic systems and products for defense, civil, and commercial markets. NSD offers inertial navigation systems, with and without embedded GPS, Identification Friend or Foe (IFF) systems, Fiber-Optic Acoustic Sensors (FOAS), integrated avionics and vetronics systems, and logistic support products and services.

Northrop Grumman Electronic Systems
Navigation Systems Division
Litton Indistries Inc
5500 Canoga Avenue, Woodland Hills, California, 91367-6698, United States
Tel: (+1 818) 715 40 40
(+1 818) 715 42 56
Fax: (+1 818) 715 20 19
e-mail: info@ngnavsys.com
■ Supplies situational awareness products for international and domestic defence and commmerical markets and offers intergrated avionics, navigation-grade and tactical-grade inertial systems.

Northrop Grumman Information Technology (IT)
Commercial, State and Local Group
13825 Sunrise Valley Drive, Suite 120, Herndon, Virginia, 20171, United States
Tel: (+1 703) 713 41 00
Web: www.it.northropgrumman.com

Northrop Grumman Space Technology (NGST)
a subsidiary business unit of Northrop Grumman Corporation
1 Space Park, Redondo Beach, California, 90278, United States
Tel: (+1 310) 812 43 21
Fax: (+1 310) 814 45 07
Web: www.st.northropgrumman.com
■ Develop leading edge technologies across a wide array of disciplines including avionics, sensors, optics, lasers, microelectronic processors, and spacecraft structures, power propulsion, deployable antennas and reflectors.

Northrop Grumman, Electronic Systems
9318 Gaither Road, Gaithersburg, Maryland, 20877-1441, United States
Tel: (+1 301) 840 15 97
(+1 301) 527 79 00
Fax: (+1 301) 216 19 87
Web: www.es.northropgrumman.com
■ Supplier of voice switching and remote control systems used in air traffic control, shipboard communication and air defence applications. Products include: the model 400/ETVS, model 3080/RDVS and the model STVS (a small VCS). The Shipboard Integrated Voice Communication System (SIVCS) for ATC and internel shipboard communications on aircraft carriers and other vessels.The RCE 2001 Remote Control Equipment enables remote monitoring and control of radios over standard grade 4 wire telephone lines.

Norut Information Technology Ltd
N-9291 Tromsø, Norway P O Box 6434, N-9294 Tromsø, Norway
Tel: (+47) 77 62 94 00
Fax: (+47) 77 62 94 01
e-mail: itek@norut.no
Web: www.norut.no/itek
■ Research and technological development.

Norwegian Defence Research Establishment (FFI)
Forsvarets forskningsinstitutt
Forsvarets Forskningsinstitutt
PO Box 25 N-2027 Kjeller, Norway
Instituttveien 25, N-2027 Kjeller, Norway
Tel: (+47) 63 80 70 00
Fax: (+47) 63 80 71 15
e-mail: infoenheten@ffi.no, ffi@ffi.no
Web: www.ffi.no www.mil.no
■ Defence research and technology establishment.
Programme areas:
Basic research
Command and control systems
Electronic warfare
Environmental measures in the armed forces
Operations and cost analysis
Protection against NBC weapons
Strategic analysis
Surveillance systems
Tactical underwater systems
Weapon effectiveness, vulnerability and protection
Weapon guidance and weapon control systems
Weapon technology

Norwegian Industrial Forum for Space Activities (NIFRO)
Middelthunsgate 27, PO Box 5250 Majorstua, N-0303, Oslo, Norway
Tel: (+47) 23 08 83 08
Fax: (+47) 23 08 80 18
e-mail: nifro@nho.no
Web: www.nifro.no
■ A grouping of 17 Norwegian research and development and manufacturing companies and research institutes promoting members' participation in space activities. Areas of interest include: ground segment and support data systems, space electronics, materials technology, manfacturing techniques, space transportation technology, remote sensing, telecommunications and quality assurance/control and consulting. Integrated remote sensing data interpretation from multispectral and non-optical sources.

Norwegian Space Centre (NSC)
Norsk Romsenter
PO Box 113 Skoyen, N-0212 Oslo, Norway, Drammensveien 165, N-0277 Oslo, Norway
Tel: (+47 22) 51 18 00
Fax: (+47 22) 51 18 01
e-mail: spacecentre@spacecentre.no
Web: www.spacecentre.no
■ Provides development, co-ordination and evaluation of national space activities.

Noshiro Testing Center (NTC)
Asani, Noshiro-city, Akita, 016-0179, Japan
Tel: (+81 185) 52 71 23
Fax: (+81 185) 54 31 89
■ Site containing solid propellant rockets ground firing test stands, multipurpose vacuum firing test cells, liquid engine vertical firing test stand, cryogenic propellant test house, warehouses, measurement and control centre and an administration building.

Novespace
15 rue des Halles, F-75001 Paris, France
Tel: (+33 1) 42 33 41 41
Fax: (+33 1) 40 26 08 60
e-mail: espace@novespace.fr
Web: www.novespace.fr
■ French business corporation with activities in transfer of space technologies and promotion of access to low gravity environments.
Publications: Mutations; Mutations Microgravity

Novintec
ZA De la Pillardiere, F-45600 Sully-sur-Loire, France
Tel: (+33 2) 38 29 57 10
Fax: (+33 2) 38 29 57 11
e-mail: contact@novintec.com
Web: www.novintec.com/
■ Design and manufacture of filtration products and sub-assembly components for aerospace, defence, space, nuclear industries. Manifold filters and accessories, differential pressure indicators for filters and accumulators, visual and electrical reservoir gauges, blocking valves and accumulators, magnetic plugs, check valves, relief valves, drain valves, sampling valves and fuel injectors.

NPA Group
Crockham Park, Edenbridge, Kent, TN8 6SR, United Kingdom
Tel: (+44 1732) 86 50 23
Fax: (+44 1732) 86 65 21
e-mail: info@npagroup.com
Web: www.npagroup.com
■ Provide services of acquire, process and distribute both optical and radar satellite image data including landsat, SPOT, RADARSAT, ERS, QUICKBIRD and IKONOS. Also provide remote sensing applications including interpretation services and the generation of Digital Elevation Models (DEMs).

NPO Lavochkin
24 Leningradskaya Street, 141400 Khimki, Moscow Region, Russian Federation
Tel: (+7 095) 573 90 56
(+7 095) 573 55 65
(+7 095) 573 86 92
Fax: (+7 095) 573 35 95
(+7 095) 573 27 86
Telex: 911721 irbis
■ Planetary and deep-space craft, science and communication satellites.

NPO Mashinostroyenia
33 Gagarin Street, 143966 Reutov, Moscow Region, Russian Federation
Tel: (+7 95) 528 50 12
(+7 95) 302 05 84
Fax: (+7 95) 302 20 01
e-mail: export@land.ru
vvs@npomit.ru
fnpc@npomash.ru
Web: www.npomit.ru
■ Development and manufacture of intercontinental ballistic missiles, heavy space launch vehicles, earth monitoring satellites, orbital manned and automatic space stations.

NPO Prikladnoi Mekhaniki (NPO PM)
Scientific Production Association of Applied Mechanics
Prikladnoi Mekhaniki, NPO 52 Lenin Street, Zheleznogorsk, 662972 Krasnoyarsk, 26, Russian Federation
Tel: (+7 39197) 346 17
(+7 39197) 280 08
Fax: (+7 39197) 226 35
(+7 39197) 258 33
e-mail: gonti@npo-pm.krasnoyarsk.su
postmaster@npo.pm.krasnoyarek.ru
Web: www.npopm.ru
■ Design of communications systems, TV broadcasting, navigation and geodetic satellites. Manufacture and operation of satellites. Manufacture of ground antennas for satellite communication.

NPO Soyuz
Ulitsa Sovetskaya 6, 140056 Dzerzhinsk, Russian Federation
Tel: (+7 095) 551 78 08
Fax: (+7 095) 551 11 44

NRAO
520 Edgemont Road, Charlottesville, Virginia, VA 22903, United States
Tel: (+1 804) 296 02 33
Fax: (+1 804) 296 02 78

Nu-Cast Inc
29 Grenier Field Road, Londonderry, New Hampshire, 03053, United States
Tel: (+1 603) 432 16 00
Fax: (+1 603) 432 07 24
e-mail: nci@grolen.com
Web: www.nu-cast.com

NZ Guide Ltd
OJT Associates
17 Trenant Close, Polzeath, Wadebridge, Cornwall, United Kingdom
Tel: (+44 1208) 86 39 29
(+44 1208) 86 32 80
Fax: (+44 870) 164 53 86
■ Motor gliding and gliding services.

O

Obs de Strasbourg
11 Rue de l'Université, F-67000 Strasbourg, France
Tel: (+33 3) 88 15 07 21
Fax: (+33 3) 88 25 01 60
e-mail: genova@astro.u-strasbg.fr

Observatoire de Paris
Section de Meudon
F-92195 Meudon, PPL Cedex, France
Tel: (+33 1) 45 07 75 65
Fax: (+33 1) 45 07 74 69

Octec Ltd
The Western Centre, Western Road, Bracknell, Berkshire, RG12 1RW, United Kingdom
Tel: (+44 1344) 46 52 00
Fax: (+44 1344) 46 52 01
e-mail: octec@octec.co.uk
sales@octec.co.uk
Web: www.octec.com
■ Automatic video target tracking and position analysis systems.
Accreditations: AS 9100; ISO 9001:2000

Odetics Inc
Communications Division
1585 South Manchester Avenue, Anaheim, California, 92802, United States
Tel: (+1 714) 780 76 80
Fax: (+1 714) 780 76 49
Web: www.odetics.com
■ Supply products for Closed-Circuit Television (CCTV) and manufacture of fully automated libraries for video and data storage.

Oerlikon Contraves AG
a division of the Rheinmetall Defence Group, Germany
Birchstrasse 155, CH-8050 Zürich, Switzerland
Tel: (+41 44) 316 22 11
Fax: (+41 44) 311 31 54
e-mail: info@ocag.ch
Web: www.rheinmetall-defence.com
■ Main products and activities:
■ Defence: Skyranger highly mobile multi-mission system, Skyshield 35-Ahead gun and missile; Skyguard twin gun and missile air defence systems; Ahead system; Gun-King multi-divergence laser sight system; Shorar/Pagoda search and acquisition radar; Gun-Star compact fire control system; Seaguard CIWS system; ADATS missile systems; training systems and simulators. Automatic cannon for air defence, vehicle and naval armament.
Subsidiaries:
Contraves Advanced Devices Sdn Bhd
ILEE AG
Singapore Pte Inc
Oerliton contraves

Office for Outer Space Affairs (UNO/OOSA)
Vienna International Centre – Room EO954, PO Box 500 A-1400 Wien, Austria
Tel: (+43 1) 260 60 49 50
Fax: (+43 1) 260 60 58 30
e-mail: oosa@unvienna.org
Web: www.unoosa.org
Publications: Annual Reports, Intergovernmental & Expert Group Meetings

Office National d'Études et de Recherches AÉrospatiales (ONERA)
Chalais-Meudon Centre
8 rue des Vertugadins, F-92190 Meudon, France
Tel: (+33 1) 46 23 50 50
Fax: (+33 1) 46 23 51 51

Office National d'Études et de Recherches AÉrospatiales (ONERA)
Headquarters
29 Avenue de la Division Leclerc, PO Box 72 F-92322 Châtillon, Cedex, France
Tel: (+33 1) 46 73 40 40
Fax: (+33 1) 46 73 41 41
Web: www.onera.fr
■ Fundamental research, applied research and contribution to development in the fields of aerodynamics, structures, energetic materials, optics, acoustics, electronics, flight mechanics, ground wind tunnel and flight testing. Computer science.
Publications: Aerospace Science and Technology

Office National d'Études et de Recherches AÉrospatiales (ONERA)
Le Fauga Mauzac Center
Noe, F-31410 Mauzac, France
Tel: (+33 5) 61 56 63 00
Fax: (+33 5) 61 56 63 63
e-mail: cfm@onera.fr

Office National d'Études et de Recherches AÉrospatiales (ONERA)
Lille Center
5 Boulevard Paul-Painlev, F-59045 Lille, Cedex, France
Tel: (+33 3) 20 49 69 00
Fax: (+33 3) 20 52 95 93
e-mail: capon@imf-lille.fr

Office National d'Études et de Recherches AÉrospatiales (ONERA)
Modane-Avrieux Center
Route Départementale n° 215, PO Box 25 F-73500 Modane, Avrieux, France
Tel: (+33 4) 79 20 21 22
Fax: (+33 4) 79 20 21 68
e-mail: becle@onera.fr

Office National d'Études et de Recherches AÉrospatiales (ONERA)
Palaiseau Center
Chemin de la Hunière, F-91761 Palaiseau, Cedex, France
Tel: (+33 1) 69 93 60 60
Fax: (+33 1) 69 93 61 61

Office National d'Études et de Recherches AÉrospatiales (ONERA)
Toulouse Center
PO Box 4025 F-31055 Toulouse, Cedex 4, France
Tel: (+33 5) 62 25 25 25
Fax: (+33 5) 61 55 71 72
Telex: 521596

Office of Naval Research (ONR)
Public Affairs Office
Ballston Centre Tower One, 800 North Quincy Street, Arlington, Virginia, 22217-5660, United States
Tel: (+1 703) 969 50 31
Web: www.onr.navy.mil

Office of Science and Technology Policy (OSTP)
Executive Office of the President, Washington, District of Columbia, 20502, United States
Tel: (+1 202) 395 73 47
e-mail: ostpinfo@ostp.eop.gov
■ Advises the President on science and technology investments and contributes to the advancement of education, science and international co-operation.

Ohio State University
Center for Mapping
1216 Kinner Road, Columbus, Ohio, 43212, United States
Tel: (+1 614) 292 16 00
Fax: (+1 614) 292 80 62
Web: www.cfm.ohio-state.edu
■ Research centre focused on spatial data technologies including remote sensing, geodesy using the Global Positioning System (GPS), Inertial Navigation Systems (INS), photogrammetry, image processing, computer vision, image understanding, spatial cognition, modeling and Geographic Information Systems (GIS). The centre performs basic research but focuses on applied research projects that yield commercially viable mapping and positioning technologies.

OIP Sensor Systems
Delft Sensor Systems
OIP Optronic Instruments & Products
Westerring 21, B-9700 Oudenaarde, Belgium
Tel: (+32 55) 33 38 11
Fax: (+32 55) 33 38 02
e-mail: sales@oip.be
Web: www.oip.be
■ Portable display systems, thermal observation equipment (handheld and vehicle mounted), night vision systems, fire control systems, for armoured vehicles: space projects, advanced and holographic optics, customised research and development projects and equipment qualification.

Oki Electric Industry Co Ltd
7-12 Toranomon 1-chome, Minato-ku, Tokyo, 105-8460, Japan
Tel: (+81 3) 35 01 31 11
Fax: (+81 3) 35 81 55 22
Web: www.oki.com/jp
■ Sonobuoys, sonars, computer systems, communications equipment and electronic devices.

Omnimed
Pr. Gagarina, 28, office 67, 603098 Nizhnyi Novgorod, Russian Federation
Tel: (+7 8312) 30 68 60
Fax: (+7 8312) 65 14 19
e-mail: omnimed@mail.ru
■ Production of equipment for medical aid.

Oologah Technologies Inc (OTI)
PO Box 2670 Bay St Louis, Mississippi, 39521, United States
Tel: (+1 228) 688 73 86
Fax: (+1 228) 688 73 86
e-mail: oti@oologahtech.com
Web: www.oologahtech.com

Optech Incorporated
Optech Systems Inc
300 Interchange Way, Vaughan, Ontario, L4K 5Z8, Canada
Tel: (+1 905) 660 08 08
Fax: (+1 905) 660 08 29
e-mail: inquiries@optech.ca
Web: www.optech.on.ca
■ Manufacture of laser radar systems including, military laser systems, lidar bathymetry systems, laser altimeters, high-accuracy laser rangefinders, laser terrain profilers and laser water depth sounders. Remote-sensing systems, lidar pollution-detection systems, remote-sensing optical scanners and wideband logarithmic amplifiers. Airborne ASW and MCM systems.

Optus Networks Pty Ltd
Cable & Wireless Optus Pty Ltd
Optus Communications Pty Ltd
PO Box 1, North Sydney, New South Wales, 2060, Australia
L29 Optus Centre, 101 Miller Street, North Sydney, New South Wales, 2060, Australia
Tel: (+61 2) 93 42 78 00
(+61 2) 92 38 78 00
Fax: (+61 2) 93 42 76 67
Web: www.cwo.com.au

Orbcomm Global LP
21819 Atlantic Boulevard, Dulles, Virginia, 20166, United States
2115 Linwood Avenue, Fort Lee, New Jersey, 07024, United States
Tel: (+1 201) 363 49 00
Freephone: (+1 800) 672 26 66
Fax: (+1 703) 433 64 00
Web: www.orbcomm.com

Orbital Commerce Project
23 Alafaya Woods Boulevard #222, Oviedo, Florida, 32765, United States
Tel: (+1 321) 244 25 50
Fax: (+1 321) 244 09 91
e-mail: orbital@orbitalcp.com
Web: www.orbitalcommerceproject.com

Orbital Sciences Corp
Orbital Image Corp
21700 Atlantic Boulevard, Dulles, Virginia, 20166, United States
Tel: (+1 703) 406 50 00
Web: www.orbital.com

Orbital Sciences Corporation

Missile Defense Systems
3380 South Price Road, Chandler, Arizona, 85248, United States
Tel: (+1 602) 899 60 00
Fax: (+1 602) 899 88 00
Web: www.orbital.com

Orbital Sciences Corporation

Space and Electronics Systems Group
2771 North Garcy Avenue, Pomona, California, 91767, United States

Orbital Sciences Corporation

21839 Atlantic Boulevard, Dulles, Virginia, 20166, United States
Tel: (+1 703) 406 50 00
e-mail: public.relations@orbital.com
Web: www.orbital.com
■ Development and manufacture of space systems for commercial, civil government and military customers. Primary products include low-orbit, geosynchronous and planetary spacecraft for communications, scientific and remote sensing missions; ground- and air-launched rockets that deliver satellites into orbit; missile defence boosters for use as target and interceptor vehicles. Provides space-related technical services to government agencies, develops and builds satellite-based transportation management systems for public transit agencies and private vehicle fleet operators.

Orbital Technologies Corporation (ORBITEC)

Space Center
1212 Fourier Drive, Madison, Wisconsin, 53717, United States
Tel: (+1 608) 827 50 00
Fax: (+1 608) 827 50 50
Web: www.orbitec.com
■ Aerospace research and product development.

Orbitale Hochtechnologie Bremen System AG (OHB System AG)

Universitätsallee 27-29, D-28359 Bremen, Germany
Tel: (+49 421) 202 08
Fax: (+49 421) 202 07 00
e-mail: ohb@ohb-system.de
■ Design and development of mechanical and electronic equipment (including MGSE); microgravity systems, experiment facilities and equipment; manned and unmanned space systems and mission operation; re-entry technology and arodynamics/aerothermodynamics; remote and in situ pollution sensors; small satellites for science, earth observation and telecommunication/reconnaissance.

Oreet International Media Ltd

Oreet-Marketing Communications
15 Kineret Street, IL-51201 Bene Barak, Israel
Tel: (+972 3) 570 65 27
Fax: (+972 3) 570 65 26
■ Public relations and marketing communications company.
Representing: BVR, Israel
Representing: Elbit Systems Electro-Optics Elop Ltd, Israel
Elbit Systems Ltd, Israel
Elisra, Israel
Plasan Sasa, Israel
Tadiran Communications Ltd, Israel
Tadiran Spectralink Ltd, Israel

Oremet – Wah Chang (WC)

Wah Chang
a subsidiary of Allegheny Technologies, US
1600 North East Old Salem Road, PO Box 460, Albany, Oregon, 97321, United States
Tel: (+1 541) 926 42 11
Freephone: (+1 888) 926 42 11
Fax: (+1 541) 967 69 94
e-mail: custservice@wahchang.com
Web: www.alleghenytechnologies.com/wahchang
Web: www.wahchang.com
■ Manufacture of speciality metals and chemicals, used in energy production, chemical and mineral processing, aerospace, medical, research and consumer products. Materials include hafnium, niobium, titanium, vanadium, zirconium, silicon tetrachloride, zirconium and hafnium chemicals.
European operations
Liechtenstein

Orion Atlantic LP

Orion Satellite Corp
2440 Research Boulevard, Boulevard 400, Rockville, Maryland, MD20850, United States
Tel: (+1 301) 258 32 22
Fax: (+1 301) 258 32 56

Orion Propulsion

105 A-4 Commerce Circle, Madison, Alabama, 35758, United States
Tel: (+1 256) 461 80 32
Fax: (+1 256) 461 87 30
Web: www.orionpropulsion.com

Oscilloquartz SA (OSA)

CH-2002 Neuchatel, 2, Switzerland Rue des Brevards 16, CH-2002 Neuchatel, 2, Switzerland
Tel: (+41 32) 722 55 55
Fax: (+41 32) 722 55 56
Cable: OSCILLOQUARTZ
e-mail: osa@oscilloquartz.com
Web: www.oscilloquartz.com
■ Atomic and quartz frequency and time standards, quartz crystal resonators and oscillators, frequency and time generating systems for telecommunications, satellite ground stations, navigation/positioning and metrology. Master reference clock systems for stratum/level 1. Tracking oscillators for timing at stratum/level 2 and 3. Products for private and public network synchronisation. GPS/Cesium/Masor/Galelio.

Oss Astron di Palermo

Piazza del Parlemento 1, I-90134 Palermo, Italy
Tel: (+39 91) 23 32 51
Fax: (+39 91) 23 34 44

Oss Astronomico di Torino

St Osservatorio 20, I-10025 Pino Torinese, Italy
Tel: (+39 11) 461 90 35
Fax: (+39 11) 461 90 30

P

PanAmSat Corporation

20 Westport Road, Wilton, Connecticut, 06897, United States
Tel: (+1 203) 210 80 00
Fax: (+1 203) 210 80 01
e-mail: corpcomm@panamsat.com
Web: www.panamsat.com
■ Provide global video and data broadcasting services via satellite. The company builds, owns and operates networks that deliver entertainment and information to cable television systems, TV broadcast for direct-to-home TV operators, internet service providers, telecommunications companies and corporations. The company currently has 20 satellites in orbit with plans to expand the fleet to 22.

Paradigm Secure Communications

an EADS joint venture
EADS Astrium Site, Gunnels Wood Road, Stevenage, Hertfordshire, SG1 2AS, United Kingdom
Tel: (+44 1438) 28 21 21
Fax: (+44 1438) 28 21 20
e-mail: contact@paradigmsecure.com
Web: www.paradigmsecure.com

Paradigm Services

an EADS joint venture
Astrium Site, Gunnels Wood Road, Stevenage, Hertfordshire, SG1 2AS, United Kingdom
Tel: (+44 1438) 28 22 44
Fax: (+44 1438) 28 25 00
e-mail: contact@paradigmservices.com
Web: www.paradigmservices.com

Parker Aerospace Group

Air & Fuel Division, Irvine Plant
16666 Von Karman Avenue, Irvine, California, 92606, United States
Tel: (+1 949) 833 30 00
Fax: (+1 949) 851 33 41
Web: www.parker.com/airfuel
■ Manufacture of aircraft fuel system components, aerial refuelling equipment, high-temperature bleed air valves, turbine clearance control valves, air turbine starters and aircraft fuel-tank inerting systems and components.
Accreditations: Certified to AS9100 and ISO 9001

Parker Aerospace Group

Control Systems Division, Irvine
14300 Alton Parkway, Irvine, California, 92618, United States
Tel: (+1 949) 833 30 00
Fax: (+1 949) 586 84 56
■ Manufacture primary flight control actuation and engine control systems, integrated fly-by-wire electrohydraulic servomodules, electrically controlled actuators, electro-hydraulic servovalves, electro-mechanical products, electro-mechanical actuators and flight control components.
Accreditations: ISO 9000, registered to ISO 9001:1994

Parker Aerospace Group

Electronic Systems Division
Gull Electronic Systems Division, Parker Aerospace
300 Marcus Boulevard, PO Box 9400 A, Smithtown, New York, 11787-9400, United States
Tel: (+1 631) 231 37 37
Fax: (+1 631) 434 81 52
■ Fluid management and quantity-guaging systems, electronic controllers and fibre-optic systems.
Accreditations: ISO 9000, registered to ISO 9001:2000

Parker Aerospace Group

Group Headquarters
Parker Hannifin
14300 Alton Parkway, Irvine, California, 92618, United States
Tel: (+1 949) 833 30 00
Fax: (+1 949) 851 32 77
e-mail: csd@parker.com
Web: www.parker.com
■ Design, manufacture and service of hydraulic, fuel and pneumatic components, systems and related electronic controls for aerospace and other high-technology markets.

Parker Hannifin Corporation

6035 Parkland Boulevard, Cleveland, Ohio, 44124-4141, United States
Tel: (+1 216) 896 30 00
Fax: (+1 216) 896 40 00
Web: www.parker.com
■ Manufacture of motion control components and systems for industrial, automotive, aviation, space and marine markets.
Operation with defence business
Parker Aerospace

Particle Physics and Astronomy Research Council (PPARC)

Polaris House, North Star Avenue, Swindon, Wiltshire, SN2 1SZ, United Kingdom
Tel: (+44 1793) 44 20 00
Fax: (+44 1793) 44 20 02
e-mail: pr.pus@pparc.ac.uk
Web: www.pparc.ac.uk
■ Promoting the UK's role in research and training in the fields of particle physics, solar system science, astrophysics and astronomy.

Pasifik Satelit Nusantara (PSN)

Kawasan Karyadeka Pancamurni, Blok-A Kav, 3 Lemah Abang, 17550 Bekasi, Indonesia
Tel: (+62 21) 89 90 81 11
Fax: (+62 21) 89 90 81 10
Web: www.psn.co.id
■ Provider of a range of satellite-based telecommunications services, including both mobile and fixed telephony systems.

Patria Aviation Oy

Patria Finavitec Oy
Valmet Aviation Industries Inc
a subsidiary of Patria Industries Oy, Finland
Lentokonetehtaantie 3, FI-35600 Halli, Finland
Tel: (+358 20) 469 33 85
Fax: (+358 20) 332 14
(+358 20) 469 33 85
(+358 20) 469 36 00
e-mail: aviation@patria.fi
Web: www.patria.fi
■ Overhaul, modification and repair of airframes and engines, final assembly. Development and support of aircraft systems. Overhaul of fast diesel engines. Structures for aircraft and spacecraft. Military systems integration.

Patria New Technologies Oy

Systems Division
Patria Finavitec Oy
Naulakatu 3, FI-33100 Tampere, Finland
Tel: (+358 20) 46 91
Fax: (+358 20) 469 26 92
e-mail: new.technologies@patria.fi
Web: www.patria.fi
■ Design and manufacture of surveillance equipment, space electronics and test systems.

Payload System Inc
247Third Street, Cambridge, Massachusetts, 02141, United States
Tel: (+1 617) 868 80 86
Fax: (+1 617) 868 66 82
Web: www.payload.com
■ Provision of test and research services for space equipment.

Paz Logistics Ltd
10 Yad Harutzim Street, PO Box 2242 IL-44425 Kfar-Saba, Israel
Tel: (+972 9) 766 15 55
Fax: (+972 9) 766 15 50
e-mail: info@pazlog.co.il
Web: www.pazlog.co.il
■ High level logistic and technical consulting. Specialise in weapons, guns, ammunition and related accessories, optical systems, night vision equipment, communication systems, GPS and satellite communication, navigation, compass systems, police and military equipment and textiles, automative and retrofit projects. Representation of foreign firms in Israel.

Pecos County/West Texas Spaceport Development Corporation
1000 Railroad Avenue, Fort Stockton, Texas, 79735, United States
Tel: (+1 915) 336 22 64
Fax: (+1 915) 336 61 14
e-mail: fortstocktontx@ftstockton.net

Pedeo Techniek NV
Martijn van Torhoutstraat 15, B-9700 Oudenaarde, Belgium
Tel: (+32 55) 31 35 61
Fax: (+32 55) 31 26 85
Web: www.pedeotechniek.be
■ Design and manufacture of molds and dyes, precise equipment and tool parts for industries including automation, medical, electronics, aviation, aerospace and defence.

Peleng JSC
23 Makayonok Street, 220023 Minsk, Belarus
Tel: (+375 17) 263 82 04
Fax: (+375 17) 263 65 42
e-mail: market@belomo.minsk.by
■ Space photographic systems, photogrammetric instruments and techniques, stabilised night vision devices and binoculars and anti-tank weapon systems.

Penn State College of Engineering
Communications & Space Sciences Laboratory
318 EE East, University Park, Pennsylvania, 16802, United States
Tel: (+1 814) 865 23 54
Fax: (+1 814) 863 84 57
Web: www.engr.psu.edu/cssl
■ Current research interests focus on the upper atmosphere, weather, lightning and atmospheres of other planets and electromagnetic related problems such as antennas, microwaves, plasma diagnostics, scattering and shielding.

Penn State College of Engineering
Pennsylvania State University
101 Hammond, University Park, Pennsylvania, 16802, United States
Tel: (+1 814) 865 75 37
(+1 814) 865 55 44
Fax: (+1 814) 863 47 49
e-mail: news@engr.psu.edu
Web: www.engr.psu.edu/news

Penn State College of Engineering
Propulsion Research Center
NASA Center for Space Propulsion Engineering
240 Research Building East, University Park, Pennsylvania, 16802, United States
Tel: (+1 814) 863 12 85
Fax: (+1 814) 865 33 89
■ The research programme emphasises five major areas: chemical propulsion, electric nuclear propulsion; advanced propulsion concepts; diagnostics and material compatibility/reliabilty.

Perkin Elmer
Data Systems Group
45 William Street, Wellesley, Massachusetts, 02481-4078, United States
2 Crescent Place, Oceanport, New Jersey, 07757, United States
Tel: (+1 781) 237 51 00
Web: www.perkinelmer.com
■ Minicomputers for simulation systems

Perkinelmer Fluid Sciences (PKI)
Belfab Products
EG&G Belfab
John Crane Belfab
a subsidiary of Perkinelmer Inc, US
305 Fentress Boulevard, Daytona Beach, Florida, 32114, United States
Fax: (+1 386) 257 01 22
■ Design and manufacture of welded metal bellows assemblies.

Perkinelmer Optoelectronics
Headquarters
EG&G Optoelectronics
a subsidiary of Perkinelmer, Inc. US
44370 Christy Street, Freemont, California, 94538-3180, United States
Tel: (+1 510) 979 65 00
Freephone: (+1 800) 775 OPTO
Fax: (+1 510) 687 11 40
e-mail: opto@perkinelmer.com
Web: www.perkinelmer.com
■ Ordnance for missile and space vehicle applications including EBW and EFI detonators, chip slapper detonators and laser initiated detonators and actuators. Electronic Safe and Arm Devices (ESAD) and fuzes for tactical missile systems. Continuous and pulsed light sources for spacecraft docking and missile targeting applications. High energy switches including spark gaps and thyratrons as well as custom transformers for radar, laser, fuzing, ESAD and other military applications.

Philips Research
Headquarters
Philips Laboratories
Briarcliff Manor, New York, NY 10510, United States
Tel: (+1 914) 945 61 95
Fax: (+1 914) 945 65 80

Photonis-DEP
Delft Electronic Products BV
PO Box 60 NL-9300 AB Roden, Netherlands
Dwazziwegen 2, NL-9301 Roden, Netherlands
Tel: (+31 50) 501 88 08
Fax: (+31 50) 501 14 56
e-mail: nightvision@photonis.com
Web: www.photonis-dep.com
■ Image intensifier tubes, (position sensitive) photon counters, auroral images, and intensified CCDs.

Photo-Sonics Inc
820 South Mariposa Street, Burbank, California, 91506-3196, United States
Tel: (+1 818) 842 21 41
(+1 818) 531 32 19
Fax: (+1 818) 842 26 10
e-mail: mail@photosonics.com
info@photosonics.com
Web: www.photosonics.com
■ 16 mm gun sight, HUD cameras and airborne video recorder for military aircraft. 16, 35 and 70 mm and HS video photo-instrumentation cameras for military and aerospace applications. Tracking mount mobile and fixed, instrumentation cameras. Automatic film and video analysis equipment.
Subsidiary
ARAN Electronics, Israel

Phototelesis Corporation
4801 North West Loop 410, Suite 203, San Antonio, Texas, 78229, United States
Tel: (+1 210) 349 20 20
Fax: (+1 210) 349 20 70
Web: www.photot.com
■ Base, vehicular, airborne, forward BASE and manpack image transceivers and associated centers, printers and scanners.

Physical Research Laboratory (PRL)
Navrangpura, Ahmedabad, 380009, India
Tel: (+91 79) 630 21 29
Fax: (+91 79) 630 15 02
e-mail: root@prl.ernet.in
Web: www.prl.ernet.in

Pioneer Aerospace Corp
Headquarters
a subsidiary of Zodiac SA, France
45 South Satellite Road, PO Box 207, South Windsor, Connecticut, 06074-0207, United States
Tel: (+1 860) 528 00 92
Fax: (+1 860) 528 81 22
e-mail: sales@pioneer.zodiac.com
Web: www.pioneeraero.com
■ Engineering and manufacture of parachutes and related items. Analysis, design, development, testing, manufacturing, integration and support of advanced recovery and retrieval systems for aerospace vehicles and components.
Accreditations: Mil-Q-9858A, Mil-45208A, ISO 9002

Planetarium Negara
53 Jalan Perdana, 50480 Kuala Lumpur, Malaysia
Tel: (+60 3) 273 54 84
Fax: (+60 3) 273 54 88

Plansee SE
A-6600 Reutte, Austria
Tel: (+43 5672) 60 00
Fax: (+43 5672) 60 05 00
e-mail: communications@plansee.com
Web: www.plansee.com
■ Manufacture of powder metallurgical products and components including refractory metal components for launchers and satellite engines; thermal protection systems for reuseable space transport systems; heavy metal materials for aerojet balance weights and radiation protection shieldings; hard metal cutting tools for the machining of structural components. Industry development and engineering services.
Accreditations: ISO 9001, BSI, GAZ

PMD (UK) Limited
Broad Lane, Coventry, CV5 7AY, United Kingdom
Tel: (+44 2476) 46 66 91
Fax: (+44 2476) 47 30 34
e-mail: sales@pmdgroup.co.uk
Web: www.pmdgroup.co.uk
■ Gold plating service for satellites and aerospace applications.

Polish Astronautical Society (PAS)
Polskie Towarzystwo Astronautyczne
Space Research Centre, Bartycka str. 18A room 137, PL-00-716 Warszawa, Poland
Tel: (+48 22) 840 37 66
Fax: (+48 22) 840 31 31
e-mail: poczta@ptastronaut.org.pl
Web: www.ptastronaut.org.pl
■ Society formed for the promotion of the studies in space sciences and space flight.
Publications: Astronautyka
Progress in Astronautics

Polish Space Office
ul. Bartycka 18a, PL-00-716 Warszawa, Poland
Tel: (+48 22) 851 18 12
e-mail: spaceoffice@cbk.waw.pl

Polspace Ltd
a member of the European Space Incubators Network (ESINET)
Ul. Bartycka 18A, PL-00-716 Warszawa, Poland
Tel: (+48 22) 841 75 35
Fax: (+48 22) 840 31 31
e-mail: bartosz@polspace.pl
Web: www.polspace.pl
■ Consultancy service in space-related policy, industrial and technology matters. Supports the development of space industry and space applications in Poland.

Polymarin BV
a subsidiary of Royal Schelde
PO Box 151 NL-1670 AD Medemblik, Netherlands
Nijverheidsweg 7, NL-1671 GC Medemblik, Netherlands
Tel: (+31 227) 54 30 44
Fax: (+31 227) 54 38 60
e-mail: dhofstra@polymarin.nl
Web: www.polymarin.nl
■ Design and manufacture of high grade, fibre re-inforced composite structures for aerospace (airframe), shipbuilding, wind energy, swimming pools, covering and roofing. Design and manufacture of mouldings in thermoplastic and thermosetting resins for the medical sector, process equipment, machinery and coachwork.

Polyot Production Association
Experimental Design Bureau
ASA Polyot Design Bureau
Polyot, PO 226 Khmelnitskogo Street, 644021 Omsk, Russian Federation
Tel: (+7 3812) 57 70 21
Fax: (+7 381) 253 88 31
(+7 3812) 57 70 21
■ Manufacture spacecraft, launch vehicles, high powered rocket engines and satellites.

Polyot Research and Production Company
GUP NPP Polyot
1 Komsomolskaya Pl, GSP 462, 603950 Nizhnyi Novgorod, Russian Federation
Tel: (+7 8312) 45 21 04
Fax: (+7 8312) 49 39 41
(+7 8312) 45 31 57
e-mail: polyot@atnn.ru
polyot@nnov.rfnet.ru
Web: www.polyot.nnov.rfnet.ru
■ Development of airborne radio communication equipment. Development and manufacture of aviation communication systems and devices, airborne and ground-based communication complexes for military and civil applications, automated receive-transmit centres for ATC purposes, antenna-feeders and aircraft-based ecological monitoring systems.

Pratt & Whitney (P&W)
Corporate Headquarters
a division of United Technologies Corp, US
400 Main Street, East Hartford, Connecticut, 06108, United States
Tel: (+1 860) 565 43 21
(+1 860) 214 75 02
e-mail: info@pratt-whitney.com
Web: www.pratt-whitney.com
www.pw.utc.com
■ Design, develop and support of market propulsion systems. Install fighter, transport and surveillance aircraft. Rocket engines and propulsion systems for various space programmes.
Subsidiary company:
Pratt & Whitney Canada Inc

Pratt & Whitney (P&W)
Military Engines
Government Engines & Space Propulsion
400 Main Street 181-36, East Hartford, Connecticut, 06108, United States
Tel: (+1 860) 565 96 00
(+1 860) 565 43 21
(+1 860) 565 01 40
Freephone: (+1 800) 565 01 40 49
(+1 800) 526 11 59
(+1 800) 233 18 49
e-mail: help24@pw.utc.cdm
Web: www.pw.utc.com
www.pratt-whitney.com/prod_space.asp

Pratt & Whitney Rocketdyne
Pratt & Whitney Space Propulsion
PO Box 109600 West Palm Beach, Florida, 33410-9600, United States
Tel: (+1 561) 796 20 00
Fax: (+1 561) 796 92 21
■ Solid and liquid rocket propulsion.

Precision Systems and Space Company Inc
BEI Technologies Inc
1100 Murphy Drive, Maumelle, Arkansas, 72113, United States
Tel: (+1 501) 851 40 00
Fax: (+1 501) 851 54 76
e-mail: sales@belprecision.com
Web: www.belprecision.com
■ Design and produce pointing and positioning components and servo controls used in space, tracking and military systems. Standard optical encoders give angular resolution of over 33 million absolute positions per turn with accuracy to 0.3 arc seconds. Manufacture and sale of linear accelerometers and inertial reference units.

Pressed Steel Tank Co Inc (PST)
1445 S 66th Street, Milwaukee, Wisconsin, 53214, United States
Tel: (+1 414) 476 05 00
Fax: (+1 414) 476 71 91
e-mail: sales@pressedsteel.com
Web: www.pressedsteel.com
■ Oxygen cylinders, rocket motor casings, deep drawn components. Pressed steel tank manufactures, all steel Dot-3AA and Dot-3HT cylinders, primarily used as replacement oxygen and nitrogen bottles for commercial and military aircraft.

PRIN
AO Prin
Prin Joint-Stock Company
4 Volokolamsk Highway, GSP-E, 125993 Moskva, Russian Federation
Tel: (+7 495) 785 57 37
Fax: (+7 495) 626 97 79
e-mail: pm@prin.ru
Web: www.prin.ru
■ Supplies and customises aero navigation equipment, navigation GPS Navstar systems, Inmarsat-C communication systems and different landing and correcting systems.
Representing: Trimble Navigation Ltd, USA

Princeton University
Peyton Hall, Princeton, New Jersey, NJ 08544, United States
Tel: (+1 609) 258 35 89
Fax: (+1 609) 258 10 20

Progress Plant Samara
18 Pskovskaya Street, 443009 Samara, Russian Federation
Tel: (+7 8462) 58 47 72
(+7 8462) 55 02 81
(+7 8462) 58 49 33
Fax: (+7 8462) 58 46 11
(+7 8462) 27 12 91
■ Production of aircraft and rocket engines, satellites.

Prototech AS
PO Box 6034 N-5892 Bergen, Norway
Tel: (+47) 55 57 41 10
Fax: (+47) 55 57 41 14
Web: www.prototech.no
■ Design, building and testing of space related mechanical equipment for use in projects such as sounding rockets, satellites, space laboratories and scientific probes, and space shuttle equipment.

PRT Advanced MAG-LEy Systems Inc
PRT Systems Corp
PRT Systems Inc
219 Hickory Street, Park Forest, Illinois, 60466, United States
Tel: (+1 708) 756 70 90
Fax: (+1 708) 747 58 11
■ Missile support, handling and test equipment. Space launch assist systems.

PT Indica Dharma Consulting Services
Golden Plaza Blok 43-44, Fatmawati No. 15, Jakarta, Indonesia
Tel: (+62 21) 750 89 86
Fax: (+62 21) 750 89 85

Puroflow Corp
10616 Lanark Street, Sun Valley, California, 91352, United States
Tel: (+1 818) 504 40 00
Fax: (+1 818) 779 39 02
e-mail: sales@puroflow.com
Web: www.puroflow.com

Q

QinetiQ
Company Headquarters
Cody Technology Park, Ively Road, Farnborough, Hampshire, GU14 0LX, United Kingdom
Tel: (+44 8700) 10 09 42
Fax: (+44 1252) 39 58 90
(+44 1252) 39 45 71
Web: www.qinetiq.com
■ Scientific, technological, research, development and evaluation organisation. Non-nuclear research, technology and test evaluation services of the UK Ministry of Defence. Operational studies and analysis from basic to applied research. Consultancy services for procurement and evaluation requirements during development and operational services. Test facilities include indoor and outdoor weapon ranges, underwater target ranges, marine testing facilites, automotive test tracks and climate testing laboratories. Health care.

Q-Par Angus Ltd (QPA)
Barons Cross Laboratories, Leominster, HR6 8RS, United Kingdom
Tel: (+44 1568) 61 21 38
(+44 1568) 61 49 13
Fax: (+44 1568) 61 63 73
e-mail: enquires@q-par.com
Web: www.q-par.com
■ Manufacture and design of microwave and RF antennas. Also, components, antenna positioners and special systems.

Quadrics
Bristol Office
One Bridewell Street, Bristol, BS1 2AA, United Kingdom
Tel: (+44 117) 907 53 75
Fax: (+44 117) 907 53 95
e-mail: sales@quadrics.com
Web: www.quadrics.com
■ Manufacture of high performance computer hardware.

Quadrics
Rome Office
Via Vittorio Veneto, 183, I-00187 Roma, Italy
Tel: (+39 06) 42 45 79 01
Fax: (+39 06) 42 01 63 03
e-mail: sales@quadrics.com
Web: www.quadrics.com
■ Manufacture high performance computer hardware.

Quantic Industries Inc (QII)
Whittaker Ordnance
2751 San Juan Road, PO Box 148 Hollister, California, 95024-0148, United States
Tel: (+1 831) 637 58 51
Fax: (+1 831) 637 10 13
Telex: TWX 910 590 0457
■ Electro-explosive devices: igniters, gas generators and starter cartridges, cable and bolt cutters, piston actuators, pin pullers, explosive valves and airbag initiators.

Queen Mary, University of London (QMUL)
Microwave Antenna Research Group
Electronic Engineering Department, Mile End Road, London, E1 4NS, United Kingdom
Tel: (+44 20) 78 82 53 39
Fax: (+44 20) 78 82 79 97
Web: www.elec.qmul.ac.uk/research/antennas

R

Régie Générale d'Annuaires (RGA)
Le Polaris, 76 Avenue Pierre Brosolette, F-92247 Malakoff, Cedex, France
Tel: (+33 1) 42 53 97 00
Fax: (+33 1) 42 53 14 66
■ Directory of French aviation and space industries including constructors, manufacturers, equipment suppliers and sub-contractors.

Racal Antennas Ltd
a wholly owned subsidiary of Cobham plc, UK
First Avenue, Millbrook Trading Estate, Southampton, SO15 0LJ, United Kingdom
Tel: (+44 23) 80 70 57 05
Fax: (+44 23) 80 70 11 22
e-mail: sales@racalantennas.com
Web: www.racalantennas.com
■ Design, develop and manufacture of base-station antennas including GSM900, PCN1800, PCS1900 and UMTS bands, supplies micro and pico cell antennas, base-station components including amplifiers, filters and towers. Also provide a range of tactical antennas for the defence market including HF/VHF and UHF frequency bands, telescopic masts suitable for all field applications.

Radarsat International (RSI)
European Office
21-22 Old Steyne, Brighton, East Sussex, BN1 1EL, United Kingdom
Tel: (+44 1273) 64 83 46
Fax: (+44 1273) 64 83 49

Radarsat International (RSI)
Headquarters
a member of Geomatics Industry Association of Canada
13800 Commerce Parkway, MacDonald Dettwiler Building, Richmond, British Columbia, V6V 2J3, Canada
Tel: (+1 604) 231 50 00
(+1 604) 244 04 00
Fax: (+1 604) 231 49 00
(+1 604) 244 04 04
e-mail: info@rsi.ca
Web: www.rsi.ca
■ Satellite remote sensing data distribution and processing.

Radarsat International (RSI)
Ottawa Office
2060 Walkley Road, Ottawa, Ontario, K1G 3P5, Canada
Tel: (+1 613) 238 54 24
Fax: (+1 613) 238 54 25

Radiall
101 rue Philibert Hoffmann, F-93116 Rosny sous Bois, France
Tel: (+33 1) 49 35 35 35
Fax: (+33 1) 49 35 35 14
Web: www.radiall.com
■ Coaxial connectors, rack and panel connectors, optical connectors and passive microwave devices.

Radio Holland Belgium NV
Sait Marine NV
a subsidiary of Euro Marine Belgium NV, Belgium
Noordersingel 17, B-2140 Borgerhout, Antwerpen, Belgium
■ Marine communication and navigation equipment.

Radio Research Instrument Co Inc
584 North Main Street, Waterbury, Connecticut, 06704-3506, United States
Tel: (+1 203) 753 58 40
Fax: (+1 203) 754 25 67
e-mail: radiores@prodigy.net
Web: www.radioresearch.com
■ Radar systems and spares, threat emitters, high-power RF sources, antennas, azimuth-elevation pedestals and microwave tubes.

Radyne Comstream Inc
China
Comstream Corporation
a subsidiary of Radyne Comstream Inc., US
Room 405, Building B, No. 8 Guanghaua Road, Chaoyang District, 100026 Beijing, China
Tel: (+86 10) 65 83 19 75
(+86 10) 65 83 19 76
(+86 10) 65 83 19 77
(+86 10) 65 83 19 78
Fax: (+86 10) 65 83 19 74
e-mail: chinasales@radn.com
Web: www.radn.com
Representing: Radyne Comstream Inc., US

Radyne Comstream Inc
European/Middle East/Africa Office
a subsidiary of Radyne Comstream Inc., US
8 Oriel Court, Alton, GU34 2YT, United Kingdom
Tel: (+44 1420) 54 42 00
e-mail: eurosales@radn.com
Web: www.radn.com
Representing: Radyne Comstream Inc., US

Radyne Comstream Inc
Indonesia
Comstream Corporation
a subsidiary of Radyne Comstream Inc., US
Jalan MT Haryono, Kav 25, 12820 Jakarta, Indonesia
Tel: (+62 21) 521 37 33
Fax: (+62 21) 252 01 42
e-mail: asiapacsales@radn.com
Web: www.radn.com
Representing: Radyne Comstream Inc., US

Radyne Comstream Inc
Latin America
7700 Congress Avenue, Suite 3206, Boca Ratón, Florida, 33487, United States
Tel: (+1 561) 988 12 10
Fax: (+1 561) 988 82 90
e-mail: lasales@radynecomstream.com

Radyne Comstream Inc
San Diego Office
6340 Sequence Drive, San Diego, California, 92121, United States
Tel: (+1 858) 458 18 00
Fax: (+1 858) 657 54 03

Radyne Comstream Inc
Singapore
Comstream Corporation
a subsidiary of Radyne Comstream Inc., US
15 McCallum Street, #12-04 Natwest Centre, 069045, Singapore
Tel: (+65) 62 25 40 16
Fax: (+65) 63 25 19 50
Web: www.radn.com
Representing: Radyne Comstream Inc., US

Radyne Comstream Inc
Comstream Corporation
3138 East Elwood Street, Phoenix, Arizona, 85034, United States
Tel: (+1 602) 437 96 20
Fax: (+1 602) 437 48 11
e-mail: sales@radn.com
Web: www.radn.com
■ Satellite-based communications technology, supply interactive and broadcast networks and digital componentry. Systems include: SCPC earth stations, modems, receivers, uplinks, network management systems and custom chips. Point-to-multipoint voice, data, imaging and fax transmission. Point-to-point, full mesh or 'star' connectivity. Uplink only for low cost network hubs. Applications include: mesh networking for private business and government services. Long distance telephone trunking terminals and voice backbone networks. LAN interconnection. Video conferencing. Data acquisition by a central site polling many remote locations (SCADA). Digital Data Applications. Real time financial market information. News/weather gathering and distribution. Computer file transfer. Photograph or image transmission. Paging networks. Digital Audio Applications. Radio programme distribution. Remote audio broadcasting. News gathering. Remote studio recording. Major systems designs and implementation of large turnkey satellite communication systems.

Rafael Advanced Defense Systems Ltd
Corporate Headquarters
Rafael Armament Development Authority Ltd
PO Box 2250 IL-31021 Haifa, Israel
Tel: (+972 4) 879 44 44
Fax: (+972 4) 879 46 81
e-mail: intl-mkt@rafael.co.il
Web: www.rafael.co.il
■ Design, development, manufacture and supply of a range of advanced defence systems. Rafael's products include: Missiles, Air Defense, Naval Systems, Target Acquisition, EW, C4ISR, Communication Networks, Data Links, EO Payloads, Trainers and Simulators, Add-on Armor; Combat Vehicle Upgrading, Mine Field Breaching, Border and Coastal Protection Systems and UAVs. The company has also formed partnerships with civilian counterparts to develop commercial applications based on its proprietary technology.
■ Products include, Python 5 and Derby air-to-air missiles; AGM-142 Popeye and SPIKE-ER air-to-surface missiles, SPICE Guidance Kits, Litening Targeting and Navigation Pod, Samson Remote-Controlled Weapon Stations, Typhoon Naval Weapon Stations, Add-on Reactive Armour, SkyLite B Mini-UAV, Simon Door Breaching Rifle Grenade and others.

Sub-divisions:
Missile Division
Ordnance Systems Division
Propulsion and Explosives Division
Systems Division

Subsidiaries:
Rafael Development Corp Ltd
Semi-Conductor Devices
Semiconductor Engineering Laboratories Ltd (SELA)

Rafael Advanced Defense Systems Ltd
Propulsion and Explosive Systems Division
Rafael Armament Development Authority Ltd
PO Box 2250/ M1 IL-31021 Haifa, Israel
e-mail: ilraf30@attglobal.net
■ Design, development and manufacture of rocket propulsion, warhead and pyrotechnic components for missile systems. Space propulsion systems and components, remote breaching devices and naval EW decoy rockets and launchers.

RAM System GmbH
Daimlerstrasse 11, D-85521 Ottobrunn, Germany
Tel: (+49 89) 608 00 30
(+49 89) 60 80 03 43
e-mail: info@ramsys.de
■ Consortium formed for the production of the RAM Rolling Airframe Missile ship-to-air missile system.

Member Companies:
Diehl BGT Defence GmbH & Co KG
Diehl Stiftung & Co KG
EADS Deutschland GmbH

Ramaer BV
Ramaer Printed Circuits BV
Ramaer Connection Technology
Vossenbeemd 101, PO Box 825 NL-5700 AV Helmond, Netherlands
Tel: (+31 492) 58 49 11
Fax: (+31 492) 55 09 75
e-mail: info@ramaer.nl
Web: www.ramaer.nl
■ Manufacture of printed circuit boards, multilayer board up to 22 layers, back panels, press-fit boards, rigid-flexible boards, rigid-flexible multilayer boards, controlled impedance boards, and boards certified to military specifications.

Raman Research Institute
Sadashivanagar, C V Raman Avenue, Bangalore, 560 080, India
Tel: (+91 80) 334 01 22
Fax: (+91 80) 334 04 92

Raufoss Belgium NV
Dammestraat 80, B-8800 Roselare, Belgium
Tel: (+32 51) 26 72 72
Fax: (+32 51) 22 63 44

Raufoss Metall GmbH
An der Schleuse 8, D-5870 Hemer-Becke, Germany
Tel: (+49 237) 29 19 75
Fax: (+49 237) 21 35 77

Raufoss Metall GmbH
Raufoss Germany GmbH
An der Schleuse 8, D-58675 Hemer-Becke, Germany
Tel: (+49 2372) 91 975
Fax: (+49 2372) 13 577
Web: www.raufoss.com/isiflo

Raufoss SA
2 rue du Quartier Targe, BP 5 F-42152 l'Horme, France
Tel: (+33 47) 729 23 40
Fax: (+33 47) 729 23 49

Raufoss VA AB
Kvartsen 3, Porfyr Vägen 10, SE-22478 Lund, Sweden
Tel: (+46 46) 33 39 90
Fax: (+46 46) 33 39 99

Raymond Engineering Operations (REO)
a subsidiary of Kaman Aerospace Corporation, US
217 Smith Street, Middletown, Connecticut, 06457, United States
Tel: (+1 860) 632 45 82
Fax: (+1 860) 632 43 29
Web: www.raymond-engrg.com
■ Electronic systems, both ground-based and airborne. Safety products for advanced tactical missiles, penetration fusing devices for major devices for major programmes such as AMRAAM, Phoenix, Hawk and Standard Missile; sensing systems for target recognition and data acquisition; complete testing facilities including sled testing and environmental laboratories for vibration, shock, temperature, firing and special purpose testing. Ruggedised mass memory products for weapons platforms, space satellites and the space shuttle.

Raytheon Company
Corporate Headquarters
870 Winter Street, Waltham, Massachusetts, 02451–1449, United States
Tel: (+1 781) 522 30 00
(+1 617) 860 24 14
(+1 781) 862 66 00
Fax: (+1 617) 860 25 20
(+1 781) 860 25 20
e-mail: corpcom@raytheon.com
Web: www.raytheon.com
www.raytheon.com/businesses/rsas

For details of the latest updates to *Jane's Space Systems and Industry* online and to discover the additional information available exclusively to online subscribers please visit
jsd.janes.com

Business Groups:
Raytheon Aircraft
Raytheon Homeland Security
Raytheon Integrated Defense Systems
Raytheon Intelligence & information Systems
Raytheon Missile Systems
Raytheon Network Centric Systems
Raytheon Space & Airborne Systems
Raytheon Technical Services Company

Raytheon Vision Systems

Raytheon Company
Santa Barbara Remote Sensing
75 Coromar Drive, Goleta, California, 93117, United States
Tel: (+1 805) 968 35 11
(+1 805) 562 41 27
Fax: (+1 805) 685 82 27
e-mail: rvsmarketing@raytheon.com
Web: www.raytheon.com

Reaction Engines Ltd (REL)

Building D5, Culham Science Centre, Abingdon, Oxon, OX14 3DB, United Kingdom
Tel: (+44 1865) 40 83 14
Fax: (+44 1865) 40 83 01
e-mail: mail@reactionengines.co.uk
■ Design and development of novel high performance space propulsion systems and lightweight heat exchanges.

Regional African Satellite Communications Organization (RASCOM)

Organisation Regionale Africaine de Communications par Satellite 2 avenue Thomasset, Abidjan, Côte d'Ivoire
Tel: (+225 20) 22 36 74
(+225 20) 22 36 83
Fax: (+225 20) 22 36 76
(+225 20) 22 36 79
e-mail: info@rascom.org
Web: www.rascom.org

Remote Sensing Instruments

2-2-18/18/3/16, Bagh Amberpet, Hyderabad, Andhra Pradesh, 500 013, India
Tel: (+91 40) 23 07 85 66
(+91 40) 23 07 97 83
Fax: (+91 40) 23 07 95 66
e-mail: contact@rsi.bz
Web: www.rsi.bz
■ Remote sensing and geographic information systems (GIS). Application of GIS and digitisation.

Remote Sensing Technology Center of Japan (RESTEC)

Roppongi First Building, 2F, 1-9-9 Roppongi, Minato-ku, Tokyo, 106-0032, Japan
Tel: (+81 3) 55 61 97 77
Fax: (+81 3) 55 74 85 15
Web: www.restec.or.jp

REOSC

Avenue de la Tour Maury, F-91280 Saint Pierre du Perray, France
Tel: (+33 1) 69 89 72 00
Fax: (+33 1) 69 89 72 32
e-mail: reossc@sfim.fr
Web: www.reosc.com
■ Design, production and integration of precision optics for space, astronomy, lasers and defence.

Research Institute for Advanced Computer Science (RIACS)

NASA Ames Research Center, Mail Stop T35B-1, Moffett Field, California, 94035-1000, United States
Tel: (+1 650) 604 54 02
Fax: (+1 650) 604 49 54
e-mail: info@riacs.edu
Web: www.riacs.edu
■ Human centered computing, high performance computing and autonomous reasoning.

Reynolds Industries Ltd

a subsidiary of Reynolds Industries Inc, US
Navigation House, Canal View Road, Newbury, Berkshire, RG14 5UR, United Kingdom
Tel: (+44 1635) 311 37
Fax: (+44 1635) 52 19 36
e-mail: rey@reynoldsindustries.ltd.uk
Web: www.reynoldsindustries.ltd.uk
■ Cable and wire products for space and aerospace. High voltage connectors, high and low voltage cable assemblies for the aerospace industry. Microwave cable assemblies (flexible and semi-rigid to 40 GHz). High voltage capacitors, transient protection devices used in military and airborne electronic equipment. Custom built electro-mechanical and electro-optical interconnect products for a wide range of space and aerospace applications.

Rhea System SA

New Tech Center, Avenue Einstein 2a, B-1348 Louvain la Neuve, Belgium
Tel: (+32 10) 48 72 50
Fax: (+32 10) 45 25 07
e-mail: salesrhea@rheagroup.com
Web: www.rheagroup.com

Rheinisch-Westfalische Technische Hochschule Aachen (RWTH)

Templergraben 55, D-52062 Aachen, Germany
Tel: (+49 241) 809 43 22
Fax: (+49 241) 809 23 24
e-mail: pressestelle@zhv.rwth-aachen.de
Web: www.rwth-aachen.de

Rheinmetall Defence Electronics GmbH

Headquarters
a part of Rheinmetall Defence, Germany
Brueggeweg 54, D-28309 Bremen, Germany
Tel: (+49 421) 457 01
Fax: (+49 421) 457 29 00
e-mail: info@rheinmetall-de.com
Web: www.rheinmetall-de.com
■ Systems for land forces include: command, control, communication, intelligence and reconnaissance systems, fire control systems and periscope and night sight units, air defence systems, artillery systems and test systems. Systems for air forces include aircraft simulators. Provide maintenance and logistics services for army, naval and air force related products. Submarine equipment trainers for diving and control room operations.

Company Divisions:
Reconnaissance
Command
Fire Control
Airborne systems
Flight simulation
Land Simulation
Maritime and process simulation

RocketPlane Kistler

4300 Amelia Earhart Lane, Oklahoma City, Oklahoma, 73159, United States
Tel: (+1 405) 488 12 00
Fax: (+1 405) 488 12 04
e-mail: info@rocketplane.com
Web: www.rocketplanekistler.com

Rockwell Collins

Kaiser Electronics
Kaiser Electronics
a subsidiary of Kaiser Aerospace & Electronics Corp, US
2701 Orchard Parkway, San Jose, California, 95134-2083, United States
Tel: (+1 408) 532 48 41
Fax: (+1 408) 954 10 42
e-mail: info@kaisere.com
Web: www.vsi-hmcs.com
■ Manufacture of advanced head-up displays, liquid crystal multifunction displays, helmet displays and display processors for combat aircraft.

Rockwell Collins Deutschland GmbH (RCD)

a division of Rockwell Collins, US
PO Box 105608 D-69046 Heidelberg, Germany
Grenzhöfer Weg 36, D-69123 Heidelberg, Germany
Tel: (+49 6221) 51 20
Fax: (+49 6221) 51 23 05
e-mail: rcd-info@rockwellcollins.com
rcd-customer@rockwellcollins.com
rcd-space@rockwellcollins.com
Web: www.rockwellcollins.de
www.rockwellcollins.com
■ Distribution, manufacturing, service and support of communication and navigation equipment for airborne and aerospace platforms, integrated communication systems for ship, ground or shelter applications, TELDIX® precision space wheels.

Rockwell Collins Inc

Commercial Systems
Air Transport Systems
Business and Regional Systems
Flight Dynamics
Kaiser Electro Precision
Passenger Systems
400 Collins Road North East, Cedar Rapids, Iowa, 52498, United States
Tel: (+1 319) 295 10 00
Freephone: (+1 800) COLLINS
Fax: (+1 319) 295 44 51
e-mail: collins@rockwellcollins.com
mktgsvcs@rockwellcollins.com
Web: www.rockwellcollins.com
■ Provide avionics systems products, and support the world's airlines, airframe manufactures, regional airlines and turbine-powered business aircraft. The company is directed at moving the aviation industry toward the satellite-based CNS/ATM free flight environment. Avionics solutions meeting those requirements include Pro Line 21 flight deck; Collins communication, navigation, situational awareness, display and automatic flight control systems; Integrated Information System (I2S); Rockwell Collins Flight Dynamics Head-up Guidance System (HGS) and Surface Guidance System (SGS) technology. Produce cabin management and in-flight entertainment systems that are now flown with over 100 airlines around the world.

Rockwell Collins Inc

Government Systems
Collins Avionics Communications Division
Government Systems
350 Collins Road, North East, Cedar Rapids, Iowa, 52498-0120, United States
Tel: (+1 319) 295 10 00
(+1 319) 295 19 48
(+1 319) 295 51 00
Freephone: (+1 800) COLLINS
Fax: (+1 319) 295 47 77
(+1 319) 295 70 00
(+1 319) 295 54 29
e-mail: collins@rockwellcollins.com
mktgsvcs@rockwellcollins.com
Web: www.rockwellcollins.com
■ Supply of defence electronics products and systems, including communications, navigation and integrated systems for airborne, ground and shipboard applications. Customers include the US Department of Defense, foreign militaries, government agencies and manufacturers of military aircraft and helicopters. Provide open systems architecture and commercial off-the-shelf technology solutions offering the growth and flexibility needed to address the emerging Global Air Traffic Management (GATM) requirements. Communication and navigation solutions meeting these mandates include the Joint Tactical Radio System (JTRS) Cluster 1 Radios, Collins Flight2 products and systems, GPS receivers, data links, flight management systems, communications systems and aviation electronics systems.

Rockwell Collins Inc

Headquarters
Collins Avionics and Communications Division
400 Collins Road North East, Cedar Rapids, Iowa, 52498, United States
Tel: (+1 319) 295 10 00
(+1 319) 295 51 00
(+1 800) 321 22 23
Fax: (+1 319) 295 47 77
e-mail: collins@rockwellcollins.com
Web: www.rockwellcollins.com
■ Design, production and support of communications and aviation electronics solutions for government and commercial customers worldwide.

Domestic service centres in
Atlanta, Georgia
Boston, Massachusetts
Cedar Rapids, Iowa
Charlotte, North Carolina
Chicago, Illinois
Dallas/Fort Worth, Texas
Detriot, Michigan
Honolulu, Hawaii
Houston, Texas
Los Angeles, California
Miami, Florida
Minneapolis, Minnesota
Newark, New Jersey
New York, New York
Philadelphia, Pennsylvania
Pittsburg, Pennsylvania
Pomona, California
Raleigh, North Carolina
St Louis, Missouri
Salt Lake City, Utah
San Francisco, California
San Jose, California
Seattle, Washington
Tulsa, Oaklahoma
Washingtron, District of Colombia
Wichita, Kansas

International service centres in
Amsterdam, Netherlands
Auckland, New Zealand

Bangkok, Thailand
Beijing, China
Brisbane, Australia
Buenos Aires, Argentina
Frankfurt, Germany
Hong Kong, China
Johannesburg, South Africa
London, UK
Manchester, UK
Manila, Philippines
Melbourne, Australia
Moscow, Russia
Narita, Japan
Osaka, Japan
Paris, France
Reading, England
Rome, Italy
San Jose, Brazil
San Juan, Puerto Rico
Santiago, Chile
Shanghai, China
Singapore
Sydney, Australia
Toulouse, France
Zurich, Switzerland
Operations in
Ann Arbor, Michigan
Bellevue, Iowa
Carlsbad, California
Cedar Rapids, Iowa
Charlotte, North Carolina
Colchester, Vermont
Coralville, Iowa
Cypress, California
Decorah, Iowa
Irvine, California
Kirkland, Washington
Manchester, Iowa
Melbourne, Florida
Mexicali, Mexico
Pomona, California
Portland, Oregon
Richardson, Texas
San Jose, California
Tustin, California
White Marsh, Maryland

Rockwell Collins Optronics

Kaiser Electro-Optics Inc
a subsidiary of Kaiser Aerospace & Electronics Corp, US
2752 Loker Avenue West, Carlsbad, California, 92008, United States
Tel: (+1 760) 438 92 55
Fax: (+1 760) 438 68 75
e-mail: mktginfo@keo.com
optronics@rockwellcollins.com
Web: www.keo.com
■ Design, development and manufacture of optical, opto-mechanical, electro-optical and display products for demanding aerospace, medical, defence, industrial, training and simulation applications. Products include: custom and commercial off-the-shelf optical systems for remote sensing, reconnaissance surveillance space instruments, space star tracker, countermeasures and laser communications; head-mounted, head-up, head-down displays for virtual flight simulators; tank sight display systems; land warrior monocular displays and complete virtual reality systems.
Accreditations: ISO 9001

Romanian Space Agency (ROSA)

21-25 Mendeleev Street, Sector 1, R-0R0 362 Bucharest, Romania
Tel: (+40 21) 212 87 23
Fax: (+40 21) 312 88 04
e-mail: mosa-hg@rosa.ro
Web: www.rosa.ro
■ The company was founded in 1991 and reorganised by a government decision in 1995 as an independent public institution.

Rotadata Ltd

Bateman Street, Derby, DE23 8JQ, United Kingdom
Tel: (+44 1332) 34 80 08
Fax: (+44 1332) 33 10 23
e-mail: sales@rotadata.com
Web: www.rotadata.com
■ Fixed and traversing aerodynamic probes for aircraft and aircraft engines. Tip clearance measurement equipment. Application of sensors to engine test components. Design and manufacture of turbo-machinery test rigs and wind tunnel models. High speed sliprings and telemetry.

Routes Astroengineering

303 Legget Drive, Ottawa, Ontario, K2K 2B1, Canada
Tel: (+1 613) 592 07 48
Fax: (+1 613) 592 65 53
e-mail: staff@routes.com
Web: www.routes.com
■ Design and development of scientific instruments for space applications. Manufacture of solar panels.

Roxel SAS

Tactical Propulsion
Celerg
Immeuble Jura, Centre d'Affaires La Boursidiere, F-92357 Le Plessis-Robinson, Cedex, France
Tel: (+33 1) 41 07 82 21
(+33 1) 141 07 82 82
Fax: (+33 1) 46 30 22 37
(+33 1) 41 07 82 03
■ Development and sale of propulsion systems for all types of rockets and missiles, including tactical and cruise missiles.

Royal Meteorological Institute of Belgium (RMIB)

3 Avenue Circulaire, B-1180 Bruxelles, Belgium
Tel: (+32 2) 373 05 08
Fax: (+32 2) 374 67 88
(+32 2) 373 05 28
e-mail: ui@oma.be
Web: www.meteo.be
■ Research and development study services.

RTG Aero-Hydraulic Inc

Niederlassung Deutschland, Handelshof 26, D-28816 Stuhr, Germany
Tel: (+49 421) 168 36 80
Fax: (+49 421) 16 83 68 77
e-mail: info@rtg.aero
Web: www.rtg.aero
■ Provide overhaul, repair, modification and acceptance testing of airliner components and systems, design, manufacture and integration of spacecraft propulsion systems. Also design and assembly of hydraulic, pneumatic and fuel test stands.
Accreditations: LBA, JAR 145, FAA 145

RUAG Aerospace

Aircraft and Defence Systems
a subsidiary of RUAG Holding, Switzerland
Seetalstrasse 175, PO Box 301 CH-6032 Emmen, Switzerland
Tel: (+41 412) 68 41 11
(+41 412) 68 41 48
Fax: (+41 412) 68 39 43
(+41 412) 60 25 88
e-mail: info.aerospace@ruag.com
aircraft.aerospace@ruag.com
military.aircraft@ruag.com
Web: www.ruag.com
■ Provide integrated solutions in aerospace core activities including research and development, manufacture and assembly of structural modules, systems integration, upgrading and maintenance of aircraft, helicopters, drones, missiles, anti-aircraft and related command and communication systems.

Russian Aerospace Agency Federal State Unitary Enterprise Research Institute for Parachute Engineering (FSUE RIPE)

State Unitary Enterprise Scientific Research Institute of Parachute Construction
Scientific Research Institute of Parachute Construction, State Unitary Enterprise Ulitsa Irkutskaya 2, 107241 Moskva, Russian Federation
Tel: (+7 095) 462 09 82
(+7 095) 462 13 19
Fax: (+7 095) 462 52 33
■ Design, research, testing and manufacture of parachutes for different purposes, including parachute systems for unmanned and manned space vehicles and booster rockets. Related equipment.

Rutherford Appleton Laboratory (RAL)

Council for the Central Laboratory of the Research Councils
Appleton Laboratory
Rutherford Lab
Chilton, Didcot, Oxfordshire, OX11 0QX, United Kingdom
Tel: (+44 1235) 44 50 00
Fax: (+44 1235) 44 58 48
e-mail: enquiries@cclrc.ac.uk
Web: www.sstd.rl.ac.uk
■ Part of the Central Laboratory of the Research Councils (CCLRC). Manages and provides technical support for CCLRC's space hardware programmes and carries out research in astronomy, space science, radio communications and earth observation. Undertakes commercial contracts. Provides facilities and support for research in particle physics, neutron beam scattering, high-power lasers, cryogenics, electron beam lithography, computing and radio wave propagation. Development of models for radio propagation; high power lasers, (dye, excimer and Ti sapphire) for studying high temperature, high density, plasma; state of the art sensors (neutron, x-ray, visible and microwave); advanced database techniques, expert systems and data visualisation; a powerful neutron source for determining the structure of materials; microelectronics design (including rad-hard components); instrumentation for space (including environmental test facilities) and the design and fabrication of advanced microstructures.

Ryerson University

Mechanical, Aerospace and Industrial Engineering
350 Victoria Street, Toronto, Ontario, M5B 2K3, Canada
Tel: (+1 416) 979 53 03
Fax: (+1 416) 979 52 65

S

Saab Avitronics

Ericsson Saab Avionics AB
Saabtech Electronics
a subsidiary of Saab AB, Sweden
Nettovägen 6, SE-175 88 Jarfalla, Sweden
Tel: (+46 8) 58 08 40 00
Fax: (+46 8) 58 03 22 44
e-mail: avitronics@saabgroup.com
info@saabtech.se
Web: www.saabgroup.com
www.saabsystems.se
■ Development and production of EW systems, cockpit displays, tactical reconnaissance systems, advanced electronics and mechanical equipment. Supply of avionics systems for the Gripen multi-role aircraft.

SAAO

Observatory 7935
Box 9 South Africa
Tel: (+27 21) 47 00 26
Fax: (+27 21) 47 36 39

Saddleback Aerospace

Issaquah Office
3611 262nd Avenue, SE, Issaquah, Washington, 98029-9119, United States
Tel: (+1 206) 391 56 67

Safe Training Systems Ltd (STS)

Company Headquarters
Holly House, Maidenhead Road, Wokingham, Berkshire, RG40 5RR United Kingdom
Tel: (+44 1344) 48 35 63
Fax: (+44 1344) 48 51 75
e-mail: sales@safetrainingsystems.com
Web: www.safetrainingsystems.com
■ Design, manufacture and sale of simulators for hazardous materials monitors – radiation, chemical warfare agents and toxic chemicals.

SAFRAN

Group Headquarters
Snecma SA
2 Boulevard du Général Martial-Valin, F-75724 Paris, Cedex 15, France
Tel: (+33 1) 40 60 80 80
Fax: (+33 1) 40 60 81 02
Web: www.safran-group.com
■ Manufacture turbojet engines and components for commercial and military aircraft, rocket engines for satellite launchers and missiles.

SAFRAN

a subsidiary of Snecma, France
10 Allée du Brévent, F-91019 Courcouronnes, Evry Cedex, France
Tel: (+33 1) 69 87 09 29
Fax: (+33 1) 69 87 09 22
Web: www.snecma.com

Saft

Defence and Space Division
Rue Georges Leelanche, BP 1039 Poitiers, Cedex 9, France
Tel: (+33 5) 49 55 56 43
Fax: (+33 5) 49 55 47 80

Saft
International Headquarters
12 rue Sadi Carnot, F-93170 Bagnolet, France
Tel: (+33 1) 49 93 19 18
Fax: (+33 1) 49 93 19 69
Web: www.saftbatteries.com
■ Design and manufacture of high-technology specialist battery solutions for industries including aeronautics, rail, aerospace, defence, automotive, energy, emergency and security installations and mobility and communications.
Subsidiary Companies:
Saft AB, Sweden
Saft AS, Norway
Saft America Inc, US
Saft Australia Pty Ltd, Australia
Saft Batterien GmbH, German
Saft Ferak AS, Czech Republic
Saft Ltd, UK
Saft Nife M E Ltd, Middle East
Saft Singapore Pte Ltd, Singapore

Saft Ltd
Alcad Ltd
1st Floor, Unit 5, Astra Centre, Edinburgh Way, Harlow, Essex, CM20 2BN, United Kingdom
Tel: (+44 1279) 77 25 50
Fax: (+44 1279) 42 09 09
Web: www.saftbatteries.com
■ Manufacture of a range of vented nickel-cadmium batteries to match specific operating conditions. Ni-Cd and Ni-Mh cells and battery packs and Lithium primary and secondary cells and battery packs.
Representing: Saft, France

SAG Alutech Nederland BV
Raufoss Nederland BV
Ambachtsweg 6, NL-2222 AK Katwijk, Netherlands
Tel: (+31 71) 402 93 20
(+31 71) 305 10 00
Fax: (+31 71) 403 25 77
(+31 71) 305 10 81
e-mail: info@sag-alutech.nl
Web: www.sag.at

Sage Laboratories Inc
11 Huron Drive, Natick, Massachusetts, 01760-1314, United States
Tel: (+1 508) 653 08 44
Fax: (+1 508) 653 56 71
e-mail: info@sagelabs.com
■ Microwave components operating DC-606 Hz (passive). Space qualified.

Sagem Défense Sécurité (SAFRAN Group)
Optronics and Defense Division
Sagem SA
a subsidiary of SAFRAN Group, France
Le Ponant de Paris, 27 rue Leblanc, F-75512 Paris, Cedex 15, France
Tel: (+33 1) 58 11 78 00
(+33 1) 40 70 63 63
Fax: (+33 1) 58 11 78 20
(+33 1) 40 70 84 36
(+33 1) 40 70 68 68
(+33 1) 40 70 66 40
(+33 1) 58 11 78 50
Web: www.sagem-ds.com
www.sagem.com
■ Design, manufacture and supply of avionics, aeronautic, inertial navigation, helicopter flight control, optronics, thermal vision equipment and air-land systems.

Saint-Gobain Performance Plastics (SGPPL)
Aerospace Components
3711 South Hudson Street, Seattle, Washington, 982118, United States
Fax: (+1 206) 723 68 69
e-mail: seattle.sales@saint-gobain.com
Web: www.plastics.saint-gobain.com
■ Dryliner, manufacture and assembly vacuum and pressure formed enclosures, composite components.

Sandia National Laboratories (SNL)
Sandia Corporation
a division of Lockheed Martin Corporation, Technology Services, US
PO Box 5800, Albuquerque, New Mexico, 87185, United States
1515 Eubank Boulevard, South East, Albuquerque, New Mexico, 87123, United States
Tel: (+1 505) 845 00 11
(+1 505) 844 25 55
(+1 505) 765 47 52
(+1 505) 844 34 41
Fax: (+1 505) 844 63 67
e-mail: webmaster@sandia.gov
Web: www.sandia.gov
■ Operates Sandia National Laboratories on behalf of the US Department of Energy.

Sandia National Laboratories
California Office
PO Box 969, Livermore, California, 94551, United States
7011 East Avenue, Livermore, California, 94550, United States
Tel: (+1 925) 294 20 65

Sanmina-SCI Corporation
Huntsville Office
SCI Systems Inc
SCI Technology Inc
8600 South Memorial Parkway, PO Box 1000 Huntsville, Alabama, 35802, United States
Tel: (+1 256) 882 48 00
Fax: (+1 256) 882 46 52
e-mail: info@sanmina-sci.com
Web: www.sanmina-scl.com
■ Design and manufacture of electronic products for the aerospace, defence, telecommunications industries and the US Government.

Sanmina-SCI Corporation
SCI Systems Inc
SCI Technology Inc
2700 North First Street, San Jose, California, 95134, United States
Tel: (+1 256) 882 45 77
(+1 408) 964 35 55
(+1 256) 882 41 99
(+1 860) 434 45 73
Fax: (+1 256) 882 46 52
(+1 408) 964 36 36
(+1 860) 434 45 99
e-mail: products@sanmina-sci.corp
Web: www.sanmina-sci.com
■ Militarised voice recognition systems, communications terminals, control panels, data switches and avionics processors.

Sanriku Balloon Center (SBC)
Yoshihama, Sanriku-cho, Kesennuma, gun, Iwate, 022-0102, Japan
Tel: (+81 192) 45 23 11
Fax: (+81 192) 43 70 01

Satcom Distribution Ltd
Unit 3, The Woodford Centre, Old Sarum Park, Lysander Way, Old Sarum, Salisbury, Wiltshire, SP4 6BU, United Kingdom
Tel: (+44 1722) 41 08 00
Fax: (+44 1722) 41 07 77
e-mail: enquiries@satcomdistribution.com
Web: www.satcomdistribution.com
■ Worldwide supplier of satellite communications equipment and airtime.

Satcom1
Headquarters
Sprogovej 6, 2.t.h., DK-2000 Frederiksberg, Denmark
Tel: (+33 145) 75 10 54
Fax: (+33 145) 78 83 67
e-mail: activation@satcom1.com
support@satcom1.com
finance@satcom1.com
info@satcom1.com
webmaster@satcom1.com
sales@satcom1.com
Web: www.satcom1.com
■ Communication provider and integrator for aero, maritime and land mobile applications.

Satel Conseil (SC)
34, rue des Bourdonnais, F-75001 Paris, France
Tel: (+33 1) 44 88 99 30
Fax: (+33 1) 44 88 99 39
Web: www.satelconseil.com

SATELLIFE
Watertown Office
30 California Street, Watertown, Massachusetts, 02472, United States
Tel: (+1 617) 926 94 00
Fax: (+1 617) 926 12 12
e-mail: hnet@usa.healthnet.org
Web: www.satellife.org/
■ Charity utilising satellite, telephone and internet communications to provide health communication and information in the third world.

Satellife
30 California Street, Watertown, Massachusetts, 02472, United States
Tel: (+1 617) 926 94 00
Fax: (+1 617) 926 12 12
e-mail: hnet@healthnet.org
Web: www.healthnet.org
■ Design and dvelopment of information services for health care workers, governments, medical schools and health researchers in the developing world.

Satellite Engineering Research Corporation
7701 Woodmont Avenue, Suite 208, Bethesda, Maryland, 20814, United States
Tel: (+1 301) 657 96 41
Fax: (+1 301) 657 96 42
Web: www.satellitecorp.com
■ Provides engineering and research services in satellite communication systems engineering, including RF link analysis, orbital mechanisms, constellation design, spacecraft design, GPS applications, theoretical analysis of high precision time transfer, and technical legal assistance. Conducts professional training courses in satellite communication.

Satellite Systems Distribution
Satellite House, Bessemer Way, Harfreys Industrial Estate, Great Yarmouth, Norfolk, NR31 0LX, United Kingdom
Tel: (+44 1493) 60 34 60
Fax: (+44 1493) 44 59 09
e-mail: sales@satellite-distribution.com
Web: www.satellite-distribution.com
■ Supplier of satellite communication equipment for use in situations where landline and GSM networks have failed.

Satlantic Inc
Richmond Terminal, Pier 9, 3481 North Marginal Road, Halifax, Nova Scotia, B3K 5X8, Canada
Tel: (+1 902) 492 47 80
Fax: (+1 902) 492 47 81
e-mail: info@satlantic.com
Web: www.satlantic.com
■ Provision of remote sensing solutions to support tactical, surveillance and environmental monitoring missions of government and commercial organisations.

Scaled Composites LLC
Head Office
1624 Flight Line, Mojave, California, 93501-1663, United States
Tel: (+1 661) 824 45 41
Fax: (+1 661) 824 41 74
e-mail: info@scaled.com
Web: www.scaled.com
■ Design, build and flight testing of prototype composite aircraft and spacecraft.

Scandinavian Avionics AS
Stratusvej 9, DK-7190 Billund, Denmark
Tel: (+45 79) 50 80 00
(+45 44) 64 88 28
Fax: (+45 79) 50 80 99
e-mail: sa@scanav.com
Web: www.scanav.com
■ Sales, installation, repair and overhaul of HF, VHF, UHF, communications and navigation, DME, transponder, radio/radar altimeters autopilots, flight directors, voice/flight data recorder, inertial navigation, GPS, VLF/Omega, instrument, gyro, ground proximity and aircraft electrical systems for civil and military purposes.

Schonstedt Instrument Co
100 Edmond Road, Kearneysville, West Virginia, 25430, United States
Tel: (+1 304) 725 10 50
(+1 304) 724 47 00
Freephone: (+1 800) 999 82 80
Fax: (+1 304) 725 10 95
e-mail: info@schonstedt.com
Web: www.schonstedt.com
■ Design, manufacture, marketing and servicing of instrumentation for domestic utilities, construction, excavation and exploration. The main lines of instrumentation include magnetic locating gradiometers capable of locating ferrous objects up to 16 feet underground, and multi-frequency pipe and cable locators used to find underground utilities for repair and damage avoidance during excavation.

Science & Technology International of Russia
1st Schipovski per. 20, 113093 Moskva, Russian Federation
Tel: (+7 095) 237 73 51

Science and Production Association of Automation and Instrument-engineering

Nauchno-proizvodstvennoe ob'edinenie avtomatiki i priborostroeniya
AP, NPO Vvedenskiy Street 1, 117342 Moskva, Russian Federation
Tel: (+7 095) 330 65 70
(+7 095) 330 30 56
(+7 095) 330 47 83
Fax: (+7 095) 334 83 80
■ Development and manufacture of satellite guidance systems.

Science Applications International Corporation (SAIC)

Headquarters
an employee-owned company
10260 Campus Point Drive, San Diego, California, 92121, United States
Tel: (+1 505) 842 78 91
(+1 858) 826 60 00
Fax: (+1 858) 546 68 00
Web: www.saic.com
■ Research and engineering services in technology development and analysis, computer systems, computer hardware and computer software. Applications in the areas of national and international security, information technology, transportation, telecommunications and environmental systems and engineering.

Scientific-Atlanta Inc

World Headquarters
5030 Sugarloaf Parkway, PO Box 465447 Lawrenceville, Georgia, 30042, United States
Tel: (+1 770) 903 50 00
Freephone: (+1 877) SFA INFO
Web: www.scientificatlanta.com

SciSys UK Ltd

Chippenham
SciSys Space and Defence
a part of the CODASciSys plc group
a subsidiary of CODA Scisys plc, UK
Methuen Park, Chippenham, Wiltshire, SN14 0GB, United Kingdom
Tel: (+44 1249) 46 64 66
Fax: (+44 1249) 46 66 66
e-mail: marketing@scisys.co.uk
Web: www.scisys.co.uk
■ Spacecraft monitor and control, flight dynamics support, simulation, mission planning, data processing, system studies and consultancy.

SEA (Group) Ltd (SEA)

Systems Engineering & Assessment Limited
Beckington Castle, 17 Castle Corner, Beckington, PO Box 800, Frome, Somerset, BA11 6TA, United Kingdom
Tel: (+44 1373) 85 20 00
Fax: (+44 1373) 83 11 33
e-mail: info@sea.co.uk
Web: www.sea.co.uk
■ A systems and software house with wide-ranging system and product development, systems engineering and software capabilities. Clients are blue-chip organisations principally in the marine, space, battlespace and transport domains. Assessed and approved to undertake design and development systems, feasibility studies and system specification, signal and image processing systems development, management systems and services, technical publications and manufacturing.
Accreditations: ISO 9001:2000 and ISO 9000-3 (TickIT)

Sea Launch Company LLC

One World Trade Center, Suite 950, Long Beach, California, 90831, United States
Tel: (+1 562) 499 47 00
Fax: (+1 562) 499 47 55
Web: www.sea-launch.com
■ Commercial satellite launch service provider.

Sea Tel Inc

European Office
Unit 1, Orion Industrial Centre, Wide Lane, Swaythling, Southampton, S018 2HJ, United Kingdom
Tel: (+44 2380) 67 11 55
Fax: (+44 2380) 67 11 66
e-mail: europe@seatel.com
Web: www.seatel.com

Sea Tel Inc

a wholly owned subsidiary of Cobham plc, UK
4030 Nelson Avenue, Concord, California, 94520, United States
Tel: (+1 925) 798 79 79
Freephone: (+1 888) 798 79 79
Fax: (+1 925) 798 79 86
e-mail: sales@seatel.com
service@seatel.com
seatel@seatel.com
Web: www.seatel.com
■ Design, development, manufacture and marketing of shipboard stabilised antennas, at-sea communications and data transfer technologies for yachts, commercial ships, drill rigs, work boats, cruise ships and military.

Seakr Engineering Inc

6221 South Racine Circle, Centennial, Colorado, 80111, United States
Tel: (+1 310) 542 93 02
Fax: (+1 310) 542 32 07
Web: www.seakr.com

SED

a subsidiary of Calian Ltd, Canada
18 Innovation Boulevard, PO Box 1464, Saskatoon, Saskatchewan, S7K 3P7, Canada
Tel: (+1 306) 931 34 25
Fax: (+1 306) 933 14 86
Web: www.sedsystems.ca
■ Space and communications systems design, engineering and integration specialising in satellite ground terminals, defence communication systems, satellite test and control and mobile satellite communication systems. Major products include: INMARSAT coast-earth stations, satellite in-orbit test systems, environmental monitoring, remote control systems, HF adaptive beam forming receiver systems.

Seimac Limited

Seimac Research Ltd
a wholly owned subsidiary of Cobham plc, UK
271 Brownlow Avenue, Dartmouth, Nova Scotia, B3B 1W6, Canada
Tel: (+1 902) 468 30 07
Fax: (+1 902) 468 30 09
e-mail: info@seimac.com
Web: www.seimac.com
■ Design and manufacture of radio systems for search and rescue products and telemetry. Tracking systems for research, military, search and rescue and the oil and gas industries.

Selex Communications GmbH

ANT Nachrichtentechnik GmbH
Bosch Telecom GmbH
Spinnerei 48, D-71522 Backnang, Germany
Tel: (+49 7191) 37 85 11
Fax: (+49 7191) 37 83 00
Web: www.selexcom.de
■ Satellite communications systems: ground and space. Microwave radio systems and antennas, mobile radio systems, fibre optic systems, multiplex transmission systems and the COMSEC system. Communications network control systems.
Divisions (Corporate HQ)
Microwave Systems
Multiplex Systems Telecommunications Cable Systems
Space Communications Systems Special Communications Systems

Sensitised Coatings Ltd

Bergen Way, North Lynn Industrial Estate, King's Lynn, Norfolk, PE30 2JL, United Kingdom
Tel: (+44 1553) 76 48 36
Fax: (+44 1553) 76 03 77
e-mail: sales@senco.co.uk
Web: www.senco.co.uk
■ Manufacture recording paper rolls for sonar, echosounder, maritime surveillance, hydrographic, oceanographic, seismic and satellite signal monitoring systems. Thermal paper rolls for facsimile equipment, non-impact printers and data terminals.

Sepura Limited

Simoco Europe Ltd
Simoco International Ltd
Radio House, St Andrews Road, Cambridge, CB4 1GR, United Kingdom
Tel: (+44 1223) 87 60 00
Fax: (+44 1223) 87 90 00
e-mail: info@supura.co.uk
Web: www.sepura.co.uk
■ Design, development and supply of digital (TETRA) radios for the public safety market and emergency services; such as police, fire, and ambulance, transportation and utility companies, government and municipal organisations. Selection of vehicle mounted and handportable radios include the SRP2000 public safety handportable.

Sequa Corporation

Corporate Headquarters
200 Park Avenue, New York, New York, 10166, United States
Tel: (+1 212) 986 55 00
Fax: (+1 212) 370 19 69
Web: www.sequa.com
■ Manufacture of aircraft engine components, solid rocket propellant motors for missiles and space applications.
Subsidiaries with defence business:
Atlantic Research Corp
Chromalloy Gas Turbine Corp

Serco Europe Ltd

5th Floor, Kempton Point, 68 Staines Road West, Sunbury-on-Thames, Middlesex, TW16 7AX, United Kingdom
Tel: (+44 1932) 73 30 00
Fax: (+44 1932) 73 30 99

Sermatech Power Solutions

PO Box 451047, Houston, Texas, 77245-1047, United States
25 South Belt Industrial Drive, Houston, Texas, 77047, United States
Tel: (+1 832) 485 80 51
Fax: (+1 713) 413 10 09
Web: www.sermatech.com

Service Argos Inc

Largo Office
1801 McCormick Drive, Suite 10, Largo, Maryland, MD 20774, United States
Tel: (+1 301) 925 44 11
Fax: (+1 301) 925 89 95
e-mail: useroffice@argosinc.com
Web: www.argosinc.com

Servo Corporation of America

123 Frost Street, Westbury, New York, 11590-5026, United States
Tel: (+1 516) 938 97 00
Fax: (+1 516) 938 96 44
Web: www.servo.com
■ UHF and VHF Direction Finder (DF) systems.

SES Global

Société Européenne des Satellites
Château de Betzdorf, L-6815 Betzdorf, Luxembourg
Tel: (+352) 710 72 51
Fax: (+352) 710 72 52 27
(+352) 710 72 53 24
Web: www.ses-global.com
■ Operate a satellite services network providing seamless communications.

SETI Institute

2035 Landings Drive, Mountain View, California, 94043, United States
Tel: (+1 650) 961 66 33
Fax: (+1 650) 961 70 99
Web: www.seti.org

Sevig Press

Rue Bellart 6, F-75015 Paris, France
Tel: (+33 1) 42 73 28 37
Fax: (+33 1) 42 73 20 95
e-mail: sevig.press@noos.fr
Publications: Russian Space Directory – published every two years in co-operation with the Russian Space Agency (RSA) and the European Space Agency (ESA).
European Space Directory – Published annually in co-operation with Eurospace and the technical assistance of the European Space Agency (ESA).

Shanghai Obseravatory

80 Nandan Road, 200030 Shanghai, China
Tel: (+86 21) 438 61 91
Fax: (+86 21) 438 46 18

Sharp Corporation

282-1 Hajikami, Shinjo-cho, Katsuragi-shi, Nara, 639-2198, Japan
Web: sharp-world.com
■ Worldwide marketing commercial LCD displays.

Sharp Corporation

Sharp Plaza, Mahwah, New Jersey, New Jersey, 04730, United States
22-22 Nagaike-cho, Abeno-ku, Osaka, 545-8522, Japan
Tel: (+81 6) 621 12 21
Web: sharp-world.com

Sheldahl Inc

1150 Sheldahl Road, Northfield, Minnesota, 55057, United States
Tel: (+1 507) 663 80 00
Fax: (+1 507) 663 85 45
e-mail: sheldahl.info@sheldahl.com
Web: www.sheldahl.com

- Supplier to the military and aerospace industries. Product line includes: flexible circuitry, passive thermal control for spacecraft and satellites, radar absorbing materials, engineered flexible structures.

Shinawatra Satellite Public Co Ltd (SSA)

a wholly owned subsidiary of Shin Corporation plc
41/103 Rattanathibet Road, Nonthaburi, 11000, Thailand
Tel: (+66 2) 591 07 36
(+66 2) 591 07 37
(+66 2) 591 07 49
(+66 2) 591 07 36
(+66 2) 591 07 36
Fax: (+66 2) 591 07 05
(+66 2) 591 07 19
(+66 2) 591 07 14
(+66 2) 591 07 19
(+66 2) 591 07 06
e-mail: mc@thaicom.net
Web: www.thaicom.net

- Launch and operation of the THAICOM satellite system, a direct-to-home broadcasting system to Asia, Europe, Australia, Africa and the Middle East. Services include turnkey operation consultancy, transponder leasing and broadcast services. The company's three satellites, THIACOM 1A, 2 and 3, have a combined capacity of 69 transponders and are capable of TV broadcasting transmission.

Shock-Tech Inc

360 Route 59, Monsey, New York, 10952, United States
Tel: (+1 845) 368 86 00
Fax: (+1 845) 368 87 99
e-mail: info@shocktech.com
Web: www.shocktech.com

- Manufacture of shock and vibration isolators, cabinets and cases.

Representing: SMAC S.A.S, France
Socitec International, France
Vibtech Ltd, Brazil

Siemens AG Österreich (SAGO-PSE)

Space Business Unit
Siemenßtraße 92, A-1210 Wien, Austria
Tel: (+43 5) 170 70
e-mail: kontakt.at@siemens.com
Web: www.siemens.com

- Activities in the fields of satellite test systems, telecommunications, mission control software, information management, software technology. Satellite traffic monitoring.

SIME sas di Paolo De Gaetano & Co

Via F S Benucci 9, I-00149 Roma, Italy
Tel: (+39 06) 55 27 08 29
(+39 06) 55 27 30 49
Fax: (+39 06) 551 56 62
Web: www.simesas.com

- Numerical simulation, virtual prototype software, antennas satellite earth stations, telescopic hangars for ships and helicopter landing systems. Shipborne helicopter handling.

Representing: Engineering Systems International, Netherlands
ESI Software Group, Paris
Indal Technologies Inc, Canada
Jered Industries Inc, US
Vertex RSI, US

Simulation and Software Technology (SISTEC)

SISTEC Braunschweig
Lilienthalplatz 7, D-38108 Braunschweig, Germany
Fax: (+49 531) 295 27 67

Simulogix

Top-Down Engineering
555 Veterans Boulevard, Suite 123, Redwood City, California, 94063, United States

Singapore Telecommunications Ltd

31 Exeter Road #28-00 Comcentre, Singapore, 239732, Singapore
Tel: (+65 838) 88 09
Fax: (+65 733) 13 50
e-mail: contact@singtel.com

SINTEF Group

Stiftelsen for industriell og teknisk forskning ved NTH
Sintef Group
Strindveien 4, N-7465 Trondheim, Norway
Tel: (+47) 73 59 30 00
Fax: (+47) 73 59 33 50
Web: www.sintef.no

- Advanced technical services and laboratory services for the exploration and production petroleum resources, nationally and internationally.

Sira Technology Ltd (SIRA)

Sira Electro-Optics Ltd
South Hill, Chislehurst, Kent, BR7 5EH, United Kingdom
Tel: (+44 20) 84 67 26 36
Fax: (+44 20) 84 68 17 71
e-mail: marketing@siraeo.co.uk
info@siraeo.co.uk

- Independent research and development organisation specialising in the design and custom manufacture of electro-optical systems. Reflectometers, thermal imager test equipment, high-speed IR spectro radiometers, helmet-mounted display test equipment, laser scanners. Special purpose CCD cameras, scanning sub-systems for laser radars, robust sensors, special purpose microscopes and telescopes. Satellite payloads including hyperspectral imagers, startrackers and optical intersatellite data link terminals.

Sirtrack Ltd

a subsidiary of Landcare Research New Zealand Ltd, New Zealand
Private Bag 1403 Havelock North, New Zealand
Goddard Lane, 4201 Havelock North, New Zealand
Tel: (+64 6) 877 77 36
Freephone: (+64 1800) 74 78 72
(+64 1800) 42 72 47
Fax: (+64 6) 877 54 22
e-mail: sirtrack@sirtrack.com
Web: www.sirtrack.com

- Design and manufacture of VHF and UHF tracking equipment, mainly for wildlife research, certified by Argos.

Accreditations: ISO 9001

Skyview Systems

Skyview Centre, 9 Church Field Road, Chilton Industrial Estate, Sudbury, Suffolk, CO10 6GT, United Kingdom
Tel: (+44 1787) 88 31 38
Fax: (+44 1787) 88 31 39
e-mail: skyview@skyview.co.uk
Web: www.skyview.co.uk

- Supply meteorological, hydrological and marine instruments to commercial, civil, avionic and military authorities worldwide.

Slovak Academy Sciences

Astronomical Institute
Dubravska 9, SK-842 28 Bratislava, Slovakia
Tel: (+421 7) 37 51 57
e-mail: astropor@savba.sk

SMC Detachment 12

3548 Aberdeen Avenue SE, Kirtland AFB, New Mexico, 87117-5778, United States
Tel: (+1 505) 846 09 49
Fax: (+1 505) 846 13 49

SME Propulsion

Saint Médard Plant
F-33160 Saint-Médard-en-Jalles, France
Tel: (+33 5) 56 70 50 50
Fax: (+33 5) 56 70 50 51
Web: www.sme-propulsion.com

- Solid propellants; military powders and lightweight materials.

Smith Semiconductor Inc

3185 Tozer Road, North Branch, Michigan, 48461, United States
Tel: (+1 810) 688 11 74
Fax: (+1 810) 688 47 38
e-mail: jeremy@smithsemi.com
Web: www.smithsemi.com

- Procurement and supply of obsolete and diminished military, aerospace and space electronic components.

Smiths Detection

Toronto
Barringer Instruments Ltd
7030 Century Avenue, Mississauga, Ontario, L5N 2V8, Canada
Tel: (+1 905) 817 59 90
Fax: (+1 905) 817 89 92
Web: www.smithsdetection.com

Snecma

Communication Department
10, Allee du Brevent, CE1420, F-91019 Courcouronnes, Evry Cedex, France
Tel: (+33 1) 69 87 09 21
Fax: (+33 1) 69 87 09 22
Web: www.snecma.com

- Manufacturer of aircraft and space engines with a wide range of propulsion systems. Also designs and builds commercial aircraft engines and military aircraft engines. Develops and produces propulsion systems and equipment for launch vehicles and satellites.

Snecma

Engine and Flight Test Department
Chemin des Bellons, PO Box 644, F-13804 Istres, Cedex, France
Tel: (+33 4) 42 35 90 00
Fax: (+33 4) 42 35 90 68

- Engine testing.

Snecma

Evry-Corbeil Plant
Snecma Moteurs
PO Box 81, F-91003 Evry, Cedex, France
Tel: (+33 1) 69 87 92 60
Fax: (+33 1) 69 87 89 28

- Production plant, quality control, purchasing, sales services, military engine support, technical training centre.

Snecma

Gennevilliers Plant
291 Avenue d'Argenteuil, PO Box 48 F-92234 Gennevilliers, Cedex, France
Tel: (+33 1) 47 60 72 06
Fax: (+33 1) 47 60 73 00

- Forge, foundry, and machining of mechanical parts.

Snecma

Le Creusot Plant
Avenue de l'Europe, PO Box 97 F-71203 Le Creusot, Cedex, France
Tel: (+33 3) 85 77 77 12
Fax: (+33 3) 85 77 77 77

- Automated production of turbine discs.

Snecma

Space Engines Division
Large Liquid Propulsion, BP 802, F-27208 Vernon, Cedex, France
Tel: (+33 2) 32 21 72 00
Fax: (+33 2) 32 21 27 01
Web: www.snecma.com
www.snecma-moteurs.com

Snecma

Villaroche Plant
Rond Point René Ravaud-Reau, F-77550 Moissy Cramayel, Cedex, France
Tel: (+33 1) 60 59 71 23
Fax: (+33 1) 60 59 71 36

- Design, development, assembly and test for aircraft engines. Spare parts warehouse. Civil engine advertising and sales.

Snecma Moteurs

Villaroche Nord Plant
Direction Propulsion et Équipements de Satellites, Aérodrome de Melun-Villaroche, F-77550 Moissy Cramayel, France
Tel: (+33 1) 64 71 46 97
Fax: (+33 1) 64 09 33 38

- Manufacture of satellite thrusters, propulsion systems and equipment. Solar array drive mechanisms, microgravity furnaces, surface tension propellant tanks, cryogenic engine testing.

SociÉtÉ Anonyme d'Études et RÉalisations Nucleaires (SODERN)

EADS Sodern
20 Avenue Descartes, PO Box 23, F-94451 Limeil Brévannes, Cedex, France
Tel: (+33 1) 45 95 70 00
Fax: (+33 1) 45 95 17 64
e-mail: com@sodern.fr
Web: www.sodern.fr

- Space: satellites and spacecraft attitude measurement, star sensing, earth observation cameras. Nucleonics: neutron emission and detection, detection of explosives, chemical materials and narcotics by neutron activation. Bulk material analysis. Optics: optical systems for space instrumentation (IR,UV), flight simulation, nuclear industry.

Société Anonyme Belge de Constructions Aéronautiques (SABCA)

1470 Chaussée de Haecht, B-1130 Bruxelles, Belgium
Tel: (+32 2) 729 55 11
Fax: (+32 2) 705 15 70
e-mail: sabca.secr@sabca.be
Web: www.sabca.com

■ Design, development and manufacture of aircraft structures and components for F-16, Mirage 2000, Mirage 5, Mirage F-1, Alpha Jet, Fokker F-27/F-50, Airbus A330-A340, Dornier 728, Dassault Business Jets, Breguet Atlantic 2 and helicopters. Design, development, manufacture, repair and overhaul of hydraulic servo-actuators for F-16 and Ariane launcher. Design and development of digital and direct drive servo-actuactors. Design, development and manufacture of electronic fire control systems for Leopard tanks and training simulators for main battle tanks. Development of low altitude air defence fire control systems. Development of advanced systems of automatic data processing for field artillery. Participation in space programmes for Ariane (parts, structures and servo-actuators), Spacelab, Eureca, ERA. Design, development and manufacture of composite structures such as glass, aramid and carbon fibre. Co-operation with Belgian universities. Final assembly, systems integration, ground and flight tests of aircraft and helicopters (F-16, Mirage, Alouette, Puma, Agusta 109, Sea King, Mirage F-1, Northrop F-5). Overhaul and repair of electrical, mechanical and hydraulic systems. Integration, testing of weapon systems and ECM devices on F-16, A-10, Mirage and Northrop F-5 aircraft. Repair, overhaul and integration of airborne electronic equipment and missile electronic devices. Study of improvements of present aircraft systems.

Société Anonyme de Télécommunication (SAT)

41 rue Cantagrel, F-75631 Paris, Cedex 13, France
Tel: (+33 1) 45 82 31 11

Société Métallurgique de Gerzat SA (SMG)

Factory and Commercial Services
a part of the Péchiney Group, France
Rue de l'Industrie, PO Box 7 F-63360 Gerzat, France
Tel: (+33 4) 73 23 64 00
Fax: (+33 4) 73 23 64 01

■ Aluminium alloy high pressure cylinders. Tubular parts with or without bottoms for missiles, rockets, projectile elements. Hydraulic containers, jacks. Liners for fibre-wound cylinders.

Society of Indian Aerospace Technologies & Industries (SIATI)

Aeronautical Society Building, Suranjandas Road, Off Old Madras Road, Bangalore, 560075, India
Tel: (+91 80) 521 99 51
(+91 80) 527 52 62
Fax: (+91 80) 529 24 40
e-mail: siati@dataone.in
Web: www.siatiaero.com

■ Provide a forum for encouragement of interaction between research, design, development and industry within India, and overseas collaborations for aerospace technologies, industries, assist potential entrepreneurs with technology to set up medium/small scale industries for manufacture/maintenance of aerospace components, systems and materials, research and development in institutions and universities, participate in national, international air shows, exhibitions, seminars, education, institute scholarships/fellowships in the area of aerospace.

Members interests:

■ Aircraft, helicopter, satellite/launch vehicle structures; design, aircraft engines, starters and APUs; aircraft seats, systems, equipment and accessories; avionics and radars; airline operators; electrical connectors, cables and batteries; electrical components and systems; precision machined parts; rubber polymer, FRP parts; standard parts and fasteners; aircraft filters; ground support equipment; jigs, fixtures, gauges and tools; metallic and non-metallic materials; castings and forgings; consumables; machines and special purpose machines; research, development and testing laboratories; training and educational institutions; government organisations; professional bodies and associations.

Sodankyla Geophysical Observatory

FI-99600 Sodankyla, Finland
Tel: (+358 16) 61 98 11
Fax: (+358 16) 61 98 75
Web: www.sgo.fi

■ Measurement and research of ionosphere, seismicity and magnetic fields. Design and production of magnetometers.

Soloy Aviation Solutions

Soloy, LLC
450 Pat Kennedy Way Southwest, Olympia, Washington, 98501-7298, United States
Tel: (+1 360) 754 70 00
Fax: (+1 360) 943 76 59
e-mail: soloy@soloy.com
Web: www.soloy.com

■ Manufacture of Soloy Dual Pac power plant, a single propeller twin engine powerplant, for general aviation aircraft. Also provide turbine conversion for fixed wing aircraft, engineering modifications and services, FAA repair station services and customised subcontract sheet metal fabrication.

Sonaca SA

Société Nationale de Construction Aérospatiale SA
Aérodrome de Charleroi, Route Nationale 5, B-6041 Gosselies, Belgium
Tel: (+32 71) 25 51 11
Fax: (+32 71) 34 40 35
Telex: 51241 sonaca
Web: www.sonaca.com

■ Design, development, manufacturing, assembly and testing of aerospace structures and their associated systems.

Accreditations: AQAP 110, ISO 9001, JAR/FAR 145, BCAR

Sonic Communications (Int) Ltd

Birmingham International Park, Starley Way, Bickenhill, Birmingham, West Midlands, B37 7HB, United Kingdom
Tel: (+44 121) 781 40 00
Fax: (+44 121) 781 44 02
(+44 121) 781 44 04
e-mail: sales@sonic-comms.com
Web: www.sonic-comms.com

■ Provide a complete range of communication solutions required in operational situations of surveillance, law enforcement, defence, fireground and other industries.

Accreditations: ISO 9001, ISO 2000

SOPEMEA

SociÉtÉ pour le Perfectionnement des MatÉriels et Équipements AÉrospatiaux
a subsidiary of Messier-Bugatti, France
Zone Aéronautique Louis Bréguet, PO Box 48, F-78142, Vélizy-Villacoublay, Cedex, France
Tel: (+33 1) 45 37 64 64
Web: www.sopemea.fr

■ Environmental testing for aerospace equipment. Ground vibration testing for aircraft, rockets and complex structures. Maintenance of test facilities. Design and construction of testing and measurement equipment. Technical assistance, training for engineers and technicians.

SORDAL Incorporated

12813 Riley Street, Holland, Michigan, 49429-9201, United States
Tel: (+1 616) 994 60 00
Fax: (+1 616) 994 61 40
e-mail: ddanver@sordal.org

■ Manufacture of polyimide foams for the next generation of space shuttle (2GRLV). Invent and develop new honeycomb cores and composite materials.

Southern Avionics Co (SAC)

PO Box 5345, Beaumont, Texas, 77726-5345, United States
5000 Belmont, Beaumont, Texas, 77707, United States
Tel: (+1 409) 842 17 17
Fax: (+1 409) 842 29 87
(+1 409) 842 13 24
e-mail: sales@southernavionics.com
Web: www.southernavionics.com

■ Manufacture of digital and analog non-directional beacons (NDB)/low and medium frequency radiobeacons, portable NDB, remote maintenance monitoring (RMM) systems, Marine Band Differential GPS (DGPS), reference station transmitters, Antenna Tuning Units (ATU), monitor alarm receivers, Remote-Control Units (RCU), shelters and battery standby systems.

Sovinformsputnik

47 Leningradskiy Prospekt, 125167 Moskva, Russian Federation
Tel: (+7 095) 943 07 57
Fax: (+7 095) 943 05 85
e-mail: common@sovinformsputnik.com
Web: www.sovinformsputnik.com

■ Provide satellite imaging services.

Sovzond

4 Khoromny tupik, 107817 Moskva, Russian Federation
Tel: (+7 495) 514 83 39
Fax: (+7 495) 623 30 13
e-mail: sovzond@sovzond.ru
mail@sovzond.ru
Web: www.sovzond.net

Space & Missile Systems Center (SMC)

2420 Vela Way, Suite 1467, El Segundo, California, 90245, United States
Tel: (+1 310) 363 00 30
(+1 310) 833 00 30

Space & Naval Warfare Systems Command (SpaWar)

Headquarters
SpaWarSysCom
4301 Pacific Highway, San Diego, California, 92110-3127, United States
Tel: (+1 619) 524 34 32
Web: www.spawar.navy.mil

Space Activities Promotion Committe

Industrial Affairs Bureau II
9-4 Ohtemachi 1-chome, Chiyoda-ku, Tokyo, 100-8188, Japan
Tel: (+81 3) 52 04 15 00
Fax: (+81 3) 52 55 62 33
Web: www.keidanren.or.jp

■ Provide launching services and space vehicles.

Space Adventures Ltd

8000 Towers Crescent Drive, Vienna, Virginia, 22182, United States
Tel: (+1 888) 85 SPACE
(+1 703) 524 71 72
Fax: (+1 703) 524 71 76
e-mail: info@spaceadventures.com
Web: www.spaceadventures.com

Space and Upper Atmosphere Research Commission (SUPARCO)

Sector 28, Gulzar-e-Hijri, Off University Road, PO Box 8402, 75270, Karachi, Pakistan
Tel: (+92 21) 814 46 67
(+92 21) 814 46 74
(+92 21) 814 49 23
(+92 21) 814 49 27
Fax: (+92 21) 814 49 28
(+92 21) 814 49 41
Telex: 25720 space
e-mail: suparco@digicom.net.pk
Web: www.suparco.gov.pk
www.sgs-suparco.org

■ Development and launch of sounding rockets and satellite applications.

Space Association of Australia Inc

Astronautical Society Melbourne
PO Box 351, Mulgrave, North, Victoria, 3170, Australia
Tel: (+61 3) 95 60 86 09
Web: www.vicnet.net.au/~saa

■ Space advocacy organization, meetings, seminars, publications, displays, technical projects. A weekly radio programme.

Space Communications Corp of Japan (SCC)

Uchu Tsushin
2-2-8 Higashi-shinagawa, Shingawa-ku, Tokyo, 140, Japan
Tel: (+81 3) 54 62 13 66
Fax: (+81 3) 54 62 13 90

Space Data Resources/Information (SDR/I)

PO Box 3438, Boulder, Colorado, 80307-3438, United States
Tel: (+1 303) 494 76 77
e-mail: newsspace@aol.com

■ Provide research services regarding the world's space programmes. Maintain databases on space applications programmes as well as space science projects. Freelance writing and press relations are also offered.

Space Dev (SPDV)

13855 Stowe Drive, Poway, California, 92064, United States
Tel: (+1 858) 375 20 00
Freephone: (+1 877) 375 10 04
Fax: (+1 858) 375 10 00
(+1 858) 375 10 50
e-mail: info@spacedev.com
sales@spacedev.com
Web: www.spacedev.com

Space Environment Center (SEC)
R/SE 325 Broadway, Boulder, Colorado, 80305, United States
Tel: (+1 303) 497 33 11
Fax: (+1 303) 497 36 45
Web: www.sec.noaa.gov
■ Provide real-time space environment monitoring and forecasting services, development of techniques for forecasting solar disturbances and their effects on the Earth's environment and research in solar-terrestrial physics.

Space Exploration Technologies
1310 East Ground Avenue, El Segundo, 90245, United States
Tel: (+1 310) 414 65 55
Fax: (+1 310) 414 65 52
e-mail: news@spacex.com
Web: www.spacex.com

Space Florida
Florida Space Authority
Spaceport Florida Authority
State of Florida Spaceport Florida Authority
Building M6-306, Room 9030, State Road 405, Kennedy Space Centre, Florida, 32899, United States
Tel: (+1 321) 730 53 01
Fax: (+1 321) 730 53 07
Web: www.spaceflorida.gov
■ Provide economic development for space-related businesses.

Space Imaging Europe SA (SIE)
13 Aegidon & Seneka Street, GR-14564 Nea Kifissia, Greece
Tel: (+30 1) 08 19 84 00
Fax: (+30 1) 06 25 37 61
e-mail: info@si-eu.com
Web: www.si-eu.com
■ IKONOS satellites.

Space Instruments Inc
4403 Manchester Avenue, Suite 203, Encinitas, California, CA 92024, United States
Tel: (+1 619) 944 70 01
Fax: (+1 619) 944 70 56

Space Machine Advisors Inc
PO Box 1317, Norwalk, Connecticut, 06856, United States
Tel: (+1 203) 857 56 67
e-mail: rguest@spacemachine.com
Web: spacemachine.com

Space Propulsion Systems
4707 140th Avenue North, Suite 303, Clearwater, Florida, 33762-3840, United States
Tel: (+1 727) 524 88 08
Fax: (+1 727) 524 16 17
Web: www.sps.aero

Space Research Centre
Department of Physics and Astronomy
Leicester University Space Research Centre
University of Leicester, University Road, Leicester, LE1 7RH, United Kingdom
Tel: (+44 116) 252 34 91
(+44 116) 252 35 52
Fax: (+44 116) 252 24 64
Web: www.src.le.ac.uk

Space Research Centre
Polish Academy of Sciences
Bartycka 18A, PL-00-716 Warszawa, Poland
Tel: (+48 22) 840 37 66
e-mail: cbk@cbk.waw.pl

Space Research Institute
84/32 Profsoyuznaya Street, 117810 Moskva, Russian Federation
Tel: (+7 095) 333 25 88
(+7 095) 333 20 45
Fax: (+7 095) 333 33 11
(+7 095) 310 70 23
e-mail: zakharov@iki.rssi.ru
Web: www.iki.rssi.ru

Space Software Italia SpA (SSI)
a joint venture between Alenia Spazio SpA, Italy and Computer Sciences Corp, US
Viale del Lavoro 101, Quatiere Paolo VI, I-74100 Taranto, Italy
Tel: (+39 099) 740 11 11
Fax: (+39 99) 470 17 77
e-mail: info@ssi.it
marketing@ssi.it
dirgen@ssi.it
risoreumane@ssi.it
Web: www.ssi.it
■ Design and develops large advanced software systems for space, civil and military applications.

Space Studies Institute (SSI)
PO Box 82, Princeton, New Jersey, 08542, United States
5 Cresent Avenue, Rocky Hill, New Jersey, 08553, United States
Tel: (+1 609) 921 03 77
Fax: (+1 609) 921 03 89
e-mail: ssi@ssi.org
Web: www.ssi.org
■ Sponsors and conducts research using space resources.

Space Systems Finland Ltd (SSF)
Kappelitie 6, Espoo, FI-02200, Finland
Tel: (+358 9) 61 32 86 00
Fax: (+358 9) 61 32 86 99
Web: www.ssf.fi
■ On-board flight software, system design, software and integration for embedded systems, small satellite technology, software simulations, EO instrument and optical system analysis and consultancy.

Space Systems Loral (SSL)
3825 Fabian Way, Palo Alto, California, 94303-4604, United States
Tel: (+1 650) 852 40 00
Freephone: (+1 800) 332 64 90
Fax: (+1 650) 852 47 88
e-mail: inquiries@ssd.loral.com
Web: www.ssloral.com
■ Produce commercial communications and weather satellites.

Space Technology Commerce & Communications
TF Associates Inc
17 Heritage Drive, Lexington, Massachusetts, 02420, United States
Tel: (+1 781) 862 71 74
Fax: (+1 781) 862 22 45
e-mail: tfassoc@tiac.net
■ Conference and exposition management; advertising and marketing.

Space Transportation Association (STA)
2800 Shirlington Road, Suite 405A, Arlington, Virginia, 22206, United States
Tel: (+1 703) 671 41 16
Fax: (+1 703) 931 64 32
e-mail: spacetra@erols.com
e-mail: sta4space@aol.com
Web: www.spacetransportation.org
■ Represents organisations engaged in developing, building, operating and using space transportation vehicles, systems and services to provide access to space for government, civil and military users.

Space Vacuum Epitaxy Center (SVEC)
University of Houston, 4800 Calhoun Road, Science and Research One, Room 724, Houston, Texas, 77204-5507, United States
Tel: (+1 713) 743 36 21
Fax: (+1 713) 747 77 24

Space Vector Corp (SVC)
9223 Deering Avenue, Chatsworth, California, 91311, United States
Tel: (+1 818) 734 26 00
Web: www.spacevector.com

Space Warfare Center (SWC)
730 Irwin Avenue Suite 83, Schriever Air Force Base, 80912, United States
Tel: (+1 719) 554 37 31
Fax: (+1 719) 554 83 47

Spacebel SA
Rue des Chasseurs Ardennais, Liege Science Park, B-4031 Angleur, Belgium
Tel: (+32 4) 361 81 11
Fax: (+32 4) 361 81 20
e-mail: info@spacebel.be
Web: www.spacebel.com
■ Study, design and realisation of software systems for scientific, technical and industrial applications in the field of aerospace and geomanagement.

Spacecoast Aeromedical Institute (SAMI)
1698B West Hibiscus Boulevard, Melbourne, Florida, 32901, United States
Tel: (+1 321) 676 32 00
Fax: (+1 321) 674 00 20

Spacecom Satellite Communication Services Ltd
Suite 1017 Twin Towers I, 33 Jabotinsky Street, IL-52511 Ramat Gan, Israel
Tel: (+972 3) 613 47 20
Fax: (+972 3) 613 47 23
e-mail: info@spacecom.co.il
Web: www.spacecom.co.il
■ Satellite communication services.

Spacehab Inc
12130 Highway 3, Building 1, Webster, Texas, 77598-1504, United States
Tel: (+1 713) 558 50 00
Fax: (+1 713) 558 59 60
Web: www.spacehab.com
■ Habitation and logistics services to NASA. Payload processing services for manned and unmanned vehicles.

Spacehab Inc
Payload Processing Facility
620 Magellan Road, Cape Canaveral, Florida, 32920, United States
Tel: (+1 321) 868 74 11

Spacenet Inc
1750 Old Meadow Road, McLean, Virginia, 22102, United States
Tel: (+1 703) 848 10 00
Fax: (+1 703) 848 10 10
e-mail: info@spacenet.com
Web: www.spacenet.com

Spaceport Systems International (SSI)
a CCSI-ITT Partner
3769-C Constellation Road, Lompoc, California, 93436, United States
Tel: (+1 805) 733 73 70
Fax: (+1 805) 733 73 72
Web: www.calspace.com

Spaceweek International Association (SIA)
14523 Sun Harbor Drive, Houston, Texas, 77062, United States
Tel: (+1 281) 461 62 45
e-mail: admin@spaceweek.org
Web: www.spaceweek.org
■ Co-ordinator of World Space Week.

Sparcent-Suparco (Suparco)
Space Science Division
PO Box 8402, 75270 Karachi, Pakistan
Tel: (+92 21) 496 56 38
Fax: (+92 21) 496 56 38
Telex: 25720 space
Cable: SUPARCO

Spearhead Machinery Ltd
Green View, Salford Priors, Evesham, Hereford and Worcestershire, WR11 8SW, United Kingdom
Tel: (+44 1789) 49 18 60
Fax: (+44 1789) 77 86 83
e-mail: info@spearhead.uk.com
enquiries@spearhead.uk.com
Web: www.spearhead.uk.com
■ Design and manufacture of a range of tractor powered mowing equipment in use with civil and military airfields.

Specac Ltd
Graseby Specac Ltd
a subsidiary of Smiths Industries plc, UK
River House, 97 Cray Avenue, St Mary Cray, Orpington, Kent, BR5 4HE, United Kingdom
Tel: (+44 1689) 87 31 34
Fax: (+44 1689) 87 85 27
Web: www.specac.com
■ Infra-red and laser optical components including polarisers and thermal imaging optics, FIR interferometers and laser beam splitters, optical filters, optical assemblies for astronomical telescopes and accessories for FT-IR spectroscopy.
Overseas agents
Specac Inc., US

Speck Systems Pvt Ltd
B-49 Electronics Complex, Kushaiguda, Hyderabad, 500062, India
Tel: (+91 40) 712 53 05
(+91 40) 712 53 06
Fax: (+91 40) 712 24 89
e-mail: internationaldesk@specksystems.com
■ Digital image processing systems and electronic visual interpretation aids.

Spectrolab Inc
a Boeing Company
12500 Gladstone Avenue, Sylmar, California, 91342-5373, United States
Tel: (+1 818) 365 46 11
Fax: (+1 818) 365 76 80
(+1 818) 361 51 02
(+1 818) 365 77 71
Telex: twx 9104961750 spectillm sylm
Web: www.spectrolab.com
■ Helicopter and aircraft searchlights, ground based searchlights, solar cells and optoelectronic products.

Spectrum Astro Inc
1440 North Fiesta Boulevard, Gilbert, Arizona, 85233, United States
Tel: (+1 480) 892 82 00
Fax: (+1 480) 892 29 49
Web: www.specastro.com
■ Active in the development of high performance, lower cost spacecraft for defence, scientific and commercial opportunities.

Sperry Marine Systems Inc
Northrop Grumman Sperry Marine
a subsidiary of Northrop Grumman Corporation, US
1070 Seminole Trail, Charlottesville, Virginia, 22901, United States
Tel: (+1 1434) 974 20 00
Fax: (+1 1434) 974 22 59
Web: www.sperry-marine.com
www.sperrymarine.northropgrumman.com
■ Design, development and manufacture of marine electronic products and systems for defence and commercial markets. Products include: integrated ship systems, ship bridge consoles, inertial navigation systems, gyrocompasses, steering systems, gyrofin stabilisers, marine surface search radars, airborne search radars, vessel traffic systems, dopplar speed logs, GMDSS communication systems, heading repeaters and marine navigation software.
Accreditations: ISO 9001

Spheris
Prospace
a subsidiary of Centre National d'Études Spatiales, France
34 rue des Bourdonnais, F-75001 Paris, France
Tel: (+33 1) 44 88 99 30
Fax: (+33 1) 44 88 99 39
e-mail: i-space@spheris-fr.com
Web: www.i-space.fr
www.spheris-fr.com
■ Provide services of space applications and telecommunications.

Spincraft
Massachusetts Plant
500 Iron Horse Park, North Billerica, Massachusetts, 01862, United States
Tel: (+1 978) 667 27 71
Fax: (+1 978) 667 38 99

Spincraft
2455 Commerce Drive, New Berlin, Wisconsin, 53151, United States
Tel: (+1 414) 784 84 40
Fax: (+1 414) 784 84 63
■ Metal spinning and fabrication for the aerospace, energy and nuclear markets.

Spot Asia Pte Ltd
73 Amoy Street, Singapore, 069892, Singapore
Tel: (+65) 62 27 55 82
Fax: (+65) 62 27 62 31
e-mail: spotasia@pacific.net.sg

Spot Image
an affiliate of CNES and Astrium, France
5 rue des Satellites, PO Box 14359, F-31030 Toulouse, Cedex 4, France
Tel: (+33 5) 62 19 40 55
(+33 5) 62 19 40 40
Fax: (+33 5) 62 19 40 11
(+33 5) 62 19 40 51
e-mail: info@sportimage.fr
Web: www.spotimage.fr
■ Commercial operator of the Spot earth observation satellites. Also offers distribution of worldwide optical and radar satellite data and services.
Representing:
Beijing Spot Image, China
Spot Asia, Singapore
Spot Image Corp, US
Spot Imaging Services, Australia
Tokyo Spot Image, Japan

Spot Image Corp (SICORP)
a subsidiary of Spot Image, France
14595 Avion Parkway, Suite 500, Chantilly, Virginia, 20151, United States
Tel: (+1 703) 715 31 00
Fax: (+1 703) 715 31 20
Web: www.spot.com
■ Marketing and distribution of satellite systems. Remotely sensed data includes high resolution imagery, rapid revist capability, worldwide coverage and digital elevation models.

Spot Imaging Services Pty Ltd
a subsidiary of Spot Image, France
Deakin Chambers, 4/14 Hannah Place, PO Box 9141, Epping, New South Wales, 2600, Australia
Tel: (+61 2) 62 32 51 71
Fax: (+61 2) 62 32 59 92
Web: www.spotimage.com.au
■ Distribution of RS data and products from Spot satellites.

Spur Electron Ltd
Hayward House, Hayward Business Centre, New Lane, Havant, Hampshire, PO9 2NL, United Kingdom
Tel: (+44 23) 92 41 71 40
Fax: (+44 23) 92 47 08 74
e-mail: pevans@spurelectron.com
Web: www.spurelectron.com
■ Procurement of EEE parts, component engineering and laboratory services. Manufacturing to ESA/NASA and IPC standards, cleanroom assembly facility. Software development and technical publishing.

SSE Technologies Inc
47823 Westinghouse Drive, Fremont, California, 94539-7437, United States
Tel: (+1 510) 657 75 52
Fax: (+1 510) 490 85 01
Web: www.sset.com

SSE Technologies Inc
791 Meacham Ave, Elmont, New York, NY 11003, United States
Freephone: (+1 800) 729 24 44
Fax: (+1 516) 872 90 74
e-mail: sse@isit.com
Web: www.ssetechnologies.com

ST Aviation Services Co Pte Ltd (SASCO)
a subsidiary of Singapore Technologies Aerospace Ltd, Singapore
8 Changi North Way, Singapore, 499611, Singapore
Tel: (+65) 65 40 56 13
(+65) 65 45 09 88
(+65) 65 40 57 39
Fax: (+65) 65 45 67 57
(+65) 65 42 07 31
Sita code: sinascr
Web: www.staero.aero
■ Specialise in maintenance, repair and overhaul of a wide range of commercial wide body and narrow body aircraft.

ST Electronics (Satcom & Sensor Systems) Pte Ltd
a subsidiary of Singapore Technologies Electronics Limited, Singapore
ST Electronics Jurong East Building, 100 Jurong East Street 21, Level 4, Singapore, 609602, Singapore
Tel: (+65) 65 67 67 91
Fax: (+65) 65 68 74 02
Web: www.stee.stengg.com
■ Design, development, manufacture and maintenance of microwave and satellite components and subsystems: VSAT ODUs, microwave sensors, radar systems and microwave digital radios.

Stanford Mu Corporation
24003 Frampton Avenue, Harbor City, California, 90710-2101, United States
Tel: (+1 310) 325 63 00
Fax: (+1 310) 325 63 52
Web: www.stanfordmu.com

Starsem
5-7 rue Francois Truffaut, F-91042 Evry, Cedex, France
Tel: (+33 1) 69 87 01 10
Fax: (+33 1) 60 78 31 99
e-mail: communication@starsem.com
Web: www.starsem.com
■ Marketing, sales and management of commercial launch services using Soyuz launchers; supervision of Soyuz commercial launcher production and development of new European and Russian joint space programs.

State Centre Priroda
45 Volgogradski Prospekt, 109125 Moskva, Russian Federation
Tel: (+7 095) 177 65 74
(+7 095) 163 11 41
Fax: (+7 095) 177 65 74
(+7 095) 164 49 07
e-mail: infoc12@telebox.telefi
■ Remote sensing acquisition and application.

State Governmental Scientific-Testing Area of Aircraft Systems (FKP (GkNIPAS)
GkNIPAS FKP, 140250 Belozersky, Voskresensky District, Russian Federation
Tel: (+7 095) 556 07 09
Fax: (+7 095) 556 07 40
Web: www.aha.ru/~leokon
■ Ground testing of aviation, technology products and armament. Research and development, study services in the field of aviation and space.

State Research and Production Space Rocket Center
GNPRKTs TsSKB-Progress
TsSKB-Progress
18 Zemets Street, 443009 Samara, Russian Federation
Tel: (+7 846) 955 13 61
(+7 846) 992 65 18
Fax: (+7 846) 992 65 18
e-mail: mail@progress.samara.ru
■ Design and manufacture of launch vehicles such as Molniya and Soyuz-2. Other products include the Photon and Resurs-DK. Development, manufacture and launch of the LV Soyuz and Soyuz-2, SC for Earth remote sensing and research into relevant biological studies.

State Scientific Research Institute of Graphite
Gosudarstvenny NII konstruktsionnykh materialov na osnove grafita
Ulitsa Elektodnaya 2, 111141 Moskva, Russian Federation
Tel: (+7 095) 176 13 06
(+7 095) 176 12 63
Fax: (+7 095) 176 12 63
Telex: 412280 inkar
e-mail: mail@niigrafit.ru
■ Research and development of graphite-based materials.

Steatite Group Ltd
Steatite Insulations Ltd
a subsidiary of Steatite Group (Holdings) Ltd
Kettles Wood Drive, Woodgate Business Park, Birmingham, B32 3DA, United Kingdom
Tel: (+44 121) 678 68 88
Fax: (+44 121) 683 69 99
e-mail: sales@steatite.co.uk
Web: www.steatite.co.uk
■ EMI/RFI/EMP components, precise time and frequency systems, VME computers, telemetry and telecommunications for military and aerospace sectors. Batteries and battery packs. EMC filters and solutions. Also sales office for Zyfer who manufacture GPS timing and synchronization systems and encryption products. Rugged computing technology.
Representing:
FEI, US
Meinberg GmbH, Germany
Trak Microwave Corporation, US
Zyfer Inc, US
Lau Technologies

Stein Seal Co
1500 Industrial Boulevard, Kulpsville, Pennsylvania, 19443-0316, United States
Tel: (+1 215) 256 02 01
Fax: (+1 215) 256 48 18
e-mail: mailbox@steinseal.com
Web: www.steinseal.com
■ Design, manufacture and test of mechanical seals for jet engines, rockets and space shuttle applications. Main product is jet engine main shaft seals.

Sternwarte Bochum – Planetarium
Castroper Strasse 67, D-44777 Bochum, Germany
Tel: (+49 234) 51 60 60
Fax: (+49 234) 516 06 51
e-mail: planetarium@bochum.de
Web: www.bochum.de
■ Public observatory and planetarium to promote space sciences and astronomy.

Stesalit AG

Fabrikstraβe 54,
CH-4234, Basel, Zullwil, Switzerland
Tel: (+41 61) 795 06 01
Fax: (+41 61) 795 06 04
e-mail: info@stesalit.com

■ Development and manufacture of prepreg systems based on phenol, epoxy, BMI, Cyanatester resins for aerospace, space, rail and ship building industries. Structural components and laminates for aerospace and space applications. Manufacture of carbon epoxy based pultrusion profiles for civil engineering applications. NATO cage code: SH034

Stop-choc Ltd

a subsidiary of The Hutchinson Group
Banbury Avenue, Slough, Berkshire, SL1 4LR, United Kingdom
Tel: (+44 1753) 53 32 23
Fax: (+44 1753) 69 37 24
e-mail: sales@stop-choc.co.uk
Web: www.stopchoc.co.uk

■ Design and manufacture of metal and elastomeric anti-vibration mountings, helicopter blade lead/lag dampers, avionics and military racking and accessories, illuminated instrument cockpit panels with integrated electronics, pushbutton controls and alpha-numeric displays.

Accreditations: ISO 9001

Stork NV

Head Office
PO Box 5004, NL-1410 AA Naarden, Netherlands
7 Amersfoortsestraatweg, NL-1412 KA Naarden, Netherlands
Tel: (+31 35) 695 74 11
Fax: (+31 35) 694 11 84
e-mail: info@storkgroup.com
Web: www.storkgroup.com

■ Supply high-level technological products and services to the industrial community.

Stratospheric Observatory For Infrared Astronomy (SOFIA)

Program Office
7600 Maehr Road, Mail Code 1261, Waco, Texas, 76705, United States
Tel: (+1 254) 867 44 53
Fax: (+1 254) 867 41 37
e-mail: frenette@sofia.waco.usra.edu

Stratospheric Observatory For Infrared Astronomy (SOFIA)

Science Office
NASA Ames Research Center, Building 144, Room 105, Mail Stop 144-2, Moffett Field, California, 94035-1000, United States
Tel: (+1 415) 604 21 09
Fax: (+1 415) 604 19 84
e-mail: becklin@sofia.astro.ucla.edu

Sumitomo Heavy Industries (Europe) Ltd

5th Floor, Bury House, 31 Bury Street, London, EC3A 5AR, United Kingdom
Tel: (+44 20) 76 21 10 00
Fax: (+44 20) 76 21 23 00
Representing:
Sumitomo Heavy Industries Ltd, Japan.

Sumitomo Heavy Industries (USA) Inc

666 Fifth Avenue, 10th Floor, New York, New York, 10103-1099, United States
Tel: (+1 212) 459 24 77
Fax: (+1 212) 459 24 90
Representing:
Sumitomo Heavy Industries Ltd, Japan.

Sumitomo Heavy Industries Ltd (SHI)

Chiba Works
731-1, Naganumahara-machi, Inage-ku, Chiba, 263 0001, Japan
Tel: (+81 43) 420 13 55
Fax: (+81 43) 420 15 80

Sumitomo Heavy Industries Ltd (SHI)

Head Office
Sumitomo Heavy Industries Building, 9-11 Kitashinagawa 5-chome, Shinagawa-ku, Tokyo, 141-8686, Japan
Tel: (+81 3) 54 88 80 00
Fax: (+81 3) 54 88 80 56
Web: www.shi.co.jp

Summa Technology Inc

140 Sparkman Drive, Huntsville, Alabama, 358051916, United States
Web: www.summa.com

■ Design, manufacture and support of commercial and military aircraft, space vehicles and defence systems to a global market.

SUPARCO

SGSC Division
G-8/1 Mauve Area, PO Box 1271, Islamabad, Pakistan
Tel: (+92 5777) 61 17 92
e-mail: sgs@comsats.net.pk
Web: www.suparco.gov.pk

Superform Aluminium

a subsidiary of Luxfer Gas Cylinders Ltd
Cosgrove Close, Worcester, WR3 8UA, United Kingdom
Tel: (+44 1905) 87 43 00
Fax: (+44 1905) 87 43 01
e-mail: sales@superform-aluminium.com
Web: www.superform-aluminium.com

■ Manufacture and distribution of superplastic components in aluminium alloys. Pressings and fabrications both complex and deep drawn.

Surrey Satellite Technology Ltd (SSTL)

Tycho House, 20 Stephenson Road, Surrey Research Park, Guildford, Surrey, GU2 7YE, United Kingdom
Tel: (+44 1483) 80 38 03
Fax: (+44 1483) 80 38 04
e-mail: info@sstl.co.uk
Web: www.sstl.co.uk

■ Design, manufacture and operation of small satellites with a flight heritage of 27 satellites launched ranging from a 6 kg nano-satellite to a 600 kg mini-satellite.

Shareholders:
University of Surrey 85%, Elon Musk (founder of SpaceX) 10% and SSTL staff 5%

Swales Aerospace

5050 Powder Mill Road, Beltsville, Maryland, 20705, United States
Fax: (+1 301) 902 41 14
Web: www.swales.com

Swedish Institute of Space Physics

Institutet för rymdfysik
Box 812, SE-981 28 Kiruna, Sweden
Fax: (+46 980) 790 50
Web: www.irf.se

■ Government research institute whose primary task is to carry out basic research, education and associated observatory activities in space and atmosphere physics. IRF has participated in several sounding rocket and satellite investigations primarily targeted at ionosphere and magnetosphere studies. A programme for atmospheric research using ground-based and balloon experiments was established in 1996. Recently developed space instruments include hot plasma and wave experiments for the Swedish satellites Viking and Freja, spherical Langmuir probe for the ESA/NASA Cassini mission and ion mass spectrometers for the two Russian Interball spacecraft. Particle and other instruments for Swedish Astrid 2, Japanese mission to Mars (Nozomi) and ESA Rosetta, Cluster 2, Mars Express and venus express missions.

Offices in:
Lund
Umea
Uppsala

Swedish National Space Board (SNSB)

Rymdstyrelsen
Box 4006, SE-17104 Solna, Sweden
Albygatan 107, Solna, Sweden
Tel: (+46 8) 627 64 80
Fax: (+46 8) 627 50 14
Web: www.snsb.se

■ State authority for space and remote sensing activities and general space programme administration.

Swedish Society of Aeronautics & Astronautics

Swedish Soc. of Aero & Astro
Flygtekniska Foreningen
c/o Rymdbolaget, PO Box 4207, SE-17154 Solna, Sweden
Tel: (+46 8) 62 76 20
Fax: (+46 8) 98 70 69
Telex: 17128 spaceco s
Web: www.flygtekniskaforeningen.org

■ Conferences, study tours/visits and lectures of aeronautics and astronautics.

Swedish Space Corporation (SSC Solna)

Airborne Systems
PO Box 4207, SE-171 04 Solna, Strandvag 86, Sweden
Solna strandväg 86, Solna, Sweden
Tel: (+46 8) 627 62 00
Fax: (+46 8) 98 70 69
e-mail: info@ssc.se
Web: www.ssc.se

■ Rocket and balloon launches, development of satellites, space vehicle subsystems, payloads for rockets and airborne systems for maritime surveillance.

Subsidiary
SSC Satellitbild, Kiruna

Swedish Space Corporation (SSC)

Esrange
PO Box 802, SE-981 28 Kiruna, Sweden
Tel: (+46 980) 720 00
Fax: (+46 980) 128 90
Telex: 8744 esrange
Web: www.ssc.se/esrange

Swiss Astronautics Association

Association Suisse d'Astronautique
Schweizerische Raumfahrt-Vereinigung
Schweizerische Vereinigung f. Weltraumtechnik (SVWT)
c/o Hesso-EIVD, Route de Cheseaux 1, CH-1400 Yverdon-les-Bains, Switzerland
Tel: (+41 24) 423 23 12
Fax: (+41 24) 423 22 05
e-mail: office@srv-ch.org
Web: www.srv-ch.org

■ The organisation's primary activity is astronautics which involves the promotion of space related activities in Switzerland. Organisation of visits, forums and meetings. Careers advice for those interested in space.

Swiss Center for Electronics and Microtechnology Inc (CSEM)

Centre Suisse d'Electronique et de Microtechnique SA
Jaquet-Droz 1, PO Box 41, CH-2007 Neuchatel, Switzerland
Tel: (+41 32) 720 51 11
Fax: (+41 32) 720 57 20
Web: www.csem.ch

■ Engineering and electronic assemblies. Research and development work and small quantity production of components and systems for industry and government agencies. Main space specialisations are: systems design and engineering, mechanical engineering instrumentation, electronic payloads, data communication, robotics subsystems, and instrumentation work for aerospace industry and for ESA including mechanisms and subsystems for optical calibration of Earth observation instruments, extreme-temperature low-power electronics, microsystems and MEMS, opto-mechanical systems for laser pointing and lubrication of ball bearings.

Swiss Space Office (SSO)

Hallwylstrasse 4, CH-3003 Bern, Switzerland
Tel: (+41 31) 324 10 74
Fax: (+41 31) 324 10 73
(+41 31) 322 78 54

■ The administrative unit charged with planning and implementing Swiss space policy, as defined by the Federal Council. The SSO is the executive body of the Federal Space Affairs Commission (CFAS) and chairs the Interdepartmental Committee for Space Affairs (IKAR). The secretariat of both these groups is located at the SSO.

■ Swiss space policy is part of Swiss science and technology policy, as well as Swiss foreign policy in and outside Europe. The national space policy is determined by the Federal Council (Government) with the advice of a 20-member Federal Space affairs commission (CFAS). The Swiss Space Office (SSO) of the Federal Department of Home Affairs has the main responsibility within the Federal Administration for the implementation of this policy. The Interdepartmental Committee for Space Affairs (IKAR) assures co-ordination of interests and activities in space matters within the seven Departments. IKAR is chaired by the SSO. The Committee on Space Research of the Swiss Academy of Sciences promotes and coordinates space research in Switzerland.

- Space research in Switzerland is carried out at various laboratories of the universities, at Federal Institutes of Technology and by Industry. Research is supported by the Swiss National Science Foundation (fundamental research), by the Commission for Technology and Innovation (applied research) as well as by the federal and cantonal governments.
- The SSO has prime responsibility for Swiss participation in ESA programmes and activities and oversees the budget. The SSO looks after Swiss interests in ESA regarding political, institutional, legal, financial and industrial aspects of scientific, applications and infrastructure programmes. Under the SSO's guidance, the Federal Office for Professional Training and Technology (OFFT), the Federal Finance Administration (AFF) and the Foreign Ministry's Political Division III (DP/DFAE) give their support on technological, financial and political aspects, respectively.
- The SSO leads the Swiss Delegation to ESA and chairs the Delegation's coordination meetings and assures that Swiss space policy is consistent among international organisations for the utilisation of satellite operations, e.g. EUTELSAT, INTELSAT, IMSO and EUMETSAT as well as in the respective bodies of the United Nations and the European Union where Switzerland is represented.
- As there is no national space programme, Swiss space involvement is almost totally routed through ESA.

Syderal SA (SYD)
Alcatel Space Switzerland SA
Compagnie Industrielle Radioélectrique SA (CIR)
a subsidiary of Alcatel Space Industries, France
Neuenburgstrasse 7, CH-2076 Gals, BE, Bern, Switzerland
Tel: (+41 32) 338 98 00
Fax: (+41 32) 338 99 34
(+41 32) 338 99 36
e-mail: info@syderal.ch
- Space and military electronic equipment.

Symmetricom
NavSymm Positioning Systems
3750 Westwind Boulevard, Santa Rosa, California, 95403, United States
Tel: (+1 707) 528 12 30
Fax: (+1 707) 527 66 40
e-mail: ttm_sales@symmetricom.com
Web: www.symmetricom.com
- Military GPS Receivers, 6 and 12 channel versions. XR5 kernel-OEM GPS boardsets. UHF datalink. RTK systems.

Symmetricom Ltd
4 Wellington Business Park, Dukes Ride, Crowthorne, Berkshire, RG45 6LS, United Kingdom
Tel: (+44 1344) 75 34 80
Fax: (+44 1344) 75 34 99
Web: www.symmetricom.com
- Production of GPS receivers for military and survey applications with centimetre accuracy.

Synthecon Incorporated
8054 El Rio, Houston, Texas, 77054, United States
Tel: (+1 713) 741 25 82
(+1 800) 853 07 40
Fax: (+1 713) 741 25 88
Web: www.synthecon.com

Sypris Data Systems
Datatape Inc
Metrum – Datatape Inc
160 East Via Verde, San Dimas, California, 91773-5120, United States
Tel: (+1 909) 962 94 00
(+1 877) 797 74 78
Fax: (+1 909) 962 94 01
Web: www.syprisdatasystems.com
- Manufacture of data storage systems, products and services for various military/aerospace and commercial applications.

Systecon AB
PO Box 5205, SE-102 45 Stockholm, Sweden
Tel: (+46 8) 459 07 50
Fax: (+46 8) 459 07 80
e-mail: systecon@systecon.se
Web: www.systecon.se
- Systems and logistics engineering consultancy in aerospace, military, civil and space. Operations research model development. Computer programs for analysis and optimisation of logistic support systems. Provision of professional training.

T

T S Space Systems
Unit A5, Rose Business Estate, Marlow Bottom, Marlow, Buckinghamshire, SL7 3ND, United Kingdom
Tel: (+44 1628) 47 40 40
Fax: (+44 1628) 56 61 55
Web: www.ts-space.co.uk
- Physics R+D and test work. Manufacture of bespoke test equipment for the space industry.

TA Mfg Co
a subsidiary of Esterline Technologies, USA
28065 West Franklin Parkway, Valencia, California, 91384, United States
Tel: (+1 661) 775 11 00
Fax: (+1 661) 775 11 55
e-mail: sales@tamfg.com
Web: www.tamfg.com
- Specialty elastomeric compounds in custom-moulded shapes; design and production of clamps and support blocks for wiring and tubing installations.

Taber Industries
Teledyne Taber
455 Bryant Street, North Tonawanda, New York, 14120-0164, United States
Tel: (+1 716) 694 40 00
Freephone: (+1 800) 333 53 00
Fax: (+1 716) 694 14 50
Telex: twx 7102621264
e-mail: transducers@taberindustries.com
Web: www.taberindustries.com
- Manufacture of testing equipment, machine tools and stainless steel bonded strain gauge pressure transducers. Users include Insat, Arabsat, tethered satellite, EOS, Landsatt and numerous commercial communications satellites and launch vehicles.

Tactical Missiles Corporation JSC
7 Ilicha Street, Korolev, 141075 Korolev, Moscow Region, Russian Federation
Tel: (+7 495) 542 57 09
(+7 095) 542 57 09
Fax: (+7 495) 511 94 39
e-mail: kmo@ktrv.ru
Web: www.korolev.ru
www.krtv.ru
- Development and production of airborne, shipborne and ground-based tactical missiles. Also manufacture civil products including aerospace, medical and fire security products.

Representing:
Zvezda-Strela State Scientific and Production Centre, Russian Federation
Avtomatika Omsk Plant JSC, Omsk
Azov Optomechanical PLant JSC, Azov, Rostove region
Detal Ural Design Bureau JSC, Kamensk-Uralskiy
Engineering Design Bureau JSC, Moscow
Gorizont JSC, Moscow
Iskra Ural Design Bureau named after Kartukov JSC, Moscow
Krasny Gidropress JSC, Taganrog
Region Scientific and Production Enterprise JSC, Moscow
Salyut JSC, Samara
Soyuz Turaevo Engineering Design Bureau JSC, Lytkarino
Vimpel State Design Bureau named after Toropov JSC, Moscow

Tadiran Spectralink Ltd
Communications and Data Link Systems
a division of Elisra, Israel
29 Hamerkava Street, PO Box 150, IL-58101 Holon, Israel
Tel: (+972 3) 557 31 02
Fax: (+972 3) 557 75 79
e-mail: info@tadspec.com
Web: www.tadiran-spectralink.com
www.tadspec.com
- Supplier of data link systems for a variety of manned and unmanned airborne platforms and ground installations including; UAVs, guided weapons, C4ISR networks, TTR&C systems for space vehicles, search and rescue systems for combat and non combat applications, data link design, software development, systems integration, testing and manufacture of hardware, technical support and global marketing. Core technologies include: digital processing techniques, direct sequence, spread spectrum, frequency hopping, error detection and correction codes, data interleave and deinterleave and burst synchronisation, communication and network management protocols, real-time video compression and expansion, frequency bands from UHF, L, S, C, X up to Ku.

Accreditations: ISO 9001:2000; CMMI Maturity Level 3

Taiyo Nippon Sanso Corp
Headquarters
Toyo Building, 1-3-26 Koyama, Shinagawa-ku, Tokyo, 142-8558, Japan
Web: www.tn-sanso.co.jp
- Manufacture of industrial gases, cryogenic and space equipment.

Tayco Engineering Inc
10874 Hope Street, PO Box 6034, Cypress, California, 90630, United States
Tel: (+1 714) 952 22 40
Fax: (+1 714) 952 20 42
e-mail: sales@taycoeng.com
Web: www.taycoeng.com
- Electrical heaters, temperature sensors, temperature controllers for satellites, infra-red devices, pressure switches and flexible cable (space qualified).

Taylor Devices Inc
90 Taylor Drive, PO Box 748, North Tonawanda, New York, 14120-0748, United States
Tel: (+1 716) 694 08 00
Fax: (+1 716) 695 60 15
e-mail: taylordevi@aol.com
Web: www.shockandvibration.com
www.taylordevices.com
- Liquid springs, shock absorbers, shock isolators, energy absorbers, single and double acting dampers and snubbers, hydraulic components produced on a custom, design to order basis.

TCI
Technology for Communications International
a subsidiary of Dielectric Communications
47300 Kato Road, Fremont, California, 94538, United States
Tel: (+1 510) 687 61 00
Fax: (+1 510) 687 61 01
Web: www.tcibr.com
- Provide solutions for communications and broadcasting applications and design direction finding technology. Produces high-power MF and HF broadcasting antennas as well as HF antennas for military and civilian communications.

Teamcom AS
TSAT A/S
PO Box 333, N-1379 Nesbru, Oslo, Norway
Tel: (+47) 66 77 44 00
Fax: (+47) 66 77 44 01
Web: www.teamcom.no

Technology Development and Aerospace Environments Program (TDAE)
Marshall Space Flight Center, Building 4487, Room A262, Alabama, 35812, United States
Tel: (+1 256) 544 21 04
Fax: (+1 256) 544 88 07
e-mail: gayle.brown@msfc.nasa.gov

Technology Service Corp (TSC)
1900 South Sepulveda Boulevard, Suite 300, Los Angeles, California, 90025-5659, United States
Tel: (+1 310) 954 22 00
Fax: (+1 310) 477 21 96
e-mail: info@tsc.com
Web: www.tsc.com
- Defence related services: radar, sonar, electro optical and electronic warfare.

Techspace Aero SA
a joint venture between SAFRAN (51 per cent), the Walloon region (30 per cent) and Pratt and Whitney (19 per cent).
a subsidiary of SAFRAN, France
121 Route de Liers, B-4041 Herstal (Milmort), Belgium
Tel: (+32 4) 278 81 11
Fax: (+32 4) 278 52 07
e-mail: info@techspace-aero.be
Web: www.techspace-aero.be
- Design, development and production of modules, components and equipment for aircraft engines, lubrication systems, boosters (low pressure compressors), critical engine components including: casings, bearing supports, disks, drums and spool assemblies; turbine blades and vanes; compressors and turbine disks. Design and production of components for space

propulsion systems, control cryogenic valves, regulators and leak control systems. Assembly and testing of complete turbo engines. Full overhaul of F100 military engines. Design and installation of test cells for aerospace propulsion.

Tecnavia SA

Via Cadepiano 28, Barbengo, CH-6917 Lugano, Switzerland
Tel: (+41 91) 993 21 21
Fax: (+41 91) 993 22 23
e-mail: info@tecnavia.com
Web: www.tecnavia.com
■ Specializes in land-based and maritime satellite reception systems for receiving and processing real-time weather data.

Tecstar Inc

15251 Don Julian Road, City of Industry, California, 91745-1002, United States
Tel: (+1 626) 934 65 00
Fax: (+1 626) 336 86 94
e-mail: busmgmt@tecstar.com

Tekes

Finnish Funding Agency for Technology & Innovation
Ministry of Trade and Industry, PO Box 69, FI-00100, Helsinki, Finland
Kyllikinportti 2, FI-00100 Helsinki, Finland
Tel: (+358 10) 605 58 56
Fax: (+358 9) 521 59 01
Web: www.tekes.fi/space
■ Governmental agency responsible for planning, promotion, financing and co-ordination of Finnish space activities.

Tekhnotkan Flux Factory

ZAO Tekhnotkan
Molitovskaya Linen Weaving Factory
Per. Motalni, 6, 603140 Nizhnyi Novgorod, Russian Federation
Tel: (+7 8312) 42 52 68
(+7 8312) 42 01 81
Fax: (+7 8312) 42 61 62
e-mail: com@ttka.nnov.ru
■ Produce linen for making of protective clothes and tarpaulins delivered to farming, steel and defence industries and to gas producing companies. Offers services of an accredited testing centre for certification purposes.

Telecommunications Company of Iran (TCI)

Dr Shariati Avenue, PO Box 15875-4415, Tehran, 15598, Iran
Tel: (+98 21) 860 10 41
Fax: (+98 21) 85 85 66
Telex: 210400
213425

Telediffusion de France

Cen Emetteur, Deleg. Lorraine Nord, F-57935 Luttange, France
Tel: (+33 03) 82 82 54 54
Fax: (+33 03) 82 82 54 55

Teledyne Brown Engineering (TBE)

Albuquerque Office
557 Windsong Lane, Corrales, New Mexico, 87048, United States
Tel: (+1 505) 792 44 18
Fax: (+1 505) 792 44 27

Teledyne Brown Engineering (TBE)

Colorado Springs Office
1330 Inverness Drive, Suite 350, Colorado Springs, Colorado, 80910, United States
Tel: (+1 719) 574 72 70
Fax: (+1 719) 574 63 29

Teledyne Brown Engineering (TBE)

Dayton Office
4027 Colonel Glenn Highway, Suite 101, Beavercreek, Ohio, 45431-1670, United States
Tel: (+1 937) 427 95 54
Fax: (+1 937) 427 12 11

Teledyne Brown Engineering (TBE)

Edgewood Project Office
1310 Woodbridge Station Way, Suite 101, Edgewood, Maryland, 21040-3830, United States
Tel: (+1 410) 676 61 95
Fax: (+1 410) 676 62 12

Teledyne Brown Engineering (TBE)

Gulf Office
51 Third Street, Shalimar, Florida, 32579, United States
Tel: (+1 850) 651 34 43
Fax: (+1 850) 651 46 48

Teledyne Brown Engineering (TBE)

Houston Aerospace Office
13100 Space Center Boulevard, Mail Stop HB3-40, Houston, Texas, 77059, United States
Tel: (+1 281) 226 42 48
Fax: (+1 281) 226 61 33

Teledyne Brown Engineering (TBE)

Knoxville Office
2508 Quality Lane, Knoxville, Tennessee, 37931-3133, United States
Tel: (+1 865) 690 68 19
Fax: (+1 201) 690 61 87

Teledyne Brown Engineering (TBE)

Los Angeles Field Office
21221 Western Avenue, Suite 220, Room 235, Torrance, California, 90501, United States
Tel: (+1 310) 320 84 22
Fax: (+1 310) 320 84 20

Teledyne Brown Engineering (TBE)

Oklahoma Office
2701 Liberty Parkway, Suite D, Midwest City, Oklahoma, 73110, United States
Tel: (+1 405) 739 09 52
Fax: (+1 405) 739 09 47

Teledyne Brown Engineering (TBE)

San Diego Office
5045 Caminito Exquisito, San Diego, California, 92130-2851, United States
Tel: (+1 858) 509 36 80
Fax: (+1 858) 259 69 26

Teledyne Brown Engineering (TBE)

Warner Robins Office
PO Box 98568, Robins AFB, Georgia, 21098, United States
Tel: (+1 478) 929 23 91
Fax: (+1 478) 922 38 66

Teledyne Brown Engineering (TBE)

Washington Operations
2111 Wilson Boulevard, Suite 900, Arlington, Virginia, 22201-3052, United States
Tel: (+1 703) 276 46 00
Fax: (+1 703) 276 40 63

Teledyne Brown Engineering (TBE)

a Teledyne Technologies company
PO Box 070007, Huntsville, Alabama, 35807-7007, United States
300 Sparkman Drive North West, Cummings Research Park, Huntsville, Alabama, 35807-7007, United States
Tel: (+1 256) 726 55 55
Freephone: (+1 800) 933 20 91
Fax: (+1 256) 726 55 56
e-mail: info@tbe.com
Web: www.tbe.com
■ Manufacture, engineer and integrate systems concerning aerospace, defence, environmental and energy programmes.
Advanced Programs:
Business Units
Systems Group
Technologies Group
Teledyne Solutions Inc

Telegenix Inc

Telegenix/Grim Corp
a subsidiary of Inductotherm Industries Inc, Australia
1930 Olney Avenue, Cherry Hill, New Jersey, 08003, United States
Tel: (+1 856) 424 52 20
(+1 800) 424 52 20
Fax: (+1 856) 424 08 89
Telex: 6851048 induct
e-mail: sales@telegenix.com
Web: www.telegenix.com
■ Manufacture of communications and control electronics systems and electronic graphic displays for use in air traffic, vessel and command and control applications. Programmable communications voice switches for internal and external communications, signalling interface to radio, telephone and intercom channels. Communications consoles for data, voice or telephone communications between central operator and remote operators. Intelligent voice and data acquisition control and transmission systems used for remote radio control using voice and data on a single telephone line. Contract manufacturing.

Telesat Canada

1601 Telesat Court, Gloucester, Ontario, K1B 5P4, Canada
Tel: (+1 613) 748 01 23
Fax: (+1 613) 748 87 12
e-mail: info@telesat.ca
Web: www.telesat.ca
■ Provide engineering consulting service to governments and private companies interested in establishing, operating or upgrading satellite systems and associated ground infrastructures. Also provide design, implementation, maintenance operation and marketing of satellite telecommunications services for North and South America.

Telespazio SpA

Headquarters
Nuova Telespazio SpA
joint venture between Finmeccanica (67 per cent) and Thales (33 per cent).
Via Tiburtina, 965, I-00156 Roma, Italy
Tel: (+39 6) 407 91
Fax: (+39 6) 40 99 99 06
(+39 06) 40 79 37 60
e-mail: sales@telespazio.com
marketing@telespazio.com
communication@telespazio.com
Web: www.telespazio.it
■ Provide satellite operational services including: management of satellites, earth observation services, satellite navigation, broadband multimedia telecommunications and defence. Programs include: Galileo, EGNOS, GMES and COSMO-SkyMed.

TELKOM

GKP PT Telekomikasi Indonesia, 4th & 7th Floor, Jalan Japati No 1, 40133 Bandung, Indonesia
Web: www.telkom.co.id

Telkom Indonesia

PT Industri Telekomunikasi Indonesia
PT Telekom
Industri Telekomunikasi Indonesia PT
Jalan Japati 1, Bandung, Indonesia
Tel: (+62 22) 250 00 00
(+62 22) 250 00 00
■ A full information and communications service and network provider of wireless, fixed wireless, cellular and data and Internet and network communication services.

Telonic Instruments Ltd

Toutley Industrial Estate, Toutley Road, Wokingham, Berkshire, RG41 1QN, United Kingdom
Tel: (+44 118) 978 69 11
Fax: (+44 118) 979 23 38
e-mail: info@telonic.co.uk
Web: www.telonic.co.uk
■ Sale of oscilloscopes, recorders, oscillators, power supplies, RF filters, attenuators, amplifiers and quartz crystal microbalances for satellite applications.
Representing:
American Reliance Inc, US
EPS (Takosago), Japan
Heinz-Gunter Lau GmbH, Germany
Kikusui Electronics Corp, Japan
QCM Research Inc, US
Sorensen Inc, US
Telonic Berkeley Inc, US

Telonics Inc

932 E Impala Avenue, Mesa, Arizona, 85204-6699, United States
Tel: (+1 480) 892 44 44
Fax: (+1 480) 892 91 39
e-mail: info@telonics.com

Telstra Corp Ltd

Melbourne Office
Level 41, 242 Exhibition Street, Melbourne, 3000, Australia
Tel: (+61 3) 96 34 64 00

Terma AS

Aerospace and Defence
Terma Elektronik AS
Mårkaervej 2, Tastrup, DK-2630 Denmark
Tel: (+45 87) 43 60 00
(+45 43) 52 15 13
Fax: (+45 43) 52 57 58
(+45 87) 43 60 01
e-mail: terma.dk@terma.com
Web: www.terma.com

■ Development of naval C3 systems that integrate, operate and control the entire weapon and sensor inventory on naval ships.

Terma AS
Aerospace and Defense
Per Udsen Co Aircraft Industry AS
Terma Industries Grenaa AS
Fabrikvej 1, DK-8500 Grenaa, Denmark
Tel: (+45 8743) 60 00
Fax: (+45 8743) 60 01
e-mail: ihr@terma.com
Web: www.terma.com
■ Design, manufacture and assembly of aircraft parts (airframe). Graphite composite parts.

Terma AS
Headquarters
Hovmarken 4, DK-8520 Lystrup, Denmark
Tel: (+45 87) 43 60 00
Fax: (+45 87) 43 60 01
e-mail: terma.dk@terma.com
acj@terma.com
Web: www.terma.com
■ Radar and display systems for communication, navigation, surveillance and velocity measurements. Satellite equipment, antennas and telecommunication systems. Countermeasure dispenser systems.
Subsidiaries:
Terma GmbH
Terma Ejendomme Skive AS
Terma BV
Terma North America Inc

Terma B V
Terma Elektronik Netherlands BV
a subsidiary of Terma A/S, Denmark
Schuttersveld 9, NL-2316 XG Leiden, Netherlands
Tel: (+31 71) 524 08 00
Fax: (+31 71) 514 32 77
e-mail: terma.nl@terma.com
Web: www.terma.com

Terma GmbH
Terma Elektronik (Deutschland) GmbH
a subsidiary of Terma A/S, Denmark
Europahaus, Europaplatz 5, D-64293 Darmstadt, Germany
Tel: (+49 6151) 86 00 50
Fax: (+49 6151) 860 05 99
e-mail: terma.de@terma.com
Web: www.terma.com

Terra-Mar Resource Information Services Inc
Terra-Mar Resource Inf. Serv. Inc
918 Oak Street, Hood River, Oregon, 95246, United States
Tel: (+1 541) 387 24 85
Fax: (+1 541) 387 24 85
e-mail: info@terra-mar.net
Web: www.terra-mar.net
■ Orthophoto maps. A selection of airborne, real-time mapping systems geared toward tactical emergency response and resource monitoring.
Representing:
Spot Image Corp, US

TerreStar Networks
One Discovery Square, 12010 Sunset Hills Road, Suite 600, Reston, Virginia, 20190, United States
Tel: (+1 877) TSTAR 01
Fax: (+1 877) 476 71 43
Web: www.terrestar.com
■ Provide integrated mobile astellite and terrestrial communications systems for government applications.

Tether Applications
1813 Gotham Street, Chula Vista, California, 91913, United States
Web: www.tetherapplications.com/

Tethers Unlimited Inc (TUI)
11711 North Creek Parkway South, Suite D-113, Bothell, Washington, 98011-8804, United States
Tel: (+1 425) 486 01 00
Fax: (+1 425) 482 96 70
e-mail: information@tethers.com
Web: www.tethers.com/

Texas Composite Inc
1281 North Main Street, Boerne, Texas, 78006, United States
Tel: (+1 830) 249 33 99
Fax: (+1 830) 249 32 75
e-mail: mail@texascomposite.com
Web: www.texascomposite.com

■ Manufacture engineered composite structures for military and commercial aerospace. Turbine engine parts and assemblies, Bypass Vanes or Fan OGV's, fan ducts and assemblies. Wing and fuselage fairings, control surfaces, winglets, radomes and gear doors. Space-rated structures for human space flight and satellite applications. Missile and launch vehicle structures.

Textron Systems Corporation (TS)
Wilmington
a wholly owned subsidiary of the Avco Corporation, which is a wholly owned subsidiary of Textron Inc
201 Lowell Street, Wilmington, Massachusetts, 01887, United States
Tel: (+1 978) 657 11 83
Fax: (+1 978) 657 66 44
Web: www.systems.textron.com
■ Provision of technological solutions for the aerospace and defence industries. Support military precision engagement and dominant manoeuvre with strike weapons, mobility and surveillance systems. Weapons, surveillance systems, speciality marine craft, Cadillac Gage armoured vehicles and turret control and stabilisation systems.

TGV Rockets
Headquarters
2420 Springer Drive, Suite 100, Norman, Oklahoma, 73069, United States
Tel: (+1 405) 366 07 79
Web: www.tgv-rockets.com

Thales
Corporate Headquarters
Thomson-CSF
45 rue de Villiers, F-92526 Neuilly sur Seine, Cedex, France
Tel: (+33 1) 57 77 80 00
(+33 1) 57 77 83 00
e-mail: press@thalesgroup.com
info@thalesgroup.com
dtri@thalesgroup.com
Web: www.thalesgroup.com
■ Design, development and implementation of high technology electronic solutions for global communication and information requirements. Major business areas include airborne systems (surveillance and combat systems, missile electronics); avionic systems (military and commercial aircraft, aircraft components, power generation); air security and missile systems (missile systems, air command and control systems, air traffic management, armament and propulsion); naval systems (surface and underwater systems); communications systems, military communications, professional applications); optronics; information systems and services (simulation, training and synthetic training and environment, TV and radio broadcasting, information processing systems, computer services engineering and security); and tubes and components (electron tubes, microelectronics).
■ Undertakes design, development, manufacture and the provision of in-service support.

Thales Aerospace Division (TAV)
Head Office
Thomson-CSF Sextant
45 rue de Villiers, BP 49, F-92 526 Neuilly sur Seine, Cedex, France
Tel: (+33 1) 57 77 80 00
e-mail: contact.info@fr.thalesgroup.com
■ Manufacture of avionics and associated components. Communication surveillance systems and inflight entertainment systems.

Thales Air Systems
Headquarters
Thales Systemes Aeroportes
Thales Airborne Systems
Thomson-CSF Detexis
a subsidiary of Thales, France
Aerospace Division, 45 Rue de Villiers, F-92526 Neuilly sur Seine, France
Tel: (+33 1) 34 81 60 00
(+33 1) 57 77 88 10
Fax: (+33 1) 30 66 79 66
(+33 1) 57 77 80 24
e-mail: customer.service-tasf@fr.thalesgroup.com
info@fr.thalesgroup.com
Web: www.thales-airborne.com
www.thalesgroup.com
■ Electronic warfare, radars, missile electronics, avionics and information technology systems.

Thales Alenia Space ETCA
Head Office
Alcatel Alenia Space ETCA
Alcatel Etca SA
a subsidiary of Alcatel Telecom
101 rue Chapelle Beaussart, B-6032 Mont sur Marchienne, Belgium
Tel: (+06 41) 51 25 74
Fax: (+06 41) 51 42 52
e-mail: info@etca.alcatel.be
Web: www.thalesaleniaspace.com
■ Electronic assemblies and subassemblies, HF modems equipment and terminals, tactical message terminals, power conditioning and distribution systems, hybrid circuits (signal and power) and ASIC's.

Thales Angénieux SA
Angénieux SA
a subsidiary of Thales, France
Boulevard Ravel de Malval, F-42570 Saint Héand, France
Tel: (+33 4) 77 90 78 00
Fax: (+33 4) 77 30 78 03
e-mail: angenieux@fr.thalesgroup.com
Web: www.angenieux.com
■ Design and manufacture of electro-optics products.

Thales ATM Navigation GmbH
Thales Air Traffic Management GmbH
Alcatel Air Navigation Systems GmbH
Lilienthalstrasse 2, D-70825 Korntal-Münchingen, Germany
Tel: (+49 711) 82 14 50 23
Fax: (+49 711) 82 13 44 05
Web: www.thalesgroup.com
■ Conventional navigation and ATC systems: VOR/DVOR/ILS Cat.I-III/DME;TACAN/ELTA/NDB; RMMC (remote maintenance and monitory concept); Mobile VOR/NDB/TACAN; Mobile Control Towers (2–4 operators). Factory and on-the-job training. New navigation and ATS systems: MLS Cat. I-III (azimuth-elevation-DME/D); DGPS (ground reference and monitor station); STREAMS (Surface Traffic Enhancement and Automation System).

Thales Avionics Electrical Motors SA (TAEM)
5 rue du Clos d'en Haut, PO Box 115, F-78702 Conflans-Ste.Honorine, France
Tel: (+33 1) 39 19 60 60
Fax: (+33 1) 39 19 96 31
■ Special motors for aerospace applications, space and defence (AC/DC BLDC and PM-VR). Stepper motors, blowers and fans.

Thales Communications Belgium SA (TCB)
Thomson-CSF Sysbel SA
a subsidiary of Thales, France
Rue des Frères Taymans 28, B-1480 Tubize, Belgium
Tel: (+32 2) 391 22 11
Fax: (+32 2) 391 23 00
e-mail: contact@be.thalesgroup.com
Web: www.thalesgroup.com
■ Development and manufacture of defence and aerospace C3 systems and navigation aids.

Thales Communications SA
Gennevilliers Industrial Centre
66 rue du Fossé-Blanc, BP 156, F-92231 Gennevilliers, Cedex, France
Tel: (+33 1) 46 13 20 00
Fax: (+33 1) 46 13 21 63
■ Mobile radiocommunications, radio surveillance and COMINT systems, information technology security, space communications systems.

Thales Communications UK
Headquarters
Thales Communications Limited
Thales Land & Joint Systems
a subsidiary of Thales, France
Newton Road, Crawley, West Sussex, RH10 9TS, United Kingdom
Tel: (+44 1293) 51 88 55
Fax: (+44 1293) 54 00 45
e-mail: info@rmel.com
Web: www.thalesgroup.com
www.thales-communications.ltd.uk
■ Thales Land and Joint Systems Division brings together the communications interests of the former Racal, Redifon and Thales Communications companies.

- The company provides ATC communications systems ranging from stand-alone single console radio installation, through to national coverage VHF systems or multisite MWARA and RDARA systems. The company carries out all activites from site surveys, system design, factory testing, installation, commissioning, and site acceptance testing.
- The company offer the following ATC related communication systems; TXM5500 voice communications switch, series 5000 HF transmitters, receivers and transceivers, emergency voice communications switch, series 900 and 9000 VHF transmitters, receivers and transceivers, landing clearance indicator system, missed approach alarm system, back up radio telephony, voice commuications simulator.

Thales Electron Devices SA
Thomson Tubes Electroniques
2 bis rue Latecoere, F-78941 Vélizy, France
Tel: (+33 1) 30 70 35 00
Fax: (+33 1) 30 70 35 35
Telex: 634280 thomtub
e-mail: info@thales-electrondevices.com
Web: www.thalesgroup.com/electrondevices
- Systems, equipment and services for aerospace.

Thales Italia SpA
Communications Division
Alcatel Teletta
Thales Communications SpA
Thomson-CSF Electronic Systems Italia SpA
a subsidiary of Thales, France
Via Vincenzo Bellini 24, I-00198 Roma, Italy
Tel: (+39 06) 85 37 22 32
Fax: (+39 06) 85 37 22 50
- Development and manufacture of tactical communication systems and equipment.

Thales Land & Joint Systems
Rhyl
Pilkington PE Ltd
Thales Optics Ltd
Unit 2, Kinmel Park Industrial Estate, Bodelwyddan, Clwyd, LL18 5TY, United Kingdom
Tel: (+44 1745) 58 98 00
Fax: (+44 1745) 58 42 08
- Solar cell coverglasses, optical solar reflectors for satellites.

Thales Land & Joint Systems
Thales Land and Joint Systems
Thales Communications
Thales, Communications Systems, Business Group
160 Boulevard de Valmy, F-92704 Colombes, Cedex, France
Tel: (+33 1) 41 30 30 00
Fax: (+33 1) 41 30 33 57
(+33 1) 41 30 41 86
e-mail: supportline@fr.thalesgroup.com
Web: www.thalesgroup.com/l-j
- Prime contractor specialising in radio communications equipment and networks, radio navigation, identification, electronic warfare, command systems, security, microwave communications and optronics equipment.

Companies in the Land & Joint Systems Division:
–Land & Joint Systems
–Broadcast Solutions

Thales Technology & Research France
Thales Laboratoire Central de Recherches
Thomson-CSF Laboratoire Central de Recherches
Domaine de Corbeville, F-91404 Orsay, Cedex, France
Tel: (+33 1) 60 33 00 00
Fax: (+33 1) 60 33 08 78
- Advanced research into new technologies, material science, optics, liquid crystal displays and rapid semiconductor components.

The Aerospace Consortium Inc
730 Gana Court, Mississauga, Ontario, L5S 1P1, Canada
Tel: (+1 905) 564 66 01
Fax: (+1 909) 670 16 95
- Aerospace and defence components, subsystems and systems: machined components and assemblies, military communications and electronics equipment, whip antennas, composite parts, repair and overhaul, programme management, meteorological equipment.

The Aerospace Corp
Albuquerque Office
PO Box 80360, Albuquerque, New Mexico, 87198, United States
Tel: (+1 505) 872 62 00
Fax: (+1 505) 872 62 13

The Aerospace Corp
Aurora Office
c/o 2SWS (Stop 69), East Crested Butte Avenue, Buckley Air Force Base, Aurora, Colorado, 80011-9518, United States

The Aerospace Corp
Chantilly Office
15049 Conference Center Drive, Suite 600, Chantilly, Virginia, 20151-3824, United States
Tel: (+1 703) 633 50 00
Fax: (+1 703) 633 50 13

The Aerospace Corp
Colorado Springs Office
1150 Academy Park Loop, Suite 136, Colorado Springs, Colorado, 80910, United States
Tel: (+1 719) 573 24 40
Fax: (+1 719) 573 24 97

The Aerospace Corp
Columbia Office
8840 Stanford Boulevard, Suite 4400, Columbia, Maryland, 21045-5852, United States
Tel: (+1 410) 312 14 00
Fax: (+1 410) 312 29 15

The Aerospace Corp
Corporate Headquarters
PO Box 92957, Los Angeles, California, 90009-2957, United States
2350 East El Segundo Boulevard, El Segundo, California, 90245-4691, United States
Tel: (+1 310) 336 50 00
Fax: (+1 310) 336 70 55
e-mail: corpcomm@aero.org
Web: www.aero.org
- An independent, non-profit corporation supporting the US government in planning, research, development, acquisition and operation of space launch and associated ground systems and in science and technology important to national security. Provide engineering support to Air Force space programmes and national security space systems. Co-operate with NASA and other civil agencies in order to apply space-related skills to critical issues.

The Aerospace Corp
Eastern Range Directorate-Cape Canaveral
Kennedy Space Center, PO Box 21205, Cape Canaveral, Florida, 32815-0205, United States
Tel: (+1 321) 853 66 66
Fax: (+1 321) 853 36 90

The Aerospace Corp
Falls Church-Skyline
5205 Leesburg Pike, Suite 907, Falls Church, Virginia, 22041-3234, United States
Tel: (+1 703) 812 21 44
Fax: (+1 703) 575 79 76

The Aerospace Corp
Houston Office
2525 Bay Area Boulevard, Ste 600, Houston, Texas, 77058, United States
Tel: (+1 281) 283 64 00
Fax: (+1 281) 280 81 88

The Aerospace Corp
Houston-Johnson Space Center Office
2101 NASA Road One, Johnson Space Center, M/C NQ112, Houston, Texas, 77058-3696, United States
Tel: (+1 281) 280 80 14
Fax: (+1 281) 280 81 88

The Aerospace Corp
Omaha Office
Defense Meteorological Satellite Program, PO Box 819, Bellevue, Nebraska, 68005-0819, United States
Tel: (+1 402) 292 50 80
Fax: (+1 402) 291 31 37

The Aerospace Corp
Peterson Air Force Base
HQ AFSPC/DR-(A), 150 Vandenberg Street, Suite 1105, Peterson, AFB, Colorado, 80914-4680, United States
Tel: (+1 719) 554 58 12
Fax: (+1 719) 554 58 76

The Aerospace Corp
Program Executive Office for Space Support
1500 Wilson Boulevard, Suite 515, Arlington, Virginia, 22209-2404, United States
Tel: (+1 703) 526 17 40
Fax: (+1 703) 526 17 56

The Aerospace Corp
Schriever Air Force Base
442 Discoverer Street, Suite 71, Schriever, AFB, Colorado, 80912-4471, United States
Tel: (+1 719) 567 25 64
Fax: (+1 719) 567 35 78

The Aerospace Corp
Suitland Office
FB4, Room 3316, 4700 Silver Hill Road, Suitland, Maryland, 20746, United States
Tel: (+1 301) 457 51 54
Fax: (+1 301) 457 57 13

The Aerospace Corp
Sunnyvale-Onizuka Air Force Station
1080 Lockheed Martin Way, Box 018, Sunnyvale, California, 94089-1232, United States
Tel: (+1 408) 744 64 84
Fax: (+1 408) 744 66 68

The Aerospace Corp
Washington D.C. Office
1000 Wilson Boulevard, Suite 2600, Arlington, Virginia, 22209-3988, United States
Tel: (+1 703) 812 06 00
Fax: (+1 703) 812 94 15

The Aerospace Corp
Western Range Directorate Vandenberg AFB
PO Box 5068, Vandenberg, AFB, California, 93437-0068, United States
Tel: (+1 805) 606 77 22
(+1 805) 606 20 03

The Association of Specialist Technical Organisations for Space (ASTOS)
c/o Nohmia Limited, 79 Larksway, Bishops Stortford, Hertfordshire, CM23 4DG, United Kingdom
Tel: (+44 870) 765 65 44
e-mail: enquiries@astos.org.uk
Web: www.astos.org.uk
- Association representing SMEs (small/medium enterprises) specialist skills and services to the space industry.

Members:
Actheric Engineering
AEA Technology
Aegis Systems
Analyticon Ltd
ARC
ComDev Europe
Exemplar Associates
Grafton Technology
IGG Component Technology
ISP International Space Prolulsion
JRA Aerospace & Technology
L-3 Storm Control Systems
Moltek Consultants Ltd
Moreton Hall Associates
Mullard space Science Labarotary
Nohmia
Polyflex
RDSL
RPC Telecommunications Ltd
Starchaser
Spur Electron Ltd
Surrey Satellite Technology Ltd
Tri Polus

The Bercha Group
Head Office
PO Box 61105, Calgary, Kensington PO, T2N 4S6, Canada
Tel: (+1 403) 270 22 21
Fax: (+1 403) 270 20 14
e-mail: bgroup@berchagroup.com

The Boeing Company
Aerospace and Defence Headquarters
Boeing Company, The 2201 Seal Beach Boulevard, PO Box 4250, Seal Beach, California, 90740, United States
Tel: (+1 310) 797 33 11
Fax: (+1 310) 797 58 28
Telex: 664465 rockwell rkwl
- Military aircraft avionics, military and civil navigation and communication products and systems for air, land and sea applications. Design, develop and manufacture information message handling and communication systems and products that support C3 for land, sea and air applications. Battlefield weapons, standoff weapons. The division also provides ICBM ground support equipment, nuclear radiation shielding, glass fibre reinforced plastic, honeycomb core structures and X-ray shielding.

The Boeing Company
Communications Department
Boeing Company, The
600 Gemini Avenue, Houston, Texas, 77058, United States
Tel: (+1 281) 336 51 14
Fax: (+1 281) 336 43 37

The Boeing Company
Corporate Offices
Boeing
Boeing Company, The
100 North Riverside Plaza, Chicago, Illinois, 60606, United States
Tel: (+1 312) 544 20 00
Web: www.boeing.com
■ Development, production and marketing of commercial and military aircraft; development and production of commercial and defence missiles and space exploration equipment; development and production of defence electronic systems and computer services and large-scale information networks.
Products:
Commercial jet aircraft
Defence and space systems
Defence electronics
Guided missiles
Military aircraft
Rotorcraft
Space vehicles

The Boeing Company
Integrated Defense Systems Headquarters
Boeing Integrated Defense Systems
Boeing Military Aircraft & Missile Systems
McDonnell Douglas Corp
Boeing Company, The
Boeing IDS
PO Box 516, St Louis, Missouri, 63166, United States
Tel: (+1 314) 232 02 32
(+1 253) 773 28 16
Fax: (+1 253) 773 39 00
(+1 314) 234 82 96
Web: www.boeing.com/ids/
■ Provision of defence, intelligence, communications and space capabilities combining sophisticated communications networks with air, land, sea and space-based platforms.
■ Ground-based Midcourse Defense, International Space Station, Joint Tactical Radio Systems and Military Aerospace Support.

The Boeing Company
Integrated Defense Systems Missile Defense Sys, AL
Boeing Defense and Space Group, Missiles and Space Divsion
Boeing Company, The
a part of Boeing Integrated Defense Systems, US
PO Box 240002, Huntsville, Alabama, 35824-6402, United States
Tel: (+1 205) 461 28 05
(+1 205) 461 21 21
Fax: (+1 205) 561 22 52
Web: www.boeing.com
■ Responsible for Boeing's partcipation in a variety of US DoD and NASA programmes, including the Avenger air defence system. International Space Station, airborne laser, PAC-3 and ground based midcourse defence.

The Boeing Company
Public Relations Office
Boeing Company, The
Mail Stop 80-RF, PO Box 3999, Seattle, Washington, 98124-2499, United States
Tel: (+1 253) 773 59 15
Fax: (+1 253) 773 19 40

The Boeing Company
Space and Communications
Boeing Company, The
PO Box 2515, Seal Beach, California, 90740-1515, United States
Tel: (+1 562) 797 56 30
Fax: (+1 562) 797 50 13
■ Space Systems is responsible for Boeing's prime and subcontracting roles on a variety of DoD and NASA space programs, including the International Space Station, the Space Shuttle, Sea Launch, Inertial Upper Stage, Global Positioning System and Delta II, Delta III and Delta IV rockets. Information and Communication Systems is responsible for multiple product areas including: information and surveillance systems such as the E-3 and B767 AWACS programs; communication and information management systems; commercial communication systems which includes Boeing's involvement in the Teledesic program; advanced integrated defense systems; satellite systems which involve the Global Positioning System; Boeing Australia Limited; airborne laser program; Tier III Minus Program; electronic products which involve commercial avionics systems, cabin management systems, phased-array antenna development and military electronics; and special projects and advanced system concepts.

The Charles Stark Draper Laboratory Inc (CSDL)
Headquarters
Draper Laboratory
Charles Stark Draper Laboratory Inc, The
555 Technology Square, Cambridge, Massachusetts, 02139-3563, United States
Tel: (+1 617) 258 10 00
Fax: (+1 617) 258 11 31
(+1 617) 258 21 21
(+1 617) 258 13 33
e-mail: busdev@draper.com
communications@draper.com
Web: www.draper.com
■ Research and development services in strategic systems, space systems, tactical systems, special operations and biomedical engineering including air, land, sea and space systems. Information integration, distributed sensors and networks, precision guided munitions, air traffic flow management, military logistics, biomedical engineering and chemical and biological defence.

The Evry Space Centre
Centre National d'Etudes Spatiales
Rond Point de l'Espace, F-91003 Evry, Cedex, France
■ Manufacture and marketing of launchers.

The George Washington University
Space Policy Institute
Elliott School of International Affairs, 1957 East Street Northwest, Washington, District of Columbia, 20052, United States
Tel: (+1 202) 994 72 92
Fax: (+1 202) 994 16 39
Web: www.gwu.edu/~spi

The Institute of Information Science and Technologies (ISTI)
CNUCE
Istituto di Scienza e Tecnologies
Consiglio Nazionale delle Ricerche, Area della Ricarca di Pisa, Via G. Moruzzi, 1, I-56124 Pisa, Italy
Tel: (+39 050) 315 24 03
Fax: (+39 050) 315 28 10
Web: www.isti.cnr.it

The Mitre Corporation
Washington Headquarters
7515 Colshire Drive, McLean, Virginia, 22102-7508, United States
Tel: (+1 703) 883 60 00
Fax: (+1 703) 883 63 08

The Polish Academy of Sciences
Nicolaus Coepernicus Astronomical Centre
Department of Astrophysics, Rabianska 8, PL-87-100 Torun, Poland
Tel: (+48 56) 621 93 19
Fax: (+48 56) 621 93 81
e-mail: basia@mcac.torun.pl

The Recorder Co
Recorder Co PO Box 8, San Marcos, Texas, 78667, United States
Tel: (+1 830) 629 14 00
Fax: (+1 572) 353 53 33
Web: www.recordercompany.com
■ CAD/CAM systems and equipment for design engineers. Strip chart recorders.

The Society of Japanese Aerospace Companies Inc (SJAC)
Society of Japanese Aerospace Companies
2nd Floor, Toshin-Tameike Building, 1-14 Akasaka 1-chome, Minato-ku, Tokyo, 107-0052, Japan
Tel: (+81 3) 35 85 05 11
Fax: (+81 3) 35 85 05 41
Web: www.sjac.or.jp
■ Manufacture of aircraft, engines, avionics, space vehicles and related equipment and products.
Associated Members:
Advanced Space Business Corporation
Aleris Aluminum Japan Ltd
ASAH Air Supply Inc
BAE Systems (International) Limited
BARCO Co Ltd
Chip One Stop Inc
Chuden CTI Co Ltd
CSP Japan Inc
Earth Remote Sensing Data Analysis Center
Explorer Consulting Japan Inc
Fuji Industries Co Ltd
Galaxy Express Corporation
High-Reliability Engineering & Components Corporation (HREC)
International Aircraft Development Fund
International Task Force Ltd
Ishida Jyuscho Inc
Itochu Aviation Co Ltd
Itochu Corporation
Japan Aerospace Corporation
Jupitor Corporation
Kanematsu Corporation
Kyokuto Boeki Kaisha Ltd
Maruben Aerospace Corporation
Maruben Corporation
Mikuni Shoko Corp
Mitsubishi Corporation
Mitsubishi Research Institute Inc
Mitsui & Co Ltd
Mitsui Bussan Aerospace Co Ltd
Morimura Bros Inc
Nippon Aircraft Supply Co Ltd
NTK International Corporation
OKI Engineering Co Ltd
Research Forum on Social Infrastructure for Advanced Positioning
SANKI Supplier Co Ltd
Shintoa Corporation
Smart Vision Co Ltd
Sojitz Aerospace Corporation
Sojitz Corporation
SORUN Corporation
Space Engineering Development Co Ltd
Sumitomo Corporation
Tedopres International BV Japan Branch
Tokyo Marine & Nichido Fire Insurance
Turbomeca Japan
Yamada Corporation
SJAC Member Companies:
Aero Asahi Corporation
All Nippon Airways Co Ltd
APC Aerospecialty Inc
Astro Research Corporation
Chiyoda Advanced Solutions Corporation
Chubu Nihon Maruko Co Ltd
Commercial Airplane Company
Daicel Chemical Industries Ltd
Daido Steel Co Ltd
Daikin Industries Ltd
Eagle Industry Co Ltd
Fuji Bellows Co Ltd
Fuji Heavy Industries Ltd
Fujikin Incorporated
Fujitsu Limited
Fujiwara Co Ltd
The Furukawa Battery Co Ltd
Furukawa-Sky Aluminum Corp
Furuno Electric Co Ltd
GS Yuasa Corporation
Hirobo Limited
Hitachi Ltd
Hitachi Kokusai Electric Inc
Hitachi Metals Ltd
Hoden Seimitsu Kako Kenkyusho Co Ltd
Honda Motor Co Ltd
IHI Aerospace Co Ltd
Ishikawajima-Harima Heavy Industries Co Ltd
Ishikawa Seisakusho Ltd
Jamco Corporation
Japanese Aero Engines Corporation
Japan Aircraft Development Corporation
Japan Airlines International Co Ltd
Japan Aviation Electronics Industry Ltd
Japan Radio Co Ltd
The Japan Steel Works Ltd
JTEKT Corporation
Kanto Aircraft Instrument Co Ltd
Kawanishi Aero-Parts Products Co Ltd
Kawasaki Heavy Industries Ltd
Kayaba Industry Co Ltd
Kobe Steel Ltd
Koito Industries Ltd
Koito Manufacturing Co Ltd
Machida Endoscope Co Ltd
Meira Corporation
Minebea Co Ltd
Mitsubishi Electric Corporation
Mitsubishi Heavy Industries Ltd
Mitsubishi Materials Corporation
Mitsubishi Precision Co Ltd
Mitsubishi Rayon Co Ltd
Mitsubishi Space Software Co Ltd
Murata Machinery Ltd
Nabtesco Corporation
NAC Image Technology Inc
Nachi-Fujikoshi Corp

NEC Aerospace Systems Ltd
NEC Corporation
NEC Toshiba Space Systems Ltd
NGK Spark Plug Co Ltd
Nihon Pall Ltd
Nikkiso Co Ltd
Nippi Corporation
Nippon Avionics Co Ltd
NOF Corporation
NSK Ltd
NTN Corporation
NTTDATA Corporation
Owari Precise Products Co Ltd
Sakura Rubber Company Limited
Shimadzu Corporation
Shinko Electric Co
Shinmaywa Industries Ltd
Shounan Seiki Co Ltd
Showa Aircraft Industry Co Ltd
Sogo Spring Mfg Co Ltd
Sts Corporation
Sumijyu Precision Forging Co Ltd
Sumitomo Electric Industries Ltd
Sumitomo Light Metal Industries Ltd
Sumitomo Metal Industries Ltd
Sumitomo Precision Products Co Ltd
Tamagawa Seiki Co Ltd
Terauchi Manufacturing Co Ltd
Toho Tenax Co Ltd
Tokimec Inc
Tokyo Aircraft Instrument Co Ltd
Toray Industries Inc
Toshiba Corporation
Toshiba Electro-Wave Products Co Ltd
Toyo Radio Systems Co Ltd
Toyo Seiko Co Ltd
Yamaha Motor Co Ltd
Yamaha Motor Co Ltd
Yokogawa Electric Corporation
The Yokohama Rubber Co Ltd
Yoshimitsu Industries Inc

The Toulouse Space Centre

18 Avenue Edouard-Belin, F-31 401 Toulouse, Cedex 4, France
Tel: (+33 5) 61 27 31 31
Fax: (+33 5) 61 27 31 79
Web: www.cnes.fr

Thomas Electronics Inc

100 Riverview Drive, Wayne, New Jersey, 07470, United States
Tel: (+1 973) 696 52 00
Fax: (+1 973) 696 82 98
e-mail: sales@thomaselectronics.com
Web: www.thomaselectronics.com

Thompson Valves

FCX Thompson Valves
NEI Thompson Valves
a subsidiary of FCX International, a UK private company
17 Balena Close, Creekmoor, Poole, Dorset, BH17 7EF, United Kingdom
Tel: (+44 1202) 69 75 21
Fax: (+44 1202) 60 53 85
Web: www.thompson-valves.com
■ Solenoid valves for cryogenic solenoid fuel control for the second stage of the Ariane Space vehicle. Also produce pressure regulators and solenoid valves for ground based support and service and test equipment. Valves for special development, high pressure air and fluids, containment and regulation and solenoid valves.

Thrane & Thrane Aalborg

Sailor
SP Radio A/S
Postbocks 7071, DK-9200 Aalborg, SV, Denmark
Porsvej 2, DK-9000 Aalborg, Denmark
Tel: (+45 39) 55 88 00
Fax: (+45 39) 55 88 88
Web: www.sailor.dk
www.thrane.com
■ Maritime radio communication equipment, including VHF, SSB radio-telephones, radio-telex and complete GMDSS equipment. Portable VHF/UHF programme. Satellite communication equipment B, C, Mini-M.

Thrane & Thrane AS (T&T)

Lundtoftegaardsvej 93 D, DK-2800 Lyngby, Denmark
Tel: (+45) 39 55 88 00
Fax: (+45) 39 55 88 88
e-mail: info@thrane.com
Web: www.tt.dk
www.thrane.com
■ Manufacture of radio and satellite communication equipment.

Thyssen Krupp VDM GmbH

PO Box 1820, D-58778 Werdohl, Germany
Tel: (+49 2392) 550
Fax: (+49 2392) 55 22 17
e-mail: info@tbs-vdm.thyssenkrupp.com
Web: www.thyssenkruppvdm.com
■ Development and production of high-performance materials for the aerospace industries: flat products (hot and cold-rolled sheet and plate, strip); rod and bars made of nickel-base alloys, cobalt alloys and demanding special stainless steels.

Ticra

Laederstraede 34, DK-1201 København, Denmark
Tel: (+45) 33 12 45 72
Fax: (+45) 33 12 08 80
e-mail: ticra@ticra.com
Web: www.ticra.com
■ Commercial foundation. Engaged in the development of antenna synthesis, analysis and measurement methods on research contracts for international organisations and private companies. Organisation has a library of advanced antenna software which is commercially available. Manufacture of dual polarised probes for near-field measurements.

Tietronix

1331 Gemini Avenue, Suite 300, Houston, Texas, 77058, United States
Tel: (+1 281) 461 92 00
Fax: (+1 281) 461 93 50
e-mail: emeimoun@tietronix.com

Timestep Weather Systems Ltd

Timestep Electronics Ltd
PO Box 2001, Dartmouth, Devon, TQ6 9QN, United Kingdom
Tel: (+44 1803) 83 33 66
e-mail: information@time-step.com
Web: www.time-step.com
■ Manufacture and integrator of weather satellite and remote sensing ground stations.

Timet UK Ltd

a division of Timet, US
PO Box 704, Witton, B6 7UR, United Kingdom
Tel: (+44 121) 356 11 55
Fax: (+44 121) 356 54 13
Web: www.timet.com
■ Titanium and titanium alloys, ingots, bars, billet, sheet, rod, tube, extrusions, wire and castings. Super-conducting composites wires and cables.

Timken Coventry

Nadella UK Sales Office
Nadella UK Ltd
a fully owned subsidiary of the Torrington Company, US
a subsidiary of the Timken Company
Progress Close, Leofric Business Park, Binley, Coventry, West Midlands, CV3 2TF, United Kingdom
Tel: (+44 24) 76 29 69 00
Fax: (+44 24) 76 29 69 93
Web: www.timken.com
■ Manufacture of Torrington and Nadella airframe needle roller bearings. Fafnir aircraft control bearings and SNR mainshaft jet engine, gearbox, auxilliary drives, spacecraft and helicopter transmission bearings.

Tiros Space Information

Astronautical Society of Western Australia Inc
92 Modillion Avenue, Riverton, Western Australia, WA6148, Australia
Tel: (+61 8) 94 57 36 20
e-mail: tirosspace@hotmail.com
Publications: News Bulletin, World Spacecraft Digest

Titan Systems Corporation (TSC)

Datron Advanced Technologies Division
Datron/Transco Inc
200 West Los Angeles Avenue, Simi Valley, California, 93065-1650, United States
Tel: (+1 805) 584 17 17
Fax: (+1 805) 526 08 85
(+1 805) 526 36 90
Telex: 4942787
e-mail: info-request@dtsi.com
Web: www.dtsi.com
■ Serves markets worldwide with antennas and products for telemetry, satellite communications and Remote Sensing Satellite (RSS) data aquisition and control. Complete RSS Ground Station design, manufacture, upgrade and integration services. Mobile Direct Broadcast Satellite (DBS) antennas for on air platforms.

TNO Defence, Security and Safety

TNO PML
PO Box 45, NL-2280 AA Rijswijk ZH, Netherlands
Lange Kleiweg 137, NL-2288 GJ Rijswijk ZH, Netherlands
Tel: (+31 15) 284 30 00
Fax: (+31 15) 284 39 91
e-mail: info-denv@tno.nl
Web: www.tno.nl
■ Research and development into chemical and biological protection.

TNO Science and Industry

PO Box 155, NL-2600 AD Delft, Netherlands
Tel: (+31 15) 269 20 00
Fax: (+31 15) 269 21 11
Web: www.tno.nl
■ The realisation of scientific instruments for use in space to observe the Earth and for other scientific purposes. The development of techniques for aperture synthesis for astronomical applications. The calibration under space conditions (spectral and radiometric) of optical instruments and diffusers for Earth observation. The realisation of precision mechanisms for use in space. The realisation of sensors for navigating and controlling the position of satellties.

TNO Space

PO Box 477, NL-2600 AL Delft, Netherlands
Tel: (+31 15) 269 21 61
Fax: (+31 15) 269 21 11
e-mail: woenzel@tpd.tno.nl
Web: www.tnospace.tno.nl

Tokyo Astronomical Observatory

NAOJ
Osawa Mitaka, Tokyo, 181, Japan
Tel: (+81 422) 34 36 45
Fax: (+81 422) 34 36 41

Toshiba Corp

Headquarters
1-1-1 Shibaura, Minato-ku, Tokyo, 105-8001, Japan
Tel: (+81 3) 33 57 84 15
(+81 3) 34 57 30 50
(+81 3) 34 57 30 89
(+81 3) 34 57 32 45
(+81 3) 34 57 49 24
Fax: (+81 3) 34 56 16 31
(+81 3) 34 57 83 85
(+81 3) 54 44 92 18
(+81 3) 54 44 93 33
Telex: 22587 toshiba
Cable: TOSHIBA
Web: www.toshiba.co.jp
■ Close range SAM (SAM-2, Keiko SAM, SAM-3) for air-to-air and shipborne applications. Short range SAM (SAM-1 TAN SAM, SAM 1-C improved TAN SAM), base air defence. Medium range SAM, hawk and main firing radar. Long range SAM, PATRIOT, module, power plant, radar set and transmitter. Avionics for P-3C, HRN 107 ILS receiver, HSN-3 INS, AN/APS-115C search radar, AN/ASQ-212(J) data processing system. Avionics for SH-60J, HRQ-109 VOR receiver, HYQ-1 HCDS (Helicopter Combat Direction System). Avionics for fighter aircraft F-15 and F-2, AN/ASN-109 INS, lead computing gyro unit, map generator. Avionics for C-1 cargo aircraft, J/AYN-1 map navigation display system, J/APN-93 navigation weather radar. Avionics for AH-1S attack helicopter, FCC Fire Control Computer, JARN-A10 VOR/ILS receiver. Avionics for U-125 rescue aircraft, J/APQ-2 search radar. Avionics for E-2C and E-767 AWACS. Field artillery systems including FADS Field Artillery Data-Processing System, JMPQ-P13 mortar locating radar, JTPS-P16 artillery locating radar. Navigation support systems, GCA Ground Controlled Approach, RIPS Radar Information Processing Systems, ILS Instrument Landing System, relocatable control tower. Weather radars: J/FPH-6; J/FPH-8 and LPN-10, NLPN-8. Advanced bridge systems. Information and communication systems, electronic devices, medical equipment, power plant systems, transportation equipment, consumer products.
Subsidiaries with defence business:
Toshiba Electro-Wave Products Corporation

Toyo Communication Equipment Co Ltd (TOYOCOM)

484 Tsukagoshi 3-chome, Saiwai-ku, Kawasaki-shi, Kanagwa, 212 8513, Japan
Tel: (+81 44) 542 65 00
Fax: (+81 3) 542 63 54

Toyo Communication Equipment Co Ltd (TOYOCOM)
Sagami Plant
1-1 Koyato 2-chome, Samukawa-machi, Koza-gun, Kanagawa, Ken, 253-0192, Japan
Tel: (+81 467) 74 11 31
Fax: (+81 467) 74 11 79

Transformational Space Corporation
11710 Plaza America Drive, Suite 2000, Reston, Virginia, 20190, United States
Tel: (+1 703) 871 51 02
Fax: (+1 703) 871 51 59
Web: www.transformspace.com/

Trimble UK
Trimble Navigation Europe Ltd
a subsidiary of Trimble Navigation Ltd, US
Trimble House, Meridian Office Park, Osborn Way, Hook, Hampshire, RG27 9HX, United Kingdom
Tel: (+44 1256) 76 01 50
Fax: (+44 1256) 76 01 48
Web: www.trimble.com
■ High accuracy navigation position finding, surveying and mapping systems based upon the GPS satellite constellation. Receivers, sensors, software suitable for all applications from guidance systems for drones through to troop movement monitoring and tracking.
Representing:
Trimble Navigation Ltd, US

TRW Astro Aerospace
Astro Aerospace Corp
a subsidiary of TRW Inc., USA
6384 Via Real, Carpinteria, California, 93013-2920, United States
Tel: (+1 805) 684 66 41
Fax: (+1 805) 684 33 72
Web: www.trw.com
■ Design and manufacture of space deployable devices such as unsurfacable antennas, solar arrays, extendible support structures and masts.

TSI Telsys
7100 Columbia Gateway Drive, Columbia, Maryland, 21046-2141, United States
Tel: (+1 410) 872 39 00
Fax: (+1 410) 872 39 01
e-mail: info@tsi-telsys.com
Web: www.tsi-telsys.com

TsNIIMASH-Export JSC (TsE)
4 Pionerskaya Street, 141070 Korolev, Moscow Region, Russian Federation
Tel: (+7 095) 516 59 10
Fax: (+7 095) 187 15 88
Telex: 411952 mcc
e-mail: corp@tse.ru
Web: www.tse.ru
■ Official dealer, co-ordinator of foreign activity and contractor on behalf of the Central Research Institute of Machine Building – TsNIIMASH. Main activities include marketing, scientific, technical and economic expertise of projects and proposals; information provision on TsNIIMASH and other aerospace organisations with space test capabilities; import/export activities; arrangement of international congresses, lecturing programs. Project cost analysis. Legal advice, including preparation of export licenses. Arrangement of direct contacts and exchange of specialists among Russian and foreign space enterprises. Advertising and publishing.
Representing:
Aerospatiale, France
Arnold Engineering Dev (AEDC), US
BAE SYSTEMS, UK
Ballistic Missile Defense (BMDO), US
Bureau Veritas, France
CNES, France
DARA, Germany
Dassault Aviation, France
Deloitte & Touche, France
DERSI, France
DLR, Germany
Draper Laboratory, US
EADS, Germany
General Applied Scientific Lab. (GASL), US
GKNPTs Khrunichev, Russian Federation
GP Agat, Russian Federation
Great Wall Industry Corp, China
Hilton Corp, US
IAI/MBT-System & Space Tech, Israel
JPL, US
NASA, US
NPO Molniya, Russian Federation
Projespace, France
Prospace, France
RKK Energiya, Russian Federation
RSA, Russian Federation
SEP, France
Sverdup Technology Inc, US
TsNiimash, Russian Federation
US Universities, US
Representing:
ESA
ESTEC

TT Electronic Manufacturing Services Ltd (TTems)
Rogerstone-South Wales
AB Electronic Assemblies Ltd
a subsidiary of TT Electronics plc, UK
Tregwilym Industrial Estate, Rogerstone, Newport, Gwent, NP10 9YA, United Kingdom
Tel: (+44 1633) 89 23 45
Fax: (+44 1633) 89 57 55
e-mail: sales@ttems.com
info@ttems.com
Web: www.ttems.com
■ Manufacture electronic and electro-mechanical assemblies and finished products. Services include NPI, design services, PCB assembly, programme management and international procurement.
Accreditations: ISO 9000:2000, ISO 14001, BSI Kitemark, TS 16949

Tupolev PSC
Aviation Scientific Technical Complex Named for A N Tupolev JSC
OKB-156
17 Academician Tupolev Embankment, 105005 Moskva, Russian Federation
Tel: (+7 095) 267 25 33
Fax: (+7 095) 267 27 33
(+7 095) 261 08 68
(+7 095) 261 71 41
e-mail: tu@tupolev.ru
Web: www.tupolev.ru
■ Design and manufacture of civil and military aircraft and space projects. Activities involve the Tu-334 passenger aircraft and the Burlak air launched space booster. Design and production of Tu-22M and Tu-160 supersonic heavy bombers.

Turbocam Europe Ltd
Unit 1, 155 Highlands Road, Fareham, Hampshire, PO15 6JR, United Kingdom
Tel: (+44 1329) 84 58 00
Fax: (+44 1329) 84 60 00
e-mail: tel@turbocam.com
Web: www.turbocam.com
■ Machining for turbomachinery, products include propellers and impellers for marine applications; aircraft engine components, impellors for spaceborne platforms and rail locomotive engine components.
Accreditations: ISO 9001; ISO 9002; AS9100A

Turbocam Inc
5 Faraday Drive, Dover, New Hampshire, 03820-4352, United States
Tel: (+1 603) 749 60 66
Fax: (+1 603) 749 24 04
e-mail: sales@turbocam.com
Web: www.turbocam.com
■ Specialist machining for turbomachinery, products include propellers and impellers for marine applications; aircraft engine components, impellers for spaceborne platforms and rail locomotive engine components.
Accreditations: ISO 9001; ISO 9002; AS9100A

U

Ueda Japan Radio Co Ltd
2-10-19 Fumiiri, Ueda-City, Nagano, 386-8608, Japan
Tel: (+81 268) 26 21 21
(+81 268) 26 21 23
Fax: (+81 268) 26 20 79
e-mail: sales@ujrc.co.jp
■ Manufacture of VHF radio equipment, radio receivers, measuring instruments and electromedical equipment.

Ueda Japan Radio Co Ltd
Tokyo Sales Office
Kouraku-building, 5th Floor, 1-1-8, Bunkyo-Ku, Tokyo, 112-0004, Japan
Tel: (+81 3) 3814 68 20
Fax: (+81 3) 3814 68 21

UK Astronomy Technology Centre (UK ATC)
The Royal Observatory Edinburgh, Blackford Hill, Edinburgh, EH9 3HJ, United Kingdom
Tel: (+44 131) 688 81 00
Fax: (+44 131) 688 82 64
Web: www.roe.ac.uk

UKWTechnik Jaeger
UKW-Technik Electronic GmbH
PO Box 60, D-91081 Baiersdorf, Germany
Tel: (+49 9133) 779 40
Fax: (+49 9133) 77 94 30
e-mail: info@ukwtechnik.com
Web: www.ukwtechnik.com
■ Production and sales of weather satellite receivers and image processing systems.

Ultra Electronics Holdings plc
Signature Management Systems Division
Dowty Magnetics
Fallow Park, Rugeley Road, Hednesford, Cannock, Staffordshire, WS12 0QZ, United Kingdom
Tel: (+44 1543) 87 88 88
Fax: (+44 1543) 87 82 49
Telex: 335004 domain
e-mail: mag@ultra.demon.co.uk
Web: www.ultra.demon.co.uk
■ Multi-influence measurement ranges for measuring the signatures of naval vessels. Magnetic sensors. Magnetic measurement instruments. Degaussing systems for ships.

Ultra Electronics Inc
BAE Systems Canada Inc
Canadian Marconi Co
CMC Electronics Inc
5990 Cote de Liesse, Ville Mont-Royal, Montréal, Québec, H4T 1V7, Canada
Tel: (+1 514) 855 63 63
(+1 514) 748 31 48
Fax: (+1 514) 748 31 00
(+1 514) 855 63 57
e-mail: info@ultra-tcs.com
marketing@ultra-tcs.com
Web: www.ultra-tcs.com
■ Supplier of tactical Line of Sight (LOS) radio relay equipment and systems.

Unisource Technology Inc
9010 Ryan Avenue, Montréal, Québec, H9P 2M8, Canada
Tel: (+1 514) 748 88 88
Fax: (+1 514) 748 07 48
e-mail: info@unisourcetechnology.com
Web: www.unisourcetechnology.com

United States Air Force (USAF)
Diego Garcia Tracking Station
FPO AP 96995-0052 Diego Garcia, British Indian Ocean Territory
■ Satellite command and control, USAF satellite control network, global positioning system.

United States Aircraft Insurance Group (USAIG)
managed by United States Aviation Underwriters, Inc.
199 Water Street, New York, New York, 10038, United States
Tel: (+1 212) 952 01 00
Fax: (+1 212) 349 82 26
Web: www.usaig.com
■ Aviation and aerospace insurance, including launch and in orbit coverages.

United Technologies Corp (UTC)
United Technologies Building, Hartford, Connecticut, 06101, United States
Tel: (+1 860) 728 70 00
(+1 860) 728 79 21
Fax: (+1 860) 493 41 69
(+1 860) 493 41 65
Telex: 99212 utchfda
Web: www.utc.com

Univelt Inc
Escondido Office
740 Metcalf Street 13, Escondido, California, 92025-1671, United States
Tel: (+1 760) 746 40 05
Fax: (+1 760) 746 31 39

Univelt Inc
PO Box 28130, San Diego, California, 92198-0130, United States
Tel: (+1 760) 746 40 05
Fax: (+1 760) 746 31 39
e-mail: sales@univelt.com
Web: www.univelt.com

■ Publisher in the field of astronautics and communications for the American Astronautical Society. Distribute books and microforms and provide services for technical associations.

Universal Avionics Systems Corporation
Corporate Headquarters
3260 East Universal Way, Tucson, Arizona, 85706, United States
Tel: (+1 520) 295 23 00
Freephone: (+1 800) 321 52 53
Fax: (+1 520) 295 23 90
(+1 520) 295 23 95
e-mail: info@uasc.com
Web: www.uasc.com
■ Products include: flight management and navigation management systems, long- and short-range navigation sensors, air-to-ground two way datalink, cockpit voice recorder and cabin display systems.

Universidade de Sao Paulo
IAG
Rua do Matao 1226, Cidade Universitaria, SP 05508-090 São Paulo, Brazil
Tel: (+55 11) 577 85 99
(+55 11) 30 91 47 62
Fax: (+55 11) 276 38 48
(+55 11) 30 91 28 01
e-mail: iag@usp.br
Web: www.iag.usp.br
www4.usp.br/index.php/home

Universität zu Köln
Institute of Air and Space Law
Albertus-Magnus-Platz, D-50923 Köln, Germany
Tel: (+49 221) 470 23 37
Fax: (+49 221) 470 49 68
e-mail: sekretariat-hobe@uni-koeln.de
Web: www.uni-koeln.de/jur-fak/instluft

Universiteit Gent
Sterrenkundig Observ
Krijgslaan 281, B-9000 Gent, Belgium
Tel: (+32 9) 264 47 99
Fax: (+32 9) 264 49 89

Universitetskij
Sternberg State Astronomical Institute
119899 Moskva, Russian Federation
Tel: (+7 095) 939 33 18
Fax: (+7 095) 939 16 61

Universities Space Research Association (USRA)
Headquarters
10227 Wincopin Circle, Suite 500, Columbia, Maryland, 21044-3432, United States
Tel: (+1 410) 730 26 56
Fax: (+1 410) 730 34 96
e-mail: info@hq.usra.edu
Web: www.usra.edu

Universities Space Research Association (USRA)
Huntsville Program Office
6700 Odyssey Drive, Suite 203, Huntsville, Alabama, 35806, United States
Tel: (+1 256) 971 02 40
(+1 256) 971 02 42
Fax: (+1 256) 971 02 41
e-mail: debra@space.hsv.usra.edu

Universities Space Research Association (USRA)
Seabrook Office
7501 Forbes Boulevard, Suite 206, Seabrook, Maryland, 20706, United States
Tel: (+1 301) 805 83 96
Fax: (+1 301) 805 84 66
e-mail: dhold@gvsp.usra.edu
donj@ssec.wisc.edu

Universities Space Research Association (USRA)
Space Operations Programs
13422 Laurinda Way, Santa Ana, California, 92705-1926, United States
Tel: (+1 714) 544 95 90
Fax: (+1 714) 544 34 71
e-mail: dean@usra.edu

Universities Space Research Association (USRA)
Space Technology Development Office
901 University Boulevard, SE, Suite 218, Albuquerque, New Mexico, 87106-4339, United States
Tel: (+1 505) 272 73 24
Fax: (+1 505) 272 72 03
e-mail: finnegan@usra.edu

Universities Space Research Association (USRA)
Washington Office
1101 17th Street NW, Suite 1004, Washington, District of Columbia, 20036, United States
Tel: (+1 202) 689 12 90
Fax: (+1 202) 689 12 97

University of Aarhus
Institute of Physics and Astronomy
DK-8000 Århus, C, Denmark
Tel: (+45 89) 42 36 14
Fax: (+45 89) 12 07 40

University of Alaska Fairbanks
Geophysical Institute
Poker Flat Research Range (PFRR), PO Box 757320, Fairbanks, Alaska, 99775-7320, United States
Mile 30 Steese Highway, Fairbanks, Alaska, 99712, United States
Tel: (+1 907) 455 21 10
(+1 907) 477 75 58
Fax: (+1 907) 455 21 20
(+1 907) 474 73 44
e-mail: info@gi.alaska.edu
Web: www.pfrr.alaska.edu
■ University owned sounding rocket launch facility. Facilities to handle complete NASA sounding rocket inventory.

University of California
Department of Astronomy
405 Hilgard Avenue, Los Angeles, CA 90024, United States
Tel: (+1 310) 825 44 34
Fax: (+1 310) 206 20 96
Web: astro.ucla.edu

University of California
Lick Observatory
Santa Cruz, California, CA 95064, United States
Fax: (+1 831) 426 31 15

University of California
National Fuel Cell Research Center
Engineering Laboratory Facility, Irvine, California, 92697-3550, United States
Tel: (+1 949) 824 19 99
Fax: (+1 949) 824 74 23

University of California
Space Sciences Lab
Astronomy Department, Berkeley, California, CA 94720-7450, United States
Tel: (+1 510) 642 16 48
Fax: (+1 510) 643 83 03

University of Canterbury
Department of Physics and Astronomy
Private Bag 4800, Christchurch, 1, New Zealand
Tel: (+64 3) 366 70 01
(+64 3) 366 75 59
Fax: (+64 3) 364 24 69

University of Cape Town
Department of Astronomy
7700 Rondebosch, South Africa
Tel: (+27 21) 650 23 94
(+27 21) 650 23 91
Fax: (+27 21) 650 37 26
(+27 21) 650 33 52
e-mail: astro@physci.act.ac.za
Web: artemisia.ast.uct.ac.za

University of Colorado
Center for Astrophysics and Space Astronomy an affiliated unit within the structure of the Astrophysical and Physical Sciences Department (APS) at the University of Colorado
389 UCB, Boulder, Colorado, 80309-0389, United States
Tel: (+1 303) 492 40 50
Fax: (+1 303) 492 71 78
Web: casa.colorado.edu
■ An astronomy and astrophysics centre that offers an academic program leading to the PhD degree in a variety of areas of astrophysics and planetary sciences.

University of Dundee
NERC Satellite Receiving Station
Department of Electronic Engineering and Physics, Perth Road, Dundee, DD1 4HN, United Kingdom
Tel: (+44 1382) 34 51 94
(+44 1382) 34 44 09
Fax: (+44 1382) 20 25 75
Web: www.sat.dundee.ac.uk
■ Data acquisition and distribution services funded by the UK Natural Environment Research Council. 12,400 students and over 3,000 staff in the entire university.

University of Houston Clear Lake
2700 Bay Area Boulevard, Houston, Texas, 77058-1098, United States
Tel: (+1 281) 283 31 22
Fax: (+1 281) 283 31 23

University of London Observatory (ULO)
Part of University College London
553 Watford Way, Mill Hill, London, NW7 2QS, United Kingdom
Tel: (+44 20) 89 59 04 21
Fax: (+44 20) 82 38 88 72
e-mail: secretary@ulo.ucl.ac.uk
Web: www.ulo.ucl.ac.uk
■ Provides research, teaching and training services in observational astronomy.

University of Maryland
Astronomy Department
College Park, Maryland, MD 20742, United States
Tel: (+1 301) 405 58 22
Fax: (+1 301) 314 90 67

University of Nottingham
Institute of Engineering, Surveying & Space Geodesy
University Park, Nottingham, NG7 2RD, United Kingdom
Tel: (+44 115) 951 38 68
Fax: (+44 115) 951 38 81
e-mail: iessg@nottingham.ac.uk
Web: www.nottingham.ac.uk/iessg
■ University research group. Postgraduate teaching and research, consultancy. Research ('blue sky' and applied) in satellite navigation and positioning, navigation systems, satellite orbit determination and satellite system simulation.

University of Oslo
Institute of Theoretical Astrophysics
Box 1029, N-0315 Oslo, 3, Norway
Tel: (+47 22) 85 65 15
Fax: (+47 22) 85 65 05

University of Punjab
Department of Space Science
1 Shahrah-e-al-Beruni, Lahore, Pakistan
Tel: (+92 42) 35 44 28
■ Astronomical observatory.

University of Tennessee
Department of Geography
Knoxville, Tennessee, 37996, United States
Tel: (+1 865) 974 10 00
Fax: (+1 615) 974 60 25
■ Research using remote sensing and geographic information systems (GIS).

University of Tokyo
Department of Astronomy
Bunkyo Ku, Tokyo, 113, Japan
Tel: (+81 358) 00 68 80
Fax: (+81 358) 13 94 39

University of Toronto (UT)
Institute for Aerospace Studies
4925 Dufferin Street, Toronto, Ontario, M3H 5T6, Canada
Tel: (+1 416) 667 77 18
Fax: (+1 416) 667 77 43
e-mail: info@utias.utoronto.ca
Web: www.utias.utoronto.ca
www.ornithopter.net
■ Research and education services in selected areas of aerospace science and engineering including. Flight mechanics, aircraft flight systems and control, flight simulation, aircraft design, aerodynamics, gasdynamics, propulsion, combustion, computational fluid dynamics, multidisciplinary optimization, structural mechanics, advanced materials, smart structures and materials, space robotics, space dynamics and control, microsatellite technology, plasmas and fusion energy and the design of environmentally friendly aircraft.

University of Victoria
Department of Physics
Box 1700, Victoria, BC, V8W 2Y2, Canada
Tel: (+1 250) 721 77 39
Fax: (+1 250) 721 77 15

University of Victoria
Department of Physics and Astronomy
Box 3055, Victoria, BC, V8W 3P6, Canada
Tel: (+1 250) 721 77 40
Fax: (+1 250) 721 77 15

University of Washington
Department of Astronomy
FM 20, Seattle, Washington, WA 98195, United States
Tel: (+1 206) 543 19 88
(+1 202) 543 77 73
Fax: (+1 206) 685 04 03

US Air Force Materiel Command (AFMC)
Directed Energy Directorate
Air Force Research Laboratory
Phillips Laboratory
3550 Aberdeen Avenue Southeast, Kirkland AFB, New Mexico, 87117-5776, United States
Tel: (+1 505) 846 19 11
Fax: (+1 505) 846 04 23
e-mail: afrl.public.affairs@kirtland.af.mil
Web: www.de.afrl.af.mil
- Conducts research in high-energy lasers, high power microwave technology, optics and beam control.

US Air Force Materiel Command (AFMC)
Headquarters
Public Affairs Office,
4375 Chidlaw Road, Building 262, Room N-152, Wright-Patterson AFB, Ohio, 45433-5001, United States
Tel: (+1 937) 257 32 37
e-mail: webmaster@wpafb.af.mil
Web: www.afmc.af.mil
- Management of research development, test, acquisition and support for Air Force systems.
- Organisationally this branch of the US Air Force is the Directed Energy Directorate at Kirtland Air Force Base in New Mexico. The branch is one of nine directorates of the Air Force Base research laboratory whose headquarters are at Wright-Patterson Air Force Base in Ohio. The laboratory is under Air Force Materiel command whose headquarters are also at Wright-Patterson AFB.

Major AFMC installations (A-K):
Arnold AFB: (Arnold Engineering Development Center) Flight simulation test complex.
Brooks AFB: (Armstrong Laboratory and Human Systems Center) Concentrates on human aspects of USAF weapon systems and facilities.
Edwards AFB: Flight test centre for all USAF and NASA aircraft.
Eglin AFB: Tests and evaluates munitions (non-nuclear), electronic combat systems and navigation/guidance systems.
Hanscom AFB: (Electronic Systems Center) Development and acquisition of C3I systems.
Hill AFB: (Ogden Air Logistics Center) Maintenance for ICBM, F-16, Maverick, GBU-15 and laser-guided bombs.
Kirtland AFB: (Phillips Laboratory) Research and development for space and missile related technology.

Major AFMC installations (L–Z):
Robins AFB: (Warner-Robins Air Logistics Center) Maintenance for F-15, C-141 and C-130 aircraft.
Tinker AFB: (Oklahoma City Air Logistics Center) Maintenance for B1-B, B-2, B-52, C-5, E-3, E-4 and KC-135 series aircraft.
Wright-Patterson AFB: (Aeronautical Systems Center and Air Force Research Laboratory) Research, development and acquisition of aerospace systems, electronics, materials and manufacturing technologies.

US Air Force Research Laboratory (AFRL/PA)
Public Affairs Office
1864 4th Street, Suite 1, Wright Patterson AFB, Ohio, 45433-7132, United States
e-mail: afrl.pa.dl.all@wpafb.af.mil

US Army Space & Missile Defense Command (USASMDC)
Public Affairs Office
PO Box 1500, Huntsville, Alabama, 35807-3801, United States
Tel: (+1 256) 955 34 12
(+1 256) 955 35 01
Fax: (+1 256) 955 10 56

US Army Space & Missile Defense Command (USASMDC)
Public Affairs Office
PO Box 15280, Arlington, Virginia, 22215-0280, United States
Tel: (+1 703) 607 20 39
(+1 703) 607 24 45
(+1 703) 607 18 74
Fax: (+1 256) 955 12 14

US Naval Observatory (USNO)
3450 Massachusetts Avenue NW, Washington, District of Columbia, 20392-5100, United States
Tel: (+1 202) 762 14 37
Fax: (+1 202) 652 05 87

V

Vacco Industries
a subsidiary of Esco Technologies Inc, US
10350 Vacco Street, South El Monte, California, 91733, United States
Tel: (+1 626) 443 71 21
Fax: (+1 626) 442 69 43
e-mail: navy@vacco.com
Web: www.vacco.com
- Specialised flow control valves, filters and filtration systems, photochemical machining.

Vaisala Oyj
Head Office
PO Box 26, FI-00421 Helsinki, Finland
Vanha Nurmijärventie 21, FI-01670 Vantaa, Finland
Tel: (+358 9) 894 91
Fax: (+358 9) 89 49 22 27
Web: www.vaisala.com
- Development and manufacture of electronic measurement systems and equipment for meteorology and the environmental sciences, traffic safety and industry. Products include: radiosondes and their ground equipment, dropsonde systems, surface weather observation instruments and systems, along with transmitters and instruments for the measurement of relative humidity, dewpoint, barometric pressure and carbon dioxide.

Vanguard Research Inc (VRI)
10400 Eaton Place, Suite 450, Fairfax, Virginia, 22030-2201, United States
Tel: (+1 703) 934 63 00
Fax: (+1 703) 273 93 98
Web: www.vriffx.com
- Research on C3I and BMDO. Engineering and technical professional services. Environmental remediation.

Vanguard Research Inc (VRI)
Colorado Springs Division
1 Gateway Plaza, 1330 Inverness Drive, Suite 450, Colorado Springs, Colorado, 80910, United States
Tel: (+1 719) 596 11 74
Fax: (+1 719) 596 01 83
- Provides support to the Strategic Defence Initiative organisation's National Test Bed Joint Program Office (NTBJPO) and provides technical and analytical support to the United States Space Command (AFSPACECOM).

VCS Aktiengesellschaft
VCS Nachrichtentechnik GmbH
Borgmannstrasse 2, D-44894 Bochum, Germany
Tel: (+49 234) 925 81 15
Fax: (+49 234) 925 81 90
Web: www.vcs.de
- High-availability solutions for satellite communications, earth observation, and media broadcasting.

Vectronix AG
Leica Vectronix AG
Max-Schmidheiny-Strasse 202, CH-9435 Heerbrugg, Switzerland
Tel: (+41 71) 726 72 00
(+41 71) 726 72 54
Fax: (+41 71) 726 72 01
e-mail: info@vectronix.ch
Web: www.vectronix.ch
- High-precision optronic instruments for military survey and artillery applications, digital compass, diode laser rangefinder. Day and night observation.

VEGA Group plc
Bristol Office
360 Bristol Business Park, Coldharbour Lane, Bristol, BS16 1EJ, United Kingdom
Tel: (+44 117) 988 00 33
Fax: (+44 117) 988 00 34
- Consulting and technology services in space, defence and government. Core competencies include capability acquisition, control and operations, simulation and modelling and information security.

VEGA Group plc
Head Office
2 Falcon Way, Shire Park, Welwyn Garden City, Hertfordshire, AL7 1TW, United Kingdom
Tel: (+44 1707) 39 19 99
Fax: (+44 1707) 39 39 09
e-mail: marketing@vega.co.uk
defence@vega.co.uk
Web: www.vega-group.com
- Provide consultancy, technology and managed solutions specialising in programme and system assurance. Areas covered include capability acquisition, human performance improvement, information security, simulation and technology solutions, technology support services and science and technology research.

Verhaert
Hogenakkerhoekstraat 21, B-9150 Kruibeke, Belgium
Tel: (+32 3) 250 14 14
Fax: (+32 3) 253 14 64
e-mail: info@verhaert.com
Web: www.verhaert.com
- Instruments and laboratories for microgravity research services in life, material and fluid sciences and biotechnology. Development, construction, launch and operation of small autonomous satellite systems from 10 to 400 kg, satellite subsystems and attached platforms, planetary research. Development and construction of satellite-related user terminals. Instruments and professional equipment for military, life science, environmental and automotive applications.

Viasat Inc
6155 El Camino Real, Carlsbad, California, 92009-1699, United States
Tel: (+1 760) 476 22 00
Fax: (+1 760) 476 47 03
(+1 760) 929 39 41
Web: www.viasat.com
- Digital communications company specialising in high-performance networking hardware and software for satellite and other wireless applications. Products for government and commercial applications. Satellite communication products include terminals and modems for UHF, Ku-, C- and Ka-band. Information security products include inline network encryptors that can secure communication on public networks up to top secret levels. Tactical communications products include MIDS/Link-16 terminals and advanced data controllers for error-free communication of images and email on military channels. Other products include multiple emitter C3 and EW simulators and satellite simulator/training terminals.

Vibro-Meter SA
Route de Moncor 4, CH-1701 Fribourg, Switzerland
Tel: (+41 26) 407 11 11
(+41 26) 407 15 22
Fax: (+41 26) 402 36 62
(+41 26) 407 15 55
Telex: zrhvm8x
e-mail: aerospace@vibro-meter.ch
Web: www.vibro-meter.com
Web: www.meggitt.com
- Transducers and associated electronic equipment for measurement, monitoring and control of physical parameters on aircraft, gas turbine engines and rocket engines, vibration, speed, ice, displacement and pressure. Helicopter rotor trim and balance systems. Central maintenance computers and engine and propellor interfaces.

Associated companies:
Vibro-Meter France SA, France
Vibro-Meter GmbH, Germany
Vibro-Meter Singapore Pte Ltd, Singapore

Videsh Sanchar Bhavan
Calcutta Office
1/18 CIT Scheme VIIM, Ultadanga, Kolkata (Calcutta), 700054, India
Tel: (+91 33) 355 40 14
Fax: (+91 33) 355 40 13
Telex: 81 21 5020
e-mail: helpdesk@giascl01.vsnl.net.in

Videsh Sanchar Bhavan
Madras Office
5 Swami Sivanada Salai, Chennai (Madras), 600 002, India
Tel: (+91 44) 536 67 40
Fax: (+91 44) 538 38 38
Telex: 81 41 01016
e-mail: helpdesk@giasmd01.vsnl.net.in

Videsh Sanchar Bhavan
Mumbai (Bombay) Office
287 Videsh Sanchar Bhavan, Mahatma Gandhi Road, Mumbai (Bombay), 400001, India
Tel: (+91 22) 262 40 86
Fax: (+91 22) 262 41 01
e-mail: customer@giasbm01.vsnl.net.in
helpdesk@giasbm01.vsnl.net.in

Videsh Sanchar Bhavan
New Delhi Office
Bangla Saheb Road, New Delhi, 110 001, India
Tel: (+91 11) 374 73 10
Fax: (+91 11) 334 73 37
e-mail: helpdesk@giasdl01.vsnl.net.in

Videsh Sanchar Nigam Ltd
Videsh Sanchar Bhavan
Thrikkakara, Ernakulam, Kakkanad, 682030, India
Tel: (+91 484) 42 17 11
(+91 484) 42 11 64
(+91 484) 42 11 65
Fax: (+91 484) 42 17 12

Vikram Earth Station
Arvi Office
Taluka Junnar Dist, Maharastra, Arvi, 412415, India
Tel: (+91 2132) 424 58
Fax: (+91 22) 262 48 06

Villafranca Satellite Tracking Station
PO Box 50727, E-28080 Madrid, Spain
Tel: (+34 91) 401 96 61
Fax: (+34 91) 402 45 58
e-mail: info@esa.int
vilspaweb@esa.int
■ Location for ESA satellite communication ground stations and antennas.

Villanova University
Astronomy Department
800 Lancaster Avenue, Villanova, Pennsylvania, PA 19085, United States
Tel: (+1 610) 519 48 23
Fax: (+1 610) 519 61 32

Virginia Economic Development Partnership (VEDP)
Central Office
901 East Byrd Street, Richmond, Virginia, 23219, United States
Tel: (+1 804) 545 57 00
Fax: (+1 804) 545 57 01
e-mail: info@yesvirginia.org
Web: www.yesvirginia.org

Virginia Economic Development Partnership
European Office
Avenue Louise 479, P O Box 55, B-B1050 Bruxelles, Belgium
Tel: (+32 2) 647 74 33
Fax: (+32 2) 647 14 63
e-mail: mduys@yesvirginia.org

Vitrociset SpA
Head Office and Plant
Via Salaria 1027, I-00138 Roma, Italy
Tel: (+39 06) 882 01
Fax: (+39 06) 88 20 22 06
e-mail: presidenza@vitrociset.it
Web: www.vitrociset.it
■ Technical servicing in Italy and overseas for automated air traffic control (ATC) centres, ATC radar systems, air navigation, radio aids (including ILS, VOR, DME), communications systems, auxiliary systems. Overhaul, repair and calibration of test and measuring instruments. In flight radio-measurements. Servicing of Nike and Hawk missile systems in addition to Sparrow missile. Test engineering for SPADA missile system. Logistic engineering and technical servicing for radars and missile systems.

VNII TransMash
2 Zarechnaya Ultisa, 198323 St Petersburg, Russian Federation
Tel: (+7 812) 135 99 26
Fax: (+7 812) 146 18 51
e-mail: rcl@rcl.spb.su

Vocality International Ltd
Lydling Barn, Lydling Farm, Puttenham Lane, Shackleford, Surrey, GU8 6AP, United Kingdom
Tel: (+44 1483) 81 31 30
Fax: (+44 1483) 81 31 31
Web: www.vocality.com
■ Manufacture of voice, fax, data and ISDN multiplexers for transportable satellite terminals, ship-based satellite multiplexers and Inmarsat multiplexers. Suitable for some encrypted systems. Satellite simulators for bench-testing satellite applications.

Volvo Aero Corp
a subsidiary of AB Volvo, Sweden
SE-461 81 Trollhättan, Sweden
Tel: (+46 520) 940 00
Fax: (+46 520) 340 10
Telex: 420 40 volfa
Web: www.volvoaero.com
■ Development, manufacture and maintenance of commercial and military aircraft engines. Activities include operations development and production of key components for the Ariane space program.

Voss Industries Inc
2168 West 25th Street, Cleveland, Ohio, 44113-4172, United States
Tel: (+1 216) 771 76 55
Fax: (+1 216) 771 28 87
e-mail: voss@vossind.com
sales@vossind.com
Web: www.vossind.com
■ Standard and special T-bolt clamps, V-band couplings and mating flanges, strap assemblies, rigid V couplings and machined flanges, CNC machining, stampings, light fabrications and hydraulic bulge-formed shapes for aerospace and defence applications. Mil-Spec painting, spot welding and heliarc welding.

Votaw Precision Technologies Inc
13153 Lakeland Road, Santa Fé Springs, California, 90670, United States
Tel: (+1 562) 944 06 61
Fax: (+1 562) 944 06 75
■ Manufacture of very large precision tooling for the aerospace industry. Performs machining fabrication and tooling of aerospace components. Shuttle, Titan, Delta II, Atlas, EELV, JSF and F-22.

Votkinsky Zavod State Production Association
2 Kirov St, Votkinsk, 427410 Udmert Republic, Russian Federation
Tel: (+7 34145) 652 01
e-mail: svet@vmz.udmurtia.su
■ Production and assembly of SS-20, SS-23 and SS-25 ballistic missiles, missile components, refrigeration systems, lathes and machine tools.

W

W L Pritchard & Co LC
4405 East West Highway, Suite 501, Bethesda, Maryland, 20814 , United States
Tel: (+1 301) 654 11 44
Fax: (+1 301) 654 18 44
e-mail: wlpco@ix.netcom.com
Web: www.wlpco.com

Wear-Cote International Inc
PO Box 4177, Rock Island, Illinois, 61204-4177, United States
101 10th Street, Rock Island, Illinois, 61201, United States
Tel: (+1 309) 793 12 50
Fax: (+1 309) 786 65 58
e-mail: webmaster@wear-cote.com
Web: www.wear-cote.com
■ Electroless coatings for ferrous/non-ferrous metals. Electroless nickel is employed in Shuttle main engine ignition and mechanical systems and is now applied to other cryogenic propellant systems and ground handling equipment.
Accreditations: ISO9001

Weir Valves and Controls France (Sebim)
Head Office
Groupe Sebim
a subsidiary of The Weir Group Plc, UK
Zone Industrielle La Palunette, F-13165 Chateauneuf-les-Martigues, France
Tel: (+33 4) 42 07 00 95
Fax: (+33 4) 42 07 11 77
e-mail: coseb@sebim.fr
Web: www.weir-valve.com
■ Safety valves and rupture discs for Ariane launching pads.

Weir Valves and Controls France
Sarasin RSBD
a subsidiary of The Weir Group plc, UK
Zone Industrielle du Bois Rigault, rue Jean Baptiste Grison, F-62880 Vendin-le-Vieil, France
Tel: (+33 3) 21 79 54 50
Fax: (+33 3) 21 28 62 00
e-mail: sarasin@weirvalvefr.com
Web: www.weir-valve.com
■ Safety valves and rupture discs for Ariane launching pads.

Welnavigate Inc
510 Spectrum Circle, Oxnard, California, 93030, United States
Tel: (+1 805) 485 59 08
Fax: (+1 805) 485 76 87
e-mail: welnav@aol.com
Web: www.welnav.com
■ Production of simulators and test equipment for the GPS satellite navigation system.

Western Electrochemical Co (WECCO)
a subsidiary of American Pacific Corp
PO Box 629, Cedar City, Utah, 84720, United States
Tel: (+1 801) 865 50 00
Fax: (+1 801) 865 50 05
■ Manufacture of ammonium perchlorate, sodium perchlorate and potassium perchlorate.

Western European Union (WEU)
Union de l'Europe Occidentale
15 rue de l'Association, B-1000 Bruxelles, Belgium
Tel: (+32 2) 500 44 12
(+32 2) 500 44 15
Fax: (+32 2) 500 44 70
e-mail: ueo.secretariatgeneral@skynet.be
secretariatgeneral@weu.int
Web: www.weu.int
■ European intergovernmental defence and security organisation created by the modified Brussels Treaty in 1954. Member States: Belgium, France, Germany, Greece, Italy, Luxembourg, Netherlands, Portugal, Spain and United Kingdom. Associate Members (also members of NATO and the EU, except for Turkey) : Czech Republic, Hungary, Iceland, Norway, Poland and Turkey. Observers (also members of the EU): Austria, Denmark, Finland, Ireland and Sweden. Associate Partners (also members of NATO AND THE EU) : Bulgaria, Estonia, Latvia, Lithuania, Romania, Slovakia and Slovenia.

Wilhelm-Foerster-Sternwarte e V mit Zeiss-Planetarium Berlin
Munsterdamm 90, D-12169 Berlin, Germany
Tel: (+49 30) 790 09 30
Fax: (+49 30) 79 00 93 12
e-mail: wfs@wfs.be.schule.de
Web: www.planetarium-am-insulaner.de
■ Information on astronomic and space research.

Willacy County Development Corporation for Spaceport Facilities
190 North 3rd Street, Raymondville, Texas, 78580, United States
Tel: (+1 956) 689 33 93
(+1 956) 689 48 17
e-mail: county.judge@willacycounty.org
Web: southtexasspaceport.org

Willis Aerospace
10 Trinity Square, London, EC3P 3AX, United Kingdom
Tel: (+44 20) 78 60 90 90
Fax: (+44 20) 78 60 91 20
(+44 20) 74 88 85 87
Cable: WILLIS
e-mail: willis.aerospace.airline@willis.com
Web: www.willis.com
■ Provides insurance broking and risk management services. Specialises in services for airports and associated activities.

Willis Inspace
Willis Corroon Inspace
6700–A Rockledge Drive, Bethesda, Maryland, 20817, United States
Tel: (+1 301) 530 50 50
Fax: (+1 301) 897 88 17
Web: www.willis.com
■ Provide international space insurance brokerage services. Provide services including customised risk management program design, risk management consulting and insurance consulting. Offers coverage for space industry companies worldwide.

Wissenschaftliche Arbeitsgemeinschaft fuer Raketentechnik und Raumfahrt (WARR)
Scientific Working Group for Rocketry and Spaceflight
c/o Lehrstuhl f. Raumfahrttechnik, Boltzmannstrasse 15, D-85748 Garching, Germany
Tel: (+49 89) 28 91 60 12
Fax: (+49 89) 28 91 60 04
e-mail: info@warr.de
Web: www.warr.de
■ Student education with special emphasis on development and testing of experimental rockets and liquid propulsion systems.

WL Gore & Associates GmbH
Pleinfeld Office
Nordring 1, D-91785 Pleinfeld, Germany
Tel: (+49 9144) 60 10
Fax: (+49 9144) 63 61
Web: www.gore-careers.eu.com
■ Space wires and cables according to ESA/SCC detail specification 3901/007, 008, 009, 014, 019. Highly flexible, low attenuation microwave cable assemblies, applicable up to 40 GHz. Dielectric waveguides from 26.5 to 110 GHz. Highly flexible, low-attenuation coaxial cables. Round and flat conductor band cables. Data transmission cables, round cables, high voltage wires and cables. Harness shielding and jacket, sealants and EMI gaskets.

WorldSpace Corporation
Headquarters
2400 North Street, Washington, 20037, United States
Tel: (+1 202) 969 60 00
Fax: (+1 202) 969 60 01

Wyle Laboratories Inc
Test Engineering and Research Group
1960 East Grand Avenue, Suite 900, El Segundo, California, 90245-5023, United States
Tel: (+1 310) 563 68 00
Fax: (+1 310) 322 36 03
(+1 310) 322 07 81
e-mail: service@wylelabs.com
Web: www.wylelabs.com
■ Engineering, testing and support services, launch support, life sciences systems and services, small payload integration, and small experiment payload engineering.

Wyman-Gordon Ltd
Cameron Forged Products
Wyman-Gordon Co
Houstoun Road, Livingston, EH54 5BZ, United Kingdom
Tel: (+44 1506) 44 62 00
Fax: (+44 1506) 44 63 00
■ Manufacture of closed-dye forgings and extruded seamless pipe and other sections in nickel-base, titanium, aluminium, copper and steel alloys, using 30,000 and 9,000 presses. Manufacture aircraft components.

X

X Rocket LLC
9609 North East 195th Circle #L-11, Bothell, Washington, 98011, United States
e-mail: marketing@x-rocket.com
Web: www.x-rocket.com

X-Ray Crystallography Shared Facility
262 Basic Health Sciences Building, Birmingham, Alabama, 35294-0005, United States
Tel: (+1 205) 934 53 29
Fax: (+1 205) 934 04 80

Xybion Corporation
240 Cedar Knolls Road, Cedar Knolls, New Jersey, 07927-1698, United States
■ Design systems, produce commercial software, manufacture electronic and mechanical equipment to conduct system engineering and integration; and perform scientific studies

Y

Ylinen Electronics Ltd
Teollisuustie 9 A, FI-02700 Kaunianen, Finland
Tel: (+358 9) 57 12 17 32
Fax: (+358 9) 505 55 47
Web: www.ylinen.fi
■ Provision of engineering services including requirements definition, prototype design and verification, design, development and manufacturing. Provision of microwave products from 1 GHz to 120 GHz and from single waveguide components to complete systems. Current activities include provision of 70 GHz radiometric receivers for the Planck/ Herschel spacecraft for low noise temperature (–30 K) and wide bandwidth (14 GHz). Provision of L-band sympathetic aperture radiometers and calibration subsystem of the SMOS earth observation spacecraft. Provision of transmit/ receive assemblies for the TerraSAR – X synthetic aperture radar spacecraft.

Yuzhnoye Design Office
SDO Yuzhnoye
3 Krivorozhskaya Street, 49 008 Dniepropetrovsk, Ukraine
Tel: (+380 56) 770 04 47
Fax: (+380 56) 770 01 25
e-mail: info@yuzhnoye.com
Web: www.yuzhnoye.com
■ Design and preparation of rocket and space complexes, launch vehicles (Cyclone, Zenit, Dnepr), spacecraft (Sich, Ocean) and production of their components. Provision of launch services. Testing: strength, pneumohydraulic, electrical, exposure to various environment conditions (temperature, humidity), radio frequency and firing tests of rotation motors.

Z

Zavod Petrovskogo
Ulitsa Turgeneva, 24, 603600 Nizhnyi Novgorod, Russian Federation
Tel: (+7 8312) 36 74 01
Fax: (+7 8312) 36 84 07
Telex: 151028
e-mail: petroff@sandy.ru
■ Production of naval instruments, industrial products and recording equipment, including special purpose aircraft tape recorders. New products include digital automated electronic stations, thermo-equipment sensors and elevator control panels.

Zavod Teploobmennik
Pr. Lenina, 93, 603600 Nizhnyi Novgorod, Russian Federation
Tel: (+7 8312) 58 99 68
(+7 8312) 58 98 09
Fax: (+7 8312) 53 09 96
(+7 8312) 53 17 76
e-mail: teplo@teplo.sinn.ru
■ Development and manufacture of air pressure regulation and air-conditioning systems for aerospace applications.

Zero Gravity Corporation (ZERO-G)
5275 Arville Street, Suite 116, Las Vegas, Nevada, 89118, United States
Web: www.gozerog.com/
■ Provide zero gravity parabolic flights for scientific, training and recreational purposes.

Zin Technologies
3000 Aerospace Parkway, Cleveland, Ohio, 44142, United States
Tel: (+1 216) 977 04 22
Fax: (+1 216) 977 06 26
Web: www.zin-tech.com

ZODIAC
AeroSafety Systems Segment
Aérazur
a part of the Zodiac Group, France
2 Rue Maurice Mallet, F-92130 Issy-les-Moulineaux, Cedex, France
Tel: (+33 1) 41 23 23 23
Fax: (+33 1) 46 48 74 79
e-mail: info@aerazur.com
Web: www.aerosafety.zodiac.com
■ Design and production of flexible aeronautical equipment for original manufacturers of aircraft, helicopters and weapons systems.

Zodiac
Deceleration and Protection Division
a subsidiary of Zodiac SA, France
2 rue Maurice Mallet, F-92130 Issy-les-Moulineaux, France
Tel: (+33 1) 41 23 23 23
Fax: (+33 1) 46 48 74 79
(+33 1) 46 48 83 41
Web: www.elastomeres.zodiac.com
■ Aircraft arresting systems, life-saving and survival equipment, parachutes, technical storage and fluid transfer systems. Flexible tanks for aircraft, helicopter and battle tanks, collapsible storage or transport tanks, Flexit electrical systems, aircraft pneumatic de-icers. Inflatable tents, pollution boom, aristock, inflatable seals and airbags.

Zodiac International
Marine Division
a subsidiary of Zodiac SA, France
2 rue Maurice Mallet, F-92130 Issy-les-Moulineaux, France
Tel: (+33 1) 41 23 23 23
Fax: (+33 1) 41 23 23 98
(+33 1) 46 48 83 87
e-mail: solassales@zodiac.com
Web: www.zodiacsolas.com
■ Manufacture of inflatable boats, hurricane semi-rigid craft and SOLAS liferafts. Design and manufacture of inflatable products. Superinsulating materials for satellite thermal control systems. Aircraft interiors.

Zvezda Joint Stock Company
123 Babushkina Strasse, 192012 St Petersburg, Russian Federation
Tel: (+7 812) 267 83 35
(+7 812) 103 00 70
(+7 812) 262 07 47
Fax: (+7 812) 267 23 64
(+7 812) 367 37 76
e-mail: zakr@zvezda.spb.ru
office@zvezda.spb.ru
press@zvezda.spb.ru
Web: www.zvezda.spb.ru
■ Design, manufacture and maintenance of high-speed lightweight diesel engines for marine, locomotive and industrial applications. Also provides marine gearboxes, gensets and power plants for various facilities.

For details of the latest updates to *Jane's Space Systems and Industry* online and to discover the additional information available exclusively to online subscribers please visit
jsd.janes.com

SPACE LOGS

CONTENTS

LOGS

Table 1: Space Records

Top Ten Single Flight Individual Space Endurance Records

World Rank	Astronaut/Cosmonaut	Mission	Flight Duration DD:HH:MM
1	Polyakov, Russian Federation	Soyuz TM-18/Mir Crews 15–17	437:17:59
2	Avdeyev, Russian Federation	Soyuz TM-28/Mir Crews 26–27	379:14:51
3	Manarov, USSR	Soyuz TM-4/Mir Crew 3	365:22:39
3	Titov, USSR	Soyuz TM-4/Mir Crew 3	365:22:39
4	Romanenko, USSR	Soyuz TM-2/Mir Crew 2	326:11:38
5	Krikalev, USSR/Russian Federation	Soyuz TM-12/Mir Crews 9–10	311:20:01
6	Polyakov, USSR	Soyuz TM-6/Mir Crews 3–4	240:22:35
7	Atkov, USSR	Soyuz T-10/Salyut 7	236:22:50
7	Kizim, USSR	Soyuz T-10/Salyut 7	236:22:50
7	Solovyov, USSR	Soyuz T-10/Salyut 7	236:22:50

Top Ten Cumulative Flight Individual Space Endurance Records

World Rank	Astronaut/Cosmonaut	Cumulative Duration DD:HH:MM	Total Flights
1	Krikalev, Russian Federation	803:09:43	6
2	Avdeyev, Russian Federation	747:14:15	3
3	Polyakov, Russian Federation	678:16:34	2
4	Solovyov, Russian Federation	651:00:04	5
5	Kaleri, Russian Federation	609:21:53	4
6	Afanasyev, Russian Federation	555:18:35	4
7	Usachev, Russian Federation	552:22:27	4
8	Manarov, USSR	541:00:30	2
9	Viktorenko, Russian Federation	489:01:36	4
10	Budarin, Russian Federation	444:01:27	3

Cumulative Individual Extra-Vehicular Activity Endurance Records

World Rank	Astronaut/Cosmonaut	Cumulative EVAs Duration HH:MM	Total EVAs
1	Solovyov, Russian Federation	77:41	16
2	Ross, United States	58:14	9
3	Smith, United States	49:48	7
4	Budarin, Russian Federation	46:01	9
5	Newman, United States	43:13	6
6	Musabayev, Kazakhstan	42:36	8
7	Onufrienko, Russian Federation	42:32	8
8	Avdeyev, Russian Federation	41:59	10
9	Krikalev, Russian Federation	41:26	8
10	Afanasyev, Russian Federation	38:33	7

Table 2: World Human Spaceflight 1996 — 2007

Mission	Launch Date	Crew	Crew Flight Duration (DD:HH:MM)
NASA Shuttle Endeavour STS-72	11.01.96	Duffy, Jett, Barry, Chiao, Scott, Wakata	08:22:01
Retrieve Japan's Space Flyer Unit; deploy/retrieve SPARTAN OAST-Flyer; 2 generic Extra-Vehicular Activities (EVAs)			
RKA Soyuz TM-23	21.02.96	Onufrienko, Usachev	193:19:08
21st Mir Space Station long stay			
NASA Shuttle Columbia STS-75	22.02.96	Allen, Horowitz, Chang-Diaz, Cheli, Hoffman, Nicollier, Guidoni	15:17:40
USMP-3; deploy/retrieve Italy's Tethered Satellite System (TSS-1R) – TSS lost: tether snapped; seventh Extended Duration Orbiter (EDO)			
NASA Shuttle Atlantis STS-76	22.03.96	Chilton, Searfoss, Lucid, Godwin, Clifford, Sega	09:05:16
Shannon Lucid remained on Mir, returning aboard STS-79 in September 1996. NASA's Clifford/Godwin exited docking adapter hatch to mount three experiments on Mir's docking module			
NASA Shuttle Endeavor STS-77	19.05.96	Casper, Brown, Bursch, Runco, A Thomas, Garneau	10:00:39
SPACEHAB-4; deploy/retrieve SPARTAN-207 (inflatable antenna experiment); deploy Passive Aerodynamically Stabilised Magnetically Damped Satellite (PAMS)			
NASA Shuttle Columbia STS-78	20.06.96	Henricks, Kregel, Helms, Linnehan, Brady, Favier, Thirsk	16:21:48
Life & Microgravity Spacelab (LMS); eighth EDO			
RKA Soyuz TM-24	17.08.96	Korzun, Kaleri André-Deshays	196:17:26 15:18:24
22nd long-stay Russian mission to Mir; French 'Cassiopeia' mission			*(5th French visit)*
NASA Shuttle Endeavour STS-79	16.09.96	Readdy, Wilcutt, Apt, Akers, Walz, Blaha (up) Lucid (down)	10:03:18 188:04:00
Delivered Blaha to Mir, returned Lucid			
NASA Shuttle Columbia STS-80	19.11.96	Cockrell, Rominger, Jernigan, Jones, Musgrave	17:15:53
Deployment and retrieval of WSF-3 and ORFEUS-SPAS-II; ninth EDO; planned EVAs by Jernigan and Jones cancelled			
NASA Shuttle Atlantis STS-81	12.01.97	Baker, Jett, Wilsoff, Grunsfeld, Ivins, Linenger (up) Blaha (down)	10:04:56 128:05:27
Delivered Linenger to Mir, returned with Blaha			
RKA Soyuz TM-25	10.02.97	Tsibilev, Lazutkin Ewald	184:22:08 19:16:35
23rd long-stay Russian mission to Mir; second German visit to the station. Accident-prone residency saw fire aboard Mir during the hand-over period with the Soyuz-TM 24 crew, collision of Progress-M 34 with Spektr module and the brief accidental disconnection of the computer system			
NASA Shuttle Discovery STS-82	11.02.97	Bowersox, Horowitz, Tanner, Smith, Hawley, Harbaugh, Lee	9:23:37
Second servicing mission for the Hubble Space Telescope: new instruments installed, older ones returned to Earth, repairs to insulation; five EVAs completed, Lee/Smith 3, Harbaugh/Tanner 2			
NASA Shuttle Columbia STS-83	04.04.97	Halsell, Still, J E Voss, Gernhardt, Linteris, D Thomas, Crounch	3:23:13
Microgravity Science Laboratory-1 (MSL-1) primary payload for 10th EDO: mission curtailed due to fuel cell problems and re-flown as STS-94			
NASA Shuttle Atlantis STS-84	15.05.97	Precourt, Collins, Clervoy, Noriega, Lu, Kondakova, Foale (up) Linenger (down)	9:05:21 132:03:59
Sixth Mir docking mission with SPACEHAB-DM. Delivered supplies to Mir, left Foale on board for medium-length stay, returned with Linenger			
NASA Shuttle Columbia STS-94	01.07.97	Halsell, Still, J E Voss, Gernhardt, D Thomas, Crouch, Linteris	15:16:45
Reflight of STS-83, MSL-1 primary payload for 11th EDO: fastest return-to-orbit for the crew members			
RKA Soyuz TM-26	05.08.97	Solovyov, Vinogradov	197:17:34
24th long-stay Russian mission to Mir: planned flight of a French astronaut delayed until Soyuz TM-27 because of power problems on Mir following the Progress-M 34 collision with Spektr			
NASA Shuttle Discovery STS-85	07.08.97	Brown, Rominger, Davis, Curbeam, Robinson, Tryggvason	11:20:28
CRISTA-SPAS-2 deployment, test of Japanese remote manipulator system			
NASA Shuttle Atlantis STS-86	25.09.97	Wetherbee, Bloomfield, Titov, Parazynski, Chretien, Lawrence, Wolf (up) Foale (down)	10:19:22 144:13:48
Seventh Mir docking; Lawrence originally planned to remain aboard Mir but replaced by Wolf because her height precludes use of Russian EVA suit. EVA with Parazynski and Titov.			
NASA Shuttle Columbia STS-87	19.11.97	Kregel, Lindsey, Chawla, Scott, Doi, Kadenyuk	15:16:35
USMP4, deployment and retrieval of SPARTAN 201-04. 12th EDO. Scott and Doi performed two EVAs, one to manually capture SPARTAN			
NASA Shuttle Endeavour STS-89	22.01.98	Wilcutt, Edwards, Reilly, Anderson, Dunbar, Sharipov, A Thomas (up) Wolf (down)	08:19:48 127:10:01
Delivered Thomas, final United States astronaut to Mir, returned Wolf			
RKA Soyuz TM-27	29.01.98	Musabeyev, Budarin Eyharts	207:12:49 20:16:36
25th long-stay aboard Mir carrying French CNES astronaut Eyharts			
NASA Shuttle Columbia STS-90	17.04.98	Searfoss, Altman, Hire, Linnehan, Williams, Buckey, Pawelczyk	15:21:51
US Neurolab mission; final Spacelab module flight. 13th and last planned EDO before ISS operations for long duration missions			
NASA Shuttle Discovery STS-91	02.06.98	Precourt, Gorie, Chang-Diaz, Lawrence, Kavandi, Ryumin A Thomas (down)	09:19:55 140:15:12
Ninth and last Mir docking, first with Discovery. Cosmonaut Ryumin conducted Mir inspection to determine suitability for sustained operations. A Thomas returned to Earth, last United States astronaut to Mir			
RKA Soyuz TM-28	13.08.98	Padalka Avdeyev Baturin	198:16:31 379:14:51 11:19:40
26th long-stay aboard Mir; with seats on the next and last manned flight to Mir pre-sold to Slovak and French cosmonaut/astronaut; flight plan constantly revised			
NASA Shuttle Discovery STS-95	29.10.98	Brown, Lindsey, Parazynski, Robinson, Duque, Mukai, Glenn	08:21:45
Life sciences mission including the World's oldest astronaut, John Glenn aged 77. Deployed and retrieved SPARTAN 201-05 and tested Hubble Space Telescope orbital systems test platform			
NASA Shuttle Endeavour STS-88	04.12.98	Cabana, Sturckow, Ross, Currie, Newman and Krikalev	11:19:53
RKA Soyuz TM-29	20.02.99	Afanasyev, Haignere Bella	188:20:16 07:21:56
27th long-stay aboard Mir; French astronaut stays with Mir crew, while Slovak cosmonaut returns; Avdeyev continues to remain aboard with Afanasyev and Haignere until 28 August 1999, when the docked TM-29 vessel returns to Earth and leaves Mir uninhabited			

Mission	Launch Date	Crew	Crew Flight Duration (DD:HH:MM)
NASA Shuttle Discovery STS-96	27.05.99	Rominger, Husband, Jernigan, Ochoa, Barry, Payette, Tokarev	09:19:14
First Shuttle ISS servicing mission to docked Zarya/Unity modules. Released Student-Tracked Atmospheric Research Satellite for Heuristic International Networking Experiment-1 (STARSHINE-1) satellite			
NASA Shuttle Columbia STS-93	23.07.99	Collins, Ashby, Coleman, Hawley, Tognini	04:22:50
Launch of Chandra X-ray observatory; first female commander (Collins)			
NASA Shuttle Discovery STS-103	19.12.99	Brown, Kelly, Clervoy, Smith, Foale, Grunsfeld, Nicollier	07:23:11
Third Hubble Space Telescope servicing mission			
NASA Shuttle Endeavour STS-99	11.02.00	Kregel, Gorie, Thiele, Kavandi, J E Voss, Mohri	11:05:38
Shuttle Radar Topography Mission (SRTM)			
RKA Soyuz TM-30	04.04.00	Zalyotin, Kaleri	72:19:52
28th Mir crew reopened uninhabited station for engineering evaluation; Zalyotin and Kaleri depart Mir and return to Earth on 16 June 2000, leaving the space station unmanned for the last time; final piloted Mir expedition before the space station's scheduled atmospheric re-entry and destruction on 23 March 2001			
NASA Shuttle Atlantis STS-101	19.05.00	Halsell, Horowitz, Weber, Williams, J S Voss, Helms, Usachev	09:20:10
Third Shuttle ISS flight (2A.2a) with double SPACEHAB module; first upgraded "glass cockpit" flight			
NASA Shuttle Atlantis STS-106	08.09.00	Wilcutt, Altman, Lu, Mastracchio, Burbank, Malenchenko, Morukov	11:19:11
Fourth Shuttle ISS flight (2A.2b) with double SPACEHAB module			
NASA Shuttle Discovery STS-92	11.10.00	Duffy, Melroy, Wakata, Wisoff, Chiao, McArthur, Lopez-Alegria	12:21:44
100th Shuttle Mission. Fifth Shuttle ISS flight (3A) delivering and installing Z1 Truss and Pressurized Mating Adaptor-3 (PMA-3)			
RKA Soyuz TM-31	31.10.00	Shepherd, Krikalev, Gidzenko	140:23:39
ISS Expedition 1			
Expedition 1 crew for ISS, returned in STS-102			
NASA Shuttle Endeavour STS-97	30.11.00	Jett, Bloomfield, Garneau, Tanner, Noriega	10:19:58
Sixth Shuttle ISS flight (4A) delivering and attaching P6 Integrated Truss Segment with solar array			
NASA Shuttle Atlantis STS-98	07.02.01	Cockrell, Polansky, Ivins, Jones, Curbeam	12:21:20
Seventh Shuttle ISS flight (5A) carrying US Destiny laboratory			
NASA Shuttle Discovery STS-102	08.03.01	Weatherbee, Kelly, A Thomas, Richards	12:19:49
ISS Expedition 2		Usachev (up), Helms (up), J S Voss (up)	167:06:41
ISS Expedition 1		Shepherd (down), Krikalev, (down), Gidzenko (down)	140:23:39
Eighth Shuttle ISS flight (5A.1) with Leonardo Multi-Purpose Logistics Module (MPLM), Expedition 2 crew			
NASA Shuttle Endeavour STS-100	19.04.01	Rominger, Ashby, Hadfield, Phillips, Parazynski, Guidoni, Lonchakov	11:21:30
Ninth Shuttle ISS flight (6A) with Rafaello MPLM and Space Station Mobile Servicer System (SSRMS)			
RKA Soyuz TM-32	28.04.01	Musabeyev, Baturin, Tito	07:22:04
Taxi flight to ISS carrying private US citizen and first space tourist, Dennis Tito; returned in Soyuz TM-31			
NASA Shuttle Atlantis STS-104	12.07.01	Lindsay, Hobaugh, Gernhardt, Reilly, Kavandi	12:18:35
Tenth Shuttle ISS flight (7A) delivers Quest Airlock module			

Mission	Launch Date	Crew	Crew Flight Duration (DD:HH:MM)
NASA Shuttle Discovery STS-105	10.08.01	Horowitz, Sturckow, Forrester, Barry	11:21:13
ISS Expedition 3		Culbertson (up), Tyurin (up), Dezhurov (up)	128:20:45
ISS Expedition 2		Usachev (down), J S Voss (down), Helms (down)	167:06:41
Eleventh Shuttle ISS flight (7A.1) with Leonardo MPLM, Expedition 3 crew			
RKA Soyuz TM-33	21.10.01	Afanasyev, Kozeev, Haignere	09:20:00
Taxi flight to ISS, returned in TM-32			
NASA Shuttle Endeavour STS-108	05.12.01	Gorie, Kelly, Godwin, Tani	11:19:36
ISS Expedition 4		Onufrienko (up), Bursch (up), Walz (up)	195:19:39
ISS Expedition 3		Culbertson (down), Dezhurov (down), Tyurin (down)	128:20:45
Twelfth Shuttle ISS flight (UF1) with Rafaello MPLM, Expedition 4 crew; STARSHINE-2 deployed; September 11 flag tribute			
NASA Shuttle Columbia STS-109	01.03.02	Altman, Carey, Grunsfeld, Currie, Massimino, Linnehan, Newman	10:22:10
Fourth Hubble Space Telescope servicing mission			
NASA Shuttle Atlantis STS-110	08.04.02	Bloomfield, Frick, Walheim, Ochoa, Morin, Ross, Smith	10:19:43
Thirteenth Shuttle ISS flight (8A) with S0 Truss and Mobile Transporter			
RKA Soyuz TM-34	25.04.02	Gidzenko, Vittori, Shuttleworth	9:21:25
Final mission of the Soyuz TM spacecraft type; taxi to the ISS carried Italian ESA astronaut, Vittori and the second space tourist, private South African citizen, Mark Shuttleworth			
NASA Shuttle Endeavour STS-111	05.06.02	Cockrell, Lockhart, Perrin, Chang-Diaz	13:20:35
ISS Expedition 5		Whitson (up), Korzun (up), Treschev (up)	184:22:14
ISS Expedition 4		Walz (down), Bursch (down), Onufrienko (down)	195:19:39
Fourteenth Shuttle ISS flight (UF-2) with Leonardo MPLM, Mobile Base System (MBS), Expedition 5 crew			
NASA Shuttle Atlantis STS-112	07.10.02	Ashby, Melroy, Wolf, Magnus, Sellers, Yurchikhin	10:19:58
Fifteenth Shuttle ISS flight (9A) with S1 Truss installation			
RKA Soyuz TMA-1	30.10.02	Zalyotin, Lonchakov, De Winne	10:20:53
First Soyuz TMA spacecraft type in piloted orbital flight; carried ESA astronaut, Frank De Winne; taxi crew returned to Earth in TM-34, marking the final flight and re-entry of the Soyuz TM type spacecraft			
NASA Shuttle Discovery STS-113	23.11.02	Wetherbee, Lockhart, Lopez-Alegria, Herrington	13:18:47
ISS Expedition 6		Bowersox (up), Budarin (up), Pettit (up)	161:01:17
ISS Expedition 5		Korzun (down), Whitson (down), Treschev (down)	184:22:14
Sixteenth Shuttle ISS flight (11A) with P1 Truss, Expedition 6 crew			
NASA Shuttle Columbia STS-107	16.01.03	Husband, McCool, Brown, Chawla, Anderson, Clark, Ramon	15:22:20
Microgravity research with SPACEHAB-Research Double Module; Shuttle destroyed and crew lost during re-entry on 1 February 2003			
RKA Soyuz TMA-2	26.04.03	Malenchenko, Lu	184:21:47
ISS Expedition 7			
ISS Expedition 7 crew, carrying one Russian cosmonaut and one US astronaut to the ISS			
RKA Soyuz TMA-1	04.05.03 (Landing Date)	Bowersox, Budarin, Pettit	161:01:17
ISS Expedition 6			
Expedition 6 crew use the docked, Soyuz TMA-1 vessel to depart the ISS following the Columbia disaster and US Shuttle programme grounding			

Mission	*Launch Date*	*Crew*	*Crew Flight Duration (DD:HH:MM)*
CNSA Shenzhou 5	15.10.03	Yang	00:21:23
Astronaut Yang Liwei commands China's first piloted, orbital space mission			
RKA Soyuz TMA-3	18.10.03	Kaleri, Foale	194:18:35
ISS Expedition 8		Duque	09:21:02
Expedition 8 crew; Duque, an ESA astronaut, departs the ISS along with the Expedition 7 crew in TMA-2; Kaleri and Foale remain aboard			
FSA Soyuz TMA-4	19.04.04	Padalka, Fincke	187:21:17
ISS Expedition 9		Kuipers	10:20:53
Expedition 9 crew; Kuipers, an ESA astronaut, departs the ISS along with the Expedition 8 crew in TMA-3; Padalka and Fincke remain			
FSA Soyuz TMA-5	14.10.04	Chiao, Sharipov	192:19:02
ISS Expedition 10		Shargin	09:21:29
Expedition 10 crew; Cosmonaut Shargin returns to Earth with Expedition 9 crew in TMA-4; Astronaut Chiao and Cosmonaut Sharipov remain			
FSA Soyuz TMA-6	15.04.05	Krikalev, Phillips	179:00:23
ISS Expedition 11		Vittori	09:21:21
Expedition 11 crew; ESA astronaut Vittori returns to Earth with Expedition 10 crew in TMA-5; Krikalev and Phillips remain			
NASA Shuttle Discovery STS-114	26.07.05	Collins, Kelly, Camarda, Lawrence, Noguchi, Robinson, A Thomas	13:21:32
First Shuttle Return to Flight test mission; seventeenth Shuttle ISS flight (LF1); Raffaello MPLM, External Stowage Platform-2 (ESP-2); in-flight shuttle repair EVA			
FSA Soyuz TMA-7	01.10.05	McArthur, Tokarev	189:19:53
ISS Expedition 12		Olsen	09:21:15
Gregory Olsen, private US citizen and the third space tourist, accompanies the Expedition 12 crew into orbit; Olsen, also a scientist and PhD holder, conducts ESA-sponsored medical experiments and departs with the Expedition 11 crew; McArthur and Tokarev remain			
CNSA Shenzhou 6	12.10.05	Fei, Nie	04:19:33
China's second piloted orbital mission, carrying astronauts Fei Junlong and Nie Haisheng; the crew performed medical experiments			
FSA Soyuz TMA-8	30.03.06	Vinogradov, Williams	182:23:43
ISS Expedition 13		Pontes	8:22:18
		Reiter	171:03:54
Brazilian AEB astronaut, Pontes launched along with the Expedition 13 crew and returned to Earth with the Expedition 12 crew, McArthur and Tokarev, on 9 April 2006; ESA astronaut, Reiter launched with the STS-121 crew (see below) in July 2006, and remained aboard the ISS with the Expedition 13 crew; Reiter stayed aboard the ISS during part of the Expedition 14 Crew's mission as well and was replaced by Sunita Williams, who accompanied the STS 116 crew (see below) to the ISS in December 2006			
NASA Shuttle Discovery STS-121	04.07.06	Lindsey, Kelly, Fossum, Nowak, Reiter, Sellers, Wilson	12:19:37
Second Return to Flight test mission, eighteenth Shuttle ISS flight; utilization and logistics flight (ULF 1.1); three EVAs; repeated flight schedule delays; ESA astronaut, Thomas Reiter remained aboard the ISS with the Expedition 13 and 14 crews (see above and below)			
NASA Shuttle Atlantis STS-115	09.09.06	Jett, Ferguson, Burbank, MacLean, Stefanyshyn-Piper, Tanner	11:19:06
Nineteenth Shuttle ISS flight; ISS port three and four (P3/P4) integrated truss segment installation; solar wing deployment; three EVAs scheduled			
FSA Soyuz TMA-9	18.09.06	Lopez-Alegria, Tyurin	215:08:22
ISS Expedition 14		Ansari	10:21:05
		S Williams	TBA
Daisuke Enomoto, the fourth space tourist in training, was expected to accompany the Soyuz TMA-9/Expedition 14 crew to the ISS, however, his participation was cancelled due to an unspecified medical condition that made him unfit to fly on the coming mission; originally Enomoto's backup, Anousheh Ansari (a member of the 'Ansari X Prize' family), participated as the fourth space tourist – she returned to Earth with the Expedition 13 crew after her flight to the ISS; Sunita Williams joined the ISS Expedition 14 crew, after launch with the Shuttle STS-116 crew (see below)			
NASA Shuttle Discovery STS-116	10.12.06	Polansky, Oefelein, Curbeam, Fuglesang, Higginbotham, Patrick, Reiter, S Williams	12:20:44
Twentieth Shuttle ISS flight (12A.1); Discovery carried the P5 Truss and the SPACEHAB module; the crew rewired the ISS electrical system; ESA astronaut Fuglesang accompanied the NASA crew, and Sunita Williams replaced ESA astronaut, Thomas Reiter, who descended with STS-116 and was aboard the ISS since July 2006; Shuttle liftoff marked its first night launch since 2002; four EVAs, of which Curbeam participated in all, setting a single Shuttle mission, single astronaut EVA record			
FSA Soyuz TMA-10	07.04.07	Yurchikhin, Kotov	TBA
ISS Expedition 15		Simonyi	13:19:00
Charles Simonyi, the fifth space tourist to visit the ISS, launched along with the Expedition 15 crew, and returned to Earth along with Expedition 14 on 21 April 2007. He reportedly paid up to USD25 million to be included in the journey. Sunita Williams, aboard the ISS since December 2006, will remain there until June 2007, due to the damage caused to the Shuttle Atlantis' external fuel tank by a hail storm, and that Orbiter's subsequent launch delay. Atlantis' STS-117 original launch date was set for April 2007			

Space Agency Acronyms:
AEB: Agência Espacial Brasileira
CNES: French Centre National d'Etudes Spatiales
CNSA: China National Space Administration
ESA: European Space Agency
FSA: Russian Federal Space Agency
NASA: US National Aeronautics and Space Administration
RKA: Russian Space Agency (1992–2004)

Table 3: US Space Shuttle Flights

Flight	Launch/Landing dates (DD.MM.YY) GMT Orbiter	Launch pad Landing site	Crew	Duration (DD:HH:MM) Inclination	Notes
01 **STS-1**	12.04.81/14.04.81 Columbia	KSC 39A EAFB	Young, Crippen	02:06:21 40.4°	*Shuttle debut mission; some thermal tiles lost*
02 **STS-2**	12.11.81/14.11.81 Columbia	KSC 39A EAFB	Engle, Truly	02:06:13 38.0°	*Tested Remote Manipulator System (RMS) arm; 5-day mission halved by fuel cell fault*
03 **STS-3**	22.03.82/30.03.82 Columbia	KSC 39A White Sands, New Mexico	Lousma, Fullerton	08:00:05 38.0°	*Third test; landing delayed by one day by storm and diverted to White Sands from EAFB*
04 **STS-4**	27.06.82/04.07.82 Columbia	39A EAFB	Mattingly, Hartsfield	07:01:10 28.45°	*Final test; first concrete runway landing. Two SRB lost; Cirris DoD payload failure*
05 **STS-5**	11.11.82/16.11.82 Columbia	KSC 39A EAFB	Brand, Overmyer Allen, Lenoir	05:02:14 28.45°	*First operational flight. SBS-C and ANIK C-3 satellites deployed. EVA cancelled*
06 **STS-6**	04.04.83/09.04.83 Challenger	KSC 39A EAFB	Weitz, Bobko Peterson, Musgrave	05:00:23 28.45°	*Challenger debut. First Tracking and Data Relay Satellite-1 (TDRS-1) deployed on IUS. First Shuttle EVA*
07 **STS-7**	18.06.83/24.06.83 Challenger	KSC 39A EAFB	Crippen, Hauck Fabian, Ride, Thagard	06:02:24 28.45°	*First US woman (Ride). First deploy/retrieve with Shuttle Pallet Satellite (SPAS-01). ANIK C-2 and PALAPA-B1 satellites deployed*
08 **STS-8**	30.08.83/05.09.83 Challenger	KSC 39A EAFB	Truly, Brandenstein Bluford, Gardner, W Thornton	06:01:07 28.45°	*First night launch/landing. India's INSAT-1B telecom satellite deployed; first African-American in space (Bluford)*
09 **STS-9**	28.11.83/08.12.83 Columbia	KSC 39A EAFB	Young, Shaw Garriott, Parker, Lichtenberg, Merbold	10:07:47 57.0°	*Spacelab-1 demonstration in partnership with ESA; 90 per cent science achieved. First 6-crew member flight; carried first ESA astronaut (Merbold)*
10 **41B**	03.02.84/11.02.84 Challenger	KSC 39A KSC	Brand, Gibson McCandless, Stewart, McNair	07:23:17 28.45°	*First Manned Maneuvering Unit (MMU) tests, untethered EVA; Solar Maximum satellite EVA repair rehearsal. WESTAR-VI and PALAPA-B2 lost. First KSC landing*
11 **41C**	06.04.84/13.04.84 Challenger	KSC 39A EAFB	Crippen, Scobee G Nelson, van Hoften, Hart	06:23:40 28.45°	*Long Duration Exposure Facility (LDEF) deploy. First satellite retrieval/repair (Solar Maximum)*
12 **41D**	30.08.84/05.09.84 Discovery	KSC 39A EAFB	Hartsfield, Coats Resnik, Hawley, Mullane, C Walker	06:00:56 28.45°	*Discovery debut. SBS-D, SYNCOM IV-2 (also known as LEASAT-2), TELSTAR satellites deployed. Solar panel deploy test. First commercial Payload Specialist*
13 **41G**	05.10.84/13.08.84 Challenger	KSC 39A KSC	Crippen, McBride Sullivan, Ride, Leestma, Garneau, Scully-Power	08:05:23 57.0°	*Office of Space and Terrestrial Applications-3 (OSTA-3), Earth Radiation Budget Satellite (ERBS), Large Format Camera. Firsts: 7 crew, US woman EVA (Sullivan), 2-woman flight (Ride/Sullivan). Second KSC landing*
14 **51A**	08.11.84/16.11.84 Discovery	KSC 39A KSC	Hauck, D Walker A Fisher, D Gardner, Allen	07:23:45 28.45°	*First satellite recovery/return: PALAPA-B2 and WESTAR-VI. TELESAT-H (ANIK) and SYNCOM IV-1 (also known as LEASAT-1) deployed. Third KSC landing*
15 **51C**	24.01.85/27.01.85 Discovery	KSC 39A KSC	Mattingly, Shriver Onizuka, Buchli, Payton	03:01:33 28.45°	*First dedicated DoD flight: Mercury SIGINT on IUS. 100th manned orbital mission. Fourth KSC landing*
16 **51D**	12.04.85/19.04.85 Discovery	KSC 39A KSC	Bobko, Williams Seddon, Hoffman, Griggs, C Walker, Garn	06:23:56 28.45°	*TELESAT-I (ANIK C-1) and SYNCOM IV- 3 (also known as LEASAT-3) deployed; EVA failed to repair LEASAT-3, first US congress member in space (Garn); Fifth KSC landing*
17 **51B**	29.04.85/06.05.85 Challenger	KSC 39A EAFB	Overmyer, Gregory Lind, Thagard, W Thornton, van den Berg, Wang	07:00:08 57.0°	*Spacelab-3 science mission; released small NUSAT satellite*
18 **51G**	17.06.85/24.06.85 Discovery	KSC 39A EAFB	Brandenstein, Creighton, Lucid Fabian, Nagel, Baudry, Al-Saud	07:01:38 28.45°	*Shuttle Pointed Autonomous Research Tool for Astronomy (SPARTAN-1) deploy/retrieve; MORELOS-A, ARABSAT-A and TELSTAR-3D satellite deployments*
19 **51F**	29.07.85/06.08.85 Challenger	KSC 39A EAFB	Fullerton, Bridges Musgrave, Henize, England, Acton, Bartoe	07:22:45 49.5°	*Spacelab-2 astronomy/science mission (no pressurised module)*
20 **51I**	27.08.85/03.09.85 Discovery	KSC 39A EAFB	Engle, Covey van Hoften, Lounge, W Fisher	07:02:18 28.45°	*ASC-1, AUSSAT-1 and SYNCOM IV-4 (also known as LEASAT-4) satellites deployed; LEASAT-3 EVA repair*
21 **51J**	03.10.85/07.10.85 Atlantis	KSC 39A EAFB	Bobko, Grabe Hilmers, Stewart, Pailes	04:01:44 28.5°	*DoD Defense Satellite Communications System-III (DSCS-III) pair on IUS. Atlantis debut*
22 **61A**	30.10.85/06.11.85 Challenger	KSC 39A EAFB	Hartsfield, Nagel Dunbar, Buchli, Bluford, Messerschmid, Furrer, Ockels	07:00:44 57.0°	*Spacelab D-1: first foreign dedicated (West German) Shuttle mission; first 8-man crew*
23 **61B**	27.11.85/03.12.85 Atlantis	KSC 39A EAFB	Shaw, O'Connor Cleave, Ross, Spring, C Walker, Vela	06:21:05 28.45°	*MORELOS-B, SATCOM KU-2, AUSSAT-2 satellites deployed; Experimental Assembly of Structures in Extravehicular Activity (EASE) and Assembly Concept for Construction of Erectable Space Structure (ACCESS) experiments conducted, marking first International Space Station (ISS) assembly demo*
24 **61C**	12.01.86/18.01.86 Columbia	KSC 39A EAFB	Gibson, Bolden G Nelson, Hawley, Chang-Diaz, Cenker, W Nelson	06:02:04 28.45°	*SATCOM KU-1 deploy; SDI experiments; second US congress member in space (W Nelson); scheduled KSC landing diverted to EAFB*
25 **51L**	28.01.86/None Challenger	KSC 39B None	Scobee, Smith Resnik, McNair, Onizuka, McAuliffe, Jarvis	00:00:01 28.5° (planned)	*Vehicle exploded after 73 seconds; crew killed; first lift-off from pad 39B*
26 **STS-26**	29.09.88/03.10.88 Discovery	KSC 39B EAFB	Hauck, Covey Hilmers, Lounge, G Nelson	04:01:00 28.5°	*First Return to Flight mission following the Challenger tragedy; TDRS-3 deployed on IUS stage; 11 mid-deck secondary payloads*
27 **STS-27**	02.12.88/06.12.88 Atlantis	KSC 39B EAFB	R Gibson, G Gardner Mullane, Ross, Shepherd	04:09:06 57.0°	*Third DoD mission; first DoD Lacrosse imaging radar satellite for all-weather day/night reconnaissance*
28 **STS-29**	13.03.89/18.03.89 Discovery	KSC 39B EAFB	Coats, Blaha Buchli, Springer, Bagian	04:23:39 28.45°	*TDRS-4 on IUS; five mid-deck + two cargo bay secondary payloads*
29 **STS-30**	04.05.89/08.05.89 Atlantis	KSC 39B EAFB	D Walker, Grabe Thagard, Cleave, Lee	04:00:58 28.85°	*Magellan Venus orbiter on IUS stage; arrived at Venus August 1990; first shuttle planetary mission*

Flight	Launch/Landing dates (DD.MM.YY) GMT / Orbiter	Launch pad / Landing site	Crew	Duration (DD:HH:MM) / Inclination	Notes
30 **STS-28**	08.08.89/13.08.89 Columbia	KSC 39B EAFB	Shaw, Richards Leestma, Adamson, Brown	05:01:00 57.00°	*Fourth DoD mission; possibly deployed Mercury satellite into Molniya-type orbit for relaying imaging reconnaissance satellite imagery. Flight similar to STS-53*
31 **STS-34**	18.10.89/23.10.89 Atlantis	KSC 39B EAFB	Williams, McCulley Lucid, Baker, Chang-Diaz	04:23:39 34.30°	*Galileo Jupiter orbiter on IUS stage; arrived at Jupiter December 1995*
32 **STS-33**	23.11.89/27.11.89 Discovery	KSC 39B EAFB	Gregory, Blaha Musgrave, K Thornton, Carter	05:00:07 28.45°	*Fifth DoD mission; possibly deployed second Mercury SIGINT satellite*
33 **STS-32**	09.01.90/20.01.90 Columbia	KSC 39A EAFB	Brandenstein, Wetherbee, Dunbar Low, Ivins	10:21:01 28.50°	*Deployed USN SYNCOM IV-F5 (also known as LEASAT 5) telecom satellite, retrieved Long Duration Exposure Facility (LDEF)*
34 **STS-36**	28.02.90/04.03.90 Atlantis	KSC 39A EAFB	Creighton, Casper Hilmers, Mullane, Thuot	04:10:18 62.00°	*Sixth DoD mission; deployed Misty/AFP-731 reconnaissance satellite*
35 **STS-31**	24.04.90/29.04.90 Discovery	KSC 39B EAFB	Shriver, Bolden Hawley, McCandless, Sullivan	05:01:16 28.45°	*Deployed Hubble Space Telescope; Shuttle altitude record (619 km)*
36 **STS-41**	06.10.90/10.10.90 Discovery	KSC 39B EAFB	Richards, Cabana Shepherd, Melnick, Akers	04:02:10 28.45°	*Ulysses solar probe on IUS stage*
37 **STS-38**	15.11.90/11.20.90 Atlantis	KSC 39A KSC	Covey, Culbertson Springer, Meade, Gemar	04:21:54 28.45°	*Seventh DoD mission; possibly deployed Mercury relay satellite; sixth KSC landing, first since 1985*
38 **STS-35**	02.12.90/11.12.90 Columbia	KSC 39B EAFB	Brand, G Gardner Lounge, Hoffman, Parker, Parise, Durrance	08:23:05 28.45°	*Carried ASTRO-1 observatory's four telescopes*
39 **STS-37**	05.04.91/11.04.91 Atlantis	KSC 39B EAFB	Nagel, Cameron Ross, Godwin, Apt	05:23:33 28.45°	*Deployed Gamma Ray Observatory; 2 EVAs (first since 1985)*
40 **STS-39**	28.04.91/06.05.91 Discovery	KSC 39A KSC	Coats, Hammond Harbaugh, McMonagle, Bluford, Veach, Hieb	08:07:22 57.0°	*Unclassified DoD Air Force Program-675 (AFP-675); Infrared Background Signature Survey (IBSS) and IBSS platforms. Seventh KSC landing*
41 **STS-40**	05.06.91/14.06.91 Columbia	KSC 39B EAFB	O'Connor, Gutierrez Bagian, Jernigan, Seddon, Gaffney, Hughes-Fulford	09:02:14 39.0°	*Spacelab Life Sciences (SLS-1); first dedicated life sciences research; 12 Get Away Special (GAS) canisters*
42 **STS-43**	02.08.91/11.08.91 Atlantis	KSC 39A KSC	Blaha, Baker, Low Adamson, Lucid	08:21:21 28.45°	*Deployment of TDRS-5 on IUS stage; Lucid becomes first woman to make three flights. Eighth KSC landing, first planned since flight 61-C (diverted to EAFB)*
43 **STS-48**	12.09.91/18.09.91 Discovery	KSC 39A EAFB	Creighton, Reightler Buchli, Brown, Gemar	05:08:28 57.0°	*Deployment of Upper Atmosphere Research Satellite (UARS)*
44 **STS-44**	24.11.91/01.12.91 Atlantis	KSC 39A EAFB	Gregory, Henricks J F Voss, Musgrave, Runco, Hennen	06:22:51 28.5°	*Unclassified DoD deployment of Defense Support Program (DSP) early warning satellite on IUS*
45 **STS-42**	22.01.92/30.01.92 Discovery	KSC 39A EAFB	Grabe, Oswald Thagard, Readdy, Hilmers, Bondar, Merbold	08:01:15 57.0°	*Spacelab International Microgravity Laboratory (IML-1)*
46 **STS-45**	24.03.92/02.04.92 Atlantis	KSC 39A KSC	Bolden, Duffy Foale, Leestma, Sullivan, Lichtenberg, Frimout	08:22:09 57.0°	*Atmospheric Lab for Applications & Science (ATLAS-1). Ninth KSC landing*
47 **STS-49**	07.05.92/17.05.92 Endeavour	KSC 39B EAFB	Brandenstein, Chilton Melnick, Thuot, Hieb, K Thornton, Akers	08:21:18 28.35°	*Endeavour debut; INTELSAT VI (F-3) recovery/repair. 4 EVAs by 4 crew, total 60.1 hours (including first three-man and two record length EVAs)*
48 **STS-50**	25.06.92/09.07.92 Columbia	KSC 39A KSC	Richards, Bowersox Dunbar, Baker, Meade, DeLucas, Trinh	13:19:31 28.5°	*US Microgravity Laboratory-1 (USML-1) Spacelab. First Extended Duration Orbiter (EDO); Shuttle duration record. Tenth KSC landing*
49 **STS-46**	31.07.92/08.08.92 Atlantis	KSC 39B KSC	Shriver, Allen Chang-Diaz, Ivins, Hoffman, Nicollier, Malerba	07:23:15 28.5°	*Deployment of ESA's European Retrievable Carrier (EURECA-1); deployment/retrieval of Italy's Tethered Satellite System (TSS)*
50 **STS-47**	12.09.92/20.09.92 Endeavour	KSC 39B KSC	Gibson, Brown, Apt, Lee, Davis, Jemison, Mohri	07:22:30 57.0°	*Japan-sponsored Spacelab-J: 43 materials/life sciences experiments. First Japanese (Mohri), first African-American woman (Jemison) and first married couple (Lee and Davis) in space*
51 **STS-52**	22.10.92/01.11.92 Columbia	KSC 39B KSC	Wetherbee, Baker Veach, Shepherd, Jernigan, MacLean	09:20:56 28.45°	*Deployment of Italy's Laser Geodynamic Satellite II (LAGEOS II). US Microgravity Payload (USMP-1)*
52 **STS-53**	02.12.92/09.12.92 Discovery	KSC 39A EAFB	D Walker, Cabana Bluford, J S Voss, Clifford	07:07:19 57.00°	*Final DoD mission; deployment of Mercury satellite, into Molniya-type orbit, possibly to relay reconnaissance satellite imagery. Flight similar to STS-28*
53 **STS-54**	13.01.93/19.01.93 Endeavour	KSC 39B KSC	Casper, McMonagle Runco, Harbaugh, Helms	05:23:38 28.45°	*Deployment of TDRS-6 on IUS. 268 minute EVA*
54 **STS-56**	08.04.93/17.04.93 Discovery	KSC 39B KSC	Cameron, Oswald Foale, Cockrell, Ochoa	09:06:08 57.00°	*ATLAS-2; SPARTAN-201 deploy/retrieve*
55 **STS-55**	26.04.93/06.05.93 Columbia	KSC 39A EAFB	Nagel, Henricks Ross, Precourt, Harris, Walter, Schlegel	09:23:40 28.46°	*Spacelab D-2: 90 materials/life sciences, Earth observation & astronomy experiments. Shuttle programme exceeded one year aggregate flight time*
56 **STS-57**	21.06.93/01.07.93 Endeavour	KSC 39B KSC	Grabe, Duffy Low, J E Voss, Sherlock, Wisoff	09:23:45 28.45°	*EURECA-1 retrieval; SPACEHAB-1 with 21 experiments. Generic EVA*
57 **STS-51**	12.09.93/22.09.93 Discovery	KSC 39B KSC	Culbertson, Readdy Bursch, Newman, Walz	09:20:11 28.45°	*Advanced Communications Technology Satellite (ACTS) deploy, German Orbiting and Retrievable Far and Extreme Ultraviolet Spectrograph-Shuttle Pallet Satellite (ORFEUS-SPAS) deploy/retrieve, generic EVA; first KSC night landing*
58 **STS-58**	18.10.93/01.11.93 Columbia	KSC 39B EAFB	Blaha, Searfoss Seddon, Lucid, McArthur, Wolf, Fettman	14:00:13 39.0°	*Spacelab Life Sciences (SLS-2) dedicated life sciences research; second EDO (record duration)*
59 **STS-61**	02.12.93/13.12.93 Endeavour	KSC 39B KSC	Covey, Bowersox Musgrave, Akers, Hoffman, K Thornton, Nicollier	10:19:59 28.47°	*First Hubble servicing mission; five EVAs for four crew totalled 70.93 hours*
60 **STS-60**	03.02.94/11.02.94 Discovery	KSC 39A KSC	Bolden, Reightler Davis, Sega, Chang-Diaz, Krikalev	08:07:10 57.0°	*SPACEHAB-2; employed Wake Shield Facility (WSF-1); first Russian aboard US craft*
61 **STS-62**	04.03.94/18.03.94 Columbia	KSC 39B KSC	Casper, Allen Thuot, Gemar, Ivins	13:23:17 39.0°	*USMP-2; OAST-2; third EDO*

Flight	Launch/Landing dates (DD.MM.YY) GMT Orbiter	Launch pad Landing site	Crew	Duration (DD:HH:MM) Inclination
62 **STS-59**	09.04.94/20.04.94 Endeavour	KSC 39A EAFB	Gutierrez, Chilton Godwin, Jones, Apt, Clifford	11:05:49 57.0°
Space Radar Lab-1 (SRL-1) mapped 20 per cent Earth, collected 133 hours of data. Endeavour made record 412 manoeuvres				
63 **STS-65**	8.07.94/23.07.94 Columbia	KSC 39A KSC	Cabana, Halsell, Hieb Chiao, D Thomas, Walz, Naito-Mukai	14:17:55 28.5°
Spacelab IML-2; fourth EDO (record duration)				
64 **STS-64**	09.09.94/20.09.94 Discovery	KSC 39B EAFB	Richards, Hammond Helms, Lee, Linenger, Meade	10:22:50 57.0°
Lidar In-space Technology Experiment (LITE-1); deploy/retrieve SPARTAN 201-2; untethered EVA				
65 **STS-68**	30.09.94/11.10.94 Endeavour	KSC 39A EAFB	Baker, Wilcutt Jones, Bursch, Smith, Wisoff	11:05:46 57.0°
SRL-2; Apollo 11 25th anniversary commemoration by US Postal Service sends 500,000 stamps into orbit				
66 **STS-66**	03.11.94/14.11.94 Atlantis	KSC 39B EAFB	McMonagle, Brown Ochoa, Parazynski, Tanner, Clervoy	10:22:34 57.0°
ATLAS-3; deploy/retrieve Germany's Cryogenic Infrared Spectrometers and Telescopes for the Atmosphere-Shuttle Pallet Satellite (CRISTA-SPAS-1)				
67 **STS-63**	03.02.95/11.02.95 Discovery	KSC 39B KSC	Wetherbee, Collins Foale, Harris, J E Voss, Titov	08:06:28 51.6°
Rendezvous with Mir; SPACEHAB 3; deploy/retrieve SPARTAN-204. First female pilot (Collins); second Russian aboard US craft (Titov); generic EVA				
68 **STS-67**	02.03.95/18.03.95 Endeavour	KSC 39A EAFB	Oswald, W Gregory Jernigan, Grunsfeld, Lawrence, Durrance, Parise	16:15:10 28.5°
ASTRO-2 UV telescope array; fifth EDO (record duration)				
69 **STS-71**	27.06.95/07.07.95 Atlantis	KSC 39A KSC	Gibson, Precourt Baker, Harbaugh, Dunbar, Solovyov, Budarin	09:19:22 51.6°
First docking with Mir, delivering station's 19th main crew (Solovyov/Budarin), returning with 18th crew (Dezhurov/Strekalov) and Soyuz TM-21's Dr Norman Thagard (aboard since March 1995). Spacelab carried				
70 **STS-70**	13.07.95/22.07.95 Discovery	KSC 39B KSC	Henricks, Kregel Currie, D Thomas, Weber	08:22:20 28.5°
Deploy TDRS-7 on IUS				
71 **STS-69**	07.09.95/18.09.95 Endeavour	KSC 39A KSC	D Walker, Cockrell J S Voss, Newman, Gernhardt	10:20:29 28.45°
Deploy/retrieve WSF 2; deploy/retrieve SPARTAN-201-3; generic EVA				
72 **STS-73**	20.10.95/05.11.95 Columbia	KSC 39B KSC	Bowersox, Rominger K Thornton, Coleman, Lopez-Alegria, Leslie, Sacco	15:21:52 39.0°
Spacelab USML-2; sixth EDO				
73 **STS-74**	12.11.95/20.11.95 Atlantis	KSC 39A KSC	Cameron, Halsell McArthur, Ross, Hadfield	08:04:31 51.64°
Second Mir docking. Delivered docking unit for future Shuttle missions and new solar arrays				
74 **STS-72**	11.01.96/20.01.96 Endeavour	KSC 39B KSC	Duffy, Jett, Barry Chiao, Scott, Wakata	08:22:01 28.45°
Retrieve Japan's Space Flyer Unit; deploy/retrieve SPARTAN OAST-Flyer; 2 generic EVAs				
75 **STS-75**	22.02.96/09.03.96 Columbia	KSC 39B KSC	Allen, Horowitz Chang-Diaz, Cheli, Hoffman, Nicollier, Guidoni	15:17:40 28.45°
USMP-3; deploy/retrieve Italy's Tethered Satellite System (TSS-1R) – TSS lost: tether snapped; seventh EDO				
76 **STS-76**	22.03.96/31.03.96 Atlantis	KSC 39B EAFB	Chilton, Searfoss Lucid, Godwin, Clifford, Sega	09:05:16 51.6°
Third Mir docking, with SPACEHAB short module. Delivered logistics and Lucid for Mir stay (returning STS-79). Godwin/Clifford EVA mounted experiments on Mir's docking module				
77 **STS-77**	19.05.96/29.05.96 Endeavour	KSC 39B KSC	Casper, Brown Bursch, Runco, A Thomas, Garneau	10:00:39 28.5°
SPACEHAB-4; deploy/retrieve SPARTAN-207 (inflatable antenna experiment); deploy Passive Aerodynamically Stabilised Magnetically Damped Satellite (PAMS)				
78 **STS-78**	20.06.96/07.07.96 Columbia	KSC 39B KSC	Henricks, Kregel Helms, Linnehan, Brady, Favier, Thirsk	16:21:48 39.0°
Life & Microgravity Spacelab (LMS); eighth EDO				
79 **STS-79**	16.09.96/26.09.96 Endeavour	KSC 39A KSC	Readdy, Wilcutt, Apt, Akers, Walz Blaha (up), Lucid (down)	10:03:18 51.6°
Fourth Mir link with SPACEHAB double module. Delivered supplies to Mir, left Blaha on board for medium-length stay, returned with Lucid (mission duration 188:04:00) from STS-76				
80 **STS-80**	19.11.96/07.12.96 Columbia	KSC 39B KSC	Cockrell, Rominger, Jernigan, Jones Musgrave	17:15:53 28.5°
Deployment and retrieval of WSF-3 and ORFEUS-SPAS-II; ninth EDO; planned EVAs by Jernigan and Jones cancelled				
81 **STS-81**	12.01.97/22.01.97 Atlantis	KSC 39B KSC	Baker, Jett, Wilsoff, Grunsfeld Ivins, Linenger (up), Blaha (down)	10:04:56 51.6°
Fifth Mir docking with SPACEHAB double module. Delivered supplies to Mir, left Linenger on board for medium-length stay, returned with Blaha from STS-79 (mission duration 128:05:27)				
82 **STS-82**	11.02.97/21.02.97 Discovery	KSC 39A KSC	Bowersox, Horowitz, Tanner, Smith Hawley, Harbaugh, Lee	9:23:37 28.5°
Second servicing mission for the Hubble Space Telescope: new instruments installed, older ones returned to Earth, repairs to insulation; five EVAs completed, Lee/Smith 3, Harbaugh/Tanner 2				
83 **STS-83**	04.04.97/08.04.97 Columbia	KSC 39A KSC	Halsell, Still, J E Voss, Gernhardt, Linteris D Thomas, Crounch	3:23:13 28.5°
Microgravity Science Laboratory-1 (MSL-1) primary payload for 10th EDO: mission curtailed due to fuel cell problems and re-flown as STS-94				
84 **STS-84**	15.05.97/24.05.97 Atlantis	KSC 39A KSC	Precourt, Collins, Clervoy, Noriega Lu, Kondakova, Foale (up), Linenger (down)	9:05:21 51.6°
Sixth Mir docking mission with SPACEHAB-DM. Delivered supplies to Mir, left Foale on board for medium-length stay, returned with Linenger from STS-81 (mission duration 132:03:50)				
85 **STS-94**	01.07.97/17.07.97 Columbia	KSC 39A KSC	Halsell, Still, J E Voss, Gernhardt, D Thomas Crouch, Linteris	15:16:45 28.5°
Reflight of STS-83, MSL-1 primary payload for 11th EDO: fastest return-to-orbit for the crew members				
86 **STS-85**	07.08.97/19.08.97 Discovery	KSC 39A KSC	Brown, Rominger, Davis, Curbeam, Robinson Tryggvason	11:20:28 57.0°
CRISTA-SPAS-2 deployment, test of Japanese remote manipulator system				
87 **STS-86**	25.09.97/06.10.97 Atlantis	KSC 39A KSC	Wetherbee, Bloomfield, Titov, Parazynski Chretien, Lawrence, Wolf (up), Foale (down)	10:19:22 51.6°
Seventh Mir docking; Lawrence originally planned to remain aboard Mir but replaced by Wolf because her height precludes use of Russian EVA suit. EVA with Parazynski and Titov, first to involve non-US crew member				
88 **STS-87**	19.11.97/05.12.97 Columbia	KSC 39B KSC	Kregel, Lindsey, Chawla, Scott, Doi, Kadenyuk	15:16:35 29.45°
USMP4, deployment and retrieval of SPARTAN 201-04. 12th EDO. Scott and Doi performed two EVAs, one to manually capture SPARTAN				
89 **STS-89**	22.01.98/31.01.98 Endeavour	KSC 39A KSC	Wilcutt, Edwards, Reilly, Anderson, Dunbar Sharipov, A Thomas (up), Wolf (down)	08:19:48 51.6°
Eighth Mir docking. A Thomas replaced Wolf				
90 **STS-90**	17.04.98/03.05.98 Columbia	KSC 39B KSC	Searfoss, Altman, Hire, Linnehan, Williams Buckey, Pawelczyk	15:21:51 39.0°
US Neurolab mission; final Spacelab module flight. 13th and last planned EDO before ISS operations for long duration missions				

Flight	Launch/Landing dates (DD.MM.YY) GMT / Orbiter	Launch pad / Landing site	Crew	Duration (DD:HH:MM) / Inclination
91	02.06.98/12.06.98	KSC 39A	Precourt, Gorie, Chang-Diaz, Lawrence, Kavandi, Ryumin	09:19:55
STS-91	Discovery	KSC	A Thomas (down)	51.6°
Ninth and last Mir docking, first with Discovery. Cosmonaut Ryumin conducted Mir inspection to determine suitability for sustained operations. A Thomas returned to Earth				
92	29.10.98/07.11.98	KSC 39B	Brown, Lindsey, Parazynski, Robinson	08:21:45
STS-95	Discovery	KSC	Duque, Mukai, Glenn	28.45°
Life sciences mission including world's oldest astronaut, John Glenn age 77. Deployed and retrieved SPARTAN 201-05 and tested Hubble Space Telescope orbital systems test platform				
93	04.12.98/16.12.98	KSC 39A	Cabana, Sturckow, Ross, Currie	11:19:53
STS-88	Endeavour	KSC	Newman, Krikalev	51.6°
Carried Unity module to Zarya Control Module in first docking to begin assembly of ISS. Ross and Newman made three EVAs to attach cables and deploy fixtures to ISS exterior				
94	27.05.99/06.06.99	KSC 39B	Rominger, Husband, Jernigan, Ochoa	09:19:14
STS-96	Discovery	KSC	Barry, Payette, Tokarev	51.6°
First Shuttle ISS servicing mission to docked Zarya/Unity modules. Released Student-Tracked Atmospheric Research Satellite for Heuristic International Networking Experiment-1 (STARSHINE-1) satellite				
95	23.07.99/28.07.99	KSC 39B	Collins, Ashby, Coleman, Hawley, Tognini	04:22:50
STS-93	Columbia	KSC		28.45°
Launch of Chandra X-ray observatory; first female commander (Collins)				
96	19.12.99/28.12.99	KSC 39B	Brown, Kelly, Clervoy, Smith, Foale	07:23:11
STS-103	Discovery	KSC	Grunsfeld, Nicollier	28.45°
Third Hubble Space Telescope servicing mission				
97	11.02.00/22.02.00	KSC 39A	Kregel, Gorie, Thiele, Kavandi, J E Voss	11:05:38
STS-99	Endeavour	KSC	Mohri	57°
Shuttle Radar Topography Mission (SRTM)				
98	19.05.00/29.05.00	KSC 39A	Halsell, Horowitz, Weber, Williams	09:20:10
STS-101	Atlantis	KSC	J S Voss, Helms, Usachev	51.6°
Third Shuttle ISS flight (2A.2a) with double SPACEHAB module; first upgraded "glass cockpit" flight				
99	08.09.00/19.09.00	KSC 39B	Wilcutt, Altman, Lu, Mastracchio	11:19:11
STS-106	Atlantis	KSC	Burbank, Malenchenko, Morukov	51.6°
Fourth Shuttle ISS flight (2A.2b) with double SPACEHAB module				
100	11.10.00/24.10.00	KSC 39A	Duffy, Melroy, Wakata, Wisoff, Chiao	12:21:44
STS-92	Discovery	EAFB	McArthur, Lopez-Alegria	51.6°
100th Shuttle Mission. Fifth Shuttle ISS flight (3A) delivering and installing Z1 Truss and Pressurized Mating Adaptor-3 (PMA-3)				
101	30.11.00/11.12.00	KSC 39B	Jett, Bloomfield, Garneau, Tanner	10:19:58
STS-97	Endeavour	KSC	Noriega	51.6°
Sixth Shuttle ISS flight (4A) delivering and attaching P6 Integrated Truss Segment with solar array				
102	07.02.01/20.02.01	KSC 39A	Cockrell, Polansky, Ivins, Jones	12:21:20
STS-98	Atlantis	KSC	Curbeam	51.6°
Seventh Shuttle ISS flight (5A) carrying US Destiny laboratory				
103	08.03.01/21.03.01	KSC 39B	Weatherbee, Kelly, A Thomas, Richards	12:19:49
STS-102	Discovery	KSC	Usachev (up), Helms (up), J S Voss (up), Shepherd (down), Krikalev, (down), Gidzenko (down)	51.6°
Eighth Shuttle ISS flight (5A.1) with Leonardo Multi-Purpose Logistics Module (MPLM), Expedition 2 crew				
104	19.04.01/01.05.01	KSC 39A	Rominger, Ashby, Hadfield, Phillips	11:21:30
STS-100	Endeavour	EAFB	Parazynski, Guidoni, Lonchakov	51.6°
Ninth Shuttle ISS flight (6A) with Rafaello MPLM and Space Station Mobile Servicer System (SSRMS)				
105	12.07.01/25.07.01	KSC 39B	Lindsay, Hobaugh, Gernhardt, Reilly	12:18:35
STS-104	Atlantis	KSC	Kavandi	51.6°
Tenth Shuttle ISS flight (7A) delivers Quest Airlock module				
106	10.08.01/22.08.01	KSC 39A	Horowitz, Sturckow, Forrester, Barry,	11:21:13
STS-105	Discovery	KSC	Culbertson (up), Tyurin (up), Dezhurov (up), J S Voss (down), Helms (down), Usachev (down)	51.6°
Eleventh Shuttle ISS flight (7A.1) with Leonardo MPLM, Expedition 3 crew				
107	05.12.01/17.12.01	KSC 39B	Gorie, Kelly, Godwin, Tani, Onufrienko (up), Bursch (up)	11:19:36
STS-108	Endeavour	KSC	Walz (up), Culbertson (down), Dezhurov (down), Tyurin (down)	51.6°
Twelfth Shuttle ISS flight (UF1) with Rafaello MPLM, Expedition 4 crew; STARSHINE-2 deployed; September 11 flag tribute				
108	01.03.02/12.03.02	KSC 39A	Altman, Carey, Grunsfeld, Currie, Massimino	10:22:10
STS-109	Columbia	KSC	Linnehan, Newman	28.45°
Fourth Hubble Space Telescope servicing mission				
109	08.04.02/19.04.02	KSC 39B	Bloomfield, Frick, Walheim, Ochoa, Morin, Ross, Smith	10:19:43
STS-110	Atlantis	KSC		51.6°
Thirteenth Shuttle ISS flight (8A) with S0 Truss and Mobile Transporter				
110	05.06.02/19.06.02	KSC 39A	Cockrell, Lockhart, Perrin, Chang-Diaz, Whitson (up)	13:20:35
STS-111	Endeavour	EAFB	Korzun (up), Treschev (up), Walz (down), Bursch (down), Onufrienko (down)	51.6°
Fourteenth Shuttle ISS flight (UF-2) with Leonardo MPLM, Mobile Base System (MBS), Expedition 5 crew				
111	07.10.02/18.10.02	KSC 39B	Ashby, Melroy, Wolf, Magnus, Sellers, Yurchikhin	10:19:58
STS-112	Atlantis	KSC		51.6°
Fifteenth Shuttle ISS flight (9A) with S1 Truss installation				
112	23.11.02/07.12.02	KSC 39A	Wetherbee, Lockhart, Lopez-Alegria, Herrington, Bowersox (up), Budarin (up), Pettit (up), Korzun (down)	13:18:47
STS-113	Endeavour	KSC	Whitson (down), Treschev (down)	51.6°
Sixteenth Shuttle ISS flight (11A) with P1 Truss, Expedition 6 crew				
113	16.01.03/None (01.02.03 scheduled)	KSC 39A	Husband, McCool, Brown, Chawla, Anderson, Clark	15:22:20
STS-107	Columbia	KSC (scheduled)	Ramon	39°
Microgravity research with SPACEHAB-Research Double Module; Shuttle destroyed and crew lost during re-entry on 1 February 2003				
114	26.07.05/09.08.05	KSC 39B	Collins, Kelly, Camarda, Lawrence, Noguchi, Robinson, A Thomas	13:21:32
STS-114	Discovery	EAFB		51.6°
First Return to Flight test mission; seventeenth Shuttle ISS flight (LF1); Raffaello MPLM, External Stowage Platform-2 (ESP-2); in-flight shuttle repair EVA				
115	04.07.06/17.07.06	KSC 39B	Lindsey, Kelly, Fossum, Nowak, Reiter (up), Sellers, Wilson	12:19:37
STS-121	Discovery	KSC		51.6°
Second Return to Flight test mission, eighteenth Shuttle ISS flight; utilization and logistics flight (ULF 1.1); three EVAs; repeated flight schedule delays; ESA astronaut, Thomas Reiter remained aboard the ISS with the Expedition 13 crew				
116	09.09.06/21.09.06	KSC 39B	Jett, Ferguson, Burbank, MacLean, Stefanyshyn-Piper,	11:19:06
STS-115	Atlantis	KSC	Tanner	51.6°
Nineteenth Shuttle ISS flight; ISS port three and four (P3/P4) integrated truss segment installation; solar array deployment; three EVAs; MacLean, a Canadian Astronaut, accompanied the US crew; sighting and identification of mysterious debris close to the Shuttle prior to the craft's return delayed its landing by one day				

Flight	Launch/Landing dates (DD.MM.YY) GMT Orbiter	Launch pad Landing site	Crew	Duration (DD:HH:MM) Inclination
117	10.12.06/22.12.06	KSC 39B	Polansky, Oefelein, Curbeam, Fuglesang, Higginbotham, Patrick, Reiter (down), S Williams (up)	12:20:44
STS-116	Discovery	KSC		51.6°
Twentieth Shuttle ISS flight (12A.1); Discovery carried the P5 Truss and the SPACEHAB module; the crew rewired the ISS electrical system; ESA astronaut Fuglesang accompanied the NASA crew, and Sunita Williams replaced ESA astronaut, Thomas Reiter, aboard the ISS since July 2006; Shuttle liftoff marked its first night launch since 2002; four EVAs, of which Curbeam participated in all, setting a single Shuttle mission, single astronaut EVA record				
118	08.06.07/22.06.07	KSC 39A	Frederick Sturckow, Lee Joseph Archambault, James Reilly II, Steven Swanson, Patrick Forrester, John D. Olivas, Sunita L. Williams, Clayton C. Anderson	13:20:11
STS-117	Atlantis	KSC		51.6°
STS-117 was the 118th Shuttle flight and 21st ISS flight. The crew delivered, installed and activated on the ISS a 17.5 ton S3/S4 starboard truss segment, which included a pair of 73 m long photo-voltaic solar arrays. The mission also involved the retraction of two solar arrays on the P6 truss prior to their relocation to the port side of the ISS in a later mission. Four spacewalks were performed by the crew, including repair to the thermal blanket on the port side of the orbital manoeuvring system pod. Clay Anderson, flight engineer, replaced ISS crew member Sunita Williams and remained on the ISS until November				
119	08.08.07/21.08.07	KSC 39A	Richard A. (Rick) Mastracchio, Barbara R. Morgan, Charles O. Hobaugh, Scott J. Kelly, Tracy E. Caldwell, Dafydd R. (Dave) Williams (CSA), and Benjamin Alvin Drew Jr.	12:17:56
STS-118	Endeavour	KSC		51.6°
The 119th shuttle flight and 22nd ISS flight. This was also Endeavour's first flight since STS-113 (2002). The crew installed their primary payload, the S5 truss spacer to the end of the S4 truss, conducting repairs to the ISS control moment gyroscopes (CMG) and removing the External Stowage Platform ESP-2 to deploy a new CMG (gyro) stowed inside the ESP-3. In preparation for the relocation of the P6 truss, they retracted the P6 radiator, and moved two Crew and Equipment Translation Aid (CETA) carts to the starboard side. Using the Station's robotic arm the astronauts upgraded the port S-band communications system and the P6 transponder was returned to Earth for upgrading and use in future missions. Four spacewalks were conducted				
120	23.10.07/07.11.07	KSC 39A	Pamela A. Melroy, George D. Zamka, Scott E. Parazynski, Douglas H. Wheelock, Stephanie D. Wilson, Paolo A. Nespoli (ESA), Daniel M. Tani, and Clayton C. Anderson (down)	15:02:23
STS-120	Discovery	KSC		51.6°
The 120th shuttle flight and 23rd flight to the ISS. The crew delivered and installed the primary payload Node 2, also referred to as Harmony, a 7.2 × 4.4 m passageway, which will connect the European and Japanese ISS modules to the US segment. Harmony was initially installed on a temporary node and then attached to the Pressurized Mating Adapter-2 (PMA-2), located at the end of Destiny once the shuttle had undocked from the PMA-2 port. Canadarm2 was used to move and install Harmony. The crew also completed the relocation, deployment and activation of the P6 Photovoltaic Power Module (PVM) or P6 truss from Z1 to Z5, after repairing a torn solar array on the truss. The S-band Antenna Structural Assembly from Z1 was returned to Earth. Four spacewalks were performed. Daniel Tani replaced ISS crew member Clayton Anderson				
121	07.02.08/20.02.08	KSC 39A	Commander Steve Frick, ESA Mission Specialist Léopold Eyharts, Pilot Alan G. Poindexter, Mission Specialists Leland D. Melvin, Rex J. Walheim, Stanley G. Love, and Hans Schlegel (ESA)	12:18:25
STS-122	Atlantis	KSC		51.6°
The 121st shuttle flight and 24th ISS flight. The crew using Canadarm2 mated the 7-m long, 12.8-tonne ESA Columbus Laboratory module, which was carrying 2.5 tonnes of science payloads, to the starboard side of Harmony (Node 2) on the ISS. The Solar observatory and the European Technology Exposure Facility were mounted externally to the Columbus External Payload Facility zenith position. Commissioning of the Columbus Laboratory continued after the Atlantis mission, mainly involving ESA astronaut Léopold Eyharts, who replaced ISS crew member Daniel Tani. The shuttle crew also replaced the Nitrogen Tank Assembly (NTA) Orbital Replacement Unit (ORU) on the International Space Station's Port 1 (P1) truss segment, which was returned to Houston for refurbishing to be used in future ISS missions. Three spacewalks were performed during the mission. Tani had been on the ISS since October 2007				

Notes
Shuttle Orbiters: *Challenger* 099, *Enterprise* 101, *Columbia* 102, *Discovery* 103, *Atlantis* 104, *Endeavour* 105.
Please also see the separate Space Shuttle entry in the Human Space Flight section.

The Base system for the ISS is placed in the payload bay of the Shuttle in readiness for STS-111 (NASA) 1047512

Italy's Leonardo module provides resupply container space for essential ISS equipment (ESA) 0137140

One of the new Block 2 Space Shuttle Main Engines to be fitted to Atlantis for STS-104 launched December 2001 (Boeing) 1047500

New and improved Block 2 Space Shuttle Main Engine elements are installed in Atlantis in the Orbiter Processing Facility at NASA's Kennedy Space Center, Florida (NASA) 0137134

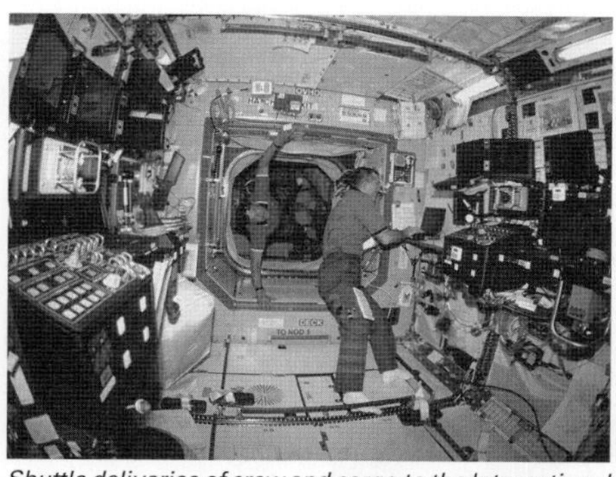
Shuttle deliveries of crew and cargo to the International Space Station facilitates extension of experiments formerly conducted from inside the Orbiter (NASA) 1047477

The Payload Changeout Room, or PCR, is seen clearly in the view of the Shuttle at Launch Complex 39 (NASA) 1047499

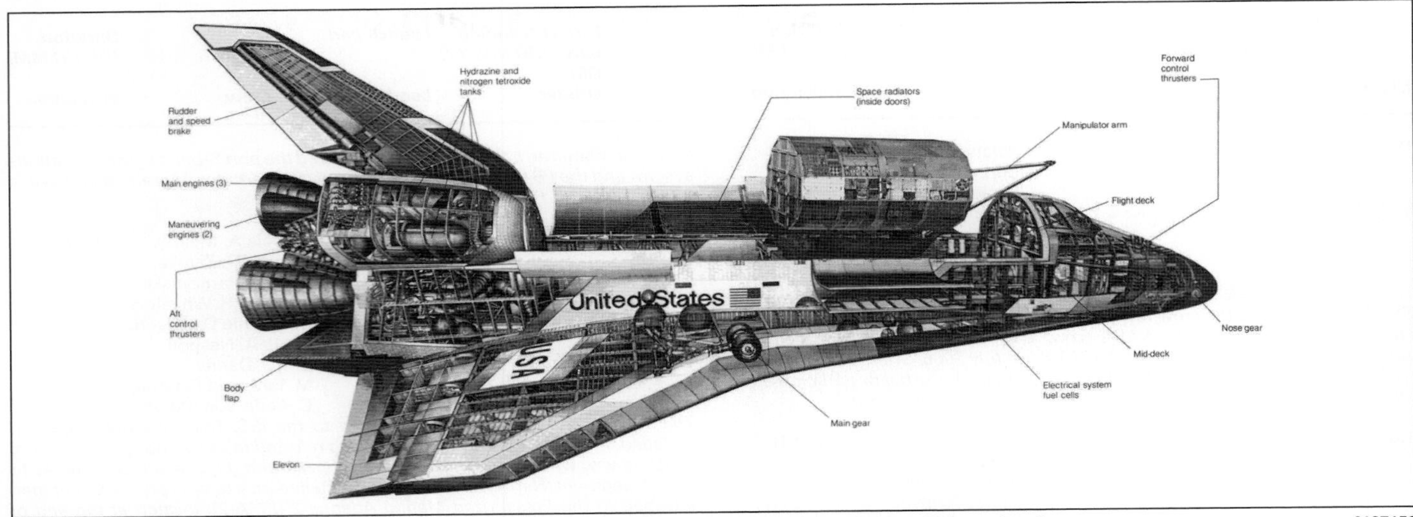

Cutaway view of the Shuttle Orbiter with remote manipulator and typical payload above the cargo bay (Rockwell International) 0137158

The Shuttle lifts off at the start of an ISS re-supply flight, one of many scheduled before the ISS is fully configured for its international crew (NASA) 0137137

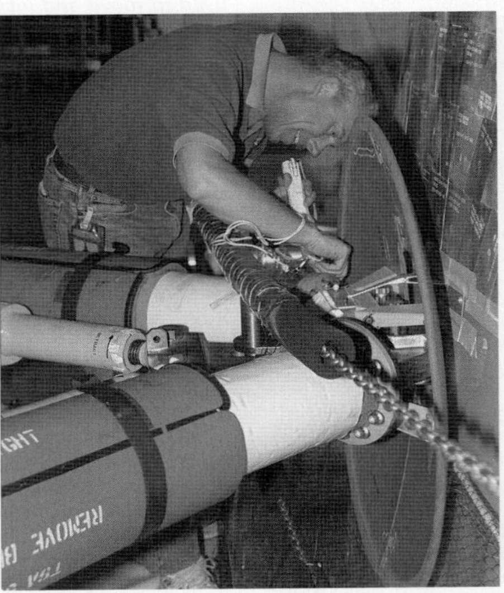

The forward bipod connecting the External Tank to the underside of the Orbiter forward fuselage (NASA) 1047478

Shuttle moves to the pad up the crawlerway ramp at Complex 39 (NASA) 1047473

Atlantis is moved on its flatbed transporter at the Kennedy Space Center (NASA) 1047504

Clearly seen in the overhead view of the Shuttle on its mobile launch platform, the two tail service umbilical shrouds through which fluids and electrical power flow to the stacked vehicle prior to lift-off (NASA) 1047474

Astronaut Sellers in the cargo bay of the Shuttle during a scheduled EVA (NASA) 0573536

Table 4.0: World Orbital Launch Log Explanatory Notes

See *jsd.janes.com* for World Orbital Launch Log 1957–1994 (Tables 4.1–4.8)

Table 4.0a: Launch log world launch sites abbreviations

Launch Site Abbreviation	*World Launch Site*
ALCA	CLA, Alcantara Space Center, Maranhao, Brazil
ALCA VLS	VLS pad, Alcantara
APC KT-408	KT No. 408, Luna-20 Site, Apollonius C, Mare Undarum B3 (Luna)
BLA	Barents Launch Area, Barentsovo More, Murmansk Oblast
BLA LP1	K-407 submarine, Barents Launch Area, Barentsovo More, Murmansk
BLAN	Black Mesa, Blanding, Utah
BRET	Zone de lancement SNLE, Golfe de Gascogne, Bretagne
CAIR	80 km WSW of Cairo, Egypt
CARD	Le Cardonnet, France
CARN	Carnarvon, Western Australia
CARN2	Carnarvon north site, Western Australia
CASS	Cassino, Rio Grande Do Sul, Brazil
CC	Cape Canaveral Air Station, Florida
CC 26A – 26B	Launch Complex 26A – B, Cape Canaveral
CC LA	Launch Area, Cape Canaveral
CC LC5 – LC6	Launch Complex 5 – 6, Cape Canaveral
CC LC11 – LC14	Launch Complex 11 – 14, Cape Canaveral
CC LC17 – LC19	Launch Complex 17 – 19, Cape Canaveral
CC LC32B – LC 46	Launch Complex 32B – 46, Cape Canaveral
CC RW30/12	Skid Strip, Cape Canaveral Air Station
CC SLC36A -SLC41	Space Launch Complex 36A – SLC41
CSG	Centre Spatial Guyanais, Kourou, French Guiana
CSG CECLES	Aire de Lancement CECLES, CSG
CSG Diamant	Aire de Lancement Diamant, CSG
CSG ELA1 – ELA3	Ensemble de Lancement Ariane 1 – 3, CSG
DES LM11-DS	LM 11 Descent Stage, Descartes Base, Descartes Plain, Theophilus D2 (Luna)
DOM	Dombarovsky (Yasny)
DZGC	Gran Canaria Drop Zone, North Atlantic Ocean
DZKW	Kwajalein Drop Zone, Pacific Ocean
DZSB	Santa Barbara Channel Drop Zone, California
DZWI	Wallops Island Drop Zone, North Atlantic Ocean
EAFB	Edwards AFB, California
EAFB RW04/22	RW04/22, Edwards AFB, California
EAFB RW18/36	RW18/36, Rogers Dry Lake, Edwards AFB, California
EAFBTL1	Tethered Launch test silo, Edwards AFB
EAFBTL2	Tethered Launch test silo, Edwards AFB
FEC KT-406	KT No. 406, Luna-16 Site, Mare Fecunditatis, Langrenus A2 (Luna)
FMR LM8-DS	LM 8 Descent Stage, Fra Mauro Base, Fra Mauro, Montes Riphaeus B1 (Luna)
GAN	Base Aerea de Gando, Gran Canaria
GAN RW03/21	RW03/21L and RW03/21R, Base Aerea de Gando, Gran Canaria
GIK-1	GIK-1, Plesetsk, Russian Federation
GIK-1 LC16/2 – LC133/3	Ploshchadka 16/2 – 133/3, GIK-1, Plesetsk, Russian Federation
GIK-1 PU11	PU 11, NIIP-53, Plesetsk, Russian Federation
GIK-2	GIK-2, Svobodniy, Amurskaya Oblast, Russian Federation
GIK-2 LC5	Ploshchadka 5, GIK-2, Svobodniy, Russian Federation
GIK-5	GIK-5, Baikonur, Kazakhstan
GIK-5 LC1 – LC250 (and variations)	Ploshchadka 1 – 250, GIK-5, Baikonur, Kazakhstan
GIK-5 PL251	Buran runway, GIK-5 Baikonur, Kazakhstan
GIK-5 PU31 – PU33	Ploshchadka Unknown 31 – 33, GIK-5, Baikonur, Kazakhstan
GNIIP	GNIIP, Plesetsk, Russian Federation
GNIIP LC158	Ploshchadka 158, GNIIP, Plesetsk, Russian Federation
GNIIP LC(T1) – (T4)	GNIIP, Plesetsk, Russian Federation
GTsP-4	GTsP-4, Kapustin Yar, Volgograd, Russian Federation
GTsP-4 LC86 (and variations)	Ploshchadka 86 GTsP-4, Kapustin Yar
GTsP-4 LC107	Ploshchadka 107 (PU 1 or 2), GTsP-4, Kapustin Yar
GTsP-4 LC107/1 & LC107/2	Ploshchadka 107/1 & 107/2, GTsP-4, Kapustin Yar
GTsP-4 Mayak-1–2	Mayak-1 – 2 silos, GTsP-4, Kapustin Yar
HAD LM10-DS	LM 10 Descent Stage, Hadley Base, Montes Apenninus, Montes Apenninus B4 (Luna)
HMG	Centre Interarmees d'Essais d'Engins Speciaux (CIEES), Algeria
HMG Brigitte	Brigitte, CIEES B2, Hammaguir, Algeria
HMG Brigitte/A	Agate launcher, Brigitte, CIEES B2, Hammaguir, Algeria
JQ	Jiuquan Space Center, Nei Monggol Zizhiqu, China
JQ LA2A	Launch Area 2 Pad 5020, Jiuquan Space Center
JQ LA2B	Launch Area 2 Pad 138, Jiuquan Space Center
JQ LA3	Launch Area 3, Jiuquan Space Center
JQ SLS	South Launch Site, Jiuquan Space Center
KASC	Kagoshima Space Center, Kagoshima, Kyushu, Nippon, Japan
KASC K	Kappa Pad, Kagoshima Space Center
KASC L	Lambda Pad, Kagoshima Space Center
KASC M	Mu Pad, Kagoshima Space Center
KASC M1	Refurbished Mu Pad, Kagoshima Space Center
KASC M-V	M-V Pad, Kagoshima Space Center
KASCTMP	Temporary pad, Kagoshima Space Center, Kagoshima, Kyushu, Nippon
KLA	Kiritimati Launch Area, Pacific Ocean launch, Sea Launch Company, LLC
KLC	Kodiak Launch Complex, Alaska
KMR OM	Omelek Island, Kwajalein Atoll, Marshall Islands, USAKA
KMR RW06/24	RW06/24, Bucholz Army Airfield (PKWA), USAKA, Kwajalein, Marshall Islands
KSC	NASA John F. Kennedy Space Center, Florida
KSC LC39A	Launch Complex 39A, NASA Kennedy Space Center, Florida
KSC LC39B	Launch Complex 39B, NASA Kennedy Space Center, Florida
KSC RW15/33	Shuttle Landing Facility, NASA Kennedy Space Center, Florida
L1011	Orbital Sciences L-1011 Stargazer air launch
LIT LM12-DS	LM 12 Descent Stage, Taurus-Littrow Base, Littrow Valley, Macrobius B1 (Luna)
MFWA	Drop Zone, Mayport, Florida, Warning Area, North Atlantic Ocean
MFWA1	Drop Zone, Mayport, Florida, Warning Area, North Atlantic Ocean
MUSU	Musudan-ri, Hwadae-gun, Hamgyong Pukdo, Choson, Korea
NOTS	Naval Ordnance Test Station, China Lake, California
NOTS RW	RW, Naval Ordnance Test Station, China Lake, California
Odyssey, POR	Odyssey launch platform, Pacific Ocean launch, Sea Launch Company, LLC
PA	Naval Missile Facility, Point Arguello, California
PALB	Israeli Air Force Test Range, Palmachim Beach, Israel
PALB SH	Shaviyt pad, Israeli Air Force Test Range, Palmachim Beach, Israel
PA LC1-1	Launch Complex 1, Pad 1, Naval Missile Facility, Point Arguello
PA LC1-2	Launch Complex 1, Pad 2, Naval Missile Facility, Point Arguello
PA LC2-3	Launch Complex 2, Pad 3, Naval Missile Facility, Point Arguello
PA LC2-4	Launch Complex 2, Pad 4, Naval Missile Facility, Point Arguello
PA LC-D	Launch Complex D, Naval Missile Facility, Point Arguello, California
PAWA	Point Arguello Warning Area Drop Zone, Pacific Ocean
SHAR	Sriharikota Range, Sriharikota, Andhra Pradesh, India
SHAR PSLV	PSLV pad, Sriharikota Range
SHAR SLV	SLV pad, Sriharikota Range
SLC17A -17B	Space Launch Complex 17A – B
SMLC	San Marco Launch Complex, Formosa Bay, Kenya
SPFL	Spaceport Florida, Cape Canaveral
SPFL SLC46	Space Launch Complex 46, Spaceport Florida, Cape Canaveral
SV3 LM6-DS	LM6 Descent Stage, Surveyor 3 Site, Mare Insularum, Montes Riphaeus A2 (Luna)
TNSC	Tanegashima Space Center, Tanegashima, Nippon
TNSC N	N Launch Complex, Osaki Launch Site, Tanegashima Space Center
TNSCY	Yoshinobu Launch Complex, Osaki Site, Tanegashima Space Center
TRAN LM5-DS	LM 5 DS, Tranquility Base, Mare Tranquillitatis, Julius Caesar C4 (Luna)
TYSC	Taiyuan Space Center, Wuzhai, Shanxi, China
TYSC LC1	Pad 1, Taiyuan Space Center, Wuzhai, Shanxi, China
V	Vandenberg AFB, California
V 75-1-1	Complex 75-1, Pad 1, Vandenberg AFB
V 75-1-2	Complex 75-1, Pad 2, Vandenberg AFB
V 75-3-4	Complex 75-3, Pad 4, Vandenberg AFB
V 75-3-5	Complex 75-3, Pad 5, Vandenberg AFB
V 576E	Complex 576-E, Vandenberg AFB
V 4300B6	Complex 4300-B6, Vandenberg AFB
V ABRESA2	Complex ABRES-A, Pad 2, Vandenberg AFB
V ABRESB3	Complex ABRES-B, Pad 3, Vandenberg AFB
V BMRSA1	Complex BMRS-A, Pad 1, Vandenberg AFB
V BMRSA2	Complex BMRS-A, Pad 2, Vandenberg AFB
V LE-6	Launch Emplacement 6, Vandenberg AFB
V PALC1-1	Point Arguello Launch Complex 1-1, Vandenberg AFB
V PALC1-2	Point Arguello Launch Complex 1-2, Vandenberg AFB

Launch Site Abbreviation	*World Launch Site*
V PALC2-3	Point Arguello Launch Complex 2-3, Vandenberg AFB
V PALC2-4	Point Arguello Launch Complex 2-4, Vandenberg AFB
V PALC-D	Point Arguello Launch Complex D, Vandenberg AFB
V RW30	RW30, Vandenberg AFB, California
V RW30/12	RW30/12, Vandenberg AFB, California
V SLC1E	Space Launch Complex 1-East, Vandenberg AFB
V SLC1W	Space Launch Complex 1-West, Vandenberg AFB
V SLC2E	Space Launch Complex 2-East, Vandenberg AFB
V SLC2W	Space Launch Complex 2-West, Vandenberg AFB
V SLC3E	Space Launch Complex 3-East, Vandenberg AFB
V SLC3W	Space Launch Complex 3-West, Vandenberg AFB
V SLC4E	Space Launch Complex 4-East, Vandenberg AFB
V SLC4W	Space Launch Complex 4-West, Vandenberg AFB
V SLC5	Space Launch Complex 5, Vandenberg AFB
V SLC6	Space Launch Complex 6, Vandenberg AFB
V SLC10W	Space Launch Complex 10-West, Vandenberg AFB
V SLF	Commercial Space Launch Facility, California Spaceport, Vandenberg AFB
WI	Wallops Flight Facility, Wallops Island, Virginia
WI LA0A	Launch Area 0A, Wallops Flight Facility
WI LA3	Mk I Launcher, Launch Area 3, Wallops Flight Facility
WI LA3A	Mk II Launcher, Launch Area 3, Wallops Flight Facility
WI RW04/22	RW04/22, WFF Research Airport, Wallops Main Base, Virginia
WOO	Woomera Instrumented Range, Woomera, South Australia
WOO LA5B	Launch Area 5B, Woomera Range, South Australia
WOO LA6A	Launch Area 6A, Woomera Range, South Australia
WOO LA8	Launch Area 8, Woomera Range, South Australia
WS WSSH	White Sands Space Harbor, New Mexico
XSC	Xichang Space Center, Sichuan, China
XSC LC1	Launch Complex 1, Xichang Space Center
XSC LC2	Launch Complex 2, Xichang Space Center

Source: Jonathan McDowell, PhD

Table 4.0b: Launch log source reference column abbreviations

Source Reference Column Abbreviation	*Original information source:*
AVM	Vladimir Agapov, internet postings
AWST	Aviation Week
Barensky	Stefan Barensky, personal communication
CCMOPS	Cape Canaveral historical records
CNES32	CNES history book
ESAPR	ESA press release
Energiya	Energiya history book
ExplnOrbit	Explorers In Orbit, MSFC, 1960
Grahn-WWW	Sven Grahn's web page
GWU	George Washington University
Henderson	B. Henderson, personal communication
ISIR	International Space Industry Report
JBIS	Journal of the BIS
JG	J. Geiger, VAFB History Office
JPL	JPL technical reports
Jupiter	Jupiter history, Redstone Arsenal
KHR-1	KSC Historical Report-1
KIAM	Vladimir Agapov's web page at KIAM
LAAFB	Los Angeles AFB history office
LR	Lompoc Record newspaper
Lissov	Igor Lissov, personal communication
McDonald-B	B. McDonald, personal communication
MIR	Chris van den Bergh's Mir News
Morton	Morton, Fire across the Desert (Woomera history)
MSFCFLASH	NASA MSFC STS flash report
NASA-TM	NASA Tech Memo
NezavB	Nezav. Baykonur
NK	Novosti Kosmonavtiki
NOTS	NOTS satellite program files, NASM
NRO	National Reconnaissance Office declassified archives
NYT	New York Times
PER	National Reconnaissance Office declassified archives
Petrov71	G. Petrov, Soviet Space Research 1967–1970
PH	Peter Hunter, personal communication
POES-WWW	POES web site
PR/AE	Arianespace press release
RAE	Launch time estimated to 0.01 day from RAE tables
Ransom	Martin Ransom, personal communication
Rosenberg	J. Rosenberg, personal communication
Rothmund	C. Rothmund, Europa et Cora.
SAC7383	SAC-7383, DMSP launch history
Seesat-L	Seesat-L email list
Sergeev96	Sergeev, Plesetsk press release
SP	NASA special publication
Stein	K. Stein, personal communication
STSJSC	NASA-JSC Shuttle records
TASS	TASS wire service
Thor	Arms, 1972, History of the Thor
Tiziou	J. Tiziou, personal communication
TLE	Estimated from orbital data
VCR	Vandenberg Launch Commanders Reports, declassified
VSA	Sergey Voevodin, internet postings
Wire	Wire stories (AP, Reuters, TASS etc)
WWW-FAST	FAST web page
WWW-NASDA	NASDA web page

Source: Jonathan McDowell, PhD

Table 4.9: World Orbital Launch Log 1995 – 1999

See *jsd.janes.com* for World Orbital Launch Log 1957–1994 (Tables 4.1–4.8)

Launch International Designation:	*Launch Date GMT:*	*COSPAR Payload International Designation:*	*Agency Post-Launch Payload Name:*	*Manufacturer Pre-Launch Payload Name:*	*SATCAT Number:*	*Launch Vehicle Type:*	*Launch Vehicle Serial Number:*	*Launch Site:*	*Source Reference:*
1995-001	1995 Jan 10 0618	1995-001A	Intelsat 704	Intelsat 704	S23461 Intelsat	Atlas IIAS	AC-113	CC LC36B	Wire
1995-U01	1995 Jan 15 1345	1995-U01	EXPRESS	EXPRESS	A04472 DARA	Mu-3S-II	M-3S2-8	KASC M1	JCM
		1995-U01	EXPRESS RV	EXPRESS RV	A04473 DARA				
1995-002	1995 Jan 24 0354:22	1995-002A	Tsikada	Tsikada	S23463 MO RF	Kosmos 11K65M	–	GIK-1 LC132/1	NK9501-24
		1995-002B	Astrid	Astrid	S23464 SSC				
		1995-002C	FAISAT	FAISAT	S23465 FAI				
1995-F01	1995 Jan 25 1926	1995-F01	Apstar 2	Apstar 2	F01134 APT	Chang Zheng 2E	CZ2E-5	XSC LC2	JCM
1995-003	1995 Jan 29 0125	1995-003A	UHF F/O F4	UFO F4 EHF	S23467 HCI	Atlas II	AC-112	CC LC36A	Wire
1995-004	1995 Feb 3 0522:04	1995-004A	Discovery	OV-103	S23469 NASA JSC	Space Shuttle	STS-63	KSC LC39B/ MLP2	STSJSC
		1995-004	Spacehab SH-03	Spacehab	A04477 Spacehab				
		1995-004B	Spartan-204	Spartan-204	S23470 NRL				
		1995-004	CGP/ ODERACS	HH-M	A04479 NASA GSFC				
		1995-004C	ODERACS 2A	ODERACS	S23471 NASA JSC				
		1995-004D	ODERACS 2B	ODERACS	S23472 NASA JSC				
		1995-004E	ODERACS 2C	ODERACS	S23473 NASA JSC				
		1995-004F	ODERACS 2D	ODERACS	S23474 NASA JSC				
		1995-004G	ODERACS 2E	ODERACS	S23475 NASA JSC				
		1995-004	ODERACS 2F	ODERACS	A04474 NASA JSC				
1995-005	1995 Feb 15 1648:28	1995-005A	Progress M-26	Progress 7K-TGM No. 226	S23477 MOM	Soyuz 11A511U	Ya15000-641	GIK-5 LC1	NK9807-46
1995-006	1995 Feb 16 1739:59	1995-006A	Foton	Foton No. 10	S23497 MOM	Soyuz 11A511U	–	GIK-1 LC43/4?	NK9721-34
1995-007	1995 Mar 2 0638:13	1995-007A	Endeavour	OV-105	S23500 NASA JSC	Space Shuttle	STS-67	KSC LC39A/ MLP1	STSJSC
		1995-007	ASTRO-2 Fwd	ASTRO-2 PLT + Igloo	A04483 NASA MSFC				
		1995-007	ASTRO-2 Aft	ASTRO-2 PLT	A04484 NASA MSFC				
		1995-007	EDO	EDO	A04486 NASA JSC				
1995-008	1995 Mar 2 1300:00	1995-008A	Kosmos-2306	Romb	S23501 MO RF	Kosmos 11K65M	–	GIK-1 LC132/1	NK9505-47
		1995-008G	Kosmos-2306 SS 1	ESO	S24781 MO SSSR				
		1995-008H	Kosmos-2306 SS 2	ESO	S24782 MO SSSR				
		1995-008J	Kosmos-2306 SS 3	ESO	S24783 MO SSSR				
		1995-008K	Kosmos-2306 SS 4	ESO	S24784 MO SSSR				
		1995-008L	Kosmos-2306 SS 5	ESO	S24785 MO SSSR				
		1995-008M	Kosmos-2306 SS 6	ESO	S24789 MO SSSR				
		1995-008N	Kosmos-2306 SS 7	ESO	S26584 MO SSSR				
		1995-008P	Kosmos-2306 SS 8	ESO	S26585 MO SSSR				
		1995-008Q	Kosmos-2306 SS 9	ESO	S26586 MO SSSR				
		1995-008R	Kosmos-2306 SS 10	ESO	S26587 MO SSSR				
		1995-008S	Kosmos-2306 SS 11	ESO	S26588 MO SSSR				
		1995-008T	Kosmos-2306 SS 12	ESO	S26589 MO SSSR				
		1995-008U	Kosmos-2306 SS 13	ESO	S26592 MO SSSR				
		1995-008V	Kosmos-2306 SS 14	ESO	S26593 MO SSSR				
		1995-008W	Kosmos-2306 SS 15	ESO	S26594 MO SSSR				
		1995-008X	Kosmos-2306 SS 16	ESO	S26595 MO SSSR				
		1995-008Y	Kosmos-2306 SS 17	ESO	S26596 MO SSSR				
		1995-008Z	Kosmos-2306 SS 18	ESO	S26597 MO SSSR				
		1995-008AA	Kosmos-2306 SS 19	ESO	S26598 MO SSSR				
		1995-008	Kosmos-2306 SS 20	ESO	A04500 MO SSSR				
		1995-008	Kosmos-2306 SS 21	ESO	A04501 MO SSSR				

Launch International Designation:	Launch Date GMT:	COSPAR Payload International Designation:	Agency Post-Launch Payload Name:	Manufacturer Pre-Launch Payload Name:	SATCAT Number:	Launch Vehicle Type:	Launch Vehicle Serial Number:	Launch Site:	Source Reference:
		1995-008	Kosmos-2306 SS 22	ESO	A04502 MO SSSR				
		1995-008	Kosmos-2306 SS 23	ESO	A04503 MO SSSR				
		1995-008	Kosmos-2306 SS 24	ESO	A04504 MO SSSR				
1995-009	1995 Mar 7 0923:44	1995-009A	Kosmos-2307	Uragan No. 65L	S23511 MOM	Proton-K/ DM-2	370-02	GIK-5 LC200/39	NK9505-47
		1995-009B	Kosmos-2308	Uragan No. 66L	S23512 MOM				
		1995-009C	Kosmos-2309	Uragan No. 67L	S23513 MOM				
1995-010	1995 Mar 14 0611:34	1995-010A	Soyuz TM-21	Soyuz 7K-STM No. 70	S23519 MOM	Soyuz 11A511U2	–	GIK-5 LC1	AVMN
1995-011	1995 Mar 18 0801	1995-011A	SFU	SFU	S23521 NASDA	H-II	H-II-3F	TNSC Y	NK9506-46
		1995-011B	Himawari-5	GMS-5	S23522 NASDA				
1995-012	1995 Mar 22 0409:03	1995-012A	Kosmos-2310	Parus	S23526 MO RF	Kosmos 11K65M	–	GIK-1 LC132/1	NK9910-34
1995-013	1995 Mar 22 0618	1995-013A	Intelsat 705	Intelsat 705	S23528 Intelsat	Atlas IIAS	AC-115	CC LC36B	Wire
1995-014	1995 Mar 22 1644:59	1995-014A	Kosmos-2311	Kobal't	S23530 MOM	Soyuz 11A511U	–	GIK-1 LC43/3	AVMN
		1995-014	Spuskaemiy Kapsula	SpK	A04510 MOM				
		1995-014	Spuskaemiy Kapsula	SpK	A04514 MOM				
1995-015	1995 Mar 24 1405	1995-015A	DMSP 24547	DMSP S-13	S23533 USAF	Atlas E	45E	V SLC3W	Wire
1995-F02	1995 Mar 28 1000	1995-F02	TECHSAT-1	Gurwin-1	F01138 AMSAT-IL	Start	–	GNIIP LC158	NK9507-24
		1995-F02	ENB	ENB UNAMSAT-1	F01139 UNAM/ SAI				
		1995-F02	EKA-2	EKA-2	F01140 MO RF				
1995-016	1995 Mar 28 2314:19	1995-016B	Hot Bird 1	Eutelsat HB1	S23537 Eutelsat	Ariane 44LP	V71	CSG ELA2	CNES32
		1995-016A	Brasilsat B2	Brasilsat B2	S23536 Telebras				
1995-017	1995 Apr 3 1348	1995-017A	Orbcomm F1	Orbcomm FM1	S23545 Orbcomm	Pegasus H	F8	V RW30/12 PAWA	Wire
		1995-017B	Orbcomm F2	Orbcomm FM2	S23546 Orbcomm				
		1995-017C	OrbView-1	Microlab 1	S23547 Orbimage				
1995-018	1995 Apr 5 1116	1995-018A	'Ofeq-3	'Ofeq-3	S23549 ISA	Shaviyt 1	3	PALB	JCM
1995-019	1995 Apr 7 2347	1995-019A	AMSC 1	AMSC 1	S23553 AMSC	Atlas IIA	AC-114	CC LC36A	Wire
1995-020	1995 Apr 9 1934:12	1995-020A	Progress M-27	Progress 7K-TGM No. 227	S23555 MOM	Soyuz 11A511U	–	GIK-5 LC1	NK9807-46
		1986-017JE	GFZ-1	GFZ-1	S23558 GFZ				
1995-021	1995 Apr 21 0144:02	1995-021A	ERS-2	ERS-2	S23560 ESA	Ariane 40	V72	CSG ELA2	CNES32
1995-022	1995 May 14 1345:00	1995-022A	USA 110	Adv ORION 1	S23567 NRO/CIA	Titan 401A/ Centaur	K-23 (45E-8)	CC LC40	Wire
1995-023	1995 May 17 0634:00	1995-023A	Intelsat 706	Intelsat 706	S23571 Intelsat	Ariane 44LP	V73	CSG ELA2	CNES32
1995-024	1995 May 20 0333:22	1995-024A	Spektr	TsM-O No. 17301	S23579 MOM	Proton-K	378-02	GIK-5 LC81/23	NK9510-30
1995-025	1995 May 23 0552:02	1995-025A	GOES 9	GOES J	S23581 NOAA	Atlas I	AC-77	CC LC36B	Wire
1995-026	1995 May 24 2010:09	1995-026A	Kosmos-2312	Oko	S23584 MOM	Molniya 8K78M	–	GIK-1 LC16/2	NK9715-48
1995-027	1995 May 31 1527:01	1995-027A	UFO 5	UHF F/O F5-EHF	S23589 USN	Atlas II	AC-116	CC LC36A	Wire
1995-028	1995 Jun 8 0443:00	1995-028A	Kosmos-2313	US-P	S23596 MO RF	Tsiklon-2	–	GIK-5 LC90/20	AVMN
1995-029	1995 Jun 10 0024:00	1995-029A	DBS 3	DBS 3	S23598 HCI	Ariane 42P	V74	CSG ELA2	CNES32
1995-F03	1995 Jun 22 1958	1995-F03	STEP M3	STEP 3	F01141 USAF STP	Pegasus XL	F9	V RW30/12 PAWA	Wire
1995-030	1995 Jun 27 1932:18	1995-030A	Atlantis	OV-104	S23600 NASA JSC	Space Shuttle	STS-71	KSC LC39A/ MLP3	STSJSC
		1995-030	External Airlock/ODS	EAL/ODS	A04519 NASA JSC				
		1995-030	Spacelab-Mir LM	Spacelab Long Module	A04521 NASA JSC				
1995-031	1995 Jun 28 1825:00	1995-031A	Kosmos-2314	Kobal't	S23601 MOM	Soyuz 11A511U	–	GIK-1 LC43/3	AVMN
		1995-031	Spuskaemiy Kapsula	SpK	A04523 MOM				
		1995-031	Spuskaemiy Kapsula	SpK	A04530 MOM				
1995-032	1995 Jul 5 0309:03	1995-032A	Kosmos-2315	Nadezhda-M	S23603 MO RF	Kosmos 11K65M	–	GIK-1 LC132/1	AVMN
1995-033	1995 Jul 7 1623:34	1995-033A	Helios 1A	Helios 1A	S23605 CNES/ DGA	Ariane 40	V75	CSG ELA2	CNES32
		1995-033B	CERISE	CERISE	S23606 CNES/ DGA				
		1995-033C	UPM/SAT 1	UPM/LBSAT	S23607 UPM				
1995-034	1995 Jul 10 1238:00	1995-034A	USA 112	TRUMPET 2	S23609 USAF/ NSA	Titan 401A/ Centaur	K-19 (45E-5)	CC LC41	Rigg
1995-035	1995 Jul 13 1341:55	1995-035A	Discovery	OV-103	S23612 NASA JSC	Space Shuttle	STS-70	KSC LC39B/ MLP2	STSJSC
		1995-035B	TDRS 7	TDRS G	S23613 NASA GSFC				

Launch International Designation:	*Launch Date GMT:*	*COSPAR Payload International Designation:*	*Agency Post-Launch Payload Name:*	*Manufacturer Pre-Launch Payload Name:*	*SATCAT Number:*	*Launch Vehicle Type:*	*Launch Vehicle Serial Number:*	*Launch Site:*	*Source Reference:*
1995-036	1995 Jul 20 0304:41	1995-036A	Progress M-28	Progress 7K-TGM No. 228	S23617 MOM	Soyuz 11A511U	–	GIK-5 LC1	NK9807-46
1995-037	1995 Jul 24 1552:10	1995-037A	Kosmos-2316	Uragan No. 80L	S23620 MOM	Proton-K/ DM-2	374-01	GIK-5 LC200/39	NK9515-28
		1995-037B	Kosmos-2317	Uragan No. 81L	S23621 MOM				
		1995-037C	Kosmos-2318	Uragan No. 85L	S23622 MOM				
1995-038	1995 Jul 31 2330:00	1995-038A	USA 113	DSCS III B-7	S23628 USAF	Atlas IIA	AC-118	CC LC36A	Rigg
1995-039	1995 Aug 2 2359:11	1995-039A	Interbol-1	SO-M2 No. 511	S23632 MOM	Molniya 8K78M	N15000-294 10M1	GIK-1 LC43-3	NK9516-38
		1995-039F	Magion-4	Magion-4	S23646 Czech				
1995-040	1995 Aug 3 2358:00	1995-040A	PAS 4	Panamsat K3	S23636 Panamsat	Ariane 42L	V76	CSG ELA2	JCM
1995-041	1995 Aug 5 1110:00	1995-041A	Mugunghwa	Koreasat 1	S23639 Korea Tel	Delta 7925	D228	CC LC17B	Rigg
1995-042	1995 Aug 9 0121:00	1995-042A	Molniya-3	Molniya-3 No. 59	S23642 MOM	Molniya 8K78M	PVB77031-674	GIK-1 LC43-3	NK9815-26
1995-F04	1995 Aug 15 2230	1995-F04	Gemstar DSS-1	Vitasat 1	F01143 CTA/VITA	LLV-1	DLV	V SLC6	JCM
1995-043	1995 Aug 29 0053:01	1995-043A	JCSAT 3	JCSAT 3	S23649 JSAT	Atlas IIAS	AC-117	CC LC36B	Wire
1995-044	1995 Aug 29 0641:00	1995-044A	N-STAR a	N-STAR a	S23651 NTT	Ariane 44P	V77	CSG ELA2	CNES32
1995-045	1995 Aug 30 1933:00	1995-045A	Kosmos-2319	Geizer No. 20L	S23653 MOM	Proton-K/ DM-2	369-02	GIK-5 LC200/39	NK9518-35
1995-046	1995 Aug 31 0649:59	1995-046A	Sich-1	Okean-O1 No. 8 (NKhM 10)	S23657 NKAU	Tsiklon-3	801	GIK-1 LC32/2	NK9518-36
		1995-046	Fasat-Alfa	Fasat-Alfa	A04532 FACh				
1995-047	1995 Sep 3 0900:23	1995-047A	Soyuz TM-22	Soyuz 7K-STM No. 71	S23665 MOM	Soyuz 11A511U2	–	GIK-5 LC1	AVMN
1995-048	1995 Sep 7 1509:00	1995-048A	Endeavour	OV-105	S23667 NASA JSC	Space Shuttle	STS-69	KSC LC39A/ MLP1	STSJSC
		1995-048C	Wake Shield Facility	WSF	S23669 SII				
		1995-048B	Spartan 201	Spartan 201	S23668 NASA GSFC				
		1995-048	GBA-8/CAPL	GBA	A04537 NASA GSFC				
		1995-048	IEH-1	IEH-1/HH-M	A04538 NASA GSFC				
1995-049	1995 Sep 24 0006:00	1995-049A	Telstar 402R	Telstar 402R	S23670 AT&T	Ariane 42L	V78	CSG ELA2	CNES32
1995-050	1995 Sep 26 1120:00	1995-050A	Resurs-F2	Resurs-F2 No. 10	S23672 MOM	Soyuz 11A511U	–	GIK-1 LC43-4	AVMN
1995-051	1995 Sep 29 0425:00	1995-051A	Kosmos-2320	Neman	S23674 MOM	Soyuz 11A511U	–	GIK-5 LC31	AVMN
1995-052	1995 Oct 6 0323:10	1995-052A	Kosmos-2321	Parus	S23676 MO RF	Kosmos 11K65M	–	GIK-1 LC132/1	NK9910-34
1995-053	1995 Oct 8 1850:40	1995-053A	Progress M-29	Progress 7K-TGM No. 229	S23678 MOM	Soyuz 11A511U	V15000-645	GIK-5 LC1	NK9807-46
1995-054	1995 Oct 11 1626:00	1995-054A	Luch-1	Gelios	S23680 MOM	Proton-K/ DM-2	386-01	GIK-5 LC81/23	NK9521-36
1995-055	1995 Oct 19 0038:00	1995-055A	Astra 1E	Astra 1E	S23686 SES	Ariane 42L	V79	CSG ELA2	CNES32
1995-056	1995 Oct 20 1353:00	1995-056A	Columbia	OV-102	S23688 NASA JSC	Space Shuttle	STS-73	KSC LC39B/ MLP3	STSJSC
		1995-056	Spacelab USML-2	Spacelab Long Module	A04548 NASA MSFC				
		1995-056	EDO	EDO	A04549 NASA JSC				
1995-057	1995 Oct 22 0800:02	1995-057A	UFO F6	UHF F/O F6-EHF	S23696 USN	Atlas II	AC-119	CC LC36A	Rigg
1995-F05	1995 Oct 23 2203	1995-F05	Meteor SM	Meteor SM	F01145 CTA	Conestoga 1620	F1	WI LA0	Wire
		1995-F05	Meteor RV	Meteor RV	F01146 EER				
1995-058	1995 Oct 31 2019:00	1995-058A	Kosmos-2322	Tselina-2	S23704 MO RF	Zenit-2	–	GIK-5 LC45L	AVMN
1995-059	1995 Nov 4 1422:00	1995-059A	Radarsat	Radarsat	S23710 CSA	Delta 7920-10	D229	V SLC2W	Rigg
		1995-059B	SURFSAT	SURFSAT	S23711 NASA/JPL				
1995-060	1995 Nov 6 0515:00	1995-060A	Milstar DFS 2	Milstar DFS 2	S23712 USAF	Titan 401A/ Centaur	K-21 (45E-7)	CC LC40	Rigg
1995-061	1995 Nov 12 1230:43	1995-061A	Atlantis	OV-104	S23714 NASA JSC	Space Shuttle	STS-74	KSC LC39A/ MLP2	STSJSC
		1995-061	External Airlock/ODS	EAL/ODS	A04555 NASA JSC				
		1995-061	Stikovochniy Otsek	DM 316GK No. 1	A04553 RKA				
1995-062	1995 Nov 17 0120:00	1995-062A	ISO	ISO	S23715 ESA	Ariane 44P	V80	CSG ELA2	CNES32
1995-063	1995 Nov 17 1425:00	1995-063A	Gals 2	Gals No. 12L	S23717 MOM	Proton-K/ DM-2	384-01	GIK-5 LC200/39	NK9523-59
1995-064	1995 Nov 28 1130	1995-064A	Asiasat 2	Asiasat 2	S23723 Asiasat	Chang Zheng 2E	CZ2E-6	XSC LC2	JCM
1995-065	1995 Dec 2 0808:01	1995-065A	SOHO	SOHO	S23726 ESA	Atlas IIAS	AC-121	CC LC36B	Wire
1995-066	1995 Dec 5 2118:00	1995-066A	USA 116	Improved Crystal	S23728 NRO/CIA	Titan 404A	K-15 (45J-3)	V SLC4E	Rigg
1995-067	1995 Dec 6 2323:16	1995-067B	Insat 2C	Insat 2C	S23731 ISRO	Ariane 44L	V81	CSG ELA2	CNES32
		1995-067A	Telecom 2C	Telecom 2C	S23730 France Tel				

Launch International Designation:	*Launch Date GMT:*	*COSPAR Payload International Designation:*	*Agency Post-Launch Payload Name:*	*Manufacturer Pre-Launch Payload Name:*	*SATCAT Number:*	*Launch Vehicle Type:*	*Launch Vehicle Serial Number:*	*Launch Site:*	*Source Reference:*
1995-068	1995 Dec 14 0610:31	1995-068A	Kosmos-2323	Uragan No. 82L	S23734 MOM	Proton-K/ DM-2	378-01	GIK-5 LC200/39	NK9525-30
		1995-068B	Kosmos-2324	Uragan No. 78L	S23735 MOM				
		1995-068C	Kosmos-2325	Uragan No. 76L	S23736 MOM				
1995-069	1995 Dec 15 0023	1995-069A	Galaxy 3R	Galaxy 3R	S23741 HCG	Atlas IIA	AC-120	CC LC36A	Wire
1995-070	1995 Dec 18 1431:35	1995-070A	Progress M-30	Progress 7K-TGM No. 230	S23744 MOM	Soyuz 11A511U	647	GIK-5 LC1	NK9807-46
1995-071	1995 Dec 20 0052:14	1995-071A	Kosmos-2326	US-P	S23748 MO RF	Tsiklon-2	–	GIK-5 LC90/20	AVMN
1995-072	1995 Dec 28 0645:18	1995-072A	IRS-1C	IRS-1C	S23751 ISRO	Molniya 8K78M	–	GIK-5 LC31	AVMN
		1995-072B	Skipper	Skipper	S23752 BMDO				
1995-073	1995 Dec 28 1150	1995-073A	Echostar 1	Echostar 1	S23754 TCI	Chang Zheng 2E	CZ2E-7	XSC LC2	WSL
1995-074	1995 Dec 30 1348:00	1995-074A	Rossi X-ray Timing Explorer	XTE	S23757 NASA GSFC	Delta 7920-10	D230	CC LC17A	Wire
1996-001	1996 Jan 11 0941:00	1996-001A	Endeavour	OV-105	S23762 NASA JSC	Space Shuttle	STS-72	KSC LC39B/ MLP1	STSJSC
		1996-001B	OAST-Flyer	Spartan 206	S23763 NASA GSFC				
		1996-001	SLA-1/GAS	SLA-1/GAS	A04565 NASA GSFC				
1996-002	1996 Jan 12 2310:00	1996-002B	Measat 1	HS376 Measat 1	S23765 Binariang	Ariane 44L	V82	CSG ELA2	CNES32
		1996-002A	PAS 3R	Panamsat K4	S23764 Panamsat				
1996-003	1996 Jan 14 1110	1996-003A	Koreasat 2	Koreasat 2	S23768 Korea Tel	Delta 7925	D231	CC LC17B	Wire
1996-004	1996 Jan 16 1533:45	1996-004A	Kosmos-2327	Parus	S23773 MO RF	Kosmos 11K65M	–	GIK-1 LC132/1	NK9602-23
1996-005	1996 Jan 25 0956:00	1996-005A	Gorizont	Gorizont No. 43L	S23775 MOM	Proton-K/ DM-2	374-02	GIK-5 LC200/39	NK9602-24
1996-006	1996 Feb 1 0115:01	1996-006A	Palapa C1	Palapa C1	S23779 Satelindo	Atlas IIAS	AC-126	CC LC36B	Wire
1996-007	1996 Feb 5 0719:38	1996-007A	N-Star b	N-Star b	S23781 NTT	Ariane 44P	V83	CSG ELA2	CNES32
1996-F01	1996 Feb 14 1901	1996-F01	Intelsat 708	Intelsat 708	F01151 Intelsat	Chang Zheng 3B	CZ3B-1	XSC LC2	JCM
1996-008	1996 Feb 17 2043:27	1996-008A	NEAR	NEAR	S23784 NASA GSFC	Delta 7925-8	D232	CC LC17B	Wire
1996-009	1996 Feb 19 0058:25	1996-009A	Gonets-D1	Gonets-D1 No. 1	S23787 RKA	Tsiklon-3	–	GIK-1 LC32/1	NK9701-40
		1996-009B	Gonets-D1	Gonets-D1 No. 2	S23788 RKA				
		1996-009C	Gonets-D1	Gonets-D1 No. 3	S23789 RKA				
		1996-009D	Kosmos-2328	Strela-3	S23790 MO RF				
		1996-009E	Kosmos-2329	Strela-3	S23791 MO RF				
		1996-009F	Kosmos-2330	Strela-3	S23792 MO RF				
1996-010	1996 Feb 19 0819:00	1996-010A	Raduga	Gran' No. 44L	S23794 MOM	Proton-K/ DM-2	383-02	GIK-5 LC200/39	NK9604-45
1996-011	1996 Feb 21 1234:05	1996-011A	Soyuz TM-23	Soyuz 7K-STM No. 72	S23798 MOM	Soyuz 11A511U	651	GIK-5 LC1	KIAM
1996-012	1996 Feb 22 2018:00	1996-012A	Columbia	OV-102	S23801 NASA JSC	Space Shuttle	STS-75	KSC LC39B/ MLP3	STSJSC
		1996-012	TSS-1R Deployer	TSS-1R Deployer	A04569 NASA MSFC				
		1996-012B	TSS-1R	TSS-1R	S23805 ASI				
		1996-012	TSS-1R MPESS	TSS-1R MPESS	A04570 NASA MSFC				
		1996-012	USMP-3 Fwd	USMP-3	A04	571 NASA MSFC			
		1996-012	USMP-3 Aft	USMP-3	A04	572 NASA MSFC			
		1996-012	EDO	EDO	A04	573 NASA JSC			
1996-013	1996 Feb 24 1124:00	1996-013A	Polar	Polar	S23802 NASA GSFC	Delta 7925-10	D233	V SLC2W	Rigg
1996-014	1996 Mar 9 0153	1996-014A	REX II	REX II	S23814 USAF STP	Pegasus XL	F10	V RW30/ 12 PAWA	Wire
1996-015	1996 Mar 14 0711:01	1996-015A	Intelsat 707	Intelsat 707	S23816 Intelsat	Ariane 44LP	V84	CSG ELA2	CNES32
1996-016	1996 Mar 14 1740:00	1996-016A	Kosmos-2331	Kobal't	S23818 MOM	Soyuz 11A511U	–	GIK-1 LC43-4	NK9701-40
		1996-016	Spuskaemiy Kapsula	SpK	A04588 MOM				
		1996-016	Spuskaemiy Kapsula	SpK	A04597 MOM				
1996-017	1996 Mar 21 0453	1996-017A	IRS-P3	IRS-P3	S23827 ISRO	PSLV	PSLV-D3	SHAR PSLV	JCM
1996-018	1996 Mar 22 0813:04	1996-018A	Atlantis	OV-104	S23831 NASA JSC	Space Shuttle	STS-76	KSC LC39B/ MLP2	STSJSC
		1996-018	External Airlock/ODS	EAL/ODS	A04578 NASA JSC				
		1996-018	Spacehab-SM	Spacehab	A04580 NASA JSC				
1996-019	1996 Mar 28 0021:00	1996-019A	Navstar GPS 33	Navstar SVN 33	S23833 USAF	Delta 7925	D234	CC LC17B	Rigg
1996-020	1996 Apr 3 2301:01	1996-020A	Inmarsat III F-1	Inmarsat III F-1	S23839 Inmarsat	Atlas IIA	AC-122	CC LC36A	Wire
1996-021	1996 Apr 8 2309:01	1996-021A	Astra 1F	Astra 1F	S23842 SES	Proton-K/ DM-2M	390-01	GIK-5 LC81/23	NK9608-29

Launch International Designation:	*Launch Date GMT:*	*COSPAR Payload International Designation:*	*Agency Post-Launch Payload Name:*	*Manufacturer Pre-Launch Payload Name:*	*SATCAT Number:*	*Launch Vehicle Type:*	*Launch Vehicle Serial Number:*	*Launch Site:*	*Source Reference:*
1996-022	1996 Apr 20 2236:00	1996-022A	MSAT-1	MSAT-1	S23846 TMI	Ariane 42P	V85	CSG ELA2	CNES32
1996-023	1996 Apr 23 1148:50	1996-023A	Priroda	TsM-I 77KSI No. 17401	S23848 RKA	Proton-K	385-01	GIK-5 LC81/23	NK9609-13
1996-024	1996 Apr 24 1227:40	1996-024A	MSX	MSX	S23851 BMDO	Delta 7920-10	D235	V SLC2W	Wire
1996-025	1996 Apr 24 1300:01	1996-025A	Kosmos-2332	Yug	S23853 MO RF	Kosmos 11K65M	–	GIK-1 LC132/1	NK9701-40
1996-026	1996 Apr 24 2337:00`	1996-026A	USA 118	MERCURY 2	S23855 USAF/ NSA	Titan 401A/ Centaur	K-16 (45E-4)	CC LC41	Rigg
1996-027	1996 Apr 30 0431:01	1996-027A	BeppoSAX	SAX	S23857 ASI	Atlas I	AC-78	CC LC36B	Wire
1996-028	1996 May 5 0704:18	1996-028A	Progress M-31	Progress 7K-TGM No. 231	S23860 MOM	Soyuz 11A511U	–	GIK-5 LC1	NK9807-46
1996-029	1996 May 12 2132:00	1996-029D	USA 122	NOSS B-4	S23862 NRO/ NRL	Titan 403A	K-22 (45F-11)	V SLC4E	Rigg
		1996-029A	USA 119	SSU	S23893 NRO/ NRL				
		1996-029B	USA 120	SSU	S23907 NRO/ NRL				
		1996-029C	USA 121	SSU	S23908 NRO/ NRL				
		1996-029E	USA 123	TIPS Ralph	S23936 NRO/ NRL				
		1996-029F	USA 124	TIPS Norton	S23937 NRO/ NRL				
		1996-029	TLD	TLD	A05013 NRO/ NRL				
1996-F02	1996 May 14 0855:00	1996-F02	–	Kometa No. 18	F01154 MOM	Soyuz 11A511U	PVB78051 -368	GIK-5 LC31	NK96-10-32
1996-030	1996 May 16 0156:29	1996-030A	Palapa C2	Palapa C2	S23864 PSN	Ariane 44L	V86	CSG ELA2	CNES32
		1996-030B	AMOS 1	AMOS 1	S23865 IAI/ Mabat				
1996-031	1996 May 17 0244	1996-031A	MSTI 3	MSTI 3	S23868 BMDO	Pegasus H	F11	V RW30/ 12 PAWA	Wire
1996-032	1996 May 19 1030:00	1996-032A	Endeavour	OV-105	S23870 NASA JSC	Space Shuttle	STS-77	KSC LC39B/ MLP1	STSJSC
		1996-032	Spacehab 4	Spacehab 4	A04601 Spacehab				
		1996-032B	Spartan 207	Spartan 207	S23871 JPL				
		1996-032C	IAE	Inflatable Antenna Expt.	S23872 JPL				
		1996-032	TEAMS	TEAMS	A04603 NASA GSFC				
		1996-032	GBA-9	GBA	A04604 NASA GSFC				
		1996-032D	PAMS	PAMS	S23876 NASA GSFC				
1996-033	1996 May 24 0109:59	1996-033A	Galaxy 9	Galaxy 9	S23877 HCI	Delta 7925	D236	CC LC17B	Rigg
1996-034	1996 May 25 0205	1996-034A	Gorizont	Gorizont No. 44L	S23880 MOM	Proton-K/DM-2	379-01	GIK-5 LC200/39	NK9611-38
1996-F03	1996 Jun 4 1234:06	1996-F03	Cluster F3	Cluster F3	F01157 ESA	Ariane 5G	V88 (501)	CSG ELA3	ESAPR
		1996-F03	Cluster F4	Cluster F4	F01158 ESA				
		1996-F03	Cluster F1	Cluster F1	F01160 ESA				
		1996-F03	Cluster F2	Cluster F2	F01161 ESA				
1996-035	1996 Jun 15 0655:09	1996-035A	Intelsat 709	Intelsat 709	S23915 Intelsat	Ariane 44P	V87	CSG ELA2	CNES32
1996-036	1996 Jun 20 1449:00	1996-036A	Columbia	OV-102	S23931 NASA JSC	Space Shuttle	STS-78	KSC LC39B/ MLP3	STSJSC
		1996-036	Spacelab LMS	Spacelab Long Module	A04611 NASA MSFC				
		1996-036	EDO	EDO	A04613 NASA JSC				
1996-F04	1996 Jun 20 1845	1996-F04	[Kosmos]	Kobal't	F01163 MOM	Soyuz 11A511U	–	GIK-1 LC16	NK96-12
		1996-F04	Spuskaemiy Kapsula	SpK	F01166 MOM				
		1996-F04	Spuskaemiy Kapsula	SpK	F01168 MOM				
1996-037	1996 Jul 2 0748	1996-037A	TOMS-EP	TOMS-EP	S23940 NASA GSFC	Pegasus XL	F12	V RW30/12 PAWA	Wire
1996-038	1996 Jul 3 0031	1996-038A	USA 125	SDS B-4	S23945 NRO/ USAF	Titan 405A	K-2 (45H-1)	CC LC40	Wire
1996-039	1996 Jul 3 1047	1996-039A	Apstar 1A	Apstar 1A	S23943 APT	Chang Zheng 3	CZ3-10	XSC LC1	JCM
1996-040	1996 Jul 9 2224:55	1996-040B	Turksat 1C	Turksat 1C	S23949 Turk Telecom	Ariane 44L	V89	CSG ELA2	CNES32
			1996-040A	Arabsat IIA	Arabsat 2A	S23948 Arabsat			
1996-041	1996 Jul 16 0050	1996-041A	Navstar SVN 40	Navstar SVN 40	S23953 USAF	Delta 7925	D237	CC LC17A	Wire
1996-042	1996 Jul 25 1242	1996-042A	UFO F7	UHF F/O F7-EHF	S23967 HCI	Atlas II	AC-125	CC LC36A	Wire

For details of the latest updates to ***Jane's Space Systems and Industry*** online and to discover the additional information available exclusively to online subscribers please visit **jsd.janes.com**

Launch International Designation:	*Launch Date GMT:*	*COSPAR Payload International Designation:*	*Agency Post-Launch Payload Name:*	*Manufacturer Pre-Launch Payload Name:*	*SATCAT Number:*	*Launch Vehicle Type:*	*Launch Vehicle Serial Number:*	*Launch Site:*	*Source Reference:*
1996-043	1996 Jul 31 2000:06	1996-043A	Progress M-32	Progress 7K-TGM No. 232	S24071 MOM	Soyuz 11A511U	–	GIK-5 LC1	NK9807-46
1996-044	1996 Aug 8 2249:00	1996-044A	Italsat F2	Italsat F2	S24208 ASI	Ariane 44L	V90	CSG ELA2	PR/ AE960809
		1996-044B	Telecom 2D	Telecom 2D	S24209 France Tel				
1996-045	1996 Aug 14 2220:59	1996-045A	Molniya-1T	Molniya-1T	S24273 MOM	Molniya 8K78M	–	GIK-1 LC43/3	NK9701-40
1996-046	1996 Aug 17 0153	1996-046A	Midori	ADEOS	S24277 NASDA	H-II	H-II-4F	TNSCY	WWW-NASDA
		1996-046B	JAS-2	JAS-2	S24278 JARL				
1996-047	1996 Aug 17 1318:03	1996-047A	Soyuz TM-24	Soyuz 7K-STM No. 73	S24280 RKA	Soyuz 11A511U	–	GIK-5 LC1	MIR.323
1996-048	1996 Aug 18 1027	1996-048A	Zhongxing 7	Chinasat 7	S24282 Chinasat	Chang Zheng 3	CZ3-11	XSC LC1	Wire
1996-049	1996 Aug 21 0947:26	1996-049A	FAST	FAST	S24285 NASA-GSFC	Pegasus XL	F13	V RW30/12 PAWA	WWW-FAST
1996-050	1996 Aug 29 0522:00	1996-050A	Victor	Microsatelite	S24291 Cordoba	Molniya 8K78M	–	GIK-1 LC43-3	AVMN
		1996-050B	Interbol-2	SO-M2 No. 512	S24292 RKA				
		1996-050C	Magion-5	Magion-5	S24293 Czech				
1996-051	1996 Sep 4 0901:00	1996-051A	Kosmos-2333	Tselina-2	S24297 MO RF	Zenit-2	–	GIK-5 LC45L	NK9701-40
1996-052	1996 Sep 5 1247:39	1996-052A	Kosmos-2334	Parus	S24304 MO RF	Kosmos 11K65M	–	GIK-1 LC132/1	NK9618-23
		1996-052B	UNAMSAT-B	UNAMSAT-B	S24305 UNAM				
1996-053	1996 Sep 6 1737:39	1996-053A	Inmarsat III F-2	Inmarsat III F-2	S24307 INMARSAT	Proton-K/ DM-2	375-01	GIK-5 LC81/23	NK9618-26
1996-054	1996 Sep 8 2149	1996-054A	GE 1	GE 1	S24315 GE Americom	Atlas IIA	AC-123	CC LC36B	Wire
1996-055	1996 Sep 11 0000:59	1996-055A	Echostar II	Echostar II	S24313 Echostar	Ariane 42P	V91	CSG ELA2	PC/Ransom 960912
1996-056	1996 Sep 12 0849	1996-056A	Navstar SVN 30	Navstar SVN 30	S24320 USAF	Delta 7925	D238	CC LC17A	Wire
1996-057	1996 Sep 16 0854:49	1996-057A	Atlantis	OV-104	S24324 NASA JSC	Space Shuttle	STS-79	KSC LC39A/ MLP1	STSJSC
		1996-057	External Airlock/ODS	Orbiter Docking System	A04628 NASA JSC				
		1996-057	Spacehab Double Module	Spacehab FU2/STA	A04630 NASA JSC				
1996-058	1996 Sep 26 1750:53	1996-058A	Ekspress	Ekspress No. 12L	S24435 AO Inform	Proton-K/ DM-2M	379-02	GIK-5 LC200/39	NK9810-25
1996-059	1996 Oct 20 0720	1996-059A	FSW-2 No. 3	FSW-2 No. 3	S24634 CASC	Chang Zheng 2D	CZ2D-3	JQ	Wire
		1996-059	FSW-2 RV	FSW-2 RV	A05014 CASC				
1996-060	1996 Oct 24 1137:00	1996-060A	Molniya-3	Molniya-3 No. 62	S24640 MOM	Molniya 8K78M	PVB71612-697	GIK-1 LC43-4	NK9701-40
1996-061	1996 Nov 4 1708:56	1996-061A	SAC-B	SAC-B	A04634 CONAE	Pegasus XL	F14	WI RW04/22? DZWI	Wire
		1996-061A	HETE	HETE	S24645 NASA-GSFC				
1996-062	1996 Nov 7 1700:49	1996-062A	Mars Global Surveyor	MGS	S24648 NASA/ JPL	Delta 7925	D239	CC LC17A	Wire
1996-063	1996 Nov 13 2240:00	1996-063A	Arabsat IIB	Arabsat 2B	S24652 Arabsat	Ariane 44L	V92	CSG ELA2	Wire
		1996-063B	Measat 2	Measat 2	S24653 Binariang				
1996-064	1996 Nov 16 2048:53	1996-064	Mars-8	M1 No. 520 (Mars-96)	A04638 RKA	Proton-K/D-2	392-02	GIK-5 LC200/39	NK9622-25
		1996-064	MAS 1	MAS No. 520/1	A04640 RKA				
		1996-064	MAS 2	MAS No. 520/2	A04641 RKA				
		1996-064	Penetrator 1	PN No. 520/4	A04642 RKA				
		1996-064	Penetrator 2	PN No. 520/5	A04643 RKA				
1996-065	1996 Nov 19 1955:47	1996-065A	Columbia	OV-102	S24660 NASA JSC	Space Shuttle	STS-80	KSC LC39B/ MLP3	MSFCFLASH
		1996-065B	ORFEUS-SPAS	ASTRO-SPAS	S24661 DLR				
		1996-065C	Wake Shield Facility	WSF	S24662 SII				
		1996-065	EDO	EDO	A04646 NASA JSC				
1996-066	1996 Nov 19 2320:38	1996-066A	Progress M-33	Progress 7K-TGM No. 233	S24663 RKA	Soyuz 1 1A511U	–	GIK-5 LC1	NK9807-46
1996-067	1996 Nov 21 2047	1996-067A	Hot Bird 2	Eutelsat HB2	S24665 Eutelsat	Atlas IIA	AC-124	CC LC36A	Wire
1996-068	1996 Dec 4 0658:07`	1996-068	MPF Cruise Stage	MPF Cruise Stage	A04654 NASA/ JPL	Delta 7925	D240	CC LC17B	Wire
		1996-068A	Sagan Memorial Station	MPF Lander	S24667 NASA/JPL				
		1996-068	Sojourner	MFEX Rover	A04656 NASA/ JPL				
1996-069	1996 Dec 11 1200:00	1996-069A	Kosmos-2335	US-P	S24670 MO RF	Tsiklon-2	–	GIK-5 LC90/19	NK96-25-41
1996-070	1996 Dec 18 0157	1996-070A	Inmarsat III F3	Inmarsat III F3	S24674 Inmarsat	Atlas IIA	AC-129	CC LC36B	Wire
1996-071	1996 Dec 20 0643:58	1996-071A	Kosmos-2336	Parus	S24677 MO RF	Kosmos 11K65M	–	GIK-1 LC132/1	NK9626-30
1996-072	1996 Dec 20 1804	1996-072A	USA 129	Improved CRYSTAL 4	S24680 NRO/ USAF	Titan 404A	K-13 (45J-5)	V SLC4E	Wire

Launch International Designation:	*Launch Date GMT:*	*COSPAR Payload International Designation:*	*Agency Post-Launch Payload Name:*	*Manufacturer Pre-Launch Payload Name:*	*SATCAT Number:*	*Launch Vehicle Type:*	*Launch Vehicle Serial Number:*	*Launch Site:*	*Source Reference:*
1996-073	1996 Dec 24 1350:00	1996-073A	Bion No. 11	Bion No. 11	S24701 RKA	Soyuz 11A511U	PVB15000-050	GIK-1 LC43-4	NK96-26-34
1997-001	1997 Jan 12 0927:23	1997-001A	Atlantis	OV-104	S24711 NASA JSC	Space Shuttle	STS-81	KSC LC39B/ MLP2	MSFCFLASH
		1997-001	External Airlock/ODS	EAL/ODS	A04670 NASA JSC				
		1997-001	Spacehab Double Module`	Spacehab DM	A04672 NASA JSC				
1997-F01	1997 Jan 17 1628	1997-F01	GPS SVN 42	GPS SVN 42	F01173 USAF	Delta 7925	D241	CC LC17A	Wire
1997-002	1997 Jan 30 2204:00	1997-002A	GE 2	GE 2	S24713 GE Americom	Ariane 44L	V93	CSG ELA2	Wire
		1997-002B	Nahuel 1A	Nahuel 1A	S24714 Nahuelsat				
1997-003	1997 Feb 10 1409:30	1997-003A	Soyuz TM-25	Soyuz 7K-STM No. 74	S24717 RKA	Soyuz 11A511U	–	GIK-5 LC1	AVMN
1997-004	1997 Feb 11 0855:17	1997-004A	Discovery	OV-103	S24719 NASA JSC	Space Shuttle	STS-82	KSC LC39A/ MLP1	MSFCFLASH
		1997-004	External Airlock	EAL	A04674 NASA JSC				
		1997-004	SAC	SAC	A04675 NASA JSC				
		1997-004	ORUC	ORUC	A04676 NASA JSC				
		1997-004	FSS	FSS	A04677 NASA JSC				
1997-005	1997 Feb 12 0450	1997-005A	Haruka	MUSES-B	S24720 ISAS	M-V	M-V-1	KASC M-V	Wire
1997-006	1997 Feb 14 0347:22	1997-006D	Kosmos-2337	Strela-3	S24728 MO RF	Tsiklon-3	–	GIK-1 LC32/1	NK9704-54
		1997-006E	Kosmos-2338	Strela-3	S24729 MO RF				
		1997-006F	Kosmos-2339	Strela-3	S24730 MO RF				
		1997-006A	Gonets-D1	Gonets-D1 No. 4	S24725 RKA				
		1997-006B	Gonets-D1	Gonets-D1 No. 5	S24726 RKA				
		1997-006C	Gonets-D1	Gonets-D1 No. 6	S24727 RKA				
1997-007	1997 Feb 17 0142:02	1997-007A	JCSAT 4	JCSAT 4	S24732 JSAT	Atlas IIAS	AC-127	CC LC36B	Wire
1997-008	1997 Feb 23 2020	1997-008A	DSP F18	DSP 20	S24737 USAF	Titan 402B/ IUS	4B-24 (K-24, 45)	CC LC40	Wire
1997-009	1997 Mar 1 0107:42	1997-009A	Intelsat 801	Intelsat 801	S24742 Intelsat	Ariane 44P	V94	CSG ELA2	Wire
1997-010	1997 Mar 4 0200:02	1997-010A	Zeya	Zeya	S24744 MO RF	Start-1.2	–	GIK-2 LC5	NK9705-11
1997-011	1997 Mar 8 0601`	1997-011A	Tempo 2	Tempo 2	S24748 TCI	Atlas IIA	AC-128	CC LC36A	Wire
1997-012	1997 Apr 4 1647	1997-012A	DMSP 5D-2 F-14	DMSP 5D-2 S-14	S24753 USAF	Titan II SLV	23G-6	V SLC4W	Wire
1997-013	1997 Apr 4 1920:32	1997-013A	Columbia	OV-102	S24755 NASA JSC	Space Shuttle	STS-83	KSC LC39A/ MLP3	MSFCFLASH
		1997-013	Spacelab MSL-1	Spacelab Long Module 1	A04684 NASA MSFC				
		1997-013	EDO	EDO	A04685 NASA JSC				
1997-014	1997 Apr 6 1604:05	1997-014A	Progress M-34	Progress 7K-TGM No. 234	S24757 RKA	Soyuz 11A511U	–	GIK-5 LC1	NK9807-46
1997-015	1997 Apr 9 0858:44	1997-015A	Kosmos-2340	Oko	S24761 MO RF	Molniya 8K78M	PVB76032-647	GIK-1 LC16-2	NK9708-30
1997-016	1997 Apr 16 2308:44	1997-016A	Thaicom 3	Thaicom 3	S24768 Shinawatra	Ariane 44LP	V95	CSG ELA2	Wire
		1997-016B	BSAT 1a	BSAT 1a	S24769 BSAT				
1997-017	1997 Apr 17 1303:21	1997-017A	Kosmos-2341	Parus	S24772 MO RF	Kosmos 11K65M		GIK-1 LC132/1	NK9708-36
1997-018	1997 Apr 21 1159	1997-018B	Celestis	CPAC	S24780 Celestis	Pegasus XL	F15	GAN RW03/21 DZGC	Wire
		1997-018A	Minisat-01	Minisat-01	S24779 INTA				
1997-019	1997 Apr 25 0549	1997-019A	GOES 10	GOES K	S24786 NOAA	Atlas I	AC-79	CC LC36B	Wire
1997-020	1997 May 5 1455:28	1997-020E	Iridium 4	Iridium SV004	S24796 Iridium	Delta 7920-10C`	D242	V SLC2W	AWST 970512-25
		1997-020D	Iridium 5	Iridium SV005	S24795 Iridium				
		1997-020C	Iridium 6	Iridium SV006	S24794 Iridium				
		1997-020B	Iridium 7	Iridium SV007	S24793 Iridium				
		1997-020A	Iridium 8	Iridium SV008	S24792 Iridium				
1997-021	1997 May 11 1617	1997-021A	Zhongxing 6	DFH-3	S24798 Chinasat	Chang Zheng 3A	CZ3A-3	XSC LC2	Stein
1997-022	1997 May 14 0033:57	1997-022A	Kosmos-2342	Oko	S24800 MO RF	Molniya 8K78M	–	GIK-1 LC43-4	NK9710-34
1997-023	1997 May 15 0807:48	1997-023A	Atlantis	OV-104	S24804 NASA JSC	Space Shuttle	STS-84	KSC LC39A/ MLP2	MSFCFLASH
		1997-023	External Airlock/ODS	EAL/ODS	A04696 NASA JSC				
		1997-023	Spacehab Double Module	Spacehab Double Module	A04698 NASA JSC				
1997-024	1997 May 15 1210:00	1997-024A	Kosmos-2343	Don	S24805 MO RF	Soyuz 11A511U	–	GIK-5 LC31	NK9710-35
		1997-024	Spuskaemiy Kapsula	SpK	A04660 MO RF				
		1997-024	Spuskaemiy Kapsula	SpK	A04661 MO RF				

Launch International Designation:	*Launch Date GMT:*	*COSPAR Payload International Designation:*	*Agency Post-Launch Payload Name:*	*Manufacturer Pre-Launch Payload Name:*	*SATCAT Number:*	*Launch Vehicle Type:*	*Launch Vehicle Serial Number:*	*Launch Site:*	*Source Reference:*
		1997-024	Spuskaemiy Kapsula	SpK	A04662 MO RF				
		1997-024	Spuskaemiy Kapsula	SpK	A04663 MO RF				
		1997-024	Spuskaemiy Kapsula	SpK	A04663 MO RF				
		1997-024	Spuskaemiy Kapsula	SpK	A04665 MO RF				
		1997-024	Spuskaemiy Kapsula	SpK	A04666 MO RF				
1997-F02	1997 May 20 0707:00	1997-F02	–	Tselina-2	F01179 MO RF	Zenit-2	–	GIK-5 LC45L	NK9711-24
1997-025	1997 May 20 2239	1997-025A	Thor II	Thor 2A	S24808 Telenor	Delta 7925	D243	CC LC17A	Wire
1997-026	1997 May 24 1700:00	1997-026A	Telstar 5	Telstar 5	S24812 Loral Skynet	Proton-K/ DM-2M	380-02	GIK-5 LC81/23	NK9711-18
1997-027	1997 Jun 3 2320:06	1997-027B	Insat 2D	Insat 2D	S24820 ISRO	Ariane 44L	V97	CSG ELA2	Wire
		1997-027A	Inmarsat III F4	Inmarsat III F4	S24819 Inmarsat				
1997-028	1997 Jun 6 1656:54	1997-028A	Kosmos-2344	11F664 No. 1	S24827 MO RF	Proton-K/ 17S40	380-01	GIK-5 LC200/39	NK9712-20
1997-029	1997 Jun 10 1201	1997-029A	Feng Yun 2	FY-2	S24834 CASC	Chang Zheng 3	CZ3-12	XSC LC1	Wire
1997-030	1997 Jun 18 1402:45	1997-030C	Iridium 9	Iridium SV009	S24838 Iridium	Proton-K/ 17S40	390-02	GIK-5 LC81/23	NK9713-34
		1997-030D	Iridium 10	Iridium SV010	S24839 Iridium				
		1997-030G	Iridium 11	Iridium SV011	S24842 Iridium				
		1997-030B	Iridium 12	Iridium SV012	S24837 Iridium				
		1997-030E	Iridium 13	Iridium SV013	S24840 Iridium				
		1997-030A	Iridium 14	Iridium SV014	S24836 Iridium				
		1997-030F	Iridium 16	Iridium SV016	S24841 Iridium				
1997-031	1997 Jun 25 2344:00	1997-031A	Intelsat 802	Intelsat 802	S24846 Intelsat	Ariane 44P	V96	CSG ELA2	Wire
1997-032	1997 Jul 1 1802:00	1997-032A	Columbia	OV-102	S24849 NASA JSC	Space Shuttle	STS-94	KSC LC39A/ MLP1	MSFCFLASH
		1997-032	Spacelab Long Module 1	Spacelab Long Module 1	A04713 NASA MSFC				
		1997-032	EDO	EDO	A04714 NASA JSC				
1997-033	1997 Jul 5 0411:54	1997-033A	Progress M-35	Progress 7K-TGM No. 235	S24851 RKA	Soyuz 11A511U	–	GIK-5 LC1	NK9807-46
1997-034	1997 Jul 9 1304	1997-034A	Iridium 15	Iridium SV015	S24869 Iridium	Delta 7920-10C	D244	V SLC2W	Wire
		1997-034B	Iridium 17	Iridium SV017	S24870 Iridium				
		1997-034D	Iridium 18	Iridium SV018	S24872 Iridium				
		1997-034C	Iridium 20	Iridium SV020	S24871 Iridium				
		1997-034E	Iridium 21	Iridium SV021	S24873 Iridium				
1997-035	1997 Jul 23 0343	1997-035A	GPS SVN 43	GPS SVN 43	S24876 USAF	Delta 7925	D245	CC LC17A	Wire
1997-036	1997 Jul 28 0115	1997-036A	Superbird C	Superbird C	S24880 SCC	Atlas IIAS	AC-133	CC LC36B	Wire
1997-037	1997 Aug 1 2020	1997-037A	OrbView-2	Seastar	S24883 Orbimage	Pegasus XL	F16	V RW30/12 PAWA	Wire
1997-038	1997 Aug 5 1535:53	1997-038A	Soyuz TM-26	Soyuz 7K-STM No. 75	S24886 RKA	Soyuz 11A511U	–	GIK-5 LC1	AVMN
1997-039	1997 Aug 7 1441:00	1997-039A	Discovery	OV-103	S24889 NASA JSC`	Space Shuttle	STS-85	KSC LC39A/ MLP3	MSFCFLASH
		1997-039B	CRISTA-SPAS	ASTRO-SPAS	S24890 DLR				
		1997-039	MFD	MFD	A04722 NASDA				
		1997-039	TAS-1	TAS-1	A04723 NASA GSFC				
		1997-039	IEH-2	IEH-2	A04724 NASA GSFC				
1997-040	1997 Aug 8 0646:00	1997-040A	PAS 6	PAS 6	S24891 Panamsat	Ariane 44P	V98	CSG ELA2	Wire
1997-041	1997 Aug 14 2049:14	1997-041A	Kosmos-2345	Prognoz	S24894 MO RF	Proton-K/ DM-2	381-01	GIK-5 LC200/39	NK9713-57
1997-042	1997 Aug 19 1750	1997-042A	Agila 2	Mabuhay	S24901 MPSC	Chang Zheng 3B	CZ3B-2	XSC LC2	Wire
1997-043	1997 Aug 21 0038:40	1997-043E	Iridium 22	Iridium SV022	S24907 Iridium	Delta 7920-10C	D246	V SLC2W	Seesat-L
		1997-043D	Iridium 23	Iridium SV023	S24906 Iridium				
		1997-043C	Iridium 24	Iridium SV024	S24905 Iridium				
		1997-043B	Iridium 25	Iridium SV025	S24904 Iridium				
		1997-043A	Iridium 26	Iridium SV026	S24903 Iridium				
1997-044	1997 Aug 23 0651:01	1997-044A	Lewis	SSTI/Lewis	S24909 TRW/ NASA HQ	LMLV-1	LM-002	V SLC6	AWST 970901-28
1997-045	1997 Aug 25 1439	1997-045A	ACE	Advanced Composition Expl	S24912 NASA GSFC	Delta 7920-8	D247	CC LC17A	Wire
1997-046	1997 Aug 28 0033:30	1997-046A	PAS 5	PAS 5	S24916 Panamsat	Proton-K/ DM-2M	387-02	GIK-5 LC81/23	NK9721-38
1997-047	1997 Aug 29 1502:22	1997-047A	FORTE	FORTE	S24920 USAF STP	Pegasus XL	019/F17	V RW30/12 PAWA	Wire
1997-048	1997 Sep 1 1400:15	1997-048A	Iridium MFS 1	Iridium MFS 1	S24925 CASC	Chang Zheng 2C-III/SD	CZ2C-15	TYSC LC1	McDonald-B-priv
		1997-048B	Iridium MFS 2	Iridium MFS 2	S24926 CASC				
1997-049	1997 Sep 2 2221:07	1997-049A	Hot Bird 3	Hot Bird 3	S24931 Eutelsat	Ariane 44LP	V99	CSG ELA2	Wire
		1997-049B	Meteosat 7	MTP 1	S24932 Eumetsat				
1997-050	1997 Sep 4 1203	1997-050A	GE 3	GE 3	S24936 GE Americom	Atlas IIAS	AC-146	CC LC36A	Wire
1997-051	1997 Sep 14 0136:54	1997-051D	Iridium 27	Iridium SV027	S24947 Iridium	Proton-K/1 7S40	391-01	GIK-5 LC81/23	NK9718-55

Launch International Designation:	*Launch Date GMT:*	*COSPAR Payload International Designation:*	*Agency Post-Launch Payload Name:*	*Manufacturer Pre-Launch Payload Name:*	*SATCAT Number:*	*Launch Vehicle Type:*	*Launch Vehicle Serial Number:*	*Launch Site:*	*Source Reference:*
		1997-051E	Iridium 28	Iridium SV028	S24948 Iridium				
		1997-051A	Iridium 29	Iridium SV029	S24944 Iridium				
		1997-051F	Iridium 30	Iridium SV030	S24949 Iridium				
		1997-051G	Iridium 31	Iridium SV031	S24950 Iridium				
		1997-051B	Iridium 32	Iridium SV032	S24945 Iridium				
		1997-051C	Iridium 33	Iridium SV033	S24946 Iridium				
1997-052	1997 Sep 23 1644:51	1997-052A	Kosmos-2346	Parus	S24953 MO RF	Kosmos 11K65M	–	GIK-1 LC132/1	NK9910-34
		1997-052B	FAISAT-2V	FAISAT-2V	S24954 FAI				
1997-053	1997 Sep 23 2358	1997-053A	Intelsat 803 (NSS 803)	Intelsat 803	S24957 Intelsat	Ariane 42L	V100	CSG ELA2	Wire
1997-054	1997 Sep 24 2130:59	1997-054A	Molniya-1T	Molniya-1T	S24960 MO RF	Molniya 8K78M	–	GIK-1 LC43-4	AVMN
1997-055	1997 Sep 26 0234:19	1997-055A	Atlantis	OV-104	S24964 NASA JSC	Space Shuttle	STS-86	KSC LC39A/ MLP2	Wire
		1997-055	External Airlock/ ODS	EAL/ODS	A04747 NASA JSC				
		1997-055	Spacehab Double Module	Spacehab Double Module	A04749 NASA JSC				
1997-056	1997 Sep 27 0123:36	1997-056A	Iridium 19	Iridium SV019	S24965 Iridium	Delta 7920-10C	D248	V SLC2W	Wire
		1997-056E	Iridium 34	Iridium SV034	S24969 Iridium				
		1997-056D	Iridium 35	Iridium SV035	S24968 Iridium				
		1997-056C	Iridium 36	Iridium SV036	S24967 Iridium				
		1997-056B	Iridium 37	Iridium SV037	S24966 Iridium				
1997-057	1997 Sep 29 0447	1997-057A	IRS-1D	IRS-1D	S24971 ISRO	PSLV	PSLV-C1	SHAR PSLV	Wire
1997-058	1997 Oct 5 1508:57	1997-058A	Progress M-36	Progress 7K-TGM No. 237	S25002 RKA	Soyuz 11A511U	–	GIK-5 LC1	NK9807-46
		1997-058C	Spoutnik-40	RS-17	S24958 RKA				
		1997-058D	X-Mir Inspector	Inspector	S25100 DASA				
1997-059	1997 Oct 5 2101	1997-059A	Echostar 3	Echostar 3	S25004 Echostar	Atlas IIAS	AC-135	CC LC36B	Wire
1997-060	1997 Oct 9 1759:59	1997-060A	Foton	Foton No. 11	S25006 RKA	Soyuz 11A511U	–	GIK-1 LC43/3	NK9721-34
		1997-060	Mirka	Mirka	A04762 DLR				
1997-061	1997 Oct 15 0843	1997-061A	Cassini	Cassini	S25008 NASA/JPL	Titan 401B/ Centaur	4B-33 (K-33, 45)	CC LC40	Wire
		1997-061	Huygens	Huygens	A04759 ESA				
1997-062	1997 Oct 16 1913	1997-062A	Apstar 2R	Apstar 2R	S25010 APT	Chang Zheng 3B	CZ3B-3	XSC LC2	Wire
1997-063	1997 Oct 22 1313	1997-063A	STEP M4	STEP M4	S25013 USAF STP	Pegasus XL	F18	WI RW04/ 22? DZWI	Wire
1997-064	1997 Oct 24 0232	1997-064A	USA 133	LACROSSE 3	S25017 NRO/ CIA	Titan 403A	4A-18 (K-18, 45)	V SLC4E	Wire
1997-065	1997 Oct 25 0046	1997-065A	USA 135	DSCS III B-13	S25019 USAF	Atlas IIA	AC-131	CC LC36A	Wire
		1997-065B	Falcon Gold	Falcon Gold	A04764 USAF				
1997-066	1997 Oct 30 1343	1997-066B	Maqsat-B	Maqsat-B	S25024 ESA	Ariane 5G	V101 (502)	CSG ELA3	Wire
		1997-066A	Maqsat-H	Maqsat-H	A04766 ESA				
		1997-066A	TEAMSAT	TEAMSAT	S25023 ESA				
		1997-066C	YES	Young Engineers Satellite	S25025 ESA				
1997-F03	1997 Nov 2 1225:00	1997-F03	SCD-2A	SCD-2A	F01181 INPE	VLS-1	V01	ALCA VLS	Wire
1997-067	1997 Nov 6 0030:00	1997-067A	GPS SVN 38	GPS SVN 38	S25030 USAF	Delta 7925	D249	CC LC17A	Wire
1997-068	1997 Nov 8 0205	1997-068A	USA 136	TRUMPET 3	S25034 USAF/ NSA	Titan 401A/ Centaur	4A-17 (K-20, 45)	CC LC41	Wire
1997-069	1997 Nov 9 0134:26	1997-069E	Iridium 38	Iridium SV038	S25043 Iridium	Delta 7920-10C	D250	V SLC2W	Wire
		1997-069D	Iridium 39	Iridium SV039	S25042 Iridium				
		1997-069C	Iridium 40	Iridium SV040	S25041 Iridium				
		1997-069B	Iridium 41	Iridium SV041	S25040 Iridium				
		1997-069A	Iridium 43	Iridium SV043	S25039 Iridium				
1997-070	1997 Nov 12 1700:00	1997-070A	Kupon	K95K	S25045 TsBank	Proton-K/ DM-2M	382-01	GIK-5 LC200/ 39	NK9723-36
1997-071	1997 Nov 12 2148	1997-071B	Cakrawarta 1	Indostar 1	S25050 Indostar	Ariane 44L	V102	CSG ELA2	Wire
		1997-071A	Sirius 2	Sirius 2	S25049 NSAB				
1997-072	1997 Nov 18 1114:59	1997-072A	Resurs F-1M	Resurs F-1M	S25059 RKA	Soyuz 11A511U	–	GIK-1	Wire
1997-073	1997 Nov 19 1946:00	1997-073A	Columbia	OV-102	S25061 NASA JSC	Space Shuttle	STS-87	KSC LC39B/ MLP1	MSFCFLASH
		1997-073B	Spartan 201	Spartan 201	S25062 NASA GSFC				
		1997-073	USMP-4 Forward	USMP-4	A04774 NASA MSFC				
		1997-073	USMP-4 Aft	USMP-4	A04775 NASA MSFC				
		1997-073	EDO	EDO	A04776 NASA JSC				
		1997-073	AERCam/ Sprint	Sprint	A04788 NASA JSC				
1997-074	1997 Nov 27 2127	1997-074A	TRMM	TRMM	S25063 NASA GSFC	H-II	H-II-6F	TNSCY	WWW-NASDA
		1997-074B	Hikoboshi	ETS-7	S25064 NASDA				
		1997-074E	Orihime	ETS-7 Target	S25424 NASDA				
1997-075	1997 Dec 2 2252	1997-075A	JCSAT 5	JCSAT 5	S25067 JSAT	Ariane 44P	V103	CSG ELA2	Wire
		1997-075B	Equator-S	Equator-S	S25068 MPE				

Launch International Designation:	*Launch Date GMT:*	*COSPAR Payload International Designation:*	*Agency Post-Launch Payload Name:*	*Manufacturer Pre-Launch Payload Name:*	*SATCAT Number:*	*Launch Vehicle Type:*	*Launch Vehicle Serial Number:*	*Launch Site:*	*Source Reference:*
1997-076	1997 Dec 2 2310:37	1997-076A	Astra 1G	Astra 1G	S25071 SES	Proton-K/ DM-2M	382-02	GIK-5 LC81/23	NK9725-35
1997-077	1997 Dec 8 0716	1997-077A	Iridium 42	Iridium SV042	S25077 Iridium	Chang Zheng 2C-III/SD	CZ2C-16	TYSC LC1	Wire
		1997-077B	Iridium 44	Iridium SV044	S25078 Iridium				
1997-078	1997 Dec 8 2352	1997-078A	Galaxy 8i	Galaxy 8-i	S25086 Panamsat	Atlas IIAS	AC-149	CC LC36B	Wire
1997-079	1997 Dec 9 0717:20	1997-079A	Kosmos-2347	US-P	S25088 MO RF	Tsiklon-2	–	GIK-5 LC90/19	Wire
1997-080	1997 Dec 15 1540:00	1997-080A	Kosmos-2348	Kobal't	S25095 MO RF	Soyuz 11A511U	–	GIK-1	TLE
		1997-080	Spuskaemiy Kapsula	SpK	A04798 MO RF				
		1997-080	Spuskaemiy Kapsula	SpK	A04821 MO RF				
1997-081	1997 Dec 20 0845:02	1997-081A	Progress M-37	Progress 7K-TGM No 236	S25102 RKA	Soyuz 11A511U	–	GIK-5 LC1	NK9807-46
1997-082	1997 Dec 20 1316	1997-082A	Iridium 45	Iridium SV045	S25104 Iridium	Delta 7920-10C	D251	V SLC2W	Wire
		1997-082B	Iridium 46	Iridium SV046	S25105 Iridium				
		1997-082C	Iridium 47	Iridium SV047	S25106 Iridium				
		1997-082D	Iridium 48	Iridium SV048	S25107 Iridium				
		1997-082E	Iridium 49	Iridium SV049	S25108 Iridium				
1997-083	1997 Dec 22 0016	1997-083A	Intelsat 804	Intelsat 804	S25110 Intelsat	Ariane 42L	V104	CSG ELA2	Wire
1997-084	1997 Dec 23 1911:42	1997-084A	Orbcomm A1	Orbcomm FM5	S25112 Orbcomm	Pegasus XL/ HAPS	F19	WI RW04/ 22? DZWI	Rosenberg
		1997-084B	Orbcomm A2	Orbcomm FM6	S25113 Orbcomm				
		1997-084C	Orbcomm A3	Orbcomm FM7	S25114 Orbcomm				
		1997-084D	Orbcomm A4	Orbcomm FM8	S25115 Orbcomm				
		1997-084E	Orbcomm A5	Orbcomm FM9	S25116 Orbcomm				
		1997-084F	Orbcomm A6	Orbcomm FM10	S25117 Orbcomm				
		1997-084G	Orbcomm A7	Orbcomm FM11	S25118 Orbcomm				
		1997-084H	Orbcomm A8	Orbcomm FM12	S25119 Orbcomm				
1997-085	1997 Dec 24 1332:52	1997-085A	EarlyBird	EarlyBird	S25123 EarthWatch	Start-1	–	GIK-2 LC5	Lissov
1997-086	1997 Dec 24 2319	1997-086A	HGS-1	Asiasat 3	S25126 Asiasat	Proton-K/ DM-2M	394-01	GIK-5 LC81/23	NK9726-45
1998-001	1998 Jan 7 0228:44	1998-001A	Lunar Prospector	LP	S25131 NASA ARC	Athena-2	LM-004	SPFL SLC46	Wire
1998-002	1998 Jan 10 0032:01	1998-002A	Skynet 4D	Skynet 4D	S25134 UK MOD	Delta 7925-9.5	D252	CC LC17B	Wire
1998-F01	1998 Jan 22 1256	1998-F01	'Ofeq-4	Ofeq-4	F01183 ISA	Shaviyt 1	4	PALB	Wire
1998-003	1998 Jan 23 0248:15	1998-003A	Endeavour	OV-105	S25143 NASA JSC	Space Shuttle	STS-89	KSC LC39A/ MLP3	MSFCFLASH
		1998-003	External Airlock/ ODS	EAL/ODS	A04809 NASA JSC				
		1998-003A	Spacehab Double Module	Spacehab Double Module	A04807 NASA JSC				
1998-004	1998 Jan 29 1633:42	1998-004A	Soyuz TM-27	Soyuz 7K-STM No. 76	S25146 RKA	Soyuz 11A511U	–	GIK-5 LC1	NK9902-45
1998-005	1998 Jan 29 1837	1998-005A	CAPRICORN	CAPRICORN	S25148 NRO	Atlas IIA	AC-109	CC SLC36A	Wire
1998-006	1998 Feb 4 2329	1998-006A	Brasilsat B3	Brasilsat B3	S25152 Embratel	Ariane 44LP	V105	CSG ELA2	Wire
		1998-006B	Inmarsat 3 F5	Inmarsat 3 F5	S25153 Inmarsat				
1998-007	1998 Feb 10 1320	1998-007D	Celestis-02	CPAC	S25160 Celestis	Taurus 2210	T2 2210	V 576E	Wire
		1998-007B	Orbcomm G1	Orbcomm FM3	S25158 Orbcomm				
		1998-007C	Orbcomm G2	Orbcomm FM4	S25159 Orbcomm				
		1998-007A	GFO	GFO	S25157 USN				
1998-008	1998 Feb 14 1434	1998-008A	Globalstar FM1	Globalstar FM1	S25162 Globalstar	Delta 7420-10C	D253	CC LC17A	Wire
		1998-008B	Globalstar FM2	Globalstar FM2	S25163 Globalstar				
		1998-008C	Globalstar FM3	Globalstar FM3	S25164 Globalstar				
		1998-008D	Globalstar FM4	Globalstar FM4	S25165 Globalstar				
1998-009	1998 Feb 17 1035:00	1998-009A	Kosmos-2349	Kometa No. 19	S25167 RKA	Soyuz 11A511U	–	GIK-5 LC31	NK9902-45
1998-010	1998 Feb 18 1358:09	1998-010D	Iridium 50	Iridium SV050	S25172 Iridium	Delta 7920-10C	D254	V SLC2W	Wire
		1998-010A	Iridium 52	Iridium SV052	S25169 Iridium				
		1998-010E	Iridium 53	Iridium SV053	S25173 Iridium				
		1998-010C	Iridium 54	Iridium SV054	S25171 Iridium				
		1998-010B	Iridium 56	Iridium SV056	S25170 Iridium				
1998-011	1998 Feb 21 0755	1998-011A	Kakehashi	COMETS	S25175 NASDA	H-II	H-II-5F	TNSCY	WWW-NASDA
1998-012	1998 Feb 26 0707	1998-012A	SNOE	SNOE	S25233 NASA GSFC	Pegasus XL	F20	V RW30 PAWA	Wire
		1998-012B	T1	BATSAT	S25234 Teledesic				
1998-013	1998 Feb 27 2238	1998-013A	Hot Bird 4	Hot Bird 4	S25237 Eutelsat	Ariane 42P	V106	CSG ELA2	Wire
1998-014	1998 Feb 28 0021	1998-014A	Intelsat 806 (NSS 806)	Intelsat 806	S25239 Intelsat	Atlas IIAS	AC-151	CC SLC36B	Wire

Launch International Designation:	Launch Date GMT:	COSPAR Payload International Designation:	Agency Post-Launch Payload Name:	Manufacturer Pre-Launch Payload Name:	SATCAT Number:	Launch Vehicle Type:	Launch Vehicle Serial Number:	Launch Site:	Source Reference:
1998-015	1998 Mar 14 2245:55	1998-015A	Progress M-38	Progress 7K-TGM No 240	S25256 RKA	Soyuz 11A511U	–	GIK-5 LC1	NK9807-46
1998-016	1998 Mar 16 2132	1998-016A	UHF F/O F8	UHF F/O F8	S25258 USN	Atlas II	AC-132	CC SLC36A	Wire
1998-017	1998 Mar 24 0146	1998-017A	SPOT-4	SPOT-4	S25260 CNES	Ariane 40	V107	CSG ELA2	Wire
1998-018	1998 Mar 25 1701:06	1998-018A	Iridium 51	Iridium SV051	S25262 Iridium	Chang Zheng 2C-III/SD	CZ2C-17	TYSC LC1	CALT-WWW
		1998-018B	Iridium 61	Iridium SV061	S25263 Iridium				
1998-019	1998 Mar 30 0602:46	1998-019A	Iridium 55	Iridium SV055	S25272 Iridium	Delta 7920-10C	D255	V SLC2W	Rigg
		1998-019B	Iridium 57	Iridium SV057	S25273 Iridium				
		1998-019C	Iridium 58	Iridium SV058	S25274 Iridium				
		1998-019D	Iridium 59	Iridium SV059	S25275 Iridium				
		1998-019E	Iridium 60	Iridium SV060	S25276 Iridium				
1998-020	1998 Apr 2 0242:39	1998-020A	TRACE	TRACE	S25280 NASA GSFC	Pegasus XL	F21	V RW30/ 12 PAWA	Wire
1998-021	1998 Apr 7 0213:03	1998-021A	Iridium 62	Iridium SV062	S25285 Iridium	Proton-K/ 17S40	391-02	GIK-5 LC81/ 23	NK9809-11
		1998-021B	Iridium 63	Iridium SV063	S25286 Iridium				
		1998-021C	Iridium 64	Iridium SV064	S25287 Iridium				
		1998-021D	Iridium 65	Iridium SV065	S25288 Iridium				
		1998-021E	Iridium 66	Iridium SV066	S25289 Iridium				
		1998-021F	Iridium 67	Iridium SV067	S25290 Iridium				
		1998-021G	Iridium 68	Iridium SV068	S25291 Iridium				
1998-022	1998 Apr 17 1819:00	1998-022A	Columbia	OV-102	S25297 NASA JSC	Space Shuttle	STS-90	KSC LC39B/ MLP2	MSFCFLASH
		1998-022A	Neurolab	Spacelab Long Module 2	A04835 NASA MSFC				
		1998-022	EDO	EDO	A04838 NASA JSC				
1998-023	1998 Apr 24 2238:34	1998-023A	Globalstar FM6	Globalstar FM6	S25306 Globalstar	Delta 7420-10C	D256	CC LC17A	Wire
		1998-023B	Globalstar FM8	Globalstar FM8	S25307 Globalstar				
		1998-023C	Globalstar FM14	Globalstar FM14	S25308 Globalstar				
		1998-023D	Globalstar FM15	Globalstar FM15	S25309 Globalstar				
1998-024	1998 Apr 28 2253	1998-024A	Nilesat 101	Nilesat 101	S25311 Nilesat	Ariane 44P	V108	CSG ELA2	Wire
		1998-024B	BSAT 1b	BSAT 1b	S25312 BSAT				
1998-025	1998 Apr 29 0436:54	1998-025A	Kosmos-2350	Prognoz	S25315 RVSN	Proton-K/ DM-2	384-02	GIK-5 LC200/ 39	NK9810-25
1998-026	1998 May 2 0916:53	1998-026A	Iridium 69	Iridium SV069	S25319 Iridium	Chang Zheng 2C-III/SD	CZ2C-18	TYSC LC1	CALT-WWW
		1998-026B	Iridium 71	Iridium SV071	S25320 Iridium				
1998-027	1998 May 7 0853:22	1998-027A	Kosmos-2351	Oko	S25327 MO RF	Molniya 8K78M	–	GIK-1 LC16-2	NK9811-14
1998-028	1998 May 7 2345:00	1998-028A	Echostar 4	Echostar 4	S25331 Echostar	Proton-K/ DM-2M	393-02	GIK-5 LC81/ 23	NK9811-15
1998-029	1998 May 9 0138:01	1998-029A	USA 139	Adv ORION 2	S25336 NRO/ NSA	Titan 401B/ Centaur	4B-25 (K-25)	CC LC40	Wire
1998-030	1998 May 13 1552:04	1998-030A	NOAA 15	NOAA K	S25338 NOAA	Titan II SLV	23G-12	V SLC4W	Wire
1998-031	1998 May 14 2212:59	1998-031A	Progress M-39	Progress 7K-TGM No 238	S25340 RKA	Soyuz 11A511U	–	GIK-5 LC1	NK9811-03
1998-032	1998 May 17 2116:56	1998-032A	Iridium 70	Iridium SV070	S25342 Iridium	Delta 7920-10C	D257	V SLC2W	Wire
		1998-032B	Iridium 72	Iridium SV072	S25343 Iridium				
		1998-032C	Iridium 73	Iridium SV073	S25344 Iridium				
		1998-032D	Iridium 74	Iridium SV074	S25345 Iridium				
		1998-032E	Iridium 75	Iridium SV075	S25346 Iridium				
1998-033	1998 May 30 1000:04	1998-033A	Zhongwei 1	Chinastar 1	S25354 China Orient	Chang Zheng 3B	CZ3B-4	XSC LC2	CALT-WWW
1998-034	1998 Jun 2 2206:24	1998-034A	Discovery	OV-103	S25356 NASA JSC	Space Shuttle	STS-91	KSC LC39A/ MLP1	MSFCFLASH
		1998-034	External Airlock/ ODS	EAL/ODS	A04854 NASA JSC				
		1998-034A	Spacehab	Spacehab FU1	A04851 Spacehab				
		1998-034A	AMS	Alpha Magnetic Spectromet	A04852 CERN				
1998-035	1998 Jun 10 0035	1998-035A	Thor III	Thor 3	S25358Telenor	Delta 7925-9.5	D258	CC LC17A	Wire
1998-036	1998 Jun 15 2258:05	1998-036A	Kosmos-2352	Strela-3	S25363 MO RF	Tsiklon-3	–	GIK-1 LC32/1	NK9814-08
		1998-036B	Kosmos-2353	Strela-3	S25364 MO RF				
		1998-036C	Kosmos-2354	Strela-3	S25365 MO RF				
		1998-036D	Kosmos-2355	Strela-3	S25366 MO RF				
		1998-036E	Kosmos-2356	Strela-3	S25367 MO RF				
		1998-036F	Kosmos-2357	Strela-3	S25368 MO RF				
1998-037	1998 Jun 18 2248	1998-037A	Intelsat 805	Intelsat 805	S25371 Intelsat	Atlas IIAS	AC-153	CC SLC36A	Wire
1998-038	1998 Jun 24 1829:58	1998-038A	Kosmos-2358	Kobal't	S25373 MO RF	Soyuz 11A511U	–	GIK-1 LC43/3	NK9814-10
		1998-038	Spuskaemiy Kapsula	SpK	A04862 MO RF				
		1998-038	Spuskaemiy Kapsula	SpK	A04867 MO RF				
1998-039	1998 Jun 25 1400	1998-039A	Kosmos-2359	Neman	S25376 MO RF	Soyuz 11A511U	–	GIK-5 LC31	NK9814-10
1998-040	1998 Jul 1 0048:01	1998-040A	Molniya-3	Molniya-3 No. 61	S25379 MOM	Molniya 8K78M	–	GIK-1 LC43/3	NK9815-26

Launch International Designation:	*Launch Date GMT:*	*COSPAR Payload International Designation:*	*Agency Post-Launch Payload Name:*	*Manufacturer Pre-Launch Payload Name:*	*SATCAT Number:*	*Launch Vehicle Type:*	*Launch Vehicle Serial Number:*	*Launch Site:*	*Source Reference:*
1998-041	1998 Jul 3 1812	1998-041A	Nozomi	Planet B	S25383 ISAS	M-V	M-V-3	KASC M-V	Wire
1998-042	1998 Jul 7 0315:00	1998-042A	Tubsat-N	Tubsat-N	S25389 TUB	Shtil'-1	–	BLA, K-407	NK9815-15
		1998-042B	Tubsat-N1	Tubsat-N1	S25390 TUB				
		1998-042C	Shtil'-1	Shtil'-1 instr. package	S25391 Makeev				
1998-043	1998 Jul 10 0630:00	1998-043A	Resurs-O1	Resurs-O1 No. 4L	S25394 RKA	Zenit-2	–	GIK-5 LC45L	NK9815-19
		1998-043B	Fasat-Bravo	Fasat-Bravo	S25395 FACh				
		1998-043C	TM-SAT	TM-SAT	S25396 FACh				
		1998-043D	Gurwin Techsat 1B	Techsat 1B	S25397 Technion				
		1998-043E	WESTPAC	WESTPAC	S25398 WPLTN				
		1998-043F	SAFIR-2	SAFIR-2	S25399 DLR				
1998-044	1998 Jul 18 0920	1998-044A	Sinosat	Sinosat	S25404 Eurasspace	Chang Zheng 3B	CZ3B-5	XSC LC2	CALT-WWW
1998-045	1998 Jul 28 0915:00	1998-045A	Kosmos-2360	Tselina-2	S25406 MO RF	Zenit-2	–	GIK-5 LC45L/1	NK9815-25
1998-046	1998 Aug 2 1624	1998-046H	Orbcomm B1	Orbcomm FM13	S25420 Orbcomm	Pegasus XL/HAPS	F22	WI DZWI	Wire
		1998-046G	Orbcomm B2	Orbcomm FM14	S25419 Orbcomm				
		1998-046F	Orbcomm B3	Orbcomm FM15	S25418 Orbcomm				
		1998-046E	Orbcomm B4	Orbcomm FM16	S25417 Orbcomm				
		1998-046A	Orbcomm B5	Orbcomm FM17	S25413 Orbcomm				
		1998-046B	Orbcomm B6	Orbcomm FM18	S25414 Orbcomm				
		1998-046C	Orbcomm B7	Orbcomm FM19	S25415 Orbcomm				
		1998-046D	Orbcomm B8	Orbcomm FM20	S25416 Orbcomm				
1998-F02	1998 Aug 12 1130:01	1998-F02	USA	MERCURY 3	F01185 USAF/NSA	Titan 401A/ Centaur	4A-20 (K-17)	CC LC41	ISIR980831-3
1998-047	1998 Aug 13 0943:11	1998-047A	Soyuz TM-28	Soyuz 7K-STM No. 77	S25429 RKA	Soyuz 11A511U	–	GIK-5 LC1	NK9902-45
1998-048	1998 Aug 19 2301:46	1998-048A	Iridium 3	Iridium SV078	S25431 Iridium	Chang Zheng 2C-III/SD	CZ2C-19	TYSC LC1	CALT-WWW
		1998-048B	Iridium 76	Iridium SV076	S25432 Iridium				
1998-049	1998 Aug 25 2307	1998-049A	ST-1	Singapore-Taiwan 1	S25460 Singapore T	Ariane 44P	V109	CSG ELA2	Wire
1998-F03	1998 Aug 27 0117	1998-F03	Galaxy X	Galaxy 10	F01225 Panamsat	Delta 8930	D259	CC LC17B	Wire
1998-050	1998 Aug 30 0031:00	1998-050A	Astra 2A	Astra 2A	S25462 SES	Proton-K/ DM-2M	383-01	GIK-5 LC81/ 23	NK9817-28
1998-F04	1998 Aug 31 0307:00	1998-F04	Kwangmyong-song 1	–	F01187 CMIK	Taepodong 1	–	MUSU	NYT980901-1
1998-051	1998 Sep 8 2113	1998-051E	Iridium 77	Iridium SV077	S25471 Iridium	Delta 7920-10C	D260	V SLC2W	Wire
		1998-051D	Iridium 79	Iridium SV079	S25470 Iridium				
		1998-051C	Iridium 80	Iridium SV080	S25469 Iridium				
		1998-051B	Iridium 81	Iridium SV081	S25468 Iridium				
		1998-051A	Iridium 82	Iridium SV082	S25467 Iridium				
1998-F05	1998 Sep 9 2029:00	1998-F05	Globalstar FM5	Globalstar FM5	F01194 Globalstar	Zenit-2 11K77.05	22D (67047801)	GIK-5 LC45L/1	Wire
		1998-F05	Globalstar FM7	Globalstar FM7	F01195 Globalstar				
		1998-F05	Globalstar FM9	Globalstar FM9	F01196 Globalstar				
		1998-F05	Globalstar FM10	Globalstar FM10	F01197 Globalstar				
		1998-F05	Globalstar FM11	Globalstar FM11	F01198 Globalstar				
		1998-F05	Globalstar FM12	Globalstar FM12	F01199 Globalstar				
		1998-F05	Globalstar FM13	Globalstar FM13	F01200 Globalstar				
		1998-F05	Globalstar FM16	Globalstar FM16	F01201 Globalstar				
		1998-F05	Globalstar FM17	Globalstar FM17	F01202 Globalstar				
		1998-F05	Globalstar FM18	Globalstar FM18	F01203 Globalstar				
		1998-F05	Globalstar FM20	Globalstar FM20	F01204 Globalstar				
		1998-F05	Globalstar FM21	Globalstar FM21	F01205 Globalstar				
1998-052	1998 Sep 16 0631	1998-052A	PAS 7	PAS 7	S25473 Panamsat	Ariane 44LP	V110	CSG ELA2	Wire
1998-053	1998 Sep 23 0506	1998-053A	Orbcomm C1	Orbcomm FM21	S25475 Orbcomm	Pegasus XL/ HAPS	F23	WI DZWI	Wire
		1998-053B	Orbcomm C2	Orbcomm FM22	S25476 Orbcomm				
		1998-053C	Orbcomm C3	Orbcomm FM23	S25477 Orbcomm				
		1998-053D	Orbcomm C4	Orbcomm FM24	S25478 Orbcomm				
		1998-053E	Orbcomm C5	Orbcomm FM25	S25479 Orbcomm				
		1998-053F	Orbcomm C6	Orbcomm FM26	S25480 Orbcomm				
		1998-053G	Orbcomm C7	Orbcomm FM27	S25481 Orbcomm				
		1998-053H	Orbcomm C8	Orbcomm FM28	S25482 Orbcomm				
1998-054	1998 Sep 28 2341:27	1998-054A	Molniya-1T	Molniya-1T	S25485 MOM	Molniya 8K78M	–	GIK-1 LC43/3	NK9821-10
1998-055	1998 Oct 3 1004:49	1998-055A	STEX	STEX	S25489 NRO	ARPA Taurus	T3 1110	V 576E	Rigg
		1998-055C	ATEX	ATEX	S25615 NRO/NRL				
1998-056	1998 Oct 5 2251	1998-056B	Sirius 3	Sirius 3	S25492 NSAB	Ariane 44L	V111	CSG ELA2	Wire
		1998-056A	Eutelsat W2	Eutelsat W2	S25491 Eutelsat				
1998-057	1998 Oct 9 2250	1998-057A	Hot Bird 5	Eutelsat HB5	S25495 Eutelsat	Atlas IIA	AC-134	CC SLC36B	Wire
1998-058	1998 Oct 20 0719	1998-058A	UHF F/O F9	UHF F/O F9	S25501 USN	Atlas IIA	AC-130	CC SLC36A	Wire

Launch International Designation:	Launch Date GMT:	COSPAR Payload International Designation:	Agency Post-Launch Payload Name:	Manufacturer Pre-Launch Payload Name:	SATCAT Number:	Launch Vehicle Type:	Launch Vehicle Serial Number:	Launch Site:	Source Reference:
1998-059	1998 Oct 21 1637:21	1998-059A	Maqsat-3	Maqsat-3	S25503 ESA	Ariane 5G	V112 (503)	CSG ELA3	Wire
1998-060	1998 Oct 23 0002	1998-060A	SCD-2	SCD-2	S25504 INPE	Pegasus H	F24/P-33	CC RW30/ 12 MFWA	Wire
1998-061	1998 Oct 24 1208:00	1998-061B	SEDSAT	SEDSAT	S25509 SEDS	Delta 7326-9.5	D261	CC SLC17A	Wire
		1998-061A	Deep Space 1	Deep Space 1	S25508 NASA/JPL				
1998-062	1998 Oct 25 0414:57	1998-062A	Progress M-40	Progress 7K-TGM No 239	S25512 RKA	Soyuz 11A511U	660	GIK-5 LC1	NK9823-01
		1998-062C	Spoutnik-41	RS-18	S25533 ACF/ RuAF				
1998-063	1998 Oct 28 2215:00	1998-063B	GE 5	GE 5	S25516 GE	Ariane 44L	V113	CSG ELA2	JCM-P
		1998-063A	Afristar	Afristar	S25515 Worldspace				
1998-064	1998 Oct 29 1919:34	1998-064A	Discovery	OV-103	S25519 NASA JSC	Space Shuttle	STS-95	KSC LC39B/ MLP2	MSFCFLASH
		1998-064A	Spacehab	Spacehab FU1	A04876 Spacehab				
		1998-064C	Spartan 201	Spartan 201	S25521 NASA GSFC				
		1998-064B	PANSAT	PANSAT	S25520 USN PGS				
1998-065	1998 Nov 4 0512:00	1998-065A	PAS 8	PAS 8	S25522 Panamsat	Proton-K/ DM-2M	395-02	GIK-5 LC81/ 23	NK9823-30
1998-066	1998 Nov 6 1337:52	1998-066A	Iridium 2	Iridium SV087	S25527 Iridium	Delta 7920-10C	D262	V SLC2W	Wire
		1998-066E	Iridium 83	Iridium SV083	S25531 Iridium				
		1998-066D	Iridium 84	Iridium SV084	S25530 Iridium				
		1998-066C	Iridium 85	Iridium SV085	S25529 Iridium				
		1998-066B	Iridium 86	Iridium SV086	S25528 Iridium				
1998-067	1998 Nov 20 0640:00	1998-067A	Zarya	77KM No. 17501	S25544 NASA	Proton-K	395-01	GIK-5 LC81/ 23	NK9901-02
1998-068	1998 Nov 22 2354	1998-068A	BONUM-1	BONUM-1	S25546 Telenor	Delta 7925-9.5	D263	CC SLC17B	Wire
1998-069	1998 Dec 4 0835:34	1998-069A	Endeavour	OV-105	S25549 NASA JSC	Space Shuttle	STS-88	KSC LC39A/ MLP3	MSFCFLASH
		1998-069F	Unity	Node 1	S25575 NASA JSC				
		1998-069F	PMA-1	PMA-1	A04890 NASA JSC				
		1998-069F	PMA-2	PMA-2	A04891 NASA JSC				
		1998-069B	SAC-A	SAC-A	S25550 CONAE				
		1998-069C	Mightysat 1	Mightysat 1	S25551 USAF				
1998-070	1998 Dec 6 0043	1998-070A	Satmex 5	Satmex 5	S25558 Satmex	Ariane 42L	V114	CSG ELA2	Wire
1998-071	1998 Dec 6 0057:54	1998-071A	SWAS	SWAS	S25560 NASA GSFC	Pegasus XL	F25	V RW30/ 12 PAWA	Wire
1998-072	1998 Dec 10 1157:07	1998-072A	Nadezhda	Nadezhda	S25567 MO RF	Kosmos 11K65M	–	GIK-1 LC132/1	SGrahn
		1998-072B	Astrid-2	Astrid 2	S25568 SSC				
1998-073	1998 Dec 11 1845:51	1998-073A	Mars Climate Orbiter	MCO	S25571 JPL	Delta 7425-9.5	D264	CC SLC17A	Wire
1998-074	1998 Dec 19 1139:44	1998-074A	Iridium 11A	Iridium SV088?	S25577 Iridium	Chang Zheng 2C-III/SD	CZ2C-20	TYSC LC1	CALT-WWW
		1998-074B	Iridium 20A	Iridium SV089?	S25578 Iridium				
1998-075	1998 Dec 22 0108	1998-075A	PAS 6B	PAS 6B	S25585 Satmex	Ariane 42L	V115	CSG ELA2	Wire
1998-076	1998 Dec 24 2002:19	1998-076A	Kosmos-2361	Parus	S25590 MO RF	Kosmos 11K65M	–	GIK-1 LC132/1	NK9902-17
1998-077	1998 Dec 30 1835:46	1998-077A	Kosmos-2362	Uragan No. 86L	S25593 MOM	Proton-K/ DM-2	385-02	GIK-5 LC200/ 39	WWW-CFSCIC
		1998-077B	Kosmos-2363	Uragan No. 84L	S25594 MOM				
		1998-077C	Kosmos-2364	Uragan No. 79L	S25595 MOM				
1999-001	1999 Jan 3 2021:10	1999-001A	Mars Polar Lander	MPL	S25605 JPL	Delta 7425-9.5	D265	CC SLC17B	Wire
		1999-001	Deep Space 2 Microprobe 1	Mars Microprobe 1	A04907 JPL				
		1999-001	Deep Space 2 Microprobe 2	Mars Microprobe 2	A04908 JPL				
1999-002	1999 Jan 27 0034:02	1999-002A	ROCSAT-1	ROCSAT-1	S25616 NSPO	Athena-1	LM-006	SPFL SLC46	Wire
1999-003	1999 Feb 7 2104:15	1999-003A	Stardust	Stardust	S25618 JPL	Delta 7426-9.5	D266	CC SLC17A	Wire
		1999-003	Sample Return Capsule	SRC	A04909 JPL				
1999-004	1999 Feb 9 0353:59	1999-004A	Globalstar FM36	Globalstar FM36	S25621 Globalstar	Soyuz 11A511U	S15000-058 ST01	GIK-5 LC1	NK9903-32
		1999-004B	Globalstar FM23	Globalstar FM23	S25622 Globalstar				
		1999-004C	Globalstar FM38	Globalstar FM38	S25623 Globalstar				
		1999-004D	Globalstar FM40	Globalstar FM40	S25624 Globalstar				
1999-005	1999 Feb 15 0512:00	1999-005A	Telstar 6	Telstar 6	S25626 Loral Skynet	Proton-K/ DM-2M	396-01	GIK-5 LC81/23	NK9904-28
1999-006	1999 Feb 16 0145:26	1999-006A	JCSAT 6	JCSAT 6	S25630 JSAT	Atlas IIAS	AC-152	CC SLC36A	Wire
1999-007	1999 Feb 20 0418:01	1999-007A	Soyuz TM-29	Soyuz 7K-STM No. 78	S25632 RKA	Soyuz 11A511U	M15000-662	GIK-5 LC1	NK9904-03
1999-008	1999 Feb 23 1029:55	1999-008A	ARGOS	P91-1	S25634 USAF SMC/TEV	Delta 7920-10	D267	V SLC2W	Wire
		1999-008B	Sunsat	Sunsat	S25635 Stellenbosch				
		1999-008C	Orsted	Orsted	S25636 DMI				
1999-009	1999 Feb 26 2244	1999-009B	Skynet 4E	Skynet 4E	S25639 UK MoD	Ariane 44L	V116	CSG ELA2	Wire
		1999-009A	Arabsat 3A	Arabsat 3A	S25638 Arabsat				

Launch International Designation:	*Launch Date GMT:*	*COSPAR Payload International Designation:*	*Agency Post-Launch Payload Name:*	*Manufacturer Pre-Launch Payload Name:*	*SATCAT Number:*	*Launch Vehicle Type:*	*Launch Vehicle Serial Number:*	*Launch Site:*	*Source Reference:*
1999-010	1999 Feb 28 0400:00	1999-010A	Raduga-1	Globus No. 14	S25642 MO RF	Proton-K/ DM-2	387-01	GIK-5 LC81/ 23	NK9904-38
1999-011	1999 Mar 5 0256	1999-011A	WIRE	WIRE	S25646 NASA GSFC	Pegasus XL	F26/M-22	V RW30/12 PAWA	Wire
1999-012	1999 Mar 15 0306:00	1999-012A	Globalstar FM22	Globalstar FM22	S25649 Globalstar	Soyuz 11A511U	–	GIK-5 LC1	NK9905-21
		1999-012B	Globalstar FM41	Globalstar FM41	S25650 Globalstar				
		1999-012C	Globalstar FM46	Globalstar FM46	S25651 Globalstar				
		1999-012D	Globalstar FM37	Globalstar FM37	S25652 Globalstar				
1999-013	1999 Mar 21 0009:30	1999-013A	Asiasat 3S	Asiasat 3S	S25657 Asiasat	Proton-K/ DM-2M	388-01	GIK-5 LC81/ 23	NK0003-17
1999-014	1999 Mar 28 0129:59	1999-014A	DemoSat	DemoSat	S25661 SeaLaunch	Zenit-3SL	–	KLA, SL Odyssey	Wire
1999-015	1999 Apr 2 1128:43	1999-015A	Progress M-41	Progress 7K-TGM No 241	S25664 RKA	Soyuz 11A511U	–	GIK-5 LC1	NK9905-01
		1999-015C	Sputnik-99	Sputnik-99	S25685 AMSAT-F				
1999-016	1999 Apr 2 2203	1999-016A	Insat 2E	Insat 2E	S25666 ISRO	Ariane 42P	V117	CSG ELA2	Wire
1999-017	1999 Apr 9 1701:00	1999-017A	DSP F19 (USA 142)	DSP 18?	S25669 USAF	Titan 402B/ IUS	4B-27 (K-32)	CC LC41	Wire
1999-018	1999 Apr 12 2250	1999-018A	Eutelsat W3	Eutelsat W3	S25673 Eutelsat	Atlas IIAS	AC-154	CC SLC36A	Wire
1999-019	1999 Apr 15 0046:00	1999-019A	Globalstar FM19	Globalstar FM19	S25676 Globalstar	Soyuz 11A511U	S15000-060 ST03	GIK-5 LC1	NK9906-12
		1999-019B	Globalstar FM42	Globalstar FM42	S25677 Globalstar				
		1999-019C	Globalstar FM44	Globalstar FM44	S25678 Globalstar				
		1999-019D	Globalstar FM45	Globalstar FM45	S25679 Globalstar				
1999-020	1999 Apr 15 1832:00	1999-020A	Landsat 7	Landsat 7	S25682 NASA/ GSFC	Delta 7920-10	D268	V SLC2W	Wire
1999-021	1999 Apr 21 0459:12	1999-021A	UoSAT-12	UoSAT-12	S25693 SSTL	Dnepr	6703542509	GIK-5 LC109/95	NK9906-16
1999-F01	1999 Apr 27 1822:01	1999-F01	Ikonos 1	Ikonos 1	F01208 Spacelm	Athena-2	LM-005	V SLC6	Wire
1999-022	1999 Apr 28 2030:00	1999-022A	ABRIXAS	ABRIXAS	S25721 DLR	Kosmos 11K65M	65036-413	GTsMP-4 LC107	NK9906-1
		1999-022B	Megsat-0	Megsat-0	S25722 MegSat				
1999-023	1999 Apr 30 1630	1999-023A	USA 143	Milstar-2 F1	S25724 USAF	Titan 401B/ Centaur	4B-32 (K-26)	CC LC40	Wire
1999-024	1999 May 5 0100	1999-024A	Orion 3	Orion 3	S25727 Loral Orion	Delta 8930	D269	CC SLC17B	Wire
1999-025	1999 May 10 0133:00	1999-025A	Feng Yun	Feng Yun 1C	S25730 CASC	Chang Zheng 4B	CZ4B-1	TYSC LC1	CGW-WWW
		1999-025B	Shi Jian 5	SJ-5	S25731 CASC				
1999-026	1999 May 18 0509:36	1999-026B	MUBLCOM	MUBLCOM	S25736 Orbcomm	Pegasus XL/ HAPS	F27	V RW30/ 12 PAWA	Wire
		1999-026A	TERRIERS	TERRIERS	S25735 NASA GSFC				
1999-027	1999 May 20 2230:00	1999-027A	Nimiq	Telesat DTH-1	S25740 Telesat	Proton-K/ DM-2M	396-02	GIK-5 LC81/ 23	NK9907-42
1999-028	1999 May 22 0936	1999-028A	USA 144	NRO X5?	S25744 NRO	Titan 404B	4B-12	V SLC4E	Wire
1999-029	1999 May 26 0522	1999-029A	Oceansat	IRS-P4	S25756 ISRO	PSLV	PSLV-C2	SHAR PSLV	Wire
		1999-029B	KITSAT-3	KITSAT-3	S25757 KAIST				
		1999-029C	DLR-TUBSAT-C	DLR-TUBSAT-C	S25758 DLR				
1999-030	1999 May 27 1049:42	1999-030A	Discovery	OV-103	S25760 NASA JSC	Space Shuttle	STS-96	KSC LC39B/MLP2	MSFCFLASH
		1999-030A	Spacehab Double Module	Spacehab Double Module	A04924 NASA JSC				
		1999-030A	ICC	Integrated Cargo Carrier	A04925 NASA JSC				
		1999-030B	Starshine	Starshine	S25769 NASA GSFC				
1999-031	1999 Jun 10 1348:43	1999-031A	Globalstar M052	Globalstar M052	S25770 Globalstar	Delta 7420-10C	D270	CC SLC17B	Wire
		1999-031B	Globalstar M049	Globalstar M049	S25771 Globalstar				
		1999-031C	Globalstar M025	Globalstar M025	S25772 Globalstar				
		1999-031D	Globalstar M047	Globalstar M047	S25773 Globalstar				
1999-032	1999 Jun 11 1715:33	1999-032A	Iridium 14A	Iridium SV090?	S25777 Iridium	Chang Zheng 2C-III/SD	CZ2C-21	TYSC LC1	CGW-WWW
		1999-032B	Iridium 20A	Iridium SV091?	S25778 Iridium				
1999-033	1999 Jun 18 0149:30	1999-033A	Astra 1H	Astra 1H	S25785 SES	Proton-K/ DM-2M	397-02	GIK-5 LC81/ 23	NK9908-12
1999-034	1999 Jun 20 0215:00	1999-034A	QuikScat	QuikScat	S25789 NASA GSFC	Titan II SLV	23G-7	V SLC4W	Wire
1999-035	1999 Jun 24 1544:00	1999-035A	FUSE	Far UV Spec. Explorer	S25791 NASA GSFC	Delta 7320-10	D271	CC SLC17A	Wire
1999-F02	1999 Jul 5 1332:00	1999-F02	Raduga	Gran' No. 45	F01213 MO RF	Proton-K/ Briz-M	389-01	GIK-5 LC81/ 24	NK9909-26
1999-036	1999 Jul 8 0845:06	1999-036A	Molniya-3	Molniya-3 No. 63?	S25847 MOM	Molniya 8K78M	–	GIK-1 LC43/3	NK9909-32
1999-037	1999 Jul 10 0845:37	1999-037A	Globalstar M032	Globalstar M032	S25851 Globalstar	Delta 7420-10C	D272	CC SLC17B	Wire
		1999-037B	Globalstar M030	Globalstar M030	S25852 Globalstar				
		1999-037C	Globalstar M035	Globalstar M035	S25853 Globalstar				
		1999-037D	Globalstar M051	Globalstar M051	S25854 Globalstar				
1999-038	1999 Jul 16 1637:33	1999-038A	Progress M-42	Progress 7K-TGM No 242	S25858 RAKA	Soyuz 11A511U	667	GIK-5 LC1	NK9909-05
1999-039	1999 Jul 17 0638:00	1999-039A	Okean-O	Okean-O No. 1	S25860 RAKA	Zenit-2	17L	GIK-5 LC45L/1	NK9909-34
1999-040	1999 Jul 23 0431:00	1999-040A	Columbia	OV-102	S25866 NASA JSC	Space Shuttle	STS-93	KSC LC39B/ MLP1	JCM-P
		1999-040B	Chandra X-ray Observatory	AXAF	S25867 NASA MSFC				

Launch International Designation:	Launch Date GMT:	COSPAR Payload International Designation:	Agency Post-Launch Payload Name:	Manufacturer Pre-Launch Payload Name:	SATCAT Number:	Launch Vehicle Type:	Launch Vehicle Serial Number:	Launch Site:	Source Reference:
1999-041	1999 Jul 25 0746:03	1999-041A	Globalstar M026	Globalstar M026	S25872 Globalstar	Delta 7420-10C	D273	CC SLC17A	Wire
		1999-041B	Globalstar M028	Globalstar M028	S25873 Globalstar				
		1999-041C	Globalstar M043	Globalstar M043	S25874 Globalstar				
		1999-041D	Globalstar M048	Globalstar M048	S25875 Globalstar				
1999-042	1999 Aug 12 2252	1999-042A	Telkom 1	Telkom 1	S25880 PT Telkom	Ariane 42P	V118	CSG ELA2	Wire
1999-043	1999 Aug 17 0437:31	1999-043A	Globalstar M024	Globalstar M024	S25883 Globalstar	Delta 7420-10C	D274	CC SLC17B	Wire
		1999-043B	Globalstar M027	Globalstar M027	S25884 Globalstar				
		1999-043C	Globalstar M053	Globalstar M053	S25885 Globalstar				
		1999-043D	Globalstar M054	Globalstar M054	S25886 Globalstar				
1999-044	1999 Aug 18 1800	1999-044A	Kosmos-2365	Kobal't	S25889 MO RF	Soyuz 11A511U	–	GIK-1 LC43/3	NK9910-32
		1999-044	Spuskaemiy Kapsula	SpK	A04938 MO RF				
		1999-044	Spuskaemiy Kapsula	SpK	A04949 MO RF				
1999-045	1999 Aug 26 1202:15	1999-045A	Kosmos-2366	Parus	S25892 MO RF	Kosmos 11K65M	–	GIK-1 LC132/1	NK9910-34
1999-046	1999 Sep 4 2234	1999-046A	Mugunghwa 3	Koreasat 3	S25894 Korea Tel	Ariane 42P	V120	CSG ELA2	Wire
1999-047	1999 Sep 6 1636:00	1999-047A	Yamal 101	Yamal 101	S25896 AO Gazcom	Proton-K/ DM-2M	388-02	GIK-5 LC81/23	NK9911-04
		1999-047B	Yamal 102	Yamal 102	S25897 AO Gazcom				
1999-048	1999 Sep 9 1800:00	1999-048A	Foton	Foton No. 12	S25902 MOM	Soyuz 11A511U	–	GIK-1 LC43/4	NK9911-08
1999-049	1999 Sep 22 1433:00	1999-049A	Globalstar M033	Globalstar FM33	S25907 Globalstar	Soyuz 11A511U	S15000-061 ST04	GIK-5 LC1	NK9911-15
		1999-049B	Globalstar M050	Globalstar FM50	S25908 Globalstar				
		1999-049C	Globalstar M055	Globalstar FM55	S25909 Globalstar				
		1999-049D	Globalstar M058	Globalstar FM58	S25910 Globalstar				
1999-050	1999 Sep 23 0602	1999-050A	Echostar 5	Echostar 5	S25913 Echostar	Atlas IIAS	AC-155	CC SLC36A	Wire
1999-051	1999 Sep 24 0621:08	1999-051A	Ikonos	Ikonos 2	S25919 SpaceIm	Athena-2	LM-007	V SLC6	Wire
1999-052	1999 Sep 25 0629	1999-052A	Telstar 7	Telstar 7	S25922 Loral Skynet	Ariane 44LP	V121	CSG ELA2	Wire
1999-053	1999 Sep 27 2230:00	1999-053A	LMI 1	LMI 1	S25924 LMI	Proton-K/ DM-2M	398-02	GIK-5 LC81/23	NK9911-20
1999-054	1999 Sep 28 1100:05	1999-054A	Resurs F-1M	Resurs F-1M	S25929 RKA	Soyuz 11A511U	–	GIK-1 LC43/4	NK9911-24
1999-055	1999 Oct 7 1251:01	1999-055A	GPS SVN 46	GPS SVN 46	S25933 USAF	Delta 7925-9.5	D275	CC SLC17A	Wire
1999-056	1999 Oct 10 0328:00	1999-056A	DirecTV-1R	DirecTV-1R	S25937 DirecTV	Zenit-3SL	–	KLA, SL Odyssey	Wire
1999-057	1999 Oct 14 0315	1999-057A	Zi Yuan 1	CBERS	S25940 CASC/ INPE	Chang Zheng 4B	CZ4B-1	TYSC LC1	Wire
		1999-057B	SACI 1	SACI 1	S25941 INPE				
1999-058	1999 Oct 18 1322:00	1999-058A	Globalstar M031	Globalstar FM31	S25943 Globalstar	Soyuz 11A511U	M15000-062 ST05	GIK-5 LC1	NK9912-10
		1999-058B	Globalstar M056	Globalstar FM56	S25944 Globalstar				
		1999-058C	Globalstar M057	Globalstar FM57	S25945 Globalstar				
		1999-058D	Globalstar M059	Globalstar FM59	S25946 Globalstar				
1999-059	1999 Oct 19 0622	1999-059A	Orion 2	Orion 2	S25949 Loral Orion	Ariane 44LP	V122	CSG ELA2	Wire
1999-F03	1999 Oct 27 1616:00	1999-F03	Ekspress A1	Ekspress A No. 1	F01219 AO Inform	Proton-K/ DM-2	386-02	GIK-5 LC200/39	Kosmodrom 99-11
1999-060	1999 Nov 13 2254	1999-060A	GE 4	GE 4	S25954 GE Americom	Ariane 44LP	V123	CSG ELA2	Wire
1999-F04	1999 Nov 15 0729	1999-F04	MTSAT	MTSAT	F01221 NASDA	H-II	H-II-8F	TNSCY	WWW-NASDA
1999-061	1999 Nov 19 2230	1999-061A	Shenzhou	Shenzhou	S25956 CASC	Chang Zheng 2F	CZ2F-1	JQ SLS	CALT-WWW
1999-062	1999 Nov 22 1620:00	1999-062A	Globalstar M029	Globalstar FM29	S25961 Globalstar	Soyuz 11A511U	#NAME?	GIK-5 LC1	NK0003-17
		1999-062B	Globalstar M034	Globalstar FM34	S25962 Globalstar				
		1999-062C	Globalstar M039	Globalstar FM39	S25963 Globalstar				
		1999-062D	Globalstar M061	Globalstar FM61	S25964 Globalstar				
1999-063	1999 Nov 23 0406	1999-063A	UHF F/O F10	UHF F/O F10	S25967 USN	Atlas IIA	AC-136	CC SLC36B	Wire
1999-064	1999 Dec 3 1622:46	1999-064A	Helios 1B	Helios 1B	S25977 CNES/DGA	Ariane 40	V124	CSG ELA2	Wire
		1999-064B	Clementine	Clementine	S25978 CNES/DGA				
1999-065	1999 Dec 4 1853	1999-065A	Orbcomm D1	Orbcomm FM30	S25980 Orbcomm	Pegasus XL/ HAPS	F28	WI DZWI	Wire
		1999-065B	Orbcomm D2	Orbcomm FM31	S25981 Orbcomm				
		1999-065C	Orbcomm D3	Orbcomm FM32	S25982 Orbcomm				
		1999-065D	Orbcomm D4	Orbcomm FM33	S25983 Orbcomm				
		1999-065E	Orbcomm D5	Orbcomm FM34	S25984 Orbcomm				
		1999-065F	Orbcomm D6	Orbcomm FM35	S25985 Orbcomm				
		1999-065G	Orbcomm D7	Orbcomm FM36	S25986 Orbcomm				
1999-066	1999 Dec 10 1432:07	1999-066A	XMM	XMM	S25989 ESA	Ariane 5G	V119 (504)	CSG ELA3	Wire
1999-067	1999 Dec 12 1738:01	1999-067A	DMSP 5D-3 F-15	DMSP 5D-3 S-15	S25991 USAF	Titan II SLV	23G-8	V SLC4W	Wire
1999-F05	1999 Dec 12 1940	1999-F05	SACI-2	SACI-2	F01223 INPE	VLS-1	V02	ALCA VLS	AEB-WWW
1999-068	1999 Dec 18 1857:39	1999-068A	Terra	EOS AM-1	S25994 NASA GSFC	Atlas IIAS	AC-141	V SLC3E	Wire
1999-069	1999 Dec 20 0050:00	1999-069A	Discovery	OV-103	S25996 NASA JSC	Space Shuttle	STS-103	KSC LC39B/ MLP2	Wire

Launch International Designation:	*Launch Date GMT:*	*COSPAR Payload International Designation:*	*Agency Post-Launch Payload Name:*	*Manufacturer Pre-Launch Payload Name:*	*SATCAT Number:*	*Launch Vehicle Type:*	*Launch Vehicle Serial Number:*	*Launch Site:*	*Source Reference:*
		1999-069	ORUC	ORUC	A04965 NASA JSC				
		1999-069	FSS	FSS	A04966 NASA JSC				
1999-070	1999 Dec 21 0713	1999-070C	Celestis-03	CPAC	A04967 Celestis	Taurus 2110	T4 2110	V 576E	Wire
		1999-070A	KOMPSAT	KOMPSAT	S26032 KARI				
		1999-070B	ACRIMSAT	ACRIMSAT	S26033 JPL				
1999-071	1999 Dec 22 0050	1999-071A	Galaxy 11	Galaxy 11	S26038 Panamsat	Ariane 44L	V125	CSG ELA2	Wire
1999-072	1999 Dec 26 0800	1999-072A	Kosmos-2367	US-P	S26040 MO RF	Tsiklon-2	801 (45082801)	GIK-5 LC90/20	NK0002-42
1999-073	1999 Dec 27 1912:44	1999-073A	Kosmos-2368	Oko	S26042 MO RF	Molniya 8K78M	–	GIK-1 LC16/2	NK0002-46

Source: Jonathan McDowell

Table 4.10: World Orbital Launch Log 2000 – 2004

See *jsd.janes.com* for World Orbital Launch Log 1957–1994 (Tables 4.1–4.8)

Launch International Designation:	*Launch Date GMT:*	*COSPAR Payload International Designation:*	*Agency Post-Launch Payload Name:*	*Manufacturer Pre-Launch Payload Name:*	*SATCAT Number:*	*Launch Vehicle Type:*	*Launch Vehicle Serial Number:*	*Launch Site:*	*Source Reference:*
2000-001	2000 Jan 21 0103	2000-001A	USA 148	DSCS III B-8	S26052 USAF	Atlas IIA	AC-138	CC SLC36A	Wire
2000-002	2000 Jan 25 0104	2000-002A	Galaxy 10R	Galaxy 10R	S26056 Satmex	Ariane 42L	V126	CSG ELA2	Wire
2000-003	2000 Jan 25 1645:05	2000-003A	Zhongxing 22	DFH	S26058 Chinasat	Chang Zheng 3A	CZ3A-4	XSC LC2	CALT-PR
2000-004	2000 Jan 27 0303:06	2000-004A	JAWSAT	JAWSAT	S26061	Minotaur	1	V CLF	ASUSAT-WWW
		2000-004E	ASUSAT	ASUSAT	S26065				
		2000-004C	OPAL	OPAL	S26063				
		2000-004B	OCS	OCSE	S26062				
		2000-004D	Falconsat	Falconsat I	S26064				
		2000-004L	JAK	Artemis	S26093				
		2000-004M	STENSAT	STENSAT	S26094				
		2000-004H	Picosat 1	MEMS 1	S26080				
		2000-004H	Picosat 2	MEMS 2	S26080				
		2000-004J	Thelma	Artemis	S26091				
		2000-004K	Louise	Artemis	S26092				
2000-005	2000 Feb 1 0647:23	2000-005A	Progress M1-1	Progress 7K-TGM No 250	S26067 RAKA	Soyuz 11A511U	A15000-669	GIK-5 LC1	NK0004-14
2000-006	2000 Feb 3 0926:00	2000-006A	Kosmos-2369	Tselina-2	S26069 MO RF	Zenit-2	45025801	GIK-5 LC45L	NK0004-17
2000-007	2000 Feb 3 2339	2000-007A	Hispasat 1C	Hispasat 1C	S26071 Hispasat	Atlas IIAS	AC-158	CC SLC36B	Wire
2000-008	2000 Feb 8 2124	2000-008A	Globalstar M060	Globalstar M060	S26081 Globalstar	Delta 7420-10C	D276	CC SLC17B	Wire
		2000-008B	Globalstar M062	Globalstar M062	S26082 Globalstar				
		2000-008C	Globalstar M063	Globalstar M063	S26083 Globalstar				
		2000-008D	Globalstar M064	Globalstar M064	S26084 Globalstar				
2000-009	2000 Feb 8 2320:00	2000-009A	Dummy satellite	–	S26086 Lavochkin	Soyuz 11A511U	A15000-079 ST07	GIK-5 LC31	NK0004-24
		2000-009	IRDT	Mission 2000	A04971 ESA				
2000-F01	2000 Feb 10 0130	2000-F01	ASTRO E	ASTRO E	F01226 ISAS	M-V	M-V-4	KASC M-V	Wire
2000-010	2000 Feb 11 1743:40	2000-010A	Endeavour	OV-105	S26088 NASA JSC	Space Shuttle	STS-99	KSC LC39A/ MLP3	Wire
		2000-010	SRL-3	SRL PLT	A04974 JPL/DARA				
		2000-010	SRTM Mast	SRTM Mast	A04978				
2000-011	2000 Feb 12 0910:54	2000-011A	Garuda 1	Garuda 1	S26089 ACES	Proton-K/ DM-2M	399-02	GIK-5 LC81/23	NK0004-32
2000-012	2000 Feb 18 0104	2000-012A	Superbird 4	Superbird 4	S26095 SCC	Ariane 44LP	V127	CSG ELA2	Wire
2000-013	2000 Mar 12 0407:00	2000-013A	Ekspress 6A	Ekspress A No. 2	S26098 AO Inform	Proton-K/ DM-2M	399-01	GIK-5 LC200/39	NK0005-20
2000-014	2000 Mar 12 0929	2000-014A	MTI	MTI	S26102 DOE	Taurus 1110	T5 1110	V 576E	Wire
2000-F02	2000 Mar 12 1449:15	2000-F02	ICO F-1	ICO-F1	F01228 ICO	Zenit-3SL	–	KLA, SL Odyssey	Wire
2000-015	2000 Mar 20 1828:30	2000-015A	Dumsat	–	S26106 Starsem	Soyuz 11A511U	–	GIK-5 LC31	NK0005-27
2000-016	2000 Mar 21 2328:19	2000-016B	Insat 3B	Insat 3B	S26108 ISRO	Ariane 5G	V128 (505)	CSG ELA3	AE-TV
		2000-016A	Asiastar	Asiastar	S26107 Worldspace				
2000-017	2000 Mar 25 2034:43	2000-017A	IMAGE	IMAGE	S26113 NASA GSFC	Delta 7326-9.5	D277	V SLC2W	Wire
2000-018	2000 Apr 4 0501:29	2000-018A	Soyuz TM-30	Soyuz 7K-STM No. 204	S26116 RKA	Soyuz 11A511U	–	GIK-5 LC1	NK0006-08
2000-019	2000 Apr 17 2106:00	2000-019A	Sesat	Sesat	S26243 Eutelsat	Proton-K/ DM-2M	397-01	GIK-5 LC200/39	NK0006-26
2000-020	2000 Apr 19 0029	2000-020A	Galaxy 4R	Galaxy 4R	S26298 Satmex	Ariane 42L	V129	CSG ELA2	Wire
2000-021	2000 Apr 25 2008:02	2000-021A	Progress M1-2	Progress 7K-TGM No 252	S26301 RAKA	Soyuz 11A511U	–	GIK-5 LC1	NK0006-16
2000-022	2000 May 3 0707	2000-022A	GOES 11	GOES L	S26352 NOAA	Atlas IIA	AC-137	CC SLC36A	Wire
2000-023	2000 May 3 1325	2000-023A	Kosmos-2370	Neman	S26354 MO RF	Soyuz 11A511U	A15000-649	GIK-5 LC1	NK0007-32
2000-024	2000 May 8 1601	2000-024A	DSP F20 (USA 149)	DSP 20	S26356 USAF	Titan 402B/ IUS	4B-29	CC LC40	Wire
2000-025	2000 May 11 0148	2000-025A	GPS SVN 51	GPS SVN 51	S26360 USAF	Delta 7925-9.5	D278	CC SLC17A	Wire
2000-026	2000 May 16 0827:41	2000-026A	Simsat-1	Simsat-1	S26365 Krunichev	Rokot	–	GIK-1 LC133	Eurockot-WWW
		2000-026B	Simsat-2	Simsat-2	S26366 Krunichev				
2000-027	2000 May 19 1011:10	2000-027A	Atlantis	OV-104	S26368 NASA JSC	Space Shuttle	STS-101	KSC LC39A/ MLP1	Wire
		2000-027	External Airlock/ ODS	EAL/ODS	A04995 NASA JSC				
		2000-027	Spacehab Double Module	Spacehab Double Module	A04999 NASA JSC				
		2000-027	ICC	Integrated Cargo Carrier	A05000 NASA JSC				
2000-028	2000 May 24 2310	2000-028A	Eutelsat W4	Eutelsat W4	S26369 Eutelsat	Atlas 3A	AC-201	CC SLC36B	Wire
2000-029	2000 Jun 6 0259:00	2000-029A	Gorizont	Gorizont No. 45L	S26372 MOM	Proton-K/ Briz-M	392-01	GIK-5 LC81/24	NK0008-16
2000-030	2000 Jun 7 1319:30	2000-030A	TSX-5	STEP M5	S26374 USAF STP	Pegasus XL	F29	V RW30/ 12 PAWA	Wire

Launch International Designation:	*Launch Date GMT:*	*COSPAR Payload International Designation:*	*Agency Post-Launch Payload Name:*	*Manufacturer Pre-Launch Payload Name:*	*SATCAT Number:*	*Launch Vehicle Type:*	*Launch Vehicle Serial Number:*	*Launch Site:*	*Source Reference:*
2000-031	2000 Jun 24 0028:00	2000-031A	Ekspress 3A	Ekspress A No. 3	S26378 AO Inform	Proton-K/ DM-2M	394-02	GIK-5 LC200/39	NK0008-23
2000-032	2000 Jun 25 1150	2000-032A	Feng Yun 2	FY-2	S26382 CASC	Chang Zheng 3	CZ3-13	XSC LC1	CALT-WWW
2000-033	2000 Jun 28 1037:42	2000-033A	Nadezhda	Nadezhda	S26384 MO RF	Kosmos 11K65M	-	GIK-1 LC132/1	NK0008-25`
		2000-033B	Tsinghua	–	S26385 Tsinghua				
		2000-033C	SNAP	–	S26386 SSTL				
2000-034	2000 Jun 30 1256	2000-034A	TDRS 8	TDRS H	S26388 NASA	Atlas IIA	AC-139	CC SLC36A	Wire
2000-035	2000 Jun 30 2208:47	2000-035A	Sirius 1	Sirius 1	S26390 Loral	Proton-K/ DM-2M	400-01	GIK-5 LC81/24	NK0008-35
2000-036	2000 Jul 4 2344:00	2000-036A	Kosmos-2371	Geizer	S26394 MO RF	Proton-K/DM-2	389-02	GIK-5 LC200/39	NK0009-24
2000-037	2000 Jul 12 0456:36	2000-037A	Zvezda	17KSM No. 128-01`	RAKA	Proton-K	398-01	GIK-5 LC81/23	NK0009-02
2000-038	2000 Jul 14 0521	2000-038A	Echostar VI	Echostar VI	S26402 Echostar	Atlas IIAS	AC-161	CC SLC36B	Wire
2000-039	2000 Jul 15 1200:00	2000-039B	CHAMP	–	S26405 DLR	Kosmos 11K65M	47136-414	GIK-1 LC132/1	NK0009-29
		2000-039A	MITA	–	S26404 ASI				
		2000-039	BIRD	–	A05007 DLR				
2000-040	2000 Jul 16 0917:00	2000-040A	GPS SVN 48	GPS SVN 48	S26407 USAF	Delta 7925-9.5	D279	CC SLC17A	Wire
2000-041	2000 Jul 16 1239:34	2000-041A	Samba	Cluster FM7	S26410 ESA	Soyuz 11A511U	A15000-069 ST09	GIK-5 LC31	NK0009-35
		2000-041B	Salsa	Cluster FM6	S26411 ESA				
2000-042	2000 Jul 19 2009:00	2000-042A	Sindri	Mightysat 2.1	S26414	Minotaur	2	V CLF	Wire
		-	Picosat 1	MEMS 1	–				
		-	Picosat 2	MEMS 2	–				
2000-043	2000 Jul 28 2242:00	2000-043A	PAS 9	PAS 9	S26451 DirecTV	Zenit-3SL	–	KLA, SL Odyssey	NK0009-45
2000-044	2000 Aug 6 1826:42	2000-044A	Progress M1-3	Progress 7K-TGM No 251	S26461 RAKA	Soyuz 11A511U	K15000-668	GIK-5 LC1	NK0010-17
2000-045	2000 Aug 9 1113:35	2000-045A	Rumba	Cluster FM5	S26463 ESA	Soyuz 11A511U	A15000-070 ST10	GIK-5 LC31	NK0010-02
		2000-045B	Tango	Cluster FM8	S26464 ESA				
2000-046	2000 Aug 17 2316:00	2000-046B	Nilesat 102	Nilesat 102	S26470 Nilesat	Ariane 44LP	V131	CSG ELA2	Wire
		2000-046A	Brasilsat B4	Brasilsat B4	S26469 Embratel				
2000-047	2000 Aug 17 2345:01	2000-047A	USA 152	ONYX 4	S26473 NRO/CIA	Titan 403B	4B-28	V SLC4E	Wire
2000-048	2000 Aug 23 1105:00	2000-048A	DM-F3	DM-F3	S26475 Boeing/HB	Delta 8930	D280	CC SLC17B	Wire
2000-049	2000 Aug 28 2008	2000-049A	Kosmos-2372	Globus	S26477 MO RF	Proton-K/DM-2	401-02	GIK-5 LC81/24	NK0010-13
2000-050	2000 Sep 1 0325	2000-050A	Zi Yuan 2	ZY-2	S26481 CASC	Chang Zheng 4B	CZ4B-2	TYSC LC1	Wire
2000-051	2000 Sep 5 0943:58	2000-051A	Sirius 2	Sirius 2	S26483 Loral	Proton-K/DM-2M	400-02	GIK-5 LC81/23	NK0011-29
2000-052	2000 Sep 6 2233	2000-052A	Eutelsat W1	Eutelsat W1	S26487 Singapore T	Ariane 44P	V132	CSG ELA2	Wire
2000-053	2000 Sep 8 1245:47	2000-053A	Atlantis	OV-104	S26489 NASA JSC	Space Shuttle	STS-106	KSC LC39B/ MLP2	Wire
		2000-053	External Airlock/ODS	EAL/ODS	A05049 NASA JSC				
		2000-053	Spacehab Double Module	Spacehab Double Module	A05053 NASA JSC				
		2000-053	ICC	Integrated Cargo Carrier	A05054 NASA JSC				
2000-054	2000 Sep 14 2254:07	2000-054A	Astra 2B	Astra 2B	S26494 SES	Ariane 5G	V130 (506)	CSG ELA3	Wire
		2000-054B	GE 7	GE 7	S26495 GE				
2000-055	2000 Sep 21 1022	2000-055A	NOAA 16	NOAA L	S26536 NOAA	Titan II SLV	23G-13	V SLC4W	Wire
2000-056	2000 Sep 25 1010:00	2000-056A	Kosmos-2372	Yenisey	MO RF	Zenit-2	–	GIK-5 LC45L	NK0011-36
2000-057	2000 Sep 26 1005:00	2000-057A	Tiungsat	–	ATSB	Dnepr	–	GIK-5 LC109/95	NK0011-40
2000-058	2000 Sep 29 0930:15	2000-058A	Kosmos-2373	Kometa No. 20?	S26552 RKA	Soyuz 11A511U	–	GIK-5 LC31	NK0011-46
2000-059	2000 Oct 1 2200	2000-059A	GE-1A	GE-1A	S26554 LMMS	Proton-K/DM-2M	401-01	GIK-5 LC81/23	Wire
2000-060	2000 Oct 6 2300	2000-060A	N-SAT-110	Superbird 5	S26559 SCC	Ariane 42L	V133	CSG ELA2	Wire
2000-061	2000 Oct 9 0538:18	2000-061A	HETE 2	HETE 2	S26561 MIT	Pegasus H	F30/P-35	KMR RW06/24 DZKMR	Wire
2000-062	2000 Oct 11 2317:00	2000-062A	Discovery	OV-103	S26563 NASA JSC	Space Shuttle	STS-92	KSC LC39A/ MLP3	Wire
		–	PMA-3	PMA-3	NASA JSC				
		–	Z1						
2000-063	2000 Oct 13 1412:45	2000-063A	Kosmos-2374	Uragan No.	S26564 MOM	Proton-K/DM-2	–	GIK-5 LC200/39?	CSIC-WWW
		2000-063B	Kosmos-2375	Uragan No.	S26565 MOM				
		2000-063C	Kosmos-2376	Uragan No.	S26566 MOM				
2000-064	2000 Oct 16 2127:06	2000-064A	Progress M-43	Progress 7K-TGM No 243	S26570 RAKA	Soyuz 11A511U	–	GIK-5 LC1	Wire
2000-065	2000 Oct 20 0040	2000-065A	USA 153	DSCS III B-11	S26575 USAF	Atlas IIA	AC-140	CC SLC36A	Wire

Launch International Designation:	Launch Date GMT:	COSPAR Payload International Designation:	Agency Post-Launch Payload Name:	Manufacturer Pre-Launch Payload Name:	SATCAT Number:	Launch Vehicle Type:	Launch Vehicle Serial Number:	Launch Site:	Source Reference:
2000-066	2000 Oct 21 0552	2000-066A	Thuraya 1	Thuraya 1	S26578 Thuraya	Zenit-3SL	–	KLA, SL Odyssey	Wire
2000-067	2000 Oct 21 2200:00	2000-067A	GE-6	GE-6	S26580 LMMS	Proton-K/ DM-2M	–	GIK-5 LC81/23	Wire
2000-068	2000 Oct 29 0559	2000-068A	Europe*Star F1	Europe*Star FM1	S26590 EuropeStar	Ariane 44LP	V134	CSG ELA2	Wire
2000-069	2000 Oct 30 1602	2000-069A	Beidou	Beidou	S26599 CNSA?	Chang Zheng 3A	CZ3A-5	XSC LC2	Wire
2000-070	2000 Oct 31 0752:47	2000-070A	Soyuz TM-31	Soyuz 7K-STM No. 205	S26603 RKA	Soyuz 11A511U	–	GIK-5 LC1	Wire
2000-071	2000 Nov 10 1714:02	2000-071A	GPS SVN 41	GPS SVN 41	S26605 USAF	Delta 7925-9.5	D281	CC SLC17A	Wire
2000-072	2000 Nov 16 0107:07	2000-072A	PAS 1R	PAS 1R	S26608 Panamsat	Ariane 5G	V135 (507)	CSG ELA3	AE-TV
		2000-072B	Amsat P3D	Amsat P3D	Amsat				
		2000-072C	STRV-1c	STRV-1c	DERA				
		2000-072D	STRV-1d	STRV-1d	DERA				
2000-073	2000 Nov 16 0132:36	2000-073A	Progress M1-4	Progress 7K-TGM No 253?	S26615 RAKA	Soyuz 11A511U	–	GIK-5 LC1	AVMN
2000-074	2000 Nov 20 2300	2000-074A	QuickBird	QuickBird	S26617 EWatch	Kosmos 11K65M	–	GIK-1 LC132/1	Wire
2000-075	2000 Nov 21 1824:25	2000-075A	EO-1	EO-1	–	Delta 7320-10	D282	V SLC2W	Wire
		2000-075B	SAC-C	–	–				
		2000-075C	Munin	–	–				
2000-076	2000 Nov 21 2356	2000-076A	Anik F1	Anik F1	S26624 Panamsat	Ariane 44L	V136	CSG ELA2	Wire
2000-077	2000 Nov 30 1959:47	2000-077A	Sirius 3	Sirius 3	S26626 Loral	Proton-K/ DM-2M	402-02	GIK-5 LC81/23	Wire
2000-078	2000 Dec 1 0306:01	2000-078A	Endeavour	OV-105	S26630 NASA JSC	Space Shuttle	STS-97	KSC LC39B/ MLP1	Wire
2000-079	2000 Dec 5 1232	2000-079A	EROS A1	EROS A1	S26631 ImageSat	Start-1	–	GIK-2 LC5	Wire
2000-080	2000 Dec 6 0247	2000-080A	USA 155	NRO satellite	S26635 NRO	Atlas IIAS	AC-157	CC SLC36A	Wire
2000-081	2000 Dec 20 0026	2000-081D	Astra 2D	Astra 2D	S26641	Ariane 5G	V138 (508)	CSG ELA3	AE-TV
		2000-081B	GE 8	–	S26639				
		2000-081C	LDREX	–	S26640				
2000-082	2000 Dec 20 1620	2000-082A	Beidou	Beidou	S26643 CNSA?	Chang Zheng 3A	CZ3A-6	XSC LC2	Wire
2000-F03	2000 Dec 27 0956:31	2000-F03	Kosmos	Strela-3	F01232 MO RF	Tsiklon-3	–	GIK-1 LC32/1	NK0102-35
		2000-F03	Kosmos	Strela-3	F01233 MO RF				
		2000-F03	Kosmos	Strela-3	F01234 MO RF				
		2000-F03	Gonets-D1	Gonets-D1 No. 7	F01235 RKA				
		2000-F03	Gonets-D1	Gonets-D1 No. 8	F01236 RKA				
		2000-F03	Gonets-D1	Gonets-D1 No. 9	F01237 RKA				
2001-001	2001 Jan 9 1700:03	2001-001A	Shenzhou	Shenzhou	S26664 CASC	Chang Zheng 2F	CZ2F-2	JQ SLS	Wire
2001-002	2001 Jan 10 2209	2001-002A	Turksat 2A	Eurasiasat 1	S26666	Ariane 44P	V137	CSG ELA2	Wire
2001-003	2001 Jan 24 0428:42	2001-003A	Progress M1-5	Progress 7K-TGM No 254	S26688 RAKA	Soyuz 11A511U	K15000-673	GIK-5 LC1	NK0103-25
2001-004	2001 Jan 30 0755	2001-004A	GPS SVN 54	GPS SVN 54	S26690 USAF	Delta 7925-9.5	D283	CC SLC17A	Wire
2001-005	2001 Feb 7 2306	2001-005B	Skynet 4F	Skynet 4F	S26695 UK MoD	Ariane 44L	V139	CSG ELA2	Wire
		2001-005A	Sicral	Sicral	S26694				
2001-006	2001 Feb 7 2313:02	2001-006A	Atlantis	OV-104	S26698 NASA JSC	Space Shuttle	STS-98	MLP2, KSC LC39A	Wire
		2001-006B	Destiny	Destiny	S26700 NASA MSFC				
2001-007	2001 Feb 20 0848:27	2001-007A	Odin	–	S26702 SSC	Start-1	–	GIK-2 LC5	Wire
2001-008	2001 Feb 26 0809:35	2001-008A	Progress M-44	Progress 7K-TGM No 244	S26713 RAKA	Soyuz 11A511U	670	GIK-5 LC1	NK0104-29
2001-009	2001 Feb 27 2120	2001-009A	USA	Milstar-2 F2	S26715 USAF	Titan 401B/ Centaur	4B-41 (K-30)	CC SLC40	
2001-010	2001 Mar 8 1142:09	2001-010A	Discovery	OV-103	S26718 NASA JSC	Space Shuttle	STS-102	MLP3,KSC LC39B	Wire
		–	Leonardo	MPLM	NASA JSC				
2001-011	2001 Mar 8 2251	2001-011A	Eurobird	Eurobird	S26719	Ariane 5G	V140 (509)	CSG ELA3	Wire
		2001-011B	BSAT 2a	–	S26720				
2001-012	2001 Mar 18 2233:30	2001-012A	XM Radio-2	XM Radio 2	S26724 XM Radio	Zenit-3SL	–	ODYSSEY, KLA	Wire
2001-013	2001 Apr 7 0347:00	2001-013A	Ekran	Ekran	S26736	Proton-M/Briz-M	535-01	GIK-5 LC81/24	NK0106-38
2001-014	2001 Apr 7 1502:22	2001-014A	2001 Mars Odyssey	Mars Odyssey	S26734	Delta 7925-9.5	D284	CC SLC17A	Wire
2001-015	2001 Apr 18 1013	2001-015A	GSAT-1	–	S26745	GSLV	GSLV-D1	SHAR PSLV	Wire
2001-016	2001 Apr 19 1840:42	2001-016A	Endeavour	OV-105	S26747 NASA JSC	Space Shuttle	STS-100	MLP1,KSC LC39A	Wire
2001-017	2001 Apr 28 0737:20	2001-017A	Soyuz TM-32	Soyuz 7K-STM No. 206	S26749 RKA	Soyuz 11A511U	674	GIK-5 LC1	NK0106-30
2001-018	2001 May 8 2210:29	2001-018A	XM Radio-1	XM Radio 1	S26761 XM Radio	Zenit-3SL	–	ODYSSEY, KLA	Wire
2001-019	2001 May 15 0111:30	2001-019A	PAS 10	PAS 10	S26766	Proton-K/ DM-2M	403-01	GIK-5 LC81/23	NK0107-22

Launch International Designation:	*Launch Date GMT:*	*COSPAR Payload International Designation:*	*Agency Post-Launch Payload Name:*	*Manufacturer Pre-Launch Payload Name:*	*SATCAT Number:*	*Launch Vehicle Type:*	*Launch Vehicle Serial Number:*	*Launch Site:*	*Source Reference:*
2001-020	2001 May 18 1745:00	2001-020A	GeoLITE	GeoLITE	S26770	Delta 7925-9.5	D285	CC SLC17B	Wire
2001-021	2001 May 20 2232:40	2001-021A	Progress M1-6	Progress 7K-TGM No 255	S26773 RAKA	Soyuz-FG	–	GIK-5 LC1	NK0107-09
2001-022	2001 May 29 1755:00	2001-022A	Kosmos-2377	Kobal't	S26775 MO RF	Soyuz 11A511U	–	GIK-1 LC43/4	NK0107-26
2001-023	2001 Jun 8 1508:42	2001-023A	Kosmos-2378	Parus	S26818 MO RF	Kosmos 11K65M	–	GIK-1 LC132/1	NK0108-21
2001-024	2001 Jun 9 0645	2001-024A	Intelsat 901	–	S26824	Ariane 44L	V141	CSG ELA2	Wire
2001-025	2001 Jun 16 0149:00	2001-025A	Astra 2C	Astra 2C	S26853	Proton-K/ DM-2M	403-02	GIK-5 LC81/23	NK0108-23
2001-026	2001 Jun 19 0441:02	2001-026A	ICO 2	–	S26857	Atlas IIAS	AC-156	CC SLC36B	Wire
2001-027	2001 Jun 30 1946:46	2001-027A	MAP	MAP	S26859	Delta 7425-10	D286	CC SLC17B	Wire
2001-028	2001 Jul 12 0903:59	2001-028A	Atlantis	OV-104	S26862 NASA JSC	Space Shuttle	STS-104	KSC LC39B	Wire
		–	Quest	–	NASA JSC				
2001-029	2001 Jul 12 2158	2001-029A	Artemis	–	S26863	Ariane 5G	V142 (510)	CSG ELA3	Wire
		2001-029B	BSAT 2b	–	S26864				
2001-030	2001 Jul 20 0017:00	2001-030A	Molniya-3	Molniya-3K	S26867 MOM	Molniya 8K78M	–	GIK-1 LC43/4	NK0109-38
2001-031	2001 Jul 23 0723	2001-031A	GOES 12	GOES M	S26871 NOAA	Atlas IIA	AC-142	CC SLC36A	Wire
2001-032	2001 Jul 31 0800:00	2001-032A	Koronas-F	AUOS-SM-KF-IK?	S26873 RKA	Tsiklon-3	–	GIK-1 LC32/1	NK0109-45
2001-033	2001 Aug 6 0728	2001-033A	DSP F21	–	S26880 USAF	Titan 402B/IUS	4B-31	CC SLC40	Wire
2001-034	2001 Aug 8 1613:40	2001-034A	Genesis	Genesis	S26884	Delta 7326-9.5	D287	CC SLC17A	Wire
2001-035	2001 Aug 10 2110:14	2001-035A	Discovery	OV-103	S26888 NASA JSC	Space Shuttle	STS-105	KSC LC39A	Wire
		2001-035B	Simplesat	–	S26889				
2001-036	2001 Aug 21 0923:54	2001-036A	Progress M-45	Progress 7K-TGM No 245	S26890 RAKA	Soyuz 11A511U	–	GIK-5 LC1	NK0110-18
2001-037	2001 Aug 24 2035	2001-037A	Kosmos-2379	–	S26892 MO RF	Proton-K/DM-2	404-01	GIK-5 LC81/24	NK0110-37
2001-038	2001 Aug 29 0700:00	2001-038A	LRE	LRE	S26898 NASDA	H-IIA 202	H-IIA-1F	TNSCY	NASDARept112
2001-039	2001 Aug 30 0646	2001-039A	Intelsat 902	–	S26900	Ariane 44L	V143	CSG ELA2	Wire
2001-040	2001 Sep 8 1525:05	2001-040A	NRO	–	S26905	Atlas IIAS	AC-160	V SLC3E	Wire
2001-041	2001 Sep 14 2334:55	2001-041A	Progress SO-1	Progress	S26908 RAKA	Soyuz 11A511U	677	GIK-5 LC1	NK0111-06
		2001-041	Pirs	–	A05160 RAKA				
2001-F01	2001 Sep 21 1849	–	Celestis-04	CPAC	Celestis	Taurus 2110	T6	V576E	Wire
		2001-F01	Orbview-4	–	F01239				
		2001-F01	QuikTOMS	–	F01240				
		–	SBD	–	–				
2001-042	2001 Sep 25 2321	2001-042A	Atlantic Bird 2	–	S26927	Ariane 44P	V144	CSG ELA2	Wire
2001-043	2001 Sep 30 0240	2001-043A	Starshine 3	–	S26929 LMA	Athena-1	LM-001	KLC	Wire
		2001-043B	Picosat	–	S26930 LMA				
		2001-043C	PCSat	–	S26931 LMA				
		2001-043D	SAPPHIRE	–	S26932 LMA				
2001-044	2001 Oct 5 2121:01	2001-044A	USA 161	Improved CRYSTAL 5	S26934 NRO/ USAF	Titan 404B	4B-34	V SLC4E	Wire
2001-045	2001 Oct 6 1645:00	2001-045A	Raduga-1	Globus	S26936 MO RF	Proton-K/ DM-2	–	GIK-5 LC81/24	NK0112-37
2001-046	2001 Oct 11 0301	2001-046A	USA 162	NRO satellite	S26948 NRO	Atlas IIAS	AC-162	CC SLC36B	Wire
2001-047	2001 Oct 18 1851:26	2001-047A	QB 2	QB	S26953	Delta 7320-10	D288	V SLC2W	spaceflightnow. com
2001-048	2001 Oct 21 0859:35	2001-048A	Soyuz TM-33	Soyuz 7K-STM No. 207	S26955 RKA	Soyuz 11A511U	672	GIK-5 LC1	NK0112-03
2001-049	2001 Oct 22 0453:00	2001-048B	TES	–	S26956 ISRO	PSLV	PSLV-C3	SHAR PSLV	www.dlr.de
		2001-049C	BIRD	–	S26959				
		2001-049B	PROBA	–	S26958				
2001-050	2001 Oct 25 1134:13	2001-050A	Molniya-3	Molniya-3	S26970 MOM	Molniya 8K78M	77013-687	GIK-1 LC43/3	NK-0112-45
2001-051	2001 Nov 26 1824:12	2001-051A	Progress M1-7	Progress 7K-TGM No 256	S26983 RAKA	Soyuz-FG	–	GIK-5 LC1	rosaviakosmos.ru
2001-052	2001 Nov 27 0035	2001-052A	DirecTV-4S	–	S26985 DirecTV	Ariane 44LP	V146	CSG ELA2	Wire
2001-053	2001 Dec 1 1804	2001-053A	Kosmos-2380	Uragan No.	S26987 MOM	Proton-K/DM-2	–	GIK-5 LC81	Wire
		2001-053B	Kosmos-2381	Uragan No.	S26988 MOM				
		2001-053C	Kosmos-2382	Uragan No.	S26989 MOM				
2001-054	2001 Dec 5 2219:28	2001-054A	Endeavour	OV-105	S26995 NASA JSC	Space Shuttle	STS-108	MLP1, KSC LC39B	Wire
2001-055	2001 Dec 7 1507:35	2001-055B	TIMED	–	S26998	Delta 7920-10	D289	V SLC2W	spaceflightnow. com
		2001-055A	Jason	Jason	S26997				
2001-056	2001 Dec 10 1718:56	2001-056A	Meteor-3M	–	S27001	Zenit-2	–	GIK-5 LC45/1	rosaviakosmos.ru

Launch International Designation:	Launch Date GMT:	COSPAR Payload International Designation:	Agency Post-Launch Payload Name:	Manufacturer Pre-Launch Payload Name:	SATCAT Number:	Launch Vehicle Type:	Launch Vehicle Serial Number:	Launch Site:	Source Reference:
		2001-056C	Badr B	–	S27003				
		2001-056D	Maroc-Tubsat	–	S27004				
		2001-056B	Kompas	–	S27002				
		2001-056E	Reflektor	–	S27005				
2001-057	2001 Dec 21 0400	2001-057A	Kosmos-2383	US-P	S27053 MO RF	Tsiklon-2	–	GIK-5 LC90/20?	www. cosmoworld.ru
2001-058	2001 Dec 28 0324:24	2001-058A	Kosmos	Strela-3	S27055 MO RF	Tsiklon-3	–	GIK-1 LC32/1?	rosaviakosmos.ru
		2001-058B	Kosmos	Strela-3	S27056 MO RF				
		2001-058C	Kosmos	Strela-3	S27057 MO RF				
		2001-058D	Gonets-D1	Gonets-D1 No. 10	S27058 RKA				
		2001-058E	Gonets-D1	Gonets-D1 No. 11	S27059 RKA				
		2001-058F	Gonets-D1	Gonets-D1 No. 12	S27060 RKA				
2002-001	2002 Jan 16 0030	2002-001A	USA 164	Milstar-DFS-5	S27168	Titan 401B/ Centaur	4B-38	CC SLC40	Wire
2002-002	2002 Jan 23 2347	2002-002A	Insat 3C	Insat 3C	S27299	Ariane 42L	V147	CSG ELA2	Wire
2002-003	2002 Feb 4 0245	2002-003A	Tsubasa	MDS-1	S27367	H-IIA 2024	H-IIA-2F	TNSC Y	Wire
		2002-003B	VEP-3	VEP-3/Upper Adapter	S27368				
		2002-003	DASH	DASH	A05184				
		2002-003	DASH Reentry Vehicle	–	A05185				
2002-004	2002 Feb 5 2058:12	2002-004A	RHESSI	HESSI	S27370 OSC	Pegasus XL	F31	L-1011, CC RW30	Wire
2002-005	2002 Feb 11 1743:44	2002-005A	Iridium 91	Iridium SV091	S27372	Delta 7920-10C	D290	V SLC2W	Spacelightnow. com
		2002-005B	Iridium 90	Iridium SV090	S27373				
		2002-005C	Iridium 94	Iridium SV094	S27374				
		2002-005D	Iridium 95	Iridium SV095	S27375				
		2002-005E	Iridium 96	Iridium SV096	S27376				
2002-006	2002 Feb 21 1243:00	2002-006A	Echostar 7	Echostar 7	S27378	Atlas 3B	AC-204	CC SLC36B	Wire
2002-007	2002 Feb 23 0659	2002-007A	Intelsat 904	Intelsat 904	S27380	Ariane 44L	V148	CSG ELA2	Wire
2002-008	2002 Feb 25 1726	1996-008A	Kosmos-2387	Kobal't	S23782	Soyuz-11A511U	–	GIK-1 LC43/4	Wire
2002-009	2002 Mar 1 0107:59	2002-009A	Envisat	Envisat PPF	S27386	Ariane 5G	V145 (511)	CSG ELA3	Wire
2002-010	2002 Mar 1 1122:02	2002-010A	Columbia	OV-102	S27388	Space Shuttle	STS-109	KSC LC39A	Wire
		2002-010	RAC	Rigid Array Carrier	A05193				
		2002-010	SAC	SAC	A05194				
		2002-010	FSS	FSS	A05195				
		2002-010A	MULE	Multi-Use Lightweight Equipment Carrier	A05196				
2002-011	2002 Mar 8 2259	2002-011A	TDRS 9	TDRS I	S27389	Atlas IIA	AC-143	CC SLC36A	Wire
2002-012	2002 Mar 17 0921:17	2002-012A	Grace-1	Grace-1	S27391	Rokot	–	GIK-1 LC133	Wire
		2002-012B	Grace-2	Grace-2	S27392				
2002-013	2002 Mar 21 2013:36	2002-013A	Progress M1-8	Progress M1 No. 257	S27395	Soyuz-11A511U	678	GIK-5 LC1	rosaviakosmos.ru
2002-014	2002 Mar 25 1415	2002-014A	Shenzhou 3	Shenzhou	S27397	Chang Zheng 2F	CZ2F-3 (66)	JQ SLS	Wire
		2002-014C	SZ-3 orbital module	Shenzhou module	S27408				
2002-015	2002 Mar 29 0129	2002-015A	JCSAT 2A	JCSAT 8	S27399	Ariane 44L	V149	CSG ELA2	Wire
		2002-015B	Astra 3A	Astra 3A	S27400				
2002-016	2002 Mar 30 1725:00	2002-016A	Intelsat 903	Intelsat 903	S27403	Proton-K/ DM-2M	–	GIK-5 LC81/23	Spaceflightnow. com
2002-017	2002 Apr 1 2206:45	2002-017A	Kosmos-2388	Oko US-KS	S27409	Molniya 8K78M	–	GIK-1 LC16/2	NK0206-34
2002-018	2002 Apr 8 2044:19	2002-018A	Atlantis (STS-110)	OV-104	S27413	Space Shuttle	STS-110	KSC LC39B	Wire
		2002-018	ISS S0 ITS	S0 ITS	A05199				
2002-019	2002 Apr 16 2302`	2002-019A	NSS-7	NSS-7	S27414	Ariane 44L	V150	CSG ELA2	Wire
2002-020	2002 Apr 25 0626:35	2002-020A	Soyuz TM-34	Soyuz TM No. 208	S27416	Soyuz 11A511U	–	GIK-5 LC1	www.energia.ru
2002-021	2002 May 4 0131:46	2002-021A	SPOT 5	SPOT 5	S27421	Ariane 42P	V151	CSG ELA2	Wire
		2002-021B	BreizhSat-Oscar-47 (BO-47)	Idefix CU-1	A05206				
		2002-021B	BreizhSat-Oscar-48 (BO-48)	Idefix CU-2	A05319				
2002-022	2002 May 4 0954:58	2002-022A	Aqua	EOS PM-1	S27424	Delta 7920-10L	D291	V SLC2W	spaceflightnow. com
2002-023	2002 May 7 1700:00	2002-023A	DirecTV-5	DirecTV-5	S27426	Proton-K/ DM-2M	404-02	GIK-5 LC81/24	NK0207-30
2002-024	2002 May 15 0150	2002-024A	Hai Yang 1	HY-1 (Hai Yang Yi)	S27430	Chang Zheng 4B	CZ4B-4 (67)	TYSC LC1	Wire

Launch International Designation:	*Launch Date GMT:*	*COSPAR Payload International Designation:*	*Agency Post-Launch Payload Name:*	*Manufacturer Pre-Launch Payload Name:*	*SATCAT Number:*	*Launch Vehicle Type:*	*Launch Vehicle Serial Number:*	*Launch Site:*	*Source Reference:*
		2002-024B	Feng Yun 1D	FY-1D (Feng Yun Yi Ding)	S27431				
2002-025	2002 May 28 1525	2002-025A	'Ofeq-5	'Ofeq-5	S27434	Shaviyt 1 11K65M	5	PALB	www.iai.co.il
2002-026	2002 May 28 1814:41	2002-026A	Kosmos-2389	Parus?	S27436	Kosmos 11K65M	–	GIK-1 LC132/1	NK0207-36
2002-027	2002 Jun 5 0644`	2002-027A	Intelsat 905	Intelsat 905	S27438	Ariane 44L	V152	CSG ELA2	Wire
2002-028	2002 Jun 5 2122:49	2002-028A	Endeavour (STS-111)	OV-105	S27440	Space Shuttle	STS-111	MLP1, KSC LC39A	Wire
		2002-028	Leonardo	MPLM-1	A05210				
		2002-028	Mobile Base System	MBS	A05211				
2002-029	2002 Jun 10 0114:00	2002-029A	Ekspress A1R	Ekspress-A No. 4	S27441	Proton-K/DM-2M	407-01	GIK-5 LC200/39	NK0208-33
2002-030	2002 Jun 15 2239:30	2002-030A	Galaxy 3C	Galaxy III-C	S27445	Zenit-3SL	–	ODYSSEY, KLA	NK0208-36
2002-031	2002 Jun 20 0933:46	2002-031A	Iridium 97	Iridium SV097	S27450	Rokot	492882034	GIK-1 LC133/3	NK0208-38
		2002-031B	Iridium 98	Iridium SV098	S27451				
2002-032	2002 Jun 24 1823:04	2002-032A	NOAA 17	NOAA M	S27453	Titan II SLV	23G-14	V SLC4W	spaceflightnow. com
2002-033	2002 Jun 26 0536:30	2002-033A	Progress M-46	Progress M No. 246	S27454	Soyuz 11A511U	–	GIK-5 LC1	www.energia.ru
2002-034	2002 Jul 3 0647:41	2002-034A	Contour	Contour	S27457	Delta 7425-9.5	D292	CC SLC17A	spaceflightnow. com
2002-035	2002 Jul 5 2322:00	2002-035A	Atlantic Bird 3	Stellat 5	S27460	Ariane 5G	V153 (512)	CSG ELA3	Wire
		2002-035B	N-Star c	N-Star c	S27461				
2002-036	2002 Jul 8 0635:41	2002-036A	Kosmos-2390	Strela-3	S27464	Kosmos 11K65M	–	GIK-1 LC132/1	www.tsenki.com
		2002-036B	Kosmos-2391	Strela-3	S27465				
2002-037	2002 Jul 25 1513:21	2002-037A	Kosmos-2392	11F664 No. 2? (Arkon)	S27470	Proton-K/17S40	–	GIK-5 LC81/24	www. cosmoworld.ru
2002-038	2002 Aug 21 2205:00	2002-038A	Hot Bird 6	Eutelsat HB6	S27499	Atlas V 401	AV-001	CC SLC41	Wire
2002-039	2002 Aug 22 0515:00	2002-039A	Echostar VIII	Echostar 8	S27501	Proton-K/ DM-2M	406-02	GIK-5 LC81/23	NK0210-19
2002-040	2002 Aug 28 2245:00	2002-040A	Atlantic Bird 1	Atlantic Bird 1	S27508	Ariane 5G	V155 (513)	CSG ELA3	Wire
			2002-040B	Meteosat 8	MSG 1	S27509			
2002-041	2002 Sep 6 0644	2002-041A	Intelsat 906	Intelsat 906	S27513	Ariane 44L	V154	CSG ELA2	Wire
2002-042	2002 Sep 10 0820	2002-042A	USERS REM	USERS REM	S27515	H-IIA 2024	H-IIA-3F	TNSC Y	Wire
		2002-042B	Kodama	DRTS	S27516				
		2002-042H	USERS	USERS SEM	S27815				
2002-043	2002 Sep 12 1023:40	2002-043A	KALPANA-1	METSAT-1	S27525	PSLV	PSLV-C4	SHAR PSLV	CurrSci83-1081
2002-F01	2002 Sep 15 1030	2002-F01	HTSTL-1	HTSTL-1	F01245	KT-1	–	TYSC	pc
2002-044	2002 Sep 18 2204	2002-044A	Hispasat 1D	Hispasat 1D	S27528	Atlas IIAS	AC-159	CC SLC36A	Wire
2002-045	2002 Sep 25 1658:24	2002-045A	Progress M1-9	Progress M1 No. 258	S27531	Soyuz-FG	E15000-003	GIK-5 LC1	www.energia.ru
2002-046	2002 Sep 26 1427:14	2002-046A	Nadezhda-M	Nadezhda-M 17F118M No. 41	S27534	Kosmos 11K65M	65021-706	GIK-1 LC132/1	NK0211-40
2002-047	2002 Oct 7 1945:51	2002-047A	Atlantis (STS-112)	OV-104	S27537	Space Shuttle	STS-112	KSC LC39B	Wire
		2002-047	ISS S1 ITS	S1 ITS	A05231				
2002-F02	2002 Oct 15 1820:00	2002-F02	Foton-M No. 1	Foton-M No. 1	F01247	Soyuz 11A511U	PVB 066	GIK-1 LC43/3	NK0212-34
2002-048	2002 Oct 17 0441:00	2002-048A	Integral	Integral	S27540	Proton-K/17S40	409-01	GIK-5 LC200/39	NK0212-40
2002-049	2002 Oct 27 0317	2002-049A	Zi Yuan 2	ZY-2B	S27550	Chang Zheng 4B	CZ4B-5 (68)	TYSC LC1	Wire
2002-050	2002 Oct 30 0311:11	2002-050A	Soyuz TMA-1	Soyuz TM No. 211	S27552	Soyuz-FG	E15000-004	GIK-5 LC1	NK0301-07
2002-051	2002 Nov 20 2239:00	2002-051A	Eutelsat W5	Eutelsat W5	S27554	Delta 4M+(4,2)	D4-1 (293) 4240	CC SLC37B	Wire
2002-052	2002 Nov 24 0049:47	2002-052A	Endeavour (STS-113)	OV-105	S27556	Space Shuttle	STS-113	KSC LC39A	Wire
		2002-052B	MEPSI	MEPSI	S27562				
		2002-052	ISS P1 ITS	P1 ITS	A05238				
2002-053	2002 Nov 25 2304:23	2002-053A	Astra 1K	Astra 1K	S27557	Proton-K/DM-2M	408-02	GIK-5 LC81/23	NK0301-50
2002-054	2002 Nov 28 0607:07	2002-054A	AlSat-1	DMC Alsat-1	S27559	Kosmos 11K65M	53779-802	GIK-1 LC132/1	NK0301-54
		2002-054B	Mozhaets-3	Mozhaets	S27560				
		2002-054	Rubin-3-DSI	Rubin-3-DSI	A05242				
2002-055	2002 Dec 5 0242	2002-055A	TDRS 10	TDRS J	S27566	Atlas IIA	AC-144	CC SLC36A	Wire
2002-F03	2002 Dec 11 2222	2002-F03	Hot Bird 7	Hot Bird 7	F01251	Ariane 5ECA	V157 (517)	CSG ELA3	Wire
		2002-F03	Stentor	Stentor	F01252				
2002-056	2002 Dec 14 0131	2002-056A	Midori-2	ADEOS-2	S27597	H-IIA 202	H-IIA-4F	TNSC Y	Wire
		2002-056B	Fedsat	Fedsat	S27598				
		2002-056C	WEOS	WEOS	S27599				
		2002-056D	Micro-LabSat	Mu-Lab-Sat	S27600				
2002-057	2002 Dec 17 2304	2002-057A	NSS 6	NSS 6	S27603	Ariane 44L	V156	CSG ELA2	Wire
2002-058	2002 Dec 20 1700:00	2002-058A	Rubin-2	Rubin-2	S27605	Dnepr	–	GIK-5 LC109/95	NK0302-34

Launch International Designation:	Launch Date GMT:	COSPAR Payload International Designation:	Agency Post-Launch Payload Name:	Manufacturer Pre-Launch Payload Name:	SATCAT Number:	Launch Vehicle Type:	Launch Vehicle Serial Number:	Launch Site:	Source Reference:
		2002-058B	Latinsat-B	Latinsat-B	S27606				
		2002-058C	Saudi OSCAR 50 (SO-50)	Saudisat 1C	S27607				
		2002-058D	Unisat 2	Unisat 2	S27608				
		2002-058E	Trailblazer Dummy	TB Dummy	S27609				
		2002-058H	Latinsat-A	Latinsat-A	S27612				
2002-059	2002 Dec 24 1220:13	2002-059A	Kosmos-2393	Oko US-KS	S27613	Molniya 8K78M	77045-865	GIK-1 LC16/2	NK0302-38
2002-060	2002 Dec 25 0737:58	2002-060A	Kosmos-2394	Uragan No. 791	S27617	Proton-K/ DM-2M	409-02	GIK-5 LC81/23	NK0302-40
		2002-060B	Kosmos-2395	Uragan No. 792	S27618				
		2002-060C	Kosmos-2396	Uragan No. 793	S27619				
2002-061	2002 Dec 29 1640	2002-061A	Shenzhou 4	Shenzhou	S27630	Chang Zheng 2F	CZ2F-4 (69)	JQ SLS	Wire
		2002-061C	SZ-4 orbital module	Shenzhou module	S27634				
2002-062	2002 Dec 29 2316:40	2002-062A	Nimiq 2	Nimiq 2	S27632	Proton-M/Briz-M	535-02	GIK-5 LC81/24	Wire
2003-001	2003 Jan 6 1419	2003-001A	Coriolis	P98-2 Coriolis	S27640	Titan II SLV	23G-4	V SLC4W	orbireport.com
2003-002	2003 Jan 13 0045:00	2003-002A	Icesat	Icesat	S27642	Delta 7320-10	D294	V SLC2W	spaceflightnow. com
		2003-002B	Chips	Chipsat	S27643				
2003-003	2003 Jan 16 1539:00	2003-003A	Columbia (STS-107)	OV-102	S27647	Space Shuttle	STS-107	KSC LC39A	Wire
		2003-003	Spacehab RDM	Spacehab RDM	A05248				
2003-004	2003 Jan 25 2013:35	2003-004A	SORCE	SORCE	S27651	Pegasus XL	F32 L1011	CC RW3	Wire
2003-005	2003 Jan 29 1806:00	2003-005A	Navstar GPS IIR-8 (USA 166)	GPS SVN 56	S27663	Delta 7925-9.5	D295	CC SLC17B	spaceflightnow. com
		2003-005B	XSS-10	XSS-10	S27664				
2003-006	2003 Feb 2 1259:40	2003-006A	Progress M-47	Progress M No. 247	S27681	Soyuz-U	E15000-680	GIK-5 LC1	www.energia.ru
2003-007	2003 Feb 15 0700	2003-007A	Intelsat 907	Intelsat 907	S27683	Ariane 44L	V159	CSG ELA2	Wire
2003-008	2003 Mar 11 0059:00	2003-008A	DSCS III A-3	DSCS III A-3	S27691	Delta 4M	D4-2 (296) 4040	CC SLC37B	Wire
2003-009	2003 Mar 28 0127	2003-009A	IGS-1a	IGS-1a	S27698	H-IIA 2024	H-IIA-5F	TNSC Y	Wire
		2003-009B	IGS-1b	IGS-1b	S27699				
2003-010	2003 Mar 31 2209:01	2003-010A	Navstar GPS IIR-9 (USA 168)	GPS SVN 45	S27704	Delta 7925-9.5	D297	CC SLC17A	spaceflightnow. com
2003-011	2003 Apr 2 0153:01	2003-011A	Molniya-1T	Molniya-1T	S27707	Molniya 8K78M	684	GIK-1 LC16/2	NK0306-30
2003-012	2003 Apr 8 1343:00	2003-012A	USA 169 (Milstar 6)	Milstar DFS-6	S27711	Titan 401B/ Centaur	4B-35/TC-23	CC SLC40	Wire
2003-013	2003 Apr 9 2252:19	2003-013A	Insat 3A	Insat 3A	S27714	Ariane 5G	V160 (514)	CSG ELA3	Wire
		2003-013B	Galaxy XII	Galaxy 12	S27715				
2003-014	2003 Apr 12 0047	2003-014A	Asiasat 4	Asiasat 4	S27718	Atlas 3B	AC-205	CC SLC36B	Wire
2003-015	2003 Apr 24 0423:17	2003-015A	Kosmos-2397	US-KMO	S27775	Proton-K/DM-2	410-02	GIK-5 LC81/24	NK0306-40
2003-016	2003 Apr 26 0353:52	2003-016A	Soyuz TMA-2	Soyuz 11F732 No. 212	S27781	Soyuz-FG	15000-006	GIK-5 LC1	NK0306-02
2003-017	2003 Apr 28 1159:57	2003-017A	GALEX	GALEX	S27783	Pegasus XL	F33 L-1011	CC RW3	Wire
2003-018	2003 May 8 1128	2003-018A	GSAT-2	GSAT-2	S27807	GSLV	GSLV-D2	SHAR PSLV	Wire
2003-019	2003 May 9 0429:25	2003-019A	Hayabusa	MUSES C	S27809	M-V	M-V-5	KASC M-V	Wire
		2003-019	Minerva	Minerva Hopper Lander	A05279				
2003-020	2003 May 14 2210	2003-020A	Hellas Sat 2	Hellas Sat 2	S27811	Atlas V 401	AV-002	CC SLC41	Wire
2003-021	2003 May 24 1634	2003-021A	Beidou 3	Beidou Navigation Test Sat	S27813	Chang Zheng 3A	CZ3A-7 (70)	XSC LC2	Wire
2003-022	2003 Jun 2 1745:26	2003-022A	Mars Express	Mars Express	S27816	Soyuz-FG	E15000-005/ ST11	GIK-5 LC31	Wire
		2003-022	Beagle 2	Beagle 2 Lander	A05294				
2003-023	2003 Jun 4 1923:52	2003-023A	Kosmos-2398	Parus?	S27818	Kosmos 11K65M	–	GIK-1 LC132/1	NK0308-28
2003-024	2003 Jun 6 2215:15	2003-024A	AMC-9	Americom 9	S27820	Proton-K/Briz-M	410-01	GIK-5 LC200/39	NK0308-29
2003-025	2003 Jun 8 1034:19	2003-025A	Progress M1-10	Progress M1 No. 259	S27823	Soyuz-U	D15000-681	GIK-5 LC1	NK0309-05
2003-026	2003 Jun 10 1355:59	2003-026A	Thuraya 2	Thuraya 2	S27825	Zenit-3SL	–	ODYSSEY, KLA	www.sea-launch.com
2003-027	2003 Jun 10 1758:47	2003-027A	Spirit (MER-A) Rover	Mars Exploration Rover 2	S27827	Delta 7925-9.5	D298	CC SLC17A	spaceflightnow. com
		2003-027	Columbia Memorial Station	MER-2 Lander	A05297				
2003-028	2003 Jun 11 2238:15	2003-028A	BSAT-2c	BSAT-2c	S27830	Ariane 5G	V161 (515)	CSG ELA3	Wire
		2003-028B	Optus and Defense C1	Optus C1/D	S27831				

Launch International Designation:	*Launch Date GMT:*	*COSPAR Payload International Designation:*	*Agency Post-Launch Payload Name:*	*Manufacturer Pre-Launch Payload Name:*	*SATCAT Number:*	*Launch Vehicle Type:*	*Launch Vehicle Serial Number:*	*Launch Site:*	*Source Reference:*
2003-029	2003 Jun 19 2000:35	2003-029A	Molniya-3	Molniya-3 No. 65?	S27834	Molniya 8K78M	689	GIK-1 LC16/2	NK0308-37
2003-030	2003 Jun 26 1853	2003-030A	Orbview-3	Orbview-3	S27838	Pegasus XL	F34 L-1011	V RW3	Wire
2003-031	2003 Jun 30 1415:26	2003-031B	Mimosa	Mimosa	S27841	Rokot	–	GIK-1 LC133/3	NK0308-42
		2003-031C	DTUSat	DTUSat (NLS-1)	S27842				
		2003-031D	MOST	MOST	S27842				
		2003-031E	Cute-I	Cubesat-TiTech	S27844				
		2003-031F	QuakeSat	QuakeSat (NLS-2)	S27845				
		2003-031G	AAU-Cubesat	AAU-Cubesat (NLS-1)	S27846				
		2003-031H	Can X-1	Can X-1 (NLS-1)	S27847				
		2003-031J	Cubesat XI-IV	XI-IV	S27848				
		2003-031A	Monitor-E GVM	Monitor-E	A05303				
2003-032	2003 Jul 8 0318:15	2003-032A	Opportunity Rover	Mars (MER-B) Rover 1	S27849 Exploration	Delta 7925H	D299	CC SLC17B	spaceflightnow.com
		2003-032	Challenger Memorial Station	MER-1 Lander	A05314				
2003-033	2003 Jul 17 2345	2003-033A	Rainbow 1	Rainbow 1	S27852	Atlas V 521	AV-003	CC SLC41	Wire
2003-034	2003 Aug 8 0330:55	2003-034A	Echostar IX (IA-13)	Echostar IX	S27854	Zenit-3SL	10	ODYSSEY, KLA	NK0310-22
2003-035	2003 Aug 12 1420:00	2003-035A	Kosmos-2399	Don	S27856	Soyuz-U	–	GIK-5 LC31	NK0310-24
		2003-035E	Spuskaemiy Kapsula	–	S28086				
		2003-035	Spuskaemiy Kapsula	SpK	A05466				
		2003-035	Spuskaemiy Kapsula	SpK	A05467				
		2003-035	Spuskaemiy Kapsula	SpK	A05468				
		2003-035	Spuskaemiy Kapsula	SpK	A05469				
		2003-035	Spuskaemiy Kapsula	SpK	A05470				
		2003-035	Spuskaemiy Kapsula	SpK	A05471				
2003-036	2003 Aug 13 0209:33	2003-036A	SCISAT-1	SCISAT-1	S27858	Pegasus XL	F35 L-1011	V RW3	KSC-PR-69-03
2003-037	2003 Aug 19 1050:45	2003-037A	Kosmos-2400	Strela-3	S27868	Kosmos 11K65M	–	GIK-1 LC132/1	NK0310-29
		2003-037B	Kosmos-2401	Strela-3	S27869				
2003-E01	2003 Aug 22 1630?					VLS-1	V03	ALCA VLS	www.iae.cta.br
2003-038	2003 Aug 25 0535:39	2003-038A	Spitzer Space	SIRTF Telescope	S27871	Delta 7920H	D300	CC SLC17B	spaceflightnow.com
2003-039	2003 Aug 29 0147:59	2003-039A	Progress M-48	Progress M No. 248	S27873	Soyuz-U	D15000-682	GIK-5 LC1	NK0310-08
2003-040	2003 Aug 29 2313:00	2003-040A	DSCS III B-6	DSCS III B-6	S27875	Delta 4M	D4-3 (301) 4040	CC SLC37B	Wire
2003-041	2003 Sep 9 0429	2003-041A	USA 171	Adv ORION 3	S27937	Titan 401B/ Centaur	4B-36/ TC-20	CC SLC40	Wire
2003-F01	2003 Sep 16	2003-F01	KT-1 Test Payload	KT-1 Test Payload	F01258	KT-1	–	TYSC	ChenLan
2003-042	2003 Sep 27 0611:44	2003-042A	Mozhaets-4	Mozhaets 4	S27939	Kosmos 11K65M	103	GIK-1 LC132/1	NK0311-36
		2003-042C	NigeriaSat-1	DMC NigeriaSat-1	S27941				
		2003-042D	UK-DMC	DMC BNSCSat-1	S27942				
		2003-042E	BILSAT-1	DMC BILSAT-1	S27943				
		2003-042F	Larets	Larets	S27944				
		2003-042G	STSat-1	KAISTSAT-4	S27945				
		2003-042G	Rubin-4-DSI	Rubin-4-DSI	S27947				
2003-043	2003 Sep 27 2314:46	2003-043A	e-Bird	e-Bird	S27948	Ariane 5G	V162 (516)	CSG ELA3	Wire
		2003-043C	SMART-1	SMART-1	S27949				
		2003-043E	Insat 3E	Insat 3E	S27951				
2003-044	2003 Oct 1 0402:59	2003-044A	Galaxy 13/ Horizons-1	Galaxy XIII/ Horizons-1	S27954	Zenit-3SL	13	ODYSSEY, KLA	NK0312-34
2003-045	2003 Oct 15 0100:03	2003-045A	Shenzhou 5	Shenzhou 5	S28043	Chang Zheng 2F	CZ2F-5 (71)	JQ SLS	Wire
		2003-045G	SZ-5 orbital module	Shenzhou module	S28049				
2003-046	2003 Oct 17 0452:08	2003-046A	IRS-P6 (Resourcesat)	IRS-P6	S28051	PSLV	PSLV-C5	SHAR PSLV	NK0312-36
2003-047	2003 Oct 18 0538:03	2003-047A	Soyuz TMA-3	Soyuz 11F732 No. 213	S28052	Soyuz-FG	D15000-007	GIK-5 LC1	www.energia.ru
2003-048	2003 Oct 18 1617	2003-048A	DMSP 5D-3 F-16	DMSP 5D-3 S-16	S28054	Titan II SLV	23G-9	V SLC4W	Wire
2003-049	2003 Oct 21 0316	2003-049A	Zi Yuan 1-2	ZY-1-2/ CBERS 2	S28057	Chang Zheng 4B	CZ4B-6 (72)	TYSC LC1	spaceflightnow.com
		2003-049B	Chuangxin-1	CX-1	S28058				

Launch International Designation:	*Launch Date GMT:*	*COSPAR Payload International Designation:*	*Agency Post-Launch Payload Name:*	*Manufacturer Pre-Launch Payload Name:*	*SATCAT Number:*	*Launch Vehicle Type:*	*Launch Vehicle Serial Number:*	*Launch Site:*	*Source Reference:*
2003-050	2003 Oct 30 1343:42	2003-050A	SERVIS-1	SERVIS-1	S28060	Rokot	4921921121	GIK-1 LC133/3	NK0312-44
2003-051	2003 Nov 3 0720	2003-051A	JB-4 No. 1?	FSW Service Module	S28078	Chang Zheng 2D	CZ2D-4 (73)	JQ SLS-2	Wire
2003-052	2003 Nov 14 1601	2003-052A	Zhongxing-20 (Feng Huo?)	Zhongxing-20	S28082	Chang Zheng 3A	CZ3A-8 (74)	XSC LC2	Xinhua
2003-053	2003 Nov 24 0622:00	2003-053A	Yamal-201	Yamal-200 No. 1	S28089	Proton-K/ DM-2M	407-02	GIK-5 LC81/23	NK0401-19
		2003-053F	Yamal-202	Yamal-200 No. 2	S28094				
2003-F02	2003 Nov 29 0433	2003-F02	IGS-2a	IGS-2a	F01260	H-IIA 2024	H-IIA-6F	TNSC Y	Wire
		2003-F02	IGS-2b	IGS-2b	F01261				
2003-054	2003 Dec 2 1004	2003-054A	USA 173	NOSS C2-1	S28095	Atlas IIAS	AC-164	V SLC3E	Wire
		2003-054C	USA 173 P/L 2	NOSS C2-2	S28097				
2003-055	2003 Dec 5 0600:00	2003-055A	Gruzomaket	Kondor-E GVM?	S28098	Strela	–	GIK-5 LC175/2	NK0402-13
2003-056	2003 Dec 10 1742:12	2003-056A	Kosmos-2402	Uragan 794	S28112	Proton-K/ Briz-M	410-03	GIK-5 LC81/24	NK0402-17
		2003-056B	Kosmos-2403	Uragan 795	S28113				
		2003-056C	Kosmos-2404	Glonass-M No. 11L (701)	S28114				
2003-057	2003 Dec 18 0230	2003-057A	UHF F/O F11 (USA 174)	UHF F/O F11	S28117	Atlas 3B	AC-203	CC SLC36B	Wire
2003-058	2003 Dec 21 0805:00	2003-058A	Navstar GPS IIR-10 (USA 175)	GPS SVN 47	S28129	Delta 7925-9.5	D302	CC SLC17A	spaceflightnow. com
2003-059	2003 Dec 27 2130:00	2003-059A	Amos 2	Amos 2	S28132	Soyuz-FG	D15000-008 084/	GIK-5 LC31	NK0402-23
2003-060	2003 Dec 28 2300:00	2003-060A	Ekspress AM-22	Ekspress AM-22	S28134	Proton-K/ DM-2M	410-04	GIK-5 LC200/39	NK0402-26
2003-061	2003 Dec 29 1906:18	2003-061A	Tan Ce 1	Double Star DSP-E	S28140	Chang Zheng 2C	CZ2C-22 (75)	XSC LC1?	Wire
2004-001	2004 Jan 11 0412:59	2004-001A	Estrela do Sul 1	Telstar 14	S28137	Zenit-3SL	12	ODYSSEY, KLA	NK0403-09
2004-002	2004 Jan 29 1158:08	2004-002A	Progress M1-11	Progress M1 No. 260	S28142	Soyuz-U	D15000-683	GIK-5 LC1	NK0403-17
2004-003	2004 Feb 5 2346	2004-003A	AMC-10	AMC-10	S28154	Atlas IIAS	AC-165	CC SLC36A	Wire
2004-004	2004 Feb 14 1850	2004-004A	DSP F22	DSP F22	S28158	Titan 402B/IUS	4B-39	CC SLC40	Wire
2004-005	2004 Feb 18 0705:55	2004-005A	Molniya-1T	Molniya-1T?	S28163	Molniya 8K78M	–	GIK-1 LC16/2	Wire
2004-006	2004 Mar 2 0717:44	2004-006A	Rosetta	Rosetta	S28169	Ariane 5G+	V158 (518)	CSG ELA3	Wire
		2004-006	Philae	Rosetta Lander	A05494				
2004-007	2004 Mar 13 0540	2004-007A	MBSAT	MBSAT	S28184	Atlas 3A	AC-202	CC SLC36B	Wire
2004-008	2004 Mar 15 2306:00	2004-008A	Eutelsat W3A	Eutelsat W3A	S28187	Proton-M/Briz-M	535-03	GIK-5 LC81/24	NK0405-23
2004-009	2004 Mar 20 1753:00	2004-009A	Navstar GPS IIR-11	GPS SVN 59	S28190	Delta 7925-9.5	D303	CC SLC17B	Wire
2004-010	2004 Mar 27 0330	2004-010A	Raduga-1	Globus No. 17?	S28194	Proton-K/DM-2	410-05	GIK-5 LC81/23	NK0405-28
2004-011	2004 Apr 16 0045	2004-011A	Superbird A2	Superbird 6	S28218	Atlas IIAS	AC-163	CC SLC36A	Wire
2004-012	2004 Apr 18 1559	2004-012A	Shiyan 1	Shiyan 1	S28220	Chang Zheng 2C	CZ2C-23 (76)	XSC LC1?	Wire
		2004-012B	Naxing 1	Nanosatellite 1 (NX-1)	S28221				
2004-013	2004 Apr 19 0318:57	2004-013A	Soyuz TMA-4	Soyuz 11F732 No. 214	S28228	Soyuz-FG	Zh15000-009	GIK-5 LC1	NK0406-10
2004-014	2004 Apr 20 1657:24	2004-014A	Gravity Probe B	GP-B	S28230	Delta 7920-10C	D304	V SLC2W	spaceflightnow. com
2004-015	2004 Apr 26 2037:00	2004-015A	Ekspress AM-11	Ekspress AM-11	S28234	Proton-K/ DM-2M	410-06	GIK-5 LC200/39	NK0406-34
2004-016	2004 May 4 1242:00	2004-016A	DirecTV 7S	DirecTV 7S	S28238	Zenit-3SL	–	ODYSSEY, KLA	www.sea-launch.com
2004-017	2004 May 19 2222	2004-017A	AMC-11	Americom 11	S28252	Atlas IIAS	AC-166	CC SLC36B	Wire
2004-018	2004 May 20 1747:03	2004-018A	ROCSAT-2	ROCSAT-2	S28254	Taurus 3210	T7 (XL)	V 576E	Wire
2004-019	2004 May 25 1234:23	2004-019A	Progress M-49	Progress M No. 249	S28261	Soyuz-U	D15000-684	GIK-5 LC1	NK0407-08
2004-020	2004 May 28 0600:00	2004-020A	Kosmos-2405	US-P	S28350	Tsiklon-2	–	GIK-5 LC90/20	NK0407-30
2004-021	2004 Jun 10 0128:00	2004-021A	Kosmos-2406	Tselina-2	S28352	Zenit-2	–	GIK-5 LC45/1	NK0408-25
2004-022	2004 Jun 16 2227:00	2004-022A	Intelsat 10-02	Intelsat 10-02	S28358	Proton-M/ Briz-M	535-06	GIK-5 LC200/39	NK0408-27
2004-023	2004 Jun 23 2254:00	2004-023A	Navstar GPS IIR-12	GPS SVN 60	S28361	Delta 7925-9.5	D305	CC SLC17B	spaceflightnow. com
2004-024	2004 Jun 29 0358:59	2004-024A	Telstar 18	Apstar 5	S28364	Zenit-3SL	11	ODYSSEY, KLA	NK0408-31

For details of the latest updates to ***Jane's Space Systems and Industry*** online and to discover the additional information available exclusively to online subscribers please visit **jsd.janes.com**

Launch International Designation:	*Launch Date GMT:*	*COSPAR Payload International Designation:*	*Agency Post-Launch Payload Name:*	*Manufacturer Pre-Launch Payload Name:*	*SATCAT Number:*	*Launch Vehicle Type:*	*Launch Vehicle Serial Number:*	*Launch Site:*	*Source References:*
2004-025	2004 Jun 29 0630:06	2004-025A	Latinsat-D	Latinsat D/ Aprizesat-2	S28366	Dnepr	–	GIK-5 LC109/95	NK0408-33
		2004-025C	Demeter	Demeter	S28368				
		2004-025D	Saudi Comsat-1	Saudi Comsat-1	S28369				
		2004-025E	Saudi Comsat-2	Saudi Comsat-2	S28370				
		2004-025F	SaudiSat-2	SaudiSat-2	S28371				
		2004-025G	LatinSat-C	Latinsat C/ Aprizesat-1	S28372				
		2004-025H	UniSat-3	UniSat-3	S28373				
		2004-025K	Amsat-Oscar-E (AO-51)	Amsat-Echo	S28375				
2004-026	2004 Jul 15 1001:59	2004-026A	Aura	EOS Chem-1	S28376	Delta 7920-10L	D306	V SLC2W	spaceflightnow.com
2004-027	2004 Jul 18 0044	2004-027A	Anik F2	Anik F2	S28378	Ariane 5G+	V163 (519)	CSG ELA3	Wire
2004-028	2004 Jul 22 1746:28	2004-028A	Kosmos-2407	Parus	S28380	Kosmos 11K65M	–	GIK-1 LC132/1	NK0409-27
2004-029	2004 Jul 25 0705	2004-029A	Tan Ce 2	Double Star DSP-P	S28382	Chang Zheng 2C	CZ2C-24 (77)	TYSC LC1	Wire
2004-030	2004 Aug 3 0615:56	2004-030A	Messenger	MESSENGER	S28391	Delta 7925H	D307	CC SLC17B	spaceflightnow.com
2004-031	2004 Aug 4 2232:00	2004-031A	Amazonas	Amazonas	S28393	Proton-M/Briz-M	535-07	GIK-5 LC200/39	NK0410-25
2004-032	2004 Aug 11 0503:07	2004-032A	Progress M-50	Progress 7K-TGM No. 350	S28399	Soyuz-U	D15000-685	GIK-5 LC1	NK0410-07
2004-033	2004 Aug 29 0750	2004-033A	FSW No 19	FSW-3?	S28402	Chang Zheng 2C	CZ2C-25 (78)	JQ SLS	xinhuanet.com
2004-034	2004 Aug 31 2317	2004-034A	USA 179	SDS C-4	S28384	Atlas IIAS	AC-167	CC SLC36A	Wire
2004-F01	2004 Sep 6 1035	2004-F01	'Ofeq-6	'Ofeq-6	F01280	Shaviyt 1	6	PALB	jpost.com
2004-035	2004 Sep 8 2314	2004-035A	SJ-6A	SJ-6A	S28413	Chang Zheng 4B	CZ4B-7 (79)	TYSC LC1	xinhuanet.com
		2004-035B	SJ-6B	SJ-6B	S28414				
2004-036	2004 Sep 20 1031	2004-036A	EDUSAT	EDUSAT (GSAT-3)	S28417	GSLV	GSLV-F1	SHAR PSLV	Wire
2004-037	2004 Sep 23 1507:36	2004-037A	Kosmos-2408	Strela-3	S28419	Kosmos 11K65M	–	GIK-1 LC132/1	NK0411-39
		2004-037B	Kosmos-2409	Strela-3	S28420				
2004-038	2004 Sep 24 1650:00	2004-038A	Kosmos-2410	Kobal't-M?	S28396	Soyuz-U	–	GIK-1 LC16/2	NK0411-40
2004-039	2004 Sep 27 0800	2004-039A	FSW No 20	FSW-3?	S28424	Chang Zheng 2D	CZ2D-5 (80)	JQ SLS	xinhuanet.com
2004-040	2004 Oct 14 0306:26	2004-040A	Soyuz TMA-5	Soyuz 11F732 No. 215	S28444	Soyuz-FG	Zh15000-012	GIK-5 LC1	NK0412-02
2004-041	2004 Oct 14 2123:00	2004-041A	AMC-15	AMC-15	S28446	Proton-M/Briz-M	535-08	GIK-5 LC200/39	NK0412-32
2004-042	2004 Oct 19 0120	2004-042A	Feng Yun 2C	FY-2 O4	S28451	Chang Zheng 3A	CZ3A-9 (81)	TYSC LC1	xinhuanet.com
2004-043	2004 Oct 29 2211:00	2004-043A	Ekspress AM-1	Ekspress AM-1	S28463	Proton-K/ DM-2M	410-08	GIK-5 LC200/39	NK0412-37
2004-044	2004 Nov 6 0310	2004-044A	Zi Yuan 2C	FY-2 O4	S28451	Chang Zheng 3A	CZ3A-9 (81)	TYSC LC1	xinhuanet.com
2004-045	2004 Nov 6 0539:00	2004-045A	Navstar GPS IIR-13	GPS SVN 61	S28474	Delta 7925-9.5	D308	CC SLC17B	spaceflightnow.com
2004-U01	2004 Nov 8 1830	2004-U01	Oblik	Oblik	A05638	Soyuz-2-1A	–	GIK-1 LC43/4	NK0501-25
2004-046	2004 Nov 18 1045	2004-046A	Shiyan 2	Shiyan 2	S28479	Chang Zheng 2C	CZ2C-26 (83)	XSC LC1?	xinhuanet.com
2004-047	2004 Nov 20 1716:00	2004-047A	Swift	Swift GRB Explorer	S28485	Delta 7320-10C	D309	CC SLC17A	spaceflightnow.com
2004-048	2004 Dec 17 1207:00	2004-048A	AMC 16	AMC 16	S28472	Atlas V 521	AV-005	CC SLC41	Wire
2004-049	2004 Dec 18 1626	2004-049A	Helios IIA	Helios 2A	S28492	Ariane 5G+	V165 (520)	CSG ELA3	Wire
		2004-049B	Nanosat 1	Nanosat	S28493				
		2004-049C	Essaim 1	Essaim 1	S28494				
		2004-049D	Essaim 2	Essaim 2	S28495				
		2004-049E	Essaim 3	Essaim 3	S28496				
		2004-049F	Essaim 4	Essaim 4	S28497				
		2004-049G	Parasol	Parasol	S28498				
2004-050	2004 Dec 21 2150:00	2004-050A	Demosat	HLV-OLDSP	S28500	Delta 4H	D4-4 4050	CC SLC37B	Wire
		2004-050	Sparky	3CS-1	A05645				
		2004-050	Ralphie	3CS-2	A05646				
2004-051	2004 Dec 23 2219:34	2004-051A	Progress M-51	Progress 7K-TGM No. 351	S28503	Soyuz-U	Zh15000-092	GIK-5 LC1	NK0502-36
2004-052	2004 Dec 24 1120:00	2004-052A	Sich-1M	Okean-O1 No. 9	S28505	Tsiklon-3	701	GIK-1 LC32/2	NK0502-21
		2004-052C	MK-1TS Mikron	KS5-MF2	S28507				
2004-053	2004 Dec 26 1353:31	2004-053A	Kosmos-2411	Uragan-M No. 712	S28508	Proton-K/DM-2	410-07	GIK-5 LC200/39	glonass-center.ru
		2004-053B	Kosmos-2412	Uragan No. 797	S28509				
		2004-053C	Kosmos-2413	Uragan No. 796	S28510				

Source: Jonathan McDowell, PhD

Table 4.11: World Orbital Launch Log 2005 – 2007

See *jsd.janes.com* for World Orbital Launch Log 1957–1994 (Tables 4.1–4.8)

Launch International Designation:	*Launch Date GMT:*	*COSPAR Payload International Designation:*	*Agency Post-Launch Payload Name:*	*Manufacturer Pre-Launch Payload Name:*	*SATCAT Number:*	*Launch Vehicle Type:*	*Launch Vehicle Serial Number:*	*Launch Site:*	*Source References:*
2005-001	2005 Jan 12 1847:08	2005-001A	Deep Impact	Deep Impact	S28517	Delta 7925-9.5	D311	CC SLC17B	spaceflightnow.com
		2005-001	Impactor	Deep Impact	A05650				
2005-002	2005 Jan 20 0300:07	2005-002A	Kosmos-2414	Parus?	S28521	Kosmos 11K65M	411	GIK-1 LC132/1	NK0503-17
		2005-002C	MGU-250	Universitetskiy/ Tatiana	S28523				
2005-003	2005 Feb 3 0227:32	2005-003A	AMC 12	Americom 12	S28526	Proton-M/Briz-M	535-09	GIK-5 LC81/24	NK0504-06
2005-004	2005 Feb 3 0741	2005-004A	USA 181	NOSS C3-1	S28537	Atlas 3B	AC-206	CC SLC36B	Wire
		2005-004C	USA-181 P/L 2	NOSS C3-2	S28541				
2005-005	2005 Feb 12 2103:01	2005-005A	XTAR-EUR	XTAR-EUR	S28542	Ariane 5ECA	V164 (521)	CSG ELA3	Wire
		2005-005C	Sloshsat-FLEVO	Sloshsat-FLEVO	S28544				
		2005-005D	Maqsat-B2	Maqsat-B2	A05666				
2005-006	2005 Feb 26 0925	2005-006A	Himawari-6	MTSAT-1R	S28622	H-IIA 2022	H-IIA-7F	TNSC Y	rocketsystem.co.jp
2005-007	2005 Feb 28 1909:18	2005-007C	TNS-0 Nanosputnik	TEKh-42	S28547	Soyuz-U	Zh15000-093	GIK-5 LC1	NK0504-27
		2005-007A	Progress M-52	Progress 7K-TGM No. 352/1	S28624				
2005-008	2005 Mar 1 0350:59	2005-008A	XM Radio 3 (Rhythm)	XM Radio-3	S28626	Zenit-3SL	17	ODYSSEY, KLA	NK0505-06
2005-009	2005 Mar 11 2142	2005-009A	Inmarsat 4 F1	Inmarsat 4 F1	S28628	Atlas V 431	AV-004	CC SLC41	Wire
2005-010	2005 Mar 29 2131:00	2005-010A	Ekspress AM-2	Ekspress AM-2	S28629	Proton-K/DM-2M	410-09	GIK-5 LC200/39	NK0505-11
2005-011	2005 Apr 11 1335	2005-011A	XSS-11 (USA 165)	XSS-11	S28636	Minotaur	3	V SLC-8	Wire
2005-012	2005 Apr 12 1200	2005-012A	Apstar 6	Apstar 6	S28638	Chang Zheng 3B	CZ3B (84)	XSC LC2	xinhuanet.com
2005-013	2005 Apr 15 0046:25	2005-013A	Soyuz TMA-6	Soyuz 11F732 No. 216	S28640	Soyuz-FG	Zh15000-014	GIK-5 LC1	NK0506-01
2005-014	2005 Apr 15 1726:50	2005-014A	DART	DART/HAPS	S28642	Pegasus XL/HAPS	F36	L-1011,V RW3	NASA-TV
2005-015	2005 Apr 26 0731:29	2005-015A	Spaceway 1	Spaceway 1	S28644	Zenit-3SL	15	ODYSSEY, KLA	www.sea-launch.com
2005-016	2005 Apr 30 0050	2005-016A	USA 182	ONYX 5	S28646	Titan 405B	4B-30	CC SLC40	Wire
2005-017	2005 May 5 0445	2005-017A	Cartosat-1	Cartosat-1	S28649	PSLV	PSLV-C6	SHAR SLP	Wire
		2005-017B	VO-52 HAMSAT	VUSat (HAMSAT)	S28650				
2005-018	2005 May 20 1022:01	2005-018A	NOAA 18	NOAA N	S28654	Delta 7320-10C	D312	V SLC2W	spaceflightnow.com
2005-019	2005 May 22 1759:08	2005-019A	DirectTV-8	DirectTV-8	S28659	Proton-M/Briz-M	535-10	GIK-5 LC200/39	NK0507-17
2005-020	2005 May 31 1200:00	2005-020A	Foton-M No. 2	Foton-M No. 2	S28686	Soyuz-U	Zh15000-091	GIK-5 LC1	Wire
2005-021	2005 Jun 16 2309:34	2005-021A	Progress M-53	Progress 7K-TGM No. 353/1	S28700	Soyuz-U	Zh15000-094	GIK-5 LC1	NK0508-16
2005-F01	2005 Jun 21 0048:37	2005-F01	Molniya-3K	Molniya-3K No. 12L	F01282	Molniya 8K78M	77046-694	GIK-1 LC16/2	NK0508-01
2005-F02	2005 Jun 21 1946:09	2005-F02	Cosmos-1	TPS Solar Sail	F01286	Volna	–	BORIS, BLA	Wire
2005-022	2005 Jun 23 1403:00	2005-022A	Intelsat Americas 8	IA-8	S28702	Zenit-3SL	14 (04L?)	ODYSSEY, KLA	NK0508-08
2005-023	2005 Jun 24 1941:00	2005-023A	Ekspress AM-3	Ekspress AM-3	S28707	Proton-K/DM-2	410-10? (07?)	GIK-5 LC200/39	NK0508-09
2005-024	2005 Jul 5 2240	2005-024A	SJ-7	SJ-7	S28737	Chang Zheng 2D	CZ2D-6 (85)	JQ SLS-2?	xinhuanet.com
2005-025	2005 Jul 10 0330	2005-025A	Suzaku	ASTRO-E2	S28773	M-V	M-V-6	KASC M-V	Wire
2005-026	2005 Jul 26 1439:00	2005-026A	Discovery (STS-114)	OV-103	S28775	Space Shuttle	STS-114	KSC LC39B	Wire
		2005-026	Raffaello	MPLM-2	A05763				
2005-027	2005 Aug 2 0730	2005-027A	FSW No. 21	FSW-3?	S28776	Chang Zheng 2C	CZ2C-27 (86)	JQ	xinhuanet.com
2005-028	2005 Aug 11 0820:44	2005-028A	Thaicom 4	iPSTAR 1	S28786	Ariane 5GS	V166 (523)	CSG ELA3	Wire
2005-029	2005 Aug 12 1143:00	2005-029A	Mars Reconnaissance Orbiter	MRO	S28788	Atlas V 401	AV-007	CC SLC41	Wire
2005-030	2005 Aug 13 2328:26	2005-030A	Galaxy 14	Galaxy 14	S28790	Soyuz-FG	Zh15000-011/ST1	GIK-5 LC31	NK0510-11
2005-031	2005 Aug 23 2109:59	2005-031A	Kirari	OICETS	S28809	Dnepr	–	GIK-5 LC109/95	NK0510-13
		2005-031B	Reimei	INDEX	S28810				
2005-032	2005 Aug 26 1834:28	2005-032A	Monitor-E No. 1	–	S28822	Rokot	–	GIK-1 LC133/3	NK0510-16
2005-033	2005 Aug 29 0845?	2005-033A	FSW No 22	FSW-3?	S28824	Chang Zheng 2D	CZ2D-7 (87)	JQ	TLE
2005-034	2005 Sep 2 0950	2005-034A	Kosmos-2415	Kometa No. 21?	S28841	Soyuz-U	–	GIK-5 LC31	NK0511-03
2005-035	2005 Sep 8 1307:54	2005-035A	Progress M-54	Progress 7K-TGM No. 354/1	S28866	Soyuz-U	P15000-095	GIK-5 LC1	NK0511-14
		2005-035C	RadioSkaf	Orlan-M No. 14	S28933				

Launch International Designation:	*Launch Date GMT:*	*COSPAR Payload International Designation:*	*Agency Post-Launch Payload Name:*	*Manufacturer Pre-Launch Payload Name:*	*SATCAT Number:*	*Launch Vehicle Type:*	*Launch Vehicle Serial Number:*	*Launch Site:*	*Source Reference:*
2005-036	2005 Sep 8 2153:40	2005-036A	Anik F1R	Anik F1R	S28868	Proton-M/ Briz-M	535-12	GIK-5 LC200/39	NK0511-06
2005-037	2005 Sep 23 0224:29	2005-037A	STP-R1	Streak	S28871	Minotaur	4	V SLC-8	spaceflightnow. com
2005-038	2005 Sep 26 0337:00	2005-038A	Navstar GPS IIR-M1	GPS SVN 57	S28874	Delta 7925-9.5	D313	CC SLC17A	spaceflightnow. com
2005-039	2005 Oct 1 0354:53	2005-039A	Soyuz TMA-7	Soyuz 11F732 No. 217	S28877	Soyuz-FG	Zh15000-017	GIK-5 LC1	NK0512-02
2005-F03	2005 Oct 8 1502:00	2005-F03	CRYOSAT	Cryosat	F01289	Rokot	–	GIK-1 LC133/3	NK0512-34
2005-040	2005 Oct 12 0100	2005-040A	Shenzhou 6	Shenzhou 6	S28879	Chang Zheng 2F	CZ2F-6 (88)	JQ SLS	xinhuanet.com
2005-041	2005 Oct 13 2232	2005-041A	Syracuse 3A	Syracuse 3A	S28884	Ariane 5GS	V168 (524)	CSG ELA3	Wire
		2005-041B	Galaxy 15	Galaxy 15	S28885				
2005-042	2005 Oct 19 1805	2005-042A	USA 186	Improved CRYSTAL?	S28888	Titan 404B	4B-26	V SLC4E	Wire
2005-043	2005 Oct 27 0652:26	2005-043A	Beijing-1	China-DMC	S28890	Kosmos 11K65M	104	GIK-1 LC132/1	NK0503-43
		2005-043B	Topsat	Topsat	S28891				
		2005-043C	UWE-1	UWE-1	S28892				
		2005-043D	Sinah-1	Sina-1	S28893				
		2005-043E	SSETI Express XO-53	SSETI Express	S28894				
		2005-043F	Cubesat XI-V	Cubesat XI-V	S28895				
		2005-043J	Rubin-5	Rubin-5	A05790				
		2005-043J	Mozhaets-5	Mozhaets-5	A05791				
2005-044	2005 Nov 8 1407	2005-044A	Inmarsat 4 F2	Inmarsat 4 F2	S28899	Zenit-3SL	23	ODYSSEY, KLA	www.sea-launch. com
2005-045	2005 Nov 9 0333:34	2005-045A	Venus Express	Venus Express	S28901	Soyuz-FG	Zh15000-010/ ST1	GIK-5 LC31	Wire
2005-046	2005 Nov 16 2346:00	2005-046A	Telkom 2	Telkom 2	S28902	Ariane 5ECA	V167 (522)	CSG ELA3	Wire
		2005-046B	Spaceway 2	Spaceway F2	S28903				
2005-047	2005 Dec 21 1838	2005-047A	Progress M-55	Progress 7K-TGM No. 355/2	S28906	Soyuz-U	095?	GIK-5 LC1	NK0511-14
2005-048	2005 Dec 21 1934	2005-048A	Gonets-D1	Gonets-D1M No. 13?	S28908	Kosmos-11K65M	232	GIK-1 LC132/1	Wire
		2005-048B	Kosmos-2416	Rodnik	S28909				
2005-049	2005 Dec 21 2233	2005-049A	Insat 4A	Insat 4A	S28911	Ariane 5GS	V169 (525)	CSG ELA3	Wire
		2005-049B	MSG 2	MSG 2	S28912				
2005-050	2005 Dec 25 0507:10	2005-050A	Kosmos-2417	Uragan-M No. 713	S28915	Proton-K/DM-2	410-12	GIK-5 LC81/23	glonass-center.ru
		2005-050B	Kosmos-2418	Uragan-M No. 714	S28916				
		2005-050C	Kosmos-2419	Uragan No. 798	S28917				
2005-051	2005 Dec 28 0519	2005-051A	Giove A	GSTB-v2/A	S28922	Soyuz-FG	Zh15000-015/ST1	GIK-5 LC31	Wire
2005-052	2005 Dec 29 0228	2005-052A	AMC 23	Americom 23	S28924	Proton-M/ Briz-M	535-13	GIK-5 LC200/39	Wire
2006-001	2006 Jan 19 1900:00	2006-001A	New Horizons	New Horizons	S28928	Atlas V 551	AV-010	CC SLC41	Wire
2006-002	2006 Jan 24 0133	2006-002A	Daichi	ALOS	S28931	H-IIA 2022	H-IIA-8F	TNSC Y	Wire
2006-003	2006 Feb 15 2334:59	2006-003A	Echostar 10	Echostar 10	S28935	Zenit-3SL	–	Odyssey, KLA	www.sea-launch. com
2006-004	2006 Feb 18 0627	2006-004A	MTSAT-2	MTSAT-2	S28937	H-IIA 2024	H-IIA-9F	TNSC Y	Wire
2006-005	2006 Feb 21 2128	2006-005A	Akari	Astro-F	S28939	M-V	M-V-8	KASC M-V	Wire
		2006-005B	SSP	Solar Sail Sub Payload	A05878				
		2006-005C	CUTE-1.7-APD	CUTE-1.7	S28941				
2006-006	2006 Feb 28 2010	2006-006A	Arabsat 4A (Badr 1)	Arabsat 4A	S28943	Proton-M/ Briz-M	535-11	GIK-5 LC200/39	NK0604-10
2006-007	2006 Mar 11 2233	2006-007A	Hotbird 7A	Hotbird 7A	S28945	Ariane 5 ECA	V170 (527)	CSG ELA3	Wire
		2006-007B	Spainsat	Spainsat	S28946				
2006-008	2006 Mar 22 1403	2006-008A	ST5-FWD	Space Technology 5	S28980	Pegasus XL	F37	V, L1011, RW30/12	NASA-TV
		2006-008B	ST5-MID	Space Technology 5	S28981				
		2006-008C	ST5-AFT	Space Technology 5	S28982				
2006-F01	2006 Mar 24 2230	2006-F01	Falconsat-2	Falconsat-2	F01299	Falcon 1	1	KMR OM	Wire
2006-009	2006 Mar 30 0230	2006-009A	Soyuz TMA-8	Soyuz 11F732 No. 218	S28996	Soyuz-FG	P15000-018	GIK-5 LC1	NK0605-01
2006-010	2006 Apr 12 2330	2006-010A	JCSAT-9	JCSAT-9	S29045	Zenit-3SL	21	Odyssey, POR	NK0606-32
2006-011	2006 Apr 15 0140	2006-011A	COSMIC 1	FORMOSAT 3 FM1 (S00-8)	S29047	Minotaur	5	V SLC8	spaceflightnow. com
		2006-011B	COSMIC 2	FORMOSAT 3 FM2	S29048				
		2006-011C	COSMIC 3	FORMOSAT 3 FM3	S29049				
		2006-011D	COSMIC 4	FORMOSAT 3 FM4	S29050				
		2006-011E	COSMIC 5	FORMOSAT 3 FM5	S29051				
		2006-011F	COSMIC 6	FORMOSAT 3 FM6	S29052				
2006-012	2006 Apr 20 2027	2006-012A	Astra 1KR	Astra 1KR	S29055	Atlas V 411	AV-008	CC SLC41	Wire
2006-13	2006 Apr 24 1603	2006-013A	Progress M-56	Progress 7K-TGM No. 356	S29057	Soyuz-U	P15000-100	GIK-5 LC1	NK0606-17
2006-14	2006 Apr 25 1647	2006-14A	EROS B	EROS B	S29079	Start-1	No. 441	GIK-2	NK0606-40
2006-15	2006 Apr 26 2248	2006-15A	Yaogan 1	Yaogan 1	S29092	CZ-4B	CZ4B-9 (89)	TYSC	xinhuanet.com

Launch International Designation:	*Launch Date GMT:*	*COSPAR Payload International Designation:*	*Agency Post-Launch Payload Name:*	*Manufacturer Pre-Launch Payload Name:*	*SATCAT Number:*	*Launch Vehicle Type:*	*Launch Vehicle Serial Number:*	*Launch Site:*	*Source Reference:*
2006-16	2006 Apr 28 1002	2006-016B	Cloudsat	Cloudsat	S29107	Delta 7420	D314	V SLC2W	spaceflightnow.com
		2006-016A	Calipso	Calipso	S29108				
2006-17	2006 May 3 1738	2006-17A	Kosmos-2420	Kobal't-M	S29111	Soyuz-U	–	GIK-1 LC16	NK0607-35
2006-18	2006 May 24 2211	2006-18A	GOES 13	GOES-N	S29155	Delta 4M+(4,2)	D4-5 (315) 4240	CC SLC37B	Wire
2006-19	2006 May 26 1850	2006-19A	COMPASS-2	Kompas-2	S29159	Shtil'-1	–	BLA, K-84	NK0607-39
2006-20	2006 May 27 2109	2006-20A	Satmex 6	Satmex 6	S29162	Ariane 5ECA	V171 (529)	CSG ELA3	Wire
		2006-20B	Thaicom 5	Thaicom 5	S29163				
2006-21	2006 Jun 15 0800	2006-21A	Resurs-DK	Resurs-DK No. 1	S29228	Soyuz-U	096	GIK-5 LC1	NK0608-01
2006-22	2006 Jun 17 2244	2006-22A	Kazsat	Kazsat 78KS	S29230	Proton-K/DM-2M	410-12?	GIK-5 LC200/39	NK0608-06
2006-23	2006 Jun 18 0750	2006-23A	Galaxy 16	Galaxy 16	S29236	Zenit-3SL	29	Odyssey, POR	NK0608-10
2006-24	2006 Jun 21 2215	2006-24A	USA 187	MITEX OSC satellite	S29240	Delta 7925-9.5	D316	CC SLC17A	spaceflightnow.com
		2006-24B	USA 188	MITEX Lockheed satellite	S29241				
		2006-24C	USA 189	NRL Upper Stage	S29242				
2006-25	2006 Jun 24 1508	2006-25A	Progress M-57	Progress 7K-TGM No. 357	S29245	Soyuz-U	S29245	GIK-5 LC1	NK0608-25
2006-26	2006 Jun 25 0400	2006-26A	Kosmos-2421	US-P	S29247	Tsyklon-2	–	GIK-5 LC90/20	NK0608-13
2006-27	2006 Jun 28 0333	2006-27A	USA 184	NROL-22	S29249	Delta 4M+(4,2)	D4-6 (317) 4240	V SLC-6	Wire
2006-28	2006 Jul 4 1838	2006-28A	Discovery (STS-121)	OV-103	S29251	Space Shuttle	STS-121	KSC LC39B	Wire
		2006-28	Leonardo	MPLM-1	A05895				
2006-F02	2006 Jul 10 1208	2006-F02	Insat 4C	Insat 4C	F01301	GSLV	GSLV-F02	SHAR SLP	Wire
2006-29	2006 Jul 12 1453	2006-29A	Genesis 1	Bigelow Genesis-1	S29252	Dnepr	–	DOM	NK0609-39
2006-30	2006 Jul 21 0420	2006-30A	Kosmos-2422	Oko US-KS?	S29260	Molniya 8K78M	–	GIK-1 LC16/2	NK0609-41
2006-F03	2006 Jul 26 1943	2006-F03	BelKa	BelKa	F01305	Dnepr	4502973804	GIK-5	NK0609-44
		2006-F03	Baumanets	Baumanets	F01306				
		2006-F03	Unisat-4	Unisat-4	F01307				
		2006-F03	PICPOT	PICPOT	F01308				
		2006-F03	ICECube-1	ICECube 1	F01312				
		2006-F03	ION	Illinois Observing Nanosat	F01309				
		2006-F03	RINCON	RINCON	F01313				
		2006-F03	AeroCube-1	AeroCube 1	F01318				
		2006-F03	CalPoly CP1	CP1	F01320				
		2006-F03	SEEDS	SEEDS	F01314				
		2006-F03	NCUBE-1	NCUBE 1	F01316				
		2006-F03	HAUSAT-1	HAUSAT 1	F01315				
		2006-F03	MEROPE	MEROPE	F01317				
		2006-F03	CalPoly CP2	CP2	F01319				
		2006-F03	KUTESat	KUTESat	F01311				
		2006-F03	SACRED	SACRED	F01310				
		2006-F03	Voyager	Mea Huaka'i	F01322				
		2006-F03	ICECube 2	ICECube 2	F01321				
2006-31	2006 Jul 28 0705	2006-31A	Arirang-2	Kompsat-2	S29268	Rokot	–	GIK-1	NK0609-52
2006-32	2006 Aug 4 2148	2006-32A	Hot Bird 8	Hot Bird 8	S29270	Proton-M/Briz-M	535-14	GIK-5 LC200/39	coopi.krunichev.ru
2006-33	2006 Aug 11 2215	2006-33A	JCSAT 3A	JCSAT-10	S29272	Ariane 5ECA	V172 (531)	CSG ELA3	Wire
		2006-33B	Syracuse 3B	Syracuse 3B	S29273				
2006-34	2006 Aug 22 0327	2006-34A	Mugunghwa 5	Koreasat 5	S29349	Zenit 3SL	–	ODYSSEY, KLA	www.sea-launch.com
2006-35	2006 Sep 9 0700?	2006-35A	SJ-8	Shi Jian 8 (FSW)	S29385	Chang Zheng 2C	CZ2C-28 (90)	JQ	TLE
2006-36	2006 Sep 9 1515	2006-36A	Atlantis (STS-115)	OV-104	S29391	Space Shuttle	STS-115	KSC LC39B	Wire
		2006-036	ITS P3/P4	P3/P4 IEA	A05918				
2006-37	2006 Sep 11 0435	2006-37A	IGS Optical-2	IGS-3a	S29393	H-IIA 202	H-IIA-10	TNSC	Wire
2006-38	2006 Sep 12 1602	2006-38A	Zhongxing-22A	Zhongxing-22A	S29398	Chang Zheng 3A	CZ3A-10 (82)	XSC	xinhuanet.com
2006-39	2006 Sep 14 1341	2006-39A	Kosmos-2423	Don	S29402	Soyuz-U	–	GIK-5 LC31	Wire
2006-40	2006 Sep 18 0408	2006-40A	Soyuz TMA-9	Soyuz 11F732 No. 219	S29400	Soyuz-FG	023	GIK-5 LC1	Wire
2006-41	2006 Sep 22 2136	2006-41A	Hinode	SOLAR-B	S29479	M-V	M-V-7	KASC	Wire
		2006-41F	HITSAT	HITSAT	S29481				
		2006-41-D	SSSAT	Solar Sail Sub Payload Sat	S29482				
2006-42	2006 Sep 25 1850	2006-42A	Navstar GPS IIR-M2	GPS SVN 58	S29486	Delta 7925-9.5	D318	CC SLC17A	spaceflightnow.com
2006-43	2006 Oct 13 2056	2006-43A	DirecTV 9S	DirecTV 9S		Ariane 5ECA		CSG ELA3	
		2006-43	Optus D1	Optus D1					
		2006-43	LDREX-4	LDREX-4					

Launch International Designation:	*Launch Date GMT:*	*COSPAR Payload International Designation:*	*Agency Post-Launch Payload Name:*	*Manufacturer Pre-Launch Payload Name:*	*SATCAT Number:*	*Launch Vehicle Type:*	*Launch Vehicle Serial Number:*	*Launch Site:*	*Source Reference:*
2006-44	2006 Oct 19 1628	2006-44A	METOP A	METOP A		Soyuz-2-1A		GIK-5 LC31	
2006-45	2006 Oct 23 1340	2006-45A	Progress M-58	Progress M-58		Soyuz-U		GIK-5 LC1	
2006-46	2006 Oct 23 2334	2006-46A	SJ-6-2A	SJ-6-2A		Chang Zheng 4B		TYSC	
		2006-46B	SJ-6-2B	SJ-6-2B					
2006-47	2006 Oct 26 0052	2006-47A	Stereo Ahead	Stereo Ahead		Delta 7925-10L		CC SLC17B	
		2006-47B	Stereo Behind	Stereo Behind					
2006-48	2006 Oct 28 1620	2006-48A	Xinnuo 2 (Sinosat-2)	Xinnuo 2 (Sinosat-2)		Chang Zheng 3B		XSC	
2006-49	2006 Oct 30 2349	2006-49A	XM-Blues	XM-Blues		Zenit-3SL		ODYSSEY, KLA	
2006-50	2006 Nov 4 1353	2006-50A	DMSP 5D-3 F-17	DMSP 5D-3 F-17		Delta 4M		V SLC6	
2006-51	2006 Nov 8 2001	2006-51A	Badr 4	Badr 4		Proton-M/ Briz-M		GIK-5 LC200/39	
2006-52	2006 Nov 17 1912	2006-52A	GPS 58	GPS 58		Delta 7925-9.5		CC SLC17A	
2006-53	2006 Dec 8 0053	2006-53A	Fengyun 2D	Fengyun 2D		Chang Zheng 3A		XSC	
2006-54	2006 Dec 8 2208	2006-54A	WildBlue 1	WildBlue 1		Ariane 5ECA		CSG ELA3	
		2006-54B	AMC-18	AMC-18					
2006-55	2006 Dec 10 0147	2006-55A	Discovery (STS-116)	Discovery (STS-116)		Shuttle		KSC LC39B	
2006-56	2006 Dec 11 2328	2006-56A	Measat 3	Measat 3		Proton-M/ Briz-M		GIK-5 LC200/39	
2006-57	2006 Dec 14 2100	2006-57A	USA-193 (NROL-21)	USA-193 (NROL-21)		Delta 7920		V SLC2W	
2006-58	2006 Dec 16 1200	2006-58A	Tacsat 2	Tacsat 2		Minotaur		WI LA0B	
		2006-58B	Genesat-1	Genesat-1					
2006-59	2006 Dec 18 0632	2006-59A	ETS-8	ETS-8		H2A 204		TNSC	
2006-60	2006 Dec 19 1400	2006-60A	SAR-Lupe 1	SAR-Lupe 1		Kosmos-3M		GIK-1 LC132/1	
2006-61	2006 Dec 21 0019	2006-55B	MEPSI-2A/2B	MEPSI-2A/2B		Discovery		LEO	
2006-55	2006 Dec 21 0156	2006-55C	RAFT	RAFT		Discovery		LEO	
		2006-55	NMARS	NMARS		Discovery			
2006-55	2006 Dec 21 1822	2006-55	ANDE-MAA	ANDE-MAA		Discovery		LEO	
		2006-55	FCAL	FCAL					
2006-61	2006 Dec 24 0834	2006-61A	Meridian No. 1	Meridian No. 1		Soyuz-2-1A		GIK-1 LC43/4	
2006-62	2006 Dec 25 2018	2006-62A	Glonass-M	Glonass-M		Proton-K/DM-2		GIK-5 LC81/24	
		2006-62B	Glonass-M	Glonass-M					
		2006-62C	Glonass-M	Glonass-M					
2006-63	2006 Dec 27 1423	2006-63A	COROT	COROT		Soyuz-2-1B		GIK-5 LC31/6	
2007-001	2007 Jan 10 0416	2007-001A	LUPAN – TUBsat	LUPAN –TUBsat	S29709	PSLV	PSLV-C7	SHAR PSLV	isro.org unoosa.org
		2007-001B	Cartosat-2	Cartosat-2	S29710				
		2007-001C	SRE-1	Space Recovery Experiment	S29711				
		2007-001D	Pehuensat-1	Pehuensat	S29712				
2007-002	2007 Jan 18 0212	2007-002A	Progress M-59[2]	Progress 7 K-TGM No.359	S29714	Soyuz-U	Zh15000-107	GIK-5 LC2	S.P. Korolev RSC Energia (energia.ru)
2007-F01	2007 Jan 20 2322	2007-F01	NSS-8	NSS-8	F01324	Zenit-3sl	24SL	Odessey, KLA	www.ses-newskies.com
2007-003	2007 Feb 02 1628	2007-003A	Beidou 2A	Beidou 2A	S30323	Chang Zheng (LongMarch) 3A	CZ3A-12	XSC LC2	xinhuanet.com
2007-004	2007 Feb 17 2301	2007-004A	THEMIS P5	THEMIS A	S30580	Delta II 7925-10C	D323	CC SLC17B	WDC SI NSSDC
		2007-004B	THEMIS P1	THEMIS B	S30581				
		2007-004C	THEMIS P2	THEMIS C	S30582				
		2007-004D	THEMIS P3	THEMIS D	S30797				
		2007-004E	THEMIS E	THEMIS P4	S30798				
2007-005	2007 Feb 24 0441	2007-005A	IGS 4A	IGS 4A	S30586	H-IIA 2024	H-IIA-12	TNSC Y	jaxa.jp and oosa.org
		2007-005B	IGS 4B	IGS 4B	S30587				
2007-006	2007 March 9 0310	2007-006A	ASTRO	Orbital Express	S30772	Atlas V 401	AV-013	CC SLC41	WDC SI – NASA NSSDC
		2007-006B	MidSTAR 1	MidSTAR 1	S30773				
		2007-006C	NEXSAT	Orbital Express	S30774				
		2007-006D	STPSat-1	STPSat-1	S30775				
		2007-006E	FalconSat-3	FalconSat-3	S30776				
		2007-006F	CFESat	CFESat	S30777				
			MEPSI Picosat 4A	MEPSI Picosat					eoportal.org/
			MEPSI Picosat 4B	MEPSI Picosat					eoportal.org/
2007-007	2007 Mar 11 2203	2007-007A	Insat 4B	Insat 4B	S30793	Ariane 5ECA	V175 (535)	CSG ELA3	
		2007-007B	Skynet 5A	Skynet 5A	S30794				

Launch International Designation:	Launch Date GMT:	COSPAR Payload International Designation:	Agency Post-Launch Payload Name:	Manufacturer Pre-Launch Payload Name:	SATCAT Number:	Launch Vehicle Type:	Launch Vehicle Serial Number:	Launch Site:	Source Reference:
2007-F02	2007 Mar 21 0110	2007-F02	Demo Flight 2	Autonomous Safety System (AFSS) and Low-Cost Tracking and Data Relay Satellite System (TDRSS) Transmitter (LCT2)	F01325	Falcon 1	2	KMR OM	nasa.gov spacex.com
2007-008	2007 Apr 7 1731	2007-008A	Soyuz TMA-10[1]	Soyuz 11 F732 No. 220	S31100	Soyuz-FG	019	GIK-5 LC1	nasa.gov
2007-009	2007 Apr 9 2254	2007-009A	Anik F3	Anik F3	S21102	Proton-M/ Briz-M	535-16	GIK-5 LC200/39	
2007-010	2007 Apr 11 0327	2007-010A	Hai Yang 1B	HY-1 (Hai Yang Yi)	S31113	Chang Zheng 2C	CZ2C-29 (96)	TYSC	unoosa.org
2007-011	2007 Apr 13 2011	2007-011A	Beidou M1	Beidou M1	S31115	Chang Zheng 3A	CZ3A-13 (97)	XSC	
2007-012	2007 Apr 17 1046	2007-012A	Eqyptsat 1	Eqyptsat 1	S31117	Dnepr		GIK-5 LC109/95	kosmotras.ru/ or cubesat.atl.calpoly.edu
		2007-012B	Saudisat 3	Saudisat 3	S31118				
		2007-012C	Saudi Comsat-7	Saudi Comsat-7	S31119				
		2007-012E	Saudi Comsat-6	Saudi Comsat-6	S31121				
		2007-012F	Aerocube 2	Aerocube 2	S31122				
		2007-012H	Saudi Comsat-5	Saudi Comsat-5	S31124				
		2007-012J	Saudi Comsat-3	Saudi Comsat-3	S31125				
		2007-012K	MAST	MAST	S31126				
		2007-012L	Saudi Comsat-4	Saudi Comsat-4	S31127				
		2007-012M	CalPoly CP3	CP3	S31128				
		2007-012N	Libertad	Libertad	S31129				
		2007-012P	CAPE 1	CAPE 1	S31130				
		2007-012Q	CalPoly CP4	CP4	S31132				
		2007-012R	CSTB 1	CSTB 1	S31133				
2007-013	2007 Apr 23 1000	2007-013A	AGILE	AGILE	S31135	PSLV	PSLV-C8	SHAR SLP	
		2007-013B	AAM	Advanced Avionics Module	S31136				
2007-014	2007 Apr 24 0648	2007-014A	NFIRE	NFIRE	S31140	Minotaur 1	7	WI LA0B	
2007-015	2007 Apr 25 2026	2007-015A	AIM	AIM	S31304	Pegasus XL	F38	L-1011, V RW3	
2007-016	2007 May 5 2229	2007-016A	Astra 1L	Astra 1L	S31306	Ariane 5ECA	V176 (536)	CSG ELA3	
		2007-016B	Galaxy 17	Galaxy 17	S31307				
2007-017	2007 May 12 0325	2007-017A	Progress M-60 [3]	Progress 7K-TGM No.360	S31393	Soyuz-U	106	GIK-5 LC2	S.P. Korolev RSC Energia (energia.ru)
2007-018	2007 May 13 1601	2007-018A	Nigcomsat 1	Nigcomsat 1	S31395	Chang Zheng 3B	CZ3B (98)	XSC	
2007-019	2007 May 25 0712	2007-019A	YaoGan WeiXing 2	Jianbing 6	S31490	Chang Zheng 2D	CZ2D-6	JQ	
		2007-019	MEMS-Pico satellite	MEMS-Pico satellite					
2007-020	2007 May 29 2031	2007-020A	Globalstar-A	Globalstar M65	S31571	Soyuz-FG	Ts15000-021	GIK-5 LC6	globalstar.com
		2007-020C	Globalstar-B	Globalstar M69	S31573				
		2007-020D	Globalstar-C	Globalstar M72	S31574				
		2007-020F	Globalstar-D	Globalstar M71	S31576				
2007-021	2007 May 31 1608	2007-021A	SinoSat 3	SinoSat 3	S31577	Chang Zheng 3A	CZ3A-14 (100)	XSC	WDC SI – NASA NSSDC
2007-022	2007 Jun 07 1800	2007-022A	Cosmos 2427	Cosmos 2427	S31595	Soyuz-U		GIK-1	WDC SI – NASA NSSDC
2007-023	2007 Jun 08 0234	2007-023A	COSMO-SkyMed 1	COSMO-SkyMed 1	S21598	Delta 2 7420-10C	D324	V SLC2W	vandenberg.af.mil
2007-024	2007 Jun 08 2338	2007-024A	Atlantis STS-117	S3/S4 Truss	S31600	Space Shuttle	STS-117	KSC LC39A	nasa.gov
2007-025	2007 Jun 10 2340	2007-025A	Ofeq 7	Ofeq 7	S31601	Shavit	7	PALB	WDC SI – NASA NSSDC
2007-026	2007 Jun 15 0214	2007-026A	Terra SAR-X	Terra SAR-X	S31698	Dnepr 1		GIK-5 LC109/95	dlr.de and kosmotras.ru/
2007-027	2007 Jun 15 1512	2007-027A	USA 194	NROL-30	S31701	Atlas V 401	AV-009	CC SLC41	nro.gov
2007-028	2007 Jun 28 1502	2007-028A	Genesis 2	Genesis 2	S31789	Dnepr 1		Yasny (Russia)	kosmotras.ru/
2007-029	2007 Jun 29 1000	2007-029A	Cosmos 2428	Cosmos 2428	S31792	Zenit-2M	1-2005	GIK-5 LC45/1	WDC SI – NASA NSSDC
2007-030	2007 July 02 1939	2007-030A	SAR Lupe 2	SAR Lupe 2	S31797	Kosmos 3M	11K65M-SL	GIK-1 LC132/1	ohb-system.de
2007-031	2007 Jul 05 1208	2007-031A	ChinaSat 6B	ChinaSat 6B	S31800	Chang Zheng 3B	CZ3B (101)	XSC LC2	alcatel-lucent.com
2007-032	2007 July 07 0116	2007-032A	DirecTV 10	DirecTV 10	S31862	Proton-M		GIK-5 LC200/39	boeing.com
2007-033	2007 Aug 02 1734	2007-033A	Progress M-61[4]	Progress M-61	S32001	Soyuz-U	Sh15000-108	GIK-5 LC1	energia.ru
2007-034	2007 Aug 04 936	2007-034A	Phoenix Mars Lander	Phoenix Mars Lander	S32003	Delta 2 7925-9.5	D325	CC SLC17A	nasa.gov

Launch International Designation:	*Launch Date GMT:*	*COSPAR Payload International Designation:*	*Agency Post-Launch Payload Name:*	*Manufacturer Pre-Launch Payload Name:*	*SATCAT Number:*	*Launch Vehicle Type:*	*Launch Vehicle Serial Number:*	*Launch Site:*	*Source Reference:*
2007-035	2007 Aug 08 2237	2007-035A	Endeavour STS-118	S5 Truss	S32008	Space Shuttle	STS-118	KSC LC39A	nasa.gov
2007-036	2007 Aug 14 2344	2007-036A	SPACEWAY 3	SPACEWAY 3	S32018	Ariane 5-ECA	V177 (537)	CSG ELA3	arianespace.com
		2007-036B	BSAT-3A	BSAT-3A	S32019				
2007-037	2007 Sep 02 1251	2007-037A	Insat 4CR	Insat 4CR	S32050	GLNV	GSLV-F04	SHAR SLP	WDC SI – NASA NSSDC
2007-F03	2007 Sep 05 2343	2007-F03	JCSAT-11	JCSAT-11	F01326	Proton M/ Breeze M		GIK-5 LC39	WDC SI – NASA NSSDC
2007-038	2007 Sep 11 1305	2007-038A	Cosmos 2429	Cosmos 2429	S32052	Kosmos 3M		GIK-1 LC132/1	WDC SI – NASA NSSDC
2007-039	2007 Sep 14 0131	2007-39A	Kaguya	SELENE	S32054	H-IIA 2022	H-IIA-13	TNSCY	WDC SI – NASA NSSDC
		2007-039B	RSAT	RSAT	S32055				
		2007-039C	VRAD	VRAD	S32056				
2007-040	2007 Sep 14 1100	2007-40A	Foton M-3	Foton M-3	S32058	Soyuz-U	098	GIK-5 LC1	WDC SI – NASA NSSDC
		2007-40B	YES2	YES2					
2007-041	2007 Sep 18 1835	2007-041A	WorldView 1	WorldView 1	S32060	Delta 2 7920-10C	D326	V SLC2W	vandenberg.af.m
2007-042	2007 Sep 19 0326	2007-042A	CBERS 2B	Zi Yuan 1-2B	S32062	Chang Zheng 4B	CZ4B (102)	TYSC	WDC SI – NASA NSSDC
2007-043	2007 Sep 27 1234	2007-043A	Dawn	Dawn	S32249	Delta 2 7925H	D327	CC LC17B	nasa.gov
2007-044	2007 Oct 05 2202	2007-044A	Optus D2	Optus D2	S32252	Ariane 5 GS	V178 (526)	CSG ELA3	arianespace.com
		2007-044B	PAS 11	Intelsat 11	S32253				
2007-045	2007 Oct 10 1322	2007-045A	Soyuz TMA-11[1]	Soyuz TMA-11 No 221	S32256	Soyuz-FG	Ts15000-020	GIK-5 LC1	nasa.gov
2007-046	2007 Oct 11 0022	2007-046A	Wideband Global SATCOM WGS F1	USA 195	S32258	Atlas 5 421	AV-011	CC LC41	ulalaunch.com/ news_WGS.html
2007-047	2007 Oct 17 1223	2007-047A	GPS 2R-17 M4	GPS 2R-17M4	S32260	Delta 27925-9.5	D328	CC LC17A	ulalaunch.com/ news_GPS_IIR-17.html
2007-048	2007 Oct 20 2012	2007-048A	Globalstar 67	Globalstar A	S32263	Soyuz-FG	Ts15000-022	GIK-5 LC31	nasa.gov
		2007-048B	Globalstar 70	Globalstar B	S32264				
		2007-048C	Globalstar 66	Globalstar C	S32265				
		2007-048D	Globalstar 68	Globalstar D	S32266				
2007-049	2007 Oct 23 0439	2007-049A	Cosmos 2430	Cosmos-Oko	S32268	Molniya-M8K78M		GIK-1 LC16/2	WDC SI – NASA NSSDC
2007-50A	2007 Oct 23 1538	2007-50A	STS120 Discovery	Node-2 Harmony	S32272	Space Shuttle	STS-120	KSC LC39A	nasa.gov
2007-051	2007 Oct 24 1005	2007-051A	Chang'e 1	Chang'e 1	S32274	Chang Zheng 3B	CZ3A-15 (103)	XSC	WDC SI – NASA NSSDC
2007-052	2007 Oct 26 0735	2007-052A	Cosmos 2433	Glonass 720	S32275	Proton-K	410-17	GIK-5 LC81/24	WDC SI – NASA NSSDC
		2007-052B	Cosmos 2432	Glonass 719	S32276				
		2007-052C	Cosmos 2431	Glonass 718	S32277				
2007-053	2007 Nov 01 0051	2007-053A	SAR Lupe 3	SAR Lupe 3	S32283	Kosmos-3M	11K65M-SL	GIK-1 LC132/1	WDC SI – NASA NSSDC
2007-054	2007 Nov 11 1550	2007-054A	DSP 23	USA 197	S32287	Delta 4H	D4-8 (329) 4050	CC SLC 37B	ulalaunch.com
2007-055 \	2007 Nov 11 2248	2007-055A	Yaogan 3	Yaogan 3	S32289	Chang Zheng 4C	CZ4C (104)	TYSC	xinhuanet.com
2007-056	2007 Nov 14 2206	2007-056A	Skynet 5B	Skynet 5B	S32293	Ariane 5 ECA	V179 (538)	CSG ELA3	arianespace.com
		2007-056B	Star One C1	Star One C1	S32294				
2007-057	2007 Nov 17 2239	2007-057A	SIRIUS 4	SIRIUS 4	S32299	Proton M		GIK-5 LC39	ils.launch.com
2007-058	2007 Dec 09 0016	2007-058A	Raduga-1M	Raduga-1M	S32373	Proton M		GIK-5 LC200/39	RIA Novosti
2007-059	2007 Dec 09 0231	2007-059A	COSMO SkyMed 2	COSMO SkyMed 2	S32376	Delta 27420-10	D330	V SLC2W	ulalaunch.com
2007-060	2007 Dec 10 2205	2007-060A	USA 198	NRO L24	S32378	Atlas 5 401	AV-015	CC SLC41	ulalaunch.com
2007-061	2007 Dec 14 1317	2007-061A	RADARSAT 2	RADARSAT 2	S32382	Soyuz-FG	Ts15000-025	GIK-5 LC31	starsem.com
2007-062	2007 Dec 20 2004	2007-062A	GPS 2R-18M	GPS 2R-18M	S32384	Delta 2 7925-9.5	D331	CC SLC17-A	ulalaunch.com
2007-063	2007 Dec 21 2142	2007-063A	RASCOM-QAF1	RASCOM-QAF1	S32387	Ariane 5 GS	V180 (530)	CSG ELA3	arianespace.com
		2007-063B	Horizons 2	Horizons 2	S32388				
2007-064	2007 Dec 23 0712	2007-064A	Progress M62	Progress M62	S32391	Soyuz-U		GIK-5 LC1	WDC SI – NASA NSSDC
2007-065	2007 Dec 25 1932	2007-065A	Cosmos 2434	Glonass-M 21	S32393	Proton-M/DM-2		GIK-5 LC81/24	
		2007-065B	Cosmos 2435	Glonass-M 22	S32394				
		2007-065C	Cosmos 2436	Glonass-M 23	S32395				

Sources: Jonathan McDowell, PhD and World Data Centre for Satellite Information (WDC SI) hosted by NASA's National Space Science Data Centre (NSSDC).
[1]This is a human spaceflight mission to the ISS. For further information please see separate entry Table 3 US Space Shuttle Flights.
[2]ISS cargo mission 24P
[3]ISS cargo mission 25P
[4]ISS cargo mission 26P

Table 5: World Orbital Launches by Location 1957 – 2006

Country:	Launch Site[1]:	Latitude:	Longitude:	1957–1994:	1995[2]:	1996:	1997:	1998:	1999:	2000:	2001:	2002:	2003:	2004:	2005:	2006:	1995–2006:	1957–2006:
National Launch Sites:																		
Brazil	Alcantara	2.3°S	44.4°W	0	0	0	1	0	1	0	0	0	1	0	0	0	3	3
China	Jiuquan	40.6°N	99.9°E	22	0	1	0	0	1	0	1	2	2	2	4	1	14	36
	Taiyuan/ Wuzhai	37.5°N	112.6°E	2	0	0	2	4	3	1	0	3	2	4	0	2	21	23
	Xichang	28.25°N	102.0°E	15	3	3	4	2	0	4	0	0	3	2	1	3	25	40
France	Hammaguir, Algeria	31.0°N	8.0°W	4	0	0	0	0	0	0	0	0	0	0	0	0	0	4
India	SHAR Centre, Sriharikota	13.9°N	80.4°E	6	0	1	1	0	1	0	2	1	2	1	1	1	11	17
Israel	Palmachim/ Yavne	31.5°N	34.5°E	2	1	0	0	1	0	0	0	1	0	1	0	0	4	6
Italy	San Marco Platform	2.9°S	40.3°E	8	0	0	0	0	0	0	0	0	0	0	0	0	0	8
Japan	Tanegashima	30.4°N	131.0°E	26	1	1	1	1	1	0	1	3	2	0	1	4	16	42
	Uchinoura/ Kagoshima	31.2°N	131.1°E	21	1	0	1	1	0	1	0	0	1	0	1	2	8	29
North Korea	Musudan	40°N	129°E	0	0	0	0	1	0	0	0	0	0	0	0	0	1	1
Russian Federation	Barents Sea	69.3°N	35.3°	0	0	0	0	1	0	0	0	0	0	0	1	1	3	3
	Dombarovsky/ Yasny	50.75°N	59.50°E	0	0	0	0	0	0	0	0	0	0	0	0	1[3]	1	1
	Kapustin Yar	48.4°N	45.8°E	83	0	0	0	0	1	0	0	0	0	0	0	0	1	84
	Plesetsk	62.8°N	40.1°E	1,413	14	11	9	7	6	5	6	10	7	6	6	5	92	1,505
	Svobodny	51.37°N	128.3°E	0	0	0	2[4]	0	0	1	1	0	0	0	0	1	5	5
Spain	Gando AFB, Gran Canaria /L-1011	27°N	15°W	0	0	0	1	0	0	0	0	0	0	0	0	0	1	1
United States	Cape Canaveral/ Kennedy Space Center	28.5°N	81.0°W	494	23	24	24	23	19	19	16	14	17	13	7	10	209	703
	Edwards AFB/B52	35°N	118°W	5	0	0	0	0	0	0	0	0	0	0	0	0	0	5
	Reagan Test Site, Omelek, Kwajalein	8.9917°N	167.72°W	0	0	0	0	0	0	1	0	0	0	0	0	1	2	2
	Vandenberg AFB	34.4°N	120.35°W	503	6	8	11	11	11	8	5	3	6	3	5	6	83	586
	Wallops Island	37.8°N	75.5°W	19	1	1	2	2	1	0	0	0	0	0	0	1	8	27
International Launch Sites:																		
Australia/United Kingdom	Woomera	31.1°S	136.8°E	2	0	0	0	0	0	0	0	0	0	0	0	0	0	2
Europe/France	Centre Spatial Guyanais/Kourou	5.2°N	52.8°W	69	11	11	12	11	10	12	8	12	4	3	5	5	104	173
Kazakhstan; Russian Federation	Baikonur/Tyuratam	45.6°N	63.4°E	968	19	16	18	17	21	30	16	15	14	17	19	17	219	1,187
Commercial Launch Sites:																		
Norway; Russian Federation; Ukraine; United States; private sector partnership	Sea Launch Company, LLC; Odyssey Launch Platform, Pacific Ocean	0°N	154°W	0	0	0	0	0	2[5]	3	2	1	3	3	4	5	23	23
United States; private sector and Alaska State partnership	Alaska Aerospace Development Corporation; Kodiak Launch Complex, Alaska	57°N	152°W	0	0	0	0	0	0	0	1	0	0	0	0	0	1	1
			Total Launches	**3,662**	**80**	**77**	**89**	**82**	**78**	**85**	**59**	**65**	**64**	**55**	**55**	**66**	**855**	**4,517**

[1]Although other sites with orbital launch capabilities may exist, only facilities that have conducted known orbital launch attempts are included in this table
[2]Data from 1995 forward includes launch failures
[3]First launch 12 July 2006
[4]First launch 4 March 1997
[5]First launch 27 March 1999

The Apollo-era Mobile Launch Platform supports External Tank, Solid Rocket Boosters and Orbiter as the Shuttle is carried to its launch pad (NASA) 0137132

Crater Endurance on Mars imaged by a Mars Exploration Rover showing layered outcrops (NASA JPL) 1047543

Reflecting the International nature of launch vehicle marketing services, models of (left to right): Ariane 4, Soyuz and Ariane 5 0084625

Separated by seven years of development, the diminutive Mars Pathfinder (launched 1996) compares in scale with the Mars Exploration Rover spacecraft (launched in 2003) (NASA JPL) 1047528

Ariane 5 is transported to its launch complex at Kourou in French Guiana (ESA) 0137120

Europe's largest science satellite, the XMM X-ray observatory launched in 1999 was the most powerful of its type yet sent into space (ESA) 0137133

NASA's High Energy Astronomy Observatory is encapsulated prior to launch (NASA MSFC) 0137154

ESA's Mars Express in Earth orbit prior to ignition of its injection stage (ESA) 1047485

INDEXES

General Index

Q

R

S

T

U

V

W

X

Z